POWER PLANT ENGINEERING

POWER PLANT ENGINEERING

POWER PLANT ENGINEERING

by

BLACK & VEATCH

Lawrence F. Drbal
Managing Editor

Patricia G. Boston
Associate Editor

Kayla L. Westra
Associate Editor

R. Bruce Erickson
Art Editor

KLUWER ACADEMIC PUBLISHERS
BOSTON/DORDRECHT/LONDON

Distributors for North, Central and South America:
Kluwer Academic Publishers
101 Philip Drive
Assinippi Park
Norwell, Massachusetts 02061 USA
Telephone (781) 871-6600
Fax (781) 871-6528
E-Mail <kluwer@wkap.com>

Distributors for all other countries:
Kluwer Academic Publishers Group
Distribution Centre
Post Office Box 322
3300 AH Dordrecht, THE NETHERLANDS
Telephone 31 78 6392 392
Fax 31 78 6546 474
E-Mail services@wkap.nl>

Electronic Services <http://www.wkap.nl>

Library of Congress Cataloging-in-Publication

Power plant engineering/by Black & Veatch;: Lawrence F. Drbal, managing editor, Patricia G. Boston, associate editor, Kayla L. Westra, associate editor.
p. cm.
Includes bibliographical references and index.
ISBN 0-412-06401-4
1. Electric power-plants--Design and construction. I. Drbal, Lawrence F. II. Boston, Patricia G. III. Westra, Kayla L. IV. Black & Veatch.
TK1191.P64839 1996 96-34590
621.31'21--dc20 CIP

Fourth Printing 2000 by Kluwer Academic Publishers

Cover photo courtesy of Black & Veatch; Mike McPheeters (photographer). Intermountain Power Plant.
Cover design: Curtis Tow Graphics

Printed on acid-free paper.

Printed in Great Britain by IBT Global, London

Contents

Foreword

The electric power generation and distribution industry has evolved during almost exactly one century into its present form. This development has been driven by economics, technological advancements, and government regulations. These diverse forces have produced generation and distribution systems that are reliable, safe, and environmentally acceptable, suitable for service throughout the world. The technologies which make up the systems have brought a highly convenient and economic form of energy to both densely populated cities and remote geographic areas. Electrical energy has become a fundamental requirement for the existence of modern society. Likewise, it is essential for economic progress in lesser developed regions.

Because the need for electricity is pervasive in our society, there is a continuing interest in the technology of electric power production and distribution. Unfortunately for those with such interests, a comprehensive source of information has not been available until now. Existing texts have tended to be single topic treatments that address power plant design and plant components (primarily pre-1950) or electrical distribution system design. *Power Plant Engineering*, however, offers a unified treatment of power system economics, planning, design, and operation that includes consideration of all of the forces which have shaped the industry.

The problem in producing a reliable book that treats all aspects of the industry has been in recruiting authors of the talent level required and in providing time and support for successful completion of the project. Producing such a book requires developing the concept and organizing the effort required, as well as managing and encouraging individuals who can effectively communicate their precise technical knowledge and experience in many diverse fields. The Black & Veatch organization is fortunate to have available within its structure the complete cast of players necessary to mount such a production. Equally important to its success, Black & Veatch management had the vision to encourage the project. The high level of talent and the effort expended on the task of producing the book are evident in the quality of the resulting product.

The book wisely concentrates on state-of-the-art technology. It emphasizes the why and the what of current practice. However, it does not neglect some of the underutilized energy sources and systems or the more promising of the developing technologies. The reader will gain a firm foundation in the technology and a good understanding of commonly encountered system configurations. Additionally, the ground work is adequately laid for the reader to be prepared for developments in the future.

I feel the book should be suitable and of strong appeal to university students (and faculty), to equipment manufacturers, to utility operators and owners, to system design engineers, and to representatives of the variety of local, state, and federal government agencies involved in planning and regulation. The book will find use as a university textbook, as a technical reference, and as a comprehensive handbook and guide to the numerous technologies that are necessary components of the electric power industry.

As a university professor and consultant who has taught and worked in this area for several decades, I am well aware of the lack of suitable texts and references available for use by students and by those employed in the power industry. I believe this book represents a vital service to that community and that it will be considered the standard work in the area.

Robert L. Gorton, PE, Ph.D.
Manhattan, Kansas

PREFACE

Blaise Pascal, scientist, mathematician, and physicist, once noted that the last thing we discover in writing a book is to know what to put at the beginning. I do not have that problem with this book, for I believe it is a book of such quality that the beginning must be an acknowledgment of the tremendous effort expended by many, many talented people.

In a manner of speaking, it has taken Black & Veatch nearly 80 years to prepare this book. It has required that length of time, not because of the magnitude of the task, but rather because we have drawn on the knowledge and experience acquired over the full history of the firm to accumulate, test, and apply the information compiled here. In one sense, the book has been authored by literally thousands of engineers, architects, technicians, scientists, and others who have contributed to the success of Black & Veatch. From that resource we recruited 32 authors, each an expert in a specific field, who could best state the firm's approach to his or her specialty. It is these people who have made this book possible.

As with any book of this magnitude, its preparation required the efforts and support of countless others who had nothing to do with the writing of chapters. Initially, the task of coordinating the contents of 26 disparate chapters necessitated a filing system for keeping track of the thousands of references, figures, tables, permissions, and pages of draft text produced for the book. And as the development of the book extended into years, people changed positions, retired, had babies, and went on international assignments—constantly changing the book's staff roster. The system for tracking was begun by Lorraine Gehring, editor turned marketer, and ended by Laura Patton, teacher turned editor turned new mother. In between, editors Dana Campbell and Elaine Rhodes cajoled engineers into using simpler, yet proper English, and Marcia Cones corrected and photocopied the revisions of revisions. The literally countless pages of drafts were typed and typed and retyped by the world's cheeriest word processing group—Cindy McDonald, Arlene Siegismund, Marsha Johnson, Lori Lange, Pat Anderson, and Becky Nanninga. The many excellent figures which illustrate the concepts discussed on these pages were constructed under the direction of Bruce Erickson, assisted by Nicole Genever-Watling, Mark Hayden, and Mark Rump. The entire effort over the years was overseen by Virgil Snell, Manager of Engineering for the Power Division.

As the project drew to a close, the person assigned to coordinate the efforts was Roy McIntosh. Also during this last year, the expert charged with technical review of the entire manuscript was Robert L. Gorton, professor emeritus from Kansas State University Engineering Department, a true gentleman whose comments were consistent, timely, and made with great affection for the profession from which he retired during the preparation of the book.

Black & Veatch wanted to prepare this book for a number of reasons. First, though many authoritative texts address the various aspects of power generation, no text exists which can be used as the definitive source book as well as a teaching tool. Second, the various technologies and processes described in this textbook are continuously changing, driven by the dynamic economic, regulatory, political, and social influences that shape the power industry. In the new global climate of world banks, turnkey contracting, information superhighways, partnering, and other concepts for which "change" seems to be the watchword, this book reflects a stable, proven approach to power plant design which, at the time of publication, reflects current technology. And finally, we believe Black & Veatch's approach to power plant design is *the* most effective approach and by sharing our experiences, we are providing a valuable service to everyone connected with the power industry.

Dr. Patrick G. Davidson
Managing Partner and Head of the Power Division
Black & Veatch

Authors

Stanley A. Armbruster, P.E.
B.S., Kansas State University, Nuclear Engineering. (*Steam Turbine Generators*)

Gordon V.Z. Beard, P.E., C.P.A.
B.S., University of Tennessee-Knoxville, Mechanical Engineering; B.S., University of Tennessee-Knoxville, Business Administration-Accounting; M.S., Virginia Polytechnic Institute and State University, Mechanical Engineering; M.B.A., Virginia Polytechnic Institute and State University. (*Resource Recovery*)

Mitchell N. Bjeldanes, P.E.
B.S., University of Minnesota, Mechanical Engineering. (*Resource Recovery*)

David J. Brill, P.E.
B.S., University of Dayton, Mechanical Engineering. (*Circulating Water Systems*)

Amy L. Carlson
B.A., University of Northern Iowa, Geology; M.S., Iowa State University, Earth Sciences/Meteorology. (*Permitting and Environmental Review Requirements*)

Kenneth E. Carlson, P.E.
B.S., University of Illinois, Mechanical Engineering; M.S., University of Illinois, Mechanical Engineering. (*Fossil Fuels*)

Richard G. Chapman, P.E.
B.S., University of Missouri-Rolla, Chemical Engineering; M.S., University of Missouri-Columbia, Civil Engineering. (*Water Treatment*)

Augustine H. Chen, P.E.
B.S., National Taiwan University, Mechanical Engineering; M.S., Oklahoma State University, Mechanical Engineering. (*Plant Control Systems*)

Lawrence F. Drbal, Ph.D., P.E.
B.S., University of Nebraska, Chemical Engineering; M.S., Kansas State University, Nuclear Engineering; Ph.D., Kansas State University, Nuclear Engineering. (*Introduction, Nuclear Power*)

Michael J. Eddington, P.E.
B.S., University of Missouri-Columbia, Mechanical Engineering. (*Cycle Performance Impacts*)

Alan W. Ferguson, P.E.
B.S., Clarkson University, Mechanical Engineering. (*Power Plant Atmospheric Emissions Control*)

Kris A. Gamble, P.E.
B.S., Iowa State University, Mechanical Engineering. (*Fans*)

Stephen M. Garrett, P.E.
B.S., Colorado State University, Mechanical Engineering. (*Power Plant Planning and Design*)

Kenneth E. Habiger, P.E.
B.S., Kansas State University, Nuclear Engineering; M.S., Kansas State University, Nuclear Engineering. (*Fluidized Bed Combustion*)

Anne F. Harris, J.D.
B.A., Florida State University, Biology; M.A.T., Emory University, Biology; M.S., University of North Carolina, Botany; J.D., University of Kansas, Law. (*Permitting and Environmental Review Requirements*)

Benjamin W. Jackson, P.E.
B.S., University of Missouri-Columbia, Mechanical Engineering. (*Steam Generators*)

Todd S. Jonas, P.E.
B.S., North Dakota State University, Mechanical Engineering. (*Combustion Processes*)

Lloyd L. Lavely
B.S., University of Kansas, Chemical Engineering. (*Power Plant Atmospheric Emissions Control*)

Richard H. McCartney, P.E.
B.S., Clarkson University, Mechanical Engineering; M.S., Clarkson University, Mechanical Engineering. (*Coal and Limestone Handling*)

Mark F. McClernon, Ph.D.
B.S., Rockhurst College, Engineering Science; M.S., University of Notre Dame, Engineering Science; Ph.D., University of Notre Dame, Engineering Science. (*Thermodynamics and Power Plant Cycle Analysis*)

Jay F. Nagori, P.E.
B.E., Osmania University, Hyderabad, India, Mechanical Engineering; M.S., University of Kansas, Mechanical Engineering. (*Steam Cycle Heat Exchangers*)

Roger M. Prewitt, P.E.
B.S., University of Missouri-Rolla, Mechanical Engineering. (*Site/Plant Arrangements*)

Lawrence J. Seibolt, P.E.
B.S., University of Missouri-Columbia, Mechanical Engineering. (*Pumps*)

Lloyd Wade Sherrill, P.E.
A.S., Arkansas Polytechnic College, Engineering; B.S., University of Arkansas, Electrical Engineering; M.S., University of Missouri-Columbia, Electrical Engineering. (*Electrical Systems*)

Jeffrey M. Smith, P.E.
B.S., University of Illinois, Mechanical Engineering. (*Gas Turbines*)

G. Scott Stallard, P.E.
B.S., University of Kansas, Mechanical Engineering; B.S., University of Kansas, Business Administration. (*Combustion Processes*)

Larry E. Stoddard, Ph.D., P.E.
B.S., South Dakota State University, Electrical Engineering; M.S., University of Missouri-Rolla, Electrical Engineering; Ph.D., University of Missouri-Rolla, Electrical Engineering. (*Emerging Technologies*)

Samuel Tarson, P.E.
B.S., University of Missouri-Rolla, Mechanical Engineering; M.S., Purdue University, Mechanical Engineering. (*Steam Cycle Heat Exchangers*)

Douglas C. Timpe
B.S., North Dakota State University, Zoology; M.S., Western Kentucky University, Biology. (*Permitting and Environmental Review Requirements*)

Kenneth R. Weiss, P.E.
B.S., University of Missouri-Rolla, Chemical Engineering. (*Liquid and Solid Waste Treatment and Disposal*)

Lesley A. Wallingford
Certificate of Paralegal Studies, Johnson County (Kansas) Community College; Associate of Applied Science, Johnson County (Kansas) Community College. (*Permitting and Environmental Review Requirements*)

John M. Wynne
B.S., Northwest Missouri State University, Economics/ Management/ Marketing; M.S., Bowling Green State University, Economics. (*Engineering Economics*)

Contributors

The authors are grateful to many people and organizations who assisted in the development of their chapters. Contributors are those who helped mold the content of the chapters and, in some cases, wrote specific sections. Acknowledgments are given to those individuals and businesses who helped the authors by sharing technical expertise or critiquing sections of a chapter. Unless otherwise noted, individuals listed are employees of Black & Veatch.

Chapter 2–Engineering Economics

Acknowledgments: Gordon V.Z. Beard; Myron R. Rollins

Chapter 3–Thermodynamics and Power Plant Cycle Analysis

Acknowledgments: Jeffrey R. Dykstra

Chapter 4–Fossil Fuels

Acknowledgments: Kristian B. Fosse; Robert L. Irvine; Richard B. Rinehart; Phil Rogers, Utilicorp United Inc.; Michael D. Sharp; Richard K. Van Meter; Steven R. Witthar

Chapter 5–Coal and Limestone Handling

Acknowledgments: Andrew J. Schwartz; United States Environmental Protection Agency

Chapter 6–Combustion Processes

Acknowledgments: John Pavlish

Chapter 7–Steam Generators

Acknowledgments: Lawrence F. Drbal; Charles J. Schutty, II

Chapter 8–Steam Turbine Generators

Contributors: Jon C. Erickson

Acknowledgments: Carl G. Granberg, Jr.; General Electric Power Systems, Schenectady, NY; Westinghouse Power Generation Business Unit, Orlando, FL

Chapter 9–Steam Cycle Heat Exchangers

Acknowledgments: David J. Brill; Mark D. Shaw

Chapter 10–Fans

Acknowledgments: Anthony L. Compaan

Chapter 11–Pumps

Contributors: John W. Kruse

Acknowledgments: Robert E. Cornman, Ingersoll-Dresser Pump Company; Jeffrey B. Galush, Ingersoll-Dresser Pump Company; Michael T. Radio, Ingersoll-Dresser Pump Company; Tim L. Wotring, Ingersoll-Dresser Pump Company

Chapter 12–Circulating Water Systems

Contributors: Michael H. Ostdiek; Mark D. Shaw; Jeffrey A. Wootton

Acknowledgments: Marley Cooling Tower Company; Kermit E. Trout, Jr.

Chapter 13–Cycle Performance Impacts

Contributors: Jon C. Erickson; Mitch L. Rackers; Jason A. Zoller

Acknowledgments: Carl G. Granberg

Chapter 14–Power Plant Atmospheric Emissions Control

Contributors: John R. Cochran; Diane M. Fischer; Mike G. Gregory; David K. Harris; Benjamin W. Jackson; Ricki L. Lausman; Kendall R. Shannon; James E. Stresewski, Jr.

Acknowledgments: Larry R. Alfred; Morgen E. Fagan

Chapter 15–Water Treatment

Contributors: Russell R. Helling; Bruce A. Larkin; James E. O'Connor; Thomas J. Shrader

Acknowledgments: Charles H. Fritz; Lester C. Webb, Jr.

Chapter 16–Liquid and Solid Waste Treatment and Disposal

Contributors: Lawrence J. Almaleh; James A. Hengel; David L. Holt

Acknowledgments: David M. Lefebvre; Gary Van Reissen; Lester C. Webb, Jr.

Chapter 17–Electrical Systems

Acknowledgments: Mike W. Kelly

Chapter 18–Plant Control Systems

Acknowledgments: Roger L. Ayers; Daniel G. Couture; Harry B. McCarl; Parke H. Woodard, Jr.

Chapter 19—Site/Plant Arrangements
Contributors: Lynn E. Brown
Acknowledgments: W. Keith Krambeck; Larry S. Newland

Chapter 20—Gas Turbines
Contributors: Michael R. Chandler; Steven M. Clark; Linus A. Drouhard; Morgen E. Fagan; John R. Hughes; Peter P. Majerle; Michael D. Morris
Acknowledgments: David L. Frieze; Barry J. Scrivner

Chapter 21—Fluidized Bed Combustion
Contributors: Kevin A. Kerschen; Ronald J. Ott
Acknowledgments: Lawrence F. Drbal; Benjamin W. Jackson

Chapter 22—Resource Recovery
Contributors: Stephen M. Garrett
Acknowledgments: Hunter F. Taylor

Chapter 23—Nuclear Power
Acknowledgments: Rebecca Jung, Westinghouse Electric Corporation; John C. Kirkland; Joe Miller, Scientech, Inc.; Dr. M. John Robinson; Dr. Craig Sawyer, GE-Nuclear Energy; D.R. Shiflett, AECL Technologies; Don Warembourg, Public Service Company of Colorado; Jan Wistrom, General Atomics

Chapter 24—Emerging Technologies
Contributors: Lawrence F. Drbal; Tracy M. Fiedler; George P. Gruber
Acknowledgments: John E. Harder; Kevin A. Kerschen; Paul G. LaHaye, Hague International; Arthur S. Seki, Hawaiian Electric

Chapter 25—Power Plant Planning and Design
Acknowledgments: Ronald D. Hubbell; Helen Z. Kennon; Richard K. Van Meter

Chapter 26—Permitting and Environmental Review Requirements
Acknowledgments: Larry R. Alfred; Marsha A. Brustad; Helen Z. Kennon; Stanley L. Rasmussen

POWER PLANT ENGINEERING

1

INTRODUCTION

Lawrence F. Drbal

One only has to experience a power outage to be reminded of how much we take electricity for granted. Our lighting, heating, and cooling systems no longer operate. Computers, televisions, videos, and other communication systems become unusable. Traffic control lights become useless; elevators no longer move people; and industries, schools, and commercial buildings become virtually inoperable.

We are clearly dependent on electricity for most of our everyday activities. This dependence also demonstrates why electricity is regarded as one of the most significant *sociological* inventions of the 20th century. Although the United States and our industrialized neighbors enjoy the full benefits of electricity, the world's developing nations do not. Supplying electricity to these areas under various economic, environmental, social, and political constraints will be one of the major challenges in the 21st century.

Power producers must consider the local social and economic conditions in the conceptual and planning phases of all projects. Their engineering decisions must reflect an awareness and knowledge of the best equipment and systems available. The final product, the electrical generation and distribution system, must reflect responsible application of economic and engineering principles and of thoughtful recognition of social and environmental concerns. The approach Black & Veatch has used to meet these challenges is described in this book.

Power Plant Engineering originated from a series of lectures prepared by Black & Veatch instructors for use in teaching steam power plant design through the University of Kansas. The material was later expanded by the firm for use in teaching its own engineers and technical specialists. As a definitive text on the *design* of coal-fueled power plants, this book was developed from the system design engineer's perspective. In other words, the book describes the approach the engineer uses to design and integrate the plant systems and to specify performance and design requirements, but it does not address the detailed design of the equipment, which is usually done by the manufacturer.

Power Plant Engineering includes both theory and practical application of the major areas of power plant related topics, ranging from permitting and engineering economics to coal and limestone handling, from design processes to plant thermal heat balances. Although the focus of the text is on coal, the book reviews all major power-generating technologies, giving particular emphasis to current approaches. It further defines and analyzes the features of various plant systems and major components, and discusses promising emerging technologies.

The text presents a full development of the design of a pulverized coal-fueled power plant from theory through the details of application.

Figure 1-1 depicts a modern pulverized coal-fueled electrical generation facility that uses the most commonly encountered Rankine-based thermodynamic cycle. This facility generates electricity by producing steam in a steam generator and expanding the steam through a turbine generator. The steam is then condensed in a condenser, and the condensed liquid is used again in the steam generator.

Chapter 2, Engineering Economics, presents the relationship between engineering design and economic costs and provides the designer with economic analysis concepts to identify the least-cost options among alternatives. Chapter 3, Thermodynamics and Power Plant Cycle Analysis, presents a method for analysis and evaluation of the modern Rankine-based cycle and procedures for a simplified heat balance analysis.

Chapter 4, Fossil Fuels, provides information on reserves, production, consumption, specific properties, and economics for oil, natural gas, and other major fossil fuels, including coal, which supplies over 50% of the fuel for electrical generation in the United States.

Coal is usually delivered to the facility by unit trains, although barges and trucks are also used. The coal handling system unloads the coal, then stacks, reclaims, crushes, and conveys it to storage silos near the steam generator. Coal and limestone handling system designs are discussed in Chapter 5. Coal is fed from the storage silos, pulverized to a powder, and blown into the steam generator. Within the steam generator, pulverized coal is mixed with air and combusted, and the combustion energy is used to produce steam. The characteristics of the combustion process, air requirements, flue gas quantities, and steam generator efficiencies are discussed in Chapter 6.

The steam generator produces, superheats, and reheats steam as it proceeds through the cycle. Chapter 7 discusses steam generator system and component design, steam temperature control, and pulverizer design.

The steam turbine generator, addressed in Chapter 8,

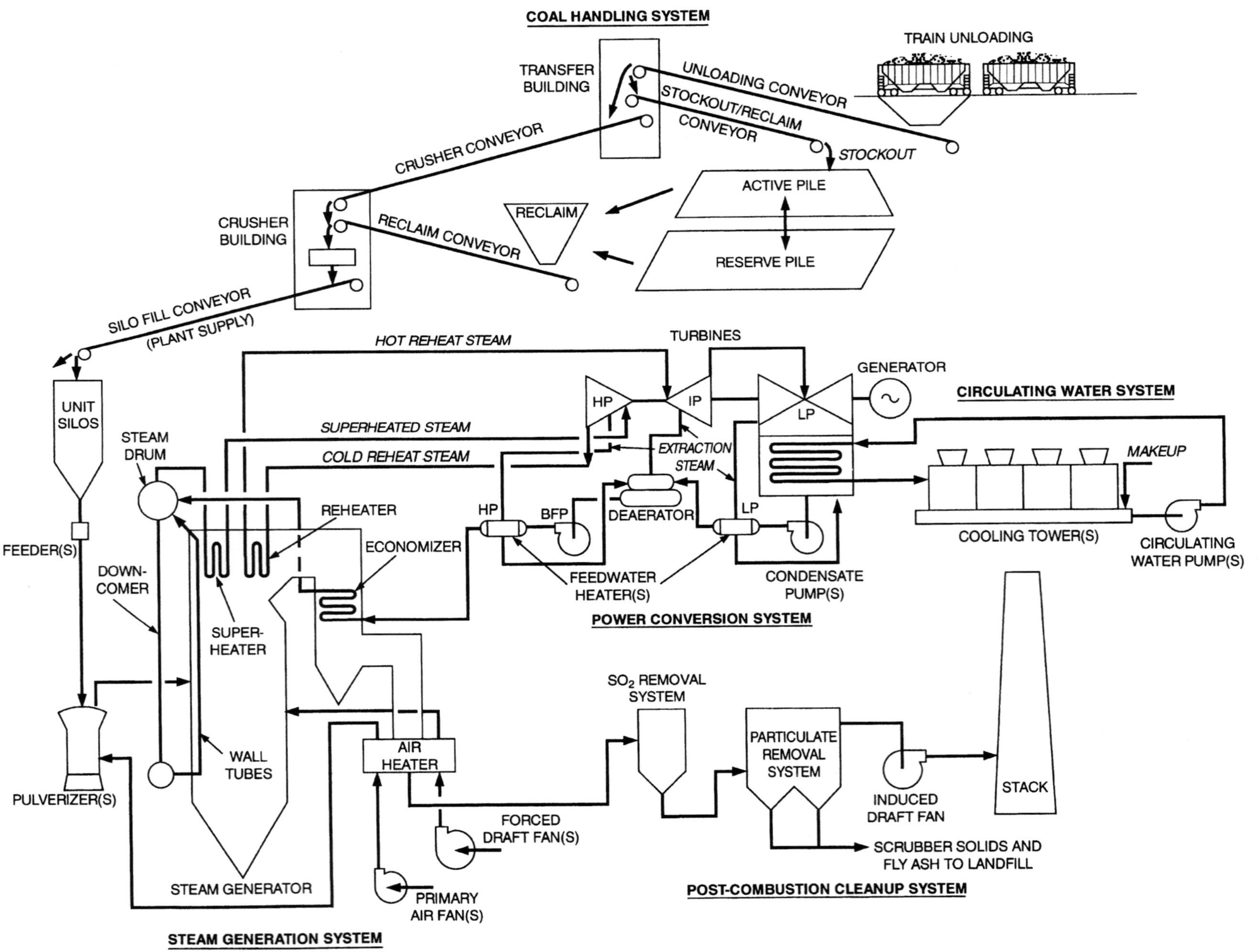

Figure 1-1. Modern pulverized coal fueled electrical generating unit.

converts the thermal energy of the superheated and reheated steam from the steam generator to electrical energy. The steam turbine converts the steam thermal energy to rotating mechanical energy, and the generator, which is coupled to the steam turbine, converts the mechanical energy to electrical energy.

Steam exhausted from the low-pressure section of the steam turbine is condensed to liquid in the condenser. The condensed liquid is moved from the condenser by condensate pumps through low-pressure regenerative feedwater heaters to a deaerator. Boiler feed pumps move the deaerated liquid through high-pressure regenerative heaters to the steam generator. Extraction steam from the steam turbine is supplied to the low- and high-pressure feedwater heaters for the liquid regenerative heating, which improves cycle efficiency. These steam cycle heat exchangers, which include the condenser, deaerator, and feedwater heaters, are discussed in Chapter 9. Design, performance, and arrangement requirements are addressed, as well as design parameters for steam turbine water protection.

Forced draft fans supply combustion air to the steam generator, and the primary air fans transport pulverized coal into the steam generator. Induced draft fans remove the flue gases from the steam generator and exhaust them to the plant stack. Chapter 10 discusses the specific fan design requirements for the combustion air and flue gas exhaust, including performance, sizing requirements, operating characteristics, and fan selection.

Just as fans move air, pumps move liquids through the plant. Chapter 11 focuses on pumps (primarily centrifugal pumps) used in power plant applications, addressing circulating water, condensate, and boiler feed pumps. Pump performance, sizing requirements, operating characteristics, and specific considerations for selection are addressed.

Cooling water for the condenser is supplied by the circulating water system, which takes the heat removed from the condenser and rejects it to cooling towers or another heat sink, such as a cooling lake. Chapter 12, Circulating Water Systems, addresses the circulating water system design, system and component design requirements, arrangements, performance requirements, material considerations, and performance optimization of cooling towers with the condenser.

Chapter 13, Cycle Performance Impacts, describes how steam cycle performance is affected by numerous design and operating parameters such as main steam and reheat steam pressure and temperature, steam turbine back pressure, number of feedwater heaters, feedwater heater design parameters, and superheat and reheat temperature spray flows. Steam cycle performance impacts are reflected in changes in steam cycle mass and energy balances. The use of a steam turbine generator thermal kit is also demonstrated.

Combustion gases exiting the steam generator require additional treatment for removal of fly ash particulates, sulfur dioxide, and nitrogen oxides by the plant pollution control systems before the gases are released through the plant exhaust stack. Environmental regulations included in the Clean Air Act of 1970 and subsequent 1977 and 1990 Amendments have established emissions limits for particulate matter, sulfur dioxide, and nitrogen oxides. Chapter 14, Power Plant Atmospheric Emissions Control, summarizes current air emission regulations, emission reduction technologies, and emission monitoring systems.

Chapter 15, Water Treatment, addresses the proper treatment and conditioning of water for the multitude of uses in power plant processes. This chapter provides basic information on water technology, principal types of treatment and conditioning equipment, and the more important design considerations and techniques applicable to steam power plants.

Coal-fueled power plants also require systems for collection, treatment, and disposal of liquid and solid wastes. Liquid wastes may include cooling tower blowdown, ash transport water, demineralizer regeneration chemicals, sulfur dioxide removal system wastewater, and coal pile runoff. Solid wastes include steam generator bottom and fly ash and combustion gas desulfurization solid wastes. Chapter 16 discusses the current federal and state regulations for liquid and solid wastes, as well as treatment and disposal systems.

Power plants not only produce energy; they also use electrical energy to control and power the various systems in the plant. Chapter 17, Electrical Systems, addresses some of the important design considerations for the auxiliary electrical systems that power the coal handling, feedwater, condensate, circulating water, combustion air, sulfur dioxide removal, and particulate removal systems. Chapter 17 also describes the transmission systems used to transmit electrical energy to the end user.

The plant control system enables the operators to direct plant operation for the reliable and efficient production of electrical energy. Chapter 18, Plant Control Systems, presents an overview of the plant control system, its functions, and the type of control equipment used in a modern coal-fueled electrical generating facility.

Chapter 19, Site/Plant Arrangement, discusses requirements for site and plant arrangements. Site arrangements show the locations of major power plant facilities, utilities and facilities entering and exiting the site, the interconnections of major areas on the site, and traffic patterns. Plant arrangements define the physical arrangement of the buildings on the site and the significant equipment within the buildings; locations and spacings of support columns; and locations of walls, floors, doors, etc.

Although coal is the source of fuel for over 50% of the electrical generation in the United States, natural gas is increasingly used with combustion turbines, in either simple or combined cycle applications. Natural gas, used in a modified Brayton thermodynamic cycle (Chapter 3), supplies about 10% of the electrical generation in the United States. Chapter 20, Combustion Turbines, discusses combustion turbine and combined cycle technology, its advantages, and current system design requirements.

Fluidized bed combustion has emerged as a feasible alternative to pulverized coal steam generators. A solid bed of fuel (such as coal) is fluidized by an upward flow of air at a velocity sufficient to expand the bed. Combustion takes place within the bed under either atmospheric or pressurized conditions. Chapter 21, Fluidized Bed Combustion, discusses this technology, its advantages and disadvantages in comparison to pulverized coal designs, and the current fluidizing bed approaches (bubbling bed and circulating bed). Important system design requirements, supporting equipment, steam temperature control, and fuel requirements are also addressed.

Chapter 22, Resource Recovery, describes the use of municipal and agricultural solid waste as fuel for electrical energy production. Technologies that convert waste to electrical energy enable resource recovery of municipal wastes, landfill volume reduction, material recycling, and energy recovery by burning waste in steam generators to produce electricity. The chapter discusses these technologies, basic designs, commercial availability, fuels, and cost and performance data.

Nuclear power, discussed in Chapter 23, supplies about 20% of the electricity generated in the United States and 19% of the electricity generated worldwide. Current boiling water and pressurized water designs are discussed, including the supporting fuel cycle and radioactive waste management. Future designs for large revolutionary reactors and smaller passive reactor designs are presented as well.

No textbook on power plant engineering would be complete without a discussion of future or emerging technologies for electrical energy production. Chapter 24, Emerging Technologies, presents this evaluation. Technologies are presented for fossil fuel (gasification combined cycle, magnetohydrodynamics, and fuel cell); renewables (solar, wind, biomass, ocean, and geothermal); energy storage (battery and compressed air); and nuclear (including fusion).

The remaining two chapters tie the design process together. Chapter 25, Power Plant Planning and Design, presents the planning and design process, which varies with specific financial, engineering, environmental, and other plant requirements. This process may involve special studies, such as fuel supply, system planning, siting evaluations, transmission planning, and other analyses, early in the project development. Design engineering includes conceptual design, which supports permitting and licensing activities, and detailed design, which determines the technical requirements for all plant systems and components and supports equipment procurement and construction specifications. Essential project control activities include critical path scheduling of engineering and plant construction, cost control, design control, and construction control.

Chapter 26 discusses essential permits and approvals and the requirements for satisfying the environmental review processes applicable to construction and operation of an electrical generation facility, a critical first step. Chapter 26 gives generic descriptions of the permits and approvals typically required for fossil-fueled units in the United States.

Power Plant Engineering is the result of over 80 years of successful utility engineering by Black & Veatch and more than 50 years of power plant engineering, design, and construction worldwide. This book may be the most complete reference source available, offering information that is practical and proven in the field. This source is intended to become the standard in the professional engineer's library as the primary source of information on steam power plant generation systems.

EDITOR'S NOTE ON THE STYLE OF THIS BOOK

As stated previously, this text was authored and supported by many experienced Black & Veatch engineers and specialists. In all cases, the styles of the individual authors were retained as much as possible, and no attempt was made to reconcile the diversity of approaches. Every effort has been made to use engineering codes and standards that are current as of publication. Readers should consult these publications directly when using information for design purposes.

2

Engineering Economics

John M. Wynne

2.1 INTRODUCTION

Understanding and maintaining the proper relationship between engineering design and economic costs should be an integral part of any engineering design project, and the engineer must have some understanding of economic concepts to make proper design decisions. This chapter discusses that relationship and introduces economic concepts that will enable the engineer to identify the least-cost option among alternatives. Centered on various net present value methodologies, the chapter is written for those who have little or no formal background in finance, yet the material advances quickly enough to be of value to those already familiar with utility economics. Most readers will find it beneficial to work through the chapter's many examples and problems, as the concepts are best understood through application.

2.2 ROLE OF ECONOMIC EVALUATION IN POWER PLANT DESIGN AND PLANNING

2.2.1 Economics of Public Utilities

Power plant design and system planning require consideration of both engineering design and economics. Engineers are usually aware of the need for reliable design, but are often less conscious of the need for cost-effective design. Cost minimization is discussed in this section from the perspective of the regulated utility, government regulators, nonutility generators, and the design engineer.

2.2.2 The Electric Utility's Market Structure

Historically, the generation, transmission, and distribution of electricity have been considered a natural monopoly. Because of the economies of scale, a monopolist can produce and deliver electricity at a lower average cost than can numerous smaller firms. For example, if several electric utilities serve a single area, multiple rows of transmission lines may be present, although, in the absence of competition, a single line from one firm may be able to serve all customers at a lower cost. Similar inefficiencies from unnecessary duplication and low utilization of facilities can be seen on the generating side when numerous suppliers exist in a natural monopoly market. Since ratepayers want to purchase electricity at the lowest possible cost, it is preferable to allow a single efficient firm to supply electricity.

On the other hand, economic theory and historical experience indicate that an unrestrained monopolist can make excess profits by reducing output and raising price. This presents a dilemma in that such a pricing strategy can offset the economy of scale benefits. From a public interest standpoint, it is desirable to allow an optimally sized monopolist to produce output efficiently, yet control the price of electricity.

2.2.3 Regulation of Public Utilities

The federal government has determined that regulation of some monopoly industries is appropriate. In the 1877 landmark case Munn vs. Illinois, the Supreme Court ruled that monopoly industries affected with a public interest can be subjected to government regulation. The concept of public interest generally refers to an industry producing a product that is an indispensable commodity or practically a necessity. In most states, firms supplying electricity, gas, water, wastewater disposal, and telephone services are under rate regulation.

2.2.4 The Regulatory Compact

Under what is known as the regulatory compact, a public utility is charged with specific responsibilities and given special rights not realized in unregulated markets. A utility has the responsibility to provide adequate, safe, and reliable service to all customers who can pay for the product. It is also obligated to provide service at reasonable rates, without undue discrimination, and must obey regulatory commission orders, which have the effect of law.

In return, a utility is granted an exclusive service territory, the right to charge reasonable prices, and an opportunity, though not a guarantee, to earn a fair return on investment, among other rights.

The compact's call for reliable power at reasonable rates provides the link between engineering and economics. Thus, although a regulated utility is allowed to recover legitimate costs from ratepayers, it has a mandate to minimize these costs while providing a reliable supply of electricity.

2.2.5 Regulatory Review

While the regulatory compact alone mandates cost-effective power plant design, the regulatory environment since the mid-1970s has made cost minimization increasingly important. Beginning in the 1970s, rapid electricity price escalations brought about by cost overruns, excess generating capacity, high interest rates, long construction lead times, and rapid fuel price increases all contributed to a general dissatisfaction with the regulatory environment. Between 1970 and 1985, for example, the average price of electricity on a kilowatt-hour basis increased nearly fourfold from 2.10 cents to 8.37 cents for the residential sector (Phillips 1988), while the Consumer Price Index (CPI) increased less than threefold. Another revealing statistic is that between 1980 and 1988, the average total cost of United States steam plants entering commercial operation increased from $504/kW to $3,273/kW (Anderson 1991). Although many of the causes of the rapid price escalations were beyond the control of utilities, regulators came under increasing pressure to develop a more efficient industry. State commissions generally responded by becoming increasingly active in the review of utility decisions and operations.

Most state commissions now translate the regulatory compact's reasonable rate requirement to mean an electric utility must provide power in the least-cost manner and make prudent decisions. Many states, through a Certificate of Public Convenience and Necessity law, require prior approval of generating facility construction and require a utility to demonstrate that a proposal is the least-cost alternative. Even more imposing are the provisions of the Energy Policy Act of 1992, which require state public utility commissions to consider mandating utilities to file an integrated resource plan which includes the consideration of cost-effective demand-side programs as well as supply-side options.

Commissions may also subject a utility to prudence reviews. Findings of imprudence or that generating capacity is not used and useful can result in cost disallowance (costs cannot be recovered from the ratepayer) or deferred cost recovery. In the previous decade, excess capacity, construction of nonoptimal generation resources, unjustified cost overruns, and unreliable capacity have all been bases for disallowances. During the 1980s, between 8% and 15% of final plant costs were disallowed (Phillips 1988). In many cases, this resulted in reduced or eliminated stock dividends, bond deratings, lower stock prices, and other adverse financial consequences.

To the extent that engineering design and planning are subject to regulatory review, engineers must plan and design to meet utility needs cost-effectively. Anything less is not consistent with the regulatory compact, and leads to higher than necessary ratepayer costs and, possibly, utility cost disallowances.

2.2.6 Competitive Generation

Utilities now face even greater pressure to design cost-effective power plants because of competition in the form of nonutility generators (NUGs), which have become a significant component of the United States power supply. In addition, exempt wholesale generators (EWGs), as created in the Energy Policy Act of 1992, and the advent of retail wheeling, whereby utility customers who have the ability to choose their electricity supplier, will result in unprecedented competition for the supply of future generation capacity.

The emergence of the NUG industry has occurred largely as a result of the rapid escalation of electricity prices in the 1970s, which fostered the view that regulation was failing to secure an efficient supply of electricity and was a poor substitute for competition. Thus, instead of maintaining the traditional view of the generation market as a natural monopoly, many came to believe that competition in the electricity generation market would increase efficiency and lower the costs to consumers.

The call for a more efficient electric utility industry resulted in the passage of the Public Utilities Regulatory Policies Act (PURPA) by Congress in 1978. Under Section 210 of PURPA, a utility is required to purchase electricity from certain nonregulated power producers, termed qualifying facilities (QFs), at a rate that does not exceed the utility's avoided cost (the capacity and energy costs that the utility would have incurred had it not made the purchase from the QF). The rules implementing PURPA required that a QF be either a cogeneration facility meeting certain efficiency requirements, or a small power producer (80 MW or less) whose energy input was primarily from waste, biomass, or renewable resources (the size limitation has since been removed).

For the franchised electric utility that prefers to supply its own capacity, a low-cost power plant design that results in a low avoided cost is necessary to ensure that the utility will be the incremental capacity supplier. Conversely, since PURPA provides a market for QF-produced electricity but only at or below a utility's avoided cost, the QF has a financial incentive to design a generating facility that can produce electricity below the utility's avoided cost.

NUGs can also be independent power producers (IPPs). An IPP is an owner of a wholesale generating facility that is permitted under the Federal Power Act to charge market-based prices for electricity. Unlike the QF, there are no fuel or efficiency constraints placed upon the IPP. Further, an IPP is not generally regulated at the state level since, unlike a franchised electric utility, it lacks market power. Although an IPP does not face some of the technical restrictions placed on a QF, a utility is not obligated to purchase power from IPPs. Therefore, the IPP project must meet the utility's criteria for lowest cost incremental supply. Thus, as with the QF, a potential IPP has tremendous incentive to design and construct a low-cost, reliable, generating facility.

The price competition between utilities and NUGs is most dramatic where potential suppliers of capacity participate in a bidding program for the right to supply incremental capacity needs. Typically, a bidding program is initiated when a utility issues an RFP (Request for Proposal), which provides information on the type and amount of capacity needed. Once bids are received and evaluated, the most cost-effective bid demonstrating viability and reliability is generally contracted to supply the capacity.

An EWG is a special type of IPP created through Title VII of the Energy Policy Act of 1992 (Act). This new type of electricity producer was created to eliminate a number of impediments to IPP development caused by the provisions of the Public Utility Holding Company Act (PUHCA). PUHCA placed restrictions on IPP ownership (particularly by public utilities). These restrictions were severe enough to curtail significantly the expansion of the IPP industry.

Under the Act, utilities, holding companies, and utility subsidiaries may be EWGs and, like IPPs, are allowed to charge market-based rates for their power. However, some restrictions apply. For example, EWG status is not permitted for facilities already in a utility's rate base when the Act became effective unless it is approved by jurisdictional commissions, is in the public interest, and is consistent with state law. Also, to prevent the potential for self-dealing, the Act restricts purchases by an electric utility from an affiliated EWG unless the jurisdictional commissions determine in advance that transaction benefits consumers and does not violate state law. Despite these and other restrictions, the creation of EWGs removed much of the legal and bureaucratic restrictions for IPP development. As a result, the number of potential suppliers for incremental generating capacity has dramatically increased and this increased competition will flow through to consumers in the form of lower energy costs.

In the United States, the most recent but potentially most significant impetus driving the industry toward increased competition is the advent of retail wheeling. Retail wheeling would generally allow consumers to shop among power suppliers for the most attractive rate and would require the traditional franchised utility to transport this power to the customer. Although many of the details of retail wheeling are still being developed, it is remarkable that in only a few years the image of retail wheeling has evolved from that of an interesting but probably unworkable concept to a viable means to minimize power costs. Although some state commissions are taking a wait and see approach toward retail wheeling, a number of other states such as California and Michigan are spearheading the drive toward implementation of retail wheeling and, it is hoped, benefits for all power consumers. Although it is too early to tell whether all power customers can gain from retail wheeling, the concept has forced utilities to recognize they are in a truly competitive industry.

The increased competition in the electricity generation market means that more than ever before, cost considerations will determine who designs and builds the industry's next generation of capacity. For a utility, NUG, or EWG attempting to be the supplier of incremental capacity, cost will be a major, if not the primary, determinant of who provides that power.

2.2.7 Engineering Design Firm

Cost-effective power plant design is also vital for the engineering design firm. Firms and engineers that have the ability and reputation for designing reliable, low-cost power plants and systems will be sought out in the increasingly competitive environment. Conversely, if a client utility suffers disallowances or cost overruns due to poor planning or design, the engineering firm may well lose a client.

Many engineering design firms have been active also in the nonutility generation market as designers and owners of potential NUGs and EWGs. The success of these facilities will be directly dependent on cost, since many utilities have a preference for building their own capacity unless there is clearly a reliable and lower cost alternative. The emergence of turnkey projects, in which the turnkey contractor is bound to cost and performance guarantees, has also meant that the turnkey contractor must be able to supply a superior product at a low and quantifiable price.

2.2.8 Future Needs

While the electric utility industry continues to move toward increased competition and cost consciousness, there are enormous opportunities for the engineers and firms who are able to compete. In addition to the new and refurbishment opportunities in the United States, there are also unprecedented international opportunities due to rapid growth in the demand for electricity and the need for design expertise. For example, the Asian and Pacific Rim countries of the Philippines, Indonesia, Malaysia, Thailand, Singapore, and even China are seen as areas of great opportunity over the next two decades. Other regions such as eastern Europe and Central and South America are also expected to experience a dramatic increase in capacity additions during this time frame and beyond.

2.3 ECONOMIC AND FINANCIAL CONCEPTS

The economic analysis methods introduced in this chapter are useful in several types of present value studies that evaluate the cost and (for nonregulated firms) the financial returns of an alternative. These studies typically include the following:

- *System analyses*, which evaluate the costs of design projects and are used to select the preferred design of major and auxiliary systems,
- *Bid analyses*, which evaluate alternative proposals for equipment or services,
- *System planning studies*, which determine the optimum generation plant or transmission additions for the utility system,
- *Nonutility analyses*, which evaluate the financial feasibility of a proposed nonutility project.

Table 2-1. Compounding and Future Values

Period (A)	Beginning Balance (B) ($)	×	Interest Rate (C)	=	Interest Earned (D) ($)	End of Period Value (E) ($)
1	1,000.00	×	0.05	=	50.00	1,050.00
2	1,050.00	×	0.05	=	52.50	1,102.50
3	1,102.50	×	0.05	=	55.13	1,157.63
4	1,157.63	×	0.05	=	57.88	1,215.51

The process of selecting the least-cost alternative is often complex; most alternatives have different design and performance characteristics as well as different costs at various times. This section develops the present worth arithmetic needed to identify the most economical alternative. Some numerical examples in this chapter contain round numbers. For simplicity, it is assumed all alternatives in this chapter are comparable in terms of reliability and the service provided. It is noted that actual recovery of utility costs through the regulatory accounting framework differs somewhat from the methods introduced in this chapter. This is because the concern of this chapter and of engineering economics is one of resource allocation, that is, selection of the least-cost resource. The intricacies of regulatory accounting are not addressed in this chapter. Finally, many of the terms introduced in this chapter are included in the glossary at the end of this chapter.

2.3.1 Time Value of Money

Power plant engineers are interested in the least-cost plant, system, or component among acceptable alternatives. Evaluation of an alternative's cost must not only measure the total installed cost, but also must consider the time value of money. Simply stated, the concept of the time value of money is that a dollar received in the future is worth less than if it were received today. Just as an individual would have financial incentive to take payment of $100 today rather than in 10 years (since interest could be earned on the payment, the $100 would hve grown to a larger amount 10 years from now), it is also appropriate to consider the time value of money when evaluating a utility's alternatives. This is appropriate because utility costs are ultimately paid by ratepayers who have a time value of money, and the regulatory compact requires a reliable supply of electricity at the minimum cost to ratepayers.

How significant is the time value of money? In 1803, France sold the 529,920,000 acres of the Louisiana Territory to the United States for $27,267,622; or about 5 cents per acre. At first glance, this seems a ridiculously low price if compared to the land that amount would buy at today's prices. But, consider the time value of money. If France had invested the $27 million, earned a 10% return annually, and kept all principal and returns invested from 1803 to 1992, France would have been able to offer a repurchase price of $3.4 million per acre in 1992; up considerably from the 5 cents per acre it originally received. If a 12% interest rate was realized, France could have offered $103.2 million per acre. This example demonstrates the significant impact of the time value of money on total cost; only by considering the time value of money can proper economic conclusions and choices be made.

2.3.2 Future Values

A fundamental time value measurement is the determination of a future value through compounding. Compounding involves the application of a rate of growth for more than one period. For example, if $1,000 is deposited in a savings account earning 5% interest annually, the future value of the $1,000 in 1 year would be $1,050 ($1,000 initial deposit + $50 interest). Now assume that the initial deposit, or principal, plus all interest remained in the savings account for 4 years. Would the amount in the account after the fourth year equal $1,200 ($1,000 + $50 interest for four periods)? The answer is no, because of compounding. As seen in Table 2-1, if the $1,050 in principal and interest after year 1 is left in the account, during the second year, interest is earned not only on the original principal of $1,000 but on the accumulated interest of $50 as well. As a result, at the end of the second year, the account value is $1,102.50. Through the compounding process, the value of the savings account after 4 years is $1,215.51.[1]

Throughout this chapter, formulas are presented to facilitate time value calculations. Without these formulas, calculations would be very time consuming and subject to error. Equation (2-1) is used to determine future values. A listing of all time value equations presented in this chapter is included in Appendix 2D.

$$FV_n = PV(1 + i)^n \tag{2-1}$$

where

FV_n = future value in period n,

PV = present value,

i = cost of money rate per period (can also represent the escalation rate),

[1]Most of the amounts shown throughout this chapter have been rounded as appropriate to illustrate the point of the calculation.

n = number of compounding periods, and

$(1 + i)^n$ = future value factor (FVF).

The future value of a present sum can be found by multiplying the present value by the future value factor (FVF). Applying Eq. (2-1) to the above example, the future value after year 3 is the following:

$$FV = PV \times FVF$$

$$FV = 1{,}000 \times (1 + 0.05)^3 = 1{,}000 \times 1.157625$$

$$FV = \$1{,}157.63$$

After year 4, the future value is $1,215.51 ($1,000 × 1.21551), which confirms the consistency of Eq. (2-1) with the results in Table 2-1.

2.3.3 Compounding and Escalation Rates

In most real-world situations, future costs and expenditures are subject to price increases or escalations brought about by inflation (an increase in the general price level) and often by other supply and demand conditions (real effects) such as resource depletion or changes in production technology. As Eq. (2-2) indicates mathematically, the total escalation rate is equal to the product of the inflation and real escalation rates, minus 1.

$$\text{TER} = (1 + \text{inflation rate})\ (1 + \text{real escalation rate}) - 1 \quad (2\text{-}2)$$

where

TER = total escalation rate.

For example, given an expected inflation rate of 5% and an expected real escalation rate of 1.5%,

$$\text{Total escalation rate} = (1.05)\ (1.015) - 1 = 0.06575 \text{ or } 6.575\%$$

Future prices and costs are calculated by the application of the (total) escalation rate to the current cost. This is nothing more than a compounding problem in which i represents the escalation rate. For example, suppose a utility wanted to determine the cost of a boiler part in 5 years, given a current cost of $2,000 and an expected total escalation rate of 10% per year. Using Eq. (2-1), the expected cost in 5 years is $3,221.02 ($2,000 × 1.61051). This future cost estimate, which takes inflation and real effects into account, is said to be stated in actual, current, or nominal dollars and reflects the actual out-of-pocket cost that one would expect to pay in a particular year.

Occasionally, it is necessary to escalate costs on a monthly rather than on an annual basis. The monthly escalation rate is calculated according to Eq. (2-3).

$$\text{Monthly escalation rate} = (1 + \text{Annual escalation rate})^{1/12} \quad (2\text{-}3)$$

For example, a component costing $1,000 currently, escalated for 2 months at an annual escalation rate of 10%, would be equal to the following:

$$\text{Escalated cost} = \$1{,}000\ (1 + 10)^{3/12}$$

$$= \$1{,}000\ (1.01601) = \$1{,}016.01$$

Using this formula to compute the cost after escalation for 12 months yields $1,100, which equals the annual escalation rate times the original cost.

When a study requires the application of an escalation rate, the assumed value of the inflation and real escalation rate components requires careful consideration since escalation directly affects an alternative's cost. For some studies, it may be appropriate to assume that real escalation effects will be zero or near zero, in which case the total escalation rate equals or approaches the expected inflation rate. On the other hand, estimates of future costs for items that have varied widely from the general inflation rate historically may require real escalation effects considerably different than zero. When decisions are being made on the value of the real escalation rate, the real historical price trends and their underlying causes should be identified. The likelihood that such factors or new influences will be present in the future must be also considered. In many cases, it may be prudent to estimate real escalation rates through an economic forecasting model.

In some studies, it is necessary to evaluate prices and costs independent of the effects of inflation. This type of analysis is said to use real or constant dollars. Constant dollar evaluations are primarily limited to forecasting models (demand for electricity, for example, may be modeled econometrically as a function of the real income) or in evaluating long-term price trends (such as the real price of coal over time). Unless otherwise directed, engineering evaluations of alternatives should use current dollar analysis; that is, apply the total escalation rate to determine future prices and costs.[2]

2.3.3.1 Future Value/Escalation Guidelines. For future values and escalations, the following can be generalized:

1. Other things being equal, the longer the compounding period, the greater the future value;
2. Other things being equal, the higher the interest rate or escalation rate, the greater the future value;
3. Other things being equal, the larger the principal or present value, the greater the future value;

[2]The principal reason is to ensure consistency. The present worth discount rate, which is usually assumed to equal the utility's weighted cost of capital, includes an allowance for expected inflation. If the real escalation rate was used to estimate future costs, the present worth discount rate would have to be adjusted to reflect the cost of capital in the absence of inflation in order to be consistent.

4. Unless otherwise stated, all future costs should be calculated as the current cost times an escalation rate. In other words, it should not be assumed that an item's future cost will equal the present cost unless special circumstances clearly justify such an assumption.

Work Problems (Answers to work problems can be found in Appendix 2A.)

Problem 1.—Compounding

1A. Assume that a utility invests $100,000 of its earnings and realizes a 9% return annually. What is the future value of the principal plus interest in 3 years?

1B. Would this future value be higher or lower if an 8% return was earned?

1C. If the compound period was 4 years?

1D. If the initial investment was $99,999? Calculate the future value for each scenario.

Problem 2.—Escalation

2A. Assume that a boiler component currently costs $3,800. If its price increases at 6% per year, what will the component cost in 10 years?

2B. Confirm the answer in 2A by calculating the escalated cost on a yearly basis.

2.3.4 Present Values

Another fundamental time value concept is present values, which are really the opposite of future values. Present value calculations deal with the notion that a sum in the future is worth less today than the future amount. For example, earlier it was determined that the future value of $1,000 compounded at 5% annual rate for 4 years is $1,215.51. This can be reversed to state the present value of $1,215.51 received 4 years into the future, with a 5% interest rate or discount rate is worth $1,000 today. How do we know this to be true? Because, as demonstrated in Section 2.3.1, $1,000 (a current or present value) invested at a 5% per year interest rate would equal $1,215.51 in 4 years. To determine the present value of a lump sum, Eq. (2-4) is used.

$$PV = FV_n \times \frac{1}{(1 + i)^n} \tag{2-4}$$

where

PV = present value,

FV_n = future value in period n,

i = Discount rate per period,

n = Number of periods, and

$\frac{1}{(1 + i)^n}$ = present worth factor.

The present value can be calculated by multiplying the future value by the present worth factor (PWF). Using Eq. (2-4), the present value of the $3,221.02 boiler part presented in Section 2.3.3, discounted 5 years at a 10% annual discount rate is as follows:

$$PV = FV \times PWF$$
$$PV = \$3{,}221.02 \times 0.620921$$
$$PV = \$2{,}000$$

2.3.4.1 Discount Rate. In most utility analyses, the discount rate is set equal to the utility's weighted cost of capital, that considers the cost of debt (bonds) and equity (preferred and common stock). For public utilities that finance by issuing only debt, the discount rate is usually assumed to equal the cost of debt. Table 2-2 calculates the weighted cost of capital given the utility's component costs of capital.

2.3.4.2 Present Value Guidelines. The following generalizations can be made about present value calculations:

1. Other things being equal, the larger the future value, the larger the present value;
2. Other things being equal, the smaller the discount rate, the larger the present value;
3. Other things being equal, the greater the number of discounting periods, the smaller the present value;
4. In most utility analyses, the present worth discount rate is set equal to the utility's weighted cost of capital.

Work Problems

Problem 3.—Compounding and Present Value

In Problem 2, the cost of a boiler component costing $3,800 currently and escalated at 6% annually for 10 years was calculated to be $6,805.22.

Table 2-2. Weighted Average Cost of Capital

Component	Capitalization Ratio (%)	×	Annual Cost (%)	=	Weighted Average Cost (%)
Debt	50	×	9	=	4.5
Preferred stock	10	×	10	=	1.0
Common stock	40	×	14	=	5.6
Weighted average cost of capital					11.1

3A. If a utility wanted to set aside an amount currently that would equal the boiler component cost in 10 years, how much should it set aside, given a 10% annual discount rate?

3B. Assuming the utility knows with certainty that it will need the component in 10 years and that there are no other risks or relevant costs (such as obsolescence or storage costs), should the utility purchase the component currently or set aside a lump sum and purchase the component in 10 years?

3C. If the escalation rate for the cost of the component and the interest rate earned by the utility were reversed, would the recommendation be the same?

2.3.5 Uniform Series (Annuities)

Although future and present values form the basis of economic evaluations, additional present value concepts better reflect real-world situations. Utilities seldom pay for large capital expenditures in one lump sum. Instead, equal periodic payments are often made over time; this involves an annuity or a uniform series.

2.3.6 Present Value of a Uniform Series

Frequently, an engineer must determine the present value of a number of equal cost payments called a uniform series of costs. Conceptually, this involves determining the present value of each uniform payment, then summing each present value to arrive at the present value for the entire uniform series. Consider a utility purchasing several fleet vehicles. The utility is offered two payment options. Option A requires a \$30,000 payment at the end of each year for the next 4 years. Alternatively, Option B requires \$39,000 payments at the end of each of the following 3 years. What is the utility's least costly option if it has an 8% annual discount rate?

Since the time value of money must be considered, it should not be assumed that Option B is the least costly even though the actual dollar outlay of \$117,000 is less than under Option A (\$120,000). The present value calculated for Option A is illustrated in Fig. 2-1 using Eq. (2-4).

The present value of Option A is calculated as the sum of the present value of each periodic payment and equals \$99,363.81. What does this figure mean? If a \$99,363.81 lump sum was set aside in period zero and earned 8% interest per period, the lump sum would be just enough to make the four \$30,000 periodic payments when due.

To determine which option is the least costly, the present value of Option B must also be determined. Fortunately, the present value of a uniform series can be calculated quickly through Eq. (2-5).

$$PV_{US} = US \times \frac{(1+i)^n - 1}{i(1+i)^n} \tag{2-5}$$

where

PV_{US} = present value of a uniform series,

US = uniform series payment or receipt,

i = discount rate per period,

n = number of periods, and

$\frac{(1+i)^n - 1}{i(1+i)^n}$ = uniform series present worth factor (USPWF).

The present value of a uniform series can be found by multiplying the uniform series payment by the USPWF. Using Eq. (2-5), the present value for Option A equals \$99,363.81. Using Eq. (2-5) for Option B yields the following:

$$PV_{US} = US \times USPWF$$

$$PV_{US} = \$39{,}000 \times 2.57710$$

$$PV_{US} = \$100{,}506.78$$

Having stated the present value of the two uniform series, it is now possible to compare costs on an equal basis. On a present value basis, Option A represents the lower cost option and this method of payment should be selected. Notice the effect of the time value of money. If one looked only at the total actual dollar payments, Option B would appear to be the least-cost option, yet Option A is the true least-cost alternative when the time value of money is considered.

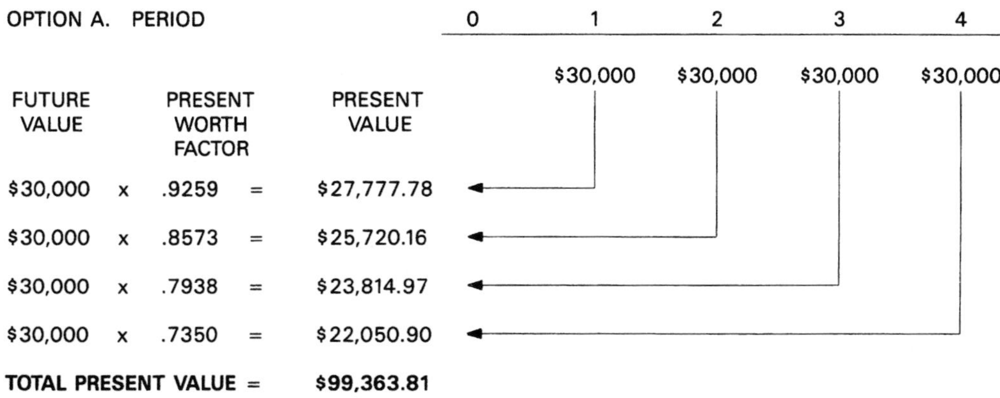

Figure 2-1. Present value of a uniform series.

This example illustrates the general approach that should be taken to decide among alternatives. Since most alternatives have different costs and timing of payments, the present value cost of each alternative must be determined, and the option with the lowest present value is the least-cost option. In the utility industry, this approach is consistent with the least-cost planning requirement. It results in the selection of the alternative that has the lowest present value cost (would require the smallest present value dollar outlay) to ratepayers.

2.3.6.1 Uniform Series Guidelines. In the above example, it was assumed that uniform series payments were made at the end of the year. The end-of-year convention for payments and expenses is a standard assumption in utility financial analysis and should generally be used. It is also standard to assume all investments are made at the beginning of the year.

Work Problems

Problem 4.—Present Value of a Uniform Series

4A. Determine the present value of 30 annual payments of \$10,000, each given a discount rate of 12% per year.

4B. How does the present value of the uniform series compare to the actual dollar outlay for all payments?

4C. What does the present value figure mean and why is it useful?

2.3.7 Uniform Series Equal to a Present Value

A second uniform series calculation involves a uniform series equal to a present value. This requires the determination of a constant amount payable at the end of each year such that the present value sum of all payments equals the present amount. Consider a home mortgage in which a lump sum is borrowed and repaid through equal monthly payments over the next 30 years. This is a process of establishing a uniform series (mortgage payment) equal to a present value (the amount of the loan). Equation (2-6) is used to determine the uniform series equal to a present value.

$$US_{PV} = PV \times \frac{i(1 + i)^n}{(1 + i)^n - 1} \qquad (2\text{-}6)$$

where

US_{PV} = uniform series equal to a present value,

PV = present value,

i = interest rate per period,

n = number of periods, and

$\frac{i(1 + i)^n}{(1 + i)^n - 1}$ = capital recovery factor (CRF).

The uniform series equal to a present value is calculated by multiplying the present value by the capital recovery factor (CRF). For example, if a utility obtains a 30-year, \$100,000 loan at a 10% interest rate to finance a replacement part, the utility will need to make yearly payments of \$10,607.92 to liquidate the loan at the end of 30 years:

$$US_{PV} = PV \times CRF$$

$$US_{PV} = \$100{,}000 \times 0.10608$$

$$US_{PV} = \$10{,}607.92$$

Work Problems

Problem 5.—Uniform Series Equal to a Present Value

5A. Assume a utility purchases 100 acres of land for a total of \$10 million. If the utility finances the purchase with a 20-year, 8% loan, what is the annual payment to liquidate the loan in 20 years?

5B. Calculate the annual payment if a 30-year loan was secured.

5C. What would the total actual dollar outlay be under 5A and 5B?

2.3.8 Levelized Values

Utilities also incur nonuniform costs, which are costs that vary from year to year. Equations (2-5) and (2-6) cannot be used for nonuniform present value calculations, so another technique is required. One common method is to levelize nonuniform costs. Levelization refers to the process of converting a series of nonuniform costs into a uniform series. Both series have the same present value; therefore, economically correct decisions may be made on the basis of levelized cost comparisons. Each uniform payment is called the levelized annual cost or the equivalent uniform annual cost.

Levelized annual costs are determined by a three-step process: (1) calculate the present worth of each annual cost, (2) sum the present worth of each annual cost to obtain the total present worth, and (3) divide the total present worth by the sum of the present worth factors. When two alternatives are being evaluated, the alternative with the lowest levelized cost has the lowest present value, provided the time frame and discount rate are the same for each alternative. The three-step levelization process is summarized in Eq. 2-7.

$$\text{Levelized annual cost} = \frac{\text{Total present worth of all annual costs}}{\text{Sum of present worth factors}} \qquad (2\text{-}7)$$

Consider a nonuniform series of 10 annual payments that begins with a \$20,000 payment in the first year and increases each year by \$5,000. Assuming a 10% discount rate per year, what is the levelized value of the nonuniform payments?

Following the three-step levelization process, the present worth of each annual cost is first determined in Column C in

Table 2-3. Determination of Levelized Costs

	Nonuniform Annual Costs			Uniform Annual Costs	
Year	Annual Cost (A) $)	Present Work Factor[a] (B)	Present Worth of Annual Cost (C) ($)	Levelized Cost (D) ($)	Present Worth of Levelized Cost (E) ($)
1	20,000	0.90909	18,181.82	38,627.30	35,115.73
2	25,000	0.82644	20,661.16	38,627.30	31,923.39
3	30,000	0.75131	22,539.44	38,627.30	29,021.26
4	35,000	0.68301	23,905.47	38,627.30	26,382.97
5	40,000	0.62092	24,836.85	38,627.30	23,984.51
6	45,000	0.56447	25,401.33	38,627.30	21,804.10
7	50,000	0.51315	25,657.91	38,627.30	19,821.91
8	55,000	0.46650	25,657.91	38,627.30	18,019.92
9	60,000	0.42409	25,445.86	38,627.30	16,381.75
10	65,000	0.38554	25,060.31	38,627.30	14,892.50
Total		6.14457	237,348.05		237,348.05

[a]Present worth factor [from Eq. (2-4)] $= \dfrac{1}{(1+i)^n}$

Levelized annual cost $= \dfrac{\text{Total present worth}}{\text{Sum of present worth factors}} = \dfrac{237{,}348.05}{6.14457} = \$38{,}627.30$

Note: Uniform series present worth factor (USPWF) for 10 years, 10% = 6.14457

Table 2-3. These present worths are then summed to arrive at the total present worth of $237,348.05. Finally, the total present worth is divided by 6.14457, the sum of present worth factors (Column B). The resulting annual levelized cost is $38,627.30.

As shown in Columns D and E, the 10 annual levelized cost figures have a present value of $237,348.05, as do the nonuniform payments in Column C. In general, it is a good practice to check the equivalence of the levelized cost and nonuniform series present values.

At first glance, levelization may seem burdensome since economically correct choices can be made by comparing the total present worth of alternatives (the first step of levelization shown in Column C in Table 2-3). However, a levelized cost figure is often more intuitive than a single large present worth figure and is usually preferred by those unfamiliar with present worth arithmetic.

In most studies, the present worth discount rate is assumed to be constant each year of the study period. In such cases, the sum of the annual present worth factors is equal to the USPWF, which in turn is the reciprocal of the CRF. This means the levelized annual cost can be determined by dividing the total present worth by the appropriate USPWF or by multiplying the total present worth by the corresponding CRF.

Work Problems

Problem 6.—Levelized Values

A utility with a 9% annual weighted cost of capital must choose between two alternatives having nonuniform annual costs. The costs of the two options are presented below.

Annual Costs of Options A and B

	Annual Costs	
Year	Option A ($)	Option B ($)
1	10,000	8,000
2	11,000	9,000
3	12,000	10,000
4	13,000	12,000
5	14,000	14,000
6	15,000	16,000
7	16,000	18,000
8	17,000	20,000
9	18,000	22,000
10	19,000	24,000

6A. Determine the levelized cost of each option. Which alternative is the least costly plan?

6B. Prove that the levelized cost of the recommended alternative has the same present value as the nonuniform series.

2.3.9 Levelized Values for Constant Percentage Cost Increases

In cases where annual costs increase by a constant percentage, levelized costs can be calculated directly by Eq. 2-8.

$$\text{Levelized cost} = (\text{Cost at end of first year})\,(\text{CRF})\,\frac{1-K^n}{i-e} \qquad (2\text{-}8)$$

where

i = present worth discount rate,

e = annual escalation rate,

n = number of years,

$k = \frac{1 + e}{1 + i}$, and

$$\text{CRF} = \frac{i(1+i)^n}{(1+i)^n - 1} \text{ [from Eq. (2.6)]}$$

Table 2-4 demonstrates the equivalence of Eq. (2-7) with Eq. (2-8) for a series of five annual payments that begin with \$5,000 and increase by 7% annually. The present worth annual discount rate is 13%. The levelized cost is first determined in Table 2-4 using Eq. (2-7). Next, the levelized cost is determined by Eq. (2-8). The levelized cost is \$5,656.70 using either method.

Equation (2-8) assumes the first annual cost in the series to be levelized is paid as of the end of the first year. This is in accordance with the end of year convention for payments and costs, although these costs are actually paid throughout the year. However, if costs are given as of the start of the first year, Eq. (2-9) can be used to calculate levelized cost when escalation is uniform. This equation differs from Eq. (2-8) by the amount of one year's escalation, or by the factor $(1 + e)$.

Levelized cost

$$= (\text{Cost at beginning of first year})\ (\text{CRF})\ (K)\ \frac{1 - K^n}{1 - K} \qquad (2\text{-}9)$$

Work Problems

Problem 7.—Levelized Value with Uniform Escalation

A utility is considering a 6-year power purchase that would cost \$15,000 the first year and escalate at 7% annually thereafter. What is the levelized cost to the utility if it has a 14% annual discount rate? Use Eq. (2-7) to determine the levelized value, then confirm the answer using Eq. (2-8).

2.3.10 Busbar Costs

Costs are frequently stated on a levelized cost per kilowatt-hour, also called a levelized busbar cost. Levelized busbar costs are obtained by dividing the levelized annual cost by the levelized annual kilowatt-hour net output. The net output is the total generator output minus the energy used for all station auxiliaries, including the main transformer but not the plant substation. For example, if a (net) 10,000 kW plant with a levelized annual cost of \$800,000 is projected to operate at a 40% capacity factor, the net annual kilowatt-hour output equals 10,000 kW (net plant capacity) times 8,760 (total hours in a year) times 0.40 (capacity factor) or 35,040,000 kWh/yr. Thus, the levelized busbar cost is \$800,000/35,040,000 kWh or \$0.0228/kWh. This can also be stated as 22.8 mills/kWh.

If the annual kilowatt-hour output is not constant, the kilowatt-hours must be levelized in the same manner as the annual costs, using the same present worth discount rate.

Table 2-4. Levlized Cost with Uniform Escalation

Year	Annual Cost at 7% Escalation (%)	Present Worth Factor	Present Worth of Annual Cost ($)
1	5,000.00	0.88496	4,424.78
2	5,350.00	0.78315	4,189.83
3	5,724.50	0.69305	3,967.37
4	6,125.22	0.61332	3,756.71
5	6,553.98	0.54276	3,557.24
Total		3.51723	19,895.93

Discount rate = 13%

Annual cost escalation = 7%

$$\text{Levelized cost} = \frac{\text{Total present worth}}{\text{Sum of present worth factors}} = \frac{\$19{,}895.93}{3.51623} = \$5{,}656.70$$

By Eq. (2-8):

$$\text{CRF} = \frac{1}{3.51723} = 0.28431$$

$$K = \frac{1+e}{1+i} = \frac{1.07}{1.13} = 0.94690$$

$$\text{Levelized cost} = \$5{,}000 \times (0.28431) \times \frac{1 - (0.94690)^5}{0.13 - 0.07}$$

$$= 1{,}421.57 \times \frac{0.23876}{0.06} = 5{,}656.90$$

2.4 UTILITY COSTS

Several equations were developed in Section 2.3 that enable present value cost evaluations. This section presents a detailed view of the types of costs incurred by a utility and will enable realistic present worth applications in Section 2.5. In particular, this section separates total utility costs into fixed and operating costs, which differ in a regulatory accounting and engineering sense. It is helpful to remember that in utility regulation, cost recovery is usually structured so that an item's cost is paid for over its useful life. In the long term, all utility costs must be recovered through rates, making total costs synonymous with revenue requirements.

2.4.1 Operating Costs

Operating costs (also called variable costs and expenses) are those costs that vary with the level of output. For generating plants, operating costs include fuel costs and variable operation and maintenance (O&M) costs. Approximately 60% of a utility's total revenue requirements are needed to offset operating costs; fuel is usually the largest single component.

Operating costs are incurred on an ongoing basis and are usually consumed or used in a year's time or less. Lubricating oils and fuel costs, for example, are typical operating costs incurred every year. To match the recovery of an operating item's cost with the benefit received from its use, these costs are recovered from revenue (through rates) on a dollar-for-dollar basis when incurred. In theory, each dollar of operating cost incurred annually is matched by a dollar of revenue.

The determination of the present value of operating costs is relatively straightforward. The yearly operating costs for each year of the investment's life are discounted to the present and can be added to the present value of fixed costs to determine the total present worth revenue requirements.

2.4.2 Fixed Charges

Fixed costs (or fixed charges) make up the remainder of total costs. They are a direct function of the level of capital investment and do not vary with production. Once an investment in a facility such as a generating plant is made, the associated fixed costs will be incurred regardless of the level of output.

Recovery of fixed costs is more complicated than for operating costs because they are not matched by revenue on a dollar-for-dollar basis when incurred. Instead, only a percentage of an investment is recovered through rates annually, and fixed cost recovery continues over the life of an investment. This method of fixed cost recovery provides a degree of equity to ratepayers since generating plants and other investments have long service lives and may provide benefits to ratepayers for 30 years or more. Fixed cost recovery spread over the investment's life causes those receiving the long-term benefits to pay for the investment. It also avoids unacceptably high rates which would occur if the total investment cost—which can be in the hundreds of millions of dollars—was immediately recovered when placed in service. Finally, the nature of a utility's fixed cost payments allows for recovery of only a portion of the total investment annually.

The components of fixed costs for investor-owned utilities and public utilities are listed in Table 2-5. The components differ primarily in that a public utility does not issue stock and its income is not taxed. The investor-owned fixed cost components are discussed below; the public utility components are discussed in Appendix 2B. Discussion of these components enable the yearly quantification of fixed cost components. These yearly fixed costs can be combined with yearly variable costs of an option on a present value basis, and then can be compared with other alternatives.

2.4.2.1 Return on Investment. Utilities seldom pay for large investments entirely from retained earnings which are profits not paid out to stockholders. Instead, large investments such as a new plant are usually financed through debt (bonds) and equity (retained earnings, common and preferred stock). Investors provide financing (purchase bonds and stock) in the expectation of future returns on their investment. Returns are in the form of interest payments to bondholders and dividends to stockholders. Appreciation of the value of the stock or bond may also occur. Since investors are willing to accept periodic returns equal to only a portion of their total investment, a utility need not recover the total cost of an investment on a dollar-for-dollar basis when incurred.

Table 2-5. Fixed Cost Components

Fixed Costs for an Investor-Owned Utility	Fixed Costs for a Public Utility
Return on investment	Amortization
Depreciation	Insurance
Income taxes	Payment in lieu of taxes
Property taxes	Renewals and replacements
Insurance	
Other administrative costs	

For example, if a utility's weighted average cost of capital is 10% and a new plant is placed into service at a total capitalized cost of $100 million, then a $10 million return on investment would provide an adequate return on capital during the first year of operation. In general, the required return on capital is calculated by applying the weighted cost of capital to the undepreciated investment. The undepreciated investment equals the original cost less accumulated depreciation.

The return on equity and debt is often called the minimum acceptable return since investors are not willing to supply additional capital if this return is not earned on a long-term basis. The actual rate of return required for debt and equity varies according to many company, industry, and market factors. For any utility, however, the cost of debt should be less than the cost of equity, since a stockholder's return depends on a utility's profits and they must be compensated for this higher investment risk. In general, debt for investor-owned utilities usually makes up from 55% to 60% of total financing.

2.4.2.2 Depreciation. Another fixed cost that is recovered by a utility is depreciation. Depreciation arises because of capital consumption; that is, because a plant wears out and becomes technologically obsolete over time. If it were not for the need to provide electricity, a plant would not be built and capital consumption could be avoided. Therefore, capital consumption represents a cost of providing service. The annual depreciation charge allows for the recovery of the plant's original cost and provides an approximate match between the yearly benefits received from the plant's use and the cost of service incurred from the consumption of capital.

An alternative view of depreciation is that ratepayers must not only provide suppliers of capital with a reasonable return, but must also repay the amount originally invested. This is done through depreciation. The sum of all annual depreciation charges should add up to the total original investment cost.

The Tax Reform Act of 1986 established the allowable depreciation recovery period for new utility property. Depreciation recovery periods are determined under either the modified accelerated cost recovery system (MACRS) or the alternative depreciation system (ADS). The MACRS is usually preferred by utilities because it allows for a shorter depreciation schedule for most types of utility property. However, the ADS is mandatory for some property including that financed with tax-exempt bonds and for property designated as imported or used outside the United States.

If the ADS is used, then the straight-line depreciation method is used over the ADS recovery period. Under straight-line depreciation, a constant percentage of the original investment is recovered over the property life. If the MACRS is used, then an accelerated depreciation method is used that allows for a greater depreciation deduction in the early years of the property's life than does the straight-line method. Under MACRS, the 200% (double) declining balance method, with switchover to straight-line method, is used for 3-, 5-, 7-, and 10-year MACRS recovery periods. A 150% declining balance method, with switchover to straight-line method, is used for 15- and 20-year MACRS recovery periods. Appendix 2C lists the depreciation recovery period for several types of utility property.

*2.4.2.3 **Federal and Local Income Taxes.*** Federal and local income taxes are also considered a fixed cost for evaluation purposes. Taxable utility income is calculated as revenue minus deductible expenses. Operating expenses and certain fixed costs including interest on debt, insurance, property taxes, and tax depreciation are deductible for tax purposes.

Under an accelerated tax depreciation method discussed in Section 2.4.2.2, a utility may take a greater depreciation deduction than if straight-line depreciation is used. Accelerated tax depreciation methods result in deferred taxes during the early years of a plant's operation compared to the straight-line method and result in present value benefits to the utility. This reduced tax obligation may be used to reduce revenue requirements and increase return to stockholders under the flow-through method of accounting. Alternatively, under the more common normalization method, revenue requirements are collected as if the straight-line method was being used and the utility may use the funds in excess of its tax obligation internally until it becomes due.

Once the taxable income is determined, the federal tax rate and local income tax rate are applied to determine revenue requirements needed to offset income taxes. Local income tax rates are usually 5% to 8% and the federal tax is usually assumed to equal 34%. Since local income taxes are deductible for federal tax computation, the total tax rate usually is near 40%, as demonstrated below, assuming a 7% local income tax rate.

$$\text{Local tax rate} = 0.07$$

$$\text{Federal tax rate} = 0.34 \times (1 - 0.07) = 0.32$$

$$\text{Total tax rate} = 0.07 + 0.32 = 0.39 \text{ or } 39\%$$

Although a utility's tax rate may stay constant, the tax collected from an individual investment generally decreases over time as the taxable income, a function of the undepreciated investment, decreases.

*2.4.2.4 **Property Taxes, Insurance, and Other Administrative Costs.*** Property taxes, insurance, and often other administrative costs are also included as fixed costs. These costs are usually based on current property values rather than on depreciated book values. As a result, the yearly cost of these items may remain fairly constant or escalate over the life of the facility. The other administrative costs component may include renewals and replacements, and general overheads. In present value analysis, the yearly cost of property taxes plus insurance and other administrative costs is typically assumed to make up about 2% of the initial investment.

2.4.3 Other Cost Considerations

Economic evaluations must include all costs associated with an alternative. Fixed costs include direct costs which are made up of items such as the cost of equipment and materials, installation, general construction, sales tax, and contingency. Two less apparent but important costs that are also often appropriate to include in fixed cost estimates are indirect costs and allowance for funds used during construction (AFUDC).

*2.4.3.1 **Indirect Costs.*** Capital cost estimates for power plants should include an item for indirect costs. Indirect costs include expenses for general costs such as equipment checkout and testing, start-up costs, operator training, taxes other than sales taxes, and miscellaneous construction expenses as well as engineering services, field construction management services, and certain owner costs.

Indirect costs are usually calculated as a percentage of escalated direct costs. The percentage varies with generating unit size, and the percentage decreases as the unit size increases. The percentage for the first unit at a site is also larger. Typical values range from 11% to 15% for a first unit, and from 8% to 12% for subsequent units (Black & Veatch 1983).

Indirect costs are usually included only in estimates for an entire project. They are not normally included in equipment estimates used in system analyses, and are not considered in bid evaluations.

*2.4.3.2 **Allowance for Funds Used During Construction.*** Another important cost component is the allowance for funds used during construction, or AFUDC. AFUDC includes interest on borrowed funds and return on equity funds used during construction. AFUDC arises because of the regulatory treatment of funds used during construction. In most states, ratepayers do not begin to pay for the cost of a plant until it is placed in service. Nevertheless, when a utility issues stocks and bonds to finance the construction of a power plant, it must make payments on these issuances during the construction period. Generally, utilities meet these costs through additional debt and equity issue or through the use of internal funds. When the plant is placed in service, the utility must be compensated for these financing costs that arose because the construction work in progress was not allowed in the rate base. Through the AFUDC process, these costs are capitalized, and when the plant is placed in service, the accumulated AFUDC is added to the book cost of the plant. The resulting total plant cost is placed in the rate base and depreciated over the life of the plant.

For the engineer making cost estimates, an AFUDC rate, usually assumed to equal the weighted cost of capital, is typically applied to the total of all other costs. This means that the AFUDC rate is the last multiplier applied in the compounding process. When a monthly AFUDC rate is needed, Eq. 2-10 may be used:

$$\text{Monthly AFUDC rate} = (1 + i)^{1/12} \qquad (2\text{-}10)$$

where

i = annual interest rate.

The total AFUDC is usually calculated by one of two methods. The first involves an estimate of the variable monthly cash payments to be made for the plant or item during construction. The financing costs for each month are then calculated from the time of payment to the commercial operation date using the AFUDC rate. The sum of the financing costs for all months is the total AFUDC for the plant or item. This method is used for bid analyses, with the bidder's payment terms reflected in the monthly cash flows. The method may also be used to calculate AFUDC for a total project.

When specific information on payment and delivery dates is not available, it is common to assume that all payments are made in a lump sum at the midpoint of the construction period and the AFUDC is calculated from the midpoint of the construction period until the plant is placed in service. This method is normally used in cost estimates for systems analyses and for preliminary total plant cost estimates. Regardless of which method is used, the resulting AFUDC amount is usually listed separately in stating the plant's total capital cost.

2.4.4 The Fixed Charge Rate

The investor-owned fixed cost components were identified in the preceding sections. These fixed costs are recovered over the life of a plant. For example, if a $100 million plant was placed in service, revenue requirements to offset fixed charges would not increase by $100 million. Instead, if the sum of all fixed charges for the first year of operation was equal to 20% of the initial investment, then $20 million would be sufficient to offset fixed costs during the first year (additional revenue requirements would be needed to recover operating costs). The summation of all fixed charges as a percent of the *initial* investment is called the fixed charge rate. In general, fixed cost revenue requirements in any given year can be determined by applying the yearly fixed charge rate to the initial investment.

It is important to remember that a return on capital is provided only for the unrecovered investment, which equals the original cost less accumulated depreciation charges. This means an investment's fixed charge rate and the yearly revenue requirements needed to offset fixed charges will generally decline over time as illustrated in Fig. 2-2.

As seen in Fig. 2-2, when the investment is first placed in service, a fixed charge rate of roughly 20% is required to offset fixed charges. During the first year of operation, some of the useful life of the plant is consumed; ratepayers pay for this through the depreciation charge. As this portion of the original investment is recovered, it requires no further return. During the second year, debt and equity returns are based on a lower unrecovered investment and this results in a lower fixed charge rate. This process continues over the life of the investment.

2.4.5 Fixed Charge Revenue Requirements

Ultimately, the present value total lifetime revenue requirements needed to offset an investment's fixed charges must be determined in an analysis and combined with present value operating costs to evaluate the total investment cost. When an annual fixed charge rate is applied to the original investment, the revenue requirements needed to offset the fixed charges for a given year are derived.

As demonstrated in this section, fixed cost revenue requirements may be stated on a total present value basis or on a levelized basis. If stated on a levelized basis, the levelized annual costs can be determined using one of two equivalent methods.

Consider the first 5 years of service of the option presented in Table 2-6. The option has a total initial cost of $10,000 and a yearly fixed charge rate of 20% for the first year, which decreases by 2% in each of the next four years. The utility's discount rate is 10%.

In year 1, the fixed charge rate of 20% is applied to the initial cost, yielding fixed revenue requirements of $2,000 (Column C). On a present value basis, $1,818.18 in revenue

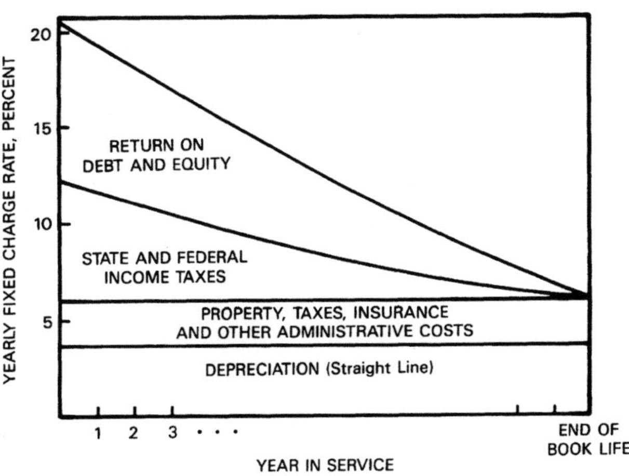

Figure 2-2. Fixed charge rates over time.

Table 2-6. Total Fixed Charge Revenue Requirements

Year	Fixed Charge Rate (A)	Initial Cost (B) ($)	Fixed Revenue Requirements (C) ($)	Present Value of Fixed Annual Revenue Requirements (D) ($)
1	0.20	10,000	2,000	1,818.18
2	0.18	10,000	1,800	1,487.60
3	0.16	10,000	1,600	1,202.10
4	0.14	10,000	1,400	956.22
5	0.12	10,000	1,200	745.11
Cumulative present worth of fixed revenue requirements				6,209.21

requirements is needed to offset fixed charges (Column D). The present values of fixed charges in years 2 through 5 are calculated in a similar fashion.

2.4.5.1 Total Annual Fixed Charges. Using the example in Table 2-6, there are three methods of stating an option's lifetime present value fixed revenue requirements. The first is to add the present values of the yearly fixed revenue requirements. This results in the total present worth revenue requirements needed to offset fixed charges.

In Table 2-6, the total present value of the yearly fixed charges is $6,209.21. If total fixed charges are added to the present value of all yearly operating costs, the cumulative present worth of total costs is obtained.

2.4.5.2 Levelized Annual Fixed Charges. Rather than state fixed costs on a total present worth basis, many utilities prefer a levelized figure. In Table 2-7, the levelized annual fixed cost is calculated as the sum of the present value of the yearly revenue requirements divided by the sum of present worth factors. The resulting levelized cost equals $1,637.97, which results in total present value fixed cost revenue requirements of $6,209.21. If added to the levelized operating costs, a total levelized cost is obtained.

2.4.5.3 Levelized Fixed Charges Using a Levelized Fixed Charge Rate. An alternative method of calculating levelized fixed charges is to use a levelized annual fixed charge rate. This is a single rate that, when applied to the initial investment, results in a constant dollar amount of fixed charges each year throughout the life of the plant. In Table 2-8, the levelized fixed charge rate is calculated to be 0.163797 which, when applied to the initial investment, results in a levelized fixed charge of $1,637.97 and a total fixed cost present value of $6,209.21. When added to the levelized operating cost, total levelized costs are derived. The levelized fixed charge rate is used most often by utilities since it is easily applied. The levelized fixed charge rate is used later in Section 2.5 to demonstrate various economic analysis methods.

Table 2-7. Levelized Fixed Costs

Year	Present Value of Fixed Revenue Requirements ($)	Present Worth Factor	Present Value of Levelized Costs ($)
1	1,818.18	0.9091	1,489.07
2	1,487.60	0.8264	1,353.70
3	1,202.10	0.7513	1,230.63
4	956.22	0.6830	1,118.76
5	745.11	0.6209	1,017.05
Total	6,209.21	3.7908	6,209.21

Levelized cost = 6,209.21/3.7908 = $1,637.97

2.4.6 Sensitivity Analyses

The previous sections presented total utility costs broken down according to their fixed and operating classifications. Knowledge of those cost components and the present worth equations of Section 2.3 will aid in understanding the economic analysis methods presented in Section 2.5. First, however, some areas need to be clarified.

Thus far, this chapter has implied that the alternative having the smallest present value should be implemented. In practice, the least costly alternative may not be chosen if it is very capital intensive and if cost effectiveness relies on narrow input assumptions for variables subject to considerable volatility or uncertainty. Before making a recommendation, therefore, it is usually appropriate to run several sensitivity analyses to determine whether the option initially identified as the least-cost option in a base case will remain the least-cost option under alternative but realistic scenarios. For system planning studies, typical sensitivity analyses include alternative assumptions about the fixed and variable cost escalations (particularly for fuel costs), growth in system demand, interest rates, and the company's cost of capital among others.

There are numerous types of sensitivity analyses. These methods vary in sophistication, time requirements, and cost. To perform a basic sensitivity analysis, input variables are changed one by one from the "best estimate" assumed in the base case. If sensitivity analysis on a particular variable does not affect the recommendation, the variable need not be considered further. If, however, the analysis indicates sensitivity to a reasonable change in assumptions, further investigation regarding probable value of that input variable is warranted. Following completion of the sensitivity analyses, if an option is cost-effective only under a narrow set of assumptions, it may be appropriate to recommend another option less sensitive to future uncertainties.

Table 2-8. Levelized Fixed Charge Rate

Year (A)	Fixed Charge Rate (B)	Present Worth of Fixed Charge Rates (C) = (B) × (D)	Present Worth Factor (D)	Levelized Cost (E) ($)	Present Worth Levelized Cost (F) ($)
1	0.20	0.18182	0.9091	1,637.97	1,489.07
2	0.18	0.14876	0.8264	1,637.97	1,353.70
3	0.16	0.12021	0.7513	1,637.97	1,230.63
4	0.14	0.09562	0.6830	1,637.97	1,118.76
5	0.12	0.07451	0.6209	1,637.97	1,017.05
Total		0.62092	3.7908		6,209.21

$$\text{Levelized fixed charge rate} = \frac{0.620908}{3.7907} = 0.163797$$

Levelized costs = 0.163797 × \$10,000 = \$1,637.97

2.5 ECONOMIC ANALYSIS METHODS

2.5.1 Introduction

The previous sections introduced present value arithmetic and described the utility's fixed and operating costs. This section combines that information and presents five present value evaluation methods used by utilities and one method used primarily by nonutility generators (NUGs) to evaluate alternatives. These economic analysis methods are the following:

- Total present worth method
- Cumulative present worth method
- Capital equivalent method
- Levelized annual cost method
- Capital recovery period method
- Pro forma analysis.

To illustrate the similarities and differences among the first five evaluation methods, a hypothetical example using three alternative investments is evaluated. The pro forma analysis is discussed separately, since it evaluates an alternative based on projected financial returns rather than from a cost minimization perspective.

2.5.2 Economic Inputs for Economic Analysis Methods

The economic inputs for the hypothetical example are presented in Table 2-9. The example assumes that capital expenditures occur midway through the 4-year construction period. Thus, escalation occurs until 2 years before the commercial operation date of 1999. The AFUDC rate of 25% approximates 2 years of interest between the expenditure date and operation date.

2.5.3 Levelized Annual Cost Method

The levelized annual cost method involves the calculation of total levelized annual costs. Total levelized costs are the sum of levelized fixed and variable costs. To determine the levelized fixed costs in Table 2-10 (line G), the 1999 total capital cost is multiplied by the levelized fixed charge rate.

Table 2-9. Economic Input for Alternatives

Economic Inputs

Economic life	30 years
Commercial operation date	January 1, 1999
Annual escalation rate	7%
Present worth discount rate	11.5
Levelized annual fixed charge rate	20.419%
AFUDC rate	25%
Construction period	4 years

Investment Alternatives

	Option A ($)	Option B ($)	Option C ($)
Capital cost, 1/1/92	570.39	1,140.78	2,281.56
Cost of operation, 1999			
Beginning of year	1,000.00	750.00	500.00
End of year	1,070.00	802.50	535.00

Present Worth Calculations

$$\text{USPWF [from Eq. (2-5)]} = \frac{(1 + i)^n - 1}{i(1 + i)^n} = 8.36371$$

$$\text{CRF} = \frac{1}{\text{USPWF}} = 0.11956$$

Equation (2-9) is used to determine the levelized operating costs.

Combining the levelized fixed and operating costs (line I), Option A has a total levelized cost of \$2,221.05; the total levelized cost for Option B and Option C is \$1,921.02 and \$1,825.19, respectively. Since Option C has the lowest levelized annual cost, it is the least cost alternative and would likely be recommended using the levelized annual cost method only.

2.5.3.1 Evaluation of the Levelized Annual Cost Method. The advantages of the levelized annual cost method are its wide use in the industry, easy application, and intuitive appeal to those not familiar with present worth arithmetic. A potential drawback is that the total levelized cost figure does not differentiate between options by capital intensity; an option with high fixed costs and low operating costs may be chosen if it has the lowest total present worth cost. Since fixed costs are "sunk" up front, the utility must incur these

Table 2-10. Levelized Annual Cost Method

Line	Capital and Operating Cost	Plan A ($)	Plan B ($)	Plan C ($)
A	Capital cost, 1/1/92	570.39	1,140.78	2,281.56
B	Escalation (5 years at 7%)	229.61	459.22	918.44
C	Direct cost	800.00	1,600.00	3,200.00
D	AFUDC (25%)	200.00	400.00	800.00
E	Capital cost, 1/1/99	1,000.00	2,000.00	4,000.00
F	Operating cost (end of 1999)	1,070.00	802.50	535.00
Line	Levelized Costs	Plan A ($)	Plan B ($)	Plan C ($)
G	Fixed charges (0.20419 × Line E)	204.19	408.38	816.76
H	Operating cost (1.88491 × Line F)	2,016.86	1,512.64	1,008.43
I	Total levelized annual cost	2,221.05	1,921.02	1,825.19

Levelized operating cost for Plan A from Eq. (2-9):

$$\text{Leveling factor} = (\text{End of year cost})\ (\text{CRF})\ \frac{1 - K^n}{1 - e}$$

$$K = \frac{1 - e}{1 + i} = \frac{1.07}{1.115} = 0.95964$$

$$\text{Levelizing factor} = (\$1{,}070.00)(0.11956)\frac{1 - (0.95964)^{30}}{0.115 - 0.07}$$

$$= (\$1{,}070.00)(1.88491)$$
$$= \$2{,}016.86$$

costs even if future conditions change and the option is no longer viable. Thus, the uncertainty of future conditions results in a potential disadvantage of using the levelized annual cost method.

The danger of relying too heavily on uncertain future conditions can be illustrated by situations that occurred in the 1970s and 1980s when utilities based their expansion plans on high load growth forecasts. When the high growth did not materialize, many utilities had already invested in large, capital intensive nuclear units. Consequently, a number of utilities were ruled to have excess capacity that was not used and useful and suffered cost disallowances.

In general, selection of a capital-intensive investment should not be made unless the option is the least-cost alternative under a reasonable range of assumptions. In addition, if the capital intensity of options being considered differs significantly, alternative evaluation methods such as the capital recovery period method (discussed later) should be used, as this method indicates the time required for an alternative to recoup additional capital investments.

Table 2-10 indicates that Plan C is the preferred alternative under the levelized annual cost method, although it has a relatively high capital cost. Plan A has the highest total levelized annual cost but is the least capital intensive. Given the contrast in the capital intensiveness of these options, additional information concerning the time required for Plans B and C to recover the additional capital investments is required to determine the preferred alternative.

2.5.4 Total Present Worth Method

The total present worth method states an option's present value cost as the present worth sum of all capital and operating costs over an option's economic life. The total present worth method will result in the same decision among alternatives as the levelized annual cost method; the difference is that costs are stated on a total cost basis. Table 2-11 calculates the total present worth for the three investment options.

The present worth of the fixed and operating costs can be derived from the existing data most quickly by multiplying the levelized cost values by the uniform series present worth factor (this follows from the discussion in Section 2.3.8), which is equal to 8.36371 in this example. Line K of Table 2-11 indicates that the total present worth cost of Plan C equals $15,265.37 and is the least-cost option.

2.5.4.1 Evaluation of the Total Present Worth Method. The total present worth method is relatively easy to apply and is commonly used in the utility industry. The total present worth method has the same disadvantages as the levelized annual cost method. In addition, the method can result in very high cost totals, which may be less meaningful than an annual levelized cost figure to those unfamiliar with present value arithmetic.

2.5.5 Capital Equivalent Cost Method

The capital equivalent cost method is used for bid analyses, usually to compare a plan with a lower operating cost to bids with higher operating costs but a lower bid (capital cost). The method expresses an option's operating costs as a capital equivalent. When added to the option's actual capital cost, a total capital equivalent cost is derived. Table 2-12 applies the capital equivalent cost method to Plans A, B, and C. Capital equivalent costs are determined by dividing the levelized operating costs by the levelized annual fixed charge rate, which is 0.20419 in this example. The total capital equivalent cost (line K) is the sum of the original

Table 2-11. Total Present Worth Method

Line		Plan A ($)	Plan B ($)	Plan C ($)
	Capital Operating Cost			
A	Capital cost, 1/1/92	570.39	1,140.78	2,281.56
B	Escalation (5 years at 7%)	229.61	459.22	918.44
C	Direct cost	800.00	1,600.00	3,200.00
D	AFUDC (25%)	200.00	400.00	800.00
E	Capital cost, 1/1/99	1,000.00	2,000.00	4,000.00
F	Operating cost (end of 1999)	1,070.00	802.50	535.00
	Levelized Cost (from Table 2-9)			
G	Fixed charges (0.20419 × Line E)	204.19	408.38	816.76
H	Operating cost (1.88491 × Line F)	2,016.86	1,512.64	1,008.43
	Total Present Worth			
I	Fixed charges (8.36371 × Line G)	1,70.79	3,415.57	6,831.15
J	Operating cost (8.36371 × Line H)	16,868.44	12,651.29	8,434.22
K	Total present worth	18,576.23	16,066.86	15,265.37

capital cost and the capital equivalent operating cost. As with the previous two evaluation methods, Plan C is the recommended alternative.

Table 2-12 can be used to demonstrate that the capital equivalent cost method is financially equivalent to the annual cost and present worth methods. The values shown under total capital equivalent cost in Table 2-12 can be multiplied by 0.20419 to yield the values under the total annual cost method in Table 2-10 (line I). The line K values of Table 2-12 can also be multiplied by the product of 0.20419 and 8.36371 to obtain the values under the total present worth method (line K) in Table 2-11.

A variation on the capital equivalent cost method is the differential capital equivalent method in which all options are compared to a base option. The base option is the plan with the lowest operating costs. Lines L through O in Table 2-12 illustrate this variation. The difference in dollars between plans is the same using either the total or differential capital equivalent cost method.

2.5.5.1 Evaluation of the Capital Equivalent Cost Method. This method is useful for evaluation of bids as it involves minimum manipulation of the bid price itself. This method has the same disadvantages as the first two methods; thus, for some bid analyses, the capital recovery period method should be used.

Table 2-12. Capital Equivalent Cost Method

Line		Plan A ($)	Plan B ($)	Plan C ($)
	Capital and Operating Cost			
A	Capital cost, 1/1/92	570.39	1,140.78	2,281.56
B	Escalation (5 years at 7%)	229.61	459.22	918.44
C	Direct cost	800.00	1,600.00	3,200.00
D	AFUDC (25%)	200.00	400.00	800.00
E	Capital cost, 1/1/99	1,000.00	2,000.00	4,000.00
F	Operating cost (end of 1999)	1,070.00	802.50	535.00
	Levelized Cost (from Table 2-9)			
G	Fixed charges (0.20419 × Line E)	204.19	408.38	816.76
H	Operating cost (1.88491 × Line F)	2,016.86	1,512.64	1,008.43
	Capital Equivalent Cost			
I	Fixed charges (Line G / 0.20419)	1,000.00	2,000.00	4,000.00
J	Operating cost (Line H / 0.20419)	9,877.37	7,408.00	4,938.68
K	Total capital equivalent cost	10,877.37	9,408.00	8,983.68
	Differential Capital Equivalent Cost Method			
L	Differential operating cost	1,008.43	504.21	Base
M	Capital cost (same as Line J)	1,000.00	2,000.00	4,000.00
N	Differential operating cost (Line L / 0.20419)	4,938.68	2,469.32	Base
O	Total capital equivalent cost (differential method)	5,938.68	4,469.32	4,000.00

2.5.6 Cumulative Present Worth Method

In the cumulative present worth method, annual present worth costs are added and stated on a year-by-year basis throughout the evaluation period. Although similar to the total cost method, alternatives can be compared throughout the horizon rather than only on the basis of total lifetime present worth costs.

The added information supplied by this method is seen through a plot of cumulative costs. Figure 2-3 plots the cumulative costs for Plans A, B, and C. The graph indicates that over the 30-year economic life, Plan C is the preferred investment. Plan B is the next most attractive plan at this year, followed by Plan A. Comparison of the total present worth costs at year 30 with Table 2-11 reveals the same end of life present worth cost as the total present worth method. Figure 2-3 also indicates that, in comparison to Plan A, the cumulative present worth cost of Plan C is higher for the first 10 years of operation and lower thereafter. Also, the cumulative present worth cost is lower for Plan B than for Plan A in all years. Comparing Plans B and C, it can be seen that the cumulative present worth cost of Plan B is lower for the first 19 years and higher thereafter. The cumulative present worth method is sometimes referred to as breakeven analysis since it can be used to identify the "breakeven" year in which the cumulative revenue requirements for a capital intensive option equals that for a lower capital cost option with higher operating costs.

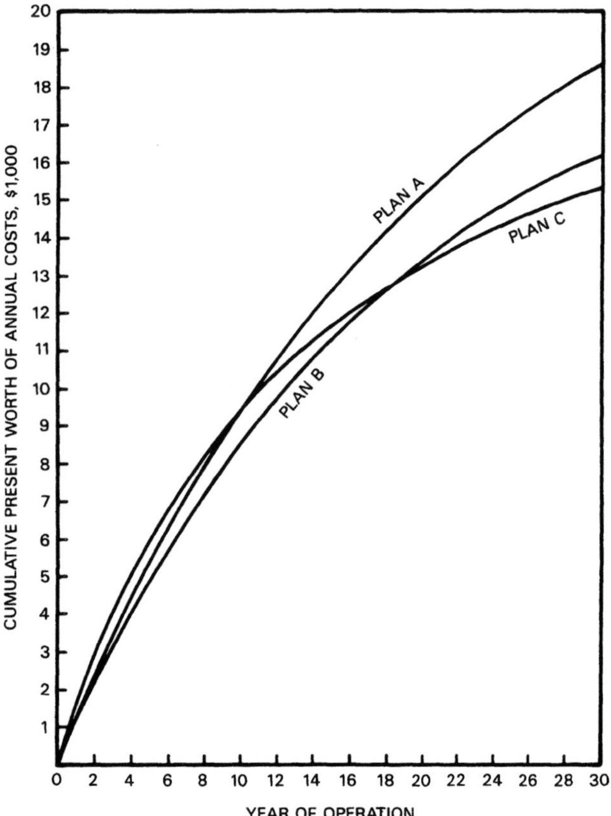

Figure 2-3. Cumulative present worth costs for Plans A, B, and C.

*2.5.6.1 **Evaluation of Cumulative Present Worth.*** The cumulative present worth method results in additional information, compared to previous methods. It identifies the year in which the cumulative present worth costs for a particular plan become lower than for another plan. This is important additional information given the uncertainties involved in predicting future costs.

Referring to Fig. 2-3, it can be seen that Plan B is preferred over Plan A, since the Plan B cumulative present worth costs are lower throughout the evaluation period. Plan C would also probably be preferred over Plan A, since the Plan C cumulative present worth costs are lower after about 10 years of operation. Comparing Plans B and C, however, it can be seen that Plan C becomes the least cost plan after 19 years of operation, well into the economic lifetime of the plant. Considering that the initial capital cost of Plan B is half that of Plan C and the uncertainties involved in predicting future costs of operation, Plan C may not be selected over Plan B. Although it is difficult to apply generalizations to such decisions, some utilities will not select a high capital cost plan unless the total cumulative present worth costs become less than those for an alternative plan with lower capital cost during the first half of the evaluation period. However, other utilities take a less conservative view, and sensitivity analysis must also be considered in such a decision. Here, the important point is that the cumulative present worth method provides additional information compared to previous methods and may lead to a different recommendation.

The cumulative present worth method is especially applicable when alternative plans involve the installation of facilities at different times, or with different lives. The major disadvantage is that computation of year-by-year cumulative present worth costs can be a time-consuming and laborious process. Computer programs are available for this purpose, but their use introduces additional steps and complexities to the process of preparing a system analysis. In many instances, the year-by-year cost breakdown provided by this evaluation method is not needed. For example, in many system analyses the differential cost of operation for the various plans is negligible and only capital costs need to be considered. In these cases, the levelized annual cost method or the total present worth method presents all the information that is needed. Also, even in those cases with differential costs of operation, there are many times when the low capital cost plan is also the low total annual cost (or total present worth) plan. In these cases, the low total annual cost plan is the obvious selection.

2.5.7 Capital Recovery Period Method

The capital recovery period method is similar to the cumulative present worth method in that year-by-year costs are generated. However, the capital recovery method charges the present worth of the total fixed costs to each option at the beginning of the evaluation period. Cumulative annual present worth costs of operation are then added to the fixed charges on a year by year basis.

Figure 2-4 illustrates this method. The three curves representing Options A, B, and C intersect the vertical axis at the point which equals the total present worth of the fixed charges for each curve. The cumulative present worths of annual operating costs are then added to generate the year-by-year costs. The intersection of curves is that year in which the cumulative savings in operating costs for a higher capital cost plan completely offset the differential (higher) capital investment committed to that plan over the life of the plant.

As with the cumulative present worth method, it is difficult to set guidelines as to how quick a recovery period justifies a higher capital cost investment. Some utilities will not adopt an alternative with a high capital cost and low operating costs unless the capital recovery years occur within the first half of the expected life or 15 years, whichever is less. Other utilities take a less conservative view, and sensitivity analyses also influence the decision.

From Fig. 2-4, it can be seen that in comparison to Plan A, the capital recovery periods of Plans B and C are about 8.2 and 13.7 years, respectively. In comparison to Plan B, the capital recovery period of Plan C is about 21 years. Comparison of Figs. 2-3 and 2-4 indicates that the curves intersect at

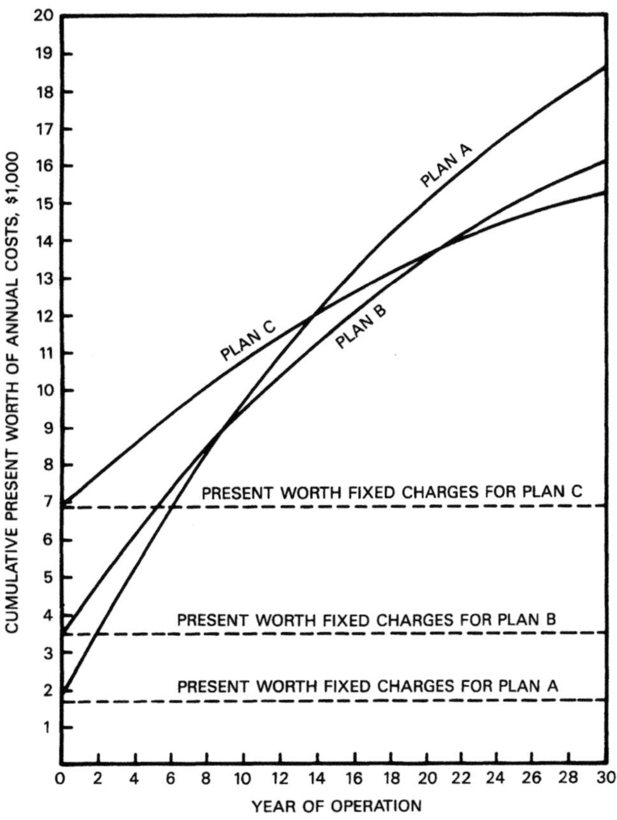

Figure 2-4. Capital recovery period for plans A, B, and C.

later years for the capital recovery period method than for the cumulative present worth method. This is because the intersection points for the capital recovery period method represent the year in which the savings for a plan offset the higher fixed charges for that plan that would accrue over the entire economic lifetime. Intersection points for the cumulative present worth method represent the year at which the cumulative fixed charges and operation costs accrued *to that point in time* are equal to the accrued costs for another plan.

2.5.7.1 Evaluation of the Capital Recovery Period Method. The capital recovery period method has many of the same advantages as the cumulative present worth method. It is especially applicable to design projects where the two plans being considered are for alternative facilities with the same expected life. In these cases, the capital recovery period clearly indicated how many years it would take for savings in operating costs to justify a higher capital cost. Also, the method is easier to apply than the cumulative present worth method.

The capital recovery period method has the same disadvantages as those listed for the cumulative present worth method.

2.5.8 Pro Forma Analysis

Under the regulatory compact, electric utilities are mandated to provide electricity at the lowest reasonable cost. This mandate justifies the use of the present value evaluation methodologies presented earlier for evaluation of investment alternatives. In contrast, a potential nonutility generator is not subject to the provisions of the regulatory compact; profits are not regulated, there is no existing obligation to serve, and it is not ensured of the opportunity to earn a reasonable return on investment. Because of this basic difference, a nonregulated business generally evaluates a potential generating facility investment based on the investment's financial returns as projected through a pro forma financial analysis.

The objective of a pro forma analysis is to reflect, as accurately as possible, the future financial cash flows and balances of a proposed project over time. Pro forma statement format varies according to the type of project being evaluated, the availability of detailed input data, and the level of detailed output desired. Typical pro forma analysis input for a generating plant includes the following (Black & Veatch 1990):

1. General operating parameters such as the construction, operating and on/off peak periods, capacity factors, and fuel specifications;
2. Plant capital costs;
3. Financing assumptions;
4. Sales and revenue data;
5. Expense data for fuel and operating and maintenance items;
6. Other cash flow parameters such as depreciation and income tax assumptions.

Once the project-specific input data are derived, the yearly and cumulative financial outputs which summarize the project's financial feasibility are calculated. Typical pro forma output for a proposed generating facility includes the following:

- Yearly cash flows and balances. It is important to measure the yearly cash flows and balances since, even if the total lifetime earnings of a project are satisfactory, ignoring period-to-period cash flows could result in cash shortages and liquidity problems.
- Debt service coverage. The debt service coverage ratio (defined as gross revenues less operating expenses divided by the period's debt service) is important because it measures how well a project can cover its required annual debt repayments. Technically, the debt coverage ratio can be as low as 1.0 to cover the annual debt service payment but should be higher to ensure an adequate after-tax return to the investor.
- Net present value (NPV). The NPV measure determines whether the financial returns will be high enough to justify the project. In pro forma analysis, a project's NPV can be calculated by discounting the after-tax cash inflows to the beginning of the operating period and then subtracting the corresponding future value of the equity investment. If the difference is greater than zero, the project is acceptable, whereas a negative NPV would mean that the project is not acceptable at the chosen discount rate. Technically, if the NPV is negative, the project would not be acceptable since the firm's opportunity cost of money is usually used as the discount rate and a negative NPV would indicate other investment opportunities were more beneficial to the firm's owners. Since the NPV is the theoretically correct measure of whether a project is financially feasible, it is often considered the single most important output of a pro forma analysis.
- The internal rate of return (IRR). The IRR is another measure of a project's overall financial feasibility. The IRR is that discount rate at which the NPV equals zero. An IRR higher than the chosen discount rate indicates the project is acceptable (caution should be used, however, since multiple IRRs may exist for a stream of cash flows if there are sign changes within that stream).
- The return on investment (ROI). The ROI measures how profitable a project is for a given year or over time and is calculated by dividing the after-tax equity cash flow by the amount of the equity investment.

Because the format of a pro forma analysis varies according to the project being evaluated, no specific pro forma model is recommended in this section. For illustration, however, an input/output summary sheet from a pro forma model developed by Black & Veatch (1990) is presented for a hypothetical project in Fig. 2-5. In this example, the NPV is positive, the debt coverage ratio is significantly greater than 1, and the IRR is greater than the discount rate. Therefore, if the year-by-year cash flows and balances are sufficient to avoid a cash flow shortage and if the project remained attractive under sensitivity analyses, it would likely be recommended.

2.5.9 Appropriate Application of the Economic Analysis Methods

Each of the evaluation methods discussed in the previous section is appropriate for particular applications, as listed below:

Application	Economic analysis method
System analyses	Levelized annual cost method or capital recovery period method (if necessary)
Bid analyses	Capital equivalent cost method or capital recovery period (if necessary)
System planning analysis	Cumulative present worth method, total present worth method (sometimes)
Nonutility feasibility study	Pro forma analysis method

2.5.9.1 System Analyses. The levelized annual cost method should normally be used for economic evaluations in system analyses. When selection of a high capital cost plan is justified on the basis of a reduction in operating costs, the annual cost analysis should be supplemented by a capital recovery period analysis. Some utilities accept higher capital cost alternatives only if the capital recovery period is less than half the expected life or 15 years, whichever is less. Those alternatives having the capital recovery periods greater than half the expected life or 15 years must be carefully analyzed before being accepted over an alternative with lower capital costs and higher operating cost. In such cases, sensitivity analysis is especially appropriate.

2.5.9.2 Bid Analyses. The capital equivalent cost method or the differential capacity equivalent cost method should be used for bid analyses. In cases where selection of a high capital cost bid is favored because of savings in operating costs, the capital equivalent cost method should be supplemented by a capital recovery period analysis and is usually subject to the same conditions for acceptance as described previously.

2.5.9.3 System Planning. Since most system planning studies normally involve installations of different size units at different times, the cumulative present worth method is most applicable for this type of study. However, the total present worth method is also occasionally used and annual costs may also be presented for consideration.

2.5.9.4 Nonutility Feasibility Analysis. Unregulated firms evaluate an investment on the basis of its ability to increase shareholder wealth. Thus, rather than undertake a project that minimizes electricity costs, these firms will normally undertake a project only if it has a positive net present value under a reasonable range of assumptions and has adequate yearly cash flows and balances over its economic life. Because of this basic difference between regulated and nonregulated firms, a pro forma analysis is normally used to evaluate the feasibility of a nonutility project.

ENERGY GROUP SHORT FORM PRO FORMA
SUMMARY REPORT

Pro Forma Training - Case 2 (CASE2)
700 MW PC Plant

MAJOR RESULTS

	Minimum	Average	P.V. Average
Financing Coverage Ratios:			
Lease	NA	NA	NA
Total Long-Term Debt	1.628	1.628	1.628

Net Present Values (1998.0 @ 20.0%, $1000):	
After-Tax Cash Flow	222,547
Less: Future Value of Equity Investment	222,547
Net Return on Equity	0
Internal Rate of Return (Project) =	20.00%
Breakeven Year (P.V.) =	2027

OPERATION PERIOD: Start = 1998 Duration (yrs) = 30
Finish = 2027

ELECTRIC GENERATION DATA:

Item	On-Peak	Off-Peak	Total
Hour fraction (%)	100.0%	0.0%	100.0%
Capacity factor (%)	80.0%	0.0%	80.0%
Operating hours	7,008	0	7,008
Base generating capacity (kW)	NA	NA	700000
Add generating capacity (kW)	NA	NA	0
Base electric sales (GWh)	4905.6	0.0	4905.6
Add electric sales (GWh)	0.0	0.0	0.0
Total electric sales (GWh)	4905.6	0.0	4905.6

STEAM DELIVERY FUEL USAGE DATA:

Item	Type of Hours: Stm/Ele	Add Stm	Total
Delivery rate (lb/hr)	0	0	NA
Capacity factor (%)	80.0%	0.0%	80.0%
Operating hours	7,008	0	7,008
Steam sales (Mlb)	0.0	0.0	0.0
Heat input rate (HHV, MBtu/hr)	7,126	0	NA
Heat input (GBtu)	49,939	0	49,939

Fuel specifications (Coal, Solid):
Heating value (HHV) = 8600 Btu/lb
Weight density = 1.00 lb/lb
Content (%) Sulfur = 0.3% Ash = 7.5%
Percent of metal recovered from solid fuel = 0.0%

Fuel burned = 2903431 Tons Fuel not burned = 0 Tons
Fuel received = 2903431 Tons

FGD CHEMICAL USAGE / SOLID WASTE PRODUCTION DATA

Required sulfur removal (%) = 90.0%
Treatment option = Non-FBC plant, dry spray absorber
FGD chemical (Lime) data:
Usage ratio = 1.50 Stoich. S removed
Purity (% Ca) = 90.0% Usage (tons) = 22,865

Solid waste production (tons):
Ash = 217,757
SO2 solids = 40,748 Total = 258,505

ANNUAL INPUT DATA

	Units	1st Year	Total Esc
Electricity Sales Rates:			
On-Peak Energy	$/kWh	0.0178	NA
Off-Peak Energy	$/kWh	0.0000	0.0%
Additional Energy	$/kWh	0.0000	0.0%
Fixed O&M Payment	$/kW-yr	32.45	NA
Capacity Charge	$/kW-yr	238.95	0.0%
Capacity Rate (@ 80.0%)	$/kWh	0.0341	0.0%
Calculated Total	$/kWh	0.0566	NA
Miscellaneous Sales Rates:			
Steam Sales	$/klb	0.000	0.0%
Tipping Fees	$/ton	0.000	0.0%
Metal Recovery	$/ton	0.000	0.0%
Ash Sales	$/ton	0.000	0.0%
Other Revenues:			
	$1000	0	0.0%
	$1000	0	0.0%
	$1000	0	0.0%
Total	$1000	0	0.0%
Fuel Price (Calc. Coal)	$/ton	20.451	4.0%
Fuel Price (Input, Coal)	$/MBtu	1.189	4.0%
Non-Fuel O&M Expenses:			
Variable:			
FGD Chem. (Lime) Price	$/ton	0.000	0.0%
Solid Waste Disposal Cost	$/ton	0.000	0.0%
Base & SCR	$1000	27,001	4.0%
CAA Allowances	$1000	1,058	4.0%
Fixed:			
Operating Labor	$1000	0	0.0%
Maintenance	$1000	0	0.0%
Total Fixed	$1000	11,514	4.0%
	$1000	0	0.0%
Other Operating Expenses:			
Property Taxes	$1000	11,201	NA
Insurance	$1000	0	0.0%
	$1000	0	0.0%
	$1000	0	0.0%
Additional Capital Investments	$1000	7,396	4.0%

TOTAL CAPITAL COST DEPRECIABLE AMOUNT ($1000): < Jan-98 >

	Capital Cost	Deprec. %	Amount $1000
Plant Construction Cost	683,865	100.0%	683,865
Contingency	20,551	100.0%	20,551
Land	1,169	0.0%	0
Total Direct Cost	705,585		
Indirect Construction Costs (ICC):			
Primary Indirect Costs	65,619		
Financing Fees	22,255		
Commitment Fees	5,564		
Total Indirect Construction Costs	93,438	100.0%	93,438
Escalation	121,793	100.0%	121,793
Initial Contribution to Debt Reserve Fund	51,356	0.0%	0
Initial Contribution to Working Capital Fund	12,200	0.0%	0
Interest During Construction (Input)	128,363	100.0%	128,363
Total Capital Cost	1,112,735		1,048,010
Allowable Depreciable Amount (100%) =	1,048,010		

PERMANENT FINANCING BREAKDOWN

Type	Percent	Amount $1000
Lease	0.0%	0
Senior long-term debt	80.0%	890,188
Subordinate long-term debt	0.0%	0
Equity	20.0%	222,547
Total	100.0%	1,112,735

FINANCING OPTIONS

Option	Lease	Senior Debt	Subord Debt
Total ($1000)	0	890,188	0
Term (years)	0	20	0
Interest rate	0.0%	9.7%	0.0%
Other fees	0.0%	0.0%	0.0%
First payment	0	1998	0
Last payment	0	2017	0
Payment ($1000)	0	102,712	0
	(1 x 1 yr)	(1 x 1 yr)	(1 x 1 yr)

CASH FLOW PARAMETERS

Treatment of tax losses option = Carryforward
Income tax rates (%)
Federal = 34.0% Municipal = 0.0%
State = 5.0% Effective = 37.3%

Interest / discount rates (%):
Annual before-tax interest rate = 0.0%
Net present value discount rate = 20.0%

Depreciation expense options:
Calculation method = 150%DB SLN
Calculation convention = Mid-year
Asset life (years) = 20

Construction financing options
Financing type = 100% debt
Input/calculate IDC = Input
Construction annual interest rate (%) = 0.0%
Construction period duration (months) = 0

05/17/93

Figure 2-5. Pro forma summary report.

ENERGY GROUP SHORT FORM PRO FORMA
DETAILED CASH FLOW REPORT

Pro Forma Training - Case 2 (CASE2)
700 MW PC Plant

Item	Units	NPV @ 20.0%	1998	1999	2000	2001	2002	2003	2004	2005	2006	2007	2008	2009	2010	2011	2012
OPERATING REVENUES:																	
Electric Sales:																	
Base Energy Sales																	
On-Peak	$1000	539,011	87,436	90,934	94,571	98,354	102,288	106,380	110,635	115,060	119,663	124,449	129,427	134,604	139,989	145,588	151,412
Off-Peak	$1000	0	0	0	0	0	0	0	0	0	0	0	0	0	0	0	0
Total	$1000	539,011	87,436	90,934	94,571	98,354	102,288	106,380	110,635	115,060	119,663	124,449	129,427	134,604	139,989	145,588	151,412
Additional Energy Sales	$1000	0	0	0	0	0	0	0	0	0	0	0	0	0	0	0	0
Fixed O&M Payment	$1000	129,030	22,715	23,253	23,812	24,393	24,998	25,626	26,280	26,960	27,668	28,403	29,168	29,964	30,791	31,652	32,547
Capacity Sales	$1000	832,799	167,264	167,264	167,264	167,264	167,264	167,264	167,264	167,264	167,264	167,264	167,264	167,264	167,264	167,264	167,264
Total Electric Sales	$1000	1,500,840	277,416	281,451	285,647	290,012	294,550	299,271	304,180	309,285	314,595	320,117	325,860	331,833	338,044	344,504	351,223
Miscellaneous Sales:																	
Steam Sales	$1000	0	0	0	0	0	0	0	0	0	0	0	0	0	0	0	0
Tipping Fees	$1000	0	0	0	0	0	0	0	0	0	0	0	0	0	0	0	0
Metal Recovery Sales	$1000	0	0	0	0	0	0	0	0	0	0	0	0	0	0	0	0
Ash Sales	$1000	0	0	0	0	0	0	0	0	0	0	0	0	0	0	0	0
Total Miscellaneous Sales	$1000	0	0	0	0	0	0	0	0	0	0	0	0	0	0	0	0
Other Revenues:																	
	$1000	0	0	0	0	0	0	0	0	0	0	0	0	0	0	0	0
	$1000	0	0	0	0	0	0	0	0	0	0	0	0	0	0	0	0
	$1000	0	0	0	0	0	0	0	0	0	0	0	0	0	0	0	0
Total Other Revenues	$1000	0	0	0	0	0	0	0	0	0	0	0	0	0	0	0	0
TOTAL OPERATING REVENUES	$1000	1,500,840	277,416	281,451	285,647	290,012	294,550	299,271	304,180	309,285	314,595	320,117	325,860	331,833	338,044	344,504	351,223
OPERATING EXPENSES:																	
Fuel Expenses (Coal)	$1000	366,039	59,377	61,753	64,223	66,792	69,463	72,242	75,131	78,137	81,262	84,513	87,893	91,409	95,065	98,868	102,823
Non-Fuel O&M Expenses:																	
Variable:																	
FGD Chem (Lime) Usage	$1000	0	0	0	0	0	0	0	0	0	0	0	0	0	0	0	0
Solid Waste Disposal	$1000	0	0	0	0	0	0	0	0	0	0	0	0	0	0	0	0
Base & SCR	$1000	166,450	27,001	28,081	29,204	30,372	31,587	32,851	34,165	35,531	36,953	38,431	39,968	41,567	43,229	44,959	46,757
CAA Allowances	$1000	6,522	1,058	1,100	1,144	1,190	1,238	1,287	1,339	1,392	1,448	1,506	1,566	1,629	1,694	1,762	1,832
Total	$1000	172,973	28,059	29,181	30,349	31,563	32,825	34,138	35,504	36,924	38,401	39,937	41,534	43,196	44,923	46,720	48,589
Fixed:																	
Operating Labor	$1000	0	0	0	0	0	0	0	0	0	0	0	0	0	0	0	0
Maintenance	$1000	0	0	0	0	0	0	0	0	0	0	0	0	0	0	0	0
Total Fixed	$1000	70,979	11,514	11,975	12,454	12,952	13,470	14,009	14,569	15,152	15,758	16,388	17,044	17,725	18,434	19,172	19,939
	$1000	0	0	0	0	0	0	0	0	0	0	0	0	0	0	0	0
Total	$1000	70,979	11,514	11,975	12,454	12,952	13,470	14,009	14,569	15,152	15,758	16,388	17,044	17,725	18,434	19,172	19,939
Total Non-Fuel O&M Expenses	$1000	243,952	39,573	41,156	42,802	44,514	46,295	48,147	50,072	52,075	54,158	56,325	58,578	60,921	63,358	65,892	68,528
Other Operating Expenses:																	
Property Taxes	$1000	58,051	11,201	11,278	11,358	11,441	11,528	11,618	11,712	11,809	11,910	12,015	12,125	12,239	12,357	12,480	12,608
Insurance	$1000	0	0	0	0	0	0	0	0	0	0	0	0	0	0	0	0
	$1000	0	0	0	0	0	0	0	0	0	0	0	0	0	0	0	0
	$1000	0	0	0	0	0	0	0	0	0	0	0	0	0	0	0	0
Total Other Operating Expenses	$1000	58,051	11,201	11,278	11,358	11,441	11,528	11,618	11,712	11,809	11,910	12,015	12,125	12,239	12,357	12,480	12,608
TOTAL OPERATING EXPENSES	$1000	668,041	110,152	114,187	118,383	122,747	127,286	132,006	136,915	142,021	147,331	152,853	158,596	164,568	170,780	177,240	183,959
CASH AVAILABLE FOR FINANCING EXPENSES	$1000	832,799	167,264	167,264	167,264	167,264	167,264	167,264	167,264	167,264	167,264	167,264	167,264	167,264	167,264	167,264	167,264

05 17 93

ENERGY GROUP SHORT FORM PRO FORMA
DETAILED CASH FLOW REPORT

Pro Forma Training - Case 2 (CASE2)
700 MW PC Plant

Item	Units	NPV @ 20.0%	2013	2014	2015	2016	2017	2018	2019	2020	2021	2022	2023	2024	2025	2026	2027
OPERATING REVENUES																	
Electric Sales:																	
Base Energy Sales																	
On-Peak	$1000	539.011	157.468	163.767	170.318	177.130	184.215	191.584	199.247	207.217	215.506	224.126	233.091	242.415	252.112	262.196	272.684
Off-Peak	$1000	0	0	0	0	0	0	0	0	0	0	0	0	0	0	0	0
Total	$1000	539.011	157.468	163.767	170.318	177.130	184.215	191.584	199.247	207.217	215.506	224.126	233.091	242.415	252.112	262.196	272.684
Additional Energy Sales	$1000	0	0	0	0	0	0	0	0	0	0	0	0	0	0	0	0
Fixed O&M Payment	$1000	129.030	33.478	34.446	35.452	36.499	37.588	38.720	39.898	41.123	42.397	43.721	45.099	46.532	48.022	49.572	51.184
Capacity Sales	$1000	832.799	167.264	167.264	167.264	167.264	167.264	167.264	167.264	167.264	167.264	167.264	167.264	167.264	167.264	167.264	167.264
Total Electric Sales	$1000	1,500.840	358.210	365.477	373.034	380.894	389.068	397.569	406.410	415.605	425.167	435.112	445.455	456.211	467.398	479.032	491.132
Miscellaneous Sales:																	
Steam Sales	$1000	0	0	0	0	0	0	0	0	0	0	0	0	0	0	0	0
Tipping Fees	$1000	0	0	0	0	0	0	0	0	0	0	0	0	0	0	0	0
Metal Recovery Sales	$1000	0	0	0	0	0	0	0	0	0	0	0	0	0	0	0	0
Ash Sales	$1000	0	0	0	0	0	0	0	0	0	0	0	0	0	0	0	0
Total Miscellaneous Sales	$1000	0	0	0	0	0	0	0	0	0	0	0	0	0	0	0	0
Other Revenues																	
	$1000	0	0	0	0	0	0	0	0	0	0	0	0	0	0	0	0
	$1000	0	0	0	0	0	0	0	0	0	0	0	0	0	0	0	0
	$1000	0	0	0	0	0	0	0	0	0	0	0	0	0	0	0	0
Total Other Revenues	$1000	0	0	0	0	0	0	0	0	0	0	0	0	0	0	0	0
TOTAL OPERATING REVENUES	$1000	1,500.840	358.210	365.477	373.034	380.894	389.068	397.569	406.410	415.605	425.167	435.112	445.455	456.211	467.398	479.032	491.132
OPERATING EXPENSES																	
Fuel Expenses (Coal)	$1000	366.039	106.935	111.213	115.661	120.288	125.099	130.103	135.308	140.720	146.349	152.203	158.291	164.622	171.207	178.055	185.178
Non Fuel O&M Expenses																	
Variable																	
FGD Chem (Lime) Usage	$1000	0	0	0	0	0	0	0	0	0	0	0	0	0	0	0	0
Solid Waste Disposal	$1000	0	0	0	0	0	0	0	0	0	0	0	0	0	0	0	0
Base & SCR	$1000	166.450	48.627	50.572	52.595	54.699	56.887	59.163	61.529	63.990	66.550	69.212	71.980	74.859	77.854	80.968	84.207
CAA Allowances	$1000	6.522	1.905	1.982	2.061	2.143	2.229	2.318	2.411	2.507	2.608	2.712	2.820	2.933	3.051	3.173	3.300
Total	$1000	172.973	50.533	52.554	54.656	56.842	59.116	61.481	63.940	66.498	69.157	71.924	74.801	77.793	80.904	84.141	87.506
Fixed																	
Operating Labor	$1000	0	0	0	0	0	0	0	0	0	0	0	0	0	0	0	0
Maintenance	$1000	0	0	0	0	0	0	0	0	0	0	0	0	0	0	0	0
Total Fixed	$1000	70.979	20.736	21.566	22.428	23.325	24.258	25.229	26.238	27.287	28.379	29.514	30.694	31.922	33.199	34.527	35.908
	$1000	0	0	0	0	0	0	0	0	0	0	0	0	0	0	0	0
Total	$1000	70.979	20.736	21.566	22.428	23.325	24.258	25.229	26.238	27.287	28.379	29.514	30.694	31.922	33.199	34.527	35.908
Total Non-Fuel O&M Expenses	$1000	243.952	71.269	74.119	77.084	80.168	83.374	86.709	90.178	93.785	97.536	101.438	105.495	109.715	114.104	118.668	123.414
Other Operating Expenses																	
Property Taxes	$1000	58.051	12.741	12.880	13.024	13.174	13.330	13.492	13.660	13.836	14.018	14.207	14.405	14.610	14.823	15.045	15.275
Insurance	$1000	0	0	0	0	0	0	0	0	0	0	0	0	0	0	0	0
	$1000	0	0	0	0	0	0	0	0	0	0	0	0	0	0	0	0
	$1000	0	0	0	0	0	0	0	0	0	0	0	0	0	0	0	0
Total Other Operating Expenses	$1000	58.051	12.741	12.880	13.024	13.174	13.330	13.492	13.660	13.836	14.018	14.207	14.405	14.610	14.823	15.045	15.275
TOTAL OPERATING EXPENSES	$1000	668.041	190.946	198.212	205.770	213.629	221.803	230.304	239.146	248.340	257.903	267.848	278.190	288.947	300.134	311.768	323.867
CASH AVAILABLE FOR FINANCING EXPENSES	$1000	832.799	167.264	167.264	167.264	167.264	167.264	167.264	167.264	167.264	167.264	167.264	167.264	167.264	167.264	167.264	167.264

05 17 93

(continued)

ENERGY GROUP SHORT FORM PRO FORMA
DETAILED CASH FLOW REPORT

Pro Forma Training - Case 2 (CASE2)
700 MW PC Plant

Item	Units	NPV @ 20.0%	1998	1999	2000	2001	2002	2003	2004	2005	2006	2007	2008	2009	2010	2011	2012
CASH AVAILABLE FOR FINANCING EXPENSES	$1000	832.799	167.264	167.264	167.264	167.264	167.264	167.264	167.264	167.264	167.264	167.264	167.264	167.264	167.264	167.264	167.264
FINANCING EXPENSES																	
Lease Expenses:																	
Lease Payment	$1000	0	0	0	0	0	0	0	0	0	0	0	0	0	0	0	0
Other Fees	$1000	0	0	0	0	0	0	0	0	0	0	0	0	0	0	0	0
Total Lease Expenses	$1000	0	0	0	0	0	0	0	0	0	0	0	0	0	0	0	0
Long-Term Debt Expenses:																	
Debt Service																	
Beginning Balance	$1000		890.188	874.181	856.614	837.337	816.181	792.966	767.489	739.531	708.849	675.179	638.230	597.682	553.185	504.353	450.765
Total Principal	$1000	129.905	16.007	17.567	19.278	21.155	23.216	25.477	27.958	30.681	33.670	36.949	40.548	44.497	48.832	53.588	58.807
Total Interest	$1000	370.258	86.704	85.145	83.434	81.557	79.496	77.235	74.753	72.030	69.042	65.762	62.164	58.214	53.880	49.124	43.905
Total Debt Service	$1000	500.163	102.712	102.712	102.712	102.712	102.712	102.712	102.712	102.712	102.712	102.712	102.712	102.712	102.712	102.712	102.712
Other Fees	$1000	0	0	0	0	0	0	0	0	0	0	0	0	0	0	0	0
Total Long-Term Debt Expenses	$1000	500.163	102.712	102.712	102.712	102.712	102.712	102.712	102.712	102.712	102.712	102.712	102.712	102.712	102.712	102.712	102.712
TOTAL FINANCING EXPENSES	$1000	500.163	102.712	102.712	102.712	102.712	102.712	102.712	102.712	102.712	102.712	102.712	102.712	102.712	102.712	102.712	102.712
BEFORE-TAX OPERATING CASH FLOW	$1000	332.636	64.553	64.553	64.553	64.553	64.553	64.553	64.553	64.553	64.553	64.553	64.553	64.553	64.553	64.553	64.553
DEPRECIATION EXPENSES																	
Initial Capital Cost	$1000	273.712	39.300	75.653	69.979	64.731	59.876	55.385	51.231	47.389	46.757	46.757	46.757	46.757	46.757	46.757	46.757
Annual Capital Investments	$1000	12.189	247	512	797	1.106	1.438	1.798	2.188	2.611	3.071	3.573	4.120	4.719	5.377	6.102	6.902
TOTAL DEPRECIATION EXPENSES	$1000	285.901	39.547	76.165	70.777	65.836	61.314	57.184	53.420	50.000	49.829	50.330	50.877	51.477	52.134	52.859	53.659
TAXABLE INCOME	$1000	176.640	41.013	5.954	13.054	19.871	26.454	32.846	39.091	45.234	48.394	51.172	54.223	57.574	61.250	65.282	69.701
ADJUSTED TAXABLE INCOME	$1000		41.013	5.954	13.054	19.871	26.454	32.846	39.091	45.234	48.394	51.172	54.223	57.574	61.250	65.282	69.701
INCOME TAXES																	
Municipal	$1000	0	0	0	0	0	0	0	0	0	0	0	0	0	0	0	0
State	$1000	8.832	2.051	298	653	994	1.323	1.642	1.955	2.262	2.420	2.559	2.711	2.879	3.062	3.264	3.485
Federal	$1000	57.055	13.247	1.923	4.216	6.418	8.545	10.609	12.627	14.610	15.631	16.529	17.514	18.596	19.784	21.086	22.513
TOTAL INCOME TAXES	$1000	65.887	15.298	2.221	4.869	7.412	9.867	12.252	14.581	16.872	18.051	19.087	20.225	21.475	22.846	24.350	25.998
AFTER-TAX OPERATING CASH FLOW	$1000	266.749	49.255	62.332	59.684	57.141	54.685	52.301	49.972	47.681	46.502	45.466	44.327	43.078	41.707	40.203	38.554
RESERVE FUND TRANSFERS																	
Debt Reserve Fund	$1000	1.340	0	0	0	0	0	0	0	0	0	0	0	0	0	0	0
Working Capital Fund	$1000	51	0	0	0	0	0	0	0	0	0	0	0	0	0	0	0
TOTAL RESERVE FUND TRANSFERS	$1000	1.391	0	0	0	0	0	0	0	0	0	0	0	0	0	0	0
ADDITIONAL CAPITAL INVESTMENTS	$1000	45.593	7.396	7.692	8.000	8.319	8.652	8.998	9.358	9.733	10.122	10.527	10.948	11.386	11.841	12.315	12.807
AFTER-TAX INTEREST INCOME	$1000	0	0	0	0	0	0	0	0	0	0	0	0	0	0	0	0
AFTER-TAX CASH FLOW	$1000	222.547	41.859	54.640	51.684	48.821	46.033	43.303	40.613	37.948	36.380	34.939	33.379	31.692	29.865	27.888	25.747
CUM PW NET CASH DISTRIBUTION	$1000		(187.665)	(149.720)	(119.810)	(96.266)	(77.767)	(63.265)	(51.930)	(43.105)	(36.054)	(30.411)	(25.919)	(22.364)	(19.573)	(17.401)	(15.730)
FINANCING COVERAGE RATIOS:																	
Lease			0.000	0.000	0.000	0.000	0.000	0.000	0.000	0.000	0.000	0.000	0.000	0.000	0.000	0.000	0.000
Long-Term Debt			1.628	1.628	1.628	1.628	1.628	1.628	1.628	1.628	1.628	1.628	1.628	1.628	1.628	1.628	1.628

05 17 93

ENERGY GROUP SHORT FORM PRO FORMA
DETAILED CASH FLOW REPORT

Pro Forma Training - Case 2 (CASE2)
700 MW PC Plant

Item	Units	NPV @ 20.0%	2013	2014	2015	2016	2017	2018	2019	2020	2021	2022	2023	2024	2025	2026	2027
CASH AVAILABLE FOR FINANCING EXPENSES	$1000	832,799	167,264	167,264	167,264	167,264	167,264	167,264	167,264	167,264	167,264	167,264	167,264	167,264	167,264	167,264	167,264
FINANCING EXPENSES:																	
Lease Expenses:																	
Lease Payment	$1000	0	0	0	0	0	0	0	0	0	0	0	0	0	0	0	0
Other Fees	$1000	0	0	0	0	0	0	0	0	0	0	0	0	0	0	0	0
Total Lease Expenses	$1000	0	0	0	0	0	0	0	0	0	0	0	0	0	0	0	0
Long-Term Debt Expenses:																	
Debt Service:																	
Beginning Balance	$1000		391,958	327,423	256,603	178,884	93,596	0	0	0	0	0	0	0	0	0	0
Total Principal	$1000	129,905	64,535	70,821	77,719	85,288	93,596	0	0	0	0	0	0	0	0	0	0
Total Interest	$1000	370,258	38,177	31,891	24,993	17,423	9,116	0	0	0	0	0	0	0	0	0	0
Total Debt Service	$1000	500,163	102,712	102,712	102,712	102,712	102,712	0	0	0	0	0	0	0	0	0	0
Other Fees	$1000	0	0	0	0	0	0	0	0	0	0	0	0	0	0	0	0
Total Long-Term Debt Expenses	$1000	500,163	102,712	102,712	102,712	102,712	102,712	0	0	0	0	0	0	0	0	0	0
TOTAL FINANCING EXPENSES	$1000	500,163	102,712	102,712	102,712	102,712	102,712	0	0	0	0	0	0	0	0	0	0
BEFORE-TAX OPERATING CASH FLOW	$1000	332,636	64,553	64,553	64,553	64,553	64,553	167,264	167,264	167,264	167,264	167,264	167,264	167,264	167,264	167,264	167,264
DEPRECIATION EXPENSES:																	
Initial Capital Cost	$1000	273,712	46,757	46,757	46,757	46,757	46,757	23,379	0	0	0	0	0	0	0	0	0
Annual Capital Investments	$1000	12,189	7,790	8,780	9,888	11,136	12,553	14,173	16,046	18,237	20,841	24,001	27,944	33,071	40,179	51,268	74,334
TOTAL DEPRECIATION EXPENSES	$1000	285,901	54,547	55,537	56,645	57,894	59,310	37,552	16,046	18,237	20,841	24,001	27,944	33,071	40,179	51,268	74,334
TAXABLE INCOME	$1000	176,640	74,540	79,837	85,626	91,948	98,838	129,712	151,218	149,027	146,423	143,263	139,320	134,194	127,085	115,996	92,931
ADJUSTED TAXABLE INCOME	$1000		74,540	79,837	85,626	91,948	98,838	129,712	151,218	149,027	146,423	143,263	139,320	134,194	127,085	115,996	92,931
INCOME TAXES																	
Municipal	$1000	0	0	0	0	0	0	0	0	0	0	0	0	0	0	0	0
State	$1000	8,832	3,727	3,992	4,281	4,597	4,942	6,486	7,561	7,451	7,321	7,163	6,966	6,710	6,354	5,800	4,647
Federal	$1000	57,055	24,077	25,787	27,657	29,699	31,925	41,897	48,844	48,136	47,295	46,274	45,000	43,345	41,049	37,467	30,017
TOTAL INCOME TAXES	$1000	65,887	27,804	29,779	31,939	34,296	36,867	48,383	56,404	55,587	54,616	53,437	51,966	50,054	47,403	43,267	34,663
AFTER-TAX OPERATING CASH FLOW	$1000	266,749	36,749	34,774	32,614	30,256	27,686	118,882	110,860	111,677	112,649	113,827	115,298	117,210	119,862	123,998	132,601
RESERVE FUND TRANSFERS:																	
Debt Reserve Fund	$1000	1,340	0	0	0	0	51,356	0	0	0	0	0	0	0	0	0	0
Working Capital Fund	$1000	51	0	0	0	0	0	0	0	0	0	0	0	0	0	0	12,200
TOTAL RESERVE FUND TRANSFERS	$1000	1,391	0	0	0	0	51,356	0	0	0	0	0	0	0	0	0	12,200
ADDITIONAL CAPITAL INVESTMENTS	$1000	45,593	13,320	13,853	14,407	14,983	15,582	16,206	16,854	17,528	18,229	18,958	19,717	20,505	21,325	22,178	23,066
AFTER-TAX INTEREST INCOME	$1000	0	0	0	0	0	0	0	0	0	0	0	0	0	0	0	0
AFTER-TAX CASH FLOW	$1000	222,547	23,429	20,921	18,207	15,273	63,460	102,676	94,006	94,149	94,420	94,869	95,581	96,705	98,536	101,819	121,736
CUM PW NET CASH DISTRIBUTION	$1000		(14,462)	(13,520)	(12,836)	(12,358)	(10,702)	(8,470)	(6,768)	(5,346)	(4,159)	(3,164)	(2,329)	(1,625)	(1,028)	(513)	0
FINANCING COVERAGE RATIOS:																	
Lease			0.000	0.000	0.000	0.000	0.000	0.000	0.000	0.000	0.000	0.000	0.000	0.000	0.000	0.000	0.000
Long-Term Debt			1.628	1.628	1.628	1.628	1.628	0.000	0.000	0.000	0.000	0.000	0.000	0.000	0.000	0.000	0.000

 05/17/93

(continued)

ENERGY GROUP SHORT FORM PRO FORMA
DETAILED ANNUAL INPUT DATA REPORT

Pro Forma Training - Case 2 (CASE2)
700 MW PC Plant

	Units	1st Year	Total Escal	(Start) 1998	1999	2000	2001	2002	2003	2004	2005	2006	2007	2008	2009	2010	2011	2012
Electricity Sales Rates:																		
On-Peak Energy	$/kWh	0.0178	NA	0.0178	0.0185	0.0193	0.0200	0.0209	0.0217	0.0226	0.0235	0.0244	0.0254	0.0264	0.0274	0.0285	0.0297	0.0309
Off-Peak Energy	$/kWh	0.0000	0.0%	0.0000	0.0000	0.0000	0.0000	0.0000	0.0000	0.0000	0.0000	0.0000	0.0000	0.0000	0.0000	0.0000	0.0000	0.0000
Additional Energy	$/kWh	0.0000	0.0%	0.0000	0.0000	0.0000	0.0000	0.0000	0.0000	0.0000	0.0000	0.0000	0.0000	0.0000	0.0000	0.0000	0.0000	0.0000
Fixed O&M Payment	$/kW-yr	32.45	NA	32.45	33.22	34.02	34.85	35.71	36.61	37.54	38.51	39.53	40.58	41.67	42.81	43.99	45.22	46.50
Capacity Charge	$/kW-yr	238.95	0.0%	238.95	238.95	238.95	238.95	238.95	238.95	238.95	238.95	238.95	238.95	238.95	238.95	238.95	238.95	238.95
Capacity Rate (@ 80.0%)	$/kWh	0.0341	0.0%	0.0341	0.0341	0.0341	0.0341	0.0341	0.0341	0.0341	0.0341	0.0341	0.0341	0.0341	0.0341	0.0341	0.0341	0.0341
Calculated Total	$/kWh	0.0566	NA	0.0566	0.0574	0.0582	0.0591	0.0600	0.0610	0.0620	0.0630	0.0641	0.0653	0.0664	0.0676	0.0689	0.0702	0.0716
Miscellaneous Sales Rates:																		
Steam Sales	$/klb	0.000	0.0%	0.000	0.000	0.000	0.000	0.000	0.000	0.000	0.000	0.000	0.000	0.000	0.000	0.000	0.000	0.000
Tipping Fees	$/ton	0.000	0.0%	0.000	0.000	0.000	0.000	0.000	0.000	0.000	0.000	0.000	0.000	0.000	0.000	0.000	0.000	0.000
Metal Recovery	$/ton	0.000	0.0%	0.000	0.000	0.000	0.000	0.000	0.000	0.000	0.000	0.000	0.000	0.000	0.000	0.000	0.000	0.000
Ash Sales	$/ton	0.000	0.0%	0.000	0.000	0.000	0.000	0.000	0.000	0.000	0.000	0.000	0.000	0.000	0.000	0.000	0.000	0.000
Other Revenues:																		
	$1000	0	0.0%	0	0	0	0	0	0	0	0	0	0	0	0	0	0	0
	$1000	0	0.0%	0	0	0	0	0	0	0	0	0	0	0	0	0	0	0
	$1000	0	0.0%	0	0	0	0	0	0	0	0	0	0	0	0	0	0	0
Total	$1000	0	0.0%	0	0	0	0	0	0	0	0	0	0	0	0	0	0	0
Fuel Price (Calc. Coal)	$/ton	20.451	4.0%	20.451	21.269	22.120	23.004	23.925	24.882	25.877	26.912	27.988	29.108	30.272	31.483	32.742	34.052	35.414
Fuel Price (Input. Coal)	$/MBtu	1.189	4.0%	1.189	1.237	1.286	1.337	1.391	1.447	1.504	1.565	1.627	1.692	1.760	1.830	1.904	1.980	2.059
Non-Fuel O&M Expenses:																		
Variable:																		
FGD Chem. (Lime) Price	$/ton	0.000	0.0%	0.000	0.000	0.000	0.000	0.000	0.000	0.000	0.000	0.000	0.000	0.000	0.000	0.000	0.000	0.000
Solid Waste Disposal Cost	$/ton	0.000	0.0%	0.000	0.000	0.000	0.000	0.000	0.000	0.000	0.000	0.000	0.000	0.000	0.000	0.000	0.000	0.000
Base & SCR	$1000	27.001	4.0%	27.001	28.081	29.204	30.372	34.587	32.851	34.165	35.531	36.953	38.431	39.968	41.567	43.229	44.959	46.757
CAA Allowances	$1000	1.058	4.0%	1.058	1.100	1.144	1.190	1.238	1.287	1.339	1.392	1.448	1.506	1.566	1.629	1.694	1.762	1.832
Fixed:																		
Operating Labor	$1000	0	0.0%	0	0	0	0	0	0	0	0	0	0	0	0	0	0	0
Maintenance	$1000	0	0.0%	0	0	0	0	0	0	0	0	0	0	0	0	0	0	0
Total Fixed	$1000	11.514	4.0%	11.514	11.975	12.454	12.952	13.470	14.009	14.569	15.152	15.758	16.388	17.044	17.725	18.434	19.172	19.939
	$1000	0	0.0%	0	0	0	0	0	0	0	0	0	0	0	0	0	0	0
Other Operating Expenses:	$1000	11.201	NA	11.201	11.278	11.358	11.441	11.528	11.618	11.712	11.809	11.910	12.015	12.125	12.239	12.357	12.480	12.608
Property Taxes	$1000	0	0.0%	0	0	0	0	0	0	0	0	0	0	0	0	0	0	0
Insurance	$1000	0	0.0%	0	0	0	0	0	0	0	0	0	0	0	0	0	0	0
	$1000	0	0.0%	0	0	0	0	0	0	0	0	0	0	0	0	0	0	0
Additional Capital Investments	$1000	7.396	4.0%	7.396	7.692	8.000	8.319	8.652	8.998	9.358	9.733	10.122	10.527	10.948	11.386	11.841	12.315	12.807
Depreciation Expense	%			3.750%	7.219%	6.677%	6.177%	5.713%	5.285%	4.888%	4.522%	4.462%	4.462%	4.462%	4.462%	4.462%	4.462%	4.462%

ENERGY GROUP SHORT FORM PRO FORMA
DETAILED ANNUAL INPUT DATA REPORT

Pro Forma Training - Case 2 (CASE2)
700 MW PC Plant

	Units	1st Year	Total Escal	2013	2014	2015	2016	2017	2018	2019	2020	2021	2022	2023	2024	2025	2026	(Finish) 2027
Electricity Sales Rates																		
On-Peak Energy	$/kWh	0.0178	NA	0.0321	0.0334	0.0347	0.0361	0.0376	0.0391	0.0406	0.0422	0.0439	0.0457	0.0475	0.0494	0.0514	0.0534	0.0556
Off-Peak Energy	$/kWh	0.0000	0.0%	0.0000	0.0000	0.0000	0.0000	0.0000	0.0000	0.0000	0.0000	0.0000	0.0000	0.0000	0.0000	0.0000	0.0000	0.0000
Additional Energy	$/kWh	0.0000	0.0%	0.0000	0.0000	0.0000	0.0000	0.0000	0.0000	0.0000	0.0000	0.0000	0.0000	0.0000	0.0000	0.0000	0.0000	0.0000
Fixed O&M Payment	$/kW-yr	32.45	NA	47.83	49.21	50.65	52.14	53.70	55.31	57.00	58.75	60.57	62.46	64.43	66.47	68.60	70.82	73.12
Capacity Charge	$/kW-yr	238.95	0.0%	238.95	238.95	238.95	238.95	238.95	238.95	238.95	238.95	238.95	238.95	238.95	238.95	238.95	238.95	238.95
Capacity Rate (@ 80.0%)	$/kWh	0.0341	0.0%	0.0341	0.0341	0.0341	0.0341	0.0341	0.0341	0.0341	0.0341	0.0341	0.0341	0.0341	0.0341	0.0341	0.0341	0.0341
Calculated Total	$/kWh	0.0566	NA	0.0730	0.0745	0.0760	0.0776	0.0793	0.0810	0.0828	0.0847	0.0867	0.0887	0.0908	0.0930	0.0953	0.0977	0.1001
Miscellaneous Sales Rates:																		
Steam Sales	$/klb	0.000	0.0%	0.000	0.000	0.000	0.000	0.000	0.000	0.000	0.000	0.000	0.000	0.000	0.000	0.000	0.000	0.000
Tipping Fees	$/ton	0.000	0.0%	0.000	0.000	0.000	0.000	0.000	0.000	0.000	0.000	0.000	0.000	0.000	0.000	0.000	0.000	0.000
Metal Recovery	$/ton	0.000	0.0%	0.000	0.000	0.000	0.000	0.000	0.000	0.000	0.000	0.000	0.000	0.000	0.000	0.000	0.000	0.000
Ash Sales	$/ton	0.000	0.0%	0.000	0.000	0.000	0.000	0.000	0.000	0.000	0.000	0.000	0.000	0.000	0.000	0.000	0.000	0.000
Other Revenues:																		
	$1000	0	0.0%	0	0	0	0	0	0	0	0	0	0	0	0	0	0	0
	$1000	0	0.0%	0	0	0	0	0	0	0	0	0	0	0	0	0	0	0
	$1000	0	0.0%	0	0	0	0	0	0	0	0	0	0	0	0	0	0	0
Total	$1000	0	0.0%	0	0	0	0	0	0	0	0	0	0	0	0	0	0	0
Fuel Price (Calc. Coal)	$/ton	20.451	4.0%	36.831	38.304	39.836	41.430	43.087	44.810	46.603	48.467	50.405	52.422	54.518	56.699	58.967	61.326	63.779
Fuel Price (Input. Coal)	$/MBtu	1.189	4.0%	2.141	2.227	2.316	2.409	2.505	2.605	2.709	2.818	2.931	3.048	3.170	3.296	3.428	3.565	3.708
Non-Fuel O&M Expenses:																		
Variable:																		
FGD Chem (Lime) Price	$/ton	0.000	0.0%	0.000	0.000	0.000	0.000	0.000	0.000	0.000	0.000	0.000	0.000	0.000	0.000	0.000	0.000	0.000
Solid Waste Disposal Cost	$/ton	0.000	0.0%	0.000	0.000	0.000	0.000	0.000	0.000	0.000	0.000	0.000	0.000	0.000	0.000	0.000	0.000	0.000
Base & SCR	$1000	27.001	4.0%	48.627	50.572	52.595	54.699	56.887	59.163	61.529	63.990	66.550	69.212	71.980	74.859	77.854	80.968	84.207
CAA Allowances	$1000	1.058	4.0%	1.905	1.982	2.061	2.143	2.229	2.318	2.411	2.507	2.608	2.712	2.820	2.933	3.051	3.173	3.300
Fixed:																		
Operating Labor	$1000	0	0.0%	0	0	0	0	0	0	0	0	0	0	0	0	0	0	0
Maintenance	$1000	0	0.0%	0	0	0	0	0	0	0	0	0	0	0	0	0	0	0
Total Fixed	$1000	11.514	4.0%	20.736	21.566	22.428	23.325	24.258	25.229	26.238	27.287	28.379	29.514	30.694	31.922	33.199	34.527	35.908
	$1000	0	0.0%	0	0	0	0	0	0	0	0	0	0	0	0	0	0	0
Other Operating Expenses:																		
Property Taxes	$1000	11.201	NA	12.741	12.880	13.024	13.174	13.330	13.492	13.660	13.836	14.018	14.207	14.405	14.610	14.823	15.045	15.275
Insurance	$1000	0	0.0%	0	0	0	0	0	0	0	0	0	0	0	0	0	0	0
	$1000	0	0.0%	0	0	0	0	0	0	0	0	0	0	0	0	0	0	0
	$1000	0	0.0%	0	0	0	0	0	0	0	0	0	0	0	0	0	0	0
Additional Capital Investments	$1000	7.396	4.0%	13.320	13.853	14.407	14.983	15.582	16.206	16.854	17.528	18.229	18.958	19.717	20.505	21.325	22.178	23.066
Depreciation Expense	%			4.462%	4.462%	4.462%	4.462%	4.462%	2.231%	0.000%	0.000%	0.000%	0.000%	0.000%	0.000%	0.000%	0.000%	0.000%

2.6 SUMMARY

An electric utility must provide a reliable supply of electricity at the lowest reasonable cost. Proper evaluation of an alternative's cost requires the consideration of the time value of money. Several present value equations have been introduced in this chapter and are included in Appendix 2D. When applied to the cost of various alternatives, these present value equations allow the engineer to determine which alternative is the least-cost option. The identification of the least-cost option allows the utility to implement the project that represents the lowest dollar outlay for consumers and is consistent with the reasonable rate mandate placed on utilities. Caution should be used when recommending an alternative, however, since future costs are always subject to uncertainty. Before an alternative is recommended, sensitivity analyses should be performed.

Since the late 1970s, the electric generation market has experienced a significant increase in the amount of nonutility generated power. Much of this increase is attributable to PURPA. The increasing importance of nonutility generation requires the engineer to be able to evaluate the financial feasibility of potential nonutility projects. As opposed to a regulated utility project which is evaluated according to its ratepayer cost, a nonutility project is typically evaluated by its financial returns as projected in a pro forma analysis.

Specific economic analysis methods appropriate for both utility and nonutility evaluation are presented in Section 2.5 of this chapter. These methods may be applied to system planning studies, bid analyses, system analyses, and for a project not subject to regulation, financial feasibility studies.

2.7 GLOSSARY

AFUDC—(allowance for funds used during construction) The method of accounting for the cost of debt and equity money used by the utility during construction, prior to the project's commercial operation date. An AFUDC rate is applied to the total of all direct and indirect investment costs and is usually set equal to the weighted cost of capital.

Annuity See Uniform series.

Capital equivalent cost method A present value evaluation method used primarily in bid analyses in which operating costs are expressed as a capital equivalent and added to an alternative's actual capital costs.

Capital recovery period method A present value evaluation method sometimes used for system and bid analyses which identifies the year in which the cumulative savings in operating costs for a higher capital cost plan completely offset the differential capital investment of that plan.

Compounding The application of a rate of growth for more than one period.

Cumulative present worth method The present value evaluation method used in some system planning studies in which the present value of an investment's yearly costs are stated, often in graphical form, on a year-by-year basis throughout the life of the investment.

Depreciation A fixed cost component through which the original cost of an investment is recovered over its useful life. Depreciation is a cost of service that arises because investments wear out and become obsolete over time.

Discount rate The rate used to state a future value or cost in present value terms. It is usually set equal to a utility's weighted cost of capital.

Energy Policy Act of 1992 Far reaching federal legislation which, among other objectives, created a new class of independent power producers called exempt wholesale generators (EWGs) in an effort to relax the restrictions of IPP ownership created by the Public Utility Holding Company Act (PUHCA). In many instances, utilities can easily be granted EWG status for new generation facilities.

Escalation rate The rate of growth applied to a present value cost to determine the future cost of the item. It is equal to the expected inflation rate times any real price effects.

Equivalent uniform annual cost See levelization.

Fixed charge rate The sum of all fixed charges as a percent of an alternative's initial investment cost. When the fixed charge rate is applied to the initial investment, the product equals the revenue requirements needed to offset fixed costs for a given year.

Fixed costs Those costs associated with an investment that do not vary with the level of output. For the investor-owned utility, fixed costs include return of investment, depreciation, income taxes, property taxes, insurance, and often, other administrative costs. For public utilities, fixed costs typically include amortization, insurance, payment in lieu of taxes, and renewals and replacements.

Future value A time value of money measurement in which a current or present value is compounded or escalated to determine its value at some future date.

Indirect costs Those total project costs not included in direct costs including startup costs, training, equipment tests, engineering fees, and other miscellaneous costs.

IPP—(independent power producer) A nonregulated generator lacking market power that is free to sell power at market-based rates under the Federal Power Act.

Levelization The process of converting a nonuniform series of payments into a uniform series having the same present value as the nonuniform series.

Levelized fixed charge rate A single rate, that when applied to the original investment cost, results in a levelized cost having the same present value as the nonuniform costs.

Levelized annual cost method A present value evaluation method usually used for system analyses in which an alternative's cost is stated on the basis of a total levelized annual cost.

NUG—(nonutility generator). A generation facility that is not a franchised utility and not directly subject to rate regulation. Composed of QFs and IPPs.

Operating costs Those costs such as fuel and variable operating and maintenance costs that vary with the level of output and are recovered on a dollar-for-dollar basis from revenue.

Present value A time value of money measure in which a future amount is discounted and stated in terms of its current or present worth.

Present value of a uniform series A time value of money measure in which a series of equal payments or receipts is discounted back to a single present value figure.

Pro forma analysis A method used primarily by NUGs to project and evaluate the profitability of an investment.

PURPA—(Public Utilities Regulatory Policy Act of 1978) United States legislation that requires, among other things, an electric utility to purchase power generated from QFs at rates that are not to exceed the utility's avoided cost.

QF—(qualifying facility) A cogenerator or qualifying power producer under PURPA that may sell power to an electric utility at a rate not to exceed a utility's avoided cost rate.

Regulatory compact The unwritten agreement between electric utilities and regulators in which electric utilities are provided an exclusive franchised territory and the opportunity to earn an adequate return on investment among other rights in exchange for providing an adequate, reliable, and reasonably priced supply of electricity to customers able and willing to pay for the service.

Return on investment The annual revenue required to adequately compensate the lenders of debt and equity money. Return on investment revenue requirements are equal to the weighted cost of capital times the undepreciated investment.

Revenue requirements The dollar amount that must be collected through rates to offset all utility costs. For an individual investment, revenue requirements must equal all operating plus fixed costs.

Sensitivity analyses Analyses in which the value of input variables are changed to different but reasonable values from the most likely scenario to determine whether a project remains cost-effective under alternative future conditions.

Time value of money A financial concept that states a dollar received in the future is worth less than if received today. Because there is a time value of money, present worth arithmetic must be used to compare the economic costs of alternatives.

Total present worth method A present value evaluation method sometimes used in system analysis or system planning in which the yearly present worth cost of an alternative over all years of operation is summed and stated as a single amount.

Uniform series A series of equal yearly costs or payments.

Weighted cost of capital The utility's weighted average cost of debt and equity. When applied to the undepreciated investment, it results in the revenue requirements needed to provide an adequate return on investment in a given year. The weighted cost of capital is usually used as the discount rate and AFUDC rate.

2.8 REFERENCES

ANDERSON, JOHN A. 1991. Are prudence reviews necessary? *Public Utilities Fortnightly* 127:3 (February).

BENESH, BRUCK K. and M. KEVIN BRYANT. 1992. *Depreciation Handbook.* Matthew Bender & Co., Inc., New York, NY. pp. 2-42 through 2-50, App. 2-6 through 2-17.

BLACK & VEATCH. 1983. *Economic Assessment Guide.* Black & Veatch, Kansas City, MO. pp. 2-15 through 2-30.

BLACK & VEATCH. 1990. *User's Manual for Program SPF, Energy Group Short Pro Forma* (Preliminary). Black & Veatch, Kansas City, MO.

EDISON ELECTRIC INSTITUTE. 1992. Nonutility generation has grown 17 percent annually since 1987, EEI says. *Independent Power Report.* McGraw-Hill. New York, NY. December.

ELECTRIC POWER RESEARCH INSTITUTE. *TAG-Technical Assessment Guide Volume 3: Fundamentals and Methods, Supply—1986.* Electric Power Research Institute, Palo Alto, CA.

ENERGY INFORMATION ADMINISTRATION. *Annual Energy Outlook 1991.* US Government Printing Office, Washington, DC. Table A5, p. 49.

PHILLIPS, CHARLES F. JR. 1988. *The Regulation of Public Utilities.* Public Utilities Reports, Inc., Arlington, VA. Table A5, p. 13.

RAO, RAMESH K. S. 1987. *Financial Management.* Macmillan, New York, NY.

STOLL, HARRY G. 1989. *Least-Cost Electric Utility Planning.* Wiley-Interscience, New York, NY.

WILLIAMS, DAN R. and LARRY GOOD. 1994. *Guide to the Energy Policy Act of 1992.* The Fairmont Press, Inc., Lilburn, GA.

Appendix 2A Solutions to Work Problems

Solution to Problem 1. Compounding

1A. Using Eq. (2-1), the future value is equal to the present value times the future value factor. Thus:

$$FV = \$100{,}000 \times 1.295029 = \$129{,}520.90$$

1B. Using the guidelines for future values, is an 8% annual return is realized, all else being equal, the future value will be lower than the $129,502.90 in 1A. The future value equals $125,971.20 ($100,000 × 1.259712).

1C. If the compounding period is increased from 3 to 4 years, the future value will be $141,158.16 ($100,000 × 1.4115816), greater than the $129,502.90 in 1A.

1D. If the present value is $99,999 rather than $100,000, the future value will be $129,501.60 ($99,999 × 1.295029), lower than the $129.502.90 in 1A.

Solution to Problem 2. Escalation

2A. Using Eq. (2-1):

$$FV = \$3{,}800 \times 1.790848 = \$6{,}805.22$$

2B. Table 2A-1 uses the year-by-year calculation.

Table 2A-1. Solution to Problem 2A

Period	Initial		Escalation		Ending
Year 1	$3,800.00	×	1.06	=	$4,028.00
Year 2	$4,028.00	×	1.06	=	$4,269.68
Year 3	$4,269.68	×	1.06	=	$4,545.86
Year 4	$4,545.86	×	1.06	=	$4,797.41
Year 5	$4,797.41	×	1.06	=	$5,085.26
Year 6	$5,085.26	×	1.06	=	$5,390.37
Year 7	$5,390.37	×	1.06	=	$5,713.79
Year 8	$5,713.79	×	1.06	=	$6,056.62
Year 9	$6,056.62	×	1.06	=	$6,420.02
Year 10	$6,420.02	×	1.06	=	$6,805.22

Solution to Problem 3. Compounding and Present Values

3A. Substituting into Eq. (2-4),

$$PV = FV \times PVF;$$

$$PV = \$6{,}805.22 \times 0.38554$$

$$PV = \$2{,}623.71$$

If the utility set aside a present value equal to $2,623.71 and earned a 10% annual return, it would have enough to pay for the boiler component costing $6,805.22 in 10 years.

3B. Since the utility has the option to purchase the component now by spending $3,800 or setting aside $2,623.71, the least-cost option is to set aside the $2,623.71 now and purchase the component in 10 years.

3C. If the cost of the component escalated at 10% annually, the future value would be [from Eq. (2-1)]:

$$FV = PV \times FVF$$

$$FV = \$3{,}800 \times 2.59374$$

$$FV = \$9{,}856.22$$

If the utility could earn only 6% per year, the amount required to be set aside currently to purchase the component in 10 years would be [from Eq. (2-4)]:

$$PV = FV \times PVF$$

$$PV = \$9{,}856.22 \times 0.55839$$

$$PV = \$5{,}503.66$$

Assuming no other costs and risks, the utility would be better off purchasing the component for $3,800 now rather than setting aside $5,503.66 and purchasing the component in 10 years for $9,856.22.

Solution to Problem 4. Present Value of a Uniform Series

4A. Utilizing Eq. (2-5), the present value is:

$$PV_{US} = US \times USPWF$$

$$PV_{US} = \$10{,}000 \times 8.05518$$

$$PV_{US} = \$80{,}551.84$$

4B. $300,000 in actual dollar outlays will be made over the 30 years of payments compared to the $80,551.84 present value figure.

4C. The interpretation of the present value figure is that if an $80,551.84 lump sum was set aside currently and earned 12% interest per year, it would be just enough to make all required payments over the next 30 years. The present value figure is useful since it enables comparisons between alternatives. The option having the lowest present value represents the least costly alternative. Thus, economically correct decisions depend on the computation and comparison of present values.

Solution to Problem 5. Uniform Series Equal to a Present Value

5A. Using Eq. (2-6), the annual payments for a $10 million loan for 20 years at 8% per year would be:

$$\$10 \text{ million} \times 0.1018522088 = \$1{,}018{,}522.09$$

5B. For a 30-year loan the annual payment would equal:

$$\$10 \text{ million} \times 0.0888274334 = \$888{,}274.33$$

5C. Under the 20-year financing option, actual dollar outlays would equal $20.370 million (20 years × $1,018,522.09). Under the 30-year financing option, actual outlays would equal $26.648 million (30 years × $888,274.33).

Table 2A-2. Levelized Values of Options A and B

		Option A		Option B	
Year	PW Factor	Annual Costs ($)	Present Worth ($)	Annual Costs ($)	Present Worth ($)
1	0.91743	10,000	9,174.31	8,000	7,339.45
2	0.84168	11,000	9,258.48	9,000	7,575.12
3	0.77218	12,000	9,266.20	10,000	7,721.83
4	0.70843	13,000	9,209.53	12,000	8,501.10
5	0.64993	14,000	9,099.04	14,000	9,099.04
6	0.59627	15,000	8,944.01	16,000	9,540.26
7	0.54703	16,000	8,752.55	18,000	9,846.62
8	0.50187	17,000	8,531.73	20,000	10,037.33
9	0.46043	18,000	8,287.70	22,000	10,129.41
10	0.42241	19,000	8,028.81	24,000	10,137.86
Totals	6.41766		88,549.35		89,928.04

Levelized annual cost $\frac{\$88{,}549.35}{6.41766}$ = $13.797.77 (Option A) $\frac{\$89{,}928.04}{6.41766}$ = $14,012.59 (Option B)

Table 2A-3. Proof of the Equality of the Levelized and Nonuniform Annual Costs For Option A

Year	PW Factor	Annual Costs ($)	Present Worth ($)
1	0.91743	13,797.77	12,658.50
2	0.84168	13,797.77	11,613.31
3	0.77218	13,797.77	10,654.41
4	0.70843	13,797.77	9,774.69
5	0.64993	13,797.77	8,967.60
6	0.59627	13,797.77	8,227.16
7	0.54703	13,797.77	7,547.85
8	0.50187	13,797.77	6,924.64
9	0.46043	13,797.77	6,352.88
10	0.42241	13,797.77	5,828.33
Total present value			88,549.35

Solution to Problem 6. Levelized Values

6A. Table 2A-2 calculates the levelized values for Options A and B. Option A has the lowest present value and is the least-cost option.

6B. Table 2A-3 proves that the levelized annual cost has the same present worth as the nonuniform annual payments. The present worth sum of the levelized payments is $88,549.35, which equals the present value sum of the annual nonuniform payments in Table 2A-3.

Solution To Problem 7. Levelized Value with Uniform Escalation

7. The levelized cost of the power purchase is calculated first by Eq. (2-7) and then by Eq. (2-8) in Table 2A-4.

Appendix 2B Public Utility Fixed Cost Components

This appendix provides a brief overview of the fixed cost components for public utilities. A public utility's fixed costs comprise amortization, insurance, payment in lieu of taxes, bond charges, and renewables and replacements. These fixed costs differ from an investor-owned utility's fixed costs since public utilities are not subject to income taxes and usually finance large projects completely through bonds. As a result of these differences, fixed charge rates are usually less than for private utilities, despite the usual 3% to 4% bond fee requirement and a reserve account equal to 1 year's debt service which reduces the risk of default.

One difference between public utilities and investor-owned utilities is the way depreciation is handled. On the other hand, an investor-owned utility (IOU) may make no payment of the principal portion of its long-term debt, making instead a provision in its books for paying off the investment by setting up a depreciation account. Often, the procedure is to pay the bond holders only the interest on the long-term debt; when the bonds mature, new bonds are sold to pay off the old bonds and to finance new capital additions. The funds earmarked for depreciation are usually not deposited into an actual depreciation fund. Instead, the IOU may use these funds to meet capital expansion requirements or for debt reduction. Thus, the depreciation account of an investor-owned utility can be looked on as an accounting technique to ensure that the true cost to the utility is considered in calculating the minimum revenue requirements and determining rates.

On the other hand, municipal and other noninvestor-owned utilities are usually required by law to repay the bond principal as well as the interest. The principal payments accomplish the same purpose for the municipal utility as

Table 2A-4. Proof of the Present Value Equivalence of Equations 2-7 and 2-8

Year (A)	Annual Cost (B)	Present Worth Factor (C)	PW of Annual Cost (d)	Levelized Cost[a] (E)	PW of Levelized Cost (F)
1	$15,000.00	0.87719	$13,157.89	$17,429.08	$15,288.66
2	$16,050.00	0.76947	$12,349.95	$17,429.08	$13,411.11
3	$17,173.50	0.67497	$11,591.62	$17,429.08	$11,764.13
4	$18,375.65	0.59208	$10,879.86	$17,429.08	$10,319.41
5	$19,661.94	0.51937	$10,211.80	$17,429.08	$9,052.12
6	$21,038.28	0.45559	$9,584.76	$17,429.08	$7,940.45
Totals		3.88867	$67,775.88		$67,775.88

[a]Proof of correct levelized value using Eq. (2-7) is shown in Columns E and F.

$$\text{Levelized cost } \frac{\$67{,}775.88}{3.88867} = \$17{,}429.08$$

By Eq. (2-8):

$$\text{Capital recovery factor } \frac{1}{3.88867} = 0.25716$$

$$K = \frac{1+e}{1+i} = \frac{1.07}{1.14} = 0.93860$$

$$\text{Levelized cost} = \$15{,}000 \times (0.25716) \times \frac{1-(0.93860)^6}{0.14-0.07}$$

$$= \$3{,}857.36 \times \frac{0.31628744}{0.07} = \$17{,}429.08$$

does the depreciation account for the investor-owned utility—that is, the return of the investment over the life of the facility.

In an economic analysis for a noninvestor-owned utility, the usual method of accounting for the cost of an investment is by amortizing it with equal annual payments over the life of the facility. Annual payments are determined by multiplying the initial investment by the appropriate capital recovery factor based on the anticipated bond interest rate and the estimated life of the facility. The rate of interest to be used for present worth calculations for a noninvestor-owned utility should be its cost of money, which is the anticipated bond interest rate.

Total fixed charges for municipal utilities consist of the capital recovery factor plus insurance and other fixed costs directly related to the investment. Many municipal utilities make payments to the city in lieu of taxes, and such payments may be included as part of the fixed charges.

It is not uncommon for public utilities to make annual payments to a special fund for capital additions. Such funds are similar in operation to the depreciation account of a private utility in that they provide a reserve fund for large expenditures in excess of normal operating expenses. These funds may be identified in various ways, such as renewals and replacements, capital reserve fund, or depreciation fund. Annual payments to such funds would normally be included as part of the fixed charges.

Another difference may be in the method of calculating AFUDC, more correctly termed interest during construction (IDC), for public utilities. Many public utilities do not have sufficient retained earnings to partially fund construction and the interest due on debt issuances during construction. When this is the case, these utilities must borrow enough to make debt payments due during construction as well as the total amount needed for actual construction. Since this can add a significant amount to the total plant cost, utility specific information needs to be collected before estimating IDC costs. In cases where the utility's borrowings include an amount to pay for IDC costs, the methods in Section 2.4.3.2 may not be appropriate.

Appendix 2C MACRS Property Classes for Steam and Electric Generation Facilities

Asset Class	Description of Assets Included	Recovery Periods (years)	
		MACRS	ADS
00.4	*Industrial steam and electric generation and/or distribution systems* • Includes assets used in production and/or distribution of electricity with rated total capacity > 500 kW. • Includes assets used in production and/or distribution of steam with rated total capacity >12,500 lb/h. • Electricity or steam for use only by taxpayer in its industrial manufacturing process or plant activity and not for sale to others. • Excludes buildings and structural components.	15	22
49.11	*Electricity utility hydraulic production plant* Includes assets used in the hydraulic power production of electricity for sale, including related land improvements, such as dams, flumes, canals, and waterways.	20	50
49.12	*Electric utility nuclear production plant* Includes assets used in the nuclear power production of electricity for sale and related land improvements. Does not include nuclear fuel assemblies.	15	20
49.121	*Electric utility nuclear fuel assemblies* Includes initial core and replacement core nuclear (i.e., the composite of fabricated nuclear fuel and container) when used in a BWR, PWR, or HTGR used in the production of electricity. Does not include nuclear fuel assemblies used in breeder reactors.	5	5
49.13	*Electric utility steam production plant* Includes assets used in the steam power production of electricity for sale, combustion turbines operated in a combined cycle with a conventional steam unit and related land improvements. Also includes package boilers, electric generators, and related assets such as electricity and steam distribution systems as used by a waste reduction and resource recovery plant if the steam or electricity is normally for sale to others.	20	28
49.14	*Electric utility transmission and distribution plant* Includes assets used in the transmission and distribution of electricity for sale and related land improvements. Excludes initial clearing and grading land improvements.	20	30
49.15	*Electric utility combustion turbine production plant* Includes assets used in the production of electricity for sale by the use of such prime movers as jet engines, conmbustion turbines, diesel engines, gasoline engines, and other internal combustion engines; their associated power turbines and/or generators; and related land improvements. Does not include combustion turbines operated in a combined cycle with a conventional steam unit.	15	20

(continued)

Appendix 2C Continued

Asset Class	Description of Assets Included	Recovery Periods (years) MACRS	ADS
49.223	*Substitute natural gas–coal gasification* Includes assets used in the manufacture and production of pipeline quality gas from coal using the basic Lurgi process with advanced methanation. Includes all process plant equipment and structures used in this coal gasification process and all utility assets such as cooling systems, water supply and treatment facilities, and assets used in the production and distribution of electricity and steam for use by the taxpayer in a gasification plant and attendant coal mining processes but not for assets used in the production and distribution of electricity and steam for sale to others. Also includes all other related land improvements. Does not include assets used in the direct mining and treatment of coal prior to the gasification process itself.	10	18
49.4	*Central steam utility production and distribution* Includes assets used in the production and distribution of steam for sale. Does not include assets used in waste reduction and resource recovery plant which are classified elsewhere.	20	28
49.5	*Waste reduction and resource recovery plants* Includes assets used in the conversion of refuse or other solid waste or biomass to heat or to a solid, liquid, or gaseous fuel. Also includes all process plant equipment and structures at the site used to receive, handle, collect, and process refuse or other solid waste or biomass to a solid, liquid, or gaseous fuel or to handle and burn refuse or other solid waste or biomass in a water wall combustion system, oil or gas pyrolysis system, or refuse-derived fuel system to create hot water, gas, stem, and electricity. Includes material recovery and support assets used in refuse or solid waste receiving, collecting, handling, sorting, shredding, classifying, and separation systems. Does not include any package boilers, or electric generators and related assets such as electricity,hot water, steam and manufactured gas production plants classified in classes 00.4, 49.13, 49.221, and 49.4. Does include, however, all other utilities such as water supply and treatment facilities, ash handling, and other related land improvements of a waste reduction and resource recovery plant.	7	10
XX.X	*Miscellaneous five-year property* Includes geothermal, ocean thermal, solar, wind energy properties; cogeneration equipment; and biomass properties that constitute qualifying small power production facilties [Section 3(17)(C) of the Federal Power Act].	5	12

Source: Benesh, Bruck K. and M. Kevin Bryant. 1992. *Depreciation Handbook.* New York: Matthew Bender & Co., Inc., New York, NY.

Appendix 2D Summary of Chapter 2 Time Value Equations

Equation (2-1): Future Value

$$FV_n = PV(1 + i)^n$$

where

FV_n = future value in period n,

PV = present value,

i = cost of money, rate per period (can also represent the escalation rate),

n = number of compounding periods, and

$(1 + i)^n$ = future value factor (FVF).

Equation 2-2: Total Escalation Rate

$$TER = (1 + \text{inflation rate})(1 + \text{real escalation rate}) - 1$$

where

TER = total escalation rate.

Equation 2-3: Monthly Escalation Rate

$$\text{Monthly escalation rate} = (1 + \text{annual escalation rate})^{1/12}$$

Equation 2-4: Present Value of a Lump Sum

$$PV = FV_n \times \frac{1}{(1 + i)^n}$$

where

PV = present value,

FV_n = future value in period n,

i = discount rate per period,

n = number of periods, and

$\frac{1}{(1 + i)^n}$ = present worth factor (PWF).

Equation 2-5: Present Value of a Uniform Series

$$PV_{US} = US \times \frac{(1 + i)^n - 1}{i(1 + i)^n}$$

where

PV_{US} = present value of a uniform series,

US = uniform series payment or receipt,

i = discount rate,

n = number of periods, and

$$\frac{(1 + i)^n - 1}{i(1 + i)^n} = \text{uniform series present worth factor (USPWF)}$$

Equation 2-6: Uniform Series Equal to a Present Value

$$US_{PV} = PV \times \frac{i(1 + i)^n}{(1 + i)^n - 1}$$

where

US_{PV} = uniform series equal to a present value,

PV = present value,

i = interest rate,

n = number of periods, and

$$\frac{i(1 + i)^n}{(1 + i)^n - 1} = \text{capital recovery factor (CRF).}$$

Equation 2-7: Levelized Annual Cost

$$\text{Levelized annual cost} = \frac{\text{total present worth of all annual costs}}{\text{sum of present worth factors}}$$

Equation 2-8: Levelized Cost With Uniform Escalation

$$\text{Levelized cost} = (\text{Cost at end of first year})\ (\text{CRF})\ \frac{1 - K^n}{i - e}$$

where

i = present worth discount rate,

e = annual escalation rate,

n = number of years,

$K = \frac{1 + e}{1 + i}$, and

$\text{CRF} = \frac{i(1 + i)^n}{(1 + i)^n - 1}$ [from Eq. (2.6)]

Equation 2-9: Levelized Cost With Uniform Escalation and Costs Given at the Start of the First Year

$$\text{Levelized cost} = (\text{Cost at beginning of first year})\ (\text{CRF})\ (K)\ \frac{1 - K^n}{1 - K}$$

where

i = present worth discount rate,

e = annual escalation rate,

n = number of years in study,

$K = \frac{1 + e}{1 + i}$, and

$\text{CRF} = \frac{i(1 + i)^n}{(1 + i)^n - 1}$ [from Eq. (2.6)]

Equation 2-10: Monthly AFUDC Rate Calculation

$$\text{Monthly AFUDC rate} = (1 + i)^{1/12}$$

where

i = annual interest rate.

3

THERMODYNAMICS AND POWER PLANT CYCLE ANALYSIS

Mark F. McClernon

3.1 INTRODUCTION

Thermodynamics grew out of a desire to convert thermal energy into mechanical energy. There has always been an abundant supply of things that would burn, but no one was quite sure how to convert this thermal energy into useful work.

As the industrial age dawned, people noticed not so much that expanding steam could do work, but that condensing steam could create a vacuum that could do useful work as it pulled something back to fill the void. It was almost 1800 before James Watt allowed the expanding steam to do work, but his steam was less than 10 psig, and his concept of the external condenser was the real key to success. By 1815, Oliver Evans was using 200 psia steam, however, and by 1881, the first steam power plants were in operation in the United States.

This chapter presents the basic thermodynamic principles necessary for understanding power plant cycle and efficiency analysis. It addresses general methods for analysis and evaluation and examines the two primary power plant cycles, the Rankine cycle and the Brayton cycle, in detail.

3.2 PROPERTIES AND THEIR UNITS

Any quantifiable parameter that describes the state of a substance is defined as a property. For example, if discussions of a "geometric state" refer to properties such as length and width, the property of color is irrelevant.

Thermodynamics is basically concerned with energy and equilibrium. The thermodynamic state of a substance is therefore described by properties that give insight to energy and transfers of energy that lead toward equilibrium.

Properties that are quantifiable must be discussed within an acceptable system of units. The system of units used in this chapter is the English Engineering system (pound, second, foot). This system has been chosen because it is the common language of the United States power industry. Table 3-1 contains a conversion table for Système International (SI) units.

In the English unit system, the unit of mass is the "pound-mass" (lbm), and the unit of force is the "pound-force" (lbf). One pound-force is the force exerted by the earth's gravitational field (32.17 ft/s^2) on 1 lbm. Examining this in terms of Newton's second law:

$$F = ma \tag{3-1}$$

where

F = force,

m = mass, and

a = acceleration (due to gravity)

It is apparent that a proportionality constant k is required such that the following occurs:

$$1 \text{ lbf} = k \times 1 \text{ lbm} \times 32.17 \text{ ft/s}^2 \tag{3-2}$$

In the English system, therefore, Newton's second law is written as follows:

$$F = kma \tag{3-3}$$

where

k = (1 lbf-s^2/32.17 lbm-ft) This constant is typically denoted "g_c."

The benefit of the English system is that weight and mass are numerically equal. The metric and SI systems of units make no pretense that these properties are in any way the same, and therefore have no requirements for proportionality constants.

The sections that follow examine some of the thermodynamic properties and the laws and relationships that govern them. Some of the properties, such as volume and mass, are familiar, and do not warrant further discussion. Other properties, such as temperature and pressure, are also familiar, but still require additional thermodynamic interpretation.

3.2.1 Temperature

Temperature is the property of matter most naturally associated with thermodynamics. When hot matter contacts cooler matter, an energy transfer occurs until a thermal equilibrium is achieved and both substances share the same temperature. A further observation can be stated as follows:

Table 3-1. Conversions for Système International Units

Length	1 inch = 2.54 centimeters (cm)
	1 foot = 0.3048 meters (m)
Area	1 inch^2 = 645.16 millimeters^2 (mm^2)
	1 foot^2 = 0.0929 meter^2
Volume	1 foot^3 = 0.028317 meter^3
Mass	1 lbm = 0.453592 kilogram (kg)
Force	1 lbf = 448222 Newtons (N)
Pressure	1 psia = 6.894757 kiloPascals (kPa)
Specific volume	1 ft^3/lbm = 0.062428 m^3/kg
Energy	1 Btu = 1.055056 kiloJoule (kJ)
Power	1 hp = 0.7457 kiloWatt (kW)
Enthalpy	1 Btu/lbm = 2.326 kJ/kg
Entropy	1 Btu/lbm-° R = 4.1868 kJ/kg-K

> Suppose Body A is in thermal equilibrium with Body C, and Body B is in thermal equilibrium with Body C. Then Body A is in thermal equilibrium with Body B, and all three bodies share the same temperature.

This observation, commonly referred to as the Zeroth Law of Thermodynamics, is really not a law, but part of the definition of the thermodynamic property of temperature.

Because thermodynamic properties must be quantifiable, numerical scales exist to define temperature. In the Fahrenheit measurement system, the melting point of ice at atmospheric pressure (14.696 psia) is 32° F and the boiling point of water is 212° F. In the Celsius (formerly Centigrade) measurement system, the melting point of ice is 0° C, and the boiling point is 100° C. The two scales are related to one another by the equation:

$$T(°\,F) = (9/5) \times T(°\,C) + 32 \tag{3-4}$$

An alternative equation to relate the two scales is as follows:

$$T(°\,C) = (5/9) \times (T(°\,F) - 32) \tag{3-5}$$

Both of these scales are relative in the sense that a temperature of zero does not imply a minimum achievable level.

A minimum level does exist, however, and is referred to as absolute zero. Absolute zero occurs at −459.67° F or −273.15° C. When the Fahrenheit and Celsius temperature scales are transposed by these values, two new temperature scales result. In the sense that both of these scales have the value 0 as the minimum achievable temperature, they are referred to as absolute temperature scales.

The absolute Rankine scale is defined in terms of the Fahrenheit scale as follows:

$$T(°\,R) = T(°\,F) + 459.67 \tag{3-6}$$

The absolute Kelvin scale is defined in terms of the Celsius scale:

$$T(K) = T(°\,C) + 273.15 \tag{3-7}$$

The two absolute scales are simply related by the equation:

$$T(K) = (5/9) \times T(°\,R) \tag{3-8}$$

It should be noted that temperature is one of the intensive thermodynamic properties. This means that the temperature of 100 pounds of matter is no different from the temperature of 1 pound of that same matter. This is not true for extensive properties such as volume, mass, and energy.

3.2.2 Specific Volume

Specific volume is a simple property. It is the volume per unit mass.

$$v = V/M \tag{3-9}$$

where

v = specific volume,

V = total volume, and

M = total mass.

Like temperature, it is an intensive property, and in the English system carries units of cubic feet per pound-mass.

Specific volume is such a simple concept, in fact, that it would hardly deserve a dedicated section in this chapter, except for the myriad of similar, but different, properties that it is often confused with.

Density, represented by the Greek letter ρ (rho), is the mass per unit volume (pounds-mass per cubic foot [lbm/ft^3]) and is the reciprocal of specific volume.

Specific gravity, when describing a solid or liquid, is the ratio of the specific volume to the specific volume of water at a standard temperature (usually 39.2° F, but sometimes 60° F; general practice is to multiply the specific volume by 62.4). Specific gravity is not commonly used to describe gases. When it is, the reference matter is generally hydrogen or air.

Specific weight is the weight per unit volume (lbf/ft^3). In the English system, it is numerically equal to density (lbm/ft^3) for all locations where acceleration due to gravity is 32.17 ft/s^2.

The density of solids and liquids can generally change with temperature, but it is fairly independent of pressure. (For example, the density of liquid water at atmospheric pressure varies by about 2.3 lbm/ft^3 over a range of 32° F to 200° F. Over a pressure range of 0.5 atm to 2.0 atm, however, the density of liquid water at 32° F varies less than 0.005 lbm/ft^3). The density of gases is highly dependent on both temperature and pressure.

3.2.3 Pressure

The mechanical concept of pressure is that it is the force exerted by matter on the surface it contacts. Mechanical and thermodynamic pressures are quantitatively the same, but reflect different views of a single property.

The thermodynamic concept of temperature is defined in terms of thermal equilibrium. Similarly, the thermodynamic concept of pressure is defined in terms of mechanical equilibrium. This can be illusrated with a simple cylinder with its two ends separated by a movable piston. If the gas on one end is at pressure 1, and the gas on the other end is at pressure 2, the piston will move to a point of balanced forces (mechanical equilibrium) where both ends are at new pressure 3. In terms of this simple image, the thermodynamic concept of pressure can be visualized as the property that is shared by two systems in mechanical equilibrium.

Although the units of pressure can always be traced back to the simple concept of force per unit area, a myriad of specific units are in common use. Many of these take the form of head, where force per unit area is displayed as a height of a liquid (at standard temperature to fix specific weight and a gravitational acceleration of 32.17 ft/s^2) necessary to cause an equivalent pressure. Table 3-2 shows equivalent quantities for pressure units.

The quantitative representation of pressure can take various forms. Absolute pressure refers to any system where zero indicates an absolute minimum (the complete absence of pressure). Pressure values in this system are stated in units followed by the word absolute (pounds force per square inch absolute or psia). This absolute pressure represents the thermodynamic property which is useful in analysis. Gauge pressure and vacuum pressure are relative pressure scales that refer to local atmospheric pressure as zero. Similar to absolute scale, gauge and vacuum pressure units are followed by the words gauge and vacuum (as in expressions such as pounds force per square inch gauge or psig). The only real difference between these two scales is their sign. This may be written as follows:

Gauge pressure = Absolute pressure − Atmospheric pressure

Vacuum pressure = Atmospheric pressure − Absolute pressure

3.2.4 Work and Heat

A thermodynamic property describes the energy equilibrium state of a substance. As such, thermodynamic properties represent information about the energy that matter possesses. Work and heat are not thermodynamic properties, but are thermodynamic quantities, phenomena that change the energy state (and thermodynamic properties) of matter.

Table 3-2. Equivalent Pressure Units

Pounds Per Square Inch	Standard Atmospheres	Inches H_2O at 39.2° F	Inches Hg at 32° F
1.0	0.68046	27.673	2.037
14.696	1.0	33.899 ft	29.921
0.0361	0.0024583	1.0	0.0735
0.491	0.033421	13.60	1.0

Work is the energy transfer that occurs when force moves through a distance (or a pressure acts through a volume). The system exerting the force gives up energy, and that energy is accepted by the system moving through the distance. This is demonstrated mathematically for a simple compressible substance:

$$\delta W = P\,dV \tag{3-10}$$

where

W = work,

P = pressure (absolute), and

V = volume.

The convention is set that the system that does the work (the one with the higher pressure) is using energy, and W is positive for this system; the system being worked on may receive energy, so W is negative.

In thermodynamic analysis, defining the exact force and distance can be challenging, but a fairly simple assessment of the work process can be made by a balancing examination of the thermodynamic systems. First law and control volume analysis offers further insight into this statement.

Power is the time rate at which work is performed. The basic unit for measurement of energy transfer as work is the foot-pound (ft-lbf), but units for power are varied, and include foot-pounds per minute (ft-lbf/min), horsepower (hp), and kilowatts (kW).

Similar to work, heat is an energy transfer. However, whereas work is a macroscopically organized transfer of energy, heat is a microscopically disorganized transfer of energy. Suppose a sealed container of gas is placed in contact with a flame. Examination of the thermodynamic properties of temperature and pressure demonstrate that the energy state of the gas has changed, even though we observed no work being done. The thermodynamic concept of heat accounts for this energy transfer. Of course, energy is energy, and once an equilibrium state is reached, there is no way of knowing whether energy was transferred in by heat or by work.

A few more topics and terms warrant attention in a discussion of heat. First, energy transfer as heat is typically indicated by the symbol Q. Second, the English units of Q are typically British Thermal Units (Btu), and are defined as the amount of energy required to raise 1 lbm of water by 1 degree F. Third, when a process involves no energy transfer as heat (e.g., it is thermally insulated), it is referred to as adiabatic.

Conventionally, heat entering a system is taken as positive and heat leaving a system is considered negative. Table 3-3 shows equivalent heat and power units.

3.2.5 Internal Energy and Enthalpy

Thermodynamic temperature and pressure are fairly meaningless on a microscopic plane. Rather, they are macroscopic

Table 3-3. Equivalent Heat and Power Units

Ft-lbf/min.	Btu/h	Watts	hp
1,000	77.1	22.6	0.0303
12,970	1,000	293.1	0.3929
44,254	3,413	1,000	1.34
33,000	2,545	746	1.0

properties in the sense that they are measurable and represent an average or net energy of billions of microscopic conditions. Internal energy is a microscopic property.

Individual molecules are capable of possessing energy, even though it is difficult to measure these directly. As the molecule vibrates, rotates, and makes random motions, it stores energy not attributable to the organized modes of kinetic energy (*KE*) and potential energy (*PE*). Internal energy (*U*) is the thermodynamic property that describes the energy stored in these disorganized modes. For a simple substance energy can be characterized as follows:

$$E = KE + PE + U \tag{3-11}$$

The specific internal energy (u) is an intensive property obtained by dividing the internal energy by the total mass (M):

$$u = U/M \tag{3-12}$$

Specific internal energy carries the units of Btu/lbm. Although internal energy is not easily measured, the macroscopic property of temperature provides a significant signpost for its determination.

When the product of pressure and specific volume is added to internal energy, a new property called enthalpy is formed. Mathematically, the specific enthalpy (i.e., enthalpy per unit mass) is given as follows:

$$h = u + Pv \tag{3-13}$$

Since h and u have units of Btu/lbm, P has units of lbf/ft^2, and v has units of ft^3/lbm, a conversion factor is needed. This conversion factor is provided as follows:

$$h = u + Pv/J \tag{3-14}$$

where

J = Joule's constant, and is equal to 778 ft-lbf/Btu.

The Pv portion of enthalpy is sometimes referred to as the flow energy, although this is sensible only for a particular type of problem. The real power of enthalpy is in its application.

3.2.6 Specific Heats

Two parameters related to internal energy and enthalpy are worthy of further discussion. The specific heat at constant pressure (c_p) and the specific heat at constant volume (c_v) are formally defined as follows:

$$c_p = \left(\frac{\partial h}{\partial T}\right)_p ; \; c_v = \left(\frac{\partial u}{\partial T}\right)_v \tag{3-15}$$

Under special conditions, these partial derivatives represent the energy transfer as heat required for a 1° temperature rise in a unit mass, and are written as follows:[1]

$$c_v = \frac{Q}{M\Delta T} \text{ (constant volume)} \tag{3-16}$$

and

$$c_p = \frac{Q}{M\Delta T} \text{ (constant pressure)} \tag{3-17}$$

where

Q = energy transfer as heat,

M = mass, and

T = temperature.

These equations assume that the specific heats remain constant over the range of temperatures in ΔT.

The units of c are Btu/lbm-°R. Representative values are shown in Table 3-4.

3.3 ANALYSIS TOOLS

3.3.1 The First Law of Thermodynamics

The laws of thermodynamics are laws in the sense that they govern our thinking processes because they have never been known to fail. The first law of thermodynamics is simply a formalization of our observation and belief that matter possesses energy, and that energy can't suddenly appear or disappear for no apparent reason.

If a change in a system's energy occurs, then that change must be accounted for. Since, in the absence of mass transfer, energy can be transferred in only two ways [work (*W*) and heat (*Q*)]; this can be stated mathematically as follows:

$$dE = \delta Q - \delta W \tag{3-18}$$

where

E = sum total of the kinetic, potential, and internal energies of the mass in a system.

[1]A constant pressure process has $dh = du + P\,dv = du + \delta W = \delta Q$. A constant volume process has $du = \delta Q$ ($P\,dv = 0$).

Table 3-4. Specific Heats of Water

Substance	c_p (BTU/lbm-°R)	c_v (BTU/lbm-°R)
Ice (32°F, 1 atm)	0.49	0.49
Water (32°F, 1 atm)	1.007	1.007
Water (100°F, 1 atm)	0.998	0.998
Steam (300°F, 1 atm)	0.474	0.357
Air	0.24	0.1714

When the effects of kinetic and potential energy are ignored, dE can be reduced to internal energy only, and the first law becomes

$$dU = \delta Q - \delta W \tag{3-19}$$

Work and heat are energy transfers from and/or to a system, and the normal convention is for positive work to imply energy transferred from the system as work, and positive heat to imply energy transferred to the system as heat.

3.3.2 Entropy and the Second Law

The first law provides a powerful tool for analyzing a thermodynamic process. However, the first law falls short in providing information on the direction of a process. If a hot block of copper is put in contact with a cold block of copper, the energy of the system will be conserved (specifically, the internal energy, if the kinetic and potential terms are not active). The first law cannot predict, however, that as time passes, energy flows as heat from the hot block to the cold block until a state of equilibrium occurs in which two warm blocks exist that have equal temperatures. The first law doesn't explain why this system will never spontaneously revert to its original state. The critical concept of entropy and its manifestation in the second law of thermodynamics provides the ability to predict equilibrium states and to determine whether a process is possible or impossible.

Entropy is a conceptual property rather than a physical one. It has central importance in areas outside of basic thermodynamics, such as information theory and statistical thermodynamics, where entropy is examined from a microscopic viewpoint, and quantitatively defined in terms of quantum state probabilities. Qualitative definitions of entropy extending from these disciplines speak of microscopic disorder and the unavailability of internal energy. Because the viewpoint of basic thermodynamics is macroscopic, these definitions may be more confusing than helpful. The following statements concerning entropy aid in understanding the concept:

1. Entropy is an extensive property. The entropy of a system is the sum of the entropies of its components.
2. All systems move toward equilibrium with time, and the entropy of a system is maximized at equilibrium.
3. The quantitative evaluation of specific enthalpy, s, is based on the Gibbs equation:

$$ds = \frac{1}{T}du + \frac{P}{T}dv \tag{3-20}$$

   Since most of second law analysis requires only information on the change in entropy, integration of the Gibbs equation provides adequate information without specifying an absolute reference point.
4. The entropy of an isolated system can only stay the same or increase.

This last statement is the basic content of the second law of thermodynamics. If the entropy of an isolated system stays the same throughout the various stages of a process, the process is said to be reversible; if the total entropy of the isolated system increases as the process progresses, the process is irreversible. The previous example of the hot and cold copper bars illustrates an irreversible process. Once the bars have reached thermal equilibrium (equal temperatures), the process will never reverse direction and go back to a hot and a cold bar.

A less obvious process might be the oscillations of a block attached to a spring. As long as the block is on a frictionless surface, sits in a vacuum, and the spring is perfectly elastic, the system will oscillate forever, constantly exchanging spring energy and kinetic energy. Given system energies at two points in time, it is impossible to know which one came first. This is a reversible process, and the entropy of the system remains constant.

In contrast to this ideal situation is a system in which the surface is not frictionless, air resistance is present, and the spring has internal friction losses. In this real-life process, the oscillations of the system gradually decay with time, and energy that began as spring or kinetic energy is slowly dissipated as the spring, air, and surface warm as a result of resistance losses.

In comparison to the ideal system, any set of energy profiles for this system can easily be time sequenced, and as the sum of the spring and kinetic energies becomes lower and lower, the energy lost to friction and resistance is unavailable to continue the mechanical operation of the system. This process is irreversible, and the entropy of the system continues to increase until it reaches its maximum at equilibrium (in this case, a complete stop).

Another aspect of entropy can aid in understanding the process. If the Gibbs Equation is rewritten as follows:

$$T\,dS = dU + P\,dV \tag{3-21}$$

then substituting the first law for dU formulates the following equation:

$$T\,dS = \delta Q - \delta W + P\,dV \tag{3-22}$$

Since $P\,dV$ equals δW for reversible work on a simple compressible substance,[2] the equation reduces to the following under these conditions:

$$dS = \frac{\delta Q}{T} \tag{3-23}$$

If the process is irreversible, and lost work makes δW less than $P\,dV$, then the following occurs:

$$dS > \frac{\delta Q}{T} \tag{3-24}$$

The entropy property is really easier to use than this discussion might imply. The units of specific entropy in the English system are Btu/lbm-°R.

3.3.3 The Control Mass

To this point, the discussion of the first and second laws has made the inherent assumption that a fixed amount of material is being considered. This description of a thermodynamic closed system constitutes a control mass (Fig. 3-1), and the first and second laws for this situation can be restated in the form previously stated:

$$dE = \delta Q - \delta W \tag{3-25}$$

and

$$dS \geq \frac{\delta Q}{T} \tag{3-26}$$

where

E = the total energy, is generally confined to internal energy, U, and kinetic or potential energy effects seldom get involved.

In discussions of a fixed amount of gas in a piston, for example, this framework would describe the system adequately, but in energy conversion work, the interest is frequently in the analysis of open systems where mass continuously flows in and out. To apply the first and second law analysis tools in this situation will require some modification.

3.3.4 The Control Volume

The analysis of an open system still utilizes the first and second laws, but additional accounting tools must be used as the mass flows through the system. These control volume problems (Fig. 3-2) generally assume that the mass flow into the system is the same as the mass flow out of the system. There is no requirement that there be only one inlet or outlet flow, but the sum of the inlet flows must equal the sum of the outlet flows (mass balance).

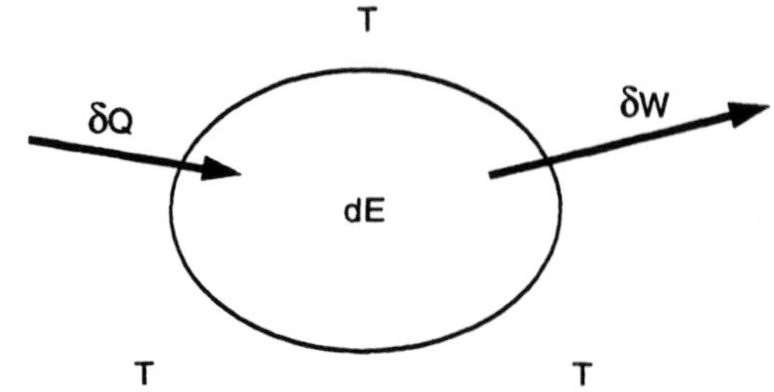

Fig. 3-1. The control mass.

Just as a mass balance provides a valuable accounting for the flow, the first and second laws take the form of a balance for energy and entropy. The first law looks at the energy in, the energy out, and the difference between them. Since time has now entered the analysis in the form of mass flow rates, energy per time, or power, is analyzed as follows:

1. Energy (Btu/lbm) associated with incoming or outgoing mass takes three forms:

$$\text{Internal: } u \qquad \text{Kinetic: } \frac{V^2}{2g_c J} \qquad \text{Potential: } \frac{zg}{g_c J} \tag{3-27}$$

where

V = velocity (ft/sec),

z = elevation (ft), and

g_c = $1/k$ of Eq. (3-3) = 32.17 lbm ft/(sec^2 lbf).

g acceleration due to gravity (ft/sec^2), and

J = Joule's constant.

Each of these quantities is specific and must be multiplied by the mass flow rate (m) to yield power.

2. The rate of energy transfer with respect to time associated with work or heat:

$$Q;\ W \tag{3-28}$$

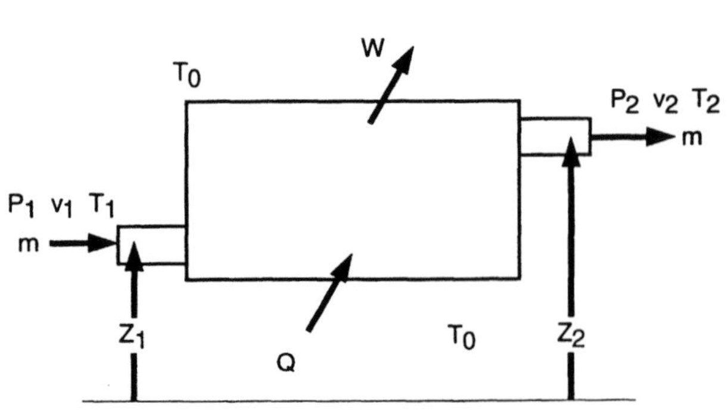

Fig. 3-2. The control volume (steady state, steady flow).

[2]A simple compressible substance means that $P\,dV$ is the only reversible work mode, and magnetism, chemical reactions, and the like can be ignored.

3. Energy per time associated with work done by the actual flow in and out of the volume can be analyzed, as always, by starting with the concept of force multiplied by distance per time. In this case, the form of the force is pressure multiplied by area, and distance per time is the velocity, so that

$$\frac{\text{Work}}{\text{Time}} = (PA)V = P(AV) = P(vm) = m(Pv) \qquad (3\text{-}29)$$

where

m = mass flow rate (lbm/sec),

P = absolute pressure (lbf/ft^2),

A = area (ft^2),

V = velocity (ft/sec), and

v = specific volume (ft^3/lbm).

Making the assumption that no energy is being stored with the control volume (steady state, steady flow), the balance of these terms at the inlet (subscript 1) and outlet (subscript 2) points of the control volume yields the following equation (for locations where $g = g_c$, these factors cancel):

$$m\left(u_1 + \frac{P_1v_1}{J} + \frac{V_1^2}{2g_cJ} + \frac{z_1}{J}\right) + Q = m\left(u_2 + \frac{P_2v_2}{J} + \frac{V_2^2}{2g_cJ} + \frac{z_2}{J}\right) + W \qquad (3\text{-}30)$$

Recognizing that $u + Pv$ is enthalpy makes it possible to simplify this equation to the following:

$$m\left(h_1 + \frac{V_1^2}{2g_cJ} + \frac{z_1}{J}\right) + Q = m\left(h_2 + \frac{V_2^2}{2g_cJ} + \frac{z_2}{J}\right) + W \qquad (3\text{-}31)$$

This is the single stream, steady flow energy equation. It can easily be extended to multiple inlet (subscript i) and outlets (subscript o) as follows:

$$\sum m_i\left(h_i + \frac{V_i^2}{2g_cJ} + \frac{z_i}{J}\right) + Q = \sum m_o\left(h_o + \frac{V_o^2}{2g_cJ} + \frac{z_o}{J}\right) + W \qquad (3\text{-}32)$$

$$\sum m_i = \sum m_o$$

The second equation, which balances the mass flow in and out of the control volume, is sometimes referred to as the continuity equation. There is no reason to confine the principles of this analysis to steady state, steady flow, but this has been done here because most of the applications will fit this format.

Similar to the first law analysis for the control volume, there is also a second law analysis. Since the thermodynamic state of the control volume is not changing (steady state, steady flow), the entropy of the control volume is constant, and the entropy flow can be balanced as follows:

$$\sum m_o s_o \geq \sum m_i s_i + \frac{Q}{T} \qquad (3\text{-}33)$$

Second law analysis for steady flow control volume problems is most often manifested in the form of availability analysis.[3] The next section discusses this important analysis technique.

3.3.5 Thermodynamic Availability Analysis

Thermodynamic availability analysis examines the control volume problem, "What is the maximum possible power output available from a control volume in contact with an environment at temperature T_0?" To answer this question, thermodynamics combines the first and second laws of thermodynamics by solving each for energy transfer with the environment as heat, Q, and then equating the resulting formats (Fig. 3-3). To simplify, use the single-stream representation displayed in the figure and ignore the effects of kinetic and potential energy as negligible. The first law, as derived above, can now be represented as follows:

$$Q = mh_2 - mh_1 + W \qquad (3\text{-}34)$$

The second law is as follows:

$$T_0ms_2 - T_0ms_1 \geqslant Q \qquad (3\text{-}35)$$

Equating the two, the following occurs:

$$W \leqslant \text{m}[(h_1 - T_0s_1) - (h_2 - T_0s_2)] \qquad (3\text{-}36)$$

The quantity $(h - T_0s)$ is represented by the Greek symbol Φ and given the special name of steady flow availability function.

The inequality demonstrates that the maximum work available from the system is given by difference in the availability function evaluated at the inlet and outlet. Because of the way the second law was used to establish the inequality, it is understood that the maximum work is available when all the processes are reversible. The process irreversibility can be quantified as the difference between the actual work and the maximum work available.

An engineer seeking to remove irreversibilities from a process will find this to be a powerful tool in locating opportunities for system improvement.

[3] Availability analysis in thermodynamics is completely unrelated to another type of availability that examines the relative time a component or system is working or broken. RAM (reliability/availability/maintainability) analysis represents an important part of power plant design and optimization. Unfortunately, the common name can be directly confused with thermodynamic availability analysis.

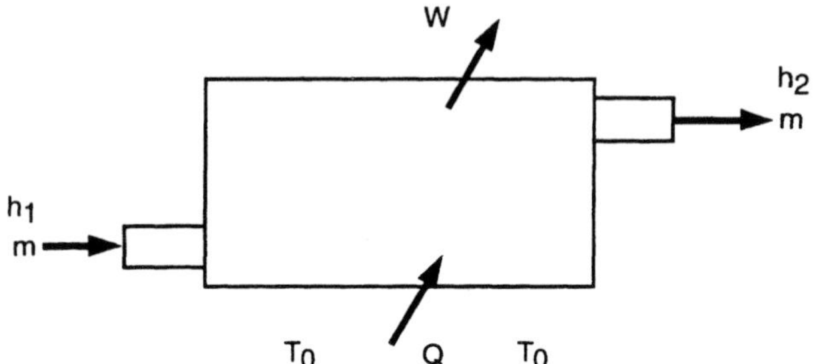

Fig. 3-3. Availability analysis control volume.

3.4 STATES OF MATTER

3.4.1 A Simple Compressible Substance

Many reversible work modes occur in nature. The compression work mode, represented by $P\,dv$, for example, has already been discussed here. Other work modes can be due to magnetic, dielectric, or elastic properties of the material. If only one work mode exists for a substance, it is referred to as a simple substance. If that work mode is compression ($P\,dv$), it is referred to as a simple compressible substance.

As thermodynamics examines various energy conversion systems,[4] the medium chosen for energy transfer, or working fluid, is generally considered to be a simple substance. Classically, the working fluid has been treated as a simple compressible substance.

When the total number of possible reversible work modes for a system or substance have been identified, the State Postulate dictates that the number of reversible work modes plus one is equal to the number of thermodynamic properties that are independently variable. For a simple substance (with one reversible work mode), only two intensive thermodynamic properties can be varied independently. Equivalently, when any two intensive thermodynamic properties are specified for a simple compressible substance, all other properties are determined, and a state has been defined. The relationship between properties is sometimes given in tables, and sometimes idealized in equations of state.

In the following sections, two important examples of property relationships will be examined.

3.4.2 States of Water

Water has been one of the most widely used simple compressible substances used as a working fluid for thermodynamic energy conversion systems. It is generally inexpensive, plentiful, nontoxic, and displays the capability for changing phase at temperatures and pressures well suited to our environment. This phase change capability creates a good opportunity for implementing common processes required for energy conversion systems.

A simple experiment will illustrate the thermodynamic properties of water.[5] A 1 lbm block of ice is isolated in a container designed to maintain a pressure of 14.7 psia (or 1 atm). The system is perfectly insulated against energy loss as heat, but energy can be added to the system as heat. The system can expand its volume to perform work ($P\,dv$), but no other energy is transferred as work to the system. As this experiment continues, energy is very slowly transferred as heat to the system—slowly because as long as it is done very slowly, all of the $P\,dv$ is work, δW, and reversible work has occurred.

Before the experiment is continued, the thermodynamic implications of this system should be considered. Since there is exactly 1 lbm of ice, it can be referred to in terms of specific, unit mass properties. Since kinetic and potential energy are not pertinent here, the first law is stated as follows:

$$du = \delta Q - \delta W \tag{3-37}$$

Since

$$dh = du + P\,dv + v\,dP \tag{3-38}$$

and $dP = 0$ (constant pressure), then

$$dh = du + P\,dv \tag{3-39}$$

Substituting Eq. (3-37) for du in Eq. (3-39):

$$dh = \delta Q - \delta W + P\,dv \tag{3-40}$$

but $\delta W = P\,dv$ (reversible work): so that

$$dh = \delta Q \tag{3-41}$$

Any energy added as heat to this system will be exactly equal to the change in enthalpy of the system.

In addition, the Gibbs equation can be written as

$$\begin{aligned} T\,ds &= du + P\,dv \\ &= (dh - P\,dv + v\,dP) + P\,dv \end{aligned} \tag{3-42}$$

or

$$T\,ds = dh \tag{3-43}$$

These equations can be used to track thermodynamic properties in the experiment. As energy as heat is added to the ice, the temperature of the ice increases, consistent with the concept of specific heat defined in Section 3.2.6. When the temperature reaches 32° F, however, continued energy

[4]Nature provides many sources of thermal energy. Many applications, however, require mechanical energy or electrical energy. An energy conversion system employs thermodynamic processes to achieve conversion of thermal energy to mechanical energy (or sometimes directly to electrical energy).

[5]This discussion actually pertains to any simple compressible substance. Water is specifically chosen here because of its familiarity and widespread use.

transfer as heat yields no additional temperature increase, but a change in "phase" from ice (solid) to water (liquid). As the enthalpy of the system increases, the ratio of ice to water decreases until a prescribed enthalpy change has been achieved, and no ice exists in the system. The amount of enthalpy required to complete this phase change is referred to as the enthalpy of melting (or the enthalpy of freezing), and denoted as h_{sf}.

When the phase change is complete, the temperature once again begins to increase, although now with a different specific heat. When a temperature of 212° F is reached, a new phase change begins, this time from water (liquid) to steam (vapor). Once again, increasing enthalpy results not in a temperature increase, but in a decrease in the ratio of water to steam. The increase in enthalpy necessary to complete the phase change is called the enthalpy of vaporization,[6] and denoted as h_{fg}. The properties for the experiment are shown in Table 3-5 and plotted on Fig. 3-4. The enthalpy and entropy are arbitrarily set to zero at the state defined by "saturated" liquid at 32° F.

Repeating the experiment for a range of pressures shows that the specific volume of liquids and solids may change with temperature; however, it is basically independent of pressure for these phases. The solid and liquid phases are basically incompressible. For liquid and vapor phases, typical results can be found in Table 3-6.

Several interesting observations can be made from the experiment. If plotting lines of constant pressure on a temperature versus entropy graph, the data would look as shown in Fig. 3-5.

The general shape depicted here is referred to as a vapor dome. States to the left of the vapor dome are subcooled liquids; states to the right of the vapor dome are superheated vapor. The actual dome itself shows the states where a phase change begins or ends, and these states are referred to as saturated liquid or saturated vapor. Thermodynamic properties at saturation are shown with the subscript "f" for saturated liquid, and the subscript "g" for saturated vapor. For example, h_f is the enthalpy of saturated liquid, and h_g is the enthalpy of saturated vapor. The enthalpy of vaporization is h_{fg}. As pressure continues to increase, the vapor dome narrows, and the difference between thermodynamic properties of saturated liquid and saturated vapor narrows. At a critical point, corresponding to 3,206.2 psia and 705.4° F and found at the very top of the vapor dome, no difference exists between saturated liquid and saturated vapor. The water is simply a fluid. Above the critical pressure, supercritical states exist. In this region, any state with $T > T_{crit}$ is arbitrarily called a superheated vapor, and any state with $T < T_{crit}$ is called a subcooled liquid.

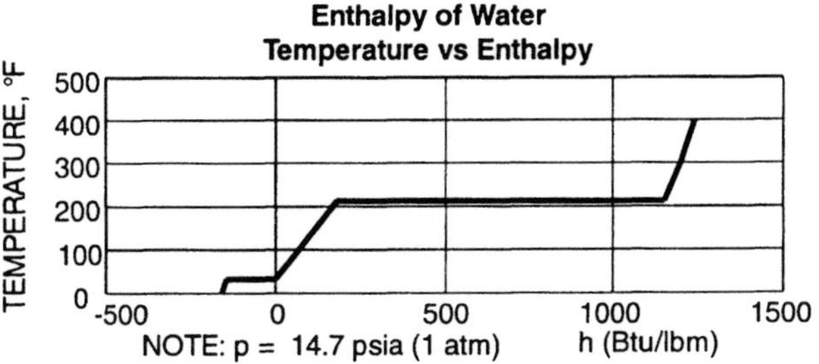

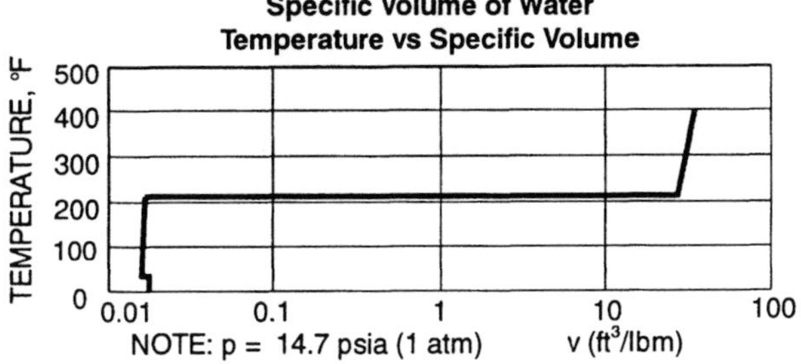

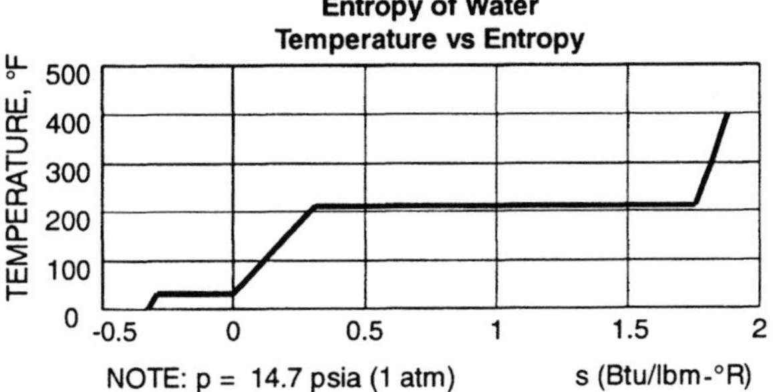

Fig. 3-4. Results of a simple experiment.

Table 3-5. Results of a Simple Experiment

Phase	T (° F)	v (ft³/lbm)	h (Btu/lbm)	s (Btu/lbm-° R)
Ice	0	0.0174	−158.98	−0.3244
Ice	32	0.0175	−143.40	−0.2916
Water	32	0.0160	0.00	0.0000
Water	212	0.0167	180.16	0.3121
Steam	212	26.8000	1,150.52	1.7567
Steam	300	30.5300	1,192.80	1.8160
Steam	400	34.6800	1,329.90	1.8743

Within the vapor dome, mixture states are defined by the thermodynamic properties at saturation and the quality, represented by the symbol x. Quality is defined as follows:

$$x = \frac{\text{lbm saturated vapor}}{\text{lbm mixture}} \tag{3-44}$$

so that for x pounds of vapor, there are $(1 - x)$ pounds of liquid. Once the quality is determined, thermodynamic properties for the mixture states are obtained from the following equations:

[6]Heat was once thought of as a property rather than a process. At this stage of thermodynamic thought, the terms latent heat and sensible heat were used. Sensible heat referred to heat transfer that resulted in an increased temperature, and latent heat referred to heat transfer that resulted in a phase change (with no temperature increase.) Unfortunately, this terminology has lived on in such expressions as "latent heat of vaporization."

Table 3-6. Results of an Extended Experiment

P (psia)	T (° F)	h (Btu/lbm)	s (Btu/lbm-° R)	v (ft³/lbm)	State
1	32.0	0.0	0.000	0.016	Sub. liq.
1	101.7	69.8	0.133	0.016	Sat. liq.
1	101.7	1,105.8	1.978	333.330	Sat. steam
1	200.0	1,150.2	2.051	392.160	Super. steam
1	400.0	1,241.8	2.172	512.820	Super. steam
14.7	32.0	0.0	0.000	0.016	Sub. liq.
14.7	212.0	180.2	0.312	0.017	Sat. liq.
14.7	212.0	1,150.5	1.757	26.800	Sat. steam
14.7	300.0	1,192.8	1.816	30.530	Super. steam
14.7	400.0	1,239.9	1.874	34.680	Super. steam
1,000	544.8	542.4	0.743	0.022	Sat. liq.
1,000	544.8	1,192.4	1.390	0.446	Sat. steam
1,000	1,000.0	1,505.9	1.653	0.831	Super. steam
2,000	636.0	671.9	0.862	0.026	Sat. liq.
2,000	636.0	1,136.3	1.286	0.188	Sat. steam
2,000	1,000.0	1,474.1	1.560	0.394	Super. steam
3,000	695.5	802.1	0.973	0.034	Sat. steam
3,000	695.5	802.1	1.157	0.084	Sat. steam
3,206.2	705.4	919.0	1.072	0.054	Critical pt.

$$h = (1 - x)h_f + xh_g$$
$$s = (1 - x)s_f + xs_g \qquad (3\text{-}45)$$
$$v = (1 - x)v_f + xv_g$$

Alternatively, of course, knowledge of the value of any thermodynamic property within the vapor dome can be used to determine the quality of the mixture.

A complete view of the temperature–entropy diagram for steam is shown in standard texts on thermodynamics as referenced at the end of this chapter. It should be noted that these figures contain not only lines of constant pressure, but also lines of constant enthalpy, specific volume, and quality. Appendix A also contains two tables of thermodynamic properties. Table A-1 (metric) and Table A-2 (English) list saturation properties for water and vapor, indexed by both temperature and pressure. (Within the vapor dome, temperature and pressure can't be varied independently. For each saturation or mixture pressure there is one saturation temperature, and vice versa.) Table A-3 (metric) and Table A-4 (English) list properties of superheated steam. These tables were generated from standard equations provided by the American Society of Mechanical Engineers (ASME). At one point, tables such as these were a mainstay reference, but they have been somewhat displaced in recent years by computer-based programs.

The temperature–entropy diagram and its partner, the Mollier Diagram (h–s plane) (Fig. 3-6), are discussed more extensively in following sections.

3.4.3 The Ideal Gas

A single equation that relates pressure, temperature, and specific volume for the full range of phases and mixtures of water and steam would be very difficult to produce. But in some cases, careful examination of laboratory data (or quan-

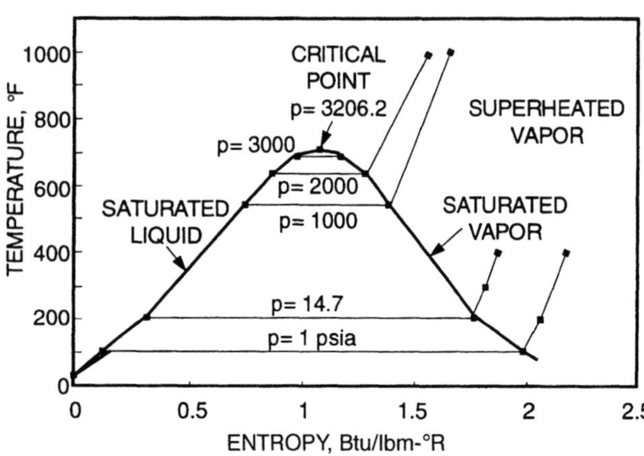

Fig. 3-5. Water and steam, temperature versus entropy.

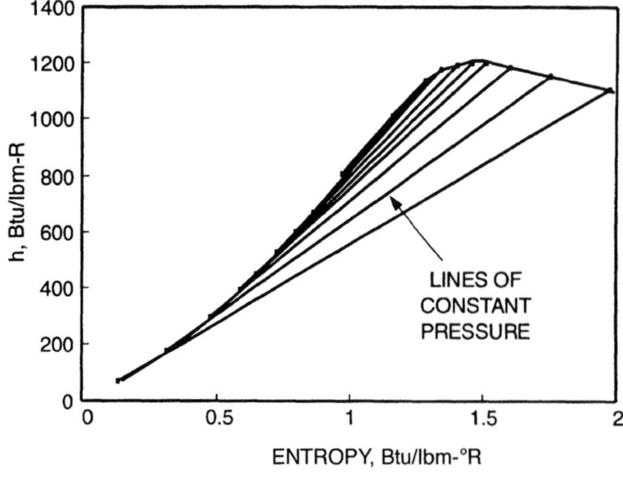

Fig. 3-6. A simplified Mollier diagram.

tum statistical thermodynamic theory) will yield such an equation. These equations of state are generally limited in range or by idealizations, but in some circumstances, they are highly accurate over a wide range of conditions and materials. The case of the ideal or perfect gas is a good example.

The concept of a perfect gas is one of gas molecules so small that they occupy a negligible part of the total volume, and exert no forces on one another except when they happen to collide. In fact, this model is fairly accurate for most gases when the pressure is low, and/or the temperature is high.

The equation of state for the perfect (or "ideal") gas is given by the following:

$$Pv = RT \tag{3-46}$$

or equivalently

$$PV = mRT \tag{3-47}$$

where

P = absolute pressure, lbf/ft^2,

v = specific volume, ft^3/lbm,

V = total volume, ft^3,

R = gas constant, ft-lbf/lbm-°R,

T = absolute temperature, °R, and

m = mass, lbm.

The gas constant R is different for different gases, but can be calculated for any gas as follows:

$$R = \frac{R^u}{M} \tag{3-48}$$

where

R^u = the universal gas constant, 1,545.33 ft-lbf/lbmol-°R, and

M = the molecular weight of the gas (lbm/lbmol).

Having a perfect gas results in several simplifications:

1. Internal energy, u, is only a function of temperature, T. Recalling the definition of c_v, this means that

$$u_2 - u_1 = \int_{T_1}^{T_2} c_v dT \tag{3-49}$$

and if c_v remains constant,

$$u_2 - u_1 = c_v(T_2 - T_1) \tag{3-50}$$

2. Since enthalpy for a perfect gas is given by

$$\begin{aligned} h &= u + Pv \\ h &= u + RT \end{aligned} \tag{3-51}$$

then h is also a function of T only, and by the definition of c_p,

$$h_2 - h_1 = \int_{T_1}^{T_2} c_p dT \tag{3-52}$$

and for constant c_p,

$$h_2 - h_1 = c_p(T_2 - T_1) \tag{3-53}$$

3. Differentiation of the enthalpy definition with respect to T gives

$$\begin{aligned} \frac{\partial h}{\partial T} &= \frac{\partial u}{\partial T} + \left(v\frac{\partial P}{\partial T} + P\frac{\partial v}{\partial T}\right) \\ c_p &= c_v + \left(v\frac{\partial P}{\partial T} + P\frac{\partial v}{\partial T}\right) \end{aligned} \tag{3-54}$$

The perfect gas law can also be differentiated with respect to T to give

$$\left(v\frac{\partial P}{\partial T} + P\frac{\partial v}{\partial T}\right) = R \tag{3-55}$$

Substituting the second equation into the first,

$$c_p = c_v + R \tag{3-56}$$

Since the units of c are (Btu/lbm-° R) and the units of R are (ft-lbf/lbm-° R), a conversion factor is required for the last equation, and it is sometimes written

$$c_p = c_v + \frac{R}{J} \tag{3-57}$$

where

J = 778 ft-lbf/Btu.

As an example, the molecular weight of dry air[7] is 28.966 lbm/lbmol. Dividing the universal gas constant by this amount yields a gas constant for air of 53.34 ft-lbf/lbm-° R. The specific heats for air are c_p = 0.24 and c_v = 0.1714. Checking the arithmetic:

$$0.24 = 0.1714 + \frac{53.34}{778} \tag{3-58}$$

[7]Dry air contains 78.08% (by volume) nitrogen, 20.95% oxygen, and 0.97% traces of 15 other gases. The composition is sometimes simplified to 20.9% (by volume) oxygen and 79.1% nitrogen. Dry air behaves well as a perfect gas over a fairly wide range of moderate to high temperatures and moderate to low pressures.

There are three more equations of interest for the perfect gas with constant specific heats, and they all relate to entropy:

$$s_2 - s_1 = c_v \ln\frac{T_2}{T_1} + R\ln\frac{v_2}{v_1}$$

$$s_2 - s_1 = c_p \ln\frac{T_2}{T_1} + R\ln\frac{P_2}{P_1} \tag{3-59}$$

$$s_2 - s_1 = c_p \ln\frac{v_2}{v_1} + c_v\ln\frac{P_2}{P_1}$$

The last of these equations leads to another important finding. For a reversible ($dS = 0$), adiabatic ($dQ = 0$) process, a perfect gas with constant specific heats exhibits the property

$$Pv^k = \text{constant} \tag{3-60}$$

where

$k = c_p/c_v$.

This representation of the perfect gas isentropic ($dS = 0$) process is, in fact, only one special case of a more general case. Any process that can be represented by the general form of the equation Pv^n = constant is called a polytropic process. Some specific cases are found in Table 3-7.

Work from a polytropic process, $Pv^n = C$, can be calculated between two states, 1 and 2, as

$$W = \int P\,dv = \int \frac{C}{v^n}dv = \frac{Cv_2^{1-n} - Cv_1^{1-n}}{1-n} \tag{3-61}$$

But $C = P_1v_1^n = P_2v_2^n$, and substituting these values into the work equation results in the following:

$$W = \frac{P_2v_2^nv_2^{1-n} - P_1v_1^nv_1^{1-n}}{1-n} \tag{3-62}$$

or simply

$$W = \frac{P_2v_2 - P_1v_1}{1-n} \tag{3-63}$$

(work of compression is negative.) An isentropic process with exponent $n = kc_p/c_v$, reduces to

$$W = \frac{R(T_2 - T_1)}{1-k} = \frac{(c_p - c_v)(T_2 - T_1)}{\frac{(c_v - c_p)}{c_v}} \tag{3-64}$$

Table 3-7. Polytropic Exponents

Process	Constant	n
Isentropic	s	$k = c_p/c_v$
Isobaric	P	0
Ishothermal	T	1
Constant volume	V	$n \rightarrow \infty$

or simply

$$W = c_v(T_1 - T_2) \tag{3-65}$$

3.4.4 Mixtures of Perfect Gases

The previous section used air as an example of a perfect gas. As mentioned previously, air is not a single gas, but a mixture of gases. This subject deserves special attention. The most important factor to remember is that perfect gas molecules exert no forces on one another. The consequence of this is that each gas behaves as if the other gases were not present. This means that in a mixture of two gases, gas 1 and gas 2, that follows the perfect gas law

$$PV = mRT \tag{3-66}$$

the law can be written

$$\begin{aligned} P_1V_1 &= m_1R_1T_1 \\ P_2V_2 &= m_2R_2T_2 \end{aligned} \tag{3-67}$$

for each gas. Based on what is known about equilibrium, the situation can be simplified by some observations about these properties:

$$\begin{aligned} T &= T_1 = T_2 \\ V &= V_1 = V_2 \\ m &= m_1 + m_2 \\ P &= P_1 + P_2 \end{aligned} \tag{3-68}$$

The last of these equations, $P = P_1 + P_2$, isn't as obvious. It's called Dalton's Law, and the pressures P_1 and P_2 are referred to as partial pressures. The ratio of any partial pressure to total pressure is equal to the mole fraction, x, of the gas component, or equivalently, the gas percent by volume.

The only question remaining is how to obtain the value of R for the gas mixture. The universal gas constant could be divided by a composite molecular weight based on the mole fractions, but this is often inconvenient when the mixture composition is known by mass flow rates.

Alternatively, solving the ideal gas equation of the mixture for R:

$$R = \frac{PV}{mT} = \frac{P_1V + P_2V}{mT}$$

$$R = \frac{m_1R_2T + m_2R_2T}{mT} \tag{3-69}$$

$$R = \frac{m_1R_1 + m_2R_2}{m}$$

This mass weighted mean can be used not only for R, but also for values of h, u, s, c_p, and c_v.

3.4.5 Real Gases

Real gases can vary from the ideal gas law as pressures increase or temperatures decrease. Van der Waals forces form attractions between molecules, and increasing density violates the premise of negligible volume.

The most common way of dealing with this variation from perfect gas behavior is through the concept of compressibility. The compressibility factor is represented by the symbol Z and changes the form of the gas equation to

$$Pv = ZRT \tag{3-70}$$

As long as Z is close to 1, the perfect gas law is a good approximation.

Although Z changes with temperature and pressure, it is essentially the same for all gases in terms of the reduced pressure, P_r, defined as

$$P_r = P/P_{crit} \tag{3-71}$$

and the reduced temperature, T_r, defined as

$$T_r = T/T_{crit} \tag{3-72}$$

where

P_{crit} = critical pressure for the substance

T_{crit} = critical temperature for the substance

In addition to the critical temperature and pressure already given for water vapor, other critical values are given in Table 3-8.

Compressibility charts can be found in any standard text on thermodynamics.

3.4.6 Incompressible Liquids

An incompressible liquid has no reversible work modes since $P\,dv$ must always equal zero. The State Postulate therefore defines the number of independent thermodynamic properties as one.

Since all processes are at constant volume,

$$c_v = c = \frac{\partial u}{\partial T} \tag{3-73}$$

Substituting this result into the differential definition of enthalpy results in

$$\begin{aligned} dh &= du + P\,dv + v\,dP \\ dh &= du + v\,dp \\ dh &= c\,dT + v\,dp \end{aligned} \tag{3-74}$$

and for a constant pressure process, c is the rate of change of h with respect to T. Recalling that this is the definition of c_p,

Table 3-8. Critical Temperature and Pressures

Gas	T_{crit} (°R)	P_{crit} (psia)
Air	235.8	547.0
Carbon dioxide	547.8	705.0
Carbon monoxide	242.2	508.2
Nitrogen	227.2	492.5
Oxygen	278.1	730.9
Sulfur dioxide	775.0	1,141.0
Water vapor	1,165.4	3,206.0

$$c_p = \left(\frac{\partial h}{\partial T}\right) = c = c_v \tag{3-75}$$

There is only one specific heat for the incompressible liquid, and assuming that it remains fairly constant, the following rules can be written for the incompressible liquid:

$$\begin{aligned} u_2 - u_1 &= c(T_2 - T_1) \\ h_2 - h_1 &= c(T_2 - T_1) + v(P_2 - P_1) \\ s_2 - s_1 &= c\ln(T_2/T_1) \end{aligned} \tag{3-76}$$

Of course, incompressible liquids are just an approximation, and the approximation degrades as the range of pressure increases. The specific volume of 200° F water may actually change by about 0.3% between atmospheric pressure and 1,000 psia, and the enthalpy may change by about 0.4%.

3.5 THERMOMECHANICAL ENERGY CONVERSION

Much of the development of thermodynamics has been motivated by a desire to convert thermal energy into mechanical energy. Any system devised to achieve this is referred to as a *heat engine*. This section examines the thermodynamic aspects of this conversion process from three basic viewpoints: the general guidelines for any successful thermomechanical process, how to engineer/analyze such a system, and how efficient a process can be.

3.5.1 An Introductory Experiment

This experiment involves a cylinder and piston arrangement containing a unit of ideal gas.[8] A valid question would be whether or not this system can convert thermal energy into mechanical efficiency with 100% efficiency. Assume that any piston movement is slow and frictionless to avoid an easy answer to the question.

At the beginning of the experiment, the following conditions exist for the ideal gas:

Pressure:	P_1
Temperature:	T_1
Specific Volume:	v_1

[8]The fact that this is an ideal gas is important only because it allows the equations of state to be used.

A small amount of energy, δQ, is added as heat to the cylinder, and at the same time, the piston is allowed to do a small amount of reversible work, $\delta W = P\,dv$. If the process is to be 100% efficient, no energy can be stored in the ideal gas, and its temperature must remain constant as T_1 (an isothermal expansion.)

The energy transfer out as work can be analyzed as follows:

$$\delta W = P\,dv = (RT/v)\,dv \tag{3-77}$$

and

$$W = RT\ln(v_2/v_1) \tag{3-78}$$

The energy transfer in as heat can be analyzed as follows:

$$\delta Q = T\,ds \tag{3-79}$$

and, using Eq. (3-59),

$$Q = T(s_2 - s_1) = TR\ln(v_2/v_1) \tag{3-80}$$

The energy transfer in and out is identical, and a 100% efficient conversion process occurs.

If this system is to produce work at a constant rate, the ratio of specific volumes must remain constant across any time step. As an example, if the specific volume were to double with each time step, the original volume would have to expand over a million times within about 20 timesteps, and any cylinder of finite size could not sustain the process.

The general type of process described here is referred to as a 1T heat engine (because only one process of energy transfer as heat is involved), and the statement that a 1T heat engine cannot operate continuously is known as the Kelvin–Planck Statement of the second law of thermodynamics.

3.5.2 Cycles

Because the 1T heat engine has limited usefulness, a more sustainable process is needed. The experiment could allow the isothermal expansion to occur for a short time and then push the piston back to its original position to start over again. As the piston is pushed back, the gas starts to increase in temperature, consistent with the first law requirement that energy added to the gas as work be either rejected as heat, or stored as internal energy in the form of increased temperature.

To return to the original conditions, heat must be rejected as the gas is compressed. This isothermal compression, of course, would only trace the original process backwards and reject the same amount of energy as heat that was absorbed. As stated in the first law, the amount of work required to recompress the gas is the same as the amount extracted from the isothermal expansion. Although the goal of creating a sustainable process has been met, no net progress has been generated.

This last experiment teaches two very important lessons:

1. Any sustainable process for thermomechanical energy conversion eventually has to return the working media to an original state. This concept is the informal statement that the processes involved must create a cycle. Just as in mathematics, a cycle is a set of repetitious processes that has the same starting and ending point in terms of state properties. The processes may be as simple or complicated as desired.
2. The net work produced is the difference between the energy absorbed as heat and the energy rejected as heat. The first law makes this point in no uncertain terms. The more done to maximize the energy absorbed and minimize the energy rejected, the more net work will result and the more efficient the process will be.

With these two lessons in mind, the cycle will be revised in the next section to achieve net work.

3.5.3 A 2T Heat Engine

The isothermal expansion provided a good start for the cycle. Since a reversible process (now a reversible cycle) is still the objective, the options are fairly limited.

The expansion is allowed to continue, but the system is isolated from the energy transfer as heat. Reversible mechanical work is occurring as follows:

$$W = \int P\,dv \tag{3-81}$$

Since no energy is being transferred as heat (adiabatic) and no irreversibilities are being generated by the work process, the entropy of gas stays constant, and the process is isentropic. Isentropic processes for ideal gases are described by the following polytropic process equation:

$$Pv^k = \text{constant} \tag{3-82}$$

If the isentropic expansion is allowed to double the cylinder volume, the end state of the gas could be defined as follows (use k of 1.4 for this example):

$$\begin{aligned} s_3 &= s_2 \\ v_3 &= 2v_2 \\ P_3 &= 0.38P_2 \\ T_3 &= 0.76T_2 \end{aligned} \tag{3-83}$$

The process is still reversible and the temperature has dropped significantly.

If an isothermal compression back to the original entropy was started now (s_1), the energy that would have to be rejected as heat ($Q_{34} = T_3(s_2 - s_1)$) would be only 76% of the original energy absorbed as heat in the isothermal expansion.

One last step must be performed to close the cycle and return the cooled, low-pressure gas to state 1. So as not to increase the entropy, no energy transfer as heat will be involved. Rather, a mechanical compression of the gas with no heat transfer and no irreversibilities will be performed.

This isentropic compression, coupled with the isothermal compression of the previous step, requires some work, called back work, but the back work will be much lower than the work extracted during the expansion processes. In fact, without any analysis of the compression processes, the first law mandates that the net work, W, will be equal to the following:

$$W = Q_{12} - Q_{34} \tag{3-84}$$

The processes just described can be viewed on a T–s diagram for the gas (Fig. 3-7).

Previously, a heat source for Q_{in} had a temperature at least slightly higher than T_{12}. Because energy as heat won't spontaneously flow from a cold mass to a hot mass, the process now requires a heat sink for Q_{out} at a temperature at least slightly lower than T_{34}. This second temperature reservoir makes the process a 2T heat engine, with energy extracted from a hot temperature reservoir and rejected to a cool temperature reservoir. If the temperatures of the reservoirs are sufficiently close to the actual process temperatures and reversible mechanical work processes are maintained, a reversible 2T heat engine exists.

There are ways to make a 2T heat engine other than the procedure described in the preceding paragraph. In fact, any cycle with the general form meets the requirements (Fig. 3-8). The following sections will describe some of these cycles. The process just described does have a special significance, however, in the fact that it is reversible and represents a special benchmark case for measurement of all other cycles. The reversible 2T heat engine bears the special name Carnot cycle, and its significance is examined in the next section.

3.5.4 The Carnot Cycle

The reversible 2T heat engine, or Carnot cycle, described previously, assumed many factors. No mechanical irreversibilities occurred as the pistons moved with frictionless effort and moved slowly enough not to create internal flow losses in the gas. The isothermal compression and expansion processes were controlled with microscopic precision as the ideal gas tracked the reservoir temperatures perfectly. As the reversible 2T engine completed its power cycle, no net increase in the entropy of the universe occurred, and at a moment's notice, the process could be reversed and return everything to the original state.

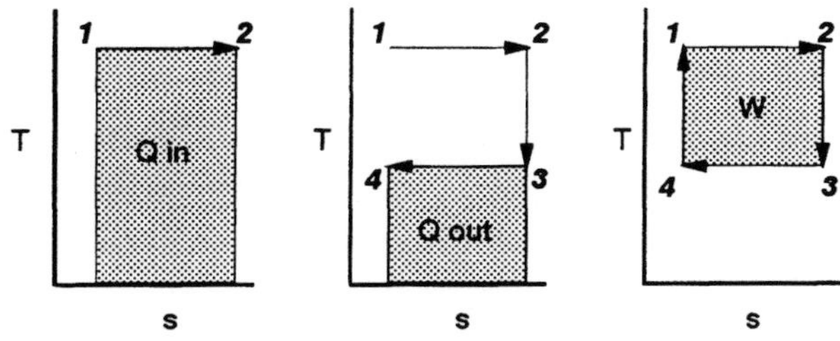

Fig. 3-7. A reversible 2T heat engine.

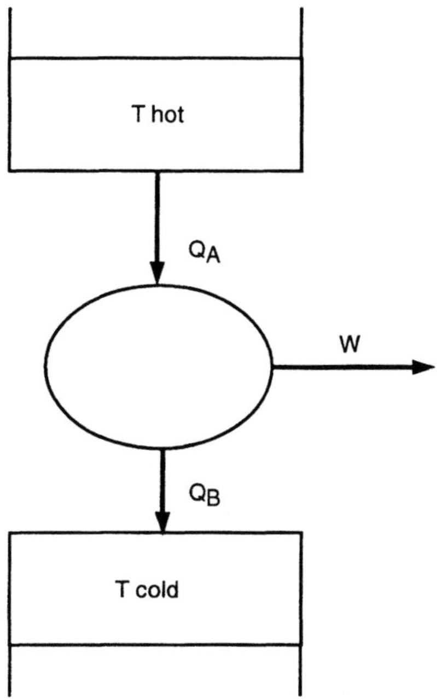

Fig. 3-8. A generic 2T heat engine.

The efficiency, η, of any 2T heat engine can be defined as follows:

$$\eta = \frac{W}{Q_A} = \frac{Q_A - Q_B}{Q_A} = 1 - \frac{Q_B}{Q_A} \tag{3-85}$$

where

Q_A = the energy absorbed as heat from the high-temperature reservoir, and

Q_B = the energy rejected as heat to the low- temperature reservoir.

If the reservoir absolute temperatures are T_A and T_B, respectively, then the efficiency of a reversible 2T engine can be written as follows:

$$\eta_{rev} = 1 - \frac{Q_B}{Q_A} = 1 - \frac{T_B \Delta s}{T_A \Delta s} = 1 - \frac{T_B}{T_A} \tag{3-86}$$

This is referred to as the Carnot efficiency.

The second law reveals valuable information about the 2T heat engine (not necessarily a reversible process) as follows. As energy leaves the high-temperature reservoir, A, the entropy of the reservoir decreases as follows:

$$\Delta s_A = -\frac{Q_A}{T_A} \tag{3-87}$$

Similarly, as energy is rejected as heat to the low-temperature reservoir, the entropy of the reservoir increases as follows:

$$\Delta s_B = -\frac{Q_B}{T_B} \quad (3\text{-}88)$$

Because no entropy is being stored in the working fluid over a period of one cycle, any irreversibilities in the process will show up as net entropy production equal to the following:

$$\frac{Q_B}{T_B} - \frac{Q_A}{T_A} \geqslant 0 \quad (3\text{-}89)$$

The equality represents the reversible 2T engine. Multiplying through by T_B/Q_A, the following is seen:

$$\frac{Q_B}{Q_A} \geqslant \frac{T_B}{T_A} \quad (3\text{-}90)$$

The equality once again represents the reversible Carnot cycle. The efficiency equation can be written as follows:

$$\eta = 1 - \frac{Q_B}{Q_A} \leqslant 1 - \frac{T_B}{T_A} = \eta_{rev} \quad (3\text{-}91)$$

The Carnot efficiency is the upper limit for any 2T heat engine operating between T_A and T_B.

3.5.5 Gas and Vapor Power Cycles

Before beginning a review of power cycle equipment, a new concept must be introduced. Although the example used to this point has been based on an all-gas power cycle where the working fluid is always in a gaseous state, many power cycles are based on operation around the vapor dome, with both liquid and gaseous phases playing an important part in the cycle. These vapor power cycles are appealing because compression is confined to the liquid phase and requires little back work, and also because heat addition and rejection occur as a phase change and therefore at a constant temperature within the vapor dome.

The vapor power cycle of primary importance in power generation work is the Rankine cycle, and the most important all-gas power cycle is the Brayton cycle. Both of these cycles are investigated more thoroughly in the following sections.

3.6 THE RANKINE CYCLE

As a matter of theory in previous sections, isothermal expansion and various other pattern processes were discussed without regard as to how to build or operate devices that could provide these services. Departures from these idealized processes occur for two principal reasons. First, it is hard to avoid the irreversibilities of flow and mechanical friction in real life. Second, although these processes can be very closely approximated in the lab, other factors must be balanced into the design of commercial equipment. Capital cost, size, capacity, and reliability cannot be ignored in the pursuit of the ideal process.

The Rankine cycle is a vapor power cycle that forms the thermodynamic basis for most steam power plants. These plants may use coal, oil, gas, or nuclear power as fuel for a high temperature source, but the basic thermodynamic operation remains fairly constant.

Although detailed examination of various power cycle components is reserved for other chapters in this book, this section briefly examines the general nature of some of the components available for construction of an operating power cycle. The operation and efficiency of these components will be addressed in comparison to the ideal process, and enough information will be provided to perform a conceptual design.

3.6.1 The Basic Cycle

Like many other cycles, the Rankine cycle relies on the isentropic expansion of high-pressure gas to produce work. The gas of choice for most Rankine cycles is steam. The system operates on the premise that the easiest way to make high-pressure steam is to start with high-pressure water and then heat that water at constant pressure. It's a concept with merit. Since water is basically an incompressible liquid, it takes relatively little energy to compress it to high pressures, and operation within the vapor dome greatly simplifies control of steam properties.

As an example, a conceptual drawing of the most basic representation of this system is shown in Fig. 3-9. This cycle is referred to as an open loop cycle because the environment closes the cycle. In effect, the environment is the device in which cycle heat rejection takes place. (In a closed cycle, a condenser takes the place of the environment.) In Fig. 3-9, water at atmospheric pressure is pumped to 200 psia. The pump is basically a control volume that adiabatically increases the pressure of a liquid. The steady-state, steady-flow energy equation for this device reduces to the following:

$$W_p = m(h_2 - h_1) \quad (3\text{-}92)$$

where

W_p = pump power, and

m = mass flow rate.

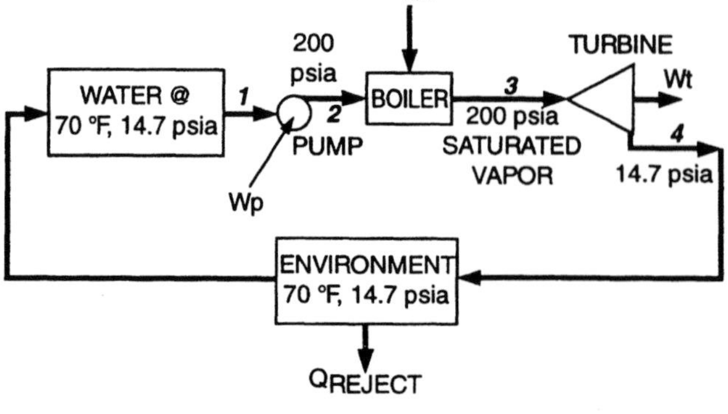

Fig. 3-9. The basic open Rankine cycle.

If the process is considered as an isentropic compression of an incompressible liquid, the Gibbs equation, $T\,ds = dh - v\,dp$, yields the following:

$$h_2 - h_1 = v(P_2 - P_1) \tag{3-93}$$

The specific volume of water is very stable for constant temperature and is available from the steam tables as the specific volume of saturated water at 70° F. (Since $s_2 - s_1 = c \ln T_2/T_1$, no entropy change dictates no temperature change.)

Pumps are really not isentropic, and friction losses are part of real life. Measuring the efficiency of pumps requires comparing their performance with the ideal, isentropic system as follows:

$$\eta_p = \frac{h_{2s} - h_1}{h_2 - h_1} \tag{3-94}$$

where h_{2s} represents the enthalpy that would result from a true isentropic compression.[9] Actual centrifugal pump efficiencies range from about 65% for a pump capacity of several hundred gallons per minute (gpm), up to about 90% for a 10,000 gpm pump. (Chapter 11 discusses this in more detail.) For this example, pump efficiency of 80% is assumed.

After the water achieves high pressure, energy is added to the water as heat in the boiler or steam generator. Entire books are written about boiler operation and Chapter 7 of this book presents a detailed discussion. For purposes of this chapter, however, it can be defined as a control volume operating at constant pressure, with a steady-state, steady-flow-energy equation:

$$Q_b = m(h_3 - h_2) \tag{3-95}$$

where

Q_b = the energy added to the boiler as heat.

Since the boiler operates in the presence of both the liquid and steam phases, the output is saturated steam. (Any attempt to increase Q_b would simply increase mass flow, not temperature.)

Finally, the saturated steam is expanded in a turbine to produce mechanical power (W). The turbine is once again a control volume controlled by a steady-state, steady-flow-energy equation:

$$W_t = m(h_3 - h_4) \tag{3-96}$$

where

W_t = the mechanical power produced by the turbine.

Like the pump, the turbine is isentropic in the ideal case. Since the steam begins as a saturated vapor, an isentropic expansion to lower pressures produces a mixture of steam and water with the following quality:

$$x = \frac{s - s_f}{s_g - s_f} \tag{3-97}$$

The following enthalpy is found:

$$h = (1 - x)h_f + xh_g \tag{3-98}$$

Because actual turbines are not isentropic, the efficiency of a turbine is measured against the ideal isentropic case as follows:

$$\eta_t = \frac{h_3 - h_4}{h_3 - h_{4s}} \tag{3-99}$$

where

h_{4s} = the enthalpy resulting from the isentropic expansion.

Real steam turbines have isentropic efficiencies of between about 60% and 90%. For this example, an efficiency of 90% is used.

The thermodynamic cycle is approximated as shown in Fig. 3-10. The energy input as heat is the area below the curve between points 2, 3a, and 3, and the energy rejected is the area below the curve made up of points between 4 and 1. The net work, which is the turbine work minus the pump work, is the difference between these two amounts. (It should be remembered that the y-axis really extends to −460° F.) In this example, it is apparent that most of the energy input is rejected from the cycle as heat.

A detailed thermodynamic evaluation of this cycle is elementary. At each state, two factors control the state properties. More properties are listed than are really needed, but they may be of interest. The properties can be obtained from

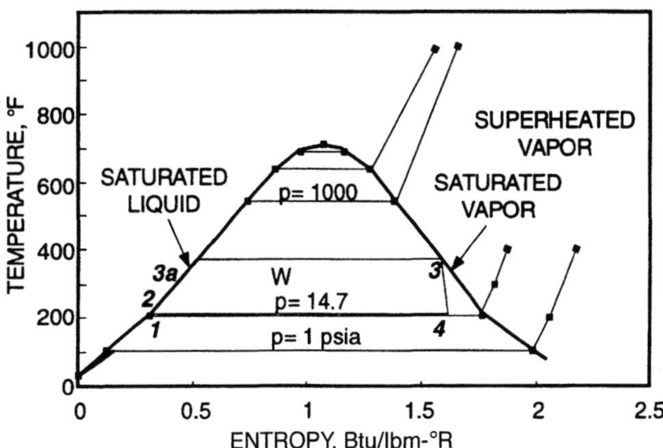

Fig. 3-10. The basic open Rankine cycle.

[9]Any enthalpy change not associated with $v\,dP$ shows up as $c\,dT$ and results in increased temperature and increased entropy.

state laws, the steam tables, the Mollier chart, or interactive steam tables on a personal computer. The chart would look similar to Table 3-9.

Key calculations are as follows:

$$Q_b = \text{heat input} = 1{,}199.3 - 38.81 = 1{,}160.49 \text{ Btu/lbm}$$

$$Q_{rej} = \text{rejected energy} = 1{,}028.3 - 38.13 = 990.16 \text{ Btu/lbm}$$

$$W_p = \text{pump power} = 38.81 - 38.13 = 0.68 \text{ Btu/lbm}$$

$$W_t = \text{turbine power} = 1{,}199.3 - 1{,}028.3 = 171 \text{ Btu/lbm}$$

$$\eta = (Q_b - Q_{rej})/Q_b = (W_t - W_p)/Q_b = 0.1468 \text{ or } 14.68\%$$

Only 14.68% of the energy into the boiler is being used to produce power. If it is arbitrarily assumed that the hot source is at 382° F (saturation temperature at 200 psia), the Carnot efficiency would be as follows:

$$\eta_{Carnot} = 1 - \frac{460 + 70}{460 + 382} = .3705 \qquad (3\text{-}100)$$

This demonstrates that the cycle is only about half as efficient as a Carnot (maximum efficiency) cycle operating between the same temperatures.

An additional note of importance is the amount of work energy used by the pump to compress the water. Of the 171 Btu/lbm produced by the turbine, only 0.68 Btu/lbm is consumed by the pump. The ratio of pump energy to work energy is referred to as the back work ratio, and has a value well below 1%.

3.6.2 The Condenser

An obvious point of concern about the basic system described previously is that the initial water temperature is 70° F, but a 212° F steam/air mixture is discharged from the turbine at point 4. This temperature was dictated by the outlet pressure of the turbine (atmospheric pressure at 14.7 psia). A concept of some merit might be to discharge into a smaller, controlled environment where the pressure could be kept low. A discharge pressure of 10 psia allows a discharge temperature of 193.2° F; a discharge pressure of 5 psia allows a discharge temperature of 162.2° F; and a discharge pressure of 1 psia allows a discharge temperature of 101.7° F.[10] If nothing but water vapor is in the environment and it is condensing (two-phase), then the pressure is dictated to be the saturation pressure at the condensate temperature. Once the steam/water exhaust is released into this low-pressure environment, something must be done with it, and the best idea is to finish condensing it to a full liquid state, and then pump it out. (Pumping an incompressible liquid is a low power activity.) Because water is being pumped to a higher pressure for the boiler, it could be reused.[11] Using this process has caused a change from an open loop cycle to a closed loop cycle.

The new component to achieve this, the condenser, is defined as a control volume at constant pressure where energy as heat is rejected from the liquid/vapor mixture to return it to a fully liquid state (possibly subcooled.) The steady-flow-energy equation for the condenser is as follows:

$$Q_{rej} = m(h_{inlet} - h_{out}) \qquad (3\text{-}101)$$

where

Q_{rej} = the rate of energy rejection.

The actual operating pressure is a function of the temperature of the thermal reservoir available for heat rejection. A continuous supply of 70° F water could theoretically produce an exhaust at 0.36 psia. In reality, however, heat transfer is associated with ΔT, and a practical exhaust pressure will be higher. For the following example, an exhaust pressure of 1 psia with a corresponding exhaust temperature of 101.7° F is assumed. The condensate is assumed to be subcooled to 80° F before it reenters the pump.

The basic closed loop Rankine cycle is now represented as shown in Fig. 3-11.

A thermodynamic analysis of the system would now be similar to the previous analysis in Section 3.1.1, with the only

Table 3-9. A Basic Open Rankine Cycle State Table

Cycle Point	T (° F)	P (psia)	h (Btu/lbm)	s (Btu/lbm-° R)	ρ (lbm/ft³)	State	Notes
1	70	14.7	38.13	0.074625	62.31	Liquid	
2s	70	200	38.68	0.074625	62.34	Liquid	
2	70.2	200	38.81	0.074862	62.34	Liquid	$\eta = 0.8$
3a	381.9	200	355.61	0.543978	54.4	Sat. liq.	
3	381.9	200	1,199.3	1.546455	0.4368	Sat. vap.	
4s	212	14.7	1,009.3	1.546455	0.0437	$x = 0.854$	
4	212	14.7	1,028.3	1.57474	0.0427	$x = 0.874$	$\eta = 0.9$

[10]Two items should be noted. First, when around the 1 psia range, pressure is usually given in "in. Hg" rather than psia (1 psia = 2.036 in. Hg). Second, the drop in temperature with pressure is not linear. The drop is about 3.5° F per psia at 14.7 psia, but over 20° F per psia at 1 psia. Small changes in discharge pressure around 1 psia can have significant impacts.

[11]This is also good from the aspect that the water used in real steam plants is generally treated to remove minerals and impurities, and reuse of the water minimizes the cost of this processing.

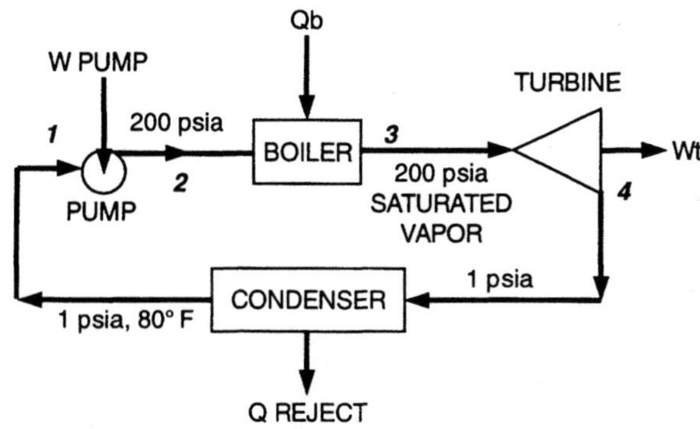

Fig. 3-11. The basic closed loop Rankine cycle.

new part being determination of the enthalpy of the subcooled water. (Enthalpy of the subcooled 80° F water is essentially the same as saturated 80° F water.) The process state table is shown in Table 3-10.

The enthalpy of the turbine exhaust has dropped from 1,028.3 Btu/lbm in the open system to 897.09 Btu/lbm in the closed system.

Key calculations are as follows:

$$Q_b = \text{heat input} = 1{,}199.3 - 48.83 = 1{,}150.47 \text{ Btu/lbm}$$

$$Q_{rej} = \text{rejected energy} = 897.09 - 48.09 = 849.00 \text{ Btu/lbm}$$

$$W_p = \text{pump power} = 48.83 - 48.09 = 0.74 \text{ Btu/lbm}$$

$$W_t = \text{turbine power} = 1{,}199.3 - 897.1 = 302.2 \text{ Btu/lbm}$$

$$\eta = (Q_b - Q_{rej})/Q_b = (W_t - W_p)/Q_b = 0.262$$

Compared with the open system, the energy input is very similar, but the net work has almost doubled, with the efficiency increasing from 14.7% to 26.2%. However, the quality of the turbine exhaust flow has dropped to about 80%. Low quality (high liquid content) can have destructive effects on a turbine.

3.6.3 Boiler Pressure

If the pressure is arbitrarily raised from 200 psia to 600 psia, the process diagram doesn't look any different, but an examination of the T–s diagram on Fig. 3-12 reveals some interesting effects. The increased pressure increases the work on the top side of the cycle,[12] but since the vapor dome is narrowing with higher pressure, a reduction in work on the right hand side of the cycle occurs. The positive effect is that a large reduction has occurred in the amount of heat rejected (area under curve 4-1 compared to curve 4x-1). These amounts can be quantified with a new process state table (Table 3-11).

Key calculations are as follows:

$$Q_b = \text{heat input} = 1{,}204.06 - 50.32 = 1{,}153.75 \text{ Btu/lbm}$$

$$Q_{rej} = \text{rejected energy} = 847.09 - 48.09 = 798.93 \text{ Btu/lbm}$$

$$W_p = \text{pump power} = 50.32 - 48.09 = 2.22 \text{ Btu/lbm}$$

$$W_t = \text{turbine power} = 1{,}204.06 - 847.09 = 356.97 \text{ Btu/lbm}$$

$$\eta = (Q_b - Q_{rej})/Q_b = (W_t - W_p)/Q_b = 0.3075$$

Despite the higher pressure and temperature, the heat input has remained almost constant. Rejected heat has dropped by about 50 Btu/lbm, and turbine work has increased by about 50 Btu/lbm. The net efficiency of the cycle has increased from 26.2% to 30.8%. The only drawback is that the quality of the exhaust flow has fallen to 75%.

If the pressure were elevated to 2,500 psia, a pressure not uncommon in modern coal-fired steam units, the efficiency would be a little over 34% and the exhaust quality would be less than 64%.

3.6.4 Superheat

Boiler output must be kept on the saturated vapor line because the boiler is a constant-pressure device, and the presence of both a liquid and gaseous phase dictates that the vapor will never go beyond saturated conditions.

The high-temperature thermal reservoir (the combustion gases) may have the capacity to increase the steam temperature further. If the saturated steam at 600 psia (486.3° F) is allowed to superheat to 1,000° F (still 600 psia), it's obvious that this new portion of the diagram (Fig. 3-13) has a higher efficiency than the earlier process. Heat rejection has been

Table 3-10. State Table for the Basic Closed Rankine Cycle

Cycle Point	T (° F)	P (psia)	h (Btu/lbm)	s (Btu/lbm-° R)	ρ (lbm/ft³)	State	Notes
1	80.00	1	48.09	0.0933	62.200	Liquid	
2s	80.00	200	48.68	0.0933	62.260	Liquid	
2	80.20	200	48.83	0.0936	62.260	Liquid	$\eta = 0.8$
3a	381.90	200	35.61	0.5440	54.390	Sat. liq.	
3	381.90	200	1,199.30	1.5460	0.437	Sat. vap.	
4s	101.70	1	863.51	1.5465	0.004	$x = 0.766$	
4	101.70	1	897.09	1.6063	0.004	$x = 0.799$	$\eta = 0.9$

[12]Heat input is the area under the curve from point 2 to point 3, energy rejected is the area under the curve from point 4 to point 1, and net work is the difference between them. Because of irreversibilities in the system, the areas on the curve are a bit stretched, but the concept is the important thing. The numbers are correct on the state table.

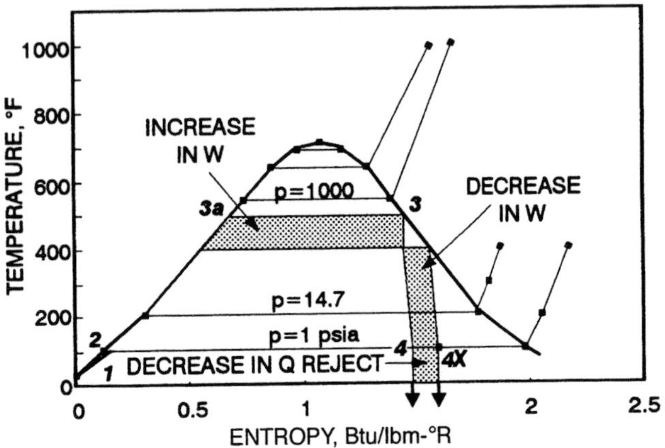

Fig. 3-12. Effects of increased pressure on the Rankine cycle.

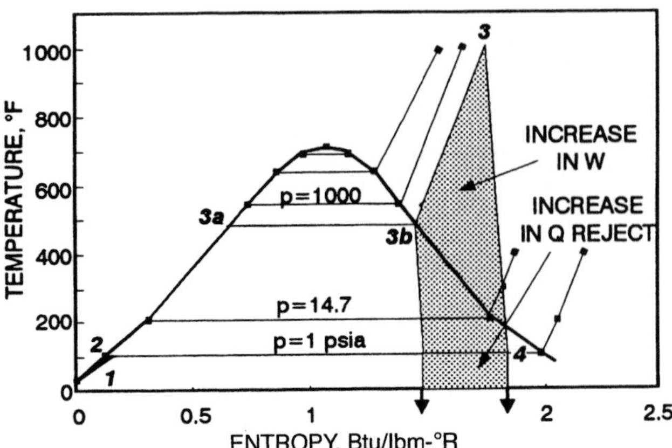

Fig. 3-13. Rankine cycle with superheat.

increased, but the heat rejection fraction of the input has not increased proportionately. Thus, a greater fraction of this input is being converted into work. The device used to accomplish this is called a superheater. The superheater is a control volume operating at constant pressure, but operates on steam only. Because the energy as heat for the superheater generally comes from the same thermal reservoir as the boiler, the energy inputs to both the boiler and the superheater are grouped as simply Q_b. It is important, however, to note that these devices are separate (Fig. 3-14).

An analysis of the superheater for thermodynamic properties can be accomplished by examining the Mollier diagram or the superheated steam tables. A process state table would look like Table 3-12, with state 3b representing the saturated vapor state between the boiler and the superheater.

Key calculations are as follows:

$$Q_b = \text{energy input} = 1{,}517.83 - 50.32 = 1{,}467.51 \text{ Btu/lbm}$$

$$Q_{rej} = \text{wasted energy} = 1{,}014.38 - 48.09 = 966.29 \text{ Btu/lbm}$$

$$W_p = \text{pump power} = 50.32 - 48.09 = 2.23 \text{ Btu/lbm}$$

$$W_t = \text{turbine power} = 1{,}517.83 - 1{,}014.38 = 503.45 \text{ Btu/lbm}$$

$$\eta = (Q_b - Q_{rej})/Q_b = (W_t - W_p)/Q_b = 0.3415$$

About 314 Btu/lbm more enthalpy is added, but over 146 Btu/lbm of that enthalpy is extracted as useful work. This means that the new superheat portion of the process has a local efficiency of almost 47% and raises the overall efficiency of the process from 30.75% to 34.15%. The quality of the turbine exhaust has also been raised from 75% to over 91%, a significant reduction in the liquid fraction of the turbine exhaust.

3.6.5 Reheat

The higher a boiler pressure and the lower a condenser pressure, the lower the quality of the turbine exhaust that results. Not even superheat may be sufficient to remedy the situation. If a final quality of 98% in the turbine exhaust was desired, the superheat temperature would have to extend to over 1,500° F.

A more practical option to achieving drier steam is the concept of reheat. In this system, as before, the steam is expanded in the turbine. Before complete expansion occurs, however, the steam is extracted and reheated to add enthalpy, and then finally taken to a low-pressure turbine for expansion to condenser pressure. A T–s (Fig. 3-15) diagram demonstrates the operation of a 600 psia system with 1,000° F superheat, and 1,000° F, 100 psia reheat steam. In this example, the steam is initially expanded in the high-pressure turbine to 100 psia (point 4a) before reheating. An interesting question (and one beyond the scope of this chapter) is what reheat conditions maximize the efficiency of the cycle. It's obvious that the situation envisioned will raise the net efficiency of the cycle. What is not obvious is that it's possible to actually reduce the efficiency of a cycle with reheat.

A process schematic of the reheat Rankine cycle is shown

Table 3-11. State Table for the Basic Closed Rankine Cycle with 600 psia Boiler Pressure

Cycle Point	T (° F)	P (psia)	h (Btu/lbm)	s (Btu/lbm-° R)	ρ (lbm/ft³)	State	Notes
1	80.00	1	48.09	0.0933	62.220	Liquid	
2s	80.00	600	49.87	0.0933	62.330	Liquid	
2	80.60	600	50.32	0.0941	62.330	Liquid	$\eta = 0.8$
3a	486.31	600	471.63	0.6722	49.670	Sat. liq.	
3	486.31	600	1,204.06	1.4464	1.299	Sat. vap.	
4s	101.70	1	807.35	1.4464	0.004	$x = 0.712$	
4	101.70	1	847.02	1.5170	0.004	$x = 0.750$	$\eta = 0.9$

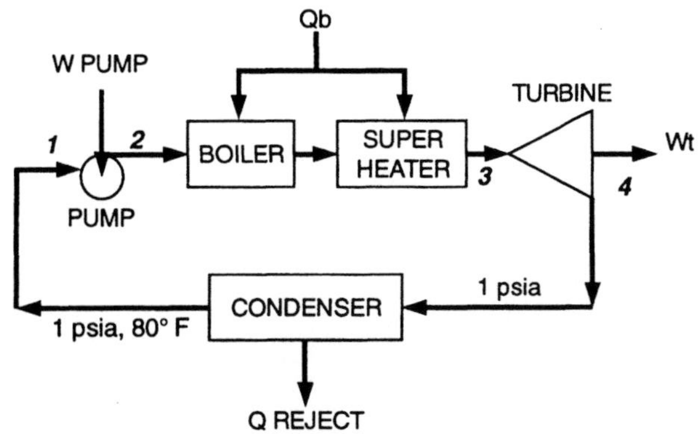

Fig. 3-14. Superheat Rankine cycle.

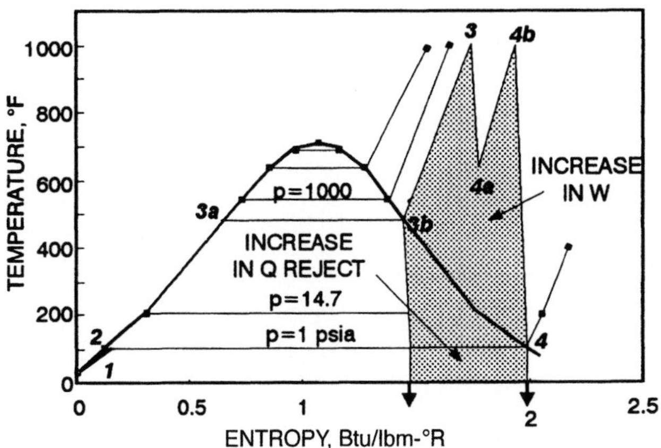

Fig. 3-15. Reheat Rankine cycle.

in Fig. 3-16. A high-pressure and a low-pressure turbine is shown. In practice, these turbines are likely to be on a single shaft (tandem compound) with a single work output, although they may be on separate shafts with separate work outputs (cross compound). Thermodynamic analysis of the reheater is similar to that for the superheater, and an updated process state table is shown in Table 3-13.

Key calculations are as follows:

Q_b = boiler/superheat = 1,517.83 − 50.32 = 1,467.51 Btu/lbm

Q_{reh} = reheater = 1,532.1 − 1,309.1 = 223 Btu/lbm

Q_{rej} = rejected energy = 1,119.33 − 48.09 = 1,071.24 Btu/lbm

W_p = pump power = 50.32 − 48.09 = 2.23 Btu/lbm

W_{t1} = turbine power = 1,517.83 − 1,309.1 = 208.73 Btu/lbm

W_{t2} = turbine power = 1,532.1 − 1,119.33 = 412.77 Btu/lbm

$$\eta = (Q_b + Q_{reh} - Q_{rej})/(Q_b + Q_{reh}) = (W_{t1} + W_{t2} - W_p)/(Q_b + Q_{reh}) = 0.3663$$

Compared with the straight superheat cycle, the energy input has gone up by 223 Btu/lbm and the energy output has gone up by 118 Btu/lbm (53% efficient). Also, the net efficiency of the cycle has increased from 34.2% to 36.6%.[13] The turbine exhaust has actually moved out of the vapor dome and is steam with 20° F of superheat.[14]

One final observation about multiple stages of reheat in a cycle should be made. A second stage of reheat would involve expanding the superheated high-pressure steam to an intermediate pressure; reheating and expanding steam to a low-pressure; and reheating and expanding steam to condenser pressure. This process could remove a good portion of the void between the superheat horns, but the effect would be smaller than the effect of the initial stage. If a third stage is added, even more of the void could be filled, but the marginal gain would be even smaller. In the limit, almost Carnot efficiency in the superheat portion of the process could be achieved. Large units with high capacity factors and/or expensive fuels frequently employ more than one stage of reheat. However, this is an expensive process and can only be justified by extensive systems analysis.

3.6.6 Regeneration

How to improve the efficiency of the Rankine cycle by modifying processes on the right-hand side of the *T*–*s* dia-

Table 3-12. Rankine Cycle with Super Heat

Cycle Point	*T* (° F)	*P* (psia)	*h* (Btu/lbm)	*s* (Btu/lbm-° R)	ρ (lbm/ft³)	State	Notes
1	80.00	1	48.09	0.0933	62.220	Liquid	
2s	80.00	600	49.87	0.0933	62.330	Liquid	
2	80.60	600	50.32	0.0941	62.330	Liquid	η = 0.8
3a	486.31	600	471.63	0.6722	49.670	Sat. liq.	
3b	486.31	600	1,204.06	1.4464	1.299	Sat. vap.	
3	1,000.00	600	1,517.83	1.7155	0.709	Steam	
4s	101.70	1	958.44	1.7155	0.003	$x = 0.858$	
4	101.70	1	1,014.38	1.8152	0.003	$x = 0.912$	η = 0.9

[13]Although it may not be obvious at this point, using the 90% efficient turbines has raised the efficiency of the cycle. The actual efficiency based on reheat only and an overall turbine efficiency of 90% would be closer to 35.9%. A later section will explain this phenomenon more clearly.

[14]If too far out of the vapor dome, the constant pressure line rises dramatically, and cycle efficiency is lost.

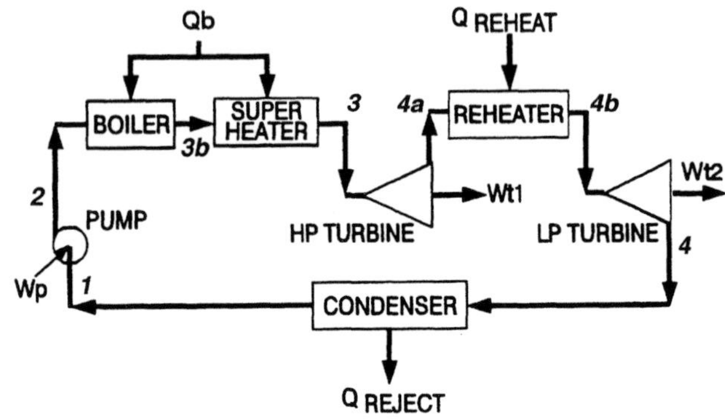

Fig. 3-16. Rankine cycle with reheat.

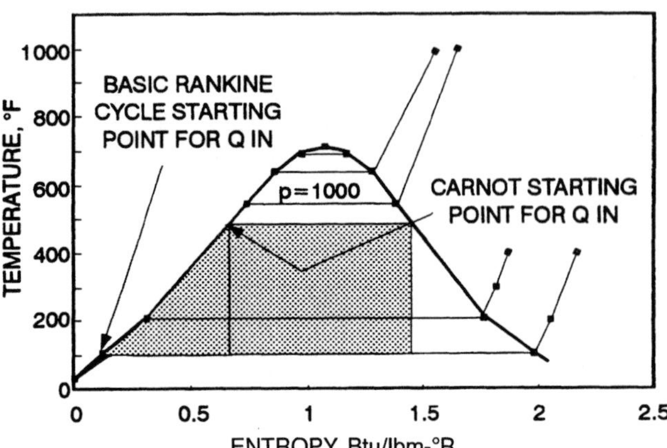

Fig. 3-17. Comparison of Rankine and Carnot cycles.

gram has been discussed. One of the biggest efficiency losses in the cycle, however, occurs on the left side.

In our example (which was 600 psia boiler pressure), a Carnot cycle would begin the heating process ($dQ = T\,ds$) with saturated liquid at 600 psia, but the Rankine cycle begins with subcooled liquid. This portion of the cycle is providing only 99 Btu/lbm of useful work even at 100% efficiency from almost 425 Btu/lbm of energy input (about 23% efficient compared with a simple Carnot efficiency of about 41%) (Fig. 3-17). Regeneration is the concept of using waste heat that would otherwise be rejected to accomplish the initial heating, and reserving actual energy input for more effective activities.

In the continuing example, superheated steam is allowed to expand in the turbine to 100 psia as before. At this point, almost 210 Btu/lbm of mechanical work has been extracted, and the steam is at 560° F with an enthalpy of 1,309 Btu/lbm. If the expansion is continued to 1 psia (no reheat), another 295 Btu/lbm of work is obtained, and the process is finished with an enthalpy of 1,014.38 Btu/lbm, which would then be rejected in the condenser. On the other hand, if part of the 100 psia steam (called extraction steam) is used to heat the subcooled boiler feedwater, all of the 1,309 Btu/lbm enthalpy (minus the feedwater enthalpy) would displace part of the ineffective Q_b used previously for heating feedwater. This is an ideal situation. Useful energy (210 Btu/lbm) is obtained from this steam, and no heat rejection occurs.

In practice, regeneration is accomplished through feedwater heaters. Feedwater heaters come in two basic varieties— open and closed. The open feedwater heater (Fig. 3-18) is a contact device where extraction steam and subcooled feedwater are mixed directly. By necessity, both must be at the same pressure. The open feedwater heater is a multiple stream control volume with continuity equation and steady-state, steady-flow-energy equation as shown in the figure. No real law says that the output must be saturated liquid, but it generally is, and this is specified in the heater description as "TTD = 0." The terminal temperature difference, or TTD, is simply the difference between the saturation temperature corresponding to pressure (P) and the exit temperature of the liquid.[15]

The closed feedwater heater is a surface heat exchanger

Table 3-13. Rankine Cycle with Reheat

Cycle Point	T (° F)	P (psia)	h (Btu/lbm)	s (Btu/lbm-° R)	ρ (lbm/ft³)	State	Notes
1	80.00	1	48.09	0.0933	62.220	Liquid	
2s	80.00	600	49.87	0.0933	62.330	Liquid	
2	80.60	600	50.32	0.0941	62.330	Liquid	$\eta = 0.8$
3a	486.31	600	471.63	0.6722	49.670	Sat. liq.	
3b	486.31	600	1,204.06	1.4464	1.299	Sat. vap.	
3	1,000.00	600	1,517.83	1.7155	0.709	Steam	
4as	513.46	100	1,285.92	1.7155	0.176	Steam	
4a	559.63	100	1,309.1	1.7388	0.168	Steam	$\eta = 0.9$
4b	1,000.0	100	1,532.1	1.9204	0.116	Steam	
4s	101.70	1	1,073.47	1.9204	0.003	$x =$ 0.9688	
4	131.80	1	1,119.33	2.0015	0.003	Steam	$\eta = 0.9$

[15]Open feedwater heaters with a TTD of zero are frequently used as a deaerator in a system. Because the condenser operates at such a low pressure, air can leak in and be absorbed by the condensate. These dissolved gases can be highly corrosive and must be removed. As the condensate is taken to saturation, the dissolved gases are released, and if the deaerator is operating at atmospheric pressure, the gases are vented directly to the atmosphere. If another pressure is chosen, the gases must be vacuum extracted.

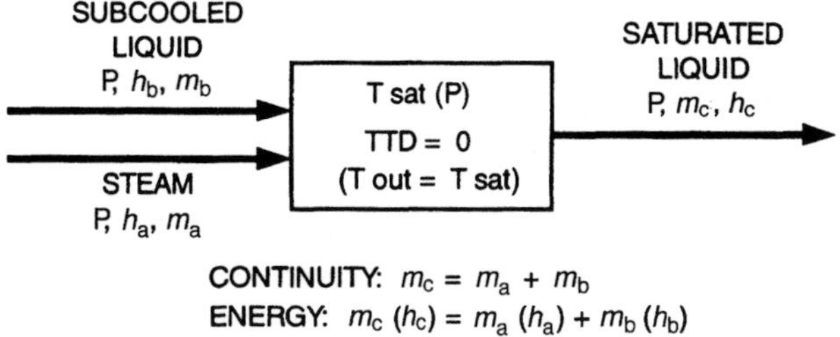

Fig. 3-18. An open feedwater heater.

with the feedwater inside the tubes, and the extraction steam condensing on the outside of the tubes. In contrast to the open feedwater heater, which is a contact heater, the closed feedwater heater can operate with different pressures for the extraction steam and the feedwater. The closed feedwater heater (Fig. 3-19) is fully specified by the three equations shown in the figure for TTD, DCA, and steady flow energy. The TTD is generally about 5° F, but may be as high as 10° F. The DCA, or Drain Cooler Approach ($t_{drain}-T_{in}$), is generally about 10° F, but may be as high as 20° F. Because the two streams are at different pressures in the closed feedwater heater, two principal alternatives exist for what to do with the condensed extraction steam from the heater drain:

1. Allow it to expand isenthalpically, and send it to a lower pressure feedwater heater, or condenser.
2. Increase its pressure with a small pump and mix it into the feedwater stream.

Both of these alternatives are frequently used.

In the continuing example, a regenerative cycle with a single, open feedwater heater operating at 100 psia (Fig. 3-20) is examined. (Reheat is left out to get a good comparison with the straight superheat configuration. In the next section it will be put back in, and closed feedwater heaters will be included.) Instead of a single boiler feed pump, a condensate pump is used to increase the condensate pressure to 100 psia, and then a second pump takes the feedwater heater discharge to 600 psia. The major thing to note here is that for the first time, the flow is not the same at all points in

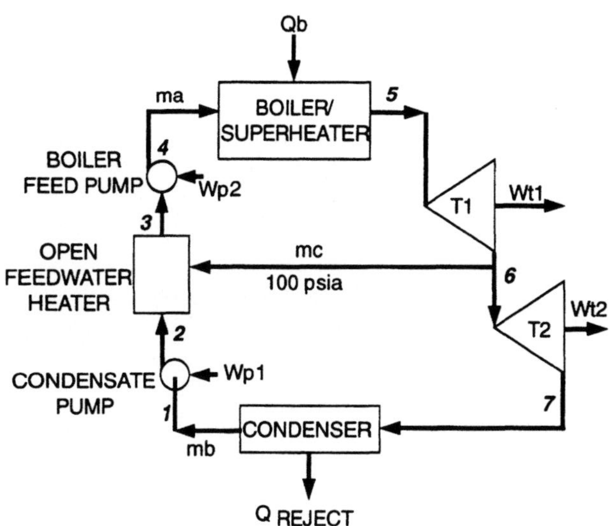

Fig. 3-20. Regenerative open feedwater heater Rankine cycle.

the cycle. The T–s diagram (Fig. 3-21) for the cycle is not very useful for this particular cycle. It is included for the sake of completeness, but only with the warning that different mass flows are present at different points, and no simple interpretation is possible. It should be noted that no external energy input to the system was needed to get from point 2 to point 3. It should also be noticed that part of the extraction steam was at a higher temperature than it needed to be, and more work could have been derived from it before using it for regeneration. That is, if two stages of regeneration were available, one could be used to get from the temperature at point 1 to some intermediate temperature, and one could be used to get from that point to the temperature of point 2. The additional steam could then be expanded to some lower intermediate pressure before it is used for regeneration. If a large number of stages of regeneration at varying pressures between 1 and 600 psia were available, the 600 psia saturation temperature could almost be attained with no external thermal energy input and minimized work loss. In fact, as the number of stages of feedwater heating increases, Carnot efficiency ($T_{high} = T_{sat}$ (600 psia)) is approached. Unfor-

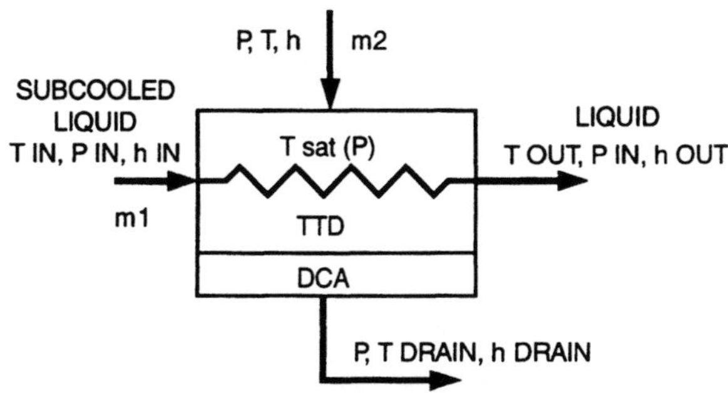

Fig. 3-19. A closed feedwater heater.

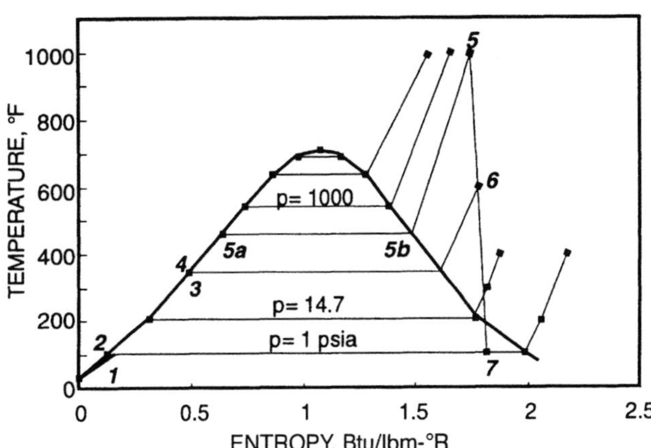

Fig. 3-21. Open feedwater heater regeneration.

Table 3-14. Regenerative Rankine Cycle with Single Open Feedwater Heater

Cycle Point	T (°F)	P (psia)	h (Btu/lbm)	s (Btu/lbm-°R)	ρ (lbm/ft³)	State	Notes
1	80.00	1	48.09	0.0933	62.220	Liquid	
2s	80.00	100	48.38	0.0933	62.240	Liquid	
2	80.10	100	48.45	0.0935	62.237	Liquid	$\eta = 0.8$
3	327.87	100	298.62	0.4744	58.380	Sat. liq.	
4s	328.63	600	300.26	0.4744	56.490	Liquid	
4	329.03	600	300.67	0.4749	56.470	Liquid	$\eta = 0.8$
5a	486.31	600	471.63	0.6732	49.670	Sat. liq.	
5b	486.31	600	1,204.06	1.4464	1.299	Sat. vap.	
5	1,000.0	100	1,517.83	1.7155	0.709	Steam	
6s	513.46	100	1,285.92	1.7155	0.176	Steam	
6	559.63	100	1,309.11	1.7388	0.168	Steam	$\eta = 0.9$
7s	101.70	1	971.51	1.7388	0.003	$x = 0.870$	
7	101.70	1	1,005.27	1.7990	0.003	$x = 0.903$	$\eta = 0.9$

tunately, the number of pumps, the piping, the extraction points, and the heat exchangers are also increased. Actual large power stations may have 8 to 12 feedwater heaters (including a deaerator.)

The process state table for the example with one open feedwater heater is shown in Table 3-14. No mass flows are required to produce this table. Key calculations are as follows:

$$m_a = m_c + m_b$$

$$m_a(h_3) = m_c(h_6) + m_b(h_2)$$
$$\Rightarrow m_a = 1.0;\ m_c = 19.84\%;\ m_b = 80.16\%, \text{ and}$$

$$Q_b = \text{boiler/superheat} = (1{,}517.83 - 300.67)1.0 = 1{,}217.16 \text{ Btu/lbm}$$

$$Q_{rej} = \text{rejected energy} = (1{,}005.27 - 48.09)0.0816 = 767.24 \text{ Btu/lbm}$$

$$W_{p1} = \text{pump power} = (48.45 - 48.09)0.8016 = 0.29 \text{ Btu/lbm}$$

$$W_{p2} = \text{pump power} = (300.67 - 298.62)1.0 = 2.05 \text{ Btu/lbm}$$

$$W_{t1} = \text{turbine power} = 1{,}517.83 - 1{,}309.11 = 208.72 \text{ Btu/lbm}$$

$$W_{t2} = \text{turbine power} = (1{,}309.11 - 1{,}005.27)0.8{,}016 = 243.55 \text{ Btu/lbm}$$

$$\eta = (Q_b - Q_{rej})/Q_b = (W_t - W_p)/Q_b = 0.3697$$

Once again, the cycle has been artificially benefitted by using 90% efficiency for all turbine stages, but this will be addressed later.

3.6.7 A Simple Heat Balance

When all the aforementioned concepts are combined, a system such as the following results.[16] To analyze a system such as this, different flows in the system as well as the thermodynamic properties must be solved. Because the continuity and steady-state, steady-flow-energy equations for the components demand a balance of mass and energy that is not apparent at first glance, the analysis leading to the full system description is called a heat balance. One new concept is demonstrated by the valve shown between point 13 and point 14 in Fig. 3-22. Because water is rejected from a high-pressure to a low-pressure region, the pressure must be reduced in a controlled manner. This isenthalpic expansion maintains all the enthalpy of the water, and it is common for some flashing of water to steam to occur during the process. The only other major points to note are that the boiler pressure has been raised to 1,000 psia, two closed feedwater heaters and one open feedwater heater are used, and extraction points off the turbines are shown without actually splitting the turbine into multiple pieces. (The next section addresses thermodynamic state of the extraction point. For now, it is assumed that the efficiency of each segment is known.)

Even though there are now seven distinct flows at various parts of the network, the process state table can be set up without knowing the specific values of the flows. (This is because at the design stage, the flows support the thermodynamic properties.) The process state table is shown in Table 3-15.

The control volume equations for the components are set up as follows:

Turbine 1: $m_a = m_b + m_c + m_d$
$m_a(h_7) = m_b(h_{12}) + m_d(h_8) + m_c(h_8) + W_{t1}$

Turbine 2: $m_c = m_e + m_f$
$m_c(h_9) = m_e(h_{11}) + \text{mf}(\text{h}_{10}) + W_{t2}$

Condenser: $m_g = m_f + m_e$
$m_g(h_1) = m_f(h_{10}) + m_e(h_{16}) - Q_{rej}$

Heater 3a: $m_g(h_3 - h_2) = m_e(h_{11} - h_{15})$

Heater 4a: $m_a(h_4) = m_d(h_8) + m_g(h_3) + m_b(h_{14})$

[16]Some practical concepts haven't been included to this point, such as pressure drops, leakages, and boiler efficiency. Although most of this is left for discussion in the chapter on heat balances, some concepts surrounding the turbine exhaust line need to be addressed. The assumption that $\eta = 0.9$ for both turbine stages has increased the overall turbine efficiency over the straight superheat case. Section 3.6.8 addresses the turbine expansion line, and it will be examined in detail in a later chapter.

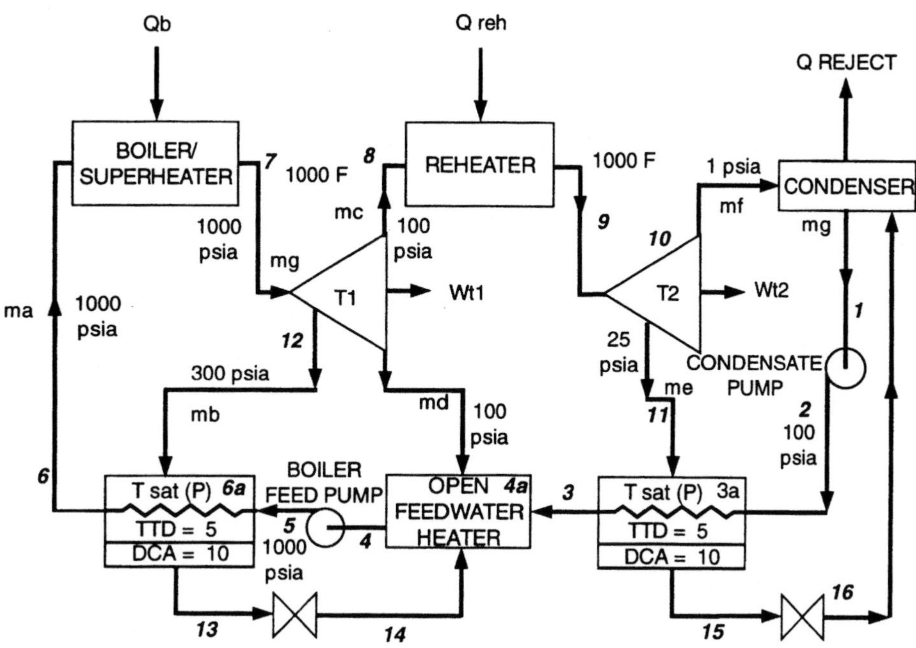

Fig. 3-22. A simple heat balance.

Heater 6a: $m_a(h_6 - h_5) = m_b(h_{12} - h_{13})$

Boiler/superheater: $m_a(h_6) + Q_b = m_a(h_7)$

Reheat: $Q_{reh} = m_c(h_9 - h_8)$

Pump power: $W_p = m_a(h_5 - h_4) + m_g(h_2 - h_1)$

Net work: $W_{net} = W_{t1} + W_{t2} - W_p$

Efficiency: $\eta = W_{net}/(Q_b + Q_{reh})$

To keep the solution general, set m_a = 1 lbm/s, and then scale up the solution to whatever power is required. The solution set is shown in Table 3-16.

The efficiency of the cycle is 41.41%.

3.6.8 The Turbine Expansion Line and Efficiencies

A detailed discussion of the turbine expansion line is in Chapter 13. The following section provides a brief discussion of this subject.

Table 3-15. A Simple Heat Balance: Process State Table

Cycle Point	T (°F)	P (psia)	h (Btu/lbm)	s (Btu/lbm-°R)	ρ (lbm/ft³)	State	Notes
1	80.00	1	48.09	0.0933	62.220	Liquid	
2s	80.00	100	48.38	0.0933	62.240	Liquid	
2	80.10	100	48.45	0.0935	62.240	Liquid	η = 0.8
3a	240.40	25	208.50/1,160.75	0.3530/1.7140	59.100/0.060	Sat. liquid/Sat. steam	
3	235.10	100	203.65	0.3461	59.245	Liquid	TTD = 5
4	327.87	100	298.62	0.4744	56.380	Sat. liq.	
5s	329.24	1,000	301.57	0.4744	56.572	Liquid	
5	329.95	1,000	302.31	0.4753	56.547	Liquid	η = 0.8
6a	417.41	300	394.06/1,203.9	0.5880/1.5110	52.920/0.648	Sat. liquid/Sat. steam	
6	412.41	1,000	389.33	0.5801	53.390	Liquid	TTD = 57
7	1,000.00	1,000	1,505.90	1.6530	1.204	Steam	
8s	402.10	100	1,228.66	1.6530	1.653	Steam	
8	455.32	100	1,256.38	1.6842	0.189	Steam	η = 0.9
9	1,000.00	100	1,532.11	1.9204	0.116	Steam	
10s	101.70	1	1,073.47	1.9204	0.003	x = 0.9688	
10	131.80	1	1,119.33	2.005	0.003	Steam	η = 0.9
11	612.58	25	1,340.59	1.9204	0.039	Steam	
11	651.96	25	1,359.74	1.9380	0.038	Steam	η = 0.9
12s	652.88	300	1,343.18	1.6530	0.471	Steam	
12	683.35	300	1,359.45	1.6674	0.456	Steam	η = 0.9
13	339.95	300	311.54	0.4899	56.010	Liquid	DCA = 10
14	327.87	100	311.54	0.4908	12.210	x = 0145	
15	90.10	25	58.24	0.1118	62.120	Liquid	DCA = 10
16	90.20	1	58.24	0.1120	62.110	Liquid	

Table 3-16. Heat Balance Solution Set

Mass Flow	Percentage of m_a	Energy Use	Btu/lbm-s
m_a (1 lbm/s)	100.00	W_{t1}	240.96
m_b	8.3	W_{t2}	320.83
m_c	83.53	W_p	3.99
m_d	8.17	W_{net}	557.80
m_e	9.96	Q_b	1,116.57
m_f	73.57	Q_{reh}	230.31
m_g	83.53	Q_{rej}	789.07

As various extraction points on the expansion line through the turbine are reviewed, a question of thermodynamic states arises. For the sake of convenience, the turbine efficiency has been assumed locally applicable to every segment of the expansion line in the examples to this point. In reality, every segment does have its own efficiency, but the combination of 90% efficient segments does not provide an overall efficiency of 90%. If the expansion line were a straight line on the h–s diagram with a 90% efficient end point, the upper portion might have an efficiency of 85%, and the lower portion might have an efficiency of 90%. Further, if an expansion line is reviewed as a series of 90% efficient segments, the overall efficiency might be 92 or 93%.

Each turbine segment has its own efficiency, and these can be used when available. The example problems were done from this perspective.

When individual segment efficiencies are unknown, the straight line approach is simply to join the inlet and exhaust points on the Mollier diagram and assume all intermediate points lie on that line. The following example uses the "straight line" approach.

3.6.8.1 The Straight Line Method. Consider a straight line (h–s) expansion from 1,000 psia, 1,000° F, to 1 psia with an isentropic efficiency of 0.9. What are the conditions at a 60 psia extraction point? A straightforward evaluation of the start and end points for the expansion is available as follows:

State 1:	$P = 1{,}000$	$T = 1{,}000$
	$h = 1{,}505.9$	$s = 1.653$
Isentropic expansion:	$P = 1$	$T = 102$
	$h = 923.3$	$s = 1.653$
State 2 ($\eta = 0.9$)	$P = 1$	$T = 102$
	$h = 981.6$	$s = 1.757$

If these points are located on a Mollier diagram and a straight line is drawn between them, the extraction point could be estimated with accuracy varying with the Mollier chart size and resolution, e.g., $P = 60$; $T = 385$; $h = 1{,}226$; $s = 1.705$. A method more suitable for automation is an iterative technique called "Gauss–Siedel Successive Iteration." Despite the long name, the method is simple.

1. Guess a value of entropy. A good place to start is the isentropic value. For the example, this would be $s = 1.653$.
2. Using computer property evaluation program, solve for the enthalpy at the desired pressure and the current guess for entropy. The example would have $h = h\ (p = 60, s = 1.653) = 1{,}184$.
3. Write a straight line equation for the expansion with enthalpy as the dependent variable. For this example: $s = 1.95105 - h/5052.75$. Now solve for a new "s" using the current estimate of h as follows: $s = 1.95105 - 1{,}184/5{,}052.75 = 1.7167$.
4. Repeat steps 2 and 3 until h does not vary significantly. The results would look like this for this example:

Iteration 1:	$h = 1{,}184$	$s = 1.7167$	
Iteration 2:	$h = 1{,}236.4$	$s = 1.7063$	
Iteration 3:	$h = 1{,}227.5$	$s = 1.708$	
Iteration 4:	$h = 1{,}228.9$	$s = 1.70779$	
Iteration 5:	$h = 1{,}228.7$	$s = 1.70782$	
Iteration 6:	$h = 1{,}228.8$	$s = 1.70787$	$T = 390.5$

It only took three iterations to get closer than the best estimate from the Mollier diagram, and it is generally true that three iterations will be very close to the final solution.

3.6.8.2 Efficiency Along the Expansion Line. An efficiency check from state 1 to the extraction point reveals an isentropic efficiency of 86%. Efficiency from the extraction point to state 2 is very close to 90%. How can the final efficiency be higher than any of its components? An analysis for a perfect gas can be performed to illustrate this. The requirement for a straight line on the h–s diagram dictates a slope as follows:

$$\frac{h_2 - h_1}{s_2 - s_1} = \frac{c_p(T_2 - T_1)}{c_p \ln\frac{T_2}{T_1} - R\ln\frac{P_2}{P_1}} = \frac{c_p(T_2 - T_1)}{c_p \ln\frac{T_2}{T_1} - c_p \ln\frac{T_{2s}}{T_1}} = \frac{(T_2 - T_1)}{\ln\frac{T_2}{T_{2s}}} \tag{3-102}$$

The definition of efficiency yields the following:

$$\eta = \frac{T_1 - T_2}{T_1 - T_{2s}} \tag{3-103}$$

Combining these to eliminate T_{2s}, the following is found:

$$\text{slope} = \frac{T_2 - T_1}{\ln(\eta T_2) - \ln(T_2 - T_1[1 - \eta])} \tag{3-104}$$

Similar to the previous experience, a 90% efficient expansion from 1,000° F to 350° F would begin with an efficiency of only about 83% and finish with an efficiency just below 90%.

The answer to this mystery is found by analyzing a two-part expansion. Steam initially at 1,000° F and 1,000 psia will be expanded to 60 psia, and subsequently to 1 psia with an overall efficiency of 90%. Proceeding as before:

State 1:	$T = 1{,}000°\text{F}$	$P = 1{,}000$ psia
	$h = 1{,}505.9$	$s = 1.653018$
State 3s:	$T = 101.7°\text{F}$	$P = 1$ psia
	$h = 923.34$	$s = 1.653018$

State 3: $T = 101.7°F$ $P = 1$ psia
$h = 981.60$ $s = 1.756784$

An isentropic expansion from the initial state to 60 psia would result in the following:

$$T = 304.9°F \quad P = 60 \text{ psia} \quad h = 1{,}184.57 \quad s = 1.653018$$

A 90% efficient expansion would change the enthalpy to $h = 1505.9 - 0.9\ (1505.9 - 1184.57) = 1216.7$, and this would yield a state 2 as follows:

$$T = 366.7°F \quad P = 60 \text{ psia} \quad h = 1{,}216.7 \quad s = 1.693445$$

If the exercise is repeated with a 90% efficient expansion to 1 psia, the following occurs:

$$T = 101.7°F \quad P = 1 \text{ psia} \quad h = 973.1$$

An overall efficiency for the expansion would be $(1{,}505.9 - 973.1)/(1{,}505.9 - 923.34) = 0.915$, which is higher than the 0.900 we wanted.

But here's the solution to the mystery. When the second expansion was performed, 90% efficiency was used as compared to a new state point entropy. The process only called for it to be 90% efficient compared to the original state entropy. That is, the corrected enthalpy at 60 psia was OK, but the correction for the second expansion could have been calculated as $h = 1{,}216.7 - 0.9\ (1{,}184.57 - 923.34) = 981.6$.

Here, 90% of the enthalpy drop at the original entropy replaces 90% of the enthalpy drop at the new entropy, and the integrity of the process is maintained. An isolated calculation of efficiency for the second expansion based on the new entropy would show about 86.9%.

3.6.8.3 The Constant Efficiency Method. Shown all at one time, the constant efficiency method looks like the following:

State 1: $T = 1{,}000°F$ $P = 1{,}000$ psia
$h = 1{,}505.9$ $s = 1.653018$
State 2s: $T = 304.9°F$ $P = 60$ psia
$h = 1{,}184.57$ $s = 1.653018$
$h_2 = 1{,}505.9 - 0.9\ (1{,}505.9 - 1{,}184.57) = 1{,}216.7$
State 2: $T = 366.7°F$ $P = 60$ psia
$h = 1{,}216.7$ $s = 1.693445$
State 3s: $T = 101.7°F$ $P = 1$ psia
$h = 923.34$ $s = 1.653018$
$h_3 = 1{,}216.7 - 0.9\ (1{,}184.57 - 923.34) = 981.6$
State 3: $T = 101.7°F$ $P = 1$ psia
$h = 981.6$ $s = 1.756784$
$\eta = (1{,}505.9 - 981.6)/(1{,}505.9 - 923.34) = 0.9$

Compared with the straight line method, the starting point and end point are the same, but the extraction point at 60 psia has a slightly lower temperature and enthalpy.

3.6.8.4 Comparison of Expansion Line Methods. The straight-line method and the constant efficiency method developed are both approximations (Fig. 3-23). The straight line method has lower efficiencies at high pressure and higher efficiencies at low pressures. The constant efficiency method has higher efficiencies at higher pressures. The straight-line method is very convenient for graphical solution. The continuous efficiency method is easy for manual solution and requires no graphs. The best method is still to have the actual expansion line, or segment efficiencies.

3.7 THE BRAYTON CYCLE

The Rankine cycle is a vapor cycle, constantly condensing and evaporating water. By contrast, the Brayton cycle is an all-gas cycle, using air and combustion gases directly as the working fluid. It is implemented in the power industry as the combustion gas turbine (CGT).[17] There are other familiar all-gas cycles, such as the Otto and Diesel cycles, but these internal combustion engines are based on piston and cylinder batch-processing of the working fluid, and generally manufactured in sizes too small to be of significance in large power systems. The Brayton cycle is a steady flow, steady state system, and the mainstay of low capital cost power supply market. Although CGTs tend to come in a prepackaged format and few power system engineers will ever have the opportunity to design one, it is important to understand the fundamentals of their operation for proper selection and implementation. This section explores the thermodynamic operation of the cycle from this aspect.

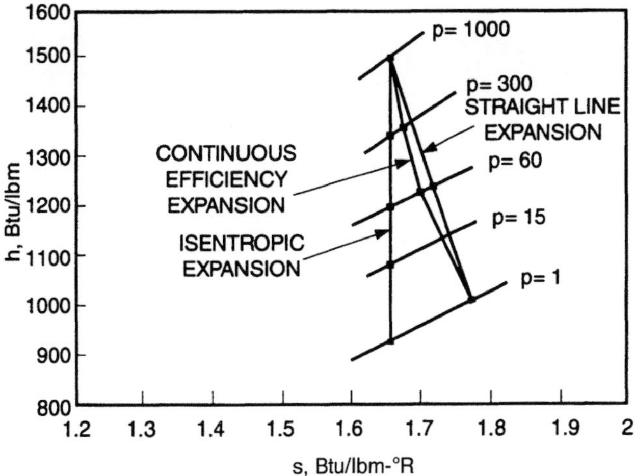

Fig. 3-23. Two turbine expansion line options ($n = 0.9$).

[17]Some people are uncomfortable with the nomenclature "combustion gas turbine," or "gas turbine," because the word "gas" may be misinterpreted as implying that natural gas is the fuel, even though it really just means that combustion gas is the working fluid. Because of this, the term "combustion turbine," or CT, has become a common substitute.

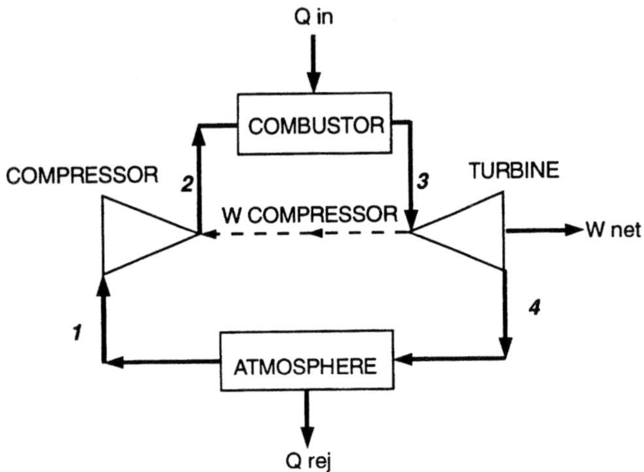

Fig. 3-24. The basic Brayton cycle.

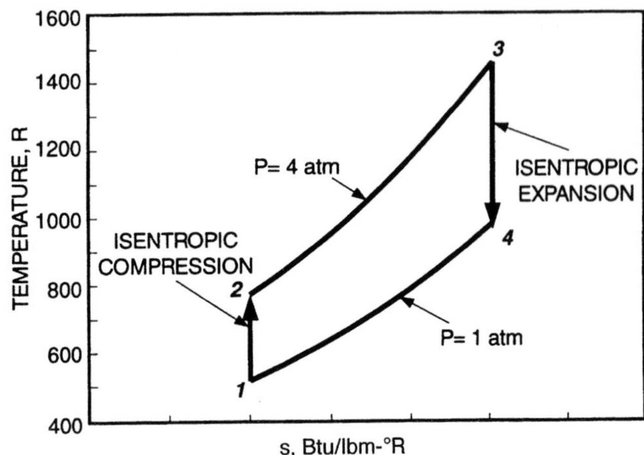

Fig. 3-25. An ideal Brayton Cycle.

3.7.1 The Basic Cycle

When all-gas cycles are analyzed, the fact that the combustion gas varies with the fuel tends to complicate the calculations. For this reason, it has become standard practice to use the air-standard cycle for analysis of these processes. The air-standard cycle replaces the combustion process with a constant pressure control volume that accepts energy as heat, and maintains air as the working fluid through the entire cycle (Fig. 3-24). Since the atmosphere closes all combustion gas processes, the inlet and exhaust states are joined by a constant pressure line. This approach can be used in visualizing a basic, ideal Brayton Cycle (Fig. 3-25). Air is isentropically compressed and then heated at constant pressure in the combustor. The high temperature gas is then isentropically expanded back to state 1 pressure through a turbine. The remaining enthalpy in the air is rejected and the air returned to its original state for reuse.

Treating the air as an ideal gas, a T–s diagram representation of the cycle is shown as follows. The paths from state 2 to state 3, and from state 1 to state 4, follow the constant pressure equation:

$$S_3 - S_2 = c_p \ln \frac{T_3}{T_2}$$
$$S_1 - S_4 = c_p \ln \frac{T_1}{T_4} \qquad (3\text{-}105)$$

Temperature and pressure at the inlet (point 1) are set by the atmosphere. T_3 is limited by the metallurgy of the machinery. Whereas all gas cycles with cylinder combustion can withstand momentary temperatures of 4,000° F, the steady state nature of the Brayton cycle demands a constant exposure to T_3, and this limits the allowable temperature to about 2,000° F. Various pressures can have the same T_3, but result in differing efficiencies and work.

The ideal cycle can be analyzed in the following example:

1. Given: T_1, P_1, T_3, P_2 and $P_3 = P_2$, $P_4 = P_1$. $c_p = 0.24$, $c_v = 0.1714$, and $k = 1.4$.

2. Isentropic compression or expansion of an ideal gas (a polytropic process with $k = 1.4$):[18]

$$\frac{T_2}{T_1} = \left(\frac{P_2}{P_1}\right)^{\frac{k-1}{k}} = \frac{T_3}{T_4} \qquad (3\text{-}106)$$

or

$$T_2 = \left(\frac{P_2}{P_1}\right)^{\frac{k-1}{k}} T_1; \quad T_4 = \left(\frac{P_1}{P_2}\right)^{\frac{k-1}{k}} T_3 \qquad (3\text{-}107)$$

3. Gross turbine work (W_t), compressor work (W_c), heat input (Q_{in}), and heat rejected (Q_{rej}) are given as follows:

$$W_t = h_3 - h_4 = c_p (T_3 - T_4); \quad W_c = h_2 - h_1 = c_p (T_2 - T_1)$$
$$Q_{input} = h_3 - h_2 = c_p (T_3 - T_2); \quad Q_{reject} = h_4 - h_1 = c_p (T_4 - T_1) \qquad (3\text{-}108)$$

Based on these equations, the solution to the example problem is as follows:

$T_1 = 520°$ R (70° F)	$P_1 = 1$ atm
$T_2 = 772.7°$ R (312.7° F)	$P_2 = 4$ atm
$T_3 = 1{,}460°$ R (1000° F)	$P_3 = 4$ atm
$T_4 = 982.5°$ R (522.5° F)	$P_4 = 1$ atm
$W_c = 60.65$ Btu/lbm	$W_t = 114.60$ Btu/lbm
$W_{net} = 53.95$ Btu/lbm	$Q_{in} = 164.95$ Btu/lbm
$bwr = 52.93\%$	$\eta = 32.70\%$

The *bwr* is simply W_c/W_T, and is the same back work ratio discussed for the Rankine cycle. For the Rankine cycle with

[18] All temperatures are absolute temperature (Rankine or Kelvin).

liquid pumping or compression, it was less than 1%. Compression of gases is an energy intensive process.

3.7.2 Inefficiencies

The analysis in the last section assumed pure isentropic compression and expansion processes. In real life, the compressor and turbine have isentropic efficiencies associated with them, just as the pump and steam turbine did for the Rankine cycle. The isentropic efficiencies for these components are given as follows:

$$\eta_c = \frac{h_{s2s} - h_1}{h_2 - h_1} = \frac{T_{2s} - T_1}{T_2 - T_1} \tag{3-109}$$

$$\eta_T = \frac{h_3 - h_4}{h_3 - h_{4s}} = \frac{T_3 - T_4}{T_3 - T_{4s}} \tag{3-110}$$

The subscript *s* refers to the isentropic state achieved with the ideal reversible machine. The cycle is extremely sensitive to these compression and expansion efficiency parameters.

Efficiencies for the turbines are good, with values of 0.90 not being unreasonable. It is easier to design an efficient turbine than it is to design an efficient axial flow compressor, however; it has only been recently that compressor efficiencies as high as 85% have been possible. If the turbine efficiency is set to 0.85 and the compressor efficiency to 0.75 the example problem is redone as follows:

1. Given: T_1, P_1, T_3, P_2 and $P_3 = P_2$, $P_4 = P_1$

   $c_p = 0.24$, $c_v = 0.1714$, and $k = 1.4$

   $\eta_c = 0.75$, $\eta_T = 0.85$

2. T_{2s} is available from:

$$T_{2s} = \left(\frac{P_2}{P_1}\right)^{\frac{k-1}{k}} T_1 \tag{3-111}$$

3. T_2 is available from:

$$T_2 = T_1 + \frac{T_{2s} - T_1}{\eta_c} \tag{3-112}$$

4. T_{4s} is available from:

$$T_{4s} = \left(\frac{P_1}{P_2}\right)^{\frac{k-1}{k}} T_3 \tag{3-113}$$

5. T_4 is available from:

$$T_4 = T_3 - \eta_T (T_3 - T_{4s}) \tag{3-114}$$

6. Work and heat are available as:

$$W_T = c_p (T_3 - T_4);\ W_c = c_p (T_2 - T_1)$$
$$Q_{\text{input}} = c_p (T_3 - T_2);\ Q_{\text{rej}} = c_p (T_4 - T_1) \tag{3-115}$$

7. The efficiency is calculated as:

$$\eta = \frac{Q_{\text{input}} - Q_{\text{rej}}}{Q_{\text{input}}} = 1 - \frac{T_4 - T_1}{T_3 - T_2} \tag{3-116}$$

Based on these updated equations, the solution is now as follows:

T_1 = 520°R (70°F)	P_1 = 1 atm
T_2 = 857.0°R (397.0°F)	P_2 = 4 atm
T_3 = 1,460°R (1000°F)	P_1 = 4 atm
T_4 = 1,054.1°R (594.1°F)	P_1 = 1 atm
W_c = 80.87 Btu/lbm	W_t = 97.41 Btu/lbm
W_{net} = 16.54 Btu/lbm	Q_{in} = 144.73 Btu/lbm
bwr = 83.02%	η = 11.43%

The efficiency has fallen from 32.7% to 11.4 percent, and the net capacity has fallen from 53.9 Btu/lbm to 16.5 Btu/lbm. Doing a set of solutions for different efficiencies shows how the combustion turbine responds. In this example, all conditions will be constant with η_T at 0.85 and η_c will vary.

η_c	η	W_{net}
0.9	0.1897	30.02
0.8	0.1441	21.59
0.7	0.0775	10.76
0.6227	0	0

At a compressor efficiency of 62.3%, all of the turbine work is consumed by compressor backwork, and the turbine will not turn.[19]

3.7.3 Pressure Ratio and Efficiency

As mentioned earlier, setting T_1 and T_3 doesn't mandate a particular pressure ratio. In fact, some question exists as to what the correct pressure ratio should be. Fig. 3-26 shows the original ideal cycle with a pressure ratio of 4, and another cycle, operating between the same temperature limits, with a pressure ratio of 6. The new cycle has a higher efficiency (40%) and a higher net work (57 Btu/lbm).

Previously, efficiency for the Brayton cycle was derived as follows:

$$\eta = 1 - \frac{T_4 - T_1}{T_3 - T_2} \tag{3-117}$$

That equation is now modified to the following:

$$\eta = 1 - \frac{T_4\left(\frac{T_4}{T_1} - 1\right)}{T_2\left(\frac{T_3}{T_2} - 1\right)} \tag{3-118}$$

It should be noted that $T_4/T_1 = T_3/T_2$ for the ideal reversible

[19]If the compressor efficiency had been at 85%, a turbine efficiency of 62.3% would stop the turbine (but the path would be different). If the turbine and compressor efficiency were equal, an efficiency of 0.7225 would reduce the output to zero.

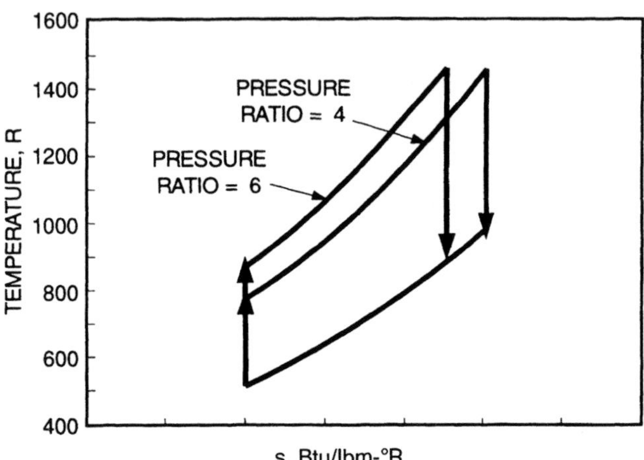

Fig. 3-26. Comparison of pressure ratios.

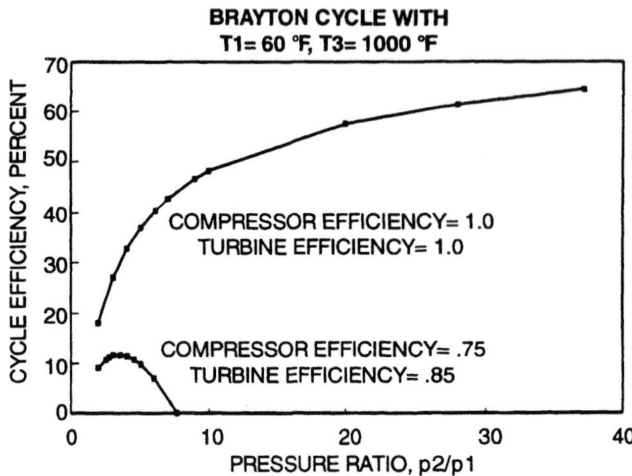

Fig. 3-27. Cycle efficiencies for the example Brayton cycle.

cycle. An expression for the efficiency of the ideal cycle can be written as follows:

$$\eta = 1 - \frac{T_1}{T_2} = 1 - \left(\frac{P_1}{P_2}\right)^{\frac{k-1}{k}} \quad (3\text{-}119)$$

This shows that the higher the pressure ratio, the higher the efficiency for the ideal cycle. As inefficiency is introduced into the cycle, this no longer holds true. Fig. 3-27 shows cycle efficiencies for the example problem with varying pressure ratios. As the equation predicts, the efficiency of the ideal cycle continues to grow with increasing pressure ratio. The cycle with inefficiencies, however, peaks with an efficiency of 11.69% at a pressure ratio of 3.5, and then descends rapidly. At a pressure ratio of 7.67, the efficiency is reduced to zero.

3.7.4 Pressure Ratio and Work

The work of the ideal cycle is given as follows:

$$c_p(T_3 - T_4) - c_p(T_2 - T_1) \quad (3\text{-}120)$$

Maximum work of the ideal cycle requires eliminating T_2 or T_4 from the equation (T_1 and T_3 are fixed), and setting the derivative equal to zero. This proceeds as follows:

$$\max\,(c_p(T_3 - T_4) - c_p(T_2 - T_1))$$
$$= \max\,((T_3 + T_1) - (T_2 + T_4))$$
$$= \max\,(-(T_2 + T_1T_3/T_2))$$

When the derivative of this quantity is taken with respect to T_2, the following occurs:

$$-1 + (T_1T_3)/T_2^2 = 0$$

or $T_1T_3 = T_2T_2$. Now when both sides are divided by T_1T_1, the following results:

$$\frac{T_3}{T_1} = \frac{T_2T_2}{T_1T_1} = \left(\frac{P_2}{P_1}\right)^{\frac{2(k-1)}{k}} \quad (3\text{-}122)$$

The optimum pressure ratio for maximum work in the ideal cycle is given as follows:

$$\text{for maximum work, } \left(\frac{P_2}{P_1}\right) = \left(\frac{T_3}{T_1}\right)^{\frac{k}{2(k-1)}} \quad (3\text{-}123)$$

Once again, the real cycle with inefficiencies will vary from this rule significantly, and Fig. 3-28 demonstrates this point for the example problem with T_3 = 1,000° F and T_1 = 60° F.

Maximum work for the ideal cycle in our example occurs at a pressure ratio of 6.09, with a corresponding work of 56.97 Btu/lbm and an efficiency of 40.3%. The same cycle with a compressor efficiency of 0.75 and a turbine efficiency of 0.85 produces maximum work at a pressure ratio of 2.77, with 19.0 Btu/lbm and an efficiency of 11.22%.

Another major issue that has developed is that whether talking about the theoretical or the actual cycle, the maximum work and the maximum efficiency don't occur at the same point.[20] The actual cycle has maximum work of 19 Btu/lbm at a pressure ratio of 2.77 (η = 11.2%). Maximum efficiency of 11.7% occurs at a pressure ratio of 3.5 (W = 18 Btu/lbm.)

3.7.5 Regeneration

In looking for ways to improve efficiency, it is hard not to notice that the exhaust gas is at 1,000° F, and it seems like there should be something useful to do with it.

[20]There truly is a pressure ratio that produces maximum efficiency for the ideal cycle. It occurs just before the work goes to zero.

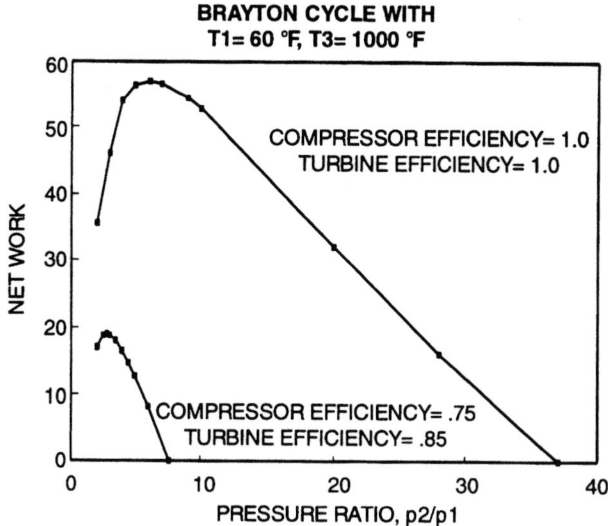

Fig. 3-28. Brayton cycle net work.

The regenerator uses the exhaust to replace part of the energy input as heat to the combustor. It is seen in Fig. 3-29 that as long as the exhaust at T_4 is hotter than T_2, heat transfer to the compressed air is possible. In theory, a counterflow heat exchanger could be used to cool the exhaust gas from T_4 to T_b and heat the compressed air from T_2 to T_a. Then only enough energy would have to be provided to the cycle to get from point a to point 3, and there would be a significant savings in fuel. A primary practical consideration associated with this concept is that heat exchangers have pressure drops, and any pressure drop increases the exhaust pressure of the turbine, thus reducing the turbine work and increasing the heat rejection.

The regenerator efficiency is defined as follows:

$$\eta_{\text{regen}} = \frac{h_a - h_2}{h_4 - h_2} = \frac{T_a - T_2}{T_4 - T_2} \tag{3-124}$$

To minimize pressure drop and maintain a reasonable size, this regenerator efficiency will be fairly low. In the following, a value of 0.65 is assumed and the example of Section 3.7.2 is reworked.

The first five steps of the solution process are still valid, but before step 6, T_a is calculated as follows:

$$T_a = T_2 + \eta_{\text{regen}} (T_4 - T_2) \tag{3-125}$$

The expressions for Q_{input} and Q_{rej} must also be modified.

$$Q_{\text{input}} = c_p (T_3 - T_a);\ Q_{\text{rej}} = Q_{\text{input}} - W_t + W_c \tag{3-126}$$

The expression for efficiency is simply the following:

$$\eta = \frac{W_t - W_c}{Q_{\text{input}}} \tag{3-127}$$

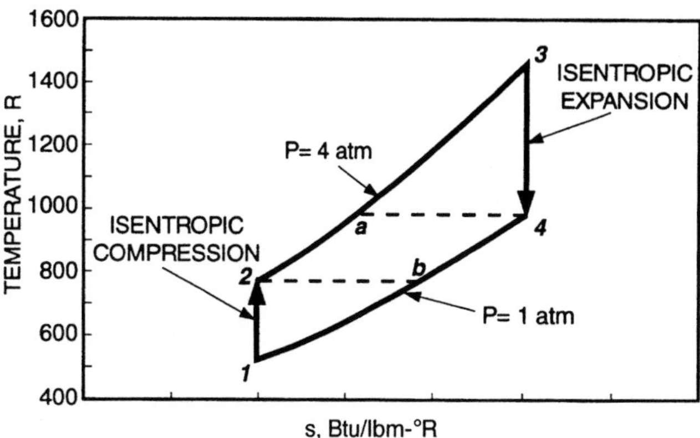

Fig. 3-29. Ideal Brayton cycle regeneration.

Based on these updated equations, the solution is now as follows:

$T_a = 985.2°\text{R}\ (525.1°\text{F})$

$W_c = 80.87$ Btu/lbm	$W_t = 97.41$ Btu/lbm
$W_{\text{net}} = 16.54$ Btu/lbm	$Q_{\text{in}} = 113.87$ Btu/lbm
$bwr = 83.02\%$	$\eta = 14.5\%$

The efficiency has increased from 11.4% to 14.5%.

3.7.6 An Updated Cycle

Additional measures for increasing the efficiency of the combustion turbine, including intercooling, reheat, and more specific information on regeneration, will be presented in later chapters. Before this discussion is ended, however, the basic calculation can be updated with numbers more representative of an actual design.

Redo the Brayton cycle with an inlet temperature of 60° F and a maximum temperature of 1,800° F.[21] Set η_T at 0.90 and η_c at 0.85 to reflect more modern design. Set the pressure ratio at 6 (even though this doesn't reflect maximum work or maximum efficiency.) Determine the following state points for this cycle:

$T_1 = 70°$ F	$P_1 = 1$ atm
$T_2 = 486.6°$ F	$P_2 = 6$ atm
$T_3 = 1{,}800°$ F	$P_1 = 6$ atm
$T_4 = 985°$ F	$P_1 = 1$ atm
$W_c = 99.98$ Btu/lbm	$W_t = 195.59$ Btu/lbm
$W_{\text{net}} = 95.61$ Btu/lbm	$Q_{\text{in}} = 315.22$ Btu/lbm
$bwr = 51.12\%$	$\eta = 30.3\%$

[21]The assumption that c_p is still .24 is getting a little shaky. As shown in Table 3-17, the c_p of air at 1,800° F is closer to 0.27. It could be modified to make it more of an average, obtained from the air tables, or the fact can just be ignored. For this example, the last one should be chosen.

Table 3-17. Specific Heats of Air

T (°R)	C_P (Btu/lb-°R)	C_V (Btu/lb-°R)	k
540	0.2397	0.1712	1.401
700	0.2414	0.1728	1.397
900	0.2457	0.1772	1.387
1,200	0.2545	0.1860	1.369
1,500	0.2638	0.1953	1.351
1,800	0.2722	0.2036	1.337
2,100	0.2790	0.2104	1.326

It is apparent that the increased value of T_3 and improved efficiencies for the components are the most effective efficiency improvement measures.

3.7.7 Mass Flow Considerations

Combustion turbines are often referred to as "mass flow devices." This can be interpreted as meaning that the power available from the cycle is directly proportional to the mass flow processed. Since the working fluid at the compressor inlet is air and basically a perfect gas, the mass flow accepted by the compressor per unit volume is inversely proportional to the temperature. As an example, the difference in mass flow between 60° F air and 100° F air is as follows:

$$\frac{T_2}{T_1} = \frac{460 + 60}{460 + 100} = 0.928 = \frac{m_1}{m_2} \tag{3-128}$$

Based on this simplistic evaluation, a 7% reduction in power can be expected. For this reason, coupled with the fact that combustion turbines are generally most needed when the temperature is the hottest and electrical demand is highest, options such as inlet cooling are growing in importance, and these will be discussed further in later chapters.

Similarly, the impact of low barometric pressure associated with increased elevation can decrease combustion turbine capacity. The barometric pressure at 5,000 ft is only 80% of that at sea level. This could cause a significant loss of capacity.

3.7.8 The Combined Cycle

The combined cycle (Fig. 3-30) uses the exhaust heat of the combustion turbine by channeling it into a heat recovery

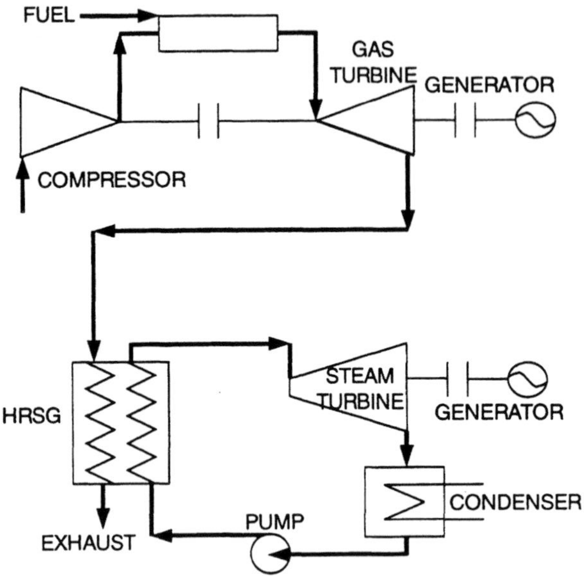

Fig. 3-30. A combined cycle.

steam generator (HRSG) to produce steam for a Rankine cycle. Although no new thermodynamic considerations are introduced, the concept will be introduced here for further discussion in later chapters. Combined cycle units can produce efficiencies in excess of 50%, and are becoming a mainstay of the power industry. In a combined cycle, the combustion turbine generally produces about 2 units of energy for each unit of energy produced in the Rankine cycle.

3.8 REFERENCES

KEENAN, JOSEPH H. and JOSEPH KEYES. 1936. *Thermodynamic Properties of Steam.* John Wiley & Sons, New York, NY.

NELSON, L. C. and E. F. OBERT. 1954. Generalized Compressibility Charts. *Chem. Engin,* 61:203.

REYNOLD, W. C. and H. C. PERKINS. 1970. *Engineering Thermodynamics.* McGraw-Hill, New York, NY.

THRELKELD, JAMES L. 1970. *Thermal Environmental Engineering.* Prentice-Hall, Englewood Cliffs, NJ.

VAN WYLEN, G. J. and R. E. SONNTAG. 1978. *Fundamentals of Classical Thermodynamics.* John Wiley & Sons, New York, NY.

4

Fossil Fuels

Kenneth E. Carlson

4.1 INTRODUCTION

Fuel represents the largest operating expense associated with electric power production. Of all the costs associated with the production of electricity, it is the only one whose control can still have a significant impact on busbar generation costs after the power plant has been constructed. Fuel selection for electric power generation plays a significant role in establishing the parameters that will be used in designing combustion and pollution control systems of the power plant.

Fossil fuels are the carbon-based fuels—coal, petroleum, and natural gas—found in fossil-bearing strata of the geologic column. In 1990, they supplied energy for 70% of all electricity generated in the United States and 64% of all electricity generated in the world (*International Energy Annual* 1991, 1992).

This chapter surveys fossil fuel usage, formation, reserves, production, processing, transportation, properties, and associated economic considerations. This should give the design and power plant operating engineer a better understanding of fuel selection and its impact on power plant design, operation, and busbar generation costs.

4.1.1 Power Generation Cost Structure

Figure 4-1 compares the relative production costs for power plants that are baseloaded. Relative costs are shown since the actual costs are site and time period dependent. Not only does the total cost of generation vary significantly, but its division between capital, nonfuel operation and maintenance, and fuel also varies significantly. Based on these costs, it is evident that some types of generation are better utilized for peaking (low utilization) than baseloaded generation. For example, a simple cycle combustion turbine burning expensive low sulfur No. 2 fuel oil is most suitable for peaking units for which the utilization of the capital investment is low. On the other hand, the fuel cost for a nuclear power plant is small when compared to its capital cost. Therefore, such units would be more suitable for satisfying baseload requirements with annual capacity factors above 70%. For such duty cycles, the large fixed capital cost can be spread over many kilowatt-hours of generation.

One of the key questions that must be answered prior to fuel and unit type selection is what the expected duty cycle is likely to be during the economic life of the unit. This is normally determined as a part of the Generation Planning Study, discussed in Chapter 25, Power Plant Planning and Design.

4.1.2 Cost of Fossil Fuels

Fossil fuels are sometimes purchased based on quantities and unit costs that are not representative of their energy value. However, the price of the fuel will almost always be adjusted to reflect its energy content and therefore its true commercial value. Table 4-1 shows the delivered cost of fossil fuels for power generation in normal units of purchase and energy-based units of cents per million British thermal units (Btu). While the costs when described in units of cents per million Btu may seem small, the economic impacts are significant as evident from the billions of dollars spent each year on fuel.

As an illustration, Fig. 4-2 shows the annual value of 1¢/MBtu[1] in fuel cost for a power plant having an electrical output capacity of 400 MW(e), a net plant heat rate[2] (on a higher heating value basis) that varies from 8,000 Btu/kWh (combined cycle plant) to 13,000 Btu/kWh (small simple cycle combustion turbine), and an annual capacity factor that varies from 15% (peaking units) to 40% to 60% (cycling plant) to 70% to 95% (baseloaded power plant). Because of the significant differences in delivered fuel prices for different types of fuel, generation type and fuel choice play a significant role in future generation costs. The figure also shows that an improvement in the net plant heat rate produces significant economic benefits, particularly for those types of generation that require high-cost fuels.

4.1.3 Fuel Cost Savings from Improving Net Plant Heat Rate

Figure 4-3 shows the reduction in annual generation cost that results from reducing the net plant heat rate by 1 Btu/kWh

[1]As used throughout this chapter, "M" means mega or million and "m" means thousand.

[2]Energy input to generate 1 kWh of electricity.

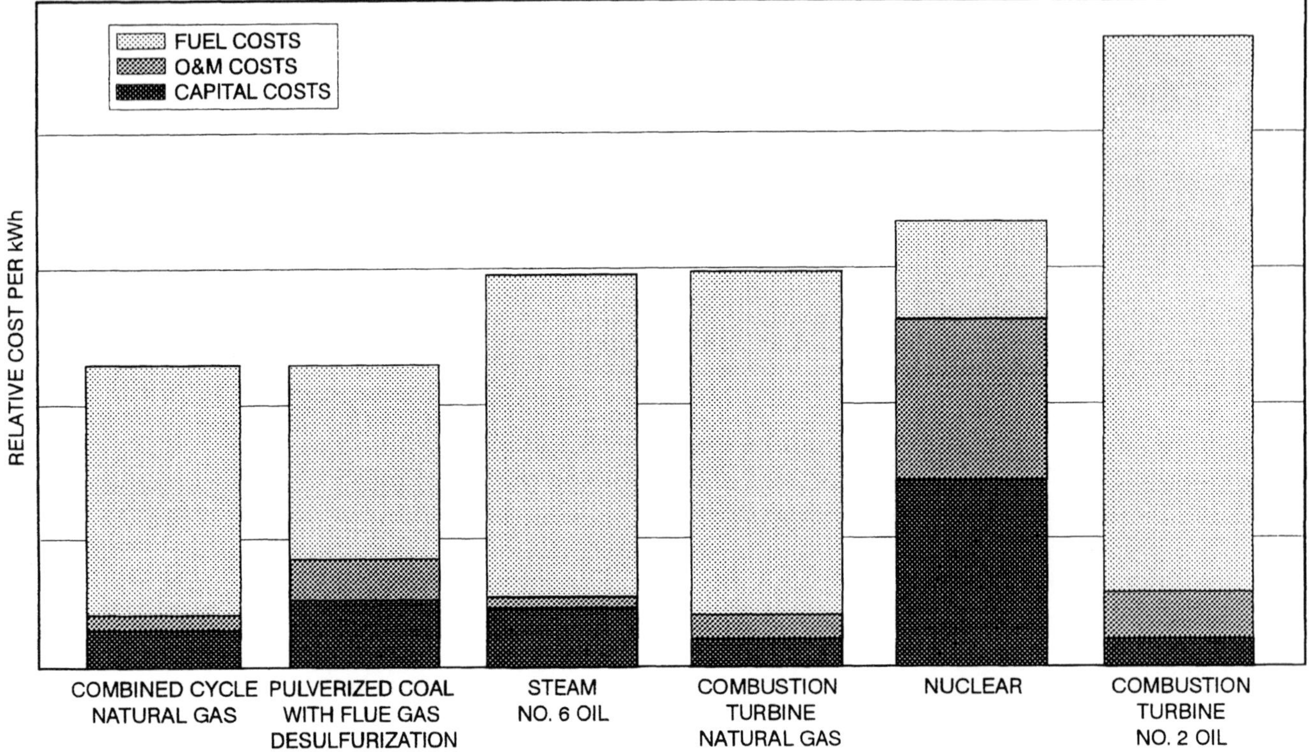

Fig. 4-1. Power plant production costs—baseloaded operation.

for a power plant with 400 MW(e) of net electrical output capability when operating at various annual capacity factors. These annual savings are based upon a fuel cost of 1¢/MBtu. A power plant designer or plant operator can equate these savings to the additional capital or operating costs required to realize the savings. For example, for a baseloaded 400 MW(e) power plant with an 80% capacity factor, an economic life of 20 years, and a 10% cost of capital, the capital recovery factor—the percentage of the initial capital cost expended each year to recover the cost of the capital investment and interest expenses—would be 11.7%.[3] Based on this capital recovery factor, Fig. 4-3 shows that an initial capital investment of about $240 per 1¢/MBtu ($28/0.117) of fuel cost can be justified for each 1 Btu/kWh reduction in net plant heat rate. For a plant with a delivered fuel cost of 200¢/MBtu, this amounts to an initial capital investment of about $48,000.

During the 1970s when fossil fuel prices were expected to continue their high rates of escalation for many years, the power plant design engineer could justify the increased capital investment for additional feedwater heaters to improve steam cycle efficiency. The design engineer could also jus-

Table 4-1. Average Delivered Cost of Fossil Fuels for Power Generation in the United States—1992

Fuel	Quantity '000	Quantity Trillion Btu	Heating Value	Delivered Cost Unit Cost	Delivered Cost ¢/MBtu	Delivered Cost $ Million
Coal	775,963 tons	16,132	10,400 Btu/lb	$29.36/ton	141	22,782
Natural gas	2,637,678 mcf	2,696	1,022 Btu/cf	$2.38/mcf	233	6,278
No. 6 Oil	138,261 bbl	878	6.35×10^6 Btu/bbl	$15.71/bbl	247	2,172
No. 2 Oil	5,853 bbl	34.1	5.82×10^6 Btu/bbl	$26.26/bbl	451	154
Petroleum coke	687 tons	19.1	13,900 Btu/lb	$20.85/ton	75	14
No. 4 oil	249 bbl	1.5	6.1×10^6 Btu/bbl	$20.60/bbl	340	5
Total/average	N/A	19,761	N/A	N/A	159	31,405

There can be significant variations depending on the power plant location, fuel source, and contractual arrangements. For example, the delivered cost of coal ranged from a low of 49¢/MBtu to 315¢/MBtu.

Source: Adapted from data in *Cost and Quality of Fuels for Electric Utility Plants 1992*, US Department of Energy, August 1993.

[3]This does not include the cost of administering the capital investment and costs such as property taxes and insurance which would be included in the annual fixed charge rate. See Chapter 2, Engineering Economics, for a more detailed discussion.

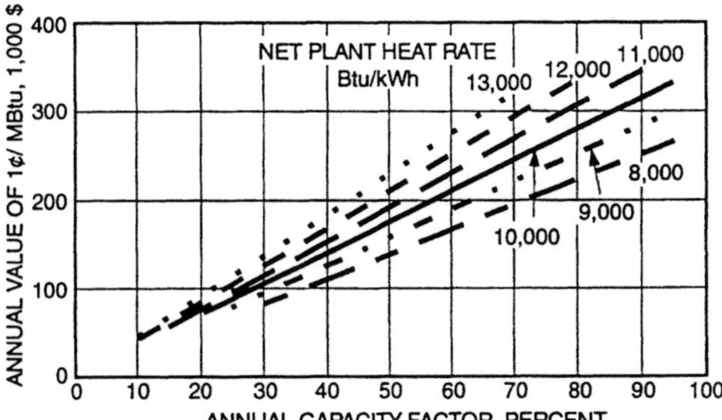

Fig. 4-2. Annual value of 1¢/MBtu in fuel cost—400 MW(e) plant.

tify the selection of natural draft instead of mechanical draft cooling towers to reduce auxiliary power requirements and hence greater consumption of increasingly expensive fuel.

4.1.4 The Need for Fuel Flexibility

Fuel properties and the delivered cost of fuel play a major role in developing design parameters and establishing design margins for fuel property dependent components and systems. Those who are involved with fuels procurement activities for electric power generation are often asked to provide "typical" fuel properties that can be used in designing fuel quality dependent components and systems. For refined products such as No. 2 light distillate and natural gas this is usually not too difficult, since their properties tend to be fairly homogeneous from one source to another. As discussed in Section 4.8, this is definitely not the case for coal, residual oil, and biomass fuels.

After the request for "typical" fuel properties, the next most often received request of the fuels procurement staff is, "What is a typical delivered cost for the fuel?" As discussed later in this chapter, delivered cost ranges widely depending upon the type of fuel, degree of refining or coal beneficiation (preparation or cleaning), and transportation requirements.

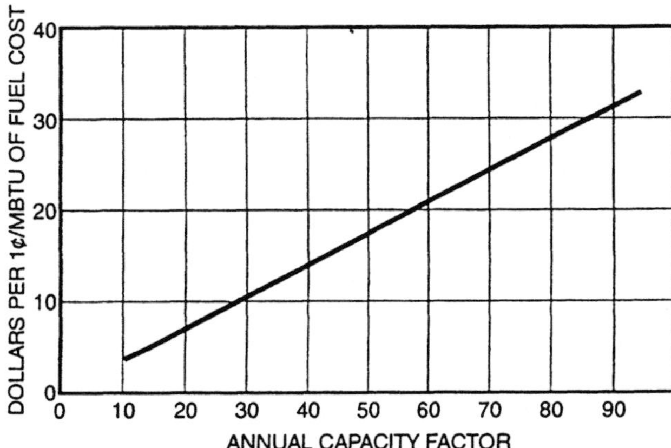

Fig. 4-3. Annual value of 1 Btu/kWh reduction in net plant heat rate—400 MW(e) plant.

Also, the future availability and price volatility of these cost components needs to be addressed.

In the 1960s, some electric utilities designed new electric power generation that was limited to burning a single fuel such as oil. These utilities then experienced the rapidly rising fuel prices in the 1970s and can testify that having flexibility in sourcing fuel and transportation can pay large dividends. Such flexibility can be utilized only if it has been planned for and designed into the electric power generating facility.

4.2 FOSSIL FUEL USE BY FUEL TYPE

To understand better the role of fossil fuels in power generation, it is good to consider their use in the major sectors of the economy such as the residential, commercial, industrial, transportation, and electric utility sectors.

In the 1800s, wood, coal, and water dominated all forms of energy usage. By the early 1900s, the United States relied almost exclusively on coal for its energy needs.

Beginning with the operation of Thomas Edison's commercial power generation station in New York City in 1882, coal became the dominant fuel for electric utility plants and was the nation's primary energy source from 1885 until the end of World War II. At that time, oil and natural gas began to make inroads into the industrial and transportation (railroad) sectors. By the time the first nuclear power plant began commercial operation in 1957 at Shippingport, Pennsylvania, the use of coal had shifted away from most sectors of the economy except the electric utility and industrial sectors, where it was dominant.

By 1992, about 80% to 90% of the coal produced in the world and in the United States was used in electricity-generating power plants. The remaining uses were primarily as a metallurgical coking coal (for use in reducing iron ore to pig iron) and to produce industrial process steam. However, in some parts of the world (such as China), coal supplies about 75% of the country's total energy needs.

While the cost of the fuel is often a major consideration in how it is used, other factors can play a significant role in how and which end-use sector the fuel is used. These factors include the following:

- Cost of fuel processing or refining and other equipment required to convert the energy of the fuel into a more useful form;
- Waste or saleable by-products produced during the combustion process;
- Costs of environmental protection associated with the use of the fuel;
- Specific energy content (Btu/lb or Btu/ft^3); and
- Convenience associated with the fuel's receipt, storage, and use.

When fuels are used for transportation, convenience, specific energy content, and minimal environmental impact are important. For those fuels that best satisfy these constraints, consumers in the transportation sector are willing to pay significant price premiums relative to the cost of other fossil fuels. For large stationary baseloaded power generating facilities for which fuel is one of the major cost components in generating electricity, its delivered cost is of primary concern.

Figure 4-4 shows world and United States electric power generation by fuel type (excluding the United States). Outside the United States, there is less dependence on coal and nuclear fuels, but greater use of oil and hydroelectric power. For the United States the market share for hydroelectric power has declined steadily from 40% in 1920 to 9% in 1992 (Historical statistics of the United States, Colonial times to 1970 and US DOE Annual Energy Review 1992).

Table 4-1 shows the quantity and cost of fossil fuels used for power generation in the United States during 1992. Of fossil fuel costs totaling 31.4 billion dollars, the majority (73%) was for coal. Since coal dominates all other fossil fuels for power generation in the United States and the world, it is the focus of discussion in the succeeding sections of this chapter.

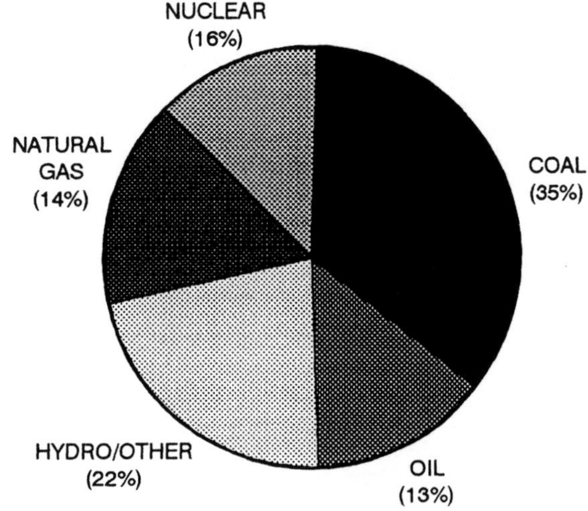

1991 TOTAL WORLD (EXCLUDING USA) = 9,210 BILLION kWh

Source: Adapted from data in Energy Statistics and Balances of Non-OECD* countries 1990-1991, International Energy Agency.

*Organization for Economic Cooperation and Development

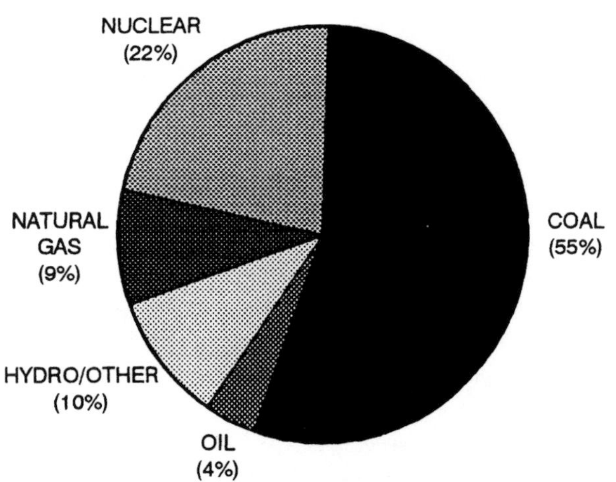

1991 TOTAL USA = 2,825 BILLION kWh

Source: Annual Energy Review by U.S. DOE 1992.

Fig. 4-4. Electric power generation by fuel type.

4.3 FORMATION OF FOSSIL FUELS

Coal, oil, and natural gas are called fossil fuels because they represent the remains of plant and animal life that are preserved in the sedimentary rocks. It is generally believed that coal was formed from plant matter and oil formed from the remains of marine organisms, although there is evidence that some oil has formed from coal (plant matter).

Coal formation (its geologic history) determines coal properties. This includes the noncombustible mineral content, or ash, associated with the coal. The ash mineralogy in turn also influences many important aspects of coal-fired power plant design and operation. Coal deposit formation also has considerable influence on mining access (surface or underground), the method of mining, and the cost of mining each deposit. Therefore, the coal formation process bears directly on the cost of mining and generating electricity. Variables such as a coal's rank, a measure of the deposit's maturation (coalification), and coal ash mineralogy demonstrate why there are no "typical" properties for coal that can be widely used in designing coal-fueled electric generating facilities.

In contrast to coal, the variation in the properties of oil and natural gas due to formation plays a less important role.

4.3.1 Geologic Setting of Fossil Fuels

Sedimentary rocks represent the accumulation of sediments (rock fragments or minerals) that generally are deposited in and by water. The resulting water-deposited sediments often exhibit a layered structure in the form of bedding planes.

The bedrock immediately below about 75% of the earth's land surface is rock of sedimentary origin. These accumulations of sedimentary rocks, known as depositional basins, have thicknesses that range from shallow sequences < 100 ft to deep sequences up to 30,000 ft (9,000 m) deep that occur in places such as the Mississippi River delta. The accumulation of sedimentary rocks in such basins account for about 5% of the earth's crust by volume, or about 100 million cubic miles (417 million km^3) of sediments.

After deposition, the deposited sediments became natu-

Fig. 4-5. Thick coal seam in the Powder River Basin. (From Peabody Holding Company. Used with permission.)

rally cemented (lithified) into rocks such as sandstone, limestone, and mud rocks (shale). Clay-sized sediments composed of rock particles < 5 microns in diameter, and silt (fine-grained sediments < 100 microns in diameter) usually lithified to form shales. Coal seams or coalbeds are often found in cyclical or repetitive sequences layered between beds of sandstone, fireclay, shale, or limestone.

4.3.2 Formation of Coal

Sedimentary rocks containing coal deposit sequences underlie about 460,000 square miles of the United States or about 13% of its land area (Energy Information Administration 1985). Thicknesses of the carbonaceous seams or coalbeds can range from a few inches to several hundred feet. Australia has brown coal (lignite) seams several hundred feet thick. The thickest coal seams in the United States are in the Powder River Basin of Wyoming and Montana, where the seams typically approach 100 ft (30 m) in thickness (Fig. 4-5). Considering that one unit of coal thickness is believed to represent at least three or more units of compacted plant matter, the original uncompressed plant material source must have been incredible. The uniformity, continuity, large areal extent of some seams, and the frequent occurrence of thin uniform in-seam partings[4] of inorganic sediment also suggest a coal formation mechanism that was quite orderly.

Coal is a combustible rock composed of a heterogeneous mixture of organic and inorganic materials. The organic materials are mainly carbon and volatile hydrocarbons within a disseminated matrix of inorganic mineral matter called ash, the noncombustible residue that remains after complete combustion of the coal. There are large variations in the type and concentration of mineral matter between and within seams. After mining, the coal contains two distinct ash components. One is inseam or inherent mineral matter disseminated within the coal seam. The other is diluent or extraneous mineral matter from the roof (top of the seam), floor (bottom of the seam), and partings within the seam recovered during mining operations.

Some of the inherent mineral matter in coal was apparently derived from inorganic compounds associated with plant life. Iron, phosphorus, calcium, potassium, and magne-

[4]Layers and lenses of mineral matter within the coal seam that vary in thickness from less than one inch to several feet.

sium are some of the elements associated with the inherent ash. This mineral matter is generally responsible for about 1% to 1.3% of the ash in coal (Stutzer and Noe 1940). Some of this inherent mineral matter occurs as clay-sized or finely disseminated inorganic sediments of < 5 microns in diameter. On the other hand, the extraneous mineral matter that tends to concentrate in the joints of the coal seam is composed of alumino silicate clays, and comprises the bulk of the ash in the coal. Some of these impurities leach into the seam from the overburden or are present as intraseam partings and clay bands. The sedimentary rock layers consisting of fireclay, slate, shale, sandstone, and limestone layered deposits that are above and below the coal seam also exist as partings or microscopic bands within the seam. As thick layers, they form the floor or the roof of the coal seam.

Another source of mineral matter is probably the result of ion-exchange processes from minerals in waters associated with the deposition of plant matter or precipitation of sulfate and carbonate compounds from these waters.

The initial deposit of plant matter can be compared to a porous sponge. The weight of the overlying sediments and water compress the plant remains. This drives out the moisture from the plant matter, thereby increasing the proportion of carbon. Thomas and Danberger (1976) have shown that coals of the Illinois Basin exhibit decreasing porosity with increasing depth of burial. The result is an increase in the rank of the coal from high-moisture lignite or brown coals to low-moisture anthracite (composed mainly of carbon). With an increase in rank, there is an associated increase in heating value.

There are two principle theories for coal formation; the Autochthonous (swamp) theory and the Allochthonous (drift) theory.

4.3.2.1 Swamp or Autochthonous Theory of Coal Formation. James Hutton (1726 to 1797) was a Scottish geologist who hypothesized that the present-day processes of erosion and sedimentation have continued in the same fashion for long periods of time to account for the sedimentary rock layers. This theory (uniformitarianism) holds that all geologic occurrences are explainable by natural physical processes that can be observed today, and that these processes and rates have continued uniformly or in the same way throughout geologic history. The theory of uniformitarianism does not allow for catastrophic processes to explain coal formation.

Hutton's theory was promoted by Charles Lyell (1797 to 1875) and others. They organized the different sedimentary rock layers according to the proportions of species of shell fossils found in each layer. These layers were later associated with what are known as geologic epochs.

The uniformitarian view explains the autochthonous or swamp theory of coal formation. It is also referred to as the in situ theory of coal formation. It states that the coalbed source material came from plants that grew in a near-coastal swamp. These source plants presumably lived and died in the same location where the coal seams are found. After dying, various mosses, leaves, and parts of trees settled to the bottom of the swamp. Following partial decomposition, the plant matter became peat. At this stage, it is important that the peat deposits be covered with a layer of sediment, otherwise it is possible they could be attacked by microbes, become oxidized, and form humus—a more fully chemically decomposed transitional material unsuitable for forming coal. The covering may come from nearby rivers or seas flooding the swamp with mud and sandy sediments.

The repetitive formation and subsequent subsidence of other superimposed swamps gave rise to the multilayered coal formations that we observe today. This process is assumed to have occurred many times during the earth's history; for example, up to 50 distinct coal seams have been identified in the Illinois Basin. In Europe, the process was assumed to have occurred during the Carboniferous (coal-making) period of the Paleozoic era. The Carboniferous period is said to comprise the Pennsylvanian and Mississippian periods 280 to 345 Ma (million years ago), which is considered to have been an epoch that contained large plants and trees that grew in a lush tropical environment. However, coal is also found in the Cretaceous (65 to 135 Ma) and Jurassic (135 to 180 Ma) periods of the Mesozoic era (65 to 230 Ma).

Some of the thickest coal seams in the geologic column did not form during the Carboniferous or coal-making period, but occur in strata interpreted as being nearly 300 million years younger. In fact, coal is found throughout the geologic column. The thick brown coal seams of the Latrobe Valley near Melbourne, Australia, were formed during the Tertiary period 600,000 years to 65 Ma. Seam thicknesses range from 200 ft (61 m) at Yallourn, to 400 ft (122 m) at Morwell, to 600 ft (183 m) at Loy Yang (Higgins and Garner 1981). These younger coals tend to be lower in rank, perhaps because there has been less overlain rock to bury and compress them.

4.3.2.2 Drift or Allochthonous Theory of Coal Formation. The Drift or Allochthonous Theory for coal formation holds that the vast amounts of plant matter that formed the coal seams did not grow in the geographic area where the seams are found. Rather, they were uprooted by vast flooding and transported to the location of the coal basins where they became waterlogged and settled out of suspension. The allochthonous view of coal source material deposition basically holds that coal was detritus (particulate woody material) that was physically transported to deposition sites in a manner similar to the other coal sequence members such as the coal floor, coal intraseam partings, and roof of the coal seams. The theory posits that the plant matter was suspended in ocean waters or floated as mats on the surface of these waters. As a result of wave agitation and autogenous grinding of the floating log mats, macerated bark and woody material became waterlogged and settled to the bottom where they were buried by layers of sediments. Agitation of the mat material would have caused the waterlogged plant matter to settle out at different times during the flood to form individual seams.

4.3.2.3 Observations. An important aspect of either theory of coal formation is that the plant matter must be buried quickly so that it has little access to oxygen, either atmospheric or water dissolved. This has led some scientists to question the gradual, in situ theory of coal formation. The swamp model alone seems inadequate to explain many coal geology features, such as the abrupt but uniform and persistent level nature of the boundary contact between the coal seam and the floor and roof sediments.

For example, research by Leonard G. Schultz (1958) has shown that the underclays (strata immediately beneath a coal seam) of coal seams in the Appalachian Region, Illinois Basin, and Mid-Continent Region lack a soil profile common to known soils. The underclays exist essentially as they were when transported into the basin as sediments. The lack of a clear soil profile has caused observers to reflect on the nature of the evidence offered in support of the swamp theory of coal formation.

Research by Hayatsu and others (1984) at Argonne National Laboratory suggests that long periods of time and high temperature and pressure do not play the essential roles in coal formation. Rather, the absence of oxygen and the presence of illite, montmorillonite, or kaolinite clays that act as chemical catalysts play the most important roles in coalification. The research at Argonne showed that when lignin (lignin and cellulose are the primary components of woody tissue) was heated at moderate temperatures of 150° C in the presence of these clays, a coal-like substance was created in a matter of months. The temperature of 150° C (302° F) was selected because it is not an unusually high temperature geologically, and because at temperatures greater than 200° C (392° F), the common clay minerals metamorphose (change from sedimentary rock to metamorphic rock). If the process was continued for up to 8 months or longer, a higher rank coal was created.

Related research has shown that a significant amount of organic material in coal is exposed to fine micron size mineral grains (Allen and VanderSande 1984). According to Schultz (1958), "These fine particles provide a large surface contact area that can have a significant impact on the organic chemistry of coal."

There is evidence that coal formation occurred contemporaneously with significant volcanic activity. Volcanic deposits, known as tuff (dust/ash), occur in the sedimentary rocks that surround some coal seams. The material ejected during eruptions is primarily composed of water, silica, alumina, iron, magnesium, calcium, sodium, potassium, titanium, manganese, phosphorus, and sulfur. These are the primary constituents of the mineral matter in coal. Silica and alumina are the basic compounds of illite, montmorillonite, and kaolinite clays. It is this same mineral matter that creates the greatest challenges in designing and operating the power plant's systems and components that are impacted by the ash residue after the combustion process. See Chapter 6, Combustion Processes, for more information. Additional evidence that supports the drift theory over the swamp theory is summarized as follows:

1. The bottom of present-day swamps consists of a mass of root structures. Such root systems must be present to support the growth of the extensive vegetation that makes up the seams. For example, some of the lycopod trees that were prevalent in the "coal forests" of the past were more than 100 ft (30.5 m) tall and several feet in diameter. They had roots up to 31 in. (79 cm.) in diameter and 46 ft (14 m) long (Andrews 1961). Such root structures are not found in the sedimentary deposits beneath coal seams, which contradicts the swamp or in situ theory.

2. In the Latrobe Valley of Australia, various pine trees have been found in the coal seams. Such pine trees are uncharacteristic of vegetation that grows in swamps.

3. Polystrate fossils (e.g., intact tree trunks that penetrate several strata) have been found that appear to be growing out of one coal seam and through many feet of overlying sediments. For example, tree trunks penetrate more than 30 ft (> 9 m) between the Lower and Upper Pilot seams near Newcastle, Australia. This indicates that the plant matter and sediments between the lower and upper pilot seams were deposited and buried rapidly. The fact that these trees consist solely of trunks (no roots or limbs) also supports the idea that they did not grow in the place where they are found but became waterlogged and sank beneath a floating mat of trees and plant material.

4. Evidence for the phenomena described in Items 2 and 3 above have been observed after the eruption of the Mt. St. Helens volcano in the state of Washington. Numerous Douglas fir trees were found floating in nearby Spirit Lake. These trees were devoid of limbs and many had lost their roots when they were uprooted. Some of them have since settled to the bottom of Spirit Lake in a vertical position. They are found resting on a 3-ft layer of peat (mostly bark) that has also settled down from the floating mat of logs. This is quite similar to the scene found near some coal seams where polystrate fossilized trees are found.

5. The thin partings of sedimentary rock within coal seams with no root systems penetrating these layers and the banded nature (bedding planes) of some coal seams could be interpreted as evidence for having been deposited by the same geologic event that deposited the sedimentary rocks that surround the seam and comprise the partings within it. (Note midseam partings in Fig. 4-5.)

6. In the Illinois Basin, cyclothems, a series of beds deposited during the same sedimentation cycle, also contain coal seams. Weller has shown that the "typical" cyclothem starts with coarse sandstone, then finer fire clay, coal, etc., and ends with shale composed of compacted clay size particles (Weller 1930). The heavy, coarse-grained particles that make up the sandstone are at the bottom because they sink

fastest, and the finer, lighter particles that compose the fire clay beneath the coal seam, mineral matter within the seam, and the shale above the seam have slower settling rates. That is, the particles settling in the water stratify by their size and weight in a manner described by Stokes' law of hindered settling [Eq. (4-1)].

$$V_{\text{settling}} \propto \frac{gD^2(\rho_p - \rho_f)}{\eta} \tag{4-1}$$

where

V_{Settling} = particle settling velocity,

g = acceleration of gravity,

D = particle diameter,

ρ_p = particle density,

ρ_f = fluid density, and

η = coefficient of viscosity for the fluid.

It may be these fine clays with their relatively large surface area settling with the organic plant matter that acted as catalysts for the coalification reaction. This is consistent with the results of experiments conducted by Hyatsu et al. (1984).

7. The juxtaposition of 4 in., 22 to 27 in., and 3 to 5 ft thick coal seams separated by thin (3/8 in. or 9.5 mm) shale partings extending for more than 15,000 square miles (39,000 km^2) unpenetrated by plant matter such as roots or trunks of growing flora (Gresley 1894) argues for their deposition from a floating mat rather than from in situ growth.

8. Some of the thickest seams, such as those found in the Powder River Basin of Wyoming, are up to 100 ft (30.5 m) thick (the amount of plant matter would be several times this thickness) and yet exhibit very uniform coalification. The uniformity of coalification is evidenced by a nearly constant moisture and ash free heating value ($< \pm 100$ Btu/lb or 56 kcal/kg) from the bottom of the seam to the top. Based on the research by Hayatsu et al. (1984), such uniform coalification would require the uniform distribution of catalyst-acting clays, that appears difficult in a swamp environment over millions of years, but not in a floating mat environment with significant quantities of fine clay sediments and fine volcanic ash[5] dispersed throughout the waters and upon the floating mats of plant matter.

9. When plant matter is buried with little access to oxygen, a composting (fermentation) type of reaction takes place providing elevated temperatures. While the temperature may not rise to a level of 150° C (the temperature of the coalification studies at Argonne National Laboratory), the coalification reaction would merely take a little longer as suggested by the Arrhenius equation for chemical reaction rates which shows that they increase with temperature.

For the growth of vegetation in the swamp environments on the earth today, coal formation is not evident, and the mechanism by which the plant matter could form a coal seam of relatively uniform thickness is difficult to imagine.

The Drift or Allochthonous Theory for coal formation helps us in understanding why we have the level and horizontally continuous seams of reasonably uniform coal deposits to work with. If the plant matter and surrounding sediments had not been deposited in the fairly uniform layers that we find them, it would be difficult to recover the coal for use as the primary fuel in generating electricity. The mechanized underground mining machinery used to recover this coal requires that the seams be fairly level and uniform in thickness to have commercial value.

4.3.3 Formation of Oil and Natural Gas

There is no clear consensus with respect to the genesis of hydrocarbons such as natural gas and petroleum. Most researchers believe that natural gas and crude oil are biogenic and derived from marine life. However, in Australia there is evidence that the source rocks for some of the oil and gas deposits were underlying coalbeds. This is particularly true in the offshore oil fields in the Bass Strait between the states of Victoria and Tasmania. It appears that the hydrocarbons forming the oil and gas originated from the pollen and spores in the underlying coal seams. They appear to have migrated from the coal seam and to have been captured in suitable overlying sediment traps. This is not unlike the mechanism employed in the manufacture of coke from coal in that methane gas, creosote, and other organic compounds are "boiled off" as the coking coal is heated in a reducing atmosphere (coke oven).

Oil is contained primarily in limestone and sandstone reservoirs. It is usually found associated with sedimentary rocks of marine origin. The general theory is that crude oil and natural gas minerals were formed by the decay of marine organisms and plants. In addition, it is believed that most oil and gas has migrated from a source rock such as shale or limestone to the present reservoir rock. Sandstone reservoirs are usually the richest oil and natural gas bearing areas because of their porous/permeable nature. The most favorable geologic structure or formation for such migration is an anticline. It forms a classic structural "trap" with a natural gas cap on top of crude oil with salt water beneath it. Sometimes the natural gas may be associated or dissolved in the crude oil; however, the majority of the gas is nonassociated with oil.

Unconventional gas may exist in coal seams (coalbed methane), geopressurized brine, tight Devonian shales, and

[5]For example, bentonite is a montmorillonite type of clay that is formed from the alteration of volcanic ash (tuffs). Tuffs are rocks of compacted volcanic fragments characterized by angular rather than rounded particles (sharp untumbled edges). Volcanic ash is often found in the sedimentary rocks that surround the coal seams.

arctic and subsea hydrates. However, these sources of gas and gas from depths > 15,000 ft (4,600 m) are very expensive to exploit. They have not been economically recoverable in recent years without tax subsidies from governments.

4.4 FOSSIL FUEL RESERVES

It is difficult to compare fossil fuel reserves among sources, whether they be individual suppliers or countries. Considerable confusion exists regarding the definition and use of such terms as resources, resource base, reserve base, in-place reserves, and economically recoverable reserves. Sometimes these terms are used interchangeably, resulting in large discrepancies between estimated resources and portions of economically recoverable, saleable fuel.

Figure 4-6 illustrates the relationship between some of these terms. For example, the United States has reported approximately four trillion tons of total coal resources in the identified and undiscovered categories. Of this total, it is estimated that 1.7 trillion tons are in the measured, indicated, and inferred categories. The demonstrated reserve base is 474 billion tons, which includes coal resources in the measured and indicated categories. Of this total, 261 billion tons of reserves, or about 7% of the total resources, are estimated to be recoverable using today's mining technology (Energy Information Administration, February 1993).

The important considerations are how much of the fossil fuel reserve base can be recovered, at what cost, and how this cost compares to current market prices for the fuel. Therefore, reserve estimates change as market prices and the exploitation technologies used for their discovery and development improve.

4.4.1 Exploration Programs

The areal extent, quantity, and quality of coal reserves are determined through exploration programs. This involves drilling and sampling the coalbeds and associated strata to determine the structure, quantity, and quality of a reserve block. Such a program establishes the depth and thickness characteristics of overlying strata, seam characteristics including floor and roof conditions, amount and extent of rock partings within the seam, mining type and methods, and coal beneficiation processes that are appropriate for developing the coal reserve.

In estimating the quantity of a fossil fuel that can be recovered from a given deposit, its density is multiplied by the planar area of the deposit times the thickness of the deposit times the estimated recovery to obtain the raw recoverable reserves. The following densities are typically used in estimating reserves when site-specific densities are unknown (Leonard et al. 1968).

- Anthracite—2,000 tons per acre-foot,
- Bituminous coal—1,800 tons per acre-foot,
- Subbituminous coal—1,770 tons per acre-foot, and
- Lignite—1,750 tons per acre-foot.

4.4.2 Coal Reserves—United States

Figure 4-7 shows the major coal fields of the conterminous United States. About 60% of United States coal reserves are located west of the Mississippi River. These reserves are equally distributed between surface and underground minable reserves. In general, coal seams in the western United States tend to be thicker than those in the east. The large seam thickness and relatively shallow depths of the major coal deposits in the Powder River Basin account for that coal's low mining cost and its ability to compete in a variety of markets.

In the east, the majority of remaining coal reserves are underground minable. The heating value of the eastern coals is significantly higher than that for the lower rank shallower surface minable coals in the west.

Figure 4-8 shows the distribution of United States re-

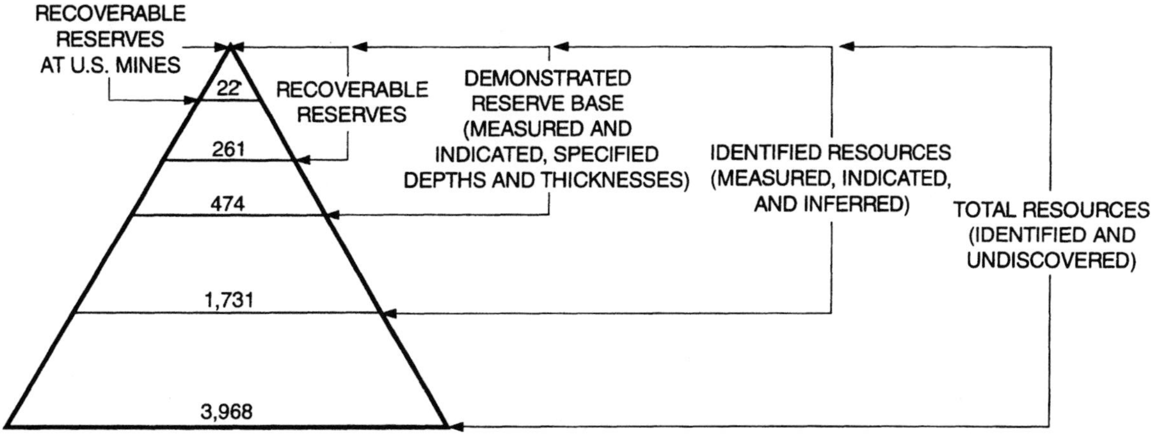

Fig. 4-6. Delineation of US coal resources and reserves (billion short tons).

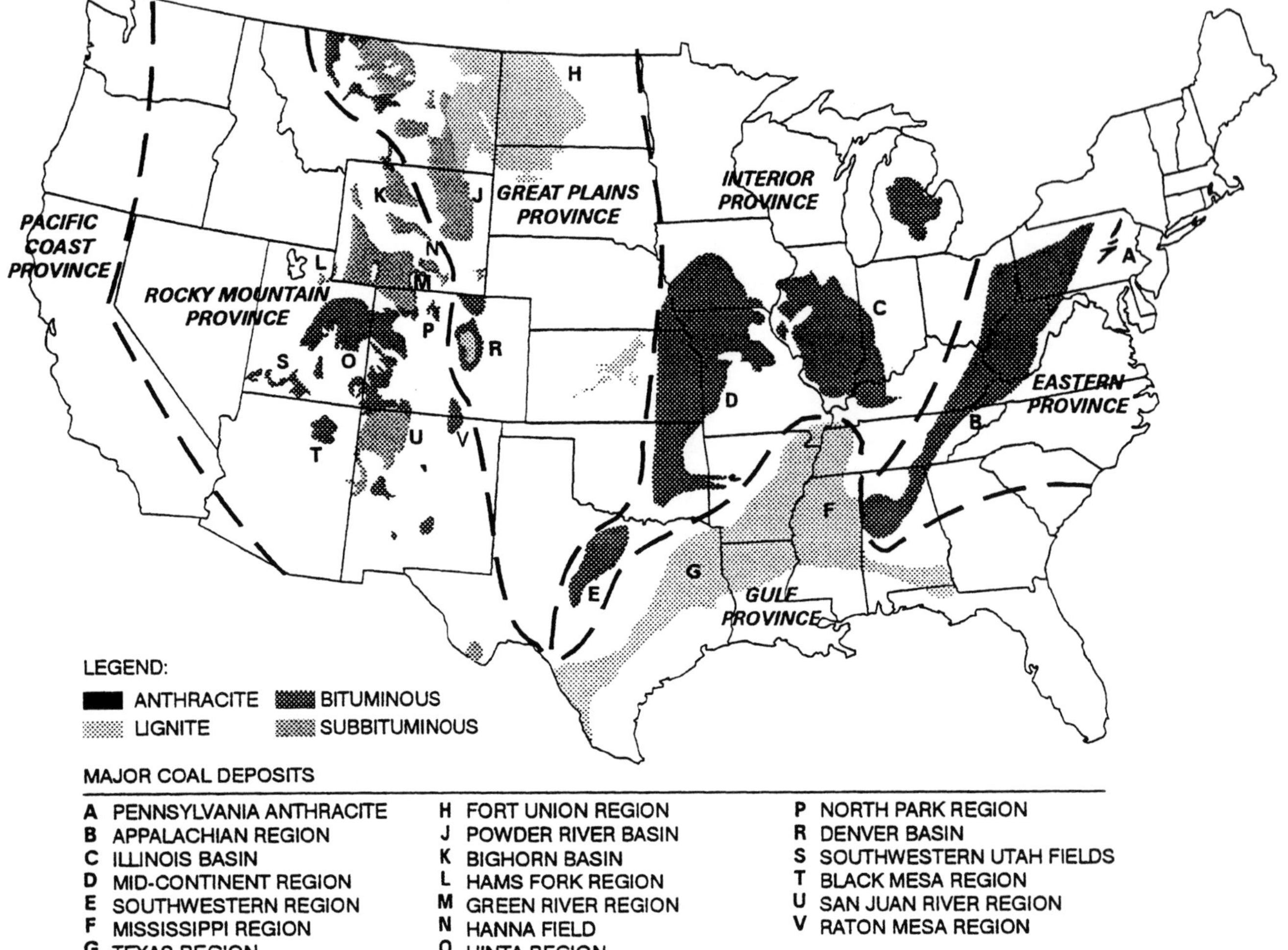

Fig. 4-7. Major coal fields of the conterminous United States.

serves by heating value and sulfur dioxide emission potential.[6] The sulfur content has been expressed as potential sulfur dioxide emissions since it is an important environmental consideration. The majority of the low sulfur coal reserves have a heating value of < 10,000 Btu/lb (5,600 kcal/kg) and primarily represent the lower rank (subbituminous and lignite) deposits west of the Mississippi River.

There are significant variations in coal properties between regions. Such differences must be considered when establishing the design basis coal quality for the power plant.

4.4.3 Coal Reserves—World

Figure 4-9 shows the coal reserves of the world by their rank and geographic area. Most areas of the world except the Middle East have significant reserves. Also, there is a fairly even distribution between low-rank (subbituminous and lignite or brown coals) and high-rank coals (anthracite and bituminous).

4.4.4 Oil Reserves—World

Figure 4-10 shows the world's proven oil reserves. While the Middle East was noticeably absent in terms of coal reserves, it dominates world oil reserves. As of 1993, the Middle East accounted for 66% of world oil reserves. Member nations of the producer cartel known as the Organization of Petroleum Exporting Countries (OPEC) account for 77% of the world's total crude oil reserves.

4.4.5 Natural Gas Reserves—World

Figure 4-11 presents the world's proven natural gas reserves. On a regional basis, Eastern Europe and the Commonwealth of Independent States (primarily Russia) and the Middle East have dominant natural gas reserve positions. In terms of natural gas, OPEC does not dominate to the extent that it does in crude oil.

[6] $\text{lb } SO_2\text{/MBtu} = \dfrac{2 \times \text{Sulfur (\%)} \times 10{,}000}{\text{Heating value (Btu/lb)}}$

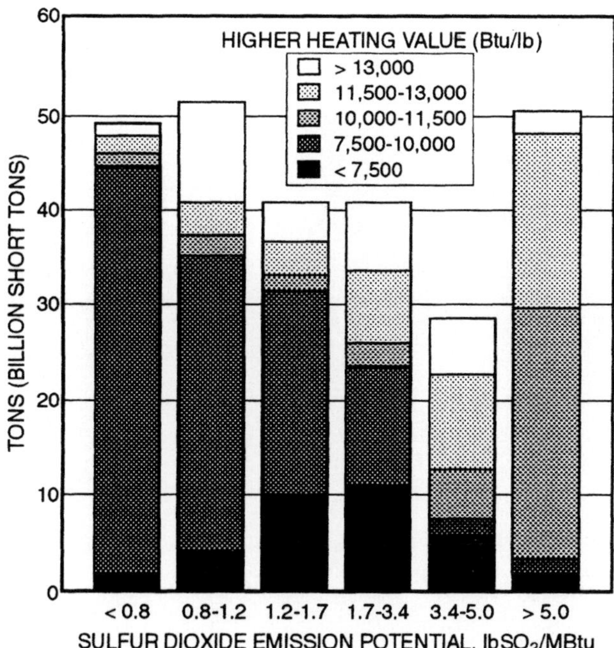

Fig. 4-8. United States recoverable coal reserves. (Adapted from *U.S. Coal Reserves: An Update by Heat and Sulfur Content*, U.S. Energy Information Administration, February 1993.)

4.4.6 Years of Remaining Reserves

It is often difficult to compare different fossil fuel producing areas based on their reserves because it is difficult to grasp their magnitude. One point of reference is to compare existing reserves with current production to determine the years of remaining reserves at current production levels *if* production is held constant.

However, this statistic can prove to be misleading, particularly if production levels are quite low because of market constraints, as is the present case for natural gas in the Middle East. In addition, for regions where exploration and finding costs are high, reserves are usually lower, reflecting the economic fact that the holding costs are higher.

With consideration of these caveats, Fig. 4-12 shows the years of remaining reserves for oil, natural gas, and coal by major geographic region. With the notable exception of the Middle East, coal has the longest reserve life and oil the least.

4.5 OTHER FOSSIL FUELS

A variety of alternate fuels are available to satisfy the increasing need for energy and to supplement the depletion of the primary fuels. At one time, coal and crude oil were viewed as alternate fuels to wood. Now their roles have reversed.

As with most alternate fuels, their properties are not well defined and often are quite variable. These fuels require special analyses of the particular sources being investigated to evaluate the typical and range values of their properties. This requires that the power plant design engineer obtain representative samples of the fuel in the condition that it will be burned. Appropriate tests need to be conducted on these samples to determine the concentration of those properties that can have an impact on the design and performance of coal quality dependent components. Significant design margins are often required to accommodate the variability in properties.

Where alternate fuels are waste products of other primary products, the demand for the primary product can have a significant impact on the supply and quality of the waste fuel.

4.5.1 Liquefied Natural Gas

Liquefied natural gas (LNG) is stored and transported at approximately $-260°$ F ($-162°$ C). Its density varies with the source of the natural gas and the amount of ethane and

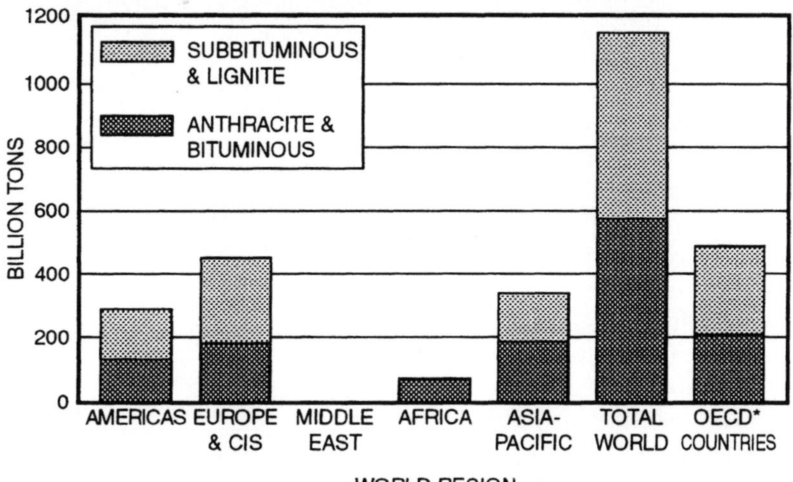

Fig. 4-9. World coal reserves 1992. (Adapted from *BP Statistical Review of World Energy*, June 1993.)

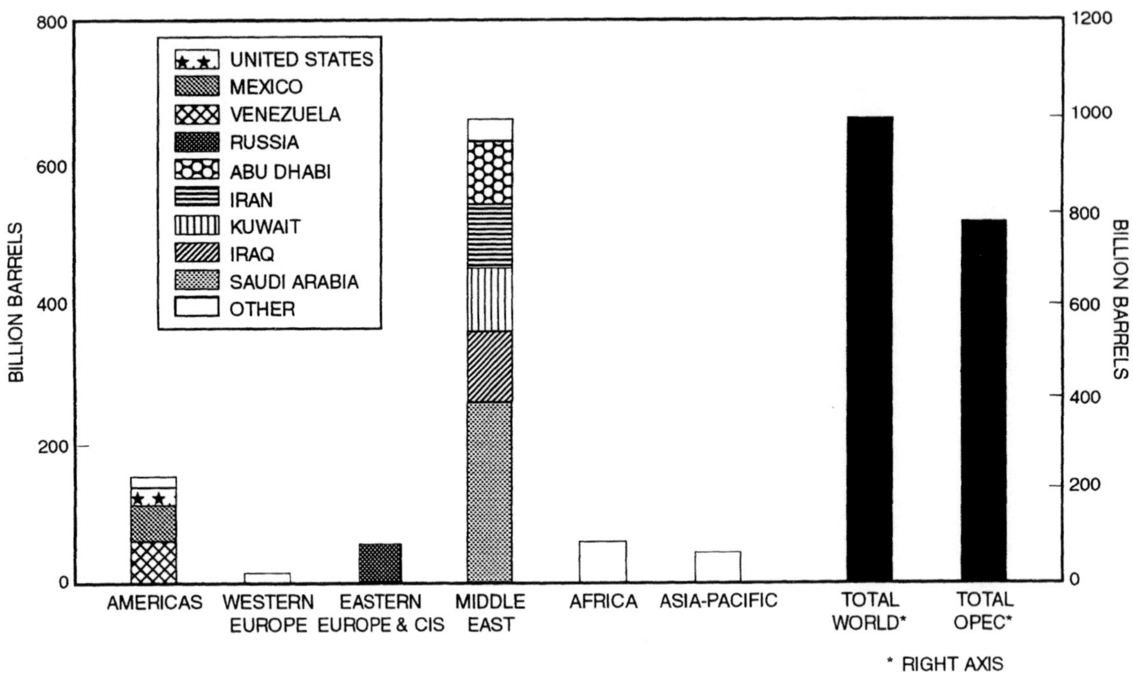

Fig. 4-10. Proven crude oil reserves—1993. (Adapted from *Oil and Gas Journal*, December 27, 1993.)

heavier components that are part of the liquid. One tonne of Algerian LNG contains approximately 52.3 million Btu on a higher heating value basis. This is equivalent to 8.3 barrels of residual fuel oil (6.3 MBtu/bbl), 52.3×10^3 ft^3 (1,480 m^3) of natural gas (1,000 Btu/ft^3), or 2.4 tons (2.2 tonnes) of coal (10,800 Btu/lb or 6,000 kcal/kg).

World LNG trading patterns can be separated into two categories that are based upon export–import shipping patterns by using the Suez Canal as a line of demarcation. Trade west of the Suez is based upon liquefaction facilities in Algeria and Libya with demand from Western Europe and the United States. LNG trade east of the Suez moves from export facilities in the Middle East, Southeast Asia, and Alaska to markets in Japan, South Korea, and Taiwan.

The first major transaction in trade west of the Suez began in 1964, when Algeria began shipping LNG from the Hassi R'Mel field to France and the United Kingdom. Algeria dominates the trade west of the Suez, with exports to Belgium, France, Spain, and the United States.

The first major LNG transaction east of the Suez occurred in 1969, with movements from Cook Inlet, Alaska, to Japan. This plant, located at Kenai, Alaska, is a relatively small facility producing about 1 million tonnes (metric tons) per year of LNG. A larger 5 million tonnes per year facility started production in 1972 at Lamut, Brunei, followed by plants at Das Island, United Arab Emirates (1977), and at Bontang and Arum, Indonesia (1977 to 1978). Indonesia is now the largest LNG producer in the world. Many of these plants have expanded their production capacity in recent years. Together with new facilities in Australia and Malaysia, these plants produced approximately 60 million tonnes (66 million tons) in 1992 with the majority being shipped to Japan (*Oil & Gas Journal*, June 1993).

LNG projects, by their very nature, are very capital intensive. Since one train of an LNG plant can produce about 2.5 million tonnes per year and at least two trains are installed to assure reliability of supply, these are world class multi-billion-dollar projects. The dedication of large gas reserves and the construction of special ships for the particular producer/end-user project are required to realize the economies of scale.

4.5.1.1 LNG Shipping. There are two types of LNG vessels. One incorporates spherical containers and the other a membrane-type system. These ships are becoming fairly standardized with six LNG storage compartments. They have a total capacity of about 59,000 tonnes of LNG. This is energy equivalent to 87.4 million m^3 (3,086 million ft^3) of natural gas at standard conditions. It is also energy equivalent to 130,000 tonnes of coal with a heating value of 10,800 Btu/lb (6,000 kcal/kg). Therefore, the energy transporting capability of a Panamax class vessel carrying LNG is nearly twice that of an equivalent size vessel carrying coal.

4.5.1.2 Regasification. Seawater is often used as the primary source of heat for revaporization. If seawater temperatures are too low, supplemental firing of natural gas may be required to revaporize the LNG.

Since regasification of LNG consumes significant amounts

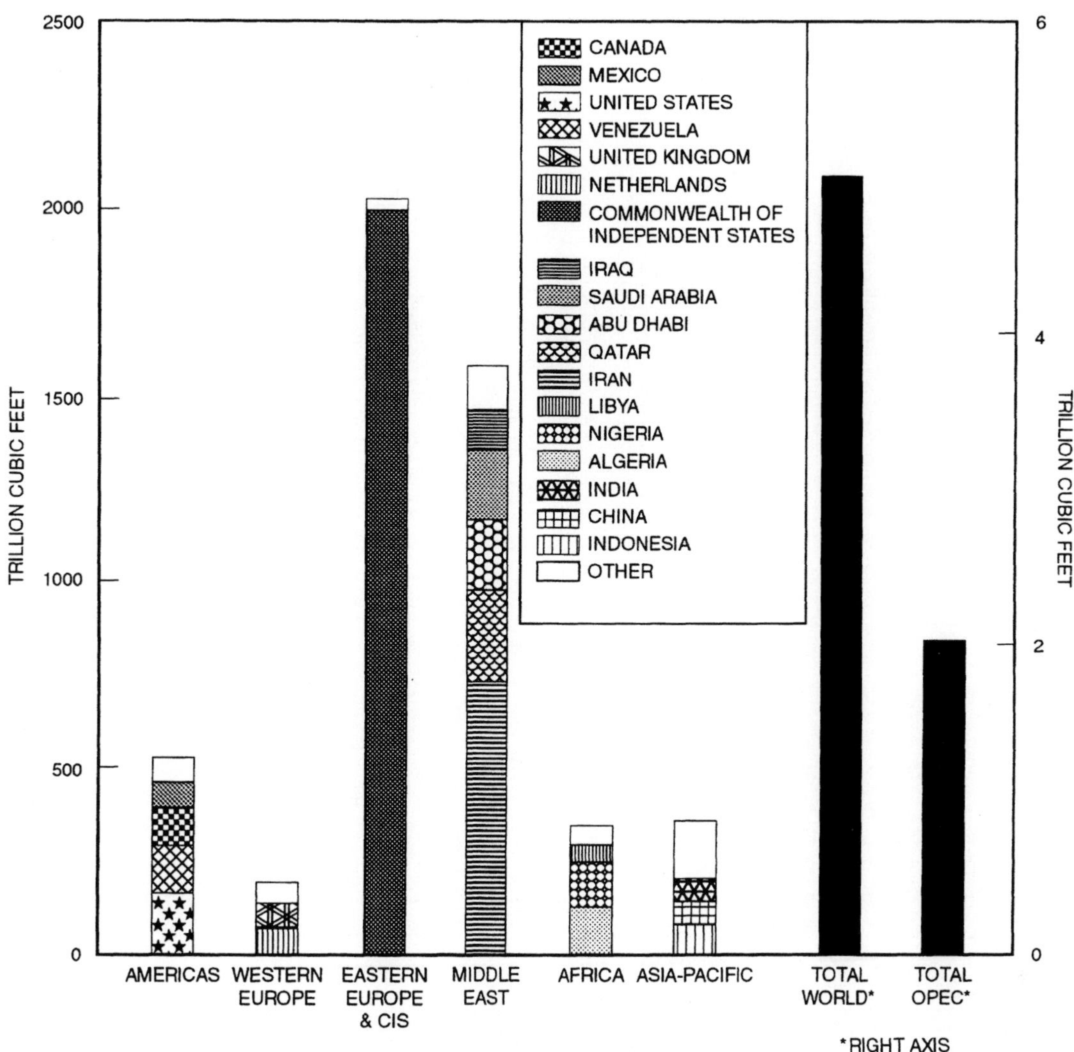

Fig. 4-11. Proven natural gas reserves—1993. (Adapted from *Oil and Gas Journal*, December 27, 1993.)

of energy, creative planning can produce ancillary economies. For example, some regasification facilities in Japan have been constructed in association with cold food storage plants. The relatively higher temperature food loses heat to the cold LNG that results in freezing the food and vaporizing the LNG. LNG regasification facilities have also been incorporated as a part of air separation plants used to produce liquid oxygen and liquid nitrogen. The air is pre-cooled by using it to revaporize the LNG.

4.5.2 Petroleum Coke

After crude oil has been distilled, the heavier fractions or "bottom of the barrel" remain. This material is often fed to vacuum distillation units or a coker. Most of these processes try to recover more of the valuable light ends (gasoline and middle distillates) that remain in the residual oil.

The coking process produces delayed coke or fluid coke. Delayed coke is sometimes referred to as shot coke or sponge coke because of its appearance as an agglomeration of small buck shot (⅛ in. diameter) or a solid with spongelike voids. Fluid coke is developed using a fluidized bed combustion process. The fluid coke consists of very hard, small spherical particles (about 200 microns in diameter) with very low volatile matter content. A description of petroleum coke production can be found in Section 4.6.10.4, Petroleum Coke.

Fuel grade coke accounted for 52% of total world coke production of 32.7 million tons in 1992. In 1992, United States refiners produced 82% of the world's fuel grade petroleum coke. If the sulfur content is quite low, petroleum coke can be used in other markets such as a graphite used in constructing anodes for electric arc furnaces. This market consumed about 45% of all petroleum coke produced in 1992 (*Oil and Gas Journal* 1994).

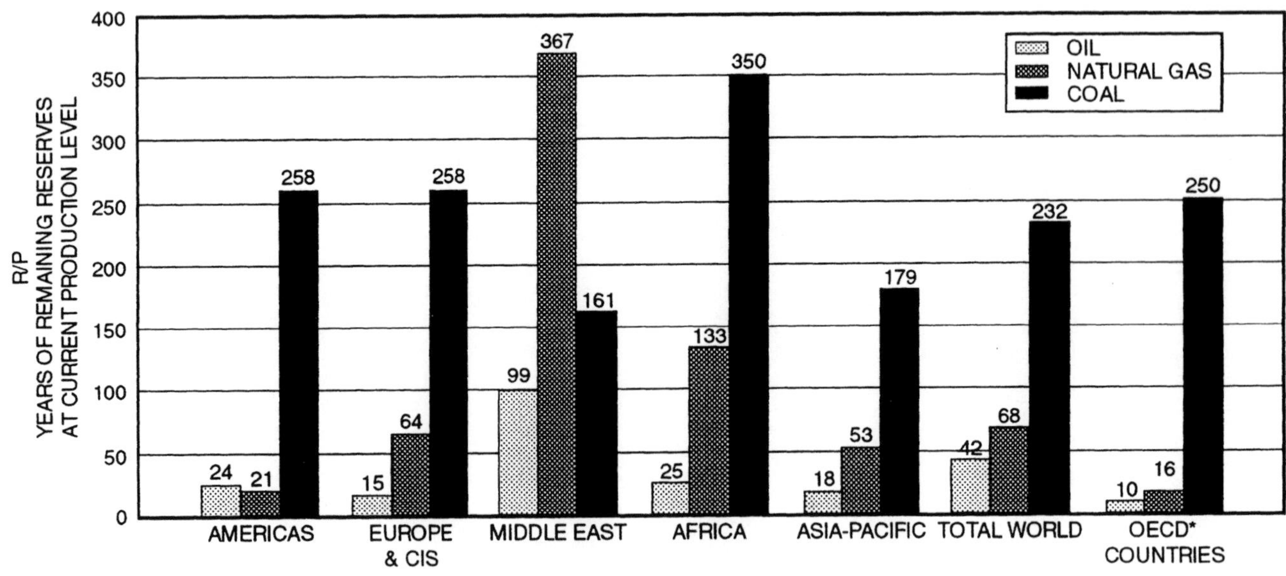

Fig. 4-12. Fossil fuel Reserves to Production (R/P) Ratio 1992. (Adapted from *BP Statistical Review of World Energy*, June, 1993.)

Petroleum coke is characterized by a fairly low ash content (< 1%), high heating value (14,000 to 15,000 Btu/lb or 7,800 to 8,300 kcal/kg), and high sulfur content (3% to 8%). It is also characterized by very low volatile matter. The low volatile matter content of petroleum coke compared to its high fixed carbon content (90% plus) leads to flame instability in the furnace unless the coke is ground to a very fine size. Unburned carbon losses can also be significant. In this regard, its combustion properties are similar to those of high-rank low-volatile coals such as anthracite. Fuel grade petroleum coke also has a high sulfur content. Because of these utilization problems, most combustion facilities burn a blend of petroleum coke and other bituminous or subbituminous coals.

4.5.3 Orimulsion®

Orimulsion® is a bitumen based fuel that has properties similar to those of residual No. 6 oil. It is produced from large bitumen reserves near the Orinoco River in Venezuela. These heavy and extra heavy oils have properties similar to those found in tar. Orimulsion® consists of about 70% Orinoco extra heavy oil and 30% water, that is combined to form an emulsified mixture, hence the name Orimulsion®.

To date, there have been several test burns of Orimulsion® in power plants originally designed to burn No. 6 oil. Most of these test burns have been accomplished with minimal modification of existing fuel handling and combustion equipment.

A primary disadvantage of Orimulsion® fuel is its very high sulfur content, that ranges up to 2.9%. For a heating value of about 12,900 Btu/lb, the uncontrolled sulfur dioxide emission potential can range up to 4.5 lb of SO_2 per million Btu. Depending on how stringent the environmental regulations are, some form of flue gas desulfurization similar to that employed on coal-fueled power plants may be required. There is also a significant amount of vanadium in Orimulsion® that can lead to the corrosion of boiler tubes. Therefore, a magnesium additive is used to inhibit such corrosion.

The production costs of Orimulsion® are quite low. On the basis of the large amount of low cost reserves, it appears that some form of bitumen-based fuel will be a competitor in traditional No. 6 oil and steam coal markets.

4.5.4 Municipal Waste and Refuse-Derived Fuels

The cost of municipal refuse received at a landfill or processing facility is negative, since the processing facility usually charges a "tipping" fee, that is the fee paid by trash haulers for dumping their load at a landfill site. Therefore, the cost of municipal refuse is fairly independent of the cost of fossil fuels but is primarily a function of the location and cost of alternative disposal sites available to trash haulers.

With high landfilling costs and increasing tipping fees for dumping at the landfill site, many communities near large urban areas have been evaluating alternatives for municipal waste disposal. Problems with municipal waste, as with most biomass fuels, include its high moisture content and the cost of processing it into a higher grade fuel. These added costs may not offset the benefits associated with a fuel having a higher heating value and lower utilization costs. Chapter 22, Resource Recovery, discusses the use of these municipal wastes and refuse-derived fuels in greater detail.

There has been some success in supplying a portion of a coal-fired unit's fuel requirements with municipal refuse. Usually such refuse amounts to < 20% of total fuel consumption to prevent undue slagging and fouling in the steam generator.

Another waste that has become popular in recent years because of its high heating value (about 14,000 Btu/lb or 7,800 kcal/kg) is used tires. Some power plants in California and Connecticut burn automobile and truck tires as their sole fuel. Used tires are also being burned at coal-burning power plants as a supplemental fuel similar to municipal refuse.

4.5.5 Wood Fuel

The three primary sources of wood waste are forest harvest residue, landfill waste, and industrial residue. Forest harvest residue can consist of slash (limbs and tree tops) that remains after harvesting trees for pulpwood or hardwood. In-forest chipping lends itself to tree thinning and improves the forest and wood quality. If the slash is removed from the harvested area, the site is better prepared for forest regeneration activities. Most of the waste wood and slash from the forest is processed by a wood chipper to make it easier to transport in "chip vans" and to facilitate drying.

The supply of waste wood available is a function of the demand for the higher end-use value products such as hardwood and pulpwood. Industrial residue can also be used by other manufacturers to make various wood products such as particle board.

The moisture content of fresh cut wood can be quite high (upward of 50%). With very high moisture percentages, the problem is similar to that of burning peat; most of the energy is used to evaporate the moisture and very little thermal energy is left to produce steam in the boiler.

Because of the low heating value of slash and wood chips obtained from forest harvesting operations, a large portion of the delivered cost of fuel is associated with truck transportation. As such, the economics of using waste wood harvested from the forest usually limit the radial distance to < 50 miles from the electric power generating plant.

While the moisture content of freshly cut wood varies from about 30% to 50%, it can be reduced to about 20% after open-air drying for about a year. However, long periods of outdoor drying and weathering can lead to decay and, therefore, loss of energy value.

One advantage of wood fuels and most biomass fuels is that their sulfur, nitrogen, and ash contents are usually low. Thus sulfur dioxide emissions are not a problem, and oxides of nitrogen are less of a problem as a result of the lower fuel-bound nitrogen and lower flame temperatures that result from combustion of the high-moisture wood.

4.5.6 Agricultural Waste

The primary agricultural waste that is being used as a fuel is bagasse. Bagasse is the portion of the sugar cane that remains after it has been compressed to extract the sugar juices. It has been used in the Hawaiian Islands as well as in other sugar-producing countries as a fuel source for generating process steam and steam for use in generating electricity.

4.5.7 Industrial Process Waste Fuels

There are many fuels, gaseous and liquid, that are produced as a part of industrial operations. These include the by-product gases from oil refinery operations. Such refinery gases usually consist of heavier hydrocarbons than natural gas and therefore have a greater heating value. These gases are generally used for steam raising purposes at the refinery.

Another industrial gas is coke oven gas that results from the manufacture of coke from metallurgical grade coking coal for reducing iron ore to make pig iron. Coke oven gas is recovered at the facility and used as fuel for raising oven temperatures and at other heat-consuming processes at an iron/steel-making facility.

4.5.8 Peat

The United States has 53 million acres of peat lands and ranks second only to Russia, which has 228 million acres of indigenous peat resources (Hamilton et al. 1981). In rankings of United States fossil fuel resources by energy content, coal ranks first and peat ranks second (Electric Power Research Institute 1983). The states with the largest peat resources are Alaska, Minnesota, Michigan, Florida, and North Carolina. Because of its low heating value, peat, like wood and lignite or brown coal, cannot be practically transported very far from where it is produced.

Russia, Finland, and Ireland have developed the most peat-fueled power plants. In Finland and Sweden, where there are no indigenous fossil fuels other than wood and peat, the local peat is used in cogeneration facilities where the steam is also used for district heating. However, they also use No. 6 oil as a backup fuel. In Russia, there was a need to avoid transporting other primary fuels great distances for electric power generation.

4.5.8.1 Peat Characteristics Peat is the partially decayed remains of plant material. The properties of peat vary widely among deposits and even within a deposit because of the type of original vegetation and the degree of its decomposition. Peat also contains inorganic minerals that have been transported as sediment load carried in by streams and flood waters that enter the bog.

One of the distinctive characteristics of peat compared to other fossil fuels is its high in situ moisture content, typically 80% to 95%. Even after draining the peat bog, the moisture content may drop to only 70% or 90%.

After air-drying in the field, the moisture content can vary from 35% to 55%. The moisture level must be reduced to $< 50\%$ to provide an acceptable fuel that burns with a stable flame and does not require the firing of supplemental fuels. Air-drying increases the as-received heating value of the peat to between 4,000 and 7,000 Btu/lb (2,200 to 3,900 kcal/kg) which is equivalent to that of lignite or brown coal. Because peat is usually found in areas with a cool climate and where precipitation exceeds the rate of evaporation, conditions are not always favorable for air-drying.

The sulfur content of peat tends to be very low, typically about 0.2%. Ash content is usually low, although this varies depending on the depositional environment of the bog.

*4.5.8.2 **Peat Production.*** Several steps are required to make a peat deposit acceptable for harvesting. The first is to develop a pattern of ditches to allow the surface water to drain from the bog. The second is to remove some of the peat and spread it on the surface of the bog for solar and wind drying. The third step is to collect the dried peat and transport it to the storage stockpile at the power plant.

Climatic conditions influence the length of the harvest season and the amount produced in one year compared to another. In areas where precipitation is minimal and evaporation rates high, more harvests are possible because of the greater thickness of the peat that can be milled during each harvest and the shorter drying time.

It is noteworthy that historically, interest in the use of peat has been high when real or constant dollar oil prices are near a peak. It is likely that during the next rise in fuel prices, interest will again develop in using peat as a power plant fuel.

*4.5.8.3 **Socioenvironmental Impacts*** At the present time, regulations in the United States are stringent regarding the loss of wetlands. The mandate to the US Army Corps of Engineers is to avoid adverse impacts, offset unavoidable adverse impacts to existing aquatic resources, and for wetlands to strive to achieve a goal of no overall net loss of values and functions (Went 1990). Therefore, under the current regulatory environment, peat development appears to be very difficult.

4.5.9 Coalbed Methane

Some coal seams contain a significant amount of methane and are referred to as gassy seams. In recent years, the United States Government has promoted the development of coalbed methane as an alternative fuel. Based on the tax credits provided, there was significant incentive to drill for and develop the methane in these coal seams. For example, production of coalbed methane in 1992 amounted to 92 billion ft^3 in the Black Warrior Basin of Alabama and 447 billion ft^3 in the San Juan Basin of Colorado and New Mexico, the two primary producing areas in the United States (Black 1994).

4.5.10 Tar Sands and Shale Oil

Near the end of the energy crisis of the 1970s, tar sands and oil shales began to be developed in the United States. This was based on forecasts of crude oil prices reaching $50 to $100 per barrel before the end of the century.

The tar sands are similar to the heavy bitumen used in Venezuela to produce Orimulsion®. The primary tar sands being developed today are in the Athabasca area of Alberta, Canada, about 75 miles north of Edmonton. Here the oily sand mixture is mined using draglines, shovels, and trucks and then heated to remove the heavy oil. While improvements continue to be made, the cost of the oil is still quite high. It also has a fairly low American Petroleum Institute (API) gravity that decreases its value as a refinery feedstock.

Shale oil in the United States exists primarily in Colorado, Utah, and Wyoming, where the estimated reserves exceed 1 trillion barrels and are about twice as large as the proven oil reserves in the Middle East. Shale oil is a fine grain rock containing organic materials called Kerogens. To remove the oil, the rock is heated to about 875° F (468° C), and a retorting process is used to recover the liquids. At 900° F (482° C), Kerogen decomposes into hydrocarbons and carbonaceous residues. When cooled, the hydrocarbons condense into oil. About 25 gallons of oil can be produced per ton of high-quality ore.

Another method for recovering shale oil consists of in situ refining. A void is created under the shale deposit. The shale is fractured using explosives and then heated in place by burning some of the oil. The remaining oil flows to the bottom of the cavity, where it is pumped out.

Shale oil can be burned in a boiler and with conditioning it can be upgraded to a feed material for oil refineries.

4.5.11 Synfuels

Synthetic or process-derived fuels represent an "ultimate" cap on the price of petroleum products. For example, coal can be processed to produce liquid or gaseous fuels as a substitute for petroleum refined products and natural gas.

The primary difference between coal, natural gas, and petroleum products is the ratio of hydrogen to carbon atoms. The lighter the refined product, such as middle distillates, the greater the hydrogen/carbon ratio compared to the heavier oils. Natural gas or methane (CH_4) has four hydrogen atoms for every carbon atom. This compares with coal, which has about 16 carbon atoms for every hydrogen atom. Therefore, the synfuel process must increase the hydrogen/carbon ratio of the coal by enriching it with hydrogen atoms. Coal gasification and liquefaction processes usually strip the hydrogen from steam or natural gas (if available at low cost) to enrich the coal with additional hydrogen.

Synfuel processes are very capital intensive, which has limited their acceptance. Because they are capital intensive, the technology has not been widely used. The use of synfuels in the form of coal gasification is discussed in Chapter 24, Emerging Technologies.

4.6 FOSSIL FUEL PRODUCTION AND CONSUMPTION

If all the oil and coal produced in the world during 1 year was placed in containers that had cross-sectional areas of 1 square mile (2.6 km^2), the height of the crude oil container would be about 0.9 mile (1.4 km) and the height of the coal container would be about 1 mile (1.6 km). While 1 cubic mile for each fuel may seem large, this is quite small when compared to the volume of more than 100 million cubic miles (420 million km^3) for the sedimentary basins in the world.

4.6.1 Comparison of Fossil Fuel Production Characteristics

Fossil fuel pricing and price volatility vary significantly among fuel types. This is primarily a result of the way a particular type of energy source is discovered, developed, produced, and marketed, as well as the degree of interfuel competition.

The finding or exploration cost for coal is low. Often the coal seams are found during drilling for oil, natural gas, or water. In comparison, a large portion of the total costs of producing oil and natural gas is associated with exploration and the success of such exploration efforts.

Although the finding costs are high for oil, the production costs are usually very low. This is usually not the case for coal, particularly for underground mines producing from thin seams.

Table 4-2 compares the primary differences between the production and market characteristics of coal and petroleum fuels. For example, though an oil field may last for several years, each individual well within that field has a relatively limited life. Oil wells have a production versus time curve that often approaches an exponential rate of decline. This is in contrast with the production level of a coal mine that tends to be fairly uniform over its reserve life, assuming steady market demand.

In terms of reserve recovery, there are significant differences between fossil fuels. For petroleum, approximately one fourth to one third of the in-place reserves are recovered. Reserve recoveries for a coal mine vary from 50% for room and pillar underground mining up to 70% to 80% for retreat (support pillar recovery) or longwall mining and 85% to 95% for surface mining methods. Enhanced oil recovery operations, such as flooding a depleted oil field with water or carbon dioxide, can enhance recoveries; however, the costs of extracting the additional oil are comparable to the cost of discovering and developing the initial reserves.

The quality of coal varies significantly among coal basins and between specific seams within each basin. The quality of coal can also vary significantly within a seam. By contrast, the variation in the quality of crude oil within a producing region tends to be minimal.

In the marketing of coal and petroleum products, there are significant differences between the relationships of the fuel producer and the end-user. With coal, for which the major portion tends to be consumed in large electric power generating plants, the relationships between producer and consumer tend to be much stronger and of a longer term than for refined products and natural gas where the consumer gives little thought to the source of the fuel.

In terms of market concentration, petroleum, particularly crude oil, is dominated by a few large producing areas whose domination and market influence have significant impacts on world prices. For example, more than 90% of the worldwide oil and gas reserves and 77% of all oil production is controlled by national oil companies (*World Oil* 1993). This is not the case for coal mining, that has a large number of producers.

Figures 4-13, 4-14, and 4-15 compare the production (left bar) and consumption (right bar) patterns for coal, crude oil, and natural gas for the world's major producing areas.

4.6.2 Coal Production and Consumption Patterns

Figure 4-13 shows the coal production and consumption patterns for geographic regions of the world. The Middle East, that is a major supplier of oil, is conspicuous by its absence in coal production. Similar to natural gas, coal is consumed primarily in the same regions where it is produced.

Table 4-2. Comparison of Coal and Petroleum Production

Category	Coal	Petroleum
Reserve life for mine or well	10–40 years	5–15 years
Production rate	Uniform	Majority of recoverable reserves produced in first few years
Reserve recovery[a]	50–90%	25+ percent
Cost of finding reserves	Low	Very high
Cost of production	Very high	Low
Variability of fuel quality	Very high	Low
Market transactions[b]	Spot and term	Spot
Primary customers	Electric utilities	All sectors of the economy
Concentration/market influence	Very competitive with a large number of independent producers	A few large producers, e.g., OPEC and large government-owned companies

[a]Portion of in-place reserves that are recovered and sold.

[b]Spot transactions involve single or multiple deliveries that are usually completed over a period of <12 months or a fiscal year. Term transactions usually involve multiple deliveries that will be completed over a period of >12 months and are usually governed by the terms and provisions of a long-term contract.

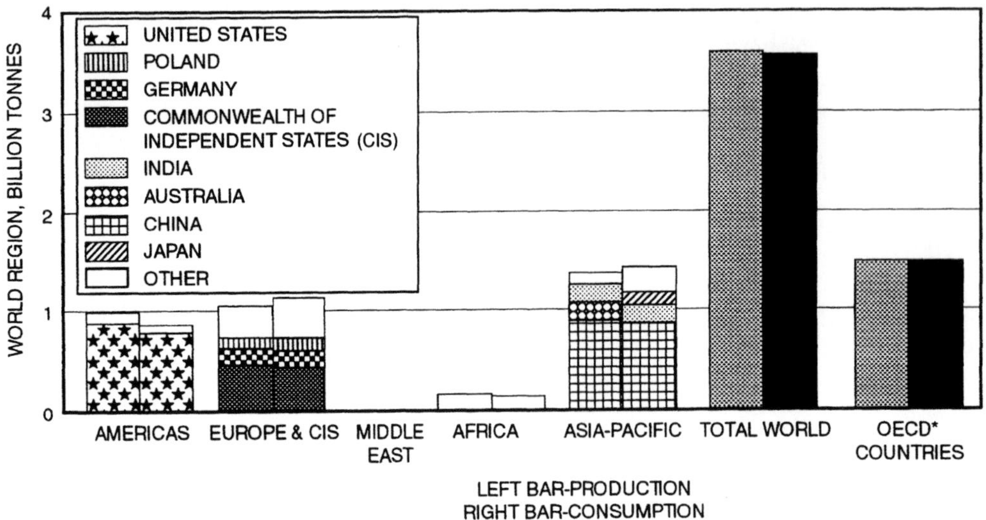

Fig. 4-13. Coal production and consumption—1992. (Adapted from *BP Statistical Review of World Energy*, June, 1993.)

4.6.3 Oil Production and Consumption Patterns

A review of the production and consumption patterns for oil (Fig. 4-14) shows that the largest producing areas are in the Americas (Western Hemisphere) and the Middle East. The largest consuming areas are in the Americas and Europe. There are large imbalances between production and consumption in the Middle East, where consumption is about 20% of production; in Africa, where consumption is about 30% of production; and in the Asia–Pacific region, where consumption is about 225% of production. World crude oil production and consumption were projected to average

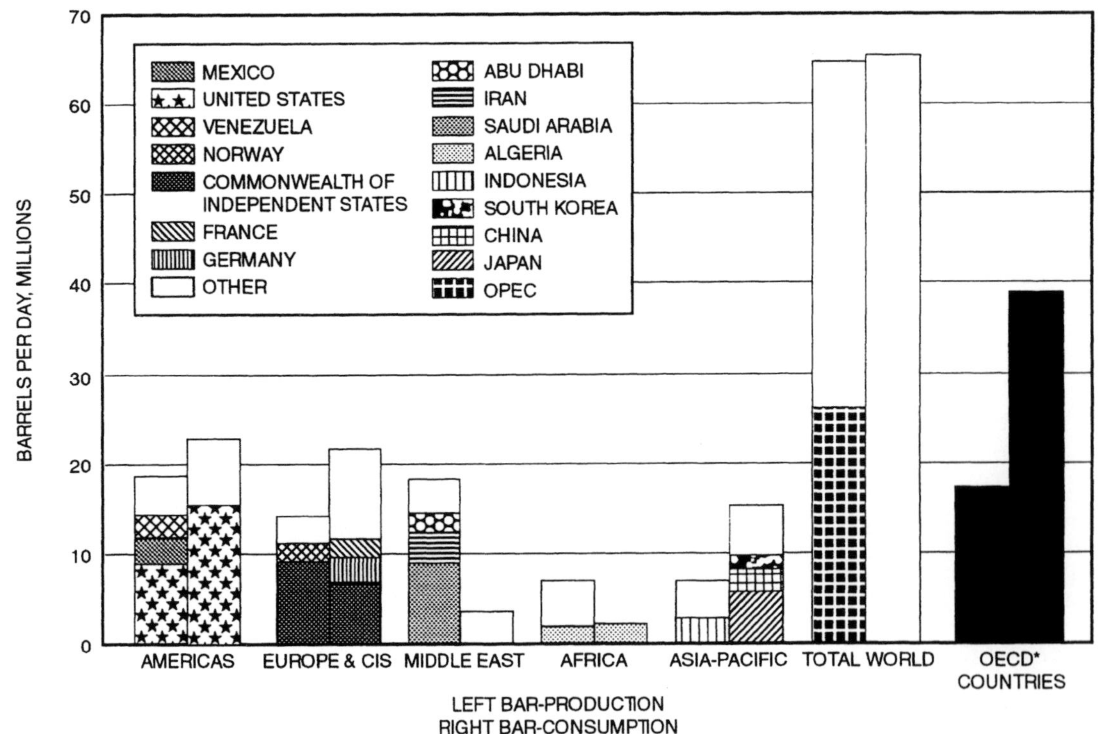

Fig. 4-14. Oil production and consumption—1992. (Adapted from *BP Statistical Review of World Energy*, June, 1993.)

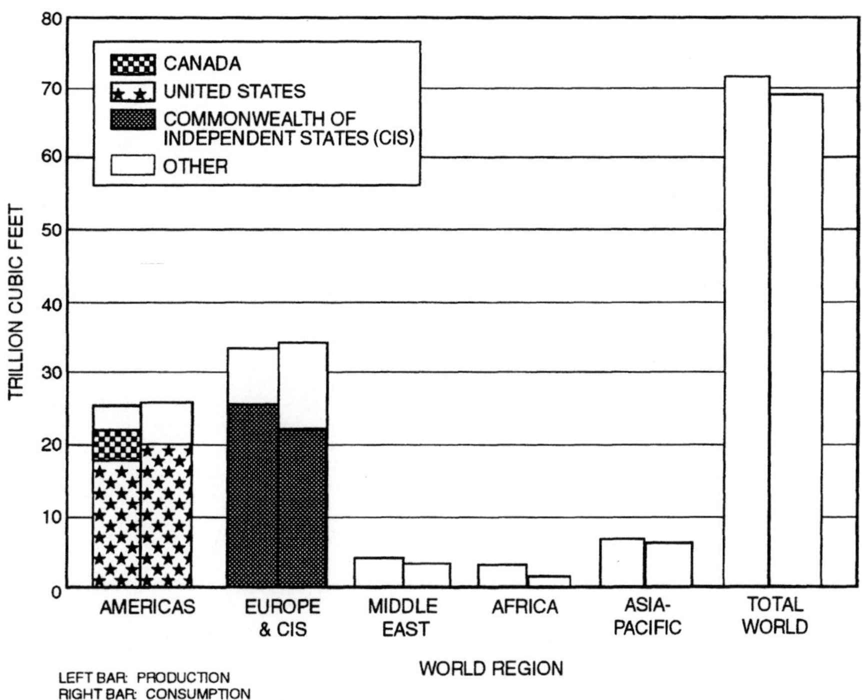

Fig. 4-15. Natural gas production and consumption—1992. (Adapted from *BP Statistical Review of World Energy*, June, 1993.)

about 60 million barrels per day in 1992 (*International Energy Statistics Sourcebook: Oil and Gas Journal Energy Database* 1991).

4.6.4 Natural Gas Production and Consumption Patterns

Figure 4-15 shows natural gas production and consumption for major regions of the world. The major producing areas tend to be Europe and the Americas. This is not because other areas such as the Middle East lack reserves, but because a key requirement for utilizing natural gas is the availability of suitable transportation, usually a pipeline. This is the reason production and consumption patterns within world regions for natural gas are nearly identical. The natural gas tends to be consumed in the regions where it is produced.

In 1992, total world natural gas consumption was about 70 trillion ft^3 (2 trillion m^3). The Americas accounted for about 25 trillion ft^3 (0.71 trillion m^3) with the majority of the remainder being consumed in Europe and the Commonwealth of Independent States. In terms of total energy consumption, the 70 trillion ft^3 of natural gas is energy equivalent to 33 million barrels per day of crude oil production or about 50% of current oil production levels.

4.6.5 Coal Production—United States

Table 4-3 compares coal production statistics for the major coal-producing regions in the United States. By comparing the number and types of mines, their average production levels, and their productivity, much can be inferred about the large differences in the geologic settings of the various coal basins and seams.

While total 1993 coal production east and west of the Mississippi River was comparable at 504 and 424 million tons,[7] respectively, the type of production, surface versus underground, and the scale of mining as measured by annual tons per mine is significantly different.

In the west, production is dominated by large surface mines. In terms of annual production, these mines are, on average, about 22 times as large as eastern surface mines (3,877,000 tons per year versus 180,000 tons per year). It is also seen that these large surface mines in only a few regions account for the majority of total western coal production. In the east, production is more evenly distributed between surface and underground mines.

In the east, where surface mining accounts for 40% of production, surface mines are about 25% more productive than underground mines. In the west, where surface mines account for about 90% of total production, they are about 130% more productive than underground mines.

Since the 1970s, United States production growth has shifted to the west because of its very favorable geology and the low sulfur content of the coals. For example, Wyoming ranked 11th in coal production in 1970, but it has ranked number one since 1988.

[7]In terms of energy produced, it was significantly greater east of the Mississippi River because of the generally higher heating value of eastern coals.

Table 4-3. Profile of United States Coal Production—1993

Region	Number of Producing Mines			Production (1,000 tons)						Productivity (tons/manday)[a]		
	Surface	Underground	Total	Surface	Average per Mine	Underground	Average per Mine	Total	Average per Mine	Surface	Underground	Total
East of the Mississippi River												
Central Appalachia	365	825	1,191	88,230	242	156,311	189	244,603	205	36	27	30
Northern Appalachia	488	158	646	47,415	97	75,486	478	122,901	190	30	29	30
Illinois Basin	103	47	150	49,724	483	54,376	1,157	104,101	694	41	30	35
Southern Appalachia	74	24	98	9,278	125	16,106	671	25,385	259	25	20	22
Pennsylvania Anthracite	90	51	141	3,643	40	396	8	4,039	29	22	11	20
Mississippi Region	2	0	2	3,113	1,557	0	—	3,113	1,557	132	—	132
East of the Mississippi River subtotal	1,122	1,105	2,228	201,403	180	302,675	274	504,142	226	35	28	30
West of the Mississippi River												
Powder River Basin	24	0	24	227,351	9,473	0	—	227,351	9,473	274	—	274
Texas Region	15	0	15	54,879	3,659	0	—	54,879	3,659	71	—	71
Uinta Region	1	22	23	4,709	4,709	28,795	1,309	33,504	1,457	63	58	59
Fort Union Region	7	0	7	32,331	4,619	0	—	32,331	4,619	140	—	140
San Juan River Region	7	1	8	26,500	3,786	134	134	26,633	3,329	69	17	68
Green River Region	5	4	9	13,005	2,601	4,591	1,148	17,596	1,955	58	58	58
Black Mesa Region	2	0	2	12,146	6,073	0	—	12,146	6,073	52	—	52
Washington	2	0	2	4,739	2,370	0	—	4,739	2,370	40	—	40
Hanna Field	1	2	3	2,614	2,614	1,581	790	4,195	1,398	97	43	66
Hams Fork Region	2	0	2	4,065	2,033	0	—	4,065	2,033	42	—	42
Mid-Continent Region	30	2	32	2,806	94	94	47	2,901	91	21	10	20
Raton Mesa Region	1	2	3	980	980	1,458	729	2,438	813	20	23	22
Alaska	1	0	1	1,598	1,598	0	—	1,598	1,598	62	—	62
Bighorn Basin	1	0	1	16	16	0	—	16	16	45	—	45
North Park Region	1	0	1	4	4	0	—	4	4	7	—	7
West of the Mississippi River subtotal	100	33	133	387,744	3,877	36,652	1,111	424,396	3,191	121	53	109
Total/average for US	1,222	1,138	2,361	589,147	482	339,328	298	928,538	393	66	29	45

Source: Adapted from US Mine Safety and Health Administration data.

[a]Productivity is based on all workers except those at central shops, coal preparation plants, and offices.

4.6.6 Coal Markets—United States

Figure 4-16 compares changes in total coal usage patterns for the United States between 1950 and 1992. Total coal usage increased about 80% during this 42-year period. However, the growth has not been uniform over all end-use sectors. In fact, for such sectors as transportation (steam locomotives), residential, and commercial, there has been a significant decline in coal's market share. The primary use of coal went from fairly uniform demand by industry, coke plants, residential and commercial properties, and electric utilities in 1950 to the electric utility sector being almost the only marKet in the 1990s, consuming nearly 90% of all production.

Some producing regions, such as the Illinois Basin and Northern Appalachia, are relatively close to their markets, while others, such as the Powder River Basin in Wyoming, are relatively distant from the markets they serve. It is the tradeoff between mining cost (very low in the Powder River Basin), transportation cost, and coal quality that determines which markets a particular producing region will be competitive in.

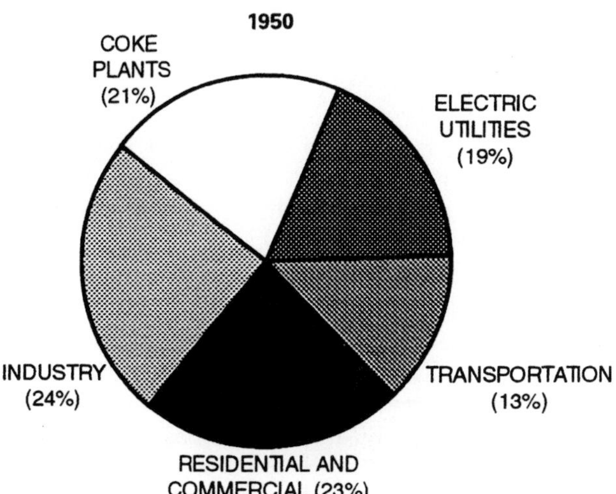

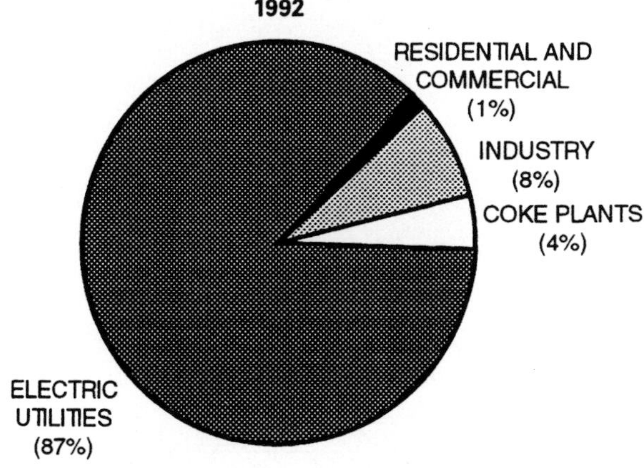

Fig. 4-16. End use patterns for coal usage in the United States in 1950 and 1992. (Adapted from *Annual Energy Review*, EIA, 1992, Quarterly Coal Report, EIA, July-September, 1993.)

By the nature of their siting and sometimes their design, power plants are often restricted in terms of the transportation modes and carriers that are available and, therefore, their fuel sources. In addition, the purchasing strategies (long term contract versus spot purchases), the provisions of the contracts that govern the price of coal, the nature of the supplier/buyer relationship, and transportation can have a significant impact on delivered cost. Based on these factors, it is not unusual to see suppliers in one coal-producing state shipping coal to power plants in another coal-producing state.

In terms of market reach, the most interesting producing region is the Powder River Basin that encompasses northeastern Wyoming and southeastern Montana. This coal is competitive in markets as distant as Georgia, which is about 1,800 rail miles away. In fact, despite its long distance from an ocean port and low heating value, this coal has been shipped to Japan, Spain, and other European countries. This is because of the very favorable geology (thick seams with minimal overburden) which leads to very low mining costs. Also, the favorable terrain for railroads encourages the use of highly efficient coal unit trains. An additional desirable attribute of this coal is its low sulfur dioxide emission potential, which is often less than 1 lb SO_2/MBtu.

4.6.7 Coal Production—International

One significant contrast to coal production in the United States is that in the past, some countries, particularly those in Europe, have subsidized domestic mining operations to boost domestic employment. In the case of Germany and the United Kingdom, the subsidies were significant, sometimes more than the cost of imported coal itself. With a more competitive world and a concern among major trading partners for fairer trade (i.e., fewer government subsidies for domestic producers), these large subsidies are being phased out or eliminated. As a result, demand for imported coals in these countries is likely to increase.

4.6.8 Coal Markets—International

Table 4-4 shows the current makeup of the international steam coal trade between exporting and importing countries. The largest supplier in the international steam coal trade is Australia. The economies of ocean shipping are very favorable compared to land transport, and in most cases, for the same cost, a ton of coal can be transported about 10 to 20 times the distance over the ocean that it can over land by rail. The economies of scale are particularly noticeable for large Capesize vessels that are capable of transporting more than 120,000 tonnes.

In the international steam coal trade, the FOBT (free on board vessel and trimmed) cost, not distance, is the primary

Table 4-4. International Steam Coal Trade (Thousand tonnes), 1992

Major Importing Country	Major Exporting Country					
	Australia	South Africa	United States	Indonesia	China	Colombia
Japan	30,662	2,530	2,561	4,343	3,863	294
Taiwan	4,826	5,197	2,840	3,306	2,339	0
Netherlands	4,167	3,523	3,911	805	178	1,168
France	2,236	2,501	3,397	0	462	1,044
Hong Kong	3,605	3,288	323	2,396	1,990	0
Germany	263	5,360	651	280	163	554
Spain	167	6,329	1,575	546	0	767
Canada	0	0	8,925	0	0	0
Denmark	1,314	505	3,476	76	57	1,899
Belgium	1,172	3,016	1,842	0	296	258
South Korea	0	2,281	674	137	3,194	0
Italy	133	3,251	2,866	0	307	602
United Kingdom	1,412	591	1,363	0	177	2,368
Other	8,337	6,817	4,155	2,066	877	2,400
Total	58,294	45,189	38,559	13,955	13,903	11,354

Source: Adapted from *Coal Year 1993* and *Coal Year 1994* by International Coal Report.

contributor to delivered cost economics. Therefore, the international steam coal trade tends to be dominated by low-cost producers with low labor costs and/or low capital cost per ton of production resulting from favorable coal geology and/or short rail or truck distances to port.

4.6.9 Coal Production Technologies

Throughout the world, mining conditions (geology) and the type of mining that is suitable for reserve development varies from one coal-producing region to another.

In most countries, the exploitation of coal reserves historically begins on a very small scale, employing low capital investment underground mining technologies to serve local markets. As production expands and producers seek more distant markets, FOB mine or FOBT ship cost competitiveness is of greater importance. This usually results in the employment of greater capital investment and a focus on larger reserve blocks that are situated near major transportation corridors. As distance to market increases, there is greater market pressure to reduce mining costs to remain competitive. This is particularly true for areas where transportation costs are relatively high because of mode, terrain, or lack of competition.

There are basically two types of coal mining: surface (i.e., open cut or open pit) mining and underground (deep) mining. These different types of surface and underground mining operations are illustrated in Fig. 4-17. The type of mining selected depends on several factors. One of the most important is the stripping ratio. For a given prospective mining area, there are various geologic and economic factors that must come together to maximize reserve recovery, minimize environmental disturbances, and allow the use of proven technologies to minimize costs and maximize profit potential.

Mining costs are influenced by the number of seams and their structure such as thickness, pitch, depth of overburden, thickness of interburden, gassy nature (methane concentration), hydrology (potential for flooding and pumping requirements), composition of overburden, roof and floor and their stability, proximity to old mine workings, and the contiguous nature of the reserve block.

In general, underground mining methods would be employed if the depth of overburden was > 200 ft (60 m). However, multiple seam open pit mines can be up to 1,000 ft (300 m) deep. Mining could also begin with a surface operation and when the depth of overburden becomes significant, convert the operation to underground mining.

4.6.9.1 Surface Mining. One of the primary determinants of the economics of surface mining is the stripping ratio. One of the reasons for the declining amount of production from surface mines east of the Mississippi River is that the reserve blocks with low stripping ratios have been depleted.

One of the major features of surface coal mining is that it is less labor intensive and can accommodate geologic unknowns and uncertainties better than when these conditions are encountered in underground mining. For example, if a fault goes undetected in an underground mining operation and production advances to this fault, it can present a significant technical and economic challenge to future mine development since the seam may be displaced several tens of feet above or below current mining operations. Even minor displacements can result in a significant increase in costs and less safe working conditions. With surface mining there are fewer impacts. Encountering the same fault usually means that a little more or less overburden must be removed.

Because of the more selective nature of surface mining, it is usually possible to recover a greater percentage of the in-place reserves (90%) and reduce contamination of the coal by roof and floor material (out-of-seam dilution). Therefore, surface mining, by its nature, lessens the need for coal beneficiation (coal washing).

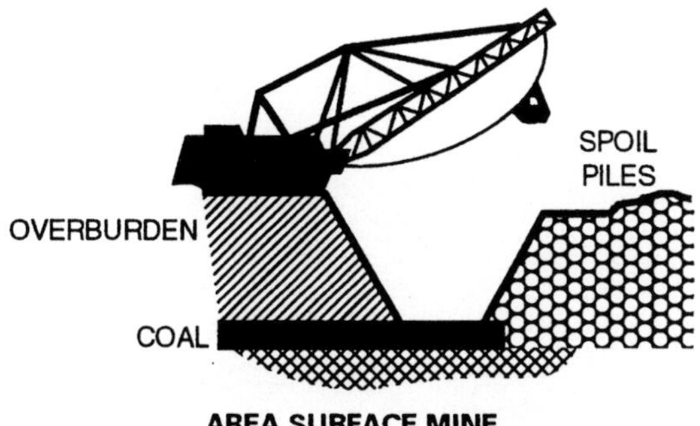

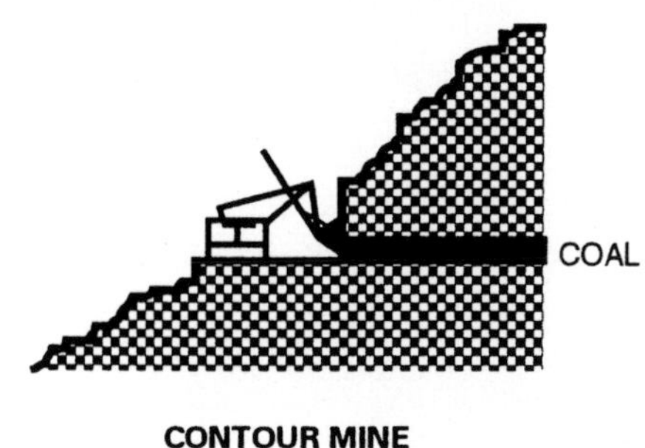

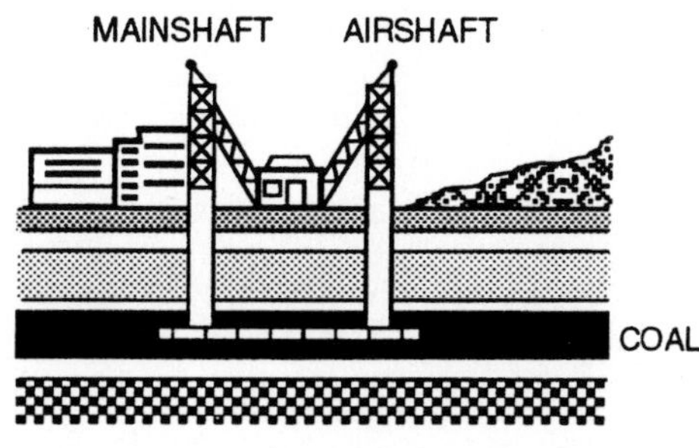

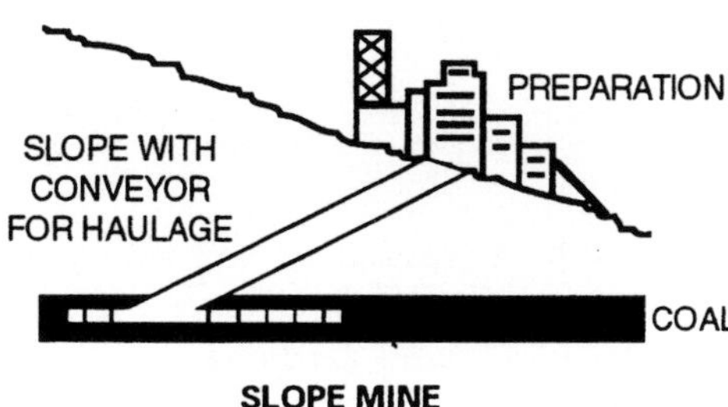

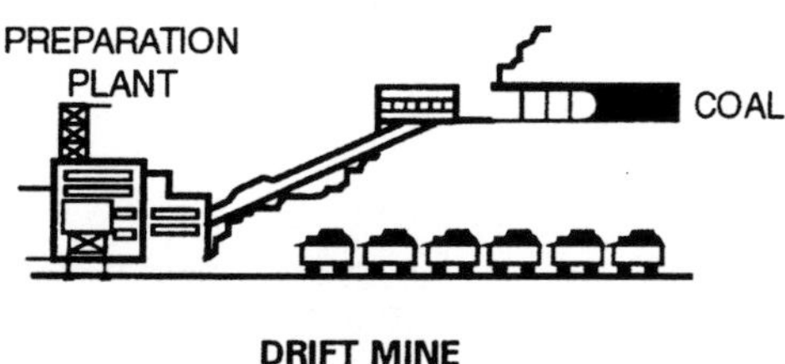

Fig. 4-17. Types of surface and underground mines. (From EIA, 1985, *Coal Data: A Reference.*)

There are five primary types of surface mining: mountain top removal, contour mining, augur mining, area mining, and open pit mining. A mountain top removal operation consists of removing all of the overburden above the coal (top of mountain) to expose the seam.

Contour mining follows the outcrop of the seam around the side (along the contour line) of a hill. When the cost of removing the overburden becomes prohibitive, augers can be used to drill into the seam about 200 ft (60 m) from the working face to remove additional coal. A small underground drift or punch mine can also be developed to remove a portion of the remaining reserves.

Area surface mining is done on relatively flat ground and consists of cuts about 100 to 200 ft (30 to 60 m) wide with the spoil being placed as fill in the previously mined-out area as shown in Fig. 4-17, Area Surface Mine. Open pit mining is used where the coalbeds have significant dip and are relatively deep or where coalbeds are greater than about 30 ft (9 m) thick.

Surface mining begins with prestripping the topsoil in layers based on their fertility. The topsoil is removed and stacked out typically using scrapers so it can later be replaced as a part of the reclamation program.

The primary production operation in surface mining is the removal of the overburden. The primary pieces of equipment used are draglines, power shovels, and bucket wheel excavators.

Figure 4-18 shows a truck and shovel operation of the type that has been popular in developing the large western surface mines, eastern contour, and mountain top removal mines. A drill, shown in the background of Fig. 4-18, drills the holes into which explosives are placed to loosen the overburden and in some cases remove the overburden (blast casting). Some large surface mines use draglines to remove the overburden. The larger draglines have bucket capacities that range from 80 to 175 cubic yards (61 to 134 m^3).

The large draglines, stripping shovels, and bucket wheel excavators are powered by electricity. Since they are the key piece of equipment in a surface mining operation, they strip ahead of coal removal to uncover several months of coal production so as to not curtail production in the event of a major component failure.

After the coal seam has been exposed, if it is fairly thick, it is first drilled and blasted to make it easier for loading operations. Thin seams may be loosened for loadout using a ripper mounted on the back end of a large bulldozer. The coal is then loaded with rubber tired front-end loaders or tracked shovels (Fig. 4-18). The coal is then loaded into diesel-powered trucks, with some having dump bed capacities ranging from 100 to 240 tons. The size of these trucks is impressive when one considers that a railroad hopper car typically holds about 100 tons of coal. Sometimes conveyors are used to transport the coal from the pit to the mine's coal preparation plant or loadout facilities.

After the coal has been removed, the disturbed land is reclaimed by restoring the surface to its approximate original

Fig. 4-18. Truck and shovel operation. (From Bucyrus Erie Co. Used with permission.)

contour and replacing the indigenous soils in the same horizon sequence from which they were removed. If the land is reclaimed properly, its productivity in terms of crop yield usually is not impaired. In some cases, the productivity of the soil has been improved through the addition of appropriate nutrients during reclamation.

During the 50-year period between 1930 and 1980, coal mining in the United States disturbed 2.7 million acres (4,200 square miles) (Johnson and Paone 1982). This compares to 4 million acres disturbed for airports and the 77 million acres in the National Park System (National Coal Association 1984).

4.6.9.2 Underground Mining. If the depth of the overburden is several hundred feet and the seam thickness is minimal (high stripping ratio), underground mining operations are usually employed. Depending on the type of opening used to reach the coal seam, the underground mine is called a drift, slope, or shaft mine. Figure 4-17 illustrates these different types of underground mines.

The drift mine is employed where the coal seam outcrops on the side of a mountain. Surface contour mining around the side of a hill or mountain may have been employed prior to the development of the drift mine. Entry to the drift mine is horizontally from the point where the coal seam outcrops to the surface.

For depths of coal up to about 300 ft (90 m), slope mines can be employed. The slope entry consists of a sloped tunnel excavated from the surface to the point where it intersects the coal seam. The angle of these slopes is usually about 17 degrees and is related to the degree of incline for the conveyor that can be used to remove the coal from the underground operations. For depths $>$ 300 ft and up to a few thousand feet, shaft mines are usually employed. A shaft consists of a vertical opening from the surface to the coal seam, with equipment and coal being transported out of the mine using a mine hoist.

After the seam has been reached, three basic mining methods are employed: conventional, continuous, and longwall mining. Both the conventional and continuous mining methods leave pillars of coal to support the weight of the overburden and prevent subsidence. The amount of in-place reserves recovered using continuous mining methods ranges from a low of about 50% to upward of 70% if some of the pillars are extracted during retreat mining operations and the surface is allowed to subside.

Continuous mining uses a machine called a continuous miner that has a rotating cutter head, as shown in Fig. 4-19. The rotating head is advanced into the coal seam and "chisels" the coal from the seam. The continuous miner loads the coal into a shuttle car that usually hauls it to a nearby conveyor system for transport to the surface.

The other method of underground mining operation is called longwall mining. It is illustrated in Fig. 4-20. This method of underground mining is favored where there are poor roof conditions, such as soft shale and unconsolidated soap stone that can be difficult to hold up using roof bolts. Since the objective of longwall mining is for the roof to collapse after the coal has been removed, the size of the room or panel of coal removed can be increased considerably. Whereas a width of 20 ft (6 m) of coal might be removed using conventional or continuous mining methods, longwall mining systems remove panels that are 400 to 1,000 ft (122 to 305 m) wide and several thousand feet long.

The longwall mining unit consists of three primary pieces of equipment: the longwall rotating drum shear, the face conveyor that transports the coal cut off the coal face to conveyors along the side of the panel being removed, and hydraulic supports called shields or chocks that temporarily support the overburden after the shear has removed the coal from under it. The shear cuts up to 36 in. (91 cm) of coal from the face during each pass. The roof material (gob) is allowed to fall behind the support shields as the longwall unit advances into the seam.

Where geologic conditions such as minimal seam dip and pitch, uniformity of thickness, and competence of the floor (to support hydraulic supports) exist and the roof caves evenly, the economics of longwall mining can be quite favorable. Coal recovery as a percentage of in-place reserves often reaches 80%. While the capital investment in the longwall section is several times that of the continuous miner section, the greater recovery and enhanced productivity of labor and capital (depending on geologic conditions) can be quite favorable.

Fig. 4-19. Continuous miner. (From Joy Technologies, Inc. Used with permission.)

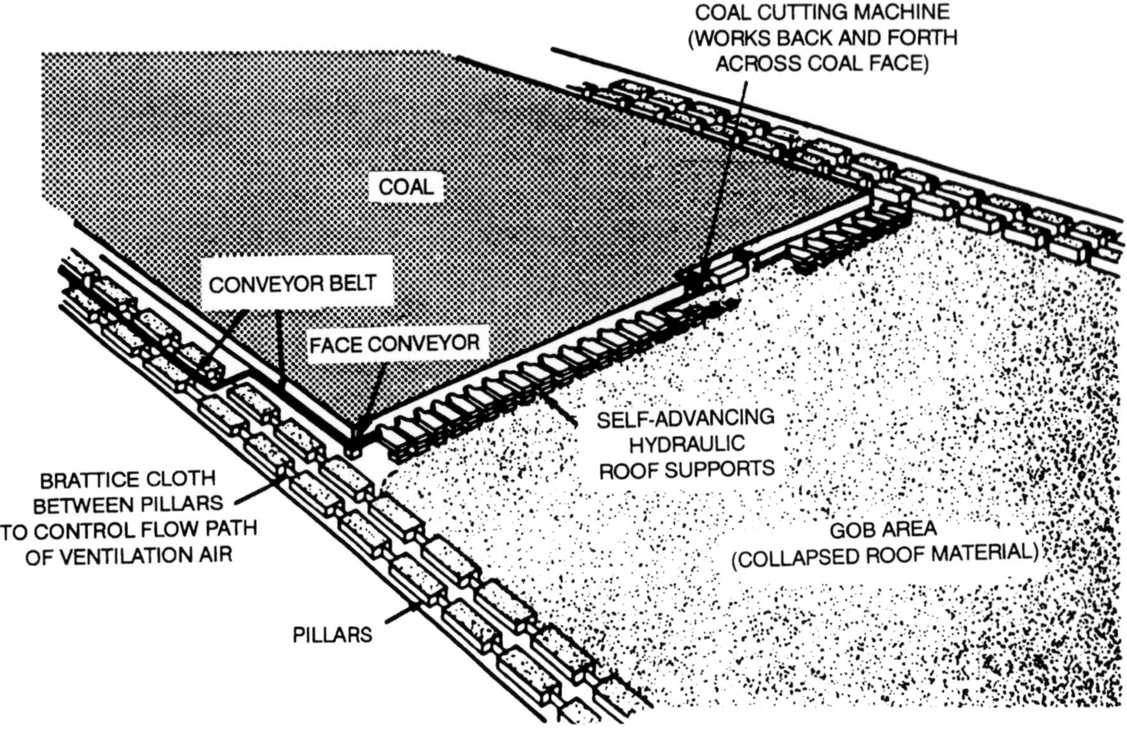

Fig. 4-20. Longwall mining system. (From EIA, 1985, *Coal Data: A Reference.*)

In the United States, about 65% of underground coal production is performed using continuous mining machines, 30% using longwall systems, and only 4% using conventional mining methods.

While both surface and underground mining present numerous challenges, the challenges are probably greater for underground mining. Not only must underground coal deal with less recovery of in-place reserves than surface mined coal, but raw or run-of-mine coal is often not salable without beneficiation to remove out-of-seam impurities (diluent) that contaminate the coal during mining operations. Offsetting these obstacles, the deeper underground mined coal may have a higher rank (less moisture and higher Btu) than surface mined coal. Where surface and underground mined coals compete in the same markets, their total cost (adjusted for the quality of the coal being sold) from the mine to the end-user must be competitive.

4.6.10 Oil and Natural Gas Production Technologies

Since the exploitation of oil began in Romania and the United States in the 1850s, significant technological advances have occurred in the exploration, development, and production of oil and natural gas.

In exploiting new reserves, the primary costs, aside from drilling success rates, are lease acquisition, geological and geophysical investigations, and drilling. These costs vary inversely with the size of the field discovered and exponentially with the depth of an actual discovery.

Natural gas, after coal, is the second most important fossil fuel used in electric power generation. Of the total natural gas produced each year in the United States, there are two primary sources: nonassociated gas, that comes from wells that produce gas only; and associated or casinghead gas, that is natural gas dissolved in the crude oil and produced with the oil. In 1989, associated gas amounted to 25% of total gas production.

Figure 4-21 shows the development of natural gas fields in the United States since the early 1900s by size and depth of field discovery. It is seen that as technology and higher market prices allow greater depths to be exploited, there is an increase in the average size of gas field discovered that peaks after a couple of decades of drilling activity at the new depth. The data show that the long-term trend has been toward the discovery of smaller natural gas fields and wells that are producing from greater depths. For example, it appears that the mean field size peaked in the 5,000 to 10,000 ft (1,500 to 3,000 m) depth interval at 672 trillion Btu. Exploration at greater depths, although significantly more costly, appears to be producing smaller payoffs. There is a general perception that with exploration at greater depths, the amount of gas discovered would be significantly greater, which would compensate for the significantly greater drilling cost.

The mean field size discovered has decreased from a high of 404 trillion Btu in the 1920s to 7 trillion Btu during the period from 1980 to 1982 (Energy Information Administration [EIA], An Economic Analysis of Natural Gas Resoures and Supply). For perspective, based on a mean field size of 7 trillion Btu, a combined cycle plant with a generating capacity of 460 MW(e) and a net plant heat rate of 7,100 Btu/kWh

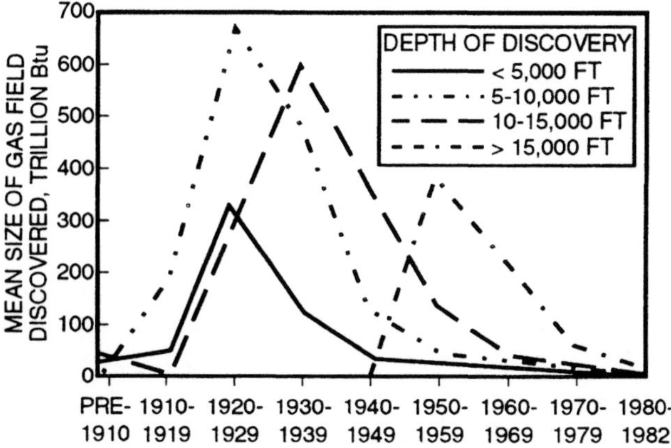

Fig. 4-21. Trends in the size of natural gas field discoveries by depth. (From EIA, 1986, *An Economic Analysis of Natural Gas Resources and Supply*.)

(7,500 kj/kWh) (higher heating value basis) operating at a capacity factor of 90% would consume the natural gas from one new field discovery in about 3 months. Even if it is assumed that the most recent mean field size of 7 trillion Btu has not accounted for a significant amount of development drilling, and a typical reserve appreciation factor[8] of about 7 is applied, the amount of exploration and drilling activity to supply the fuel needs of such a facility is considerable.

There is a dynamic interaction between field size, depth, and geology of new gas field discoveries and the technologies and market price levels under which the discovery takes place. This interaction is related to the uncertainty regarding the future price of gas required to support the probable cost of discovering and developing a new field. There is a constant struggle to develop new technologies such as three-dimensional seismic and horizontal drilling to offset the rising cost impacts of resource depletion.

In 1978, onshore nonassociated natural gas supplied nearly 80% of the total gas needs of the United States and offshore nonassociated natural gas wells supplied about 5%. The remaining 15% was associated with onshore crude oil production. By 1988, only about 57% of the total natural gas supply came from onshore while 20% came from offshore sources (EIA, Natural Gàs Production Responses to a Changing Market Environment 1978 to 1988). The switch from onshore to offshore and unconventional deep gas supplies is important, since it affects trends in average well life and deliverability characteristics.

Deep wells, because of their higher bottom hole pressure and the need to maximize return on investment, tend to have very high initial rates of production that decline exponentially such that within a 3- to 5-year period, well production rates are < 50% of the initial rates of the first 2 years. This phenomenon also characterizes Gulf Coast offshore wells, that tend to have short production lives and rapid declines in deliverability.

4.6.10.1 Deliverability. The term deliverability refers to the quantity of gas that can be produced from a well, reservoir, or natural gas field over a given period of time with consideration of the restrictions imposed by pipeline capacity and regulations regarding well flow rates. For a given well, the flow rate is determined by bottom hole pressure, size of the controlling orifice, and back pressure as determined by compression capability on the receiving pipeline system.

4.6.10.2 Life of Natural Gas Wells The typical gas well tends to be about 6 years old when it is abandoned. Wells less than 3 years old usually supply 40% of the delivery production capability (EIA, March 1993). More than 80% of the offshore wells have service lives of < 5 to 10 years, with the typical offshore well being a major producer for only 2 or 3 years. This compares to wells in the Oklahoma portion of the Anadarko Basin which have a productive life of 15 to 20 years. In general, decline rates range from 5 to 30% of the previous year's production. Just to maintain the United States natural gas reserve-to-production ratio at its current level of about 9 years, nine units of additional reserves have to be added for each additional unit of annual production required.

4.6.10.3 Production of Refined Products. The type and quantity of the refined products produced will vary with the type of crude refined and the complexity of the refinery. Heavy crudes (API gravity[9] < 25 degrees) will produce fewer light fractions such as gasoline than will lighter crudes (API gravity > 25 degrees). The principal products produced by the refinery are gasoline (25% to 50%), jet fuel or kerosene (5% to 20%), middle distillates (15% to 25%), and residual oil (20% to 50%). The remaining products consist of motor oils, waxes, asphalt, and petroleum coke. The primary revenue producers for the refinery are gasoline and middle distillates such as highway diesel fuel and heating oils.

The first step in the refining process is to heat the crude in a distillation column. The hydrocarbons in crude oil boil at different temperatures. By taking advantage of the different boiling temperatures and by condensing these vapors, the barrel of crude oil can be divided into a series of products (fractions) that range from the lighter butane through the heavy oils, the material that boils off at > 800° F (> 427° C).

The heavy oils are also referred to as asphaltines, residuum, bottom of the barrel, residual oil, No. 6 oil, or bunker C oil. The heavy oil is usually sold to electric utilities, large industrial users, or as a bunker fuel for ships. Because of their high density (low API gravity) and relatively high pour

[8]The reserve appreciation factor is the ratio of the total gas ultimately recovered from a field divided by the recoverable reserves estimated at the time of initial field discovery.

[9]
$$°\mathrm{API} = \frac{141.5}{\text{Specific gravity } 60/60°\,\mathrm{F}} - 131.5$$

point, heavy oils must be heated to pump them from onsite storage tanks to the burner.

4.6.10.4 Petroleum Coke. After crude oil has been distilled, the heavier fractions or "bottom of the barrel" remains. This material is often fed to vacuum distillation units or a coker. Most of these processes try to recover more of the valuable light ends (gasoline and middle distillates) that remain in the residual oil.

The heavy oil is heated in the coker to a temperature of 900 to 1,000° F (482 to 538° C). The lighter fractions rise to the top of the coker and are drawn off. The heavier product remains as a solid black coal-like substance. A high-pressure water jet is then used to cut the solid coke from the drum. The process just described produces a delayed coke that is sometimes referred to as a shot coke or sponge coke based on its appearance as an agglomeration of shot pellets (1/8 inch diameter) or a solid with spongelike voids.

Another form is fluid coke, which is produced in a fluidized bed combustion process. The fluid coke consists of very hard, small spherical particles (about 200 microns in diameter) with very low volatile matter content.

4.7 FUEL TRANSPORTATION SYSTEMS

Several modes of transportation are used in transporting fuel from the mine, natural gas field, or refinery to the power plant. For land movements, the main transportation technologies are conveyor, truck, rail, and natural gas or petroleum products pipeline. For water movements, barges, lake vessels, and oceangoing vessels are used. The focus in this chapter is on the transportation of coal rather than natural gas or petroleum products because of the greater number of logistical variables that must be dealt with.

Figure 4-22 shows the modes used to transport coal to end-users. As can be seen, nearly 60% of the movements in the United States are by railroad. For the international steam coal trade, the majority of movements are by large oceangoing dry bulk carriers called colliers.

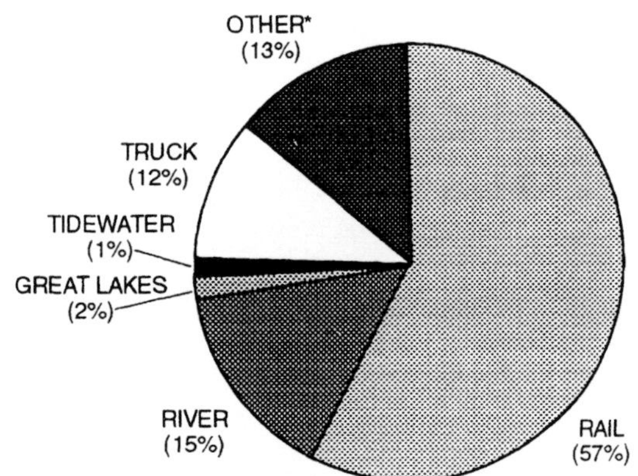

Fig. 4-22. Modes of coal transportation in the United States, 1992. (From EIA, Quarterly Coal Report, Oct-Dec, 1992.)

Fuel transportation costs are subject to significant variation, depending on the distance, terrain, annual volume, degree of equipment utilization, potential for backhaul movements, type of fuel, energy density, and the number of competitive alternatives. Governmental regulations may also affect operations, costs, and the tariff structure for the movement. In general, water movements are less expensive than rail, and rail is less expensive than truck. These cost differences are accentuated as distance and annual volume increase.

4.7.1 Rail

Coal is the most important commodity by tonnage that is carried by the railroads. In 1992, coal accounted for 40% of railroad tonnage and 23% of total revenues (*Railroad Facts* 1993).

As the size of power plants and mines increased and environmental regulations limited potential sulfur dioxide emissions, the sourcing of coal for power generation took on greater significance. In the mid-1970s, the average coal rail shipment was a few hundred miles long. However, with the development of the low-cost western coal reserves since then, rail transport distances have increased significantly. As distances increased, ways were sought to reduce costs. This usually resulted in bigger movements and ultimately led to dedicated unit train operations.

For large annual volumes and long distances, unit trains provide significant economies of scale compared to single- or multiple-car movements of coal that are part of a mixed freight train and are transferred from one way-train to another until the shipment reaches its destination. Unit trains were also motivated by the interest that electric utilities showed in developing nuclear power plants in the late 1960s and 1970s. If the coal and rail transportation industries were going to maintain their fair share of the electric power generation market, they had to become more competitive. A unit train consists of 90 to 115 open-top coal cars that are dedicated to a single cargo type such as coal and a continuous transportation cycle between a single origin and destination.

Each coal car in the unit train contains 90 to 115 tons of coal. Steel hopper cars have less capacity than aluminum rotary dump cars. Such a train is usually pulled with three to five locomotives, depending on the terrain, and has an overall length that exceeds 1 mile (1.6 km).

Figure 4-23 is a photograph of a unit train being unloaded at a power plant site. When the unit train reaches the power plant site, there are two primary means of unloading the cars. One method requires that each car be rotated 135 degrees about its coupling axis to discharge the coal into a hopper that then feeds the coal to a conveyor belt for stockout in the coal yard. The rotary car dumper is shown in Fig. 4-23 in the left corner below the oil storage tanks. Another method is for the train to pass over a trestle where the hopper doors be-

Fig. 4-23. Unloading a coal unit train at a power plant.

neath each car are actuated to dump the coal into a surge bin or storage pile. Trestle facilities are usually found in warmer climates where there are few problems with coal frozen to the sides of the hopper cars. If the coal particles tend to freeze together or stick to the sides of the cars, it is sometimes necessary to spray them with de-icers or for the train to pass through a thaw shed.

A number of factors impact unit train costs. Two congressional acts were passed to promote the fiscal and physical rehabilitation of the nation's railroads. These were the Railroad Revitalization and Regulatory Reform Act of 1976 and the Staggers Rail Act of 1980. The Staggers Rail Act also gave railroads much greater rate-making freedom. For example, it allowed them to enter into long-term contracts with shippers and to negotiate rates reflective of the economies realized through large movements of goods and bulk materials such as coal in dedicated unit train movements.

When coal cars are purchased for a particular dedicated movement, consideration is given to the route—particularly distance, curvature, gradient, traffic density, and annual quantities to be transported. The type of coal, particularly its sulfur content, density, and size consist are also important. These parameters can play a role in coal car design, such as the benefit of steerable trucks for trains routed through mountainous terrain with high curvature and gradient and the type of car material, whether aluminum or steel. Since the total weight of the car on the track is limited, lightweight aluminum cars have greater coal-carrying capacity than do heavy steel cars. Depending on the relative cost of aluminum cars versus steel cars, this can reduce the cost of coal transportation since the unit train can transport more coal with the same number of cars and locomotives.

One of the most important parameters in a unit train operation is the cycle time the railroad proposes for the movement. In developing the cycle time, a key component is the free time allowed for loading the coal at the origin and unloading it at the power plant before demurrage (an equipment detainment charge) is assessed by the rail carrier. The cycle time establishes the amount of coal that can be transported annually by the unit train and hence the productivity of the unit train (train set). This impacts the number of train sets and spare cars required to transport the coal. This has a direct impact on the required capital investment and some impact on the maintenance cost associated with the movement.

Maintenance cost is usually low during the first few years of operating a new train set but increases significantly after the coal cars attain an age of about 10 years. Poor load distribution in the car and overloads can lead to excessive wheel wear. Other contributors to excessive wheel, coupling, and draft gear[10] wear are excessive train handling forces and sharp curvature of the rail in locations where the railroad has no wheel flange lubrication systems.

4.7.2 Barge

Coal transportation by barge is common on all of the inland waterways, but particularly on the Ohio and Mississippi Rivers and their tributaries. The main constraint to barge transportation on the Ohio River and the upper Mississippi (northward of Cairo, Illinois) has been the limitation of the locks to handle barges and tow boats in a timely manner. Some key congestion points are being renovated and capacity throughput is being improved to reduce locking times.

The typical river barge has a capacity of about 1,500 tons. These barges are gathered together as a flotilla to form a "tow" that is pushed by a tow boat. The size of the tow constructed is primarily constrained by the river's channel width and the ability to accommodate the tows in the locks. In some cases, it is necessary to break the tow to pass through the lock. The lower Mississippi River, which has no locks, can accommodate tows with up to forty 1,500-ton barges. On the other hand, the upper Mississippi, Illinois, Ohio, and Tennessee rivers would be restricted to tows with about 15 barges.

4.7.3 Truck

Trucks are usually competitive with railroads for distances up to about 50 or 100 miles (80 to 160 km), depending on the origin and destination points and the availability of rail transportation between them. However, even for short movements of under 50 miles and large annual volumes, "optimization" of short shuttle trains of about 25 cars with efficient loading and unloading operations in a dedicated unit train movement would be more economic than truck transportation. Truck transportation is typically utilized where power plants are located in close proximity to a number of small mining operations.

Trucks are also used to haul the coal from small mining operations, particularly in the Appalachian region, to a central coal washing and processing plant that would be equipped with a unit train loadout facility. In such applications, the trucks provide an initial "gathering" service as an extension of the mining operation.

For minemouth power plants, off-highway trucks can be used to transport the coal from the mine to the power plant coal storage area. These trucks have capacities significantly greater than the 22 to 25 tons of highway semitrailer trucks that are limited to a gross weight of 80,000 lb on the United States interstate highway system.

4.7.4 Conveyors

Conveyor systems are usually associated with minemouth power plant operations, particularly those that are served by large nearby underground mines. The conveyors are used to transport the coal directly to the power plant or more often to a coal beneficiation plant where some of the diluent and pyritic sulfur can be removed. Conveyors have been used over significant distances, up to 15 miles (24 km), where the terrain between the mine and the central coal processing facility or power plant is quite rugged.

4.7.5 Lake Vessels

As shown in Fig. 4-22, some coal moves on the Great Lakes, that are usually open for shipping between April 1 and December 15. This is about the same navigation season that is available on the upper Mississippi River.

Virtually all of the coal shipped on the Great Lakes moves in self-unloading vessels. Such vessels usually consist of a conveyor belt that runs beneath the cargo holds of the vessel. The coal is fed onto the conveyor belt and then conveyed to shore via a boom conveyor. Unloading rates as high as 5,000 tons per hour are possible. These vessels may also be equipped with bow thrusters that enable them to moor without the need for an attending tug. A typical Great Lakes self-unloader has a draft of about 26 to 28 ft (7.9 to 8.5 m) and a cargo capacity of about 25,000 to 30,000 tons.

4.7.6 Ocean Vessels

Ocean vessels are commonly categorized by size classes or cargo carrying capacity. As used in the shipping industry, the deadweight tonnage (dwt) carrying capacity of a ship includes cargo, bunkers (fuel), and stores that a vessel is capable of carrying when floating at its load line. It is usually stated in long tons of 2,240 lb (1,016 kg) or in metric tonnes of 2,205 lb (1,000 kg).

There are four common size classes: Handysize, Panamax, Capesize, and Large Size. Handysize vessels are usually those with a cargo capacity of about 35,000 dead weight tons (dwt). The Panamax class, as its name implies, is the maximum size vessel that can be accommodated in the Panama Canal. This limits vessels to about 65,000 dwt. Capesize vessels are too large to transit the Panama Canal and must therefore be routed around the Cape Horn of South America. These vessels generally range in size from 75,000 to 150,000 dwt. Large Size vessels are usually those > 150,000 dwt. The vessel sizes considered for a movement depend on

[10]Draft gear is the double-acting "shock absorber" that cushions draft (pull) and buff (compression) forces.

the characteristics of the ports of origin, the restrictions enroute, and limitations at the port and terminal serving the power plant.

The variation in fully loaded draft with vessel size would tend to be as follows:

Class	Size	Approximate draft
Handysize	35,000 dwt	11.1 m
Panamax	60,000 dwt	12.8 m
Capesize	125,000 dwt	15.0–17.5 m

The required under-keel clearance and channel width in port areas are a function of the ship's size and port authority restrictions.

The self-unloading vessel uses an onboard conveying system to discharge its bulk cargo. It can moor alongside a few mooring dolphins and discharge its cargo into a relatively inexpensive hopper that feeds a high-speed conveyor. Alternatively, it can discharge the coal directly onto the coal storage pile. As with coal unit trains, the amount of time required to load and unload the vessel can be a significant factor in the economics, particularly for short movements. The self-unloading vessel also has an economic advantage over the larger dry bulk carrier when the transport distances are relatively short and annual coal requirements do not justify the large capital investment in a berth and ship unloaders that are idle for a great deal of the time. For example, the initial development of a power plant project might employ self-unloaders for the first and second units, and then with full-scale development of the project, revert to gearless, larger dry bulk carriers to provide greater fuel sourcing flexibility and lower cost transport from more distant sources. The viability of the use of self-unloaders depends on the need for diversity of supply, transportation distances, and the general availability of self-unloading vessels committed to such movements.

The cost distribution between capital, fuel, and crew varies with vessel size since the economies of scale are significant. One factor that influences the cost of ocean transportation is the amount of time required to unload the vessel. Nonfuel vessel costs including capital charges continue whether the vessel is in port or at sea. If the time required to unload the vessel can be reduced, it should be reflected in the quoted rate. The amount of this credit is dependent on the level of demand for dry bulk carriers versus available supply. For long-term charters, vessel utilization and hence cycle time also influences the number of vessels required to transport the coal.

The desire to reduce the amount of idle time in port for short movements spurred the development of dry-bulk self-unloaders with their very high rates of discharge. These high discharge rates can make self-unloading vessels very competitive for short movements.

The design of the planned port, terminal, and vessel unloading equipment is influenced by the balance between the initial cost of dredging and then maintaining a deeper and wider channel, heavier berth structures, greater capacity of the unloading equipment, and the longer reach for the unloader boom for larger deeper draft vessels versus the reduced transportation costs that can be realized from the use of larger vessels. In addition, prevailing winds, currents, and sea conditions, including tidal variation, will affect the number and size of tugs and the design of the berth and ship unloaders.

Computer models are often used to evaluate the large number of variables in transocean shipping. The major ones include vessel operating costs, unloading rates and costs, onshore handling, and inventory costs associated with the use of different types and sizes of vessels.

4.7.7 Pipelines

Pipelines have been used primarily to transport natural gas and refined products. However, they have also been used to transport other commodities such as coal in a slurry form. Two coal slurry pipelines have been built: one was operated between Cadiz and Cleveland, Ohio, a distance of 108 miles by the Consolidation Coal Company from 1957 to 1963; and the current operating coal slurry pipeline from a coal mine at Black Mesa, Arizona, to the Mohave Power Plant near Laughlin, Nevada. This pipeline is 275 miles long and has been operating successfully since 1970.

Coal slurry pipelines are most economical for long distances and very large annual volumes over which the large fixed costs can be spread. Therefore, they require long-term contractual commitments among coal producers, end-users, and the pipeline owner/operator. Coal slurry pipelines also limit transportation flexibility.

Natural gas, also transported by pipeline, is somewhat similar to electricity in that it is difficult to store. There are underground storage caverns near major usage centers that are used to accommodate the large swings in short period demands. For example, the annual average daily demand for natural gas in the United States is about 52 billion cubic feet. The average daily demand in December is 70 billion ft^3 and the peak daily demand has been as much as 104 billion ft^3 (Energy Information Administration, March 1993).

Natural gas and petroleum product pipelines are capital intensive. The major portion of the transportation cost is associated with capital recovery and return on the invested capital. This is a fixed annual cost that is recovered through reservation fee (demand charge) payments by the shippers of the natural gas or petroleum whether or not all of their allocated capacity is utilized.

When gas is transported in a pipeline system, a certain percentage of the gas is removed from the lines to operate the compressors that are powered by natural gas-fueled combustion turbines. The gas that is used for compression is considered to be part of the payment for its transport. Depending on the transportation distance, 3 to 10% of the gas may be used to fuel compression facilities associated with its transport.

4.8 FOSSIL FUEL PROPERTIES

For purposes of combustion facility design, the properties of liquid and gaseous fossil fuels are fairly uniform. For example, the quality of natural gas that producers and gatherers are allowed to put into interstate pipelines is specified by the operator of the pipeline to a fairly tight tolerance. This reduces the amount of variation in fuel properties that must be considered in the power plant design process. However, even for natural gas, the variation in its calorific value can be considerable depending on its nitrogen content.

The quantity of fuel to be handled in the power conversion process will vary considerably as a function of the fuel's energy density. Table 4-5 shows the specific energy contents for different types of fuels used in power generation. Since each of these fuels has a different process conversion efficiency (kilowatt-hours of generation per unit of thermal energy input), the variation in the quantities required per kilowatt-hour of generation is even greater.

For coal, there is no such thing as typical properties. This is in large part due to the way in which it was formed. Of all the fossil fuels, coal has the most variable quality and delivered cost economics. Since coal dominates all other fossil fuels for electric power generation and its properties are so variable, most of the discussion on fossil fuel properties is directed toward the properties of coal.

Table 4-5. Comparison of Specific Energy Content of Various Fossil Fuels

Fuel	Heating Value	"Specific Energy" Content
Coal	4,000–13,000 Btu/lb	77–250 lb/MBtu
No. 6 oil	18,000 Btu/lb	55 lb/MBtu
No. 2 oil	19,500 Btu/lb	51 lb/MBtu
Natural gas	22,750 Btu/lb	44 lb/MBtu
Natural gas	1,030 Btu/ft^3	970 ft^3/MBtu

4.8.1 Coal Properties

Table 4-6 lists the primary coal properties and the numerous power plant components or systems that are impacted by them.

Coal is an admixture of organic material and mineral matter. The organic matter is responsible for the energy content of the fuel. However, it is the mineral matter that presents significant challenges in the design and operation of a power plant.

Table 4-7 shows the variation in primary coal properties such as heating value, sulfur, and ash for various coals in the United States. There are significant variations among the various coal producing regions of the United States, as well as within regions and between seams within these regions.

Several types of analyses are performed to evaluate those coal properties that impact the design and operation of power plant components and systems. These analyses are commonly referred to as Higher Heating Value or Gross Calorific Value, Proximate Analysis, Short Proximate, Ultimate Analysis, Sulfur, Forms of Sulfur, Equilibrium Moisture, Mineral Analysis of Ash, Ash Fusion Temperatures, Grindability (Hardgrove Grindability Index), Free Swelling Index (caking characteristics), and Rank. In addition, other physical characteristics of the coal may be determined, such as bulk density, size consist, and friability.

All coal analyses begin with the requirement for a representative sample. Section 4.9 covers this subject in detail.

The methods for performing the various tests on fossil fuels have been developed by various standards organizations such as the American Society for Testing and Materials (ASTM) and the International Standards Organization (ISO). The ASTM Standards have been cited. The complete citation for the referenced ASTM Standard cited can be found in the reference section. Since these standards are under continuous review and are often updated, refer to the latest revision of the Standard for current information.

Table 4-6. Primary Impacts of Coal Properties on the Design of a Pulverized Coal-Fueled Unit

Component or System	Heating Value	Moisture and Hydrogen	Volatile Matter	Grindability	Total Sulfur	Total Nitrogen	Total Ash	Ash Properties
Coal handling	X	X		X				
Coal feeders	X	X		X				
Forced draft fans	X	X						
Pulverizers	X	X	X	X				X
Burners	X	X	X					
Furnace	X	X	X		X		X	X
Convection passes	X	X	X				X	X
Sootblowers							X	X
Air heaters		X			X		X	X
Precipitators	X	X			X		X	X
Induced draft fans	X	X			X			
Air quality control	X				X	X		
Waste handling	X				X	X	X	X

Table 4-7. Coal Classification—Source and Analyses of United States Coals

Coal Rank	State/County	Bed	Proximate (%)				Ultimate (%)					Higher Heating Value, Btu/lb
			Moisture	Ash	Volatile Matter	Fixed Carbon	Sulfur	Hydrogen	Carbon	Nitrogen	Oxygen	
Meta-anthracite	Rhode Island	Middle	13.2	18.9	2.6	65.3	0.3	0.4	64.2	0.2	2.8	9,310
Anthracite	Pennsylvania	Clark	4.3	9.6	5.1	81.0	.8	2.4	79.7	.9	2.3	12,880
Semianthracite	Arkansas	Lower	2.6	7.5	10.6	79.3	1.7	3.5	81.4	1.6	1.7	13,880
Low-volatile bituminous coal	West Virginia	Pocahontas	2.9	5.4	17.7	74.0	.8	4.3	83.2	1.3	2.1	14,400
Medium-volatile bituminous coal	Pennsylvania	Upper	2.1	6.1	24.4	67.4	1.0	4.8	81.6	1.4	3.0	14,310
High-volatile A bituminous coal	West Virginia	Pittsburgh	2.3	5.2	36.5	56.0	.8	5.2	78.4	1.6	6.5	14,040
High volatile B bituminous coal	Kentucky	No. 9	8.5	10.8	36.4	44.3	2.8	4.5	65.1	1.3	7.0	11,680
High-volatile C bituminous coal	Illinois	No. 5	14.4	9.6	35.4	40.6	3.8	4.2	59.7	1.0	7.3	10,810
Subbituminous A coal	Wyoming	No. 3	16.9	3.6	34.8	44.7	1.4	4.1	60.4	1.2	12.4	10,650
Subbituminous B coal	Wyoming	Monarch	22.2	4.3	33.2	40.3	.5	4.4	53.9	1.0	13.7	9,610
Subbituminous C coal	Colorado	Fox Hill	25.1	6.8	30.4	37.7	.3	3.4	50.5	.7	13.2	8,560
Lignite	North Dakota	Unnamed	36.8	5.9	27.8	29.5	.9	2.8	40.6	.6	12.4	7,000

Source and analysis of coal were selected to represent the various ranks of the specifications for classification of coals by rank adopted by the American Society for Testing and Materials.

Source: Adapted from *Coal Data: A Reference*, US Department of Energy, Energy Information Administration.

4.8.1.1 Classification of Coals By Rank. The rank of a coal is a general measure of the degree of coalification or metamorphism. The primary properties used to establish the rank of the coal are its gross calorific value, moisture, volatile matter content, and agglomerating character. The standard for classifying coals by rank is ASTM Standard D388. The calorific value used in establishing the rank of a coal is its gross calorific value on a moist, mineral-matter-free basis using the Parr formulas. These formulas are as follows:

Dry, mineral-matter-free fixed carbon

$$= \frac{FC - 0.15S}{100 - (M + 1.08A + 0.55S)} \times 100, \%$$

Dry, mineral-matter-free volatile matter

$$= 100 - \text{Dry, mineral-matter-free FC, } \%$$

Moist, mineral-matter-free calorific value

$$= \frac{\text{Btu} - 50S}{100 - (1.08A + 0.55S)} \times 100, \text{ Btu per lb}$$

where

FC = fixed carbon (%).

A = ash (%).

M = moisture (%).

S = total sulfur (%).

Btu = gross calorific value (Btu/lb).

(All of the above properties are calculated on an inherent or equilibrium moisture basis.)

The moisture level used in establishing rank is that level that duplicates the in situ (inherent) or bed moisture of the coal seam. ASTM Standard D1412 is used to measure this.

The agglomerating nature of the coal is determined by ASTM Standard D720. This standard specifies the method for determining the caking characteristics of the coal as it loses moisture and volatile matter. The test determines if the sample fuses together (agglomerates) or if it remains a powder. Depending on the cross-sectional area of the coke button that is formed, Free Swelling Indices of 1 to 9 are determined. A Free Swelling Index of 1 or greater indicates that the coal is agglomerating; 0.5 or less indicates that it is nonagglomerating.

Figure 4-24 shows the trend in moisture, volatile matter, and fixed carbon when expressed on a moist, mineral-matter-free basis from low-rank coals through high-rank anthracites. The general trend with increasing rank is an increase in the as-received heating value and fixed carbon and a corresponding decrease in moisture and volatile matter. This trend is so pronounced that a classification system based on the fuel ratio (ratio of fixed carbon to volatile matter) has been used as a rough indicator of the coal's rank.

Lower rank coals (lower fuel ratio) are also characterized by a greater oxygen content, that aids ignition and enhances combustibility and flame stability. Flame stability is primarily a function of the volatile matter content and the particle size distribution of the fuel. High combustibility improves carbon burnout (reduces carbon carryover) and hence boiler efficiency, and for pulverized coal-fired units, this allows the coal to be ground to a coarser size. Low-rank coals (high moisture content) produce a "self-pulverization" of the coal particles during combustion. As the inherent moisture in the pore structure of the coal is heated and rapidly expands, its volume increases (as water flashes to steam at atmospheric pressure, the volume expansion is 1,600 to 1), thus fragmenting the coal particles. This increases the exposed surface for combustion.

4.8.1.2 Gross Calorific Value (Higher Heating Value). Of all the properties of a fuel, gross calorific value or higher

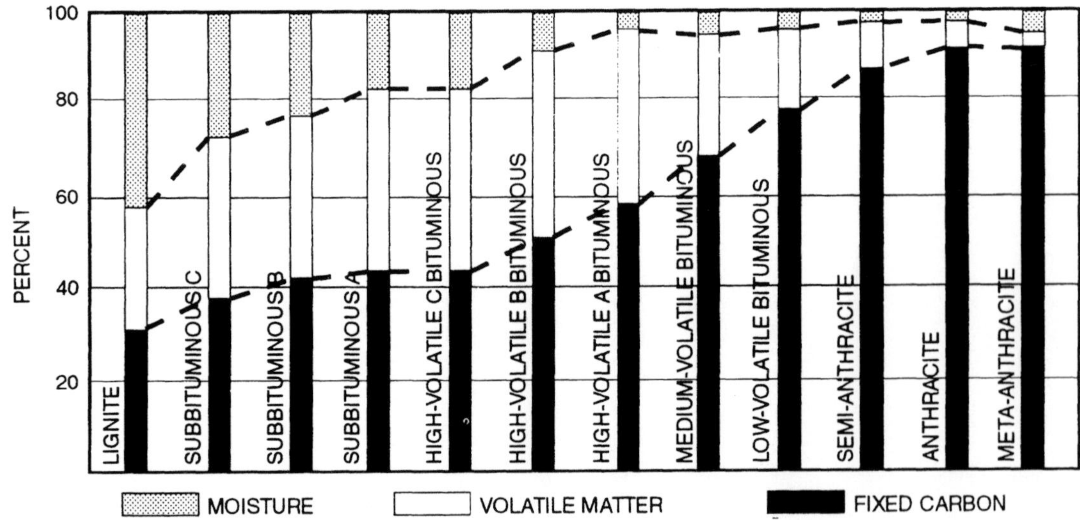

Fig. 4-24. Coal rank compared to proximate analysis (ash-free basis). (From: EPRI, 1978, *Coal Preparation for Combustion and Conversion.*)

heating value is the most important from a commercial perspective. It is the primary determinant of how much a buyer is willing to pay for one fuel compared to another.

The heating value is expressed in Btu/lb, kcal/kg (international steam coal trade), kJ/kg, or MJ/kg.[11] The primary contributors to heating value are carbon, hydrogen, and sulfur. The heating value is usually determined according to ASTM Standard D2015 or D3286 using a bomb calorimeter. The gross or higher heating value is the total heat liberated by burning the coal as measured from the time of initial combustion to the time when the temperature of the combustion products is equivalent to the initial temperature of the oxygen and fuel that were combusted.

The net or lower heating value (LHV) is the heat produced by a pound of coal when all the water in the products of combustion remains as vapor. It is calculated from the gross or higher heating value (HHV) as follows:

$$\mathrm{LHV} = \mathrm{HHV} - 10.30(\mathrm{H} \times 9 + \mathrm{M}),\ [\mathrm{Btu/lb}]$$

or

$$\mathrm{LHV} = \mathrm{HHV} - 5.72(\mathrm{H} \times 9 + \mathrm{M}),\ [\mathrm{kcal/kg}]$$

where

LHV = net calorific value at constant pressure, Btu/lb or kcal/kg;

HHV = gross calorific value at constant volume, as-received basis, Btu/lb or kcal/kg;

H = hydrogen, as-received basis, percent, and

M = total moisture, as-received basis, percent.

4.8.1.3 Approximation of Gross Calorific Value The gross calorific value can be approximated by using formulas such as Dulong's and Francis and Lloyd's (Francis and Lloyd 1983).

The Dulong formula has been used for many years and is as follows:

$$\mathrm{HHV\ (Btu/lb)} = 146 \times \mathrm{C} + 620(\mathrm{H} - 0/8) + 40.5 \times \mathrm{S}$$

The Francis and Lloyd formula is as follows:

$$\mathrm{HHV\ (Btu/lb)} = 149 \times \mathrm{C} + 530(\mathrm{H} - 0.835 \times 0/8) + 26.7 \times \mathrm{S}$$

where

C = carbon,

H = hydrogen,

O = oxygen, and

S = total sulfur.

(All measured in percent on an as-received basis.)

While these formulas give results that only approximate the higher heating value of the coal, they provide a quick method for ascertaining whether or not the gross calorific value reported in the analysis is consistent with the properties shown in the ultimate analysis.

4.8.1.4 Proximate Analysis. The proximate analysis includes the total moisture, volatile matter, fixed carbon, and ash of the sample that has been analyzed in the laboratory. Total moisture is determined by ASTM Standards D2013, D3173, and D3302; ash is determined by ASTM Standard D3174; and volatile matter is determined by ASTM Standard D3175. The fixed carbon is the difference between the other constituents and 100%.

The determination of moisture and volatile matter involve heating the coal sample at various temperature levels for different periods of time and noting the weight loss in the sample. As such, the determination of moisture and volatile matter involves a rather subjective procedure.

The amount of moisture as reported by the laboratory impacts the contribution of the various coal constituents to the total and the calorific value of the coal. It has been a source of considerable disagreement when the seller and buyer cannot agree on the quality of a particular coal. This is particularly true since various coal producing countries and testing standards organizations have adopted different conventions for determining and reporting properties. For example, some report properties on an as-received basis and others on an air-dried basis.

The total moisture in coal is determined using ASTM Standards D2013 (air-dry moisture loss) and D3173 (residual moisture in the analysis sample). When analytically combined, the two moisture losses represent the total moisture for the sample. The purpose of air-drying the coal is to bring its moisture level into equilibrium with ambient conditions in the laboratory. Air-drying also reduces the potential for unaccounted moisture loss because of the heat generated when grinding the sample to 60-mesh.

Ash is the noncombustible residue left after complete combustion of a coal sample. As such, it usually weighs less than the initial mineral matter content of the coal since some of the mineral constituents are driven off as gasses. Ash may increase for coals high in calcium that react with the sulfur in the coal to form calcium sulfates or sulfites that retain some of the oxides of sulfur that would normally appear in the gases during combustion. This is why the Parr formulas have correction factors applied to the ash to account for this type of loss of mineral matter.

The volatile matter of coal consists of hydrocarbons, including tars such as creosote, light oils such as benzene, and other gases that result from distillation and decomposition of the coal as it is heated. The determination of volatile matter is based on ASTM Standard D3175.

Fixed carbon is the difference between 100% and the sum of moisture, ash, and volatile matter.

[11]kcal/kg = (Btu/lb)/1.8; MJ/kg = (Btu/lb)/429.92.

4.8.1.5 Short Prox. The short proximate (short prox), an abbreviated proximate analysis, consists of an analysis of heating value, ash, sulfur, and moisture content of the coal.

4.8.1.6 Ultimate Analysis. The ultimate analysis is very important since it provides information regarding the primary combustible and noncombustible components that are used to compute combustion air requirements, flue gas volumes, and the losses associated with the combustion of hydrogen whose latent heat of evaporation is lost with the flue gas.

ASTM Standard D3176 is used in the determination of the ultimate analysis of coal. The ultimate analysis consists of moisture, ash, carbon, hydrogen, nitrogen, sulfur, and oxygen. Some laboratories also include chlorine as a part of the ultimate analysis. If not, it shows up in the oxygen content, that is determined by difference.

One problem with the interpretation of ultimate analyses is that they may be reported on several different bases. The difference primarily relates to the manner in which moisture is treated. Sometimes the ultimate analysis is reported on a moisture-free (dry) basis. Sometimes moisture is not shown, but its hydrogen and oxygen components are included with the hydrogen and oxygen values. Therefore, it is very important to first determine the basis of the analysis. It is usually less confusing if the ultimate analysis includes the total moisture and total ash for the coal and the hydrogen and oxygen gases are shown separately.

SULFUR IN COALS. Sulfur in the coal exists in three forms: organic, inorganic, and sulfate sulfur. Inorganic sulfur occurs primarily as pyrite or marcasite. The pyrite (iron disulfide [FeS_2] or fools' gold) has an isometric crystal structure. Marcasite is in the chemical form of FeS_x, is gray in color, and has an orthorhombic crystal structure.

Sulfate sulfur usually exists in the form of calcium sulfate or iron sulfate and usually is a small contributor to total sulfur content.

Whereas the organic sulfur is combined within the carbon matrix, the pyritic sulfur is deposited in the fractures and joints of the coal seams where it forms a lenticular or tabular kind of deposit, as small nodules distributed throughout the coal, or as dendritic or treelike veins that radiate from a central point. No one is exactly sure of the origin of pyritic sulfur, but there are indications that some of the material accumulated at the bottom and top of the coal seams from volcanic activity while that deposited in the joints of the seams may have leached in from sources in the overburden. A majority of this sulfur tends to be in the pyritic form, particularly for high-sulfur coals. The sulfur content in coal is usually determined using ASTM Standard D3177 or D4239.

CHLORINE IN COALS. The amount of chlorine in the coal is determined by ASTM Standard D2361 or D4208. Coals in the Illinois Basin and in England with relatively small amounts of chlorine (0.3%) can exhibit severe fouling tendencies. This is not due to the chlorine itself, but is caused by the chemically associated presence of "free volatile" sodium, for which the chlorine is an indicator. The volatile or active sodium provides the "glue" for ash particles to fuse to each other and to the tubes in the convection passes as the gas and particles are cooled below their melting points. Tests have been developed to measure the types of sodium and other active and inactive alkalies in coal that contribute to fouling. These tests consist of using a dilute acetic acid to leach the coal. The alkalies removed during the leaching process are those that can be expected to volatilize during combustion.

Although some of the Illinois Basin coals may be high in chlorine content, they may or may not be high in alkalies that easily volatilize.

4.8.1.7 Mineral Analysis of Ash. The ash mineralogy affects fuel selection and the design and sizing of the furnace and other coal quality dependent components. One of the primary factors that contributes to coal being a heterogeneous nonfungible commodity is the variability in the total ash and the constituents that compose it. Such variability exists within and between seams. The amount of this variability can be reduced by coal beneficiation or washing.

The mineral matter consists of ash inherent within the coal seam, partings within the seams, and out-of-seam dilution (diluent) from the sedimentary rock and clays that compose the roof and floor of the coal seam.

The mineral constituents in the ash can promote a number of boiler ash problems including slagging, the running deposits within the furnace (those areas of the steam generator subjected to radiant heat transfer). These mineral constituents also produce fouling, the accumulation of sintered deposits, on the closely spaced tubes in the convection passes. Convection passes are those areas of the steam generator where convection heat transfer occurs. The mineral matter in the ash plays a significant role in the slagging, fouling, erosion, and corrosion of components exposed to the combustion gases.

ASTM Standard D3682 or D4326 is usually used to determine the major elements in the ash and express them as oxides on an ignited basis. The primary mineral matter constituents analyzed for the coal are silicon dioxide (SiO_2), aluminum oxide (Al_2O_3), titanium dioxide (TiO_2), ferric oxide (Fe_2O_3), calcium oxide (CaO), magnesium oxide (MgO), potassium oxide (K_2O), sodium oxide (Na_2O), sulfur trioxide (SO_3), phosphorus pentoxide (P_2O_5), and an undetermined fraction by difference. Some laboratories also include strontium oxide (SrO), barium oxide (BaO), and manganese oxide (Mn_3O_4).

The mineral matter in the ash can be subdivided into basic and acidic components. The basic components consist of ferric oxide, lime, magnesia, potassium oxide, and sodium oxide. The acidic components consist of silica, alumina, and titania. The major mineral constituents in the ash tend to be silica and alumina that come from the clays that are distributed throughout the coal matrix, and iron, the majority of which for bituminous coals comes from the pyritic sulfur in the coal (FeS_2). For low-rank western United States coals, the calcium content is also quite high.

Quartz is the hardest mineral found in coal; for large particle sizes, high temperatures are required for vitrification. Therefore, the amount and shape of the post combustion quartz particles are a function of flame temperature. Low-rank coals such as lignites and subbituminous coals that burn with low flame temperatures are likely to produce an ash having greater amounts of quartz that is quite erosive because of its hardness and sharp edges.

It should be noted that the abrasiveness and erosiveness of a coal or its ash are not the same. Abrasion results from the wear caused by particles sliding over the surface, whereas erosion results from impact of the particles on the boiler tube surface. Both quartz and pyrite are major contributors to abrasion and erosion. It should also be noted that a coal that may not be significantly abrasive or erosive prior to combustion may be quite abrasive and erosive after combustion when hard oxides have formed from the cooled and resolidified ash.

The composition of the ash that is analyzed in the laboratory is not necessarily representative of those deposits that will ultimately form as slag on the furnace walls or the fly ash to be collected by the electrostatic precipitator or bag house. The actual combustion environment (oxidizing or reducing atmosphere) and the higher temperatures at which it can take place (up to 3,000° F or 1,650° C in a pulverized coal fueled unit compared to about 1,500° F or 815° C in the laboratory) mean that the weight of the laboratory ash is likely to be greater than the weight of the actual boiler ash.

The mineral constituents reported in the ash analyses as oxides may in fact not be the constituents that contribute to the particles that make up the fly ash or bottom ash after the coal has been combusted at laboratory rather than furnace temperatures and eutectic recrystallization has occurred. There is a natural classification of ash particles by size and weight that differentiate the chemical composition of the bottom ash from the fly ash and that obtained from the laboratory ash sample. Therefore, such laboratory analyses of the ash may not be appropriate for designing power plant components such as the electrostatic precipitator that is exposed only to the fly ash.

4.8.1.8 Ash Fusion Temperatures. Ash fusion temperatures can be viewed as the melting stages that the ash passes through in a reducing or oxidizing atmosphere while being heated. The results are primarily used to predict the degree of slagging on the furnace walls.

The procedures outlined in ASTM Standard D1857 are used for determining the fusion temperatures of the ash. The temperature at which the ash melts is determined in both a reducing atmosphere (composed of hydrogen, hydrocarbons, or carbon monoxide) and an oxidizing atmosphere (composed of oxygen, carbon dioxide, and water vapor or air), depending on whether a gas fired or electric furnace is used for the test. The different atmospheres are used to simulate coal combustion in an oxygen rich or oxygen deficient zone in the furnace.

To perform the test, small 3/4 in. (19 mm) high cones of ash are formed and placed in a furnace and the temperature of the environment is increased until initial deformation of the pyramid tip is observed. This temperature is called the initial deformation (ID) temperature. The relevant temperatures that are recorded when the cone is heated in an oxidizing or reducing atmosphere are the following:

- Initial deformation (ID)—the temperature at which the tip of the cone begins to deform;
- Softening temperature ($H = W$)—the temperature at which the height of the cone equals its width at the base;
- Hemispherical temperature ($H = W/2$)—the temperature at which the height equals half of the cone's width; and
- Fluid temperature—the temperature at which the melted cone spreads into a flat layer with a maximum height of 1/16 in. (1.6 mm).

The ash fusion temperatures in an oxidizing atmosphere are generally higher than those reported in a reducing atmosphere.

When different coals are blended, it is difficult to predict the resultant ash fusion temperature for the blend. This is particularly true when one of the coals has a lignitic type ash (calcium plus magnesium is greater than iron) and the other coal has a bituminous type ash (calcium plus magnesium is less than iron). In such cases, minerals form eutectics such that the resultant ash fusion temperatures for the blend may be lower than those for either coal.

Another important temperature for ash is its T_{250} temperature. This is the temperature at which the viscosity of the ash is 250 poise. It is significant for units with cyclone burners which require that the ash melt and flow at a relatively low temperature so that it can be tapped at the bottom of the furnace into slag tanks. For such units, the majority of the ash (70% to 80%) ends up as molten slag in the bottom of the furnace and only 20% to 30% ends up as fly ash.

4.8.1.9 Grindability of Coal. The ease of pulverizing a coal is determined by its Hardgrove Grindability Index (HGI). The procedures for this test are provided in ASTM Standard D409.

4.8.1.10 Size Consist. The size consist of a coal is the percentage of the total coal by weight that exists between selected sieve sizes. Each size range is designated by a top size and a bottom size (for example, 1¼ in. × ¼ in.). If no bottom size is designated, the nomenclature −1¼ in. or 1¼ in. × 0 is used.

The majority of all coal shipped in the United States has a top size of about 2 in. (51 mm) as defined by ASTM Standard D4749, "Sieve Analysis of Coal and Designating Coal Size." The top size of the coal is the sieve that defines the upper limit and is the smallest of a series of sieves upon which is retained < 5% of the total sample by weight. In contrast, the bottom size of the coal is defined as the largest sieve through which no more than 15% of the coal passes.

In recent years with the development of lower rank western coals that are considered to be somewhat dusty, attempts have been made to ship a larger top size coal to reduce the amount of fines. The western low-rank coals are highly slaking. Because of the large inherent moisture content of these coals, when they lose some of this moisture they also lose some of their physical integrity and the particles begin to disintegrate.

Although there are many opinions of what size range constitutes the "fines" or smallest particle sizes in a coal, the maximum is probably 1/8 in. (3.2 mm), although from the standpoint of fugitive dust, it is the −28 mesh (< 600 microns) fraction that creates the biggest problems. If fugitive dust becomes a significant problem, commercially available dust suppressing agents, heavy oils, and calcium chloride can be sprayed on the coal to reduce dusting.

4.8.1.11 Spontaneous Combustion. Spontaneous combustion can present problems during transportation, handling, and storage. Combustion (oxidation) can take place rapidly as in a furnace or slowly on a stockpile. If it takes place slowly in a stockpile, there is a degradation or loss of energy content and hence value of the fuel. Although the factors that influence spontaneous combustion are not completely known, there is enough experience to indicate that the following play a significant role:

- Rank of the coal, with low-rank coals being more susceptible than high-rank coals. Because of the porosity of low-rank coals, they seem to be affected more by the "heat-of-wetting" (release of heat during the condensation of water vapor).
- Amount of surface area exposed to air;
- Ambient temperature, with high solar insolation aiding spontaneous combustion;
- Oxygen content of the coal;
- Free moisture content of the coal;
- Pyritic sulfur content of the coal; and
- Configuration of the coal stockpile. Steep conical piles with coarse coal around the edges and fines near the top are more susceptible because they promote natural convection (chimney effect) and good air flow through the pile to support combustion as it develops.

To prevent spontaneous combustion, it is important to maintain a dry pile. Compaction at regular intervals while building the storage pile seems to offer the best solution in most cases to preventing spontaneous combustion.

The critical temperature for initiating spontaneous combustion seems to be about 150 to 170° F (66 to 77° C). Once the coal reaches this temperature, further increases are likely and the process eventually heads towards "thermal runaway," with progression from smoldering to total combustion likely.

4.8.2 Fuel Oil Properties

Petroleum products, other than No. 6 fuel oil, are usually not sold based on their heating value, but rather on properties such as their API gravity, flash point, pour point, viscosity, sulfur content, and cetane number.

4.8.2.1 API Gravity of Oils The gravity or weight of an oil is measured in degrees API, which is a standard formulated by the American Petroleum Institute. The API gravity is related to the specific gravity as shown in the following equations:

$$\text{Specific gravity at } 60^\circ\text{F} = 141.5/(^\circ\text{API} + 131.5)$$

$$^\circ\text{API} = 141.5/\text{SpGr}_{60/60^\circ\text{F}} - 131.5$$

The specific gravity at 60° F (15.5° C) means that the fluid whose density is being measured as well as the reference water is at a temperature of 60° F.

The term fuel oil applies to a range of refined products commonly referred to as No. 1, 2, 4, 5, and 6 oils. This is in decreasing order of API gravity that ranges from about 50 to 40 for No. 1 light distillate and 40 to 30 for No. 2 light distillate, down to 20 to 0 for heavy No. 6 residual oil. The heavy oils, Nos. 5 and 6, are referred to as ash-bearing fuels because they leave an ash after combustion. The lighter No. 1 and No. 2 oils are referred to as light distillates and leave no ash when burned.

The specifications for the heavier oils are often based on their API gravity, viscosity, as well as their pour point, the temperature at which the fuel barely flows. For example, the pour point of No. 6 oil can be as high as 70° F (21° C). Such a high pour point means that the oil must be heated, particularly in the wintertime, to reduce its viscosity so that it can be pumped from the storage tanks to the burners.

Table 4-8 shows typical analyses and properties of fuel oils. It also shows typical trade names for the various products and types of use and handling requirements.

4.8.2.2 ASTM Standards Related to the Properties of Oils and Their Uses. ASTM Standard D396, "Standard Specifications for Fuel Oils," describes the key properties and the minimum and maximum values for each type of fuel that satisfies the specifications. With the heavier oils such as No. 6, the specifications are less stringent since these oils are sold primarily based on their energy content, with price premiums and penalties depending on sulfur content.

4.8.2.3 Heating Values of Fuel Oils. The actual heating value of a fuel oil is determined in a fashion similar to that used for coal using a bomb calorimeter. ASTM Standard D240 is used in performing this test. The heating value of oils can be stated on a Btu/gal or Btu/lb basis. The Btu/gal basis should also have associated with it a specific temperature, usually 60° F (15.5° C) since the density of the fuel changes with temperature as does the Btu/gal.

Heating values are not normally specified for No. 1 and No. 2 fuel oils. These fuels are primarily sold as transportation fuels in which parameters such as cetane number are

Table 4-8. Illustrative Analyses for Fuel Oils

Grade: Type:	No. 1 Distillate (Kerosene)	No. 2 Distillate (Diesel or Heating Oil)	No. 4 Distillate or Low Viscosity Residual	No. 5 Residual	No. 6 Residual (Heavy fuel oil or Bunker C)
°API	40	32	21	17	12
Specific gravity 60/60° F (15.5° C)	0.825	0.865	0.928	0.953	0.986
Pounds/U.S. gallon, 60° F (15.5° C)	6.87	7.21	7.73	7.94	8.21
Kinematic viscosity, centistokes (cs) at 100° F	1.6	2.7	15	50	360
ASTM[a] maximum kinematic viscosity, cs	2.1 (104° F)	3.4 (104° F)	24 (104° F)	12 (212° F)	50 (212° F)
Pumping temperature, minimum° F	Ambient	Ambient	15 (−10° C)	35 (2° C)	100 (38° C)
Atomizing temperature, ° F	Ambient	Ambient	25 (−4° C)	130 (54° C)	200 (93° C)
ASTM water and sediment, max. vol. %	0.05	0.05	0.5	1.0	2.0
Carbon residue, wt%	Trace	Trace	2.5	5.0	12.0
Ash, wt%	Not applicable	Not applicable	0.05	0.08	0.10
Vanadium, ppm by wt	Not applicable	Not applicable	Varies with crude source	Varies with crude source	Varies with crude source
Gross heating value[b]	19,926 (11,070)	19,489 (10,827)	19,066 (10,592)	18,887 (10,493)	18,644 (10,358)
Net heating value[b]	18,711 (10,395)	18,320 (10,178)	17,971 (9,984)	17,810 (9,894)	17,677 (9,821)
Sulfur, wt%	Less than 0.3 wt% S	Less than 0.5 wt% S[c]	1.0[d]	1.9[d]	2.7[d]
Oxygen, wt%	Less than 0.04	Less than 0.1	0.46	0.54	0.62
Nitrogen, wt%	Less than 0.04	Less than 0.1	0.24	0.28	0.32
Hydrogen, wt%	13.2	12.7	11.9	11.7	10.5
Carbon, wt%	86.5	86.6	86.4	85.5	85.7

Actual properties will be dependent on the crude source and the degree of refining.

[a]ASTM standard specifications for fuel oils, ASTM D396-92.

[b]Btu/lb (kcal/kg) @60° F (15.5° C).

[c]In the United States <0.05% for highway diesel fuel.

[d]Actual value depends on application, environmental restrictions, and crude source.

important for diesel engine use. Their heating value will usually be about 139,000 Btu/gal (5.8 MBtu/bbl).

*4.8.2.4 **Sulfur Content.*** For most fuel oils, environmental regulations control sulfur and the potential for sulfur dioxide emissions, that are felt to be a precursor to the formation of acid rain. As of October 1, 1993, the United States Environmental Protection Agency requires that all diesel fuels sold for highway use are limited to a maximum sulfur content of 0.05% by weight. This compares with the ASTM Standard D396 value for the maximum sulfur in No. 2 oil of $< 0.5\%$. However, federal and state environmental regulations had already limited the maximum sulfur content of these fuels to 0.3% or 0.2% in many areas of the United States.

SULFUR CONTENT OF NO. 6 OIL. The sulfur content of residual oil and other lighter fractions is a function of the sulfur in the crude oil feedstock and the process used to refine the crude oil. The "normal" sulfur content of residual No. 6 oil can range from less than 1% (if sweet low sulfur crudes are refined) up to 3 or 4%.

The sulfur content of No. 6 oil can be reduced by blending it with "cutter stock" that has a much lower sulfur level. Cutter stock can be off-specification light or middle distillate oils, or other streams in the refinery that are low in sulfur, but have limited market value.

Another means of reducing the sulfur content of liquid oil products is to use hydrotreating. Hydrogen under high temperature and pressure is forced to react with the sulfur in the fuel oil to produce hydrogen sulfide. The addition of hydrogen to the heavy oil also increases its hydrogen-to-carbon ratio that further increases the amount of light ends produced from a barrel of crude. The hydrogen required for the process is produced in a steam and/or methane gas reforming operation. The gas stream containing the hydrogen sulfide is scrubbed with an amine solution. The hydrogen sulfide and amine are then heated in an amine regeneration tower to separate the amine from the hydrogen sulfide gas. The hydrogen sulfide gas is then processed in a Claus plant where oxygen and hydrogen sulfide react to produce sulfur dioxide which further reacts with the remaining hydrogen sulfide to produce liquid sulfur.

*4.8.2.5 **Viscosity of Fuel Oils.*** The viscosity of fuel oils can be measured in several different units including centistokes, Saybolt, Saybolt universal, and Saybolt furol. These viscosity determinations are usually made in conformance with ASTM Standard D445. In this standard, the kinematic viscosity, which is a measure of the resistive flow of a fluid under gravity, is measured in centistokes or 1/100th of a stoke which is equal to 1 m^2/s. The viscosity is measured as the time for a given volume of fluid to flow under the force of gravity through a calibrated glass capillary viscometer at a controlled temperature.

*4.8.2.6 **Flash Point.*** The flash point is the lowest temperature at standard atmosphere pressure at which the vapor of a fuel ignites when a flame is brought near the surface of the fuel. It is determined by procedures given in ASTM Standard D93.

*4.8.2.7 **Storage of Fuel Oils.*** One problem that can occur with the storage of fuel oils is the growth of microorganisms that live in the fuel storage tanks. These microorganisms form slimes that later plug fuel filters and burner nozzles. These organisms usually live at the moisture–fuel interface. To reduce the opportunity for their growth, the potential for moisture or condensation forming in the tank must be minimized. Usually by keeping the tanks full, this problem can be minimized. If the microorganisms become a problem, the fuel can be treated with a biocide (compatible with combustion system) to limit their growth. For periods of long-term storage, the oil can be circulated from one tank to another and the contaminants such as microorganisms removed using appropriate filtering processes. In any event, there should be an adequate filtering system between the storage tank and the combustion burner.

4.8.3 Natural Gas Properties

The primary components of natural gas are methane (CH_4) and ethane (C_2H_6). Since hydrogen sulfide is corrosive, it is usually removed at the natural gas processing plant to eliminate corrosion of pipelines and compression equipment. Therefore, the gas received from a local distribution pipeline is fairly free of sulfurous compounds. However, in remote areas, the processing plant may not remove the hydrogen sulfide from the gas. When the natural gas contains significant amounts of hydrogen sulfide, it is often referred to as "sour gas" because of its odor.

The heating value of the natural gas is primarily impacted by contaminants such as carbon dioxide and nitrogen. The gross or higher heating value of interstate pipeline quality natural gas is usually about 1,000 Btu/ft^3. Natural gas is normally purchased on the basis of mcf (1,000 ft^3). One thousand cubic feet would contain about 1 million Btu or a decatherm of energy. A therm is equivalent to 100,000 Btus. If the constituents of the natural gas and their relative contributions are well known, the approximate higher heating value of the natural gas can be calculated fairly accurately. However, the best means of determining the calorific value is by using a calorimeter.

4.8.4 Properties of Waste Fuels

The sulfur content of municipal refuse is usually negligible, which would appear to make it a fairly desirable fuel for power plant use. However, while the sulfur is low, municipal refuse and refuse-derived fuels are high in moisture content and experience a considerable decline in their heating value during storage. In addition, municipal refuse usually contains significant quantities of polyvinyl chloride (PVC). The chlorine content of PVC can be as high as 50%. This can aggravate corrosion of low-alloy steels.

Landfill methane has also received some attention as a fuel, but it contains large amounts of CO_2, therefore reducing its effective heating value.

The quality of wood waste is a function of the wood species and type of debarking (hydraulic debarking leads to high moisture content) and the type of the waste (sawdust, shavings, bark).

Since the quality of waste fuels can vary significantly, they must be adequately sampled and the properties analyzed to assess their importance in the design and operation of the power plant.

4.9 COAL SAMPLING AND REPORTING OF LABORATORY RESULTS

The three primary steps involved in analyzing coal properties are coal sampling, sample preparation, and laboratory analysis of prepared samples. The heterogeneity of coal properties makes it very difficult to sample coal. Yet, if the sample is not representative of the shipment, the results of the laboratory analysis will not be representative.

About 60% to 80% of the variance in analyzing the ash content of a coal is due to sampling, while the variance of the laboratory analysis accounts for $< 4\%$ of the total (Aresco and Orning 1965). The importance of obtaining a representative sample is obvious.

For a unit train shipment lot of 10,000 tons, the analyses upon which the properties of the coal in the shipment are based involve a final sample of only 1 g. It is no wonder that disputes sometimes arise between the coal supplier and buyer regarding the quality of coal shipped from the mine and the quality of coal received at the power plant when a sample of only 1 g from a shipment of 10,000 tons or 0.00000001% of the total weight (about 0.1 part per billion) is actually used in determining these properties.

Assume there is a consistent error in that the reported heating value is 50 Btu/lb (28 kcal/kg) greater or less than the actual value. The total annual cost of such an error to the buyer or seller for shipments to a 600-MW power plant consuming 1.7 million tons of coal per year of an Appalachian Basin bituminous coal with a delivered price of 150 ¢/MBtu would amount to about $300,000 per year.

4.9.1 Sample Quantity

The amount of sample taken from the shipment lot must be adequate to be representative of the shipment. The basic procedures for collecting such a sample are described in ASTM Standard D2234.

To properly collect a representative sample, consideration must be given to the physical character and variations in the properties of the coal, the number and size of the sample increments composing the gross sample, and the required precision.

A good sampling system affords every particle in the shipment lot an equal opportunity of being selected and ultimately analyzed in the laboratory. If this occurs, a representative sample of the shipment lot has been obtained. To obtain the gross sample, several increments or small portions of the shipment lot are collected, usually using an automatic sampling device. The number of increments taken depends on whether or not the shipment of coal has been cleaned (processed in a beneficiation plant), or represents run-of-mine raw coal. More than twice the number of increments are required for raw coal because of the larger variation in properties as compared to beneficiated or washed coal.

The maximum size of a shipment lot of coal that can be represented by a gross sample should not exceed about 10,000 tons. This is equivalent to the usual unit train shipment or the quantity of coal contained in one hold of a Panamax size vessel.

The minimum weight for each increment of sample taken depends on the top size of the coal. With increasing top size, the required minimum weight for each sample increment increases. If the required number of increments and minimum weight for each increment becomes prohibitive, the gross sample can be divided to reduce its weight. For example, the large increment can be reduced in quantity by secondary sampling, that is, taking at least six secondary increments from each primary increment. Prior to taking the secondary increments from the primary increment, the primary increment may be crushed to a smaller top size. The number of these sample reductions can vary between sampling systems.

Each of the cutters that cuts the full stream of coal has an aperture opening that is at least three times the diameter of the maximum top size of the coal stream being sampled.

After the sample has been collected and sealed in a container to prevent moisture loss or gain, it is shipped to the laboratory and prepared for analysis.

4.9.2 Preparation of a Sample for Analysis

When received at the laboratory, gross samples can weigh from 20 or 30 lb for top sizes of 8-mesh or up to several thousand pounds if the top size is 6 in. or greater. For most laboratory analyses, the top size of the coal must be reduced. ASTM Standard D2013 presents the procedures for preparing the gross sample into representative splits with a top size of No. 60 sieve size (250 microns) and having sample weights of 50 to 100 g. A process similar to that used in obtaining the gross sample is used to obtain a representative laboratory sample. That is, the top size of the coal is reduced and the resultant material is divided (or riffled) into two equally representative portions, one of which is retained for further processing, and the other discarded.

4.9.3 Accuracy and Reproducibility of Laboratory Results

Good sampling technique requires not only accuracy but also precision and reproducibility of results. For mechanical sampling systems, bias tests are performed periodically to ensure that the gross sample is representative.

As defined by the ASTM Test Procedure, precision is used to indicate the capability of a person, an instrument, or a method to obtain repeatable results. Accuracy is a term used to indicate the reliability of a sample, a measurement, or an observation. It is a measure of the closeness of agreement between the experimental result and the true or actual value. Reproducibility is the difference in results obtained by two different laboratories employing the same procedures but different equipment when analyzing splits of the same gross sample.

Table 4-9 shows ASTM reproducibility standards between laboratories analyzing different splits from the same gross sample. This is the kind of variation one could expect between the three samples that are usually taken during the loadout of a coal unit train. One sample is used to analyze the key properties of the shipment such as moisture, ash, sulfur, and heating value, which form the basis for the commercial transaction. The other sample is sent to the buyer's laboratory for analysis. A third sample is held as a "referee" sample and is used should a dispute over the coal's quality result between the buyer and seller. Therefore, the differences between these results as shown in Table 4-9 would apply to such a situation.

It is seen that the variation between laboratory results for the same property can be significant. For example, the table shows that a difference of 100 Btu/lb (55 kcal/kg) between two laboratories may actually be no difference at all when using ASTM approved coal analysis methods.

4.9.4 Bases of Analyses

The primary difference in the way analyses are reported is based on how moisture is treated in the analysis. For example, in some Asian countries, the sample is air-dried on reaching the laboratory to attain equilibrium conditions with the laboratory environment. These analyses are then reported on an air-dried basis. In the United States, it is typical to report the results on an as-received basis, that is, with the moisture in the sample as it was when received at the laboratory. Furthermore, the air-dried moisture is often referred to as the inherent moisture for the coal. For ASTM test methods, inherent or equilibrium moisture represents the moisture retained in the sample when it is subjected to a relative humidity of 95% to 97% in a desiccator.

Figure 4-25 illustrates the relationship between these different types of moistures. The difference between as-received moisture and air-dried moisture is relatively small for high-rank bituminous coals. However, as more low-rank coal deposits are developed (such as in Asia), the difference between as-received and air-dried moisture can become significant.

Table 4-9. ASTM Reproducibility Requirements Between Laboratories for Coal Quality Analyses

Type of Analysis	Variation Allowed Between Analyses
Heating value, dry basis (Btu/lb)	100 (55.5 kcal/kg or 233 kJ/kg)
Proximate analysis (%)	
Moisture	0.65
Ash	0.3–1.0[a]
Volatile matter	0.6–2.0[a]
Fixed carbon	—[b]
Ultimate analysis (%)	
Moisture	0.65
Carbon	0.3[c]
Hydrogen	0.07[c]
Nitrogen	—[a]
Chlorine	0.06
Sulfur	0.1–0.2[a]
Ash	0.3–1.0[a]
Oxygen	—[b]
Sulfur forms (%)	
Pyritic	—[a]
Sulfate	0.04
Organic	—[b]
Mineral analysis of ash (%)	
Silicon dioxide, SiO_2	2.0
Aluminum oxide, Al_2O_3	2.0
Titanium dioxide, TiO_2	0.25
Ferric oxide, Fe_2O_3	0.7
Calcium oxide, CaO	0.4
Magnesium oxide, MgO	0.5
Potassium oxide, K_2O	0.3
Sodium oxide, Na_2O	0.3
Sulfur trioxide, SO_3	0.2–1.0[a]
Phosphorus pentoxide, P_2O_5	0.15
Undetermined	—[b]
Fusion temperature of ash, °F	
Reducing atmosphere	
Initial deformation	125 (70°C)
Softening ($H = W$)	100 (55°C)
Hemispherical ($H = W/2$)	100 (55°C)
Fluid	150 (85°C)
Oxidizing atmosphere	
Initial deformation	100 (55°C)
Softening ($H = W$)	100 (55°C)
Hemispherical ($H = W/2$)	100 (55°C)
Fluid	100 (55°C)
Moisture	
Equilibrium moisture (%)	0.5–1.5[a]
Hardgrove grindability index	3 Index points
Size consist (% by weight)	0.1
Free swelling index	2 Index points

[a]Variation allowed depends on amount of this and/or other constituents present in the sample. See ASTM Volume 5.05, *Gaseous Fuels: Coal and Coke* for details.

[b]Not applicable. Amount of constituent is determined by subtraction from total of all other constituents in the analysis.

[c]Repeatability within the same laboratory. Reproducibility requirement is greater.

Values shown as percentage points unless otherwise stated.

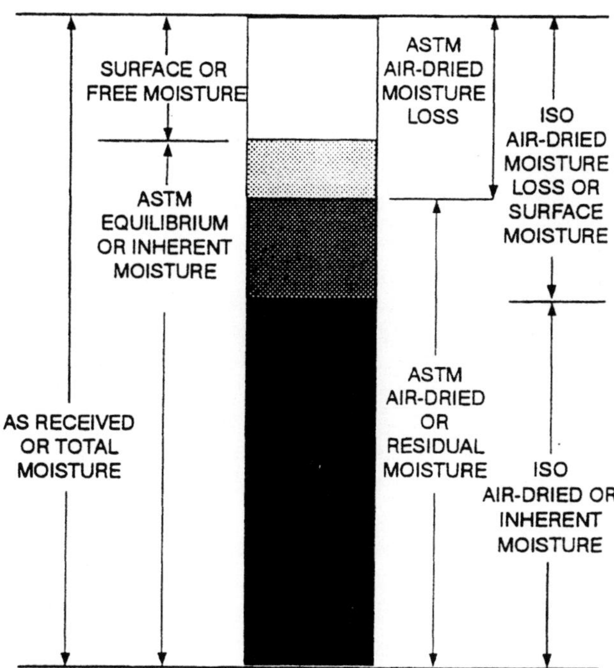

Fig. 4-25. Forms of moisture in coal.

The power plant design engineer must design the plant for the properties of the coals that are likely to be received at the power plant, not those that have come into "equilibrium with laboratory conditions" which can change depending upon the environmental control in the laboratory.

Coal properties may also be reported using different bases. The relationships between properties reported on an as-received, air-dried, dry, ash-free, and dry and ash-free basis are shown in Table 4-10.

4.9.5 Differences Between Coal Testing Standards

The ASTM standards for determination of coal properties are used primarily in the United States but have also been adopted by other countries. However, Australia, Japan, Germany (DIN), South Africa, and the United Kingdom have their own standards that are sometimes employed. The International Standards Organization (ISO) has also developed coal testing standards. The intent is that over time, countries will adopt the ISO standards as their national standards.

For example, to determine volatile matter content, the ISO standards heat the coal sample at a temperature of 1,652° F (900° C) for 7 minutes while the ASTM standards require heating the sample at 1,742° F (950° C) for 7 minutes. In a test conducted by the Consolidation Coal Company on an eastern United States high-sulfur coal, the following differences were noted (Technical Bulletin TB-4 1983):

- Ash—Procedures were very similar (as were the results).
- Volatile matter—ASTM was 1% to 3% higher than other test methods depending upon the volatile matter level.
- Gross calorific value—All used similar procedures and gave similar results. The results may be reported on a lower or net heating value basis, while in the United States, they are reported on a gross or higher heating value basis.

Since coal properties may be analyzed and reported differently depending on the standard used for analysis, some of the engineering design practices used in sizing furnaces and convection passes, such as fuel ratio, a measure of the coal's combustibility, may be different when using a standard that is different from the one on which the original design practice was based.

When an analysis of coal properties is being interpreted, two important questions must be answered: "How representative was this sample of the coal that will be burned at the power plant?" and, "What standards were used in performing the sampling and analysis?"

4.9.6 Automated Sampling and Analysis Systems

With more stringent environmental restrictions requiring better control of coal quality (particularly its sulfur dioxide emission potential), some electric utilities have installed coal blending systems at their power plants. To facilitate these blending operations, on-line coal analyzers are used. These on-line performance analyzers use Prompt Gamma Neutron

Table 4-10. Converting Analyses to Different Bases

	Desired Basis			
Given Basis	As-Received (ar) Multiply by:	Air-Dried or As-Determined (ad) Multiply by:	Dry (d) Multiply by:	Dry, Ash-Free (daf) Multiply by:
As-received (ar)	—	$\frac{100 - M_{ad}}{100 - M_{ar}}$	$\frac{100}{100 - M_{ar}}$	$\frac{100}{100 - M_{ar} - A_{ar}}$
Air-dried or as-determined (ad)	$\frac{100 - M_{ar}}{100 - M_{ad}}$	—	$\frac{100}{100 - M_{ad}}$	$\frac{100}{100 - M_{ad} - A_{ad}}$
Dry (d)	$\frac{100 - M_{ar}}{100}$	$\frac{100 - M_{ad}}{100}$	—	$\frac{100}{100 - A_{d}}$
Dry, ash-free (daf)	$\frac{100 - M_{ar} - A_{ar}}{100}$	$\frac{100 - M_{ad} - A_{ad}}{100}$	$\frac{100 - A_{d}}{100}$	—

M = moisture, percent; A = ash, percent; ar = as-received basis; ad = air-dried or as-determined basis; d = dry basis; daf = dry, ash-free basis.

Activation Analysis techniques for measuring the elemental content of the coal.

Some of the neutrons emitted from a source are absorbed by a nucleus of an atom in the coal being analyzed. The extra energy acquired by the atom in absorbing the neutron causes it to become unstable. The unstable nucleus emits this excess energy in the form of gamma rays. The energy level of the gamma rays is related to the nucleus of the atom (element) that emitted them. The elements and the degree of their presence in the coal can then be determined. The primary properties of the coal such as its calorific value, volatile matter, ash fusion temperatures, and sulfur dioxide emission potential (lb SO_2/MBtu) are then estimated by calculation based on the elemental analysis of the coal. Some elements can be measured fairly accurately, while others do not activate well in a neutron flux. This impacts the accuracy of the estimate of properties for the coal.

4.9.7 Automated Analyzers

ASTM Standard D5142 covers the standard test methods for proximate analysis of the analysis sample of coal and coke by instrumental procedures. There are automatic analyzers, particularly those that measure sulfur, that may not necessarily follow ASTM standards. While they may not be "ASTM Approved," if properly calibrated, they appear to perform fairly well.

The important thing is that both the buyer and seller agree on the standards and devices that are to be used in sampling and analysis and that they come to some agreement regarding their use and the frequency and procedures for their calibration.

4.10 COAL BENEFICIATION

Beneficiation refers to the removal or separation of the desirable constituents from the undesirable constituents in coal. Some undesirable constituents that exist in coal are sulfur, ash, and moisture. These constituents contribute very little to the useful energy of the coal and are detrimental to power generation, thereby increasing the as-fired cost of utilizing the fuel. There is also the additional cost of transporting the undesirable constituents and increased utilization costs from slagging, fouling, and erosion of heat transfer surfaces, and the cost of disposing combustion wastes.

Many processes have been developed for the removal of undesirable constituents from coal before it is burned in the furnace. Beneficiation is a generic term that is applied to this broad category of coal cleaning technologies. Other terms that are used synonymously with beneficiation are coal cleaning, coal washing, and coal preparation. Beneficiation primarily refers to the removal of sulfur, ash, and moisture from the coal. However, it is generally not used to refer to processes that alter the coal's structure, including those that occur in processes such as coal gasification and coal liquefaction.

Beneficiation processes are broadly categorized as chemical, physical, and biologic. The discussion focuses on physical beneficiation processes that are the most prevalent. The level of beneficiation, or the degree to which undesirable properties such as ash and sulfur can be removed from the coal, is dependent on the coal's properties and the new market opportunities for the upgraded clean coal product.

4.10.1 Evaluating a Coal's Beneficiation Potential

When the beneficiation potential of a particular coal is evaluated, the form and structure of the coal must be analyzed. Whether beneficiation is technically and economically feasible depends on the form of the undesirable constituents and how they are distributed throughout the coal, that is, the washability characteristics of the coal. For example, sulfur can exist in several different forms. Organic sulfur is chemically bound to the carbon, that is the most desirable constituent of the coal. The inorganic sulfur (such as marcasite and pyrite) is usually concentrated as mineral particles and dispersed throughout the coal seam. It is this type of sulfur that physical beneficiation processes attempt to remove. Sulfate sulfur may also be distributed throughout the coal, but this usually represents a very small percentage of the total sulfur and most beneficiation processes do not attempt to remove it.

Mineral matter (ash) may exist as fine material disseminated throughout the seam, as distinct partings within the seam, and out-of-seam dilution composed of roof and floor material. The out-of-seam dilution and in-seam partings recovered during mining operations are usually independent of the carbon matrix (they are "liberated") and are more easily removed. The distribution of organic and pyritic sulfur, as well as the ash, may vary greatly between seams and can depend on the type of mining operation used in recovering the coal. Therefore, the beneficiation potential can vary significantly between mines and between seams.

Table 4-11 shows the specific gravity distribution of various coal constituents. As shown in this table, a significant variation exists between the specific gravity of desirable constituents (pure coal) and the undesirable constituents,

Table 4-11. Specific Gravity of Coal Constituents

Coal Constituent	Specific Gravity
Pure bituminous coal	1.12–1.35
Middlings (bone coal)	1.35–1.70
Carbonaceous shale[a]	1.6–2.2
Shale	2.0–2.6
Clay	1.8–2.2
Pyrite (iron disulfide)	4.8–5.2

[a]High in ash, very little heating value.

Source: *Coal Preparation for Combustion and Conversion*, Electric Power Research Institute, EPRI AF-791, May 1978, p. 2–9.

such as ash (S.G. ≅ 2.2) and pyritic sulfur (S.G. ≅ 4.8 to 5.2). If these constituents are liberated from carbonaceous material, a gravimetric separation process could be employed to separate the desirable low specific gravity constituents from the undesirable high specific gravity constituents. Most physical coal beneficiation processes use gravimetric separation methods. For sulfur, it must first be ascertained how much of the sulfur is in the pyritic (inorganic) form and therefore available for removal using physical coal cleaning methods.

Figure 4-26 is a photograph of several pieces of coal taken from various mines. Piece No. 1 shows a spotted pattern in the carbon that represents small nodular particles of pyrite. This piece of coal would have to be crushed to a very fine size to enhance the density difference between the particles containing nodular pyrite and those containing mostly carbon. All of these fine particles would complicate the beneficiation process. On the other hand, piece No. 2 contains pyrite in a dendritic (treelike) form. A large portion, particularly by weight, of this particle is associated with the heavy pyrite. Therefore, this large heavy particle of undesirable coal could be removed fairly easily using a simple gravimetric process. Piece No. 3 is virtually 100% pyrite and is the kind of material that sometimes collects in the joints or on the top and bottom portions of the coal seam. This large piece of heavy material is very easily removed which would have a significant impact on reducing the sulfur dioxide emission potential of the coal.

If the sulfur is not in the pyritic form, as shown in the photographs, but rather is organically bound to the carbon matrix, little can be done using gravimetric separation processes to reduce the sulfur dioxide emission potential of the coal.

For some coal seams, the organic sulfur level is fairly low and uniform with variations in total sulfur being primarily attributable to the variability in pyritic sulfur. This characteristic, combined with pyrite which is easily liberated from the carbon matrix without requiring significant crushing, leads to good potential for technical and economic feasibility using traditional physical coal beneficiation processes.

Fig. 4-26. Distribution of pyritic sulfur in coal.

4.10.2 Float/Sink Washability Tests

The most commonly used test to determine the suitability of a coal for physical beneficiation is the float/sink washability test. This test also determines how much carbon (Btu) will be lost during the beneficiation process. The loss in carbon has a significant impact on the economic feasibility of the process.

The results of the float/sink washability test show the quantity and quality of the coal that can be recovered at various separation gravities. The results are used to develop process flow sheets. The flow sheets show the type and arrangement of coal sizing, gravity separation, and dewatering equipment.

4.10.3 Coal Beneficiation Processes

Coal beneficiation processes try to duplicate what was done in the laboratory float/sink tests. Two beneficiation processes are the most prevalent: the baum jig and the heavy media cyclone. Both use water as the primary separation medium.

In a coal-cleaning plant using a jig, the raw coal is fed into a coal jig or wash box such as the one shown in Fig. 4-27. As the coal enters the jig, vertical pulses of water and air begin to stratify it by particle size, shape, and weight. The lighter carbon-dominated particles migrate toward the top and the heavier pyrite- and ash-dominated particles settle toward the bottom. The intensity and frequency of the pulsations are the primary determinant of the effective separation gravity. By the time the coal reaches the end of the jig, it is segregated into a top layer that is dominated by lighter carbonaceous material and a bottom layer that is dominated by the heavier particles containing significant quantities of ash and pyrite. These fractions are divided into clean coal, refuse, and sometimes middlings (neither clean coal nor reject) streams.

The heavy media cyclone uses a medium that has a specific gravity close to the specific gravity of the desired separation. The principal medium used today is magnetite or iron oxide (Fe_3O_4) suspended in water. This medium can be used in heavy media separations involving vessels, baths, or cyclones of the type shown in Fig. 4-28. The heavy media cyclone can produce a fairly precise separation with little misplaced material. That is, particles with a specific gravity greater than the separation gravity have a high probability of ending up in the refuse stream and particles with a specific gravity lower than the separation gravity have a high probability of ending up in the clean coal stream.

The heavy media cyclone relies on centrifugal force and the addition of magnetite in a water slurry to effect the separation. Its operating principle is similar to that of the mechanical dust collectors installed at saw mills, factories, and in power plants to remove particulate matter from the exhaust or flue gases. Those particles with a specific gravity lower than the effective specific gravity of the magnetite/water slurry, float off in the overflow from the cyclone. The magnetite is reclaimed using magnetic separators to reduce makeup quantities.

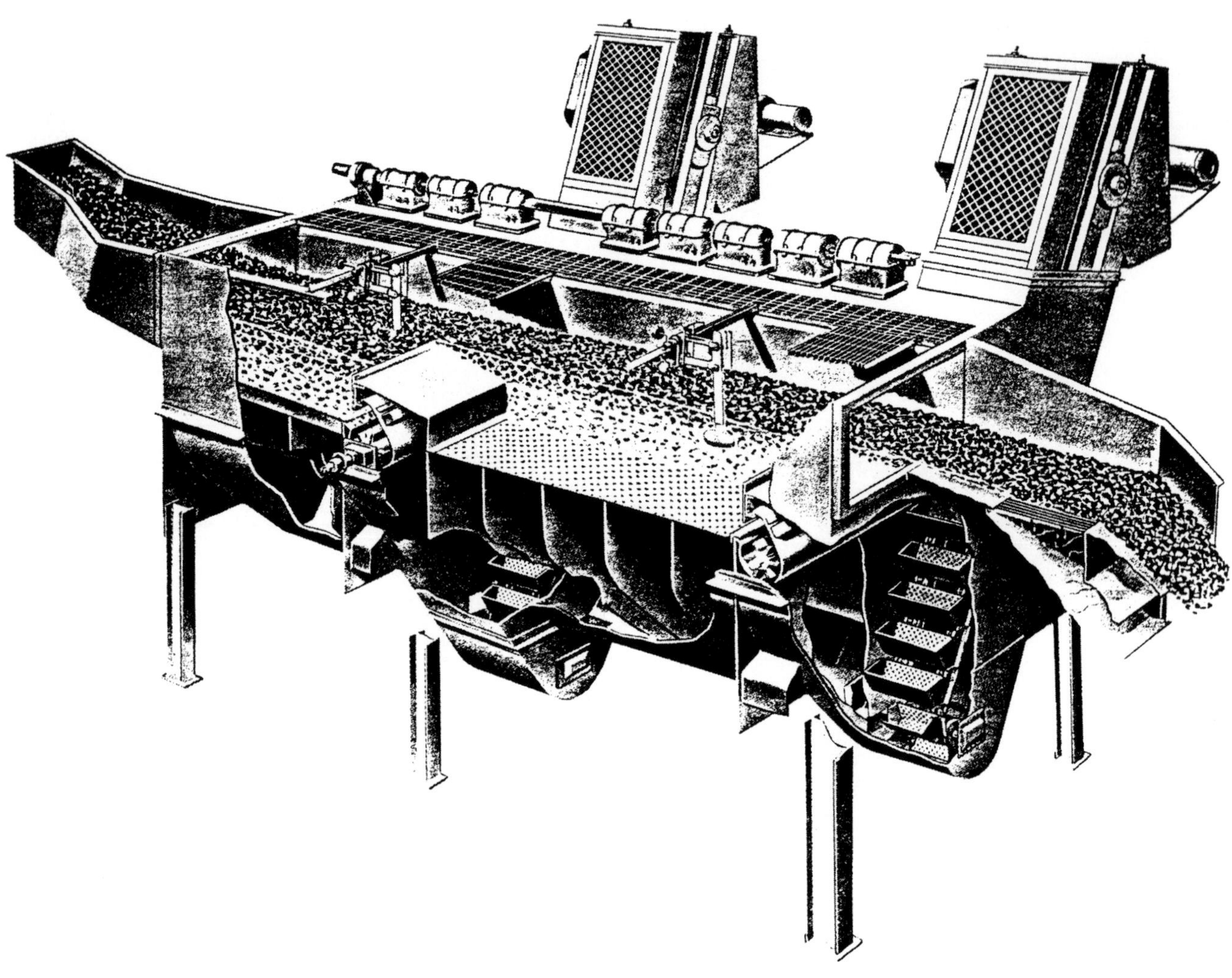

Fig. 4-27. Coal jig. (From Jeffrey Division of Indresco. Used with permission.)

4.10.4 Economics of Coal Beneficiation

In terms of the cost of coal beneficiation, the primary costs are the capital investment associated with the cleaning plant, operating costs, including the cost of disposing refuse, and the amount of energy in the raw coal that is lost to the refuse stream and must be replaced with additional mined coal.

Current environmental regulations in the United States require such low sulfur dioxide emission levels for new power plants that some form of post-combustion sulfur dioxide removal system is almost always required. Therefore, the cost savings from having to remove less sulfur dioxide and ash using post-combustion cleanup systems and the benefits of utilizing a coal with potentially lower slagging, fouling, erosion, and reduced transportation cost must be compared with the added cost of coal beneficiation.

Therefore, one of the first steps in evaluating the use of a coal beneficiation process is to compare the cost of the process versus the total benefits derived by the consumer in using the upgraded fuel.

4.11 FOSSIL FUEL ECONOMICS

Black & Veatch has spent considerable effort during the past 15 years in trying to understand and evaluate the underlying factors that influence fuel prices. This has included a study of coal and crude oil prices back to the mid-1800s. The results of this effort have shown that fossil fuel prices, when expressed in constant dollars, exhibit cyclical price behavior. Constant or real dollars are computed by taking the nominal dollar prices and adjusting them for general inflation in the economy since that time as measured by the broadest measure of inflation, the Gross National Product Implicit Price Deflator (GNPIPD). For example, to convert nominal, actual, or current dollars to real or constant 1993 dollar prices, Eq. (4-2) is used:

$$\begin{gathered}\text{(Price in nominal dollars for 19XX)}\\ \times\ \text{GNPIPD for 1993/GNPIPD for 19XX}\end{gathered} \tag{4-2}$$

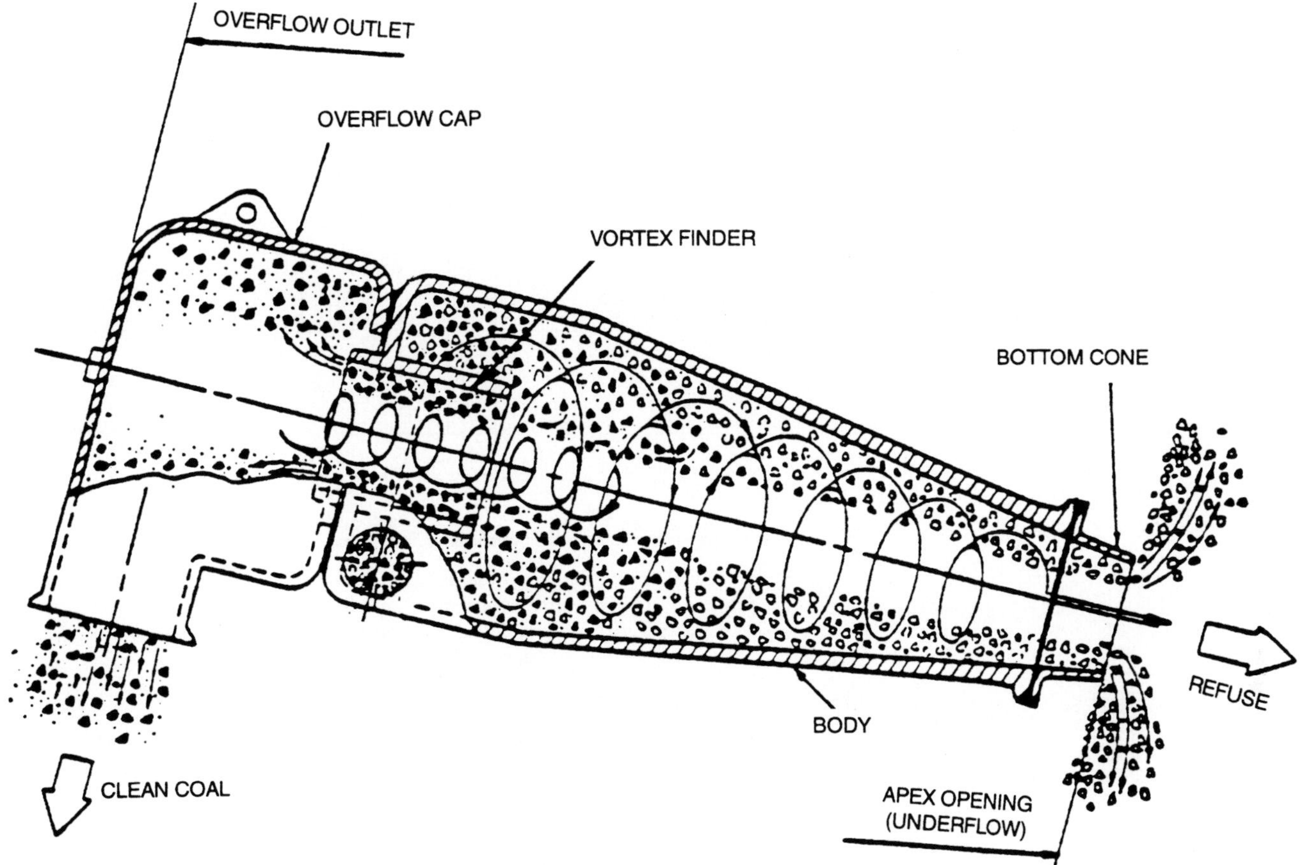

Fig. 4-28. Heavy media cyclone. (From Roberts & Schaefer Co. Used with permission.)

The following discussion will present the reason for this cyclical price behavior and some evidence as to why it is likely to continue in the future.

4.11.1 What We Learn from History

Figure 4-29 shows the price of imported oil since 1968. The prices are shown in 1993 constant dollars with past prices adjusted for inflation using the GNPIPD. This figure also shows how forecasts for the price of oil have varied since 1981. These forecasts by the Energy Information Administration (EIA) of the Department of Energy are similar to those made by many private forecasters during the same period.

While we often speculate about the future, the past is something we do not have to guess about. Figure 4-30 shows the price of United States coal and crude oil for the last 100 years expressed in 1993 constant dollars. The prices are shown on a logarithmic scale, such that equal distance in the vertical direction represents equal percentage change in the price of the fuel (crude oil or coal). While most current forecasts call for gradually rising prices through the year 2000 and beyond, the actual historical experience has been quite different. Rather than constant rates of change over long periods of time (that are represented by straight lines on a logarithmic graph), fuel price changes have demonstrated significant volatility. Figure 4-31 shows that price volatility, as measured in terms of annual price changes, has been significant with the volatility in the price of oil being about twice that of coal.

The low prices for each cycle throughout this long period of history are remarkably similar. Over this long period of time, one would have expected that depletion[12] effects would have led to gradually rising real or constant dollar prices. The fact that they do not do this suggests that advancements in exploration, development, and production technologies have counteracted the effects of increased scarcity.

Throughout history, there has been periodic concern about the diminishing availability of a given resource. However, it is likely that we will never run out of any scarce resources, provided a free market exists and supply and demand forces are allowed to operate freely. In other words, scarcity of a fuel or commodity leads to an increase in price, that leads to an increase in supply of that particular commodity or substitutes that are often better than the original in satisfying primary needs. Some examples are kerosene for

[12]Resource depletion is the upward pressure on market-clearing fuel prices over time, relative to what the price would normally be if there were an infinite supply of the fuel under geological conditions similar to current conditions.

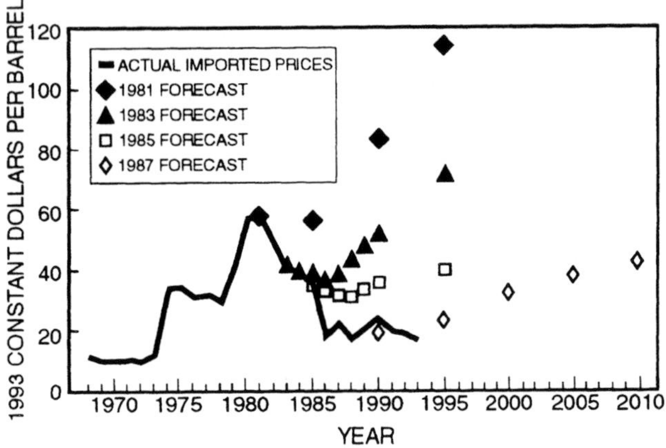

Fig. 4-29. Oil price forecasts. (From Annual Energy Outlook by EIA.)

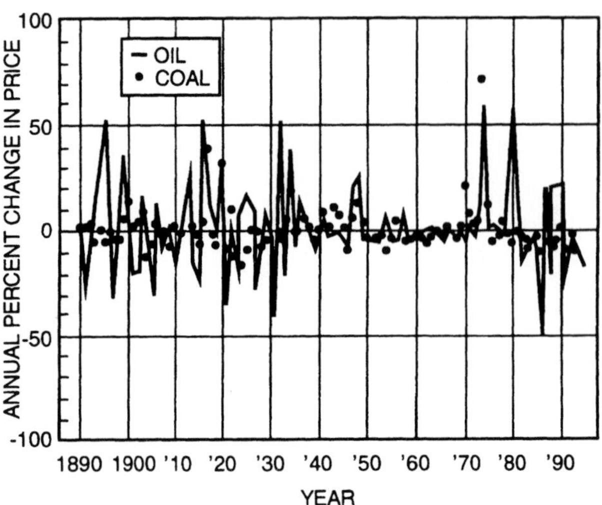

Fig. 4-31. Price volatility.

whale oil, synthetic rubber for natural rubber, and fiberoptics for copper wire.

4.11.2 Consumer and Supplier Responses to Changing Prices

The general scenario is that during a decline in prices, productivity and performance at coal- and oil-producing facilities improves. Technical innovations that were so expensive during the boom years are now priced more reasonably, and producers are more interested in using them to remain competitive in a marketplace of falling real prices.

Under the backdrop of falling prices, fuel consumers start reducing inventories; fuel is no longer a good investment. The need for a 90-day coal inventory policy is reduced to 60, 40, and 30 days, and then "just in time." The money saved from a policy of buying less expensive fuel each month, having a less expensive fuel in inventory, and less inventory, is considerable. However, there comes a time when prices have been reduced to such low levels that there is little incentive for the producer to invest additional capital to purchase marginally better technology to improve the productivity of existing operations. Rather, the less competitive suppliers begin to leave the marketplace. This has been the recent trend.

This trend accelerates as many large coal mines that were predicated on 20-year reserve blocks cannot afford to develop new reserves since the market prices are not sufficient to cover the cost of the new capital investment and the required investment returns.

When a new price cycle begins, the time required to purchase new reserves, design the mine, obtain the required permits and licenses, arrange financing, and construct the mine can be several years. Also, the fuel consumer's response to rising energy prices is not immediate. For example, it usually takes several years of rising prices for consumers to realize that they are not experiencing a normal aberration in the long-term downward trend in prices that they have experienced for many years. It is often near the top of the fuel price cycle where most consumers convert plans into action. For example, the article "Eastern Utilities Get Started on Conversions Back to Coal" appeared in the August 1980 issue of *Electric Light and Power*, one year before oil prices peaked in 1981.

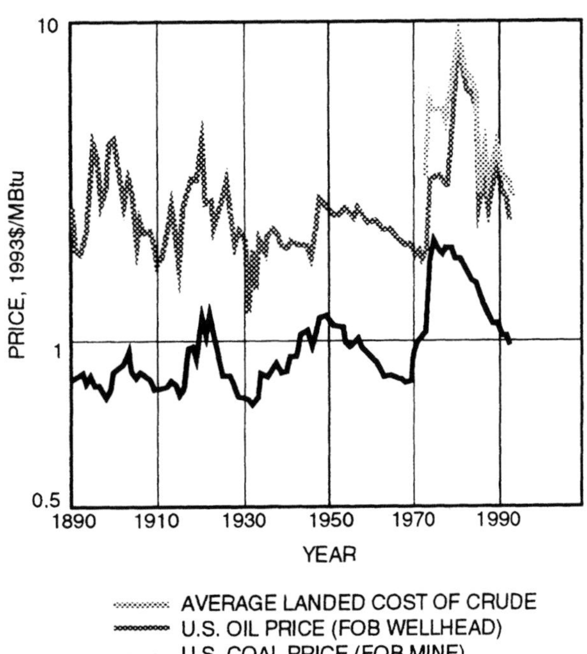

Fig. 4-30. Historical US oil and coal prices.

4.11.3 Regulatory Influences on Supply and Pricing

The free movement of prices adjusts supply and demand, allocates resources, and promotes substitution to maximize economic efficiencies. However, if we look at the energy markets, prices have often been subject to controls for long periods of time. For natural gas, regulation began in 1938 with the Natural Gas Act that controlled the interstate pipeline transportation rates and required that these pipelines have at least 20 years of gas reserves under long-term contract prior to constructing a new pipeline. The end-user purchased the gas from the local distribution company who purchased it from the interstate pipeline who had long-term contracts with the gas producers.

In 1954, a Supreme Court decision interpreted the Natural Gas Act as requiring the Federal Power Commission to regulate natural gas wellhead prices (Phillips Petroleum Company 1954). The Federal Power Commission adopted a policy of low-wellhead gas prices to decrease the price for residential consumers. During the rise in energy prices during the 1970s, this led to large price disparities not only between oil and natural gas, but between different natural gas supplies. For example, gas wells serving unregulated intrastate markets had prices that were several times that of a nearby controlled well that supplied the interstate markets.

Price controls were extended to crude oil in 1971, and as the historical price data on Fig. 4-30 shows, there was a large disparity between United States domestic crude oil prices and the landed cost of imported crude oil. With such an environment, it is difficult to get investors to invest funds in oil and gas drilling prospects.

The United States Congress recognized these limitations and passed the Natural Gas Policy Act of 1978 that mandated the phased decontrol of natural gas wellhead prices. In this same year, the Fuel Use Act was passed that stated that all electric utilities should quit using natural gas in generating electricity after 1990. Deregulation of the wellhead price of crude oil began in January 1981. In May of 1988, the Fuel Use Act was amended to allow electric utilities to again utilize natural gas.

4.11.4 Interfuel Competition

The degree of interfuel competition depends on the type of market into which the fuel is being sold. For example, natural gas has a hierarchy that is a function of the marginal utility of usage among end-user categories. These categories include residential, commercial, industrial, and electric utility. Obviously, the residential consumer places a higher value on utilizing natural gas for heating than he would on coal. The electric utility providing baseload power does not put as much marginal utility on the ease of use as would the commercial or residential consumer.

Prior to World War II, coal markets were regional, serving smaller local power plants, industrial and retail users (for heat sources), and railroads for fueling steam driven locomotives. Today, coal is usually used to serve large stationary energy needs, natural gas for unattended stationary energy needs, and light oil products are the premier fuels in the transportation sector.

There can be interfuel competition only if there is excess production capacity for both competing fuels. Once excess production capacity has been depleted, additional supply to satisfy a given level of demand can come only from the development of new fuel sources.

4.11.5 Cost and Price Structure of Fossil Fuels

For each type of fossil fuel, there are significant differences between the geologic settings, production technologies, markets, and producer–consumer relationships. The primary costs incurred in providing fuel to markets include production of the unprocessed fuel, refining or coal beneficiation, marketing, and transportation. On the other hand, the pricing of the fuel is determined by its cost of production, the degree of competition, its uniqueness (environmental acceptability, handling characteristics, storage, and usage characteristics), and the degree of interfuel competition or potential substitution. Pricing parity with other end-user alternatives is also an important factor.

During the downward portion of the "price cycle," competition within each primary producing region forces prices down to the variable cost of production. During the recent oil price decline since 1981, the cost of operating wells has also decreased. Near the bottom of the cycle, only the variable or cash costs of production are covered. At this low point, only the large low-cost suppliers with production from low-cost reserves survive.

As fuel prices are forced down to near the cash cost of production near the bottom of the cycle, this is usually when the oil minus coal price differential also tends to bottom. At this time, some electric power generators begin to consider premium fuels such as oil and natural gas instead of coal. These fuels usually have a low capital cost associated with their use in electric power generation. As the marginal high-cost fuel producers leave the market, supply concentration increases. This is opposite to what occurs during periods of expanding price, which encourages many new entrants into the market and a decreasing concentration of producer power.

During the upside portion of the "price cycle," the reverse phenomenon takes place. As the price of oil begins to increase, there is fuel switching from oil and natural gas to coal, that causes an incremental increase in demand that cannot be satisfied without developing new mines because existing mines have depleted their reserves. Throughout the resource supply chain, suppliers of goods and services begin marking up the price of those goods and services compared to what had been charged in the past. This changing cost structure of the industry is illustrated in Fig. 4-32, which shows drilling cost as measured in dollars per foot for gas wells during the period 1960 to 1991.

In addition, if new oil and gas fields are to be developed, higher risk prospects must be drilled which requires higher prices to cover higher dry hole costs. The net result is a rapid price rise in the early years of the next "price cycle."

Meanwhile, there is a lag between market price increases and market responses. For example, there may have been

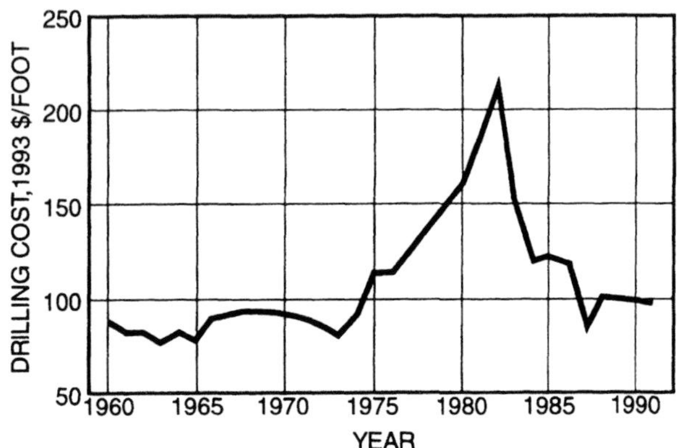

Fig. 4-32. Drilling cost for natural gas wells. (From Annual Energy Review, 1992, EIA.)

several years during the previous price decline when prices rose significantly for a year or two but ultimately led to a continuation in the decline in real prices. However, eventually a market price response takes place that causes end-users to increase inventories and make long-term commitments that allow the producers to invest in new production capability. In addition, new mine expansion brings a need for new workers, training programs, and a demand for higher wages that was foregone during the previous years of declining prices. This also leads to a decrease in productivity, that adds cost and price pressure.

As with other goods and services in a free market, the price of oil and gas is established by the interaction of the forces of supply and demand. The market supply curve represents the quantity supplied at any given price. In the short term, supply will be influenced by the supplier's ability to cover his variable costs of production. As long as the supplier can cover all variable costs and make a contribution toward fixed costs, he will have incentive to continue production in the short term. In the long term, however, all costs including fixed costs, must be covered for the supplier to remain in the industry. The factors affecting the long-term supply costs include depletion; level of technology employed in the discovery, development, and production of the resource; and the cost and productivity of labor and capital.

4.12 DESIGN BASIS FUEL QUALITY AND PROCUREMENT PROCESS

There are many steps in the fuel procurement process that ultimately lead to the successful design, construction, and operation of a power plant.

- Fuel supply study,
- Fuel procurement strategy,
- Bidder prequalification,
- Request and evaluation of fuel supply proposals,
- Request and evaluation of transportation proposals,
- Evaluation of ownership and maintenance alternatives for transportation equipment,
- Selection of finalists,
- Possible visits to mines and transportation facilities,
- Evaluation of intangible factors,
- Selection of candidates for negotiations, and
- Negotiation of fuel supply and transportation contracts.

Some of these steps may require studies or investigations to provide input to the decision making process. As listed above, the process usually begins with a fuel supply study. The primary tasks associated with the fuel supply study are as follows:

- Identify potential fuel types.
- Identify potential fuel sources.
- Evaluate reserves (quantity and quality).
- Determine production capability and quality.
- Identify current and future markets.
- Evaluate current and future FOB mine or FOBT ship costs.
- Identify potential transportation modes.
- Evaluate potential transportation systems.
- Evaluate current and future transportation costs.
- Estimate current and future delivered cost of fuel.
- Determine fuel handling and storage/inventory requirements.
- Evaluate combustion performance and costs (e.g., capital, operation and maintenance, availability).
- Determine combustion products disposal requirements and costs.
- Compare the busbar generation economics of each alternative.
- Evaluate intangible factors.
- Establish preliminary design basis fuel quality.

The three primary outputs of the fuel supply study are the most competitive fuel supply and transportation alternatives, the preliminary design basis fuel quality, and input for use in developing a fuel and transportation procurement strategy.

The preliminary design basis fuel quality would be confirmed before awarding contracts for major coal quality dependent equipment such as the steam generator and air quality control equipment. Confirmation would result from the solicitation of fuel supply proposals from those potential bidders identified in the fuel supply study. The detailed fuel quality data submitted with the proposals would be compared with the preliminary design basis fuel quality established during the fuel supply study. The results of this comparison would also be used to confirm that sufficient design margin has been included to provide some flexibility in the procurement of fuel and transportation.

4.13 COMMERCIAL CONSIDERATIONS

The larger, more efficient power plants developed in the late 1950s and early 1960s to obtain economies of scale also encouraged long-term contractual commitments for fuel and the construction of minemouth power plants. Also during the 1960s, competitive threats from low oil prices delivered to east coast power plants and the threat of utilities switching to nuclear power put market pressure on coal prices. Therefore, coal companies also preferred long-term contracts. Today, project financing for independent power generation projects often requires a fuel supply contract for the period of the loan or 15 to 20 years.

The greater environmental awareness in the 1950s and 1960s led to the Clean Air Act of 1955 that was amended in 1963, 1966, and 1970. These regulations, combined with low oil and natural gas prices relative to coal, caused utilities to switch from coal fueled alternatives to relatively clean oil burning facilities in the late 1960s and early 1970s. This was particularly true for electric utilities located near natural gas fields or distant from the coal fields.

This is similar to the experience of the late 1980s and early 1990s when the price of electricity generated by oil and natural gas fueled plants became more competitive with electricity generated from coal-fueled plants. In some respects, we may now be replaying the experience of the late 1960s and early 1970s, prior to the last energy crisis in the 1970s. If so, the importance of maintaining adequate fuel flexibility in designing the power plant will pay large dividends in the future. The objective of fuel flexibility is to maintain competitive alternatives at each link in the fuel supply chain.

4.13.1 Commercial Relationships

The corporate entity with whom a purchaser is signing the contract is very important. As the buyer moves away from organizations that own and produce the reserves, the strength of the relationships begins to deteriorate.

4.13.2 Purchase of Natural Gas

The traditional method of purchasing natural gas was from the local distribution company (LDC) under a tariff schedule. The gas consumer would purchase the gas from the LDC who in turn had long-term contracts for gas and transportation with the interstate pipeline. The interstate pipeline performed a merchant role (sold gas) and also provided for its transport. The interstate pipeline had long-term contracts with the gas producers to assure an adequate and reliable supply. As new gas fields were discovered and end-user demand increased, the interstate pipeline would work with the regulatory authorities such as the Federal Energy Regulatory Commission (FERC) and its predecessor, the Federal Power Commission (FPC) to arrange for planning, licensing, establishing tariffs, and constructing and operating the new pipeline. Transportation services were provided on an interruptible and firm basis. That is, as long as there was excess capacity in the line, one could transport gas with a negotiated rate that at times was only slightly greater than the marginal cost of operating the pipeline. On the other hand, natural gas moving under firm transportation contracts was required to pick up most of the costs associated with the return of capital and return on investment, commonly known as the demand charge. In addition, those shippers or buyers who transported natural gas through the pipeline were required to supply the natural gas that was used to fuel the compressors that move the gas.

With the implementation of FERC Order 636 during the early 1990s, the interstate pipeline transportation companies became independent of the natural gas consumer and the producer. Their primary role now is that of a transporter of natural gas.

In the future, interruptible transportation may no longer be available from the interstate pipeline. The shipper will have to deal with other shippers who have purchased firm transportation to determine if they are willing to resell some of their pipeline capacity at that time or during particular seasons of the year.

As a result of FERC Order 636, the end-user can deal directly with the gas producer and transporter and arrange for the purchase and transport of the gas he needs. He can also deal with a broker or a marketer who will arrange for the purchase of gas and its transport to his facilities.

4.13.3 Provisions of Long-Term Coal Supply Agreements

The provisions of the long-term coal supply agreement govern the long-term relationship between the buyer, seller, and third parties such as transporters. While the number of provisions can be quite extensive, as a minimum they usually include the following articles:

- Parties—Description of the parties to the contract;
- Term—The length of the agreement and options for termination or extension;
- Quantity—The amount of fuel to be shipped during the term, year, and other shorter periods of time and how it might vary;
- Quality—Specifications and agreement as to where and how the coal will be sampled and analyzed;
- Price adjustment—Initial price and how it will be adjusted in the future;
- Weighing—How and where the quantity of coal shipped will be determined. The coal should be weighed at the same time it is sampled.
- Transportation responsibilities of buyer and seller;
- Notices;
- Billing and payment—When, where, and how payments for coal shipments will be made;
- Force majeure—The response of the parties to acts beyond the control of either party that affect the buyer and/or seller's ability to perform under the contract;
- Dispute resolution—How (mediation, arbitration, or litigation) and what forum will be used to resolve disputes between the buyer and seller.

There are five primary types of long-term contracts: fixed price, cost plus, base price plus escalation, annual price nomination (evergreen), and market price. Moving from the first to the last type in this list, market forces play an increasingly greater role in the determination of the price of coal.

The annual price nomination or evergreen contract removes some of the problems encountered in the cost plus or base price plus escalation type of contractual arrangement by allowing market conditions to affect the price paid for the coal. Such agreements have been problematic in that often the market where the price is to be determined has not been adequately defined. For example, during some periods in the fuel price cycle, different grades of coal (such as low sulfur compliance versus high sulfur noncompliance) may be escalating at different rates or in opposite directions. The buyer will obviously bring to the annual negotiations evidence that favors a lower price and the seller evidence that favors a higher price. In many cases, failure to agree on a price has led to termination of the contract, lawsuits, or arbitration before a fair market price could be determined.

As a result of past problems with long-term contracts, there are those who advocate the purchase of fuel solely in the spot market. This is particularly true during the downward portion of the fuel price cycle when market prices are weak and contract prices tend to be above spot prices.

Whether a procurement strategy that favors spot prices or long-term contracts is adopted will depend on the answers to many questions regarding the sourcing, transport, and utilization of the fuel. Ultimately, it is not a question of whether long-term contracts are good or bad, but whether their provisions are. Those provisions that do not satisfy the needs of the three parties to the contract—buyer, seller, and the marketplace—will usually promote a deterioration of the relationship between the buyer and the seller.

4.14 REFERENCES

ALLEN, R. M. and J. B. VANDERSANDE. 1984. Analysis of Submicron Mineral Matter in Coal Via Scanning Transmission Electron Microscopy. *Fuel*. Vol. 63, January, pp. 24–29.

AMERICAN PETROLEUM INSTITUTE (API). 1984. *Facts about Oil*.

ANDREWS, HENRY N. JR. 1961. *Studies in Paleobotany*. John Wiley & Sons, New York, pp. 236, 241.

ARESCO, S. J. and A. A. ORNING. 1965. A Study of the Precision of Coal Sampling, Sample Preparation and Analysis. *Transactions*. Society of Mining Engineers. September, pp. 258–264.

BLACK, HERBERT T. 1994. U.S. Coalbed Methane Production. *Natural Gas Monthly*. Energy Information Administration. January, pp. 1–11.

BP Statistical Review of World Energy. 1993. British Petroleum Company. June. London, England.

BUCYRUS ERIE COMPANY. P.O. Box 500, South Milwaukee, WI 53172.

ELECTRIC POWER RESEARCH INSTITUTE (EPRI). 1978. Coal Preparation for Combustion and Conversion. EPRI AF-791. May, pp. 1–6 and 2–9.

ELECTRIC POWER RESEARCH INSTITUTE. 1983. *Evaluation of Peat as a Utility Boiler Fuel*. EPRI Report CS-2913. March.

ENERGY INFORMATION ADMINISTRATION (EIA). *Annual Energy Outlook*. US Department of Energy. DOE/EIA-0383.

ENERGY INFORMATION ADMINISTRATION (EIA). 1990. *Natural Gas Production Responses to a Changing Market Environment 1978–1988*. US Department of Energy. DOE/EIA-0532.

ENERGY INFORMATION ADMINISTRATION (EIA). 1985. *Coal Data: A Reference*. US Department of Energy. DOE/EIA-0064(84). January.

ENERGY INFORMATION ADMINISTRATION (EIA). 1986. *An Economic Analysis of Natural Gas Resources and Supply*. US Department of Energy. DOE/EIA-0481.

ENERGY INFORMATION ADMINISTRATION (EIA). 1992. *Annual Energy Review*. June 1993, p. 197.

ENERGY INFORMATION ADMINISTRATION (EIA). 1993. *US Coal Reserves: An Update by Heat and Sulfur Content*. US Department of Energy. DOE/EIA-0529. February, pp. 6, 8.

ENERGY INFORMATION ADMINISTRATION (EIA). 1993. *Natural Gas Productive Capacity for the Lower 48 States 1982–1993*. US Department of Energy. DOE/EIA-0542(93). March, pp. 8–9.

ENERGY INFORMATION ADMINISTRATION (EIA). 1993. *Cost and Quality of Fuels for Electric Utility Plants 1992*. US Department of Energy. DOE/EIA-0191. August.

ENERGY INFORMATION ADMINISTRATION (EIA). 1993. *Coal Production 1992*. US Department of Energy. DOE/EIA-0118. October.

ENERGY INFORMATION ADMINISTRATION (EIA). 1993. *Quarterly Coal Report*. US Department of Energy. DOE/EIA-0121. October–December 1992.

ENERGY INFORMATION ADMINISTRATION (EIA). 1994. *Quarterly Coal Report*. US Department of Energy. DOE/EIA-0121. July–September 1993.

FRANCIS, HENRY E. and WILLIAM G. LLOYD. 1983. Predicting Heating Value from Elemental Composition. *Journal of Coal Quality*. Spring, pp. 21–25.

GRESLEY, WILLIAM STUKELEY. 1894. The Slate Binders of the Pittsburg Coal-Bed. *American Geologist*. December, pp. 356–365.

HAMILTON, T. B., ET AL. 1981. Peat as a fuel for steam generation. Paper presented at the American Power Conference. Chicago, IL, April 27–29.

HAYATSU, RYOICHI, ROBERT L. MCBETH, ROBERT G. SCOTT, ROBERT E. BOTTO, and RANDALL E. WINANS. 1984. Artificial Coalification Study: Preparation and Characterization of Synthetic Materials. *Organic Geochemistry*, Vol. 6, pp. 463–471.

HIGGINS, R. S. and L. J. GARNER. 1981. Developments in the Use of Low-Rank Coals in Australia. Proceedings of the Eleventh Biennial Lignite Symposium, June 15–17.

INDRESCO—JEFFREY DIVISION. P.O. Box 387, Woodruff, SC 29388.

INTERNATIONAL COAL REPORT. 1993. *Coal Year 1993*. FT Business Information Limited, London, England.

INTERNATIONAL COAL REPORT. 1994. *Coal Year 1994*. FT Business Information Limited, London, England.

INTERNATIONAL ENERGY AGENCY. 1993. *Energy Statistics and Balances of non OECD Countries 1990–1991*. Organization for Economic Cooperation and Development/International Energy Agency. Paris, France.

International Energy Annual 1991. 1992. US Energy Information Administration. December, p. 87.

International Energy Statistics Sourcebook: Oil & Gas Journal Energy Database. 1991. PennWell Publishing Company. Tulsa, OK.

JOHNSON, WILTON and JAMES PAONE. 1982. *Land Utilization and Reclamation in the Mining Industry, 1930–1980*. US Bureau of Mines. Information Circular 8862. p. 17.

JOY TECHNOLOGIES, INC.—MINING MACHINERY DIVISION. Meadowlands Facility, 2101 West Pike Street, Houston, PA 15342-1154.

LEONARD, JOSEPH W., ET AL. 1968. *Coal Preparation*, American Institute of Mining, Metallurgical and Petroleum Engineers. p. 4–18.

NATIONAL COAL ASSOCIATION. 1984. *Facts About Coal 1984/1985*. Public and Media Affairs Group. Washington, D.C., pp. 4–63.

Oil & Gas Journal. 1993. World LNG Trade to Soar to 2010 if Prices, Funds Line Up. June 28, pp. 23–30.

Oil & Gas Journal. 1993. PennWell Publishing. Tulsa, OK. December 27.

Oil & Gas Journal. 1994. Strong Growth Seen For World Coke Market. January 31, p. 38.

PEABODY HOLDING COMPANY, INC. 701 Market Street, Suite 700, St. Louis, MO 63101.

Phillips Petroleum Company vs. Wisconsin et al. 347 US 672(1954).

Railroad Facts. 1993. Association of American Railroads. Washington, D.C.

ROBERTS & SCHAEFER COMPANY. 120 South Riverside Plaza, Chicago, IL 60606-3986.

SCHULTZ, LEONARD G. 1958. Petrology of Underclays. *Bulletin of the Geological Society of America*. April, pp. 363–402.

SMOCK, ROBERT W., Managing Editor. 1980. Eastern utilities get started on conversion back to coal. *Electric Light and Power*. August.

STUTZER, OTTO and ADOLPH NOE. 1940. *Geology of Coal*. University of Chicago Press, p. 12.

Steam, Its Generation and Use. 1978. Babcock & Wilcox. 39th edit., p. 7-2.

Technical Bulletin TB-4. 1983. Comparison of ASTM, ISO, DIN, and BSI Methods of Coal Analysis. Coal Research Division. CONOCO INC. September.

THOMAS, J., JR. and H. H. DANBERGER. 1976. Internal Surface Area, Moisture Content and Porosity of Illinois Coals: Variations with Coal Rank. Illinois State Geological Survey Circular 493.

US DEPARTMENT OF COMMERCE, US BUREAU OF CENSUS. 1976. *Historical Statistics of the United States Colonial Times to 1970*, Bicentennial Edition. Government Printing Office.

US MINE SAFETY AND HEALTH ADMINISTRATION. US Department of Labor. Room 601, 4015 Wilson Boulevard, Arlington, VA 22203.

WELLER, J. M. 1930. Cyclical Sedimentation of the Pennsylvanian Period and Its Significance. *Journal of Geology*. Vol. 38 (1930), pp. 97–135.

WENT, W. L. 1990. *Law of Wetlands Regulation*. Clark Boardman Co., New York. Appendix 16.

World Oil. 1993. ICEED (International Research Center for Energy and Economic Development) Shapes Pursestring Attitudes, Too. June, p. 31.

American Society for Testing and Materials Standards

Designation	Description
D93	Test methods for flash point by Pensky–Martens closed tester
D240	Test method for heat of combustion of liquid hydrocarbon fuels by bomb calorimeter
D388	Classification of coals by rank
D396	Specification for fuel oils
D409	Test method for grindability of coal by the Hardgrove–Machine method
D445	Test method for kinematic viscosity of transparent and opaque liquids (and the calculation of dynamic viscosity)
D720	Test method for free-swelling index of coal
D1412	Test method for equilibrium moisture of coal at 96% to 97% relative humidity and 30° C
D1857	Test method for fusibility of coal and coke ash
D2013	Method of preparing coal samples for analysis
D2015	Test method for gross calorific value of coal and coke by the adiabatic bomb calorimeter
D2234	Test methods for collection of a gross sample of coal
D2361	Test method for chlorine in coal
D3173	Test method for moisture in the analysis sample of coal and coke
D3174	Test method for ash in the analysis sample of coal and coke from coal
D3175	Test method for volatile matter in the analysis sample of coal and coke
D3176	Practice for ultimate analysis of coal and coke
D3177	Test methods for total sulfur in the analysis sample of coal and coke
D3286	Test method for gross calorific value of coal and coke by the Isoperibol bomb calorimeter
D3302	Test method for total moisture in coal
D3682	Test method for major and minor elements in coal and coke ash by atomic absorption
D4208	Test method for total chlorine in coal by the oxygen bomb combustion/ion selective electrode method
D4239	Test methods for sulfur in the analysis sample of coal and coke using high-temperature tube furnace combustion methods
D4326	Test method for major and minor elements in coal and coke ash by X-ray fluorescence
D4371	Test method for determining the washability characteristics of coal
D4749	Test method for performing the sieve analysis of coal and designating coal size
D5142	Test methods for proximate analysis of the analysis sample of coal and coke by instrumental procedures

5

Coal and Limestone Handling

Richard H. McCartney

5.1 INTRODUCTION

The coal handling facility is the lifeline of a coal-fueled power plant. Modern plants have high coal demands because of the ever-increasing sizes of turbine generator units and the economic advantages of a single coal (fuel) handling facility serving a multi-unit power plant. Thus, coal handling facilities have had to become more flexible, more reliable, and capable of handling larger quantities of coal in less time than ever before. The limestone handling facilities for the scrubber additive for these power plants have also grown in importance to support the larger plants and the ever increasing environmental concerns.

Coal handling facilities normally require large ground areas, and an ideal area for siting these facilities is seldom available. When a plant must be sited on irregular terrain, additional conveyor length may be required solely to gain the necessary elevations to transfer the coal from system to system. Consequently, the contour, size, shape, and availability of plant site property profoundly affect the configuration, size, and cost of a coal handling facility. The limestone handling facility requires less ground area and is usually developed with the coal handling facility, as a secondary system, for best use of the available plant site property.

In planning and determining the design parameters for a coal handling facility, it might be assumed that the maximum fuel burn rate of the power plant is the ruling criterion. However, although the burn rate is extremely important, other elements are equally essential. The plant and coal source locations, the physical extent of the system, the environmental considerations, the redundancy desired, and the type and properties of the coal to be burned must be considered as well. All of these factors must be collectively evaluated in planning the basic design for the coal handling facility. The limestone handling facility has many of the same design parameters as the coal handling facility. Since it handles a much smaller quantity of material, however, the consideration of these factors is less critical and they are balanced and coordinated with the coal handling facility.

The major systems of a coal handling facility from coal delivery to the generating units are interrelated as shown in Fig. 5-1, a block diagram of the coal flow pattern. The type of unloading system used is dependent on the location of the coal source(s) and the mode of transportation. The stockout system capacity supports the unloading capacity for a single continuous process. The stockout, storage, and reclaim systems are coordinated to balance the plant supply with the plant consumption and the site considerations. The crushing and silo fill system is coordinated between the reclaim system and the maximum daily plant consumption. The secondary systems of a coal handling facility include the conveying, sampling, weighing, and dust control systems. These subsystems provide information, transportation, and material control for and between the major systems.

The major systems of a limestone handling facility from limestone delivery to the preparation area are interrelated as shown in Fig. 5-2, a block diagram of the limestone flow pattern. These major systems are coordinated as stated previously for the coal handling facility, but the complexity is reduced. The secondary systems of a limestone handling facility include the conveying, weighing, and dust control systems. As with the coal handling, these subsystems provide information, transportation, and material control for and between the major systems.

5.2 COAL AND LIMESTONE TRANSPORTATION AND UNLOADING SYSTEMS

The mode of transporting coal and limestone from the mines or quarries to the power plant is determined by the location of the plant relative to the location of the mines or quarries and also by the available practical methods of transportation such as railroads, waterways, highways, and conveyors. The year-round availability of each must be evaluated to determine the most reliable and economical transportation.

More than half of the United States coal-fueled power plants are supplied by coal train delivery. A typical coal unit train consists of about 100 railcars, 3 to 5 locomotives (depending on grades and curves), with each railcar carrying 100 tons (91 tonnes) of coal, for a total capacity of 10,000 tons (9,070 tonnes). The train functions as a unit, continuously shuttling between the coal source and the plant. Power plants receive limestone deliveries by train also, if the distance to the quarry is too great for truck deliveries. The delivery is usually by a small number of railcars, each carrying 70 tons (64 tonnes) of limestone.

For long distances and where ocean or river delivery is

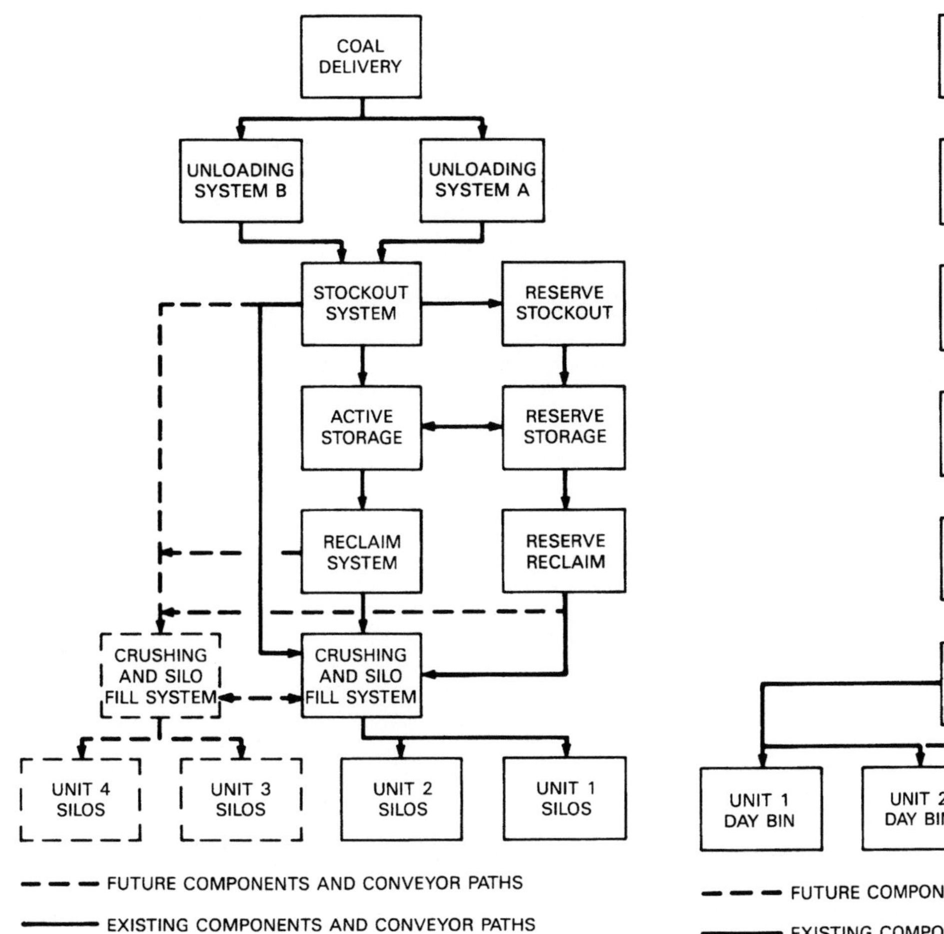

Fig. 5-1. Coal flow diagram.

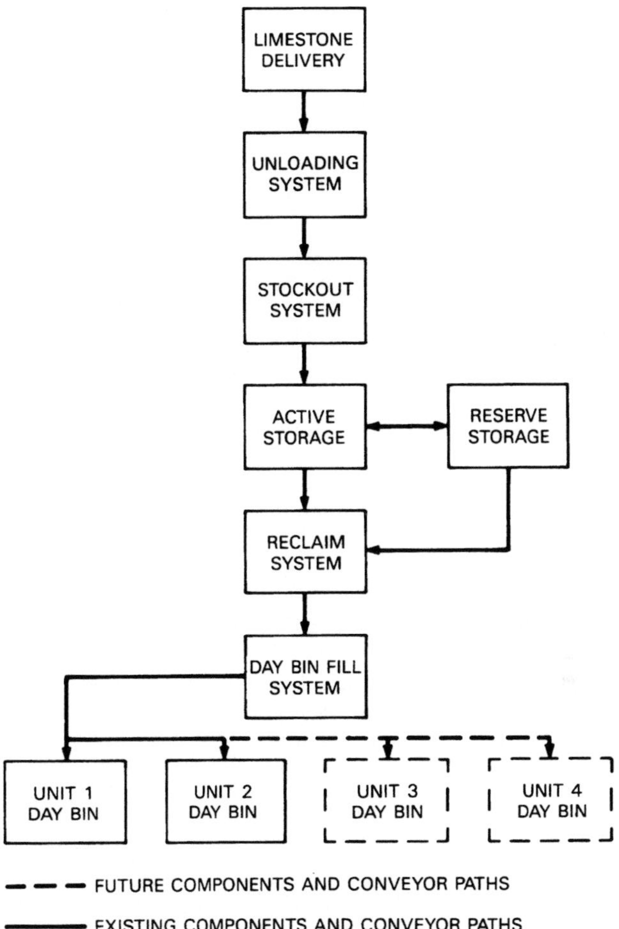

Fig. 5-2. Limestone flow diagram.

possible, barge or ship delivery of coal and limestone is usually the most practical mode, with the lowest cost per ton-mile. A typical river barge for coal transportation is an open hopper barge loaded with 1,500 tons (1,360 tonnes) of coal or 1,000 tons (910 tonnes) of limestone. The normal practice is to use a tugboat to push 15 barges—3 barges wide by 5 barges long—in an arrangement called a unit-tow, carrying 22,500 tons (20,410 tonnes) of coal or 15,000 tons (13,610 tonnes) of limestone. An ocean barge for coal and limestone transportation is a covered hopper barge loaded with between 9,000 and 15,000 tons (8,160 and 13,610 tonnes) of coal. These barges are pushed one at a time by a tugboat. An ocean ship for coal and limestone transportation is a cargo ship with covered holds loaded with 20,000 to 100,000 tons (18,140 to 90,720 tonnes) of coal, depending on the ship size. For limestone deliveries by ocean barge or ship, usually one hold is loaded with limestone and the rest with coal. The plant limestone handling system normally does not receive a complete shipment of limestone at one time due to the small quantities used.

Truck and overland conveyor transportation are used primarily for delivery of coal to plants near the mines or coal mine-mouth plants. Truck delivery is used extensively for small power plants in the eastern United States. Truck delivery is the most common method of receiving limestone because of the small quantities used and the local availability of limestone. A typical coal or limestone truck holds 20 to 25 tons (18 to 23 tonnes) with a single trailer or 40 tons (36 tonnes) with tandem trailers. An overland conveyor commonly moves coal up to 10 miles (16 km) from the mine to the plant. Longer routes are technically feasible but may be limited by the economics of long overland conveyors compared to truck hauls or short-haul rail transport. Limestone normally is not conveyed directly from the quarry to the plant, since they are not usually located adjacent to each other. The quantity of coal required also greatly affects the relative economics. Conveyor capacity is based on both mine output capacity and power plant consumption.

As with most major components of the coal and limestone handling systems, the capacity of both the coal and limestone transportation and unloading systems are based on the projected annual consumption of all generating units (final plant size) for the peak years of generation. Annual coal consumption is based on the heating value of the design coal in Btu per pound and the annual capacity factor. This annual coal consumption is divided into monthly requirements and then increased by 10% to 15% to account for fluctuation in transportation schedules and generation unit consumption. If the coal is not to be received on a year-round basis, the monthly totals are increased accordingly.

Annual limestone consumption is based on the coal properties (requiring scrubbers) and the limestone quality. Limestone may be delivered on a prorated daily basis or in large monthly shipments.

For a generating plant with multiple coal transportation methods available, the yearly capacity of each system must be determined. Usually, the primary mode of transportation is designed to handle the total yearly requirements. The secondary mode of transportation is designed to handle a percentage of the total based on the anticipated use of this mode.

The capacity of the coal unloading system is based on the monthly receiving requirements and specific design criteria. The system must be available when the coal arrives to maintain the specified turnaround time for the selected mode of transportation. For example, United States tariff agreements for most coal unit train operations specify that the turnaround time onsite not exceed a fixed period, usually 4 h, to avoid incurring monetary penalties. An unloading rate of 3,500 tons (3,180 tonnes) per hour allows a 10,000-ton (9070 tonne) train to be unloaded in approximately 3 h, allowing 1 h for train positioning, for a total onsite turnaround time of 4 h. River barges, ocean barges, and ships usually have similar tariff agreements allowing specified turnaround times. These times vary depending on the vessel size and the unloading method. Truck shipment agreements are usually based on the trucks being unloaded when received, with only minimal delays. The capacity of the limestone unloading system is usually based on the quantity of limestone in the shipment and the economics of the unloading system with minimal usage.

To permit proper maintenance and makeup of any out-of-service time, the equipment associated with the coal and limestone unloading systems should not operate in excess of 12 h per day during normal operation. This can be illustrated using the example of a 2-day outage on 8-h and 16-h operating cycles. If the unloading system is normally required to operate 8 h per day, a 2-day outage can be made up on the third day, by working three 8-h segments that third day. If the unloading system is normally required to operate 16 h per day, a 2-day outage will not be back on schedule until the end of the sixth day, since only 8 h per day is available for makeup time.

5.2.1 Railcar Types and Unit Trains

For coal and limestone delivery by rail, the unloading system is designed to accommodate the specific type of railcar that will deliver the material. Railcars are built to two basic designs: the bottom dump car, in which doors in the car bottom are opened for unloading, and a top dump car, a solid bottom car unloaded by turning the car over. The type of car used depends on the haul distance combined with the climate conditions. Railcars subject to freezing are easier to unload

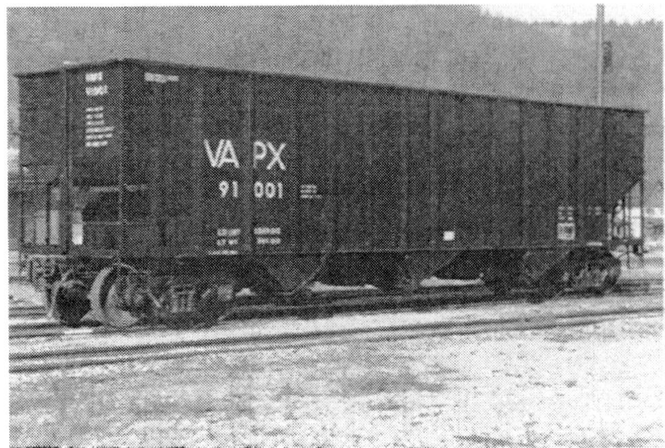

Fig. 5-3. Conventional bottom dump railcar. (From Johnstown America Corporation. Used with permission.)

when top dumped. A unit train consists of a specific number of railcars (usually 100) that are all of one type and the train, as a unit, travels continuously from the mine to the plant and back.

5.2.1.1 Bottom Dump Railcar Unloading. The bottom dump railcars are either the conventional bottom dump hopper car with three small manual doors or the air-door bottom dump type in that virtually the entire railcar bottom is automatically opened (by multiple large doors) for quick unloading. The conventional bottom dump car is shown in Fig. 5-3 and the air-door bottom dump car is shown in Fig. 5-4. The conventional bottom dump car is used for coal in bulk-rate or small multicar shipments to small power plants and for limestone shipments to large or small power plants. The railcars are positioned individually over an unloading hopper and the doors are manually opened. The unloading rate for each 70- to 100-ton (64- to 91-tonne) car varies from 2 to 5 min with good material flow conditions, or from 10 min to over an hour with frozen or poor material flow conditions. Manually opening and closing of the railcar doors requires 1 to 2 min. The plant locomotive usually positions the railcars through the unloading area in 5- to 10-car segments. All

Fig. 5-4. Air-door bottom dump railcar. (From Trinity Industries. Used with permission.)

of these operations combine for a maximum unloading rate of 15 cars per hour with an average of 8 to 10 cars per hour.

A car shaker is required for faster unloading times, especially when the coal or limestone is wet or frozen. The car shaker is a heavy vibrating mass that, when placed in contact with the railcar, causes the car to vibrate. This loosens the material in the car and improves the flow out of the car. Car shakers are positioned either on top of the car or against the side of the car. A top-mounted car shaker, as shown in Fig. 5-5, is supported on a monorail hoist and lowered onto the car as required. Usually, the side-mounted car shaker is hydraulically positioned against the side of the car.

For the larger coal-fueled power plants that use unit trains for coal shipments, air-door bottom dump railcars are used. Limestone shipments are not large enough to justify the use of unit trains with air-door bottom dump railcars. This coal unloading usually is performed while the train is in motion and powered by the railroad locomotives. When the cars pass over the unloading area, as shown in Fig. 5-6, the car doors are automatically opened and closed. The advantage of this type of unloading is the speed at which the coal cars are

Fig. 5-5. Top-mounted railcar shaker. (Black & Veatch.)

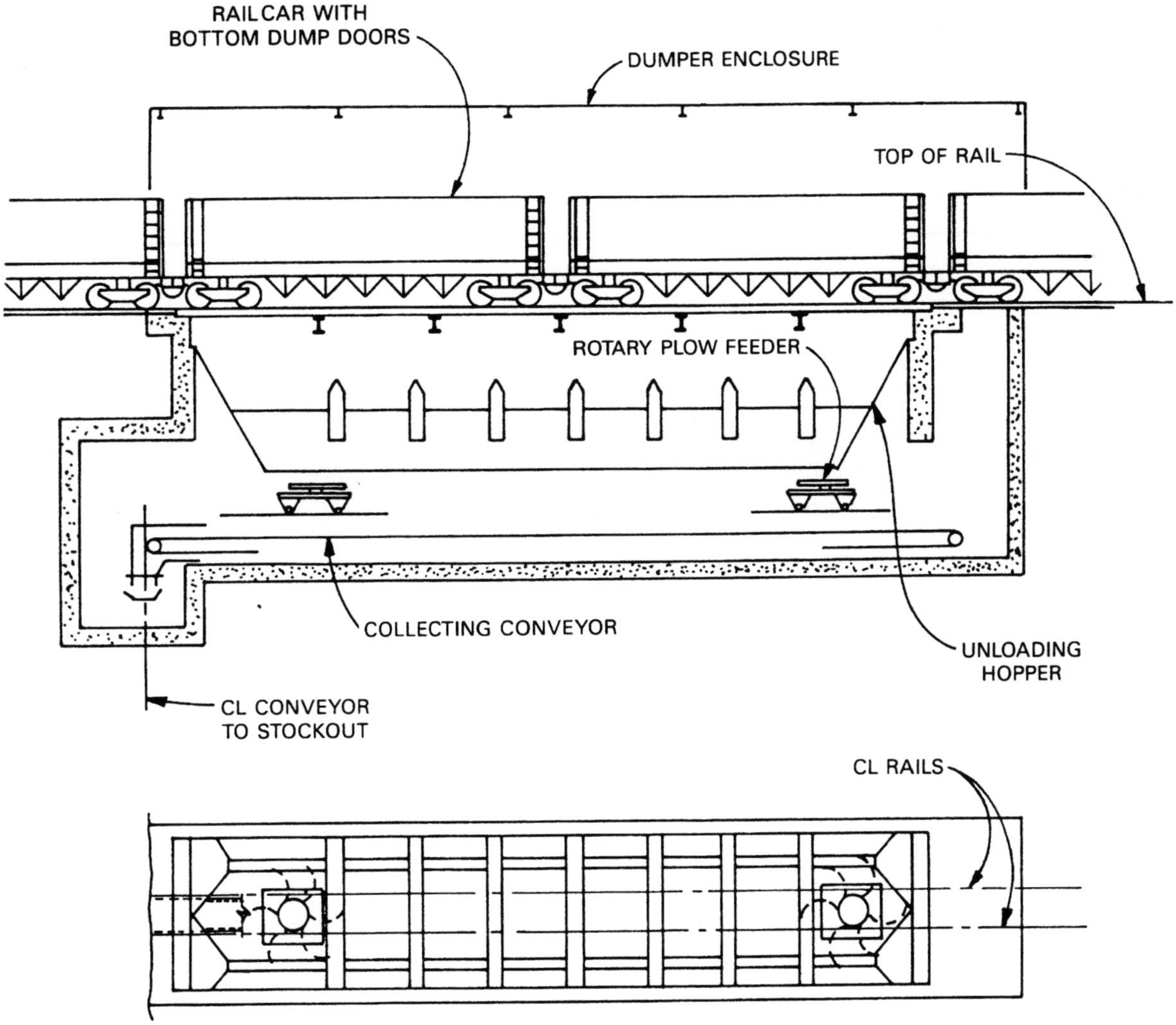

Fig. 5-6. Train unloading system with air-door bottom dump railcars.

unloaded; a 100-ton (91-tonne) car can be unloaded in 20 s. Where the unloading rate is not limited by hopper capacity, as with an elevated trestle, a 100-car unit train can be unloaded in approximately 30 min at a train speed of 2 miles (3.2 km) per hour. Using a 225-ft (69-m) long hopper, 8,000 tons (7,260 tonnes) per hour of coal can be unloaded dumping four cars at a time and maintaining a train speed of 0.8 mile (1.3 km) per hour.

Using a 125-ft (38-m) long hopper, 3,500 tons (3,180 tonnes) per hour can be unloaded dumping two cars at a time and maintaining a train speed of 0.35 mile (.56 km) per hour. Hoppers such as those discussed previously typically do not have a capacity of 10,000 tons (9,070 tonnes) each (a 100-car unit train) and therefore require a take-away rate that matches the unloading rate. The hopper discharge equipment is discussed in Section 5.3.2.

For unloading air-door bottom dump railcars in motion, the railroad locomotives are equipped with a pace setter, that allows the locomotives to travel at consistent speeds below 1 mile (1.6 km) per hour. The railcars have a "hot shoe" located at the car sill level on diagonal corners of the car. When contacted by a 24 volt dc current, the shoe causes the air system to either open or close the car doors, depending on the polarity. A system of rail segments along one side of the track at the car shoe elevation contacts the car shoe, and if a rail segment is energized, either opens or closes the car doors.

The air-door bottom dump railcar unloading rate is dependent only on the design of the receiving area (hopper or trestle) and the take-away rate. In areas where freezing weather is not a major concern, unit trains of air-door bottom dump railcars are usually used.

5.2.1.2 Rotary Car Dumper Unloading. The rotary car dumper system uses the top dump method with either gondola (flat bottom) or conventional hopper cars. Some small power plants use a rotary car dumper to unload individual, uncoupled railcars that are positioned in the dumper by a switch engine. Large coal-fueled power plants that require the use of unit trains use rotary car dumpers and train positioning equipment.

As previously discussed, unit train operations normally have a 4-h turnaround—3 h for unloading and 1 h for train positioning. For 100 cars of 100 tons (91 tonnes) each, the 3 h equates to an unloading rate of 3,500 tons (3,180 tonnes) per hour, or 103 s per car. Rotary dumper systems can achieve this rate with both coupled and uncoupled railcars when utilizing the proper car positioning equipment.

For unloading unit trains with the railcars coupled, the rotary dumper arrangement typically used is like that shown in Fig. 5-7. The system uses unit trains consisting of either high-side gondola (flat bottom) or conventional hopper railcars, each with a rotary coupler on one end. The train is unloaded, without uncoupling the locomotives or individual cars, by inverting each railcar 140 to 160 degrees (2.4 to 2.8 rad) as it is positioned in the dumper. The cars are rotated about the center line of the coupling, as shown in Fig. 5-8 (railcar shown uncoupled for clarity). An automatic, electric, or hydraulic powered train positioner moves the coupled unit train through the car dumper and automatically positions the

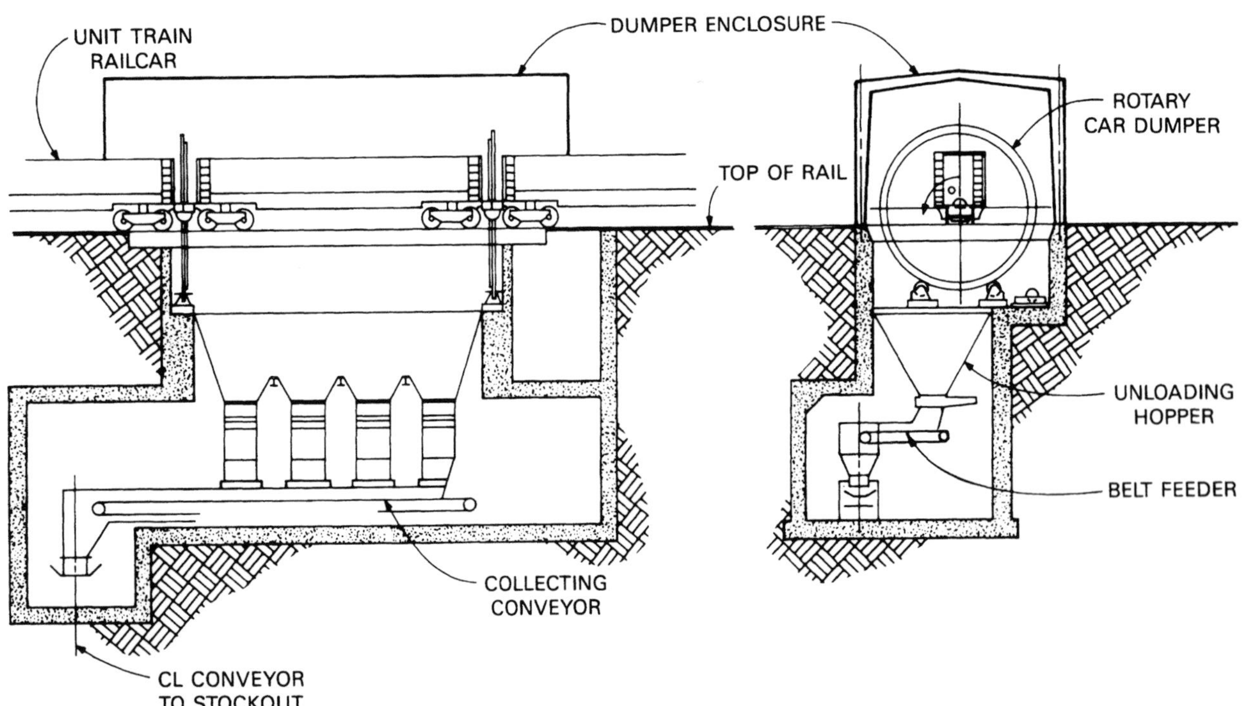

Fig. 5-7. Train unloading system with rotary car dumper.

Fig. 5-8. Rotary car dumper during unloading. (From Heyl & Patterson, Inc. Used with permission.)

individual cars for dumping. The train positioner arm, shown in the up position in Fig. 5-9, is lowered over the railcar coupling to position the train. A holding device, either wheel chocks or a holding arm that engages the coupling, locks the cars in position in front of and behind the dumper.

While the positioner is operating, the unit train is completely controlled by the positioner and holding devices. This system requires only one operator stationed in an elevated operating cab, located at the dumper.

For unloading unit trains consisting of varying sizes and types of cars or short trains of random cars, the cars must be uncoupled during the unloading operation. With the proper positioning equipment, an unloading rate of 3,500 tons per hour can still be maintained. Two types of rotary car dumpers available for this service are the solid-end ring type and the open-end ring type. The solid-end ring dumper has circular end rings and is combined with a car positioner located in front of the dumper and car retarders on the dumper to stop the car in the correct position. The railroad locomotives position the train at the positioner and disconnect from the train. The track gradient on the outbound side of the dumper incorporates a 1% grade for a short distance to move cars, by gravity, away from the dumper.

Fig. 5-9. Automatic train positioner. (From Heyl & Patterson, Inc. Used with permission.)

The equipment and operating sequence are similar in concept to that of the rotary car dumper for coupled unit trains, with variations to accommodate the random car sizes and the dumped car being uncoupled. The unloading operation for the solid end ring dumper requires a positioner operator, a dumper operator, a coupling knuckler, and a plant locomotive engineer. Two control stations are provided, one at the positioner and one at the dumper.

The open-end ring dumper has C-shaped end rings and is combined with a car indexer located in front of the dumper and a side arm charger at the dumper to position the car. The railroad locomotives position the train at the indexer and disconnect from the train. The outbound trackage is sloped as previously discussed. The unloading operation for the open-end ring dumper requires an indexer/charger/dumper operator, a plant locomotive engineer, and one control station to position cars.

Any of these types of rotary car dumpers can be equipped with car vibrators to assist with the unloading of wet or slightly frozen coal. The vibrators are either mounted in the car clamps that hold the car during rotation, or counterweighted to pivot and contact the upper side of the railcar as it is rotated.

5.2.1.3 Frozen Coal. A major problem for coal unloading systems is the problem of frozen coal, and depending on the location and winter conditions at the plant site, provisions must be made to deal with the problem.

Several provisions are made for power plant locations where frozen coal lumps on the unloading hopper grillage are a problem. For only occasional problems, a ramp down to the hopper grillage can be provided to allow a "bobcat" access to break up the lumps. When frozen coal on the grillage is a major problem, however, a traveling hammermill can be used. The traveling hammermill consists of pinned, hammer-shaped steel bars spaced around the shaft support that spans the hopper grillage, as shown in Fig. 5-10. As the shaft rotates, the hammermill assembly travels across the grillage and reduces the frozen coal masses to a size that will allow it to pass through the grillage.

Fig. 5-10. Traveling hammermill. (From Heyl & Patterson, Inc. Used with permission.)

A thaw shed can be used with either bottom dump or top dump railcars during freezing weather conditions. For top dump railcars, the purpose of the thaw shed is to break the bond between the car and the coal. For bottom dump cars, the thaw shed is also used to thaw the area of the car doors. The thaw shed is not designed to totally thaw all the coal in the railcar. Though thaw sheds initially used gas or oil heaters to thaw the cars, they now use electric infrared heaters with deflectors to concentrate the heat on the railcar. The heating elements are arranged as shown in Fig. 5-11, in a U shape, beneath and up the sides of the railcar. The amount of heat applied and the length of time in the thaw shed are controlled so that damage to the railcar does not occur.

A thaw shed is arranged with car length sections, normally with decreasing heating capacity in each section. The last section is a soaking section with minimal or no heaters, in that the heat is allowed to soak into the railcar. A typical thaw shed with two heating sections and one soak section would have between 2 and 4 megawatts of total heater capacity.

5.2.2 River Barges

The river barges are made up into a "unit-tow" of 15 barges or shipped as part of a mixed consignment. Each 1,500-ton (1,360-tonne) coal barge measures 195 ft (59.4 m) long by 35 ft (10.7 m) wide by 12 ft (3.7 m) deep. The barge has a draft of about 9 ft (2.7 m) loaded and 3 ft (0.9 m) empty. Two primary types of river barge unloaders are the clamshell bucket unloader and the continuous bucket ladder unloader.

5.2.2.1 Clamshell Bucket Unloader. The clamshell bucket unloader, as shown in Fig. 5-12, uses a bucket suspended from a set of hoisting cables, that is opened and closed by a set of control cables. Unloading is cyclical, about one bucketful of material every 30 to 40 s. The clamshell bucket unloader for river barges usually has a low to moderate capacity. A moderately sized clamshell bucket unloader has a maximum unloading rate, or free digging rate, of 1,500 tons (1,360 tonnes) per hour when starting to unload the barge. This rate decreases considerably as the barge is emptied and a full bucket is harder to obtain. For a complete barge, including clean-up and barge positioning, the average unloading rate is about 50% of the free digging rate, or 700 to 800 tons (635 to 725 tonnes) per hour.

Fig. 5-11. Thaw shed heater arrangement. (From Spectrum Infrared, Inc. Used with permission.)

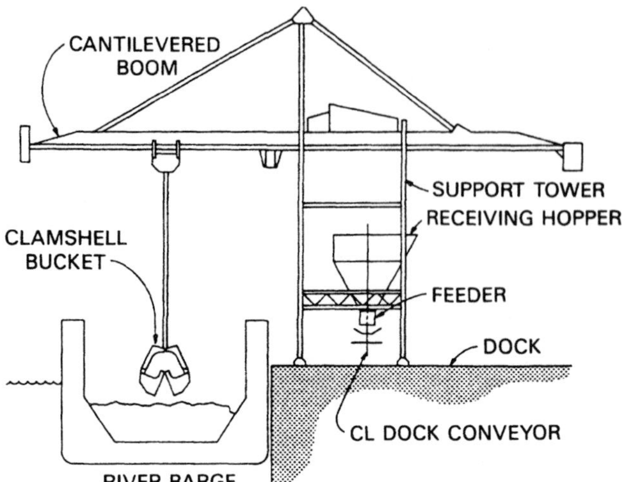

Fig. 5-12. River barge unloading system with clamshell bucket unloader.

The clamshell bucket unloader for river barge unloading is normally a stationary machine with a support tower with cantilevered boom. The support tower straddles the dock conveyor and supports the receiving hopper. The cantilevered boom supports the clamshell bucket trolley and cable systems. The hoist, travel, discharge, travel, and lower motion is a continuous arc when the machine is operated properly.

The travel distance of the bucket is varied to unload the material from the full barge width. The barge is positioned by a barge haul system to unload material from the full barge length. The barge haul system for the clamshell unloader is a simple haul system since the barge movement is intermittent and the exact position of the barge is not critical. The barge haul system consists of a steel cable and hoist system connected directly to the barge being unloaded. The barges are positioned from the unit-tow to the clamshell unloader barge haul system and back by a tugboat operated by power plant personnel.

5.2.2.2 Continuous Bucket Ladder Unloader. The continuous bucket ladder unloader employs a series of buckets supported between two strands of roller chain running in a continuous loop (digging head assembly), as shown in Fig. 5-13. The buckets are filled as they are dragged through the material in the barge and are emptied as they pass over a discharge sprocket. The continuous bucket ladder unloader for river barges usually has a moderate to high capacity. A moderate-capacity continuous bucket ladder unloader has a maximum free digging rate of 1,500 tons (1,360 tonnes) per hour. For a complete barge, including cleanup and barge positioning, the average unloading rate is about 65% of the

Fig. 5-13. Continuous bucket ladder unloader—first unloading pass. (From McNally Wellman. Used with permission.)

free digging rate, or 950 to 1,000 tons (860 to 910 tonnes) per hour.

The continuous bucket ladder unloader for river barge unloading is normally a stationary machine. The structure includes a support tower with fixed or movable cantilevered boom, depending on the water level fluctuations of the river. The boom supports the digging head assembly and a transfer conveyor. If the boom is fixed, the digging head assembly pivots to change the height of the digging buckets. The barge is unloaded by a two-pass operation, as shown in Fig. 5-14. On the first pass, the digging head digs a trough down the centerline of the barge, as the barge passes beneath the unloader. On the second pass, the digging head is oscillated back and forth across the width of the barge to unload the remaining coal. A three-pass operation also can be used. The first pass is the same as described previously and on the second and third passes, the digging head is positioned on one side and then on the other side of the barge. The barge is returned to the starting position between each pass. The loose catenary at the bottom of the bucket chain loop allows the buckets to contact the bottom of the barge without damage for good cleanup. The high-capacity continuous bucket ladder unloaders use two digging head assemblies.

The barge haul system for the continuous bucket ladder unloader employs a small shuttle barge, that is permanently attached to the barge haul cables. The barges to be unloaded are attached to the shuttle barge. This barge haul system provides a continuous smooth movement of the barge during unloading. In addition, to reduce barge hookup time, while one barge is being unloaded, a second barge is being attached to the other end of the shuttle barge. The barges are positioned at the shuttle barge by a tugboat operated by power plant personnel.

5.2.3 Ocean Barges and Ships

The ocean barge for coal and limestone transportation is a covered hopper barge with separate holds. The barge has a capacity normally varying from 9,000 to 15,000 tons (8,160 to 13,610 tonnes) of coal. A 15,000-ton (13,610-tonne) capacity coal barge measures about 470 ft (143.3 m) long by 80 ft (24.4 m) wide by 34 ft (10.4 m) deep. The barge has a draft of about 20 ft (6.1 m) loaded and 10 ft (3.0 m) empty. These barges are pushed one at a time by a tugboat.

The ocean ship for coal and limestone transportation is a cargo ship with separate covered holds. Ocean ships vary in capacity from under 20,000 tons (18,140 tonnes) to over 100,000 tons (90,720 tonnes) of coal. A 40,000-ton (36,290-tonne) capacity ocean ship, for example, is longer than two football fields, and measures about 640 ft (195.1 m) long by 100 ft (30.5 m) wide by 38 ft (11.6 m) deep. For limestone delivery by ocean barge or ship, usually one hold is loaded with limestone and the rest with coal.

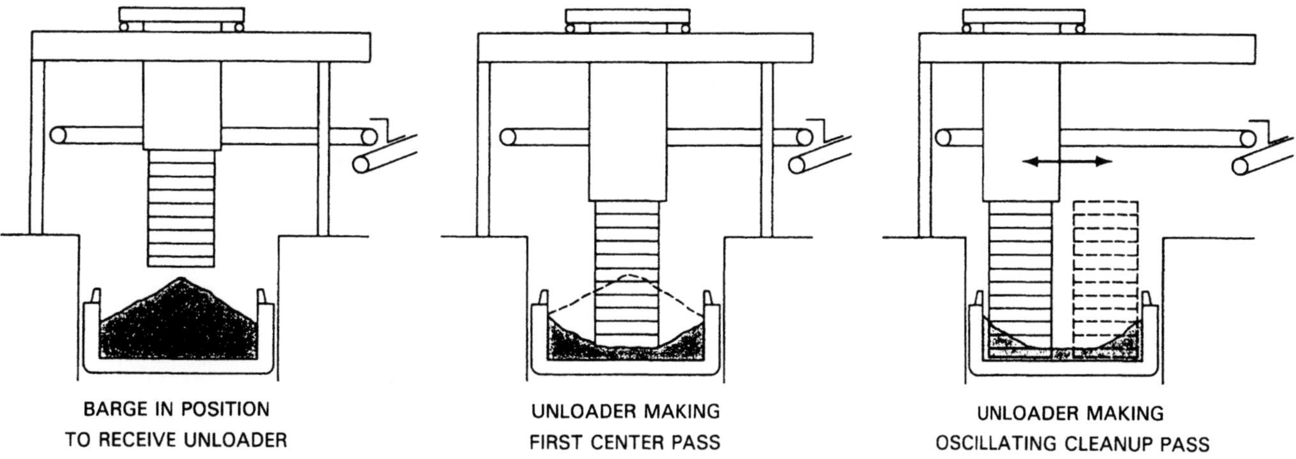

Fig. 5-14. Continuous bucket adder unloader—unloading procedure. (From McNally Wellman. Used with permission.)

Ocean barges and ships use similar unloading equipment. The primary types of unloaders are the clamshell unloader, continuous bucket ladder unloader, and the vertical screw unloader. In addition, ocean ships can be self-unloading.

The clamshell bucket unloader for ocean barges or ships is similar in design and operation to the unloader described for river barges, except that the machine travels on a set of rails on the dock, as shown in Fig. 5-15. This allows the ocean barge or ship, because of its size, to remain stationary while the unloader repositions from hold to hold. In addition, the boom is hinged at the front of the support tower and can be raised to allow the ships to dock.

The clamshell bucket unloader for ocean barges and ships can have a maximum free digging rate of 3,000 tons (2,720 tonnes) per hour, falling to about 50% of the free digging rate for the complete unloading operation.

The continuous bucket ladder unloader for ocean barges or ships may have design and operation similar to that for the unloader described for river barges, or it may use the bucket chain assembly "L" shaped at the bottom to improve the reach in the hold. The machine travels on a set of rails on the dock as does the clamshell unloader for ocean barge or ship, as shown in Fig. 5-16. In addition, the boom and digging head assembly can be rotated to improve the machine reach. Also, the boom is hinged or pivoted at the support tower to clean the ship during docking.

The continuous bucket ladder unloader, for ocean barges and ships, can have a maximum free digging rate of 5,000 tons (4,540 tonnes) per hour when starting to unload the vessel, or a finished unloading rate of about 65% of the free digging rate.

Fig. 5-15. Clamshell bucket unloader with ocean ship. (From McNally Wellman. Used with permission.)

The third unloader type for ocean barges and ships is the vertical screw unloader. This unloader is used for coal and is similar in concept to the continuous bucket ladder unloader in that the coal is continuously discharged from the barge or ship. The lower end of the vertical screw may have a counter-

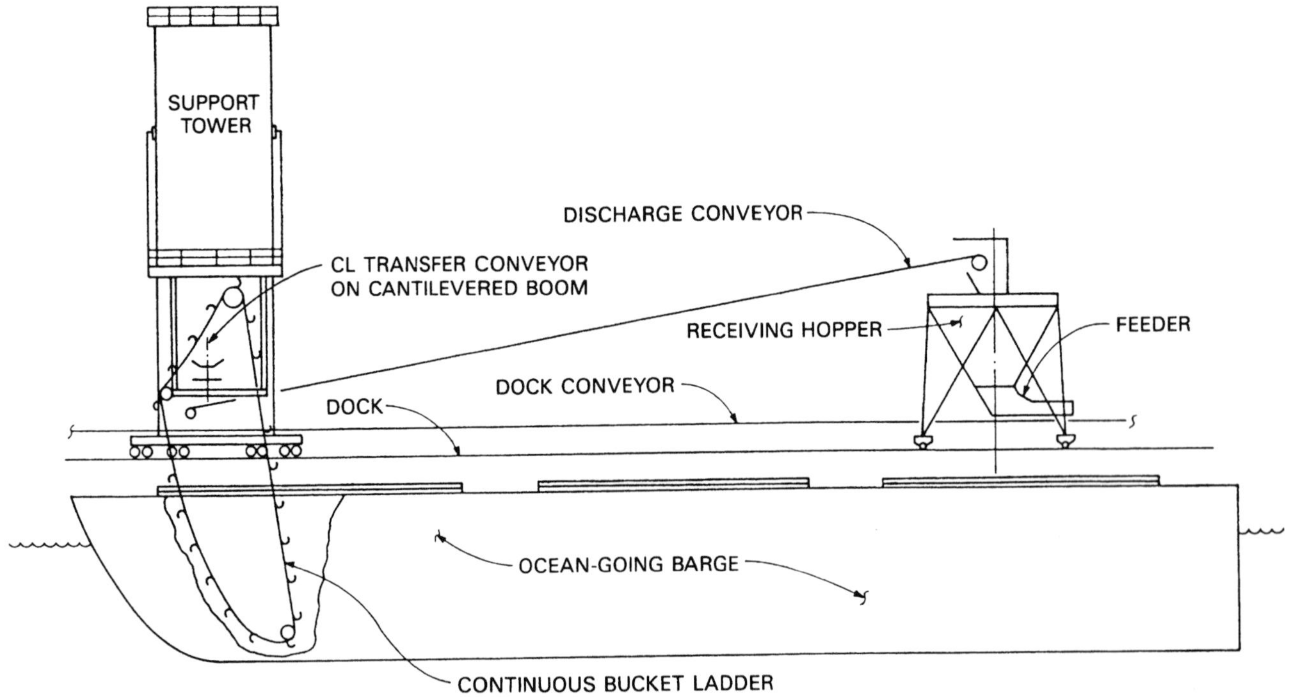

Fig. 5-16. Ocean barge unloading system with continuous bucket ladder unloader.

rotating feeder or a horizontal feeder screw to feed the coal to the vertical screw. The vertical screw conveyor is supported from a cantilevered boom. The travel of the machine, the rotation of the boom, and the tilt of the vertical screw permit the unloading of coal from all areas of the hold.

The vertical screw unloader can have an unloading rate of 2,000 tons (1,810 tonnes) per hour and an average unloading rate for the complete vessel of 60% of the unloading rate.

The self-unloading ship has a conveyor system beneath the hold's hopper bottom to discharge the material from the cargo holds and elevate it to the boom conveyor. The boom conveyor can be rotated perpendicular to the ship and discharged either into a hopper or onto the ground, as shown in Fig. 5-17. Self-unloading ships, in the popular size range, can vary in capacity from 35,000 to 75,000 tons (31,750 to 68,040 tonnes) of coal, depending on design and size. The self-unloading ship can have an unloading rate from 3,000 up to 10,000 tons (2,720 to 9,070 tonnes) per hour.

5.2.4 Trucks

Highway trucks for coal or limestone transportation are either rear dump or bottom dump type. Either type can unload into a hopper or directly onto the storage pile. A typical coal or limestone truck has a capacity of 20 to 25 tons (18 to 23 tonnes) with a single trailer, or as shown in Fig. 5-18, a total of 40 tons (36 tonnes) with tandem trailers. Without frozen or sticky material problems, these trucks usually require less than a minute to unload.

For coal shipments, highway trucks are normally used for moderate to small plants with coal mines within a 50-mile (81-km) radius. They are also used for plants of all sizes as a secondary mode of transportation. In some western states, truck transportation up to 120 miles (193 km) one way is economical for spot purchases of coal. For limestone shipments, highway trucks are the most common method of transportation, determined by the availability and required quantities of limestone.

Off-road trucks for coal transportation are also rear dump or bottom dump type, and either type can be unloaded into a hopper or directly onto the storage pile. A coal capacity of 80 to 100 tons (73 to 91 tonnes) is common for these trucks, which serve only power plants adjacent to coal mines. These trucks have a normal haul distance of less than 10 miles (16 km), since the haul roads for these large trucks are nonpublic roadways.

5.2.5 Overland Conveyors

Overland conveyors transport coal to all sizes of power plants that have adjacent coal mines. Overland conveyors are also used for both coal and limestone when the barge or ship terminal is several miles from the plant. Overland conveyors commonly move material up to 10 miles (16 km) from the mine or terminal to the plant. Longer routes are technically feasible, but may be limited by the economics of long overland conveyors compared to truck or rail transport. The straightness of the route and the roughness of the terrain also play a large part in the economics.

Overland conveyor systems employ a variety of conveyor profiles and belt configurations. Normal troughed belts, similar to the conveyors used at the power station, use transfer points at each location the belt must change direction. They also use conveyor covers to protect the material on the belt from the weather. Several types of conveyor systems allow the belt to convey material around a horizontal curve and also completely enclose the material within the belt. One of these systems rolls the belt edges up and pinches them together into a tear-drop shape as shown in Fig. 5-19. These systems use an idler configuration that completely surrounds the belt and maintains its shape.

The capacity of these conveyor systems for mine-mouth plants is dictated by the mining operations and the location of active storage, at either the mine or the plant. Their capacity may be sized to match the plant silo fill rate of 300 to 1,000 tons (270 to 910 tonnes) per hour or the maxi-

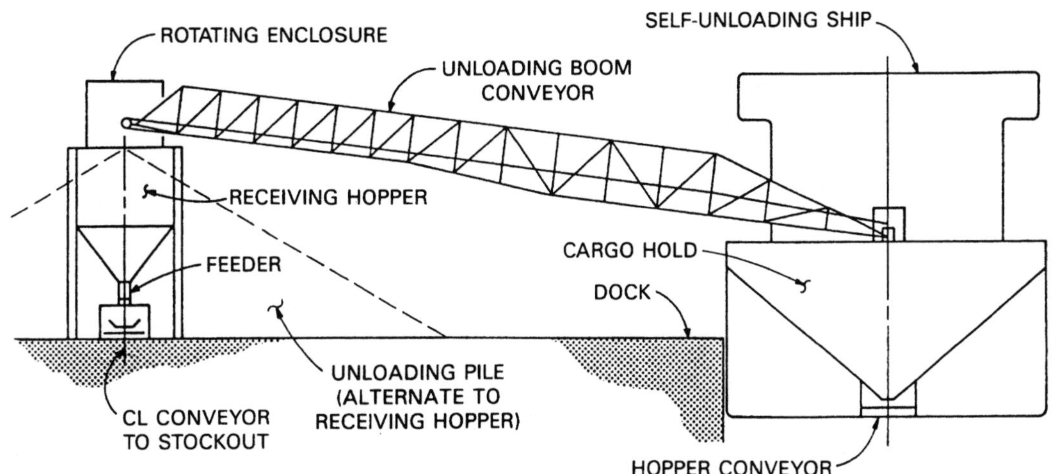

Fig. 5-17. Ship unloading system with self-unloading ship.

Fig. 5-18. Tandem trailer truck at unloading hopper. (Black & Veatch.)

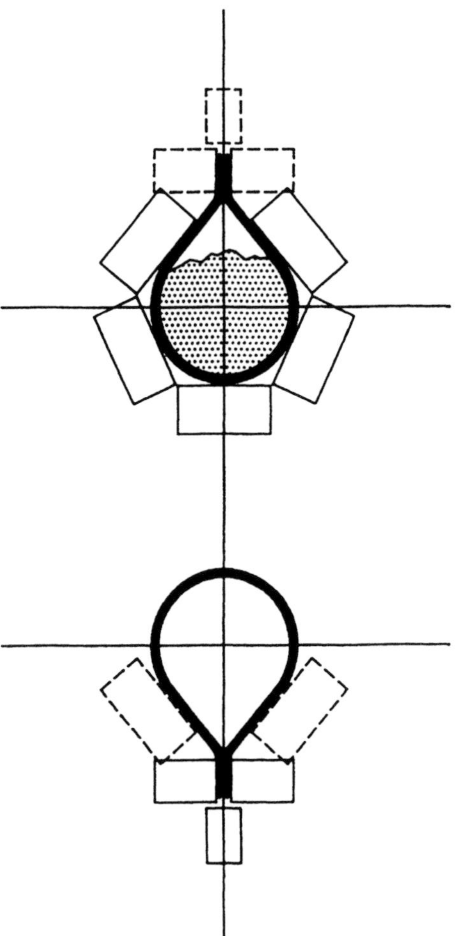

Fig. 5-19. Conveyor system, tear-drop belt shape. (From Sumitomo Heavy Industries Ltd., Willis & Paul Group Engineers Construction. Used with permission.)

mum conveying rate of up to 4,000 tons (3,630 tonnes) per hour. The capacity of the overland conveyor systems, in conjunction with a terminal, is usually the upper end of the range to maximize the unloading rate at the terminal.

5.3 STOCKOUT, RECLAIM, AND STORAGE SYSTEMS

Stockout is the term for the method and equipment used to place the material in the storage area, whereas *reclaim* is the term for the method and equipment used to retrieve the material from the storage area. *Storage* is the area where the material is held from the time it is received to the time it is required for plant consumption.

The stockpiling and unloading of coal or limestone are closely related functions, in that continuous unloading operations require the simultaneous withdrawal and conveying of the material from the unloading point to the plant storage areas. Rates of conveying material to storage must correspond with the maximum unloading rates so that both operations can be carried out as a single continuous process. For a generating plant with multiple coal receiving systems, the stockout system must be capable of supporting each unloading system. This can be done with separate stockout systems, or the secondary unloading system can be designed to "hold" the coal during unloading operations at the primary unloading system.

The capacity of the coal and limestone reclaim systems, as well as the coal crushing and silo fill system and the limestone day bin fill system, are based on the maximum daily consumption of all generating units. These total capacities may be designed initially, or phased in as required.

The coal and limestone storage areas include both active

and reserve storage. The active areas coordinate delivery with plant usage. The reserve areas accommodate a disruption in deliveries.

The equipment associated with the stockout and reclaim system should also not operate in excess of 12 h per normal day. As with the unloading system, this criterion applies to the complete coal handling facility.

5.3.1 Stockout Equipment

Selection of the most appropriate stockout system is based on the unloading system selected, blending requirements, enclosed storage requirements, and economic considerations. Alternate types of outdoor stockout equipment for coal or limestone include a fixed boom conveyor, radial stacker, traveling stacker, bucket wheel stacker–reclaimer, and an elevated reversing shuttle conveyor. This equipment is normally used to stockpile material on the ground in nonenclosed areas. Alternate types of indoor stockout equipment for coal and limestone include a traveling tripper and portal stacker reclaimer. This equipment usually stockpiles material in enclosed areas. For specific applications, any piece of stockout equipment can be used outdoors or in an enclosed area. All of these types of equipment differ in concept and capacity, and each type has multiple variations in design and layout.

5.3.1.1 Fixed Boom Conveyor. A fixed boom conveyor is a stationary inclined conveyor with the discharge end cantilevered to allow the formation of a conical shaped pile beneath the conveyor discharge. A fixed boom conveyor is shown in Fig. 5-20. The conveyor discharge has a telescopic chute that is raised as the pile is formed, minimizing the free-fall distance of the material.

A variation of the fixed boom conveyor with a telescopic chute is an inclined conveyor discharging into a lowering well. The lowering well is a small-diameter [about 12 to 20 ft (3.7 to 6.1 m)] tube with a series of doors staggered around the perimeter at various elevations. The conveyor discharges into the lowering well and the material flows out of the doors (lowest ones first) to form a conical shaped pile centered around the lowering well.

A silo, used for enclosed active storage, requires only an inclined loading conveyor as the stockout equipment.

The fixed boom conveyor method of stockout is most commonly used for reserve coal stockout and active limestone stockout at large power plants. It is also used for active coal stockout at small power plants. A typical reserve coal stockout pile has a capacity of 10,000 tons (9,070 tonnes) (one unit train) with a pile height of 60 ft (18.3 m). The active limestone stockout pile has a capacity varying between 2,000 and 20,000 tons (1,810 and 18,140 tonnes), based on the limestone delivery method and plant usage. The pile height varies from 30 to 65 ft (9.1 to 19.8 m).

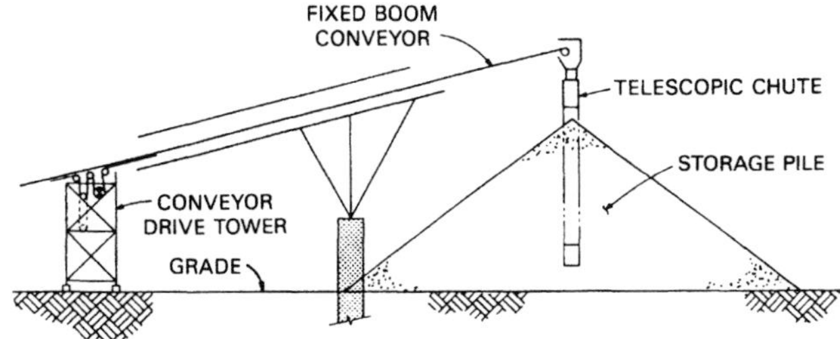

Fig. 5-20. Stockout system with fixed boom conveyor.

5.3.1.2 Radial Stacker. A radial stacker is similar to the fixed boom conveyor except that the conveyor can slew (rotate horizontally) in a semicircle to form a kidney-shaped pile beneath the conveyor discharge. The radial stacker conveyor can either be fixed at a set incline and equipped with a telescopic chute or capable of a luffing (raising and lowering) motion, as shown in Fig. 5-21. Either method minimizes the free-fall distance of the material.

The radial stacker rotational capability varies from less than 90 degrees (1.5 rad) if the tail end of the radial stacker is within a building, and up to 270 degrees (4.6 rad) if the radial stacker is in a open area and limited only by avoiding proximity to the elevating conveyor feeding the radial stacker.

This method of stockout is similar to the fixed boom conveyor for both coal and limestone, but is one step up from the fixed boom conveyor in both capacity and flexibility. Since the discharge point is variable, it can stockout both coal and limestone in separate piles for a small power plant. A typical radial stacker with a 120 ft (36.6 m) boom and 180 degrees (3.1 rad) of rotation can stockout a 30 ft (9.1 m) high pile. The total pile capacity is either 12,500 tons (11,340 tonnes) of coal or 21,000 tons (19,050 tonnes) of limestone.

5.3.1.3 Traveling Stacker. A traveling stacker is similar to the radial stacker except that the conveyor travels linearly instead of rotating. The cantilevered boom conveyor is supported on a gantry that travels on a set of rails as shown in Fig. 5-22. A long pile, with a triangular cross-section, is formed beneath the conveyor discharge as the stacker travels. The traveling stacker conveyor can either be fixed at a set incline and equipped with a telescopic chute or capable of a luffing motion to minimize the free-fall distance of the material during pile construction. The boom conveyor can also slew 180 degrees (3.1 rad) so that a pile can be stocked out on both sides of the traveling stacker track.

The traveling stacker pile length is limited only by the travel distance of the stacker. Unless confined by the site conditions, a travel distance of up to 1,200 ft (366 m) is common. The traveling stacker rails are usually elevated on an earth berm to minimize machine height while maximizing pile height.

This method of stockout is most commonly used for active coal stockout at large power plants. Segregated coal piles can be stockpiled or the traveling stacker can be used for both coal and limestone stockout. A typical traveling stacker with a 100 ft (30.5 m) boom can stockout a pile 40 ft (12.2 m) high. It will stockpile 53 tons/ft (158 tonnes/m) of coal or 90 tons/ft (268 tonnes/m) of limestone.

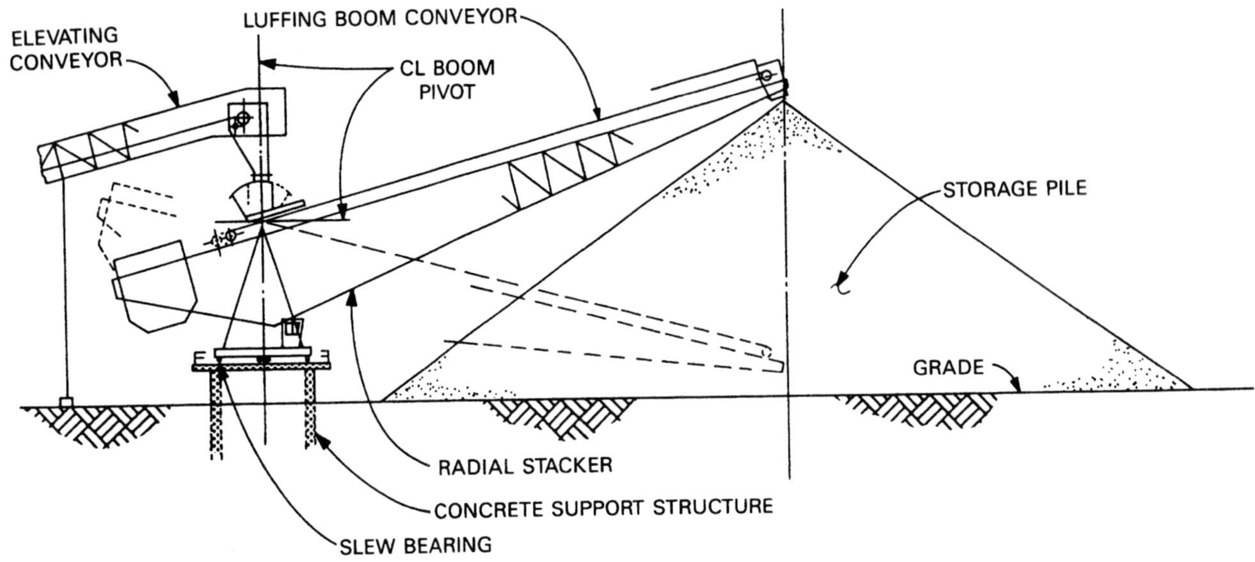

Fig. 5-21. Stockout system with radial stacker.

5.3.1.4 Bucket Wheel Stacker–Reclaimer. A bucket wheel stacker–reclaimer is very similar to the traveling stacker in the stockout mode. The stacker–reclaimer is equipped with a bucket wheel at the end of the boom for the reclaim mode, discussed in Section 5.3.2.1. The boom conveyor is also reversible for stockout versus reclaim. The counterweighted, cantilevered boom conveyor is supported on a gantry that travels on a set of rails. The stacker–reclaimer boom is capable of a luffing motion and is utilized in both the stockout and reclaim modes.

The two types of bucket wheel stacker–reclaimers are the trench type and the slewing type. As shown in Figs. 5-23 and 5-24, the trench type machine forms a long triangular cross-sectional pile beneath the conveyor discharge as the machine travels. The boom luffs to minimize the material free-fall distance as the pile is constructed. The trench type boom is normally 50 ft (15.2 m) long. The trench type stacker–reclaimer boom is slewed only when the machine is required to stockpile material on the other side of the track. As shown in Fig. 5-25, the slewing machine forms a long flat-topped pile beneath the boom conveyor discharge as the machine travels and the boom slews. The boom also luffs to minimize the material free-fall distance as the pile is constructed. The slewing type boom length varies from 80 to 200 ft (24.4

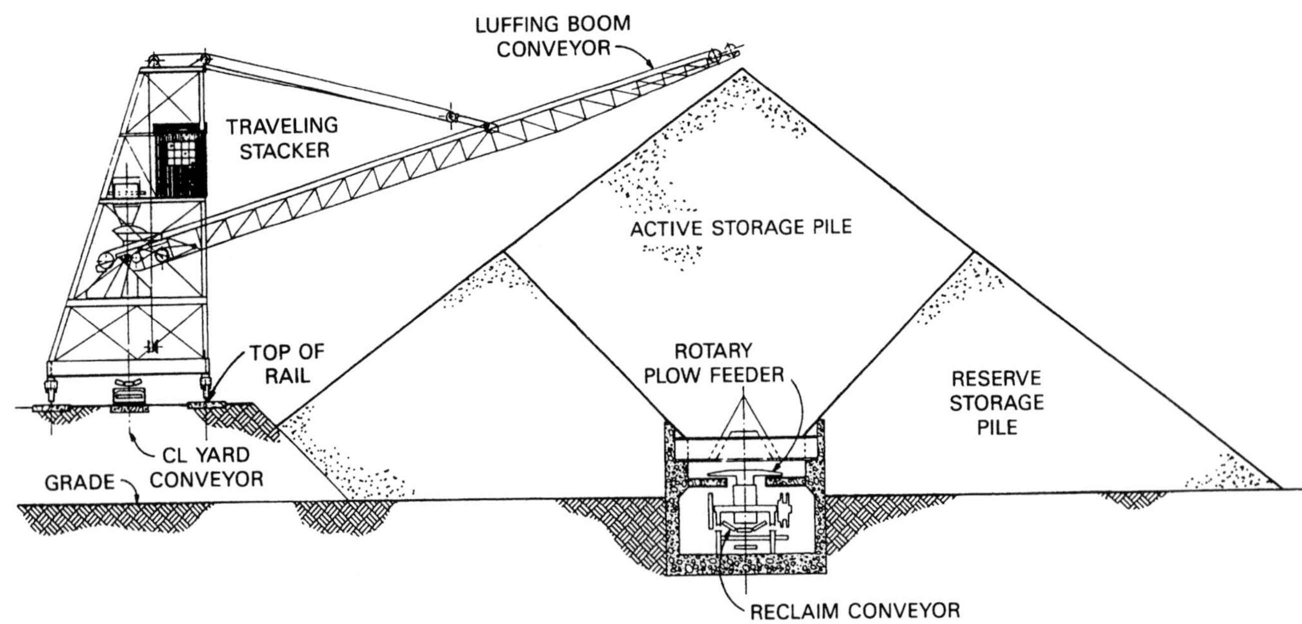

Fig. 5-22. Stockout system with traveling stacker and rotary plow feeder reclaim.

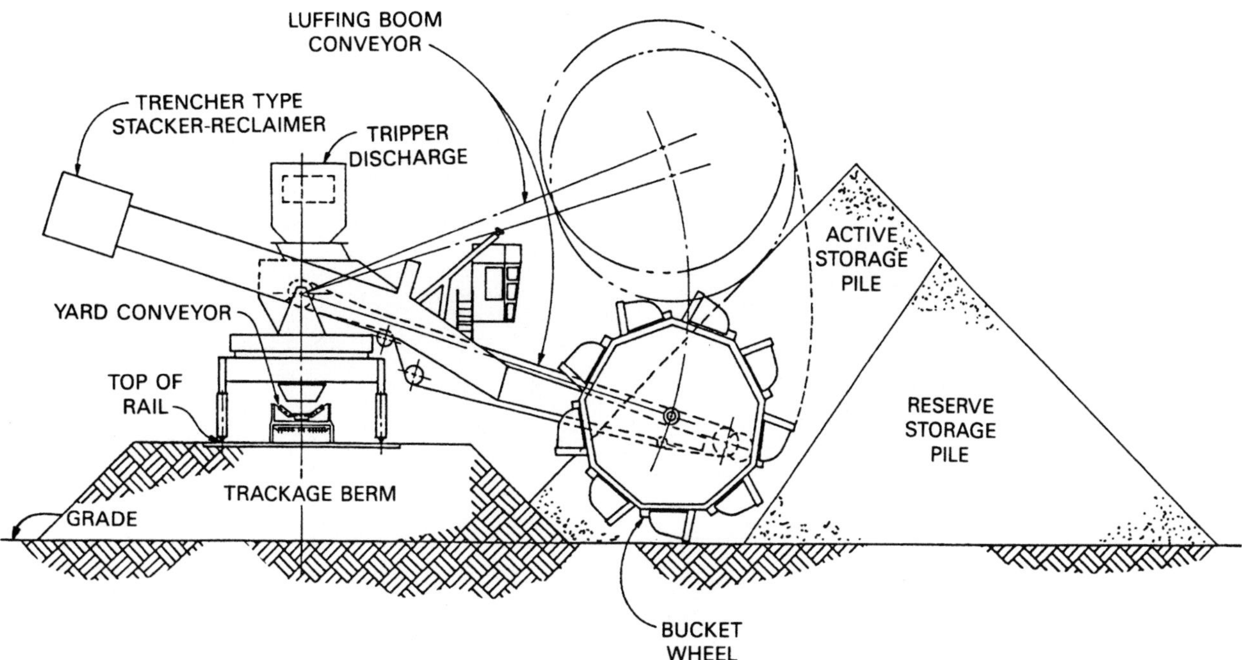

Fig. 5-23. Stockout/reclaim system with trench type bucket wheel-stacker–reclaimer.

to 61.0 m). The slewing type stacker–reclaimer boom can also be slewed 180 degrees (3.1 rad) to stockpile material on the other side of the track.

Similar to that of the traveling stacker, the bucket wheel stacker–reclaimer pile length is limited only by its travel distance, and a pile length of up to 1,200 ft (366 m) is common. The trench machine travel rails are usually elevated 12 to 15 ft (3.7 to 4.6 m) on an earth berm to maximize the pile height because of the short boom length. The slewing machine travel rails are usually elevated only several feet since the longer booms accommodate the required pile heights.

This method of stockout is most commonly used for active coal stockout at large power plants to form one or segregated coal piles. A bucket wheel stacker–reclaimer is not normally used for both coal and limestone stockout because of the one machine being used for both stockout and reclaim operations. A typical trench-type stacker–reclaimer with a 50-ft (15.2-m) boom can stockout a 40-ft (12.2-m) high pile of coal with a total capacity of 40 tons/ft (119 tonnes/m). A small size slewing-type stacker–reclaimer with a 95-ft (29.0-m) boom can stock out a 50 ft (15.2 m) high pile of coal with a total capacity of 110 tons/ft (327 tonnes/m). A large-size slewing type stacker–reclaimer with a 185-ft (56.4-m) boom can stockout a 70 ft (21.3 m) high pile of coal with a total capacity of 310 tons/ft (923 tonnes/m).

Fig. 5-24. Stacker–reclaimer, trencher type-stockpiling coal. (Black & Veatch.)

5.3.1.5 Elevated Reversing Shuttle Conveyor. An elevated reversing shuttle conveyor, as shown in Fig. 5-26, is an elevated horizontal conveyor with a reversing belt drive. The complete conveyor also has transverse (shuttle) motion for a fixed travel distance. A long pile, with a triangular cross-section, is formed beneath the conveyor discharge as the conveyor shuttles. A second pile is formed when the belt is reversed and the conveyor shuttles in the opposite direction. Both discharge ends of the shuttle conveyor have a telescopic chute that is raised as the pile is formed. This minimizes the free-fall distance of the material.

The total length of the shuttle conveyor piles is limited by the length of the shuttle conveyor and the elevated structure. The shuttle conveyor can vary in length from 200 to 400 ft (61.0 to 121.9 m), with a pile height of 40 to 70 ft (12.2 to 21.3 m). The combination of pile length and height establishes the stockout capacity. These dimensions are optimized based on the required active storage capacity and the type of reclaim equipment to be used. This method of stockout, similar to the traveling stacker, is most commonly used for active coal stockout at large power plants. Two segregated coal piles or one coal pile and one limestone pile can be

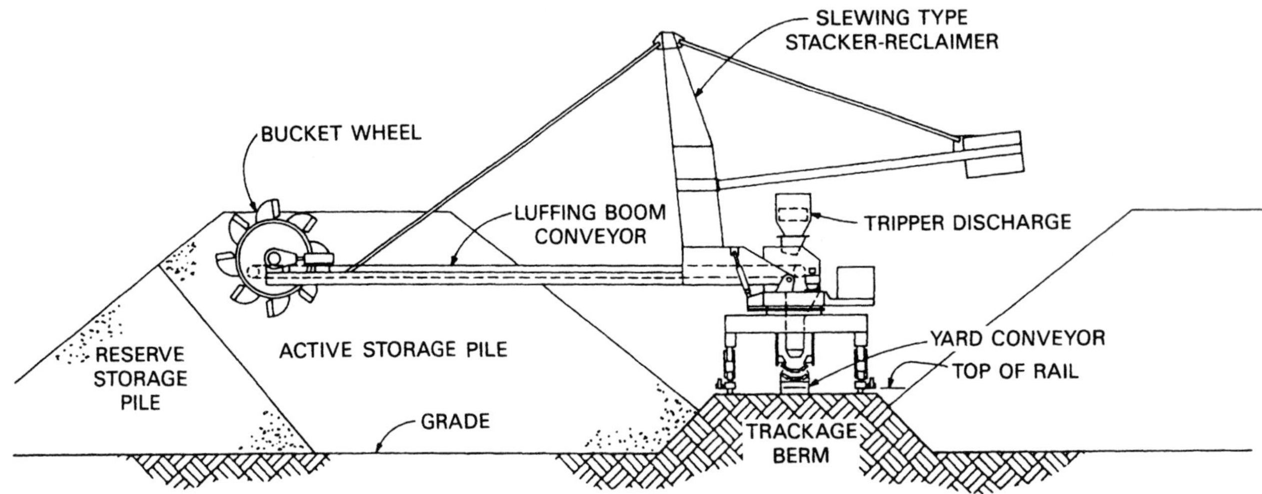

Fig. 5-25. Stockout/reclaim system with slewing type bucket wheel stacker–reclaimer.

stockpiled. A typical shuttle conveyor 300 ft (91.4 m) long forms two piles, each 280 ft (85.3 m) long at the pile tip. With a pile height of 60 ft (18.3 m), the shuttle conveyor stockpiles 120 tons/ft (357 tonnes/m) of coal or 200 tons/ft (595 tonnes/m) of limestone.

5.3.1.6 Traveling Tripper. A traveling tripper is a movable discharge point on an elevated horizontal conveyor. The self-propelled tripper travels on a set of rails located on either side of the conveyor. The tripper elevates the belt and discharges the material into a pant-leg chute that diverts the material around the conveyor and down into the enclosed storage structure, as shown in Fig. 5-27. A long pile, with a triangular or diamond cross-section, is formed beneath the tripper discharge as the tripper travels in segmented lengths.

The length of the traveling tripper pile is defined by the length of the enclosure, normally 400 to 600 ft (121.9 to 182.9 m). The combination of pile height [normally 40 to 70 ft (12.2 to 21.3 m)], length, and shape establish the stockout capacity. This method of stockout is used mostly for enclosed active coal storage at large power plants. For a typical traveling tripper the coal pile capacity of a 60 ft (18.3 m) high pile will be 120 tons/ft (357 tonnes/m) for a triangular pile cross-section and 40 tons/ft (119 tonnes/m) for a diamond-shaped pile cross-section.

5.3.1.7 Portal Stacker–Reclaimer. A portal stacker–reclaimer, shown in Fig. 5-28, is similar to the traveling stacker in the stockout mode. The boom is supported from a portal structure that spans the storage pile and travels on a set of rails, one on either side of the pile. A long pile, with a triangular cross-section, is formed as the portal stacker–

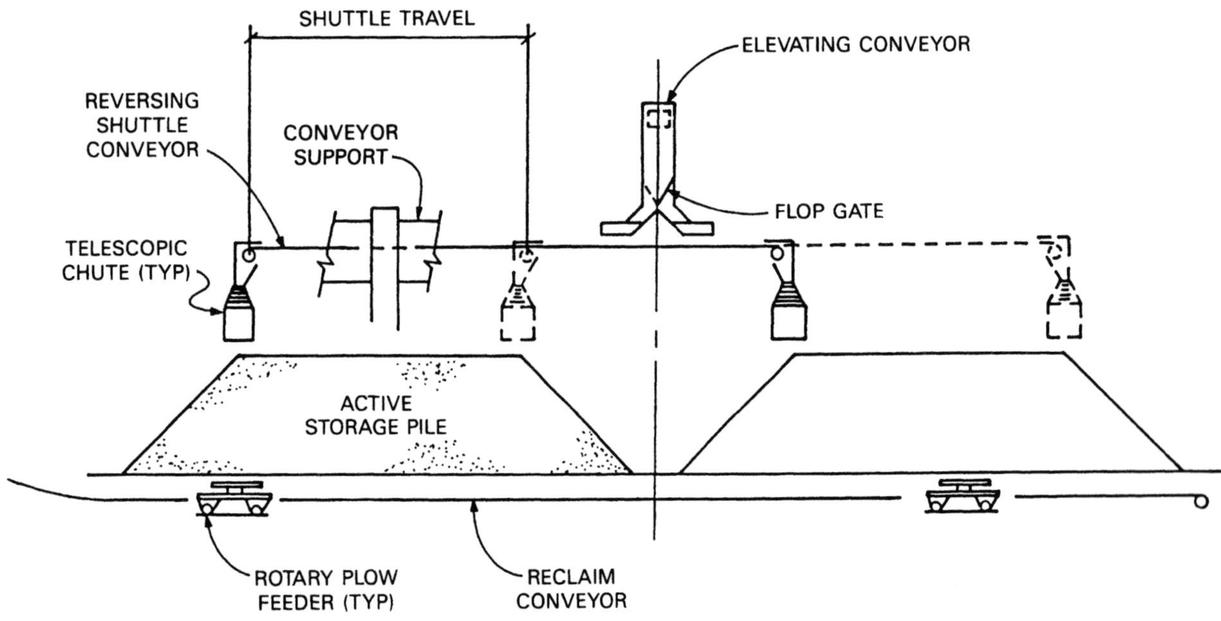

Fig. 5-26. Stockout system with elevated reversing shuttle conveyor and rotary plow feeder reclaim.

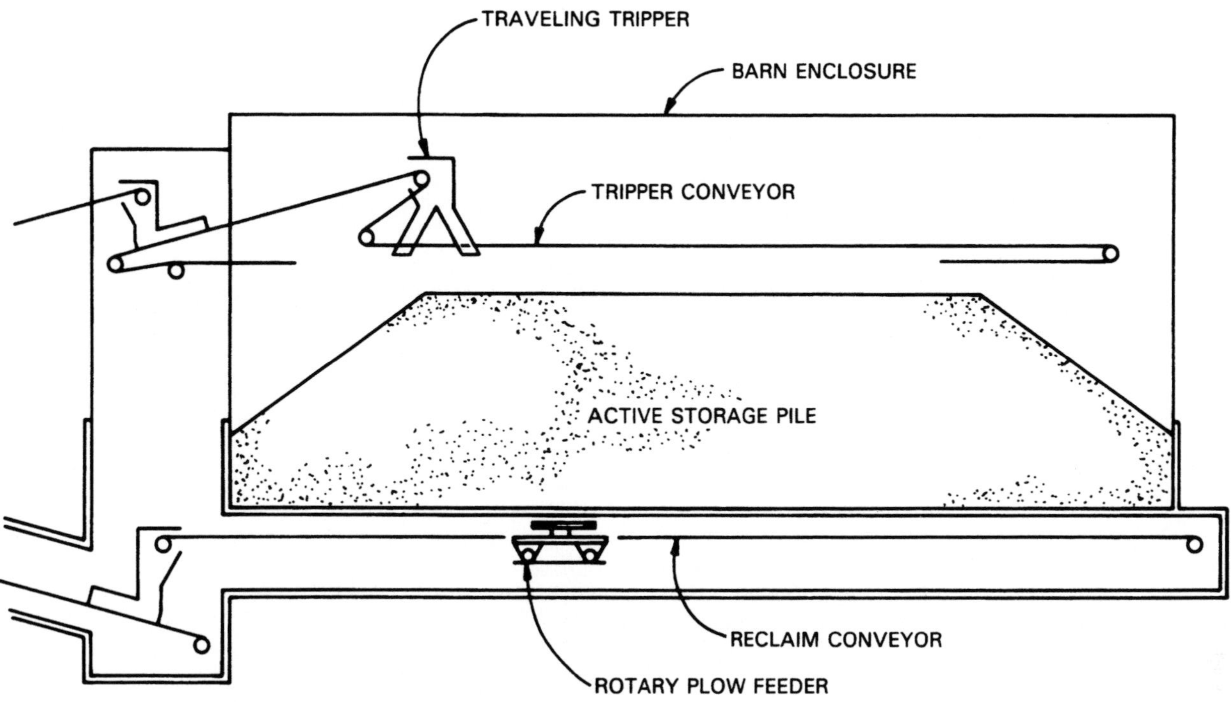

Fig. 5-27. Stockout system with traveling tripper and rotary plow feeder reclaim.

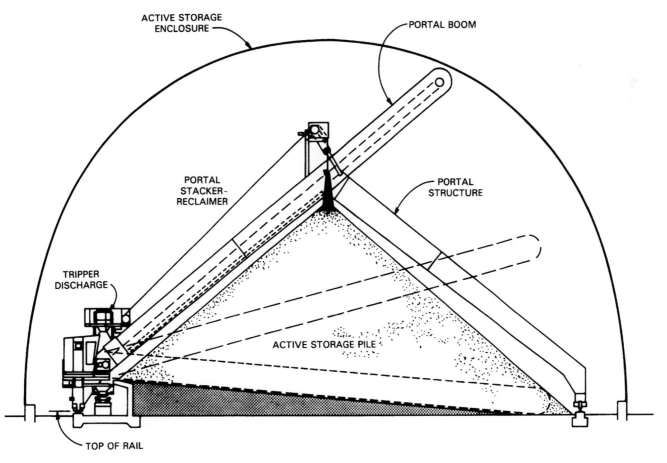

Fig. 5-28. Stockout/reclaim system with portal stacker–reclaimer.

reclaimer travels. The boom supports the chain and scraper blade assembly that is reversible for both stockout and reclaim operations.

Stockout of the material is accomplished by the boom also being equipped with a conveyor that elevates the material to the discharge point of the boom. The material then free falls onto the top of the pile. The boom is raised as the pile is formed to minimize the free fall height. This method of stockout is most commonly used for moderate to small power stations where enclosed storage is required or for a small plant for both coal and limestone. A typical portal stacker–reclaimer with a 100-ft (30.5-m) boom can stockpile a pile 40 ft (12.2 m) high. It stockpiles 50 tons/ft (149 tonnes/m) of coal or 85 tons/ft (253 tonnes/m) of limestone.

5.3.1.8 Stockout Summary. The stockout capacities for the stockout equipment discussed previously are summarized in Table 5-1 for per unit length and specified pile size.

Table 5-1. Stockout and Reclaim Capacities

Equipment	Material	Pile Height (ft)	Per Unit Length[a] (ton/ft)	Pile Length[b] (ft)	Total Capacity[c] (tons)
Fixed boom conveyor	Coal	60			10,000
	Limestone	30			2,000
	Limestone	65			20,000
Radial stacker	Coal	30	62.5 ton/deg	180 deg	12,500
	Limestone	30	106.3 ton/deg	180 deg	21,000
Traveling stacker	Coal	40	53	600	34,700
	Limestone	40	90	600	59,000
Bucket wheel					
Stacker–Reclaimer					
Trench type	Coal				
Stockout		40	40	600	27,000
Reclaim 90 deg			25	600	15,000
Reclaim 90/45 deg			30	600	18,000
Slewing type—small	Coal				
Stockout		50	100	600	76,000
Reclaim			75	600	45,000
Slewing type—large	Coal				
Stockout		70	310	600	210,000
Reclaim			245	600	147,000
Elevated reversing					
Shuttle conveyor	Coal	60	120	600	86,400
	Limestone	60	200	600	146,000
Traveling tripper					
Triangular pile	Coal	60	120	600	84,000
Diamond-shaped pile	Coal	60	40	600	24,600
Portal stacker–reclaimer					
Stockout and Reclaim	Coal	40	50	300	17,400
	Limestone	40	85	300	30,000

Equipment	Material	Reclaim Capacity (tons/h)
Drum reclaimer	Coal	Complete pile
Stationary feeders		
Vibratory—24 in.	Coal	225
	Limestone	300
Vibratory—72 in.	Coal	1,200
	Limestone	1,700
Stationary feeders		
Belt—24 in.	Coal	120
	Limestone	210
Belt—72 in.	Coal	2,400
	Limestone	4,080
Rotary flow feeder	Coal	200–2,000

[a]Per unit length in tons per foot, except as noted.

[b]Pile length in feet, except as noted.

[c]Volumes for stockout piles based on the following:

- Angle of repose, coal or limestone—37 degrees
- Density, coal—50 lb/ft^3
- Density, limestone—85 lb/ft^3

5.3.2 Reclaim Equipment

The capacity of the coal reclaim system, as well as the coal crushing and silo fill system, is based on the maximum daily consumption of all generating units (final plant size) for the peak years of generation. This total capacity may be designed initially or in phases as additional units are added. The capacity of the limestone reclaim system is based on the maximum daily requirements for limestone. Except for small power stations, the coal and limestone reclaim systems are independent systems.

In general, reclaiming systems may be classified as either "above-grade" or "below-grade" systems, depending on whether the material is removed from the top or bottom of the storage pile. Alternate types of above-grade reclaim systems include the bucket wheel stacker–reclaimer, portal stacker–reclaimer, and drum reclaimer. The principal form of above-grade reclaiming in use is the bucket wheel stacker–reclaimer.

The tunnel conveyor is the principal form of below-grade reclaiming and is used with alternate feed systems including stationary feeders (vibratory and belt) and rotary plow feeders. Mobile equipment, such as bulldozers, frontend loaders, or scrapers, is used in conjunction with some of the above equipment for a complete reclaim system. The above equipment is utilized to reclaim coal or limestone from either open or enclosed storage depending on the type of stockout equipment it is associated with. All of these types of equipment differ in concept and capacity, and each type has multiple variations in design and layout.

5.3.2.1 Bucket Wheel Stacker–Reclaimer. A bucket wheel stacker–reclaimer is used for both stockout (discussed previously in Section 5.3.1.4 and shown in Figs. 5-23 and 5-25) and reclaim operations. The stacker–reclaimer is equipped with a bucket wheel at the end of the boom for the reclaim mode. The bucket wheel, or digging wheel, is used for reclaiming material from the active storage pile and discharging it onto the boom conveyor. The digging wheel has a series of buckets arranged around the circumference of the wheel. The buckets are scoop-shaped with two open faces, one at the leading edge, for digging, and one where the bucket is attached to the wheel for discharge to the boom conveyor, as shown in Fig. 5-29.

The two types of bucket wheel stacker–reclaimers are the trench type and the slewing type. The trench type machine reclaims by continuously traveling while the boom and rotating bucket wheel remain at 90 degrees (1.5 rad) to the track. The bucket wheel reclaims the material in benches as the boom is lowered after each set travel distance. The slewing type machine reclaims by continuously slewing the boom from 90 to 30 degrees (1.5 to 0.5 rad) to the track and back while traveling in short increments between each slew motion. The bucket wheel reclaims the material in benches similar to the trencher.

A typical trench type stacker–reclaimer with a 50-ft (15.2-m) boom can reclaim 25 tons/ft (74 tonnes/m). The newer trenchers also slew the boom to 45 degrees (0.8 rad) from the track and increase the reclaim capacity to 30 tons/ft (89 tonnes/m), by reclaiming additional coal next to the track. This reclaim rate of 25 to 30 tons/ft (74 to 89 tonnes/m) is from the total stockout capacity of 40 tons/ft (119 tonnes/m). During normal stockout and reclaim operations, the unreclaimed coal is either compacted and stays in place or is dozed to the stacker–reclaimer for total reclaim of the stocked out coal. A small size slewing type stacker–reclaimer with a 95-ft (29.0-m) boom can reclaim 75 tons/ft (223 tonnes/m). A large-size slewing type stacker–reclaimer with a 185 ft (56.4 m) boom can reclaim 240 tons/ft (714 tonnes/m). These reclaim rates of 75 and 240 tons/ft (223 to 714 tonnes/m) are from the total stockout capacity of 110 and 310 tons/ft (327 and 923 tonnes/m), respectively.

Fig. 5-29. Stacker reclaimer bucket—reclaiming coal. (From McNally Wellman. Used with permission.)

5.3.2.2 Portal Stacker–Reclaimer A portal stacker–reclaimer is used for both the stockout (discussed previously in Section 5.3.1.7 and shown in Fig. 5-28) and reclaim operations. The portal stacker–reclaimer is equipped with the reversible chain and scraper blade assembly on the boom for the reclaim mode. The boom is lowered to the top of the pile, and as the portal stacker–reclaimer travels, the scraper blades drag the material down the face of the pile to the reclaim conveyor. The scraper blades reclaim the material in benches as the boom is lowered after each set travel distance.

A typical portal stacker–reclaimer with a 100-ft (30.5-m) boom can reclaim 50 tons/ft (149 tonnes/m) of coal or 85 tons/ft (253 tonnes/m) of limestone. The reclaim rate per foot matches the total stockout rate since the portal stacker–reclaimer can reclaim all of the stocked out material, without the assistance of mobile equipment.

5.3.2.3 Drum Reclaimer. The drum reclaimer is similar to the stacker–reclaimer in the reclaim mode. As shown in Fig. 5-30, the drum reclaimer is a horizontal drum 10 to 12 ft (3.0 to 3.7 m) in diameter that spans the base of the stockpile. The drum has staggered rows of buckets mounted on its

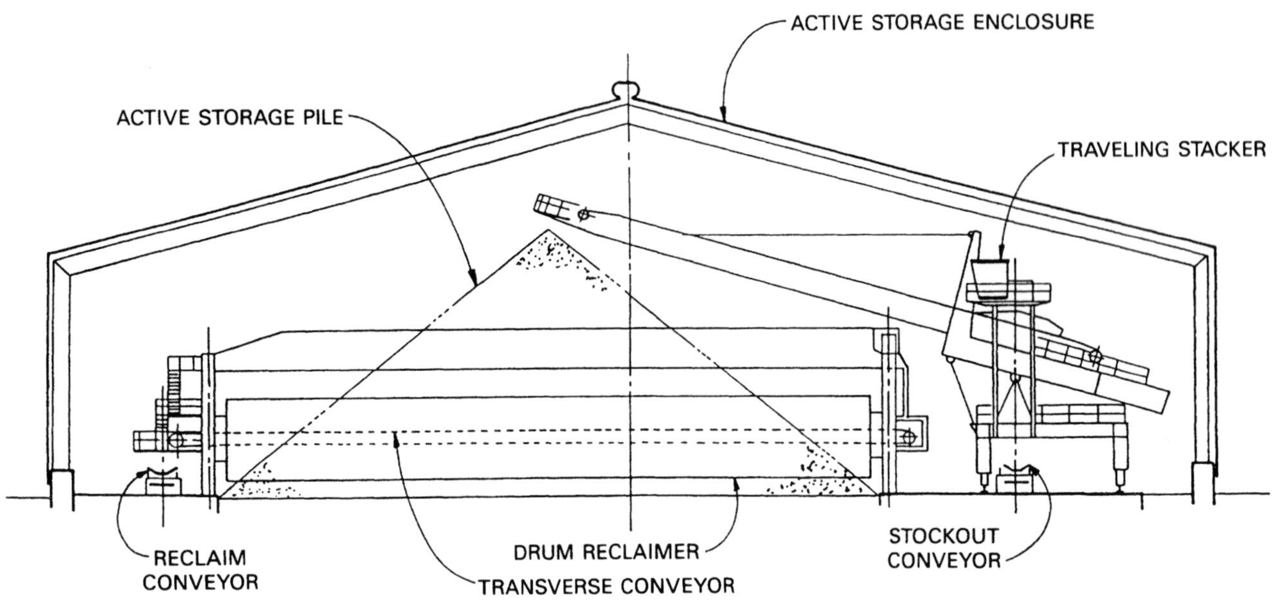

Fig. 5-30. Reclaim system with drum reclaimer and traveling stacker stockout.

circumference to scoop up the material as the drum reclaimer travels into the pile. The buckets discharge the material to a transverse conveyor in the center of the drum. The drum reclaimer also employs a rake to keep the face of the pile sliding down to the drum.

The drum reclaimer is excellent for use where blending is required and the coal types were layered in the stockout operation, since it simultaneously reclaims material from the complete cross-section of the pile as the material flows down to the buckets. In addition, the drum reclaimer can reclaim all of the stocked out material without the assistance of mobile equipment.

*5.3.2.4 **Stationary Feeders.*** Stationary feeders reclaim material from a hopper and the portion of the storage pile above the hopper that free flows into the hopper. These feeders have variable speed drives to control and regulate the flow of material out of the hopper and onto the reclaim conveyor. Stationary feeders are usually used in pairs for each hopper. A conical shaped storage pile for reserve coal storage or active limestone storage is normally reclaimed with one pair of feeders beneath a hopper. A long storage pile with a triangular cross-section for active coal storage is normally reclaimed by a series of feeders and hoppers spaced to maximize economically the amount of reclaimed material.

Two types of stationary feeders are the vibratory feeder and the belt feeder. The vibratory feeder generates the vibratory force from a relatively small centrifugal counterweight connected to the drive motor mounted on a secondary mass. The secondary mass is supported from the main feeder frame by a spring system that amplifies the vibration, as shown in Fig. 5-31. The feeder is mounted horizontally or declined to approximately 10 degrees (0.17 rad). A vibratory feeder moves material across the pan at 30 to 60 ft/min (0.15 to 0.30 m/s). A 24 in. (0.6 m) wide feeder has a midrange

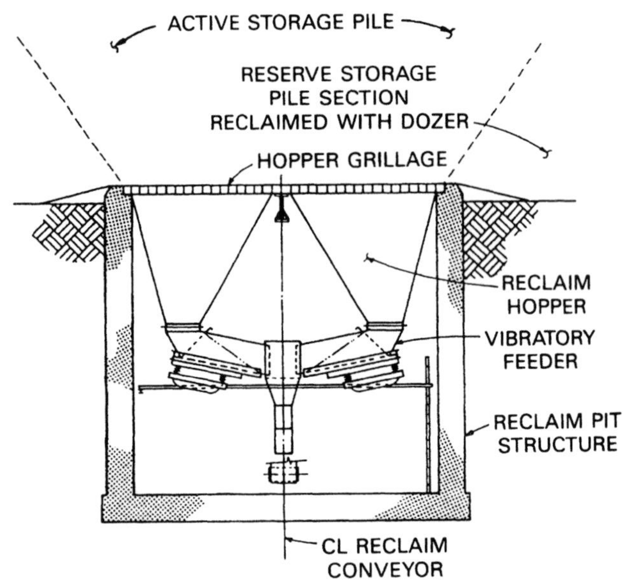

Fig. 5-31. Reclaim system with vibratory feeder.

capacity of 225 tons (204 tonnes) per hour of coal and 300 tons (272 tonnes) per hour of limestone. A 72 in. (1.8 m) wide feeder has a midrange capacity of 1,200 tons (1,090 tonnes) per hour of coal and 1,700 tons (1,540 tonnes) per hour of limestone. The reclaim capacity is varied by varying the amplitude of vibration. The reclaim rate is usually more accurate in the upper half of the range, and changes in the material moisture content and partial size also vary the capacity.

The belt feeder is a short belt conveyor with pulley centers in the range of 8 to 20 ft (2.4 to 6.1 m), as shown in Fig. 5-32. The feeder is mounted horizontally or inclined to approximately 14 degrees (0.24 rad). A belt feeder moves the

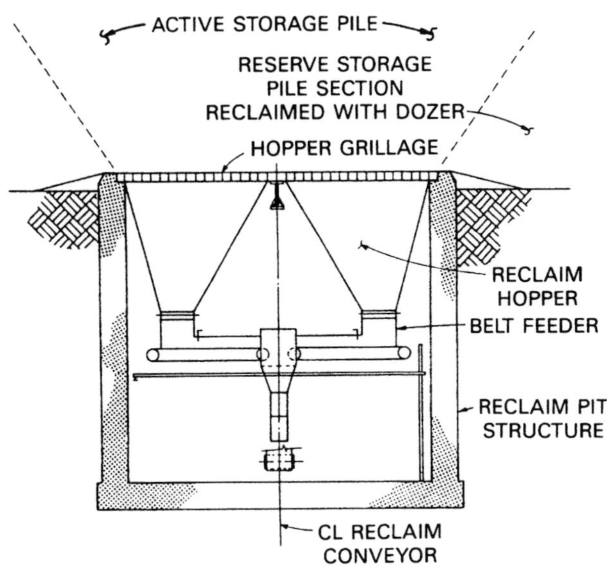

Fig. 5-32. Reclaim system with belt feeder.

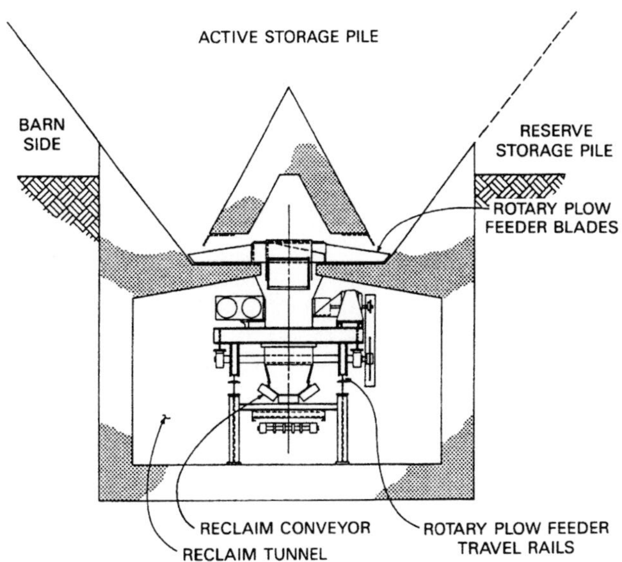

Fig. 5-33. Reclaim system with rotary plow feeder.

material with a belt speed of up to 100 ft/min (0.5 m/s), the preferred maximum speed. A 24-in. (0.6-m) belt feeder with a skirt width of 14 in. (0.4 m) has a capacity (at 100 fpm) of 120 tons (109 tonnes) per hour of coal and 210 tons (190 tonnes) per hour of limestone. A 72 in. (1.8 m) wide feeder with a skirt width of 62 in. (1.6 m) has a capacity (at 100 fpm) of 2,400 tons (2,180 tonnes) per hour of coal and 4,080 tons (3,700 tonnes) per hour of limestone. The reclaim capacity is varied by varying the belt speed. The optimum material bed depth is 60% of the material width. The reclaim rate is directly proportional to the belt speed for a constant-density material.

Stationary feeders, either vibrating or belt, are also used as the hopper discharge equipment in the unloading system. Rotary car dumper hoppers normally use belt feeders, as previously shown in Fig. 5-7, or vibrating feeders in the same configuration. Bottom dump hoppers two-car lengths or shorter also use stationary feeders.

5.3.2.5 Rotary Plow Feeder. A rotary plow feeder reclaims material from a slotted hopper and the portion of the storage pile above the hopper that will free flow into the hopper as shown in Fig. 5-33. A storage barn, with a rotary plow feeder reclaim system, will have sides inclined at 55 degrees (0.96 rad) or steeper so that all of the material stored in the barn will be reclaimable by the rotary plow feeder(s). The self-propelled rotary plow feeder travels on a set of rails, one on either side of the reclaim conveyor. The rotary plow feeder reclaims material with a nominal 10-ft (3-m) diameter blade that rotates horizontally and removes the material from the reclaim shelf of the hopper. The blade is equipped with six curved arms to move the material from the shelf inward to the feeder chute and down onto the tunnel (reclaim) conveyor. The rotary plow feeder travels at a fixed speed in the range of 10 to 20 ft/min (0.05 to 0.10 m/s). The reclaim capacity is varied by varying the rotational speed of the blade. A rotary plow feeder with a 10-ft (3-m) diameter blade can reclaim coal at a variable rate of 200 to 2,000 tons (180 to 1,810 tonnes) per hour.

Rotary plow feeders are also used as the hopper discharge equipment in the unloading system. Bottom dump hoppers of two-car lengths or longer generally use rotary plow feeders, as previously shown in Fig. 5-6.

5.3.2.6 Reclaim Summary. The reclaim capacities for all of the reclaim equipment discussed in the previous sections are summarized in Table 5-1 for per unit length and specified pile size.

5.3.3 Material Storage

5.3.3.1 Coal Storage—Active and Reserve. The coal storage area includes both active and reserve coal storage. Active coal storage implies the reclaiming and combustion of coal that has been stored for only a relatively short time, usually less than a week. Coal from an active coal storage pile is usually reclaimed without the use of mobile equipment. Normally, a power plant has an active storage capacity of 3 days at the maximum burn rate of all units. The 3 days are increased or reduced depending on coal sources, transportation modes and distance, climate, and plant availability requirements. For a plant with multiple coal sources and delivery modes, a short haul distance, and average climate conditions, the active storage requirement relates to a capacity of 3 days or less. A single coal source and delivery mode and high baseload availability requirements equate to a capacity of 3 days or more of active storage. To permit flexibility in the selection of coal sources and to attend to the increased environmental considerations, the design criteria for the storage area should also include consideration of segregated storage of multicoals.

Enclosed storage in silos or barns should be considered

for extremely cool areas and in congested areas when coal handling dust is a major problem.

Active coal storage silos have been 70 ft (21.3 m) in diameter and constructed of concrete. The silo capacity ranges from 6,000 tons (5,440 tonnes) at 120 ft (36.6 m) tall to 18,000 tons (16,330 tonnes) at 240 ft (73.2 m) tall. The silo is fed by an elevating belt conveyor and has four hopper outlets with stationary feeders and a reclaim conveyor, as shown in Fig. 5-34. The reclaim feeders are the vibrating or belt type, as previously shown in Figs. 5-31 and 5-32.

The current design is toward smaller diameter silos with a single outlet for better discharge flow characteristics and silo stability. A 54 ft (16.5 m) diameter silo has a capacity of 6,000 tons (5,440 tonnes) at 190 ft (57.9 m) tall.

Active coal storage barns are normally constructed with a diamond-shaped cross-section storage area. The lower half is concrete or reinforced earth with the top half steel framing, as shown in Fig. 5-35. The coal is fed to the barn by a tripper conveyor, also shown in Fig. 5-27, and reclaimed with rotary plow feeders, also as shown in Fig. 5-33. The capacity is based on the barn's cross-sectional shape and length, normally in the range of 20,000 to 60,000 tons (18,140 to 54,430 tonnes). A storage barn can also be a free-standing building that houses the storage pile with stockout and reclaim equipment, as previously shown in Fig. 5-30. These barn capacities are typically in the range of 10,000 to 20,000 tons (9,070 to 18,140 tonnes).

The reserve storage pile is a long-term storage area that provides an emergency supply of coal for the power plant in the event of an interruption of coal shipments. The reserve storage pile usually holds a 60- to 90-day supply at 65% to 80% of maximum burn rate. The number of days of supply is

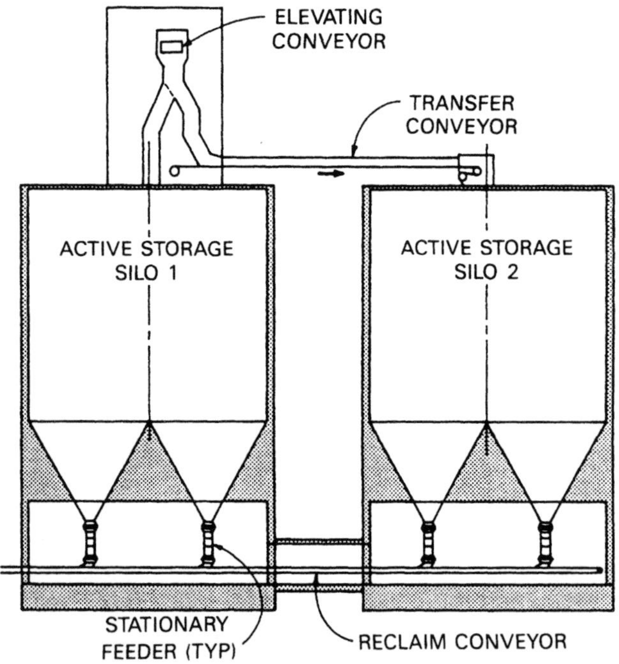

Fig. 5-34. Silos for enclosed storage.

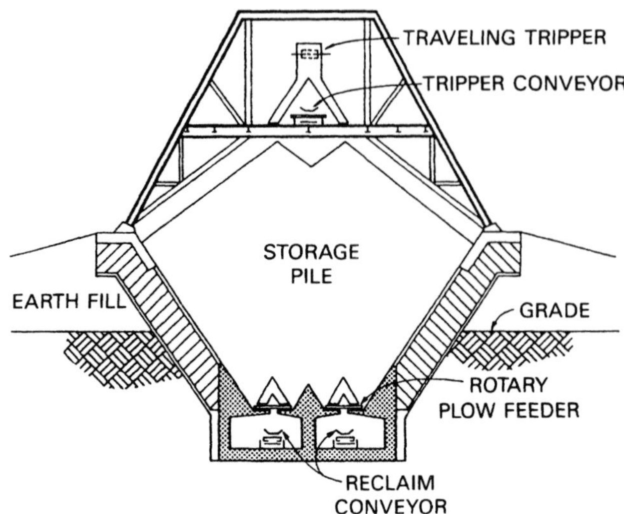

Fig. 5-35. Barn for enclosed storage.

usually based on the reliability of the coal sources and modes of transportation. The percent of maximum burn rate is typically the annual capacity factor or the annual capacity factor while operating for all generating units.

5.3.3.2 Coal Pile Management and Arrangement. The active coal storage pile is stocked out and reclaimed by the equipment described in Sections 5.3.1 and 5.3.2, with the pile configuration based on the stockout and reclaim equipment used. The pile is noncompacted, with a nominal density of 50 lb/ft^3 (800 kg/m^3). The reserve storage pile configuration is normally rectangular, but it can be an irregular shape to suit the plant site or equipment arrangement. It is a compacted pile about 40 ft (12.2 m) high. The pile side slopes are 4 to 1 (horizontal to vertical) to enable mobile equipment to also compact the pile sides. The pile is compacted in 1-ft (0.3-m) layers by rubber-tired mobile equipment to a density of 70 lb/ft^3 (1,120 kg/m^3). The pile top is crowned at 30 to 1 to facilitate drainage, then sealed to reduce the effect of wind and water erosion, as discussed further in Section 5.7.3.

The reserve storage pile is located next to or as near to the active storage pile as practical. It is stocked out and reclaimed by mobile equipment. The reserve storage system arrangement should be such that coal directed to the reserve stockout pile can be either reclaimed for plant supply or moved to reserve storage by use of mobile equipment. The reserve storage pile may be divided into two piles, inactive reserve and ready reserve, with the greater portion in the inactive reserve pile. Mobile equipment should not be allowed to disturb this pile after it is constructed and sealed. The ready reserve pile size should be large enough to accommodate fluctuations in active coal caused by nonuniform deliveries, but should be small enough that dust emissions can be controlled. If there are large seasonal fluctuations in coal delivery, a different amount of coal in each of the reserve coal piles may be required. Successful dust control on a coal pile depends on proper pile construction to provide a stable base for dust suppressants and to reduce the possibility of spontaneous combustion. Coal from the inactive reserve pile should be reclaimed from the sides, cutting from top to bottom and leaving a 4 to 1 slope. Properly stockpiled, the reserve storage pile should have only minimal problems with spontaneous combustion, and in combination with a surface dust suppressant, should have adequate resistance to erosion from wind or rain.

The reserve storage area for the Rawhide Energy Station of Platte River Power Authority, located in north central Colorado, was designed as shown in Fig. 5-36 to protect the piles from a strong, constant west wind. The reserve storage piles start at the reserve stockout pile and extend westward. They are divided into three piles oriented in a direction normal to the prevailing winds and are about 35 ft (10.7 m) deep. This configuration allows the top surface of the piles to be maintained below the level of the built up earth berms, that reduces windborne emissions. In addition, the top of the piles can be sealed easily to minimize spontaneous combustion. This is a good example of tailoring the reserve storage area to the site conditions.

5.3.3.3 Mobile Equipment. Regardless of the number of coal piles and types of coal in storage, adequate space must be provided to facilitate the plant requirements as well as the use of mobile equipment in the area. Mobile equipment for use in handling and stock piling coal can be divided into five types:

- Track-type tractors—good for high capacity over short distances. Good tractive capabilities aid in pile forming.
- Wheel-type dozers with articulated design—good stability and maneuverability for moving the coal over greater distances. They have a lower coefficient of traction than track-type tractors.
- Wheel-type front-end loaders—versatile machines for dozing, digging, and pile management.
- Wheel-type scrapers with tandem drives and self-loading—good for long-haul and large-capacity operations. Best suited for top-loading systems and drive-over reclaim hoppers.
- Water wagons using a baffled tank and spray nozzles—provide dust suppression on the storage pile and haulage roads, and spray surfactant to seal inactive reserve storage areas.

Fig. 5-36. Reserve coal storage pile arrangement (Black & Veatch.)

The wheel-type equipment is best for pile compaction.

5.3.3.4 Coal Reserve Reclaim. Coal from the reserve storage area can be reclaimed by the reserve reclaim hopper or the active reclaim equipment working in conjunction with the mobile equipment.

Three positions are workable for the location of the reserve reclaim hopper when located in the area of the reserve stockout conveyor (fixed boom conveyor):

- On the center line of the stockout pile, providing 25% to 30% reclaim of the stockout pile without dozing.
- Half way between the stockout pile center and edge, providing 15% to 18% reclaim without dozing.
- At the edge of the stockout pile, requiring dozing for all reclaim.

The advantage of the first two locations is that a portion of the stockout pile can be reclaimed without the use of mobile equipment. However, the actual tonnage reclaimed from a 10,000-ton (9,070-tonne) capacity pile is relatively small. There are also two major disadvantages of these first two locations. First is the possibility of fire from spontaneous combustion in the coal over the hopper, because of the chimney effect of the hopper. The second disadvantage is the possible flow of water into the hopper during storms if the "doughnut" or funnel hole is allowed to remain after initial automatic reclaim. In the winter, this can also cause problems with wet coal freezing in the hopper and reclaim equipment. The location of the coal reclaim hopper adjacent to the stockout pile greatly reduces the disadvantages related to the other two locations and offers more flexibility for the hopper when it is used primarily for reserve reclaim.

5.3.3.5 Limestone Storage—Active and Reserve. Limestone is usually stored in one common pile with an active area near the reclaim system. The capacity of the total pile varies from a 30-day supply to a 6-month supply, depending on the transportation, shipment size, and plant usage rate. Both the active and reserve areas have a noncompacted nominal density of 85 lb/ft^3 (1,360 kg/m^3). Limestone does not present the problem of spontaneous combustion, and the amount in reserve storage does not justify compaction for reduced area.

The pile configuration is based on the stockout equipment used. The fixed boom conveyor, as previously shown in Fig. 5-20, is the stockout equipment used most commonly, with either one large conical pile or a small active conical pile with a dozed reserve pile adjacent. The previous discussion related to the location of a coal reclaim hopper also applies to limestone, except that there are no problems with spontaneous combustion. In addition, since the reclaim hopper is used for active reclaim and the pile is used and managed daily, the hopper is normally positioned under the pile to minimize the requirement for mobile equipment. The same mobile equipment used on the coal piles is also used for the limestone storage area, except that normally scrapers are not required.

5.3.4 Combined Stockout, Reclaim, and Storage Systems

Various methods and equipment have been discussed for coal and limestone (Stockout Equipment, Section 5.3.1, Reclaim, Section 5.3.2, and Material Storage, Section 5.3.3). These three systems are integrated into a complete facility for each power plant to handle the coal and limestone from unloading to silo or day bin fill. Several pieces of equipment are capable of both stockout and reclaim, whereas others are designed specifically for one function or the other. The equipment is also sized and arranged for either outdoor or enclosed storage. These three systems are combined in many variations to handle the specific requirements for each power plant. The common combinations of stockout, reclaim, and active storage for coal and limestone are summarized in Table 5-2.

5.4 CRUSHING AND PLANT SUPPLY SYSTEMS

The capacities of the coal crushing and silo fill system and the limestone day bin fill system are based on the maximum daily consumption for the final plant size for the peak years of generation. Where multiple units occupy a plant site, a coal crushing and silo fill system is usually developed for each two units. For these plant sites, the limestone day bin fill system is expanded as required to handle all of the units.

5.4.1 Coal Crushing and Silo Fill Parameters

For the multi-unit plant sites, the arrangement would be as follows. The crushing and silo fill system would handle the requirements of two units. The conveying path is to the first unit, with connecting conveyors to the second unit. Additional crushers, conveyors, and other equipment to service the future units are installed as required for each additional pair of units, as shown in Fig. 5-37. With one crushing and plant supply conveyor system serving three or more units, its reliability becomes critical. A separate crushing and plant supply conveyor system for each two units would mean that only two units would be dependent on each plant supply system. If one entire plant supply system became inoperable, only two of the units would be affected while the other units would still operate using the other plant supply system(s).

A four-unit site with two crushing and plant supply conveyor systems could also be interconnected with a reversing conveyor at the crusher buildings or plant transfer areas. This would permit one entire plant supply conveyor system to become inoperable and still supply coal to all four units. These criteria are based on units in the size range of 400 megawatts (MW) and above. A site developed for six 100-MW units would probably use a single crushing and plant supply conveyor system with all of the unit silos arranged in a single row.

The total capacity of the crushing and silo fill system for the two units is determined on the basis of the daily maximum burn rate for both units times approximately 2.4. This service factor allows for maintenance and flexibility of operation, requiring approximately 10 h per day to supply the units with 24 h worth of coal. The conveyor path usually consists of dual conveyors with each conveyor handling half of the total capacity (1.2 times maximum burn rate). The use of dual conveyors and the above capacity criteria enable the operation of both units at full capacity, even with one conveyor out of service for maintenance or repair.

For example, two 450-MW units with a maximum burn rate of 250 tons (227 tonnes) per hour (tph) each will have the following dual conveyor system. The system will handle: 250 tph $\times$ 2 units $\times$ 2.4 = 1,200 tph. Each conveyor will, therefore, have a capacity of 600 tons (544 tonnes) per hour.

Table 5-2. Stockout, Reclaim, and Active Storage Combined Systems

Stockout	Reclaim	Active Storage	Figure
Fixed boom conveyor	Stationary feeder	Conical pile	
	Mobile equipment	Conical pile	5-20
	Stationary feeder	Silo	5-34
Radial stacker	Stationary feeder	Semicircular pile	
	Rotary plow feeder	Semicircular pile	
	Mobile equipment	Semicircular pile	5-21
Traveling stacker	Rotary plow feeder	Long triangular pile	5-22
	Stationary feeder	Long triangular pile	
	Drum reclaimer	Barn	5-30
Bucket wheel stacker–reclaimer	Bucket wheel stacker–reclaimer	Long triangular pile	5-23
	Bucket wheel stacker–reclaimer	Flat topped pile	5-25
Elevated reversing shuttle conveyor	Rotary plow feeder	Long triangular pile	5-26
Traveling tripper	Rotary plow feeder	Barn	5-27 5-35
Portal stacker–reclaimer	Portal stacker–reclaimer	Barn	5-28

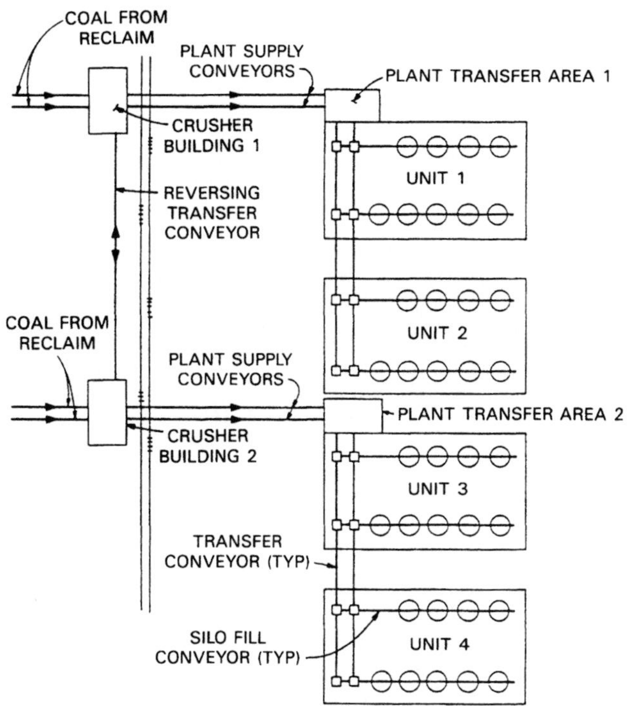

Fig. 5-37. Four-unit plant supply systems.

For a single large baseload unit (400 MW or above) plant site, the dual plant supply conveyor system is still recommended, as determined by the utility's dependence on that unit. The dual conveyor system is sized for 2.4 times the daily maximum burn rate, or each conveyor is sized for 1.2 times the daily maximum burn rate, so that the above criteria apply. For a single small unit (250 MW or below) plant site, the plant supply conveyor system uses one conveyor, based on economics. This one conveyor is sized for 2.4 to 8 times the daily maximum burn rate. The larger capacity decreases the required silo filling time and increases the time allowed for maintenance and repairs. Also, small units usually have silo capacities of up to 24 h at the maximum burn rate, allowing silo filling only once per day.

5.4.2 Coal Crushing

In general, the size of the coal received at power plants is typically 4 in. (0.10 m) or smaller. The size required for the pulverizers used with most pulverized coal steam generators is 1¼ to 1½ in. (0.03 to 0.04 m). This requires the reclaimed coal to be reduced by crushing before it is transported to the unit silos. Crushing capacity for frozen lumps should also be considered if a nonenclosed active storage system is utilized. The ring granulator, reversible hammermill, and bradford breaker type crushers have all been used for this application. The ring granulator crusher is normally selected for power plant application, since its maximum product size can be controlled while minimizing the amount of fines and dust produced. In addition, a ring granulator crusher can handle coal capacities from 50 tons (45 tonnes) per hour up to 3,000 tons (2,270 tonnes) per hour. As shown in Fig. 5-38, the ring granulator consists of alternate plates mounted on the crusher main shaft. Smaller shafts are supported by these plates at the periphery of the plates. On these smaller shafts are small round cutting rings that perform the crushing or grinding action. This assembly is mounted in a steel housing, the bottom of that is a curved grate. As the rings rotate they force coal through the grates (cage bars) in the bottom. The spacing in the grates is adjustable to compensate for wear. Uncrushable material is deposited in a tramp iron area without damage to the crusher. After the coal is crushed, it is conveyed directly to the unit(s) to fill the silos.

5.4.3 Coal Silo Fill

Older plants with multiple small units used a long coal bunker across all the units. The bunkers were large-capacity, rectangular-shaped bins with multiple square hopper-type outlets. Because of problems with hang-up of the coal on the slopes and hopper valleys of the bunkers, steel silos are now commonly used with all sizes of units. The steel silos are vertical cylinders, 24 to 30 ft (7.3 to 9.1 m) in diameter. They have conical outlets and sometimes use special equal percentage decrease shaped bottom hoppers to promote mass flow. These hoppers are constructed in sections with varying slopes such that the rate of change in material flow area is nearly constant (Snell 1982). The capacity of in-plant coal silos is usually 8 to 12 h at maximum burn rate (all silos).

There are a variety of methods and equipment to feed the coal silos. The major types of silo fill equipment for coal silos include a fixed tripper conveyor, traveling tripper, reversible stationary conveyors, and en masse chain conveyors.

5.4.3.1 Fixed Tripper Conveyor. The fixed tripper conveyor consists of a series of inclined conveyor sections with

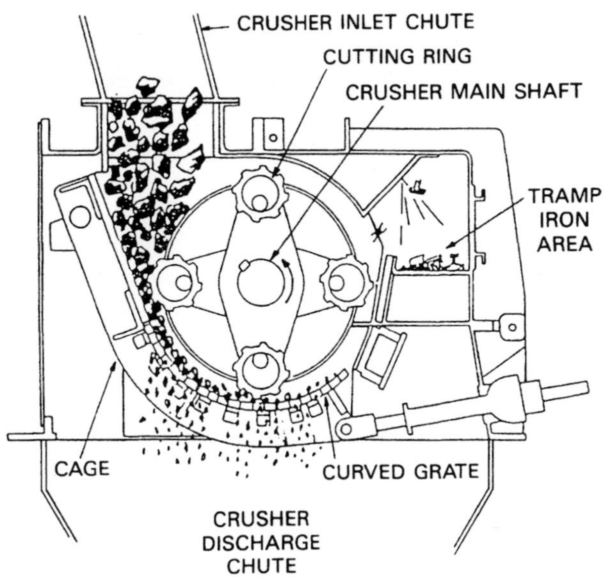

Fig. 5-38. Coal crusher—ring granulated type. (From Pennsylvania Crusher Corporation. Used with permission.)

each conveyor section discharging over the center of a silo, as shown in Fig. 5-39. Each discharge chute except the one over the last silo includes a diverter gate and pant-leg chute to direct the coal flow into the silo or back onto the next inclined conveyor section, as shown in Fig. 5-40. Plant silos are loaded with the fixed tripper conveyor by the correct positioning of the respective diverter gate at each silo. The silo fill operation proceeds from silo to silo. If one silo is not to be filled, the diverter gate over that silo is positioned to bypass the coal to the next silo without any disruption in the coal flow from the plant surge hopper or plant supply conveyors. The fixed tripper conveyor is better suited for a plant arrangement with one or two short rows of silos per unit (four or fewer silos per row). The capacity of this system is based on the belt conveyor capacity, that can easily handle the normal required silo fill rate of up to 1,000 tons (910 tonnes) per hour.

Dust control for the fixed tripper conveyor consists of a bag filter unit dust collector with pick-up points at the load chute and discharge chute of each inclined conveyor section. There is also a pick-up point at each silo.

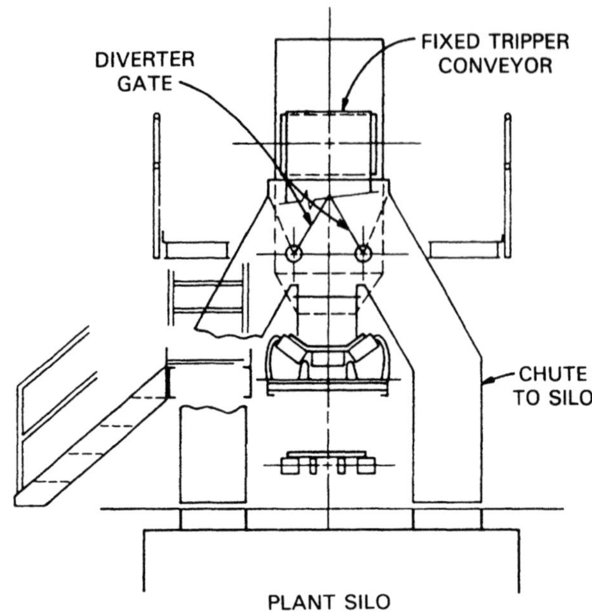

Fig. 5-40. Fixed tripper conveyor cross-section.

5.4.3.2 Traveling Tripper. The traveling tripper consists of a single horizontal conveyor above the row of silos. The conveyor has a traveling tripper that elevates the conveyor belt and discharges the coal to the silos through a pant-leg chute, as shown in Fig. 5-41. Plant silos are loaded in sequence, with the tripper stopping over the center of each silo. If one silo is not to be filled, the coal flow to the tripper must be stopped while the tripper travels over that silo. This can be accomplished by either turning off the feeder at the plant surge hopper or stopping the flow of coal to the plant supply conveyor. Either method requires purge capacity in the surge hopper or silo, as well as a more involved control system. A traveling tripper equipped with a gate, similar to the fixed tripper, as shown in Fig. 5-40, could divert the coal back onto the conveyor and into the last silo. This would allow the tripper to travel over a silo without purging the system, but the gate requires additional tripper height and complicates the tripper system arrangement.

The traveling tripper is better suited for a plant arrangement with a single long row of silos per unit (six or more). Considerable additional structural costs are required for the traveling tripper since this system requires an additional bay in front of the silo row. This additional space is necessary for proper loading of the tripper conveyor. As with the fixed tripper conveyor, the capacity of this system is based on the belt conveyor capacity, that can easily handle the normal required silo fill rate of up to 1,000 tons (910 tonnes) per hour.

Dust control for the traveling tripper consists of a seal belt over the two slots in the silo floor. The seal belt is lifted by the tripper discharge chute at the tripper discharge location. The bag filter unit dust collector has pick-up points at the load chute of the conveyor and at each silo. It is not practical to have dust pick-up points on the traveling tripper.

5.4.3.3 Reversible Stationary Conveyors. The reversible stationary conveyors consist of a supply conveyor to discharge the coal to two reversible conveyors staggered such that coal is delivered to alternating silos by each conveyor, as shown in Fig. 5-42. The conveyors are reversible and require the coal flow from the supply conveyor to be interrupted before the direction of the belt can be reversed. The coal flow is interrupted by actuating the supply conveyor gate, diverting the coal to the other reversible conveyor. The reversed belt is then ready to fill the next silo. The capacity of this

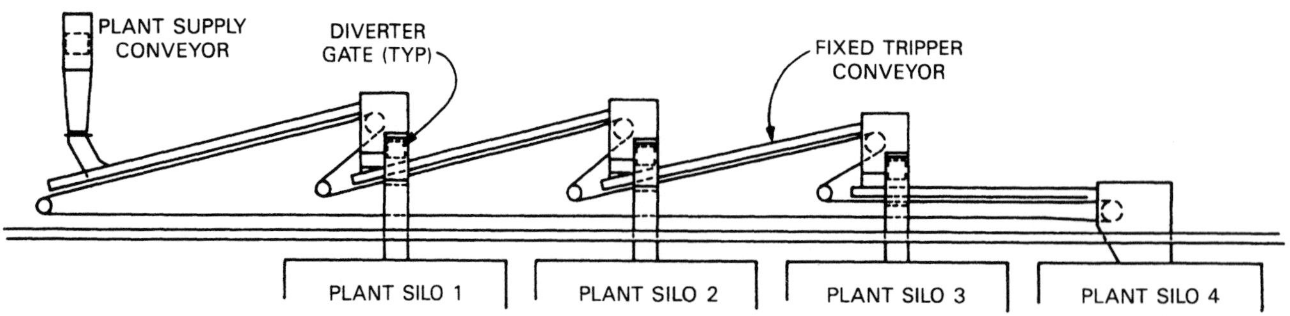

Fig. 5-39. Silo fill system with fixed tripper conveyor.

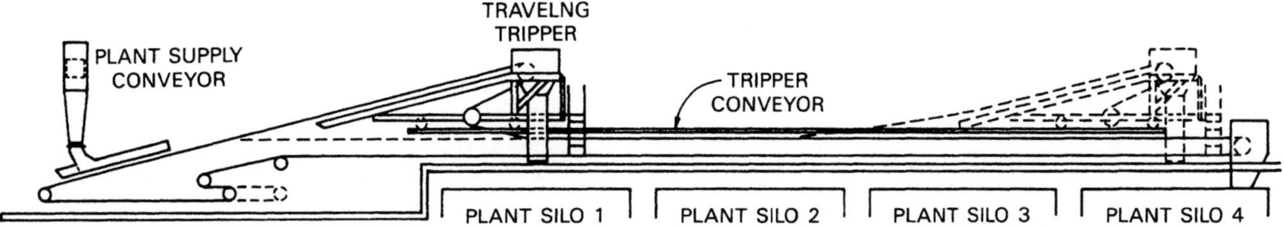

Fig. 5-41. Silo fill system with traveling tripper.

system is based on the belt conveyor capacity, that can easily handle the normal required silo fill rate of up to 1,000 tons (910 tonnes) per hour.

The reversible stationary conveyors are suited for a plant arrangement with one or two rows of four silos per unit. A row of five silos can be accommodated by chutework to the fifth (center) silo, but increased transfer height will be required.

Dust collection for the two reversible stationary conveyors is similar to that for the fixed tripper system. It consists of a bag filter unit dust collector with pick-up points at the load chute and discharge chute of the supply conveyor and each reversible conveyor, as well as at each silo.

5.4.3.4 En Masse Chain Conveyor. The en masse chain conveyor consists of a totally enclosed continuous chain with flight bars to move the coal en masse along the bottom of the casing, as shown in Fig. 5-43. At the center of each silo, a discharge gate is located in the bottom of the conveyor casing. Plant silos are loaded by the en masse chain conveyor by the correct positioning (open or closed) of the discharge gates at each silo. The silo fill operation proceeds from silo to silo. If one silo is not to be filled, the discharge gate over that silo remains closed to bypass the coal to the next silo without any disruption in the coal flow from the plant surge hopper or plant supply conveyors.

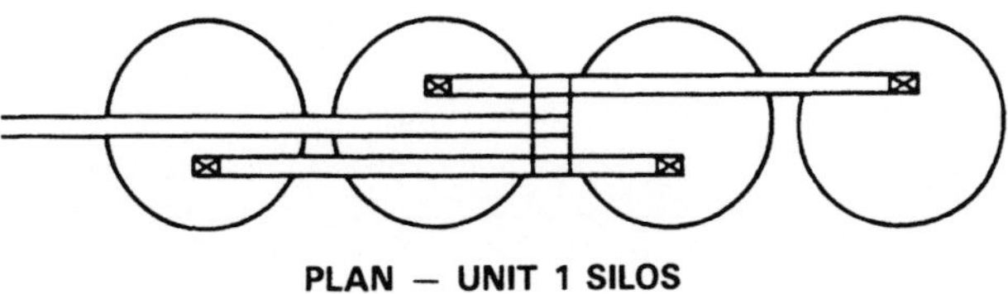

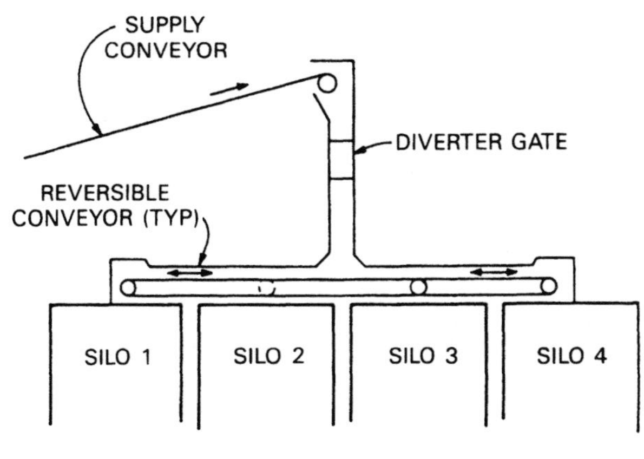

Fig. 5-42. Silo fill system with reversible stationary conveyors.

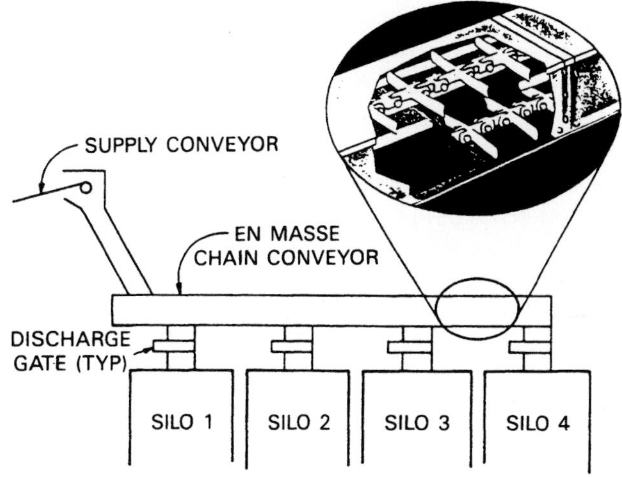

Fig. 5-43. Silo fill system with en masse chain conveyor.

The en masse chain conveyor is better suited for a plant arrangement with a single long row of silos per unit (four or more). The capacity of this system is limited by chain tension and at present is in the maximum range of 500 tons (455 tonnes) per hour. Large plants with silo fill rates of up to 1,000 tons (910 tonnes) per hour would require dual conveyor systems over each silo row.

Dust control for the en masse chain conveyor consists of a bag filter unit dust collector. The collector has pick-up points at several locations on the conveyor casing and at each silo.

5.4.4 Limestone Day Bin Fill

For multiple-unit plants, the day bin fill system usually handles the requirements of all of the units. This is based on the low usage rate [usually 5 to 15 tons (4.5 to 13.6 tonnes) per hour] and the high conveying rate, usually 200 to 600 tons (181 to 544 tonnes) per hour. Therefore, all of the in-use day bins can be filled in only a few hours each day.

Limestone is typically purchased sized properly for the limestone preparation facility, normally 1½ in. (0.04 m) or smaller top size. Limestone generally is not crushed in the reclaim area, except in colder climates to reduce frozen lumps to the purchased size.

In the limestone preparation facility, the normal practice is to have two day bins for a single unit plant site and one day bin for each unit of a multiple-unit plant site plus one spare. In this way, the plant site always has one spare day bin and associated limestone processing equipment. The day bin normally has a capacity equal to the limestone required for 24 h of each unit operation at maximum output.

Each day bin is a steel vertical cylinder with a conical outlet and a gyratory bin bottom to aid the limestone flow from the day bin to the limestone preparation equipment. For a one- to four-unit plant site, there are normally two to five day bins in a single row, with one per unit in use at any one time. Alternate types of day bin fill equipment for limestone day bins include chutework, chutework and conveyor, and a fixed tripper conveyor. These types of equipment differ in concept and can service different day bin arrangements.

A two-day-bin arrangement can be fed by the reclaim conveyor discharging between the two day bins with chutes to each day bin. They can also be fed by the reclaim conveyor discharging into the first bin and onto a short conveyor discharging to the second day bin. For three or more day bins in a row, a fixed tripper conveyor is typically used. The fixed tripper functions as previously described in Section 5.4.3.1 and shown in Fig. 5-39, for coal silo filling. With the last section of the fixed tripper conveyor elevated, the conveyor can easily be extended as other day bins are added with additional generating units.

5.5 CONVEYOR SIZING, DESIGN, AND SUPPORT SYSTEMS

Conveyor belts are used within and between the coal or limestone unloading, stockout and reclaim, and silo fill or day bin fill systems. The sizing, design, and support of the conveyors for both coal and limestone are very similar.

5.5.1 Conveyor Sizing

The proper design of a belt conveyor begins with an accurate appraisal of the characteristics of the material to be conveyed. The important characteristics are as follows:

1. Angle of repose—the angle that the surface of a normal, freely formed pile makes to the horizontal, typically 35 to 38 degrees for coal or limestone.
2. Angle of surcharge—the angle to the horizontal that the surface of the material assumes while it is at rest on a moving conveyor belt, typically 20 to 25 degrees for coal or limestone.
3. Flowability of material—determined by the material size, shape, roughness, moisture content, etc. Maximum belt incline angle commonly used is 14 degrees for coal and limestone.
4. Material density—for coal, density is 45 to 50 lb/ft^3, and for limestone, density is 85 lb/ft^3.

These material characteristics are defined in greater detail for all forms of coal and limestone in Chapter 3 of *Belt Conveyors for Bulk Material* (CEMA 1988).

The parameters of the belt conveyor and the determination of the belt width are calculated as follows:

1. Idler shape (troughed belt)—normally 35 degrees (idler side roll slope) for a coal or limestone conveyor.
2. Belt speed—recommended maximum for coal or limestone conveyors is 700 ft/min, based on belt width and material characteristics. Slower speeds are recommended for silo and day bin filling.
3. Capacity in cubic ft per hour—convert the capacity to be conveyed in tons per hour to cubic ft per hour

$$\text{ft}^3/\text{h} = \frac{\text{ton/h} \times 2{,}000 \text{ lb/ton}}{\text{material denstiy (lb/ft}^3)} \tag{5-1}$$

4. Capacity (equivalent)—convert the capacity in ft^3/h to the equivalent capacity at a belt speed of 100 ft/min.

$$\text{Capacity (equivalent)} = (\text{ft}^3/\text{h}) \times \left(\frac{100 \text{ ft/min}}{\text{actual belt speed (ft/min)}}\right) \tag{5-2}$$

The belt width can now be determined from Table 5-3 by selecting the proper belt width for the corresponding surcharge angle and capacity (equivalent). These belt conveyor parameters and selection of the belt width are defined in greater detail and for all troughed belt configurations in Chapter 4 of *Belt Conveyors for Bulk Material* (CEMA 1988).

The following example problem illustrates the selection of the correct belt size (width) and actual belt speed for a coal stockout conveyor with a required capacity of 3,500 ton/h and a maximum belt speed of 600 ft/min.

Assume 35-degree idlers, 25-degree angle of surcharge, and a coal density of 50 lb/ft^3 and using Eq. (5-1):

$$\text{Capacity (ft}^3/\text{h}) = \frac{3{,}500 \text{ ton/h} \times 2000 \text{ lb/ton}}{50 \text{ lb/ft}^3} = 140{,}000 \text{ ft}^3/\text{h}$$

The capacity and maximum belt speed are then used in Eq. (5-2).

$$\text{Capacity (equivalent)} = 140{,}000 \text{ ft}^3/\text{h} \times \left(\frac{100 \text{ ft/min}}{600 \text{ ft/min}}\right) = 23{,}333 \text{ ft}^3/\text{h}$$

Table 5-3 is based on 35-degree idlers. Using the column for a 25-degree surcharge angle, the capacity (equivalent) must be larger than 23,333 ft^3/h to keep the belt speed below 600 fpm. A belt width of 72 in. with a capacity (equivalent) of 27,196 ft^3/h is the first belt width above 23,333 ft^3/h.

The actual belt speed for the 72 in. wide conveyor will be

$$\frac{23{,}333}{27{,}196} \text{ ft}^3/\text{h} \times 600 \text{ ft/min} = 515 \text{ ft/min}$$

Table 5-3. 35-Degree Troughed Belt—Three Equal Rolls at Standard Edge Distance

Belt Width (in.)	A_t—Cross-Section of Load (ft²)							Capacity at 100 fpm (ft³/h)						
	Surcharge Angle							Surcharge Angle						
	0°	5°	10°	15°	20°	25°	30°	0°	5°	10°	15°	20°	25°	30°
18	0.144	0.160	0.177	0.194	0.212	0.230	0.248	864	964	1,066	1,169	1,274	1,381	1,492
24	0.278	0.309	0.341	0.373	0.406	0.440	0.474	1,668	1,857	2,048	2,241	2,438	2,640	2,847
30	0.455	0.506	0.557	0.609	0.662	0.716	0.772	2,733	3,039	3,346	3,658	3,975	4,300	4,636
36	0.676	0.751	0.826	0.903	0.980	1.060	1.142	4,058	4,508	4,961	5,419	5,886	6,364	6,857
42	0.940	1.044	1.148	1.254	1.361	1.471	1.585	5,644	6,266	6,891	7,524	8,169	8,830	9,511
48	1.248	1.385	1.523	1.662	1.804	1.949	2.099	7,491	8,312	9,138	9,974	10,825	11,698	12,598
54	1.599	1.774	1.950	2.128	2.309	2.494	2.686	9,598	10,646	11,700	12,768	13,855	14,969	16,118
60	1.994	2.211	2.429	2.651	2.876	3.107	3.345	11,966	13,269	14,580	15,906	17,257	18,642	21,058
72	2.913	3.229	3.547	3.869	4.197	4.532	4.879	17,484	19,378	21,285	23,215	25,182	27,196	29,275
84	4.007	4.440	4.876	5.317	5.766	6.226	6.701	24,043	26,641	29,256	31,902	34,597	37,360	40,210
96	5.274	5.842	6.415	6.994	7.584	8.189	8.812	31,645	35,058	38,490	41,966	45,506	49,134	52,876

From Belt Conveyors for Bulk Materials by Conveyor Equipment Manufacturers Association. 1988. Used with permission.

The actual belt speed should be increased by a 10% (minimum) safety factor because of the unknowns of the actual coal properties. The result is therefore a 72 in. wide belt with a belt speed of approximately 570 ft/min.

5.5.2 Conveyor Drive Design (Horsepower)

The horsepower (hp) required at the conveyor drive is derived from the effective tension (T_e) in pounds required at the drive pulley to propel the loaded conveyor at the belt velocity (V) in ft/min.

$$\text{hp} = \frac{T_e \times V}{33{,}000} \tag{5-3}$$

T_e is the summation of the belt tensions produced by gravitational load and frictional resistance of the conveyor components and the material conveyed.

T_e is determined as follows:

$$\text{Effective belt tension, } T_e = T_1 - T_2 \tag{5-4}$$

where

$T_1 = T_x + T_y + T_z$, tight side tension at the drive pulley, lb, (5-5)

where

$T_x + T_y$ = resistance due to friction, lb (belt on idlers, bearings, belt scrapers, etc.)

where

T_z = lift tension, lb (height of material elevated) × (weight of material per foot on belt).

$T_2 = C_w \times T_1$, slack side tension at the drive pulley, lb, (5-6)

where

C_w = wrap factor (amount of belt wrap on drive pulley).

Weight of material per foot on belt (W_m) is equal to the capacity in pounds per minute divided by belt speed in feet per minute.

$$W_m = \frac{\text{Capacity}\left(\frac{\text{tons}}{\text{h}}\right) \times \frac{2{,}000 \text{ lb}}{\text{ton}} \times \frac{1 \text{ h}}{60 \text{ min}}}{\text{Belt speed (ft/min)}} = \frac{\text{lb}}{\text{ft}} \tag{5-7}$$

Conveyor drive horsepower is defined in greater detail for all conveyor profiles in Chapter 6 of *Belt Conveyors for Bulk Materials* (CEMA 1988).

The following example problem illustrates the selection of the correct conveyor drive horsepower for the coal stockout conveyor in the previous example. The conveyor has a horizontal length of 400 ft and is inclined at 14 degrees. The total frictional resistance is 3,000 lb and the wrap factor is 0.38.

1. Calculate W_m using Eq. (5-7):

W_m = (3,500 ton/h × 2,000 lb/ton × 1 h/60 min)/570 ft/min

W_m = 205 lb/ft

2. Calculate H (height of material elevated):

H = 400 ft × tan 14 degrees = 100 ft

3. Calculate T_e using Eqs. (5-4), (5-5), and (5-6):

$T_e = T_1 - T_2$

$T_1 = T_x + T_y + T_z$

$T_x + T_y$ = 3,000 lb

T_z = 100 ft × 205 lb/ft

T_z = 20,500 lb

T_1 = 20,500 lb + 3,000 lb = 23,500 lb

$T_2 = C_w \times T_1$ = 0.38 (23,500 lb)

T_2 = 8,930 lb

T_e = 23,500 lb − 8,930 lb = 14,570 lb

4. Calculate hp using Eq. (5-3):

$$\text{hp} = \frac{T_e \times V}{33{,}000} = \frac{14{,}570 \text{ lb} \times 570 \text{ ft/min}}{33{,}000}$$

$$\text{hp} = 252$$

The required drive horsepower at the drive pulley is 252 hp. The actual motor horsepower is based on the drive efficiency (normally 0.97) and the motor service factor (nominally 1.15).

$$\text{hp motor} = 252 \times \frac{1}{0.97} \times 1.15 = 299 \text{ hp}$$

The selection of the next larger standard motor results in a 300-hp motor.

5.5.3 Conveyor Support and Enclosure Systems

Conveyor supports and enclosures vary depending on the location of the conveyor at the plant site and the plant area climate conditions.

Below-grade conveyors are usually supported on the pit or tunnel floor with stringer legs at 10-ft (3.0-m) centers. These conveyors are enclosed by the pit or tunnel and do not require further enclosures or conveyor covers. At-grade conveyors are usually supported on ties or piers with the stringer legs at 10- or 20-ft (3.0- or 6.1-m) centers. These conveyors have covers to protect the conveyed material from the elements. Elevated conveyors are supported and enclosed by a variety of methods. The supports include open box trusses beneath the conveyor, enclosed walk-through galleries (box trusses), and structural tubes. The support bent spacing for elevated conveyors varies based on the economics of the support span and site facilities to be avoided.

There are two arrangements for open box trusses. The conveyor stringer legs and the walkway are supported on the top chord of the box truss for the first arrangement as shown in Fig. 5-44. For the second, the conveyor idlers are supported directly on the top chord, and the walkway is cantilevered from the side of the box truss. Both conveyor arrangements use covers to protect the conveyed material from the elements. The covers are normally either ¾ fixed or full-hinged types. The ¾ fixed type allows the operator to view the belt and material without opening the cover, but the cover must be unbolted for maintenance other than idler lubrication. The full-hinged cover completely covers the belt and idlers and must be unlatched to view the belt and material or for maintenance. Walk-through galleries, as shown in Fig. 5-45, are large box trusses with the conveyor stringer legs and the walkway supported on the bottom chord. The complete box truss can be enclosed with wall paneling and a roof to protect both the conveyed material and the plant personnel from the elements. Structural tubes are round reinforced steel tubes with the conveyor stringer legs and the walkway supported on the lower section of the tube. As shown in Fig. 5-46, the steel tube furnishes both the conveyor support and the enclosure for the conveyed material and the plant personnel.

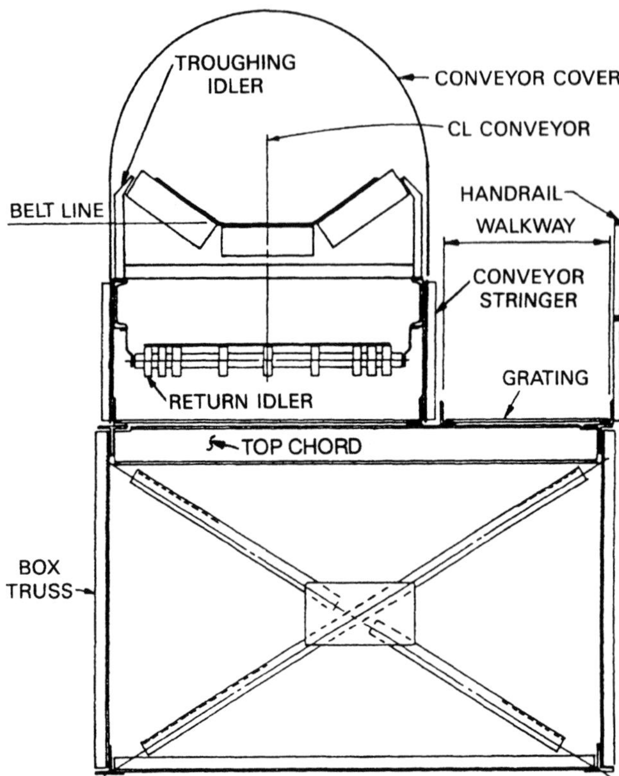

Fig. 5-44. Conveyor support and enclosure—box truss and cover.

5.6 SAMPLING AND WEIGHING SYSTEMS

Two important subsystems of a coal handling facility are the sampling system, including as-received and as-fired, and the weighing system, including belt, truck, and train scales. Normally, a limestone handling facility uses only a weighing system.

Coal samples are taken from the coal stream to determine and monitor plant efficiency and to verify the heating value of the coal delivered to the plant. As a minimum, the plant should have an as-fired sampling system for the determination of plant performance. It could also have an as-received sampling system for verification of mine point quality analysis and data. The as-fired system would be at the coal crusher building or the plant transfer area, whereas the as-received system would be in the first transfer building after the unloading building.

Weighing scales at various points in the coal and limestone handling systems are essential for checking material quantities and equipment handling rates. As a minimum, scales should be placed to determine the amount of coal and limestone delivered to the plant site, coal sent to the plant silos for each unit, and limestone sent to the day bins. Belt scales are a short section of conveyor idlers supported on

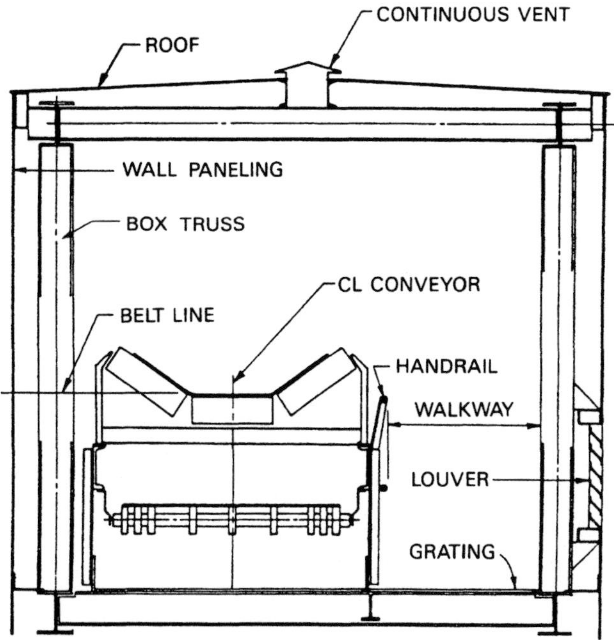

Fig. 5-45. Conveyor support and enclosure—walk-through gallery.

load cells to weigh the material on the belt. The belt scales for determining the total coal or limestone delivered to the plant should be positioned on each of the conveyors leaving the unloading systems. The belt scales for determining the coal sent to the plant silos should be located in front of each generating unit, and the belt scales for determining the limestone conveyed to the day bins should be located in front of the limestone preparation building.

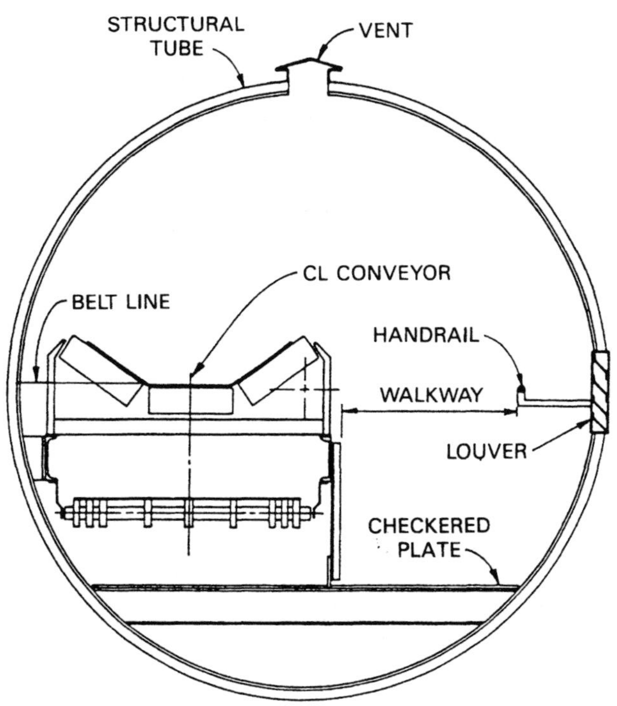

Fig. 5-46. Conveyor support and enclosure—structural tube.

Train and truck scales can also be used to determine the amount of coal and limestone delivered to the plant site. If the material is to be weighed by the plant weigh system for the purpose of determining material and/or transportation costs, train or truck scales are often used instead of belt scales.

5.6.1 Coal Sampling

The objective of collecting a sample of coal is to obtain a small portion that is representative of the whole of a shipment or consignment and then analyze it to determine characteristics such as heating value, and ash, sulfur, and moisture contents. At most utilities, coal samples are collected in accordance with American Society for Testing and Materials (ASTM) Standard D-2234, "Standard Test Methods for Collection of a Gross Sample of Coal" (ASTM 1993). Collected samples are prepared for analysis by reduction and division according to ASTM Standard D-2013, "Standard Method of Preparing Coal Samples for Analysis" (ASTM 1993). The analysis of the collected coal sample is then made by a lab analysis group either within or outside the power plant. The maximum shipment or consignment size from which one sample is taken should usually not exceed 10,000 tons (9,070 tonnes).

The basic process includes removing an increment of coal from the main coal stream and crushing it to the proper size, then reducing the sample again before it is collected for lab analysis. The sampling system employs two or three stages of sampling, depending on the consignment size, coal lump size, and the primary sampler type. The two types of primary samplers differ based on the location from which the sample is taken.

The first primary sampler type extracts a sample from the free-falling coal stream after it has been discharged by the conveyor. A three-stage sampling system is required with this type of primary sampler, due to the large amount of coal collected, if the conveyor capacity exceeds 2,000 tons (1,810 tonnes) per hour and/or the coal size exceeds 3 in. (0.08 m). The second primary sampler type extracts a sample from the coal stream while it is still being conveyed by the belt conveyor. A two-stage sampling system is used with this type of primary sampler, due to the smaller amount of coal collected, if the conveyor capacity does not exceed 3,500 tons (3,180 tonnes) per hour and/or the coal size does not exceed 3 in. (0.08 m). The first primary sampler (cutter) type and the components for a three-stage sampling system are shown in Fig. 5-47. The second primary sample cutter type and the components for a two-stage sampling system are shown in Fig. 5-48. Either primary sample cutter type can be used with a three-stage or a two-stage sampling system.

In a three-stage coal sample system, the primary sampler (cutter) is designed to extract samples on a consignment basis in accordance with published standards. A representative of all sizes of coal present is obtained from a stream of moving coal by passing the primary cutter at uniform speed

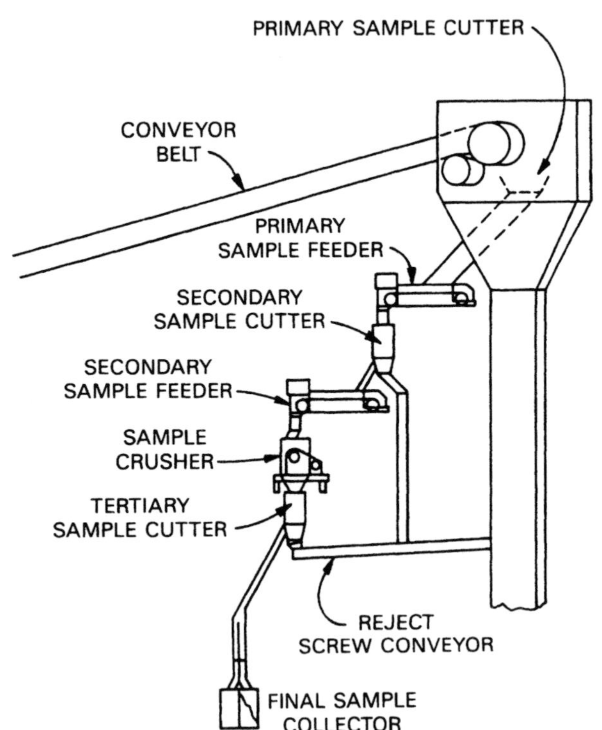

Fig. 5-47. Coal sampling system—three-stage with primary sample from free falling coal stream.

across the entire width of the coal stream. The cutter opening width should be at least three times the size of the largest lump of coal but not less than 1¼ in. (0.03 m). The sample is then transported by the primary sample belt feeder to the secondary sample cutter, where a number of increments must be obtained at the minimum weights prescribed by ASTM D-2234. As specified, the number of increments for the

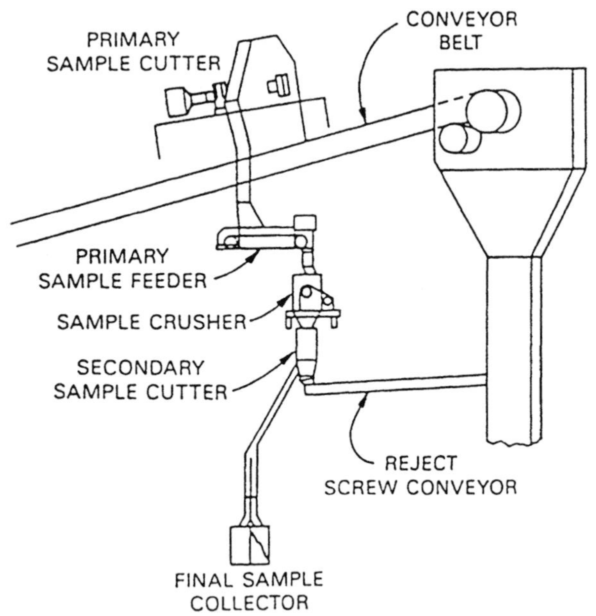

Fig. 5-48. Coal sampling system—two-stage with primary sample from conveyor belt.

secondary sample cutter should be six times the number of primary sample cutter increments, and the cutter width is also three times the largest coal lump. From the secondary sample cutter, the coal is metered by a secondary sample belt feeder or chutework to the crusher and reduced to 100% smaller than 4-mesh with 95% smaller than 8-mesh. The coal is then fed to the tertiary sample cutter with a minimum cutter width of 1¼ in. (0.03 m), where final extraction of the sample is accomplished. A minimum of 60 total increments (cuts) for the total consignment is required. The travel speed of each sample cutter should be in the range of 6 to 18 in./s (0.15 to 0.46 m/s). The final sample of approximately 40 to 50 lb (14.9 to 18.7 kg) is then collected in a suitable sample collector where it is retained in a dust- and moisture-tight container until required for analysis.

In the two-stage sampling system, the secondary sample cutter is deleted and the primary sample proceeds directly to the crusher. The tertiary (or final) sample cutter is the secondary sample cutter in a two-stage system. All other aspects of the system are the same as for the three-stage system.

Air circulation through the sampling equipment should be held to a minimum with no dust collection taking place on the sample process. Loss of fines and moisture is to be kept to a minimum. Chutes and equipment should allow no rust or contamination to mix with the coal samples. To avoid contamination, 304 stainless steel chutework is often used.

Coal rejected from the sample reduction process is returned automatically to the coal conveying system. The sampling system is designed so that failure in no way interferes with the normal operation of the remainder of the coal conveying system.

5.6.2 Coal and Limestone Weighing

The weighing system determines the quantity of material delivered to the plant site, the rate of material consumption, the quantity of material in storage, and the equipment handling rates. The design of the belt, truck, or train scales is the same for either coal or limestone.

5.6.2.1 Belt Scale. A conveyor belt scale, as shown in Fig. 5-49, is located on the conveyor receiving material for the power plant. This conveyor originates at the unloading system. In addition, belt scales are on conveyors feeding the plant silos or day bins, and should be used anywhere in the conveyor system where rate control is required. The quantity difference between the received material and the consumed material is the amount of material in active and reserve storage. Belt scales should not be located on conveyor belts near their takeups, terminal point, vertical curves, or loading points, since fluctuations in belt tensions affect belt scale accuracy.

Belt scales are typically the digital electronic type, with an accuracy range of 0.25% to 1.00% when the belt is from 25% to 100% loaded. Belt scales with an accuracy of 0.25% and used for commercial weighing are certified in accordance with National Institute of Standards and Technology

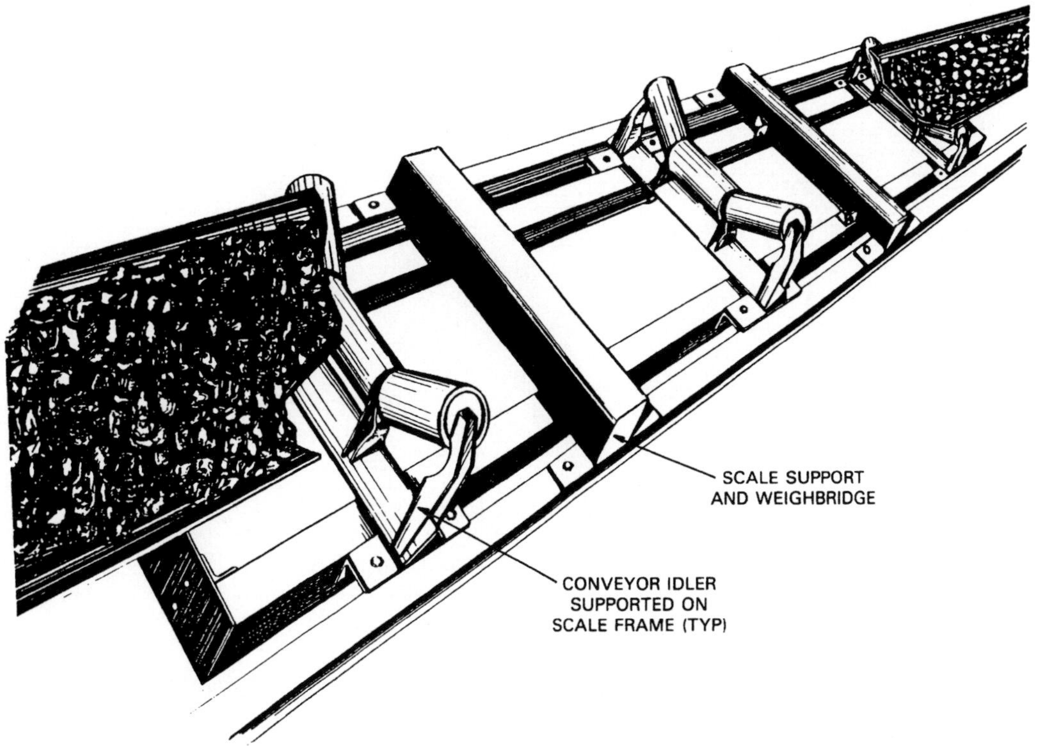

Fig. 5-49. Conveyor belt scale. (From Ramsey Technology, Inc. Used with permission.)

(NIST) Handbook 44 (U.S. Department of Commerce 1993) and state requirements. Belt scales with an accuracy of 0.25% to 0.5% are used for accurate weight and flow control, but are not certifiable. Belt scales with an accuracy of 0.5% to 1.0% are used for general-purpose flow control.

*5.6.2.2 **Truck Scale.*** Truck scales used today are the fully electronic, load-cell type for static weighing of trucks, as shown in Fig. 5-50. This type of scale has minimum effect on truck unloading time and is the most accurate. A second scale may be used after unloading if truck traffic or the traffic pattern is not suitable for use of one scale weighing both loaded and unloaded trucks.

Present truck scales are the pitless type with a minimum capacity of 120% of total truck weight and a minimum length of 120% of total truck length. For best accuracy, each truck is weighed individually with the total truck on the scale.

The truck scale is generally designed and constructed to meet the American Association of State Highway Officials (AASHO) bridge loading specification governing the type of trucks to be used, with the scale certified in accordance with NIST Handbook 44 and State requirements. The scale should be certified for proof of accuracy and can then be used for commercial weighing (pay scale). In most cases, the scale is used to check the material supplier's scale or for payment of the shipping costs. A microprocessor-based digital display unit displays net weight, tare weight, gross weight, and truck identification number.

Fig. 5-50. Truck scale for stationary weighing. (Black & Veatch.)

*5.6.2.3 **Train Scale.*** Newer train scales are the fully electronic, load-cell type for in motion one-direction unattended, coupled, two-draft railcar use. These scales are commonly known as coupled-in-motion scales, as shown in Fig. 5-51. This type of scale does not affect the unloading time of the train. A second train scale for tare weighing may be used if

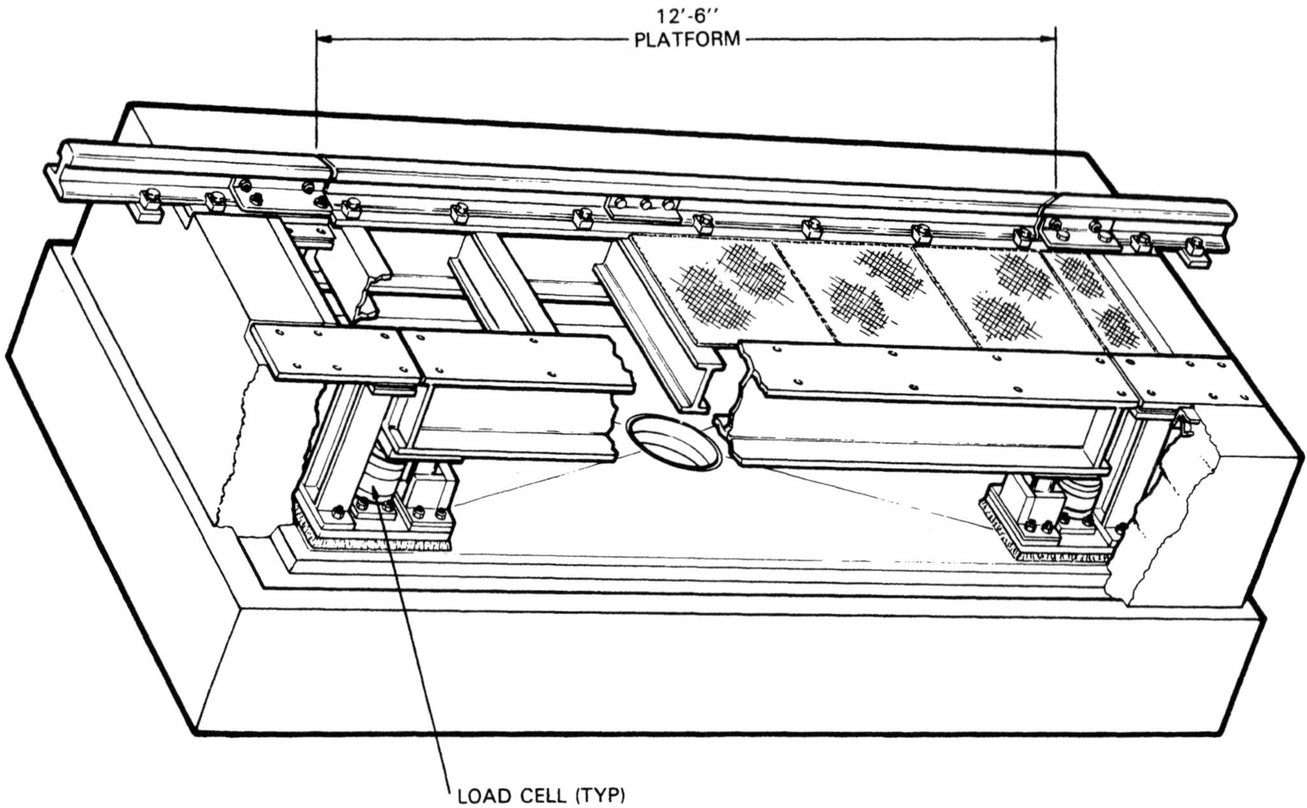

Fig. 5-51. Train scale for coupled-in-motion weighing. (From Toledo Scale. Used with permission.)

the scale system is used commercially. If not, the tare weights may be obtained from the cars, eliminating a second scale. Railcar tare weights, if used, should be checked periodically.

The typical train scale has a minimum capacity of 180 tons (163 tonnes) and a platform length of 12.5 ft (3.8 m). It is a pit-type scale with access provisions for load cell maintenance, that maintains its accuracy better over the life of the plant.

The scale is designed and constructed to meet the American Railway Engineering Association (AREA) bridge loading specification for a Cooper rating of E-80, or as specified by the railroad, and is certified in accordance with NIST Handbook 44 and State requirements. Similar to the truck scale, the train scale should be certified for proof of accuracy, and can then be used for commercial weighing (pay scale). In most cases, the scale is used to check the material supplier's scale. A microprocessor based, digital display unit displays net weight, tare weight, gross weight, and car identification number (optional).

Coupled-in-motion scales require 1,000 ft (305 m) of straight track, 500 ft (152 m) in front of and 500 ft (152 m) behind the scale. Scale approaches should have 80 ft (24 m) of concrete supported rail before and after the scale. Grade should be constant for 500 ft (152 m) on either side of the scale. It can then change, but should not go from uphill to downhill or vice versa.

Other types of scales are available for railcar weighing. They are static scales that require the railcar to be located on the scale platform and decoupled. Today, static scale use is limited to spot checking or for small plants, because the weighing time is too long for them to be considered for plants using unit train coal shipments.

5.7 DUST CONTROL SYSTEMS

The coal and limestone handling dust control systems minimize fugitive dust by confining it and minimizing its rehandling. The proper combination of dry dust collection and wet dust suppression is based on the type of material, plant location, and the site climate conditions. Dry dust collection reduces rehandling of dust by properly returning it to the system, processing it, or transporting it directly to the plant silos or day bins. With wet dust suppression, rehandling the dust is minimized, but excess moisture and freezing problems must be carefully monitored.

A combination of wet dust suppression and dry dust collection can be used at the coal unloading system. The extent of each depends on the unloading method and equipment selected. Since clogging problems associated with wet dust suppression systems may hamper coal handling operations, a dry dust collection system should be employed at coal storage structures and transfer points. Limestone unloading and handling systems typically use only dry dust collection systems since water and limestone dust causes material buildup problems. Supplemental dust control for exposed conveyors and transfer points consist of conveyor covers, enclosed conveyor galleries, lowering walls, or telescopic chutework. A wet spray system can be used in conjunction with a telescopic chutework to control coal dust emissions during stockout operations.

Reserve coal piles require supplemental dust control such as sealing or stabilizing to reduce the climate effects, such as wind and rain erosion. Feasible methods include vegetative or chemical stabilization and capping. Reserve limestone piles do not require supplemental dust control.

5.7.1 Dust Collection

Dry dust collection for coal or limestone involves removal of dust-laden air from enclosed transfer points or other dusty areas into a duct work system. A typical dust pick-up arrangement and ductwork are shown in Fig. 5-52. The duct system transports the dust-laden air to a fabric filter unit. The collection process creates a slight negative pressure within enclosed transfer points, hoppers, chutes, etc. This negative pressure causes an indraft of ambient air that is effective in reducing the expiration of dust along skirt seals, at points of loading impact in conveyor systems, and other points of potential dust leakage.

Dust collectors are manufactured by a large number of companies in the United States. Designs are highly varied,

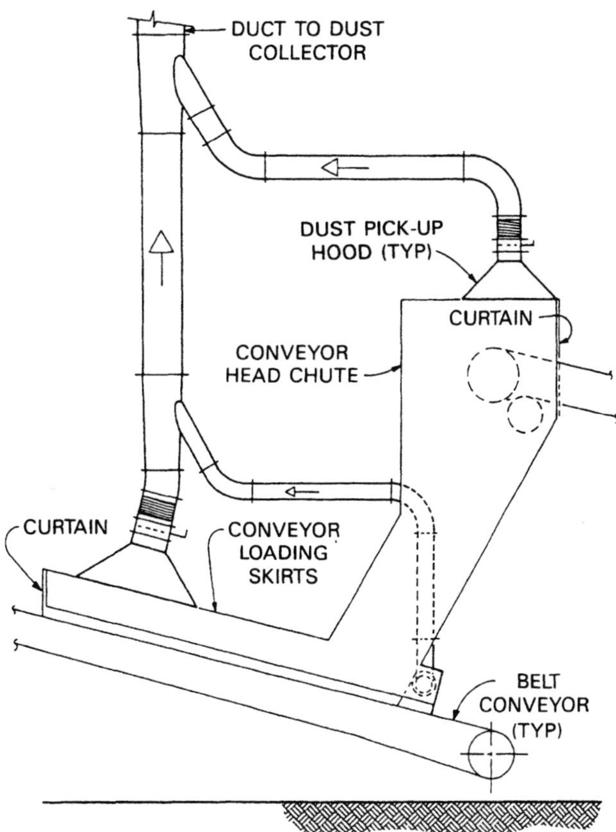

Fig. 5-52. Dust collection—pick-up arrangement.

often employing one or more of the known principles of dust separation from an air stream. Two basic types of dry dust collectors that are effective in similar material handling installations are the centrifugal and bag filter units.

The high-efficiency centrifugal unit uses a highly specialized refinement of the principle of cyclonic collection and increases the centrifugal forces available for dust separation. By minimizing dust re-entrainment, good collection efficiencies can be maintained on particles as small as the 10 to 20 micron range, with an overall dust collection efficiency of approximately 85%.

Fabric filter units operate by passing dust-laden air through a fabric at low velocity, as shown in Fig. 5-53. Good collection efficiencies can be maintained for dust particle sizes to 1 micron and smaller with overall dust collection efficiencies exceeding 99% for properly maintained units.

Based on the present-day environmental requirements for dust collector efficiency, the fabric filter units are most commonly employed and are the only type discussed further for coal and limestone handling systems.

The fabric in fabric filters is arranged in envelope (pillowcase style)- or tubular-shaped bags within the collection areas. Dust removal is obtained by building up a cake or mat of dust on the outside (dirty air) surface of the bag. As dust is collected on the fabric, resistance to air flow increases. Periodically, the fabric is cleaned by an air-pulse vibration down the inside of the bag to drop the bulk of the adhering material to an underlying collection hopper. Automatic differential pressure controls and timers for each fabric filter unit control the cleaning cycle frequency during unit operation. Collection efficiencies of greater than 99% by weight can be expected with a resulting pressure drop ranging from 2 to 6 in. (500 to 1,500 Pa) water gauge. Air-pulse units typically have air-to-cloth (cubic feet per minute of air flow to square feet of cloth collecting dust at a given time) ratios of 6 or 7 to 1. Fabric material for dust collection is usually 16 oz per square yard (0.5 kg/m^2) minimum, and can be polypropylene, polyester, acrylic, or other acceptable synthetic fiber, woven cloth, or felt. Bags from coal handling service are flameproofed and grounded to dissipate static electric charges.

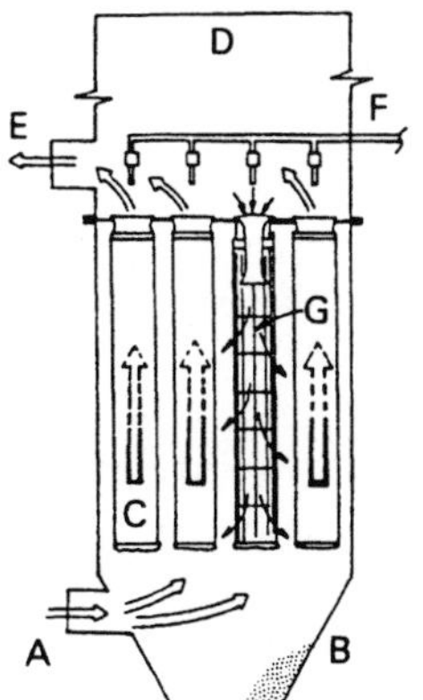

Fig. 5-53. Dust collection—air pulse type fabric filter.

The air-pulse fabric filters, such as the one shown in Fig. 5-53, require an external compressed air system for bag cleaning. Typical air volumes and pressures range from 4 to 12 ft^3/min (0.002 to 0.006 m^3/s) (compressed air) at 80 to 100 lb/in.^{2} (552,000 to 689,500 Pa) per 1,000 ft^3/min (0.5 m^3/s) (dust-laden air flow). Variation exists between models and manufacturers with respect to air requirements, and this aspect requires close evaluation in final system designs.

All fabric filters are limited to air conditions dry enough to prevent condensation of free moisture on the fabric. Bonding tendencies of moist dust particles gradually seal the fabric and lead to premature bag replacement. When both wet and dry dust control systems are used, care must be taken to ensure that free moisture is not entrained in the air entering the fabric filter units.

Fabric filter dust collectors present a potential explosion hazard within the collection housing and adjacent ductwork when handling coal dust. Locating the dust collector in exterior or low-hazard areas should be a primary consideration. Blowout doors or panels should be provided with the collector to safely vent the pressure. If the dust collectors are located within a building, the explosion vents should be extended outdoors. Fabric filter dust collectors handling limestone dust do not require the installation of explosion vents. For all dust collectors, a fire protection system should be installed to contain and extinguish fires.

For material transfer points, as previously shown in Fig. 5-52, dust is collected at the head of the discharging conveyor and the rear and top front of the skirtboard of the receiving conveyor. The dust pick-up air flow requirements for coal or limestone fall within ranges. For the top of the discharge chute the air flow is based on the belt width. The range is from 600 to 1,400 ft^3/min (0.28 to 0.66 m^3/s) for 24 to 72 in. (0.61 to 1.83 m) belt widths. For the rear of the loading skirtboard, the air flow is 700 ft^3/min (0.33 m^3/s) for belt widths less than 36 in., and 1,000 ft^3/min (0.47 m^3/s) for belt widths 36 in. and larger. For the top front of the loading skirtboard, the air flow is based on induced air, displaced air, entrained air, and control air. These air factors are based on the skirtboard area, belt width, capacity, and speed. The range is from 1,500 to 6,000 ft^3/min (0.71 to 2.83 m^3/s) for 24 to 60 in. (0.61 to 1.52 m) and above belt widths.

5.7.2 Wet Suppression

One method of controlling dust is to suppress it by confining it to the dust-producing area with a curtain of moisture, wetting it through direct contact between dust particles and moisture droplets, and combining small dust particles with each other and with the moisture droplets to form agglomerates too heavy to become or remain airborne.

A typical wet suppression system applies a spray of moisture to the coal stream at hopper discharge, feeders, or transfer points. The turbulent motion of the coal stream at these locations exposes many faces of the coal to the spray so that thorough wetting is achieved. Under some conditions, the effect of the spray application is carried over the length of the conveyor to transfer points downstream of the spray point. The carryover effect depends greatly on the physical condition of the coal being sprayed.

The amount and type of spray applied and the timing of the spray application will also affect the system's ability to suppress dust. Spray actuation and control are important, especially when the flow of coal is intermittent. Sensing controls actuate the individual spray header only when coal is detected on the belt where the application will take place.

Two common types of suppressants are water suppressants and chemically treated water suppressants. Descriptions of each and their proper applications are as follows.

*5.7.2.1 **Water Suppressant.*** From the standpoint of economy, availability, and safety, water is an ideal fluid. However, water is not an efficient dust suppressant. As previously stated, wet suppression is achieved through wetting of the dust by direct contact between the dust particles and moisture droplets. The dust particles and moisture droplets combine, becoming too heavy to remain airborne.

Fluids of high surface tension exhibit poor wetting, spreading, and penetrating qualities. The high surface tension of water keeps dust particles on the exterior of the water droplets. The required combining of particle and droplet does not readily occur, thereby reducing the effectiveness of the suppression system.

When water is used, large volumes are required to adequately suppress the dust; approximately 5% to 8% of moisture by weight is needed. It has limited effectiveness, further limited by ice deposits under freezing conditions for which freeze prevention is necessary throughout the system in unheated areas. Because of the problems associated with its use, water is seldom used alone to treat the main coal stream and is not considered a viable alternative for this type of application.

*5.7.2.2 **Chemically Treated Water Suppressant.*** In general, adding moisture to coal is undesirable. Excessive moisture increases the likelihood of spontaneous combustion, accentuates freezing of coal during winter months, and reduces the effective heating value of the coal. To overcome some of the disadvantages of water, a small percentage of specially formulated dust control compounds or surface active agents can be introduced. These agents work chemically in the water to reduce the surface tension by as much as 50%.

Treated water, because of its low surface tension, is more readily atomized; more droplets are produced per unit volume, increasing the available surface area and improving the potential for contact with dust particles. The moisture is more effectively used so less moisture is required to do the job. The amount of solution used is 0.5% to 1.0% of moisture by weight for treated water as compared to 5% to 8% for untreated water.

Wet suppression systems using chemically treated water usually consist of the chemical compound, the proportioning equipment to mix the chemicals and water, a distribution system, and actuating controls to automatically control application of the chemical solution to the coal stream. A general schematic for this type of system is shown in Fig. 5-54. Typical spray solutions contain 3,000 to 4,000 parts water to 1 part chemical agent, and the rate of spray application is about 1 to 2 gallons per ton (0.004 to 0.008 m^3/tonne) of coal. The total water use is much less for this wet suppression system than for a water-only suppressant system.

The piping and spray systems in areas not reliably heated require freeze protection. Nozzle pluggage may occur from buildup of ice, lime, mineral deposits, or corrosion. Special precautions may be required in the proportioning and pumping equipment area, as well as at spray discharge points, because of the corrosive nature of some chemicals. These precautions include acid-resistant coatings and finishes on equipment, special drainage and curbing for spill containment, and exterior wall exhaust ventilation of equipment enclosures. Other disadvantages of chemically treated water suppressant systems include higher initial cost and a high degree of maintenance.

5.7.2.3 Foam and Fog Suppressants. Two additional forms of suppressants are foam and fog. In a foam system, dust suppression is achieved when a small dust particle contacts the surface of a foam bubble. The bubble breaks, thereby wetting the particle, and effectively destroying the foam. The wetted particles then combine to form agglomerates too heavy to remain airborne.

In a fog system, the water spray particles are so small that they are classified as a fog (less than 100 microns). The fog particle size range closely matches the size range of the fugitive dust that results in an easier union between the water and coal particles.

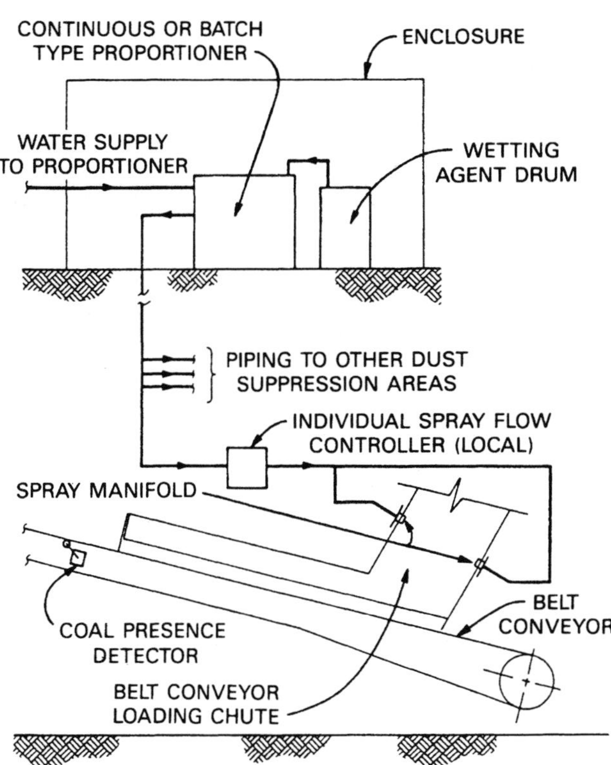

Fig. 5-54. Wet suppression piping arrangement.

Both of these systems decrease the amount of water required, but application is more difficult and the initial operating and maintenance costs are considerably higher.

5.7.3 Dust Control for Reserve Coal Storage Pile

Wind erosion contributes most significantly to fugitive dust problems. Therefore, a reduction in surface wind speed across the dust source reduces dust emissions.

The minimum velocity required to start particle movement varies with grain size. The range of threshold velocities for most grain sizes is from 13 to 30 mi (21 to 48 km) per hour at a height of 1 ft (0.3 m) above the surface. The following compares particle size and the susceptibility to erosion.

Particle size	Wind susceptibility
0- to 40-mesh	Highly erodible
40- to 20-mesh	Difficult to erode
20-mesh to ¼-in.	Usually not erodible
¼ in. and larger	Not erodible

As can be seen from the preceding, coals high in fines (20 mesh and smaller), such as most western and lignite coals, will be susceptible to erosion.

To reduce surface wind speeds across the coal piles, natural or manmade windbreaks can be used. Based on agricultural applications, windbreaks on the windward side of a field protect the field from wind erosion for a distance equal to 10 times the height of the windbreak. Applying this concept, the coal piles should be arranged such that the reserve coal storage pile shields the active storage pile whenever possible. Since the reserve coal pile is subject to minimal disturbances if coal shipments coincide with plant requirements, the exposed pile surfaces can be sealed or stabilized for dust control.

Three methods of dust control for the reserve coal pile storage are vegetative stabilization, chemical stabilization, and capping, as described in the following sections.

5.7.3.1 Vegetative Stabilization. Vegetative techniques have been used to control dust emissions from tailing piles and agricultural applications. Vegetative techniques need a soil that supports growth, and their efficiency in reducing wind erosion is dependent on the density and type of vegetation that can be grown. For tailing piles, vegetative stabilization decreases emissions by approximately 65%. Combined with a chemical stabilizer, the efficiency increases to approximately 90%.

This technique can be used for coal if the piles are not subjected to frequent disturbances from coal handling operations. Sod can be used in this application, requiring a 4- to 6-in. (0.10- to 0.15-m) layer of soil on top of the coal pile and frequent irrigation, that would be a major disadvantage.

Although effective, this technique is not practical for most coal handling operations. Handling the sod is cumbersome and expensive during reclaiming operations, and the soil layer contaminates the upper layer of the coal pile.

*5.7.3.2 **Chemical Stabilization.*** A variety of chemicals have been effective in reducing fugitive dust emissions. These chemicals control dust by different methods and are classified by composition—petroleum based, polymers, resin, surface active agent, and latex.

From the many chemical dust suppressants on the market, a product can be selected for optimum dust control. Some of the products will reform if the treated surface is disturbed, but most will not. Weathering abilities of these products depend on the application and chemical composition.

Chemical dust suppressants must be selected by several criteria:

- Effect on personnel, animals, and vegetation in the surrounding areas;
- Contamination of the material being treated (combustion affects);
- Capability for proper application;
- Effect on water quality from pile runoff;
- Effect on coal heating values, short and long term.

Two basic methods of chemical dust suppression are wet dust suppression throughout receiving and handling operations and chemical binders that form a crust on the surface of the stockpile.

Chemical wetting agents provide better wetting of fines and longer retention of moisture when compared with plain water. These treatments protect stockpiled materials until the added moisture has been removed by heat and wind. Depending on local conditions, some of these agents remain effective for weeks or months without rewatering the piles.

Crusting procedures involve the use of Bunker C crude oil, No. 6 crude oil, water-soluble acrylic polymers, or organic binders. The compounds are sprayed on the surface of the stockpile, coating the top layer of particles with a thin film. This film causes the particles to adhere to one another, forming a tough, durable crust resistant to wind and rain. As long as the crust remains intact, the stockpile is protected from wind losses. Depending on the material and climate, some crusting agents offer up to 1 year of protection. Some of these agents are concentrated and are diluted with water before application. Unless care is taken in proportioning, the spray adhesive may be ineffective.

Crusting agents of the polymer and latex type have been tested on coal piles. No major problems were encountered in the tests, although it was noted that the protective crust on the pile slopes tended to break up during heavy rains. It should be noted that sealing a coal pile with these crusting agents may increase the chances of spontaneous combustion. Any openings in the seal act as orifices for relatively high velocity air movement. This movement of air into and out of the pile may result in localized hot spots. This is especially true for piles subject to frequent disturbances due to stockout and reclaim operations. If a hot spot is found, extra compaction in the affected area may cut airflow enough to discontinue the reaction. If this does not work, the affected area should be removed, without disturbing the rest of the pile and spread out on the ground to cool. Water should not be used since it is likely to induce a recurrence of the heating within a few days.

*5.7.3.3 **Capping.*** Capping procedures involve asphalt compounds, road oil, earth, fly ash, and polyethylene tarpaulins to cover the surface area of the pile.

One method employs a mixture of wood pulp and asphalt in a slurry that is sprayed over the surface of the storage pile. Also, the oil produce normally used to treat the road surface before applying asphalt is an economical product that can be sprayed over the surface of the storage pile. This product has the advantage that it will allow limited trespass without totally destroying the dust suppression layer. In addition, when reclaimed, it will not affect the boiler performance because of the small quantity mixed with the coal and because it is a petroleum-based product. These methods offer approximately six months to one year of protection.

Another method of sealing stockpiles involves covering the piles with a straw mulch in a hydrocarbon binder. This material is an emulsified hydrocarbon material that contains a special mulch of cut straw. It is sprayed on in a layer approximately 5 in. (0.125 m) thick and then compacted.

Covering the storage pile with earth can also control coal dust emissions but creates a dusting problem of its own and requires soil stabilization. This method is not considered practical because of the additional dusting problem and contamination of the outer layer of the storage pile.

A mixture of fly ash and water sprayed over the coal pile and built up to a ½ to 1 in. (0.013 to 0.025 m) thick crust has proven successful. Feasibility of this method depends on the chemical properties of the fly ash that causes it to harden when water is added. The crust remains stable for a year or so, if not disturbed. When reclaimed, the minimal amount of fly ash has no effect on the coal quality.

Polyethylene tarpaulins can effectively control dust, but this method is not practical in areas of high wind speeds and large storage piles. Handling these tarpaulins is also a cumbersome operation for coal yard personnel.

All of these capping methods are subject to the orifice action described earlier and may increase the potential for hot spots in the coal pile.

5.8 COMPOSITE FACILITY ARRANGEMENTS

To illustrate the application of the various equipment and arrangements that make up coal and limestone handling facilities, an example of one complete coal handling facility is presented in some detail. In addition, the characteristic features of this facility and four others are summarized in Table 5-4. The tabulated data are for both the coal and limestone handling facilities of each power plant. This table

Table 5-4. Characteristic Features of Coal and Limestone Handling Facilities

Item	Intermountain	Crystal River	Coal Creek	Rawhide	Old Dominion
Utility	Intermountain Power Agency	Florida Power Corporation	Cooperative Power Assn./ United Power Assn.	Platte River Power Authority	Old Dominion Electric Coop.
Station location	Utah	Florida	North Dakota	Colorado	Virginia
Present units/MW	2/825 each	1/380 1/490 2/700 each	2/520 each	1/250	2/393 each
Ultimate/MW	3,300	2,270	2,200	690	786
Coal	Bituminous	Bituminous and Subbituminous	Lignite	Subbituminous	Bituminous
Limestone used	Yes	No	No	No	Yes
Coal Handling Facilities					
Coal transportation	Train/truck	Train/ocean barge	Overland conveyor	Train	Train
Unloading and stockout rate (TPH)	6,000/1,000	2,500/1,500	2,500	3,500	3,500
Reclaim and silo fill rate (TPH)	2 × 1,000	1 × 600 2 × 800	2 × 1,000	1 × 500	2 × 800
Unloading	Air-door bottom dump/bottom dump	Air-door bottom dump/clamshell bucket unloader	Conveyor discharge	Rotary dumper	Air-door bottom dump
Stockout	Traveling stacker	Slewing S-R Two-trench S-R	Conveyor Slewing S-R	Conveyor	Conveyor
Active storage (ton)	4 × 30,000 Piles	1 × 24,000 Pile 4 × 25,000 Piles	2 × 16,000 Silos 2 × 62,000 Piles	2 × 6,000 Silos	2 × 17,000 Silos
Reclaim	Rotary plow feeders	Slewing S-R Two-trench S-R	Vibratory feeders Slewing S-R	Vibratory feeders	Vibratory feeders
Silo fill	En-masse chain conveyors	Fixed tripper conveyors	Fixed tripper conveyor	Traveling tripper	Traveling trippers
Limestone Handling Facilities					
Limestone transportation	Truck	N/A	N/A	N/A	Train/truck
Unloading and stockout rate (TPH)	600	N/A	N/A	N/A	2,000
Reclaim and day bin fill rate (TPH)	600	N/A	N/A	N/A	400
Unloading	Bottom dump	N/A	N/A	N/A	Bottom dump/bottom or back dump
Stockout	Fixed boom conveyor	N/A	N/A	N/A	Fixed boom conveyor
Reclaim	Belt feeders	N/A	N/A	N/A	Vibratory feeder

Note: S-R is bucket wheel stacker–reclaimer.

shows that each facility is unique and that the facilities have been developed to match the requirements of the specific power plant.

The coal handling facility presented in detail is the Intermountain Power Agency's Intermountain Generating Station, located about 100 miles (160 km) south of Salt Lake City, Utah. Unit 1 started commercial operation in June 1986, and Unit 2 in July 1987. The generating plant site and much of the coal handling facility was designed to support four 825-MW generating units. For the coal handling facility, an important design criterion was availability, since the facility must handle 4.4 million tons (4.0 million tonnes) of coal per year.

The coal for the plant comes from several underground mines in east-central Utah. The coal is transported primarily by eighty-four 105-ton (95-tonne) railcar unit trains and secondarily by tandem 40-ton (36-tonne) trucks. Obtaining coal from several mines and the use of alternate transportation modes help to ensure fuel availability.

The coal handling facility is highly flexible and redundant, as shown in Fig. 5-55, the coal handling flow diagram. Coal can be delivered directly to the generating units from the train unloading system, the truck unloading system, the active reclaim system, and the reserve reclaim system. Any one of these paths is capable of handling the maximum coal requirements of Units 1 and 2 operating simultaneously at full load. Coal conveying within the units is with en masse chain conveyors arranged to provide complete redundancy in the coal path to each silo. Coal dust control is furnished to confine the coal dust and to minimize rehandling. The coal handling facility shown in Fig. 5-56 can be expanded for Units 3 and 4 with Crusher Building 2 directly south of

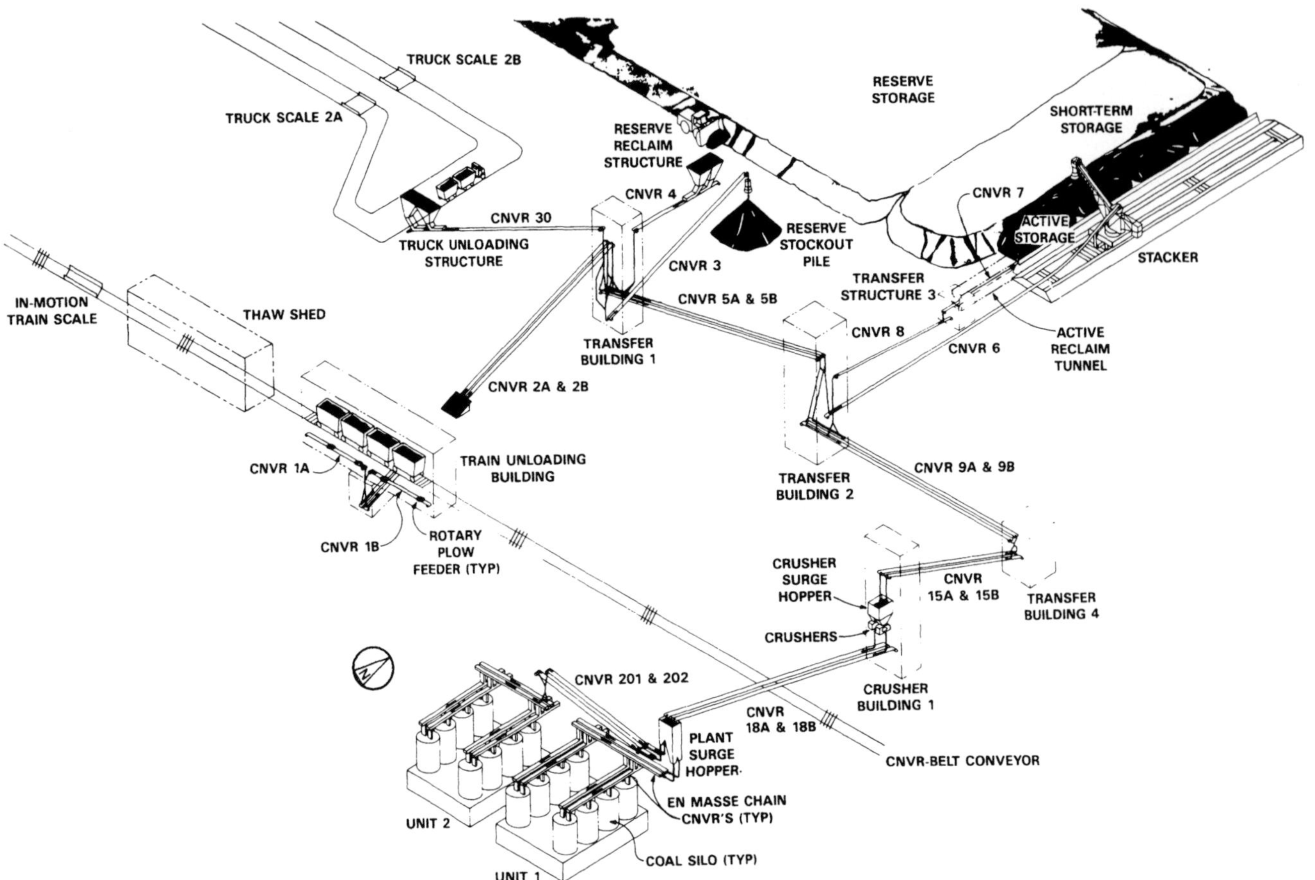

Fig. 5-55. Coal handling flow diagram.

Transfer Building 2 and an additional active storage system directly north of Transfer Building 4.

For the coal train transportation and unloading system, the trains in service are based on a 24-h turnaround time. The maximum of eighty-four 105-ton (95-tonne) cars that make up a unit train is based on the severe railroad grades and curves in the mining areas. The coal handling facility and railroad criteria prescribed a bottom dump unloading system capable of handling up to three trains per day with an unloading time of less than 2 h per train. The air-door bottom dump railcars are unloaded at 6,000 tons (5,440 tonnes) per hour by moving the train across the unloading hopper and dumping the coal by remote control operation of the car doors. Coal is reclaimed from the hopper by four rotary plow feeders. The unloading system also includes a coal car thawing shed and an in-motion train scale.

For the coal truck transportation and unloading system, tandem trailer bottom dump trucks with a total capacity of 40 tons (36 tonnes) are available for coal hauling in central Utah. A large-capacity unloading hopper system rated at 1,000 tons (910 tonnes) per hour and capable of handling up to 100 trucks per day was selected. The tandem bottom dump trucks are unloaded by positioning one trailer at a time over the hopper. Coal is reclaimed from the hopper by two vibratory feeders. The unloading system also includes two stationary truck scales with 100 ft (30.5 m) long weigh bridges to accommodate the tandem trucks.

Coal can be simultaneously unloaded from both train and truck systems, with the train coal going to active stockout and the truck coal being held in the hopper or sent to reserve stockout.

The stockout system capacity is presently 6,000 tons (5,440 tonnes) per hour, and a dual-conveyor path system, with each conveyor rated at 4,000 tons (3,630 tonnes) per hour, was furnished. The stockout rate and the need for segregated coal storage resulted in the selection of a large-capacity traveling stacker for the active stockout equipment. The stacker discharges the coal into flat-topped segregated piles in the active storage area. The stacker boom conveyor has a telescopic chute and wet suppression to help control fugitive dust during coal discharge. A reserve stockout system consisting of a stockout conveyor with telescopic chute and a capacity of 4,000 tons (3,630 tonnes) per hour was provided. The stockout system also includes a three-stage as-received train sampling system, a two-stage as-received truck sampling system, both at Transfer Building 1, and conveyor belt scales.

The coal storage system contains two coal storage areas. Active coal storage provides a continuous coal supply to the units by handling daily variations in coal consumption and

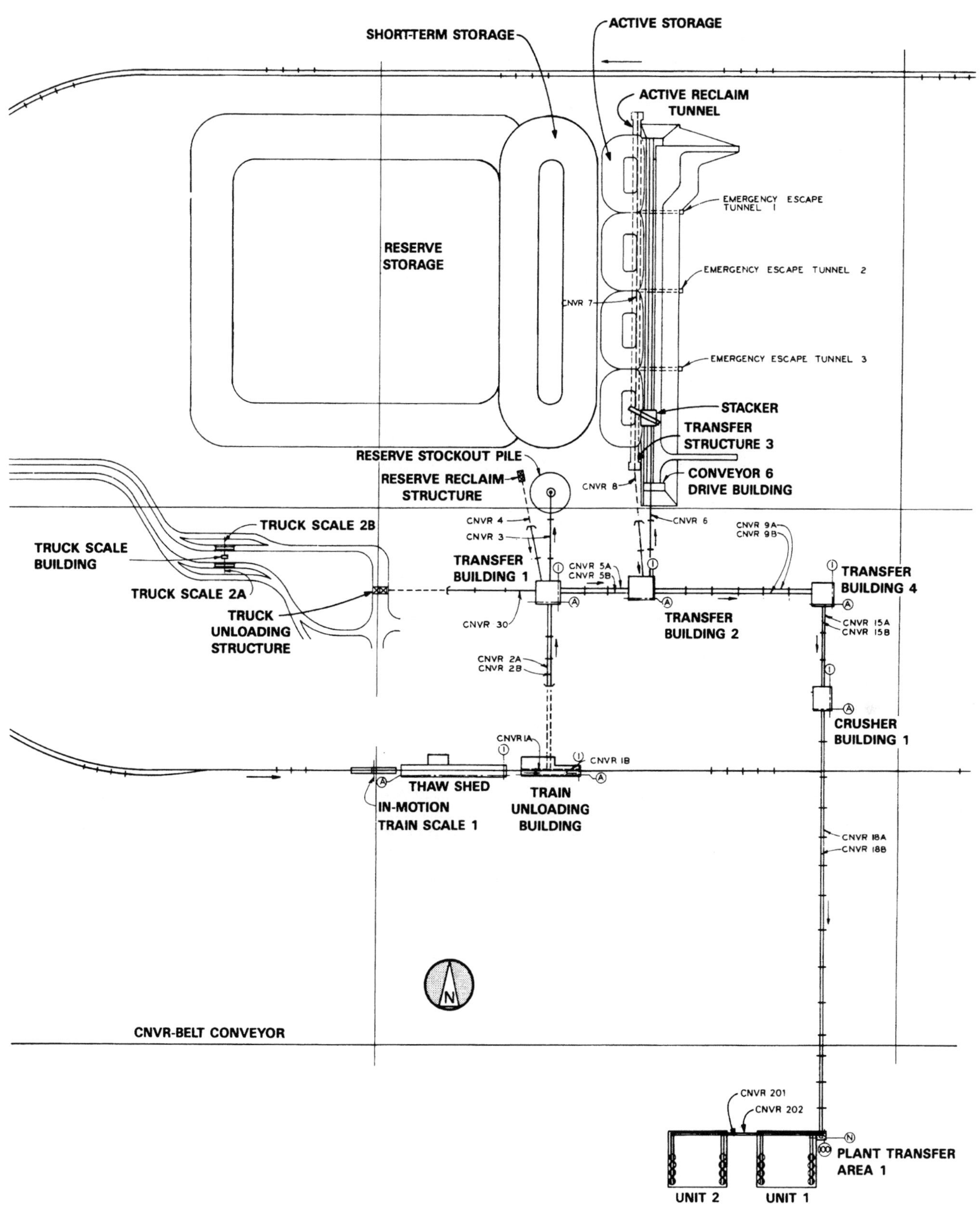

Fig. 5-56. Coal handling plot plan.

coal delivery schedules. The 3-day active storage capacity is approximately 30,000 tons (27,200 tonnes) per unit. Reserve coal storage provides a supply of coal to the units in the event of a long-term interruption in coal deliveries. The 60-day reserve storage capacity is approximately 440,000 tons (399,200 tonnes) per unit.

The reclaim system rate for each unit is based on burning blended coal at a maximum burn rate of 400 tons (360 tonnes) per hour, with a 2.5 service factor to allow for maintenance and flexibility of operations. The total reclaim rate of 2,000 tons (1,810 tonnes) per hour can supply the maximum 24-h coal requirements of both units in approximately 10 h. Four variable-capacity (2,000 tons/h maximum) rotary plow feeders were selected for the active reclaim equipment and are located in the Active Reclaim Tunnel beneath the storage piles. Coal can be reclaimed from the reserve storage area and transferred by mobile equipment to either the active reclaim system or the reserve reclaim system, depending on the location of the coal and the desired blend. The capacity of the reserve reclaim hopper and two vibratory feeders is 2,000 tons (1,810 tonnes) per hour.

The coal crushing and silo fill system capacity is 2,000 tons (1,810 tonnes) per hour. The system consists of redundant conveyor paths following the criteria of the reclaim system. To handle these requirements, dual 1,000 tons (910 tonnes) per hour crusher and conveyor paths are supplied from the crusher surge hopper to the plant surge hopper. Dual 500 tons (450 tonnes) per hour en masse chain conveyor paths transfer the coal from the plant surge hopper to each silo. The feeding of coal to each silo is controlled by the positioning of the en masse chain conveyor discharge gates below the silo fill conveyors. These dual conveyors and crossovers provide a reliable flow of coal to each silo. The silo fill system also includes a two-stage as-fired sampling system at Plant Transfer Area 1 and conveyor belt scales.

The dust control system confines the coal dust and minimizes handling the same dust at each transfer point. The system selected consists of minimal wet dust suppression and maximum dry dust collection. Dust collection serving the coal handling facility is furnished by induced draft filter bag dust collection units. The collected coal dust is processed by granulation equipment before it is returned to the coal stream. Wet dust suppression controls dust at the rotary plow feeders, at the conveyor transfer points after the rotary plow feeders, and at the discharge of the stacker telescopic chute.

5.9 REFERENCES

ASTM. 1993. Standard test methods for collection of a gross sample of coal; standard method of preparing coal samples for analysis. *1993 Annual Book of ASTM Standards*, Volume 05.05. ed. ASTM Committee D-5. ASTM, Philadelphia, PA.

BEHLING, R. A. and S. M. COULTER. 1989. Alternatives for fuel receiving system selection based on plant parameters. Paper read at Power-Gen '89 Conference, 5–7 December 1989. New Orleans, LA.

CEMA. 1988. Characteristics and conveyability of bulk materials. Capacities, belt widths, and speeds. Belt tension, power, and drive engineering. *Belt Conveyors for Bulk Materials*, 3rd edit., ed. Engineering Conference, Conveyor Equipment Manufacturers Association.

DUTTON, R. W. and R. H. MCCARTNEY. 1986. Intermountain Generating Station's coal handling facility—flexibility, reliability, and expandability. Paper read at Joint Power Generation Conference, 19–23 October 1986, at Portland, OR.

HEYL & PATTERSON, INC. 250 Park West Drive, P.O. Box 36, Pittsburgh, PA 15230.

JOHNSTOWN AMERICA CORPORATION. FREIGHT CAR DIVISION. 17 Johns Street, Johnstown, PA 15901.

MCCARTNEY, R. H., J. H. PLANNER, and P. S. WHITTAKER. 1990. A comparison of United States and Australian practices for power station coal handling facilities. Paper read at 1990 International Coal Engineering Conference, 19–21 June 1990, at Sydney, Australia.

MCNALLY WELLMAN COMPANY. 4800 Grand Avenue, Neville Island, Pittsburgh, PA 15225.

PENNSYLVANIA CRUSHER CORPORATION. 600 Abbott Drive, Box 100, Broomall, PA 19008-0100.

RAMSEY TECHNOLOGY, INC. 501 90th Avenue N.W, Minneapolis, MN 55433.

SNELL, V. H. 1982. Design and construction—flow dependable coal silos. Paper read at American Power Conference, 26–28 April 1982, at Chicago, IL.

SPECTRUM INFRARED, INC. 246 East 131 Street, Cleveland, OH 44108.

SUMITOMO HEAVY INDUSTRIES (USA), INC. Park Avenue Tower, 65 East 55th Street, Suite 2303, New York, NY 10022.

TOLEDO SCALE. 60 Collegeview Road, Westerville, OH 43081.

TRINITY INDUSTRIES. FREIGHT CAR DIVISION. 600 East 9th Street, Suite 5, Michigan City, IN 46360.

U.S. DEPARTMENT OF COMMERCE. 1993. NIST Handbook 44—Specifications, tolerances, and other technical requirements for weighing and measuring devices, 1994 edition. U.S. Government Printing Office, Washington, D.C.

6

Combustion Processes

G. Scott Stallard and *Todd S. Jonas*

6.1 INTRODUCTION

At the heart of fossil-fueled power plant operation is the combustion process. Through the combustion process, modern power plants burn fuel to release the energy that generates steam—energy that ultimately is transformed into electricity. Yet, while the combustion process is one of a power plant's most fundamental processes, it is also one of the most complex.

Combustion, or the conversion of fuel to useable energy, must be carefully controlled and managed. Only the heat released that is successfully captured by the steam is useful for generating power. Hence, the ability of the steam generator to successfully transfer energy from the fuel to steam is driven by the combustion process, or more precisely, the characteristics of the combustion process.

6.2 FUNDAMENTALS OF COMBUSTION

Combustion of fuels must be considered both from theoretical and practical perspectives. From the theoretical perspective, combustion can be defined as the rapid chemical reaction of oxygen with the combustible elements of a fuel. From a practical standpoint, the engineer concerned with boiler design and performance might define combustion as the chemical union of the fuel combustibles and the oxygen of the air, controlled at a rate that produces useful heat energy. Both of these definitions implicitly consider many key factors. For complete combustion within a furnace, four basic criteria must be satisfied:

1. Adequate quantity of air (oxygen) supplied to the fuel,
2. Oxygen and fuel thoroughly mixed,
3. Fuel–air mixture maintained at or above the ignition temperature, and
4. Furnace volume large enough to give the mixture time for complete combustion.

Quantities of combustible constituents available within the fuel vary by fuel type. Figure 6-1 illustrates the significant change in combustion air requirements for various fuels, resulting from changes in fuel composition. This figure illustrates the minimum combustion air theoretically required to support complete combustion.

In an ideal situation, the combustion process would occur with the appropriate proportions of oxygen and a combustible, based on underlying chemical principles (the stoichiometric quantities). However, since complete mixing of air and fuel within the furnace is virtually impossible, excess air must be supplied to the combustion process to ensure complete combustion. The amount of excess air that should be provided varies with the fuel, boiler load, and type of firing equipment. As shown in Table 6-1, air is a mixture of nitrogen, oxygen, water vapor, and other gases with oxygen comprising only about 20% of the mixture. Therefore, the quantity of excess air must be properly controlled to minimize the quantity of air and flue gas conveyed throughout the power plant.

Ultimately, the efficiency of the mixing process and the quantity of excess air supplied determine whether the exhaust gases contain the products of both complete and incomplete combustion. If combustion is incomplete, in addition to loss of the heat available in the unburned fuel, undesirable atmospheric pollutants such as carbon monoxide are often produced. Excessive turbulence and availability of combustion air can also increase production of nitrogen oxide, another atmospheric pollutant.

6.2.1 Combustion Reactions

Combustion reactions involving oxygen and fuel combustibles are consistent with the following chemical principles:

- Specific compounds form in fixed combinations when two or more reactants combine (stoichiometric ratio). Table 6-2 lists combustible elements and compounds formed in typical combustion reactions.
- The mass of any element in the reactants must equal the mass of that element in the products (conservation of matter).
- Chemical compounds form from elements combining in fixed weight relationships (law of combining weights).
- Formation of a compound either produces heat (exothermic reaction) or requires heat (endothermic reaction), based on the free energy change for the reaction.

For hydrocarbon fuels, the principal sources of energy are elemental carbon, hydrogen, and their compounds; in the combustion process, these produce carbon dioxide and water vapor. Small quantities of sulfur are also present in most fuels. Although sulfur is a combustible and contributes en-

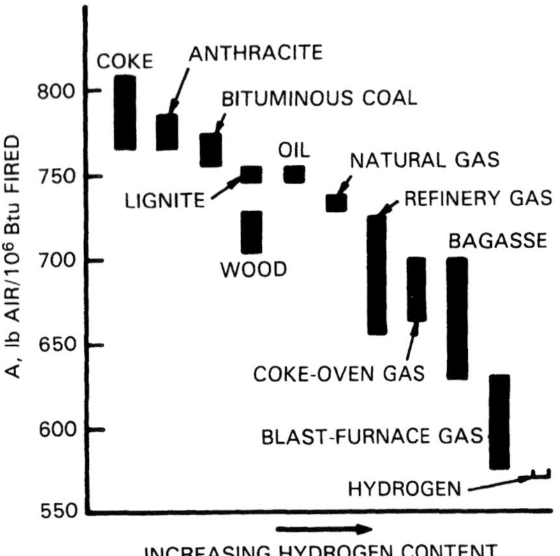

Fig. 6-1. Combustion air requirements for various fuels at zero excess-air—a range of values as an approximate function of hydrogen content. (From *Combustion*, ed. Joseph G. Singer, Copyright © 1991, used by permission from Combustion Engineering, Inc.)

Table 6-1. Composition of Combustion Air

Dry atmospheric air

The volumetric composition of dry atmoshperic air guide is given in the NACA Report 1235 (Standard Atmosphere—Tables and Data for Altitude, November 20, 1952). Volumetric composition and molecular weights of these constituents are as follows:

	Volume (%)	Molecular Weight
Nitrogen	78.09	28.016
Oxygen	20.95	32.000
Argon	0.93	39.944
Carbon dioxide	0.03	44.010

(Neon, helium, krypton, hydrogen, xenon, ozone, are <0.003%)

Source: Elliott, T., *Standard Handbook of Powerplant Engineering*. ©1989. Used with permission of McGraw-Hill.

ergy, its presence is not desired because its compounds are corrosive in nature. Like nitrogen oxides, sulfur dioxide (SO_2) emissions must be closely monitored and controlled.

Combustion constants for hydrocarbon-based fuels are shown in Table 6-3. The heat of combustion corresponds to the heat released during the reaction. This table also presents the stoichiometric, volumetric, and mass ratios for combustion air and flue gas flow rates for each fuel. A typical furnace mass balance is shown in Fig. 6-2.

6.2.2 Fuels and Their Characteristics

Hydrocarbon fuels generally are classified according to their phase—solid, liquid, or gas. Each fuel possesses its own unique combustion and handling requirements and, as such, requires specialized equipment to mix reactants properly and to transfer the heat of combustion efficiently. Gaseous fuels typically are burned in premix or diffusion burners. Liquid fuel burners atomize or vaporize the fuel prior to mixing with air. Solid fuel combustion equipment is necessarily more complex. Fuel preparation systems crush the fuel to expose more surface area and heat the fuel to vaporize sufficient volatiles for ignition and sustained combustion. In addition, the furnace must be large enough to give sufficient residence time for complete combustion.

The type of fuel selected is a function of the availability and dependability of the fuel supply, transport and storage issues, uncontrolled pollutant emission rates, and, most importantly, economics. Combustion characteristics of fuels selected must then be factored into the design of the steam generator and fuel preparation and handling systems. Key fuel-specific factors are discussed briefly below.

6.2.2.1 Natural Gas. Natural gas historically has been a popular source of energy. Natural gas is very clean, burns efficiently, and can be easily transported. In the 1970s, economic and political factors eliminated gas from serious consideration as a principal fuel for new power facilities. However, natural gas as a clean burning fuel has enjoyed renewed interest in the 1990s. Such interest typically has focused on the use of natural gas in combustion turbines and combined cycle plants; however, it can be a potentially viable fuel source for certain existing steam generators as well. In addi-

Table 6-2. Combustion Equations

Combustible	Molecular Weight	Reaction	Heat Release (Btu/lb)
Carbon	12	$C+O_2 \rightarrow CO_2$	14,100
Hydrogen	2	$H_2 + 0.5O_2 \rightarrow H_2O$	61,000
Sulfur	32	$S + O_2 \rightarrow SO_2$	4,000
Hydrogen sulfide	34	$H_2S + 1.5O_2 \rightarrow SO_2 + H_2O$	7,100
Methane	16	$CH_4 + 2\,O_2 \rightarrow CO_2 + 2H_2O$	23,900
Ethane	30	$C_2H_6 + 3.5O_2 \rightarrow 2CO_2 + 3H_2O$	22,300
Propane	44	$C_3H_8 + 5O_2 \rightarrow 3CO_2 + 4H_2O$	21,500
Butane	58	$C_4H_{10} + 6.5O_2 \rightarrow 4CO_2 + 5H_2O$	21,300
Pentane	72	$C_5H_{12} + 8O_2 \rightarrow 5CO_2 + 6H_2O$	22,000

Source: *Combustion*, edited by Joseph G. Singer, © 1991. Used by permission from Combustion Engineering, Inc.

Table 6-3. Combustion Constants

No.	Substance	Formula	Molecular Weight	Lb/ft³	Ft³/lb	Sp gr air 1,000	Heat of Combustion Btu/ft³ Gross	Btu/ft³ Net	Btu/lb Gross	Btu/lb Net	Moles/Mole or ft³/ft³ Combustible: Required for Combustion O_2	N_2	Air	Flue products CO_2	H_2O	N_2	Lb/lb Combustible: Required for Combustion O_2	N_2	Air	Flue products CO_2	H_2O	N_2
1	Carbon[c]	C	12.01	—	—	—	—	—	14.093	14.093	1.0	3.76	4.76	1.0	—	3.76	2.66	8.86	11.53	3.66	—	8.86
2	Hydrogen	H_2	2.016	0.0053	187.723	0.0696	325	275	61.100	51.623	0.5	1.88	2.38	—	1.0	1.88	7.94	26.41	34.34	—	8.94	26.41
3	Oxygen	O_2	32.000	0.0846	11.819	1.1053																
4	Nitrogen (atm.)	N_2	28.016	0.0744	13.443	0.9718																
5	Carbon monoxide	CO	28.01	0.0740	13.506	0.9672	322	322	4.347	4.347	0.5	1.88	2.38	1.0	—	1.88	0.57	1.90	2.47	1.57	—	1.90
6	Carbon dioxide	CO_2	44.01	0.1170	8.548	1.5282																
Paraffin series																						
7	Methane	CH_4	16.041	0.0424	23.565	0.5543	1,013	913	23.879	21.520	2.0	7.53	9.53	1.0	2.0	7.53	3.99	13.28	17.27	2.74	2.25	13.28
8	Ethane	C_2H_6	30.067	0.0803	12.455	1.0488	1,792	1,641	22.320	20.432	3.5	13.18	16.68	2.0	3.0	13.18	3.73	12.39	16.12	2.93	1.80	12.39
9	Propane	C_8H_8	44.092	0.1196	8.365	1.5617	2,590	2,385	21.661	19.944	5.0	18.82	23.82	3.0	4.0	18.82	3.63	12.07	15.70	2.99	1.63	12.07
10	*n*-Butane	C_4H_{10}	58.118	0.1582	6.321	2.0665	3,370	3,113	21.308	19.680	6.5	24.47	30.97	4.0	5.0	24.47	3.58	11.91	15.49	3.03	1.55	11.91
11	Isobutane	C_4H_{10}	58.118	0.1582	6.321	2.0665	3,363	3,105	21.257	19.629	6.5	24.47	30.97	4.0	5.0	24.47	3.58	11.91	15.49	3.03	1.55	11.91
12	*n*-Pentane	C_6H_{12}	72.144	0.1904	5.252	2.4872	4,016	3,709	21.091	19.517	8.0	30.11	38.11	5.0	6.0	30.11	3.55	11.81	15.35	3.05	1.50	11.81
13	Isopentane	C_6H_{12}	72.144	0.1904	5.252	2.4872	4,008	3,716	21.052	19.478	8.0	30.11	38.11	5.0	6.0	30.11	3.55	11.81	15.35	3.05	1.50	11.81
14	Neopentane	C_6H_{12}	72.144	0.1904	5.252	2.4872	3,993	3,693	20.970	19.396	8.0	30.11	38.11	5.0	6.0	30.11	3.55	11.81	15.35	3.05	1.50	11.81
15	*n*-Hexane	C_6H_{14}	86.169	0.2274	4.398	2.9704	4,762	4,412	20.940	19.403	9.5	35.76	45.26	6.0	7.0	35.76	3.53	11.74	15.27	3.06	1.46	11.74
Olefin series																						
16	Ethylene	C_2H_4	28.051	0.0746	13.412	0.9740	1,614	1,513	21.644	20.295	3.0	11.29	14.29	2.0	2.0	11.29	3.42	11.39	14.81	3.14	1.29	11.39
17	Propylene	C_3H_4	42.077	0.1110	9.007	1.4504	2,336	2,186	21.041	19.691	4.5	16.94	21.44	3.0	3.0	16.94	3.42	11.39	14.81	3.14	1.29	11.39
18	*n*-Butene	C_4H_8	56.102	0.1480	6.756	1.9336	3,084	2,885	20.840	19.496	6.0	22.59	28.59	4.0	4.0	22.59	3.42	11.39	14.81	3.14	1.29	11.39
19	Isobutene	C_4H_8	56.102	0.1480	6.756	1.9336	3,068	2,869	20.730	19.382	6.0	22.59	28.59	4.0	4.0	22.59	3.42	11.39	14.81	3.14	1.29	11.39
20	*n*-Pentene	C_6H_{10}	70.128	0.1852	5.400	2.4190	3,836	3,686	20.712	19.363	7.5	28.23	35.73	5.0	6.0	28.23	3.42	11.39	14.81	3.14	1.29	11.39
Aromatic series																						
21	Benzene	C_6H_6	78.107	0.2060	4.852	2.6920	3,751	3,601	18.210	17.480	7.5	28.23	35.73	6.0	3.0	28.23	3.07	10.22	13.30	3.38	0.69	10.22
22	Toluene	C_7H_8	92.132	0.2431	4.113	3.1760	4,484	4,284	18.440	17.620	9.0	33.88	42.88	7.0	4.0	33.88	3.13	10.40	13.53	3.34	0.78	10.40
23	Xylene	C_8H_{10}	106.158	0.2803	3.567	3.6618	5,230	4,980	18.650	17.760	10.5	39.52	50.02	8.0	5.0	39.52	3.17	10.53	13.70	3.32	0.85	10.53
Miscellaneous gases																						
24	Acetylene	C_2H_6	26.036	0.0697	14.344	0.9107	1,499	1,448	21.500	20.776	2.5	9.41	11.91	2.0	1.0	9.41	3.07	10.22	13.30	3.38	0.69	10.22
25	Naphthalene	$C_{10}H_3$	128.162	0.3384	2.955	4.4208	5,584	5,654	17.298	16.708	12.0	45.17	57.17	10.0	4.0	45.17	3.00	9.97	12.96	3.43	0.56	9.97
26	Methyl alcohol	CH_3OH	32.041	0.0846	11.820	1.1052	868	768	10.259	9.078	1.5	5.65	7.15	1.0	2.0	5.65	1.50	4.98	6.48	1.37	1.13	4.98
27	Ethyl alcohol	C_2H_3OH	46.067	0.1216	8.221	1.5890	1,600	1,451	13.161	11.929	3.0	11.29	14.29	2.0	3.0	11.29	2.08	6.93	9.02	1.92	1.17	6.93
28	Ammonia	NH_3	17.031	0.0456	21.914	0.5961	441	365	9.668	8.001	0.75	2.82	3.57	—	1.5	3.32	1.41	4.69	6.10	—	1.59	5.51
														SO						SO		
29	Sulfur[c]	S	32.06	—	—	—	—	—	3.983	3.983	1.0	3.76	4.76	1.0	—	3.76	1.00	3.29	4.29	2.00	—	3.29
30	Hydrogen sulfide	H_2S	34.076	0.0911	10.979	1.1898	647	596	7.100	6.545	1.5	5.65	7.15	1.0	1.0	5.65	1.41	4.69	6.10	1.88	0.53	4.69
31	Sulfur dioxide	SO_2	64.06	0.1733	5.770	2.264																
32	Water vapor	H_2O	18.016	0.0476	21.017	0.6215																
33	Air	—	28.9	0.0766	13.063	1.0000																

[a]From American Gas Association.

[b]All gas volumes corrected to 60° F and 30 in. Hg dry.

[c]Carbon and sulfur are considered as gases for molal calculations only.

Source: Perry, R. and C. Chilton. *Perry's Chemical Engineer's Handbook.* © 1973. Reproduced with permission of McGraw-Hill.

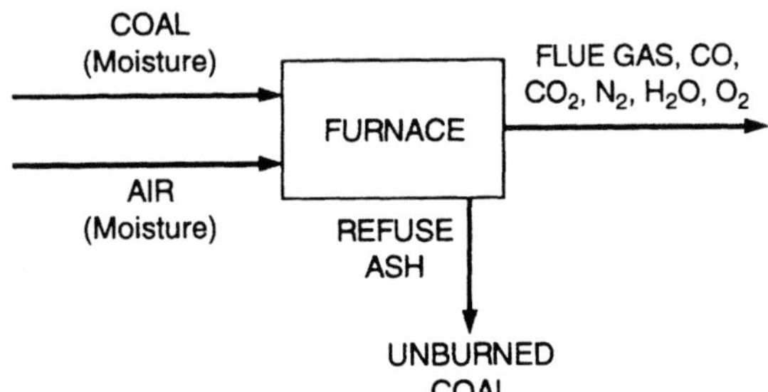

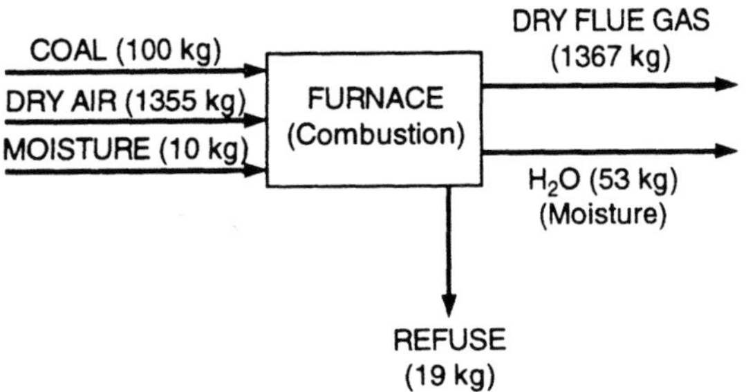

Fig. 6-2. Typical furnace mass balance. (From J. Weisman and L. E. Eckart, *Modern Power Plant Engineering*, 1985. Used with permission of Prentice-Hall, Englewood Cliffs, NJ.)

tion, natural gas is often fired in existing gas/oil fueled steam generators and can be fired in concert with coal in existing, coal-fueled units.

Natural gas is a mixture of methane (55% to 95%), more complex hydrocarbons (primarily ethane), and noncombustible gases. Typical heating values range from 950 to 1,100 Btu/ft^3 (35.4 to 41.0 MJ/m^3) at standard conditions.

Because the gaseous fuel can rapidly disperse in air, combustion is rapid once ignition temperature is reached and proper turbulence is provided. Two type of burners are primarily used: premix burners and nozzle-mix burners. Premix burners are principally used in forced draft boiler application. Nozzle-mix burners can be configured in a variety of ways. However, in each case, the gas and air are mixed at the burner tip.

6.2.2.2 Fuel Oil. Fuel oil is a liquid fuel from a family of hydrocarbon products derived from crude petroleum oil. Other liquid fuels include liquified petroleum gases (LPGs), gasoline, kerosene, jet fuel, diesel fuels, and light heating oils. The level of refinement determines fuel composition, ignition temperature, flash point, viscosity, and heating value.

Specifications for various grades of fuel oils are based on requirements of the target equipment burners. Per ASTM Fuel Specification D 396, fuel oils are classified as distillate oils (lighter petroleum products) and residual fuel oils (heavier oils) as follows:

- Grade No. 1: A light distillate with high volatility, used in vaporizing-type burners; highest in cost/gallon.
- Grade No. 2: A distillate oil heavier in viscosity and API gravity than No. 1, used in pressure atomizing burners; in common use domestically and in medium-capacity industrial burners.
- Grade No. 4: Light residual oil or heavy distillate used in burners designed to atomize oils of higher viscosities.
- Grade No. 5L (Light): A residual oil heavier than No. 4; may require preheating for pumping and burning.
- Grade No. 5H (Heavy): A residual oil more viscous than No. 5L; requires preheating.
- Grade No. 6: Also known as Bunker C oil; frequently used in industrial applications.

Fuel oil is composed principally of paraffins, isoparaffins, aromatics, napthenes, and related hydrocarbon derivatives of hydrogen, sulfur, and oxygen not removed by the refining process. Fuel oil can also contain traces of other impurities or pollutants such as vanadium- and nickel-based compounds. Although the percentage of residues or ash is small (typically less than 0.5%), such residues can form the basis of troublesome deposits on heat transfer surfaces. Such deposits are often controlled by adding chemical agents or additives to the fuel oil prior to combustion.

As in the case with all fuels, proper combustion of fuel oil is dependent on developing an appropriate mix of air and fuel within the furnace. In most utility installations, No. 6 fuel oil is burned using atomizing oil burners. Air or steam atomization is required to break the fuel oil into sufficiently small droplets and provide adequate mixing of the fuel with the combustion air to ensure effective combustion. Burner problems can cause impingement of the flame on the boiler surface, resulting in hardened carbon deposits, refractory spalling, washout, or slagging. In addition, No. 6 fuel oil typically must also be preheated to reduce its viscosity to ensure that it can be successfully transported and atomized.

6.2.2.3 Coal. As discussed in Chapter 4, coal is a quite complex fuel. Coal is a product of nature and time, derived from the decay of ancient trees, bushes, ferns, mosses, vines, and other forms of plant life that flourished millions of years ago in humid, tropical climates. Time, coupled with pressure, heat, and chemical and bacterial decay, has transformed the plant remains into coal. ASTM D 388 establishes categories or ranks of coal on the basis of measurable properties that relate to its metamorphism or its degree of transformation while buried.

Coal is heterogeneous. Unlike natural gas or fuel oil, composition or quality of coal can vary significantly depending on the degree of metamorphosis, type of vegetation, location in seam, surrounding materials, etc. Noncombustible material or ash in a furnace can pose significant operational problems because of the tendency of the ash to deposit and stick on boiler heat transfer surfaces.

Proper steam generator design must consider the specific characteristics of the coal, including its rank, composition, and volatility. In addition, coal characteristics directly impact the design and operation of the coal handling system, the fuel preparation and firing system, ash handling, and particulate removal systems. The complexities of coal combustion are explored in greater detail in Section 6.3. Coal-fueled steam generator design is discussed in Chapter 7.

6.2.3 The Thermodynamic View

Combustion produces heat; this heat is then transferred to water and steam within the steam generator to produce the steam necessary to drive the steam turbine. Efficient transfer of heat is important; any heat not transferred to the steam within the steam generator system is effectively lost. Therefore, steam generator system design must determine where and how such heat is transferred to meet design specifications and provide maximum efficiency.

The steam generator system comprises the fuel preparation and firing system, the steam generator itself, and the air heaters. Figure 6-3 illustrates how a thermodynamic control volume can be drawn around this equipment. Heat input to this control volume is principally derived from the fuel; some heat may be available in the combustion air as well. Within the steam generator, heat is transferred to the two steam circuits for main steam and reheat steam. Any heat remaining in the flue gas at the exit of the air heater is lost. In addition, other minor heat losses occur as a function of radiation from the steam generator, unburned combustibles, and other unaccounted losses. A full discussion of steam generator heat input and loss calculations is included in Section 6.3.

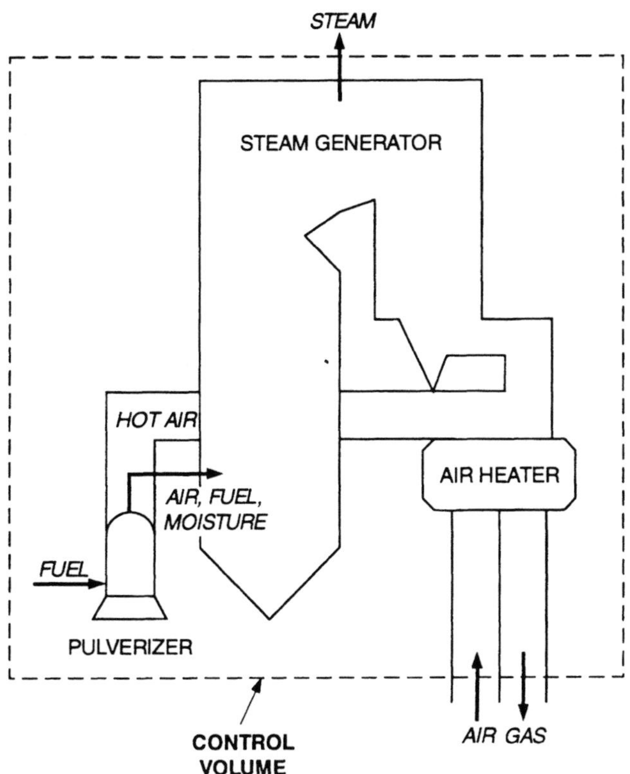

Fig. 6-3. Thermodynamic control volume drawn around equipment.

Steam generator, or boiler, heat input can be defined as the quantity of heat released within the steam generator. The total heat input must meet the needs of the turbine cycle (heat input required to match main and reheat steam duty) and heat losses. Hence, turbine cycle demands and boiler efficiency determine steam generator heat input.

An important measure of efficiency for a generating unit or power plant is its net plant heat rate (NPHR). Net plant heat rate, expressed in Btus per kilowatt-hour (Joules per kWh), is the amount of fuel energy or boiler heat input required to generate a kilowatt-hour and deliver it to the transmission lines leaving the plant. Net plant heat rate is used to determine the amount of fuel required and serves as the basis for determining the fuel costs used in economic assessments of alternative plans.

For a coal-fueled steam electric generating unit, determination of the NPHR at a particular value of turbine output involves net turbine heat rate, auxiliary power, and boiler efficiency, as follows:

$$\mathrm{NPHR} = Q_B/\mathrm{NPO} \qquad (6\text{-}1)$$

$$\mathrm{NTHR} = Q_T/\mathrm{NTO} \qquad (6\text{-}2)$$

$$\mathrm{NPO} = \mathrm{NTO} - \mathrm{AP} \qquad (6\text{-}3)$$

$$\eta_B = Q_T/Q_B \qquad (6\text{-}4)$$

where

NPHR = net plant heat rate, Btu/kWh (J/kWh);

NTHR = net turbine heat rate, Btu/kWh (J/kWh);

NPO = net plant output, kW;

NTO = net turbine output, kW (as measured at electric generator terminals minus motor-driven boiler feed pump power);

AP = auxiliary power, kW (exclusive of motor-driven boiler feed pump power);

Q_B = heat input to boiler from fuel, Btu/h (J/h) = fuel burn rate;

Q_T = heat input to turbine cycle, Btu/h (J/h); and

η_B = boiler efficiency (fuel efficiency).

From these relationships, a more useful expression for net plant heat rate may be derived, as follows:

From Eq. (6-2), $Q_T = \mathrm{NTO} \times \mathrm{NTHR}$

From Eq. (6-4), $Q_B = Q_T/\eta_B$

Therefore, $Q_B = \dfrac{\mathrm{NTO} \times \mathrm{NTHR}}{\eta_B}$

Substituting in Eq. (6-1), $\mathrm{NPHR} = \dfrac{\mathrm{NTO} \times \mathrm{NTHR}}{\eta_B(\mathrm{NPO})}$

From Eq. (6-3), $\text{NPHR} = \dfrac{\text{NTO} \times \text{NTHR}}{\eta_B(\text{NTO} - \text{AP})}$

Finally, dividing through by NTO, we arrive at the following expression:

$$\text{NPHR} = \frac{\text{NTHR}}{\eta_B\left[1 - \dfrac{\text{AP}}{\text{NTO}}\right]} \qquad (6\text{-}5)$$

Equation (6-5) is normally used to determine net plant heat rate.

The term "gross plant heat rate" (GPHR) is sometimes used to refer to the amount of heat input required to generate a kilowatt-hour at the generator terminals, neglecting the effect of auxiliary power. GPHR may be determined by the following relationships:

$$\text{GPHR} = Q_B/\text{NTO} = \text{NTHR}/\eta_B \qquad (6\text{-}6)$$

Since 1 kilowatt-hour is equal to 3,413 Btu (3.598 MJ), net plant heat rate may also be used to determine the thermal efficiency of a generating unit by the following relationship:

$$\text{Thermal efficiency} = \frac{3{,}413\ (3.598\ \text{MJ}) \times 100}{\text{NPHR}} \qquad (6\text{-}7)$$

Therefore, based on the preceding, for a typical net plant heat rate of approximately 10,000 Btu/kWh (10.542 MJ/kWh), the thermal efficiency would be 34.13%.

Heat transferred within the steam generator must meet three discrete turbine cycle requirements: heat must convert feedwater into steam, superheat this steam, and reheat the steam returning from the high-pressure turbine. Therefore, the design of a steam generator must focus on arranging appropriate heat transfer surface throughout the unit to ensure that the correct quantity of heat is transferred for each requirement. Waterwall surface within the furnace collects heat to boil the feedwater. Economizer surface located at the exit of the convective pass is also frequently used to preheat the feedwater prior to introduction into the waterwalls. Superheat surface is normally located above the furnace and in the backpass. Suspended reheat surface is located in a similar manner; radiant reheat panels may also be located on front/side walls of the upper furnace. The actual location of surface varies depending on boiler manufacturer, type of fuel or fuels, and pressure class. Figure 6-4 illustrates a potential surface arrangement.

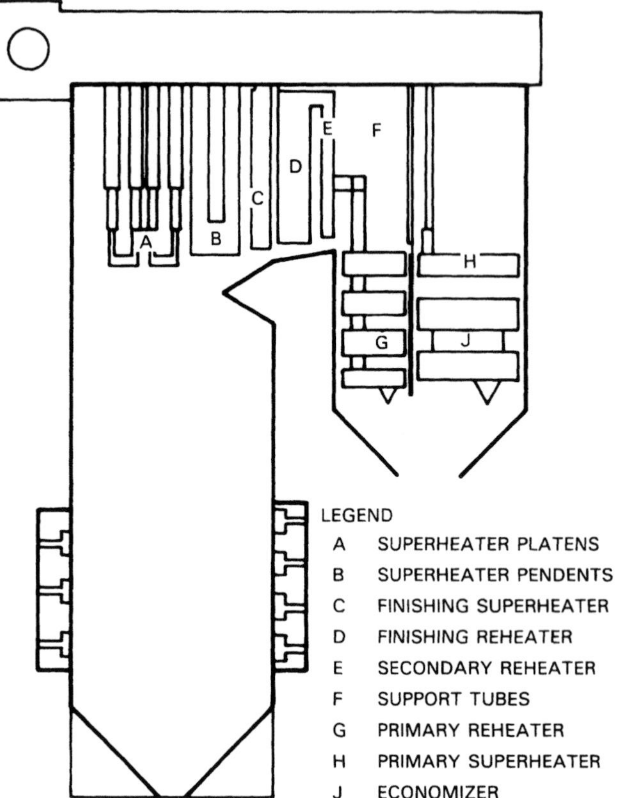

Fig. 6-4. Example of steam generator surface arrangement.

6.3 COMBUSTION CALCULATIONS

6.3.1 Key Concepts

6.3.1.1 Moles. Combustion calculations involving gaseous mixtures can be simplified by the use of the mole. Since equal volumes of gases at the same pressure and temperature contain the same number of molecules (Avogadro's law), the weights of equal volumes of gases are proportional to their molecular weights. If "M" is the molecular weight of the gas, 1 mole weighs "M" units of measure. This chapter uses the term lb-mole in which the unit of weight (mass) measure is the pound (lb). For example, from the Periodic Table of Elements, oxygen (O_2) has an approximate molecular weight of 32 (2 × 16); thus, a lb-mole of O_2 weighs 32 lb-mass. Similarly, a kg-mole of O_2 weighs 32 kg. Therefore, for purposes of discussion, this chapter uses lb/lb-mole or lb/100 lb fuel to represent equivalent relationships of kg/kg-mole and kg/100 kg. The number of molecules in a lb-mole and a kg-mole are, of course, different. However, this relationship between mass and volume is a valuable tool in determining the products of combustion.

6.3.1.2 Heating Value. In boiler practice, the heat of combustion of a fuel is the amount of heat, expressed in Btu, generated by the complete combustion, or oxidation, of a unit weight (1 lb in the United States) of fuel. Calorific value or "fuel Btu value" are also used to describe the unit heat of combustion of a fuel.

In the determination of heating value, several distinctions must be made. First, two types of heating values exist: higher heating value (HHV) and lower heating value (LHV). Higher heating value includes the heat given off from condensation of moisture in the combustion products. Lower heating value does not include this latent heat. Second, the manner in that heating value is determined and used must be recognized. Heating value determined from a bomb calorimeter is the heating value under constant volume. Combustion in a boiler occurs at constant pressure. Therefore, if heating value determination is via a bomb calorimeter, correction of heating value to constant pressure should be performed before using heating value in the equations presented in this chapter.[1]

Regardless of the type of heating value being considered, the amount of heat generated by complete combustion is a constant for any given combination of combustible elements and compounds, and is not affected by the manner of combustion, provided it is complete.

6.3.1.3 Stoichiometric Air. To support combustion of the fuel, oxygen must be present. The oxygen is provided through combustion air. The minimum amount of air required to theoretically combust a fuel completely is called *stoichiometric combustion air*. The stoichiometry of combustion is determined from the constituent analysis of the fuel. During the perfect combustion process, the fuel is oxidized by the following exothermic reactions:

$$C + O_2 \rightarrow CO_2$$

$$2H_2 + O_2 \rightarrow 2H_2O$$

$$S + O_2 \rightarrow SO_2$$

An unfortunate result of the sulfur combustion reaction is that the combustion product SO_2, when combined with the moisture in the gas, forms sulfuric acid that is very corrosive to equipment and the environment. Therefore, the benefit of the heat release from oxidation of sulfur is greatly outweighed by the negative impact on generation costs resulting from emissions control and equipment maintenance.

The other constituents of the fuel, as well as the nitrogen and moisture in the air, do not provide useful heat. They are, however, constituents of the flue gas produced by combustion and must be accounted for when determining the efficiency of the boiler. Some of the nitrogen in the fuel and the air combine with oxygen to form oxides of nitrogen (NO_x). These compounds normally are not considered in combustion calculations but are of concern in considerations of boiler emissions.

6.3.1.4 Stoichiometry. Combustion reaction equations determine molar relationships between reactants and products. A mole of substance has a mass equal to its molecular weight; mass can be measured in grams, kilograms, or pounds. Based on the carbon combustion reaction, the amount of oxygen required to combust carbon is 1 mole of oxygen per mole of carbon. Since one lb-mole of carbon weighs approximately 12 lb and one lb-mole of oxygen (O_2) weighs approximately 32 lb, each pound of carbon in the fuel requires 2.67 (32/12) lb of oxygen. Similarly, each pound of hydrogen in the fuel requires 8.00 (32/4) lb of oxygen, and each pound of sulfur in the fuel requires 1.00 (32/32) lb of oxygen for combustion. Also, given the composition of atmospheric air, for each pound of oxygen required, 4.32 lb of air must be provided. From Table 1, the derivation of this relationship for combustion of coal is as follows; note that relationships for natural gas and fuel oil are the same, given fuel constituents present:

	Volume fraction	Mol. Wt.	lb per lb-mole Air[a]
Nitrogen	0.7809	28.016	21.878
Oxygen	0.2095	32.000	6.704
Argon	0.093	39.944	0.371
Carbon dioxide	0.003	44.010	0.013

Molecular weight of air: 28.966 lb/lb-mole

	Weight (%)
Nitrogen[b]	76.86
Oxygen	23.14

Pounds of air per pound of oxygen = 100/23.14
= 4.32

[a]Calculated by multiplying volume fraction by mol. wt.

[b]Nitrogen weight percent includes argon and carbon dioxide.

However, the oxygen in the fuel is used during combustion so the actual amount required from the combustion air is reduced by the amount of oxygen in the fuel. The following example illustrates the calculation of stoichiometric air (dry) required for combustion:

Fuel constitutent, lb/100 lb of fuel	Multiplier	Stoichiometric oxygen, lb/100 lb of fuel
C = 55.37	32/12 = 2.67	+147.84
H_2 = 3.69	32/4 = 8.00	+29.52
S = 0.38	32/32 = 1.00	+0.38
N_2 = 0.72	Not applicable	Not applicable
O_2 = 15.66	32/32 = 1.00	−15.66
H_2O = 14.58	Not applicable	Not applicable
Ash = 9.60	Not applicable	Not applicable
Total = 100.00		162.08

Stoichiometric air = 4.32 × stoichiometric oxygen
= 4.32 × 162.08
= 700.19 lb of dry air/100 lb of fuel

The combustion air utilized in the boiler contains moisture in the form of water vapor. Figure 6-5 illustrates the relationship between moisture content in air as a function of dry bulb temperature and relative humidity. For example, the moisture contained in combustion air at 80° F (26.7° C) and 60% relative humidity is 0.0132 lb of moisture per pound of dry air. This number is referred to as the humidity ratio and can be found from Fig. 6-5. From the preceding stoichiometric air calculation, the moisture content of the air is as follows:

Moisture in Air = 0.0132 lb of moisture/lb of dry air × stoich. air
= 0.0132 × 700.19
= 9.2 lb of moisture/100 lb of fuel

[1]For calculation, refer to ASME PTC 4.1, Steam Generating Units.

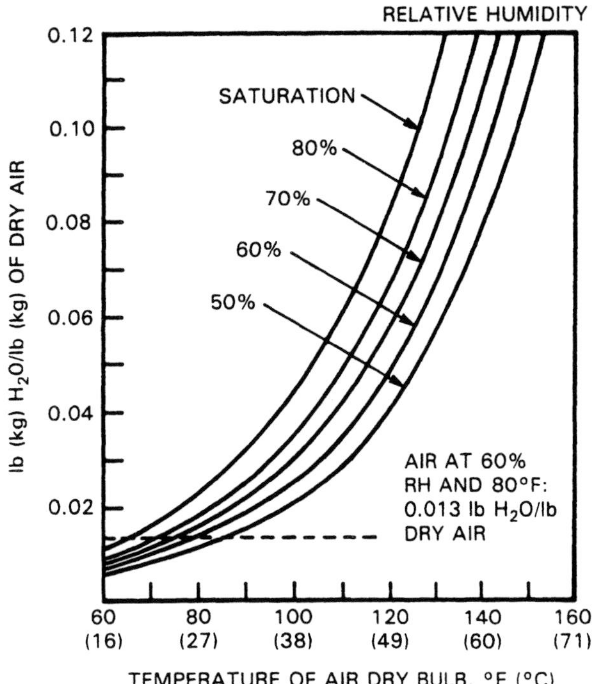

Fig. 6-5. Moisture content of dry air as a function of dry-bulb temperature and relative himidity. (From T. Elliott, *Standard Handbook of Power Plant Engineering*, 1989. Used with permission of McGraw-Hill.)

The wet stoichiometric air is equal to the dry stoichiometric air plus the moisture in the air. For this example, the wet stoichiometric combustion air is equal to 709.4 lb/100 lb of fuel (700.2 + 9.2). This moisture in the air does not participate in the combustion process but must be accounted for in calculating flue gas flow rates and boiler efficiency.

6.3.1.5 Excess Air. The amount of air calculated in the preceding example represents the requirement for theoretical complete combustion. In other words, if every molecule of fuel came into contact with the right number of molecules of air, all the fuel would be combusted and no "excess" air would remain. This is impossible to achieve from a practical standpoint. Consequently, additional air must be supplied to the furnace to ensure that all fuel comes in contact with enough air to support complete combustion. This amount of additional air is referred to as "excess air" and is represented as a percentage of stoichiometric air.

The amount of excess air required is dependent on a number of parameters including fuel fired, boiler design, burner design, and load. Typically, the excess air levels for pulverized coal boilers range from 15% to 30% excess air. Gas- and oil-fueled boilers require less excess air, typically 5% to 10% for gas and 3% to 15% for fuel oil. The liquid or gaseous form of these fuels facilitates contact with the air in the furnace more readily than the solid particles of coal.

Load level has a dramatic effect on excess air requirements. Furnace cross-sectional area and burner design must support full load gas flow rate at acceptable velocities. However, when load is decreased, gas flow rates and associated velocities drop off and proper mixing of the fuel with the combustion air becomes more difficult. Therefore, additional excess air is introduced as load decreases to maintain effective combustion. For a pulverized-coal boiler, the excess air level at 50% load may be controlled to twice that of 100% load.

Excess air level is determined by measuring the oxygen content of the flue gas, typically at the exit of the economizer. If combustion was complete and no excess air was required, the oxygen content of the flue gas would be zero. The oxygen content of the flue gas can therefore indicate the excess air level in the furnace. However, on balanced-draft boilers where the furnace is under slight negative pressure, air inleakage is a concern. Air inleakage cools the flue gas so that less heat is transferred to the steam/water and increases the loading on the induced draft fans.

Air inleakage can impact the accuracy of measured excess air levels. For practical reasons, excess air is measured at the economizer rather than in the furnace (at point of combustion). Without knowledge of air inleakage, it must be assumed that the excess air level in the furnace is identical to that at the economizer. However, if air is leaking through the boiler skin between the furnace and the economizer, the actual excess air level in the furnace will be less than the level measured at the economizer. Thus, the furnace may be "starved" of air while the excess air level at the economizer may appear perfectly normal. Carbon monoxide (CO) monitors are often used in parallel with oxygen monitors to verify adequate air levels in the furnace. A rise in the CO level indicates incomplete combustion in the furnace, possibly due to insufficient air.

Excess air, although necessary for complete combustion, has a negative impact on boiler efficiency. As excess air increases, the heat lost from the boiler envelope via the flue gas increases. Therefore, from an efficiency perspective, the lowest possible excess air level should be maintained. This desire to lower excess air levels must be balanced against the potential for increased CO level (incomplete combustion). Typically, CO measurements are not used for excess air control. Rather, O_2 at the economizer exit is typically controlled based on a curve relating O_2 level to load. However, the optimum level of excess air can be determined by operating the boiler at decreasing levels of excess air and monitoring the CO level of the flue gas. Once the CO level begins to rise, the excess air level should be increased an appropriate amount for margin. In this manner, complete combustion and efficient operation can be achieved simultaneously.

6.3.2 Calculating Unit Air/Flue Gas Flow Rates

6.3.2.1 Combustion Air. From the prior example, the stoichiometric combustion air requirements were 700.2 lb/100 lb of fuel (dry) and 709.4 lb/100 lb of fuel (wet). By recognizing excess air requirements, the total combustion air flow rates per 100 lb of fuel can be determined. Given an excess air level of 20%, the total combustion air flow rates are calculated as follows:

Dry combustion air = 1 + (excess air/100) × dry stoichiometric air (6-8)
= (1 + 20/100) × 700.2 lb/100 lb of fuel
= 840.2 lb/100 lb of fuel

Moisture in dry air = Humidity ratio × dry combustion air (6-9)
= 0.0132 lb of moisture/lb of dry air × 840.2 lb/100 lb of fuel
= 11.1 lb/100 lb of fuel

Wet combustion air = Dry combustion air + moisture in dry air (6-10)
= 840.2 lb/100 lb of fuel + 11.1 lb/100 lb of fuel
= 851.3 lb/100 lb of fuel

6.3.2.2 Flue Gas. To calculate the flue gas flow rate from the economizer, the various constituents must be identified. Flue gas calculations consider the following components:

- CO_2—combustion product from oxidation of carbon in fuel, assuming all carbon is oxidized to CO_2.
- H_2O—combustion product from oxidation of hydrogen in fuel and from moisture in fuel and in combustion air.
- SO_2—combustion product from oxidation of sulfur in fuel.
- N_2—from nitrogen in combustion air and in fuel.
- O_2—from combustion air (excess air).

Continuing with the previous example, the following illustrates the calculation of flue gas flow from the economizer:

Constituent[a]	Economizer flue gas, Fuel + air	lb/100 lb of fuel
Carbon dioxide	55.37 + 147.84	203.21
Nitrogen		
Combustion air	0 + 645.78[b]	645.78
Fuel	0.72 + 0	0.72
Sulfur dioxide	0.38 + 0.38	0.76
Excess oxygen	0 + 32.40[c]	32.40
Total dry flue gas		882.87
Moisture		
From H_2 content in fuel	3.69 + 29.30	32.99
From fuel moisture contact	14.58 + 0	14.58
From combustion air	0 + 11.10	11.10
Total wet flue gas		941.54

[a]In some cases, fly ash is induced in the determination of the flue gas flow rate.

[b]Nitrogen in combustion air is determined from percent by weight as previously stated (76.86%). Value is found by multiplying total dry combustion air (840.2 lb/100 lb of fuel) by nitrogen weight percent (76.86).

[c]Excess oxygen in combustion air is determined from percent by weight as previously stated (23.14%). Value is found by multiplying excess air (840.2 − 700.2) by oxygen weight percent (23.14).

6.3.2.3 Air Heater Leakage. Combustion air and flue gas flow rate calculations presented previously do not recognize air heater leakage. Air heater leakage can be defined as the quantity of air "leaking" from the higher-pressure air stream past the mechanical seals of the air heater into the flue gas stream. Leakage rates vary for different types of air heaters. Typical values for regenerative air heaters range from 6% percent to 15%. Tubular and heat-pipe air heaters have no mechanical seals, and therefore leakage have minimal leakage (less than 3%).

Air heater leakage is expressed as a percentage of the flue gas entering the air heater. For calculational purposes, it is assumed that leakage does not pass through the air heater and absorb heat.

For this example, the air heater leakage is assumed to be 10%. The flue gas flow from the air heater can be calculated as follows:

Flue gas from air heater = Flue gas from economizer × (1 + leakage/100) (6-11)
= 941.5 × (1 + 10/100)
= 1,035.7 lb/100 lb of fuel

6.3.3 Boiler Efficiency

Boiler efficiency is defined as the percentage of the heat input to the boiler, in the form of fuel, that is absorbed by the working fluid. This working fluid is the steam and/or water that flows on the inside of the boiler tubes. Practical design considerations limit the boiler efficiency that can be achieved. Typically, boiler efficiency ranges from 75% to 95%. Boiler efficiency is dependent on many parameters, including boiler design, operation, and fuel.

To determine boiler efficiency, one of two methods is generally employed. These methods, the Input–Output Method and the Heat-Loss Method, are both recognized by ASME. This organization has formed committees from members of the industry to analyze and standardize the method by that boiler performance is determined and evaluated. The committee on steam generators has developed the Performance Test Code (PTC) 4, Steam Generating Units, that details the method of determining boiler efficiency by the two methods stated. The following sections are summaries of PTC 4 and are not meant to be substitutes for the code.

6.3.3.1 Input–Output Method. Perhaps the easiest method of calculating boiler efficiency is the Input–Output Method. In this method, the heat input from the fuel and the heat absorbed by the working fluid are directly measured. Boiler efficiency from the Input–Output Method is calculated as follows:

$$\eta_b = 100 \times Q_{abs}/Q_{fuel} \tag{6-12}$$

where

η_b = boiler efficiency, percentage;

Q_{abs} = heat absorbed, Btu/h

= $\Sigma m_o h_o - \Sigma m_i h_i$;

$m_o h_o$ = mass flow-enthalpy products of working fluid streams leaving boiler envelope, including main steam, hot reheat steam, blowdown, soot blowing steam, etc.;

$m_i h_i$ = mass flow-enthalpy products of working fluid streams entering boiler envelope, including feedwater, cold reheat steam, desuperheating sprays, etc.;

Q_{fuel} = heat from fuel and other heat credits, Btu/h

= $m_{fuel} \times H_{fuel}$;

m_{fuel} = mass flow of fuel into boiler, lb/h; and

H_{fuel} = higher heating value of fuel, Btu/lb.

In the previous equations, the other heat sources, or credits, are ignored since fuel efficiency is desired. For determing gross efficiency, these credits should be included in the denominator of the equation. These credits include the sensible heat in the entering fuel and air. ASME PTC 4 defines appropriate procedures for calculating these credits.

The previous calculation of boiler efficiency requires relatively few measurements, many of that are often readily available. However, the accuracies of these measurements, especially for fuel flow and heating value, are sometimes an issue. For gas- and oil-fueled boilers, the preceding method produces acceptable levels of accuracy because of the relatively easy measurement of liquid or gaseous fluid flow and the consistency of fuel heating value. For coal-fueled boilers, however, measurements of fuel flow and heating value present a challenge. Today's gravimetric feeders represent a large improvement over the older volumetric feeders for measuring fuel flow. However, inaccuracy still exists in determination of a representative heating value for the large, often variable quantities of fuel consumed by a modern boiler. Generally, the best method for determining boiler efficiency in a coal fueled boiler is the Heat-Loss Method.

6.3.3.2 Heat-Loss Method. Instead of measuring the actual heat absorbed relative to the heat supplied in the Input–Output Method, the Heat-Loss Method concentrates on determining the heat lost from the boiler envelope, or the heat not absorbed by the working fluid. Derivation of the Heat-Loss Method is as follows:

$$\eta_b = 100 \times Q_{abs}/Q_{fuel} \qquad (6\text{-}13)$$

$$Q_{abs} = Q_{fuel} - Q_{loss} + Q_{credit}$$

$$\eta_b = 100 \times (Q_{fuel} - Q_{loss} + Q_{credit})/Q_{fuel}$$
$$= 100 \times (1 - (Q_{loss} - Q_{credit}/Q_{fuel}$$
$$= 100 - 100 \times (Q_{loss} - Q_{credit})/Q_{fuel}$$

where

Q_{loss} = heat loss across steam generator control volume, and

Q_{credit} = heat credit across steam generator control volume.

The measurement of fuel flow can be eliminated from the preceding equation by dividing Q_{loss}, Q_{credit}, and Q_{fuel} by fuel flow. The equation for boiler efficiency then becomes the following:

$$\eta_b = 100 - 100 \times (\Sigma H_{loss} - \Sigma H_{credit})/H_{fuiel} \qquad (6\text{-}14)$$

where

ΣH_{loss} = summation of individual loss terms divided by fuel flow, and

ΣH_{credit} = summation of individual heat credits divided by fuel flow.

From the preceding derivation, it can be seen that the Heat-Loss Method concentrates on determining the lost heat—the heat that prevents boiler efficiency from being 100%. Generally, boiler efficiency is between 75% and 90% for large utility coal-fueled boilers.

Accuracy or uncertainty in boiler efficiency calculations is a function of what is measured and what quantities are determined. Using the Input–Output Method, these quantities are related to overall efficiency. For example, if the accuracy of all measurements was assumed equal at 1%, the minimum expected error of the Input–Output Method would be 0.8% (equivalent to 1% of the measured boiler efficiency of 80%). However, for the Heat-Loss Method, the measured and determined parameters are related to net losses. Therefore, for our example for the heat-loss method, the maximum expected error would be 0.2% (1% of the measured losses of 20%). As a result, the Heat-Loss Method is inherently more accurate than the Input–Output Method for coal-fueled boilers.

The losses and credits considered in determining Heat-Loss boiler efficiency include the following:

- Losses
  —Dry gas sensible
  —Fuel moisture latent
  —Fuel hydrogen latent
  —Combustion air moisture latent
  —Other losses
- Credits
  —Heat in entering air
  —Heat from auxiliary equipment
  —Sensible heat in fuel

With the presence of fluidized-bed boilers in industry, the calculation of heat losses and heat credits becomes somewhat more complex because of the injection of lime or limestone into the furnace to control emissions. Additional chemical reactions affect the efficiency of the boiler. Two of these reactions include calcination and sulfation. Calcination is an endothermic reaction whereby limestone (calcium carbonate and magnesium oxide) forms calcium oxide and magnesium oxide and carbon dioxide. This reaction, since it is endothermic, is considered a loss. Sulfation is an exothermic reaction whereby calcium oxide, oxygen, and sulfur dioxide combine to form calcium sulfate. This reaction, since it is exothermic, is considered a credit. In 1993, ASME provided an industry review of the new code, PTC 4, that considers these reactions in determining boiler efficiency.

HEAT LOSSES. The following illustrate the calculation of each individual heat loss using the previous numerical example. In addition to the previous data, the following data are assumed:

- Ambient air temperature: 80° F (26.7° C)
- Air heater gas outlet temperature (with leakage): 280° F (137.8° C)
- Higher heating value of fuel: 10,457 Btu/lb (24.3 MJ/kg)

Dry Gas Heat Loss. The dry gas heat loss represents the sensible heat lost from the flue gas leaving the air heater. From our example, the measured, or observed, gas temperature at the air heater exit is 280° F (137.8° C). However, to calculate the sensible heat loss, the "no-leakage" air heater exit gas temperature must be determined because it is important to isolate performance from air heater performance. Leakage from the air side to the gas side of the air heater artificially depresses the gas temperature and could, without investigation, be a misleading indicator of good boiler performance. Therefore, the no-leakage gas temperature is higher than the observed gas temperature. The no-leakage gas temperature can be approximated by the following:

$$t_{nl} = t_g + (\mathrm{AHL}/100) \times (t_g - t_{ahi}) \qquad (6\text{-}15)$$

where

t_{nl} = no-leakage air heater gas outlet temperature
= 300° F (148.9° C),

t_g = observed air heater gas outlet temperature
= 280° F (137.8° C),

AHL = air heater leakage, percentage
= 10%, and

t_{ahi} = air heater air inlet temperature
= 80° F (26.7° C).

The preceding equation is an approximation in that the specific heat capacity of the air and gas, C_p, is assumed equal. In actuality, they are not equal, but for ease of calculation, the error introduced is acceptable.

With the no-leakage gas temperature determined, the dry gas sensible heat loss is found as follows:

$$L_{dg} = (W_{dg}/100) \times C_p \times (t_{nl} - t_{ref}) \qquad (6\text{-}16)$$

where

L_{dg} = dry gas sensible heat loss, Btu/lb
= 466.2 Btu/lb (1.1 MJ/kg),

W_{dg} = dry flue gas flow
= 882.87 lb/100 lb of fuel (882.87 kg/100 kg of fuel),

C_p = specific heat of flue gas integrated over temperature range of t_{nl} to t_{ref}
= value from ASME PTC 4 curve or estimated value of 0.24 Btu/lb/° F (0.24 cal/g/° C),

t_{nl} = no-leakage gas temperature
= 300° F (148.9° C), and

t_{ref} = reference temperature, typically ambient air or air heater inlet air upon that properties of streams entering or leaving the control volumes are based.
= 80° F (26.7° C).

Determination of this loss can be with either specific heat capacities and gas temperatures, or enthalpies of flue gas. Enthalpies, as with specific heat capacities, can be determined from published curves. If enthalpy curves are used, the reference temperature for the calculation of boiler efficiency must be the reference temperature of the curves. However, if specific heats are used, it is common to use the air inlet temperature.

Fuel Moisture Latent Heat Loss. The moisture in the fuel absorbs heat and leaves the air heater at the gas temperature, t_{nl}. The change in enthalpy of the water vapor due to moisture in the fuel constitutes a heat loss. This heat loss is determined as follows:

$$L_{mf} = (H_2O/100) \times (h_{wv} - h_{wref}) \qquad (6\text{-}17)$$

where

L_{mf} = fuel moisture latent heat loss, Btu/lb (MJ/kg)
= 167.3 Btu/lb (389.1 kJ/kg),

H_2O = fuel moisture, lb/100 lb of fuel (kg/100 kg of fuel)
= 14.58 lb/100 lb of fuel (14.58 kg/100 kg of fuel),

h_{wv} = enthalpy of water vapor at t_{nl} and 1 psia, Btu/lb (MJ/kg)
= 1,195.8 Btu/lb (2.8 MJ/kg), and

h_{wref} = enthalpy of water at t_{ref}, Btu/lb (MJ/kg)
= 48.0 Btu/lb (111.7 kJ/kg).

Fuel Hydrogen Latent Heat Loss. Combustion of hydrogen yields moisture as a product. As with moisture in the fuel, heat is lost as a result of a change in enthalpy of the water vapor leaving the air heater. The heat loss corresponding to fuel hydrogen content is determined as follows.

$$L_h = (8.936 \times H_2/100) \times (h_{wv} - h_{wref}) \quad (6\text{-}18)$$

where

L_h = fuel hydrogen latent heat loss, Btu/lb (kJ/kg)
= 378.5 Btu/lb (880.4 kJ/kg), and

H_2 = fuel hydrogen, lb/100 lb of fuel (kg/100 kg of fuel)
= 3.69 lb/100 lb of fuel (3.69 kg/100 kg of fuel).

Combustion Air Moisture Latent Heat Loss. The moisture carried into the boiler envelope by the combustion air absorbs heat as well. The heat loss corresponding to moisture content in the combustion air is determined as follows.

$$L_{ma} = (W_{ma} \times W_{ca}/100) \times (h_{wv} - h_{vref}) \quad (6\text{-}19)$$

where

L_{ma} = combustion air moisture latent heat loss, Btu/lb (kJ/kg)
= 11.0 Btu/lb (25.6 kJ/kg),

W_{ma} = humidity ratio (kg of moisture/kg of dry air)
= 0.0132 lb of H_2O/lb of dry air (0.00599 kg of H_2O/kg of dry air).

W_{ca} = dry combustion air, lb/100 lb of fuel (kg/100 kg of fuel)
= 840.2 lb/100 lb of fuel (840.2 kg/100 kg of fuel), and

h_{vref} = enthalpy of saturated vapor at t_{ref}, Btu/lb (MJ/kg)
= 1,096.6 Btu/lb (2.6 MJ/kg).

Other Losses. Other losses include unburned carbon, incomplete combustion, radiation and convection, and unaccounted losses. The following paragraphs describe these losses.

Unburned carbon loss, or Loss On Ignition (LOI), occurs when not all of the available carbon comes in contact with combustion air. This loss can be caused by insufficient grinding of the fuel. If the fuel particles are too large, the surface area is reduced and not all of the particles are completely burned. Samples of the ash from various locations in the boiler are analyzed for heating value and the unburned carbon loss is determined as follows:

$$L_{uc} = (\text{ASH}/100) \times \Sigma S_a H_a \quad (6\text{-}20)$$

where

L_{uc} = unburned carbon loss, Btu/lb (J/kg),

ASH = ash content of the fuel, lb/100 lb of fuel (kg/100 kg of fuel),

S_a = split fraction from each collection point (bottom ash, fly ash, air heater ash, etc.) fraction, and

H_a = higher heating value of ash from each collection point (bottom ash, fly ash, air heater ash, etc.), Btu/lb (J/kg).

Incomplete combustion occurs when insufficient oxygen is available to completely oxidize the carbon in the fuel. Carbon monoxide is formed instead of carbon dioxide. When carbon dioxide is formed, the heat of reaction is 14,540 Btu (15.3 MJ). When carbon monoxide is formed, the heat of reaction is only 4,380 Btu (4.6 MJ). Therefore, heat is lost by not oxidizing all of the carbon to carbon dioxide. The heat lost because of incomplete combustion is determined as follows:

$$L_{co} = (\text{CO}/[\text{CO} + \text{CO}_2]) \times C_b/100 \times 10{,}160 \quad (6\text{-}21)$$

where

L_{co} = incomplete combustion loss, Btu/lb (J/kg),

CO = carbon monoxide content of flue gas, percentage by volume,

CO_2 = carbon dioxide content of flue gas, percentage by volume,

C_b = carbon burned in fuel, lb/100 lb (kg/100 kg),
= $C - 100 \times L_{uc}/14{,}500$ (15.3 MJ), and

C = carbon content of fuel, lb/100 lb of fuel (kg/100 kg).

Radiation and convection loss terms represent the heat losses from the boiler skin to the surrounding environment, due to each heat transfer mechanism. Since these losses are quite difficult to measure, the combined loss term is typically estimated from a curve supplied from the American Boiler Manufacturers' Association (ABMA). This loss is a function of boiler heat release and type of cooling of furnace walls.

Unaccounted losses include those not considered explicitly in the design calculations of the boiler. Typically, these losses are specified by the boiler manufacturer. Many times these losses include sensible heat in the ash, pulverizer rejects, etc.

HEAT CREDITS. Heat credits represent useful heat available to the boiler in a form other than fuel. These credits include heat in entering air, heat added by auxiliary equipment, and sensible heat in the fuel. The following paragraphs illustrate these heat credits.

Heat in Entering Air. If the temperature of entering air is different from the reference air temperature, a credit must be calculated to account for this heat crossing the control volume. Heat sources include air preheat coils, combustion air fans, and high ambient air temperatures. This credit is calculated as follows:

$$B_a = (W_{ca}/100) \times C_p \times \Delta t_{air} \quad (6\text{-}22)$$

where

B_a = heat credit in entering air, Btu/lb;

W_{ca} = combustion air flow (wet), lb/100 lb of fuel;

C_p = specific heat of air, Btu/lb/° F,
≈ 0.24 Btu/lb/° F (0.24 cal/g/° C); and

Δt_{air} = temperature differential between temperature of air entering air heater and t_{ref}.

Heat from Auxiliary Equipment. The heat generated by auxiliary equipment within the boiler envelope must be accounted for. Generally, the energy added by the draft fans must be considered. The calculation of the heat credit from draft fans can be determined as follows:

$$B_{fan} = \Sigma(W_{fan}/100) \times C_p \times \Delta t_{fan}) \tag{6-23}$$

where

B_{fan} = heat credit from draft fans (Btu/lb),

W_{fan} = air flow through fan, lb/100 lb (Btu/lb) of fuel, and

Δt_{fan} = temperature rise across fans, °F.

Sensible Heat in Fuel. If the fuel enters the boiler envelope at a temperature other than the reference temperature, typically ambient dry bulb, a credit must be determined to account for the heat entering the envelope. This credit is determined as follows:

$$B_{fuel} = 100 \times C_{pfuel} \times (t_{fuel} - t_{ref}) \tag{6-24}$$

where

B_{fuel} = sensible heat in fuel, Btu/lb;

C_{pfuel} = specific heat of fuel, Btu/lb/°F,
= 0.3 Btu/lb/°F (coal); and

t_{fuel} = temperature of fuel, °F.

6.3.4 Total Air/Gas Flows

To determine the total combustion air and flue gas flows, the fuel flow must be measured or calculated. With the Input–Output Method, fuel flow is measured. With the Heat-Loss Method, fuel flow is calculated by the following equation:

$$W_{fuel} = \frac{Q_{abs}}{(\eta_b/100)H_f)} \tag{6-25}$$

Once fuel flow is found, the total combustion air and flue gas flows can be found by multiplying the unit air/gas flows by fuel flow.

6.3.4.1 Fan Requirements. While boiler requirements for combustion air and the corresponding flue gas flows are on a mass basis, lb/h, fans are volumetric machines, for which the requirements must be on a volumetric basis, actual cubic feet per minute (ACFM).

Since the combustion air fans generally handle air prior to the air heaters, the amount of air they are required to handle must include the mass of air determined from air heater leakage. To determine the volumetric requirements of the combustion air fans, the following equations apply:

$$Q_{ca} = (W_{ca}/100 + \text{AHL}/100 \times W_g/100) \times W_{fuel}/(\rho_{ca} \times 60\ \text{min/h}) \tag{6-26}$$

where

Q_{ca} = combustion air volumetric flow rate, acfm;

W_{ca} = combustion air flow (wet), lb/100 lb of fuel,

AHL = air heater leakage, percentage;

W_g = flue gas flow from economizer, lb/100 lb of fuel;

W_{fuel} = fuel flow, lb/h; and

ρ_{ca} = density of combustion air, lb/ft^3 (from air tables).

Fans handling flue gas must consider leakage. The requirements of the induced draft (ID) fans are calculated as follows:

$$Q_g = (1 + \text{AHL}/100) \times W_g \times W_{fuel}/(\rho_g \times 60\ \text{min/h}) \tag{6-27}$$

where

Q_g = flue gas volumetric flow rate to ID fans, acfm;

ρ_g = density of flue gas at ID fan inlet, lb/ft^3 (kg/m^3); and

AHL = air heater leakage, or in cases where hot precipitators or baghouses are used, leakage is sum of air heater and dust collection device leakage, percentage.

The first step in determining flue gas density is to calculate the molecular weight of flue gas. This is done by determining the number of moles of flue gas by constituent from the combustion calculations: carbon dioxide, sulfur dioxide, nitrogen, oxygen, and moisture. Once the number of moles has been determined, the molecular weight is found by the following:

$$M_g = W_g/N_g \tag{6-28}$$

where

M_g = molecular weight of flue gas, lb/lb-mole (kg/kg-mole);

W_g = flue gas mass flow, lb/100 lb fuel (kg/100 kg); and

N_g = moles of flue gas, lb-mole/100 lb (kg-mole/100 kg).

From the ideal gas law, the density can be determined.

$$P = \rho RT \tag{6-29}$$

or

$$\rho = P/(RT) \tag{6-30}$$

where

ρ = density, lb/ft^3;

P = pressure, lb/ft^2, absolute;

R = gas constant, ft-lbf/lbm/deg R,

= 1,545 ft-lbf/lb-mole/deg R/M_g;

T = temperature, deg R; and

M_g = molecular weight of gas, lb/lb-mole (kg/kg-mole).

The preceding density calculation was for gas at the economizer exit. To determine the density of gas after the air heater, convert the density of flue gas to air heater conditions via the ideal gas law, and then use a mass weighted average of flue gas and air due to leakage. The density of gas at the fan can then be determined and used to calculate the volumetric flow of flue gas.

6.3.4.2 Gas Velocity. From the volumetric gas flow rates, the velocity of gas throughout the boiler can be determined by dividing the gas flow by the cross-sectional flow area. Velocities in the boiler, ductwork, and precipitator are important design parameters. High flue gas velocity enhances heat transfer but also increases erosion because of the ash suspended in the gas. Therefore, it is important to check velocities throughout the boiler, ductwork, and precipitator to verify that design criteria are not being exceeded.

6.4 COMPLEXITIES OF COMBUSTION

6.4.1 Relationship of Combustion to Design and Operation

Combustion characteristics may alter where and how the heat is transferred to the water or steam. Heat transfer imbalances occur when the ratio of heat passing into waterwalls versus that passing into the superheat surface fails to meet design conditions. In that case, one of the following will occur:

- If excessive heat is being transferred into the waterwalls, the fuel heat input must be reduced to match total steam demand. Less heat will be available for superheat, thus main steam temperatures will be low. In most cases, insufficient heat will also be available for reheat duty and therefore reheat steam temperatures will typically be low as well.
- If heat transfer to the waterwall is too low, the unit will tend to "overfire" fuel to match waterwall demand. Excessive heat will typically be transferred to superheat and reheat surface. Steam temperature must then be controlled with additional attemperation spray.

For a steam generator to operate properly, a synergy must exist between fuel properties, combustion control parameters, steaming requirements, and boiler design/surface arrangement. Changes in any of these categories can drastically impact capacity and performance of the boiler. Hence, it is important to understand the relationship between these parameters and boiler operation/performance. Table 6-4 illustrates some of these relationships.

6.4.2 Complexities of Oil-Fueled Combustion

Satisfactory combustion performance from firing a heavy residual oil requires optimization or trade-offs among a number of parameters (EPRI 1988).

- Carbon burnout/utilization,
- Flue gas losses (sensible and latent heat),
- Flame shape and heat release distribution,
- Flame stability,
- Stack opacity (visible smoke),
- Gaseous and particulate emission (CO, NO_x, SO_x, ash, condensible particulate),
- Load following capability and turndown, and
- Flame detection and safety.

Most of these parameters are directly involved in two important boiler performance activities: optimizing boiler efficiency and minimizing emissions.

Optimum combustion performance is strongly dependent on effective atomization, that requires effective control of fuel oil firing temperature/viscosity and atomizing pressure. Effective control of the excess air level over the boiler load range is necessary for efficient combustion with complete burnout of the fuel. Operating at high excess air levels wastes fuel by heating and conveying additional air not essential for complete combustion. High excess air levels can also negatively impact unit operation by increasing SO_3 emissions or creating a plume visibility problem with some fuels. In contrast, an operating excess air level that is too low can lead to formation of smoke and soot, violation of stack opacity limits, flame impingement and deposits, and carryover of unburned fuel into the particulate control equipment (if any) or out the stack. In addition to lost energy, fires and safety problems can also result from carryover of unburned fuel.

Degradation in fuel quality can also be a major cause of poor steam generator performance. Oil-fueled power plants normally fire residual fuel oil. Although this fuel costs less and has higher heat content per gallon than distillate fuel, it usually has higher sulfur content plus sodium, vanadium, nickel, and other ash-forming ingredients. In and near the combustion zone, molten ash can cause corrosion and deposits; in other areas below 350° F (176.7° C), water and sulfur compounds condense into corrosive acid solutions.

As in the case of coal ash, during combustion, the ash-forming materials are converted to oxides that interact to form a variety of chemical compounds. If they cool and solidify before striking a solid surface, the ash particles are likely to pass through the boiler. However, if such particles strike boiler surface while above the ash fusion temperature, slag deposits may collect. In addition to inhibiting heat transfer, the presence of such deposits may lead to significant corrosion problems. Vanadium compounds are particularly corrosive when molten.

As ashes are heterogeneous, and contain mixtures of compounds with different sintering and softening temperatures, the actual fusion of the ash particles may occur over a range

Table 6-4. Relationship Between Steam Generator Performance and Fuel, Operations, and Design Parameters

Source	Parameter	Impact
Fuel	Moisture content	Higher moisture contents in fuel tend to lower the flame temperature and reduce the radiant heat transfer to waterwall surface.
	Heating value	High heating value fuels tend to burn with a hot flame, increasing radiant heat transfer.
	Ash composition	Type and quantity of minerals can strongly influence the potential for ash to create deposition problems.
	Ash content	Quality of ash affects the radiation heat transfer characteristics by contributing to gas emissivities, and altering wall deposit emissivities. Ash quality also impacts bottom ash generation rates and storage capacity of bottom ash hoppers.
	Volatile matter	Fuels with high-volatile matter burn quickly and require less furnace residence time. Alternatively, fuels with low-volatile matter contents tend to burn poorly (incomplete combustion) and may require use of supplemental fuels to assure proper flame stability.
Operations	Excess air level	Higher excess air levels tend to cool the flame and raise the temperature of flue gas passing into the backpass. For coal-fueled units burning bituminous coal, raised excess air levels can be used to mitigate furnace deposition problems. However, higher excess air will increase quantity of heat lost in flue gas leaving the steam generator and lower boiler efficiency.
	Flue gas tempering or recirculation	Some steam generators use recirculated flue gas to cool the flame and help push additional heat into the convective pass.
	Burner tilt	Tangentially fired boilers often tilt burners up or down to move the flame within the furnace. Movement of the flame alters the effective split between heat transferred to the waterwalls and heat leaving the furnace. Hence, burner tilt can be used to control steam temperature.
Design	Heat input per plan area	The heat input per plan area defines the cross-sectional area available within the furnace. Lower heat input per plan areas correspond to large funace volumes; larger furnace volumes are used for lower-rank fuels to assure adequate radiant surface area is available and that ash deposition problem can be effectively controlled.
	Height of boiler	Like the heat input per plan area, the height of the boiler alters the volume of surface available for radiant heat transfer; similarly, taller boilers allow the flue gas to more fully cool prior to entering the convective pass. Taller units are also necessary for fuels that require additional residence time for complete combustion.
	Use of platens and pendants	In most boilers, heat transfer surface is suspended above the furnace in the form of platens or pendants; this surface can collect radiant heat from the flame and convective heat from the local flue gas. This type of surface is not used if the lower quality of the fuel (high risk for slugging or fouling problems) dictates a more conservative surface arrangement.
	Burner and wind box design	Burner and wind box configuration alter the characteristics of the flame including degree of mixing, speed of combustion, degree of combustion, and potential for accumulation of deposits in the furace. Burners designed to reduce the formation of NO_x tend to spread out and cool the flame, potentially altering both the heat transfer and ash deposit profiles. Low NO_x burners must be carefully tuned for the unit to realize both efficient combustion and actual low NO_x emission rates.

of temperatures. Oil ash fusion temperatures range from below 1,000° (537.8° C) to over 2,000° F (1,075.6° C), depending on the relative concentrations of fluxes (principally sodium) and refractory compounds (such as silica, magnesia, and alumina). As a rule, vanadium corrosion usually occurs at temperatures above 1,250° F (676.7° C), and sulfidation (attack of nickel alloys by sulfates) above 1,650° F (898.9° C) (Perry and Chilton 1973).

Chemical additives are commonly used to control ash deposition problems. As stated in the *Chemical Engineer's Handbook*, fuel suppliers should be consulted for possible adverse reactions between the additive and fuel, and manufacturer claims for additives should be evaluated cautiously, but their potential usefulness for specific performance problems should not be overlooked (Perry and Chilton 1973). Magnesia, epsom salts, and other inexpensive magnesium compounds are added at M_g/V weight ratios of 3 or 3.5 to 1 to prevent corrosion and deposition by raising the ash fusion temperature. There is disagreement over the value of alumina as a co-additive to overcome the slight tendency of magnesia to form deposits. Calcium compounds have been widely tested but are now considered undesirable because they form hard, adherent, insoluble deposits. Manganese compounds and, to a lesser extent, lead and copper compounds are being used as combustion catalysts to reduce soot and smoke formation. Aside from additives designed to modify the ash or combustion performance, many proprietary additives are sold to benefit the fuel handling system. These may contain solvents or dispersants to combat sludge deposits, emulsifiers or de-emulsifiers for water in the fuel, corrosion inhibitors, and other specific functional ingredients.

6.4.3 Complexities of Coal-Fueled Combustion

Coal characteristics or quality affect nearly every aspect of plant operation including unit capability, plant heat rate, equipment failure rates, and waste disposal requirements. Each of these impacts is either directly or indirectly a function of the combustion process. The most troublesome—the creation of runny, sticky, hardened deposits on furnace or convective surface—relates to the combustion process itself.

The accumulation of these deposits is referred to as slagging or fouling. Slagging refers to the deposition of molten or partially fused ash particles on the walls and suspended surface of the furnace. The deposits are normally confined to the radiant heat transfer surfaces of the steam generator.

Wall blowers and retractable soot blowers keep the heat transfer surfaces sufficiently clean to allow the unit to operate at full load. However, because of variations in the ash deposition rate and ash deposit tenacity, the wall and soot blowers may not be capable of adequately removing ash deposits.

Fouling refers to the deposition of volatile constituents in ash on the convective surface of the steam generator. Certain ash constituents, particularly active sodium, volatilize at relatively low temperatures and then condense on other ash particles and on metal surfaces as heat is absorbed and gas temperatures are lowered. The condensed matter provides a binding matrix for ash particles to fuse and build up on the tube surface. Fouling deposits can become extremely difficult to remove, rapidly plugging convective suspended surface and forcing unit shutdown for removal.

Slagging and fouling can affect net plant heat rate, maximum achievable load, and maintenance and availability.

6.4.3.1 Net Plant Heat Rate. Slagging and fouling can impact net plant heat rate in different ways. Switching to a higher slagging coal increases the resistance to heat transfer of the steam generator walls, necessitates a higher fuel burn rate to achieve the same steaming rate, and therefore decreases boiler efficiency. The increased fuel burn rate causes a higher temperature in the furnace, that can further aggravate slagging. The higher gas volumes and temperatures entering the convective pass usually necessitate an increase in steam attemperation flows that, in turn, increases the net turbine heat rate.

Higher fouling tendency, on the other hand, lessens the ability of the convective surface to absorb heat, causing high economizer gas outlet temperatures, thus reducing boiler efficiency and potentially derating downstream equipment such as fans. Reduced heat absorption can result in lower attemperation spray levels or, in more severe cases, the inability to meet target main or reheat steam temperatures.

With either slagging or fouling problems, increased use of soot blowing steam for deposit removal also increases the turbine heat rate. Slagging and fouling also have a secondary effect on heat rate. The increased fuel burn rate, due to a decreased boiler efficiency, requires additional auxiliary power for air, gas, coal, and ash handling systems. Further increases in auxiliary power can be associated with additional use of compressed air for soot blowing.

6.4.3.2 Maximum Achievable Load. Variations in design philosophies among utilities, consulting engineers, and manufacturers make it difficult to predict maximum achievable load for a particular unit. Some steam generators are conservatively designed and can burn a wide variety of fuels with a broad range of slagging or fouling characteristics. Other units do not have such fuel flexibility and are able to accommodate problematic coal supplies only by lowering the heat input (permanent derate) or modifying steam generator operating parameters such as excess air level, flue gas recirculation rates, and soot blowing frequency.

The most commonly used indicator of a steam generator's susceptibility to slagging problems is heat input per furnace plan area (HIPA). HIPA relates to slagging propensity in two ways. First, a lower HIPA or larger steam generator cross-sectional area decreases the probability for an ash particle to impinge on the waterwall surface and therefore lessens the potential for slagging to cause adverse effects on steam generator operation. Second, additional wall surface will absorb more heat and reduce combustion zone gas temperatures.

Fouling is particularly sensitive to gas temperature and tube spacing. Therefore, if gas temperatures are too high entering a particular tube bank, excessive fouling may result. To control fouling, it is often necessary to increase soot blower cycle length and frequency or, in some cases, to reduce load (derate) to lower the temperature of the gas entering the particular zone. Often, furnace exit gas temperature (FEGT) is used to characterize the fouling susceptibility of a particular unit. Other factors being equal, the lower this temperature, the more likely the unit is to accommodate a more severely fouling coal. Fouling can be aggravated further by slagging conditions; as discussed above, slagging tends to increase the temperature of the gas entering the convective pass.

Lowering load to control deposition problems effectively derates the unit. Whether or not a unit is derated, the burning of an alternative coal likely affects the average thickness and composition of the various ash deposits, changing the heat transfer characteristics and temperature profile of the steam generator. This, in turn, affects many aspects of plant operation, maintenance, and availability.

6.4.3.3 Maintenance/Availability. Boiler maintenance costs and availability are directly affected by combustion characteristics because of several factors. Variations in flame temperature and deposition profiles can significantly alter local gas/metal temperatures and deposit chemistry.

Temperature and chemistry of deposits impact fireside corrosion rates and corresponding maintenance/availability impacts. Similarly, tube erosion from additional wall blower operation to combat slagging can reduce tube life. For suspended surface, higher gas velocities and temperatures lead to higher erosion, corrosion, and overheating of tubes, that ultimately results in increased convective pass tube failures and replacement costs. Availability can also be negatively impacted if unit load must be lowered at night for slag shedding or if the additional tube failures cause more forced outages.

Fouling can cause plugged tube banks, higher localized gas velocities and thus higher erosion rates, increased tube corrosion, and eventual unit shutdown for cleanup.

Secondary effects of increased slagging or fouling on unit maintenance costs and availability are imposed on the unit auxiliary systems. Because of the higher fuel burn rate, ash generation rate, air requirements, etc., nearly all equipment experiences increased usage and wear.

6.4.4 NO_x Formation and Control

Nitrogen monoxide (NO) and nitrogen dioxide (NO_2) are byproducts of the combustion process of fossil fuels. Historically, the quantity of these inorganic compounds in the products of combustion was not sufficient to affect boiler performance, and their presence was largely ignored. In recent years, nitrogen oxides have been linked to the formation of acid rain and smog. Based on such findings, the presence of NO_2 and NO (collectively referred to as NO_x) is now regulated by both state and federal authorities. In 1990, the United States Clean Air Act was amended to require the installation of NO_x combustion control equipment on the bulk of existing coal-fueled units. Hence, NO_x emission rates have become an important consideration in both design of fuel-firing equipment and in the modification of existing coal-fueled units.

The formation of NO_x in the combustion process is related to the source of nitrogen required for the reaction. NO_x formed from the nitrogen in the combustion air is referred to as "thermal NO_x." NO_x derived from the organically bound nitrogen components in all coals and fuel oils is called "fuel NO_x."

6.4.4.1 Thermal NO_x. The mechanisms involving thermal NO_x were first described by Zel'dovich (1946) and later modified to what is referred to as the extended Zel'dovich mechanism.

$$N_2 + O \leftrightarrow NO + N$$

$$N + O_2 \leftrightarrow NO + O$$

$$N + OH \leftrightarrow NO + H$$

As the equilibrium values predicted by this mechanism are higher than those actually measured, it is generally assumed that the $N_2 + O$ relationship is rate determining, due to its high activation energy of 317 kJ/mol (Singer 1991). The rate of conversion is governed by the gas temperature (exponentially), residence time of the gas in high temperature regions, and excess air levels (affecting oxygen availability). Formation of thermal NO_x is also related to combustion kinetics (dissociation rates) and the flow patterns (residence time at temperature) for the specific steam generator in question.

6.4.4.2 Fuel NO_x. Although the kinetics involved in the conversion of organically bound nitrogen compounds found in fossil fuels are not yet well understood, numerous investigators have shown fuel NO_x to be an important mechanism in NO_x formation from fuel oil, and the dominant mechanism in NO_x generated from the combustion of coal (Singer 1991). Bench-scale tests burning fuel oils in a mixture of oxygen and carbon dioxide (to exclude thermal NO_x) have shown a remarkable correlation between the percentage N_2 in the fuel oil versus NO_x (Habelt 1977). The percentage of fuel-nitrogen conversion is not constant, but decreases with increasing fuel nitrogen.

The availability of oxygen for conversion of fuel nitrogen, while in its gaseous state, affects the design of fuel-firing equipment. Compounds that evolve from a coal particle such as NCH and NH_3 are relatively unstable and are reduced to harmless N_2 under fuel-rich conditions, or to NO under air-rich conditions (Singer 1991). The technique used to demonstrate this is staging.

In staging, a portion of the total air required to complete combustion is withheld initially and the balance of the air is mixed with the incomplete products of combustion only after the oxygen content of the first-stage air is consumed. By varying the quantity of air introduced as first-stage air, the combustion of a coal particle or oil droplet can be interrupted at different stages of the reaction because of lack of oxygen, and allowed to proceed further at such time as the balance of air (second-stage air) is introduced. Staging is an important element in the design of low NO_x burners. For NO_x control, the ideal quantity of first-stage air would be that that is sufficient to only generate the temperatures necessary to drive the gaseous state, but insufficient to provide sufficient oxidant to complete a reaction to NO.

6.4.5 Using Computer Models to Predict As-Fired Costs

Even though a boiler is designed for a certain fuel, circumstances influenced by economics or politics can occur that dictate the combustion of fuels other than the design fuel. Combustion of alternate fuels will alter both unit performance and associated economics as follows:

- Loss of unit load capability,
- Increased maintenance,
- Higher auxiliary load,
- Higher heat rate,
- Increased consumables, and
- More waste disposal.

Fuel-related impacts on the cost of electricity tend to increase with the associated increase in the complexity of the fuel source. Thus, the highest risk for combustion and performance problems is associated with the most complex of fuels—coal. To facilitate the evaluation of coal quality effects on power plant performance and costs, utilities and architecture and engineering firms have developed coal quality evaluation computer tools. An example of these computer tools is the Coal Quality Impact Model (CQIM™) developed by the Electric Power Research Institute (EPRI). This model performs a rapid, systematic evaluation of coal quality related impacts and associated costs for each alternative fuel analyzed, on a plant basis. CQIM can therefore be used to illustrate the fundamental performance and cost considerations and modelling techniques.

To determine fuel-related costs, coal quality impacts in the following areas are considered:

- *Equipment capability.* Derates, if any, due to lack of equipment capability affect maximum achievable load and associated day-to-day replacement power costs.
- *Auxiliary power requirements.* Changes in auxiliary power of fuel-related equipment can either be related directly to power sales lost (due to internal use) or incremental fuel consumption required to increase overall gross unit output to compensate for increased auxiliary power.
- *Consumable requirements.* Consumables include fuel, scrubber additive, and waste fixative. Requirements are primarily a function of plant heat rate. Heat rate is dependent on auxiliary power requirements, turbine heat rate, and boiler efficiency, all of that are dependent on coal quality. Consumable requirements affect delivered fuel costs as well as operating costs.
- *Waste generation.* Waste streams include bottom ash, fly ash, and scrubber sludge. These quantities are also related to plant heat rate and impact waste disposal costs and/or ash sales.
- *Long-term maintenance/availability costs.* Maintenance/availability predictions consider failure rates for equipment, lost generation (if any), and the cost to repair.

Determination of overall costs for alternative fuels can offer significantly different outcomes, depending on that cost components are considered. Figure 6-6 illustrates the difference in "cost" of burning alternative coals, on a total cost basis versus simply a delivered cost basis.

6.4.5.1 Equipment Capability. Specific equipment or systems directly affected by coal quality are as follows:

- Coal receiving, preparation, and handling systems;
- Pulverizer train equipment;
- Steam generator;
- Air heater(s);
- Primary air, forced draft, induced draft, and scrubber booster fans, as appropriate;
- Particulate removal systems;
- Fly ash handling system;
- Bottom ash handling system;
- Flue gas desulfurization system, as appropriate; and
- Waste disposal system, as appropriate.

For all of the above, system performance requirements are directly driven by the handling requirements for the coal itself, ash, combustion air, or flue gas. Duty for each system varies as a function of the demand placed on the system by the fuel, as a function of either the air/gas/ash/coal flow rates, properties of coal or ash, or both.

6.4.5.2 Overall Unit Derate Analysis. Changes in system demand or duty required to accommodate an alternative coal

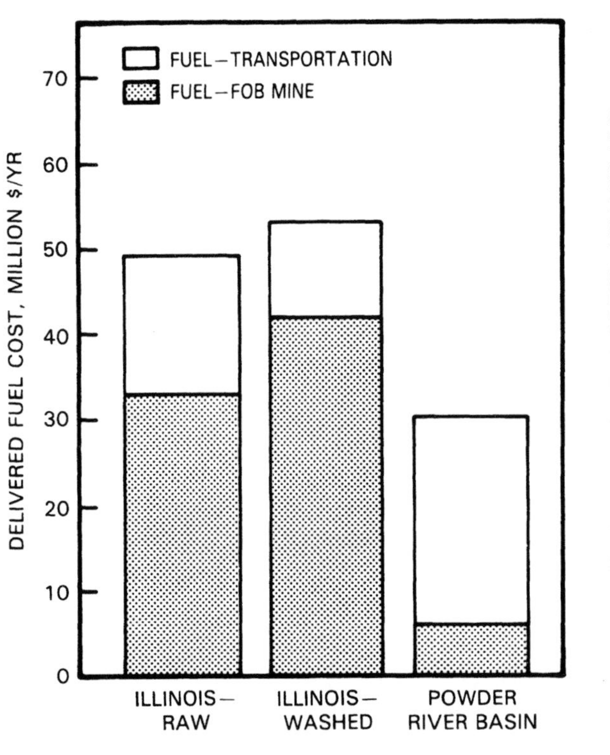

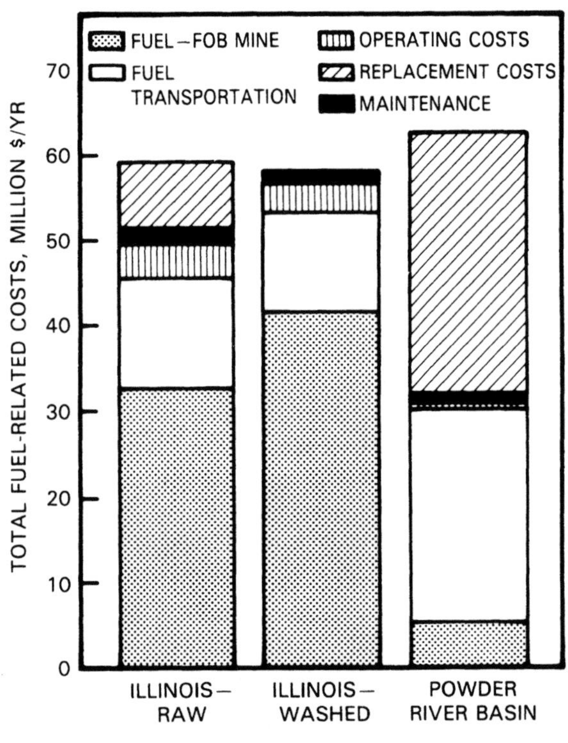

Fig. 6-6. Delivered fuel costs differ from fuel-related costs. (From Galluzzo, N.G., G. S. Stallard, M. E. Boushka, A. K. Mehta. 1987. Model predicts as-fired cost of changing coal sources. Used with permission of *POWER*.)

may exceed current system capacity. If this is the case, unit capability will decrease. The lost generation may be either "replaced" by another power source within the utility system or recovered by installation of new retrofit equipment.

The derates predicted for each system are a function of specific coal quality parameters. Since properties vary for any particular coal, the level of component derate similarly varies depending on day-to-day changes in such properties and their impact on unit capability. Thus, derate levels cannot be calculated merely considering typical properties but must be based on alternative, more complex approaches that consider variability as a function of time. The CQIM employs a Monte Carlo treatment to simulate the expected variability of the coal supply, thus constructing a probability distribution of expected system and unit capability, as illustrated in Fig. 6-7.

6.4.5.3 Auxiliary Power Calculations. Changes in auxiliary power requirements are calculated for each fuel evaluated. These changes are used to determine the amount of lost generation (replacement power costs) assuming that gross power output remains fixed. Alternatively, the effect of changes in auxiliary power requirements can be related to the additional fuel burn rate requirements (operating costs), assuming that net unit power output remains fixed.

6.4.5.4 Consumable Requirements and Waste Generation. The fuel burn rate, scrubber additive requirements, ash and scrubber solids generation rates, and fixative requirements, as appropriate, are evaluated for each fuel. Fuel burn rates reflect calculated boiler efficiencies, turbine heat rate adjustments, and auxiliary power requirements, and ultimately,

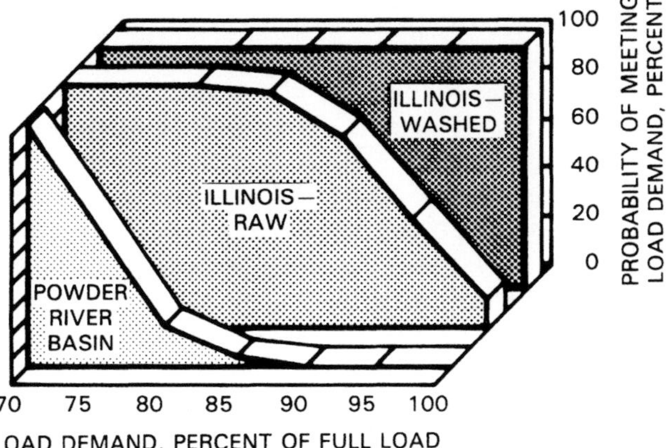

Fig. 6-7. Probability distribution of expected system and unit capability as a function of coal supply. (From Galluzzo, N.G., G. S. Stallard, M. E. Boushka, A. K. Mehta. 1987. Model predicts as-fired cost of changing coal sources. Used with permission of *POWER*.)

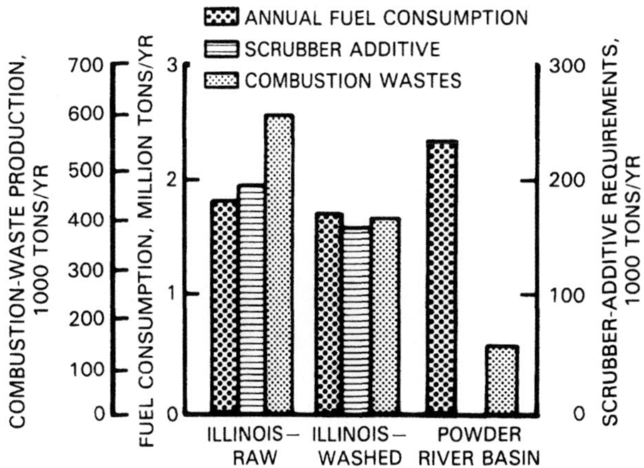

Fig. 6-8. Consumable requirements by coal supply. (From Galluzzo, Model predicts as-fired cost of changing coal sources. Used with permission of *POWER*.)

overall unit heat rate impacts. Consumable requirements can vary significantly by coal supply, as illustrated in Fig. 6-8.

6.4.5.5 Long-Term Maintenance/Availability Costs. Maintenance and availability impacts are a function of changes in equipment duty or parts life. Equipment failures tend to increase, with increased run time or change in abrasive characteristics of the fluid. The CQIM employs a detailed maintenance/availability treatment that models each of these factors as follows:

1. The generating unit is subdivided into multiple systems, each system subdivided into subsystems, and each subsystem divided into components.
2. Statistical data are provided within the CQIM for each component. These data, that were developed under the CQIM research project (EPRI RP2256), include mean times between failure (MTBF); a determination of whether the failure leads to a forced outage or can be repaired without causing a forced outage; and the time, man-hours, and material costs to repair. The MTBF typically are functions of coal quality and therefore vary by coal. Data are included for some components where the number of spare components, such as mills and scrubber modules, may vary by coal.
3. To obtain annual maintenance costs, the number of annual component failures—whether or not those failures cause forced outages—is calculated for each component for each coal and multiplied by the total cost to repair each failure. The total cost to repair each failure is the sum of material costs to repair and the labor costs for the man-hours required per failure.
4. To obtain unavailability costs, the annual forced outage rates for the various components are considered by a UNIRAM[2] model. The UNIRAM model calculates the probability of meeting load as a function of load. As with the Monte Carlo derate analysis, the probability of meeting load

[2]The UNIRAM methodology for availability analysis was developed by ARINC Research Corporation for EPRI. This method involves the partition of a power plant into independent systems, definition of plant states (percent capacity) if various systems are unavailable, development of subsystem fault trees, calculation of probabilities of subsystem group states, and evaluation of overall unit availability.

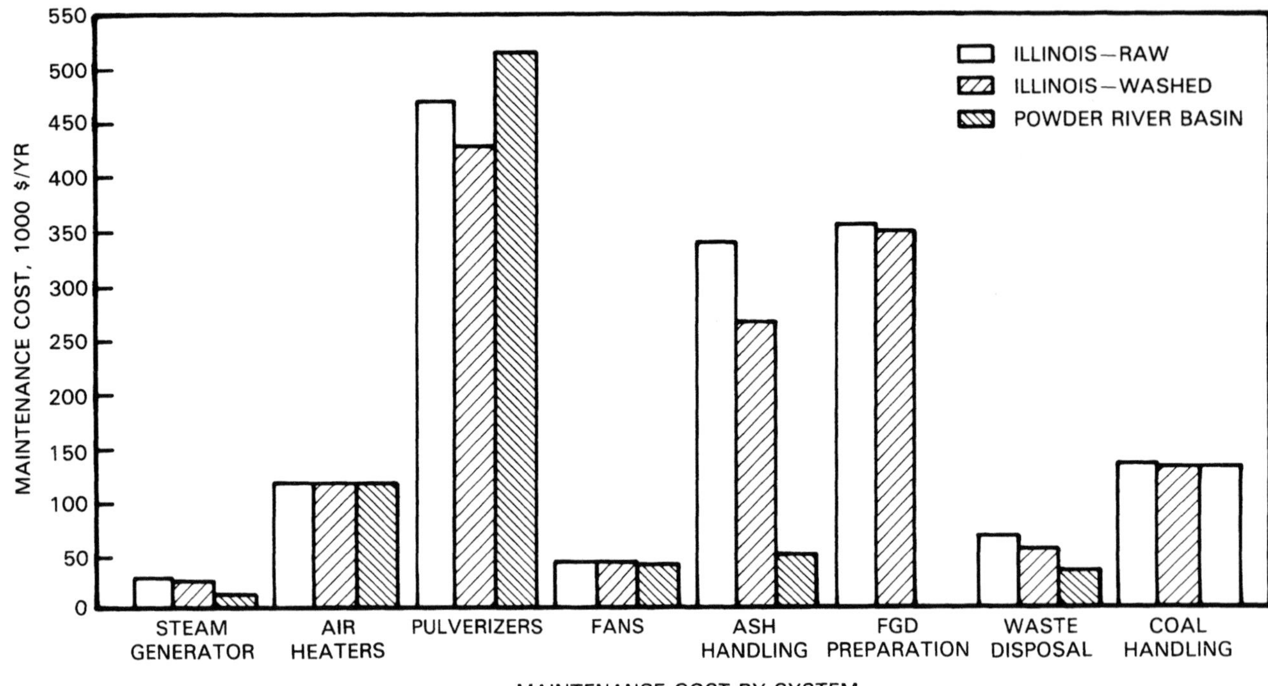

Fig. 6-9. Relative maintenance costs among alternate coals. (From Galluzzo, N.G., G. S. Stallard, M. E. Boushka, A. K. Mehta. 1987. Model predicts as-fired cost of changing coal sources. Used with permission of *POWER*.)

for each coal is multiplied by the unit load demand (hours per year as a function of load) to obtain the expected annual generation in MWh/yr for each coal. The calculated differential annual generation is multiplied by an input replacement power cost to yield the differential unavailability cost for each fuel.

An example of relative system maintenance costs is illustrated by Fig. 6-9.

6.4.5.6 Overall Fuel-Related Costs. The bottom line of a coal quality computer model is to determine the overall fuel-related costs associated with alternative fuel supplies.

The fuel-related costs include the following:

- FOB mine fuel costs;
- Fuel transportation costs for rail, barge, ship, truck, or other transportation methods;
- Operating and maintenance costs for waste disposal, scrubber fixative, scrubber additive, railcar maintenance, and differential maintenance costs discussed earlier;
- Replacement power costs for derates, differential availability, and differential auxiliary power, as appropriate;
- Capital costs for equipment modifications and railcar ownership, as appropriate;
- Other costs, specific to the station in question, including staffing, parts inventory, etc.

Only by considering overall fuel-related costs can fuels of differing quality be satisfactorily compared on an economic basis; such evaluations can play an important role in utility planning and operation.

6.5 REFERENCES

ASME. Performance Test Code (PTC) 4. Steam Generating Units. American Society of Mechanical Engineers.

BARRETT, R. E., R. C. TUCKFIELD, and R. E. THOMAS OF BATTELLE LABORATORIES DIVISION. 1987. *Slagging and Fouling in Pulverized Coal Fired Boilers, Volume 1: A Survey and Analysis of Utility Data*, EPRI CS-5523. Electrical Power Research Institute, Palo Alto, CA.

ELLIOTT, T. C. 1989. *Standard Handbook of Powerplant Engineering*. McGraw-Hill, New York, NY.

EPRI. 1988. *Residual Fuel Oil User's Guidebook, Volume 2*. Residual oil-fired boilers. Electric Power Research Institute Report AP-5826, Palo Alto, CA.

GALLUZZO, N. G., G. S. STALLARD, M. E. BOUSHKA, and A. K. MEHTA. June 1987. Model predicts as-fired cost of changing coal sources. *POWER*. New York, NY **131**: pp. 53–56.

HABELT, W. W. 1977. The influence of coal oxygen to coal nitrogen ratio on NO_x formation. Paper presented at the 70th Annual AiChE Meeting. 13–17 November 1977. New York. Published as Combustion Engineering Publication TIS-5140. Windsor, Connecticut.

PERRY, ROBERT H. and C. CHILTON. 1973. *Chemical Engineers' Handbook*, 5th Edit. McGraw-Hill, New York, NY.

SINGER, JOSEPH G., editor. 1991. *Combustion*. Combustion Engineering, Inc. Windsor, CT. p. 4-3, 4-6.

WEISSMAN, JOEL and L. E. ECKART. 1985. *Modern Power Plant Engineering*. Prentice-Hall, Englewood Cliffs, NJ. p. 112.

ZEL'DOVICH, YA. B. 1946. The oxidation of nitrogen in combustion and explosions. *Acta Physicochim. U.S.S.R.* **21**:577–628 (C–E Combustion).

7

STEAM GENERATORS

Benjamin W. Jackson

7.1 INTRODUCTION

In essence, a steam generator is a machine that safely, reliably, and efficiently transfers heat released during the combustion of fuel to both feedwater and steam.

The development of steam generators has been a progression of design that, within the constraints of available materials, construction methods, and economics, has increased the area for transfer of heat from the combustion process to the cycle working fluid. The increase in heat transfer surface has been accomplished not only by increases in the size of the steam generator, but also by changing the arrangement and configuration of the heat transfer surface.

The earliest steam generators consisted of a water-filled vessel heated by fire. The heat transfer surface of this arrangement was increased by placing tubes through the vessel and passing hot flue gas from the combustion process through the tubes. As this firetube design reached size limitations, the watertube design was developed. The watertube design is based on water and/or steam within the heat transfer tubes and the combustion process and flue gas outside the tubes. The watertube design is the basic arrangement currently used in large-utility steam generators.

A second steam generator design progression has occurred in the evolution of the combustion process. Efficient combustion is based on thoroughly mixing fuel with air to achieve complete burnout or utilization of the fuel heat content. In the early boilers which used solid fuels, fuel–air mixing was achieved by placing the fuel on a stationary grate and passing air through and above the grate. As grate technology improved, the stationary grate was modified into a traveling grate that provided continuous removal of solid combustion products and other unburned material from the combustion zone.

As traveling grate stoker designs reached size limitations, a replacement process was needed that would allow larger steam generators and the firing of coal in suspension in a manner similar to the firing of oil and gas.

Mixing air with liquid or gaseous fuels is somewhat easier to accomplish than with solid fuels; gas can be directed into the combustion chamber through burner nozzles, mixed with air at the front of the burner, and fired. Similarly, oil can be sprayed into the combustion chamber by mechanical, steam, or compressed air atomization processes which create small oil droplets and increase its surface area. The atomized oil mixes readily with air at the burner prior to combustion.

Pulverization of coal, compared to unprocessed coal, yields a fuel with a much larger surface area for combustion. The small particle size and increased surface area allows coal to be transported from the pulverizer, distributed into the furnace, and fired in suspension in an arrangement similar to gas and oil furnaces.

During the period 1918 to 1920 at the Oneida Street Station, owned by the Milwaukee Electric Railway and Light Company (Singer 1991), suspension firing of pulverized coal was developed and accomplished for the first time in utility service.

The most recent evolution of steam generator design involves the improvement of cycle efficiency through incorporation of a reheat steam system and increases in steam temperature and pressure.

Early steam generators produced saturated steam. To increase the cycle efficiency, provisions for increasing the steam temperature above saturation temperature were added. Superheating of the steam was accomplished through the addition of a heat transfer surface downstream and separate from the evaporation section of the boiler.

Reheating the exhaust from the high-pressure turbine at pressure levels typical of current power station design produces approximately a 5% improvement in cycle efficiency. Further improvement of the reheat steam cycle was achieved through the addition of a second reheat system (double reheat) in conjunction with supercritical main steam pressure.

Current cycle development efforts focus on designs for main steam pressures of 4,500 psi and higher and steam temperatures of 1,050° F, 1,100° F, and above.

This chapter focuses on suspension firing of pulverized coal in a reheat steam generator, which continues to be the principal process for release and recovery of heat from coal fuels for the majority of large utility installations in the United States.

The function of a steam generator is to provide controlled release of heat in the fuel and efficient transfer of heat to the feedwater and steam. The transfer of heat produces main steam at the pressure and temperature required by the high-pressure turbine. Heat is also transferred through the reheater to increase the temperature of the high-pressure turbine exhaust, or cold reheat steam, to the conditions required by the intermediate-pressure turbine.

The steam generator receives fuel from the coal silos, combustion air from the forced draft fans and primary air fans, feedwater from the boiler feed pump, and cold reheat steam from the high-pressure turbine exhaust. The boiler uses these to produce steam at the required operating conditions for main steam flow to the high-pressure turbine and for hot reheat steam flow to the intermediate-pressure turbine.

The combustion process also produces flue gas (including fly ash), which is routed through pollution control equipment to the chimney, as well as bottom ash and boiler hopper ash, which are removed by ash handling equipment.

Any discussion of steam generators relies on an understanding of the principal components of the steam generator, which is also frequently referred to as the boiler. Throughout this chapter, the terms appearing in the glossary at the end of the chapter are used in descriptions of the component function. Selected components are illustrated in Figs. 7-1, 7-2, and 7-3.

7.2 STEAM GENERATOR SYSTEM DESIGN

This section discusses design criteria and considerations for the Steam Generator System. The component design considerations are presented later in this chapter. This distinction facilitates the discussion since the system design criteria affect the overall system and typically affect two or more of the steam generator components. The system design categories are as follows:

- Turbine cycle heat balance,
- Fuel characteristics,
- Pulverizer system,
- Excess air,
- Steam temperature control methods,
- Minimum load capability,
- Variable pressure operation,
- Flue gas emissions,

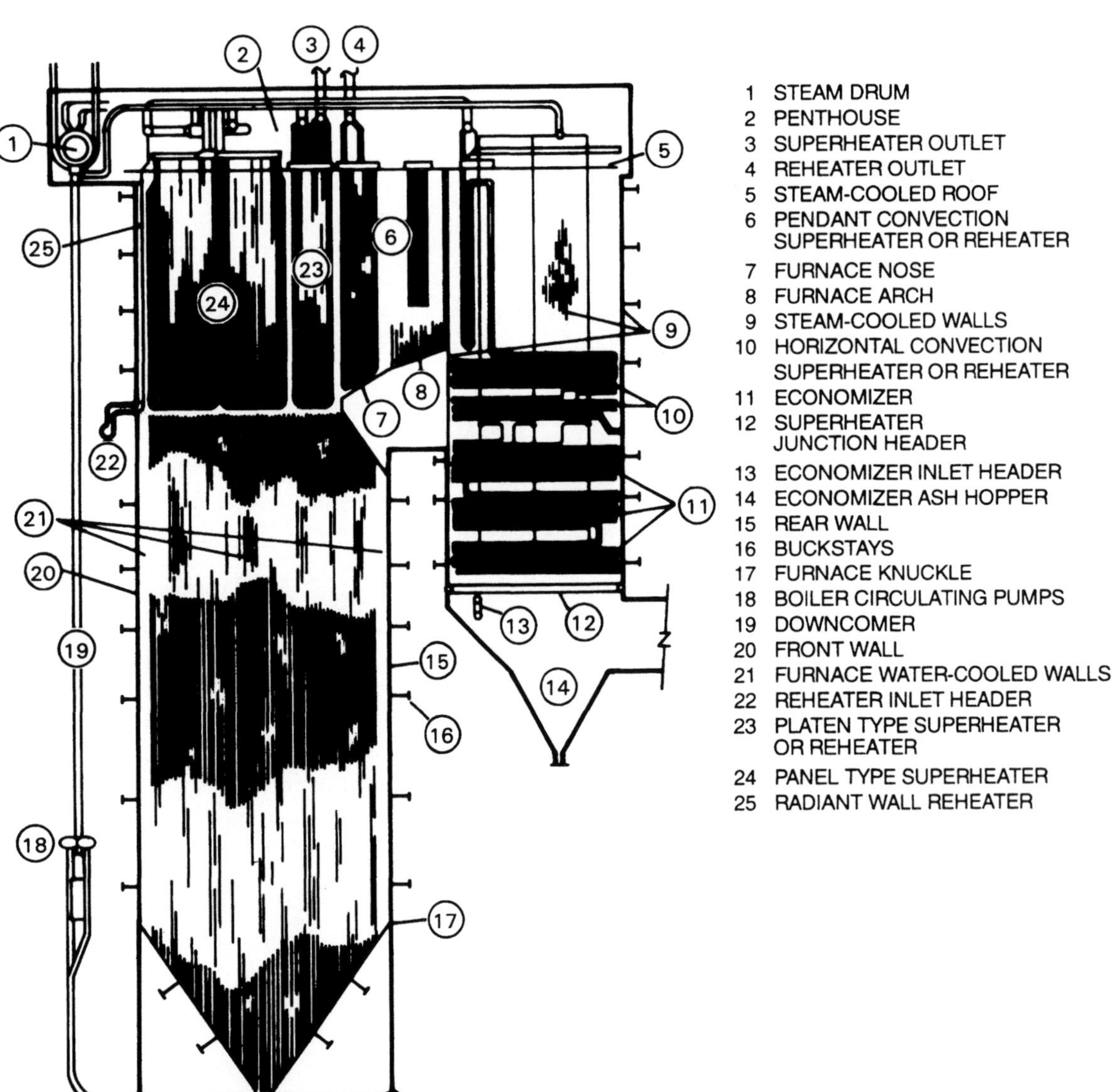

Fig. 7-1. Boiler nomenclature.

- Feedwater quality,
- Steam purity,
- Steam generator arrangement, and
- Startup systems.

The discussion in this chapter is based on subcritical drum type boilers. However, the fireside design criteria and the majority of the boiler components are common to both subcritical and supercritical units. Additional information on steam generator systems and components can be found in two excellent references: *Combustion Fossil Power Systems* (Singer 1991) and *Steam: Its Generation and Use* (Stultz and Kitto 1992).

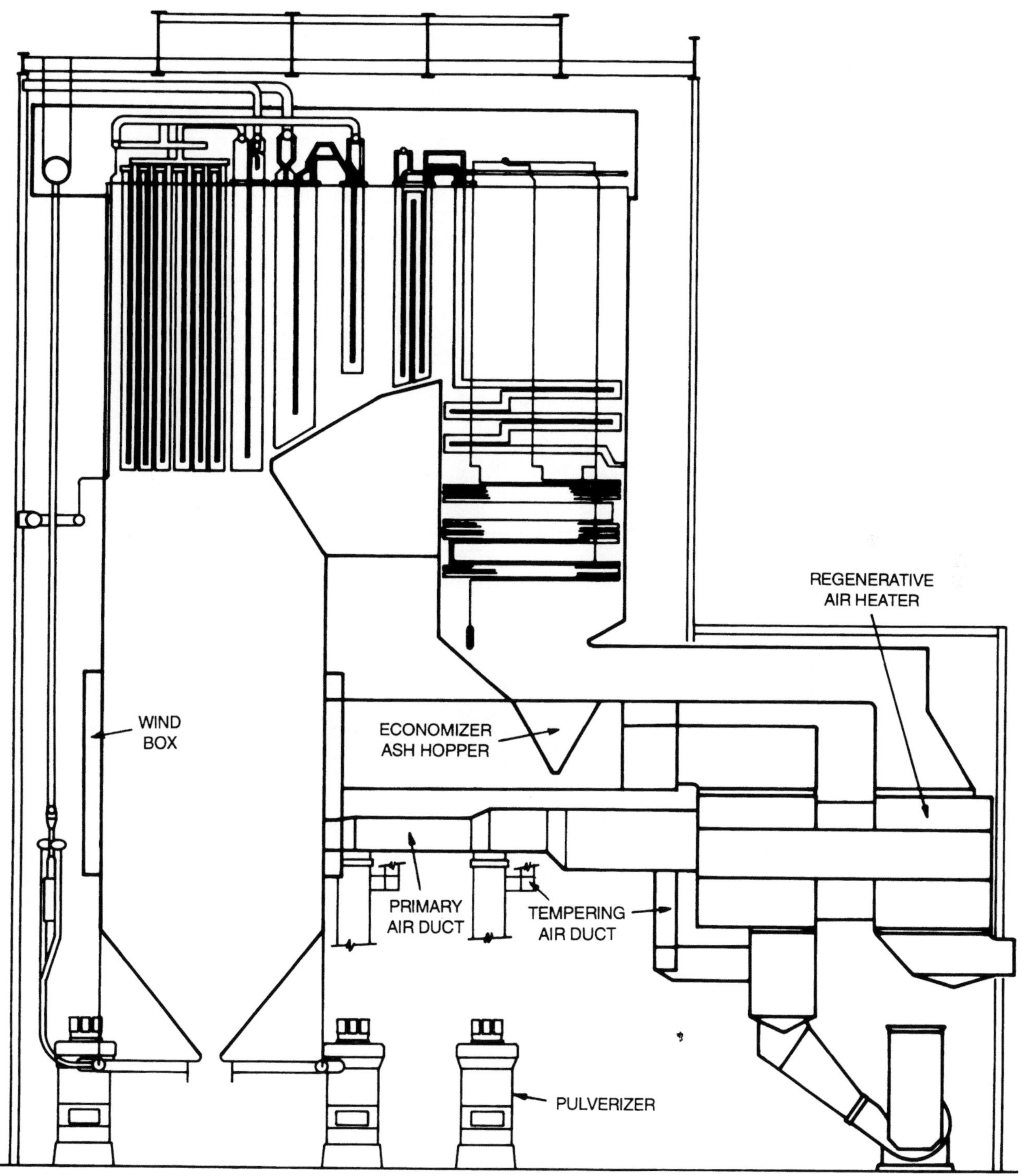

Fig. 7-2. Typical corner fired boiler cross-section with series convection pass. (From Salt River Project. Used with permission.)

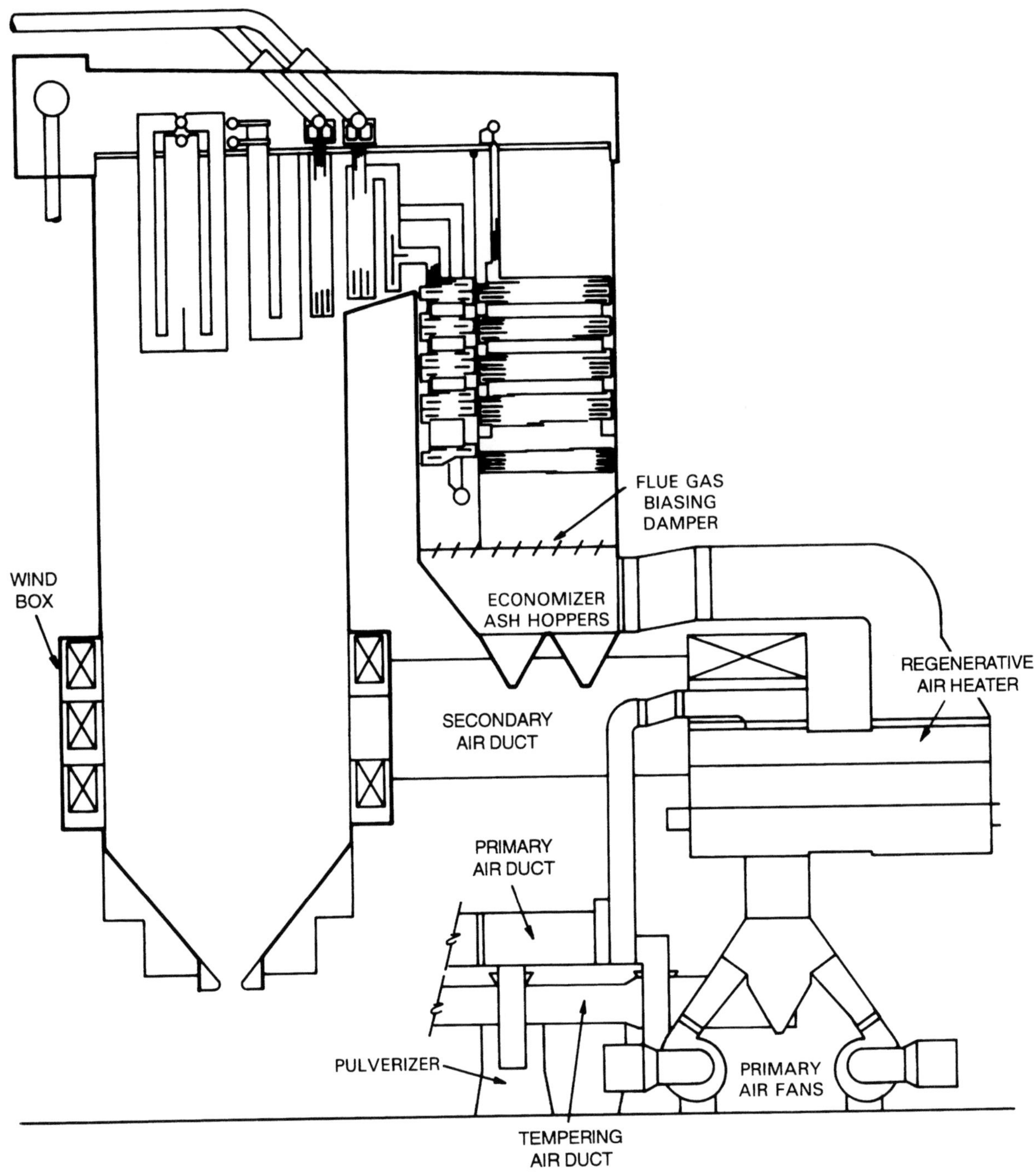

Fig. 7-3. Typical wall fired boiler cross-section with divided convection pass. (From Babcock & Wilcox. Used with permission.)

7.2.1 Turbine Cycle Heat Balance

The steam production requirements of a steam generator used in conjunction with a turbine for electrical power generation are based on a turbine cycle heat balance. Turbine cycle heat balances are discussed in Chapter 13, Plant Heat Balances. The following discussion identifies the relationship of the turbine heat balance to the steam generator design.

The heat balance defines steam and feedwater boundary conditions which are basic requirements for the steam generator performance. These boundary conditions include the following:

- Main steam flow, pressure and temperature;
- Feedwater flow and temperature;
- Hot reheat steam flow, pressure and temperature;
- Cold reheat steam flow, pressure and temperature (or enthalpy); and
- Desuperheating spray water temperature—superheat and reheat.

The operation of a given turbine is defined by a set of six or more heat balances. The turbine guarantee conditions are defined by a guarantee or 100% load heat balance which is based on rated throttle pressure, typically 2,400 psig (16.5 MPa) or 1,800 psig (12.4 MPa). A heat balance is also supplied for a valves-wide-open, 5% overpressure condition of 2,520 psig (17.4 MPa) or 1,890 psig (13.0 MPa) throttle pressure. If operation with a high pressure feedwater heater out of service is anticipated, a heat balance must be provided to define those conditions. The turbine performance represented by the "overpressure" heat balance is not guaranteed by the turbine manufacturer, although it typically is achieved. However, it does represent the normal maximum heat duty required of the boiler, and typically the boiler guaranteed maximum continuous rating (MCR) is based on the flows and conditions defined on the valves-wide-open, 5% overpressure turbine heat balance.

An adaptation of the heat balance information must be applied regarding the main steam and reheat steam conditions. The turbine heat balance shows throttle pressure, temperature, and reheat steam conditions at the turbine. However, the boiler manufacturer needs to know steam conditions at the respective boundaries (e.g., the superheater outlet, reheater inlet, and reheater outlet). The following approximations typically are used for main steam:

	1,800 psig (12.4 MPa) Turbine cycle	2,400 psig (16.5 MPa) Turbine cycle
Rated throttle pressure, psig	1,800 (12.4 MPa)	2,400 (16.5 MPa)
Overpressure throttle pressure, psig	1,890 (13.0 MPa)	2,520 (17.4 MPa)
Main steam pressure drop, psi	100 (0.7 MPa)	120 (0.8 MPa)
Superheater outlet pressure, psig	1,990 (13.7 MPa)	2,640 (18.2 MPa)

The turbine generator heat balance is typically based on a 10% pressure loss between the high-pressure turbine outlet (cold reheat piping inlet) and the intermediate-pressure turbine inlet (hot reheat piping outlet). That is, 10% of the high-pressure turbine exhaust pressure is allowed for the pressure drop in the cold reheat piping, reheater, and hot reheat piping. Once the reheater pressure drop is determined, as discussed later in this chapter, the remaining allowable pressure drop is allocated between the hot and cold reheat piping. Thus, by allowing for the pressure losses in the reheat piping, the designer can calculate the pressures at the respective steam generator terminals from the heat balance values.

Steam temperatures indicated on the turbine heat balance also require adaptation. In the main steam piping, the steam undergoes an essentially constant enthalpy pressure loss between the superheater outlet and the turbine throttle. The resulting temperature loss is generally less than 5° F (2.8° C) and is approximated by specifying the superheater outlet temperature 5° F (2.8° C) higher than the turbine throttle temperature. Based on a design turbine throttle temperature of 1,000° F (538° C), this yields a superheater outlet temperature of 1,005° F (540.6° C). The same type of correction is applied to the reheat steam temperature to obtain a reheater outlet steam temperature of 1,005° F (540.6° C).

The feedwater conditions indicated on the turbine heat balance do not require modification since the change in conditions is negligible between the outlet of the last feedwater heater and the inlet of the economizer.

An additional modification of heat balances may be required in the definition of steam generator part load conditions. The turbine generator part load heat balances are based on percentages of the guaranteed turbine load, which may not be the same as the guaranteed steam generator load. The key steam generator part load conditions—the control point or the lowest load at which the steam generator is designed to maintain the rated main and reheat steam temperature, and the minimum load while firing coal only—typically are described as round number percentages of MCR. Typically, obtaining the values for the heat balance corresponding to a round number percentage of MCR requires interpolation between two of the turbine heat balances. The factors influencing the selection of the control point and the minimum stable coal load are discussed later in this chapter.

7.2.2 Fuel Characteristics

In addition to the turbine heat balance, the design fuel analysis is an integral part of the basic information contained in the steam generator design and specification. The fuel analysis for the design coal should include at least the proximate analysis, ultimate analysis, ash analysis, Hardgrove Grindability Index, and oxidizing and reducing atmosphere ash fusion temperatures.

Both typical and range values should be provided for all fuel analysis values. Typical United States fuels are discussed in Chapter 4, Fossil Fuels.

A single coal must be selected as the "performance coal" or the design fuel. The design coal is the basis for determining heat transfer surface allocation, and ideally, is the coal fired during performance testing. Unless the steam generator is a minemouth unit with a dedicated fuel source sufficient for the life of the boiler, the ability to fire coals other than the design coal is desirable. However, designing the steam generator for a wide range of fuel flexibility degrades the performance of the unit. A steam generator design that has maximum fuel flexibility evidenced by a wide range of individual fuel properties compromises performance when some of the individual coals within the design fuel range are fired. Alternatively, a steam generator designed for minimal fuel flexibility and that fires the design fuel is more likely to achieve optimum boiler efficiency and steam temperature control.

The typical performance deficiencies from firing off design (not the performance fuel, but among the specified fuels) or nondesign fuels (not among the specified fuels) are failure to achieve superheat and/or reheat steam temperature, or excessive superheat and/or reheat desuperheating spray flow. In extreme cases, main steam flow may be deficient. These performance deficiencies result from a fuel-induced departure from the design balance between radiant heat transfer in the furnace and convective heat transfer. The most common fuel related causes of performance deficiencies are shown in Table 7-1.

Operating remedies for deficient steam temperature or excessive desuperheating spray can be applied, such as increasing excess air or selective soot blowing, respectively. However, increasing excess air, for example, increases the steam generator stack loss and reduces boiler efficiency. Operators must determine an optimum balance between reduced steam temperature and increased excess air.

Typical flue gas quantities and moisture content for several United States coals are shown in Table 7-2. Changes in flue gas quantity will change the heat transfer to convection pass surfaces. Likewise, changes in flue gas moisture content will change the radiant and convection heat transfer. The differences in flue gas quantity and moisture content among the fuels indicate why differing quantities and proportions of furnace and convection heat transfer surface are required for different coals.

The following guidelines are suggested for the specification of design fuels:

- One fuel must be designated as the design fuel, upon which performance guarantees will be based.
- The design fuel should be the fuel that is actually fired the majority of the time. This practice is not always possible, given the complexities of coal procurement, but it offers the best chance for achievement of long-term design boiler performance.
- If multiple fuels must be specified, complete individual analyses of each fuel must be provided.
- Except for the design of the pulverizer system, designers should avoid inventing a composite design fuel from several likely fuel analyses. Since the resulting design fuel does not exist, the best opportunity for achieving optimum boiler performance does not exist either.

Table 7-2. Typical Combustion Product Quantities

	Combustion Products			
	Flue Gas[1]		H_2O in Flue Gas	
	lb/MBtu	g/MJ	lb/MBtu	g/MJ
Eastern Bituminous	1,065	458	57	24.5
Western Subbituminous	1,019	439	87	37.4
Northern Plains lignite or Texas lignite	1,030	443	95	40.9

[1]Inclues normal operating excess air.

7.2.3 Pulverizer System

Five primary interrelated aspects must be considered in the specification of the pulverizer system design: number of spare pulverizers, pulverizer design coal, unit turndown, product fineness, and wear allowance. Pulverizers are discussed in more detail in Section 7.3.11 and are illustrated in Figs. 7-20 through 7-23.

7.2.3.1 Number of Spare Pulverizers. United States design practice has generally been to provide one installed spare pulverizer based on firing the steam generator design coal. Large units of the 700-megawatt (MW) class may even have two installed spare pulverizers. This differs from recent European design in which 900-MW units have been built with no installed spare pulverizers. The difference in design practice is due to differences in the value assigned to the capital cost of additional pulverizers and the forced outage and availability of the resulting design.

Table 7-1. Fuel Effects on Convection Pass Performance

Fuel-Related Cause of Performance Degradation	Result	Effect on Performance
Radiant heat transfer is adequate to evaporate the design steam quantity but less flue gas is produced.	Less convective heat transfer to superheater and reheater	Superheat and/or reheat steam temperatures less than design
Radiant heat transfer is adequate to evaporate the design steam quantity, but fuel moisture content is less than design and flue gas moisture content is reduced.	Less convective heat transfer to superheater and reheater	Superheat and/or reheat steam temperatures less than design
Radiant heat transfer is adequate to evaporate the design steam quantity, but flue gas quantity is higher than design.	Increased convective heat transfer to superheater and reheater	Superheat and/or reheat attemperating spray flows too high
Radiant heat transfer is adequate to evaporate the design steam quantity, but the fuel moisture content is higher than design and flue gas moisture content is increased.	Increased convective heat transfer to superheater and reheater	Superheat and/or reheat attemperating spray flows too high
Radiant heat transfer is insufficient to evaporate the design steam quantity, typically because of excessive slag accumulation on the furnace walls and/or entrance to the convection pass. Furnace exit gas temperature is high due to reduced heat transfer in the furnace.	Increased convective heat tansfer to superheater and reheater	Superheat and/or reheat attemperating spray flows are high; however, the main problem is a unit derating caused by excessive slagging and inability to generate design steam flow.

*7.2.3.2 **Pulverizer Design Coal.*** As discussed in the section on design fuel, fuel properties vary. The pulverizer system should therefore be designed to accommodate the fuel with the worst combination of properties that still allow the steam generator to achieve the design steam flow. Three fuel properties affect pulverizer fuel processing capacity: moisture, heating value, and Hardgrove Grindability Index. The fuel rank determines the pulverized fuel fineness requirement, which in turn affects the fuel processing capacity of a particular pulverizer. The following tabulation lists the effects of fuel properties on pulverizer capacity:

Change in fuel property or product fineness	Resulting change in pulverizer fuel processing capacity
Moisture content increase	Decrease
Heating value increase	Increase[1]
Hardgrove Grindability Index increase	Increase
Pulverizer product fineness increase	Decrease

The worst-case fuel properties, although they may not occur simultaneously, are used to define the pulverizer design coal. The pulverizer design coal is therefore specified as follows:

Fuel moisture	Highest range value
Fuel heating value	Lowest range value
Hardgrove Grindability Index	Lowest range value
Pulverizer product fineness	Corresponding to fuel rank

The range values used for pulverizer design fuel selection should be the range values of the steam generator design fuel. If additional fuels are included in the specification, the pulverizer system capacity should be checked with these coals as well. However, to avoid oversizing the pulverizer system, it is not recommended that the worst characteristics of the entire fuel list be used in defining the pulverizer design coal.

The application of two pulverizer selection criteria—one spare pulverizer with typical coal, and all pulverizers in service with the pulverizer design coal—may result in two different answers for the number of pulverizers. This may occur if the extreme values of the properties that affect pulverizer capacity (moisture, Hardgrove Grindability Index, fuel heating value) cause a significant pulverizer derating when the pulverizer design coal is fired.

For example, the steam generator design coal may require four pulverizers in service at full load. With the addition of one spare, the system then requires five pulverizers. However, use of the worst or pulverizer design coal properties may yield a coal that requires six pulverizers to achieve full load. In this case, the range values used with the steam generator design coal may be too wide, and thus indicate an unrealistic pulverizer design coal. However, if the range values are determined to be realistic, a system with two spare pulverizers (when firing the normal steam generator design coal) may be the best design solution.

*7.2.3.3 **Unit Turndown.*** The design of the pulverizer system determines the turndown capability of the steam generator. The minimum stable load for an individual pulverizer firing coal is about 50% of the pulverizer rated capacity. Normal utility minimum load operating practice is to operate at least two pulverizers. Thus, the minimum steam generator load when firing coal without supporting fuel is equal to the full capacity of one pulverizer. A minimum of two pulverizers in operation is preferred, since a loss of one of two pulverizers will not require tripping the steam generator because of loss of fuel and loss of flame.

The selection of the pulverizer size, therefore, determines the steam generator turndown. A pulverizer system composed of a few large pulverizers results in poorer turndown (higher minimum load) than a system composed of a large number of smaller pulverizers. The capital cost, however, typically is less for a system with a few large pulverizers than for a system composed of a larger number of small pulverizers. See Section 7.2.6 for additional discussion of unit minimum load capability.

An additional consideration in the selection of pulverizer number and size is the number of pulverizers of the same model already in operation within the owner's system. The use of the same pulverizer model number at other units or stations within the system offers the potential for sharing spare parts, particularly during unplanned pulverizer outages.

The total number of a particular pulverizer model sold and operating should also be considered. This aspect is especially important if the pulverizer selection is a new model with relatively few sold and operating. The improved performance and improved maintenance features of a new pulverizer model should be evaluated as part of the risk assessment of purchasing a new pulverizer model.

*7.2.3.4 **Pulverizer Product Fineness.*** Fineness of the pulverized fuel is one of the pulverizer sizing criteria. Fineness is expressed as the proportions of the pulverized product that will pass through a fine screen, 200-mesh [0.0029-in. (0.074-mm) diameter screen openings] and a coarse screen, 50-mesh [0.0117-in. (0.297-mm) diameter screen openings]. The typical fineness criteria for the most common coal ranks are as follows:

	Percentage passing 200-mesh	Percentage passing 50-mesh
Subbituminous and lignite	65	98
Bituminous	70	98

The fineness requirement for subbituminous coal and lignite is less stringent due to the higher reactivity of the lower rank fuels.

The use of low-NO_x burners is likely to result in more stringent fineness criteria. Current pulverizers can improve the coarse mesh fineness from 98% to 99.7%. The reduction

[1]Fuel processing quantity does not change but heat content delivered to the burners does increase.

of coarse material in low-NO_x burner applications is expected to reduce the unburned carbon content of the fly ash and improve burner performance.

7.2.3.5 Pulverizer Wear Allowance. A final factor affecting pulverizer system design is a capacity margin to account for loss of grinding capacity as a result of wear between overhauls. A 10% capacity loss due to wear is a typical mill sizing criterion. Some pulverizer designs do not lose grinding capacity between overhauls, but do consume more power as the grinding surfaces wear. For these designs, the increase in power required should be considered in the sizing criteria for the pulverizer drive motor.

7.2.4 Excess Air

Excess air is quantified by measurement of the oxygen content of the flue gas. To calculate excess air, the volume percentage of oxygen on a dry gas basis is entered in the following formula, which is given in the ASME Power Test Code 4.1 Steam Generators (ASME 1974).

$$\text{Percent excess air} = \frac{100 \times (O_2 - CO/2)}{0.2682 \times N_2 - (O_2 - CO/2)} \qquad (7\text{-}1)$$

where

N_2 = nitrogen volume percentage,

O_2 = oxygen volume percentage, and

CO = carbon monoxide volume percentage.

If the nitrogen content of the flue gas was not measured or not reported, it can be approximated with a value of 81% with little loss of accuracy. If carbon monoxide was not measured, it can be approximated as zero in the typical pulverized coal combustion case.

The oxygen content of the flue gas is usually measured at the economizer outlet for several reasons. The flue gas temperature is lower there than at other positions in the convection pass nearer the furnace. Also, significant air in-leakage in the flue gas stream occurs just downstream in the rotary regenerative air heater. Air heater air leakage increases the proportion of oxygen in the flue gas, and, therefore, makes the concentration of oxygen at the regenerative air heater flue gas outlet a poor indicator of the furnace flue gas oxygen concentration and the corresponding furnace excess air level.

Two different excess air levels are addressed in a specification: the operating excess air, and the equipment excess air capability. The operating excess air is the excess air level used for performance calculations, and is a function of the coal rank as follows: subbituminous coal and lignite, 20% operating excess air; bituminous coal, 25% operating excess air. The excess air percentages are nominal values, and individual units may operate in a band defined by the nominal value ± 3%.

The difference in excess air level between the low rank coals and bituminous coal is based on furnace slag control. The formation of molten ash or slag is dependent largely on the ash fusion temperature. An ash fusion temperature of 2,500° F (1,371 °C) or higher indicates a coal that is less likely to form slag deposits. Conversely, ash fusion temperatures of 2,100° F (1,149° C) or lower are more likely to form slag.

Coal ash fusion temperatures are measured and reported in both oxidizing (oxygen-rich) and reducing (oxygen-poor) atmospheres. Oxidizing atmosphere ash fusion temperatures are higher than reducing atmosphere ash fusion temperatures. Eastern United States bituminous coal ash is often characterized by iron oxide content of 15% to 25%, which in turn significantly spreads the oxidizing and reducing atmosphere ash fusion temperatures. The difference may be 300 ° F (167° C) or higher and can affect the slagging tendency of the coal. Higher excess air is, therefore, specified to ensure that the entire combustion zone is oxygen rich (oxidizing atmosphere). The ash in the furnace would then exhibit the higher oxidizing atmosphere ash fusion temperature and reduced slagging tendency.

Western subbituminous coals and lignites typically exhibit low iron content and a reduced differential ash fusion temperature between oxidizing and reducing atmosphere conditions. The increased reactivity, characterized by higher volatile matter and oxygen content of the lower rank coals, also favors a reduced excess air requirement.

The equipment excess air capability is specified to set a maximum capability for the entire combustion air and flue gas system. The maximum flow capability is typically set at 30% excess air. The 30% excess air requirement requires that burners function properly with as much as 30% excess air, that normal operating ductwork pressure losses are not excessive, and that the ductwork is not susceptible to flow-induced vibration at the increased flow rate.

7.2.5 Steam Temperature Control Methods and Control Range

One of the basic functions of a steam generator is to produce steam at design temperature. However, as the firing rate and steam flow decrease from design full load conditions, the main and reheat steam temperatures decrease. Reduced steam temperature reduces turbine cycle efficiency. To minimize this reduction at part load conditions, the steam generator is designed to maintain the steam temperature over a specified temperature control range. The *control point* is the lowest load at which design steam temperatures can be produced.

A control point of 60% of MCR during constant pressure operation is typical for United States utility installations. The control point should be specified with consideration of the load model expected for the steam generator. For example, if the steam generator is to operate principally at high load with few hours of part load operation, capital cost can be reduced by specification of a higher control point and a smaller temperature control range. Unfortunately, the load model for a new unit may not be known with confidence beyond the initial 5 to 10 years of operation. Therefore, a control point at 60% load represents a good general service value.

The steam generator incorporates both primary and secondary systems for reheat steam temperature control, but uses only a primary system for superheat steam temperature control. The systems used for steam temperature control are as follows:

	Primary	Secondary
Superheat steam temperature control	Desuperheating spray	—
Reheat steam temperature	Flue gas Proportioning	Desuperheating spray
control	Tilting burners	Desuperheating spray

Superheat steam temperature control is achieved by spraying feedwater into the superheater steam at an intermediate position between superheater heat transfer tube banks. Desuperheaters typically are located at the primary superheater outlet. Two-stage desuperheaters may be located at the outlets of both the primary and intermediate superheater tube banks. The water source for desuperheating can be either the boiler feed pump discharge or a feedwater connection between the highest pressure feedwater heater outlet and the economizer inlet.

Using feedwater from the economizer inlet for superheat desuperheating yields a small turbine cycle benefit compared to the use of feedwater directly from the boiler feed pump discharge. Feedwater taken directly from the boiler feed pump discharge bypasses the high-pressure feedwater heaters, which decreases the turbine cycle efficiency, or increases the turbine heat rate (expressed as heat energy per kilowatt-hour). However, the use of economizer inlet feedwater for desuperheating may require a throttling valve on the economizer feedwater inlet to ensure adequate pressure drop and flow through the desuperheating system at all load conditions.

The available head or pressure to force feedwater flow through the desuperheating system is determined by the pressure drop through the portion of the feedwater-steam circuitry which forms a parallel flow path to the desuperheating water flow path. If desuperheating water is taken from the boiler feed pump discharge, the available head includes the feedwater pressure drop across the high-pressure feedwater heaters and the economizer, and the steam pressure drop across the primary superheater. However, if desuperheating water is taken from the economizer inlet, the feedwater/steam pressure drop across the economizer and the primary superheater may not be adequate to force sufficient flow under all load conditions. Although the desuperheating water requirements typically decreases with decreasing load, the pressure drop decreases in proportion to the square of the flow. Therefore, throttling of the main feedwater flow path may be required if desuperheating water is taken from the economizer inlet.

An evaluation of the alternative desuperheating water sources may be appropriate, depending on the economic benefit of improved turbine cycle efficiency.

Superheater heat transfer surface is designed such that desuperheating spray is required for all firing conditions in the upper half of the load range. Ideally, the superheater is "oversurfaced" so that operation of the reheat steam temperature control system affects only the amount of superheat desuperheating spray and not the superheater outlet steam temperature.

As listed previously the principal reheat steam temperature control primary systems are flue gas flow proportioning and tilting burners. Gas flow proportioning systems are used for wall-fired burner systems, and tilting burners are used on corner or tangentially fired systems.

A divided convection pass, shown in Fig. 7-3, is used in a flue gas flow proportioning system. The front parallel convection pass passage contains reheat heat transfer surface, and the rear pass contains superheat heat transfer surface. Economizer surface may be arranged within the parallel passes, or may be downstream from the two gas passes. Flue gas flow through the passes is biased from one pass to the other by a multiple leaf damper at the parallel pass outlet.

Thus, if reheat steam temperature is too low, the flue gas proportioning damper is repositioned to increase the gas flow through the reheat pass and to reduce the flow through the superheat pass. This action reduces heat transfer to the superheater. However, this should reduce only the superheater desuperheating spray, not the final superheater steam temperature.

Conversely, if reheat steam temperature is too high, flue gas flow is biased away from the reheat pass, more heat is transferred to the superheater, and superheater desuperheating spray flow is increased.

Tilting burners control the reheat steam temperature by changing the elevation of the fireball within the furnace, which changes the heat absorption of the furnace, and changes the quantity of heat delivered to the superheater and reheater in the convection pass. A typical furnace arrangement with corner mounted tilting burners is shown in Fig. 7-2. To increase reheat steam temperature, the burners are tilted upward to deliver more heat to the reheater, which also increases the need for superheater desuperheating spray. Conversely, tilting the burners downward reduces the reheat steam temperature and the superheat desuperheating spray flow requirement.

The secondary system of reheat steam temperature control is desuperheating spray. The reheat desuperheating spray nozzle is in the cold reheat piping at the reheater inlet rather than the reheater outlet to avoid the possibility of spray water induction into the intermediate pressure turbine. Reheat spray water is taken from an intermediate stage of the boiler feed pump since the spray water pressure requirement is lower for the reheat steam than for the superheat steam. The reheat desuperheating spray system, unlike the superheat desuperheating spray system, is designed for use in emergency and abnormal situations only. Reheat spray flow bypasses the entire high-pressure turbine, degrading turbine cycle efficiency.

Additional operating practices and systems affect superheat and reheat steam temperature. These include biased firing of burners, excess air, and flue gas recirculation.

Biased firing is similar to tilting burners and involves firing upper level burners preferentially to raise steam temperature and firing lower level burners to reduce steam temperature. Since pulverizers are dedicated to an individual level or row of burners, the application of biased firing would involve bringing the upper and lower burner levels and pulverizers in and out of service based on steam temperature criteria. Biased firing has limited effectiveness and, therefore, is not used as a principal steam temperature control system.

Increasing the excess air increases the flue gas velocity over the tubes, the heat transfer to the convective heat transfer surfaces and, therefore, increases both the superheat and reheat steam temperature. However, increasing the excess air also increases the stack loss, which reduces the boiler efficiency. Therefore, excess air is not used as a principle steam temperature control system.

Flue gas recirculation has been used as a steam temperature control method on coal-fueled boilers. Flue gas is drawn from the economizer outlet by a fan and returned to the furnace. This arrangement allows control of the convection pass gas velocity without increasing the stack loss. However, the hot ash-laden flue gas which passes through the flue gas recirculation fan is very abrasive and represents a difficult fan application, with high maintenance requirements. In addition, during upset conditions, 2,500 to 3,000° F (1,371 to 1,649° C) flue gas from the furnace can backflow through the gas recirculation fan and overheat the fan rotor. Rotors thus weakened have failed in service. For these reasons, flue gas recirculation on pulverized coal-fueled boilers is not recommended for new steam generator designs.

7.2.6 Minimum Load Capability

The minimum load capability, or turndown, of the steam generator when firing coal only is an important consideration for units that are not base loaded but are kept in operation during off-peak periods to provide "rolling reserve" for the system. For such units, good turndown capability is essential to avoid the increased fuel costs associated with natural gas or fuel oil firing for flame stabilization.

Typically, a single pulverizer-burner system produces a stable flame at half of its full load, corresponding to a turndown ratio of 2:1. Technically, a single pulverizer can be operated alone, although operation of a single pulverizer is not recommended. Conservative minimum load operating practice dictates no less than two pulverizers in operation, preferably firing on adjacent burner levels to enhance the flame stability. The use of two pulverizers prevents unnecessary emergency shutdowns of the boiler in response to individual pulverizer operating problems. If one of two pulverizers trips off due to a malfunction, the remaining pulverizer prevents a subsequent boiler trip.

As discussed in Section 7.2.3, the pulverizer system should be designed such that full boiler load can be achieved with one or two pulverizers out of service. If the numbers of pulverizers and spare pulverizers are known, the minimum load while firing only coal with two pulverizers, each operating at half capacity, can be calculated as follows:

$$\text{Minimum load} = \frac{100}{N - S} \tag{7-2}$$

where

minimum load is in percent,

N = total number of pulverizers, and

S = number of spare pulverizers when firing design coal.

7.2.7 Variable Pressure Operation

Constant pressure operation consists of operation at a constant throttle steam pressure over the entire load range. *Variable pressure operation* is defined as controlling the boiler steam production by controlling and adjusting the boiler steam pressure. Essentially, the turbine throttle valves are left in an open position and steam flow is increased or decreased by increasing or decreasing the throttle steam pressure.

Variable pressure operation is further divided into two types: *pure variable pressure* and *hybrid variable pressure*. In pure variable pressure, the turbine throttle valves are wide open and any load reduction below maximum continuous rating is achieved by steam pressure reduction. Hybrid variable pressure operation consists of a combination of constant pressure operation over the upper load range and variable pressure operation over the lower load range.

The benefits of variable pressure operation are improved turbine cycle efficiency and improved boiler steam temperature control range. Part-load turbine cycle efficiency improves as a result of reduction of high-pressure turbine throttling valve losses. The reduction of throttle valve losses results in higher steam temperatures throughout the high-pressure turbine, higher high-pressure turbine exhaust (cold reheat) steam temperature, and therefore an improvement in reheat steam temperature control range, since the cold reheat steam is hotter than it would otherwise be with constant pressure operation at the same steam flow.

Variable pressure operation can be achieved either by control of firing rate during operation, or through the use of large division valves in the intermediate superheater flow connecting links. (Figure 7-4 is a flow diagram that includes division valves.) Variable pressure operation through control of the firing rate is accomplished by changing the firing rate prior to changing the load, or steam flow. If a load reduction is desired, then the firing rate is reduced without moving the

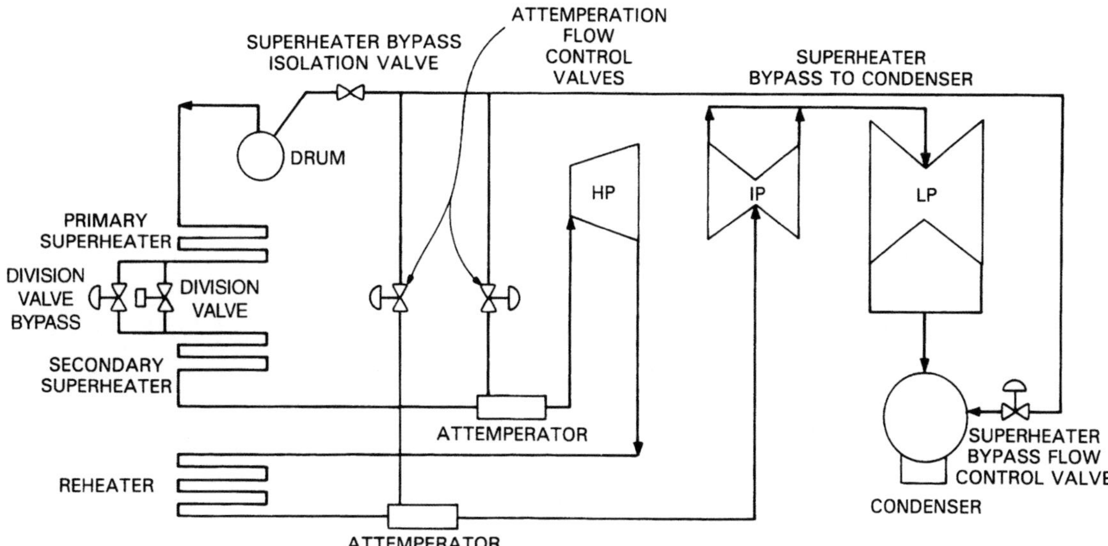

Fig. 7-4. Full startup system.

turbine throttle. As energy delivery to the steam system is reduced, the drum temperature, steam pressure, and steam flow decrease. As the desired lower load is achieved, the firing rate is stabilized, drum temperature and pressure stabilize, and a reduced steam flow is established. For an increase in load, the firing rate is increased and the effects are reversed, which results in an increase in steam flow.

Control of steam pressure by control of the firing rate is the least expensive and most widely used variable pressure system. Control by firing rate has the added benefit of a reduction of boiler feed pump power, since the drum pressure and therefore the static pressure requirement of the boiler feed pump are reduced during variable pressure operation. Since the drum operates at full pressure with a division valve system, the boiler feed pump power is not reduced.

The division valve system is somewhat more responsive to load changes than a firing rate variable pressure control system. With a division valve system, the firing rate is controlled to maintain drum pressure. Load reduction is achieved by throttling the main steam flow with the division valve. As the division valves are closed, the superheater outlet pressure and flow are reduced. The drum pressure increases as the division valves close and the firing rate is decreased, returning the drum pressure to its original value. When the boiler stabilizes at a reduced load, the drum pressure returns to its original value, the division valve is in a throttled position, superheater outlet pressure and flow are reduced, and the firing rate is reduced to establish a balance with steam production energy requirements. The division valve action and effects are reversed to increase the steam generator load. The load change capabilities for the two variable pressure systems are 2% of full load per minute with firing rate control, and 10% of full load per minute with a division valve system.

An additional benefit of variable pressure operation and higher steam temperatures is the ability to trip the turbine at a higher temperature on shutdown. If the turbine is subsequently restarted from hot conditions, the resulting thermal cycle of the turbine components is reduced.

Because of the turbine heat rate and steam temperature control range improvements that can be obtained, variable pressure capability typically is recommended as a design feature for new utility steam generation units.

7.2.8 Flue Gas Emissions

Flue gas emissions from pulverized coal boilers generally can be divided into two categories:

- Combustion products (NO_x, carbon monoxide, and hydrocarbon emissions) which are controlled through manipulation of the combustion process.
- Combustion products (NO_x, sulfur dioxide, fly ash, or particulate matter) which are controlled through a post-combustion cleanup process.

The design practices for both combustion control methods and postcombustion cleanup processes for flue gas emission control are discussed in Chapter 14.

7.2.9 Feedwater Quality and Steam Purity

Feedwater quality and steam purity are interrelated in that the feedwater quality affects the steam purity. Boiler feedwater quality is addressed in a boiler specification by specifying the steam purity requirements, which in turn are based on turbine manufacturer recommendations. Current basic steam purity criteria which encompass the requirements of three major United States and European turbine manufacturers are as follows:

Sodium as Na, g/L	3.0
SiO_2, micro g/L	10.0
Cation conductivity, mho/cm	0.2

Using these requirements for steam purity, the boiler manufacturer must submit the corresponding feedwater quality requirements as part of his proposal. The feedwater treatment equipment required for a given boiler design can then be evaluated as part of the overall proposal.

7.2.10 Steam Generator Arrangement

Steam generator arrangement issues include pulverizer arrangement, regenerative air heater location, use of a fan room, and boiler enclosure.

Pulverizers may be located either in a row across the boiler front or on one or both sides of the boiler as shown in Fig. 7-5. The arrangement of pulverizers across the boiler front simplifies the coal handling system, since a single row of pulverizers and silos minimizes transfer points in the coal conveyor system. However, pulverizers across the front generally are between the boiler and the turbine room, and a bay for pulverizer maintenance must be provided. The extra bays between the steam generator and the turbine result in longer runs of high-pressure piping for main steam, reheat steam, and feedwater.

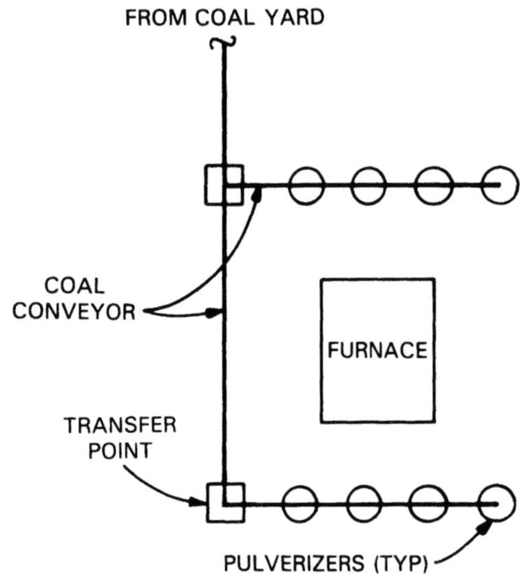

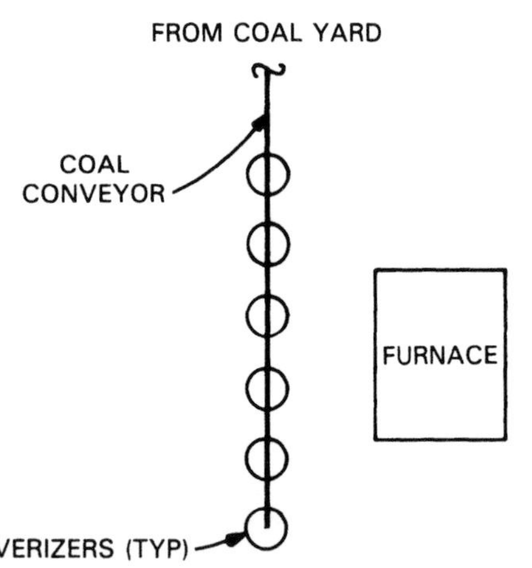

Fig. 7-5. Pulverizer arrangements.

Pulverizers along both sides of the boiler may be required for large boilers, because the number of pulverizers may exceed the space available across the front of the boiler. Access to the outside of the boiler building from the pulverizer bay can be provided easily through roll-up overhead doors in the side of the building adjacent to each pulverizer. However, arrangement of the pulverizers on both sides of the boiler means a more complex coal conveyor arrangement.

The principal issue on the regenerative air heater (RAH) arrangement is its location. The preferred location of the RAH is behind the rear boiler column line rather than underneath the convection pass. The preferred arrangement is shown in Figs. 7-2 and 7-3. This arrangement behind the rear boiler column line both enlarges and increases the cost of the boiler building. However, the preferred arrangement allows a sharper change in the flue gas flow direction and enhances the separation of economizer ash from the flue gas stream. This arrangement also offers better access and routing of economizer ash piping from the economizer hoppers. In addition, during construction, this arrangement allows better access to the RAH.

The use of a room for enclosure of the primary and secondary air fans is a design decision that must be evaluated for the project. A fan room protects the fans and drives, and facilitates maintenance during inclement weather. A fan room also reduces the transmission of noise generated by the fans to other areas of the plant. However, the room must be designed to accommodate fan maintenance and fan rotor removal.

The enclosure of the boiler also must be evaluated on a project basis. Northern United States sites are good candidates for a complete boiler enclosure, which facilitates wintertime maintenance and operation and protects water and steam systems. A semi-enclosed boiler building is an alternative for utilities that prefer open construction. Semi-enclosed boilers are enclosed below the top burner level, are open at the platform over the top burner level and above, and are provided with a roof over the top of the boiler, above the penthouse. In general, enclosed boilers are preferred because of the expected additional operation and maintenance productivity and the protection of components from weather-induced deterioration.

7.2.11 Startup Systems

Steam generator startup involves firing with natural gas or fuel oil to heat up the boiler. When combustion air temperatures are sufficiently high, coal firing is initiated. The principal firing rate constraints during startup are as follows:

- Heat transfer surface metal temperature,
- Heatup rate of thick-walled pressure parts such as the drum and steam headers,
- Steam pressure control, and
- Temperature mismatch between steam and turbine metal.

During startup, when the steam flow is zero or very low and the cooling effect of the steam is limited, the superheater metal temperature is essentially equal to the flue gas temperature entering the superheater tube bank. During this early stage of the startup, the metal temperatures are monitored indirectly by measurement of the flue gas temperature with an air-cooled flue gas temperature probe in the upper furnace. The firing rate is controlled to limit the flue gas temperature to approximately 1,000° F (538 ° C).

The heatup rate of thick walled pressure parts is monitored with thermocouples. Typical drum heatup rates range from 100° F (56° C)/h on older boilers to 400° F (222° C)/h on a modern installation. Pressure part temperatures must be monitored and the firing rate controlled to achieve an acceptable heatup rate.

Several specialized systems are available to aid in steam temperature and pressure control during startup. These systems include the superheater bypass, the reheater outlet steam attemperator, and the superheater outlet steam attemperator with secondary superheater stop valve and stop valve bypass.

7.2.11.1 Superheater Bypass. The superheater bypass is a startup system that provides a method of releasing steam and relieving drum pressure during boiler startup without the loss of feedwater. Figure 7-6 shows a typical superheater bypass system. On hot starts, including starts following overnight shutdowns, the turbine metal temperature is high prior to rolling the turbine, and high steam temperatures are desirable to match the turbine metal temperature. As the turbine is rolled, the firing rate and the gas temperature leaving the furnace have to be kept high to maintain high main steam and reheat temperatures. This generally results in a rapid rise in throttle pressure which is undesirable because of the resulting large temperature drop when steam is throttled across the turbine control valves.

The superheater bypass to the condenser provides a means to maintain throttle pressure in the range of 500 to 1,000 psi (3.45 to 6.90 MPa), and offers several advantages for hot restarts and starts following overnight shutdown:

- An optimum match of steam conditions and turbine metal temperatures at the maximum heat input to the boiler within the limits dictated by metal protection of the reheater and superheater;
- Reductions in startup time from improved turbine metal to steam temperature matching and increased heat input; and
- Turbine starting with a greater opening of the turbine control valves as a result of the lower throttle pressure, thus easing the temperature shock of the turbine first-stage (impulse stage) shell metals.

7.2.11.2 Reheater Outlet Steam Attemperator—Partial Startup System. Before steam is admitted to the turbine, the reheater is without flow. The reheater metal absorbs heat from the flue gas and eventually reaches the temperature level of the flue gas, which can be as high as 1,000° F (538° C). When steam is first admitted to the turbine and passes through the reheater, reheat outlet steam temperature rises very rapidly to the gas temperature level, resulting in a poor match with the intermediate-pressure turbine metal temperatures for cold starts and starts following a weekend shutdown. A reheat steam attemperation system, including a superheater bypass (Fig. 7-7), provides a method of startup reheat steam temperature control.

Steam attemperation, using saturated steam at the reheater outlet, limits the rise of reheat steam temperature when steam is first admitted to the turbine and offers positive reheat steam temperature control up to about 10% of full load. This system extends the hot start advantages described

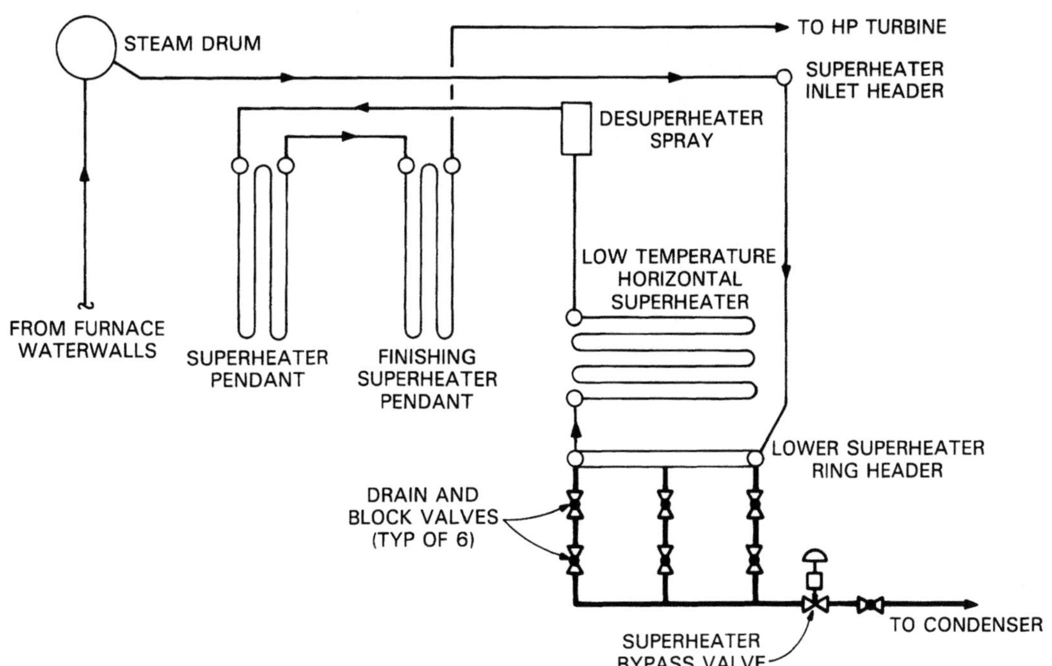

Fig. 7-6. Superheater bypass system.

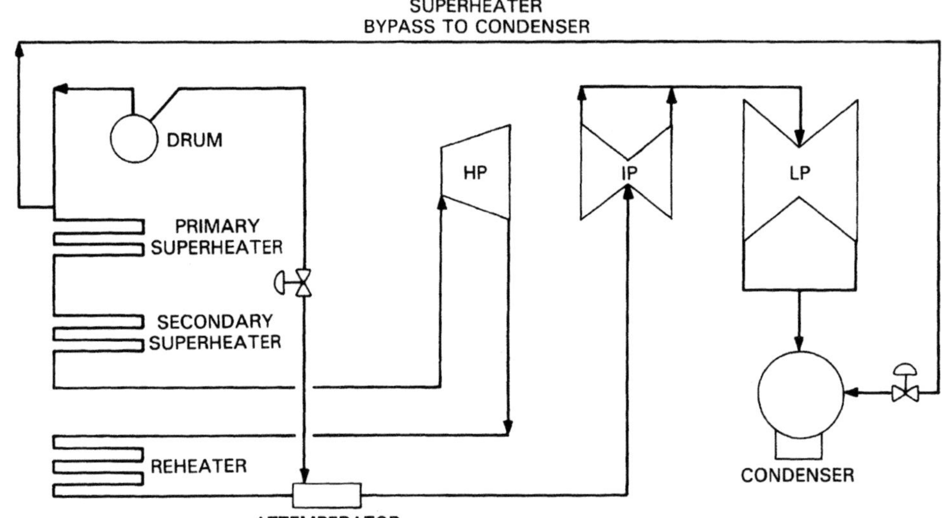

Fig. 7-7. Partial startup system—reheat steam attemperation system.

for the superheater bypass to cold starts and starts following a weekend shutdown.

7.2.11.3 Superheater Outlet Steam Attemperator with Secondary Superheater Stop Valve and Stop Valve Bypass—Full Startup System. For cold starts and starts following weekend shutdowns, the requirements early in the startup of low steam temperature for a good temperature match and high heat input for a quick startup are generally not compatible. With low steam flows up to about 10% to 15% of full load, the main steam temperature approaches the gas temperature in the final tube passes of the secondary superheater. To keep the steam temperature low, the gas temperature must be kept low. Low furnace exit flue gas temperature is incompatible with a firing rate and heat input high enough to generate sufficient steam for rolling and initial loading of the turbine.

Separate systems for main and reheat steam temperature control facilitate steam-to-turbine metal temperature matching during cold startups. A partial startup system consists of reheat steam attemperation with drum steam. A full startup system adds attemperation of superheater steam with drum steam and superheater division valves or stop valves to the partial system. A full boiler startup system is shown in Fig. 7-4.

The superheater outlet steam attemperator, using saturated steam from the drum, permits startup steam temperature control independent of heat input to the boiler. To bypass saturated steam from the drum to the superheater outlet, it is necessary to install stop valves and a stop valve bypass between the primary and secondary superheater to provide flow resistance and control flow through the superheater. Without the stop valve, the superheater pressure drop during startup is insufficient to induce adequate drum steam flow for superheat temperature control.

The secondary superheater stop valve bypass is a smaller control valve that provides pressure and flow control through the secondary superheater to the turbine under low-flow startup conditions.

The superheater outlet steam attemperator valve is used at loads less than 10% to introduce saturated steam from the drum to the superheater outlet for rolling the turbine. Initial rolling of the turbine may be started with saturated steam through the superheater outlet steam attemperator valve, mixed with a limited quantity of steam passing through the stop valve bypass, and through the secondary superheater to control the high-pressure turbine inlet temperature.

On hot restarts with very high turbine metal temperature, steam temperature is controlled by furnace exit flue gas temperature with the superheater outlet steam attemperator valve closed. For moderately high metal temperature [700° F (371° C)], the attemperator valve may be used for steam temperature control to achieve a close match between entering steam temperature and first-stage shell and shaft temperatures. The following advantages are gained for cold starts and for starts following a weekend shutdown:

- Variable pressure operation during startup by throttling through the stop valve bypass with the turbine valves at or near the wide-open position,
- Optimum control of main steam temperature during startup for turbine metal temperature matching, and
- Reduced startup time because of the improvement in steam temperature control.

7.3 STEAM GENERATOR COMPONENT DESIGN

The principal steam generator components described in this section include the following:

- Furnace;
- Drum;
- Boiler circulating pumps;
- Convection pass
  Superheater,

Reheater,
Economizer;

- Air heater;
- Air preheat coils;
- Soot blowers;
- Coal feeders;
- Pulverizers;
- Coal piping;
- Burners;
- Ignitors and warmup burners;
- Ductwork; and
- Insulation and lagging.

7.3.1 Furnace

The furnace serves as an enclosure for the combustion process and is shown in Fig. 7-1. The furnace walls are formed by water-filled tubes or waterwalls that contain the upward flow of water and steam.

The size of the furnace is determined by the required steam capacity and the characteristics of the fired fuel. The principal design methods used to control furnace size are as follows:

- Fuel heat input per furnace plan area;
- Heat available to the furnace per effective projected radiant surface area, or furnace release rate; and
- Furnace exit gas temperature.

The fuel heat input per furnace plan area is calculated as the product of the fuel flow and fuel higher heating value divided by the horizontal cross-section or plan area of the furnace. The specification of the heat input per plan area is used as a method of preventing slag formation in the combustion zone. The value specified for heat input per plan area is a function of the fuel properties. The ranges of typical furnace design values for United States coal ranks are as follows:

Fuel Heat Input per Furnace Plan Area		
	Btu/h-ft²	MJ/h-m²
Bituminous coal	1,600,000–1,800,000	18,200–20,400
Subbituminous coal	1,500,000–1,700,000	17,000–19,300
Lignite	1,200,000–1,500,000	13,600–17,000

Values in the low range are used for coals known to be slagging fuels through previous firing, test burns, or the calculation of slagging indices based on fuel properties. Low-range values may also be specified if fuel blending is planned, since a blend coal may have poorer slagging characteristics than any of its component coals. Conversely, the higher values can be specified if the fuel supply is known to be lower slagging type within its coal rank.

The heat available to the furnace is a modification of the fuel heat input. The heat available to the furnace includes the fuel heat content based on the higher heating value plus the heat provided by the combustion air compared to a datum temperature of 80° F (27° C). This heat quantity is reduced by the latent heat required to evaporate water in the fuel and water formed by combustion of hydrogen, the heat loss due to radiation from the furnace and convection pass exterior walls, and the heat loss due to unburned combustible material in the ash.

The effective projected radiant surface (EPRS) is the total projected area of the planes that pass through the centers of all furnace wall, roof, and floor tubes, plus the area of a plane that passes perpendicular to the gas flow where the furnace gases reach the first convection superheater or reheater surface. In calculating the EPRS, the projected area of both sides of the superheater and reheater platens extending into the furnace is included (Singer 1991).

The ratio of heat available divided by the EPRS, or furnace heat release rate, is a furnace design parameter that determines the amount of heat transfer surface available to cool the flue gas prior to its exposure to the tube banks of the convection pass. Cooling of the gases in the open furnace is essential to prevent molten ash (slag) accumulation at the furnace exit. The ranges of typical values of heat available per EPRS are as follows:

Furnace Heat Release Rate		
	Btu/h-ft² EPRS	MJ/h-m² EPRS
Bituminous coal	70,000–90,000	795–1,020
Subbituminous coal	60,000–80,000	680–910
Lignite	40,000–60,000	450–680

The furnace exit gas temperature is determined primarily by the ratio of heat available per effective projected radiant surface (Stultz and Kitto 1992). Although the two parameters are interdependent, most steam generator specifications include both values. Ranges of typical furnace exit gas temperatures for United States coals are as follows:

	Furnace exit gas temperature, °F	Furnace exit gas temperature, °C
Bituminous coal	1,900–2,000	1,038–1,093
Subbituminous coal	1,850–1,950	1,010–1,066
Lignite	1,800–1,900	932–1,038

7.3.2 Drum

The drum encloses the steam–water interface in a subcritical boiler, and provides a convenient point for addition of chemicals and removal of dissolved solids from the feedwater–steam system. The drum also contains equipment for removal of liquid from the steam as the steam leaves the drum and enters the connecting links to the primary superheater.

The following drum design parameters typically are specified [based on a nominal 2,400 psig (16.547 MPa) cycle]:

Drug design pressure	2,800–2,950 psig (19.3 MPa–20.3 MPa)
Drum inside diameter, as a function of main steam flow	
6,000,00 lb/h (2,722,000 kg/h) and higher steam flow	72 in. (1,829 mm)
3,000,000 lb/h (1,361,000 kg/h) to 6,000,000 lb/h (2,722,000 kg/h)	66 in. (1,676 mm)
Less than 3,000,000 lb/h (1,361,000 kg/h)	60 in. (1,524 mm)

7.3.3 Boiler Circulating Pumps

Subcritical boilers may be designed with natural circulation through the furnace waterwalls or forced circulation with boiler circulating pumps. Natural circulation systems must be designed with low flow resistance in the water circuit which consists of the drum, downcomers, lower headers, furnace waterwalls, upper headers, and connecting links back to the drum. The forced circulation design allows the use of smaller diameter tubing in the furnace walls, since the higher pressure drop in the smaller tubing can be offset through pump circulation. The smaller diameter also allows thinner tube walls.

The principal design consideration for the boiler circulating pump system is to specify an installed spare that allows full load operation of the boiler with one boiler circulating pump out of service. In addition, suction gate valves and discharge stop check valves should be specified for each pump. The isolation valves allow pump removal without the need to drain the boiler after a shutdown.

7.3.4 Superheater

The superheater heat transfer surface may be radiant surface in the furnace or convective surface in the convection pass. Typical maximum superheater steam side pressure drops are as follows:

	Maximum superheater presure drop, psi	MPa
Less than 2,000,000 lb/h (907,000 kg/h) steam flow	150	1.03
Greater than 2,000,000 lb/h (907,000 kg/h) steam flow	170	1.17

Tube spacing may be specified as a function of flue gas temperature. Specification of tube spacing is of interest in projects where a high level of conservatism in convection pass design is desired. Specification of tube spacing is not particularly desirable from the manufacturer's standpoint, since the specified spacing probably will not conform with the individual standards established by the boiler manufacturers. The following values are conservative tube spacing criteria for a fouling coal:

		Mininum center-to-center	
Gas temp., °F	Gas temp., °C	Spacing, in.	mm
2,450–2,181	1,343–1,194	54	1,372
2,180–1,901	1,193–1,038	24	610
1,900–1,750	1,037–954	12	305
1,749–1,450	953–788	8	203
1,449–800	787–427	4.5	114

If the fuel is known as a severe fouling coal, increased tube spacing may be required.

Tube materials typically are selected by the boiler manufacturer based on the temperature of the tube surface during operating conditions. United States steam generator manufacturers have developed temperature limits for tube materials based on metal oxidation characteristics. The manufacturers have developed similar criteria, although there are individual differences. In instances where uniform metal selection criteria and additional conservation are desired, the following maximum allowable metal temperatures have been specified:

Material	ASME tubing specification	Maximum allowable metal temp., °F	°C
Carbon steel	SA178, SA192, SA210	775	413
Low-alloy steel	SA209T1A	875	468
Low-alloy steel	SA213T2	900	482
Low-alloy steel	SA213T11	1,000	538
Low-alloy steel	SA213T22	1,075	579
High-alloy (stainless) steel	SA213TP304H	1,300	704
High-alloy (stainless) steel	SA213TP321H	1,300	704
High-alloy (stainless) steel	SA213TP347h	1,300	704

The superheater tube banks typically are designed with a maximum 48 in. (1,219 mm) to 72 in. (1,829 mm) bank depth, with a minimum of 36 in. (914 mm) between banks. The maximum depth allows reasonable access, by tube spreading, to a tube leak in the middle of the tube bank. The minimum interbank spacing allows access for maintenance.

The flue gas velocity through the superheater is based on the amount of ash in the fuel and the amount of silica and alumina in the ash. Typical maximum flue gas velocities are as follows:

Silica and alumina content, lb/10^6 Btu	g/MJ	Erosion potential	Maximum flue gas velocity, ft/s	m/s
0–10	0–4.3	Low	60	18.3
10–15	4.3–6.5	Medium	55	16.8
>15	76.5	High	50	15.2

7.3.5 Reheater

Like the superheater, the reheater heat transfer surface may be composed of either radiant or convective surface. Radiant reheater surface can be either radiant wall heat transfer surface or pendant heat transfer surface. A radiant wall reheater can be mounted on the front and/or side walls of the upper furnace. A radiant wall reheater is shown on the front wall in Fig. 7-1. Pendant type heat transfer surface is suspended from the roof of the furnace.

The maximum reheater steam side pressure drop ranges from 20 psi (0.14 MPa) to 25 psi (0.17 MPa). Higher reheater pressure drops are undesirable because of the resulting turbine cycle efficiency loss. Lower reheater pressure drops are insufficient to ensure steam flow distribution to adequately cool all the reheater steam circuits, particularly at intermediate and low loads.

The reheater tube spacing, material selection, tube bank

spacing criteria, and gas velocity criteria are as discussed in Section 7.3.4.

7.3.6 Economizer

The economizer is composed of low-temperature convection pass surface. The economizer tubes should be specified as bare tubes, because finned tubes plug with ash when firing with all but the best coals. Similarly, the tubes should be arranged in line rather than staggered to allow passage of large ash chunks through the tube bank. The minimum clear space between economizer tubes ranges from 2½ to 4 in. (63.5 to 102 mm).

The economizer material selection, tube bank spacing criteria, and gas velocity criteria are as discussed in Section 7.3.4.

7.3.7 Air Heater

The air heater for utility installations typically consists of a rotary regenerative air heater(s) of the Ljungstrom[2] design as manufactured by Asea Brown Boveri (ABB) Air Preheater, Inc., or a Rothemuhle[2] design manufactured by Babcock & Wilcox. The Ljungstrom and Rothemuhle designs are shown in Figs. 7-8 and 7-9, respectively. Tubular, heat pipe, and plate and frame air heaters have been used on small (100-MW range) boilers. However, the rotary regenerative air heater design using either rotating heat transfer surface or rotating air distribution hoods predominates for utility air heating applications. The air heater arrangement may consist of one, two, three, or four air heaters, depending on the size of the unit and the degree of fuel flexibility desired. Two

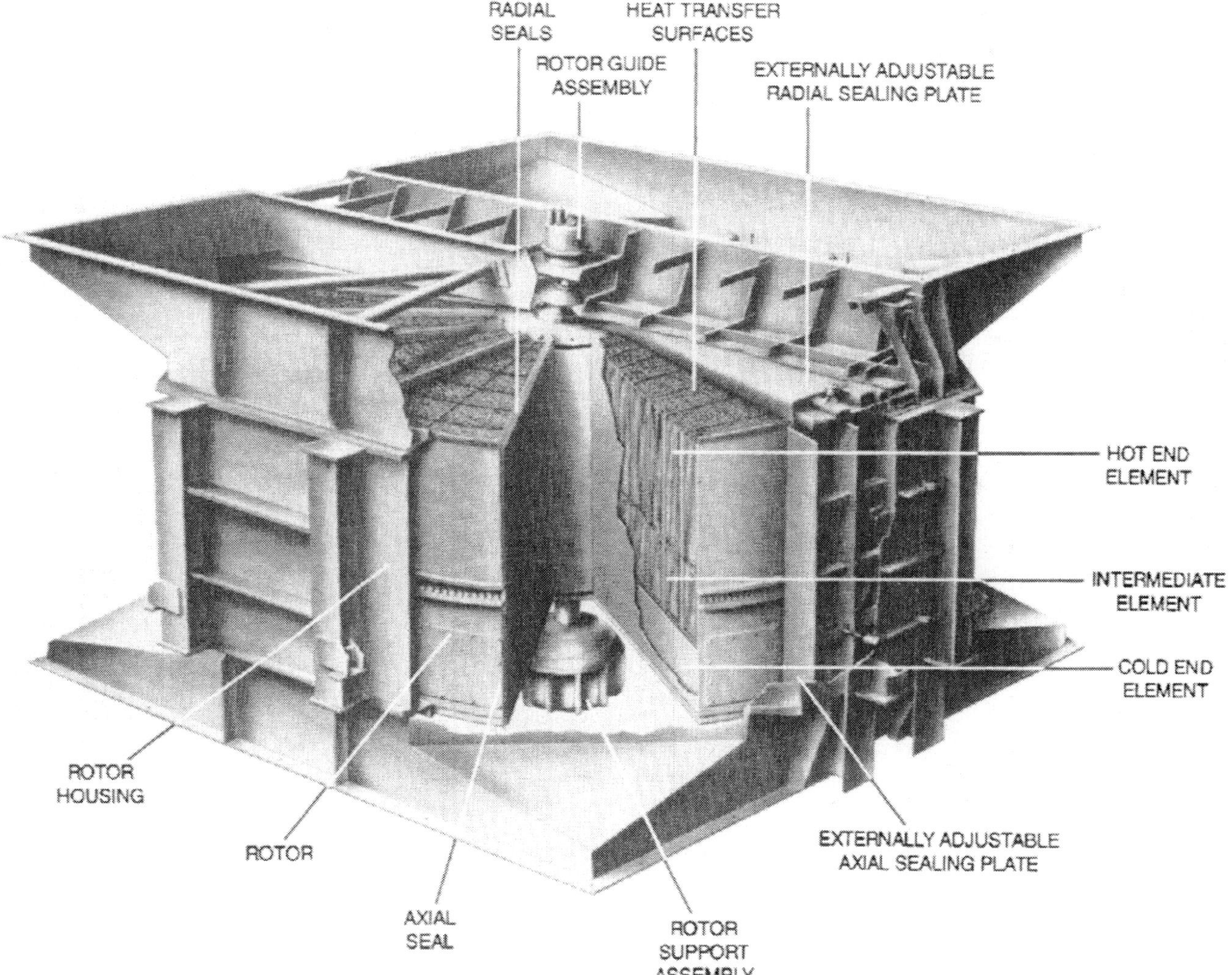

Fig. 7-8. Bisector Ljungstrom air preheater isometric cut-away view. (From ABB Air Preheater, Inc. Used with permission.)

[2]Ljungstrom and Rothemuhle are registered names for designs marketed by ABB Air Preheater, Inc., and Babcock & Wilcox, respectively.

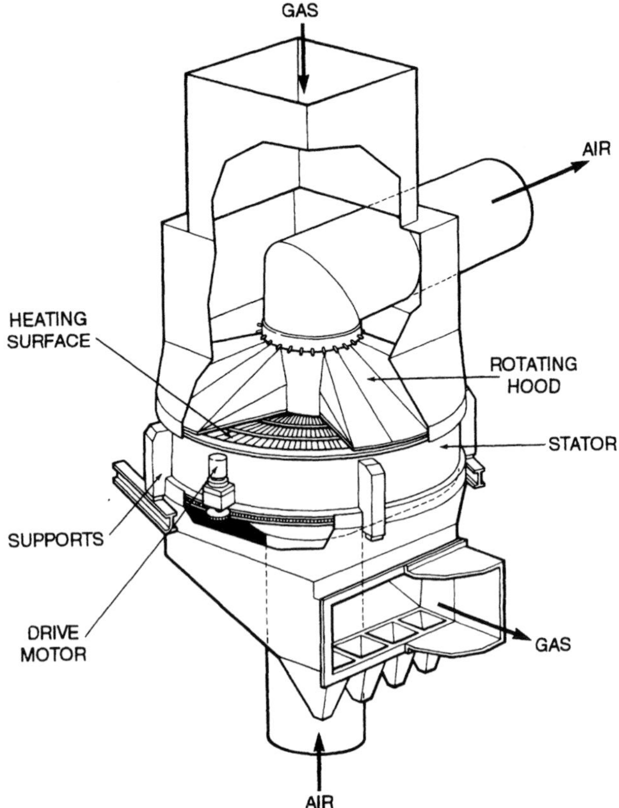

Fig. 7-9. Rothemuhle air heater isometric cut-away view. (From Babcock & Wilcox, Brochure PS113R. Used with permission.)

types of regenerative air heaters are used: bisector (one air stream per air heater) and trisector (two air streams per air heater).

The trisector design incorporates both a primary and a secondary air heater within one housing. The bisector and trisector terminologies are typically associated with the Ljungstrom design. However, the configuration is available with either Ljungstrom or Rothemuhle equipment.

The following tabulation shows typical air heater arrangements for pulverized coal units. These arrangement variations are illustrated in Figs. 7-10 and 7-11.

Number of air heaters and types		
1	Bisector	500 MW and smaller units using exhauster pulverizers or hot primary air fans
2	Bisectors	Units larger than 500 MW using exhauster pulverizers or hot primary air fans
3	Bisectors	Any size, two secondary air heaters and one primary air heater
4	Bisectors	Large units, two secondary air heaters and two primary air heaters
1	Trisector	Units less than 500 MW
2	Trisectors	Units larger than 500 MW.

The arrangement of two bisectors in which one is a primary air heater and the other is a secondary air heater is not recommended. The flue gas temperature profile leaving a one-primary, one-secondary heater arrangement is stratified from side to side. The uneven temperature profile is not a favorable design condition for the downstream air quality control equipment. The use of two bisectors in combination with exhauster pulverizers or hot primary air (PA) fans exists, but is not common. Two bisectors would be used on units of 500 MW and larger. Large units typically have pressurized pulverizers, atmospheric suction PA fans, and,

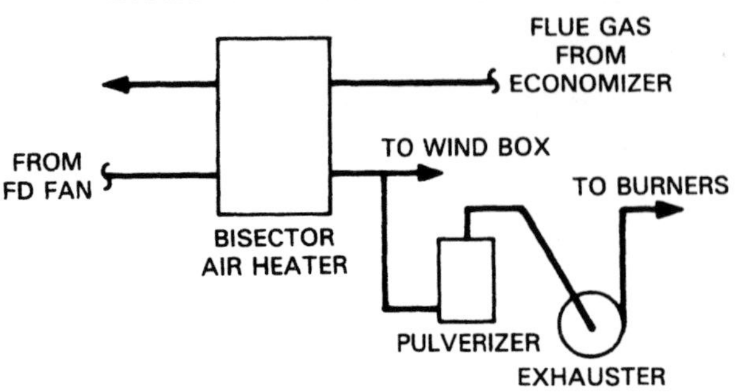

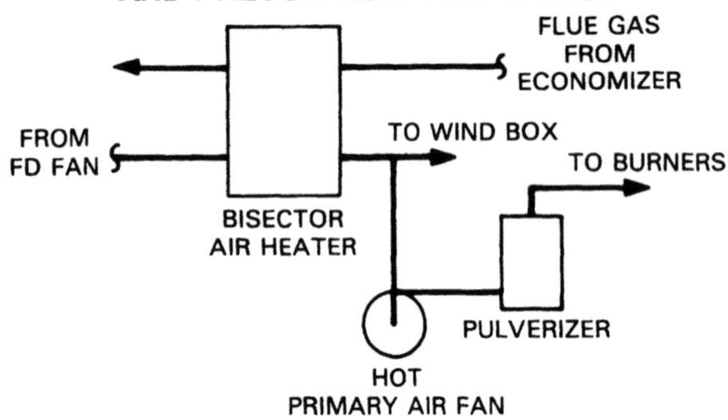

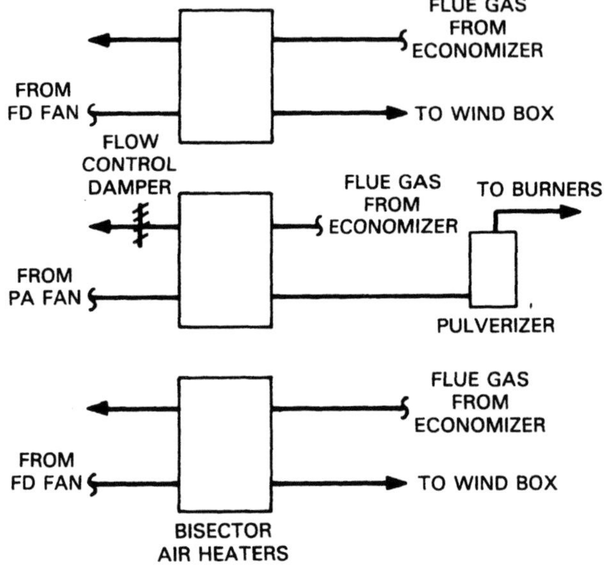

Fig. 7-10. Bisector air heater arrangements.

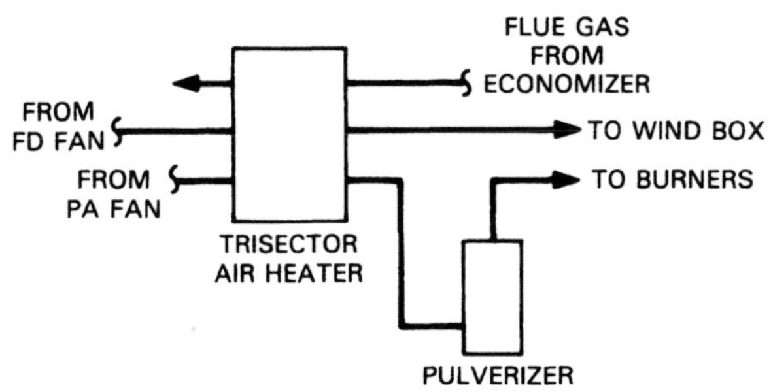

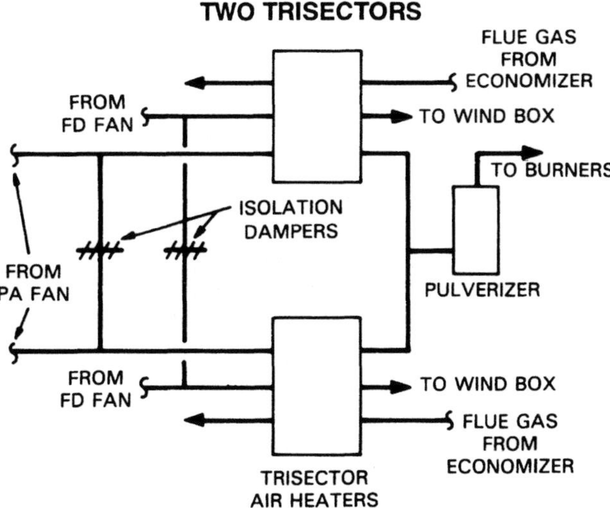

Fig. 7-11. Trisector air heater arrangements.

therefore, a primary air system separate from the secondary air system.

The three- or four-bisector arrangement with two secondary air heaters and one or two primary air heaters yields the best fuel flexibility. Because the primary air is used to dry the fuel during pulverization, the amount of primary air heating required is proportional to the fuel moisture content. The heat transfer in the separate primary air heater can be controlled with a damper in the flue gas stream at the primary air heater outlet. Opening the damper provides more flue gas to the air heater and hotter primary air to accommodate higher moisture coal. For lower moisture coal, the damper is closed to reduce the primary air temperature.

Air heater systems consisting of a single trisector are used frequently on boilers smaller than 500 MW. Two trisectors are used frequently on 500 MW and larger boilers. The capital and operating costs of a one- or two-trisector air heater arrangement often are more favorable than the costs of a three-bisector arrangement. However, if fuel flexibility is weighed heavily in the evaluation, the three- or four-bisector arrangement may be advantageous.

The fuel flexibility of either type of air heater can be improved by specifying that the air heater be designed to accommodate an increase in heat transfer surface depth of up to 12 in. (305 mm) without significant housing or rotor modifications. The addition of heat transfer surface increases the capacity of the primary air system and reduces the exit gas temperature from the air heater. Also, the additional surface can be installed as a remedial action if the final flue gas temperature is higher than design.

The two-trisector and three- or four-bisector air heater designs appear to have some capacity for operation with one of two trisectors out of service or one of the two secondary air or primary air bisectors out of service. However, this capability cannot be realized unless isolation dampers are installed in the air and flue gas streams to stop the flow of air and flue gas through the idle air heater. Depending on the arrangement, cross-connection ducts may be required on the air and gas suction and discharge streams. Because of their clean air environment, the air dampers pose no significant operation or maintenance problems. However, the exposure of the flue gas dampers to hot dust-laden flue gas containing sulfur dioxide necessitates frequent maintenance to prevent corrosion. Seal deterioration, ash deposits, binding, and eventual inoperability of the damper will result if the damper is not cycled periodically. A damper system can be designed to allow part-load capability if one of two air heaters is inoperable; however, the owner must recognize that periodic cycling of the damper and intensive maintenance are required to ensure long-term operability of the system.

In addition to provision of multiple air heaters, a second method of addressing rotary regenerative air heater failure is to provide an installed spare electrical drive motor. The spare drive motor is designed to start automatically on failure of the operating drive motor. The spare drive motor design improves the reliability of the air heater system.

Air-driven motors are also provided for rotary regenerative air heaters. The air motor drive is designed for use during initial construction and maintenance of the air heater. The air motor typically will rotate the air heater at a lower than normal speed. The air motor can be used for short-term unit operation at load, but typically is not suitable for long-term operation.

7.3.8 Air Preheat Coils

Air preheat coils are installed upstream from the regenerative air heater. Although their use increases the boiler efficiency, their primary purpose is to prevent corrosion of the regenerative air heater by increasing heat transfer surface temperatures. The increased heat transfer surface temperatures are less likely to cause condensation of acids from the flue gas stream.

The design and operation of the air preheat system are based on maintaining an average cold end temperature (ACET) in the regenerative air heater. The ACET is defined as follows:

$$\text{ACET} = \frac{T_{\text{air in}} + T_{\text{gas out}}}{2} \tag{7-3}$$

where

$T_{\text{air in}}$ = the air temperature entering the regenerative air heater, and

$T_{\text{gas out}}$ = the uncorrected (no air to flue gas leakage) flue gas temperature leaving the regenerative air heater.

Recommendations for design ACET as a function of coal sulfur content are shown in Fig. 7-12. These design guidelines are published by ABB Air Preheater, Inc., a manufacturer of rotary regenerative air heaters.

The selection of the design ambient or inlet temperature for the air preheat coils involves a check of the meteorological records for the plant site. The minimum recorded ambient temperature often is used as a design requirement for air preheat coils. Obviously, coils using water or steam as a working fluid must be designed so that freezing will not occur under the coldest possible ambient temperature. However, the use of a design ambient temperature 10 to 20° F (6 to 11° C) higher than the minimum recorded should be considered for sites where extreme cold is of short duration. The use of a higher minimum ambient temperature value reduces the size and capital cost of the coils, and operation at lower than design ACET for a few hours per year does not significantly shorten the life of the regenerative air heater.

The air preheat coils may be arranged as first- or second-stage coils or a combination of both types. First-stage coils are upstream from the fans. Second-stage coils are between the fan and the regenerative air heater. Typically, an analysis is required to determine the most advantageous arrangement. Two generalities can be made regarding the system design:

- A first-stage coil system can be considered if a fan room encloses the secondary air and primary air fans or if the draft fans have inlet ductwork. The first-stage coils are typically designed to maintain a fan room or fan inlet temperature of 80° F (27° C). As a result of the large space available in a fan room wall, first-stage coils may have a lower air side pressure drop than second-stage coils with comparable heating duty.
- Second-stage air preheat coils are a typical arrangement, particularly if the fans are located outdoors. Second-stage coils also offer an operating benefit in that the draft fan inlet temperature is kept as low as possible, thus minimizing the volume of air and the fan power requirement.

The arrangement of the air preheat coils is affected by the regenerative air heater system. If bisector air heaters are used, each air stream will require an air preheat coil to protect each regenerative air heater. If trisector air heaters are used, coils are provided for the secondary air streams only.

The secondary air comprises about 80% of the total air flow through the air heater. Therefore, preheating of the secondary air treats the majority of the air flow, and is sufficient for corrosion protection of the air heater.

A second air preheat system design consideration is the choice of heat transfer medium. Alternative methods include steam, water, and ethylene glycol. As with the coil arrangement, a complete analysis is often required for the selection of the heat transfer medium, but the following generalities apply:

- A steam heat transfer system (Fig. 7-13) has the lowest capital cost and the lowest operating power requirement. A water system (Fig. 7-14) will have a higher capital cost and higher operating power requirement than a steam coil system. The ethylene glycol system (Fig. 7-15) has the highest capital cost and highest power requirements because of the intermediate heat exchangers and pumps required for the system.
- The ethylene glycol system has the best resistance to freezing, followed by the water system. Although a steam system is equally or only slightly more susceptible to freezing than a water system, a properly designed and operated steam coil

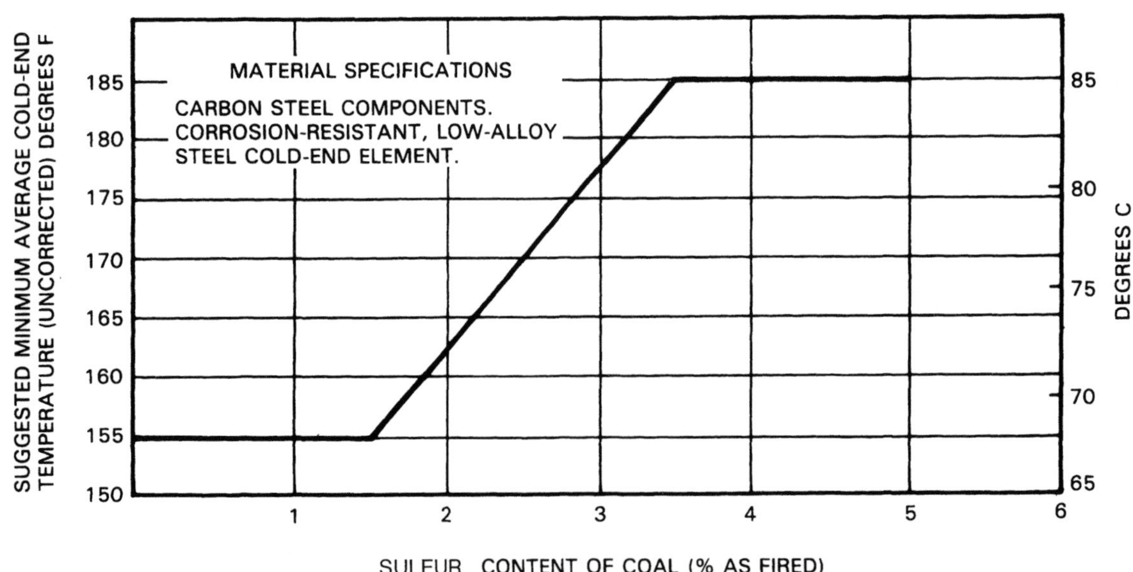

Fig. 7-12. Average cold end temperature design guide. (From ABB Air Preheater, Inc. Used with permission.)

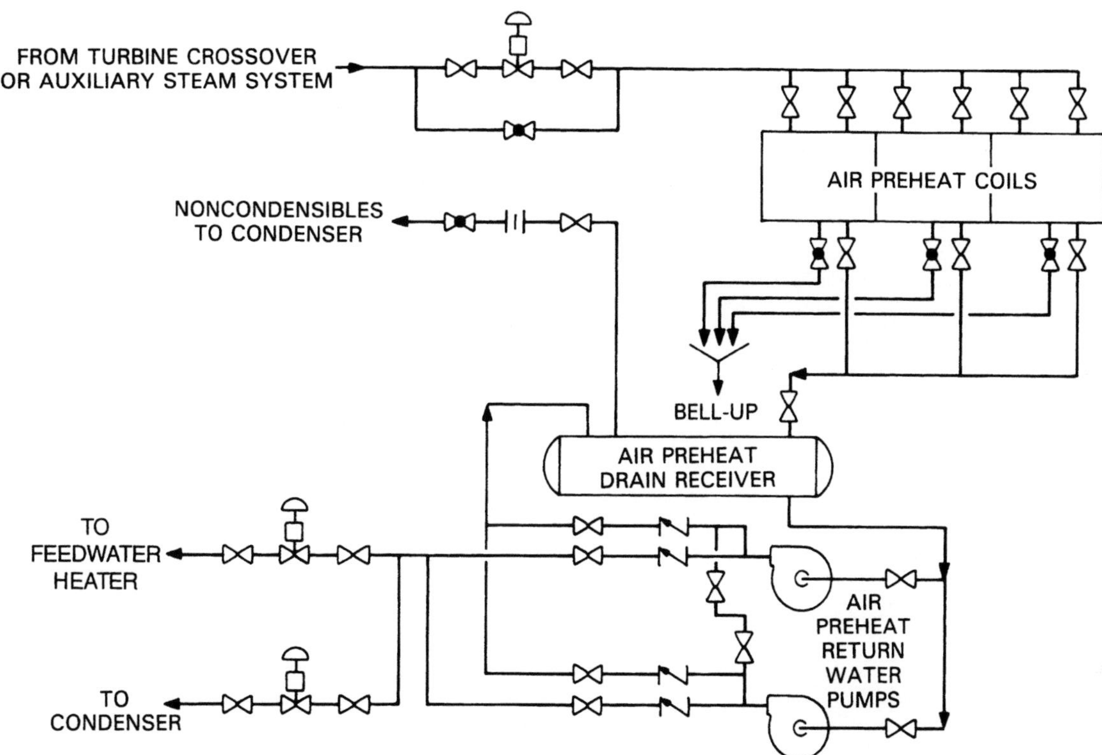

Fig. 7-13. Steam air preheating coils flow diagram.

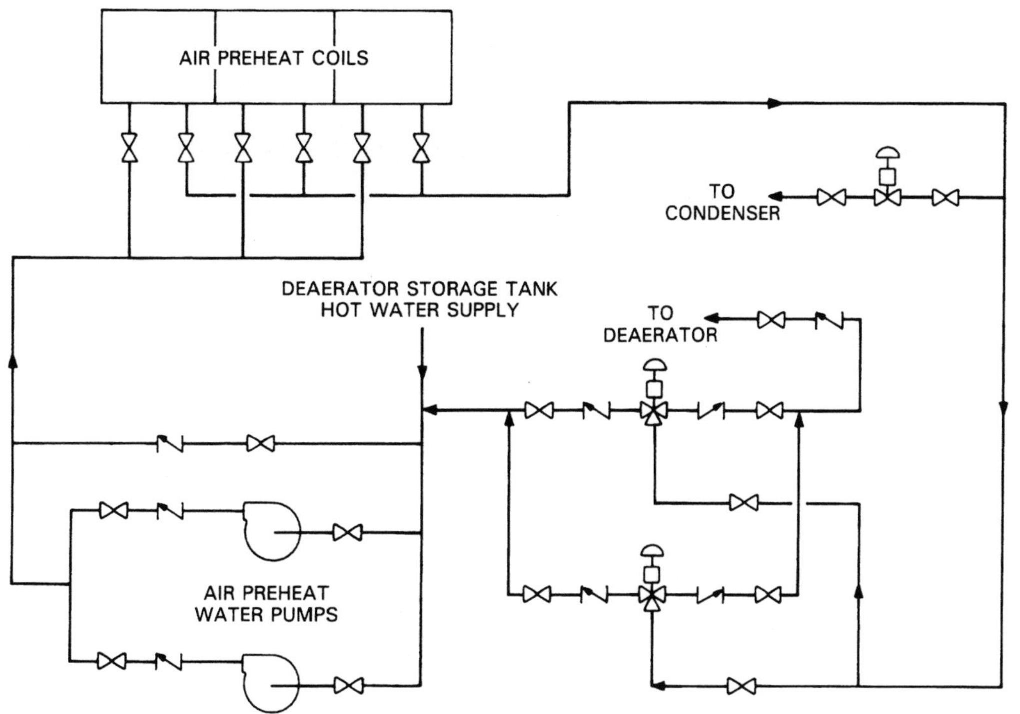

Fig. 7-14. Hot water air preheating coils flow diagram.

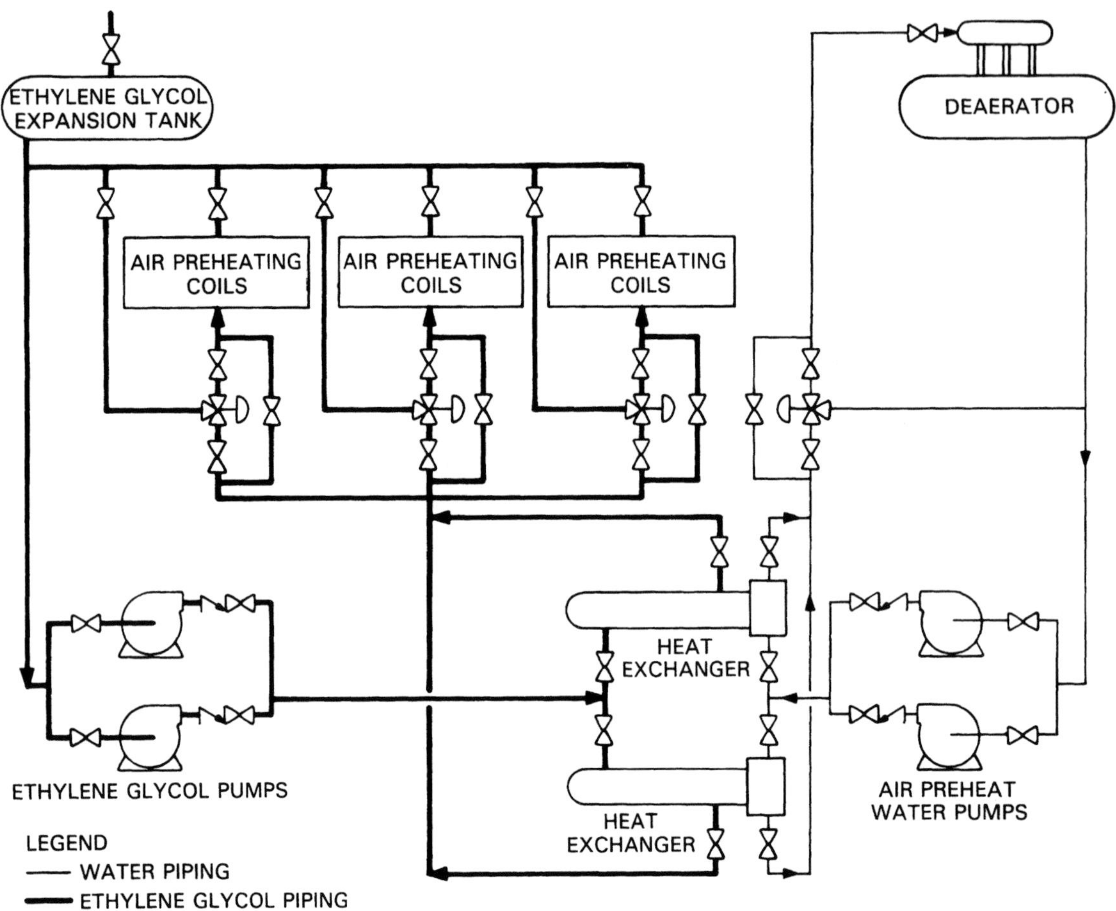

Fig. 7-15. Ethylene glycol air preheating coils flow diagram.

system can be operated reliably without freezing if the coils are drained properly.

- The operating pressures in an ethylene glycol system must be designed so that any leakage in the intermediate heat exchanger will be from the water side to the glycol side. Ethylene glycol contamination of the steam cycle is detrimental to the steam system and is very difficult to clean up.

Steam coils should be designed to drain by gravity through a level controlled drain tank. Pumping the coil drains without a collection tank is a very difficult pump application because of the rapid and large pressure variations inherent in the operation of steam air preheat coils.

The heat source for air preheating systems typically is intermediate pressure turbine exhaust or "crossover" steam to the low-pressure turbine (Fig. 7-13). This extraction typically serves as the deaerator heating steam. Therefore, a steam system uses crossover steam directly in the coils.

As shown in Fig. 7-14, a water system uses deaerator water in the coils; therefore, the deaerator must be designed with extra water heating capability if a water air preheating system is used.

The water temperature to the coils is controlled by the three-way control valves shown in Fig. 7-14. If an increase in the air temperature from the preheat coils is needed, more of the coil discharge water is directed to the deaerator and more hot water is drawn from the deaerator. Conversely, if less air preheating is needed, the three-way valve directs less water to the deaerator, and more water bypasses the deaerator directly to the air preheat water pump suction, thus reducing the water temperature to the air preheat coils.

A glycol system (Fig. 7-15) uses water from the deaerator circulating through intermediate water/glycol heat exchangers. The heated glycol circulates through the air preheat coils. The energy delivered by the water system is controlled as described for a water–air preheat coil system.

7.3.9 Soot Blowers

Soot blowers are used for removal of ash deposits from the fireside of heat transfer surfaces. Several types of soot blowers are used in utility steam generators. Typical locations of the various types are shown in Fig. 7-16.

Wall blowers are used for furnace walls. Wall blowers have a very short lance with a nozzle on the tip. The lance rotates as it moves into the furnace, and the nozzle directs the soot blowing medium onto a circular area of the furnace wall.

Retractable water lances are used in difficult applications

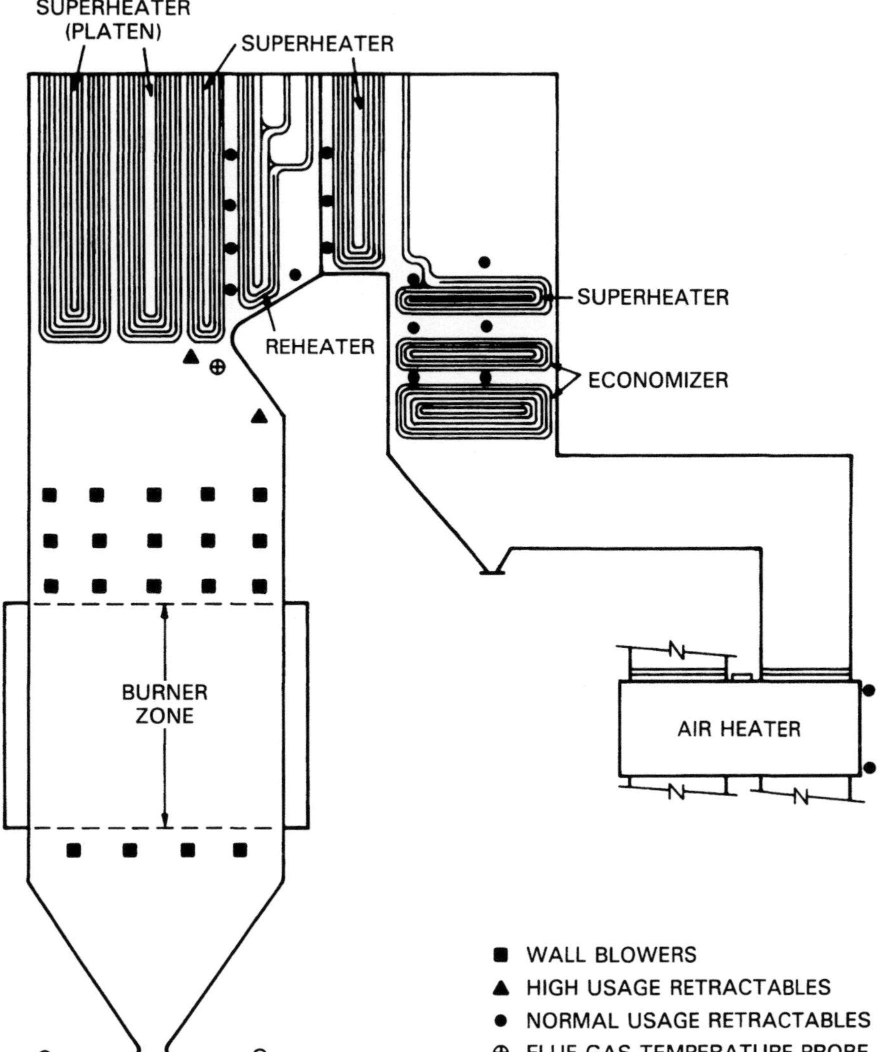

Fig. 7-16. Typical soot blower locations.

for spot removal of heavy slag. Water lances are typically located at the furnace knuckle or in the furnace throat, where heavy slag may accumulate.

Retractable soot blowers (Fig. 7-17) are used for cleaning of furnace pendant surface and convection pass tube banks. Retractables may be fully retractable or partially retractable, depending on the temperature zone. Retractable soot blowers have a nozzle at the end of a lance that rotates as the lance travels along the tube surface, perpendicular to the glue gas flow (Fig. 7-18).

Fully retractable soot blowers are required in high flue gas temperature zones, where the soot blower materials could not withstand the flue gas temperature without cooling flow through the soot blower. Partially retractable soot blowers, including half-tract or one-third-tract soot blowers, can be used where the soot blower metals can withstand the flue gas temperature without cooling flow. Half-tract soot blowers retract or move laterally only half of the lance length, and provide full cleaning coverage with two nozzles on the lance. One-third-tract soot blowers retract only a distance of one-third of the lance length and provide full cleaning coverage with three equally spaced nozzles on the lance.

Rotary soot blowers are used for low-temperature convection pass surface. Rotary soot blowers have multiple nozzles along the shaft of the blower, are fixed axially, and rotate to distribute the soot blowing medium over the adjacent tubes.

Air heater soot blowers are typically found in the flue gas side of the air heater on both the inlet and the outlet. Selection of the air heater soot blower depends on the size of the air heater, as follows:

Air heater size	Soot blower type
Small, package type	Stationary multinozzle
Medium	Nozzle mounted on pivoted arm
Large	Retractable

The soot blower types listed presume the use of a fluid soot blowing medium. Acoustical soot removal systems

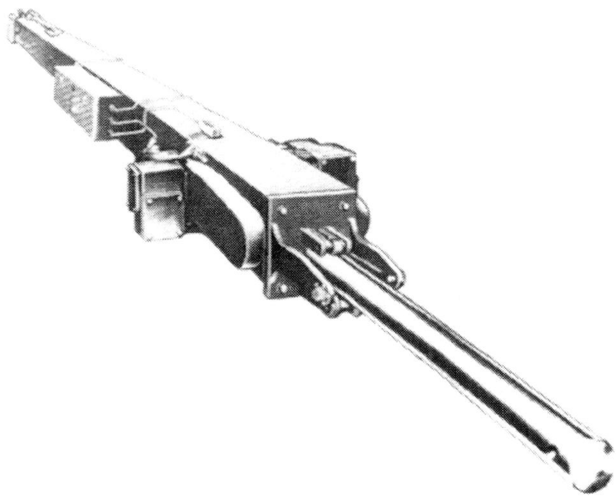

Fig. 7-17. Retractable soot blower. (From Copes Vulcan. Used with permission.)

have been used in utility and industrial service. Acoustical systems apply low-frequency sound waves to induce vibration of the tubes and dislodge the ash. Such systems are not widely used, but remain an alternative to conventional systems.

Mechanical rapping of tubes has also been used on small installations; however, this arrangement is not considered practical for large utility systems.

Conventional soot blowing media include two alternatives: steam and air. Although water lances were listed among the soot blower types, water has not been widely used because of the potential for thermal shocking of tubes. However, heavy slag accumulations that are not removed by air or steam can be effectively removed by water soot blowers. The potential for thermal shock can be reduced by avoiding water blowing on clean heat transfer surface and minimizing the blowing frequency. Water blowers, like conventional soot blowers, should be operated only when surfaces are dirty, rather than on a fixed interval.

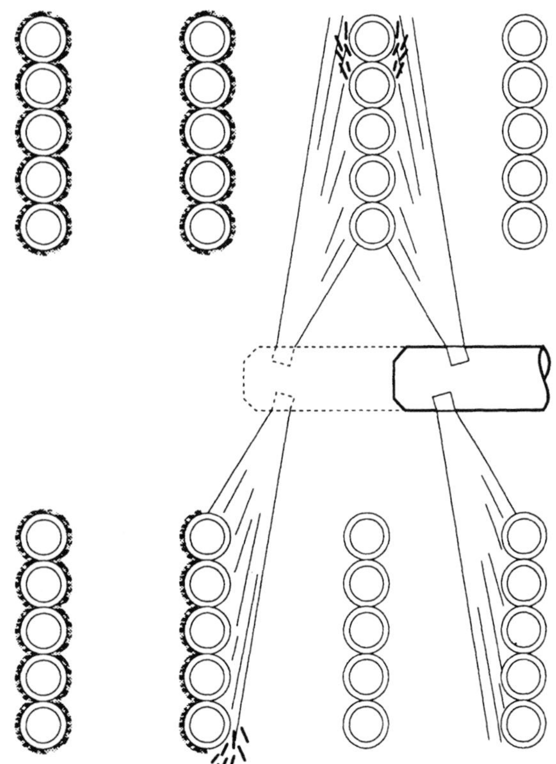

Fig. 7-18. Soot blowing flow pattern. (From Copes Vulcan. Used with permission.)

A clear case for ash removal superiority of steam over air probably cannot be made. Each medium has its advocates among the operating community based on individual experience. Beyond ash removal effectiveness, each medium offers some advantage.

Air systems have much simpler piping system requirements than steam systems. Steam piping systems must be sloped to allow condensate drainage, and must have a valve system to remove condensate from the soot blowing piping. The maintenance on an air system is less than that required for the numerous valves in a steam system.

The steam system has a clear advantage in the capability for system expansion. Additional soot blowers can be accommodated since the steam supply is essentially unlimited. The soot blowing capacity of a compressed air system is dependent on the number of compressors, compressor capacity, air receiver capacity, and soot blower flow requirement. Because the air supply system capacity is finite, the extension of an existing air soot blower system may be capacity limited.

Project requirements must be individually evaluated for selection of a soot blowing medium. However, the ease of expansion of a steam system often is cited as a deciding factor in the selection of a steam soot blowing system. If plant water consumption is to be minimized, an air soot blowing system may be advantageous.

7.3.10 Coal Feeders

Coal feeders are located between each coal silo and its respective pulverizer. The principal function of a coal feeder is to control the flow of coal to the pulverizer, thus matching the fuel flow to the steam demand.

The feeder design commonly used for large-scale power plants is the horizontal belt type shown in Fig. 7-19. Coal flows onto the moving belt from a vertical feed pipe and is discharged from the end of the belt into the vertical pulverizer feed pipe. The belt speed is varied to control the coal flow. The feed pipe is typically constructed of 304 stainless steel to enhance coal flow.

Two variations of the belt feeder are the volumetric feeder and the gravimetric feeder.

Volumetric belt feeders typically use a fixed position leveling bar in combination with a variable speed belt to control the coal flow.

The gravimetric feeder is equipped with a belt scale that

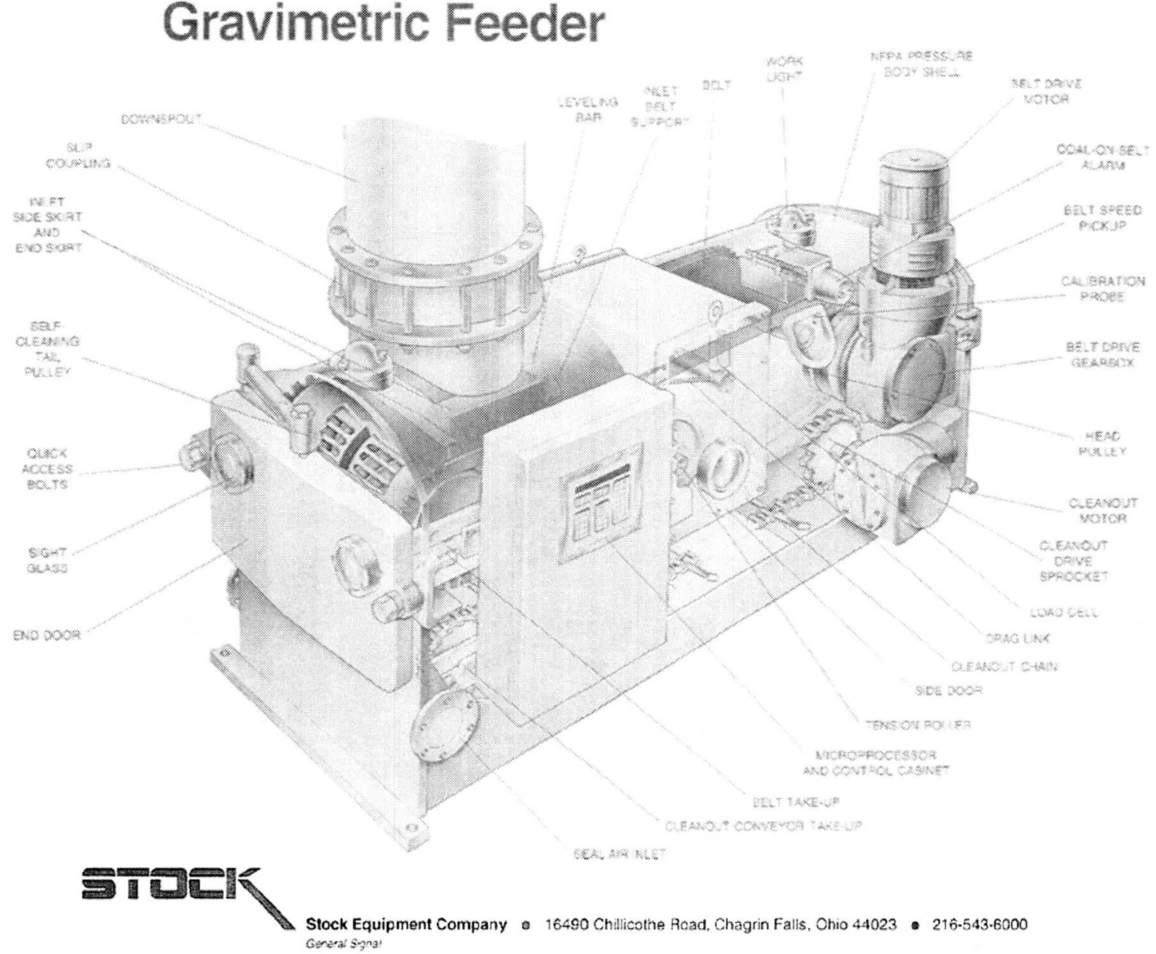

Fig. 7-19. Stock gravimetric feeder. (From Stock Equipment Company. Used with permission.)

weighs the coal as it passes through the feeder. The associated feeder control system measures and records both instantaneous coal feed rate and the cumulative weight of coal fed. The gravimetric feeder is preferred for utility installations because of its ability to sense and respond to changes in coal density. As the heat content of coal is generally more dependent on the weight of the coal being fed than on the volume, the gravimetric feeder should yield a better control of the energy flow to the furnace. Also, the gravimetric feeder offers a method of measuring coal consumption if feeder maintenance and scale calibration are performed regularly.

The feeder system is pressurized with seal air to provide a seal against hot air flow from the pulverizer to the silo in pressurized pulverizer systems. (Note that in pulverizers equipped with exhauster fans, the pulverizer operates at negative pressure and air flow is from the silo, through the feeder, and into the pulverizer.) With pressurized pulverizers, which operate at a positive pressure of 10 to 15 in. (254 to 381 mm) of water, air (if unrestricted) flows from the pulverizer through the feeder and up into the silo. To counter this effect, the vertical coal pipe at the feeder inlet serves as a seal against gross air movement. The inlet pipe typically is 24-in. (610-mm) diameter, but may be as large as 36 in. (914 mm) if the coal flow properties are difficult. The seal pipe sizing criterion is 4 in. (102 mm) of vertical coal pipe per inch (23.4 mm) of water gauge pressure in the pulverizer.

In lieu of a round feeder inlet, rectangular slot feeder inlets have also been used for coals with clay content and difficult flow characteristics. Slot feeder inlets are not common among United States utility installations, but they appear to be an effective alternative design.

7.3.11 Pulverizers

Three types of pulverizers have been used on coal-fueled utility boilers: the ball tube mill, vertical spindle mill, and attrition mill.

The low-speed ball tube mill (Fig. 7-20) is characterized by very low maintenance costs but high power consumption. Wear part maintenance consists of replacing wear liners on a 10- to 15-year frequency and replenishing the ball charge several times a year.

The medium-speed vertical spindle mill can be either a bowl-and-roller or ball-and-race mill. The bowl-and-roller type (Figs. 7-21, 7-22, and 7-23) is used predominantly in power installations and is expected to be used in virtually all

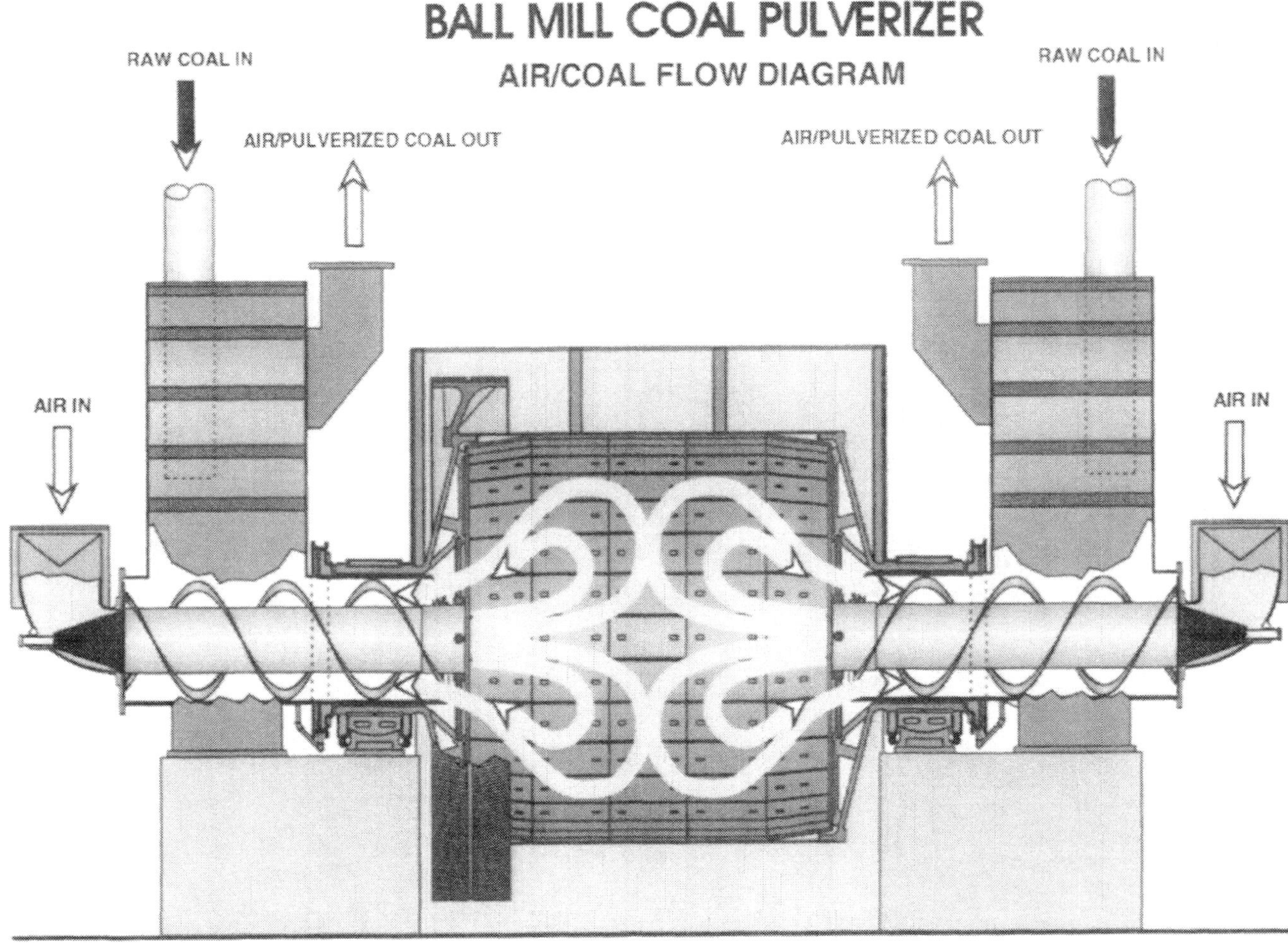

Fig. 7-20. Ball tube mill. (Illustration courtesy of Foster Wheeler Corporation.)

large boilers in the future. The bowl-and-roller pulverizer is characterized by medium to high maintenance and low power consumption. Pulverizer overhauls for replacement or renewal of roller wear surfaces are required on a 2- to 5-year frequency, depending on the abrasion characteristics of the coal. Ball-and-race pulverizers (Fig. 7-24) typically have been used on small unit sizes, although some large installations do have ball-and-race mills.

The high-speed attrition mill (Fig. 7-25) typically has been used on small installations. It is characterized by high power consumption and high maintenance, and mill overhauls are required approximately every year.

Several variations of vertical spindle mill arrangements are available (Fig. 7-10). The most common variation is the pressurized pulverizer with atmospheric suction (cold) primary air fans. This pulverizer arrangement operates with a pressure of about 10 to 15 in. (254 to 381 mm) of water gauge above the grinding zone.

A second arrangement pressurizes the pulverizer with hot PA fans. The hot PA fans are usually arranged with one fan for each pulverizer. Hot PA fans have typically been used on 200-MW and smaller boilers.

The third arrangement uses exhauster fans, one exhauster per pulverizer, and the pulverizer operates at about −5 to −10 in. (−127 to −254 mm) of water gauge above the grinding zone.

Pressurized mills with cold PA fans are the most common arrangement for large boiler service. Hot PA fans may be economical for small boilers, and individual manufacturers will determine if a hot fan arrangement is the most competitive arrangement for the unit size and the coal type. Exhauster pulverizers are used on units as large as 900 MW. As with hot PA fans, the manufacturer determines if the exhauster arrangement is the most economic selection.

7.3.12 Coal Piping and Burners

Coal piping conveys the pulverized coal–primary air mixture to the burners. The pulverized coal piping is typically steel with ½ in. (12.7 mm) wall thickness. Ceramic linings are recommended on coal pipe bends if the pulverized coal is particularly abrasive. A basic ceramic liner arrangement includes the mill outlet area, the first two pipe bends out of the pulverizer, and 5-ft straight sections beyond these bends. For severe erosion problems, all coal pipe bends may require ceramic lining.

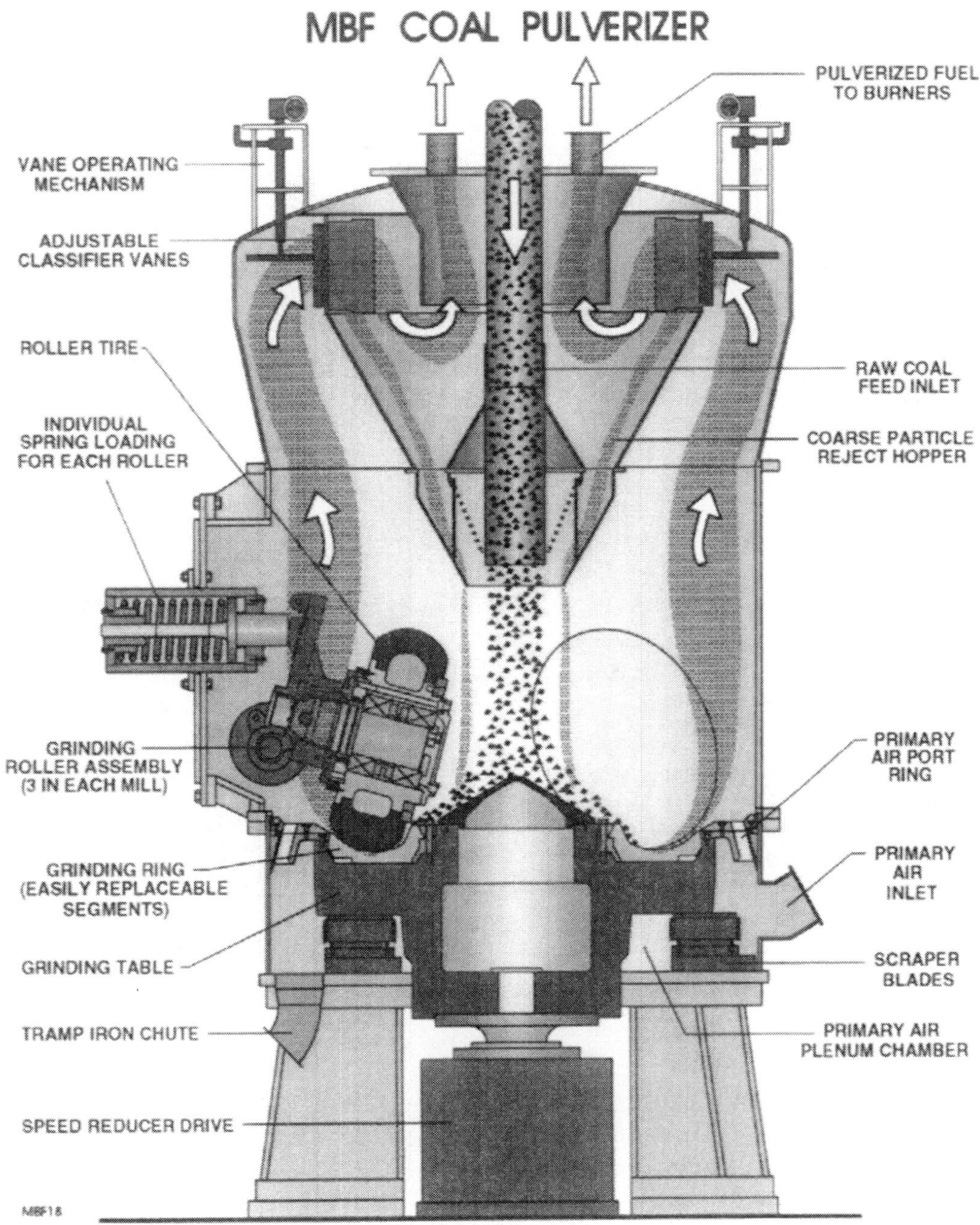

Fig. 7-21. Vertical spindle mill. (Illustration courtesy of Foster Wheeler Corporation.)

The burner pipe arrangement and accessories are determined by the burner arrangement in the furnace and the type of pulverizer. Two types of burner arrangements are used on large utility boilers: wall-fired (Fig. 7-3) and corner- or tangentially fired (Fig. 7-2). The wall-fired furnaces may be front wall-fired, rear wall-fired, or the burners may be front- and rear-wall opposed. The typical wall-fired burner arrangement is configured such that one pulverizer feeds all the burners on one level on one wall. Pressurized pulverizers have outlets equal in number to the burners fed by the pulverizer. Exhauster pulverizers have a single outlet stream from the exhauster fan. The exhauster fan discharge is divided with coal pipe splitters to obtain an individual coal pipe for each burner served by the pulverizer. Exhauster pulverizers are typically used in conjunction with tangentially fired furnaces. The exhauster mill and wall-fired furnace combination is unusual, but does exist on some older boilers.

For tangentially fired furnaces, modern designs are arranged to fire from four furnace corners. Older tangential furnace designs may be arranged with a divided furnace with two fire balls, one in each furnace side. The divided furnace design has been built both with and without a center division wall. The divided furnace has eight burners per burner eleva-

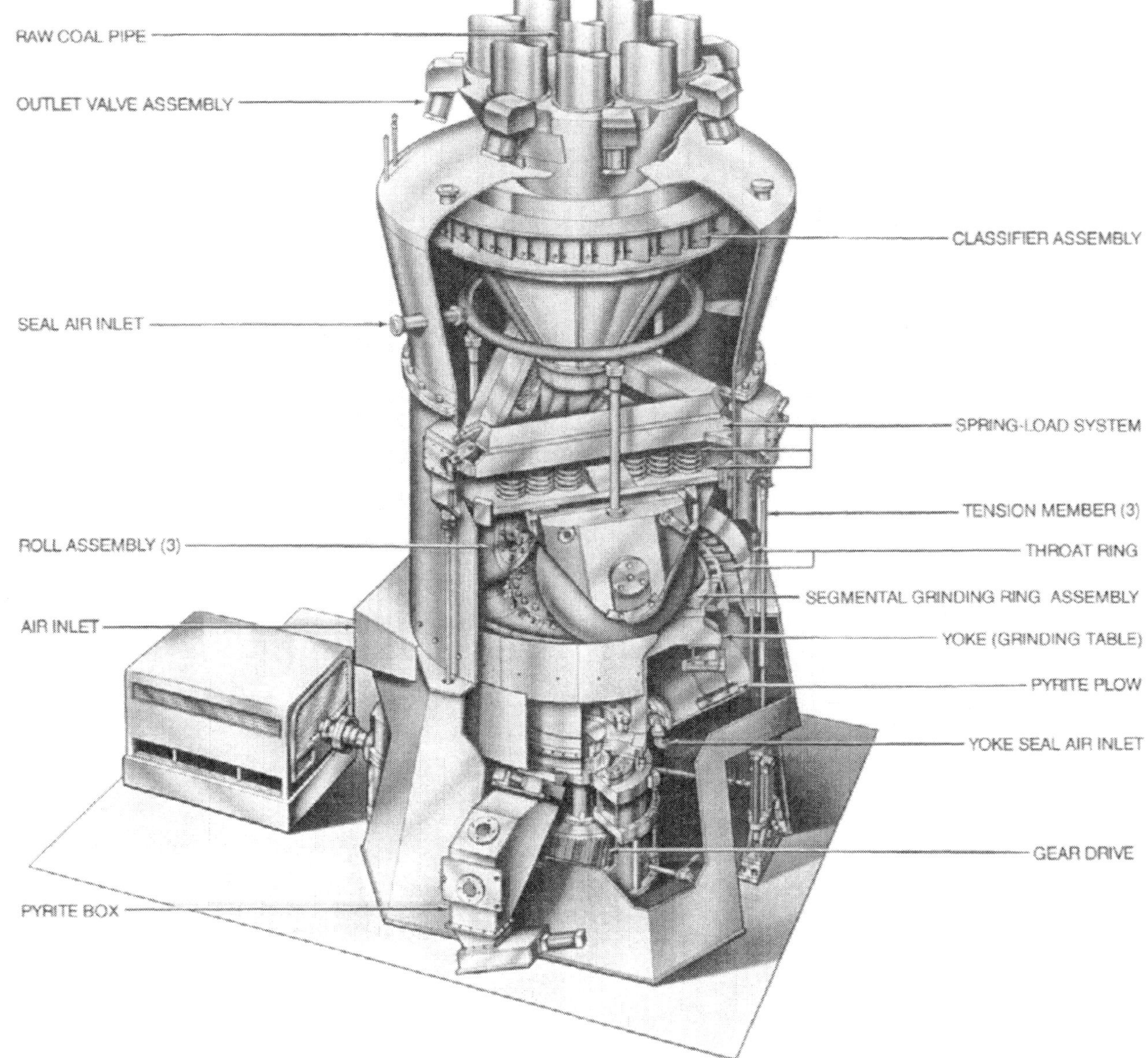

Fig. 7-22. Vertical spindle mill. (From Babcock & Wilcox. Used with permission.)

tion. One pulverizer feeds all eight burners on one elevation of a divided furnace or all four burners per elevation of a "single fire ball" tangential furnace.

7.3.13 Ignitors and Warmup Burners

Ignitors and warmup burners are necessary for flame initiation and low load stabilization. The type of equipment provided and even the terminology varies depending on the boiler manufacturer.

All manufacturers use the term ignitor except for Babcock & Wilcox, which uses the term "lighter." Warmup burners or warmup guns are used only by ABB Combustion Engineering Services in tangentially fired furnace designs. The warmup gun is a high heat input gas or oil burner that performs the functions of an ignitor. As an example, the heat input of one level of four warmup guns typically equals or exceeds the heating capacity of all ignitors on all elevations, both front and rear, of a comparably sized wall-fired boiler.

The following example for a 5,200,000 lb/h (2,360,000 kg/h) 750 MW boiler illustrates the burner, ignitor, and warmup gun variations that may be encountered:

	Manufacturer designation		
	A	B	C
Number of coal burners	56	32	28
Burner rating, MBtu/h	120	210	240
MJ/h	126,000	221,000	253,000
Number of ignitors	56	28	28
Ignitors rating, MBtu/h	12	10	25
MJ/h	12,600	10,500	26,400
Number of warmup guns	—	4	—
Warmup gun rating, MBtu/h	—	245	—
MJ/h		258,000	

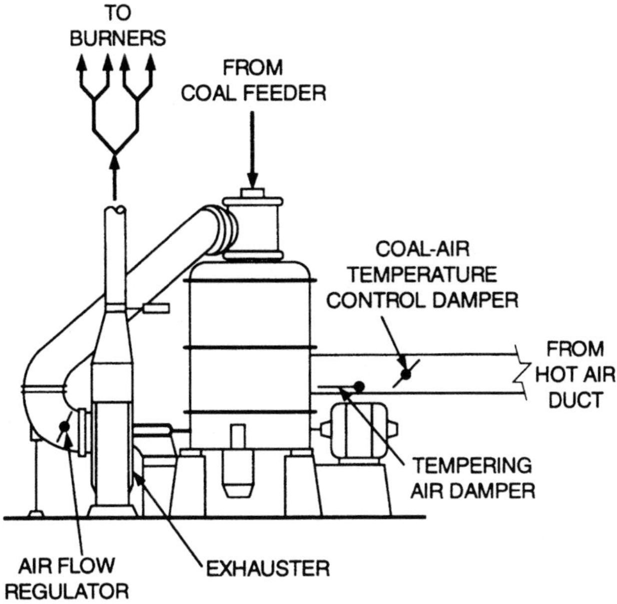

Fig. 7-23. Vertical spindle mill. (From Asea Brown Boveri. Used with permission.)

The ignitor itself requires an ignition source. For large pulverized coal units, the initial ignition source is a high-energy electrical spark that starts the ignitor flame.

Fossil-fueled ignitors typically are fueled with natural gas or any grade of fuel oil. Propane can be used, but typically is not the most economic ignitor fuel for a utility. Electrically powered "plasma torch" ignitors, which use neither natural gas nor fuel oil, are also commercially available.

The plasma torch direct ignition system (PTDIS) is a direct-fired coal ignitor fuel system offered by Babcock & Wilcox. The plasma torch converts electrical energy to thermal energy by establishing an electrical arc between a pair of electrodes. The plasma torch is located on the axis of the B&W coal burner and replaces the conventional oil-fired ignitor. Virtually no other changes in the burner/pulverizer system are required compared to a conventional ignitor/burner/pulverizer system. However, the plasma torch system does require additional secondary unit substation breakers, conductor, and raceway. Auxiliary power transformer sizing is not affected since the plasma torch system operates when other auxiliary loads are low.

If fuel oil is used, oil atomizing and ignitor purge methods must be selected. Mechanical atomization, or compressed air or steam atomization mediums have been used. Mechanical atomization has been used for main burner service and igni-

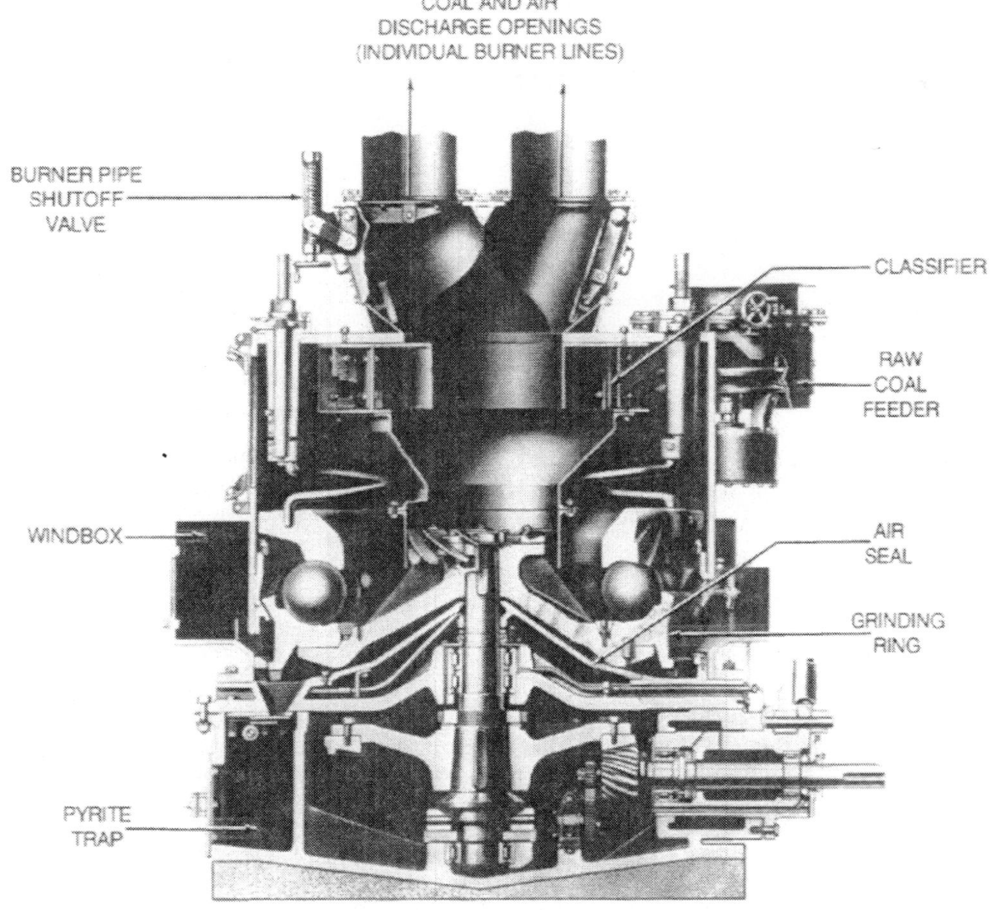

Fig. 7-24. Ball and race mill. (From Babcock & Wilcox. Used with permission.)

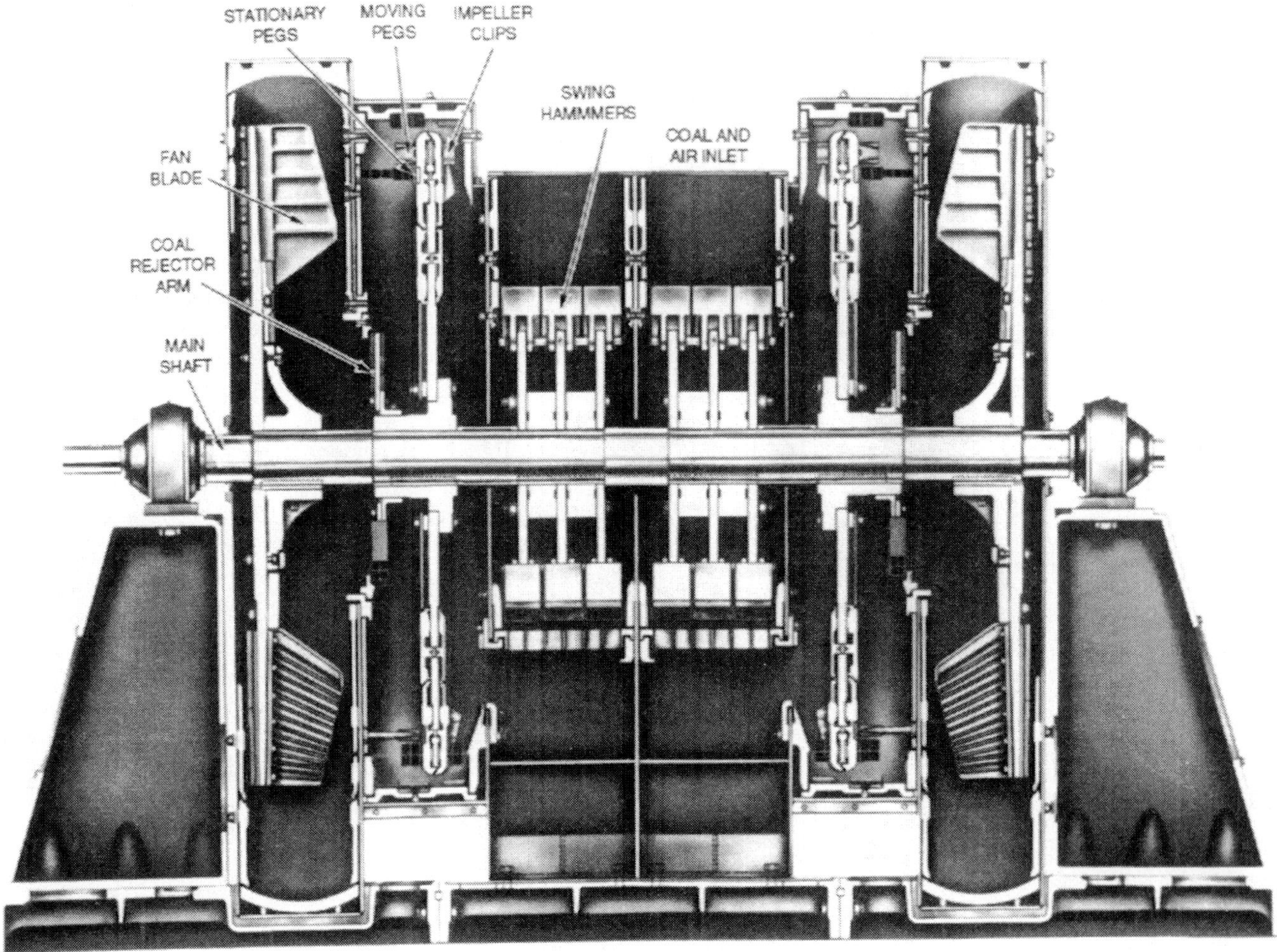

Fig. 7-25. High-speed attrition mill. (From Riley Stoker. Used with permission.)

tor service in older units. However, compressed air or steam typically provides better atomization of the oil and more complete combustion. Therefore, modern utility boilers typically are specified with compressed air or steam atomization.

The type of oil used also influences the atomization medium selection. Light oils such as No. 2 are effectively atomized with compressed air. Heavy oils such as No. 6 can be atomized with compressed air, but since No. 6 oil must be maintained at 200° F (93° C) for atomization, the heat energy of the steam is considered to be an inherent advantage. Steam is recommended as the atomizing medium for No. 6 fuel oil.

The ignitor purge medium is the same as the selected atomizing medium.

Ignitor classes are defined for different types of service in the National Fire Protection Association (NFPA) Code 85C. The ignitor class definitions are as follows (NFPA 1993):[3]

- Class 1 (Continuous Ignitor). An ignitor applied to ignite the fuel input through the burner and to support ignition under any burner light-off or operating conditions. Its location and capacity are such that it provides sufficient ignition energy (generally in excess of 10% of full load burner input) at its associated burner to raise any credible combination of burner inputs of both fuel and air above the minimum ignition temperature.

- Class 2 (Intermittent Ignitor). An ignitor applied to ignite the fuel input through the burner under prescribed light-off conditions. It is also used to support ignition under low load or certain adverse operating conditions. The range of capacity of such ignitors is generally 4% to 10% of full load burner fuel input. It shall not be used to ignite main fuel under uncontrolled or abnormal conditions. The burner shall be operated under controlled conditions to limit the potential for abnormal operation, as well as to limit the charge of fuel to the furnace in the event that ignition does not occur during light-off. Class 2 ignitors may be operated as Class 3 ignitors, but not to extend turn-down range.

[3]Reprinted with permission from NFPA 85C, *Prevention of Furnace Explosions/Implosions in Multiple Burner Boiler-Furnaces*, Copyright © 1991, National Fire Protection Association, Quincy, MA 02269. This reprinted material is not the complete and official position of the National Fire Protection Association, on the referenced subject which is represented only by the standard in its entirety.

- Class 3 (Interrupted Ignitor). A small ignitor applied particularly to gas and oil burners to ignite the fuel input to the burner under prescribed light-off conditions. The capacity of such ignitors generally does not exceed 4% of the full load burner fuel input. As a part of the burner light-off procedure, the ignitor is turned off when the timed trial for ignition of the main burner has expired. This is to ensure that the main flame is self-supporting, is stable, and is not dependent on ignition support from the ignitor. The use of such ignitors to support ignition or to extend the burner control range shall be prohibited.
- Class 3 Special (Direct Electric Ignitor). A special Class 3 high energy electrical ignitor capable of directly igniting the main burner fuel. This type ignitor shall not be used unless supervision of the individual main burner flame is provided.

Class 1 ignitors are typically specified for utility service.

The combustion air for the ignitors may be supplied either by the forced draft fans or by separate ignitor air fans. If separate ignitor air fans are furnished, two full-capacity fans should be specified to enhance the reliability of the ignitor system.

7.3.14 Ductwork, Ash Hoppers, and Dampers

The steam generator includes both air and flue gas ductwork. The following ductwork systems typically are included in the steam generator specification:

- Air heater secondary air outlet to windbox,
- Primary air fan discharges to air heater,
- Air heater primary air outlet to pulverizers,
- Tempering air to pulverizers,
- Seal air ducts,
- Flame detector cooling air ducts,
- Ignitor air ducts,
- Steam generator flue gas outlet to air heater, and
- Air heater gas outlet to a specified terminal point.

If the forced draft (FD) fans are included as part of the steam generator scope of work, the ductwork between the FD fan and the air heater should be included as part of the steam generator. If the FD fan is purchased separately from the steam generator, the FD fan discharge ductwork may be purchased separately from the steam generator for convenience.

For utility steam generators, ductwork is specified to be steel plate, ¼ in. (6 mm) thick. Ductwork deflection criteria are as follows:

- Duct platework designed to limit deflection to 1/100 of the span of plate between stiffeners or the plate thickness, whichever is less, under normal operating conditions.
- Duct stiffeners designed to limit the deflection to 1/240 of the stiffener span for normal operating conditions.

Duct air and gas velocities at full load are specified as follows:

Air ducts, maximum velocity	3,500 ft/m (17.8 m/s)
Flue gas ducts, minimum velocity	3,500 ft/m (17.8 m/s)

For air, the velocity is given as a maximum to limit the duct pressure drop and the potential for vibration. For flue gas, a minimum velocity is used to minimize deposition of fly ash in the ductwork.

Both air and gas ducts may be constructed of ASTM A36 carbon steel. Ash hoppers are provided under the convection pass for collection of economizer ash. The following criteria are used in the specification of economizer ash hoppers:

Ash storage capacity, hours of operation	12
Slope of hopper sides, minimum degrees	60
Design ash density	
Volume	45 pcf (721 kg/m^3)
Weight	120 pcf (1,922 kg/m^3)

A low ash density is used for volume to ensure the hopper is adequately sized. A high ash density is specified to ensure that the hopper structure support is sufficient.

Ash hoppers may be provided under the regenerative air heater. Often these hoppers are designed for collection of wash water. A shallow hopper design may be used that collects water, but is "self-sweeping" with respect to fly ash.

The following dampers are typically purchased with the steam generator:

- Primary air fan inlet vanes;
- Primary air fan discharge isolation;
- Primary air heater gas outlet (when separate primary air heaters are provided);
- Pulverizer air flow, temperature control, and isolation;
- Sealing, cooling, pressurizing, and ignitor air control and isolation; and
- Burner air registers.

Isolating and control dampers are specified as multileaf opposed blade type. Typical damper technical specifications follow:

Maximum allowable bending stress, percentage of yield	60
Maximum combined axial compressive and bending stress, percentage of yield	60
Maximum deflection of frame components	Length/360

In addition, damper blade deflection at the maximum design differential pressure is not to impair damper operation or sealing capability.

Damper materials are typically specified as ASTM A36 carbon steel.

Typical damper blade criteria:	
two blades, maximum duct area	40 ft^2 (3.7 m^2)
three blades, duct area	40 to 80 ft^2 (3.7 to 7.4 m^2)
four blades, duct area	80 or more ft^2 (7.4 m^2)
Maximum blade width	30 in. (762 mm)

7.3.15 Insulation and Lagging

Insulation both reduces the heat loss from the boiler and protects operations and maintenance personnel from contact with high-temperature surfaces. Lagging or jacketing protects the insulation from physical damage.

Insulation typically is specified for surfaces that operate at a temperature exceeding 130° F (54° C). Coal piping between the pulverizers and the burners exceeds that temperature, but typically is not insulated. Insulation on coal piping would obscure the source of pulverized coal leaks as the coal piping wears.

Insulation is typically specified to limit the cold face temperature to 120° F (49° C) maximum based on an ambient air temperature of 80° F (27° C) with an air velocity of 60 ft (18.3 m) per minute. This requirement is typically met with 4 to 6 in. (102 to 152 mm) of insulation on boiler flat surfaces.

Steam generator surfaces, hot air, and gas ducts are insulated with mineral fiber block insulation. Rigid calcium silicate molded insulation may be used on the penthouse roof and on the top surface of ducts. The rigid material is expected to withstand occasional foot traffic.

Lagging may be specified as either aluminum or steel, although aluminum is used most often.

The following are typical lagging types and thicknesses for aluminum lagging:

Location	Type	Thickness	
		in.	mm
Steam generator	Ribbed or fluted aluminized steel for roof deck	0.040	1.016
Ducts	Ribbed or fluted	0.040	1.016
Equipment	Flat aluminum	0.050	1.27
		0.080 in areas subject to foot traffic	2.032
Piping	Aluminum sheet	0.020 up to 13 in. OD	0.508 mm up to 330 mm OD
		0.024 all other sizes	0.6096 mm

Steel is an alternate lagging material for equipment. Steel lagging is specified with a minimum 20-gauge (0.95-mm) thickness.

7.4 GLOSSARY

Air heater. Heat transfer device that transfers heat from the flue gas to the combustion air.

Air preheat coils. Heat transfer device that transfers heat from a heating medium (steam or hot water or other heated fluid) to the combustion air, but prior to entry of the air into the air heater.

Ash collection. Hoppers for collection and removal of ash from the flue gas stream.

Boiler circulating pumps. Pumps that force water circulation through the furnace water walls.

Buckstays. See Structural Members.

Burners. Equipment that directs the primary air/coal mixture and the secondary air such that mixing and efficient combustion occurs.

Coal feeders. Feeders receive coal from the coal silos and meter the coal flow to the pulverizers.

Coal piping. Piping that directs the coal/air mixture from the pulverizers to coal burners.

Coal pulverizers. Equipment that reduces the fuel particle size, mixes the fuel with hot primary air, and thereby dries the fuel.

Coal silos. Silos that provide ready storage of fuel within the boiler structure.

Column lines. See Structural Members.

Critical pressure. At or above the critical pressure of 3,208.2 psi (22.12 MPa), addition of heat to water causes a continuous transition from water to steam, with a continuous increase in temperature as heat is added.

Drum. Vessel that encloses the steam–water interface within the steam generator.

Drum type steam generator. A steam generator that incorporates a drum in the steam and feedwater circuitry. The alternative to a drum steam generator design is a once-through steam generator design. The drum type boiler can be used with subcritical steam cycles. The drum type boiler can be of either reheat or nonreheat design.

Ductwork. Duct system that directs the flow of combustion air and flue gas.

Economizer. Heat transfer surface used to increase the temperature of the feedwater prior to its entry into the drum.

Flame scanners. Radiation sensing devices that provide indication of flame in the furnace.

Forced draft or secondary air fans. Fans that provide air flow for completion of combustion in the furnace. The terms forced draft and secondary air are used interchangeably in the industry and throughout this chapter.

Furnace. Enclosure for combustion of fuel and heat transfer to the feedwater.

Hardgrove Grindability Index. The Hardgrove Grindability Index (HGI) was developed to quantify the difficulty of grinding a coal sample. The HGI is derived from the quantity of 200-mesh [0.0029 in. (0.074 mm)] coal sample particles produced by 60 revolutions of the grinding spindle of a Hardgrove grindability machine. The procedure for determining HGI is given in ASTM Standards D409, Grindability of Coal by the Hardgrove-Machine Method.

Ignitors or lighters and warmup burners or warmup guns. Equipment that ignites and stabilizes the coal flame. Warmup burners or warmup guns and ignitors or lighters are burners used specifically for ignition and heatup. Warmup guns are used in corner-fired boilers; ignitors or lighters (one for each main burner) are used in wall-fired arrangements.

Induced draft fans. Fans that exhaust the flue gas from the furnace such that a slight negative pressure is maintained within the furnace.

Insulation and lagging. Insulation prevents heat loss to ambient and protects personnel from contact with the hot surfaces. Lagging is a metal sheet that protects the insulation from damage.

Once-through steam generator. A steam generator designed with no fixed steam–water interface. The alternative steam generator design incorporates a steam drum that contains the steam–water interface. A once-through design may operate at either subcritical or supercritical steam pressure, with a reheat or nonreheat steam cycle.

Overfire air ports. Air ports that direct secondary air into the furnace above the combustion zone, thus reducing NO_x formation.

Penthouse. The penthouse encloses the drum, high-pressure headers, and connecting links on top of the steam generator. The floor of the penthouse is formed by the roof tubes of the steam generator.

Primary air fans. Fans that provide the air flow quantity required to dry the coal within the pulverizers and convey the pulverized coal to the burners.

Reheat steam generator design. A single reheat steam generator includes a heat transfer surface for heating the high-pressure turbine exhaust steam to a temperature equal or above the main steam temperature. The reheat steam is subsequently expanded in the intermediate- and low-pressure turbines.

A double reheat steam generator includes heat transfer surface for reheating the intermediate-pressure turbine exhaust steam to a temperature equal or above the main steam temperature. The double-reheat cycle typically is used with 3,500 psig (22.12 MPa) and higher pressure steam cycles.

Reheater. Heat transfer surface used to increase the temperature of the high-pressure turbine exhaust steam.

Soot blowers. Devices that direct a high-pressure medium (steam, water, or air) onto heat transfer surface to remove ash or slag deposits.

Stack loss. Stack loss is the sensible heat in the flue gas leaving the last heat recovery section of the boiler. The last useful heat recovery device typically is an economizer in small boilers and a flue gas to combustion air heater in large utility boilers.

Structural members. The principal structural members consist of vertical structural steel columns, a top support grid, and beams (buckstays) which provide stiffness to the furnace, combustion air, and flue gas flow enclosures. In a plan view, the vertical columns that support the steam generator form a rectangular grid system. The lines thus formed are called column lines.

Subcritical steam cycle. A subcritical steam cycle is designed with a main steam pressure that is less than the critical pressure.

Superheater. Heat transfer surface used to increase the temperature of the steam from the drum.

Superheat/reheat desuperheaters. Devices used in both the superheater and reheater to mix water with steam, and thus reduce the steam temperature.

Windbox. A secondary air distribution plenum attached to the furnace which distributes secondary air among the burners.

7.5 REFERENCES

ABB AIR PREHEATER, INC. P.O. Box 372, Andover Road, Wellsville, NY 14895.

ASME. 1974. *ASME Performance Test Code 4.1—Steam Generating Units*. American Society of Mechanical Engineers, New York, NY.

ASTM. 1993. *ASTM Standard D409—Grindability of Coal by the Hardgrove Machine Method*. American Society for Testing and Materials. Philadelphia, PA.

BABCOCK & WILCOX. 20 S. Van Buren Avenue, P.O. Box 351, Barberton, OH 44203-0351.

COPES-VULCAN, INC. Martin and Rice Avenues, Lake City, PA 16423.

FOSTER WHEELER CORPORATION. Perryville Corporate Park, Clinton, NJ 08809-4000.

NFPA. 1993. *NFPA 85C—Standard for Prevention of Furnace Explosions/Implosions in Multiple Burner Boiler-Furnaces*. National Fire Protection Association, Quincy, MA.

RILEY STOKER CORPORATION. 5 Neponset Street, P.O. Box 15040, Worcester, MA 01615-0040.

SALT RIVER PROJECT. P.O. Box 52025, Phoenix, AZ 85072-2025.

SINGER, JOSEPH G., editor. 1991. *Combustion Fossil Power Systems*. Combustion Engineering, Inc., Windsor, CT.

STOCK EQUIPMENT COMPANY. 16490 Chillicothe Road, Chagrin Falls, OH 44023-4398.

STULTZ, S. C. and J. B. KITTO, editors. 1992. *Steam: Its Generation and Use*. Babcock & Wilcox Company, New York, NY.

8

STEAM TURBINE GENERATORS

Stanley A. Armbruster

8.1 INTRODUCTION

The steam turbine generator is the primary power conversion component of the power plant. The function of the steam turbine generator is to convert the thermal energy of the steam from the steam generator to electrical energy. Two separate components are provided: the steam turbine to convert the thermal energy to rotating mechanical energy, and the generator to convert the mechanical energy to electrical energy. Typically, the turbine is directly coupled to the generator.

This chapter provides a general overview of steam turbine generators. The detailed design of individual components is not addressed since typically the power plant designer is not involved in the detailed design of the steam turbine generator. The power plant designer specifies performance and general design requirements for a turbine generator manufacturer to meet in the production of the turbine generator. The turbine generator manufacturer performs the detailed design of the unit. The knowledge required by the power plant designer is an understanding of the basic operating principles; familiarity with the types, components, and auxiliaries of the turbine generator; and an understanding of the unit performance with the capability to predict overall turbine generator performance. This required information is presented in this chapter. In-depth information on steam turbine generators can be obtained from the additional references noted at the end of this chapter.

The principal medium to large steam turbine manufacturers in the United States are General Electric and Westinghouse. The major European manufacturers who supply turbine generators to the United States are ASEA Brown Boveri, Siemens, MAN, and GEC Alsthom. The major Japanese manufacturers are Hitachi, Mitsubishi, Toshiba, and Fuji Electric.

8.2 OPERATING PRINCIPLES

The operation of the steam turbine generator involves the expansion of steam through numerous stages in the turbine, causing the turbine rotor to turn the generator rotor. The generator rotor is magnetized, and its rotation generates the electrical power in the generator stator.

8.2.1 Turbine Stage Types

The thermal energy of the steam is converted to mechanical energy by expanding the steam through the turbine. The expansion of the steam occurs in two types of stages: impulse and reaction. The impulse stage can be compared to a water wheel on which a stream of water strikes the paddles, causing the wheel to turn. The reaction stage can be compared to a rotating sprinkler in that the jet of water from the sprinkler causes the arms to rotate.

8.2.1.1 Impulse Stages. An impulse stage consists of a stationary nozzle with rotating buckets or blades (Fig. 8-1). The steam expands through the nozzle, increasing in velocity as a result of the decrease in pressure. The steam then strikes the rotating buckets and performs work on the rotating buckets, which in turn decreases the steam velocity. The impulse stages can be grouped together in *velocity compound stages* or *pressure compound stages* (Fig. 8-2).

The velocity compound stage involves a stationary nozzle followed by several rotating and stationary buckets. The nozzle has a large pressure drop with a resulting increase in velocity. The first set of rotating buckets partially decreases the velocity as a result of the work performed on the buckets. The steam then passes through a set of stationary buckets in that the steam direction is changed back to the original direction. The steam then enters a second set of rotating buckets where the steam velocity is completely dissipated by performing work on this row of buckets. The velocity compound stage can consist of the stationary nozzles and many rotating and stationary buckets; however, there usually are only two rotating bucket rows and one stationary bucket row. The velocity compound stage typically is used as the first stage of a turbine because of its ability to withstand high-pressure reductions and the resultant efficiency in quickly reducing pressure and minimizing the requirements for high-pressure casings. The velocity compound stage is also called a *Curtis stage*.

Also shown in Fig. 8-2 is a pressure compound series of impulse stages. Rather than involving a large pressure drop in the one nozzle set, the pressure compound stages involve several sets of nozzles with small pressure drops through each set of nozzles and complete velocity dissipation in each row of rotating buckets. The pressure compound stages are also called *Rateau impulse stages*. The pressure compound

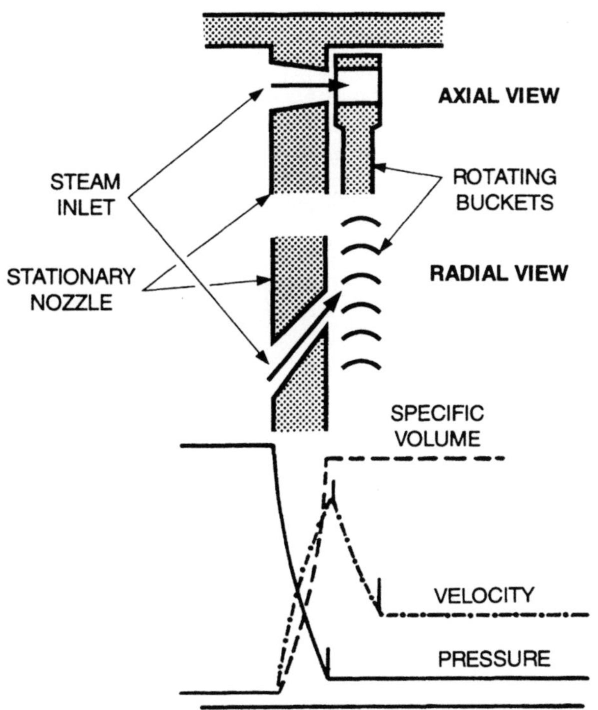

Fig. 8-1. Simple impulse turbine stage. (From GP Publishing. Used with permission.)

stage is typically used for all stages except for older design reheat and nonreheat turbines. These older designs often used a Curtis stage as the first or governing stage. In addition, some controlled extraction pressure applications use a Curtis stage immediately following the controlled extraction pressure zone.

8.2.1.2 Reaction Stages. Ideal reaction stages would consist of rotating nozzles with stationary buckets. However, it is impractical to admit steam to rotating nozzles. In practical applications, the reaction stage consists of both rotating and stationary nozzles. Reaction staging in a turbine is shown in Fig. 8-3. The steam expands through the stationary nozzles with an increase in velocity. The steam then enters the rotating nozzles where it expands further. The velocity force from the initial expansion and the expansion in the rotating blades is imparted to the rotating nozzles.

The expansion of the steam in the stationary nozzles of the reaction turbine is an impulse action. Therefore, the reaction stage in actual turbine applications is a combination of impulse and reaction principles. The reaction turbine is classified as percent reaction by the amount of energy conversion in the rotating nozzles. The term reaction stage generally implies a stage where 50% of the pressure drop occurs in the rotating blade and 50% occurs in the stationary nozzles.

The reaction stage has pressure drop across the rotating blades, inducing axial thrust in the rotor. To offset this thrust, reaction turbines have balancing pistons (refer to Section 8.4.2.1) in high-pressure zones of single-flow turbines.

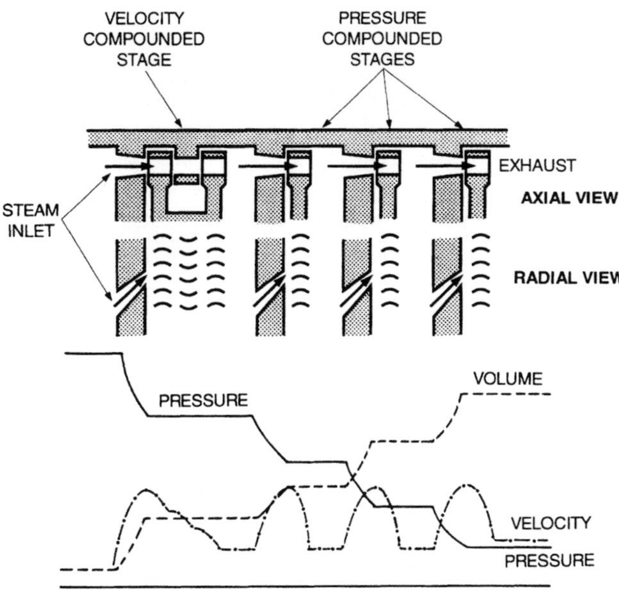

Fig. 8-2. Velocity compounded and pressure compounded turbine. (From GP Publishing. Used with permission.)

8.2.1.3 Impulse Versus Reaction Comparison. Many differences exist between impulse and reaction turbines. Three significant differences related to the nature of the expansion process are the number of stages, the bucket design, and the stage sealing requirements.

Peak efficiency is obtained in an impulse stage with more work per stage than in a reaction stage, assuming the same bucket diameter. Relative to an impulse turbine, this results in a reaction turbine requiring either 40% more stages, 40% greater stage diameters, or some combination of the two to obtain the same peak efficiency. This contrast is more prevalent in the high- and intermediate-pressure turbines. The contrast is less in the low-pressure turbines where the long bucket lengths significantly increase velocity of the bucket from the root to the tip and require both impulse and reaction design features in the blades.

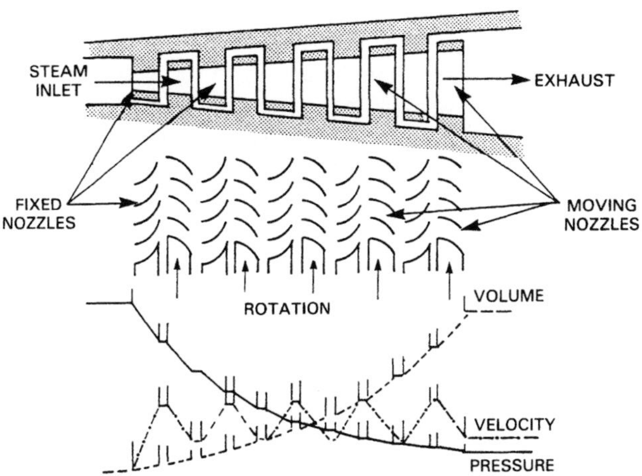

Fig. 8-3. Reaction turbine stages. (From GP Publishing. Used with permission.)

The pressure drop in an impulse turbine occurs across the stationary nozzle, whereas the reaction turbine has pressure drop across the stationary nozzle and the rotating bucket. Pressure drop across a rotating bucket causes thrust in the rotor. To minimize the thrust load to the rotor, the high- and intermediate-pressure sections of reaction turbines have the rotating blades mounted directly on the rotor, resulting in a small overall diameter and the need for a large number of stages. Impulse turbines do not have this thrust concern, and the buckets are mounted on disk extension of the rotor (wheels), resulting in larger overall diameters, smaller rotor diameters, and fewer stages than reaction turbines, as noted previously.

Turbines have internal sealing systems between the rotating buckets and the stationary casing and between the stationary nozzles and the rotor. The rotating bucket to stationary casing seal is more critical in a reaction turbine than in an impulse turbine since the reaction turbine has pressure drop across the buckets. The stationary nozzle to rotor seal is more critical in the impulse turbine because of the higher pressure drop across the impulse stationary nozzle. However, the impulse turbine has a smaller rotor and thus a smaller sealing diameter, offsetting the effects of the higher pressure drop. In addition, the wheel design of the impulse turbine rotor allows the installation of rotor seal leakage steam passages (holes) in the wheels, minimizing leakage steam interference with the main steam flow from the stationary nozzle to the rotating bucket. The rotor-mounted bucket design of the reaction turbine does not allow the installation of these leakage passages. However, because of the higher reaction at the base of the rotating blade, the reaction stage is less affected by re-entering leakage from the stationary row.

The above discussion is generally applicable to impulse and reaction turbines. However, it should be noted that true impulse stages having 0% reaction and reaction stages that always have at least 50% reaction do not exist in practical turbine design. The amount of reaction in a blade varies to accommodate the natural variation of reaction with the blade height. Impulse stages typically have 3% to 5% reaction at the base of a rotating blade in order to avoid zero or negative reaction that results in efficiency loss and may lead to flow separation in the rotating blade or bucket. For long reaction stage blades, the reaction percentage at the mean diameter may be as low as 40%. Thus, impulse and reaction turbines in the classical definition do not exist in practical power plant applications.

With regard to overall efficiency comparisons, the reaction stage has a higher aerodynamic or profile efficiency than an impulse stage. However, leakage losses are higher on the reaction stages as noted previously. Consequently, on stages with small blade heights, the impulse stage has higher efficiency because the difference in leakage losses offsets the higher profile of the reaction stage. As the blade height increases, the influence of leakage losses decrease and a point is reached where the reaction stage is more efficient.

8.2.2 Steam Expansion

As noted previously, the expansion of the steam through the turbine results in the heat (enthalpy) and pressure energy conversion to kinetic energy in steam jets. The kinetic energy is changed to mechanical energy in the rotating buckets or blades. This expansion process can be examined by plotting it on a Mollier chart (*h–s* diagram). The expansion of the steam in a steam turbine for a typical thermal power plant is shown in Fig. 8-4. The cycle represented has main steam conditions of 2,414.7 psia (16.65 MPa) and 1,000° F (537.8° C) with a reheat temperature of 1,000° F (537.8° C). The steam exhausts to a condenser operating at a pressure of 1.5 in. mercury absolute (HgA) (5.08 kPa).

As shown in Fig. 8-4, the steam enters the turbine at 2,414.7 psia (16.65 MPa), 1,000° F (537.8° C), and 1,460.15 Btu/lb (3,396.31 J/g). The steam expands through the high-pressure turbine and exhausts the turbine to the cold reheat lines at 550 psia (3.79 MPa), 640° F (337.8° C), and 1,318.54 Btu/lb (3,066.92 J/g). The steam then flows to the reheater of the steam generator where it is reheated and returned to the inlet of the intermediate-pressure turbine at 500 psia (3.45 MPa), 1,000° F (537.8° C), and 1,520.74 Btu/lb (3,537.24 J/g). The steam then expands through the intermediate-pressure turbine and the low-pressure turbine, exhausting to the condenser at a pressure of 1.5 in. HgA (5.08 kPa) with an expansion line end point enthalpy of 1,010.00 Btu/lb (2,349.26 J/g).

The energy extracted from each pound of steam can be determined for each turbine section by subtracting the exhaust enthalpy conditions ($h_{\text{out-act}}$) from the inlet enthalpy conditions ($h_{\text{in-act}}$). Dividing this quantity by the amount of energy that would have been extracted if the process were isentropic or ideal ($h_{\text{in-act}} - h_{\text{out-ideal}}$) provides the efficiency of the turbine section. This is expressed by the following equation.

$$\text{Efficiency} = \frac{h_{\text{in-act}} - h_{\text{out-act}}}{h_{\text{in-act}} - h_{\text{out-ideal}}} \times 100\% \qquad (8\text{-}1)$$

For the turbine section shown in Fig. 8-4, the high-pressure section of the turbine has an efficiency of 78.65%, and the combined intermediate- and low-pressure turbine sections have an efficiency of 90.20%.

8.2.3 Electrical Energy Generation

The rotational mechanical energy is converted to electrical energy in the generator by the rotation of the rotor's magnetic field.

The rotation of the turbine turns the rotor of the generator, producing electrical energy in the stator of the generator. The generator rotor consists of a steel forging with slots for conductors that are called the *field windings*. An electrical direct current is passed through the windings, causing a magnetic field to be formed in the rotor, as shown in Fig. 8-5. This magnetic field is rotated by the turbine. The rotor is

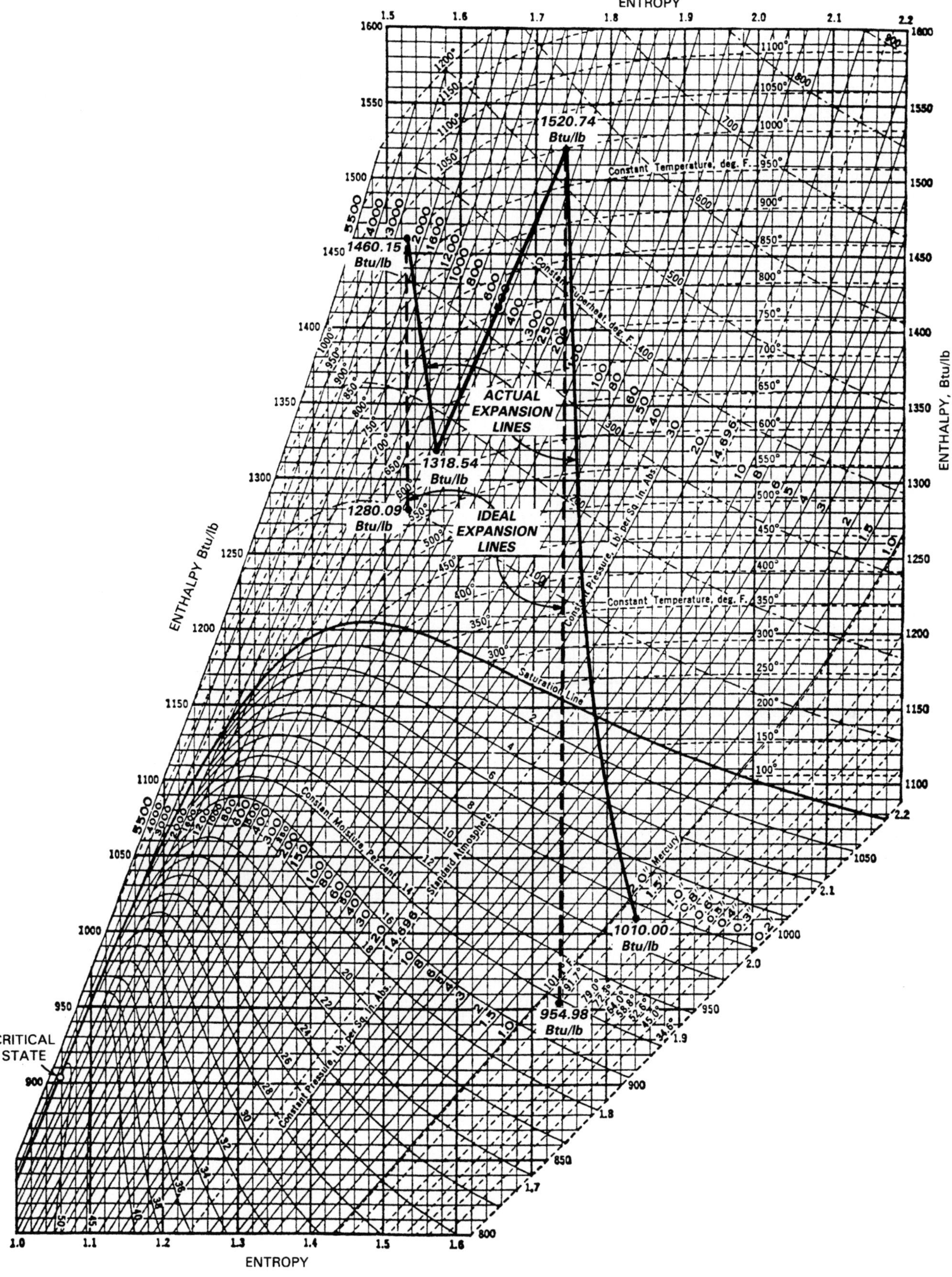

Fig. 8-4. Example steam expansion line curve.

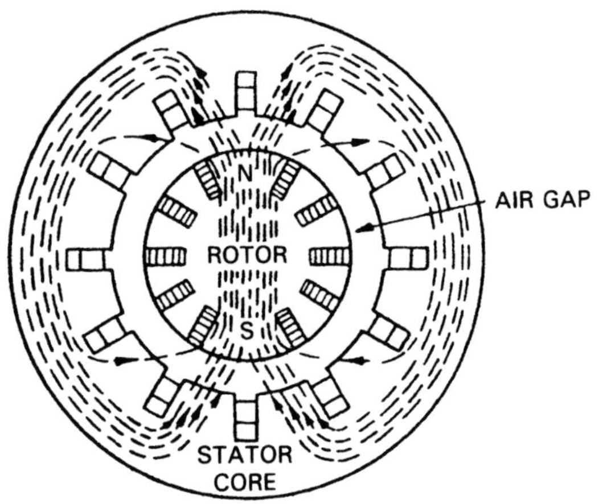

Fig. 8-5. Cross-section of generator showing magnetic circuit. (From General Electric Company. Used with permission.)

surrounded by the generator stator that includes copper conductors. The magnetic field of the rotor passes through the stator, setting electrons in the stator conductor in motion. The flow of electrons is called *current*. As the rotor's north pole passes through the stator conductors, the current flows in one direction. When the south pole of the rotor's magnetic field passes through the same conductor, the current flows in the opposite direction. This type of electrical current is called *alternating current* (*ac*) and is the type of current produced in most power plants.

The stator contains three groups of separate conductors in that the voltage and current are generated. The conductors are arranged at 120 degrees from each other and are connected in series to form three sets of coils called phases. The electrical power is generated in each of the three separate phases. Since the coils are displaced in space by 120 mechanical degrees, and there is only one rotating magnetic field, the voltage and current produced in each of the three phases are displaced in time by 120 electrical degrees. Thus, the power produced is called balanced three-phase power.

The components of power produced in the generator are shown in Fig. 8-6. The real power generated is represented by kilowatts (kW). The kilovolt amps reactive (kVAR) component is the reactive component, that is a result of the current that is out of phase with the generated voltage from the rotor field. The magnitude of the kW component is determined by the amount of mechanical power supplied by the turbine. The magnitude of the kVAR component is a function of the electrical system load to which the generator is connected and the amount of dc current supplied to the field winding. The total generator output is represented by kilovolt amperes (kVA) and is calculated as follows:

$$\text{kVA} = (\text{kW}^2 + \text{kVAR}^2)^{0.5} \qquad (8\text{-}2)$$

The power factor of a generator is defined as the ratio of the kW to kVA. The power factor also represents the cosine of the angle describing the angular difference between the

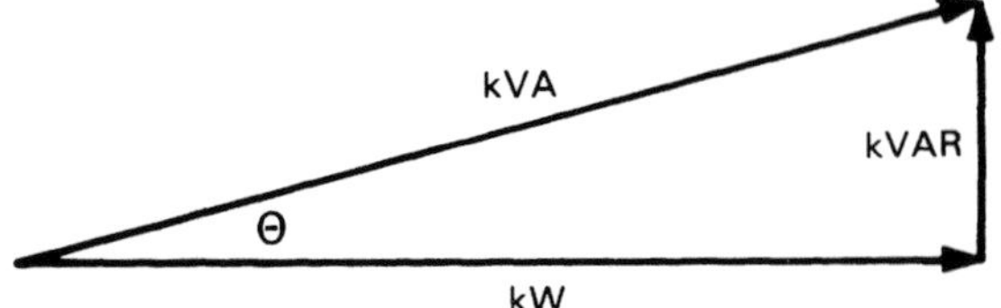

Fig. 8-6. Components of power.

generator phase current and the phase voltage. Additional discussion of generator operation is contained in Chapter 17.

8.3 TURBINE TYPES

Steam turbines are divided into many types with various designations. The designations may indicate the various combinations of turbine types that make up a turbine as well as the turbine size. Figure 8-7 shows various representative turbine types. The commonly used turbine types are described in the following sections.

8.3.1 Pressure–Reheat Designation

The designation of a turbine by a pressure may also involve the cycle arrangement with regard to reheat. The benefits of incorporating reheat into the cycle are discussed in Chapter 3. For small units without reheat, the steam turbine may consist of a single turbine with the steam flow entering the turbine, expanding through the turbine, then exhausting either to a condenser or to a process line. This is a straight-flow turbine as shown in Figs. 8-7(A) and (B). For a large unit without reheat, the steam may expand through an initial section and then exhaust to another turbine. This later turbine may then exhaust to a condenser or to a process. In this arrangement, the initial turbine is designated the high-pressure turbine and the second turbine the low-pressure turbine, as shown in Fig. 8-7(C).

For a single reheat cycle, the steam from the boiler flows to the high-pressure turbine where it expands and is exhausted back to the boiler for reheating. The reheat steam coming from the boiler flows to the intermediate-pressure or reheat turbine where it expands and exhausts into a crossover line that supplies the steam to the low-pressure turbine. The steam expands through the low-pressure turbine and exhausts to a condenser. Thus, the single-reheat cycle has high-, intermediate-, and low-pressure turbine sections as shown in Fig. 8-7(D). The designation is based on the location of the turbine in the cycle relative to inlet pressure. The intermediate-pressure turbine also is called the reheat turbine since it receives the reheated steam.

For a double-reheat cycle, the steam from the boiler flows to the high-pressure turbine where it expands and is exhausted back to the boiler for reheating. The reheat steam flows to the intermediate-pressure turbine where it expands and is exhausted back to the boiler again for reheating. The

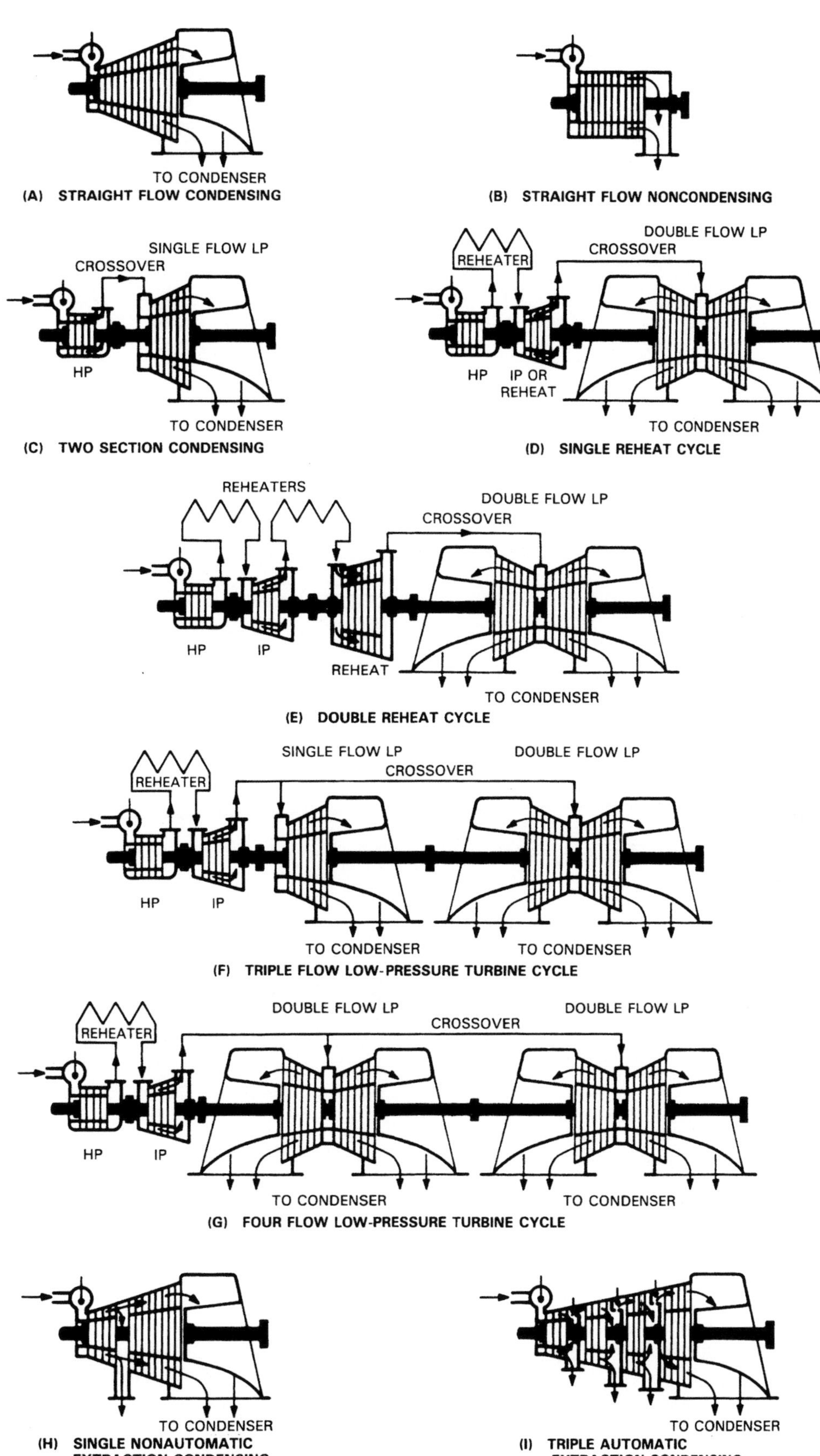

Fig. 8-7. Turbine types. (Adapted from *Power*, June 1989. Used with permission of McGraw-Hill.)

second stage of reheated steam flows from the boiler to the reheat turbine where it expands and exhausts to the crossover line that provides the steam to the low-pressure turbine. The steam expands through the low-pressure turbine and exhausts to the condenser. Thus in the double-reheat cycle, high-, intermediate-, and low-pressure turbine designations are used as for the single-reheat cycle, with the addition of the reheat turbine designation for the turbine required for the second stage of reheat. This arrangement is shown in Fig. 8-7(E).

8.3.2 Exhaust Conditions

Two designations exist based on the turbine exhaust conditions: condensing and noncondensing. These two designations are shown in Figs. 8-7(A) and (B). The condensing turbine exhausts to a condenser where the steam is condensed at subatmospheric pressure (vacuum). The low-pressure turbines of a typical power plant cycle are condensing turbines in that they exhaust to a steam surface condenser or to a direct condensing air-cooled condenser. The condensing turbines have large exhaust areas since the steam is expanded to low pressures, extracting as much of the useful energy as reasonably possible prior to being exhausted. The low pressures result in a large volume of steam, requiring a large exhaust area to minimize energy loss in the exhausting process.

The noncondensing turbine exhausts the steam above atmospheric pressure into a line for supply to the boiler, another turbine, or a process. The high- and intermediate-pressure turbines and the reheat turbines in the single- and double-reheat cycles described previously are noncondensing turbines. Turbines used in processes that first expand the steam through the turbine with the exhaust steam supplied to a heat exchanger or other process function are of the noncondensing type. Because of the higher exhaust pressure, the noncondensing turbine exhaust area is significantly smaller than for a condensing turbine.

8.3.3 Flow Designation

The turbine can also be described by the number of directions steam flows to exhaust from the turbine. The number of directions (paths) required depends on the amount of steam and the specific volume (pressure) of the steam. A single-flow turbine has the steam flowing in one direction and exhausting at one end of the turbine. The steam enters the turbine and expands in one direction as shown in Fig. 8-7(A). Small nonreheat turbines, mechanical drive turbines, reheat cycle high-pressure, and reheat cycle intermediate-pressure turbines typically employ single direction flow.

Double-flow turbines have two steam flow paths. The steam enters the center of the turbine and flows in two opposite directions as shown for the low-pressure sections in Figs. 8-7(D) and (E). This type of turbine is also called an opposed-flow turbine. Large reheat cycle high-pressure, intermediate-pressure, reheat, and low-pressure sections typically are double-flow turbines. Low-pressure turbines in power plant cycles of 150 megawatts (MW) and larger are typically double-flow turbines. Large unit low-pressure turbines are usually double-flow because of the low pressure, resulting in a large steam volume.

Flow designations of triple-flow, four-flow, six-flow, and eight-flow are also used. These designations typically apply to the low-pressure sections of power plants. A triple-flow designation indicates the use of one double-flow low-pressure turbine in combination with a single-flow low-pressure turbine, as shown in Fig. 8-7(F). Four-flow designation indicates the use of two double-flow low-pressure turbines as shown in Fig. 8-7(G). Six- and eight- flow designations indicate the use of three and four double-flow low-pressure turbines, respectively.

8.3.4 Extraction Types

Turbines are also designated by the type of extraction involved, if any. During the expansion of the steam through a turbine, removing steam from an intermediate stage of the turbine is called an *extraction*. In most power plants, some steam is extracted from the turbine expansion process and supplied to heat exchangers for feedwater heating. This type of extraction varies in pressure and flow as a function of load. This variation is acceptable and therefore no effort is made to regulate the pressure. This type of extraction is called uncontrolled, simple, or nonautomatic extraction, and is shown in Fig. 8-7(H).

Turbines supplying steam for process applications typically must supply steam at a constant pressure. Since the pressure available varies with load unless controlled, valving is included in the turbine steam expansion path to control the pressure. This valving restricts the flow to the downstream stages as required to maintain the pressure of the extraction stage. This type of turbine is called a controlled or automatic extraction turbine and is shown in Fig. 8-7(I). If several controlled extraction points are required in any one turbine, multiple internal control valves are provided. One controlled extraction would be called a single automatic extraction turbine; two would be called a double automatic extraction turbine.

Some turbines have no extraction points and therefore are called nonextraction turbines. This type of turbine is shown in Fig. 8-7(A).

8.3.5 Shaft Orientation

The overall steam turbine generator arrangement of a power plant is designated as *tandem-compound* or *cross-compound* on the basis of the shaft orientation. These two arrangements are shown in Fig. 8-8. The tandem-compound unit has all turbines and the generator in-line, connected to the same shaft. The turbines all drive the same generator and thus operate at the same speed.

The cross-compound unit has two turbine generator alignments. This type of arrangement is used to increase turbine

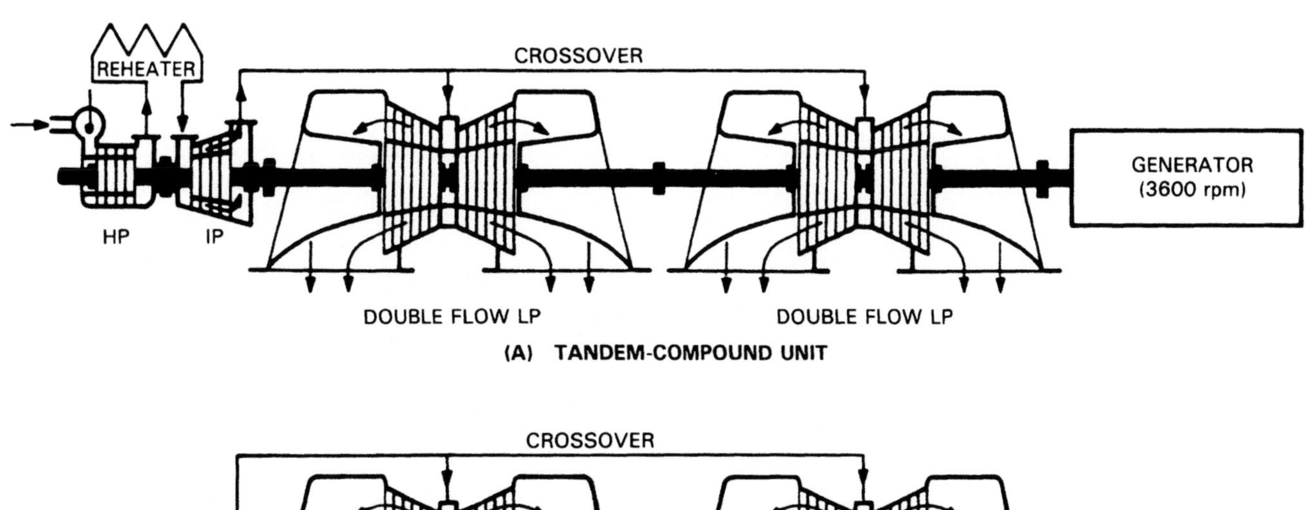

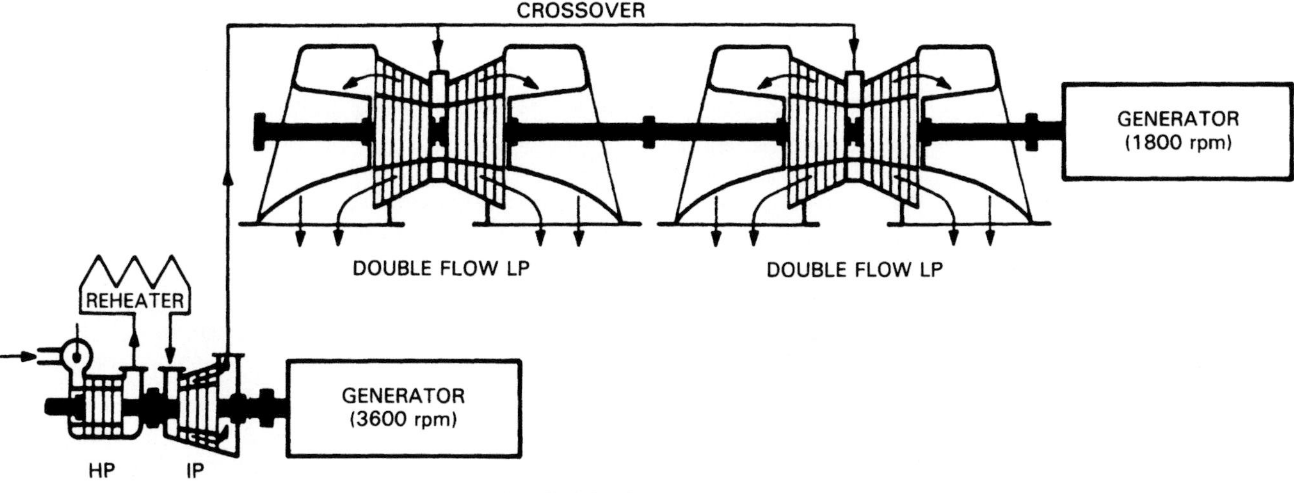

Fig. 8-8. Turbine shaft orientations. (Adapted from *Power*, June 1989. Used with permission of McGraw-Hill.)

efficiency. The cross-compound arrangement typically consists of high-pressure and intermediate-pressure turbines operating at 3,600 rpm driving a generator. The exhaust steam of the intermediate turbine crosses over to a low-pressure turbine that operates at 1,800 rpm, driving a separate generator. The low-pressure turbine operating at the slower speed allows the use of longer last-stage turbine blades with expansion to higher moisture percentages and less exhaust losses, as discussed later. These characteristics result in higher overall turbine efficiencies.

Large cross-compound units in the 1,300-MW range with both shaft orientations operating at 3,600 rpm have been built. The dual-shaft arrangement was used to minimize shaft length and any single generator size.

8.3.6 Designations

Power plant steam turbines are typically designated by shaft orientation, number of low-pressure turbine steam flow paths, and the last-stage blade length of the low-pressure turbine. A turbine designated as TC4F30, for example, indicates a unit that is tandem-compound (TC) having two double-flow (4F) low-pressure turbines with 30 in. (76.2 cm) last-stage blade length. A CC2F23 indicates a unit that is cross-compound (CC) having one double-flow (2F) low-pressure turbine with 23 in. (58.4 cm) last-stage blade length.

The last two parts of the designations are related to the low-pressure turbine since low-pressure turbines are standard designs. The manufacturers custom design or adapt existing designs of the high-pressure, intermediate-pressure, and reheat turbines for each power plant project. A range of standard low-pressure designs exist and the optimum standard low-pressure design is used with the custom designed higher-pressure turbines. The low-pressure designs are designated by the length of the last-stage blades.

The Japanese designation also includes the number of casings in the arrangement. A designation of TC2C2F23 indicates a tandem-compound (TC) unit having two casings (2C) and a double-flow (2F) low-pressure turbine with 23 in. (58.4 cm) last-stage blade length. If the unit is a reheat unit, the two casing designation indicates that the high-pressure and intermediate-pressure turbine are in the same casing (representing one casing designation) with the low-pressure turbine being the second casing.

For small units in the United States, the casing designation is also included. The designation SCSF17 indicates a single casing turbine (SC) with a single-flow path (SF) and last-stage blade length of 17 in. (43.2 cm).

8.3.7 Turbine Arrangements

Present day electric power generation is largely by fossil and nuclear power plants. The general turbine arrangements used in modern day power plants are summarized in Table 8-1 by listing the number of reheat stages, steam pressures, and turbine configurations as a function of the unit output. Figure 8-9 shows the typical steam expansion conditions for nuclear and fossil plants.

8.3.7.1 Fossil Power Plant Turbine Arrangement. For fossil plants, the turbines typically operate at 3,600 rpm in countries with 60 hertz (Hz) electrical systems and 3,000 rpm for 50 Hz electrical systems. The high main steam pressures allow the use of compact high-speed designs. Units typically are tandem-compound because the cross-compound arrangement cannot be justified economically. The cross-compound units cost significantly more than tandem-compound units because of the need for separate generators and larger foundations. The use of reheat and number of turbines increases with the plant size. Main steam and reheat steam temperature are typically 1,000° F (537.8° C). Steam expansion in the turbine cycle is mainly in superheated steam with only the last one or two stages of the low-pressure turbine expanding in saturated steam (Curve C in Fig. 8-9). Typically, the expansion in the low-pressure turbine is to a moisture content of 6% to 8%. For sizes up to 600 MW, the high-pressure and intermediate-pressure sections may be in a single casing with an opposed flow arrangement. Last-stage blade lengths vary from 17 to 40 in. (43.2 to 101.6 cm).

8.3.7.2 Nuclear Power Plant Turbine Arrangement. For light water reactor nuclear plants, the turbines are tandem-compound units operating at 1,800 rpm in 60-Hz electrical systems and 1,500 rpm in 50-Hz electrical systems. To take advantage of economies of scale, nuclear plants are usually large in size. This large size, in combination with low operating pressures and temperatures of 1,000 psia (6.89 mPa) and 560° F (293.3° C), results in large steam flows. To pass these large flows, long rotating blades (large turbine flow passage area) are required. The 1,800 and 1,500 rpm speeds can accommodate the longer blades while maintaining acceptable blade tip speeds.

The nuclear steam cycles start with steam at saturated conditions (Point M in Fig. 8-9) with the initial expansion in the high-pressure section increasing moisture content. The exhaust steam exiting the high-pressure turbine (Point X in Fig. 8-9) is passed through a moisture separator to remove the condensed moisture prior to further expansion in the low-pressure turbines (from Point Z along Curve A in Fig. 8-9). The low-pressure turbine expansion is almost entirely in the saturated steam with expansion to 10% to 14% moisture content before exhausting to the condenser. Grooves on the low-pressure turbine blades channel moisture to the extraction ports, reducing the moisture content during the expansion through the low-pressure turbine. To improve efficiency, some of the main steam or steam extracted for the high-pressure turbine is used in a heat exchanger to reheat the high-pressure turbine exhaust steam (from Point X to Point R in Fig. 8-9) prior to expansion through the low-pressure turbines (Curve B in Fig. 8-9). Steam is not returned to the nuclear steam generator for reheating. Based on the low initial pressures and the limited or no reheating, only high- and low-pressure turbines are required for typical nuclear units. Low-pressure turbine last stage blade lengths vary from 38 to 54 in. (96.2 to 137.2 cm).

8.4 MAJOR STEAM TURBINE COMPONENTS

The power plant designer must have an understanding of the components that compose a steam turbine so that interfacing

Table 8-1. Typical Steam Turbine Arrangements[a]

Gross Ouput (MW)	Number of Reheat Stages	Steam Pressure psig (MPa)	Turbine Pressure Sections			
			HP	IP	RH	LP
Fossil						
50–150	0 or 1	1,450 (10.1)	1SF	—	—	1SF
150–250	1	1,800 (12.5)	1SF	1SF	—	1DF
250–450	1	2,400 (16.6)	1SF	1SF	—	1DF
450–600	1	2,400 or 3,500 (16.6 or 24.2)	1SF	1SF	—	2DF
600–850	1	2,400 (16.6)	1DF	1DF	—	2DF
	1 or 2	3,500 (24.2)	1SF	1SF	1DF	2DF
850–1100	1	2,400 (16.6)	1DF	1DF	—	3DF
	1 or 2	3,500 (24.2)	1SF	1SF	1DF	3DF
Nuclear						
600–900	—	1,000 (7.0)	1DF	—	—	2DF
900–1300	—	1,000 (7.0)	1DF	—	—	3DF

[a]All listings are for tandem-compound units.

1SF = one single-flow turbine; 1DF = one double-flow turbine; 2DF = two double-flow turbines; 3DF = three double-flow turbines

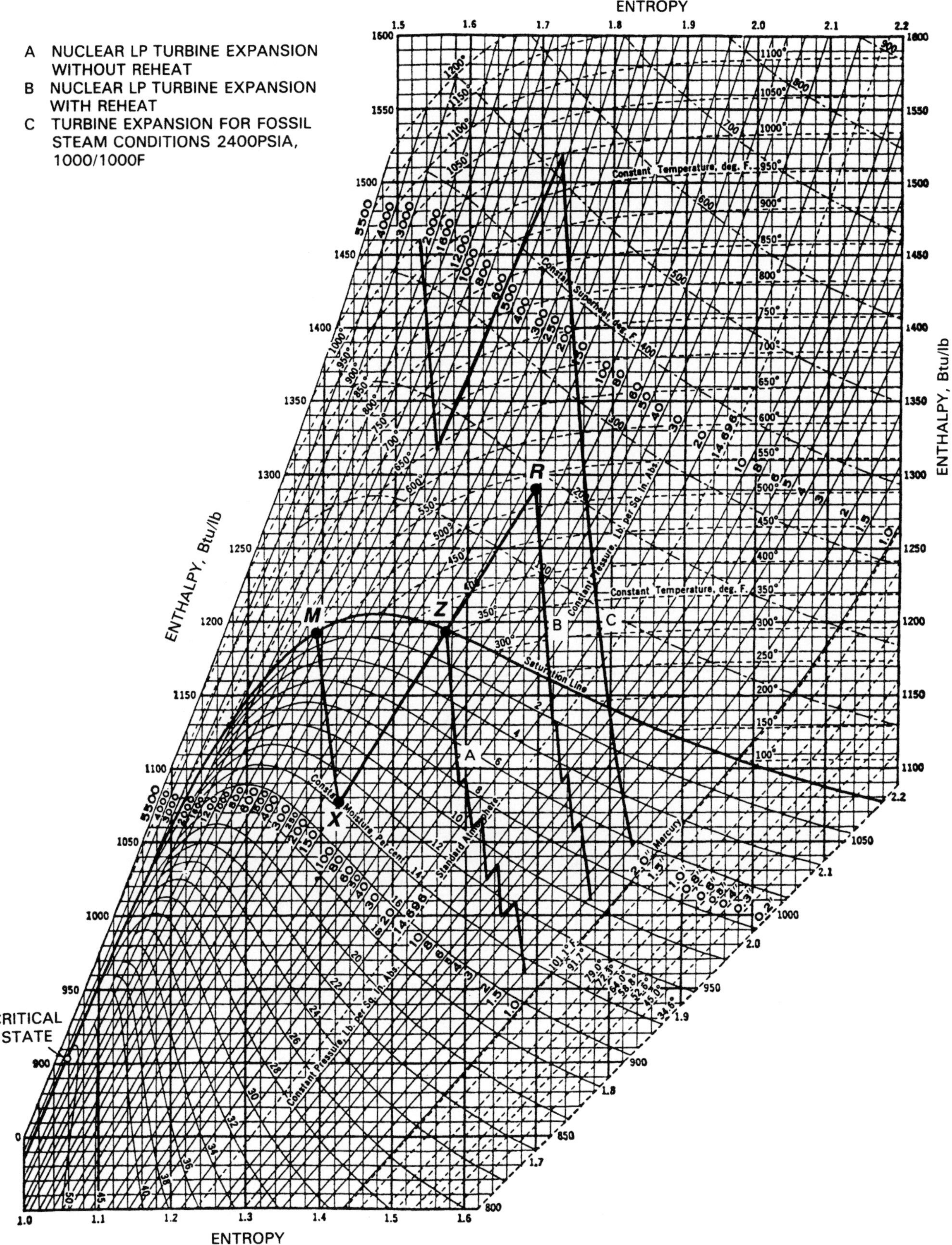

Fig. 8-9. Typical expansion lines for nuclear and fossil turbines. (From General Electric Company. Used with permission.)

systems and plant controls can be properly designed for optimum plant operation without affecting component life. The details and names of components of a steam turbine vary by manufacturer and type (impulse or reaction). For this chapter, when several terms are used for a particular component or item, the most commonly used terms are given. The arrangement and cross-section drawings used here are for a reaction turbine. A typical 400-MW reheat, tandem-compound steam turbine generator perspective arrangement is shown in Fig. 8-10, with side and plan views shown in Figs. 8-11 and 8-12, respectively.

8.4.1 Valves

Major valves associated with the steam turbine are shown in Fig. 8-13. The functions and significant features of the valves are discussed in this section as they occur in the steam path flow.

8.4.1.1 Main Steam Stop (Throttle) Valves. The steam from the steam generator flows to the main steam stop or throttling valves. The primary function of the stop valves is to provide backup protection for the steam turbine during turbine generator trips in the event the main steam control valves do not close. The energy in the main steam and steam generator can quickly cause the turbine to reach destructive overspeed on loss of the generator load. The main steam stop valves close from full open to full closed in 0.15 to 0.5 s. The stop valves are also closed on unit normal shutdown after the control valves have closed.

A secondary function of the stop valves is to provide steam throttling control during startup. The stop valves typically have internal bypass valves that allow throttling control of the steam from initial turbine roll to loads of 15% to 25%. During this startup time, the main steam control valves are wide open and the bypass valves are used to control the steam flow. Some recent and current designs do

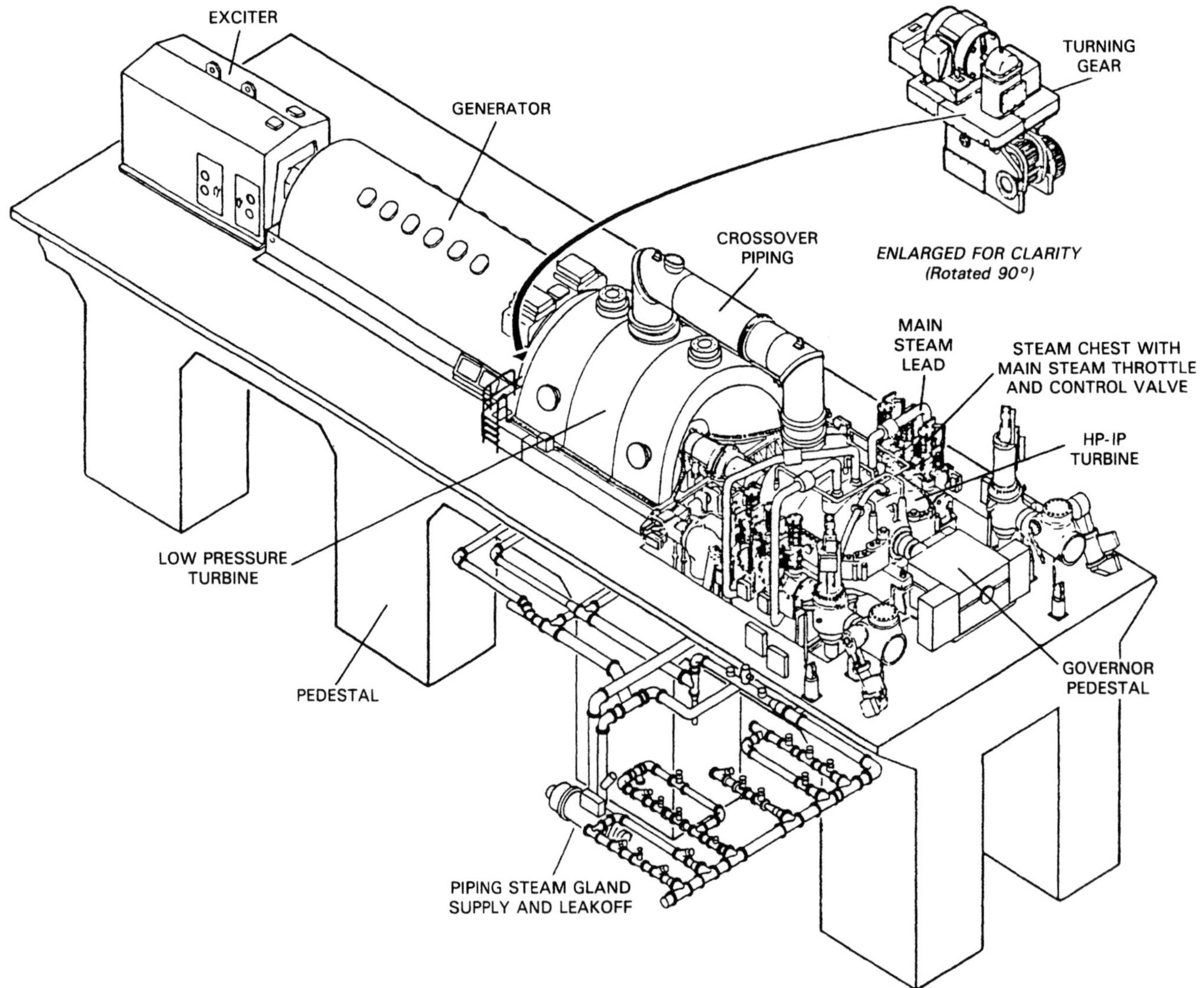

Fig. 8-10. Overall steam turbine generator arrangement. (From Westinghouse Electric Corporation. Used with permission.)

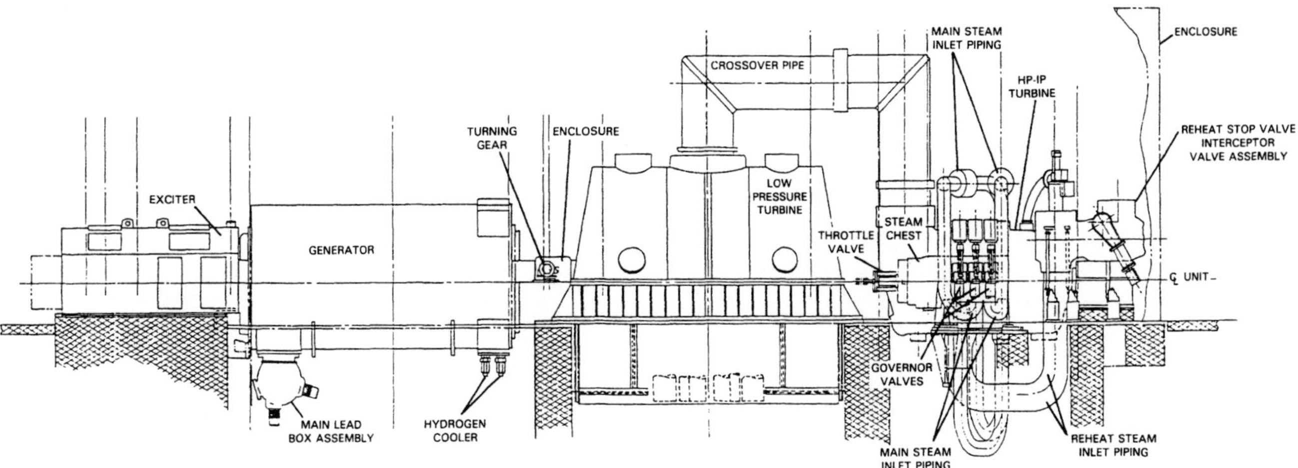

Fig. 8-11. Steam turbine generator general arrangement (side view). (From Westinghouse Electric Corporation. Used with permission.)

not have these bypass valves. Initial turbine speed runup is controlled by the main steam stop valves.

The main steam stop valves can be mounted separately or on a steam chest. For the steam turbine shown in Figs. 8-11 and 8-12, there are two stop or throttle valves, one on the side of each steam chest. The design of the valves enables positive sealing by the main steam pressure. The valves are opened by hydraulic actuators with springs to assist in closing the valve. Steam strainers in the valves prevent foreign material from being carried into the turbine. The valves also have before and after seat drains to prevent water induction into the turbine.

8.4.1.2 Main Steam Control (Governor) Valves. The steam flows from the stop valves to the main steam control or governor valves. The primary function of the control valves is to regulate the steam flow to the turbine and thus control the power output of the steam turbine generator. The modes of steam flow control are discussed in Section 8.7, Steam Flow Control. The control valves also serve as the primary shutoff of the steam to the turbine on unit normal shutdowns and trips.

The control valves are normally mounted on a steam chest that receives steam from the stop valves. For some units, the stop and control valves are directly connected, one stop valve providing steam to one control valve. For the unit shown in Figs. 8-11 and 8-12, there are two steam chests, each having three control or governor valves mounted on top. Each control valve supplies steam to one section or arc of the high-pressure turbine first stage. The unit shown has six control valves, and therefore six steam lines (leads) from the steam chest to the high-pressure turbine section. Each lead supplies steam to one arc (a segment of the full 360 degrees; 60-degree arc, in this example) of the turbine's first stage.

Control valves are of the plug or poppet type with venturi

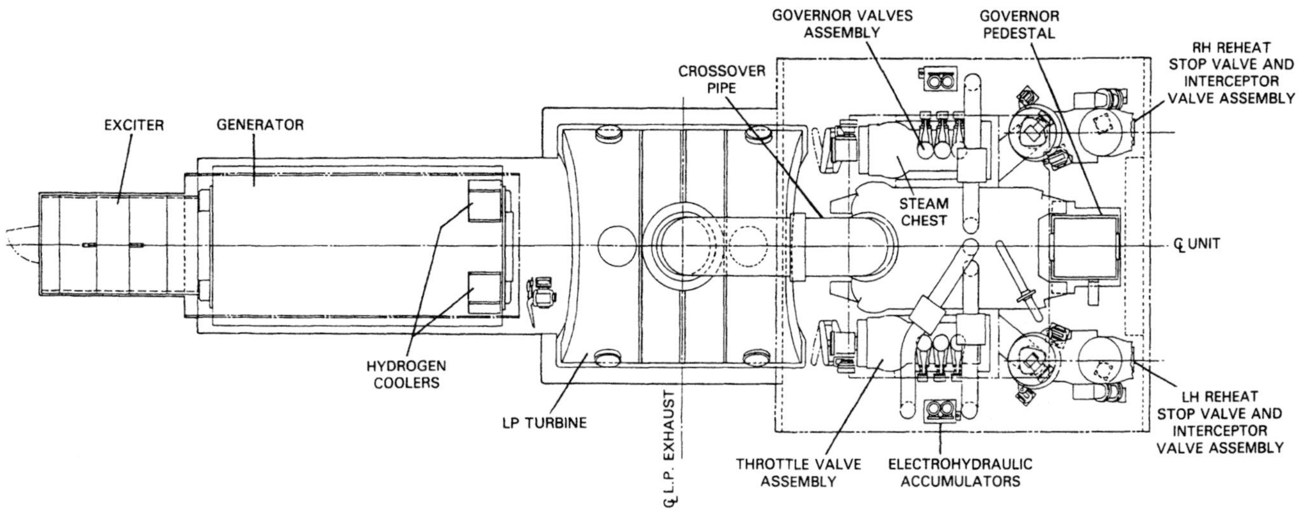

Fig. 8-12. Steam turbine generator general arrangement plan view. (From Westinghouse Electric Corporation. Used with permission.)

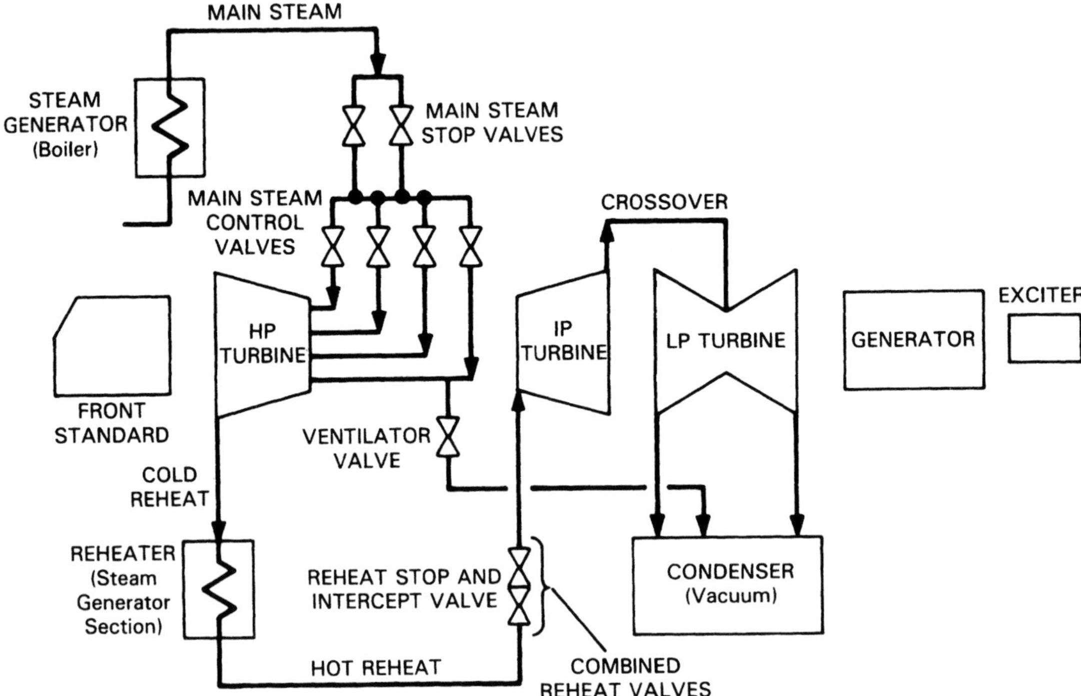

Fig. 8-13. Typical power plant steam flow diagram.

seats. The valves are spring-closed with hydraulic actuators to open the valves. The valves can be of the balanced or unbalanced disk type. Like the stop valves, the control valves have before-seat drains for water induction protection.

8.4.1.3 Reheat Stop and Intercept Valves. As shown in Fig. 8-13, the steam exhausted from the high-pressure turbine flows through the cold reheat lines to the reheater in the steam generator. The reheated steam then flows through the hot reheat piping to the reheat stop and intercept valves. The reheat stop and intercept valves function similarly to the main steam stop and control valves. The reheat stop valves offer backup protection for the steam turbine in the event of a unit trip and failure of the intercept valves to close. The intercept valves control unit speed during shutdowns and on large load swings, and protect against destructive overspeeds on unit trips.

The need for these valves is a result of the large amount of energy available in the steam present in the high-pressure turbine, the hot and cold reheat lines, and the reheater. On large load changes, the main steam control valves start to close to control speed; however, energy in the steam present after the main steam control valves may be sufficient to cause the unit to overspeed. The steam after the main steam control valves could expand through the intermediate- and low-pressure turbines to the condenser, supplying more power output than is required, causing the turbine to overspeed. The intercept valves are used to throttle the steam flow to the intermediate-pressure turbine in this situation to control turbine speed. During unit shutdowns, a similar situation could occur, and the intercept valves are used to control speed under these conditions as for the trip condition. During unit trips, both the reheat stop and the intercept valves close, preventing the reheat-associated steam from entering the intermediate-pressure turbine. During normal unit operation, the reheat stop and intercept valves are wide open with load control only by the main steam control valves.

The reheat stop and intercept valves are positioned as close to the intermediate-pressure turbine as possible to minimize the amount of steam available to expand through the turbines on unit trips. The valves are typically located in the same assembly or in the same valve body with two separate disks. Even though the valves are placed together, each valve has its own operating mechanism. The valves are spring-closed with hydraulic actuators for opening. The intercept section has an internal steam strainer to prevent foreign materials from entering the turbine. The valves have before-seat drains and, depending on the arrangement, after-seat drains for turbine water induction protection.

The unit shown in Figs. 8-11 and 8-12 has two reheat stop and intercept valve assemblies. Each assembly is connected to the intermediate-pressure turbine through a separate steam lead that supplies steam to half the intermediate-pressure turbine first stage.

8.4.1.4 Ventilator Valves. During a unit trip, the closure of the main steam stop and control valves and of the reheat stop and intercept valves traps steam in the high-pressure turbine. During the turbine overspeed and subsequent coast-down, the high-pressure turbine blades are subject to windage losses from rotating in this trapped steam. The windage losses cause the blades to be heated. This heating, in combination with the overspeed stress, can damage the high-pressure turbine. To prevent this, a ventilator valve is pro-

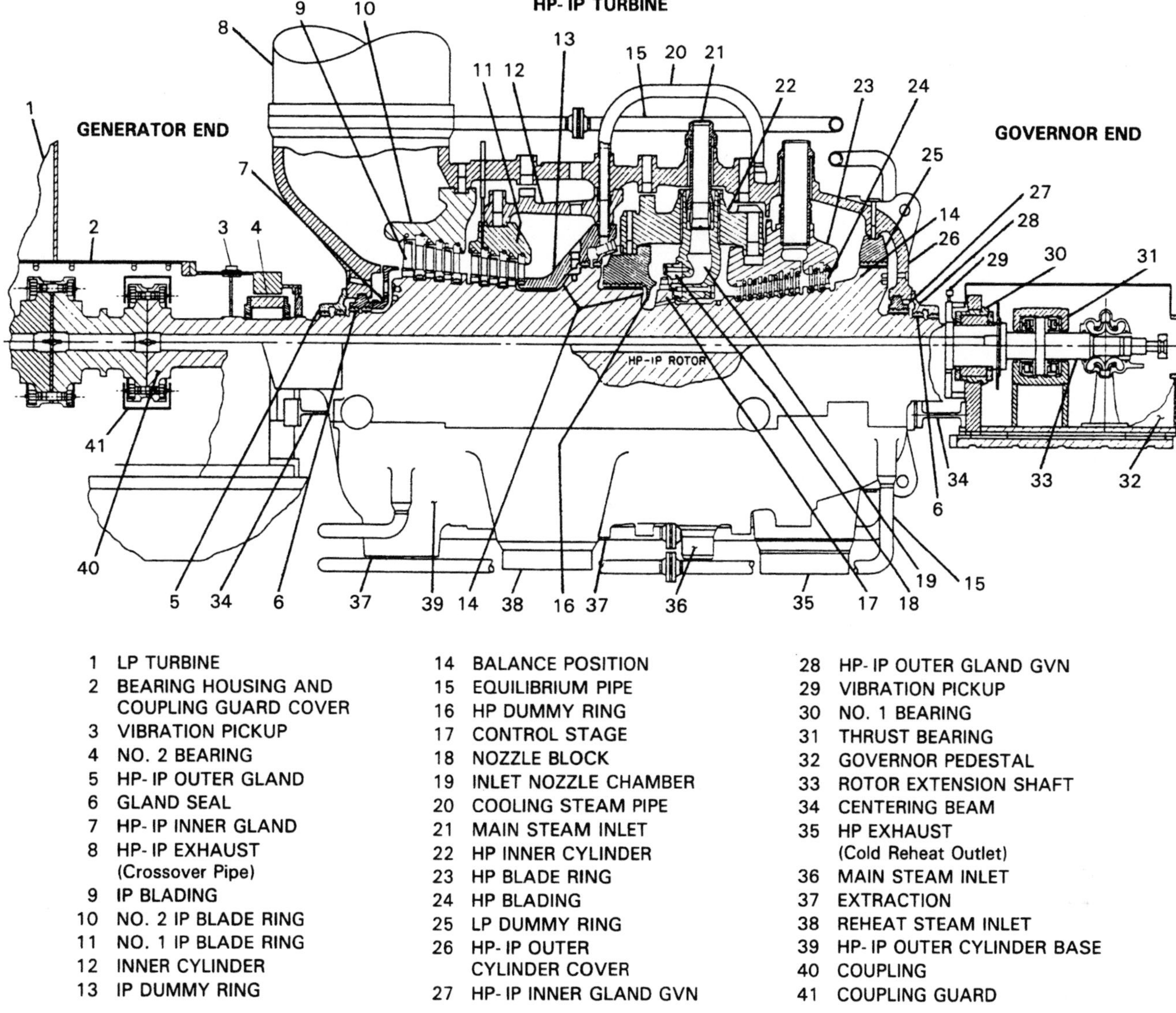

Fig. 8-14. Combined high-pressure/intermediate-pressure turbine longitudinal section. (From Westinghouse Electric Corporation. Used with permission.)

vided to bleed the trapped steam to the condenser. The ventilator valve is connected to one of the main steam leads between the main steam control valves and the high-pressure turbine, as shown in Fig. 8-13. On some combined high-pressure–intermediate-pressure turbines, the ventilator valve is connected to the steam seal between the two pressure sections. On a unit trip, the valve automatically opens, bleeding the trapped steam to the condenser. This bleeding action causes the trapped steam to flow through the high-pressure turbine, maintaining the high-pressure turbine temperature within acceptable limits by preventing heat buildup from the windage losses.

8.4.2 Turbine Stationary Parts

The major stationary parts of the steam turbine are mainly those associated with the casing of the various turbine sections. The combined high-pressure–intermediate-pressure turbine of the 400 MW unit is shown in Fig. 8-14, with the low-pressure turbine shown in Fig. 8-15. Stationary parts of these two turbines are discussed in the following sections.

8.4.2.1 High-Pressure–Intermediate-Pressure Turbine. Figure 8-14 shows an opposed flow, combined high-pressure–intermediate-pressure turbine. This type of arrangement is used on units up to 600 MW to save space and cost. Separate high-pressure and intermediate-pressure shells are used on larger units because of the larger steam volumes requiring the use of separate high-pressure and intermediate-pressure sections. The use of a single shell instead of two separate shells eliminates the intermediate casing seals, shaft couplings, and bearings. The opposed flow arrangement also helps in balancing axial thrusts.

The steam flows to the high-pressure and intermediate-pressure turbine sections are high-pressure, high-temperature flows. Outer and inner cylinders or casings contain the

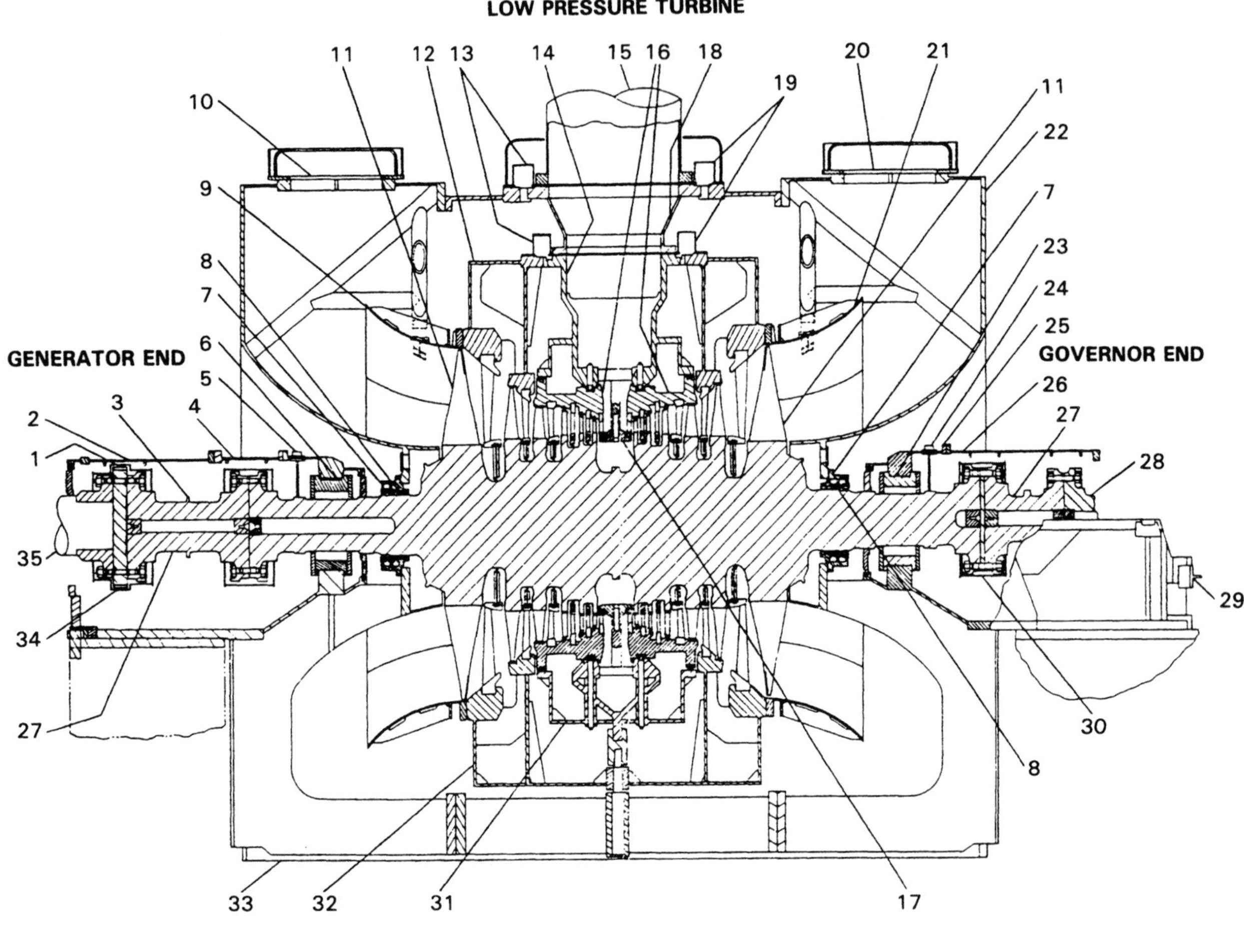

1 SPEED SENSING PICKUP
2 COUPLING COVER
3 DIFFERENTIAL EXPANSION PICKUP LOCATION
4 BEARING COVER
5 VIBRATION PICKUP
6 NO. 4 BEARING
7 GLAND
8 GLAND SEAL
9 EXHAUST FLOW GUIDE
10 BREAKABLE DIAPHRAGM
11 BLADING
12 INNER CYLINDER COVER NO. 2
13 TOP SEALING DIAPHRAGMS
14 INNER CYLINDER COVER NO. 1
15 TURBINE STEAM INLET
16 BLADE RINGS
17 INLET FLOW GUIDE
18 CROSSOVER PIPE ADAPTER
19 TOP SEALING DIAPHRAGMS
20 BREAKABLE DIAPHRAGM
21 EXHAUST FLOW GUIDE
22 OUTER CYLINDER COVER
23 NO. 3 BEARING
24 VIBRATION PICKUP
25 BEARING AND COUPLING COVER INNER
26 BEARING AND COUPLING COVER OUTER
27 COUPLING
28 HP- IP ROTOR
29 CENTERING BEAM
30 COUPLING GUARD
31 INNER CYLINDER BASE NO. 1
32 INNER CYLINDER BASE NO. 2
33 OUTER CYLINDER BASE
34 TURNING GEAR COUPLING SPACER
35 GENERATOR ROTOR

Fig. 8-15. Low-pressure turbine longitudinal section. (From Westinghouse Electric Corporation. Used with permission.)

steam. The cylinders are castings. The outer cylinder acts as an overall pressure vessel and provides the support for the inner parts. The outer cylinder has a horizontal joint with precision-machined bolted flanges. This joint separates the outer cylinder into upper and lower halves. The two halves are bolted together without any gaskets to form the outer pressure vessel. The lower half rests on the foundation, transferring the turbine weight to the building structure. The outer cylinder lower half is supported by two paws on each end of the cylinder. The paws are slotted and rest on slide plates with bolts to prevent vertical movement while still allowing longitudinal and transverse movement for thermal expansion. The high-pressure–intermediate-pressure turbine expands approximately 1 in. (2.54 cm) from cold to hot conditions. The upper half of the outer cylinder is removable to allow access to the turbine internals for maintenance.

The inner cylinder also acts as a pressure barrier and supports the blade rings. Like the outer cylinder, the inner cylinder consists of upper and lower halves.

Two cylinders (outer and inner) are provided instead of

one to allow faster startups. The arrangement of the two nested cylinders results in the higher pressure inside the inner cylinders, with lower pressure on the outside of the inner cylinder that is the pressure on the outer cylinder. This two-pressure arrangement allows use of two thinner cylinders as opposed to one thick cylinder. The two thinner metal cylinders can be heated faster with less stress and distortion than can one thick cylinder.

Some European designs with the high-pressure turbine in a separate shell use a barrel-type outer casing. The outer casing does not have a horizontal joint. Maintenance on the high-pressure turbine requires removal of an end retaining ring. The inner casing and rotor can then be withdrawn from one end of the barrel. The inner casing is split as on other units.

The inner cylinder supports the blade rings. The blade rings support the stationary blades. The stationary blades are directly in the steam path of the rotating blades. The stationary blades have a pressure drop, increasing the steam velocity and directing the steam to the next row of rotating blades. The end of the stationary blades adjacent to the rotor has a labyrinth-type seal to minimize steam leakage between the rotor and the stationary blade. The high-pressure blade ring has steam extraction ports to provide steam for feedwater heating. The intermediate-pressure blade ring has two separate sections with the space between the two rings being a steam extraction port. For an impulse turbine, the blade ring is called a *nozzle diaphragm.*

The steam enters the high-pressure turbine from the main steam control valves. The steam enters the turbine nozzle chamber, expands through the nozzle block, strikes the control stage blades, is turned 180 degrees (this turn not provided for all turbines), and flows to the remainder of the high-pressure turbine stages. Each nozzle chamber is supplied steam from one control valve. The first stage of steam expansion is a severe-duty, high-pressure, high-temperature process with steam supplied only to some of the stationary blades at lower loads when not all control valves are open. The first set of stationary blades that are of special design is called a *nozzle block.* The first stage of expansion is typically an impulse stage, even in reaction turbines. The impulse expansion has the severe high-pressure drop across the stationary blades with little or no pressure drop on the rotating blades. The rotating blades of the control (first) stage are subjected to extreme duty, passing in and out of flow several times per revolution when only a portion of the control valves are open.

The steam is turned 180 degrees after the first stage to allow installation of a thrust balancing piston on the rotor to offset the pressure drop through the high-pressure rotating blades. The high-pressure balancing piston is provided under the stationary high-pressure dummy ring. Two balancing pistons offset the axial thrust of the intermediate-pressure turbine rotating blades. One is located at the high-pressure exhaust under the low-pressure dummy ring and the other is positioned at the intermediate-pressure inlet, under the intermediate-pressure dummy ring. The dummy rings separate various pressure zones and act as flow guides to direct steam flow within the high-pressure and intermediate-pressure sections. The dummy rings form a labyrinth path between the casing and rotor to minimize steam leakage between the pressure zones.

Where the rotating shaft extends through the outer cylinder, casing to shaft seals are provided. The seals are labyrinth-type with backing springs to ensure tight clearances between the seal and the rotating shaft. The seals are in several stages to allow higher-pressure leakage steam to be extracted for reinsertion into a lower-pressure section of the steam turbine, for use in feedwater heating, or for use as seal steam supply to the low-pressure turbine casing seals. The outer seal port is at a slight vacuum to prevent steam leakage out of the unit. The seal system is described in more detail in Section 8.8.1, Turbine Seal System.

At the high-pressure end of the turbine is the governor pedestal or front standard. This compartment contains much of the turbine mechanical protective devices and the main oil supply pump for lubrication oil. The major protective device in the governor pedestal is the mechanical overspeed trip.

8.4.2.2 Low-Pressure Turbine. Figure 8-15 shows a double-flow low-pressure turbine. Like the high-pressure–intermediate-pressure turbine, the low-pressure turbine has an outer cylinder, inner cylinder, blade rings, and casing seals. Outer and inner cylinders are provided because of thermal considerations. The steam temperature in the low-pressure turbine drops 600 to 700° F (316 to 371° C) during its expansion through the low-pressure turbine. The outer and inner cylinders are of fabricated construction. The outer cylinder supports the weight of the low-pressure turbine on a support flange around the entire outside of the cylinder, that is bolted to the foundation. The outer cylinder also includes the bearing supports on both ends. The outer cylinder of the low-pressure turbine is exposed to the vacuum pressure of the condenser. A section of the outer cylinder is the exhaust flow guide or hood, that directs the steam from the last stage of rotating blades to the condenser. The guide must be aerodynamically designed to minimize pressure losses. A 1 in. (2.54 cm) Hg pressure loss in this guide can result in a 1% increase in plant overall heat rate. To prevent overpressurization of the low-pressure shell on loss of condenser vacuum with steam still flowing to the condenser, breakable diaphragms or rupture discs are provided on top of the shell. These breakable diaphragms should be designed to vent to an area that will not cause injury to personnel in the event the diaphragms relieve the pressure. During low loads and high condenser pressures, windage action of the last stage blades may cause overheating of the flow guide and the blades. An exhaust hood water spray system cools the exhaust zone.

The inner cylinders support the blade rings. The inner cylinders have ports or are in sections to allow extraction of steam for feedwater heating. The inner cylinders rest on supports in the outer cylinder that allow the inner cylinders to move with thermal expansion.

The casing to shaft seals prevent leakage of air into the condenser. The seals are labyrinth-type with two pressure zones. The zone nearest the turbine is supplied low-pressure steam that leaks into the condenser and to the other zone that is at a slight vacuum. This arrangement prevents drawing cool ambient air over the shaft, causing thermal stresses in the shaft and reducing the condenser vacuum.

8.4.3 Rotor Assembly

The major components of a turbine rotor assembly are the shaft (rotor), the wheels, and the buckets or blades. The rotor assembly is supported on both ends by two stationary bearings.

The shaft of each turbine section is a single high-quality forging machined to provide the required contours and functioning parts. Each end contains an integral coupling, gland seal area, and bearing area. For high-pressure and intermediate-pressure reaction turbines, serrated, axial grooves are machined into the rotor for the blades as in the case in Fig. 8-16. Other designs incorporate different types of roots and grooves to attach the blades to the rotor. In some instances, the blades are pinned to the rotor. In the case of Fig. 8-16, the grooves are sized and contoured to match the blade root designated for each row. The blades are mounted directly in the rotor to minimize the diameter and thus the resulting blade thrust on the rotor from the pressure drop across the reaction blading. For impulse turbine high-pressure and intermediate-pressure sections and for low-pressure turbines, wheels are machined or shrink-fitted onto the rotor with the blades mounted in serrations or dovetail grooves at the end of the wheels (Fig. 8-16). The impulse high-pressure and intermediate-pressure turbines and the low-pressure turbines have considerably less pressure drop across the blading; therefore, the blading can be mounted on larger diameters (wheels), providing a larger steam flow path without inducing large thrust forces into the rotor. The rotors of the high-pressure–intermediate-pressure turbine may have a center bore machined in the shaft to remove impurities that occur during the forging process and to allow periodic internal inspection of the rotor.

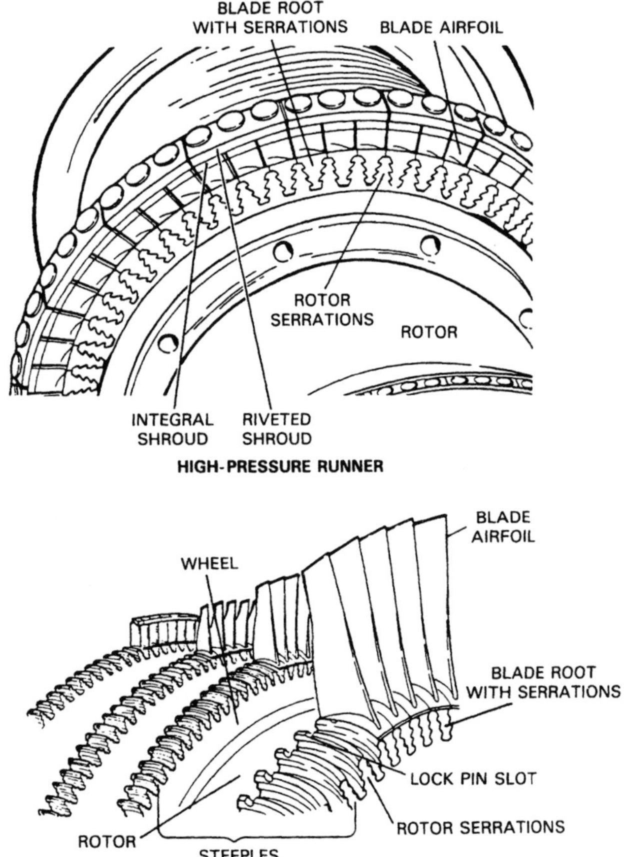

Fig. 8-16. Rotor blading grooves and blading. (From Westinghouse Electric Corporation. Used with permission.)

Turbine blades are machined from solid bars or forging of chrome–iron alloys. Each blade row has a unique profile to optimize the efficiency. The blades are mounted and locked into the rotor or wheels. The high-pressure, intermediate-pressure, and initial stages of the low-pressure turbines have shrouds. The shroud can be integral to the blades or fastened to the blades by integral tenons on the ends of the blades. The tenons act as retaining devices for the shroud. The shrouds dampen blade vibration and help direct and contain the steam flow around the blade tips. The shrouds are fitted over the tenons, that are peened over to hold the shrouds as shown in Fig. 8-16. The shrouds are final-machined to the rotor blade diameter necessary for proper running clearances. The last stages of the low-pressure turbine typically have free-standing blades (no shrouds) or blades that are continuously coupled at the ends. Some designs include tie wires at the blade center length or other grouping devices to minimize vibration. The last rows operate in steam that has high moisture content. To minimize blade erosion from the moisture, the leading edges of the blades in the later stages are induction hardened or covered with stellite. *Stellite* is a binary alloy of cobalt and chromium that is very hard and has high resistance to erosion. To aid in removing water from the back of the blades, a series of grooves is also often used.

A seal at the blade tip prevents steam leakage around the blade path between the blade tip and the casing. The seal is mounted in the casing and varies from a strip-type seal to a spring labyrinth-type seal. The seal design is dependent on the pressure drop across the blades. Reaction turbines that have high pressure drops across the blading require the labyrinth type seal, whereas impulse turbines with small pressure drops across the blading require only spill strips.

8.4.4 Bearings

Steam turbines have journal bearings and thrust bearings. Journal bearings are at each end of each rotor to support the weight of the rotor. One thrust bearing typically is provided for the entire steam turbine to maintain the axial position of the rotor.

8.4.4.1 Journal Bearings. The journal bearings are constructed of two halves that enclose the shaft. The inside of the bearing adjacent to the shaft is lined with babbitt metal. *Babbitt* is an alloy of tin, copper, and antimony that has antiseizing qualities and a natural oiliness. The journal bearings are oil-pressure lubricated. Oil flow is controlled to limit oil temperature rise to a set value that is typically 30° F (17° C). The two types of journal bearings commonly used are shell and tilting pad.

The shell bearing is used in heavily loaded areas. The bearing shell is spherical, resting in a corresponding spherical seat for ease of alignment. The bearings are oil-pressure lubricated to maintain an oil film between the shaft and the bearing. The oil is pumped into holes with axial grooves in the center of the bearing. The oil flows toward the outer edges of the bearing, where it is collected in radial grooves and drained back to the oil storage reservoir.

The tilting pad bearing is used in lightly loaded areas where rotor instability (oil whip) could occur. The tilting pad bearing is made of a series of individual babbitted segments (usually four) or shoes upon which the shaft rides. During normal operation, the pads tilt in the direction of rotation. This provides a converging passage necessary for hydrodynamic lubrication. The bearing compartment is flooded with oil entering the space between the pads. Oil outflow is control by the seals at the ends of the bearings. Oil flows back to the oil storage reservoir.

8.4.4.2 Thrust Bearings. The thrust bearing consists of babbitt metal lined, stationary shoes that run against the rotor thrust runner. Shoes on both sides of the runner prevent movement in either axial direction. The thrust bearing compartment is oil-pressure flooded with the oil introduced near the shaft and flowing outward by the centrifugal action of the runner. Oil flow is controlled to limit oil temperature rise to a set value, usually 30° F (17° C). Two shoe arrangements are typically used: tapered land or Kingsbury (tilting pad).

The tapered land thrust shoe consists of a flat, fixed surface with radial grooves cut into the thrust face to pass the oil. The segments or shoes created by the grooves are tapered in circumferential and radial directions so that the motion of the runner wipes oil into the contacting wedge-shaped area and builds up load-carrying oil pressures. Although the tapered land thrust plates have a high load-carrying capacity, they are sensitive to misalignment. Misalignment causes uneven loading and thermal distortion due to uneven temperature distribution around the thrust plate. This distortion can destroy the oil film thickness. To alleviate this, copper-backed plates are used to distribute temperature differences between lands. This procedure can double the load-carrying capability of a thrust bearing.

The tilting pad thrust bearing is a self-equalizing design with individual shoes or pads. Each shoe is free to tilt about a pivot. Normally, the pivot takes the form of a hardened spherical surface inserted behind each shoe. The shoe is free to tip in both circumferential or radial directions. This allows accommodation of any misalignment that may exist between the shoes and the thrust runner. When the runner is stationary, the shoes are parallel to the runner. As the bearing starts to rotate, an oil film is created between the shoes and the runner and the shoes tilt to an angle that generates the proper distribution of oil film pressure. In addition to the shoe pivot feature, the shoe pivot points are supported by leveling plate bases that equalize the thrust loading around the entire thrust bearing. The tilting-shoe type of thrust bearing is capable of carrying the greater load of the two thrust-bearing types.

8.4.5 Turning Gear

A turning gear rotates the turbine rotor at slow speeds before startup and just after shutdown. This minimizes the bowing of the rotor, that can cause uneven temperature distribution in the covers and bases of the turbine. The turning gear consists of an electric motor that gear drives the turbine generator rotor.

8.4.6 Pedestal

The steam turbine generator support foundation of large units is called a *pedestal.* Larger units have low-pressure turbines that exhaust downward and have numerous steam lines connected to the turbines. This requires the installation of the steam turbine generator on an elevated pedestal. The condenser can be located under the low-pressure turbine and access is provided to the underside of the turbines for steam lines.

Two types of turbine pedestals are used: heavy reinforced concrete and steel. The prevalent type in the United States is the massive concrete foundation, as shown in Fig. 8-10. A few of the steel foundations have been installed. The pedestal is supported on a based mat that is isolated from the remainder of the building to prevent any turbine generator vibration from being transmitted to the building. This isolation is also applied to any floor above the ground floor.

8.5 MAJOR GENERATOR COMPONENTS

The major generator components are the generator frame, stator, and rotor. A cross-section of a typical hydrogen-cooled large steam turbine driven generator is shown in Fig. 8-17. Additional information on generators is provided in Chapter 17, Electrical Systems.

8.5.1 Generator Frame

The generator frame supports the weight of the stator and rotor and acts as a containment vessel for the hydrogen cooling gas. The frame is a fabricated structure supported by a frame foot along its length on foundation seating plates. Foundation bolts are provided to resist short-circuit torques

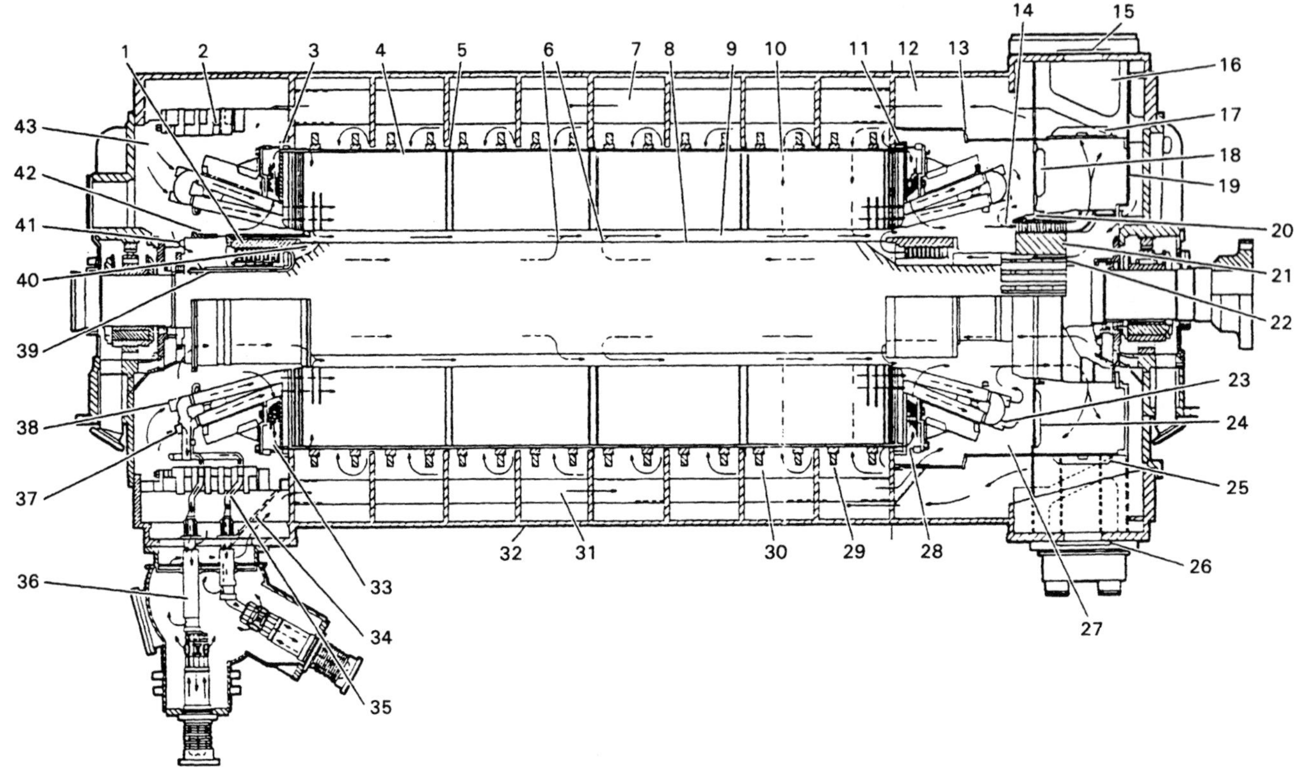

1 ROTOR END WINDING RETAINING RING
2 PARALLEL RING HOLLOW CONDUCTOR (Typical)
3 FRAME RING END ZONE SEAL (EE)
4 STATOR CORE
5 FRAME RING SEAL (3 Places)
6 ROTOR WINDING HOT GAS EXIT PORTS
7 COOL GAS DUCTS TO EXCITER END
8 ROTOR
9 AIR GAP (LP Zone)
10 STATOR CORE LAMINATION GAS FLOW
11 FRAME RING END ZONE SEAL (TE)
12 FRAME RING HP ZONE
13 WINDING DRUM BARRIER
14 HOT GAS RETURN TO BLOWER ASSEMBLY
15 MANHOLE COVER
16 HYDROGEN COOLERS
17 BAFFLE ASSEMBLY TOP COVER
18 BAFFLE ASSEMBLY INNER COVERS
19 BAFFLE ASSEMBLY OUTER COVER
20 BLOWER SHROUD ASSEMBLY
21 ROTOR HUB AND BLADES
22 COLD GAS VENTILATION SLOTS FOR ROTOR WINDING
23 STATOR COIL VENT TUBE (Hot Gas Exit)
24 BAFFLE ASSEMBLY INNER COVERS
25 BAFFLE ASSEMBLY BOTTOM COVER
26 MANHOLE COVER
27 LP ZONE
28 STATOR CORE END SHIELD
29 BORE RING (Typical)
30 STATOR CORE OUTER PERIMETER HP ZONE (4 Places)
31 HOT GAS RETURN DUCTS FROM LEAD BOX
32 GENERATOR FRAME
33 STATOR END SHIELD VENT PATH
34 FRAME FABRICATION DUCT AND COVER
35 PHASE LEAD CONNECTOR
36 LEAD CONNECTOR
37 PARALLEL RING VENT PORT (Cool Gas Entrance)
38 STATOR WINDING VENT TUBES (Cool Gas Entrance)
39 J STRAP CONDUCTOR
40 ROTOR END WINDING HOT GAS EXIT (LP Zone)
41 COOL GAS ENTRANCE (EE) TO ROTOR WINDING
42 AIR GAP BAFFLE
43 HP ZONE

Fig. 8-17. Typical generator cross-section. (From Westinghouse Electric Corporation. Used with permission.)

applied to the frame. Axial anchors at the frame foot and at the seating plates allow for thermal expansion of the generator in width. At each end of the frame is a journal bearing to support the rotor. Transverse anchors at the bearing bracket on each end of the generator allow expansion lengthwise.

The frame contains cooler baffles that create a flow path for the circulating hydrogen gas to cool the generator. The gas is circulated by a blower on the rotor shaft that discharges to the hydrogen coolers. The cooled gas from the water to hydrogen coolers flows through ducts and ventilation passages throughout the generator. The reverse side of the blower creates a low-pressure zone that pulls the heated gas back to the blower. The hydrogen gas is pressurized from 45 to 90 psia (308 to 618 kPa), depending on the generator size and rating. A gas–oil seal prevents gas leakage from the end of the frame at the shaft. Labyrinth-type seals on the gas side of the seal reduce the gas flow and pressure at the seal. Oil is injected in a center cavity of the seal and flows toward both the air and gas sides of the seal. Collection chambers are on each side of the oil injection point. The gas side collects any oil and gas leakage and routes it to a defoaming tank. This hydrogen seal oil system is described in more detail in Section 8.8.4.1, Generator Seal Oil System.

The main lead box, lead bushings, and current transformers are located under the generator at the exciter end. The main lead box is bolted to the underside of the generator frame. Six lead bushings are bolted to the main lead box. The bushings are high-voltage, gas-tight conductors that carry the generator output through the frame. Three of the bushings are bolted to the bottom of the lead box. The other three bushings are bolted to the slant of the lead box and connected together by a neutral bus in a wye-connected generator. Bushing-type current transformers are mounted on the main lead box. These transformers monitor the generator output for control room instruments and protective relays. The transformers also supply inputs to the voltage regulator.

8.5.2 Stator

The stator consists of a stator core and stator windings. The stator core has two major functions: to support the stator winding coils and to provide the flux path that links the stator coils and the rotating magnetic field.

The stator core is made of many high-permeability, low-loss silicon steel laminations. These laminations are coated on both sides with a high-temperature insulation. The laminations minimize the transformer-type losses caused by eddy currents. Hydrogen gas flowing through the ventilation passages between the lamination groups removes the heat generated in the stator.

Bolts at the outer diameter of the core and insulated through-bolts hold the lamination together to form a solid tubular cylinder. The through-bolts spread their clamping force over the entire face of the core through nonmagnetic plates that extend over the end portion of the core. A magnetic end shield is clamped outside the plates to shield the core from axial magnetic flux that creates radial eddy currents and heats the lamination. At the air gap surface, equally spaced slots run the entire length of the stator core. These slots extend into the core to a depth that allows installation of two separately wound coil sections. The portion of the core between the slots is called the tooth area.

Flexible elements connect the stator core to the generator frame. These elements help reduce the core vibratory forces transmitted to the generator frame.

The stator winding provides the generator output voltage and current. The stator winding consists of stationary high-voltage coils arranged in slots in the stator core to form a three-phase winding that is usually wye-connected. Each phase is made of two phase groups connected in parallel and containing an equal number of top and bottom coils connected in series. These coils are arranged in the phase groups spaced at intervals of 120 electrical degrees. Half-circle parallel rings connect the phase groups to the respective lead bushings. The three main lead bushings are connected to the main power transformer. The other three bushings are connected together to form the generator neutral.

Two stator coils are arranged in each stator slot. Each coil consists of mica-insulated strands of copper plus ventilation tubes surrounded by insulation. The ventilation tubes are open at each end of the coil, allowing the blower to circulate gas through the tubes to cool the windings. The coils are held firmly in place by a system of high-strength wedge filler and prestressed driving strips. The ends of the coils that extend beyond the core are mechanically braced to withstand vibration and abnormal forces. The bracing and blocking restrain both radial and tangential movement. Filler blocks between the end winding structure and the support brackets allow for normal axial thermal expansion of the coils.

8.5.3 Rotor

The rotor provides the magnetic field of the generator. The rotor consists of the shaft and the windings.

The rotor shaft is a one-piece, high-strength forging. The machined shaft contains the winding slots, end-turn ventilation slots, pole face slotting, bearing journals, and the couplings. The couplings are typically shrunk-fit and keyed onto the shaft.

The rotor winding is a multiple-turn, continuous-series winding. The winding is connected to the exciter through slip rings to receive power for generation of the magnetic field. The winding is constructed of both straight sections and end turns by use of two U-shaped bars per conductor. These bars are placed concave-to-concave in direct electrical contact. This forms a hollow conductor to allow hydrogen cooling.

The straight-section rotor windings are placed in variable width radial slots. The slots are tapered to give maximum winding space in the rotor. Molded insulation is used to line the slots. This insulation electrically insulates the windings from the shaft. Inter-turn insulation is provided by strips of insulation bonded to the conductors. Dovetail-shaped wedges along the tops of the slots over the entire length of the outer conductors secure the windings against centrifugal forces. The center portion of the wedge has holes for the discharge of hydrogen into the air gap between the rotor and stator. The end-turn winding ventilation path is separate from the straight section.

Blocks are fitted between the coil straight sections and between the coil arc parts to maintain alignment stability of the end turns. The rotor winding end turns are held against centrifugal force by retaining rings.

8.6 TURBINE GENERATOR PERFORMANCE AND RATING

The power plant designer must be able to estimate steam turbine generator performance, set the required ratings, and analyze the effects of various power plant operating modes on the purchased steam turbine generator. Prior to the purchase of a steam turbine generator, the designer must determine the plant steam cycle. This is usually accomplished through a series of optimization studies. These optimization studies require estimating the performance of the steam turbine generator. Once the steam cycle has been optimized, the steam turbine generator rating can be established and specified for purchasing of the unit. Following purchase of the unit, further plant optimization can occur and off-design performance evaluated through the use of information provided to the purchaser by the manufacturer in the form of a thermal kit.

8.6.1 Predicting Steam Turbine Performance

The prediction of steam turbine generator performance for development of the steam cycle is a long and complicated effort that is best performed by a computer simulation. A method of predicting performance is presented in the publication by Spencer, Cotton, and Cannon (1974). This publication is a revision of ASME Paper No. 62-WA-209 presented originally in 1962. The basic method presented in the paper is summarized in this section. This method can be used to approximate performance prior to purchase of the steam turbine. Once the unit is purchased, the turbine manufacturer will provide data that will allow more accurate prediction of performance.

The prediction of the turbine generator performance involves consideration of the following parameters:

- Expansion line efficiency,
- Exhaust loss,
- Packing and valve steam leakage flows,
- Mechanical losses, and
- Generator losses.

8.6.1.1 Expansion Line Efficiency. Expansion line efficiency is used to develop the turbine expansion line for plotting on a Mollier chart. This expansion line provides the enthalpies available at any stage in the turbine for purpose of estimating reheating requirements, turbine power output, and extraction flows for feedwater heating. The expansion efficiency is defined in Eq. (8-1), with typical expansion curves shown in Figs. 8-4 and 8-9. The expansion through a turbine section is actually made up of several stage expansions. However, prediction of a proposed (not purchased) turbine's performance on a stage-by-stage basis is a lengthy effort for which actual calculated efficiency would be no more meaningful than evaluating the entire turbine section (a section relating to high, intermediate, or low pressures). Thus, the development of an expansion line curve for a particular proposed turbine section should be based on an overall section efficiency.

The expansion line efficiency is a function of volume flow, pressure ratio, initial pressure and temperature, and governing stage design. At flows other than design, the efficiency of the stages, except of the first stage of the high-pressure turbine and the last few stages of the low-pressure turbine, is the same as at design because the pressure ratios and volume flows do not change. Thus, the efficiency of the intermediate-pressure turbine section does not change with throttle flow. The efficiency of the high-pressure section is primarily a function of throttle flow ratio because of the influences of the governing stage. The efficiency of the low-pressure condensing section is primarily a function of the annulus exhaust velocity that relates to the exhaust losses.

The high-pressure expansion lines for reheat units are drawn as straight lines. Partial-flow expansion lines are drawn parallel to the design expansion line through the calculated expansion line end point. The expansion line end point for the high-pressure section with governor stages can be obtained by multiplying the internal efficiency by the available energy and subtracting this value from the enthalpy of the steam ahead of the main steam stop valves. The available energy is the difference between the energy from ahead of the main steam stop valves to the energy at the section exhaust for an ideal expansion [refer to Eq. (8-1)].

The expansion lines for the intermediate-pressure sections of double reheat units are drawn as straight lines. The upper end of the line is drawn from the bowl enthalpy and entropy to the calculated expansion end point. Partial flow expansions are drawn in the same manner. The expansion line end point for the intermediate-pressure section can be obtained by multiplying the internal efficiency by the available energy and subtracting this value from the enthalpy of the steam at the inlet bowl. The available energy is the difference between the energy from the inlet bowl to the energy at the section exhaust for an ideal expansion. The 2% pressure drop from the intercept valve to the inlet bowl must be considered in the determination of the entropy.

The expansion lines for the low-pressure sections are drawn as slight concave curves (curving from the top toward the bottom to the right). The line is drawn from the inlet bowl enthalpy and entropy to the calculated expansion line end point. Partial flow expansion lines are drawn in the same manner. The expansion line end point for the low-pressure section can be obtained by multiplying the internal efficiency by the available energy and subtracting this value from the enthalpy of the steam at the inlet bowl. The available energy should be based on the difference between inlet bowl energy and the ideal expansion line end point energy. The actual expansion line end point is calculated based on 1.5 in. HgA (5.080 kPa) exhaust pressure with a correction to the actual end point as a function of incremental turbine efficiency and available energy.

For single-reheat units, the intermediate-pressure section is combined with the low-pressure section to develop the expansion line with the intermediate-pressure inlet bowl used as the starting point and the efficiency and expansion line end point calculated based on both sections being the low-pressure section. This also applies for the reheat section of a double reheat unit.

Nonreheat expansion lines are drawn in a manner similar to the low-pressure section starting at the throttle valve enthalpy. The internal efficiency calculation accounts for the effects of any governing stage.

8.6.1.2 Exhaust Loss. The exhaust loss is a kinetic energy loss and enthalpy increase associated with the steam exhausting from the last stage blades of the low-pressure section. The exhaust loss consists of two components: hood loss and leaving loss. The hood loss is due to the pressure drop associated with the exhaust hood. The leaving loss is that associated with the kinetic energy of the steam. The steam leaving the last stage blades of the low-pressure section is travelling at a high velocity, resulting in pressure drop. As it turns down and slows in the exhaust hood, most of its kinetic energy is converted into enthalpy.

The exhaust loss is a function of the exhaust area and the steam velocity. Typical exhaust loss curves are shown in Fig. 8-18. The exhaust loss is added to the expansion line end point (ELEP) enthalpy to determine the actual enthalpy of the steam leaving the low-pressure section. This actual expansion end point is called the used energy end point (UEEP) or the turbine end point (TEP). The actual end point should be used in determining the power generated by the low-pressure section of the turbine.

8.6.1.3 Packing and Valve Stem Leakages. In determining the overall power output, the designer should account for steam leakages in establishing steam flows. The leakage occurs from the main and reheat steam valves and from the shaft packings. The shaft packings include the seal between pressure steam and atmosphere and the leakages associated with seals between two turbine sections in one casing such as an opposed flow high-pressure–intermediate-pressure turbine. The leakages are used for steam seal, feedwater heating, or expansion in a lower pressure section of the turbine. The leakages are a function of the turbine type, size, and steam pressures. Typical leakages are shown in the heat balance in Chapter 13.

8.6.1.4 Mechanical Losses. These losses include all the mechanical losses of the turbine such as the turbine bearing losses and the shaft-driven bearing oil pump. The mechanical losses also include the bearing losses of the generator and exciter. Mechanical losses are estimated as a function of unit size and shaft speed. These losses are subtracted from the turbine output.

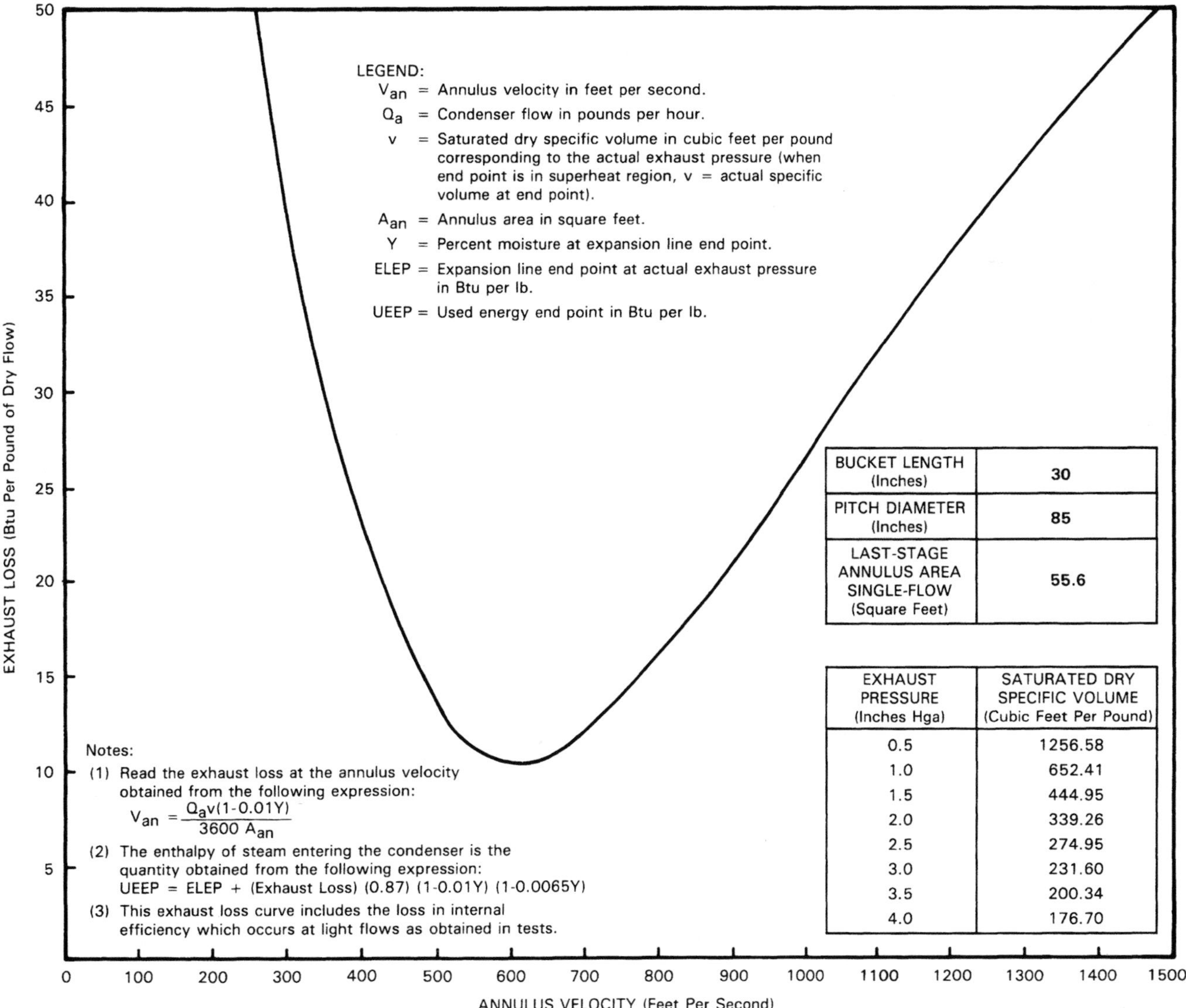

Fig. 8-18. Typical condensing low-pressure turbine exhaust loss curve. (From General Electric Company. Used with permission.)

8.6.1.5 Generator Losses. The generator losses include all mechanical and electrical losses of the generator, except for bearing losses, that are included with the turbine mechanical losses. The electrical losses include the resistive (I^2R) heating losses and the excitation power. The generator losses are a function of unit size, speed, and generator cooling method. These losses are subtracted from the turbine output to obtain the final generator output.

8.6.1.6 Predicted Unit Output. By use of the expansion lines, exhaust loss, and leakages, the power generated by the expansion of the steam can be calculated. Subtracting the mechanical losses from this provides the power supplied to the generator by the turbine. Subtracting the generator losses from the turbine output yields the final generator output to the main transformer.

8.6.1.7 Thermal Kit. The turbine manufacturer should be required to supply various calculated unit performance cases (heat balances) and information to enable the designer to predict the unit performance. This information is typically contained in a document called a thermal kit. Typical information contained in the thermal kit is as follows:

- Heat balances,
- Expansion line diagrams,
- Exhaust loss curve,
- Expansion line end point versus reheat flow and condenser pressure,
- Stage, bowl, and extraction pressures as a function of the downstream stage flow,
- Backpressure correction curves,
- Throttle and reheat temperature and pressure correction curves,
- Leakage information as a function of throttle flow,
- Generator loss curves, and
- Generator characteristic curves.

This information can be used to estimate changes in unit performance at off-design conditions. These estimations can be performed by hand. However, some calculations can be lengthy, and if several conditions are being evaluated, a detailed computer model is typically used with this information to predict the performance of the actual turbine purchased.

8.6.2 Turbine Ratings

The rating of a steam turbine is dependent on the overall power plant steam cycle. The steam cycle should be established by the power plant designer before specification of the steam turbine generator. The basic conditions that must be determined are those necessary for development of a steam cycle heat balance as discussed in Chapter 13. Once the basic steam cycle has been established, the designer can specify the turbine rating by indicating the required generator output, rated steam conditions, low-pressure turbine exhaust pressure, special extraction flows, and the percent makeup to the condenser. These items, along with the steam cycle layout, enable the turbine generator manufacturer to establish the steam turbine rating.

8.6.2.1 Turbine (Generator) Output. The required generator output should be established by the designer based on the plant net output plus the plant auxiliary loads. Once the turbine generator manufacturer knows the generator output, he establishes the steam turbine output, taking into account the generator electrical losses and turbine generator mechanical losses.

To reduce cost, the plant designer may establish the steam turbine generator rating considering design margins inherent to steam turbine design. Historically, turbine designers have included 5% margins above required rated steam flows and pressure to provide for manufacturing tolerances and variations in flow coefficients. The steam flow margin is included by designing the turbine to achieve a steam flow with the main steam control valves wide open (VWO) that is 5% greater than that required for rated output with rated steam pressure (normal pressure). Under the valves-wide-open, normal pressure condition (VWO-NP), the turbine generator output is expected to be approximately 104% of rated.

The pressure margin is included by designing the turbine to operate safely and continously at 105% of rated pressure (overpressure) with the control valves wide open (VWO-OP). Under the VWO-OP condition, the turbine generator output is approximately 109% of rated and the main steam flow is 110% to 111% of rated.

The designer can include some or all of these margins in the calculation of the turbine generator rating. Recent trends indicate that some manufacturers are not including all of the 5% steam flow margin as they have in the past. The designer may want to include only a part of the steam flow margin with consideration of the full over-pressure operation margins. The designer should specify that the turbine be capable of operation at VWO-OP because operators typically attempt to operate at those conditions. The designer needs to design the steam generator and balance of plant equipment to support the VWO-OP conditions if he has included them in the establishment of the steam turbine generator rating. Designing for the VWO-OP condition is recommended even if not included in the rating definition, since significant output can be achieved at little cost by being capable of operating at VWO-OP.

8.6.2.2 Rated Steam Conditions. During the conceptual design of the plant, the designer determines the steam cycle temperatures and pressure. These are the conditions at the inlet to the main steam stop valves (pressure and temperature) and the reheat stop valves (temperature). These temperatures and pressures should be those specified to the manufacturer, who lists these on the heat balances as rated pressure, main steam temperature/reheat steam temperature; for example, 2,415 psia (16.65 MPa), 1,000° F/1,000° F (537.8° C/537.8° C); and 1,815 psia (12.51 MPa), 1,000° F/1,000° F (537.8° C/

537.8° C). Overpressure conditions would show the pressures on the main steam flow as 2,535 psia (17.48 MPa) and 1,905 psia (13.13 MPa) for these two examples.

8.6.2.3 Low-Pressure Exhaust Pressure. The low-pressure exhaust pressure is the condenser pressure under the conditions that the designer wants the required plant output. This may be a peak summer condition, a peak winter condition, or an annual average condition.

8.6.2.4 Extraction Flows. In the establishment of the steam cycle, the number of feedwater heaters and the sources of their extraction steam are determined. The extraction flows to these heaters are determined by the turbine manufacturer based on the turbine design. The extraction steam source for the turbine drive of the boiler feed pump (if provided) is also selected by the designer and typically is the intermediate-pressure turbine exhaust for moderate to large plants. In addition, special steam extractions may be required for items such as air preheating, flue gas reheating, or process steam. The turbine manufacturer must be given the required steam flows, temperatures, and pressures of these special steam extractions at the rated plant conditions. This is used in establishing the turbine rating. The manufacturer should also be given the required conditions over the entire operating range to ensure that the turbine performs satisfactorily over the entire operating range.

8.6.2.5 Makeup Flows. In setting the turbine rating, designers should specify the amount of makeup flow to the steam cycle. This makeup is typically introduced into the condenser and therefore must be heated in the feedwater cycle on the way to the boiler. The makeup typically replaces steam losses in the cycle and losses in the boiler associated with boiler blowdown and steam soot blowing. Typical amounts specified are from 1% to 3% of the throttle flow. Consideration should also be given to process extractions that involve less than 100% return of condensate.

8.6.3 Generator Ratings

The establishment of the generator rating involves defining the size (power rating), power factor, short circuit ratio, and the response ratio.

8.6.3.1 Power Rating (kVA). The generator must be capable of producing the maximum peak load power requirement established for the plant. Generator power ratings are given as the total power that is the vectorial sum of the active (or real) power (kW) and reactive power (kilovars). The generator must be sized to handle the maximum shaft power output of the driving turbine and provide reactive power required by the electric system. The generator nameplate rating is specified as net terminal kVA with maximum hydrogen pressure (if hydrogen cooling is employed) and rated power factor. The generator rating (kVA) is established by dividing the generator maximum real power output (kW) by the power factor.

8.6.3.2 Power Factor. The power factor is an important design parameter and is defined as the ratio of the active power (kW) to total power (kVA). For a given active power, the lower the rated power factor, the higher the armature and field currents (reactive power), which causes greater resistive (I^2R) losses. In addition, larger generators are needed, which increases capital costs and other losses such as windage and core losses.

In general, the power factor used in specification of generators should be based on expected electrical transmission system needs, with a reasonable design margin to give stability in the event of transient system disturbances. A power factor of 0.9 is adequate for most instances and is more or less an industry standard. In addition, experience has shown it is generally less expensive to adjust system reactance by capacitive and inductive devices installed in the transmission system than by employing generators of greater capability.

8.6.3.3 Short-Circuit Ratio. The generator short-circuit ratio is a measure of the generator stability characteristics. It is defined as the ratio of field current required to produce no-load rated armature voltage at rated frequency to the current required to produce rated armature current at rated frequency with the generator terminals short-circuited. For a given change in load, the change in excitation required to maintain constant terminal voltage is inversely proportional to the short-circuit ratio. Therefore, for a given change in load, a machine with a low short-circuit ratio requires a larger incremental change in excitation than a machine with a higher short-circuit ratio. High short-circuit ratios increase the cost of the generator. Other factors affect stability such as turbine-generator inertia, transmission system stability, and response characteristics of the voltage regulator and excitation equipment.

Over the past few years, generator specifications have called for lower short-circuit ratios, as present transmission systems tend to be more stable. More importantly, the response characteristics of voltage regulator and excitation equipment have improved significantly; recent developments of solid-state control equipment have made the equipment more cost effective. A short-circuit ratio of 0.58 range is typically recommended, unless system stability studies indicate otherwise.

8.6.3.4 Exciter Response Ratio. The exciter response ratio (or nominal response) is the primary variable involved in the specification of an excitation system and indicates the average speed with that an excitation system responds to transients. More exactly, the response ratio is defined as the average rate of rise in exciter voltage for the first half second after change initiation, divided by the generator rated voltage. A standard value for exciter response is 0.5 for a brushless-type exciter. Static-type excitation systems have much higher response ratios (1.0 to 6.0) since they do not have mechanical inertia.

The proper exciter response ratio is typically determined on a case-by-case basis. Higher response ratios become more advantageous when the generator is connected to a relatively large transmission network, as commonly experienced in the western United States, where transmission lines running several hundred miles are not uncommon. Often, distribution pools have specific requirements as to the response characteristics of member utilities' plants. Selection of a higher response ratio than necessary does not pose any operation problems; however, an insufficiently rapid exciter can prevent the unit from remaining on-line when transient disturbances occur on the system. Except for long transmission systems, a response ratio of 0.5 is typically adequate for most installations.

8.7 STEAM FLOW CONTROL

The steam flow to a power plant steam turbine is controlled by the main steam control valves. These valves regulate the steam flow to the high-pressure turbine, that also controls the steam flow to remaining turbine sections. The operation of these valves in conjunction with pressure variation in the steam generator provides for four methods of controlling steam turbine generator output:

1. Throttling control,
2. Governing control,
3. Variable-pressure control, and
4. Hybrid variable-pressure–governing control.

These four methods of control and comparisons of the heat rates for each mode are presented in the following section. The control methods are also referred to as admission modes.

8.7.1 Throttling Control

Throttling control is the simultaneous operation of all main steam control valves at the same time. The main steam pressure is typically held constant at rated conditions and the control valves are all opened the same amount. The steam turbine output increases as the valves are opened and full load is reached when the valves are wide open. This type of operation minimizes the mechanical loadings on the steam turbine control stage as a result of the reduced pressure to the stage and the equal loading on all stage sections. At partial loads, throttling operation is the least efficient of all control modes because of the throttling process that reduces the energy available in the steam expansion process.

Depending on the turbine design, this mode of operation may also be referred to as full arc admission because of the steam admission to all portions of the control stage. The term typically is used in conjunction with the startup of a turbine. During the startup of some units, the control valves are wide open. Steam is initially admitted to the turbine by throttling the steam flow by use of bypass valves internal to the main steam stop valves. This flow control method is used up to 15% to 25% load. Above this load, the main steam control valves are used to control the steam flow and the main steam stop valves are wide open.

8.7.2 Governing Control

Governing control is the sequential operation of the main steam control valves. Governing control varies the output of the steam turbine by increasing or decreasing the arc of admission of steam flow to the turbine control (first) stage. Each control valve feeds a particular section of the control stage, and the amount of arc in use is determined by the number of valves open. The valves are opened in a particular order that is determined by the allowable stresses on the control stage. Typically, a minimum of 25% of the control valves are opened simultaneously when initially starting a unit to minimize stresses on the control stage. When these valves are fully open, the remaining valves are opened in series. This control mode is also referred to as partial arc admission. Rated throttle conditions are used throughout the load range to the extent allowed by the steam generator. This control mode is more efficient than throttling control because the throttling process loss is minimized by reducing the number of control valves throttling at any one time.

8.7.3 Variable-Pressure Control

Variable-pressure control is steam flow control by varying the steam generator pressure with the main steam control valves in a fixed position. In this control mode, the main steam control valves typically are wide open. The steam flow to the turbine is controlled by the pressure of the main steam from the steam generator. Main steam pressure is controlled by the steam generator firing rate. The main advantage of variable pressure operation is that the turbine first-stage temperature remains relatively constant across the load range which shortens startup times and increases turbine rotor life. The disadvantages of variable pressure operation are poorer thermodynamic performance and limited load response capability. The lower main steam pressures of this mode result in less available energy than in the governing mode, but more than in the throttling mode. This reduced efficiency still exists on an overall cycle basis even when the reduced power requirements of the boiler feed pump (due to reduced discharge pressure) are included. The response time is limited to the rate the steam generator firing can be increased since the main steam pressure is controlled by the firing rate. In the other modes of control, steam turbine load can be increased more rapidly by opening the closed main steam control valves and utilizing pressure and thermal energy stored in the steam generator.

8.7.4 Hybrid Variable-Pressure–Governing Control

The hybrid variable-pressure–governing control mode uses the low-load high-temperature, no throttling loss advantages of variable-pressure operation and the thermodynamic and load response advantages of governing control. In addition, hybrid operation provides improvements in part load control stage efficiency. At low loads, some of the main steam control valves are wide open and steam flow is controlled by variable pressure operation of the steam generator. Increasing main steam pressure increases steam turbine load until the steam pressure reaches rated conditions. Load is increased further by maintaining the rated main steam pressure and sequentially opening the remaining main steam control valve as in the governing control.

8.7.5 Heat Rate Comparisons

A comparison of the steam cycle overall heat rates for the four control modes is shown in Fig. 8-19. The throttling control has the worst overall heat rate as a result of throttling losses. The variable-pressure control mode has a slightly better heat rate than throttling control, but still has poor performance because of the low steam pressures. The governing control has the best heat rate at higher loads because of the high steam pressures and minimized throttling losses resulting from throttling with only one valve at a time. The hybrid control mode takes advantage of the governing high load performance while reducing throttling losses and boiler feed pump power at low load with the use of variable-pressure operation. This hybrid mode results in the most efficient operation while extending unit life by maintaining high temperatures at low loads, reducing cycling effects.

Governing heat rates are typically plotted as a "locus of valve points," that is, as a smooth curve passing through the heat rate points where any valves open are wide open. The governing heat rate is shown by the solid line in Fig. 8-19. The actual heat rate curve is represented by a valve loop that incorporates the throttling losses associated with a valve throttling between full closed and full open. This actual heat rate is represented by the dashed line. As can be seen from the heat rate curve, the more control valves, the smaller the valve loop, the greater the possibility of operating without throttling, and the better the heat rate.

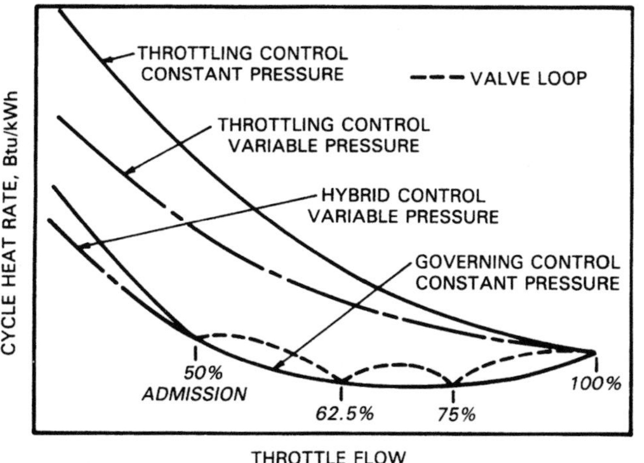

Fig. 8-19. Control mode thermal performance curve. (From Westinghouse Electric Corporation. Used with permission.)

8.8 AUXILIARY SYSTEMS

Steam turbine generators require numerous ancillary systems to support their operation. These systems are normally supplied with the units. The function and scope of the auxiliary systems are described in the following subsections. Also included as part of the description of the turbine drains system are turbine interfacing recommendations and requirements with regard to water induction prevention. The design of the auxiliary systems varies by manufacturer. Although the information provided here is generic, it still may be more applicable to one manufacturer in some cases.

8.8.1 Turbine Seal System

The turbine seal system prevents seal leakage from pressurized packings of the turbine and valves, and prevents air leakage into zones under a vacuum. The basic concept of the seal system is shown in Fig. 8-20. The turbine shaft seals and the operation of the system are described below.

8.8.1.1 Shaft Seals. At the end of each turbine casing is a shaft seal (shaft packing) that prevents steam leakage to atmosphere and air infiltration into the turbine casing. The shaft seals consist of labyrinth-type packing rings with glands. There are two types of shaft packings: pressure packings and vacuum packings. Figure 8-20 shows a schematic of typical pressure and vacuum packings. Both seal systems incorporate labyrinth packing rings with glands and connections to two sources that have constant pressure. One source of constant pressure is the steam seal header, that is typically held to 3 to 4 psig (21 to 28 kPa gauge). The other source is the steam packing exhauster vent header, for that pressure is typically maintained at about 3 to 5 in. (746 to 1,244 Pa) of water vacuum.

The pressure packings are the shaft seals that seal against a positive pressure at full load. This type of seal is typically found on the exhaust end of high- and intermediate-pressure turbines, where the pressures can be on the order of 815 and 195 psia (5.62 and 1.34 MPa), respectively. The steam pressure is broken down to the steam seal header pressure in stages. At each stage there is a leakoff that discharges steam to a turbine extraction point or to the steam seal header. At full load, a portion of the leakage flow goes to the steam seal header. The remaining leakage steam flows to the condenser or a feedwater heater. During startup, when there is a vacuum condition throughout the turbine, the steam seal header supplies steam to the pressure packings to ensure a seal to prevent air inleakage.

The vacuum packings are the shaft seals that always seal against a vacuum. Steam is supplied from the steam seal header. A portion of this flows to the vacuum end of the turbine. The remaining portion of steam flows to the steam packing exhauster vent header.

8.8.1.2 Seal System. During startup and low-load operation, steam from the main steam supply system or auxiliary

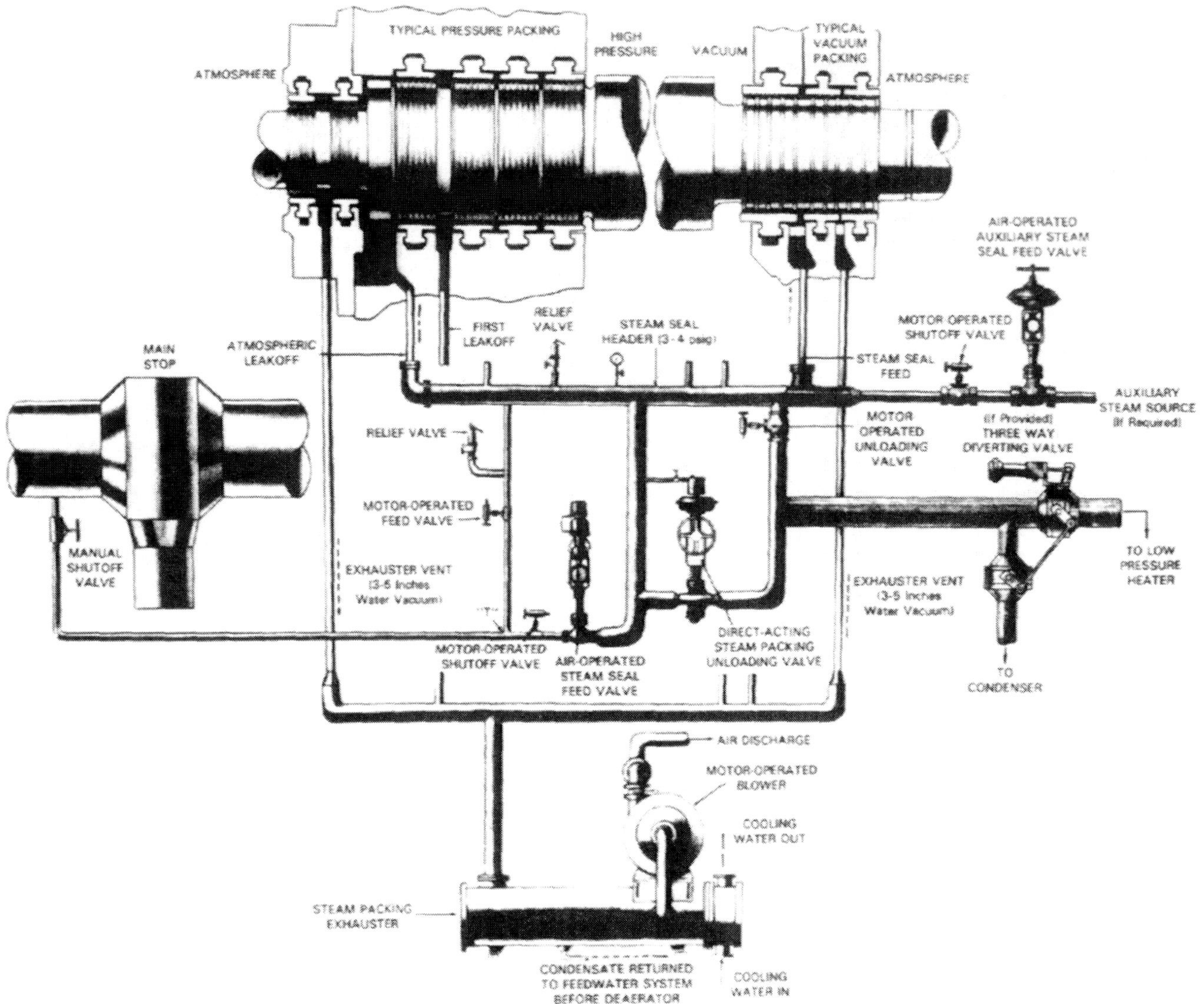

Fig. 8-20. Diagram of a typical steam seal system. (From General Electric Company. Used with permission.)

steam supply system supplies the steam seal header. Pressure in the steam seal header is maintained by diaphragm-actuated feed valves. The main steam and auxiliary steam supply are desuperheated with condensate as required.

The steam seal header is supplemented by leakage from the pressure packings on the high-pressure turbine, intermediate-pressure turbine, and boiler feed pump turbines. As the pressure packing leakage increases with load, the steam supplied through the main steam feed valve or the auxiliary steam seal feed valve is decreased. Above approximately 50% turbine load, the leakage from the pressure packing exceeds the steam seal requirements of the low-pressure turbine vacuum packings. As the vacuum packing steam seal requirements are more than satisfied by the pressure packing leakage, the excess steam is discharged through the steam packing unloading valve. The excess steam is piped from a steam dump valve to a feedwater heater or to the condenser.

The outer packings of the stop valves, control valves, and turbine shaft are maintained under vacuum by a steam packing exhauster that consists of a steam condenser and motor driven blower (exhauster). Air and leakoff steam are drawn from the packings into the steam condenser where the steam is condensed and the air is discharged to the atmosphere. Because of the critical nature of this system, a spare backup blower is often installed.

8.8.2 Turbine Drains System

The introduction of water into a steam turbine can cause significant damage to the turbine. Water can be inducted into the turbine through many sources. The primary sources of water are condensation in steam lines, valves and casings during startups, desuperheating water sprays, and water from feedwater heaters on unit trips. The potential for water induction is great unless the plant designer takes special precautions when designing systems that connect to the turbine. The extent of the concern has resulted in an ANSI/ASME Standard No. TDP-1, *Recommended Practices for the Prevention of Water Damage to Steam Turbines Used for Electric Power Generation*, to provide design guidelines for the prevention of water induction into the turbine.

8.8.2.1 Turbine Induction Prevention Measures. Turbine water induction prevention measures for each potential induction point should include the following techniques:

- Detection of the presence of water external to the turbine before the accumulation of water is sufficient to cause damage.
- Where possible, isolation of the turbine from the water source after the water has been detected. Where practical, isolation should be automatic.
- Automatic disposal of the accumulated water after it has been detected and isolated, where practical.

During the implementation of the above techniques, the designer should consider failure modes. The design should be reviewed to ensure two failure considerations.

First, where a water source is particularly hazardous, no single failure of equipment associated with that source should in itself be a cause of water entering the turbine.

Second, the failure mode of the various devices used to prevent water induction should be reviewed to ensure that a single failure of the signals from or to the device will not allow water to enter the turbine.

Many of the water induction prevention measures involve the collection and draining of condensed steam in steam lines. The design of the drain valves and piping should include the following general requirements:

- All drain valves should be power-operated.
- All automatic actuated power-operated isolation and drain valves should be remotely operated from the control room.
- Each drain pipe should be continuously sloped, or a downward trend maintained, in the direction of flow to the discharge. When a low point cannot be avoided, the downstream end of the horizontal run should always be lower than the upstream end. No part of a drain pipe should be lower than the final discharge.
- Intersecting drain lines should be joined by wye or lateral connections instead of tees. Drain lines should use radius bends instead of elbows.

The following are areas of a power plant that should receive special attention in the design of water induction prevention measures:

- Main steam line low-point drains,
- Main steam stop valves before seat drains,
- Desuperheating water spray systems (main and reheat steam),
- Cold reheat line low-point drains,
- Intermediate-pressure turbine cooling steam pipe connection location,
- Hot reheat line low-point drains,
- Reheat stop and intercept valve seat drains,
- Steam lead and turbine drains as set by the manufacturer,
- Intercept valve bypass valves and piping,
- Steam seal system,
- Feedwater heater level control system,
- Feedwater heater extraction piping system, and
- Steam supply systems to auxiliary steam turbine drives.

8.8.2.2 Drains System. The turbine manufacturer supplies a turbine drain system with the unit. The system includes drains required downstream of the main steam stop valves and the reheat stop valves. Drains at the low points of the steam leads drain any condensate that collects between the turbine control valves and the high-pressure turbine. The turbine design includes provisions for supplying warming steam to the steam leads that are isolated by their corresponding control valves during turbine operation with partial arc admission. The drains from all the steam leads are manifolded into a common header that contains a power-actuated shutoff valve and discharges to the condenser.

The main stop valves and turbine control valves have drains to remove condensate that collects in the valves. The auxiliary drive steam turbine stop valves also have drains. The drains have power-operated shutoff valves and are piped to the condenser or blowdown tank.

The reheat stop valves have after-seat drains that drain the valve bodies and the piping between the valves and the intermediate-pressure turbine. The drains have power-operated shutoff valves and are piped to the condenser.

Drains to remove condensate from the turbine casing may be required. The drains are piped to the condenser and may be either continuous drain or isolated drains. Isolated drains have power-operated shutoff valves. The wastewater and oil drains from the turbine, main stop valves, and combined reheat valves are piped to waste.

8.8.3 Turbine Lubrication Oil System

Before discussing turbine lubricating oil systems, it is important to state that the hydraulic control fluid system used in actuating the main steam stop valves, main steam control valves, reheat intercept valves, etc., may be part of the turbine lubricating oil system. If the turbine lubricating oil system supplies control oil in addition to lubricating oil, then the control system can be identified as a mechanical–hydraulic control system. If the hydraulic control system is kept separate from the bearing lubricating oil system, the control system is called an electrohydraulic control system. In general, large- and medium-sized turbines use an electrohydraulic control system. The descriptions of the turbine lubrication oil system given in this text are for turbines with an electrohydraulic control system. Oil purification systems are also described.

8.8.3.1 Lubrication System. A diagram of a typical turbine lube oil system is shown in Fig. 8-21. The oil tank (reservoir) is physically below the turbine, often on the ground floor. The turning gear oil pump (TGOP), emergency

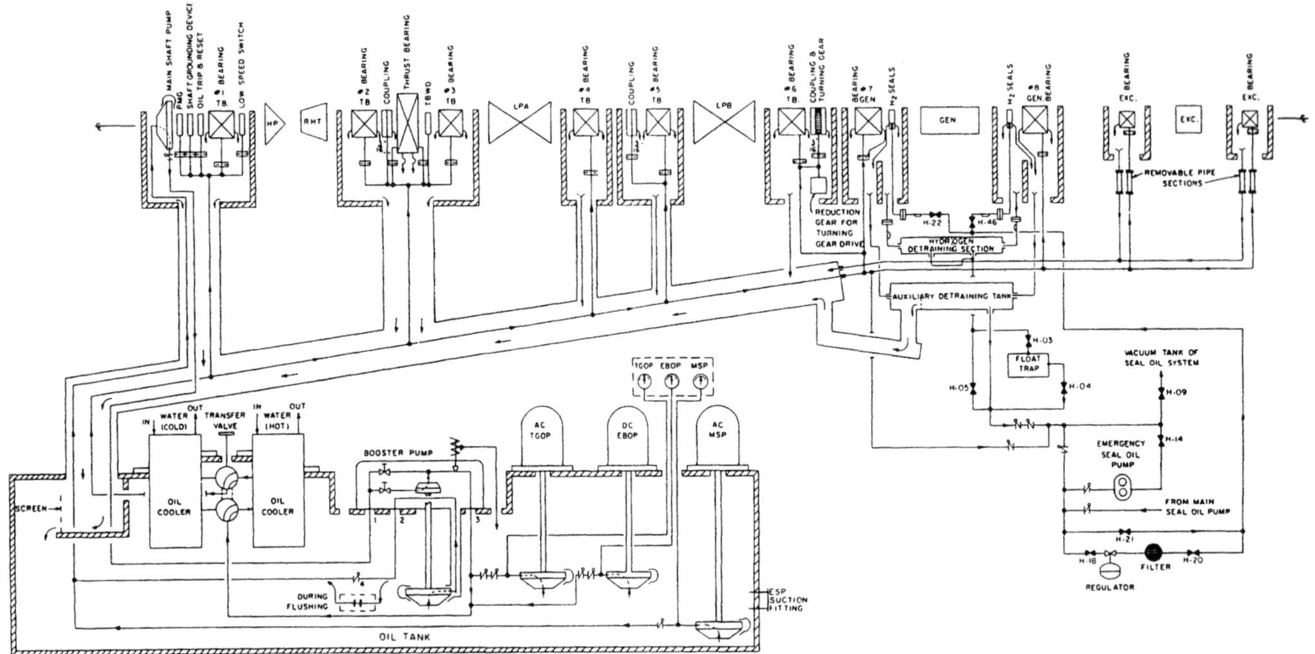

Fig. 8-21. Diagram of a turbine lubricating oil system. (From General Electric Company. Used with permission.)

backup oil pump (EBOP), main suction pump (MSP), booster pump, oil coolers, and control gauges are mounted on top of the tank. The main oil pump is shaft-driven and is located in the front standard. The pressure lube oil supply pipe to the bearing is located in the return pipe, that acts as a guard. In some cases, supply pipes are separately guarded. A guard reduces the possibility of oil contacting superheated-steam piping (that is a fire hazard) in the event of a pressure line leak. In this system, the return pipe acts as a guard for the supply pipe. Not shown in the diagram are the turbine lube oil vapor extractors on the oil tank.

When the unit is on-line and in normal operation, the booster pump supplies oil to the main shaft-driven pump, that supplies oil to the bearings. The booster pump is powered by an oil turbine that receives oil from the main oil pump discharge. During startup of the turbine, the main suction pump supplies the main oil pump until the main turbine reaches approximately 90% of rated speed. The turning gear oil pump is used when the unit is on turning gear. On designs that do not employ this pump, turbine bearing lifting pumps are used in conjunction with the turbine bearing pumps when the turbine is on turning gear.

8.8.3.2 Oil Purification. An Electric Power Research Institute study concluded that over half of the forced outage hours related to bearings, journals, and oil systems were a result of contaminated oil (Rippel and Colsher 1982). A lube oil purification system should remove from the turbine lubricating oil the scale, sediment, water, sludge, and insoluble oxides that are foreign to the oil or that may be chemically or mechanically formed in the oil, and should restore the purity, viscosity, and other desirable properties of new oil. Water is a very common contaminant and is introduced to the lube oil at the bearing and oil return lines. The source of the water is from the turbine steam seals.

Lube oil conditioning may be performed in either continuous or batch purification modes. In the batch mode, lube oil is periodically replaced with clean oil. ASME Standard LOS-1M, *ASTM-ASME-NEMA Recommended Practices for the Cleaning, Flushing, and Purification of Steam and Gas Turbine Lubrication Systems*, along with turbine manufacturing standards, recommend continuous purification methods over batch purification methods.

The standard also recommends that the continuous purification system be capable of cleaning 10% to 20% of the total lube oil capacity. Most turbine manufacturers recommend 20%. Solid particulate matter, down to 10 microns or less, along with all free water should be removed.

Lube oil purification systems used for purification of steam turbine lubricating oil in central power stations include the mechanical filter, the centrifuge, the vacuum dehydrator, and the coalescer types. The mechanical filter and centrifuge types are the most commonly used. For all types, oil is taken from the lube oil reservoir, conditioned, and returned to the reservoir. In addition to the purification systems, an in-line bearing filtration system may also be employed. The in-line bearing filtration system is located in the pressure lines to the turbine bearings. This system improves lube oil quality but imposes a small risk to the bearing in the event of a filter failure.

The mechanical filter (or precipitator) type lube oil conditioner has three compartments: a precipitation compartment with water and particulate removal screens, a filtration compartment with filter bags, and a storage compartment with polishing filters. A large portion of the free water in the lube oil is removed in the precipitation compartment by the use of screens and gravity. Filter bags in the filtration compartment remove a majority of solid contaminants. Final filtering is accomplished in the storage compartment through the use of polishing filters.

A centrifuge-type lube oil conditioner consists of a centrifuge in parallel with pressure filters. The centrifuge spin bowl causes the heavy contaminants and water to flow outward and the oil to flow inward, allowing separation. Both the centrifuge and the pressure filter are capable of removing particulate down to at least 10 microns. Normal operation involves continuous filtering with periodic operation of the centrifuge for water removal. One may be used for oil purification while the other is serviced.

Vacuum dehydration systems are seldom used in today's large power plants. The principle of the vacuum dehydrator unit is to heat the oil enough to vaporize and remove contaminants. The remaining contaminants are removed by filtering.

The coalescer-type unit employs a fiberglass element. As contaminated oil flows through the element, small water particles become trapped. The increasing differential pressure across the element causes the particles to consolidate and forces the relatively large particles to the downstream side. A water-repellent screen collects the water drops and routes them to a drain by gravity. Nearly all free and emulsified water is removed. Polishing filters remove the remaining solid contaminants.

8.8.4 Generator Cooling and Purge

Heat is produced in a generator as a result of resistive losses caused by current flow in the stator and field windings, stator core magnetic losses, and windage losses. This heating action, that is primarily a function of load and power factor, is the primary limiting factor in generator rating. By providing forced cooling of the rotating and stationary components, the generator rating may be increased and the physical size of the components made smaller.

Large generators typically are cooled with hydrogen. The thermal properties of hydrogen are superior to those of air and allow for reduced windage and better cooling. Windage and ventilating losses are lower because of the low density of hydrogen gas. Generators can be built smaller because of the high specific heat capacity, thermal conductivity, and heat transfer coefficients of hydrogen. Hydrogen must be supplied to the generator under carefully controlled conditions of purity and pressure. The hydrogen in the generator is circulated by fan blades mounted on the rotor. The hydrogen gas is circulated past the stator coil and core to remove heat. The rotor is cooled by a self-pumping action that draws hydrogen gas into passages in the rotor. Heat is removed from the hydrogen by gas to water coolers mounted in the generator casing. The hydrogen coolers are connected to the plant cooling water system.

For large generators, the stators may also be water cooled as well as hydrogen cooled. This system consists of a closed deionized water loop.

8.8.4.1 Generator Seal Oil System. The hydrogen leakage out of the generator casing is minimized by a seal oil system. The seal oil system consists of two segments: the hydrogen-side seal oil segment, that provides seal oil to the hydrogen side of the generator gland seals, and the air-side seal oil segment, that supplies seal oil to the air side of the gland seals.

The hydrogen-side seal oil segment consists of a seal oil pump, seal oil cooler, oil filters, defoaming tank at each seal, and a drain regulator. The oil pump takes suction from the hydrogen-side drain regulator and discharges through the oil cooler and filters to the hydrogen side of the generator seals. This seal oil drains to the defoaming tanks where most of the hydrogen is evacuated from the oil. The oil drains by gravity from the defoaming tanks to the hydrogen-side drain regulator.

The air-side seal oil system consists of seal oil pumps, backup seal oil pumps, seal oil coolers, filters, a vapor extractor, and a drain loop seal. The seal oil pump and seal oil backup pump take suction from the generator bearing oil drain line loop seal and discharge through an oil cooler and filter to the air side of the generator gland seals. The oil returns to the loop seal from the gland seals, where a vapor extractor removes hydrogen gas from the oil and provides a negative pressure on the return side of the gland seals. Seal oil makeup is supplied by oil from the turbine bearings via emergency seal oil backup pumps.

8.8.4.2 Generator Compressed Gas Equipment (or Generator Purging). The generator compressed gas equipment purges the generator and monitors and maintains the hydrogen purity. Generator purging is required whenever the generator hydrogen in installed or removed. Hydrogen is not combustible or explosive if its purity is greater than 95%. To prevent an explosive hydrogen–air mixture in the generator, an inert gas, typically carbon dioxide, is used to purge the air or hydrogen in the generator casing. When hydrogen is being removed from the generator, the hydrogen is vented off while carbon dioxide is introduced into the generator by a manifold at the bottom. Once all the hydrogen has been removed from the generator, ambient air is introduced and the unit is ready for maintenance. On refilling the generator, carbon dioxide is again introduced in the generator through the bottom manifold while the air is vented off. Once all of the air is removed, the hydrogen is introduced through a manifold at the top of the generator and the carbon dioxide is drained off.

In addition to the purging equipment required, the generator compressed gas equipment consists of a gas dryer, water detectors, and a hydrogen control panel containing gas purity and pressure monitoring equipment. The gas dryer is connected across high- and low-pressure areas of the generator so that a portion of the hydrogen gas is routed through the dryer on a continuous basis during operation. An absorbent material in the dryer removes any moisture in the gas. The gas purity equipment continuously monitors the hydrogen gas quality to detect any gas degradation that would indicate a potential problem. Leakage of hydrogen from the generator is made up from hydrogen storage to maintain the required generator pressure.

8.8.4.3 Stator Winding Cooling System. Depending on the manufacturer and unit size, a generator stator winding cooling system may be provided. With this cooling system, the stator windings are made of hollow conductors through which a cooling liquid is pumped. Usually, high-purity water is the stator coolant. By using liquid cooling in conjunction with hydrogen cooling, the amount of copper used in the generator can be significantly reduced for a given rating, as compared to that used for hydrogen cooling alone.

The stator winding cooling water system is supplied by the turbine manufacturer as a complete deionized water circulating and cooling unit (for water units). The complete system includes two full-capacity cooling water pumps, two stator water coolers, deionizer, and flowmeter; and associated piping, valves, and controls. The pumps pump the water through the generator stator passage. The heated water is cooled in the coolers by the plant cooling water system.

8.8.5 Controls and Instrumentation

The steam turbine generator is controlled and monitored by several interrelated systems:

- Turbine governor system—automatically controls turbine speed, acceleration, and load;
- Trip system—provides protection through trips and runbacks;
- Supervisory instrumentation system—provides past and present operating data through parameter sensing, indicating, and recording; and
- Excitation system—controls generator voltage (described in Chapter 17).

8.8.5.1 Turbine Governor. The turbine governor system for medium and large steam turbines typically is a hybrid electrical–hydraulic system. This type of control system is significantly more advanced and preferable to the older mechanical–hydraulic designs in that the linkages and cams of the mechanical designs have been replaced with electrical logic and fast-acting hydraulic servomotors. The hydraulic fluid is supplied to the stop and control valve servomotors by a high-pressure power pumping unit. The fluid is flame retardant to minimize fire hazards in the event of a leak. The system also includes controls and instrumentation.

The turbine governor facilitates control of the turbine over the full operational range by positioning the turbine control valves and the interceptor valves to control turbine speed, load, and throttle pressure.

During manual startups, control is based on selected turbine speeds using full-arc admission. The governor is designed to offer a selection of operating speed and acceleration rates when the operation is on speed control. Turbine generator synchronization speed is approached either manually or by a speed-matching circuit in the governor system. The speed-matching circuit maintains rotor speed at or near synchronous speed.

Automated turbine startup, shutdown, and load change operations are directed by the turbine automation program incorporated in the governor system. All acceleration, heat soaking, and speed control are executed by the automated startup system. Automated startup is an operator-selected and -initiated option and includes all startup operations to bring the turbine generator to synchronization.

After synchronization of the turbine generator, turbine operation under load control uses load demand as the basis of control. Loading rates may be operator-controlled or automatically controlled. When the operator controls load rate, the governor system alarms out-of-limit rotor thermal stresses. In general, partial arc admission is employed when the system is operating under load control. The governor system includes the capability of transferring from full-arc to partial-arc admission (and vice versa) at any load from minimum to rated, or automatically at preselected load or valve position, with the actual transfer rate selected manually.

8.8.5.2 Trip System. The trip system initiates protective tripping of the turbine by sensing potentially damaging operating conditions and automatically tripping the main and reheat steam valves closed while opening the drain valves. Typical parameters sensed for initiation of turbine protective functions include the following. (Items followed by R are recommended to have redundant sensors.)

- Overspeed governor trip,
- Manual trip device (in turbine front standard),
- Generator trip—R,
- Generator protection trips (loss of coolant, high stator temperature, etc.),
- High differential expansion—R,
- Solenoid trip—R,
- Turbine overspeed—R,
- Thrust bearing failure—R,
- Low lubricating oil or hydraulic fluid pressure—R,
- Operator manual trip—R,
- Governor system protective trips—R,
- Condenser low vacuum trip—R,
- High exhaust temperature trip—R,
- High vibration trip—R.

To permit testing and prevent false tripping while still ensuring a true trip is executed, the system has redundant sensors (independent channels from signal sensor to the tripping device) to achieve an appropriate logic philosophy. The trip system also protects the turbine from potential overspeed conditions in the event of a loss of load.

8.8.5.3 Supervisory Instrumentation System. The supervisory instrumentation system includes devices to sense, indicate, and record parameters necessary to monitor the operation of the machine. The following parameters are monitored, as a minimum:

- Shaft speed;
- Governor or control valve position;
- Shaft eccentricity;
- Radial (X–Y) shaft vibration at all turbine, generator, and exciter bearings;
- Shell, rotor, and differential expansion;
- Shell and valve chest temperatures;
- Water induction thermocouple temperatures;
- Vibration phase angles;
- Bearing metal temperature, including the thrust bearing;
- Generator winding temperatures;
- Generator gas temperatures;
- Generator cooling water temperatures; and
- Exciter temperatures.

These parameters are measured, recorded, displayed, and alarmed by hard-wired monitors in the system cabinets. High differential expansion, turbine overspeed, and thrust bearing wear alarms are provided to the trip system. Rotor vibration alarms are displayed in the main control room. All parameters monitored can be displayed as graphics or advisory messages on a monitor or dedicated printer.

8.9 REFERENCES

ANSI/ASME PTC 6-1976. 1976. *Performance Test Codes for Steam Turbines*. ASME, New York, NY.

ANSI/ASME PTC 6A-1982. 1982. *Appendix A to Test Code for Steam Turbines*. ASME, New York, NY.

ANSI/ASME STANDARD NO. TDP-1-1985. 1985. *Recommended Practices for Prevention of Water Damage to Steam Turbines Used for Electric Power Generation*. ASME, New York, NY.

ASTM-ASME-NEMA Committee on Turbine Lubrication Systems. 1980. *ASTM-ASME-NEMA Recommended Practices for the Cleaning, Flushing, and Purification of Steam and Gas Turbine Lubrication Systems*. ASME Standard LOS-1M-1980. ASME, New York, NY.

BARTLETT, R. L. 1958. *Steam Turbine Performance and Economics*. McGraw-Hill, New York, NY.

GP PUBLISHING, INC. 5727 South Lewis Avenue, Suite 727, Tulsa, OK 74105.

GENERAL ELECTRIC COMPANY. Industrial and Power Systems. One River Road, Schenectedy, NY 12345.

MOORE, J. H. 1990. *Steam Turbines for Utility Application*. General Electric. GER-3646.

RIPPEL, H. C., AND R. COLSHER. 1982. *Failure-Cause Analysis: Turbine Bearing Systems*. Paper read at the Proceeding of Turbine Bearings and Rotor Dynamics Workshop. Electric Power Research Institute. September 1982, Palo Alto, CA.

SALISBURY, J. K. Reprinted 1974. *Steam Turbines and Their Cycles*. Robert E. Krieger Publishing Company, New York, NY.

SPECIAL SECTION. 1989. Steam Turbines. *Power*. June. McGraw-Hill, New York, NY.

SPENCER, R. C., K. C. COTTON, and C. N. CANNON. 1974. *A Method for Predicting the Performance of Steam Turbine-Generators ... 16,500 kW and Larger*. Based on ASME Paper No. 62-WA-209, July 1974.

WESTINGHOUSE ELECTRIC CORPORATION. The Quadrangle. 4400 Alafaya Trail, Orlando, FL 32826-2399.

9

STEAM CYCLE HEAT EXCHANGERS

Jay F. Nagori and *Samuel Tarson*

9.1 INTRODUCTION

Two classifications of heat exchangers are associated with the steam cycle. These are the condenser located at the exhaust of the steam turbine, and feedwater heaters located in the condensate/boiler feedwater cycle that utilize extracted steam from the steam turbine. There are many other heat exchangers in the power plant, even some associated with the steam turbine or condensate cycle, but only the condenser and feedwater heaters are considered "steam cycle heat exchangers" and discussed in this chapter.

9.2 CONDENSER

The function of the condenser is to condense the steam leaving the turbine, collect the condensate, and lower the turbine exhaust pressure. A turbine with inlet steam conditions of 850 pounds per square inch (psi) [5,861 kiloPascals (kPa)] at 900° F (482° C) discharging at atmospheric pressure converts about 300 British Thermal Units (Btu) per pound [698 kilojoules per kilogram (kJ/kg)] of steam to electricity. The same unit operating with a steam condenser at 2 in. of mercury absolute pressure (HgA) (50.8 mm HgA) converts about 450 Btu per pound (1,047 kJ/kg) of steam to electricity.

The condensing of the steam requires the condenser to remove the heat of vaporization from the steam and reject it. Condensers are designed to reject this energy directly into cooling water or directly into the atmosphere. Condensers using cooling water are the norm, except in a few locations where water is very expensive, in that case air-cooled condensers may be used. This chapter addresses primarily the water-cooled condenser, and only generally discusses the air-cooled condenser.

The Heat Exchange Institute (HEI) has published the *Standards for Steam Surface Condensers*, which covers recommended standards of definitions, performance, and construction, and it is recommended that the designer become familiar with this publication.

9.2.1 Condenser Arrangement

The condenser is a steam-to-water, tube and shell heat exchanger with the cooling water, normally called circulating water, passing through the tubes and the steam in the shell. Because of the large volume of steam to be condensed, the specific volume of steam at 2 in. HgA (50.8 mm HgA) is 339 ft^3/lb (21.2 m^3/kg). The condenser is a relatively large piece of equipment. In very small sizes, the condenser shell may be round, but in most cases, the size of the condenser shell requires it to be constructed of reinforced flat plate in a box shape. Figure 9-1 shows a typical condenser with water boxes for inlet and outlet circulating water, tubes, tube support plate, shell, and hotwell where the condensate collects. Most condensers are mounted in the plant directly below the turbine exhaust so that the steam passes down through the condenser neck into the condenser. Other physical arrangements may locate the condenser at the side or end of the turbine.

On very large turbines that have two or three low-pressure casings, a separate condenser shell is mounted under each low-pressure casing. Each shell is in effect a separate condenser. Figure 9-2 represents various circulating water systems and condenser shell arrangements.

On Figure 9-2, the top row of condensers indicates the circulating water system for one-pass condensers. In one-pass condensers, the circulating water enters the inlet water box at one end of the shell, passes through the straight tubes, and exits from the outlet water box at the other end of the shell.

The second row of condensers indicates the circulating water system for two-pass condensers. In two-pass condensers, the circulating water enters the inlet water box at one end of the shell and passes through the tube bundle consisting of half the total number of tubes. It then exits the tubes into the return water box, from that the water passes through the other half of the tubes exiting from the outlet water box at the same end of the shell as the inlet water box. If there are two or three condenser shells, all are arranged identically.

The third row of condensers indicates an arrangement that utilizes two and three shell installations where the circulating water passes through the shells in series. The water exiting the first shell enters the second shell, and so on. As the water entering the second shell is warmer than the water entering the previous shell, the performance of the second shell is less than that of the first shell, resulting in a higher pressure in the second shell than in the first shell. The pressure in the third

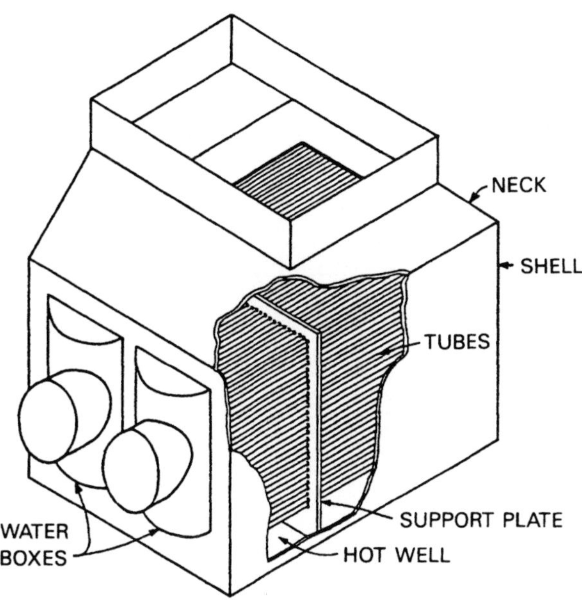

Fig. 9-1. Typical condenser.

shell is even higher than in the second shell. This condition is referred to as multipressure and is used where the lower water flow and its economics exceed the loss in turbine cycle efficiency resulting from higher condenser pressures.

Figure 9-3 gives standard nomenclatures for the various parts of a condenser.

9.2.2 Condenser Sizing

Condensers are basically steam to water heat exchangers and are sized according to normal heat exchanger design technology. Figure 9-4 indicates some of the terms used in calculating the size of a condenser, including the following:

Turbine exhaust steam flow, pounds per hour (lb/h) (Fl_s)
Turbine exhaust steam pressure, in. HgA (P_s)
Turbine exhaust steam enthalpy, Btu per pound (Btu/lb) (h_s)
Saturation temperature of exhaust steam, °F (T_s)
Circulating water inlet temperature, °F (T_1)
Circulating water outlet temperature, °F (T_2)
Circulating water flow, gallons per minute (gpm) (Fl_{cw})
Effective tube length, ft (L).

Additional condenser sizing information includes the following:

Number of passes (NP)
Tube diameter, in., and gauge, Birmingham wire gauge (BWG)
Condensate temperature, °F (T_c)
Condensate enthalpy, Btu/lb (h_c)

From these definitions the following are derived:

Initial temperature difference = $T_s - T_1$ (ITD)
Terminal temperature difference = $T_s - T_2$ (TTD)
Temperature rise = $T_2 - T_1$ (TR).

9.2.2.1 Design Heat Load. The heat load to be used in designing a condenser is that energy that must be removed from the exhaust steam to cause it to condense. Turbine manufacturers normally list the exhaust steam enthalphy on their heat balance, as shown in Fig. 9-5. These are listed as ELEP and UEEP. ELEP stands for expansion line end point, that would be the enthalpy of the exhaust steam if there were no losses at the outlet of the turbine. However, some losses do occur at the outlet of the turbine, that reheats the steam, resulting in the UEEP, usable energy end point. This is the enthalpy of the steam entering the condenser and should be used in determining design heat load.

By subtracting the enthalpy of the condensate leaving the condenser (h_c) from the steam enthalpy (h_s) and multiplying by the exhaust steam flow (Fl_s), the design heat load is obtained. As $\Delta h = (h_s - h_c)$ is normally approximately 950 Btu/lb (2,210 kJ/kg), this number is frequently used in the computation of design heat load.

In addition to the heat load created by the turbine exhaust steam, other flows into the condenser may add further heat load. These flows may include steam flow from a boiler feed pump drive turbine, other drive turbines, heater drains, turbine leak-offs, and other sources. All these must be added to the turbine exhaust steam load to determine total condenser design heat load.

9.2.2.2 Condenser Operating Pressure Optimization. The optimum condenser operating pressure must be determined in conjunction with the steam turbine. Limitations imposed by circulating water parameters of flow, inlet temperature, and temperature rise establish lowest possible condenser pressure. This lowest possible pressure may not be the most

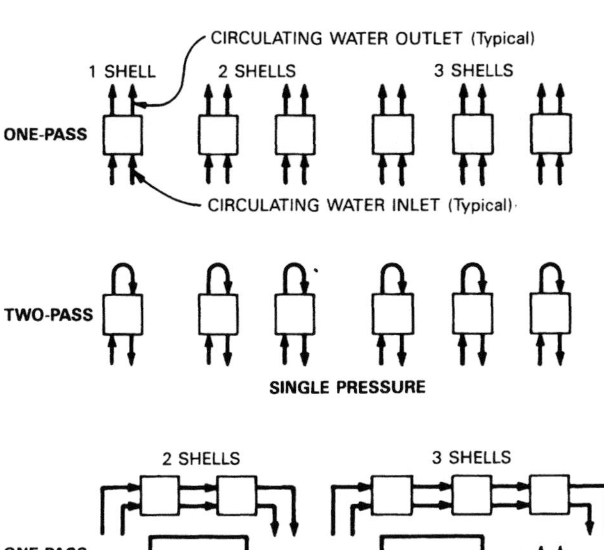

Fig. 9-2. Circulating water flow alternatives.

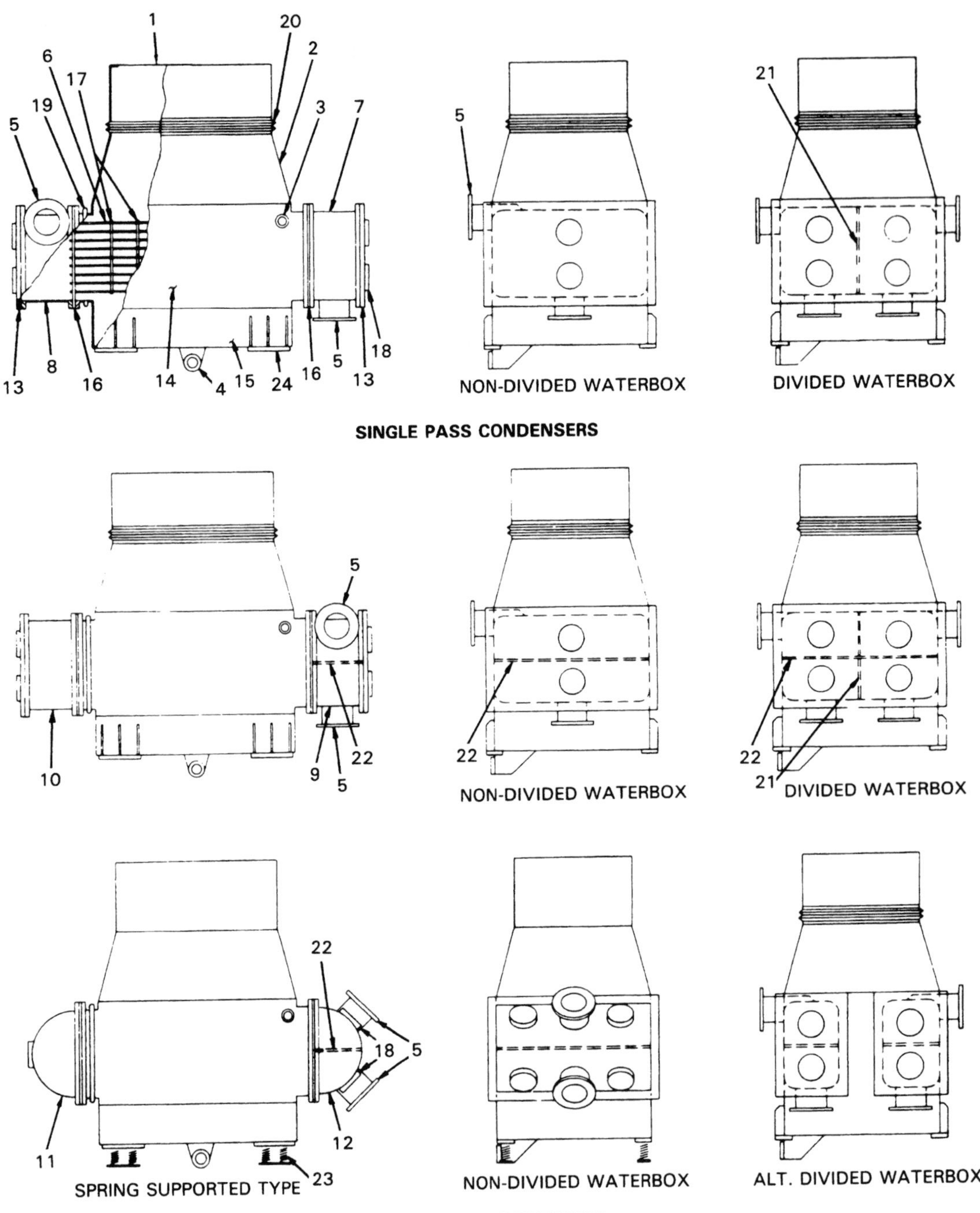

1. EXHAUST CONNECTION
2. EXHAUST NECK
3. VENTING OUTLET
4. CONDENSATE OUTLET
5. CIRCULATING WATER INLET OR OUTLET
6. TUBES
7. INLET WATERBOX
8. OUTLET WATERBOX
9. INLET-OUTLET WATERBOX
10. RETURN WATERBOX
11. BONNET TYPE RETURN WATERBOX
12. BONNET TYPE INLET – OUTLET WATERBOX
13. WATERBOX COVERS
14. CONDENSER SHELL
15. HOTWELL
16. TUBE SHEETS
17. TUBE SUPPORT PLATES
18. HANDHOLES OR MANHOLES
19. SHELL EXPANSION JOINT
20. EXHAUST NECK EXPANSION JOINT
21. WATERBOX DIVIDING PARTITION
22. WATERBOX PASS PARTITION
23. SPRING SUPPORTS
24. SUPPORT FEET

NOTE: MORE THAN ONE OF THE VARIOUS PARTS MAY BE INCLUDED IN A GIVEN CONDENSER. EXACT LOCATIONS OF CONNECTIONS WILL VARY FROM ONE CONDENSER TO ANOTHER.

Fig. 9-3. Standard nomenclature for condenser parts. (From *Standards for Steam Surface Condensers, Eighth Edition*, Heat Exchange Institute, Inc. Used with permission.)

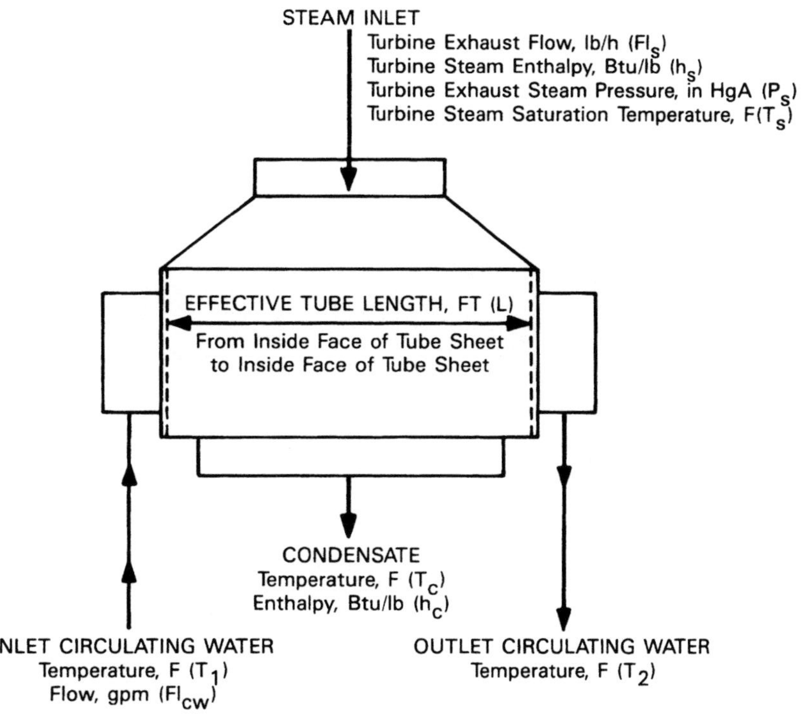

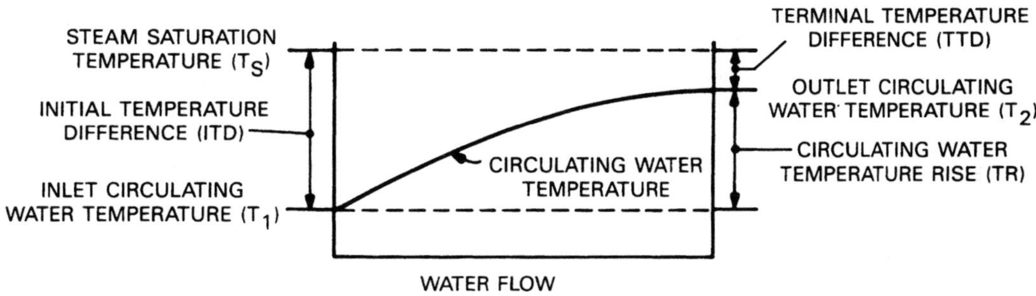

Fig. 9-4. Condenser sizing terminology.

economical from the viewpoint of first cost, operating cost, and efficiency. An economic analysis is required to determine the optimum condenser operating design pressure.

Figures 9-6 and 9-7 from the HEI Condenser Standards give recommended minimum operating design pressures for varying cooling water inlet temperatures. It is recommended that the operating pressure be limited by not allowing the design terminal temperature difference to be less than 5° F (2.8° C).

9.2.2.3 Condenser Sizing. The basic heat transfer equation used in sizing a condenser is listed below:

$$q = UA\Delta T \tag{9-1}$$

where

q = design heat load in Btu/lb,

ΔT = log mean temperature difference defined as

$$\Delta T = \frac{\text{ITD} - \text{TTD}}{\ln\left(\frac{\text{ITD}}{\text{TTD}}\right)} \tag{9-2}$$

A = tube surface area in square feet, and

U = Function of the heat transfer coefficient and the actual construction and operation of the unit, and can be expressed as,

$$U = C \times \sqrt{V} \times f \times m \times Cl \tag{9-3}$$

where

C = heat transfer coefficient,

V = velocity of cooling water through the tubes,

f = cooling water inlet temperature correction factor,

m = tube material and thickness correction factor, and

Cl = tube overall cleanliness factor.

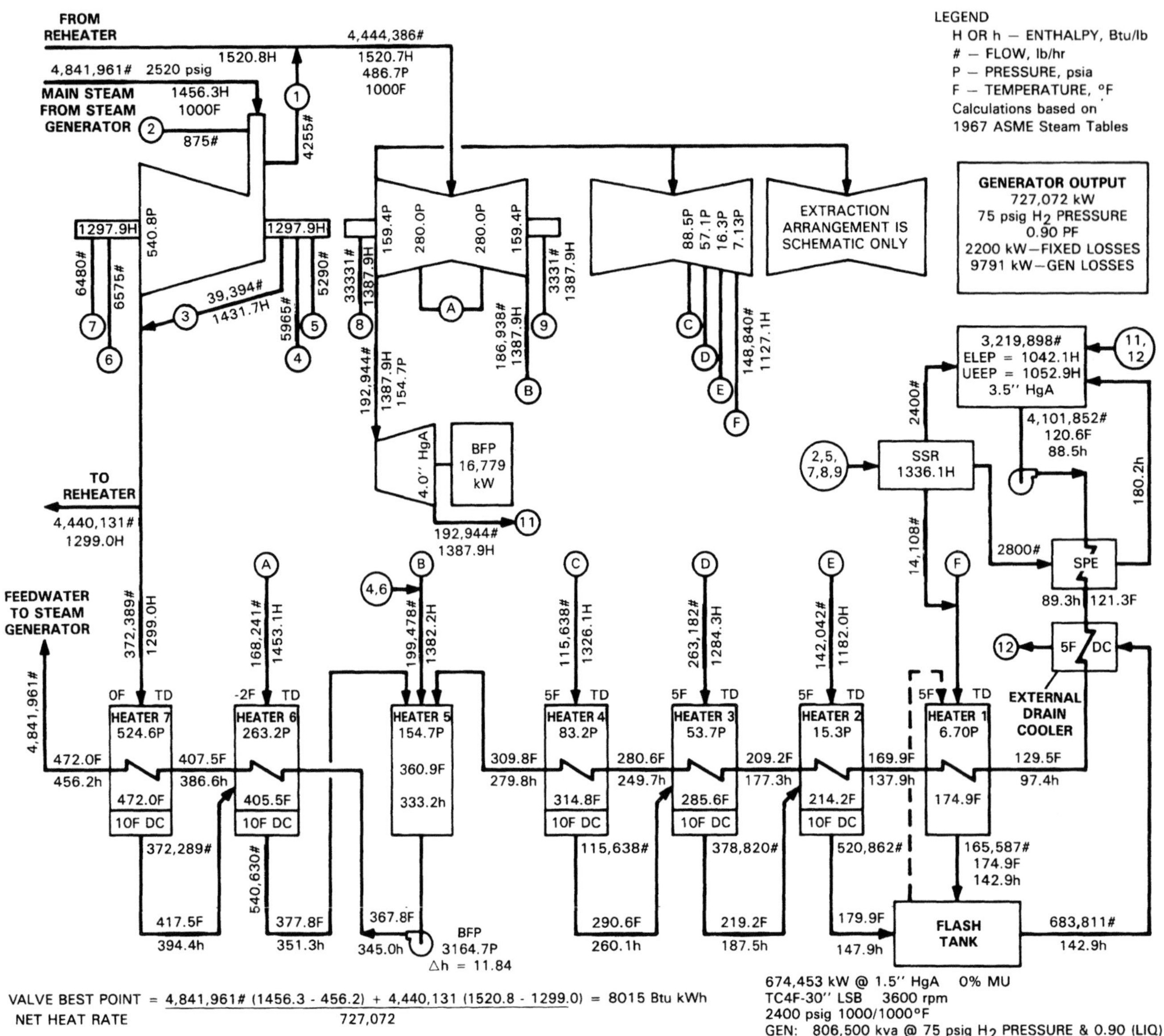

Fig. 9-5. Typical heat balance for a large turbine generator. (From General Electric Co. Used with permission.)

The Heat Exchange Institute has performed extensive tests to establish reasonable heat transfer coefficients. These are published in their *Standards for Steam Surface Condensers*. Table 9-1, taken from that source, gives values of C for various tube diameters. These are based on admiralty metal tubes of 18 BWG with inlet cooling water temperature of 70° F. Figure 9-8, also from the HEI Standards, gives correction factors for other inlet water temperatures, listed above as Factor f. Table 9-2 gives correction factors for several other tube materials and gauges, listed above as Factor m. Table 9-3 lists tube characteristics. Cleanliness factor provides for the loss of heat transfer with time attributable to scale or other buildup and is established by the designer based on the selected tube materials and cooling water conditions.

In the design of a condenser, the majority of design parameters must be established by the designer. These normally include the following:

- Turbine exhaust steam flow and enthalpy,
- Turbine exhaust steam pressure,
- Circulating water inlet temperature,
- Tube diameter and gauge,
- Tube material,
- Number of passes,
- Water velocity through tubes, and
- Cleanliness factor.

In addition to the above parameters, the effective tube length or the circulating water temperature rise is normally

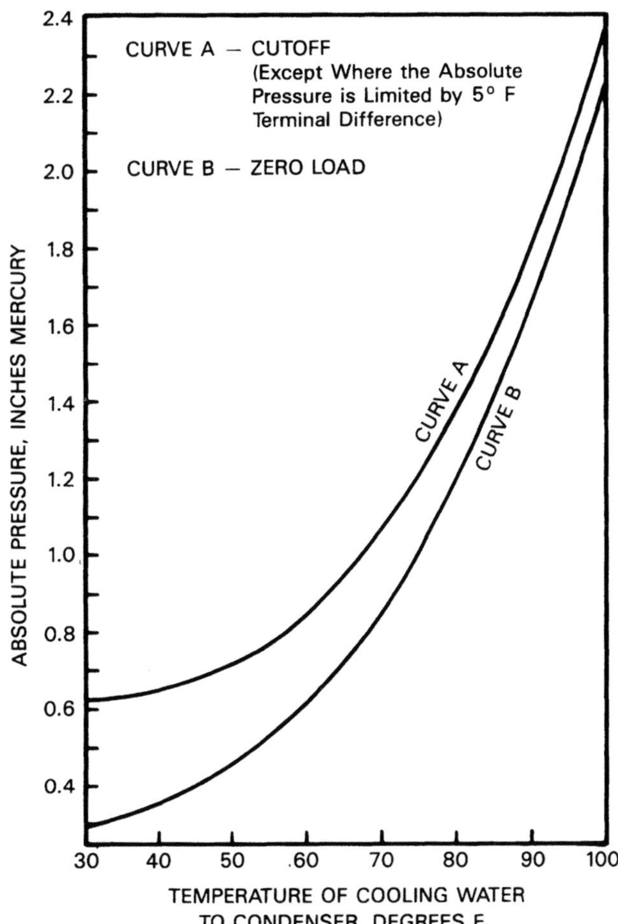

Note: Performance is based on air removal equipment having a capacity at one inch mercury absolute suction pressure, of not less than that recommended by HEI and the air and non-condensibles being removed from the system not exceeding 50 percent of those values.

Fig. 9-6. Absolute pressure limit curves for steam turbine service. (From *Standards for Steam Surface Condensers, Eighth Edition*, Heat Exchange Institute, Inc. Used with permission.)

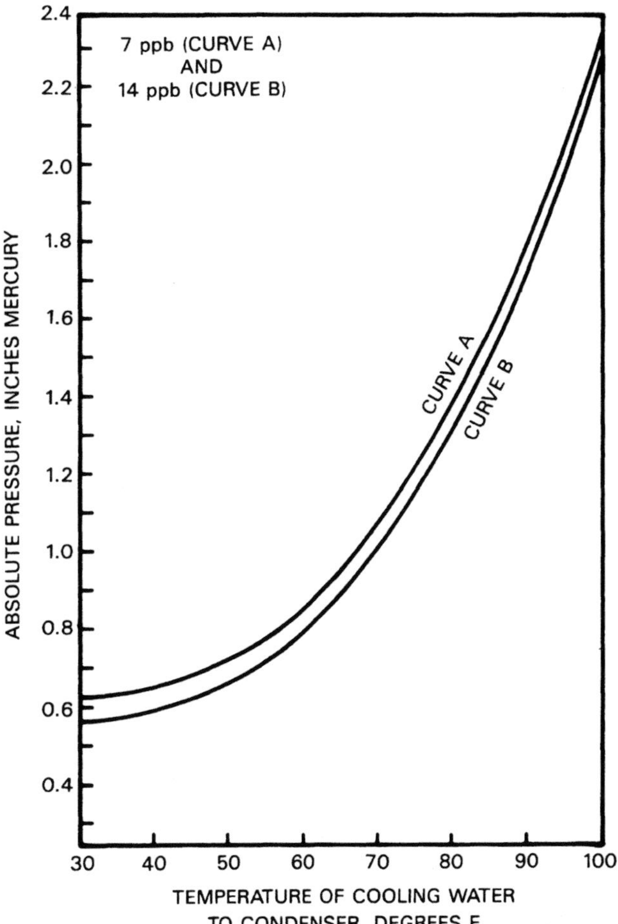

Fig. 9-7. Absolute pressure limit curves for steam turbine service for oxygen contents. (From *Standards for Steam Surface Condensers, Eighth Edition*, Heat Exchange Institute, Inc. Used with permission.)

established. With this information, the other parameters of the condenser, surface area, and circulating water flow can be calculated. The following example illustrates this.

Example 9-1

Given:

Steam flow = Fl_s = 800,000 lb/h
Steam turbine exhaust enthalpy = h_3 = 1,038.6 Btu/lb
Condenser pressure = P_s = 3.5 in. HgA
Circulating water inlet temperature = T_1, = 85° F
Circulating water velocity through tubes = V = 7.0 fps
Temperature rise = TR = 27° F
Tube size = D = 1.0 in. OD; BWG = 22 gauge
Tube material = 304 SS
Cleanliness factor = 90%
Number of tube passes = 2

Calculate:

Circulating water outlet temperature, $T_2 = T_1 + \text{TR} = 85 + 27 = 112°\text{F}$

Saturated steam temperature at 3.5 in. HgA, $T_s = 120.56°\text{F}$

Initial temperature difference, ITD = $T_s - T_1$, = 120.56 − 85 = 35.56° F

Terminal temperature difference, TTD = $T_s - T_2$ = 120.56 − 112 = 8.56° F

Table 9-1. Heat Transfer Coefficient for Steam Surface Condensers

Tube Diameter, OD, inches	Coefficient (*C*)
⅝ and ¾	267
⅞ and 1	263
1⅛ and 1¼	259
1⅜ and 1½	255
1⅝ and 1¾	251
1⅞ and 2	247

Note: Based on new clean Admiralty metal tubes of 18 BWG gauge and with inlet cooling water temperature of 70° F

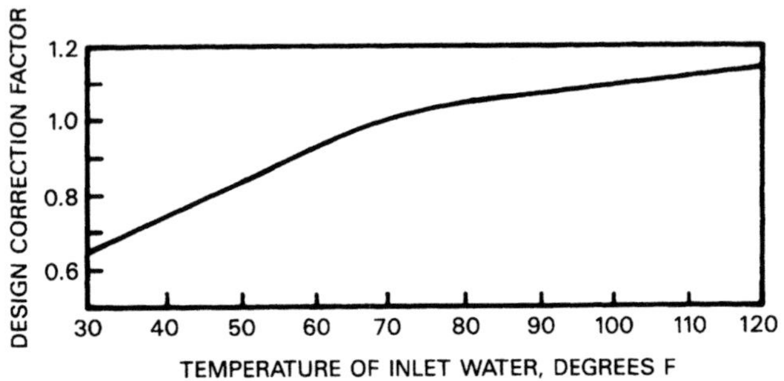

Fig. 9-8. Inlet water temperature correction factor. (From *Standards for Steam Surface Condensers, Eighth Edition*, Heat Exchange Institute, Inc. Used with permission.)

LMTD, $\Delta T = (35.56 - 8.56) \div \ln(35.56 \div 8.56) = 18.96$
Heat input, q = steam flow $\times \Delta h = 800{,}000 \times (1{,}038.6 - 88.6) = 760 \times 10^6$ Btu/h

From the Heat Exchange Institute Standard, the following is used:

Reference heat transfer coefficient, C, for 1.0-in. tubes = 263 (Table 9-1)
Inlet temperature correction, f, for 85° F = 1.06 (Fig. 9-8)
Tube material and thickness correction, m, for 304 SS, 22 BWG = 0.87 (Table 9-2)
Tube characteristics for 1.0-in. 22 BWG
External surface, square feet per foot of length = 0.2618 ft^2/ft (Table 9-3)
Length for 1 ft^2 of surface = 3.817 ft/ft^2
gpm water flow at 1 ft/s (fps) velocity = 2.182 gpm/fps (Table 9-3)

With this information, the condenser physical parameters can be determined.

$$q = UA\Delta T \quad (9\text{-}4)$$

$$A = \frac{q}{U\Delta T}$$

Corrected heat transfer

$$U = C \times \sqrt{V} \times f \times m \times Cl \quad (9\text{-}5)$$
$$= 263 \times \sqrt{7} \times 1.06 \times 0.87 \times 0.90$$
$$= 577.5$$

Condenser surface

$$A = \frac{760 \times 10^6}{577.5 \times 18.96} = 69{,}410 \text{ ft}^2$$

Circulating water flow

$$Fl_{cw} = \frac{q}{TR} \quad (9\text{-}6)$$
$$= \frac{760 \times 10^6}{27} = 281.5 \times 10^5 \text{ lb/h}$$
$$= \frac{281.5 \times 10^5}{500} = 56{,}300 \text{ gpm}$$

(1lb/h = 1 gpm × [60 min/h/7.48 gal/ft^3/0.016 ft^3/lb]
= 1 gpm × 500 lb/h/gpm)

Number of tubes per pass

$$= \frac{\text{Total gpm}}{\text{gpm per tube per fps} \times \text{velocity, fps}} \quad (9\text{-}7)$$
$$= \frac{56{,}300}{2.182 \times 7.0}$$
$$= 3{,}686 = \text{tubes/pass}$$

Total number of tubes 2 × 3,686 = 7,372

Table 9-2. Tube Material Correction Factor of Heat Transfer Coefficient

	Tube Wall Gauge—BWG						
Tube Materials	24	22	20	18	16	14	12
Admiralty metal	1.06	1.04	1.02	1.00	0.96	0.92	0.87
Arsenical copper	1.06	1.04	1.02	1.00	0.96	0.92	0.87
Copper iron 194	1.06	1.04	1.02	1.00	0.96	0.92	0.87
Aluminum brass	1.03	1.02	1.00	0.97	0.94	0.90	0.84
Aluminum bronze	1.03	1.02	1.00	0.97	0.94	0.90	0.84
90–10 Cu–Ni	0.99	0.97	0.94	0.90	0.85	0.80	0.74
70–30 Cu–Ni	0.93	0.90	0.87	0.82	0.77	0.71	0.64
Cold-rolled low carbon steel	1.00	0.98	0.95	0.91	0.86	0.80	0.74
Stainless steels[a] type 304/316	0.91	0.87	0.83	0.76	0.70	0.63	0.55
Titanium[a]	0.91	0.87	0.83	0.76	0.70	0.63	0.55

[a]The user is specifically cautioned that these noncopper bearing tube materials are more susceptible to biofouling than tubes with high copper content. This may call for selection of design cleanliness factors that suitably reflect the probable operating condition the tubes will be subject to in service.

Source: *Standards for Steam Surface Condensers*, Heat Exchange Institute, Inc. Used with permission.

Table 9-3. Tube Characteristics

OD of Tubing Inches	BWG	Thickness (inches)	Inside Diameter (inches)	Surface External (ft²/linear ft)	Length in Feet for 1 ft² Surface	Water-gpm at 1 ft/s Velocity
⅝	12	0.109	0.407	0.1636	6.112	0.406
	13	0.095	0.435	0.1636	6.112	0.463
	14	0.083	0.459	0.1636	6.112	0.516
	15	0.072	0.481	0.1636	6.112	0.566
	16	0.065	0.495	0.1636	6.112	0.600
	17	0.058	0.509	0.1636	6.112	0.634
	18	0.049	0.527	0.1636	6.112	0.680
	19	0.042	0.541	0.1636	6.112	0.716
	20	0.035	0.555	0.1636	6.112	0.754
	21	0.032	0.561	0.1636	6.112	0.770
	22	0.028	0.569	0.1636	6.112	0.793
	23	0.025	0.575	0.1636	6.112	0.809
	24	0.022	0.581	0.1636	6.112	0.826
¾	12	0.109	0.532	0.1963	5.094	0.693
	13	0.095	0.560	0.1963	5.094	0.768
	14	0.083	0.584	0.1963	5.094	0.835
	15	0.072	0.606	0.1963	5.094	0.899
	16	0.065	0.620	0.1963	5.094	0.941
	17	0.058	0.634	0.1963	5.094	0.984
	18	0.049	0.652	0.1963	5.094	1.041
	19	0.042	0.666	0.1963	5.094	1.086
	20	0.035	0.680	0.1963	5.094	1.132
	21	0.032	0.686	0.1963	5.094	1.152
	22	0.028	0.694	0.1963	5.094	1.179
	23	0.025	0.700	0.1963	5.094	1.200
	24	0.022	0.706	0.1963	5.094	1.220
⅞	12	0.109	0.657	0.2291	4.367	1.057
	13	0.095	0.685	0.2291	4.367	1.149
	14	0.083	0.709	0.2291	4.367	1.231
	15	0.072	0.731	0.2291	4.367	1.308
	16	0.065	0.745	0.2291	4.367	1.359
	17	0.058	0.759	0.2291	4.367	1.410
	18	0.049	0.777	0.2291	4.367	1.478
	19	0.042	0.791	0.2291	4.367	1.532
	20	0.035	0.805	0.2291	4.367	1.586
	21	0.032	0.811	0.2291	4.367	1.610
	22	0.028	0.819	0.2291	4.367	1.642
	23	0.025	0.825	0.2291	4.367	1.666
	24	0.022	0.831	0.2291	4.367	1.690
1	12	0.109	0.782	0.2618	3.817	1.497
	13	0.095	0.810	0.2618	3.817	1.606
	14	0.083	0.834	0.2618	3.817	1.703
	15	0.072	0.856	0.2618	3.817	1.794
	16	0.065	0.870	0.2618	3.817	1.853
	17	0.058	0.884	0.2618	3.817	1.913
	18	0.049	0.902	0.2618	3.817	1.992
	19	0.042	0.916	0.2618	3.817	2.054
	20	0.035	0.930	0.2618	3.817	2.117
	21	0.032	0.936	0.2618	3.817	2.145
	22	0.028	0.944	0.2618	3.817	2.182
	23	0.025	0.950	0.2618	3.817	2.209
	24	0.022	0.956	0.2618	3.817	2.237
1⅛	12	0.109	0.907	0.2944	3.397	2.014

(Continued)

Table 9-3. Continued

OD of Tubing Inches	BWG	Thickness (inches)	Inside Diameter (inches)	Surface External (ft^2/linear ft)	Length in Feet for 1 ft^2 Surface	Water-gpm at 1 ft/s Velocity
	13	0.095	0.935	0.2944	3.397	2.140
	14	0.083	0.959	0.2944	3.397	2.251
	15	0.072	0.981	0.2944	3.397	2.356
	16	0.065	0.995	0.2944	3.397	2.424
	17	0.058	1.009	0.2944	3.397	2.492
	18	0.049	1.027	0.2944	3.397	2.582
	19	0.042	1.041	0.2944	3.397	2.653
	20	0.035	1.055	0.2944	3.397	2.725
	21	0.032	1.061	0.2944	3.397	2.756
	22	0.028	1.069	0.2944	3.397	2.797
	23	0.025	1.075	0.2944	3.397	2.829
	24	0.022	1.081	0.2944	3.397	2.861
1¼	12	0.109	1.032	0.3271	3.057	2.607
	13	0.095	1.060	0.3271	3.057	2.751
	14	0.083	1.084	0.3271	3.057	2.877
	15	0.072	1.106	0.3271	3.057	2.994
	16	0.065	1.120	0.3271	3.057	3.071
	17	0.058	1.134	0.3271	3.057	3.148
	18	0.049	1.152	0.3271	3.057	3.249
	19	0.042	1.166	0.3271	3.057	3.328
	20	0.035	1.180	0.3271	3.057	3.409
	21	0.032	1.186	0.3271	3.057	3.443
	22	0.028	1.194	0.3271	3.057	3.490
	23	0.025	1.200	0.3271	3.057	3.525
	24	0.022	1.206	0.3271	3.057	3.560
1⅜	12	0.109	1.157	0.3600	2.778	3.272
	13	0.095	1.185	0.3600	2.778	3.433
	14	0.083	1.209	0.3600	2.778	3.573
	15	0.072	1.231	0.3600	2.778	3.705
	16	0.065	1.245	0.3600	2.778	3.790
	17	0.058	1.259	0.3600	2.778	3.876
	18	0.049	1.277	0.3600	2.778	3.987
	19	0.042	1.291	0.3600	2.778	4.075
	20	0.035	1.305	0.3600	2.778	4.164
	21	0.032	1.311	0.3600	2.778	4.202
	22	0.028	1.319	0.3600	2.778	4.254
	23	0.025	1.325	0.3600	2.778	2.292
	24	0.022	1.331	0.3600	2.778	4.331
1½	12	0.109	1.282	0.3927	2.546	4.018
	13	0.095	1.310	0.3927	2.546	4.196
	14	0.083	1.334	0.3927	2.546	4.351
	15	0.072	1.356	0.3927	2.546	4.496
	16	0.065	1.370	0.3927	2.546	4.589
	17	0.058	1.384	0.3927	2.546	4.683
	18	0.049	1.402	0.3927	2.546	4.806
	19	0.042	1.416	0.3927	2.546	4.902
	20	0.035	1.430	0.3927	2.546	5.000
	21	0.032	1.436	0.3927	2.546	5.042
	22	0.028	1.444	0.3927	2.546	5.099
	23	0.025	1.450	0.3927	2.546	5.141
	24	0.022	1.456	0.3927	2.546	5.183
1⅝	12	0.109	1.407	0.4254	2.351	4.840
	13	0.095	1.435	0.4254	2.351	5.035

(Continued)

Table 9-3. Continued

OD of Tubing Inches	BWG	Thickness (inches)	Inside Diameter (inches)	Surface External (ft²/linear ft)	Length in Feet for 1 ft² Surface	Water-gpm at 1 ft/s Velocity
	14	0.083	1.459	0.4254	2.351	5.205
	15	0.072	1.481	0.4254	2.351	5.363
	16	0.065	1.495	0.4254	2.351	5.465
	17	0.058	1.509	0.4254	2.351	5.567
	18	0.049	1.527	0.4254	2.351	5.701
	19	0.042	1.541	0.4254	2.351	5.806
	20	0.035	1.555	0.4254	2.351	5.912
	21	0.032	1.561	0.4254	2.351	5.958
	22	0.028	1.569	0.4254	2.351	6.019
	23	0.025	1.575	0.4254	2.351	6.065
	24	0.022	1.581	0.4254	2.351	6.111
1¾	12	0.109	1.532	0.4581	2.183	5.738
	13	0.095	1.560	0.4581	2.183	5.950
	14	0.083	1.584	0.4581	2.183	6.135
	15	0.072	1.606	0.4581	2.183	6.306
	16	0.065	1.620	0.4581	2.183	6.417
	17	0.058	1.634	0.4581	2.183	6.528
	18	0.049	1.652	0.4581	2.183	6.673
	19	0.042	1.666	0.4581	2.183	6.786
	20	0.035	1.680	0.4581	2.183	6.901
	21	0.032	1.686	0.4581	2.183	6.950
	22	0.028	1.694	0.4581	2.183	7.016
	23	0.025	1.700	0.4581	2.183	7.066
	24	0.022	1.706	0.4581	2.183	7.116
1⅞	12	0.109	1.657	0.4909	2.037	6.713
	13	0.095	1.685	0.4909	2.037	6.942
	14	0.083	1.709	0.4909	2.037	7.141
	15	0.072	1.731	0.4909	2.037	7.326
	16	0.065	1.745	0.4909	2.037	7.445
	17	0.058	1.759	0.4909	2.037	7.565
	18	0.049	1.777	0.4909	2.037	7.721
	19	0.042	1.791	0.4909	2.037	7.843
	20	0.035	1.805	0.4909	2.037	7.966
	21	0.032	1.811	0.4909	2.037	8.019
	22	0.028	1.819	0.4909	2.037	8.090
	23	0.025	1.825	0.4909	2.037	8.143
	24	0.022	1.831	0.4909	2.037	8.197
2	12	0.109	1.782	0.5236	1.910	7.764
	13	0.095	1.810	0.5236	1.910	8.010
	14	0.083	1.834	0.5236	1.910	8.224
	15	0.072	1.856	0.5236	1.910	8.422
	16	0.065	1.870	0.5236	1.910	8.550
	17	0.058	1.884	0.5236	1.910	8.679
	18	0.049	1.902	0.5236	1.910	8.845
	19	0.042	1.916	0.5236	1.910	8.976
	20	0.035	1.930	0.5236	1.910	9.107
	21	0.032	1.936	0.5236	1.910	9.164
	22	0.028	1.944	0.5236	1.910	9.240
	23	0.025	1.950	0.5236	1.910	9.298
	24	0.022	1.956	0.5236	1.910	9.354

Source: *Standards for Steam Surface Condensers*, Heat Exchange Institute, Inc. Used with permission.

Surface per tube

$$= \frac{\text{Total surface}}{\text{No of tubes}} \qquad (9\text{-}8)$$

$$= \frac{69{,}410}{7{,}372}$$

$$= 9.42 \text{ ft}^2 \text{ per tube}$$

Length of tube

$$= \text{Surface per tube} \times \text{length for 1 square foot} \qquad (9\text{-}9)$$

$$= 9.42 \times 3.817$$

$$= 36.0 \text{ ft}$$

It should be noted that the quantity of noncondensible gases leaking into a condenser affects its performance. Normally, the effect of this leakage is negligible, but prediction of performance is less precise when the terminal difference, TTD, is less than 5° F. For this reason, condensers should not be designed with TTD less than 5° F.

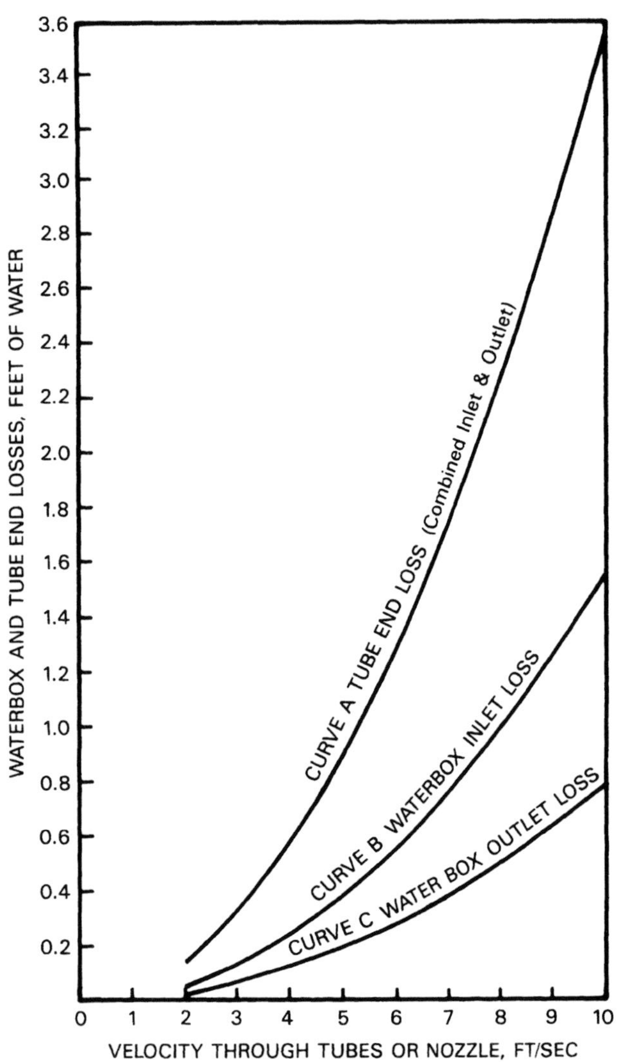

Fig. 9-9. Waterbox and tube end losses two pass condensers. (From *Standards for Steam Surface Condensers, Eighth Edition*, Heat Exchange Institute, Inc. Used with permission.)

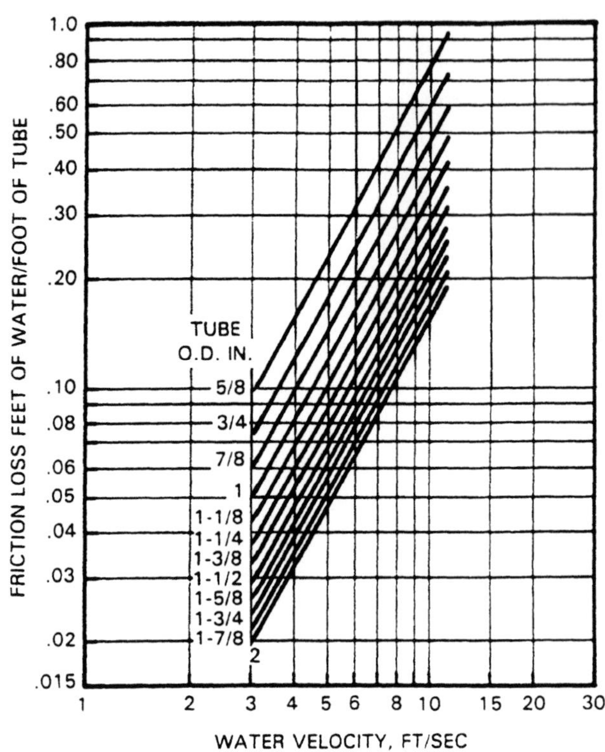

GAUGE CORRECTION FACTOR FOR FRICTION LOSS

Tube O.D. In.	12 BWG	14 BWG	16 BWG	18 BWG	20 BWG	22 BWG	24 BWG
0.625	1.38	1.21	1.10	1.00	0.94	0.91	0.89
0.750	1.28	1.16	1.06	1.00	0.95	0.93	0.90
0.875	1.25	1.13	1.06	1.00	0.96	0.94	0.92
1.000	1.19	1.11	1.05	1.00	0.96	0.94	0.93
1.125	1.16	1.09	1.04	1.00	0.97	0.95	0.94
1.250	1.14	1.08	1.04	1.00	0.97	0.96	0.94
1.375	1.13	1.07	1.03	1.00	0.97	0.96	0.94
1.500	1.12	1.06	1.03	1.00	0.97	0.96	0.95
1.625	1.10	1.05	1.02	1.00	0.97	0.96	0.95
1.750	1.10	1.05	1.02	1.00	0.98	0.97	0.96
1.875	1.09	1.05	1.02	1.00	0.98	0.97	0.96
2.000	1.08	1.04	1.02	1.00	0.98	0.97	0.96

Fig. 9-10a. Friction loss for water flowing in 18 BWG tubes. (From *Standards for Steam Surface Condensers, Eighth Edition*, Heat Exchange Institute, Inc. Used with permission.)

To determine circulating water pressure drop, use Figs. 9-9, 9-10a, and 9-10b, with the following:

ΔP = friction loss per foot for water flowing in 1-in. 18 BWG tubes at 7 fps (Fig. 9-10a) × correction factor for 22 BWG tubes (Fig. 9-10a) × correction factor for average water temperature (85 + 27/2 = 98.5° F) (Fig. 9-10b) × tube length × number of passes + combined inlet and outlet tube end loss (Fig. 9-9) + water box inlet loss (Fig. 9-9) + water box outlet loss (Fig. 9-9).

Using Figs. 9-9, 9-10a, and 9-10b, results in the following:

$\Delta P = 0.22 \times 0.94 \times 0.949 \times 36.0 \times 2 + 1.74 + 0.76 + 0.38 =$ 17.0 ft of water

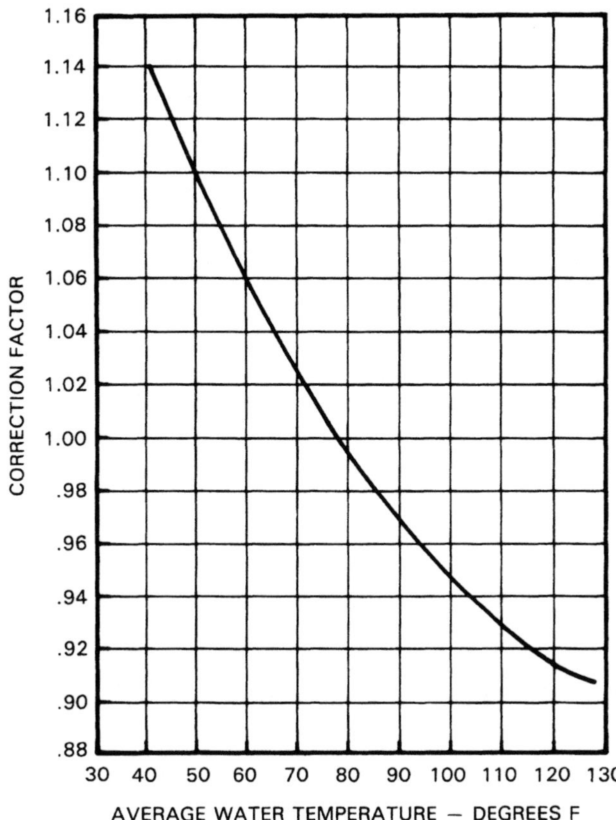

Fig. 9-10b. Temperature correction for friction loss in tubes. (From *Standards for Steam Surface Condensers, Eighth Edition,* Heat Exchange Institute, Inc. Used with permission.)

Calculations of condensers and their performance, with a knowledge of other parameters, can be preformed in a similar manner. Most condenser manufacturers publish an empirical method of calculating condenser size that is based on the preceding approach.

9.2.3 Materials of Construction

Condensers are normally fabricated with a welded steel shell and tube support plates. The tube sheets, tubes, and water boxes are constructed of material compatible with the circulating water. Material specifications and details of construction are recommended in the HEI *Standards for Steam Surface Condensers.*

When fresh water is used as circulating water, a typical condenser uses admiralty tubes, muntz metal tube sheet, and cast iron or carbon steel waterboxes. The air-cooling section of the tube bundle uses 90–10 copper–nickel (CuNi) tubes.

The selection of tube, tube sheet, and water box material for brackish water or seawater must carefully consider the type and amount of impurities in the water, including the amount of sand and dirt being carried by the water.

Tube materials for these applications normally vary from 90–10 CuNi to stainless steel to titanium. Tube sheet materials are selected to be compatible with the tube material. Waterboxes normally are made of fabricated steel and are protected with a coating and cathodic protection.

As different materials normally have different life spans, an economic analysis should be performed to evaluate the various materials.

9.2.4 Corrosion Protection

To protect the waterboxes and tube sheets from corrosion, a combination of cathodic protection and protective coating is used. Cathodic protection is provided by fastening sacrificial anodes to the waterboxes. Protective coatings must be provided with cathodic protection. These coatings vary with the circulating water analysis and may range from a coal tar enamel to an epoxy.

9.2.5 Air Extraction

Because the condenser operates at below atmospheric pressure, air leaks into the system. This must be continuously removed to maintain the design vacuum.

The air removal equipment for small condensers usually consists of a steam jet air ejector. This ejector uses steam from the main steam header as the operating medium. A two-stage dual unit is normally provided. The steam is condensed out of the exhaust stream by a small after-condenser, and the remaining gasses are discharged to the atmosphere.

On larger condensers, mechanical evacuators are used. These are basically air compressors that raise the pressure of the air from condenser vacuum to atmospheric. They are specially designed for this service.

The HEI *Standards for Steam Surface Condensers* discusses air removal equipment and its capacity. Table 9-4 lists recommended capacities from that source (HEI 1984). To use Table 9-4:

1. Determine the total steam flow of the unit by adding the main turbine exhaust flow and the auxiliary turbine exhaust flow entering all shells of the condenser.
2. Determine the total number of MAIN turbine exhaust openings for all shells. Do not include auxiliary turbine exhaust openings.
3. Divide flow obtained in Step 1 by exhaust opening number obtained in Step 2. The resultant number is the EFFECTIVE STEAM FLOW FOR EACH MAIN EXHAUST OPENING.
4. Enter the appropriate section of Table 9-4 based on whether the unit is a single-shell, twin-shell, or triple-shell condenser and locate the flow obtained in Step 3 in the left vertical column.
5. Determine TOTAL NUMBER OF EXHAUST OPENINGS for all shells by adding the total number of main turbine exhaust openings to the total number of auxiliary turbines exhausting into the condenser. Split auxiliary turbine exhaust ducts coming from one auxiliary turbine count as one auxiliary turbine exhaust.
6. Locate the appropriate column and capacity using the number obtained in Step 5.

Table 9-4. Venting Equipment Capacities

A. One Condenser Shell

Effective Steam Flow Each Main Exhaust Opening (lb/h)		Total Number of Exhaust Openings								
		1	2	3	4	5	6	7	8	9
Up to 25,000	SCFM[a]	3.0	4.0	5.0	5.0	7.5	7.5	7.5	10.0	10.0
	Dry air (lb/h)	13.5	18.0	22.5	22.5	33.8	33.8	33.8	45.0	45.0
	Water vapor (lb/h)	29.7	39.6	49.5	49.5	74.4	74.4	74.4	99.0	99.0
	Total mixture (lb/h)	43.2	57.6	72.0	72.0	108.2	108.2	108.2	144.	144.0
25,001–50,000	SCFM[a]	4.0	5.0	7.5	7.5	10.0	10.0	10.0	12.5	12.5
	Dry air (lb/h)	18.0	22.5	33.8	33.8	45.0	45.0	45.0	56.2	56.2
	Water vapor (lb/h)	39.6	49.5	74.4	74.4	99.0	99.0	99.0	123.6	123.6
	Total mixture (lb/h)	57.6	72.0	108.2	108.2	144.0	144.0	144.0	179.8	179.8
50,001–100,000	SCFM[a]	5.0	7.5	10.0	10.0	12.5	12.5	15.0	15.0	15.0
	Dry air (lb/h)	22.5	33.8	45.0	45.0	56.2	56.2	67.5	67.5	67.5
	Water vapor (lb/h)	49.5	74.4	99.0	99.0	123.6	123.6	148.5	148.5	148.5
	Total mixture (lb/h)	72.0	108.2	144.0	144.0	179.8	179.8	216.0	216.0	216.0
100,001–250,000	SCFM[a]	7.5	12.5	12.5	15.0	17.5	20.0	20.0	25.0	25.0
	Dry air (lb/h)	33.8	56.2	56.2	67.5	78.7	90.0	90.0	112.5	112.5
	Water vapor (lb/h)	74.4	123.6	123.6	148.5	175.1	198.0	198.0	247.5	247.5
	Total mixture (lb/h)	108.2	179.8	179.8	216.0	251.8	288.0	288.0	360.0	360.0
250,001–500,000	SCFM[a]	10.0	15.0	17.5	20.0	25.0	25.0	30.0	30.0	35.0
	Dry air (lb/h)	45.0	67.5	78.7	90.0	112.5	112.5	135.0	135.0	157.5
	Water vapor (lb/h)	99.0	148.5	173.1	198.0	247.5	247.5	297.0	297.0	346.5
	Total mixture (lb/h)	144.0	216.0	251.8	288.0	360.0	360.0	432.0	432.0	504.0
500,001–1,000,000	SCFM[a]	12.5	20.0	20.0	25.0	30.0	30.0	35.0	40.0	40.0
	Dry air (lb/h)	56.2	90.0	90.0	112.5	135.0	135.0	157.5	180.0	180.0
	Water vapor (lb/h)	123.6	198.0	198.0	247.5	297.0	297.0	346.5	396.0	396.0
	Total mixture (lb/h)	179.8	288.0	288.0	360.0	432.0	432.0	504.0	576.0	576.0
1,000,001–2,000,000	SCFM[a]	15.0	25.0	25.0	30.0	35.0	40.0	40.0	45.0	50.0
	Dry air (lb/h)	67.5	112.5	112.5	135.0	157.5	180.0	180.0	202.5	225.0
	Water vapor (lb/h)	148.5	247.5	247.5	297.0	346.5	396.0	396.0	445.5	495.0
	Total mixture (lb/h)	216.0	360.0	360.0	432.0	504.0	576.0	576.0	648.0	720.0
2,000,001–3,000,000	SCFM[a]	17.5	25.0	30.0	35.0	40.0	45.0	50.0	55.0	60.0
	Dry air (lb/h)	78.7	112.5	135.0	157.5	180.0	202.5	225.0	247.5	270.0
	Water vapor (lb/h)	173.1	247.5	297.0	346.5	396.0	445.5	495.0	544.5	594.0
	Total mixture (lb/h)	251.8	360.0	432.0	504.0	576.0	648.0	720.0	792.0	864.0
3,000,001–4,000,000	SCFM[a]	20.0	30.0	35.0	40.0	45.0	50.0	55.0	60.0	65.0
	Dry air (lb/h)	90.0	135.0	157.5	180.0	202.5	225.0	247.5	270.0	292.5
	Water vapor (lb/h)	198.0	297.0	346.5	396.0	445.5	495.0	544.5	594.0	643.5
	Total mixture (lb/h)	288.0	432.0	504.0	576.0	648.0	720.0	792.0	864.0	936.0

B. Two Condenser Shells

Effective Steam Flow Each Main Exhaust Opening (lb/h)		Total Number of Exhaust Openings												
		2	3	4	5	6	7	8	9	10	11	12	13	14
100,001–250,000	SCFM[a]	15.0	20.0	20.0	20.0	25.0	25.0	30.0	30.0	35.0	35.0	40.0	40.0	40.0
	Dry air (lb/h)	67.5	90.0	90.0	90.0	112.5	112.5	135.0	135.0	157.5	157.5	180.0	180.0	180.0
	Water vapor (lb/h)	148.5	198.0	198.0	198.0	247.5	247.5	297.0	297.0	346.5	346.5	396.0	396.0	396.0
	Total mixture (lb/h)	216.0	288.0	288.0	288.0	360.0	360.0	432.0	432.0	504.0	504.0	576.0	576.0	576.0
250,001–500,000	SCFM[a]	20.0	20.0	25.0	30.0	30.0	35.0	40.0	40.0	50.0	50.0	50.0	60.0	60.0
	Dry air (lb/h)	90.0	90.0	112.5	135.0	135.0	157.5	180.0	180.0	225.0	225.0	225.0	270.0	270.0
	Water vapor (lb/h)	198.0	198.0	247.5	297.0	297.0	346.5	396.0	396.0	495.0	495.0	495.0	594.0	594.0
	Total mixture (lb/h)	288.0	288.0	360.0	432.0	432.0	504.0	576.0	576.0	720.0	720.0	720.0	864.0	864.0
500,001–1,000,000	SCFM[a]	25.0	25.0	30.0	35.0	40.0	50.0	50.0	50.0	60.0	60.0	70.0	70.0	70.0
	Dry air (lb/h)	112.5	112.5	135.0	157.5	180.0	225.0	225.0	225.0	270.0	270.0	315.0	315.0	315.0
	Water vapor (lb/h)	247.5	247.5	297.0	346.5	396.0	495.0	495.0	495.0	594.0	594.0	693.0	693.0	693.0
	Total mixture (lb/h)	360.0	360.0	432.0	504.0	576.0	720.0	720.0	720.0	864.0	864.0	1,008.0	1,008.0	1,008.0
1,000,001–2,000,000	SCFM[a]	30.0	35.0	40.0	40.0	50.0	50.0	60.0	60.0	70.0	70.0	80.0	80.0	90.0
	Dry air (lb/h)	135.0	157.5	180.0	180.0	225.0	225.0	270.0	270.0	315.0	315.0	360.0	360.0	405.0
	Water vapor (lb/h)	297.0	346.5	396.0	396.0	495.0	495.0	594.0	594.0	693.0	693.0	792.0	792.0	891.0
	Total mixture (lb/h)	432.0	504.0	576.0	576.0	720.0	720.0	864.0	864.0	1,008.0	1,008.0	1,152.0	1,152.0	1,296.0
2,000,001–3,000,000	SCFM[a]	35.0	40.0	40.0	50.0	60.0	60.0	70.0	70.0	80.0	80.0	90.0	100.0	100.0
	Dry air (lb/h)	157.5	180.0	180.0	225.0	270.0	270.0	315.0	315.0	360.0	360.0	405.0	450.0	450.0
	Water vapor (lb/h)	346.5	396.0	396.0	495.0	594.0	594.0	693.0	693.0	792.0	792.0	891.0	990.0	990.0
	Total mixture (lb/h)	504.0	576.0	576.0	720.0	864.0	864.0	1,008.0	1,008.0	1,152.0	1,152.0	1,296.0	1,440.0	1,440.0
3,000,001–4,000,000	SCFM[a]	40.0	50.0	50.0	60.0	70.0	70.0	80.0	80.0	90.0	100.0	100.0	110.0	120.0
	Dry air (lb/h)	180.0	225.0	225.0	270.0	315.0	315.0	360.0	360.0	405.0	450.0	450.0	495.0	540.0
	Water vapor(lb/h)	396.0	495.0	495.0	594.0	693.0	693.0	792.0	792.0	891.0	990.0	990.0	1,089.0	1,188.0
	Total mixture (lb/h)	576.0	720.0	720.0	864.0	1,008.0	1,008.0	1,152.0	1,152.0	1,296.0	1,440.0	1,440.0	1,584.0	1,728.0

(Continued)

Table 9-4. (Continued)

Effective Steam Flow Each Main Exhaust Opening (lb/h)		C. Three Condenser Shells — Total Number of Exhaust Openings											
		3	4	5	6	7	8	9	10	11	12	13	14
250,000–500,000	SCFM[a]	30.0	30.0	37.5	37.5	37.5	45.0	52.5	52.5	60.0	60.0	75.0	75.0
	Dry air (lb/h)	135.0	135.0	168.8	168.8	168.8	202.5	236.3	236.3	270.0	270.0	337.5	337.5
	Water vapor (lb/h)	297.0	297.0	371.4	371.4	371.4	445.5	519.9	519.9	594.0	594.0	742.5	742.5
	Total mixture (lb/h)	432.0	432.0	540.2	540.2	540.2	648.0	756.2	756.2	864.0	864.0	1,080.0	1,080.0
500,001–1,000,000	SCFM[a]	30.0	37.5	45.0	45.0	52.5	52.5	60.0	75.0	75.0	75.0	90.0	90.0
	Dry air (lb/h)	135.0	168.8	202.5	202.5	236.3	236.3	270.0	337.5	337.5	337.5	405.0	405.0
	Water vapor (lb/h)	297.0	371.4	445.5	445.5	519.9	519.9	594.0	742.5	742.5	742.5	891.0	891.0
	Total mixture (lb/h)	432.0	540.2	648.0	648.0	756.2	756.2	864.0	1,080.0	1,080.0	1,080.0	1,296.0	1,296.0
1,000,001–2,000,000	SCFM[a]	37.5	45.0	52.5	52.5	60.0	75.0	75.0	75.0	90.0	90.0	105.0	105.0
	Dry air (lb/h)	168.8	202.5	236.3	236.3	270.0	337.5	337.5	337.5	405.0	405.0	472.5	472.5
	Water vapor (lb/h)	371.4	445.5	519.9	519.9	594.0	742.5	742.5	742.5	891.0	891.0	1,039.5	1,039.5
	Total mixture (lb/h)	540.2	638.0	756.2	756.2	864.0	1,080.0	1,080.0	1,080.0	1,296.0	1,296.0	1,512.0	1,512.0
2,000,001–3,000,000	SCFM[a]	45.0	52.5	60.0	75.0	75.0	75.0	90.0	90.0	105.0	105.0	120.0	120.0
	Dry air (lb/h)	202.5	236.3	270.0	337.5	337.5	337.5	405.0	405.0	472.5	472.5	540.0	540.0
	Water vapor (lb/h)	445.5	519.9	595.0	742.5	742.5	742.5	891.0	891.0	1,039.5	1,039.5	1,188.0	1,188.0
	Total mixture (lb/h)	648.0	756.2	864.0	1,080.0	1,080.0	1,080.0	1,296.0	1,296.0	1,512.0	1,512.0	1,728.0	1,728.0
3,000,001–4,000,000	SCFM[a]	52.5	60.0	75.0	75.0	90.0	90.0	105.0	105.0	120.0	120.0	135.0	135.0
	Dry air (lb/h)	236.3	270.0	337.5	337.5	405.0	405.0	472.5	472.5	540.0	540.0	607.5	607.5
	Water vapor (lb/h)	519.9	594.0	742.5	742.5	891.0	891.0	1,039.5	1,039.5	1,188.0	1,188.0	1,336.5	1,336.5
	Total mixture (lb/h)	756.2	864.0	1,080.0	1,080.0	1,296.0	1,296.0	1,512.0	1,512.0	1,728.0	1,728.0	1,944.0	1,944.0

[a]14.7 psia at 70° F

Note: These tables are based on air leakage only and the air vapor mixture at 1 in. Hg abs and 71.5° F.

Source: *Standards for Steam Surface Condensers*, Heat Exchange Institute, Inc. Used with permission

7. If independent venting systems are utilized for each shell of a multishell condenser, the capacity of each system is determined by dividing the total capacity obtained from the appropriate table by the number of independent venting systems.

When starting a turbine, it is desirable to rapidly evacuate the air from the condenser and reduce its pressure to some lower value, after that the main ejector takes over. To do this, a starting air (hogging) ejector is installed in addition to the regular steam air ejector. It has a much higher capacity but normally will not produce the low vacuum desired during normal operation. The mechanical evacuators also have a method to permit rapid evacuation of the condenser during startup.

Table 9-5 lists recommended capacities.

9.2.6 Performance

Condensers are very large heat exchangers, with thousands of tubes handling very large volumes of steam. In addition to providing the required tube surface area, the manufacturer must ensure that the design provides for the proper distribution of the steam throughout the shell so that all of the surface is effective. Manufacturers employ various tube location patterns to permit steam access to all the tubes while maintaining minimum pressure drop. If properly designed, the condenser will perform as intended, provided that air leakage is within limits and that the tubes are maintained within the cleanliness factor allocated.

9.2.6.1 Leak Detection. Air leakage into the condenser can be detected by a measuring device provided with the air removal equipment that measures the quantity of air discharged. Air leakage into the condenser during plant startup can come from incomplete welds, connections not plugged, and other similar construction errors. Other sources that may contribute at startup or later during operation include leaking valves or valve packing on lines leading to the condenser, leaking or broken gauge glasses, weld cracks, and similar sources.

Large leaks, such as incomplete welds, are usually located by physical inspection or by flooding the steam space prior to startup. There are several methods of detecting a small leakage. One of the most common is by installing a freon detector on the air removal equipment outlet. A small stream of freon directed at the suspected leak point will show up on the leak detector if a leak actually exists.

A defective condenser tube results in leakage of circulating water into the shell where it combines with the condensate. This is caused by lower pressure in the shell than inside the tubes. This leakage, unless very large, does not affect condenser performance. The circulating water, because of its dissolved chemicals, has a much higher conductivity than the condensate. Leaking circulating water mixing with the condensate raises the conductivity of the condensate and is detected by a conductivity meter in the condensate outlet.

Table 9-5. Hogger Capacities

Total Steam Condensed (lb/h)	SCFM[a]—Dry Air at 10 in. Hg abs Design Suction Pressure
Up to 100,000	50
100,001–250,000	100
250,001–500,000	200
500,001–1,000,000	350
1,000,001–2,000,000	700
2,000,001–3,000,000	1,050
3,000,001–4,000,000	1,400
4,000,001–5,000,000	1,750
5,000,001–6,000,000	2,100
6,000,001–7,000,000	2,450
7,000,001–8,000,000	2,800
8,00,0001–9,000,000	3,150
9,000,001–10,000,000	3,500

[a]SCFM—14.7 psia at 70° F

Note: In the range of 500,000 lb/h steam condensed and above, the above table provides evacuation of the air in the condenser and LP turbine from atmospheric pressure to 10 in. Hg abs in about 30 min if the volume of condenser and LP turbine is assumed to be 26 ft^3/1,000 lb/h of steam condensed.

Source: *Standards for Steam Surface Condensers*, Heat Exchange Institute, Inc. Used with permission.

9.2.6.2 On-Line Cleaning. Lowering of the performance of a condenser may be a result of scaling of the inside surface of the tubes, accumulation of sediment in the tubes, or an accumulation of leaves or other debris on the tube sheet, shutting off the inlet to the tubes. Several methods are available for on-line cleaning of the condenser.

To remove an accumulation of material from the face of the tube sheet, the condenser can be "back flushed" by reversing the flow of the circulating water through the condenser. This is performed by adding additional valves and piping to the circulating water system. It is relatively simple for a two-pass condenser, but more complicated for a single-pass condenser. Figure 9-11 shows the pipe and valve arrangement required to permit backwash. To switch flow on-line requires that the flow be restricted for a very short time. To accomplish this, the circulating water valves are all motorized and programmed to function in the proper order on start of the program.

Backwashing removes accumulation of material from the tube sheet and to a lesser extent may remove some sediment in the bottom of the tubes. Two proprietary systems have been developed to help remove sediment settling in the tubes and may help in removing scaling. One system uses small plastic shuttles; the other uses sponge balls.

In the shuttle system, a small plastic shuttle is placed in each tube and a cage is placed on the end of each tube. The shuttle normally rests in the cage at the outlet of the tube. When the condenser is back-washed, the reverse flow causes the shuttle to move down the tube, pushing the sediment ahead of it, and to a lesser extent, scraping the scale from the tube. The shuttle is captured in the cage on the inlet of the tube where it stays until the direction of the water is reversed

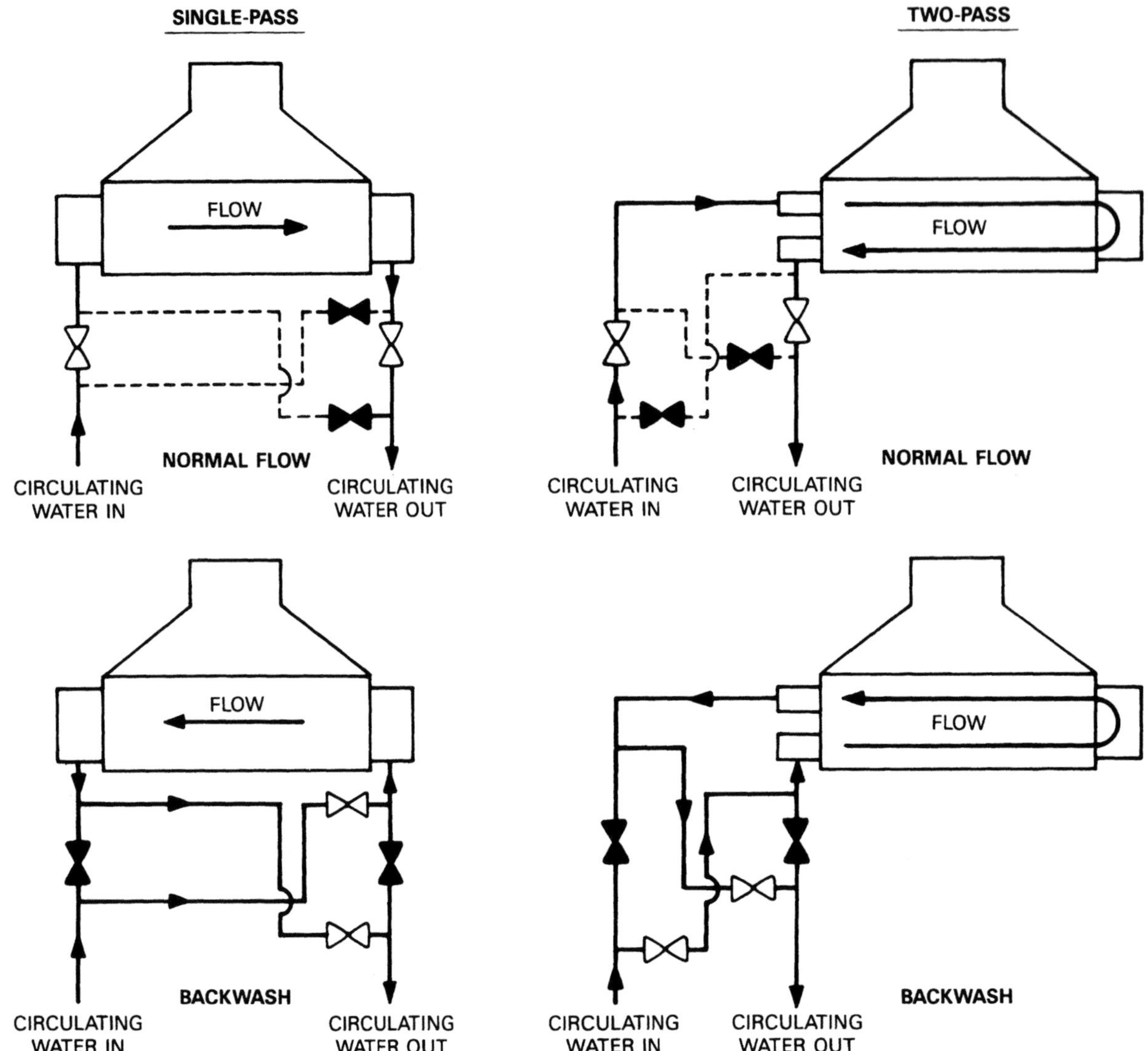

Fig. 9-11. Backwash piping arrangements.

to normal, when it again travels through the tube to be captured by the outlet cage.

The other system employed to clean the condenser on-line uses sponge balls of the same diameter as the tube's inside diameter. These balls are injected into the circulating water upstream of the condenser, pass through the tubes, and are collected downstream of the condenser. A pump returns the balls to the upstream injection point. In passing through the tubes, the balls push the sediment out of the tube. Balls can be provided with an abrasive coating that helps remove scale from the tube.

9.2.6.3 Performance Testing. Performance testing of a condenser requires an accurate measurement of the various parameters. These measurements can best be taken if arrangements for them are incorporated into the original design.

ANSI/ASME *Performance Test Code*, PTC 12.2, Code on Steam Condensing Apparatus outlines the recommended method for measuring condenser performance.

Circulating water flow is normally determined by taking pitot tube readings across the diameter of the circulating water pipe. These readings should be taken in a straight run of pipe. As in any flow measurement, the pitot tube should be located properly from any upstream or downstream pipe fitting.

Circulating water inlet and outlet temperatures must be measured accurately. With a condenser designed for a temperature rise of 10° F (5.6° C), a one-degree error in reading inlet or outlet temperature results in a 10% error in the calculated temperature rise. The circulating water does not leave each condenser tube at exactly the same temperature. To obtain an average outlet temperature, the measurement point should be located some distance down the pipe, away from the condenser to allow time for uniform mixing of the water.

Steam flow is almost impossible to measure directly. Indirectly, the steam flow can be determined by performing a complete heat balance on the turbine and feedwater system at the same time as the condenser test is being performed. The other method is to measure condensate flow leaving the condenser and correct this flow for other flows entering the condenser, such as heater drains.

Condenser vacuum should be determined from connections on the turbine exhaust casing or on the condenser shell close to the connecting point with the turbine. The measuring instrument should be accurate to 0.01 in. Hg (0.254 mm Hg).

With these readings and the design parameters of the condenser, calculations can be performed, similar to those for the original design, to determine expected vacuum, that is then compared with actual. Alternatively, by using actual vacuum the cleanliness factor can be calculated to determine tube status.

9.2.7 Air-Cooled Condensers

The previous paragraphs have discussed water-cooled steam condensers. The great majority of the condensers installed in the United States are of this type. The other type of condenser is the air-cooled condenser.

The air-cooled condenser has both advantages and disadvantages. Among the advantages are that it minimizes water make-up requirements and eliminates cooling tower blowdown disposal problems, cooling tower freeze-up, tower vapor plume, and circulating water pollution restrictions. The disadvantages of the air-cooled condenser include higher condenser operating pressure (lower cycle efficiency), higher first cost, larger site, higher noise levels, and higher operating cost.

There are two basic types of air-cooled condenser systems, as shown in Fig. 9-12. These are the jet condenser with dry cooling tower arrangement, and the direct air-cooled condenser system. In the jet condenser with dry cooling tower, part of the steam condensate is cooled in a dry cooling tower. It is then returned to the condenser where it is sprayed into the steam flow, causing the steam to condense and collect in the bottom of the condenser.

In the direct air-cooled condenser, the steam is piped from the turbine exhaust direct to air-cooled steam coils. The steam condenses in the coils, the condensate draining to the bottom collection tank.

The jet condenser/dry cooling tower has several advantages. The most prominent advantage is that the cooling tower may be located farther from the plant than the air-cooled condenser. The large size of the steam duct and the need to maintain a low-pressure drop in this duct dictates that the air-cooled condenser be located as close to the turbine as possible. It also eliminates the cost of the large steam duct from turbine to condenser.

The advantages of the air-cooled condenser are that it eliminates the cost of the jet condenser and the cost of the circulating water pumps and piping. It normally also produces slightly better condenser vacuum. In the air-cooled condenser, there is only one approach temperature involved, from air to steam. In the jet condenser/dry cooling tower, two approach temperatures are involved—from air to cooling water and from cooling water to steam, although the cooling water to steam approach is very small (less than 1° F).

Although the use of air-cooled condensers in this country has been very limited, the scarcity of water, the need for zero water discharge, and the better modern designs of this equipment indicate that more of these types of installations will be used in the future, particularly in the northeast United States.

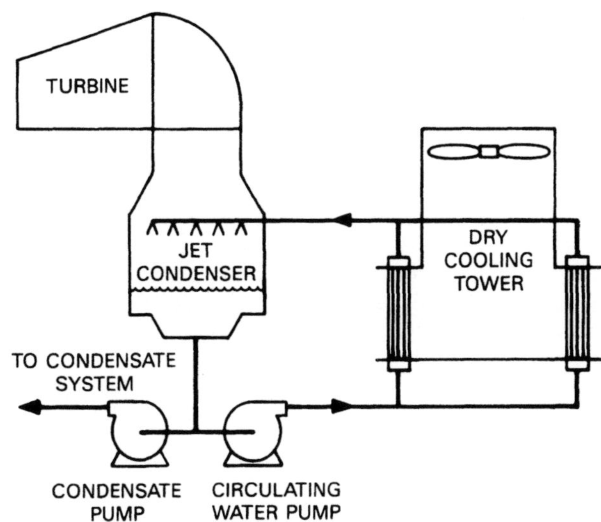

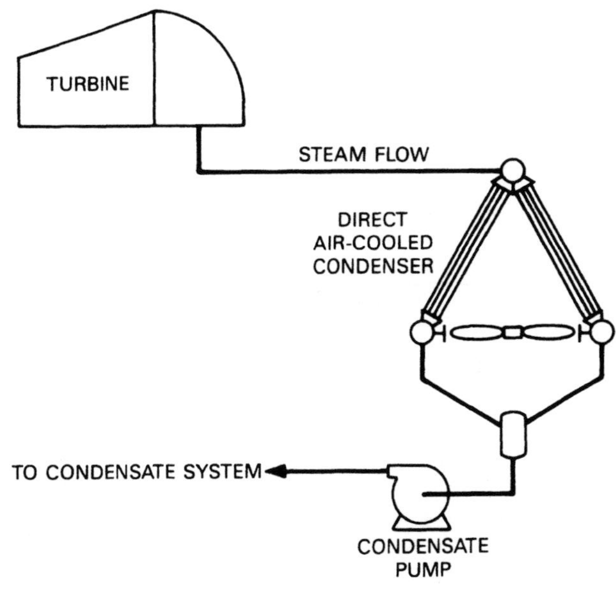

Fig. 9-12. Air-cooled condensers.

9.3 FEEDWATER HEATERS

9.3.1 Regenerative Cycle

In the regenerative cycle, steam is extracted from the steam turbine at various stages and used to heat the feedwater. This results in a higher cycle efficiency by increasing the temperature of the feedwater and by reducing the amount of energy lost in the condenser.

Heat balances for two representative turbine generators will illustrate the method of using these heaters in the cycle. The first, shown in Fig. 9-13, is a heat balance for a small, low-pressure turbine generator producing 26,860 kW (26.86 MW). This turbine operates with inlet steam at 850 psig (5,861 kPa), 900° F (482° C), with a condenser back pressure of 1.5 in. HgA (38.1 kPA Hg). It utilizes four feedwater heaters and has a heat rate of 9,605 Btu/kWh (10,124 kJ/kWh).

Figure 9-5 (p. 254) is a heat balance for a large, high-pressure turbine generator producing 727,072 kW (727.072 MW). This turbine operates with inlet steam at 2,520 psig (17,375 kPa), 1,000° F (538° C) with 1,000° F (538° C) reheat, and with a condenser back pressure of 3.5 in. HgA (88.9 mm HgA). It uses seven feedwater heaters and has a heat rate of 8,015 Btu/kWh (8,448 kJ/kWh).

Obviously, with the unit depicted in Fig. 9-5, the higher initial steam pressure and temperature, along with the steam reheat, contribute to the lower cycle heat rate, as compared to the unit shown in Fig. 9-13. However, the use of seven feedwater heaters instead of four also contributes to this lower heat rate.

The number of feedwater heaters used in any specific cycle depends mainly on the size of the turbine, the inlet and exhaust steam conditions, to some extent the overall plant cycle, and certainly the economic considerations.

The overall plant cycle may help determine the total number of feedwater heaters. In a stoker-fired unit, only a minimum of combustion air heating is normally desired. It

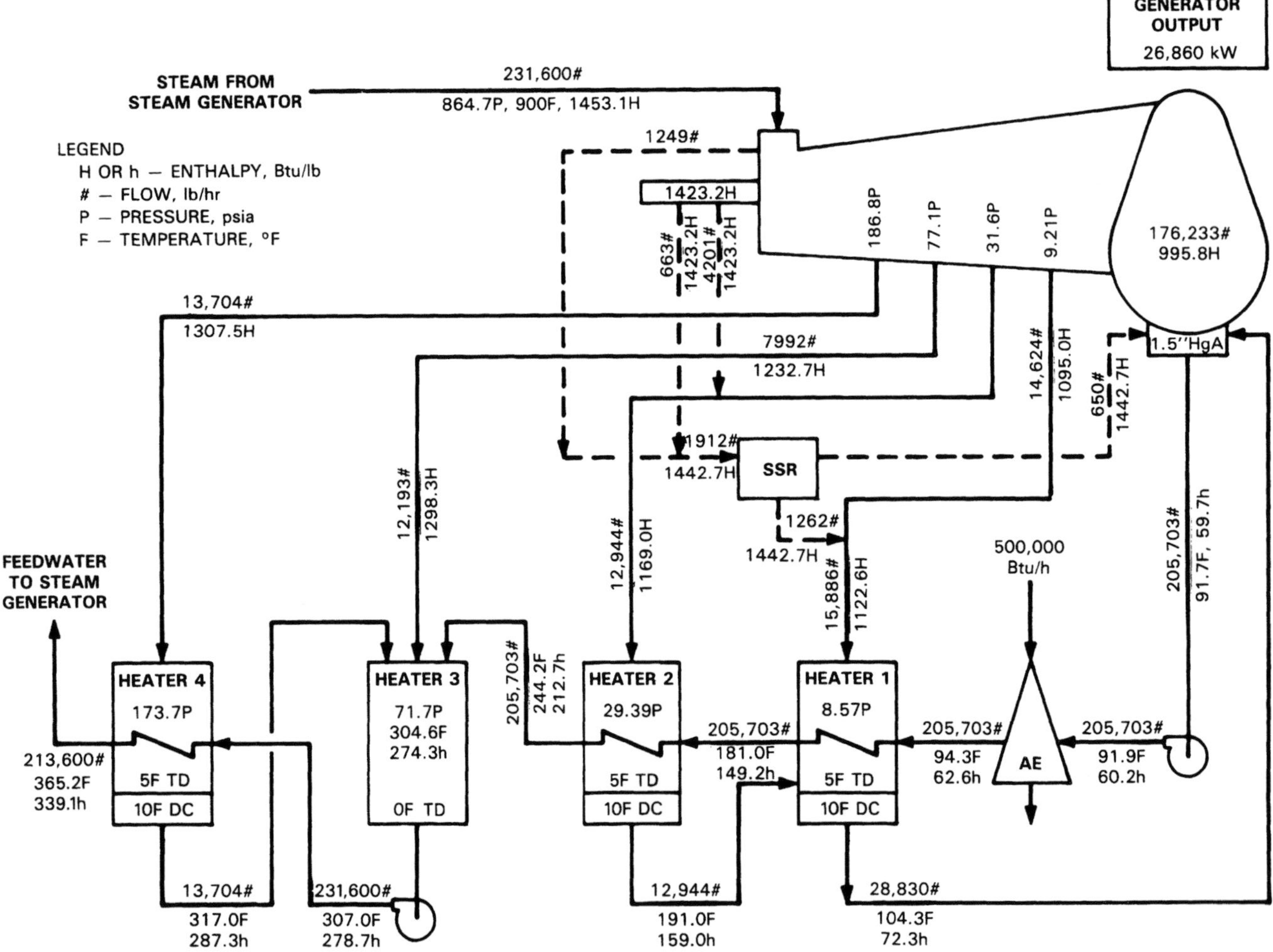

$$\text{Gross Heat Rate} = \frac{231{,}600(1453.1-339.1)}{26{,}860} = 9605 \text{ Btu/kWh}$$

Fig. 9-13. Typical heat balance for a small turbine generator. (From General Electric Co. Used with permission.)

then can be advantageous to eliminate the highest pressure feedwater heater, thus reducing the temperature of the feedwater to the steam generator economizer and allowing it to recover more of the energy from the steam generator exhaust gasses. In cogeneration plants, the entire cycle must be analyzed carefully to establish the optimum plant cycle and number of feedwater heaters.

Overall plant economics always affect plant design. Each feedwater heater added to the plant results initially in higher costs. These higher costs result from the first cost of the heater; piping and valving systems for extraction steam, feedwater, and drains; controls and instrumentation; and space requirements. In addition, there are costs for maintenance and upkeep as well as additional pumping cost due to pressure drop across the heater. On the other hand, fuel costs are lower because of the improved cycle efficiency due to the additional feedwater heater.

An economic analysis should be performed as part of the design to determine the optimum number of heaters to be used in any particular plant. Obviously, fuel cost plays a major role in this analysis. High fuel cost normally justifies additional heaters.

9.3.2 Heater Construction

As part of the cycle analysis, consideration must be given to the type of feedwater heaters to be used, their design parameters, and the method for disposal of the condensed steam from the heater.

Feedwater heaters are classified as "open" or "closed" heat exchangers. The open heater directly mixes the extraction steam with the feedwater to be heated. The extraction steam is condensed and becomes part of the feedwater leaving the heater.

The closed heater maintains a separation of the extraction steam and the feedwater to be heated. The extraction steam is condensed in its chamber of the heater and leaves the heater separately from the feedwater.

The open heater is designed to deaerate the incoming condensate.[1] This action liberates the dissolved, noncondensible gasses consisting mainly of oxygen, nitrogen, ammonia, and carbon dioxide, from the condensate. They are present as a result of leaks and chemical reactions. From this action, the open heater has received the name of "deaerator." Deaeration of the feedwater is essential to the proper operation of the plant and is discussed in more detail in Chapter 15, Water Treatment. In addition to deaeration, this heater provides proper suction conditions for the boiler feed pump and a natural break between high-pressure and low-pressure closed heater design. The disadvantages to open heaters are that they are large and heavy, and require a pump on the feedwater outlet to move the feedwater forward in the cycle.

The closed heater is normally referred to as the feedwater heater. It is physically smaller than the open type of heater and is easier to control. One pump can be used to move the feedwater through a series of heaters, thereby saving the cost and complications of a series of pumps.

A plant could be designed using all open heaters, but then a pump would be required with each open heater.

A plant could also be designed to use all closed heaters. Some plants in the past have been designed in this manner, employing the condenser to perform the required deaeration. However, deaeration was found to be not as complete, and control of the boiler feed pump was more difficult. Most modern cycles using high-pressure steam generators are designed with one deaerator and with the remaining heaters of the closed variety.

9.3.2.1 Open Heater Construction. The open heater (deaerator) is constructed of three sections: the heater section, the vent condenser, and the storage section. Figure 9-14 shows a cross section of a typical deaerator. Although this configuration is typical, other configurations such as those shown in Fig. 9-15 are also used.

In the heater section, the incoming condensate comes into contact with the extraction steam, heating the condensate, and condensing the steam. Deaeration of the condensate is based upon Dalton's and Henry's laws. These laws combine to state that the quantity of a gas that dissolves in a liquid decreases as the temperature of the liquid rises, and if the liquid is raised to the boiling point all the dissolved gases will be liberated. To heat the incoming condensate to the boiling point and release the dissolved gases, the incoming water must be broken down into a fine spray or thin sheets of water and permitted to come into contact with the extraction steam. Deaerators are made using both principles, some using a spray system, but most using a stack of small trays that allow the condensate to cascade over the edges of the trays, falling from one layer of trays to the next. The steam flows upward through the falling condensate, heating it and causing liberation of the dissolved gases, and, of course, condensing the extraction steam. A tray-type deaerator is shown in Fig. 9-14, and a spray-type unit in Fig. 9-16.

An opening in the top of the heater section allows the released gasses and some steam to be vented from the heater section. To prevent loss of this venting steam, a "vent condenser" uses the incoming condensate to condense the steam being vented with the gases. The vent condenser can be built as a simple tube and shell heat exchanger mounted on top of the heater section, or as is done in most modern large units, can be constructed as a spray section located in the top of the shell of the heater section. Figure 9-17 shows a typical deaerating heater section using a tray distribution system and an internal spray vent condenser. Figure 9-15 shows the shell and tube-type vent condenser as a dashed circle at the top of the heater section.

The heated and deaerated condensate, along with the condensed steam, falls to the bottom of the heater section and

[1]The condensed steam leaving the condenser and passing through the closed heaters to the deaerator is normally called "condensate." The mixture of "condensate" and extraction steam leaving the deaerator is called "feedwater."

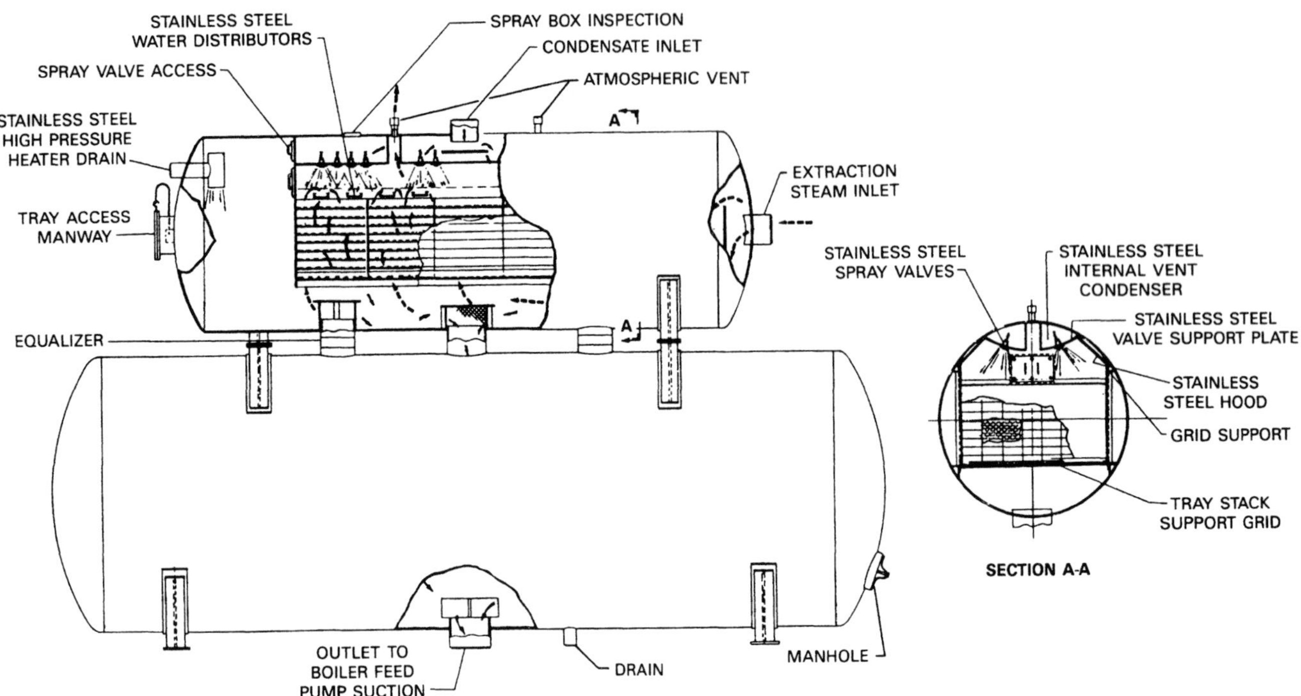

Fig. 9-14. Typical tray-type deaerator.

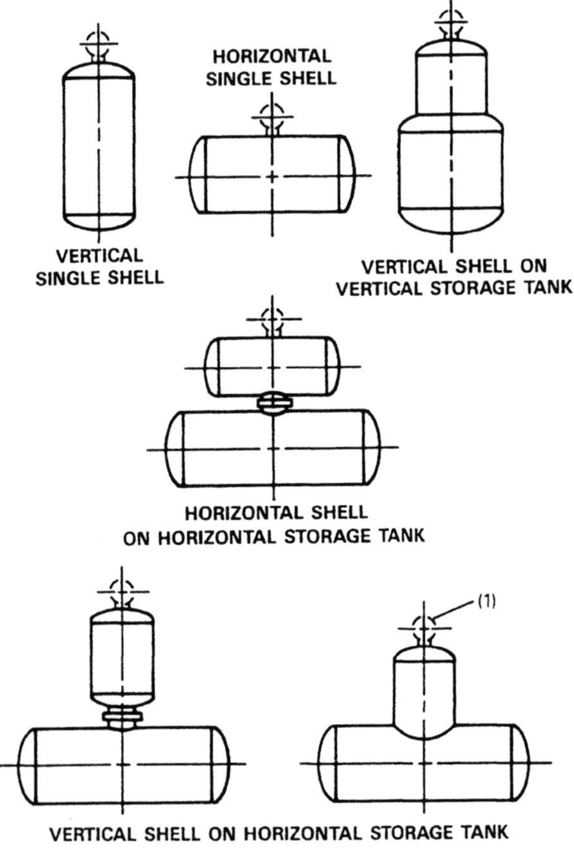

Fig. 9-15. Typical deaerator configurations. (From *Standards for Steam Surface Condensers, Eighth Edition*, Heat Exchange Institute, Inc. Used with permission.)

then passes into the storage section. For small heaters, the storage section can be located within the same shell as the heater section. However, for most large heaters, the storage section is a separate vessel with the heater section mounted on top of it to permit gravity discharge of the condensate into the storage section.

The deaerating capability of a unit is measured by the amount of dissolved oxygen in the feedwater leaving the storage section. Modern plants normally specify a dissolved oxygen guaranteed maximum of 0.005 cc/L. In specifying

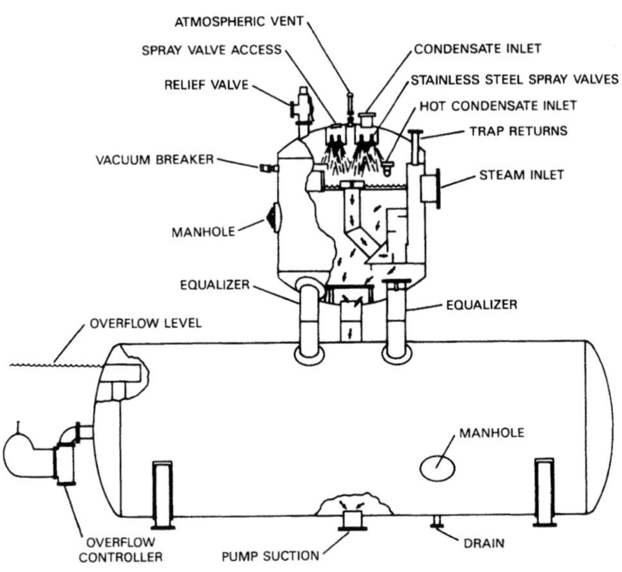

Fig. 9-16. Typical spray-type deaerator.

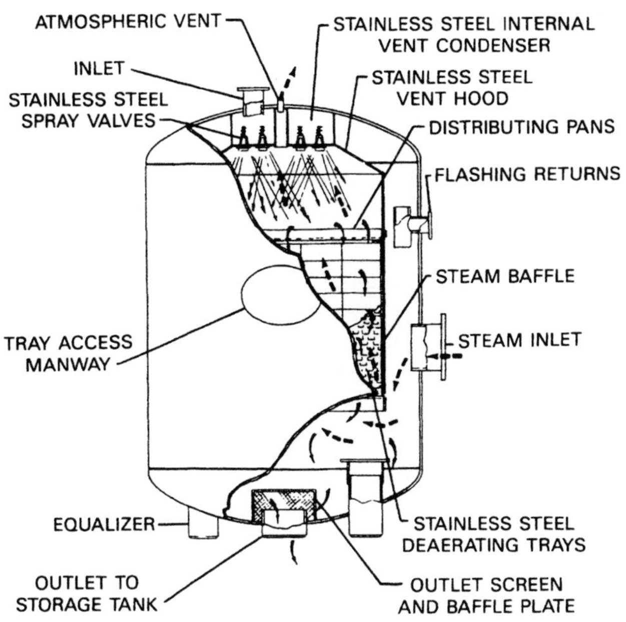

Fig. 9-17. Typical deaerating heater section with internal spray vent condenser.

the capacity of the unit, the reference is to the quantity of feedwater leaving the storage section. The unit can be designed to operate at any pressure, although subatmospheric operation requires additional facilities to provide for removal of the released gasses from the shell. Most units are designed for positive pressure only, and the shells are specified for a pressure exceeding the maximum pressure of the turbine extraction steam. Water storage capacity of the storage section is measured as that volume below the overflow level. It is normally specified as the equivalent to 10 min of rated capacity flow, but on very large units this may be reduced somewhat because of limits in physical size. This storage capacity, working together with the condenser hotwell storage capacity, provides for sudden changes in condensate or feedwater flow in the system. It also assists in flash protection of the boiler feed pump, as discussed later.

The deaerator is also used for several other purposes. Drains from the high-pressure heaters are normally cascaded into the deaerator. If an air preheater is installed on the steam generator, it normally uses extraction steam with the condensate returned to the deaerator. As an alternative, the air preheater may use hot water from the deaerator, with the cooled water being returned to the deaerator for reheating.

Discharges from high-pressure traps throughout the plant are also normally piped to the deaerator. Because these drains come from sources of higher pressure than the deaerator shell design pressure, a safety valve is provided on the shell to relieve a possible overpressure by these drains. If an auxiliary steam supply is connected to the heater that can provide a large capacity of high-pressure steam, the unit must be adequately protected from this additional source. The unit must also be protected from a loss of steam supply, that can create a subatmospheric pressure in the unit. A "vacuum breaker" valve is installed on the unit for this protection.

The unit should be constructed in accordance with the ASME *Boiler and Pressure Vessel Code*, Code for Unfired Pressure Vessels, Section VIII.

9.3.2.2 Closed Heater Construction. The closed heaters (feedwater heaters) are of the tube and shell design, with the condensate or feedwater in the tube side and the extraction steam and resulting condensed steam (heater drains) in the shell side.

Figure 9-18 shows a cross-section through a straight tube feedwater heater. In this type of construction, the feedwater enters the divided channel, passes through the tubes to the reverse channel, reverses direction, passes through return tubes to the divided channel, and exits the heater. A partition

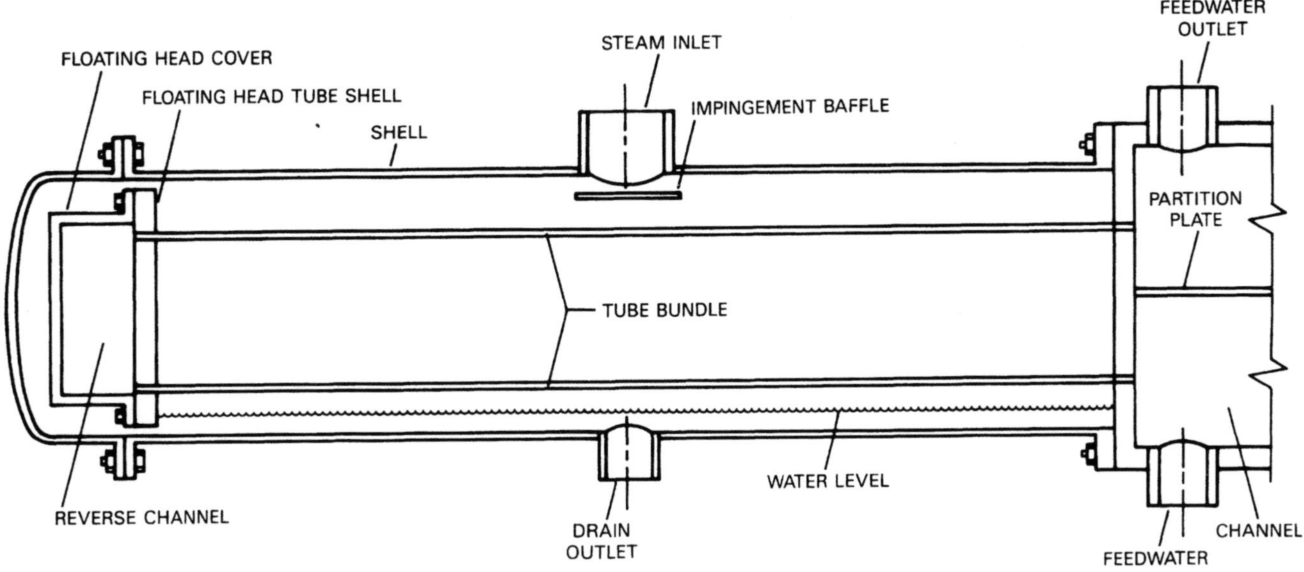

Fig. 9-18. Straight tube feedwater heater with floating reverse channel.

plate in the divided channel separates the incoming feedwater from the outgoing. To accommodate the differential expansion between the tubes and the heater shell, the reverse channel is "floated" in the heater shell. This type of construction permits access to both ends of the tubes and allows for mechanical cleaning of scale from the inside of the tubes. With the advent of better water chemistry and chemical cleaning, however, the floating head design has given way to the "U" tube construction shown in Fig. 9-19.

In a typical, straight condensing feedwater heater, such as that shown in Fig. 9-19, the extraction steam condenses on the tubes, drains to the bottom of the heater, and exits through the drain outlet. Since the steam condenses on the outside of the tubes, the outside tube temperature is the steam saturation temperature at the prevailing pressure. Therefore, the feedwater temperature can only approach the saturation temperature even when the extraction steam entering the heater is superheated. An economical design is one in which the unit is designed for a 5° F (2.8° C) terminal difference between the saturated steam temperature and the outlet feedwater temperatures.

In the straight condensing heater, the drains falling to the bottom of the shell pass through the steam atmosphere, resulting in their leaving the heater at approximately saturation temperature.

Thermodynamically, it is often advantageous if the feedwater outlet temperature can be raised above the steam saturation temperature or the drains cooled below the steam saturation temperature.

Figure 9-20 illustrates a typical two-zone feedwater heater constructed with both condensing and subcooling zones. In this design, the tubes containing the inlet and coldest feedwater are enclosed in such a manner that the drains must pass over these tubes before exiting. This zone now becomes a water-to-water heat exchanger, and the outlet drain temperature can approach the inlet feedwater temperature. An economical design for this zone is one in that the drain to feedwater approach temperature is 10° F (5.6° C).

Figure 9-21 illustrates a typical two-zone feedwater heater constructed with both desuperheating and condensing zones. In this design, the tubes containing the outlet and hottest feedwater are enclosed in such a manner that the inlet steam must first pass across these tubes before entering the main portion of the shell. In this zone, the temperatures and velocities are such that the steam is cooled but not condensed. The zone becomes a gas-to-liquid heat exchanger, and the outlet feedwater temperature can be raised above the steam saturation temperature. Obviously, this zone is used only when the extraction steam contains a considerable amount of superheat. An economical design for this zone is one in that the final feedwater temperature is 2 to 3° F (1 to 1.7° C) above the steam saturation temperature. As approach temperature is normally referring to approach to the steam saturation temperature, a temperature of 2° F (1° C) above saturation, is

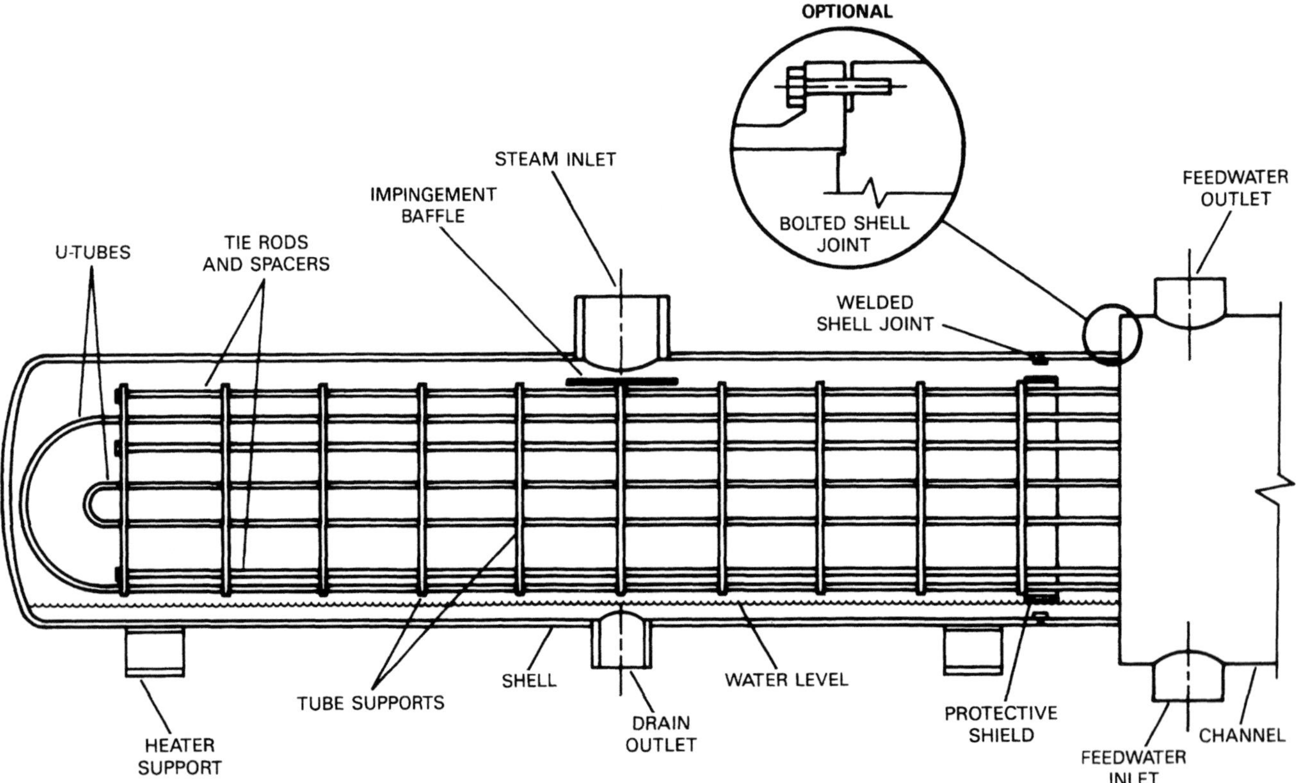

Fig. 9-19. Typical straight-condensing "U" tube feedwater heater. (From *Standards for Steam Surface Condensers, Eighth Edition*, Heat Exchange Institute, Inc. Used with permission.)

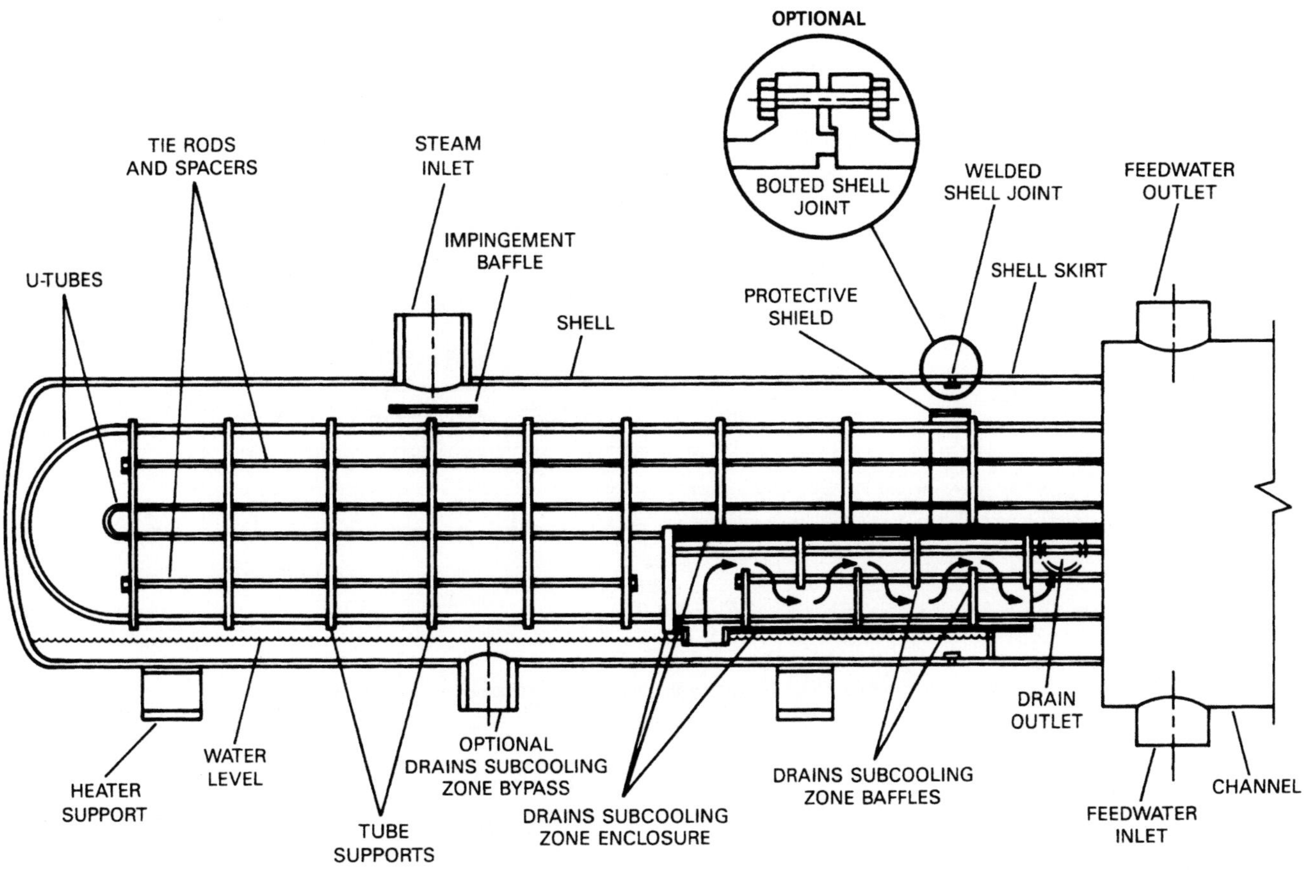

Fig. 9-20. Typical two-zone feedwater heater (subcooling and condensing zones). (From *Standards for Steam Surface Condensers, Eighth Edition*, Heat Exchange Institute, Inc. Used with permission.)

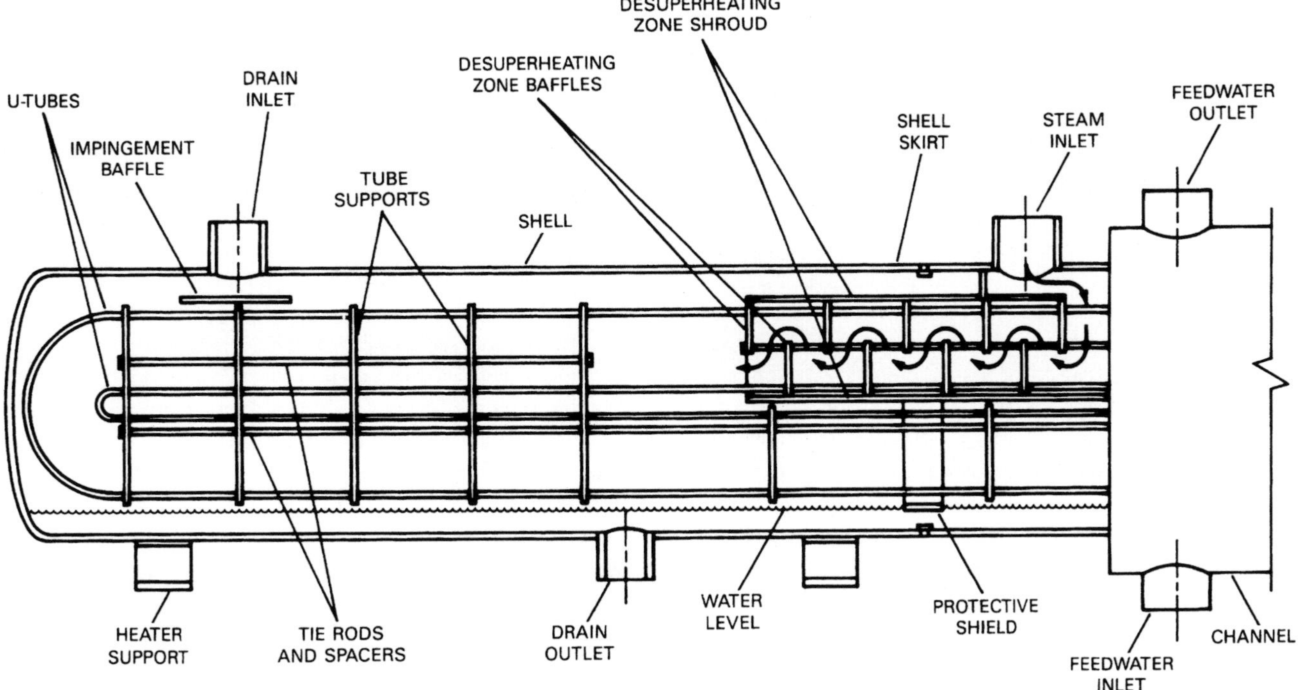

Fig. 9-21. Typical two-zone feedwater heater (desuperheating and condensing zones). (From *Standards for Steam Surface Condensers, Eighth Edition*, Heat Exchange Institute, Inc. Used with permission.)

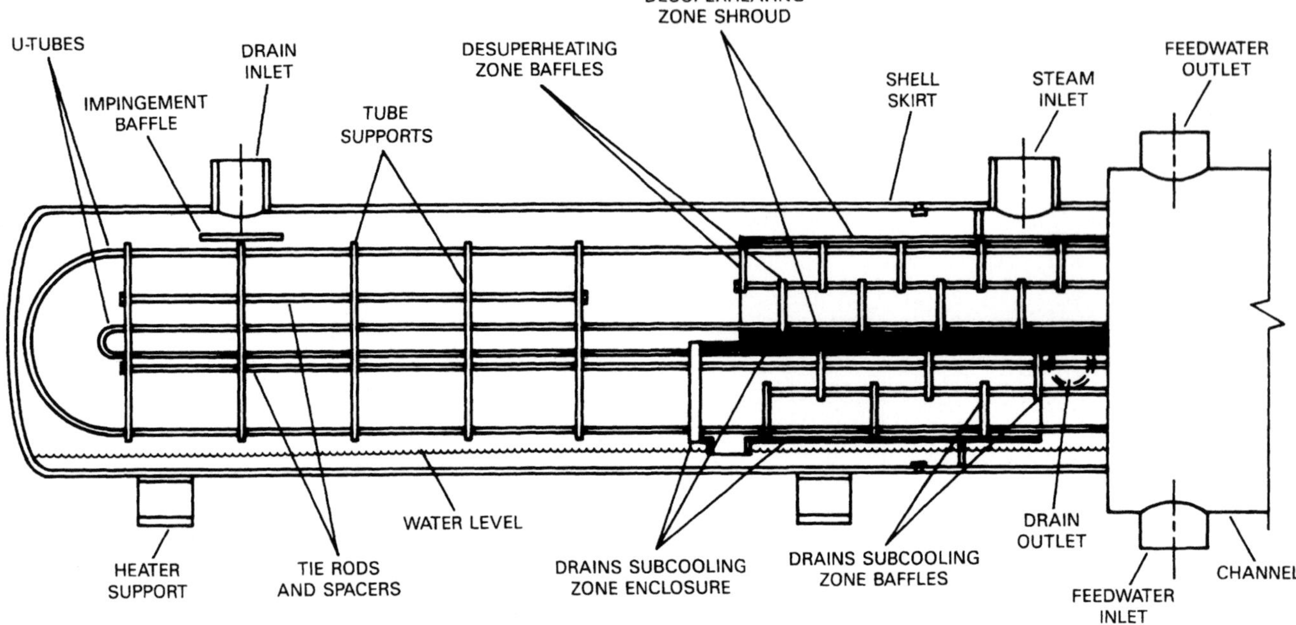

Fig. 9-22. Typical three-zone feedwater heater (desuperheating, condensing, and subcooling zones). (From *Standards for Steam Surface Condensers, Eighth Edition*, Heat Exchange Institute, Inc. Used with permission.)

referred to as a minus two degree (−2° F) (−1° C) approach. Figure 9-22 illustrates a typical three-zone feedwater heater constructed with desuperheating, condensing, and subcooling zones.

To perform maintenance on the tube bundle, it must be removed from the heater shell. Figures 9-19 and 9-20 show the two types of construction available, flanged and bolted or all welded. In the all welded design, the shell must be cut to separate it from the tube bundle. A protective shield protects the tubes during the cutting process. Most larger heaters are constructed all welded, as it is less expensive than flanges and eliminates the problems associated with leakage on a large-diameter, high-pressure flanged joint.

To expose the tube bundle, the designer can arrange to "shell pull," in that the shell is pulled off the tube bundle, or "bundle pull," in that the bundle is pulled out of the shell. With the present use of all-welded piping systems, it is necessary to cut the piping connections on the shell or on the channel before the two parts can be pulled apart. Most designers feel that it is easier for the maintenance personnel to cut and reweld the thinner walled piping associated with the shell than the heavy walled piping associated with the feedwater. Therefore, in most installations, the units are designed for shell pull. When specified, the manufacturer can provide wheeled shell supports to facilitate pulling of the shell. In either design, the designer should provide adequate space for disassembly of the unit.

Figure 9-23 illustrates construction of the channel for a low-pressure feedwater heater. The channel is that portion of the heater in that the feedwater enters and leaves. It consists of the tubesheet, barrel with inlet and outlet connections, cover, and pass partition to separate the inlet water from the outlet water and force it to pass through the tubes. This type of construction is common for low-pressure heaters with diameters of less than 48 in. (121.9 cm). To eliminate the gasketing of the pass partition in the main channel cover, an internal pass partition cover such as that shown in Fig. 9-23(b), is used. The pressure differential across the pass partition and pass partition cover is only the pressure drop through the heater tubes, normally only a few pounds per square inch. The partition and cover are relatively lightweight and the gasketing is relatively simple.

Figure 9-24 illustrates various types of construction of the channel for a high-pressure feedwater heater. Because of the high pressure of the feedwater, a full-size flanged cover, as shown in Fig. 9-23, is impractical. The design utilizing a hemispherical head is usually the most efficient but presents difficulty in fitting nozzles on the hemisphere's periphery. The larger size low-pressure heaters are often constructed in the same manner.

The feedwater tubes are "rolled" into the tubesheet. Rolling is a process in that a tool is inserted into the tube and it expands the tube beyond its elastic limit, forcing it out tight against the tube sheet. On some high-pressure heaters, the hole in the tube shell may be serrated to provide further resistance to pull out of the tube. On some high-pressure heaters, the tubes are rolled into place and then a small weld bead is placed around the tube end that welds the tube to the tube sheet, providing further protection against leaks between the tube and tube sheet.

Feedwater heaters can be constructed for mounting in the horizontal or vertical position as illustrated in Fig. 9-25. While most heaters are installed in the horizontal, some smaller plants have the heaters mounted vertically. The verti-

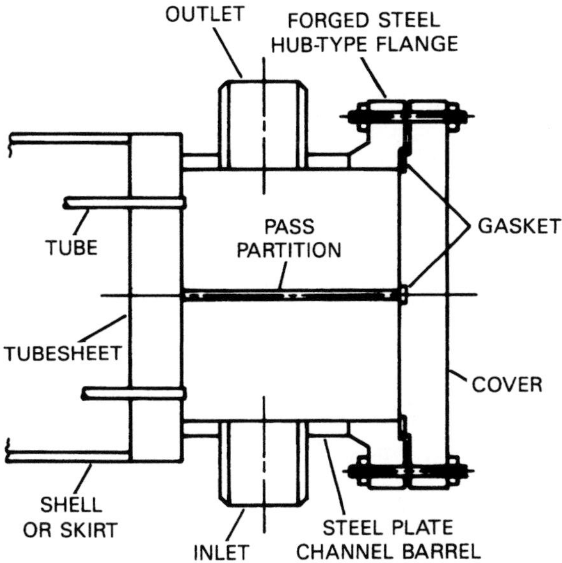

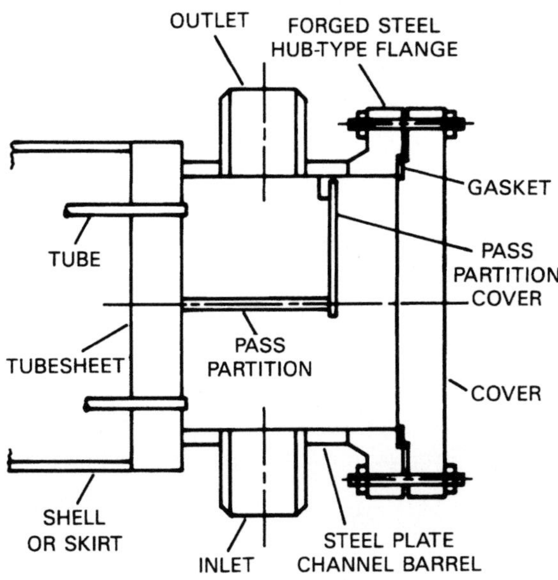

Fig. 9-23. Low-pressure feedwater heater channel construction. (From *Standards for Steam Surface Condensers, Eighth Edition*, Heat Exchange Institute, Inc. Used with permission.)

cal heaters are normally installed in the turbine room, where the overhead crane can be used to lift the top part of the heater when disassembling for maintenance. The heaters can be designed for either channel-up or channel-down installation, as shown in Fig. 9-25. The internal configuration of a straight condensing heater is not complicated by being designed for vertical installation, although with channel-down construction some of the tubes below the water level are submerged and become ineffective for condensing purposes, as shown in Fig. 9-26. When a subcooling and/or desuperheating zone is added to a vertical heater, it can complicate the design. The various manufacturers have different approaches in designing this type of heater, but in most of these designs, some of the tube surfaces are ineffective. Figure 9-27 shows a typical two-zone vertical, channel-up heater with subcooler zone. It can be seen that the inlet to the subcooler section is brought down to near the bottom of the heater. This results in some of the tubes at the bottom being submerged below water level and may result in oversurfacing of the subcooler section.

Horizontal heaters are placed in various locations in the plant where there is adequate room and where piping costs will be minimized. On large turbines, the lowest pressure heaters are sometimes located in the neck of the condenser. This permits the extraction steam piping (which may be as many as four large-size pipes) to be kept short and completely confined within the condenser neck.

Feedwater heaters are normally constructed with carbon steel shell and channel, the shell being rolled plate or pipe. The channel of low-pressure heaters is fabricated steel, and of forged steel for high-pressure heaters. Tubes for heaters can be stainless steel, monel, inconel, copper nickel in amounts of 70–30, 80–20, or 90–10; admiralty; copper; or carbon steel. Most small plants use admiralty tubes in the low-pressure heaters and copper nickel in the high-pressure heaters. On large units using very high-pressure and high-temperature steam, a phenomenon called "copper pickup" was encountered where minute amounts of dissolved copper were found in the feedwater. This collected in the steam generator, causing problems. To combat this, the steam, condensate, and feedwater systems are now designed of material containing no copper. Originally, the tube material of choice for this application was carbon steel. However, it later was found that preventing corrosion of these tubes when the heater was out of service required that the heater be drained, dried out, and sealed with an inert gas. Because of these problems, the present tube of choice for these units is normally stainless steel.

Feedwater heaters should be constructed in accordance with the *ASME Boiler and Pressure Vessel Code*, Code for Unfired Pressure Vessels, Section VIII. They should also comply with *Standards for Closed Feedwater Heaters*, published by the Heat Exchange Institute, Inc. (HEI). This publication was developed by the leading manufacturers of feedwater heaters and contains much information, including definitions, performance, and design recommendations. The following are some of the recommendations of this publication that are important for the design engineer:

- A minimum fouling resistance of 0.0002 h $\times$ F $\times$ ft^2/Btu (7.0 $\times$ 10^{-9} s $\times$ C $\times$ m^2/J) should be applied to the tubeside surface. An additional fouling resistance of 0.0003 h $\times$ F $\times$ ft^2/Btu (10.5 $\times$ 10^{-9} s $\times$ C $\times$ m^2/J) should be applied for the outer tube surfaces in the desuperheating and drain subcooling zone.

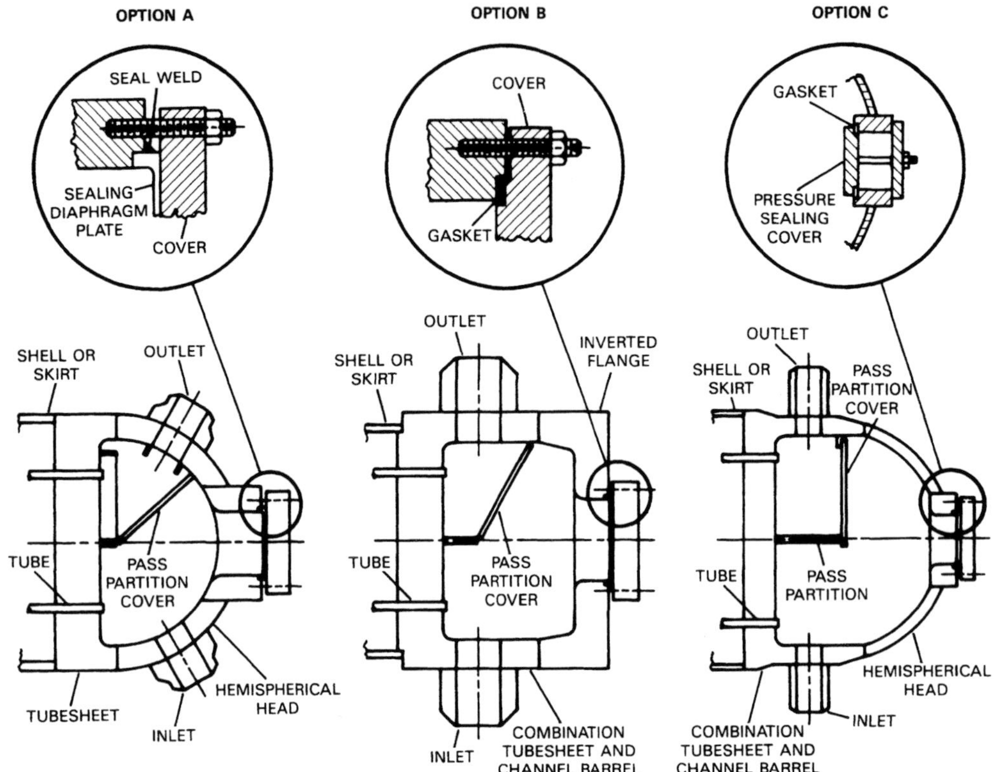

Fig. 9-24. High-pressure feedwater heater channel construction. (From *Standards for Steam Surface Condensers, Eighth Edition*, Heat Exchange Institute, Inc. Used with permission.)

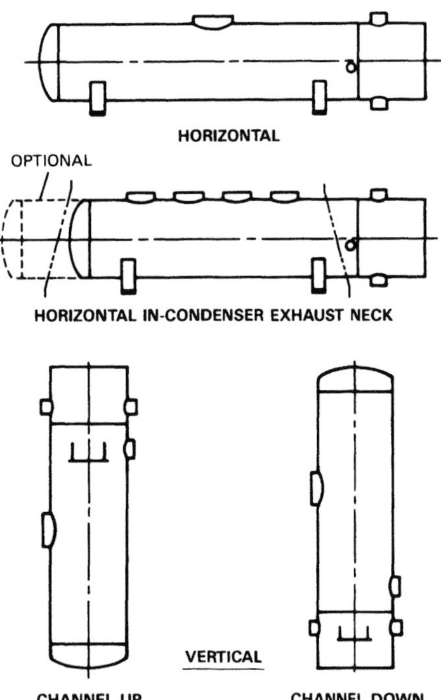

Fig. 9-25. Feedwater heater mounting positions. (From *Standards for Steam Surface Condensers, Eighth Edition*, Heat Exchange Institute, Inc. Used with permission.)

- Terminal temperature difference, without desuperheating zone, should not be less than 2° F (1.1° C).
- Drain to feedwater approach when subcooler is provided should not be less than 10° F (5.6° C).
- Maximum tube-side velocities should be the following:

Stainless steel, monel, inconel	10.0 ft/s (304.8 cm/s)
Copper nickel	9.0 ft/s (274.3 cm/s)
Admiralty and copper	8.5 ft/s (259.1 cm/s)
Carbon steel	8.0 ft/s (243.8 cm/s)

- Method of determining design temperature for both shell and tubes is as follows:

  Shell side—Enter the Mollier diagram at the normal operating steam temperature and pressure and follow a constant entropy line to the maximum operating pressure. Read the temperature at that point and round off to the next higher 10° F (5.6° C). For a heater with a desuperheating zone, only the shell skirt need be designed for this temperature. The design temperature of the main shell barrel of such heater shall be at least equal to the saturation temperature at the design pressure.

  Tube Side—The maximum design temperature shall be the saturated steam temperature corresponding to the shell side design pressure. Where a desuperheating zone is employed, the temperature of the straight lengths of tubes in the desuperheating zone shall be considered to be 35° F (19.4° C) higher than the saturated steam temperature corresponding to the shell side design pressure.

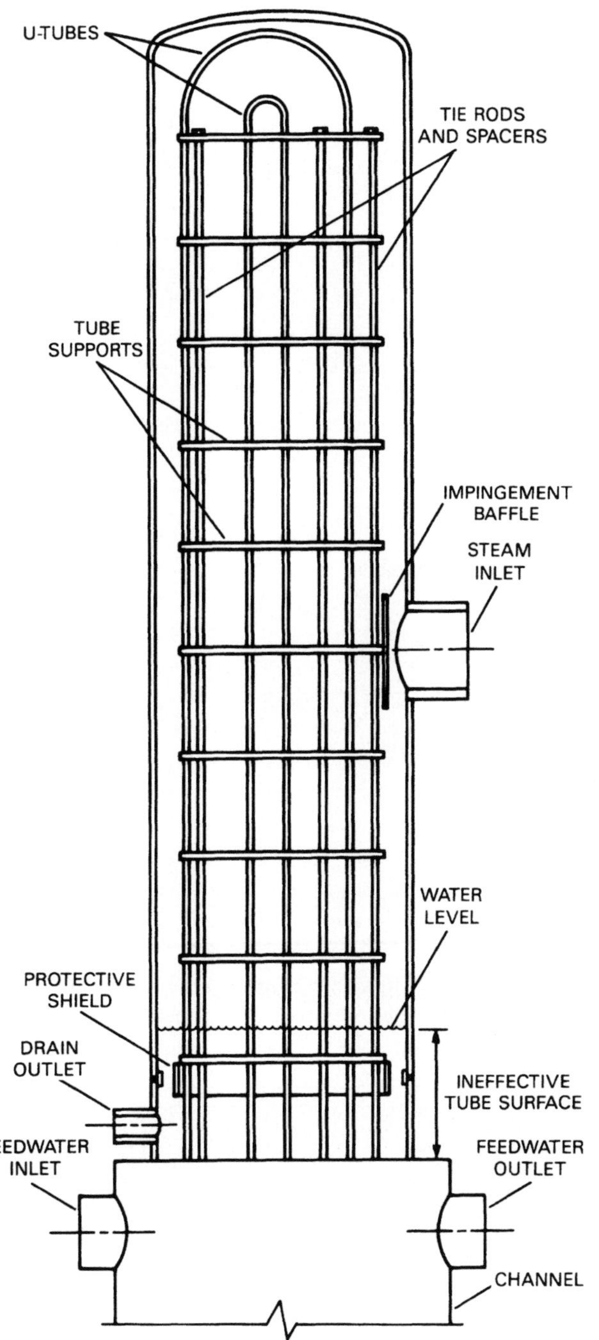

Fig. 9-26. Typical straight condensing vertical feedwater heater with channel down.

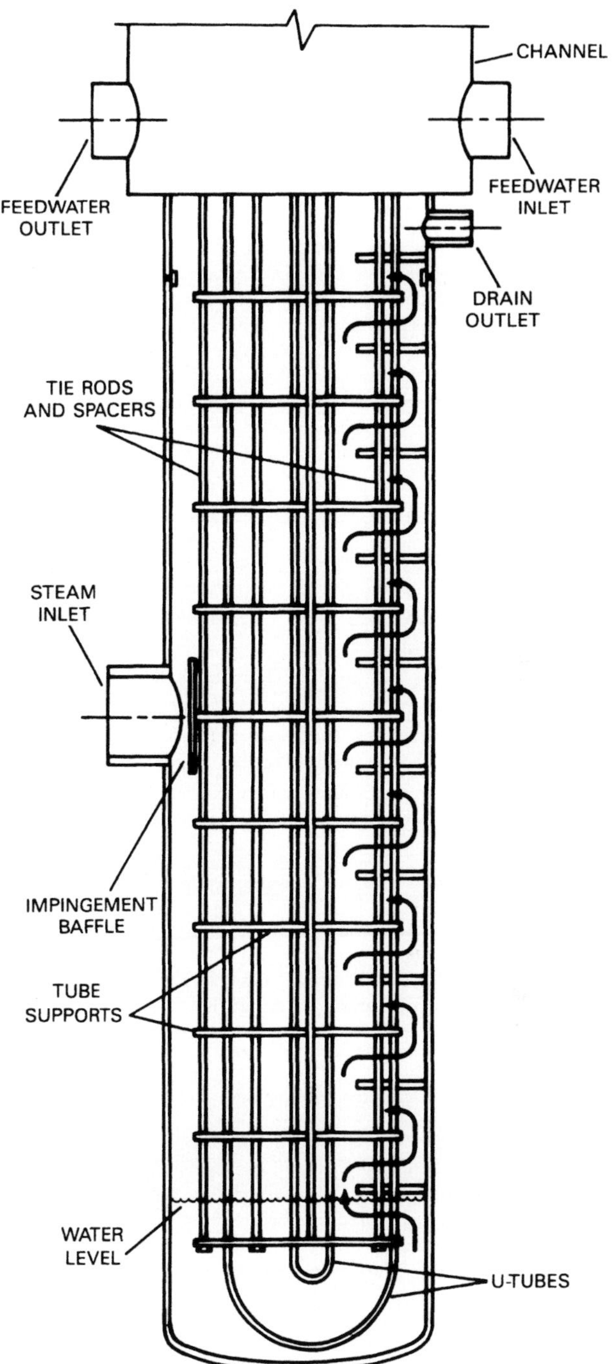

Fig. 9-27. Typical two-zone vertical feedwater heater with channel up.

- Maximum metal temperature for the various tube materials should be as shown in Table 9-6.
- Maximum metal temperature of expanded tube joints should be as shown in Table 9-7.
- Minimum tube wall thickness should be as shown in Table 9-8.
- U-tube minimum wall prior to bending should be

$$t = \frac{P \times d}{2S + 0.8\ P} \times \left(1 + \frac{d}{4R}\right) \qquad (9\text{-}10)$$

where

t = tube wall minimum wall before bending, in.;

P = design pressure, psi;

d = outside diameter of tube, in.;

S = allowable design stress, psi; and

R = radius of bend at center line of tubes, in.

- There should be a ¾ in. tube-side relief valve to protect against

Table 9-6. Maximum Metal Temperatures for Tube Materials

Material	Temperature (° F)
Arsenical copper	400
Admiralty metal	450
90–10 Copper–nickel	600
80–20 Copper–nickel	700
70–30 Copper–nickel (annealed)	700
70–30 Copper–nickel (stress relieved)	800
70–30 Nickel–copper (annealed)	900
70–30 Nickel–copper (stress relieved)	800
Carbon steel	800
Stainless steel	800

Source: *Standards for Closed Feedwater Heaters*, Heat Exchange Institute, Inc. Used with permission.

overpressure from water expansion when the water inlet and outlet valves are closed.

- There should be a relief valve to protect the shell from overpressure in case of tube failure, when the shell design pressure is less than the tube-side design pressure. The relief valve to be sized to pass the larger of the following flows at 10% accumulation with water at T_v, average tube-side temperature at the normal operating conditions, ° F:
  —10% of the maximum overload feedwater flow.
  —Flow based on the clean rupture of one heater tube resulting in two open ends discharging as orifices.

$$Q = 54d^2\sqrt{P_t - P_s} \qquad (9\text{-}11)$$

where

Q = flow of water, gpm at 70° F;

d = nominal inside diameter of tubes, in.;

P_t = tube-side design pressure, psi; and

P_s = shell side design pressure, psi.

Table 9-7. Maximum Temperatures of Expanded Tube Joints

Material	Temperature[a] (° F)
Arsenical copper	350
Admiralty metal	350
90–10 Copper–nickel	400
80–20 Copper–nickel	450
70–30 Copper–nickel (annealed)	500
70–30 Copper–nickel (stress relieved)	500
70–30 Nickel–copper (annealed)	550
70–30 Nickel–copper (stress relieved)	550
Carbon steel	650
Stainless steel	500

[a]For this purpose, the temperature of the tube joint shall be considered to be the outlet temperature of the feedwater at the specified operating conditions. Welded tube joints should be used when temperatures range from the values in Table 9-7 up to the maximum metal temperatures given in Table 9-6.

Source: *Standards for Closed Feedwater Heaters*, Heat Exchange Institute, Inc. Used with permission.

Table 9-8. Design Minimum Tube Wall Thickness

Material	Average Wall Thickness, inches
Copper and copper alloy	0.049
Nickel alloy	0.049
Stainless steel (U-tubes)	0.035
Stainless steel (straight tubes)	0.028
Carbon steel	0.050

Source: *Standards for Closed Feedwater Heaters*, Heat Exchange Institute, Inc. Used with permission.

- There should be venting and draining of all high and low points of both shell and tubes. The shell side should have continuous operating vents and they should be individually piped to the condenser or deaerator as applicable. Manifolding or cascading of vents is not recommended.

9.3.2.3 Heater Drain Disposal. Laws of thermodynamics prove that it is never as efficient to cool and then reheat a medium as it is to keep it at the original temperature. In feedwater cycle design, the most efficient cycle would be to eliminate the heater drain coolers, keeping the drains as hot as possible, and then injecting the drains into the feedwater where the feedwater is about the same temperature. The heater drain cycle would then look as shown in Fig. 9-28. Unfortunately, the pressure in the heater shells is always lower than the pressure in the feedwater system. A pump would therefore be needed in each heater drain loop, as indicated in dashed outline in Fig. 9-28. Such a series of pumps would increase the first cost and maintenance cost of the plant and adversely affect availability.

To eliminate all the heater drain pumps and still maintain a high efficiency (although not as high as the pumping cycle), drains from the highest pressure heater are cascaded to the next lower pressure heater, where it cools and joins the drains from that heater. This cycle is repeated until the drains enter the deaerator or condenser, where it joins the main

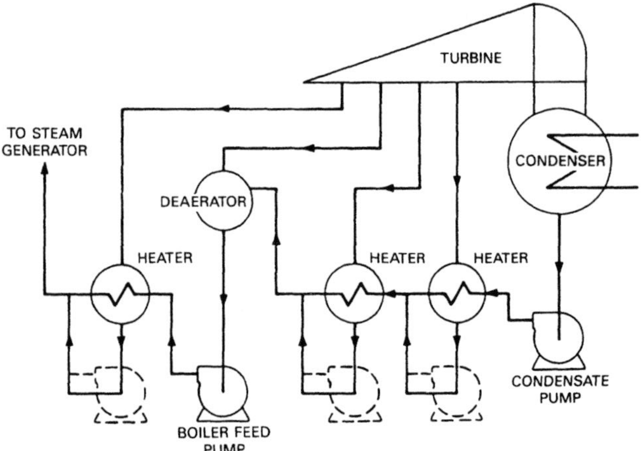

Fig. 9-28. Heater drain cycle with drains injected downstream of heater.

stream of feedwater or condensate. This cycle utilizes the differential pressure existing between the two heaters to force the drains into the lower pressure heater. A drain subcooler in each heat exchanger further increases the efficiency of this cycle. When the differential pressure between the lowest pressure heater and the condenser is very small and physical arrangements prevent use of gravity flow, it is common to employ a heater drain pump to pump the drains out of the lowest pressure heater.

Another variation is the external drain cooler. The reasons for using an external drain cooler rather than an integral subcooler zone are discussed under Turbine Overspeed Protection (Section 9.3.5.1). In this design, the lowest pressure heater does not contain an integral drain cooler. The drains from this heater are cascaded to the condenser, but go through an external drain cooler. This is a water-to-water heat exchanger with condensate in the tubes and heater drains in the shell.

Figure 9-13 shows a configuration for a small (26 MW) unit. It utilizes two low-pressure closed heaters, a deaerator, and one high-pressure closed heater. All three of the closed heaters contain condensing and subcooling zones. Terminal temperature difference between the outlet feedwater temperature and the steam saturation temperature is 5° F (5° F TD). Drain to feedwater approach temperature is 10° F (10° F DC). Drains from High-Pressure Heater 4 are cascaded to the deaerator. Drains from Low-Pressure Heater 2 are cascaded to Heater 1, and from Heater 1, to the condenser.

Figure 9-5 shows a configuration for a large (727 MW) high-pressure unit. The unit uses four low-pressure closed heaters, a deaerator, and two high-pressure closed heaters. High-Pressure Heaters 6 and 7 are three-zone heaters including desuperheating and subcooling zones. Terminal temperature difference for Heater 7 is 0° F(0° F TD) and −2° F (−2° F TD) for Heater 6. Low-Pressure Heaters 1 through 4 do not have desuperheating zones and have terminal temperature differences of 5° F (5° F TD). Heaters 2, 3, 4, 6, and 7 have internal drain coolers with drain to feedwater approach temperature of 10° F (10° F DC). Heater 1 has an external drain cooler with approach temperature of 5° F (5° F DC).

Drains from High-Pressure Heater 7 are cascaded to High-Pressure Heater 6, and from High-Pressure Heater 6 to the deaerator. Drains from Low-Pressure Heater 4 are cascaded to Low-Pressure Heater 3, from Low-Pressure Heater 3 to Low-Pressure Heater 2, from Low-Pressure Heater 2 to Low-Pressure Heater 1, from Low-Pressure Heater 1 through the drain cooler to the condenser.

9.3.3 Cycle/Heater Calculations

Cycle heat balance calculations determine the flows, temperatures, and pressures through the feedwater cycle. Heater calculations determine the design of the heater components such as tube surface, tube diameter, tube wall thickness, design temperatures, and pressures.

9.3.3.1 Cycle Calculations. Cycle heat balance calculations may be performed to establish a number of parameters such as heat rates for evaluation of various cycles: flows, pressures, and temperatures for heater calculations; and part load and full load performance.

The normal starting point for a cycle heat balance calculation is similar to the information shown in Fig. 9-29. Turbine performance data that would be known would include inlet steam flow, pressure, and temperature; extraction pressures and enthalpies; and condenser vacuum pressure. The heater arrangement would have been established, including the number of low-pressure and high-pressure heaters, location of the deaerator in the cycle, use of drain subcoolers and desuperheater zones, and the heater drain arrangement. If the purpose of the calculations is to evaluate various heater arrangements, each of these variations must be set up and separate calculations must be performed.

In Fig. 9-29, the turbine inlet flow and conditions are given, as well as the turbine extraction point pressure and enthalpies. The cycle is similar to that shown in Fig. 9-5, with high-pressure heaters, a deaerator, and four low-pressure heaters. The two high-pressure heaters have desuperheating zones designed for terminal temperature difference of 0° F for Heater 7 and −2° F for Heater 6. All the other low-pressure closed heaters are designed for a terminal temperature difference of 5° F. All six closed heaters have drain coolers and are designed with drain to inlet feedwater approach of 10° F.

Example 9-2

The first step in cycle calculations is to determine heater shell pressure. While the pressure of the extraction steam at the turbine is known, a pressure drop will occur in the extraction steam piping between the turbine and heater and through the desuperheating zone in the heater, if one is present. This pressure drop is a function of items such as pipe size, length, configuration, and number and type of valves. Usually, these values have not been determined at this stage of the design. However, experience has shown that this drop should be between 3% and 6% of the turbine extraction pressure. For this example in Fig. 9-29, a 5% drop is assumed. The heater shell pressures are then calculated as follows:

Heater 7	540.0 × 0.95	=	513.0 psia
Heater 6	280.0 × 0.95	=	266.0 psia
Deaerator	160.0 × 0.95	=	152.0 psia
Heater 4	90.0 × 0.95	=	85.5 psia
Heater 3	60.0 × 0.95	=	57.0 psia
Heater 2	15.0 × 0.95	=	14.7 psia
Heater 1	7.0 × 0.95	=	6.7 psia

A steam table can be used to determine the saturation temperature in the heater shells that corresponds to the shell pressure. By use of the established terminal temperature difference, the temperature and enthalpy of the feedwater exiting each heater can be determined. The second-stage cycle heat balance then contains the information as shown in Fig. 9-30. Note that the deaerator raises the inlet condensate to the saturation temperature.

Continuing the example, the temperature of the condensate (or

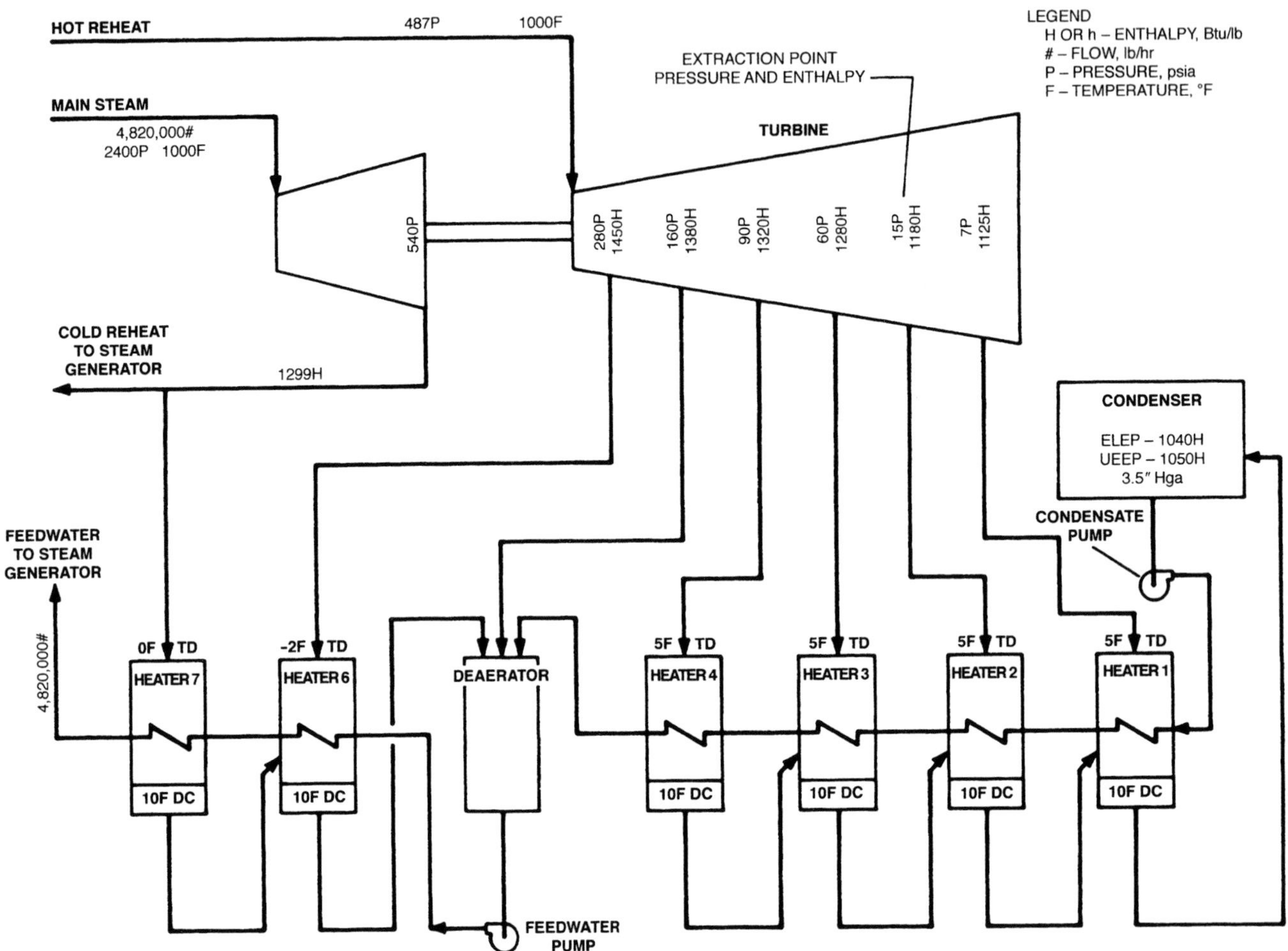

Fig. 9-29. Normal starting point information for a cycle heat balance.

feedwater) entering and leaving all the heaters is now known, except for that entering Heaters 1 and 6. Heater 1 condensate comes from the condenser, from that it leaves at the saturation temperature corresponding to the condenser pressure. The condensate pump does increase the condensate temperature, but this is minor and normally ignored. Other heat exchangers are sometimes located between the condenser and Heater 1, and if they are, the corrected temperature should be used in Heater 1 calculations. For this example, it is assumed there are no such heat exchangers.

The boiler feed (BF) pump also increases the boiler feedwater temperature and enthalpy as it passes through the pump. This temperature rise is enough that it should not be ignored, particularly on large high-pressure units. The rise in enthalpy through the pump can be estimated using the following formula:

$$\Delta h = v_f(144)(P_o - P_i) \div (778 \times E_p) \qquad (9\text{-}12)$$

where

Δh = increase in enthalpy, Btu/lb;

v_f = specific volume of liquid at pump inlet conditions, ft³/lb;

P_o = BF pump outlet pressure, psia;

P_i = BF pump inlet pressure, psia; and

E_p = efficiency of BF pump.

BF pump outlet pressure should be actual pump outlet pressure, or turbine inlet pressure plus all the pressure drops of the high pressure heaters, economizer, superheater, piping, controls, and so on. As these pressure drops are seldom known at this stage of plant design, the use of 125% of 5% overpressure throttle pressure is a reasonable estimate for the BF pump outlet pressure in the following example:

$$\Delta h = 0.01811 \times 144 \times (1.25 \times 2535 - 152) \div (778 \times 0.80) = 12.6$$

The temperature of the feedwater entering Heater 6 can be determined from its enthalpy using the compressed water portion of a standard steam table.

With the condensate or feedwater inlet temperature known for each of the heaters, the temperature of the drains from the drain coolers and their enthalpies can be established, as shown in the third-stage heat balance in Fig. 9-31. In this figure, Heater 7 heat balance is complete except for extraction flow and drain flow. As

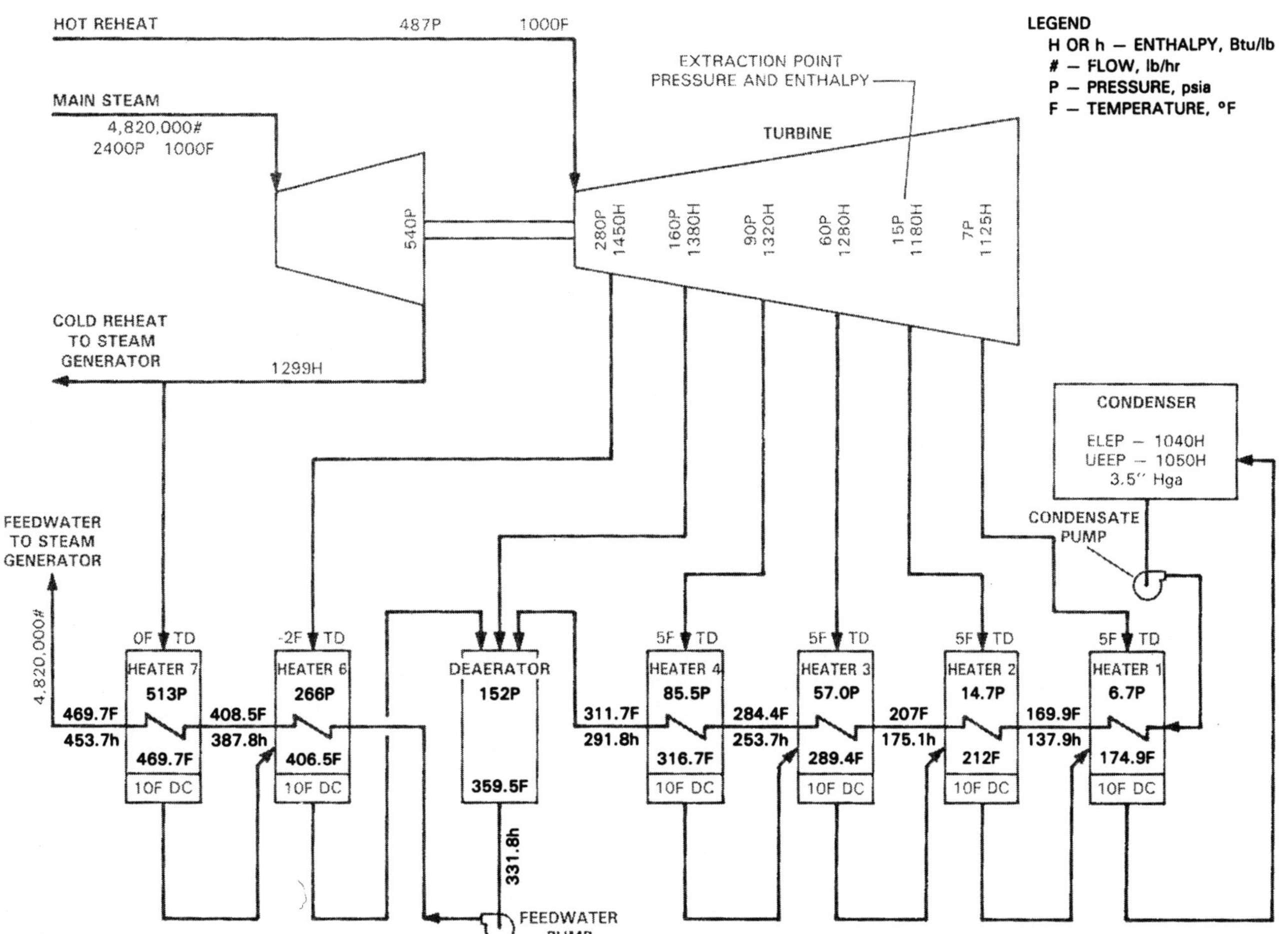

Fig. 9-30. Second stage information for a heat balance.

these two flows are equal, only one unknown remains. Using the conservation of energy theory, a heat balance can be performed on this heater and the extraction flow determined as follows:

Energy in = Energy out

Feedwater inlet flow × feedwater inlet enthalpy + extraction flow × extraction enthalpy = feedwater outlet flow × feedwater outlet enthalpy + drain flow × drain enthalpy

4,820,000 lb/h × 387.8 Btu/lb + extraction flow lb/h × 1,299 Btu/lb = 4,820,000 lb/h × 453.7 Btu/lb + drain flow lb/h × 395.2 Btu/lb

Since drain flow is the same as the extraction flow, exsubstituting extraction flow for drain flow, the equation can be solved for extraction flow:

Extraction flow = 351,400 lb/h

With the extraction and drain flows determined for Heater 7, a similar heat balance can be performed for Heater 6. In this case, drain flow from Heater 6 equals extraction flow to Heater 6 plus drains entering Heater 6 from Heater 7. Continuing the example as follows:

4,820,000 lb/h × 344.4 Btu/lb + 351,400 lb/h × 395.2 Btu/lb + extraction flow lb/h × 1,450 Btu/lb = 4,820,000 lb/h × 387.8 Btu/lb + (351,400 lb/h + extraction flow lb/h) × 350.3 Btu/lb

Extraction flow = 175,900 lb/h

Drain flow = 175,900 lb/h + 351,400 lb/h = 527,300 lb/h

Proceeding to the deaerator, the extraction flow and the condensate flow from Heater 4 are unknowns. However, by setting up an energy balance and a mass balance, two equations with two unknowns can be established that can be solved simultaneously giving both flows. The mass balance is as follows:

Mass in = Mass out

Condensate from Heater 4 lb/h + extraction steam lb/h + drains from Heater 6 lb/h = Feedwater out lb/h

Condensate lb/h + extraction steam lb/h + 527,300 lb/h = 4,820,000 lb/h

The energy balance is as follows:

Energy in = Energy out

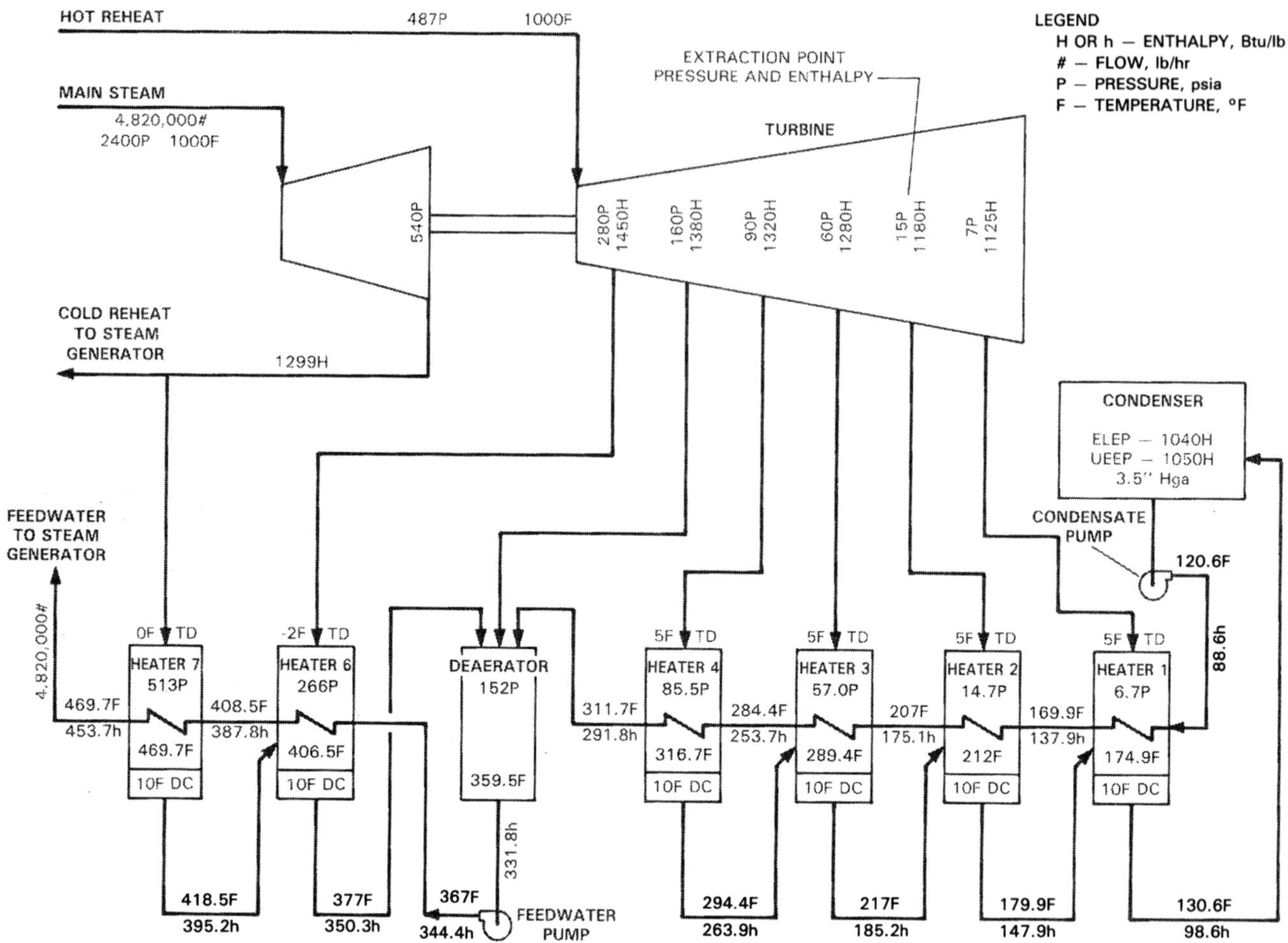

Fig. 9-31. Third stage information for a heat balance.

Condensate from Heater 4 lb/h × condensate enthalpy Btu/lb + extraction steam lb/h × extraction steam enthalpy Btu/lb + drains from Heater 6 lb/h × drain enthalpy Btu/lb = Feedwater out lb/h × Feedwater enthalpy Btu/lb.

Condensate lb/h × 291.8 Btu/lb + extraction steam lb/h × 1,380 Btu/lb + 527,300 lb/h × 350.3 Btu/lb = 4,820,000 lb/h × 331.8 Btu/lb

Solving these two equations simultaneously gives the following:

Extraction flow = 148,900 lb/h

Condensate flow from Heater 4 = 4,143,800 lb/h

Once the condensate flow is known, extraction flow can be calculated for Heaters 4, 3, 2, and 1.

Based on extraction flows to each heater, the resulting steam flow through the turbine from extraction point to extraction point can then be calculated. The turbine output can then be calculated as described in Chapter 8, Steam Turbine.

The cycle heat rate is the thermal energy input to the cycle divided by the electric output. The heat rate then is as follows:

([Turbine inlet steam flow lb/h] × [turbine inlet steam enthalpy Btu/lb − final feedwater enthalpy Btu/lb] + [reheater steam flow lb/h] × [hot reheat enthalpy Btu/lb − cold reheat enthalpy Btu/lb]) ÷ (turbine kW output).

The complete heat balance and net heat rate are shown in Fig. 9-32.

9.3.3.2 Heater Calculations. Once a heat balance has been completed, the design of the heater, including heater tube surface, can be performed. The determination of required surface for a single zone, condensing heater is very similar to the procedure used for determining required surface for the steam condenser, as discussed earlier in this chapter. The basic equation remains

$$A = \frac{Q}{U(\text{LMTD})} \tag{9-13}$$

where

A = required surface area (ft^2), (m^2);

Q = heat exchanged per hour (Btu/h), (kJ/s);

U = heat transfer coefficient $\left(\frac{\text{Btu}}{\text{ft}^2\text{h}F}\right)\left(\frac{\text{kJ}}{\text{m}^2\text{s}C}\right)$, and

LMTD = log mean temperature difference (° F), (° C).

The Q and LMTD can be calculated from the heat balance. The heat transfer coefficient is a function of the tube material, tube size and thickness, water velocity, and heat transfer coefficients. After selection of the tube material, an analysis similar to that for the condenser design can be performed to establish tube size, velocity, surface, number, and length. The desuperheating zone and the drain cooler zone surfaces can be computed in a similar manner, considering them as separate heat exchangers with their related Q, U, and LMTD. Once the condensing section is designed and tube material, tube size, gauge and water velocity are established, these must be used in the design of the desuperheater and drain cooler sections.

Because the physical design of the interior of a closed feedwater heater affects the thermal design, it is normal to leave the physical design of the heater to the heater manufacturer, with the plant designer specifying the design pressures, temperatures, flows, tube material, and maximum overall length. The manufacturer then determines the physical arrangement, tube size, tube wall thickness and length, and surfaces for each zone.

9.3.4 Physical Arrangement in Plant

To help protect against flashing and cavitation at the boiler feed pump, the deaerator is normally located at a high point in the plant. This may be in the auxiliary bay or elsewhere in the boiler area where space and supporting steel are available. The boiler feed pump suction piping from the deaerator outlet should be as short as practical, and a deaerator location approximately above the pumps is desirable.

The physical size of the deaerator makes installation difficult after the building is enclosed, so the deaerator is nor-

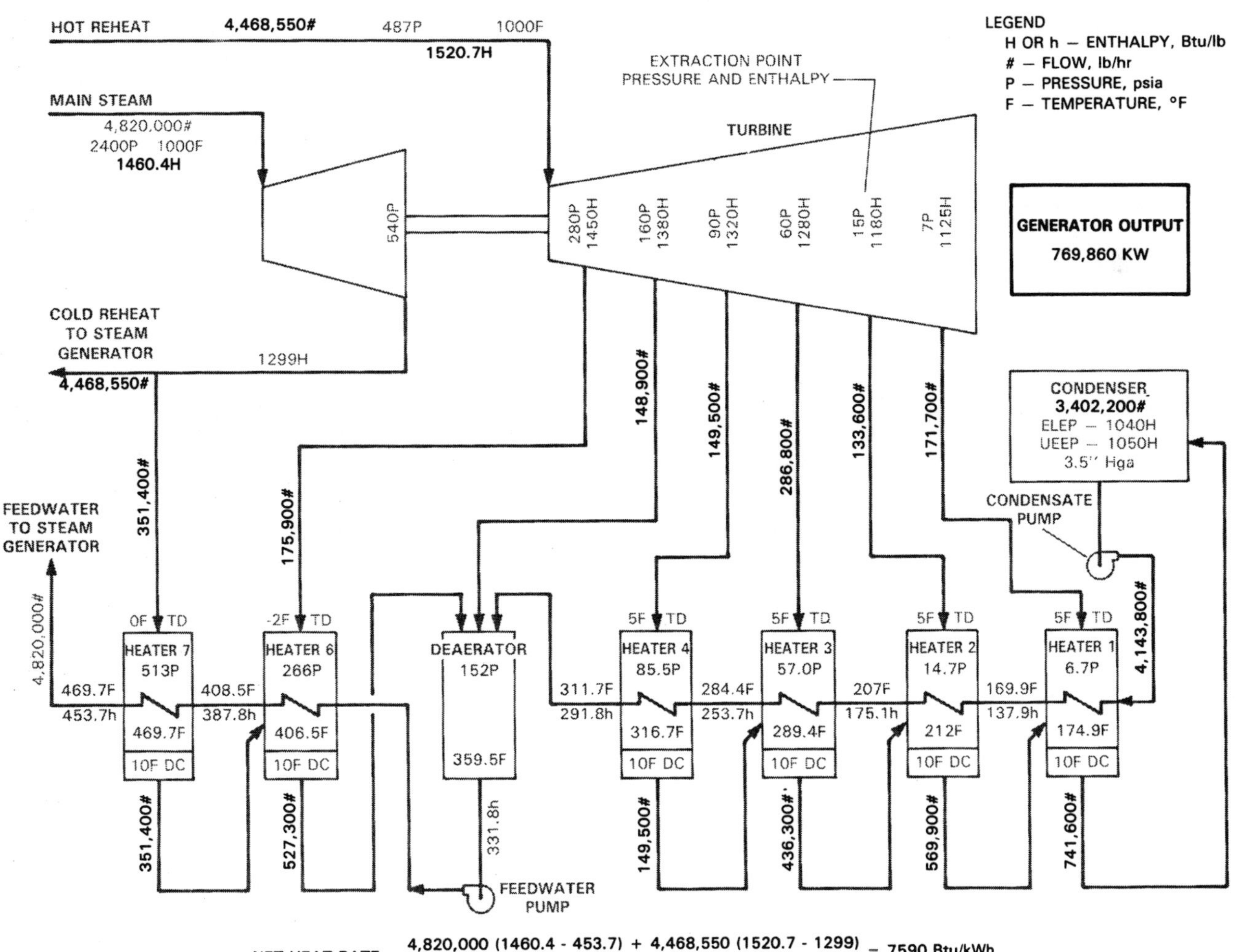

Fig. 9-32. Complete heat balance and net heat rate.

mally erected during construction of the building or boiler steel. After erection, maintenance is usually performed only on internal parts of the deaerator, that are nominal in size and weight.

The closed feedwater heaters can be located in the turbine room, auxiliary bay, or boiler bay to minimize piping expense, considering all piping including extraction steam, condensate, feedwater, and drains. Other considerations are pressure drops in piping systems and maintenance space, including space for shell or bundle removal. Differential pressure available for cascading drains from heater to heater should be considered in determining the location of heaters relative to their next lower heater. This is particularly important for low-pressure heaters. During part load operation, the differential pressure between the heater and the condenser is very low, making it difficult to cascade the drains to the condenser without the assistance from static head. Lacking static head, a drain pump may be required. As previously indicated, the lowest pressure heaters on large units are frequently located in the neck of the condenser. On occasion, both the lowest pressure and next to lowest pressure heaters are so located.

Closed heaters are a relatively large piece of equipment normally erected during building erection. Replacement of a tube bundle is rare but not unheard of. Thus, provisions for tube bundle replacement should be considered.

9.3.5 Piping Systems

Design of the connecting piping systems for the deaerator and feedwater heaters involves resolution of many unique problems, such as turbine overspeed protection, turbine water induction protection, and boiler feed pump cavitation protection.

9.3.5.1 Turbine Overspeed Protection. Turbines have emergency governors that close off the inlet steam supply when an overspeed is detected, protecting them from damage. However, there is another potential source of steam to a turbine that has been operating normally and suddenly loses all or most of this load. The following scenario describes this source:

- The turbine is at a normal load, all heaters are in service and contain saturated drains at the bottom of the shell at a temperature corresponding to the extraction pressure.
- Upon sudden loss of load, the steam governor valves on the turbine close, and the pressures throughout the turbine, including the pressure at each extraction point, drop to, or near, condenser pressure.
- The pressure in the heaters drops to near condenser pressure. Part of the saturated drains in the bottom of the heater shell flash to steam, that flows backward through the extraction pipe into the turbine and continues through the turbine to the condenser, driving the turbine.
- If this flow is considerable, it can overspeed the turbine, causing damage. The amount of steam formed by this flashing action depends on the pressure, temperature, and quantity of the heater drains.

The highest-pressure heater will have drains at the highest pressure and temperature, but the quantity of drains in the shell is usually small. The deaerator, however, has a high quantity of water in its storage section, and is actually the possible source of the largest quantity of flashed steam.

To prevent this backward flow of steam from the heaters or deaerater to the turbine, a check valve is installed in each extraction steam line. As check valves can malfunction and fail to close on reverse flow, redundant valves in series are in the extraction line to the deaerator. In addition, these valves are provided with an electric-actuated, compressed-air piston that assists the valve in closing. These pistons are actuated by the turbine emergency governor when it senses overspeed conditions.

Low-pressure heaters in the neck of the condenser have short, direct piping from the turbine to the heater shell, all within the condenser shell. Check valves in these lines are not practical. As the pressure and temperature of these drains are very low, the amount of flash steam can normally be held to an acceptable quantity by controlling the quantity of drains maintained in the heater shell. The use of an external drain cooler in this arrangement can also help minimize the number of drains maintained in the heater shell.

The turbine manufacturer provides the designer with the maximum quantity of flash steam permissible at the various extraction points.

9.3.5.2 Turbine Water Protection. The cause of most turbine accidents is water entering the steam passages of the turbine. One of the sources for such water is a backflow through the extraction steam lines from the heaters or deaerator. This water can come from a failure of the heater drain system, a flooding of the heater shell by broken tubes in the heater, or even improper draining of the extraction piping prior to placing the heater into service.

The problems of water damage to turbines at one time became so acute that American National Standards Institute and the American Society of Mechanical Engineers developed and published ANSI/ASME Standard TDP-1, *Prevention of Water Damage to Steam Turbines Used for Electric Power Generation* (1985). This standard recommends that the feedwater heaters and extraction systems be designed so that "no single failure of equipment should result in water entering the turbine." Sections 3.7 and 3.8 of Standard TDP-1 cover the recommendations for Feedwater Heaters and Extraction Systems and Direct Contact (DC) Feed-water Heaters (deaerators). As with all standards, TDP-1 may be revised occasionally, so the designer should be certain to use the current standard for designing the extraction, heater drain, and feedwater systems.

Two items of equipment required by this standard are automatic shutoff valves in the extraction line from the turbine to the feedwater heater, and a primary and emergency drain line for the heater drains.

The automatic shutoff valve in the extraction steam line normally is an electric operated valve interlocked to close on high water level in the heater. For heaters located in the neck of the condenser where automatic valves cannot be installed in the extraction piping, an alternate solution is to add automatic stop valves on both inlet and outlet condensate connections that close on high water level and prevent flooding of the heater due to a tube rupture.

Primary and emergency drain lines are installed on heaters. The emergency drain line usually takes the heater drains directly to the condenser and functions during emergencies when the water level of the heater rises above the normal level.

9.3.5.3 Boiler Feed Pump Cavitation Protection. Much has been written about boiler feed pumps and the cavitation problems they experience by flashing of the inlet feedwater upon sudden unit load reduction.

Several actions can be taken to alleviate this problem, including raising the location of the deaerator and increasing the storage capacity in the deaerator storage section. These requirements and the method of designing the various components are discussed in detail in Chapter 11, Pumps.

9.3.6 Operation

This chapter has discussed design andoperation of the feedwater heaters at design conditions, but in actual operation the units must also perform at maximum turbine overload, startup low-load, and with one or more heaters out of service. Comprehensive design of the cycle requires consideration of each of these conditions.

9.3.6.1 Maximum Turbine Overload. Turbine manufacturers normally design their units about 5% larger than the guaranteed capacity. The turbine can be operated at this higher design load. The manufacturer may also permit operation with a 5% increase in throttle inlet pressure, that increases unit capacity another 5%. Thus, maximum load on the turbine may be 10% greater than guaranteed. At this condition, the extraction pressure and the feedwater flow will increase, increasing extraction steam flow, heater drain flow, and exit feedwater temperature. Generally, this mode of operation is not considered normal. The closed heater surfaces are designed thermodynamically for the guaranteed normal load, but mechanically for this maximum overload. This results in a slight loss of efficiency at this overload condition, but this is not considered serious. However, the heater shell pressure and temperature, and the tube side pressure and temperature must be designed for this maximum service. The extraction steam line normally handles the extraction flow with additional pressure drop. The feedwater heater and drain system must be designed for the higher drain flow at the maximum overload condition. The increased pressure drop across the drain subcooler zone must be considered. Calculations can be performed to determine drain flow under these conditions, although a reasonable estimate considers drain flow change as directly proportional to feedwater flow change.

9.3.6.2 Heater Out of Service. The bypassing of a heater or series of heaters causes the feedwater to enter the next higher operating heater at a much lower temperature. This results in a greatly increased extraction steam and drain flow for this heater. To determine extraction steam flow, a heat balance can be performed on this heater using the new inlet feedwater temperature and the same feedwater exit temperature. Although actual feedwater exit temperature will be somewhat less than design, this assumption gives a conservative estimate of extraction steam flow under these conditions. As is the case with turbine overload, the main concern to the designer is the ability to handle the heater drains. With the heater immediately below out of service, the drains cannot be cascaded to that heater. Usually, the emergency drain, required for turbine water induction protection, is relied on to handle these drains. To alleviate the high-pressure drop through the subcooler zone created by this large drain flow, it is common to install the emergency drain connection on the bottom of the heater shell, up-stream of the entrance to the subcooler, as shown in Fig. 9-21. This lowers the heater performance, but is not considered critical under this abnormal operating condition.

9.3.6.3 Startup Low-Load. Heater shell pressure and extraction flow decline with turbine load reduction, approximately directly proportional to load. The differential pressure between heaters used to cascade the drains from heater to heater also declines with turbine load. This can create a situation in that the most stringent design conditions for a system occur at some load other than full load. This condition normally occurs when the lower-pressure heater is positioned higher physically in the plant than the higher-pressure heater.

Such a condition almost always exists between the lowest-pressure high-pressure heater and the deaerator. This condition may suggest that the primary drain or an alternate drain be routed into the highest pressure low-pressure heater.

The designer must review the various operating conditions from startup to maximum load, including the possibility of heaters being out of service, to ensure that the systems perform safely and properly under all conditions.

9.4 REFERENCES

ANSI/ASME. 1983. *Code on Steam Condensing Apparatus Performance Test.* PTC 12.2, Section 7. American National Standards Institute and American Society of Mechanical Engineers.

ANSI/ASME. 1985. *Prevention of Water Damage to Steam Turbines Used for Electric Power Generation, Standard TDP-1.* American National Standards Institute and American Society of Mechanical Engineers.

GENERAL ELECTRIC COMPANY. 1 River Road, Schenectady, NY 12345.

HEI. 1992. *Standards for Closed Feedwater Heaters*, 5th edit. Heat Exchange Institute, Inc., Cleveland, OH.

HEI. 1992. *Standards and Typical Specifications for Deaerators*, 5th edit. Heat Exchange Institute, Inc., Cleveland, OH.

HEI. 1984. *Standards for Steam Surface Condensers*, 8th edit. Heat Exchange Institute, Inc., Cleveland, OH.

10

FANS

Kris A. Gamble

Fans are provided throughout the steam electric generating unit to supply or exhaust air or flue gas to meet the needs of a variety of systems. In addition, fans are used for building heating, ventilation and cooling, equipment pressurization to prevent contamination due to inleakage, and cooling for a wide variety of equipment from lubricating oil coolers to mechanical draft cooling towers. This chapter discusses primarily the specific fan requirements for fossil fuel combustion air and flue gas exhaust, but the general considerations presented for fan performance and operating characteristics are also representative for other fan applications.

Fans are provided to control the flow of air or gas through equipment or through a series of components interconnected with a system of ductwork. The fan increases the pressure of a flow stream to offset the pressure losses that result from system resistance.

10.1 TYPES OF FANS

Draft fans are classified as either centrifugal or axial, according to the direction of air or gas flow through the fans. Centrifugal fans move air or gas perpendicular to the impeller shaft. Axial fans move air or gas parallel to the impeller shaft.

10.1.1 Centrifugal Fans

Centrifugal fan blades are mounted in an impeller (or rotor) that rotates within a spiral housing. Figures 10-1 and 10-2 illustrate the construction and components of centrifugal fans. Centrifugal fans can be designed with either one inlet, called single-width single-inlet (SWSI) or with two inlets, called double-width double-inlet (DWDI).

The performance characteristics of centrifugal fans are highly dependent on the type of blade used. Basically, three types of blades are used on centrifugal fans in power plant applications: backward curved, straight, and radial tip.

These types of blades, along with their performance curves, are shown in Fig. 10-3.

10.1.1.1 Backward Curved Blades. Backward curved blades are normally the airfoil type, as shown in Fig. 10-3. The airfoil blade is most widely used for power plant applications and has several advantages over other types of blades:

- Highest efficiency, over 90%,
- Very stable operation,
- Low noise level,
- Ideal capabilities for high-speed service, and
- Nonoverloading horsepower characteristic.

The rising cost of auxiliary power makes the high efficiency of airfoil-bladed centrifugal fans attractive for most power plant applications. Airfoil blades are used for almost all forced draft and induced draft centrifugal fan applications. The use of backward curved blades should be avoided on applications where significant levels of large or adhesive particles are present.

10.1.1.2 Straight Blades. Straight blades, sometimes called radial blades, have limited applications for use in power plants, but they offer several operating advantages:

- Good abrasion resistance;
- Simplified maintenance, particularly blade replacement; and
- Wide range of capacities.

The disadvantage of straight blades is the relatively low operating efficiency and an overloading horsepower characteristic. This characteristic is illustrated in Fig. 10-3. The power requirements at maximum flow conditions (runout) are significantly higher than power requests during operation at the point of highest efficiency.

10.1.1.3 Radial Tip Blades. Radial tip blades are normally used for moderately erosive gas applications. Radial tip blades have several advantageous operating characteristics:

- Nonoverloading horsepower,
- High capacity for size,
- Excellent abrasion resistance,
- Very stable operation, and
- Essentially self cleaning capabilities.

The disadvantage of using radial tip blades is that efficiency is not as high as that of backward curved blades.

10.1.2 Axial Fans

Single-stage (one rotor and one set of blades) axial fans (Fig. 10-4) are typically used in forced draft service on a balanced draft steam generator. When axial fans are designed for induced draft service, the higher pressure requirements nor-

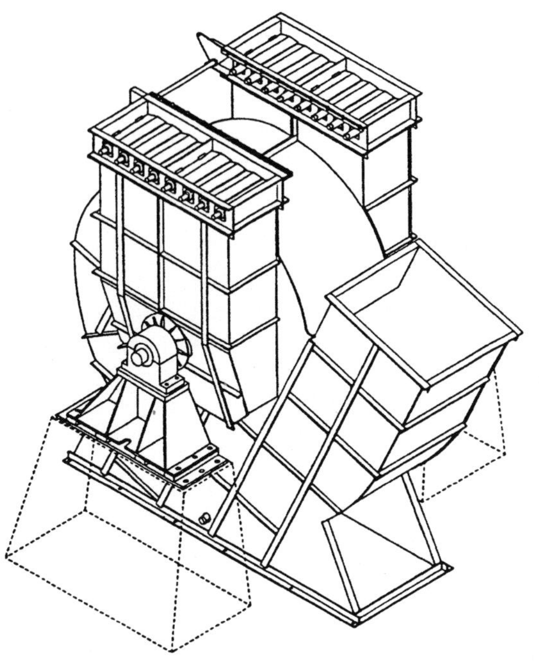

Fig. 10-1. Double-width, double-inlet centrifugal fan.

mally dictate the use of a two-stage fan. Figure 10-5 shows a typical two-stage axial fan. Axial fans are driven by single-speed or two-speed motors. The flow and pressure output is controlled by adjusting the pitch of the fan blades. These are called variable pitch blades. A typical axial fan performance curve is shown in Fig. 10-6.

Axial fans can maintain higher efficiencies at various steam generator loads than can constant-speed centrifugal fans controlled with inlet dampers or vanes. As fuel costs continue to rise, the higher capital cost of axial fans over centrifugal fans frequently can be offset by operating cost savings over the life of the plant. However, a detailed evaluation is required to show what type of fan and drive system should be used in a given situation.

Either a mechanical or a hydraulic mechanism is used to adjust blade pitch while the fan operates at the design speed. Mechanical pitch change mechanisms, however, are usually insufficient to properly control fans of the size required for power plant applications.

Unlike centrifugal fans that consist of one moving part, the fan rotor weldment, axial fan rotors contain many moving parts. A rotating union in the blade pitch control mecha-

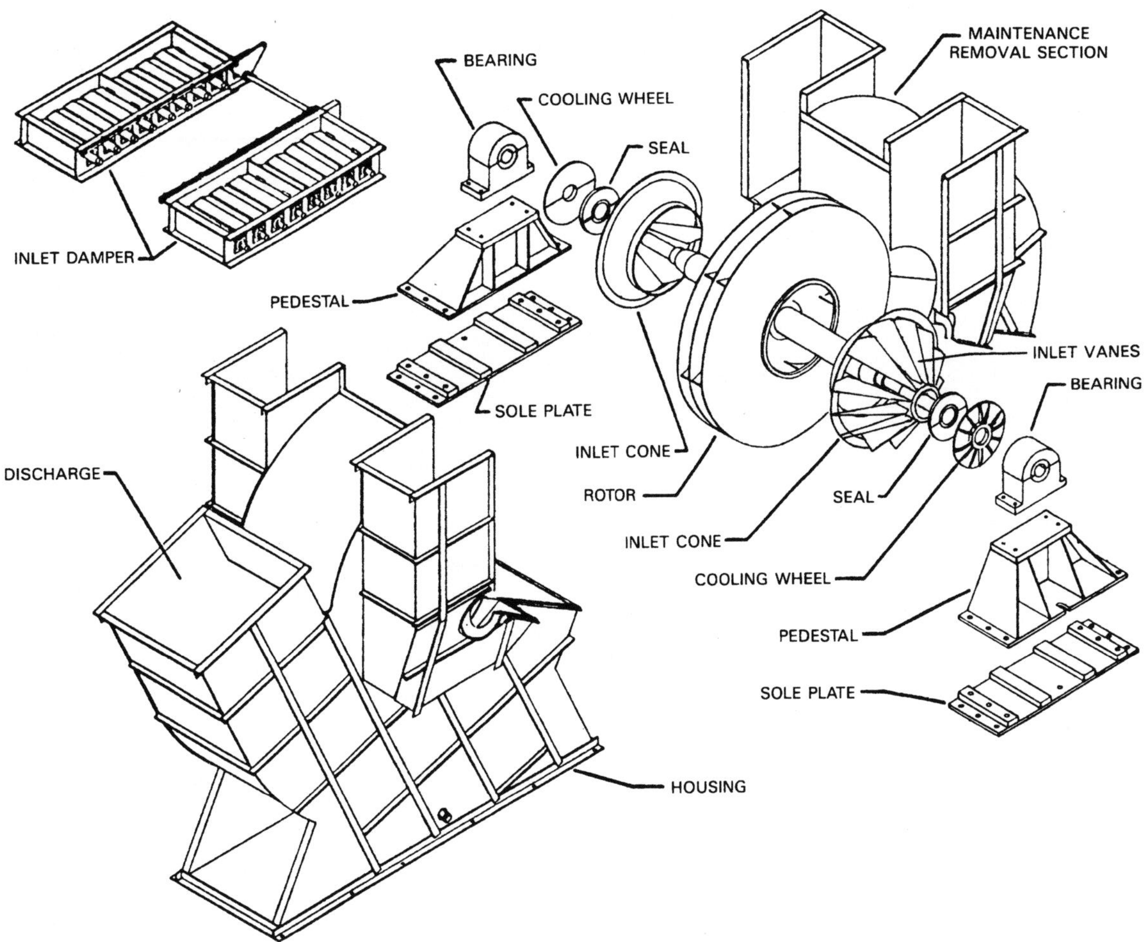

Fig. 10-2. Centrifugal fan components.

FAN BLADE TYPES

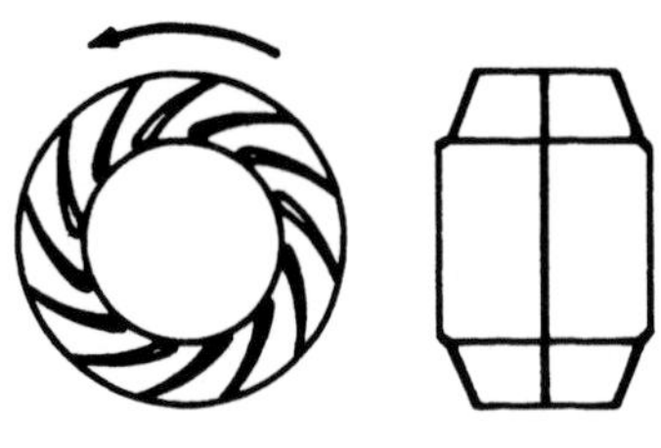

(a) BACKWARD INCLINED (Airfoil)

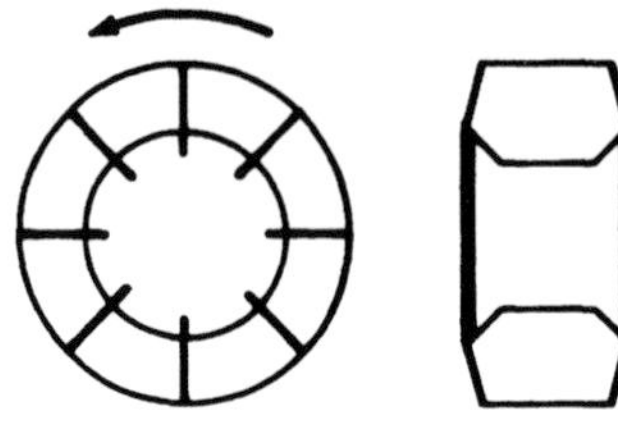

(c) FORWARD CURVED BACKWARD INCLINED

PERFORMANCE CURVES

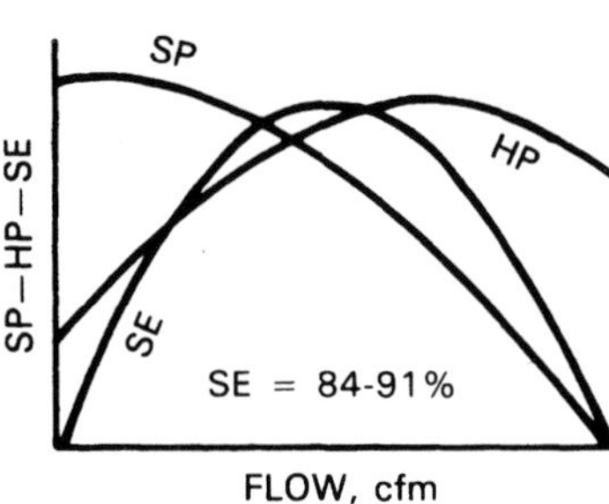

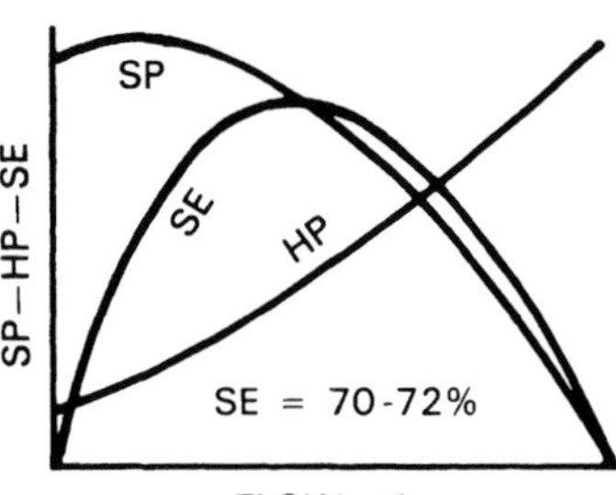

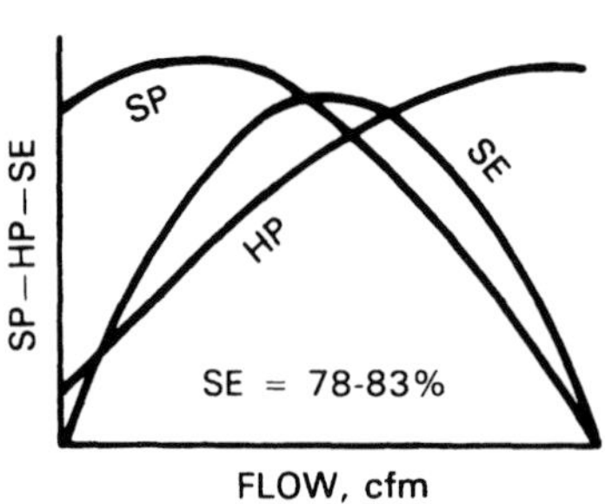

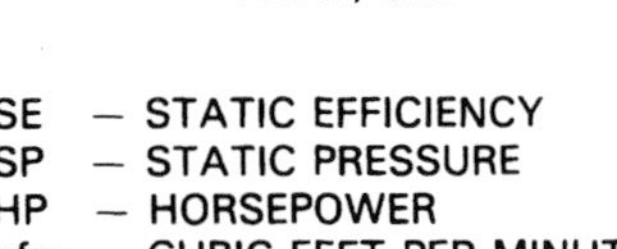

Fig. 10-3. Fan blade types and performance curves. (From The Green Fan Company. Used with permission.)

nism enables hydraulic control oil flow between the stationary oil pump and the blade pitch control mechanism. The blade pitch regulating disk inside the fan hub moves in response to hydraulic oil pressure and is connected to each blade crank arm. Blade pitch regulating disk movement enables blade pitch adjustment during operation. The fan blades are supported on the fan hub with antifriction thrust bearings.

The use of lightweight fan blade materials such as aluminum or magnesium greatly reduces the fan hub strength requirements and the fan blade thrust bearing loads. By reducing the hub strength and bearing thrust load requirements, the axial fan cost can be significantly reduced. The disadvantage of the lightweight fan blade materials is the lower erosion resistance of aluminum or magnesium com-

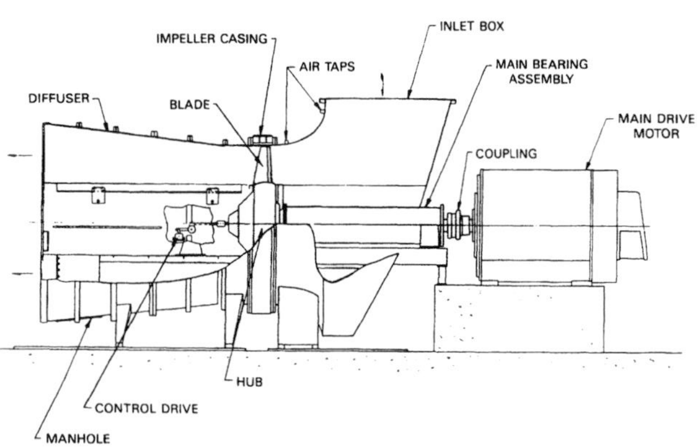

Fig. 10-4. Single-stage axial fan. (From Novenco, Inc. Used with permission.)

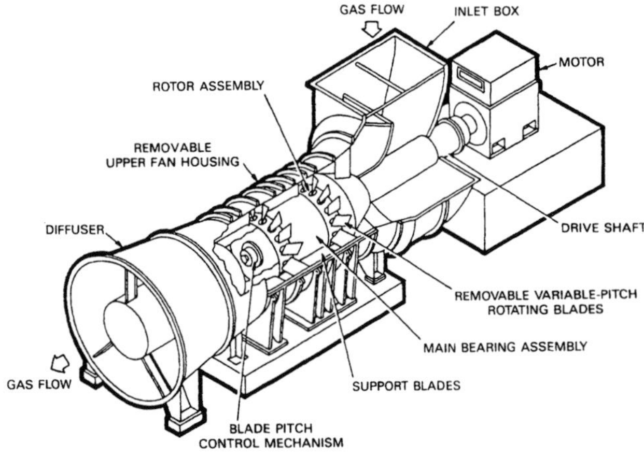

Fig. 10-5. Two-stage axial fan. (Reprinted from *Combustion Fossil Power Systems*, 3rd edition. Joseph Singer, editor, with permission from Combustion Engineering, Inc.)

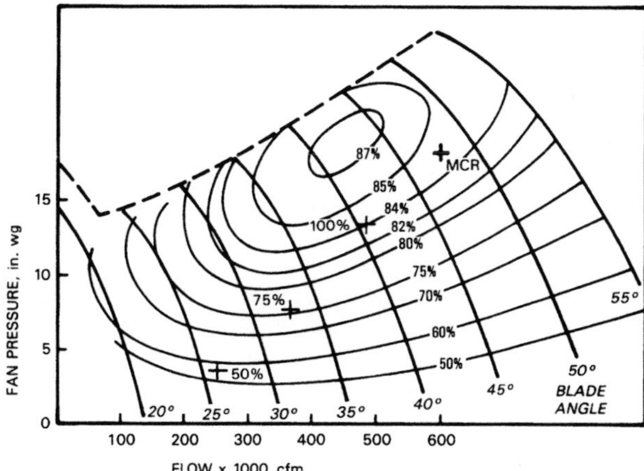

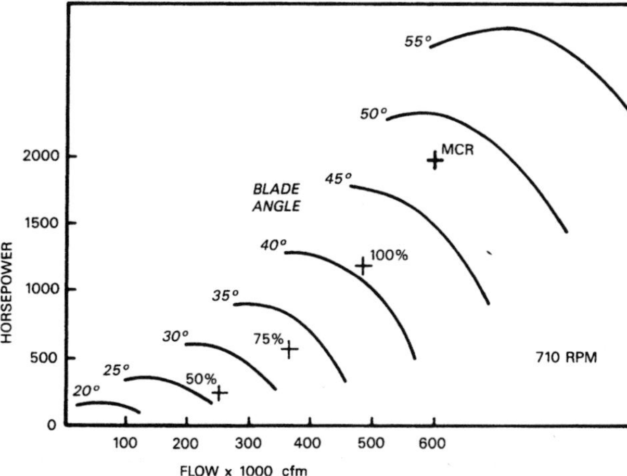

Fig. 10-6. Typical axial fan performance curve.

pared to steel. Both types of blades can be protected with erosion-resistant coatings. In applications where the fans are handling particulate laden gas, aluminum or magnesium blades are also protected by a hardened steel strip on the blade leading edge.

Because of the additional complexity of the adjustable pitch axial fans compared to centrifugal fans, axial fans require more frequent maintenance than centrifugal fans of equivalent capacity and must be maintained at a higher level of balance due to lower tolerance for operating vibration. When used in induced draft applications, axial fans are more sensitive than centrifugal fans to performance degradation caused by fan blade erosion from fly ash and thrust bearing failure due to contamination with fly ash and corrosive flue gas components.

Axial fans used in particulate free applications such as forced draft fan service and in induced draft fan service downstream of high-efficiency particulate removal equipment have exhibited reliable and efficient operating performance. In non-corrosive and particulate-free applications, the use of aluminum fan blades reduces capital costs compared to more corrosion or erosion-resistant blade materials.

The variable pitch axial fan design was developed in Europe and used extensively prior to its introduction in the United States during the 1970s. Since that time, variable pitch axial fans have been used regularly in United States power generating facilities primarily in clean air, forced draft fan applications. With the enforcement of more stringent emission control requirements that will result in lower particulate emission levels, axial fans are expected to be used more often in induced draft applications.

10.2 SYSTEM RESISTANCE

When a gas is forced through a duct system, a loss in pressure occurs. This loss in pressure is called system resistance. System resistance is composed of two components: friction losses and dynamic losses. Friction losses occur at the walls of the duct system and can be quantified by the following equation:

$$\Delta P \text{ (friction)} = \frac{f\rho v^2 L}{D g_c} \qquad (10\text{-}1)$$

where

ΔP = frictional pressure loss, lb/ft^2;

f = a dimensionless friction factor that is a function of relative roughness and Reynolds number;

ρ = gas density, lb/ft^3;

V = gas velocity, ft/s;

L = duct length, ft;

D = duct diameter, ft; and

g_c = unitary constant, 32.2 $\dfrac{\text{lbm}}{\text{lbf}}\dfrac{\text{ft}}{\text{s}^2}$.

Dynamic losses occur at changes of direction in gas flow and at sudden duct enlargements and contractions. Dynamic losses can also be called velocity pressure losses. Provided that the gas flow is in the turbulent range, as is the case for nearly all combustion air and flue gas handling systems and equipment, dynamic losses can be quantified by the following equation:

$$\Delta P \text{ (dynamic)} = \tfrac{1}{2}\frac{\rho K V^2}{g_c} \qquad (10\text{-}2)$$

where

ΔP = pressure loss, lb/ft^2;

ρ = gas density, lb/ft^3;

v = gas velocity, ft/s;

g_c = unitary constant, 32.2 $\dfrac{\text{lbm}}{\text{lbf}}\dfrac{\text{ft}}{\text{s}^2}$; and

K = system constant based on the geometry of the duct system most accurately determined by actual testing.

By adding the friction losses to the dynamic losses, the total system resistance can be determined.

Once the system resistance is determined for one set of flow conditions, gas density, and average duct velocity (or volumetric flow rate), the system resistance can be predicted for any other flow rate or gas density. As shown in the friction and dynamic loss equations, when the system geometry is not modified, that is, when f, L, D, g_c, and K are held constant, only changes in gas density and flow rate cause the system resistance to change. As shown by the equations, system resistance changes directly with changes in gas density and is directly proportional to the square of volumetric flow changes.

Once the pressure test has been made at a known gas flow rate and density, and as long as the system geometry does not change, the system resistance can be predicted for any flow condition.

Example 10-1

A duct system with multiple elbows, dampers, sudden enlargements, and contractions has been tested to have a system resistance of 2 in. of water gauge (in. wg) when the flow through the system is 50,000 actual cubic feet per minute (acfm) and the gas density is 0.075 lb/ft³.

How does this system resistance change as the flow rate changes from 50,000 to 75,000 acfm with gas densities of 0.075 and 0.06 lb/ft³?

Solution:

Refer to Eq. (10-2) to calculate the effects of flow and density on system resistance. With constant density (0.075 lb/ft³), as the system flow changes from 50,000 acfm to 75,000 acfm, the system resistance will increase from 2 in. wg to 4.5 in. wg (2.0 in. wg × $\left(\frac{75{,}000 \text{ acfm}}{5{,}000 \text{ acfm}}\right)^2$ = 4.5 in. wg). With constant volumetric flow (50,000 acfm), as the gas density changes from 0.075 lb/ft³ to 0.06 lb/ft³, the system resistance will be reduced from 2 in. wg to 1.6 in. wg (2 in. wg × $\left(\frac{0.060 \text{ lb/ft}^3}{0.075 \text{ lb/ft}^3}\right)$ = 1.6 in. wg).

System resistance curves are normally presented with static pressure (in. wg) on the y-axis and gas flow, acfm, on the x-axis. The effects of changing flow rates and gas densities are illustrated below in tabular form and also graphically in Fig. 10-7.

	System resistance, in. wg				
System flow, 1,000 acfm	0	25	50	75	100
Gas density, lb/ft³					
0.075	0	0.5	2	4.5	8.0
0.06	0	0.4	1.6	3.6	6.4
0.04	0	0.3	1.1	2.4	4.3

In some cases, it may be more useful to plot a system resistance curve as a function of mass flow rather than volumetric flow. In systems that include heat transfer equipment where the air or gas undergoes significant temperature change, a system resistance curve based on volumetric flow is applicable to only one location in the system. In this case, a system resistance curve based on mass flow is applicable to the entire system.

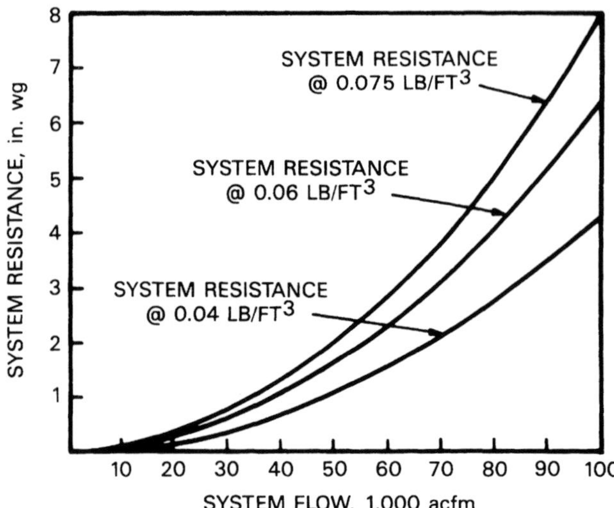

Fig. 10-7. System resistance.

An additional item to be considered with systems that undergo significant temperature change is that in predicting the effects of temperature change on total system resistance, the temperature (and density) change through each pressure drop component must be evaluated individually.

Example 10-2

A combustion air system consists of the following equipment, each component operating at its respective temperatures and pressure drops with a forced draft (FD) fan flow of 500,000 acfm:

	Temperature, °F		
	Entering	Leaving	ΔP, in. wg
FD fan inlet silencer	100	100	0.5
Ducts to air heater	110[a]	110	0.5
Air heater	110	700	5.0
Ducts to wind box	700	700	1.0
Wind box dampers	700	700	2.0
Burners	700	700	4.0
Total			13.0

[a]Assumes a 10° F rise across FD fans.

How does this system resistance change with a 60° F ambient temperature reduction?

Solution:

Assume the 60° F temperature reduction will occur through each component and the FD fan flow of 500,000 acfm will be maintained.

The component pressure drops will all be increased because of the higher air density. The density change will result in a linear increase in each component pressure drop. To calculate the change in pressure drop through the air heater, use the average $\frac{(T_{\text{in}} + T_{\text{out}})}{2}$ air heater temperature for both ambient temperature conditions. The density change will be equivalent to the inverse of the change in absolute temperature (R).

	Temperature, °F		
	Entering	Leaving	ΔP, in. wg
FD fan inlet silencer	40	40	$\left(\frac{T_2}{T_1}\right)\left(\frac{100 + 460}{40 + 460}\right) \times 0.5 = 0.56$
Ducts to air heater	50	50	$\left(\frac{T_2}{T_1}\right)\left(\frac{110 + 460}{50 + 460}\right) \times 0.5 = 0.56$
Air heater	50	640	$\left(\frac{T_2}{T_1}\right)\left(\frac{405^a + 460}{345^a + 460}\right) \times 5.0 = 5.37$
Ducts to wind box	640	640	$\left(\frac{T_2}{T_1}\right)\left(\frac{700 + 460}{640 + 460}\right) \times 1.0 = 1.06$
Wind box dampers	640	640	$\left(\frac{T_2}{T_1}\right)\left(\frac{700 + 460}{640 + 460}\right) \times 2.0 = 2.11$
Burners	640	640	$\left(\frac{T_2}{T_1}\right)\left(\frac{700 + 460}{640 + 460}\right) \times 4.0 = 4.22$
Total			13.88

[a]Average Air Heater Temperature $\frac{(T_{in} + T_{out})}{2}$

10.3 SYSTEM GEOMETRY CHANGE

In most cases, the total system resistance must be determined during the design phase of a project to enable timely construction and delivery of equipment. System component manufacturers are required to predict the equipment pressure drop data to enable the system designer to compile the total system resistance. Once all the component pressure drop data have been established, the designer can construct a system resistance curve and can also evaluate the impacts of changes in temperature and flow rate.

In some cases, the effects of system geometry changes can also be predicted. When parallel equipment trains are included in a system, which is often the case with air quality control equipment, the impact of taking equipment in and out of service can be predicted.

Example 10-3

An induced draft system consists of the steam generator, air heater, electrostatic precipitator, wet scrubber, chimney, and the interconnecting ductwork. Flue gas from the steam generator leaves the common air heater outlet plenum and flows to two parallel electrostatic precipitators and four parallel wet scrubber modules. Typical pressure loss data supplied by the component suppliers are tabulated below. A system resistance curve is shown in Fig. 10-8.

Component	Flue gas flow rate, 1,000 lb/h	Total pressure loss, in. wg
Steam generator	5,000	8
Air heater	5,000	6
Precipitator, each	2,500	2
Wet scrubber, per module	1,750	5
Chimney	5,000	2
Total		23

How does this total system resistance change if one precipitator and one scrubber module are removed from service and the total system flow at 5,000,000 lb/h is held constant?

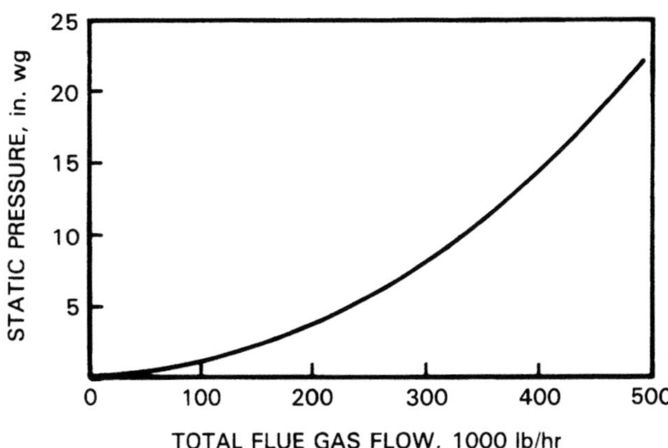

Fig. 10-8. System resistance curve (all equipment in service).

Solution:

Removal of the precipitator and scrubber modules from service will not impact the pressure drop through the steam generator and air heater. The flow through the remaining precipitator will double, increasing the precipitator pressure drop by a factor of $(2)^2$, or 4. Since only three of the scrubber modules will remain in service, the flow through each module will increase by a factor of $4/3$. This will increase the scrubber pressure drop by a factor of $(4/3)^2$ or 1.78.

The chimney flow will remain unchanged so the chimney pressure drop will not be affected.

The calculated pressure loss data for the revised operating conditions are tabulated below:

Component	Flue gas flow rate, 1,000 lb/h	Static pressure loss, in. wg
Steam generator	5,000	8
Air heater	5,000	6
Precipitator, each	5,000	8
Wet scrubber, per module	2,333	8.9
Chimney	5,000	2
Total		32.9

The effects of removing the precipitator and scrubber module from service are further illustrated on the revised system resistance curve in Fig. 10-9. The system resistance curve is constructed by plotting the point at 500,000 lb/h and 32.9 in. wg and constructing a curve with $\Delta p \alpha$ $(\text{lb/h})^2$. This example assumes the flue gas density through each component will not change significantly over the system flow range.

10.4 FAN PERFORMANCE

Fans are used to provide the pressure necessary to overcome system resistance. Fan performance characteristics are developed from test data and are typically illustrated on flow (acfm) vs. static pressure curves similar to those used to show system resistance. A typical fan performance curve for a centrifugal fan operating at a given speed and gas density is shown in Fig. 10-10.

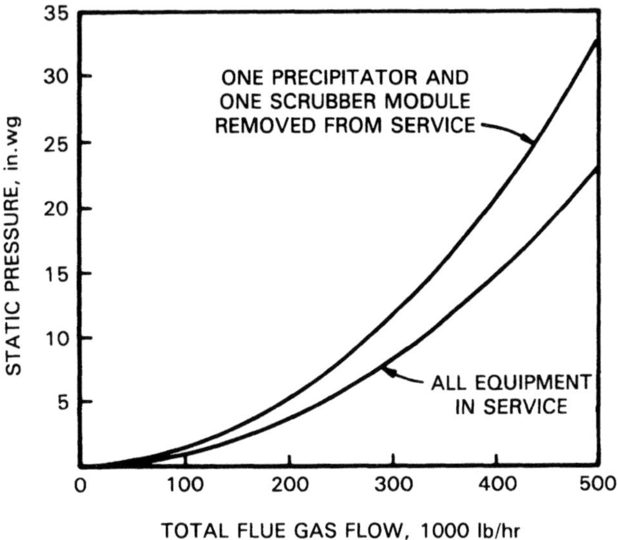

Fig. 10-9. System resistance curve (alternative equipment in service).

In addition to the static pressure capability, the fan power requirements are also shown on the fan performance curve.

Once the fan is installed into a ductwork system, the intersection of the system resistance curve and the fan characteristic curve defines the system operating point. By plotting the system resistance curve on the fan performance curve, this operating point (flow, pressure, and power requirement) can be determined (Fig. 10-11).

An alternative to showing fan brake horsepower (bhp) requirement is to show fan efficiency as a function of flow. A typical fan efficiency curve is shown in Fig. 10-12.

As shown on the efficiency curve, fan efficiency approaches zero when the fan operates at zero flow and shut-off head, and also approaches zero at maximum flow conditions where the static pressure approaches zero (fan runout conditions). Fan efficiency (η_{FAN}) is calculated from test data and is a ratio of the air horsepower (ahp) to the actual shaft bhp requirement.

$$\text{ahp} = \frac{\text{acfm} \times \Delta P \text{ in. wg}}{6{,}356} \tag{10-3}$$

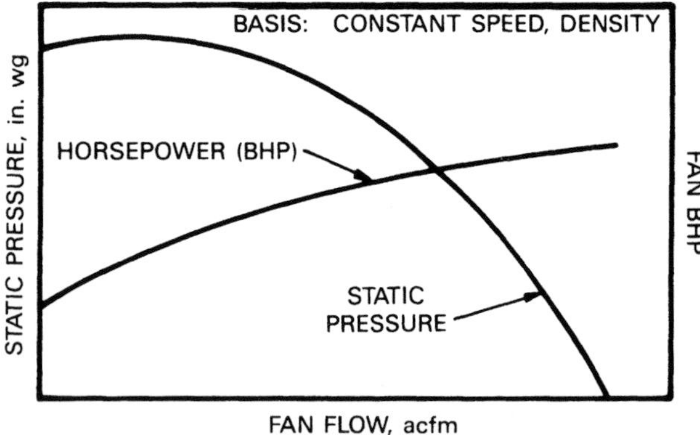

Fig. 10-10. Typical centrifugal fan performance curve.

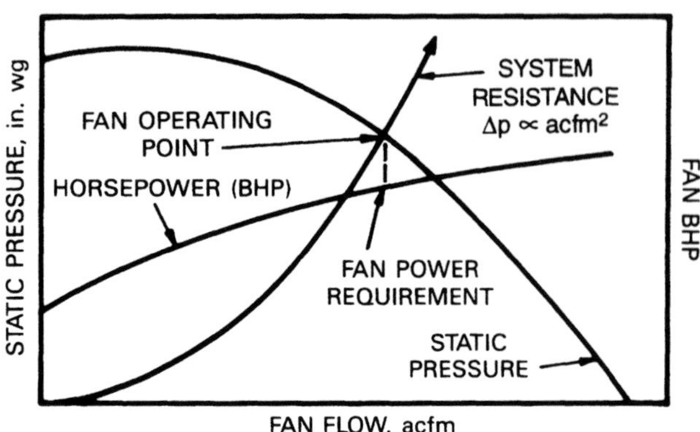

Fig. 10-11. Typical installed centrifugal fan performance curve.

$$\text{bhp} = \frac{\text{acfm} \times \Delta P \text{ in. wg}}{6{,}356 \times \eta_{FAN}} \tag{10-4}$$

$$\eta_{FAN} = \frac{\text{ahp}}{\text{bhp}} \tag{10-5}$$

Fan performance curves are developed by testing model fans. The results of model fan tests are used as a basis for determining the performance capabilities of full-size geometrically similar fans. Three basic relationships between fan size (SIZE), fan speed (RPM), and gas density (LB/FT^3) are the bases for predicting full-size fan performance. The three relationships are referred to as Fan Laws, described below for Fans A and B.

1. $$\text{CFM}_a = \text{CFM}_b \left(\frac{\text{SIZE}_a}{\text{SIZE}_b}\right)^3 \times \left(\frac{\text{RPM}_a}{\text{RPM}_b}\right) \tag{10-6}$$

2. $$\Delta P_a = \Delta P_b \left(\frac{\text{SIZE}_a}{\text{SIZE}_b}\right)^2 \times \left(\frac{\text{RPM}_a}{\text{RPM}_b}\right)^2 \times \left(\frac{\text{LB/FT}_a^3}{\text{LB/FT}_b^3}\right) \tag{10-7}$$

3. $$\text{HP}_a = \text{HP}_b \left(\frac{\text{SIZE}_a}{\text{SIZE}_b}\right)^5 \times \left(\frac{\text{RPM}_a}{\text{RPM}_b}\right)^3 \times \left(\frac{\text{LB/FT}_a^3}{\text{LB/FT}_b^3}\right) \tag{10-8}$$

Fan efficiency = $\eta_a = \eta_b$

Fan A is geometrically similar to Fan B.

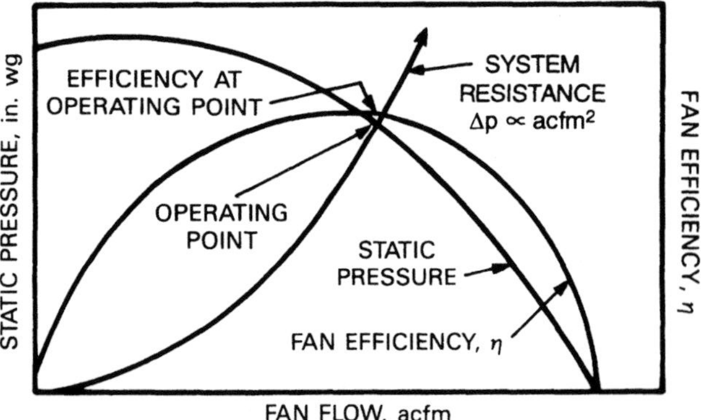

Fig. 10-12. Fan efficiency curve.

These fan laws apply to both centrifugal and axial flow fans and are based on incompressible flow. Correction factors due to the effects of compressibility, surface boundary conditions, and flow stream Reynolds Number can also be used to fine tune the use of fan laws to predict more accurately the performance of full-size fans from model fan performance. As a general rule, if correction factors are not used, the full-size fan performance capabilities of flow, pressure, and efficiency are understated by a relatively small amount.

As listed above, geometric similarity must be maintained between the model fan and full size fan. In addition, proper use of these fan law relationships (without the use of correction factors) will result in equal efficiency between Operating Point A (CFM_a, ΔP_a, HP_a) and Operating Point B (CFM_a, ΔP_b, HP_b).

10.4.1 Fan Law 1—Fan Flow

Fan Law 1 states that as the fan size increases, flow increases as the cube of the size changes and increases directly as rpm changes. A change in fan size consists of an equivalent change in each linear dimension. A 50% increase in fan size would result in a 50% addition to the diameter and width, as well as all fan component linear dimensions.

Consider the fan rotor to be a cylinder with diameter equal to the rotor diameter and length equal to the rotor width. Each rotation of the rotor produces a certain number of air changes in the fan rotor. The volume of each air change is equivalent to the volume of the rotor cylinder. As the volume of the rotor cylinder increases, so does the capacity of the fan.

The cylinder volume increases directly with a change in length and with the square of a diameter change. Since an equivalent change in each linear dimension (length and diameter) occurs with a fan size change for geometrically similar fans, the volume of the fan rotor increases as the cube of the size changes. The rotor volume or the volume of one complete air change through the rotor would increase by a factor of $(2)^3$ if the rotor diameter was double.

As the rpm of the fan rotor changes, the number of air changes for every revolution remains constant, but the number of air changes through the rotor per minute changes. If the fan speed changes, there will be a linear change in fan flow.

10.4.2 Fan Law 2—Fan Pressure

Fans generate pressure by accelerating gas flow through the inlet of the fan wheel and regaining this velocity pressure in the discharge of the fan housing. This velocity pressure regain in the fan housing is the opposite of the dynamic or velocity pressure loss component of the system resistance and can be quantified by the following equation.

$$P_{\text{velocity}} = \tfrac{1}{2}\frac{\rho V^2}{g_c} \tag{10-9}$$

where

P = lbf/ft^2;

ρ = gas density, lb/ft^3;

V = velocity, ft/s; and

g_c = unitary constant $\dfrac{\text{lbm}}{\text{lbf}}\dfrac{\text{ft}}{\text{s}^2} = 32.17$.

As the gas density increases, the velocity pressure increases linearly, and as the gas velocity increases, the velocity pressure increases with the square of the velocity increase.

A change in fan size changes the rotor tip speed. The flow velocity leaving the fan rotor is a direct function of tip speed. As shown in the velocity pressure equation (10-9), the pressure changes with the square of the velocity. In terms of a fan size, the pressure changes with the square of the size change. A change in fan rpm causes a linear change in rotor tip speed and a corresponding change in flow velocity leaving the fan rotor. This velocity change results in a change in velocity pressure by the square of the rpm change.

Unlike Fan Law 1, Fan Law 2 is dependent on gas density. Since a change in gas density directly affects velocity pressure, a density change also has a direct effect on fan pressure capability.

10.4.3 Fan Law 3—Fan Power

As defined in Eqs. (10-3) and (10-4), fan BHP is a product of the fan flow (acfm) and fan pressure (in. wg). Fan Law 3 (for BHP) is simply a product of Fan Law 1 for fan flow (acfm) and Fan Law 2 for fan pressure.

Fan Law 3 can be derived by multipying Eq. (10-6) by Eq. (10-7), noting that $HP_b = ACFM_b \times \Delta P_b$ results in the following:

$$HP_a = HP_b\left(\frac{SIZE_a}{SIZE_b}\right)^5 \times \left(\frac{RPM_a}{RPM_b}\right)^3 \times \left(\frac{LB/FT^3_a}{LB/FT^3_b}\right)$$

10.4.4 Fan Law Example 1

Convert model fan (b) performance to that of a full-size fan (a) with different speed and operating temperature as indicated below. Assume that the inlet pressure and gas molecular weight are the same for the model and full size fan.

	Model Fan (b)	Full-Size Fan (a)
Diameter, in.	20	80
rpm	1,200	900
Temperature	60° F (520° R)	320° F (780° R)

The model fan performance curve is shown in Fig. 10-13.

The ratio of model fan performance to that of the full-size fan is calculated by using the three fan laws:

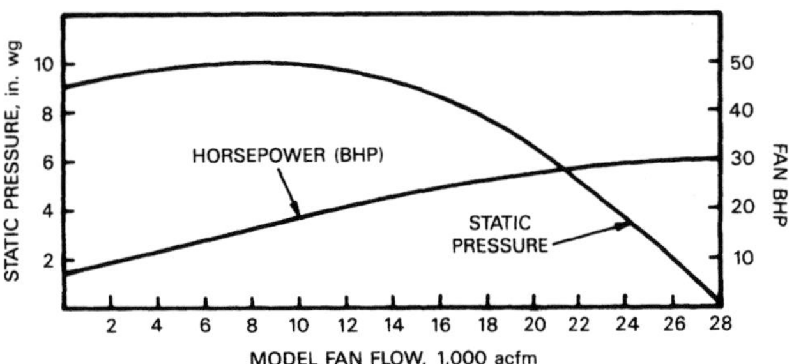

Fig. 10-13. Model fan performance curve.

$$CFM_a = CFM_b\left(\frac{80}{10}\right)^3 \times \left(\frac{900}{1,200}\right) = CFM_b \times 48$$

$$\Delta P_a = \Delta P_b\left(\frac{80}{20}\right)^2 \times \left(\frac{900}{1,200}\right)^2 \times \left(\frac{520}{780}\right) = \Delta P_b \times 6$$

$$HP_a = HP_b\left(\frac{80}{20}\right)^5 \times \left(\frac{900}{1,200}\right)^3 \times \left(\frac{520}{780}\right) = HP_b \times 288$$

(*Note*: Gas density ratios are equivalent to the inverse of absolute temperature ratios whenever the molecular weight of the gas is constant for both the model fan and the full-size fan and the inlet pressure to both the model fan and full-size fan are equivalent. The ideal gas law below can be used to determine the actual density ratio when the molecular weight or inlet pressure is not equivalent for the model and full-size fans.)

$$\rho\ (\text{lbm/ft}^3) = \frac{P\ (\text{absolute pressure lbf/ft}^2)}{R\left[\frac{\text{ft-lbf}}{\text{lbm-}^\circ\text{R}}\right] \times T\ (\text{Absolute temperature, }^\circ\text{R})} \qquad (10\text{-}10)$$

where

$$R = \text{gas constant} = \frac{R_u}{M}$$

$$R_u = \text{universal gas constant} = 1,545\ \frac{1,545\ \text{ft–lbf}}{\text{mole }^\circ\text{R}}$$

M = molecular weight, lbm/lb-mole

°R = absolute temperature, degrees Rankine

The following model fan performance points have been extracted from the fan performance curve and the efficiency has been calculated using Eq. (10-5) for each point:

Model fan performance

Flow	ΔP	bhp	η
0	9	7	0
6,000	10	16	79
12,000	8.6	25	87
18,000	5.2	28	64
24,000	~0	30	0

By using the ratios of model fan performance to full-size fan performance, the following full-size fan performance points can be calculated:

Full-size fan performance

Flow	ΔP	bhp	η
0	54	2,016	0
384,000	60	4,608	79
768,000	51.6	7,200	87
1,056,000	31.2	8,064	64
1,344,000	~0	8,640	0

Since model and full-size fan efficiencies must be the same, proper application of the fan law equations can be checked by calculating the efficiency of the full-size fan performance points and comparing them to the model fan efficiencies. The full-size fan performance curve is presented in Fig. 10-14.

10.4.5 Fan Law Example 2

From the full-size fan performance curve developed in the previous example, predict the impact on fan performance of a 40° F temperature increase. Assume the gas temperature is constant throughout the entire system. When operating at 320° F, the fan performance curve intersects the system resistance curve at a flow of 768,000 acfm and a pressure of 51.6 in. wg. What are the operating point and power requirement when the system operates at 360° F? Plot the system resistance for both temperatures on the fan performance curve.

Fan Law Example 2—Solution:

The graphical solution is shown in Fig. 10-15. A mathematical solution based on fan laws is as follows:

Step 1. Plot the 320° F system resistance curve on the 320° F fan curve. Generate some data points based on $\Delta P \alpha$ (acfm)^2.

ΔP	51.6	60	45	30	15	10
acfm	768,000	828,000	717,000	586,000	414,000	338,000

Step 2. At 360° F, the system resistance will be reduced by the absolute temperature ratio at equivalent system flows in action: ΔP at 320° F $\times \left(\frac{320 + 460}{360 + 460}\right) = \Delta P$ at 360° F. Generate some data points:

ΔP	49.1	57.1	42.8	28.5	14.3	9.5
acfm	768,000	828,000	717,000	586,000	414,000	338,000

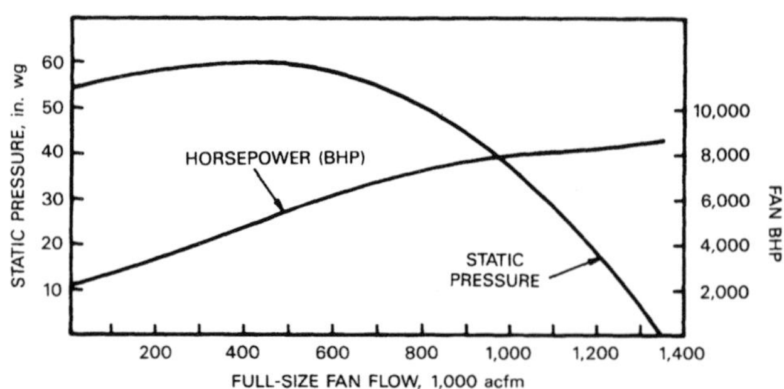

Fig. 10-14. Full-size fan performance curve.

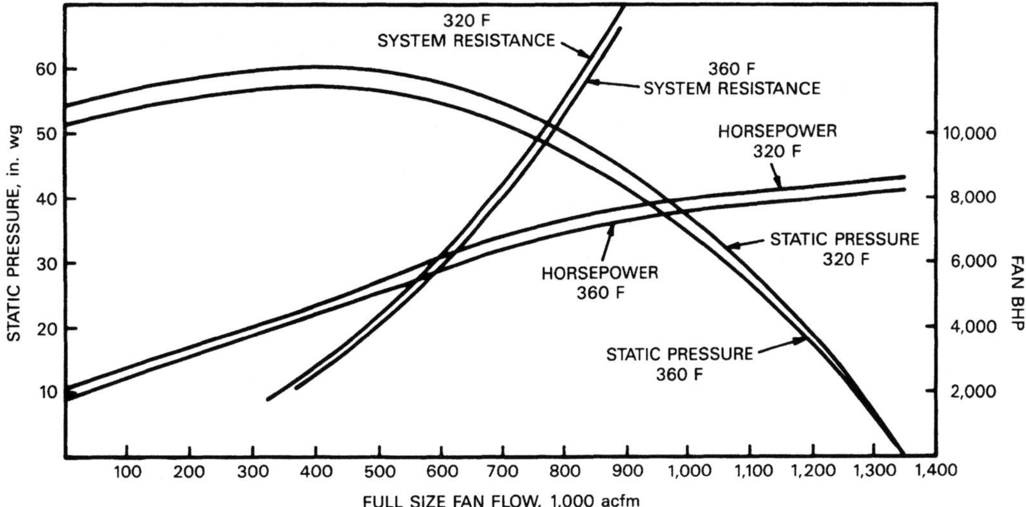

Fig. 10-15. Fan performance temperature impacts.

Step 3. Modify the fan performance curve for operation at 360° F. Fan Law 1 indicates density change will not affect fan flow capacity. Fan Law 2 indicates density change will affect fan pressure capacity. Fan Law 3 indicates density change will affect fan power requirement. Generate new performance points by modifying the 320° F full-size fan performance data by the ratio of the absolute temperatures.

Fan flow	Performance at 320° F		Performance at 360° F	
acfm	ΔP	bhp	ΔP	bhp
0	54	2,016	51.4	1,918
834,000	60	4,608	57.1	4,383
768,000	51.6	7,200	49.1	6,849
1,056,000	31.2	8,064	29.7	7,671
1,344,000	0	8,640	0	8,220

Step 4. Plot the 360° F system resistance curve and fan performance data. Locate the fan operating points with the system at 320° F and 360° F. For fan operation at 768,000 acfm (review Fan Law 1; fan flow is not dependent on temperature), fan bhp at 320° F will be 7,200 hp, and 6,849 hp at 360° F.

When fan laws are being used to predict fan performance for various operating conditions and fan sizes, changes in fan size must be geometric (all linear dimensions change equally).

The following examples illustrate methods to maintain constant-efficiency between given and desired operating conditions.

10.4.5.1 Maintaining Constant Efficiency with Changing Fan Speed. When fan laws are used to predict fan performance for changes in operating speed, fan size, and gas density, the efficiencies for the known and predicted conditions must be equal. To ensure this requirement is maintained, the following procedure can be used:

Example 10-4

From the fan performance curve presented in Fig. 10-16, determine the operating speed and power required to operate at the reduced load operating point of 600,000 acfm flow at 40 in. wg.

Solution:

The first step in solving this type of problem is to construct a constant-efficiency operating line from the desired operating point to the fan static pressure curve. The intersections of this operating line with the fan static pressure curves for both the known and desired operating points must be at equal fan efficiency conditions. To ensure equal fan efficiency at both points, the following relationship between fan flow and pressure is established to enable construction of a constant efficiency operating line:

- Fan Law 1:

$$\mathrm{CFM_a} = \mathrm{CFM_b}\left(\frac{\mathrm{SIZE_a}}{\mathrm{SIZE_b}}\right)^3 \times \left(\frac{\mathrm{RPM_a}}{\mathrm{RPM_b}}\right) \qquad (10\text{-}11)$$

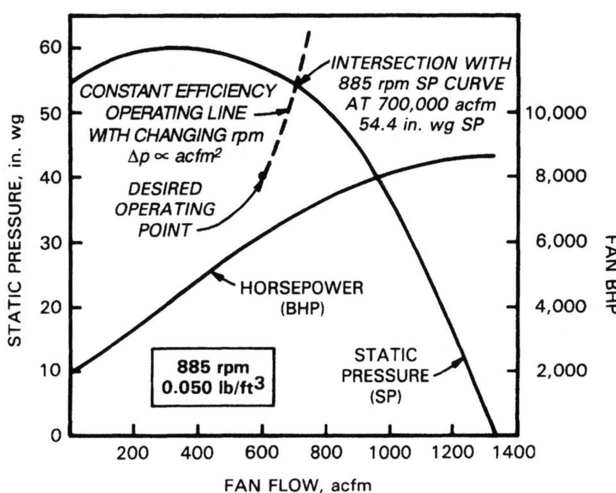

Fig. 10-16. Fan performance with changing fan speed.

Expressed in terms of rpm changes, Fan Law 1 is revised to:

$$\left(\frac{\text{RPM}_a}{\text{RPM}_b}\right) = \left(\frac{\text{CFM}_a}{\text{CFM}_b}\right) \times \left(\frac{\text{SIZE}_b}{\text{SIZE}_a}\right)^3 \quad (10\text{-}12)$$

- Fan Law 2:

$$\Delta P_a = \Delta P_b \left(\frac{\text{SIZE}_a}{\text{SIZE}_b}\right)^2 \times \left(\frac{\text{RPM}_a}{\text{RPM}_b}\right)^2 \times \left(\frac{\text{LB/FT}_a^3}{\text{LB/FT}_b^3}\right) \quad (10\text{-}13)$$

Expressed in terms of rpm changes, Fan Law 2 is revised to:

$$\frac{\text{RPM}_a}{\text{RPM}_b} = \left(\frac{\Delta P_a}{\Delta P_b}\right)^{\frac{1}{2}} \left(\frac{\text{LB/FT}_b^3}{\text{LB/FT}_a^3}\right)^{\frac{1}{2}} \left(\frac{\text{SIZE}_b}{\text{SIZE}_a}\right) \quad (10\text{-}14)$$

- Substituting Eq. (10-12) for $\frac{\text{RPM}_a}{\text{RPM}_b}$ in Eq. (10-14) results in the following:

$$\left(\frac{\text{CFM}_a}{\text{CFM}_b}\right)\left(\frac{\text{SIZE}_b}{\text{SIZE}_a}\right)^3 = \left(\frac{\Delta P_a}{\Delta P_b}\right)^{\frac{1}{2}} \left(\frac{\text{LB/FT}_b^3}{\text{LB/FT}_a^3}\right)^{\frac{1}{2}} \left(\frac{\text{SIZE}_b}{\text{SIZE}_a}\right) \quad (10\text{-}15)$$

Expressing Eq. (10-15) in terms of ΔP changes results in the following:

$$\left(\frac{\Delta P_a}{\Delta P_b}\right) = \left(\frac{\text{CFM}_a}{\text{CFM}_b}\right)^2 \times \left(\frac{\text{SIZE}_a}{\text{SIZE}_b}\right)^{-4} \times \left(\frac{\text{LB/FT}_a^3}{\text{LB/FT}_b^3}\right) \quad (10\text{-}16)$$

Provided the fan size and gas density remain constant for the alternative operating speeds, Eq. (10-16) can be further simplified to the following:

$$\Delta P_a = \Delta P_b \times \left(\frac{\text{CFM}_a}{\text{CFM}_b}\right)^2 \quad (10\text{-}17)$$

This relationship described in Eq. (10-17) between ΔP and CFM defines the constant-efficiency operating line for alternative operating speeds. To construct this line, start with the desired operating condition (600,000 acfm, 40 in. wg) and calculate a series of points with the pressure changing with the square of the flow, so the intersection with the fan curve can be determined. Sample points are 650,000/46.9, 700,000/54.4, and 750,000/62.5. By connecting the points to develop an operating line, the intersection with the known fan static pressure curve and horsepower curve can be determined (700,000 acfm/54.4 in. wg/6,750 bhp). Once determined, the ratio of this flow to 600,000 acfm is equivalent to the ratio of the speed reduction required to operate at 600,000 acfm (600,000/700,000) × 885 rpm = 759 rpm [Eq. (10-12)]. The power required at the 600,000 acfm, 40 in. wg operating point is equivalent to the cube of the speed ratio times the power requirement at the 885 rpm operating point $(759/885)^3 \times 6{,}750 = 4{,}258$ (Fan Law 3).

10.4.5.2 Maintaining Constant Efficiency with Changing Fan Size In accordance with Fan Laws 1 and 2, a change in size produces a change in flow capacity with the cube of the size change and a change in pressure capacity with the square of the size change. To predict the required size for a desired operating condition, this relationship of the flow increasing as the cube of the size increase and the pressure capacity increasing with the square of the size increase must be maintained.

Similar to changes in fan speeds, when using fan laws to predict changes in fan size, the efficiency of the known and predicted conditions must be equal. To ensure this requirement is maintained, the following procedure can be used.

Example 10-5

From the model fan performance curve presented in Fig. 10-17, determine the fan size and power required to operate at the full-size operating conditions of 40 in. wg and a flow of 700,000 acfm. Both the model fan and the full-size fan operate at 900 rpm with inlet air conditions of 29.92 in. HgA and 70° F. The wheel size of the model fan is 50 in.

Solution:

The first step in this type of problem is to construct an operating line of constant efficiency from the desired operating point to the model fan performance curve.

The intersections of this constant-efficiency operating line with the fan static pressure curves for both the model and full-size fan must be at equal fan efficiency conditions. The following relationship between fan flow and pressure is established to enable construction of a constant-efficiency operating line. To evaluate the effects of changing fan size, Fan Law 1 and Fan Law 2 are expressed in terms of size changes:

- Fan Law 1:

$$\text{CFM}_a = \text{CFM}_b \left(\frac{\text{SIZE}_a}{\text{SIZE}_b}\right)^3 \times \left(\frac{\text{RPM}_a}{\text{RPM}_b}\right) \quad (10\text{-}18)$$

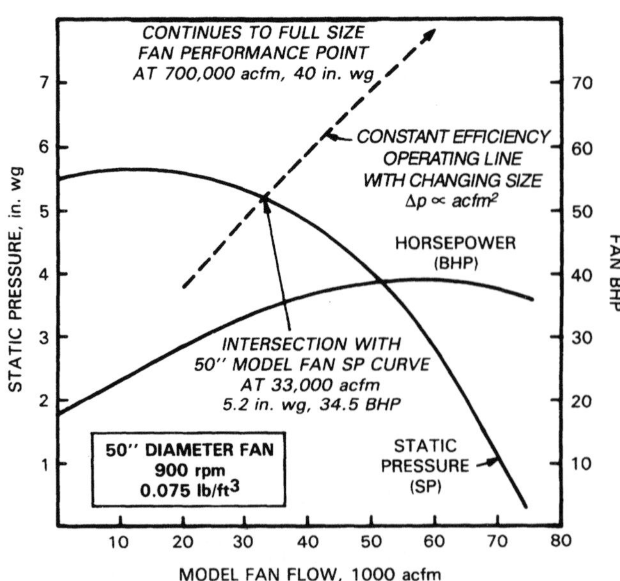

Fig. 10-17. Fan performance with changing fan size.

Expressed in terms of size changes, Fan Law 1 is revised as follows:

$$\left(\frac{\text{SIZE}_a}{\text{SIZE}_b}\right) = \left(\frac{\text{CFM}_a}{\text{CFM}_b}\right)^{\frac{1}{3}} \times \left(\frac{\text{RPM}_a}{\text{RPM}_b}\right)^{-\frac{1}{3}} \quad (10\text{-}19)$$

- Fan Law 2:

$$\Delta P_a = \Delta P_b \left(\frac{\text{SIZE}_a}{\text{SIZE}_b}\right)^2 \times \left(\frac{\text{RPM}_a}{\text{RPM}_b}\right)^2 \times \left(\frac{\text{LB/FT}^3_a}{\text{LB/FT}^3_b}\right) \quad (10\text{-}20)$$

Expressed in terms of size changes, Fan Law 2 is revised as follows:

$$\frac{\text{SIZE}_a}{\text{SIZE}_b} = \left(\frac{\Delta P_a}{\Delta P_b}\right)^{\frac{1}{2}} \left(\frac{\text{RPM}_a}{\text{RPM}_b}\right)^{-1} \left(\frac{\text{LB/FT}^3_a}{\text{LB/FT}^3_b}\right)^{-\frac{1}{2}} \quad (10\text{-}21)$$

- Substituting Eq. (10-19) for $\frac{\text{SIZE}_a}{\text{SIZE}_b}$ in Eq. (10-21) results in the following:

$$\left(\frac{\text{CFM}_a}{\text{CFM}_b}\right)^{\frac{1}{3}} \left(\frac{\text{RPM}_a}{\text{RPM}_b}\right)^{-\frac{1}{3}} = \left(\frac{\Delta P_a}{\Delta P_b}\right)^{\frac{1}{2}} \left(\frac{\text{RPM}_a}{\text{RPM}_b}\right)^{-1} \left(\frac{\text{LB/FT}^3_a}{\text{LB/FT}^3_b}\right)^{-\frac{1}{2}} \quad (10\text{-}22)$$

- Expressing Eq. (10-22) in terms of ΔP changes results in the following:

$$\left(\frac{\Delta P_a}{\Delta P_b}\right) = \left(\frac{\text{CFM}_a}{\text{CFM}_b}\right)^{\frac{2}{3}} \times \left(\frac{\text{RPM}_a}{\text{RPM}_b}\right)^{\frac{4}{3}} \times \left(\frac{\text{LB/FT}^3_a}{\text{LB/FT}^3_b}\right) \quad (10\text{-}23)$$

Provided that fan speed and gas density remain constant for the alternative fan sizes, Eq. (10-23) can be further simplified to the following:

$$\Delta P_a = \Delta P_b \times \left(\frac{\text{CFM}_a}{\text{CFM}_b}\right)^{\frac{2}{3}} \quad (10\text{-}24)$$

This relationship described in Eq. (10-24) between ΔP and CFM defines the constant-efficiency operating line for the alternative fan sizes. To construct this line, start with the desired operating point (40 in. wg, 700,000 acfm) and calculate a series of points so the intersection with the fan curve can be determined.

Sample points would be 70,000/8.6; 50,000/6.9; and 30,000/4.9.

By connecting the points to develop a constant-efficiency operating line, the intersection with the model fan static pressure curve and horsepower curve can be determined (33,000 acfm, 5.2 in. wg, 34.5 bhp). Once determined, the cube root of the ratio of 700,000 acfm to this flow is equivalent to the ratio of the sizes of the full-size model and fan $(700{,}000/33{,}000)^{1/3} \times 50$ in. = 138.4 in. The full-size fan power required at the 700,000 acfm operating point is equivalent to the ratio of fan sizes raised to the fifth power times the model fan power requirement (Fan Law 3), or $(138.4 \text{ in.}/50 \text{ in.})^5 \times 34.5$ bhp = 5,606 bhp.

10.4.6 Multiple Fan Operation

For added reliability or operating versatility, fan systems regularly have two to four fans operating in parallel, supplying or exhausting gas from a common source. Fan performance for these parallel operating conditions can be graphically presented on a total system flow basis or on a flow per fan basis. Note that system resistance is dependent on total system flow. The following exmple illustrates fan performance on a flow per fan basis. Since fan suppliers normally furnish fan performance curves on a flow per fan basis, this method of presentation is normally the least complex of the two and would be recommended in most cases.

Example 10-6

Multiple Fan Operation Example:

Four fans operate in parallel supplying air to a common duct. The fan performance curve on a flow per fan basis and system resistance with all four fans in service is presented in Fig. 10-18. Assuming constant fan speed and density, what would be the total system flow, system resistance, and total fan power requirement with four, three, two, and one fans in service?

Solution:

Plot the system resistance curves on a flow per fan basis for the alternative cases based on the pressure drop through the system remaining constant for constant total system flow. Determine the fan power requirements from the fan performance curve at the operating point where the system resistance curves intersect the fan static pressure curve.

Multiple Fan Operation Example—Graphical Solution (Fig. 10-19):

Step 1. Plot the alternative system resistance curves on the fan performance curve based on the total system resistance remaining constant for the same total system flow. For example: the flow per fan at constant pressure will double when two of four fans are removed from service. Flow per fan at constant pressure will increase by a factor of four when three of four fans are removed from service. Flow per fan at constant pressure will increase by a factor of 4/3 when one of four fans is removed from service.

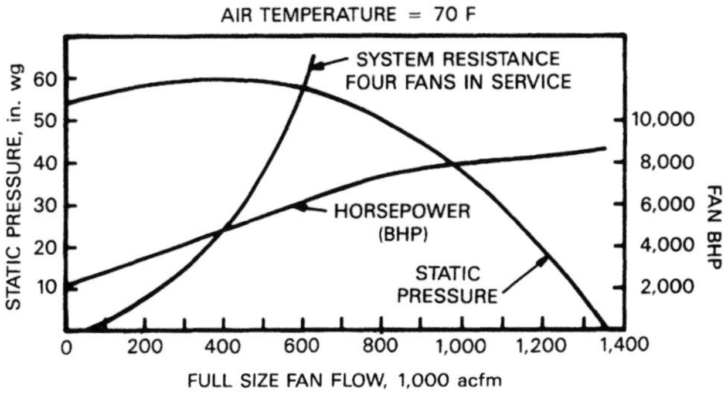

Fig. 10-18. Multiple-fan operation curve.

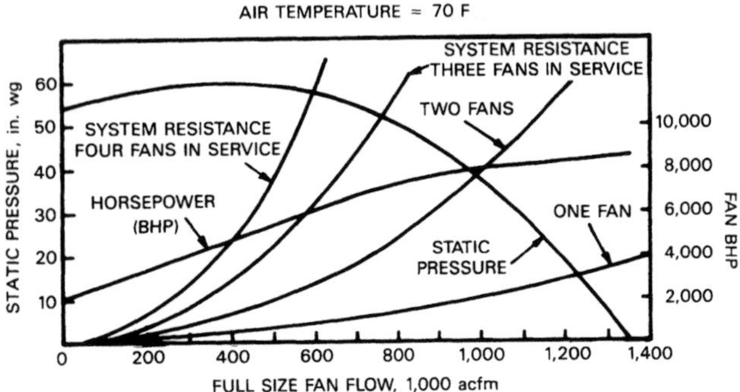

Fig. 10-19. Multiple-fan operation.

Step 2. Locate the intersections of the alternative system resistance curves with the fan static pressure and bhp curves.

No. of fans in service	Flow/fan 1,000 acfm	bhp/ fan	Total flow 1,000 acfm	Total bhp
4	600	6,200	2,400	24,800
3	765	7,150	2,295	21,450
2	985	8,000	1,970	16,000
1	1,230	8,400	1,230	8,400

10.5 FAN OPERATING CHARACTERISTICS

In applications that require fans to maintain variable and controlled pressure or flow conditions, the fans are equipped with flow or pressure regulating capability. With centrifugal fans, this capability is normally provided by either inlet dampers, outlet dampers, two-speed control, or variable-speed control. With axial flow fans, this capability is normally provided by adjustable pitch blade control.

10.5.1 Centrifugal Fan Operating Characteristics

Inlet damper and/or inlet vane centrifugal fan control are used when the fan is to be driven with a single- or two-speed motor. Inlet dampers are shown in Fig. 10-2. Inlet vanes (when supplied) are located in the fan inlet cone, also shown in Fig. 10-2. The dampers or vanes are full open for full flow, and as the dampers or vanes close, the fan flow and pressure capacities are reduced. Figure 10-20 illustrates typical impacts of inlet dampers or inlet vanes on centrifugal fan performance. The fan provides flow for the system ABC. With full open inlet dampers or inlet vanes, the fan operates at point A with bhp of A′. With three-quarter-open inlet dampers or inlet vanes, the fan operates at Point B with bhp of B′, and with half-open inlet dampers or inlet vanes, the fan operates at Point C with bhp of C′.

The inlet dampers or inlet vanes, when partially closed, impart a preswirl on the flow stream entering the fan in the direction of rotor rotation. This preswirl reduces the relative velocity of the fan blades to the air streams, thus reducing the fan flow and pressure capacities. This preswirl also reduces

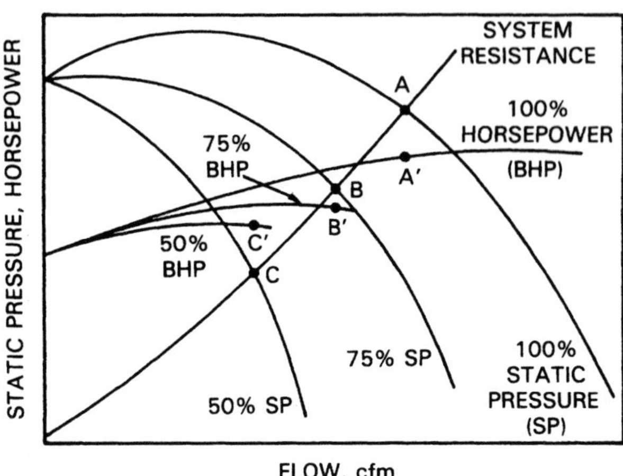

Fig. 10-20. Inlet vane control of centrifugal fans.

the fan power requirement. The majority of fan manufacturers state that inlet vane control is slightly more efficient than inlet damper control, since the inlet vanes are more effective at imparting a preswirl to the flow stream entering the fan rotor. Figure 10-20 shows a typical performance curve, and indicates that the actual percentage of damper position versus flow may vary significantly depending on the specific fan and damper design. Fan flow versus damper position is typically nonlinear, with initial damper opening from the closed position generating relatively large changes in flow. When the dampers are nearly full open, changes in damper positions result in relatively small changes in flow.

Determination of fan performance with inlet dampers at partially closed positions is accomplished by fan performance testing. These tests are normally performed by the manufacturer on reduced-size scale model fans and are then estimated for the full-size fan performance.

Discharge damper control could be used instead of inlet damper or inlet vane control when a fan is to be driven with a single-speed or two-speed motor. Figure 10-21 illustrates the typical impacts of discharge damper control on centrifugal fan performance. The fan provides flow for the system DEF. With full-open discharge dampers, the fan operates at point D with bhp of D′. With three-quarter-open discharge damper, the fan operates at point E″ with bhp of E′. The pressure differential between points E and E″ represents the pressure drop across the discharge damper. With a half-open discharge damper, the fan operates at point F″ with bhp of F′.

Since discharge damper control simply adds resistance to the system rather than altering the aerodynamic characteristics of the fan, discharge damper control results in higher part load power requirements than inlet damper or inlet vane control. Similar to operating with inlet damper control, fan flow versus discharge damper position is typically nonlinear.

The above power savings is illustrated by comparing Figs. 10-20 and 10-21. With inlet damper or inlet vane control, a specific bhp versus flow characteristic curve exists for each inlet vane position. With discharge damper control, the bhp versus flow characteristic for the entire fan operating

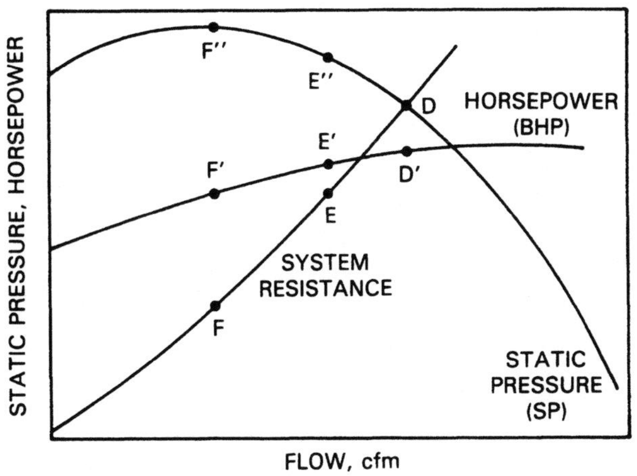

Fig. 10-21. Discharge damper control of centrifugal fans.

range is equivalent to the bhp curve representative of full-open inlet vane position.

Variable-speed operation provides the ultimate in power savings at reduced load operating conditions. When the fan is operating at full speed, the fan performance is identical to that of damper-controlled fans operating with full-open dampers. When variable-speed is used to regulate flow, the fans are not required to overcome the pressure drop across partially open dampers. This results in fans that can maintain maximum efficiency across the entire system operating range. Figure 10-22 illustrates the typical impacts of variable-speed control on centrifugal fan performance. The fan provides flow for system HIJ. During full-speed operation, the fan operates at point H with bhp of H′. With three-quarters speed operation, the fan operates at point I with bhp of I′; and with half-speed operation, the fan operates at point J with bhp of J′. Provided that the flow versus pressure requirement of system HIJ is a parabola with the pressure varying with the square of the flow, a reduction in fan speed will result in a linear flow reduction equal to the speed ratio.

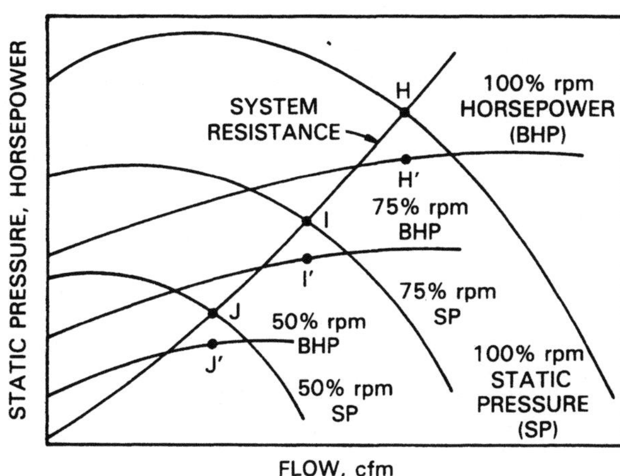

Fig. 10-22. Variable-speed control of centrifugal fans.

The fan laws presented in the preceding sections can be used to determine the impact of speed changes on fan operating conditions and power requirements.

Even though variable-speed operation results in significant power savings at reduced flow and pressure conditions, the system resistance must be analyzed across the expected operating range to ensure variable-speed control provides acceptable operating characteristics. Figure 10-23 illustrates a typical system that would not be appropriate for a variable-speed fan installation. The fan provides flow for system KLM. This type of system resistance curve is typical for primary air fan systems that must maintain a relatively constant pressure across a wide flow range.

As illustrated in Fig. 10-23, if the fan was controlled with variable speed, operation at point M would not be stable. At the flow conditions close to point M, the fan static pressure curve is nearly parallel to the system resistance curve KLM. With this characteristic, small fluctuations in system resistance result in large fluctuations in system flow. Stable operation is ensured when the system resistance curve intersects the fan static pressure curve at a point to the right of the peak in the fan static pressure curve and also when the curves intersect at an acute angle that provides a well-defined operating point. Small fluctuations in system resistance at a constant flow should not produce large changes in fan flow capacity.

10.5.2 Axial Fan Operating Characteristics

Rather than being controlled with inlet or outlet dampers or speed control devices, axial flow fans are normally controlled by adjusting the blade pitch. A characteristic performance field for a typical constant-speed, variable-pitch axial flow fan is shown in Fig. 10-6. As shown in the figure, each blade pitch angle results in a specific flow versus head characteristic curve. As the blade pitch angle is increased, the pressure and flow capacity of the fan also increase. Once installed in a system, the flow through the fan can be con-

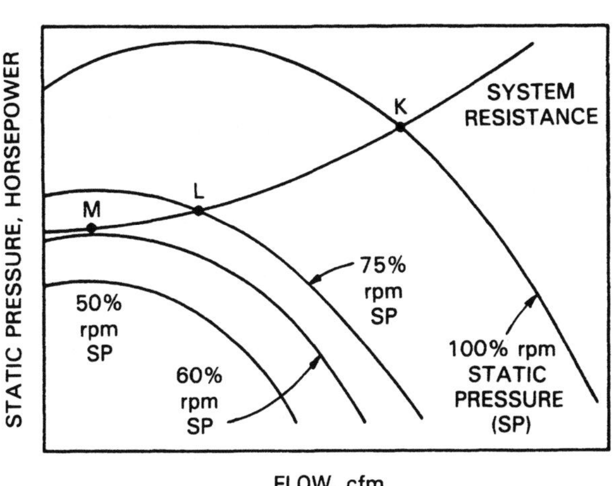

Fig. 10-23. Centrifugal fan performance characteristics with a system not suitable for variable speed operation.

trolled to any condition in its performance field, provided the fan drive motor is of sufficient horsepower. Axial fans are much more prone to operation in a stall condition than are centrifugal fans. Stall is the aerodynamic phenomenon that occurs when fan operation is attempted at a condition that is beyond the fan pressure capacity. When the fan blades are required to provide more static pressure rise than they are designed to produce, flow separation occurs around the blade. When this occurs, the fan becomes unstable and no longer operates on its normal performance curve. Operation in this stall condition is accompanied with increased noise and vibration. Continuous operation in this condition leads to fatigue failure of the rotating blades because of the high-frequency vibrations the blades encounter during stall conditions.

An illustration of this typical stall phenomenon is presented in Fig. 10-24. The curves in Fig. 10-24 marked "N" are the normal fan performance curves for a constant-blade angle. Each blade angle characteristic curve has its individual stall point, marked S on the figure. The curve "SL" connects the stall points of the individual blade angle characteristic curves and is referred to as the stall line. The dotted curves "U" are the characteristic stall curves for individual blade pitch angles. The curves show the path the fan follows when operating in the stalled condition.

A typical axial fan performance curve is presented in Fig. 10-25 to illustrate further axial fan stall conditions. If the normal operating system resistance increases for any reason (damper closure, fuel trip, system pluggage), the normal operating point "X" will change to overcome the higher system resistance. For a given blade pitch angle, the fan operating point will move along the static pressure curve to meet the new pressure requirements. If the pressure increases to the stall point "S," the fan will stall. Once the fan enters the stall condition, performance will remain unstable

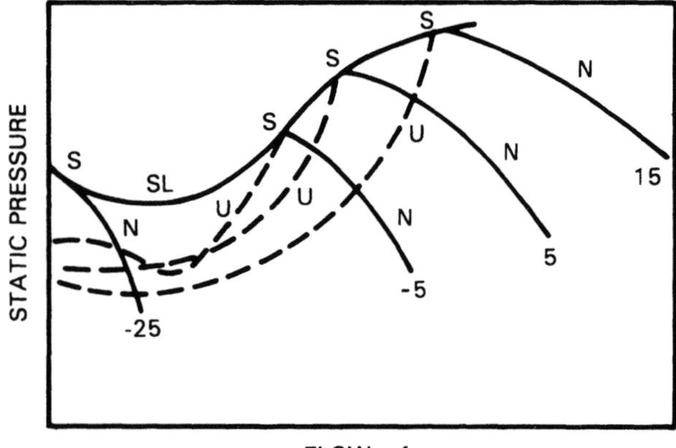

Fig. 10-24. Axial fan performance stall characteristics.

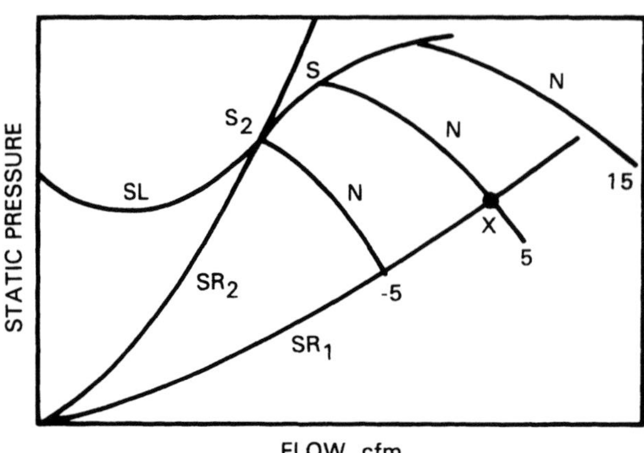

Fig. 10-25. Axial fan performance during stall conditions with changing system resistance.

until the system resistance returns to normal or the blade pitch angle is reduced until the fan regains stability. The fan will be stable when the static pressure capability at the reduced blade pitch angle provides a stall point S_2 that is higher than the pressure required by the new system resistance (SR_2). If stabilization cannot be attained by blade adjustment, the fan must be shut down to avoid damage to the fan rotating components.

When axial fans are sized properly and the normal system resistance line is a reasonable distance from the stall line, the probability of stall is low. This probability increases if the fan is oversized, or if the system resistance increases significantly.

To prevent stall, either the equipment operator or the control system must have sufficient operating data to monitor operation of the fan in relation to the stall line. This can be accomplished by instrumentation to monitor fan flow, pressure, and blade position, or by equipment that monitors and alarms on occurrence of flow and pressure instability directly upstream of the fan blades at the onset of stall conditions.

Stall conditions can also result when a second fan of a two-fan system is brought in service. As shown in Fig. 10-26, if a single fan in a two-fan system is operating at P_1 and the second fan is started against the system pressure that exceeds the minimum point or saddle point in the stall line, the second fan will not be capable of stable operation since it will go into stall conditions as the blades are opened. To enable the second fan to be brought into service, the operating conditions of the first fan must be reduced to Point P_2 where the operating pressure is below the pressure that corresponds to the saddle point in the stall line. Once both fans are providing stable operation with nearly equivalent blade pitch

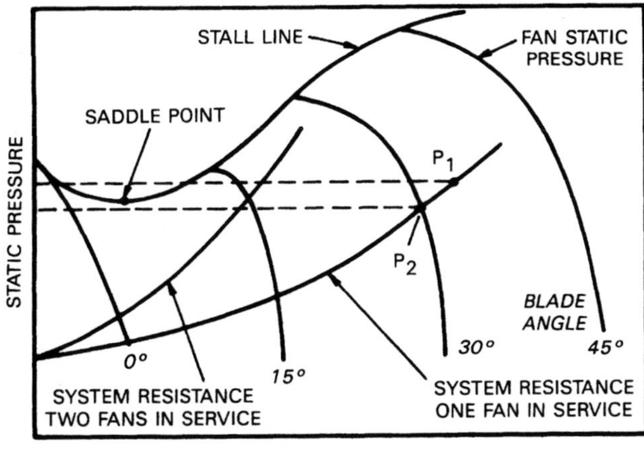

Fig. 10-26. Potential complications of bringing a second axial fan in service.

angle, the blades of both fans can be opened in unison to provide the system flow and pressure requirements.

Even with the higher complexity of axial fan performance characteristic resulting from the potential for stall and the higher mechanical complexity resulting from the blade pitch adjustment mechanism, axial fans have been accepted in most instances as the standard for relatively low-pressure, high-flow applications for clean air duty. Major advantages over single-speed centrifugal fans are improved controllability, especially at low flow conditions, and lower operating power requirements.

10.6 ADDITIONAL FAN DESIGN CONSIDERATIONS

10.6.1 Fan Vibration

The primary cause of fan vibration is fan rotor imbalance. This results from eccentricity between the center of mass and the center of rotation of the fan rotor. The level of vibration can be reduced by placing a weight on the outer diameter of the rotor opposite the mass center such that weight times its distance from the center of rotation equals the unbalance. Perfect balance is unattainable by practical methods. Some imbalance always remains. If the fan rotor and bearing system natural frequency are designed to be above the fan speed, the vibration sensitivity of the fan rotor to the remaining imbalance can be reduced to acceptable levels.

High sensitivity to imbalance can be avoided by tuning the fan–foundation system natural frequencies such that the fan operating speed is far from the natural frequencies. Tuning to achieve the desired system natural frequencies is performed during the design phase of the fan–foundation system.

10.6.2 Fan-Induced Duct Vibrations

Fan-induced duct vibrations are attributed to internal static pressure pulsations rather than to fan rotor imbalance that results in fan rotor and bearing vibration. Fan-induced duct vibrations normally cause an entire duct system to shake with a magnitude ranging from annoying rattles to large-scale destructive movement. If both the bearing and the ducts vibrate, two problems exist that require independent solutions.

Pressure pulsations of sufficient magnitude to damage ductwork systems are normally attributable to the associated fan and the fan inlet flow conditions. The fan inlet box, inlet cone, and fan blades each can be the location of unstable flow conditions that have been found to be the source of pressure pulsations with distinctive characteristics. In addition, the style of fan and fan control method also determines the potential source of pressure pulsation for a particular installation.

Most occurrences of fan-induced pressure pulsations can be categorized as inlet cone vortex, inlet box vortex, or rotating stall areas of flow instability.

The inlet cone vortex category of flow instability and pressure pulsation has been found to exist on large centrifugal fans controlled with radial inlet vanes. The instability occurs with inlet vane positions between approximately 30% and 70% open. Vibration levels peak at 50% vane opening and reduce as the vanes are further opened or closed. Normal pressure pulsations emitted from centrifugal fans are in the range of 1 to 2 in. wg. Pulsations higher than 40 in. wg have resulted from the inlet cone vortex phenomenon.

The most common method of eliminating an inlet cone vortex induced pressure pulsation is with the addition of radial dorsal fins downstream of the radial inlet vanes. The dorsal fins consist of stationary fins attached to a pipe sleeve that is attached to the inlet vane hub assembly. The fan impeller shaft passes through the pipe sleeve or alternatively can be attached directly to the inlet vanes perpendicular to the vanes. The dorsal fins partially straighten or spoil the inlet air prespin generated by the partially closed inlet vanes. Dorsal fins have the disadvantages of reducing prespin and increasing fan power requirements at reduced loads. Dorsal fin heights of approximately 1/3 the height of the inlet vane height have been found to be effective in reducing inlet cone vortex pulsations with minimal impact on fan efficiency.

The inlet box vortex category of flow instability and pressure pulsation occurs on both inlet vane and inlet damper controlled centrifugal fans. The pulsations occur both upstream and downstream, increase with increasing damper opening, and normally have a magnitude of less than 4 in. wg. The most common method of eliminating the problem is to install a splitter plate in the bottom of the inlet box. The splitter plate is mounted in the bottom of the inlet box parallel with the fan shaft to partially straighten the gas flow entering the inlet cone. The height of the splitter plate varies for alternative fan designs from less than 1 ft to extending all the way from the bottom of the inlet box to directly below the fan shaft. Fan efficiency with full-open control vanes or dampers is normally improved with the installation of splitter plates, and efficiency is reduced during operation with partially open dampers since the splitter plates reduce the effectiveness of the prespin provided by the inlet vanes or inlet dampers.

The rotating stall category of flow instability and pressure pulsation occurs on inlet damper controlled centrifugal fans and, in most cases, on fans with backward curved airfoil blades. The pulsations occur during operation with control dampers from zero to approximately 30% open and have magnitudes less than 10 in. wg. Once the fan dampers exceed 30% open, pulsations return to normal levels.

The most common methods of eliminating the rotating stall problem are either to replace the inlet damper control equipment with inlet vane controls or to throttle the fan with outlet dampers instead of inlet dampers during extended operating periods with inlet dampers less than 30% open.

10.6.3 Fan Noise Generation

Fan noise consists of two separate components. Single-tone noise results from the flow leaving the blades passing by stationary objects such as flow straightening devices or the nose cut-off of the centrifugal fan housing. The sound intensity changes inversely with the clearance between the rotating blades and stationary object. The blade passing frequency and its first harmonic are normally the most predominant. Broad band noise is produced by the high-velocity air stream passing through the stationary objects in the ductwork, dampers, and fan housing.

The noise generated by both of these components travels out of the inlet box, through the discharge duct, and also through the fan housing.

Forced draft (FD) and primary air (PA) fan inlets normally have absorptive-type inlet silencers to lower the noise being emitted from the inlets to acceptable levels. Absorptive-type silencers normally consist of several rows of panels in the airstream that have perforated plate skin and are filled with acoustically absorptive material. Absorptive silencers are economically constructed, produce small pressure drop, and work well in clean air or gas applications. The installation of absorptive insulation and acoustic lagging on fan housings and ductwork is the normal method of reducing noise levels emitted from the exterior surfaces of the fans and ducts. In most cases, the noise transmitted through the fan housing is a very small portion of the total noise generated by the fan.

The noise emitted from FD and PA fan outlets is normally absorbed by ductwork insulation, pulverizer, and steam generator components.

Induced draft (ID) fans located between particulate removal and desulfurization equipment normally do not require additional noise reduction equipment because of the noise absorbing characteristics of these particulate and SO_2 control devices. In certain applications where high-pressure ID fans are located between particulate removal systems and the exhaust stack, discharge silencers have been employed to reduce the noise level emitted from the stack outlet. For ID applications in that fly ash and SO_2 removal equipment reaction products may exist in the flue gas stream, absorptive-type silencers have been found to be prone to material buildup and reduced acoustical performance. In these applications, resonant-type silencers designed specifically to reduce the noise emissions at specific frequencies have been found to perform reliably with minimal maintenance requirements.

10.6.4 Fan Inlet Duct Design

Uniform flow conditions at the fan inlet are required if the fan is to operate at design capacity. Fan installations that result in fan operation significantly below expected capacity are most commonly hampered by uneven flow distribution into the fan inlet box or inlet cone. An elbow or a 90-degree duct turn directly upstream of the fan inlet box prevents uniform inlet conditions and results in uneven flow distribution at the fan impeller. If the inlet duct produces a vortex or prespin in the air entering the impeller in the same direction as impeller rotation, fan performance will be similar to that from throttling with a set of inlet control vanes or dampers. If the prespin cannot be eliminated, full fan performance cannot be expected. When the inlet conditions result in a prespin in the opposite direction of impeller rotation, a slight increase in the pressure volume curve may result, but the fan power requirement will increase significantly.

When inlet elbow connections cannot be avoided, inlet conditions can normally be improved by the use of turning vanes in the elbow or splitter plates in the fan inlet box to reduce the tendency for prespin at the impeller inlet.

10.6.5 Turning Gear Applications

Turning gear auxiliary fan drive systems significantly reduce centrifugal fan vibration levels during hot or warm restarts of fans installed in hot gas applications. In cases where vibration levels are high during initial operation following hot or warm restarts, but gradually smooth out as operation continues, slow continuous rotation of the fan rotor prior to full speed operation can eliminate the startup vibration excursions.

Thermal stratification in the idle fan housing causes excessive startup vibration levels. Following the unit shutdown and fan damper closure, heat loss through the fan housing cools the gas trapped in the fan housing at the housing surfaces, and the cooler gas migrates to the bottom of the fan housing. As this thermal stratification continues, the gas temperature at the housing bottom can become significantly cooler than at the top. The cooler gas at the bottom causes more rapid cooling of the impeller at the bottom than at the top, resulting in thermal distortion. As this thermal distortion continues, the center of mass of the impeller displaces upward, away from the center of rotation. Relatively small amounts of thermal distortion, although not permanent, have the same effect as large balance weights placed on the fan impeller.

A turning gear, operated to prevent thermal distortion, has been found to be the best solution to the problem. Turning gear normally consists of a motor, a speed reducer, and a single-direction engagement clutch attached to the fan or main drive motor shaft through a flexible coupling. Turning gear operation normally rotates the fan at 25 to 50 rpm.

10.7 FAN SELECTIONS

Power plant applications that require the largest fans for a steam generator fall into four categories: forced draft (FD), primary air (PA), induced draft (ID), and gas recirculation.

10.7.1 Forced Draft

FD fans supply combustion air to the steam generator. The fans must have a pressure capability high enough to overcome the total resistance of inlet silencers, air preheat coils, air ducts, air heaters, wind boxes, burner registers, and any other resistance between the air intake of the fan and the furnace.

The FD fans must supply the total stoichiometric combustion air less credit for air supplied by the primary air fans, plus the excess air needed for proper burning of the design fuel. In addition, FD fans supply air to make up for air heater leakage and for some sealing air requirements.

10.7.2 Primary Air

PA fans normally handle relatively low flows and very high-pressure differentials. This usually requires centrifugal fans of large diameter and relatively narrow width impellers operating at high speeds, sometimes as high as 1,800 rpm. PA fans are generally driven by single-speed or two-speed motors. There are basically three types of primary air (PA) fans: hot PA fans, cold PA fans, and exhauster PA fans.

Hot and cold PA fans that use centrifugal fans usually have airfoil blades. In situations where fly ash or coal dust enters the PA fan with the combustion air, radial tip blades are preferred because of their improved abrasion resistance. Exhauster PA fans are normally constructed with straight radial blades.

10.7.3 Induced Draft

ID fans exhaust combustion products from the steam generator. ID fans control furnace pressure, that is normally set at −0.5 in. of water relative to atmosphere. The condition is known as balanced draft operation. The ID fans must be designed with sufficient pressure capability to overcome the total resistance of the convection passes of the steam generator, gas ducts, air heater, fabric filter or precipitator, wet or dry scrubber, stack, and any other resistance between the furnace and the stack outlet.

The ID fans must handle all of the flue gas from the furnace plus any infiltration caused by the negative pressure in equipment downstream of and including the furnace and any air heater seal leakage.

ID fans are normally downstream of particulate removal equipment and are in a moderate to low erosion potential service. In most instances, an airfoil centrifugal fan is specified. The airfoil design can develop efficiencies of more than 90% and have capacities greater than 1.5 million actual cubic feet per minute (acfm). The shape of the airfoil blade minimizes turbulence and noise. The blades and center plate may be fitted with wear pads and replaceable nose sections to simplify replacement. Structural strength, particularly important in fans of larger sizes, is excellent with this design.

Variable-pitch axial flow fans with steel blades are sometimes specified for induced draft service where the potential for blade erosion is minimal. The axial flow fan has a higher potential for erosion and has higher maintenance costs than an airfoil-bladed centrifugal fan. Where greater erosion resistance is needed because of particulate loading or where a very conservative approach is desirable, a radial tip centrifugal fan design is used. Without unduly sacrificing efficiency, the shape of this blade minimizes dust buildup and reduces downtime for cleaning.

Selection of an ID fan arrangement must consider plant size and layout, as well as the client's preference. Recent developments in adjustable-speed motors make the variable-speed centrifugal fans economically advantageous for many power plants. Other ID fan arrangements include the following:

- Centrifugal fans with variable-speed fluid couplings and single-speed electric motor drives,
- Centrifugal fans with variable-speed steam turbine drives,
- Centrifugal fans with two-speed electric motor drives, and
- Axial flow fans with variable-pitch blades and single-speed electric motor drives.

10.7.4 Gas Recirculation

Gas recirculation (GR) fans draw gas from a point between the economizer outlet and the air heater inlet and discharge it into the bottom of the furnace or near the furnace outlet. Recirculated gas introduced in the vicinity of the initial burning zone of the furnace is used for steam temperature control, while recirculated gas introduced near the furnace outlet is used for control of gas temperature. A high-efficiency, high draft loss mechanical dust collector must be installed upstream of the fan for coal-fueled units. If the recirculation fan on a coal-fueled unit is used only for standby emergency oil firing, the dust collector is omitted.

GR duty provides the most severe test of a power plant fan. The combination of heavy dust loads and rapid temperature changes demands the use of a rugged, reliable fan design. Particularly important is how the fan hub is mated with the shaft; conventional shrink-fit may not be adequate. To cope with temperature excursions, fans with an integral hub are preferable. Straight or radial tip blades are normally used. Turning gears are sometimes required to prevent thermal distortion of the rotor under cool-down conditions.

10.8 FAN SIZING

The flow and pressure capacities of boiler draft fans are determined by analyzing the fan system requirements at the maximum continuous rating (MCR) point of the steam generator, then adding design margins for additional fan flow and pressure capabilities to develop the fan design point or the test block point. The fans are designed with flow and pressure capacities above the MCR system requirements to allow for changes in combustion air or flue gas flow rates resulting from changes in system leakage rates, system flow obstructions, and the variable combustion characteristics of alternative fuels.

10.8.1 Forced Draft Fan Flow Margins

The standard flow margins applied to FD fans to establish the fan test block flow rate are illustrated in Table 10-1. A stoichiometric combustion air flow of 5,000,000 lb/h used in Table 10-1 is for illustration only. The actual amount of stoichiometric combustion air for a specific generating unit would be based on the MCR fuel burn rate and the combustion characteristics of the design basis fuel.

The excess air flow requirement depends on the fuel used in the steam generator. Table 10-2 lists the excess air requirements for various fuels. The 20% excess air requirement shown in Table 10-1 is based on burning a western subbituminous coal with lignitic ash and is recommended for complete and stable combustion of this type of fuel.

The 10% flow margin shown in Table 10-1 is included only in the test block flow requirement to ensure a conservative fan design. This margin is 10% of the combined stoichiometric combustion air and excess air.

The PA fan credit shown in Table 10-1 is the amount of air supplied to the steam generator by the PA fans in a cold PA fan arrangement. For steam generators firing a western subbituminous coal under normal operating conditions at the MCR load point, PA flow represents approximately 20% to 25% of the combined stoichiometric combustion air and excess air. PA flow requirements for steam generators firing a high moisture coal such as a North Dakota or Texas lignite may represent as much as 35% to 40% of the combined stoichiometric combustion air and excess air. To allow for the variations in flow due to the different types and quality of coals available for use in power plants, the PA fan credit is based on the number of pulverizers in service and the air flow requirements per pulverizer. The MCR PA fan credit in Table 10-1 is based on the air flow required for the pulverizers to supply the average design coal and operate the steam generator at MCR. The test block PA fan credit in Table 10-1 is based on the air flow required for the pulverizers to supply the best design coal and operate the steam generator at MCR. This will require the least number of pulverizers in service, the lowest primary air flow at the MCR load, and the highest secondary air flow from the FD fans. This variation in coal quality normally reduces the PA fan flow by approximately 25% compared to design conditions.

The total secondary air heater leakage listed in Table 10-1 is the net amount of air that leaks past the air heater seals and into the flue gas stream. The secondary air heater leakage is discussed in Section 10.7.3.

Table 10-2. Excess Air Requirements

Fuel	Excess Air (percent—range)
Eastern bituminous coal	25
Western subbituminous coal	20
Lignite coal	20
Oil	3–15
Natural gas	5–10
Refinery gas	8–15
Blast-furnace gas	15–25
Coke-oven gas	5–10

10.8.2 Induced Draft Fan Flow Margins

The standard flow margins applied to ID fans to establish the fan test block flow rate are illustrated in Table 10-3. The flow of stoichiometric combustion products at the MCR steam generator load point depends on the type and moisture content of the fuel. A typical value of 10 lb of combustion air per ash-free pound of fuel has been used for illustrative purposes.

The excess air flow through the ID fans is the same as the excess air flow through the FD fans, since excess air passes

Table 10-1. Standard Criteria for Forced Draft Fan Flow Conditions

Item	System Flow, lb/h: MCR Conditions	System Flow, lb/h: Test Block Conditions
Stoichiometric combustion air	5,000,000	5,000,000
Excess air	1,000,000	1,000,000
10% margin (test block only)	—	600,000
PA fan credit	(1,200,000)	(900,000)
Subtotal	4,800,000	5,700,000
Total secondary air heater leakage	325,000	536,250
Total FD fan flow	5,125,000	6,236,250

Table 10-3. Standard Criteria for Induced Draft Fan Flow Conditions

Item	System Flow, lb/h: MCR Conditions	System Flow, lb/h: Test Block Conditions
Combustion products	5,500,000	5,500,000
Excess air	1,000,000	1,000,000
10% margin (test block only)	—	650,000
Subtotal	6,500,000	7,150,000
Total gas side air heater leakage	650,000	1,072,500
Total ID system flow	7,150,000	8,222,500

through the steam generator without reacting with the fuel constituents.

The 10% flow margin listed in Table 10-3 is based on 10% of the total of the stoichiometric combustion products and excess air flows.

The total gas side air heater leakage, listed in Table 10-3, is the amount of air that leaks into the flue gas stream from the primary and secondary air.

10.8.3 Air Heater Leakage

Air heater leakage occurs through seals on the hot and cold ends of the air heater. The majority of the leakage occurs on the cold end, where corrosion due to the relatively low air and gas temperatures results in larger clearances between sealing surfaces.

Design allowance for air heater leakage is based on a percentage of the flue gas that enters the air heater. The percentage is based on data provided by the air heater manufacturer. Typical air heater leakage percentages are 6% to 10%. If the seals and metallic surfaces of the air heater are not maintained, air heater leakages of 15% or more can be expected. The air heater leakage, for study purposes, is normally assumed to be 10% of the flue gas entering the air heater.

Table 10-4 shows the impact of air heater leakage on fan sizing, based on the example in Table 10-3. Air heater leakage is divided into four parts: primary air to flue gas, secondary air to flue gas, primary air to secondary air, and air heater leakage margin. Figure 10-27 shows the direction of air heater leakage for the two types of air heaters: bisectors (primary and secondary) and trisectors.

Table 10-4. Impact of Air Heater Leakage on Fan Sizing

	System Flow, lb/h	
Item	MCR Conditions	Test Block Conditions
Total flue gas entering air heater	6,500,000	7,150,000
Air heater leakage		
Primary to flue gas	325,000	357,500
Secondary to flue gas	325,000	357,500
Total leakage to flue gas (10%)	650,000	715,000
Primary to secondary[a]	—	—
Impact on FD fan sizing		
Secondary to flue gas leakage	325,000	357,500
Primary to secondary leakage	(—)	(—)
Margin	—	178,750
Total leakage allowance	0,325,000	0,536,250
Impact on PA fan sizing		
Primary to flue gas leakage	325,000	357,500
Primary to secondary leakage[a]	—	—
Margin	—	178,750
Total leakage allowance	0,325,000	0,536,250
Impact on ID fan sizing		
Primary to flue gas leakage	325,000	357,500
Secondary to flue gas leakage	325,000	357,500
Margin	—	357,500
Total leakage allowance	0,650,000	1,072,500

[a]Applies only to trisector air heaters. With a trisector air heater, primary air leakage to secondary air is approximately equal to primary air leakage to flue gas.

[b]Applies only to trisector air heaters. Leakage is a credit for FD fan sizing.

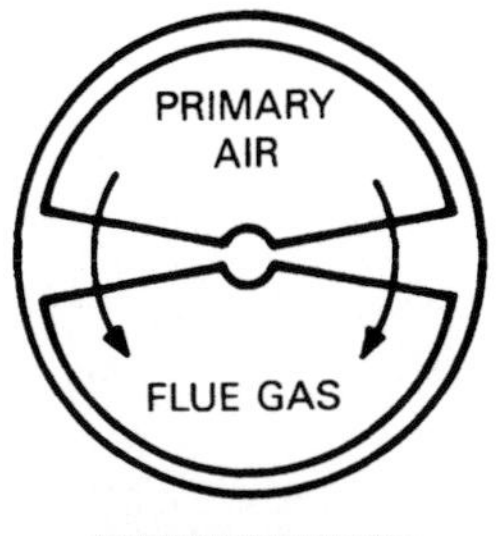

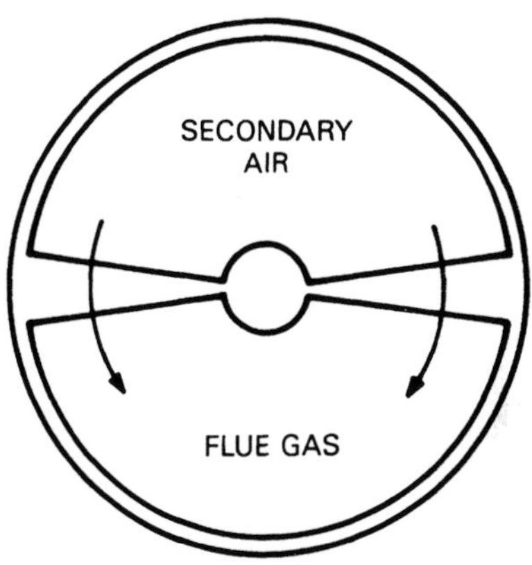

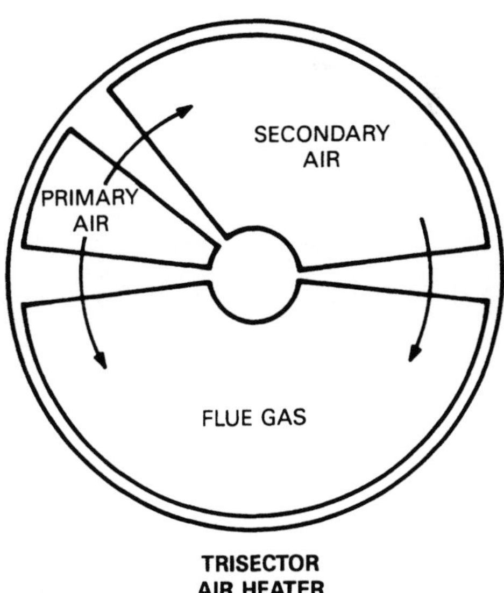

Fig. 10-27. Air heater leakage.

10.8.3.1 Primary Air to Flue Gas. This is the amount of primary air that leaks into the flue gas stream. Experience has shown that approximately 50% of the total air heater leakage into the flue gas stream is from primary air. The actual leakage rate is based on the guaranteed leakage rate proposed by the air heater manufacturer. This leakage affects only the PA and ID fan sizing and has no effect on the FD fan sizing.

10.8.3.2 Secondary Air to Flue Gas. This is the amount of secondary air that leaks into the flue gas stream. This leakage is the remainder, approximately 50%, of the total air heater leakage into the flue gas stream. The actual leakage rate is based on the guaranteed leakage rate proposed by the air heater manufacturer. Secondary air to flue gas leakage affects only the FD and ID fan sizing and has no effect on the PA fan sizing.

10.8.3.3 Primary Air to Secondary Air. Leakage of primary air to secondary air occurs only on trisector air heaters. The leakage from primary air to secondary air is approximately equal to the leakage from primary air to flue gas. The actual leakage rate is based on the guaranteed leakage rate proposed by the air heater manufacturer. This leakage affects only the PA and FD fan sizing and has no effect on the ID fan sizing.

10.8.3.4 Air Heater Leakage Margin. Air heater leakage margins are applied to the fan test block conditions to ensure adequate fan capacity if air heater seals are not properly maintained. The air heater leakage margin used for FD fans is 50% of the expected secondary air to flue gas leakage. The air heater leakage margin used for ID fans is 50% of the total primary air plus secondary air expected leakages into the flue gas stream. The air heater leakage margin used for PA fans is 50% of the expected primary air to flue gas leakage, except for trisector air heaters. If a trisector air heater is used, the PA fan leakage margin is 50% of the total expected leakage from primary air to secondary air plus primary air to flue gas.

10.8.4 Static Pressure Requirements

The fan test block pressure capabilities are established by tabulating the pressure drops for each of the components in the FD and ID systems at the MCR boiler load point, as shown in Tables 10-5 and 10-6. Sufficient pressure margins are then added to the MCR pressure requirement to ensure sufficient fan pressure capability to pass the test block flow through each of the fan system components. Pressure drops through components designed to operate with flow in the fully developed turbulent range increase in direct proportion to the second power of the flow through each component for flows above the MCR flow rate.

In addition to the component pressure drops at the test block flow rate, the fan test block static pressure includes design margins for air heater pluggage and for a control allowance. The air heater pluggage margin is based on 50% of the pressure drop through the air heater at the test block flow rate. A control allowance pressure margin of 1.0 in. wg is added as a component of the test block pressure requirement to allow for control system fluctuations at the test block flow.

Table 10-5. Criteria for Forced Draft Fan Design Pressure Conditions

	System Pressure Drop, in. wg	
Item	MCR Conditions	Test Block Conditions
Inlet silencers	0.5	0.7
Air preheat coils (if required)	—	—
Normal air heater	3.5	4.9
Air heater margin	—	2.5
Burners and windboxes	6.0	8.5
Dampers, ducts, and venturis	2.0	2.8
Total drop	12.0	19.4
Control allowance	—	1.0
Total static pressure	12.0	20.4

Table 10-6. Criteria for Induced Draft Fan Design Pressure Conditions

	System Pressure Drop, in. wg	
Item	MCR Conditions	Test Block Conditions
Furnace pressure	−0.5	−0.5
Components		
Furnace to economizer outlet	5.5	6.7
Ducts to air heater	1.0	1.2
Normal air heater	5.0	6.1
Air heater margin	—	3.0
Ducts to precipitator[a]	1.0	1.3
Precipitator[a]	1.5	2.0
Ducts to scrubber	1.0	1.3
Scrubber	5.0	6.6
Reheater (if required)	—	—
Ducts to stack	1.0	1.3
Stack	1.0	1.5
Total system pressure drop	22.5	31.5
Control allowance	—	1.0
Total static pressure	22.5	32.5

[a]Substitute data for fabric filter if applicable.

10.8.5 Primary Air Fan Sizing

Under normal circumstances the PA fans are purchased with the steam generator because the PA fans are such an integral part of the coal flow and combustion process for the steam generator. It is a good policy to specify in the steam generator contract that the PA fans must be sized to supply the required PA flow to operate the steam generator at 70% load with only one PA fan operating.

For conceptual design studies, the PA fans are sized with a 10% margin on the PA fan flow credit used for FD fan sizing. PA fan static pressure requirements should be based on manufacturer-supplied information and on the designer's experience with each of the components in the PA system.

10.9 FAN DRIVES

Steam generator draft fans are normally driven with one of five drive systems:

- Single-speed induction motor,
- Two-speed induction motor,
- Adjustable-speed electric motor,
- Steam turbine (either condensing or noncondensing), or
- Variable-speed fluid drive.

10.9.1 Single-Speed Induction Motor

The single-speed induction motor is the most common fan drive and has the lowest capital cost of the alternative drive arrangements. Single-speed motors are widely used because of their rugged construction and the absence of moving electrical contacts, that makes motor maintenance requirements very low. Steam generator draft fans driven with single-speed motors have been selected to operate at speeds as high as 1,800 rpm and as low as 450 rpm. The speed at that induction motors are designed to operate depends on the number of poles in the motor field winding. Motor operating speeds and the associated number of motor field winding poles, based on a 60 hertz power supply, are tabulated below.

Motor operating speed (rpm)	Number of poles in field winding
1,800	4
1,200	6
900	8
720	10
600	12
514	14
450	16

10.9.2 Two-Speed Induction Motor

Two-speed induction motors operate at either of two constant speeds. The higher speed is typically selected to satisfy test block conditions, and the lower speed is selected to satisfy the system requirements at the MCR load point. In actual operation, the fan normally operates at the lower speed, using the higher speed only if air heater leakage, system cleanliness, or other factors necessitate additional fan capacity.

Some installations in the past used two-speed motors with a low speed selected to satisfy the system requirements at the load where the maximum operating hours occur and a high speed selected to satisfy all higher load points, including test block. Some of these installations have experienced problems with motor speed changes. As a result, many operators were reluctant to change speeds and, therefore, the advantages of two-speed motors were seldom realized.

Speeds are typically changed by switching the internal winding connections at six lead points in the motor terminal box. The usual method is to use a nonload break, oil-filled switch (five pole) in conjunction with a standard three-pole circuit breaker. In transferring from low to high speed, the breaker is first opened. After a delay of about 2 s, the switch is activated to reconnect the motor windings and the breaker is then reclosed. On transferring from high to low speed, the motor is usually allowed to coast after the breaker is opened and the speed is transferred until a speed monitor senses the low speed and the breaker is reclosed. During the speed change cycle, there should be ample time for the steam generator control system to reposition the fan control dampers to avoid pressure transients in the furnace. An automatic control sequencer would normally be provided to perform the speed switching operation. The control sequencer would be manually initiated.

10.9.3 Adjustable Speed Motor

Motor manufacturers have recently developed adjustable speed control systems for large motors that use solid-state electronic components to control motor speed. These control systems permit the motor efficiency to be maintained at very high levels across the entire motor operating range and are available to control either wound rotor motors or synchronous motors.

Typically, adjustable speed synchronous motors are specified because of their lower maintenance requirements and slightly higher efficiency. The variable-speed synchronous motor is controlled with an adjustable frequency control system. The motor speed is adjusted from rated speed to a very low speed by reducing the frequency and line voltage imposed on the motor terminals. The AC amperage, line voltage, and frequency are controlled through the use of static electronic control converters. This power conversion system consists of two three-phase thyristor bridges with their DC terminals connected through a reactor. The first-stage three-phase bridge is connected to the 60 hertz power supply and operated as a phase-controlled rectifier to supply power to a DC link. The second bridge is connected to the synchronous motor and inverts the power from the DC link to the stator of the motor.

10.9.4 Steam Turbine

Basically, two types of steam turbines are available for draft fan drives: condensing and noncondensing.

A condensing turbine normally drives fans through zero speed disconnect couplings and speed reduction gears. The condensing turbine receives steam from the main turbine low-pressure crossover piping and exhausts to a condenser directly below the fan drive turbine. Condensate is returned to the main condenser hot well.

A noncondensing turbine is arranged in the same manner as a condensing turbine except for the steam supply and exhaust. Supply steam is taken from the cold reheat line or the main turbine high-pressure extraction, and exhaust steam is routed to a feedwater heater. The noncondensing turbine does not provide as great a heat rate advantage as does the condensing turbine.

10.9.5 Variable-Speed Fluid Drive

Three basic types of variable-speed fluid couplings are available for draft fan service: hydrokinetic, hydroviscous, and two-speed fluid coupling. Any of the three are driven with single-speed electric motors.

10.9.5.1 Hydrokinetic. Figure 10-28 shows a typical example of a hydrokinetic fluid coupling. This coupling consists of a runner attached to the driven shaft that receives a vortex of oil from the impeller attached to the driving shaft. There is no mechanical connection between the runner and the impeller, and the two are almost identical in shape. Kinetic energy is imparted to the oil by the impeller, and the oil flows radially outward and into the vanes of the runner. The oil then flows through the runner and transfers its energy to the runner, similar to steam imparting energy to a turbine blade. The speed of the runner and the energy transferred to the runner by the impeller are controlled by the amount of oil flowing from the impeller. The oil flow from the impeller is regulated by a controllable scoop tube that regulates the depth of oil in the impeller assembly.

10.9.5.2 Hydroviscous. Figure 10-29 shows a typical hydroviscous fluid coupling. The hydroviscous fluid coupling operates on the principle of shearing an oil film to transmit torque. The amount of torque that can be transmitted varies with the number of disc surfaces over that the shearing action occurs. To increase the number of working surfaces, a hydroviscous disc pack is built up by alternately stacking discs splined to the input shaft between discs splined to the output shaft. All discs are free to move axially but must rotate with the member to that they are splined. The controllable clamp force is applied to one end of the disc pack and distributed to the remainder of the discs due to the freedom of the disks to slide axially. The working oil is introduced to each set of

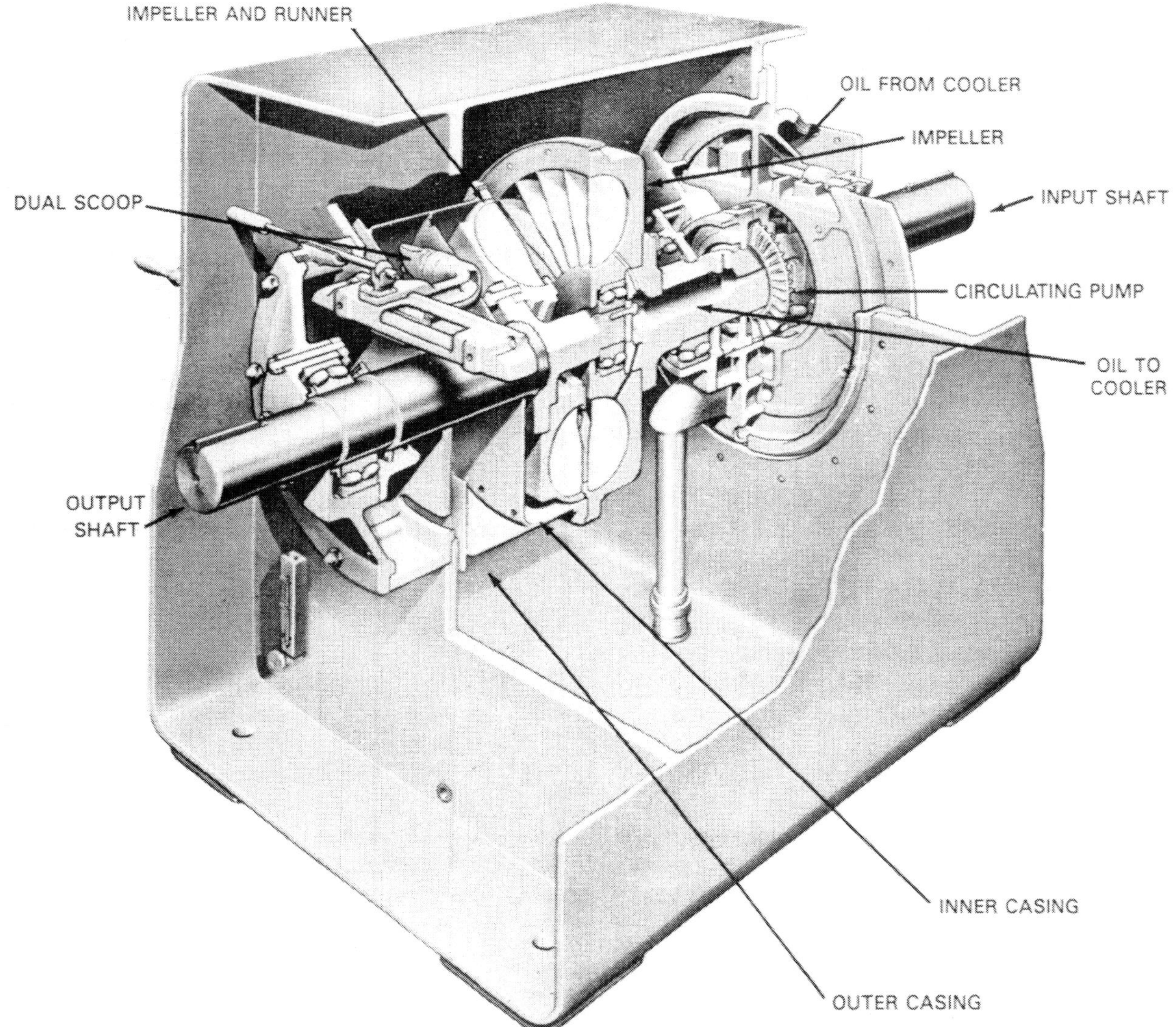

Fig. 10-28. Hydrokinetic fluid coupling. (From Howden Sirocco, Inc., Gyro® Fluid Drive. Used with permission.)

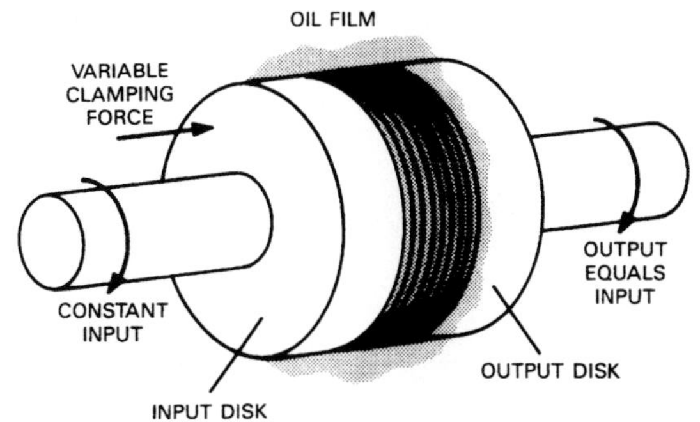

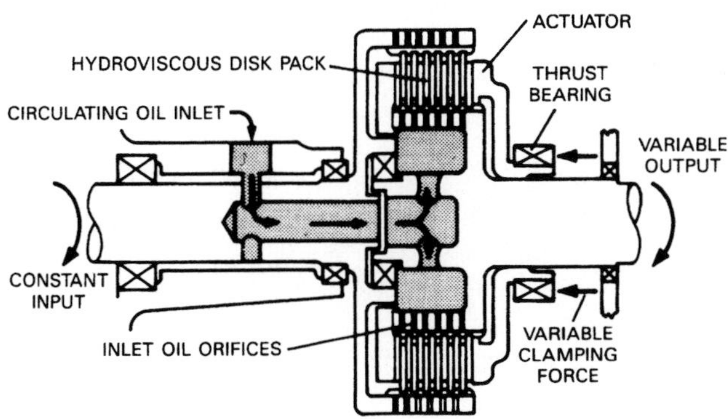

10-29. Hydroviscous fluid coupling.

surfaces to set up the hydrodynamic film between each of the discs. The working oil is continuously circulated through the disc pack to dissipate the heat generated by the shearing action. By varying the clamping force with hydraulic pressure on one end of the disc pack, the input-to-output speed ratio can be controlled from very low output speed to full no-slip output speed. The major advantage of the hydroviscous fluid drive over the hydrokinetic fluid drive is this no-slip, lockup capability.

By varying the speed of the fan rather than controlling with inlet vanes or dampers, peak fan efficiency can be maintained over the operating range. The fan efficiency remains high, but efficiency losses due to fixed losses and slip losses in the fluid coupling are introduced into the fan system. The amount of slip loss in the fluid drive is proportional to the ratio of the in-put speed over the output speed and can be determined by the following equation:

$$\text{HP slip losses} = (\text{HP fan})\left(\frac{\text{RPM}_{\text{in}}}{\text{RPM}_{\text{out}}} - 1\right) \qquad (10\text{-}25)$$

10.10 REFERENCES

Asea Brown Boveri, ABB Power Plant Systems. 1000 Prospect Hill Road, Post Office Box 500, Windsor, CT 06095-0500.

AMCA. 1973. Fans and Systems. *Fan Application Manual.* AMCA Publication 201. Air Movement and Control Association, Inc., Arlington Heights, IL.

Hoffman, J. H. 1976. Turning gear is solution to vibration problems when restarting hot-gas fans. *POWER* 120(2):55–56.

Howden Sirocco, Inc. One Westinghouse Plaza, Suite 300, Hyde Park, MD 02136.

Jorgensen, R., editor. 1983. *Fan Engineering, an Engineer's Handbook on Fans and Their Applications.* Buffalo Forge Company, Buffalo, NY.

Joy Technologies, Inc. Joy/Green Fan Division. 338 South Broadway, P.O. Box 5000, New Philadelphia, OH 44663.

Novenco, Inc. 685 Martin Drive, South Elgin, IL 60177-5700.

Osborne, William C. 1966. *Fans.* Pergamon Press, Oxford.

Rodgers, J. D. and C. H. Gilkey. 1975. A Summary of Experience with Fan Induced Duct Vibrations on Fossil Fueled Boilers. *Fossil Power Systems.* Combustion Engineering, Inc., Windsor, CT.

Singer, Joseph G., editor. 1981. *Combustion Fossil Power Systems,* Third edit. Combustion Engineering, Inc., Windsor, CT.

Stultz, S. C. and J. B. Kitto, editors, 1992. *Steam: Its Generation and Use.* Stacks, Fans and Drafts (Chapter 17). Babcock & Wilcox Company, New York, NY.

11

Pumps

Lawrence J. Seibolt

11.1 INTRODUCTION

A pump is a machine that imparts energy into a liquid to lift the liquid to a higher level, to transport the liquid from one place to another, to pressurize the liquid for some useful purpose, or to circulate the liquid in a piping system by overcoming the frictional resistance of the piping system.

The pump is one of the oldest machines invented by man. The earliest pumps devised for raising water, such as the Persian and Roman waterwheels and the more advanced Archimedean screw, were invented to raise water from the hold of a ship. Since that time, many variations and applications of pumps have been developed, which in most cases can be classified into two basic methods used to impart energy to a liquid: volumetric or positive displacement, and addition of kinetic energy. Volumetric (positive) displacement of a fluid can be accomplished either mechanically by the action of a screw or a plunger, or by the use of another fluid. Kinetic energy can be added to a fluid by rotating the fluid at high speeds, in a device typically known as a centrifugal pump, or by providing an impulse in the direction of flow.

Because of the many and varied uses and applications for a pump, it is easy to understand why so many types of pumps have evolved, as shown in Fig. 11-1.

This chapter focuses on pumps in power station applications and those characterized as kinetic, or more commonly known as centrifugal pumps. This narrowing of the focus permits a more detailed examination of power station pump applications and their associated pumping systems, and therefore positive displacement type pumps are not discussed further here. A good starting source for a basic understanding of these pumps is the *Hydraulic Institute Standards*. Table 11-1 illustrates some of the typical power station pump applications and the type of pump commonly used in each application.

11.2 KINETIC PUMP TYPES AND CHARACTERISTICS

Kinetic pumps can be divided into two classes: centrifugal and regenerative. As listed in Table 11-1, the most common type of kinetic pump used in a modern central power plant is the centrifugal pump. Although the first centrifugal pump was developed in the late 1600s, most of the development of this pump type has occurred in the present century. Centrifugal pumps include radial, axial, and mixed flow types, with the radial flow volute type used for the bulk of power plant applications. Regenerative pumps are typically referred to as vortex, peripheral, or turbine pumps.

11.2.1 Volute Type Centrifugal Pumps

The operating principle of a volute pump is shown in Fig. 11-2. Liquid enters the pump at the impeller eye and is thrown radially outward through an expanding pump casing. This action creates a low pressure at the impeller eye which draws more liquid into the impeller. The velocity head imparted to the liquid by the impeller is converted into static pressure by the widening spiral pump casing.

11.2.2 Diffuser Type Centrifugal Pumps

Another type of radial flow centrifugal pump is the diffuser pump, shown in Fig. 11-3. As with the volute centrifugal pump, the liquid enters a diffuser type pump at the impeller eye and is thrown outward along the impeller vanes. After the liquid has left the impeller, it is passed through a ring of stationary guide vanes that surround the impeller and diffuse the liquid to provide a controlled flow and efficient conversion of velocity head into static pressure.

11.2.3 Axial Flow and Mixed Flow Centrifugal Pumps

In an axial flow pump, the impeller acts as a propeller. Liquid flows parallel to the axis or shaft of the pump, as shown in Fig. 11-4. Pressure is generated by the propelling and lifting action of the impeller vanes on the liquid. Normally, diffusion vanes are located on the discharge side of the pump to eliminate rotation and radial velocity of the liquid imparted by the impeller.

In mixed flow pumps, liquid is discharged both radially and axially into a volute type of casing. Pressure is developed both by centrifugal force and by the lift of the impeller vanes on the liquid. The action of a mixed flow impeller is shown in Fig. 11-5.

11.2.4 Regenerative Pumps

A regenerative type of centrifugal pump has an impeller with vanes on both sides of a rim that rotates in a channel in the pump's casing, as shown in Fig. 11-6. Liquid enters through a

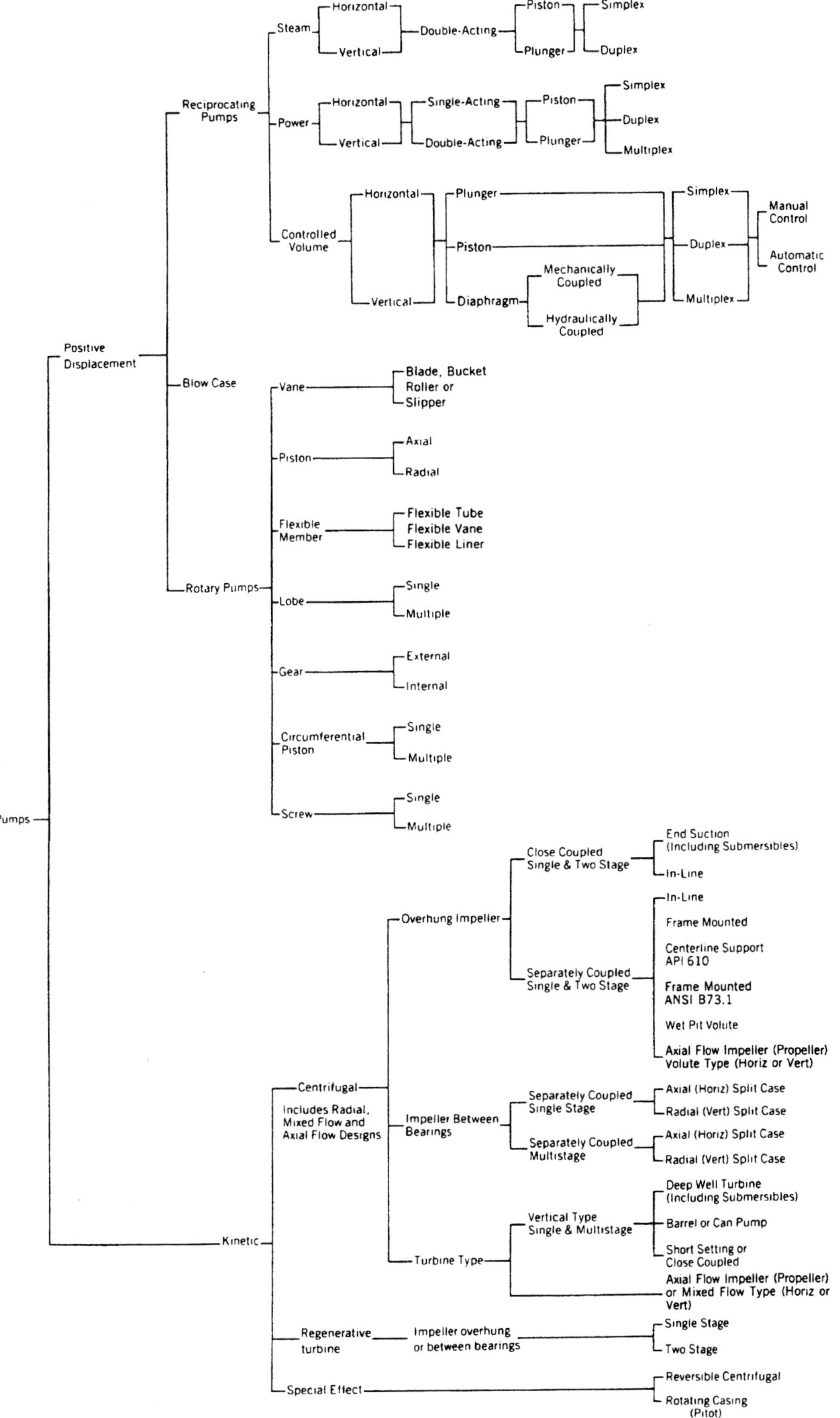

Fig. 11-1. Types of pumps. (From Hydraulic Institute. Used with permission.)

Table 11-1. Power Station Pump Applications

Application	Pumped Fluid	Typical Pump Type
Steam cycle heat rejection		
Circulating water	Surface water (seawater, lake water, treated sewage effluent, etc.)	Mixed flow vertical or HSCP
Auxiliary cooling water	Surface water (seawater, lake water, treated sewage effluent, etc.)	Mixed flow vertical or HSCP
Screen wash	Surface water (seawater, lake water, treated sewage effluent, etc.)	Turbine-type, vertical, multistage
Cooling tower makeup	Surface water (seawater, lake water, treated sewage effluent, etc.)	Turbine-type, vertical, multistage
Condenser vacuum	Condensate and noncondensible gases	Positive displacement (rotary)
Preboiler–boiler feed/condensate		
Condensate	Condensate/feedwater	Centrifugal (turbine type, vertical, multistage, can type)
Boiler feed	Condensate/feedwater	Centrifugal (horizontal, diffuser or volute type, double can barrel)
Boiler feed booster	Condensate/feedwater	Centrifugal (horizontal, diffuser or volute type, double can barrel)
Heater drains	Condensate/feedwater	Centrifugal (CMVSP)
Bearing injection water	Condensate/feedwater	Centrifugal (HSCP or FMESP)
Air preheater water	Condensate/feedwater	Centrifugal (CMVSP)
Flue gas reheater water	Condensate/feedwater	Centrifugal (CMVSP)
Cycle makeup and storage		
Condensate makeup	Demineralized water	Centrifugal (FMESP or HSCP)
Demineralized water	Demineralized water	Centrifugal (FMESP or HSCP)
Condensate miscellaneous services	Demineralized water	Centrifugal (FMESP or HSCP)
Fuel oil pumps		
Fuel oil unloading	Fuel oil distillate	Centrifugal (FMESP or HSCP) or positive displacement (rotary)
Fuel oil transfer	Fuel oil distillate	Centrifugal (FMESP or HSCP) or positive displacement (rotary)
Fuel oil supply	Fuel oil distillate	Centrifugal (FMESP or HSCP) or positive displacement (rotary)
Injector oil	Fuel oil distillate	Centrifugal (FMESP or HSCP) or positive displacement (rotary)
Plant service pumps		
Firewater	Service/firewater	Centrifugal (HSCP or turbine type, vertical, multistage)
Firewater booster	Service/firewater	Centrifugal (HSCP, FMESP, or turbine type, vertical, multistage)
Firewater jockey	Service/firewater	Centrifugal (HSCP, FMESP, or turbine type, vertical, multistage)
Service water	Service water	Centrifugal (HSCP or FMESP)
Closed cycle cooling water	Treated condensate	Centrifugal (HSCP)
Chilled water	Treated condensate	Centrifugal (CCMP)
Hot water heating	Condensate	Centrifugal (CMVSP)
Well water	Well water	Centrifugal (turbine type, vertical, multistage, deep-well type)
Slurry pumps		
Ash transfer	Ash/water slurry	Centrifugal, single-stage, single-suction, volute, vertically split
Ash sludge	Ash/water slurry	Centrifugal, single-stage, single-suction, volute, vertically split
Scrubber spray	Limestone/water slurry	Centrifugal, single-stage, single-suction, volute, vertically split
Thickener underflow	Limestone/water slurry	Centrifugal, single-stage, single-suction, volute, vertically split
Chemical pumps		
Ammonia feed	Ammonia solution	Positive displacement, diaphragm
Caustic feed	Concentrated (50%) sodium hydroxide	Positive displacement, diaphragm
Acid feed	Sulfuric or hydrochloric acid	Centrifugal, vertical (sulfuric acid), positive displacement, diaphragm (hydrochloric acid)
Hydrazine feed	Hydrazine solution	Positive displacement, diaphragm
Phosphate feed	Phosphate solutions	Positive displacement, diaphragm
Demineralizer	Demineralized water	Centrifugal (FMESP or HSCP)
Regeneration water	Demineralized water	Centrifugal (FMESP or HSCP)

FMESP = frame-mounted end suction pump; HSCP = horizontal split-case pump; CCMP = close-coupled motor pump; CMVSP = centerline mounted vertically split pump.

nozzle into an impeller vane and is forced outward by centrifugal force. The liquid impacts the pump casing and is turned inward to reenter the impeller at a different vane. This cycle is repeated throughout the rotation of the impeller, generating pressure until the liquid is forced out of the pump at the discharge nozzle. Because of close running tolerances, the regenerative pump is suitable only for clear liquids with relatively low viscosities (less than 250 SSU). These pumps are useful in pumping liquids containing vapors and gases because of their resistance to cavitation.

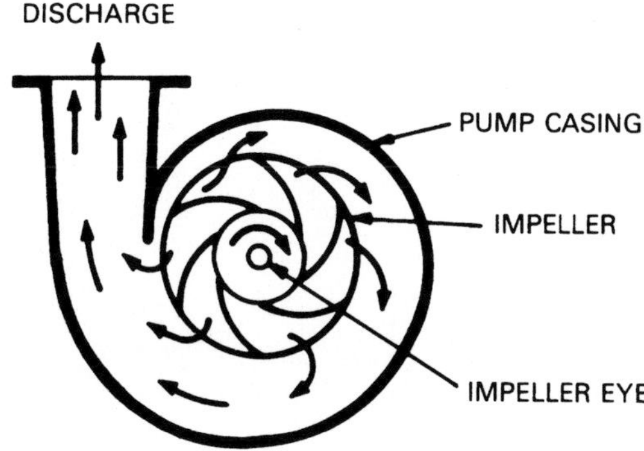

Fig. 11-2. Volute centrifugal pump.

11.2.5 Specific Speed and Impeller Configurations

A dimensionless index of pump type known as specific speed has been developed for pump design and selection to show the relationship between pump capacity, head, and impeller speed. Specific speed of an impeller is defined as the speed in revolutions per minute at which a geometrically similar impeller would operate to deliver 1 gpm at a developed head of 1 ft. Specific speed is algebraically defined in Eq. 11-1.

$$N_s = \frac{n\sqrt{Q}}{h^{\frac{3}{4}}} \tag{11-1}$$

where

N_s = specific speed;

n = pump speed, rpm;

Q = pump flow at best efficiency point (BEP), gpm; and

h = pump developed head at BEP, ft.

Specific speed characterizes the shape and configuration of an impeller. Since the ratios of the major impeller dimensions vary uniformly with specific speed, specific speed is useful to the pump designer in determining impeller proportions and dimensions required, as well as to the application engineer in checking suction limitations of the pump.

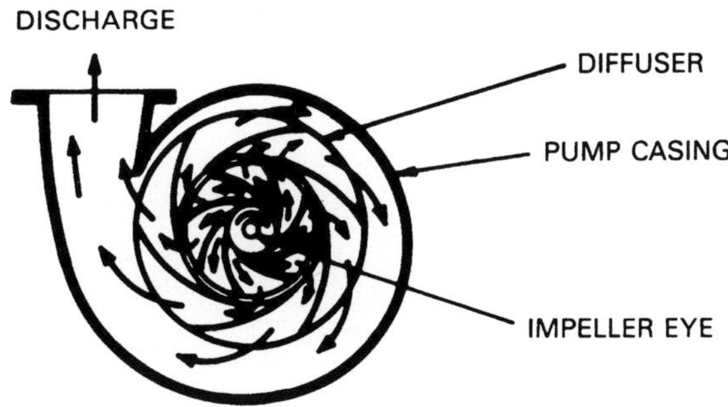

Fig. 11-3. Diffuser centrifugal pump.

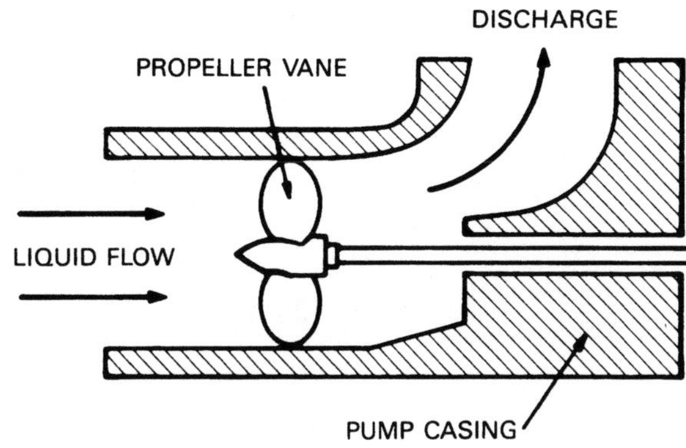

Fig. 11-4. Axial flow pump.

It should be noted that Eq. (11-1) for specific speed is written for single-stage pump applications. For multistage pumps, the head per stage is used to calculate specific speed. Generalizing, the head per stage is found by dividing the total head of the pump by the number of stages of the pump.

The following example illustrates the specific speed calculation for a first-stage impeller of a five-stage pump. The pump is designed to operate with the following conditions:

Pump flow, Q = 5,000 gpm
Pump total head, h = 7,500 ft
Pump speed, n = 5,450 rpm

For the first stage impeller:

$$N_s = \frac{5{,}450 \text{ rpm } (5{,}000 \text{ gpm})^{\frac{1}{2}}}{\left(\frac{7{,}500 \text{ ft}}{5 \text{ stages}}\right)^{\frac{3}{4}}} = 1{,}600$$

Impeller form and proportions vary with specific speed, as shown in Fig. 11-7.

Each type of impeller has been found to have a range of specific speeds in which it will give the best performance. Radial-vane area impellers have specific speeds up to about 1,000, whereas Francis-vane area impeller specific speeds go up to 4,000 to 4,500. The specific speeds of mixed flow area impellers range from that of the Francis-vane impellers up to about 9,500 to 10,000. The specific speeds of axial flow area

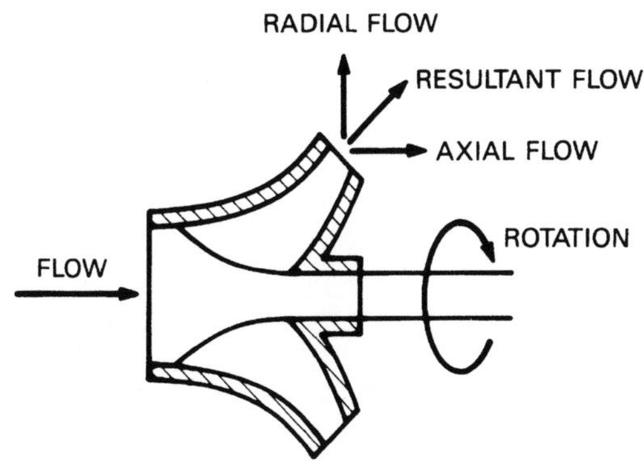

Fig. 11-5. Mixed flow impeller.

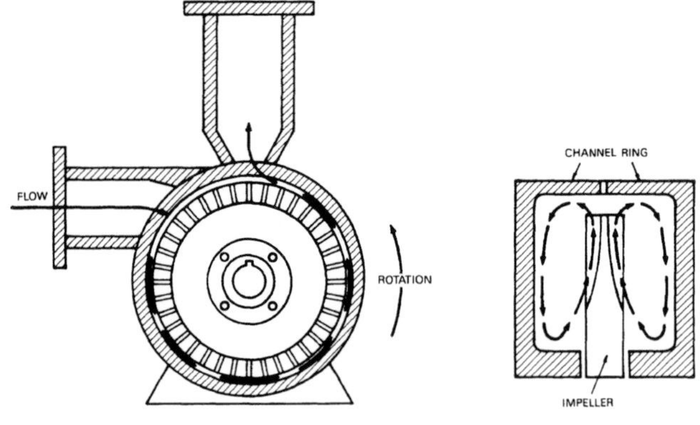

Fig. 11-6. Regenerative turbine pump.

impellers range from 10,000 to 14,000. Generally, as flow goes up and pump head decreases, specific speed increases.

For power station centrifugal pump applications, the following specific speeds and impeller profiles are generally seen:

Application	Typical impeller profile	Specific speed, N_s
Boiler feed	Radial-vane area or Francis-vane area	1,000–1,800
Condensate	Francis-vane area	1,500–2,500
Circulating water	Mixed flow area	4,500–7,500

In addition, the impeller configuration and specific speed characterize the shape of the pump operating head-capacity curve, as shown in Fig. 11-8.

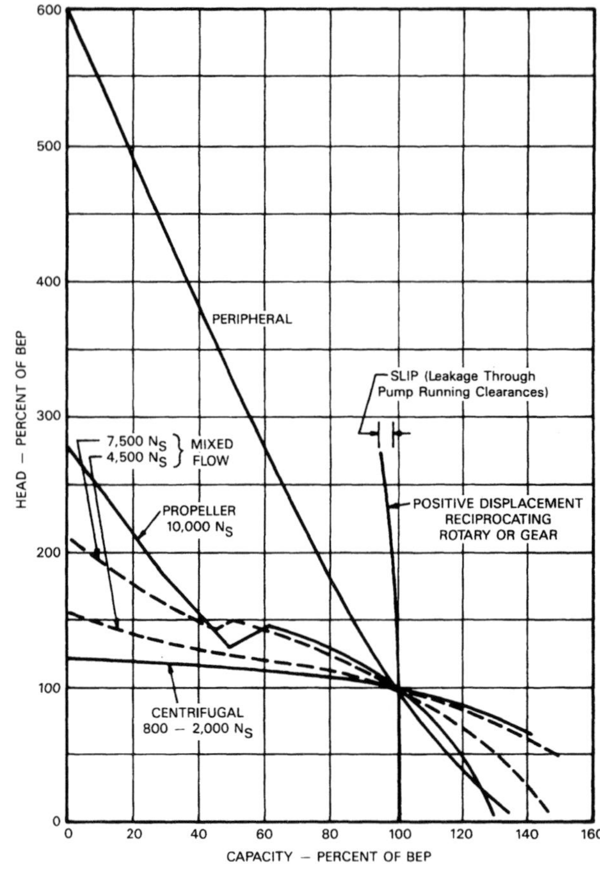

Fig. 11-8. Head capacity curves for various pump impeller types.

11.3 PUMP HYDRAULIC NOMENCLATURE AND DEFINITIONS

11.3.1 Terminology

Accurate terminology is essential to those who make pump applications and selections. Many pumping terms are confusing to both experienced and inexperienced pump designers and application engineers. One of the best sources of current terminology, nomenclature, and definitions is the *Hydraulic Institute Standards*. The definitions and terminology used herein are based on today's practice among pump manufacturers.

Pressure. Figure 11-9 illustrates the relationship between gauge and absolute pressures. Perfect vacuum cannot exist on the surface of the earth; however, it makes a convenient datum for pressure measurement.

Depending on the designer's or application engineer's preference, calculations regarding pressure may be made in any unit of pressure. Some calculations are simplified by using either gauge or absolute pressure terms, but this is left to the designer's experience.

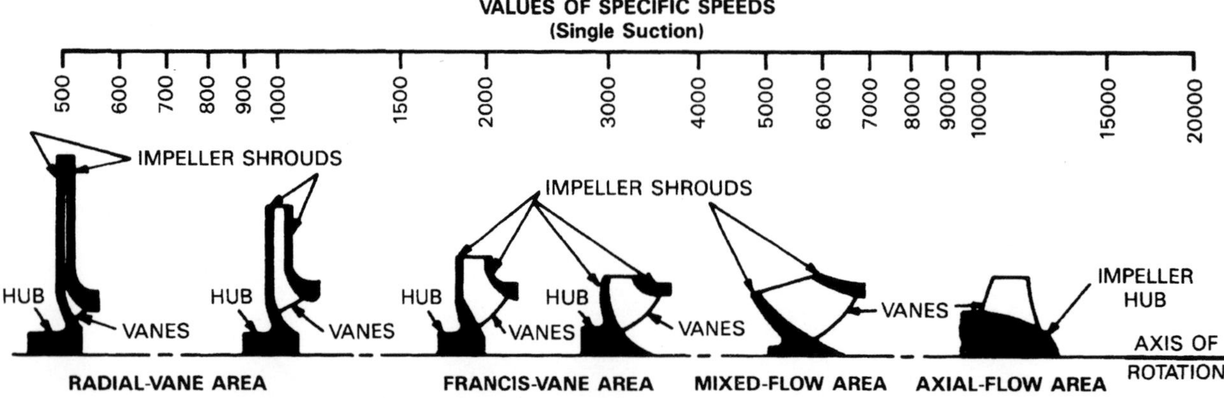

Fig. 11-7. Comparison of pump profiles. (From Hydraulic Institute. Used with permission.)

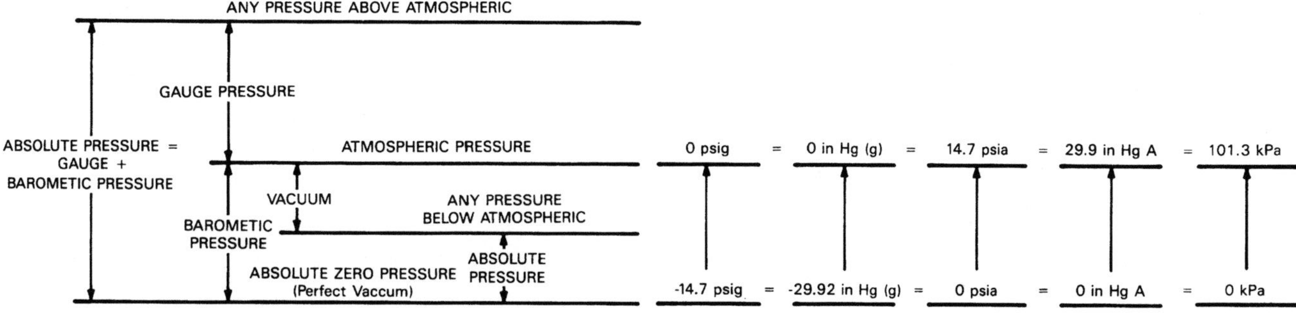

Fig. 11-9. Relationship between various pressure terms used in pumping.

Barometric pressure is the level of the atmospheric pressure above perfect vacuum. It is also the pressure at the site being investigated, and changes with weather condition and altitude.

"Standard" atmospheric pressure is 14.696 lb/in.^{2} absolute (101.326 kilopascals, kPa).

Gauge pressure is measured above atmospheric pressure, whereas **absolute pressure** always refers to perfect vacuum as the datum point.

Vacuum, usually expressed in inches of mercury, is the depression of pressure below the atmospheric level.

Head. The pressure at any point in liquid contained in a vertical column of that liquid is caused by the weight of that liquid in the column above the point in question. The height of the column of liquid is called **static head** and is normally expressed in terms of feet (meters) of the liquid.

The static head which corresponds to any pressure is dependent upon the weight of the liquid and is calculated using Eq. (11-2).

$$\text{Static head, ft} = \frac{\text{Pressure, psig} \times 2.31}{\text{Specific gravity of the liquid (SG)}} \qquad (11\text{-}2)$$

Suction lift. The vertical distance in feet (meters) from the liquid supply level to the pump center line with the pump physically located above the liquid level supply is defined as **suction lift.** A suction line may run several feet before entering the pump and may include horizontal piping runs. These horizontal piping runs are not considered as part of the suction lift, as shown in Figs. 11-10a and c.

Suction head. The vertical distance in feet (meters) from the liquid supply level to the pump center line with the pump physically located below the liquid level supply is defined as **suction head** (Fig. 11-10b). Again, horizontal piping runs are not considered as part of the suction head.

Static discharge head. The vertical distance in feet (meters) between the pump center line and the point of free discharge on the surface of the liquid in the discharge tank is defined as static discharge head (Figs. 11-10a, b, and c).

Total static head. The vertical distance in feet (meters) between the liquid level of the supply and the point of free discharge on the surface of the liquid in the discharge tank is defined as **total static head** (Figs. 11-10a, b, and c).

Friction head. The head required to overcome the resistance to flow in the pipe and fittings is defined as **friction head.** It is dependent on the size and type of pipe, flow rate, and liquid prop-Erties.

Velocity head. The energy of a liquid as a result of its motion at some velocity is defined as **velocity head.** Velocity head is the equivalent head in feet (meters) through which a liquid would have to fall to acquire the velocity, or the head, necessary to accelerate the liquid.

Velocity head is calculated from Eq. 11-3.

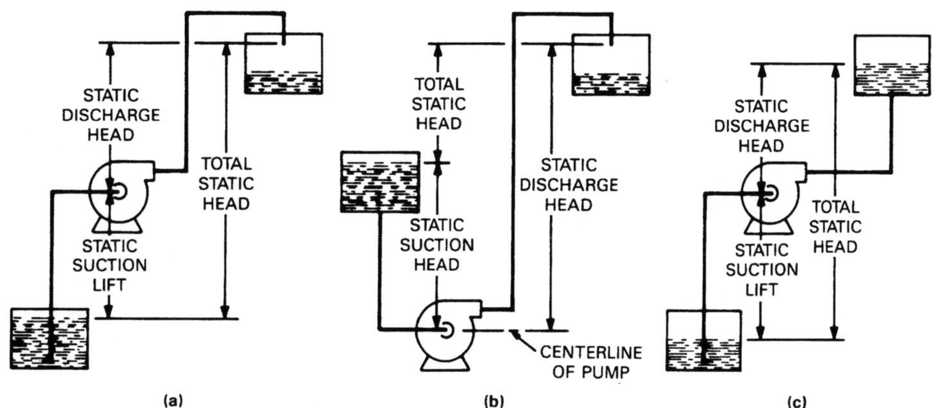

Fig. 11-10. Head terms used in pumping.

$$h_v = \frac{V^2}{2g} \tag{11-3}$$

where

h_v = velocity head, ft (m);

V = liquid velocity, ft/s (m/s); and

g = acceleration due to gravity = 32.2 ft/s^2 (9.81 m/s^2).

Normally, the velocity head is insignificant and can be ignored in most high-head system applications such as a boiler feed system. However, in a low-head system, such as a circulating water system, velocity head can be a large factor and must be considered in the design.

Pressure head. If a pumping system begins or ends in a tank in which internal pressure is above or below atmospheric pressure, the **pressure head** of the pumping system must be included in the head calculation of a pump. The pressure in such a tank must be first converted to feet (meters) of the liquid. The vacuum in a suction tank or a positive pressure in a discharge tank must be added to the pumping system head; on the other hand, a positive pressure in a suction tank and a vacuum in a discharge tank must be subtracted.

Total dynamic suction lift (h_s) is the static suction lift plus the velocity head at the pump suction flange plus the total friction head in the suction pipeline.

Total dynamic suction head (h_s) is the static suction head minus the velocity head at the pump suction flange minus the total friction head in the suction pipeline.

Total dynamic discharge head (h_d) is the static discharge head plus the velocity head at the pump discharge flange plus the total friction head in the discharge line.

Total head or total dynamic head (total developed head—TDH) is the total dynamic discharge head minus the total dynamic suction head or plus the total dynamic suction lift.

$\text{TDH} = h_d - h_s$ (for a suction head)

$\text{TDH} = h_d + h_s$ (for a suction lift)

Capacity is the throughput of the pump normally expressed in gallons per minute (gpm), (cubic meters per hour, m^3/h). Since liquids are essentially incompressible, there is a direct relationship between the capacity in a pipeline and the velocity of the flow, as shown in Eq. (11-4).

$$Q = A \times V \text{ or } V = \frac{Q}{A} \tag{11-4}$$

where

Q = capacity, ft^3/s (m^3/sec);

A = area of the pipeline, ft^2 (m^2); and

V = liquid velocity, ft/s (m/s).

The conversion from ft^3/min to gpm can be accomplished if the liquid specific volume is known.

Power and Efficiency. The pump output is normally expressed as liquid horsepower (water horsepower), as shown in Eq. (11-5). The pump input or power delivered to the pump shaft is normally expressed as brake horsepower, as shown in Eq. (11-6).

$$\text{Water horsepower (whp)} = \frac{Q(\text{TDH})\ (\text{SG})}{3{,}960} \tag{11-5}$$

$$\text{Brake horsepower (bhp)} = \frac{Q(\text{TDH})\ (\text{SG})}{3{,}960 \times \text{Pump efficiency}} \tag{11-6}$$

The constant 3,960 is determined by dividing the number of foot-pounds per minute for one horsepower (33,000) by the weight (pounds) of water per gallon (8.33).

The brake horsepower is greater than the water horsepower due to the mechanical and hydraulic losses in the pump. The pump efficiency is expressed as the ratio of these two horsepower values as shown in Eq. (11-7).

$$\text{Pump efficiency} = \frac{\text{whp}}{\text{bhp}} = \frac{Q(\text{TDH})\ (\text{SG})}{3{,}960\ (\text{bhp})} \tag{11-7}$$

11.3.2 Affinity Laws

Another important tool used by pump designers and application engineers is Affinity Laws. These laws express the mathematical relationship and illustrate the effect of changes in pump operating conditions or pump performance variables such as pump head, flow, speed, horsepower, and pump impeller diameters. The affinity laws are summarized by Eqs. (11-8), (11-9), (11-10), and (11-11), where the subscript 1 refers to the original conditions and 2 refers to the new conditions.

$$\text{Flow } Q_2 = Q_1 \frac{n_2}{n_1} \frac{D_2}{D_1} \tag{11-8}$$

$$\text{Head } h_2 = h_1 \left(\frac{n_2}{n_1}\right)^2 \left(\frac{D_2}{D_1}\right)^2 \tag{11-9}$$

$$\text{Horsepower } hp_2 = hp_1 \left(\frac{n_2}{n_1}\right)^3 \left(\frac{D_2}{D_1}\right)^3 \tag{11-10}$$

where

Q = pump flow, gpm (m^3/h);

n = pump speed, rpm;

D = impeller diameter, in. (m);

h = pumphead, ft (m);

hp = pump brake horsepower, bhp (kw); and

$$\text{Efficiency } \varepsilon_{p2} = \varepsilon_{p1} \tag{11-11}$$

(Nearly constant for pump speed changes and small changes in impeller diameter)

where

ε_p = pump efficiency, %.

For example, assume that a variable-speed pump is delivering 3,000 gpm of water with a total head of 200 ft when operating at 1,550 rpm and an efficiency of 83%. The brake horsepower is 182 bhp. The outside diameter of the impeller is 12 in. Calculate the performance of the pump if the impeller outside diameter is increased to 12.5 in. and the speed increased to 1,600 rpm.

$$Q_2 = 3{,}000 \text{ gpm}\left(\frac{1{,}600 \text{ rpm}}{1{,}550 \text{ rpm}}\right)\left(\frac{12.5 \text{ in.}}{12.0 \text{ in.}}\right) = 3{,}226 \text{ gpm}$$

$$h_2 = 200 \text{ ft}\left(\frac{1{,}600 \text{ rpm}}{1{,}550 \text{ rpm}}\right)^2\left(\frac{12.5 \text{ in.}}{12.0 \text{ in.}}\right)^2 = 231 \text{ ft}$$

$$\text{hp}_2 = 182 \text{ bhp}\left(\frac{1{,}600 \text{ rpm}}{1{,}550 \text{ rpm}}\right)^3\left(\frac{12.5 \text{ in.}}{12.0 \text{ in.}}\right)^3 = 226 \text{ bhp}$$

$$\varepsilon_{p2} = \varepsilon_{p1} = 83\% \text{ (Efficiency)}$$

More typically, either the pump speed or the pump impeller outside diameter is held constant and pump characteristic curves such as those shown in Figs. 11-11a and 11-11b developed accordingly. This method of presentation allows the application engineer to determine the best pump, pump speed, and impeller diameter size for the system in which the pump will be used. Normally, when selecting a pump to fit the system head and flow parameters, the pump characteristic curve is chosen that defines the most efficient pump performance for an impeller outside diameter that is mid-range for the pump casing size. This is done to allow modifications to the pump impeller outside diameter (if necessary) making the impeller either larger or smaller, without changing the pump casing size. If a larger impeller should be required, another consideration is the horsepower available from the pump driver. Since the larger impeller will provide more flow and total head for the same pump casing size, the driver may be undersized to provide the horsepower needed to improve the pump's performance.

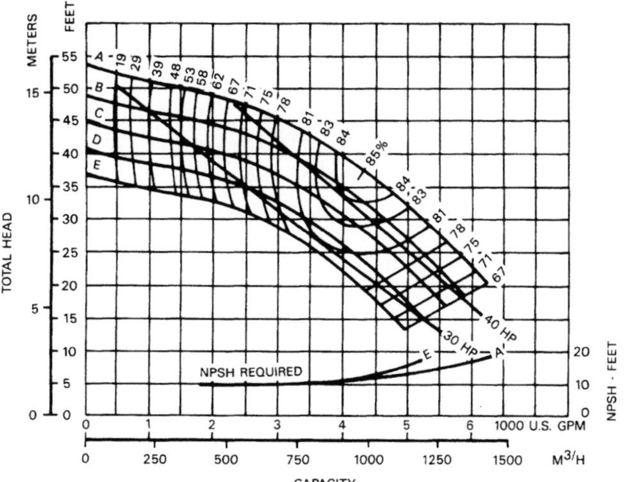

Fig. 11-11a. Typical pump rating chart—constant speed with various impeller diameters.

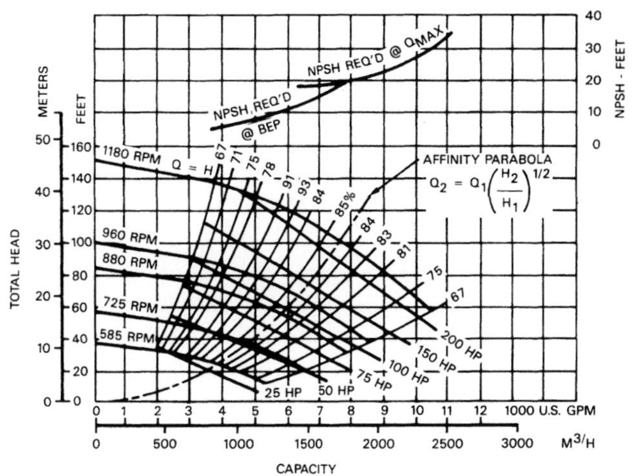

Fig. 11-11b. Typical pump rating chart—constant-diameter impeller with various speeds.

11.3.3 Cavitation

In the handling of incompressible liquids, all pumps have a tendency to cavitate when the pressure in the pump inlet regions approaches the vapor pressure of liquids. The total energy at any point in the flowing liquid consists of pressure and velocity (kinetic) heads. Therefore, until the impeller vanes begin to add energy to the liquid being pumped, the pressure decreases as the velocity increases. So, the suction pressure necessary to prevent cavitation increases with an increase in the pump output. Cavitation is a two-stage phenomenon consisting of the formation of vapor cavities resulting from low pressure and their collapse as they move out of the low-pressure into higher pressure regions. The higher pressure region causing the vapor cavity to collapse can be immediately following the formation of the vapor cavity or some distance downstream from the impeller inlet, depending on the downstream pressure conditions and the quantity of vapor formed. The formation and collapse of the vapor cavities occur in a very short time interval. It is at the point of collapse where the physical damage to the impeller occurs. Therefore, when selecting a pump for a piping system, it is necessary to avoid cavitation in the pump operating range. The pump must be located so that there is sufficient pressure above the vapor pressure at the pump inlet to ensure no cavitation.

The cavitation characteristics of each pump design are determined by testing on the basis of handling water at or near room temperature. However, the pump application may be for another fluid of varying viscosities or temperatures. In each case, the pump manufacturer should be consulted for accompanying changes in the cavitation characteristics of the pump design.

At the inception of cavitation there may not be a measurable decrease in pump performance although pump damage may be taking place. As cavitation increases, there is a reduction in pump output because the vapor cavities occupy some of the impeller water passage area. This reduces developed head and efficiency. The point at which suction lift begins to affect head and efficiency is the critical condition shown on the pump rating charts (characteristic curves).

Net positive suction head (NPSH) is the term used by the pump industry for describing pump cavitation characteristics. NPSH is defined as the pressure (head) in excess of the saturation pressure of the fluid being pumped. NPSH is expressed as NPSHA (available) and NPSHR (required). NPSHA is the NPSH available or existing at the pump installed in the system. NPSHR is a performance characteristic of a pump and is established through closed loop or valve suppression tests conducted by the pump manufacturer. These tests consist of lowering the NPSHA provided to the test pump until the pump head, power, or efficiency noticeably decreases. At this point, the pump is cavitating. The NPSHR is then established based on a predefined percentage reduction in head, power, or efficiency. Usually NPSHR is established as 3% head reduction in single-stage pumps or 3% first-stage head reduction in multistaged pumps.

NPSHR is defined for each flow rate shown on the pump rating chart and is the pumping system NPSH required to prevent cavitation at the pump impeller inlet.

Typically, the NPSHR shown on the pump rating chart was determined by the pump manufacturer based on a 3% or greater reduction in pump discharge head. Piping system designers should keep this in mind when allowing for the pump suction conditions in a particular pumping system. First, the pump manufacturer should be consulted to determine the basis of his stated values of NPSHR. Second, the pumping system designer should provide some margin above the stated NPSHR value when designing for the pump suction conditions. Typical margins over the published NPSHR values may run from 10% to 50% for a simple cold water pumping system to 50% to 100% for a complex boiler feed pumping system with transient suction condition operations. Particular attention to NPSH margin should be given to boiler feedwater systems where load rejection and system transients are expected.

The NPSH provided by the piping system design and configuration is called the net positive suction head available or NPSHA. NPSHA must be at least equal to or greater than NPSHR by some margin as discussed above to prevent cavitation in the pump impeller inlet area.

The NPSH available is expressed as follows:

For suction lift,

$$\text{NPSHA} = P - (V_p + L_s + h_f) \tag{11-12}$$

For suction head,

$$\text{NPSHA} = P + L_h - (V_p + h_f) \tag{11-13}$$

where

P = absolute pressure at the surface of the liquid, ft (m) (barometric pressure in open suction tank);

V_p = vapor pressure (absolute) of the liquid at maximum pumping temperature, ft (m);

L_s = maximum static suction lift, ft (m);

L_h = maximum static suction head, ft (m); and

h_f = friction loss in suction pipe at the required capacity, ft (m).

NPSH available is a function of the system in which the pump operates. Figure 11-12 shows several typical suction systems with the NPSH available formula applicable to each.

11.3.4 Suction Specific Speed Required

Hydraulic Institute Standards have defined suction specific speed required, S, as an index number descriptive of the suction characteristics of a given pump design as follows in Eq. (11-14).

$$S = \frac{n\sqrt{Q}}{(\text{NPSHR})^{\frac{3}{4}}} \tag{11-14}$$

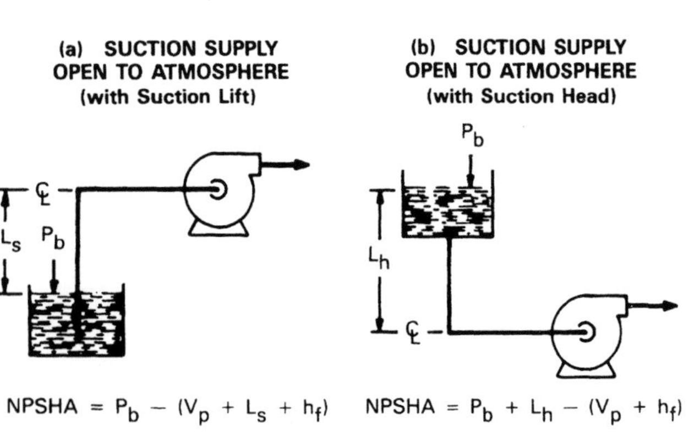

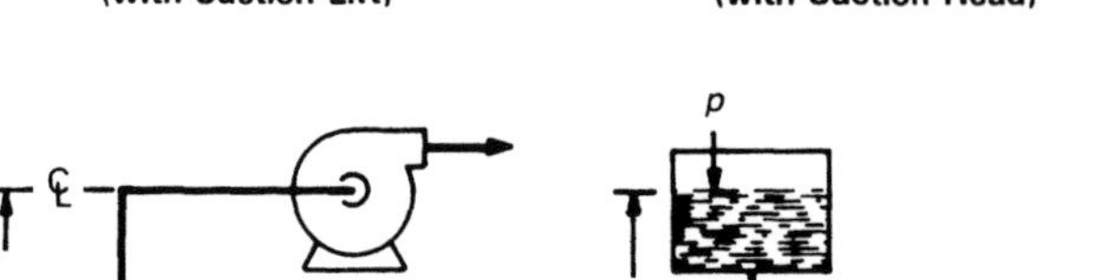

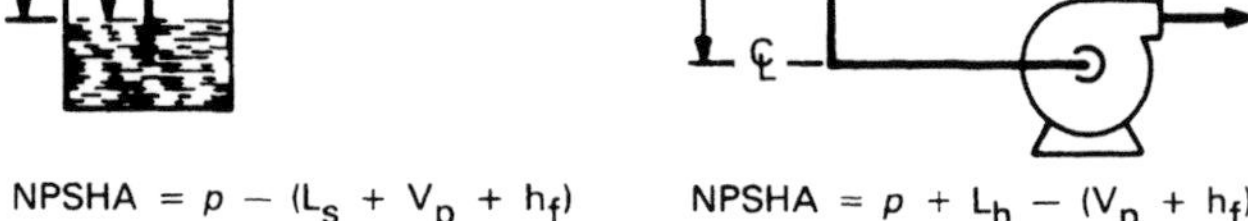

P_b = BAROMETRIC PRESSURE, IN FEET ABSOLUTE.
V_p = VAPOR PRESSURE OF THE LIQUID AT MAXIMUM PUMPING TEMPERATURE, IN FEET ABSOLUTE.
p = PRESSURE ON SURFACE OF LIQUID IN CLOSED SUCTION TANK, IN FEET ABSOLUTE.
L_s = MAXIMUM STATIC SUCTION LIFT IN FEET.
L_h = MINIMUM STATIC SUCTION HEAD IN FEET.
h_f = FRICTION LOSS IN FEET IN SUCTION PIPE AT REQUIRED CAPACITY.

Fig. 11-12. NPSH available for various pump suction arrangements.

where

S = suction specific speed required;

n = pump speed, rpm;

Q = pump flow, gpm (for double-suction pumps, use $Q/2$); and

NPSHR = pump net positive suction head required, ft (based on 3% head reduction as the manufacturer's standard).

Normally, the highest value of S is at or near the capacity corresponding to the optimum or best efficiency point (bep). Special pump designs may cause the highest value of S to shift away from the pump best efficiency point (bep).

Higher values of S are associated with better pump suction capabilities. The numerical value of S is mainly a function of the impeller inlet and suction inlet design. For pumps of normal design, S varies from 6,000 to 12,000. In special designs, including inducers, higher values of S can be found. However, other special pump features for these pumps, including materials, may be necessary.

11.3.5 Suction Specific Speed Available

Hydraulic Institute Standards have defined suction specific speed available, SA, as an index number descriptive of the available suction conditions of the pumping system from which the pump is receiving suction, as shown in Eq. (11-15).

$$SA = \frac{n\sqrt{Q}}{(\text{NPSHA})^{\frac{3}{4}}} \tag{11-15}$$

where

SA = suction specific speed available;

n = pump speed, rpm;

Q = pump flow, gpm (for double-suction pumps, use $Q/2$); and

NPSHA = system net positive suction pressure available, ft.

The suction specific speed required, S, must equal or exceed the suction specific speed available, SA, to prevent pump cavitation. The difference between S and SA is the safety margin. Some of the factors that affect the degree of margin necessary are pump size, power consumption, intake design, range and mode of operation, type of service, and materials of construction.

Increased pump speeds without proper suction conditions will result in abnormal wear and possible failure from excessive vibration, noise, and cavitation damage. Suction specific speed available, SA, is a valuable criterion in determining the maximum permissible pump speed. The curves shown in Figs. 11-13 and 11-14 are based on a suction specific speed of 8,500, which represents a practical value. Values may be lower.

On special applications, SA may exceed 8,500. In cases where the characteristics of the pump are based on the manufacturer's experience and test data, the values may be exceeded.

Given an NPSHA value of 8,500, the maximum permissible pump speed can be determined as follows in Eq. (11-16):

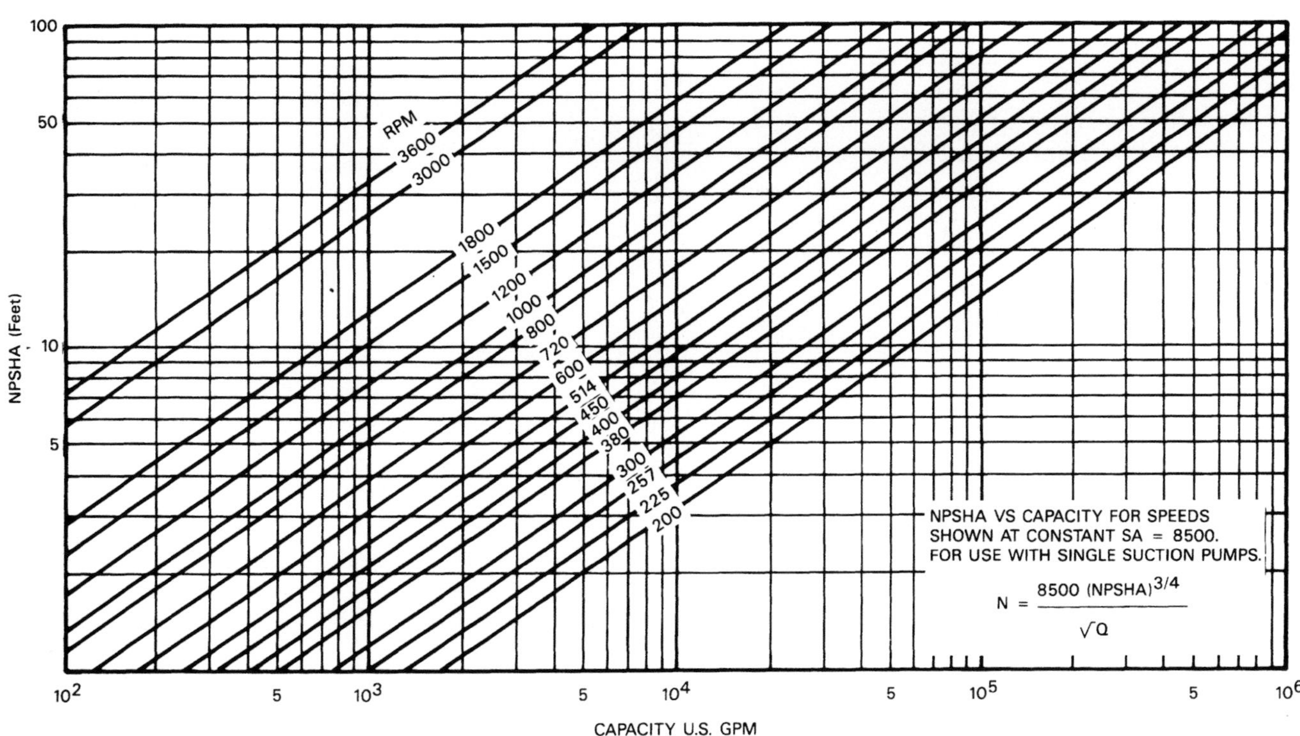

Fig. 11-13. Recommended maximum operating speeds for single-suction pumps. (From Hydraulic Institute. Used with permission.)

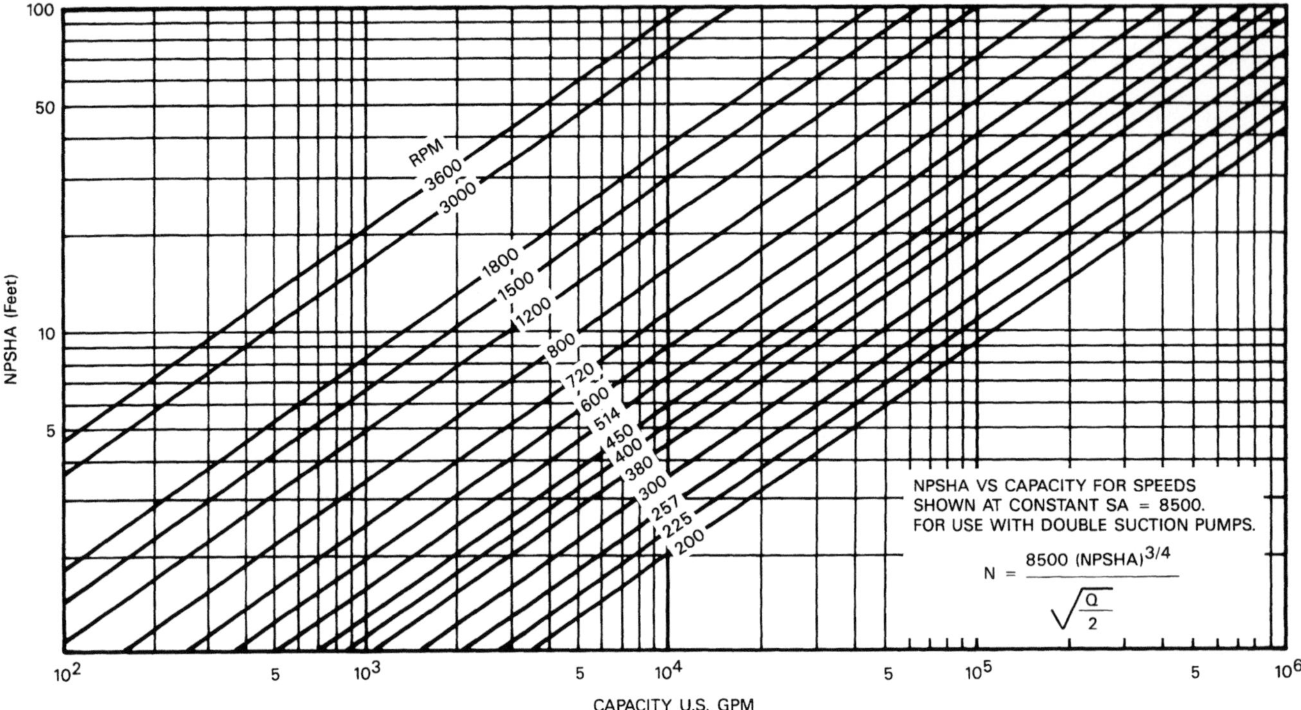

Fig. 11-14. Recommended maximum operating speeds for double-suction pumps. (From Hydraulic Institute. Used with permission.)

$$n = \frac{8{,}500\ (\text{NPSHA})^{\frac{3}{4}}}{\sqrt{Q}} \tag{11-16}$$

The curves shown in Figs. 11-13 and 11-14 are graphical representations of the suction specific speed formula for single- and double-suction centrifugal pumps, respectively.

For example, determine the maximum permissible pump operating speed for a pump operating at an elevation of 5,000 ft and 140° F water and for the same pump operating at sea level with 85° F water. Assume that in both cases, the pump flow rate is 700 gpm, the static suction lift is 10 ft, and the suction pipe friction is 1.8 ft. Determine the maximum permissible speed for both single-suction and double-suction pumps.

The NPSHA value at 5,000 ft elevation is the following:

$$\text{NPSHA} = (12.2\ \text{psia} - 2.89\ \text{psia}) \times \frac{2.31}{0.983} - (10 + 1.8)$$

$$\text{NPSHA} = 10.07\ \text{ft absolute}$$

where

Barometric pressure = 12.2 psia,

Vapor pressure = 2.89 psia for 140° F water, and

Specific gravity = 0.983 for 140° F water.

The NPSHA value at sea level is the following:

$$\text{NPSHA} = (14.7\ \text{psia} - 0.6\ \text{psia}) \times \frac{2.31}{0.996} - (10 + 1.8)$$

$$\text{NPSHA} = 20.9\ \text{ft absolute}$$

where

Barometric pressure = 14.7 psia,

Vapor pressure = 0.6 psia for 85° F water, and

Specific gravity = 0.996 for 85° F water.

The pump speed at 5,000 ft elevation is the following:

$$n = \frac{SA(\text{NPSHA})^{\frac{3}{4}}}{\sqrt{Q}} = \frac{8{,}500\ (10.2)^{\frac{3}{4}}}{\sqrt{700}}$$

$$= 1{,}833\ \text{rpm (for single-suction pump)}$$

or

$$n = \frac{8{,}500(10.2)^{\frac{3}{4}}}{\sqrt{700/2}} = 2{,}593\ \text{rpm (for double-suction pump)}$$

Similarly, the pump speed at sea level is 3,117 rpm for a single-suction pump and 4,409 rpm for a double-suction pump.

In conclusion, a double-suction pump at 3,600 rpm could be used without cavitation at sea level, but not at 5,000 ft elevation. A single-suction pump could be operated without cavitation at both sea level and 5,000 ft elevation at 1,800 rpm, but not at 3,600 rpm.

The Hydraulic Institute recommended maximum operating speeds for centrifugal pumps are the upper limits and are for average pumps with optimum efficiencies. Special designs are made that exceed the Hydraulic Institute limits at some sacrifice in efficiency.

11.3.6 System Head Curves

Each pump with a specific impeller diameter and operating at a given speed exhibits a fixed performance characteristic curve. The point on this characteristic curve where a pump operates is dependent on the characteristics of the pumping system in which the pump is operating. The pumping system characteristic curve is typically referred to as the system head curve. By imposing the pump characteristic curve on the system head curve, one can determine where the pump will operate (head and capacity).

The development of the system head curve, as shown in Fig. 11-15, is dependent on the determination of the pumping system resistance to flow at various system flow rates, known as system friction losses and including so-called "minor" losses due to the various pipe fittings and valves in the system, and the pumping system fixed losses or gains, known as static heads. It should be noted that system friction losses vary approximately as the square of the liquid flow in turbulent systems and vary directly with liquid flow in laminar systems as determined by the magnitude of the Reynolds Number.

The pumping system designer uses graphical plots of the characteristics of a pump system as an aid in the analysis of pump operation in the pumping system. Although the details of design are left to the pumping system designer, the following simple pumping system configurations, plotted representation of the pump characteristic curves and system head curves, shown in Figs. 11-16 through 11-19, are examples of how various pumping system configurations affect a system head curve.

11.3.7 Parallel and Series Pump Operation

In many central power station pumping applications, more than one pump is aligned in a pumping system to meet the operating, economic, or reliability considerations of the system. For example, in the boiler feed pumping system, two pumps—one booster pump and one main boiler feed pump—may be arranged in series. In this configuration, the booster pump is aligned in series ahead of the main boiler feed pump to provide enough head to overcome the main boiler feed pump NPSH requirements. In this same pumping system, two identical lines of boiler feed pumping equipment may be aligned in parallel to provide the operator system operating flexibility to start and stop pumping equipment based on system load or demand, thus reducing the plant fuel and auxiliary power costs. Again in the same system, a reduced capacity startup boiler feed pump may be installed to operate alone to provide the plant startup capability and also to operate in parallel with either of the main boiler feed pumping equipment lines in case one of the lines is out of service, thus providing additional pumping system on-line reliability.

In designing a pumping system for either parallel or series pumping operation, the system head curve transposed by the pump characteristic curve must be determined. In a parallel pumping system application, as always, the system head consists of the total static head and the pumping system friction losses. The pumps operating in parallel have a com-

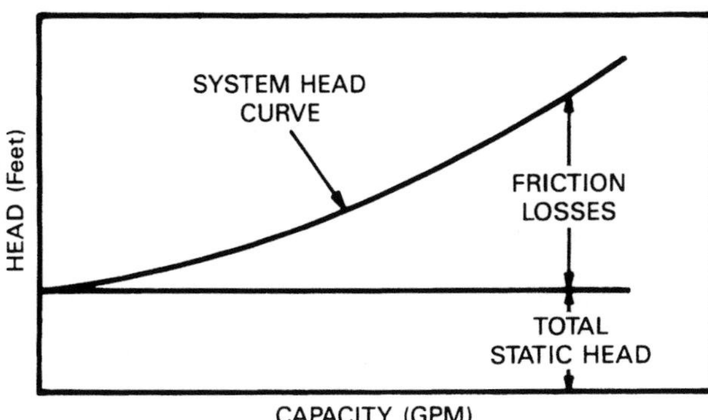

Fig. 11-15. System head curve.

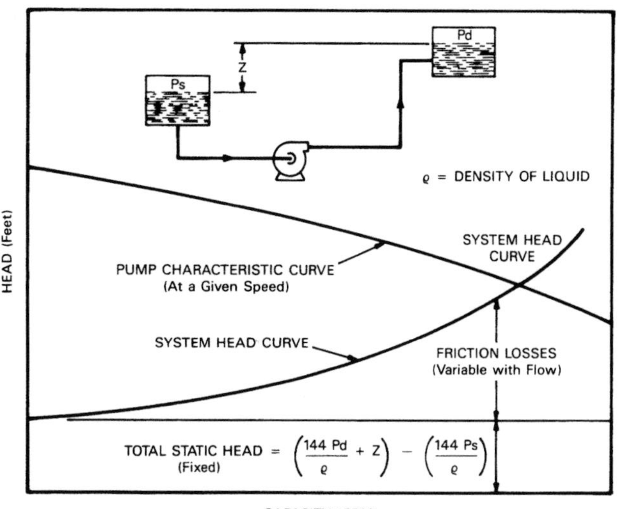

Fig. 11-16. Pumping system with fixed flow, pump speed, system static head, discharge above suction datum.

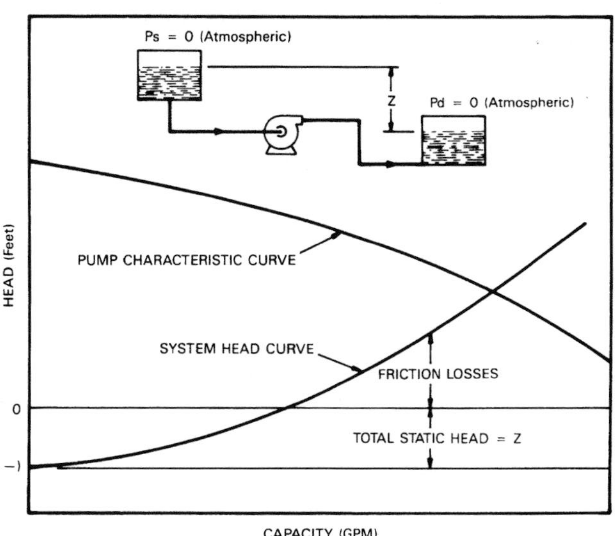

Fig. 11-17. Pumping system with fixed flow, pump speed, system static head, discharge below suction datum.

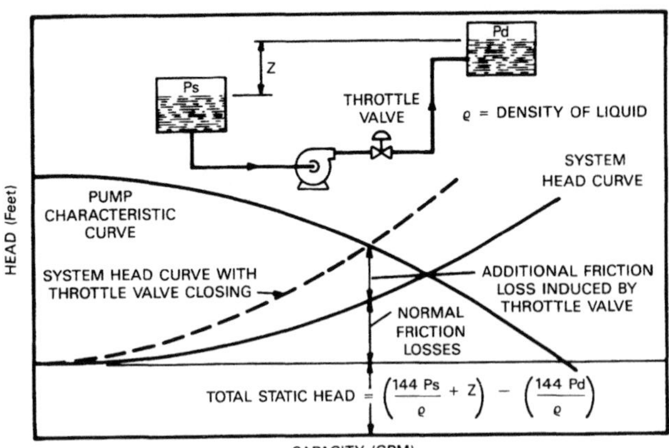

Fig. 11-18. Pumping system with throttle flow, fixed pump speed and system head, discharge above suction datum.

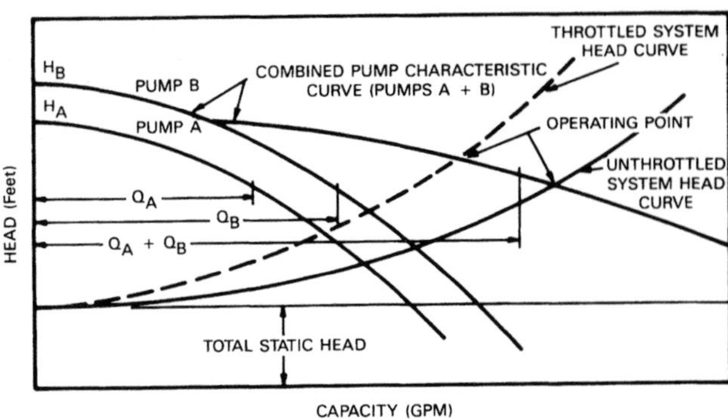

Fig. 11-20. Pump characteristic curves for pumps operating singly or in parallel.

bined system characteristic curve determined by adding the pump capacities at each value of pump head. The intersection of the pumps' combined system characteristic curve with the system head curve is the operating point. Figure 11-20 shows the pump characteristic curve for two pumps operating singly and the pumps' combined characteristic curve for two pumps operating in parallel in the same pumping system.

For series pumping applications, the pumps have a combined system characteristic curve determined by adding the pump heads at each value of capacity. The intersection of the pumps' combined system characteristic curve with the system head curve is the operating point. Figure 11-21 shows the pump characteristic curve for two pumps operating singly and the pumps' combined characteristic curve for two pumps operating in series in the same pumping system.

11.3.8 Pump Capacity Control

A pump operates within a pumping system to provide flow and head in accordance with the requirements of the pumping system. These pumping system requirements may be either continuous or only periodic. In some cases, a pump could be sized and selected to provide a "design" flow and head to overcome the worst case scenario of a pumping system. These pumps may never be required to operate at the design point flow and head and would require some type of flow or head control to adequately satisfy the instantaneous demands of the pumping system.

In power station pumping applications, pump capacity is accomplished by changing a pump's head or speed, or in some systems, by changing both. Normally, capacity regulation is by throttling the discharge of the pump, pump speed regulation, and bypass regulation. Other means of capacity control are pump suction throttling, adjustable guide vanes at the pump impeller inlet, air admission, and "cavitation" control. These latter methods of pump control are described in greater detail in many pump handbooks and references and are not discussed further herein.

11.3.8.1 Discharge Throttling. This is the most common and most economical (with respect to the capital costs of the capacity control devices required) means of capacity control for most power station applications for pumps with low- to

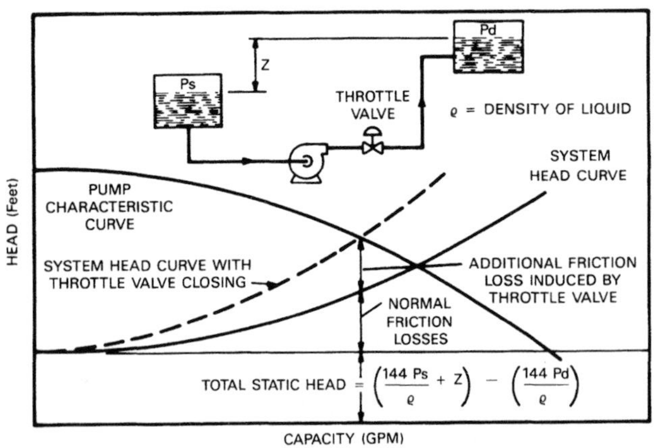

Fig. 11-19. Pumping system with variable static head, fixed pump speed discharge above suction datum.

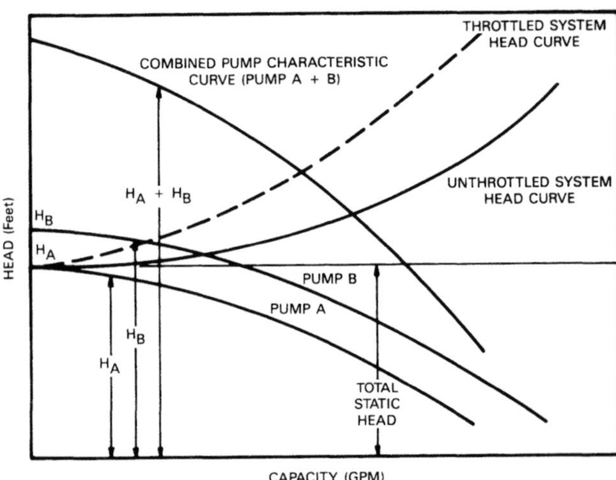

Fig. 11-21. Pump characteristic curves for pumps operating single or in series.

medium-range specific speeds. With this type of pump capacity control, the partial closure of a discharge piping throttle valve will increase the pumping system friction losses, causing the system head curve to move up and intersect the pump characteristic curve at a lower capacity, as shown previously in Fig. 11-18. However, by inducing more friction losses in the system, the pump operating point moves to one of lower efficiency and increased horsepower loss. This may be satisfactory in a lower flow and head pumping system, but other methods of capacity control should be considered to determine the most economical method of capacity control for each installation. For example, continuous modulation of a throttle valve in a 2,400 psig boiler feed system would be prohibitive not only because of the fuel costs and auxiliary power costs but also because of severe service requirements on the internal design and materials of the throttle valve.

11.3.8.2 Speed Regulation. In power station applications, capacity control by raising or lowering the pump speed and thus changing the intersection of the pump characteristic curve with the system head curve normally is economically justified for use in the major power plant pumping system. These systems normally include the boiler feed, condensate, and circulating water systems where substantial fuel cost and auxiliary power cost savings can be realized. In each case, the pumping system operation, plant load model, and utility economics should be considered to determine the justification for such capacity control. Several methods of varying pump speeds have been successfully used in recent power station installations. These include steam turbines and variable speed drive units/speed increasers (fluid drive couplings) for boiler feed pumping equipment; fluid drive couplings and variable frequency drive equipment for condensate pumping equipment; and variable frequency and multiple (two- to three-speed)-speed drive motors for circulating water pumping equipment.

11.3.8.3 Bypass Regulation. All or part of a pump flow may be diverted from the pump discharge through a bypass line back to the pump's suction source to control the pump flow or head to the pumping system. The bypass system normally contains metering orifices or flow nozzles and controlling valves to regulate the bypassed flow. This type of pump capacity control typically is used in boiler feed and condensate pumping systems to provide flow control and prevent overheating of the pump. This control is also used on long supply systems, such as a fuel oil supply system to multiple supply points, to provide a constant oil supply pressure regardless of which supply points are in operation and taking suction from the main fuel oil supply line.

11.3.9 Minimum Recirculation

Centrifugal pumps are usually required to operate over a flow range from minimum flow to beyond the design rated capacity of the pump to runout. Pump impellers are usually designed for best performance at or near the rated condition. The efficiency of the pump decreases as the flow deviates from this point at constant speed. Most of the power difference between pump input (brake horsepower) and work performed in pumping (water horsepower) is converted into heat energy and is transferred to the fluid being pumped.

Operation of a centrifugal pump at very low flows increases the temperature of the pumped fluid, which may result in flashing of the water within the pump. Sustained operation at this condition damages the pump. Therefore, a minimum permissible flow rate must be maintained to limit the temperature of the fluid in the pump below the saturation temperature. The allowable temperature rise across the pump varies with the system conditions (fluid pumped, suction pressure, and temperature and head developed by the pump).

The temperature rise in a pump may be calculated from Eq. (11-17).

$$\Delta t = \frac{(1 - \varepsilon_p)h}{778C_p\varepsilon_p} \tag{11-17}$$

where

Δt = temperature rise, °F;

ε_p = overall pump efficiency;

C_p = specific heat of pumped fluid $\left(C_p = 1.0\frac{\text{Btu}}{\text{lb-°F}} \text{ for water}\right)$;

h = total pump head, ft; and

Conversion constant = 778 ft-lb equals 1 Btu.

Further, after establishing an allowable temperature rise, an approximate minimum safe continuous flow efficiency can be determined by Eq. (11-18), which then may be used with the pump characteristic curve information to determine a corresponding minimum safe continuous flow.

$$\varepsilon_p = \frac{h_{so}}{778C_p(\Delta t)} \tag{11-18}$$

where

h_{so} = pump head at shutoff, ft.

Overheating is not the only factor related to the minimum flow. Operation of the pump at low capacities causes noise, hydraulic pulsation, and rapid deterioration of the impeller. These effects are caused by internal recirculation in the impeller as a result of a mismatch between the inlet vane angle and the liquid flow direction at the low flow velocity.

The minimum flow rate for a pump depends on these and other design factors and must be determined by the manufacturer. Minimum flow required to prevent overheating does not always determine the minimum pump recirculation requirement.

11.3.10 Recirculation Applications

Each pump application, system control, pipe configuration, and method of operation should be reviewed to determine necessity for a minimum flow recirculation system and the type of minimum flow recirculation control. Consideration should be given to operation during system and plant startup and shutdown, as well as during normal conditions.

Pumps operating with constant flow that will not result in operation at low flows for substantial periods need not be provided with a recirculation system. Operation of motor-operated valves at discharge of pumps must be coordinated with pump controls to prevent prolonged pump operation at low flows. Pumps in this category typically include pumps for the circulating water, auxiliary cooling, and closed cycle cooling water systems.

Variable speed pumps, such as condensate pumps, should have individual minimum flow recirculation systems to prevent operation of a pump at speeds that produce insufficient head to provide flow to the system.

11.3.10.1 Arrangement. One of the functions of recirculation is to prevent overheating of the pump; therefore, the recirculation flow should be cooled prior to returning to the pump. Usually this is accomplished by returning the flow to the suction source tank, where the fluid is tempered by mixing with the stored volume prior to repumping.

The recirculation flow may be returned to the suction of the pump if a suction source tank or other means of cooling is not available, but only for very short durations, such as for checkout of the pump.

11.3.10.2 Minimum Flow Rate. The minimum flow rate is normally determined by the pump manufacturer. The recommended minimum flow is usually between 20% and 50% of the flow at the pump's best efficiency point.

11.3.10.3 Type of Control. The following types of systems are used to provide minimum flow through the pump:

1. *Modulating*. The modulating control automatically recirculates only the flow required to maintain the minimum flow through the pump regardless of flow to the main system. This type of control conserves pumping power as recirculation flow is minimized. Modulating control should be used in applications where the pump is expected to operate below the minimum flow rate a large portion of the time, in order for the energy savings to justify the cost of the required equipment.

Modulating control requires a flow element to measure pump flow upstream of the pump recirculation line. A flow controller modulates the recirculation control valve, which is the pressure breakdown device.

2. *Continuous*. This system provides for continuous recirculation flow whenever the pump is operating. Since this type of control requires continuous expenditure of pump power for recirculation, it should be used for low-flow or head pumps that operate only for short periods of time, or where the pump has excess capacity as a result of the unavailability of an appropriately sized pump. The rated capacity of the pump must include the recirculation flow at the corresponding pump head. A pressure breakdown orifice is normally used as the flow limiting device.

3. *Automatic On–Off*. The automatic on–off control recirculates a constant flow rate to maintain pump flow above the required minimum. An on–off power-operated valve is actuated to the full open position by a flow sensing switch whenever the pump is operating below the minimum flow rate. The valve normally remains open until the pump flow exceeds twice the minimum flow to ensure adequate pump flow after the recirculation valve is closed. The excessive recirculation flow increases the pumping power above that of the modulating type.

A pressure breakdown orifice is used as the flow limiting device with the power-operated valve providing on–off control.

A flow element or elbow tap may be used for the flow sensing device for the switch instead of the flow element and controller required for the modulating control. The flow should be measured upstream of the pump recirculation line.

Care should be used in the selection of automatic on–off recirculation control. Depending on the pump flow rate and pump head, the transition from no recirculation to full recirculation may not be smooth and may upset other pumping system controls.

4. *Manual On–Off*. This system has a manual on–off valve to control the recirculation at a constant flow rate. The flow rates at which the manual valve must be opened and may be closed are the same as for the automatic on-off control system.

This system should be used on pumps that normally do not require recirculation except for infrequent pump checking, such as during initial startup or after pump maintenance.

A pressure breakdown orifice should be used as the flow-limiting device.

Table 11-2 provides some general guidelines that can be used to select between a modulating or continuous type of pump recirculation control system, depending on the pump horsepower requirement and hours of pump operation. These guidelines are based on typical power plant economic factors and conservative engineering design practices. Automatic on–off control and manual on–off control are not included in the table, but could be used as determined by the system designer as described previously.

Table 11-2. Guidelines for Selection of Pump Recirculation Control System

Single or Two Half-Capacity Pumps		
Annual Operating Hours per Pump	Type of Recirculation Control: Continuous	Modulating
2,000	200 hp and below	250 hp and above
4,000	100 hp and below	125 hp and above
6,000	60 hp and below	75 hp and above
8,000	50 hp and below	60 hp and above
Two-Full Capacity Pumps		
Annual Operating Hours[a]	Type of Recirculation Control: Continuous	Modulating
2,000	300 hp and below	350 hp and above
4,000	150 hp and below	200 hp and above
6,000	100 hp and below	125 hp and above
8,000	75 hp and below	100 hp and above

[a]Total annual operating hours assuming one pump always idle as a backup.

11.4 PUMP SELECTION—GENERAL GUIDELINES

One of the most critical decisions a power plant system designer faces is the proper application and selection of pumps for installation and operation in a piping system. As previously discussed, there are many different sizes, types, classes, and configurations of pumps that make the pump selection difficult. In addition to the sheer number of pumps available to the designer, the design must also consider the system hydraulic design, plant economics, plant spaces and maintenance area required, structural support, electrical/control requirements, noise, vibration, and in the end, the procurement and actual installation of the pumps.

Although not completely a science, the selection process for a system pump usually takes place over a reasonable period of time and follows the same general procedures for each pump application to allow the designer to make an accurate pump selection. The process is also aided by the availability of many fine references prepared both by independent sources and by the pump manufacturers themselves. These references should always be reviewed by the designer during the pump selection process. Further, the designer must rely on his experience and the experience of the pump manufacturer to make proper pump selections. The manufacturer should always be consulted for special or custom pump operations.

The discussion that follows should give a power plant system designer some useful general guidelines in the selection of a pump for a particular system application. Pumps normally are selected by the designer by one of the following general methods:

- Preparation of a pump performance specification used by the pump manufacturer to suggest a pump to the designer for a particular application;
- Selection of a pump by the designer using the hydraulics required and the pump performance curves and data provided by the pump manufacturer; or,
- Most typically, using a combination of both of the above to provide an accurate fit to the application requirements.

In any case, the following steps are the usual procedure for selection of each pump:

1. *Determine the need for pumping.* This step is very basic. The piping system is analyzed in the design to determine if, in fact, a pump or a number of pumps are required. The need may be to provide a regulated flow to move a liquid from one place to another, lift a liquid to a higher level, pressurize the liquid, or circulate a liquid in a system by overcoming system friction losses.

2. *Prepare a piping system flow schematic.* The system must be conceptually configured with all liquid flow paths identified and all system equipment positioned. This allows the designer to understand and confirm the logical operation of the system and to determine what calculations he must make to determine the hydraulic requirements of the pumps.

3. *Determine the physical arrangement of the piping system.* Once the designer has established the need for pumping and has a clear understanding of how the system must operate, he can physically locate the pump, piping equipment, and piping system instrumentation and control devices within the plant to most logically satisfy the requirements of the system. This step is necessary to give the designer important information, such as lengths of pipe and static head requirements, so that he can begin the detailed analysis to determine the hydraulic requirements of the pump.

4. *Calculate the hydraulic requirements of the pumps.* In this step, the designer determines the pump flow, head, NPSH available, suction requirements, and ultimately the system head curve on which the pump will operate. Important data, such as the liquid, liquid specific gravity, liquid temperature, and liquid pH will be determined and used to calculate the pump hydraulic requirements. This information will also be given to the pump manufacturer to use to provide a pump to meet the performance requirements determined during this step or to verify that the designer has selected the proper pump.

5. *Determine the pump materials of construction.* Consultation with the pump manufacturer or other users during this step is recommended. With the information determined in the preceding steps, the manufacturer can provide very useful information regarding the materials of construction available and recommended for an application. The manufacturer can provide recent experience with a material selection in similar applications. Material costs, material durability, manufacturing experience, and predicted length of service are available from the pump manufacturer. Designers may also rely on their own experience or the experience of the end user/operator to determine the proper materials of construction.

6. *Determine the pump trim/construction requirements.* The details of this step are available in numerous handbooks that address various aspects the designer must determine, such as the pump type, impeller design, shaft seal type, shaft sleeves, bearings, couplings, and other accessories necessary for a successful pump application. Again, experience in the selection of these components is important. Consultation with the pump manufacturer, designer's experience, and user/operator preferences are important.

7. *Determine the pump driver arrangement.* This is a critical step and involves much consideration by the designer. Most general service pumps are driven by a single-speed electric motor; however, many pump drivers are available for the designer's choosing. Drivers may include simple squirrel cage induction motors, synchronous motors, wound rotor induction, steam/gas turbines, variable frequency drives, diesel engine drivers, and others.

The proper selection of a drive for pumping equipment requires a careful study of all system parameters concerned, and includes the following important topics:

- Portion of time the pump/drive is in use;
- Acceptable duration for lost or partial pumping capacity;
- Need for variable pumping rates to the piping system;
- Importance and type of remote or automatic starting and/or speed control;
- Availability and dependability of electric power or other fuels;
- Pump characteristics, including horsepower, torque, and speed under all operating conditions, and availability of matching drives;
- First costs, including building and other equipment costs;
- Total annual cost of consumables, including fuel, cooling water, cooling air, and for electricity, the combination of energy, demand, and power factor charges;
- Cost of any additional operating/inspection personnel;

- Cost of maintenance parts, materials, and labor; and
- Dependability of types and extent of equipment.

8. *Prepare a comprehensive pump performance/selection specification.* All of the important pump design/selection data should be consolidated in a comprehensive pump performance/ selection specification for presentation to the pump manufacturer for a tendered proposal. The specification should include but not necessarily be limited to the following sections:

- Code requirements,
- Description of pump types,
- Pump arrangement,
- Pump hydraulic characteristics,
- Characteristic curves,
- Pump guarantees,
- Pump tests (shop and/or field),
- Pump construction,
  —Materials of construction,
  —Type of shaft seals,
  —Shaft sleeve construction,
  —Casing and connections,
  —Bearings,
  —Shaft couplings,
  —Miscellaneous trim (recirculation orifices, vibration detector mountings, safety guards, etc.)
- Pump driver requirements, and
- Pump hydraulic specification summary sheet/system head curve.

An example of a miscellaneous general service pump specification and hydraulic specification summary sheet is shown in Fig. 11-22.

9. *Select and prequalify pump manufacturers for bidding.* Generally, the completed specification should be issued to four to six prequalified pump manufacturers as available for tendering, in order to receive an acceptable number of complete and proper proposals for evaluation. The bidders should be given 4 to 6 weeks to prepare their proposals, depending on the complexity of the pumps or number of pumps specified. In prequalifying pump manufacturers for bid, the following items should be considered:

- Experience and manufacturing capabilities;
- Financial stability;
- Number of similar pump installations with same pump size (e.g., five pumps each with 5 years of successful operating experience);
- Extent of engineering department necessary to support pump design, design documentation and drawings, testing facilities;
- Ability to provide installation/startup services;
- Extent of service representative network;
- Ability to supply spare parts quickly;
- Extent of user references; and
- QA/QC program and procedures.

10. *Receive and evaluate pump proposals.* After bids are received and deemed acceptable, a thorough analysis of the bid should be undertaken. A bid analysis consists of an evaluation of alternative proposals for the pumps specified and usually includes a recommendation as to which proposal/bidder should be selected. Items that should be considered in the bid analysis should include proposal price, terms of payment, general conditions (including legal issues), delivery of equipment, escalation, user/operator allowance for funds (cost of money) during equipment installation and plant construction (AFUDC), differential operating costs, and cost of electrical/fuel power losses. Annual operating costs associated with the pump proposals are usually expressed as capital equivalent costs. In some cases, the equivalent capital cost of power or fuel losses that will be used in the proposal evaluation will be stated in the specification, for example, dollars per foot of head of pumping power.

11.5 POWER PLANT PUMP APPLICATIONS

The following sections provide more detailed information regarding what are typically the main pump applications in a power station. Although only the main power station pumping applications are presented here, much of the theory and general guidelines are relevant to any pumping application.

Most of the information given here is general, and should be used as guideline information only. The pump and pumping system designer should be aware of the site-specific and installation-specific requirements that could affect the successful operation of the pump installation.

Power station pump applications discussed in this section are as follows:

- Circulating water pumps,
- Condensate pumps,
- Boiler feed pumps, and
- General service/slurry pumps.

11.5.1 Circulating Water Pumps

Circulating water pumps are high-capacity low-head pumps that provide the cooling water flow for the circulating water system (Chapter 12). Because of their large size and continuous operation, circulating water pumps must be carefully selected for economical and reliable operation over the lifetime of the plant.

Circulating water pumps are typically selected from one of three pump designs: vertical wet pit, horizontal dry pit, and vertical dry pit pumps. These pump types are shown in Fig. 11-23. For once-through circulating water systems (systems without cooling towers), vertical wet pit pumps are most commonly used, followed by horizontal dry pit pumps and vertical dry pit pumps, in that order. For closed cycle circulating water systems (systems with cooling towers), vertical wet pit and horizontal dry pit pumps are used about equally, with vertical dry pit pumps used less frequently. Electric Power Research Institute (EPRI 1978) data on circulating water pump selection for United States electric utilities are presented in Table 11-3.

11.5.1.1 Vertical Wet Pit Pumps. Vertical wet pit pumps, as shown in Figs. 11-24a and 11-24b, are typically of the mixed

GENERAL SERVICE CENTRIFUGAL PUMP SPECIFICATION SHEET

HORIZONTAL PUMP

PUMP DESCRIPTION

Pump Designation ________________ Pump Tag No(s) ________
Type Designation ________________ ________
Number to be Furnished ________________ ________
____

PERFORMANCE REQUIREMENTS

Pumped Fluid ________________ Specific Gravity: ______ @ ______ F
Temperature: ______ F Max ______ F Min ______ F at Design Conditions pH Range: ______
Capacity: ________ GPM Total Head, H: ______ Ft Maximum RPM: ________
Total Suction Pressure, H_S: ________ Ft (Positive)
Net Positive Suction Head Available, NPSH: ____________ Ft
Suction Pressure Range: ________ Ft Min to ________ Ft Max Plant Elevation: ________ Ft
Parallel Operation Capability Required: ________

CONSTRUCTION REQUIREMENTS

Casing Material: ________________ Design Pressure:____ PSI Design Temperature:_____ F
Shaft Material: ________________ Shaft Sleeves: ________________
Impeller Material: ________________ Impeller Type: ________ No. of Stages: ___
Wearing Rings: Impeller: ________________ Casing: ________________
Suction Can (VWPCP Type Only) Material: ________________ Coating: ________________
Connection Location (VWPCP Type Only) Suction: ____________ Floor, Discharge: ____________ Floor
Shaft Seals: Stuffing Box: ________________ Mechanical Seal: ________________
Seals for Pressure: ____________ Vacuum: ____________ Pressure and Vacuum: ____________
Bearing Type: ________________________________
Pump Location: ____________ Baseplate: ____________ Coupling: ____________
Suction Strainer Required: ________________

MOTOR REQUIREMENTS

Motor Required with Pump: ________________ Motor Supplied by: ________________
Voltage: ____________ No. of Phases: ____________ Frequency: ____________ CPS
Motor Type: ________________ Enclosure Type: ________________
Insulation Class: ____________ Sealed: ____________ Encapsulated: ____________
Ambient Temperature: ________ F Temperature Rise: ________ F Service Factor: ____________
Bearings: Type: ____________ Lubrication: ____________ Average Life (AFBMA): ______ Yrs
Conduit Box: ________________ Space Heaters: ________________

REMARKS

____
____
____
____
____

Fig. 11-22. General service centrifugal pump specification sheet—horizontal pump.

flow, single-stage, single-suction type for circulating water service. Axial flow pumps are occasionally used for very-low-head applications. Vertical wet pit pumps use a vertical, internally lubricated shaft to drive the pump impeller. The pump is partially submersed in a wet pit with the motor mounted directly over the pump above the water level. Location of the motor directly above the pump column minimizes horizontal space requirements.

Vertical wet pit pumps may be of pull-out or non-pull-out design. Pull-out design allows the rotating elements and critical nonrotating components such as the impeller shroud and pump bowl/diffuser/volute to be quickly removed without removing the column or disconnecting the pump discharge. Non-pull-out design has a 20% to 25% lower capital cost; however, pump disassembly is more difficult and requires a longer pump outage.

Another design variable for vertical wet pit pumps is the location of the discharge relative to the baseplate. An above-floor or aboveground discharge indicates that the pump discharge is above the baseplate, whereas a below-floor or belowground discharge refers to the opposite. The below-floor discharge is more difficult to disconnect since access to

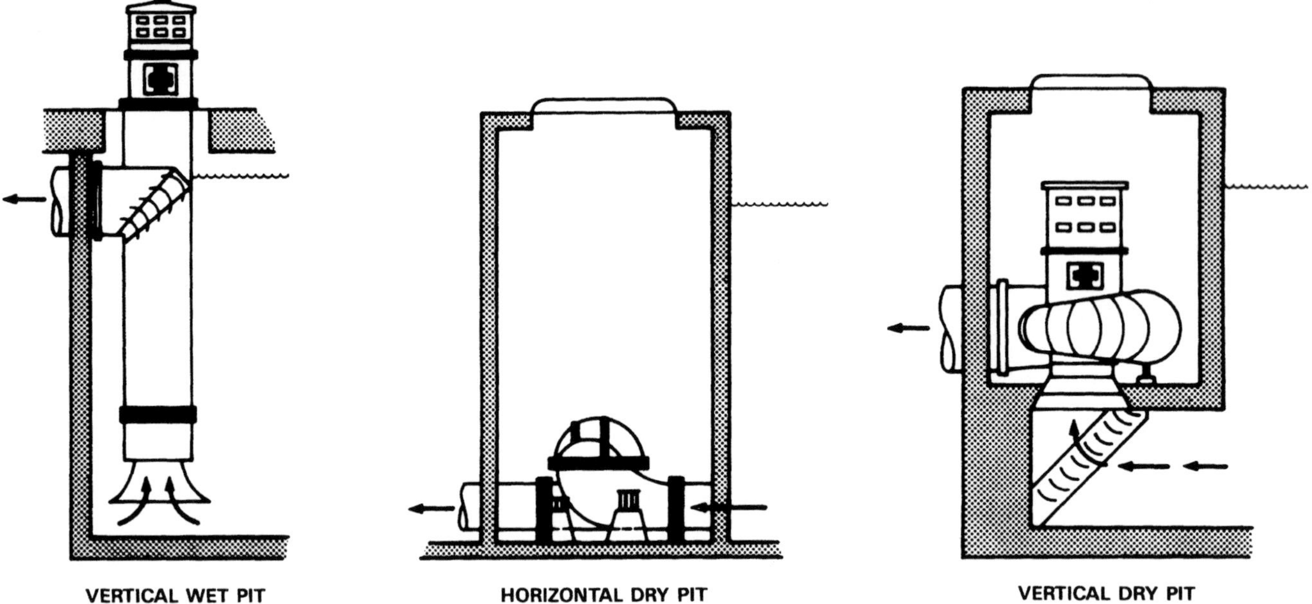

Fig. 11-23. Circulating water pump types.

the discharge is usually limited. Because disconnecting the discharge is required for disassembly of non-pull-out pumps, below-floor discharge combined with non-pull-out design may create maintainability problems.

In applications where large variations in water level exist, short column vertical wet pit pumps can be used to avoid the use of pumps with long column lengths. Short column vertical wet pit pumps, as shown in Fig. 11-25, place the pump discharge and motor in a dry pit below the high water level. Seals are placed around the pump baseplate and discharge pipe to prevent water leakage into the dry pit. The resulting pump has fewer bearings and is less susceptible to vibration. However, failure of the seals could flood the dry pit and damage the motor. In addition, isolation of the pump through use of stop logs or slide gates is difficult because of the intake structure design.

Table 11-3. Circulating Water Pump Selection for United States Electrical Utilities

Pump Type	Once-Through System	Recirculating Systems	Total All Systems
Vertical wet pit			
Number of units	253	22	275
Number of pumps	541	53	594
Horizontal dry pit			
Number of units	32	26	58
Number of pumps	63	54	117
Vertical dry pit			
Number of units	24	1	25
Number of pumps	50	4	54
Total number of units	309	49	358
Total number of pumps	654	111	765

 Size of unit is not specified.

11.5.1.2 Horizontal Dry Pit Pump. For circulating water service, horizontal dry pit pumps are typically split-case, single-stage, double-suction type with either centrifugal or mixed flow design. Horizontal dry pit pumps, as shown in Fig. 11-26, use a horizontal shaft with external bearings on each side of the pump casing. The pump is located in a dry pit below the water level with suction taken through horizontally mounted suction piping that connects to the sump. The pump motor is adjacent to the pump on the floor of the dry pit.

Where horizontal space is limited, horizontal dry pit pumps can be installed in a vertical position with the motor above the pump. This arrangement, however, is unusual and requires redesign of the bearings.

11.5.1.3 Vertical Dry Pit Pumps. Vertical dry pit pumps, as shown in Fig. 11-27, are of the mixed flow, single-stage, single-suction type. Vertical dry pit pumps use a vertical shaft with bearings external to the pump casing. The pump is located in a dry pit below the water level, with suction taken from a suction tunnel located below the pump. The pump motor is above the pump, thus minimizing horizontal space requirements.

As shown in Fig. 11-28, turning vanes are used with vertical dry pit pumps to improve flow uniformity into the suction bell. The turning vanes are typically made of fabricated steel embedded in the concrete walls of the suction tunnel. A draft tube suction can be used as an alternate to turning vanes.

Vertical dry pit pumps normally have a lower design point efficiency than either vertical wet pit or horizontal dry pit

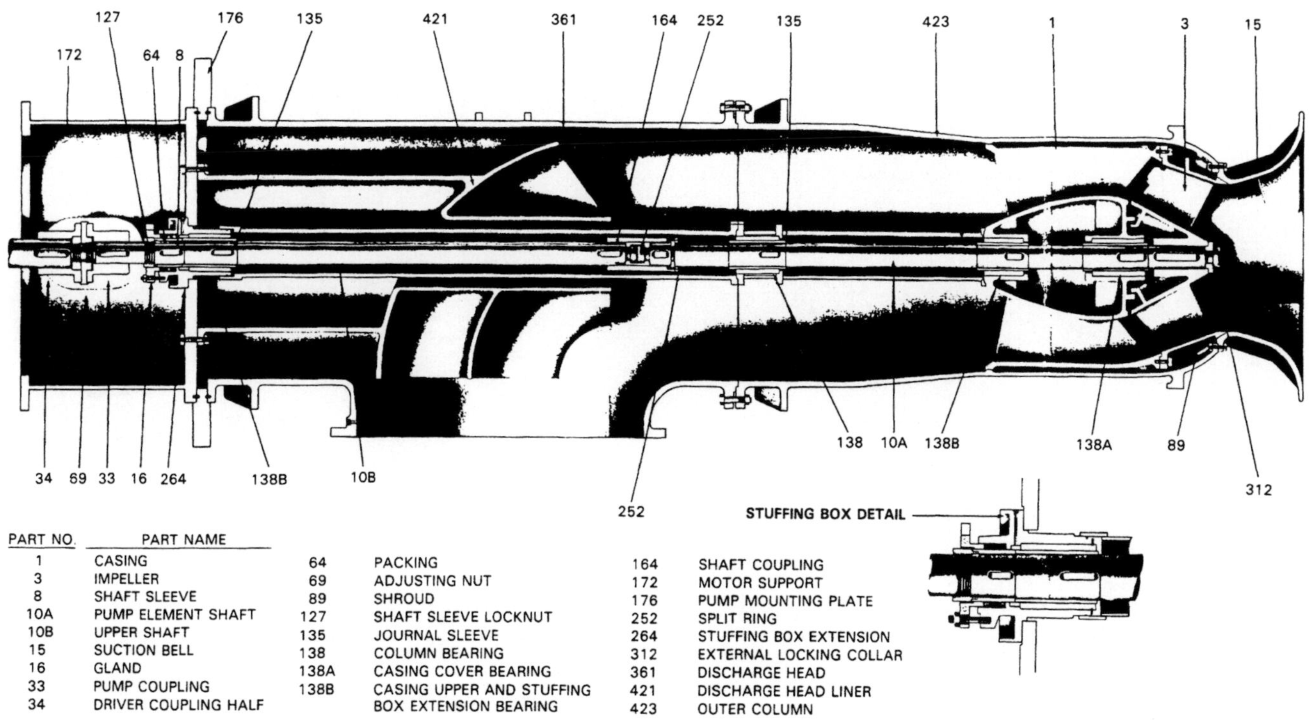

PART NO.	PART NAME
1	CASING
3	IMPELLER
8	SHAFT SLEEVE
10A	PUMP ELEMENT SHAFT
10B	UPPER SHAFT
15	SUCTION BELL
16	GLAND
33	PUMP COUPLING
34	DRIVER COUPLING HALF
64	PACKING
69	ADJUSTING NUT
89	SHROUD
127	SHAFT SLEEVE LOCKNUT
135	JOURNAL SLEEVE
138	COLUMN BEARING
138A	CASING COVER BEARING
138B	CASING UPPER AND STUFFING BOX EXTENSION BEARING
164	SHAFT COUPLING
172	MOTOR SUPPORT
176	PUMP MOUNTING PLATE
252	SPLIT RING
264	STUFFING BOX EXTENSION
312	EXTERNAL LOCKING COLLAR
361	DISCHARGE HEAD
421	DISCHARGE HEAD LINER
423	OUTER COLUMN

Fig. 11-24a. Vertical wet pit pump components—pullout design.

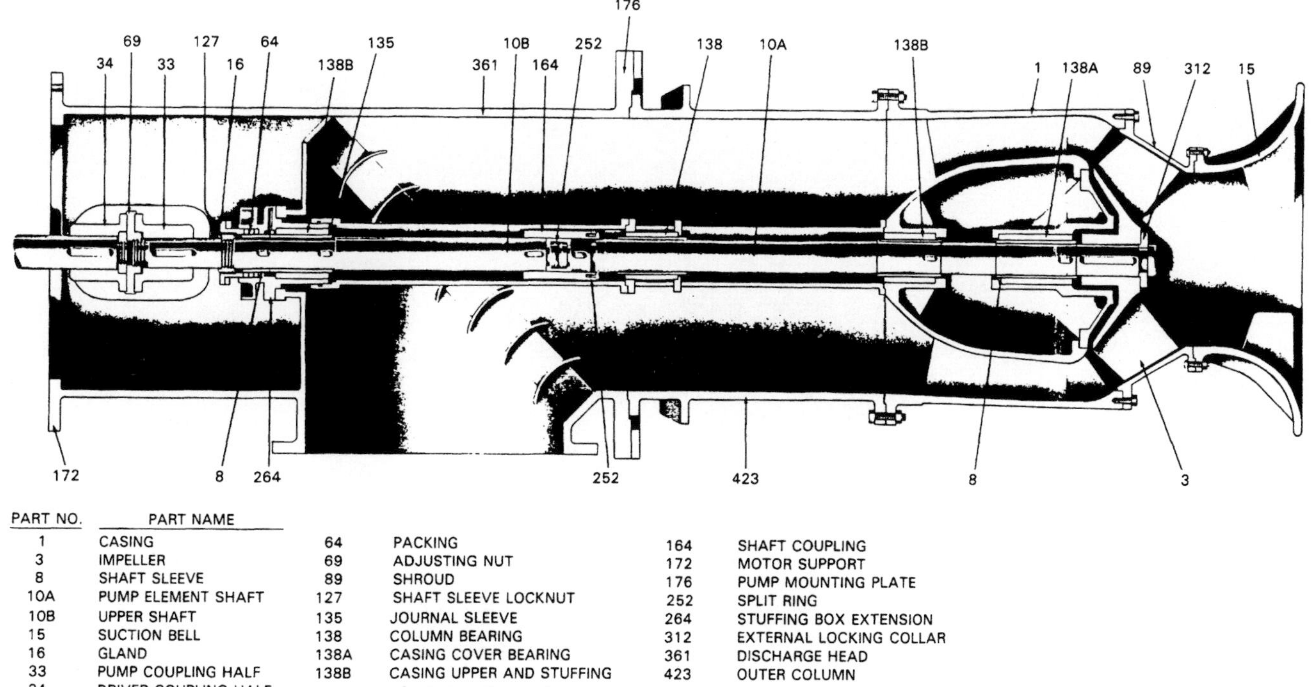

PART NO.	PART NAME
1	CASING
3	IMPELLER
8	SHAFT SLEEVE
10A	PUMP ELEMENT SHAFT
10B	UPPER SHAFT
15	SUCTION BELL
16	GLAND
33	PUMP COUPLING HALF
34	DRIVER COUPLING HALF
64	PACKING
69	ADJUSTING NUT
89	SHROUD
127	SHAFT SLEEVE LOCKNUT
135	JOURNAL SLEEVE
138	COLUMN BEARING
138A	CASING COVER BEARING
138B	CASING UPPER AND STUFFING BOX EXTENSION BEARING
164	SHAFT COUPLING
172	MOTOR SUPPORT
176	PUMP MOUNTING PLATE
252	SPLIT RING
264	STUFFING BOX EXTENSION
312	EXTERNAL LOCKING COLLAR
361	DISCHARGE HEAD
423	OUTER COLUMN

Fig. 11-24b. Vertical wet pit pump components—nonpullout design.

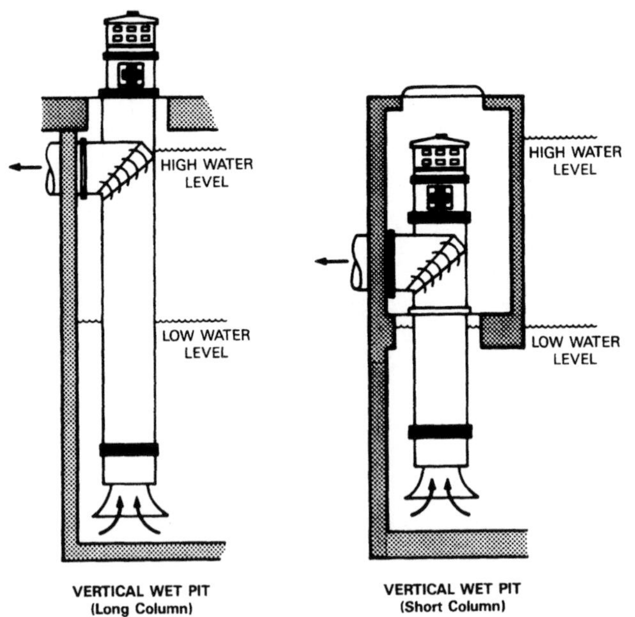

Fig. 11-25. Comparison of long-column and short-column vertical wet pit pumps.

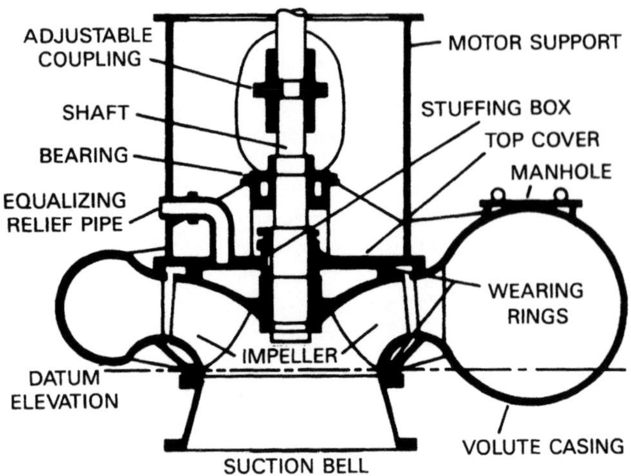

Fig. 11-27. Vertical dry pit pump components.

pumps. In addition, United States electric utilities have substantially less operating experience with vertical dry pit pumps than with either vertical wet pit or horizontal dry pit pumps.

11.5.1.4 Circulating Water Pump Design Criteria. The selection of a circulating water pump for a specific circulating water system application requires an evaluation of several design criteria. Pump design criteria include pump capacity and total developed head, net positive suction head, submergence, suction specific speed, and rotative speed. The recommended evaluation procedure for each design criterion is as follows.

DESIGN POINT. The design point for circulating water pumps is specified by the design flow rate (capacity) and total developed head (TDH). Design capacity per pump is determined based on the design circulating water flow rate, including main condenser flow and auxiliary cooling water flow, and the number of pumps. The design circulating water flow rate is determined based on the maximum circulating water requirement and the number of pumps being provided. For two half-capacity circulating water pumps, the design capacity of each pump is determined as follows. Each circulating water pump should be designed to pass 50% of the total circulating water flow required to maintain the design turbine exhaust pressure at the steam turbine low-pressure exhaust connection to the condenser neck, with the maximum cooling water body temperature, with a condenser tube cleanliness factor of 90%, and with a steam turbine exhaust flow, boiler feed pump turbines exhaust flow, and other

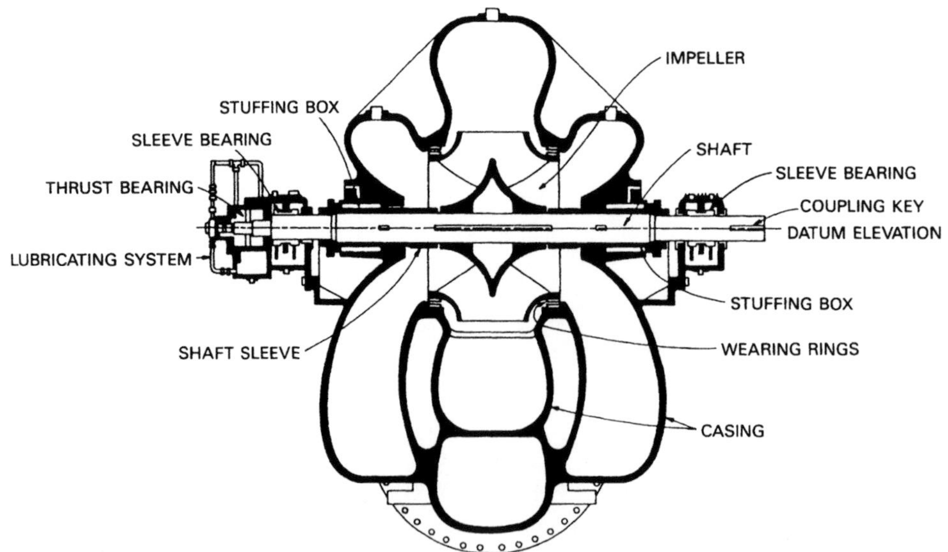

Fig. 11-26. Horizontal dry pit pump components.

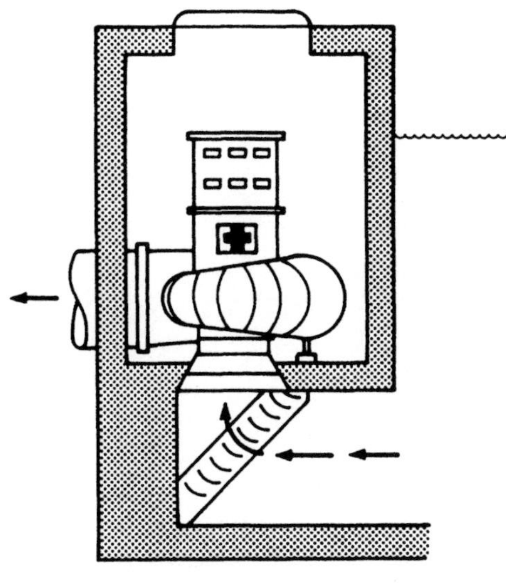

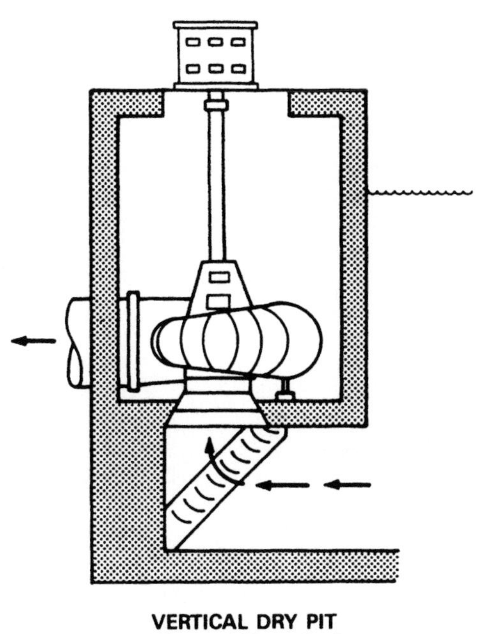

Fig. 11-28. Comparison of integral and elevated mounting for vertical dry pit pump motors.

reclaimable steam flows associated with plant operation at maximum load.

If auxiliary cooling water is to be provided by the circulating water pumps, 50% of the required auxiliary cooling water flow at the maximum cooling water body temperature and at maximum plant load is added to the above circulating water flow for each pump.

The Hydraulic Institute Standards require circulating water pump manufacturers to meet the design capacity with margins of plus 10% and minus 0%. For this reason, no flow margin is included in the pump design capacity.

The design total developed head is calculated as shown in Table 11-4 for once-through cooling systems and as shown in Table 11-5 for recirculating cooling systems. No head margin is included in the design TDH due to conservatism in the head calculation. Also, the Hydraulic Institute Standards require manufacturers of low-head pumps to meet the design TDH with margins of plus 5% and minus 0%.

NET POSITIVE SUCTION HEAD. As was previously discussed, net positive suction head (NPSH) represents the total pump suction head referenced to the datum elevation of the pump less the vapor pressure of the liquid. The datum elevations for the three types of circulating water pumps are shown in Figs. 11-24, 11-26, and 11-27. NPSH is characterized by the net positive suction head available (NPSHA) and the net positive suction head required (NPSHR). NPSHA represents the NPSH provided to the pump whereas NPSHR represents the NPSH needed by the pump to prevent cavitation. To prevent cavitation, NPSHA must be equal to or greater than NPSHR for the entire range of pump operation.

For circulating water pumps, NPSHA is primarily a characteristic of the sump design. Higher NPSHA values generally indicate better suction conditions.

Good design practice requires the NPSHA to exceed the NPSHR at the pump design point and at each appropriate runout point including one-pump runout conditions. The basis for the NPSHR value is dependent on the pump operating scenario or runout conditions. Table 11-6 lists conservative recommended NPSHR basis for various operating and runout conditions. Most pump manufacturers would recommend using the 3% head reduction value with appropriate margin included. Satisfying NPSH criteria at the pump design point and appropriate runout points generally ensures that NPSH requirements will be met at all points of operation. However, the entire range of pump operation must be checked for NPSH acceptability.

SUBMERGENCE. Submergence is a measure of the water depth in the pump sump above the pump suction. Submergence is defined in Fig. 11-29 for each type of circulating water pump. Submergence must be adequate to meet NPSH requirements and to prevent vortexing. The Hydraulic Institute has provided guidance on minimum submergence to prevent vortexing for vertical wet pit pumps. These recommendations, along with recommendations provided by the pump manufacturer, should be considered when designing the pump intake structure for vertical wet pit pumps. As a general rule for horizontal pumps, submergence should be about 1 ft for each foot per second velocity at the bell mouth.

Submergence may also be set by the sump depth required to meet traveling screen or trash rack approach velocities. The minimum acceptable submergence is established based on meeting all the above criteria.

Table 11-4. Circulating Water Pump Design Total Developed Head Determination: Once-Through Systems

Item	Head (ft)	Reference/Comments
Difference between low water level and circulating water system discharge water level	a	Based on expected minimum level of cooling water source and water level at circulating water system discharge, at design flow
Trash rack and traveling screen loss	1.0–2.0	Conservative design head loss to allow for rack and screen pluggage
Head loss between screen and pump intake	0	No losses in structure between screen and pump
Friction drop in circulating water pump suction piping (including entrance loss)	a	Friction loss in suction piping up to pump suction flange, at design flow (suction losses not applicable for vertical wet pit and vertical dry pit pumps)
Circulating water pump internal losses	N/A	Not included in total head calculation. Prospective pump manufacturer must adjust the total head to include the losses
Pump discharge velocity head	a	Based on design flow and pump discharge connection size
Friction drop between circulating water pump discharge connections and condenser water box connections	a	Friction loss in discharge piping from pump discharge flange to condenser inlet connections, at design flow
Friction drop through condenser	a	Calculated based on the latest Heat Exchange Institute Standards for Steam Surface Condensers or manufacturer's design information
Friction drop between condenser water box connections and outfall structure	a	Friction loss from condenser outlet connections to circulating water system discharge, at design flow
Total developed head	a	Sum of above items

[a]Specific installation data.

PUMP INTAKE SUMP DESIGN. In addition to pump shop performance tests discussed herein, the designer should consider model testing on any circulating water pump intake structure/piping installations. These tests, as determined by the designer and pump manufacturer, would consist of a number of simulated flow conditions within the circulating water pump intake structure to confirm the intake structure design. Normally, these tests are included in the pump purchase specifications and sometimes become part of a combined pump and intake structure arrangement guarantee.

Table 11-5. Circulating Water Pump Design Total Developed Head Determination: Recirculating Systems

Item	Head (ft)	Reference/Comments
Difference between normal and low basin levels	0	Basin will be maintained at or above normal water level
Head loss between cooling tower basin and pump suction screens	0	No head loss between basin and screens
Pump suction screen loss	1.0–2.0	Conservative design head loss to allow for screen pluggage
Friction drop in circulating water pump suction piping (including entrance loss)	a	Friction loss in suction piping up to pump suction flange, at design flow (suction losses not applicable for vertical wet pit and vertical dry pit pumps)
Circulating water pump internal losses	N/A	Not included in total head calculation. Prospective pump manufacturer must adjust the total head to include the losses
Pump discharge velocity head	a	Based on design flow and pump discharge connection size
Friction drop between circulating water pump discharge connections and condenser water box connections	a	Friction loss in discharge piping from pump discharge flange to condenser inlet connections, at design flow
Friction drop through condenser	a	Calculated based on the latest Heat Exchange Institute Standards for Steam Surface Condensers or manufacturer's design information
Friction drop between condenser water box connections and cooling tower inlet connections	a	Friction loss from condenser outlet connections to cooling tower inlet connections, at design flow
Cooling tower pumping head requirement (includes all friction losses, nozzle head losses, and static head)		Coolig tower manufacturer's design information
Total developed head	a	Sum of above items

[a]Specific installation data.

Table 11-6. NPSHR Head Reduction Criteria

Pump Operation	Head Reduction (%)
Two half-capacity pumps	
2 of 2 pumps	0
1 of 2 pumps	3
Three one-third capacity pumps	
3 of 3 pumps	0
2 of 3 pumps	0
1 of 3 pumps	3
Four one-fourth capacity pumps	
4 of 4 pumps	0
3 of 4 pumps	0
2 of 4 pumps	3
1 of 4 pumps	3

These simple, inexpensive tests could preclude limited pump operation, pump damage, and costly redesign of the intake structure after installation of the facility.

There are many good references for pump intake piping and structure design that a designer should consult during pump selection and detailed design. The Hydraulic Institute offers many good recommendations and the specific pump manufacturers should also be consulted for their recommendations.

SUCTION SPECIFIC SPEED. Suction specific speed is another parameter used to predict pump performance and is characterized by the suction specific speed required (S) and the suction specific speed available (SA) as described previously. To prevent pump cavitation, S must equal or exceed SA for the entire range of pump operation.

ROTATIVE SPEED LIMITATIONS. As stated in the *Hydraulic Institute Standards* 1983, "Increased pump speed without proper suction conditions can result in abnormal wear and possible failure from excessive vibration, noise, and cavitation damage. Suction specific speed available, SA, has been found to be a valuable criterion in determining the maximum permissible speed."

The Hydraulic Institute has a maximum SA value of 8,500 to limit pump speeds. The Institute does, however, allow higher values of SA "where the characteristics of the pump are based on the manufacturer's experience and test data." In special applications, SA values of up to 12,000 may be allowable. However, the pump manufacturer must have proven successful experience with pumps at installations with similar designs.

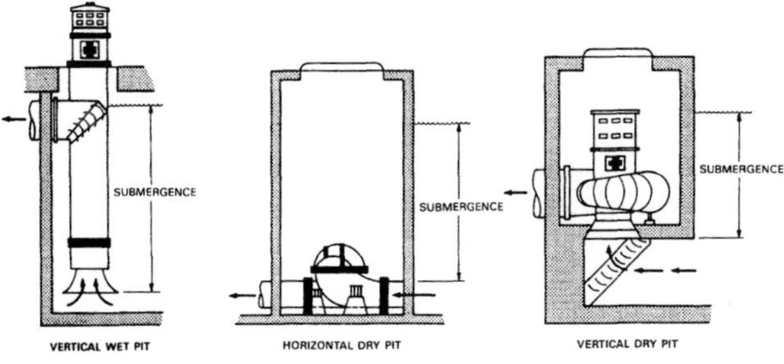

Fig. 11-29. Circulating water pump submergence.

Given the SA value of 8,500, the rated pump capacity, and the NPSHA, Eq. (11-16) can be used to calculate the maximum pump speed.

Pump speed may be additionally limited to prevent high erosion rates of the impeller and impeller bowl and to reduce vibration-related problems in vertical wet pit pumps. The maximum pump speed should be set based on specific installation considerations, namely the impeller and impeller bowl materials, the total suspended solids in the circulating water, the vertical wet pit pump shaft length, and manufacturer's recommendations. However, it should be recognized that artificially lowering pump speed to prevent high erosion rates may not achieve that result because the exit tip speeds of a slower, larger impeller pump and a faster, smaller impeller pump providing the same head are comparable. Pump speeds must also be selected based on commercially available motor speeds.

11.5.1.5 Circulating Water Pump Selection. Several factors must be considered in the selection of the type of circulating water pump to be used for a particular power plant application. During the preliminary design of the circulating water system, all pertinent factors must be evaluated in detail to ensure proper pump selection. Some typical factors considered in the evaluation are discussed below. Each factor must be weighed against the others to determine the optimal circulating water pump.

OWNER/OPERATOR EXPERIENCE. Perhaps the most important factor in selecting a circulating water pump type is the owner's or operator's preference. Generally, this preference is supported by reliability and maintenance records of existing pumps. Equipment operating records for similar circulating water system configurations are of great value and should be investigated as part of the selection process.

CIRCULATING WATER SYSTEM CONFIGURATION/SITE CONDITIONS. The configuration of the circulating water system—whether it is closed cycle with cooling towers or once-through—is a major consideration in selecting the right circulating water pump. A closed cycle system typically allows more flexibility in the pump selection because the suction hydraulics are relatively constant. The once-through system is generally more restrictive because of varying pump suction conditions. In the selection of a pump for a once-through system, the variance in the level of the cooling water body dictates the required depth of the pump suction to satisfy submergence requirements at the low water level. The required depth of the pump suction may make the use of a dry pit pump impractical because of the required depth and complexity of the intake structure. Also, a deep pump suction may require a very long vertical pump column, and may lead to consideration of a short column vertical pump. Equally important, the possibility of flooding must be considered if a dry pit or short column vertical pump is selected, in which case the pump drives cannot be mounted above flood level. Each of these factors must be weighed and analyzed for the specific site conditions.

COSTS. An economic evaluation must be conducted that compares the capital cost of the pumps and drives and the associated pump intake structure. Capital costs vary with pump manufacturer, type, material requirements, and the specific site requirements. Pump operating and maintenance (O&M) costs are also evaluated based on manufacturer supplied data and industry experience. O&M costs include energy costs, performance penalties, and parts and labor costs for each pump type and are typically evaluated as a levelized annual cost. The economic evaluation often includes a comparison between various pumping arrangements as well. For example, the costs associated with four one-fourth capacity pumps are compared to that of three one-third capacity pumps. In this case, the operating flexibility provided by four pumps lowers the energy costs because of the more efficient operation at reduced plant load, and this cost is evaluated against the increased capital cost of four pumps compared to three.

SPACE CONSIDERATIONS/ACCESSIBILITY. The available space for the pump intake structure and the degree of access required by the owner must be considered. Vertical wet pit pumps require the least amount of space; however, the rotating elements must be pulled out of the pit for bearing or impeller access. Vertical pumps with a pull-out design do allow easier maintenance, but at an increased capital cost. Horizontal dry pit pumps require the most space, but are also the most accessible for maintenance. Bearings are easily accessed by removing connecting lubricating oil piping and bearing covers. Impellers are accessed by removing the top half of the pump casing.

WATER QUALITY/MATERIAL CONSIDERATIONS. Water quality can be the determining factor in the proper selection of a circulating water pump type. High concentrations of suspended solids may preclude the use of pump discharge for water-lubricated bearings in vertical wet pit pumps. Depending on the availability and cost of treated water for bearing lubrication, this may make a vertical wet pit pump less attractive, and may favor the use of a dry pit pump that uses oil or grease-lubricated bearings. Also, water chemistry may dictate the use of a special material or coating for vertical wet pit pump columns and shafts. Depending on the required materials and associated costs, a dry pit pump may be desirable because of the minimal wetted surface associated with the dry pit designs. In general, poor water quality can shift the selection in favor of dry pit designs.

11.5.1.6 Circulating Water Pump Specification. Once the circulating water pump type is selected, a purchase specification can be prepared for bidding by the pump manufacturers. To properly define the pump requirements, the pump type and arrangement, system design conditions, operational requirements, guarantees, and pump construction requirements should be included in the specification. The following items provide an overview of the criteria that are suggested for inclusion in any circulating water pump specification:

PUMP TYPE AND ARRANGEMENT. The specification should define the desired pump type and quantity, the location of the pumps (indoors, outdoors, partially covered, etc.), the arrangement of the pump structure, the piping arrangement, and any other physical attributes that may affect the pump design. A sketch of the pump structure and associated pump and piping arrangement is often included.

SYSTEM DESIGN CONDITIONS. The following design conditions should be provided to the pump manufacturer:

- Design capacity for each pump during parallel operation, at minimum water level.
- Total head, based on the design capacity and low water level.
- Net positive suction head available. Minimum value is provided based on low water level and maximum water temperature. Because pump details usually are not known during the development of the specifications, a minimum water level below the vertical pump soleplate or horizontal pump baseplate and a water temperature range may be provided instead of an NPSHA value. The NPSHA value can then be calculated and compared to the NPSHR values after pump data are received from the manufacturer.
- Maximum pump speed, based on a suction specific speed of 8,500 and the NPSHA value provided above. A conservatively estimated NPSHA value may be used for the specification and the maximum pump speed verified after pump data are received from the manufacturer. The maximum design speed may be increased based on the manufacturer's successful operating history at higher speeds.
- Maximum shutoff head, based on circulating water system and condenser design requirements. It should be noted that maximum shutoff head is normally not an issue for circulating water pump applications since these pumps do not run at this point under normal operating conditions.
- Water chemistry. The specification should include anticipated concentrations of all constituents in the circulating water, including circulating water pH.

OPERATIONAL REQUIREMENTS. The specification should include the single or parallel operation requirements, anticipated startup and shutdown cycles, requirements for passing foreign objects through the pump, minimum expected pump life, and any other operational constraints that may affect pump operation. System head curves for single pump operation and all parallel-pump operation modes should be provided.

GUARANTEE. The guarantees required from the pump manufacturer should be stated. Guarantees vary for each application and may be negotiated during contract development. Some typical guarantees are as follows:

- Design point capacity and head;
- Submergence and NPSH requirements at design and runout conditions;
- Pump efficiency at design;

- Power required at design and maximum power requirement, generally at one pump runout conditions;
- Maximum shutoff head;
- Parallel- and single-pump operation in accordance with certified pump curves, including operation during the starting or stopping of a parallel pump;
- No critical speeds near the pump operating speed; and
- Sound level.

TESTING. Pump shop testing requirements should be provided and should include tests to verify design capacity and head, NPSH required, power requirements at design and runout, efficiency at design, and minimum flow, runout head and flow, and shutoff head. For high-speed or high-energy pumps, the manufacturer should be committed for testing at the pump shutoff head conditions. Sufficient operating data (eight to ten test points) should be collected to allow plotting of an accurate performance curve. In some cases, a pump manufacturer may be unable to shop test his pump due to limitation of the pump test stand. Should this occur, empirical data or field testing data may be available to verify or prove the pump design and operating characteristics.

MATERIALS AT CONSTRUCTION. Materials should be specified for each component of the pump. Material selection varies with site conditions. The technology of engineering materials is constantly evolving, offering new material developments for circulating water applications. Some typical materials are presented in Tables 11-7 and 11-8; however, each circulating water pump application must be carefully studied to determine the best combination of materials.

11.5.2 Condensate Pumps. Condensate pumps are medium-flow medium-head pumps that pump condensate from the condenser hot well generally through a series of low-pressure feedwater heaters to a deaerating heater. These pumps operate at a very low pressure at their suction with cold condensate, which can lead to pump damage because of cavitation if the pumps are not carefully selected to allow for these severe operating conditions.

Both horizontal and vertical condensate pumps can be used. Years ago, typical condensate pumps used in central power stations were horizontal multistage centrifugal pumps because of their ease of installation, maintainability, availability, reliability, and operability. However, with the trend toward larger power stations, larger condensers, and lowered hot wells to reduce building costs, the NPSH available to the condensate pumps, already low, was reduced considerably more. As a result, the horizontal centrifugal condensate pumps have been replaced with vertical, multistage pumps, mounted in suction "cans" beneath the plant ground floor. The suction "can" lengths are determined so that adequate NPSH is provided to the centerline of the first-stage impeller for proper operation of the condensate pump.

Condensate pumps, as shown in Figs. 11-30(a) and 11-30(b), are motor-driven vertical, multistage, wet suction, "can" type pumps with either single- or double-suction first-stage impellers. The pumps are located near the condenser with the pump baseplates at approximately ground floor elevation and arranged for below-floor suction from the condenser hot well and for above-floor discharge. The drive motor is mounted directly above the condensate pump to minimize

Table 11-7. Typical Materials of Construction for Vertical Wet Pit Pumps

Pump Component	Circulating Water Quality		
	Fresh Water	Brackish Water	Seawater
Impeller	Bronze ASTM B584 Alloy C92200 or stainless steel ASTM A276 Type 316 Condition A	Stainless steel ASTM A351 Grade CF3M	Stainless steel ASTM A351 Grade CF3M
Shaft	Stainless steel ASTM A276 Type 410 Condition T	Stainless steel ASTM A276 Type 316 Condition A	Nitronic 50 ASTM A351 Grade CG6MMN
Shaft sleeves	Stainless steel ASTM A276 Type 410 Condition H	Stainless steel ASTM A276 Type 316 Condition A	Nitronic 50 ASTM A351 Grade CGMMN
Shaft enclosing tube	Stainless steel ASTM A276 Type 316 Condition A	Stainless steel ASTM A276 Type 316L	Stainless steel ASTM A276 Type 316L
Radial shaft bearings	Bronze backed cutless rubber	Bronze backed cutless rubber	Bronze backed cutless rubber
Suction bell	Cast iron ASTM A48 or carbon steel ASTM A283	Stainless steel ASTM A276 Type 316L	Ni-Resist ASTM A439 Type D-2 or stainless steel ASTM A276 Type 316L
Column and discharge head	Carbon steel ASTM A283 or ASTM A516 (cold climates)	Stainless steel ASTM A276 Type 316L	Ni-Resist ASTM A439 Type D-2 or stainless steel ASTM A276 Type 316L
Impeller bowl	Cast iron ASTM A48 or carbon steel ASTM A283	Stainless steel ASTM A351 Grade CF3M	Stainless steel ASTM A276 Type 316L
Wearing rings	Bronze ASTM B584 Alloy C93200 or stainless steel 17-4 PH	Stainless steel ASTM A276 Type 316 Condition A	Nitronic 50 ASTM A351 Grade CGMMN and Nitronic 60 (Use alternately between stationary and rotating elements to prevent galling.)

Table 11-8. Typical Materials of Construction for Horizontal Dry Pit Pumps

Pump Component	Circulating Water Quality		
	Fresh Water	Brackish Water	Seawater
Impeller	Bronze ASTM B584 Alloy C92200	Stainless steel ASTM A351 Grade CF3M	Stainless steel ASTM A351 Grade CF3M
Shaft	Carbon steel AISI 1045	Stainless steel ASTM A276 Type 316 Condition A	Stainless steel ASTM A276 Type 316 Condition A
Shaft sleeves	Stainless steel ASTM A582 Type 416 Condition H	Stainless steel ASTM A743 Grade CF8M	Stainless steel ASTM A276 Type 316 Condition A
Casing	Cast iron ASTM A48	Ni-Resist[a] ASTM A439 Type D-2	Ni-Resist[a] ASTM A439 Type D-2
Wearing rings	Bronze ASTM B584 Alloy C93200 (impeller) Cast iron ASTM A48 (casing)	Stainless steel 17-4 PH	Nitronic 50 ASTM A351 Grade CGMMN and Nitronic 60 (Use alternately between stationary and rotating elements to prevent galling.)

[a]Ni-Resist may not be a practical selection for the casing material for larger (i.e., above approximately 100,000 gpm) horizontal circulating water pumps because of casting limitations.

horizontal space requirements. The number and size of condensate pumps vary with each installation.

11.5.2.1 Condensate Pump Design Criteria. The selection of a condensate pump requires the proper application and evaluation of several important condensate pump operating parameters and general guidelines. These include the pump capacity and total head requirements, net positive suction head, pump/suction can length, suction specific speed, peripheral velocity of the outer tip of the impeller inlet vanes, and piping system layout.

DESIGN POINT. The design point for condensate pumps is specified by the design capacity and total developed head. The design capacity for each condensate pump is determined by summing the condensate flow requirements for the following items:

- Condensate flow requirements from the condenser hot well with the unit operation at maximum turbine heat balance conditions; typically 5% overpressure conditions with the turbine control valves wide open (VWO);
- Condensate cycle makeup flow, typically 1% to 3% of the maximum condensate flow;

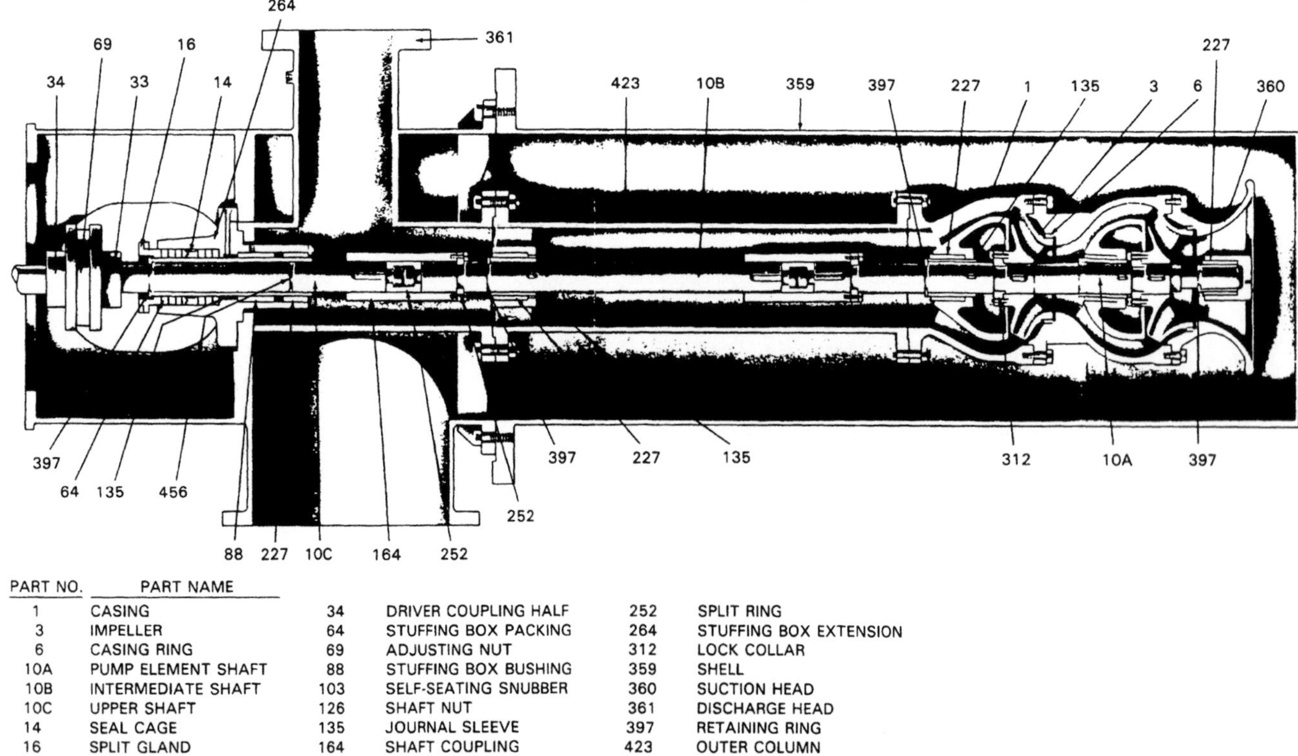

Fig. 11-30a. Condensate pump—vertical, multistage, wet suction can type single-suction first-stage impeller.

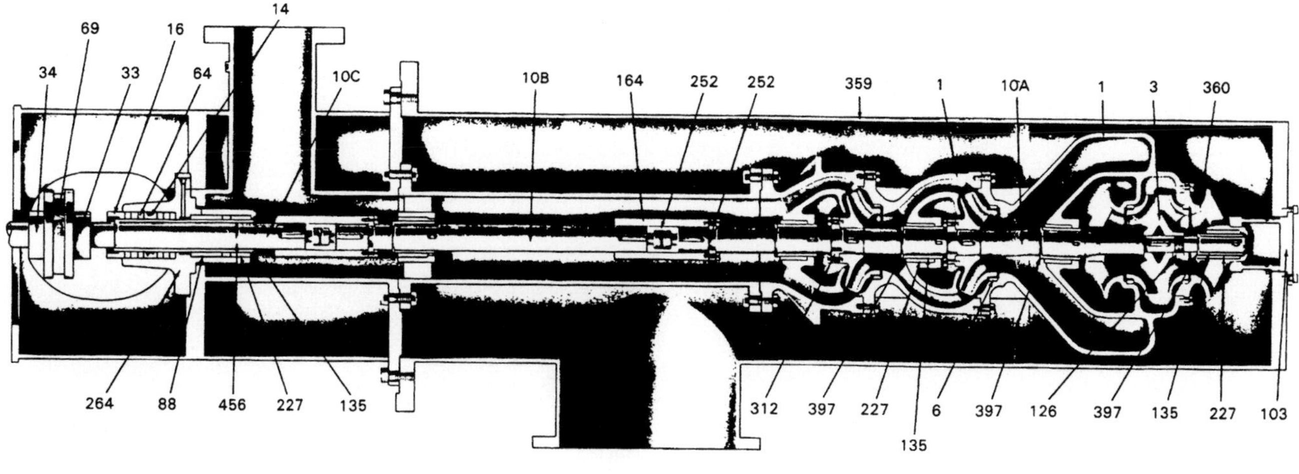

PART NO.	PART NAME				
1	CASING	34	DRIVER COUPLING HALF	252	SPLIT RING
3	IMPELLER	64	STUFFING BOX PACKING	264	STUFFING BOX EXTENSION
6	CASING RING	69	ADJUSTING NUT	312	LOCK COLLAR
10A	PUMP ELEMENT SHAFT	88	STUFFING BOX BUSHING	359	SHELL
10B	INTERMEDIATE SHAFT	103	SELF-SEATING SNUBBER	360	SUCTION HEAD
10C	UPPER SHAFT	126	SHAFT NUT	361	DISCHARGE HEAD
14	SEAL CAGE	135	JOURNAL SLEEVE	397	RETAINING RING
16	SPLIT GLAND	164	SHAFT COUPLING	423	OUTER COLUMN
33	PUMP COUPLING HALF	227	BEARING	456	"O"-RING

Fig. 11-30b. Condensate pump—vertical, multistage, wet suction can type double-suction first-stage impeller.

- Soot blowing steam flow, if applicable;
- Auxiliary steam flow, if applicable;
- Secondary air preheating flow, if applicable; and
- Boiler feed pump seal water injection flow, if applicable.

A flow margin of 5% of the total flow determined above is typical to provide latitude for future increase in the flow requirements due to any of the contributing factors. The design total developed head for each condensate pump is calculated by determining the difference between the condensate pump discharge head and the suction head as follows and as shown in Tables 11-9 and 11-10. A head margin of 5% of the total developed head determined is typical.

The shape of the condensate pump characteristic curve is important in vertical condensate pump designs. For stable parallel pump operation, the minimum rise in the pump curve from the design point should be 20% to 25%. The maximum rise should be approximately 40%. Lower rises to shutoff are acceptable if an independent minimum flow recirculation system is provided for each pump to guard against possible operation of a pump at or near shutoff conditions.

Excessively high rises to shutoff can be inefficient at off design operation of the pump during the plant life.

FIRST-STAGE IMPELLER DESIGN. Proper design of the condensate pump first-stage impeller is a critical consideration for reliability in a condensate pump to eliminate cavitation problems.

Foremost in the design of a condensate pump is the pump suction specific speed required (S). The Hydraulic Institute suggests for normal pump design that S varies from 6,000 to 12,000, and allows higher suction or specific speeds for special pump designs. For condensate pump service, the suction specific speed allowance of 12,000 (based on total pump flow) has been found acceptable. Another important parameter established for good condensate pump design is the limitation of the peripheral velocity of the outer tip of the first-stage impeller inlet vanes to 65 ft/s. Most condensate pumps operate without cavitation if they meet these two separate but related criteria.

Adherence to these two limits normally ensures stable

Table 11-9. Condensate Pump Total Suction Head Determination

Item	Head (ft)	Reference/Comments
Static head between the condenser hot well water level and the center line of the condensate pump first-stage impeller	a	Based on the low condenser hot well water level
Condenser pressure	a	Based on the condenser operating pressure at maximum turbine heat balance condition
Friction drop between the condenser and the condensate pump suction connection (the friction losses in the pump should be accounted for by the pump manufacturer)	a	Friction loss in the pump suction piping from the condenser to the condensate pump at maximum turbine heat balance condition
Total suction head	a	Sum of the above items

[a]Specific installation data.

Table 11-10. Condensate Pump Design Total Developed Head Determination

Item	Head (ft)	Reference/Comments
Static head between the deaerator level and the inlet of the condensate pump first-stage impeller discharge	a	Based on the water level maintained in the deaerator storage tank
Friction drop between the condensate pump discharge connection and the deaerator inlet (the friction losses in the pump should be accounted for by the pump manufacturer)	a	Friction loss in the pump discharge piping from condensate pump to the deaerator at maximum turbine heat balance conditions
Feedwater heat losses	a	Generally limited by feedwater heater design and specifications to approximately 50 psi or less, depending on number of feedwater heaters
Regulating valve pressure drop	a	Based on the required pressure drop of the control valve for the required flow control. Typically 25–50 psi
Condensate polisher pressure drop (if applicable)	a	Generally limited by condensate polisher design and specification to approximately 80–100 psi
Deaerator pressure	a	Based on deaerator operating pressure at maximum turbine heat balance conditions
Total discharge head	a	Sum of the above items
Total developed head	a	Total discharge head minus total suction head
Design margin	a	Typically 5% of the subtotal of total developed head above
Design total developed head	a	Total developed head plus margin

[a]Specific installation data.

condensate pump operation from minimum flow to pump runout. Suction specific speeds required (S) over 12,000 and/or inlet peripheral velocities over 65 ft/s are permissible, but increase the minimum flow for stable pump operation. The pump manufacturer should be consulted in these instances. Sensitivity to low flow operation results from separation of fluid flow at the periphery in the inlet area. Suction specific speeds over 14,000 and/or peripheral velocities over 75 ft/s may require minimum pump flow in the 50% to 60% flow range.

With the suction specific speed and peripheral velocity limitations, the NPSH, and in turn, the condensate pump and suction "can" lengths can be established. Typically, both are established at the maximum pump flow operating at runout conditions. Most conservatively, the NPSH required to prevent any reduction in the pump first-stage head at pump runout operation is used to determine the length of the condensate pump. However, it is important to note that the deeper the suction "can" setting, the longer the cantilever for the pump. Problems have surfaced because pump elements are so long that pulsations resulting from cavitation or critical reed frequencies have caused pump and pump bearing failures. Other alternatives to long condensate pump designs are slower operating speeds, double-suction first-stage impellers, or first-stage impellers specially developed by the pump manufacturer.

11.5.2.2 Condensate Pump Selection. Generally, the selection of a condensate pump is highly influenced by the configuration and overall space availability of the power station. As a result, condensate pumps are usually vertical multi-staged units with below-floor suction cans and are located near the condenser. The two important condensate pump selection factors that remain are the number of condensate pumps required and the type of pump flow control to be used.

These two factors are determined by the plant designer during a detailed investigation and evaluation of the condensate system. Information regarding the planned plant operating characteristics (base loaded or peaking), pumping flexibility, and operator requirements are considered as well as the plant economics, equipment capital cost, and operation and maintenance costs.

CONDENSATE PUMP FLOW CONTROL. Constant-speed and variable-speed condensate pumps are the two types of flow control typically used in modern power station installations. A summary of the design features and operating characteristics of each type follows.

CONSTANT-SPEED CONDENSATE PUMPS WITH THROTTLE VALVE CONTROL. This arrangement consists of condensate pumps connected directly to constant-speed motors. Flow control is accomplished by control valves located downstream of the condensate pumps. This type of arrangement is shown in Fig. 11-31. The system resistance curve shown on this figure represents the sum of the deaerator pressure, static head, and pressure losses due to the piping, valves, and equipment. The friction loss across the control valve is shown as a separate item to aid in explaining one of the differences between constant-speed and variable-speed pump operation. The design head that the pump must develop at the design flow rate is the sum of the system resistance and the pressure drop through the control valve at the design flow rate.

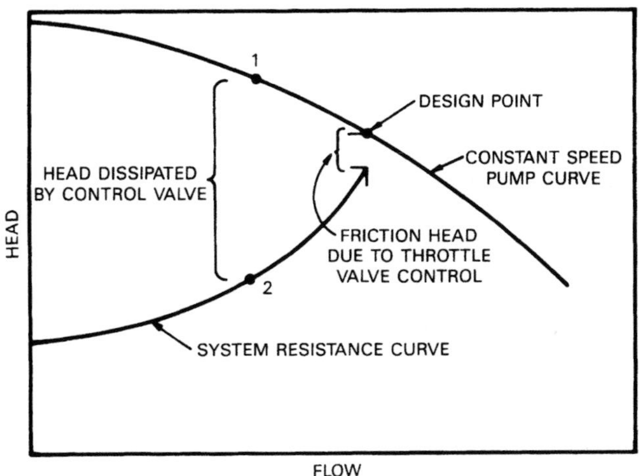

Fig. 11-31. System resistance curve, constant-speed condensate pump with throttle valve control.

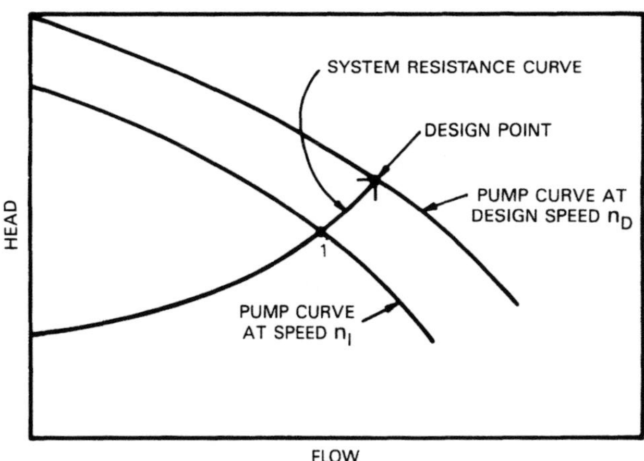

Fig. 11-32. System resistance curve, variable-speed condensate pump.

Those arrangements that use variable-speed pump drives for flow control do not require control valves. The design head for a variable-speed pump is less than the design head for a constant-speed pump by the amount of pressure loss across the control valve.

The method of controlling flow for the constant-speed arrangement is also illustrated in Fig. 11-31. At all flows, the static head produced by the pump is greater than the system resistance. For example, Point 1 on this figure is the head developed by the pump at a flow rate less than the design flow rate. Point 2 is the system resistance at the same flow rate. To achieve this flow rate, the difference in pressure between Points 1 and 2 must be dissipated by the control valve. The amount of pressure that must be dissipated by the control valve represents a loss (or inefficiency) associated with this type of control.

On units designed several years ago, the major disadvantage to the use of control valves in the condensate system was that valve cavitation sometimes occurred at low loads, resulting in noise and destructive vibration. However, control valve design innovations in recent years have minimized this problem.

The constant-speed pumping arrangement is relatively simple in design, easy to control, and highly reliable. The capital investments to install this arrangement are less than the cost of a variable-speed pump arrangement, but the energy and demand costs are higher.

VARIABLE-SPEED CONDENSATE PUMPS. The effect of varying the pump speed on the pump head curve is shown in Fig. 11-32. As the speed of the pump is reduced, the pump head curve shifts to a lower position, but the shape of the curve remains the same as the shape of the curve at the design speed. Two pump head curves (for the same pump but for two different speeds) are shown in Fig. 11-32. Also shown is the system resistance curve for the condensate system. This is the same resistance curve as shown in Fig. 11-31. The design point for the variable-speed pumping arrangement does not include the additional friction head margin required by a control valve since control valves are not required.

An advantage of the variable-speed arrangement is shown in Fig. 11-32. If Point 1 is the desired flow rate, the pump speed is reduced to speed n_1, such that the head and flow rate of the pump match the system resistance. No excess head is developed and no energy is dissipated by a throttle valve; hence, the horsepower required at this flow rate for the variable-speed arrangement is less than it would be for a constant-speed pump with a control valve.

Variable-speed condensate pumps can be driven by variable-speed motors, by constant-speed motors through variable-speed couplings, or by variable frequency drive arrangements.

The variable-speed pumping arrangement generally has higher capital costs, but has lower energy and demand costs. For a peaking unit, variable-speed condensate pumps would be very attractive and provide the operator with additional operating flexibility.

NUMBER OF CONDENSATE PUMPS. Typically, the number and size of condensate pumps is selected from the following arrangement:

- Two half-capacity pumps,
- Three half-capacity pumps, or
- Two full-capacity pumps.

Other than the obvious capital costs and annual operating costs associated with each pumping arrangement listed above, the reliability of the pumping system should be considered when designers are selecting the number of condensate pumps for the power station.

Additional condensate pump installation reliability may be provided by adding redundancy to the pumping system. The most apparent means of adding redundancy would be to install two full-capacity or three half-capacity condensate pumps instead of two half-capacity pumps so that one pump would remain in a standby mode at all times. Failure of one pump would not result in any temporary derating of the unit. The standby pump would immediately take over for the failed pump and maintain the unit load at the level prior to the failure.

The preferred approach for an arrangement with only two pumps is to provide one complete, spare internal pump assembly. Should a failure occur, the failed pump assembly could be removed and the new internal assembly installed. The estimated time to perform this replacement would be on the order of one to two shifts. However, the labor and materials costs associated with this replacement are considered small compared to the replacement power costs associated with the loss in unit capacity.

CONDENSATE PUMP HEAD PER STAGE. The head produced by an impeller is an important condensate pump parameter. Condensate pumps that produce high head per stage are more likely to have problems or actual damage resulting from higher vibrations, cavitation, and erosion because of higher energy input to the condensate. Modern condensate pumps have inputs as high as 600 to 700 horsepower per stage, which translates to approximately 365 to 400 ft of head per stage in a typical installation. A good conservative head per stage allowance is 100 to 225 ft. Any values higher than this should be proven by the condensate pump manufacturer.

CONDENSATE PUMP SHAFT DESIGN. A well-designed shaft is essential to reliable condensate pump operation. Recommended shaft design parameters include a maximum stress level of 7,500 psi, replaceable shaft sleeves under each bearing, keyed and locked shaft couplings, and bottom-supported shafts. Although the stress level of 7,500 psi is considerably less than the strength of the shaft material, this limit ensures long shaft life.

Keyed shaft couplings generally are superior to screwed shaft couplings, in that alignment is more precise and disassembly is facilitated. Screwed type couplings sometimes tend to "freeze," and the attempt to force the coupling loose can damage the pump shaft. Renewable shaft sleeves are typically specified and required under each pump interstage bearing, shaft column bearing, and uppermost bearing. Replaceable shaft sleeves facilitate removal and replacement due to bearing wear. Other shaft sleeve bearing arrangements are available that may be adequate for the specific installation. Because of the length of vertical multistage condensate pumps, together with the fact that the pump is cantilevered from its base, flow disturbances can cause vibration problems, that in turn decrease bearing life. Several condensate pump manufacturers overcome this problem and offer bottom shaft support designs that should be considered in any condensate pump specification requirements.

11.5.2.3 Condensate Pump Specifications. The following suggested condensate pump requirements that affect the design, manufacturing, and operation of a condensate pump should be included in any condensate pump procurement specification.

PUMP TYPE AND ARRANGEMENT. The specification should include the desired pump type, number of pumps required, location of the pumps, and plant arrangement. For current power stations, the condensate pump installation is typically a vertical, wet suction, multistage "can" type pump. These pumps are normally located near the condenser hot well, at ground level, arranged for a below-floor suction from the condenser hot well and for above-floor discharge.

SYSTEM DESIGN CONDITIONS. The following design conditions should be provided to the pump manufacturer:

- Design capacity for each pump during parallel operation, at minimum condenser hot well level.
- Total developed head, based on the design capacity and low condenser hot well level.
- Net positive suction head available, based on design capacity typically referred to the pump support elevation and in accordance with the worst case condenser vacuum operating condition. Plant arrangement and condensate pump setting must be considered.
- Condensate temperature, typically specified for both the design conditions and at maximum required capability of the pump.
- Allowable shutoff head for both the maximum and minimum allowable.
- Maximum suction specific speed for the first-stage impeller. See previous discussions.
- Maximum pump speed, typically specified at 1,200 rpm or 1,800 rpm, depending on the pump fit with hydraulic operating conditions included in the specification.
- Water chemistry, including the anticipated concentration of each constituent of the condensate, including pH.

OPERATIONAL REQUIREMENTS. The specification should include the single or parallel operation requirements and a description of the type of flow control that will be designed into the condensate system (regulating valve or variable speed operation). Requirements regarding how many pumps are included in parallel as well as a short description of the anticipated startup/shutdown operation of the pumps should be given. System head curves for single-pump operation and all parallel pump operating modes should be provided.

GUARANTEES. The guarantees required from the pump manufacturer should be included in the specification. Guarantees will vary for each application and may be negotiated during contract development. Some typical guarantees are as follows:

- Design point capacity and head;
- Pump speed at design capacity and head;
- NPSH requirements at design and single-pump runout operating conditions;
- Pump efficiency at design capacity and head;
- Power required at design point, and maximum power requirement, generally at one-pump runout conditions;
- Maximum shutoff head;

- Parallel- and single-pump operation in accordance with certified pump curves, including operation during the starting and stopping of a parallel pump;
- No critical speeds near the pump operating speed; and
- Sound level of operation.

TESTING. Pump shop testing requirements should include tests to verify design capacity and head, NPSH required, power requirements, efficiency, runout flow and head, and shutoff head. Sufficient operating data (8 to 10 test points) should be collected to allow plotting of an accurate performance curve. Most pump manufacturers have adequate test facilities to handle the testing of condensate pumps. Should this not be the case, then prototype or field performance data should be required to verify the condensate pump design and operating characteristics.

PUMP CONSTRUCTION. Key condensate pump construction requirements that should be specified include the following:

- First-stage and remaining stage impeller type;
- Bowl construction;
- Column construction;
- Bearing type, life span, and lubrication requirements;
- Baseplate, soleplate, discharge head/drive support requirements;
- Shaft coupling requirements;
- Shaft seal requirements;
- Manufacturing tolerances;
- Maximum sound level; and
- Vibration monitoring requirements (if any).

MATERIALS OF CONSTRUCTION. Materials of construction should be specified for each pump component. The pump material included in the specification should be the minimum requirements based on the application with the allowance that the pump manufacturer may suggest alternate materials that are equivalent to or exceed the strength of corrosion–erosion properties of the materials specified. Some typical materials are presented in Table 11-11.

Table 11-11. Typical Materials of Construction for Condensate Pumps

Pump Component	Condensate Water Quality Typical Material Selections
Suction can, column, and discharge head	Carbon steel, ASTM A36, or ASTM A283 Grade C for fabricated parts and ASTM A216 Grade WCB for castings
Bowls	Cast carbon steel, ASTM A216 Grade WCB
Impellers	12–14% chromium stainless steel, ASTM A743 or A217, Grades CA-15 or CA-6NM
Shaft	11.5–13.5% chromium stainless steel, ASTM A276, Type 410, Condition T
Shaft sleeves	11.5–13.5% chromium stainless steel, ASTM A276, Type 410, Condition H
Wearing rings	12–14% chromium stainless steel, ASTM A582, Type 416, Condition H
Bearings	Carbon or graphalloy
Bowl bolting	12% chromium stainless steel, ASTM A193, Type 416

11.5.3 Boiler Feed Pumps

At the heart of the power station preboiler systems is the boiler feed pump. This is a pump that must be reliable and rugged to withstand not only continuous normal high-pressure pumping operation but also transient system upset conditions and possible frequent starting and stopping to match the plant's load output requirements (cycling operation).

In general terms, a boiler feed pump is any pump that supplies feedwater to a steam generator for the production of steam either for energy conversion (supply to a steam turbine) or for plant or other industrial uses. Boiler feed pumps are available in many pump configurations and sizes and supply the flow and head required for any application. However, this chapter describes only boiler feed pumps that are diffuser or volute, horizontal, double-case barrel, single- and double-suction first-stage impeller, multistage centrifugal pumps. This is the most common type of boiler feed pump used in central power station applications today. A typical boiler feed pump configuration is illustrated in Fig. 11-33.

11.5.3.1 Boiler Feed Pump Systems. Many boiler feed pump system configurations have been used in power station applications. The basic system usually includes a deaerating heater and storage tank at some elevation above the suction of the boiler feed pump to provide a reservoir of heated, deaerated condensate to the boiler feed pump and available suction head for the pump. The suction pipeline to the boiler feed pump is amply sized to maintain the required suction velocity into the boiler feed pump, as well as limit the friction drop between the deaerator and the boiler feed pump.

A booster boiler feed pump may be included to provide suction head to the main boiler feed pump. The booster pump may be motor-driven or driven by the main boiler feed pump driver through an extended shaft off the driver (usually a steam turbine) through reduction gearing. A single-suction first-stage impeller is usually used in the main boiler feed pump in this system.

A conservative boiler feed pump system would also include a motor-driven startup boiler feed pump. This pump would be used primarily for plant startup activities, but could also be used to supplement the reliability and availability of the boiler feed system by operating in parallel with the main boiler pump at reduced plant loads or even at design load, depending on the capacity and head available from the startup boiler feed pump.

The discharge of the boiler feed pump should include pump recirculation instrumentation, valving, and piping back to the deaerator storage tank. Downstream, several stages of regenerative feedwater heating may be used, depending on the plant economics and cycle efficiency required. Some cycling plants also include a boiler feed system recir-

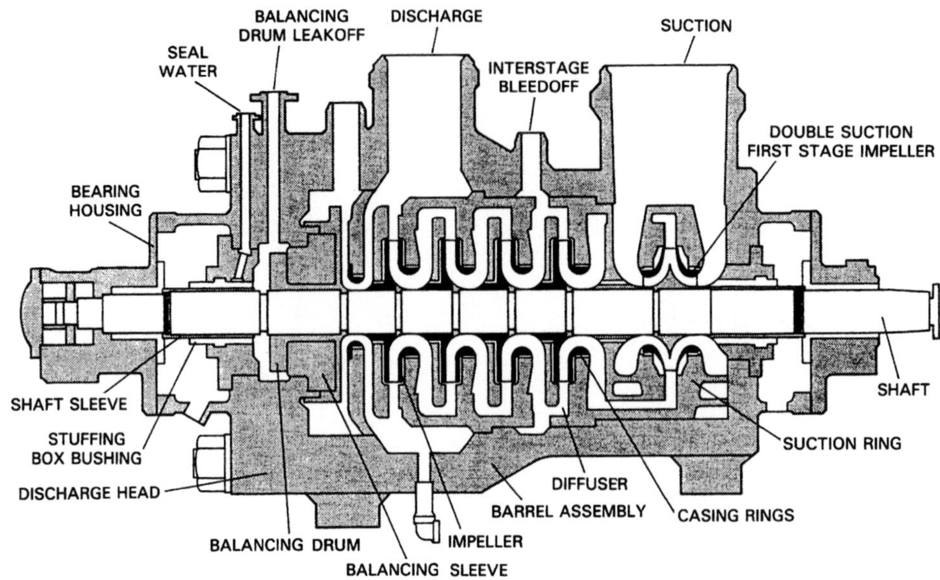

Fig. 11-33. Boiler feed pump—double case barrel pump, double-suction first-stage impeller, multistage centrifugal type.

culation to the condenser for use during off-peak/standby operation to keep the boiler feed system warmed and poised for operation.

Some typical boiler feed pump systems are shown in Figs. 11-34 through 11-38.

11.5.3.2 Boiler Feed Pump Design Criteria. The selection of a boiler feed pump is dependent on the requirements and constraints imposed by the boiler feed pump system configuration and operation. Each of the boiler feed pump system configurations previously described offers varying requirements that must be provided by the boiler feed pumps. In addition to the development of the system configuration, pump hydraulics, including the boiler feed pump capacity,

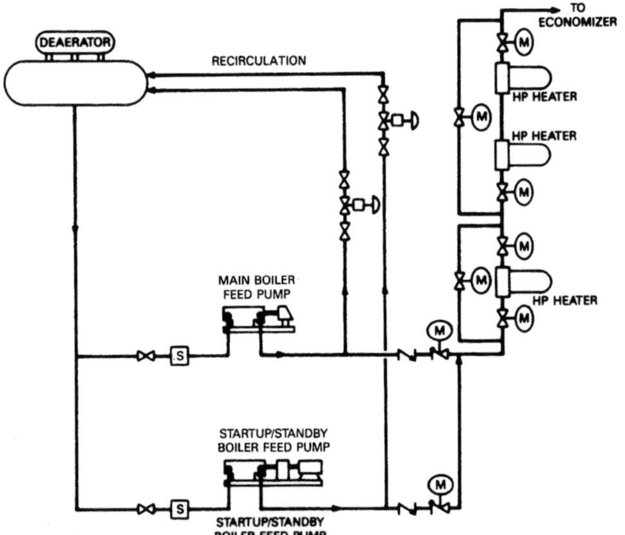

Fig. 11-34. Boiler feed pump system. One full-capacity boiler feed pump and startup/standby boiler feed pump.

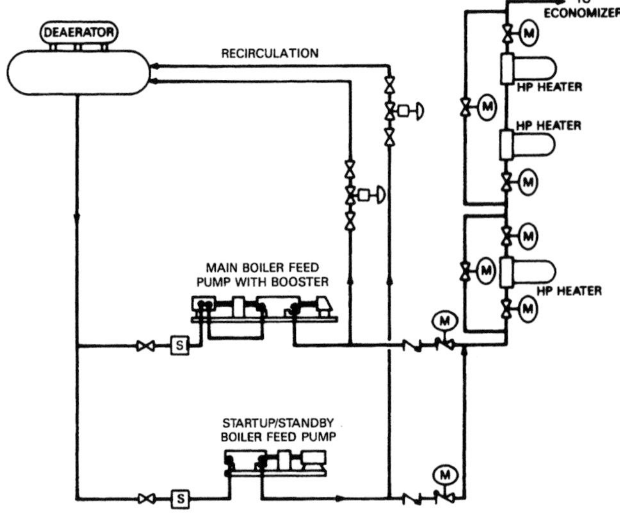

Fig. 11-35. Boiler feed pump system. One full-capacity boiler feed pump with booster and startup/standby boiler feed pump.

total head, net positive suction head, transient operation, specific speed, and pump physical construction must be considered.

DESIGN POINT. The design point for boiler feed pumps is specified by the design capacity and head. The design capacity is based on the plant's maximum steam flow requirements. For a large power station, this would be the steam turbine throttle flow at valves-wide-open and 5% overpressure, plus any steam or feedwater removed from and not returned to the flow path between the boiler feed pump suction and the turbine throttle. The following are contributing factors in the determination of the boiler feed pump design capacity:

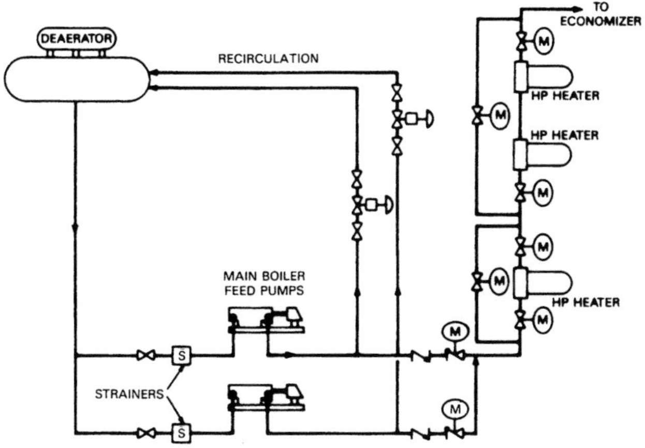

Fig. 11-36. Boiler feed pump system. Two 50% capacity boiler feed pumps.

- Turbine throttle steam flow typically at valves-wide-open and 5% overpressure conditions;
- Steam cycle makeup flow, typically 1% of the turbine throttle flow;
- Soot blowing steam flow, if applicable;
- Auxiliary steam flow, if applicable; and
- Reheat steam desuperheater flow, if applicable.

A flow margin upwards of 5% of the total flow minus innerstage bleedoff requirements is typically added to provide latitude for future increases in the flow requirement.

The total developed head for the boiler feed pump is calculated as shown in Tables 11-12 and 11-13 by determining the difference between the boiler feed pump discharge head and suction head. A head margin of 5% of the total developed head determined is typical.

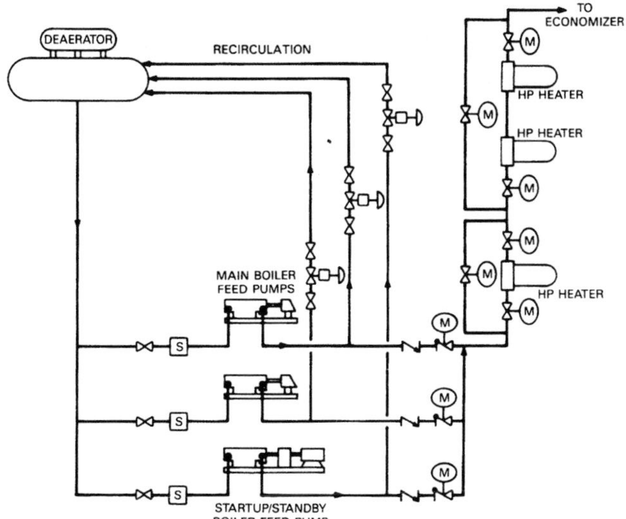

Fig. 11-37. Boiler feed pump system. Two half-capacity boiler feed pumps and startup/standby boiler feed pump.

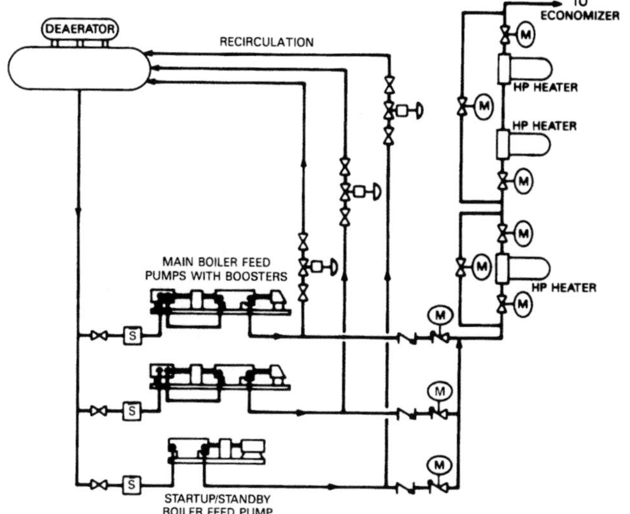

Fig. 11-38. Boiler feed pump system. Two half-capacity boiler feed pumps with boosters and startup/standby boiler feed pump.

HYDRAULIC CONSIDERATIONS. A number of hydraulic considerations influence the life and performance of the boiler feed pump.

The available suction head is the most important system parameter in relation to protection of the boiler feed pump itself and the preservation of stable pump operation at high and low loads and during transient system occurrences. Inadequate suction head can permit localized vaporization of the water when accelerated through the impeller. Cavitation can lead to pump instability and structural damage and should be avoided in high-head, high-speed applications such as boiler feed service.

Cavitation in warm water pumps is manifested by the formation of vapor bubbles created in regions of fluid acceleration where the pressure is lowered to the fluid vapor pressure. As these bubbles pass into regions of higher pressure, they collapse with a sudden release of energy that can

Table 11-12. Boiler Feed Pump Total Suction Head Determination

Item	Head (ft)	Reference/Comments
Deaerator pressure	a	Based on deaerator operating pressure at maximum turbine heat balance conditions
Static head between the deaerator level and the boiler feed pump impeller center line	a	Based on the low deaerator storage tank water level
Friction loss between the deaerator and the boiler feed pump suction connection	____	Friction loss in the pump suction piping from the deaerator to the pump suction connection at the design point capacity (a negative value)
Total suction head	a	Sum of the above items

[a]Specific installation data.

Table 11-13. Boiler Feed Pump Design Total Developed Head Determination

Item	Head (ft)	Reference/Comments
Turbine throttle pressure	a	Based on turbine throttle pressure at maximum turbine heat balance conditions
Main steam line friction loss	a	Based on maximum turbine heat balance conditions. Typically, main steam line economically sized to render a 100–200 psi pressure drop
Superheater friction loss	a	Determined by the steam generator manufacturer. Generally limited by the plant designer to an economically feasible value by specification
Economizer friction loss	a	Determined by the steam generator manufacturer and limited by specification
Static head between steam drum water level and economizer water inlet	a	Based on steam generator configuration
Static head between the economizer water inlet and the center line of the boiler feed pump impeller	a	Based on location of the economizer inlet elevation and the boiler feed pump elevation
High-pressure feedwater heater pressure drop	a	Generally limited by feedwater heater design and specifications to 50 psi or less depending on the number of feedwater heaters
Friction drop between the boiler feed pump discharge connection and the economizer inlet	a	Based on the feedwater flow at maximum turbine heat balance conditions
Total discharge head	a	Sum of the above items
Total developed head	a	Total discharge head minus total suction head
Design margin	a	Typically 5% of the subtotal of total developed head above
Design total developed head	a	Sum of the above items

[a]Specific installation data.

cause vibration of the pump shaft, damage to close internal seal ring clearances, and pitting of the impeller and casing surfaces. In addition to the vibration and noise created, cavitation also creates a drop in total developed head.

The parameter used to define the amount of suction head required above the vapor pressure is termed NPSH and is the total suction head less the vapor pressure. For a typical power plant installation, it is estimated that the maximum NPSH available (NPSHA) to the boiler feed pumps is approximately 130 to 145 ft depending on the elevation of the deaerator/deaerator storage tank. This value represents a maximum value, in that the length of the static column between the deaerator storage tank level and the boiler feed pump center line is at essentially the maximum practical length permitted by plant arrangement.

The desirable maximum deaerator height is limited by the boiler building height. The boiler feed pumps are normally installed at the operating floor level, thereby fixing the lower end of the static column. Locating the boiler feed pumps on either a mezzanine or ground floor level would result in significant complexity of the plant pipe routing. These locations may also necessitate the use of a separate condenser or a side or top exhaust boiler feed pump turbine, which could complicate maintenance and overhaul. A bottom exhaust feed pump turbine is possible if the boiler feed pump turbine is located on the mezzanine, but even this arrangement complicates maintenance. If the turbine room crane is to be used for removing the equipment, an operating floor access hatch must be included in the design. This requires that the piping that is normally located in that area of the mezzanine be rerouted. Also, experience has shown that this arrangement increases the noise level, since installing a sound enclosure becomes more difficult.

Pump manufacturers typically specify the net positive suction head required (NPSHR) for their pump based on the value observed in pump tests corresponding to a 3% reduction in first-stage developed head. The NPSHR is specified in this manner because of the inability of the pump manufacturer to determine at exactly what point the initial head reduction begins. The specified NPSHR at 3% head reduction is, therefore, not an adequate design basis since the pump will experience some cavitation under these conditions.

Consequently, a boiler feed pump should be selected to ensure ample margin between the pump NPSHR and the NPSHA of the system. This margin should be specified not only to include extra NPSH to protect the pump during foreseeable transient occurrences, but also to provide the NPSH margin between the 3% and the 0% head reduction points. Industry experience has shown that a margin of at least 50% above the manufacturers' stated 3% head reduction point is usually adequate. Additional margin above and beyond this value is justified by the additional reliability provided as indicated in the EPRI *Survey of Feed Pump Outages* (1978), in which a margin of 80% is recommended. Insufficient margin, on the other hand, may produce intense cavitation and accelerated impeller wear. In every case, the pump manufacturer should be consulted for appropriate NPSH margins.

Another hydraulic phenomenon that is important in boiler feed pump design is suction recirculation. As the flow through the impeller is decreased, a point is reached where vortices form in the suction eye of the impeller. These vortices are unstable and result in pressure pulsations at the pump discharge. The net results of this phenomenon are pump instability, noisy operation, and vibration of both the pump and the suction piping. The chances of boiler feed pump suction recirculation are much less with a booster boiler feed pump or with double-suction pumps.

Additional protective constraints should be applied to ensure hydraulically conservative designs. A parameter developed in the pump industry to describe pump similarity is specific speed, as shown in Eq. (11-1).

Specific speed allows a similar comparison of performance and is often used as a specification guideline. Customarily, boiler feed pumps have low specific speeds (1,000 to 1,800) and therefore define impeller passages that are essentially radial (Fig. 11-7). In general, lower values of N_s imply more stable pumps with smoother, more uniform head capacity curves, as shown in Fig. 11-39. For reliable partial load operation and stable operation of matched pumps in parallel, the specific speed for a boiler feed pump should be 1,600 or less.

Boiler feed pumps should be designed with a head capacity characteristic which rises continuously from high to low capacities without interruption. The specified requirement for head rise from design flow to shutoff is recommended to be not less than 120% and not more than 130% of the head at design flow. The continuously rising characteristic curve to shutoff guarantees that the boiler feed pump will be able to operate in parallel from minimum flow to runout with equal load sharing, without excessive vibration, hunting, pressure pulsations, or cavitation. This also enables the pump for unlimited, continuous operation at minimum flow.

Recalling our definition of suction specific speed required in Eq. (11-14), the suction specific speed required expresses the suction capability of the pump impellers. As with specific speed (N_s), the suction specific speed required (S) is an applicable basis for comparison only at the pump design point. For boiler feed water applications, a properly designed normal stage, other than the first-stage impeller, has an S value of about 8,000 to 9,500. A suction impeller has higher values, typically 10,000 to 12,000. However, the higher the S value, the lower the stage efficiency, and the stronger the tendency for hydraulic instabilities at low flow operation. Also for high S applications, a larger than normal impeller eye must be used. This produces more sensitivity to flow recirculation at the impeller eye, resulting in accelerated cavitation damage potential at flows other than the best efficiency operating point.

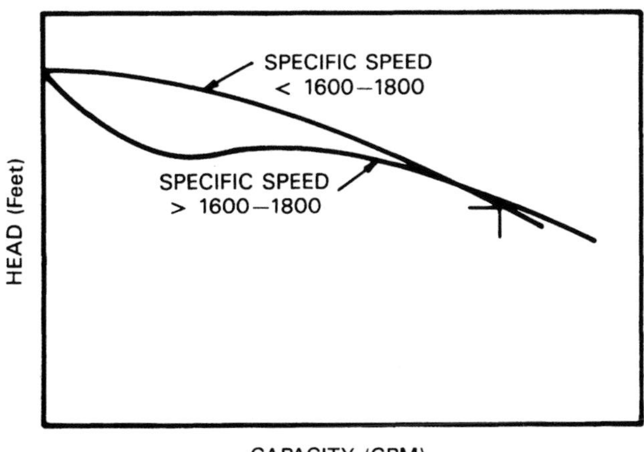

Fig. 11-39. Boiler feed pumps—pump characteristic curves for different specific speeds.

Field experience indicates that boiler feed pump reliability can decrease when developed head per stage exceeds 2,200 feet. Pumps with more than 2,200 feet developed head per stage require special designs and should be considered only where prior experience has been demonstrated by the pump manufacturer. Therefore, pumps should be specified to have a maximum of 2,200 ft of developed head per stage, unless the pump manufacturer can exhibit successful, proven experience.

OTHER BOILER FEED PUMP CONSIDERATIONS. Several references provide excellent recommendations and guidelines for boiler feed pumps and boiler feed pump installations. The reference of most significance for larger central power station boiler feed pump installations is *Recommended Design Guidelines for Feedwater Pumps in Large Power Generating Units* (1980), prepared by the Energy Research Consultants Corporation for EPRI. Another EPRI reference is *Survey of Feed Pump Outages* (1978), which also presents interesting boiler feed pump data that should be consulted by any boiler feed pump system designer or application engineer.

Recommendations regarding the following boiler feed pump design and major components that affect the pump reliability are noted in these references:

- Efficiency and hydraulic instability,
- Shop tests,
- Field instrumentation,
- Pump seals,
- Axial balancing device,
- Journal bearings,
- Internal pump clearances/tolerances,
- Minimum pump recirculation,
- Suction condition margins,
- Head per stage,
- Shaft breakage, and
- Feedwater control.

11.5.3.3 Boiler Feed Pump Specifications. After the system design engineer has determined the optimum boiler feed pump system and the requirements for the boiler feed pump are known, a purchase specification for the boiler feed pumps may be prepared for quotation by the qualified boiler feed pump manufacturers. The specification should include the pump type, arrangement, system design requirement, operational requirements, guarantees, pump construction details, and other requirements necessary for the pump manufacturer to satisfy the application requirements. The following items provide an overview of the criteria suggested for inclusion in any boiler feed pump specifications. Other guidelines can be found in EPRI *Recommended Design Guidelines for Feedwater Pumps in Large Power Generating Units* (1980).

PUMP TYPE AND ARRANGEMENT. The specification should define the boiler feed pump type and quantity, location of the pumps, and any significant plant arrangement requirements. Typically, for central power station applications, the boiler feed pumps are diffuser or volute, horizontal, double case barrel, double-suction first-stage impeller, multistage centrifugal pumps. The pumps should be designed for continuous service, and if more than one pump is to be supplied, they should be specified as "identical" as defined in the *Hydraulic Institute Standards* (1983).

Most boiler feed pumps are physically located in the plant in close proximity to the main turbine generator unit. The type and arrangement requirements of the pump(s) driver should be specified.

SYSTEM DESIGN CONDITIONS. The following system design conditions should be provided to the boiler feed pump manufacturer:

- Water chemistry, with the anticipated constituents of the boiler feed water, as well as pH range, to confirm or select the pump materials of construction.
- Maximum developed head per stage. This is typically about 2,200 feet per stage, based on operating experience and EPRI recommendations.
- Boiler feed pump suction conditions:
  —Design operating pressure at design conditions,
  —Minimum feedwater suction temperature,
  —Design feedwater operating temperature,
  —Design steady-state NPSH available,
  —Maximum NPSH available for sudden load reduction, and
  —Total boiler feedwater suction flow.
- Boiler feed pump discharge conditions:
  —Discharge design operating capacity,
  —Total head at design operating capacity,
  —Interstage bleedoff flow (if required), and
  —Interstage bleedoff discharge pressure.
- Boiler feed pump speed range at design.
- Maximum pump specific speed, typically 1,600 to 1,800.
- Maximum suction specific speed required of first stage, recommended to be approximately 10,000 to 12,000.
- Seal water temperature range.

The actual NPSH available is generally determined with all low-pressure feedwater heaters in service, with no margin or contingency included. It is important that a boiler feed pump be selected with conservative margin between the NPSH required by the pump and the NPSH available from the pump suction system. A boiler feed pump with low NPSH requirements meeting the following criteria should be considered:

1. The NPSH required by the boiler feed pump at the design operating conditions, based on 3% reduction in first-stage total head, should not be greater than the calculated NPSH available, divided by 1.8.
2. The NPSH required during sudden load reduction, based on 3% reduction in first-stage total head, should not be greater than the calculated NPSH available during sudden load reduction with the highest pressure low-pressure feedwater heater out of service, divided by 1.3.

OPERATION REQUIREMENTS. The specification should include a description of the operating requirements and scenarios that the boiler feed pump could undergo during its lifetime. A description of how the feedwater flow will be regulated should be presented. For turbine-driven boiler feed pumps, this is by pump speed regulation. For motor-driven pumps, this is accomplished by throttle valve control or variable-speed drive control.

Requirements regarding pump overspeed should be included, as well as restrictions on the avoidance of first, second, and third critical speed occurrences within the pump speed range. Typically, these critical speeds would be prohibited within a pump speed range defined by at least 125% of the pump maximum operating speed and 75% of the pump lowest operating speed.

Reliability of the boiler feed pump during all operating modes should be emphasized. Therefore, adequate pump clearances should be selected considering not only pump efficiency, but also the severe load and temperature swings the boiler feed pump could experience, including the following:

- Boiler feed pump turning gear operation,
- Startup when the feedwater suction temperature may rise from a minimum to a system warm-up operation temperature instantaneously,
- Hot restarts,
- Rapid load pickups, and
- Rapid load rejections.

Boiler feed pump internal warping, buckling, misalignment rubbing, or other damaging effects are not acceptable for long-term reliability and operations.

System head curves for single and all parallel pump operating modes should be provided. Curves indicating the pump suction temperature, steady-state NPSH available, and NPSH available after sudden load reduction versus feedwater flow at all plant loads should also be provided in the specification.

GUARANTEES. The pump guarantees required from the pump manufacturer will vary for each pump design and application. In addition to general guarantees such as those for pump operation without pitting, cavitation, damaging vibration, or excessive noise, the following should also be considered:

- Capacity at design operating point head and speed;
- Total head at the design operating point;
- Maximum NPSH required at design operating point and at the manufacturer's recommended minimum continuous flow with the suction temperature and pressure specified;
- Efficiency;
- Time required to start the pump from cold condition to design operating point flow and head;
- Horsepower input at pump design operating point;
- Minimum recommended continuous operation flow;
- Ability of the boiler feed pump to operate singly up to, and including, the intersection of the pump characteristic curve with the system head curve (runout);
- Ability of the main boiler feed pumps to operate in parallel with each other;
- Maximum shutoff head;
- Maximum permissible speed;
- Sound level;
- Ability of the main boiler feed pump to operate satisfactorily with the drive turbine on turning gear;
- Maximum vibration of the shaft with respect to the pump bearing housings as specified in *Hydraulic Institute Standards* (1983); and
- Maximum pump shaft deflection.

TESTING. Shop testing of the boiler feed pumps is recommended, as well as field testing after the pumps are installed. Some suggested shop tests are as follows.

COLD TESTS. The pumps should be tested over the complete pump head-capacity range from minimum flow to one-pump runout capacity with inlet water at not over 212°F to determine the pump characteristic curve, horsepower, and efficiency.

In addition, during cold tests the actual NPSH requirements should be determined on the basis of a 3% reduction of the first-stage pressure from minimum flow to the maximum capability of the pump (one-pump runout flow). The tests should be performed at least at flow rates including minimum recommended continuous flow, design point flow, and one-pump runout flow. Continuous vibration data should be recorded during these tests.

HOT TESTS. Following the cold running tests, the pumps should be hot tested using water at the operating temperature at the pump design conditions specified. In addition to determining the pump characteristic curve, horsepower, and efficiency, each pump should be subjected to the following tests, all of which should be performed at the pump designed speed.

NPSH Requirements Test. Each pump should be tested for its NPSH requirements on the basis of a 3% reduction of the first-stage pressure from minimum flow to one-pump runout flow. The tests should be run to prove the manufacturer's guarantee at both the design point and the minimum recommended continuous flow condition of the pump.

The test should be performed at least at flow rates including minimum recommended continuous flow, design point flow, and one-pump runout flow.

Transient Cavitation Test. Each pump should be tested to demonstrate its ability to operate under system transient cavitation conditions. With the pump operating at design flow, head, and temperature, the pump suction pressure should be reduced until the pump horsepower drops 15% to 20% of the design rated horsepower. This condition should be maintained for a minimum of 2 to 3 min before full suction pressure is restored.

Loss of Injection Seal Water Test. Each pump should be subjected to a 3-min loss of injection water to the pump seals. The pump should be tested at design speed and design flow, head, and temperature. Following the 3-min operation without injection water, the injection water should be restored to demonstrate continuous operation.

Vibration and Dynamic Pressure Pulsation Test. Shaft vibration recordings and dynamic monitoring of the pump suction and discharge pressure pulsations should be made during the pump hot running performance tests and the transient cavitation test.

Overspeed Test. Each steam turbine-driven pump should be tested at an overspeed equivalent to 115% of the design operating speed at design flow to demonstrate the ability of the pump to withstand this speed during a boiler feed pump turbine overspeed condition.

PUMP CONSTRUCTION. Key boiler feed pump construction requirements that should be specified include the following:

- Pump casings,
- Impellers and wearing rings,
- Shafts,
- Shaft seals,
- Bearings,
- Lubrication system,
- Shaft couplings,
- Maximum sound level, and
- Vibration monitoring (if any).

Materials of Construction. Materials of construction should be specified for each pump component. The pump materials included in the specification should be the minimum required based on the application, with the allowance that the pump manufacturer may suggest alternate materials that are equivalent to or exceed the strength and corrosion–erosion properties of the materials specified. Same typical materials are presented in Table 11-14.

11.5.4 Slurry Pumps

Several power plant systems require the pumping of liquids with suspended solids. The proper selection of slurry pumps

Table 11-14. Typical Materials of Construction for Boiler Feed Pumps

Pump Component	Boiler Feedwater Quality Typical Material Selections
Casing barrel	Forged carbon steel, ASTM A266, Class 1 or 2, or ASTM A105
Diffusers, volute cases, diaphragm, and stage pieces	12–14% chromium stainless steel, ASTM A743 or A487, Grade CA-15 or CA-6NM
Discharge head	Forged carbon steel, ASTM A266, Class 1 or 2, or ASTM A105, or ASTM A516 Pressure Vessel Plate
Shaft	12–14% chromium stainless steel, ASTM A276, Type 410, Condition T or ASTM A479, Type 410, Class 2
Impellers	12–14% chromium stainless steel, ASTM A743, Grade CA-15 or CA-6NM
Wearing rings	12–14% chromium stainless steel, ASTM A743, Grade CA-40, CA-6NM, or ASTM A276, Type 416 HT, AISI Type 410, 416, or 420

for the various slurry pumping applications is a major factor in the overall reliability of the plant. Slurry pumps are designed to minimize the effects of abrasion and moderate corrosion present in most power plant slurries.

Slurries are present in a number of industrial applications, and slurry pumps are available in several designs to satisfy particular requirements. Slurry pumps are used for dredging, mining, paper pulp, concrete, municipal solid waste, and various other solids handling applications. Depending on the service requirements, slurry pumps can be centrifugal, reciprocating, diaphragm, or jet pump.

This discussion concentrates on horizontal centrifugal slurry pumps as used in ash handling and scrubber slurry recirculation systems in a central power station application.

11.5.4.1 Slurry Pump Construction. A typical slurry pump (Fig. 11-40) has many features that set it apart from the typical centrifugal pump used for clear liquids. Wall thicknesses of wetted-end parts (casing, impeller, etc.) are greater than in conventional centrifugal pumps or are equipped with hard metal or rubber liners. Slurry pumps are typically ver-

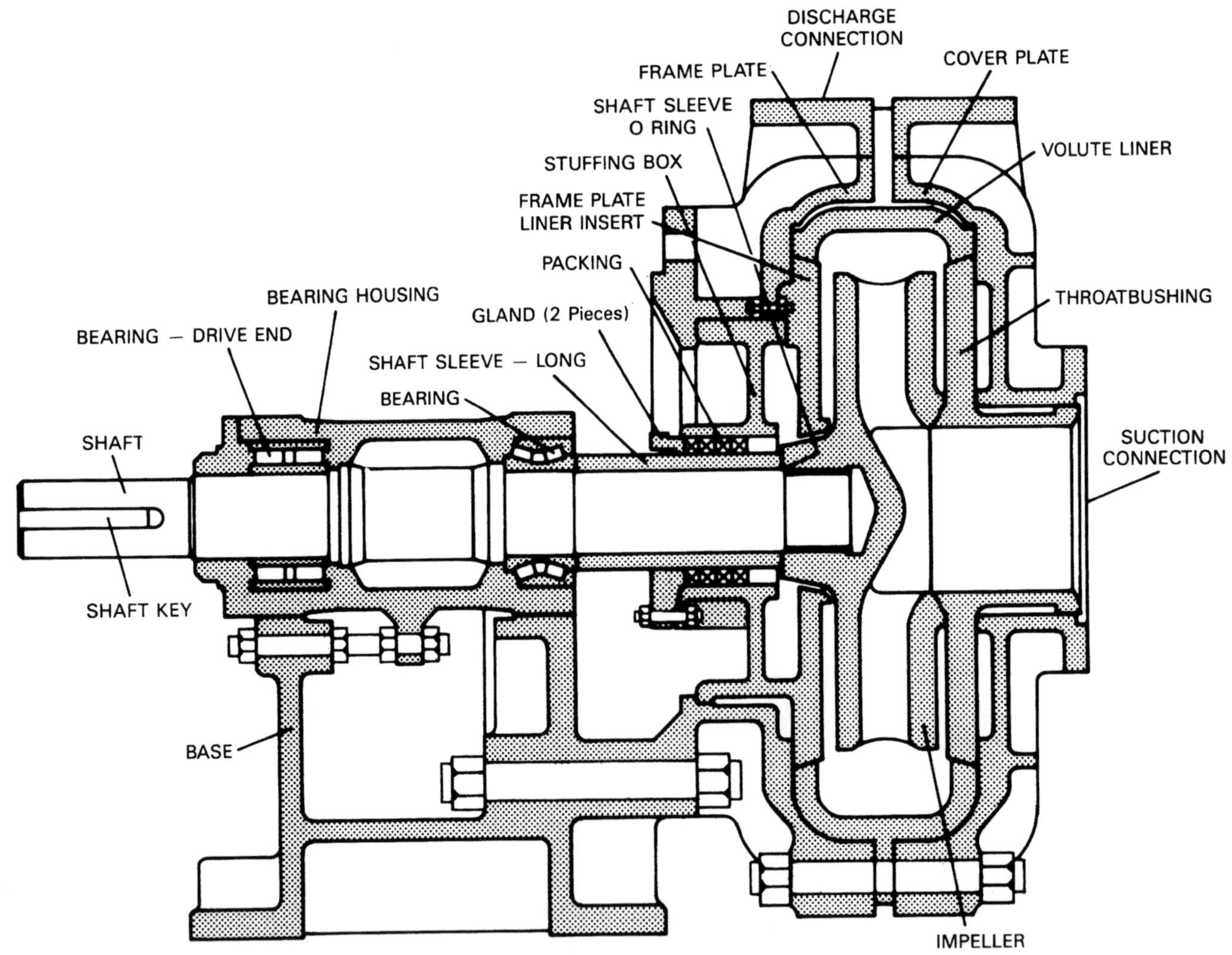

Fig. 11-40. Typical slurry pump.

tically split to allow for the replacement of casing and impeller liners. The cutwater, or volute tongue (the point on the casing at which the discharge nozzle diverges from the casing), is less pronounced in order to minimize the effects of abrasion. Flow passages through both the casing and impeller are large enough to permit solids to pass without clogging the pump. Since the gap between the impeller face and suction liner increases with wear, the rotating assembly of the slurry pump must be capable of axial adjustments to maintain the manufacturer's recommended clearance. This is critical if design heads, capacities, and efficiencies are to be maintained.

Other specialized features include extra large stuffing boxes and replaceable shaft sleeves. Some designs incorporate impeller back vanes that keep solids away from the stuffing box. Although the impeller back vanes also reduce axial thrusts by lowering stuffing-box pressures, these vanes can wear considerably in abrasive services. Therefore, both the radial and the axial-thrust bearings on the slurry pump are heavier than those on standard centrifugal pumps.

11.5.4.2 Materials of Construction. Because ash and scrubber recirculation pumps handle abrasive slurry, their design involves special considerations, many of them related to the selection of materials. Slurry pumps are available in a variety of materials of construction to handle the abrasion, corrosion, and impact requirements of the solids-handling application. Since the pump parts in contact with the slurry are subjected to abrasive–corrosive action, the "wetted" parts must be constructed of a corrosion-resistant material that is either harder than the slurry solids or resilient. The pump casings and impellers are typically made of cast iron lined with rubber or of cast iron alloys, also called hard iron. Since pump manufacturers will not accept responsibility for selection of material for these wetted parts, the design engineer must know which materials are suitable.

RUBBER. Molded rubber is the material specified most often for wetted parts of scrubber recirculation pumps. Both natural rubber and synthetic rubber of about ¼ in. thickness are used. In applications where fly ash may be present in the slurry or where the slurry contains a high level of chlorides, rubber-lined pumps are the scrubber industry's standard. Although rubber is resilient in abrasive service and resistant to corrosion, rubber parts have some disadvantages. One disadvantage is the limitation of pump discharge head to about 100 ft. In addition, the entry of tramp metal (welding rods, bolts, etc.) into the pump can destroy the rubber lining. At several facilities, strainers have been installed in the pump suction to protect the lining. When a strainer is plugged, however, the pump can cavitate, stripping the lining from the casing. Once the lining rubber is damaged, it cannot be repaired and must be replaced with new factory-supplied parts. Hence, if strainers are to be used, they should be accompanied by efficient cleaning devices.

HARD IRON. Cast iron alloys containing nickel (4%) and chrome (1.4% to 3.5%), or chrome alone (28%) hardened by heat treatment are used in a variety of slurry pump applications. These alloys are very hard, brittle materials (550 to 650 Brinell) that can be finished only by grinding. Hard iron materials have been used successfully in scrubber applications where good pH control is achieved, and hard iron is the preferred material for the wetted parts of ash slurry pumps. Hard iron should not be subjected to a pH below 4. Compared to rubber liners, hard iron allows higher heads and is not as vulnerable to damage by tramp metal. It is more subject to erosion and corrosion deterioration and should not be applied in high chloride concentration service.

11.5.4.3 Slurry Pump Design Criteria

SLURRY DESCRIPTION. To properly select a slurry pump, a comprehensive service description must be developed. This necessitates detailed analysis of the slurry composition, pH, specific gravity, and viscosity. Knowledge of the following factors is necessary in selection of pump material.

COMPOSITION. The fluid to be pumped is a slurry containing many solid and dissolved species. The solids typically found in power plant slurries are limestone, fly ash, calcium sulfite, and calcium sulfate, all of which are erosive. Solids levels normally range from 5% to 20% by weight. The dissolved species include calcium, magnesium, sodium, sulfite, sulfate, chloride, and carbonate ions, together with the ion pairs, such as hydrogen and hydroxide ions. Knowledge of the chemical analysis of the specific slurry is particularly important with closed-loop operation, since species such as chloride ions present only in trace amounts in the coal and makeup water can build up to critical levels of 10,000 ppm or more, thus dictating the use of highly corrosion-resistant materials. In addition, the nature, concentration, and size distribution of these solids should be determined. Information about these all important elements is necessary to determine abrasion–corrosion resistance and mechanical strength required of the pump.

PH. The slurry pH is an important consideration in the selection of materials for a scrubber recirculation pump. The recirculation pump normally takes suction from the reaction tank; therefore, it is typically exposed to a pH of about 5.5 or more. On systems without a reaction tank or that have a pump both before and after the reaction tank, such as some lime-based systems, the pH at the recirculation pump inlet is lower. Thus, the pump location determines the pH to which the pump material will be exposed.

The pH of ash slurries should be determined, but is typically between 6.0 and 8.0 and is not a major factor in the selection of materials.

SPECIFIC GRAVITY. The specific gravity of the slurry must be determined. Specific gravities of scrubber recirculation and ash slurries usually range between 1.05 and 1.15. Figure 11-41 is a graphic representation of specific gravity as a function of the solids content of the slurry. In systems that incorporate automatic solids control, the specific gravity of the slurry is relatively constant. Some systems, however, do

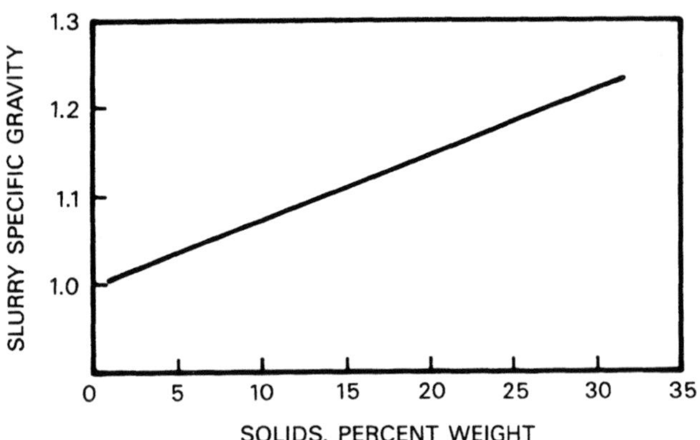

Fig. 11-41. Approximate specific gravity of scrubber recirculation and ash slurry.

not control solids content, and the specific gravity could vary over a wide range.

VISCOSITY. Knowledge of the slurry characteristics makes it possible to evaluate the reduction in pump performance due to the viscosity of the mixture. The added slip between the fluid and the solid particles as the mixture accelerates through the pump impeller decreases hydraulic efficiency. This slip is greater in mixtures with higher settling velocities, which may necessitate a higher pump speed to achieve the required head.

Each of the above factors must be analyzed by the designer and the pump manufacturer to ensure proper selection of pump materials. The slurry composition and pH define the degree of erosion and corrosion to which the pump will be subjected. The slurry-specific gravity and viscosity contribute to the required pump head and pump speed which may preclude the use of rubber-lined impellers.

PUMP CAPACITY. The required system flow for slurry systems is generally a function of the mass flow of solids to be handled and the solids percentage desired in the slurry. For scrubber recirculation pumps, this is determined by the design liquid-to-gas ratio of the system. Ash slurry flow rates are determined by the required ash removal rate and by the percent solids, which is maintained between 10% and 15% by weight.

Once the total system flow is determined, the required pump capacity is established based on the number of pumps and the operating scheme to be employed. The number of pumps required is determined based on an economic and reliability analysis for operation over the entire range of expected flows. If frequent low-capacity operation is expected, a greater number of lower capacity pumps may be economically and operationally desirable instead of two full-capacity pumps.

PUMP HEAD. Low-speed operation is one of the most important wear-reducing features of a slurry pump. Pump abrasive wear increases proportionally to the third power of pump speed. Impeller tip speed of rubber-lined pumps is limited to 3,500 to 5,000 ft/min to prevent disintegration of the impeller lining. This typically limits the rpm of slurry pumps with rubber linings to 400 to 600 rpm, which corresponds to a maximum discharge head of about 100 ft.

If a higher head is required for the application, hard-metal impellers can be specified with tip speeds up to 7,500 ft/min. This allows pump speeds of up to 900 rpm or a discharge head of approximately 225 ft. However, the corrosiveness must be carefully analyzed before metal liners are specified. Heads above 225 ft are achievable by increasing the pump speed and decreasing impeller diameter to maintain the lower tip speed. This decreases pump capacity and may require that additional pumps be added to the system. Impeller tip speeds above the levels stated above may be used for high head applications at the cost of substantially increased pump wear and maintenance.

Determining the required pump head for slurry systems is more difficult than for clear liquid systems. Slurries can be homogeneous or nonhomogeneous, and can exhibit Newtonian or non-Newtonian properties. Homogeneous slurries have a near-even distribution of solids across the flow area. This is achieved by maintaining adequate flow velocity in the slurry pipeline, typically between 5 and 10 ft/s. Non-Newtonian flow is commonly found in slurries with high solids concentrations and small particle size (less than 0.002 in.).

For typical power plant slurries, homogeneous Newtonian flow can be assumed to estimate system friction loss, as follows in Eq. (11-19);

$$\frac{H_f - H_{fw}}{H_{fw}\phi} = SG_s - 1 \tag{11-19}$$

where

H_f = head loss due to friction for the slurry, ft;

H_{fw} = head loss due to friction for water at the same flow velocity, ft;

ϕ = volume fraction of solids; and

SG_s = specific gravity of the solid (ash or scrubber solids).

A 10% design margin is typically applied to the friction loss calculation to ensure adequate pump head.

NPSH. As with all centrifugal pumps, NPSH is an important factor in the specification and selection of a slurry pump. For ash and scrubber recirculation systems, the available NPSH is generally more than adequate because of the system configuration. Most slurry pumps require a NPSH of 15 to 30 ft.

PUMP EFFICIENCY AND SPEED. In the selection of a slurry pump, the efficiencies of pumps from different manufacturers vary depending on how well the particular pump fits the design conditions. In the evaluation of the operating costs of a slurry pump, the efficiency, or energy cost associated with the pump, must be evaluated along with the pump speed, or wear rate.

The pump wear rate is approximately proportional to the pump speed to the third power. Therefore, a pump that operates at 1,000 rpm would be expected to wear approximately eight times faster than a pump operating at 500 rpm. Because of this, the pump speed is an important consideration, along with the pump efficiency, in determining the operating cost of the pump. In general, when the efficiency varies by less than 10%, a pump with a 10% lower speed will compensate for the increased energy costs with reduced wear.

11.5.4.4 Slurry Pump Specification. The following are suggested slurry pump requirements that affect the design, manufacturing, and operation of a slurry pump and that should be included in the slurry pump procurement specification.

PUMP TYPE AND ARRANGEMENT. The specification should include the desired pump type, number of pumps required, and the location of the pumps (indoor or outdoor). The typical slurry pump for an ash or scrubber recirculation application is the horizontal, centrifugal, heavy-duty material handling type with vertically split casings, cantilevered shafts, end suction, and replaceable wear parts.

SYSTEM DESIGN CONDITIONS. Each of the design criteria previously discussed should be included in the specification. These include a slurry description, design flow, design head, NPSH available, and maximum allowable rotational and/or tip speed.

OPERATIONAL REQUIREMENTS. Unique operational requirements should be specified. These may include a requirement for operation at increased slurry solids concentration or particle size due to slurry settling in tanks during inactive periods. Also, requirements for flow regulation should be specified. If flow regulation is required, it should be accomplished with a variable-speed drive instead of throttling the pump discharge.

GUARANTEES. The guarantees required from the pump manufacturer should be included in the specification. Guarantees vary for each application and may be negotiated during contract development. Some typical guarantees are as follows:

- Design point capacity and head;
- Pump speed at design capacity and head;
- NPSH requirements at design and single-pump runout operating conditions;
- Pump efficiency at design capacity and head;
- Power required at the design point, and maximum power requirement;
- Parallel- and single-pump operation in accordance with certified pump curves, including operation during the starting and stopping of parallel pumps;
- No critical speeds near the pump operating speed;
- Sound level of operation; and
- Ability to pass minimum particle size specified.

TESTING. Pump shop testing requirements should be provided and should include tests to verify design capacity and head, NPSH required, power requirements, efficiency, and runout flow and head. Sufficient operating data (8 to 10 test points) should be collected to allow plotting of an accurate performance curve. Shop tests should be conducted using a test slurry with a solids ratio, particle size, and specific gravity equivalent to the actual slurry to be pumped.

PUMP CONSTRUCTION. Key slurry pump construction requirements that should be specified include the following:

- Casing liners—rubber or hard metal, minimum thickness, attachment requirements;
- Impellers—shaft attachment, liner material (if required), one-piece construction, provisions for clearance adjustment;
- Shaft seals—type, seal water requirements;
- Connections—suction and discharge connection type;
- Bearings—radial and thrust-bearing type, lubrication requirements, service life rating, seal requirements;
- Drives—variable-speed drive type (if required), coupling or V-belt requirements, motor requirements; and
- Baseplate requirements.

MATERIALS OF CONSTRUCTION. Materials of construction should be specified for each pump component. The materials included in the specification should be the minimum requirements based on the specific slurry application. Allowance should be made for the pump manufacturers to propose alternate materials that are equivalent to or exceed the strength and corrosion–erosion properties of the materials specified.

Table 11-15. Typical Materials of Construction for Slurry Pump

Pump Component	Ash Pump	Scrubber Recirculation Pump
Casing	Ductile iron ASTM A536 Grade 60-45-12	Ductile iron ASTM A536 Grade 60-45-12
Casing liner	ASTM A532 550 minimum BHN	Natural rubber
Impeller	ASTM A532 550 minimum BHN	Ductile iron ASTM A536 Grade 60-45-12
Impeller liner	None	Natural rubber
shaft	Carbon steel ASTM A576, Grade 1045	Carbon steel ASTM A576, Grade 1045
Shaft sleeves	Stainless steel ASTM A276, Type 420 hardened	Stainless steel ASTM A276, Type 420 hardened
Bolting	Stainless steel ASTM A193, Type 410	Stainless steel ASTM A193, Type 410

Table 11-15 lists typical materials for ash slurry and scrubber recirculation pumps. Although these are common material selections, the engineer must work with the pump manufacturer to ensure that the materials selected fit the specific slurry application.

11.6 REFERENCES

ELECTRIC POWER RESEARCH INSTITUTE. 1978. FP-754. *Survey of Feed Pump Outages*. Prepared by Energy Research and Consultants Corporation, Morrisville, Pennsylvania. Electric Power Research Institute, Palo Alto, CA.

ELECTRIC POWER RESEARCH INSTITUTE. 1980. CS-1512. *Recommended Design Guidelines for Feedwater Pumps in Large Power Generating Unit*. Prepared by Energy Research and Consultants Corporation, Morrisville, Pennsylvania. Electric Power Research Institute, Palo Alto, CA.

Hydraulic Institute Standards, 14th ed. 1983. Hydraulic Institute. 30200 Detroit Road, Cleveland, OH 44145.

STONE & WEBSTER. 1982. *Failure Cause Analysis—Condenser and Associated Systems*, Volume 2. Electric Power Research Institute CS-2378, Boston, MA, pp. I-1 through I-34.

12

Circulating Water Systems

David J. Brill

12.1 INTRODUCTION

In accordance with the second law of thermodynamics, operation of a power cycle requires that heat be rejected at the low temperature of the cycle. The efficiency of the power cycle typically improves as the temperature at which the heat is rejected is lowered. This is illustrated by reference to the Carnot cycle for a two-phase working fluid as shown in Fig. 12-1. The power cycle thermal efficiency is generally defined as follows:

$$\eta_{Th} = \frac{\text{Work produced by the cycle}}{\text{Heat supplied to the cycle}}$$

For the Carnot cycle, the thermal efficiency becomes the following:

$$\eta_{Th} = \frac{T_h - T_l}{T_h} \quad (12\text{-}1)$$

where

T_h = absolute temperature of the high-temperature heat source; and

T_l = absolute temperature of the low-temperature heat reservoir, or heat sink.

Equation (12-1) indicates that the Carnot cycle efficiency can be improved by either increasing the temperature T_h or lowering the temperature T_l.

Since most power cycles depend on relatively warm ambient temperatures for heat rejection, the magnitude of waste heat is large. For each kilowatt-hour (kWh) of electricity generated by a conventional steam power cycle, approximately 2 kWh of waste heat must be rejected by some manner to the ambient environment. Through a wide variety of processes and equipment, most power cycles use a cooling system to reject waste heat.

Most power plants use a circulating water system as the mechanism by which steam cycle waste heat is transferred from the steam cycle to the ambient environment (low-temperature reservoir) (Fig. 12-2). Many factors determine the size and design of power plant circulating water systems. Of primary importance is determining the desired level of circulating water system performance (how efficiently the waste heat is transferred to the low-temperature reservoir). A power plant steam condenser that condenses the turbine exhaust steam at the lowest possible temperature and corresponding pressure achieves maximum steam cycle efficiency, which in turn minimizes the amount of waste heat to be rejected. A typical power plant heat rate curve is shown in Fig. 12-3. As the steam turbine exhaust temperature and pressure are reduced, the power cycle efficiency typically improves. At extremely low condensing temperatures and pressures, further improvements in cycle efficiency are diminished as a result of other physical operating limitations of the steam turbine and condenser.

However, designing a more efficient circulating water system to achieve lower condenser temperatures and pressures typically results in higher system capital and operating costs. During the power plant conceptual design phase, a detailed economic analysis is performed to determine the optimum balance between power cycle efficiency and circulating water system capital expenditures and operating costs.

12.1.1 Once-Through Cooling

In a once-through circulating water system, water is taken from a body of water such as a river, lake, or ocean, pumped through the plant condenser, and discharged back to the source. A schematic of this type of circulating water system is shown in Fig. 12-4. The circulating water flowing through the condenser is heated in the process of condensing the turbine exhaust steam. The temperature rise (TR) of the circulating water in the condenser may be calculated from the following equation:

$$\text{TR} = Q/(k_1)(\text{GPM})(C_p) \quad (12\text{-}2)$$

where

Q = condenser duty (Btu/h);

k_1 = constant = 500;

GPM = circulating water flow rate (gpm); and

C_p = specific heat of circulating water (Btu/lb/° F) = ≈1.0 for most fresh water circulating water sources.

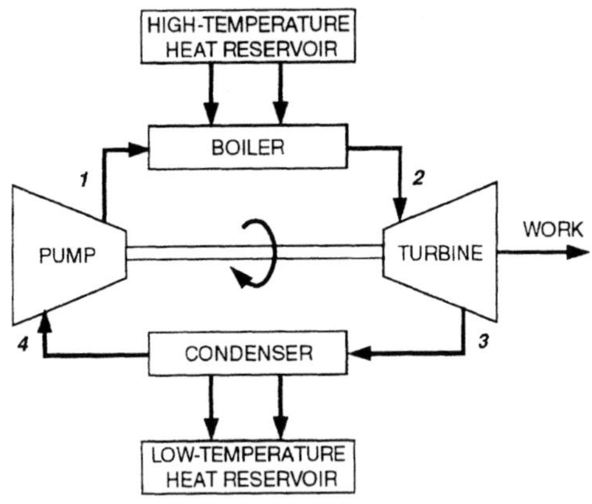

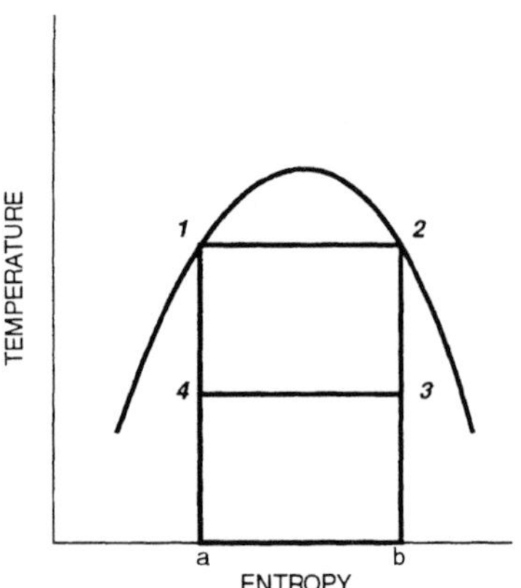

Fig. 12-1. A Carnot cycle utilizing a two-phase fluid as the working substance. (From *Power Plant System Design*. Kam W. Li and A. Paul Priddy. Copyright © 1985. Reprinted by permission of John Wiley & Sons, Inc.)

For many years, the once-through circulating water system was the most popular arrangement for power plant cycle heat rejection systems. A once-through circulating water system has two significant advantages. First, the relatively low temperature of most water sources used for once-through cooling makes this the most efficient cycle heat rejection system design. Second, the simple system arrangement typically makes once-through cooling the cycle heat rejection system design with the lowest capital and operating costs.

The disadvantage of this system is that the heated water is discharged back to the original water source, where the added heat is gradually dissipated to the earth's atmosphere. However, it may take a long time for the source water temperature to return to normal, or a new equilibrium temperature may be reached at a level higher than the normal temperature as long as the plant is in operation. Before construction of a once-through cooling system, environmental permits typically require determination of flow and temperature patterns around the circulating water discharge to estimate the thermal impact.

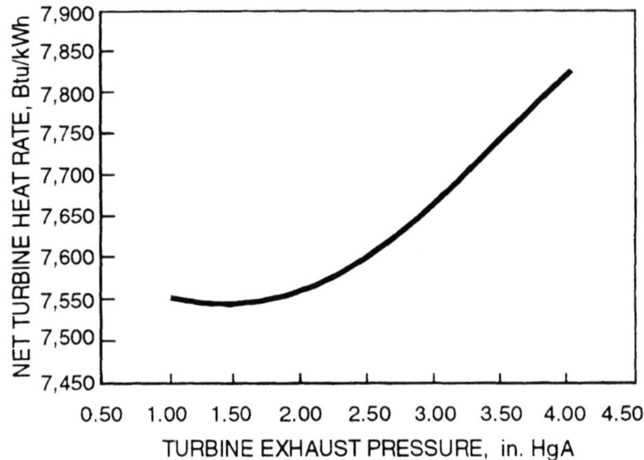

Fig. 12-3. Typical power plant heat rate curve.

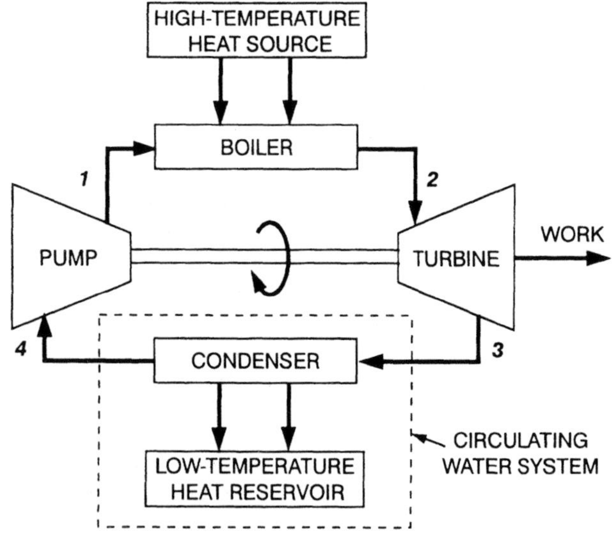

Fig. 12-2. Circulating water system. (From *Power Plant System Design*. Kam W. Li and A. Paul Priddy. Copyright © 1985. Reprinted by permission of John Wiley & Sons, Inc.)

Water quality regulations recently issued by federal and local environmental agencies have made the once-through

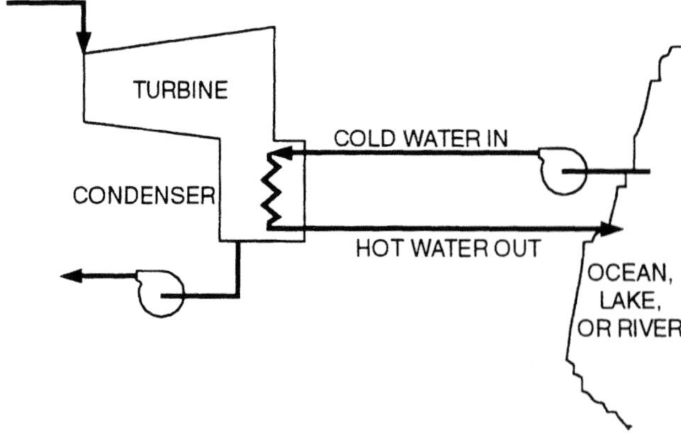

Fig. 12-4. Once-through circulating water system.

cooling system very difficult, if not impossible, to implement in domestic new power plant construction. Also, the large cooling water requirement of a once-through system limits potential plant sites to locations near large rivers, lakes, and oceans.

For existing plant locations with a once-through circulating water system that may face revised environmental regulations on the heated circulating water discharge, a helper cooling tower system may be required. As shown in Fig. 12-5, helper cooling towers can be retrofitted to the existing once-through system to reject a portion of the circulating water heat load directly to the atmosphere rather than to the water source.

12.1.2 Cooling Pond

Given the scarcity of environmentally acceptable large bodies of water available for once-through cooling, alternative cooling system arrangements must be developed economically. A cooling pond can be the simplest and least expensive alternate method for providing plant circulating water. The key to the economic viability of cooling pond construction is the availability of sufficient land at reasonable cost and geographic conditions suitable for pond or lake construction.

A cooling pond is a body of water into which the circulating water rejects the power cycle waste heat. The cooling pond then ultimately rejects the acquired heat to the atmosphere. The plant circulating water system arrangement is similar to that of the once-through circulating water system. Cooling ponds require large surface areas because they have low heat transfer rates. The temperature of the circulating water returned to the power plant depends on the heat transfer capacity of the pond. If the cooling pond does not have sufficient surface area to reject the acquired heat from the circulating water before it is recirculated, the pond temperature will rise until an equilibrium is established between the heating and cooling processes in the pond. The heat transfer mechanisms between the water surface and the atmosphere are complex. The processes involved in the heating and cooling of the cooling pond water mass can be summarized as follows:

Heating processes	Cooling processes
• Absorption of shortwave radiation from the sun and the sky • Absorption of longwave radiation from the atmosphere • Heat rejected to the pond from the plant power cycle	• Emission of shortwave solar radiation by the water • Emission of longwave radiation by the water • Convection between the water and the ambient air • Evaporation of pond water into the atmosphere

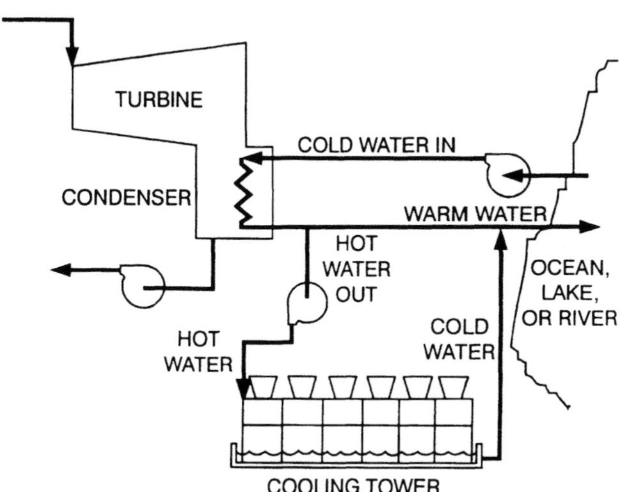

Fig. 12-5. Helper tower cooling system.

12.1.3 Recirculating Cooling System

Waste heat eventually finds its way to the earth's atmosphere. In the once-through cooling water system, heat is removed from the steam turbine exhaust and transferred to water bodies such as oceans, rivers, and lakes. The heat is then gradually transferred to the atmosphere by evaporation, convection, and radiation. However, the body of water may be negatively affected in this waste-heat transfer process. In a recirculating cooling system, the circulating water serves as an intermediate heat transfer medium from which the waste heat is directly rejected to the atmosphere.

The most common recirculating cooling water system arrangement incorporates an evaporative cooling tower as shown in Fig. 12-6. In this arrangement, waste heat removed from the steam turbine exhaust is carried by the circulating water to the cooling tower, which rejects the heat to the atmosphere. Because of this direct path to the atmosphere, surrounding water bodies typically do not suffer adverse thermal effects. Cooling towers have been used for many years at power plants in locations where some water is available for cooling system use, but where once-through cooling is not viable.

12.2 CIRCULATING WATER SYSTEM COMPONENTS AND ARRANGEMENTS

Only a few major equipment components comprise most circulating water systems. Basic descriptions of the compo-

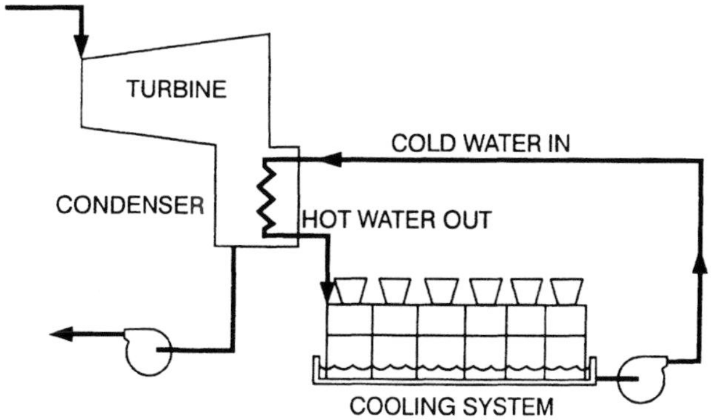

Fig. 12-6. Recirculating cooling water system.

nents are provided in this section, including the most common equipment arrangements used in power plant circulating water systems. Detailed descriptions of condensers are included in Chapter 9, Steam Cycle Heat Exchangers, and pumps are discussed in Chapter 11, Pumps.

12.2.1 Condenser

The power plant condenser is essentially a shell-and-tube heat exchanger (Fig. 12-7). The function of the condenser is to condense exhaust steam from the main power cycle steam turbine and boiler feed pump turbine. The condenser has other functions, as well:

- Recover condensed steam as condensate.
- Provide for short-term storage of condensate.
- Provide a low-pressure collection point for condensate drains from other systems in the plant.
- Provide for deaeration of the collected condensate.

Steam from the turbine is condensed by rejecting the heat of vaporization to the circulating water. The condensate is collected in the condenser hot well from which the condensate pumps take suction.

Condensers are classified as single pressure or multipressure, depending on whether the circulating water flow path creates one or more turbine back pressures. The condenser is described further by the number of shells (which is dependent on the number of low-pressure turbine casings) and as either a single-pass or a two-pass type depending on the number of parallel water flow paths through each shell. Flow schematics of common single- and multipressure condenser circulating water flow arrangements are shown in Figs. 12-8 and 12-9. Most power plant condenser arrangements have tubes running perpendicular to the longitudinal axis of the turbine generator.

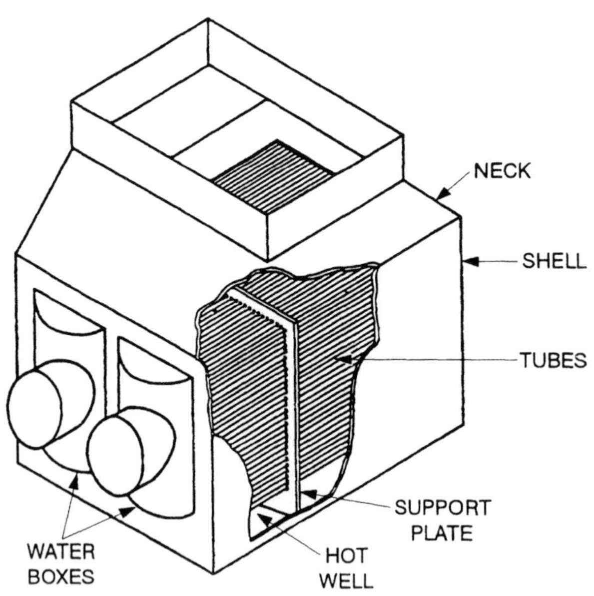

Fig. 12-7. Condenser.

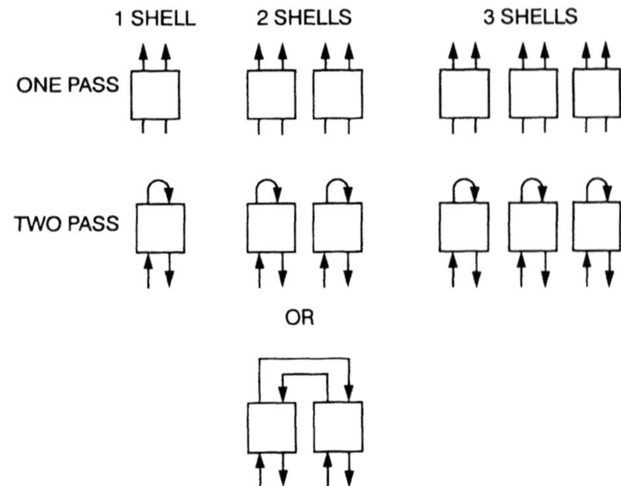

Fig. 12-8. Single-pressure condenser arrangements.

12.2.2 Cooling Tower

The cooling tower rejects waste heat from the steam cycle to the atmosphere. The use of cooling towers has become increasingly widespread as the availability of large water sources required for once-through systems has decreased and as environmental restrictions on thermal discharges associated with once-through systems have increased.

The conventional wet cooling tower transfers heat principally through evaporation. Because of the evaporative heat transfer process, water losses are actually higher for this recirculating cooling system arrangement than for a once-through system. However, withdrawals of water from the plant water source are reduced dramatically and no significant thermal impact to the water source occurs.

In areas where water conservation is a top priority, air-cooled condensers and dry cooling towers, using only sensible heat transfer, are available to reduce water usage to near zero. Also, hybrid wet–dry cooling tower designs are available that can be custom configured to merge the respective benefits of wet and dry cooling as may be required for a particular plant site.

Cooling towers are discussed in more detail throughout this chapter.

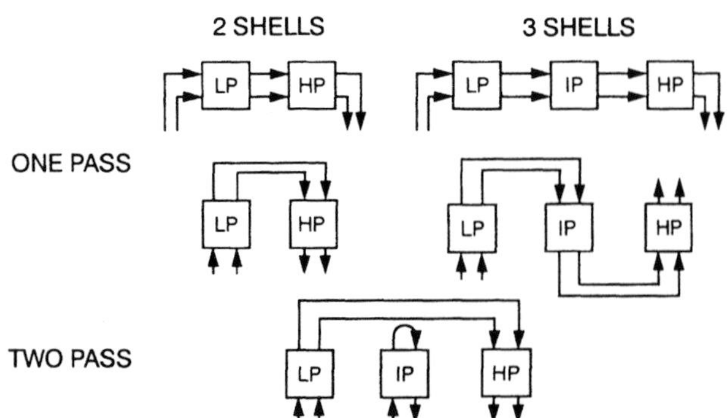

Fig. 12-9. Multipressure condenser arrangements.

12.2.3 Circulating Water Pumps

Circulating water pumps supply cooling water at the required flow rate and head to the power plant condenser and the plant auxiliary cooling water heat exchangers. The circulating water pumps must operate economically and reliably over the lifetime of the plant.

The three types of pumps most commonly used for circulating water service are vertical wet pit, horizontal dry pit, and vertical dry pit pumps. Typical arrangements of these pump types are shown in Fig. 12-10. For once-through cooling systems, vertical wet pit pumps are most commonly used, with occasional use of horizontal dry pit pumps and vertical dry pit pumps. For recirculating cooling systems, vertical wet pit and horizontal dry pit pumps are used about equally, with occasional use of vertical dry pit pumps.

12.2.4 Circulating Water Piping

Circulating water piping carries the cooling water from the circulating water pumps to the condenser and returns the water to the cooling tower (or discharge structure, in the case of once-through cooling). The routing of this piping includes above- and below-grade piping between the following components:

- The circulating water pumps and auxiliary cooling water pumps to the condenser inlet water boxes,
- The condenser outlet water boxes to the cooling tower riser (or discharge structure), and
- The circulating water pump suction piping (where required).

The large flow rates associated with circulating water systems typically require the use of large-diameter piping. Common pipe sizes range from 36 to 96 in. (900 to 2,400 mm) in diameter, although much larger pipe diameters have been used.

The design of the circulating water piping system must consider the environment internal to the pipe, as well as the external environment. Common materials used for circulating water piping, in order of increasing capital cost, include carbon steel, reinforced concrete embedded cylinder pipe, and fiberglass-reinforced plastic pipe. The large water requirement associated with circulating water systems typically makes it uneconomical to use high-quality water sources. The source of circulating water usually depends on the location of the power plant. Coastal plant sites typically use seawater or brackish water as the circulating water source, whether withdrawn by pumping directly from the body of water or by installing wells to extract nearby ground water. Inland plant sites usually are located near a lake, river, or other natural water source. Again, water is typically either withdrawn directly from the source or indirectly through ground wells.

The water from these sources can contain high concentrations of corrosive contaminants. Any pipe materials considered must include measures to protect the pipe for the service life of the plant. For example, carbon steel pipes in seawater service usually require an internal coating or an impressed current cathodic protection system, or both. Concrete pipe may require a special dense concrete mix to withstand chloride attack. These protective measures significantly increase the capital cost of an installation such that it can be as economical to install fiberglass-reinforced plastic pipe to obtain the same service life. As existing water sources become more strained and new water sources more scarce and expensive to develop, the quality of circulating water in future power plants is expected to decline further. This will likely further the trend toward corrosion-resistant piping materials.

12.2.5 Liquid Transient Analysis

The large flow rates of a typical circulating water system with flow velocities ranging from 6 to 12 ft/s introduce the

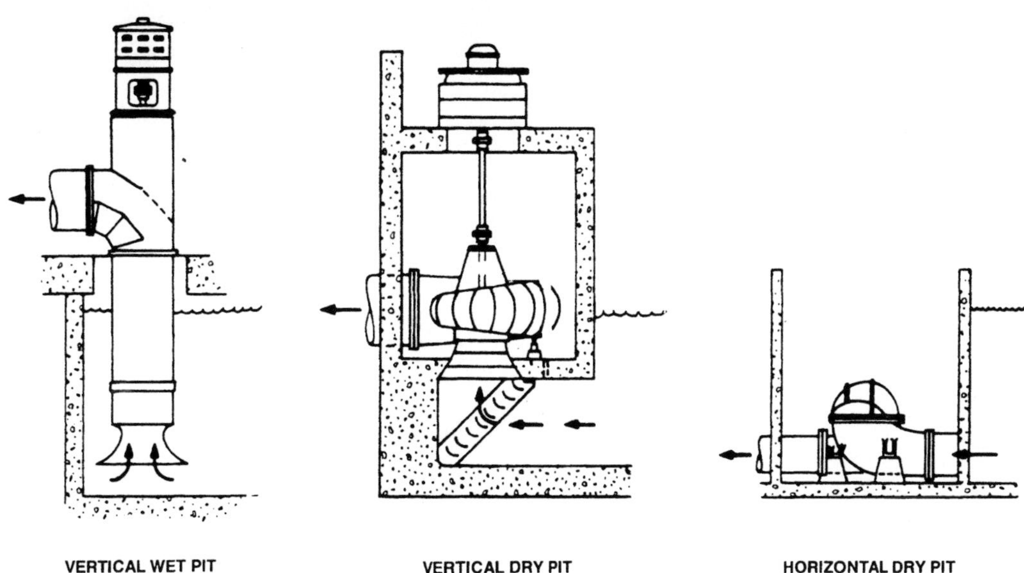

Fig. 12-10. Typical circulating water pumps.

potential for significant pressure transients that could be damaging. Liquid transient analyses are performed on circulating water systems to prevent costs associated with damage to system components from waterhammer and other unsteady flow phenomena. The potential risk of destruction of the large capital investment of a circulating water system and the threat of forced shutdown of the power plant are compelling reasons to perform such an analysis.

A conservative selection of design pressure for all circulating water system components should provide a sufficient margin to protect against damage from waterhammer. However, these design margins are not always sufficient and each system should be reviewed for waterhammer potential. The modeling techniques available are accurate and the results are often used in the preparation of system operating procedures as well as the selection of surge protection devices, if needed.

The typical analysis reviews normal valve and pump operating scenarios and foreseeable abnormal operating scenarios. The following is a nonexhaustive list of mechanical equipment operating scenarios that warrant study:

- The tripping of one or more large circulating water pump,
- The starting of one or more circulating water pump,
- The closure or opening of a valve within a circulating water system, and
- The filling of a circulating water system with velocities greater than 1 to 2 ft/s.

For more information on this subject, refer to the references at the end of this chapter.

12.2.6 Intake and Outfall Structures

Intake pumping structures are required for most wet type cycle heat rejection systems. Intake structure applications fall into two basic categories: intake structures located on a river, lake, ocean, or cooling pond for use in a once-through cooling system, and intake structures for cooling tower systems. The circulating water intake structure provides the following:

- Housing for the circulating water pumps,
- Proper hydraulic flow conditions to the pump suction, and
- Environmentally acceptable withdrawal conditions from the water source.

An outfall structure typically is included as part of a once-through circulating water system as a means to return the discharge circulating water flow to the waterway. Many once-through system piping arrangements can incorporate siphon recovery to reduce the system pumping head requirement. A siphon piping arrangement works on the principle that no pumping head is lost in a pumping system between two points having the same elevation because of elevation differences that may occur between the points. When pumping from one point to another at the same elevation, the only losses are caused by friction and valves, elbows, etc. A pipe that rises and then falls is called a siphon. A pipe that falls then rises is called an inverted siphon. On rising, the pressure head is transformed into elevation head and the reverse takes place on falling. Most once-through circulating water systems can use a siphon to some degree because the condenser elevation is usually well above the water level of the water source.

The siphon principle holds true provided that the circulating water piping flows full and is free of vapor and air. These requirements impose a limiting height for an effective siphon. The pressure in a siphon is a minimum at the highest point in the system (typically the top of the outlet condenser water box). To prevent vaporization of the liquid at the highest point, the pressure must exceed the vapor pressure of the water. Thus, the siphon may need to be broken in the circulating water discharge piping ahead of the outfall structure by installation of a seal well. The seal well exposes the circulating water flow to atmospheric pressure and the elevation is based on maintaining the back pressure needed to prevent flashing of the circulating water in the outlet condenser water box. A discharge water channel typically is constructed to carry the circulating water from the seal well to the outfall structure.

12.2.6.1 Once-Through Cooling System Intake Structure. Plan and elevation views of a typical intake structure for vertical wet pit pumps are shown in Figs. 12-11 and 12-12. Two major design parameters are considered for the once-through intake structure. The first parameter is the hydraulic/hydrologic characteristics of the water source. For a river or lake intake structure, the water level may vary on a seasonal basis. For an ocean intake structure, the water level varies hourly depending on tidal influences. In either case, the intake structure must be designed to provide adequate pump submergence and net positive suction head at low water levels, while assuring that the pumps and motors will not be flooded at high water level.

The second design parameter is the environmental regulations applicable to each plant location. An intake structure located on a natural body of water is usually subject to regulations concerning the water approach velocity to pro-

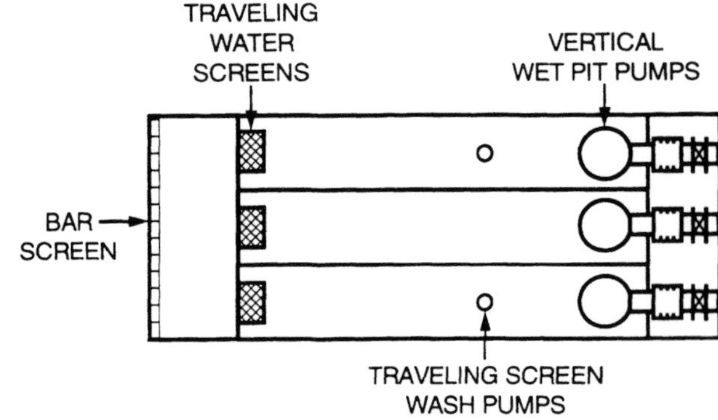

Fig. 12-11. Plan view of an intake structure for a vertical wet pit pump.

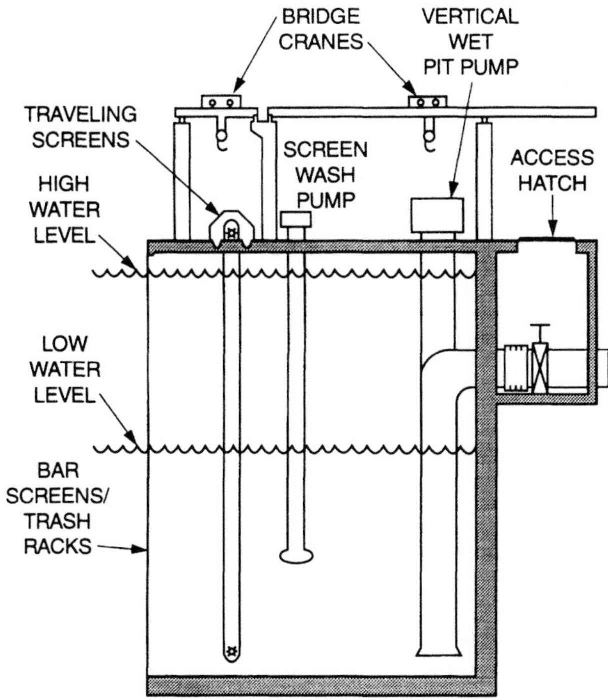

Fig. 12-12. Elevation view of an intake structure for a vertical wet pit pump.

tective intake screens, intake structure location, and other site-specific environmental concerns.

12.2.6.2 ***Cooling Tower Intake Structure.*** The cooling tower intake structure receives water from the cooling tower basin. The design of the cooling tower intake structure is similar to the once-through system. However, since the cooling tower system is a closed system, the pump structure water level varies over a narrow range, keeping the operating region well-defined, which allows for a much more shallow structure. Also, the intake structure does not interface with a river, lake, or ocean, so most of the associated concerns with environmental regulations are avoided.

12.2.6.3 ***Intake Structure Screening Systems.*** Intake structures located on a lake, river, ocean, or cooling pond require a screening system for debris removal. To be considered practical for use in protected power plant circulating water pumps, a screening system must meet the following criteria:

- Effectively screen the required amount of water without clogging or allowing bypass of screened materials.
- Be cost effective.
- Use proven mechanical components for reliable long-term performance in typically severe operating conditions.
- Be maintainable without interfering with the requirements of the cooling water supply.
- Use corrosion-resistant materials and protective coatings as required for long-life operation with the expected circulating water quality.
- Be effective in screening and protecting aquatic life.

Screening systems that satisfy the above requirements typically are classified in two categories: traveling screen systems and passive screen systems. The most commonly used traveling screen is the vertically rotating, single-entry, single-exit, through-flow screen mounted facing the water source. A through-flow traveling screen is shown in Fig. 12-13.

The vertical traveling screen consists of screens attached to a continuous belt that travels in the vertical plane between two sprockets. The screen, which is typically 3/8 in. (10 mm) mesh, is usually supplied in individual removable panels referred to as baskets or trays. The entire screen assembly is supported by two or four steel posts. The drive for the screen system is typically a two-speed drive, with normal operation at low speed to reduce equipment wear, and high-speed operation during periods of high debris loading. Screens are available in vertical lengths up to 100 ft (30 m) and widths up to 14 ft (4 m). Operation of the system can be performed manually at regular intervals or automatically based on continuously monitored differential pressure drop across the screen. A typical differential head to initiate operation of the screens is 6 to 10 in. (150 to 250 mm) of water.

Another type of vertical traveling screen used in power plant intake facilities is the dual-flow screen. Dual-flow screens are installed with the screens in parallel to the direc-

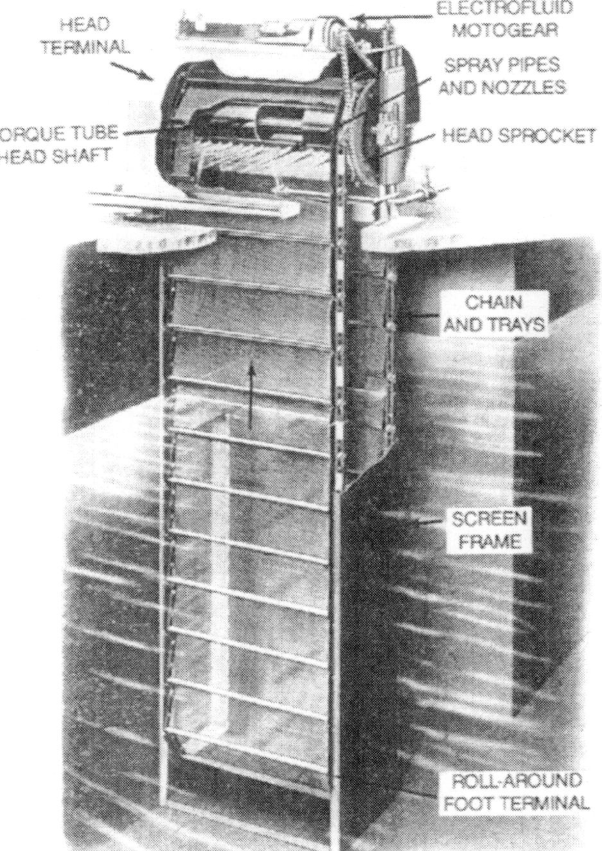

Fig. 12-13. Through-flow traveling water screen. (From FMC Corporation. Used with permission.)

tion of water flow to the intake structure. A typical arrangement is shown in Fig. 12-14. The advantage of the dual-flow screen is that twice as much screen area per screen is available as compared to the through-flow arrangement.

There are several disadvantages associated with the vertical traveling screens. Experience has shown that the screens require high maintenance, especially if located in a high debris or sediment-laden environment, or if located in a severely corrosive environment such as seawater. In through-flow arrangements there is sometimes debris carryover to the pump side of the screen. Also, the screens may be environmentally unacceptable because of damage to aquatic life. A fish handling and bypass system is available to attempt to save fish impinged against the screens.

The major difference between traveling screens and passive screens is that passive screens have no moving parts. A typical passive screen arrangement is shown in Fig. 12-15. Passive screens are typified by low approach velocities, low through-screen velocities, and minimum debris impingement and blockage. Any debris or material that may become impinged on the screens is removed by a periodic air or water backwash. The screens typically are designed for a maximum intake velocity of 0.5 ft/s (fps) (0.15 m/s) through the screen. The low velocity reduces the flow forces that cause the entrainment and impingement of debris and aquatic life. A typical arrangement for river and lake applications is shown in Fig. 12-16.

12.2.6.4 Intake Structure Sizing. The intake structure must provide acceptable pump suction hydraulic operating conditions at all possible water levels, accommodate the selected screening system, and satisfy applicable environmental regulations. Standards commonly used for intake structure sizing are the British Hydromechanics Research Association Standards and the Hydraulic Institute Standards. General criteria considered for the design of the intake structure are presented below:

- The flow approaching the pumps should be uniform across the width of the pump cell.
- Kinetic energy due to changes in level should be dissipated well in advance of the pumps.
- All flow obstructions should be streamlined to minimize flow separation near the intake structure.

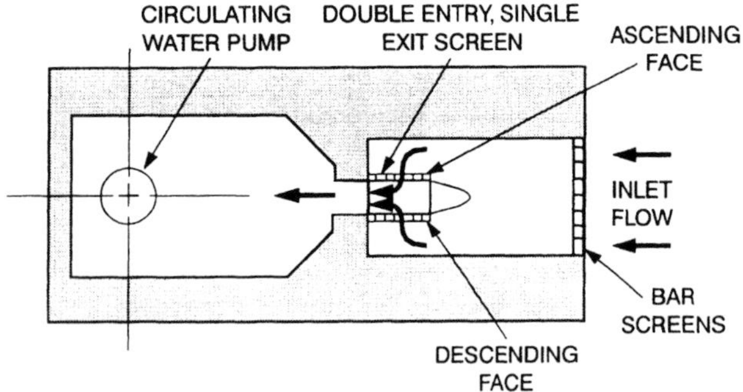

Fig. 12-14. Dual-flow traveling screen schematic.

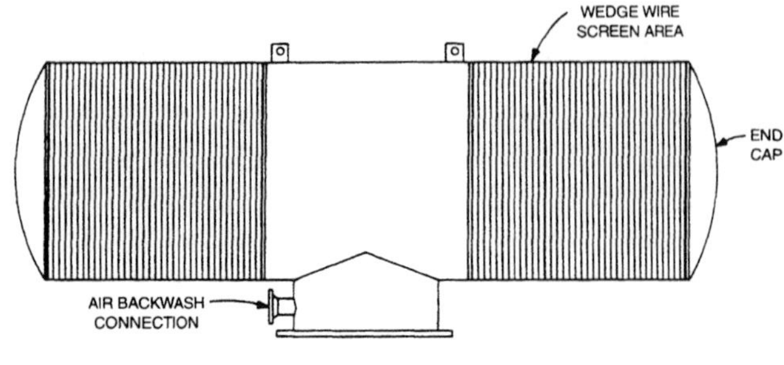

Fig. 12-15. Passive screen arrangement.

- Areas where stagnant water could occur should be filled in.
- Average velocities must be kept below 2 fps (0.6 m/s) on the approach to the pump sump, and 1 fps (0.3 m/s) or below on the approach to the pump bellmouth.
- Trash racks and screens should be located so they can also act as flow straighteners.

12.2.6.5 Outfall Structures. Outfall structures have two components: the seal well and the discharge channel. The seal well provides a minimum discharge elevation for the circulating water to maintain a siphon and prevent flashing in the outlet condenser water box and circulating water piping. The required seal well elevation is based on maintaining the back pressure needed to prevent flashing of the circulating water at the lowest absolute pressure point in the system

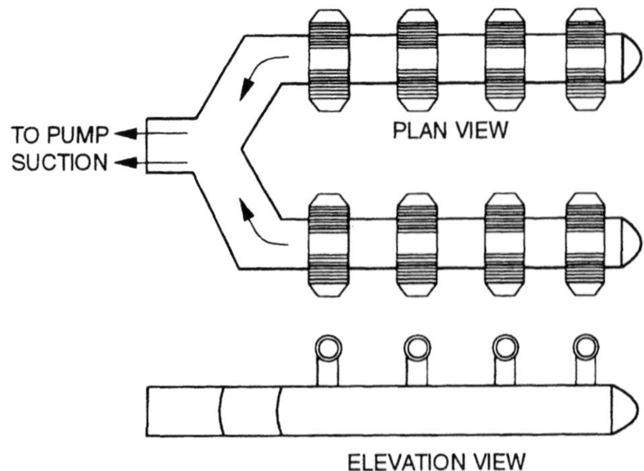

Fig. 12-16. Typical passive screen river and lake application.

(usually the top of the outlet condenser water box). A minimum 6 ft (1,800 mm) of water absolute pressure head at any point in the circulating water system is recommended. The elevation of the water in the seal well is typically maintained by a sharp-crested weir formed with adjustable stop logs. The discharge channel provides the flow path of the circulating water flow from the seal well to the waterway. In some cases, it is necessary to specially design the discharge channel to impart an optimal discharge direction and velocity to the discharge circulating water for proper dispersion into the waterway.

12.2.6.6 Model Testing. Model testing is recommended for intake structures using pumps with capacities greater than 50,000 gpm (4,700 L/s), and for smaller pumps if there are conditions in the intake approach that generate unusual circulation. Intake structure modeling offers several benefits:

- May reduce costs by optimizing the size of the pump cells.
- Identifies the need for intake structure flow baffles, fillets, and other pump structure modifications required to eliminate pump damaging surface and subsurface vortex formations. The model also provides physical verification of the effectiveness of these modifications.
- Helps reduce maintenance and improve efficiency for the pumps.

12.3 COOLING TOWERS

The cooling towers described in this section are evaporative towers that derive their primary cooling effect from the evaporation that takes place when air and water are brought into direct contact. The efficiency of the cooling process depends on the ambient conditions and the heat rejection load within which the tower must operate. Other factors that influence efficiency, and over which the cooling tower manufacturer has some measure of control, are the amount of heat transfer surface area, the time duration that the water surface is exposed to the airstream, and the ratio of airflow to water flow in the cooling tower. These factors can be balanced in an infinite array of combinations that produce the same end result of cooling the water to the design conditions.

Cooling towers can be characterized in several ways. The most common ways are by type of draft (mechanical or natural) or by the relationship between the air and water flows (crossflow or counterflow). They can also be classified according to the main mode of heat transfer (evaporative, or wet, tower versus sensible, or dry, tower). This section describes the salient features of evaporative cooling tower classifications. Portions of this section were compiled from documents published by the Marley Cooling Tower Company.

12.3.1 Mechanical Draft

A mechanical draft cooling tower uses large fans to produce airflow through the tower fill. Water is distributed over the tower fill, and heat transfer takes place by evaporation and convection as a result of water to air interface.

There are two types of mechanical draft towers: forced draft and induced draft (Fig. 12-17). Although Fig. 12-17 shows a counterflow cooling tower design, both fan arrangements are also applicable for crossflow cooling towers.

Forced draft cooling towers have the fans mounted at the base of the tower. Air is forced in at the bottom and discharged through the top of the tower. This arrangement has the advantage of locating the fan and drive motor outside the tower, where it is convenient for inspection, maintenance,

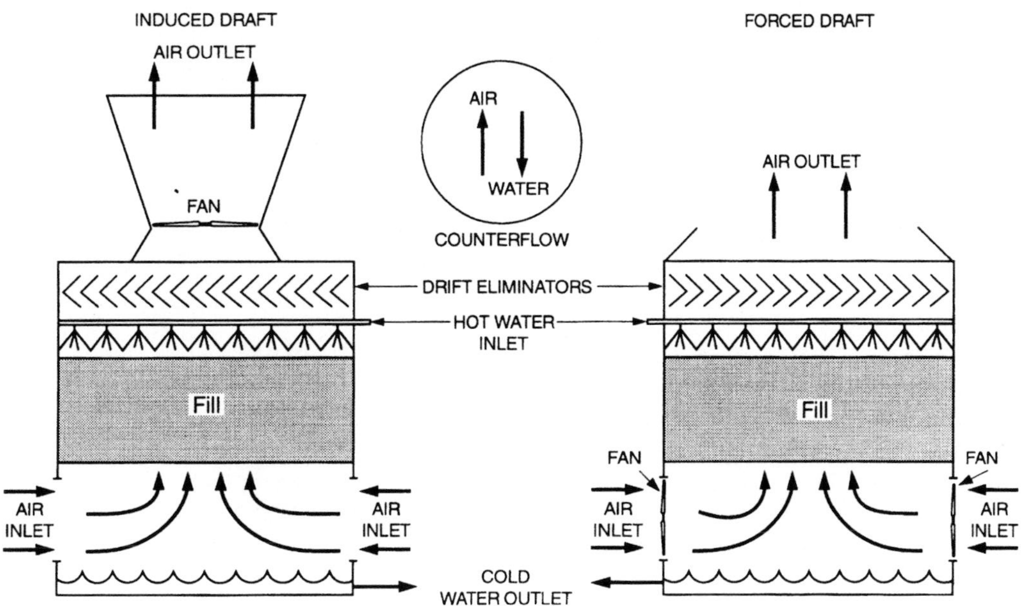

Fig. 12-17. Induced draft and forced draft counterflow cooling towers.

and repairs. Since the equipment is out of the hot, humid area of the tower, the fan is not subjected to corrosive conditions. The distinguishing airflow characteristic is a high inlet air velocity and a low exit air velocity. This characteristic can lead to excessive recirculation and decreased stability of thermal performance when compared to the induced draft tower. Recirculation describes the phenomenon of portions of the warm, moist air of the tower exit air plume being drawn back into the tower air intake, raising the inlet wet-bulb temperature. The location of the tower fans in the entering airstream also makes them subject to potential icing problems. Icing can be aggravated in periods when the forced draft tower is experiencing recirculation, because of the high moisture content of the recirculated air.

The induced draft tower has the fans located downstream of the tower fill section so that the air is pulled through the fill section. As opposed to the forced draft arrangement, the distinguishing airflow characteristic of this induced draft arrangement is a lower inlet air velocity and a higher exit air velocity. Exit air velocities three to four times higher than inlet velocities are common. The higher exit velocity reduces the tendency for a reduced pressure zone to be created at the air inlet by the action of the fan. This greatly reduces the potential for excessive recirculation. When recirculation occurs on an induced draft tower, it is usually the result of external conditions such as ambient wind conditions. Icing of the mechanical equipment is also reduced because the fans are located in the warm airstream. The induced draft tower is by far the most common type of mechanical draft used in utility power plant service.

There are two primary types of mechanical draft tower arrangements: rectangular, in-line fan arrangement, and round, clustered fan arrangement.

The in-line fan arrangement (Figs. 12-18 and 12-19) is commonly referred to as a rectangular cooling tower and is the most common type of mechanical draft tower arrangement. Rectangular towers are constructed in modular fashion; the tower length and number of cells are increased as necessary to accomplish a specified thermal performance. Each cell of the rectangular tower is nearly identical in design. This modularity allows standardization in the design of the structure and mechanical components, and in determining the thermal ratings. The rectangular tower design is usually the lowest capital cost cooling tower arrangement.

A drawback of the rectangular arrangement is the need for a long narrow strip of land on which to locate the tower. On large, multiple-unit power plant sites, siting of multiple rows of long rectangular cooling towers can be nearly impossible. Also, when multiple rectangular towers are located next to each other, the hot moist air exit plume from one tower can be drawn into the air inlet of another tower. This phenomenon is called interference. As with recirculation, tower interference affects the thermal performance of the tower by raising the average wet-bulb temperature of the tower inlet air.

Rectangular tower structures are typically wood or concrete construction. Recent advances in the design and fabrication of fiberglass structural members have enabled the commercial introduction of large-scale fiberglass structure cooling towers to the market.

As shown in Fig. 12-20, the round cooling tower arrangement is nearly circular in plan arrangement, with fans clus-

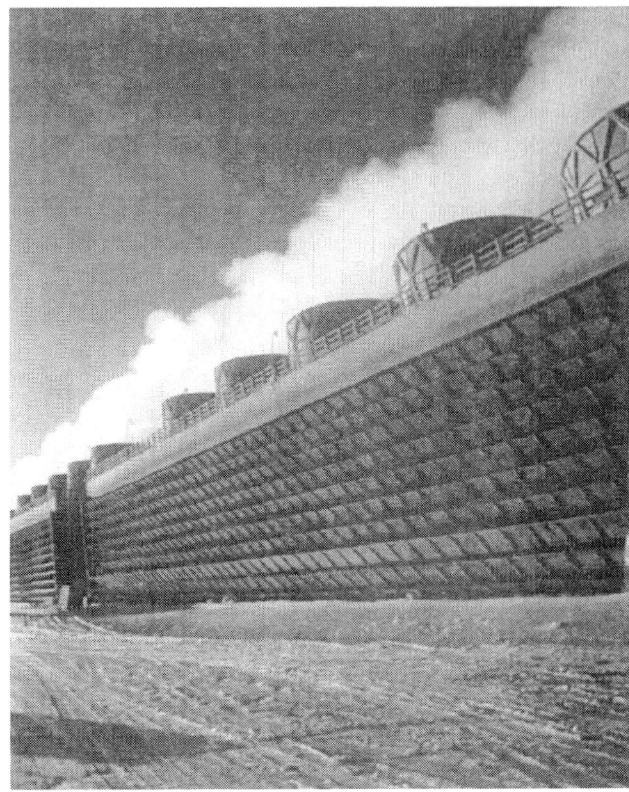

Fig. 12-18. Wood, rectangular mechanical draft cooling tower manufactured by the Marley Cooling Tower Company. (From Northern States Power. Used with permission.)

Fig. 12-19. Concrete, rectangular mechanical draft cooling tower manufactured by the Hamon Cooling Tower Company. (From Florida Power Corporation. Used with permission.)

Fig. 12-20. Round, concrete, mechanical draft cooling tower manufactured by Balcke-Dürr. (From Iowa-Illinois Gas & Electric. Used with permission.)

tered as close as practical around the centerpoint of the tower. The round tower arrangement allows cooling of very large water flows with less plant site area than required by multiple-cell rectangular cooling towers. The round tower arrangement offers other advantages over the rectangular cooling tower:

- Reduced ground level effects of fog resulting from a consolidated, more buoyant plume;
- Reduced circulating water piping lengths; and
- Less thermal performance degradation caused by recirculation.

The round, clustered fan tower generally costs about 35% to 50% more than the rectangular tower. The round, clustered fan towers usually can be located closer to the plant and closer to one another (on a multiple-unit site). This spacing consideration saves piping costs and pumping power costs. When total system costs are considered, the round tower arrangement generally has a comparable, if not lower, total evaluated cost than the rectangular arrangement for units 500 megawatts (MW) and larger. For units 250 MW and smaller, the cost for the round tower can be nearly double that of a rectangular tower and is generally not competitive. For units between 250 and 500 MW, project and site-specific criteria determine if a round tower arrangement is cost effective.

12.3.2 Natural Draft

The natural draft cooling tower produces airflow through utilization of the stack, or chimney, effect. The density difference between the warm (less dense) air inside the stack and the relatively cool (more dense) ambient air outside produces the required airflow.

These towers tend to be very large, both in height and in the amount of water flow each tower can cool. The hyperbolic natural draft cooling tower derives its name from the geometric shape of the shell. Early natural draft towers were nearly cylindrical in shape and made of wood. As the design of natural draft towers evolved and the towers became very tall and large, the hyperbolic shape of today's reinforced concrete towers was found to offer superior strength and resistance to ambient wind loadings. The hyperbolic shape of the tower shell has little to do with the heat transfer process occurring within the tower.

Figures 12-21 and 12-22 show schematics of crossflow and counterflow hyperbolic natural draft towers. As the figures indicate, the fill is located within the lower portion of the shell, with the upper 85% to 90% of the shell empty and used solely to achieve the chimney effect. A ring-shaped foundation supports the tower. Columns rising from the foundation provide an open air inlet base with the hyperbolic curved shell above. A counterflow natural draft cooling tower is shown in Fig. 12-23, and a crossflow natural draft cooling tower is shown in Fig. 12-24.

As mentioned previously, the airflow through the natural draft tower is induced by the density difference between the warm (less dense) air inside the stack and the relatively cool (more dense) ambient air. The driving force developed by the density differential is balanced by the pressure drops within the tower as the air passes through. The driving force (ΔP, lbf/ft^2) is given by the following:

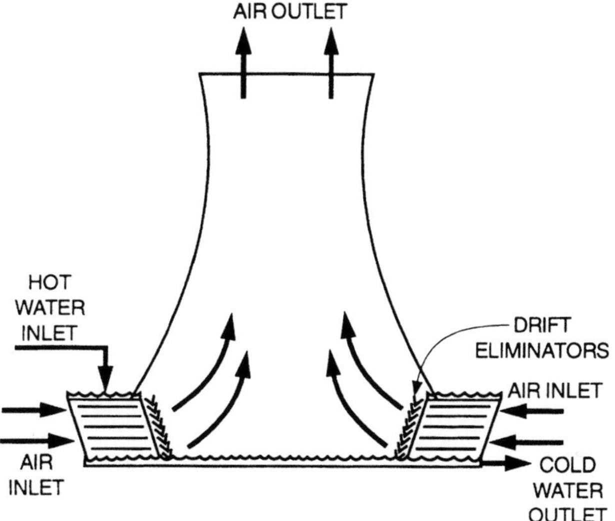

Fig. 12-21. Crossflow hyperbolic natural draft cooling tower.

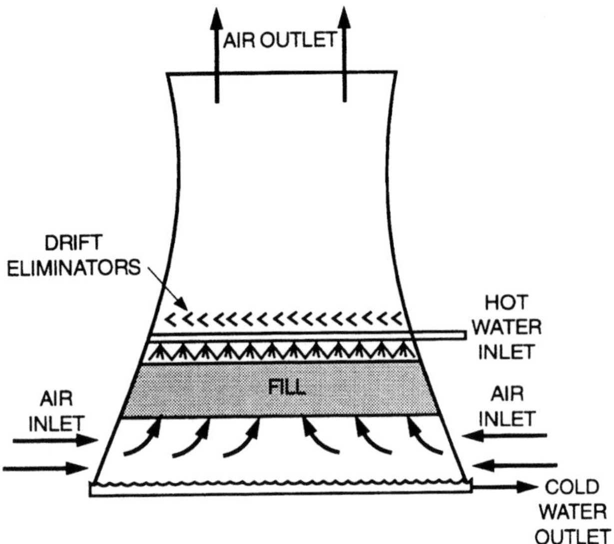

Fig. 12-22. Counterflow hyperbolic natural draft cooling tower.

$$\Delta\rho = (\rho_o - \rho_i)Hg/g_c \qquad (12\text{-}3)$$

where

ρ_o = density of outside air, lbm/ft³;

ρ_i = density of inside air, typically taken at exit of the fill, lbm/ft³;

H = height of tower above the fill, ft;

g = gravitational acceleration, ft/s²); and

g_c = gravitational constant = 32.2 lbm × ft/lbf × s².

Fig. 12-23. Counterflow natural draft cooling tower at the Stanton Energy Center in Orlando, Florida, manufactured by the Marley Cooling Tower Company. (From Orlando Utilities Commission. Used with permission.)

Fig. 12-24. Crossflow natural draft cooling tower. (From Marley Cooling Tower Company. Used with permission.)

Since $\rho_o - \rho_i$ is relatively small, H must be large to generate the ΔP necessary to produce the required airflow.

Hyperbolic towers are generally quite expensive. Stack heights are occasionally in excess of 500 ft (150 m). Compared to an equivalent capacity rectangular mechanical draft tower, costs for hyperbolic towers can range from two and a half times as expensive for a large plant to five times as expensive for a small plant. However, the hyperbolic tower has an economic advantage over the mechanical draft tower in that no energy is required to operate fans. In addition, the high stack outlet precludes most concerns of recirculation, interference, and fogging. As opposed to conventional mechanical draft towers, natural draft towers can be located nearer to the plant and to each other.

The hyperbolic natural draft tower operates most effectively in areas of higher relative humidity, since the ratio of sensible to latent heat transfer is greater than at lower relative humidities. The greater the proportion of sensible heat transfer, the greater the density differential between the stack air and the ambient air, and thus the greater the airflow for a given stack height. Because of this characteristic, and because of the high first cost of hyperbolic cooling towers, their application generally has been limited to areas of higher relative humidity. In arid or high-altitude areas, mechanical draft towers are the dominant choice.

12.3.3 Fan-Assisted Natural Draft

The fan-assisted natural draft cooling tower is a hybrid design (Fig. 12-25). The intent of the design is to augment the airflow produced by fans with airflow produced by the stack effect of a natural draft tower. Because the fans assist in producing the required airflow, the cost impact of a large stack is reduced. Because the stack augments the airflow, fan horsepower requirements are reduced. If designed properly in temperate climates, the fans may need to be operated only during relatively short periods of high ambient wet-bulb temperatures and high wind loads.

The fan-assisted natural draft cooling tower has many of the same advantages as the hyperbolic natural draft, such as the high air discharge which precludes potential problems with recirculation and interference. The tower is also smaller than the hyperbolic natural draft cooling tower, resulting in less visual impact on the surrounding area. Figure 12-26 shows a schematic of an induced draft counterflow fan-assisted natural draft cooling tower.

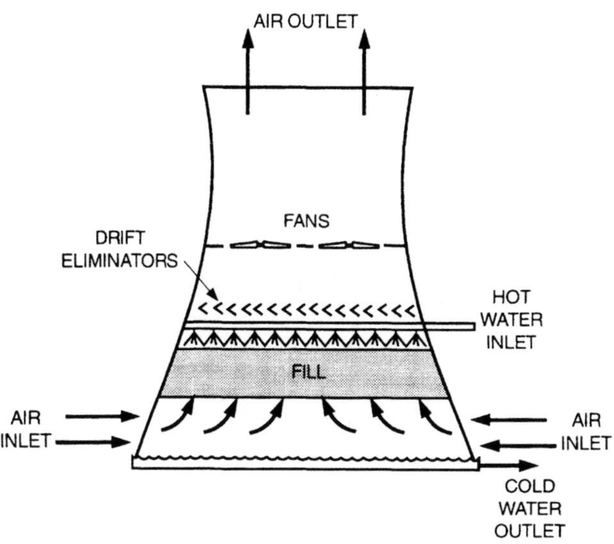

Fig. 12-26. Induced draft counterflow, fan-assisted natural draft cooling tower.

12.3.4 Crossflow

The crossflow cooling tower has a fill configuration through which the air flows horizontally, across the downward fall of water. Figure 12-27 shows a typical cross-section of a rectangular, mechanical draft crossflow cooling tower. Figure 12-28 shows a round mechanical draft, crossflow cooling tower. As these figures show, the fill is contained in the outer perimeter of the tower. The air enters the tower through louvers, passes through the fill and drift eliminators, enters the central air plenum, and is drawn up through the fan stack(s). Hot water is delivered to open hot water basins above the fill areas. The hot water is distributed to the fill by gravity through orifices in the floor of the hot water basin.

Fig. 12-25. Fan-assisted natural draft cooling tower. (From Hamon Cooling Tower Company. Used with permission.)

Crossflow tower fill typically is a splash bar type. Splash bar fill functions to impede the progress of the falling water by breaking it into tiny droplets. The bars are typically arranged in staggered rows. Heat and mass transfer occur at the surface of the droplets. As the water is broken into more small droplets, more heat and mass transfer can take place because increased droplet surface area is in contact with the air. Figure 12-29 shows a common arrangement of a splash bar crossflow fill. The splash bars can be aligned with the longitudinal axis either perpendicular or parallel to the airflow through the fill. The bars are supported horizontally. Long-term performance reliability requires that the splash bars be supported on close centers to prevent damage from long-term thermal loadings, high winds, and ice loads. The major advantage of splash bar fill is the ability to maintain thermal

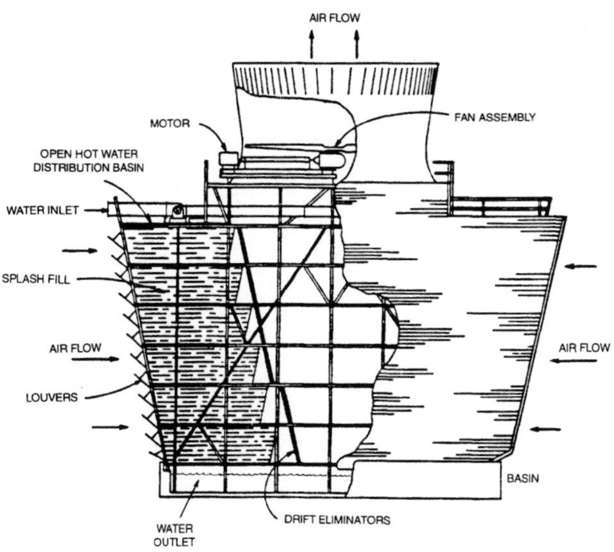

Fig. 12-27. Rectangular, mechanical draft crossflow cooling tower.

Fig. 12-28. Mechanical draft, crossflow cooling tower manufactured by the Marley Cooling Tower Company. (From Cooperative Power. Used with permission.)

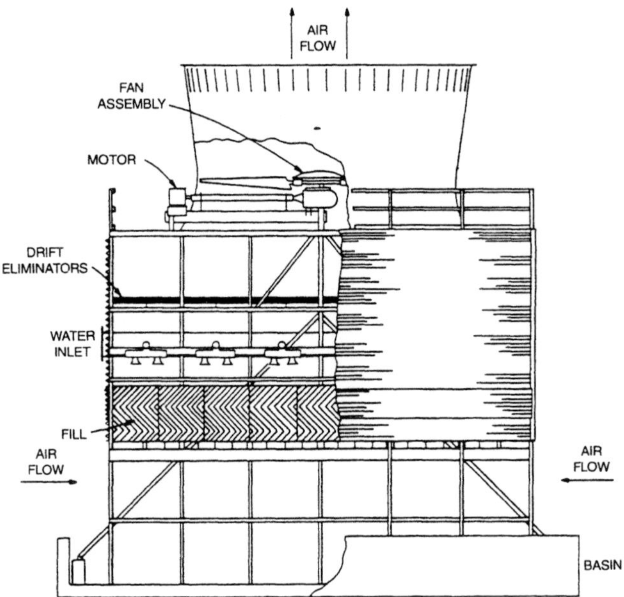

Fig. 12-30. Cross section of typical mechanical draft counterflow cooling tower.

performance capability when circulating water quality is poor. This fill is nearly impossible to plug.

In recent years, new crossflow fills have been developed that use closely spaced film type fill sheets similar to those typically found in counterflow cooling towers.

12.3.5 Counterflow

The counterflow tower has a fill configuration through which air flows vertically upward, counter to the falling water. Figure 12-30 is a cross-sectional schematic diagram of a mechanical-draft counterflow cooling tower. The fill is arranged over the entire tower plan area rather than just at the outer perimeter as in the crossflow tower. The air enters the tower through the openings in the lower portion of the tower, turns 90 degrees, and passes upward through the fill section, where heat and mass transfer between the air and the water take place. The air then passes through the drift eliminators above the fill, enters the tower plenum space, and passes out through the fan stack.

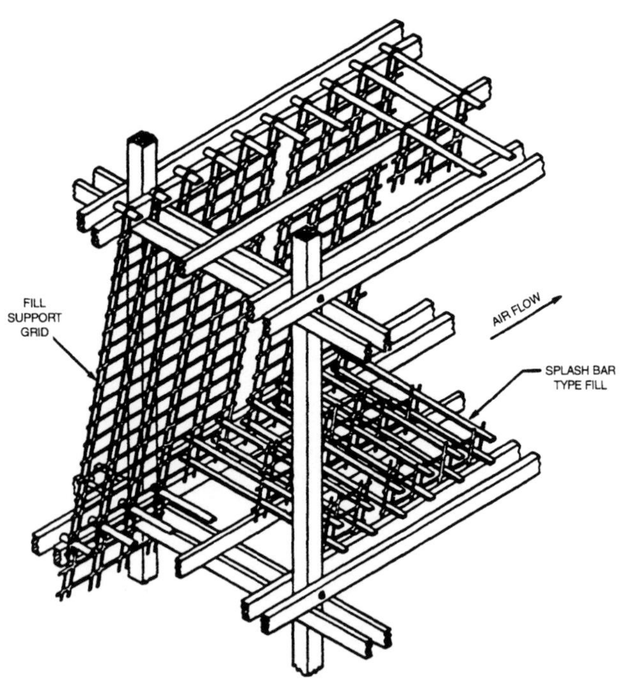

Fig. 12-29. Crossflow fill design.

The hot water distribution system in a counterflow tower typically is a closed-pipe distribution system with risers feeding a header that in turn feeds distribution piping containing nozzles for uniform distribution of water on the fill. Alternatively, the hot water is pumped up to open flumes at the water distribution level that feed the hot water to the distribution piping and nozzles by gravity.

The fill in the counterflow tower is typically a film type fill. Figure 12-31 shows a typical film fill arrangement. The fill consists of densely packed, vertically oriented sheets of material [usually polyvinyl chloride (PVC)], and functions by causing the hot water to flow down the surfaces of the fill in a thin continuous film. As air passes over the water film, heat and mass transfer occur at the surface of the water film. This type of fill promotes maximum exposure of the water to the airflow. It has the capability to provide maximum effective cooling capacity in a small amount of space. However, the thermal performance of this type of fill is extremely sensitive to poor water distribution, as well as to the air blockage and turbulence that a poorly designed fill support system can perpetuate. The overall tower design must assure uniform air and water flow throughout the entire fill area. Because the fill sheets are closely spaced, the quality of the circulating water is of critical importance to keep the fill

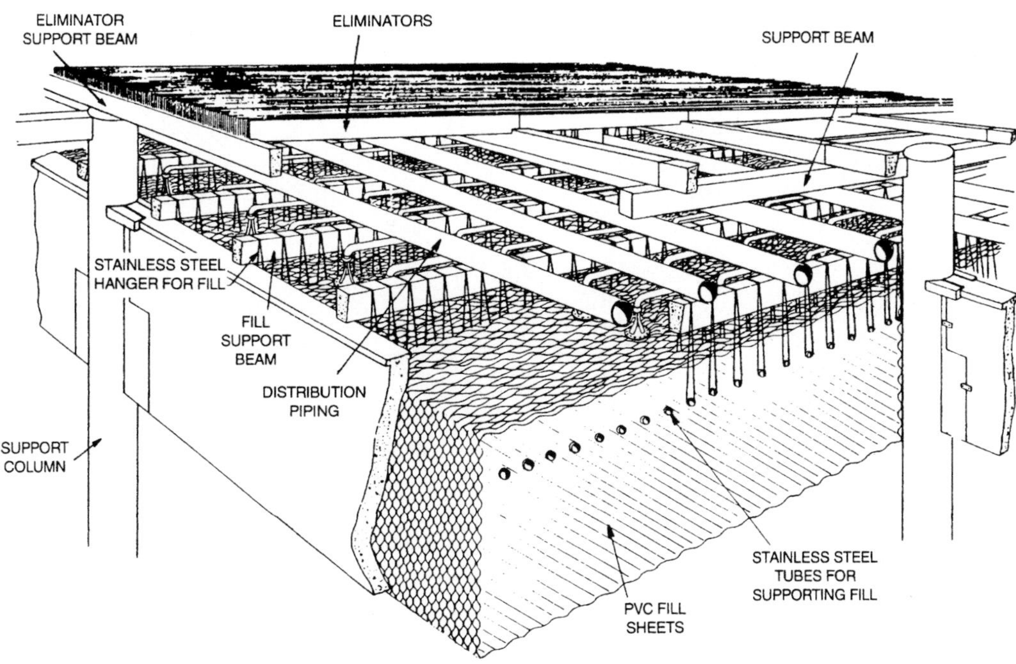

Fig. 12-31. Typical film fill design.

from fouling and plugging the fill with debris accumulation or biological growth.

The major operating differences for a counterflow tower as compared to a crossflow tower are reduced pumping head and slightly increased fan horsepower. These differences can be explained by comparing Figs. 12-27 and 12-30. As these figures show, the fill section of the counterflow tower covers the entire plan area, whereas the fill for the crossflow tower is confined to the outer perimeter of the tower. Because of this, the height of the fill section is greater and, as a result, the pumping head is greater on the crossflow tower for a given duty.

The crossflow tower also has a larger air inlet area than the counterflow tower as a result of the fill configuration. This leads to a lower entering air velocity and a lower pressure drop through the air inlets. The counterflow fill is also more densely packed, which often leads to a greater pressure drop through the fill section, even though the air travel distance through the fill section is less than in the crossflow tower. The increased pressure drop in the counterflow tower air inlets and fill section increases the counterflow tower's fan horsepower requirement when compared to a crossflow cooling tower. This is mitigated somewhat by the counterflow heat transfer process, which is more efficient than the crossflow process, and the crossflow tower, therefore, requires a greater air mass flow for given duty than does the counterflow tower. Some of the comparative advantages and disadvantages of the two tower types are as follows:

- Crossflow advantages:
  —Open hot water distribution deck allows easy inspection and cleaning of nozzles.
  —Crossflow splash fills can operate and maintain design thermal performance even with very poor circulating water quality.
  —Lower air static pressure losses.
  —Crossflow design accepts wide range of the circulating water flows without significantly affecting tower pumping head or distribution nozzle spray pattern.
- Counterflow advantages:
  —Maximum thermal efficiency,
  —Smallest tower,
  —Lowest capital cost,
  —Typically creates lower tower pumping head than crossflow tower,
  —Minimum exposure of fill and water distribution system to sunlight reduces the potential for algae growth, and
  —Design allows more effective ice control in freezing climates.

12.4 COOLING TOWER SIZING

For a new power plant project, the condenser and cooling tower arrangements are optimized during the conceptual design of the cycle heat rejection system. Condenser design parameters are determined based on an economic optimization study that evaluates and compares the numerous potential condenser arrangements and sizes, which are evaluated in combination with various cooling tower efficiencies (approaches), to select the most cost-effective condenser arrangement, size, and back pressure. The optimization is based on the client's preferences as well as the project economic criteria and projected unit load model. Discussion on the optimization of condenser size and back pressure is presented in Chapter 9, Steam Cycle Heat Exchangers. An economic evaluation of alternative cooling tower designs and arrangements is performed based on the results of the condenser optimization. This section discusses the factors to consider when sizing and optimizing cooling tower arrangements.

12.4.1 Design Heat Load

The plant condenser and cooling tower are sized to handle the maximum expected heat load from the turbine–generator manufacturer's valve-wide-open, 5% overpressure heat balance. This is also commonly referred to as the maximum continuous rating (MCR) load point. During the course of the condenser and cooling tower optimization, sufficient design margins are verified to ensure that plant capacity is never restricted by insufficient cooling system capacity. For example, many large steam turbines are limited in maximum steam exhaust pressure to about 5.0 in. HgA. Appropriate design margins would typically eliminate evaluation of any condenser and cooling tower arrangement that would produce a back pressure of greater than 4.0 in. HgA (100 mm) under peak load and the site 1% wet-bulb temperature, or 4.5 in. HgA (150 mm) when the wet-bulb temperature is at the historical maximum for the site. These design conditions allow adequate margin for unusually warm weather and ensure that unit capacity is not limited by excessive back pressures during peak load and peak ambient conditions. Maximum steam turbine back pressure limitations are specific to turbine–generator manufacturer and can vary as a function of steam turbine size.

12.4.2 Design Wet-Bulb Temperature

An important cooling tower design parameter is the design wet-bulb temperature. Wet-bulb temperature is that saturation temperature to which air can be cooled adiabatically by the evaporation of liquid water at the same temperature. Thus, the wet-bulb temperature represents the lowest temperature to which the circulating water can theoretically be cooled in the tower. Since there are inefficiencies in the cooling tower evaporative cooling process, the cooling tower can cool only to within an approach of the wet-bulb temperature. Selection of the design wet-bulb temperature must be made on the basis of historical ambient conditions at the plant site.

Since, for a given cooling tower, the cold water temperature will increase as the wet-bulb temperature increases, and, since condenser pressure is a direct function of the cold water temperature, the design wet-bulb temperature is usually selected at a peak summer temperature. As discussed above, this design method limits peak condenser pressures so that plant generating capacity is not restricted during the summer, which is the peak electricity demand period for most utilities. It is common to select the design inlet wet-bulb temperature as the 5% ambient wet-bulb temperature at the site plus an allowance for recirculation (and interference if the site has more than one tower). The 5% wet-bulb temperature is the wet-bulb temperature that has historically been exceeded an average of only 5% of the hours in the four warmest summer months (typically June through September). The hours in which peak wet-bulb temperatures exceed the 5% level are seldom consecutive hours and usually occur in periods of relatively short duration. The 5% wet-bulb temperature is a common design parameter for industrial cooling applications, and historical weather data is usually readily available for most proposed plant sites.

In some special cases, it may be necessary to perform a comprehensive study to determine the appropriate design wet-bulb temperature for a project. The necessary site weather data to perform such a study are usually publicly available in some form. Historical weather data are vital to many industrial cooling applications. Air temperatures (dry-bulb as well as coincident wet-bulb temperatures), wind direction and velocity, and other data are routinely measured and recorded by the U.S. Weather Bureau, worldwide U.S. military installations, airports, and various other organizations to whom anticipated weather patterns and specific air conditions are a vital concern.

12.4.3 Recirculation and Interference

Cooling tower recirculation, as shown in Fig. 12-32, describes the phenomenon of portions of the warm, moist air of the tower exit plume being drawn back into the air intake of

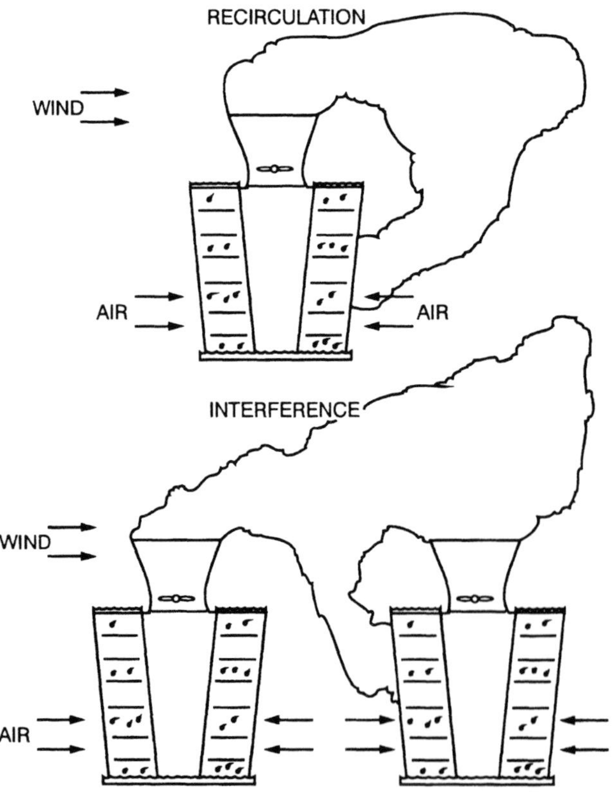

Fig. 12-32. Cooling tower recirculation and interference.

the cooling tower, thus raising the inlet wet-bulb temperature. Interference is a similar phenomenon, but is defined as the discharge from one tower that enters another nearby tower (also shown in Fig. 12-32). Since the heated discharge air that recirculates or interferes is essentially saturated, it cannot contribute to the heat transfer process because the predominant cooling mechanism of a wet cooling tower is evaporation.

Recirculation and interference increases the wet-bulb temperature of the entering air, negatively affecting tower performance. The cold water temperature to the condenser is increased, which in turn increases the turbine back pressure. That, in turn, reduces the turbine–generator output and the plant overall efficiency. Thus, the design of mechanical draft cooling towers must include an appropriate design margin to compensate for recirculation or interference effects. The most common design margin is to add a recirculation allowance to the design ambient wet-bulb temperature that mimics the average increase in wet-bulb temperature expected from recirculation effects. The typical mechanical draft cooling tower specification provides a design inlet wet-bulb temperature defined as follows:

$$\begin{pmatrix}\text{Design inlet wet}\\ \text{bulb temperature}\end{pmatrix} = \begin{pmatrix}\text{Design ambient wet}\\ \text{bulb temperature}\end{pmatrix} + \begin{pmatrix}\text{Recirculation}\\ \text{allowance}\end{pmatrix}$$

Rectangular towers have the most severe recirculation problems. As discussed earlier, the exit air outlet is close to the air inlet, and there is no benefit of plume reinforcement except in certain wind conditions. Also, the long, low tower profile creates a low-pressure zone on the downwind side of the tower which tends to draw the discharge air to that side and subsequently back into the tower air inlet. A recirculation allowance of 2° F (1.1° C) for a rectangular mechanical draft tower is typically recommended.

Round mechanical draft cooling towers have the advantage of a consolidated plume, which increases buoyancy, and the round shape helps reduce the size of the low-pressure zone on the downwind side of the cooling tower. However, the air inlets are still relatively close to the exit airstream and recirculation does occur. A recirculation allowance of 1° F (0.6° C) is typically recommended for a round mechanical draft cooling tower.

Natural draft cooling towers have no significant recirculation and interference problems, although multiple towers located close together can cause unstable inlet air conditions. No recirculation allowance is recommended for this type of cooling tower.

The magnitude of actual recirculation effects on mechanical draft cooling towers is a subject of considerable debate within the cooling tower industry. Most recommendations on recirculation allowance have been based on wind tunnel testing or limited site data. Operating experience has shown that actual recirculation effects appear to be larger than discussed above for both rectangular and round mechanical draft cooling towers. The Cooling Tower Institute has started a program to obtain tower recirculation data when field thermal testing of cooling towers is performed. It is hoped this program will allow gathering of consistent field data from which useful recommendations can be drawn.

12.4.4 Cooling Tower Flow, Range, and Approach

The condenser optimization determines the design circulating water flow rate, cold water temperature, and cooling range based on the turbine–generator heat load at the maximum continuous rating load point plus a 10% allowance for the heat load of the auxiliary cooling system. On some projects, an additional margin is included to allow for deterioration in turbine–generator performance. Recalling Eq. (12-2), the calculation of the temperature rise of the circulating water in the condenser is as follows:

$$\text{TR} = Q/500(\text{FLOW, gpm})$$

Since the cooling tower rejects the total heat load from the condenser and since the design flow rate of the circulating water pumps is fairly constant, the temperature range of the circulating water flow in the tower must be equal to that of the condenser. The relationships between the cooling tower range, approach, and wet-bulb temperature are shown in Fig. 12-33. The design tower cold water temperature is based on the cooling tower approach to the design inlet wet-bulb temperature. Cooling tower approach is defined as the temperature difference between the cold water temperature leaving the tower basin and the wet-bulb temperature of the air entering the tower.

To illustrate further how the condenser and cooling tower perform together as a system, and how heat rejection system optimization studies must consider the performance and costs of all system components, Fig. 12-34 shows the condenser and cooling tower design performance parameters for a typical power plant cycle heat rejection system that ultimately determine the turbine–generator back pressure for a given heat load:

- Wet-bulb temperature, WBT = 75° F (23.9° C);
- Cooling tower approach, APP = 15° F (8.3° C);

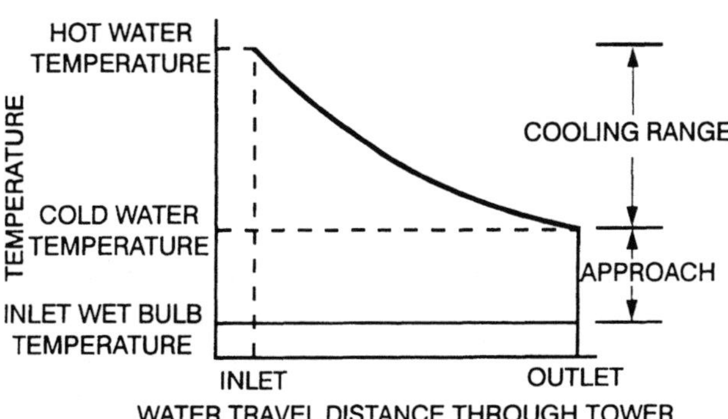

Fig. 12-33. Relationship between cooling range, approach, and wet-bulb temperature.

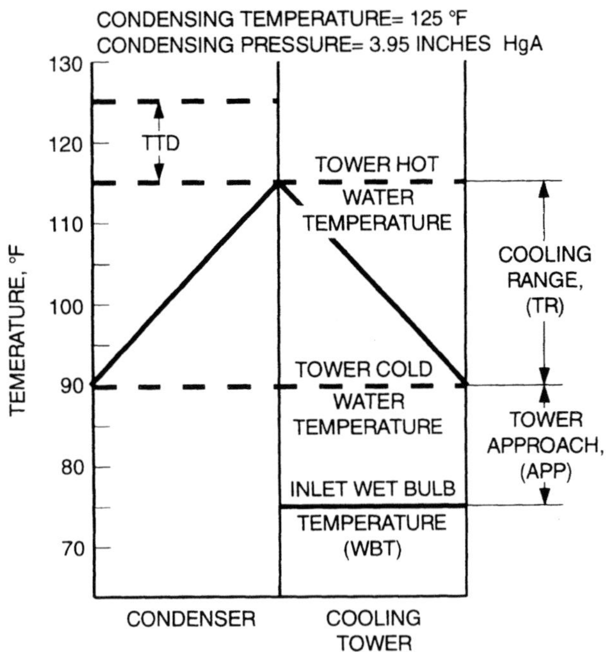

Fig. 12-34. Condenser and cooling tower performance parameters.

- Condenser/cooling tower temperature range, TR = 25° F (13.9° C); and
- Condenser terminal temperature difference, TTD = 10° F (5.6° C).

While determination of the above parameters defines the performance and efficiency of the cycle heat rejection system, each of these parameters also affects the cost of the system components. For the cooling tower, reducing the tower approach temperature reduces the turbine back pressure, but it also increases the capital and operating costs of the cooling tower. Figure 12-35 shows how the design approach temperature affects cooling tower size. The tower cost usually increases in proportion to the size of the tower. Note the asymptotic movement of the curve toward a zero approach. Tower approaches from 5 to 9° F (2.8 to 5.0° C)

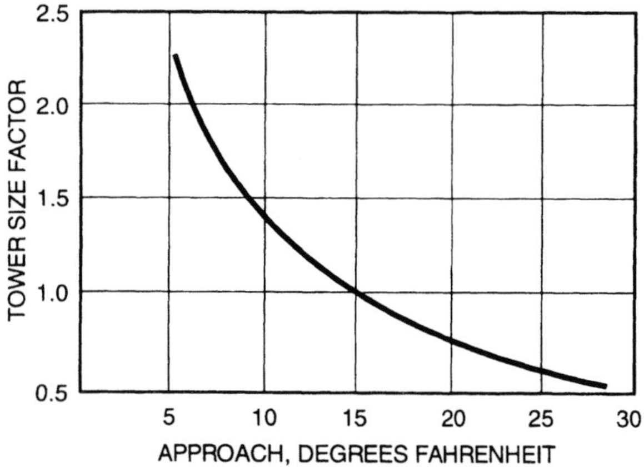

Fig. 12-35. Effect of design approach temperature on cooling tower size. (From Marley Cooling Tower Company. Used with permission.)

generally are not cost effective for power plant use, and approaches below 5° F (2.8° C) usually cannot be guaranteed by cooling tower manufacturers. The role of the cycle heat rejection system optimization is to determine the optimum size and arrangement of the system equipment according to the project economic criteria.

12.4.5 Dry-Bulb Temperature and Relative Humidity

Dry-bulb temperature is not usually an essential criterion for sizing wet mechanical draft cooling towers. However, the design dry-bulb temperature or relative humidity is used by the tower manufacturer to determine design inlet air properties so an accurate determination of evaporation can be made. Also, for natural draft, plume abatement, water conservation, and dry or wet–dry cooling towers, dry-bulb temperature is essential in sizing the cooling tower.

The site elevation is also essential for determining the size of the cooling tower because moist air properties are dependent on atmospheric pressure. Generally, as elevation increases, the required tower size and/or fan horsepower decreases if all other thermal variables remain constant.

12.4.6 Sample Cooling Tower Optimization

The objective of a cooling tower optimization is to establish the most advantageous type of cooling tower for the power plant. A typical cooling tower optimization is presented in the example below, conducted for the following design criteria:

- Power plant net turbine output = 450 MW,
- Five percent wet-bulb temperature = 74° F (23.3° C),
- Circulating water flow rate = 190,000 gpm (12,000 L/s),
- Condenser/cooling tower range = 24° F (13.3° C), and
- Condenser/cooling tower heat load [from Eq. (12-2)] = (190,000)(24)(500) = 2.28×10^9 Btu/h.

The following cooling tower types will be evaluated:

- Plan A1—rectangular crossflow tower of wood construction,
- Plan A2—rectangular crossflow tower of concrete construction,
- Plan B1—rectangular counterflow tower of wood construction,
- Plan B2—rectangular counterflow tower of concrete construction,
- Plan C1—round crossflow cooling tower of partial wood and concrete construction,
- Plan C2—round crossflow cooling tower of concrete construction, and
- Plan D—round counterflow cooling tower of concrete construction.

Table 12-1 lists the preliminary design parameters for each type of tower as provided by cooling tower vendors. As discussed previously, a recirculation allowance of 2° F (1.1° C) is used to determine the design inlet wet-bulb temperature

Table 12-1. Design Parameters for Cooling Tower Alternatives

Parameter	Plan A1—Rectangular, Wood Crossflow Tower	Plan A2—Rectangular, Concrete Crossflow Tower	Plan B1—Rectangular, Wood Counterflow Tower	Plan B2—Rectangular, Concrete Counterflow Tower	Plan C1—Round, Partial Wood Crossflow Tower	Plan C2—Round, Concrete Crossflow Tower	Plan D—Round, Concrete Counterflow Tower
Tower width and length or diameter, ft (m)	54 × 529 (17 × 161)	66 × 432 (20 × 132)	54 × 762 (17 × 232)	60 × 477 (18 × 145)	320 (98)	300 (91)	250 (76)
Number of fans	12	8	14	9	12	12	12
Brake horsepower per fan (kW)	125 (93)	188 (140)	72 (54)	191 (142)	125 (93)	110 (82)	100 (75)
Pump head, ft (m)	41 (12.5)	42 (13)	19 (6)	23 (7)	40 (12)	39 (12)	30 (9)

for the rectangular cooling tower alternatives, and a recirculation allowance of 1° F (0.6° C) is used for the round mechanical draft cooling tower alternatives.

12.4.6.1 Evaluation. The alternative cooling tower plans are evaluated based on an economic comparison. Economic criteria, typical of what may be applicable for an electric utility company, are shown in Table 12-2. The first step in the analysis is to estimate the comparative capital costs for installation of each of the alternate plans (Table 12-3). Estimated escalation and interest charges are added to the capital costs to determine the total comparative capital costs.

Differences in the operating characteristics of the equipment required by each plan result in differing annual operating costs. The differences in projected annual operating costs need to be estimated and considered in the economic analysis. The differential capital costs are annualized with the fixed charge rate to allow comparison with the differential annual operating costs.

After the capital and operating costs are calculated, a capital recovery analysis is performed to determine if a plan with higher capital costs but lower annual operating costs is economically advantageous. The payback period is the number of years required for the cumulative present worth savings of the reduced operating costs to equal the lifetime present worth of the additional fixed charges. A typical acceptable payback period for a 30-year plant life is 5 to 10 years.

12.4.6.2 Development of Capital Costs. The comparative capital costs associated with each tower type evaluated are shown in Table 12-3. All costs are based on the design conditions in Table 12-1 and are determined as follows:

Table 12-2. Economic Criteria

Fixed charge rate, percent levelized	16.0
Present worth discount rate, percent	11.0
Evaluation period, years	30
First year fuel cost, $/MBtu	2.6
Average plant capacity factor, percent	70.0
Equipment escalation rate, percent	4.0
Allowance for funds used during construction (AFUDC) rate, percent	9.0
Energy cost, mills/kWh levelized	52.00
Annual demand charge, $/kW/yr	112.0

- *Cooling tower.* Cooling tower capital costs can be obtained directly from cooling tower manufacturers. Their cost quotes are usually based on supply and erection of all tower equipment and materials. The cold water basin typically is not within the cooling tower manufacturer's standard scope of supply and is addressed as a separate cost item.
- *Circulating water piping and valves.* Large-diameter circulating water piping must be installed to transfer circulating water between the cooling tower and condenser. Variations in piping costs result from variations in tower location, tower length, and tower riser arrangements among the various plans. Towers that are located farther from the plant and towers that require individual risers to each cell require more circulating water piping, resulting in higher capital costs.

  Rectangular crossflow towers usually have one or two risers at the end of the tower closest to the plant. Rectangular counterflow towers usually require a pipe header along the base of the tower, with a riser pipe for each individual tower cell. Round cooling towers typically have one or two large riser pipes at the center of the tower. Piping costs should be calculated as total installed costs. Usually the circulating water piping between the cooling tower and the plant is buried; thus, installation cost estimates should include excavation and backfill costs. Rectangular counterflow cooling towers require additional valves for each of the individual riser pipes. These valves typically are motor operated.
- *Circulating water pumps and motors.* The circulating water pump and motor capital costs include differential material and installation costs for pumps and motors. Cooling towers with higher tower pumping heads and towers requiring longer lengths of circulating water pipe add to the system resistance of the circulating water piping. An increase in pumping system resistance requires larger and more expensive circulating water pumps and motors. For this example, total circulating water pump heads were calculated for each alternative based on circulating water pipe friction losses, condenser friction losses, and static and dynamic cooling tower pump head requirements. Differential circulating water pump and motor costs were then estimated using pump manufacturers' quotes. The circulating water pump and motor costs for this example are based on single-speed vertical wet pit pumps.
- *Cooling tower basin.* The cooling tower basin costs in-

Table 12-3. Comparative Costs for Cooling Tower Alternatives

Item	Plan A1—Rectangular, Wood Crossflow Tower	Plan A2—Rectangular, Concrete Crossflow Tower	Plan B1—Rectangular, Wood Counterflow Tower	Plan B2—Rectangular, Concrete Counterflow Tower	Plan C1—Round, Partial Wood Crossflow Tower	Plan C2—Round, Concrete Crossflow Tower	Plan D—Round, Concrete Counterflow Tower
	$1,000	$1,000	$1,000	$1,000	$1,000	$1,000	$1,000
Capital costs							
Cooling tower	300	1,160	Base	240	4,300	5,550	4,800
Circulating water piping and valves	90	90	720	550	Base	Base	70
Circulating water pumps and motors	160	160	Base	80	160	160	80
Cooling tower basin	80	190	220	290	20	Base	350
Electrical connection	160	Base	240	40	160	160	160
Fire protection	200	Base	270	Base	200	Base	Base
Subtotal	990	1,600	1,450	1,200	4,840	5,870	5,460
Escalation	313	505	458	379	1,529	1,855	1,725
Subtotal	1,303	2,105	1,908	1,579	6,369	7,725	7,185
AFUDC	176	284	258	213	860	1,043	970
Comparative capital cost	1,479	2,389	2,166	1,792	7,229	8,768	8,155
Differential capital cost	Base	910	687	313	5,750	7,289	6,676
Levelized annual cost of operation							
Energy costs	455	473	Base	244	439	375	197
Demand costs	153	158	Base	82	147	126	66
Maintenance costs	169	Base	154	Base	157	Base	Base
Comparative annual cost of operation	777	631	154	326	743	501	263
Differential annual cost of operation	623	477	Base	172	589	347	109
Levelized total annual costs							
Differential fixed charges	Base	149	113	51	943	1,195	1,095
Differential annual cost of operation	623	477	Base	172	589	347	109
Comparative total annual cost	623	626	113	223	1,532	1,542	1,204
Differential total annual cost	510	513	Base	110	1,419	1,429	1,091

clude the costs associated with the construction of a concrete basin for each tower type. For this example, the smaller footprint of the crossflow towers would result in lower basin costs.

• *Electrical connection.* The electrical connection costs are for connecting the cooling tower fans and lighting to the power plant auxiliary power system. For this example, the plans with the least number of fans had the lowest electrical connection costs.

• *Fire protection.* The towers of wood construction should have a fire protection system. The fire protection costs are for installation of the complete cooling tower fire protection system, with fire water supply from the plant fire water loop.

• *Escalation.* Escalating the capital costs from the time of the study to the midpoint of the plant construction period approximates the cash flow of the costs through the construction period. Escalation is computed from the date that the capital cost estimate is prepared to the estimated midpoint of plant construction.

• *AFUDC.* The allowance for funds used during construction (AFUDC) approximates the interest that would be paid on the construction costs through the construction period. AFUDC is computed from the midpoint of plant construction to the date of commercial operation. It is calculated by multiplying the differential escalated capital costs by the AFUDC rate in Table 12-2, compounded by the number of years between the midpoint of plant construction to the commercial operation date. For this example, the number of years from the midpoint of construction to the commercial operation date is 1.5 years.

12.4.6.3 Development of Operating Costs. The comparative levelized annual operating costs associated with each plan evaluated are shown in Table 12-3. Throughout the life of the project, the annual costs vary as a result of changes in the loading of the plant and escalation effects on fuel and energy charges. The cumulative present worth of the lifetime annual costs of operation, divided by the sum of the present worth factors, gives the levelized annual costs of operation. All costs are based on the design conditions in Table 12-1 and are determined as follows:

- *Energy costs.* Energy costs are associated with the differential circulating water pump and cooling tower fan horsepower requirements and are evaluated to account for differences in energy requirements per year between the plans. Plans that have lower energy requirements per year have lower energy costs. For this example, the fan horsepower requirements from Table 12-1 and the calculated circulating water pump horsepower requirements were multiplied by the levelized energy cost in Table 12-2.
- *Demand costs.* Annual demand costs are used to economically equalize the net plant output of the alternative plans at full load. Thus, if the auxiliary power consumption of one tower arrangement uses more auxiliary power than another tower arrangement, the demand charge approximates the incremental cost to the client to build additional capacity into the plant. For this example, the fan horsepower required at full load and the circulating water pump horsepower required at full load were multiplied by the annual demand cost in Table 12-2.
- *Maintenance costs.* Cooling tower maintenance costs are difficult to predict and the magnitude of these costs tends to be very site specific. The following factors greatly impact maintenance costs, but are difficult to quantify accurately:
  - —The quality of materials and workmanship during construction of the cooling tower;
  - —The quality of the preventative maintenance program at the plant, especially as applicable to the cooling towers;
  - —Severity of the climate extremes at the plant site;
  - —Ability of the plant to maintain consistent control of circulating water quality with an effective water treatment system; and
  - —Annual thermal loading capacity on the cooling towers.

While these factors greatly affect the magnitude of annual maintenance costs, the differential costs among alternate plans may not be significant. For this example, the differential maintenance cost estimates were calculated based on the assumption that the only major difference in tower maintenance would be useful service life of the wood portions of the wood towers. Other maintenance costs associated with fan adjustments, gearbox lubrication and oil changes, and motor lubrication were assumed to be the same for all tower plans. The material cost for the wood portion of the rectangular cooling towers is assumed to be approximately 34% of the total material costs of the tower. Assuming that the wood has a service life of 15 years, the annual replacement cost is 2.3% per year of the initial material cost over the life of the plant.

12.4.6.4 Capital Recovery Results. A capital recovery period can be determined when a plan that has low capital costs and high operating costs is compared to a plan with high capital costs and low operating costs. As can be seen in Table 12-3, Plan A1, a rectangular crossflow tower of wood construction, has the lowest total differential capital costs. Plan B1, a rectangular counterflow tower of wood construction, has the lowest differential annual costs. A capital recovery analysis shows that Plan B1 has a capital recovery period of 3 years. After 3 years, the cumulative present worth savings of the reduced operating costs equal the lifetime present worth of the additional fixed charges.

12.5 COOLING TOWER SPACING AND ORIENTATION

Mechanical draft cooling tower spacing and orientation can significantly affect cooling tower performance because recirculation and interference effects are dependent on wind direction, particularly for rectangular cooling towers. Cooling tower placement requires examination of the winter and summer wind roses. A typical wind rose is shown in Fig. 12-36.

Rectangular cooling towers should be oriented so that the longitudinal axis is parallel to the prevailing wind coincident with the highest ambient wet-bulb temperature. This allows the exit air to be blown clear of the tower for shorter towers and allows concentration of the separate cell plumes into one of greater buoyancy for longer towers. Even though this plume consolidation occurs on longer towers, considerations should be given to splitting towers longer than 500 ft (150 m) into multiple units to reduce the impact of wind conditions.

On multiple units, rectangular cooling towers should be placed so that the effluent from one tower will not significantly affect the air inlet of a second tower under predomi-

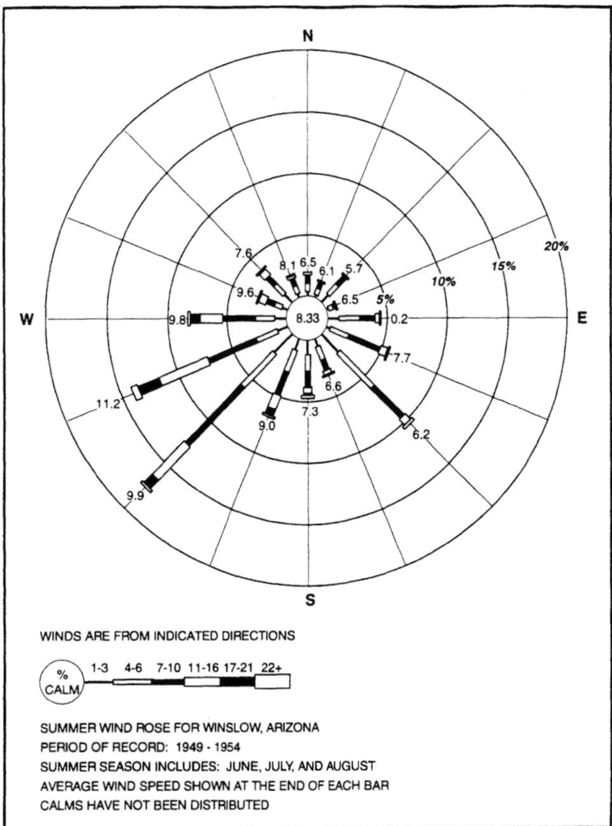

Fig. 12-36. Typical wind rose.

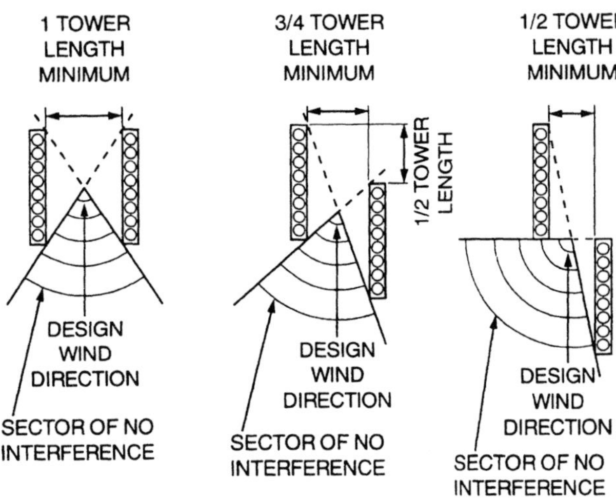

Fig. 12-37. Acceptable rectangular cooling tower spacing configurations. (From Marley Cooling Tower Company. Used with permission.)

nate summer wind directions. Several acceptable spacing configurations are shown in Fig. 12-37.

Wind direction need not be considered in siting round cooling towers to reduce recirculation tendencies. However, wind direction must be considered for other possible problems such as fogging and icing from the tower plume, and tower interference with multiple tower arrangements. To minimize interference on sites with two round towers, the cooling towers should be located a minimum of one tower diameter apart, with the centerline passing through both towers oriented perpendicular to the predominate summer wind direction. For four tower cluster arrangements, minimum tower spacing of 1 and 1/2 tower diameter should be used.

Every effort should be made to provide the least possible restriction to the free flow of air to the cooling tower on a plant site. Other factors to consider when siting cooling towers include the distance from the tower to the plant and electrical switchgear, distances between the tower and plant boundaries, and distances between the tower and surrounding roadways. The resulting effect of distance on piping and electrical wiring costs, and the offsite impact of noise from the tower must be considered. Where a large enough plant site is available, mechanical draft cooling towers should be placed no closer than 800 ft (240 m) from main plant structures and no closer than 1,000 ft (300 m) from electrical substations to minimize cooling tower drift impact. Also, the tower should be located such that prevailing winter winds do not blow plume and fog into the plant or substation.

Natural draft towers typically are not affected by wind direction, recirculation, or interference. Likewise, ground fogging and icing usually are not problems. This allows more flexibility in siting a natural draft tower versus a mechanical draft tower. However, because of the way natural draft towers develop airflow through the tower, natural draft towers are sensitive to restrictions to the free flow of air into the tower. Also, natural draft cooling tower drift deposition on the main plant buildings and plant substation can be a problem. Natural draft cooling towers should be located at least 500 ft (150 m) from main plant buildings and plant substations. For multiple natural draft tower arrangements, spacing between towers should at least equal the length of one tower base diameter.

12.6 COOLING TOWER MATERIALS

Cooling tower construction materials are a prime factor in determining the cost of a cooling tower, and the engineering characteristics of the materials are critical in determining the useful life of the tower. This section discusses materials typically used in utility cooling tower construction.

12.6.1 Tower Structural Materials

The primary load-carrying structural materials used in cooling towers are wood and reinforced concrete. From a historical standpoint, wood is the predominant material of choice; however, reinforced concrete has been increasingly used in large power plant cooling towers. Both materials are satisfactory in terms of providing a suitably durable structure. Care must be taken in the specification stage to identify possible environmental factors that may cause untimely deterioration of either material, and adequate quality control must be maintained during construction. As a new alternative, fiberglass structure towers are now available on the large cooling tower market.

The following subsections discuss these primary structural materials as well as materials used for hardware, air inlet louvers, casings, and fan cylinders.

12.6.1.1 Wood. Because of its availability, workability, relatively low cost, and durability under the very severe operating conditions encountered in cooling towers, wood is the predominant structural material used. The dominant wood used in cooling tower construction is West Coast Douglas fir, which began replacing California redwood in the early 1960s. Basic structural standards are specified in the "National Design Specification for Wood Construction (NDS)," published by the National Forest Products Association.

To augment the NDS standards, the Cooling Tower Institute (CTI) has issued standards pertaining to the design of cooling towers with Douglas fir lumber (STD-114) and redwood lumber (STD-103). The natural flexibility of wood gives it certain advantages over concrete, particularly in areas where seismic loads or freeze–thaw cycling must be considered. Douglas fir is a mechanically stronger lumber than redwood; however, it has almost no natural resistance to certain types of deterioration. Among these are chemical surface attack, biological internal attack, and iron rot. These types of wood deterioration are outlined in CTI Bulletin WMS-104, "Wood Maintenance for Water-Cooling Towers."

Most specifications require all lumber used in cooling tower construction be pressure treated with a preservative to inhibit fungal attack. Douglas fir is a very dense wood that is very difficult to penetrate with preservative chemicals. CTI Bulletin WMS-104 identified that the best preservatives are acid copper chromate (ACC), creosote, and chromated copper arsenate (CCA). Creosote and chromated copper arsenate are not used often because of environmental concerns and difficulties in handling the treated wood. The preservative treatments are typically applied to the wood by immersion in a pressure vessel. The pressure is maintained either until a specified amount of preservative is retained by the wood, or until the wood refuses to accept further treatment.

Wood has been the most popular cooling tower structural material for many years because of the availability of relatively low-cost lumber. However, in the future alternate materials are likely to receive more attention. The long-term supply of low-cost Douglas fir is in doubt because of dwindling domestic and worldwide supplies of available forest land for harvesting the lumber. Also, the use of chemical preservative treatments is receiving more environmental scrutiny because of the leaching of the chemicals into the circulating water. Disposal of dismantled cooling tower structures can be expensive if a significant concentration of residual chemical preservatives remains in the lumber. This condition can cause the lumber to be classified as a hazardous waste.

12.6.1.2 Concrete. Concrete construction offers the promise of a long-life, low-maintenance cooling tower structure. Concrete construction typically is more expensive than a comparable wood structure tower. Concrete cooling tower structural members may be a combination of precast and cast-in-place construction, with design varying according to applicable loads and tower configuration. Complete precast construction is becoming a popular choice for concrete construction of mechanical draft cooling towers. The use of standardized, modular cooling tower cell designs of precast structural concrete members has reduced tower erection costs significantly, narrowing the price differential between wood and concrete construction. The higher initial cost of concrete construction sometimes can be justified by its reduced fire risk and higher load-carrying capability.

For natural draft cooling towers, reinforced concrete construction is currently the dominant choice. Natural draft cooling towers require customized tower designs, and high-lift pumping of concrete is required for the hyperbolic shell construction.

Concrete cooling towers are not necessarily maintenance free. The extreme operating conditions existing in many cooling towers require that much attention be paid to the specified concrete mixes, reinforcing steel requirements, minimum concrete coverage over the reinforcing steel, and concrete additives.

The phenomena that can contribute to the deterioration of concrete often do so by inducing internal expansion. These phenomena include frost action, sulfate attack, and chloride attack on reinforcing steel. Deterioration of concrete also can occur from nonuniform contraction, leaching of the soluble constituents of the cement paste, and as a secondary consequence of structural overloading. In a cooling tower, thermal expansion forces must be adequately allowed for in the structural design since uncontrolled temperature variations can increase the concrete's vulnerability to sulfate attack and frost action.

Design philosophies for concrete cooling tower construction typically coincide with those espoused by the American Concrete Institute (ACI). Generally a rich concrete mix (high cement content) is specified, with a low free water-to-cement ratio. In addition, if high sulfate levels are present, a sulfate-resistant portland cement is specified (ASTM C150 Type V or a modified Type II). If high chloride levels are present, consideration should be given to specifying a minimum 2 in. (50 mm) cover over the rebar. If sulfate levels are greater than 6,000 ppm, consideration should be given to coating concrete located in immersed service, such as the cooling tower basin.

12.6.2 Miscellaneous Structure Materials

Miscellaneous structural materials include materials used for the air inlet louvers on crossflow towers, the tower casing, and the fan cylinders.

Air inlet louvers retain circulating water within the cooling tower as well as help equalize airflow into the tower. Although louvers are necessary on crossflow cooling towers, they are seldom required for counterflow towers. Air inlet louvers must be capable of supporting snow and ice loads. In the past, the predominate material for air inlet louvers was asbestos cement board. Because of environmental concerns and Federal Occupational Safety and Health Standards (OSHA) regulations regarding the handling of asbestos-carrying materials, this material is no longer used. Common materials currently used on large utility cooling towers are fiberglass-reinforced polyester and treated Douglas fir plywood. Advantages of fiberglass-reinforced polyester include its fire-retardant characteristics and inert character, which precludes chemical and biological attack. Precast, prestressed concrete louver panels are typically used on concrete cooling towers.

The cooling tower casing is the exterior enclosing wall of a cooling tower, exclusive of the air inlet louvers. Its function is to contain water within the tower, provide an air plenum for the fan, and transmit wind loads to the tower framework. It should have diaphragm strength, be watertight and corrosion resistant, and have fire-retardant qualities. As with louvers, the traditional material for tower casings was asbestos cement board, which is no longer used. The most common material used today is corrugated fiber reinforced polyester. The panels are overlapped and sealed to prevent air leakage. On concrete towers, the casing is typically concrete panels.

Fan cylinders are extremely important because they directly affect the proper flow of air through the tower. Fan efficiencies can be severely reduced by a poorly designed fan cylinder or significantly enhanced by a well-designed one. A well-designed fan cylinder will have an eased inlet to promote smooth flow of air to the fan, a minimal fan blade tip clearance, a smooth profile above and below the fan, and sufficient structural strength to maintain a stable plan and profile. Fan cylinders with a gradual increase in cross-sectional area beyond the fan (called velocity recovery stacks) serve to reduce the air exit velocity. This shape partially converts velocity pressure to static pressure, allowing an increase in airflow over that accomplished with a straight stack at the same horsepower. Typically, fan cylinder materials are either fiberglass-reinforced polyester or precast concrete. Concrete fan cylinders are more expensive than fiberglass-reinforced polyester cylinders, but they allow the use of smaller fan tip clearances, which increases fan efficiency.

12.6.3 Hardware and Structural Connectors

Materials typically used for hardware and structural connectors include carbon steel, stainless steel, fiberglass-reinforced plastic, silicon bronze, and aluminum bronze. Fiberglass-reinforced plastic, stainless steel, or silicon bronze connectors are typically used in wooden structures as shear and diagonal connectors. Fiberglass-reinforced plastic is advantageous because the inert nature of the material eliminates reactions with chemical species in the circulating water and deposition of metallic salts on the wood which can lead to accelerated biological attack of the wood.

The type of metallic hardware used depends on the circulating water chemistry. Bolting for freshwater service is typically hot-dipped galvanized or stainless steel. For brackish water and seawater applications, chloride-resistant stainless steel (usually Type 316), silicon bronze, or aluminum bronze are typically used. Care is required with silicon bronze since it is subject to impingement attack. Exposed portions of the bolting should be capped with an inert material such as polyvinyl chloride. Precast concrete connectors for freshwater applications are typically carbon steel if grouted or stainless steel if exposed. Silicon bronze or aluminum bronze precast concrete connectors are typically used for brackish water or seawater applications.

12.6.4 Materials for Cooling Tower Internals

Cooling tower internals include all components related to the basic heat transfer function of the cooling tower. These include the fill and fill support system, hot water distribution system, and drift eliminators.

12.6.4.1 Fill. Cooling tower fill accelerates the dissipation of heat from the circulating water. To function efficiently, the fill should promote a high rate of heat transfer, provide low resistance to airflow, and provide uniform water and air distribution throughout the tower. The ideal fill should be highly resistant to deterioration, able to withstand moderate ice loading, and able to maintain efficient heat transfer capabilities for many years.

The crossflow cooling tower typically uses a splash bar or open-type of fill system, whereas the counterflow cooling tower generally uses a film or closely packed fill system. However, film type fills are available for crossflow cooling towers, and splash type fills are available for counterflow cooling towers.

Before the advent of plastics, the predominate materials for splash bar fill materials were asbestos cement board and treated wood lath. These materials are used infrequently today because of environmental restrictions and OSHA regulations in the use of asbestos cement, and because of the refinement of plastics technology and availability. Today, the predominate material on new cooling towers is polyvinyl chloride, which has advantages over wood of a low flame spread rating and relative inertness to biological and chemical attack.

Film fill is a more recent development. For a film fill, it is extremely important that fill sheets are spaced uniformly since air has a tendency to take the path of least resistance, and "channeling" of air and water flows can take place in a poorly designed system. Because of the close spacing of the sheets, film fill should be avoided in situations where the circulating water has high solids concentrations, high fouling or scaling potential, or can become contaminated with debris. Early film fill materials were closely spaced sheets of asbestos cement board. Today, the predominate material is polyvinyl chloride because of its low flame spread rate, adequate strength, and formability to desired shapes. Maximum research and development effort has gone into the various types and shapes of film fill.

Another material that has a long history in air conditioning towers but has also been used in large utility cooling towers is vitrified ceramic tile fill. This fill material is used exclusively in counterflow cooling towers. The fill is made of individual ceramic tiles or bricks, with several large holes to allow passage of air and water. The tiles are stacked in a pattern that allows uniform air and water distribution. This type of fill is sometimes called a combination splash/film fill since it is more open than a typical film fill. The ceramic tile has strength, durability, and is not subject to destruction by fire. Its major drawback is its heavy weight and lower thermal efficiency as compared to other types of counterflow fill. Advantages include long life, low maintenance, and cleanability. This fill can be considered for poor water quality applications where film fill may be inappropriate.

12.6.4.2 Fill Support System. The fill support system is critical, since it supports the heat transfer surface. The ideal system has minimal impact on air and water distribution and is strong, durable, and fire-retardant.

For crossflow cooling towers, the fill should be supported on close centers since sagging of the splash bars can lead to channeling of the air and water into separate flow paths, with a resultant decrease in tower capability. Currently, the predominant materials for splash bar fill support are fiber-reinforced plastic, 300 series stainless steel, and polyvinyl chloride coated steel wire grids. The disadvantage of the coated steel wire grids is that the coating on the wire grids can be subject to abrasion and subsequent corrosion of the wire.

Two basic types of support systems are in common usage for counterflow cooling towers: a bottom-supported system and a top-supported system. Bottom-supported fill systems typically use precast concrete beams, fiberglass-reinforced polyester beams, or cast iron lintels on precast concrete support beams for concrete towers. For wood cooling towers, fiberglass-reinforced polyester beams or wood girts are used for bottom-supported fill. Top-supported systems typically use Type 304 or 316 stainless steel tubes or wires hung from concrete or wood beams above the fill. The fill is attached by various methods to the stainless steel tubes or wires.

12.6.4.3 Hot Water Distribution System. Crossflow and counterflow towers typically have different types of hot water distribution systems. In a crossflow tower, hot water is elevated to a distribution basin above the fill, where it flows by gravity over the fill through orifices in the basin floor. The counterflow tower normally uses a pressure system of closed pipe and spray nozzles.

Common cooling tower riser pipe materials are concrete, steel, or fiberglass-reinforced plastic. Typically, the cooling tower manufacturer provides risers on a round tower, but not on a rectangular tower. Crossflow rectangular cooling towers are usually served by one or two risers at one end of the tower, with distribution headers running the length of the tower at the fan deck level. On a rectangular counterflow cooling tower, an individual riser usually serves each cell, with the risers served by a common ground level distribution header running the length of the tower. A round tower typically has one or two risers depending on the manufacturer's design. Concrete risers are almost exclusively used on round concrete towers.

In crossflow towers, distribution flumes or headers feed the hot water distribution basins above the fill through adjustable weirs or flow control valves. The hot water basin material is usually treated Douglas fir plywood for wooden towers and concrete for concrete towers. In counterflow towers, the distribution flumes or headers feed branch piping which distributes the water to the entire fill area through nozzles. Typical distribution piping materials are polyvinyl chloride and fiberglass-reinforced plastic.

12.6.4.4 Drift Eliminators. Drift emissions are circulating water droplets entrained in the cooling tower exhaust air. Because drift has nearly the same water chemistry as the circulating water, it can be troublesome if the tower is located upwind of the plant boundary, power lines, substations, or the electrical switchgear. The function of the drift eliminators is to remove as much drift as practical from the tower exhaust without adding a large airside pressure drop.

Drift eliminators remove entrained water from the discharge air by causing it to make sudden changes in direction. The resulting centrifugal force separates the drops of water from the air, depositing them on the eliminator surface, from which they flow back into the tower. Eliminators are normally classified by the number of directional changes the air must make when passing through the eliminators or the general shape of the airflow passages through the eliminator.

Since drift eliminators are subject to impingement of the circulating water droplets, they must be as corrosion-resistant as the fill. Asbestos cement and treated wood were used for many years. Today, most drift eliminators are constructed of polyvinyl chloride.

12.6.5 Mechanical Equipment

The majority of the mechanical equipment in a cooling tower operates in a corrosive environment, and materials selection is very important. The mechanical equipment in a cooling tower includes the fans, speed reducers, and drive shafts.

12.6.5.1 Fans. Cooling tower fans are propeller-type fans designed to produce air velocities that are as uniform as possible across the effective area of the fan. Blades have an airfoil cross-section and are tapered and twisted. Fiberglass-reinforced plastics, with polyester or epoxy resin, are the most commonly used blade materials today because of their light weight and exceptional corrosion resistance. Fiberglass-reinforced plastic blades construction typically includes a means of protecting against ultraviolet damage.

Fan hubs should be of a material that is structurally compatible with blade weight and loading. Typical materials include galvanized steel and ductile iron for freshwater applications, and epoxy coal tar-coated steel for brackish water or seawater applications.

12.6.5.2 Speed Reducers and Drive Shafts. The function of the speed reducer is to reduce the motor speed, that is usually 1,800 rpm, down to a speed that allows an acceptable fan tip speed [typically limited to about 12,000 fpm (60 m/s)]. The speed reducer is mounted directly below the fan hub, and the casing materials must be corrosion resistant. The casing material is typically epoxy-coated cast iron.

Since the motors are mounted outside the fan cylinders, the fan drive shaft can be very long, depending on the diameter of the fan. Almost the entire length of the drive shaft is exposed to the saturated tower exhaust air. Materials that are suitable for the drive shaft are stainless steel and newly developed carbon fiber composites.

Typical applications use 18-8 stainless steel for freshwater applications and 316L stainless steel for brackish water and seawater applications. The main disadvantage of stainless steel drive shafts is their heavy weight, which for large-diameter fans has necessitated a two-piece drive shaft with an intermediate bearing coupling the two pieces. Historically this intermediate bearing has been a high maintenance item.

In recent years, carbon fiber composite drive shaft tubes

have been developed and successfully used in cooling tower service. Carbon fiber compositions can be used with practically all circulating water applications. Advantages of the carbon fiber composite shafts include their light weight and natural resistance to corrosion. The composite shafts are light enough to be installed without an intermediate bearing even for the largest typical cooling tower fans, which are 40 ft (12 m) in diameter.

12.7 TOWER PERFORMANCE CRITERIA AND DESIGN FEATURES

12.7.1 Manufacturer's Performance Curves

Cooling tower thermal performance is determined by the basic parameters of airflow, water flow, type and quantity of heat transfer surface, inlet wet-bulb temperature, and heat load imposed on the cooling tower. The design water flow, wet-bulb temperature, heat load, and cold water temperature are specified. The cooling tower manufacturer designs the cooling tower to obtain the specified design cold water temperature by selecting a design airflow rate and heat transfer surface area combination. Once these have been selected, the tower's performance at off-design conditions is fixed and performance curves can be drawn.

Figure 12-38 shows typical cooling tower performance curves for a mechanical draft tower. The performance curves serve two important functions:

- Provide plant engineers with information as to how the cold water temperature of the circulating water from the cooling tower is affected by off-design tower conditions.
- Are required for field testing of cooling tower performance since the design cooling tower conditions rarely exist during actual tower operation or testing. These curves allow verification of contractual thermal performance guarantees.

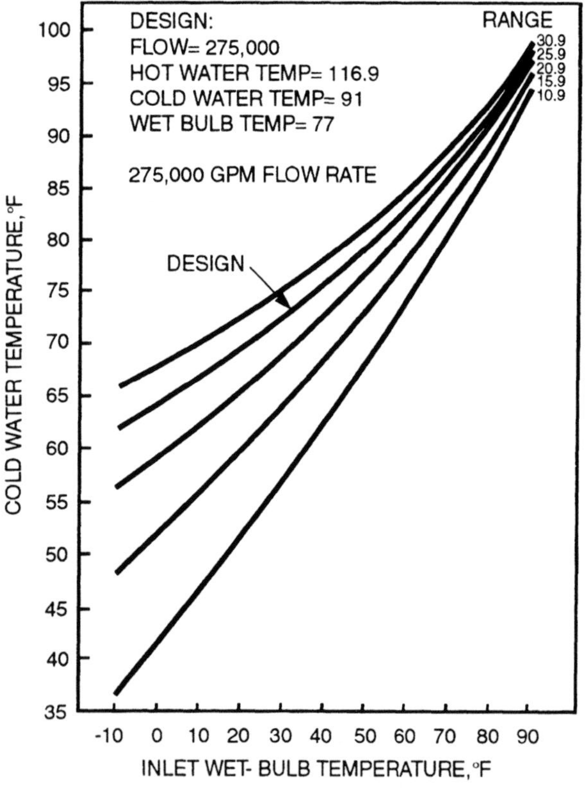

Fig. 12-38. Typical cooling tower performance curve for a mechanical draft tower.

Performance curves are drawn with the wet-bulb temperature on the abscissa and the cold water temperature on the ordinate. In addition to the performance curve at the design cooling range, the tower specification should require performance curves at cooling ranges above and below the design range to allow for off-design plant heat loads. Minimally, the additional ranges requested should be 20% above and 20% below the design range. This encompasses the variance in range from the design point which is allowed by the CTI test code. In addition, curves for cooling ranges that correspond to expected plant operating points should be obtained.

Each set of performance curves is drawn for a specific circulating water flow rate. In addition to the design flow rate, a set of performance curves for off-design flow rates should be requested. Minimally, curves requested should be for flow rates 10% above and 10% below the design point to encompass the variance in flow allowed by the test code. In addition, curves should be requested for expected operating flow rates corresponding to circulating water pump runout with a pump, or pumps, out of service. Each set of performance curves should be drawn for the design fan blade pitch setting, which reflects constant volumetric airflow through the cooling tower. If the tower is to have two-speed fans, curves should be requested for low-speed fan operation.

For natural draft cooling towers, performance curves should also be drawn for various off-design values of relative humidity. A typical natural draft tower performance curve is shown in Fig. 12-39. As the relative humidity decreases for a given wet-bulb temperature, the airflow rate decreases and the cold water temperature increases. As described for mechanical draft cooling towers, additional sets of curves should be requested to show the effects of off-design range and circulating water flow on cold water temperature.

12.7.2 Cooling Tower Performance

The cooling tower manufacturer designs the cooling tower to obtain the desired cold water temperature at the specified conditions by selecting a design airflow rate and heat transfer surface area. The tower manufacturer has many design alternatives available that affect tower performance and cost. The manufacturer may offer a relatively small tower with a high airflow rate and high tower pump head, balancing low capital costs against larger fan horsepower and circulating water pump operating costs. Alternately, the manufacturer can

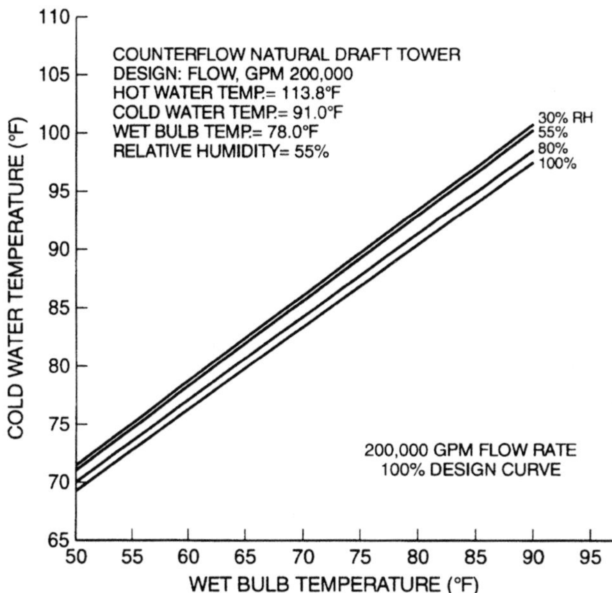

Fig. 12-39. Typical cooling tower performance curve for a natural draft tower.

adjust the tower configuration to decrease any combination of total airflow, airflow resistance, or pumping head, all leading to lower operating costs. The tower manufacturer generally attempts to optimize the cooling tower design to obtain the least total evaluated cost impact, as may be determined from a cooling tower optimization similar to that discussed in Section 12.4.6. Where possible, the cooling tower specification should contain operating cost evaluation factors for the tower manufacturer's optimization expressed as capital equivalent dollars per fan horsepower and per foot of cooling tower pumping head.

A tower that does not perform as specified results in higher than expected circulating water temperatures to the plant condenser. This in turn leads to higher turbine back pressures, lower plant output, and higher plant heat rates. The end results are significant economic penalties to the plant as a result of higher fuel costs and the costs of replacing lost plant generating capacity from other sources. The effect of deficient cooling tower performance on the overall plant performance can be estimated, and the potential economic damages quantified. The potential damages should be specified as liquidated damages to the cooling tower manufacturer if performance guarantees are not achieved.

12.7.3 Cooling Tower Testing

The cooling tower specification should provide for cooling tower acceptance testing to verify that tower performance guarantees have been achieved. It is recommended that tower acceptance testing be performed by an independent third party testing agency. The agency should be specially trained in cooling tower performance testing of large utility cooling towers. The two major cooling tower performance test codes are the CTI ATC-105 and the ASME PTC-23. The CTI has a licensing program to certify independent testing contractors qualified to perform cooling tower acceptance testing. Utilities and other plant operators can solicit competitive bids for testing services.

12.7.4 Winter Operation

At plant sites where winter temperatures often drop below freezing, special cooling tower design considerations are required. Cooling towers are designed to promote the maximum possible contact between air and water for the maximum possible time period. This design requirement results in the highest thermal efficiency in the summertime, but increases the tendency of the tower to develop damaging ice formations during winter operation. If left uncontrolled, ice formations can build quickly and cause substantial damage to the cooling tower. The design of the plant's cycle heat rejection system must include means for plant operators to limit the development of damaging ice formations.

Acceptable ice formations may be defined as ice of relatively thin cross-section that forms on the louvers or air inlet area of the tower structure as shown in Fig. 12-40. If anticipated in a tower's design loading, such ice is not normally a structural concern and, in many cases, retards airflow through the tower, achieving passive control of further ice formation. Unacceptable ice formations can be categorized as either significant amounts of ice that have formed on the fill, jeopardizing the operation and existence of the heat transfer surface, or excessive ice formations on the tower structure. Figure 12-41 illustrates an excessive ice buildup.

Methods of ice control vary somewhat with types of towers, water distribution systems, and mechanical equipment arrangements. The following guidelines apply to most situations:

Fig. 12-40a. Acceptable counterflow ice formation. (From Marley Cooling Tower Company. Used with permission.)

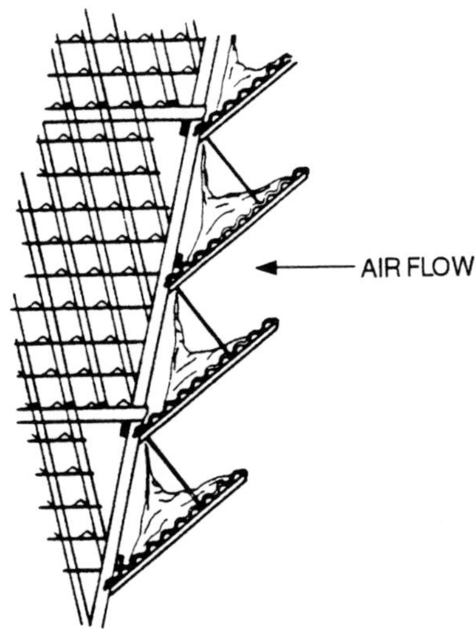

Fig. 12-40b. Acceptable crossflow ice formation. (From Marley Cooling Tower Company. Used with permission.)

- The potential for ice varies directly with the quantity of airflowing through the tower. Reducing the airflow retards the formation of ice.
- Where airflow is uncontrolled (as in the case of hyperbolic towers), the potential for ice formation varies inversely with the heat load imposed on the tower. A reduced heat load increases the probability that unacceptable ice will form.
- The potential for ice varies inversely with the amount of water flowing over the fill. A reduced pumping rate increases the likelihood of unacceptable ice formation.

All mechanical draft towers afford some degree of air-side control, the variability of which depends on the number of fans with which the tower is equipped and, most importantly, the speed-change capability of the motors. Towers operated in cold climates can also include water-side control measures. In mechanical draft towers, both air-side and water-side controls are mutually supportive.

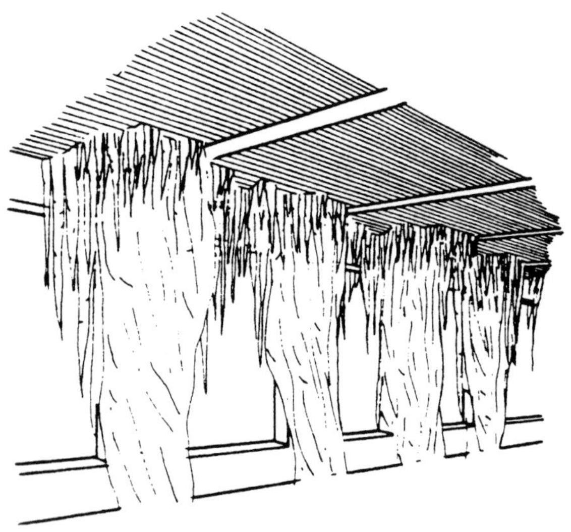

Fig. 12-41a. Unacceptable counterflow ice formation. (From Marley Cooling Tower Company. Used with permission.)

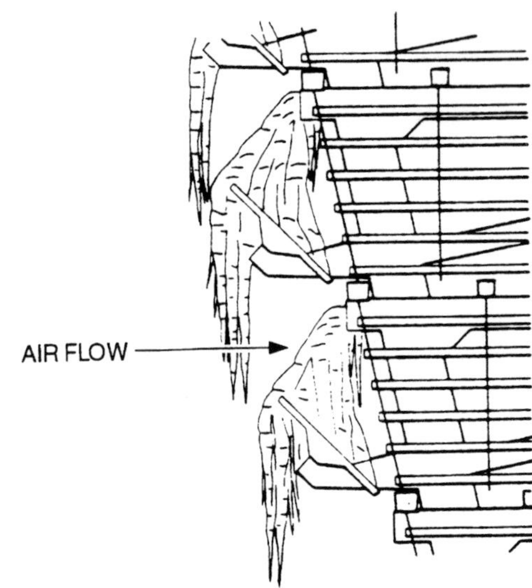

Fig. 12-41b. Unacceptable crossflow ice formation. (From Marley Cooling Tower Company. Used with permission.)

12.7.4.1 Air-Side Control. Manipulation of the airflow is an invaluable tool, not only in the retardation of ice formation, but also in the reduction or elimination of ice already formed. In addition to bringing less cold air into contact with the circulating water, reducing the entering airflow velocity alters the path of the falling water. This allows more water to fall along the tower perimeter to melt ice previously formed around the air inlets by random droplets or wind gusts.

Single-speed fans afford the least opportunity for airflow variation. Towers so equipped may require significant attention from plant operators to determine the proper cyclic fan operation that best controls ice. Two-speed fan motors offer improved operating flexibility at a relatively small increase in capital cost. Their use is recommended in freezing climates. Fans may be individually cycled back and forth between full-speed and half-speed as required to balance cooling effect and ice control. An abnormal number of speed changes per hour may cause the motor insulation temperature to rise. Care must be taken during cyclic operation not to exceed the maximum allowable motor insulation temperature. The plant distributed control system can be programmed to make some of these changes automatically.

Complete air-side control can be achieved through the use of variable-speed fan motors and switchgear. Such systems allow infinite flexibility to prevent the tower water temperature from dropping below recommended limits. These features are expensive and are typically used only in extremely cold climates. The additional capital cost, a complicated control scheme, and the decreased reliability of the variable-speed fan drives are significant disadvantages when compared to a slight increase in operating flexibility.

Round cooling towers may or may not be equipped with a separate plenum for each tower fan. If plenum isolating partitions are specified, individual fans may be shut off, providing increments in operating flexibility similar to rectangular towers. Where two or more fans operate in a common plenum, the fans should be brought to the off position in unison to prevent a down draft of cold, moisture-laden air from icing up the mechanical equipment of an inoperative fan. Most round cooling towers incorporate the necessary plenum partitions to at least segment the overall tower plenum into quadrants.

12.7.4.2 Water-Side Control. Since the potential for icing of the fill depends so much on the incoming water temperature, a provision for total water bypass directly to the cold water basin is advisable. During cold weather startup, the basin water may be at a temperature very near freezing. A tower bypass allows the total circulating water return flow from the condenser to be directed back into the cold water basin without passing through the fill. This bypass mode is typically continued until the cold water temperature to the condenser reaches an acceptable temperature level [usually about 80° F (49° C)], at which time the bypass may be closed to allow total flow over the fill.

Bypass mode is the primary freeze protection measure for natural draft cooling towers during startup and operation. If damaging ice begins to form on the cooling tower, it may be necessary to revert to total bypass flow in order to maintain a reasonable basin water temperature. Partial implementation of the bypass, whereby a portion of the water flow is allowed to continue over the fill area, should not be used. The reduced water loading on the fill could lead to rapid ice formation and significant damage could occur.

12.7.5 Fire Protection

Wood cooling towers typically require a complete fire protection sprinkler system. Fire protection systems for cooling towers are defined and governed by the National Fire Protection Association (NFPA) Bulletin 214. A fire protection system normally consists of an arrangement of piping, nozzles, valves, and sensors or fusible heads which cause the tower to be automatically deluged with water soon after the start of a fire. Piping within the tower is usually free of water to prevent freezing. Water at a prescribed residual pressure is available at an automatic valve, located within a nearby heated space. Operation of the valve is initiated either by thermostatic type sensors that react to an abnormal rate of temperature rise, or by fusible heads that cause pressure loss with a pneumatic control system.

Partition walls between cells of a rectangular tower are designed to act as fire walls to prevent or delay the spread of fire.

12.7.6 Mechanical Equipment

Mechanical equipment includes gear reducers, fans, and drive shafts. This equipment must provide safe, reliable operation for many years of nearly continuous operation. For this reason, several areas of concern exist regarding the design of mechanical equipment.

Cooling tower gear reducers historically have been one of the least reliable pieces of equipment on the cooling tower. The gear reducers are typically located in the moist exit airstream directly below the fans, and routine scheduled maintenance is often neglected since the tower is frequently a significant distance from the main plant structure. Inadequate lubrication, inappropriate lubricant viscosities, infrequent oil changes, and general neglect will eventually lead to bearing failure. For ease of maintenance, the oil level sight glass, oil fill and operational venting, and all necessary valving should be located outside the fan cylinder. To guard against gear reducer operation with inadequate lubrication, each gear reducer should be equipped with a low oil level switch.

Other problems can arise from inadequate service factors and bearing life rating. CTI Standard 111 recommends a minimum service factor for cooling tower gear reducers of 2.0 and a minimum output shaft bearing L-10 life expectancy of 100,000 h. The L-10 life expectancy rating is defined as the life expectancy in hours during which 90% or more of a given group of bearings under a specific loading condition will still be in service. To provide design margins on these standards and to enhance reliability, a minimum service factor of 2.5 and minimum L-10 rated life of 130,000 h is recommended.

In large cooling towers, fans have large diameters to handle the large volumes of air more efficiently and to reduce fan horsepower requirements. For many years, this has necessitated a two-piece drive shaft with an intermediate bearing coupling between the two pieces. The intermediate bearing historically has been a high-maintenance item. For this reason, the use of an intermediate bearing in cooling tower fan drive shafts is not desirable. Avoiding the problems associated with the intermediate bearing requirement is much easier since the successful introduction of lightweight carbon fiber composite drive shafts.

12.8 DRY AND WET–DRY COOLING TOWERS

Wet or evaporative cooling towers typically evaporate about 2% of the circulating water. A 200-MW power plant with a circulating water flow of 100,000 gpm (6,300 L/s) over the cooling tower, operating at a capacity factor of 65%, requires an annual makeup requirement of approximately 2,000 acre-ft (2.5×10^6 m^3) of water. In areas of water scarcity, necessity and economics have dictated the use of dry and wet–dry cooling towers.

12.8.1 Dry Cooling Tower

Dry cooling towers transfer heat by convection and radiation instead of by evaporation as the wet towers do. The circulating water makeup requirements are reduced to zero, making this an ideal arrangement from an environmental standpoint, particularly in areas of water scarcity. However, the dry tower typically results in higher plant fuel costs. This is because the heat transfer process in the dry tower is totally sensible, and the controlling variable is the ambient dry-bulb temperature. This is always a higher value than the ambient wet-bulb temperature, which is the cooling variable in the evaporative tower. This results in higher operating turbine back pressure, decreased turbine efficiency, and higher fuel costs when compared to a typical evaporative design.

There are two types of dry cooling systems: direct and indirect. The direct type, shown schematically in Fig. 12-42, involves condensing the turbine exhaust steam in air-cooled heat exchanger bundles called an air-cooled condenser. The indirect type, shown schematically in Fig. 12-43, uses a traditional steam surface condenser and circulating water system to transfer the waste heat to cooling towers using air-cooled heat exchanger bundles. Alternatively, the indirect system can utilize a direct contact (jet) steam condenser. Both types of dry cooling can be applied with either mechanical or natural draft cooling tower designs. Advantages of the direct system over the indirect system include lower capital cost and theoretically lower exhaust pressures, since the use of an intermediate cooling fluid and a second heat exchanger with its associated thermal resistance are not required. The large exhaust duct required to transport the steam from the turbine to the air-cooled condenser at an acceptable pressure drop is a drawback to the direct system. The direct air-cooled condenser must be located near the Turbine Building to minimize the steam exhaust pressure drop. Air-cooled condensers are discussed in more detail in Chapter 9.

In an indirect system, circulating water flow is cooled in fin-tubed, air-cooled heat exchange bundles. The air-cooled cooling water flows back to a surface condenser to condense the turbine exhaust steam. Because the indirect system requires the use of two heat transfer processes—the steam condenser and the air-cooled heat exchangers—the logarithmic mean temperature difference (LMTD) for the overall heat transfer process is lower than that of the direct system (Fig. 12-44). This can be compensated only by a larger cooling surface, increased cooling airflow, or both. The indirect system represented in the figure includes a cooling surface increase of about 30% as compared to the air-cooled condenser, based on equal cooling airflow and heat transfer surface performance.

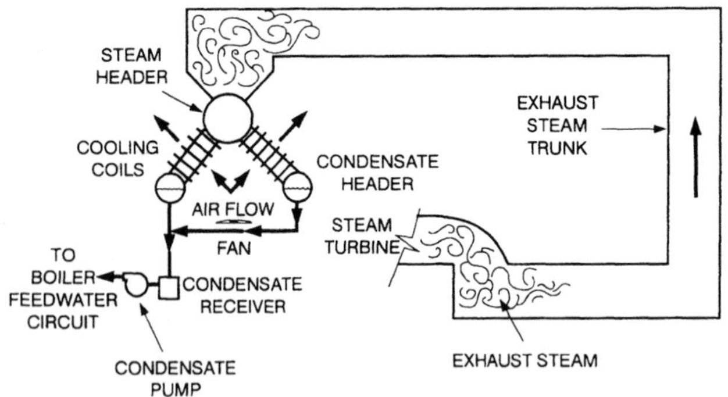

Fig. 12-42. Direct dry type cooling.

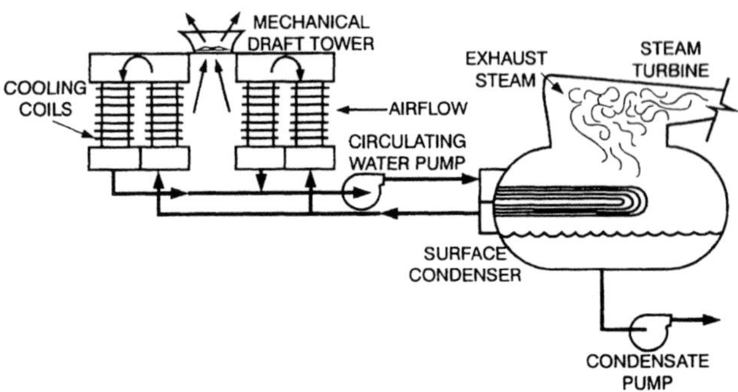

Fig. 12-43. Indirect dry type cooling tower.

12.8.2 Wet–Dry Cooling Tower (Plume Abatement Tower)

The use of a wet–dry cooling tower arrangement can be another solution to the problem of low water availability. This type of cooling tower typically has an air-cooled section as well as a conventional evaporative section. To conserve water, only the air-cooled sections are used during most of the year. The evaporative sections are used only in periods of high ambient temperatures. This type of system greatly reduces the annual makeup water requirements and allows the use of conventional, low back pressure turbines, which a totally dry system usually does not allow. Fuel costs are reduced and plant capacity increased as compared to the dry system since lower turbine back pressures are obtained in the summer months.

In addition to water conservation, wet–dry cooling towers have an application where visible plumes are undesirable. In this capacity, it is often referred to as a plume abatement tower.

Plume formation from a cooling tower occurs most often during the cooler months of the year. Figure 12-45 is a simplified psychometric chart that shows typical summer and winter operations of a wet cooling tower. Line 1–2 shows the winter mode of operation, and Line 3–4 shows the summer mode. As evaporation occurs in the tower, the moisture content of the air increases. The dry-bulb temperature also increases, and the air is discharged to the atmosphere in a saturated state (Points 2 and 4). The exit air then mixes with the ambient air, and mixing occurs along a straight line (either 2–1 or 4–3). In the winter, this mixing outside the tower occurs in the supersaturated region (above the saturation curve) of the psychometric chart and a plume forms. A plume does not form for the summer case (4–3) since mixing occurs in the superheated region of the psychometric chart.

There are two basic types of wet–dry cooling towers. The first type is the series path wet–dry tower as shown in Fig. 12-46 with an accompanying psychrometric process diagram. Ambient air is heated as it passes through the dry section (Process Line 1–2), undergoing an increase in dry-bulb and wet-bulb temperatures, and a decrease in relative humidity. The air then passes through the wet section (Process Line 2–3), bringing the condition to the saturation curve. Once the air exits the tower, it mixes (M) with the ambient air along Line 3–1, which is totally outside the fog

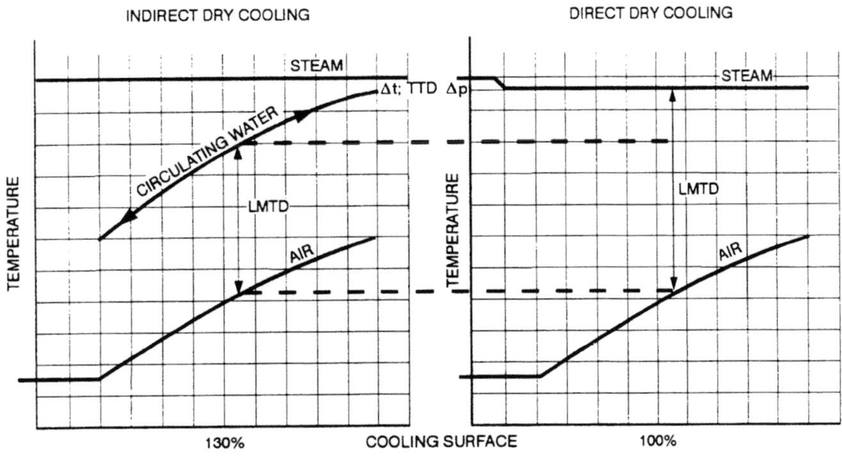

Fig. 12-44. LMTD comparison for direct dry cooling versus indirect dry cooling.

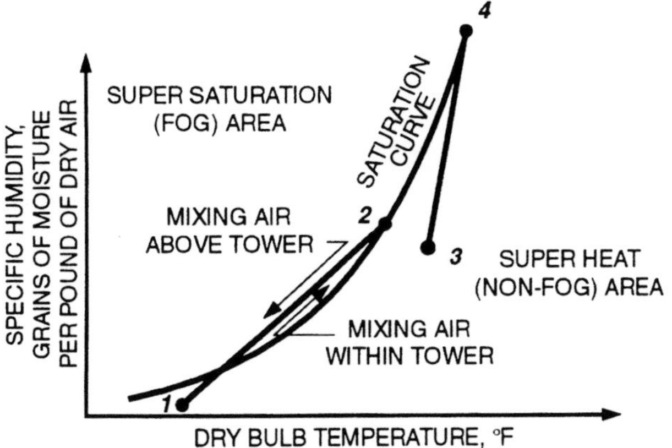

Fig. 12-45. Psychrometric chart for conventional wet cooling tower under winter and summer conditions. (From Marley Cooling Tower Company. Used with permission.)

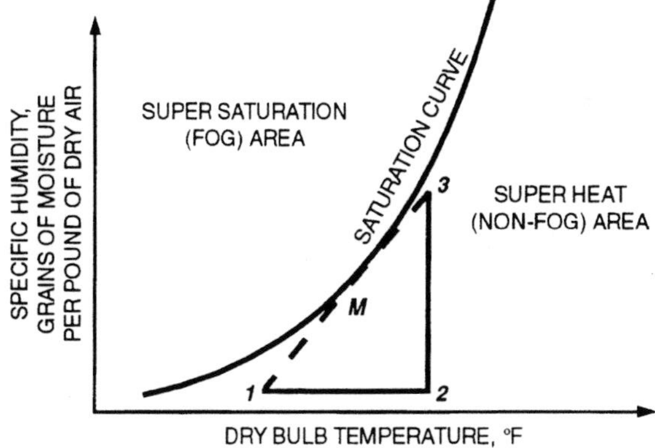

Fig. 12-46b. Psychrometric chart for series air path wet–dry with coils upstream of fill. (From Marley Cooling Tower Company. Used with permission.)

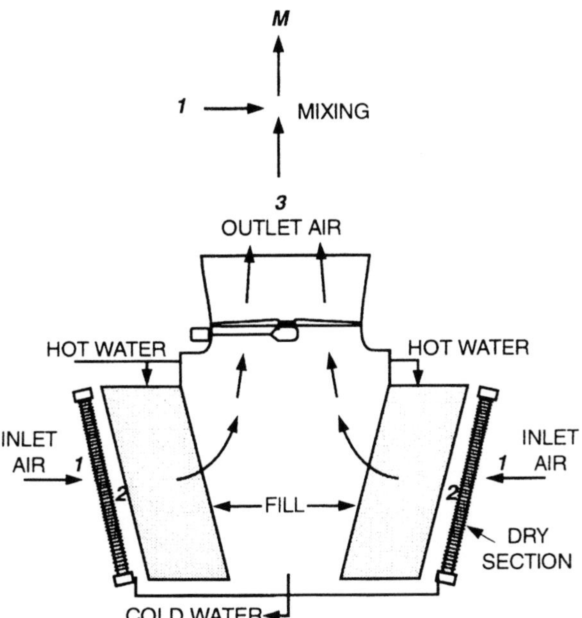

Fig. 12-46a. Series path crossflow wet–dry cooling tower. (From Marley Cooling Tower Company. Used with permission.)

region (below the saturation curve). The dry section can also be placed after the wet section, with the same result.

The second type of wet–dry tower is the parallel path cooling tower shown in Fig. 12-47 with an accompanying psychrometric process diagram. In this system, the air passes through both the wet (Process Line 1–2) and dry (Process Line 1–3) sections in parallel paths, mixing together as the streams go through the fan (Process Line 2–4–3). Above the fan, the exit air mixes (M) with the ambient air along Line 4–1. Again, this is totally outside the fog region. The parallel path wet–dry cooling tower is the preferred wet–dry cooling tower because the dry section is not located in close proximity to the evaporative fill section. In this arrangement, circulating water impingement, which could result in scaling and restricted airflow, is also minimized. A parallel path wet–dry mechanical draft cooling tower is shown in Fig. 12-48.

Capital costs for wet–dry and dry cooling systems are quite high when compared to a conventional evaporative tower. This factor and excessive unit fuel and energy costs have made these towers practical only where extreme environmental conditions have necessitated their use.

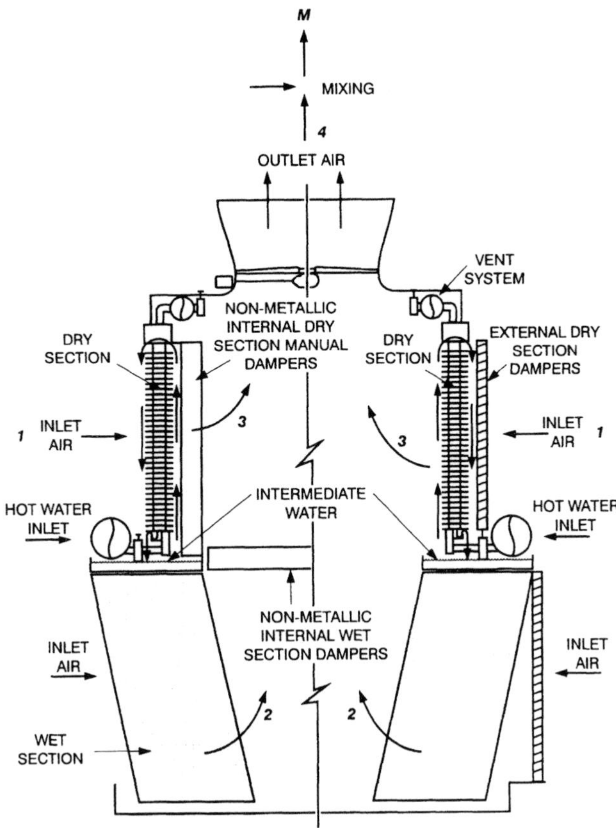

Fig. 12-47a. Parallel air path, series water path, crossflow wet–dry cooling tower. (From Marley Cooling Tower Company. Used with permission.)

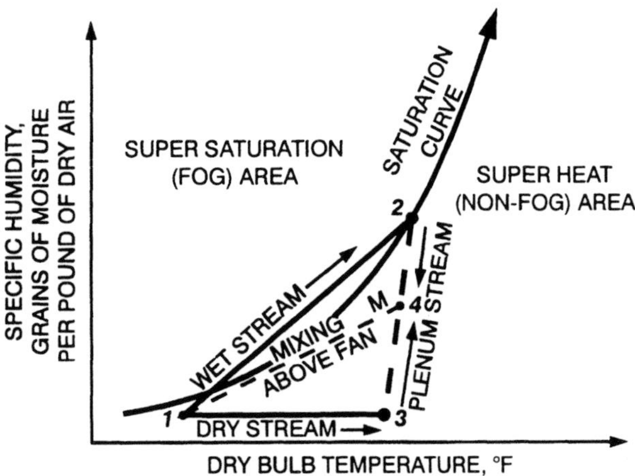

12-47b. Psychrometric chart for parallel air path, series water path, wet–dry tower. (From Marley Cooling Tower Company. Used with permission.)

Fig. 12-48. Parallel path wet–dry mechanical draft cooling tower manufactured by the Hamon Cooling Tower Company. (From Greater Detroit Resource Recovery. Used with permission.)

12.9 REFERENCES

CHAUDHRY, M. H. 1979. *Applied Hydraulic Transients.* Van Nostrand Reinhold. New York, New York.

COOPERATIVE POWER. Coal Creek Station. P.O. Box 780, Underwood, ND 58576.

FMC CORPORATION. 400 Highpoint Drive, Chalfont, PA 18914.

FLORIDA POWER CORPORATION. 3201 34th St. S., St. Petersburg, FL 33711.

HAMON COOLING TOWER COMPANY. 245 US Hwy. 22 W., Bridgewater, NJ 08807.

HENSLEY, JOHN C., EDITOR. 1985. *Cooling Tower Fundamentals,* 2nd edit. Marley Cooling Tower Company. Mission, KS.

IOWA-ILLINOIS GAS & ELECTRIC. 206 E. 2nd Street, Davenport, IA 52801.

LI, KAM W. and A. PAUL PRIDDY. 1985. *Power Plant System Design.* John Wiley & Sons. New York, NY.

MARLEY COOLING TOWER COMPANY. 5800 Foxridge Drive, P.O. Box 2912, Mission, KS 66201-9875.

MICHIGAN WASTE-TO-ENERGY. 5700 Russell Street, Detroit, MI 48211-2545.

NORTHERN STATES POWER. 414 Nicollet Ave., Minneapolis, MN 55402.

ORLANDO UTILITIES COMMISSION. P.O. Box 3193, Orlando, FL 32802.

STREETER, V. L. and E. B. WYLIE. 1982. *Fluid Transients.* FEB Press, Ann Arbor, MI.

13

CYCLE PERFORMANCE IMPACTS

Michael J. Eddington

13.1 INTRODUCTION

Steam cycle performance is impacted by numerous design and operating parameters. The operating parameters include, among others, main steam pressure and temperature, reheater system pressure drop, reheat temperature, and turbine back pressure. Cycle performance is also impacted by numerous design decisions made for a new power plant. This chapter evaluates the performance impacts of the design and operating parameters on power station steam cycles and describes how these impacts can be reflected in changes in steam cycle mass and energy balances.

13.2 THE STEAM POWER CYCLE

To understand better the thermodynamic impacts of various steam cycle parameters, a brief review of the ideal Carnot cycle is helpful. The thermal efficiency of any power cycle is maximized if the heat supplied to the cycle is supplied at the highest possible temperature and the heat rejected from the cycle is rejected at the lowest possible temperature. Figure 13-1 shows a Carnot cycle plotted on a temperature–entropy (T–S) diagram. On the T–S diagram, areas may be interpreted as proportional to heat transfer amounts. The Carnot cycle is made up of two reversible constant temperature (isothermal) processes and two reversible adiabatic (isentropic) processes. The working fluid is isentropically compressed from a two-phase liquid–vapor mixture at state 1 to saturated liquid at state 2. The fluid is evaporated at constant pressure and temperature from state 2 to state 3. The fluid then expands isentropically from state 3 to state 4. The fluid is then partially condensed at constant pressure and temperature from state 4 back to state 1. The area bounded by points 1–2–3–4 is proportional to the heat converted into work in this cycle. The area bounded by a–1–4–b is the heat rejected to the surroundings from this cycle. The cycle efficiency can be expressed by the ratio of the areas on the T–S diagram representing the heat converted into work and the total heat supplied.

$$\eta = \frac{\text{Heat converted into work}}{\text{Total heat supplied}} \tag{13-1}$$

$$= \frac{(T_2 - T_1)(S_4 - S_1)}{T_2(S_4 - S_1)}$$

In addition, given that the temperatures are in terms of absolute temperatures, $T_a = 0°$ R. Simplifying by canceling the entropy terms yields the classical expression for Carnot efficiency.

$$\eta_{\text{Carnot}} = \frac{T_2 - T_1}{T_2} = 1 - \frac{T_1}{T_2} = 1 - \frac{T_{\text{LOW}}}{T_{\text{HIGH}}} \tag{13-2}$$

The Carnot cycle is a theoretical cycle only. It is not mechanically practical to partially condense steam to a particular quality from state 4 to state 1 and then compress the wet steam from state 1 to state 2. For these reasons, actual steam cycles are based on a modified version of the Carnot cycle called the Rankine cycle. Figure 13-2 shows the equipment schematic and the T–S diagram for a simple steam power plant operating on the Rankine cycle. The major difference between the Carnot cycle and the Rankine cycle is that in the Rankine cycle the steam is fully condensed to saturated liquid in the process from state 4 to 1 and the liquid is pumped to the boiler pressure (state 2). Heat is applied at constant pressure until the compressed liquid becomes saturated at state "X." More heat is applied at constant pressure to vaporize the liquid until it is saturated vapor at state 3. The vapor is then expanded from state 3 to 4 and the cycle is repeated.

The improvements in cycle efficiency for changes in various cycle parameters can be easily seen on the T–S diagram. Several modifications to the basic Rankine cycle are explored as follows.

13.2.1 Superheating

Although Eq. (13-2) applies only to the Carnot cycle, the guidance it offers is true in general. From Eq. (13-2), it is seen that the thermal efficiency of a power cycle is increased if the heat supplied to the cycle is supplied at a higher temperature. One way to increase the temperature at which heat is supplied is to superheat the steam above the saturation temperature. The increase in thermal efficiency is shown in Fig. 13-3. The additional work (additional area "within" the cycle diagram boundaries) due to superheating is shown by

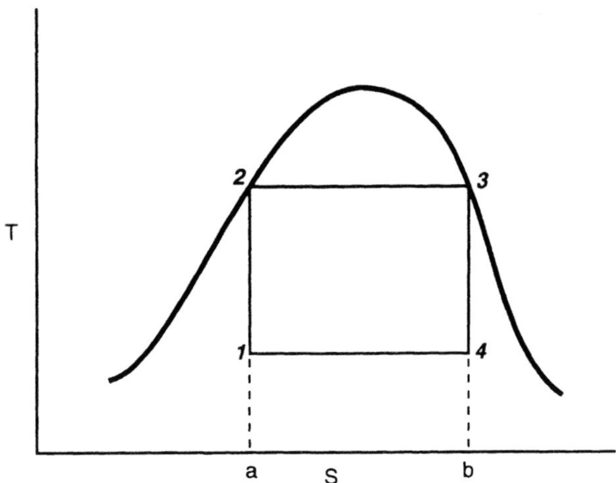

Fig. 13-1. Carnot cycle.

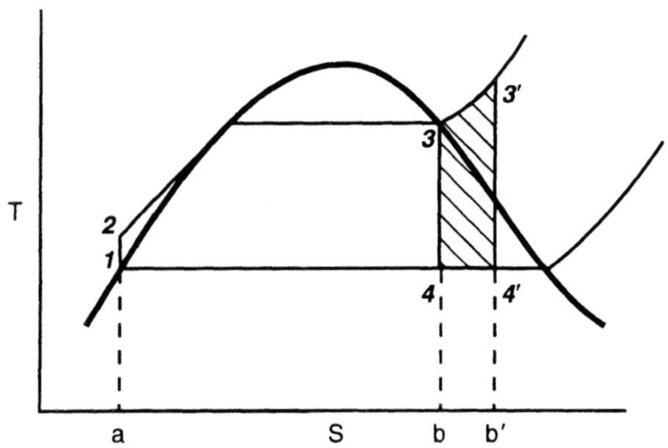

Fig. 13-3. Effect of superheating on Rankine cycle efficiency.

the shaded area bounded by points 3–3′–4′–4. The additional heat that must be rejected is shown by the area bounded by b–4–4′–b′. The increase in cycle efficiency can be seen by noting that the ratio of areas 3–3′–4′–4 to b–4–4′–b′ is larger than the ratio of net work to heat rejected for the original cycle (1–2–3–4/a–1–4–b).

13.2.2 Increasing Pressure

Increasing the pressure at which the boiler evaporates steam increases the saturation temperature which results in an increase in the average temperature of heat addition. This increase in temperature results in an increase in thermal efficiency (Fig. 13-4). The additional work due to the increased pressure is shown by the shaded area 2–2′–3′–c. The double shaded area c–3–4–4′ represents work of the lower pressure cycle that is lost by increasing the pressure. The area 2–2′–3′–c is slightly larger than area c–3–4–4′ resulting in a small net gain in work. The major gain in cycle efficiency comes from a reduction in total heat rejected from the cycle while the heat input remains the same. This reduction in heat rejection is represented by area b′–4′–4–b.

13.2.3 Lowering Exhaust Pressure

The increase in cycle efficiency as a result of lowering the pressure (and the temperature) at which the steam is condensed can be seen in Fig. 13-5. The shaded area 1′–2′–2–1–4–4′ represents the increase in available work from the cycle. This area also represents a decrease in the total cycle heat rejection. Lowering the condenser pressure captures some of the previously unavailable work, thereby increasing

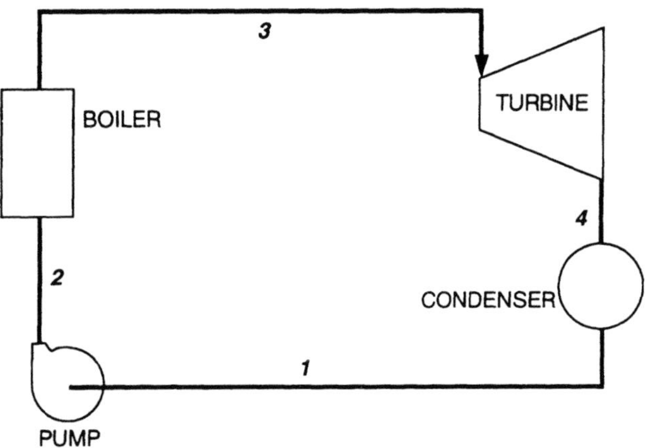

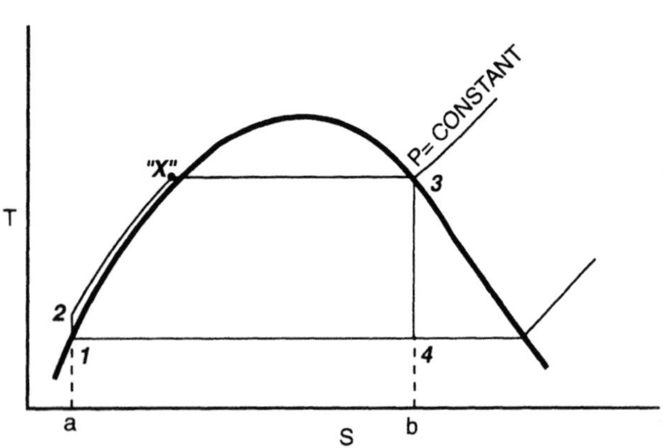

Fig. 13-2. Rankine cycle equipment schematic and T–S diagram.

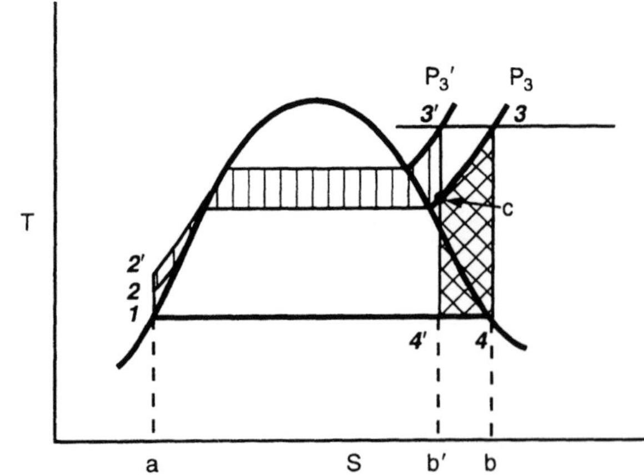

Fig. 13-4. Effect of boiler pressure on Rankine cycle efficiency.

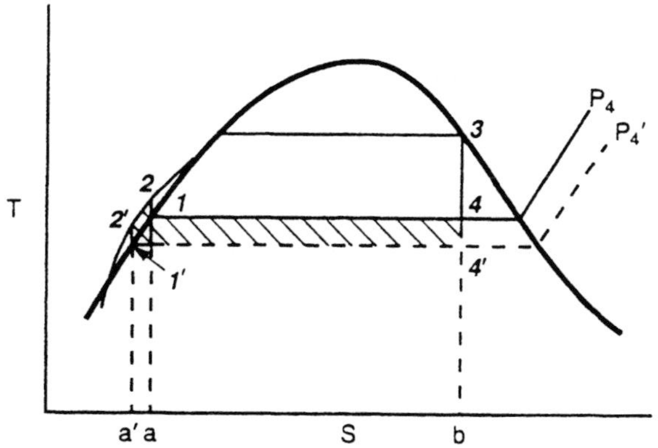

Fig. 13-5. Effect of exhaust pressure on Rankine cycle efficiency.

the cycle efficiency. Lowering the exhaust pressure also results in a very small additional heat input 2′–2–a–a′.

13.2.4 Reheat

Increasing the average temperature of heat addition increases the cycle efficiency. Reheating the steam after it has partially expanded through the turbine increases the average temperature of heat addition. Fig. 13-6 shows the equipment arrangement and the T–S diagram for the single reheat cycle.

The additional work as a result of reheating is shown by the shaded area bounded by points 4–5–6–6′. The additional unavailable heat that must be rejected is shown by the area bounded by b′–6′–6–b. The increase in cycle efficiency can be seen by noting that the ratio of areas 4–5–6–6′ to b′–6′–6–b is larger than the ratio of net work to heat rejected for the original cycle (1–2–3–6′/a–1–6′–b′). Additional reheating will continue to increase the cycle efficiency; however, the incremental gain for each additional reheat will decrease. It should be noted that an additional benefit of reheating is to provide drier steam in the last stages of the turbine, point 6 compared to 6′.

13.2.5 Regenerative Feedwater Heating

Increasing the average temperature of heat addition can also be accomplished by increasing the temperature of the feedwater entering the boiler. To realize a gain in efficiency, heat from within the cycle is used to elevate the feedwater temperature. This can be done by extracting a portion of the partially expanded steam from the turbine and directing it to a heat exchanger that heats the feedwater to the boiler. This process is called regenerative feedwater heating. Figure 13-7 shows the equipment arrangement and T–S diagram for the regenerative Rankine cycle. Steam enters the turbine at state

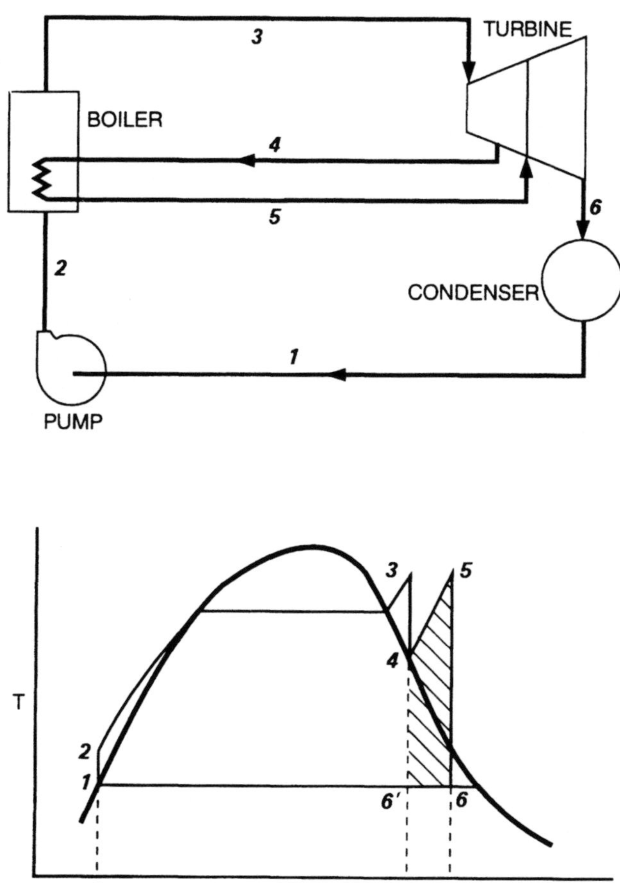

Fig. 13-6. Effect of reheat on Rankine cycle efficiency.

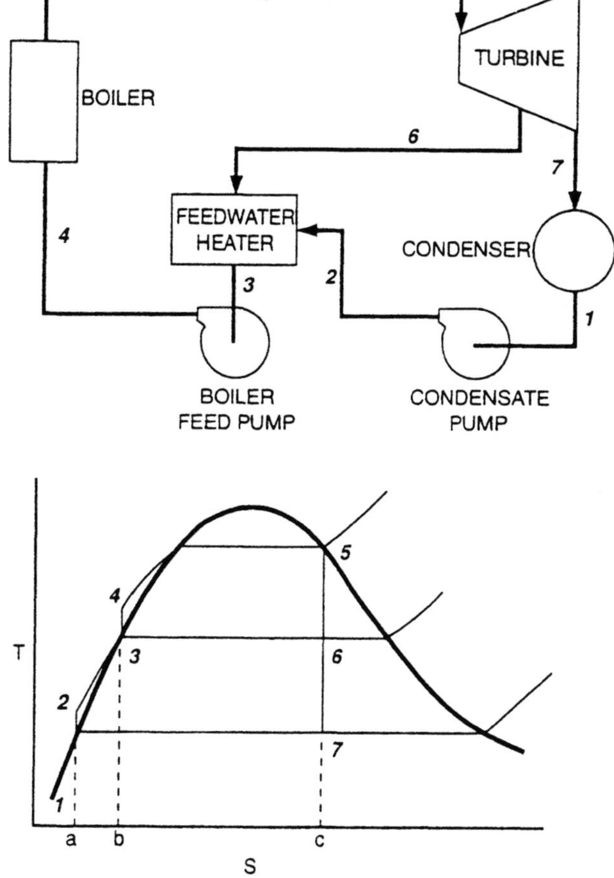

Fig. 13-7. Regenerative cycle with open feedwater heater.

5 and is partially expanded to state 6. A portion of the steam is extracted at state 6 and sent to a feedwater heater operating at state 3. The remainder of the steam expands through the steam turbine to state 7. Heat is rejected as the turbine exhaust steam is condensed in process 7–1. The condensate is pumped to the feedwater heater and mixed with the turbine extraction steam to become saturated liquid at state 3. The feedwater is pumped to the boiler pressure (state 4), heated to saturation, and evaporated in the boiler to reenter the turbine at state 5. The low temperature heat addition into the cycle (2–3) is avoided and the improvement in efficiency comes from the increase in the average temperature of heat addition. This is difficult to show graphically on the *T–S* diagram. Because the flow rates are not equal at all of the state points on the *T–S* diagram, the areas do not represent the total work and heat rejected of the cycle. Rather they represent the work and heat rejected per pound of steam.

13.3 DESIGN STEAM CONDITIONS

As discussed in the previous section, increasing the power cycle steam pressure and temperature increases the cycle efficiency. There are practical limits to the values of pressure and temperature that can be used in an actual application. Materials considerations currently limit the maximum pressure and temperature to 5,000 psig and 1,200° F. The most common rated steam pressures used in large fossil fuel steam power plants are 3,500 psig for supercritical plants and 2,400 psig for subcritical plants. The most common steam temperature (main steam and reheat) is 1,000° F although 1,050° F steam is sometimes used. Figure 13-8 shows the net turbine heat rate relationships for various turbine steam conditions.

13.4 NUMBER OF FEEDWATER HEATERS

Reversible heat transfer and an infinite number of feedwater heaters would result in a cycle efficiency equal to the Carnot cycle efficiency. The greater the number of feedwater heaters used, the better the cycle efficiency. However, each additional heater results in lower incremental heat rate improvement because of the decreasing benefit of approaching an ideal regenerative feedwater heating cycle. Because of the diminishing improvement in cycle efficiency, increasing capital costs, and turbine physical arrangement limitations, the economic benefit of additional heaters is limited. The typical number of feedwater heaters is shown in Table 13-1 for various plant size ranges.

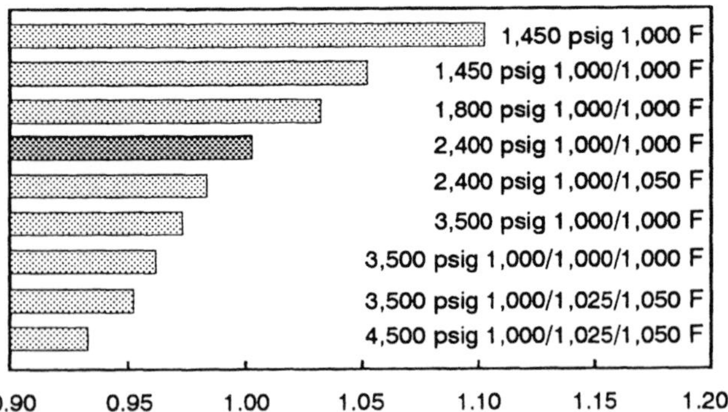

Fig. 13-8. Relative effect of steam conditions on net turbine heat rate. (From General Electric. Used with permission.)

Table 13-1. Typical Number of Feedwater Heaters

Unit Size (MW)	Number of Heaters
0–50	3–5
50–100	5 or 6
100–200	5–7
over 200	6–8

13.5 FEEDWATER HEATER DESIGN PARAMETERS

A closed feedwater heater is a heater in which the feedwater and the heating steam do not directly mix. Low-pressure (LP) and high-pressure (HP) feedwater heaters are typically of the closed type. Open feedwater heaters directly mix the feedwater and heating steam. Deaerators are of this type.

A closed feedwater heater may consist of three zones: the desuperheating zone, the condensing zone, and the drain cooling zone. All closed heaters have a condensing zone where the feedwater is heated by the condensation of the heating steam. Feedwater heaters that receive highly superheated steam require a desuperheating zone to reduce the steam temperature to approximately 50° F above saturation temperature before it enters the condensing zone. A desuperheating zone may not be required for heaters that receive heating steam with less than 100° F superheat. Usually, a drain cooler is also included in a feedwater heater to recover heat from the drains before the drains leave the heater. Figure 13-9 shows the temperature profile through a three-zone closed feedwater heater.

The feedwater heater performance is specified by the drain cooler approach (DCA) and terminal temperature difference (TTD). Both are shown in Fig. 13-9. The DCA is the difference between the temperature of the drains leaving the heater and the temperature of the feedwater entering the heater. The TTD is the difference between the saturation temperature at the operating pressure of the condensing zone and the temperature of the feedwater leaving the heater.

A drain cooler is used to recover heat from feedwater heater drains. The recovery of heat from the drains enhances cycle efficiency. By decreasing the DCA of a heater, cycle efficiency is improved while the heater surface area is increased, resulting in higher capital cost. The practical minimum DCA for an internal drain cooler is 10° F. For an external drain cooler, the minimum practical limit is 5° F.

The TTD shown in Fig. 13-9 is for a heater with a de-

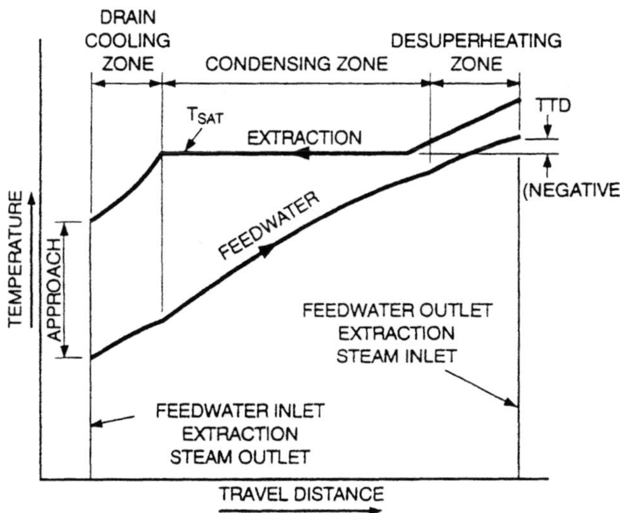

Fig. 13-9. Temperature profile for a closed feedwater heater.

superheating zone. Because of the desuperheating zone, the feedwater temperature leaving the heater may be higher than the saturation temperature of the condensing zone. Therefore, the heater may have a negative TTD as shown in the figure. If the desuperheating zone of the heater was removed, the feedwater outlet temperature would be less than the saturation temperature, which results in a positive TTD. The practical lower TTD limit on a heater without a desuperheating zone is +2° F. The negative TTD limit for a heater with a desuperheating zone depends on the amount of superheat in the extraction steam entering the heater.

The lower the TTD and DCA, the more efficient the cycle and the larger the heater surface area. The more efficient cycle results in a lower heat rate and reduced fuel costs, while the larger surface area results in a higher capital cost. Figures 13-10, 13-11, and 13-12 show the effect of various TTDs and DCAs on cycle heat rate for the HP and LP heaters for a seven-heater 500 MW cycle. A detailed engineering study using applicable economic parameters must be completed to determine the optimum feedwater heater parameters.

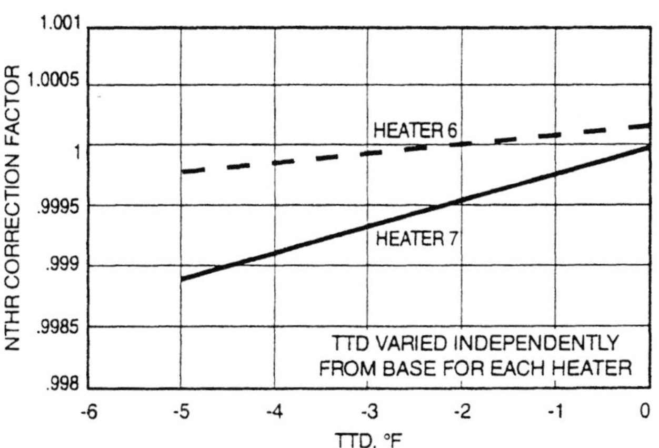

Fig. 13-10. Effect of TTD on net turbine heat rate—Heater 7 (500 MW) cycle, HP heaters 6 and 7.

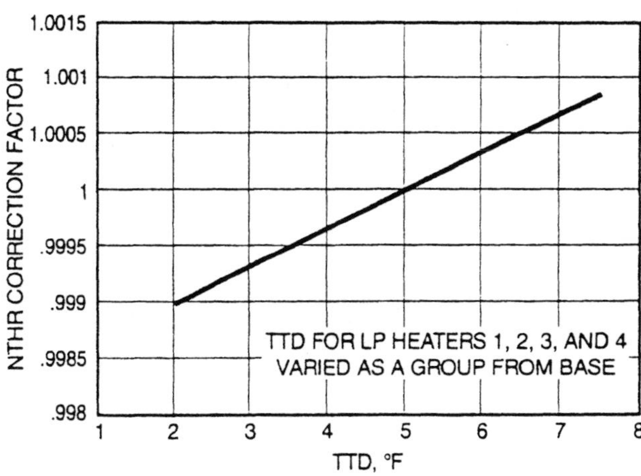

Fig. 13-11. Effect of TTD on net turbine heat rate—Heater 7 (500 MW) cycle, LP heaters 1, 2, 3, and 4.

13.6 IMPACT OF REHEATER SYSTEM PRESSURE DROP

The total reheat system pressure drop includes the pressure drop associated with the cold reheat piping from the HP turbine exhaust to the reheater section of the boiler, the reheater section of the boiler itself, and the hot reheat piping from the reheater to the intermediate-pressure turbine intercept valves. A typical design value for total reheater system pressure drop is 10% of the HP turbine exhaust pressure. Figure 13-13 shows the relative impact of changes to reheater system pressure drop on turbine output and heat rate. For a 1% decrease in reheater pressure drop, the heat rate and output improve approximately 0.1% and 0.3%, respectively.

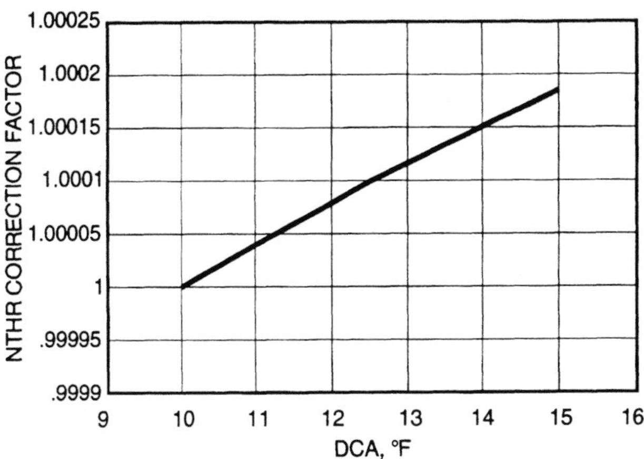

Fig. 13-12. Effect of DCA on net turbine heat rate—Feedwater Heater 7 (500 MW) cycle.

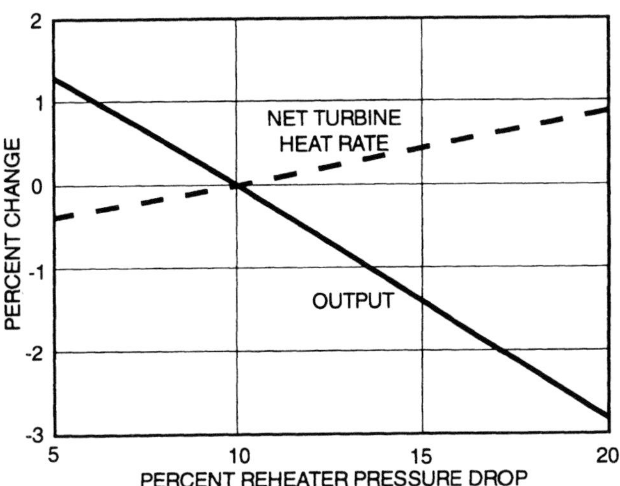

Fig. 13-13. Reheater pressure drop correction factors.

13.7 IMPACT OF EXTRACTION LINE PRESSURE DROP

The extraction line pressure drop is the pressure drop between the turbine stage pressure and the heater shell pressure. For extractions not at turbine section exhausts (HP exhaust and intermediate-pressure exhaust), 6% of the turbine stage pressure is a typical design pressure drop. Three percent is the drop across the extraction nozzle, and 3% is for the extraction piping and valves. For extractions at the turbine exhaust section, no extraction nozzle loss occurs and the total drop is 3%. The higher the extraction line pressure drop, the worse the cycle heat rate. For a 2% increase in extraction line pressure drop for all of the heaters (from 6% to 8%), the change in output and heat rate would be approximately 0.09% poorer.

13.8 IMPACT OF CYCLE MAKEUP

Cycle makeup is necessary to offset cycle water losses, the most significant of which is boiler blowdown. Boiler blowdown is necessary to maintain proper boiler water chemistry and is most commonly expressed as a percentage of throttle flow. Values between 0% and 3% are typical. The makeup water is normally supplied to the condenser hot well.

The makeup water flows through the condensate and feedwater systems, increasing the total flow through the heaters and pumps. This additional flow results in higher feedwater heater thermal duties and therefore higher extraction flows, and higher pump horsepower requirements. The energy that was taken from within the cycle to pump and heat the additional makeup water flow is then wasted in the boiler blowdown. This results in a negative impact on cycle performance. The impact of makeup on net turbine heat rate is approximately 0.4% higher per percent makeup. The impact of makeup on output is approximately 0.2% lower per percent makeup. These values are based on boiler blowdown at saturated conditions at the boiler drum pressure.

13.9 IMPACT OF TURBINE EXHAUST PRESSURE

Except for choked turbine exhaust conditions, lowering the turbine exhaust pressure increases the cycle efficiency as discussed in Section 13.2.3. Figure 13-14 indicates the percent change in turbine heat rate due to variations in steam turbine exhaust pressure. It should be noted that it is difficult to develop an accurate rule-of-thumb for turbine exhaust pressure impacts on performance that is valid for all cycles and turbines. Actual turbine characteristics (last stage blade design and exhaust area) and unit size both affect the impact of changing exhaust pressure on performance.

13.10 IMPACT OF AIR PREHEAT

Prior to entering the steam generator, combustion air enters the air heater and is heated by flue gas leaving the steam generator. The air heater improves boiler efficiency by lowering the flue gas exit temperature. Preheating of the combustion air prior to the air heater is used to keep the flue gas air heater exit temperature above its dewpoint temperature.

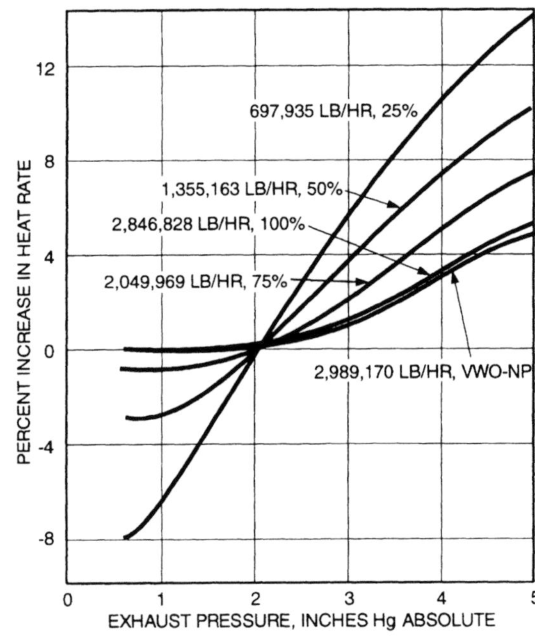

VALUES NEAR CURVES ARE FLOWS AT 2400 PSIG, 1000 F.
THESE CORRECTION FACTORS ASSUME CONSTANT CONTROL VALVE OPENING.
APPLY CORRECTIONS TO HEAT RATE AND kW LOADS AT 2.0 IN. Hg ABS. AND 0.0 PERCENT MU.

THE PERCENT CHANGE IN kW LOAD FOR VARIOUS EXHAUST PRESSURES IS EQUAL TO (MINUS PERCENT INCREASE IN HEAT RATE) 100/(100 + PERCENT INCREASE IN HEAT RATE).

THESE CORRECTION FACTORS ARE NOT GUARANTEED.

THESE FACTORS GIVE CHANGE IN NHTR.

Fig. 13-14. Exhaust pressure correction factors. (From General Electric. Used with permission.)

If the temperature of the flue gas falls below the dewpoint temperature, sulfuric acid forms which will damage the air heater and duct work. LP extraction steam or hot water from the turbine cycle is often used as the preheating source. These heating sources are readily available and minimize the impact on the turbine cycle since the thermodynamic availability of the supply source is low.

The air preheater steam supply is often supplied from the deaerator extraction point which is normally the IP/LP turbine crossover point. If the combustion air preheater has steam coils, crossover steam is used directly and condenses in the preheater. If the air preheater uses hot water, saturated liquid from the deaerator is supplied to the air preheater. The condensate is either pumped back to the deaerator, returned to the condenser, or returned to an intermediate LP feedwater heater point such as the flash tank. The 500 megawatt (MW) plant considered in this chapter for the determination of performance impacts uses saturated water from the deaerator storage tank and pumps this water through a water-to-glycol heat exchanger and back to the deaerator.

Figure 13-15 shows the normalized net turbine heat rate impact versus combustion air preheating duty at the 100% and 50% loads for a 500-MW unit. Also, shown in the figure is the heat rate impact with the air preheater water returned to the condenser. Pump impacts and condenser heat impacts are minor and were not considered.

13.11 IMPACT OF CONDENSATE SUBCOOLING

Condensate subcooling is the cooling of the cycle condensate in the condenser hot well below the saturation temperature corresponding to the turbine exhaust pressure. Condensers are normally specified to provide condensate at the condenser saturation temperature (0° F subcooling). When subcooling occurs, the duty on the first feedwater heater increases, causing the extraction flow to the heater to increase. This decreases the turbine output and increases the turbine heat rate. Figure 13-16 shows the effect on net turbine heat rate at various throttle flows for a 5° F subcooling. Figure 13-17 shows the effects of various amounts of subcooling on net turbine heat rate at different turbine loads.

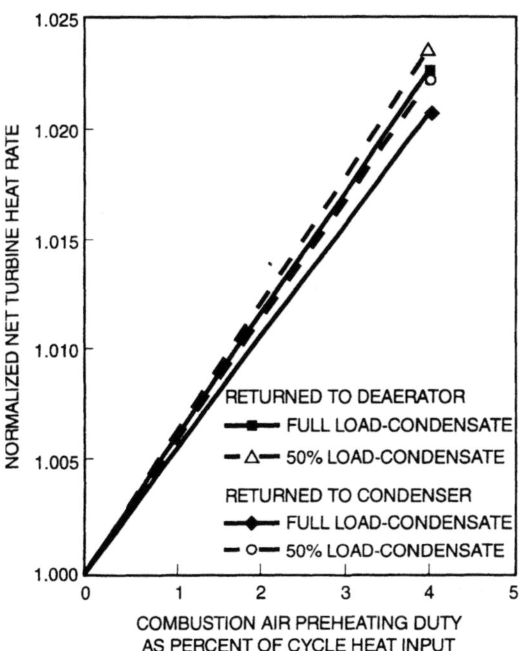

Fig. 13-15. Normalized net turbine heat rate versus combustion air preheating duty.

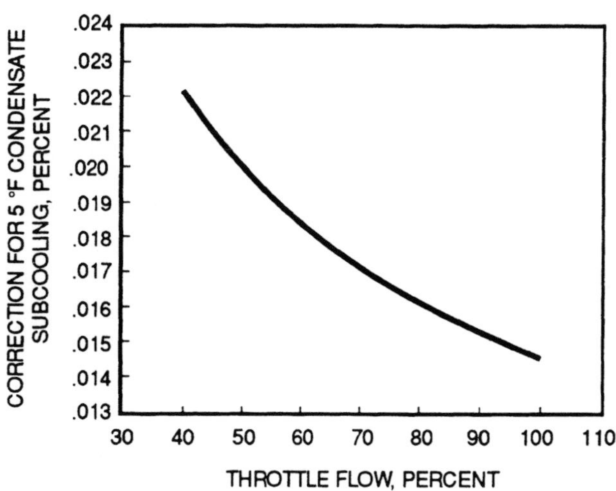

Fig. 13-16. Effect of condensate subcooling on net turbine heat rate.

13.12 IMPACT OF SUPERHEAT AND REHEAT SPRAY FLOWS

One method used to control the main steam and reheat steam temperatures is the use of desuperheaters in the boiler. The source of spray water is typically boiler feed pump discharge for main steam spray, and an interstage bleed off the boiler feed pump for reheat spray. Alternatively, the spray water is taken from after the final feedwater heater. Both main steam

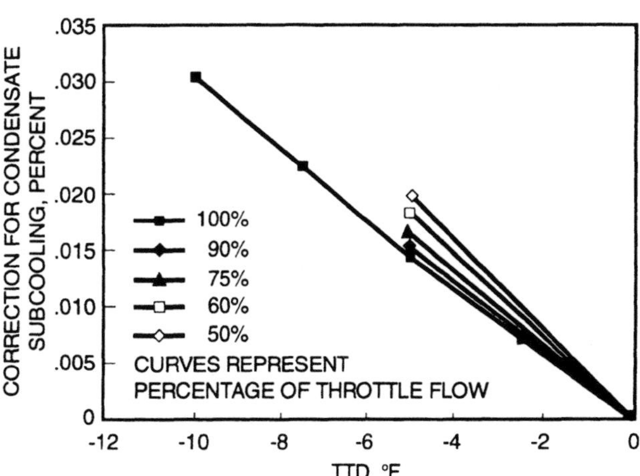

Fig. 13-17. Effect of condensate subcooling for various loads on net turbine heat rate.

and reheat steam spray flows have an adverse impact on turbine heat rate when taken from the boiler feed pump discharge. The reason for this, in the case of main steam spray, is that the spray flow evaporates in the boiler and becomes part of the main steam flow. However, it bypasses the HP feedwater heaters. For this fraction of the main steam flow, the cycle was less regenerative (using only five feedwater heaters). In the case of reheat spray, the impact on cycle heat rate is worse. This is because not only does the cycle become less regenerative, but also the reheat spray flow bypasses the HP turbine and expands only through the reheat turbine section; thus, for the portion of the steam flow that is reheat spray, the cycle is nonreheat. Figure 13-18 shows the effect on turbine output and heat rate for 1% main steam and reheat spray flows at various throttle flows.

13.13 IMPACT OF WET-BULB TEMPERATURE

The ambient wet-bulb temperature has an indirect effect on the cycle output and heat rate. For plants with evaporative cooling towers for cycle heat rejection, the circulating water temperature to the condenser increases as the ambient wet bulb increases. This results in an increase in condenser back pressure that normally adversely affects the turbine output and heat rate. Changes in ambient wet-bulb temperature are very difficult to quantify into a set of performance impacts relationships that are accurate for all plants. In addition to the turbine design features that impact performance versus back pressure, condenser and cooling tower design details also affect performance in ways that cannot be generalized to cover all possible cases. Figure 13-19 shows the impact of changes in wet-bulb temperature to turbine output and heat rate for a specific 500-MW reheat power plant. Impacts of up to 1.5% to 2.0% can occur depending on specific plant design conditions.

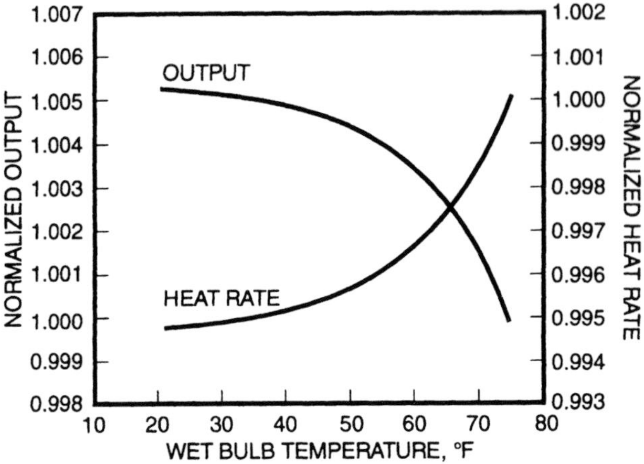

Fig. 13-19. Effect of wet bulb temperature turbine output and heat rate vs. wet bulb temperature.

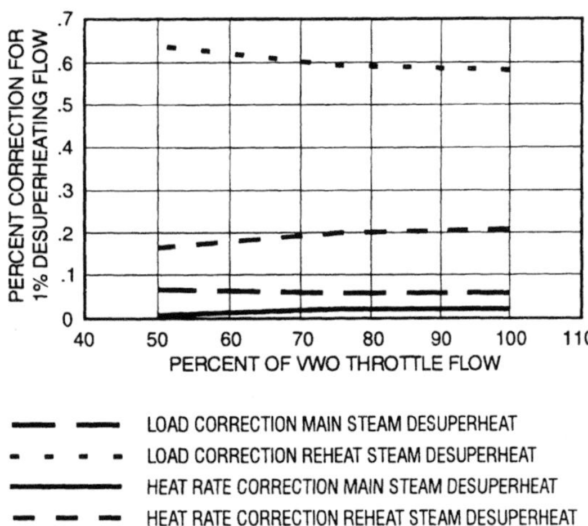

Fig. 13-18. Corrections for main steam and reheat steam desuperheating flow.

13.14 IMPACT OF REMOVING TOP HEATERS

Feedwater heaters may need to be removed from service due to tube leaks. Removing the top heater(s) from service eliminates turbine extraction for these heaters and increases steam flow through the remaining sections of the turbine. For a given throttle flow, turbine output increases because of the greater flow through the turbine and turbine cycle heat input increases because of the lower final boiler feedwater temperature. The turbine and plant heat rates are poorer when removing the top heaters from service.

Some power plants are designed for removal of the top feedwater heaters to increase net plant output. In this case, the boiler is specified to produce the same throttle steam rate (maximum continuous rating) with the lower final feedwater temperature and higher heating duty. The turbine would need to be designed to accommodate the higher HP turbine exhaust pressure, increased shaft power requirements in the IP and LP turbines, increased electrical power generation, and increased steam flow in the LP turbine last stage. The traditional industry limit on steam loading on the last row of blades in the LP turbine has been 15,000 lb/h/ft^2 (the area, ft^2, is the total annulus area of the last stage). The recent trend has been to relax this criterion to as high as 18,000 lb/h/ft^2 for specific applications. If the turbine specification requires increased output with removal of the top feedwater heaters, the manufacturer may have to select a larger last stage blade than optimal. For existing units, the steam loading limit on the last row of blades may prohibit increased output. The engineer or operator should check with the turbine manufacturer's literature or contact the manufacturer directly for limitations on operation with heaters removed from service. Unless the turbine was designed for greater capacity with heaters removed from service, the steam turbine manufacturers will have load restrictions when operating with feedwater heaters removed from service.

Table 13-2 shows the results of an analysis that removed the top feedwater heater (Heater 7) of a 500-MW unit. The circulating water inlet temperature is assumed constant in the analysis. With the heater removed from service, the HP shaft power decreases because turbine expansion is reduced as a result of the higher exhaust pressure caused by the larger cold reheat flow. However, the IP and LP turbine flow increases significantly. The turbine output increased from 522,316 kW to 550,129 kW, which is an increase of 27,813 kW. The net turbine heat rate increased 135 Btu/kWh (8,136 − 8,001 Btu/kWh). The final feedwater temperature decreased from 481.8 to 412.9° F, which contributed to the 297 MBtu/h (from 4,179 to 4,476 MBtu/h) increase in the turbine cycle heat input requirements. The increased steam flow through the IP and LP turbines results in an increased cycle heat rejection of 201 MBtu/h (from 2,373 to 2,574 MBtu/h). Turbine steam loading on the last row of LP blades increased from 14,233 lb/h/ft^2 to 15,459 lb/h/ft^2.

13.15 IMPACT OF HP HEATER DRAIN PUMP

For a typical seven feedwater heater turbine cycle, Heater 7 drains cascade to Heater 6. Heater 6 drains cascade to the deaerator. At high loads, enough pressure difference exists between Heater 6 and the deaerator for flow to overcome the elevation difference between the two heaters (Heater 6 and the deaerator). However, as the throttle flow decreases, the pressure difference between the two heaters also decreases. Typically, at around 50% load, the pressure difference between the heaters is not enough to drive the drains to the deaerator. An HP heater drains pump is needed to pump the heater drains back to the deaerator. One alternative is to dump heater drains to the condenser. Another alternative is to flash the heater drains to the next lower pressure closed feedwater heater. The net turbine cycle heat rate improvement with using an HP heater drains pump is 60 Btu/kWh at 50% load compared to the case of returning the drains to the condenser. At the 50% load condition, a HP heater drain pump improves net turbine cycle heat rate by 10 Btu/kWh compared to flashing the drains to the next lower pressure heater.

Table 13-2. Effect on Turbine Cycle Performance with Removal of Top Feedwater Heater from Service

Parameter	Case: All Heaters in Service	Case: Heater 7 Out of Service
High-pressure turbine shaft power, kW	151,440	142,823
Intermediate- and low-pressure turbine shaft power, kW	379,583	416,512
Generator and mechanical losses, kW	8,707	9,206
Net turbine output, kW	522,316	550,129
Net turbine heat rate, Btu/kWh	8,001	8,136
Final feedwater temperature, °F	481.8	412.9
Turbine cycle heat input, MBtu/h	4,179	4,476
Turbine cycle heat rejection, MBtu/h	2,373	2,574
Steam loading on last row of LP blades, lb/h/ft^2	14,233	15,459

Notes: (1) Assumes constant circulating water temperature and constant throttle flow.

(2) Assumes valves-wide-open, normal pressure throttle flow. In actuality, the throttle flow passing capability when removing Heater 7 from service would be decreased by approximately 0.65% because of an increase in the turbine first-stage pressure. This decrease in throttle flow would result in a corresponding decrease in turbine output.

13.16 FEASIBLE TYPES OF BOILER FEED PUMP DRIVES AND FLOW CONTROLS

The turbine cycle is impacted by the type of boiler feed pumps (BFPs) and method of flow control. The four alternate pump drive and control concepts typically considered for the boiler feedwater pumping system are as follows:

- Constant-speed, motor-driven BFPs with throttle valve control,
- Motor-driven BFPs with variable-speed coupling control,
- Variable-speed, motor-driven BFPs, and
- Turbine-driven BFPs.

Another alternative pump arrangement is a main turbine shaft-driven pump arrangement that uses an extension of the main turbine shaft connected to the pump. The shaft extension can be either on the HP side of the turbine through the front standard or on the LP side through the generator and exciter. Both arrangements have been used in the past but increase the cost and complexity of the main turbine generator unit. Because of the speed differences between the turbine and pump, a gear and coupling are required. The shaft-driven arrangement does not have the lowest capital costs or the lowest operating costs and will not be considered further.

13.16.1 Constant-Speed, Motor-Driven Boiler Feed Pump with Throttle Valve Control

This arrangement consists of a BFP connected to a constant-speed motor. If the pump has a design speed greater than the motorspeed, a gear is required. Flow control is accomplished by a control valve. A typical pump head curve and a typical system resistance curve for this type of arrangement are shown in Fig. 13-20. The system resistance curve represents the sum of the static head and friction losses of the piping, valves, and equipment. The friction loss across the control valve is shown as a separate item to aid in explaining one of the differences between constant-speed and variable-speed pump operation. The design head that the pump must develop at the design flow rate is the sum of the system resistance and the pressure drop through the control valve at the design flow rate.

Those arrangements that use variable-speed pump drives

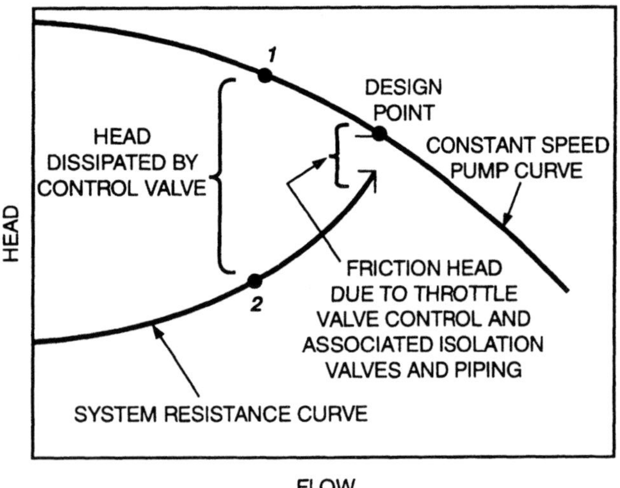

Fig. 13-20. Throttle valve control.

for flow control do not require control valves other than a start-up control valve. The design head for a variable-speed pump is less than the design head for a constant-speed pump by the amount of pressure drop across the control valve.

The method of controlling flow for the constant-speed arrangement is illustrated in Fig. 13-20. At flows less than the design flow, the static head produced by the pump is greater than the system resistance. For example, Point 1 on this figure is the head developed by the pump at a flow rate less than the design flow. Point 2 is the system resistance at the same flow. To achieve this flow rate, the difference in pressure between Points 1 and 2 must be dissipated by the control valve. The amount of pressure that must be dissipated by the control valve represents a loss (or inefficiency) associated with this type of control.

The constant-speed pumping arrangement is relatively simple in design and relatively easy to control. The capital investment required to install this arrangement is less than the cost of a variable-speed arrangement, but the operating costs are higher, particularly for low-load operation.

13.16.2 Motor-Driven Boiler Feed Pump with Variable-Speed Coupling Control

This arrangement consists of a BFP driven by a constant-speed motor via a variable-speed coupling (and integral speed increasing gear, if required). Flow control is accomplished by varying the amount of slip of the variable-speed coupling, which in turn varies the pump speed.

Two basic types of variable-speed fluid couplings are normally considered for the BFPs: hydraulic (hydrokinetic) and hydroviscous. The hydraulic coupling has the largest service record of the two types. In the hydraulic type fluid coupling, a runner attached to the driven shaft receives a vortex of oil from the impeller that is attached to the driving shaft. No mechanical connection exists between the runner and the impeller, and the two are almost identical in shape. Kinetic energy is imparted to the oil by the impeller, and the oil flows radially outward and into the vanes of the runner. The oil then flows through the runner and transfers its energy to the runner, similar to the process of steam imparting energy to a turbine blade. The output speed and the energy transferred are controlled by the amount of oil in the circuit. The oil flow from the impeller is regulated by a controllable scoop tube that extracts or feeds oil into the working circuit.

The hydroviscous fluid coupling operates on the principle of shearing an oil film to transmit torque. This coupling uses fluid shear instead of the momentum of the fluid (as in the hydraulic type) over most of the operating range to transmit torque. The amount of torque that can be transmitted varies with the number of disc surfaces over which the shearing action occurs. To increase the number of working surfaces, a hydroviscous disc pack is built up by alternatively stacking discs splined to the input shaft between discs splined to the output shaft. All discs are free to move axially but must rotate with the member to which they are splined. The controllable clamp force is applied to one end of the disc pack and is distributed to the remainder of the discs because of their freedom to slide axially. The working oil is introduced to each set of surfaces to set up the hydrodynamic film between each of the discs. The working oil is continuously circulated through the disc pack to dissipate the heat generated by the shearing action. By varying the clamping force on one end of the disc pack, the input-to-output speed ratio can be controlled from very low output speed to full no-slip output speed.

The major difference between the two types is that the hydroviscous coupling is more efficient at the design speed because it can be locked up with no slip losses. The hydraulic coupling has a definite amount of slip at rated speed/load (typically 2% to 3%), and therefore some slip losses are incurred.

The hydraulic coupling will be used for comparative purposes because of its more extensive service record.

The effect of varying the pump speed on the pump head curve is shown in Fig. 13-21. As the speed of the pump is

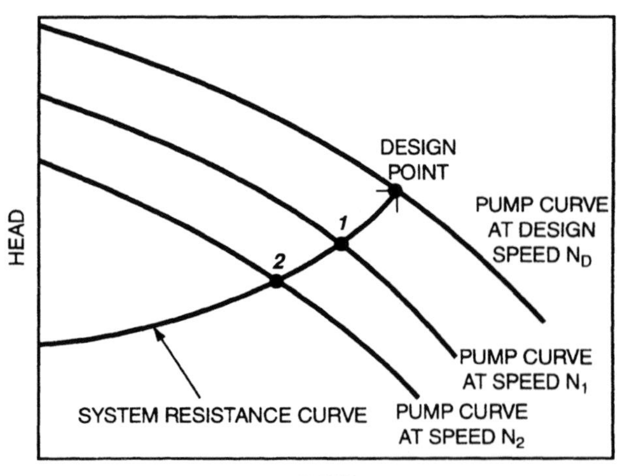

Fig. 13-21. Variable speed control.

reduced, the pump head curve shifts to a lower position, with the shape of the curve remaining generally the same as the shape of the curve at the design speed. Three pump head curves for the same pump operating at three different speeds are shown in Fig. 13-21. Also shown is the system resistance curve for the feedwater system. This is the same resistance curve as shown in Fig. 13-20.

Two flow conditions which illustrate the advantage of the variable-speed arrangement are shown in Fig. 13-21. If Point 1 (or Point 2) is the desired flow, the pump speed is reduced to speed N_1 (or N_2), such that the head of the pump matches the system resistance. No excess head is developed and no energy is dissipated by a throttle valve; hence, the horsepower required at this flow rate for the variable-speed arrangement is less than it would be for a constant-speed pump with a throttle valve. In addition, the pump efficiency is closer to the design value with variable-speed operation than with constant-speed operation. Not all of this savings in pumping power is realized because the variable-speed coupling incurs power losses. However, the net horsepower required for the variable-speed arrangement is lower at partial loads than it is for a constant-speed pumping arrangement with throttle valve control.

13.16.3 Variable-Speed Boiler Feed Pump with Adjustable-Frequency Motor

This arrangement consists of a variable-speed BFP driven by an adjustable-frequency motor. Flow control is accomplished by controlling the frequency and line voltage supplied to the motor, which in turn adjusts the motor speed. The motor is connected to the pump through a resilient coupling which reduces torsional harmonics transmitted by the pump. The effect of varying the pump speed on the pump head curve is shown in Fig. 13-21 and the discussion regarding pump head versus pump speed of Section 13.16.2 is applicable to this type of drive.

This type of pump control offers reduced electrical consumption at all loads compared to throttle valve and variable-speed coupling control. As in the case of variable-speed coupling control, pump speed regulation allows the pump to operate with little change in efficiency across the unit load range in comparison to constant-speed pumps.

The adjustable-speed motors used are synchronous motors regulated by adjustable frequency control systems. The synchronous motor speed is varied from rated speed to a very low speed by reducing the frequency and line voltage imposed on the motor terminals. The ac current, line voltage, and frequency are controlled and changed to any value desired through the use of static electronic control converters.

This power conversion system consists of two three-phase thyristor bridges with the dc terminals connected through a reactor. The first stage three-phase bridge is connected to the ac power supply and operates as a phase-controlled rectifier to supply power to a dc link. The second bridge is connected to the synchronous motor and inverts the power from the dc link to the stator of the motor. A power circuit diagram of a 12-pulse synchronous motor drive system is presented in Fig. 13-22.

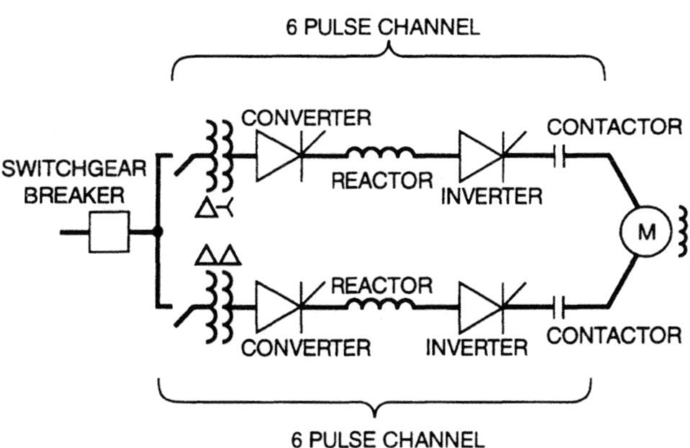

Fig. 13-22. Adjustable speed electric motor drive power circuits.

By using solid-state electronic components to control motor speed, the motor efficiency can be maintained at high levels across the entire motor operating range.

13.16.4 Turbine-Driven Feed Pump

A condensing boiler feed pump turbine (BFPT) takes extraction steam from the IP/LP turbine crossover point or a nearby extraction point and exhausts to the main condenser or an auxiliary condenser. Since some crossover steam flow is diverted to the condensing BFPT, the amount of steam flow through the last stages of the LP turbine is reduced. At rated load, lower flows to the last stage in the LP turbine can reduce overall turbine exhaust losses and therefore improve turbine heat rate. The more heavily loaded the low pressure turbine, the more beneficial it is to have a condensing BFPT. Improvements of 60 Btu/kWh or more are possible when utilizing a condensing BFP turbine. This arrangement allows for greater electrical output from a maximum LP turbine end-loaded machine.

For the purposes of comparison with other BFP drives, the condensing BFP turbine will take supply steam from the main turbine crossover and will exhaust to the main condenser. This is the most common BFPT arrangement; however, other arrangements are possible, including noncondensing arrangements that exhaust to feedwater heaters.

At low-load operation, main steam is required to supplement the crossover steam. To accommodate both steam supplies, the BFPT has a dual inlet (admission). Crossover steam is delivered to the LP admission which has a stop valve and dedicated governor control valves to control steam flow to the BFPT. Main steam is delivered to the HP admission, which has its own stop and control valve to supplement the LP admission supply, as required.

As unit load is decreased from rated load, additional volumetric flow passing capability in the LP admission is

Table 13-3. BFP Brake Horsepower Requirements and the Net Electric Power Consumption for Various BFP Drive Types

	100% Throttle Flow		75% Throttle Flow		50% Throttle Flow	
Type of Pump Drive and Control	bhp	kWe	bhp	kWe	bhp	kWe
Constant-speed motors with throttle control	17,484	14,270	15,970	13,174	14,459	12,057
Constant-speed motors with variable-speed hydraulic couplings	14,924	14,016	10,916	11,355	9,185	10,239
Variable-speed BFPs with adjustable-frequency motors	14,924	12,557	10,916	9,683	9,185	8,703
Turbine-driven BFPs	14,930	10,737	10,917	8,573	9,182	8,145

required. The LP admission has a larger admission area and larger and possibly more governing control valves. Typically, the BFPT is specified to be capable of meeting the boiler feed pump power requirements on LP steam down to 40% load. When decreasing the unit load from rated load during constant pressure operation, the main turbine valves are closing, and the BFPT LP admission valves are opening. When the LP admission valves are completely open, main steam is used to supplement the BFPT motive steam supply requirements.

13.16.5 Comparison of Alternative Pump Drive and Control Concepts

A comparison of the four alternative pump drive and control concepts for the 500-MW unit has been developed. The plant has two 50% capacity BFPs.

The analysis assumes the same 5,150 rpm design pumps are used. Therefore, a gear speed increaser with an efficiency of 98.5% is assumed for the alternatives which use 1,800 rpm synchronous motors.

Table 13-3 shows the BFP brake horsepower requirements and the net electric power consumption for three turbine throttle flow rates. Net electrical power consumption includes all associated generator, transformer, motor, frequency converters, hydraulic coupling, and gear losses as appropriate.

The constant-speed motor with throttle control alternate has the highest brake horsepower and electrical power consumption. This is because of the throttling losses across the control valve and pump inefficiencies associated with operating off the design pump efficiency point. The BFP brake horsepower requirements for the constant-speed motors with hydraulic couplings alternative, variable-speed BFP with adjustable frequency motors alternative, and the auxiliary condensing turbine drives alternative are essentially identical. The second highest net electrical power consumption is the constant-speed motors with hydraulic coupling alternative due to the losses in the hydraulic couplings. The variable-speed BFP with adjustable frequency motors alternative has the lowest net electrical power consumption of the electric motor-driven feed pump alternatives.

Table 13-3 shows net electrical power consumption for the BFPT. This is represented as a reduction in turbine power output calculated as the difference in the gross turbine electrical output between the variable-speed motor-driven pump option and the condensing BFPT alternative. The BFPT has the lowest net electrical power consumption because of the lower turbine exhaust losses in the main turbine and the absence of transformer, motor, frequency converter, coupling, and gear losses.

Table 13-4 shows the net turbine heat rate and net turbine output at three turbine throttle flow rates for the four alternatives. The constant speed motor with throttle control alternate has the highest net turbine heat rate (NTHR) and lowest net turbine output (NTO) at loads less than 100%. This is because of the throttling losses across the control valve and pump inefficiencies associated with operating off the pump design efficiency point. The NTHR is the second highest for the constant-speed motor with hydraulic coupling alternative at loads less than 100%. This alternative is the second highest because of the losses in the hydraulic coupling for the 75% and 50% throttle flow cases. The net turbine output is higher with the variable-speed coupling

Table 13-4. Net Turbine Heat Rate and Net Turbine Output for Various BFP Drive Types

	100% Throttle Flow		75% Throttle Flow		50% Throttle Flow	
Type of Pump Drive and Control	NTHR (Btu/kWh)	NTO (kW)	NTHR (Btu/kWh)	NTO (kW)	NTHR (Btu/kWh)	NTO (kW)
Constant-speed motors with throttle control	8,005	522,839	8,043	407,544	8,320	274,712
Constant-speed motors with variable-speed hydraulic couplings	8,010	522,607	8,030	408,396	8,302	275,587
Variable-speed BFPs with adjustable-frequency motors	7,988	524,066	7,997	410,068	8,256	277,123
Turbine-driven BFPs	7,960	525,886	7,975	411,178	8,239	277,681

than with the throttling control, but is not as high as the difference in net electrical power consumption (Table 13-3). The reason for this is that some of the losses due to BFP inefficiencies are recovered as additional heat input to the feedwater, which reduces feedwater heating requirements. The net turbine heat rate at 100% throttle flow is slightly better for the constant-speed motor with throttle control than for the variable-speed hydraulic coupling because the drive power consumption is lower. This is because the pump efficiency is not far from the design for the throttle control case and the inefficiencies heat the feedwater and are not losses to the surroundings which is the case for losses in the hydraulic coupling.

The second best heat rate at rated load is with variable-speed BFPs with adjustable frequency motors. As the load decreases, the heat rate with this arrangement approaches the turbine-driven BFP arrangement since the advantage of the BFPT alternative's reduced main turbine exhaust losses decreases with load (the turbine becomes less loaded). In addition, the efficiency of the adjustable frequency motors is relatively constant over the load range.

The turbine-driven boiler feed pump arrangement offers the best net turbine cycle heat rate and turbine output.

13.17 DEFINITION/DESCRIPTION OF THE CYCLE HEAT BALANCE

Steam cycle performance is typically represented in a heat balance diagram. This diagram shows the steam/condensate flow streams associated with the power generation cycle. In addition, pressures, temperatures, and enthalpies of the various flow streams are also represented. A complete heat balance provides enough information to balance the energy distribution within the power station steam cycle.

Heat balances also typically indicate the turbine expansion line end point and used energy end point of the steam as it exhausts to the condenser. Generator losses and net generator electrical output are normally indicated. On the basis of this information, the engineer can perform an energy balance for the major equipment associated with the turbine, feedwater, condensate, and heat rejection systems.

The indicated flows, pressures, and temperatures are also used to determine equipment design conditions. Typical design criteria for feedwater heaters, feed pumps, condensate pumps, and condensers are discussed in other chapters. In general, the equipment is designed to the valves-wide-open, 5% overpressure load condition. However, given that power cycles can include process steam flows which may alter the mass and energy distribution within the steam cycle, careful selection of the particular heat balance condition needs to be made to ensure equipment sizing appropriate for all anticipated power station operating modes.

Heat balances typically are provided by the turbine manufacturer. In addition, numerous heat balance computer programs are commercially available to permit the evaluation of steam power cycles. If adequate information is available to characterize the steam turbine performance at various load conditions, the programs can predict the performance of power station output with good accuracy.

Heat balances can also be performed by hand. Hand calculations can be very time consuming given the need to iterate to a solution and determine various enthalpies and other steam properties throughout the cycle. However, a hand calculation is instructive because it permits the engineer to gain an understanding of the interrelationships of the various pieces of equipment within a power cycle. In addition, hand calculations are often the most convenient method of estimating performance impacts. These calculations can later be refined through consultation with the turbine manufacturer or through the use of computer modeling. An example of a hand calculation is provided in Appendix 13A. This calculation presents the use of the thermal kit data to develop the thermal performance of the power cycle.

13.18 TURBINE THERMAL KIT DESCRIPTION AND USE

The turbine thermal kit is provided by the turbine manufacturer and consists of numerous characteristic curves and values that are used to determine the performance of the steam turbine for various steam cycle conditions. These curves are used during the development of a computer-based steam cycle performance model or to perform hand calculations of steam turbine performance.

In addition, the turbine thermal kit includes correction curves that can be used to adjust actual turbine test data to design or guaranteed turbine performance conditions. These correction curves facilitate the comparison of actual performance to guaranteed performance. The turbine manufacturer should supply a complete set of these curves to permit the adjustment of all cycle parameters that may vary between guarantee conditions and actual operating conditions. These correction curves should be obtained and their use understood prior to conducting the performance test. In addition, turbine test procedures should be developed and agreement reached on their use prior to testing. These procedures should illustrate methods of adjustment to reference conditions.

A typical turbine thermal kit is shown in Figs. 13-23 through 13-40. These curves are based on a 415-MW General Electric steam turbine. The presentation of the data is typical for General Electric. Other manufacturers' thermal kits may vary; however, the type of information and its presentation are representative. The turbine is a reheat, tandem compound, four-flow condensing design with 26-in. last stage buckets and rated steam conditions of 2,400 psig main steam pressure and 1,000° F temperature with reheat to 1,000° F (2,400 psig/1,000° F/1,000° F). The steam cycle arrangement consists of seven stages of feedwater heating and turbine-driven BFPs. The thermal kit consists of the following characteristic and correction curves:

- Characteristic curves:
  —Extraction stage shell pressures versus flow to the following stage,
  —Gland leakage and mechanical losses,
  —Expansion lines,
  —HP turbine internal efficiency,
  —HP turbine expansion line end points,
  —Reheat turbine internal efficiency,
  —Reheat turbine expansion line end points,
  —Correction to expansion line end points,
  —Exhaust loss curve,
  —Generator losses, and
  —First-stage shell pressure versus throttle flow.
- Correction curves:
  —Initial pressure correction,
  —Initial temperature correction,
  —Reheat pressure drop correction,
  —Reheat temperature correction, and
  —Exhaust pressure correction factors.

13.18.1 Characteristic Curves

The characteristic curves and related performance characteristic values are discussed as follows.

To determine the extraction stage shell pressure, multiply the following factors by:

$$\frac{\text{Flow to the Following Stage}}{10^6}$$

Extraction No.	1	2	3(XO)	4	5	6	7
Factor	210.3	110.1	59.3	23.87	8.92	4.99	2.59

Note:

1. Apply the factor for Extraction No. 1 to the flow to the first reheat stage to determine the pressure ahead of the intercept valve. High-pressure turbine exhaust flange pressure equals pressure ahead of the intercept valve divided by 0.9.
2. Flow to following stage equals throttle flow minus leakages and all extractions from preceding stages and stage in question plus any steam returned to the turbine ahead of the stage in question.
3. Use 3 percent pressure drop between drop between shell stage pressure and turbine shell flange pressure except where extractions are at the end of a shell casing.
4. These factors are to be applied for heat balance calculations at rated steam temperatures only.
5. Extraction pressures during normal operation can be greater than those calculated at maximum expected throttle flow because of manufacturing tolerances on nozzle areas, variations in flow coefficients, deposits in the steam path, etc. To allow for this, the General Electric Company recommends that extraction piping and feedwater heaters be designed for pressures at least 15 percent greater than those calculated at maximum throttle flow conditions. It is likely that there will be abnormal operating conditions which will result in pressures greater than those defined above, which should be used as the actual extraction piping and feedwater heater design pressures.

Fig. 13-23. Extraction stage shell pressure. (From General Electric. Used with permission.)

13.18.1.1 Extraction Stage Shell Pressure Extraction stage shell pressures are shown tabulated in Fig. 13-23. These factors represent the proportionality of shell pressure and flow to the following turbine stage. Extraction stage shell pressures can also be drawn as curves. For reheat units, these curves are typically straight lines passing through the origin. However, extractions that are followed by turbine stages that do not operate at a constant pressure ratio are exceptions. An example is the lowest pressure turbine extraction, which is influenced by LP turbine exhaust pressure. Another limitation of these proportionality constants, or extraction pressure versus flow to the following stage curves, is their assumption of rated steam temperatures.

A constant-pressure flow relation exists that can be used to account for deviations in steam temperatures. This may be derived from the manufacturer's heat balances. This stage flow coefficient, or flow factor, is also limited by its inability to represent pressure feedback effects because of variations in stage group pressure ratio. However, it is noted that small deviations in heater pressure for the lowest feedwater heater are usually insignificant with respect to overall plant performance. Situations when the assumption of a constant extraction flow factor becomes invalid include extractions upstream of a feedwater heater extraction that has been removed from service or locations where significant admission or extraction flows disrupt the normal mass distribution within the steam turbine. The turbine flow factor is calculated using the following equation:

$$C = \frac{m}{\sqrt{\frac{P}{v}}} \tag{13-3}$$

where

C = flow factor constant;

P = extraction stage shell pressure, psia;

v = specific volume of steam at shell enthalpy and pressure, ft³/lb; and

m = mass flow past the extraction, lb/h.

13.18.1.2 Gland Leakage and Mechanical Losses Gland leakage packing constants are shown tabulated in Fig. 13-24. The description on the determination of the leakage flow and its enthalpy are also provided. Determination of the leakage from the control stage shell to the IP turbine (via the midspan N2 packing) involves the calculation of the first-stage enthalpy and pressure. These values may also be determined from inspection of the heat balances and through the use of the first-stage pressure curve. The first-stage pressure must be corrected for variation in throttle temperature as indicated in Fig. 13-24.

Figure 13-24 also includes additional information regarding the pitch diameter of the first stage. The pitch diameter is the diameter of a turbine stage measured to the midpoint of

To determine the leakage flow in lb/h, multiply the following packing constants by the $\sqrt{P/V}$ (ahead of the leakage). The leakage through the first packing in a series will be the sum of the flows to the packing leakoffs following it.
Refer to heat balance diagram for identification of leakages.

Leakage No.	1+2	2	3	4+5	5	6
Packing Constant	56	50	500	620	970	600

When expansion line end points from Figures 13-28 and 13-30 are used in heat balance calculations, the leakages shown on this sheet must be used in the feedwater heating cycle.
The 1st of 2 leakages can be determined by subtracting the 2nd leakage from the sum of both leakages and the pressure ahead of the 2nd leakage is the extraction stage pressure to which the 1st of the 2 leakages connects into.

Use throttle enthalpy for leakages 1 and 2.
Use high-pressure turbine exhaust enthalpy for leakages 4 and 5.
Enthalpy for leakage 6 can be determined from turbine exapansion lines at crossover pressure.
5,200 lb/h are required by the steam seals. When leakages 2, 5, and 6 combined are less than 5,200 lb/h, throttle steam must be used. (Steam seal flow to steam packing exhauster is 2,800 lb/h at all loads.)

To determine the first stage enthalpy and pressure used to calculate leakage 3, use the following table:

Throttle Flow, lb/h	2,989,170	2,846,829	2,051,529	1,356,349	698,889
1st Stage Enthalpy, Btu/lb	1,433.2	1,427.9	1,404.6	1,389.7	1,382.1
1st Stage Shell Pressure, psia	1,845	1,750	1,237	807	413

Throttle flow ratio at first admission = 0.35
First stage pitch diameter = 41
For other than rated initial temperature, read the first stage shell pressure curve drawn through these points at a flow corrected for $\sqrt{\frac{\text{V DESIRED}}{\text{V RATED}}}$ where V is initial specific volume cu. ft/lb.
Turbine Generator Mechanical Losses
1. 2,264 kW fixed losses are not included in the 3,600 rpm generator loss curve.

Fig. 13-24. Gland leakage and mechanical losses. (From General Electric. Used with permission.)

the buckets. This value can be used to determine first stage wheel velocity by use of the following relation:

$$W = (\text{PDIA} \times \text{RPM} \times \pi)/(12 \times 60) \qquad (13\text{-}4)$$

where

W = wheel speed, ft/s;
PDIA = first-stage pitch diameter, in.; and
rpm = revolutions per minute.

Figure 13-24 also provides the equivalent throttle flow ratio at the first admission. This flow ratio represents the amount of valves-wide-open throttle flow that occurs when the first admission to the turbine is fully open. The equivalent throttle flow ratio is calculated by use of the following equation:

$$\text{TFR}_{eq} = [m_{\text{off-des}}/m_{\text{design}}]\sqrt{\frac{\left(\frac{P_{\text{design}}}{v_{\text{design}}}\right)}{\left(\frac{P_{\text{off-des}}}{v_{\text{off-des}}}\right)}} \qquad (13\text{-}5)$$

where

m_{design} = design throttle flow, lb/h;
$m_{\text{off-des}}$ = throttle flow, lb/h;
P_{design} = design turbine throttle pressure, psia;
$P_{\text{off-des}}$ = throttle pressure, psia;
v_{design} = specific volume at design throttle pressure and temperature, cu. ft^3/lb; and
$v_{\text{off-des}}$ = specific volume at throttle pressure and temperature, ft^3/lb.

Figure 13-24 also provides the turbine generator fixed (mechanical) losses. These losses account for bearing friction, main shaft driven oil pump losses, and other miscellaneous losses. Since the steam turbine rotational speed is constant, these losses are constant throughout the turbine operating range.

13.18.1.3 Expansion Lines Turbine expansion lines are drawn on Mollier diagrams as shown in Fig. 13-25. These expansion lines depict the thermal state of the steam as it expands through the turbine. These lines may be developed

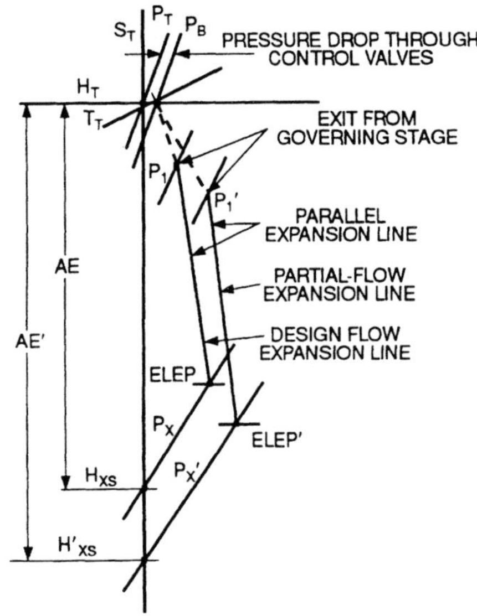

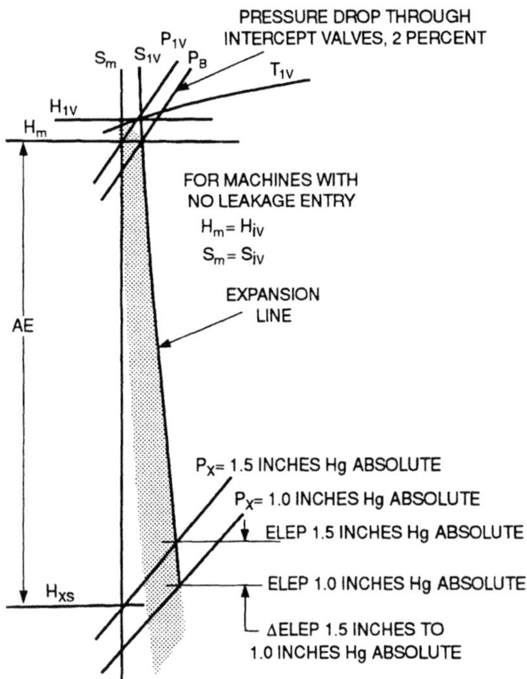

Fig. 13-25. Steam turbine expansion lines. (From ASME. Used with permission.)

based on the throttle, governing stage, and reheat conditions to determine the steam enthalpy at the various extraction points on the turbine. These lines are used in conjunction with a heat balance and the extraction stage shell pressure curves or constants to establish the extraction pressure at which to read the expansion line enthalpy for a given extraction point in the turbine.

Care should be taken to ensure reading the expansion lines at the turbine shell pressure, not the extraction nozzle pressure or feedwater heater operating pressure. Extraction nozzle pressures are typically 2% to 3% lower than the shell pressure. Heater operating pressures are typically 3% to 5% lower than the nozzle pressure. If extraction steam is supplied from the exhaust of a HP or IP turbine, a nozzle pressure drop is typically not included.

In a thermal kit, the expansion line for the HP turbine is typically shown only for the steam expansion downstream of the first stage. The expansion that takes place in the first stage can also be shown on the Mollier diagram. A pressure drop from turbine throttle conditions of approximately 3% is usually indicated to depict pressure losses between the main steam stop valve and the HP turbine bowl conditions. The assumption of 3% is typical for turbine operation in partial arc admission mode. The steam entering the turbine then expands from the HP turbine bowl conditions to the exhaust conditions of the first stage. Turbine heat balances developed on the basis of this assumption are considered to be on a "locus-of-valve point" basis. These heat balances depict heat rates assuming an infinite number of small valves admitting steam to the turbine where each small valve has a 3% pressure drop.

Actual turbine cycle performance is shown on a "valve loop" basis heat rate curve. This curve reflects the steam throttling effect as the steam passes through a partially closed steam admission. The throttling pressure drop reduces the available energy of the steam as the throttled admission steam expands across the control stage. Depending on the steam turbine manufacturer, curves of heat rate effect due to control valve position are provided in the thermal kit.

An alternative method of representing turbine heat rate impact due to turbine valve losses at part load is by a "mean-of-valve-loop" method. This method is an approximation of the heat rate impact illustrated on the "valve loop" basis curve and represents a mean of the turbine heat rate and passes through the valve loop curve. The curve is a smooth curve. Typical heat rate versus output curves for each of these methods are shown in Fig. 13-26.

For units operating with constant throttle pressure in partial arc admission mode, the pressure ratio through the control stage is not constant. As a result of the variation in pressure ratio, the available energy across the stage and the

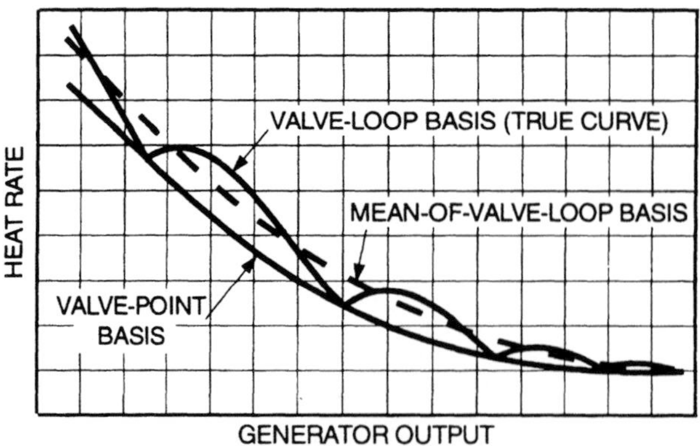

Fig. 13-26. Typical heat rate curve showing various methods of representing the control-valve throttling losses.

control stage efficiency vary with throttle steam flow and conditions. Therefore, expansion lines at different flow conditions for the control stage are not parallel to one another. However, HP turbine stages downstream of the control stage operate with essentially constant pressure ratio, and their expansion efficiency is essentially constant. Therefore, at lower steam flows, the expansion line of the HP turbine stage group downstream of the control stage is typically depicted as a straight line that is drawn parallel to the valves-wide-open expansion line.

The expansion line for the reheat turbine (IP and LP turbine sections) typically includes a 2% pressure drop between the reheat stop valve inlet and IP turbine bowl to account for the pressure drop across the intercept/stop valves. In addition, for combined HP/IP turbines, the steam leakage from the HP turbine is mixed with the hot reheat steam to determine the reheat bowl enthalpy. The steam then expands through the IP turbine to the LP turbine. Steam is often conveyed to the LP turbine through a crossover pipe. A 2% allowance for crossover pipe pressure drop is typically included by the turbine manufacturer to determine the LP turbine inlet conditions. The steam then expands through the LP turbine to the condenser.

On the Mollier diagram, the turbine expansion line is drawn to a point which is termed the expansion line end point (ELEP). This point is plotted at the turbine back pressure used as the basis of the heat balance and represents a complete expansion of the steam to the condenser pressure before the additional losses associated with the turbine exhaust. The steam leaving the LP turbine never actually reaches this steam condition since there is a leaving loss, or exhaust loss, associated with the steam as it exhausts the turbine. The actual exhaust condition, referred to as the used energy end point (UEEP), is calculated as the sum of the ELEP and the exhaust loss. Since the stages upstream are unaffected by the exhaust losses, the expansion line, which depicts the steam condition in the IP and LP turbine stages, is drawn to the LP turbine ELEP. This permits determination of the steam conditions for the reheat turbine extractions. The expansion line for the IP turbine is essentially a straight line. However, the steam expansion in the LP turbine exhibits a curvature or varying slope section. This variation in the expansion line represents efficiency degradation due to the presence of moisture.

13.18.1.4 HP Turbine Internal Efficiency The HP turbine efficiency curve is shown in Fig. 13-27. Turbine efficiency is plotted as a function of equivalent throttle flow ratio. This permits the use of this curve for various turbine inlet steam conditions and flows. Although not precise, small variations in the reheater pressure drop are typically neglected.

The turbine internal efficiency represents the overall efficiency of the HP turbine and is applied to the available energy from the turbine throttle conditions to the HP turbine exhaust pressure. A composite of the effect of throttle valve pressure drop, first-stage efficiency, and HP turbine stage group efficiency are represented by this curve. Since the

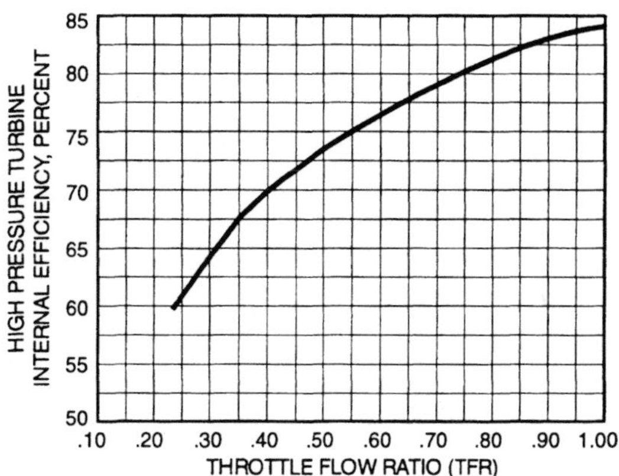

THIS CURVE IS ON A VALVE BEST POINT BASIS

TFR = (THROTTLE FLOW AT ANY STEAM CONDITIONS)/(VALVES WIDE OPEN THROTTLE FLOW AT SAME STEAM CONDITIONS)

APPLY THE EFFICIENCY FROM THIS CURVE TO THE AVAILABLE ENERGY FROM THE TURBINE STOP VALVES TO THE HIGH PRESSURE TURBINE EXHAUST

BREAK IN CURVE IS FIRST ADMISSION POINT. THROTTLING CONTROL OCCURS AT ALL LOWER THROTTLE FLOW RATIOS.

FOR OFF-RATED STEAM CONDITIONS USE EQUIVALENT TFR WHERE

$$\text{EQUIV TFR} = \frac{\text{(OFF-RATED FLOW)}}{\text{(DESIGN FLOW)}}\sqrt{\frac{\text{(P/V) RATED}}{\text{(P/V) OFF-RATED}}}$$

Fig. 13-27. High-pressure turbine internal efficiency. (From General Electric. Used with permission.)

curve is drawn using the assumption of locus-of-valve point, the curve does not reflect the loss in HP turbine efficiency due to throttling losses of partially open control valves for flows above the throttle flow ratio of the first admission. These throttling losses are small at high load because of the relatively small portion of flow that is throttled compared to the flow that is passing through the valves that are fully open. However, as load is decreased, a greater portion of turbine flow becomes throttled, further impacting turbine efficiency.

Turbines are provided with a limited numbers of valves, some of which are operated in unison. The actual number of valves and turbine admissions is a function of the mode of operation intended for the unit, the size of the unit, and the manufacturer's design practice. Steam flow control at throttle flow ratios below the first admission point is accomplished by throttling the turbine valve(s). In this mode, all flow to the turbine is throttled which results in a decrease of efficiency. This decrease in HP turbine efficiency is shown in Fig. 13-27. For this example, the throttle flow ratio corresponding to the first admission point is 0.35. Table 13-5 shows approximate throttle flow ratios at the first admission for various numbers of steam turbine admission arrangements. Because of throttling losses at lower loads, the throttle flow ratio at the first admission has a significant impact on performance at loads below this point.

13.18.1.5 HP Turbine Expansion Line End Points Fig. 13-28 illustrates the HP turbine expansion line end point plotted as a function of reheat turbine intercept valve pressure. The curve is limited to applications with rated steam conditions

Table 13-5. Throttle Flow Ratio at First Admission Point

The throttle flow ratio at first admission point with respect to the number of admissions is approximately as follows:

1 admission (full throttling)	1.0
2 admissions	0.85
3 admissions	0.60
4 admissions	0.35
8 admissions	0.30

Source: General Electric. Used with permission.

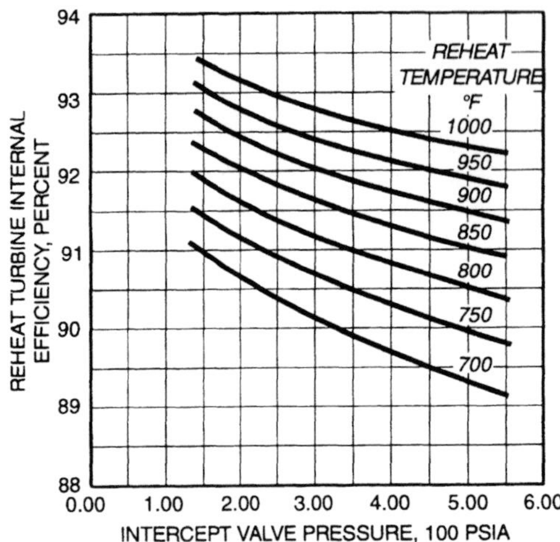
REHEAT TURBINE INTERNAL EFFICIENCY, PERCENT
94
93
92
91
90
89
88
0.00
1.00
2.00
3.00
4.00
5.00
6.00
INTERCEPT VALVE PRESSURE, 100 PSIA
REHEAT TEMPERATURE °F
1000
950
900
850
800
750
700
APPLY EFFICIENCY TO AVAILABLE ENERGY BETWEEN INTERCEPT VALVE PRESSURE AND BOWL ENTHALPY (INCLUDING PACKING LEAKAGE MIXTURE IF ANY) AND 1.5 IN. Hg ABS.
ELEP1.5= HBOWL - EFF (AE)
USE FIGURE 13-31 TO CORRECT ELEP AT 1.5 IN. Hg ABS. TO OTHER EXHAUST PRESSURES
SEE ASME PAPER 62WA209 FOR CONSTRUCTION OF EXPANSION LINE

Fig. 13-29. Reheat turbine internal efficiency. (From General Electric. Used with permission.)

and is further based on the assumption of a 10% reheater pressure drop from HP exhaust to intercept valve. The curve is provided as a quick reference for the leaving enthalpy of the steam exhausting the HP turbine. The HP turbine efficiency curve discussed previously provides a more versatile method for determining expected HP turbine exhaust conditions.

13.18.1.6 Reheat Turbine Internal Efficiency The overall IP and LP turbine internal efficiency curves are shown in Fig. 13-29. The curves are presented for various reheat turbine inlet temperatures. Like the HP turbine end point curves, the reheat internal efficiency curves are plotted as a function of pressure ahead of the intercept valve. Intercept valve pressure is a reasonable parameter to plot efficiency against since it relates to steam flow.

For turbines that have packing inleakage to the IP turbine bowl, the mixed bowl enthalpy must be calculated. The turbine efficiency is then applied to the available energy between the intercept valve pressure and mixed bowl enthalpy and a reference LP turbine exhaust pressure of 1.5 in. of mercury absolute (HgA). The exhaust pressure of 1.5 in. HgA has been selected by the turbine manufacturer as a reference pressure convention. The ELEP for the reheat turbine at a 1.5 in. HgA back pressure can then be calculated.

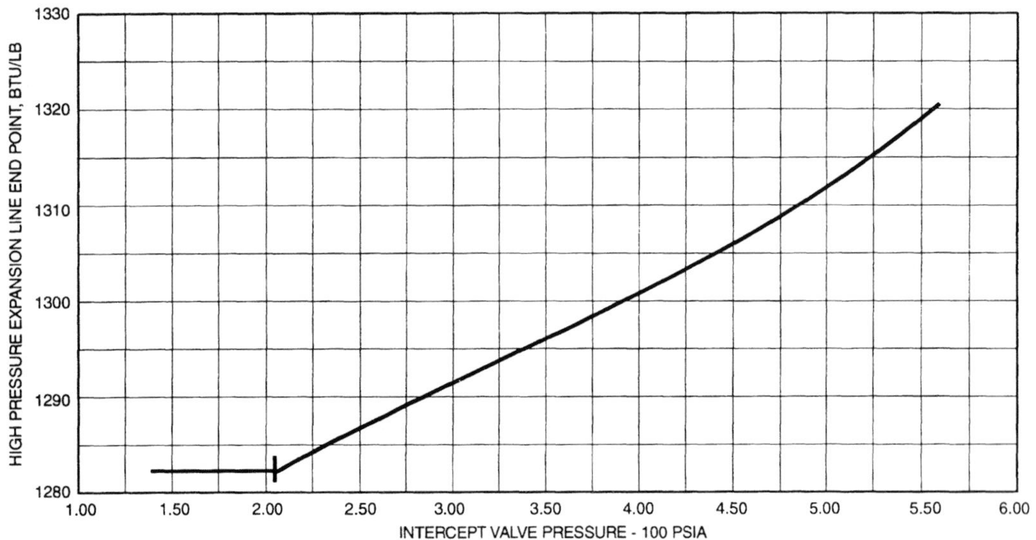
HIGH PRESSURE EXPANSION LINE END POINT, BTU/LB
1330
1320
1310
1300
1290
1280
1.00
1.50
2.00
2.50
3.00
3.50
4.00
4.50
5.00
5.50
6.00
INTERCEPT VALVE PRESSURE - 100 PSIA
THIS CURVE ASSUMES A PRESSURE DROP OF 10 PERCENT FROM HIGH PRESSURE TURBINE EXHAUST TO INTERCEPT VALVE. THESE HIGH PRESSURE EXPANSION LINE END POINTS CANNOT BE USED FOR OTHER PRESSURE DROPS.

Fig. 13-28. High-pressure turbine expansion line end points. (From General Electric. Used with permission.)

Expansion line end points at other exhaust pressures can also be determined as discussed in the following sections.

*13.18.1.7 **Reheat Turbine Expansion Line End Points*** Figure 13-30 provides the reheat turbine ELEP based on a reheat steam temperature of 1,000° F. The end points are plotted as a function of intercept valve pressure and are based on a turbine exhaust pressure of 1.5 in. HgA. This curve conveniently provides the end point conditions for rated reheat steam temperature; however, it is not as flexible as the internal efficiency curve previously discussed because end points are valid only for 1,000° F reheat steam. The end point calculations can be corrected to other exhaust pressures based on the use of the correction to expansion line end points for exhaust pressure curve.

*13.18.1.8 **Correction to Expansion Line End Points for Exhaust Pressure*** The curve shown in Fig. 13-31 is a typical exhaust pressure correction curve provided by General Electric. This curve permits the correction of turbine end points determined from the previous two sets of curves to exhaust pressures other than 1.5 in. HgA. The directions for the proper use of this curve are provided on the figure. This curve permits the determination of ELEPs from 0.5 to 5.0 in. HgA exhaust pressure.

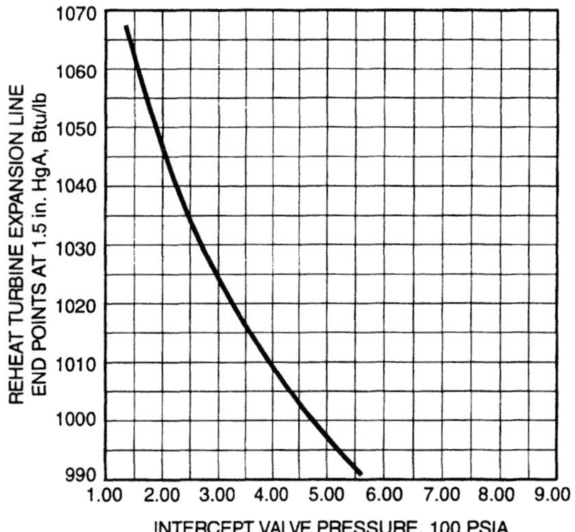

THESE EXPANSION LINE END POINTS ARE FOR HEAT BALANCE CALCULATIONS IN WHICH GLAND LEAKAGE STEAM IS USED IN THE FEEDWATER HEATING CYCLE.

TO OBTAIN THE ENTHALPY OF THE STEAM ENTERING THE CONDENSER READ THE CURVE AT 1.5 IN. Hg ABS. AND CORRECT TO THE DESIRED EXHAUST PRESSURE USING FIGURE 13-31, AND CORRECT FOR EXHAUST LOSS USING FIGURE 13-32.

Fig. 13-30. Reheat expansion line end points at 1.5 in. HgA. (From General Electric. Used with permission.)

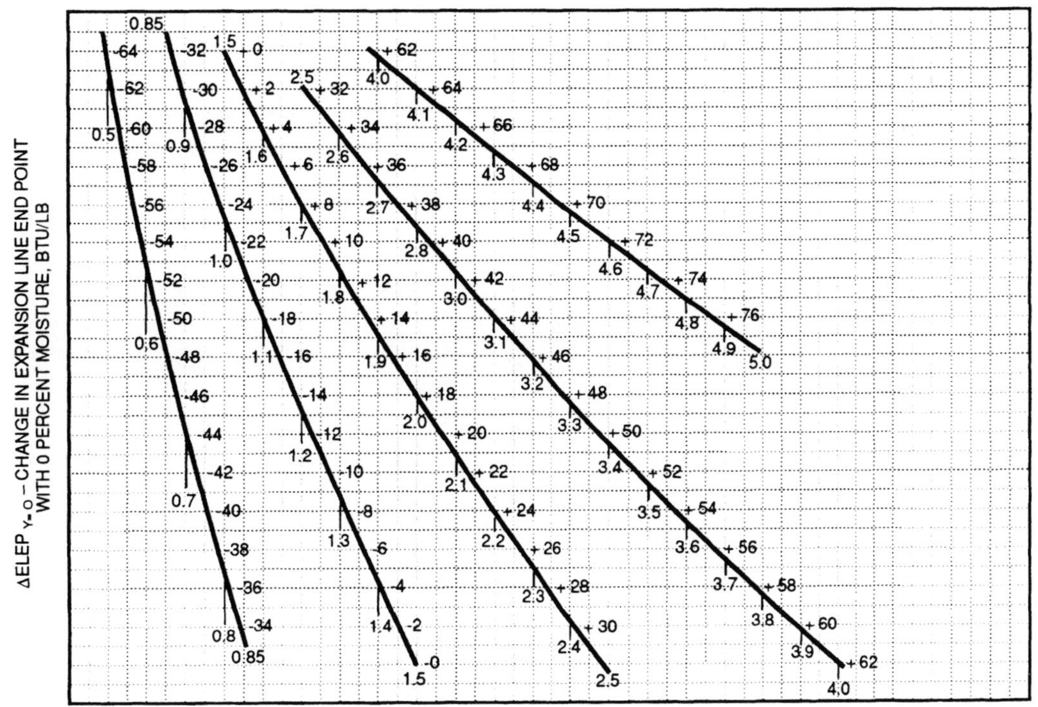

1. WHEN EXPANSION LINE END POINT AT 1.5 IN. Hg ABS. IS IN THE MOISTURE REGION, USE THE FOLLOWING EQUATION:
   $\Delta ELEP = \Delta ELEP_{Y=0}\ (0.87)\ (1\text{-}0.01Y)\ (1\text{-}0.0065Y)$
   Y= PERCENT MOISTURE AT ELEP AT 1.5 IN. Hg ABS.

2. WHEN THE EXPANSION LINE END POINT AT 1.5 IN. Hg ABS. IS IN THE SUPERHEAT REGION, MULTIPLY THE CHANGE IN EXPANSION LINE END POINT AT 0 PERCENT MOISTURE BOTH BY THE FACTOR 0.87 AND THE RATIO OF SPECIFIC VOLUME AT THE EXPANSION LINE END POINT UNDER CONSIDERATION TO THE DRY SATURATED SPECIFIC VOLUME AT 1.5 IN. Hg ABS.

Fig. 13-31. Correction to expansion line point for exhaust pressure. (From General Electric. Used with permission.)

13.18.1.9 Exhaust Loss Curve The turbine exhaust loss curve is shown in Fig. 13-32. The curve is unique to a given last stage blade length. The exhaust loss curve is used to determine the dry exhaust loss. The dry exhaust loss is found by entering the exhaust loss curve at the exhaust velocity or the exhaust volume flow depending on manufacturer preference. Equations, given on the figure, are used to correct the dry loss to the actual exhaust loss. This corrected exhaust loss is then added to the expansion line end point corresponding to the exhaust pressure under consideration. The exhaust loss plus the end point defines the used energy end point (UEEP) in General Electric nomenclature, or turbine end point (TEP) in Westinghouse nomenclature. The UEEP (TEP) is the actual leaving enthalpy of the LP turbine steam and includes losses associated with the turbine exhaust hood, the steam leaving velocity (kinetic energy), and the restriction loss or turn up loss. These losses are shown schematically in Fig. 13-33. Note that the work produced by the turbine is determined by performing an energy balance with the use of the UEEP, not the LP turbine ELEP.

The value of moisture (Y) as used in the correction equations for General Electric have different reference conditions for the expansion line end point correction and the exhaust loss correction. The moisture for the correction to expansion line end point is based on the percent moisture at the ELEP at 1.5 in. HgA exhaust pressure. The percent moisture for the calculation of exhaust loss is at the ELEP at the actual exhaust pressure.

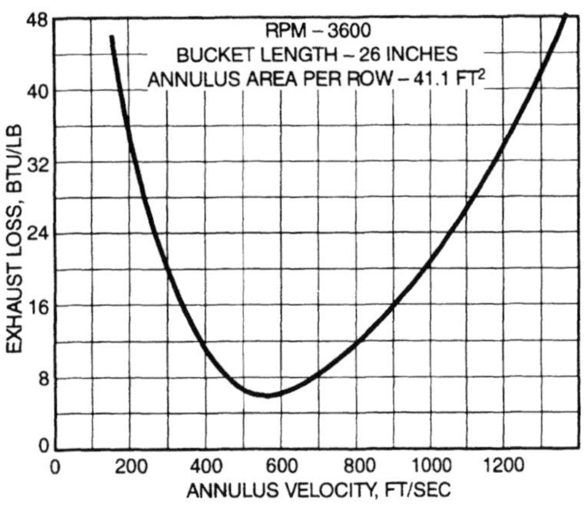

1. READ THE EXHAUST LOSS CURVE AT THE ANNULUS VELOCITY OBTAINED FROM THE FOLLOWING EXPRESSION:

$$VAN = \frac{QA\,(V)\,(1.0-0.01Y)}{3600\,(AA)}$$

2. THE ENTHALPY OF THE STEAM ENTERING THE CONDENSER IS THE QUANTITY OBTAINED FROM THE FOLLOWING EXPRESSION:

$$UEEP = ELEP + (EXHAUST\ LOSS)\,(0.87)\,(1.0-0.01Y)\,(1.0-0.0065Y)$$

3. THE EXHAUST LOSS CURVE INCLUDES THE LOSS IN INTERNAL EFFICIENCY WHICH OCCURS AT LIGHT LOADS AS OBTAINED IN TESTS.
4. THIS CURVE IS BASED ON THE 1967 ASME STEAM TABLES.

LEGEND:

VAN – ANNULUS VELOCITY IN FEET PER SECOND
QA – CONDENSER FLOW IN POUNDS PER HOUR
V – SATURATED DRY SPECIFIC VOLUME IN CUBIC FEET PER POUND CORRESPONDING TO THE ACTUAL EXHAUST PRESSURE (WHEN THE END POINT IS IN THE SUPERHEAT REGION V = THE ACTUAL SPECIFIC VOLUME AT THE END POINT.)
AA – ANNULUS AREA IN SQUARE FEET
Y – PERCENT MOISTURE AT THE EXPANSION LINE END POINT
ELEP – EXPANSION LINE END POINT AT THE ACTUAL EXHAUST PRESSURE
UEEP – USED ENERGY END POINT

Fig. 13-32. Exhaust loss curve. (From General Electric. Used with permission.)

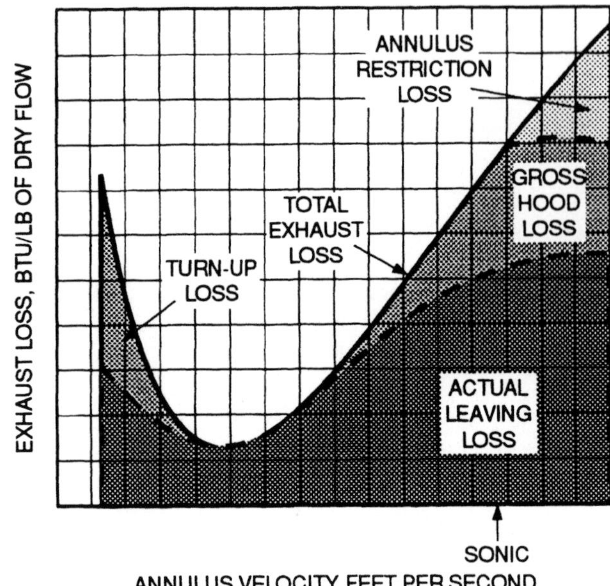

Fig. 13-33. Typical exhaust-loss curve showing distribution of component losses. (From Bartlett R.: *Steam Turbine Performance and Economics*. Copyright © 1958. Reprinted with permission of McGraw-Hill Inc.)

13.18.1.10 Generator Losses Figure 13-34 shows generator electrical losses for various turbine generator outputs at rated power factor. In addition, an equation is provided to permit the calculation of generator losses at other than rated hydrogen pressure. The generator losses are a function of generator kVa (not kW). Therefore, if generator losses at a power factor other than design is required, the curve should be read at desired output (in kilowatts) multiplied by the rated power factor divided by the desired power factor. Generator losses do not include the turbine generator fixed or mechanical losses. As discussed previously, mechanical losses should be accounted for separately and do not vary with unit load.

13.18.1.11 First-Stage Shell Pressure versus Throttle Flow Figure 13-35 shows the first-stage pressure as a function of throttle flow. The curve is presented based on rated steam conditions. To estimate the first stage shell pressure at other than rated initial temperature, the following equation should be used to determine an equivalent throttle flow before entering the curve.

$$m_{\text{th corrected}} = m_{\text{th actual}} \times \sqrt{\frac{v_{\text{actual}}}{v_{\text{design}}}} \qquad (13\text{-}6)$$

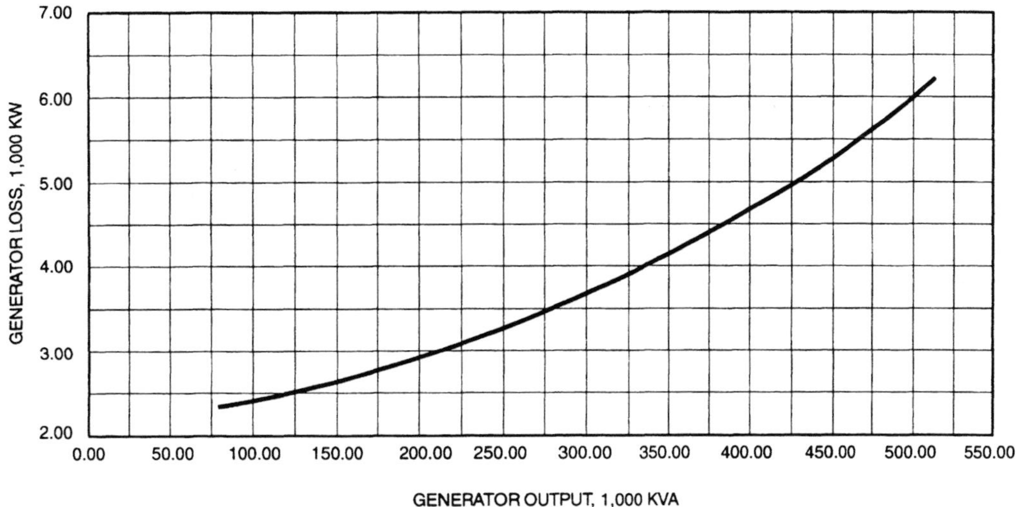
7.00
6.00
5.00
4.00
3.00
2.00
GENERATOR LOSS, 1,000 KW
0.00 50.00 100.00 150.00 200.00 250.00 300.00 350.00 400.00 450.00 500.00 550.00
GENERATOR OUTPUT, 1,000 KVA
511,000 kVA AT 45 PSIG H2 PRESSURE
CONDUCTOR COOLED 3600 RPM
NOTES:
GENERATOR LOSSES ASSUME RATED HYDROGEN PRESSURE AT ALL LOADS.
GENERATOR LOSS AT REDUCED HYDROGEN PRESSURE (P) = LOSS AT RATED HYDROGEN PRESSURE - 11.2 (PRATED - P).
USE GENERATOR REACTIVE CAPABILITY CURVE TO DETERMINE GENERATOR CAPABILITY AT REDUCED HYDROGEN PRESSURE.
TURBINE GENERATOR MECHANICAL LOSSES ARE NOT INCLUDED IN THE GENERATOR LOSS CURVE.
IF HYDROGEN AND STATOR LIQUID COOLERS ARE LOCATED IN THE CONDENSATE LINE, THE LOSS TRANSFERRED TO THE COOLERS IS 474
kW LESS THAN THE GENERATOR LOSS AT ALL LOADS.

Fig. 13-34. Generator loss curve. (From General Electric. Used with permission.)

where

$m_{\text{th corrected}}$ = corrected turbine throttle flow, lb/h;

$m_{\text{th actual}}$ = actual turbine throttle flow, lb/h;

v_{actual} = actual throttle steam specific volume, ft^3/lb; and

v_{design} = design throttle steam specific volume, ft^3/lb.

13.18.2 Correction Curves

The correction curves typically found in a thermal kit are intended to give approximate output and heat rate corrections and are often provided for correcting turbine test data to guaranteed conditions. These correction curves are discussed in the following sections.

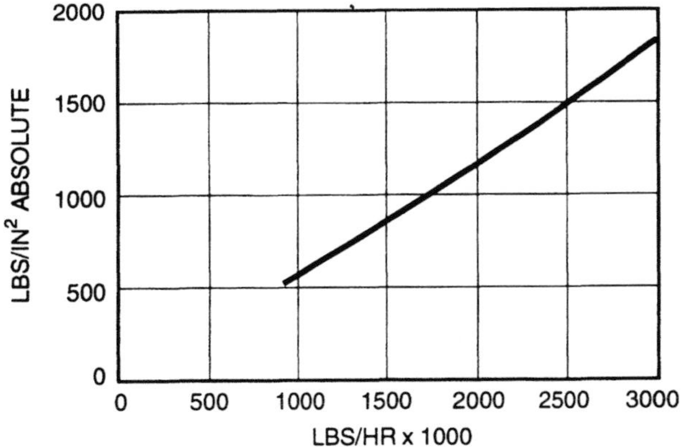
2000
1500
1000
500
0
LBS/IN2 ABSOLUTE
0 500 1000 1500 2000 2500 3000
LBS/HR x 1000

Fig. 13-35. First-stage shell pressure vs. throttle flow. (From General Electric. Used with permission.)

13.18.2.1 Initial Pressure Correction A typical standard initial pressure correction curve is shown in Fig. 13-36. The curve allows for the correction of turbine output and heat rate for changes in throttle pressure. The curve is developed based on holding a constant control valve opening while varying the turbine throttle pressure. The corrections are to be applied to kilowatt and heat rate at rated steam conditions. A similar curve is provided in the ASME Power Test Code (PTC) 6.1, Interim Alternative Test Procedure for Steam Turbines. The manufacturer may provide a curve specifically developed for the turbine. Increased throttle pressure at constant valve opening increases the mass flow to the turbine, which increases the output of the unit. The increased throttle pressure improves the turbine cycle efficiency as a result of increased available steam energy.

13.18.2.2 Initial Temperature Correction The initial temperature correction curve is shown in Fig. 13-37. The curve is also based on the assumption of fixed control valve position. Increasing the throttle temperature results in an increased specific volume, a decreased mass flow, and decreased generation. The increase in throttle temperature increases the available energy to the turbine, increases the turbine cycle efficiency, and reduces heat rate.

13.18.2.3 Reheat Pressure Drop Correction As shown in Fig. 13-38, a decrease in reheater pressure drop results in an increase in turbine output and a decrease in turbine heat rate. HP turbine output increases, and the HP turbine expansion end point decreases. As a result, the reheater duty increases.

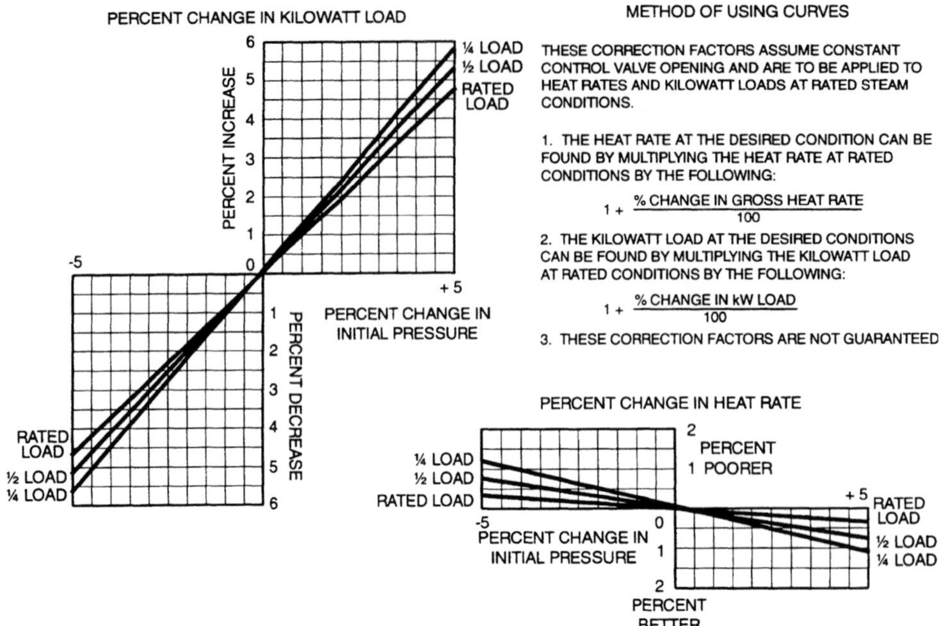

Fig. 13-36. Initial pressure correction factors. (From General Electric. Used with permission.)

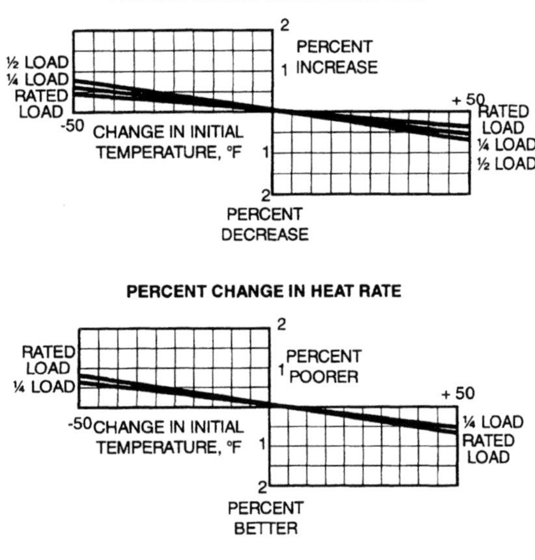

METHOD OF USING CURVES

THESE CORRECTION FACTORS ASSUME CONSTANT CONTROL VALVE OPENING AND ARE TO BE APPLIED TO HEAT RATES AND KILOWATT LOADS AT RATED STEAM CONDITIONS.

1. THE HEAT RATE AT THE DESIRED CONDITION CAN BE FOUND BY MULTIPLYING THE HEAT RATE AT RATED CONDITIONS BY THE FOLLOWING:

$$1 + \frac{\text{\% CHANGE IN GROSS HEAT RATE}}{100}$$

2. THE KILOWATT LOAD AT THE DESIRED CONDITIONS CAN BE FOUND BY MULTIPLYING THE KILOWATT LOAD AT RATED CONDITIONS BY THE FOLLOWING:

$$1 + \frac{\text{\% CHANGE IN kW LOAD}}{100}$$

3. THESE CORRECTION FACTORS ARE NOT GUARANTEED.

Fig. 13-37. Initial temperature correction factors. (From General Electric. Used with permission.)

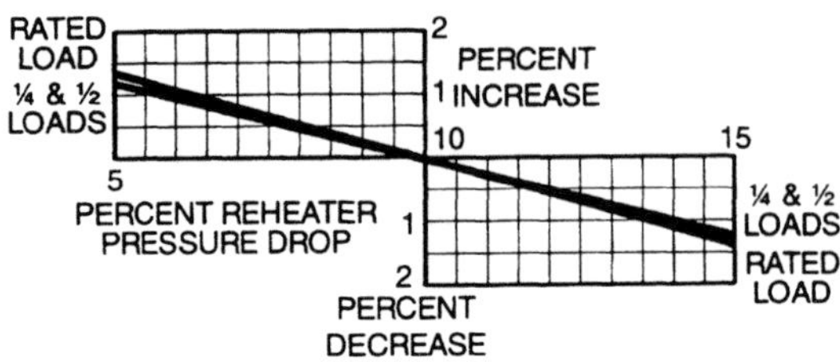

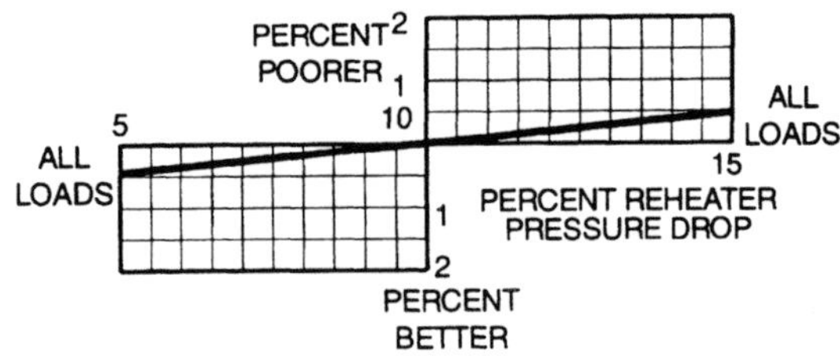

Fig. 13-38. Reheater pressure drop correction factors for single reheat units. (From General Electric. Used with permission.)

13.18.2.4 Reheat Temperature Correction As shown in Fig. 13-39, increasing reheat temperature results in increased turbine output and decreased turbine heat rate. Increased reheat temperature results in increased heat duty in the reheater. Increased reheat temperature does not impact the amount of main steam flow; however, it does increase the intercept valve pressure and decreases the amount of work produced by the HP turbine due to greater HP turbine exhaust pressure. However, the reduced HP turbine output is offset by the greater available steam energy supplied to the reheat turbine and its increased turbine output.

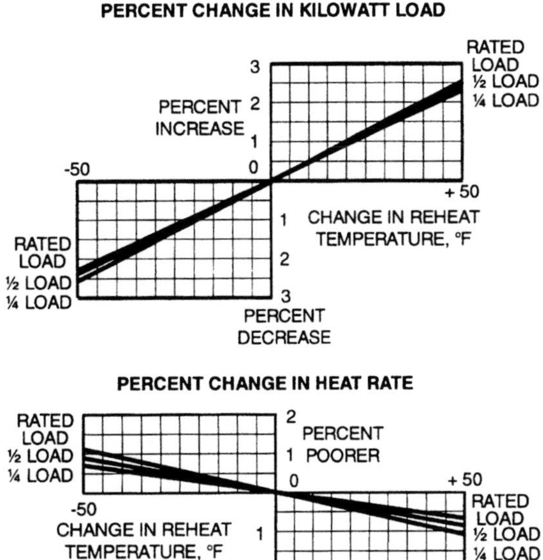

Fig. 13-39. Reheat temperature correction factors. (From General Electric. Used with permission.)

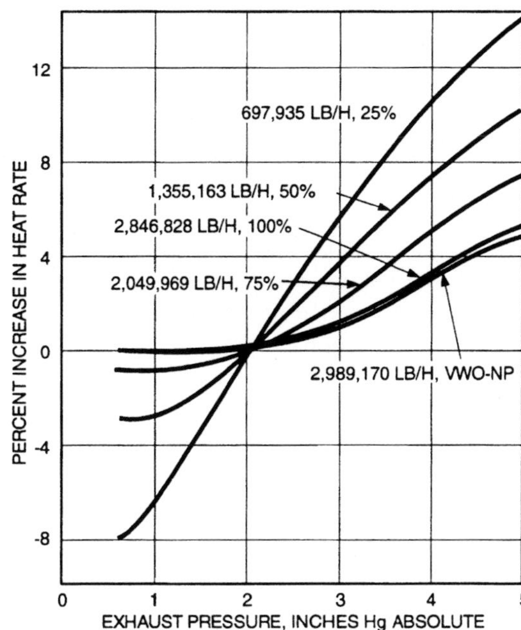

VALUES NEAR CURVES ARE FLOWS AT 2400 PSIG, 1000 F.
THESE CORRECTION FACTORS ASSUME CONSTANT CONTROL VALVE OPENING.
APPLY CORRECTIONS TO HEAT RATE AND kW LOADS AT 2.0 IN. Hg ABS. AND 0.0 PERCENT MU.

THE PERCENT CHANGE IN kW LOAD FOR VARIOUS EXHAUST PRESSURES IS EQUAL TO (MINUS PERCENT INCREASE IN HEAT RATE) 100/(100 + PERCENT INCREASE IN HEAT RATE).

THESE FACTORS GIVE CHANGE IN NTHR.

Fig. 13-40. Exhaust pressure correction factors. (From General Electric. Used with permission.)

13.18.2.5 Exhaust Pressure Correction A change in LP turbine exhaust pressure results in a change in available energy. In addition, exhaust losses and LP turbine extraction flow are impacted. Figure 13-40 indicates the percent change in turbine heat rate due to variations in steam turbine back pressure.

Figure 13-40 provides an equation to correct the turbine output. This equation is based on constant heat input and indicates that turbine heat rate is inversely proportional to turbine output. For the case of decreasing exhaust pressure, the condensate temperature to the first LP heater decreases because it is typically the saturation temperature corresponding to the turbine exhaust pressure. The colder condensate results in an increased extraction flow to the LP heater and reduced exhaust flow. These effects would tend to decrease turbine output. In addition, lower exhaust pressure typically results in higher exhaust loss because of increased exhaust velocity due to the higher specific volume at the exhaust pressure. However, these effects are offset by the increased available energy. Therefore, output and efficiency typically improve with reduced exhaust pressure.

However, as back pressure continues to decrease, the turbine may become choked. Below the choking pressure, the turbine no longer benefits from increased available energy. The condensate continues to cool further and heater extraction increases. Therefore, the output and efficiency can decrease. This can be seen by examining Fig. 13-40. The turbine becomes choked at approximately 2.0 in. HgA for the maximum flow condition. As turbine exhaust flow decreases, the turbine becomes choked at lower exhaust pressures. The point at which the turbine becomes choked depends on turbine design, exhaust flow, and exhaust pressure.

A detailed example illustrating the use of these concepts is found in Appendix 13A.

13.19 REFERENCES

American Society of Mechanical Engineers. 345 East 47th Street, New York, NY 10017-2392.

American Society of Mechanical Engineers (ASME). 1976. ASME Power Test Code (PTC) 6.1. Interim Alternate Test Procedures for Steam Turbines. ASME, New York, NY.

Bartlett, Robert L. 1958. *Steam Turbine Performance and Economics*. McGraw-Hill, New York, NY.

General Electric Company. 1 River Road, Schenectady, NY 12345.

Spencer, R. C., K. C. Cotton, and C. N. Cannon, General Electric Company. 1962. A method for predicting the performance of steam turbine generators ... 16,500 kW and larger. ASME Paper No. 62-WA-209. Contributed by the Power Division of the American Society of Mechanical Engineers for presentation at the Winter Annual Meeting. November 25–30. New York, NY.

APPENDIX 13A

Turbine Heat Balance Calculation

I. Establishing a heat balance.

The following procedure illustrates a method for establishing a turbine cycle heat balance by hand calculation.

A. Setting up a heat balance

1. Establish cycle configuration, steam conditions (throttle pressure and temperature) and unit size (megawatts required) at a specific back pressure.
2. Establish feedwater terminal temperature differences and drain cooler approaches.
3. Specify assumptions to be used for the heat balance. Refer to Table 13A-1, "Typical Assumptions Made in Establishing a Heat Balance."

B. Use thermal kit

1. Estimate flows. Refer to Table 13A-2, "Approximate Flow Distribution for Typical Regenerative/Reheat Cycles." Alternatively, and perhaps preferably, a similar heat balance diagram can be used to establish the initial flow distribution in the cycle. The approach taken in this example is to estimate flows since it provides a more generalized approach to establishing flow distribution.
   a. Reheat flow,
   b. BFPT flow,
   c. Turbine exhaust flow, and
   d. Makeup flow.
2. Calculate leakages and remaining flow distribution.
3. Estimate extraction pressures.
4. Determine the expansion lines.
5. Establish heater and feedwater/condensate conditions.
   a. Determine enthalpy rise across BFP.
   b. Determine BFPT flow rate.
   c. Determine feedwater and condensate flow rates.
6. Determine final turbine flow distribution by iteration.
7. Calculate the turbine shaft work and turbine output.
8. Calculate net turbine heat rate.

II. Example Heat Balance Calculation

Materials contained in figures in this Appendix are for example purposes. These data were derived from a Thermal Kit for a particular turbine and the numerical values given are *not* valid for general use.

A. Given a thermal kit and the throttle steam flow and steam conditions, determine unit output. Throttle flow is 2,846,829 lb/h at 2,400 psig and 1,000° F.

The assumptions used in establishing the heat balance are shown in Fig. 13A-1. These include cycle configuration, turbine back pressure, heater parameters, steam line pressure drops, and the efficiency for the BFP and BFP turbine.

B. Establish the initial assumed flow distribution.

1. Estimate flows

From Table 13A-2 on approximating the first guess flow distributions, the following can be found:

a. Reheat flow = 0.9 (2,846,829)
= 2,562,146 lb/h

b. BFPT flow = 0.04 (2,846,829)
= 113,873 lb/h

c. Turbine exhaust = 0.65 (2,846,829)
= 1,850,439 lb/h

d. Makeup flow = 0.02 (2,846,879)
= 56,937 lb/h

Table 13A-1. Typical Assumptions Made in Establishing a Heat Balance

1. The boiler feed pump suction conditions will be the temperature and pressure of the deaerator. Boiler feed pump discharge pressure is 125% of the turbine throttle pressure.

2. The boiler feed pump efficiencies will vary with load as follows:

Condition	BFP Efficiency
VWO-OP, VWO-NP, and rated load	84%
75% of rated	83%
50% of rated	67%
25% of rated	40%

3. For a turbine cycle with a motor-driven boiler feed pump, the variable speed coupling efficiency will vary with load as follows:

Condition	Coupling Efficiency
VWO-OP	85%
VWO-NP and rated load	82%
75% of rated	76%
50% of rated	73%
25% of rated	68%

The combined motor and transmission efficiency will vary with load as follows:

Condition	Motor and Transmission Efficiency
VWO-OP, VWO-NP, and rated load	94%
75% of rated	93%
50% of rated	92%
25% of rated	89%

4. For a turbine cycle with a turbine-driven boiler feed pump exhausting to the main condenser, the boiler feed pump turbine (BFPT) will operate at an exhaust pressure 0.5 in. HgA greater than the exhaust pressure of the main turbine. BFPT expansion efficiencies will vary with load as follows:

Condition	BFPT Efficiency
VWO-OP, VWO-NP, and rated load	80%
75% of rated	78%
50% of rated	77%
25% of rated	77%

The presure drop in the extraction line to the BFPT is 3% of the inlet pressure.

At low loads, the BFPT will require steam from a source of higher pressure than is available in the crossover line. Below approximately 0.35 throttle flow ratio (TFR), the BFPT takes steam as required from the main steam line.

5. There is a pressure drop between the turbine stage and the extraction flange. This value is typically 3% of stage presure. A pressure drop also occurs from the turbine extraction flange to the heater. This value is usually 3% or 5% of the extraction flange pressure. For extractions at a turbine exhaust or at the crossover pipe, no pressure drop due to an extraction flange exists, only an extraction line pressure drop.

6. There is a pressure drop from the high-pressure turbine exhaust to the intercept valves of the intermediate-pressure turbine because of hot and cold reheat piping and the reheater. This value is normally taken to be 10% of the high-pressure turbine exhaust pressure.

7. The condensate leaving the condenser will be at the saturation temperature corresponding to the turbine exhaust pressure.

8. The condensate will be considered to be saturated liquid at the heater inlet and outlet temperatures.

9. Calculations will consider the feedwater downstream of the boiler feed pump as compressed liquid.

Table 13A-2. Approximate Flow Distribution for Typical Regenerative/Reheat Cycles

1. The reheater flow is approximately 90% of the throttle flow.
2. The boiler feed pump turbine extraction flow from an IP to LP turbine crossover is 4% to 6% of the throttle flow.
3. The turbine exhaust flow is 65% to 75% of the throttle flow, with the remaining flow being taken for heating the feedwater and driving the boiler feed pump turbine.
4. Rules of thumb when the temperature rise across a heater is known:
   a. For the low-pressure (LP) heaters and deaerator, the extraction flow is approximately 1% of throttle flow for each 14° F temperature rise.
   b. For high-pressure (HP) heaters, the extraction flow is approximately 1% of throttle flow for each 10° F temperature rise.
5. If a heat balance is available for other than the desired load, ratio the extractions by the ratio of the throttle flows for a first guess. However, these may differ up to 30% from the final calculations.

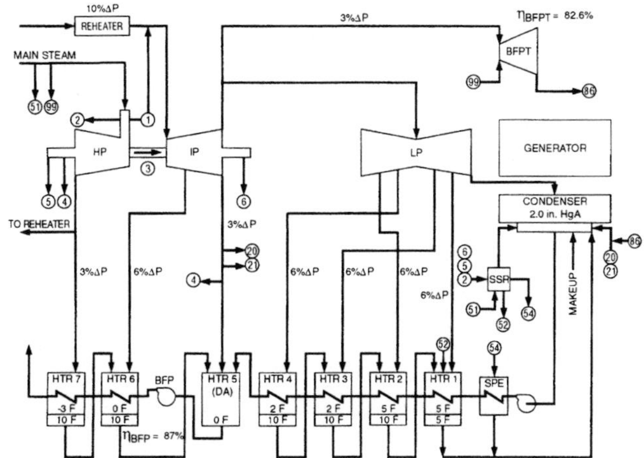

Fig. 13A-1. Cycle schematic. (From General Electric. Used with permission.)

2. Calculate leakages and remaining flow distribution
   a. Calculate combined leakage for valve stems. From Fig. 13A-2 of the thermal kit

$$\text{VSLO } (1 + 2) = C\sqrt{\frac{P}{\upsilon}}$$

where

$$C = 56$$

$$P = 2{,}415 \text{ psia}$$

and υ corresponds to the throttle pressure and enthalpy.

$$\upsilon = 0.3191 \text{ ft}^3/\text{lb}$$

$$\text{VSLO } (1 + 2) = 56\sqrt{\frac{2{,}415}{0.3191}} = 4{,}871 \text{ lb/h}$$

To determine the leakage flow in lb/h, multiply the following packing constants by the $\sqrt{P/V}$ (ahead of the leakage). The leakage through the first packing in a series will be the sum of the flows to the packing leakoffs following it.

Refer to heat balance diagram for identification of leakages.

Leakage No.	1+2	2	3	4+5	5	6
Packing Constant	56	50	500	620	970	600

When expansion line end points from Figures 13-A4 and 13-A5 are used in heat balance calculations, the leakages shown on this sheet must be used in the feedwater heating cycle.

The 1st of 2 leakages can be determined by subtracting the 2nd leakage from the sum of both leakages and the pressure ahead of the 2nd leakage is the extraction stage pressure to which the 1st of the 2 leakages connects into.

Use throttle enthalpy for leakages 1 and 2.

Use high-pressure turbine exhaust enthalpy for leakages 4 and 5.

Enthalpy for leakage 6 can be determined from turbine exapansion lines at crossover pressure.

5,200 lb/h are required by the steam seals. When leakages 2, 5, and 6 combined are less than 5,200 lb/h, throttle steam must be used. (Steam seal flow to steam packing exhauster is 2,800 lb/h at all loads.)

To determine the first stage enthalpy and pressure used to calculate leakage 3, use the following table:

Throttle Flow, lb/h	2,989,170	2,846,829	2,051,529	1,356,349	698,889
1st Stage Enthalpy, Btu/lb	1,433.2	1,427.9	1,404.6	1,389.7	1,382.1
1st Stage Shell Pressure, psia	1,845	1,750	1,237	807	413

Throttle flow ratio at first admission = 0.35

First stage pitch diameter = 41

For other than rated initial temperature, read the first stage shell pressure curve drawn through these points at a flow corrected for $\sqrt{\frac{\text{V DESIRED}}{\text{V RATED}}}$ where V is initial specific volume cu. ft/lb.

Turbine Generator Mechanical Losses

1. 2,264 kW fixed losses are not included in the 3,600 rpm generator loss curve.

Fig. 13A-2. Gland leakage and mechanical losses. (From General Electric. Used with permission.)

The separate flows for VSLO (1) and VSLO (2) cannot be calculated until the reheat pressure is known since VSLO (1) flows to the hot reheat line, changing the value of P and υ for use with packing constant VSLO (2) = 50.

b. Calculate governor stage leakage to IP turbine bowl. The governor stage leakage occurs after the steam has expanded across the governor stage. These conditions, as a function of turbine throttle flow, can be found from Fig. 13A-2. The first stage enthalpy = 1,427.9 Btu/lb. The first stage shell pressure = 1,750 psia. At these conditions, υ = 0.4185 ft³/lb. The packing constant for leak 3 is 500 from Fig. 13A-2.

$$\text{GSLO} = 500\sqrt{\frac{1{,}750}{0.4185}}$$

$$= 32{,}333 \text{ lb/h}$$

c. Calculate combined HP shaft leakages.

For this calculation, it is required to know the HP turbine exhaust pressure. Refer to Fig. 13A-3 for extraction stage shell pressure factors. Extraction No. 1 factor will give the pressure at the intercept valve of the IP turbine.

To determine the extraction stage shell pressure, multiply the following factors by:

Flow to the Following Stage / 10^6

Extraction No.	1	2	3(XO)	4	5	6	7
Factor	210.3	110.1	59.3	23.87	8.92	4.99	2.59

Note:

1. Apply the factor for Extraction No. 1 to the flow to the first reheat stage to determine the pressure ahead of the intercept valve. High-pressure turbine exhaust flange pressure equals pressure ahead of the intercept valve divided by 0.9.
2. Flow to following stage equals throttle flow minus leakages and all extractions from preceding stages and stage in question plus any steam returned to the turbine ahead of the stage in question.
3. Use 3 percent pressure drop between shell stage pressure and turbine shell flange pressure except where extractions are at the end of a shell casing.
4. These factors are to be applied for heat balance calculations at rated steam temperatures only.
5. Extraction pressures during normal operation can be greater than those calculated at maximum expected throttle flow because of manufacturing tolerances on nozzle areas, variations in flow coefficients, deposits in the steam path, etc. To allow for this, the General Electric Company recommends that extraction piping and feedwater heaters be designed for pressures at least 15 percent greater than those calculated at maximum throttle flow conditions. It is likely that there will be abnormal operating conditions which will result in pressures greater than those defined above, which should be used as the actual extraction piping and feedwater heater design pressures.

Fig. 13A-3. Extraction stage shell pressure. (From General Electric. Used with permission.)

Neglecting VSLO (1) flow for now since it is small, the IP turbine bowl flow is the reheat flow plus the governing stage leakage.

$$M_{\text{IP bowl}} = M_{\text{Rht}} + M_{\text{GSLO}}$$

$$= 2{,}562{,}146 + 32{,}333$$

$$= 2{,}594{,}479 \text{ lb/h}$$

The pressure at the IP turbine intercept valve is

$$P_{\text{IV}} = 210.3\frac{2{,}594{,}479}{10^6} = 545.6 \text{ psia}$$

Since the reheat system pressure drop is 10% of HP exhaust pressure the HP exhaust pressure can be found.

$$P_{\text{HPexh}} = \frac{545.6}{(1-0.1)} = 606.2 \text{ psia}$$

We still need the HP exhaust enthalpy, or HP ELEP. Since we know the intercept valve pressure, we can find the ELEP for the HP turbine from Fig. 13A-4.

$$\text{HP ELEP} = 1{,}318.0 \text{ Btu/lb}$$

From the steam tables at P = 606.2 psia and h = 1,318.0 Btu/lb,

$$\upsilon = 0.9931 \text{ ft}^3\text{/lb}$$

From Fig. 13A-2, the combined packing constant for leaks 4 and 5 is 620.

$$\text{SLO}_{(4+5)} = 620\sqrt{\frac{606.2}{0.9931}}$$

$$= 15{,}318 \text{ lb/h}$$

d. Calculate VSLO (1) and VSLO (2).

The calculation of the VSLOs can now be done since the pressure to which VSLO (1) leaks to is now fairly well established.

Consider this diagram which is schematically more correct than on the heat balance.

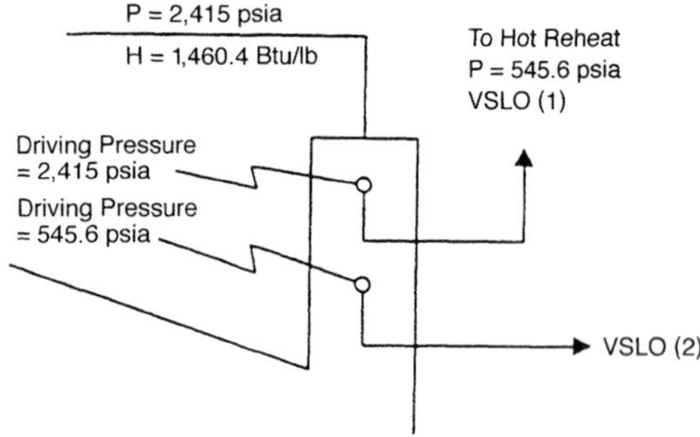

So we find υ at P = 545.6 psia and h = 1,460.4 Btu/lb.

$$\upsilon = 1.4258 \text{ cf/lb}$$

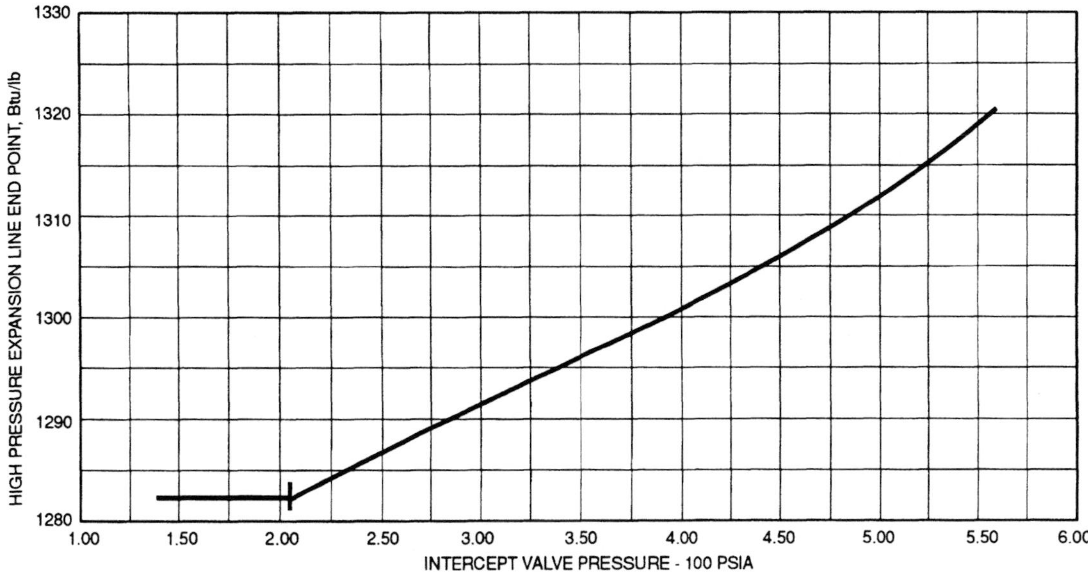

Fig. 13A-4. High-pressure turbine expansion line end points. (From General Electric. Used with permission.)

The packing constant for VSLO (2) is 50 as found from Fig. 13A-2, thus

$$\text{VSLO (2)} = 50\sqrt{\frac{545.6}{1.4258}} = 978 \text{ lb/h}$$

By difference, calculate VSLO (1) flow.

$$\begin{aligned}\text{VSLO (1)} &= \text{VSLO (1 + 2)} - \text{VSLO (2)}\\ &= 4{,}871 - 978 = 3{,}893 \text{ lb/h}\end{aligned}$$

e. Calculate IP intercept valve flow and IP bowl flow.

$$\begin{aligned}M_{\text{IV}} &= M_{\text{Rht}} + M_{\text{VSLO(1)}}\\ &= 2{,}562{,}146 + 3{,}893 = 2{,}566{,}039 \text{ lb/h}\end{aligned}$$

$$\begin{aligned}M_{\text{IP Bowl}} &= M_{\text{IV}} + M_{\text{GSLO}}\\ &= 2{,}566{,}039 + 32{,}333\\ &= 2{,}598{,}372 \text{ lb/h}\end{aligned}$$

f. Determine extraction flow to Heater 7.
Perform a mass balance around HP turbine and reheater.

$$\begin{aligned}M_{\text{Htr7}} &= M_{\text{Th}} - M_{\text{VSLO}} - M_{\text{GSLO}} - M_{\text{SLO (4+5)}} - M_{\text{Rht}}\\ &= 2{,}846{,}829 - 4{,}871 - 32{,}333 - 15{,}318 - 2{,}562{,}146\end{aligned}$$

$$M_{\text{ext Htr7}} = 232{,}161 \text{ lb/h}$$

g. Estimate remaining flows.
This is the extent of being able to calculate an estimated first iteration flow distribution. The remaining extraction flows must be estimated. With a bit of luck and possibly some reliance on a similar although not necessarily identical heat balance, reasonably close estimates of extraction flows can be assumed.

A tabulation of the calculated and estimated flows is shown in Table 13A-3. The tabulation is based on conservation of mass. Since the turbine exhaust flow had been estimated and the IP bowl flow had been

Table 13A-3. Assumed Turbine-Flow Distribution

Flow Designation	Flow (lb/h)	Pressure (psia)
Throttle flow (given)	2,846,829	
VSLO (1) (calc)	−3,893	
VSLO (2) (calc)	−978	
Flow to governing stage	2,841,958	
GSLO (calc)	−32,333	
SLO (4 + 5) (calc)	−15,267	
HP extraction flow	2,794,358	
Heater 7 (calc)	−323,212	606.2
Reheater flow (calc)	2,562,146	
VSLO (1) returned	+3,893	
IP intercept valve flow	2,566,039	
GSLO returned	+32,333	
IP bowl flow	2,598,372	
Heater 6 (assumed)	−128,400	271.9
Heater 5 (D.A.) (assumed)	−159,100	130.3
BFPT (assumed)	−113,873	
SLO (6) (calc)	−3,063	
Heater 4 (assumed)	−129,200	49.4
Heater 3 (assumed)	−72,500	17.8
Heater 2 (assumed)	−66,200	9.63
Heater 1 (assumed)	−75,597	4.80
Exhaust flow (calc)	1,850,439	

calculated, the remainder is the one leakage flow, the heater extraction flows, and the boiler feed pump turbine flow.

3. Estimate extraction pressures.
   a. The extraction to Heater 7 is at the HP turbine exhaust pressure of 606.2 psia. The remaining extraction pressures are found from the extraction factors given in Fig. 13A-3.

Heater 6

Flow to stage after extraction 6
$$= 2{,}598{,}372 - 128{,}400 = 2{,}469{,}972 \text{ lb/h}$$

$$P_{ext6} = 110.1\left(\frac{2{,}469{,}972}{10^6}\right) = 271.9 \text{ psia}$$

Extraction factors are also provided for other extraction points.

Heater 5

Flow to following stage
$$= 2{,}469{,}972 - 159{,}100 - 113{,}873 = 2{,}196{,}999 \text{ lb/h}$$

Note that SLO (6) has been neglected. This is calculated after P_{ext} and h extraction are found since we need to know the specific volume.

$$P_{ext5} = 59.3\left(\frac{2{,}196{,}999}{10^6}\right) = 130.3 \text{ psia}$$

Heater 4

Flow to following stage $= 2{,}196{,}999 - 129{,}200$
$$= 2{,}067{,}799 \text{ lb/h}$$

$$P_{ext4} = 23.87\left(\frac{2{,}067{,}799}{10^6}\right) = 49.4 \text{ psia}$$

Heater 3

Flow to following stage $= 2{,}067{,}799 - 72{,}500$
$$= 1{,}995{,}299 \text{ lb/h}$$

$$P_{ext3} = 8.92\left(\frac{1{,}995{,}299}{10^6}\right) = 17.8 \text{ psia}$$

Heater 2

Flow to following stage $= 1{,}995{,}299 - 66{,}200$
$$= 1{,}929{,}099 \text{ lb/h}$$

$$P_{ext2} = 4.99\left(\frac{1{,}929{,}099}{10^6}\right) = 9.63 \text{ psia}$$

Heater 1

Flow to following stage $= 1{,}929{,}099 - 75{,}597$
$$= 1{,}853{,}502 \text{ lb/h}$$

$$P_{extl} = 2.59\left(\frac{1{,}853{,}502}{10^6}\right) = 4.80 \text{ psia}$$

As stated, the above calculation ignored the leakage from the shaft of the IP turbine. This value is calculated after the fact and input to Table 13A-3. The flow is taken from the assumed Heater 1 extraction. These deviations will not significantly affect the calculation since it is primarily based on assumptions and rules of thumb at this stage. These deviations should be accounted for in following iterations.

   b. Determine flow values for SLO (4) and SLO (5). Now that the extraction pressure to Heater 5 is known, the flow rate for leakage 4 can be found. First calculate leakage 5.

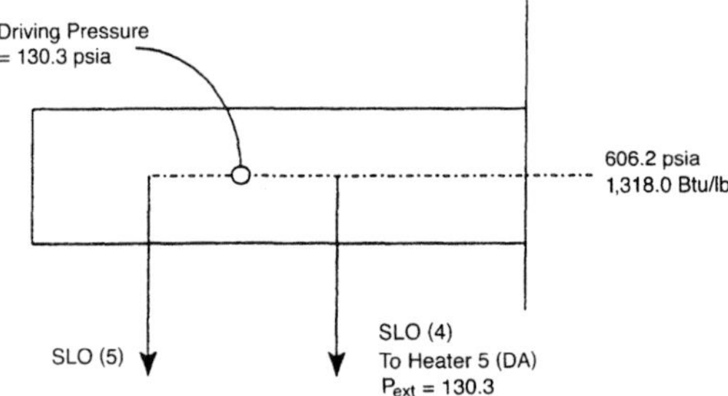
Driving Pressure
= 130.3 psia
606.2 psia
1,318.0 Btu/lb
SLO (5)
SLO (4)
To Heater 5 (DA)
P_{ext} = 130.3

From Figure 13A-2, packing constant SLO (5) = 970
For P = 130.3 psia and h = 1,318.0 Btu/lb

$$\upsilon = 4.6675 \text{ ft}^3\text{/lb}$$

$$\text{SLO}(5) = 970\sqrt{\frac{130.3}{4.6484}} = 5{,}125 \text{ lb/h}$$

By difference SLO (4) = SLO (4 + 5) − SLO (5)
$$= 15{,}267 - 5{,}125 = 10{,}142 \text{ lb/h}$$

4. Determine the expansion lines.
   a. The HP turbine efficiency has already been determined since the HP's ELEP has been found and initial and

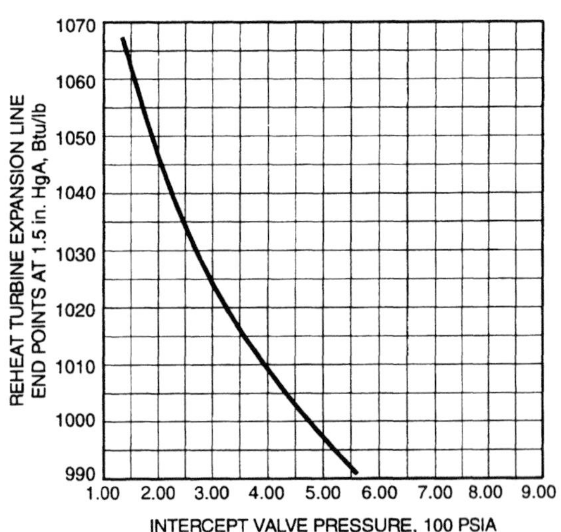
REHEAT TURBINE EXPANSION LINE END POINTS AT 1.5 in. HgA, Btu/lb
1070
1060
1050
1040
1030
1020
1010
1000
990
1.00 2.00 3.00 4.00 5.00 6.00 7.00 8.00 9.00
INTERCEPT VALVE PRESSURE, 100 PSIA
THESE EXPANSION LINE END POINTS ARE FOR HEAT BALANCE CALCULATIONS IN WHICH GLAND LEAKAGE STEAM IS USED IN THE FEEDWATER HEATING CYCLE.
TO OBTAIN THE ENTHALPY OF THE STEAM ENTERING THE CONDENSER READ THE CURVE AT 1.5 IN. Hg ABS. AND CORRECT TO THE DESIRED EXHAUST PRESSURE USING FIGURE 13A-6, AND CORRECT FOR EXHAUST LOSS USING FIGURE 13A-8.

Fig. 13A-5. Reheat expansion line end points at 1.5 in. HgA. (From General Electric. Used with permission.)

first-stage conditions are known. The expansion line could therefore be drawn, but is not actually needed for this calculation.

The IP-LP ELEP can be found for a 1.5 in. HgA exhaust pressure from Fig. 13A-5. For our $P_{IV} = 545.6$ psia, the ELEP at 1.5 in. HgA is 992.5 Btu/lb. Correct this to 2.0 in. HgA.

From the steam tables at 1.5 in. HgA;

$$X = (h - h_f)/h_{fg} = (992.5 - 59.74)/1{,}041.8$$

$$= 0.8953$$

$$Y = (1 - 0.8953)100 = 10.47\%$$

From Fig. 13A-6;

$$\Delta ELEP_{Y = 0,\ 2.0 \text{ in. HgA}} = 17.8 \text{ Btu/lb}$$

$$\Delta ELEP_{2.0} = \Delta ELEP_{Y=0}(0.87)(1 - 0.01Y)(1 - 0.0065Y)$$

$$= 17.8(0.87)[1 - (0.01)10.47][1 - (0.0065)10.47]$$

$$= 12.9 \text{ Btu/lb}$$

$$ELEP_{2.0} = ELEP_{1.5} + \Delta ELEP_{2.0}$$

$$= 992.5 + 12.9 = 1{,}005.4 \text{ Btu/lb}$$

Now calculate the bowl conditions to establish the top point of the expansion line. There is a 2% pressure drop across the intercept valve.

$$P_{IP\ Bowl} = (1 - 0.02)545.6 = 534.7 \text{ psia}$$

There is leakage into the bowl from the HP turbine. Determine mixed enthalpy of the bowl.

The exit conditions of the reheater are 545.6 psia and 1,000° F; therefore,

$$h_{Rht} = 1{,}519.0 \text{ Btu/lb}$$

$$h_{IP\ Bowl} = [M_{Rht}\, h_{Rht} + M_{VSLO(1)}\, h_{VSLO(1)} + M_{GSLO}\, h_{GSLO}]/(M_{Rht} + M_{VSLO(1)} + M_{GSLO})$$

$$= [2{,}562{,}146(1{,}519.0) + 3{,}893(1{,}460.4) + 32{,}333(1{,}427.9)]/(2{,}562{,}146 + 3{,}893 + 32{,}333)$$

$$= 1{,}517.8 \text{ Btu/lb}$$

The following points are now drawn on a Mollier chart. See Fig. 13A-7.

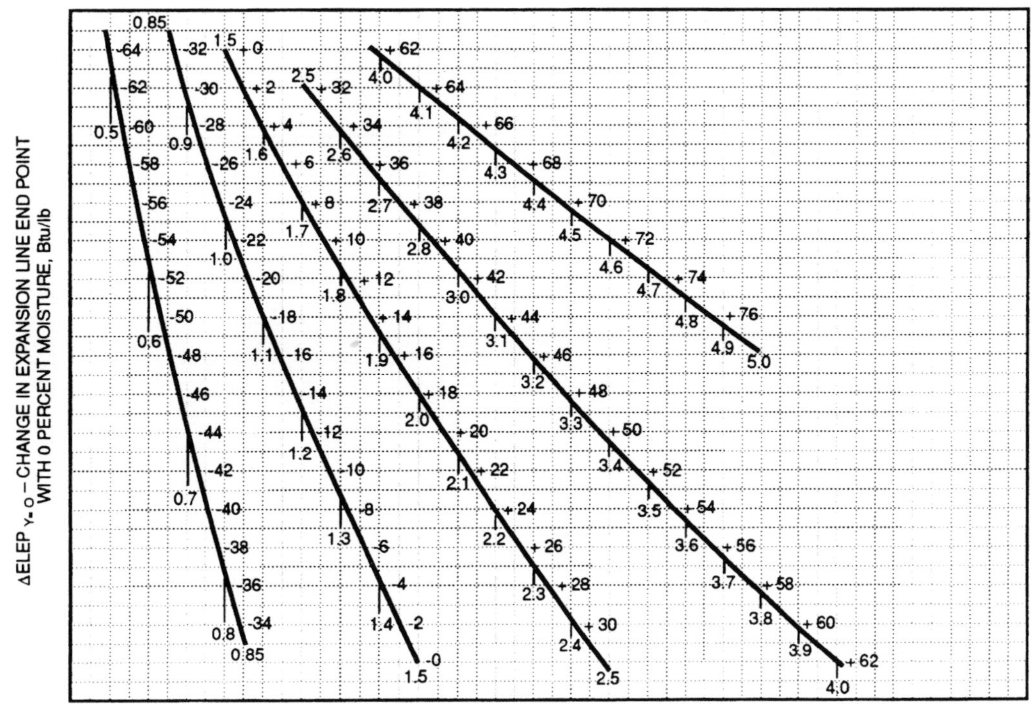

1. WHEN EXPANSION LINE END POINT AT 1.5 IN. Hg ABS. IS IN THE MOISTURE REGION, USE THE FOLLOWING EQUATION:
   ΔELEP= ΔELEPY= O (0.87) (1-0.01Y) (1-0.0065Y)
   Y= PERCENT MOISTURE AT ELEP AT 1.5 IN. Hg ABS.
2. WHEN THE EXPANSION LINE END POINT AT 1.5 IN. Hg ABS. IS IN THE SUPERHEAT REGION, MULTIPLY THE CHANGE IN EXPANSION LINE END POINT AT 0 PERCENT MOISTURE BOTH BY THE FACTOR 0.87 AND THE RATIO OF SPECIFIC VOLUME AT THE EXPANSION LINE END POINT UNDER CONSIDERATION TO THE DRY SATURATED SPECIFIC VOLUME AT 1.5 IN. Hg ABS.

Fig. 13A-6. Correction to expansion line point for exhaust pressure. (From General Electric. Used with permission.)

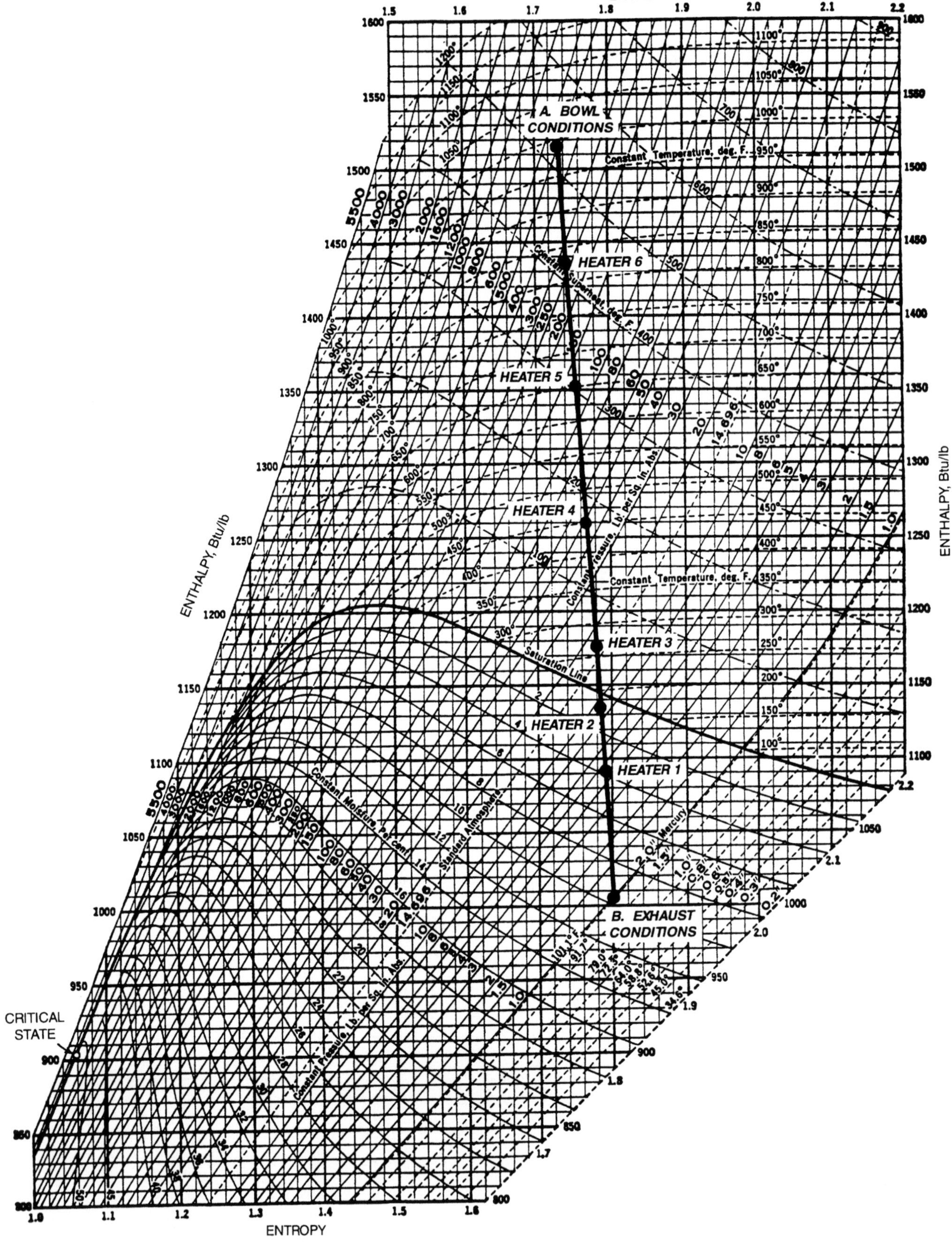

Fig. 13A-7. IP-LP expansion line.

Bowl conditions—534.7 psia and 1,517.8 Btu/lb

Exhaust conditions—2.0 in. HgA and 1,005.4 Btu/lb (ELEP)

b. Calculate IP shaft leakage, SLO(6). From the Mollier chart, the enthlapy at the leakage pressure can be found. The leakage pressure is the same as Heater 5 extraction pressure, which equals 130.3 psia. The enthalpy is 1,352.7 Btu/lb. These conditions correspond to a specific volume of υ = 4.9997 ft³/lb

From Fig. 13A-2, the packing constant for SLO(6) = 600.

$$\text{SLO(6)} = 600\sqrt{\frac{130.3}{4.9997}} = 3{,}063 \text{ lb/h}$$

5. Establish heater and feedwater/condensate conditions.

The easiest way to keep track of these values is to set up a table as shown in Table 13A-4. The extraction enthalpies are taken from the expansion line on the Mollier chart (Fig. 13A-7).

Since work is done on the feedwater by the boiler feed pump, the enthalpy rise must be accounted for.

a. Determine the enthalpy rise across the boiler feed pump

The total developed head required by the boiler feed pump is TDH = Discharge head − Suction head

The conversion from psi to feet of water is as follows.

$$\text{ft/psi at } 345.2°\text{ F} = \text{ft/psi at } 60°\text{ F} \times \frac{\text{Sp. vol. at } 345.2°\text{ F}}{\text{Sp.vol. at } 60°\text{ F}}$$

$$= 2.308\ (0.01793/0.01604)$$

$$= 2.580 \text{ ft/psi}$$

For typical heat balance purposes, the static head of the deaerator water level relative to the boiler feed pump suction center-line and the friction loss in the suction line are neglected.

The boiler feed pump discharge head is calculated using the "Typical Assumptions Made in Establishing a Heat Balance," Table 13A-1.

$$\text{Discharge head} = \text{Turbine throttle pressure} \times 1.25$$

$$= 2{,}415 \times 1.25 = 3{,}018 \text{ psia}$$

The deaerator pressure is found from P_{ext5} − 3%/100 (P_{ext5}) where 3% is the pressure drop from Fig. 13A-1.

$$P_{DA} = (1 - 0.03)\ 130.3 = 126.4 \text{ psia}$$

$$\text{Total developed head (TDH)} = 3{,}018 \text{ psia} - 126.4 \text{ psia}$$

$$= 2{,}892 \text{ psia (2.580 ft/psi)}$$

$$= 7{,}462 \text{ ft}$$

To calculate the enthalpy rise across the pump;

$$\Delta h = (\text{Head in ft/J})/(\eta_p)$$

From the heat balance diagram Fig. 13A-1, η_p = 87.0% at this load condition.

$$\Delta h = (7{,}462/778)/(0.870) = 11.0 \text{ Btu/lb}$$

This enthalpy rise is added to the saturated liquid enthalpy leaving the deaerator (h_f = 316.7 Btu/lb at 126.4 psia).

$$h_{\text{BFP exit}} = 316.7 + 11.0 = 327.7 \text{ Btu/lb}$$

Treating the feedwater as compressed liquid at 3,018 psia with an enthalpy of 327.7 Btu/lb;

$$T_{\text{BFP exit}} = 351.3°\text{ F}$$

b. Determine BFPT flow rate.

The work required by the boiler feed pump must be produced by the boiler feed pump turbine. Neglecting friction losses in the connections between the BFP and the BFPT, the work can be considered equal.

$$W_{BFP} = W_{BFPT}$$

$$W_{BFP} = M_{FW}\ \Delta h_{BFP} = M_{BFPT}\ \Delta h_{BFPT} = W_{BFPT}$$

Table 13A-4. Heater and Feedwater/Condensate Conditions

Extraction to Heater	Extraction Pressure (psia)	Extraction Enthalpy (Btu/lb)	Extraction Line Pressure Drop (%)	Heater Pressure (psia)	Heater Saturation Temp. (°F)	Terminal Diff. (°F)	Feedwater Temp. Leaving Heater (°F)	Feedwater Enthalpy[b] Leaving Heater (Btu/lb)	Drain Cooler Approach (°F)	Heater Drain Temp. (°F)	Heater Drain Enthalpy (Btu/lb)
7	606.2	1,318.0	3	588.0	484.0	−3	487.0	472.8	10	412.9	389.1
6	271.9	1,436.0	6	255.6	402.9	0	402.9	381.3	10	361.3	333.8
Condition leaving boiler feed pump							351.3	327.7			
5 (D.A.)	130.3	1,352.7	3	126.4	345.2	—	345.2	316.7	—	—	—
4	49.4	1,260.0	6	46.4	276.3	2	274.3	243.6	10	226.5	194.8
3	17.8	1,178.0	6	16.7	218.5	2	216.5	185.1	10	193.5	161.6
2	9.63	1,136.0	6	9.05	188.5	5	183.5	152.0	10	162.9	130.9
1	4.80	1,091.0	6	4.51	157.9	5	152.9	121.3	5	108.0	76.0

[a]Extraction enthalpies are found from expansion line or HP turbine exhaust conditions.

[b]Feedwater downstream of the boiler feed pump is considered to be a compressed liquid.

From the efficiency of 82.6%, the enthalpy drop across the BFPT can be calculated.

The pressure at the BFPT throttle can be calculated from the extraction 5 pressure which is the same as the IPT exhaust pressure. Note there is a 3% pressure drop in the BFPT supply line.

$$P_{ext5} = 130.3 \text{ psia} \qquad h_{ext5} = 1{,}352.7 \text{ Btu/lb}$$

$$P_{BFPT\ th} = (1 - 0.03)130.3 = 126.4 \text{ psia}$$

$$h_{BFPT\ th} = 1{,}352.7 \text{ Btu/lb}$$

At these conditions, the entropy is as follows:

$$S = 1.7512 \text{ Btu/lb-}^\circ\text{R}$$

Assuming isentropic expansion to the BFPT exhaust pressure of 2.5 in. HgA, the ideal exhaust enthalpy can be determined.

$$P = 2.5 \text{ in. HgA}$$

$$S = 1.7512 \text{ Btu/lb-}^\circ\text{R}$$

$$h_{exh\ ideal} = 991.7 \text{ Btu/lb}$$

The actual exhaust enthalpy can be calculated from the turbine efficiency.

$$\begin{aligned} h_{exh\ act} &= h_{th} - \eta_{BFPT}(h_{BFPT\ th} - h_{exh\ ideal}) \\ &= 1{,}352.7 - 0.826(1{,}352.7 - 991.7) \\ &= 1{,}054.5 \text{ Btu/lb} \end{aligned}$$

The BFPT flow rate can then be calculated.

$$M_{FW} = \text{Turbine throttle flow} + \text{makeup} = 2{,}903{,}765 \text{ lb/h}$$

$$M_{BFPT} = \frac{M_{FW}\Delta h_{BFP}}{\Delta h_{BFPT}} = \frac{2{,}903{,}765(11.0)}{(1{,}352.7 - 1{,}054.5)} = 107{,}114 \text{ lb/h}$$

c. Feedwater and condensate flow rates.

The feedwater flow rate equals the turbine throttle flow plus makeup flow.

The condensate flow rate is the sum of the bottom four feedwater heater extractions plus the excess flow from the steam seal receiver (SSR) plus the 2,400 lb/h from the LP turbine seals (shown schematically in Fig. 13A-1 as coming from the SSR and discussed in Fig. 13A-2) plus the LP turbine exhaust flow plus the BFPT flow plus the makeup flow.

6. Calculate first iteration—determine final turbine flow distribution.

Having established the initial mass flow distribution and the resulting pressures and enthalpies for extractions and the heater inlet and outlet conditions, the calculation of the turbine heat balance can begin.

While working through the iterations required to balance the heat transfer, do not correct the flow assumptions until a complete trial heat balance has been calculated. After working through the balance, new extraction flows will have been found. At this point, the newly calculated extraction flows should be used to correct the previously assumed turbine-flow distribution. New extraction pressures can then be found. If the new flow distribution results in new extraction pressures within ±0.5%, they can typically be considered acceptably accurate. If the pressures exceed this tolerance, the new extraction pressures should be used and the heater conditions corrected to correspond with their new operating pressures.

7. Calculate the turbine shaft work and turbine output. To calculate the shaft work, write the first law around the turbine.

$W_{HP_{sh}}$ = High-pressure turbine shaftwork, kW

$$W_{HP_{sh}} = \frac{[(M \times h)_{Throttle} - (M \times h)_{VSLO(1+2)} - (M \times h)_{GLSO} - (M \times h)_{SLO(4+5)} - (M \times h)_{exh}]}{3{,}412.14 \text{ Btu/kWh}}$$

$W_{IP\text{-}LP\ shaft}$

= Intermediate- and low-pressure turbine shaft work, kW

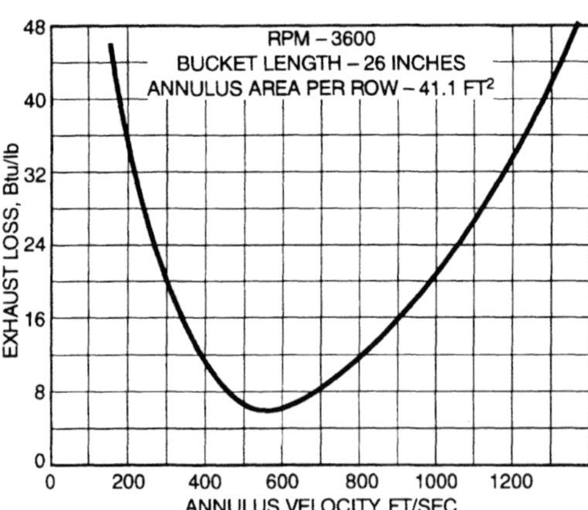

1. READ THE EXHAUST LOSS CURVE AT THE ANNULUS. VELOCITY OBTAINED FROM THE FOLLOWING EXPRESSION:

$$VAN = \frac{QA\ (V)\ (1.0\text{-}0.01Y)}{3600\ (AA)}$$

2. THE ENTHALPY OF THE STEAM ENTERING THE CONDENSER IS THE QUANTITY OBTAINED FROM THE FOLLOWING EXPRESSION:

UEEP = ELEP + (EXHAUST LOSS) (0.87) (1.0-0.01Y) (1.0-0.0065Y)

3. THE EXHAUST LOSS CURVE INCLUDES THE LOSS IN INTERNAL EFFICIENCY WHICH OCCURS AT LIGHT LOADS AS OBTAINED IN TESTS.

4. THIS CURVE IS BASED ON THE 1967 ASME STEAM TABLES.

LEGEND:

VAN – ANNULUS VELOCITY IN FEET PER SECOND
QA – CONDENSER FLOW IN POUNDS PER HOUR
V – SATURATED DRY SPECIFIC VOLUME IN CUBIC FEET PER POUND CORRESPONDING TO THE ACTUAL EXHAUST PRESSURE (WHEN THE END POINT IS IN THE SUPERHEAT REGION V = THE ACTUAL SPECIFIC VOLUME AT THE END POINT.)
AA – ANNULUS AREA IN SQUARE FEET
Y – PERCENT MOISTURE AT THE EXPANSION LINE END POINT
ELEP – EXPANSION LINE END POINT AT THE ACTUAL EXHAUST PRESSURE
UEEP – USED ENERGY END POINT

Fig. 13A-8. Exhaust loss curve. (From General Electric. Used with permission.)

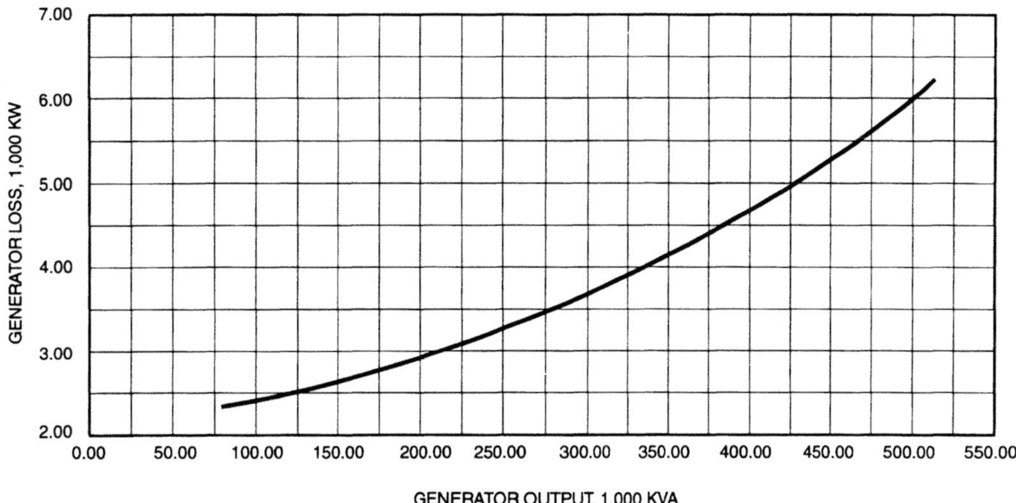
GENERATOR LOSS, 1,000 KW
7.00
6.00
5.00
4.00
3.00
2.00
0.00
50.00
100.00
150.00
200.00
250.00
300.00
350.00
400.00
450.00
500.00
550.00
GENERATOR OUTPUT, 1,000 KVA
511,000 kVA AT 45 PSIG H2 PRESSURE
CONDUCTOR COOLED 3600 RPM
NOTES:
GENERATOR LOSSES ASSUME RATED HYDROGEN PRESSURE AT ALL LOADS.
GENERATOR LOSS AT REDUCED HYDROGEN PRESSURE (P) = LOSS AT RATED HYDROGEN PRESSURE - 11.2 (PRATED - P).
USE GENERATOR REACTIVE CAPABILITY CURVE TO DETERMINE GENERATOR CAPABILITY AT REDUCED HYDROGEN PRESSURE.
TURBINE GENERATOR MECHANICAL LOSSES ARE NOT INCLUDED IN THE GENERATOR LOSS CURVE.
IF HYDROGEN AND STATOR LIQUID COOLERS ARE LOCATED IN THE CONDENSATE LINE, THE LOSS TRANSFERRED TO THE COOLERS IS 474
kW LESS THAN THE GENERATOR LOSS AT ALL LOADS.

Fig. 13A-9. Generator loss curve. (From General Electric. Used with permission.)

$$W_{\text{IP-LP shaft}} = \frac{[(M \times h)_{\text{IV}} + (M \times h)_{\text{GSLO}} - \Sigma (M \times h)_{\text{ext's to heaters}} - (M \times h)_{\text{BFPT}} - (M \times h)_{\text{SLO(6)}} - M_{\text{exh}}\,(\text{UEEP})]}{3{,}412.14 \text{ Btu/kWh}}$$

Note that the work for the IP-LP shaft is based on the used energy end point (UEEP). The used energy end point is the sum of the ELEP and the turbine exhaust loss. The turbine exhaust loss can be found by the method described in Fig. 13A-8.

To calculate the turbine output, deduct the mechanical losses and the generator losses.

$$W_{\text{send out, kW}} = W_{\text{HPsh}} + W_{\text{IP-LPsh}} - \text{Mechloss} - \text{Genloss}$$

The mechanical losses or fixed losses, as they are also called, are found in Fig. 13A-2. These losses do not vary with load.

The generator losses vary with load and are given in Fig. 13A-9.

To calculate the generator losses, reduce the shaft work by the fixed losses and a first guess generator loss. Divide the remainder by the generator power factor, which for this unit is 0.9, to calculate the generator output in kVa. Enter curve in Fig. 13A-9 and read the generator loss. If the initial guess is close, stop. If not, recalculate kVa and repeat.

8. Calculate net turbine heat rate.

   Net turbine heat rate (NTHR) equals total heat input (Q) divided by the net useful work of the turbine cycle after accounting for boiler feed pump work (W).

$$\text{NTHR} = Q \text{ in/W send out of turbine cycle}$$

For the turbine driven feed pump arrangement, this equation becomes

$$\text{NTHR} = \frac{M_{\text{main stm}}(h_{\text{Th}} - h_{\text{FW}}) + M_{\text{Rht}}(h_{\text{Hot Reheat}} - h_{\text{HP exh}})}{\text{kW send out}}$$

14

Power Plant Atmospheric Emissions Control

Lloyd L. Lavely and *Alan W. Ferguson*

14.1 INTRODUCTION

Fossil-fueled electric-generating plants emit combustion gases to the atmosphere that may contain pollutants in the gas phase or as suspended solid or liquid particulates. Environmental laws and regulations define and restrict the amounts of the gas, solid, or liquid pollutants that can be emitted to the atmosphere.

This chapter summarizes emission reduction technologies and emission monitoring technologies for combustion gases. Regulatory programs and associated requirements pertaining to air and water emissions, as well as other facility design considerations, are discussed in Chapter 26. Potential air pollutant combustion gas emissions can be controlled prior to combustion by switching, blending, or cofiring with low sulfur and ash containing fuels; or using other reduced pollutant content fuels. In addition, cleaning coal by beneficiation can reduce pollutants in the fuel. Pollutants can also be reduced by energy conservation measures or by increasing plant thermal efficiency. The combustion gas emission control technologies described in this chapter are currently available commercial processes located within or after the combustor.

Emissions associated with coal and limestone handling, or liquids and solids waste treatment, are covered in Chapters 5 and 16, respectively. Some combustor pollutants are converted to harmless gases, some are recovered or converted to useful byproducts, and the remainder are recovered, treated, and disposed of as described in Chapter 16, Liquids and Solids Waste Treatment.

The atmospheric emissions control design, including balance-of-plant impacts, should be integrated into the power plant design so that regulatory requirements are reliably achieved at the lowest overall cost.

Potential future emission regulations or regulatory programs should also be considered in new or modified existing plant designs. Atmospheric emission laws and regulations are constantly evolving to control more pollutants or to reduce currently regulated emissions to lower levels. For example, the US Department of Energy (DOE) recently received proposals regarding an integrated demonstration power plant for technologies to reduce NO_x and SO_2 emissions to 0.06 lb/MBtu, particulates to 0.003 lb/MBtu, and to meet future hazardous pollutant limits, while increasing overall thermal efficiency and lowering total generating costs. The emission limits for this program are about 10 times more stringent than existing emission limits. The United States trend to regulate more pollutants and further lower emission levels as technology develops will continue in part because of the various control technology provisions of the Clean Air Act and its associated amendments. Ideally, a new emissions control design would meet (or be compatible with retrofit additions that will meet) future regulations over the plant's life.

14.2 PARTICULATE CONTROL

After combustion, most fossil fuels, excluding natural gas or liquified petroleum gas, yield residual particulate. The primary particulate constituents are ash and unburned carbon. The ash consists of silica, alumina, and other noncombustible compounds. The amount of ash varies widely, sometimes exceeding 50 wt %. These particulates leave the boiler as bottom ash, economizer ash, or fly ash. Bottom ash falls to the bottom of the boiler and is removed periodically. Economizer ash generally has larger particle size, separates from the flue gas, and drops into hoppers for removal in the economizer area. Fly ash is relatively small and leaves the boiler entrained with the flue gas.

The following table lists typical ash distributions for a variety of boiler types:

Boiler type	Bottom ash (%)	Fly ash (%)	Economizer ash (%)
Pulverized coal	10–30	70–90	0–10
Stoker	60–80	20–40	0–5
Cyclone	50–80	15–50	0–10
Circulating fluidized bed	5–90	10–95	0–5

The ash split is a function of the fuel particle size; boiler temperatures; ash fusion and deformation temperatures; ash particle sizes, densities, and shapes; and boiler flue gas velocities. Stoker-fired units produce the largest particles. Wall- and corner-fired pulverized coal (PC) fueled boilers produce smaller, spherical shaped particles. Particles from cyclone-fired units are also mostly spherical and are smaller than ash from PC boilers. Fluidized bed units produce a wide range of particles that are asymmetrical, nonspherical, and shaped more like crystals (Stultz and Kitto 1992).

Knowledge of the boiler type, fuel characteristics, the resultant fly ash mineral analysis, particle loading, and particle sizes is essential in the selection and design of an appropriate particulate removal control technology. Electrostatic precipitators and fabric filters are the most prevalent choices. The final selection is generally based on economics and compatibility with the design fuel, operating conditions, regulatory requirements, and other pollution control technologies required for the unit. Guaranteed outlet emissions of particulates in the 0.01 to 0.015 lb/MBtu range are typically available with either choice.

PARTICULATE MEASUREMENT TECHNIQUES. The US EPA provides reference test methods (40 CFR 60, Appendix A, Methods 5 and 17) that are adequate for determining emission rates down to 0.001 grains per dry standard cubic foot (gr/dscf). These test methods are accurate to very low levels if the test equipment operator uses the equipment properly and for a sufficient amount of sampling time. However, longer tests (required by very low emission limits) also make maintaining steady-state conditions more difficult. Test methods to determine very low emission rates require continued attention and should be appropriately planned and configured to furnish accurate and repeatable data.

14.2.1 Electrostatic Precipitators

Electrostatic precipitators (ESP) sized for the efficiency required to meet current particulate emission limits are operating on the full range of utility power plant applications. Accordingly, ESPs are the most widely installed utility particulate removal technology.

ESPs use transformer–rectifiers (T–Rs) to energize discharge electrodes and produce a high-voltage direct current (dc) electrical field between the discharge electrodes and grounded collecting plates. Particulate matter entering the electrical field acquires a negative charge and migrates to the grounded collecting plates. This migration can be expressed in engineering terms as an empirically determined effective migration velocity, but takes place in the turbulent flow regimen with the particulate entrained within the turbulent gas patterns. Thus, the charged particles are actually captured when the combined effect of electrical attraction and gas flow patterns moves the particulate matter close enough so that it can attach to the collecting surfaces. A layer of collected particles forms on the collecting plates and is removed periodically by mechanically rapping the plates. The collected particulate drops into hoppers below the precipitator and is removed by the ash handling system. Some particulate is also reentrained and either collected in subsequent electrical fields or emitted from the precipitator.

The physical size of an ESP is determined by the required particulate removal efficiency, the expected electrical resistivity of the fly ash to be collected, and the expected electrical characteristics of the energization system. Many parameters determine the ESP's capability for particulate collection. The following sections discuss the most important items.

14.2.1.1 Specific Collection Area. The ESP size is often measured in terms of specific collection area. The specific collection area (SCA) of a precipitator has units of square feet per 1,000 actual cubic feet per minute (acfm) of flue gas flow. The SCA for the required performance can be determined by using the Deutsch–Anderson equation, which relates the collection efficiency (E) to the unit gas flow rate, the particulate's effective migration velocity (see previous section), and the collection surface area (Stultz and Kitto 1992):

$$1 - E = e\frac{-wA}{V} \tag{14-1}$$

or

$$A = \left[\ln\left(\frac{1}{1 - E}\right)\right]\frac{V}{w} \tag{14-2}$$

where

E = ESP removal efficiency, %

$= 100\left(\frac{\text{Inlet dust loading} - \text{Outlet dust loading}}{\text{Inlet dust loading}}\right)$,

w = effective migration velocity, ft/min (m/s);

A = collection surface area, ft^2 (m^2), and

V = gas flow, ft^3/min (m^3/s).

Because of the assumption about an effective migration velocity to make use of this sizing equation, the empirical nature of ESP design is obvious. The more application experience and performance tests for a given ESP design, the better the suppliers can accurately predict the required effective migration velocity for a new application. ESP suppliers size ESPs according to their test experience for their specific design and to meet the specified parameters. It is nearly impossible to provide effective migration velocities without an abundance of application-related information. In general, sizing an ESP is still more of an art than a science.

Comparison of the SCA between designs requires adjustment for any varied collecting plate spacing (typically ranging from 9 in. to 16 in. for coal-fueled ash applications). For example, two ESPs with the same treatment time or volume may have different collection plate spacing. For the SCA comparisons to reflect this size equivalency, they must be put on an equivalent basis. This basis is often referenced to an equivalent SCA with a plate spacing of 12 in., even though there is really not an industry-wide standard. To do this, multiply the calculated collection plate area by the actual gas passage width and divide by the referenced gas passage width which will be the equivalent basis. Use this calculated equivalent area to calculate the equivalent SCA.

Larger plate spacing with less plate area per unit volume requires higher voltage (larger TRs) for attaining similar particulate removal performance. However, the overall power consumption and efficiency of the ESP may improve because current flows and associated spark rates are lower.

ESPs installed in the 1970s and 1980s used collecting plate spacings of 9 or 12 in. This spacing defines the gas path width between collecting plates. The vertical discharge electrodes are placed in the horizontal center of this space. Equivalent performance might be obtained with ESPs that have collecting plates spaced 14, 15, 16, 18, or 21 in. apart. Other ESP parameters must be adjusted with wider plate spacing and care must be taken to avoid high dust loading "space charge" problems in the inlet fields.

14.2.1.2 Treatment Time. The treatment time is the amount of time that the flue gas is within the electric and collection fields of the ESP. High efficiency ESPs typically have treatment times between 7 and 20 seconds. Shorter times apply to high-sulfur fuels with higher permissable emissions. Longer times are required for low-sulfur fuels with low sodium oxide fly ash contents or lower permissable emissions.

14.2.1.3 Flue Gas Velocity. The speed with which the flue gas moves through the ESP is important in the design and sizing of the ESP. Higher design velocities result in smaller ESP cross-sectional area inlets but could result in an undersized ESP with high ash reentrainment losses and low treatment time. The most practical design gas velocity with coal fuel fly ash is around 4 ft/s. ESPs on oil fueled boilers generally have velocities in the 3 to 4 ft/s range. Some instances exist where a higher velocity may be appropriate but typically, higher velocities raise concern about reentrainment of the collected ash. Lower velocities are suggested for ash with a high potential for reentrainment, such as particulate with a high carbon content. However, the effects of too low a velocity as would be experienced during low boiler load operation must also be considered. Too low a velocity in the ESP could result in the flue gas meandering and possibly allowing ionization of the gas, which leads to sparkover and performance degradation.

14.2.1.4 Aspect Ratio. Another way to combat reentrainment difficulties is to install an ESP with a higher aspect ratio. The aspect ratio is the ratio of the treatment length (effective ESP length excluding walkways) to the collection plate height. Higher aspect ratios allow the particulate collected at the top of the inlet collecting plates to reach the hoppers before exiting the ESP. Many existing ESPs have aspect ratios of around 0.8 to 1.2. These low aspect ratios may be appropriate for applications where a problem with reentrainment is not expected or where regulatory requirements are not severe. However, meeting contemporary particulate emission limits generally necessitates aspect ratios in the range of 1.2 to 2.0.

14.2.1.5 Gas Distribution. Optimum particulate removal requires uniform gas velocity throughout the entire precipitator treatment volume with minimal gas bypass or "sneakage" around the discharge electrodes or collection plates. If the flue gas distribution is not even, the particulate removal will decrease and reentrainment losses will increase in high-velocity areas. This will reduce overall collection efficiency. The International Conference of Clean Air Companies (ICAC, formerly IGCI) developed standards and test methods to define the minimum acceptable gas flow distribution within an ESP. These standards are described in IGCI Publication EP-7. It may be appropriate to require that in addition to meeting IGCI EP-7 requirements, the flue gas exhibit a root mean square (RMS) velocity deviation of $<15\%$. Physical scale models with gas flow measurement studies are necessary to locate and design flow correction devices that will permit meeting these standards. The correct fabrication and installation of these devices within the ESP is also critical. These are necessary requirements when high efficiencies are needed.

14.2.1.6 Electrostatic Precipitator Control. The objective of the ESP control system is to create and maintain an electrical field so that the desired collection of fly ash occurs. An additional control objective is to minimize power consumption consistent with performance. This is accomplished by T–Rs, current-limiting reactors, and a supervisory control system. The T–Rs step up the voltage from 480 V to between 25 kV and 125 kV, depending on plate spacing and particulate electrical resistivity, and convert alternating current (ac) to direct current (dc). An ESP with wide plate spacing requires T–Rs with a higher voltage rating and more current capability than an ESP with narrow spacing.

The ESP control system maximizes the effective voltage and current. Electrical sparks or arcs cause draining of the input power and can damage the ESP internals. The control system reduces power to the affected electrical bus section when an arc or spark is sensed. Thus, the "weak link" controls the power input. The "ideal" bus section would have a single discharge electrode. Since this would be too costly, the number of electrically separate bus sections is often limited to 1 per 25,000 ft^2 of collection area. Many ESPs are meeting emission requirements with fewer and larger bus sections, but more bus sections improve removal efficiencies. The tradeoff is the cost increase.

Analog controls sense sparks or arcs and reduce power input. Newer computerized digital control systems react much faster. They sense the change in the shape or slope of the power input curve and react rather than waiting for a spark to occur. They allow the operator to control the set back voltage, set back time, power up ramp rate, and duration of set back and power up. They also allow the ESP to operate at the most efficient particulate removal by using the highest voltage and current possible with automatic phase back if the spark rate is exceeded and interrupt/pedestal/ramp or phase back if an arc occurs.

Discharge opacity can also be an input power limiting parameter with a computer control system. An opacity control system can limit power use by changing the apparent current limits of the controlled T–Rs. As the opacity dips below the set point, power is stepped back so that energy is saved. Conversely, as the opacity increases beyond a set point, the input power is allowed to seek up to the level determined by the spark or arc characteristics.

Intermittent energization uses the ESP's large capacitor behavior and shuts off power for part of the ac half-cycles. The input power to the ESP is reduced because the ac power to the T–Rs is stopped for short, adjustable periods measured in fractions of a second. During these short time periods, the ESP acts as a large capacitor, and the removal efficiency is not decreased because the ESP voltage is maintained and the particle charging rate is not significantly impacted.

14.2.1.7 Rapping Systems. Collecting and discharge electrodes retain particulate on their surface. If this material gets too thick, the electrical characteristics become unstable (spark or arc over, back-corona, or decrease in corona generation) and collection decreases. Periodic rapping cleans the collecting and discharge electrodes to ensure the proper functioning of the ESP.

During rapping, groups of these electrodes receive short and generally intense intermittent energy inputs via rotating hammers, vibrators, or dropped weights. The resultant shearing acceleration forces cause large sheets of fly ash to fall from the collecting electrodes into the hoppers. If the acceleration forces perpendicular to the plate surfaces are too large, the fly ash is excessively reentrained into the gas stream. The rapping energy is imparted to either the top, bottom, or side (edge) of the collection plates and discharge electrodes. All three locations are successful. The online capability to adjust the rapping frequency, sequence, intensity, or combinations of these parameters to match changes in load or fly ash properties, may improve overall collection or electrical efficiency. This capability is helpful in increasing fuel or operating flexibility. However, many ESPs with internal and unadjustable online rapping intensity or sequence designs operate successfully.

Advances in rapper control technology may increase ESP efficiency by enhancing the plate and discharge electrode cleaning to match the rate at which particulate is collecting in any given mechanical section. The rapper control system customizes the rapper sequence, frequency, and intensity by comprehensive measurements of ESP operating parameters.

14.2.1.8 Resistivity. Resistivity is a measure of how easily the ash or particulate acquires an electric charge. Typical values for resistivity range from 1×10^8 ohm-cm to 1×10^{14} ohm-cm. High-resistivity particulate is difficult to remove because it does not accept a charge easily. Conversely, with very low resistivity particulate, the collected particles can rapidly lose their charge and become reentrained. High-resistivity particles require a larger ESP than those with a lower resistivity. The effective migration velocity, w, is lower for high resistivity. The ideal resistivity range for electrostatic precipitation is 5×10^9 to 5×10^{10} ohm-cm.

Operating resistivity varies with the flue gas moisture and SO_3 content, ash chemical composition, and temperature. The SO_3 content is the largest single factor affecting the operating resistivity. The amount of SO_3 condensed on the fly ash is a function of the sulfur content of the coal, the amount of oxidation of SO_2 to SO_3, and the operating temperature of the ESP. For medium- and high-sulfur fuels, the particulate's surface SO_3 resistivity controls precipitation. For low-sulfur fuels, the volume resistivity, as determined by fly ash chemical composition, controls precipitation. For these, sodium oxide above 0.5 wt % has the biggest influence. The resistivity of fly ash follows a bell-shaped curve with maximum resistivity at a temperature of about 300° F. The resistivity is lower at higher or lower temperatures. Thus, low sulfur fuels, which can form only small amounts of SO_3, produce particulate that is difficult to precipitate at the typical 300° F air heater outlet temperature. High-sulfur coals with corresponding high SO_3 generation produce ash that is easily precipitated.

Designing an ESP to operate on the hot side of the air heater with temperatures between 700 and 750° F takes advantage of low-volume resistivities. Volumetric gas flow rates are much higher at hot-side temperatures, but the drastically lower resistivities for low-sulfur coals more than offset this disadvantage. During the 1970s precipitators were purchased for installation on the hot side of the air heater. However, for most of these installations, under the influence of high temperature and intense electric field, the charge-carrying sodium ions apparently migrated to the outside of the ash layer and were removed during rapping. This left a highly resistive, tenacious ash layer on the plates, which resulted in high particulate emissions. Because of this problem and other maintenance-related problems, many of these units were converted to cold-side operation; currently, hot-side precipitators usually are not offered commercially.

14.2.2 ESP Design Characteristics

ESP suppliers have specific major component design features. These include discharge and collection electrodes; gas flow distribution devices; casing and structural framework; electrical power supply including controls, high voltage dc distribution, and insulators; and rapper cleaning systems and controls. The ESP internal components are top supported to facilitate thermal expansion.

The major design types are generally characterized by the discharge electrode and support design. The early ESP discharge electrodes were wires up to 30 ft long, with alignment weights on the bottom end and the top end suspended from horizontal frames. These early day US ESP designs, built through the 1970s, became labeled the "American" or "weighted wire" type. The wires eventually exhibited breakage due to electrical erosion or fatigue. A single broken wire can ground a bus section, significantly affecting the performance and ultimately the precipitator's reliability.

The wire breakage problem was addressed in Europe by

using relatively short (15 ft or less) discharge electrode strips (or wires) stretched within vertically oriented structural support frames. These designs were labeled "pipe frame," "bed spring," or "mast" types and were characterized as the "rigid frame" or "European" type. The Babcock & Wilcox/Rothemule rigid frame precipitator shown in Fig. 14-1 is typical of this design.

The reliability advantage of the "rigid frame" with no marked wire breakage versus the lower cost of the "weighted wire" led to the rigid discharge electrode design. The rigid discharge electrodes replaced the "weighted wires" in the "American" design. They are pipes or other light structural members with points, scallops, or other sharp edges for corona generation. This design permitted taller and less expensive precipitators with collection plate heights and discharge electrode lengths up to 50 ft. The Enelco precipitator with Rigitrode discharge electrodes shown in Fig. 14-2 is typical of the current American type design.

14.2.2.1 ESP Size Variance for Different Coals. ESP design parameters for boilers depend on the different possible fuels. The wide variance in ESP size, performance, and cost reflect the sensitivity of ESPs to the properties of the coal burned. Calculation of the required SCA or treatment volume (time) depends on the design fuels, required outlet emissions, and the ESP performance parameters. Certain coals have ash properties that make collection in an ESP very difficult, if not impossible. Silica (SiO_2) and alumina (AlO_2), which are

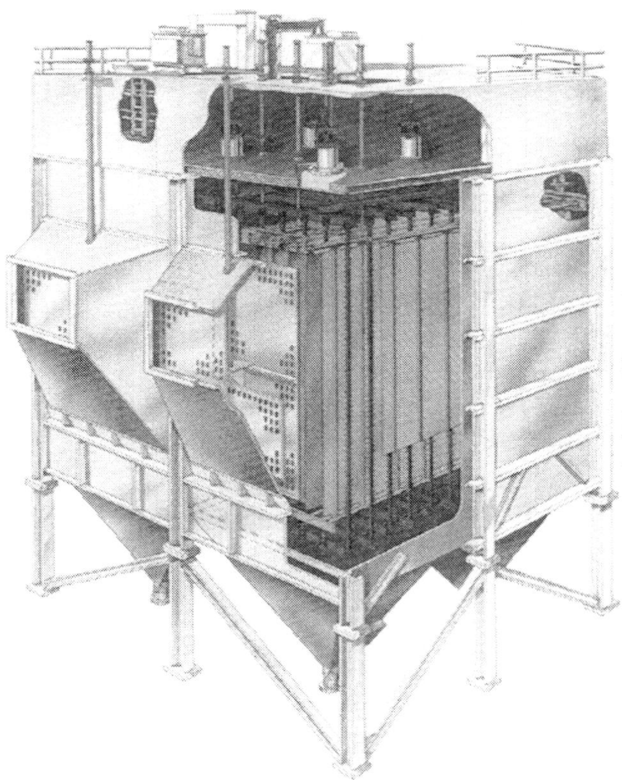

Fig. 14-2. Enelco precipitator with Rigitrode discharge electrodes. (From Environmental Elements Corporation. Used with permission.)

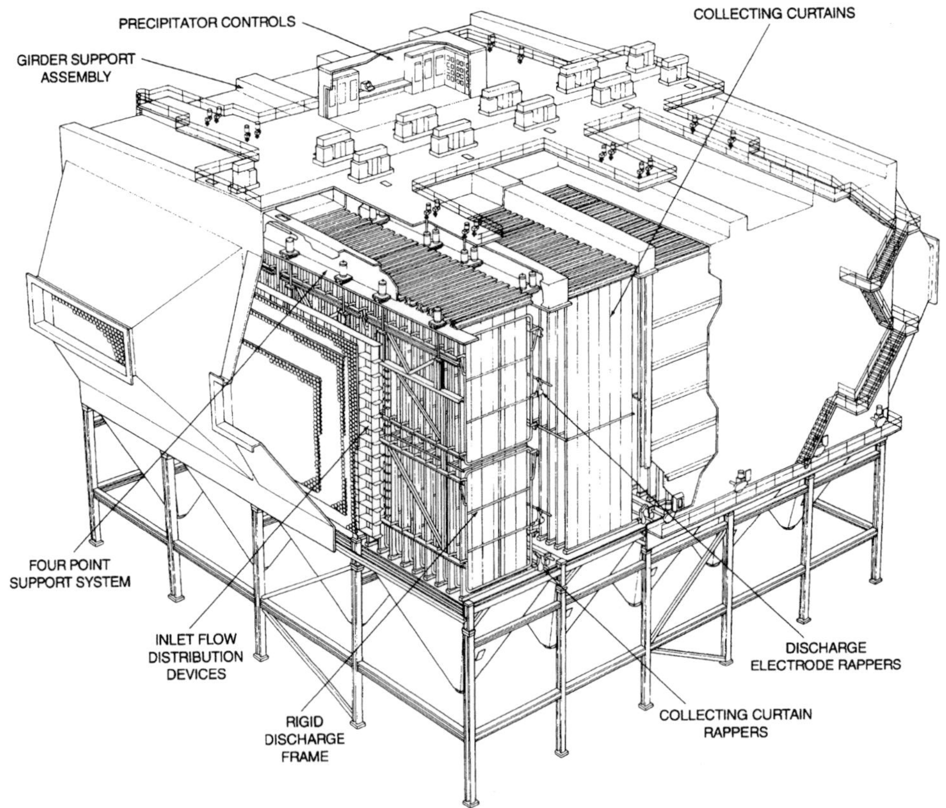

Fig. 14-1. B&W/Rothemule rigid frame electrostatic precipitator. (From Babcock & Wilcox. Used with permission.)

used as electrical insulators, do not easily acquire an electric charge, and therefore are difficult to precipitate because of the high ash resistivity. The high silica and alumina content of the ash in some coals (such as Australian) requires a very large ESP to adequately collect the fly ash.

14.2.2.2 Fly Ash Conditioning. The operating resistivity of the ash affects the effective migration velocity and, therefore, the ESP size and cost. Fly ash conditioning systems alter the effective ash resistivity and make the ash easier to collect. They use various conditioning agents to improve the fly ash's resistivity or cohesiveness.

Most conditioning systems reduce the surface resistivity by condensing the additive on the ash particle. However, higher moisture levels in the flue gas lower ash resistivity even when the gas is above the water dewpoint temperature. As temperatures fall below the 30° F approach to water saturation, resistivity may become so low that particle reentrainment may result in increased emissions.

The most popular conditioning agent is SO_3. Gaseous SO_3 is most commonly created for larger plants by burning sulfur to yield SO_2 and then catalytically oxidizing the SO_2 to make SO_3.

Surface resistivity of low-sulfur coal fly ash, at or below the acid dew point temperatures, frequently can be reduced by the injection of trace amounts of SO_3 into the flue gas ahead of the ESP. This allows a smaller ESP for a new plant. It allows an existing plant to reduce emissions or burn a coal with a lower sulfur content or higher inlet ash content than those for which the ESP was originally designed. Since nearly all the SO_3 condenses on the particulate matter and is subsequently collected in the precipitator, the emission levels of SO_3 are not significantly affected.

Very small particles may not be collected or will be reentrained when the ESP is rapped. Ammonia injection often causes the particles to agglomerate and be collected. Ammonia also enhances ash particle adhesion to reduce reentrainment.

Dual conditioned units benefit from the reduced resistivity from SO_3 injection and enhanced particle agglomeration and cohesion from the ammonia (NH_3) addition. With gas conditioning, an ESP may be 20% to 30% smaller.

14.2.2.3 Wet Electrostatic Precipitators. With the use of wet ESPs, the reentrainment potential is all but eliminated. This is because instead of rapping, the plates and discharge electrodes are washed with water to remove collected particulate.

Many wet ESPs use vertical flow through cylindrical, hexagonal, or rectangular collecting "tubes" around each discharge electrode so that the entire surface can be washed without disrupting the ESP performance. The collecting and discharge electrodes must be evenly wetted to avoid solid buildups and maintain electrical clearances and collection characteristics. The flue gas is saturated with moisture before entering the wet ESP so that water evaporation does not occur. The water cleaning systems typically use various combinations of upflow and downflow spray nozzles (continuous or intermittent) and overflow irrigation. There are also a few horizontal flow plate wet ESPs in commercial operation.

A wet ESP can be used downstream of a wet scrubber as a polishing unit for particulate control and as a mist eliminator. Since many gaseous emissions condense and form a fine liquid mist at or above the water saturation temperature, the wet ESP can also remove these compounds. The wet ESP is also very effective for fine particulate (PM_{10} and $PM_{2.5}$) emissions control.

One problem with wet ESPs is the resulting particulate slurry and the subsequent dewatering required for suspended solids disposal. In addition, soluble components may require water treatment before liquid is discharged from the plant.

14.2.3 Fabric Filters

Fabric filters or bag houses are media filters that the flue gas passes through to remove the particulate. Reduced particulate emissions limits and the selection of low-sulfur fuels made fabric filters the best technical and economic choice for many new United States utility boilers during the last 10 years.

Media flue gas velocity (air-to-cloth [A/C] ratio); fabric and dust layer pressure drop; media life and cleanability; media, flue gas, and dust cake properties; and cleaning system type affect the design and subsequent performance of fabric filters. Certain membrane-type fabrics or ceramics offer significantly finer pore structure than conventional woven or felted fabrics. These media effectively take the place of the particle cake's filtering properties and may provide lower emissions and pressure differentials. The cost of membrane fabric bags or ceramics can be significantly higher than conventional fabrics.

The media flue gas velocity (A/C ratio) refers to the velocity of the flue gas through the collection media. This critical A/C ratio is determined by dividing the actual flue gas volumetric flow by the filtration area. As the filtration area decreases, the capital cost decreases. The flue gas pressure drop across the media and cake increases as the A/C ratio increases. A higher A/C ratio also increases the potential for higher emissions and media failure. Therefore, the A/C ratio is critical to both emissions and the proper balance of capital and operating costs.

Cloth filter media is typically sewn into cylindrical tubes called bags. Each fabric filter may have thousands of these filter bags. The filter unit is typically divided into compartments which allows online maintenance or bag replacement. The quantity of compartments is determined by maximum economic compartment size, total gas volume rate, A/C ratio, and cleaning system design. Extra compartments for maintenance or offline cleaning increase the cost but increase the reliability. Each compartment includes at least one hopper for temporary storage of the collected fly ash.

When all compartments are online, the A/C ratio is called

"gross." With one compartment offline for maintenance or cleaning, the A/C ratio increases and is called the "net" A/C ratio. If two compartments are taken offline, one each for maintenance and cleaning, the A/C ratio again increases and is called a "net–net" condition.

Fabric bags vary in composition, length, and cross-section (diameter or shape). Bag selection characteristics vary with cleaning technology, emissions limits, flue gas and ash characteristics, desired bag life, capital cost, A/C ratio, and pressure differential. Fabric bags are typically guaranteed for 3 years but frequently last over 5 years.

Media cleaning is initiated by a high flue gas pressure differential signal, a timer, or both. Cleaning decreases the pressure differential which gradually increases as the dust cake builds. With typical nonmembrane bags, the accumulated cake is the primary filtration media so that clean bags have higher emissions. As particulate loading increases, the cleaning frequency typically increases. A balance between emissions, low operating costs (resulting from frequent cleaning with low pressure differentials), and maximizing the bag life must be achieved. Filter bag life is dependent on media properties, cleaning frequency and force, along with operating temperature, pressure, flue gas composition, and particulate properties.

14.2.3.1 Fabric Filter Types. Fabric filters are categorized by their cleaning methods. Utility coal fired boiler applications primarily use reverse gas (also called reverse air) or pulse jet cleaning. Initially, United States power plants used reverse gas cleaning with few exceptions. In the last decade, as pulse jet experience, bag life, bag lengths, and compartment sizes increased, United States utilities frequently selected pulse jet fabric filters for economic and site layout reasons. As utilities worldwide adopted fabric filtration, they have predominately used pulse jet cleaning. Successful Australian and South African utility pulse-jet installations have accelerated the trend.

REVERSE GAS FABRIC FILTERS. With the reverse gas fabric filter (RGFF), the flue gas enters horizontally near the top of the hoppers and just below the compartment's tube sheet floor (a horizontal plate with rows of vertical thimbles to which the bags are attached). The flue gas passes through the thimbles into the bottom of the bags. As the particulate laden flue gas moves up and passes through the bag, the particulate deposits on the inside of the bag (Fig. 14-3). The RGFF bags are typically 8 or 12 in. in diameter with lengths up to 24 or 35 ft, respectively. The filtering cloth velocity is about 2.3 acfm/ft^2 of cloth area in the net–net condition. The RGFF bags are usually made of woven fiberglass that is treated to protect the glass fibers from abrasion and provide them with chemical resistance to the acids in the flue gas. The cleaned flue gas exits the compartments through poppet valves to the exit duct manifold and ID fans.

Reverse gas fans and ducts recycle part of the clean flue gas from the exit manifold back through the compartment during the cleaning cycle. This temporary reversal of the flue gas direction flexes the fabric to loosen the filter cake and gently blows the dust cake off the filter media. This dust cake then falls to the hopper for removal. The reverse gas exits the compartment being cleaned and flows into the dirty gas inlet manifold. Anti-collapse rings are sewn into the bag to prevent the reverse gas from collapsing the bag and trapping the particulate. The bottom of the bags are attached to the thimbles with snap rings. They are suspended through a tensioning device from the top of the compartment housing.

Cleaning always occurs with the compartment offline. A reverse gas cleaning system is typically sized to clean only one compartment at a time. Proper tension of the bags is necessary to enhance the cleaning process. This is accomplished with springs or eccentric weights. Too much bag tension impairs the bag cleaning and life. Too little tension allows folding during cleaning, which decreases life and cleaning effectiveness.

When fabric cleaning is initiated, the total gas flow that

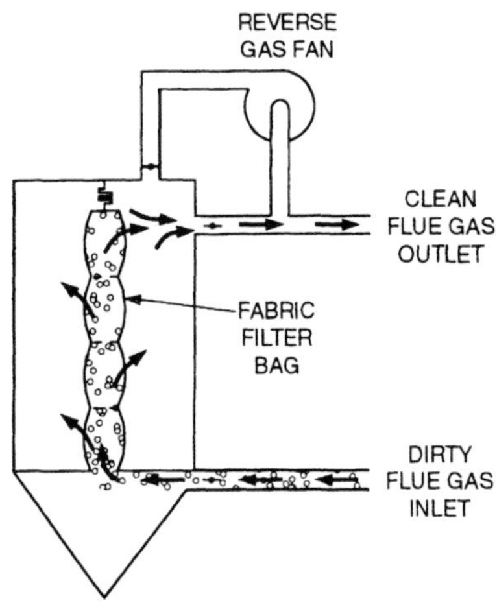

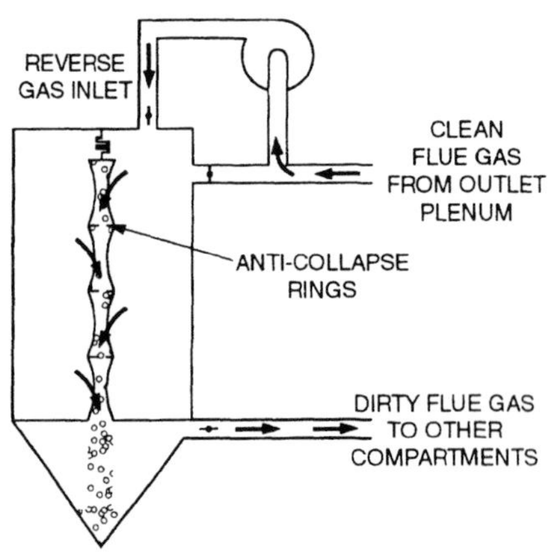

Fig. 14-3. Operating schematic reverse gas fabric filter.

the balance of the compartments must clean is increased. The additional flow comes as a result of the clean flue gas that is recirculated in the cleaning process. The clean flue gas flow rate is typically 1.5 to 2 acfm/ft2 of the fabric surface area being cleaned. RGFFs use this relatively low cloth velocity for gentle cleaning of the bags. Gentle cleaning is important because the whole dust cake should not be removed from the media surface with RGFFs. If the bags are overcleaned, the emissions may actually increase due to particle migration through the fabric. Another effect of overcleaning may be an increase in the pressure differential due to fine particulate becoming embedded in the fabric. This is called "bag blinding." Bag blinding can also be caused by sticky particulate from operation below the dew point, oil firing residuals, or deliquescence and particle adhesion due to high calcium chloride contents.

If reverse gas cleaning is inadequate, sonic horn assisted cleaning may be installed. Sonic horns may help when a high-pressure differential develops across the bags because of "bag blinding" or an excessively thick or low permeability dust cake. Excessive dust cake can strain both the fabric and the tensioning system. The horn's sound frequencies vibrate the bag material and release additional dust cake.

PULSE JET FABRIC FILTERS (PJFF) Flue gas typically enters the PJFF compartment hopper and passes from the outside of the bag to the inside, depositing the particulate on the outside (Fig. 14-4). To prevent collapse of the bag, a metal cage is installed on the inside of the bag. The flue gas then passes up through the center of the bag into the outlet plenum. The horizontal tube sheet for the PJFF is located at the top of the compartment (as compared to the bottom of the compartment for the RGFF). The bags and cages are suspended from the tube sheet. Cleaning is performed by initiating a downward pulse of air into the top of the bag. This pulse causes a rippling effect along the length of the bag. This ripple releases the dust cake from the bag surface, where it falls to the hopper. This cleaning may occur with the compartment online or offline.

Since the compartment may be cleaned online, care must be taken in the design of the inlet plenum. If the flue gas is introduced through the hopper, it flows upward to the bags. Upward flow can reentrain the particulate that has just been released from the bag surface. Upward velocity between the bags is called "can velocity" and is minimized with proper bag spacing and, in some designs, by introducing the flue gas at the compartment sides, by using particulate impingement baffles. The PJFF cleans bags in sequential and usually staggered rows within the compartment. If online cleaning is used, part of the dust cake from the row of bags being cleaned may be captured by adjacent rows. Despite this apparent shortcoming, online cleaning is successfully used on many large PJFF units.

The relatively high cleaning forces on the PJFF bags give acceptable life because of the difference in filter material. PJFF bags typically are made of felted materials that do not rely as heavily on the dust cake's filtering capability and thus can be cleaned more vigorously. The felted material is also a major reason the PJFF can operate at a cloth velocity in the net condition of approximately 4 acfm/ft^2 of cloth area, nearly twice that of the RGFF.

Three categories further differentiate the PJFF cleaning technology due to differences in cleaning pressures and volumes.

- Low-pressure/high-volume (LP/HV) — <15 psig
- Intermediate-pressure/intermediate-volume (IP/IV) — 15 to 40 psig
- High-pressure/low-volume (HP/LV) — > 40 psig

The current philosophy is that the cleaning energies of the low- and intermediate-pressure pulses carry effectively further down the bag than does the energy of the high-pressure pulses. Therefore, high-pressure systems typically have shorter bags than the low- or intermediate-pressure cleaning technologies. Low- and intermediate-pressure designs are currently in operation with bags 7 and 8 m long, compared to the typical maximum high-pressure length of 5 m. This bag length and cleaning differentiation will be under further scrutiny as results with fabric filter installations utilizing high-pressure pulsing and 7 m lengths become available.

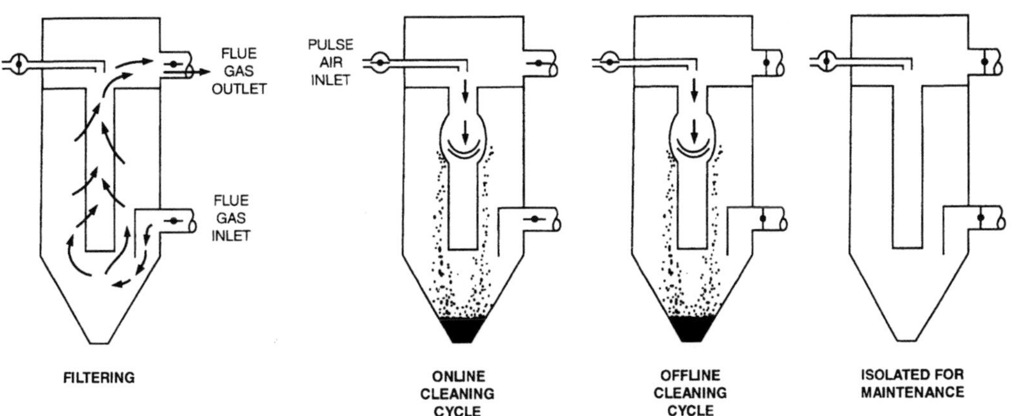

Fig. 14-4. Operating schematic pulse jet fabric filter.

The PJFF outlet plenum design has two types, the top door and the walk-in plenum (Fig. 14-5). The walk-in plenum has one small door per compartment, typically 2 ft by 4 ft, which is used to gain access to the tube sheet upper surface from which the bags are hung. This walk-in plenum is tall enough to remove the cages and bags. The plenum height is set by the length of the cage. With the advent of the long bags came two- and three-part interlocking cages. These cages are made into short sections to minimize the required plenum height and cost.

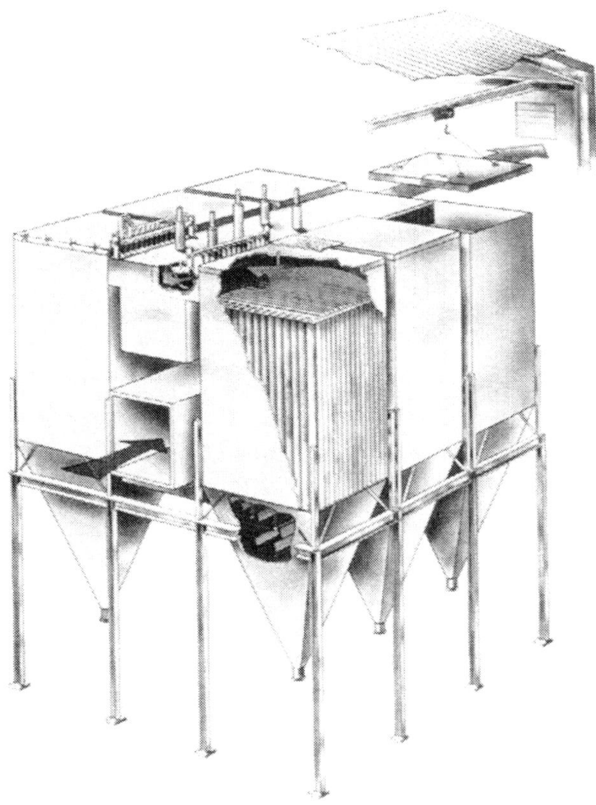

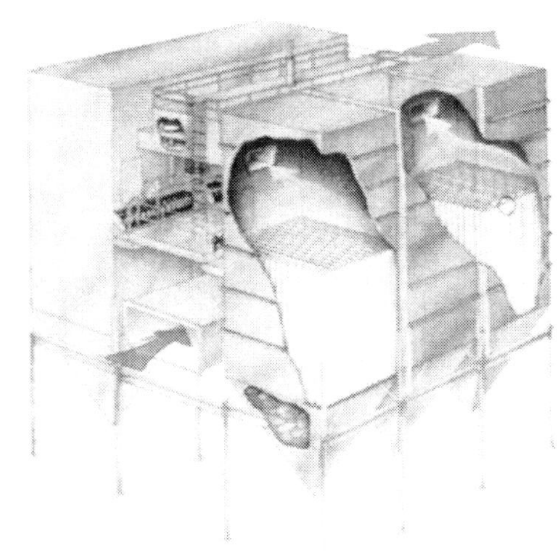

Fig. 14-5. Top door and walk-in plenum. (From environmental Elements Corporation. Used with permission.)

The top-door or roof top hatch outlet plenum design has a short outlet plenum height. Access to the bags is accomplished by lifting a cover that typically covers the top of the entire compartment. This cover is then set aside and the bags can be inspected or replaced.

The choice of outlet plenum technology must carefully be considered. Each has its advantages and disadvantages. The large doors on the top-door design result in more seal maintenance and inleakage. Inspection of the bags is easier with the walk-in plenum design. However, the working conditions when the bags are being replaced are better with the top-door technology. Most manufacturers indicate that both outlet plenum designs cost the same.

14.2.3.2 Comparison of PJFF to RGFF. A higher A/C ratio, online cleaning design, and no internal walkways make a PJFF smaller than a RGFF. Typically, the PJFF footprint can be up to 50% smaller than the RGFF footprint, even though the RGFF can be taller due to longer bags. The capital cost of the PJFF can be less than 80% of the RGFF's capital cost.

In contrast, the RGFF's bag costs are significantly lower than those for the PJFF. Woven fiberglass costs less than felted materials and the PJFF's bags also require support cages. The RGFF woven fiberglass bags have higher design, operating, and excursion temperatures than those for the felted materials typically used with PJFFs. (Bag materials can physically fail with excessively high operating temperatures.) RGFF guaranteed and expected operating bag lives are longer than those for a PJFF.

14.2.3.3 Comparison of Fabric Filters to ESPs. The emissions preference of fabric filters (FFs) over ESPs is primarily a result of the FF's consistent emissions independent of the fuel ash characteristics or inlet particulate load. The performance of an ESP is highly dependent on fuel changes with resultant variation in the ash resistivity and inlet ash load. With boiler soot blowing or ESP rapping, emission spikes may occur with the ESPs. Fabric filter emissions are seldom affected by these activities. Infrequent emission spikes because of overcleaning or due to installation of new bags may occur with fabric filters. Fabric filter pressure drops increase with higher inlet particulate load but may be compensated for with more frequent cleaning.

ESPs have a relatively constant flange-to-flange pressure differential of typically less than 2 in. wg versus the pressure differential of approximately 6 in. wg for a fabric filter. The fabric filter pressure differential variation is primarily a function of the dust cake and cleaning frequency. The higher fabric filter pressure differential results in higher ID fan energy consumption than that used by the ESP. The operating cost savings that occur as a result of the lower ID fan energy level of the ESP are often offset by the ESP's T–Rs power requirements.

Fabric filters typically have higher maintenance costs than ESPs due to their bag replacement costs. Current ESP designs typically have low maintenance costs because they often operate for many years without major work.

14.2.4 Alternate Particulate Control Technologies

14.2.4.1 Cyclone Collectors. Cyclone collectors are centrifugal collectors that rely on the particle density and velocity to separate the fly ash from the flue gas. As seen in Fig. 14-6, the particulate-laden flue gas enters the top or the side of the cyclone. Vanes impart a rotational velocity to the flue gas, driving the fly ash to the edge of the cylinder. The flue gas then exits the center of the cyclone out the top, leaving the fly ash to fall out the bottom. At pressures near one atmosphere and 2 to 5 in. wg pressure differential, this technology can effectively remove particles >20 microns in size; particles <10 microns are usually unaffected and not removed.

Cyclone collectors are currently used on circulating fluidized bed boilers to return the bed, unburned fuel, and limestone (if used) to the boiler. The finer particles that pass through the cyclone collector are then removed with an ESP or fabric filter.

14.2.4.2 Wet Venturi Scrubber. Wet venturi scrubbers use liquids to capture the fly ash. Flue gas is accelerated to between 12,000 and 18,000 fpm in the venturi throat. Liquid is typically introduced just upstream or within the throat. As the high-velocity flue gas passes the flow of liquid, the water is sheared into droplets. Droplet size is a function of pressure drop. Impacts among the fly ash and the droplets occur as a function of droplet and fly ash particle size and their initial differential velocity. The liquid droplets with captured fly ash are removed by cyclonic action, impingement trays, or interception with liquid sprays, mist eliminators, or both.

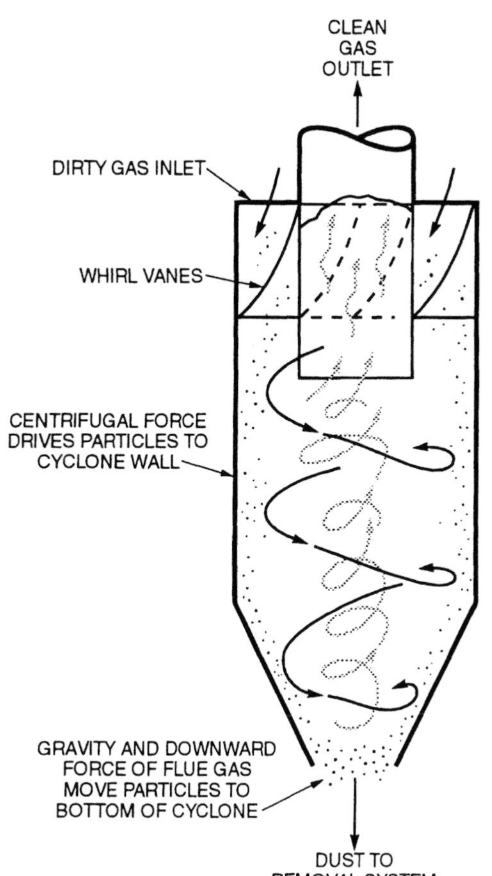

Fig. 14-6. Principle of cyclone-separator operation.

The flue gas pressure drop required to meet current regulations for particulate emissions and the resultant increased ID fan auxiliary power costs make this technology uneconomic for new electrical power plants. Some older installations still have operating wet venturi scrubbers.

Venturi scrubbers are effective particulate removers across the size spectrum, but they require 30 to 40 in. wg differential to achieve the currently required removal efficiencies for new plant's fly ash size particulates. The resultant fly ash slurry adds to water treatment, SO_2 scrubber complications, and operating expenses.

14.3 NITROGEN OXIDES EMISSIONS CONTROL

Reduction of nitrogen oxides (NO_x) emissions has become a prevalent requirement for new and existing power plants. This is primarily in response to the link between NO_x emission and ambient ozone concentrations. Therefore, in the United States, the 1990 CAAA Title I and Title IV and BACT for new units will increasingly require reduced NO_x emissions from power plants. Stringent NO_x emission controls are common in Europe and Japan. NO_x in flue gas consists of 90% to 95% nitric oxide (NO) with the balance occurring as nitrogen dioxide (NO_2). Although NO_2 is by far the minority constituent, industry practice is to express emissions based on the molecular weight of NO_2.

NO_x in flue gas is a result of oxidizing either nitrogen in the combustion air (thermal NO_x) or nitrogen in the fuel (fuel NO_x). Generally, when burning coal, less than 25% of the NO_x produced is thermal NO_x and the balance is fuel NO_x (Stultz and Kitto 1992). Boiler thermal NO_x emissions are minimized by reducing combustion zone temperatures; however, this is less significant to reduce total NO_x than firing techniques that limit fuel NO_x formation.

Fuel NO_x formation reactions are not well understood. The principal reaction process appears to be devolatilization of fuel nitrogen-forming intermediate compounds such as NO, NO_2, NH_3, and HCN (Blair and Wendt 1981). Subsequent N_2 or NO formation depends strongly on the fuel/air ratio. Therefore, most low NO_x burner designs carefully introduce reduced oxygen levels for fuel combustion.

Fuel NO_x conversion reactions appear to be dependent on the volatile reactants produced by the fuels. For example, bench scale combustion tests reveal the lack of a correlation between the fuel-bound nitrogen and fuel-generated NO_x (Singer 1981). In general, a fuel with lower nitrogen content typically produces less total NO_x than a fuel with higher nitrogen content. However, research has indicated that fuel-bound nitrogen conversion to NO_x is relatively insensitive to typical flame temperatures (Singer 1981).

Through the use of the combustion NO_x emission reduction methods, NO_x emissions for a boiler burning coal, oil, or gas can be limited to approximately 0.3, 0.2, and 0.1 lb/MBtu, respectively.

There are two ways to accomplish this reduction: combustion control and postcombustion control. Combustion control methods are the primary choices to reduce NO_x and include low-NO_x burners, air staging, fuel staging, operational modifications, and combustion turbine NO_x controls. Postcombustion controls include selective catalytic reduction (SCR) and selective noncatalytic reduction (SNCR).

14.3.1 Combustion Control

Combustion control methods are applied to reduce the formation of NO_x, carbon monoxide, and unburned hydrocarbon (generally regulated as nonmethane hydrocarbons or volatile organic compounds [VOC]) emissions, as opposed to postcombustion processes which reduce the concentration of these pollutants in the flue gas. Unburned combustibles in the ash have not been a regulatory concern, but unburned carbon is a plant efficiency and economic concern, and at certain levels, affects the salability of the fly ash and the performance of an electrostatic precipitator.

The combustion control processes that reduce the formation of NO_x are reduced oxygen concentration, reduced combustion temperature, and reduced reaction time at oxygen-rich, high-temperature conditions. However, these conditions will generally result in increased formation of CO and VOC in the flue gas and unburned carbon in the ash. Therefore, a first design criterion for combustion control of NO_x is to maintain acceptably low levels of CO, VOC, and unburned carbon while reducing NO_x emissions.

The principal NO_x reduction combustion control methods for new equipment and retrofit applications include low-NO_x burners, air staging, fuel staging, and operational and design modifications. These principal control methods are discussed in detail later.

14.3.1.1 Low-NO_x Burners. Low-NO_x burners are available for new and retrofit applications on boilers and combustion turbines burning nearly any fuel. Although the principles of operation are nearly identical, the following discussion concentrates on low-NO_x burners installed on pulverized coal (PC) boilers.

NO_x control on circulating fluidized bed (CFB) boilers does not include the use of low-NO_x burners. By maintaining low combustion temperature and intimate mixing within the combustion chamber, NO_x emissions from CFB boilers can be as low as 0.2 lb/MBtu. See Chapter 21 for further discussion of CFB boilers.

Low-NO_x burner systems on PC boilers are readily divided into wall-fired and corner-fired systems.

Wall-mounted burners are typically a multiple-register (damper) type with two separate secondary air flow paths through the burner and into the furnace. Common features include dedicated total secondary air flow control dampers and separate dedicated dampers or vanes to control the flow and spin of the individual secondary air flows through the burner. The vanes that control spin or flame shape are typically set during initial startup and then locked in place.

Control and balancing of the secondary air, primary air, and coal distribution among the burners is a basic requirement of all manufacturers. Typical allowable flow deviations from the mean are 10% for individual burner air and coal flows. This requirement may necessitate changes in operating procedures related to individual burner level turn down at part load. Conversely, additional control provisions and flow monitoring capability may be required to preserve the capability to operate with unbalanced firing at part load.

In retrofit situations, manufacturers typically suggest coal fineness of 65% to 70% through 200-mesh (0.074 mm) and 98% to 99% through 50-mesh (0.297 mm). Manufacturers agree that the coal fineness has little or no effect on NO_x production, but the burners function better with the finer coal grind. Unburned carbon in the ash will increase if coal fineness degrades.

The basic NO_x reduction principles for wall-mounted burners are to control and balance the fuel and air flow to each burner, and to control the amount and position of secondary air in the burner zone so that fuel devolatilization and high-temperature zones are not oxygen rich. Figures 14-7, 14-8, and 14-9 illustrate selected wall-mounted low NO_x burners.

Corner-fired low NO_x burner systems further divide into

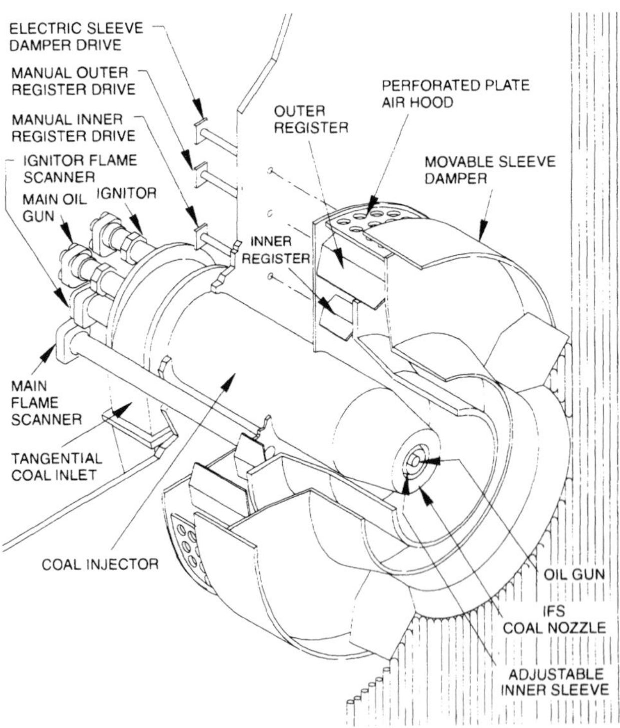

Fig. 14-7. Internal fuel-staged low-NO_x burner™. (From Foster Wheeler Corporation. Used with permission.)

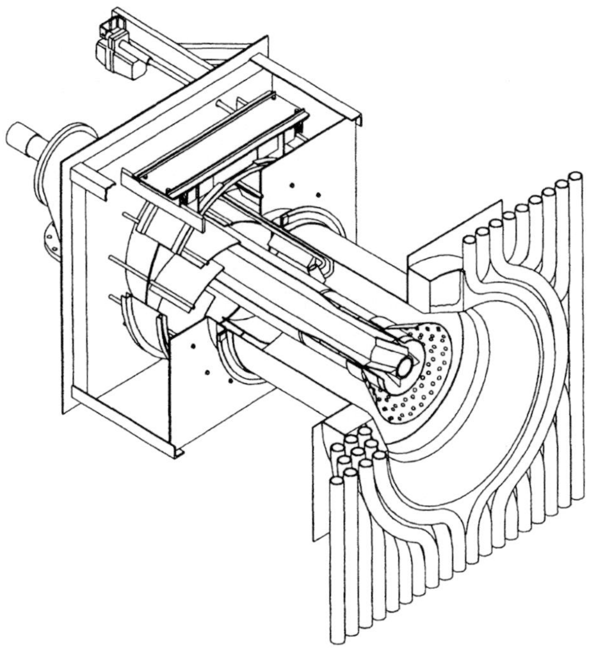

Fig. 14-8. Riley low-NO_x controlled combustion Venturi (CCT™) burner. (From Riley Stoker Corporation. Used with permission.)

two types. The Low NO_x Concentric Firing System[1] (LNCFS)™ is principally used on retrofit situations, while the Pollution Minimum[2] (PM)™ System is typically used on new construction. The technology selection for a given situation is also influenced by the required NO_x emission limit and the specific fuel to be fired.

The LNCFS™ differs from conventional tangential firing in that a portion of the secondary air, designated as auxiliary air, is directed into the furnace along a line that forms a 25° angle with the primary air/coal stream as shown in Fig. 14-10. Redirecting auxiliary air toward the furnace wall reduces the air available to the fuel stream during the devolatilization process, and thus reduces NO_x production.

Overfire air (OFA) can be used in conjunction with LNCFS™. OFA provides vertical air staging over the furnace height, while the redirection of the auxiliary air nozzles provides horizontal staging over the length and width of the furnace (Collette 1985).

In field retrofit demonstrations, LNCFS™ has been shown to provide approximately 20% to 50% NO_x reduction compared to conventional tangential firing systems.

The PM™ system for coal firing uses both OFA and local stoichiometry control during the early part of the combustion process to reduce NO_x production.

The PM™ burner system involves splitting the fuel/primary air stream into fuel-rich and fuel-lean streams to produce fuel/transport air ratios that produce less NO_x. The split burner pipe produces fuel-rich and fuel-lean streams through centrifugal separation. To accentuate the stoichiometry and separation effect, the fuel nozzles are arranged vertically in pairs so that two fuel-rich streams are adjacent, followed by two elevations of fuel-lean nozzles, and two more levels of fuel-rich burners.

The PM™ windbox height is taller than a conventional tangential windbox height, and therefore the PM system is not a "plug in" system for retrofit situations. By comparison with a conventional tangentially fired boiler, the PM system will reduce NO_x by about 60% (Collette 1985).

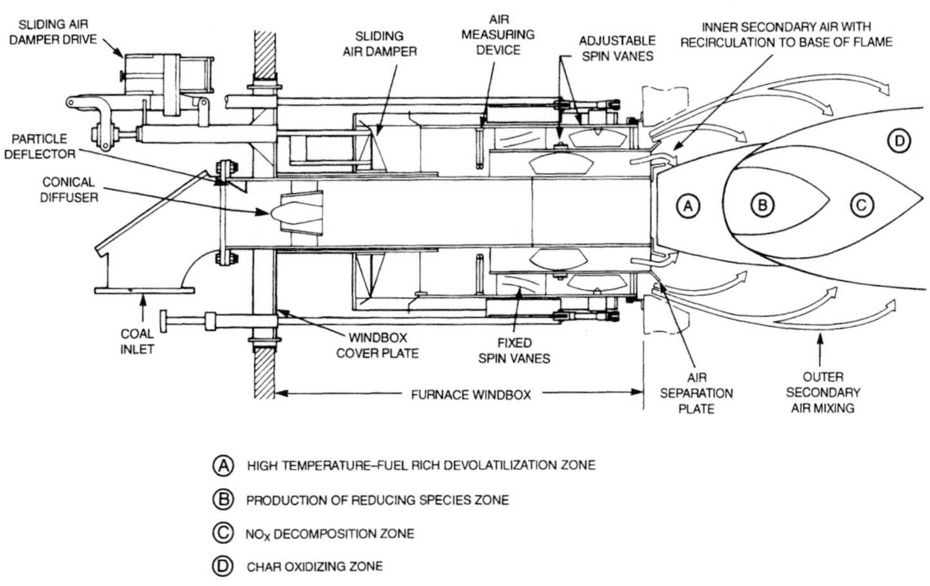

Fig. 14-9. DRB-XCL™ low-NO_x burners, pulverized coal-fired. (From Babcock & Wilcox. Used with permission.) (A) High-temperature fuel-rich devolatilization zone. (B) Production of reducing species zone. (C) NO_x decomposition zone. (D) Char oxidizing zone.

[1]Low NO_x Concentric Firing System is a registered name for a design marketed by ASEA Brown Boveri Inc., formerly Combustion Engineering, Inc.
[2]Pollution Minimum is a registered name for a design marketed by ASEA Brown Boveri, Inc., formerly Combustion Engineering, Inc.

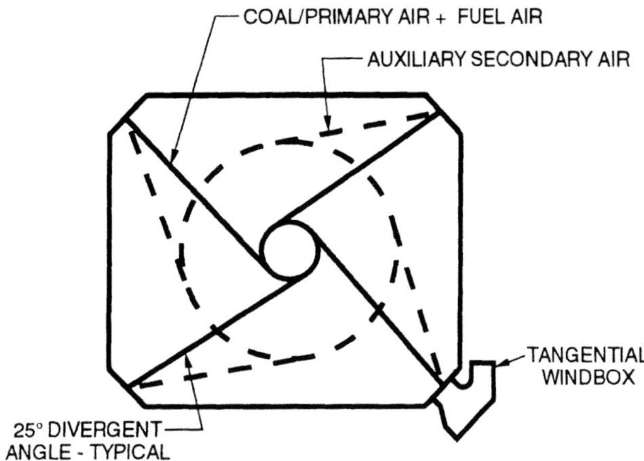

Fig. 14-10. Furnace plan view—LNCFS. (From Asea Brown Boveri. Used with permission.)

14.3.1.2 Air Staging. Air staging external to the burner consists of OFA ports. OFA may or may not be used in conjunction with low-NO_x burners, depending on the fuel and the NO_x emission requirement. The typical effect of OFA addition to a new or retrofit low-NO_x burner system is a further 10% to 20% reduction of NO_x emissions. However, unburned carbon and combustible materials may increase as a result of the addition of OFA.

OFA systems reduce NO_x formation by creating a fuel-rich combustion zone. The OFA is introduced above the main combustion zone where fuel burnout can be completed at a lower temperature with fewer volatile nitrogen-bearing combustion products.

Ducts connected to the top of the windbox, with flow control dampers, represent the simplest OFA system. However, side wall and rear wall OFA ports (on a front wall fired unit) located above the top burner elevation have been proposed. For retrofit situations with unbalanced windbox flow, a separate secondary air to OFA duct connection upstream of the windbox inlet may provide better balancing and control of both the windbox and the OFA system.

14.3.1.3 Fuel Staging. Staging fuel introduction into the furnace is a method to control NO_x formation. Reburning technology is the most common form of fuel staging.

The reburn process employs three separate combustion zones to reduce NO_x emissions, as shown in Fig. 14-11. The first zone consists of the normal combustion zone in the lower furnace, which is formed by cyclone burners or pulverized coal burners, depending on the furnace type. In this zone, 70% to 80% of the total fuel heat input is introduced. The first zone burners are operated with about 10% excess air (a 1.10 stoichiometric ratio). A second combustion zone (the reburn zone) is created above the lower furnace by operating a row of conventional pulverized coal or conventional natural gas burners at a stoichiometric ratio much less than 1.0.

The reburn zone stoichiometric ratio (mixing of reburn fuel and lower furnace combustion products) is between 0.80 and 0.95. The substoichiometric reburn zone causes NO produced in the lower furnace units to be reduced to molecular nitrogen and oxygen because the oxygen stripped from the NO_x molecules is combined with more active carbon monoxide molecules to form carbon dioxide. Fuel burnout is completed in the burnout zone by the introduction of OFA. Sufficient OFA is introduced to complete combustion of the unburned materials in the upper furnace with an overall excess air for the boiler of 15% to 20%. Reburn technology has demonstrated NO_x removal efficiencies of 40% to 65%.

Reburning offers a method of load regain for boilers that have been derated as a result of limits on fuel processing or firing capacity. The addition of the reburn zone burners provides additional heat input while substantially reducing NO_x emissions. However, successful retrofit of this technology requires space within the boiler to allow adequate residence time for both the additional burning zone and the associated OFA burnout zone. When this space is available, reburning can be effective, but a low residence time will limit system performance.

14.3.1.4 Operational Modifications. Operational modifications have been implemented on existing boilers to achieve NO_x reductions of 5% to 15%. Operational modifications typically have been applied in situations where only marginal NO_x reductions are required. Operational modifications are attractive because capital investment is either minimal or not required, although operating costs may increase. Operational modifications include flue gas recirculation, reduced air preheat, water injection, reduced excess air, biased firing, burners out of service, and fuel switching or dual fuel firing.

Reduction of excess air flow rate is one of the simplest operational modifications for NO_x reduction, particularly if the baseline excess air includes a sizable margin above the level required for complete combustion. Reduction of excess air also reduces the stack loss and improves boiler

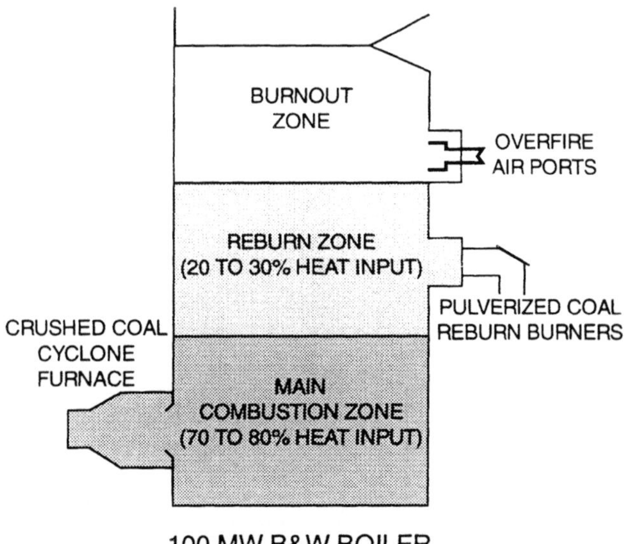

Fig. 14-11. Three zones for cyclone reburn system. (From Babcock & Wilcox. Used with permission.)

efficiency. NO_x reduction with reduced excess air is typically <10%.

Biased firing also is a relatively simple operational modification. Biased firing consists of operating lower burner levels fuel rich and upper burner levels air rich, thus achieving limited fuel staging through the furnace. NO_x reduction with biased firing is typically limited to 8% to 10%. Coal mill capacity limitations may limit the amount of fuel biasing achievable, and control system limitations on existing units may make consistent biased firing difficult (Miller 1985). Excessive biasing may result in localized reducing atmosphere conditions in the lower furnace, with increases in unburned combustible and waterwall corrosion.

14.3.1.5 Combustion Turbine NO_x Controls. Injecting water or steam in the combustion zones of a combustion turbine unit can limit NO_x formation. Thermal NO_x formation is avoided because of the lower combustion temperatures resulting from water or steam injection. The degree of reduction in NO_x formation is somewhat proportional to the amount of water or steam injected into the turbine.

In recent years, combustion turbine units have improved tolerance to the water or steam necessary to control the NO_x emissions. However, a point still exists at which the amount of water or steam injected into the turbine seriously degrades the turbine's reliability and operational life.

Advanced combustion turbine designs are capable of achieving low uncontrolled NO_x emissions without the use of water or steam injection. These dry low NO_x burners are a NO_x abatement technology capable of reducing NO_x emissions from combustion turbines from 42 to 9 ppmvd. The technology is an inherent part of the combustor design and is not an add-on technology. Advantages of this system include proven NO_x reduction capability and general similarity to conventional gas turbine combustor hardware. Disadvantages include increased carbon monoxide and unburned carbon emissions.

This combustion system is based on the principles of lean premixed combustion. Lean premixed combustion reduces the conversion of atmospheric nitrogen to NO_x by preventing high, near stoichiometric flame temperatures from occurring within the primary combustion zone. Since NO_x formation rates are exponentially dependent on temperature, lowering the temperature is extremely effective in reducing NO_x emissions.

Any high temperature regions can contribute disproportionately to overall NO_x emissions. High temperatures can occur locally if high fuel/air ratios exist within the flame zone. This is typical in a conventional combustor where fuel is injected into the primary zone and the fuel/air mixing and combustion processes occur simultaneously. In lean premixed combustion, the mixing and combustion processes are separated. The fuel and primary combustion air are mixed upstream of the combustion zone. Premixing produces a more uniform flame temperature and prevents high NO_x production zones within the combustor. Chapter 20 also addresses NO_x control for combustion turbines.

14.3.2 Postcombustion Control—Selective Catalytic Reduction Systems

Selective catalytic reduction (SCR) is a postcombustion NO_x emission reduction system. In SCR systems, vaporized ammonia injected into the flue gas stream acts as a reducing agent in the presence of a catalyst, achieving NO_x emission reductions as high as 95%. The NO_x and ammonia (NH_3) reagent react to form nitrogen and water. The reaction mechanisms are very efficient, with a reagent stoichiometry of approximately 1.0 (moles of NH3 per mole of NO_x reduced) and with very low ammonia slip (unreacted ammonia emissions).

Design ammonia slip values range from 2 to 10 ppm. With adequate design, SCR systems can be installed on new plants or retrofitted onto PC, CFB, or combustion turbine units fueled by coal, oil, or natural gas with very little effect on balance-of-plant equipment or unit availability. The following describes the basic process reactions, alternative SCR systems available, system configuration, catalyst considerations, and developmental status of SCR systems.

14.3.2.1 Process Reactions. An exothermic reaction occurs as ammonia and NO_x flow over the catalyst forming nitrogen and water vapor. The following are the predominant process reactions.

$$4NO + 4NH_3 + O_2 \xrightarrow{\text{Catalyst}} 4N_2 + 6H_2O + \text{heat}\uparrow \quad (14\text{-}3)$$

$$2NO_2 + 4NH_3 + O_2 \xrightarrow{\text{Catalyst}} 3N_2 + 6H_2O + \text{heat}\uparrow \quad (14\text{-}4)$$

Unfortunately, the following undesirable reactions can also take place:

$$2SO_2 + O_2 \xrightarrow{\text{Catalyst}} 2SO_3 \text{ (}SO_2\text{ oxidation)} \quad (14\text{-}5)$$

$$NH_3 + SO_3 + H_2O \rightarrow NH_4HSO_4 \text{ (ammonium bisulfate formation)} \quad (14\text{-}6)$$

$$2NH_3 + 2SO_3 + H_2O + 0.5O_2 \rightarrow 2NH_4SO_4 \text{ (ammonium sulfate formation)} \quad (14\text{-}7)$$

The potentially most troublesome reactions listed are the oxidation of SO_2 to SO_3 (Reaction 14-5) and the potential subsequent reaction of SO_3 with unreacted ammonia to form ammonium bisulfate (Reaction 14-6). Conversion of SO_2 to SO_3 is greatly increased at temperatures above approximately 700° F. Therefore, for applications using sulfur-containing fuels, ammonia injection (SCR operation) should be limited to temperatures below 660 to 700° F. With the appropriate catalyst composition, operation in this temperature window should limit the oxidation of SO_2 to SO_3 1% or less. Ammonia reagent for SCR systems should be injected at temperatures above approximately 570° F for effective ammonia reactivity and to help prevent the formation of ammonium salts.

Sulfur trioxide (SO_3) in the presence of ammonia forms

ammonium sulfate and ammonium bisulfate salts. The precipitation temperature of these salts from the flue gas is dependent on the relative concentration of ammonia and SO_3 in the flue gas as shown in Fig. 14-12. Ammonium bisulfate is a sticky substance that can deposit on catalysts, air heater baskets, heat recovery steam generators, and other downstream equipment.

Resultant ammonium bisulfate and sulfate salt particle diameters are on the order of 1 to 3 microns and thereby could contribute to PM_{10} (particulate matter <10 microns) emissions. These particles of ammonium sulfate and bisulfate could foul the micropore structure of the catalyst, thereby limiting catalyst reactivity. However, this has not been found to be a problem at SCR operating temperatures >570° F. On boiler units, these ammonium salts could deposit on air heater surfaces, necessitating more frequent soot blowing and offline washes. Depending on the design fuel's sulfur content, it may be necessary to limit ammonia slip to below 5 ppm to minimize the potential for air heater fouling.

The oxidation of SO_2 to SO_3 could also necessitate a change in the minimum air heater outlet temperature since the acid dew point temperature of the flue gas is directly related to SO_3 concentration. As the SO_3 concentration increases, the acid dew point of the flue gas increases, potentially increasing corrosion in downstream equipment. An alternative to changing the air heater outlet temperature is to install enameled baskets that are corrosion resistant.

For coal-fueled boilers, a significant quantity (up to approximately 80%) of the ammonia slip that exits the SCR system condenses onto the fly ash collected in particulate removal equipment (electrostatic precipitator or fabric filter). The ammonia content of the fly ash can have an impact on waste disposal or the potential for waste product sales. At elevated pH, ammonia in the fly ash is released, possibly leading to odorous emissions. In addition, fly ash used for fixation of FGD reaction products or as admixture for cement manufacturing or other pozzolonic uses could result in spontaneous ammonia releases. In extreme instances, this can be a safety problem.

German experience has indicated that fly ash ammonia concentrations of >80 to 100 ppm (mass basis) can result in disposal or reuse problems. For alkaline ashes, ammonia slip emissions may need to be limited to <2 ppm for coal ash contents of approximately 10% (based on bituminous coal) to avoid potential ash contamination. As coal ash contents increase, allowable ash ammonia contents also increase, since there is a greater volume of material for the ammonia to condense onto.

14.3.2.2 Reagent Considerations. SCR systems can use either anhydrous (pure) or aqueous (20% to 30% NH_3 in water mixture) ammonia reagents.

Anhydrous ammonia is clear in the liquid state and boils at −28° F. Liquid anhydrous ammonia must be stored under pressure at ambient temperatures. Accidental atmospheric release of anhydrous ammonia vapor can be hazardous; therefore, stringent requirements for safety and environmental protection are required.

When ammonia is diluted with water to 20% to 30% volume (aqueous ammonia), evaporation of ammonia gas from the fluid becomes negligible at ambient conditions. However, an aqueous ammonia storage facility is three to four times larger than an anhydrous ammonia storage facility. In addition, use of aqueous ammonia has the double disadvantage of increased transportation costs and the evaporation of water.

Although safety considerations are significant with onsite storage and use of anhydrous ammonia, a safe and reliable system can be designed and constructed. The majority (>95%) of current SCR systems use anhydrous ammonia. Most of these plants are located in heavily populated areas and, as such, safety is a primary consideration.

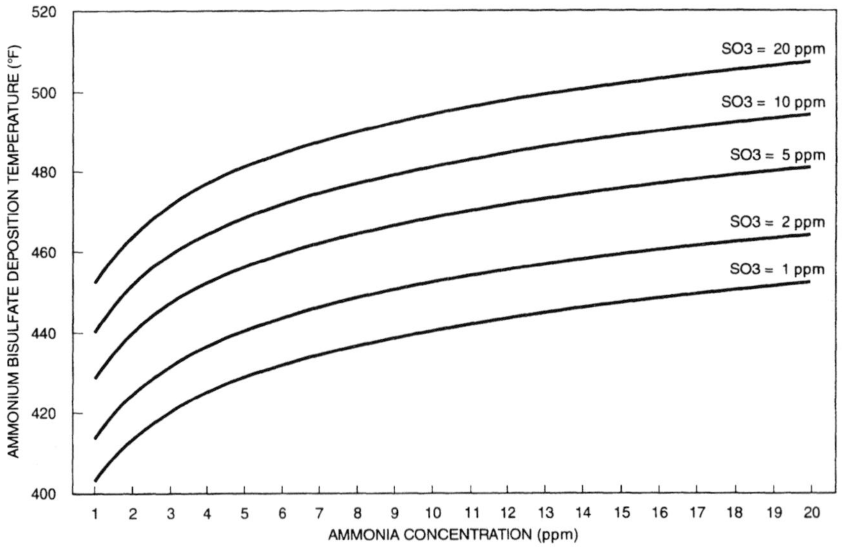

Fig. 14-12. Deposition temperature of ammonium bisulfate.

Anhydrous ammonia is the preferred reagent because it is less expensive than aqueous ammonia and is a more concentrated chemical requiring less storage volume. In addition, the energy and equipment requirements for vaporization of anhydrous ammonia are significantly less than those required to vaporize aqueous ammonia. However, in several areas of the United States, the impetus is for use of aqueous ammonia.

14.3.2.3 SCR System Configurations. Based on the previously described temperature limitations for SCR system use, three potential system arrangements should be considered for installation on a PC boiler:

- High dust—Catalyst located at the outlet of the economizer and upstream of the air heater (Fig. 14-13).
- Low dust—Catalyst located at the outlet of a hot-side ESP and upstream of the air heater (Fig. 14-14).
- Tail end—Catalyst located at the outlet of particulate removal and FGD systems and upstream of the stack (Fig. 14-15).

All three arrangements have been successfully used in coal-fueled applications in Japan and Europe. For tail-end applications, catalyst life is maximized and catalyst cost is minimized by more reactive catalyst formulations since the predominant catalyst degradients and poisons (particulate, sulfur compounds, and trace elements) have been removed from the gas stream. However, reheating the flue gas is relatively expensive, requiring a gas–gas heat exchanger as well as a supplemental heat source such as natural gas or steam from the boiler.

For low- and high-dust SCR arrangements, catalyst lives are similar, but the low-dust catalyst can have a smaller catalyst pitch (size of catalyst open areas) leading to a more cost-effective catalyst installation. Retrofit of either low- or high-dust SCR systems requires adequate space proximate to the boiler area.

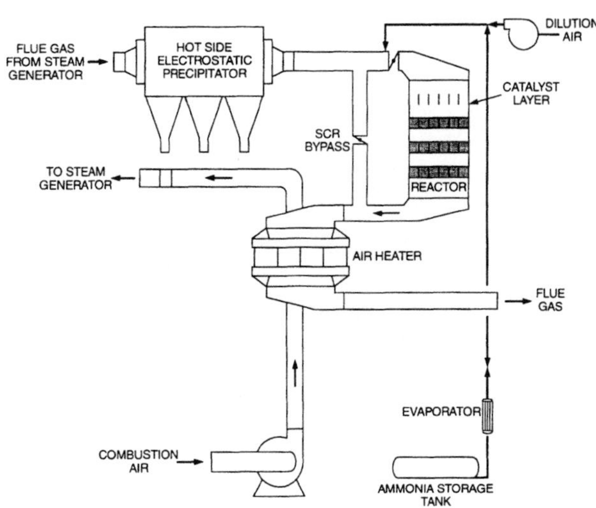

Fig. 14-14. Low dust—SCR arrangement.

SCR catalyst for oil- and gas-fueled installations is generally placed in the high-dust position at the outlet of the boiler economizer. However, more reactive catalyst with smaller pitch is allowable considering the relative cleanness of the flue gas.

Catalyst beds are oriented in vertical or horizontal flow arrangements. Vertical flow is preferred for high-dust systems because it allows effective online sootblowing to avoid plugging the catalyst bed. Low-dust and tail end arrangements can use either vertical or horizontal flow arrangements.

Two potential arrangements for SCR systems on combustion turbines (CT) are possible: within the HRSG (horizontal flow) or at the CT outlet. The maximum operating temperature of standard vanadium/titanium catalyst is 780° F. The temperature of the flue gas exiting the CT is rarely lower than 900° F, so this type of catalyst must be installed within the HRSG unit. Alternatively, the use of zeolyte catalyst allows

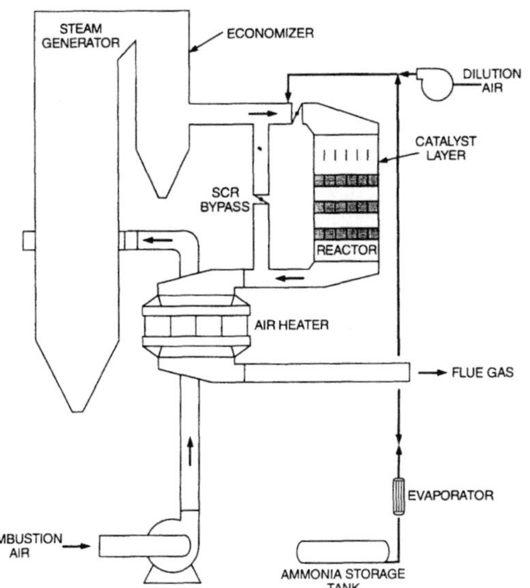

Fig. 14-13. High dust—SCR arrangement.

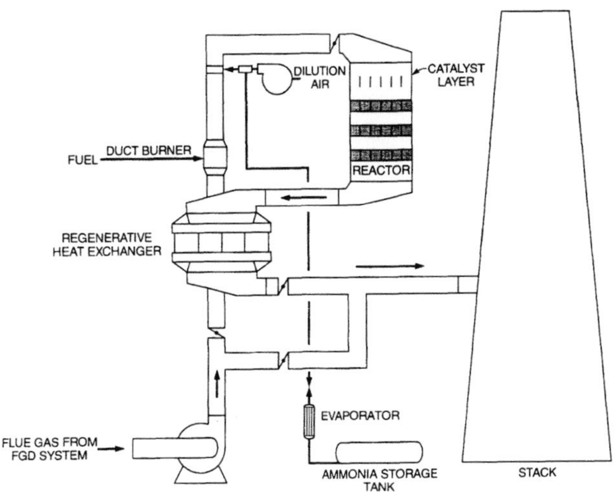

Fig. 14-15. Tail end—SCR arrangement.

SCR installation at the CT exhaust because zeolyte can operate at temperatures over 1,150° F.

14.3.2.4 Ammonia Storage and Vaporization. Anhydrous ammonia is stored in a pressure vessel rated for 250 to 300 psig. Aqueous ammonia is generally stored in a tank rated for 25 to 30 psig. The storage area should include a diked foundation with appropriate drainage to contain spills from the storage tanks. The storage vessel should have high and low temperature and pressure alarms, pressure relief valves, and liquid level monitors. Ammonia in either form can be delivered by truck, rail, or barge to the storage facility.

Ammonia must be vaporized to cause a phase change from liquid to gas. The flow of ammonia vapor is regulated by a control valve. The gaseous ammonia is diluted with air to approximately a 2% to 5% mixture. The air provides a relatively constant flow rate for more efficient distribution of ammonia in the flue gas.

14.3.2.5 Ammonia Injection and Control. SCR ammonia flow rates are controlled based on inlet NO_x concentrations, unit load or gas flow, and the required outlet NO_x concentration. High reaction efficiencies and low ammonia slip requirements result in nearly ideal SCR system stoichiometries. The inlet total NO_x concentration and flue gas flow indication are used for feedforward control to regulate the ammonia flow control valve. An outlet total NO_x concentration feedback signal is sent to the reagent feed control valve for trimming of the ammonia injection rate. This very precise control philosophy minimizes ammonia slip emissions.

Two basic alternatives exist for the uniform injection of vaporized, diluted ammonia in the flue gas stream: an ammonia injection grid (AIG) consisting of manual throttling valves on the injection lines, or a simpler matrix of injection headers with reliance on a flow mixing device (static mixer) to distribute reagent. Use of AIGs is generally favored to preserve "cause and effect" tuning capabilities and to minimize overall system pressure drops. In general, experience has indicated that adjustments to the AIG are infrequent (once per year or less).

14.3.2.6 Catalyst Design and Deactivation. The cost of a catalyst is typically a significant portion (20% to 30%) of total SCR system capital cost. Therefore, selection of catalyst materials and catalyst design are of great concern. The type of catalyst and catalyst volume required are dependent on the following parameters:

- Flue gas temperature,
- Inlet NO_x concentrations,
- NO_x emission limits,
- Allowable ammonia slip,
- Required catalyst life,
- Oxygen content of flue gas,
- Allowable SO_2 to SO_3 conversion rate,
- Inlet dust loading,
- Flue gas and NO_x distribution,
- Flue gas catalyst poison concentrations, and
- Allowable pressure drop across the catalyst.

Most catalysts used in utility service primarily consist of a vanadia–titania mixture. However, the final catalyst composition can consist of many active metals and support materials. Titanium dioxide (TiO_2) is typically used as the base material which disperses and supports vanadium pentoxide (V_2O_5), the active catalyst material. Vanadium pentoxide is widely used in the SCR industry because of its resistance to sulfur poisoning. The vanadium content controls the reactivity of the catalyst and, unfortunately, catalyzes the oxidation of SO_2 to SO_3. Therefore, in moderate- to high-sulfur fuel applications, it is necessary to minimize the vanadium content to reduce SO_2 conversion.

Tungsten oxide (WO_3) is also used in a number of catalyst compositions to reduce excess surface oxygen and thereby inhibit the oxidation of SO_2 to SO_3. Tungsten oxide also provides thermal and mechanical stability to the catalyst. The concentrations of vanadium pentoxide, titanium dioxide, and tungsten oxide can be fluctuated to meet the specific requirements for each SCR installation.

The NO_x reduction rate is not only a function of the ammonia injection rate and catalyst composition, but is also directly related to criteria regarding gas hourly space velocity, catalyst pitch, area velocity, and flue gas temperature.

The gas hourly space velocity, the inverse of residence time (1/h), is the volumetric flow rate of the flue gas (ft^3/h) divided by the catalyst volume (ft^3). Therefore, as the gas hourly space velocity increases, the contact time between the gas and catalyst decreases, and the overall NO_x reduction potential decreases (increasing NO_x emissions, potential ammonia slip, or both).

The catalyst pitch is defined as the cell wall thickness plus the flow channel width. The pressure drop through the catalyst is directly related to the catalyst pitch. As the catalyst pitch decreases, the system pressure drop increases. Typically, the fuel properties and application type (high-dust, low-dust, or tail-end) determine the catalyst pitch. Coal-burning plants with high-dust applications require the largest pitch to accommodate the large amount of fly ash present in the flue gas. High-dust SCR catalysts installed on coal-fueled plants generally have a catalyst pitch of 6 to 7.5 mm. Low-dust or tail-end catalysts installed on coal-fueled plants generally have a catalyst pitch of 3.3 to 5 mm. The catalyst pitch in applications with very low particulate (fuel oil or natural gas) can be as low as 2 to 3 mm. Site-specific issues must be considered when optimizing catalyst pitch based on enlarging the frontal area versus minimizing the catalyst bed depth for low-pressure drop.

Area velocity (ft/h) is obtained by dividing the volumetric flow of the flue gas (ft^3/h) by the surface area of the catalyst (ft^2) and can be a more representative measure of catalyst design because the pitch is reflected in this value. Conversion of area velocity to space velocity can be made by dividing the space velocity by the specific surface area of the catalyst. The specific surface area is the area of catalyst surface divided by total catalyst volume (ft^2/ft^3). Note that the surface area is the geometric value, not the surface area of the microporous structures on the surface of the catalyst.

The optimum temperature for NO_x emission reduction for vanadium–titanium catalyst is generally between 600 and 700° F; however, it can range from 550 to 780° F. Given a flue gas composition, each type of catalyst has a specific optimum temperature range. As temperatures are encountered below the optimum temperature range, NO_x reduction efficiency decreases rapidly. As temperatures increase past the optimum temperature range, SO_2 to SO_3 conversion rates increase dramatically. As a result, economizer design and operation may be affected by considerations related to effective SCR operation.

The activity of a catalyst degrades over time. Ultimately, this degradation leads to a requirement to add or replace catalyst in the reactor vessel. The initial charge of catalyst in SCR systems is typically guaranteed by the manufacturer for 2 to 4 years.

A number of alkali metals and trace elements poison the catalyst, significantly affecting reactivity and life. Known poisons are listed below:

- Arsenic
- Beryllium
- Cadmium
- Calcium
- Chromium
- Copper
- Lead
- Manganese
- Mercury
- Nickel
- Thorium
- Uranium

Arsenic, the major poison when in the form of gaseous arsenic oxide, can be deposited on catalyst surfaces, clogging small pores of the catalyst. This limits the transport of the ammonia–NO_x mixture to the active catalyst sites. This effect can be somewhat minimized by enlarging the pores of the catalyst to limit the blocking of active sites. However, arsenic, as well as the other poisons listed, can chemically attack (neutralize) the active sites on the catalyst surface and reduce catalyst effectiveness over time. These poisoning effects do not occur suddenly; they are a continual process with catalyst exposure to flue gas. As the catalyst becomes deactivated, more ammonia must be injected to maintain NO_x emission limits. This results in an increased amount of ammonia slip for a given level of performance.

14.3.2.7 Catalyst Management. As stated previously, approximately 20% to 30% of the SCR system cost is for the catalyst. Therefore, catalyst life and associated addition or replacement schedules have a significant impact on the relative economics of the SCR. The initial charge of catalyst is capable of meeting design basis performance objectives. This initial catalyst charge deactivates at a fairly uniform rate, with augmentation or replacement necessary when catalyst activity has decreased to the point that outlet NO_x emission requirements cannot be achieved consistent with ammonia slip limits. Since the activity levels are still significantly high (50% to 75%) in the initial catalyst charge, it is best to add, rather than fully replace, catalyst to more completely utilize remaining catalyst activity. Accordingly, it is advisable to equip the reactor casing with the flexibility to accommodate the addition of future catalyst layers. This flexibility substantially lowers life cycle costs for an SCR system.

A catalyst management plan (CMP) can have a significant impact on the life cycle cost of the SCR system. For example, a unit with an SCR system that requires 1,000 ft^3 of catalyst (with a guaranteed life of 3 years) and a design plant life of 20 years will require a total 5,500 ft^3 of catalyst if a CMP is not used, and 3,000 ft^3 if space for a spare layer is included in the design and a CMP is used. Including a CMP in the SCR design is cost effective whenever the physical constraints allow the installation of a spare layer.

Addition of a future catalyst layer increases the overall SCR system pressure drop. Therefore, there is a tradeoff between the replacement catalyst cost and the increase in overall power consumption that accompanies additional pressure drop. Typically, this tradeoff strongly favors a CMP.

14.3.2.8 Catalyst Disposal/Recycling. New, fresh catalyst is generally not considered to be a hazardous material although some locales may classify it as hazardous because of the high concentration of vanadium. However, after years of operation on a coal- or oil-fueled boiler in the presence of flue gas, the catalyst may absorb or collect enough of the trace pollutants in the flue gas to be classified as hazardous. Potential hazardous materials include most of the elements listed previously as poisons. Preliminary tests indicate that spent catalyst does not generally meet the definition of a hazardous waste.

Major catalyst suppliers in both the United States and Germany have indicated that catalyst recycling is the preferred method for the ultimate removal of spent catalyst. It is normal for the user to form an agreement with the catalyst supplier stipulating that when new catalyst is ordered, the catalyst supplier will accept the spent catalyst for final disposition.

14.3.3 Postcombustion Control—Selective Noncatalytic Reduction Systems.

Selective noncatalytic reduction is another commercially available technology to control NO_x emissions from fossil-fueled boilers. SNCR systems rely on an appropriate reagent injection temperature, good reagent–gas mixing, and adequate reaction time rather than a catalyst to achieve NO_x reductions. SNCR systems can use either ammonia (marketed as Thermal DeNO_x systems) or urea (marketed as NO_xOUT™ systems) as reagents. Ammonia or urea is injected into areas of the steam generator where the flue gas temperature ranges from 1,500 to 2,200° F. SNCR systems are capable of achieving NO_x reduction efficiencies as high as 70% to 80% in optimum situations (adequate reaction time, temperature, and reagent–flue gas mixing) with ammonia slips of 10 to 50 ppm. However, potential performance is *very* site specific and varies with fuel type, steam generator size, and steam generator heat transfer characteristics. With current SNCR technology, these parameters can limit potential SNCR effectiveness to as low as 20% to 30% NO_x reduction for coal fueled steam generators larger than 100 to 150 megawatts (MW). Alternatively, SNCR systems have proven very effective on circulating fluidized bed (CFB) applications (50% to 70% NO_x reduction) where the presence of the hot cyclone ensures adequate retention time at temperatures nearly ideal for NO_x reduction.

The following describes the basic process reactions, system capabilities, and development status of SNCR systems.

14.3.3.1 Process Reactions. SNCR systems reduce NO_x emissions based on a thermally based exothermic reaction between reagent and NO_x. The following are the predominant reactions for ammonia and urea based systems:

$$4NO + 4NH_3 + O_2 \rightarrow 4N_2 + 6H_2O + \text{heat}\uparrow \quad \text{(ammonia)} \tag{14-8}$$

$$4NO + 2CO(NH_2)_2 + O_2 \rightarrow 4N_2 + 2CO_2 + 4H_2O + \text{heat}\uparrow \quad \text{(urea)} \tag{14-9}$$

With the exception of the oxidation of SO_2 to SO_3 (Reaction 14-5), all of the undesirable reactions that can occur for SCR systems occur for SNCR systems (Reactions 14-6 and 14-7). In addition, the following reactions can significantly affect the performance of SNCR.

$$4NH_3 + 5O_2 \rightarrow 4NO + 6H_2O \quad \text{(ammonia oxidation to NO)} \tag{14-10}$$

$$4NH_3 + 3O_2 \rightarrow 2N_2 + 6H_2O \quad \text{(ammonia thermal decomposition)} \tag{14-11}$$

$$2NH_3 + 2O_2 \rightarrow N_2O + 3H_2O \quad \text{(nitrous oxide emission)} \tag{14-12}$$

It is anticipated that the NO_x emissions from most boiler applications would typically consist of approximately 95% NO and 5% NO_2. SNCR systems only remove NO from the flue gas (Reactions 14-8 and 14-9) and do not remove NO_2 emissions from the flue gas stream. This limits the overall NO_x reduction capability of SNCR systems by requiring higher NO emission reductions to account for the lack of NO_2 reduction.

Ammonia slip from SNCR is not time dependent. The potential for ammonia slip from SNCR systems is immediate once initial operation begins. Ammonia slip from SNCR is highly variable (10 to 50 ppm). In applications with sulfur-bearing fuel, higher ammonia slip emissions increase the likelihood of ammonium bisulfate and sulfate fouling (Reactions 14-6 and 14-7) in the air heater. As a result, more frequent online soot blowing and offline cleaning of the air heater may be required which may increase overall unit forced outage rates.

Injection of SNCR reagent (urea or ammonia) at temperatures in excess of approximately 2,200° F increases NO_x emissions (Reaction 14-10). Injection of SNCR reagent at temperatures greater than approximately 1,700° F results in reagent decomposition without a comparable reduction of NO_x emissions (Reaction 14-11). Accordingly, NO_x reductions and overall reaction stoichiometry are very sensitive to the temperature of the flue gas at the reagent injection point.

Nitrous oxide (N_2O) can be a byproduct of SNCR operation. The oxidation of ammonia to N_2O (Reaction 14-12) is undesirable because it wastes reagent and results in the emission of a significant greenhouse effect gas (N_2O). Also, the conversion of NO to N_2O can take place through a series of intermediate reactions (Muzio et al. 1991) which can account for 10% to 15% of overall SNCR apparent NO_x reductions. (Currently, N_2O emissions are not required to be measured or reported.)

14.3.3.2 System Configuration/Capabilities. The NO_x reduction potential of SNCR systems is limited by (1) boiler geometry and temperature profile (varies as a function of load) which affect reagent and flue gas mixing and (2) ammonia slip.

The ideal SNCR temperature window ranges from 1,500 to 2,200° F based on the inlet concentration of NO_x and type of reagent. Injection above the high end of the temperature range results in increased NO_x emissions. Hydrogen can be injected along with ammonia (or other additives with the urea reagent) to extend the effective range of the SNCR process down to 1,300° F. Boilers typically operate at a temperature of between 2,500 and 3,000° F. Therefore, without substantial boiler modifications, the optimum SNCR temperature window in a boiler usually occurs somewhere at the top of the furnace or in the tubed backpass of the boiler.

For larger boilers, the design challenge is to achieve appropriate residence times in the optimum temperature range. Distribution of the reagent can be difficult because of the long injection distance required to cover the relatively large cross-section of a boiler larger than 100 to 200 MW. In the past, injection of SNCR system reagent has normally been achieved through the use of wall injectors. These wall injectors have limited effectiveness for large boiler expanses, so steam-cooled injection lances are required to cover a larger area inside the boiler.

Residence times in excess of 1 second at optimum temperatures would yield high NO_x reduction levels even under less than ideal mixing conditions. However, a residence time of at least 0.3 seconds is required to assure minimal levels of SNCR performance. Computer-based flow/temperature modeling is generally conducted to estimate optimum ammonia injection locations and flow patterns. New boilers could include provisions for an SNCR system with increased space between the pendent superheat sections to facilitate installation of the injection lances in the optimum temperature range. For an existing boiler, water wall and steam piping modifications would probably be necessary to accommodate the installation of SNCR injector lances.

To accommodate SNCR reaction temperature and boiler

turndown requirements, three to four levels of injection points would typically be needed. SNCR system suppliers have the capabilities to estimate boiler temperature profiles to support the selection of injector locations; however, modification of the injectors or installation of additional injectors may be necessary after initial SNCR system operation.

Reagent utilization is a predominant concern with SNCR use on steam generators larger than 100 to 200 MW. Stoichiometries can range from 3.0 to 4.0 (moles of NH_3 per mole of NO_x removed) to attain 30% to 50% NO_x reductions. The increased reagent consumption is a result of reagent thermal decomposition, varying temperatures, and the lack of a true steady-state controlled environment, which tends to increase ammonia slip emissions.

From a process chemistry perspective, the common reagent constituent that influences NO_x reduction for either ammonia or urea is nitrogen. Therefore, an accurate measurement of relative reagent cost is dollars per ton of nitrogen. The costs of aqueous ammonia and urea nitrogen are approximately two and three times, respectively, that for anhydrous ammonia. These comparisons are based on historical data and are influenced by factors such as freight distance and local demand.

14.3.3.3 Reagent Storage and Vaporization. The use of ammonia requires the same safety systems, storage tanks, and vaporizers as those described for SCR installations in Section 14.3.2.4.

Urea is a nonhazardous material with no special storage or usage limitations. A urea reagent based process has a pumping skid instead of a vaporizer or blower. The urea is pumped and injected as a liquid. One system supplier catalytically converts the urea-based solution to ammonia prior to injection into the flue gas stream.

14.3.3.4 Reagent Injection and Control. Ideally, reagent injection control would be precise to minimize reagent consumption as well as ammonia slip. Accurate reagent injection is critical for successful SNCR NO_x reduction performance. Limiting ammonia slip is important to avoid deleterious side effects when burning sulfur-bearing fuels. Unfortunately, there are no monitors currently available that can accurately measure ammonia slip emissions as low as the required 2 to 10 ppm. Therefore, SNCR reagent control must be based on NO_x emission feedback. As a result, SNCR NO_x reduction capabilities can be significantly limited if low overall ammonia slip limits are to be met reliably. In addition, without accurate ammonia slip feedback it is not possible to precisely tune reagent injection.

14.3.4 Postcombustion Balance-of-Plant Impacts.

The following sections describe possible design and operational impacts for installation of postcombustion NO_x reduction systems.

14.3.4.1 Steam Generator Impacts. The effectiveness of SNCR is directly dependent on gas temperature and residence time considerations. The majority of an existing boiler convection pass is somewhat congested for optimum SNCR system operation. Without modification, this could lead to either limited NO_x reduction capabilities or excessive ammonia slip. Therefore, effective use of retrofit SNCR may require extensive boiler modifications with corresponding balance-of-plant impacts. The design of a new boiler can incorporate the installation of SNCR through the use of free space in areas corresponding to appropriate temperatures.

14.3.4.2 Air Heater Impacts. Sulfur trioxide (SO_3) in the presence of ammonia forms ammonium sulfate and ammonium bisulfate salts. Ammonium bisulfate deposits could foul air heater surfaces, leading to increasing pressure drops, and possibly create the need for periodic offline cleanings.

Experience has indicated that deposits in the baskets generally tend to sweep out of the air heater without significant incident. However, if the condensation temperature occurs between layers, tenacious deposits tend to build up. Cold-end soot blowers generally are not effective in reaching and removing these deposits online. To remove any deposits that might occur, offline water washing may be necessary.

Operation of both SCR and SNCR can result in ammonium bisulfate fouling the air heater. However, with SNCR systems this is not a time-dependent situation. With SNCR, the full range of ammonia slip can occur anytime the system is operating. With SCR catalyst, ammonia slip is very low with fresh catalyst and increases slowly as the catalyst is deactivated. Without tight control of ammonia slip emissions, the potential for air heater fouling and increased forced outages is much higher for an SNCR system.

An advantage of SNCR is that there is no increase in SO_3 emissions. However, whenever a sulfur bearing fuel is combusted, 1% to 3% of the SO_2 will be oxidized to SO_3 in the combustor. Therefore, considering the relative availability of ammonia slip from SNCR, substantial concern remains.

14.3.4.3 Draft System Impacts. An SCR system installed in a high-dust location can add 3 to 5 in. wg of pressure drop to the boiler system. During normal operation, the SNCR system should not increase the pressure drop more than 1 in. wg. However, with SNCR there is an increased potential for fouling the air heater with ammonium bisulfate deposits. A draft design margin of 3 to 6 in. wg would not be unreasonable to cover worst-case SNCR conditions.

14.3.4.4 Precipitator Impacts. Properly operating SCR or SNCR systems should have no detrimental effects on the precipitator. In boilers with low SO_3 concentrations, the ammonia slip that does not react with SO_3 acts to enhance the collectability of fly ash in the precipitator.

14.3.4.5 Waste Solids Handling and Disposal Impacts. Ammonia slip from the SCR or SNCR system condenses onto the fly ash collected in the electrostatic precipitator, is absorbed in the FGD system, or exits the system through the stack. The ammonia content of the fly ash can have an impact on waste disposal or reuse practices. At elevated pH, ammonia in the fly ash is released, possibly leading to odorous emissions. Experience has indicated that ammonia (NH_3)

concentrations >100 ppm (mass) in the fly ash result in noticeable odor and may result in rejection of the ash for use in the cement industry. Testing has indicated that for a coal with seven percent ash, ammonia slip must be limited to below approximately 2 ppm to avoid any potential problem with fly ash sales. Alkaline fly ash contaminated with ammonia could lead to disposal problems.

14.3.4.6 Visible Plume. The potential exists for the formation of a detached plume caused by ammonium chloride condensing as the flue gas cools. Although a threshhold concentration has not been established, it should be considered if both ammonia and chlorides exceed 15 ppm. The ammonia source is the postcombustion NO_x system, and chlorides can come from either the fuel or the FGD system.

Acceptable accuracy of ammonia slip analyzers has not been demonstrated. Therefore, these analyzers are not widely used. An ammonia slip monitor may not be required if ammonia concentrations in the fly ash are monitored on a regular basis. Ash sampling may be an adequate surrogate measurement.

14.4 SULFUR DIOXIDE EMISSION CONTROL

This section covers current commercially available combustion and postcombustion, dry and wet sulfur dioxide (SO_2) removal processes. SO_2 is an acid gas formed by combustion of sulfur in the fuel with oxygen. Generally, removal of this acid gas requires reaction with an alkali material. There are new processes that catalytically convert the SO_2 to SO_3 and condense the sulfuric acid formed ($H_2O + SO_3$ to form H_2SO_4). The alkali type, preparation, different methods of addition, and reactions with the SO_2 differentiate flue gas desulfurization (FGD) technologies.

14.4.1 Trends in SO_2 Emission Control

Several current trends are occurring in SO_2 emission control.

- Significant use of low sulfur fuel switching to achieve the United States CAAA Title IV Phase I and Phase II compliance.
- Retrofit scrubbers purchased to meet United States CAAA Title IV Phase I compliance were exclusively wet designs.
- When scrubbers are purchased, SO_2 removal levels are generally 95% or better.
- The guaranteed scrubber availability range is 96% to 98% without spare modules.
- Lower cost calcium reagent designs dominate, many with gypsum byproduct production.
- Many wet lime scrubbers incorporate magnesium enhancement.
- Semidry scrubbers achieve 95% or better SO_2 removal with low- or high-sulfur fuels.
- Single module size ranges are 600 to 1,000 MW for wet scrubbers and 200 to 250 MW for semi-dry scrubbers.
- Fluid bed boilers use low-cost limestone beds for SO_2 removal; some of the newest contract awards use quick lime generated in the boiler for use in downstream semi-dry scrubbers.
- Processes combining SO_2 removal with NO_x reduction or particulate removal are also available.

14.4.2 Combustion Processes

14.4.2.1 Furnace Sorbent Injection. Dry furnace sorbent injection (FSI) is sometimes referred to as limestone injection multistage burners (LIMB) when it is used in conjunction with low-NO_x burners. It involves the injection of a calcium-based reagent (such as powdered limestone) into the furnace. Limestone calcines to form calcium oxide (CaO) which reacts with SO_2 and oxygen to form calcium sulfate ($CaSO_4$). The reagent is pneumatically injected through the burners, overfire air ports, or other ports installed in the furnace, or through wall jets or lances that penetrate into the furnace. Transport air requirements range from 1% to 5% of the total combustion air.

Limestone, dolomite, lime, and hydrated lime are possible reagents. With any of these reagents, lime (CaO) (also called quick lime) reacts with sulfur dioxide (SO_2) at approximately 1,600 to 2,200° F in the furnace's oxidizing atmosphere to form calcium sulfate ($CaSO_4$). The basic reactions are:

Limestone

$$CaCO_3 \rightarrow CaO + CO_2 \text{ (calcination)} \tag{14-13}$$

$$CaO + SO_2 + \tfrac{1}{2}O_2 \rightarrow CaSO_4 \text{ (sulfation)} \tag{14-14}$$

Dolomitic limestone

$$CaCO_3{\cdot}MgCO_3 \rightarrow CaO{\cdot}MgO + 2CO_2 \text{ (calcination)} \tag{14-15}$$

$$CaO{\cdot}MgO + SO_2 + \tfrac{1}{2}O_2 \rightarrow CaSO_4{\cdot}MgO \text{ (sulfation)} \tag{14-16}$$

Quick lime

$$CaO + SO_2 + \tfrac{1}{2}O_2 \rightarrow CaSO_4 \text{ (sulfation)} \tag{14-17}$$

Hydrated lime (calcium hydroxide)

$$Ca(OH)_2 \rightarrow CaO + H_2O \text{ (calcination)} \tag{14-18}$$

$$CaO + SO_2 + \tfrac{1}{2}O_2 \rightarrow CaSO_4 \text{ (sulfation)} \tag{14-19}$$

Hydrated dolomitic lime (dolomitic hydroxide)

$$Ca(OH)_2{\cdot}Mg(OH)_2 \rightarrow CaO{\cdot}MgO + 2H_2O \text{ (calcination)} \tag{14-20}$$

$$CaO{\cdot}MgO + SO_2 + \tfrac{1}{2}O_2 \rightarrow CaS0_4{\cdot}MgO \text{ (sulfation)} \tag{14-21}$$

In the case of dolomitic limestone or lime, the magnesium oxide fraction does not react with SO_2 at the temperatures and pressures in the furnace and requires more material for a given calcium/sulfur (Ca/S) ratio.

FSI SO_2 capture is strongly influenced by temperature, residence time, reagent distribution within the furnace, Ca/S ratio, and reagent surface area and type. Good reagent mixing with the flue gas is essential. Good mixing is difficult to achieve when injecting reagent above the burners in the largest furnaces. The highest reagent surface areas yield the greatest utilization. Reducing the sorbent particle size to 2 microns increased SO_2 capture, but further reduction did not help. Energy costs to produce finer particles may offset increased utilization. More expensive hydrated lime appears more effective than limestone in furnace SO_2 removal. This is probably because of increased specific surface areas that result when hydrated lime is heated.

FSI is commercially demonstrated on boilers ranging from 15 MW to 700 MW. The majority of this experience is with reagents injected into the upper furnace region away from the burners. Two United States FSI demonstrations began in 1990: Richmond Power & Light's 60-MW Whitewater Valley tangentially fired Unit 2, and Ohio Edison's 105-MW Edgewater Station wall fired Unit 4. Preliminary results indicate SO_2 removals of 40% at Whitewater and 50% at Edgewater with a calcium-to-sulfur ratio of 2.0.

Tampella Power Corporation's limestone injection furnace activation of calcium (LIFAC) process (Fig. 14-16) began operation at Whitewater Valley Unit 2 in 1990 and in October 1992 at Saskatchewan Power Corporation's 300-MW Shand Station. LIFAC achieved 10% to 15% SO_2 removal at full load and up to 25% at reduced loads in the Shand furnace; 25% was expected. An additional 85% SO_2 removal was achieved in a downstream humidification activation reactor on half the exit gas flow (Ball and Ewald 1993).

FSI significantly increases the particulate loading to the convection section. The increased deposition of solids requires more frequent soot blowing and may require installation of additional soot blowers. Particulate collector inlet loadings will increase and ash resistivity will go up.

Babcock & Wilcox offers a combined limestone injection and dry scrubbing (LIDS®) process to provide low cost and high SO_2 removal for medium- and high-sulfur coals. Figures 14-17 and 14-18, respectively, show the FSI and LIDS® systems. Babcock & Wilcox advises that the FSI technology provides some initial SO_2 reduction but, more importantly, calcines the lower cost limestone to provide fresh calcium oxide (CaO) for the downstream semidry scrubber.

14.4.2.2 Fluid Bed Combustion Scrubbing. Fluid bed combustors, particularly the circulating fluid bed type, are an economical SO_2 removal choice. Chapter 21, Fluidized Bed Combustion, discusses limestone reagent sizes, feeding systems, bed composition, flow diagrams, chemical reaction equations, and module sizes. Tables 21-2 and 21-3 in Chapter 21 provide a quick overview of SO_2 removal parameters.

For higher sulfur fuels with low SO_2 emissions, combustor fluid bed Ca/S ratios can be reduced by supplemental postcombustion semidry scrubbing (Lavely and Mastalio 1994). An overall Ca/S range of 1.3 to 1.7 is indicated versus the higher ratios in Chapter 21. Low-cost limestone ($CaCO_3$) reagent is calcined in the fluid bed combustor to quick lime (CaO). Some of the CaO reacts with SO_2 to form calcium sulfate ($CaSO_4$) in the combustor. The unreacted CaO is hydrated to calcium hydroxide $Ca(OH)_2$ in downstream semidry scrubbers which reacts to form calcium sulfite ($CaSO_3$) and $CaSO_4$.

14.4.3 Postcombustion—Wet Scrubbing

Wet scrubbing is the most common FGD method in use throughout the world. These processes are used for boiler applications with high- to low-sulfur fuels. Even with the refinement of semidry scrubber processes, wet scrubbers

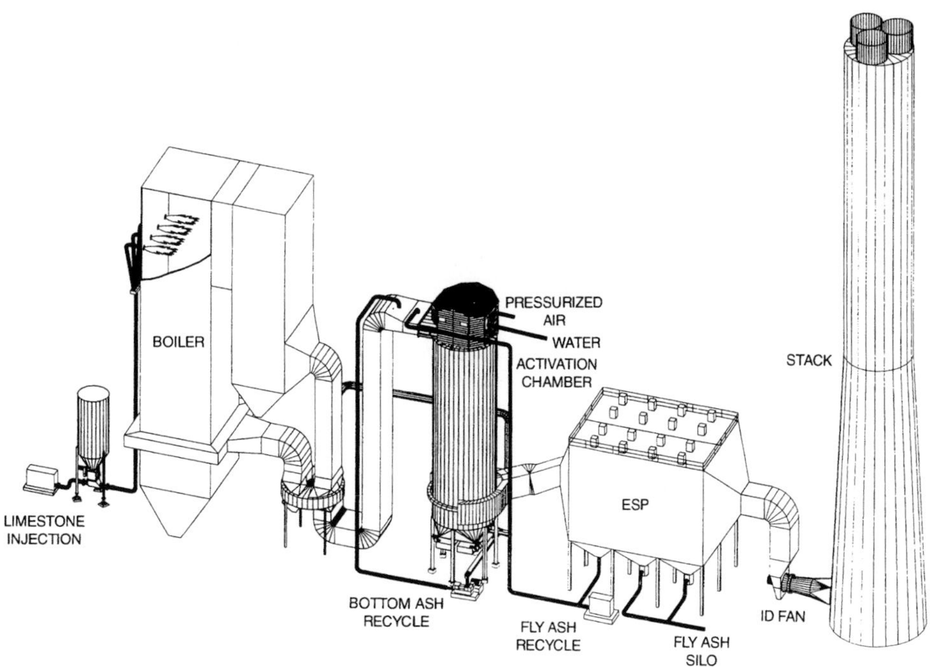

Fig. 14-16. The LIFAC dry FGD process developed by Tampella LTD's boiler division. (From Tampella Power Corporation. Used with permission.)

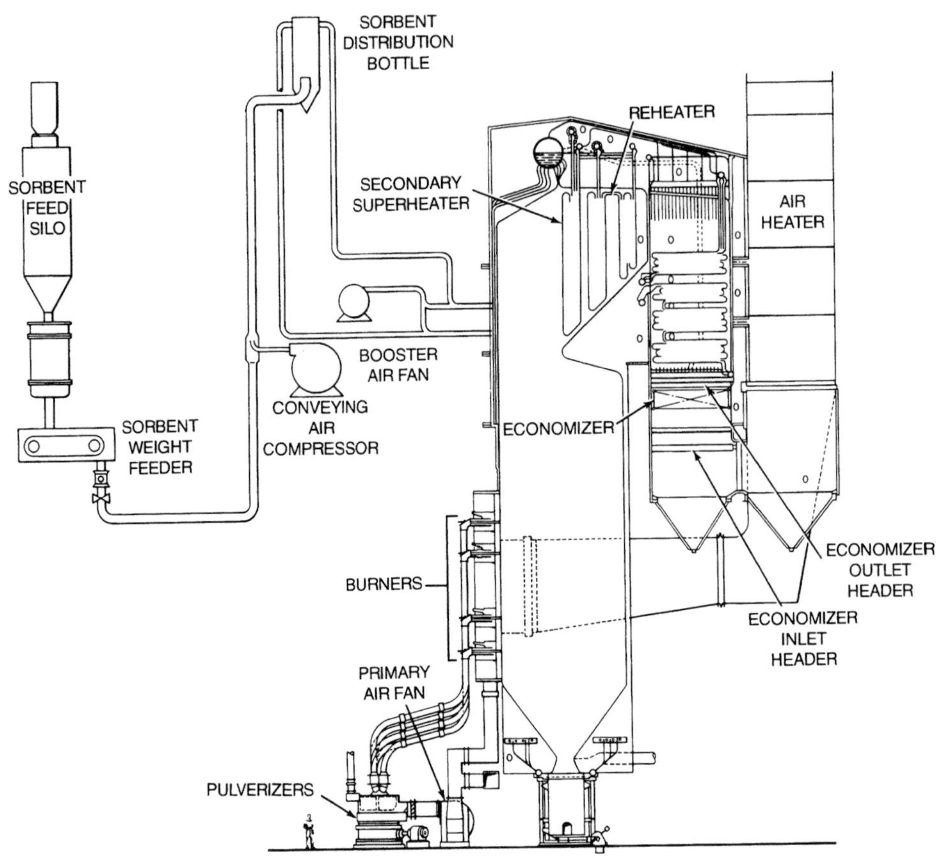

Fig. 14-17. Furnace sorbent injection system. (From Babcock & Wilcox. Used with permission.)

remain competitive, particularly for larger units with medium to higher sulfur fuel.

Better knowledge and control of wet scrubber process chemistry has increased operating availability and design reliability. By the early 1990s, these improvements led to major cost reductions, including the use of single-absorber modules without spares for very large units. A valuable source for wet scrubbing process chemistry and analytical methods is the Electric Power Research Institute's *FGD Chemistry and Analytical Methods Handbook.*

A wet scrubbing FGD process contacts the flue gas with an alkaline liquid or slurry within the absorber. A cleaned,

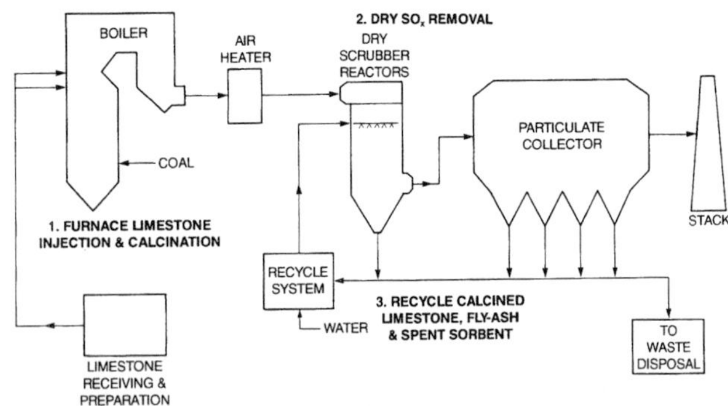

Fig. 14-18. LIDS system. (From Babcock & Wilcox. Used with permission.)

moisture-saturated exit flue gas and a liquid or slurry by-product containing the sulfur dioxide removed from the gas, are discharged from the absorber.

The gaseous SO_2 is absorbed into the liquid and forms sulfurous acid, H2SO_3, which disassociates into hydrogen (H^+) and bisulfate (HSO_3^-) ions. The latter in turn also disassociate into sulfite (SO_3^{2-}) and another hydrogen (H^+) ion according to the following equilibrium reaction.

$$SO_2 + H_2O \rightarrow H_2SO_3 \rightarrow H^+ + HSO_3^- \rightarrow SO_3^{2-} + 2H^+ \qquad (14\text{-}22)$$

Liquid phase reactions to remove or pair up with either H^+, HSO_3^-, or SO_3^{2-} ions result in more SO_2 absorption. For absorption of SO_2 to occur, the SO_2 gas phase partial pressure must be greater than the vapor pressure of SO_2 in the liquid. Therefore, as gas phase SO_2 concentration and partial pressure become lower, as the gas nears the gas exit, absorption decreases. The SO_2 absorption also decreases as the H^+ and HSO_3^- ion concentrations increase. (This would happen if the lack of reagents stopped the removal or pairing up with H^+, HSO_3^-, or SO_3^{2-} ions.) This would cause the equation's reactions to proceed predominantly to the left, which would increase the liquid's SO_2 concentration and vapor pressure.

The major scrubber alkaline ions that are more reactive than the alkaline HSO_3^- ion and that remove or pair up with

H^+ ions are hydroxide(OH^-), sulfite (SO_3^{2-}), carbonate (CO_3^{2-}), and bicarbonate (HCO_3^-). Therefore, highly soluble sodium, ammonia, or magnesium solutions containing these alkaline ions and lime or limestone (calcium) slurries which can dissolve to provide a source of these alkaline ions are effective SO_2 removal reagents. In addition, positive ions such as calcium (Ca^{2+}), magnesium (Mg^{2+}), sodium (Na^+), potassium (K^+), and ammonium (NH_4^+) react or pair up with HSO_3^- or SO_3^{2-} ions to reduce the number of their free ions in solution which, as above, increases SO_2 absorption. The presence of chloride (Cl^-) ions negatively impacts SO_2 absorption since they displace carbonate (CO_3^{2-}) ions from solution.

The FGD scrubber liquid phase is a nonideal concentrated solution containing water and all soluble ions that come from the flue gas, reagent, or makeup water sources, as well as FGD reaction products. The resulting reactions between water and all of the ions depend on the temperature, pressure, the ionic concentrations and activity coefficients, the solubility coefficients or relative saturation of solid compounds that may dissolve into or precipitate from the liquid, and the partial or vapor pressure of gasses that may absorb into or desorb from the liquid. Computerized species distribution modeling can be used to help predict scrubber properties, which include SO_2 absorption capacity. These models work best when the specific absorber design has been calibrated with actual field operating test results.

The previous SO_2 absorption and liquid phase disassociation equilibrium reaction (14-22) and the following similar one for CO2 provide buffering reactions as a function of pH (H^+ ion concentration).

$$CO_2 + H_2O \rightarrow H_2CO_3 \rightarrow HCO_3^- + H^+ \rightarrow 2H^+ + CO_3^{2-} \quad (14\text{-}23)$$

These equilibrium reactions plus the following key oxidation equilibrium reaction of sulfite to sulfate:

$$2SO_3^{2-} + O_2 \rightarrow SO_4^{2-} \quad (14\text{-}24)$$

which is a function of pH and concentration of elemental oxygen, also significantly influence scrubber absorption sulfur byproduct formation and recovery.

Current utility commercial wet scrubber SO_2 removal reagents include limestone, lime, magnesium oxide, soda ash, fly ash, ammonia, and sea water. The four current utility economic wet scrubber processes are listed in Table 14-1 with a comparison of attributes. These four processes are discussed in this section. Selection of an FGD process or reagent for a specific site is dependent on many factors that affect the overall system cost.

14.4.3.1 Forced Oxidation Wet Limestone FGD. Forced oxidation wet limestone FGD is currently the most common wet process selected because of high SO_2 removal, high unit availability, and limestone's utilization and low cost. Reactive limestones are also generally available worldwide. Because of these factors, the forced oxidation wet limestone FGD system is the focus of this section.

A typical limestone system with exit flue gas reheat is shown in Fig. 14-19. The reagent used determines the type and final design of auxiliary equipment required, but similar subsystems are required for other reagents, including the reagent preparation, absorber, dewatering, and flue gas handling. Forced oxidation wet limestone FGD processes can consistently achieve 95% to 98% SO_2 removal for a wide range of coals.

REAGENT PREPARATION. The reagent preparation system consists of the equipment necessary to unload, store, transport, and prepare the reagent for use in the system. Onsite reagent handling and storage facility requirements depend on the frequency of deliveries, potential for delivery interruption, site space available, climate, storage requirements, and costs. To obtain the surface area and reactivity required to meet typical reaction design rates, limestone should be crushed to a size such that 90% to 95% will pass a 325-mesh screen. Larger grind material may be used to save grinding costs but these savings are offset by additional costs because of lower limestone utilization, larger reaction tank size requirements, and lower reactivity. Processes designed by some suppliers may even require a smaller particle size.

Ball mills are used to grind the limestone to the required size. The ball mill system crushes the limestone and mixes it with water to form a fine slurry which is then classified for the proper particle size by hydroclones. Dry grinding and air classification of limestone is used by a few scrubber sup-

Table 14-1. Wet FGD Systems—Comparison of Attributes[a]

	Forced Oxidized Wet Limestone	Magnesium-Enhanced Wet Lime	Seawater	Ammonium Sulfate
Capital cost	Medium	Low	High	High
Industry experience	High	High	Low	Very low
Reagent cost, \$/lb SO_2 removed	Medium	High	Low	High
L/G ratio	50–150	20–70	30–130	Not available
Byproduct dewatering	Low	High	None	Medium
Byproduct management	Disposal to revenue producer	Disposal only	Disposal only	Revenue producer
Maximum removal efficiency achievable, %	95–98	98–99	To 98	To 99
Maintenance requirements	Medium	Medium	Low	To be determined

[a]Information is general in nature based on vendor and project data. These FGD system attributes should be comparatively evaluated in regard to the specific site and the total plant design.

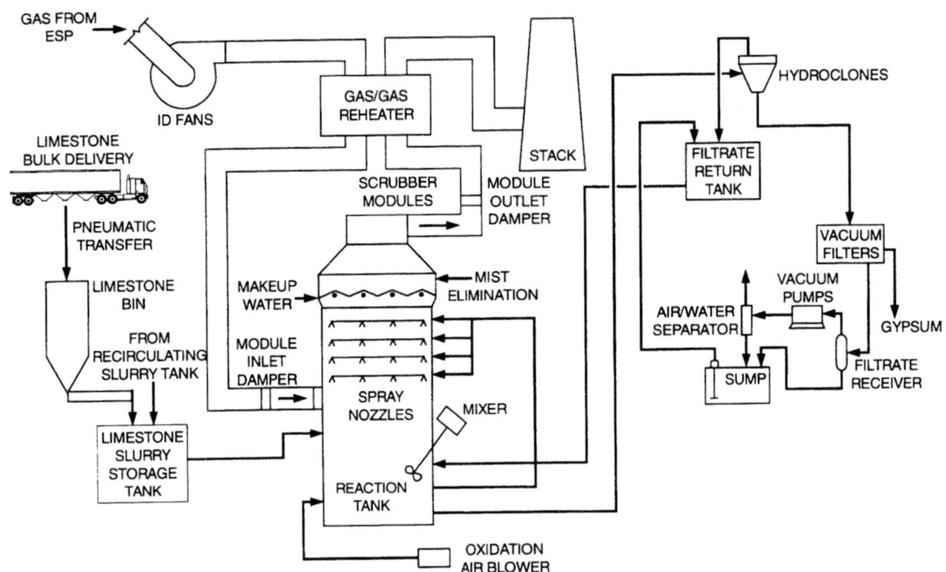

Fig. 14-19. Wet limestone FGD system.

pliers for calcium addition. The ball mill supplier typically provides a complete grinding and sizing system as shown in Fig. 14-20. The work index, or hardness, of the limestone; grind size; limestone flow rate required; and desired ball mill operating period determine the size of the ball mill system. Careful review of potential limestone supplies is necessary because the limestone work index, which may vary from 2 to 13, is proportional to the energy that must be provided to the ball mill system.

ABSORBER. The flue gas contacts the scrubbing liquid in the absorber. Many forced oxidation wet FGD processes use an open spray tower absorber with countercurrent flue gas flow as depicted in Fig. 14-21. A mixture of fresh limestone and reaction products is continuously recycled by pumps from an integral reaction tank through the absorber to contact

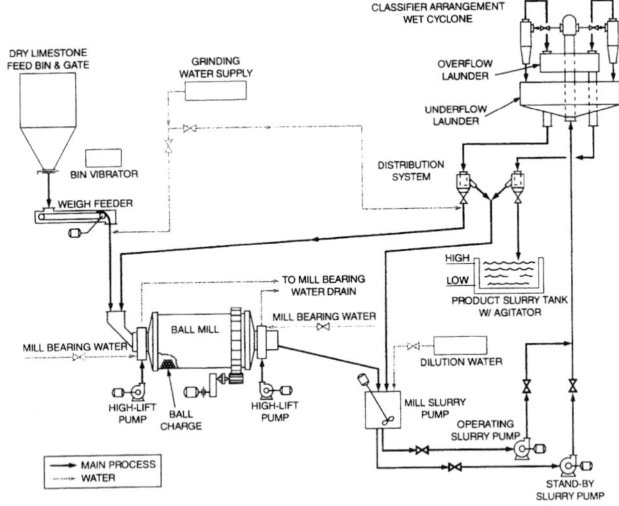

Fig. 14-20. Typical flow diagram for wet limestone grinding systems. (From Svedala Industries. Used with permission.)

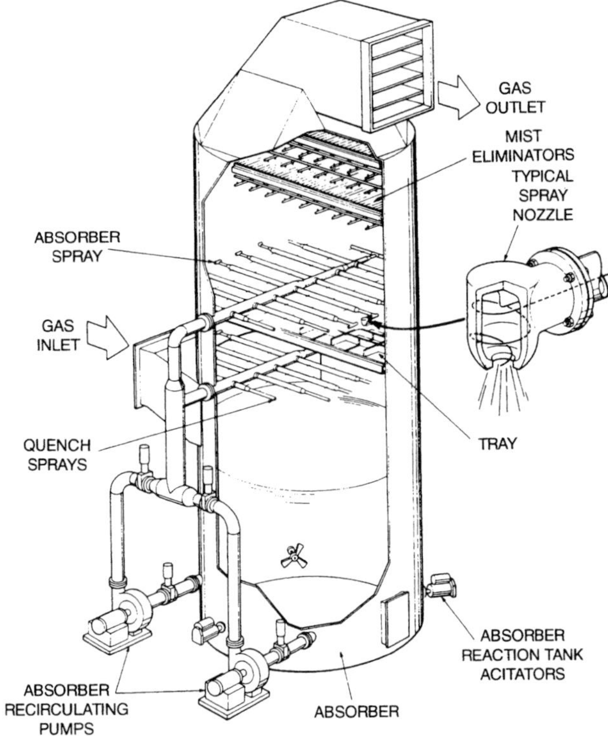

Fig. 14-21. Absorber cutaway view. (From American Electric Power Service Corporation and Babcock & Wilcox. Used with permission.)

the flue gas. This slurry is distributed by pipe headers to nozzles that spray the liquid downward through the absorber. The nozzles generate the required liquid surface area by breaking the slurry into droplets in the size range of 2,000 microns. Both tangential spray and pigtail type nozzles are successful.

The reaction tank catches the falling recycle slurry, allows sufficient retention time for any liquid phase reactions and gypsum particle growth to occur, and acts as a supply point for the recycle pumps. Limestone slurry, oxidation air, and makeup water are added to the reaction tank to maintain the pH, water level, and slurry solids content to achieve the optimum reaction rates and byproduct gypsum characteristics.

Sulfur dioxide is absorbed from the flue gas into the slurry liquid in the absorber where the following overall reactions occur:

$$SO_2 + CaCO_3 + \tfrac{1}{2}H_2O \rightarrow CaSO_3 \cdot \tfrac{1}{2}H_2O + CO_2 \quad (14\text{-}25)$$

$$SO_2 + CaCO_3 + \tfrac{1}{2}O_2 + 2H_2O \rightarrow CaSO_4 \cdot 2H_2O + CO_2 \quad (14\text{-}26)$$

Intermediate reactions occur, but generally sulfur dioxide from the flue gas is absorbed into the slurry liquid. Then the liquid phase chemical reactions occur between the dissolved alkalinity from the reagent, the oxidation air, and the absorbed acidic SO_2 component to produce gypsum ($CaSO_4 \cdot 2H_2O$). Many factors determine the rate at which these overall reactions occur and the equilibrium between them. These factors include the pH, temperature, amount of oxidation air, and the liquid concentration of chemical species such as sodium, magnesium, calcium, potassium, carbonates, sulfates, chlorides, and fluorides.

Current forced oxidation wet limestone designs inject air directly into the reaction tank to improve overall process chemistry and convert most of the calcium sulfite ($CaSO_3$) to gypsum ($CaSO_4 \cdot 2H_2O$). This conversion to gypsum serves several purposes. It minimizes hard calcium sulfate scaling on the absorber internal surfaces, which occurs if the amount of oxidation (conversion of sulfite to sulfate) is in the intermediate range of 15% to 85%. (The scaling of the absorber can be minimized by controlling the oxidation conversion either below or above these thresholds.) In fully oxidized systems, the dissolved $CaSO_4$ precipitates predominantly onto the suspended $CaSO_4$ crystals, which have a very large relative surface area, thus reducing the scaling on the absorber surfaces. The byproduct gypsum crystal shape and size is easier to dewater. The resultant lower moisture and solid properties aid landfill disposal or sales to wallbaord and cement markets. Alternately, oxidation inhibitors such as thiosulfate or liquid sulfur are introduced. This keeps oxidation to <15% which also minimizes scaling. The resulting $CaSO_3$ forms flat platelet crystals that grow into rosettes and retain water in their structure. They are thixotropic and do not have desirable physical properties for landfilling. Fly ash or lime or both must be added to obtain properties suitable for landfill disposal.

Slurry droplets entrained in the flue gas are removed by mist eliminators near the absorber outlet. A vertical disengagement zone is typically provided prior to the mist eliminators to allow large slurry droplets to fall back into the absorber. The mist eliminators, usually of a chevron design (Fig. 14-22), remove the remaining entrained slurry to as low as 0.0001 gpm/ft^2 of eliminator area. The mist eliminator flue gas velocity is approximately 15 ft/s; new designs allow up to 20 ft/s velocities with the same performance.

CROSSFLOW

COUNTERFLOW

Fig. 14-22. METAdek—Chevron mist eliminator. (From Munters: The Incentive Group. Used with permission.)

Wash systems consisting of a nozzle array are located between the stages of the mist eliminator to prevent solid scale buildup. The washing systems are split into sections. The sections are normally washed separately on an intermittent timed basis. The wash water quality can greatly affect the amount of plugging. New designs use fresh makeup wash water with limited alkaline species to prevent solids formation.

The specific FGD design, inlet SO_2 flow rate, and required SO_2 removal efficiency determine the amount of reagent required. The stoichiometric ratio (the ratio of moles of reagent feed required per mole of SO_2 removed) ranges between 1.02 and 1.06 for wet limestone scrubbers. The required SO_2 mass transfer volume (including L/G ratio), the water saturated flue gas volumetric flow rate, and maximum flue gas velocity determine the absorber cross-sectional area or diameter. These three values, the flue gas inlet and outlet duct configuration, the mist eliminators and wash system design, reaction tank height, and internal reheaters, if any, determine absorber height. The reaction tank liquid residence time (typically 3 to 10 minutes), and the suction requirements of the recycle pumps, establish the reaction tank size and configuration. The reaction tank geometry, agitator and oxidation air designs, limestone feed, inlet SO_2, liquid recycle rate, and absorber blowdown rate determine the gypsum byproduct's particle size range.

The liquid to gas ratio (L/G) is the recycle slurry volumetric flow divided by the saturated flue gas volumetric flow and is expressed as gallons per thousand cubic feet (gal/1,000 ft^3) or liters per cubic meter (L/m^3). The L/G ratio, the flue gas velocity, and gas treatment time reflect the number of SO_2 mass transfer units required to achieve a given removal efficiency. Typical L/G ratios for a limestone system range from 80 to 120. The L/G ratio is greater for high inlet sulfur dioxide concentrations and removal efficiencies. The large pumps required to recycle the slurry within the absorber represent the major power consumer for wet FGD systems. The L/G ratio provides a good indication of the auxiliary power difference between wet processes.

As previously noted, the concentration of some chemical species can influence the overall reactions that occur in the absorber. This is especially true with chlorides. The SO_2 removal efficiency for a forced oxidation system drops dramatically as the chloride concentration increases. Increasing chloride levels in either the fuel or water supply may prevent the FGD design efficiency from being achieved. Other dissolved solids such as fluorides or carbonates need to be considered in both the process design and materials selection. The resulting scrubber water chemistry is important to consider when evaluating the plant's fuels or water supply.

DEWATERING. The FGD byproducts formed in a forced oxidization wet limestone process are discharged from the absorber through a blowdown slurry stream. The blowdown stream contains primarily gypsum byproduct solids and minor solid quantities of calcium sulfite, unreacted calcium carbonate, and limestone inerts along with substantial amounts of dissolved solids in the liquid phase. The recycle of the liquid permits lower limestone and water makeup rates due to the dissolved alkalinity and water recovery.

Two dewatering steps are normally used. Hydroclones recycle the smaller unreacted limestone solids with the primary liquid overflow and concentrate the larger gypsum solids in the underflow. This underflow then goes to a vacuum filter or centrifuge for final liquid separation and recycle. This two-step approach reduces the overall dewatering size and cost. The oxidized systems produce larger, easier to settle gypsum solids that allow hydroclones as the primary step. In the second step, belt filters and centrifuges allow washing the chlorides from the gypsum byproduct to meet specifications and further reduce the moisture. Chloride levels below 100 parts per million and moisture levels of $<10\%$ (90+ percent solids) are needed to comply with salable gypsum contract specifications.

FGD BYPRODUCT MANAGEMENT. The disposal costs for FGD solids can vary greatly depending on the method of disposal and the plant location. Types of disposal methods include sale of a marketable product, landfilling, ponding, or mine reclamation. Gypsum's physical characteristics permit landfilling without any special pretreatment. However, if the FGD byproduct leachate contains high levels of total dissolved solids (TDS), landfilling or ponding designs may require special liners, or leachate collection and treatment, or both.

FLUE GAS HANDLING. With large, single absorbers without spares or flue gas bypass, the flue gas handling system is simplified. Multiple absorber systems require inlet and outlet ducts and dampers for each absorber module per unit to distribute and control the flue gas. Dampers and multiple flues are not required for a single module. Fluid modeling of the ductwork and absorber is typically done to ensure even gas distribution. Poor gas distribution among multiple or within absorbers can reduce performance.

14.4.3.2 Wet Scrubbing Design Considerations. Wet FGD scrubbing presents several major impacts to the power plant design and operation. These impacts include wet flue gas discharge or reheat, power consumption, water balance, and waste disposal. Excluding byproduct management, design and installation issues are common among most wet processes, and therefore are discussed collectively in this section.

MANAGEMENT OF SATURATED FLUE GAS. The flue gas handling design from the exit of the absorber to the stack discharge is a very important consideration in wet scrubbing systems. Flue gas leaves the absorber at the adiabatic saturation temperature and includes some entrained liquid or slurry. The entrained liquid contains dissolved chlorides, SO_2, and SO_3, all of which contribute to a very corrosive low pH environment. Because the flue gas is already saturated, cooling causes additional condensation on downstream equipment. To prevent corrosion, a facility can either reheat the flue gas to evaporate some of the entrained liquid and minimize downstream condensation or use appropriate corrosion-resistant materials downstream of the absorber.

Alternatives for flue gas reheating include a regenerative type heat exchanger between the scrubber inlet and exit flue gas ductwork, mixing treated flue gas with hot bypassed flue gas, or tubular (steam) type reheaters. All of these have advantages and disadvantages that need to be evaluated for each project.

The operation of the duct and stack in a wet mode (with entrained slurry and condensation in the liquid phase without any flue gas reheat) has become a favored method in the United States. The advantages include lower capital costs, no degradation of plant performance, and lower operating costs. These systems may have operating problems such as acidic droplet fallout from the stack exit gas or more corrosive environments than anticipated when the corrosion-resistant material or liners were selected.

POWER CONSUMPTION. The power consumption of a wet FGD can range from 1% to 2% of the total plant electric output without the inclusion of flue gas reheat. The recycle pumps are the major power consumer.

WATER USAGE/WASTE WATER. An FGD system can impact the plant water balance by being both a water consumption point and a waste water source. To minimize the FGD water consumption, other plant water discharges, such as cooling water blowdown, are used as the FGD makeup water. As previously noted, the water quality used in an FGD can impact the system performance; therefore, the effect of other available water sources must be evaluated.

The potential for net wastewater FGD production exists as a result of lower moisture FGD discharge solids either to meet byproduct specifications or to reduce disposal quantities and costs. If enough water is not entrained within the solid byproduct, the concentration of dissolved solids (such as chlorides) in the system increases. High levels of dissolved solids could cause material corrosion or increase the SO_2 emissions. A purge blowdown stream may be cost effective. The quantity and quality of the liquid blowdown need to be included in the plant's waste water treatment design.

MATERIAL SELECTION. A large number of materials are available for use in wet flue gas desulfurization systems, and the selection of a material is based on a number of factors including operating conditions, corrosion potential, erosion potential, temperature, initial cost, and maintenance philosophies. Materials used in FGD service can be classified in three categories: metallic, organic, and inorganic.

Metals include a wide range of materials including carbon steel, stainless steels (316 series, 317 series, etc.), nickel-based alloys (625, C-276, C-22, etc.), duplex steels, titanium, and other corrosion-resistant alloys. As the demand increases for minimal water discharge, closed loop systems have resulted in higher chloride levels which present significant corrosion problems for many metals and require more exotic alloys. Because of high costs, many alloys are applied as a 3/16 in. cladding or "wallpaper" over a carbon steel base.

Organic lining materials, applied over a carbon steel base, include epoxy, vinyl ester, polyester, or rubber. These linings are used for corrosion protection. Field-wound construction of vessels and tanks with fiberglass-reinforced plastic has recently become competitive. These materials can often be the low initial capital cost alternative but normally require higher periodic maintenance. When in the flue gas path, these organic materials require protection against high temperatures caused by air heater failures coupled with the loss of absorber recycle pumps. Therefore, an emergency quench system should be installed upstream of the absorber gas inlet module.

Inorganic materials consist of prefired bricks, ceramic tiles, and hydraulically or chemically bonded concretes and mortars. Typical uses of these materials are as linings for sumps, silos, tanks, and stacks. The adhesion methods used for attaching these materials and the mortar choice are crucial since these are the most common failure modes.

The installation quality is an important issue for all of these materials, especially for substrate preparation and lining application. The best materials have failed as a result of poor quality assurance/quality control during installations.

Mist eliminators are most commonly constructed of fiberglass-reinforced plastic (FRP); however, they can also be constructed of fiberglass polypropylene, stainless steel, or high nickel alloys. Stainless steel mist eliminators are not suitable for many FGD systems because of high chloride levels. The main cause of mist eliminator damage comes during maintenance outages when high-pressure water washing or mechanical cleaning is required to remove solids plugging or scale.

14.4.3.3 Wet Lime FGD Systems. Many FGD suppliers offer wet lime FGD using slaked lime as the scrubbing reagent in a spray tower scrubber. The thiosorbic (magnesium enhanced) lime process is the most commonly used lime process in the United States. Therefore, the process is discussed in the paragraphs below because it has better performance and lower costs than those for wet lime only and it competes more strongly with the forced oxidation wet limestone process. The Dravo Lime Company provided Black & Veatch with valuable information on the magnesium-enhanced wet lime process.

PROCESS DESCRIPTION. In the magnesium-enhanced wet lime process, quick lime (CaO), containing 4 to 7 wt % magnesium oxide (MgO), is slaked to produce a calcium hydroxide ($Ca[OH]_2$) and magnesium hydroxide ($Mg[OH]_2$) slurry. As in the equilibrium reactions in Eq. (14-22), when SO_2 is absorbed, sulfite ions (SO_3^{2-}) are formed. With high alkalinity (high pH) values, highly soluble magnesium sulfite exists. The magnesium-enhanced lime scrubbing reagent reacts with SO_2 to form magnesium bisulfite ($Mg[HSO_3]_2$), magnesium sulfite ($MgSO_3$), and calcium sulfite hemihydrate ($CaSO_3 \cdot \frac{1}{2}H_2O$), according to the following basic reactions:

$$MgSO_3 + SO_2 + H_2O \rightarrow Mg(HSO_3)_2 \quad (14\text{-}27)$$

$$Mg(OH)_2 + SO_2 \rightarrow MgSO_3 + H_2O \quad (14\text{-}28)$$

$$Ca(OH)_2 + SO_2 \rightarrow CaSO_3 \cdot \tfrac{1}{2}H_2O + \tfrac{1}{2}H_2O \quad (14\text{-}29)$$

After the SO_2 gas is absorbed into the liquid, it reacts with the dissolved magnesium ions to form soluble compounds. There are also some solids dissolution and precipitation reactions.

The slurry of calcium hydroxide and soluble magnesium compounds has significantly greater reactivity and solubility than a calcium carbonate slurry. The fresh magnesium enhanced lime slurry has a pH that ranges from 10.0 to 12.0, which is much higher than a limestone slurry's pH. Consequently, SO_2 removal efficiencies in the 98% to 99% range are possible. For the same FGD requirement, the greater SO_2 reactivity significantly lowers the required liquid to gas (L/G) ratio as compared to limestone scrubbers.

The magnesium-enhanced lime slurry may be prepared in

detention, paste, or ball mill slakers. Prepared slurry is stored in a slurry feed tank, ready for automatic injection into the scrubber module's reaction tank as required to maintain the tank's desired pH between 6.0 and 7.0. In this pH range, the pH is sensitive to relative concentrations of HSO_3^- and SO_3^{2-} ions and indicates, in conjunction with a boiler load signal, when more fresh lime should be added. The reaction tank reagent regeneration reactions are:

$$Mg(HSO_3)_2 + Ca(OH)_2 \rightarrow MgSO_3 + CaSO_3 \cdot \tfrac{1}{2}H_2O + 1\tfrac{1}{2}H_2O \quad (14\text{-}30)$$

$$Mg(HSO_3)_2 + Mg(OH)_2 \rightarrow 2MgSO_3 + 2H_2O \quad (14\text{-}31)$$

Liquid residence time in the reaction tank is typically about 2 minutes. Oxidation in the reaction tank is not used or wanted as magnesium sulfate ($MgSO_4$) would precipitate to cause scale formations.

Spray towers for magnesium-enhanced wet lime processes are geometrically similar to those used in wet limestone FGD processes. There are three distinct advantages due to the slurry's higher reactivity: (1) the absorber vessel height is reduced, (2) pumping heads are lower, and (3) the required L/G is less. The lime spray tower may be 10 to 30 ft shorter than the limestone scrubbers for the same SO_2 removal efficiency because of the reduced number of spray levels (lower L/G). The L/G ratio may vary from 20 to 70 gallons per 1,000 acfm of flue gas depending on the application. The stoichiometric ratio typically ranges between 1.01 and 1.05.

The higher pH of the slurry results in levels of natural oxidation of sulfites to sulfates that is less than that achieved in a wet limestone process. Therefore, magnesium-enhanced wet lime FGD processes are less susceptible to formation of hard gypsum scale. However, formation of soft sulfite scale is likely. Since sulfite scale does not harden, it is more easily washed from the mist eliminators and other surfaces in the magnesium-enhanced wet lime spray tower. Typically, the sulfite scale is not a serious problem. Research is continuing into the forced oxidation of the magnesium-enhanced lime systems to allow the production of gypsum as a salable byproduct.

Slurry blowdown from the reaction tank can be pumped to a pond for settling or dewatered using thickeners and vacuum filters. Recovered water is returned to the FGD process for reuse. The solids content of the thickener underflow is typically <35% by weight. The maximum solids content upon discharge of the vacuum filter is in the range of 50% to 65% for these byproducts. Therefore, there is more liquid blowdown in the solids discharged than with wet limestone gypsum solids.

DESIGN CONSIDERATIONS. The magnesium-enhanced lime system design considerations for the management of saturated gas, water usage/waste water, and material selection are very similar to those described for a wet limestone system. Differences in other design considerations are described below.

POWER CONSUMPTION. Power consumption for a magnesium-enhanced lime system is 1% or lower of the total plant electrical generation, which is significantly less than the 1% to 2% power consumption of a wet limestone system.

BYPRODUCT DISPOSAL. The solid waste generated by a magnesium-enhanced lime system presents additional handling complications. First, the solid byproduct is generally more difficult to handle because the solids have a higher moisture content and are thixotropic. The higher moisture levels and thixotropic nature are a result of the flat platelet $CaSO_3$ crystals. These individual crystals grow into rosettes and retain water between the platelets. When handled, the fragile rosettes break and the entrapped water acts as a lubricant for the individual platelets, which results in the thixotrop behavior. To achieve the desired physical properties for landfilling, these FGD byproducts should be mixed with fly ash and lime.

Research into the improvement of the magnesium-enhanced lime system, such as the oxidation of the $CaSO_3$ to $CaSO_4$, is evolving into process modifications that will allow the generation of easier to handle solids with the potential for producing a salable solid byproduct.

14.4.3.4 Wet Scrubbing Emerging Technologies. Wet forced oxidation limestone, wet magnesium-enhanced lime, and semidry lime systems have extensive utility FGD experience. Emerging technologies for utility SO_2 removal continue to develop. Two new wet FGD scrubbing technologies are available: seawater and ammonia scrubbing.

SEAWATER SCRUBBING. Seawater has high natural alkalinity and can effectively control SO_2 emissions without supplemental reagents. This process has removed SO_2 on industrial applications for 15 years. In 1988, the first utility seawater scrubber started up on a 500-MW power plant application near Bombay, India. It controls SO_2 emissions from 25% of the flue gas, with another 25% scheduled for service in late 1994. Sulfur dioxide emissions from two other power plants in Spain, each producing 160 MW of electricity, will be controlled by seawater scrubbers starting in 1995.

Process Description. Figure 14-23 shows a seawater FGD process flow diagram. This process uses a once-through seawater spray tower. Seawater contains significant concentrations of alkaline ions including sodium, magnesium, potassium, calcium, carbonates, and bicarbonates. It also contains significant concentrations of chloride and sulfite ions.

Flue gas passes through an electrostatic precipitator or fabric filter and induced draft fans before entering the bottom of the seawater absorber. Seawater is pumped into the top of the absorber through overflow nozzles and flows counter to the flue gas through wetted saddle-type packing. The cleaned gas passes through a mist eliminator to remove entrained liquid before discharging to the atmosphere through the chimney.

Seawater for the process typically uses only part of the

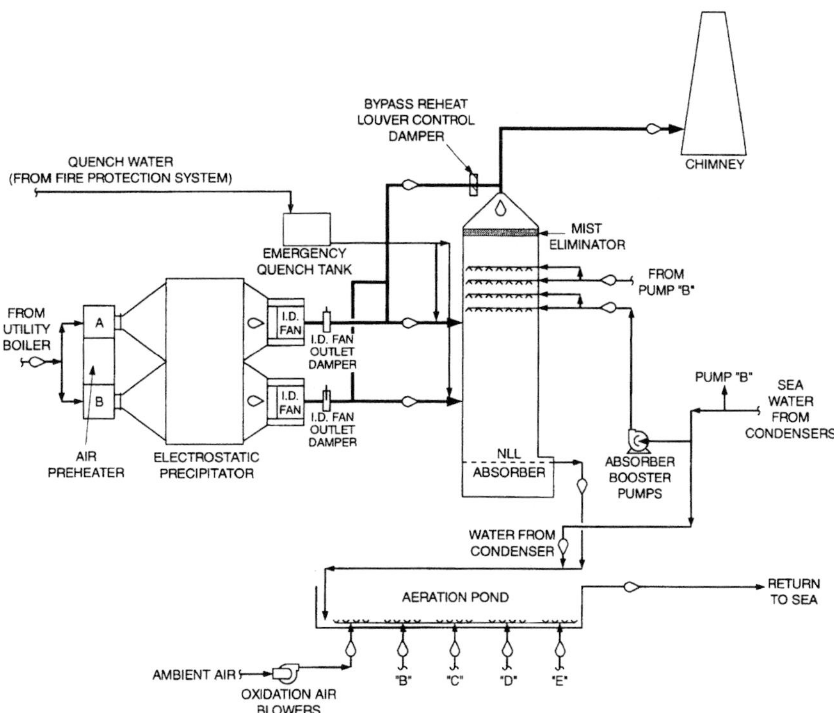

Fig. 14-23. Process flow diagram of a seawater scrubber.

condenser cooling water. It is pumped from the cooling water system at the condensor's discharge to the scrubber on a once through basis. The L/G ratio is dependent on the condenser's discharge temperature, SO_2 concentration in the flue gas, and required SO_2 removal efficiency. The L/G ratio range is 30 to 130 gpm/1,000 acfm. There is no recycle to the absorber.

The absorber's liquid effluent is pumped to the aeration pond. The remaining condenser seawater is directly bypassed to the aeration pond to dilute the absorber's liquid effluent. The absorber's diluted effluent is aerated using a fan to strip out the CO_2. This reduces acidic content and pH is raised to between 6 and 7 to oxidize the HSO_3^- and SO_3^{2-} to HSO_4^- and SO_4^{2-} before returning the stream to the sea. This conversion to SO_4^{2-} continues as pH increases to around 8 in the ocean mixing zone.

Design Considerations. Several factors that impact the suitability of the process for a specific application are upstream particulate collection efficiency, a proper ocean mixing zone, and the amount, if any, of volatile air toxics in the flue gas.

The effect on the marine environment by this system's effluent is a frequent concern. The sulfur removed from the flue gas is transferred directly to the ocean. Natural seawater contains approximately 1 kg of sulfate per tonne of seawater. The aeration pond discharge has a sulfate concentration about 3% greater than that in seawater. Another concern is introducing the acidic liquid effluent with a pH of 6 to 7 into the main body of alkalinic seawater where the pH is about 8. However, when mixed, the alkalinity of the seawater quickly raises the liquid scrubber effluent pH to that of natural seawater and converts most of the remaining HSO_4^- to SO_4^{2-}.

Potential air toxic and heavy metal releases into the seawater are another major concern. Because the seawater may remove some air toxics from the flue gas, those removed would end up in the effluent liquid to the ocean. Therefore, high-efficiency particulate removal is required upstream of the seawater scrubber to minimize particulate concentration and removal of air toxics and heavy metals in the FGD system.

Mercury and other air toxics in vapor or gaseous form may be a problem. Because elemental mercury is a vapor at temperatures found in the particulate control device, very little collection is achieved. Therefore, when some of the mercury does condense at the lower temperatures found in the scrubber, it can be absorbed. One alternative is to inject activated carbon upstream of the particulate removal device, but its effectiveness at a typical air heater exit temperature of 300° F remains to be demonstrated. Additional liquid effluent treatment might be necessary.

AMMONIA SCRUBBING. A new ammonia wet scrubber process has very high SO_2 removal efficiencies and produces a salable ammonium sulfate fertilizer byproduct. Field experience consists of a 3-MW pilot plant that operated in North Dakota from September 1992 to January 1993. A commercial ammonia scrubber contract for a 300-MW plant was awarded to General Electric Environmental Systems by the Dakota Gasification Company, with commercial operation scheduled by March 1997.

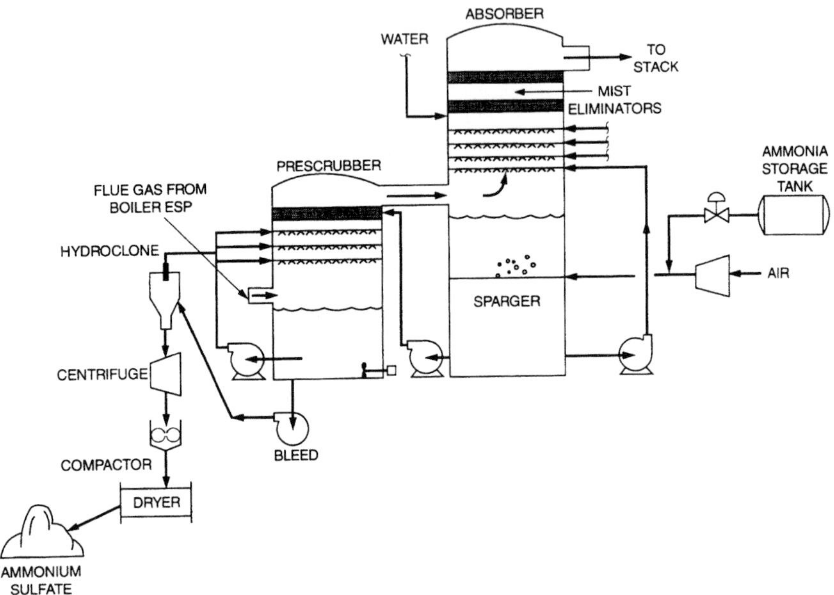

Fig. 14-24. Ammonium sulfate process simplified flow diagram. (From General Electric Environmental Systems, Inc. Used with permission.)

Process Description. Figure 14-24 shows the flow diagram for this process. Hot flue gas enters a prescrubber where it is sprayed with saturated ammonium sulfate liquor. The prescrubber cools the flue gas to the adiabatic saturation temperature and evaporates water from the saturated ammonium sulfate solution to produce ammonium sulfate crystals.

Flue gas leaving the prescrubber passes through a bulk entrainment separator and into the SO_2 absorber. Flue gas flows countercurrent to the scrubbing liquor, which is pumped from the reaction tank and sprayed through nozzles. Ammonia, diluted in air, is introduced to the reaction tank to form ammonium compounds. The ammonia reacts with and the dilution air oxidizes the absorbed SO_2 to form ammonium sulfate ($[NH_4]_2SO_4$). Makeup water is added to the absorber to keep the scrubbing liquor diluted. Finally, the flue gas passes through a mist eliminator to remove any droplets.

The absorber liquor blowdown is fed to the prescrubber where water is evaporated and ammonium sulfate crystallizes. The prescrubber tank liquor blowdown, containing crystallized ammonium sulfate, passes to the primary dewatering hydroclones. Secondary dewatering occurs in a centrifuge. The ammonium sulfate cake from this equipment has $<2\%$ moisture. This ammonium sulfate cake is processed in a drying/compacting system to yield granular ammonium sulfate with $<0.5\%$ moisture.

Design Considerations. The main benefit of this system is the salable byproduct. Ammonium sulfate prices range between $75 and $130 per ton, depending on shipping costs and demand. Therefore, a market evaluation needs to be performed before selection of this process. This evaluation is especially important because of the high ammonia reagent cost.

Capital costs for an ammonia scrubber are higher than for a wet limestone system because of the costs for the prescrubber/crystallizer and equipment needed to dry, compact, and granulate the product.

One main disadvantage is that anhydrous ammonia must be handled and stored on-site with extreme care, since it is a hazardous material. Storage tank rupture or ammonia leakage from transport piping presents a potential health and safety hazard to project site personnel.

14.4.4 Postcombustion—Semidry Scrubbing

For these processes, $Ca(OH)_2$ contacts the gas as a slurry or as a dry solid. The water added is dried before the solids exit the scrubber vessel. Semidry scrubbing processes operate above the flue gas water saturation temperature. There are no liquid discharge streams from or liquid levels in the SO_2 absorber. The fly ash and SO_2 reaction products are discharged from the particulate collector as free flowing solids. Dewatering is not required.

Quick lime (CaO) and hydrated lime ($Ca[OH]_2$) are the current commercial reagents. The CaO is hydrated to $Ca(OH)_2$ either within or before addition to the absorber. This SO_2 removal process is upstream of either an ESP or FF for byproduct particulate removal. If the price for fly ash is high, particulate collection could occur both upstream and downstream of the process. Semidry scrubbing processes include circulating dry scrubbers (CDSs) and spray dryers.

Introduction of reagent independent of the humidification water achieves a major technical improvement. In the CDS, lime is usually introduced as a dry hydrated solid. In the spray dryer, supplemental dry hydrated lime can be added upstream of the spray dryer vessel and lime slurry atomizing devices. With these changes, the required reagent can be introduced to treat high inlet SO_2 concentrations and achieve high SO_2 removals. Competition with wet scrubbing is limited only by economic considerations.

The overall process reactions are:

$$CaO + H_2O \rightarrow Ca(OH)_2 \quad (14\text{-}32)$$

$$SO_2 + Ca(OH)_2 \rightarrow CaSO_3{\cdot}\tfrac{1}{2}H_2O + \tfrac{1}{2}H_2O \quad (14\text{-}33)$$

$$SO_2 + Ca(OH)_2 + \tfrac{1}{2}O_2 \rightarrow CaSO_4{\cdot}2H_2O \quad (14\text{-}34)$$

FGD byproducts from a semidry lime absorber process contain fly ash and significantly more calcium sulfite than calcium sulfate. The process requires the evaporation of all liquid entrained in the flue gas before exiting the absorber modules. However, since the process relies on liquid phase chemical reactions, the SO_2 must react with the reagent before evaporation is complete. Reaction rates are dependent on the amount of liquid present and the drying rate.

Semidry scrubbing offers the following advantages over wet FGD processes:

- Lower energy usage,
- Lower water consumption,
- FGD byproducts collected dry,
- Reduced slurry pumping requirements,
- Flue gas discharge temperature above saturation,
- Less complex and less costly equipment,
- No exotic metals or coating materials required for scrubber or ductwork, and
- Fewer operator and maintenance hours.

The disadvantages are greater reagent quantities needed and more solids produced than for conventional wet calcium reagent spray tower processes. The unreacted lime and fly ash within the semidry scrubber byproducts may aid fixation reactions which would result in decreased permeability and increased unconfined compressive strength properties for the disposal byproduct. These properties may make landfill disposal easier or enhance byproduct use.

14.4.4.1 Circulating Dry Scrubber (CDS). The Enelco/Lurgi circulating dry scrubber process is described in this subsection. Enelco/Lurgi has a commercially available process and similar processes are being developed by Graf-EPE GmbH (Graf et al. 1993) and AirPol Inc., a Division of FLS Miljo (Burnett et al. 1993).

CDS/ESP OR FF PROCESS. The circulating dry scrubber/electrostatic precipitator or fabric filter process can achieve SO_2 removal >98%. The lime addition is independent of the evaporative capacity of the flue gas and is essentially constrained only by utilization and cost. There is no apparent constraint to the CDS's practical SO_2 removal other than reagent cost, even on high sulfur coal-fueled combustors.

Flue gas is directed to the CDS for scrubbing of acid gases and then cleaned of particulate matter by an ESP or fabric filter (FF). It is a dry process that normally produces a dry, free-flowing disposal product. Hydrated lime is introduced as a dry, free-flowing powder. Lime utilization is improved by evaporatively cooling (low-quality water can be used) the flue gas and recirculating the calcium in the process. This utilization improves as the CDS is operated closer to the adiabatic saturation temperature. Typically, the outlet approach temperature is maintained at a minimum of 30° F above adiabatic saturation to improve material handling, minimize cold spots and corrosion, and permit low-cost carbon steel construction.

Over 90% of the solids, which contain unreacted lime, discharged from the ESP or FF are recirculated to the CDS by gravity using air slides which require a higher than typical ESP or FF elevation. Fresh hydrated lime is fed into the air slides by a rotary feeder. The air slides have low operating and maintenance costs and high capacity.

The CDS/ESP or FF requires a small plan area; this is advantageous for retrofit projects as well as new installations with significant space constraints. Figure 14-25 shows relative sizes for the Black Hills Power and Light Company's Neil Simpson Unit 2 turbine–generator, boiler, and CDS–ESP configuration. The space below the ESP was used to enclose equipment such as lime hydration, boiler water treatment systems, and air compressors and pumps.

Figure 14-26 shows the simplified CDS absorber with the feed points for hydrated lime, recirculated dust, and cooling water. The absorber operates as an evaporator and as a chemical reactor for absorbing gaseous pollutants, including air toxics.

Heat and mass transfer are improved by maximizing the slip velocity (differential velocity) between the solid particles and the flue gas.

Typical characteristics are as follows:

- High mass transfer rates within the lime particles because of the hydrated lime particle structure and a very small average particle diameter of 5 to 10 μm.
- Extremely long solids retention time which allows high absorption of gaseous pollutants and improves lime utilization.

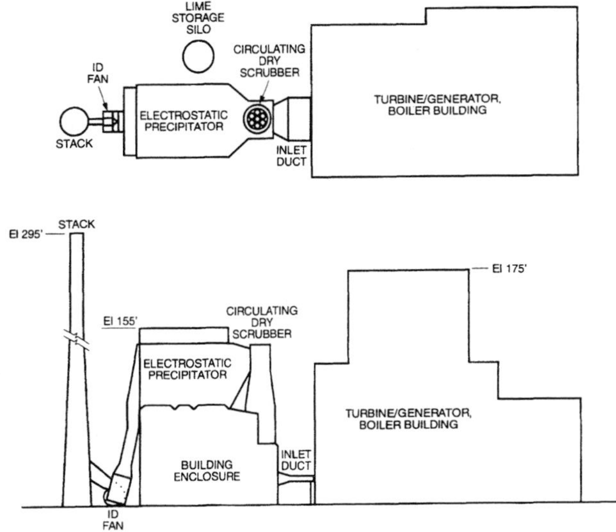

Fig. 14-25. Turbine–generator, boiler, and AQCS layout and elevation views.

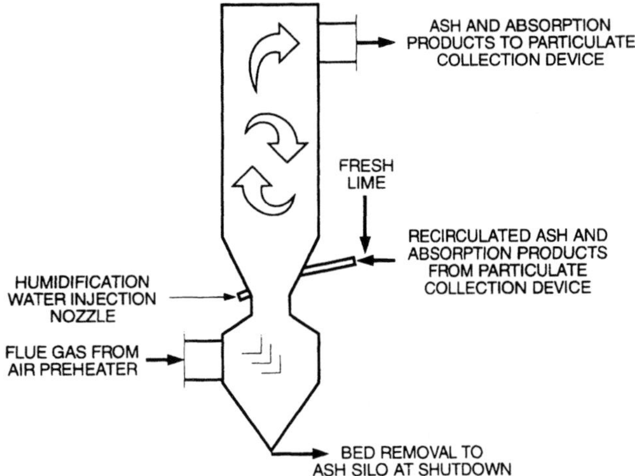

Fig. 14-26. Circulating dry scrubber.

- Continuous abrasion at the lime surface removes the $CaSO_3$ and $CaSO_4$ surface that covers active lime particle cores.
- Operating temperatures with a close approach to the adiabatic saturation temperature can be achieved which gives a higher reagent utilization but may require alloy materials.
- Wastewater can be used for flue gas cooling water injection.
- The turndown is between 100% and 30% of the boiler load. Lower turndowns are possible by recycling a portion of the exit gas with a booster fan.

PROCESS CONTROL. Figure 14-27 shows the process control, which consists of three major independent control loops for fully automatic operation. This process is, in principle, very simple and reliable. The process SO_2 and temperature control concepts are similar for other semidry scrubber processes.

- SO_2 control—The feed rate of hydrated lime is determined by the amount of SO_2 in the inlet flue gas and an outlet flue gas SO_2 signal.
- Temperature control—The gas temperature leaving the absorber directly controls the cooling water injection rate through high-pressure single fluid flow nozzles.

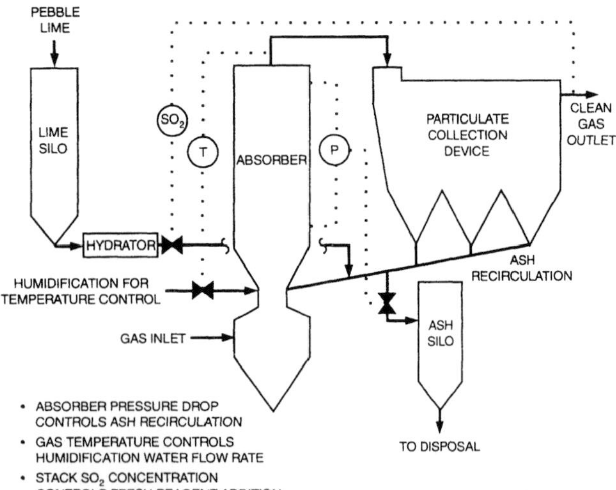

Fig. 14-27. CDS/particulate collector control schematic.

- Solids discharge—The solids discharged from the system are controlled by the solids loading in the absorber as measured by the differential pressure across the absorber height. Solids are discharged from the system at the same rate that hydrated lime, fly ash, and SO_2 enter the system. Except for startup or shutdown, this maintains a constant mass of solids in the system.

14.4.4.2 Lime Spray Dryer/Absorber. Spray dryer/absorber SO_2 removal processes use lime slurry for scrubbing the SO_2 from the flue gas and produce a dry byproduct. Joy/Niro, ABB Environmental Systems, Environmental Elements Corp., General Electric Environmental Systems, and Babcock & Wilcox incorporated the gas absorption feature of this process for use in utility FGD applications. Today, spray dryer/absorbers treat flue gas from over 17,000 MW of coal-fueled boilers (Gleiser and Felsvang 1994).

A schematic diagram of a lime spray dryer/absorber process is presented in Fig. 14-28. Lime slurry is atomized into fine droplets as it is injected into flue gas flowing through the absorber module. Atomization may be accomplished by a high-speed rotary atomizer or spray nozzles. In utility applications, rotary atomizers typically are used because they have better abrasion resistance, are less prone to plugging, and use a lower pressure slurry feed system.

Rotary atomizers typically turn at 12,000 to 17,000 rpm and use a carefully balanced wheel with replaceable ceramic inserts to atomize the lime slurry. A spare atomizer is used to permit quick removal and replacement of the rotary atomizer in an operating module. Ceramic nozzle maintenance is typically required every 2,000 to 4,000 hours of operation because of abrasion.

As shown in Fig. 14-28, flue gas enters the absorber module and contacts the atomized slurry. SO_2 is absorbed into the atomized slurry droplets from the flue gas and reacts with the calcium in the slurry.

Proper operation yields a dry, free-flowing byproduct. Normally there is minimal buildup of byproducts on the walls of the module.

Parameters requiring control include slurry feed rate, gas and slurry distribution, slurry droplet size, outlet gas temperature, droplet/particle residence time (5 to 12 seconds), and lime content in the slurry feed.

Water evaporation from the slurry reduces the flue gas temperature to 20 to 40° F (11 to 22° C) above the saturation temperature of the gas. Higher approach temperatures require higher calcium-to-sulfur ratios since the reaction rate is limited by the presence of liquid moisture, which is being evaporated. Most manufacturers recommend a minimum approach temperature of 18 to 20° F to prevent moisture condensation in the spray dryer module. However, to avoid reheat and to prevent condensation in downstream particulate removal equipment, many systems operate at 30 to 40 degrees above the gas saturation temperature. Even greater approach temperatures are required to avoid corrosion and plugging if hydroscopic calcium chloride (from hydrogen chloride removal) is present.

The FGD byproducts are usually collected with fly ash

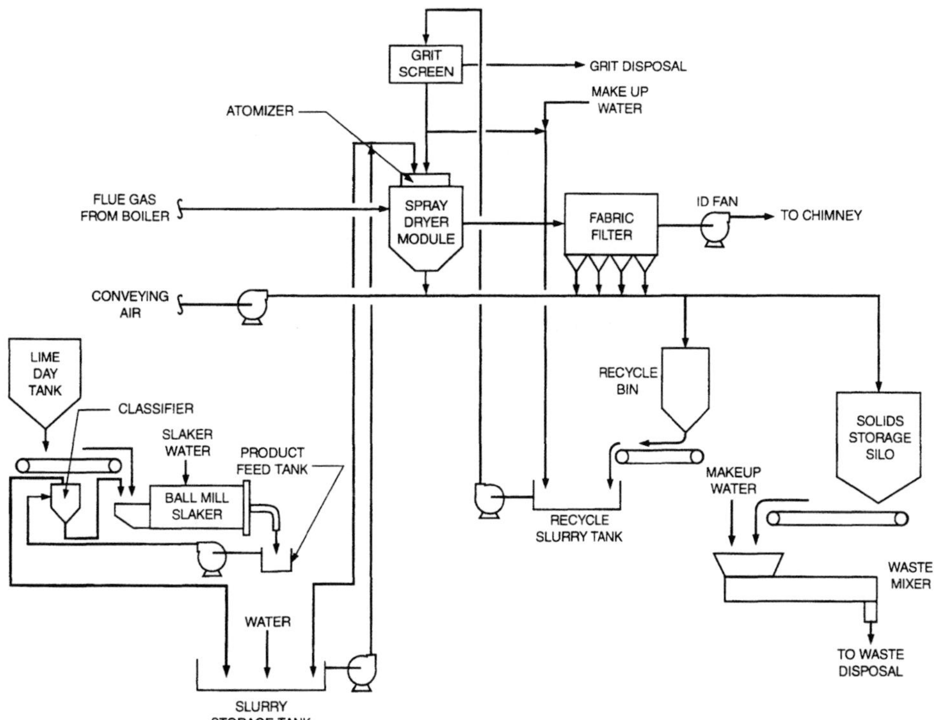

Fig. 14-28. Spray absorber/dryer process.

in a fabric filter. In some cases, part of this material, which contains unreacted lime, is recycled to improve reagent utilization. The balance, which typically contains 1% to 3% moisture, goes to storage and disposal. In a fabric filter, the flue gas passes through a dust cake of unreacted lime and reaction products. This additional contact may result in as much as 20% additional SO_2 removal. Less supplemental SO_2 removal occurs in an electrostatic precipitator. However, an electrostatic precipitator is less susceptible to solids pluggage from condensation and solids agglomeration, as compared with a fabric filter's potential for bag binding.

A typical utility sized spray drier showing quick lime storage area and ball-mill slaked lime feed system to the spray dryer atomizers is shown in Fig. 14-28.

Current generation lime spray dryer/absorber FGD processes are capable of 95% SO_2 removal on coals containing up to 3.0% sulfur. For additional control of vapor phase emissions such as mercury and dioxins, the spray dryer/absorber system can be augmented with an activated carbon injection system. This consists of a dry additive system for injection of additives, preferably upstream of the spray dryer/absorber (Gleiser and Felsvang 1994).

14.5 COMBINATION NO_x REDUCTION/SO_2 REMOVAL

Recent developments in air quality control systems include processes that remove both SO_2 and NO_x from the flue gas. At present, there are very few commercial installations of these systems. The Haldor Topsoe/ABB SNOX™ process and Mitsui-BF Activated Coke System have commercial utility installations. Both of these generate sulfuric acid or elemental sulfur as byproducts. A long-term market for these byproducts must exist. Other processes under development and near commercial availability include the Babcock & Wilcox $SO_xNO_xRO_xBO_x$ (SNRB) and the NOXSO™ Corporation NOXSO™ process.

14.5.1 Haldor-Topsoe/ABB SNOX Process

The SNOX™ Process, a totally catalytic process for reduction of sulfur oxides and nitrogen oxides from gaseous streams, was developed in Europe by Haldor–Topsoe A/S and its affiliate, Snamprogetti. The process is offered under license in North America by ABB Environmental Systems. In this process, nitrogen oxides are decomposed to elemental nitrogen and water vapor with ammonia. Sulfur dioxide (SO_2) is catalytically converted to sulfur trioxide (SO_3). The SO_3 and water from the flue gas are then condensed as sulfuric acid in a falling film condenser.

The SNOX technology consists of five key process areas: particulate collection, NO_x reduction, SO_2 oxidation, sulfuric acid (H_2SO_4) condensation, and acid conditioning. Heat addition, heat transfer, and heat recovery represent a significant part of the SNOX™ system. A process flow diagram integrating these individual steps is shown in Fig. 14-29.

The flue gas from the boiler air preheater is treated in a particulate control device and passes through the primary side of a gas/gas heat exchanger (GGH) which raises the gas temperature to above 700° F (370° C). An ammonia and air mixture is then added to the gas prior to the selective cata-

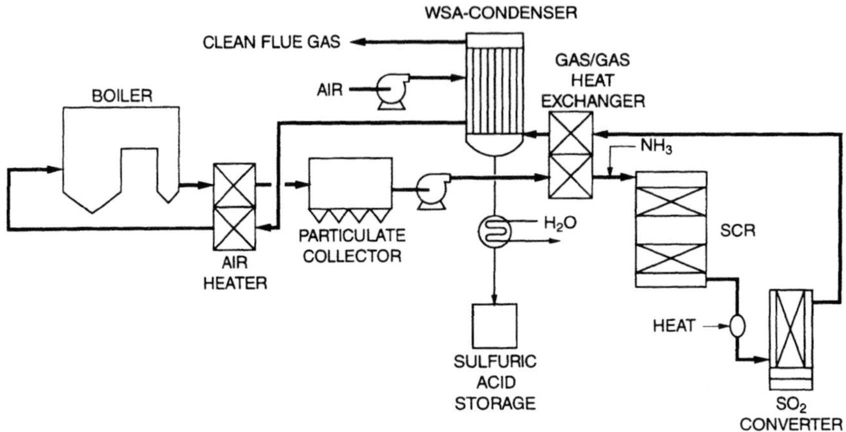

Fig. 14-29. SNOX™ integrated process flow. (From ABB Environmental Systems. Used with permission.)

lytic reactor (SCR) where nitrogen oxides are reduced to free nitrogen and water. The flue gas leaves the SCR, its temperature is adjusted slightly, and enters the SO_2 converter which oxidizes SO_2 to sulfur trioxide (SO_3). The SO_3-laden gas passes through the secondary side of the GHH where it is cooled as the incoming flue gas is heated. The cooled flue gas passes through a falling film condenser (Wet-gas Sulfuric Acid Condensor [WSA-Condenser]) where it is further cooled with ambient air to below the sulfuric acid dewpoint. Acid condenses out of the gas phase on the interior of borosilicate glass tubes and is subsequently collected, cooled, diluted, and stored. Cooling air leaves the WSA-Condenser at about 400° F (200° C), and a small amount is used for process support; the remainder is sent to the boiler's air preheater and used for combustion air.

14.5.2 Mitsui-BF System (Activated Coke Process)

The Mitsui-BF $DeSO_x/DeNO_x$ system is a dry process for the integrated removal of SO_2 and NO_x. General Electric Environmental Systems, Inc. (GEESI) is the United States licensee.

Figure 14-30 shows the equipment required for this system. All process equipment is located downstream of the air heaters. The fly ash is removed from the flue gas in a fabric filter or ESP before the gas enters the activated coke reactor. The activated coke bed adsorbs the SO_2 and serves as the catalyst for converting the NO_x to nitrogen and water. A sulfur or sulfuric acid byproduct stream can be produced.

The removal of the SO_x and NO_x is accomplished in two sequential process steps using a two-stage, activated coke adsorber. Unlike many other technologies, the SO_x and NO_x removal steps can be located in the lower temperature area following the air heater (212 to 392° F). The SO_x must be adsorbed from the flue gas prior to conversion of the NO_x. The flue gas first passes in a crossflow manner through the lower section of an activated coke bed that is moving downward in the adsorber. Sulfur dioxide and small additional amounts of fly ash are removed in this stage of the process. On leaving the first stage carbon bed adsorber, the flue gas is injected with ammonia and then flows crossflow through the second activated coke bed which serves as the catalyst for

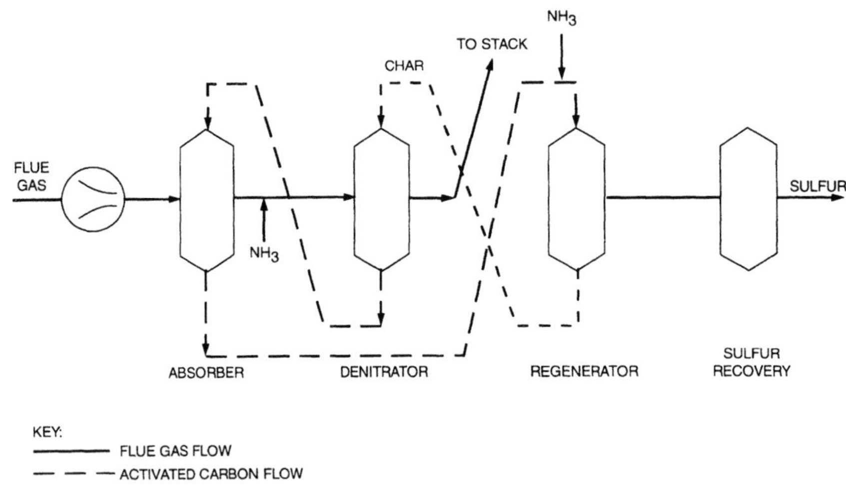

Fig. 14-30. Mitsui/BF activated coke process.

the NO_x decomposition reaction with the ammonia to form water and nitrogen.

As the activated coke bed moves through the two sections, the coke becomes saturated with SO_x, reducing its adsorption capacity. The coke is regenerated by removing the SO_x and other impurities by heating the coke to between 570 to 932° F. A SO_2-rich gas stream released from the coke is removed from the regeneration/desorption module by an induced draft blower. This SO_2 stream can be treated in a Claus unit to produce sulfur or in a scrubbing system to produce high-quality sulfuric acid.

The regenerated coke is screened to remove degraded coke fines which can be used as fuel for the boiler. Acceptable regenerated and supplemental new coke are returned to the moving bed.

14.6 HAZARDOUS EMISSIONS CONTROL

The United States enactment of the CAAA included the provisions of Title III, Hazardous Air Pollutants (HAPs), as described in Chapter 26. The core of Title III is a list of 189 chemicals (pollutants) that may cause potential hazards to human health and the environment when emitted.

The United States EPA is not currently required to promulgate standards for HAPs emitted by electric utility steam generating units. However, the EPA is directed to perform a study of the public health hazards that can be reasonably anticipated as a result of emissions by electric utility steam generating units. The EPA has indicated that the report will be delayed until November 1995 to include emissions data currently being generated at various sites by the EPRI and the United States Department of Energy (DOE). If, after considering the results of the study, the EPA finds that HAP emissions regulations for electric utility steam generating units are appropriate and necessary, then these units will be regulated.

The CAAA also required the EPA to conduct a study of mercury emissions including those from electric utility steam-generating units and submit the study report to Congress by November 1994. An update to this report that specifically addresses utility emissions is currently scheduled for submittal to Congress in November 1995. If deemed necessary and appropriate by the EPA, mercury emissions from these sources would also be regulated.

14.6.1 Emission Standards and MACT

The emission standards for new and existing sources to be established by the EPA require the maximum degree of reduction in HAPs emissions, while taking into consideration the cost of achieving such emission reductions, nonair quality impacts, and energy requirements. The emission standards associated with the required maximum degree of reduction are referred to as maximum achievable control technology (MACT). For new sources in a category, the MACT standard will not be less stringent than the emission control that is currently achieved in practice by the best controlled similar source. The MACT standard for existing sources in a category may be less stringent, but must be at least as stringent as the average emission limitation achieved by the best performing 12% of the existing sources.

14.6.2 Potential Control Technologies

The removal efficiencies of different control technologies in controlling HAPs, including mercury, are currently not well defined because of a lack of accurate testing information and operating experience on fossil-fuel-fired utility steam generators. The concentrations of trace metals in fuels can differ dramatically by fuel source region and even for coal within a single mine. The partition of trace metals between bottom ash and fly ash as a result of coal combustion can vary by element and furnace operating conditions. Therefore, HAP control technologies will likely require meeting stringent outlet emission levels over a wide range of inlet HAP concentrations.

Two important characteristics of HAPs emitted by power plants must be considered in assessing potential control devices. The first is that the HAPs, such as vanadium, nickel, cadmium, lead, and most other heavy metals, are concentrated in the particulate fraction with size less than 10 microns (PM_{10}). Control of the HAPs identified as occurring in the fine particulate fraction (PM_{10}) will be related to control of those particulate emissions. The second is that volatile HAPs, including elemental mercury and selenium, can result in emissions in the gas phase. These gaseous pollutants would then be relatively unaffected by conventional particulate control devices. Control of those HAPs occurring as gases is typically related to cooling and sublimation/condensation and removal as particulates or absorption/adsorption and removal with the recovery liquid or solid.

Water-based scrubbers, such as semidry spray absorbers and wet scrubbers, are potential control options. They both involve cooling the flue gas. Although both are in common use on coal-fueled steam generators to reduce SO_2 emissions, limited testing and data reporting have documented the actual performance of these devices in controlling HAPs.

Semidry absorber scrubbers may be suitable for the control and removal of fine particulate matter associated with certain characteristic HAPs, especially trace metals. The cooling function of the dry spray absorber provides conditions thought beneficial in controlling a wide variety of HAPs (Brna and Kilgore 1987; Getz et al. 1992; White et al. 1992; and Felsvang and Brown 1992). Major factors influencing the removal of mercury from the flue gas are the chemical form of the mercury and the carbon content of all particulate present in the flue gas. Mercury salts can be removed as particulates. The injection of carbon into a dry spray absorber can also be an effective method of increasing the volatile elemental mercury removal by adsorption. One new, 80-MW power plant located in the United States has included the ability to add carbon injection into the fluidized bed lime scrubber should future air toxic controls require it.

Wet scrubber control systems used as quenchers and particulate control devices may be fairly efficient in controlling HAPs and mercury emissions (Getz et al. 1992). The particulate and SO_2 removal efficiencies of wet scrubber control systems are well known, and the associated lowering of flue gas temperatures may be very efficient in controlling condensible HAPs, including mercury (Brna and Kilgore 1990; Jordan 1987; and Brna 1988). Additive injection systems for wet scrubbers using materials such as sodium hypochlorite to solubilize mercury can improve removal efficiencies significantly. Reports from both pilot plant and full-scale incinerator installations indicate that removal efficiencies of >95% are possible (Neme 1991).

Fabric filters are well suited for the control and removal of fine particulate associated with certain characteristic HAPs, especially trace metals. Pulse jet or reverse air fabric filters are expected to be in excess of 98% efficient in controlling particulate HAPs (Laudall et al.). The control of mercury from fabric filters alone is not expected to be highly effective. Fabric filters can, however, be very efficient in controlling mercury when used in conjunction with additives such as activated carbon or sodium sulfate injected upstream of the control device.

High-efficiency ESPs, if designed effectively, can provide control efficiencies of more than 98% for most particulate HAPs (Tumati and Devito). Flue gas conditioning with dual injection of sulfur trioxide/ammonia upstream of the ESP is being evaluated to determine its effect on control of HAPs. Current data and experience are limited, however.

Wet ESPs are particularly suited for applications downstream of wet scrubbers or other processes where the flue gas has been quenched with moisture to provide the saturated conditions. A wet ESP provides ideal conditions for mercury and other gas phase toxics to be condensed and recovered as particulates or absorbed into the water stream. Other advantages of a wet ESP include smaller size because of the reduced gas volume of the wet gas, low reentrainment, and avoidance of ash resistivity problems. In early 1995, one United States utility installed a retrofit wet ESP downstream of the mist eliminators inside an 80-MW size wet scrubber to aid in the control of fine particulate emissions. This EPRI sponsored demonstration project is notable since the wet ESP is located within the space formerly used for SO_2 absorption. Prior to the wide spread application of wet ESP technology in conjunction with utility scrubber applications, design issues concerning the wet ESP flushing water, corrosion, and wastewater treatment need to be fully addressed.

Condensing heat exchangers are a new utility application being investigated as a means of controlling HAPs while increasing plant performance. The condensing heat exchangers are placed at the end of the process where they remove heat from the flue gas and condense water, acid, and other volatile compounds. Condensation also occurs on a submicron level, enhancing the removal of fine particulate matter. The cooling medium may be either process water or inlet combustion air.

14.6.3 Impacts of Potential HAP Controls

The design and installation of air quality control systems to control the emissions of HAPs, including mercury, from utility combustion sources must consider environmental, operational, and economic impacts on the sources. This is especially true for retrofit installations. Potential control technologies may require extensive plant modifications, including ductwork, fans, water treatment, and solids stabilization systems.

In summary, the power generation industry should develop strategies to deal with hazardous air pollutants. These strategies should include assessment of potential control systems that may be required as MACT for combustion sources.

14.7 CONTINUOUS EMISSIONS MONITORING

14.7.1 Gaseous Emissions Monitoring

This section describes the continuous emissions monitoring (CEM) technologies, analytical techniques, and CEM system types used for continuous emissions monitoring. These monitors are required by United States regulations for all new units and many existing units.

14.7.1.1 Analytical Techniques. CEM utilizes various analytical techniques. These techniques include absorption and luminescence spectroscopy, and electroanalytical methods. Figure 14-31 shows the relationship between CEM system type and analysis technique. The following subsections describe these techniques.

ABSORPTION SPECTROSCOPY. Absorption or scattering of incident light by solids or gases is the basis for analyses using absorption spectroscopy. The underlying principle is explained by the Beer–Lambert law, which states that the transmittance of light is equal to the ratio of the intensity of the light leaving the gas stream to the intensity of the light entering the gas stream. An exponential relationship exists between transmittance and the attenuation coefficient of the pollutant, the concentration of the pollutant, and the distance the light travels through the gas. This relationship applies to both light absorption and light scattering. The technique is sensitive to concentration and path length. This principle is used for monitoring opacity and for various gases as described in the following paragraphs.

OPACITY MONITORING. Opacity monitors use light in only the visible spectrum because opacity must correlate to measurements made with the human eye. The light attenuation coefficient is dependent on particle size distribution and the reflectivity of the particulate. The relationship of these parameters to the measurement of particulate mass emissions is

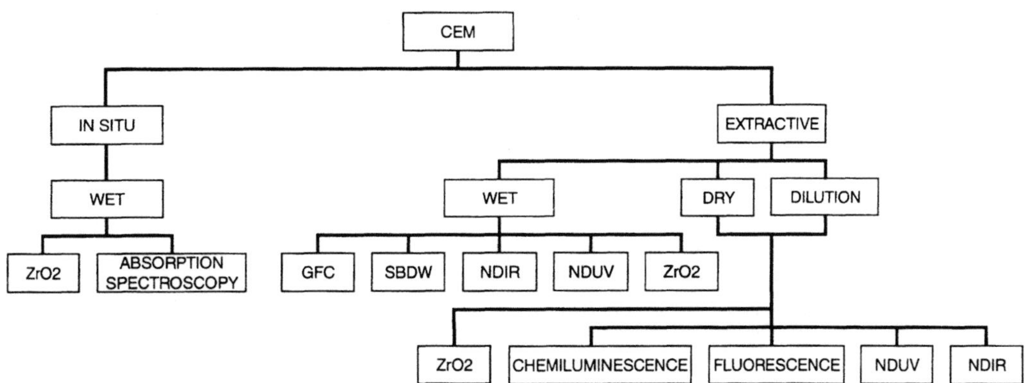

Fig. 14-31. Emission monitoring technologies.

very complex and site specific. Therefore, opacity monitors are not capable of accurate mass emission measurements.

GAS MONITORING. Nondispersive infrared analysis is a technique that uses band pass filters or diffraction gratings to select the appropriate wavelengths from an infrared source. Nondispersive means that the infrared light is not separated into its component wavelengths and only part of the infrared spectrum is used for analysis. The wavelengths sensitive (light is absorbed) to the gases of interest with the fewest interferences are selected. Nondispersive infrared analysis is sensitive to SO_2, NO, NO_2, CO, CO_2, and H_2O, although the technique is not equally sensitive to all of these gases.

Nondispersive ultraviolet analysis is another absorption spectroscopy technique used for gas analysis. Band-pass filters or diffraction gratings are used to select particular wavelengths in the ultraviolet part of the spectrum. This technique is sensitive to NO and SO_2. To determine the total NO_x, an NO_2 monitor would also be required. However, since NO_2 is generally <5% of the total NO_x from pulverized coal boilers, the NO_2 fraction is sometimes ignored in determining the approximate NO_x concentration.

The differential absorption analysis technique compares an absorbed wavelength of light to a nonabsorbed wavelength of light to determine gas concentrations. These measurements can be made only over limited path lengths, and therefore these analyzers are single-pass type or limited path length type. Differential absorption is used for NO and SO_2 measurements using nondispersive ultraviolet light and for CO_2 and H_2O measurements using nondispersive infrared light.

The gas filter correlation analysis technique compares a split beam of light in the analyzer. After passing through the measurement cell, the light beam is split. One beam passes through a gas filter correlation cell containing a high concentration of the subject pollutant. After passing through the gas filter correlation cell, nearly all the light in the wavelengths of interest will be absorbed. The difference in energy levels between these beams in the wavelengths of interest is related to the pollutant concentration. This technique is limited to analyzers measuring SO_2, NO, and CO_2 concentrations.

The single-beam dual-wavelength analysis technique uses narrow-band optical filters to select infrared wavelengths. One wavelength is selected so that it is absorbed by the gas to be measured, and the other wavelength is not absorbed by the gas. The ratio between the two energy levels is proportional to the concentration of gas.

Second-derivative spectroscopy uses ultraviolet light to measure SO_2 and NO concentrations (NO_x is determined from the NO concentrations), and is based on the varying spectral absorption of gas molecules. A diffraction grating is used to select a range of wavelengths, centered about a wavelength that is absorbed most readily by the subject pollutant. As the range of wavelengths selected by the diffraction grating change, the absorption measured by the detector varies. The absorption is greatest at the wavelengths closest to the central wavelength and as the band selected moves away from the central wavelength, and a sinusoidal signal is generated. The amplitude of this signal is proportional to the concentration of the gas. This amplitude is determined by the second derivative (slope) of the sinusoidal curve.

LUMINESCENCE SPECTROSCOPY. Gas monitoring techniques using luminescence spectroscopy are based on specific gas molecules giving off light when excited. Three types of luminescence spectroscopy techniques are fluorescence, chemiluminescence, and flame photometry.

Fluorescence techniques use light to excite the gas molecules so that the difference in the wavelength of the exciting light and the emitted light can be determined. Fluorescence is often used to determine SO_2 concentration. However, this technique has interference from H_2O, CO_2, and O_2, which quenches the emitted light. These interferences can be calibrated out with some sacrifice in accuracy. This technique requires a clean, particulate free, and essentially moisture free gas stream for analysis and is limited to the extractive systems.

Chemiluminescence uses a chemical reaction to stimulate molecules to give off light. NO_x analyzers apply this technique using the chemical reaction between NO and ozone. For determination of total NO_x, NO_2 must first be reduced to

NO catalytically. O_2, N_2, and CO_2 quench the reaction but can be compensated for by excess dilution with ozone. Because the technique requires a clean, dry sample and the reduction of NO_2 to NO, this technique is limited to extractive systems.

Flame photometry is a technique that excites the gas sample to luminescence by introduction into a hydrogen flame. The spectra given off by the hot gases are analyzed in the ranges of interest. This technique is sensitive to all forms of sulfur and thereby yields total sulfur data unless sample conditioning is used to absorb sulfur species that are not of interest. This technique requires a clean, dry sample and is limited to extractive systems. Flame photometry has been most widely used for ambient level monitoring.

ELECTROANALYTICAL TECHNIQUES. The previously described methods of analysis rely on spectroscopic or electro-optical techniques for measuring particulate or gas concentrations. Electroanalytical instruments measure electrical impulses from various chemical reactions. This technique is most often used to determine O_2 concentrations using zirconium oxide and paramagnetic cells.

The zirconium oxide cell uses a reference solution of air with 21% O_2 which is fed to one side of the zirconium oxide cell, while the other side of the zirconium oxide is exposed to flue gas. The current flow through the zirconium oxide as a result of the O_2 difference from one side to the other is measured. Zirconium oxide analyzers are relatively insensitive to vibrations in the measurement area and as such are frequently used for in situ monitoring, especially for excess air monitoring in boiler control systems.

Paramagnetic analyzers are also capable of accurately measuring oxygen concentrations. A magnetic field in the measurement chamber is changed by the presence of oxygen which is bipolar, and concentrations can be determined by the displacement of a platinum ribbon in a permanent magnetic field. This technique requires a sample stream that is clean and free of water droplets. Paramagnetic analyzers are highly susceptible to vibration and are limited to extractive type systems.

14.7.1.2 Monitoring Alternatives. Two basic types of CEM systems (classified by sampling system type) are available:

- In situ systems—Measure the flue gas at the conditions present in the stack at the monitoring location.
- Extractive systems—Draw a gas sample from the stack or duct to a remote location for sample conditioning and analysis. This group includes wet extractive, dry extractive, and dilution extractive systems.

IN SITU CEM SYSTEMS. In situ systems measure gas concentrations directly in the stack or duct without extracting samples for external analysis and can make real-time gas measurements. In situ analyzers widely available today for gas concentration measurements use two different methods of gas analysis: differential absorption and second derivative spectroscopy.

The differential absorption type is an across-flue system that analyzes the gas that passes through a specific "line of sight" of the monitor, typically ranging from a few feet to the full distance across the interior diameter of the flue.

The second type of in situ system is a point measurement instrument that analyzes the gas at one specific point or along a short path (a few inches) in the flue or duct in a cavity, protected by a ceramic filter. Particulate matter cannot pass through the filter. NO_x and SO_2 measurements typically are made using an ultraviolet light source and second-derivative spectroscopy. CO_2 and H_2O typically are measured using infrared energy in a separate instrument with the same technique. O_2 can be measured in situ using a zirconium oxide probe.

Calibration is accomplished by flooding the measurement cavity with the appropriate calibration gas at a slightly higher pressure than the measurement location so that no flue gas remains or can diffuse into the cavity. After calibration, the flue gas returns to the cavity when the calibration gas flow is turned off.

One type of across-flue in situ analyzers uses a Xenon bulb light and, when coupled with a spectrum analyzer, is sensitive to SO_2, NO_x, CO_2, and H_2O. This sensitivity can be expanded to include other gases (up to 30) with minor modifications. The transmitter shines a light across the stack or duct to the receiver which sends this light to the analyzer unit where the concentrations are determined by analyzing the spectrum of the light from the receiver. This sample path is illustrated in Fig. 14-32. Also in Fig. 14-32 is a simplified installation in a power plant application. Similar systems using other light sources are available.

Calibration of a cross-flue analyzer is accomplished by flooding part of a second, sealed tube (which also traverses the flue) with calibration gas. The light source is also directed across this calibration gas cell to the detector. The same light source is used and mirrors accomplish the path splitting. Since the tube or a cell within the tube can be flooded with fresh calibration gas each time calibration occurs, this approach meets EPA requirements for instrument calibration.

EXTRACTIVE CEM SYSTEMS. There are three types of extractive CEM systems. One type measures pollutants on a dry basis, another measures pollutants on a wet basis, and the third dilutes the sample prior to analysis.

DRY EXTRACTIVE CEM SYSTEMS. Dry extractive CEM systems use analysis techniques or analyzers that require a particulate-free, moisture-free, low-temperature sample stream. Therefore, the sample must be removed from the flue and conditioned before reaching the analyzers. As such, dry extractive CEM systems consist of sample extraction, transport, conditioning, and analysis subsystems.

Particulate is removed by filtering at the probe tip. Dust-free sample gas is transported in heat traced lines to a remotely located analysis enclosure. Moisture is removed by refrigeration condensation and by permeation tube dryers

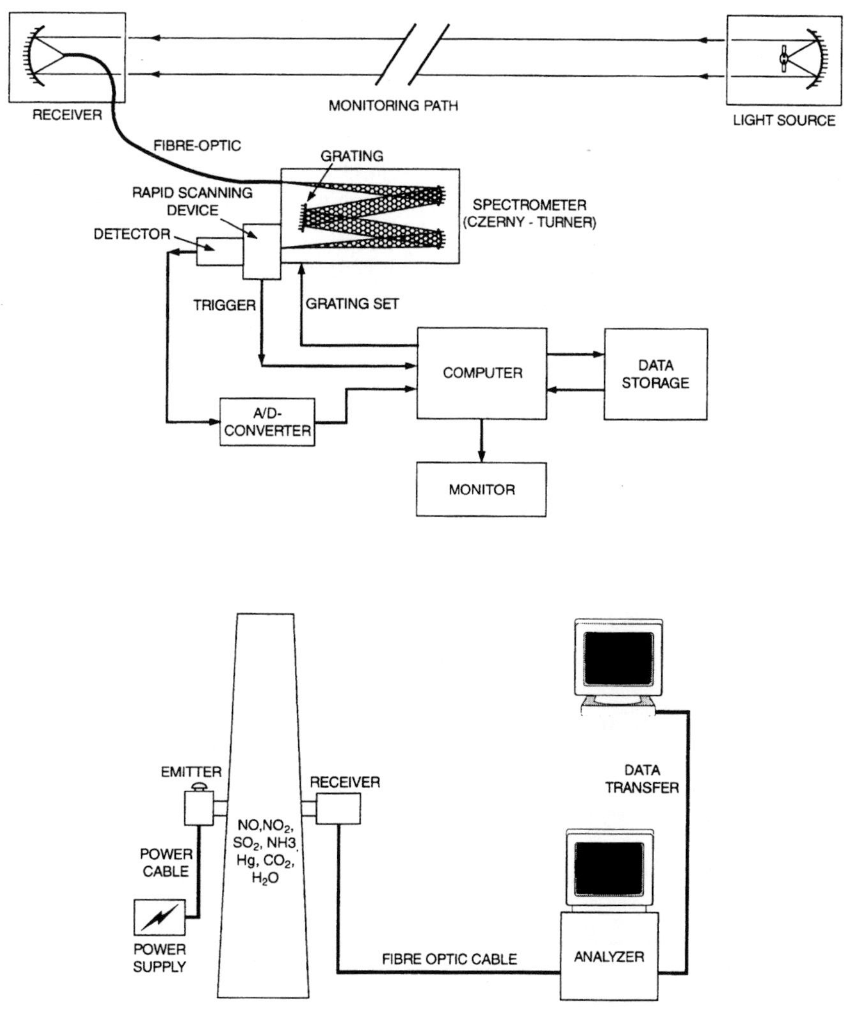

Fig. 14-32. Typical installation of across-stack in situ CEM system. (From Asea Brown Boveri. Used with permission.)

that pass the sample through membranes that do not allow water to pass. This sample extraction/conditioning is the critical component of dry extractive gas analysis. Poorly designed sample conditioning systems require significant maintenance, prevent accurate measurement by the analyzers, and lead to low monitor and system reliability.

The analysis techniques used in dry extractive CEM systems vary from one system supplier to another. Typically, these techniques include spectral absorption (nondispersive infrared, nondispersive ultraviolet, and gas filter correlation) or luminescence spectroscopy (fluorescence techniques for SO_2 and chemiluminescence for NO_x). Material selection for potential corrosion occurrences is critical in this subsystem.

WET EXTRACTIVE CEM SYSTEMS. Wet extractive CEM systems remove the particulate at the probe then maintain the sample in a hot condition. The moisture is retained throughout gas concentration measurements. Accordingly, this technique requires that the gas sample be maintained above the acid dew point temperatures (280 to 320° F) throughout the entire CEM system, including the analyzer.

Infrared analyzers commonly used with this technique determine NO_x using absorption spectroscopy techniques in a wavelength area that contains interference from moisture. In fact, infrared energy absorption by water vapor is a broad band (multiple wavelength) phenomenon. As a result, such infrared analyzers also measure water vapor concentration and correct the NO_x concentration data for water vapor. In certain situations of high moisture content in the flue gas, this has caused accuracy problems with NO_x measurements.

The critical component of a wet extractive CEM system is the heat tracing used throughout the system. Manufacturers of this system maintain sample temperatures between 360 and 480° F. A temperature of 360° F would be above most acid dew point temperatures encountered in fossil-fueled applications.

DILUTION EXTRACTIVE CEM SYSTEMS. Dilution extractive CEM systems eliminate the need for heat tracing and conditioning of extracted samples by precisely diluting the sample stream with very dry (with a moisture dew point as low as −100° F), clean, instrument air.

Measurement of pollutant concentrations is accomplished

with analyzers designed for ambient level pollutant measurements (at or below 1 part per million [ppm] accurately).

Particulate is filtered out prior to dilution. Precise dilution of the sample gas stream is accomplished using critical orifice venturis to define both sample and dilution air flow quantities. Based on the undiluted concentration of the subject pollutant in the flue gas stream, a critical orifice can be selected so that the diluted sample fits within the sampling ranges of the ambient concentration monitors. Standard critical orifice sizes are available that can dilute flue gas anywhere from 12 to 1 to upwards of 700 to 1.

Dilution extractive systems require the use of dilution air that is free from SO_2, NO_x, CO_2, CO, or moisture to avoid inaccuracies. An air cleanup system is equipped with filters and scrubbers that remove these interferents from the dilution air.

Dilution extractive sample conditioning is based on choked flow (constant mass flow) flue gas sampling with a sonic (critical) orifice and precise dilution by an air-driven aspirator. The aspirator creates a high vacuum that ensures choked flow through the orifice.

Maintaining sonic velocity through critical orifices is dependent on operation at an absolute pressure downstream of the critical orifice that is <0.53 of the upstream absolute pressure. A pressure ratio >0.53 may compromise the accuracy of a dilution extractive system. A correction must be made to the dilution ratio if the gauge pressure at the orifice inlet is negative (with respect to atmosphere).

14.7.1.3 Technical Comparison. The following discussion describes advantages and disadvantages of the previously discussed CEM technologies. Table 14-2 summarizes the advantages and the disadvantages of the various CEM technologies.

IN SITU CEM SYSTEMS. The primary advantage of in situ analysis is that the gas sample that is analyzed is essentially unchanged from the balance of the gas stream, and the integrity of the sample is very high. Some in situ analyzers are designed with relatively short path lengths over which gas concentration measurements are made. This limited path length can lead to diminished accuracy and sensitivity capabilities at low concentrations, and lack of representativeness where stratification of gases is present. Diminished accuracy capabilities are especially of concern considering 40 CFR 75 requirements.

Since the gas is analyzed at the sampling location, electronic signals from the monitors are sent to the remote data recorder. No walk-in analyzer enclosure is necessary. However, monitor maintenance can be difficult because convenient and easy access to the monitor location (typically on the stack) is required, but not always available. In addition, some end-users are reticent to install sensitive electronic measurement devices on stacks that may be struck by lightning. Weather conditions, ambient temperature changes, and vibration at the monitoring location can all impact monitor accuracy, repeatability, and maintenance requirements.

EXTRACTIVE CEM SYSTEMS. Since promulgation of early EPA CEM system quality assurance requirements (40 CFR 60 Appendix F), extractive CEM systems have been used at a large number of installations. However, these systems are

Table 14-2. Alternate CEM Technology Advantages and Disadvantages

Technology	Advantage	Disadvantage
In situ	Proven technology Measures on a wet basis. Real-time analysis No sample lines	Limited manufacturer support Electronics installed in potentially vulnerable location Poor maintainability for electronic equipment located on the stack or duct Affected by particulate and water droplets in the flue gas stream Limited path lengths for low-concentration measurements
Full extractive dry	Proven technology Wide manufacturer support	Measures on a dry basis, so it requires additional method for determining flue gas moisture content for calculation of pollutant mass flow. Potential for acid condensation in high-sulfur applications Sample conditioning system requires high maintenance Heated sample line High sampling rates decrease filter life.
Full extractive wet	Proven technology Measures on a wet basis Low maintenance	Heated sampling line and hot analyzer Moisture interference on NO_x measurements. Limited manufacturer support High sampling rates decrease filter life.
Dilution extractive	Proven technology Measures on a wet basis. Low maintenance Does not require actively heat traced lines. Low sampling rates prolong filter life. Wide manufacturer support	Requires a method to correct for changes in pressure and temperature conditions. Dilution air cleanup essential for accurate measurement Often requires freeze protection for sample lines and heating of probes.

somewhat complex and are generally more expensive than in situ monitors. This disadvantage is somewhat offset by the accuracy of extractive system measurements. However, these systems suffer the same stratification of gas limitation that point or short path in situ systems suffer; they are of questionable accuracy if gas stratification is present in the stack or flue.

Because of the relatively large sampling volumes for wet extractive and dry extractive systems (approximately 300 ft^3/day), filter pluggage is a potential problem. However, back purging the filter regularly minimizes filter pluggage difficulties.

DRY EXTRACTIVE CEM SYSTEMS. Dry extractive systems have been installed on numerous coal-fueled boilers. The accuracy of these systems is high because the moisture and particulates have been removed from the sample.

A disadvantage of dry extractive systems for determining pollutant mass flow (lb/h) is that sample conditioning systems remove the moisture in the flue gas prior to sample analysis. This does not affect the calculation of lb/MBtu emissions using EPA Method 19. However, it adds a potential source of error into the calculation because the gas moisture content must be used to correct the dry basis concentration measurement to wet basis.

WET EXTRACTIVE CEM SYSTEMS. The advantages of wet extractive CEM systems are high sample integrity, since only particulate is removed before analysis, and that sample analysis is made on a wet basis, allowing uncorrected conversion to mass flow (lb/h).

The main disadvantages of wet extractive systems are the requirement for heat tracing throughout the system and potential moisture interference for NO_x measurements. The heat tracing must be capable of maintaining a sample temperature that is above the acid dew point temperature of the gas sample. Moisture interference for NO_x measurements limits the applicability of this system where a high moisture content is present in the gas stream or when very accurate NO_x determinations are required.

DILUTION EXTRACTIVE CEM SYSTEMS. Potential advantages for dilution extractive CEM systems are the relative simplicity of operation and low cost. Accordingly, these systems have been used at a large number of recent installations. There may be an accuracy problem because of sample dilution, particularly at low pollutant concentrations such as at the outlet of an SO_2 scrubber or on combustion turbine applications. However, this technique is widely used, certifiable, accurate, and acceptable to the EPA.

Another advantage of the dilution extractive system is that the sample conditioning does not remove the water prior to analysis. As discussed earlier, analysis of a wet basis sample allows direct correlation of the measured pollutant concentration to the wet basis mass flow without the error inherent in assuming or measuring a moisture content for correction.

14.7.2 Gas Flow Monitoring

The 1990 CAAA requires measuring and recording SO_2 concentration (in ppm) and flue gas volumetric flow in standard cubic feet per hour (scfh) to determine SO_2 mass emissions in pounds per hour (lb/h). If the CEM system measures the CO_2 concentration in the flue gas, the volumetric gas flow monitor can also be used to report CO_2 emissions (in lb/day).

14.7.2.1 Measurement Techniques. Three categories of flue gas flow rate measurement techniques can be used: differential temperature using heated temperature sensors, differential pressure using annubar or pitot tube type probes, and differential frequency/transit time using ultrasonic sound (Doppler principle). Essentially, these flow measurement devices indirectly measure the flue gas velocity (ft/min), then calculate the flow rate (ft^3/min) based on the given geometry (cross-sectional flow area, ft^2) of the measurement location. Some of the monitors now available use more than one technique.

DIFFERENTIAL TEMPERATURE. This technique measures a temperature differential between heated and the unheated RTDs located in the gas flow at multiple points. This temperature differential decreases as the heated RTD is cooled by increasing flue gas flow. Direct changes in flow can incrementally affect the extent to which heat is dissipated from the sensor probe. Consequently, the magnitude of the temperature differential between the matched RTDs can be calibrated to flue gas flow rate. Figure 14-33 shows the typical installation of thermal mass flow probes which utilize differential temperature measurements.

DIFFERENTIAL PRESSURE. Differential pressure measurement of gas flow is based on the pressure difference between the upstream (total pressure) and the downstream (almost static pressure) sides of the probe which is correlated to flow. Differential pressure measurements can be made with either an annubar or pitot type probe. Annubar probes (Fig. 14-34) generally span the cross-section of the flue, providing an average reading from multiple points along the probe. Pitot probes (Fig. 14-35) provide point measurements of the velocity within the flue and are typically grouped to mitigate the effects of stratified flow. For either measurement configuration, back purge systems are required to periodically clear obstructions from the pressure sensing orifices.

TRANSIT TIME. The transit time technique (often referred to as ultrasonic flow monitoring) is based on measuring the transit time across the flue from the source to the detector (the transducers) for sound waves, which varies with the relative velocity of the gas stream. Ideally, the transducers should be mounted at a 45-degree angle across the flue. Each transducer acts alternately as a transmitter or receiver of ultrasonic sound. When a tone burst is sent through the gas stream from one transducer to the other, the movement across the gas stream alters the time required for the tone to

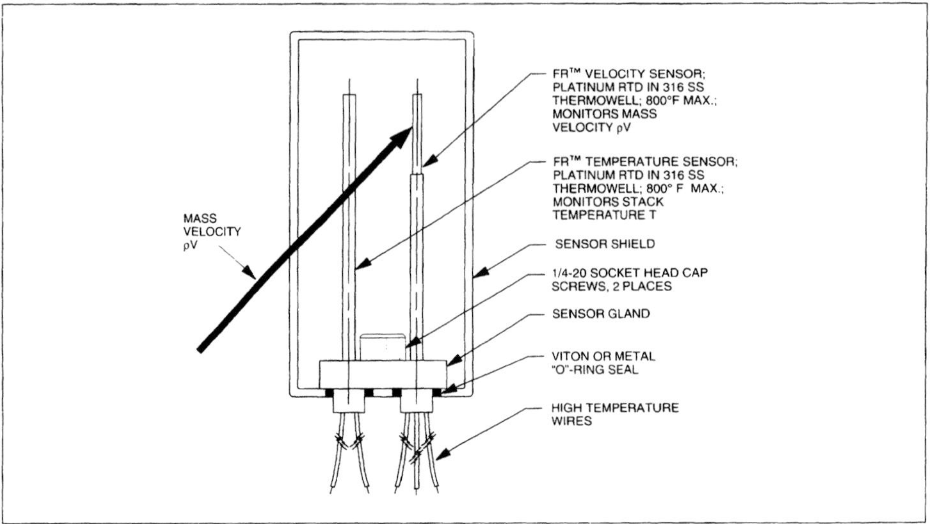

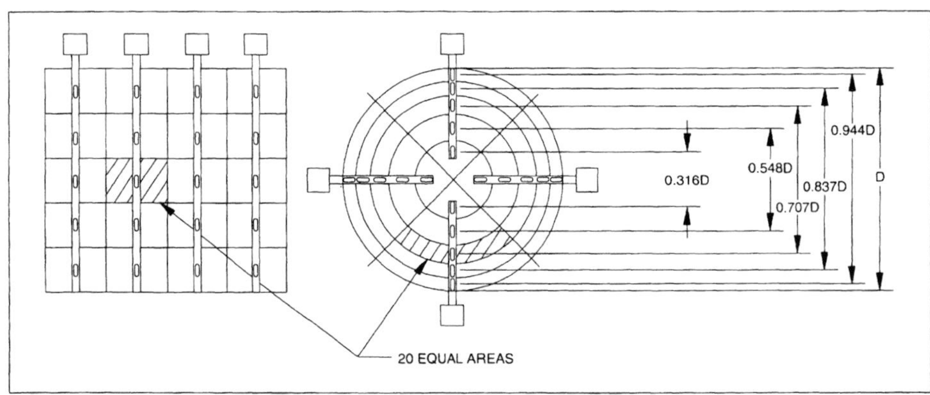

Fig. 14-33. Thermal mass flow probe and typical installation. (From Sierra Instruments. Used with permission.)

traverse the distance across the flue. If the tone burst is traveling with the gas stream, the time is reduced. If the tone burst is traveling against the gas stream, the time is increased. The flow velocity is directly related to the difference in the time required for the tone bursts to traverse the stack in both directions.

As shown in Fig. 14-36, Differential Frequency Typical Installation, ultrasonic monitors are not directly located in the flue gas flow stream. Like opacity monitors, they are located on either side of the flue gas flow stream. However, purge airflow and careful selection of the materials of construction are still required. This type of flow rate monitor requires the two transducers to be installed at an angle of approximately 45 degrees (upstream/downstream) to the direction of flow. Therefore, one of the transducers must be located on a different platform.

14.7.2.2 Technical Comparison. A measurement difficulty arises for all the flow rate monitors because, in addition to gas flow, other factors such as the gas density (affected by temperature and pressure), direction of the flow vector (affected by cyclonic flow), and flue gas molecular weight also affect the measured parameters. However, only the flue gas temperature and pressure, which affect gas density, can be measured continuously to correct flow rates. The flow vector angle and flue gas molecular weight cannot be measured continuously. Flow monitors typically operate in a corrosive and erosive environment. If the stack is wet (following a wet scrubber), the low temperature allows the condensation of acid mist onto the probe. Also, any particulate carryover from the scrubber is likely to stick to the probe, potentially blocking the sample ports. In dry stack applications, particulate can collect on the probe as a result of electrostatic forces. Because of the potential severity of the operating environment, high-nickel alloys are often selected as the materials of construction for the probe for both wet and dry installations.

DIFFERENTIAL TEMPERATURE. Thermal flow monitors are not recommended for wet stack applications because of impingement of water droplets. These droplets cause significant increases in heat transfer from the heated RTD. This

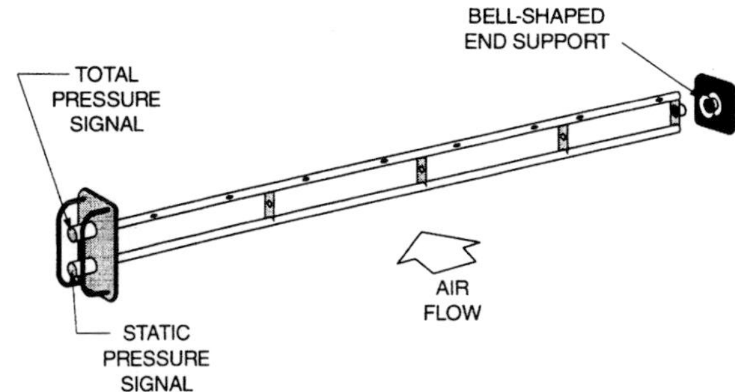

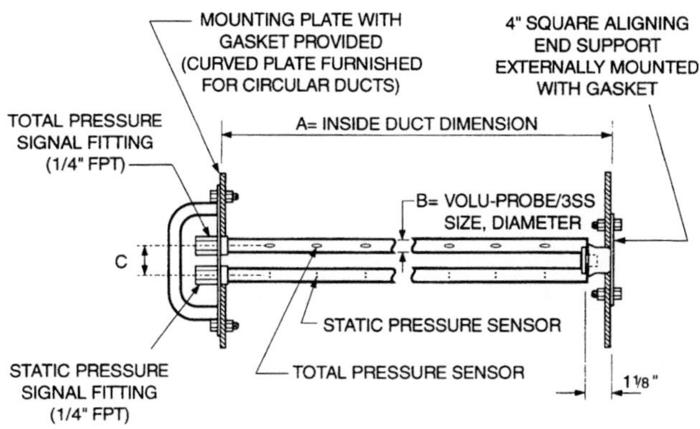

Fig. 14-34. Volu-probe differential pressure flow sensing element and typical installation. (From Air Monitor Corporation. Used with permission.)

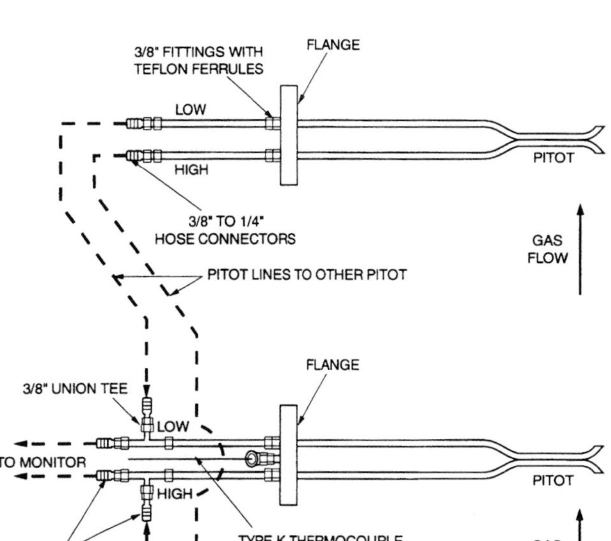

Fig. 14-35. Differential pressure pitot tube. (From EMRC. Used with permission.)

leads to erroneous readings of the temperature differential between the two RTDs.

DIFFERENTIAL PRESSURE. Differential pressure measurement is an established technology and has been widely used in measurement of flue gas flow rates on a variety of power plants and is the EPA reference method for gas flow measurement. A back purge system is required for differential pressure flow monitors to prevent probe pluggage. Differen-

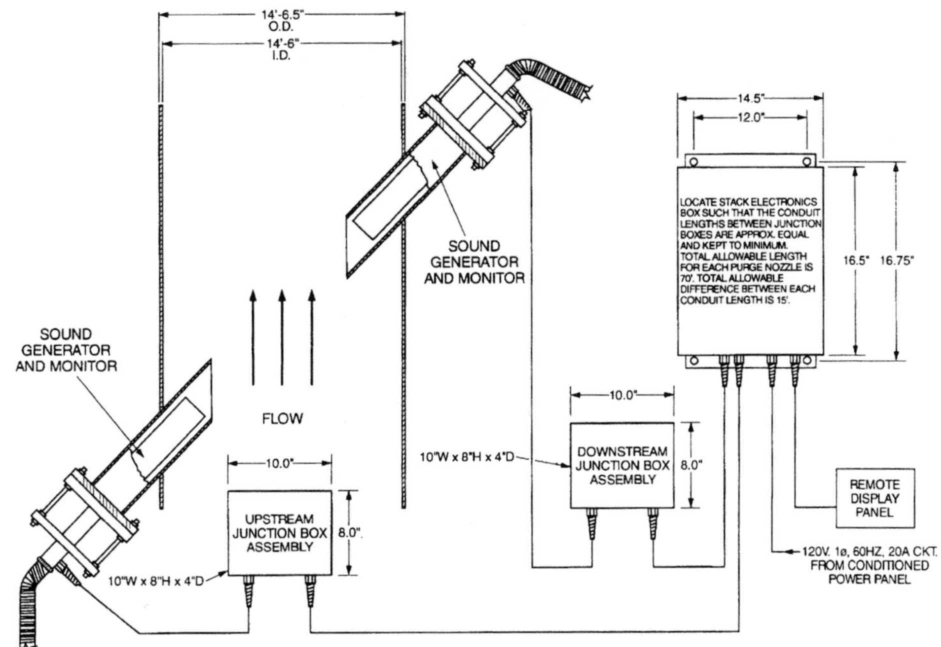

Fig. 14-36. Transit time monitor typical installation. (From KVB/Analect. Used with permission.)

tial pressure point measurements are pneumatically averaged by manifolding all total pressure lines together and all static pressure lines together to eliminate a potential inaccuracies caused by electronic averaging of the measured values.

TRANSIT TIME. Ultrasonic systems generally are more expensive than differential temperature or differential pressure systems, but may be less susceptible to errors due to gas flow conditions or characteristics. A differential frequency system often requires the installation of a new platform, and sensitive electronic components would be located in stack- or duct-mounted enclosures. Also, a differential frequency system provides line averaging in only one direction. If the flue gas flow is stratified along another axis within the flue, a single set of monitors might not be able to account for the stratification. In these situations, another set of transceivers is required to produce a cross-beam system for accurate measurement across both axis (multiple directions) of stratification.

14.8 REFERENCES

ABB POWER PLANT CONTROLS. Combustion Engineering, Inc., 2 Waterside Crossing, Windsor, CT 06095.

AIR MONITOR CORPORATION. 5 Coldhill Road, Suite 11, Mendham, NJ 07945.

AMERICAN ELECTRIC POWER SERVICE CORPORATION. One Riverside Plaza, Columbus, OH 43215.

BABCOCK & WILCOX. Power Generation Group, 20 S. Van Buren Avenue, P. O. Box 351, Barberton, OH 44203-0351.

BALL, M. E., SASKATCHEWAN POWER CORPORATION and T. A. ENWALD, TAMPELLA POWER INC. 1993. Installation and initial operation of LIFAC at Shand Power Station. Paper presented at the 1993 SO_2 Control Symposium. August 24–27, 1993, Boston, MA.

BLAIR, D. W. and J. O. L. WENDT. 1981. Formation of NO_x and other products from chemically bound nitrogen in coal combustion. DOE/ET/11314-T1. June 1981.

BRNA, T. G. ET AL. 1988. Cleaning of municipal waste incinerator flue gas in Europe. USEPA Report No. EPA/600-D-88/15. January 1988.

BRNA, THEODORE G. and JAMES D. KILGORE. 1987. The impact of particulate emissions control on the control of other MWC air emissions. *Journal of the Air & Waste Management Association.* Volume 40, No. 9, April 1987.

BURNETT, T. A., V. M. NORWOOD, E. J. PUSCHAVER, TENNESSEE VALLEY AUTHORITY; F. E. HSU and B. M. BHAGAT BHAGAT, AIRPOL, INC.; and S. K. MERCHANT and G. W. PUKANIC., USDOE. 1993. 10 MW demonstration of the AirPol gas suspension absorption flue gas desulfurization process. Paper presented at the 1993 SO_2 Control Symposium. August 24–27, 1993, Boston, MA.

CHRISTIANSEN, OVE B. and BERT BROWN. 1992. Control of heavy metals and dioxins from hazardous waste incinerators by spray dryer absorption systems and activated carbon injection. Paper presented at 85th Annual Air & Waste Management Meeting. June 21–26, 1992, Kansas City, MO.

Code of Federal Regulations (CFR). Published by the Office of the Federal Register National Archives and Rewards Administration. Washington, D.C. 20408. Mail order sales by Superintendent of Documents. Attn. New Orders, P.O. Box 371954, Pittsburgh, PA 15250-7954. Title 40—Protection of the Environment, Chapter I—Environmental Protection Agency (15 volumes) Subchapter C—Air Programs, Parts 53–60 (one volume), parts 61–80 (one volume), parts 81–85 (one volume), and parts 86–99 (one volume).

COLLETTE, R. J. 1985. 1985 update on NO_x emission control technology at combustion engineering. Paper presented at the Joint Symposium on Stationary Combustion NO_x Control. May 1985.

DIGHE, DR. A. S., A. K. KAUL, and SVEIN OLE STROMMEN. 1992. The Flakt–Hydro process SO_2 removal by seawater. Paper presented at the International Conference and Exhibition on Environmental Protection and Control Technology. October 28–31, 1992. Kuala Lumpur, Malaysia.

DRAVO LIME COMPANY. 1990. FGD capabilities and experience. Paper prepared for Black & Veatch, December 6, 1990.

EMRC. Energy and Environmental Measurement Corporation (EEMC). 3925 Placita de la Escarpa, Tucson, AZ 85715.

ELECTRIC POWER RESEARCH INSTITUTE. 1984. Operation and Design of FGD Dampers. EPRI CS-3709. Electric Power Research Institute. Prepared by Black & Veatch Engineers–Architects, Palo Alto, CA.

ELECTRIC POWER RESEARCH INSTITUTE. 1989. FGD Damper Guidelines: Volumes 1 and 2. EPRI GS-6569. Electric Power Research Institute. Prepared by Black & Veatch Engineers–Architects, Palo Alto, CA.

ELECTRIC POWER RESEARCH INSTITUTE. 1990. FGD Chemistry and Analytical Methods Handbook, Volume 1: Process Chemistry. EPRI CS-3612. Electric Power Research Institute. Prepared by Radian Corporation, August 1990, Austin, TX.

ELLIOTT, THOMAS C., ED. 1988. *Standard Handbook of Powerplant Engineering.* McGraw-Hill, New York, NY.

ENVIRONMENTAL ELEMENTS CORPORATION. P. O. Box 1318, Baltimore, MD 21203.

FELSANG, KARSTEN AND BERT BROWN. 1992. High SO_2 and mercury removal by dry FGD systems. Paper presented at the 1992 American Power Conference, Chicago, IL.

FELSVANG, KARSTEN, NIRO-USA; RICK GLEISER, JOY ENVIRONMENTAL TECHNOLOGIES, INC.; GARY JUIP, NORTHERN STATES POWER CO.; and KIRSTEN KRAGH NIELSEN, NIRO-DENMARK. 1993. Air toxics control by spray dryer absorption. Paper presented at the 1993 SO_2 Control Symposium, August 24–27, 1993, Boston, MA.

FGD and DeNO$_x$ Newsletter. 1993. The McIlvaine Company. No. 184. August 1993, Northbrook, IL.

FORSYTHE, R. C. 1990. The advanced thiosorbic flue gas desulfurization process. Paper presented at the AREGC Conference. June 26, 1990, Baton Rouge, LA.

FOSTER WHEELER CORPORATION. Perryville Corporate Park, Clinton, NJ 08809-4000.

GENERAL ELECTRIC ENVIRONMENTAL SYSTEMS, INC. (GEESI). 200 North Seventh Street, Lebanon, PA 17042.

GETZ, NORMAN ET AL. 1992. Demonstrated and innovative control technologies for lead, cadmium and mercury for municipal waste combustors. Paper presented at the 85th Annual Air &

Waste Management Association Meeting. June 21–26, 1992, Kansas City, MO.

GLEISER, RICHARD, JOY ENVIRONMENTAL TECHNOLOGIES, INC., and KARSTEN FELSVANG, NIRO A/S. 1994. Mercury emission reduction using activated carbon with spray dryer flue gas desulfurization. Paper presented at American Power Conference. April 25–27, 1994, Chicago, IL.

GRAF, R. E., B. HUCKRIEDE, H. KESSLER, and S. ZIMMER, GRAF-EPE GMBH. 1993. Commercial operating experience with advanced design, circulating fluid bed scrubbing. Paper presented at the 1993 SO_2 Control Symposium. August 24–27, 1993, Boston, MA.

INTERNATIONAL CONFERENCE OF CLEAN AIR COMPANIES (ICAC), formerly Industrial Gas Cleaning Institute (IGCI). 700 N. Fairfax Street, Suite 304, Alexandria, VA 22314. Publication Numbers EP-1 and EP-7.

JORDAN, ROBERT J. 1987. The feasibility of wet scrubbing for treating waste-to-energy flue gas. *Journal of the Air Pollution Control Association.* Volume 34, No. 4. April 1987.

KAWAMURA, TOMOZUCHI and D. J. FREY. 1980. Current Developments in LOW NO_x Firing Systems, Paper read at Joint Symposium on Stationary Combustion NO_x Control, October 6–9, 1980. Denver, CO.

KBV/ANALECT. 9420 Jeronimo Avenue, Irvine, CA 92718.

LAUDALL, DENNIS, STANLEY MILLER, and RAMSEY CHANG. Enhanced fine particulate control for reduced air toxics emissions. Undated technical paper.

LAVELY, LLOYD AND KIM MASTALIO. 1994. Circulating dry scrubber (CDS): cost effective FGD for clean coal plants. Paper presented at American Power Conference. April 25–27, 1994, Chicago, IL.

MARTINELLI, R., T. GOOTS, and K. REDINGER, BABCOCK & WILCOX. 1993. Least cost environmental protection: comparison of emerging SO_2 control technologies. Paper presented at Power-Gen Americas. November 17–19, 1993, Dallas, TX.

MARTINELLI, ROBERT, THOMAS R. GOOTS, and PAUL S. NOLAN, BABCOCK & WILCOX. 1993. Economic comparisons of energy SO_2 control technologies. Paper presented at the 1993 SO_2 Control Symposium. August 24–27, 1993, Boston, MA.

MASTALIO, K. I., L. L. LAVELY, and T. M. OHLMACHER. 1993. A fully flexible air quality control system for stringent SO_2 and particulate emission limitation. Paper presented at Power-Gen Americas. November 17–19, 1993, Dallas, TX.

MILLER, M. J., ED. 1985. *SO_2 and NO_x Retrofit Control Technologies Handbook.* Prepared by Electric Power Research Institute. EPRI CS-4277-SR. October 1985, Palo Alto, CA.

MUNTERS: THE INCENTIVE GROUP. 1205 Sixth Street Southeast, Fort Myers, FL 33907.

MUZIO, L. J., ET AL. 1991. N_2O Formation is Selective Non-Catalytic NO_x Reduction Processes. Paper presented at EPA/-EPRI 1991 Joint Symposium on Stationary Combustion NO_x Control. March 25–28, 1991, Washington, D.C.

NEME, C. 1991. *Electric Utilities and Long Range Transport of Mercury and Other Toxic Air Pollutants.* Center for Clean Air Policy. November 1991, Washington, D.C.

NISCHT, W., D. W. JOHNSON, AND M. G. MILOBOWSKI. 1991. Economic comparison of materials of construction of wet FGD absorbers and internals. Paper presented at the 1991 SO_2 Control Symposium. December 3–6, 1991, Washington, D.C.

NYMAN, GORAN B. G., ARVID TOKERUD, and NILS MOEVIK. 1991. Scrubbing by seawater, a simple method of removing of SO_2 from flue gases. Paper presented at the National Petroleum Refiners Association 1991 Annual Meeting. March 17–19, 1991, San Antonio, TX.

RILEY STOKER CORPORATION. 5 Neponset Street, P. O. Box 15040, Worcester, MA 01615-0040.

ROSENBERG, HARVEY S., GERALD O. DAVIS, BARRY HINDIN, PAUL RADCLIFFE, and BARRY SYRETT. 1991. Guidelines for FGD materials selection and corrosion protection. Paper presented at the 1991 SO_2 Control Symposium. December 3–6, 1991, Washington, D.C.

SALEEM, A. SR. VICE PRESIDENT, GENERAL ELECTRIC ENVIRONMENTAL SERVICES, INC.; KENT E. JANSSEN, VICE PRESIDENT AND COO, DAKOTA GASIFICATION COMPANY; PAUL A. IRELAND, CHIEF ENGINEER—AIR POLLUTION CONTROL, RAYTHEON ENGINEERS AND CONSTRUCTORS. 1993. Ammonia scrubbing of SO_2 comes of age with in situ forced oxidation. Paper given at the 1993 SO_2 Control Symposium. August 24–26, 1993, Boston, MA.

SIERRA INSTRUMENTS. 5 Harris Court, Building L, Monterey, CA 93940.

SINGER, JOSEPH G., ED. 1981. *Combustion Fossil Power Systems.* Combustion Engineering, Inc., Windsor, CT.

STULTZ, S. C. and J. B. KITTO, EDS. 1992. *Steam: Its Generation and Use.* Babcock and Wilcox Company, New York, NY.

STURN, BERNARD J. 1981. *Fate of Trace Elements in Coal Used in Power Plants.* Black & Veatch Special Report. August, 1981.

SVEDALA INDUSTRIES, INC. 240 Arch Street, P. O. Box 15312, York, PA 17405-7312.

TAMPELLA POWER CORPORATION. 2600 Reach Road, P. O. Box 3308, Williamsport, PA 17701-0308.

TUMATI, P. R. and M. S. DEVITO. Retention of condensed/solid phase trace elements in an electrostatic precipitator. Consolidated Coal Company. Undated technical paper.

WHITE, DAVID M. ET AL. 1992. Parametric evaluation of powdered activated carbon injection for control of mercury emissions from a municipal waste combustor. Paper presented at 85th Annual Air & Waste Management Association Meeting. June 21–26, 1992, Kansas City, MO.

15

WATER TREATMENT

Richard G. Chapman

15.1 INTRODUCTION

Water is a basic requirement for steam-generating power plants, and its availability and quality[1] are primary considerations in siting major steam-generating facilities. Water is used for a multitude of purposes in power plant processes, including equipment cooling, maintenance cleaning, air pollution control (scrubbing), solids conveying, and as the working fluid for the steam cycle. Proper treatment and conditioning of water to avoid scaling and corrosion in modern, efficient high-pressure power cycles is imperative to avoid economic losses caused by decreased production capability and increased operating costs. In a broad sense water treatment is any physical or chemical process that improves the usability of the water treated.

Management of plant water uses to meet increasingly stricter quality requirements as well as more stringent environmental restrictions poses complex challenges for both the design and operation of power plants. This chapter provides basic information on water technology, principal types of treatment and conditioning equipment used, and the more important design considerations and techniques applicable to large steam power plants. Many excellent references are available to the reader interested in additional information on water chemistry theory or general industrial water treatment. Several of these references are listed at the end of this chapter.

15.2 WATER SOURCES

Water supplies generally are divided into two major categories: surface water and ground water. Surface water is available from rivers, lakes and other impoundments, and the sea. Ground water is present below the earth's surface, confined within strata in aquifers, and withdrawn through springs or wells. Only about 3% of the earth's water is considered fresh, and 75% of this is fixed as ice in glaciers and polar ice caps. Of the remaining 25%, more than 24% is groundwater, with surface and atmospheric water making up the remaining 1% (Moffat).

All freshwater supplies are the result of precipitation from the atmosphere, which is part of the continuous evaporation and condensation process called the hydrologic cycle. A simplified schematic of the hydrologic cycle is presented in Fig. 15-1 (Pontius 1990).

Rainfall or snow, on reaching the land surface, either evaporates, transpires, runs off, or infiltrates the soil. Depending on atmospheric conditions and topography, approximately 25% of precipitation runs off to surface waters, less than 10% infiltrates, and the remainder returns to the atmosphere by evapotranspiration. Only a small amount of the infiltrating waters reaches the ground water aquifers. Most of this eventually reappears at the surface to contribute to streamflow. Aquifers near the surface (surficial aquifers) collect almost all of the infiltrating water. Water contained in deep aquifers, separated from the surficial aquifers by bedrock, usually originates in areas other than the location of the aquifer. Some very deep aquifers contain water that collected at a remote location and in a previous geologic age.

Naturally occurring waters have a wide range of qualities. If water was free from dissolved and suspended matter, treatment would be relatively simple. Water has been called the "universal solvent" because it is capable of dissolving, in whole or in part, so many substances. Thus, except as initially produced by condensation, it is never entirely free of dissolved materials.

15.3 WATER QUALITY

Precipitation, as it falls through the atmosphere, absorbs gases, principally oxygen and carbon dioxide, which add greatly to the dissolving power of water. As rain falls on the earth, runs over it, and soaks into the ground, the rainwater entrains and dissolves a variety of substances from the surfaces it contacts. The nature of these substances and the amounts dissolved are functions of the characteristics of the terrain on which the rain falls and of the underground formations into which the rainwater seeps.

Surface runoff carries varying amounts of silt, sand, leaves, decayed vegetation and organic debris, and dissolved

[1]As used here, water quality refers to water purity; the characteristics imparted to water from the contaminants or impurities contained in the water generally cause a decrease in water quality.

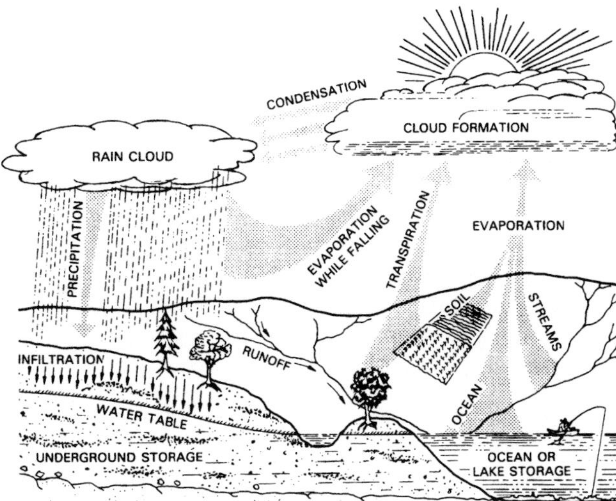

Fig. 15-1. Hydrologic cycle. (From Pontius, F. W.: *Water Quality and Treatment*, 4th edit. Copyright © 1990. Used with permission of McGraw-Hill, Inc.)

mineral constituents from rock formations that it contacts. Surface waters from rivers, lakes, and impoundments, therefore, almost invariably contain turbidity, suspended matter and color, in varying amounts as well as moderate amounts of dissolved solids. Surface waters gradually increase in solids content as a result of concentration caused by evaporation. High dissolved solids concentrations are evident in the oceans because of the large surface areas available for evaporation.

Impounded surface waters such as ponds and lakes usually contain less matter in suspension due to natural settling of the suspended constituents. Dissolved solids, gases, and organic matter may remain in impounded waters. Organic growths such as algae also may be present.

The portion of rainfall that percolates into the soil and then into underground formations, and that is the source of supply of springs and wells, is "strained" or "filtered" in the process so that water from these sources is usually clear and colorless. However, because of its longer and more intimate period of contact with the geological formations through which it passes, water from wells and springs typically contains moderate to large amounts of dissolved solids of various kinds.

15.3.1 Inorganic Constituents

The dissolved inorganic constituents in water are classified in Table 15-1 (Davis and DeWiest 1966).

The concentrations of the major, secondary, minor, and trace inorganic constituents in water are controlled by the availability of the elements in the soil and rock through which the water has passed, by geochemical constraints such as solubility and absorption, by the rates (kinetics) of geochemical processes, and by the sequence in which the water has come into contact with the various minerals in the geologic materials along the flow paths.

Table 15-1. Classification of Dissolved Inorganic Constituents in Water

Major constituents (1.0–1,000 ppm)		
Sodium	Magnesium	Sulfate
Calcium	Bicarbonate	Chloride
	Silica	
Secondary constituents (0.01–10.0 ppm)		
Iron	Potassium	Nitrate
Strontium	Carbonate	Fluoride
	Boron	
Minor constituents (0.0001–0.1 ppm)		
Antimony[a]	Copper	Nickel
Aluminum	Germanium[a]	Phosphate
Arsenic	Iodide	Rubidium[a]
Barium	Lead	Selenium
Bromide	Lithium	Titanium[a]
Cadmium[a]	Manganese	Uranium
Chromium[a]	Molybdenum	Vanadium
Cobalt	Niobium	Zinc
Trace constituents (generally less than 0.001 ppm)		
Beryllium	Lanthanum	Thallium[a]
Bismuth	Niobium[a]	Thorium[a]
Cerium[a]	Platinum	Tin
Cesium	Radium	Tungsten[a]
Gallium	Ruthenium[a]	Ytterbium
Gold	Scandium[a]	Yttrium
Indium	Silver	Zirconium[a]

[a]These elements occupy an uncertain position on the list.
Source: Davis and DeWiest, *Hydrogeology*. Used with permission of John Wiley & Sons.

Water is an excellent solvent for many ionic compounds. The ionic compounds that are dissolved in natural waters generally are major contributors to the contaminants present in the water. Typically, these contaminants are mainly the cations (positively charged ions) of calcium (Ca^{2+}), magnesium (Mg^{2+}), and sodium (Na^{+}), and the anions (negatively charged ions) of bicarbonate (HCO_3^-), sulfate (SO_4^{2-}), and chloride (Cl^-). The total concentrations of these six major constituents normally make up more than 90% of the total dissolved solids in natural water, regardless of whether the water is dilute or has salinity greater than seawater.

Human activity influences the concentrations of the dissolved inorganic constituents. Contributions from manmade sources can increase some of the minor or trace constituents to concentration levels that are orders of magnitude above the normal ranges indicated in Table 15-1.

The term "hardness" has been assigned to the calcium and magnesium present in the water and is of interest in industrial uses because of the tendency of the calcium and magnesium compounds to form scale, especially when the water is heated.

15.3.2 Organic Constituents

Organic compounds have carbon and usually hydrogen and oxygen as the main elemental components in their structural framework. By definition, carbon is the key element. However, the carbonate species—carbonic acid (H_2CO_3), carbon dioxide (CO_2), bicarbonate (HCO_3^-), and carbonate (CO_3^{-2}), which are important constituents in most natural waters—are not classified as organic constituents.

Dissolved organic matter is ubiquitous in natural water, although the concentrations are generally low compared to the inorganic constituents. Investigations of soil-contacted water suggest that most dissolved organic materials in subsurface flow systems are composed of fulvic and humic acids. The molecular weights of organic compounds present in water may range from a few thousand to many thousand grams per gram-mole. Carbon is commonly about half of the formula weight. Analyses of the total organic carbon (TOC) content are becoming a typical part of water investigations. Concentrations in the range of 0.1 to 10 mg/L are most common, but in some areas values are as high as several tens of milligrams per liter. The analyses for individual species is typically not warranted for most treatment applications, but may be necessary to meet regulatory requirements, especially for potable water services or in cases where characterization of the organics is required for wastewater discharge.

15.3.3 Dissolved Gases

The most abundant dissolved gases in water are nitrogen (N_2), oxygen (O_2), carbon dioxide (CO_2), methane (CH_4), and hydrogen sulfide (H_2S). The first three make up the earth's atmosphere, and therefore, it is not surprising that they occur in surface water. The gases CH_4 and H_2S often exist in water in significant concentrations as a result of the product of biogeochemical processes that occur in nonaerated surface zones.

Dissolved gases can have a significant influence on the usefulness of water and, in some cases, can even cause major problems or hazards. For example, because of its odor, H_2S at concentrations greater than about 1 mg/L renders water unfit for human consumption. The dissolution of CH_4 can result in accumulation of this highly flammable gas in wells or buildings and can cause explosion hazards. Gases coming out of solution can form bubbles in wells, screens, or pumps, reducing well productivity or efficiency. Radon-222, a common constituent of water because it is a decay product of radioactive uranium and thorium, common in rock or soil, can accumulate to undesirable concentrations in unventilated spaces. Decay products of radon-222 can be hazardous to human health.

15.3.4 Alkalinity

A basic understanding of alkalinity relationships is important in the study of water chemistry. The alkalinity of water is the capacity of that water to accept protons. It is a measure of the capacity of the water to neutralize acids. The alkalinity of natural waters is due primarily to the salts of weak acids. Alkalinity affects hardness, other ion solubility, and pH.

Although several bases can contribute to the alkalinity of a water, in most natural waters the alkalinity is described by the following relationship:

$$\text{Total alkalinity} = [HCO_3^-] + 2[CO_3^{-2}] + [OH^-] - [H^+] \tag{15-1}$$

Bicarbonates (HCO_3^-) represent the major form of natural alkalinity, because they are formed in considerable amounts from the action of carbonic acid (carbon dioxide plus water) upon basic materials in the soil. At low CO_2 concentration (higher pH), carbonate alkalinity can be present in natural waters. Hydroxide alkalinity is normally not present in natural waters that have not been treated or otherwise altered.

Carbon dioxide and the three forms of alkalinity are all part of one system that exists in equilibrium, as can be seen by the following equations. The symbol $\rightleftharpoons$ indicates a reversible reaction. M is a metallic ion.

$$CO_2 + H_2O \rightleftharpoons H_2CO_3 \rightleftharpoons H^+ + HCO_3^- \tag{15-2}$$

$$M(HCO_3)_2 \rightleftharpoons M^{2+} + 2HCO_3^- \tag{15-3}$$

$$HCO_3^- \rightleftharpoons CO_3^{-2} + H^+ \tag{15-4}$$

$$CO_3^{-2} + H_2O \rightleftharpoons HCO_3^- + OH^- \tag{15-5}$$

A change in the concentration of any one member of the system causes a shift in the equilibrium, alters the concentration of the other ions, and changes the pH. Conversely, a change in pH shifts the relationships.

The traditional analysis for alkalinity is by two-step titration with sulfuric acid. In the first step, phenolphthalein indicator turns the solution from pink to colorless at the end point. In the second step, methyl orange indicator is added to the first-step liquid and the acid titration continues until a pink color develops. At this point, the pH is about 4.5 and all of the alkalinity has been neutralized. The first end point is noted as the "P" alkalinity and the second end point as the "M" or total alkalinity.

From these data, alkalinity relationships as shown in the following table are established.

Titration Result	Hydroxide (OH^-) alk. as $CaCO_3$	Carbonate (CO_3^{-2}) alk. as $CaCO_3$	Bicarbonate (HCO_3^-) alk. as $CaCO_3$
P = 0	0	0	M
P < ½M	0	2P	M − 2P
P = ½M	0	2P	0
P > ½M	2P − M	2(M − P)	0
P = M	M	0	0

Figure 15-2 (Pontius 1990) is a curve of the carbonate alkalinity relationship compared to pH for an aqueous solution at 25° C (77° F) and ionic strength (I) of zero.

The preceding table and Fig. 15-2 do not quite agree because the table is theoretical and does not take temperature into account. Although the figure is more an indication of actual alkalinity relationships, the table can be used for most purposes.

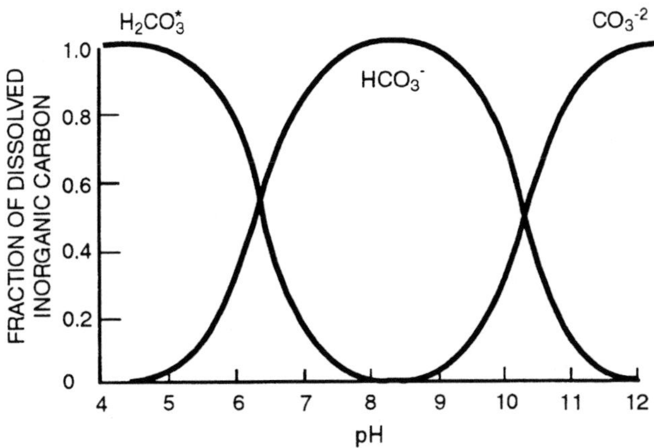

Fig. 15-2. Relationship among dissolved inorganic carbonate species in water at 25° and ionic strength = 0. (From Pontius, F. W.: *Water Quality and Treatment*, 4th edit. Copyright © 1990. Used with permission of McGraw-Hill, Inc.)

15.3.5 Water Analysis

Water treatment and water treating equipment deal with the removal or replacement of these suspended and dissolved substances to make the water suitable for specific uses.

Table 15-2 (Betz 1980) lists characteristics and contaminants commonly analyzed in natural waters to evaluate treatment requirements for industrial use of the waters. In addition to the treatment methods noted in Table 15-2, other treatment technologies are continually being developed.

The kinds and amounts of various dissolved and suspended solids in a water sample are determined by standardized laboratory testing methods such as those provided in *Standard Methods for the Examination of Water and Wastewater* (Greenborg et al. 1992). For each constituent, the results are reported in units (mg/L, ppm, or gr/gal) of that constituent stated as ions such as Ca^{2+}, Mg^{2+}, SO_4^{2-}, and Cl^-, or alternatively stated as hypothetical chemical compounds such as $Ca(HCO_3)_2$, $CaSO_4$, $MgCl_2$, and NaCl.

The ions, radicals, and combinations thereof most commonly involved in water and water treatment problems, along with their chemical symbols, are shown in Table 15-3. In many cases, water analyses are reported in terms of $CaCO_3$ equivalents. One reason for using $CaC0_3$ as a standard is that it has a molecular weight of 100, simplifying the conversion arithmetic. By using the conversion factors given in Table 15-3, all constituents determined by analysis can easily be converted to $CaCO_3$ equivalents; for example, Ca concentration × 2.5 = Ca concentration expressed as $CaCO_3$ equivalents, and Mg concentration × 4.12 = Mg concentration expressed as $CaCO_3$ equivalents. Expression of the analysis in terms of $CaCO_3$ equivalents is a convenient means of checking the analysis for equivalents. In a precisely correct analysis, the sum of the cations should be equal to the sum of the anions when both are expressed as CaCO3 equivalents.

The following example demonstrates the conversion of a raw water analysis, originally reported ions as such, to an analysis expressed in $CaCO_3$ equivalents.

Example 15-1

Given the following raw water analysis, convert the analysis to $CaCO_3$ equivalents and check the cation/anion balance:

Constituent	Concentration (mg/L as such)
Calcium (Ca)	71
Magnesium (Mg)	24
Sodium (Na)	79
Potassium (K)	8
Bicarbonate (HCO_3)	244
Sulfate (SO_4)	210
Chloride (Cl)	24
Nitrate (NO_3)	6

Solution

Constituent	Original analysis, as such (mg/L)	$CaCO_3$ conversion factor	Analysis expressed as $CaCO_3$ equivalents mg/L
Ca	71	× 2.50 =	178
Mg	24	× 4.12 =	99
Na	79	× 2.17 =	171
K	8	× 1.28 =	10
Total cations	182		458
HCO_3	244	× 0.82 ×	200
SO_4	210	× 1.04 ×	218
Cl	24	× 1.41 =	34
NO_3	6	× 0.81 =	5
Total anions	484		457

In actual practice, the sum of the cations is seldom precisely equal to the sum of the anions because of normal analytical errors experienced with the analysis of each individual constituent. A variation of 2% to 5% between the cation and anion balance in actual analyses is generally considered acceptable. In some cases, the analysis may not be complete and a constituent such as sodium may be determined by the difference between the cation and anion balance; the final adjusted analysis then presents an exact balance. The analysis above is reported in concentration units of mg/L, which are essentially equivalent to ppm in solutions and suspensions with specific gravities approximately equal to 1.0. There are a number of other conventions for expression of concentrations; conversion factors for many of these are shown in Table 15-4.

Although pure water is never found in nature, the quality of surface waters typically varies from the high quality of a pristine mountain stream to highly saline water such as that contained in the Great Salt Lake. Similarly, ground water may vary from a high-quality, pristine aquifer to highly mineralized aquifers several times more saline than seawater. Tables 15-5 and 15-6 present analyses of several different surface water and ground water supplies, respectively, and show the highly variable water characteristics and constituents encountered among various water sources.

Table 15-2. Common Characteristics of Natural Waters

Constituent	Chemical Formula	Difficulties Caused	Means of Treatment
Turbidity	None; expressed in analysis as units	Imparts unsightly appearance to water. Deposits in water lines, process equipment, etc. Interferes with most process uses.	Coagulation, settling and filtration
Hardness	Calcium and magnesium salts expressed as $CaCO_3$	Chief source of scale in heat exchange equipment, boilers, pipelines, etc. Forms curds with soap, interferes with dying, etc.	Softening. Demineralization. Internal boiler water treatment. Surface-active agents.
Alkalinity	Bicarbonate $(HCO_3)^-$ carbonate $(CO_3)^{2-}$, and hydrate $(OH)^-$, expressed as $CaCO_3$	Foaming and carryover of solids with steam. Embrittlement of boiler steel. Bicarbonate and carbonate produce CO_2 in steam, a source of corrosion in condensate lines	Lime and lime–soda softening. Acid treatment. Hydrogen zeolite softening. Demineralization. Dealkalization by anion exchange.
Free mineral acidity	H_2SO_4, HCl, etc. expressed as $CaCO_3$	Corrosion	Neutralization with alkalies.
Carbon dioxide	CO_2	Corrosion in water lines and particularly steam and condensate lines	Aeration. Deaeration. Neutralization with alkalies.
pH	Hydrogen ion concentration defined as: $pH = \log \frac{1}{[H^+]}$	pH varies according to acidic or alkaline solids in water. Most natural waters have a pH of 6.0–8.0.	pH can be increased by alkalies and decreased by acids.
Sulfate	SO_4^{-2}	Adds to solids content of water, but in itself is not usually significant. Combines with calcium to form calcium sulfate scale.	Demineralization
Chloride	Cl^-	Adds to solids and increases corrosive character of water.	Demineralization
Nitrate	NO_3^-	Adds to solids content, but is not usually significant industrially. High concentrations cause methemoglobinemia in infants. Useful for control of boiler metal embrittlement.	Demineralization
Fluoride	F^-	Cause of mottled enamel in teeth. Also used for control of dental decay. Not usually significant industrially.	Adsorption with magnesium hydroxide, calcium phosphate, or bone black. Alum coagulation.
Sodium	Na^+	Adds to solids content of water. When combined with OH^-, causes corrosion in boilers under certain conditions.	Demineralization
Silica	SiO_2	Scale in boilers and cooling water systems. Insoluble turbine blade deposits due to silica vaporization.	Hot process removal with magnesium salts. Adsorption by highly basic anion-exchange resins in conjunction with demineralization.
Iron	Fe^{2+} (ferrous) Fe^{+3} (ferric)	Discolors water on precipitation. Source of deposits in water lines, boilers, etc. Interferes with dying, tanning, papermaking, etc.	Aeration. Coagulation and filtration. Lime softening. Cation exchange. Contact filtration. Surface-active agents for iron retention.
Manganese	Mn^{2+}	Same as iron	Same as iron
Aluminum	Al^{3+}	Usually present as a result of floc carryover from clarifier. Can cause deposits in cooling systems and contribute to complex boiler scales.	Improved clarifier and filter operation
Oxygen	O_2	Corrosion of water lines, heat exchange equipment, boilers, return lines, etc.	Deaeration. Sodium sulfite. Corrosion inhibitors.
Hydrogen sulfide	H_2S	Cause of "rotten egg" odor. Corrosion.	Aeration. Chlorination. Highly basic anion exchange.
Ammonia	NH_3	Corrosion of copper and zinc alloys by formation of complex soluble ion.	Cation exchange with hydrogen zeolite. Chlorination. Deaeration.
Dissolved solids	None	"Dissolved solids" is measure of total amount of dissolved matter, determined by evaporation. High concentrations of dissolved solids are objectionable because of process interference and as a cause of foaming in boilers.	Various softening processes, such as lime softening and cation exchange by hydrogen zeolite, will reduce dissolved solids. Demineralization.*

(Continued)

Table 15-2. *(Continued)*

Constituent	Chemical Formula	Difficulties Caused	Means of Treatment
Suspended solids	None	"Suspended solids" is the measure of undissolved matter, determined gravimetrically. Suspended solids cause deposits in heat exchange equipment, boilers, water lines, etc.	Subsidence. Filtration, usually preceded by coagulation and settling.
Total solids	None	"Total solids" is the sum of dissolved and suspended solids, determined gravimetrically. See "Dissolved solids" and "Suspended solids."	

Source: Betz Laboratories, Inc. Used with permission.

*Black & Veatch notes that dissolved solids reduction using cation exchange by hydrogen zeolite typically requires additional treatment processes, for example, decarbonation.

15.3.6 Water Resources Data

Surprisingly, it is often difficult to obtain reliable water data for a given area. The U.S. Corps of Engineers usually can supply comprehensive flow data on all major and most minor rivers and streams, but quantity data on ground water and quality data for both surface and ground water are usually very limited and often unreliable. Among the sources of data that may have to be explored are the following:

- U.S. Corps of Engineers;
- United States Geological Survey;
- State and local geological, health, and water resources agencies;
- Local public utility companies;
- Adjacent industries;
- State or local agricultural agencies;
- Local well drillers; and
- Environmental and pollution control organizations.

If time and the scope of the project allow, samples of surface water supplies can be collected and analyzed, or, for prospective ground water supplies, test wells drilled and ground water samples collected and analyzed.

For surface water, especially rivers or streams, multiple analyses are required to ascertain seasonal and annual variations. Ground water usually maintains a relatively constant quality if the aquifer is not overpumped or fed by a surface water supply. When multiple quality data are used, the analysis having the highest total solids content is often used for design. If a typical analysis is required, it should be an actual analysis that is nearest in composition to the average analysis of the available data. The use of an arithmetic average of the constituents for a design analysis should be discouraged as it may not provide the proper relationship of one constituent to another, which can adversely impact treatment design considerations.

15.4 PLANT WATER REQUIREMENTS

Water requirements for a fossil-fueled plant cannot be estimated from the size of the plant alone. Water quality, plant location, fuel characteristics, steam generator design pressure, and local regulatory requirements can all have a significant effect on the quantity requirements.

Although a typical power plant has a multitude of water uses, the primary uses can be categorized for simplicity as cooling water, service water, and high-purity water. This section describes these uses and shows how to approximate the amount of water required for each use under varying conditions.

15.4.1 Description of Usage

Cooling water includes the water used for condenser cooling in the turbine heat rejection system and for the cooling of auxiliary equipment. The condenser cooling system can be a once-through system with the discharge returned to the river or surface impoundment which is serving as the source, a recirculating cooling system using a wet cooling tower, a closed system employing dry cooling, or a hybrid system using both wet and dry cooling for heat rejection.

Once-through systems require high flow rates but consume relatively small quantities of cooling water, whereas cooling tower designs have significant consumption rates as a result of the high evaporation rates and blowdown requirements. However, once-through cooling from a river is heavily regulated and permitted only in unusual circumstances. Dry cooling systems are relatively expensive and inefficient and are applicable only when water is in very limited supply. Consequently, recirculating cooling systems using a wet cooling tower are most often used for main steam cycle heat rejection. A completely closed cooling water system is often used for auxiliary equipment cooling.

Service water is used for general plant services such as sanitary water, washdowns, ash transport, flue gas desulfurization (FGD) system makeup, and pump seal water. However, many of these services may use reclaimed wastewater, depending on specific plant water and wastewater management design.

A portion of the service water is demineralized to obtain a high-purity water for use as cycle makeup, as chemical solution water, as makeup to the closed auxiliary cooling system, and in the laboratory.

15.4.2 System Requirements

Proper plant operation minimizes maintenance and protects plant equipment, and requires that certain water quality

Table 15-3. Calcium Carbonate ($CaCO_3$) Conversion Factors

Chemical Name	Symbol or Formula	Molecular Weight	Equivalent Weight	To Convert to $CaCO_3$ Equivalent Multiply By
Aluminum (element)	Al	27.0	9.0	5.56
Aluminum hydroxide	$Al(OH)_3$	78.0	26.0	1.92
Aluminum sulfate	$Al_2(SO_4)_3 \cdot 18H_2O$	666.4	111.1	0.45
Aluminum chloride	$AlCl_3$	133.5	44.5	1.12
Bicarbonate	HCO_3	61.0	61.0	0.82
Calcium (element)	Ca	40.1	20.0	2.50
Calcium bicarbonate	$Ca(HCO_3)_2$	162.1	81.1	0.62
Calcium carbonate	$CaCO_3$	100.1	50.0	1.00
Calcium sulfate	$CaSO_4$	136.1	68.1	0.74
Calcium chloride	$CaCl_2$	111.0	55.5	0.90
Calcium hyroxide	$Ca(OH)_2$	74.1	37.1	1.35
Calcium oxide	CaO	56.1	28.0	1.79
Carbonate	CO_3	60.0	30.0	1.67
Carbon dioxide	CO_2	44.0	44.0	1.14
Chloride	Cl	35.5	35.5	1.41
Ferric chloride	$FeCl_3$	162.3	54.1	0.92
Ferric hydroxide	$Fe(OH)_3$	106.8	35.6	1.40
Ferric sulfate	$Fe_2(SO_4)_3$	399.9	66.6	0.75
Ferrous hydroxide	$Fe(OH)_2$	89.8	44.9	1.11
Ferrous sulfate	$FeSO_4$	151.9	76.0	0.66
Hydrogen (element)	H	1.0	1.0	50.00
Iron (ferric)	Fe	55.8	18.6	2.69
Iron (ferrous)	Fe	55.8	27.9	1.79
Magnesium (element)	Mg	24.3	12.2	4.12
Magnesium bicarbonate	$Mg(HCO_3)_2$	146.4	73.2	0.68
Magnesium carbonate	$MgCO_3$	84.3	42.2	1.18
Magnesium sulfate	$MgSO_4$	120.4	60.2	0.83
Magnesium chloride	$MgCl_2$	95.3	47.6	1.05
Magnesium hydroxide	$Mg(OH)_2$	58.3	29.2	1.72
Magnesium oxide	MgO	40.3	20.2	2.48
Manganese (manganic)	Mn	54.9	18.3	2.73
Manganese (manganous)	Mn	54.9	27.5	1.82
Nitrate	NO_3	62.0	62.0	0.81
Phosphate	PO_4	95.0	31.7	1.58
Potassium	K	39.1	39.0	1.28
Silica	SiO_2	60.1	60.1	0.83
Sodium (element)	Na	23.0	23.0	2.17
Sodium bicarbonate	$NaHCO_3$	84.0	84.0	0.60
Sodium carbonate	Na_2CO_3	106.0	53.0	0.94
Sodium chloride	$NaCl$	58.5	58.5	0.85
Sodium hydroxide	$NaOH$	40.0	40.0	1.25
Sodium sulfate	Na_2SO_4	142.1	71.0	0.70
Sulfate	SO_4	96.1	48.0	1.04
Sulfuric acid	H_2SO_4	98.1	49.0	1.02
Hydrochloric acid	HCl	36.5	36.5	1.37
Nitric acid	HNO_3	63.0	63.0	0.79
Water	H_2O	18.0	9.0	5.56

Table 15-4. Various Conventions for Expression of Concentrations of Constituents in Water

	Milligrams per Liter	Grams per Liter	Grains per US Gal	Grains per Imp Gal	Clark Degree	French Degree	German Degree
ppm	1.0	0.001	0.0584	0.07	0.07	0.10	0.056
mg/L	1.0	0.001	0.0584	0.07	0.07	0.10	0.056
French degrees	10.0	0.010	0.583	0.70	0.70	1.00	0.56
Grains per US gal	17.1	0.0171	1.000	1.2	1.2	1.71	0.958
Grains per imp gal	14.3	0.0143	0.829	1.0	1.0	1.43	0.80

Table 15-5. Selected River Water Analyses[a]

Constituent	Expressed as	Analysis River Identification[b] VA	OH	IA	MO	WA
Cations						
Calcium	$CaCO_3$	28	280	88	178	15
Magnesium	$CaCO_3$	16	105	67	98	9
Sodium	$CaCO_3$	13	165	24	171	15
Potassium	$CaCO_3$	3	NR	NR	10	NR
Total cations	$CaCO_3$	60	550	179	457	39
Anions						
Bicarbonate	$CaCO_3$	43	112	115	200	23
Carbonate	$CaCO_3$	0	0	12	0	0
Sulfate	$CaCO_3$	10	200	26	218	6
Chloride	$CaCO_3$	8	240	21	34	8
Nitrate	$CaCO_3$	Nil	NR	NR	5	2
Total anions	$CaCO_3$	61	552	174	457	39
Carbon dioxide	CO_2	NR	18	NR	NR	NR
Iron	Fe	NR	2	NR	1	NR
Silica	SiO_2	12	5	10	16	15
pH	Std units	7.3	NR	9.3	8.3	NR

[a]All values are in mg/L except as noted. Some analyses are adjusted for ionic balance.

[b]River identification: VA, Roanoke River, Clover, VA; OH, Muskingum River, Conesville, OH; IA, Mississippi River, Muscatine, IA; MO, Missouri River, Watson, MO; WA, Skookumchuck River, Lewis County, WA. NR, not reported.

guidelines be established. Some of these are essential, but most are flexible. The following is a discussion of recommended guidelines and the importance of each.

15.4.2.1 Main Steam Cycle Cooling Water. For a once-through cooling system, it can normally be assumed that trash racks and screens on the suction side of the circulating water pumps adequately remove suspended materials from the cooling water and that additional treatment for suspended solids removal is not required. With proper design considerations and materials selections, waters with up to approximately 50,000 mg/L of total dissolved solids can be used. Organic matter and microorganisms in the water present the biggest problem and biological growth is most commonly controlled by shock chlorination. Mechanical tube cleaning systems can be used where marine growth tends to adhere to the tube surface. Since suspended sand and grit can erode tubes, condenser tubes must be protected from erosion by proper design of the intake structure, reasonably low velocities, and selection of condenser tube materials to match the service conditions. The pH of once-through cooling water normally ranges between about 6 and 8.5; pH adjustment is rarely required.

Treatment required for cooling towers can be much more complex. The evaporation of water as the cooling flow is recirculated across the tower results in high concentrations of the dissolved species. As these materials begin to reach solubility limits and the dissolved solids levels are increased, the potential for deposition, fouling, and corrosion of materials in the cooling circuit is dramatically increased. Chemicals are added for biological control and to mitigate the effects of the higher solids concentrations, and blowdown is employed to limit the cooling water cycles of concentration to control the dissolved solids within prescribed control limits. Blowdown is also used to control the suspended solids in the circulating water and minimize the accumulation of sediments in the cooling tower basin. Typical control limits commonly employed for a recirculating cooling water system are presented in Section 15.6, Treatment Applications.

The chemical composition of the cooling water is also

Table 15-6. Selected Well Water Analyses[a]

Constituent	Expressed as	Analysis Identification[b] OH	FL	IA	MO	CO	AZ	CA
Cations								
Calcium	$CaCO_3$	314	131	63	282	330	172	107
Magnesium	$CaCO_3$	140	15	25	220	145	128	91
Sodium	$CaCO_3$	80	11	8	69	292	277	77
Potassium	$CaCO_3$	NR	1	NR	5	11	NR	NR
Total cations	$CaCO_3$	534	158	96	576	778	577	275
Anions								
Bicarbonate	$CaCO_3$	233	126	50	535	205	196	141
Carbonate	$CaCO_3$	0	0	0	0	0	0	0
Sulfate	$CaCO_3$	215	2	26	22	475	136	2
Chloride	$CaCO_3$	90	29	20	19	95	245	104
Nitrate	$CaCO_3$	NR	Nil	NR	NR	3	NR	20
Total anions	$CaCO_3$	538	157	96	576	778	577	267
Carbon dioxide	CO_2	NR	13	NR	NR	NR	NR	NR
Iron	Fe	NR	2	NR	1	NR	NR	NR
Silica	SiO_2	10	10	15	30	15	12	46
pH	Std units	7.1	7.3	7.5	7.0	7.3	7.6	7.5

[a]All values are in mg/L except as noted. Some analyses are adjusted for ionic balance.

[b]Well identification: OH, Athens County, OH; FL, DeBary, FL; IA, Louisa County, IA; MO, Watson, MO; CO, Colorado Springs, CO; AZ, Holbrook, AZ; CA, Sacramento, CA. NR, not reported.

important from the standpoint of tube material selection. If the water is of brackish or seawater quality, copper alloy tubes have traditionally been used; however, in recent years, other materials, such as ferritic stainless steels and titanium, have gained in popularity. Copper alloy tubes are also satisfactory for use with freshwater, although stainless steel tubes are often used as part of a copper-free condensate feedwater cycle, as described in Section 15.7, Cycle Chemistry and Control

15.4.2.2 Auxiliary Cooling Water. A completely closed loop auxiliary cooling water system is normally preferred for oil coolers, air compressors, bearing water, etc. This permits the use of condensate quality water with corrosion inhibitors added for corrosion protection, which results in very low corrosion rates in this system. The heat in the closed cooling system is typically rejected to the condenser cooling water. If the condenser cooling water is of exceptionally high quality and low in suspended solids, it is sometimes used directly for cooling auxiliary equipment.

15.4.2.3 Service Water. Service water is used for pump and instrument seal water, hose supply, fire water, demineralizer supply, sanitary water, and makeup to ash and flue gas scrubbing systems. Service water should be essentially free of suspended solids, turbidity, and color. The pH will typically be between 6.0 and 8.5, and total dissolved solids preferably limited to less than 1,000 mg/L. If the service water is also used for potable water, it must be chlorinated and conform to applicable drinking water standards. Residual chlorine must be minimized or removed before the water is used for demineralizer supply to avoid degrading the demineralizer ion exchange resins.

Unless a very limited quantity of potable quality water is needed, it is generally preferable to design the service water system to meet drinking water standards. This avoids dual piping systems and eliminates the necessity of tagging valves and lines as potable or nonpotable.

The water quality requirements for the potable water system are normally established by regulatory standards. Federal, state, and local regulations must be investigated during the design phase. If the service water system is supplying potable water, it must meet these requirements, as well as any more stringent requirements established by the end use. If a nonpotable service water system is being used, the service water quality requirements of each of the end uses must be reviewed.

15.4.2.4 High-Purity Water. High-purity water is required for makeup to the condensate–feedwater cycle, as solution water for condensate–feedwater chemicals, and for various laboratory uses. The quality of the water required is dictated by the quality necessary in the condensate–feedwater system to prevent scaling and corrosion and to prevent carryover of solids with the steam from the steam generator.

These water quality requirements are dependent on the operating pressure of the boiler, with a higher quality water required at higher pressures. Water quality required for cycle makeup is discussed further in Section 15.7.5, Water Quality Control.

15.4.3 Quantity Requirements

15.4.3.1 Main Steam Cycle Cooling Water A once-through cooling system is a simple heat exchanger, and the amount of cooling water required can be calculated if the amount of heat to be rejected, the condenser size, and the cooling water temperature are known. For a very rough estimate, a flow of 1,000 gpm (0.063 m^3/s) per megawatt (MW) of plant electrical output can be used for a single-pressure condenser and 750 gpm (0.047 m^3/s) per MW of plant electrical output for a dual-pressure condenser.

For an all-wet, recirculating cooling system using cooling towers, the water quantity is dependent on the water quality and atmospheric conditions. The amount of makeup water is determined by a mass balance that can be expressed as follows.

$$\text{Blowdown} = \frac{\text{Evaporation}}{\text{Cycles of concentration} - 1} - \text{Drift} \quad (15\text{-}6)$$

$$\text{Makeup} = \text{Blowdown} + \text{Evaporation} + \text{Drift} \quad (15\text{-}7)$$

Evaporation is the quantity of water that enters the atmosphere as a result of vaporization caused by the heat. A heat and mass balance, such as that described in ASME Technical Paper 69-WA/PWR-3 (Leung and Moore, 1969), can be made for accurate determination, but an "order of magnitude" determination can be made using the following criteria:

Evaporation rate = 1,000 gpm (0.063 m^3/s) per 100 MW of plant electrical output

@ Sea level

@ 50% relative humidity

@ 65° F (18.3° C) wet bulb temperature

@ 40 lb/h (0.005 kg/s) of circulating water flow rate per 1,000 Btu/h (293 joules/s) of heat load

Corrections

Plus 0.5% per 1,000 ft altitude

Plus 3% per each 10% decrease in relative humidity

Minus 3% per each 10% increase in relative humidity

Plus 5% per each 10° F (5.6° C) increase in wet-bulb temperature

Minus 5% for each 10° F (5.6° C) decrease in wet-bulb temperature

Drift is the water lost when entrained water is carried out with the air. The drift contains the same dissolved solids concentration as the tower water. Modern cooling tower mist eliminator design minimizes drift losses. The drift will typically be 0.005% or less of the circulating water flow.

The cycles of concentration are defined as the multiple increases in concentration of dissolved solids in the circulating water compared to the makeup water. For example, cycles of concentration of 4.0 indicate that the dissolved constituent concentrations in the circulating water are four times greater than the concentrations in the makeup water. In most cases, the limiting constituents are calcium, alkalinity, and silica content of the circulating water. Total dissolved solids, chloride, and sulfate can also be limiting by presenting a corrosive condition, but proper selection of construction materials can often alleviate this problem.

Blowdown is the water purged from the system to control the concentration of the dissolved solids in the circulating water.

Chemical treatment of the makeup water can reduce the calcium and alkalinity contents and, in some cases, the silica. An economic evaluation may be required to determine if treatment is justified to decrease circulating water blowdown and makeup water requirements. In some cases, if treatment is required to remove suspended solids or high iron content, calcium alkalinity reduction can be incorporated at a very low incremental cost.

The preceding discussion is for turbine condenser cooling only, using a 100% wet system. Usually the circulating water rate is increased by about 10% to cover cooling for auxiliary equipment.

15.4.3.2 Service Water. In the absence of more specific plant design criteria, service water quantities can be estimated from the following:

General service	1% of maximum steam flow
Potable sanitary	50 gal (189 L) per day per person
Cycle makeup	1.5% of steam flow plus steam soot blowing requirements[2]
FGD system	
Lime	1 gal (3.785 L) per pound (0.45 kg) of lime[3]
Limestone	0.5 gal (1.89 L) per pound (0.45 kg) of limestone[3]
Makeup	Dependent on degree of flue gas saturation[3]
Fire protection	No regular usage

If the water will be chemically pretreated, about 5% to 10% should be added to the total for treatment losses, dependent on the type of pretreatment used.

The general service water usage is mainly for plant washdown and miscellaneous low-volume usages and is not directly related to plant load.

Potable water usage can be based on the normal number of plant personnel plus an allowance for visitors and special work crews.

The cycle makeup quantity is based on boiler blowdown and evaporative losses plus steam used for soot blowing and fuel oil atomization. Losses due to demineralizer regeneration must be accounted for in the demineralizer design and supply water requirements. The cycle makeup quantity estimate given does not include the water requirements for chemical cleaning.

Flue gas desulfurization (FGD) system requirements should take into consideration the use of recycled water for meeting some of the makeup demand.

Additional service water may be required if wet sluicing is used for bottom ash handling. Ash transport system makeup is based on recycling as much water as possible and is about 2% of the flow in a 20% by weight ash sluice system.

Allowance must be made in water storage and pumping capacity for fire water requirements.

15.4.3.3 High-Purity Water. In the context of this discussion, high-purity water is synonymous with cycle makeup. The service water system allows for an amount equal to 1.5% of the steam flow plus soot blowing requirements. This is a maximum number; the actual normal boiler makeup in a well-maintained facility should be less than 0.5% plus soot blowing requirements. A small amount of high-purity water is also required for makeup to the closed auxiliary cooling system and for use in the laboratory; however, these requirements are insignificant when considering overall demand.

During preoperational alkaline and acid cleaning of the preboiler and the boiler and in subsequent periodic boiler acid cleanings, there is a short-term demand for large quantities of high purity water. This can amount to as much as 500,000 gal (1,890 m^3) in 1 or 2 days. A significant portion of such demands must be met from storage or by outside purchase.

15.4.4 Water Usage Variations

Except for general service water and potable water, all plant water usage should be more or less directly proportional to unit load. But in actual operation, certain usages such as the following are not proportional because of plant operating procedures:

- Circulating water pumps and ash sluice pumps are generally not throttled with load; flow variations tend to be step changes as the number of pumps in service are varied.
- Boiler blowdown may be at a constant rate.
- Circulating water evaporation rate shows significant variation with meteorological conditions.
- The quality of the raw water can significantly affect water usage in the condenser cooling system.
- The degree of plant wastewater reuse can have a significant effect on plant water requirements.

These variations must be considered in the plant design for all of the parameters and conditions specific to that particular design.

15.4.5 Water Mass Balance

One of the initial project design activities should include development of the plant water mass balances. Water mass balances incorporate both water supply and wastewater reuse and disposal considerations. Water mass balances are

[2]This criterion assumes no steam is used in process or exported without full condensate return.

[3]These do not reflect any reuse of other system wastewaters such as circulating water blowdown.

normally developed for various conditions such as operating loads and meteorology, and to satisfy both design and permitting needs. Plant water mass balances are very helpful for the sizing of water supply and treatment facilities and are essential to developing integrated plant water and wastewater systems. These water mass balances can become complex with the integration of wastewater reuse, especially for zero discharge design plants. Development of water mass balances is discussed further in Chapter 16, Liquid and Solid Waste Treatment.

15.5 TREATMENT PROCESSES

Water treatment processes range in complexity from simple sedimentation to multiple-bed ion exchange demineralization. This section defines various treatment processes, describes the basic chemical and physical reactions involved, and discusses their applicability to power plants.

15.5.1 Sedimentation and Clarification

15.5.1.1 Settling. For many power plant uses, it is essential the water be practically free from suspended matter. This may be achieved by simple settling (sedimentation) in relatively large bodies of water, such as lakes or reservoirs that allow significant detention times.

Comparatively few water users, however, have the space required for effective settling by such means. Moreover, periodic weather disturbances and organic growths such as algae usually necessitate supplementary provisions to ensure clean water at all times. This has led to the development of equipment that clarifies the water more rapidly and with greater assurance of the end result by chemical coagulation and the use of equipment to facilitate separation and removal of coagulated solids.

15.5.1.2 Coagulation. Chemical coagulation is the gathering of suspended matter present in a finely divided or colloidal state into larger and more rapidly settling particles. To effect such chemical coagulation, various compounds are used, the most common being aluminum sulfate, $Al_2(SO_4)_3$. Others are sodium aluminate, $Na_2Al_2O_4$; ferrous sulfate, $FeSO_4$; ferric sulfate, $Fe_2(SO_4)_3$; and ferric chloride, $FeCl_3$. When ferrous sulfate is used as a coagulant, oxygen or an oxidizing agent such as chlorine must be present to convert the ferrous salt to the ferric form.

These coagulants produce a jelly-like spongy mass of floc with enormous surface area per unit of volume, which entraps and binds together the small particles of silt, organic matter, and even bacteria. The enlarged precipitate thus formed has many times the settling rate of the finer particles it has entrapped, thus accelerating fine particle separation. In this process, only suspended solids are affected; hardness is not removed.

Alum, as aluminum sulfate is generally called, is most effective as a coagulant in the pH range of 5.7 to 8.0. However, it is often used satisfactorily in higher pH ranges, especially in conjunction with lime softening. Coagulation with alum, ferrous or ferric sulfates, and ferric chloride consumes alkalinity. The alkalinity required may be naturally present in the water as bicarbonates and carbonates. When alkalinity is not present in sufficient amount, alkaline chemicals such as lime [$Ca(OH)_2$], soda ash [Na_2CO_3], or caustic soda [NaOH], must be added. When sodium aluminate is used, free carbon dioxide is reduced and alkalinity is increased.

The dissociation of alum to form gelatinous aluminum hydroxide liberates sulfuric acid which reacts with the alkalinity to form $CaSO_4$, the overall reaction being as follows:

$$Al_2(SO_4)_3 + 3Ca(HCO_3)_2 \rightarrow 2Al(OH)_3 +\downarrow 3CaSO_4 + 6CO_2\uparrow \quad (15\text{-}8)$$

The iron coagulants usually function best in the higher pH ranges, where hydroxide is present. Typical coagulation reactions for ferric chloride, $FeCl_3$, and ferric sulfate, $Fe_2(SO_4)_3$, in lime-treated water are the following:

$$2FeCl_3 + 3Ca(OH)_2 \rightarrow 2Fe(OH)_3\downarrow + 3CaCl_2 \quad (15\text{-}9)$$

$$Fe_2(SO_4)_3 + 3Ca(OH)_2 \rightarrow 2Fe(OH)_3\downarrow + 3CaSO_4 \quad (15\text{-}10)$$

The sodium aluminate reaction is the following:

$$Na_2Al_2O_4 + 2CO_2 + 4H_2O \rightarrow 2Al(OH)_3\downarrow + 2Na(HCO_3) \quad (15\text{-}11)$$

Activated silica, bentonite clays, powdered activated carbon, and organic polymers (polyelectrolytes) are examples of additives used in the clarification process to enhance suspended solids removal. Clays and activated carbon are weighting agents used to enhance the sedimentation process by producing heavier floc particles, thus improving the settling characteristics of the floc. Coagulant aids, such as polyelectrolytes, are used to further agglomerate the coagulated particles to increase the size, the settling rate, and the stability of floc produced by aluminum or iron coagulants. Polyelectrolytes can be either synthetic or natural and are typically classified by their charge as either anionic, cationic, or nonionic.

To realize their full potential and to be effective, these coagulating chemicals and coagulant aids must be used in properly designed treating equipment. The equipment must provide rapid and thorough mixing of coagulant and coagulant aid with the water, subsequent slow stirring to afford particle growth, and a zone of quiescence to permit coagulated solids to separate from the water. These functions may be accomplished in separate steps by the use of a flash mixing tank, a flocculation tank, and a clarifier unit as shown schematically on Fig. 15-3; or they may be combined in a single unit, as in a solids contact unit. Figures 15-4 and 15-5 illustrate some of the different types of clarification units commonly used. For most power plant applications, the slurry recirculation type of unit is preferred for the treatment flexibility it offers, especially for softening applications.

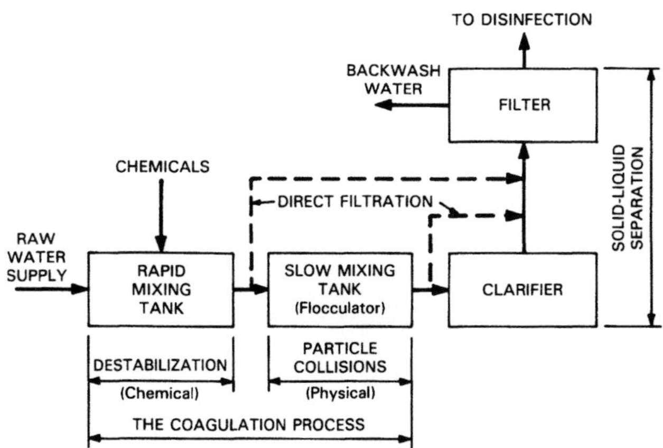

Fig. 15-3. Schematic diagram of coagulation process in a water treatment plant. (From Sanks, R. L. 1978. *Water Treatment Plant Design*. Ann Arbor Science Publishers, Inc., Ann Arbor, Michigan.)

Unlike other forms of chemical reactions where chemical dosage requirements can be calculated with considerable accuracy, coagulant requirements cannot always be accurately predicted because of the many variables in water characteristics such as kind and amount of suspended solids, color variations, and similar physical considerations. However, past experience with similar applications will usually serve to establish basic design concepts and equipment sizing. Laboratory tests ("jar tests") performed on each water supply under consideration are often beneficial to establish the most effective and economical coagulant dosages.

15.5.2 Aeration

When applied to water treatment, "aeration" is the mixing of air and water so gaseous substances are transferred into or out of the water. For example, oxygen is added to water by aeration for oxidation of the iron, manganese, and to a lesser extent, organic matter in the water. Aeration of water supersaturated with carbon dioxide, hydrogen sulfide, or other volatile compounds allows the concentration of these chemicals to be reduced to the equilibrium point.

Aeration can be accomplished by a variety of methods. The methods most commonly used for treatment of freshwater are waterfall aeration and spray aeration. Waterfall aeration is usually accomplished by directing water downflow countercurrent to air upflow. The efficiency of waterfall aeration is improved by causing the water to be spread into thin sheets or films, increasing the contact time and the area of water exposed per unit volume of air.

Spray aeration consists of spraying water through nozzles to effect contact with the air. Nozzle design and spray pressure are important parameters to maximize the surface area of the water available for air contact. Spray aeration is frequently used in conjunction with waterfall aeration.

A variation of the waterfall aerator employs a design that creates a jet action effect by discharging the raw water through multiple distribution orifices into an aspirating chamber, resulting in an excess of air being sucked into contact with the water.

Another method of aeration is the bubble or diffusion method, which can be accomplished by pumping air into the water through distribution devices such as perforated pipes or porous plates. This method of aeration offers the potential for high efficiency because bubbles of air rising through the water are continually exposed to fresh liquid surfaces, maximizing water surface per unit of air.

Mechanical aeration may also be used. One approach employs motor-driven impellers that stir the surface of the water being treated to increase the amount of water exposed to the atmosphere. Another variation of mechanical aeration is the use of a submerged mixer located directly above an air diffuser, which is typically located at the bottom of an aeration tank. Mechanical aerators have their greatest application in the treatment of wastewater.

Aeration as a treatment process should be evaluated critically to determine what can be accomplished and at what cost. The economics of alternative treatment programs to accomplish the same objective should be compared. In general, ground water sources are much more apt to be treatable by aeration than surface water sources.

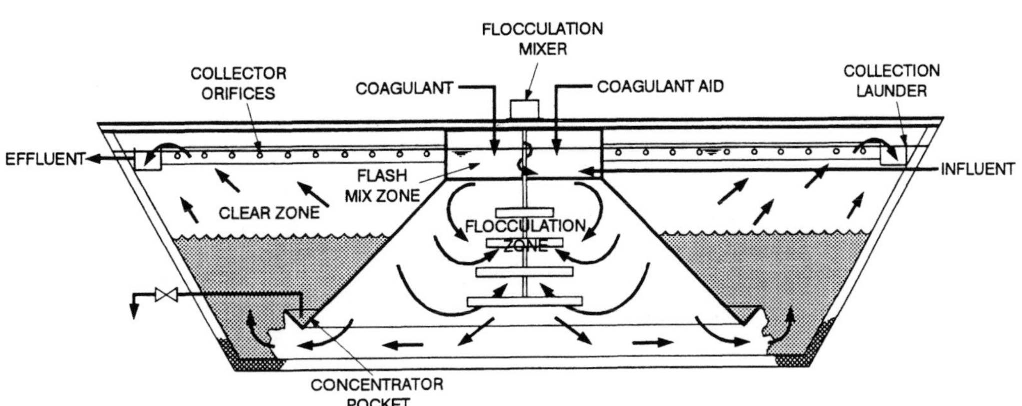

Fig. 15-4. Typical sludge blanket type solids contact unit.

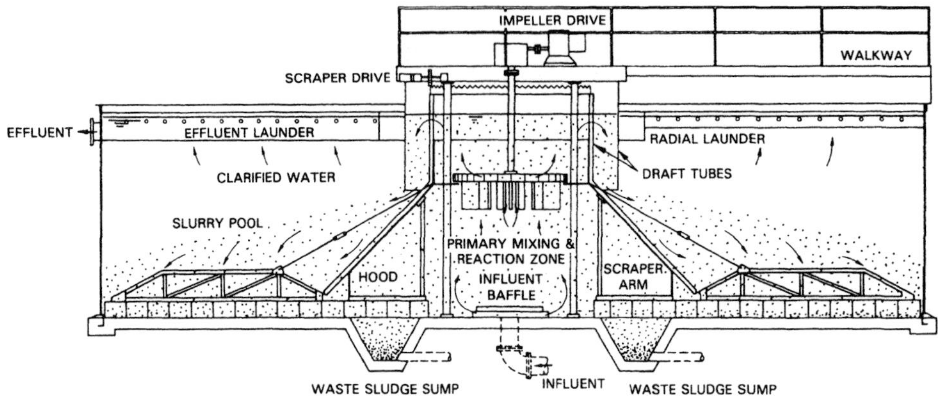

Fig. 15-5. Typical slurry recirculation type solids contact device. (From Infilco Degremont, Inc. Used with permission.)

15.5.3 Softening

Most Ca and Mg compounds form scale or sludge when water is heated or concentrated by evaporation. Because water is frequently used as a heat exchange medium, the importance of Ca and Mg removal for such water uses is obvious.

Softening of water is the process of reducing or removing the Ca and Mg content. This is accomplished by two basic methods—chemical treatment and ion exchange.

In chemical treatment softening, the reagents most commonly used are lime [in the form of quicklime, CaO, or hydrated lime, $Ca(OH)_2$], soda ash [Na_2CO_3], and to a lesser extent, sodium hydroxide [NaOH]. Ion exchange softening uses cation-type resin.

15.5.3.1 Lime Softening. The main purpose of lime softening is to reduce the Ca and Mg hardness associated with carbonate and bicarbonate alkalinity. In addition, lime softening removes any carbon dioxide dissolved in the raw water. These reductions can lower the total dissolved solids and the alkalinity of the treated water. The chemical reactions associated with lime softening are as follows:

$$CO_2 + Ca(OH)_2 \rightarrow CaCO_3\downarrow + H_2O \quad (15\text{-}12)$$

$$Ca(HCO_3)_2 + Ca(OH)_2 \rightarrow 2CaCO_3\downarrow + 2H_2O \quad (15\text{-}13)$$

$$Mg(HCO_3)_2 + Ca(OH)_2 \rightarrow CaCO_3\downarrow + MgCO_3 + 2H_2O \quad (15\text{-}14)$$

$$MgCO_3 + Ca(OH)_2 \rightarrow CaCO_3\downarrow + Mg(OH)_2\downarrow \quad (15\text{-}15)$$

15.5.3.2 Lime–Soda Ash Softening. When water has less M-alkalinity expressed as $CaCO_3$ than the sum of calcium and magnesium expressed as $CaCO_3$, the water contains noncarbonate hardness: calcium and/or magnesium ions associated with nonalkalinity anions such as chlorides, sulfates, or nitrates. Noncarbonate hardness is not reduced by lime softening; however, lime–soda ash softening reduces both carbonate and noncarbonate hardness. The following equations show the reactions associated with noncarbonate hardness reduction. Carbonate hardness reduction equations are as shown in the preceding section.

$$CaSO_4 + Na_2CO_3 \rightarrow CaCO_3\downarrow + Na_2SO_4 \quad (15\text{-}16)$$

$$CaCl_2 + Na_2CO_3 \rightarrow CaCO_3\downarrow + 2NaCl \quad (15\text{-}17)$$

$$MgSO_4 + Ca(OH)_2 + Na_2CO_3 \rightarrow Mg(OH)_2\downarrow + CaCO_3\downarrow + Na_2SO_4 \quad (15\text{-}18)$$

$$Mg(Cl)_2 + Ca(OH)_2 + Na_2CO_3 \rightarrow Mg(OH)_2\downarrow + CaCO_3\downarrow + 2NaCl \quad (15\text{-}19)$$

It is notable that although removal of noncarbonate hardness does reduce calcium and magnesium levels, the effect on total dissolved solids will vary based on the original water composition because sodium is being added in an amount equivalent to the noncarbonate hardness being removed.

15.5.3.3 Caustic Soda (NaOH) Softening. Caustic soda is a common name for sodium hydroxide (NaOH). Caustic soda softening is a variation of the combined lime–soda ash softening treatment process. The basic reactions are as follows:

$$Ca(HCO_3)_2 + 2NaOH \rightarrow CaCO_3\downarrow + Na_2CO_3 + 2H_2O \quad (15\text{-}20)$$

$$Mg(HCO_3)_2 + 4NaOH \rightarrow Mg(OH)_2\downarrow + 2Na_2CO_3 + 2H_2O \quad (15\text{-}21)$$

The sodium carbonate formed in the above reactions is available for further reaction for noncarbonate hardness removal as shown in Eqs. (15-16) through (15-19).

The choice between lime–soda ash softening versus caustic soda softening is usually based on differential chemical costs versus capital costs. The capital costs of the chemical storage and feed equipment for lime–soda ash softening are typically higher. However, it is not uncommon for the daily chemical costs of caustic soda softening to be three to four times higher than the daily chemical costs of lime–soda ash softening. The smaller the treatment flow rate, the smaller the chemical dosage necessary, the more likely that caustic soda softening is cost effective.

15.5.3.4 Sodium Cycle Ion Exchange. Sodium cycle ion exchange consists of passing raw water through a sodium ion exchange resin bed that has been regenerated with sodium chloride (NaCl). Sodium ion exchange resins are insoluble granular substances that have sodium radicals in their molecular structure. These resins exchange sodium ions for the calcium and magnesium ions in the solution in contact with them without apparent modification in their physical appearance and without deterioration. The ion exchange softening reactions are characterized by the following:

$$\underset{\text{soluble}}{\text{Ca[Salts]}} + \underset{\text{insoluble}}{\text{Na}_x\text{–[Resin]}} \rightarrow \underset{\text{soluble}}{\text{Na}_x\text{[Salts]}} + \underset{\text{insoluble}}{\text{Ca–[Resin]}} \tag{15-22}$$

$$\underset{\text{soluble}}{\text{Mg[Salts]}} + \underset{\text{insoluble}}{\text{Na}_x\text{–[Resin]}} \rightarrow \underset{\text{soluble}}{\text{Na}_x\text{[Salts]}} + \underset{\text{insoluble}}{\text{Mg–[Resin]}} \tag{15-23}$$

Essentially, all the hardness in the influent water is replaced with sodium ions. A hardness effluent of < 2 mg/L is common. To regenerate the ion exchange resin, an 8% to 10% solution of NaCl is passed over the resin, which converts the resin back to the sodium form as follows:

$$\text{Mg–[Resin]} + \text{Ca–[Resin]} + \text{NaCl} \rightarrow \text{Na}_x\text{–[Resin]} + \text{MgCl}_2 + \text{CaCl}_2 \tag{15-24}$$

15.5.4 Filtration

Water filtration is the process of separating suspended and colloidal impurities from the water by passage through a porous medium. A bed of granular filter material or media is used in most power plant applications. A number of mechanisms are involved in the filtration process. Some of these mechanisms are physical, and some are both chemical and physical.

A filter may be defined simply as a device consisting of a tank, a means of supporting a working filter bed within the tank, suitable filter media, and necessary piping, valving, and controls. Filters are designed for the following:

- Gravity flow, with natural head of water above the filter bed and low point of discharge at the filter bottom providing the pressure differential needed to move the water through the filter bed.
- Pressure units, which, as their name implies, are operated on line, under service pressure, filtering the water as it flows through the tank on its way to service or storage.

Gravity units are the simplest in design of the filters, with pressure filters being more sophisticated in engineering design and operation.

Filters are classified in a number of different ways. A common classification is by type or kind of filter medium to be employed, such as sand, anthracite coal, activated carbon, dual- or even multilayered media, diatomaceous earth, fabric, or porous membranes. Another distinction is direction of flow—upflow or downflow. Also, they are sometimes described hydraulically as the "slow rate" or "rapid rate" type.

Additional discussions in this section address filtration classified as granular media, activated carbon, cartridge, and ultrafiltration.

15.5.4.1 Granular Media Filtration. As the water to be treated is passed through a filter bed of granular media, the suspended solids are collected or retained in the voids within the media. The retained suspended solids decrease the void volume, which helps remove additional suspended solids but increases the pressure loss across the filter bed.

The selection of a granular media filtration system is normally determined by which filter type provides the desired performance at the least cost. Table 15-7 lists most of the basic variations available.

The quality of filter effluent water is a function of the filter medium type, size, and depth. In general, the finer the filter medium size, the better the water quality produced, but the head loss increases. The most commonly used granular filter media are silica sand and anthracite coal. Garnet sand is also used as a bottom layer in some mixed media filter designs.

When a single-filter medium such as sand or coal is backwashed, the bed becomes graded with the finest material on top and the coarser material at the bottom. For a downflow filter, this causes the influent water to contact the finest grained material first. Upflow operation allows the influent water to contact the coarsest grade material first, resulting in a more even loading of the entire filter bed and longer service runs.

To provide downflow filters a coarse-to-fine graded bed effect similar to that from upflow filters, dual- or multimedia beds are utilized. Dual-media filters usually consist of silica sand and anthracite coal media. The coal is selected so that the coarser coal material is still less dense than the silica sand. Therefore, after backwashing, the coarser coal medium remains on top of the sand medium. The addition of a layer of garnet sand, which is more dense than silica sand, gives an even more distinct coarse-to-fine flow path. The use of dual- and triple-media filters typically increases run lengths and produces better quality water than single-medium filters.

Filters may be designed for either manual or automatic operation. Automatic filters receive their signal for backwash from controllers actuated by pressure differential across the filter bed, by timers, or by turbidity breakthrough determined by a turbidimeter placed in the filter effluent line.

GRAVITY FILTERS. Gravity filters are usually employed for municipal applications and on some industrial applications

Table 15-7. Basic Granular Media Filter Variations

	Pressure Filters		Gravity Filters
	Upflow	Downflow	Downflow
Types of operation			
Automatic	X	X	X
Manual	X	X	X
Types of media			
Sand	X	X	X
Coal		X	X
Multimedia		X	X
Types of backwash			
Water Wash		X	X
Air–water wash	X	X	X

where large quantities of filtered water are required. Public health agencies tend to prefer gravity equipment since gravity filters cannot be overloaded as easily as can pressure filters, and gravity filters are open at the top so the operator can observe the filter during operation, including backwash.

Gravity filters normally use 8 to 12 ft (2.4 to 3.7 m) high tanks with definite preference for the higher shell heights of 10 to 12 ft (3.1 to 3.7 m) to prevent operation of any part of the filter bed under negative head. Negative head operation can have the disadvantage of releasing dissolved air in the form of bubbles that may accumulate in the filter bed and cause air binding.

Gravity filter tanks are constructed of steel or concrete. Concrete tanks are usually rectangular; steel tanks are round or rectangular. Figure 15-6 represents a typical rectangular gravity filter.

Conventional filter rates are usually 2 to 4 gpm/ft^2 of filter bed cross-sectional area (1.4 to 2.7 L/s per m^2) with 3 gpm/ft^2 (2.0 L/s per m^2) being a commonly selected criterion for power plant service. For potable water service, regulatory requirements may establish the filtration rate limits.

The end of gravity filter runs is usually determined by head loss or pressure loss developed across the filter bed, and backwashing is initiated on this basis. Water quality deterio-

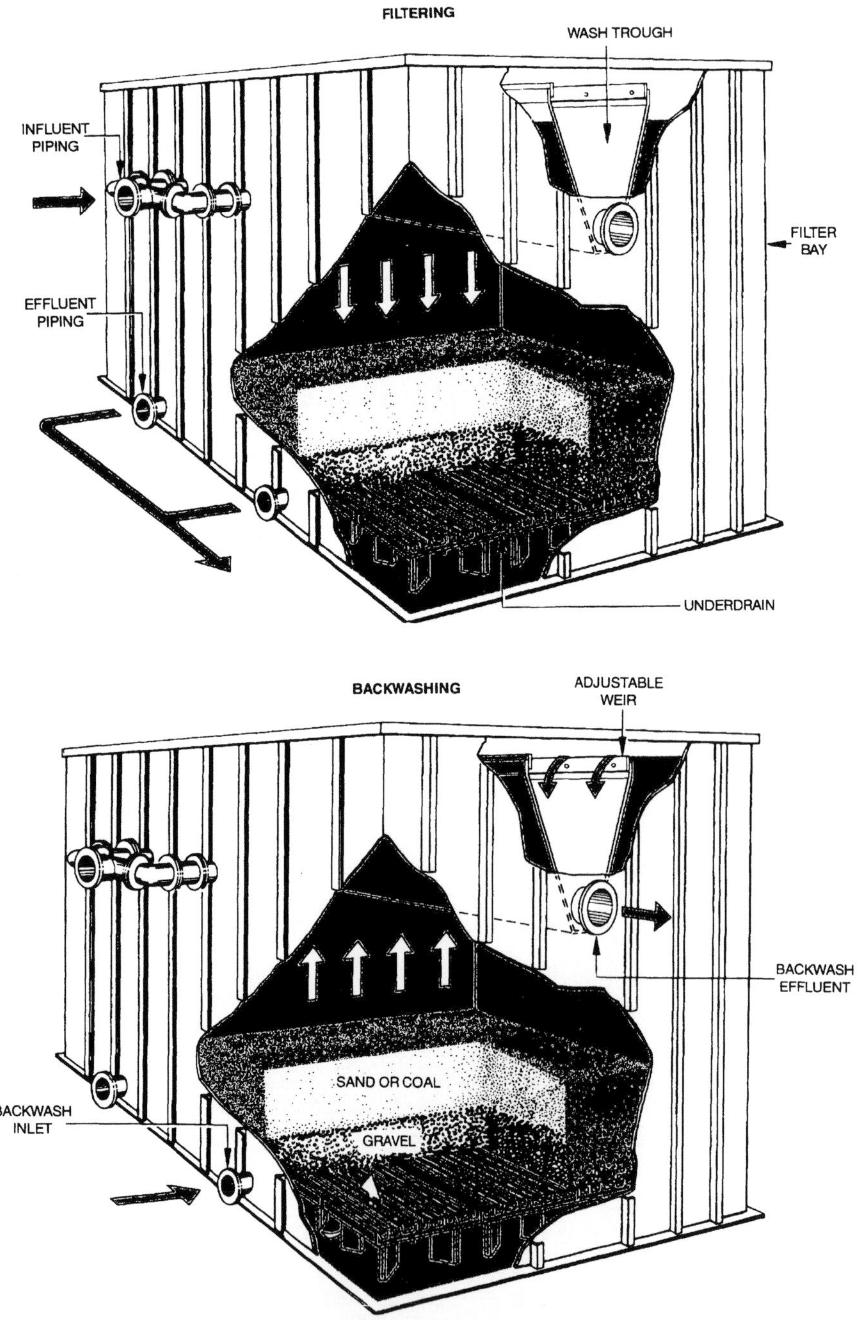

Fig. 15-6. Typical rectangular gravity filter. (From Infilco Degremont, Inc. Used with permission.)

ration (turbidity breakthrough) can also be used. When pressure loss is the determining criterion, it is usually recommended that filters be backwashed when the pressure loss builds up to about 8 to 10 ft of water (24 to 30 kPa) above the original pressure differential across the filter at the beginning of the run. Backwash should be adequate in quantity and time to ensure good bed cleansing.

PRESSURE FILTERS. Three basic types of pressure filters are vertical downflow, horizontal downflow, and vertical upflow. Each has its advantages and particular merits for general filtration purposes. Vertical and horizontal downflow pressure filter units are the "work horses" of the filtration field and typically provide reliable service.

In a vertical downflow filter, unfiltered water enters the top of the filter tank and flows downward through the bed of filter medium, wherein the suspended matter is removed. The filtered water is collected at the bottom of the tank and delivered to service. A typical vertical downflow pressure filter is shown in Fig. 15-7.

As the name implies, a horizontal downflow filter deploys the filter tank in a horizontal position. Otherwise, the operation of the filter is identical to that of the vertical type. By using the filter tank in a horizontal position, a larger bed area is obtained, thus increasing the flow rate available from a given tank size. The horizontal filters are normally 7 to 10 ft (2.1 to 3.1 m) in diameter and are 8 to 30 ft (2.4 to 9.1 m) long.

BACKWASHING. Proper filter backwashing is perhaps the most important part of good filter operating procedure, since good bed cleansing is required for continuously successful service cycles. Backwashing should expand the filter bed about 20% to 40%. The backwash flow rates employed to obtain this expansion vary with temperature; in the winter when water is colder and more viscous, lower flow rates lift the filter beds more effectively.

Surface washers, either of the rotary design or stationary type, are used in many plants. The jets, directed at high velocity down onto the surface of the filter medium, break up hardened mats and help prevent mud balls from building up. Rotary washers, to be suitably applied, require filter shapes that minimize space not reached by the jets. With stationary washers, this design consideration is of less concern.

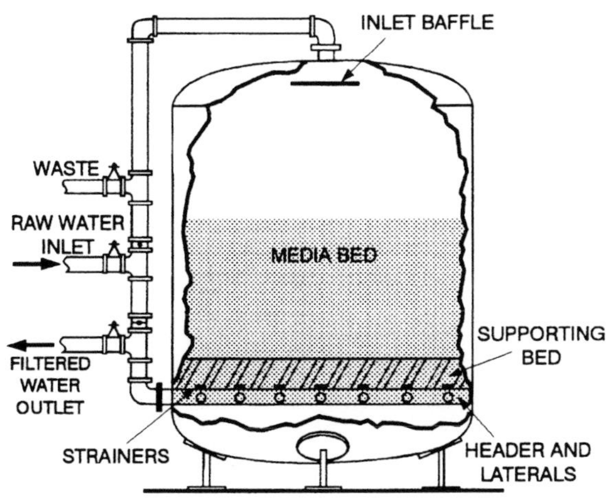

Fig. 15-7. Typical vertical downflow pressure filter.

Some manufacturers have employed the use of a separate air distribution manifold in the lower section of a filter tank to bubble air through a filter bed during the backwash and bed cleansing operation. The use of air with water is known as an air–water system. Air–water systems have been effective in improving operations at some plants. Air–water wash has been most effectively used where the quantity of precipitate is large, and good agitation of the bed is needed to dislodge coatings that have built up on the medium.

15.5.4.2 Activated Carbon. The use of activated carbon pressure filters is typically warranted when adsorption is needed to remove impurities such as chlorine, organics, hydrogen sulfide, or constituents causing tastes and odors.

Typical applications for activated carbon filter media for a power plant include potable water polishing filtration to improve taste and odor and polishing filtration in front of a cycle makeup treatment demineralization system to remove chlorine or organics. The importance of reducing chlorine and organics in the demineralization system influent water is that both these impurities degrade resin performance, reducing ion exchange capacity. Some reverse osmosis membranes are also sensitive to chlorine.

It is generally preferred to avoid the use of activated carbon media for roughing filtration. As the activated carbon medium surface and interior pores are coated with suspended solids, the medium's adsorption capability and, therefore, its ability to remove impurities, is reduced significantly. The activated carbon medium removal efficiency should be monitored to note any removal capability loss. As a general rule, the activated carbon medium should be replaced every 12 to 24 months.

15.5.4.3 Cartridge Filtration. Cartridge filtration typically uses filter elements or cartridges mounted in a pressure vessel. The filter cartridges are sheet or wound fiber material supported by screens or perforated plate made of stainless steel or plastic. Cartridge filters are usually restricted to polishing service because they foul quickly in roughing service and the replacement cost is high. A typical application is the essentially complete removal of suspended solids in cycle makeup demineralizer influent water that has already been filtered using granular media. Wound cotton or melt-blown polypropylene fiber cartridges are good choices for this service. Some of the newer cartridges are being developed with the fiber porosity graded large to small. This design offers longer service life than single-size fiber cartridges.

15.5.4.4 Ultrafiltration. Ultrafiltration (UF) is a filtration process where the water being filtered is passed through a molecular sieve-type membrane. UF membranes are very similar to reverse osmosis membranes. One notable difference is that reverse osmosis membranes reject the major portion of dissolved solids. UF membranes reject colloidal and high-molecular-weight dissolved organic solids such as humic and fulvic acids, viruses, and bacteria, but do not reject dissolved ionic constituents such as calcium, magnesium, sulfate, and chloride.

Reverse osmosis membrane materials have been used to develop UF membranes. Other polymeric film-forming substances have also been used. Membrane configurations are similar as well. Spiral wound, tubular, and hollow fiber UF systems are available.

Pretreatment requirements to reduce suspended solids loading to UF membranes are similar to those required for the reverse osmosis membranes.

15.5.5 Iron/Manganese Removal

Iron and manganese exist in nature in both the soluble (Fe^{2+} and Mn^{2+}) and relatively insoluble (Fe^{3+} and Mn^{4+}) oxidation states. The insoluble or precipitated iron and manganese are readily removed like other suspended solids by both clarification and filtration. Iron and manganese in either oxidation state can become chelated with organic matter, which makes removal more difficult.

The basic treatment methods used for iron and manganese removal are oxidation/filtration and chemical precipitation/ filtration, using combinations of the treatment processes described previously. The selection of the basic process and sequence is dependent on the concentrations and characteristics of the iron and manganese contaminants.

Ion exchange materials remove iron and manganese in the water being treated. The process is limited to handling waters from which air has been excluded so that oxidation does not occur. Some oxidized iron and manganese almost invariably precipitate, impairing the function of the ion exchange resin.

Specially processed greensand zeolite resin with a manganese dioxide coating can be used to remove iron and manganese by oxidation. In this process, the raw water is passed through special zeolites that are contained in a vessel and operate as a filter. The zeolites are regenerated with potassium permanganate, which may be fed continuously to the inlet water to keep the manganese–zeolite continuously regenerated. This process is limited by economics to applications where the water usage is small or where the iron and manganese contents are low but essentially complete removal is required.

15.5.6 Demineralization

Demineralization is the removal of dissolved ionic impurities that are present in water. Demineralized water is commonly produced by one or a combination of the following processes:

- Ion exchange
- Membrane desalination
- Thermal desalination.

The method selected to produce demineralized water depends on the quality of the influent water, the required quality of the effluent water, the availability of resources such as regenerant chemicals, and wastewater treatment and disposal requirements. The economics of the processes that produce acceptable effluent quality must be evaluated to determine the most cost-effective method for a specific application.

Makeup water quality recommendations developed by the Electric Power Research Institute (EPRI 1990) provide guidelines for determining the required quality of cycle makeup.

15.5.6.1 Ion Exchange Process. Ion exchange demineralization is one of the most important and widely applied processes for the production of high-purity water for power plant services, and it is accomplished using resins that exchange one ion for another. Cation resins are solid spherical beads with fixed negatively charged sites and exchangeable positively charged sites. Anion resins are solid spherical beads that have fixed positively charged sites and exchangeable negatively charged sites. In their regenerated state for demineralization applications, cation resins are in the hydrogen form and anion resins are in the hydroxide form. The reactions of the resin beads with the dissolved impurities in the water are represented by the following:

Cation resin: $R^-H^+ + C^+ \rightleftharpoons R^-C^+ + H^+$

$$2R^-H^+ + C^{2+} \rightleftharpoons R_2^-C^{2+} + 2H^+$$

Anion resin: $R^+OH^- + A^- \rightleftharpoons R^+A^- + OH^-$

$$2R^+OH^- + A^{2-} \rightleftharpoons R_2^+A^{2-} + 2OH^-$$

where

R = resin matrix and fixed charge site;

C = cations such as Ca^{2+}, Mg^{2+}, and Na^+; and

A = anions such as HCO_3^-, Cl^-, and SO_4^{-2}.

The hydrogen ions (H^+) displaced from the cation resin react with the hydroxide ions (OH^-) displaced from the anion resin. The net effect is that dissolved ions are removed from the water and replaced by pure water (H_2O).

The ion exchange resins are contained in ion exchange pressure vessels. The ion exchange resin in the vessels is referred to as the resin bed. This process of exchanging dissolved impurities is cyclic. When a resin bed site is exchanged with a dissolved ion, the site becomes "exhausted" and cannot remove other impurities without releasing an impurity. Exhausted resins must be regenerated to return the resin beads to the original hydrogen form for cations and hydroxide form for anions before further ion exchange can take place.

Cation resins are commonly regenerated with a strong acid solution of either sulfuric or hydrochloric acid. Sulfuric acid does not present the fuming problems associated with concentrated hydrochloric acid and is easier to handle (material selection). Consequently, sulfuric acid is frequently the recommended regenerant for cation resins. Anion resins are commonly regenerated with a sodium hydroxide solution. As can be seen from the regeneration reactions listed below, regeneration is the reverse reaction to the impurity exchange reactions.

Cation resin regeneration:

$$2R^-C^+ + H_2SO_4 \rightleftharpoons 2R^-H + C_2^+SO_4^{2-}$$

$$R_2^-C^{2+} + H_2SO_4 \rightleftharpoons 2R^-H^+ + C^{2+}SO_4^{2-}$$

Anion resin regeneration:

$$R^+A^- + NaOH \rightleftharpoons R^+OH^- + Na^+A^-$$

$$R_2^+A^{2-} + 2NaOH \rightleftharpoons 2R^+OH^- + Na_2^+A^{2-}$$

Dissolved gases such as oxygen or free carbon dioxide are usually removed by degasification. Oxygen is removed by vacuum degasification. Carbon dioxide can be removed by either vacuum degasification or forced draft degasification. A vacuum degasifier is a packed tower in which the water is sprayed and the gases are removed to a low level by maintaining a vacuum in the tower. Figure 15-8 is a schematic representation of a typical two-stage vacuum degasifier. A forced draft degasifier is a packed tower in which the water is sprayed down the column and the carbon dioxide is removed by air that is blown up the column. A typical forced draft degasifier is shown schematically in Fig. 15-9.

Determination of whether a degasifier is to be used and the type of degasifier used depends on the amount of CO_2 produced following cation exchange and the effluent water quality (dissolved oxygen content) required. For moderate to large concentrations of carbon dioxide, it is more economical to remove the CO_2 by degasification than by anion exchange.

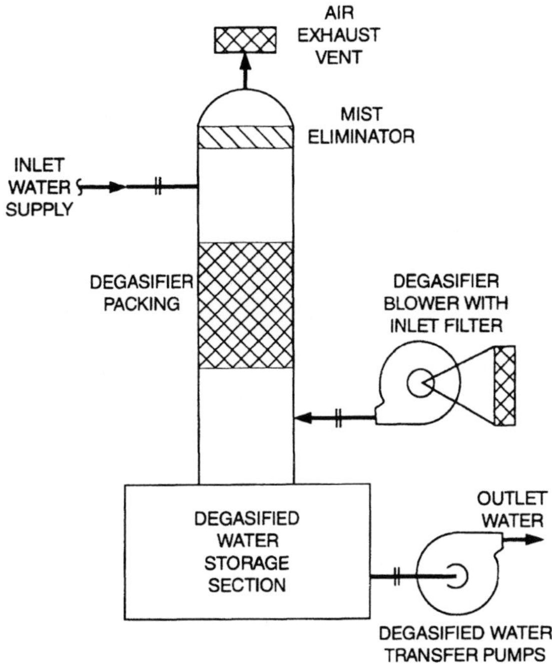

Fig. 15-9. Schematic of typical forced draft degasifier.

ION EXCHANGE RESINS. Cation resins are commercially available in strong-acid and weak-acid forms. Strong-acid cation resins can remove all cations in the influent water whereas weak-acid cation resins generally are restricted to the removal of hardness associated with carbonate alkalinity.

Anion resins are commercially available in strong-base

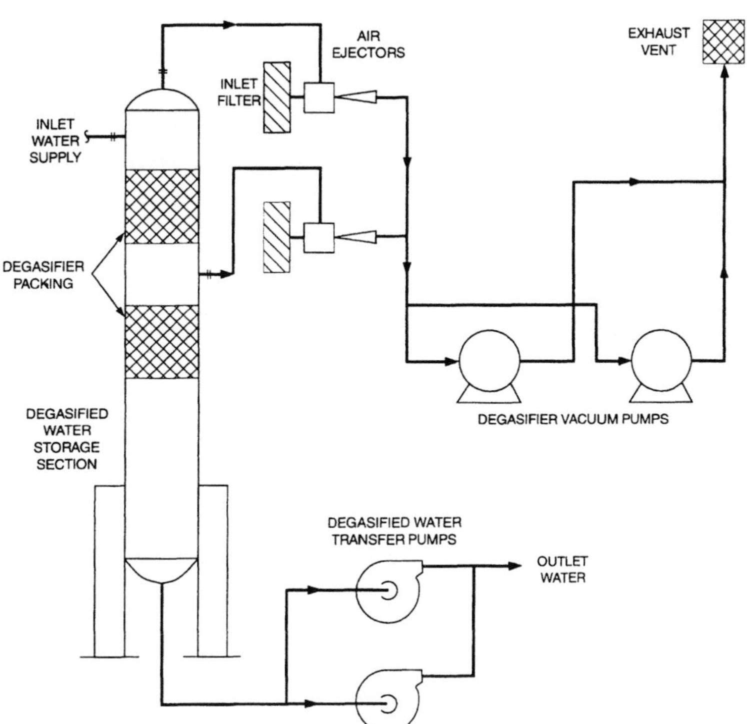

Fig. 15-8. Schematic of typical two-stage vacuum degasifier.

and weak-base forms. Strong-base anion resins are capable of removing both weakly dissociated and strongly dissociated acids. Weak-base anion resins primarily provide removal of the strongly dissociated acids (such as hydrochloric, sulfuric, and nitric) and have only limited capability for removal of weakly dissociated acids.

Cation and anion resins have different affinities for the different dissolved ions. Consequently, the cation and anion resins display a selectivity series for dissolved ions. Listed below is a representative selectivity series for commonly used strong-acid cation resins and strong-base anion resins with the ion having the greatest affinity for the resin at the top and the ion having the least affinity for the resin at the bottom of the list.

Selectivity of Strong-Acid Cation and Strong-Base Anion Resins

Cations	Anions
Calcium (Ca^{2+})	Sulfate (SO_4^{2-})
Magnesium (Mg^{2+})	Chloride (Cl^-)
Ammonium (NH_4^+)	Bicarbonate (HCO_3^-)
Potassium (K^+)	Silica (exchanges as $HSiO_3^-$)
Sodium (Na^+)	Hydroxide (OH^-)
Hydrogen (H^+)	

The relative selectivity of the ions determines the exchange chemistry of the resin bed. Figure 15-10 represents a cation and anion resin bed during a service cycle. As seen in the "partially exhausted" resin bed, since the affinity of the cation resin for the calcium ion is greater than for the other cations shown, the upper layer of cation resin will be occupied predominantly by calcium ions. Since magnesium is next in the selectivity series, the layer of resin below the calcium layer will be occupied predominantly by magnesium ions. The lowest layer will be occupied predominantly by sodium ions, for which the resin has the least affinity.

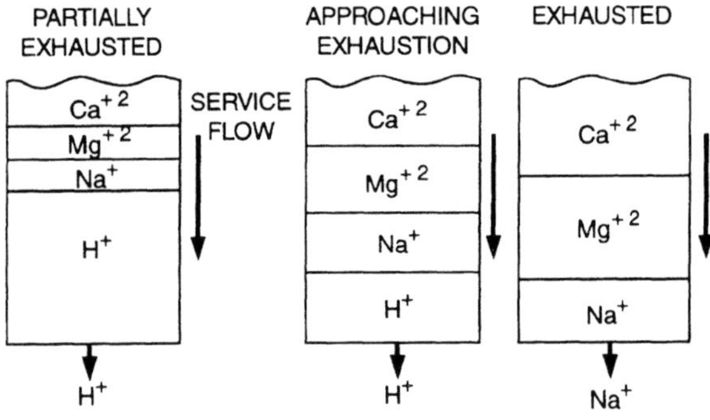

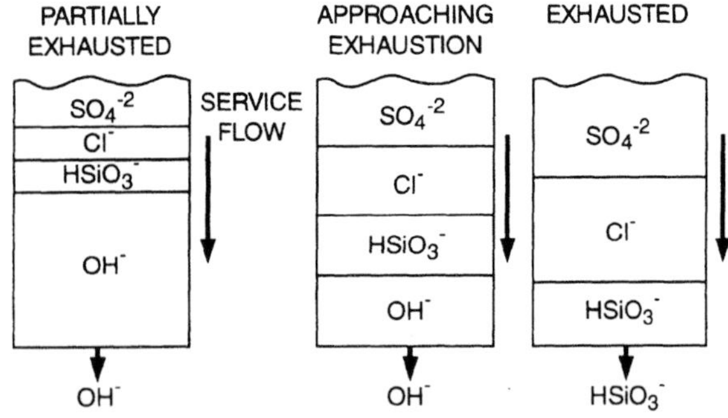

Fig. 15-10. Cation and anion resin bed during a service cycle.

As the service run continues, the incoming calcium, which has the greatest affinity, replaces the first available magnesium-held site. The displaced magnesium and the incoming magnesium displace sodium-held sites. The sodium ions move farther down the resin bed and are exchanged by the unexhausted hydrogen form resin. As illustrated in the "approaching exhaustion" view, the band of each ion grows in depth during the service run in proportion to the relative concentration of the ion in the influent water. In the "exhausted" view, all of the available hydrogen exchange sites of the resin have been exchanged and the resin bed must be regenerated. At the point of exhaustion, ions, primarily sodium, break through into the effluent as they are exchanged for an ion of higher affinity. This breakthrough of undesirable ions is called "leakage."

This same description of the service runs applies to the anion resin bed with the selectivity series determining the position of the layers of ions. Since bisilicate has the lowest affinity, it is the predominant ion to break through when an anion exchanger exhausts.

A wide variety of ion exchange resins have been designed, with different resin structures offering different advantages. For example, some resins have increased cross-linkage to improve resistance to oxidation and osmotic shock. Other resins have greater porosity to lessen irreversible organic fouling from waters containing organic materials. The type of resin selected must be determined based on the constituents in the water. Table 15-8 is a compilation of some of the resins available from various resin manufacturers illustrating differences in resin characteristics.

As a rule, water containing strong oxidizers, such as free chlorine, damages ion exchange resins. If these oxidizers are present in high concentrations, treatment is required to convert or remove the oxidizers prior to ion exchange.

The capacity of a resin is a rating of the ion removal capability of the resin and is reported in kilograins of ions per cubic foot (kilograms per cubic meter) of resin. The capacity that can be obtained from a resin is a function of the type of resin selected, the constituents in the influent water, the regenerant dosage [pounds of regenerant per cubic foot (kilograms per cubic meter) of resin], and for anion resins, the temperature of the regenerant solution when silica is to be regenerated from the bed. The regenerant dosage rate is a function of the leakage rate that can be tolerated from the ion exchange bed. Each resin manufacturer has characteristics curves for each resin product used to determine resin capacity and regenerant dosages based on a specific water analysis. Many water treatment design engineers use proprietary procedures that incorporate additional factors into the resin

Table 15-8. Comparison of Resin Types and Applications

Matrix Structure	Remarks	Dow (Dowex)	Mitsubishi (Diaion)	Purolite	Rohm & Haas (Amberlite)	Sybron (Ionac)
Strong-Acid Cation Resins						
Styrene-DVB 8% crosslinkage	Premium quality strongly acidic cation exchange resin. The quality of this resin is such that it can be used in standard operation as well as where there is a need for premium-grade gel cation exchange resin.	HCR-S HCR-W2	SK-1B	C-100	IR-120 plus IR-130	C-249 C-298
Styrene-DVB 10% crosslinkage	This is a sulfonic acid cation exchange resin, which gives greater resistance to oxidation than the 8% DVB-type cation resins.	HGR HGR-W2	SK 110	C-100-10	IR-122 IR-132	C-250 C-299
Styrene-DVB 20% crosslinkage	Superior physical stability and resistance to oxidation. Can be expected to provide at least three times greater resistance to oxidation than conventional gel-type cation-exchange resins.	—	—	C-150	IR-200	CFP-100
Macroporous DVB	Provides excellent osmotic shock and oxidative degradation resistance in water treatment applications, including hot process softeners.	MSC-1	PK-228	—	—	—
Weak-Acid Cation Resins						
Acrylic-DVB	Recommended for dealkalization of water for steam flooding or cooling tower applications as well as adjunct to strong-acid cation in deionization processes. Very high capacity.	CCR-2	WK-10	C-105	IRC-84	CC
Strong-Base Anion Resins						
Styrene-DVB	Standard crosslinkage; usually used for the treatment of waters that are essentially free of organic materials that might be irreversibly held by the resin.	SBR	SA 10A	A-600	IRA-400	ASB-1
Styrene-DVB	Chemically the same as the above resin but lower crosslinkage to give a better diffusion rate with large organic molecules.	SBR-P	SA-12A	A-400	IRA-402	ASB-1P
Styrene-DVB	Best organic fouling resistance. Recommended for highest purity water production in mixed beds.	MSA-1	PA-312	A-500	IRA-900	A-641
Acrylic-DVB	Differs from styrene-DVB resins in that it has an acrylic structure and is more hydrophilic, hence more resistant to organic fouling. It is recommended for treating water having a record of producing severe organic fouling problems for conventional strongly basic anion-exchange resins.	—	—	A-850	IRA-458	—
Intermediate or Weak-Base Anion Resins						
Acrylic-DVB	Has an unusually high capacity for large organic molecules. Used for deacidification, deionization, and desalination of water when the removal of strong mineral acids and sorption of organics is desired.	—	—	—	IRA-68	A-375
Styrene-DVB	Has excellent resistance to oxidation. Also has high exchange capacity and exceptional resistance to organic fouling. Often used because of the long operating life that can be expected.	MWA-1	WA-30	A-100	IRA-94	AFP-329
Epoxy amine	High capacity, high regenerant utilization efficiency. Stability to oxidation provides long-term low rinse requirements, making resin ideal in three bed systems. Good organic acid removal.	WGR-2	—	—	IRA-47	A-305

rating process to more accurately reflect plant operating conditions and experience. Operating experience has shown that use of these procedures often results in a more reliable long-term estimation of resin capacities, especially for some anion resins. However, the manufacturer's curves may be used to compare products and to develop preliminary design information.

Figures 15-11 and 15-12 show the manufacturer's performance curves for Rohm & Haas IR-120 strong-acid cation resin and IRA-402 strong-base anion resin, respectively, at selected regenerant dosages. Using Figs. 15-11 and 15-12, the manufacturer's ratings for ion exchange capacity and ion leakage rate can be calculated for these resins for a given water analysis and regenerant dosage. The method of determining the resin capacity using these curves is provided in Example 15-2.

Example 15-2

Ion exchange resin is needed to demineralize well water for steam cycle makeup. The following water analysis has been provided for the plant site:

Cations	Expressed as	Design analysis	Percentage of cations
Calcium	mg/L $CaCO_3$	86	31
Magnesium	mg/L $CaCO_3$	64	23
Sodium	mg/L $CaCO_3$	126	46

Anions	Expressed as	Design analysis	Percentage of anions
Bicarbonate	mg/L $CaCO_3$	98	34
Sulfate	mg/L $CaCO_3$	68	24
Chloride	mg/L $CaCO_3$	110	38
Silica	mg/L $CaCO_3$	10	4
pH	Standard units	7.6	—

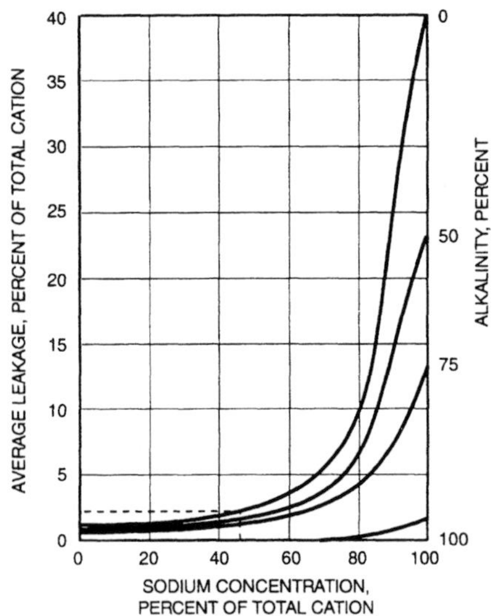

Fig. 15-11, 1 of 3. Amberlite IR-120 leakage data regeneration. (From Rohm & Haas. Used with permission.)

Determine the appropriate cation and anion resin ratings based on the above water analysis, regenerant dosages of 5 lb H_2SO_4/ft^3 (80 kg/m^3) of cation resin, 6 lb NaOH/ft^3 (96 kg/m^3) of anion resin, and Figs. 15-11 and 15-12.

Solution

In this example, the regenerant dosages are predetermined. Normally the ion leakage rate is the controlling factor and various regenerant dosages must be considered to satisfy the process requirements. Generally, the higher the regenerant dosage, the lower the average leakage rate.

With the use of Fig. 15-11, the average ion leakage can be determined based on the raw water quality. Use of the graph requires the following information which is available from the above water analysis:

Sodium concentration (percentage of total cations) = 46%.
Alkalinity (percentage of total anions) = 34%.

Locating these coordinates in Fig. 15-11 and reading to the left scale results in a leakage determination of approximately 2% of total cations, as shown by the broken line in Fig. 15-11, 1 of 3.

The following water quality information from the above analysis is used to determine the cation resin capacity:

Calcium = 31%
Magnesium = 23%
Sodium = 46%

Locating these coordinates on the triangle graph results in a capacity determination of approximately 13.9 Kgr/ft^3 (31.8 kg/m^3), as shown in Fig. 15-11, 2 of 3.

To correct the cation resin capacity for alkalinity impacts, the following information is applied to the third graph in Fig. 15-11:

Sodium concentration (percentage of total cations) = 46%
Alkalinity (percentage of total anions) = 34%

From the graph, the alkalinity correction factor is approximately 1.07.

$$\text{Cation resin capacity} = \left(\frac{13.9 \text{ Kgr}}{\text{ft}^3}\right) \times 1.07 = \frac{14.9 \text{ Kgr}}{\text{ft}^3}$$

$$\left(\frac{31.8 \text{ kg}}{\text{m}^3}\right) \times 1.07 = \frac{34.0 \text{ kg}}{\text{m}^3}$$

Figure 15-12 is used to determine the silica ion leakage and the capacity of the anion resin.

It is good practice to regenerate strong-base anion resins with warm water to aid in the elution of silica from the resin and reduce the resultant leakage rate during service runs. For this example, the regeneration temperature is predetermined to be 120° F (49° C).

From the above water analysis, the silica concentration is about 4% of the total anions. Locating the regeneration temperature and silica concentration on the first graph of Fig. 15-12 allows a silica leakage of less than 0.01 mg/L to be estimated.

The following water quality information from the above analysis is used to determine the anion resin capacity:

H_2SiO_3 = 4%
H_2CO_3 = 34%
TMA = 62%

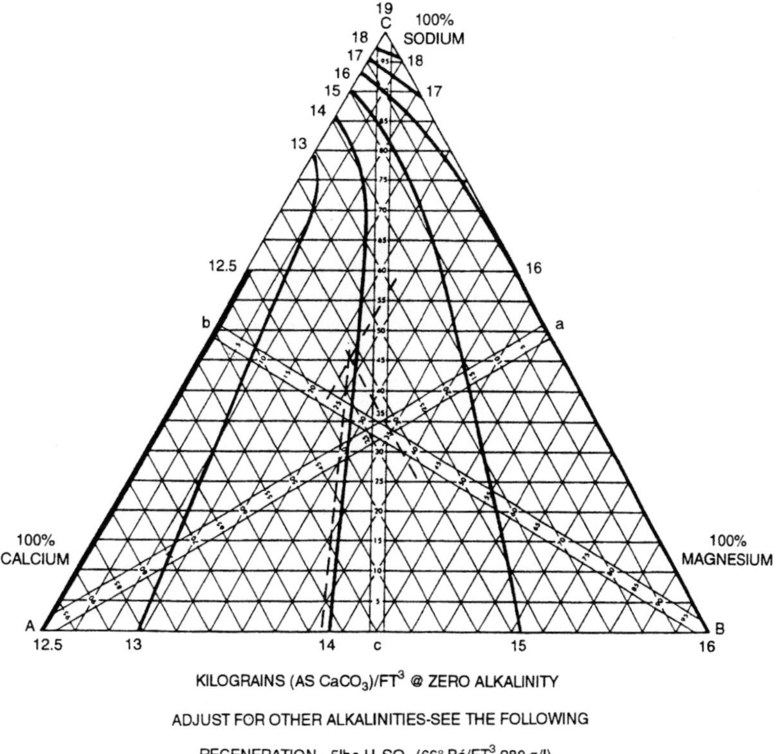

Fig. 15-11, 2 of 3. Amberlite IR-120 ISO capacity data. (From Rohm & Haas. Used with permission.)

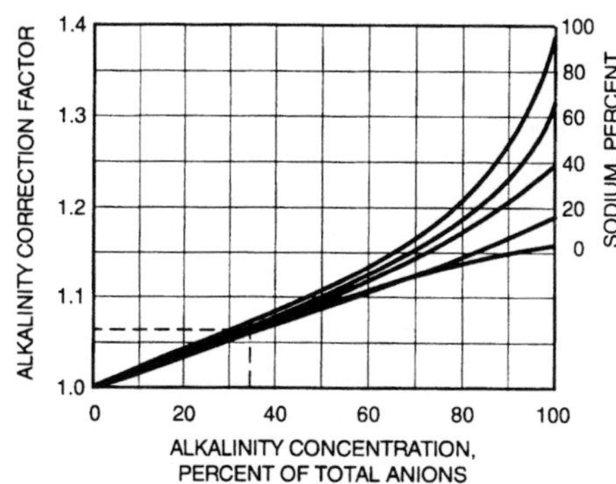

Fig. 15-11, 3 of 3. Amberlite IR-120 capacity correction for alkalinity. (From Rohm & Haas. Used with permission.)

Locating these coordinates on the triangle graph results in a capacity determination of approximately 14.3 Kgr/ft^3 (32.7 kg/m^3).

To correct the anion resin capacity for the concentration of chlorides, the following information is applied to the third graph in Fig. 15-12:

$$\text{Chloride/chloride + sulfate} = \frac{110}{178} = 0.62$$

From the third graph in Fig. 15-12, the correction factor = 0.96.

$$\text{Anion resin capacity} = \left(\frac{14.3 \text{ Kgr}}{\text{ft}^3}\right) \times 0.96 = \frac{13.7 \text{ Kgr}}{\text{ft}^3}$$

$$\left(\frac{32.7 \text{ kg}}{\text{m}^3}\right) \times 0.96 = \frac{31.4 \text{ kg}}{\text{m}^3}$$

If the required ion leakage rate cannot be achieved or can be achieved only by using excessive regenerant levels (lb of regenerant/ft^3 of resin), the required ion leakage is obtained by demineralizing the water in series. This is accomplished in "multibed" configurations, where a primary cation and anion exchanger are followed by a secondary cation and anion exchanger. The leakage from the primary cation and anion exchangers is removed by the secondary pair of ex-

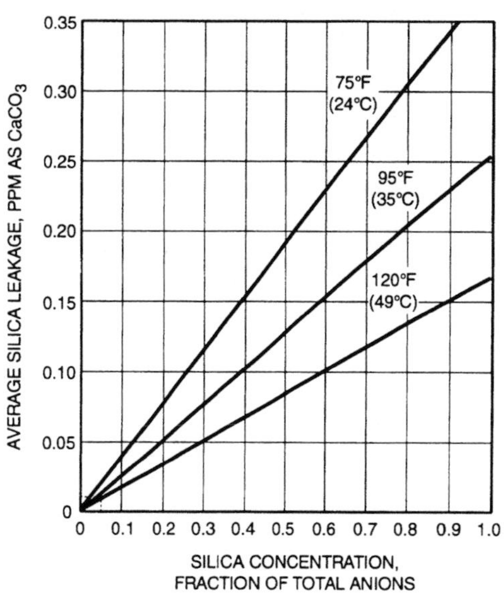

Fig. 15-12, 1 of 3. Amberlite IRA-402 capacity curves. Silica leakage at 6 lb of 100% NaOH/ft^3 (96 g/L). (From Rohm & Haas. Used with permission.)

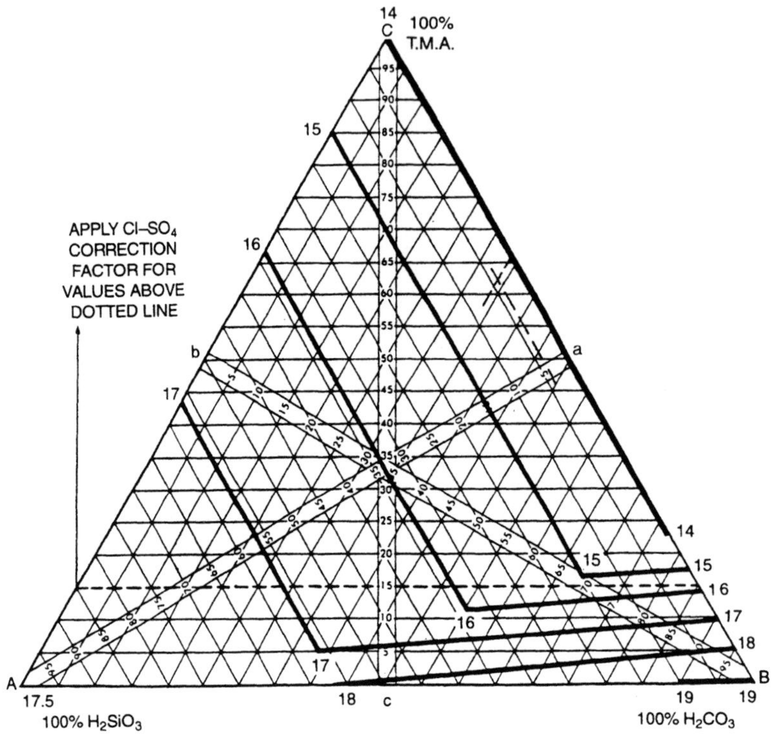

Fig. 15-12, 2 of 3. Amberlite IRA-402 capacity curves. (From Rohm & Haas. Used with permission.)

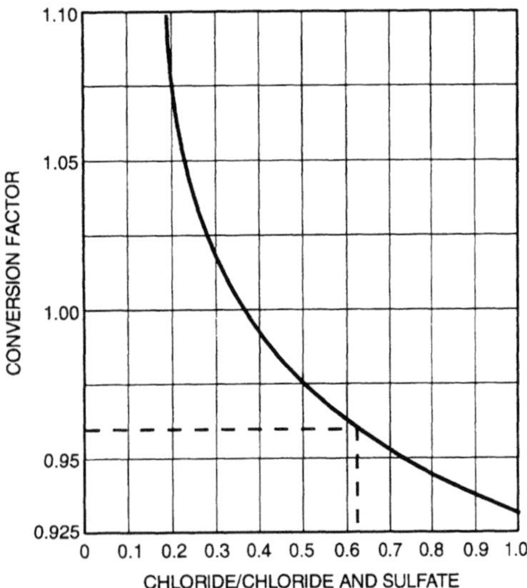

Fig. 15-12, 3 of 3. Amberlite IRA-402 chloride-sulfate correction factor curve. Use only in area above that designated on the amberlite IRA-402 capacity curve (Figure 15-12, 2 of 3). (From Rohm & Haas. Used with permission.)

changers. The number and type of exchanger vessels required to be operated in series is a function of the supply water ionic loading and the demineralization system effluent quality required.

ION EXCHANGE METHODS. Three primary methods are used to accomplish the ion exchange operation: fixed bed, fluidized bed, and continuous bed.

With the fixed bed method, the inlet solution to be treated flows through the vessel. The ion exchange resin is not moved during the exhaustion; therefore, the resin remains a compact (unexpanded) bed or column during the service run. Following exhaustion, the resins are regenerated.

With the fluidized bed method, the inlet solution to be treated flows upward in the vessel. The ion exchange resin bed is fluidized by the upward flow. The fluidized bed allows passage of suspended solids and results in less efficient contact. This process is used when suspended solids in the inlet solution are not removed. For high-purity cycle makeup treatment systems, the influent water is treated to ensure low suspended solids and fluidized beds are not employed.

The continuous bed method is similar to the fixed bed method in that the solution to be treated flows down and the resin bed is compacted. However, for the continuous method, a main vessel and a regeneration vessel are required. Small slugs of exhausted portions of the bed from the main vessel are removed to the regeneration vessel, and simultaneously, a slug of regenerated resin is returned to the main vessel. Although the resin slugs are transferred on an intermittent basis, the transfer is frequent and of short duration, so that the vessel service cycle is considered continuous. The continuous method is applicable for water treatment. However, compared to the fixed bed method, the continuous method is more complex, the capital costs for the control system are higher, and the ion exchange resin is subject to greater attrition or wear and tear because of the frequent resin transfers. The fixed bed is predominantly selected as the preferred method of ion exchange, and the discussions that follow are based on the fixed bed method.

When the resin in the fixed bed process becomes exhausted, it must be regenerated. Normally, the regeneration process is performed in the exchanger vessel; however, sluicing of the resin to a separate vessel(s) for regeneration is possible. Typically, "in-place" regeneration is employed with a demineralizer providing cycle makeup, and "external" regeneration is employed with condensate polishing systems. The two predominant in-place regeneration methods are cocurrent and countercurrent regeneration. With cocurrent regeneration, both the service (process) water and the regenerant solution are applied to the resin bed in a downflow direction. With countercurrent regeneration, the direction of the service (process) water flow is opposite to the direction of the regenerant solution flow. The service flow in a countercurrent regenerated vessel can be either downflow or upflow. Downflow service flow is the most commonly encountered technique, and for the sake of simplicity, is assumed in further discussions of countercurrent regenerated exchanger vessels in this chapter.

In theory, countercurrent operation offers benefits over cocurrent regeneration operation. Figure 15-13 illustrates an important difference between the ionic distribution of cocurrent and countercurrent cation resin beds. For both cocurrent and countercurrent regeneration processes, the distribution of ions upon bed exhaustion is similar. The difference lies in the regeneration. For cocurrent regeneration systems, hydrogen ions displace calcium, magnesium, and sodium ions from the top to the bottom of the resin bed. Complete removal of the ions is accomplished only if excessive levels of regenerant are used. In the service run, some of the residual sodium ions left at the bottom of the bed following regeneration are displaced by hydrogen ions produced in the upper bed during the exchange process. Consequently, some sodium leakage occurs.

After countercurrent regeneration, the residual ions are in the top of the bed, with the bottom of the bed being fully converted to hydrogen form. Consequently, there is little residual sodium present in the bottom of the bed to allow

COCURRENT REGENERATION

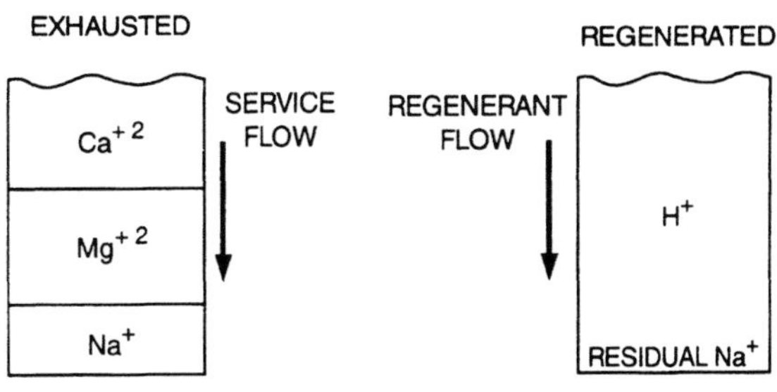

COUNTERCURRENT REGENERATION

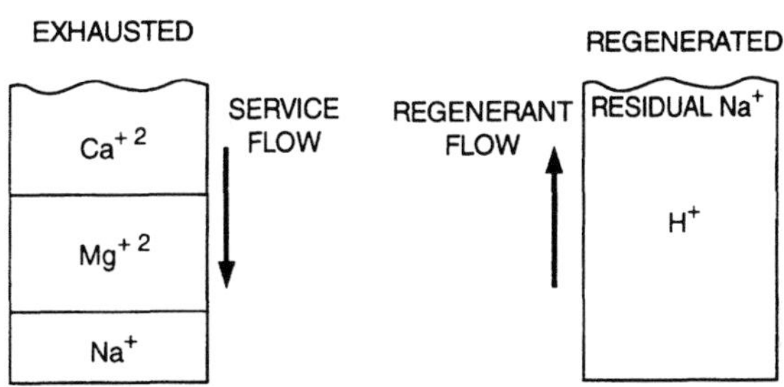

Fig. 15-13. Difference between typical ionic distribution for cocurrent and countercurrent regeneration processes (simplified for illustrative purposes).

sodium leakage to occur during the service cycle, which creates the potential for countercurrent regeneration to offer reductions in ion leakage at lower regenerant levels compared to cocurrent regeneration. In practice, however, the performance advantages of countercurrent regeneration depend on the quality of the slow rinse water and how well the resin bed is held in place during the upflow regeneration and rinse process. If the resin bed is not maintained as a packed bed during the regeneration process, channelling can occur, resulting in improper contact between the regenerant solution and the resin bed. Another problem with the resin bed becoming fluidized is that movement of the resins can occur within the bed; consequently, the resin at the bottom which is the most highly regenerated becomes mixed with resin from the top, thus negating the advantage of countercurrent regeneration. Several methods have been developed by equipment suppliers to minimize these problems.

The regeneration sequence for in-place cocurrent regeneration of either a cation or an anion exchanger is basically a four-step process consisting of backwashing, regenerant application, slow rinse, and fast rinse. Regeneration of a mixed bed exchanger follows essentially the same basic steps; however, the regenerant process is more complicated due to having both cation and anion resins in the same vessel.

The regeneration sequence for in-place countercurrent regeneration of either a cation or anion exchanger is basically a four-step process consisting of partial backwashing, regenerant application, slow rinse, and fast rinse. As discussed earlier, the successful operation of the countercurrent design is dependent on holding the resin bed in place during the upflow regenerant and slow rinse application. Several methods, as listed below, have been developed in the attempt to maintain a packed bed during the countercurrent regeneration of a downflow service vessel:

- Water block—blocking downward flow of water in an upflow regeneration cycle.
- Air block—blocking air flow to hold the bed in place during the regenerant and slow rinse steps.
- Inert granule packed column—inert granules fill the freeboard of the vessel, thus allowing no bed movement. With this design, to perform a full backwash when the pressure drop across the bed becomes excessive, the resin must be sluiced from the exchanger. This requires resin sluice equipment and a resin backwash vessel.
- Water bag—a water-filled rubber bag to hold the resin in place during regeneration.

As previously indicated, the cation resin can be regenerated using either sulfuric acid or hydrochloric acid. If sulfuric acid is used, precautions must be taken to prevent calcium sulfate precipitates from being trapped in the resin bed. The solubility of calcium sulfate is low; this, coupled with the presence of excess acid and the common sulfate anion, can result in calcium sulfate precipitation during regeneration. To minimize the effects of this precipitation on the resin bed, the acid solution flow rate must be sufficiently high to ensure that the spent regenerant is removed from the vessel while the precipitated calcium sulfate particles are still minute and can be readily removed from the bed. To minimize further the potential for trapping these precipitates in the resin bed, application of the dilute sulfuric acid regenerant solution is often designed such that concentration of the sulfuric acid solution progressively increases during regeneration. Typically, the sulfuric acid solution is applied in three steps: first at 2%, then at 4%, and finally at 6%.

The use of hydrochloric acid for regeneration avoids the calcium sulfate precipitation problem because the chlorides eluted from the bed are very soluble. However, hydrochloric acid is more expensive and the handling problems are much greater with hydrochloric acid. As a result, sulfuric acid is the more commonly used acid for cation regeneration.

The contact time for the cation regenerant step is typically 20 to 30 min. This contact time is important to ensure that the cation resin is exposed to the regenerant acid for enough time to allow the regenerating exchange process to approach completion.

For the anion resin regeneration, silica elution is difficult and causes the most frequent regeneration problems. If improperly regenerated, only a portion of the silica will be eluted during the regeneration, with an eventual silica accumulation in the resin that decreases the resin capacity and increases the silica leakage. Proper silica elution from strong-base anion resins can usually be achieved by applying the caustic solution at 120° F (49° C) and by providing long regenerant contact times. The contact time for the dilute caustic solution should be from 60 to 90 min and typical caustic solution strengths are 3% to 5%; consequently, the caustic solution flow rate is low.

Normally, cation resins and anion resins are contained in separate vessels. However, mixed bed exchanger vessels can be provided to contain a mixture of cation and anion resins. Mixed bed exchangers are generally limited to "polishing" applications downstream of a series of individual cation and anion exchangers. In this service, the mixed bed provides additional ion exchange of any ionic leakage from the upstream exchangers. The mixed bed exchangers generally are designed to produce high-quality effluent and are not designed for a large ion loading.

ION EXCHANGE EQUIPMENT. Ion exchange vessels are typically vertical cylindrical pressure vessels with either dished or elliptical heads. Ion exchanger vessels are constructed of carbon steel and are of welded construction.

The vessel interiors are lined for corrosion resistance. Several lining materials can be used, but natural rubber is frequently selected because of its long history of successful use. The internals for the ion exchange vessels are dependent on the design of the regeneration process. Except when hydrochloric acid is the regenerant, vessel internals of 316 stainless steel give good performance. Polyvinyl chloride (PVC) internals are also often supplied and have given satisfactory performance. If hydrochloric acid is used for regeneration, Hastelloy-B(TM) is a suitable construction material for internals and distributors.

To prevent channeling and ensure uniform contact of the inlet water and the regenerants with the resin bed, the design and placement of the distributors in the vessel are critical. For in-place cocurrent regeneration, an inlet distributor, regenerant distributor, and underdrain system are required. If the vessel is a mixed bed exchanger, a regenerant interface collector is also required.

Figure 15-14 shows a typical cocurrent regenerated demineralizer exchanger vessel. The inlet distributor at the top of the exchanger evenly distributes the inlet flow in the service mode and collects the backwash flow and distributes the fast rinse flow in the regeneration mode. The regenerant distributor is located slightly above the resin bed and evenly distributes the regenerant solution and the slow rinse water during regenerant application. The underdrain system collects the treated water flow in the service mode and distributes the backwash water and collects the regenerant solution, slow rinse water, and fast rinse water in the regeneration mode. For mixed bed exchangers, a regenerant interface collector is installed to collect the regenerant solution flows and the slow rinse flow instead of these flows being collected by the underdrain system.

For in-place countercurrent regeneration of a downflow service demineralizer exchanger, an inlet distributor, regenerant collector, and an underdrain system are required. The inlet distributor is located at the top of the exchanger, evenly distributes the inlet flow in the service mode, and collects the

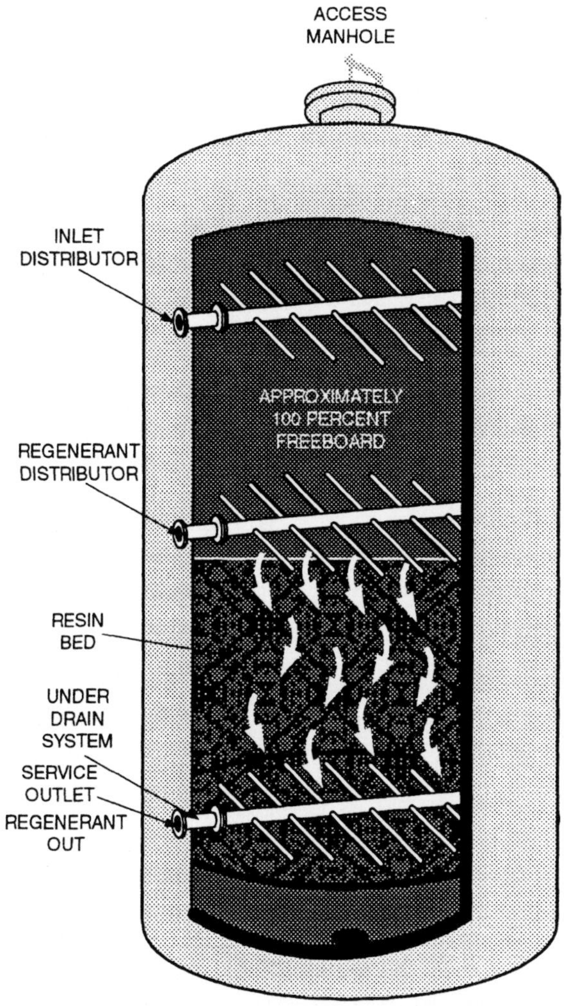

Fig. 15-14. Typical cocurrent regenerated demineralizer exchange vessel.

backwash flow and distributes the fast rinse flow in the regeneration mode. The regenerant collector is just below the top of the resin bed and collects the regenerant solution and the slow rinse water, and distributes the partial backwash water in the regeneration mode. The underdrain system collects the treated water flow in the service mode and distributes the regenerant solution, the slow rinse water, and the full backwash water (when required) in the regeneration mode.

Ion exchange vessels are sized based on both hydraulics and the quantity of resin required for a service run. Selection of the vessel dimensions is an iterative process. The minimum allowable diameter of the vessel can be determined based on hydraulic loading [gpm/ft^2 (L/sec $-$ m^2)]. The quantity of resin can be determined based on the ionic loading of the water, the capacity of the resin, the gross throughput required from a service run, and the volumetric loading [gpm/ft^3 (L/sec $-$ m^3)]. Once the volume of resin is determined, the height of the resin bed is calculated based on the vessel diameter, and must be greater than the minimum bed depth. The height of the exchanger is determined based on the design bed expansion (percentage of the resin bed depth) required for the resin. The required hydraulic loading, volumetric loading, minimum bed depth, and minimum bed expansion values should be selected based on the resin manufacturer's data.

Table 15-9 lists ranges for these design criteria, based on fixed bed exchangers regenerated cocurrently. Actual design should consider the specific resin manufacturer's recommended design criteria.

When sizing a demineralization system, the net capacity required for use must be considered. Since some of the demineralized water is actually used to regenerate the demineralizer, that portion of the production must be deducted from the gross water produced per regeneration to obtain the net production. An additional factor in system sizing is the service/regeneration cycle. The hydraulic capacity of the system must consider the regeneration period during which the exchanger is not producing demineralized water. System design criteria often include a 24-h cycle, with 3 to 6 h of the cycle in regeneration mode and the remainder of the cycle in service mode.

15.5.6.2 Membrane Desalination Process. Three membrane processes are commercially available for desalination of brackish water: reverse osmosis, electrodialysis, and electrodialysis reversal.

REVERSE OSMOSIS. As related to water solutions, osmosis occurs when two solutions of different ionic concentrations are separated by a semipermeable membrane. The membrane allows only water molecules to cross the membrane barrier. Water will flow from the lower concentration side to

Table 15-9. Guidelines for Sizing Ion-Exchange Vessels

Type of Ion Exchanger	Hydraulic Loading[a] (gpm/ft^2)	Volumetric Loading (gpm/ft^3)	Minimum Bed Depth[b] (ft)
Cation Exchanger			
Weak acid resin	5–10	2–4	3
Strong acid Resin	5–15	2–5	3
Anion Exchanger			
Weak base resin	5–10	2–4	3
Strong base Resin	5–15	2–5	3
Mixed bed exchanger	5–20	2–5	4[c]

[a]Excessive hydraulic loading can result in channeling through the resin bed, increasing leakage and consequently reducing service run lengths. The minimum hydraulic loading is a general guideline for economical design; however, many acceptable designs for high TDS waters will result in hydraulic loadings of less than 5 gpm/ft^2.

[b]The maximum bed depth preferred is 6 ft. The resin manufacturer should be consulted if bed depths in excess of 6 ft are considered.

[c]At least 2 ft each of cation and anion resin are required. If an inert resin is supplied, the inert resin volume is in addition to the 4-ft minimum cation and anion bed depth requirement.

Note: The bed depth expansion area, called vessel freeboard, allows the resin to expand during backwash. It is usually recommended that all exchangers be designed with 100% freeboard even though manufacturers may permit a lower bed expansion rate.

the higher concentration side until equilibrium is achieved. The resulting difference in the height of the liquid columns represents the osmotic pressure. Reverse osmosis occurs when sufficient pressure is applied to the higher concentration side to reverse the flow of water. Figure 15-15 provides a schematic representation of osmosis and reverse osmosis.

The osmotic pressure required to reverse the direction of water flow increases as the concentration of the brine increases. During the reverse osmosis process, water flows to the area of lower concentration and the high-concentration solution becomes more concentrated. To overcome the increased concentration, additional pressure is required to continue the reverse osmosis process. For example, normal osmotic pressure of seawater is about 385 psia (2,654 kPa), but to achieve reasonable product flow rates, about 765 psia (5,274 kPa) is required to achieve a 50% conversion. Additional pressure beyond 765 psia (5,274 kPa) will further increase the process flux (the amount of product discharged per unit of membrane area). In the reverse osmosis process, the dissolved ions concentrate on the high-pressure side of the membrane, and purified water passes through the membrane to create a product stream on the low-pressure side.

The semipermeable membranes are very sensitive to contaminants and impurities, and proper pretreatment of reverse osmosis feedwater is essential to prevent membrane fouling. A reverse osmosis system, therefore, usually consists of two parts: pretreatment equipment to filter and chemically condition the water, and a group of reverse osmosis modules to reduce the concentration of dissolved solids.

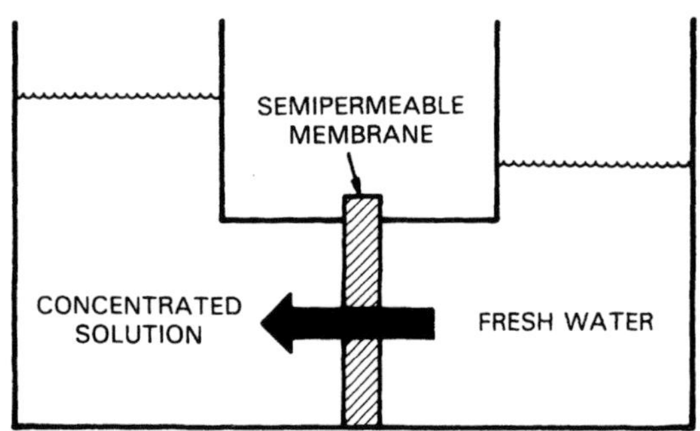

A. OSMOSIS — NORMAL FLOW FROM LOW-CONCENTRATION SOLUTION TO HIGH-CONCENTRATION SOLUTION.

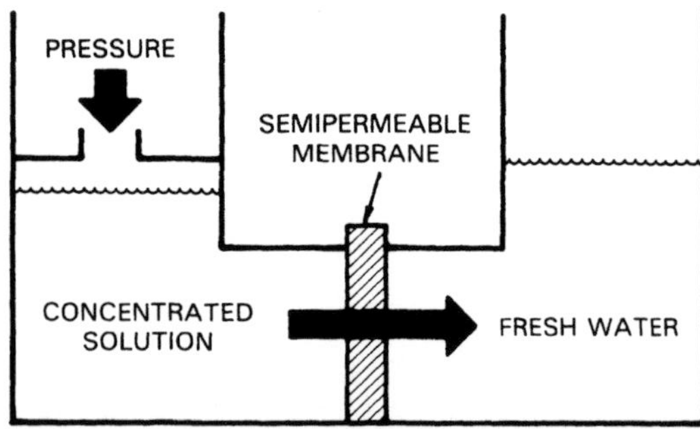

B. REVERSE OSMOSIS — FLOW REVERSED BY APPLICATION OF PRESSURE TO HIGH-CONCENTRATION SOLUTION.

Fig. 15-15. Theory of osmosis and reverse osmosis.

Typical dissolved solids rejection rates range from 90% to 98%, depending on such factors as the water temperature, ionic concentration, ionic distribution, and the system operating pressure.

The two basic types of reverse osmosis membrane materials are asymmetric and thin-film composite. Asymmetric membranes consist of a very thin, dense surface layer with a microporous substructure. The substructure is designed to provide support for the surface skin without impeding permeate flow. Cellulose acetate and aromatic polyamides are the most common asymmetric materials. Cellulose acetate membranes are susceptible to annealing and a reduction in flux if operated at higher temperatures. The membranes are also prone to hydrolysis at extreme pH but are relatively insensitive to chlorine. Aromatic polyamide membranes are more resistant to hydrolysis, but more sensitive to chlorine than cellulose acetate membranes. Both are subject to compaction.

Like asymmetric membranes, thin-film composite membranes consist of a very thin, dense surface layer with a microporous substructure. In the asymmetric membranes, these layers are created simultaneously out of the same polymer. In a composite membrane, these layers are produced separately, which increases both the flexibility and complexity of the membrane design and construction.

The best reverse osmosis membrane would offer high flux, high rejection rates, high chlorine resistance, high fouling resistance, and a strong, durable composition, as well as allow for a wide range of variation in operating temperature, pressures, and pH. Membrane advances continue to broaden the acceptable operating limits, prolong membrane life, lower installation and operating costs, and improve the recovery and rejection rates for reverse osmosis systems.

Proper pretreatment of reverse osmosis feedwater is essential in the prevention of membrane fouling. Membrane cleaning frequency serves as a general guide for evaluating the effectiveness of the pretreatment system. A cleaning frequency of more than once a month indicates inadequate pretreatment. Types of fouling that can be prevented or reduced by pretreatment include the following:

- Membrane scaling,
- Metal oxide fouling,
- Plugging,
- Colloidal fouling, and
- Biological fouling.

Once fouling occurs, the membranes must be cleaned. Table 15-10 lists a number of cleaning solutions. Thorough cleaning will remove most foulants; however, membranes must be cleaned before irreversible fouling occurs.

The four types of reverse osmosis element configurations are plate and frame, tubular, hollow fiber, and spiral wound.

Figure 15-16 illustrates a plate and frame module which

Table 15-10. Typical Chemical Cleaning Solutions

	Chemical				
Foulant	Acid	NaOH	NH_4OH	Phosphate Detergents	Sodium Bisulfite
$CaCO_3$	X	—	X	—	—
SO_4 scales	X	X	X	—	—
Silica	—	X	—	X	—
Metal oxides	X	—	X	—	—
Inorganic colloids	X	X	X	X	—
Biological	—	—	—	X	X
Organics	X	—	—	X	—

consists of a plate and frame assembly enclosed in a sealed pressure vessel. The plate and frame is one of the oldest reverse osmosis devices. Although it is conceptually simple, the high capital costs involved in construction contribute to making the plate and frame configuration most suitable for process applications characterized by low flow rates and high-value products.

Figure 15-17 shows a tubular membrane device. Also one of the earliest reverse osmosis configurations, tubular arrangements use a design much like a shell and tube heat exchanger, with the semipermeable membrane lining the tube walls. The membrane packing density is the lowest of the four configurations. By allowing turbulent flow down the tubes, the device offers a high resistance to fouling. For this reason, tubular devices have been used in several applications to process wastewaters with difficult suspended and colloidal solids.

The hollow fiber configuration, depicted in Fig. 15-18, consists of a bundle of porous hollow fibers. The external wall of each fiber is lined with a semipermeable membrane, with the fiber providing the necessary support for the membrane. The configuration is also similar to a shell and tube design. Feed water flows through the shell side and the system pressure drives water molecules through the membrane and into the fibers. The product stream is retrieved from the tube side, and the reject stream continues out the shell side. Hollow fiber devices offer the highest membrane packing densities of the four configurations. Hollow fiber applications require extensive pretreatment to remove all suspended and colloidal solids in the feedstream because the devices are particularly susceptible to fouling, and once fouled, are difficult to clean.

The spiral wound configuration consists of sheets of brine transport material, product transport material, and membrane material, sealed on three of four sides and rolled around a perforated tube. As the brine flows longitudinally down the element, the system pressure causes the water to pass through the membrane into the product transport layer and flow spirally toward the center of the element. As the product reaches the center, it flows out the unsealed side and through the tube perforations, allowing the product flow to exit the element out either end of the tube. The brine continues to flow longitudinally down the element through spacers that form the brine transport layer. A unique advantage of spiral wound systems is the ability to place several elements together in a single-pressure vessel. Spiral wound elements are

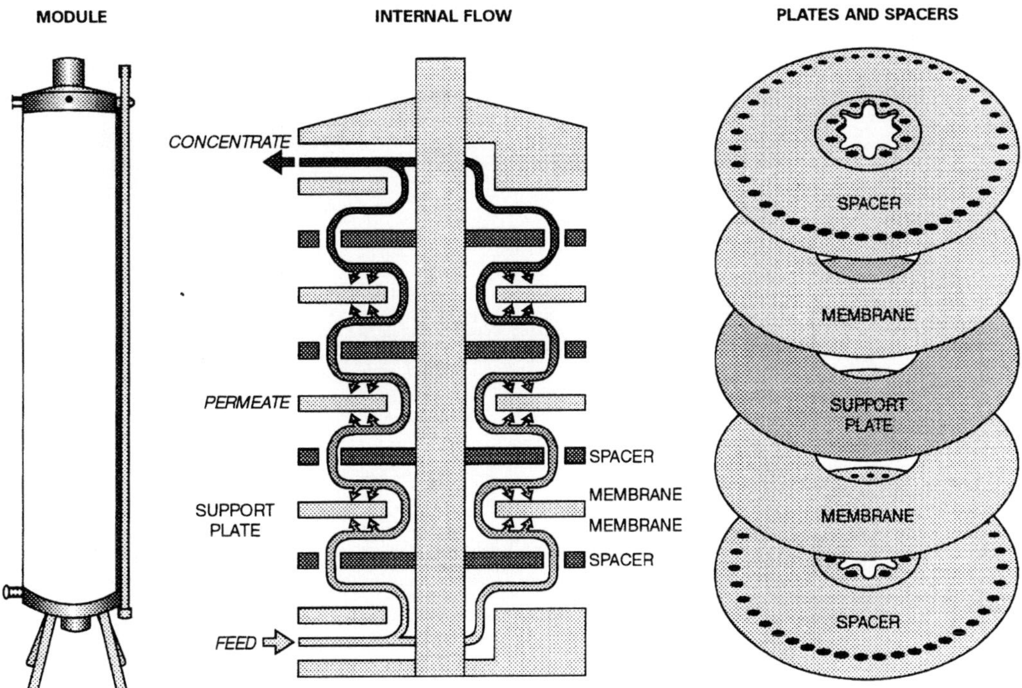

Fig. 15-16. RO plate and frame configuration.

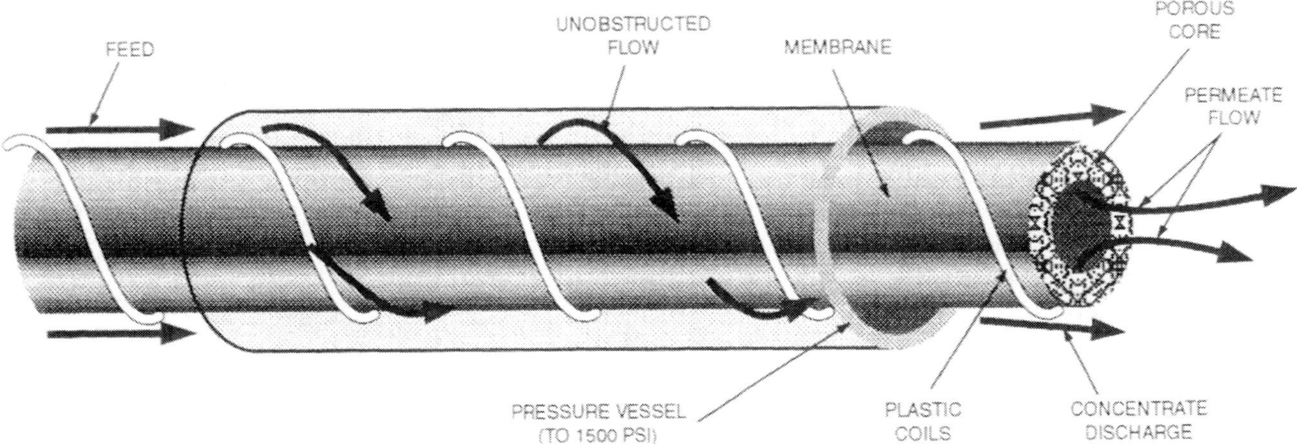

Fig. 15-17. Tubular RO configuration.

more resistant to fouling and are easier to clean than hollow fiber elements. Somewhat more space is required for spiral wound systems than for corresponding hollow fiber applications. A typical spiral wound system configuration is shown in Fig. 15-19.

Reverse osmosis has been used on its own or in combination with other treatment systems in many applications for effective, economical water treatment.

ELECTRODIALYSIS. Electrodialysis is a membrane process employed to selectively remove dissolved ionized acids, bases, and salts from one stream into another using ion-selective membranes in conjunction with an applied direct current electric field. Two types of thin membranes are used: anode-selective and cathode-selective. The anode-selective membrane is selectively permeable to anion and low-molecular-weight anionic organics. The cathode-selective membrane is selectively permeable to cations and low-molecular-weight cationic organics. The membranes are alternately placed in an electrodialyzer which is configured like a plate and frame heat exchanger.

An electrodialyzer is shown schematically in Fig. 15-20. The electrodialyzer consists of alternating series of cationic and anionic membranes separated by flow distribution gaskets. The electrodialyzer is bounded on one end by an anode compartment and anode electrode, and the other end

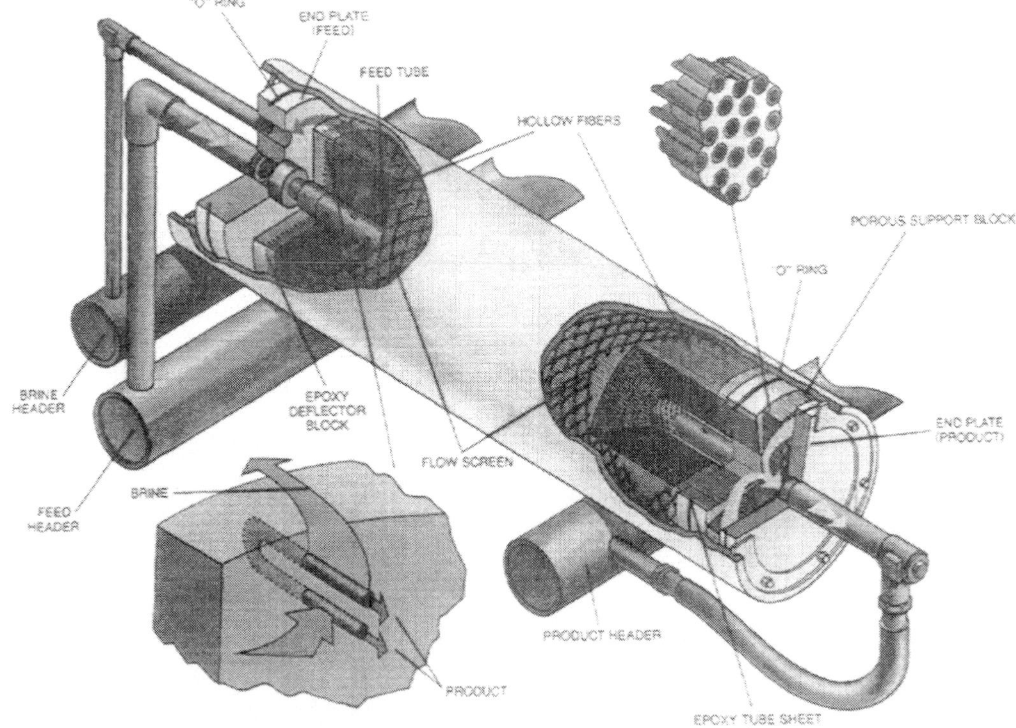

Fig. 15-18. Hollow fiber configuration. (From DuPont. Used with permission.)

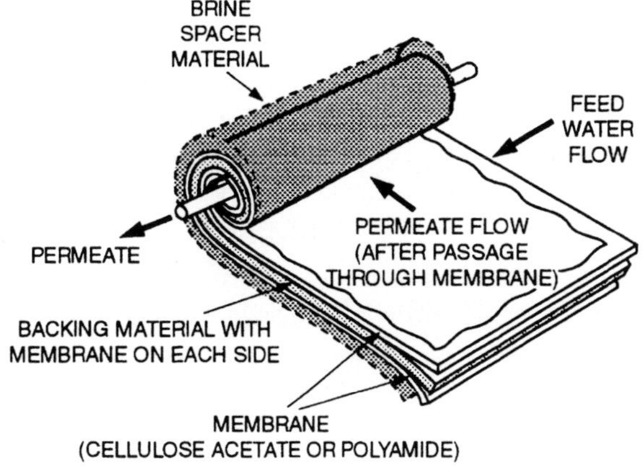

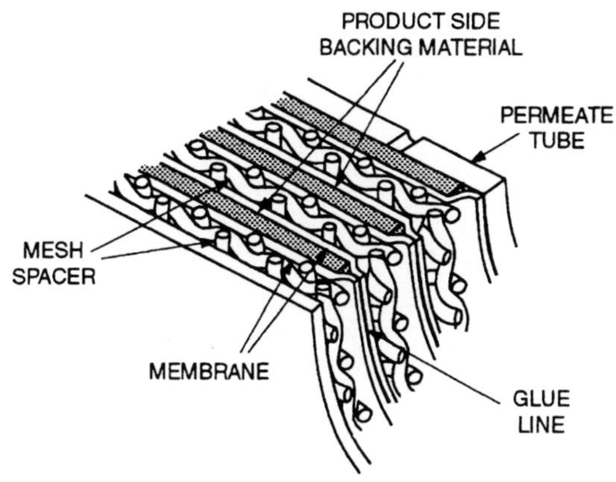

Fig. 15-19. Spiral wound element configuration. (From Fluid Systems Corporation, a member of Anglian Water Group. Used with permission.)

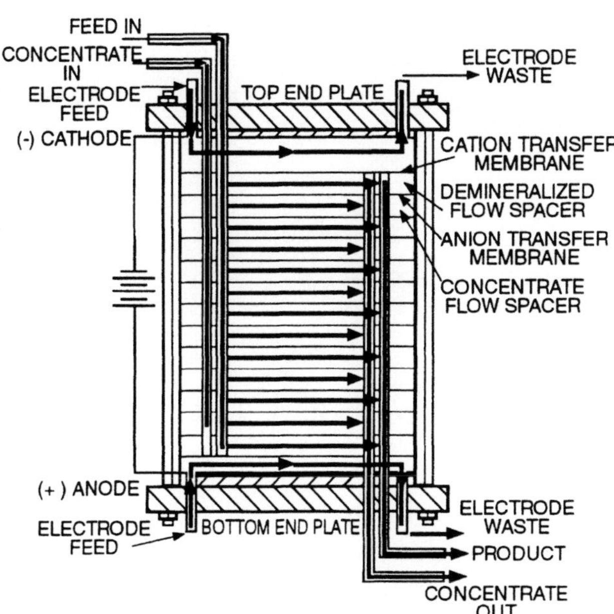

Fig. 15-20. Electrodialyzer flow schematic. (From Ionics, Inc. Used with permission.)

by a cathode compartment and cathode electrode. Alternating gaskets are hydraulically connected by flow distribution headers such that flow channels bound by these gaskets form alternately diluting or concentrating compartments.

By application of an electric current to the electrodes, feed solutions containing dissolved ions are alternately stripped or concentrated, depending on the location of the cationic and anionic permeable membranes relative to the anode and cathode.

The diluting makeup stream is fed to the diluting compartments of the electrodialyzer, and the concentrate makeup stream is fed to the concentrating compartments. The diluting stream becomes deionized, with the dissolved solids being moved to the concentrate stream.

The electric field in the electrodialysis process is unidirectional. As a result, the membranes in this process are subject to scaling and fouling from the precipitation of calcium salts. Pretreatment of the feedwater and the concentrate makeup streams is required to prevent membrane scaling and fouling.

ELECTRODIALYSIS REVERSAL. The electrodialysis reversal process is the same as the electrodialysis process and requires the same equipment, but the electrodialysis reversal process features an automatic periodic reversal of current and flow every 15 to 20 min. The reversal of the current causes a reversal in the direction of ion transport, which redissolves precipitated salts from the membrane surface. Consequently, the electrodialysis reversal process does not require the addition of chemicals to prevent membrane fouling and scaling.

COMPARISON OF MEMBRANE PROCESSES. The electrodialysis and electrodialysis reversal processes are applicable only for brackish water. Generally, waters with a TDS above 10,000 ppm are not candidates for either electrodialysis or electrodialysis reversal. These processes remove ionized salts from nonionized components but do not remove nonionized components such as organic substances, colloids, and silica. Power consumption by the electrodialyzers is about 5 kWh/1,000 gal (1.3 kWh/m^3) of product water per 1,000 ppm total dissolved solids (TDS) reduced. Typical pumping power requirements are about 3 kWh/1,000 gal (0.8 kWh/m^3) of demineralized water. Feedwater temperature and pressure limits for both processes are 115° F (46° C) and 50 psig (345 kPa), respectively. Typical dissolved solids removal efficiency of both electrodialysis processes is around 90% to 92% of the total dissolved solids in the feedwater. The effluent from these processes requires further treatment by ion exchange to meet power plant makeup water quality requirements.

The reverse osmosis process is applicable for both brackish water and seawater. The process removes both ionized and nonionized compounds. Power consumption ranges from 6 to 8 kWh/1,000 gal (1.6 to 2.1 kWh/m^3) of product water for brackish waters and from 35 to 40 kWh/1,000 gal (9.2 to 10.6 kWh/m^3) product water for seawater. Installation of hydroturbines or similar energy recovery equipment to recover the energy of the high-pressure reject water can reduce the energy consumption by up to 30%. Dissolved gases such as carbon dioxide permeate through the membrane; consequently, the reverse osmosis product water may require degasification. Typical dissolved solids removal efficiency of reverse osmosis membranes ranges from 95% to 98% of the total dissolved solids in the feedwater, depending on the type of membrane and the feedwater constituents. The effluent from the reverse osmosis process typically requires further treatment by ion exchange to meet power plant makeup water quality requirements.

Since the electrodialysis processes do not remove organics or silica and the reverse osmosis process does, reverse osmosis is generally the more frequently applied membrane process for upstream treatment of ion exchange equipment supply water.

REVERSE OSMOSIS DESALINATION. Reverse osmosis plant configurations vary with feedwater quality and desired product water recovery. A single-stage system is shown in Fig. 15-21. Feedwater passes through a cartridge filter, and a high-pressure pump boosts it to the desired pressure. The feedwater enters a manifold and is evenly distributed to all membranes operating in parallel. Deionized water is produced on the free side of the membrane and the resulting concentrated brine is rejected from the membranes for disposal.

Reverse osmosis systems can be brine staged to increase product recovery. With brine staging, brine from the first stage is collected and piped to the second-stage modules. More than two stages can be used in brine staging.

Reverse osmosis systems can also be product-staged to increase product water purity. With product staging, the product water from the first stage is collected and repumped to a second stage for additional deionization. The brine from the second stage can be passed back to the first stage to increase overall water recovery.

Selection of a system configuration is based on raw water quality and desired product water quality. Design of the system must be determined by economic evaluation. Design of system energy efficiency and water recovery rate must be optimized. The operation of the reverse osmosis feedwater high-pressure pumps accounts for the majority of the energy consumption of the process. Typically, these pumps are electric driven. However, if a source of steam is available, steam turbines can be coupled to the pump shafts to drive the pump. Reverse osmosis systems are custom designed for each installation.

15.5.6.3 Thermal Desalination Process. Thermal desalination is a process based on using heat to vaporize a portion of the fluid (brackish water or seawater) treated. The vapor is subsequently condensed as pure water. Basically, there are three commercially available technologies by which thermal desalination can be accomplished: multiple effect distillation, multistage flash evaporation, and vapor compression. The thermodynamic principles of operation for the processes are the same; the processes vary in the method and operating conditions used to accomplish the desalination.

The heat of vaporization of water varies with the liquid temperature. The relationship of water temperature to the heat of vaporization and the pressure at which water boils is shown in Fig. 15-22. At atmospheric pressure, water remains in the liquid state below 212° F (100° C); however, if the pressure is reduced, water vaporizes at a lower temperature.

The information in Fig. 15-22 is for pure water; adjustments to the boiling point curve must be made for dissolved solids content. Figure 15-23 shows the boiling point curve for water–sodium chloride solutions. Seawater averages 3.5% salt; therefore, at atmospheric conditions, seawater boils at 213° F (100.5° C). The salts in seawater do not volatilize at these low temperatures; consequently, only water vaporizes. The difference of 1° F (0.5° C) between pure water and seawater is called the boiling point rise. The boiling point rise must be used to adjust the boiling point curve provided in Fig. 15-22.

Evaporation processes rely on the physical state change of water to separate the water from the salt. All commercial thermal desalination processes operate under a vacuum to reduce the energy required for vaporization. The seawater

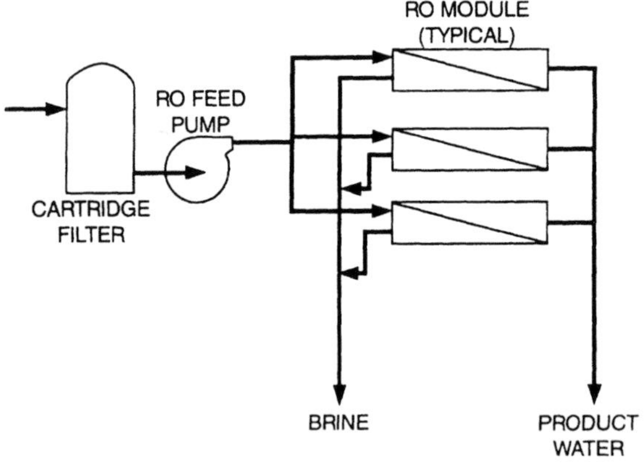

Fig. 15-21. Single-stage reverse osmosis system.

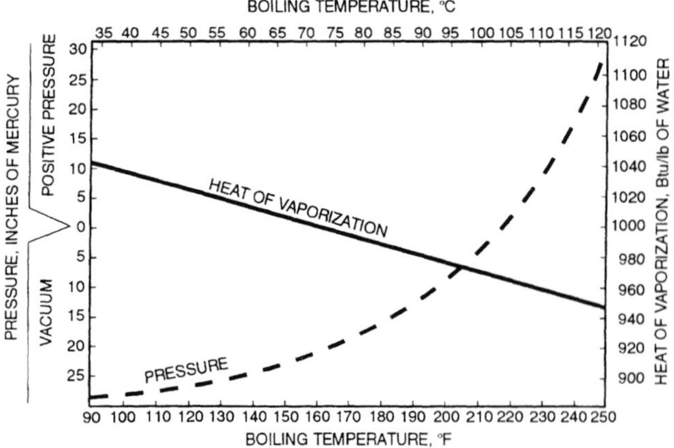

Fig. 15-22. Water boiling point versus heat of vaporization and pressure.

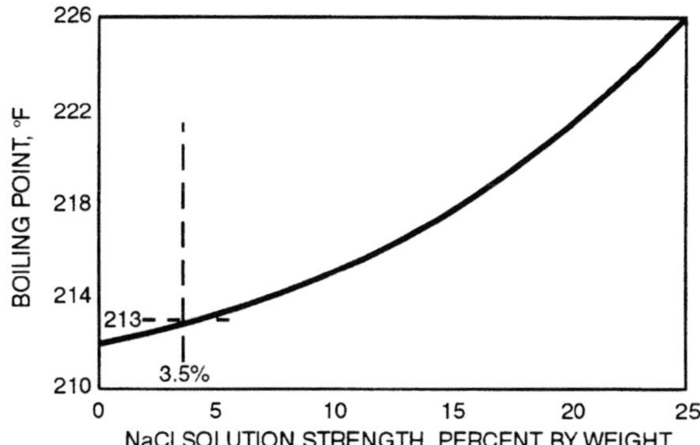

Fig. 15-23. NaCl concentration versus boiling point of water.

boiling point temperature is defined by the evaporator vacuum pressure. An inherent characteristic of evaporation is that the principal form of energy input is thermal energy, usually steam. The performance ratio of an evaporation process is the measure of the efficiency of heat use. The performance ratio is defined as the pounds of product water produced per 1,000 Btu of heat input.

THERMAL DESALINATION PRETREATMENT REQUIREMENTS. During evaporation, the salt content of the water being evaporated increases. As the salt content increases, the potential for scale formation also increases. Two types of scale are nonalkaline, or hard scale, and alkaline, or soft scale. For seawater, the nonalkaline scale that usually forms is calcium sulfate, which is difficult to remove. The most common alkaline scales formed are calcium carbonate and magnesium hydroxide, which can be removed chemically. Scale formation is also a function of solution temperature. As seawater temperature increases, the solubilities of both hard- and soft-scale-forming constituents decrease, resulting in increased scale formation potential. Figure 15-24 shows the relationship of hard scale to seawater temperature and concentration.

Soft scale formation can be controlled to some degree with chemicals. Calcium sulfate scale cannot be effectively controlled chemically. The chemicals do not eliminate scale, but do inhibit scale formation on the tube surfaces. Selection of operating conditions at points below the scale saturation point can also be used to minimize scale formation. Since neither chemical feed nor careful control of the operating point completely limits the formation of scale, periodic chemical cleaning of the heat transfer surfaces is required to maintain the design heat transfer coefficient.

MULTIPLE-EFFECT DISTILLATION. Multiple-effect distillation units consist of a heat rejection section (condenser) and heat recovery (evaporative) sections. Each heat recovery section is called an effect and is essentially a shell and tube heat exchanger. In each effect, energy is recycled to boil water. Multiple-effect distillation units receive heat from an outside source and utilize energy from the vapor produced in the preceding effect to vaporize water in the subsequent

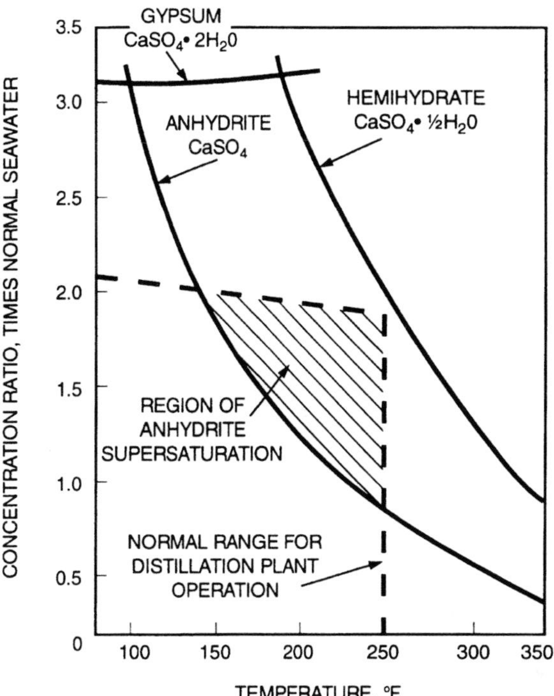

Fig. 15-24. Solubility of calcium sulfate in seawater.

effect. A thin film of feedwater is formed by spraying the feedwater over the heat exchange tubes. This device is called a falling film evaporator. The evaporator tubes can be arranged either horizontally or vertically. A flow schematic of a horizontal falling film multiple effect distillation process is shown in Fig. 15-25.

MULTISTAGE FLASH EVAPORATION. A flow schematic of a once-through multistage flash (MSF) evaporation process is shown in Fig. 15-26. Steam is used to heat the feedwater to the required operating temperature and to power a steam jet air ejector. The steam jet air ejector is an aspirating device that creates a vacuum within the evaporation chambers and removes any noncondensable gases that may collect in the system. Each stage in the MSF unit is operated at a different pressure and temperature.

The feedwater to the multistage flash process must be chemically treated for scale control to operate at high temperature. Because of the high cost of chemicals for scale control, a brine-recirculating MSF plant was developed and is shown in Fig. 15-27.

A multistage flash plant consists of a long rectangular-shaped metal vessel. The vessel is divided with vertical plates to form a series of individual stages. At the top of each stage is a tube bundle, beneath which is a product collection tray. The flashed steam condenses on the tube bundle and drops into the product collection tray that is sloped to the end of the stage, where the product water discharges into a collection trough that runs the length of the vessel. The collection trough is sealed at each stage to prevent vapor passage.

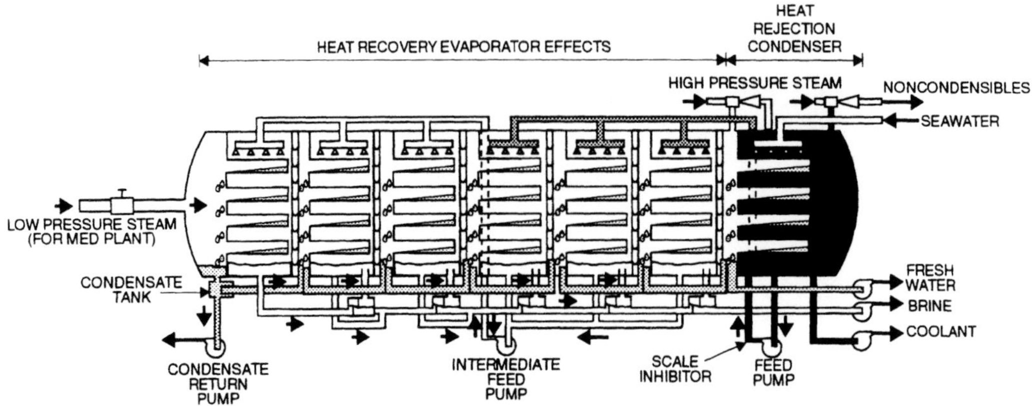

Fig. 15-25. Horizontal falling film multiple-effect distillation. (From IDE Technologies. Used with permission.)

MECHANICAL VAPOR COMPRESSION. The mechanical vapor compression (MVC) process operates on the "heat pump" principle and requires a tube bundle, a mechanical vapor compressor, an inlet feedwater heat exchanger, and a vacuum pump. Figure 15-28 shows a schematic of a MVC unit. The incoming feedwater is chemically pretreated for scale control and then passes through an inlet feedwater heat exchanger, where the heat in the discharged brine and product water streams is recovered. From the inlet heat exchanger, the feedwater is piped to an auxiliary condenser, where noncondensable gases are removed by means of a vacuum pump.

The feedwater is mixed with recirculated brine and sprayed on the outside of a bundle of tubes at a rate sufficient to create a thin continuous liquid film. The compressor provides, through its suction, a pressure lower than the equilibrium pressure of the brine. Consequently, part of the brine evaporates. The vapor passes through a mist eliminator and is compressed by the compressor and discharged to the inside of the tubes. There it condenses, supplying the latent heat required for the evaporation of the brine film flowing over the tubes.

The tube bundles may be arranged horizontally or vertically. Up to three stages can be supplied for operation with a single vapor compressor. For multiple-stage MVC units, the compressor withdraws the vapor from the last stage, compresses it, and delivers it to the first stage where it is condensed. The vapor from the first stage flows into the second stage and condenses, delivering its latent heat to the brine film flowing over the tubes in this stage.

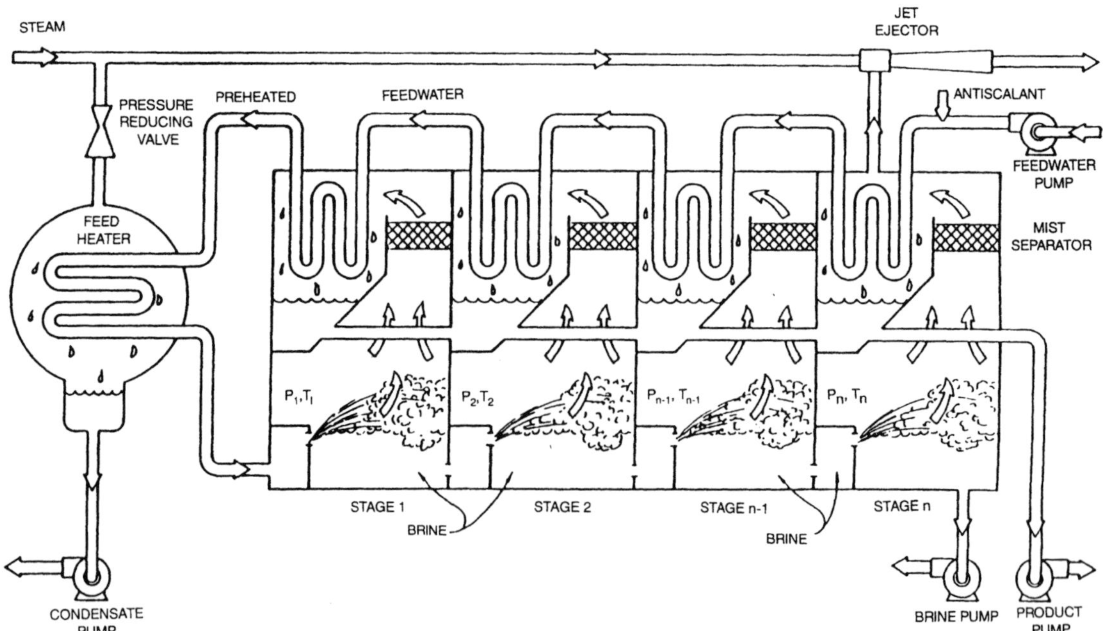

Fig. 15-26. Typical flow schematic of a once-through multiple stage flash evaporator (MSF) plant. (From Sasakura Engineering Company. Ltd. Used with permission.)

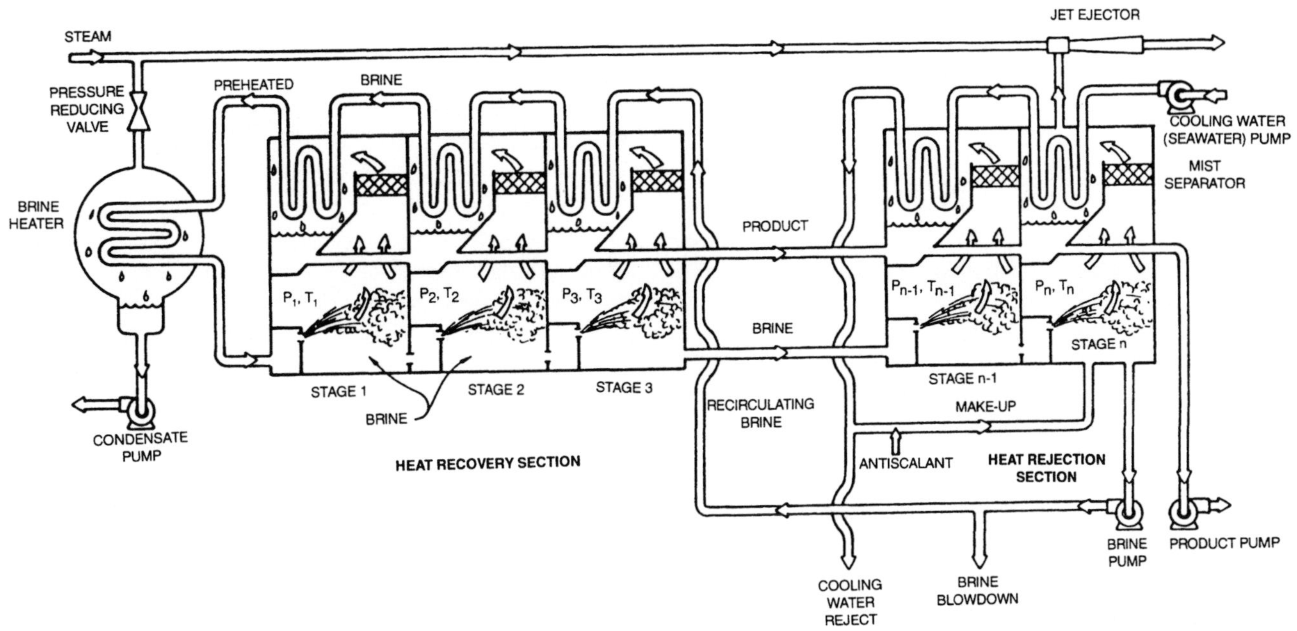

Fig. 15-27. Typical flow schematic of brine recirculating multiple-stage flash evaporator (MSF) plant. (From Sasakura Engineering Company, Ltd. Used with permission.)

The MVC production capacity is limited by vapor compressor design technology. The largest MVC units that have been installed are 0.5 million gal per day (0.02 m^3/s); consequently for large water production requirements, MVC is not considered because the number of units required is too great.

As stated earlier, the efficiency of thermal desalination processes is measured in terms of a performance ratio which is defined as the pounds of product water produced per 1,000 Btu of heat input. For a given application, the performance ratio can be chosen over a wide range, with low-efficiency

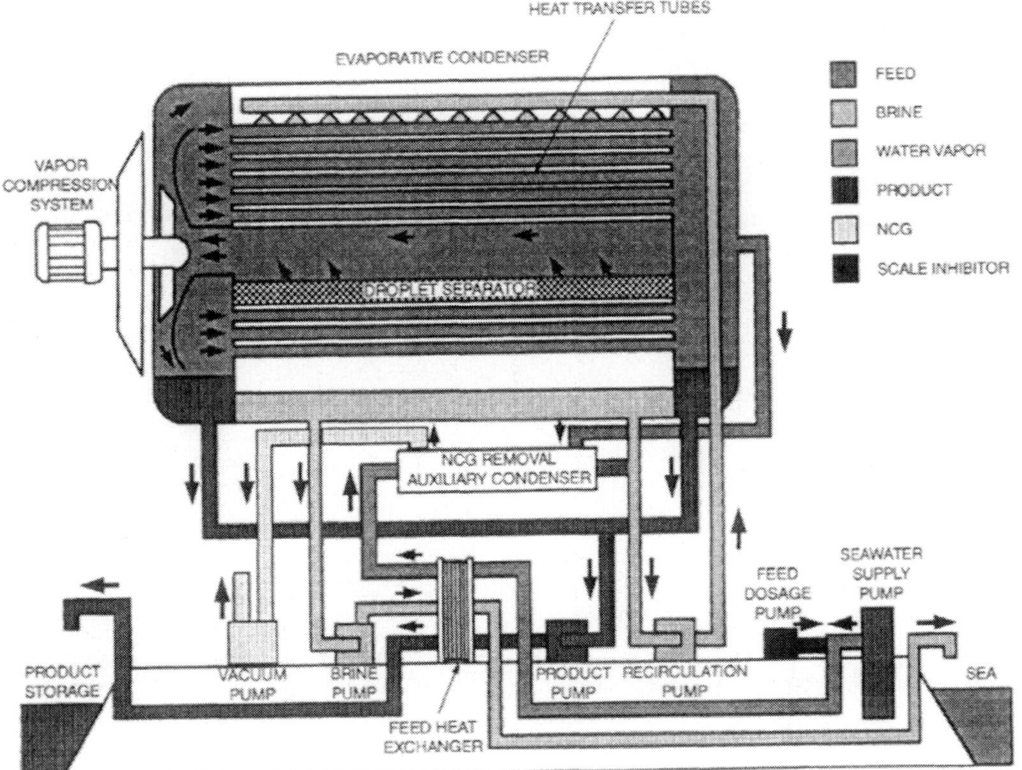

Fig. 15-28. Typical flow schematic of mechanical vapor compression (MVC) plant. (From IDE Technologies. Used with permission.)

plants at low capital cost and high-efficiency plants at high capital cost. An economic evaluation of feasible designs is required to determine the optimum design. Typical effluent quality from thermal desalination facilities is 20 mg/L of TDS or less.

Typically, thermal desalination is cost competitive with reverse osmosis only when seawater must be desalinated. For brackish water supplies, unless a very cheap source of steam is available, thermal desalination is not cost competitive.

15.6 TREATMENT APPLICATIONS

15.6.1 Circulating Water

The cooling water systems for the vast majority of power plants can be divided into two categories: once-through or open-recirculating. Once-through cooling water systems were common in the past, but most power plants designed today use the open-recirculating cooling water system, which uses a cooling tower that is open to the atmosphere. With either the once-through or the open-recirculating system, corrosion monitoring (corrosion coupons or in-line corrosion monitors) should be used to monitor the protection of the materials used within the circulating water system.

Water treatment process applications for the once-through systems are minimal and generally consist only of biocide injection for control of biofouling. Open-recirculating cooling systems with cooling towers also require biological control, but, in addition, they usually require pH adjustment and inhibitor feed for control of scaling. The most widely used treatment is to minimize scaling by feeding an alkaline cooling water scale inhibitor, to control alkalinity by feeding acid, and to limit the calcium concentration by blowdown. In some instances, makeup pretreatment is beneficial in providing a higher quality makeup resulting in reduction in blowdown and chemical feed requirements. Makeup to the open-recirculating cooling system is required to replace losses in the cooling tower resulting from evaporation, blowdown, and drift. Evaporation and drift can be estimated using the information provided in Section 15.4, Plant Water Requirements. The following discussions are based on an open-recirculating cooling system.

15.6.1.1 Circulating Water Quality Control Limits. Blowdown control is used to limit the concentration of chemical constituents in the circulating water system such that excess corrosion or damaging amounts of precipitates do not occur. Calcium carbonate, calcium sulfate, and magnesium silicate precipitates are of primary concern. Iron or manganese precipitates may also be formed if iron or manganese is present in sufficient concentration. These scales, particularly magnesium silicate scale, are difficult to remove once deposited on heat transfer surfaces.

The maximum allowable levels of these constituents vary somewhat with the chemical conditioning program being utilized at the power plant. Several computer programs have been developed by various entities, including chemical suppliers, consultants, and the Electric Power Research Institute (EPRI), to model the equilibrium chemistry present in a circulating water system. These computer tools can be helpful in predicting scaling tendencies when specific water quality conditions are defined. Additional sources of information used in predicting scaling tendencies of a particular water are available through the major water treatment chemical suppliers. In the absence of detailed design data, preliminary design estimates may be made or the following general guidelines used for circulating water limits for systems employing alkaline cooling water scale inhibitors:

Calcium as $CaCO_3$	900 mg/L
Silica as SiO_2	150 mg/L
Iron as Fe	10 mg/L
pH	7.5–8.3

The upper limit of the alkalinity range is typically 150 mg/L to 200 mg/L as $CaCO_3$.

If the magnesium and silica concentrations are controlled so that the following solubility product is not exceeded, magnesium silicate deposition is generally controlled (Drew Chemical Corporation 1984).

$$(\text{Mg, mg/L as } CaCO_3) \times (SiO_2\text{, mg/L as such}) < 35{,}000 \quad (15\text{-}25)$$

While the <35,000 solubility product guideline is useful, it is important to note that control of magnesium silicate deposition is sensitive to the circulating water pH. When the pH is at or below 7.5, the solubility product defined above could approach 100,000 with only limited concern for excessive magnesium silicate deposition.

Other chemical constituents in makeup water that commonly have an effect on circulating water system design and operation include TDS, chlorides, sulfates, and manganese. High TDS or especially high concentrations of chloride in circulating water affect the selection of construction materials for components in the circulating water system, such as condenser tubes. If high levels of sulfate are present in the makeup water, consideration should be given to including a limit on sulfate to avoid problems with calcium sulfate scaling or sulfate attack on concrete structures contacted by circulating water.

The levels of total suspended sediment (TSS) in the circulating water should also be considered. High levels of suspended sediment may result in solids settling out in the circulating water system. Settled solids increase the potential for fouling problems and contribute to bacteriological and corrosion problems. Settled solids also increase maintenance costs and reduce unit availability if the cooling tower must be removed from service periodically to allow the tower basin to be cleaned.

The maximum allowable cycles of concentration in the circulating water system are determined based on the limits discussed above and are directly affected by the makeup water analysis.

15.6.1.2 Biological Control. The condenser cooling water, whether once-through or open-recirculating with a cooling tower, usually contains organic matter and microorganisms which tend to foul condenser tubes and reduce the heat exchange efficiency of the condenser. Although other biocides are available, and in a limited number of cases a better choice, chlorine is the most common biocide used today. Chlorine is typically the most cost-effective biocide for treating circulating water. In some cases, where silty river water is used for once-through cooling, abrasion may alleviate the requirement for biocide treatment.

Shock chlorination, whereby chlorine is added periodically, is usually preferred to continuous chlorine feed. To be effective, several times as much chlorine per day would be required for continuous feed as for shock feed. Also, the federal EPA discharge regulations limit the time that discharge of any chlorine is allowed. Therefore, if once-through cooling is used or an open-recirculating system blows down to a receiving steam, continuous chlorination cannot be used without dechlorinating. Even with shock chlorination, depending on design particulars and regulatory restrictions, dechlorination may be required.

Chlorine is normally purchased as liquid sodium hypochlorite or as compressed liquid. Liquid sodium hypochlorite is fed to the circulating water system with feed pumps. Compressed liquid chlorine is typically vaporized and metered through a gas chlorinator, wetted with water, and diffused into the cooling tower basin close to the suction of the circulating water pumps.

Feed rate is usually 1 to 3 mg/L as chlorine for 15 to 30 min, two to three times a day. These are design figures, and the equipment should allow for adjustments in rate and time. Actual feed is adjusted to allow addition until a free available chlorine residual exists at the condenser outlet, assuming this keeps the cooling tower clean as well. There are situations where the chlorine feed rate must be increased to keep the tower itself functional.

15.6.1.3 Alkalinity Adjustment. As stated previously, an alkalinity limit is typically established for the water within the circulating water system. The alkalinity of the circulating water is reduced by feeding acid. Sulfuric acid is used in almost all cases. The following two equations show how alkalinity reacts with sulfuric acid.

$$2(HCO_3)^- + H_2SO_4 \rightarrow 2CO_2\uparrow + 2H_2O + SO_4^{2-} \qquad (15\text{-}26)$$

$$CO_3^{2-} + H_2SO_4 \rightarrow CO_2\uparrow + H_2O + SO_4^{2-} \qquad (15\text{-}27)$$

To determine the amount of alkalinity reduction required, three values are necessary:

A = circulating water alkalinity limit, mg/L as $CaCO_3$;

B = makeup water alkalinity before acid feed, mg/L as $CaCO_3$; and

C = cycles of concentration.

When A (circulating water alkalinity limit) is divided by C (cycles of concentration), the maximum allowable makeup water alkalinity is established. If A/C is equal to or greater than B (makeup water alkalinity before acid feed), no acid is needed. If A/C is less than B, sufficient acid to reduce B to A/C is required.

Example 15-3

Determine the amount of makeup water alkalinity reduction required given the following:

$$A = 150 \text{ mg/L as } CaCO_3$$

$$B = 70 \text{ mg/L as } CaCO_3$$

$$C = 5$$

Solution

Alkalinity reduction required

$$\text{mg/L as } CaCO_3 = B - \frac{A}{C} = 70 - \frac{150}{5} = 40 \text{ mg/L}$$

To determine the gallons per hour of 93% sulfuric acid to be fed for any particular application, multiply the gallons per minute of makeup times the desired alkalinity reduction expressed as mg/L as $CaCO_3$, times 3.47×10^{-5}. To determine the liters per hour of 93% sulfuric acid required, multiply the liters per second of makeup times the desired alkalinity reduction expressed as mg/L as $CaCO_3$, times 2.07×10^{-3}.

Example 15-4

Determine the amount of 93% sulfuric acid needed to reduce alkalinity by 40 mg/L, expressed as $CaCO_3$ from a circulating water makeup flowrate of 2,000 gpm (126 L/s).

Solution

$$\text{CWMU} \times \text{Alkalinity reduction} \times (3.47 \times 10^{-5}) = \text{gallons per hour (gph) acid feed rate}$$

$$2{,}000 \times 40 \times (3.47 \times 10{-}5) = 2.8 \text{ gph of 93\% sulfuric acid}$$

$$\text{CWMU} \times \text{Alkalinity reduction} \times (2.07 \times 10^{-3}) = \text{acid feed rate in liters per hour}$$

$$(126)\,(40)\,(2.07 \times 10^{-3}) = 10.5 \text{ L/h of 93\% sulfuric acid}$$

Metering pumps are used to feed acid to the circulating water system. Control of these pumps is typically based on circulating water makeup flow rate and circulating water pH. Because the ability to continuously monitor the circulating water alkalinity is limited, pH measurement is a convenient substitute. The correlation between circulating water alkalinity and pH should be checked periodically and the pH set point of the acid feed control adjusted as necessary.

The acid can be fed to the makeup line, the circulating water line, or the cooling tower basin. Care must be exercised wherever the concentrated acid is introduced into the circulating water system; it must be diluted rapidly regardless of the point of feed. Adequate mixing of the acid and water is necessary to prevent low pH extremes which can accelerate corrosion attack on most of the materials used in the circulating water system. Location of a mixing trough in the tower basin in an area where the makeup water line sweeps below the trough discharge point has been shown to be successful.

15.6.1.4 Circulating Water Chemical Feed. One of the earliest chemicals for treating open-recirculating cooling water systems was inorganic polyphosphate. This chemical is still used today in many operating systems. However, a problem with this approach is the potential reversion of polyphosphate to orthophosphate. Excessive orthophosphate can combine with calcium to form tricalcium phosphate scale.

Chromate was the next cooling water chemical used and was quite effective. However, increasing environmental concerns have caused the use of chromates to be drastically reduced.

Currently, the most common chemical treatment incorporates the use of an alkaline scale inhibitor. The pH of the operating system is raised above 7.5 and an organic phosphate scale inhibitor is fed to minimize the formation of calcium carbonate and other scale-forming precipitates. The scale inhibitor feed may also contain polymers for improved dispersion. Unlike polyphosphate, the organic phosphates do not readily revert to orthophosphate at normal cooling tower conditions. Blends of organic phosphates and copper corrosion inhibitors such as tolyltriazole should be considered for cooling water systems containing copper because some organic phosphates can be aggressive to copper and copper alloys. The type and treatment level of scale inhibitor are based on site-specific parameters and should be investigated thoroughly for the specific application.

15.6.1.5 Blowdown Control. The makeup water analysis, the desired circulating water analysis, the desired cycles of concentration, and the amount of heat dissipated at the cooling tower are the major factors in establishing the correct blowdown flowrate. Accordingly, typical blowdown valve control is proportional to unit load and biased with circulating water conductivity. The proportional multiplier and the conductivity set point should be monitored periodically and adjusted, as necessary, to prevent excessive scale or corrosion within the circulating water system and at the same time minimize the quantity of blowdown.

15.6.1.6 Makeup Treatment. If the water supply is unable to satisfy either the quantity or quality requirements for makeup water, pretreatment should be considered. In addition, pretreatment of the makeup source is often justified by the capital and operating cost reductions associated with the following benefits provided by makeup pretreatment:

- Decreased water demand,
- Decreased blowdown and therefore decreased capacity of wastewater treatment before discharge,
- Decreased circulating water chemical usage, and
- Improved maintainability and availability.

A number of treatment processes are applicable for circulating water makeup pretreatment. Clarification can be used to reduce TSS, and, in some cases, organics, but does not reduce the levels of dissolved constituents such as calcium, magnesium, and silica in the makeup. Reduction of TSS reduces the potential for the settling of solids in the circulating water system. This improves maintainability and availability because the fouling of heat transfer surfaces and the contributions to microbiological attack decrease as the amount of sediment accumulation is reduced.

If iron or manganese levels are dictating the cycles of concentration, one of the iron/manganese removal methods presented previously should likely be employed.

Lime or lime–soda ash softening is frequently applied to the pretreatment of water supplies for use as cooling tower makeup and service water. The process can be an economical method of increasing the utilization of water with high carbonate hardness content. Lime or lime–soda ash softening can be used to reduce TSS, calcium, magnesium, alkalinity, silica, iron, and manganese in the makeup. The benefits of the TSS reduction are the same as described above. The potential for operation at higher cycles of concentration is increased by reducing calcium, magnesium, and silica. In most instances, the alkalinity reduced by lime softening permits a reduction in the amount of acid required to control alkalinity. Increased cycles of concentration reduce the amount of makeup needed and the amount of blowdown that must be discharged.

15.6.1.7 Sidestream Treatment. Sidestream treatment is a term used to describe the practice of treating small portions of the circulating water flow and then recycling this water. Sidestream treatment flow rates are typically in the range of 1% to 5%, but can be higher.

Sidestream treatment may be as simple as filtration of circulating water with high levels of suspended solids to reduce fouling. Where "zero discharge" is desired or where water resources are inadequate, treated cooling tower blowdown is returned to the circulating water system. This type of sidestream treatment typically uses lime–soda ash softening and filtration to reduce suspended solids, hardness, alkalinity, and silica. Reverse osmosis and brine concentrators are other processes possibly applicable to sidestream treatment.

15.6.2 Plant Service/Potable Water

High levels of TSS, iron, or organic matter are problems typically encountered with the raw water supply that must be dealt with to satisfactorily use the supply for service water.

Clarification or lime softening followed by filtration is the most common approach to reduce TSS. Iron and organic matter are also reduced by lime softening. If the water supply is essentially free of TSS but has a high iron or manganese content, manganese–zeolite filters can be used. Chlorination of the potable water is typically required.

Potable water regulations have continually become more stringent. Analytical testing and reporting, especially with the enactment of the Safe Drinking Water Act, are becoming appreciably more exhaustive even for the small noncommunity type of potable water system associated with the power plant. The option of purchasing potable water from a local utility should not be overlooked, if that option is available.

15.6.3 High-Purity Water

Power plant service water or potable water supplies are often used as source water for further treatment to produce high-purity water. The resulting high-purity water is then typically used as makeup water to steam generators. Another use of high-purity water in contemporary power plants is as injection water for combustion turbines. The high-purity water is injected into the turbine combustors to control the formation of environmentally undesirable nitrogen oxides (NO_x) in the turbine exhaust gas.

The following discussions address how this high-purity water supply is produced with the treatment methods described in Section 15.5, Treatment Processes.

15.6.3.1 Cycle Makeup Water. The production of cycle makeup water of the typical qualities required is feasible only with the use of demineralization. Demineralizing water to the purity required for cycle makeup usually requires multiple ion exchange vessels in series. A very straightforward demineralizer configuration that has proven to be very reliable is the four-bed demineralizer followed by a polishing pair of exchangers. This configuration has been referred to as a six-bed demineralizer; however, only four of the exchanger vessels are designed to be working vessels. The final two vessels provide assurance that the effluent water quality will remain within design constraints over a very wide range of operating conditions.

Figure 15-29 illustrates the process flow path for a six-bed demineralizer with a vacuum degasifier. The figure shows a primary pair of exchangers, followed by a secondary pair of exchangers, followed by a polishing pair of exchangers. The primary and secondary exchangers are designed to perform the bulk of the ion exchange operation. The polished pair scavenge any remaining ion leakage from the previous exchangers. This exchanger configuration can reliably produce treated water with a specific conductance less than 0.1 micromho. With appropriate sizing and design considerations, this system can be arranged in various operating configurations to permit operational flexibility in the event of equipment outage.

Ion exchange is the most frequent process used for production of high-purity water; however, combinations of reverse osmosis and ion exchange have also proven to be cost-effective depending on the dissolved solids concentrations of the supply water to the demineralization system and the project economic criteria.

15.6.3.2 NO_x Injection Water. The production of high-purity water for NO_x injection is also typically performed with demineralization as the final treatment process. The various combustion turbine manufacturers in the marketplace today require different quality levels for injection water. The allowable solid contaminants in the injection water supplies is usually tied to a combination of the injection water, the fuel, and the combustion air.

Depending on the combustion turbine manufacturer and the chemical composition of the demineralization process supply water, it may be possible to adequately purify the water supply without the need for ion exchange. Since combustion turbine manufacturers do not specify low silica concentrations for NO_x injection water, non-ion exchange final treatment processes become a more feasible alternative. Ion exchange is the most effective process for removal of silica. Without the need to remove silica to the parts per billion levels, processes such as reverse osmosis and electrodialysis become more feasible as the final treatment process to meet NO_x injection water requirements. Continuing improvements in combustion turbine design and water treatment processes render the optimum selection of treatment for NO_x

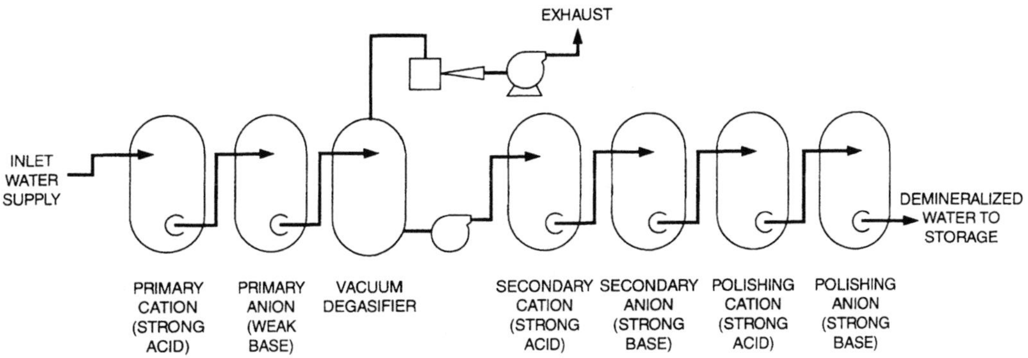

Fig. 15-29. Process flow path for a six-bed demineralizer with vacuum degasifier.

injection water a dynamic process. The process selection should be made on a case-by-case basis using the most current information available regarding combustion turbine requirements and the feasible water treatment alternatives available.

15.6.4 Auxiliary Equipment Cooling

Heat rejected from various plant auxiliary equipment heat exchangers is transferred to an auxiliary cooling water system. Some of the typical power plant heat exchangers that reject heat to the auxiliary cooling water are as follows:

- Air compressor aftercoolers, intercoolers, jacket coolers, and oil coolers;
- Condenser exhauster seal water heat exchangers;
- Exciter coolers;
- Stator cooling water coolers;
- Hydrogen coolers;
- Generator seal oil coolers;
- Sampling and analysis system coolers;
- Turbine lube oil coolers;
- Boiler feed pump turbine lube oil coolers;
- Booster boiler feed pump lube oil cooler and seal coolers;
- Startup boiler feed pump lube coolers and pump seal coolers;
- Boiler circulating pump coolers;
- Primary air preheat water pump seal coolers;
- Secondary air preheat water pump seal coolers; and
- High-pressure heater drains pump seal cooler.

Auxiliary equipment cooling is typically performed by one of two different methods. An "open" system uses circulating water to cool the various plant heat exchangers. A "closed" system uses circulating water to cool an auxiliary heat exchanger, accepting the heat rejected by an intermediate or "closed cycle cooling water" that is used to cool the various plant heat exchangers. Within the closed system, a condensate-quality water is used as makeup. The ability to chemically condition the closed cycle cooling water allows the use of more economical materials in the plant heat exchangers. A major advantage of the "closed" system is reduced potential of corrosion and fouling of the various heat exchanger surfaces and, therefore, reduced maintenance requirements to keep them clean and functional. With either the open or closed system, corrosion monitoring should be used to monitor heat exchanger material protection.

15.7 CYCLE CHEMISTRY AND CONTROL

Cycle chemistry is impacted by many factors, all of which must be controlled to achieve satisfactory operation of the cycle. Important factors include proper selection of cycle materials, use of high-purity cycle makeup, proper steam generator and cycle chemical conditioning, and proper noncondensables control. Chemical conditioning is the addition of chemicals to provide a selected chemical environment in that portion of the system being conditioned. Chemical conditioning is practiced for both the steam generator water and the water in the steam–condensate–feedwater system. Maintenance of proper cycle chemistry additionally requires surveillance of cycle parameters by the use of a water quality control system that is a part of an established water quality control program for the entire power plant.

One very important aspect of cycle chemistry is to minimize the amount of contaminants that enter the cycle. Typical means of contaminant ingress include condenser leaks (both air and water) and poor-quality makeup. Other contaminant ingress can occur from maintenance activities (paints, solvents, cleaning compounds) and from chemical feed contaminants.

In modern-day power cycles, the portion of the cycle most sensitive to contamination is the steam turbine. In general, this is because contaminants concentrate in certain areas of the turbine because the solubilities of the contaminants in the steam decrease as the steam temperature and pressure decrease. Usually, if chemistry and cycle fluid purity are acceptable for the steam turbine, the remainder of the cycle components are also protected. Exceptions are those contaminants, primarily corrosion products (crud), that deposit in the steam generator. Typical turbine steam purity requirements recommended by various sources are given in Table 15-11.

Figure 15-30 shows the basic steam cycle and describes many of the steam cycle areas that are impacted by water chemistry. A multitude of different forms of damage can occur to a steam cycle as a result of inappropriate cycle chemistry. For the boiler, these include caustic attack, hydrogen damage, and stress corrosion cracking, just to name a few. Chemical attack in the feedwater system also includes stress corrosion cracking, and erosion corrosion.

Chemical conditioning is divided into two different areas: steam generator conditioning and cycle conditioning. Steam generator conditioning can be accomplished by the addition of nonvolatile chemicals. Cycle conditioning includes the turbine, the condensate system, and the feedwater system up to the steam generator, and generally requires the use of volatile chemicals.

Table 15-11. Typical Turbine Steam Purity Requirements Recommended by Various Sources

Constituent	General Electric[a]	Westinghouse	Electric Power Research Institute
Cation Conductivity, μmho/cm (μs/cm)	<0.2	<0.3	<0.15
Sodium μg/L	<3	<5	<3
Chloride, μg/L	—	<5	<3
Silica, μg/L	—	<10	<10
Sulfate, μg/L	—	—	<3

[a]General Electric recognizes modification of the above limits for units using coordinated phosphate control.

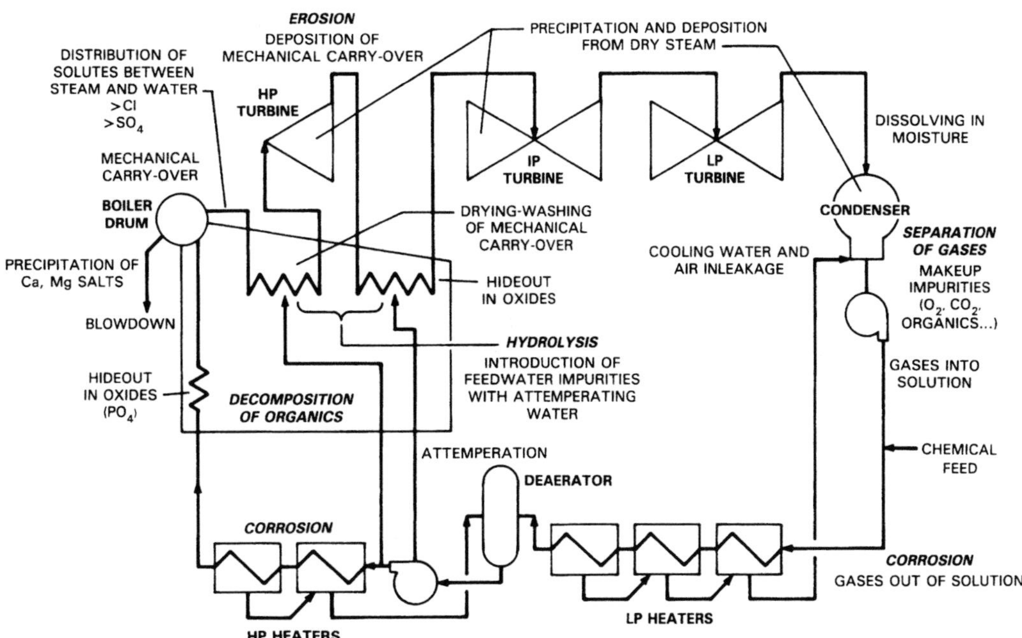

Fig. 15-30. Chemical transfer in the conventional drum boiler system. [From Sargent & Lundy for the Electric Power Research Institute (EPRI). Used with permission of EPRI.]

15.7.1 Steam Generator Conditioning

Steam is generated in the steam generator circuits and routed to the turbine. The purity of this steam is dependent on the appropriate conditioning of the steam generator water and the quality of the feedwater. Feedwater is routed to the steam generator to be converted to superheated steam which is then ready for admittance to the turbine.

The approach to steam generator conditioning varies depending on the type of steam generator used within the cycle. The approach must minimize corrosion of steam generator materials, minimize deposition of feedwater contaminants, and provide steam of adequate purity for the turbine.

Any solids in the feedwater are concentrated in the steam generator. Concentration beyond the solubility of the solids results in deposition in the high heat flux areas of the steam generator. The most common deposits are corrosion products from the condensate and feedwater systems. Hardness constituents (calcium and magnesium), if present in the feedwater, are particularly troublesome because of their property of inverse solubility; that is, these materials are less soluble at higher temperatures than at lower temperatures. In addition, calcium forms particularly hard scales. Scaling from any of these materials insulates the inside of the tubes in the steam generator. This insulation reduces heat transfer and can result in overheating and failure of the tubes in the steam generator.

Corrosion of steam generator materials is undesirable because of the associated material loss and the heat transfer resistance of the corrosion product oxides. The insulating effect of the corrosion products on the steam generator tubes can result in tube failure due to overheating. In addition to insulating heat transfer, these deposits also participate in caustic attack and other under deposit phenomena, all of which are detrimental to long-term service.

Maintaining steam purity affects both the levels of chemical additions allowed to the steam generators and the required quality of feedwater to the steam generator. Steam generator conditioning for both drum units and once-through steam generators is discussed further below.

15.7.1.1 Drum-Type Units. A primary function of the steam generator drum is to separate vapor from liquid. Feedwater routed to the steam generator is heated to the point of vaporization within the steam generator. For drum-type units, a vapor and liquid mixture is generated in the tubes, collected, and separated in the steam drum; the separated vapor is routed to the superheater and then to the turbine. This two-phase separation causes concentration of the dissolved impurities in the remaining drum water because of the preferential solubility of most impurities for the liquid phase.

Blowdown from the steam generator is used to control and remove solids. With proper feedwater purity, blowdown can be limited to less than 1% of the feedwater flow rate. It is desirable to minimize blowdown to limit both the heat loss and the use of high-purity water.

Another function of the drum is to provide proper steam purity to the turbine. The purity of the steam is affected by both mechanical and vaporous carryover. Mechanical carryover is the physical entrainment of drum water in the steam leaving the drum. This results in contamination of the steam by the less pure drum water. Excessive mechanical carryover results in unacceptable contamination of the steam since this carryover contains the same concentration of contaminants as the drum water.

Vaporous carryover is a result of the volatilization of the chemical species in the drum water. This volatilization of impurities contaminates the steam. The extent of volatilization depends on the chemical species making up the contaminant, the drum pressure, and the concentration of the contaminants in the drum water. Silica volatilization is a particularly difficult problem because of its relatively high solubility in steam at intermediate steam generator pressures. Steam turbine manufacturers have imposed steam purity limits that typically restrict the maximum silica concentration in the steam to 10 μg/L or less. Figure 15-31 provides guidelines for allowable silica in water.

The concentrations of solids in the steam from both volatilization and mechanical carryover must be limited to meet steam turbine manufacturers' limitations. Since both are affected by steam generator drum water, it is obvious that the concentrations in the drum water must be controlled.

Chemical conditions within the steam generator must be controlled to minimize corrosion within the steam generator. This means maintaining pH at a minimum of 9.0 for high-pressure units and higher values for some of the lower pressure units. For drum units, this is often achieved with the addition of a sodium phosphate salt. The sodium phosphate salt accomplishes two desirable objectives for boiler water: properly selected sodium phosphate solutions result in alkaline solutions that elevate the pH of the steam generator water, and phosphate chelates or combines with calcium to minimize the potential for the calcium to deposit on the heat transfer surfaces.

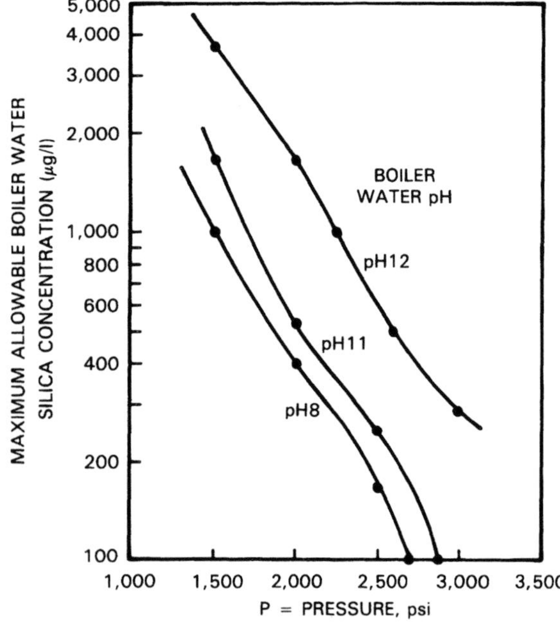

Fig. 15-31. Maximum boiler water silica (SiO_2) concentration versus drum pressure at different values of pH (10 μg/L SiO_2 limit in steam). [From J. Rios (Bechtel Power Corporation) and F. X. Blood (Union Electric Company). Used with permission.]

Several different types of phosphates can be used to add phosphate ion to the steam generator water. Sodium orthophosphates vary in sodium-to-phosphate molar ratio. The sodium orthophosphates are monosodium phosphate (1.0 Na/PO_4 ratio), disodium phosphate (2.0 Na/PO_4 ratio), and trisodium phosphate (3.0 Na/PO_4 ratio). A popular type of phosphate treatment is called coordinated phosphate control, a method that involves maintaining the sodium-to-phosphate molar ratio between 2.4 and 2.7. This type of treatment was developed to reduce the potential of caustic formation in the steam generator. Figure 15-32 shows the steam generator water pH versus phosphate residual for various sodium-to-phosphate molar ratios. Note that higher pressure units use lower levels of phosphate. This is because higher pressure units transmit higher levels of sodium phosphate to the turbine via volatilization; thus, lower levels are required for steam purity targets. Recent information indicates that strict adherence to the sodium-to-phosphate molar ratios used for coordinated phosphate control may lead to steam generator metal wastage. EPRI guidelines are currently being reviewed to consider this recent information.

There are several other types of steam generator drum conditioning. Caustic is sometimes used instead of or in addition to phosphates to maintain alkaline conditions in the steam generator. A concern with the use of caustic is the potential for caustic gouging or caustic underdeposit corrosion. EPRI guidelines are currently under review to consider the conditions under which caustic should be used in drum type steam generators. Chelants are sometimes used to maintain cleanliness in the steam generator. Some difficulties have been experienced with control of the amount of chelant available and the chelants themselves sometimes cause corrosion damage to the steam generators. All-volatile treatment, which is discussed in the following subsection, is also applicable to drum type steam generators.

15.7.1.2 Once-Through Steam Generators. The schematic representation of the steam generator in Fig. 15-30 is based on a drum-type steam generator. Steam generators may also be of the once-through type, without a drum. Steam is generated within the tubes of the steam generator and collected in headers before being routed to the superheaters and then to the turbines. There is no area for collection of a vapor and liquid mixture, and thus no opportunity to concentrate solids or contaminants in a liquid phase for discharge. Some of these once-through steam generators operate at supercritical conditions, meaning that water has exceeded its critical point and once beyond this point the water no longer exhibits a phase change from liquid to vapor. In either the subcritical or the supercritical steam generator, any impurities that enter the steam generator either deposit on the steam generator surfaces or are passed through the steam generator to the

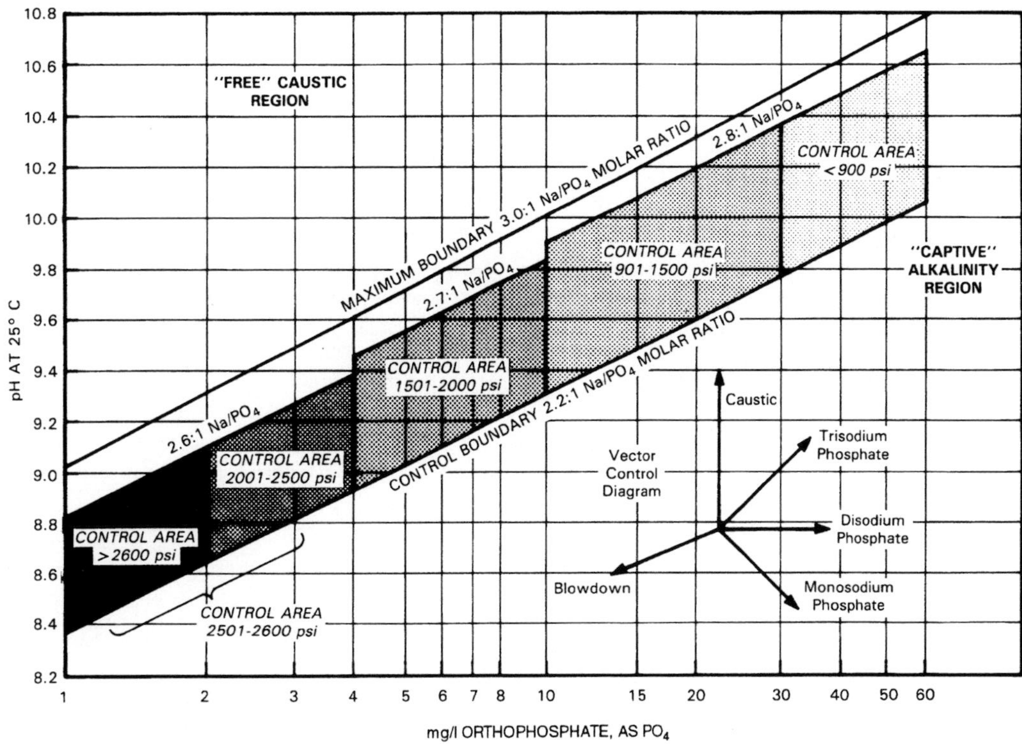

Fig. 15-32. Boiler water pH versus phosphate residual for various sodium to phosphate molar ratios. (From Betz Laboratories, Inc. Used with permission.)

turbine. Because there is no separation of phases in the once-through units, the feedwater to the steam generator must be of very high purity to meet turbine requirements.

For preservation of the steam generator materials in high-purity water, a high pH is required to minimize corrosion. Because there is no drum or phase separation, the use of non-volatile solid chemicals such as phosphates for steam generator conditioning is not acceptable. The solid chemicals would deposit in the tubes in the high heat flux zones, resulting in tube failures. Fortunately, an elevated pH can be maintained in the once-through steam generator with the amine (generally ammonia) that is used to condition the cycle. This type of treatment is called all-volatile treatment (AVT).

Because of the need for high-purity feedwater and because no additional conditioning chemicals can be fed to the boiler, the use of condensate polishing is required with once-through units. Condensate polishing is described later in this chapter.

All-volatile treatment may also be used with drum-type steam generators. As with the once-through steam generator, the all-volatile treated drum steam generator cycle should use condensate polishing.

15.7.2 Cycle Conditioning

Feedwater purity and conditioning are important factors in the maintenance of good cycle chemistry. Cycle conditioning minimizes corrosion and subsequent transport of corrosion products to downstream components and ultimately the steam generator. The most important function of these corrosion control mechanisms is to minimize the transport of iron from erosion corrosion in copper-free cycles and to minimize the transport of both iron and copper in cycles with copper alloys.

15.7.2.1 Conventional Treatment. In the United States the typical approach to minimize erosion corrosion has been to elevate cycle pH in a reducing environment. To maintain a reducing environment, major efforts are made to eliminate oxygen. Oxygen removal, as well as removal of other non-condensables, occurs in both the condenser and the deaerator. In addition, reducing agents or oxygen scavengers such as hydrazine are fed to eliminate any oxygen that may have eluded physical methods used in both the condenser and the deaerator. This reducing environment results in the formation of a protective layer of magnetite over steel materials.

To minimize the erosion corrosion of magnetite, it is necessary to select conditions that minimize the solubility of magnetite in water. Figure 15-33 indicates the relative solubility of magnetite as a function of pH. From this figure, it is apparent that minimum solubility and thus minimum magnetite transport occur when the cycle pH is maintained at around 9.5. This is typically accomplished with the addition of an amine such as ammonia. Other amines such as morpholine or cyclohexylamine can be used in lower pressure cycles. These other amines break down in cycles operating above 1,500 psig (10,340 kPa) and thus are not recommended

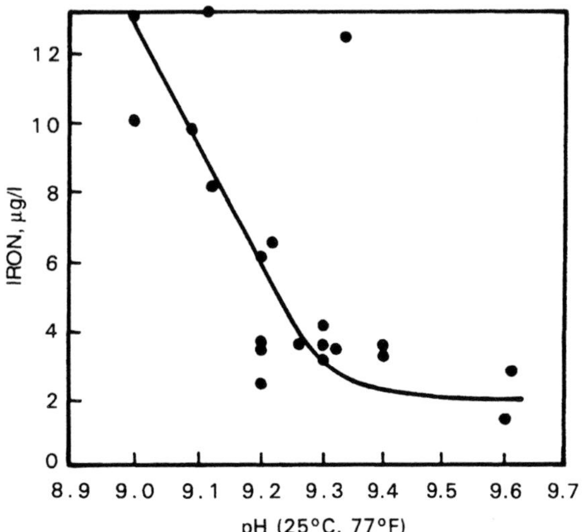

Fig. 15-33. Effect of pH on iron concentration. (Electric Power Research Institute, 1994. Reprinted with permission.)

for use in higher pressure cycles. Ammonia is also convenient to use in cycle conditioning because ammonia is formed as one of the breakdown products of hydrazine. Thus, regardless of which amine is fed for cycle conditioning, ammonia is present if hydrazine is used.

Copper is also sometimes used as a cycle material, primarily in condenser tube alloys, but may also be used in feedwater heater tubes. Figure 15-34 shows the relative solubility of copper alloys as a function of pH. To minimize the solubility of copper, and thus the transport of copper, cycle pH should be maintained around 8.5.

There is an inherent conflict with the optimum cycle conditioning of a cycle that contains both steel and copper. The cycle pH cannot be both 8.5 for limiting copper transport and 9.5 for limiting iron transport. Generally, a cycle containing copper is conditioned at a pH of around 9.0, with a control range of from 8.8 to 9.2. This compromise range typically provides satisfactory, although not ideal, conditioning. It is generally considered more desirable to use an all-ferrous-metal (copper-free) system and avoid the difficulties of conditioning a system containing copper.

15.7.2.2 Oxygenated Treatment (OT). An alternate treatment was developed in Germany in the early 1970s for treatment of once-through steam generators. Although the primary history of OT is for once-through units, OT has also seen limited use in drum units.

The basis of the treatment is the feed of oxygen or other oxidizer to an all-steel cycle for cycle conditioning. In an oxidizing environment, a different protective layer is formed over the steel materials. This protective corrosion layer, ferric hydrate oxide, covers a base layer of magnetite. Figure 15-35 shows the relative solubilities of both magnetite and ferric hydrate oxide. Reports from Germany and the former Soviet Union indicate that plants using this type of treatment have operated for significantly longer periods than all-volatile treated steam generators without the necessity of

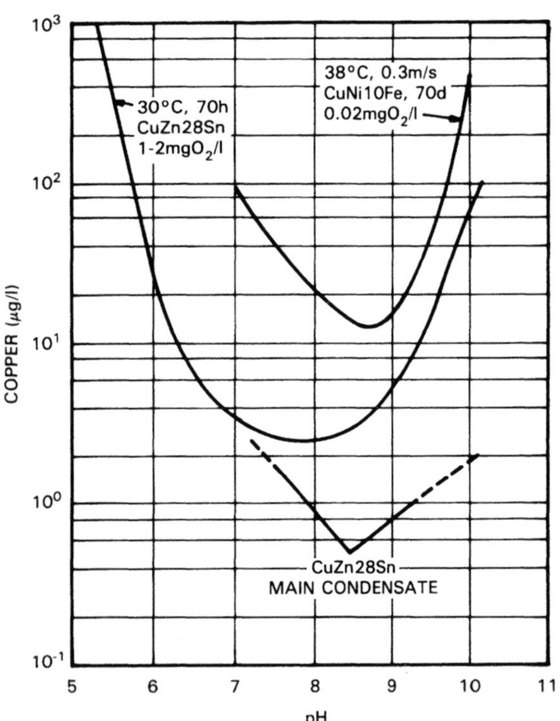

Fig. 15-34. Ammonia corrosion of copper-base alloys in high-purity water. (From Allianz Berichte. Used with permission.)

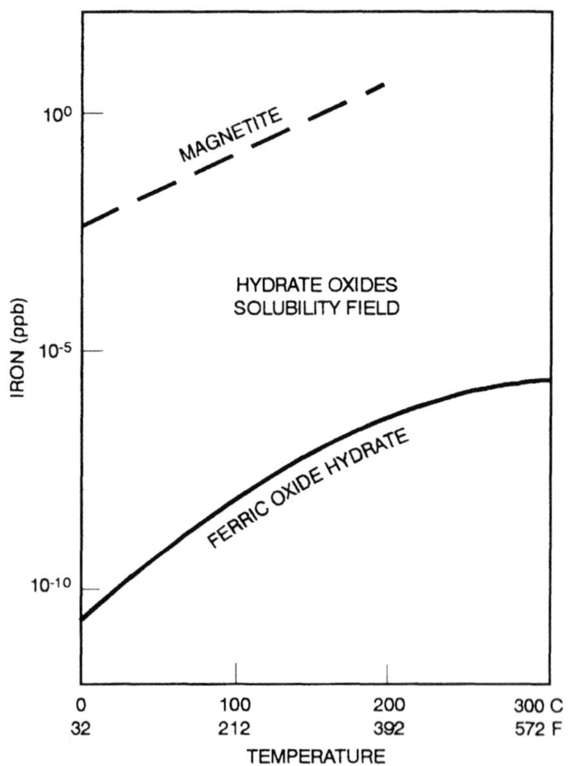

Fig. 15-35. Solubility of iron hydrate oxides. (Electric Power Research Institute, 1994. Reprinted with permission.)

chemical cleaning to remove iron deposits in the steam generator.

In oxygenated treatment, oxygen is fed either to the condensate system downstream of the condensate polisher, in the boiler feed pump suction, or both. Oxygenated treatment can be practiced satisfactorily over a large pH range. An all-ferrous condensate and feedwater system is required. Copper materials are acceptable for the condenser tubing. Any copper materials downstream of the condensate polisher will result in an unacceptably high rate of copper transport in the cycle.

Oxygenated treatment, like AVT, requires very-high-purity water for successful operation. This high purity is achievable in well-designed and well operated plants with condensate polishing.

Figure 15-36 shows a comparison between AVT conditioning and oxygenated treatment conditioning.

15.7.3 Nuclear Steam Cycles

Nuclear cycles present a different set of problems, even though feedwater heaters, condensers, condensate pumps, and boiler (reactor or steam generator) feed pumps are all basically the same as in fossil cycles.

15.7.3.1 Boiling Water Reactors. In boiling water reactors, the feedwater is routed to the reactor, where the water is exposed directly to the nuclear fuel rods in the core. Any chemicals used for conditioning the water would be subject to activation in the reactor. To avoid this activation and the subsequent potential radioactivity buildup, no conditioning chemicals are used and neutral water treatment is practiced.

Oxygen may or may not be fed directly to the cycle. Oxygen is generated within the reactor as a result of radiolytic decomposition of the reactor water. The majority of this oxygen is removed later in the cycle. Frequently, enough oxygen is present in the feedwater to obtain the desired passivation of the ferrous materials.

Stainless steel materials are used extensively within the reactor and are also used for feedwater heater tubing. Full-flow condensate polishing is also practiced with boiling water reactor cycles. Copper alloys have been used in the condensers.

The presence of oxygen, susceptible materials (stressed 300 series stainless steels), contaminant anions (generally chlorides), and high temperatures can lead to stress corrosion cracking. Stress corrosion cracking has been a problem with some of the sensitized stainless steel materials used in the reactor and associated systems. The required quality of reactor water has strictly limited the acceptable concentrations of most common anions, such as chloride and sulfate. This strict limitation has not been universally successful in preventing stress corrosion cracking.

An alternate approach is being taken in a few boiling water reactors that involves minimization of the oxygen generation in the reactor. This is accomplished by feeding hydrogen gas, thus suppressing the generation of oxygen in the reactor. With this system, oxygen must be added to the feedwater to achieve passivation of the feedwater system. Oxygen is also added to the steam downstream of the reactor to allow proper stoichiometry for offgas recombiner operation.

15.7.3.2 Pressurized Water Reactors. In pressurized water reactors, reactor water is maintained under pressure and circulated through heat exchangers that function as steam generators. Feedwater is routed to the steam generators and the steam is routed to the turbine. Like fossil-fueled steam generators, pressurized water reactor steam generators may be either once-through or recirculating. Blowdown is used on the recirculating steam generators.

All-volatile treatment is used in nearly all of the pressurized water reactor cycles. Phosphate treatment of recirculating steam generators was practiced in earlier years of pressurized water reactor steam generator treatment.

Several alternate amines are now used for cycle conditioning in many systems. Boric acid is also used in some stations as a corrosion inhibitor.

The systems with once-through steam generators were originally equipped with condensate polishing equipment. Condensate polishing has been retrofitted to some pres-

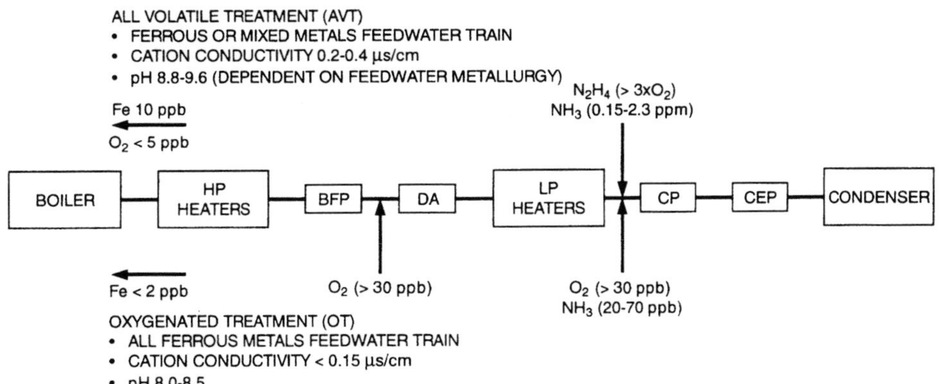

Fig. 15-36. Comparison between AVT conditioning and OT conditioning (once-through units). (Electric Power Research Institute, 1994. Reprinted with permission.)

surized water reactors with recirculating steam generators to aid in maintaining increasingly stringent steam generator purity requirements.

15.7.4 Condensate Polishing

Condensate polishing was discussed earlier as a method of improving feedwater quality and as a virtual requirement for once-through steam generators and steam generators using all-volatile treatment. Condensate polishing also finds frequent application on units subjected to cycling or peaking operation, or where the condenser cooling water contains high concentrations of dissolved solids. Condensate contains minor levels of contaminants which consist of both dissolved and suspended materials. Removal of these low levels of contaminants reduces deposition in the steam generator and also reduces transport of deposit-prone and potentially corrosive contaminants to the turbine.

For modern high-pressure steam cycles, the removal of these minor levels of contamination is highly desirable. Because the contaminants are relatively low level, the term polishing describes the removal treatment.

The level of contaminants present in condensate varies with the operation of the unit. High levels of suspended solids are present during and shortly after the startup of a unit. The suspended solids result from corrosion that occurs to materials of construction in the cycle. Suspended solids are removed by filtration. Startup practices for once-through units include recirculations through the feedwater cycle and steam generator to allow the condensate polishing system to remove the suspended solids during startup.

Dissolved solids are present due to contaminant ingress. The source of this ingress is generally either from low levels of contamination in the makeup or from condenser leakage. Other possible sources of contamination include maintenance chemicals left on wetted materials following a maintenance action. Dissolved solids can be removed by ion exchange.

In United States practice, condensate polishing is accomplished by two different kinds of systems: mixed bed system and powdered resin systems. Mixed bed systems use mixed cation- and anion-exchange resins contained in a pressure vessel. Condensate is passed through the bed of ion exchange resin for treatment of the condensate. Early development indicated that the ion exchange resin satisfactorily removed suspended solids within the bed of ion exchange beads when the flow rate was maintained above 25 gpm/ft^2 (17 L/s per m^2). This is contrasted with normal makeup demineralization, which is carried out in beds typically operating at rates of 5 to 15 gpm/ft^2 (3.4 to 10.2 L/s per m^2). Further development and application resulted in vessels sized to operate at design flow rates of about 50 gpm/ft^2 (34 L/s per m^2), with the peak flow rates up to 75 gpm/ft^2 (51 L/s per m^2). The high vessel flow rates were beneficial because they allowed relatively small vessels to treat a large condensate flow. The relative purity of the condensate means that a relatively small amount of ion exchange capacity is required, also making treatment feasible at these high flow rates. However, the high treatment flow rate of condensate polishing does require excellent quality resin and challenges the resin's kinetic capabilities.

Regeneration of the ion exchange resin is generally accomplished in external vessels. The resin is transferred hydraulically from the service vessels or polishers to the external regeneration vessels. The external regeneration vessels are capable of cleaning the resins, separating the resins, regenerating the cation resin in one vessel and the anion resin in another vessel, and recombining the cleaned and regenerated resins. Frequently, a spare resin charge is kept in the external regeneration vessels. This spare charge is transferred to the service vessel to allow it to go back into service quickly, while the exhausted charge is being regenerated. A typical mixed bed condensate polishing system is depicted schematically in Fig. 15-37.

Service vessel bed depths are generally about 3 ft (0.9 m). Deeper beds have been used for seawater-cooled plants. Deeper bed depths allow for greater ion exchange capacity, but have the drawback of increased pressure losses at the high treatment rates used for condensate polishing. Because bed depths are usually 3 ft (0.9 m), the volume of ion exchange resin per polisher vessel is determined by the condensate polisher vessel diameter, which is selected to satisfy the design flow criteria. The table below gives the approximate internal cross-sectional areas of typical 150 psi pressure vessels by vessel diameter:

Vessel diameter							
in.	72	84	96	102	108	120	132
m	1.8	2.1	2.4	2.6	2.7	3.0	3.4
Vessel cross-sectional area							
ft^2	27.4	37.1	48.7	55.1	61.9	76.0	89.3
m^2	2.55	3.45	4.52	5.12	5.75	7.06	8.30

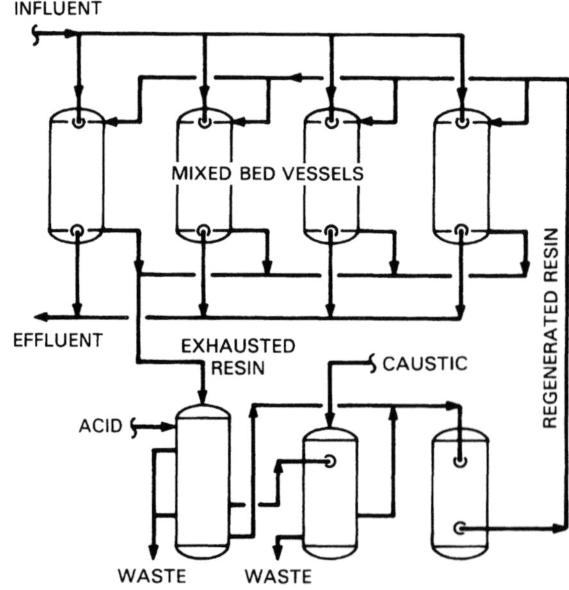

Fig. 15-37. Mixed bed condensate polishing system.

Condensate polisher vessels are often of higher pressure design, thus the above figures should be ajusted to account for the actual vessel metal thickness associated with the design pressure for the specific application.

Generally, condensate polishing systems are selected to be full-flow systems. A full-flow system includes sufficient vessels to treat the entire condensate flow. A limited number of partial flow systems have been installed. These systems assist in reducing suspended solids during startup. These partial flow systems are not effective in operating with significant condenser tube leakage because of the amount of condensate that bypasses untreated.

The other major type of condensate polishing is the powdered resin system. This system evolved from filtration equipment. The powdered resin system uses basically a precoat filter that is precoated with ground ion exchange resin. The filter vessel consists of many tubular filter elements attached to a tubesheet. The ground resin is applied as a precoat to the bare tubular element. The flow through the vessel holds the precoat on the filter element. The precoated filter elements in the powdered resin filter demineralizer remove suspended solids from the condensate and the ion exchange capability of the ground resin removes dissolved ions. A powdered resin filter demineralizer vessel is depicted in Fig. 15-38.

A relatively small amount of ground ion exchange resin is applied to the vessels. The amount of precoat applied results in a relatively thin layer of ion exchange resin on the element. Thus, the ion exchange capacity of the powdered resin filter demineralizer is very limited and is significantly less than that available with the mixed bed polisher. The fine size of the particles resulting from the resin grinding gives very good filtration capability. The vessel diameter limits the number of filter elements that can be installed. Vessels are generally sized based on the amount of filter surface area available within the vessel. During the early application of filter demineralizers, a flow rate of 4 gpm/ft^2 (2.7 L/s per m^2) of filter element area was used. Subsequent installations have tended to use lower flow rates to improve performance and a rate of 3 gpm/ft^2 (2.0 L/s per m^2) is now more typical. Typical vessel sizes and filter areas for filter demineralizer equipment are given below. Actual equipment sizes and filter area capabilities will vary from one vendor to another.

Vessel diameter					
in.	48	54	6	66	72
m	1.2	1.4	1.5	1.7	1.8
Filter area					
ft^2	512	689	872	1,060	1,280
m^2	47.6	64.0	81.0	98.5	118.9

The precoat is removed by backwashing the filter demineralizer vessel. This process is accomplished by flushing water and air in the reverse direction through the filter elements to remove the precoat. Following discharge of the precoat, a new precoat is applied with the precoating system. The precoating system generally consists of one or two tanks

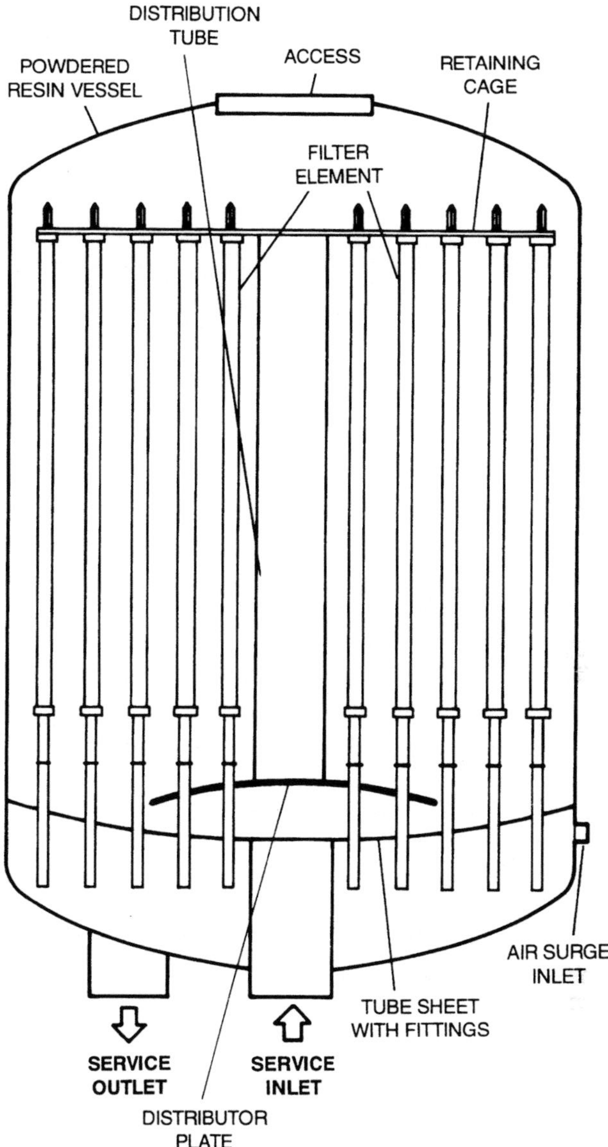

Fig. 15-38. Powdered resin filter demineralizer vessel. (From E. Salem and M. J. O'Brien. Used with permission of Graver Water.)

and a pump to apply the powdered resin to the filter demineralizer vessel. The precoat is mixed in one of the tanks starting either with two separate resin components or a single premixed precoat mixture. A recirculating flow is established through the vessel. Precoat is metered into the recirculating stream and forms a coating on the filter elements. Some earlier precoat systems applied the precoat very rapidly from the precoat mixing tank; however, this high precoat application rate has been found to be less effective than the system described previously. Fiber may also be used with the resin in premixed precoat or can be used as an overlay. A typical powdered resin condensate polisher system is depicted schematically in Fig. 15-39.

Both types of condensate polishing systems use mixtures of cation and anion resins. Generally, the resin ratio for mixed bed condensate polishing is two parts cation resin to

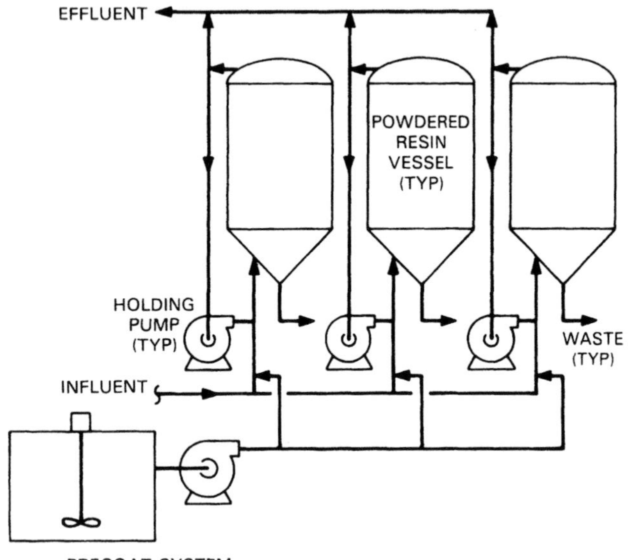

Fig. 15-39. Powdered resin condensate polisher system.

one part anion. This ratio has evolved for a number of reasons. As discussed earlier, the condensate treated for all but the boiling water reactor units is conditioned with a neutralizing amine, frequently ammonia. A majority of cation resin in the mixed bed provides for a longer operating run before the cation portion of the bed exhausts. In addition, iron removal is accomplished best by the cation resin. Finally, cation resin costs about half as much as anion resin. In a few limited applications, a 1:1 cation-to-anion resin ratio is used for plants with seawater cooling to provide greater anion exchange capacity to accommodate cooling water inleakage. For boiling water reactors, a 2:3 (by volume) cation-to-anion resin ratio is frequently used. This ratio is considered to be chemically equivalent, which is appropriate for the neutral chemistry employed in boiling water reactor units.

Similar resin ratio considerations are applicable to the powdered resin filter demineralizer. Resin cost is a more important consideration for the filter demineralizer because the resin is discarded following a single use. Often a cation-to-anion resin ratio of 3:1 is used.

As discussed above and in the cycle chemistry section, ammonia is often used to condition the condensate and feedwater of power plants. This ammonia is removed by the cation resin. When the resin is exhausted to ammonia, it may be regenerated to restore its capability to again remove ammonia. Typically, for a unit using an all-ferrous feedwater system, the cation resin in a mixed bed unit is exhausted in 3 to 7 days. An alternative for the mixed bed unit is to allow the cation resin to exhaust to ammonia and continue to operate saturated with ammonia. In this form, the resin has significantly reduced capacity for certain monovalent cations, specifically sodium. However, if the condensate remains relatively pure, it is not necessary to remove a large amount of sodium. Such operation is referred to as operating beyond ammonia break, or operating in the ammonia cycle. An advantage of this operation is that it allows greatly increased run lengths for the condensate polisher. Condensate polisher runs of about 30 days are commonly achieved.

Because of their limited ion exchange capacity, powdered resin units typically exhaust to ammonia within hours of being placed on-line. This powdered resin is often purchased in the ammonia form prior to precoat to avoid the transition from hydrogen form to ammonia form.

Alternate condensate polishing systems have limited use in the United States. One combination system consists of a powdered resin filter demineralizer used as a prefilter for a mixed bed system. This system has both the superior filtration capability of the powdered resin filter demineralizer and the superior ion exchange capability of the mixed bed demineralizer. The performance of the mixed bed resin is improved because the upstream filtration minimizes fouling of the mixed bed resins.

The obvious drawback of this combination arrangement is the capital cost of the system. However, most European condensate polishing installations are combination systems. A typical system may also consist of a cation bed in front of a mixed bed demineralizer. This particular configuration has advantages when operating with ammonia in that the cation bed can be used for ammonia removal while the mixed bed continues to operate in the hydrogen cycle. Other filtration methods such as cartridge filtration or precoat filtration are used also. Some other systems may also include individual three-bed systems consisting of cation–anion–cation beds.

The following example problem illustrates how to select vessel sizes for both mixed bed and powdered resin condensate polishing systems.

Example 15-5

For a new station design, both mixed bed polishing and powdered resin polishing are being considered. Peak condensate flow is 6,000 gpm (378 L/s). Select polishing vessel sizes for both types of polishing. For simplicity, assume cross-sectional areas of 150 psi vessels are applicable. Start with the use of three half-capacity (one standby) exchangers.

Solution

Mixed bed

$$6{,}000 \text{ gpm}/(50 \text{ gpm/ft}^2) = 120 \text{ ft}^2 \; [378 \text{ L/s}/(34 \text{ L/s m}^2) = 11.1 \text{ m}^2]$$

$$120 \text{ ft}^2/2 \text{ vessels} = 60 \text{ ft}^2 \text{ per vessel } (11.1 \text{ m}^2/2 = 5.55 \text{ m}^2)$$

From the table of vessel cross-sectional areas, select 108-in. (2.7-m) diameter vessels, each with a cross-sectional area of 61.9 ft2 (5.75 m^2). Therefore, the system consists of three, one-half capacity vessels, 108 in. diameter.

Powdered Resin

$$6{,}000 \text{ gpm}/(3 \text{ gpm/ft}^2) = 2{,}000 \text{ ft}^2 \; [378 \text{ L/s}/(2.0 \text{ L/s m}^2) = 189 \text{ m}^2]$$

$$2{,}000 \text{ ft}^2/2 \text{ vessels} = 1{,}000 \text{ ft}^2 \; (189 \text{ m}^2/2 = 94.5 \text{ m}^2)$$

From the table of vessel sizes and filter areas, select 66-in. (1.7 m) vessels, each with a filter area of 1,060 ft^2 (98.5 m^2). Therefore, the system consists of three, one-half capacity vessels, 66-in. diameter.

15.7.5 Water Quality Control

Impurities in the feedwater can cause corrosion, reduce heat transfer, lower turbine blade efficiency, and cause malfunction of valves, piping, and seals. Chemical sampling and monitoring are important functions in the operation of power generating plants. An outage of a large high-pressure steam electric power plant—especially an unplanned, forced outage—is extremely costly. Many forced outages and premature equipment failures can be prevented by application of correct cycle chemistry, monitored and controlled by properly designed sampling and analysis systems and chemical feed systems. The avoided costs of a single forced outage can sometimes more than offset the initial capital cost of a sampling and analysis system costing several hundred thousand dollars. The benefits derived from these systems when the data they provide are properly interpreted and used include lower operating costs through improved heat rates, lower maintenance costs, and higher steam turbine efficiency.

The higher operating pressures and temperatures of the modern steam power plant make it more susceptible to damage caused by dissolved and suspended constituents in the feedwater. Where lower pressure power plants of the past could tolerate several milligrams per liter (parts per million) of impurities in the feedwater, modern high-pressure plants can tolerate no more than a few micrograms per liter (parts per billion).

To increase the reliability and availability of modern power plants, several organizations have established limits on various constituents in the feedwater and boiler water. These organizations include the Electric Power Research Institute, the American Society of Mechanical Engineers Research Committee on Water in Thermal Power Systems, the Metals Properties Council, and steam turbine and steam generator equipment manufacturers.

Continuous sampling and analysis of samples drawn from critical points in the steam–condensate–feedwater cycle are employed to monitor the impurities in the system. Figure 15-40 shows typical sample locations in a drum type steam generator. Variations are dependent on cycle chemical conditioning practices. The sampling points in the feedwater cycle are intended to monitor the purity of the incoming water and to detect the entrance of contaminants from condenser leakage. Further discussion of the rationale for selection of the specific cycle locations is given in Section 15.7.5.2. Additional control over the cycle water chemistry can be obtained by using a condensate polishing system for the removal of dissolved and suspended solids. The effluent from the polisher should be monitored to verify that the system is operating properly and removing contaminants as designed.

In addition to monitoring impurities in the cycle, the sampling system allows verification that the chemical conditioning that is practiced to minimize the corrosiveness of the water is effective.

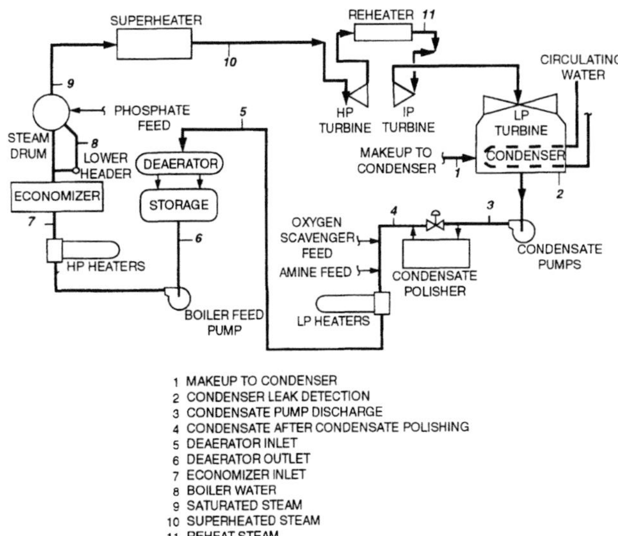

Fig. 15-40. Typical primary sampling points in cycle with drum type steam generator.

15.7.5.1 Water Quality Control Limits. The Electric Power Research Institute (EPRI) has established guidelines in EPRI CS-4629 for water quality control limits for various steam generators (EPRI 1986). These guidelines are based on strict cycle chemistry requirements necessary to minimize losses in unit efficiency and availability caused by deposition and corrosion. EPRI has developed cycle chemistry guidelines for various types of steam generators, both with and without reheat, on various chemical treatment programs.

Table 15-12 shows typical control limits for the recommended sample points for a high-pressure drum type steam generator with reheat using conventional feedwater chemical conditioning and phosphate treatment for steam generator water conditioning.

15.7.5.2 Rationale for Sample Points. Guidelines, such as those established by EPRI, have been developed to help utilities minimize losses in unit efficiency and availability caused by deposition and corrosion. Recommended sample points are established based on optimum cycle chemistry. The following describes typical recommended sample points and discusses the rationale for each:

- *Makeup to condenser.* Monitors quality of makeup water to the cycle. Also serves to monitor the performance of the makeup demineralizer to remove dissolved solids from the makeup water.
- *Condenser leak detection.* Allows detection of condenser tube leaks. Multiple sample points can be used to determine the particular tubesheet or area on a tubesheet where a leak has occurred. Cation conductivity is normally used to detect the increase in total dissolved solids that results from cooling water inleakage.

Table 15-12. Normal Chemistry Limits in High-Pressure Drum-Type Steam Generators with Phosphate Treatment and Conventional Feedwater Treatment

	Sodium (μg/L)	Chloride (μg/L)	Sulfate (μg/L)	Silica (μg/L)	Specific Conductance (μs/cm)	TOC (μg/L)	Oxygen (μg/L)	Hydrazine (μg/L)	Ammonia (μg/L)	Cation Conductivity (μs/cm)	pH	Iron (μg/L)	Copper (μg/L)	Phosphate (mg/L)
Makeup to condenser	⩽5	⩽3	⩽3	⩽10	⩽0.1	⩽300								
Condenser leak detection	a													
Condensate pump discharge						⩽200	⩽20							
Plants with polisher	⩽10									⩽0.3				
Plants without polisher	⩽5									⩽0.2				
Condensate polisher effluent (if applicable)	⩽5			⩽10						⩽0.2				
Deaerator inlet							⩽20	⩾20 or ⩾3 × O_2						
Deaerator outlet							⩽7							
Economizer inlet					b		⩽5		b	⩽0.2		⩽10	⩽2	
All ferrous metallurgy											9.0–9.6			
Mixed Fe-Cu metallurgy											8.8–9.3			
Boiler water	c	c	c	c	d						e			e
Superheat/reheat steam	f	<3	<3	<10		⩽100				⩽0.3 (degassed)				

Specific conductance, cation, conductivity, and pH all referenced to 25° C.

[a]Monitored to detect deviations from normal values.

[b]Consistent with pH.

[c]Consistent with curves of maximum allowable concentration versus pressure.

[d]Monitored as general indication of boiler water dissolved solids content.

[e]See phosphate/pH curves.

[f]Generally <5 μg/L, but may be lower depending on steam generator manufacturer's guidelines.

- *Condensate pump discharge.* Monitors carryover of contaminants and treatment chemicals in the steam, detects contaminants from cooling water inleakage at the condenser, and identifies the amount of oxygen inleakage. The pH or specific conductance may be measured at this point for a cycle without condensate polishing to allow adjustment of ammonia or other amine feed in sufficient concentration necessary to minimize corrosion in the condensate/feedwater systems. For a cycle with condensate polishing, the location for amine chemical feed control is downstream of the condensate polishing system.
- *Combined condensate polisher effluent.* Monitors the performance of the condensate polisher. Monitored parameters are similar to those at the condensate pump discharge to determine the extent of contaminant removal.
- *Deaerator inlet.* Aids in monitoring the performance of the deaerator by allowing comparison of deaerator inlet and outlet samples. This sample may serve as an alternate control point for hydrazine or other oxygen scavenger chemical feed.
- *Deaerator outlet.* Monitors the performance of the deaerator to remove dissolved oxygen from the feedwater.
- *Economizer inlet.* Allows monitoring of total contaminants entering the steam generator to determine whether the feedwater meets the chemistry limitations required by the steam generator manufacturer. This point can serve as the control point for hydrazine or other oxygen scavenger chemical feed.
- *Boiler water (drum boilers only).* Monitors boiler drum water chemistry to minimize deposits and corrosion in the boiler tubes. Monitoring this sample point allows the boiler water chemistry to be controlled by proper chemical feed and boiler blowdown. The chemistry of the boiler water directly affects steam chemistry; thus, this point is a primary control point for steam purity.
- *Saturated steam (drum boilers only).* Provides verification of compliance with the boiler manufacturer's performance guarantee for steam purity. This point also aids in monitoring mechanical entrainment and vaporous carryover of impurities into the turbine.
- *Superheat steam/reheat steam.* Indicates whether the steam purity limits are being maintained to prevent deposition and corrosion in the steam turbine. Aids in determination of whether the steam meets the turbine manufacturer's chemistry limitations. Also allows determination of the extent of steam contamination by attemperating water.

Successful control of the cycle water chemistry requires collection of samples that can be analyzed to supply useful information on the condition of the cycle. Without effective sampling and analysis techniques, it is impossible to ensure that the desired cycle chemistry is being achieved.

15.7.5.3 Sampling and Analysis Systems. Sampling systems are used to condition continuously flowing samples of condensate, feedwater, boiler water, and steam so the sample properties will be safe for manual sampling or be compatible with the requirements of automatic analytical instruments.

It is common practice to route most sample lines to a central sampling area near the laboratory for convenience of plant operating personnel. This central location is normally air-conditioned to provide a temperature and humidity suitable for obtaining optimum performance from automatic sample analyzers and instruments. Because these systems are located in a central area, the sample conditioning and analyzer system can be purchased as a prefabricated package with most equipment mounted on racks and panels. Figure 15-41 shows a typical shop-fabricated sample panel.

The main objective of the sampling system is to transport and condition a sample without altering the constituents in the sample. System parameters that need to be controlled are velocity, temperature, and pressure. Although most of today's on-line analyzers include temperature compensation devices that allow some deviation in sample temperature, it is common practice in the design of sample panels for large high-pressure units to condition the sample to a temperature of 77° F ± 1° F (25° C). As described in the following subsections the basic components of a sampling system are sample lines (through-the-plant), primary and secondary coolers, pressure reducing valves, pressure and flow regulators, isolation valves, and instruments. Figure 15-42 shows a schematic of pressure–temperature conditioning for a typical sample.

15.7.5.4 Sample Piping. An important consideration in any sampling system is the time lag between taking the sample from the bulk fluid and analyzing it at a central sample panel. Information about the cycle chemistry should be as current as possible. Sampling systems should be located and the sample piping sized appropriately to ensure that samples generally reach the sample panel within 3 to 5 min. Sample velocities resulting from the recommended lag times should also be considered. Usually a 5 to 6 ft/s (1.5 to 1.8 m/s) velocity in the sample line for liquid samples meets this requirement. Velocity considerations may require that lag times be increased or decreased as required to obtain acceptable pressure losses or to prevent suspended solids from settling.

Sample lines are typically fabricated from Type 304 or Type 316 stainless steel. The Power Piping Code (ANSI/ASME B31.1-1986) contains guidelines applicable to the design of sample system piping for power stations. The Code guidelines should be followed for the design of sample piping for the various piping pressure–temperature conditions

Fig. 15-41. Typical power plant sample panel. (From Johnson March Systems, Inc. Used with permission.)

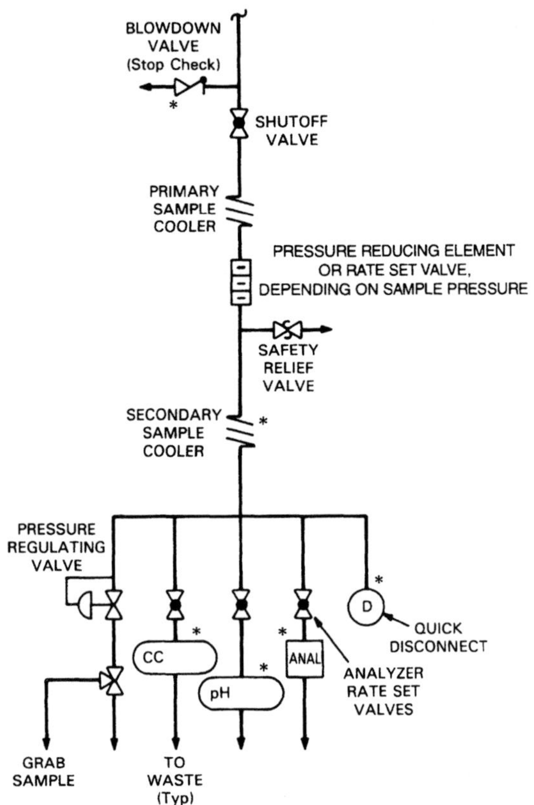

Fig. 15-42. Temperature sample conditioning schematic.

normally present in a sample system. The Code definition of "pipe" uses the terms "pipe" and "tube" interchangeably.

Considerations for sample line sizing and material selection for sample lines running between the process takeoff point and the water quality control system include support requirements and mechanical strength factors appropriate for long runs of unprotected piping.

For long service life, a ½-in. double-extra strong pipe with a 0.252-in. bore has been used for through-the-plant sample lines. This pipe is much more durable than tubing and is capable of withstanding the abuse that can occur in a power plant environment. Tubing, such as ¼ in. OD with a 0.166-in. bore or ⅜ in. OD with a 0.245-in. bore, is a less costly alternative providing a reasonable pressure drop while still maintaining desirable sample velocities. If tubing is used, it is often continuously supported to offer structural and physical protection.

15.7.5.5 Primary and Secondary Coolers Sample cooling is required for accurate performance of most analyzers. In most cases, sample cooling is conducted in two stages. Primary sample cooling is necessary to reduce sample temperatures to 100 to 120° F (38 to 44° C) or less. Generally, because of their compactness, primary sample coolers are tube coils in a shell. The cooling coil should be stainless steel, whereas the shell side may be carbon steel or stainless steel. Typically, primary coolers are included in the sample conditioning panel. Figure 15-43 shows a typical stainless steel shell and tube sample cooler.

The purpose of secondary coolers is to bring the sample to its final analysis temperature, usually 77° F ± 1° F (25° C). Three basic types of systems are used. The first is similar to primary sample cooling, as a shell and tube heat exchanger is used. In this system, the sample cooler is sized so that the approach temperature is very low and constant-temperature chilled water is used as the cooling medium.

Another type of secondary sample cooling system utilizes an isothermal, 76.5° F ± 0.5° F (25° C) water bath. The sample is passed through coils immersed in the recirculating water bath. The coils are sized such that an approach temperature of 0.5 to 1° F (0.25 to 0.5° C) can be achieved. This type of system generally is designed to incorporate hot gas bypass of the chiller compressor to permit continuous operation of the chiller and has a precision controller to adjust refrigerant temperature.

The third type of system employs a sample splitter that directs a portion of the sample flow through a cooling coil in a refrigerated water bath. Figure 15-44 shows a flow schematic for a secondary sample cooling system with sample splitting and a cooling bath maintained at a temperature well below 77° F (25° C). The remainder of the sample bypasses

Fig. 15-43. Typical shell and tube sample cooler. (From Sentry Equipment Corporation. Used with permission.)

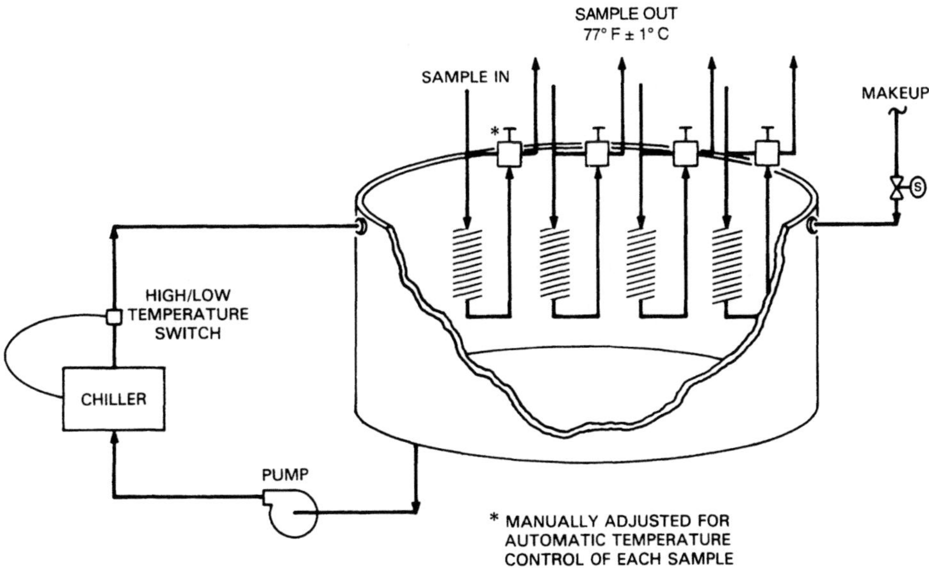

Fig. 15-44. Secondary sample cooling system with sample splitting and cooling bath. (From Waters Equipment Co. Used with permission.)

the coil and is discharged to the outlet of the coil. A thermal control element in the splitter valve proportions the two flows to achieve the set point temperature of 77° F ± 1° F.

15.7.5.6 Pressure Reducing Devices. Three types of pressure reducing devices are commonly used: rod-in-tube, labyrinth passage (drag valves), and capillary tubing. The primary problem associated with these three types of pressure reducing devices is pluggage by corrosion products. Simple capillaries and labyrinth passage devices cannot be cleaned without shutting down the system, unless special provisions are made. The same is also true for the fixed rod-in-capillary designs. However, adjustable rod-in-tube designs can be cleaned by simply retracting the rod. Figure 15-45 shows an example of a rod-in-tube pressure reducing valve. Based on ease of blockage removal, adjustable rod-in-tube pressure reducing devices are generally preferred for pressure reduction following the primary cooler.

15.7.5.7 Pressure and Flow Regulation. It is desirable to maintain a constant velocity in the sample line because many analyzers are sensitive to flow. Many analyzers also require a stable inlet pressure [5 to 20 psig (34.5 to 138 kPa)]. To satisfy this requirement, it is necessary to maintain a constant pressure in the sample line supplying the instruments. This can be accomplished by using an adjustable backpressure regulating device.

Orifices installed in sample lines have also been used to reduce pressure and control maximum flow rate. However, this is not recommended since orifice size cannot be adjusted, and the orifice is susceptible to pluggage and wear.

15.7.5.8 Isolation Valves and Relief Valves. Generally, a sample line includes two isolation valves. One, the root

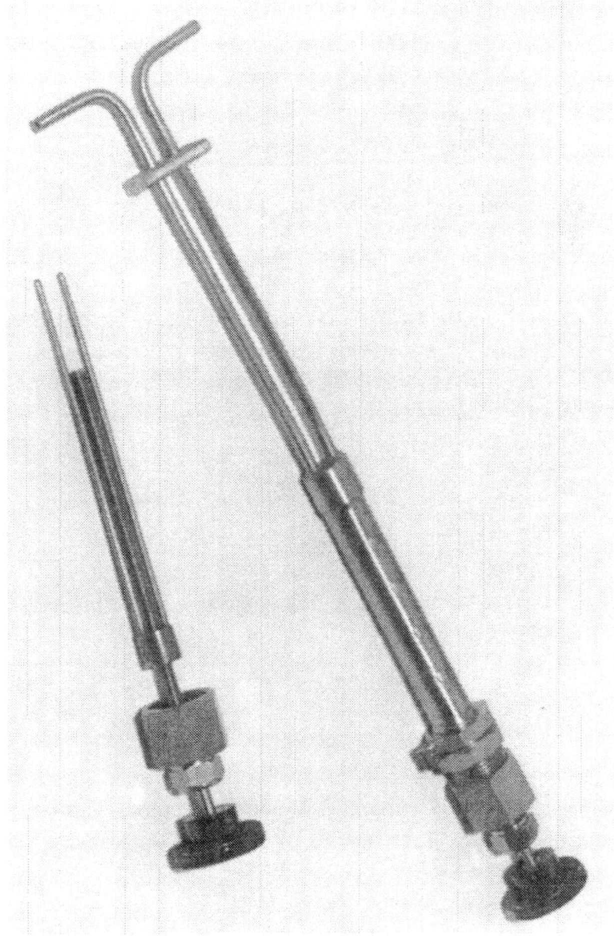

Fig. 15-45. Rod-in-tube type pressure reducing valve. (From Sentry Equipment Corporation. Used with permission.)

valve, is located at the sample source. The other is the sample panel inlet valve. On high-pressure samples (greater than 675 psig), double root valves are normally recommended. A root valve at the sample source provides a means of shutting off flow in the case of a leak in the sample line or at the sample panel.

During operation, the root valve and sample panel isolation valve should be either fully opened or fully closed. A throttling shutoff valve installed in the sample panel should be used to adjust the flow rate to each analyzer.

15.8 WATER STORAGE PROVISIONS

Adequate storage facilities for the water produced by the various treatment systems in a power station are an important consideration for reliable power plant operation. An abundant supply of demineralized water, for instance, permits continued operation of the power plant during regeneration of the ion exchange resin or while emergency maintenance is performed on the demineralization system. The need for sufficient potable and fire water storage is obvious. The sizing of the storage system is as important as the sizing of the water treatment system to meet plant needs. Likewise, the design of the storage tank for the protection of its contents is also crucial. The following subsection addresses the sizing fundamentals of storage tanks, general tank design standards, and design details necessary for tanks to be properly coated with a protective coating system.

15.8.1 Sizing Fundamentals

Bulk storage of water in a power station is generally handled by closed roof, vertical, cylindrical steel tanks supported on the ground. There are no universal sizing requirements that apply to the various storage tanks. Each project is unique when it comes to establishing sizing criteria. Several factors influence the sizing criteria, including the following:

- Reliability of the raw water source and pipeline,
- Treatment rate of the system supplying the water,
- Water usage rate,
- Space availability, and
- The value to the project of having additional operating margin due to water storage.

As part of the decision process, the design engineer must consider the impact to the power plant if water needs are not met.

Ground storage tank heights are usually most economical when selected at 8-ft intervals, to match commonly available steel plate width. Tank heights of 24 to 40 ft are frequently used for water service.

Before a storage tank can be sized, the plant water requirements need to be determined. This includes demineralized water, service water, potable water, and fire water requirements. Good engineering judgment, factoring in such

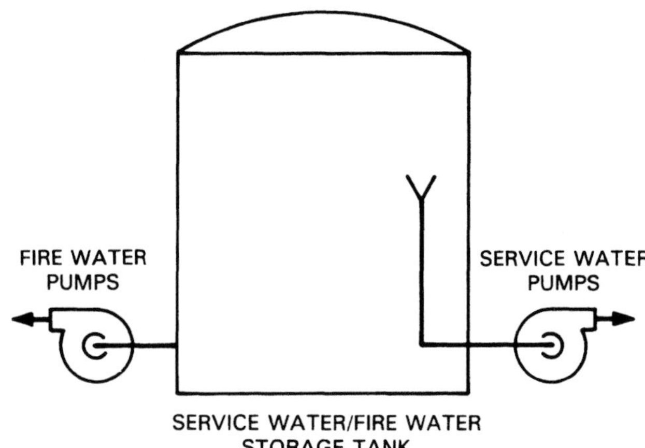

Fig. 15-46. Internal standpipe arrangement for fire water storage.

variables as the reliability of the raw water supply, water quantity requirements, the importance of the generating unit to the utility, and the degree of comfort with the established sizing criteria usually establish the tank size or the need for multiple tanks.

15.8.1.1 Service Water/Fire Water Storage In the case of fire water storage tanks, a minimum quantity of dedicated storage is usually a fire code requirement. Frequently, service water and fire water are stored in a common tank with an internal standpipe used to supply the service water so as to maintain a minimum dedicated supply of fire water, as shown schematically in Fig. 15-46. The following is an example of the approach to sizing a tank with this arrangement.

Example 15-6

A particular plant design requires that sufficient service water be stored for up to a 6-h interruption of the raw water supply. The maximum service water usage rate has been determined to be 420 gpm (26.5 L/s), so the minimum service water storage requirement is as follows:

$$(420 \text{ gpm})(6 \text{ h})(60 \text{ min/h}) = 151{,}200 \text{ gal}$$

$$\left[26.5 \text{ L/s } (6 \text{ h})(3{,}600 \text{ s/h})\left(\frac{\text{m}^3}{1{,}000 \text{ L}}\right) = 572.4 \text{ m}^3\right]$$

The component requiring the greatest amount of fire water in the event of a fire has been determined to be the main transformer. It was estimated that the minimum amount of fire water needed to bring the fire under control is 1,200 gal/min (75.7 L/s) for 2 h. The minimum fire water storage then is as follows:

$$1{,}200 \text{ gpm } (2 \text{ h})(60 \text{ min/h}) = 144{,}000 \text{ gal}$$

$$\left[75.7 \text{ L/s } (2 \text{ h})(3{,}600 \text{ s/h})\left(\frac{\text{m}^3}{1{,}000 \text{ L}}\right) = 545.1 \text{ m}^3\right]$$

The combined minimum requirement then is 151,200 gal + 144,000 gal = 295,200 gal (1,117.5 m^3). In this instance, allowing only a small margin for water level fluctuation, a storage tank with a diameter of 36 ft (11 m) and height of 40 ft (12.1 m) would provide a gross capacity slightly in excess of 300,000 gal (1,135 m^3), satisfying the minimum criteria.

15.8.1.2 Cycle Makeup Storage. Demineralized water storage tanks and/or condensate storage tanks must provide adequate storage volume to allow continued operation of the plant when the demineralizer is being regenerated. The condensate storage tank should be of adequate volume to receive the largest expected condensate dump volume. According to EPRI GS-6699 (EPRI 1990), the minimum recommended working storage capacity for treated makeup water for conventional steam plants is 500 gal/MW (1.89 m^3/MW) of generating capacity. A capacity of 1,000 gal/MW (3.78 m^3/MW) is preferred, especially in areas where water supply is short or where emergency mobile water treatment equipment is not readily available. For plants using nonrecovered process steam or demineralized water for emissions control, the amount of cycle makeup storage must be adjusted accordingly.

The following example illustrates the sizing of cycle makeup storage tanks for a combustion turbine installation.

Example 15-7

A project to install five 85-MW combustion turbines is being planned. Each combustion turbine requires 61,200 lb/h (7.71 kg/s) of demineralized water to meet NO_x emissions limits. The demineralizer can produce 260 gal/min (16.4 L/s) of demineralized water operating for 18 h per day. For regeneration of the demineralizer, 23,000 gal (87,065 L) of demineralized water per day are required. Size a tank(s) to satisfy the following two criteria, assuming that the normal minimum storage level is 75% full.

Criterion 1 12 hours of combustion turbine operation per day for 10 consecutive days at full load operation, with demineralization system operating to provide makeup.

Criterion 2 12 hours of combustion turbine operation per day for 3 consecutive days at full load operation, with no demineralization system makeup.

Solution

For Criterion 1, the amount of water required by the combustion turbines (CT) is determined as follows:

$$(5\text{ CTs})\left(\frac{61{,}200\text{ lb}}{\text{h}}\right)\left(\frac{12\text{ h}}{\text{day-CT}}\right)(10\text{ days})\left(\frac{\text{gal}}{8.34\text{ lb}}\right) = 4{,}403{,}000\text{ gal}$$

$$\left[(5\text{ CTs})(7.71\text{ kg/s})\left(\frac{12\text{ h}}{\text{day-CT}}\right)(10\text{ days})\left(\frac{\text{L}}{1.0\text{ kg}}\right)\left(\frac{3{,}600\text{ s}}{\text{h}}\right)\left(\frac{\text{m}^3}{1{,}000\text{ L}}\right) = 16{,}650\text{ m}^3\right]$$

Water produced by the demineralizer after allowance for regeneration requirements is determined as follows:

$$\left(\frac{260\text{ gal}}{\text{min}}\right)\left(\frac{60\text{ min}}{\text{h}}\right)\left(\frac{18\text{ h}}{\text{day}}\right)(10\text{ days}) - (23{,}000\text{ gal/day})(10\text{ days}) = 2{,}578{,}000\text{ gal}$$

$$\left[(16.4\text{ L/s})\left(\frac{3{,}600\text{ s}}{\text{h}}\right)(18\text{ h/day})(10\text{ days})\frac{\text{m}^3}{1{,}000\text{ L}} = 87.065\text{ L/day (10 days)}\left(\frac{\text{m}^3}{1{,}000\text{ L}}\right) = 9{,}760\text{ m}^3\right]$$

Net requirement:

$$4{,}403{,}000 - 2{,}578{,}000 = 1{,}825{,}000\text{ gal}$$

$$[16{,}650 - 9{,}760 = 6{,}890\text{ m}^3]$$

For Criterion 2, the amount of water required by the combustion turbines is as follows:

$$(5\text{ CTs})\left(\frac{61{,}200\text{ lb}}{\text{h}}\right)\left(\frac{12\text{ h}}{\text{day-CT}}\right)(3\text{ days})\left(\frac{\text{gal}}{8.34\text{ lb}}\right) = 1{,}321{,}000\text{ gal}$$

$$\left[(5\text{ CTs})(7.71\text{ kg/s})\left(\frac{12\text{ h}}{\text{day-CT}}\right)(3\text{ days})\left(\frac{\text{L}}{1.0\text{ kg}}\right)\left(\frac{3{,}600\text{ s}}{\text{h}}\right)\left(\frac{\text{m}^3}{1{,}000\text{ L}}\right) = 4{,}996\text{ m}^3\right]$$

Thus, the tank(s) should be sized for Criterion 1, or about 1,825,000 gal (6,890 m^3). For 75% full tanks then, the gross capacity of the storage should be 1,825,000 gal/0.75 = 2,433,000 gal (6,890 m^3/0.75 = 9,190 m^3). Two tanks, each with a diameter of 72 ft (21.9 m) and height of 40 ft (12.2 m) would be sufficient to satisfy the criteria.

15.8.2 General Design Standards

The majority of all steel potable water storage tanks and other types of water storage tanks constructed in the United States utilize the AWWA Steel Tank Standards. These standards are developed and maintained by special standards committees comprising utilities, engineers, and fabricators. AWWA Standards D100-84, Welded Steel Tanks for Water Storage (AWWA 1984) and D103-80, Factory Coated Bolted Steel Tanks for Water Storage Tanks (AWWA 1987) have been approved as standards by the American National Standards Institute (ANSI).

AWWA D100-84 includes specific requirements for the construction of welded steel tanks. AWWA D100-84 broadly classifies water storage tanks as elevated tanks, reservoirs, or standpipes. Elevated tanks have a supporting structure that elevates the lower operating level of the tank to an elevation that provides a predetermined minimum pressure. Reservoirs and standpipes are water storage tanks that are supported at ground level and are generally referred to as ground storage tanks. Most of the large water storage tanks used in the power industry are ground storage tanks.

A reservoir is defined by AWWA D100-84 as a ground-supported tank which has a shell height less than or equal to the tank diameter. A standpipe is defined as a ground supported tank whose shell height exceeds its diameter. Both of these ground tanks have three basic components: a flat plate bottom, a cylindrical shell, and a roof structure. Figure 15-47 depicts the four most common types of ground storage tank roofs.

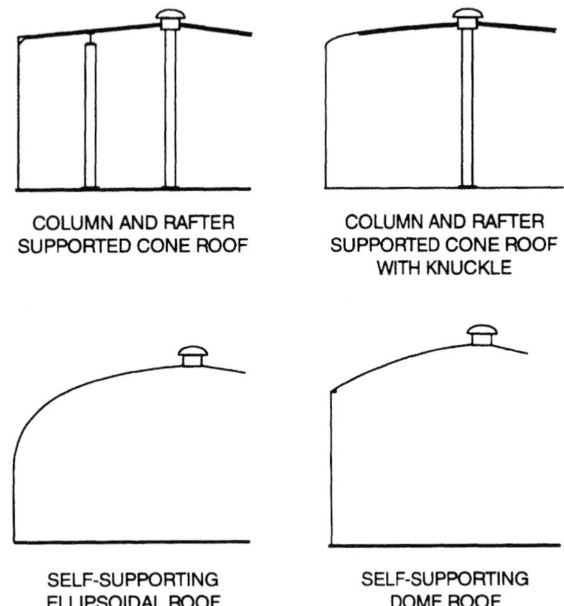

Fig. 15-47. Four common storage tank roofs. (From Steel Plate Fabricators Association. Used with permission.)

AWWA D100-84 has minimum requirements for shell manways, pipe connections, ladders, roof openings, and vents. The design engineer is responsible for specifying the size, orientation, and type of all piping connections, pumping and discharge rates, special vent requirements, ladder safety requirements, and other accessories required for the tank. Additional accessories such as obstruction lights and safety devices may be required by federal and/or state regulations. Adequately sized overflow piping and vacuum breakers must be included in the design of the tanks to prevent damage due to overfilling or collapse by vacuum created during sudden level drops. The minimum operating level in the tank must be considered to provide the minimum net pump suction head required by a pump taking suction from the tank.

Fixed roof tanks must be vented to the atmosphere to prevent overpressurizing the tank during filling or producing a vacuum during draining. If the stored liquid is nearly saturated with oxygen, then oxygen transfer from the air space to the liquid is of little concern. If the stored liquid has been deaerated and it is desirable to minimize contamination of the water with oxygen or carbon dioxide, as is often the case with cycle makeup, water should be added to or withdrawn from the tank beneath the minimum tank level.

Small tanks, typically those having a capacity of up to about 500,000 gal (1,893 m^3), are usually constructed with a height-to-diameter ratio of approximately 1. In larger tanks, this ratio decreases with volume, since the tank height is limited to about 40 to 48 ft (12.2 to 14.6 m). Tank height is economically limited by shell thickness requirements of the bottom course. AWWA D100-84 provides guidelines for calculating the minimum shell thickness for each course of the tank depending on tank height, tank diameter, and specific gravity of the liquid to be stored.

The tank interior may be protected by a cathodic protection system that utilizes a flow of direct electric current from anodes suspended in the water to the surfaces of the tank. The need for cathodic protection should be evaluated based on the quality of the interior coating system to be used, the physical and chemical characteristics of the water contained therein, and the cost of maintaining the system.

15.8.3 Coating Considerations

Storage tank design should allow for easy application of coating to protect the tank and minimize contamination of the tank contents. The National Association of Corrosion Engineers (NACE) offers recommended design practices for coating tanks in NACE Standard RP-01-78, *Design, Fabrication and Surface Finish of Metal Tanks and Vessels to Be Lined for Chemical Immersion Service* (NACE 1978). All interior surfaces of the tank should be readily accessible for surface preparation and coating application.

Thin-film coatings are most frequently used for tank interior coatings in water immersion applications. Careful design of the tank construction details and cleanliness of the surface prior to coating application will yield dividends in the performance and life of the chemical resistant coating applied.

In addition to NACE guidelines for design, fabrication, and surface preparation of tanks to be coated, the Steel Structures Painting Council (SSPC 1982) also offers guidelines for surface preparation prior to coating tanks. Table 15-13 provides a description and comparison of the NACE and SSPC surface preparation guidelines.

For immersion service, a "White Metal" abrasive blast of all steel surfaces in accordance with SSPC-SP5 or NACE No.1 is highly preferred for thin film coatings. Some coating manufacturers allow a "Near-White Metal" blast finish for surfaces that will be used in immersion services. However, the premise that the cleanliness of the surface relates directly to the performance and service life of the coating makes a "White Metal" blast surface highly preferable for immersion service.

Coatings used for protecting carbon steel tanks used to store potable water and high-purity water are generally thermosetting, resin-based materials. Epoxies and vinyls have been the most commonly employed coatings for interior surfaces of water storage tanks; however, vinyls generally have lost favor due to environmental concerns related to the high concentrations of volatile organic compounds that can be generated during the application process. The determining factor in selecting a particular coating is the type of fluid that will be contained in the tank. Coating manufacturers should be consulted to obtain their recommendations for coating systems to be used in the various applications.

NACE Publication TPC-2, "Coatings and Linings for Immersion Service," and the various coating manufacturers

Table 15-13. The Four Grades of Blast Cleaning

Designation		
NACE	SSPC	Description
No. 1	SP5	White metal blast. Defined as a gray-white (uniform metallic) color, slightly roughened to form a suitable anchor pattern for coatings. This surface is free of all oil, grease, dirt, mill scale, rust, corrosion products, oxides, paint, and other foreign matter.
No. 2	SP10	Near-white metal blast. Defined as a surface from which all oil, grease, dirt, mill scale, rust, corrosion products, oxides, paint or other foreign matter have been removed except for light shadows, streaks, or slight discolorations (of oxide bonded with metal). At least 95% of any given surface area has the appearance of NACE No. 1, and remainder of the area is limited to light discolorations.
No. 3	SP6	Commercial blast. Defined as a surface from which all oil, grease, dirt, rust scale, and foreign matter have been completely removed and all rust, mill scale, and old paint have been removed except for slight shadows, streaks, or discolorations caused by rust stain or mill scale oxide binder. At least two thirds of the surface will be free of all visible residues and the remainder will be limited to light discoloration, slight staining, or light residues mentioned above.
No. 4	SP7	Brush-off blast. Defined as a surface from which oil, grease, dirt, loose rust scale, loose mill scale, and loose paint are removed, but tightly adhering mill scale, rust, paint and coatings are permitted to remain if they have been exposed to the abrasive blast pattern.

provide recommendations for paint systems materials and their application to the interior and exterior of potable water storage tanks. In addition to NACE TPC-2, interior painting systems for potable water storage tanks must meet the minimum requirements of the National Sanitation Foundation Standard 61 (NASF 1992), with regard to acceptable contaminant levels contributed by the materials used.

Coatings used to protect the interior surfaces of high-purity water storage tanks must be carefully selected to protect the contents of the tank, that is, the coating system should not contribute contaminants to the high-purity water.

An interior coating must withstand constant immersion in the lower areas of the tank, be able to resist alternate wetting and drying in the upper portion of the operation range and high humidity above the high water level, be resistant to the actions of ice abrasion in cold climates, and adequately withstand temperature extremes. The system must also be receptive to application conditions for it to be cost effective. The integrity of the interior coating system of a tank is only as good as the quality of surface preparation, application, and inspection. Annual inspection of the interior surfaces permits minor coating deficiencies to be corrected, thus preventing the propagation of coating failures. The maximum interval for periodic inspection should be no longer than 3 years. The only effective way to ensure long coating life is through proper periodic cleaning and maintenance.

15.9 REFERENCES

Allianz Berichte. Alliance Center for Technology. P.O. Box 24, D 8000. Munich 44, Germany.

American National Standards Institute. *American National Standard Code for Pressure Piping, Power Piping.* ANSI/ASME B31.1, 1986 edition, including all addenda through ANSI/ASME B31.1b-1987.

American Society for Testing and Materials. 1988. *Annual Book of ASTM Standards*, Volume 01.01, Specification A213. Philadelphia, PA.

AWWA. 1984. *Welded Steel Tanks for Water Storage.* AWWA D100-84. American Water Works Association, Denver, CO.

AWWA. 1987. *Factory Coated Bolted Steel Tanks for Water Storage Tanks.* AWWA D103-87. American Water Works Association, Denver, CO.

Bell, M. J., J. H. Hicks, A. Banweg, and J. A. Lux. 1981. A manufacturer's view of sampling protocol in high-pressure utility steam generating plants. Paper read at American Power Conference, 27–29 April 1981, Chicago, IL.

Betz. 1980. *Betz Handbook of Industrial Water Conditioning.* Betz Laboratories, Inc., Trevolse, PA.

Bussert, B. W., R. M. Curran, and G. C. Gould. 1978. The effect of water chemistry on the reliability of modern large steam turbines. Paper read at the ASME/IEEE Joint Power Generation Conference, 10–14 September 1978, Dallas, TX.

Callahan, F. J. 1985. *Swagelok Tube Fitting and Installation Manual.* Markad Service Company, Solon, OH.

Cheremisinoff, N. P. and D. S. Azbel. 1983. *Liquid Filtration.* Ann Arbor Science Publishers, Woburn, MA.

Cull, D. G. 1986. *Specifying AWWA Steel Water Storage Tanks.* Tank Industry Consultants, Inc., Speedway, IN.

Davis, S. N. and R. J. M. DeWiest. 1966. *Hydrogeology.* John Wiley & Sons, New York, NY.

Degrémont. 1991. *Water Treatment Handbook*, Volumes 1 and 2. Lavoisier Publishing, Paris, France.

Drew Chemical Corporation. 1984. *Principles of Industrial Water Treatment.* Drew Chemical Corporation, Boonton, NJ.

DuPont. "Permasep" Products. Building 200, Glasgow Site, Wilmington, DE 19898.

Electric Power Research Institute. 3412 Hillview Avenue. P. O. Box 10412, Palo Alto, CA 94303.

Electric Power Research Institute. 1986. *Interim Consensus Guidelines on Fossil Plant Cycle Chemistry*, CS-4629. Electric Power Research Institute, Palo Alto, CA.

Electric Power Research Institute. 1987. *Guideline Manual on Instrumentation and Control for Fossil Plant Cycle Chemistry*, CS-5164. Electric Power Research Institute, Palo Alto, CA.

Electric Power Research Institute. 1990. *Guidelines for Makeup Water Treatment*, GS-6699. Electric Power Research Institute, Palo Alto, CA.

Electric Power Research Institute. 1994. *Cycle Chemistry Guidelines for Fossil Plants: Oxygenated Treatment*, TR-102285. Electric Power Research Institute, Palo Alto, CA.

FILMTEC Corporation. 1985. *FILMTEC Membranes FT 30 Membrane Description Technical Bulletin*, 500AA F.s. FILMTEC Corporation, Minneapolis, MN.

FILMTEC Corporation. 1989. *FILMTEC Reverse Osmosis Systems Analysis Manual*. FILMTEC Corporation, Minneapolis, MN.

Freeze, R. A. and J. A. Cherry. 1979. *Groundwater*. Prentice-Hall, Englewood Cliffs, NJ.

Fluid Systems Corporation. A member of Anglian Water Group. 10054 Old Grove Road, San Diego, CA 92131.

Greenborg, A. E., L. Clesceri, and S. Eaton. 1992. *Standard Methods for the Examination of Water and Wastewater*, 18th Edit. American Public Health Association, American Water Works Association, and Water Environment Federation, Washington, DC.

Hydranautics. 8444 Miralani Drive, San Diego, CA 92126.

IDE Technologies. Ambient Technologies, Inc., 2999 NE 191st Street, Suite 407. North Miami Beach, FL 33180.

Infilco Degremont, Inc. 2924 Emerywood Parkway, P.O. Box 71390, Richmond, VA 23255-1390.

Ionics, Inc. Water Systems Division. 65 Grove Street, Watertown, MA 02172.

Johnson March Systems, Inc. 220 Railroad Drive, Ivyland, PA 18974.

Kemmer, F. N. 1979. *The NALCO Water Handbook*. McGraw-Hill, New York, NY.

Kremen, S. 1984. Membrane systems. Paper read at the International Symposium—Synthetic Membranes, 20–24 August 1984, Michigan Molecular Institute, Millend, MI.

Larkin, B. A., et al. 1991. Optimum cycle chemistry for once-through units. Paper presented at the EPRI International Conference on Cycle Chemistry, 4–6 June. Baltimore, MD.

Leung, P. and R. E. Moore. Water consumption determination for steam power plant cooling towers: a heat-and-mass balance method. Paper read at the ASME Winter Annual Meeting, 16–20 November 1969, Los Angeles, CA.

Moffat, W. S. Groundwater—an introduction. *Developing World Water*, WEDC (ed.), 57 pp. Grosvenor Press International, Hong Kong.

NACE. 1972. *NACE Coatings and Linings for Immersion Service*, NACE Publication TPC-2. National Association of Corrosion Engineers, Houston, TX.

NACE. 1978. *NACE Recommended Practice: Design, Fabrication, and Surface Finish of Metal Tanks and Vessels to Be Lined for Chemical Immersion Service*, NACE Standard RP-01-78. National Association of Corrosion Engineers, Houston, TX.

NSF. 1992. *Drinking Water System Components—Health Effects*, Standard 61-1992. American National Standards Institute/National Sanitation Foundation International, Ann Arbor, MI.

Pontius, F. W., editor 1990. *Water Quality and Treatment*. McGraw-Hill, New York, NY.

Powell, T. 1954. *Water Conditioning for Industry*. McGraw-Hill, New York, NY.

Rios, J. and F. X. Blood. 1981. Design and operation of sampling and analysis system for high-pressure utility steam generating plants. Paper read at the American Power Conference, 27–29 April 1981, Chicago, IL.

Rohm & Haas. Independence Mall, W. Philadelphia, PA 19106.

Salem, E. and M. J. O'Brien. 1991. Powdered resin equipment design improvements. Paper presented at the EPRI Conference on Filtration of Particulates in LWR Systems, 10–12 September, King of Prussia, PA.

Sanks, R. L. 1978. *Water Treatment Plant Design*. Ann Arbor Science Publishers Inc., Ann Arbor, MI.

Sasakura Engineering Company, Ltd. 7-32, Takeshima 4-Chome. Nishiyodogawa-ky, Osaka 555, Japan.

Sentry Equipment Corporation. P.O. Box 127, Oconomowoc, WI 53066.

SSPC. 1982. *Steel Structures Painting Council*, 3rd edit., Volume 2, Chapter 2, and as amended by the supplement to the Third Edition, 1985. Steel Structures Painting Council, Pittsburgh, PA.

Strauss, S. D. 1981. Control of Turbine-Steam Chemistry. *Power*. 125(3):33–42.

Steel Plate Fabricators Association. 2400 S. Downing Avenue, Westchester, IL 60154.

Stumm, W. and J. J. Morgan. 1970. *Aquatic Chemistry*. John Wiley & Sons, New York, NY.

Tank Lining Guide. 1989. By the Carboline Co., 5940 Reeds Road, Shawnee Mission, KS 66202.

U. S. Filter/Permutit. 1994. *Water and Waste Treatment Data Book*. United States Filter Corporation, Warren, NJ.

Wagner, J. H. 1977. Solids contact devices—past, present and the state of the art. Paper read at 15th Annual Liberty Bell corrosion course, May 1977, Philadelphia, PA.

Waters Equipment Co. P.O. Box 576, Lansdale, PA 19446.

16

Liquid and Solid Waste Treatment and Disposal

Kenneth R. Weiss

16.1 INTRODUCTION

In 1972 the Federal Water Pollution Control Act Amendments were passed, and in 1977 the Clean Air Act Amendments were passed. These laws profoundly changed the way power generation facilities handled their waste byproducts. Various updates to those acts and other legislation, both federal and state, have further regulated the release of power generation byproducts to the environment.

The process of electric power generation is not completely efficient. The best way to begin examining waste issues facing the power generation industry is with an overall mass balance. Figure 16-1 illustrates a typical modern coal-fueled power plant overall mass balance. Entering the power plant is coal for fuel; limestone for sulfur dioxide control; water supplies for steam production, cooling, and other uses; and the various chemicals used for water treatment and conditioning, maintenance, and other uses. The coal contains the carbon and other organic material used for fuel, but also a small amount of various sulfur compounds plus a wide variety of other organic and inorganic materials, including heavy metals. Limestone contains the calcium carbonate used to react with sulfur in the coal but may also contain traces of other substances.

Depending on its source, the water supply contains various dissolved and suspended constituents, mostly inorganic in nature. Chemicals are used in power plant water treatment, such as lime and coagulants for water pretreatment; sulfuric acid, hydrochloric acid, and sodium hydroxide for ion exchange resin regeneration; ammonia, hydrazine, and sodium phosphates for steam cycle water conditioning; and chlorine, sulfuric acid, and scale and corrosion inhibitors for cooling water conditioning. Other potential substances imported include various acids, bases, and organic compounds used in maintenance, including periodic chemical cleaning of the boiler waterside surfaces.

As first stated by Lavoisier in 1785, the sum of the masses of substances entering into a chemical reaction or process must be equal to the sum of the masses of the products. At a power plant, therefore, all materials imported must be exported or retained. Various chemical reactions take place in a power plant that change the characteristics of the materials imported. These reactions result in waste products that must either be released to the environment or retained on site.

The principal reaction in a coal-fueled power plant is the coal combustion reaction, which converts carbon and other compounds in the coal, including sulfur, to their oxides, thereby generating heat. In addition to the gaseous pollution concerns addressed in Chapter 14, the combustion reactions and supporting plant systems produce various liquid and solid products. These products are the subject of this chapter.

In terms of quantity, the major solid wastes produced at a power plant are the solid ash remaining from coal combustion and the solid calcium–sulfur compounds created during the flue gas desulfurization process.

Liquid wastewater originates from many places throughout the power plant. One major source is the water treatment systems that purify the water used in power generation. Wastewater from the water treatment systems includes reaction products and sludge from water pretreatment processes, and acid and caustic demineralizer regeneration wastes. Another major source of wastewater is from the water-related systems used in power generation, including condenser and equipment cooling system wastewater, boiler blowdown, and miscellaneous floor drains. A third major source is the various gaseous pollution control systems installed in modern power plants, such as wastewater from flue gas desulferization processes, and bottom ash and fly ash sluicing systems.

Wastewater also results from general plant upkeep. After all, power generation is a large industrial process requiring continuing maintenance. Various lubricating fluids can be inadvertently or purposely released to floor drain systems. Wastewater results from various equipment cleaning operations. Also, because a large staff is needed to operate and maintain a power plant, sanitary wastewaters result. Lastly, rainfall water from disturbed areas of the site can dissolve and pick up constituents that may be undesirable for release to the environment.

16.2 FEDERAL REGULATORY REQUIREMENTS

Any discussion of power plant wastes and how to handle them must be viewed from the perspective of regulatory requirements. The regulatory requirements applicable to power plants have undergone various changes and refinements in the past 20 years, and these refinements are expected to continue. The primary impetus has been the better

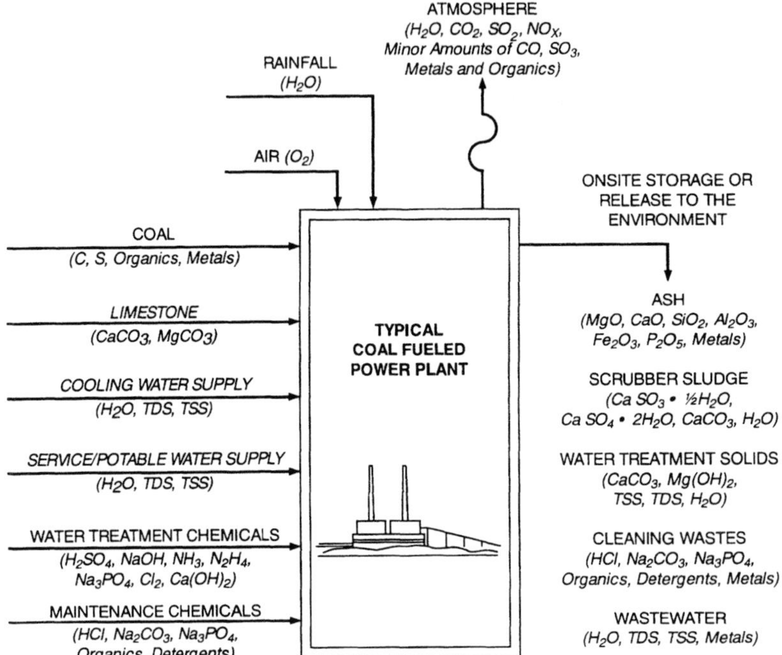

Fig. 16-1. Power plant overall chemical balance.

scientific understanding, as well as an increased public awareness of environmental risks, reflected by various legislative acts, particularly on the federal level.

Various federal legislative acts impacting power plant liquid and solid waste handling are listed in Table 16-1. Two federal acts that influence the power generation industry significantly from the standpoint of liquid and solid wastes treatment and disposal are described in the following sections.

Table 16-1. Federal Legislative Acts Impacting Power Plant Waste Handling

Act	Impact on Power Plants
Water Pollution Control Act or Clean Water Act (33 U.S.C. 1251 et seq.)	Regulates wastewater discharges to waters of the US and publicly owned treatment plants (POTW)
Resource Conservation and Recovery Act (42 U.S.C. 6901 et seq.)	Regulates generation and disposal of hazardous waste
Safe Drinking Water Act (42 U.S.C. 300 f et seq.)	Regulates potable water supply quality
Toxic Substances Control Act (15 U.S.C. 2601 et seq.)	Regulates discharge of toxic substances, including PCBs and asbestos
Natural Environmental Policy Act (42 U.S.C. 4341 et seq.)	Requires filing of an Environmental Impact Statement prior to major federal actions; i.e., issue of a NPDES permit

16.2.1 Federal Water Pollution Control Act

In 1972 Congress passed the Federal Water Pollution Control Act Amendments (FWPCA) and in 1977 enacted substantial amendments to the FWPCA. The revised act is better known as the Clean Water Act. In 1974, the Environmental Protection Agency (EPA) established effluent regulations for existing and new sources. Also established were pretreatment requirements for discharges into publicly owned treatment works (POTW). These requirements have been expanded and refined over time. Table 16-2 summarizes current regulation discharge limits applicable to power plants.

The effluent limitations for steam electric power generating units appear in 40 Code of Federal Regulations (CFR) Part 423. Part 423 is divided into three subparts to deal with three categories of generation facilities. Categories include effluent limitation guidelines representing the degree of effluent reduction attainable by the application of the best practicable control technology currently available (BPT), effluent limitation guidelines representing the degree of effluent reduction attainable by the application of the best available technology economically achievable (BAT), and the new source performance standards (NSPS).

The three subparts contain effluent limitations for three categories of pollutants:

- Conventional pollutants (biochemical oxygen demand, total suspended solids, pH, fecal coliform, and oil and grease) are subject to effluent limitations based on best conventional pollutant control technology (BCT).
- Toxic pollutants identified by the regulations are subject to BAT effluent limitations.

- All other pollutants are called nonconventional pollutants and are subject to BAT effluent limitations.

16.2.2 The Resource Conservation and Recovery Act

The Resource Conservation and Recovery Act (RCRA) was enacted in 1976 to control hazardous waste. These regulations impact coal fueled power plants in several ways. Presently excluded from hazardous waste designation are fly ash, bottom ash, and flue gas desulfurization waste products. However, certain other wastes generated at power plants may meet the criteria for hazardous waste designation.

RCRA mandates that all solid waste be designated as

Table 16-2. Discharge Limits for Steam Electric Plants[a,b]

	BPT[c] Limit, mg/L		BAT[d] Limit, mg/L		NSPS[e], mg/L	
Stream, pollutant	Max[f]	Avg[g]	Max	Avg	Max	Avg
All streams						
pH (except once-through cooling)	6.0–9.0		6.0–9.0		6.0–9.0	
PCBs	No discharge		No discharge		No discharge	
Low-volume waste streams						
Total Suspended Solids (TSS)	100	30	100	30	100	30
Oil and grease	20	15	20	15	20	15
Bottom-ash transport water						
Total Suspended Solids (TSS)	100	30	100[f]	30[h]	100[g]	30[i]
Oil and grease	20	15	20[h]	15[h]	20[i]	15[i]
Flyash transport water						
Total Suspended Solids (TSS)	100	30	100	30	No discharge	
Oil and grease	20	15	20	15	No discharge	
Chemical metal-cleaning wastes						
Total Suspended Solids (TSS)	100	30	100	30	100	30
Oil and grease	20	15	20	15	20	15
Copper (total)	1	1	1	1	1	1
Iron (total)	1	1	1	1	1	1
Boiler blowdown						
Total Suspended Solids (TSS)	100	30	100	30	100	30
Oil and grease	20	15	20	15	20	15
Copper (total)	1	1	1	1	1	1
Iron (total)	1	1	1	1	1	1
Once-through cooling water						
Free available chlorine[j]	0.5	0.2	0.2	—	0.2	—
Cooling tower blowdown						
Free available chlorine[j]	0.5	0.2	0.5	0.2	0.5	0.2
Zinc[k]	N/A	N/A	1	1	N/D	N/D
Chromium[k]	N/A	N/A	0.2	0.2	N/D	N/D
Other corrosion inhibitors	Limits determined on a case-by-case basis					
Materials-storage runoff[k]						
Total Suspended Solids (TSS)		50		50		50

NA = not applicable; ND = none detectable.

[a]Except where specified otherwise, allowable discharge equals flow multiplied by concentration limitation. Where waste streams from various sources are combined for treatment or discharge, quantity of each pollutant attributable to each waste source shall not exceed the specified limitation for that source.

[b]All sources must meet state water quality standards by 1977 [Section 301 (b)(1)(c)].

[c]Best practicable control technology (BPT) currently available.

[d]Best available technology (BAT) economically achievable.

[e]New source performance standards (NSPS).

[f]Maximum for any one day.

[g]Average of daily values for 30 consecutive days.

[h]Allowable discharge equals flow multiplied by concentration divided by 12.5.

[i]Allowable discharge equals flow multiplied by concentration divided by 20.0.

[j]Limits given are maximum and average concentrations. Neither free available chlorine nor total residual chlorine may be discharged from any unit for more than 2 hours in one day; not more than one unit of any plant may discharge free available or total residual chlorine at the same time unless the utility can demonstrate that the units in a particular location cannot operate at or below this level of chlorination. BAT and NSPS limits for plants with a total rated generating capacity of <25 megawatts are 0.5 mg/L maximum, 0.2 mg/L average.

[k]Not applicable for BPT; no detectable discharge for new sources.

[l]Runoff from detention pond must comply with these limits except overflow from a detention pond designed, constructed, and operated to hold the runoff from a 10-year, 24-hour rainfall event.

either hazardous or nonhazardous. The regulations specify that a solid waste is considered hazardous if it contains a material listed as hazardous in the regulations or if it is corrosive, ignitable, reactive, or toxic. The regulations also specify that a generator of solid waste must determine whether the waste is hazardous. This can be accomplished by either testing the waste or applying knowledge of the characteristics of the waste in light of the materials or processes used. The regulations also specify that any test results or waste analysis performed must be retained for at least 3 years.

If the solid waste contains a listed hazardous waste or is determined or conceded to be a hazardous waste, the generator must obtain an identification number from the EPA and file the appropriate forms. The generator must also file annual reports with the EPA. Furthermore, if the hazardous waste is treated, stored, or disposed of offsite, the generator is subject to the manifest and packaging requirements of the RCRA regulations. Power plant wastes and the potential impact of RCRA on their handling and disposal are discussed in the following sections.

16.2.2.1 Fly Ash, Bottom Ash, and Desulfurization Solid Waste. The RCRA regulations currently state that fly ash, bottom ash, and scrubber sludge are not hazardous wastes. Consequently, these wastes can usually be disposed of in a facility meeting the EPA standards for a sanitary landfill. (Note that most states and many localities have detailed regulations on solid wastes that can affect landfill plans.) The most significant requirements of the EPA standards for sanitary landfills are to either use natural soil or synthetic liners to provide protection against ground water contamination or to demonstrate the leachate does not pose a significant danger of contaminating ground water. In any case, monitoring of the ground water quality will almost certainly be required.

16.2.2.2 Acid and Caustic Waste. Wastewater is defined as corrosive, thus hazardous, if the pH is <2 or >12.5 or if the waste corrodes SAE steel at a rate >6.35 mm (0.250 in.) per year. Typical power plant wastes with these characteristics are makeup demineralizer regeneration wastes, condensate demineralizer regeneration wastes, and chemical metal cleaning wastes.

The impact of RCRA on these wastes depends on the method used to store and treat them. One treatment option is to pipe the wastes directly to a neutralization tank or basin. The advantage of this option is that this type of facility can be designed to be a "totally enclosed treatment facility." Facilities that are directly connected to an industrial production process and that are constructed and operated in a manner that prevents the release of any hazardous waste are exempted from the RCRA permit requirements. However, the utility is still subject to the generator notification and waste testing requirements.

Another option is to dispose of these wastes by transporting them to an ash pond or recycle pond that recycles all water except that which evaporates. RCRA regulations exempt users or treatment and storage facilities that recycle or reuse wastes.

16.2.2.3 Waste PCBs and Other Chemicals. The RCRA regulations contain a long list of chemical compounds deemed to be hazardous. Listed compounds that might be encountered at power plants include polychlorinated biphenyls (PCBs), spent solvents, pesticides, and herbicides. Wastes that contain these listed substances must be stored, treated, and disposed of in accordance with the hazardous waste management regulations.

Consequently, if a utility expects to handle waste PCBs from old transformers or waste herbicides, pesticides, or any other chemical compound listed as a hazardous waste, the utility should obtain an identification number and comply with the manifest and packaging requirements for offsite disposal or with the permit requirements for onsite disposal. The regulations allow onsite storage of hazardous waste for 90 days without a permit if the utility has an identification number and all the waste is disposed of offsite and is packaged in accordance with US Department of Transportation regulations.

16.3 STATE REGULATORY REQUIREMENTS

In company with federal regulations, state regulations for discharge of liquid and solid wastes have become more strict. Separate regulations to supplement the federal regulations have been promulgated by all states. Usually, the method of liquid and solid waste disposal is analyzed and regulated as part of the state power plant licensing and permit renewal process.

16.3.1 Liquid Wastes

States have jurisdiction over wastewater discharges within their boundaries. Nearly all states have or are in the process of developing standards for maintaining the quality of surface waters and ground waters. The standards are determined based on the present condition of the stream, the indigenous species, and the desired use for the stream. State water quality standards often necessitate more stringent limitations on a power plant's wastewater discharge than otherwise might be required by the federal regulations.

16.3.2 Combustion Wastes

In many states, the permit for operation of the combustion waste disposal landfill is issued separately from the overall permits for station construction or certification.

In general, state regulations for combustion waste disposal landfills have evolved from regulations for municipal sanitary waste landfills. These regulations provide specific limitations and requirements for the siting, design, and operation of a combustion waste disposal landfill. The submittals to secure an operation permit require extensive site studies and generally complete engineering design, including detailed methods for operation of the disposal landfill. To ensure final closure of waste disposal landfills without adding a financial burden to the state, the site operator is often required to provide sufficient financial assurance of the operator's ability to complete disposal landfill closure at the start of disposal operations.

Combustion waste disposal permits usually include some or all of the following siting requirements:

- The site must be outside the 100-year floodplain.
- The site must be consistent with existing land uses.
- The site must have sufficient buffer zones from site boundaries or existing facilities.
- A minimum separation between the base of the disposal site and ground water must be provided.
- Surface water entering the disposal site must be diverted or controlled.
- Aerial photography of the disposal site and immediate vicinity must be performed before development.
- Topographic surveys of the disposal site and immediate vicinity must be performed.
- Hydrogeologic studies must be performed, including aquifer descriptions, water levels and quality, geologic description, and geotechnical engineering parameters of subsurface materials.

A state typically requires a detailed description of the utility's plans for site development be submitted and approved prior to development. The following aspects of site development must be addressed in detail:

- Site grading,
- Liner and leachate collection systems,
- Storm water management systems,
- Sedimentation control systems,
- Disposal area containment berms,
- The system of haul roads,
- Disposal area fencing and security, and
- Detailed construction specifications.

In addition, the state generally requires submittal and approval of a detailed operations plan. The operations manual should address the following:

- Storm water control procedures,
- Fugitive dust control procedures,
- Transportation methods and traffic control,
- Stage development of disposal site,
- Leachate collection and disposal procedures,
- Ground water monitoring program,
- Safety program,
- Landfill maintenance procedures,
- Combustion waste placement and compaction requirements,
- Emergency action plans,
- Training programs, and
- Documentation and recordkeeping procedures.

Generally, the state requires detailed closure provisions to be included in the permit application. This documentation includes detailed grading drawings, cap and closure details, financial assurance documents, maintenance requirements and procedures, leachate control provisions, and security arrangements. In addition, the documentation must describe the proposed ultimate land use and post-closure monitoring provisions.

The state generally requires development of a ground water monitoring plan. The plan includes monitoring well locations, construction specifications, sampling procedures, laboratory test procedures, and quality control procedures.

The site studies, engineering design, operation manuals, and supporting documentation must be submitted to and reviewed by the appropriate state agency. During this process, changes may be requested. The permit process for a disposal site may take a year or more before initial site development is initiated.

16.4 LIQUID WASTE TREATMENT AND DISPOSAL

16.4.1 Description of Wastewater Sources

Various processes take place in a power plant that generate wastewater. In this section, the major power plant processes and operations that result in wastewater are outlined. The interactions between plant design and the types and quantities of wastewaters are considered.

16.4.1.1 Once-Through Cooling Water. As discussed in Chapter 12, heat must be rejected from the power plant condenser to maintain the proper cycle thermal characteristics. Generally, a large flow of water is used to condense the spent steam from the turbine.

The transfer of heat from the condensing stream raises the temperature of the cooling water. For a given amount of heat transfer, the temperature rise is inversely proportional to the flow.

This condenser cooling water is the largest flow of water within the power plant. Assuming a 10-degree temperature rise within the condenser, a 400 megawatt (MW) power plant requires approximately 200,000 gallons per minute (gpm), or about 760,000 liters per minute (L/min), of cooling water.

Once-through cooling involves taking water from a source, routing the water through the power plant condenser, and discharging the water. For once-through cooling to be practical, a very large source of water must be available. This source is typically available only on sites near the ocean, a large lake, or a major river.

Because of the very large flow, it is not practical to reuse any significant part of the once-through cooling water in the power plant. The heated water must therefore be discharged, probably back to the source from which it came.

Since increased temperature often has a significant effect on fish, plants, and other organisms within the receiving waters, a waiver under Section 316(a) of the Clean Water Act is required to discharge once-through cooling water. Such a waiver will be granted only if the thermal discharge does not endanger the protection and propagation of a balanced, indigenous population of shellfish, fish, and wildlife in and on the receiving water body.

In practice, obtaining a 316(a) waiver has been very difficult, and most recent power plants, including those adjacent to large bodies of water, have been built with cooling towers or cooling lakes.

Shock chlorination is often used on once-through cooling water to reduce biological fouling in the condenser. To meet federal standards for discharge of total residual chlorine, chemical dechlorination using sulfur dioxide or other means of chlorine residual minimalization in the discharge is typically required.

16.4.1.2 Cooling Tower Blowdown. An alternative condenser cooling process is to circulate the large condenser cooling flow through a cooling tower. In the cooling tower, warm cooling water from the power plant condenser is cooled by the evaporation of a portion of the water flow as it passes through the cooling tower. In addition to evaporation, a very small amount of entrained water, called drift, is also lost. As evaporation continues, salts dissolved in the remaining cooling liquid become more concentrated. When the concentrations of the dissolved salts near their solubility limit, scale formation may occur on the condenser tubes and hinder heat transfer. Although addition of certain chemicals can inhibit scale formation, a portion of the cooling water, called blowdown, must be removed. The blowdown can be calculated in the following manner:

$$\text{Blowdown} = \frac{\text{Evaporation}}{\text{Cycles} - 1} - \text{Drift} \qquad (16\text{-}1)$$

The cycles, or "cycles of concentration," relate how much the dissolved solids are allowed to concentrate in the cooling water system. For instance, if the cooling water makeup calcium ion concentration is 100 mg/L and the control limit to ensure scaling in the condenser does not occur is 900 mg/L, the number of cycles of concentration is nine.

$$\frac{900 \text{ mg/L as } CaCO_3}{100 \text{ mg/L as } CaCO_3} = 9.0 \text{ cycles} \qquad (16\text{-}2)$$

The blowdown stream contains all constituents in the makeup stream, multiplied by the cycles of concentration. Adjustment must be made if sulfuric acid is added for alkalinity control as sulfates are added.

Several factors may be of concern in a cooling tower blowdown discharge to the environment. If the blowdown is taken downstream of the condenser prior to the cooling tower, temperature could be of concern. Chlorination, used for control of biofouling, results in a chlorine residual in the cooling tower blowdown stream. Since the cooling tower circuit concentrates dissolved solids that exist in the makeup water, high salinity may affect the ability to discharge or reuse the blowdown.

16.4.1.3 Ash Transport Water. Coal-fueled electricity generating plants produce massive amounts of ash as a waste product of combustion. In a pulverized coal unit, three kinds of ash are produced: bottom ash, economizer ash, and fly ash. Bottom ash is the residue that accumulates at the bottom of the furnace. Economizer ash accumulates in hoppers under the economizer section of the boiler. Fly ash is carried in the flue gas stream. A typical breakdown is 20% bottom ash, 4% economizer ash, and 76% fly ash. The amount of ash produced is directly related to the amount of inert material present in the fuel and the quantity of fuel burned.

In a coal furnace, a wet bottom seal is generally used, requiring continual replenishment with water. The water at the bottom of the furnace also cools the ash. Periodically, the ash must be wet-sluiced from the furnace to either its permanent or an intermediate disposal point.

Economizer and fly ash is generally collected from the flue gas in mechanical separators or electrostatic precipitators in a dry form. Economizer and fly ash can either be wet sluiced or pneumatically blown to a final or intermediate disposal point. Dry handling of this ash is more common in new facilities than is wet sluicing.

Pollution concern in ash transport water includes high levels of suspended solids, high or low pH levels, and pickup of heavy metals that remain in the ash after combustion.

For many plants, the ash is transported to ponds, where it is allowed to settle and the water discharged or reused. Representative water quality data for ash ponds are given in Table 16-3.

16.4.1.4 Flue Gas Desulfurization Wastewater. Chapter 14 describes various types of air pollution control systems, including flue gas desulfurization (FGD) systems. The quantity of solids produced in an FGD system is related to the quantity of fuel burned, the amount of sulfur in the fuel, and the FGD process used. The solid waste products generally consist of a combination of the following:

- Calcium sulfate dihydrate ($CaSO_4 \cdot 2H_20$);
- Calcium sulfite hemihydrate ($CaSO_3 \cdot 1/2H_20$);
- Unreacted scrubber additive [usually $Ca(OH_2)$ or $CaCO_3$];
- Inert material contained in scrubber additive;
- Coal ash not removed by upstream electrostatic precipitators or baghouses; and
- Minor amounts of other precipitated solids, such as $Mg(OH)_2$.

The amount of each solid material produced must be determined on an individual basis for each unit, based on the design coal, design scrubber additive, and the FGD process used.

Most modern FGD systems are designed, to some extent, to reuse wastewaters that result from the desulfurization process. However, in many cases, complete reuse cannot be achieved and a wastewater discharge is necessary from the

Table 16-3. Representative Quality of Ash Pond Discharges[a]

Parameter	Fly Ash Pond			Bottom Ash Pond		
	Minimum	Average	Maximum	Minimum	Average	Maximum
Flow (gpm)	3,100	6,212.5	8,800	4,500	16,152	23,000
Total alkalinity (as $CaCO_3$)	—	—	—	30	85	160
Phen. alkalinity (as $CaCO_3$)	0	0	0	0	0	0
Conductivity (μmhos/cm)	615	810	1,125	210	322	910
Total hardness (as $CaCO_3$)	185	260.5	520	76	141.5	394
pH	3.6	4.4	6.3	4.1	7.2	7.9
Dissolved solids	141	508	820	69	167	404
Suspended solids	2	62.5	256	5	60	657
Aluminum	3.6	7.19	8.8	0.5	3.5	8.0
Ammonia (as N)	0.02	0.43	1.4	0.04	0.12	0.34
Arsenic	<0.005	0.010	0.023	0.002	0.006	0.15
Barlum	0.2	0.25	0.4	<0.10	0.15	0.30
Beryllium	<0.01	0.011	0.02	<0.01	<0.01	<0.01
Cadmium	0.023	0.037	0.052	<0.001	0.0011	0.002
Calcium	94	136	180	23	40.12	67
Chloride	5	7.12	14	5	8.38	15
Chromium	0.012	0.067	0.17	<0.005	0.009	0.023
Copper	0.16	0.31	0.45	<0.01	0.065	0.14
Cyanide	<0.01	<0.01	<0.01	<0.01	<0.01	<0.01
Iron	0.33	1.44	6.6	1.7	5.29	11
Lead	<0.01	0.058	0.2	<0.01	0.016	0.031
Magnesium	9.4	13.99	20	0.3	5.85	9.3
Manganese	0.29	0.48	0.63	0.07	0.16	0.26
Mercury	<0.0002	0.0003	0.0006	<0.0002	0.0007	0.0026
Nickel	0.06	1.1	0.13	0.05	<0.059	0.12
Total phosphate (as P)	<0.01	0.021	0.06	<0.01	0.081	0.23
Selenium	<0.001	0.0019	0.004	<0.001	0.002	0.004
Silica	10	12.57	15	6.1	7.4	8.6
Silver	<0.01	<0.01	<0.01	<0.01	<0.01	<0.01
Sulfate	240	357.5	440	41	48.75	80
Zinc	1.1	1.51	2.7	0.02	0.09	0.16

[a]All constituents in mg/L, except as indicated.

scrubber system. For instance, if the scrubber is constructed of an alloy such as Type 316L or 317L stainless steel, it is usually necessary to blow down a waste stream from the scrubber process to prevent excessive buildup of chlorides, which at high levels could cause corrosion of scrubber surfaces.

Scrubber wastewaters typically are high in dissolved and suspended solids and are saturated with calcium sulfate. Many coals contain chlorine and, to a lesser extent, fluorine. FGD systems remove these substances from the flue gas. The substances build up as dissolved solids in the recirculating liquid scrubber liquors. In addition, heavy metals released in the combustion of coal and removed by the scrubber may be present. Representative quality of FGD wastewater is shown in Table 16-4.

16.4.1.5 Coal Pile Runoff. To maintain a 30-day supply of coal, approximately 130,000 tons (118,000 tonnes) of coal are required for a typical 400-MW coal-fueled power plant. The coal is usually stored outdoors, in a coal pile area that

Table 16-4. Representative Quality of FGD Wastewater

Constituent	Range mg/L (except pH)	Constituent	Range mg/L (except pH)	Constituent	Range mg/L (except pH)
Aluminum	0.03–2.0	Lead	0.01–0.52	Chloride	420–33,000
Arsenic	0.004–1.8	Magnesium	4.0–2,750	Fluoride	0.6–58
Beryllium	0.002–0.18	Mercury	0.0004–0.07	Sulfate	600–35,000
Cadmium	0.004–0.11	Potassium	5.9–100	Sulfite	0.9–3,500
Calcium	180–2,600	Selenium	0.0006–2.7	Chemical oxygen demand	1–390
Chromium	0.015–0.5	Sodium	10.0–29,000	Total dissolved solids	2,800–92,500
Copper	0.002–0.56	Zinc	0.01–0.59	pH	4.3–12.7

allows stockout and reclaim of coal as needed. Coal pile runoff is a wastewater that includes the leachate that percolates through the coal and the runoff from the surface of the coal pile.

Coal consists primarily of carbon, mostly in a complex organic matrix that is nearly insoluble in water. Coal also contains a large variety of inorganic minerals, each with its individual solubility characteristics.

Most trace elements in coal are inorganic. These minerals are generally insoluble in neutral water. However, some of the inorganic coal minerals, such as pyrites, are unstable relative to oxidation by air or other weathering processes.

Limited field data exist on coal pile leachates and runoff. However, runoff from stored coal with a high pyritic content tends to be acidic and has been studied more than other types of coal pile runoff. Some coal with little or no pyrite produces an alkaline runoff. Coal pile runoff also can contain dissolved organic matter, which raises its chemical oxygen demand. The soluble organics may include phenols.

Pyrite in coal reacts with air and moisture to form ferrous and ferric ions and sulfuric acid. Hydrolysis of these iron sulfates yields additional sulfuric acid. These acid conditions lower the pH of waters in contact with the coal. Below a pH of 5.0, *Thiobacillus ferrooxidans*, *Ferrobacillus ferrooxidans*, and other similar species of bacteria contribute to the reactions. The resulting acid solutions attack inorganic mineral matter in coal, causing runoff to pick up trace elements.

The degree of acidity of runoff from such a coal pile depends on the concentration of pyrite in the coal, the size of the coal pile, the intensity and length of the rainfall event, and the frequency of rainfall. Temperature also affects runoff acidity.

Some field studies of runoff from pyritic coal have shown local pH values as low as 1.5, iron concentrations as high as 23,000 mg/L, and manganese concentrations as high as 45 mg/L. Total dissolved solids may be in the thousands of milligrams per liter.

Metals that may exist in coal pile leachate and runoff include aluminum, copper, lead, nickel, selenium, and mercury. Representative coal pile runoff data are presented in Table 16-5.

Federal effluent limitations for coal pile runoff currently require that the pH and solids be controlled or treated until the runoff volume reaches the equivalent of a 10-year, 24-hour storm. Rainfall runoff from a pond designed, constructed, and operated to hold and treat runoff from a 10-year, 24-hour storm need not be treated.

16.4.1.6 Metal Cleaning Wastewater. Steam generators and their associated flue gas facilities require periodic maintenance to maintain proper operation. Wastewaters from such maintenance include furnace fireside washings, stack cleaning wastes, economizer and air heater rinses, and boiler waterside cleaning solutions.

Drum-type boilers are usually chemically cleaned every 3 to 4 years. Chemical metal cleaning solutions used in power plant waterside cleaning include hydrochloric acid, citric and formic acids, ammoniated bromate, and hydroxyacetic formic acid. As an alternative to an acid solution, a chelant solution may be used. Chelants are organic substances that chemically bond to metal oxides, allowing them to be removed from heat transfer surfaces by flushing. Chelants used for chemical cleaning include ammoniated citric acid or ammoniated ethylenediamine tetraacetate (EDTA).

Before plant startup, the interior of condensate and feedwater piping surfaces is usually alkaline cleaned with a

Table 16-5. Coal Pile Runoff Analysis at Selected Plants[a]

Plant:	A	B	C	D	E	F	G
Alkalinity	6	0	—	—	14.32	36.41	—
Total solids	1,330	9,999	—	—	—	—	6,000
TDS	720	7,743	—	28,970	—	—	5,800
TSS	610	22	—	100	—	—	200
Ammonia	0	1.77	—	—	—	—	1.35
Nitrate	0.3	1.9	—	—	—	—	1.8
Phosphorus	—	1.2	—	—	—	—	—
Turbidity	505	—	—	—	2.77	6.13	—
Acidity	—	—	—	21,700	10.25	8.84	—
Total hardness	130	1,109	—	—	—	—	1,851
Sulfate	525	5,231	6,837	19,000	—	—	861
Chloride	3.6	481	—	—	—	—	—
Aluminum	—	—	—	1,200	—	—	—
Chromium	0	0.37	—	15.7	—	—	0.05
Copper	1.6	—	—	1.8	—	—	—
Iron	0.168	—	0.368	4,700	1.05	0.9	0.06
Magnesium	0	89	—	—	—	—	174
Zinc	1.6	2.43	—	12.5	—	—	0.006
Sodium	1,260	160	—	—	—	—	—
pH, pH units	2.8	3	2.7	2.1	6.6	6.6	4.4

[a]All data in mg/L except as indicated.

solution of sodium phosphates and acid cleaned with hydroxy-acetic-formic acid to ensure removal of oils, greases, and mill scale. Boiler surfaces are usually alkaline cleaned before startup using a sodium phosphate solution. For boiler acid cleaning, similar to the operational cleaning described previously, an inhibited hydrochloric solution or a chelant solution is used.

The total amount of wastewater, including rinses, generated during preoperational chemical cleaning can run from 500,000 to 1,000,000 gal (1,892,000 to 3,785,000 L) for a 400- to 600-MW pulverized coal plant.

Discharge solutions of chemical cleaning wastewater can contain strong concentrations of acids, alkaline wastes, and metals. The iron concentration in the wastewater may be as high as 5,000 to 10,000 mg/L. High concentrations of copper may also be present during operational cleanings of units with copper alloy feedwater heaters. A problem with handling chemical cleaning wastewater is that such wastes are generated infrequently in large batches. Installing waste treatment equipment with capacity large enough to meet these flows may not be economically feasible.

Various solutions to disposing of chemical cleaning solutions have been attempted. Some boiler cleaning solutions, such as organic chelants, are considered combustible and can be incinerated or evaporated in the boiler. A temporary treatment system using leased equipment can also be set up to adjust the pH of the cleaning solution and to precipitate and remove metals. At some plants, the chemical solutions are hauled off by a licensed contractor.

"Nonchemical" metal cleaning wastes resulting from operation of a power plant include wastes generated during fireside boiler surface cleaning, air preheater cleaning, and stack cleaning. Cleaning takes place on an as-needed basis, often once per year during the scheduled plant shutdown. Large hoses are often used to flush equipment surfaces to dislodge and remove ash and other residue that adhere. A high suspended solids wastewater results. Depending on the coal characteristics, this wastewater may be acidic. If acidic, various metals may be leached from the ash and can be present in the wastewater. The wastewater usually ends up in the plant floor drain system and must be addressed in the plant drains treatment system.

16.4.1.7 Low-Volume Wastewater Sources. Low-volume wastewaters include boiler blowdown, water treatment system wastes, and miscellaneous floor drains wastes.

Large, modern, high-pressure boilers typically have blowdown rates of 0.1% to 1.0% of the steam flow. Blowdown from such units is relatively high-purity water and usually contains only small amounts of suspended solids and phosphates, and traces of iron, copper, nickel, and chromium.

Water treatment wastewater includes sludge removed from pretreatment systems, filter backwash, and ion exchange resin regeneration wastes. The pretreatment sludge contains suspended solids from the raw water being treated, plus the reaction products from coagulation and treatment chemicals added to the water. In the case of lime softening, the solids are predominately calcium carbonate.

Ion exchange processes can be used for both cycle makeup water treatment and for polishing of condensate for return to the boiler as feedwater. Such ion exchange systems generally use sodium hydroxide for anion resin regeneration and a strong acid, such as sulfuric or hydrochloric acid, for cation resin regeneration. The regeneration wastewater includes unreacted sodium hydroxide and acid, plus the sulfate, or chloride, and sodium salts of the ions removed by the ion exchange process.

Polishing of condensate also can be accomplished with powdered ion exchange resin applied to a precoat filter. This ion exchange medium is not regenerated but is periodically removed using backwash water and air. The wastes consist of the powdered resin material, which contains suspended solids removed from the condensate, plus the ions that have been chemically removed from the condensate by the ion exchange media.

Floor drains are a low-volume wastewater source resulting from plant maintenance operations or from equipment leakage. This wastewater can contain suspended solids, oil, and grease.

16.4.2 Water Management Criteria

An organized process should be used in developing a power plant water management plan. This section describes the process recommended in design of new plants, but the principles also apply to existing plants. A power plant wastewater management system should be designed for optimum overall plant availability and reliability, ease of operation and maintenance, and compliance with environmental regulations.

The overall plant water and wastewater systems design considerations may include water ponds and various containments or collection devices, wastewater treatment systems, surface water discharge, ground water discharge, and wastewater reuse, depending on regulatory requirements, site conditions, and economic considerations.

A detailed water and wastewater management plan should be developed early in the design process. The design of all plant systems and equipment selection should consider this plan. Four important principles should be considered in development of the water management plan:

- Select the water source most appropriate for each service. Some plant services are more conducive to using a lower quality source than others.
- Select water treatment processes recognizing waste treatment considerations and reuse concepts to be employed.
- Ensure plant systems design is compatible with the designated water source.
- Use design factors for liquid waste treatment systems which adequately allow for changing operating conditions.

How these principles are used is explained in detail in the following subsections.

16.4.2.1 Plant Water Balance. The detailed plant water balance is the most basic tool used in developing a plant water and wastewater management plan. The water mass balance diagram illustrates the principal water and wastewater routings for the plant and defines the flow rates. In developing the water balance, one should consider all possible water sources. Typical sources of raw water include well water; surface supplies from rivers, lakes, and the ocean; and collected rainfall runoff from site areas. Effluent from municipal wastewater treatment plants may also be an option.

Reuse of plant wastewater should be considered when feasible. Cooling tower blowdown, material storage areas runoff, boiler blowdown, and plant drains can often be reused in compatible plant systems.

Major users of water at a coal-fueled power plant include the following:

- Condenser and other equipment cooling water;
- Boiler makeup water;
- Service water (pump seals, wash down, etc.);
- Flue gas scrubber makeup water;
- Ash conveying; and
- Potable water.

To develop the water mass balance, one must match the appropriate source to the use and determine the appropriate treatment concepts. Where no discharge of wastewater is permitted (i.e., a zero discharge plant), water reuse is extensive and the water mass balance becomes quite complex. Figure 16-2 is an example of a water mass balance diagram prepared for a zero discharge power plant.

Certain plant systems require a consistently good quality water supply. Potable water must meet federal and state drinking water standards, and the demineralizer supply must be of consistent quality, free of organics that foul resins or contaminate the steam cycle.

The highest volume water consumer at a power plant is the condenser cooling water system. The quality of makeup water to a cooling tower or a once-through cooling water must be relatively consistent to ensure that efficient heat transfer is maintained in the condenser, cooling tower fill (if a cooling tower is used), and other plant heat exchangers.

Other major users of water at the plant include the flue gas scrubber system, the bottom ash system, and plant washdown. As these systems can be more easily designed to use a lesser quality water, a wastewater recycle system may be designated to supply these systems.

The degree of water reuse is influenced by specific site conditions, such as availability of water supplies, whether discharges are allowed, and specific permit requirements. Wastewater reuse reduces the amount of water needed to supply the plant as well as reducing the amount of wastewater to be discharged or treated.

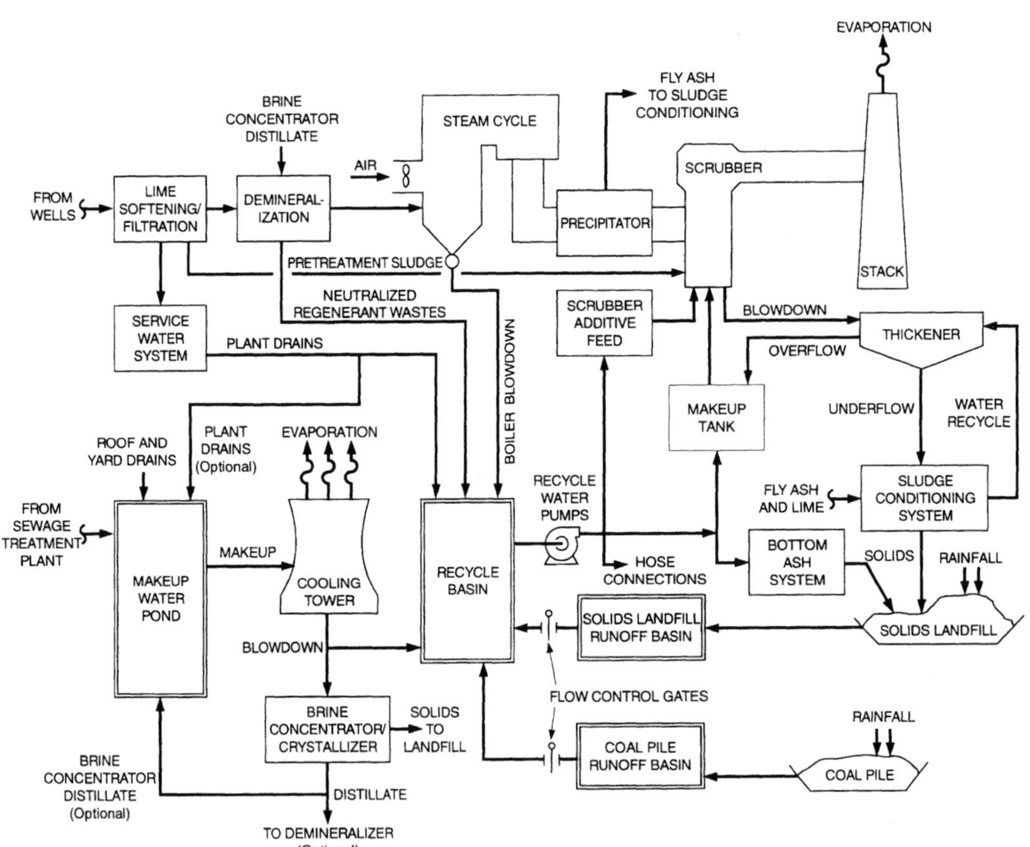

Fig. 16-2. Zero discharge water balance.

Following initial matching of water sources and uses, the wastewater management plan can be refined further to include specific features such as ponds and containments as necessitated by site conditions.

16.4.2.2 Use of Ponds and Containments. Economical and reliable operation of zero discharge plants and plants that recycle a significant amount of water can be ensured by using a conservatively designed wastewater storage and equalization system. Ponds and containments are incorporated into the plant water balance to provide the water source and use combinations identified in the initial concept stages of the design process.

The pond containment system can be designed to segregate wastewaters by quality as required by conceptual design matches of source and use. For example, high-quality wastewaters, such as rainfall runoff from uncontaminated plant areas and roof drains, and possibly cooling tower blowdown, can be collected, contained, and treated; then reused or discharged, depending on site conditions and regulatory requirements. Low-quality wastewater can be collected and routed to ponds for reuse, treatment and reuse, or treatment and discharge. Typical low-quality wastewater sources include coal pile runoff, runoff from actively worked combustion waste landfills, demineralizer resin regenerant wastewater, bottom ash sluice water, and scrubber blowdown.

Wastewaters generated from rainfall are intermittent and can occur in short, very intense storm events. A pond system allows for equalization of flows, which reduces costs, since wastewater treatment equipment can be sized to more nearly match average wastewater flows. A secondary benefit is that pond capacity allows for equalization of water quality. In addition, conservatively sized plant makeup water supply ponds may help provide a consistent, reliable supply of water from variable sources, such as rivers or municipal wastewater subject to low flow periods.

16.4.2.3 Water Reuse Concepts. Water reuse offers the dual benefit of reducing the quantity of raw water supply required and reducing or eliminating discharge requirements. The condenser cooling system is a high-volume water consumer. With appropriate design features and materials selection for components such as condenser tubes, cooling tower shell material, cooling tower fill material, and circulating pump materials, wastewater can be used as the supply source or supplemental supply source of makeup water. In water-short areas of the country, treated municipal wastewater increasingly is preferred as a cooling water makeup source.

Cooling systems can be designed with construction materials suitable for lower quality makeup sources. For instance, titanium or high-alloy stainless steel condenser tubes may be operated with wastewaters high in total dissolved solids (TDS) so long as scaling does not occur.

Low-quality wastewaters may be reused in plant processes designed to use a lesser quality water. The scrubber system can be designed to use lower quality makeup water. A plant recycle basin can be designed that collects demineralizer wastes, coal pile runoff, and other low-quality wastewater for recycle. The scrubber system can be designed to use this low-quality water by using corrosion-resistant materials, including rubber-lined modules and pumps, fiberglass piping, and coated tanks. The scrubber system pumps can be designed to allow use of recycled wastewater as seal water. Recycled wastewater can be further used for reagent preparation, such as wet limestone grinding or lime slurry dilution, pump and pipeline flush operations, and equipment cooling.

16.4.2.4 Surface Water Discharge. Surface water discharge of wastewater is usually the most economically attractive and, in practice, the most frequently used disposal method for plant wastewater. The ability of a plant to discharge to a surface water source depends on site-specific criteria, including the quality and quantity of the wastewater discharged, the wastewater treatment concept employed, the quality and quantity (flow or volume) of the receiving water body, and the environmental sensitivity of the receiving water body.

A sampling and analysis program, material balance calculations, or both may be required to characterize the process wastewater to be discharged.

As discussed earlier, plant wastewater being discharged must be treated to meet applicable federal and state categoried limits.In addition, applicable surface water quality standards for the receiving stream must be met before wastewaters can be discharged.

Technologies employed for treatment of power plant wastewaters to allow surface water discharge are often the same as those used to allow reuse as plant makeup water.

Surface water discharge may be accomplished with an outfall structure in the receiving stream. Environmental regulations may require dispersion and mixing of the wastewater by means of a diffuser or diffusion equipment.

16.4.2.5 Ground Water Discharge. Although used less extensively than surface water discharge, ground water discharge of wastewaters may be a viable disposal method. The ability of the plant to discharge to ground water depends on site-specific criteria, including the quality and quantity of the wastewaters discharged, the quality and quantity (flow or volume) of the receiving aquifer, and the environmental sensitivity of the aquifer in the region surrounding the plant site. Geologic conditions of the receiving aquifer and use of the aquifer in the plant area are important considerations.

To predict the impact to the aquifer, the wastewater should be characterized by a sampling and analysis program, material balance calculations, or both.

Ground water discharge is generally accomplished using percolation ponds or by deep well injection, and its use is limited by ground water quality standards in much the same way that surface water quality standards limit surface water discharge.

Percolation ponds are shallow containments constructed

of earthen berms. The pond bottom consists of a permeable material such as sand or gravel. The differential head of liquid between the percolation pond surface elevation and the ground water elevation provides the driving force for wastewater flow into the ground water. Application of percolation ponds is generally limited to areas where the soil is predominately sandy, such as coastal regions where the aquifer that receives the water is brackish.

Deep well injection is accomplished by pumping wastewater into wells drilled into aquifers below those used for potable water supplies. Injection well capacity is limited by the porosity of the geologic formations of the aquifer. If the aquifer structure is not porous, the well capacity is low and injection may not be practical. Expensive test wells are usually required to determine technical feasibility and to demonstrate environmental acceptability of deep well injection. Deep well injection is usually limited to low-volume wastewaters, with such injection more and more limited by regulatory authorities. Because of the high costs and technical and permitting uncertainties, use of deep well injection as a power plant wastewater disposal method is very limited.

16.4.2.6 Zero Discharge. The most direct way of ensuring that effluent regulations and water quality standards are met is not to discharge an effluent at all. If a receiving stream is not available or accessible for surface water discharge and ground water discharge is environmentally unacceptable or uneconomical, zero discharge must be employed. Zero discharge is the evaporation or onsite retention of all water imported to the site.

Generally, two methods of zero discharge treatment are used: evaporation ponds or volume reduction processes.

Evaporation ponds are an economical method of wastewater disposal in areas where annual average evaporation significantly exceeds precipitation. Evaporation ponds require large land areas and leave salt residue from evaporation of the water. Figure 16-3 indicates the areas of the United States where average annual lake evaporation rates exceed the annual average precipitation rates by more than 10 in. per year, a criterion generally considered necessary to meet if successful use of an evaporation pond is expected.

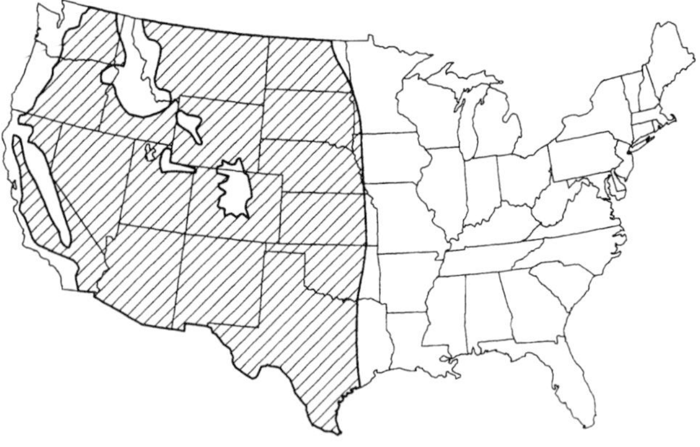

Fig. 16-3. Areas of the United States where average annual lake evaporation exceeds average annual precipitation by 10 in.

Volume reduction processes can be used to treat excess wastewater and have the advantage of not only treating the wastewater, but also recycling treated water and reducing overall plant makeup requirements. "Volume reduction" processes may not actually reduce the volume of wastewater, but rather separate it into a large low TDS stream suitable for direct reuse and a small concentrated high TDS stream, which can be concentrated further or stored onsite.

Commercially available volume reduction processes include membrane processes that usually require extensive pretreatment of the wastewater to reduce membrane scaling potential. Three of these membrane processes, reverse osmosis (RO), electrodialysis (ED), and electrodialysis reversal (EDR), are discussed later in the chapter. Other volume reduction processes discussed in this chapter include brine concentrators/evaporators and crystallizers or spray dryers.

One or more or a combination of these processes is used to concentrate the dissolved and suspended solids in the excess plant wastewater to an increasingly small wastewater flow. The end product of these treatment processes is a relatively pure, reusable product water and either a very concentrated brine that can be routed to an evaporation pond or a dry salt that requires landfill disposal in an encapsulated form.

16.4.3 Wastewater Treatment Technologies

Numerous technologies can be used for the treatment of power plant wastewater to meet permitted discharge limitations or allow reuse of the wastewater in other power plant processes. The various treatment technologies must often be used in combination to provide the results desired. For instance, treatment for suspended solids removal must be done before using a membrane process.

Some of the more common treatment technologies are described in the following sections.

16.4.3.1 Wastewater Suspended Solids Removal. Perhaps the oldest method of suspended solids reduction is the use of settling ponds or lagoons. Settling ponds can be used with or without the addition of coagulants and coagulant aids to promote settling of solids. Settling pond sizing must allow for the gravity settling of particles contained in the waste stream. Coagulants and coagulant aids increase the effective size and settling rate of the particles, reducing the amount of time needed for settling.

Jar testing can be used to determine the settling time required for an actual sample of the wastewater to be treated, as well as measure the effect of various coagulants and coagulant aids. Jar tests are laboratory bench tests using samples of the specific water to be treated. Jar tests simulate possible treatment chemical combinations and unit operations. They also can be used to help determine what size settling pond is needed.

Finely divided particles often form colloidal suspensions in wastewater, which makes settling very difficult. The addition of a coagulant such as alum or ferrous sulfate causes a reaction between these coagulants and alkalinity in the wastewater. In the coagulant reaction process, the charges on colloidal particles cause them to be attracted to the precipitates being formed. The addition of coagulant aids such as polymers assists in the settling of suspended solids and the precipitates formed in coagulation reactions. When used in conjunction with settling chambers or settling ponds, coagulant and coagulant aids are mixed and flocculated with the incoming wastewater to promote the coagulation reaction before routing the wastewater to the settling chamber.

Limitations to using settling basins include requirements for a large land area, possible contamination of ground water, and filling of the basin with suspended solids, gradually decreasing the detention time and, therefore, performance.

Clarifiers are steel or concrete basins specifically designed for suspended solids removal. Typically, a clarifier has an entry well to receive the influent wastewater and dissipate its energy, a top launder collection system to provide equal collection of the treated water across the entire surface area, and a bottom scraper and solids collection system.

Solids contact basins (SCBs) are specific types of clarifiers with separate mixing and clarification sections. The mixing section provides for mixing and flocculation of the influent wastewater with treatment chemicals and previously precipitated solids. The previously precipitated solids promote a more rapid reaction and coagulation. As the wastewater and precipitated solids pass out of the mixing section to the clarification section, the velocity decreases, promoting flocculation.

Liquid–solids separation occurs in the clarification section. The treated water rises through the clarified water zone and is collected by a system of launders at the top of the basin. Settled solids are collected and periodically blown down to a sump where they can either be reused in another power plant process, routed to a storage pond, or further dewatered prior to landfilling of the solids. Reaction time and rise rate (flow divided by the area of the clarification section) are important factors in the design of a solids contact basin and often must be determined based on experience for the process being utilized. A typical solids contact basin is illustrated in Fig. 16-4.

Jar tests can be used to help determine the optimum treatment chemical combinations. The dosages of coagulant, coagulant aid, and pH can be varied from jar to jar. In addition, paddle mixers are used to simulate the flash mixing and slow mixing operation of a solids contact unit. By observing the settling rates and the characteristics of both the flocculated solids and the clarified liquid, the optimum combination of treatment process and chemical dosage can be approximated.

Table 16-6 indicates chemicals and additives typically used in wastewater treatment. Depending on the wastewater characteristics, the treatment required, and the economics, a particular combination of chemicals will be required.

An important sizing criterion for a solids contact basin is the rise rate. Rise rate is defined as the influent flow in gallons per minute (gpm) (liters per minute [L/min]) divided

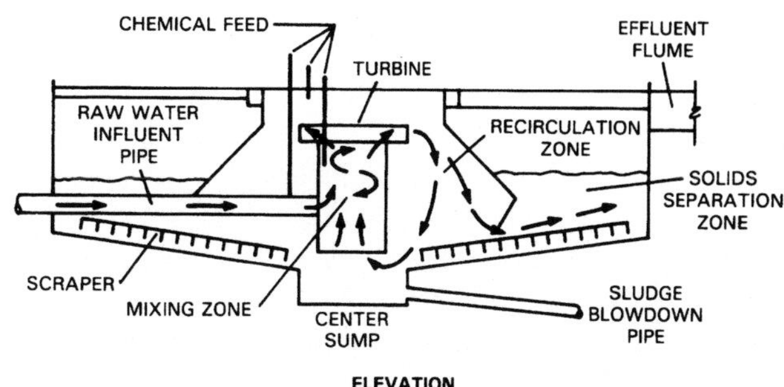

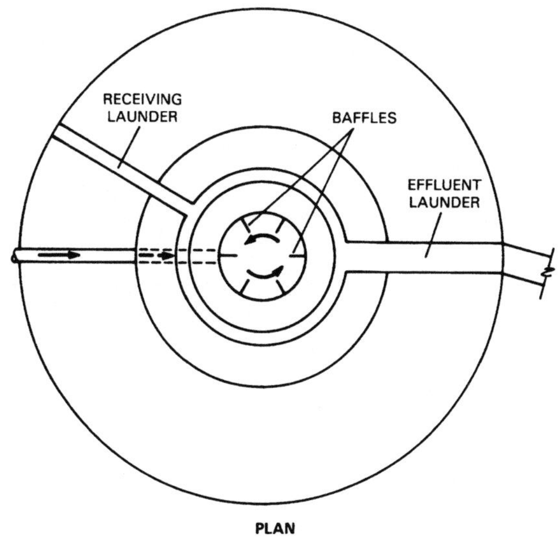

Fig. 16-4. Typical solids contact basin.

by the water surface area at the solids/liquid separation elevation, excluding the draft tube area. Depending on the density of the sludge formed and the effluent turbidity requirements, typical rise rates for solids contact basins treating wastewater are between 1.0 and 2.0 gpm (3.8 and 5.6

Table 16-6. Chemicals Typically Used in Wastewater Treatment

Type	Common Name	Chemical Formula	Mol wt
Coagulants	Alum	$Al_2(SO_4)_3 \cdot 14H_2O$	594
	Ferric sulfate	$Fe_2(SO_4)_3$	400
	Ferric chloride	$FeCl_3$	162
	Ferrous sulfate	$FeSO_4 \cdot 7H_2O$	278
	Sodium aluminate	$Na_2Al_2O_4$	164
Coagulant aids	Anionic polyelectrolytes	Varies	Varies
	Cationic polyelectrolytes	Varies	Varies
	Nonionic polymers	Varies	Varies
Oxidizing agents	Chlorine	Cl_2	70.9
	Chlorine dioxide	ClO_2	67.5
	Potassium permanganate	$KMnO_4$	158.0
	Calcium hypochlorite	$Ca(OCl)_2$	143.1
	Sodium hypochlorite	$NaOCl$	74.5
Alkalies	Hydrated lime	$Ca(OH)_2$	74.1
	Magnesium oxide	MgO	40.3
	Sodium hydroxide	$NaOH$	40.0
	Quicklime	CaO	56.1

L/min per ft^2). Chapter 15 discusses solids contact basins in more detail.

Solids contact basins have several advantages over the use of settling chambers or ponds. The primary advantage is the compact nature of treatment equipment, which results in significant space savings and, in many cases, significant savings in capital costs. A second advantage of solids contact basins is that they can be operated as a continuous process, whereas settling ponds must often be periodically dredged to remove settled material.

An alternative sometimes attempted to reduce the size of the clarifier facility is to use inclined plate or tube-type settlers. This type of settling device is illustrated in Fig. 16-5. The plates or tubes are designed to improve solids separation in the settling zone. The plates or tubes typically extend from the water surface through the clarification section. The angle and spacing of the plates or tubes must be selected for the particular application, based on the characteristics of the flocculated solids, the desire for continuous operation, and space considerations. Several types of plate and tube arrangements and flow arrangements (concurrent flow, cocurrent flow, and crossflow) have been used. The inlet flow distribution, as well as the adequate collection of the clarified effluent, are important factors in the design of such a separator.

A small amount of solids carryover is expected from any type of solids settling unit operation. In wastewater treatment, postfiltration may or may not be needed depending on the allowable effluent turbidity or suspended solids limitation.

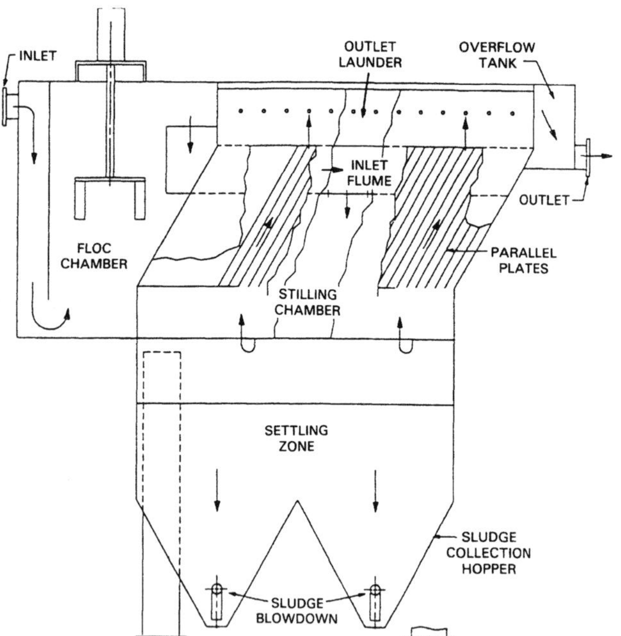

Fig. 16-5. Plate-type clarifier.

16.4.3.2 Wastewater Filtration. Filtration can be used when the wastewater suspended solids loading is relatively light and a low suspended solids effluent is required. Wastewater filtration is generally used downstream of another technology such as a settling pond or a clarification system. Table 16-7 indicates four types of filters that have been used for wastewater suspended solids reduction, and the advantages and limitations of each.

Cartridge filters use disposable filter cartridges enclosed in pressure vessels for removal of suspended solids. Through proper design selection, nearly any removal efficiency and effluent suspended solids limit can be achieved. Cartridge filters are typically used for polishing operations or are used ahead of other unit operations, such as ion exchange or membrane processes in which all but very fine particles (<5 or 10 microns) would cause plugging. When solids loadings would require cartridge replacement more often than every week or so, one of the other types of filters should be considered for use prior to or in lieu of cartridge filters. A cartridge filter is illustrated in Fig. 16-6.

Pressure filters typically use graded sand, anthracite, or both materials in a completely enclosed pressure vessel. Flow is downward, under pressure, through gradually finer materials. A typical pressure filter is shown in Fig. 16-7. As

Table 16-7. Wastewater Filtration

Type	Advantages	Limitations
Cartridge filters	Provides good control of particles to one specific size. Compact size Relatively inexpensive capital investment	Relatively low solids capacity High filter cartridge replacement costs
Pressure filters	Less expensive than gravity filters Effectively removes particles over large size range. Clearwell and secondary pumping not required.	Cannot observe filter bed. Backwash water must be disposed. Relatively high capital costs
Gravity filters	Effectively removes particles over large size range. Greater solids capacity Filter can be easily operated and maintained. Can observe filter bed.	Clearwell and secondary pumping required Backwash water must be disposed. Relatively high capital costs
Upflow filters	Coagulants can be used upstream to enhance solids capture. Greatest solids loading capacity since full volume of filter media is used in filtration.	Unproven in many applications Limited number of suppliers Backwash water must be disposed. Relatively high capital cost

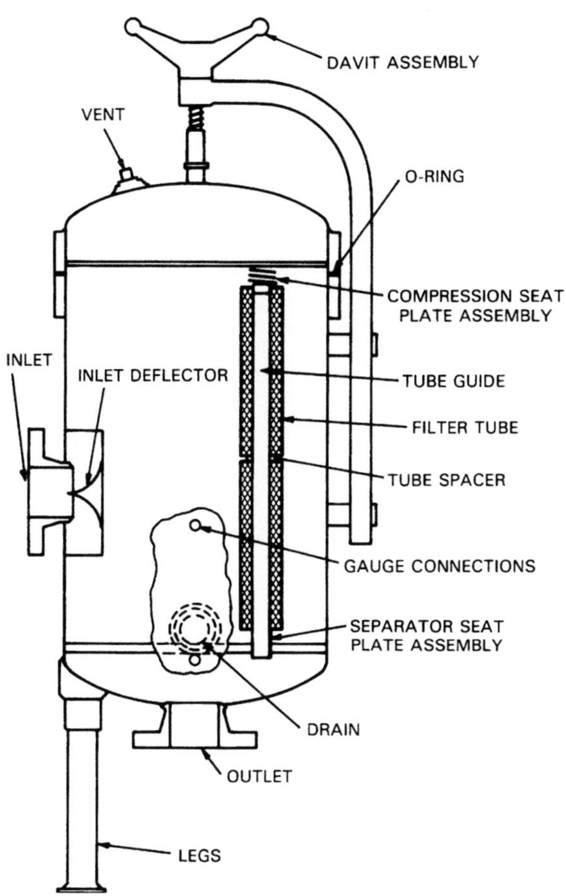

Fig. 16-6. Typical cartridge filter.

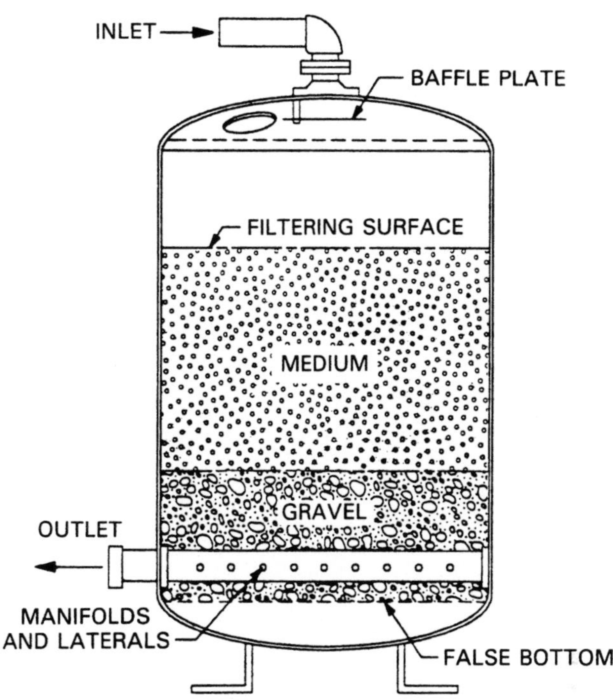

Fig. 16-7. Typical vertical pressure filter.

the filter medium becomes loaded with solids, the differential pressure across the filter increases. When the pressure drop reaches a predetermined level, the filter medium is backwashed by taking the filter out of service and passing water up through the medium to wash the solids to waste. Pressure filters are typically sized to provide adequate volume to contain the filter medium, plus 50% to 100% freeboard for bed expansion during backwash. Cross-section area for wastewater filtration is usually 3 to 5 gpm per square foot (1.05 to 1.76 L/m per m^2) of filter area.

Gravity filters are similar in concept to pressure filters, except that open top atmospheric tanks are used to hold the media. A typical gravity filter is illustrated in Fig. 16-8. Gravity filters are commonly used in association with upstream clarifiers or solids contact basins, as are pressure filters, with the overflow from the upstream basin flowing by gravity into the filter. Gravity filters may be partitioned into cells to allow individual backwash of one cell. Partitioning of a gravity filter has the added advantage of reducing the size of the backwash pump. Gravity filter backwash is handled similarly to that described for pressure filters. In certain gravity filter designs, the effluent from the other cells in operation is used to backwash the out-of-service cell, thus eliminating the need for a backwash pump.

An alternative filter sometimes used in wastewater treatment is an upflow filter. Upflow filters are designed so that water is pumped upward through the graded bed of sand and gravel, traveling in the direction of decreasing grain size. The entire bed is available as a solids trap, with the larger particles remaining in the lower coarse-channel strata and the finest particles being polished out of the top fine-grain layer. It is usually necessary to inject a polyelectrolyte to the influent water to create or increase the electrokinetic attraction between the suspended solids and the filter bed.

The filter bed is cleaned by expansion by an airstream and high rate water flushing, with the water flowing in the filtering direction. Figure 16-9 shows a typical upflow filter arrangement.

The filtered solids loading capacity of the bed is much greater than that of a conventional gravity sand filter, and higher filtration rates can be achieved.

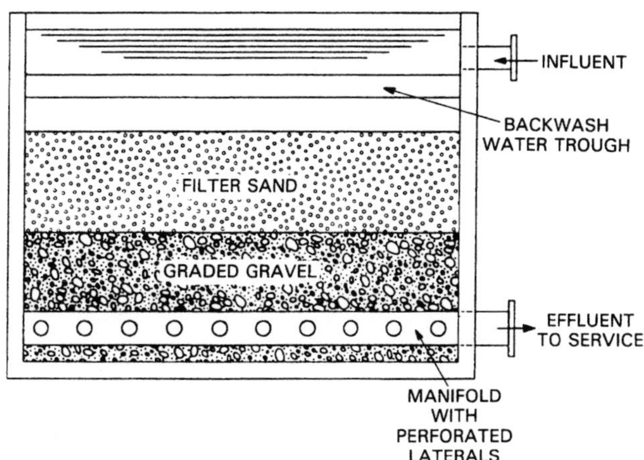

Fig. 16-8. Typical gravity filter.

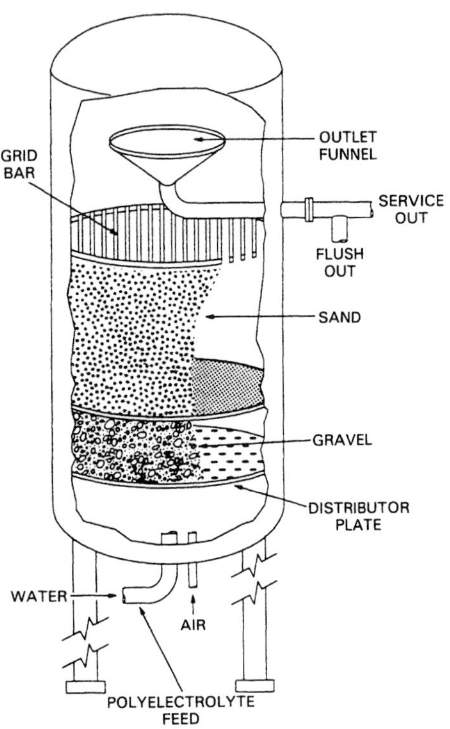

Fig. 16-9. Typical upflow filter.

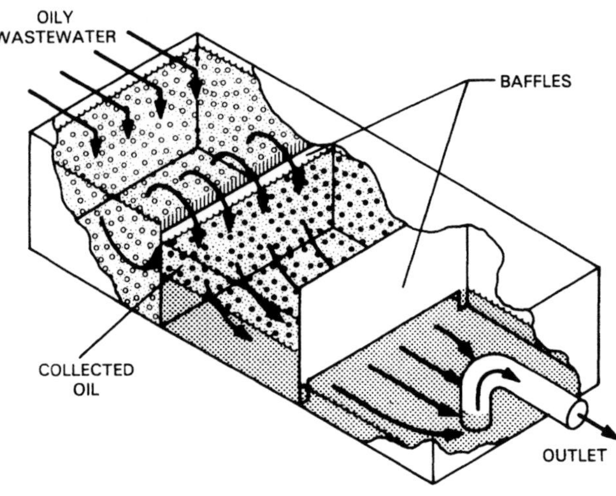

Fig. 16-10. Typical gravity-type oil separator.

Gravity or pressure filter backwash water is itself a wastewater that must be disposed of. Filter backwash recovery, in which the backwash water is pumped back into the clarifier or solids contact basin upstream of the filter inlet, is often practiced to recover this wastewater and assure proper disposal of the suspended solids.

16.4.3.3 Oil Separation and Removal. Lubricating oils and greases from various mechanical equipment and maintenance operations can be present in water from general plant drains. In addition, fuel oil storage tank fill stations and oil-filled transformers are other possible sources of drainage oil contamination. All drainage that has the potential for oil contamination can be collected and routed through an oil separator for treatment.

Since most oils have a lower specific gravity than water, oil separators typically rely on detention time to allow the oil to separate and collect as a floating layer on top of the water. Figure 16-10 illustrates a gravity-type oil separator that uses baffles and an inverted discharge to prevent the release of oil while allowing the wastewater to leave the separator. The collected oil is periodically pumped out of the separator using portable equipment and either burned onsite or disposed of or reclaimed by a licensed contractor.

Coalescing-type oil separators are gravity separators that have coalescing chevrons or other elements that permit small particles of oil to contact each other and agglomerate until the oil particles are large enough in diameter to easily separate and float. Coalescing-type separators are used when the oil present is very dispersed and there are stringent limitations for the effluent. Other types of oil separators use flotation to achieve oil removal. Chemicals can be added to break up emulsions to aid in removal.

16.4.3.4 Neutralization. Certain power plant wastewaters are considered hazardous or otherwise unsuitable for discharge or reuse in other plant systems solely or in part because of their pH. Typical power plant liquid waste streams that often require neutralization include acid or caustic regenerant wastewater from an ion exchange process, chemical feed and storage area drains, chimney drains, laboratory drains, and battery room drains.

Wastewaters can often be self-neutralizing to some extent, such as the acid and caustic waste streams that result from cycle makeup or condensate demineralizer regenerations. To take advantage of the self-neutralizing characteristics of various streams, a neutralization tank or basin is used to collect the wastewater. A basin allows gravity collection of the wastewater. The basin is normally sized to collect at least the volume from one complete demineralizer regeneration operation or two boiler volumes, whichever is greater. The ability to contain two boiler volumes allows the use of the neutralization basin during chemical cleaning of the boiler.

Neutralization basins are equipped with capability for mixing the contents of the basin and for adding acid or caustic to adjust the pH as necessary. A typical neutralization basin is shown in Fig. 16-11.

A neutralization basin can be constructed from reinforced concrete with a protective lining system or from fiberglass-reinforced plastic, if the sizing permits. One of the most reliable lining systems for a concrete neutralization basin includes a membrane liner protected by a layer of acid brick and furan mortar cement.

A "permit by rule" program has been established by the EPA for such an elementary neutralization system. This eliminates the need for a utility to file an RCRA Part A application for hazardous wastes for the acid and caustic wastewaters.

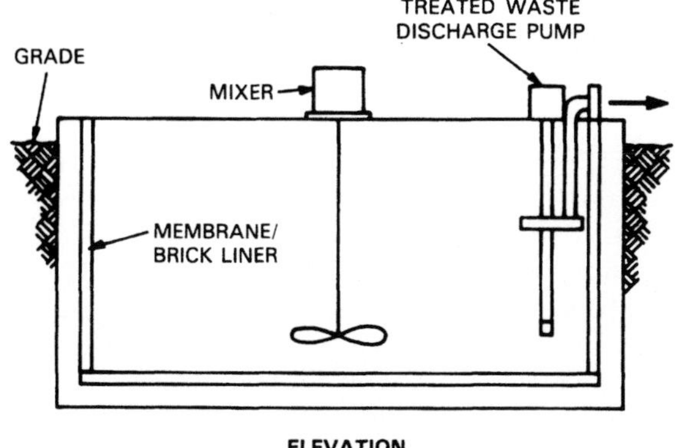

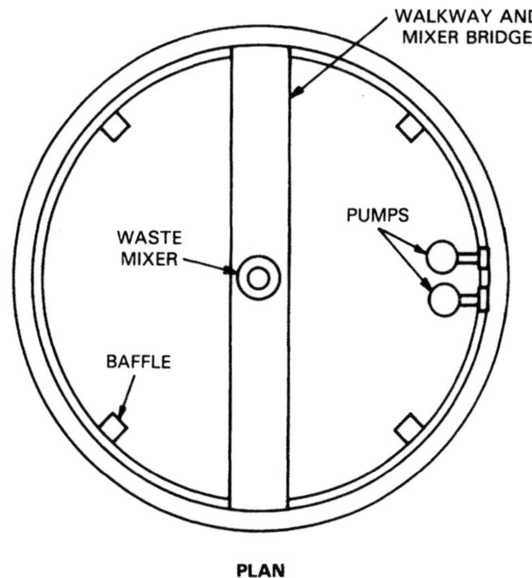

Fig. 16-11. Typical neutralization basin.

In coal-fueled power plants, where water is used to convey ash as a slurry to an ash disposal pond, the pond water may be acidic, alkaline, or essentially neutral depending on coal characteristics. In many cases, alkaline water-soluble constituents in the ash such as CaO cause the pond water to develop a higher pH than is permittable for discharge. In some cases, sulfuric acid has been used to neutralize the effluent from such a pond before discharge. This practice is difficult to control and potentially hazardous to personnel handling the sulfuric acid. An alternative method of treating this wastewater is to use carbon dioxide. Liquid carbon dioxide is available for bulk delivery in tank trucks from various suppliers. The carbon dioxide is unloaded and stored in an onsite tank, then vaporized and fed as a gas or a gas solution to the wastewater stream.

Carbon dioxide first dissolves in the wastewater to produce carbonic acid:

$$CO_2 + H_2O \rightarrow H_2CO_3$$

Then the carbonic acid neutralizes the alkaline agent in the water:

$$H_2CO_3 + NaOH \rightarrow NaHCO_3 + H_2O$$

A carbon dioxide neutralization system is illustrated schematically in Fig. 16-12. A major advantage of carbon dioxide neutralization is that overfeeding will not depress the pH below 6.0, which is commonly the lower limit for discharge to surface waters (Carpenter and Shanks 1977).

16.4.3.5 Solids Dewatering. Flue gas desulfization processes, as well as other power plant operations, usually

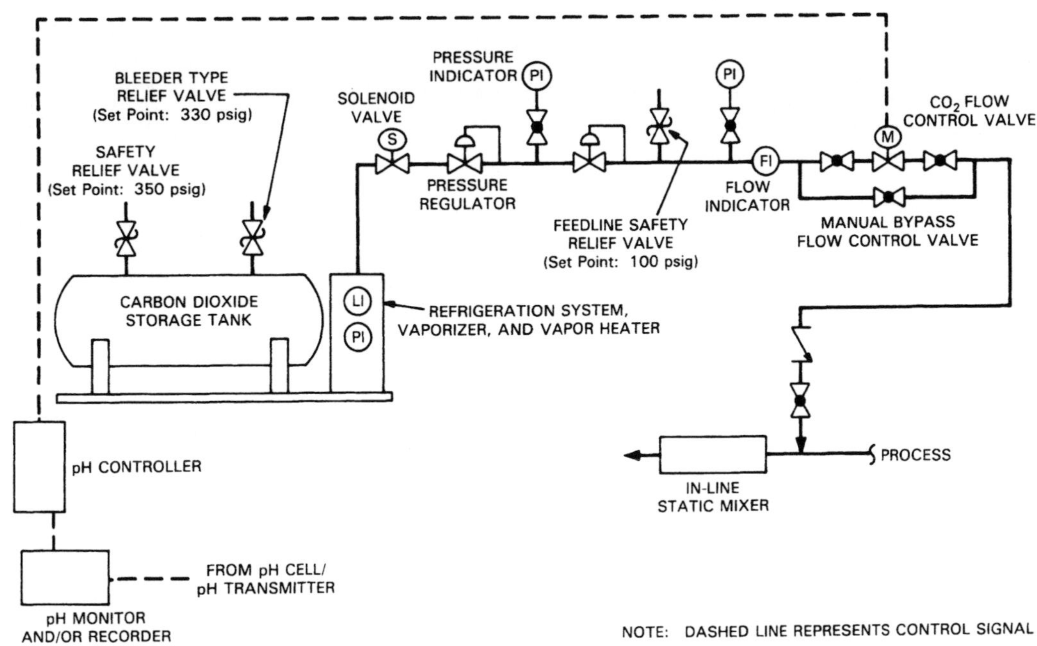

Fig. 16-12. Typical schematic—carbon dioxide storage and direct gas feed system.

Table 16-8. Primary Solids Dewatering Processes

Process	Advantages	Limitations
Cyclone separator	Low capital cost Small space requirements	Significant erosion Questionable effectiveness with certain solids
Thickener	Good capability to concentrate sludge Good separation characteristics with a variety of solids Polymers useful to improve performance	Large space requirements Higher capital costs Sludge must be pumped.

require removal of water from a high suspended solids product to allow for solids transportation and ultimate disposal. Cyclone separators and thickeners are two commonly used processes for primary sludge dewatering. The advantages and limitations of each are described in Table 16-8.

In cyclone separators, or hydroclones, centrifugal force causes the solids particles to collide with each other and with the walls of the cyclone. Solids then settle to the bottom of the cone while the centrate is discharged from the side of the separator. A typical cyclone separator is illustrated in Fig. 16-13.

Thickeners are collection tanks equipped with slow moving rakes or bottom scrapers that provide extended contact and settling of solids. Clarified water rises to the top to be recycled or reused. The thickener underflow usually must be routed to a secondary solids dewatering system for further concentration. Scrubber thickeners are often designed to provide a minimum of 3 days' storage of sludge to allow for weekend outages of secondary sludge dewatering. A typical thickener is shown in Fig. 16-14.

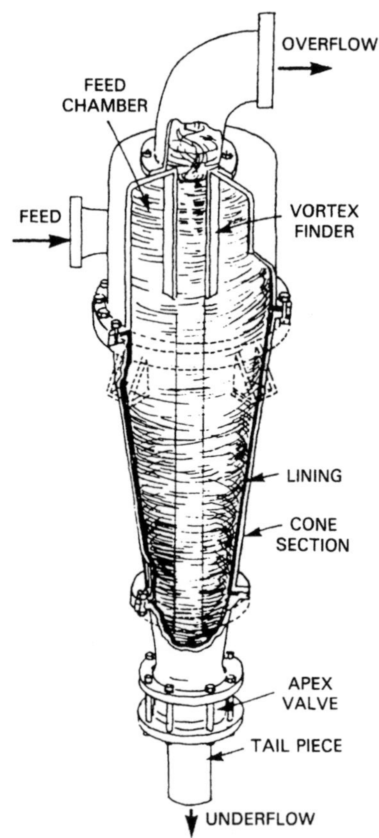

Fig. 16-13. Typical cyclone separator.

Secondary sludge dewatering further concentrates the solids from the primary sludge dewatering operation. Two secondary sludge dewatering processes commonly used at power plants are vacuum filters and filter presses. The advantages and limitations of each are summarized in Table 16-9.

Vacuum filters use belts or drums with a vacuum on one side to force drain the filtrate through a filter mat, leaving the dewatered solids on the belt or drum. A scraper then continuously removes the collected solids. A typical vacuum filter is shown in Fig. 16-15.

Filter presses are perhaps the simplest of the dewatering equipment options and are adaptable to a wide range of wastewater treatment processes. A typical filter press is shown in Fig. 16-16. Sludge is pumped into the cavities between the filter press plates. The filter retains the solids and the filtrate is collected and either discharged or returned to the front of the treatment system. Periodically the solids build up in the press and the system must be shut down to allow for removal. During cleaning, the plates are separated, and paddles or scrapers are used to remove the dewatered solids. Depending on the rate of solids production, the frequency of cleaning may be burdensome or impractical and a vacuum filter should be considered.

16.4.3.6 Dechlorination. Dechlorination is the removal of all or part of the total residual chlorine remaining after chlorination. This includes removal of both free available chlorine (HOCl) and chlorine compounds such as chloramines. In power plant wastewaters, total residual chlorine may be present in cooling tower blowdown, once-through cooling water, and onsite sewage treatment plant effluent.

In recent years, regulators have begun to place stringent limits on the discharge of wastewaters containing total residual oxidants such as chlorine. Sewage treatment plant wastewaters from power plants are not normally required to be dechlorinated because of the relatively small flow and the need to main the chlorine residual to control pathogens. However, it is becoming common practice for discharge regulations to require the complete dechlorination of cooling tower blowdown and once-through cooling water.

Dechlorination is accomplished by the addition of a reducing agent such as sulfur dioxide, sodium sulfite, or sodium metabisulfite. Activated carbon is sometimes used for dechlorination. Because of the relatively large cooling water discharge flows from power plants, less expensive sulfur dioxide is normally used as the dechlorinating agent.

Sulfur dioxide reacts with free chlorine and chloramine through the following reactions:

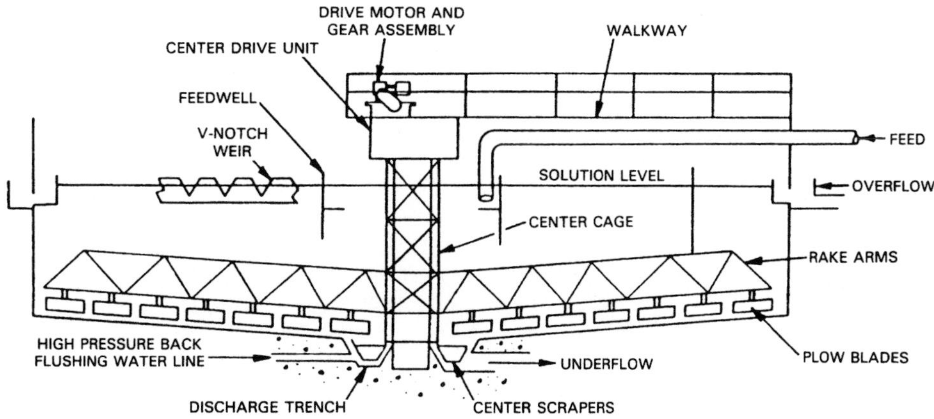

Fig. 16-14. Typical gravity thickener.

$$SO_2 + H_2O \rightarrow H_2SO_3 + HOCl \rightarrow H_2SO_4 + HCl$$

$$NH_2Cl + H_2SO_3 + H_2O \rightarrow NH_4HSO_4 + HCl$$

Both reactions are essentially instantaneous. The reaction stoichiometry requires 0.9 ppm SO_2 to dechlorinate 1 ppm chlorine.

Sulfur dioxide is a colorless, odorless gas formed in the combustion of elemental sulfur, the smelting of sulfide ores, and in the reaction of liquid sulfur with sulfur trioxide. Table 16-10 summarizes some of the physical properties of sulfur dioxide.

The equipment used to feed sulfur dioxide is very similar to that used to feed chlorine gas. Liquified sulfur dioxide gas is available in the United States in 150-lb cylinders, 1-ton containers, and in bulk deliveries using tank trucks and rail cars. The required sulfur dioxide feed rate determines the most practical and cost-effective type of container for a particular application. Generally, gas withdrawal rates in excess of 40 to 50-lb in 24 hours for 150-lb cylinders, and in excess of 400 to 450-lb in 24 hours from 1-ton containers, are impractical. Withdrawal rates at near the limits indicated above can result in frosting of the container and may lead to some reliquification of the gas in the piping system (White 1986).

Enclosing the containers and heating the room to 70 or 80° F can help maintain high withdrawal rates. Another method to support high feed rates is to manifold several containers together to reduce the net usage per container below the point where frosting becomes a problem. Liquid sulfur dioxide can also be withdrawn from the alternate connections on the cylinders and routed to an evaporator. Sulfur dioxide evaporators use a heated water bath external to a pressure vessel to vaporize the liquid continuously. By adding additional evaporators and containers or bulk storage, any required feed rate can be accommodated. Because of the

Table 16-9. Secondary Solids Dewatering Processes

Process	Advantages	Limitations
Vacuum filter	Continuous process	High maintenance
	High capacity	Vacuum pumps require seal water.
		Filter cloth can blind with certain solids.
Filter press	Lower capacity	More operator intensive
	Adaptable to large variety of dewatering operations and solids	High maintenance
		Noncontinuous; requires shutdown to clean the press.

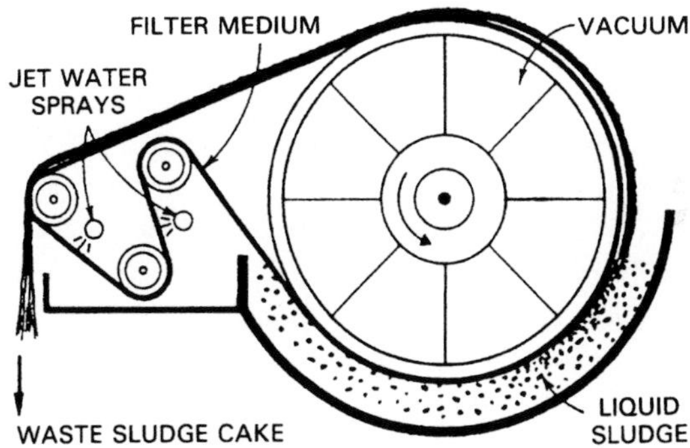

Fig. 16-15. Typical vacuum filter.

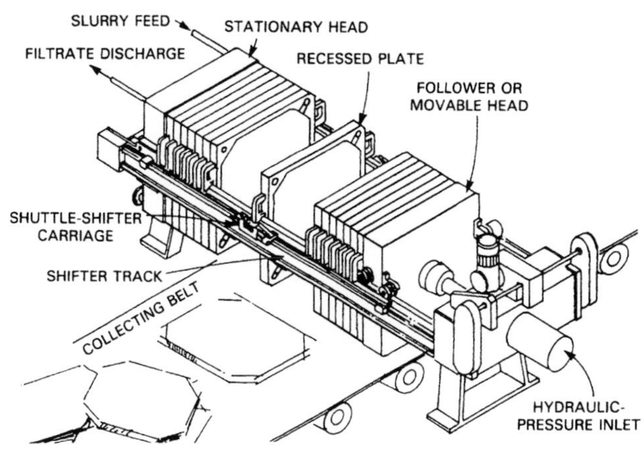

Fig. 16-16. Typical filter press.

Table 16-10. Physical Properties of Sulfur Dioxide

Molecular symbol	SO_2
Molecular weight	64.06
Boiling point at 1 atm	14° F
Freezing point at 1 atm	−104.6° F
Latent heat of vaporization at 70° F	155.5 Btu/lb
Specific gravity of vapor compared to dry air at 32° F and 1 atm	2.2638
Specific gravity of liquid at 32° F and 23.7 psia	1.436

Source: Compressed Gas Association 1988.

thermal input required and their relative efficiency, evaporators are typically used where continuous dechlorination is required.

A flow diagram for a typical continuous dechlorination system using ton cylinders and a sulfur dioxide evaporator is shown in Fig. 16-17.

Sulfonators are used to regulate the sulfur dioxide gas flow. An injector is used to develop a vacuum and draw the gas through the sulfonator into the injection water. The sulfonated injection water is then diffused into the wastewater stream using a distribution such as a perforated pipe spool to ensure distribution and mixing of the sulfurous acid solution in the discharge stream.

The feed rate of sulfur dioxide is often controlled in proportion to the waste stream flow and biased by a measured chlorine residual before sulfonation. Two chlorine residual analyzers are typically required, one before sulfonation to control the feed rate of SO_2, and one after the feed point to monitor the discharge chlorine residual.

Sulfur dioxide is a potential health hazard, and the design of the facilities housing the feed equipment must take certain requirements into account. Consideration should be given to the separation of the pressurized portion of the sulfur dioxide feed system from the vacuum portion and the system control panel. Separate rooms are advisable. The ventilation requirements for sulfur dioxide feed facilities are the same as those for chlorine feed facilities. The recommended unoccupied exhaust rate is four air changes per hour, with the recommended exhaust rate boosted to 15 air changes per hour when the room is occupied. Sulfur dioxide is heavier than air, so exhaust fan inlets must be near the floor.

Sulfur dioxide leak detectors are also recommended in storage areas and in any area where a sulfur dioxide leak might occur. The sample inlet to the detector must be near the floor (White 1986).

16.4.3.7 Heavy Metals Reduction. As stated previously, heavy metals can enter the power plant through coal, limestone, and water supplies. Heavy metals can also result from corrosion of various power plant components, particularly during maintenance activities such as boiler waterside cleaning and air heater washdown. Power plant wastewater streams that often contain heavy metals include coal pile runoff, ash storage area runoff, flue gas desulfurization wastewater, and metal cleaning wastewater.

Metals can be removed from wastewater by a variety of processes, including alkaline precipitation, sulfide precipitation, and ion exchange. Alkaline precipitation uses an alkaline reactant such as lime (CaO) or sodium hydroxide (NaOH) to precipitate metals as the hydroxide. The alkaline process generally can target only some of several metals, since metal hydroxides have minimum solubility at a particular pH, as shown in Fig. 16-18. This process will have limited applicability to power plant wastewaters in which a large variety of metals are present.

A recent process that has been demonstrated to effectively remove a large variety of metals to very low limits in power plant wastewaters is the iron coprecipitation process. Figure 16-19 schematically illustrates this process. Influent water enters the first stage basin where pH is adjusted and an iron salt added. The second stage basin is the reaction tank. Iron

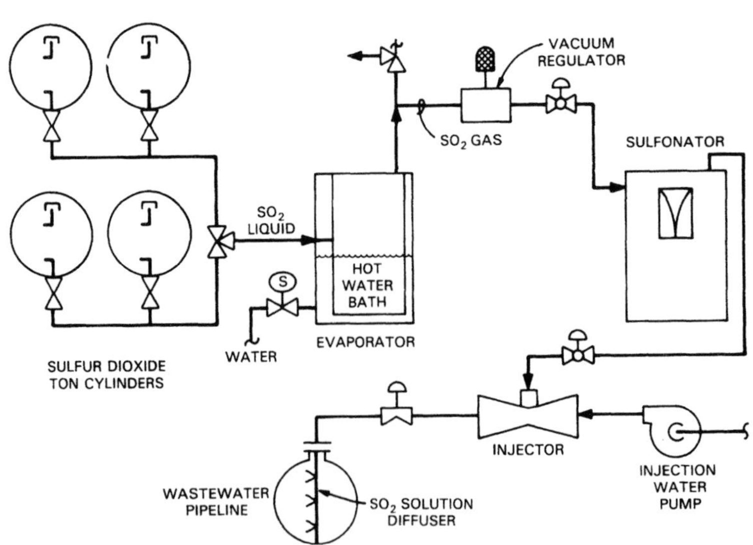

Fig. 16-17. Flow diagram of sulfur dioxide feed system.

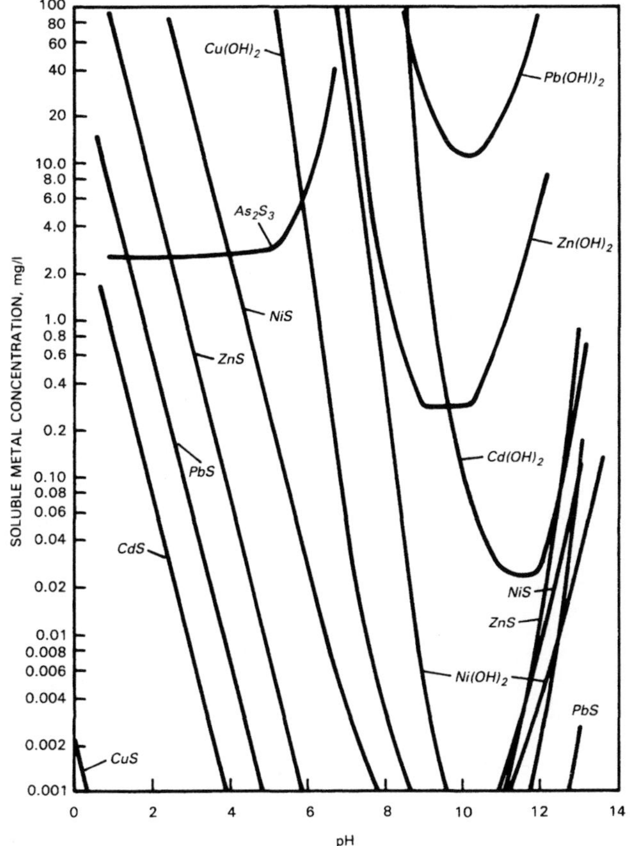

Fig. 16-18. Calculated metal hydroxide/sulfide solubilities.

salts used include ferric chloride and, in conjunction with an oxidation step, ferrous sulfate. Compounds to adjust the pH and flocculating agents to promote precipitation also are added.

In the wastewater, dissolved and suspended metals are adsorbed onto and trapped within the iron precipitate, which then settles out, leaving a clear effluent. The efficiency of the process depends on a number of factors, including the oxidation state and initial concentration of the trace elements, iron dosage, the extent of solids recirculation and contact, retention within the reaction and flocculation zones, and pH. A sulfide, such as sodium hyrosulfide, also can be used in the process to enhance metals reduction.

The treatment costs for removing metals from wastewater when using iron coprecipitation is usually less than competing processes, such as lime or lime/soda ash precipitation, ion exchange, or RO (Chow 1987).

16.4.3.8 Wastewater Volume Reduction. Wastewater volume reduction processes are becoming increasingly more important in power plant water management designs. Concentration of dissolved solids from wastewater into a smaller wastewater stream allows reuse of the relatively pure portion of the treated effluent in other plant systems, thus reducing overall plant makeup and discharge requirements. Solids removed are concentrated to a slurry or dry solids form.

Power plant wastewater volume reduction is accomplished by separation processes. These separation technologies include membrane processes, evaporation/distillation processes, and crystallization processes. These separation processes are distinguished from each other by their energy requirements, capital costs, the purity of the product water obtained, and the moisture content of the reject or waste solids removed.

The separation technologies used most frequently at power plants are discussed in the following paragraphs.

16.4.3.9 Membrane Processes. Membrane processes typically considered for volume reduction of power plant wastewater include RO, ED, and EDR. These processes produce a relatively clean product water stream that is low in total dissolved solids (TDS), and they produce a concentrated liquid waste stream very high in TDS. The liquid waste stream must be disposed of or treated to further reduce volume before disposal.

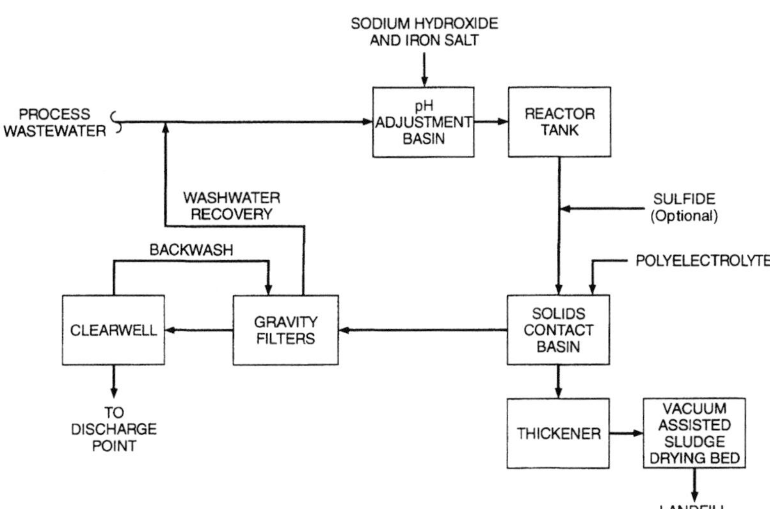

Fig. 16-19. Iron coprecipitation process for metals reduction.

RO is a commercially available membrane process used more frequently to treat plant raw water makeup waters, such as well waters or surface waters, and used less extensively to treat high TDS wastewaters. The membranes are permeable to water but generally not permeable to dissolved solids. This process allows a separation of the dissolved solids from the water and results in a relatively low solids product water and a relatively high solids waste stream. High system operating pressures are required to overcome the osmotic pressures of the brine stream. The concentration of dissolved solids in the brine stream is limited by the potential for scaling of the dissolved solids on the membranes. Frequently, silica is a controlling constituent, which requires increased brine flow rates and, therefore, decreased recovery rates. RO must be used in conjunction with pretreatment such as filtration or lime softening and filtration to prevent fouling or sealing of membranes and to optimize recovery rates. Feed of antiscalents is often required. Additional treatment of the RO reject stream using brine concentrators and crystallizers is necessary if a solid waste product or further volume reduction is required. Additional description of the RO process may be found in Chapter 15, Water Treatment.

ED and EDR are membrane technologies that use ion-selective membranes to reduce the volume of wastewater and recycle recovered water. Water is routed between alternating stacks of cation and anion membranes. The dissolved cations pass through the cation membranes and the dissolved anions pass through the anion membranes. The remaining relatively pure water is the product water. Electricity powers the process. Silica is not removed by this process; therefore, the product water produced may not be suitable for recycle to a system in which silica must be limited such as the condenser cooling system.

16.4.3.10 Evaporation/Distillation Processes. Evaporation/distillation processes for wastewater treatment use steam or mechanical energy to vaporize wastewater and a condenser to recover the purified liquid for reuse. Wastewater treatment by evaporation/distillation is energy intensive and, as such, distillation processes for power plant wastewater treatment must achieve maximum efficiency. The evaporation/distillation process most widely applied in power plant wastewater treatment is the brine concentrator.

Brine concentrators use a vapor compression process to evaporate wastewater and condense the vapor formed to provide a relatively pure product water that can be recycled to other plant systems, such as makeup to the circulating water system. In the evaporation process, the vapor from the evaporated wastewater is compressed to raise the condensing temperature of the vapors. Heat recovery is accomplished by using the compressed vapors to heat the evaporating fluid. The solids present in the supply water to the evaporator are concentrated in a waste stream.

A brine concentrator system consists of a long tube, vertical falling film, vapor recompression evaporator; a feed tank; a feedwater deaerator; heat exchangers; pumps; and controls. A typical brine concentrator system is shown schematically in Fig. 16-20. Cooling tower blowdown or other plant wastewater is transferred to the feed tank. Sulfuric acid is added to the feed tank to convert any alkalinity to carbon dioxide. The wastewater is then pumped from the feed tank through the feed strainer and heated to the boiling point in the feed heaters by recovering the product water (condensate) heat. Dissolved carbon dioxide and dissolved oxygen are removed from the preheated supply water in a deaerator to minimize downstream corrosion. The deaerated water is directed to the brine concentrator vapor body, where it is mixed with the concentrated slurry. The slurry is continuously recirculated to the inside wall of the tubes as a thin film. As the thin film falls down the inside of the tubes, a portion is evaporated and the remainder returns to the vapor body. The resulting vapor is demisted, compressed to raise the condensation temperature above the boiling point of the brine film, and returned to the shell side of the brine concentrator to provide heat for further evaporation inside the tubes. The condensate is collected as relatively pure water in the product tank. The brine concentrate is continuously blown down to maintain the solids concentration in the evaporator sump within the desired range. Brine concentrator blowdown is still mostly in the liquid form and must be directed to a crystallizer or spray dryer for further treatment if a solid waste is desired. A brine concentrator installation is shown in Fig. 16-21.

Multi-effect distillation (MED) or multistage flash evaporation (MSF) can be used to treat power plant wastewater. MED has been mostly used in the treatment of sea water and is discussed briefly in Chapter 15.

16.4.3.11 Crystallizer Processes. A crystallizer uses a boiling process to treat very high TDS fluids, such as brine concentrator blowdown. The equipment is designed to handle the suspended solids resulting from evaporation of a supersaturated liquid.

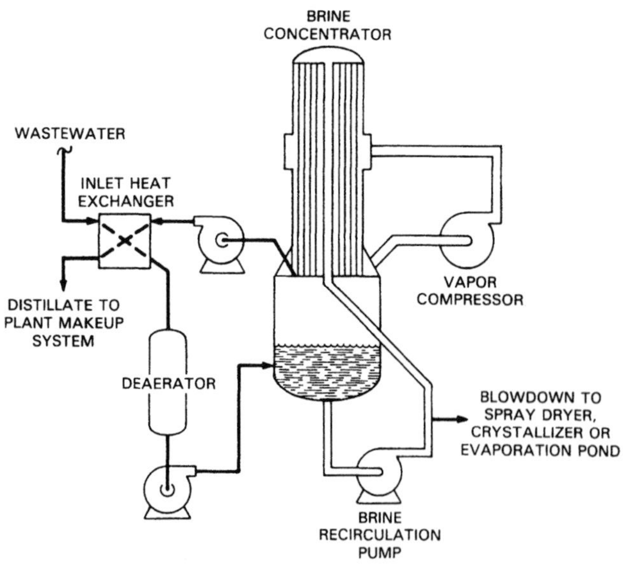

Fig. 16-20. Typical brine concentrator.

Fig. 16-21. Brine concentrator installation. (From Resources Conservation Company. Used with permission.)

Crystallization systems can be vapor recompression processes or utilize an external steam source. A vapor recompression crystallizer system is shown schematically in Fig. 16-22. Concentrated blowdown from the brine concentrator is collected in the brine return tank. The brine enters the suction side of the brine circulation pump and is pumped to the crystalizer, from where it is pumped by the brine recirculation pump through the heat exchanger to bring the feed stream to the boiling point. The heated feed stream is then directed to the crystallizer vessel, where it flashes, releasing vapors from the supersaturated liquor. The vapors leaving the crystallizer vessel pass through an entrainment separator to remove any liquid present in the product stream. The vapor from the entrainment separator is then compressed and used to heat the feed stream in the crystallizer heat exchanger. The vapor condenses on the shell side of the crystallizer heat exchanger, and the condensate is collected as product water. Blowdown from the crystallizer is routed to the centrifuge to separate liquids and solids. Solids are then collected and landfilled. Liquid from the centrifuge is returned to the brine feed tank for further treatment.

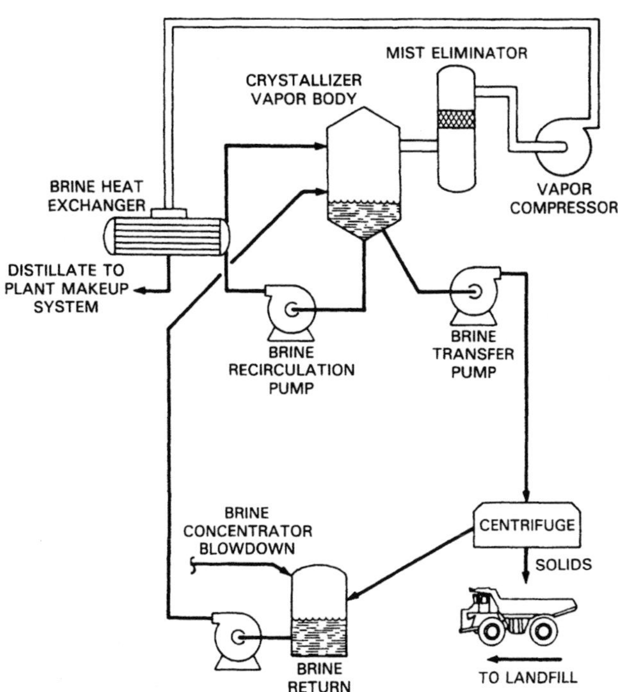

Fig. 16-22. Typical vapor compression crystallizer.

If an external steam source is used instead of vapor recompression, steam from an external source will be used in the brine heater and vapors from the crystallizer vessel entrainment separator will be condensed in an external condenser.

16.5 SOLIDS TREATMENT AND DISPOSAL

16.5.1 Description of Solid Waste Sources

The majority of the solid wastes generated at a power plant come from the coal combustion and flue gas desulfurization processes. More than 10% of the coal usually remains as ash after combustion. This waste has, to date, not been considered hazardous by the EPA and generally can be landfilled if state leachate control regulations are met.

Flue gas desulfurization (FGD) leaves a solid waste product of calcium–sulfur compounds that must be dewatered before handling as a solid waste. FGD wastes are not considered hazardous and generally can be landfilled.

Other solid wastes include those from wastewater treatment dewatering processes. In power plants, these wastes are generally nonhazardous. However, consideration must be given in landfilling to the leachability of the solids. Often, encapsulation of wastewater treatment solids in a more impervious solid, such as combustion wastes, has been employed.

16.5.2 Disposal System Selection Criteria

The development of a power plant solid waste disposal system requires the evaluation of site- and fuel-specific plant operational requirements, waste properties, site constraints, environmental impacts, and regulatory requirements. The selection of the disposal system should include the development of a set of criteria or operational requirements the disposal system must meet to successfully support station operation.

16.5.2.1 Power Plant Operational Requirements. Obviously, existing power plants need to determine their present and possible future operational requirements to select a solids disposal system, and new facilities need to define their anticipated operational requirements. In general, evaluation of the following operational factors is required:

- Types of solid wastes generated. The types generated may change over the life of the plant as treatment systems change, perhaps required by change in regulatory requests.

- Production rates of each solid waste generated. The rate of production may vary over the operational life of the power plant as a result of variability in fuel quality, plant load factor, and changes in environmental control systems.
- The location of each solid waste discharge point and locations for temporary storage,
- Plant water balance to completely define all interfaces with water supplies and wastewater discharges,
- Plant operational procedures, operating hours, and capacity factors, and
- Projected life of the facility.

16.5.2.2 Waste Properties. Information on the engineering and chemical properties of the solid wastes generated should be collected. The properties and types of solid wastes generated are determined by the type of fuel burned, method of combustion, environmental and pollution control systems, plant load factor, and water and wastewater treatment systems. The material properties may change as the operation of the power plant or fuel source changes over the life of the plant. The present and anticipated future properties of the generated solid wastes should be considered when selecting a disposal system. Physical properties to be considered include the following:

- Moisture–density relationships,
- Compatibility,
- Shear strength,
- Moisture content,
- Erodibility,
- Fugitive dust capability,
- Particle size,
- Permeability, and
- Pozzolinic properties.

Chemical properties to be considered include the following:

- Corrosiveness,
- Toxicity,
- Leachability, and
- Gas generation potential.

The potential for offsite sales or disposal is another consideration.

16.5.2.3 Site Constraints. Disposal of solid wastes by landfilling or ponding requires adequate land area. Potential site evaluations must address the following questions:

- What is the existing land use?
- Is the site in a floodplain?
- Is the site a wildlife habitat, particularly for an endangered species?
- What zoning regulations exist?
- What is the volume capability for the site?
- What is the site topography and is it conducive to development?
- What is the surface drainage pattern?
- What climatic conditions exist that can affect its use?
- What height restrictions exist?
- What are existing ground water conditions and how will they be impacted?
- Are there any impacts to wetland?
- What are the costs to transport the wastes to the site?
- What are the requirements for site closure?
- Are aesthetics important at the site?
- What buffer zones are required?

16.5.2.4 Environmental Regulatory Requirements. A complete review of all environmental regulatory requirements must be performed during selection of a solids disposal system, as described earlier in this chapter. The environmental regulatory requirements will, in many cases, define the constraints around which the disposal facility will be sited and designed.

If an onsite disposal site is selected as a plant disposal system, early discussions with the reviewing agencies are recommended to ensure an orderly development and review. An adversarial relationship with the reviewing agencies should be avoided.

16.5.3 Onsite Transportation and Disposal of Solid Wastes

The onsite disposal of solid wastes consists of two primary steps: transportation of solid wastes from the point of generation or temporary storage to the disposal site, and disposal at the ultimate storage or disposal area for the solid wastes. The following sections describe the various systems that can be used.

16.5.3.1 Transportation of Solids Wastes. Various methods have been developed for transporting solid wastes. They can be generally grouped as follows:

- Wet slurry method, using a water slurry to transport the material to the disposal area;
- Pneumatic method, using air to "slurry" and transport solid wastes to the disposal area;
- Trucking, using either over the road, highway weight-limited vehicles, or off-road, high-capacity vehicles. Roads can be existing public roads or dedicated haul roads.
- Rail transport, using existing or dedicated railroad alignments and equipment;
- Conveyor, using fixed or movable belt conveyor systems; and
- Barge, using waterways to transport waste materials.

Each system can be used for the entire transport cycle or can be combined with other systems, depending on specific plant or site conditions. For example, a fixed conveyor system could be used to transport solid wastes to a central discharge point at a distant disposal site. A truck system could be used to distribute the solids at the disposal site.

The selection of a suitable transport system should include evaluation of the following technical factors:

- Compatibility with plant operations,
- Compatibility with solid wastes,
- Regulatory restrictions,

- Site conditions,
- Length of transport,
- Capacity of system,
- Compatibility with disposal system, and
- Durability.

Economic factors that should be evaluated are capital equipment costs, such as equipment, roads, or railroad truckage costs; and operating costs, including maintenance, equipment replacement, personnel, power, and fuel costs. Support equipment costs such as maintenance buildings also merit attention. Table 16-11 provides a comparison of the transport methods.

16.5.3.2 Onsite Land Disposal. The objective of a land disposal system is to provide an environmentally safe, permittable, and economic method of disposal of solid waste. Hazardous wastes are not included, since the requirements for permanent disposal of these materials are not within the scope of normal power plant operation. Disposal system design should address all or some of the following concerns:

- Storm water management,
- Solid waste placement and control,
- Leachate control,
- Disposal site closure, and
- Monitoring.

The specific methods of addressing each of these concerns depends on the solid waste properties, site and climatic conditions, and the regulatory requirements for the facility.

Land disposal systems are separated into two general

Table 16-11. Comparative Evaluation of Transport Methods

Method	Advantages	Disadvantages
Wet slurry	High capacity	Wet disposal may not be compatible with a dry landfill.
	Generally low operating costs	Large volume of water required
	Relatively low capital costs except for long distances or high grades	Wet disposal may not be compatible with regulatory requirements.
		Self-hardening of pozzolinic combustion wastes in lines possible during long transport or temporary interruptions in flow
		Low flexibility at discharge point
		High capital and operating costs for long distances
		Collection/disposal of transport water which may be contaminated by the solid waste itself
Pneumatic	Large volumes of water not required	Not suitable for transport of wet solid wastes.
Trucking	High flexibility	Limited transport distance
	Low capital cost using contracted haulers	Low flexibility at discharge point
	Large off-road vehicle can haul high capacity (up to 120 tons per truck).	Potential fugitive dust problems
		Existing roads may not be suitable; costs of dedicated haul roads can be high.
		Spillage a concern, particularly on public roads
		High personnel and operating costs
		Not compatible with wet material
Railroad	High capacity	Low flexibility
	Lower operating costs than trucks for long haul	Fixed alignment and discharge require additional mobile equipment
	Compatible with returning combustion wastes to coal mine where coal is rail supplied	Potential fugitive dust problems
		High capital costs for dedicated equipment
		Combustion wastes may require preprocessing, such as pelletizing.
Conveyor	Low operating costs for long distances on level grades	Low degree of flexibility
		Additional equipment may be required at discharge point
		Potential fugitive dust problem
		High capital costs
		High power costs
		Abrasive material may result in high maintenance costs.
		Not compatible with wet material
Barge	Low operating costs for long hauls	Requires existing waterways
	Compatible with returning combustion wastes to coal mine with barge-supplied coal	High capital costs for dedicated equipment
		Self-hardening combustion wastes may cause problems without preprocessing.
		Not suitable for short hauls.

groups: wet ponding/sedimentation systems and dry landfill systems. The disposal systems in general correspond with the method of transport of the solid wastes. The wet system is used with slurried solid wastes, while the dry landfill system disposes of dry or moisture conditioned solids.

In many older plants built before 1980, disposal of solid wastes was performed using the wet, ponded sedimentation system. Because of environmental concerns, many modern facilities now make use of the dry landfill method.

16.5.4 Solid Waste Disposal Technologies

The design of a land disposal site must use a combination of conventional civil, geotechnical, and increasingly sophisticated environmental control engineering in the development of stormwater drainage systems, earthfill or dam embankment, and leachate control and monitoring.

16.5.4.1 Wet Disposal System Operation. A wet disposal system requires the construction of a water impoundment to collect the slurried wastes, provide for sedimentation of the solids, and decant the transport water for discharge or reuse. The impoundment can be constructed in one of the following fashions:

- Dam construction—a dam is constructed across the outlet of a valley. The resulting impoundment is used as the disposal site.
- Pond construction—a pond is created by constructing the confining perimeter dikes, excavating a hole, or using an existing depression below surrounding grade. For example, a borrow pit for site fill can be used, or a depression can be excavated and the removed fill used to build the perimeter dikes.

Stormwater control for a pond with above-grade dikes is simplified since only direct rainfall can enter the impoundment. Exterior drainage can be routed around the pond by perimeter ditches. Below-grade ponds and dam construction must contend with runoff from the surrounding area. Diversion of runoff is generally preferred to prevent uncontrolled discharges of water from the solid waste disposal area.

The solids slurry is deposited in one end of the disposal impoundment. Sufficient water is kept in the impoundment to permit sedimentation of the solids. If the sedimentation properties of the solids are compatible, it is a good practice to deposit the slurry in a "delta" above the water surface. This deposit pattern allows higher densities to form in the deposited solids, reducing storage volume requirements. In addition, this deposit pattern may provide other advantages to the plant water balance. Its flexibility allows the system to absorb additional discharges from other compatible plant systems, and it reduces evaporative losses. However, additional water surface area can be used to increase evaporative loss if advantageous for the plant mass balance.

Solid slurry transport water can be from a once-through system for a simple bottom ash or fly ash transport system, or from a water recycle system tied in with the scrubber system.

The waste delta expands as solids are added to the impoundment. However, the usable volume of the impoundment must be less than the total volume of the diked interior. Sufficient freeboard must be left to prevent overtipping of the containment dikes or dam as a result of storm inflow.

Additional volume must be left to provide sufficient detention time to permit settling of suspended solids from the decanted water. The required volume depends on the solids slurry production rate, particle size, velocity of the water through the pond, length of the flow path, depth of the water, and density of the transport water.

Because wet solids transport and disposal is not an exact science, testing of solids settling rates should be performed before construction to help determine the proper impoundment size and design.

16.5.4.2 General Dry Disposal System Operation. A dry disposal system requires the development of a landfill. The configuration of the landfill is controlled by the site topography and the drainage system. The objective of a modern waste landfill is to compact the maximum amount of solids in a minimum area and to minimize the entrance of water into the landfill solids. Water management considerations are the major reasons why modern power plants prefer a dry landfill over ponding.

Dry landfills can be subdivided into two general categories: heaped fill above grade and valley or pit fills below grade.

A dry landfill, if properly sited, constructed, and operated, has a lower potential for impact on the environment, primarily because leachate generation is reduced. Reduced leachate results because dry landfills offer several advantages:

- Greater compaction of solid wastes results in low permeabilities in the landfill, reducing infiltration of rainfall.
- Less water is introduced to the landfill because the slurry transport system is removed. Solids are dewatered to a solids content of 85% or higher prior to landfilling.
- Rapid surface drainage of rainfall is achieved by using proper grading.
- Leachable properties of materials can be reduced further by fixation or stabilization of solid wastes. When mixed, fly ash, FGD solids, and lime take on pozzolonic properties, which increases compressive strength and decreases permeability. This mixed waste material has been accepted in several states as being sufficiently stabilized to delete liner requirements.

Storm water management in a dry disposal landfill should meet the following objectives:

- Minimization of stormwater infiltration into the landfill site. Surface water runoff from outside the landfill is diverted by ditches or diversion dikes around the landfill and off the site. Rapid drainage of direct rainfall off the site is accomplished by proper grading of the landfill as it is developed.
- Containment of runoff from active disposal areas. A system of containment dikes surrounding the active landfill area directs runoff to sedimentation ponds for removal of solids eroded from landfill areas.

Management of the physical handling and disposal operation for the solid waste materials should include the following as the bases of the system design:

- Maximize storage volume in the available area.
- Minimize transport and placement costs, compatible with proper landfill operation. This operation includes proper grading of haul roads for vehicle access and economical compaction of landfill waste, as well as providing proper structural stability for the landfill.
- Control fugitive dust.
- Minimize site development costs, compatible with proper landfill design and site conditions.

A properly designed landfill integrates water management and solids management requirements.

Landfills can be described as valley (buried) or heaped fills. Some landfills are combinations of the two types, being more of one or the other depending on the landfill's stage of development in relation to the surrounding topography.

A heaped fill is a landfill mounded above the surrounding grade. Figure 16-23 shows a general arrangement of a heaped fill. A valley, or buried, fill is constructed in a topographic depression. The depression may be a natural valley or a manmade excavation, such as a quarry, mine, or earth barrow pit. A very effective form of dry landfill disposal is reclamation of a coal surface strip mine. Figure 16-24 shows a general arrangement of a natural valley fill.

In both arrangements, the waste solids are dewatered to a solids content of approximately 85% or higher before landfilling. This solids content may vary depending on the physi-

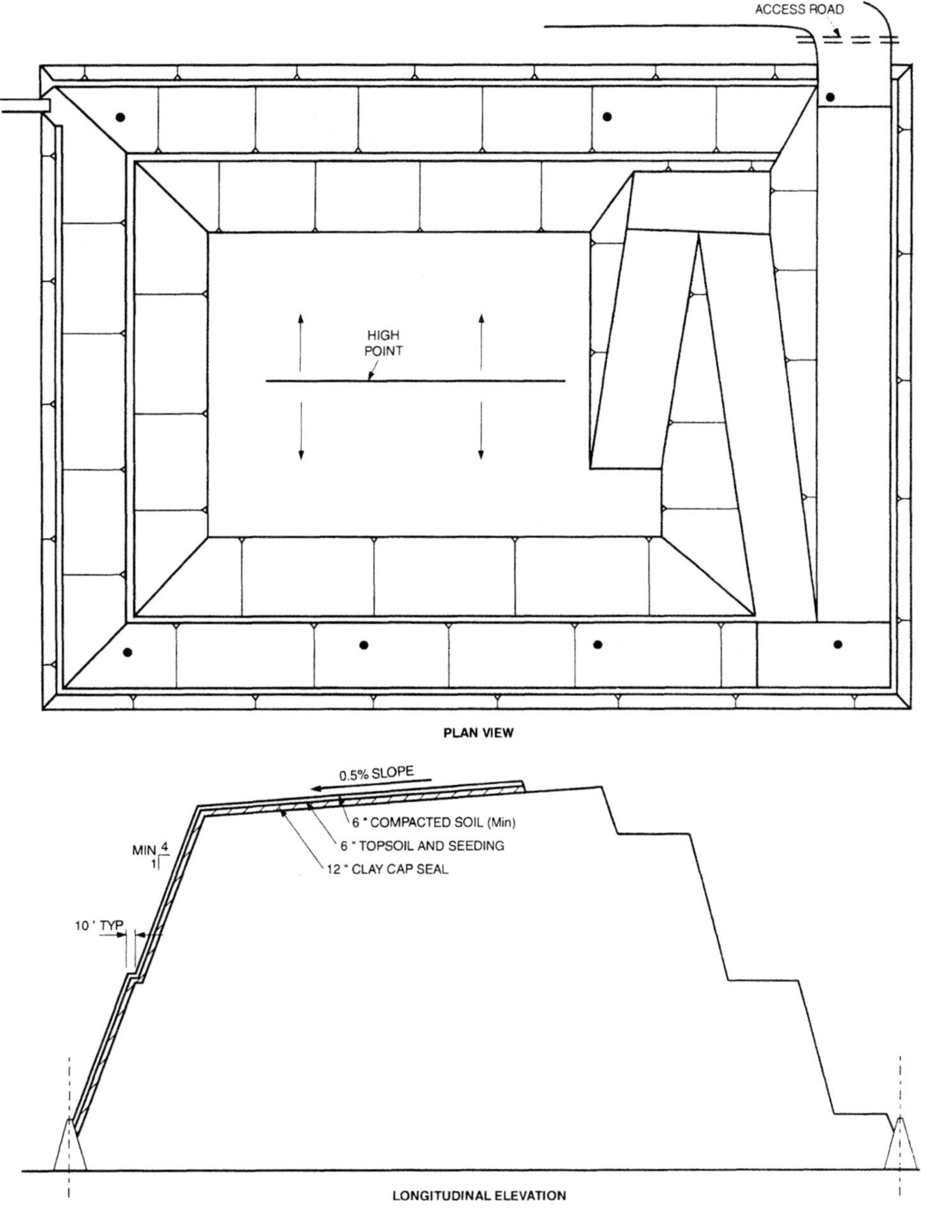

Fig. 16-23. Heaped fill solid waste landfill.

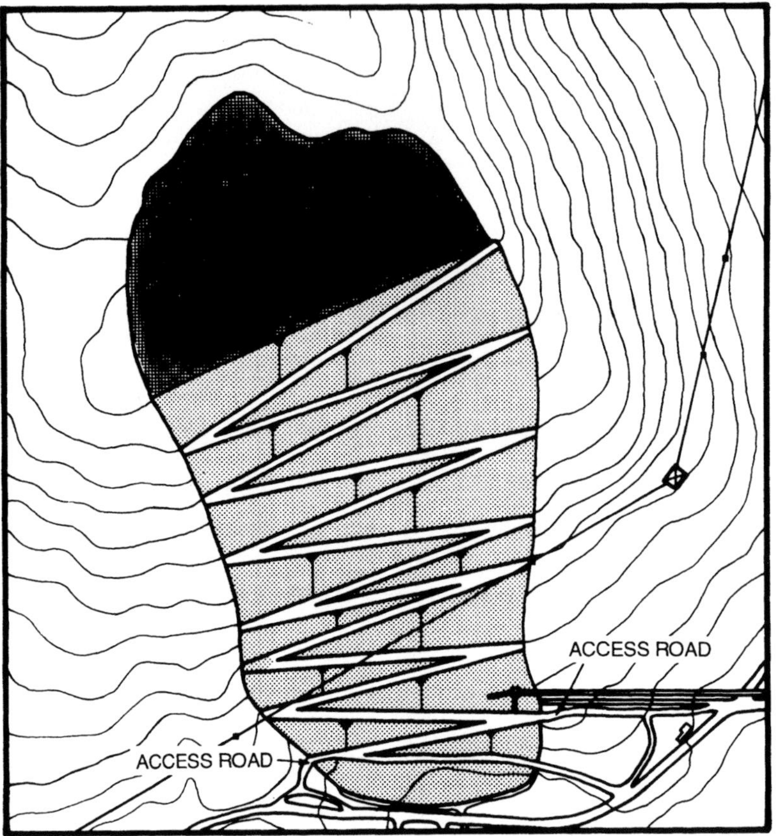

Fig. 16-24. Natural valley fill solid waste landfill.

cal properties of the waste material. Dewatering the solids simplifies transport of the material and allows mobile equipment to spread and compact the waste materials. Compaction densities of 85% to 95% of maximum density, as determined in accordance with ASTM D698, can be achieved with normal spreading equipment if proper control of moisture content and thickness of the lift or layer placed is maintained. Lift thickness should be determined based on material properties and the size of landfill equipment. Layers of 6 to 24 in. in uncompacted thickness are typical. This control of compaction is one of the items that separates a modern landfill from the much maligned "dumps" of the past. Proper compaction combined with stabilization of the waste materials can provide sufficient strength to the waste materials to construct landfills up to several hundred feet in height.

Fugitive dust control can be a major landfill problem, particularly on haul roads and in active landfill areas. Where high winds are common, dust generation can be severe. This problem can be compounded if the waste is fine grained and noncementateous, as are some fly ashes and FGD wastes. Dust generation can be controlled by performing the following operations:

- Minimize active exposed landfill areas.
- Constantly wet exposed areas and haul roads using water trucks.
- Cover daily when necessary with soil or coarse-grained materials, such as bottom ash. The use of sanitary landfill techniques such as applying foams or chopped tires may be necessary in extreme conditions.
- Minimize spillage from transport equipment.
- Minimize tracking of equipment over compacted wastes that have hardened or have been sealed.

16.5.4.3 Leachate Control. The potential generation of leachate that can degrade the quality of either ground water or surface water supplies is a major concern that must be addressed by the disposal site design. Leachate control is accomplished by minimizing leachate volume or stabilizing wastes to prevent leaching of materials that will degrade water quality.

Testing for leachate quality is generally performed using the leaching method, or "shake" test method, specified in the EPA Toxicity Characteristic Leachate Procedure (TCLP) toxicity test. The water in contact with the solids is tested to determine the materials that can be dissolved out of the waste materials and compared with surface and ground water quality standards.

An alternative method used in leachate testing is a column test. A column is constructed of waste and selected clay liner or natural subgrade soils to approximate the landfill cross section. Water is added at the top of the column and allowed to percolate through the column and exit at the bottom, where it is captured for analysis. The captured water is analyzed in a similar manner to the EPA leach test for dissolved materials. The water from the column is sampled and tested versus time or quantity of flow through the column. Often the flow through the column is accelerated so that a sufficient quantity of leachate is produced in a reasonable time frame. A large positive head greater than the anticipated head in the landfill is applied to the column to accelerate acquisition of test samples. The leachate chemistry is often plotted against the number of pore volumes of fluid that have passed through the soil.

The chemistry of the leachate generally changes with increasing pore volumes. Often the concentration of the dissolved materials is initially high and then decreases or stabilizes with time. However, this phenomenon varies depending on the constituents that leach out of the waste solids.

16.5.4.4 Liner Systems. Liner system designs are often controlled by state regulatory requirements. In general, the need for a liner system for the base of the landfill and landfill closure is determined by the type of waste material to be landfilled or the results of leach tests performed on the waste materials. If it is determined that a liner system is necessary, one or a combination of the following types is used:

- Clay liners consisting of natural or recompacted clay soil. Permeability of these liners generally ranges from 1×10^{-6} cm/s to 1×10^{-8} cm/s. The thickness of these liners generally ranges from 2 to 5 ft.
- Synthetic liners called geomembranes or flexible membrane liners (FML).[1] These types of liners are available in many forms and materials.

  Reinforced liners
    Chlorinated polyethylene (CPER)
    Chlorosulfonated polyethylene (CSPE)
    Proprietary liners, such as XR-5 or "Hypalon"™
- Unreinforced liners
    High Density Polyethylene (HDPE)
    Very Low Density Polyethylene (VLDPE)
    Polyvinyl Chloride (PVC)

  In general, the geomembrane should be chemically compatible with the wastes and physically strong enough to withstand stresses associated with liner installation, pond or landfill operations, and climatic condition at the disposal site. However, although strength is important, the synthetic liner must function as a hydraulic barrier, not a structural member.
- Modified natural soils or waste materials. These liners are often substitutes for clay liners. One common liner is a soil–bentonite clay mixture, which uses a natural or chemically treated very low permeability sodium montmorillonite clay added to the natural soil to decrease the soil's permeability. Another common liner uses lime, cement, or both to stabilize the waste material for the same purpose.
- Geosynthetic clay liners (GCL). These are premanufactured clay liners formed by encapsulating a very low permeability sodium mantmarihlinite clay between the geotextiles.

The design of the liner system often is associated with a leachate collection system immediately above the liner to collect leachate generated by the waste and transport it to a sump or collection pond. The leachate collection system minimizes the height of the leachate on the liner system, thereby reducing seepage through the liner. Leachate height limitations are often required by state regulations. The leachate collection system, in general, consists of 12 to 24 in. of granular material and a system of collection pipes. In some newer landfills, synthetic drainage nets of polyethylene webs are used. These nets are thinner, take up less space, and can be installed more quickly at a lower cost. However, extra care must be exercised to properly install these systems.

To evaluate the effectiveness of the liner and leachate collection system, a landfill water balance must be performed. The EPA has developed a program called "The Hydrologic Evaluation of Landfill Performance," or HELP, to perform this analysis. The HELP program has been modified as additional information about the rapidly changing liner technology has developed.

In general, liner systems for combustion wastes have consisted of a single liner with an overlying leachate collection system. However, for wastes with a greater potential to impact ground water quality, double liner systems or composite liner systems are used. A double liner system consists of two liners separated by a leachate collection system. A composite liner consists of a synthetic liner overlying a clay, GCL, or soil–bentonite liner. A composite liner combines the advantage of a very low permeability synthetic liner with a backup clay liner to "plug" accidental small leaks in the synthetic liner that are due to tears or minor seam failures. The greater cost of these systems is justified if there is a serious potential for degradation of the ground water system from the waste disposal site.

16.5.4.5 Closure. After completion of the landfill, final closure of the landfill will be required. This includes the following steps:

- Final grading for drainage,
- Placement of an infiltration barrier, and
- Placement of a soil erosion control cover.

The type of infiltration barrier will be dependent on the climatic conditions and the regulatory requirements. This may consist of a natural clay layer, or a combination of synthetic liners such as clays or GCLs.

To protect the infiltration barrier and prevent erosion of the placed combustion solids, an erosion protection layer

[1]A complete description of the synthetic liner types is beyond the scope of this book. Suggested references are *Designing with Geosynthetics* by R. M. Koerner, National Sanitation Foundation Standard No. 54, and the Environmental Protection Agency's EPA/600/2-88/052 "Lining of Waste Containment and Other Impoundment Facilities."

will be required. This is often a soil layer covered with a vegetated layer to prevent soil erosion. In arid climates, where vegetation would be difficult to establish and maintain, a graveled surface may be used. The thickness of this soil cover will be dependent on the climatic conditions and the slopes of the landfill.

16.5.4.6 Monitoring. Landfills will require monitoring of the underlying ground water and any discharges of storm water from active landfill areas. The monitoring requirements should be addressed with the appropriate regulatory agencies.

16.6 REFERENCES

ASTM. 1991. Test Method for Laboratory Compaction Characteristics of Soil Using Standard Effort. ASTM D698-91.

BRENNAN, J. E. and G. R. MACE. 1980. How to Treat Chemical Cleaning Waste from Your Power Plant. *The 1980 Electric Utility Generation Planbook.* McGraw-Hill, New York, NY, pp. 150–153.

CARPENTER, J. K. and V. J. SHANKS. 1977. CO2 for pH Adjustment of Wastewater. *Power Engineering 87.* October 1977, pp. 49–51.

CHOW, W. 1987. Removing Trace Elements from Power Plant Wastewater. *Pollution Engineering*, March.

Code of Federal Regulations. Steam Electric Power Generating Point Source Category. 40 CFR Part 423. Updated yearly.

COMPRESSED GAS ASSOCIATION. 1988. *Sulfur Dioxide.* Compressed Gas Association, Arlington, VA.

EPA. 1977. *Development Document for Effluent Limitations Guidelines and NewSource Performance Standards for the Steam Electric Power Generating Point Source Category.* EPA/440/1-74/029a. United States Environmental Protection Agency.

EPA. 1984. *Hydrologic Evaluation of Landfill Performance.* Vol. 1: EPA/530-SW-84-009, 1984. Vol. II: EPA/530-SW-84-010. United States Environmental Protection Agency.

EPA. 1988. *Lining of Waste Containment and Other Impoundment Facilities*, EPA/600/2-88/052. United States Environmental Protection Agency.

EPA. EPA Publication SW-846. *Test Methods for Evaluating Solids Wastes, Physical Chemical Methods.* United States Environmental Agency.

EPA. 1992. *Test Methods for Evaluating Solid Wastes, Method 1311: Toxicity Characteristics Leachate Procedure.* EPA/SW-846. United States Environmental Protection Agency. Vol. 1, Part C, Chapter 8, July 1992.

GARING, C. D., L. C. WEBB, and K. R. WEISS. 1990. Treatment of McIntosh Power Plant process wastewaters by the iron coprecipitation process. Paper given at the 51st International Water Conference. October 21–24, 1990, Pittsburgh, PA.

KOERNER, R. M. 1985. *Designing with Geosynthetics.* National SanitationFoundation Standard No. 54, November 1985.

RESOURCES CONSERVATION COMPANY. 3006 Northrup Way, Bellevue, WA 98004.

RICE, J. K. 1987. The Clock Is Running on Discharge for Steam/Electric Power Plants. *Power.* 131. October 1987, pp. 91–96.

RICE, J. K. and S. D. STRAUSS. 1977. Water-Pollution Control in Steam Plants. Power. 121. April 1977, pp. S1–S20.

WEISS, K. R., J. A. HENGEL, AND F. F. HADDAD, JR. 1989. A water management/monitoring plan for a true zero discharge power plant. Paper given at the American Power Conference. April 24–26, 1989, Chicago, IL.

WHITE, G. C. 1986. *Handbook of Chlorination.* Van Nostrand Reinhold, New York, NY.

YARD, K. D. 1991. Regulation of Utility Waste Disposal Practices. *Power Engineering.* 95. November 1991, pp. 83–85.

17

Electrical Systems

Lloyd Wade Sherrill

17.1 INTRODUCTION

A power generation plant converts other energy sources into an electrical form of energy that is convenient for transmission over long distances to many users. These sources include the potential energy contained in fossil fuels such as oil, gas, and coal; natural energy, encompassing water and wind sources; and nuclear fission sources (and perhaps fusion sources in the near future).

By burning fuel (fossil or nuclear) to make steam, the potential energy contained in the fuel is converted into heat energy, which in turn is converted into rotating mechanical energy in a steam turbine. In the case of a combustion turbine, the steam cycle is bypassed and the compressed combustion gases are allowed to expand directly in a heat engine which is used to convert the heat energy into rotating mechanical energy. This rotating mechanical energy, from whatever source, is then converted to electrical energy in an electrical generator.

The electrical systems in a power generation plant also use energy in this same electrical form to control and power the other systems in the plant. These systems include the fuel handling systems, boiler feedwater and makeup systems, and chemical waste treatment systems, among others.

This chapter addresses some important design considerations of the auxiliary electrical systems and gives brief descriptions of the transmission systems that are used to transmit electrical energy to the end-user.

It is not possible in the space of this chapter to cover all of the fundamentals of electrical engineering or the complete requirements for designing all of the electrical systems involved in a typical power plant or substation. Nevertheless, some definitions and basic fundamentals are necessary. Some of the basic concepts of electrical engineering which are used in everyday work in power systems design follow.

17.2 FUNDAMENTALS

17.2.1 Ohm's Law

The basic law of all electrical engineering is Ohm's Law. It simply states that the amount of current flowing in an electrical circuit (in amperes, symbol I) is directly proportional to the ratio of the electromotive force across the circuit (symbol E) or volts, (symbol V) to the resistance (or impedance) of the circuit (in ohms, for resistance, symbol R or Z for impedance), that is:

$$I = E/R \tag{17-1}$$

where

I = current in amperes,

E = electromotive force in volts, and

R = resistance in ohms (or Z for impedance).

This formula may, of course, be rearranged algebraically in any manner to solve for the other components.

As an example, if a power source having a terminal voltage of 100 Volts is connected to a load having a resistance of 20 ohms, the resulting current in the circuit would be 5 amperes.

$$I = \frac{100 \text{ Volts}}{20 \text{ ohms}}$$

$$I = 5 \text{ amperes}$$

Impedance is a term used to describe the alternating current equivalent of resistance in ohms of a circuit element that is not purely resistive, such as a coil of wire, called a *reactor*; or an array of parallel plates in the same circuit which are termed *capacitors*.

17.2.2 Kirchhoff's Law

The current portion of this law of electrical engineering states that the algebraic sum of all the currents flowing into and away from a given point in an electrical circuit must be zero. Stated simply, "whatever goes in—must come out." This is equivalent to a statement of the principle of conservation of mass. The other portion of this law, or voltage part, states that the algebraic sum of all the electromotive driving forces in a circuit and the potential differences (voltage drops) must also be zero. This is equivalent to a statement of the principle of conservation of energy.

17.2.3 Power

Power is the rate of doing work. In electrical systems, *power* is defined by the product of voltage and current. If Ohm's Law is multiplied by voltage and algebraically manipulated several ways, power can be determined by several equations:

$$P = E \times I$$
$$P = I^2 \times R \qquad (17\text{-}2)$$
$$P = E^2/R$$

where

P = power in watts,

I = current in amperes,

E = electromotive force in volts, and

R = resistance in ohms.

Since most electrical power work involves balanced three-phase systems, power is almost always expressed in three-phase terms:

$$P(\text{3-Ph}) = 3 \times P(\text{1-Ph}) \qquad (17\text{-}3)$$

where

P(3-Ph) = three-phase power, and

P(1-Ph) = single-phase power.

Since three-phase power systems are arranged so that the voltages and currents are electrically out of phase with each other by 120 degrees, it can be shown mathematically that the total power in a three-phase circuit is

$$P = \sqrt{3} \times E \times I \times \text{cosine}\,\Theta \qquad (17\text{-}4)$$

where

cos Θ is the cosine of the angle between the voltage waveform and the current waveform in a single phase; it is called the *power factor*.

For example, if a three-phase power line having a line-to-line voltage of 13.8 kV has a line current of 150 amperes and a power factor of 0.8, the total power the line is delivering would be

$$P = (\sqrt{3}) \times (13{,}800) \times (150) \times 0.8$$
$$= 2{,}868{,}276 \text{ watts, or } 2{,}868 \text{ kW}$$

Figure 17-1 shows the relationship between these quantities.

The hypotenuse of the triangle is called the *apparent power* or volt-amps. The vertical component of the triangle is called VARS (for volt-amps-reactive).

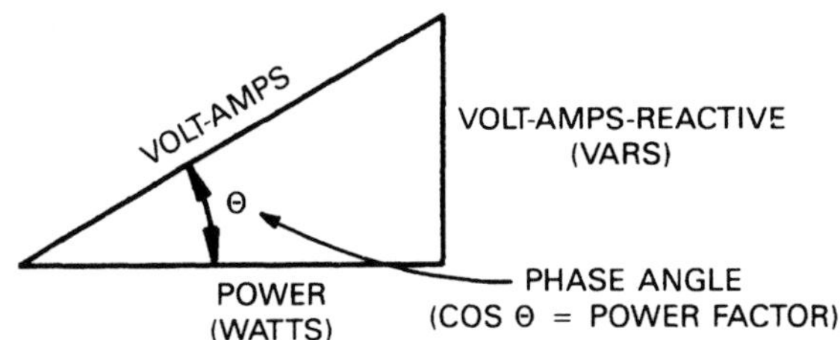

Fig. 17-1. AC power relationship diagram.

The voltages in a three-phase power system are electrically out of phase with each other by 120 degrees because the electrical power generators are mechanically arranged so that the planes of the coils are physically separated by 120 geometric degrees. Therefore, when the single rotating magnetic flux wave crosses the coils in sequence, the voltages generated are 120 degrees apart on the circular traverse of the rotating generator field.

Three sets of coils—ab, cd, and ef—are shown schematically in Fig. 17-2a, and are representative of the three windings in a three-phase generator. The generated sinusoidal voltage waveforms would be as shown in Fig. 17-2b with respect to a time axis. A vector representation of the voltages is also shown in phasor form on Fig. 17-2c. The phasor diagram is by convention assumed to be rotating counterclockwise.

As shown, the generator winding coils are independent of each other and may be connected in any manner desired. The system of voltages is said to be balanced whenever all three of the waveforms are of equal magnitude and are exactly 120 degrees apart (one third of a cycle).

The coils may be connected in delta as shown in Fig. 17-3, and the line terminals would be represented as X, Y, and Z, as shown.

Note that in the delta connected case, the line-to-line voltage is the same as the voltage across each coil or phase voltage; however, the line current is the summation of the currents in two of the coils. Mathematically (by vector algebra), it can be shown that the line current in the delta configuration is $\sqrt{3}$ times the phase current.

If the coils are connected in wye, as shown in Fig. 17-4, it is apparent that the line current is the same as the current in each coil and the line-to-line voltage is the sum of two phase to neutral voltages. In this case, by vector algebra the line-to-line voltage is equal to $\sqrt{3}$ times the line to neutral voltage.

Electrical loads may also be connected in delta or wye to the terminals of a three-phase power system. The relationships between the phase and line currents and voltages are the same as described above.

The phase sequence is defined as the order in which each of the three voltages reaches its positive maximum. Since it is always assumed that phasors or vector diagrams are rotating counterclockwise, it follows that a positive phase sequence is defined as the voltages X, Y, and Z reaching their positive maximum in that order. Interchanging any two terminals, say y and z, shows that the vector diagram would

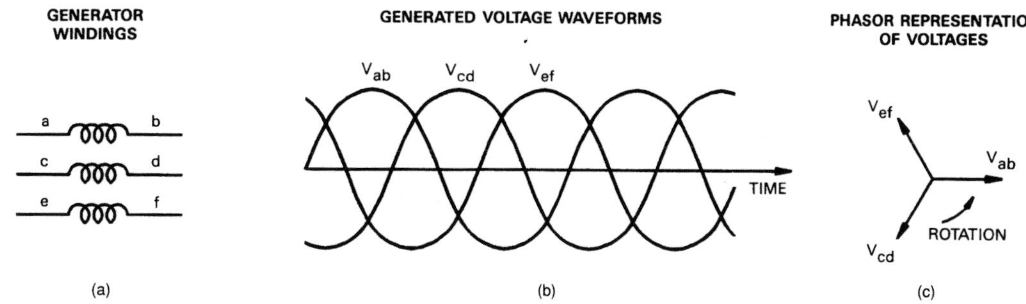

Fig. 17-2. Three-phase voltage representations.

have to rotate in the reverse direction for the voltages to reach their maximum in the same sequence as before. Conversely, if the vector diagram continues to rotate in the counterclockwise direction, the voltage peaks occur in reverse order, that is, XZY instead of XYZ. This is said to be reverse phase sequence.

17.2.4 The Per Unit System

Calculations in electric power work are most often carried out in per unit form, that is, with all pertinent quantities expressed as decimal fractions of appropriately chosen base values. All the usual computations are then carried out in these per unit values instead of in volts, ohms, and amperes. Voltage, current, impedance, and kVA (kilovolt-amperes) or MVA (mega volt amperes) are interrelated by Ohm's Law, so that by specifying any two of the quantities, the other variables are determined. In practice, it is usually most convenient to select voltage and kVA as the base quantities. Since almost all electric power work is done at the three-phase level, line-to-line voltage is usually selected as the base voltage and total kVA is given as the base kVA. Then three-phase power calculations are carried out in the normal manner. The following formulas relate the various quantities:

$$kVA_b = \sqrt{3} \times I_b \times KV_b \qquad (17\text{-}5)$$

where

kVA_b = three-phase base kVA (kilovoltamperes),

kV_b = line-to-line base voltage (kilovolts), and

I_b = line base current per unit (amperes).

$$kVA_{pu} = \frac{kVA}{KVA_b} \qquad (17\text{-}6)$$

where

kVA_{pu} = kVA per unit (unitless).

$$kV_{pu} = \frac{kV}{kV_b} \qquad (17\text{-}7)$$

where

kV_{pu} = kV per unit (unitless).

$$Z_b = \frac{(KV_b)^2 \times 1{,}000}{kVA_b} \qquad (17\text{-}8)$$

where

Z_b = base impedance (ohms).

Note that Z_b is usually a complex number consisting of resistance and reactance. The algebraic relationship of these quantities is shown as follows:

$$Z = R + jX \qquad (17\text{-}9)$$

where

Z = impedance in ohms,

R = resistance in ohms,

X = reactance in ohms, and

$j = \sqrt{-1}$ (mathematical operator).

(a)

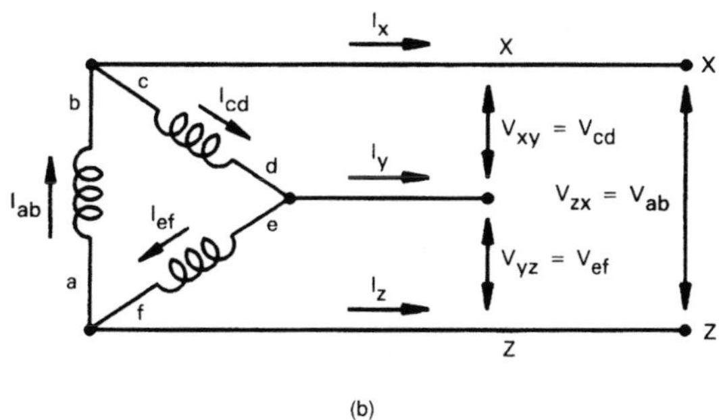

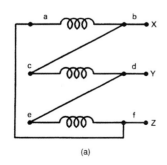

Fig. 17-3. Delta connection.

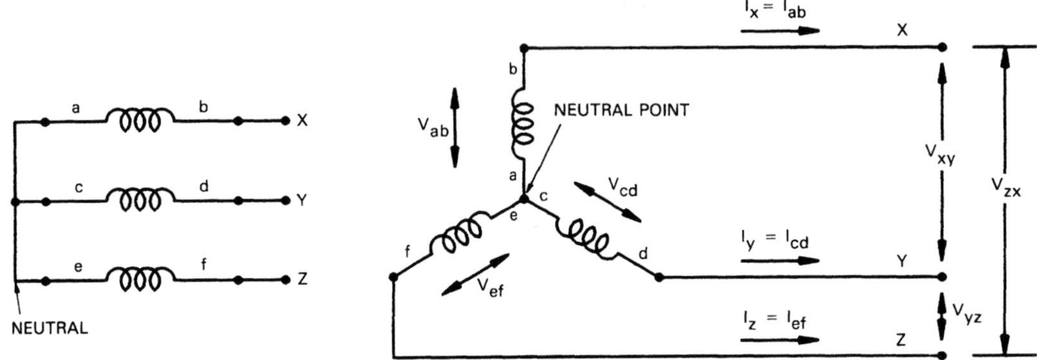

Fig. 17-4. Wye connection.

Impedance can also be stated in polar form as follows:

$$Z = Z\underline{/\Theta} \tag{17-10}$$

where

Z = magnitude of the impedance in ohms, and

$\Theta = \tan^{-1}(X/R)$.

It is sometimes desirable to change bases when working with per unit quantities. The new base quantities and the old base quantities are related to each other according to the following formula:

$$\frac{Z_{\text{new pu}}}{Z_{\text{old pu}}} = \left(\frac{\text{base } kV_{\text{old}}}{\text{base } kV_{\text{new}}}\right)^2 \times \frac{\text{base } kVA_{\text{new}}}{\text{base } kVA_{\text{old}}} \tag{17-11}$$

17.2.5 Symmetrical Components

Almost all calculations involving power systems are carried out in balanced three-phase terms. To use any other method is extremely complicated. However, in the real world of power systems, the voltages, currents, etc. are almost never exactly balanced. Most homes are supplied with only one phase of power from a three-phase system. Almost all home appliances are single-phase devices. Distribution engineers try to keep the phases on the three-phase distribution system balanced, and on the average, they do a reasonably good job. However, lightning, broken conductors, and system short circuits are almost never balanced three-phase events. For this reason, engineers have developed a set of operators to artificially allow an unbalanced set of vectors to be described in balanced three-phase terms. Three equal terms of *positive*, *negative*, or *zero sequence* can be used to describe all unbalanced conditions that may be present in a three-phase system.

The positive sequence terms are three vectors equal in magnitude and equally spaced 120 degrees from each other in polar form. The negative sequence components are also equal in magnitude and spaced 120 degrees from each other in polar space. Similarly, the zero sequence vectors are defined as three vectors equal in magnitude; however, they have no angular difference between them. This component is sometimes called the "residual" quantities. All three-phase sequence components rotate in the conventional, counterclockwise direction. The sequence diagrams are shown in Fig. 17-5.

A mathematical operator a is defined as $1\underline{/120 \text{ degrees}}$. Therefore, a^2 would be $1\underline{/240 \text{ degrees}}$ or $1\underline{/-120 \text{ degrees}}$.

Any three vectors (balanced or unbalanced, equal or unequal) may be resolved into a system of three sets of balanced three-phase components—positive sequence, negative sequence, and zero sequence. In using symmetrical components voltages and currents, it is important to note that the sequence quantities are always "wye connected," single-

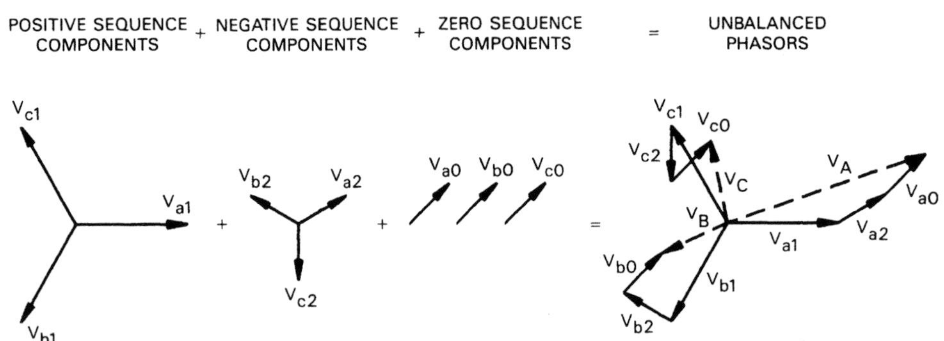

Fig. 17-5. Three-phase symmetrical components.

phase quantities. That is, voltage is line-to-ground, current is line current, and impedances are single-phase, line-to-ground quantities. Note that voltage on a power system is normally thought of as phase-to-phase voltage, whereas in symmetrical components calculations, it is always a phase-to-ground quantity. However, this is really unimportant if the per unit system of quantities is being used.

Any unbalanced three-phase voltage can be defined from the following fundamental equations using operator notation:

$$\begin{aligned} V_a &= V_1 + V_2 + V_0 \\ V_b &= a^2V_1 + aV_2 + V_0 \\ V_c &= aV_1 + a^2V_2 + V_0 \end{aligned} \tag{17-12}$$

Using vector notation, the same equations are as follows:

$$\begin{aligned} V_a &= V_1 + V_2 + V_0 \\ V_b &= V_1\angle{-120°} + V_2\angle{120°} + V_0 \\ V_c &= V_1\angle{120°} + V_2\angle{-120°} + V_0 \end{aligned} \tag{17-13}$$

The current equations are in the same form.

Within each set of sequence quantities, all components must exist. That is, $V_{a1} = V_{b1} = V_{c1}$, and $V_{a2} = V_{b2} = V_{c2}$, and $V_{a0} = V_{b0} = V_{c0}$. V_{a1} cannot exist without V_{b1}, and so on. Thus, it is necessary to define only one of the phasors.

The system of symmetrical components is one of the languages of electric power engineering. Many devices, particularly relays, are designed to operate from the symmetrical component quantities. Thus, it is an extremely important concept. However, it must always be remembered that symmetrical components are only a mathematical tool. The individual values, in themselves, are *not* real.

17.3 TRANSFORMERS

Power transformers are used to raise or to lower the voltage in electric power systems to move electrical energy in an efficient manner around the system. Before the invention of power transformers about 1885, direct current was the only method of transmission of electrical energy. The highest voltage that could be used was that of the generator itself. Consequently, electrical energy could only be transmitted over fairly short distances because of the voltage drop in the conductors. With the invention of the power transformer, the voltage could be raised to much higher levels and power could be transmitted over much longer distances using the same magnitude of current. The power transformer makes it possible to use the optimum voltage for the generator itself, then transmit the power at the most economical transmission voltage, and use the electrical energy at the most convenient voltage at the point of use. The power transformer has allowed the electric power industry to develop into the major industrial contributor it is today.

17.3.1 Basic Transformer Theory

A transformer consists of two or more coils of wire (windings) which are linked together by a magnetic field. If one of the windings—the primary—is connected to an alternating current, an alternating magnetic field will be produced within the coil. The magnitude of the field will be determined by the number of turns, the magnitude of the alternating current, and the permeability of the material within the winding. The formula for this is as follows:

$$V = 4.44\, n\Phi f \tag{17-14}$$

where

V = applied voltage (rms),

n = number of turns in the winding,

Φ = the total magnetic flux (webers), and

f = frequency of the applied voltage (hertz).

A voltage is induced in the other winding (the secondary) by the alternating magnetic field which links the two windings (Φ). The voltage in this winding is produced using the same formula, with the voltage magnitude proportional to the ratio of the number of turns between the two windings, since the flux (Φ) and the frequency (f) are the same. This magnetic coupling action is true even if the medium linking the two windings is air. However, the magnetic coupling is much more efficient if the linking medium is iron or another ferromagnetic material, because most of the magnetic flux is then confined to a definite path linking both of the windings. Almost all power transformers use iron or other ferromagnetic material in the core and are thus called "iron core" transformers. To reduce losses from eddy currents in the transformer iron, the core usually consists of stacked sheets of steel laminations. Two types of transformers are used in the electric power industry. These are called the "core form" type and the "shell form" type. Figure 17-6 shows the basic difference between the two in a single-phase transformer.

In the core form type, the two windings are wound around separate iron cores. The top and bottom of the cores are connected together with another stack of laminations called the yoke. In the shell form type, the windings are more like pancake coils, and the iron core is basically wrapped around the windings. The high-voltage and low-voltage windings are usually made up of several pancakes coils that are interlaced with each other and connected together in series to make up the phase winding.

Figure 17-7 shows a diagram of a three-phase core form transformer and a three-phase "five-legged" shell form transformer. Figure 17-8 shows a photograph of a 12-MVA, 69-kV delta to 13.8-kV wye power transformer before installation in its oil-filled tank. Figure 17-9 shows a similar size transformer in its tank after installation in a substation.

The ability of the transformer to carry a certain load is a

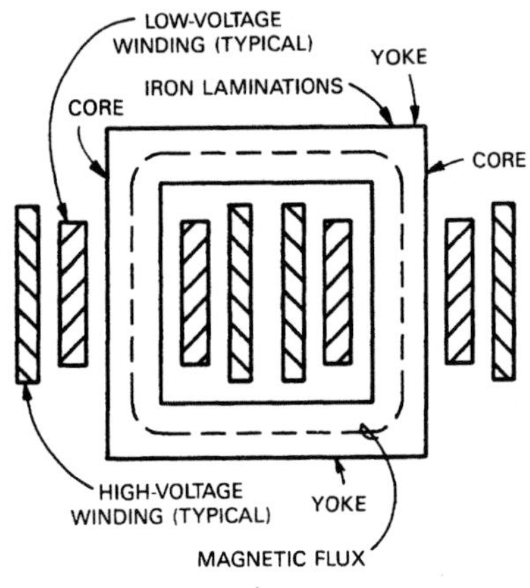

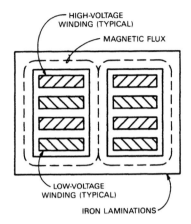

(a) CORE FORM TRANSFORMER

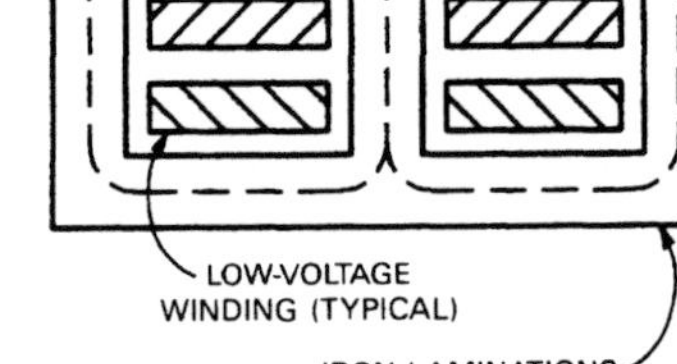

(b) SHELL FORM TRANSFORMER

Fig. 17-6. Single-phase transformers.

function of the size of the current-carrying conductors in the windings, the permeability and area of the iron core and allowable flux density (lines of flux per unit area), and the ability of the entire assembly to transfer away the heat generated inside the transformer. Transformer cooling is discussed further in Section 17.3.4.

17.3.2 Application and Sizing

Several factors are important in the selection of a substation or power transformer other than its voltage ratio and capacity. The voltage ratio is usually obvious; the capacity depends on the transformer application. If the transformer is to serve a defined load, its kVA rating may also be fairly well set. If the transformer is to serve as a bus tie transformer interconnecting two electric systems, the capacity must be determined from load flow studies. If the transformer is to be used as a generator stepup transformer, it should be capable of carrying all the power the generator can deliver. However, it should not be significantly larger than necessary; otherwise, the excess will be wasted capacity. It is common practice to rate the generator stepup transformer for its high-voltage capacity only. That is, the generator transformer nameplate is equal to the generator maximum nameplate minus generator transformer losses. If a station has two or more main auxiliary transformers, it may be assumed that only one of the main auxiliary transformers will be out of service when the generator is called upon to deliver full load. In that case, the kVA rating of one of the main auxiliary transformers may also be deducted from the generator transformer size.

Generator transformers are sized as tabulated below:

Generator MVA size (17-15)
− Auxiliary system use at maximum unit output[1]
− Generator transformer loss at maximum unit output
+ Auxiliary system use which may pass through generator transformer[1]
= Generator transformer minimum rating

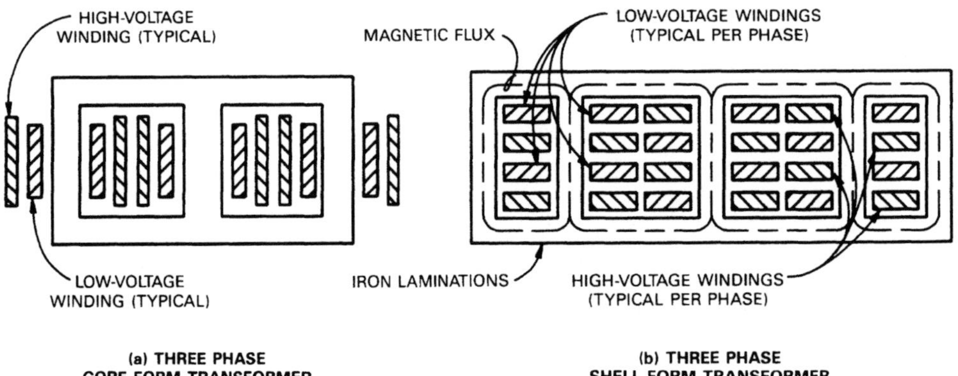

(a) THREE PHASE CORE FORM TRANSFORMER

(b) THREE PHASE SHELL FORM TRANSFORMER

Fig. 17-7. Three-phase transformers.

[1]These values are the same when there is only one main auxiliary transformer. If a generating station is to have more than one main auxiliary transformer, the generator transformer sizing may be based on only one unit auxiliary transformer being out of service at a time. If a full-capacity reserve auxiliary transformer is available, however, it should be assumed that *all* the auxiliary load will, at some time, pass through the generator transformer.

Transformer losses cannot be accurately predicted at the time the station designer is specifying the transformer because it has not been sized at that point. However, total losses may be *approximated* as equal to the product of the generator MVA and the following empirical multipliers shown in Table 17-1.

Actual transformer per unit MVAR losses are equal to the product of per unit transformer reactance times transformer rated MVA.

The following is an example of sizing a large generator transformer at a station having two main auxiliary transformers and two reserve auxiliary transformers:

	MW	MVAR		MVA
Generator rating (at 0.90 *pf*)	855	414	=	950 at 0.90 *pf*
− Aux system use (2 aux trans)	−74	−55.5	=	92.5 at 0.80 *pf*
− Generator transformer loss	−3	−139		(at 345 KV)
+ One aux trans out of service	+37	+27.75		
Generator transformer rating	815 + *j*	247.55	=	851.7 at 0.957 *pf*

pf is the power factor
j is $\sqrt{-1}$

In practice, a generator transformer for this example could be specified at 800/896 MVA 55/65° C (131/149° F). This would permit the transformer to operate normally at 808 MVA, which is close to the 55° C (131° F) rating when both main auxiliary transformers are in service and still allows a slight margin of safety in its top 65° C (149° F) rating. Note that at this stage of design, the actual auxiliary load is still estimated. Therefore, the station designer should allow some contingency in sizing all loads. In addition, it should be

Fig. 17-9. 18-MVA, 69-kV wye to 13.8-kV delta power transformer after installation in an oil-filled tank. (From City of Vero Beach. Used with permission.)

Fig. 17-8. 12-MVA, 69-kV delta to 13.8 kV wye power transformer before installation. (From Cooper Power Systems. Used with permission.)

Table 17-1. Empirical Multipliers

	Multiplier
Transformer MW loss:	0.003
Transformer MVAR loss:	
Transformer high voltage (kV)	
69	0.09
115	0.10
138	0.11
161	0.12
230	0.13
345	0.14
500	0.17

remembered that the allowable temperature rise of a transformer is based on a 30° C (86° F) average ambient temperature. If the average ambient temperature of the hottest month at the generating station is higher than 30° C (86° F), the rating of the generator transformer should be decreased by 1% for each degree C increase in the ambient temperature above 30° C, in accordance with the recommended NEMA TR 98 and American National Standards Institute (ANSI) C57.92 standards.

17.3.3 Basic Insulation Level

The basic insulation level (BIL) of transformer windings is usually selected on the basis of the system operating voltage and the protective level established by the surge arresters to be used. Full BILs are usually specified for system voltages of 69 kV and below. BILs may be reduced one or more steps below the full value for system voltage above 69 kV. Commonly used BILs and ranges of maximum conventional surge arrester rating to provide transformer protection are shown in Table 17-2. (Complete listings of BIL and system voltage are found in ANSI C57.12.00, Table 4.)

Surge arrester technology has advanced to the point that the values listed below are considered as guides only, since different manufacturers use slightly different methods to rate their arresters, and there is considerable overlap in arrester ratings. Arrester protective characteristics are a continuous function defined over a wide spectrum of time. However, insulation withstand of a transformer is defined at only three points using industry standard tests called the switching surge test, full wave test, and chopped wave tests. The amount of protection a given arrester offers the transformer is called the protective margin. This protective margin is defined as follows:

$$\% \text{ Margin} = \frac{(\text{Transformer withstand} - \text{arrrester protective level}) \times 100}{\text{Arrester protective level}} \qquad (17\text{-}23)$$

where

the transformer withstand rating is the rating at one of the above-defined test values.

Obviously, the higher the percentage, the better the transformer is protected. However, if the percent margin is too high, the arrester may fail in normal service because of the power system voltage fluctuations. Conversely, if the margin of protection is set too low, the extra cost of buying an arrester is not warranted, since it will probably offer very little protection.

The manufacturer's data should be consulted for each application. Metal Oxide Varistor (MOV) arresters with nonlinear conductive characteristics may be applied to meet the specific conditions for each application. Surge arresters are continuously exposed to normal line-to-ground voltage. For each rating, there is a recommended limit to the maximum continuous voltage that may be applied. Another considera-

Table 17-2. Transformer Basic Insulation Levels

Nominal System Voltage (kV)	Transformer Winding BIL Steps		Range of Surge Arrestor kV Rating	
	kV	Reduced	Underground	Grounded
2.5	60	0	3	2.7
5.0	75	0	6	5.1
8.7	95	0	9	6
15	110	0	15	10
25	150	0	25	18
34.5	200	0	37	27
46	250	0	50	39
69	350	0	72	54
115	350	2	108	90
	450	1	120	96
	550	0	120	108
138	450	2	132	108
	550	1	144	120
	650	0	144	120
161	550	2	168	120
	650	1	168	132
	750	0	168	144
230	650	3	192	144
	750	2	228	180
	825	1½	240	192
	900	1	240	228
	1,050	0	240	240
345	900	3	276	264
	1,050	2	312	288
	1,175	1½	312	294
	1,300	1	312	300
	1,425	0	—	312
500	1,300	4	420	396
	1,425	3½	444	420
	1,550	3	492	444
	1,675	2½	516	—
700	1,800	5	—	—
	1,925	4½	—	—
	2,050	4	636	588
	2,175	3½	—	—
	2,300	3	678	—

tion, particularly at extra high voltage (EHV) levels, is the energy dissipation (in watt-seconds) that the arrester must handle. Very long transmission lines can have a substantial charge buildup, and if the arrester is required to discharge the entire charge, it may fail dramatically (explode) if called upon to discharge energy above its rating.

The difference between grounded and ungrounded transmission lines in systems may appear to be obvious; however, in some instances it is not so simple. If a transformer is located at a great enough distance from the generation source—for example, in a stepdown substation several hundred miles from the source—it may need to be considered as a high-impedance grounded transformer, and the maximum continuous voltage rating value of the protective arrester should be made higher than it would need to be for a transformer located at the generation source. The extreme in this case would be a delta connected winding that could operate continuously with one phase grounded. In that case, the maximum continuous line-to-ground operating voltage would also be the phase-to-phase voltage. Line-to-ground short-circuit (fault) studies allow the designer to determine the proper arrester rating for the location within the power system where the application is being made.

17.3.4 Transformer Cooling

The kVA capacity of a transformer is dependent on its ability to remove heat from the windings. A power transformer may be selected to provide the required capacity for a particular application either with or without some form of forced cooling. High dielectric strength oil is the usual insulating and cooling medium for transformers which are located outdoors. Liquid containment areas and oil separators should be used to control possible oil spills or leaks.

Liquid immersed transformers are designed for a 55° C temperature rise above a 30° C average ambient temperature at full load operation or a final top oil temperature of 85° C. A self-cooled transformer (one without forced cooling) would require a rating equal to the actual power flow requirement. The equivalent kVA can be reduced by the addition of forced cooling to sustain the required kVA within the design temperature rise. In other words, the transformer capacity of a self-cooled transformer may be increased by the addition of forced cooling. Modern transformer insulation can withstand a temperature rise of 65° C without loss of insulation life. This additional 10° C of temperature rise increases the capacity of a 55° C transformer from 100% to 112% of nameplate rating with a final top oil temperature of 95° C.

The operating cost of a forced-cooled transformer is, of course, higher than that of a self-cooled transformer because of the higher effective impedance and the cost of the cooling energy. On the other hand, the initial capital cost of a self-cooled transformer to handle the same load is significantly higher than that of a forced-cooled transformer. The evaluated cost of the transformer should be considered in the selection of forced cooling systems. Load factor may dictate the most economical choice of transformer cooling equipment.

Transformer cooling systems are designated as one or more of the following combinations:

AA—Dry type, self-cooled;
OA—Liquid immersed, self-cooled;
FA—Forced air-cooled;
OW—Liquid immersed, water-cooled;
OW/A—Liquid immersed, water-cooled, with air-cooled rating (self);
FOA—Liquid immersed, forced oil, forced-air cooled; and
FOW—Liquid immersed, forced oil, forced-water cooled.

Liquid-filled transformers rated 10,000 kVA and above may have three ratings. A single stage of forced cooling such as OA/FA or OA/FOA increases the transformer OA rating from 100% to 133⅓% (at 55° C rise). An OA/FA rating would mean that the transformer was self-cooled but had one stage of fan cooling. For example, a 12-MVA transformer with one bank of cooling would be rated 12/16 MVA OA/FA. Two stages of forced cooling such as OA/FA/FA or OA/FA/FOA or OA/FOA/FOA increases the transformer rating from 100% to 166⅔% (at 55° C rise). For example, in the 12-MVA transformer, the rating would be 12/16/20 MVA. The OA/FA/FA rating would mean self-cooled at 12 MVA, 16 MVA with one bank of fans operating, and 20 MVA with both banks of fans operating. Two stages of cooling are the most that are practical and are all that the industry recognizes. An OA/FA/FOA rating would mean self-cooled with one bank of fans and one set of forced oil circulating pumps.

Many large generator stepup transformers are rated at FOA or FOW only and do not carry an OA (self-cooled) rating at all because the generating unit is intended to be loaded to 100% of its capacity at all times (in the case of a base load power plant). If the unit is not at full load, it will not be operating at all; therefore, the transformer need not have any intermediate rating. This is the most economical transformer rating because the FOA only or FOW only cooling increases the transformer pricing rating the full 166⅔% (at 55° C rise). It must be emphasized, however, that the cooling equipment must have more than one source of auxiliary power; otherwise, the entire generating plant could be forced off line because of the loss of a single small 480-V feed to the transformer cooling equipment. In the case of a generating plant that may be used in swinging load situations, the designer may wish to consider having intermediate cooling ratings.

If a transformer is specified to have both a 55° C and a 65° C rating, the top rating of the transformer should have a nameplate rating of 112% of the highest 55° C rating. In the example used above, the 12/16/20 MVA, OA/FA/FA 55° C transformer's top rating at 65° C would be 22.4 MVA.

Liquid-filled transformers are very rugged devices and can be subjected to severe overloads for short periods of time without significantly reducing their normal life. For example, the National Electrical Manufacturer's Association (NEMA) Standard TR98, which is a guide for loading liquid-filled power transformers, indicates that a liquid-filled transformer can carry 155% of its full load rating for up to 4 hours following continuous full-load operation at 30° C (86° F) ambient with an estimated loss of life of only 4% provided it is allowed to cool back to its 55° C rise temperature within 12 hours following the overload condition. This loading guide is very important in designing an auxiliary electric system because it allows auxiliary load centers to be automatically transferred between transformer sources in emergencies and to continue to operate all the auxiliaries by allowing a transformer to be loaded above its nameplate rating during the emergency. This type of temporary overload operation is also very important in utility distribution systems because it allows the utility to continue to serve all its customers in an emergency by permitting a substation transformer to operate in an overloaded condition until the utility emergency crews have time to switch distribution feeders over to alternate sources.

17.3.5 Indoor Transformers

Transformers for indoor application include secondary unit substation transformers for 480-V loads, as well as special applications such as rectifier transformers for variable-speed ac drive motors.

Oil-filled transformers should not be used indoors unless they are placed in a fireproof vault. Transformer insulation and cooling media which may be considered for indoor use are air-ventilated dry type, high-fire-point insulating liquid-filled, Freon-filled, and cast resin insulated transformers.

Evaluations of the most suitable type of insulation and cooling medium for a particular application should consider the cost of ventilating for heat dissipation and structural costs for heavy floor loading. Additional costs for installation and fire protection may also be required for high-fire-point insulating liquid transformers, depending on the particular fluid chosen.

Table 17-3 shows the fan-cooled ratings and allowable temperature rise for transformers below 2,500 kVA with various types of insulating mediums and fan- cooled ratings with an average ambient temperature of 30° C (86° F).

Air-ventilated dry-type transformers are virtually fireproof and are considered safe for operation indoors. Because the cooling air partially acts as the dielectric medium, it should be free of dust, water, and chemical contaminants. Transformer life expectancy is greatly reduced if the cooling air ducts are not kept clean and dry. When the transformer is deenergized, space heaters should be kept energized to prevent condensation of moisture in the windings.

High-fire-point liquid-filled transformers may be used indoors but should have a containment basin to control insulating medium leaks or spills. This type of transformer may be used in dusty or wet areas. Because the sealed tank is impervious to water, dust, and chemical contaminants of the cooling liquid, silicone and highly saturated paraffin oil is the most common type of insulating liquid used. Because of suspected health hazards, polychlorinated biphenol (PCB) liquids, which were once common, are no longer used as transformer insulating and cooling liquids.

Gas-filled transformers are rarely used in power generating station applications; however, they are sometimes found in industry. These transformers, depending on the manufacturer, do not all carry an additional fan-cooled rating. They can be used indoors, however, because they are nonflammable and explosion resistant. In addition, they can be used in wet or dry locations because the tank is sealed. The major disadvantage, other than cost, is the risk of failure due to an undetected loss of gas coolant. Recent US Environmental Protection Agency (EPA) rulings may result in elimination of this type of transformer cooling until a suitable replacement for Freon gas is available. Freon has been linked to upper atmospheric depletion of ozone.

The cast resin or cast coil transformers are manufactured with the coils cast in epoxy resin. Some manufacturers vacuum-impregnate the coils before casting to eliminate any possible air entrainment in the windings. The epoxy acts as both a dielectric and a heat transfer medium to transmit the heat from the conductors to the air cooling ducts. The epoxy-encapsulated coils are impervious to dust, water, and chemical contaminants and can be used indoors or out. Maintenance requirements are extremely limited.

17.3.6 Bushing Arrangement

Power transformer manufacturers follow a standard physical arrangement. This allows some degree of interchangeability and eliminates confusion when equipment is being arranged during the design of a generating station or substation. The American National Standards Institute (ANSI) Standard C57.12.10 has established preferred configurations for the location of transformer bushings and other appendages.

Figure 17-10 illustrates a transformer top view segmented into four sections as defined in ANSI standards.

17.3.7 Winding Terminal Markings

Standard (ANSI) winding terminal markings identify the voltage levels of each winding. Windings and their terminals are identified as H, X, Y, or Z in descending order of their voltage magnitude, with H being the highest voltage winding, X being the next highest, and so on. The windings on two-winding transformers are identified as H (high-voltage) and X (low-voltage). Where multiwinding transformers have two or more windings that are identical in voltage and kVA rating, their identifications are arbitrarily assigned. Where multiwinding transformers have two or more windings identical in voltage but different in kVA, the identification sequence is the highest to lowest kVA winding.

Neutral terminals are identified by a subscript 0 identify-

Table 17-3. Small Transformer—Fan-Cooled Ratings and Allowable Temperature Rise

	Self-Cooled Percent	Temp Rise (° C)	Fan-Cooled (%)	Typical kVA Rating
Dry type	100	100	133	1,000/1,333 AA/FA
		150	115	1,533 FA
High fire	100	55	115	1,000/1,150 OA/FA
Point liquid and oil-filled		65	112	1,288 FA
Freon-filled	100	45	133	1,000/1,333 OA/FA
Cast resin	100	80	150	1,000/1,500 AA/FA

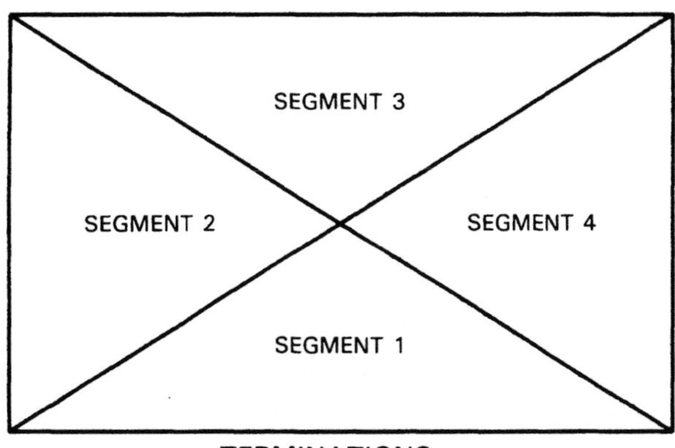

TERMINATIONS

	Segment High Voltage	Segment Low Voltage
Bushings, Cover Mounted		
Line	3	1
Neutral	2	2
*Junction Box (110 kV BIL and above)		
Cover	3	1
Tank Wall	2	4
Terminal Chamber with disconnect switch (110 kV BIL and below)		
Tank Wall	2	4
*Throat		
Cover	3	1
Tank Wall	2	4

*Only one junction box or throat may be cover mounted (High or Low Voltage).

Fig. 17-10. Transformer top view.

ing the respective winding, such as H_0. A neutral terminal common to two windings reflects both winding designations such as H_0, X_0.

Winding terminals are marked with a numerical subscript 1, 2, 3, etc., to the letter designating the winding voltage such as H_1, H_2, H_3, X_1, X_2, X_3, X_0, etc.

Terminal H_1 appears on the *right* when viewing the high-voltage terminals. Related terminals of other windings will be oriented such that Terminal 1 is opposite the H_1 terminal, Terminal 2 is opposite the H_2 terminal, and so forth. Figure 17-11 shows a three-phase two winding and a three-phase three winding transformer.

17.3.8 Angular Displacement

Transformations in polyphase systems often introduce a voltage phase shift between transformer windings. Such a phase shift in a transformer or transformer bank must be considered for the successful interconnection of two electric systems. The following comments apply only to three-phase (three- or four-wire) systems whose phase voltage magnitude reaches peak values in any definite sequence (assume A, B, C).

Although, for simplicity, this discussion covers two winding transformers only, each winding of a multiwinding transformer (including autotransformers) can be considered cou-

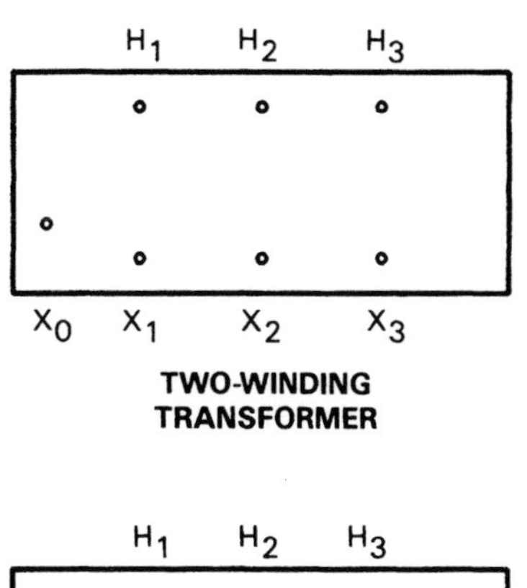

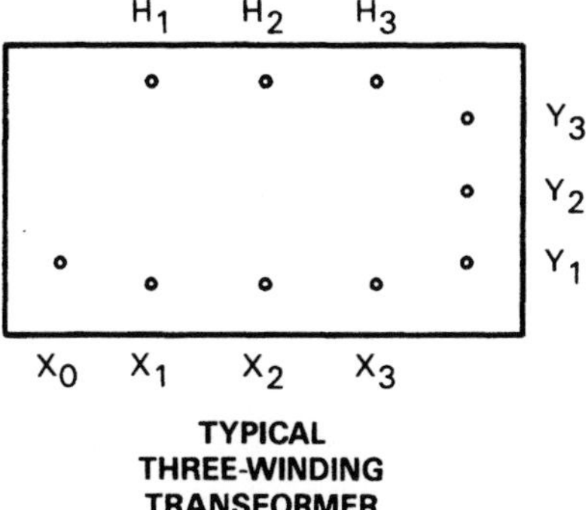

Fig. 17-11. Plan view, transformer terminal locations.

pled to any other winding and treated as a two-winding transformer.

17.3.8.1 Delta–Delta or Wye–Wye Windings. Where two windings are each connected identically in deltas or connected identically in wye (including autotransformer windings), no voltage phase shift occurs between windings. That is, X_1, X_2, and X_3 reach their peak voltage magnitudes simultaneously and in the same phase sequence as H_1, H_2, and H_3, respectively.

17.3.8.2 Delta–Wye or Wye–Delta Windings. A "standard" three-phase transformer with one winding connected delta and the other winding connected wye has an angular displacement in its line-to-neutral and line-to-line voltage vectors of 30 degrees. The low-voltage *always* lags the high-voltage when the phase rotation A, B, C is connected to terminals H_1, H_2, H_3 or X_1, X_2, X_3. This relationship is true regardless of which winding is delta or wye. This also applies to a wye-connected autotransformer main winding relationship with respect to its delta-connected tertiary winding.

The standard transformer connections are shown in Fig. 17-12 in three-line form. Its accompanying symbolic diagram is correct for phase sequence A, B, C.

17.3.9 Other Angular Displacements

It is possible to have phase shifts other than that shown in Fig. 17-12. A standard delta–wye or wye–delta transformer may also be used to achieve a low-voltage which leads the high-voltage by 30 degrees by reversing the system phase sequence at the transformer terminals. That is, if phase se-

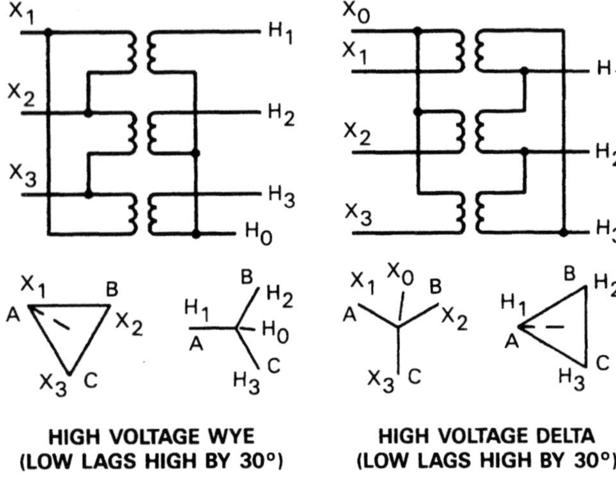

Fig. 17-12. Standard transformer connections.

quence A, B, C is connected to terminals H_3, H_2, H_1 or X_3, X_2, X_1, the phase-to-neutral low-voltage will lead the phase-to-neutral high-voltage by 30 degrees. This is shown in Fig. 17-13.

Many other angular displacements are possible in delta–wye, wye–delta, delta–delta, and wye–wye configurations by various interconnections of the windings within the transformer. Special configurations should be achieved during manufacture. The capital cost of standard transformers is always less than an equivalent capacity specially connected transformer and should be used in all but the most extraordinary circumstances. However, it should be noted that standard transformers can also be used to make many different phase relationships by redefining the terminal connections. For example, A, B, C on terminals H_1, H_2, H_3 would amount to a 150-degree shift if the corresponding low-voltage phases were defined as C, A, B on terminals X_1, X_2, X_3. This arrangement is shown in Fig. 17-14.

Many other phase relationships are possible using a standard transformer by connecting the terminals to different phases and phase sequences. The phasing diagrams may be drawn using the standard transformer phasing diagram or the power system phasing diagram. However, it is important to be consistent. It is usually much less confusing and less likely to contain a phasing error if the power system phasing diagram is always used and the delta's and wye's are drawn

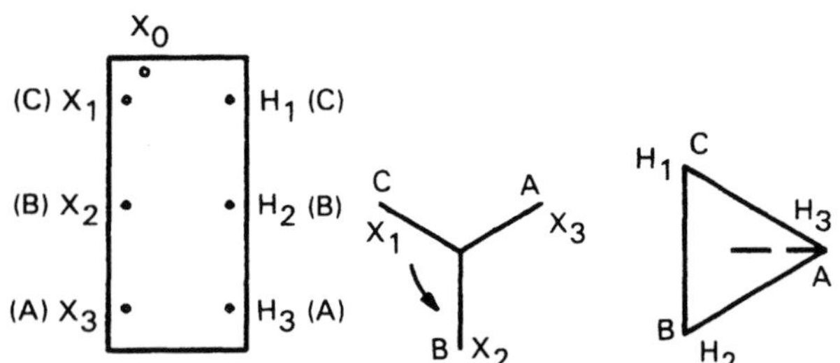

Fig. 17-13. Low voltage leading by 30 degrees using standard transformer.

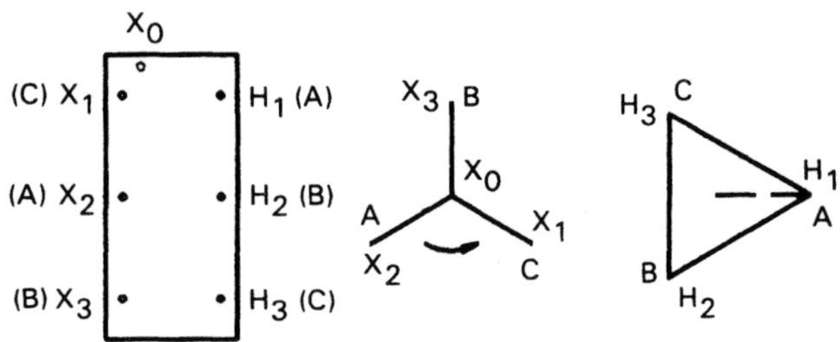

Fig. 17-14. 150-Degree phase shift using standard transformer.

in the proper orientation. This also makes one-line diagrams easier to follow between a set of drawings. The transformer manufacturer always uses the NEMA standard vector relationship. However, the transformer may eventually be connected to a variety of system diagrams. The phasing relationship is most important on the power system, regardless of the way the transformer is connected.

It should be remembered that the definition of a "standard" delta–wye transformer is one in which the voltages at the low-voltage terminals lag the voltages at the high-voltage terminals by 30 degrees *if* the phase sequence of the voltages connected to the high-voltage terminals reach their maximum positive values in the order of terminals H_1, H_2, and H_3.

17.4 MOTORS AND GENERATORS

Mechanically, there is almost no difference between a motor and a generator. Both machines have two windings, one winding on a rotating shaft which is called the rotor, and the other winding on a stationary frame surrounding the rotor, called the stator.

Figure 17-15 is a photograph of a 1,200-horsepower induction motor. Chapter 9 contains illustrations of much larger generators.

Fig. 17-15. 1,200-Horsepower induction motor. (From City of Vero Beach. Used with permission.)

Each device may be thought of as a "rotating transformer;" therefore, it is possible to have relative motion in each device between the "stator" and the "rotor."

Both devices, motors and generators, operate because of the laws of electromagnetism. The only difference is which way the energy flows. In the case of a motor, the electrical energy flows into the machine from the power system and mechanical energy is delivered to the rotating shaft, whereas in a generator, the electrical energy flows out of the machine and mechanical energy is furnished to the rotating shaft from some type of driving engine.

When an electric current is caused to flow at right angles to a magnetic field, a mechanical force is produced in a direction which is perpendicular to the plane formed by the electric current and the magnetic field. The variable can be the current, the magnetic field, or the mechanical force. If the rotating device is called a generator, then the magnetic field is usually constant and the mechanical force is the variable and is supplied by some variable driver such as a steam turbine. The magnetic field is usually supplied by an electromagnet whose strength is determined by the excitation source. This would be called a synchronous generator since it is capable of supplying its own excitation.

If the synchronous generator were to be called a motor, the magnetic field would still be provided by the electromagnet and the stator current would be the variable causing the rotating device to produce mechanical torque. Both devices still rotate in the same direction.

Synchronous motors and generators are very common in electric power generating stations. Obviously, the main generator is a synchronous type. The only reason for the existence of all the other power plant hardware is to drive the rotating field of the generator.

Generator sizes range from small 1 or 2 kilowatt size to the large 1,000-megawatt (MW) variety. All have the same function, that is, to convert mechanical energy to electrical energy in as efficient a manner as possible.

Some of the more important parts of the generator are discussed in the sections that follow and in Chapter 9.

17.4.1 Stator

The stator, as the name implies, is the stationary portion of the generator (or motor). It contains the stator windings which are connected to the electric power system. The stator windings in a three-phase machine consist of three identical coils of conductor wound into the face of high-permeability steel laminations. Each coil is wound longitudinal at 120 mechanical degree intervals around the circumference of the stator so that when the two-pole electromagnet is rotated inside the stator, voltages are produced which reach their maximum value at 120 electrical degrees apart. Three balanced phase voltages are thus produced.

The above description is for a two-pole machine. A four-pole machine would have two sets of coils connected in series and arranged at 60 mechanical degrees apart. The rotating electromagnet would then have two north poles and two south poles in an X arrangement. Thus, to produce the same frequency of voltage as the two-pole machine, the four-pole machine would have to rotate at only half the speed as a two-pole generator. Other multiple-pole arrangements are possible.

The speed of a synchronous machine is directly related to the number of poles by the following formula:

$$S = \frac{120f}{p} \qquad (17\text{-}16)$$

where

S = speed in rpm,

f = frequency (hertz), and

p = number of poles (always an even number).

The loading capacity of a generator is limited by the saturation flux density of the iron stator magnetic laminations, the ability of the conductors to remove heat from the machine, and the mechanical strength of the windings to resist the electromagnetic forces imposed on them by the rotating magnetic field and induced currents.

Stator iron flux density is controlled by the selection of the material to be used in the stator laminations. In power generation design, manufacturers use high-silicon steel to allow very high magnetic flux densities with relatively low hysteresis loss.

The stator heating problem has several solutions. On small machines (lower than about 50 MW), conventional air cooling may be used, particularly if the generator is to be used in a cycling mode. This approach may be taken one step further to use a closed cycle air-to-air heat exchanger to avoid the problem of dirt contamination inside the windings. As machine size increases, air cooling becomes less practical because of the diminishing returns caused by the windage loss in the machine and the resultant volume of air required to circulate through the machine in order to cool it. In such cases, hydrogen gas is used to cool the machine. Hydrogen offers a very high coefficient of heat capacity and a very low specific gravity. Therefore, hydrogen conducts heat away from the stator at high loads while producing low windage losses. In addition, hydrogen will not support combustion if the purity is maintained above 95%, and it is not explosive if the purity is maintained above 75%. Hydrogen coolers are located on the outside of the generator, and water, glycol, or air heat exchangers are used to cool the hydrogen, thus removing the heat from the generator.

In very large machines, the stator conductors may be hollow to allow very pure deionized water to be circulated in the windings to also assist in removing the heat. This water must remain without conductive contaminants to maintain the insulation value of the generator. The water is introduced into the stator at the generator bushings. The stator must also be insulated to withstand the induced voltage on the machine. Most modern generators are now insulated with epoxy-impregnated materials, which also provide very high mechanical strength.

17.4.2 Rotor

The rotor of the generator carries the rotating field. Most large steam turbine drives rotate at 3,600 rpm for 60-Hz systems and at 3,000 rpm for 50-Hz systems; therefore, the rotor in such a generator would be a two-pole electromagnet, usually of cylindrical design. Salient pole rotors are used in slower machines such as those used for hydroelectric applications.

Direct current (dc) is introduced into the rotating electric field by means of slip rings and sliding brushes. This dc current then produces the electromagnetic field, which is used to convert mechanical energy to electrical energy in the stator. By varying the amount of current in the field, the voltage on the terminals of the generator can be varied. However, if the generator rating is small compared to the electric system to which it is connected, then the terminal voltage of the generator may not change significantly when the generator field current is changed. The only parameter that can then change is the generator torque angle. If the machine is overexcited, the torque angle will be reduced to a small value and the machine will be very stable (in other words, the turbine drive will not be able to drive the rotating field past the pole slip point). The machine also will be operating in the lagging power factor range and therefore will "produce" VARS (reactive volt-amperes) for the power system. If the machine is underexcited, the torque angle will increase because the field strength will be reduced, and the generator will be operating in a less stable condition, with the possibility that the turbine drive could cause the machine to "slip" a pole, that is, go past its magnetic maximum torque position. The machine will be operating as a leading power factor machine under this condition, and the power system will be obliged to provide external excitation to the generator in order to maintain the proper terminal voltage. This condition is known as "taking" VARS from the power system.

In practice, generators are almost always operated with a lagging power factor for two reasons. First, nearly all power systems operate with a lagging power factor, that is, the phase current lags the phase voltage by some angle. This requires the generators in the system to furnish VARS to maintain the system voltage. (It is also common practice to install fixed capacitors in some parts of the distribution system to also provide VARS and help hold up the system voltage.) Second, good operating practice dictates that the generator should be operated with high-field current to produce a strong magnetic field in the machine air gap, which will reduce the torque angle and help maintain machine stability during system disturbances.

The subject of VARS is not magic; however, it causes a great deal of confusion among many people in the utility industry. Remember that the operating power factor of a generator is a function only of the excitation of the machine. The amount of power produced by the generator is related only to the input power of the turbine. It is not possible to change the amount of "real power" delivered to the system except by increasing or decreasing the amount of power provided by the turbine, and to increase the power delivered by the turbine, more fuel must be burned. Nothing is free. Changing the power factor of a generator requires only increasing or decreasing the field current, which does not require additional fuel to be burned (except for the minuscule amount used in the additional I^2R losses of the field winding) and thus can produce only "imaginary power," or VARS.

Nevertheless, it is important to specify the power factor of a generator when purchasing a turbine–generator set. The VAR requirements of the power system to which the generator will be connected should be considered before determining the specified power factor of the machine. However, a specified power factor smaller than 0.9 is not normally required unless the system has an unusual amount of inductive load, and in this case, the system designer may consider installing capacitors in remote substations within the system.

The specified power factor of a generator determines the size of its exciter, the rotor winding current carrying capability, the stator kVA, and the overall operating capability of the generator. Figure 17-16 shows these parameters for a typical generator. These capability curves are usually a series of curves for different hydrogen cooling pressures. The generator may be operated at any point within the capability envelope of the machine. The diagram in Fig. 17-1 also shows the relationship between VARS, MVA, and real power.

17.4.3 Excitation Systems

Several excitation systems discussed in the following paragraphs are in use in the industry today, and each manufacturer has its own preferred system. Regardless of the type employed, the purpose of all excitation systems is to provide direct current to the rotating electromagnetic field.

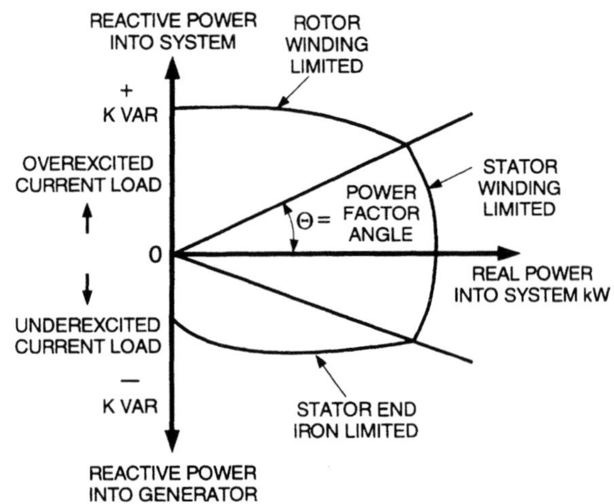

Fig. 17-16. Typical generator capability curve.

In the static excitation system, a power potential transformer is connected directly to the terminals of the generator, and the field on the generator is "flashed" with a station battery while the field is rotating. The resulting small voltage generated by the initial field is then fed to the terminal transformer, which is connected to a set of solid-state rectifiers that convert the alternating current to direct current. This is sort of a bootstrap operation, because now the new dc current is fed back to the field brushes and the process continues until either the generator iron is saturated and the voltage doesn't get any higher or a control loop in the feedback circuit limits the terminal voltage. The generator may then be connected to the power system. Now, however, without any control mechanism on the field current, the generator load current through the stator causes the terminal voltage on the power potential transformer to drop, thus causing less field current, with resulting lower terminal voltage, and so on. To correct this problem, a saturable current transformer (SCT) is used to provide supplementary current to the power rectifiers to compensate for the voltage drop caused by the stator load current.

The SCTs are connected in series with the stator line terminals; therefore, the higher the load current, the more excitation current is available to be fed to the power rectifiers to provide direct current to the field. The amount of current to be used from this current source is controlled by the bias winding on the SCT. Of course, the control loop in the feedback circuit is used to adjust the field current and, consequently, the power factor of the machine after it is online.

The terminal voltage cannot be adjusted significantly after the machine is online. Only the power factor can be changed because the capacity of a single generator is almost always small compared to the entire power system to which it is connected.

In the separately excited generator system, the excitation current is provided by a separate generator connected to the same generator shaft. The excitation generator may be either a three-phase generator with separate external rectifiers that provide the necessary direct current or it may be a dc generator itself. In either case, this is a very simple method of providing the generator excitation, although it may not necessarily be the most economical or desirable.

In the "brushless" type of excitation system, a set of rotating diodes is installed on the rotor and the voltages induced by separate windings built into the rotor are rectified by these rotating diodes and supplied directly to the field circuit. Direct current is applied to another set of coils on the stator section of the generator to provide the necessary relative motion of conductor, magnetic field, and mechanical force.

17.4.4 Induction Machines

The stator of an induction machine is built exactly the same as a synchronous machine stator. The rotor, however, is different. Whereas the rotor of the synchronous machine is a dc electromagnet, the rotor of an induction machine is another closed loop winding that receives its magnetic field by "induction" from the stator winding. This rotor is usually of the "squirrel cage" design, with the rotor consisting of copper bars running lengthwise of the rotor and connected together on the ends with a copper ring which is brazed to the rotor bars. The entire cage is embedded in a steel core made up of laminated sheets of steel. The shaft of the machine is connected to the laminations with a slotted keyway to transmit the torque from the rotor cage to the shaft. The machine is thus a transformer with the stator acting as the primary winding and the rotor cage as the secondary winding. The secondary winding is, of course, short circuited by the rotor end rings. Electrical currents are induced in the rotor by electromagnetic transformer action. These currents then react with the magnetic forces in the iron to produce torque at right angles to the rotor currents. Since the rotor is free to move in either direction, the rotor spins in the same direction as the rotating magnetic field.

However, the faster the rotor spins, the less the relative motion speed of the rotor cage with respect to the rotating magnetic field and thus the lower the frequency and magnitude of the rotor currents. As the rotor current magnitude decreases, the torque will also decrease until, if the rotor were going at exactly the same speed as the rotating magnetic field, the rotor cage would have no torque at all. Thus, an induction machine cannot run at synchronous speed, since there is no driving torque. The synchronous speed is determined by the number of poles in the stator and the frequency of the driving voltages. The actual speed is determined not only by the synchronous speed but also by the "slip" speed of the rotor. The slip speed of the rotor is the difference between the synchronous speed and the actual rotor speed.

There is a slip speed, depending on the design of the rotor bars, at which the machine develops maximum torque. This point is called the pullout speed of the motor since it will not drive a load requiring a larger torque. The typical speed torque curves of various classes of motors are shown in Fig. 17-17.

The various classes of motors can generally be given as follows:

- Class A—Normal starting torque, normal starting current, high pullout torque, low slip, high efficiency. This is the basic standard motor design.
- Class B—Normal starting torque, low starting current, low pullout torque, low slip, slightly lower efficiency. This motor is used where it is desirable to keep the starting current down and where starting torque is not severe, such as for fans.
- Class C—High starting torque, low starting current, high slip, lower efficiency. This class would be used for loads such as conveyors.
- Class D—High starting torque, low starting current, high slip, low efficiency. Applications for this motor are high impact loading, such as for crushers or punch presses, where a flywheel can be used to reduce the pulsations on the power system.

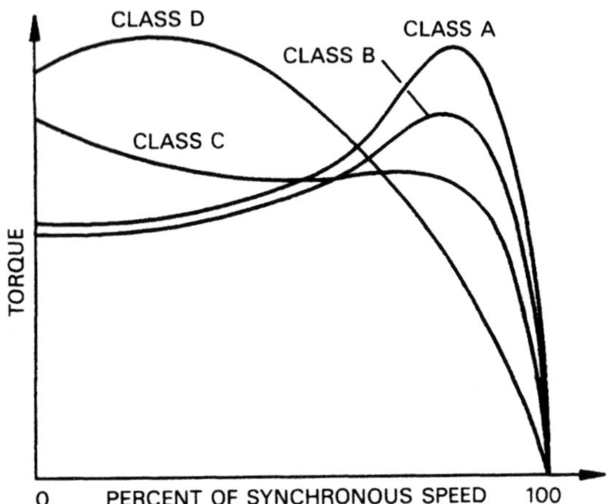

Fig. 17-17. Typical motor speed torque curves.

Motors must be sized to drive a given load at the maximum output of the load plus allow for transient overloads and for the losses in the mechanical drive as well as the motor itself. It is common practice to size a motor to be capable of handling at least 15% over its nameplate. The motor may be specified to have a 1.15 service factor while driving a load equal to its nameplate, or it may be allowed to have a load of 85% of its nameplate if it has only a 1.0 service factor. When a motor has a service factor greater than 1.0 it means that the motor will be allowed to carry a load equal to its nameplate times the service factor, but the temperature rise of the motor will be 10° C (50° F) greater than the normal temperature rise for the insulation class of the motor.

An induction machine can also be used as a generator. If the driven load, such as a fan, becomes a driver device, for example, a windmill or water turbine, which is capable of driving the rotor faster than synchronous speed, the rotor will have induced currents in it the same as it would have if it were running slower than synchronous speed. Thus the rotor will transfer power to the stator by electromagnetic action. It should be noted that an induction generator must have its excitation supplied by the system to which it is connected, because the rotor has no magnetic field of its own as does a synchronous machine. Induction generators are normally used only in small applications such as small hydroelectric dams and windmill generators. Most applications for induction machines are as motors driving large pumps or fans.

17.4.5 Insulation Systems

Motors and generators must be insulated to withstand the electrical stress on the windings without breaking down and arcing to either the adjacent turns of wire or to the steel frame to which the windings are connected. Modern insulation systems are classified into types A, B, F, and H. These classifications refer to the allowable hot-spot temperature in the motor. The standard classes as related to temperature are listed in Table 17-4.

The selection of an insulating system for a motor must include consideration of the electrical and mechanical design, the normal and abnormal operating requirements, and the environmental conditions to which the motor will be subjected.

Table 17-4. Insulation System Classification

Class	Hot-Spot Temperature
A	105° C (221° F)
B	130° C (130° F)
F	155° C (311° F)
H	180° C (356° F)

A motor insulation's greatest enemy is heat. Most large motors are Class B insulated, with a trend toward Class F. Although a motor may not require a high hot-spot rating, insulation systems that can operate at a higher temperature may provide greater reliability for a longer time because they can more effectively withstand the heat of short time overloads. Thus, many users specify Class F insulation systems with a Class B allowable temperature rise.

17.4.6 Bearings and Lubrication

Bearings are designed to support and control the motion of rotating shafts, while providing very low friction. All bearings can be classified as plain sleeve (sliding type bearings) or as antifriction bearings. Antifriction bearings include roller and ball bearings. Each type of bearing has advantages and disadvantages, and there is considerable overlap on their suitability for many applications.

The majority of bearings in rotating machinery are of the plain or sliding type. This type of bearing develops a tapered, hydrodynamic oil film which does not permit any actual contact between the shaft and the support. All the load is carried by the lubricating oil film. Properly applied, lubricating oil can provide very large load-carrying capabilities. In general, sliding type sleeve bearings and plate bearings ("Kingsbury") are more suitable for handling very large loads and shock loads than antifriction bearings. In addition, sliding type bearings are capable of operating at very high speeds. However, fluid friction increases as the square of the speed, and in high-speed motors, the heat removal system of the lubricating oil must be given due consideration.

Oil may be supplied to the bearings in various manners. It is usually supplied to horizontal sleeve bearings by small oil rings which are much larger in diameter than the shaft of the motor. The oil rings ride on the shaft of the rotor and the lower part of the ring is immersed in the lubricating oil. The turning action of the rotor causes the oil rings to ride up over the shaft and provide a continuous supply of oil. Heat is carried away from the shaft by the oil and returned to the reservoir. Heat exchangers may be required in the oil reservoir. Oil can also be provided by forced lubrication systems which include pumps, both shaft driven and motor driven, to force oil to the sliding surfaces. In large rotating devices such as generators (and turbines), a machine may have several forced lubricating pumps including a last resort dc-operated oil pump. The main bearing oil pump is usually shaft driven by the prime mover itself with one or more backup ac driven motors. In the event that all the shaft driven and ac motor-operated pumps fail, the dc pump should provide lubrication for the bearings to allow the machine to coast down to a stop without severe damage to the bearings or the machine itself.

Antifriction bearings operate on a rolling contact action between the shaft and the housing. They are capable of providing both thrust loads and radial loads at the same time. Sliding bearings do not have this capability. Rolling bearings require very little lubricant and are usually grease lubricated. Specifications for power plant applications usually require "regreasable" bearings. However, too much grease on rolling type bearings produces excessive fluid friction and may cause the bearing to fail from overheating. Antifriction bearings are best applied in slow-speed applications where sliding type bearings are not suitable. They are sensitive to dirt, corrosion, vibration, and poor fit. These type of bearings are not suitable for very high speeds because they are difficult to lubricate, and at high speeds, the fluid friction may cause excessive heating problems. Antifriction bearings are sometimes referred to by manufacturers as so-called "lifetime permanently lubricated" bearings. However, it is true that, when properly applied, they require almost no maintenance and do have a very long life.

17.4.7 Machine Terminals

Motors and generators are connected to the power system through the machine terminals. Motors usually are small enough that they require only a terminal box large enough to connect the stator cables to the supply cables with provision for grounding of the motor frame and the ground wires of the supply cables. Occasionally, it may be desirable to furnish surge protection of the motors, and this can be done with surge arresters and capacitors at the terminals. The capacitors provide a wave shaping characteristic to reduce the rate of change of the voltage wavefront (dv/dt). The surge arresters provide arc over and resealing capability for very large transient waveforms to reduce the internal electrical stress on the insulation system. Modern surge arresters may actually be nonlinear resistors that have continuous discharge characteristics which are inversely proportional to the applied voltage. Capacitors and surge arresters are normally used only in motor applications where relatively long overhead lines are used to connect a motor to the power system. Motor applications in power plants normally do not have capacitors and surge arresters, since they are rarely supplied from sources that are subject to high transient voltage values, and the arresters or capacitors may be considered needlessly expensive and subject to failures which reduce the overall reliability of the plant. However, there may occasionally be some special applications where consideration may be given to using such devices.

Generator applications are almost always designed with surge capacitors and arresters even though it may be difficult to prove mathematically that they are required. Both can be considered "insurance devices" and are very inexpensive when compared to the generator itself, which may be worth millions of dollars.

17.4.8 Bus Duct

Large generators are normally connected to the low-voltage terminals of the generator stepup transformer with bus duct, although some machines of smaller size may use multiple insulated cables per phase.

Bus duct is classified into three categories: nonsegregated phase bus, segregated phase bus, and isolated phase bus.

17.4.8.1 Nonsegregated Phase Bus Duct. Nonsegregated phase bus duct is one in which all phase conductors are in a common metal enclosure without barriers between the phases. Normally, the phase conductors are bars of copper or aluminum that are covered with insulation material throughout and the bars are supported and braced inside the bus duct with insulating boards. Plastic insulating boots are used at phase conductor joints. This type of bus is most commonly used in connecting small generators and transformers to switchgear and in applications where greater reliability is required than can be provided by power cables.

17.4.8.2 Segregated Phase Bus Duct. Segregated phase bus duct is bus duct in which all phase conductors are in a common metal enclosure, but are separated by metal barriers between phases. The phase conductors are insulated and supported from the enclosure by standoff insulators, made of either porcelain or plastic insulating material. This type of bus has greater reliability than nonsegregated phase bus and may be used in applications where space is not available for isolated phase bus.

17.4.8.3 Isolated Phase Bus Duct. Isolated phase bus duct is one in which each phase conductor is enclosed by an individual metal housing separated from adjacent conductor housings by an air space. The bus may be self-cooled, or forced-cooled by means of circulating air inside the phase enclosures. A typical self-cooled isolated phase bus installation is shown in Fig. 17-18.

If force cooling is used, the cooling air is introduced inside the two outer phases of bus near the center of the bus run between the generator and the stepup transformer. Cooling air flows in opposite directions toward the generator and transformer. At the ends, the air is directed to the center phase through deionizing baffles to maintain separation between the phases, then the air returns through the center phase back to the heat exchanger at the center of the bus run. This configuration allows the hotter air from the two outside phases to flow at considerably more velocity (and volume) through the center phase so as to provide an approximately equivalent cooling capacity. The cooling air should be in a closed cycle to avoid dust contamination. Air-to-air or

Fig. 17-18. Typical self-cooled isolated phase bus installation. (From City of Vero Beach. Used with permission.)

air-to-water heat exchangers remove the heat from the cooling air.

The conductors in segregated and isolated phase bus are normally made of aluminum although copper conductors are occasionally used. The enclosure of the phases constitutes a one-turn transformer that is shorted; therefore, the induced current in the enclosure contributes to the heat generated in the bus duct. The outer enclosure of all types of bus duct is grounded for personnel protection. Inspection ports are normally provided near the insulator supports. The number and size of the conductor insulating supports are determined by the magnitude of the available fault current. Bus duct can be designed for extremely high fault currents by bracing the interior conductor in several directions and at close spacing. Typical values for nonsegregated phase bus are up to 80,000 amperes. Segregated phase bus short circuit withstand values may be up to 240,000 amperes. It is not uncommon for isolated phase bus to be designed for short-circuit currents in excess of 350,000 amperes and a continuous current rating of up to 50,000 amperes.

17.5 GROUNDING

Power plants and substations are built upon ground mats that are designed to provide a reference value for the generated voltages within the station as well as to furnish safe working conditions for both personnel and equipment. These system ground mats are also used as the current source for protective relaying which is discussed later in this chapter.

17.5.1 Safety Grounding

The goal of safety grounding is to reduce the potential for electric shock to both personnel and sensitive electronic equipment. Two terms that are widely used in the industry are "touch potential" and "step potential." Touch potential refers to the voltage experienced when a person "touches" an energized conductor or piece of equipment while standing on another piece of equipment which may be at a different voltage level. Step potential is the term used to describe the difference in the voltage between the feet of a person walking on a surface which may have a voltage gradient along it. The object of safety grounding practice is to eliminate dangerous voltages in both these conditions by connecting all parts of the power station to the earth through heavy conductors. The object of system grounding is to connect the generating sources to the earth through heavy conductors which reduces the station's "ground rise potential" to values that can be tolerated with the available system equipment. Both aims are overlapping, and if one condition is satisfied, usually the other desirable condition is also enhanced.

The earth has fairly low resistance to electrical currents, and indeed the earth is an extremely large generator itself since it rotates within a very powerful magnetic field. There are very large electrical currents flowing within the earth at all times. At different locations throughout the world, the earth's resistivity may vary widely, and to make intelligent grounding calculations, the designer may be required to measure the ground resistivity at the specific location where the power plant or substation is to be built.

Many of the calculations involved in determining ground fault, step, touch, and rise potentials are empirically derived. Complete details for these determinations are discussed in Institute of Electrical and Electronic Engineers (IEEE) Standard 80.

In substation grounding, it is desirable to have a visible grounded platform or mat for the operator to stand upon with a visible solid ground connection between the handle of manually operated switches. This ensures that the operator has a zero touch potential while operating high-voltage switchyard equipment.

All equipment frames within the ground mat area should be connected to the ground grid to avoid electrostatic buildup of dangerous potential. This includes fences, motors, cabinets, steel structures, conduit, trays, pumps, pipes, and similar equipment. This also includes the circuit neutrals of potential transformers, current transformers, and relaying circuits. However, each circuit should have only one ground connection, because more than one will cause the circuit to be in parallel with the station ground grid between the two ground locations, and a portion of any fault current flowing through the ground grid will also flow through this parallel portion of the relay circuit. This can cause misoperation of the control circuits as well as severe damage to electronic equipment. The location of this ground connection is usually made at the control room or a central location to avoid multiple grounds on control circuits.

Some specialized electronic equipment requires an "isolated ground" to assure that their equipment will never be subjected to heavy fault currents from power-generating equipment. In this case, the equipment requiring an isolated ground should be grounded with insulated conductors directly from the equipment to a single point within the ground mat. Multiple equipment requiring an isolated ground may be connected to the same ground location. It is common practice to have ground disconnecting links in the ground junction box which will allow the sensitive equipment to be disconnected from ground for testing purposes. It is almost always preferable to connect the isolated ground to the station ground at a single location so as to prevent ground potential rise between two ground grids that may be within the same proximity. Occasionally, some specialized equipment cannot be grounded for one reason or another. This includes communications channels such as solid wire pilot channels and older telephone circuits. In this case, the equipment must be fully isolated so that personnel are not accidentally subjected to hazards during faults at remote locations.

17.5.2 Generator and System Grounding

Generators are almost always wye connected and are electrically connected to the power systems in various ways, usually depending on the size of the machine. Small generators (30 MW and less) may be connected directly to a power system distribution system. In this case, the generator almost always experiences some phase unbalance in its normal load, and therefore must be solidly grounded to the system. However, it can be shown mathematically that the short-circuit current in a single line-to-ground fault on a generator is larger in that faulted phase than the short-circuit current in any phase of a bolted three-phase fault. Generator manufacturers normally brace their machines to withstand the mechanical forces developed in a bolted three-phase fault. Therefore, in an application in which a generator must be solidly grounded, it should be grounded through a reactor whose impedance is large enough to limit the line-to-ground fault to a value no larger than the magnitude of the three-phase fault current. A reactor is used in this instance which also serves to eliminate the power losses that would result if a resistor were used instead. Such an arrangement is shown in Fig. 17-19.

It should be emphasized that there is a difference between a power system ground and the system neutral. Electrically, both points should be at the same potential and are normally connected to the earth. However, a system ground is not expected to carry a constant current, whereas the system neutral is designed to carry some constantly varying unbalanced current which makes up the difference in the phase currents. The neutral conductor in a three-phase four-wire system normally is designed to carry about one-third the

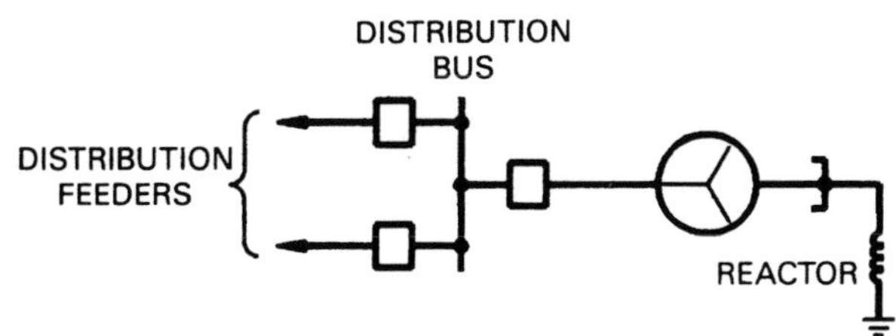

Fig. 17-19. Generator grounding with neutral reactor.

amount of the phase current. Of course in a perfectly balanced three-phase system, the neutral current is zero. In an application such as the above, involving a generator that is connected directly to a three-phase, four-wire distribution system, an insulated neutral conductor, capable of carrying about 33% of the phase conductors ampacity, should be connected from the ground side of the reactor to the neutral of the distribution bus. The actual amount of continuous current the neutral will experience depends on the amount of unbalance in the load.

When more than one generator is connected to the same distribution bus, only one of the generators need be connected to the neutral at a time. However, all of the machines may be connected to neutral and ground if the system switchgear is designed to handle the resulting available fault current. The more machines that are connected to the ground and neutral bus, the higher the available fault current will be since their transient impedances will be electrically in parallel. For this reason, a neutral breaker or switch is sometimes installed where multiple machines are connected to a distribution bus. It is necessary to have the neutral bus of at least one of the machines connected to the ground bus at all times in order to have a ground path for fault currents. It should be noted, however, that the machine which is grounded to the neutral bus is the machine that will carry any unbalanced load current.

In larger generator installations, the generator is normally connected to the power system through a generator stepup transformer and high-voltage transmission lines. Although the generator itself is connected to the delta side of the stepup transformer, the power system still must have a ground source from the generating plant to be able to clear ground faults on the transmission system. For this reason, the high-voltage side of the generator transformer is normally wye connected and the neutral of the wye is connected solidly to ground.

In rare cases in which a generating station is remotely connected to the load centers through long transmission lines, it may be necessary to connect the generator transformer to ground through a resistor for system stability reasons. A fault near the terminals of a generator will unload the generator, and consequently the machine would tend to overspeed, with the possibility of the torque angle of the machine increasing past the point where the machine could slip a pole and become unstable. The ohmic value of the resistor normally is low (around 8 to 15 ohms) and is selected so that a line-to-ground fault will not unload the generator during the fault.

This relationship is shown as follows:

$$(I_f)^2 R = 1/3 P_{gen} \tag{17-17}$$

where

I_f = transmission system line-to-ground fault current,

R = ohmic value of transformer neutral resistor, and

P_{gen} = MVA power rating of the generator.

The power rating of the resistor depends on how long the fault current is expected to last. That time depends on the coordination of the system's protective relaying equipment. The resistor itself normally is protected by a high-speed grounding switch that shorts out the resistor some time after the system protective relays and circuit breakers should have removed the fault. Power system stability studies must be performed by the interconnected utilities to determine if transformer grounding resistors are required for a particular generating station.

In the case where a generator is unit connected to a power system through a delta wye stepup transformer, the generator still should be grounded at its neutral terminals. However, since it will never see unbalanced currents in normal operation, there is no need to solidly ground its neutral terminal or to use a reactor.

It is common practice in this application to use a high-resistance ground for the generator. Since a large high-voltage resistor may be impractical, the principle of transformer action can be used to reflect an impedance through a transformer as the square of the turns ratio. Thus, a fairly small resistance value at a low voltage can be used and connected on the secondary side of a single-phase distribution transformer. The high-voltage winding of the transformer is then connected between the neutral connection of the generator wye and ground. The value of the resistor/transformer combination is selected to limit the machine line-to-ground fault current to some very small value, usually about 5 amperes. The station designer then has the option of alarming or tripping the generator unit upon detection of generator phase-to-ground faults. The design philosophy of most generating stations is to trip the generator on sustained ground faults, since the resulting fireworks from a fault on a second phase of the generator while the first phase was already grounded would be quite dramatic.

An example of a unit connected generator configuration is shown in the one-line diagram in Fig. 17-20.

A third type of generator grounding application is the unit-connected generator with a low side generator breaker. In this application, the high-resistance grounded wye connection of the generator is still applicable; however, another consideration must be examined such as the times when the generator is not connected to the system, but the generator transformer remains connected. During this condition, the low side of the generator transformer, all the bus duct up to the open generator breaker, and the high side of the main auxiliary transformer are not grounded by any direct connection. Nevertheless, the delta of the transformer and the bus duct will be grounded at this time through the capacitance of the bus duct, the capacitance of the transformer windings, and any cables connected to this energized system. The windings of the transformer also are inductive because they are coils of wire wrapped around the core steel of the transformer. When inductors and capacitors are connected in parallel, there is some frequency at which this connection is resonant. Electrical theory refers to this parallel connection as a "tank" circuit. At the resonant frequency, the voltage across the tank circuit would be infinite if there were no parallel resistance. A low value of parallel resistance reduces the Q of the tank circuit (i.e., the ratio of the energy stored to the energy dissipated), and although the circuit may still be resonant, the resultant high voltage will be greatly reduced. In practice, this resonant frequency is very difficult to determine. However, an arcing fault on the delta-connected windings generates various random frequencies of voltages and one of these voltage waveforms may be at the resonant frequency of the tank circuit. This condition is also sometimes referred to as "ferroresonance." To avoid these very high voltages, a set of "ballast resistors" can be installed on the delta bus. The resistors can be installed either on the open corner of a "broken delta" set of potential transformers or installed as a set of wye connected resistors. The values have been empirically determined as shown in Table 17-5.

The location of the ballast resistors is shown in the one-line diagram on Fig. 17-21.

17.6 PLANT AUXILIARY ELECTRIC SYSTEMS

Depending upon the type of fuel used and the environmental control systems required, a coal-fueled electric generating plant may consume as much as 10% or more of its total generation for auxiliary power. A conventional steam electric generating plant using natural gas or liquid fuel may

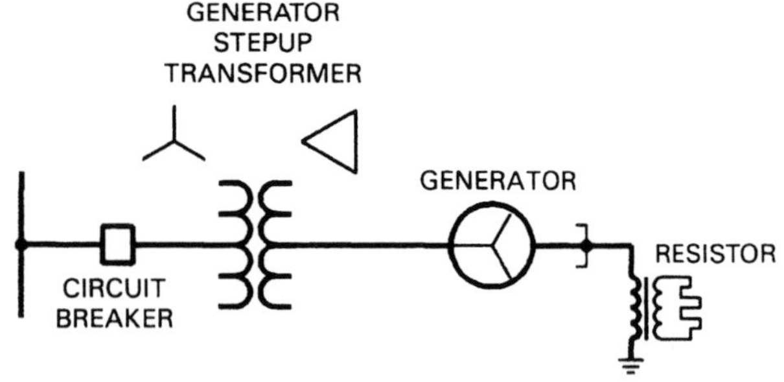

Fig. 17-20. High-resistance grounded unit-connected generator.

Table 17-5. Ballast Resistor Values

Voltage Class Approximate Ohms Across	Watt Loading (Primary) Broken Delta	Approximate Ohms on PT Secondary	per Phase (120 V)
>5,000 V	200	72	216
7.2 kV–15 kV	500	29	87
25 kV and up	750	20	60

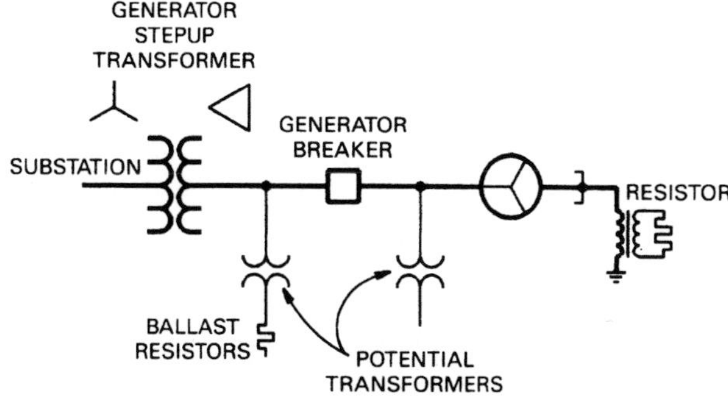

Fig. 17-21. Ballast resistors with unit-connected generator and generator breaker.

require only 5% or 6% auxiliaries, whereas a combustion turbine firing natural gas may require less than 1% auxiliary electric power while operating.

Some very large, high-speed loads such as boiler feed pumps may be driven by variable-speed steam turbines, particularly where the turbine exhaust steam may be used in the feedwater heating cycle. Such drives reduce the auxiliary electric system requirements, although they still must be accounted for in the total plant efficiency.

Some new power plant projects use large variable-speed electronic drives for large fan loads. These drives use large silicon-controlled rectifier stacks (SCRs) to produce direct current. Another set of SCRs "invert" the dc current back into alternating current (ac) at a frequency that is controlled by the power plant operator. This adjustable frequency power is then used to drive a synchronous motor. Many applications use two such SCR assemblies in parallel to drive a synchronous motor having two sets of stator windings that are mechanically displaced from each other by 60 mechanical degrees. Such an arrangement substantially reduces the harmonic content of the auxiliary electric system over that produced when only one set of SCR assemblies is used. Remember that the speed of a synchronous motor is determined exactly by the frequency of the power supply. Since the volume of air delivered by a fan is proportional to its speed, the variable-speed electronic drive system eliminates the need for control dampers in the fan discharges, thereby increasing the efficiency of the fan system, since the pressure drop across a set of dampers is eliminated. Such an arrangement is shown in one-line form in Fig. 17-22.

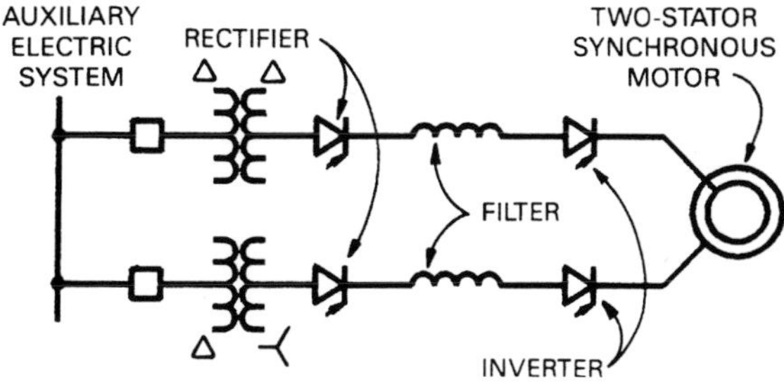

Fig. 17-22. Variable-speed electronic motor drive.

17.6.1 AC Systems

In general, the larger the power generating plant, the higher the voltage selected for the ac auxiliary electric system. For some very small generating stations, the highest plant auxiliary voltage may be 480 V. Slightly larger stations, say 10 MW to 100 MW, may have a 2,400 V auxiliary electric system. Most stations in the 75 to 500 MW range utilize 4,160 V as the base auxiliary system voltage. Large generating stations 500 MW to 1,000 MW and more have forced the electrical designer to investigate voltage levels above 4,160 V. It is not uncommon for these large stations to have multiple buses using 6,900 V as the auxiliary system voltage. Occasionally, some single dedicated loads, such as electric driven boiler feed pumps in the 10,000- to 15,000-horsepower class are supplied by a separate 13,800-V bus.

It must be emphasized that the choice of voltage level for an auxiliary electric system cannot be generalized. Each situation should be studied on its own merits to determine the most favorable voltage selection.

The following areas must be considered in the design of an auxiliary electric system.

*17.6.1.1 **Motor Starting Requirements.*** Most motors in power plant service are started by contactors which place the motor terminals directly across the incoming auxiliary electric system line. This means that full system voltage is applied to the motor windings while the motor is at rest. If the motor has been applied properly, the starting torque will be larger than the driven load's starting inertial requirements, and the motor will start to rotate. However, in this starting mode, the motor inrush current will be very large, and consequently, the motor terminal voltage will drop, due to the voltage drop in the auxiliary electric system and the leads connecting the motor to the system. Typically, the system voltage will not drop below 90% of the normal operating voltage during start, and no special provisions will be required. However, if the motor is physically remote from the main auxiliary system bus and requires long leads to connect it, the motor terminal voltage may drop excessively to the point that motor overheating may be a concern. In such cases, reduced voltage starting devices may be required or the motor may be designed to operate at 80% or even less terminal voltage during starting. The tradeoff in this case is reduced starting torque and slightly higher operating losses at normal loads.

*17.6.1.2 **Voltage Regulation Requirements.*** The auxiliary electric system bus voltage varies with load changes. The ratio of the difference between bus voltage at no load and the bus voltage at full load to the no-load bus voltage is called voltage regulation. Regulation is referred to in percentage as follows:

$$\text{Percent regulation} = \frac{(V_{\text{full load}} - V_{\text{no load}}) \times 100}{V_{\text{no load}}} \qquad (17\text{-}18)$$

The bus voltage regulation may be improved by using a larger capacity source having less impedance. Low regulation percentage is desirable because it reduces lighting flicker during system disturbances such as motor starting. However, while an attempt is being made to keep the auxiliary electric system bus voltage from dropping during motor starting, the short-circuit duty of the switchgear will increase.

17.6.1.3 Short-Circuit Duty Requirements. As an alternative to purchasing special motors that start and operate at reduced voltage, the auxiliary electric system may be designed to have a very large capacity so that large inrush currents do not cause a significant voltage drop in the auxiliary electric system. Motor starting ability is always enhanced by providing a high level of short-circuit MVA at the motor bus. This can be accomplished by prudent selection of the auxiliary transformer capacity and its impedance. However, the designer must be aware that excessive short-circuit currents may be encountered and thus require heavier mechanical bracing in the switchgear as well as high-capacity interrupting circuit breakers which supply the system. This high-capacity equipment may also be expensive or not commercially available. Excessive short-circuit current magnitude may be decreased by using more than one unit auxiliary transformer and associated switchgear buses, by using three-winding transformers, or both. Current limiting reactors might also be considered, but only if there is no reasonable alternative, since they are physically bulky, vulnerable to failure, and affect voltage regulation.

17.6.1.4 Load Current Duty Requirements. As more and more loads are added to the auxiliary electric system, the current carrying capacity of the supply switchgear and its conductors must be examined. At some point, it may become desirable to develop multiple auxiliary system buses to better utilize commercially available hardware and distribute the loads evenly throughout the plant. This may require more supply transformers, as well.

17.6.1.5 Economic Considerations. All of the previous technical considerations must be balanced with cost considerations. Motors capable of starting under 70% of normal terminal voltage will cost more than motors that are designed for a minimum of 90% terminal voltage during starting. Switchgear capable of interrupting 50,000 amperes of fault current is more expensive than switchgear that is designed to interrupt 30,000 amperes. A single bus lineup of switchgear costs less than a lineup having two buses because the single bus requires only one main feed circuit, whereas two buses require a main feed circuit for each bus. However, the two-bus lineup may have reduced bus ampacity requirements since its required capacity for each bus will be smaller. Usually, a technically well-designed system is also the most economical to purchase, operate, and install. However, no hard rules can be made for all applications. Each case should be examined for its own merits. Both the capital and operating costs of all final system designs should be seriously examined.

17.6.1.6 Future Load Requirements. The electrical power plant designer cannot anticipate what future mechanical or electrical devices needing electric power may be required by the plant owner. However, the auxiliary electric system should not be designed so close as to disallow any future load growth on the auxiliary electric system. At the same time, excessive future capacity will be expensive and may subject the designer to justifiable criticism. As in any system design, good judgment is essential and comes only with experience.

17.6.1.7 Reliability and Alternate Sources. Many loads on the auxiliary electric are essential to the operation of the plant. Loss of all the boiler feed pumps, condensate pumps, or boiler draft fans, for example, requires immediate shutdown of the steam generator and consequent loss of electric production. The mechanical designer is aware of this and usually has a system design using two 100% capacity pumps, or three 50% or 60% capacity pumps or fans. In this case, the electrical system designer should also supply half of the equipment such as the boiler feed pumps, condensate pumps, fans, etc., from different supply buses. Or, if the plant is small and has only one bus, that bus should have at least two sources of supply. In some cases, regulatory and licensing bodies may have design review responsibility over the plant and may require the design to have additional reliability considerations. Such is the case for nuclear power plants, for example, that require three sources of electric supply for the reactor cooling pumps.

The voltage level for an auxiliary electric system must be selected after careful consideration of all the criteria discussed above. Table 17-6 gives the recommended nominal system voltages and the corresponding switchgear and motor rated voltages.

The voltages listed are included in ANSI standards as acceptable voltages for systems and equipment including motors. These voltage levels are widely used by utilities in both fossil-fueled and nuclear power plants.

Occasionally, a motor with a voltage rating corresponding to the selected auxiliary electric system voltage may not be commercially available. For example, one steam generator manufacturer uses a boiler water circulating pump with a self-contained "wet rotor" pump motor. This motor is designed only for a 4,000-V supply feed. If the remainder of the auxiliary electric system is designed for 6,600 V, the de-

Table 17-6. Auxiliary Electric System Voltage Levels

Nominal System Voltage	Switchgear Nominal Voltage Class	Motor Nameplate Voltage
13,800	15,000	13,200
6,900	7,200	6,000
4,160	5,000	4,000
2,400	5,000	2,300
480	600	460

signer could utilize a "captive" or "unit" transformer installed between the 6,600-V main bus and the 4,000-V motor. This transformer–motor combination is then switched like a single high-voltage motor. The transformer also serves as a series starting reactor to reduce the starting current. However, this lower starting current is accomplished at the expense of lower motor starting voltage and torque during the starting period. Motor acceleration time must also be investigated.

In the design of an auxiliary electric system as shown in Fig. 17-23, every parameter is a variable for the electric system designer. Bus voltage is a selection. Ampacity of the bus is a function of the selected voltage and total load. Switchgear capability (MVA, ampacity, and capacity) can be selected by varying the transformer size and impedance, and the number and size of motors and other loads connected to the bus. Transformer size (and quantity) is determined by the total size of the load allowed on the bus. Motor size is determined largely by the mechanical system requirements, but the electrical designer can serve a very large mechanical load—say 12,000 horsepower—by using tandem 6,000-horsepower motor drives on the same shaft. The mechanical requirements of the load can be served by more pumps or fans to serve the particular process system requirements. The total kVA load on the bus can be controlled by using more or fewer buses. If one or more parameters are fixed by the designer, all the rest become dependent variables. And of course, all these decisions must be tempered by economic considerations.

17.6.2 System Configuration

Auxiliary electric systems may be arranged in several ways. Small units (< 50 MVA) may be bus connected directly to a distribution bus, usually with a low-voltage generator breaker and one or more bus tie transformers to a high-voltage grid system, as illustrated in Fig. 17-24.

Larger units are usually unit connected to a dedicated generator stepup transformer with one or more main auxiliary transformers and a separate set of reserve auxiliary (or startup) transformers, which may serve two adjacent units, as shown in Figs. 17-25 and 17-26.

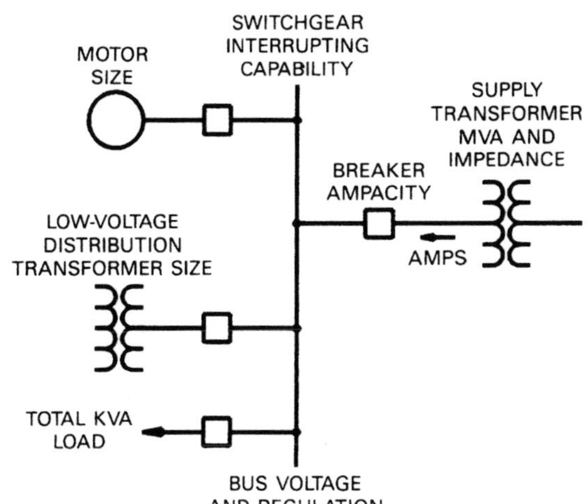

Fig. 17-23. Auxiliary electric system variables.

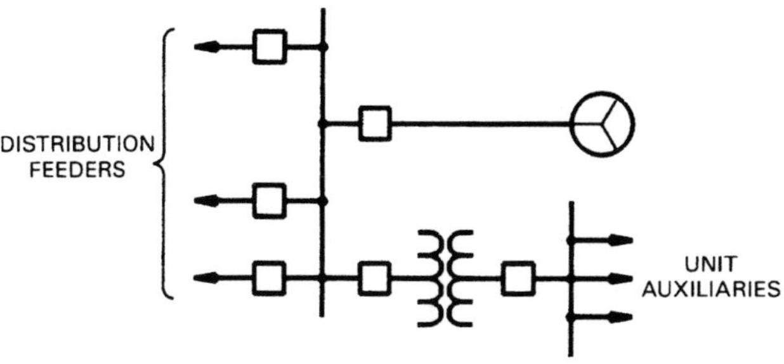

Fig. 17-24. Bus-connected small generating unit.

Also available are indoor isolated phase circuit breakers that may be installed in the generator leads. Although quite expensive, for a large generating unit (> 200 MW), these devices offer the following benefits:

- Provides through fault protection of unit connected generator transformers by disconnecting the generator from the fault, rather than allowing it to continue to supply fault current during its field decay period following a unit trip.
- Furnishes an auxiliary power source to the unit auxiliary load via the backfed generator transformer and the unit auxiliary transformer(s). This may be particularly useful in a nuclear plant design in which two separate offsite power sources are required.
- Eliminates the need for auxiliary load transfer during startup and shutdown operations.

Large generators may or may not have a low-side generator breaker. One of the disadvantages of the low-side generator breaker system is the lack of redundancy of auxiliary electric supply sources if the reserve auxiliary transformers are not in place.

17.6.3 High-Voltage Auxiliary Electric System

The high-voltage auxiliary electric system is sized to handle the majority of the large motor loads directly from metalclad high-voltage switchgear. The bus voltage for this system is

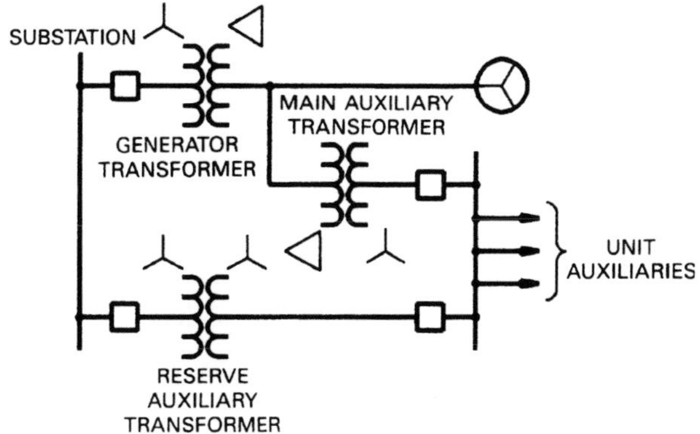

Fig. 17-25. Unit-connected large generating unit.

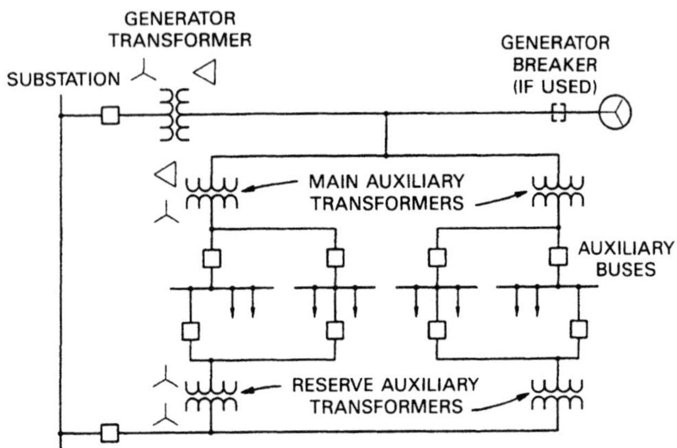

Fig. 17-26. Large generating unit, multiple auxiliary buses.

determined after considering the alternatives discussed previously. It is normally chosen at 2,400 V, 4,160 V, or 6,900 V.

It is a good practice to divide the loads among several switchgear buses to enhance the entire plant's reliability. If the generating unit has two redundant fans or pumps for a mechanical system, each fan or pump motor for that system should be fed with auxiliary electrical power from an electrically operated circuit breaker in separate switchgear buses. For instance, large loads that require repetitive starting duty, such as train or barge unloading conveyors and coal crushers, should be served from motor starting contactors instead of heavy-duty circuit breakers because contactors usually offer longer life (as measured by number of nonfault operations) than heavy-duty circuit breakers. The motor contactors, however, are not capable of interrupting fault current, and therefore must be protected by an upstream circuit breaker. The economics of this dual-series equipment arrangement should be considered for each application.

Whenever possible, designers should arrange unit loads so that the buses are approximately equally loaded. This allows unit auxiliary transformer sizes to be equal in those cases where multiple auxiliary transformers are used. In addition, large motors are normally grouped together on a "large motor" bus so that the bus voltage during motor starting does not cause "flicker" on smaller loads such as motor control centers and the plant lighting systems. The voltage on the large motor buses during starting may then be permitted to drop to a much lower level, say 80%. The voltage regulation on the "small motor buses" can be kept to about a 90% drop. Loads such as Secondary Unit Substations, which serve the 480-V system, should not be connected to the large motor buses because low-voltage motor starters could otherwise have their starting coils "drop out" during starting of large loads.

The high-voltage auxiliary electric system buses do not need to be solidly grounded if all loads are three-phase motors or delta-connected secondary unit substations. If all loads are balanced three-phase loads, normally there is no neutral current; therefore, the system need only be capable of detecting ground faults. The auxiliary electric system may then be designed as a grounded three-wire system with wye-connected transformers or as an ungrounded three-wire system with delta-connected transformers.

With the three-wire grounded system using wye-connected transformers, the system ground fault current required to detect a ground fault inside the main auxiliary transformers (using harmonic restraint transformer differential relaying) may be fairly high, possibly on the order of 1,000 or 1,500 amperes. Since there is no continuous ground or neutral current, the transformers may be resistance grounded. The ohmic value of the resistor is easily calculated by the line-to-neutral voltage and the required ground fault current. The power rating of the resistor is again dependent on the speed of the protective relaying system and how the relays are coordinated. Normally the resistor may be rated for 1 to 5 seconds. Each transformer neutral normally has its own neutral grounding resistor, although the designer may choose to use a common neutral grounding resistor for more than one transformer (or winding in the case of dual secondary transformers).

The maximum allowable ground fault current should be coordinated with the shield wire capability of the high-voltage conductor cable system to ensure that the cable shields can handle the ground fault current long enough for the relaying system to clear a fault.

The auxiliary electric system designer must also be aware of any loads that may require unbalanced load currents. If large loads from equipment such as auxiliary electric boilers occur, which may result in continuous neutral currents, the high-voltage auxiliary electric system may be required to be designed as a four- wire system with solid grounding and a full neutral. These unusual loads may also be served from a separate four-wire auxiliary electric system.

The ungrounded delta system is not as common in power system design, although it can be used with success. In this case, the ground fault current is zero unless the second phase becomes grounded, in which case a line-to-line fault will be present. Ground detection schemes using potential transformers with wye connected alarm relays or a "broken delta" connection with alarm relays may be used. Special consideration should be given to insulation coordination of ungrounded delta systems. In addition, arcing ground faults may also create very high voltages in some circumstances, as discussed previously.

In general, the high-voltage auxiliary system furnishes power directly to motors about 250 to 350 horsepower and larger. In addition, large dedicated station loads such as auxiliary electric boilers, coal handling centers, or other specialized loads may require dedicated feeds from the high-voltage auxiliary electric system. Special remote loads such as makeup well water fields, storage lake pumps, remote fuel handling facilities, or other remote dedicated plant loads may require dedicated subtransmission line feeds directly from the plant substation to be properly served.

17.6.4 480-V Auxiliary Electric System

The 480-V auxiliary electric system consists of Secondary Unit Substations (SUSs), Motor Control Centers (MCCs), and 480-V panelboards. In addition, heating, ventilating, and air conditioning equipment (HVAC), freeze protection panelboards, and station lighting loads are normally supplied with power from this intermediate source.

The 480-V secondary unit substations supply motor loads directly from electrically operated low-voltage circuit breakers in the SUS of about 100 to 300 horsepower. Large 460-V motors that require repetitive starting duty, such as conveyor motors, cooling tower fans, etc., should be served from a motor starter rather than a breaker because of the lower life duty (measured as number of operations) of circuit breakers as compared to motor starters.

Secondary unit substations nearly always are designed to have more than one power source from separate high-voltage buses. The secondary unit substations in this case are double ended, using two transformers, each having a secondary main circuit breaker which serves two 480-V buses interconnected by a bus tie circuit breaker.

Secondary unit substation transformer loading must be controlled where a bus tie is provided to prevent sustained overloading if the tie should be closed upon failure of the opposite side SUS transformer. Good practice is to design each bus for loading to approximately 80% of the transformer self-cooled rating. Each SUS transformer in that case has sufficient capacity at the fan cooled rating to carry both SUS buses for a reasonable length of time.

Transformer sizes and impedances should be selected to limit the available fault currents to within the SUS switchgear circuit breaker ratings. When this is not feasible for some reason, current limiting fuses, current limiting reactors, or a combination of both may be required in the MCCs. Reactors are not a very desirable method of limiting the fault current because they are bulky, must be cooled due to the internal losses, and are just one more device subject to failure in the auxiliary electric system.

Low-voltage (480-V) motor control centers for small motors ($<$ 100 horsepower) are generally served by radial feeders from the secondary unit substations unless a particular MCC is considered a critical bus, in which case automatic throwover contactors may be used to provide a second source.

17.6.5 480-V System Grounding

The types of 480-V system ground configurations that may be considered are solidly grounded, reactance grounded, ungrounded, and resistance grounded.

17.6.5.1 Solidly Grounded Wye Systems. Solidly grounded systems offer the greatest control of over voltages during phase-to-ground faults. The neutral driving voltage is 58% of the phase-to-phase voltage. Solidly grounded systems using delta–wye SUS transformers with the neutral directly connected to the station ground have probably been the most widely used. This method has been used both with and without ground fault protection. However, since this system depends on high fault current capability for detection and clearing of faults, considerable damage will occur from an arcing fault if not immediately cleared. Some delay is always required to properly coordinate between main feeder and branch feeder circuit breaker clearing times, and a certain amount of conductor and/or enclosure vaporization may be unavoidable, depending on the location of the fault.

Historically, special ground fault protection of 480-V systems has been largely ignored, and the phase fault detection and tripping has been allowed to provide limited coordination of ground fault protection. However, recent editions of the National Electric Code (NEC) have required ground fault protection for solidly grounded wye electrical services rated above 1,000 amperes. While installations under the exclusive control of electric utilities for the purpose of generation and transformation are not subject to the NEC, it is generally recognized as good engineering practice to follow the NEC recommendations where practical to do so.

Solid grounding of the neutral provides the maximum ground current for detection and clearing of a ground fault. However, proper coordination of ground fault relaying throughout the 480-V auxiliary electric system is difficult to achieve, particularly at the smaller motor control center motor starters. Ground faults near the source can exceed the three-phase fault current magnitude, while at the ends of long feeders the fault current may be much lower. High-impedance arcing ground faults may be considerably below the setting of the SUS main breaker phase overcurrent trip devices since these main breakers must be set high to coordinate with the branch circuits and to avoid tripping on motor starting inrush currents in motor control centers.

If ground fault protection is applied only to the large SUS main breakers and feeder breakers above 1,000 amperes capacity as required by the NEC, selective tripping of downstream devices such a motor starters and panel board feeder breakers cannot be ensured. This may result in a fault on a relatively small downstream feeder causing a trip of the entire SUS. Reliability of the generating plant should not be compromised with this type of partial ground fault protection. If ground fault protection is to be applied to the larger feeders, it must be carried out on the entire 480-V system. Selective tripping is essential to minimize the affected equipment and to maintain power plant reliability.

Solidly grounded systems are directly compatible with 277-V fluorescent lighting schemes; however, ground fault coordination of lighting circuits with upstream SUS main breakers must be carefully considered. Otherwise, a ground fault in a lighting circuit may trip the main breaker of a 480 V SUS and result in a forced outage of the entire power plant. Delta–wye isolation transformers are recommended for the lighting panelboards, freeze protection circuits, and other single phase 277-V loads.

17.6.5.2 Reactance Grounded Systems. Reactance grounded systems use reactors to reduce the available ground fault current to a level that is compatible with the low-voltage switchgear. This type of system is used only where the SUS transformer capacity must be very large to accommodate high three-phase loads in applications where the number of transformers must be limited. If the transformer MVA capacity and impedance can be properly selected, neutral reactors are not normally required.

17.6.5.3 Ungrounded Delta Systems. The ungrounded system uses delta–delta SUS transformers with no intentional connection to the station ground grid. Since this system has no ground fault current, ground fault interrupters (along with their accompanying difficult coordination problems) are not required. However, a ground fault detection scheme should be installed to alarm a single phase-to-ground fault.

The ungrounded system is actually a high-reactance, capacitively grounded, system because of the capacitance coupling to ground of every energized conductor. The operating advantage for the ungrounded system is that a single line-to-ground fault does not result in an outage of the grounded feeder. Of course, if the second phase becomes grounded, a phase-to-phase fault will be present.

For the duration of the single line-to-ground fault on one conductor, the other two-phase conductors throughout the entire system are subjected to 73% overvoltage. Therefore, it is prudent for the operations and maintenance staff to locate the faulted circuit promptly and repair or remove it before abnormal voltage stresses produce breakdown of other circuits or machines.

Because of the capacitance coupling to ground, the ungrounded system is subject to overvoltages as a result of intermittent contact ground faults (arcing faults). The magnitude of the overvoltage is almost impossible to predict in a given application because of the difficulty of determining the actual capacitance and inductance of the system, but may be several times the phase-to-phase voltage rating.

The ungrounded delta system is not directly compatible with 277-V single-phase loads such as lighting circuits. However, the use of 480 V delta to 480/277 V wye isolation transformers eliminates this problem.

17.6.5.4 Resistance Grounded Systems. The resistance grounded system consists of delta–wye transformers with the transformer neutral connected to the station ground grid through a resistor. The primary feature of the resistance grounded system is that the phase-to-ground fault current is limited to a very low magnitude.

Since most of the loads on a 480-V auxiliary electric system in power plant applications consist of three-phase motors and other balanced three-phase loads, neutral currents are zero. Therefore, as in the ungrounded system, a neutral connection is not required to serve these loads. The resistance grounded system offers a compromise between the adverse overvoltage conditions that can occur in the ungrounded delta system and the very high ground fault currents which are prevalent on solidly grounded systems.

As with the ungrounded system, the main purpose of high-resistance neutral grounding is to avoid automatic tripping of the faulted circuit at the first ground fault occurrence. However, the resistance grounded neutral system does control the limits of the transient overvoltages to approximately a maximum of the line-to-line voltage, due to the reduced neutral driving voltage and because the phase-to-ground capacitance is paralleled by the neutral resistance. Of course, this will not prevent the 73% overvoltage on the other two ungrounded phases during a sustained ground fault.

If the value of the resistor is selected so that the maximum sustained ground fault current is a very low value, say 5 to 10 amperes, the fire and shock hazard will be significantly reduced from the several thousand amperes available on a solidly grounded system.

As with the ungrounded delta system case, the high-resistance grounded system is not directly compatible with 277-V single-phase loads such as lighting circuits. Nevertheless, a delta–wye isolation transformer can provide service to these single-phase loads separately from the main 480-V auxiliary electric system. The isolated 480/277-Volt lighting panelboards or other single-phase loads must be provided with their own neutral conductors and ground fault protection schemes. Fortunately, ground fault interrupting circuit breakers for these low-level service requirements are readily available, and at modest cost.

17.7 DIRECT CURRENT SYSTEMS

The direct current (dc) system is one of the most important systems to be considered in the design of power-generating stations and substations. The power system's alternating current (ac), which is being generated at the generating stations and distributed to customers via the transmission system, is the controlled quantity. Therefore, that power is subject to outages, however momentary, and thus cannot be considered reliable enough to be used as the power supply to control itself.

Batteries provide power for many generating station and substation functions, including tripping and closing of circuit breakers, energizing emergency bearing oil pumps, emergency lighting, reliable ac power system inverters (uninterruptible power supplies, or UPSs), critical boiler controls, and for motor-operated valves and electrical disconnect switches. In addition, the supervisory control and data acquisition system (SCADA), which monitors the status of the entire power system and displays this information at central control centers, must also have its power supply furnished by dc batteries. In short, the dc system is called on to provide power in emergencies to protect costly equipment from damage and to ensure the safety of the power system's operating personnel. Figure 17-27 shows a typical lead acid battery installation.

Fig. 17-27. Typical lead acid power plant battery installation. (From City of Vero Beach. Used with permission.)

17.7.1 DC System Configuration

Usually, dc systems are operated in the "floating" configuration, as shown in Fig. 17-28.

In this configuration, the battery charger and all loads are in parallel with the battery. When the dc loads exceed the capacity of the battery charger, or if the charger can no longer deliver power, as is the case when the ac system is deenergized, the battery compensates to supply the load requirements. How long the battery can continue to supply the load is, of course, a function of the battery's size and capacity.

Dc systems may be designed to provide either single- or dual-voltage capabilities. Several standard nominal system voltages are in common use, such as 12, 24, 48, 125, and 250 V. Single-voltage systems may be any of these nominal system voltages. Occasionally, dual-voltage systems such as 24/48-V and 125/250-V are used. Typically, large loads such as motor-operated valves or emergency oil pumps are fed with 250-V power, and other smaller system loads, such as switchgear tripping and spring charging motors or emergency lights, are balanced on the other two 125-V legs of the system. Figure 17-29 is a diagram of a typical dual-voltage system.

Due to the fairly large loads and long distances encountered in most modern generating stations and substations, 125-V and 250-V systems are the most widely used voltages. Selection of a single-voltage system versus a dual-voltage system requires the judgment of the designer. In general, dc systems are designed as single-voltage systems within a given power station for simplicity, unless there are several large loads that might best be served at a higher voltage to reduce their current requirements. In addition, some turbine manufacturers have control systems that require multiple-voltage 24/48-V power supplies.

Station dc systems are normally operated as ungrounded systems; that is, the negative terminal, or ground reference terminal, is not connected to the station ground grid. Equipment enclosures, however, should be connected to ground. The ungrounded type of system offers greater reliability than the grounded type because, if one polarity of the ungrounded system becomes grounded, the system will remain operational (unless of course, the opposite polarity also becomes grounded). Ground detection systems are designed to alert plant operating personnel of one polarity becoming grounded so that the situation can be corrected before a fault develops.

A simple ground detection scheme is shown in Fig. 17-30. The two incandescent lamps are in series with their center connection grounded. Normally, both bulbs will glow with low brilliance because the voltage across each of them is 62.5 V dc. If either polarity of the battery becomes grounded, the bulb on the grounded side will go out since there will be no voltage across it. The bulb on the ungrounded side will then glow at full brilliance because the full 125 V will be applied across its terminals. The light bulbs in an unattended location could, of course, be replaced with relays having contacts that would alarm at a remote annunciator when one of the battery polarities becomes grounded.

This ground detection scheme works only on one side of a dual-voltage battery scheme. In the case of a dual-voltage system such as 125/250 V, the battery center may be con-

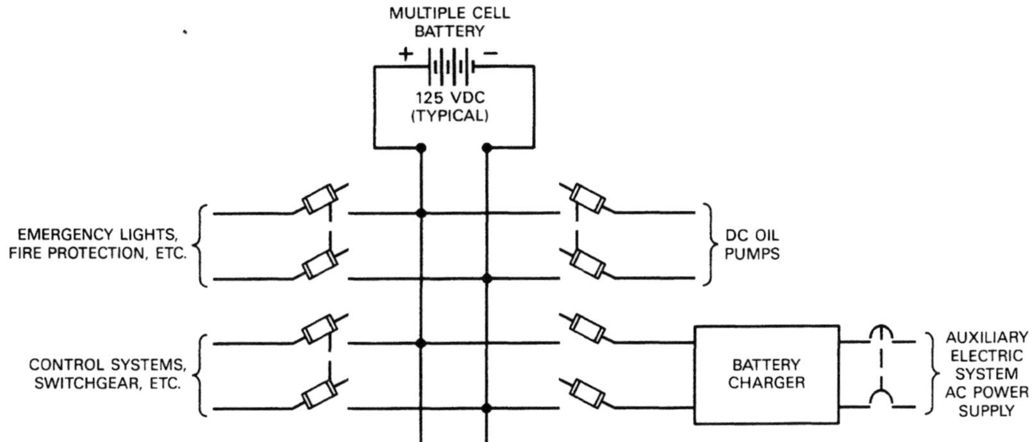

Fig. 17-28. Floating battery configuration.

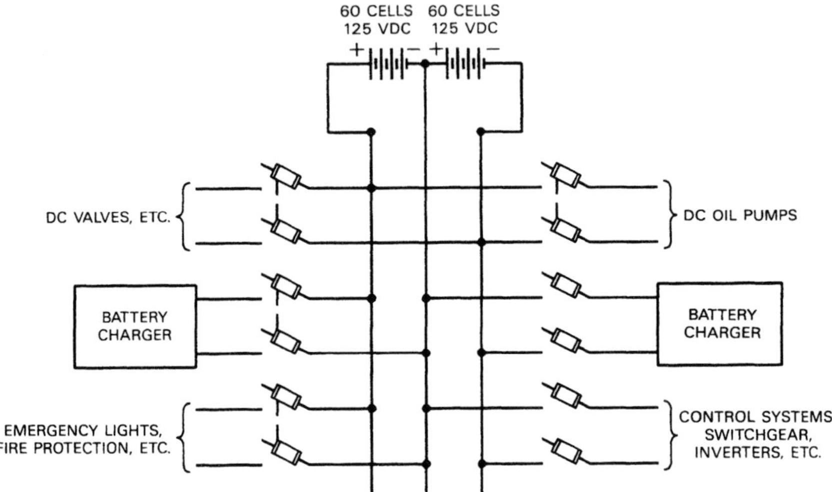

Fig. 17-29. Typical dual 125/250-V dc battery system.

nected to the center ground detection point, or one side of the battery may have the ground detection scheme discussed above while the other half of the battery floats above ground voltage. This scheme is shown in Fig. 17-31.

The incandescent bulbs of the previous example have been replaced with auxiliary relays AX-1 and AX-2. These relay contacts are wired in parallel and to a remote annunciator alarm. As shown in the figure, AX-1 has a normally closed[2] cutoff contact in series with the operate coil of AX-2. This is to keep the full 240 V from being impressed across coil AX-2 in the event that the positive terminal of the ungrounded battery becomes temporarily grounded.

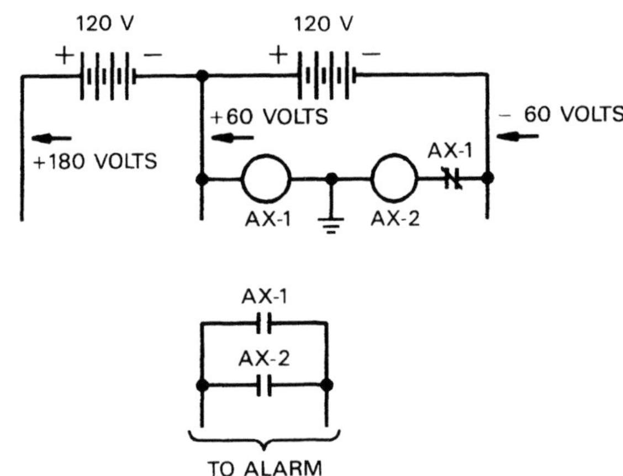

Fig. 17-31. Battery ground detection system with dual-voltage battery.

17.7.2 Battery Sizing

The station battery, either in a power plant or a substation, is sized to carry its load for several hours plus provide a safe margin for overloads, future loads, and contingencies. Normally, batteries are sized by the required duty cycle and the lowest voltage permitted on the cells before recharging, typically 1.75 V.

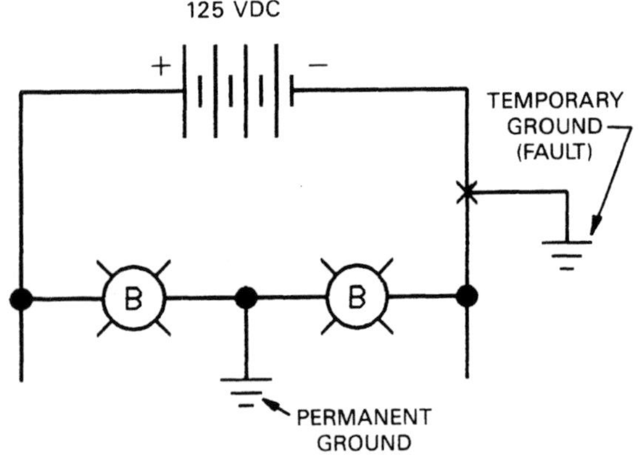

Fig. 17-30. Battery ground detection system.

The actual battery internal design is determined by the manufacturer. However, the station designer should be able to determine the number of positive plates required from the duty cycle so that adequate space in the power plant may be reserved for the battery room.

A load model should be set up by the station designer for the required sizing period. Typical loads are emergency lighting, inverters, control and indication monitors and computers, motor-operated valves, emergency oil pumps, circuit breaker operations, communication systems, and fire protection panels. Momentary loads such as circuit breaker operations should be assumed to last at least a minute for sizing calculations. The starting current for motor-operated valves and emergency oil pumps should be used at the actual load demand since these values are usually about five to six times the normal running current.

[2]A "normally closed" contact means that the contact is closed when the relay coil (or sensing device) is below its pickup voltage.

The number of positive plates would be determined from curves supplied by the battery manufacturer. IEEE Standard 485 has examples of battery calculations that are typical of some manufacturers.

A fully charged lead acid wet cell battery normally has a terminal voltage of 2.12 V per cell. Thus, a fully charged 60-cell battery would have a terminal voltage of 127.2 V.

Batteries are also given a rating called an Ampere Hour rating. This is the discharge current a given battery may supply for an 8-hour period to discharge the cells to 1.75 V per cell.

Station batteries are typically recharged to 2.33 V per cell. Also, it is usually desirable to recharge the battery in a reasonable time, say 8 hours. Therefore, the battery chargers should be at least big enough to carry the normal floating load of the battery, plus provide enough extra current to recharge the battery in the specified time.

Periodically, lead acid batteries should be "equalized." That is, a higher than normal charging voltage is applied to the battery to temporarily overcharge the cells and equalize the voltage across each cell. As the equalized battery is allowed to return to its normal voltage when the battery charger voltage is reduced, the cells tend to discharge equally and each cell's terminal voltage will return to approximately the same as that of all the other cells. This helps ensure that each cell in the battery stack shares its capacity when needed to deliver emergency power.

A word of caution should be given about excessive battery voltage. Many control systems have electronic components that are sensitive to overvoltage on their power supply circuits and may fail when subjected to overvoltages during battery charging and especially during equalization. The plant designer should be aware of this and either supply special protection to these circuits or reduce the normal station battery voltage by using fewer than 60 cells in the station battery stack. A battery made up of 58 cells with a fully charged terminal voltage of 119.96 V may be acceptable as a station battery.

Large dc motor-operated valves, pulsing dc motor-operated damper motor operators, dc governor motor circuits, inverter circuits, or other devices that use across-the-line starting contactors may also cause very large inrush currents and excessive "noise" on the battery system. These uneven waveforms and spikes may be seen on an oscilloscope trace when the scope is placed across the battery terminals. Many electronic circuits are sensitive to harmonics and electronic noise and may cause sporadic misoperation. Sometimes these spikes can be eliminated by placing a local reversed biased diode across the terminals of the offending source. However, in order to avoid these kinds of problems with sensitive electronic control systems, the station designer may elect to have a separate inverter battery to supply power to the station instruments. The station battery then supplies power to the electrical switchgear, substation circuit breakers, dc oil pumps, and other heavy-duty dc power requirements.

17.8 PROTECTIVE RELAYING

Protective relays are expensive online computer devices that require careful calibration, special handling, and controlled environments. At the same time, they are completely unnecessary for the normal operation of a power generation plant, substation, transmission or distribution line, or any other part of an electric power system.

However, if any part of the power system should experience abnormal operation or a short circuit, the entire power system could collapse—with enormous consequences—if it were not for the protective relays. These protective devices continuously monitor, detect, and provide the logic to cause prompt removal from service of any element of the power system if it suffers a fault or starts to operate abnormally in a manner that might allow damage or otherwise interfere with the effective operation of the rest of the system. The relaying equipment is aided in this task by circuit breakers that are capable of disconnecting the faulty element when required to do so by the relaying equipment. The relaying equipment supplies the intelligence (brains), and the circuit breakers provide the action element (brawn, or muscle).

Many types of protective relays are in use throughout the industry. All types require current and potential information from the power system they are trying to protect. This is accomplished by the use of current and potential transformers. The most common types of protective relays are overcurrent, differential, and impedance.

Figure 17-32 shows a typical terminal of relaying using solid-state protective relays.

ANSI standards list various numbers that are used throughout the power industry to designate various types of relays. These numbers are used in this section where applicable.

17.8.1 Overcurrent Relays

As the name implies, overcurrent relays measure the current in a circuit and send a signal or contact output to the proper circuit breaker whenever the current value in the circuit being monitored gets "over" a preset value.

An ordinary fuse is a time overcurrent device which, when properly applied, provides much the same protection as a relay and circuit breaker. Fuses operate only one time before they must be replaced, whereas a relay and circuit breaker can provide multiple protective operations on the same circuit. However, fuses are sometimes the only devices that are capable of providing protection on circuits having very high available fault currents.

Overcurrent relays may be time overcurrent type (Device Number 51), instantaneous type (Device 50), or a combination of both (50/51).

An instantaneous overcurrent relay is one which operates with no "intentional" time delay. This is typically a simple solenoid which picks up a magnetic plunger when the current through the coil is large enough to overcome the tension on the spring used to hold up the solenoid. The spring holds the

Fig. 17-32. Typical protective relay panels. (From City of Vero Beach. Used with permission.)

plunger out of the solenoid when the relay has no current flowing through it.

A time overcurrent relay operates faster on large currents than it does on small currents. These devices were originally built as induction disk devices that operated similar to a watt-hour meter whose disk had limited travel, and the shaft operated contacts that were used in the "trip circuit" of a circuit breaker. The higher the current, the faster the induction disk moved. Thus, the time overcurrent relay has an "inverse time" characteristic. Over the years, a number of shapes of inverse time curves have been adopted by the industry which approximately fit some specific purpose, such as a motor starting characteristic or a distribution fuse melting curve. Although mechanical induction disk time overcurrent relays continue to be used today, many time-overcurrent relay devices use electronic circuits to create the same characteristic shape inverse time curves.

Figure 17-33 shows typical inverse time–overcurrent relay curves. Note that the basic shape of the curve is fixed for a given relay type. However, the settings of the relay can cause the curve to shift right or left by changing the "tap" setting. The curve may be moved up and down by changing the "time" setting. A larger "tap" setting causes more current to be required to "pick up" the relay. "Pick up" is the terminology relay application engineers use to define the current required to just balance the opening spring on the induction disk relay. A current just larger than the pickup value would start the disk rotating toward the trip position (although very slowly). The higher the current, the faster the disk travels. The "Time" dial setting determines how far the disk must travel and thus the time required for the relay to trip at a given current value.

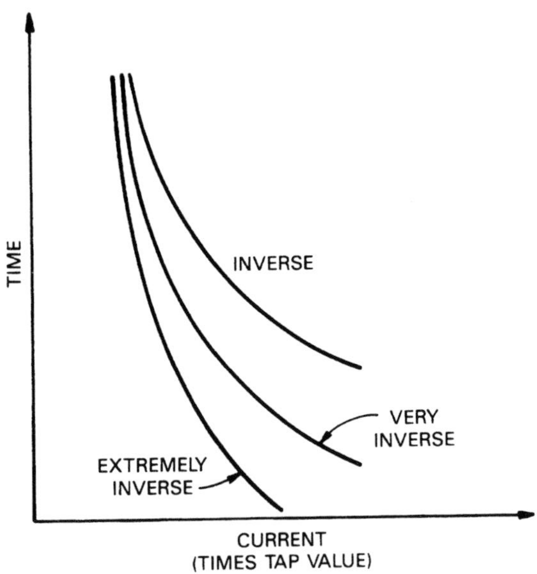

Fig. 17-33. Typical time overcurrent relay curves.

Figure 17-34 shows how overcurrent relays can be used to protect a large motor.

The current transformers in each phase of the power leads feeding the motor provide the overcurrent relay with a cur-

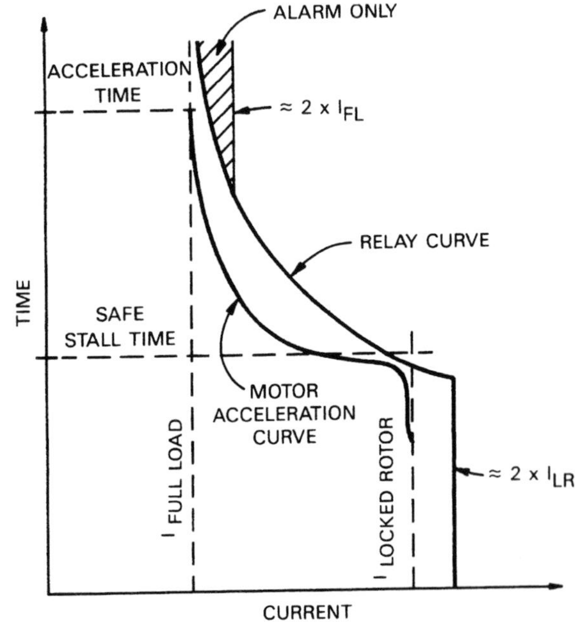

Fig. 17-34. 50/50/51 Motor overload protective relaying.

rent signal which is proportional to the motor line current. Typically, the turns ratio of the current transformer is selected so that the relay secondary current is about 5 amperes or less at full load. The time–overcurrent element (Device 51) is designed to have a time–current shape which approaches the motor acceleration curve from start to full load. The curve is set slightly above this normal motor startup curve so that if the motor has a normal start, no relay action will occur.

From just slightly higher than full load to approximately 150% to 200% full load current, the instantaneous element in series with the time–overcurrent element provides an alarm contact to alert the station operator to the overload condition so that the plant operator can take action on the driven equipment, if he deems it necessary, to reduce the motor overload before the motor or its driven process are damaged. Continuous current above this overload level will eventually cause the high dropout instantaneous element to allow the overcurrent element to send a signal to trip the motor supply circuit breaker. The higher the overload condition, the faster the trip signal occurs. If the motor current is larger than the motor's locked rotor current value, the parallel high current instantaneous element will trip the motor circuit breaker with no intentional delay. The logic for the high value instantaneous element action is that the motor or its leads are experiencing a complete short circuit and that nothing can be gained from delaying a trip, since the motor or its supply cables are probably on fire and will not operate anyway.

17.8.2 Differential Relays

Differential relays (Device 87) make use of Kirchhoff's current law of electrical circuit theory. That is, the algebraic sum of the currents entering a device must be zero. Another way of describing this type of relaying is to say that, "Whatever goes into the circuit must also come out at the proper location." Otherwise, there is a "leak," which in electrical engineering terms is called a fault.

A simple differential circuit monitoring one phase of a generator (or motor) is shown in Fig. 17-35.

Note that if a fault occurs inside the protected zone, the currents entering the first current transformer will not be equal to the current leaving the second transformer. Since Kirchhoff's current law must be satisfied, the difference between the two currents will flow through the relay operating coil and cause the relay to "trip." This signal can be used to operate a circuit breaker to clear the fault.

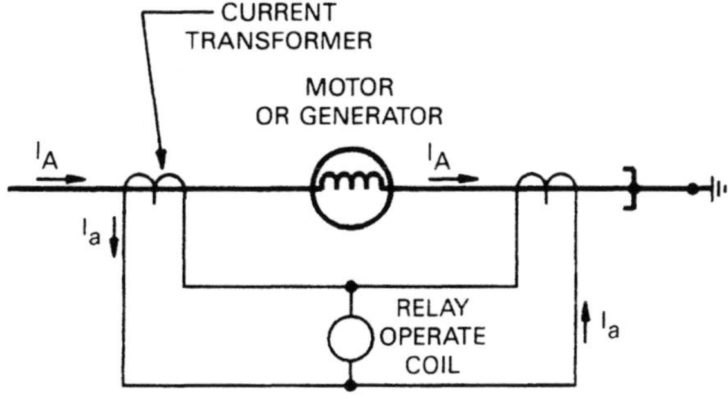

Fig. 17-35. Current differential relaying.

This type of relaying can also be used to protect a transformer, even though the actual current value on the high side of the transformer is different from the magnitude of the current on the low side. Since this difference in normal current value is inversely proportional to the voltage turns ratio of the transformer, the current transformers supplying the differential relay need only be adjusted to provide a current value to the relay which compensates for the power transformer turns ratio. If the power transformer is connected delta–wye, the current transformers should be connected wye–delta to compensate for the phase shift. The current connections for a standard delta–wye transformer are shown in Fig. 17-36.

Ideally, an entire power system could be protected by current differential relaying. All elements of the system could be divided into zones and each zone would have its own set of differential relays that would monitor the currents entering and leaving its zone. If at any time the algebraic sum did not add up to zero, the relay would act to remove only the minimum number of circuit breakers to isolate the fault. Such a "zoned" power system is shown in Fig. 17-37, with each zone overlapping the adjacent zone so that no part of the power system is left unprotected.

17.8.3 Impedance Relays

Impedance relays (Device 21) are devices which continuously monitor the ratio of the voltage of a circuit to the current in the circuit [defined in Eq. (17-1) as impedance Z] and trip the supply feeders if this ratio gets below a predetermined value. This type of relay is also called a "distance relay," since the impedance of a circuit (Z) is also a function of its physical length (or distance).

Transmission lines are particularly suited to this type of relaying, since the length of the transmission line is usually known quite accurately, and therefore its impedance can be calculated. If the impedance of the circuit becomes less than that which is attributable only to the length of the line, the relay logic reasons that the circuit must be faulted (be experiencing a short circuit), and therefore the relay acts to remove the line from service (trip).

Impedance relays are usually "directional." That is, the relay monitors a line and supplies information to trip its appropriate circuit breaker when power is flowing in a certain direction only. This characteristic is very useful when several lines emanating from a substation bus will be monitored individually with distance relaying. Figure 17-38 illustrates the advantage of directional impedance relaying when only one of the circuit breakers trips for a fault, even though the other impedance relays are also monitoring voltage-to-current ratios, which could cause them to trip if they were not directional.

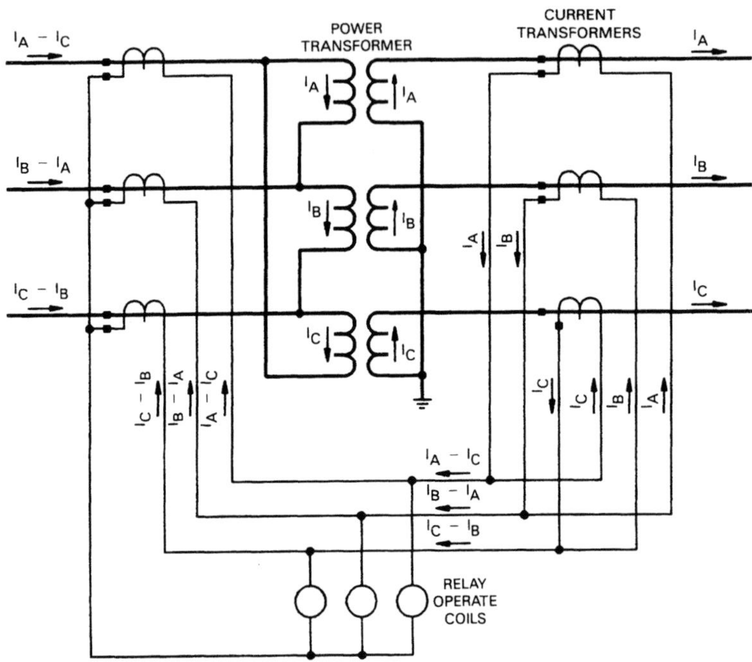

Fig. 17-36. Transformer current differential relaying.

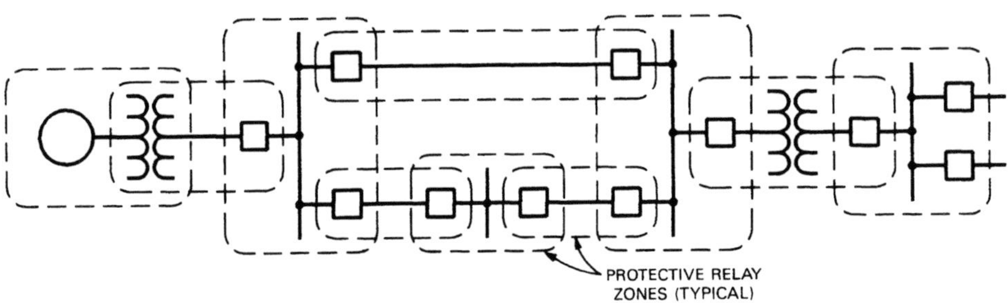

Fig. 17-37. Zoned differential relaying as applied to a power system.

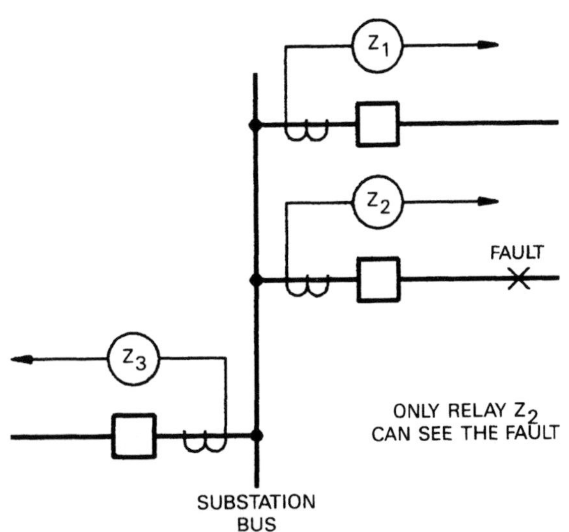

Fig. 17-38. Directional impedance relaying at a substation bus.

17.8.4 Generator Protection

In reality, many types of relays are used throughout a power system to protect the generators, motors, transmission lines, substations, and all the various elements of the system. An example of the devices that would normally be used in the protection of a synchronous generator are shown in Fig. 17-39.

The ANSI device numbers are used in the diagram. The general description of the relays and their device numbers are described below:

- *(87G) Generator Differential Relay.* This relay is the primary generator protection. It detects internal three-phase and phase-to-phase faults. This is a nondirectional current device.
- *(87T) Transformer Differential Relay.* This relay detects three-phase and phase-to-phase faults in the low-voltage delta transformer windings, and three-phase, phase-to-phase, and phase-to-ground faults in the wye high side winding of the generator stepup transformer. This relay should be equipped

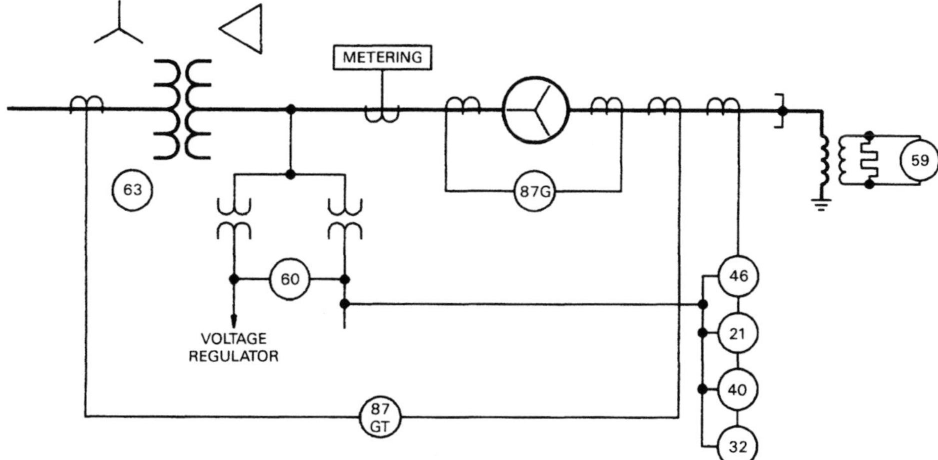

Fig. 17-39. Synchronous generator typical protective relaying.

with harmonic restraint circuits to avoid tripping on generator no-load magnetizing current.

• *(87GT) Generator and Transformer Differential Relay.* This relay provides backup protection to the generator differential (87G) and the generator stepup transformer (87T). It should have some harmonic restraint to prevent tripping on initial transformer energization. This relay also provides a second level of protection to the generator and primary protection of the isolated phase bus duct connecting the generator to its unit stepup transformer.

• *(59) Neutral Ground Relay.* This relay senses the voltage which is developed across the generator neutral grounding transformer secondary resistor to detect ground faults in the generator, its isolated phase bus duct, the generator transformer low-voltage windings, the high-voltage windings of the main auxiliary transformer, and the surge protection and potential transformer equipment. It is not necessary to trip a generator for the occurrence of a single ground on any of these devices. The relay can merely indicate an alarm condition. However, the consequences of a second ground occurring while the first fault is still present would be quite dramatic. Therefore, most utilities choose to initiate a trip of the generator upon detection of any ground fault by the neutral ground relay.

• *(46) Negative Sequence Relay.* This sensitive relay protects against sustained operation of the generator with unbalanced line currents or open phases. Generators have an $I_2{}^2t$ limit which is determined by the manufacturer. I_2 is the negative sequence component current. Its limit is a function of the effective cooling capability of the end turns on the generator stator winding.

• *(21) Distance Relay.* This impedance relay provides backup protection for the generator and the power system from damage caused by three-phase, phase-to-phase, or transmission system phase-to-ground faults which are not cleared by the transmission line primary relays. A separate timer is normally installed to provide the necessary delays required to coordinate with the primary relays, thereby allowing the proper breaker to clear the fault.

This relay is sometimes used to initiate a trip of the unit without transferring the auxiliary electric supply to the reserve auxiliary transformer. The theory for not transferring the auxiliary electric system is that if the 21 relay requires a trip of the generator because of a system disturbance, the entire power system external to the generating station is in serious trouble, and therefore probably cannot sustain the load of the generating plant's auxiliary electric system.

Many plants use this relay to allow the generating unit to trip off of the power system and still maintain its own auxiliaries for a short time by a bootstrap operation or using the residual heat from its own boiler without additional firing. In one test operation of a large 900-MW coal-fueled station, the station was able to carry its own auxiliaries for a period of over an hour on boiler residual heat, following a full load rejection and subsequent boiler trip.

• *(40) Loss-of-Field Relay.* If a generator rotor loses its magnetizing field current while the turbine is driving it at heavy load, the generator will lose synchronism and the turbine will slip a pole. The loss of field relay is designed to detect this condition and protect the generator from thermal damage and the power system itself from instability due to loss of excitation in the generator. A timer is normally used to override a stable power system swing which may appear to the relay as a partial loss of field.

• *(63GT) Sudden Pressure Relay.* This is a mechanical differential pressure device that detects sudden building of pressure in a transformer tank caused by electrical arcing, which vaporizes oil. The relay provides sensitive detection of power level faults within the transformer by sensing this abnormal gas pressure condition.

• *(32) Reverse Power Relay.* The reverse power relay is similar to a watt-hour meter and provides logic action to trip when power is flowing into the generator. This relay provides overspeed protection to the turbine. Reverse power is the safe way to remove a turbine generator from the line. Proof of reverse power into the generator means the turbine will not overspeed when the generator is disconnected from the power system.

• *(60) Voltage Balance Relay.* The voltage balance relay monitors the voltage from both sets of the generator potential transformers. One set of potentials serves the protective relays, and the other set normally serves the turbine controls such as the voltage regulator, turbine stress evaluator, and electrohydraulic

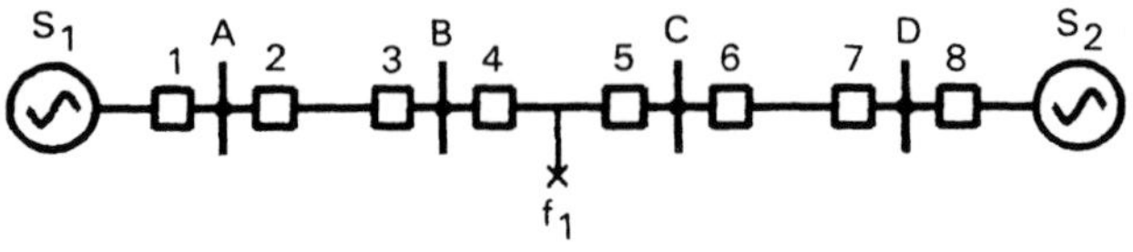

Fig. 17-40. Transmission line segments.

governor. If both sets of potential transformers are operating properly, all controls are in the normal position. If one set of potential transformers has blown fuses so that its voltages are zero, while the other set is providing voltage, the voltage balance relay detects this abnormal condition and disables the devices served from the dead potential bus to avoid misoperation of the relays and control functions. Loss of one side potential causes the voltage balance relay to disable the protective relays served by that potential. Loss of the other potential initiates a transfer from automatic to manual voltage regulator control.

- *(59/81) Volts-per-Hertz Relay.* The excitation system should be monitored by at least one volts-per-hertz relay to prevent the generator transformer and main auxiliary transformer from overexcitation at low turbine speeds before synchronizing the unit to the power system.

17.8.5 Transmission Line Protection

Transmission lines can also be protected with the same overcurrent, differential, or impedance type of relays that protect generators, motors, and transformers.

17.8.5.1 Line Protection with Overcurrent Relays. If a series of transmission lines between two power sources, S_1 and S_2, is represented as shown in Fig. 17-40, with Substations A, B, C, and D at intermediate points along the line, each circuit breaker should trip, in the event of a fault, in such a sequence that the minimum number of substations would be without service. Therefore, if the fault were at F_1, Circuit Breakers 4 and 5 should operate before 2 and 7. Likewise, if a fault were between Substations C and D, Breakers 6 and 7 should operate before Breakers 4, 2, and 8.

Normally, the relays for the breakers on each side of the bus should be of the directional overcurrent type, meaning that they should operate for faults away from their respective buses. This avoids having to coordinate adjacent breakers on the same bus.

A set of overcurrent relay curves to protect the line configuration with these busses is shown in Fig. 17-41a and Fig. 17-41b. Notice that Breakers 3, 5, and 6 in Fig. 17-41a do not trip at all since they are directional and do not see the fault at F_1. Also note the different times that Breakers 1, 2, and 4

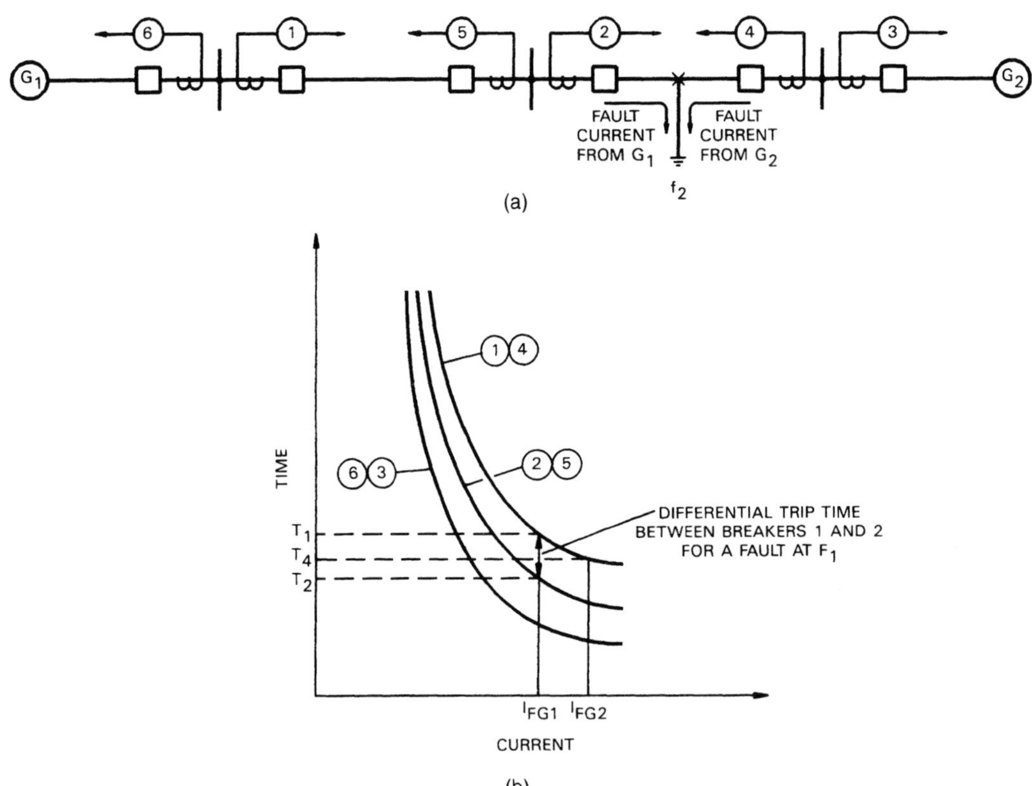

Fig. 17-41. Line segments protected with directional overcurrent relays.

would trip as shown in Fig. 17-41a. Breaker 4 "sees" a different current than Breakers 1 and 2 because Breaker 4 will be activated by the fault current supplied by Generator 2, which is higher than the current supplied by Generator 1, since the fault is closer to Generator 2 than it is to Generator 1.

Since electromechanical overcurrent relays have inertia in the disk and a definite reset time, the curves cannot be made too close or they may not properly coordinate. A reasonably good rule of thumb is that the curves should be about 0.2 to 0.3 seconds apart at the highest fault level the breakers will normally "see." While solid-state digital relays can be made to overcome the reset time and over-travel problem, 0.2 to 0.3 seconds is still a reasonable coordinating time. The coordinating time is shown in Fig. 17-41 as the difference between the trip times T_1 and T_2 for the fault at F_1.

The above scheme works fairly well for two or three lines; however, if multiple infeed from more than two sources becomes possible, the coordination problem becomes very complicated. In addition, time overcurrent relays do take a definite time to clear the fault, and this may be avoided with other types of relays.

17.8.5.2 Line Protection with Impedance Relays. It has already been mentioned that impedance relays can be used to protect a transmission line, since the impedance of a line can be calculated and the relay can be made to trip if the measured impedance becomes less than that of the line alone. In Fig. 17-42, Z_1, Z_2, and Z_3 represent three impedance relays located at Substation A which have three different settings (zones) for the impedances at which they pick up (will operate).

If the fault was at F_1 between Substation B and C, both Relays Z_2 and Z_3 would "see" the fault, while Relay Z_1 would not. This logic then could be made to trip only Breaker 4, with no intentional time delay, if a communication path was made available between Substations A and B.

Similar impedance relays would be installed at Substation D and would look in the opposite direction toward source S_1.

If the fault was between Substations C and D such that only Relay Zone Z_3 could "see" the fault, the logic could be made to trip Breaker 6 only. Similarly, if all three zones picked up on the fault, only Breaker 2 would be the proper breaker to open.

Notice that if a fault was just outside the reach of Zone Z_1 but not quite in the line segment between Breakers 4 and 5, it would create a problem because the above logic would still trip Breaker 4, although this would be the incorrect breaker to trip in that case.

17.8.5.3 Directional Comparison Impedance Protection. If a communications link is established between all substations, impedance relaying can be used to develop logic to determine the location of any fault just by knowing in which direction all relays "look" to "see" it. In Fig. 17-43, the directional impedances Zones Z_1, Z_2, and Z_3 are shown for Breakers 2, 4, and 6, respectively. Zones Z_4, Z_5, and Z_6 are for Breakers 7, 5, and 3.

If the fault is at F_1, both Zones Z_2 and Z_5 will "see" it and are the only two zones on adjacent breakers that can "see" the fault at the same time. Therefore, Breakers 4 and 5 should operate to clear the fault. Breakers 2 and 7 also "see" the fault, but because Breakers 3 and 6 do not "see" it in their forward direction, the line segments between Substations A and B and also between Substations C and D will not be opened.

This scheme is called "directional comparison, overreaching, transfer trip" and is reliable in that it will always clear a fault. Its disadvantage is that it requires a highly secure communications channel to transmit the necessary intelligence between relay terminals to prevent unnecessary tripping of breakers.

A slight modification can be made to the preceding scheme so that the impedance zones do not reach past their opposite end terminals but always overlap. In that case, when a relay "sees" a fault, it trips its own circuit breaker immediately and also transmits a tripping signal to the opposite end terminal. This scheme is referred to as "directional comparison, direct underreaching, transfer trip." This scheme permits high-speed tripping but also requires a highly secure communications channel to ensure that faults near the ends of the line are cleared by the opposite end breakers.

17.8.5.4 Blocking. This type of relaying is one of the oldest and most widely used schemes in power line protec-

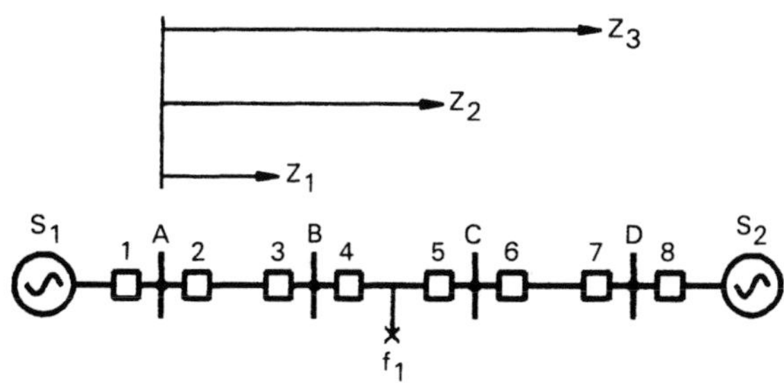

Fig. 17-42. Line segments protected with impedance relays.

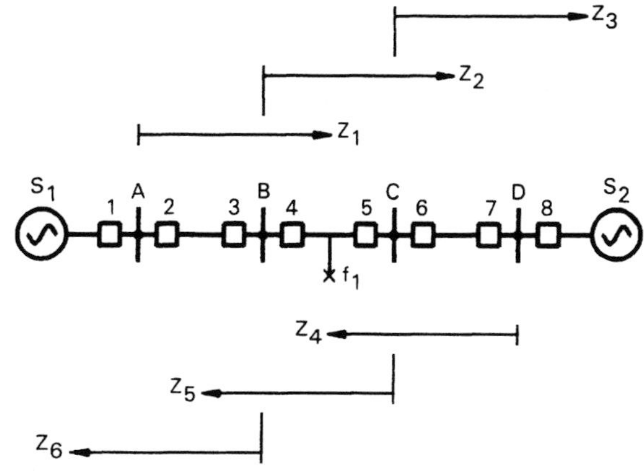

Fig. 17-43. Line segments protected with directional comparison impedance relays.

tion. It does not depend on the communications channel to transmit a tripping signal, and thus a faulted communications link will not cause a failure to clear a fault. The system monitors the power flow into both ends of a line, and tripping is permitted at all times by the impedance relays. If a fault occurs in the middle of a line, a tripping signal is not required to be transmitted from the other end of the line through the faulted portion of the line. If either set of relays at either end of the line detects a fault in its tripping direction, that relay set is permitted to trip its circuit breaker unless it gets a blocking signal. If a set of relays detects a fault in the opposite direction from its tripping direction, that relay set transmits a blocking signal to the set of relays on the opposite end of the protected line to prevent false tripping of the line.

Thus, a blocking scheme of relays does not require the communications link to transmit a tripping signal through a possible faulted communication path.

17.8.5.5 Unblocking. The unblocking scheme also uses directional comparison impedance relays. However, this scheme normally transmits a continuous blocking signal to its opposite end set of relays. In the event of an internal fault, each relay set "unblocks" the opposite end relays and permits a high-speed trip. In this case, not only do the relays unblock each other, but the internal fault itself may interrupt the blocking signal as well, depending on the communications method, thus permitting a trip.

17.8.6 Relaying Communications Methods

The principle of directional comparison is fairly simple and is a very positive method of high-speed fault detection. The problem lies in attempting to reliably communicate the logic between substation terminals at both ends of a line simultaneously during the actual fault occurrence. Four methods used to facilitate this communication are wire pilots, power line carriers, fiber optic communications, and radio communications.

17.8.6.1 Wire Pilots. If the line terminals are not too far apart, a hard metallic wire communication circuit between them may permit high-speed automatic comparison of fault directions. This is most commonly used on short lines and provides a good method of utilizing simple current differential relaying. The wire pilot can also be used as a communications link for directional comparison impedance relaying. However, if the line sections are several hundred miles apart, wire communication channels may be both expensive and unreliable. In addition, ground potential rise differences between terminals must also be considered. A fault at one station may cause that terminal's relaying equipment to have several thousand volts of potential difference between the equipment at the opposite end. Heavy-duty insulating transformers and remote grounds for the pilot wire equipment may be required to prevent direct metallic connection between remote ends of the relaying equipment.

17.8.6.2 Power Line Carrier. Since the power transmission lines physically go between substations themselves, it is possible to use this metallic link as a communications line also. Figure 17-44 shows a line between substations with a wave trap at each end.

The wave trap is nothing more than a parallel tuned inductor–capacitor "tank" circuit made to be resonant at the desired communication frequency. A coupling capacitor installed on the line side of each wave trap allows the communications signal to be superimposed onto the power conductor. If the frequency of the communications signal is 250 kilohertz (kHz), the tuned wave traps on each end will effectively block the signal from passing through them onto the substation bus and into transformers and other transmission lines.

A typical wave trap installed in a line takeoff tower in a substation is shown in Fig. 17-45.

Several lines between substations can have their own communications frequency and therefore do not interfere with each other. The impedance of the parallel tank circuit is almost nil at the power line frequency of 60 Hz and, therefore, very little power loss is experienced in the wave trap. On the other hand, the impedance of the coupling capacitor at 60 Hz is very high, so the power frequency current does not cause a phase-to-ground fault. At the same time, the impedance of the coupling capacitor at the communications frequency (250 kHz in this example) is very low; therefore, it does not take a very powerful transmitter to place the signal on the overhead power conductor.

One disadvantage of power line carrier applications is that a fault inside the protected zone may also short circuit the communications channel. If a communications channel itself becomes faulted, the impedance relays may operate incorrectly during a fault or even operate when there is not a

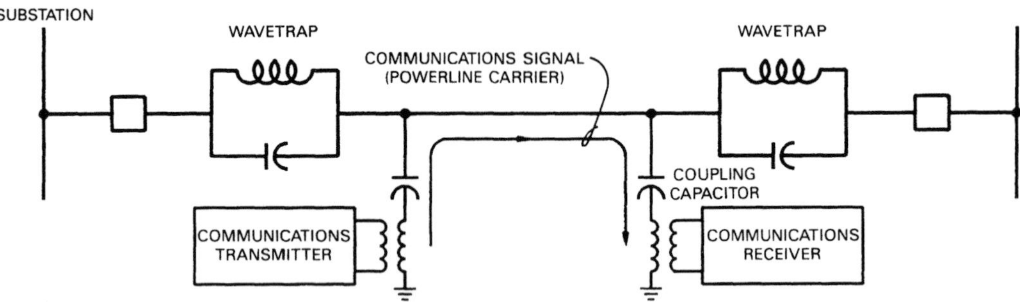

Fig. 17-44. Power line carrier communications.

Fig. 17-45. Typical wave trap installation in a line takeoff tower. (From City of Vero Beach. Used with permission.)

fault. The blocking and unblocking schemes discussed previously may use power line carrier communications channels. The relay applications engineer must try to account for this problem during system design.

17.8.6.3 Fiberoptic Communications. Recent technological advances have produced communications links using light waves in glass fibers which are encased in a plastic tube. The fiberoptic "cable" can be installed underground or lashed to overhead messenger cables that are attached to the power line structures. The fiberoptic link has several advantages over other communications media in that ground potential difference between terminals is not a problem since there is no metallic path. In addition, each fiber is capable of a very large bandwidth; thus, each fiber can have many channels. It has become fairly common on new transmission lines to install a fiberoptic cable inside the shield wires on overhead lines.

17.8.6.4 Radio Communications. A communications link between terminals may use almost any medium, including the air waves. A radio link can be a good communications method whether it is high-frequency AM, FM, or microwave. Amplitude Modulation (AM) suffers from static interference and is subject to fading. Frequency Modulation (FM) does not have the static interference problem, but its range is limited and it is also subject to fading. High-frequency radio links are not often used for relaying communications. Microwave channels are used frequently; however, they require direct line-of-sight communication paths, and implementation usually requires a feasibility and radio mapping study to ensure reliable communication paths.

17.9 PRIMARY POWER SUPPLY SYSTEM

The primary power supply system is defined here as that external electrical system to which the power plant is connected. The voltage of this system, to a large degree, is determined entirely by the interconnected utilities and previous system-wide or area-wide load flow studies that support the utility's decision to build the power plant in the first place. In the case of an independent power producer (IPP) or a cogenerator, the choice of system voltage still rests with the utility to which the power plant will supply energy. The electrical designer of the power plant must interconnect the power plant to the utility grid in such a manner that operation of the new power plant contributes to the reliability and safety of the interconnected system. Each utility has its own unique system of protective relaying practices and standard substation configurations with which the electrical designer should become familiar with in order to intelligently make design decisions that are acceptable to all participants.

17.9.1 Switchyard Configurations

The terms "substation" and "switchyard" are used interchangeably. Strictly speaking, a "switchyard" is the location where numerous transmission lines of the same voltage are interconnected. A "substation" is the location of one or more transformers which change the power system voltage to another level of the interconnected system. The configurations discussed in the sections that follow are some of the ones most commonly encountered by power plant designers. They may be used with or without transformers and, indeed, more than one configuration may be connected to each other at the same location with one or more transformers.

17.9.1.1 Radial Bus. The radial substation is one of the simplest configurations. As shown in the one-line diagram in Fig. 17-46, it consists of a single bus to which all transmission lines and generators are connected. Although operationally very simple, its biggest disadvantage is that to perform maintenance on one of the circuit breakers, the transmission line or generator that the circuit breaker is serving must also be out of service.

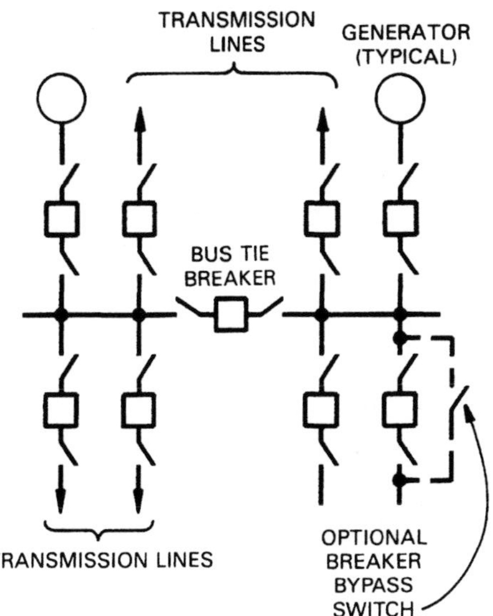

Fig. 17-46. Radial substation.

Circuit breakers are the devices used to switch the transmission lines, generators, and transformers in the substations. These electrical devices come in various sizes and shapes depending on the rating, voltage class, and insulating medium. Figure 17-47 shows a 138-kV oil-insulated circuit breaker. Figure 17-48 shows a 69-kV gas-insulated circuit breaker.

Occasionally, a breaker bypass switch is designed into a radial substation to permit the line or generator to remain in service during breaker maintenance. The obvious disadvantage of this arrangement is that the protective relaying and circuit breaker is also out of service, and in the event of a fault on the unprotected line or generator, serious system damage could result. This bypass switch is rarely used except in a distribution system for reclosers on distribution feeders where another circuit breaker or fuse device is available to provide a backup to the bypassed recloser.

Each circuit breaker in Fig. 17-46 is shown with isolation breaker disconnect switches that can be opened to permit maintenance on a circuit breaker. The configuration may also include one or more bus tie breakers which break up the substation into two or more mini substations. Some of the reasons for including bus tie breakers are to isolate a portion of the substation for future expansion, facilitate maintenance, reduce the available fault current during certain operating conditions, and provide isolation to prevent loss of the entire substation bus in the event of a breaker failure situation.

Radial substations are most often used in distribution substations at voltages of 69 kV and below.

Fig. 17-47. 138-kV oil-insulated circuit breaker. (From City of Vero Beach. Used with permission.)

Fig. 17-48. 69-kV gas-insulated circuit breaker. (From City of Vero Beach. Used with permission.)

The ampacity of the main bus must be sized to handle any scenario of load flows of the lines connected to it.

17.9.1.2 Double Bus. The double bus configuration is shown in Fig. 17-49. As the name implies, there are two main buses. Each transmission line, generator, or transformer is connected to each of the buses with a circuit breaker. This configuration solves the breaker maintenance problem, but it is approximately twice as expensive as the radial configuration because it has twice as much hardware and occupies more space.

17.9.1.3 Ring Bus. The ring bus configuration shown in Fig. 17-50 has the same number of circuit breakers as a radial bus design having the same number of lines. It also takes care of the breaker maintenance problem in that any single circuit breaker may be taken out of service without loss of any of the generators or transmission lines that are connected to it. Each transmission line or generator shares the adjacent circuit breaker with the next transmission line or generator in the ring. In this configuration, each line or generator should have a separate disconnect switch to allow the ring to be reclosed after a transmission line is taken out of service.

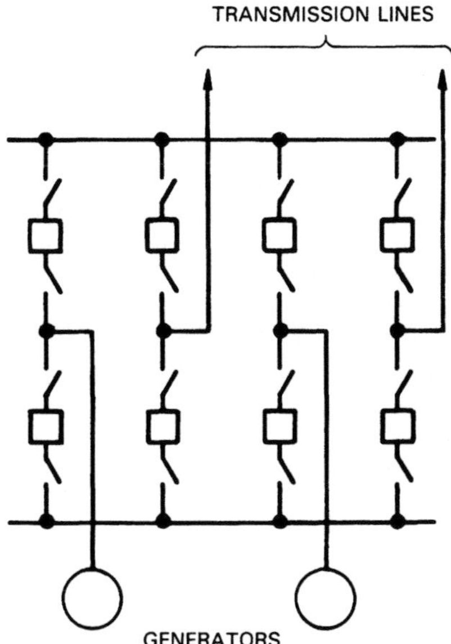

Fig. 17-49. Double bus configuration.

Without the separate line disconnect switch, a fault on another transmission line in the ring could cause a loss of several lines if the ring was not complete. The operation of the line disconnect switch must be interlocked with the adjacent circuit breakers so that the switch cannot be operated (either opened or closed) unless both adjacent circuit breakers are open.

The ring bus configuration is widely used. Its major disadvantage is that it is fairly complicated to add another line or generator at a later date. It is not an easy arrangement to expand after its initial design. In addition, the bus must be sized to handle all of the current for all the lines connected to it. For this reason, it is desirable to alternate sources on the bus—that is, to alternate generators with transmission lines or transformers around the ring to try to distribute the current in the bus.

*17.9.1.4 **Main and Transfer.*** The main and transfer configuration is an attempt to gain the benefits of a double bus configuration without the expense of the extra breakers. This configuration is shown in Fig. 17-51.

This configuration takes only one additional breaker over the radial design but offers serviceability and breaker maintenance outages without sacrificing breaker protection on the line being serviced. It does require one additional switch for each line position. The transfer breaker serves as the line breaker when one of the main breakers is out of service for maintenance. One of its disadvantage is that the protective relaying for a line must also be switched to the transfer breaker during maintenance.

*17.9.1.5 **Breaker-and-One-Half.*** This configuration is one of the most popular at voltages above 161 kV. It uses three breakers for two lines, as shown in Fig. 17-52, thus the name breaker-and-one-half. Each of the bus breakers is dedicated to the line or generator to which it is connected; however, the center breaker in each lineup is shared by the line on each side of it.

This configuration is easy to expand if the original designer has allowed spare ampacity in each of the main buses and if the short-circuit capacity of the breakers and switches is adequate. Occasionally, a bus tie breaker is included in the main buses' design to break up the switchyard during certain operating conditions or breaker failure events. The line disconnect switch may be omitted if the utility does not wish to reclose the bus ties in the event of a maintenance outage of a transmission line or generator. It is most common to include the disconnect switch for a generator position. Again, if the line disconnect switch is used, it must be interlocked with the adjacent breakers so that it may not be operated either opened or closed unless the adjacent breakers are open. It is common practice to position the switch so that the blade of the switch will be deenergized when the switch is open. Of course, this is always the case if a double break switch is used.

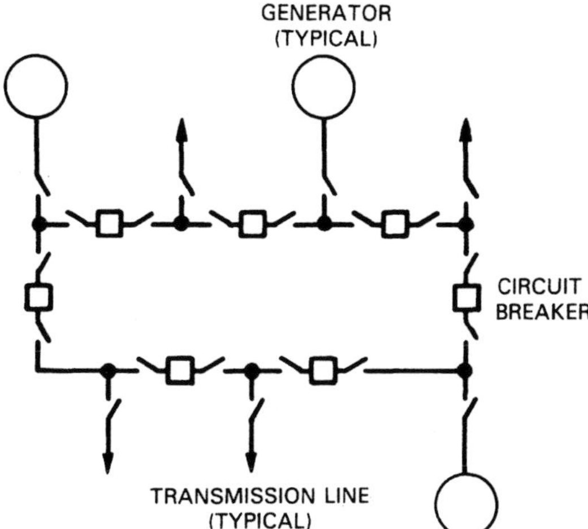

Fig. 17-50. Ring bus configuration.

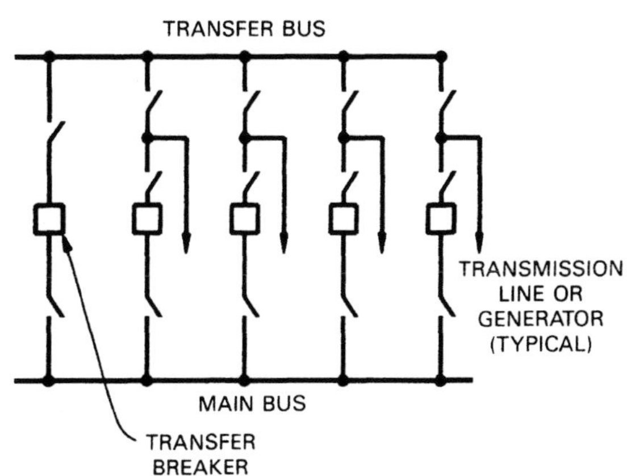

Fig. 17-51. Main and transfer configuration.

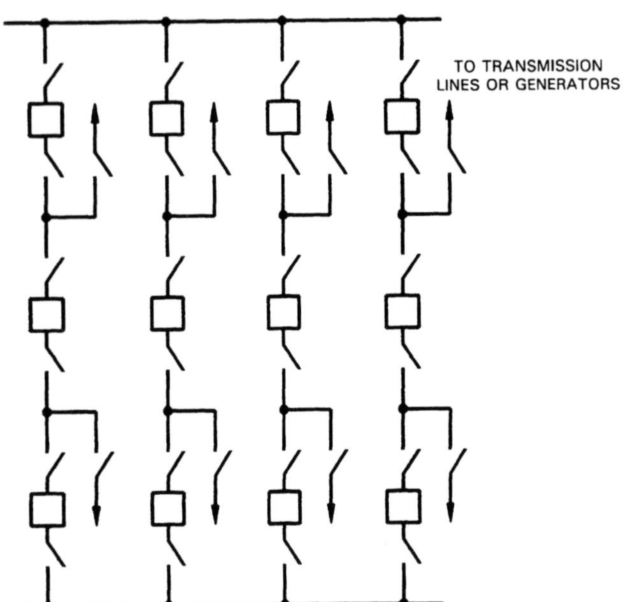

Fig. 17-52. Breaker-and-one-half configuration.

17.9.2 Substation Clearances and Spacings

Table 17-7 lists the basic electrical clearances as they appear in the standards of the National Electrical Manufacturers Association (NEMA), American National Standards Institute (ANSI), and Institute of Electrical and Electronic Engineers (IEEE), as well as the National Electrical Safety Code (NESC).

Table 17-7 may be used to assist the switchyard designer in the selection of the proper spacings for various applications. The values listed are mandatory minimums. Design practices of utilities vary in different areas of the world. It is normal practice to design for somewhat greater clearances than the absolute minimums recommended by the various standards, and there may occasionally be an exception to the recommendations. The appropriate standards should be consulted before the design is finalized to ensure conformance to the latest revisions of the codes and standards. It is good engineering practice, however, to observe these minimum standards both from a human safety and equipment rating standpoint.

Typical design clearances shown in Table 17-8 may be used for preparing switchyard space allocation arrangements. The typical electrical clearances for switchyard equipment shown in Table 17-8 are for rigid bus, low-profile designs. The locations of the clearances listed in Table 17-8 are shown in Fig. 17-53. These clearances may be used as a guide for preparing switchyard layouts. The switchyard area space requirements are a serious consideration in the arrangement of a generating facility and must be given proper consideration.

The spacings listed in Table 17-8 may be modified to meet specific client preferences or unusual circumstances. However, the minimum dimensions listed in Table 17-7 and recommended by ANSI standards should not be modified.

If the substation is to be built at elevations above 3,300 ft, the clearances and spacings must be increased from those listed in Tables 17-7 and 17-8. As the altitude increases, the air surrounding the equipment ionizes more readily. Equipment that depends upon air for its insulating and cooling medium has a higher temperature rise and a lower dielectric strength when operated at higher altitudes.

Table 17-9 gives altitude correction factors that should be used to adjust the spacings and clearances listed in Tables 17-7 and 17-8. The correction factors may be used to derate

Table 17-7. Mandatory Minimum Substation Clearances

Nom kV	Max kV	BIL kV	Min Mtl to Mtl	ϕ–ϕ Rec V.B. D.S	ϕ–ϕ Rec S.S.B. D.S.	ϕ–ϕ Rec H.G. D.S.	ϕ–ϕ Min Rigid Parts	ϕ–*G* Rec Rigid Parts	ϕ–*G* Rec O.H. Part	Min Vert Clear to UG Parts	Min Horz Clear to UG
14.4	15.5	110	12″	24″	30″	36″	7″	10″	9′	9′–0″	3′–5″
23	25.8	150	15″	30″	36″	48″	10″	12″	10′	9′–3″	3′–9″
34.5	38	200	18″	36″	48″	60″	13″	15″	10′	9′–6″	4″–0′
46	48.3	250	21″	48″	60″	72″	17″	18″	10′	9′–10″	4′–4″
69	72.5	350	31″	60″	72″	84″	25″	29″	11′	10′–5″	4′–11″
115	121	550	53″	84″	108″	120″	42″	47″	12′	11′–7″	6′–1″
138	145	650	63″	96″	132″	144″	50″	52.5″	13″	12′–2″	6′–8″
161	169	750	72″	108″	156″	168″	58″	61.5″	14′	13′–4″	7′–10″
230	242	900	89″	132″	192″	192″	71″	76″	15′	14′–0″	8′–6″
230	242	1,050	105″	156″	216″	216″	83″	90.5″	16′	—	—
345	362	1,050	—	—	—	—	—	—	—	15′–6″	10′–0″
345	362	1,300	119″	174″	—	240″	104″	105″	18′	17′–2″	11′–8″
500	550	1,550	—	—	—	—	—	—	—	18′–10″	13′–4″
500	550	1,800	—	—	—	—	—	—	—	20′–6″	15′–0″
765	800	2,050	—	—	—	—	—	—	—	—	—
1,100	1,200	—	—	—	—	—	—	—	—	—	—

The values in the above table are accumulated from the following standards: NEMA SG6, ANSI C37.32, ANSI C92.2, and NESC.

Abbreviations: BIL = basic insulation level; Min Mtl to Mtl = minimum metal to metal; ϕ–ϕ Rec V.B.D.S. = phase-to-phase recommended, vertical break disconnect switch; ϕ–ϕ Rec S.S.B.D.S. = phase-to-phase recommended, single side break disconnect switch; ϕ–*G* Rec O.H. = phase-to-ground recommended to overhead; Min Vert (Horz) Clear to UG Parts = minimum vertical (or horizontal) clearance to undergrounded parts.

Table 17-8. Typical Switchyard Spacings

System kV	Max Rated	BIL kV	Horz φ–φ C_L (ft)	Low Bus Height (ft)	High Bus Height (ft)	Dim A (ft)	Dim B (ft)	Dim C (ft)	Dim D (ft)	Dim E (ft)	Dim F (ft)
69	72.5	350	5	14	17.5	10	12	10	10	—	24
115	121	550	7	14	19	12	16	10	10	15	28
138	145	650	8	14	20	14	18	12	12	18	30
161	169	750	9	17	24	15	23	12	12	23	36
230	242	900	11	18	26	17	25	17	17	25	45
230	242	1,050	13	20	29	23	30	21	21	30	55
345	362	1,300	14.5	22	34	30	35	25	25	35	64
500	550	1,550	20	24	40	35	40	30	30	40	72
500	550	1,800	25	25	45	40	45	35	35	45	80
765	765	2,050	30	28	50	45	50	40	40	50	90

the voltage withstand and current ratings, or they may be used as a divisor into the dimensions to increase the spacings to achieve the same rating as the given dimensions at lower altitudes.

17.9.3 Bus Forces and Considerations

The design of the switchyard bus depends not only on the ampacity required of the conductors, but also on the mechanical forces to which the bus may be subjected. These forces include ice and wind loading, seismic considerations, and the

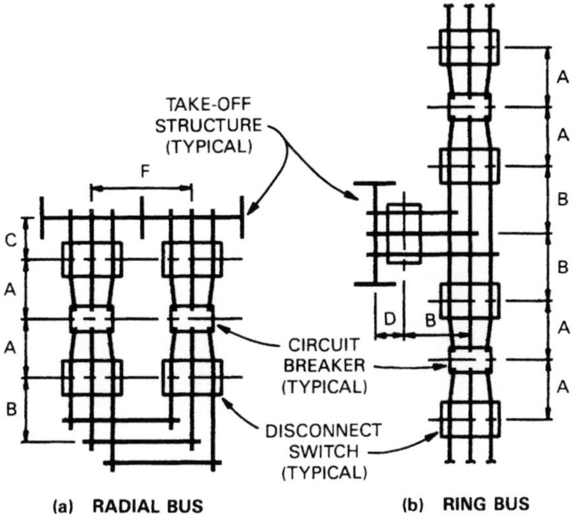

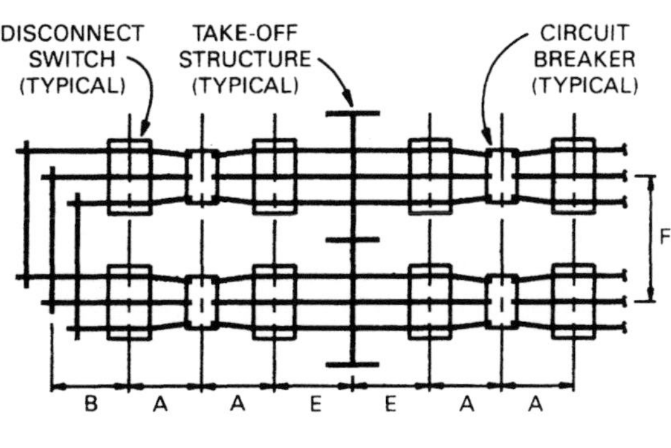

Fig. 17-53. Substation electrical clearances for switchyard equipment.

Table 17-9. Altitude Correction Factors

Feet	Meters	Rated Withstand Voltage	Current Rating[a]	Ambient Temperature[b]
3,300	1,000	1.00	1.00	1.00
4,000	1,200	0.98	0.995	0.992
5,000	1,500	0.95	0.99	0.98
6,000	1,800	0.92	0.985	0.968
7,000	2,100	0.89	0.98	0.956
8,000	2,400	0.86	0.97	0.944
9,000	2,700	0.83	0.965	0.932
10,000	3,000	0.80	0.96	0.92
12,000	3,600	0.75	0.95	0.896
14,000	4,200	0.70	0.935	0.872
16,000	4,800	0.65	0.925	0.848
18,000	5,400	0.61	0.91	0.824
20,000	6,000	0.56	0.90	0.80

[a]For maximum ambient of 40° C for nonenclosed switches and outside the enclosure for enclosed switches.

[b]For operation at continuous current rating.

Source: ANSI 19.

forces developed as a result of high currents during fault conditions. The forces will be transmitted from the bus into the support insulators, and, for rigid bus systems, will determine the cantilever strength requirements of the insulators and the maximum bus support spacing. The equation for the force between two parallel conductors due to an asymmetrical phase-to-phase fault is given below:

$$F_{sc} = \frac{5.4(1.6\sqrt{2}I_{sc})^2}{D = 10^7} \qquad (17\text{-}19)$$

where

F_{sc} = short circuit current force (lb/ft),

I_{sc} = symmetrical RMS short circuit current (amperes), and

D = conductor spacing center-to-center (inches).

This formula has been adopted by various substation standardizing committees who have concluded that the actual offset peak of the asymmetrical wave has a maximum of 1.6 times the symmetrical wave instead of the theoretical value of 2.0.

Other considerations in substation bus design include dampening of vibrations, appearance due to sag in the bus conductors; smoothness of the bus and fittings to avoid corona, especially at voltages above 230 kV; expansion due to temperature variations; and drain holes to prevent accumulation of water in moisture pockets so as to avoid freeze damage in low points of the bus.

17.10 TRANSMISSION LINES

Although the design of long cross-country transmission lines is not normally the responsibility of the power plant or switchyard designer, these lines almost always terminate at a source of generation. Therefore, it is appropriate that certain aspects of transmission line design be discussed here so that proper appreciation is given to the outgoing transmission lines and their relationship to the power plant and switchyard equipment in relation to the overall power system.

Transmission lines are the interconnecting link between the power-generating facilities and the distribution substations that feed electric energy to the utility customers. These large-capacity lines make possible the economics of scale of large generating plants and permit the lowest possible energy costs to the ultimate consumer.

The location of transmission lines is an important consideration in interconnecting large generating plants with urban load centers. Power plants are normally built in remote locations where large spaces are available near sources of water, fuel, and rail transportation.

Transmission lines also require fairly large right-of-way areas. Development of new transmission corridors should always consider such land use requirements as future highway networks, gas and oil lines, sewage and drainage systems, and recreational usage. Such planned use of land areas requires cooperation among utilities, state and local governments, industry, and the public. The planning and building of any major transmission line requires that all ecological, social, and technical aspects be carefully considered. Such considerations as archeological and historical sites, wetlands, agricultural needs, aesthetics, and public acceptance are of equal importance with the technical considerations of transmission line design. Nevertheless, this short section on transmission line design focuses entirely on the technical considerations of transmission lines as they affect power plant and switchyard design.

17.10.1 Load Flow Studies

The need for transmission lines is determined by electrical load flow studies, which are usually made several years in advance to determine the appropriate configuration and timing for the necessary construction of new transmission lines. However, new regulations recently adopted by the US Congress in the Public Utility Regulatory Policy Act (PURPA) have allowed many new sources of generation to be built by cogenerators and independent power producers without regard for local need for power or transmission system line considerations. This has made the job of planning for transmission line facilities very difficult for utility transmission system planners.

Transmission planning studies apply digital computer techniques to solve simultaneous loop equations of Kirchhoff's current and voltage laws for various scenarios of generation and transmission line configurations. Each generator and each load center is connected by an electrical model of various configurations of transmission lines. The loadings of the generators and load centers under varying operating scenarios are studied to determine the appropriate configuration for building the lines. Although analog computers called "Network Analyzers" were formerly used, particularly for small networks, load flow studies today are almost always accomplished by using digital computer simulation.

Since a transmission line consists of one or more conductors per phase stretched overhead for several miles, it can be represented as a distributed resistance and reactance that can be modeled on network circuit analysis programs as a lump sum resistor in series with a reactor. This works reasonably well for short lines, but for long lines, the capacitance should also be represented. Since a transmission line consists of metallic conductors in parallel with the earth (also a conductor), and the line usually has one or more shield wires, the line also has distributed capacitance throughout its length. The value of this capacitance may be quite difficult to determine since the configuration is usually fairly complicated. Most models make an attempt at representing a power transmission line as a pi (π) network with resistance and reactance in series, and with lumped sum capacitance on each end.

Where transmission lines are in parallel, such as on double circuit towers for appreciable distance, the mutual reactance coupling of the lines should also be modeled if accurate studies are to be made. However, mutual coupling is probably of more importance in short-circuit studies prepared for protective relaying coordination than it is for load flow studies.

After the system voltage and power handling requirements have been determined by the load flow studies and models, the transmission line physical and electrical considerations including insulation coordination, corona, conductor size, conductor sag and tensions, structure spacing, structural loading, phase spacing, right-of-way width, shielding, and tower grounding requirements must then be examined.

17.10.2 Overhead Conductors

The current-carrying capacity requirements of a conductor also determine its physical diameter. Most modern transmission lines use aluminum conductors due to their relatively high current-carrying capacity and strength. Aluminum conductor, steel-reinforced (ACSR) is the most common type of conductor used in transmission line construction. It consists of one or more strands of high-strength galvanized steel core surrounded by one or several layers of hard-drawn aluminum strands. The many different stranding combinations make it possible to vary the current-carrying capacity and strength of the conductor to fit the mechanical strength best suited for each application. For example, a long river crossing span would likely require a heavier, stronger steel core than the shorter average spans for the rest of a cross-country line over flat terrain.

The electric power industry has adopted the names of birds as a means of distinguishing various combinations of sizes of ACSR conductors from each other. For example, 795 MCM 26/7 ACSR is known as "Drake" conductor. The "795 MCM" refers to the thousand circular mils (also probably more correctly referred to as kcmil) of cross-sectional area of conductor. The 26/7 means that there are 26 strands of aluminum wrapped around seven strands of galvanized steel. More examples of popular ACSR conductors for high-voltage and extra-high-voltage lines include "Cardinal" (954 MCM 54/7) and "Finch" (1114 MCM 54/19), among several dozen others.

Other popular types of conductors are all-aluminum conductors (AAC), which are given the names of flowers; all-aluminum-alloy conductor (AAAC), using either 5005 alloy or 6201 alloy; and aluminum conductor alloy-reinforced (ACAR). Tables giving the mechanical and electrical characteristics of these and many more conductors are given in reference materials published by aluminum conductor manufacturers, the Electric Power Research Institute, and others.

17.10.3 Conductor Spacing

Conductor phase spacing in transmission lines is determined by a combination of many variables. Many conventions have been developed by various utilities for use within their own systems. Some approximate conductor phase spacing for tower takeoffs may be inferred from the switchyard spacing examples given in Table 17-8, and these may be used as a starting guide for the first span away from the switchyard. However, the first span is usually a reduced tension span, normally very short.

Typical horizontal phase spacings for cross-country lines are shown in Table 17-10. The actual phase spacing of a line depends on the span length, as discussed later. At longer span lengths, the phase spacing must be wider to avoid flashovers because of conductor swings.

Since air is used as the insulator between phases, the minimum distance between conductors is dependent on the insulating value of the air itself under various weather conditions. If a line is to be built at altitudes above 3,300 ft, the phase spacings should be increased as suggested in Table 17-9.

Table 17-10. Typical Transmission Line Conductor Phase Spacing

Voltage Level of Line (kV)	Phase Spacing Range (ft)
69	7–9
138	10–16
230	12–18
345	20–30
500	25–40

Another consideration for conductor phase spacing of a transmission line is that wind blowing on the conductors may cause them to swing and approach each other during normal operating conditions. A frequently used method for determining this is to use the sag-tension analysis to determine a Lissajous Ellipse, which gives a theoretical envelope of how close conductors may approach each other during "galloping." This is a phenomenon that may occur when blowing freezing rain builds up on a conductor in the shape of an airfoil and allows the wind to cause the conductor to rise until the airfoil loses its lift, at which time the conductor falls again. This is an extremely serious condition which can destroy the conductor supporting towers because of the shock load created by the galloping conductor weight.

Another type of vibration is Aeolian Vibration, a high-frequency low-amplitude oscillation generated by low-velocity wind blowing across the conductor, causing it to vibrate like a tuned string. This type of vibration over long periods can cause conductor fatigue and breakage at the supports.

For phase spacing of vertical conductor configurations, designers must consider other conditions, such as ice-covered conductors which may drop their ice loads at different times. Conductors that are closer to the ground have the tendency to drop their ice load first and jump upward toward the upper conductor. Even if this does not cause a phase-to-phase fault, the unloaded conductor could remain in close proximity to the ice-loaded conductor above it for extended periods and could result in a flashover at a later time.

17.10.4 Sag and Tension Relationships

The longer the span, the larger the phase spacing should be, because the sag is greater. Also, if the tension is increased, the sag is less. There is a minimum allowable distance above the earth, roads, fences, railroads, buildings, and other structures. This distance is determined by the requirements of the National Electrical Safety Code (NESC) and is administered by the American National Standards Institute (ANSI Code C2) and the Institute of Electrical and Electronic Engineers (IEEE). Electric utilities in the United States are obligated to follow minimum clearances given by this code.

In general, the minimum distances are determined by the voltage of the overhead conductor, the altitude, and the loading district. The loading district is defined by NESC according to the amount of ice that is likely to build up on conductors (thereby increasing their effective weight and resulting sag). These loading districts are very general terms referred to as "Light" (no ice), "Medium" (up to ½ in.), and "Heavy" (½ in. or more of radial ice).

For a given span length, as the sag is increased, the tension at the support decreases until a point is reached where the tension starts to increase because of the weight of the conductor.

This point occurs when the sag is equal to 0.337 times the span length as given by the formula:

$$S = 1.33R/W \qquad (17\text{-}20)$$

where

R = resultant tension at the support (lb),

W = resultant load on conductor (lb/ft), and

S = span length (ft).

This formula can be used to define the maximum possible span length between supports for a given conductor. This is most useful for large spans such as river crossings, rail yard spans, and so on.

For the relatively short spans in and around power plants and their switchyards, the approximate "parabolic" methods are usually acceptable for determining sags and tensions of conductors. This formula considers that conductors behave according to the geometric shape of a parabola. The formula is as follows:

$$D = \frac{(W \times S^2)}{(8 \times H)} \qquad (17\text{-}21)$$

where

D = sag at center of span (ft),

S = span length (ft),

W = weight of conductor (lb/ft), and

H = horizontal tension (lb).

A more exact formula using hyperbolic trigonometry formulas should be used for spans 1,000 ft and greater. It is given by the following formula:

$$D = (H/W) \times (\cosh(WS/2H) - 1) \qquad (17\text{-}22)$$

where

D = sag at center of span (ft),

S = span length (ft),

W = horizontal tension (lb),

H = weight of conductor (lb/ft), and

cosh = hyperbolic cosine.

The tension of a conductor varies from a maximum value at the point of support to a minimum value at the midpoint of the span. The tension at the point of support is given by the following formula:

$$T = H + WD \qquad (17\text{-}23)$$

or

$$T = H\cosh(WS/2H) \qquad (17\text{-}24)$$

17.10.5 Right-of-Way

The width of the transmission line corridor is determined in large part by the horizontal clearance requirements of the National Electric Safety Code. However, after determining the horizontal clearances, the designer can minimize the width requirements by using vertical conductor configurations or restraining the insulator swing at the towers. Of course, using a vertical conductor configuration requires that the support structures be taller and perhaps more expensive. Right-of-way width can be reduced by making the spans shorter, thus reducing conductor blowout, but making the spans shorter requires more support structures. Road right-of-way is also sometimes used to allow conductors to overhang the road, thus reducing right-of-way width. Nothing is free, and every transmission line consideration is a variable, just as it was in the auxiliary electric system design.

17.10.6 Conductor Support Towers

There are as many good ways of supporting transmission line conductors as there are engineers who design them, and all must meet minimum NESC code requirements. Three common types of transmission line towers are shown in Fig. 17-54. These are among the infinite variety of possibilities.

The familiar "H-Frame" structure is shown in Fig. 17-54a. This structure is usually framed using two treated wood poles with wood crossarms and wood X-braces. The conductors are suspended between the crossarms as shown. The X-braces give the structure great horizontal overturning resistance capacity. Structures without X-bracing would be no better than two single poles. The X-bracing makes the configuration a "structure" instead of just two poles connected by a crossarm at the top.

The familiar lattice steel tower is another widely used transmission line support method. A typical double circuit design structure with vertical phase conductor configuration is shown in Fig. 17-54b. This type of tower may be designed for average spans of 1,000 to 1,500 ft or more. The configuration shown with the center phases on longer arms is used in areas that are subject to icing. The offset vertical spacing helps to alleviate the ice jump problem discussed in Section 17.9.3.

In both of these structures, right-of-way width requirements could be reduced, or the span length increased, by using "V-string" insulators to restrain the conductor blowout at the support tower. However, this increases the cost of the structure, and consequently, the cost per mile of the transmission line.

Single-pole structures may be either wood, concrete, or steel. Single-pole structures permit less right-of-way space

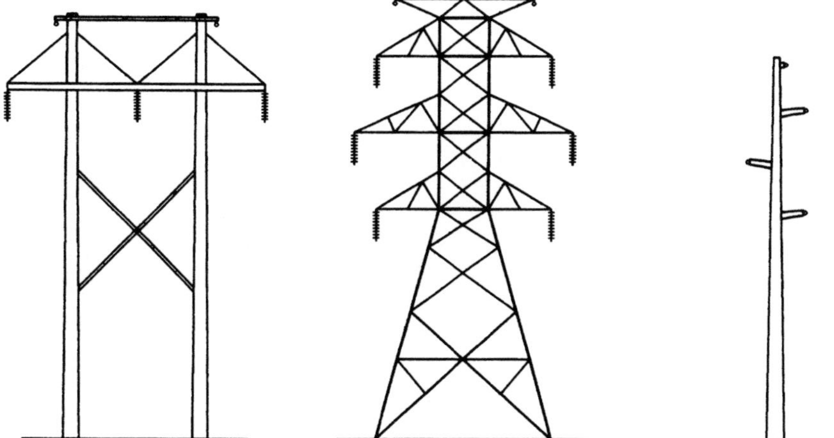

Fig. 17-54. Typical transmission line structures. (a) H-Frame. (b) Lattice double-circuit tower. (c) Single-pole horizontal post.

but require shorter spans. One configuration using horizontal line post-insulators is shown in Fig. 17-54c. The conductors may be set in a triangular arrangement, as shown, or all be on one side to overhang a roadway right-of-way if some other requirement dictates reduced width.

The voltages and currents in a flat configuration transmission line will become unbalanced because of the unbalanced mutual reactance coupling between phases. If a transmission line is to run several hundred miles, consideration may be given to "transposing" the phases to achieve better balanced impedance. This consideration does not carry as great an importance in transmission line system engineering as it once did because of the extremely high cost of transposition structures and better calculation techniques available for system relay calculations using modern high-speed digital computers.

17.10.7 Electromagnetic Considerations

There has been much publicity in recent years about the biological effects on humans from exposure to electric and magnetic fields. The US Environmental Protection Agency and other responsible scientific study groups have not yet determined any link between electrical or magnetic fields and any health hazards. Many millions of dollars are being spent on research to arrive at definitive answers. However, the modern world cannot, and will not, do without electric power. It does too much good. The benefits that electric power provides in refrigeration of foods alone are indispensable to the existence of the present-day world population. If medical researchers ultimately determine that any health hazards exist, power engineers will redesign the entire world's electric power supply system. Until then, it is the job of engineers to design power generation, transmission, and distribution facilities to supply electric power to the general public in as safe and practical a manner as possible and to cooperate with public officials in educating the public in the safe use of electrical power.

An electric field is created by any energized conductor, whether it be a table lamp, television cathode ray tube (CRT), a penlight flashlight battery, or a 765,000-V three-phase cross-country transmission line. The electric field emanates radially from the energized conductor(s). The field intensity, measured in volts per meter, is determined by the voltage magnitude to the reference (earth), the effective diameter of the conductor, and the distance from the conductor to where the measurement is made.

A magnetic field is also generated around any current-carrying conductor, whether it be a 12-V lead to an automobile fan motor, the conductors within an electric blanket, the heating elements of an electric kitchen range, the electrodes of a shop welding machine, or the 20,000-ampere bus conductors from the terminals of a 500-MW generator. That magnetic field is circumferential around the conductor, and the total magnetic strength (Φ) is directly proportional to the current flowing in the conductor. The magnetic flux density (B) measured in Webers per square inch is a function of the distance away from the conductor where the magnetic field is measured.

The electric field (E) and the magnetic field (Φ) surrounding a single conductor are shown in Fig. 17-55. Notice that both the magnetic and electric field strength intensities increase exponentially near the conductor.

If the electric current flowing in the conductor is an alternating current, the polarity of the magnetic field and the electric field reverses every half cycle depending on the direction of flow of the current. If the frequency of the alternating current is high enough, the conductor is called an "antenna." Radio waves start at about 100 kHz. The frequency of electric power transmission is normally between 50 and 60 Hz.

In areas of very high electric field intensity, human perception is very subjective. The most common first sensations are hair nerve stimulation due to hair erection and tingling between the body (arms) and clothes. This threshold of perception varies with the individual but is in the neighborhood of 10 to 15 kV/m. At much higher levels of electric field intensity (40 to 60 kV/m), corona discharges from body

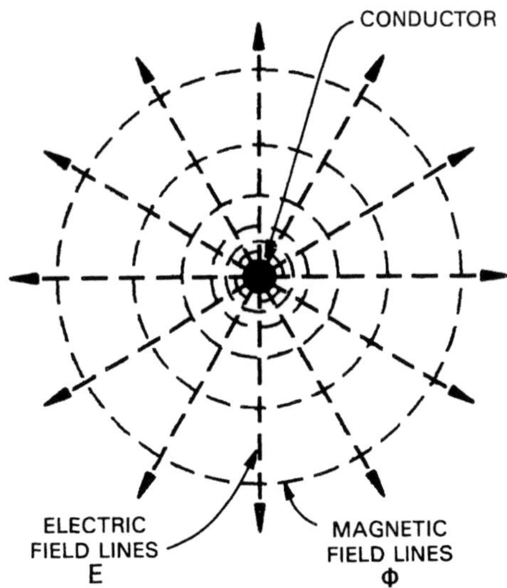

Fig. 17-55. Electric and magnetic fields surrounding a conductor.

parts, such as head to hat, ears, outstretched fingers, etc., and causes unpleasant sensations. Of more practical concern is the electrical charge that can build up on ungrounded objects such as automobiles or school buses which may happen to be parked beneath transmission lines.

The safe level of induced voltages in objects is defined by the NESC as that voltage which, if the largest anticipated truck, vehicle, or equipment under the line were shorted to ground, the resulting current would not exceed 5 milliamperes rms (root-mean-square). The level of perception in humans varies between 1.5 to 3 milliamperes (ma) with the level of discomfort about 4 ma. The electric field intensity to which the general public may be exposed should be controlled in a design application by locating the conductors at great enough distance from any possible public access so that high field intensities are not experienced.

Air, which is used as an insulator in most electrical applications (particularly in substations and transmission line designs), breaks down and ionizes in an electrical field strength of about 31,000 V/cm. At higher altitudes, or even with more conductive pollutants such as salt spray near the coasts, air may ionize at even lower field strengths. A late summer afternoon thunderstorm creates extremely high electrical fields between the clouds and the earth or between the clouds themselves and causes lightning flashes because of the high electrical field strength. These spectacular discharges are the result of millions of volts of charge buildup in the electrical field. When air ionizes, it produces electrical discharges and corresponding electrical radio waves, not unlike the early Tesla coils which Marconi used to discover radio transmission.

This ionization around an energized electrical conductor is called corona. The effects of corona are radio and television interference similar to the effects of a summer afternoon thunderstorm, although not nearly as dramatic. Corona also causes audible noise which may be objectionable, especially near residential areas. In addition, corona is a source of electrical energy loss on the transmission line.

One way of reducing the corona caused by electrical field intensity around a conductor is to make the conductor larger so that the radial electric lines of force are not concentrated near a point source. This is sometimes accomplished in transmission line design by using "bundled" conductors. Figure 17-56 shows the effects of a two-conductor bundle where the electrical field intensity and the density of the magnetic lines of force are reduced near the conductor because the effective electrical diameter of the bundled conductor is increased. Bundling of conductors also has the added effect of increasing the ampacity or load carrying capacity of a transmission line.

Corona is controlled in EHV substations by corona bells on bus terminations, corona rings on surge arresters and switch jaws, and at other areas using fittings that have large-radius smooth surfaces.

17.10.8 Shielding and Grounding

It is usual practice to attempt to protect transmission lines and substations from lightning strokes by shielding them with grounded conductors. These overhead shield wires are placed above the lines. The angle between the vertical plane of the shield wire and the conductor is called the shielding angle. The smaller the shielding angle, the better the protection for the conductor. For transmission lines, a 30-degree shielding angle is considered acceptable. For rigid bus substations, 45 degrees is about the maximum that would be considered good design. However, the actual shielding pro-

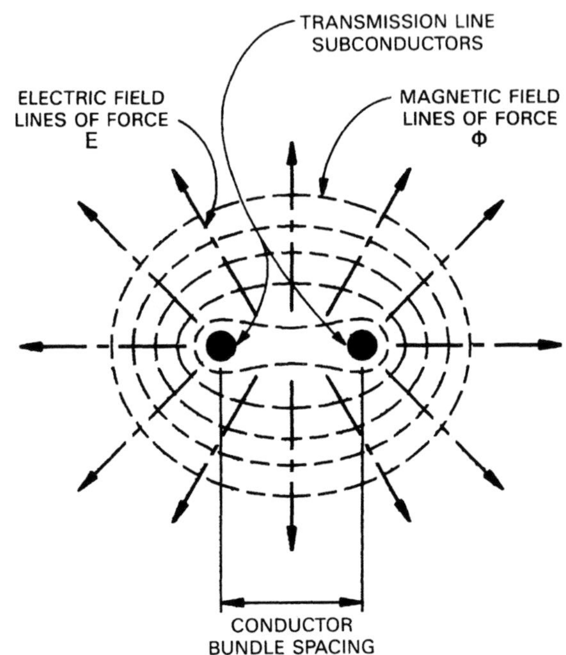

Fig. 17-56. Electromagnetic fields near bundled conductors.

tection of a substation or transmission line is based on probability and statistics. A line with a 0-degree shielding angle may receive a lightning stroke, whereas an unshielded line may never be damaged by lightning. Nevertheless, good design requires shielding. An isokeraunic map showing mean annual days of thunderstorm activity within the continental United States shows central Florida to have the most, with over 100 thunderstorm days per year, whereas the Pacific coastal areas have fewer than 10. Therefore, transmission line shielding in Florida would be more critical than on the California coast.

Lightning strokes on phase conductors of lower voltage transmission lines almost always result in flashovers of the phase conductor to ground at the next tower structure, whereas EHV and UHV lines usually sustain some lightning strokes without flashovers because the basic insulation level is so much higher.

Lightning performance of shielded transmission lines depends very heavily on how well the overhead shield wire is grounded. Therefore, tower grounding is of the utmost importance. The most common way of grounding transmission line towers is the use of driven ground rods. These rods are usually 8- to 12-ft copper-clad steel rods that are driven at the base of each structure. The shield wire is then connected to these rods at each tower. Rod diameters are usually less than 1 in.

The resistance of a single rod to ground may be approximated as follows:

$$R = \frac{\rho}{2\pi l} \times \ln\frac{(4l)}{a} - l \qquad (17\text{-}25)$$

where

ρ = ground resistivity, ohm-meters;

l = length of rod, meters;

a = radius of rod, meters;

R = resistance of rod in ohms; and

ln = natural logarithm of base ϵ.

The resistance of a rod does not decrease directly with length; therefore, there is a limit to the practical distance ground rods can be driven. Ground rods are rarely driven more than 25 ft.

17.10.9 High-Voltage Cable

Transmission lines may also be built underground. In special cases in high-density urban areas, it may not be possible to obtain the necessary right-of-way to build a high-voltage transmission line. In these cases and in other special cases, there may be no alternative to building the line underground.

Underground transmission lines are very expensive to build. Costs are measured not in dollars per mile, but in dollars per foot. In addition, losses are a serious consideration. It has been estimated that a 345-kV uncompensated cable line over 25 miles long will be thermally overloaded by its own charging current. A decision to build a transmission line underground should not be made frivolously. Underground lines are expensive, suffer performance restrictions, and are not as reliable as overhead lines. Nevertheless, underground transmission lines are an important part of the technological capability of the electrical power system designer.

Solid dielectric insulated cables and oil-filled pipe type cables are two of the high-voltage cable types used for underground transmission lines.

Solid dielectric insulated cable is common for applications of 115 kV and below. The insulation systems most commonly used are high-density crosslinked polyethylene (XLPE) and ethelythene propylene rubber (EPR). Figure 17-57 shows the makeup of a typical 69-kV underground cable.

One of the requirements of cable insulation design is to avoid excessive electric field strength discontinuity locations which would otherwise cause corona and breakdown of the insulation and result in failure of the cable. In the previous example, the extruded conductor semiconducting strand shield provides a smooth electrostatic transition between the conducting energized conductor and the nonconducting insulating medium. This extrusion is necessary so that there are no voids where corona may start in a high electrical stress environment.

Both XLPE and EPR are very good insulating materials which have good resistance to moisture and acceptable aging properties.

The insulation shield is also an extruded semiconducting material that provides a smooth electrical transition between the insulating material and the cable electrical shield and prevents air voids which could lead to corona and subsequent cable failure.

The shielding system usually consists of both helically wound copper tapes that provide continuous contact with the insulation semiconducting material and the grounded shield, and wound stranded shielding wires to achieve the necessary fault current carrying capability. The overall polyvinyl-

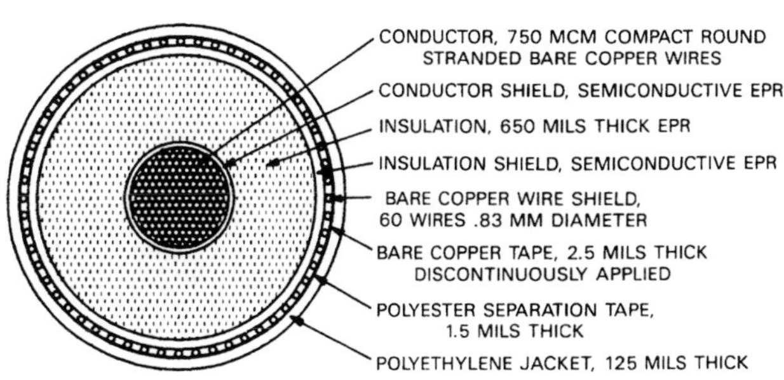

Fig. 17-57. Typical 69-kV underground high-voltage cable.

chloride (PVC) jacket provides physical protection and moisture protection to the insulation systems.

Higher voltage underground cable circuits usually use paper-insulated conductors installed in oil-filled pipes. This construction uses the same insulation principle as high-voltage oil-filled transformers. In this type of installation, the three-phase cables are usually built with paper insulation and wrapped with copper "skid wires," which are then pulled into previously installed pipes. The skid wires prevent damage to the paper insulation during installation. After installation, the pipes are filled with insulating oil. The oil then provides both cooling and insulation. Oil may be force circulated with heat exchangers at various locations to give the cable greater ampacity.

Another, less utilized insulation system is sulfur hexafluoride gas insulated bus (SF-6). This type of "cable" is normally limited to very high ampacity installations near power plants and in limited access locations such as in underground urban applications in conjunction with gas-insulated substations where space is extremely limited.

17.11 REFERENCES

American National Standards Institute. ANSI C57.92. ANSI/IEEE C57.12.00.

City of Vero Beach. Director of Power Resources—Municipal Power Plant. 100-17th Street. P.O. Box 1389, Vero Beach, FL 32961-1389.

Cooper Power Systems. P.O. Box 2850, Pittsburgh, PA 15230.

Institute of Electrical and Electronic Engineers. 1986. Guide for safety in alternating current substation groundings. IEEE Standard 80.

Institute of Electrical and Electronic Engineers. 1983. Recommended practice for sizing large storage batteries for generating stations and substations. IEEE Standard 485.

National Electric Code (NEC). 1993. National Fire Protection Association.

National Electrical Manufacturer's Association (NEMA). TR98.

National Electrical Safety Code (NESC). 1993. Institute of Electrical and Electronics Engineers.

18

Plant Control Systems

Augustine H. Chen

18.1 INTRODUCTION

In other chapters of this text, the various categories of mechanical, electrical, and chemical equipment installed in a typical coal-fueled power plant were discussed. For the power plant to run and produce electricity, the equipment in each category must be placed into operation to perform its intended function. The operation of the equipment must be coordinated to meet the demand of the electricity production process, and the production process must be regulated so that the cycle and the equipment are operated within design conditions. The plant control systems provides the necessary tools to enable the operators to orchestrate plant operation for the reliable and efficient production of electricity.

This chapter provides an overview of the plant control system, its functions, and the type of control equipment used in a modern pulverized coal-fueled power plant.

18.2 BASIC CONTROL FUNCTIONS

In a modern power plant, the majority of the control actions required to operate the processes are automated. There are numerous reasons for plant automation. First, automation reduces human error in plant operation and thus provides greater safety for plant personnel. Second, automation reduces the number of operators required to run the plant, which reduces labor costs. Third, an automated process is more efficient than a manual process because automatic controllers are capable of responding to changes in operating conditions faster and more accurately than human operators. The result is tighter control of plant operating parameters which keeps the processes running within design conditions.

The control functions used in operating a power plant can be classified in many ways. For the purpose of this discussion, the functions of the plant control system are classified by type of control action used. In this respect, control functions can be categorized into two distinctively different categories: on–off control and modulating control. Both types of control functions can be found in a power plant. The on–off control is also known as digital control, binary control, discrete control, sequential control, or motor interlocks. Modulating control is sometimes called analog control, continuous control, or closed loop control. No standard industry-wide terminology exists for these two types of control functions.

On–off control produces a control action that varies in discrete states. An example of this type of control is starting and stopping a motor. The control system produces either a start command or a stop command, and the controlled equipment reacts to the command with two distinct outcomes: motor running or motor stopped. No intermediate states exist between the "running" and the "stopped" state.

Modulating control produces a control output signal with a magnitude that varies smoothly from one value to another. An example of this type of control is regulating water flow to a tank by adjusting the position of a water supply valve. The control action is continuously adjusting the valve position to maintain the water level in the tank at a desired point. The state of the valve is continuous in that the position of the valve can be anywhere between the fully closed and the fully open position.

Using on–off and modulating control to operate a power plant may be compared to a person driving an automobile. To start the car, the driver turns on the ignition key and the engine starts. This is on–off control. Once the engine is started and the car is running, the driver controls the speed of the car by continuously adjusting the accelerator position whenever the car's speed deviates from the desired speed. This is modulating control.

A power plant operator goes through similar steps in operating the power plant. The operator starts a pump motor, a fan motor, or some such equipment. Typically, the operator waits until a pump is started and then increases or decreases the opening of a valve at the pump discharge to regulate the water flow through the pump. On some occasions, both actions can be applied to the same piece of equipment. A case in point is a pump driven by a variable-speed motor. The on–off action is used to get the motor started, and the modulating action is applied to regulate the pumping rate by adjusting the motor speed. The on–off and the modulating control action complement each other. In the case of an emergency or a desired shutdown, the on–off control can shut down equipment and put the process in a safe condition ready for a restart.

The following paragraphs describe the basic elements of on–off and modulating control functions and the degree of automation commonly encountered in the power generation process.

18.2.1 On–Off Control Functions

Figure 18-1 shows the essential elements of a basic on–off control system. In power plant applications, the equipment controlled by on–off action can be divided into the following major categories:

- Motor-driven rotating equipment such as pumps, fans, compressors, and conveyors;
- Motor-operated shutoff valves and dampers; and
- Solenoid-operated equipment such as pneumatic shutoff valves.

For motor-operated equipment, the result of an on–off action is a switching contact wired to a circuit breaker operator or a motor starter, and the opening and closing of the contact trigger the necessary circuit switching actions to connect or disconnect the electric power supply to the motor. In solenoid-operated equipment, the solenoid is either energized or deenergized through the action of a single contact opening or closing to direct compressed air into or release air out of the pneumatic actuator of the valve, thus driving the valve to either the fully open or fully closed position.

As the control action is executed, the control system requires feedback intelligence from the controlled equipment to ascertain whether the action has brought about the intended result. For motor-operated equipment, the feedback is typically derived from contacts in the circuit breaker or the motor starter that are indicative of the position of the circuit switching device. Only in rare cases is the rotation of the motor used as the feedback signal. For valves and dampers in general, contacts on the limit switches installed at the open and the closed end of the valve or damper are wired to the control system to provide the necessary feedback.

The control action of an on–off control system may be initiated manually by a command from an operator or initiated automatically by the system based on the intelligence received from the process. In all cases, the command and the intelligence received by the system vary in discrete steps such as a manual motor start/stop command from a control switch located on a control panel, or a high or low level condition from a level switch mounted on a water tank. The control logic is the part of the system that receives input data, applies Boolean algebra on the basis of the state of the input intelligence, and produces an output to start or stop a piece of equipment.

Figure 18-1 shows the control logic for a condenser exhauster pump pulling vacuum on the condenser. The exhauster is installed in parallel with other exhausters. This exhauster can be started manually by the operator using a control switch, or by putting the control in the standby mode and letting the pump start automatically whenever the condenser absolute pressure rises above 6 in. Hg. In either case, the control logic will withhold the "start" command if the seal water temperature, exhauster discharge air temperature, or lube oil pressure of the exhauster is not adequate for safe operation of the exhauster. Similar logic can be developed for the exhauster "stop" command.

This example is relatively simple and requires a few

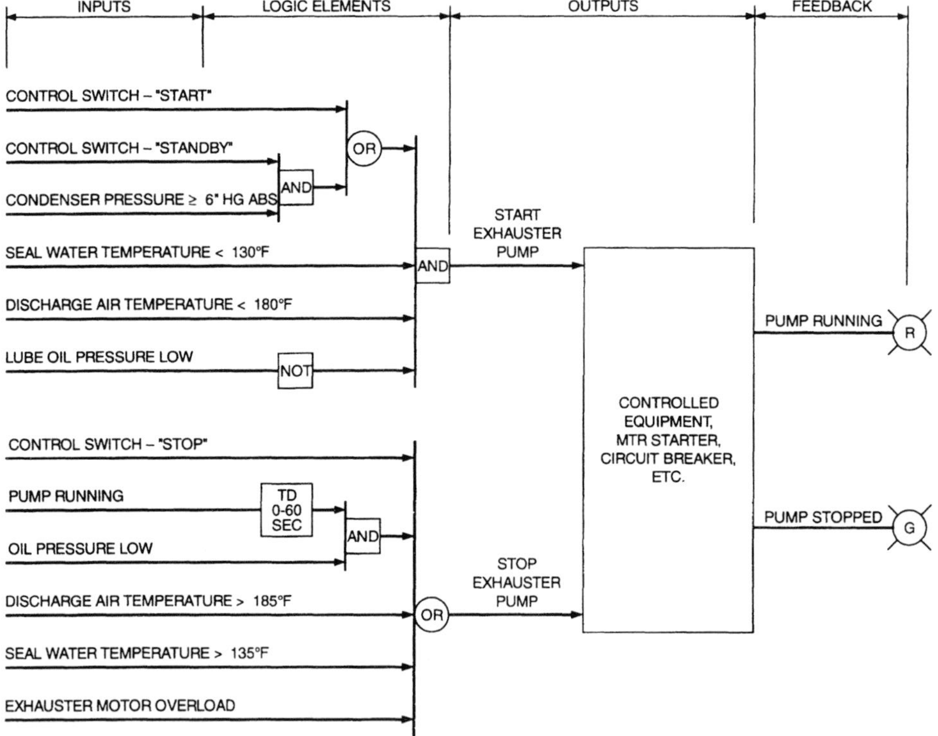

Fig. 18-1. Basic on–off control logic. TD = time delay.

pieces of feedback intelligence from field instruments such as pressure switches and level switches. The logic can be complex for large and expensive plant equipment. Regardless of the complexity of the logic, its main purpose is to protect the equipment against unsafe operating conditions and to enforce proper execution of equipment operating procedures.

18.2.2 Modulating Control Functions

Figure 18-2 shows a modulating control loop in its simplest form. The example shown is a temperature control loop. The water temperature is controlled at a desired value (set point) by automatically and continuously modulating the control valve supplying the heating steam to the heat exchanger. The basic elements of the loop are as follows:

- Controlled variable—Process parameter (water temperature) that is controlled by the control loop to a desired value (set point);
- Controller—Element that compares the value of the controlled variable to the set point value and produces a control action to correct the set point deviation (error) to zero;
- Manipulated variable—Parameter (heating steam flow) that is varied as a result of the control action of the controller so as to change the value of the controlled variable toward the set point value; and
- Final control element—Device that changes the value of the manipulated variable to correct the set point deviation according to the control action of the controller.

As shown in Fig. 18-2, the water temperature is the controlled variable. The control valve is the final control element. The heating steam flow to the heat exchanger is the manipulated variable. If the water temperature drops below the temperature set point, the controller applies a control action to the temperature deviation (error) and produces a control signal to increase the opening of the control valve. This increases the steam flow into the heat exchanger, thus bringing the water temperature back to the set point value. The control loop is said to be "closed" in that the controlled variable is measured and the measured value is used to verify the result of the control action.

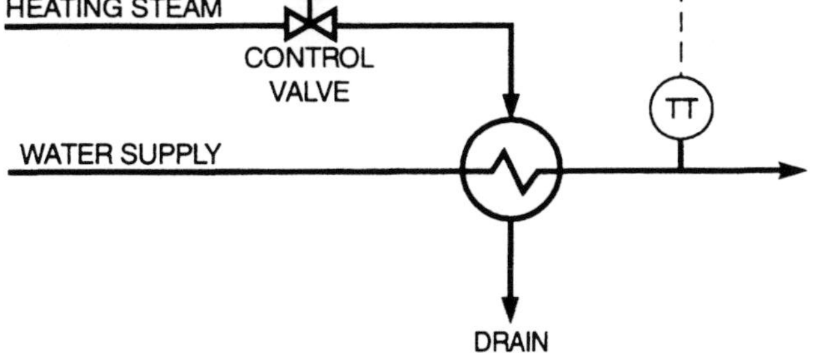

SP
TC
H/A
HEATING STEAM
CONTROL VALVE
TT
WATER SUPPLY
DRAIN

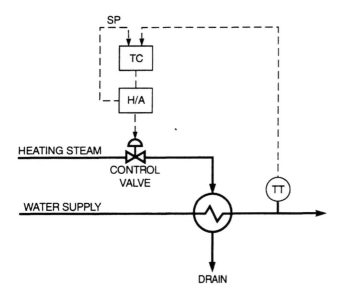

Fig. 18-2. Modulating (closed loop) control. SP = set point; TC = temperature controller; TT = temperature transmitter; H/A = hand/auto control station; controlled variable—water temperature measured by TT; manipulated variable—heating steam flow; final control element—heating steam control valve.

A Hand/Auto control station (sometimes referred to as a Manual/Auto station) can be included to provide a means of manual control. It acts like a transfer switch between the automatic and the manual mode of operation for the control loop. When in the automatic mode, the final control element is positioned by the output signal from the controller. When in the manual mode, the controller output is blocked and the output signal to position the final control element is controlled by the operator from the Hand/Auto station. A Hand/Auto station is not a necessity for every control loop. It is usually included in a control loop if manual intervention is needed during operation.

The type of control action most commonly used for process control is what is known as three-element control action. The three elements are proportional action, integral action, and derivative action. All three elements are related to the relationship between the set point error and the controller output to the final element.

In proportional control, the magnitude of the controller output is proportional to the magnitude of the set point error. A fixed relationship between the input(error) and the output exists:

$$P = (K_p)E \qquad (18\text{-}1)$$

where

P = output

E = error (set point deviation), and

K_p = proportional gain.

The proportional gain is an adjustable parameter of the controller, the magnitude of which is a measure of the sensitivity of the controller.

For example, in the control loop shown in Fig. 18-2, a given temperature deviation from set point (error) produces a control signal that drives the control valve to a corresponding position depending on the proportional gain. Ideally, this relationship keeps the temperature at the desired set point as long as the water flow and temperature entering the heat exchanger and the pressure and temperature of the supply steam remain constant. If any one of these conditions changes, the controller must make necessary corrections to maintain the exit water temperature at the set point. These changes are called load changes. If the load change is large, the controller will not be able to produce an output to bring the temperature back precisely to the set point. The difference between the new water exit temperature and the original set point is called offset. The offset is inherent in the applications of controllers with proportional action control only.

Integral control is also known as reset control. With integral control, the output of the integral controller changes whenever there is a set point error. The integral action produces an output that changes at a rate proportional to the magnitude of the error:

$$dP/dt = (1/T_i)E \tag{18-2}$$

where

T_I = integral (or reset) time.

To put the relation in terms of the output, the equation becomes as follows:

$$p = (1/T_i)\int E\,dt \tag{18-3}$$

As seen in Eq. (18-3), the integral control produces an output that keeps changing for as long as there is an error in the process. The output stops changing once the error is reduced to zero. In the example given in Figure 18-2, when a temperature error exists, the integral action of the controller keeps moving the control valve as long as the exit temperature is not at the set point. The valve keeps moving until the temperature is returned to the set point.

The integral time, like the proportional gain, is an adjustable parameter of the integral controller. In theory, the integral control action can eliminate the offset resulting from the proportional action, and the amount of integral action determines the time required to accomplish this. In practice, it is not always possible to adjust the integral action to perfection. The controller either overcorrects or its action lingers on to the point where the system becomes unstable. However, a controller with both proportion and integral action, when properly tuned, corrects the error resulting from load changes in a timely and acceptable manner for most purposes.

Derivative control is also known as rate control. The controller produces an output whenever there is a set point error and the magnitude of the output is proportional to the rate of change of the error.

$$P = (T_d)dE/dt \tag{18-4}$$

where

T_d = rate (or derivative) time.

With derivative control, the faster the error changes, the greater the controller output. The output goes to zero as soon as the error stops changing. The rate time, like the reset time, is an adjustable parameter of the controller which can be used to adjust the sensitivity of the controller. Derivative action provides an anticipatory action by responding to changes in set point error and supplements the proportional and integral action to enhance the response time of the controller.

An automatic controller can be made with any combination of the three control elements. The most commonly used are proportional, proportional plus integral (PI), and proportional plus integral plus derivative (PID) controllers. A closed control loop using one controller (such as the loop shown in Fig. 18-2) is also called a single-element control loop because only one variable (the controlled variable) is measured and used by the controller.

A single-element controller using any combination of the three control elements produces a control action only when it detects an error in the controlled variable. It applies corrective action only after the error has occurred. This is called feedback control, and the feedback is the controlled variable. To improve system response to process changes, a technique known as feedforward control is sometimes added to the feedback controller. The feedforward control employs a second element, a signal that represents a load change in the process, and passes the signal forward as a corrective action in anticipation of the set point deviation that follows as a result of the load change. The purpose of this signal is to provide a control action before a set point deviation occurs and thus to minimize any disturbance in the controlled variable caused by a change.

Figure 18-3A shows the feedforward signal added to a feedback control loop. In this example, the water flow entering the heat exchanger is measured and used as the feedforward signal. An increase in the water flow immediately produces a feedforward action to increase the heating steam supply by increasing the opening of the steam valve before the effect of the load change is reflected in a disturbance to the exit water temperature. The feedforward signal in such a loop should be calibrated to approximate the opening of the steam valve under varying load conditions (typically from 0% to 100% load). The calibration may not be precise because the required valve opening is not only a function of the water flow entering the heat exchanger, but is also a function of other factors such as water temperature and steam pressure. However, a signal calibrated in this manner is often sufficient to put the valve in the approximate correct position as soon as the change has occurred, thus minimizing set point disturbance when the system load changes from one flow to another. The feedback input to the controller can then be used to eliminate the set point deviation.

If the controller employs only a feedforward control signal, the technique is termed open loop control because the result of the control action is not verified by any measure of the controlled variable. It becomes closed loop control only when used in combination with a feedback controller as the example shown in Fig. 18-3A.

Another method often used to improve system response is cascade control. Cascade control employs two single-

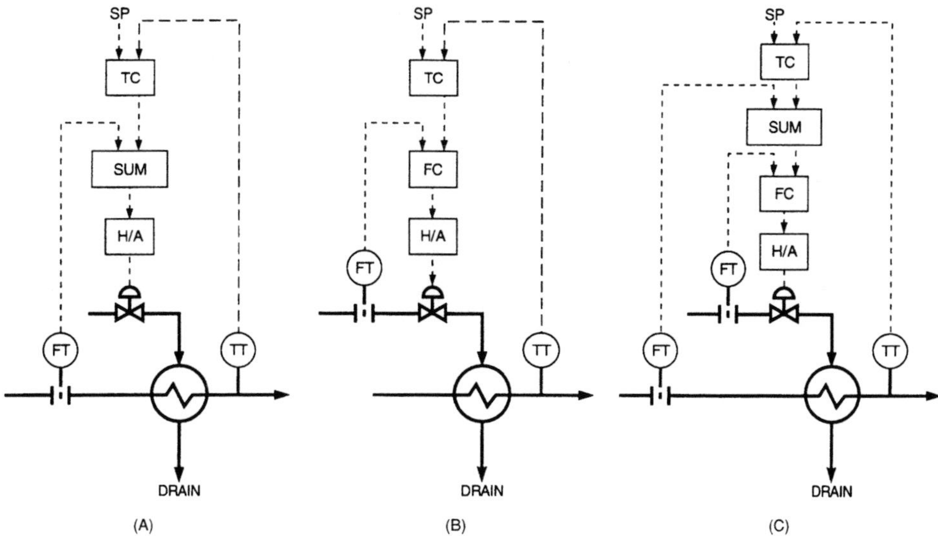

Fig. 18-3. Control loops. (A) Feedforward control. (B) Cascade control. (C) Feedforward and cascade control.

element controllers arranged so that the output of one controller becomes the set point of the other. The first and the second controllers are called primary and secondary controllers, respectively.

A typical cascade control loop (Fig. 18-3B) consists of a water exit temperature (primary) controller and a heating steam flow (secondary) controller. The output of the temperature controller becomes the set point input to the steam flow controller. The output of the flow controller is used to position the steam flow control valve. In a process such as this, a disturbance in the steam supply pressure causes a change in the steam flow to the heat exchanger that eventually results in an error in the exit water temperature. With the cascade control arrangement, the flow controller instantly reacts to any changes in the steam supply pressure, providing corrective action before the change in steam pressure develops into an error in the water exit temperature. For this application, cascade control (Fig. 18-3B) is better than the single-element control loop (Fig. 18-3A) because the simpler system is not able to change the steam flow rate fast enough to overcome variations in pressure in the steam line. Such changes would be sensed by the temperature controller only after the change has passed through the process. The addition of the flow controller to form a cascade control loop eliminates this delay and the entire system responds more quickly with respect to changes in steam supply pressure.

The feedforward control and cascade control can be combined into one loop (Fig. 18-3C). In such a scheme, the feedforward part of the loop makes the system responsive to changes in water flow, and the flow controller of the cascade portion of the loop makes the system responsive to changes in steam pressure. Lastly, the feedback (temperature) controller monitors any residual set point deviation and ensures that the water temperature is maintained precisely at the desired value. In this configuration, the feedback control does not produce the major corrective responses. It serves as a monitoring loop to trim the actions of the feedforward elements.

Numerous parameters in the steam generation process need to be controlled, such as steam temperature, steam pressure, drum level, and condenser hotwell level. The parameters react to load changes or system disturbances. Feedback, feedforward, and cascade control techniques are the basic control functions used to build a control system that will maintain plant parameters within design conditions and keep the plant running safely and efficiently.

18.2.3 Monitoring Functions

Monitoring functions are another aspect of the plant control functions involved in the electricity production process. Many variables in the process are not automatically controlled. The power plant operator must monitor equipment status and process conditions while the plant is in operation. Indicating lights, gauges, indicators, alarm annunciators, cathode-ray tube (CRT) screens, and other similar instruments and devices are used to monitor plant operation.

For example, consider operating an automobile. While driving the automobile, the driver must keep an eye on the instruments on the dashboard, watching the speed, gas tank level, water temperature, and most importantly, the red light on the alarm panel because a red light often means trouble associated with car system performance. The power plant operator goes through the same routine while the plant is in operation. Equipment status and process conditions must be checked. For example, the operator must know which one of the three boiler feed pumps is running, which draft fan is down for maintenance and not available for service, what the generator output is, where the drum level is, and the like. The tools available to keep the operator informed are indicating lights, gauges, indicators, alarm annunciators, CRT screens, and other similar instruments and devices.

In addition, the operator needs to keep a running record of what has happened for either troubleshooting or equipment maintenance functions. The operator collects plant operating data from the indicators, gauge readings, recorder charts, and computer printouts to compile reports for future reference. These activities are considered alarm monitoring and data acquisition functions. It is important to recognize that both the alarm monitoring and the data acquisition functions, while not directly related to the control functions, play an important part in the safe and efficient operation of the power plant equipment.

18.2.4 Implementation of Control and Monitoring Functions

Digital computers have revolutionized the implementation of control functions for process operations. Before digital computers, the functions were executed by mechanical or electrical devices. Until the early 1960s, on–off controllers used contact closures, timers, electromechanical relays, and similar devices to implement the logic functions. Modulating controllers employed analog devices such as mechanical or pneumatic type controllers to perform the computations necessary to produce the controller output. In the late 1960s and the early 1970s, electronic devices such as transistors and diodes gained acceptance and replaced most of the mechanical and pneumatic control devices for both the on–off and the modulating control functions. However, the basic principle of using discrete hardware to perform the control functions did not change until the advent of programmable digital computers.

The digital computer uses a processor operating with a stored program rather than mechanical or electrical devices to perform control functions. The functions performed by the computer are written in terms of algorithms or instructions and are stored in the computer memory as binary bits (ones and zeroes).

A group of algorithms or instructions constitutes a program. This is the software of the computer. The computer performs its function by executing the algorithms one at a time and one program after the other. Thus, the computer is timeshared by several functions rather than being devoted to a single function.

The computer's functions can be on–off control, modulating control, or monitoring functions such as alarm annunciation, equipment status display, or indicating and recording of operating parameters. The software can solve complicated control and monitoring problems much easier than the traditional mechanical and electrical devices. A program, once written, can be altered to accommodate changes in system functions without making physical alterations to the system hardware. Furthermore, the data stored in the computer can be shared between programs without using any physical connections between programs. The advantages of using the computers are numerous and well known in the industry. Digital computers have replaced the vast majority of traditional control hardware in power plants.

18.3 ON–OFF CONTROL APPLICATIONS

On–off control applications in a power plant can be classified into three general categories: protective interlocks, sequential control logic, and unit protection logic.

18.3.1 Protective Interlocks

Protective interlocks are on–off control logic in its simplest form. The logic is used for the operation of individual pieces of power-operated equipment. An example of this is the logic for a motor-driven boiler feed pump. The pump is operated from the central control room. Before the pump is started, sufficient water must be in the deaerator storage tank, the water path in the suction line must be open, the pump motor must be started against a closed discharge valve, and finally, lube oil must be available at the pump bearings for the pump to run. The starting logic checks the permissives using the signals from a tank level switch, suction valve limit switch, discharge valve limit switch, and oil pressure switch. If the contacts from the switches show the correct position, the pump motor starts when the operator issues the starting command. For emergency trip during pump operation in the event of very low tank level, low oil pressure, or high bearing vibration, the signals from the sensing devices can be used to develop the logic to stop the motor as soon as the condition occurs or after a time delay if the danger to the equipment is not imminent.

In this example, the logic does not provide an automatic start function for the pump. The lube oil pumps associated with the boiler feed pump may be used as an example for automatic start. If the boiler feed pump is equipped with two lube oil pumps, normally only one will be running and the other will be kept on standby. In such an arrangement, the standby pump is automatically started if the running pump stops. The logic for such operation can be implemented by using a contact closure in the motor starter of the lube oil pump to signify that the running pump has stopped.

It is important to realize that the examples given show relatively simple control logic. Permissive checking, emergency trip, and auto start functions can be used for a piece of equipment with as many inputs as the designer deems necessary. The objective of the protective interlocks for the motors is to ensure that the equipment will be started only if all starting conditions for safe operation are satisfied, and that any conditions that will expose the equipment and the related process to potential damage will automatically shut down the equipment.

18.3.2 Sequential Control Logic

Sequential control logic goes one step beyond the control logic for individual motors. It is a composition of control logic for a group of equipment operating in a predetermined sequence. Sequential control logic is used for components that must be started and stopped in a proper sequence for the system to function properly. Sequential logic is used to enforce the correct procedure and reduce the number of operational steps that must be initiated by the operator. The following sections describe some of the sequential control applications in fossil-fueled power plants.

18.3.2.1 ***Burner Control System.*** The most critical consideration in operating the fuel burning equipment of a boiler is the hazard of furnace explosion. A furnace explosion can cause catastrophic damage to the boiler equipment and endanger the safety of plant operating personnel.

The most likely cause of a furnace explosion is the spontaneous and rapid combustion of any accumulated unburned combustibles in the furnace or along the gas path of the boiler. The combustion is likely to occur when an excessive quantity of unburned combustibles is mixed with air and the mixture reaches an explosive proportion in the presence of sufficient ignition energy from electric charges, hot slag, or sparks inadvertently introduced into the furnace. Operating experience shows that such an accumulated mixture is often the result of repeated unsuccessful attempts to light off a burner, fuel leakage into the furnace through idle burners, sudden loss of burner flame during operation, or any other situations that leave the fuel and air mixture unburned in the furnace. Thus, any operation that can possibly result in such a situation must be prevented and any combustibles present must be burned or completely purged from the furnace. Every burner control system is designed with these considerations in mind.

The burners in a multiple-burner furnace are grouped for operation according to their physical location in the furnace. The boiler manufacturer's particular design of a furnace will dictate whether the burners in a burner group will be operated singly, in pairs, or as a group. Operating a burner involves manipulating the fuel valve, air register (damper), igniter, ignition fuel valve, and other accessories associated with the burner.

At the group level, operation involves starting and stopping the equipment common to the group. For a pulverized coal-fueled boiler, the group equipment includes the feeder(s), pulverizer, primary air fans, and air dampers around the pulverizer. The operation of the various components in the group must be coordinated to ensure that fuel and air are admitted into the furnace through the burners in proper order and in correct proportion to avoid the accumulation of any unburned combustibles and without any unwanted ignition source. The procedure must be repeatable and identical for all the burners, which makes it very conducive to the application of a sequential control logic.

A typical burner control system consists of three parts: burner control, master fuel trip, and furnace purge operation.

The burner control portion of the system involves the starting and stopping of individual burners or burner groups. For a pulverized coal-fueled boiler, typical burner start operation is as follows:

- Check that the secondary air flow is adequate.
- Put the air register in the light-off position.
- Check that the ignition fuel supply is available.
- Initiate the spark or other ignition source.
- Place the igniter into service by opening the ignition fuel shutoff valves.
- Check the ignition flame and shut down the igniter if the flame is not detected within a specified time period (initiate the master fuel trip if these are the burners of the first pulverizer placed into service).
- Check that the primary air flow is adequate.
- Open the coal valve on the coal pipelines from the pulverizer to the burners being started.
- Start the pulverizer.
- Start the coal feeder.
- Check the coal flame and shut down the burner and igniter if the flame is not detected within a specified time period.

The operation is considered successful when the main (coal) flame is detected within the preset time period. In general, the igniters are kept in operation during low load operation to help maintain flame stability in the furnace. They can be shut down at the operator's discretion when the boiler load reaches a certain minimum level and there is sufficient turbulence to enhance flame stability.

The burner control system is designed for remote operation from the central control room of the plant. The sequence can be operated in either manual or automatic mode. In the manual mode, the operator initiates each step of the sequence. In the automatic mode, the operator needs only to initiate the first step and the system logic automatically cycles the equipment through the remaining steps. In either case, the logic checks the permissives for each step before it is allowed to proceed to the next step, and the sequence is stopped automatically if any prescribed unsafe operating conditions are detected. Similarly, a shutdown sequence is also provided.

Typically, an operator panel with pushbuttons and indicating lights or a CRT-based control is provided for the operator to select the manual or automatic mode, and to initiate each step when in the manual mode (with all feedback information displayed to show the status of the components and the system).

The key to safe burner operation is to shut off the burner immediately at the slightest indication of any accumulation of unburned combustibles or if ignition energy is insufficient to sustain the fuel burning process. No practical means exists to monitor the accumulation of unburned combustibles or the ignition energy. Its existence can only be inferred from the lack of burner flame when the fuel is being admitted into the furnace through the burners. A "flame detected" or "no flame detected" condition for a burner is an indication of whether the fuel admitted is being burned or not. For this reason, flame detection plays a vital role in the functioning of the burner control system. A variety of techniques exist for flame detection. The detectors used for utility boilers are those sensing the light emitted by the burning fuel (ultraviolet and infrared detectors) or the flickering frequency of the light produced by the burning process (flicker detectors). The selection is largely dependent on the type of fuel burned, burner design, and the particular design consideration of the boiler and the burner control system.

In general, igniter flame is monitored on an individual basis. The monitoring philosophy for main flame varies depending on the firing pattern of the boiler. In a wall-fired furnace, a detector is typically installed for each burner. In a tangentially fired furnace, the detectors are shared between burners on adjacent elevations. The main flame is monitored either on an individual or group basis depending on the manufacturer's arrangement of the burners.

The purpose of the master fuel trip (MFT) logic is to protect the boiler by automatically stopping all fuel supplies to the furnace on detection of any unsafe operating condition. The National Fire Protection Association (NFPA) has specific guidelines for the tripping operation for utility boilers (NFPA Code 85C, "Standard for the Prevention of Furnace Explosion/Implosions in Multiple Burner Boiler-Furnaces"). This code is accepted by designers and manufacturers throughout the power industry as establishing minimum standards for operation and for systems design of multiburner boilers. The Institute of Electrical and Electronics Engineers (IEEE) also provides recommendations for tripping in the IEEE "Guide For Protection, Interlocking and Control of Fossil Fueled Unit-Connected Steam Stations." In addition, each boiler manufacturer has standard recommendations for boiler operation that address the specific characteristics of their boilers. The NFPA code recommends that a running boiler be tripped on the occurrence of any unsafe conditions. Examples of these conditions are:

- Loss of all forced draft fans,
- Loss of all induced draft fans,
- Combustion air flow below minimum,
- Excessive (high or low) furnace pressure,
- Loss of all flames, and
- Partial loss of flame introducing hazardous operating conditions.

Other tripping conditions may also cause an automatic master fuel trip. These conditions are determined by the station designer on a case-by-case basis: low boiler drum level, high drum level, loss of control power to the control system, low fuel supply pressure condition for gas- or oil-fired boilers, low circulation pump pressure differential for forced-circulation boilers, low feedwater flow for once-through boilers, and other conditions as required by the boiler manufacturer's design. To enhance the reliability of the tripping logic, redundant sensing devices are used where practical to detect unsafe operating conditions that will automatically activate the master fuel trip.

On a boiler trip, all fuel supply is immediately shut off. This is accomplished by closing all fuel valves (main fuel gas or oil supply valves, gas and oil burner valves, igniter valves, and coal burner line valves), and stopping pulverizers and feeders. Air flow control is rejected to manual mode and the air flow remains constant for post trip furnace purge.

The purge logic is used to ensure an adequate air flow supply through the furnace to get rid of any unburned combustibles in the furnace following a master fuel trip and before lighting off the first burner(s) after a trip. This is generally accomplished by opening all air and flue gas dampers in the boiler and running the forced draft and the induced draft fans for a specific time period (typically 5 minutes) to ensure that the air volume in the furnace is turned over several times during the purge operation.

The prevention of furnace implosions is addressed in Section 18.4.1.5.

18.3.2.2 Boiler Soot Blowing Control. Modern coal-fueled and heavy oil fueled boilers are equipped with a large number of soot blowers for cleaning the heating surface in the furnace and the convection pass (see Chapter 7 for more information). The soot blowers are operated in groups. Each group has a predetermined operating sequence to obtain the most effective boiler cleaning operation. The soot blowing operation is typically centralized and executed from the central control room of the plant.

The system has identical control logic for each type of blower. When a blower is placed into operation, the logic first checks the starting permissives such as the availability of the blowing medium (steam or air) and the power supply for the blower motor. Once operation is initiated, the logic opens the blowing valves, runs through the full blowing cycle in a preset time period, closes the blowing valves, and shuts itself off. For sequential operation, each group can be started by the operator and the logic will operate each blower in the group according to the predetermined sequence. Provisions are made for the operator to skip a blower during a sequential operation, to stop a blower while the sequence is in progress, or to start the blowers individually.

Although overall boiler cleaning requirements can often be predicted during the design stage, the actual slagging patterns that occur in practice and the resultant cleaning needs of the various sections of the boiler are very unpredictable. This is complicated further by variations in the characteristics of the fuel burned. This necessitates flexibility in blowing operation and sequencing to meet the changing requirements of slagging pattern and fuel characteristics. The earliest soot blower control system hardware was of the electromechanical relay type which was later replaced by solid-state logic elements. With the solid-state elements, the blowing routines could be changed by changing the wiring connections between the solid-state components. The control logic of state-of-the-art soot blower control systems is implemented by digital computers that enable the operator to change the blowing sequence and blower grouping to suit the prevailing operating conditions.

18.3.2.3 Demineralizer Operation. The operation of the demineralizer system is also a good candidate for sequential logic application. The system includes a number of cation and anion exchangers, water pumps, and acid and caustic feed systems (see Chapter 15 for further details). Each exchanger is a vessel equipped with a set of solenoid-operated valves which must be opened and closed in predetermined sequences depending on the mode of operation such as on-stream-service, regeneration, and recycle mode.

Sequential control is developed in two levels. First, each exchanger has a regeneration sequence with prescribed steps for backwash, acid or caustic application, and slow and fast rinse of the exchanger. Second, at the system level, the sequential control includes system service run, system shutdown, regeneration involving two or more exchangers, and recycling exchanger effluent before system startup, following system shutdown, or automatic recycling on high-effluent conductivity. All sequential operations can be manually or automatically initiated on the occurrence of a predetermined condition.

Some of the sequential operation steps are timed. Provisions can be made for operator intervention during a timed step to prolong or repeat the step if necessary (such as the acid or the caustic application step during the regeneration cycle). A demineralizer system is typically equipped with a control panel or workstation located near the demineralizer equipment for local operation of the system and, if desired, the same control functions for initiating the operating sequence can be duplicated on the main control panel in the central control room.

18.3.2.4 Coal Handling System Control. The coal handling system at a plant site typically serves all units in the plant. The system has many modes of operation that involve the routing of coal from one point to another, and uses a variety of material handling equipment such as feeders, conveyors, and crushers (see Chapter 5 for further information). The coal handling system is suitable for the application of sequential control logic. Typically, most of the coal handling operation is controlled from a centralized location which may be a local control panel in a building near the coal yard or in a central control room of the plant.

To start the system, the operator first selects the coal flow path and the components along the path with a starting and a finishing point (for example, stock coal from the unloading facility to the storage pile or feed coal to the silos of a particular unit from the storage pile). The operator then issues the start command from the control panel to start the sequence. If a component trips while the system is in operation, all equipment upstream of the tripped equipment is stopped automatically and the downstream equipment continues to operate until the coal is purged from the equipment. The logic also monitors the status of the sequential operation and alarms any conditions that inhibit the starting of the system. Emergency stop functions are also provided for manual stop of the system under emergency conditions.

18.3.3 Unit Protection Logic

The unit protection logic mainly refers to the protective interlocks associated with the operation of the three most important components in the power plant: the boiler, the turbine, and the generator.

The objective of the interlocks is to ensure maximum safety to plant personnel and to guard against catastrophic damage to equipment during all phases of plant operation.

The boiler, turbine, and generator are tripped individually on the occurrence of any unsafe or abnormal operating conditions. The boiler trip is the master fuel trip (MFT) and is usually incorporated as a part of the burner control system logic (Section 18.3.2.1). On a master fuel trip, the logic immediately stops all fuel flows to the furnace. The forced draft and induced draft fans will be kept running if they are in operation at the time of the trip to purge the furnace of any unburned combustibles.

The turbine trip logic hardware is an integrated part of the turbine's control system which initiates a turbine trip upon detection of any unsafe operating conditions (see Chapter 8 for more information). When the turbine trips, the turbine control system immediately dumps the hydraulic fluid from the main and reheat steam stop valve actuators and closes the stop valves. A pressure switch sensing the low-pressure condition of the hydraulic fluid or limit switches sensing the closed position of the stop valves is usually used as a verification that the turbine has tripped.

All steam extraction line air-operated nonreturn valves are closed on a turbine trip. The turbine oil system has a dump relay valve which is installed on the pneumatic line to the steam extraction line nonreturn valves. When the turbine trips, the dump relay valve is actuated, thus venting the air and causing the nonreturn valves to close. Simultaneously, all turbine and steam extraction line drain valves are opened, and all superheater and reheat spray valves are closed to minimize the possibility of water induction into the turbine.

Generator and system electrical faults are monitored by protective relays (see Chapter 17 for more information). A fault that can cause damage to the generator will trigger a generator trip. The action immediately following the initiation of a generator trip is to open the circuit breaker that connects the generator to the grid, consequently taking the generator offline.

The operation of the boiler, the turbine, and the generator are closely coupled. The boiler supplies steam to the turbine and the turbine drives the generator to produce power. A trip of any one of the three has a direct impact on the continued operation of the other two and causes an interruption to the power production process. The trips must be coordinated to effect an orderly and safe shutdown of the equipment without aggravating the system malfunctions that caused the trip in the first place.

A boiler trip requires a turbine trip because the boiler trip removes the steam supply to the turbine and there is no need to continue turbine operation without the boiler. It should be noted that the turbine control system is equipped with a low throttle pressure governor which is designed to close the turbine steam admission valves when the turbine throttle pressure drops below a preset value. However, with large boilers and the potential of water carryover from the boiler as the steam pressure decays following a boiler trip, the logic typically trips the turbine directly without waiting for the throttle pressure governor to close the turbine valves. A boiler trip will not directly trigger a generator trip.

On the other hand, a turbine trip also requires a boiler trip since the turbine valves are closed and the boiler steam has no place to go. However, on units equipped with a 100% turbine steam bypass system to the condenser, it would not be necessary to trip the boiler because the bypass valves can be opened to provide a flow path for the boiler steam. Units with a 100% turbine steam bypass system are common in Europe but are not often used in the United States.

There are two major concerns about turbine and generator tripping: turbine overspeed and generator motoring. If the generator is tripped simultaneously with the turbine, the turbine is exposed to the danger of excessive overspeed because tripping the generator in effect completely removes the load from the turbine, and the steam entrapped in the turbine flow passage keeps the turbine running. To limit turbine overspeed to a safe level under such a condition, the generator trip should be delayed until the entrapped steam has run its course. This is typically implemented by a reverse power relay. The generator is tripped only on detection of reverse power into the generator following the turbine trip. The detection of reverse power in the generator signifies that no steam is in the turbine to drive the generator, and thus there will be no danger of turbine overspeeding. Such a tripping arrangement is known as sequential tripping of the turbine and the generator.

Generator motoring occurs when the steam flow to the turbine is reduced during low load operation or when the flow is shut off as in the case of a turbine trip such that the power developed is below the no-load loss while the generator is still online. The generator will operate like a motor driving the turbine and cause overheating of the turbine blading. To prevent motoring, the turbine and the generator should be tripped simultaneously on the occurrence of either a turbine or generator fault. However, doing so runs the risk of excessive overspeed because of the entrapped steam in the turbine.

Sequential trip and simultaneous trip are at cross purpose with each other. In determining which one to use under the different fault conditions, the station designer must make a choice between the two. In general, sequential trip is used for turbine faults and for those generator faults where the risk of a few seconds delay in tripping the generator is slight. Simultaneous trip is used for all generator faults where the risk of causing damage to the generator outweighs the probability of excessive overspeeding.

On some units, the turbine is kept in operation with reduced steam flow when the breaker that connects to the grid is opened on the occurrence of certain electrical system faults in the grid. This protects the generator on a system fault. If the plant can be operated under such a condition, and if the cause of the fault can be identified and rectified quickly, it may be possible to quickly resynchronize the generator. However, this operation definitely increases the possibility of a turbine overspeed as opposed to simultaneously tripping the turbine and generator, and should be adopted only if there is an advantage in keeping the turbine in operation under such circumstances.

Boiler, turbine, and generator manufacturers have standard recommendations for the tripping logic for their equipment and how a trip of one component should be related to the other two. However, it must be emphasized that it is up to the plant designer to ensure that the tripping functions are coordinated with one another and the logic implemented truly meets the protection requirements of the boiler/turbine/generator unit.

18.4 MODULATING CONTROL APPLICATIONS

The most important applications of modulating control in the power plant are in the area of boiler control. The objective is to regulate the fuel input to the boiler in response to the load demand, and at the same time keep critical variables such as steam pressure, steam temperature, and drum level within design conditions. Other applications involve controlling pressure, temperature, level, and flow variables in the turbine cycle and plant auxiliary systems.

18.4.1 Boiler Control System

A power plant is essentially an electricity factory. Fuel is burned in the boiler to generate steam which in turn drives the turbine. The turbine drives the generator to produce electricity which is transported through the transmission lines for use by the consumers. This factory is unique in that its product is consumed as soon as it is produced and it can produce only as much electricity as the consumers demand—no more and no less. Since the factory has no control over the consumers' demands, it must at any given minute of the day keep up with the demand as it changes. The consumers can increase or decrease the demand any time they wish, and the power plant must generate sufficient power to match the demand as quickly as possible.

In a given area served by a utility company or companies, any imbalance between the generation and the consumption of electricity is reflected in a change in system frequency. The system frequency decreases as the demand picks up and increases as the demand goes down. As the frequency changes, the speed governors of the turbines automatically adjust the turbine governor valve position to change the steam flow to the turbines to support the new demand. Thus, the initial response of the system to a demand change is nearly instantaneous. It takes place automatically as soon as a change in system frequency is detected. The response is also random in that all generating units in the system simultaneously contribute to the change, and the sensitivity of each individual turbine speed governor dictates the proportionate share each unit contributes to meet the new demand. The system frequency will settle at a new level once the balance between generation and consumption has been restored.

Because the generators meet the demand in a random fashion, the system's load dispatcher will, after the initial response is over, reallocate the generation output of the individual units in the system and bring the system frequency back to normal. The load dispatcher accomplishes this by issuing demand signals to the individual generating units in the system. It is important to recognize that this signal is the load demand signal often referred to in the discussions of the boiler control system.

18.4.1.1 Load Demand Control. The load demand control of the boiler control system utilizes the load demand signal from the load dispatcher and converts the signal into demand signals for fuel flow, air flow, and feedwater flow. Load demand control can be categorized in three basic types: Boiler Follow, Coordinated Control, and Turbine Flow.

BOILER FOLLOW. The Boiler Follow scheme has been used since the early days of steam power plants. In this scheme, the boiler control and the turbine control are two separate control systems (Fig. 18-4). The load demand signal is sent directly to the turbine control system which positions the turbine governor valves in response to changes in load demand. The subsequent valve movement brings about a change in steam flow to the turbine. The change in the steam flow induces a steam pressure disturbance (increase or decrease) at the turbine throttle (turbine valve inlet). The underlying principle of this scheme is that the throttle pressure is an indication of the balance between the supply and the demand. The throttle pressure goes up when the steam supply exceeds the demand, and goes down when the supply lags behind the demand. Using the pressure error signal, a pressure controller in the system generates a control signal to adjust the boiler firing rate demand, thus changing the steam production rate of the boiler. The throttle pressure is brought back to normal when the new steam rate meets the steam demand from the turbine.

This scheme provides quick response to the load demand for two reasons. First, the hydraulically operated turbine valves can be positioned quickly to respond to demand changes. Second, the boiler itself is an energy reservoir with heat energy stored in the metal parts at the same level as the steam generated in the boiler. On a load increase, the drop in throttle pressure releases the stored energy to generate steam to make up for the shortfall before the increase in fuel input takes effect. Conversely, on a load decrease as the turbine valve opening is reduced, the energy stored in the excess steam is deposited into the boiler reservoir. The borrowing of energy from and depositing of energy into the boiler take place because of the change in throttle pressure. As the load change is taking place, the level of the stored energy must be restored to the same level at the new load condition. This is done by tuning the pressure controller to provide a certain amount of overfiring and underfiring during transients. The main drawback of the Boiler Follow scheme (which contains a single-element pressure controller) is that it is reactive. The control mechanism starts to respond to the demand for steam flow only after the occurrence of the pressure disturbance.

To improve the response of this scheme, a feedforward signal indicative of the load demand of the turbine can be added to the single-element pressure control loop. The typical signal used for this purpose is the turbine first-stage pressure signal. The first-stage pressure is an indication of the turbine steam flow and the pressure changes as soon as the turbine governor valves move to increase or decrease the steam flow to the turbine. With the addition of this feedforward element, the firing rate directly increases or decreases with the feedforward signal. The change takes place as soon as the first-stage pressure changes without waiting for the change in the throttle pressure to occur.

With the two-element scheme, the responsibility for matching the steam supply to the steam demand falls mainly on the feedforward signal. The pressure controller is used for minor adjustment of pressure error and to provide overfiring and underfiring during transient conditions. The two-element Boiler Follow scheme is more responsive to load changes. Its only disadvantage is that the system is slow in bringing the throttle pressure back to its set point. The control action starts with a change in turbine valve position followed by changes in steam flow, steam pressure, and boiler firing rate. Because the boiler and the turbine are directly connected to each other, a change in any of these variables causes a disturbance in all others and can result in unstable boiler operation.

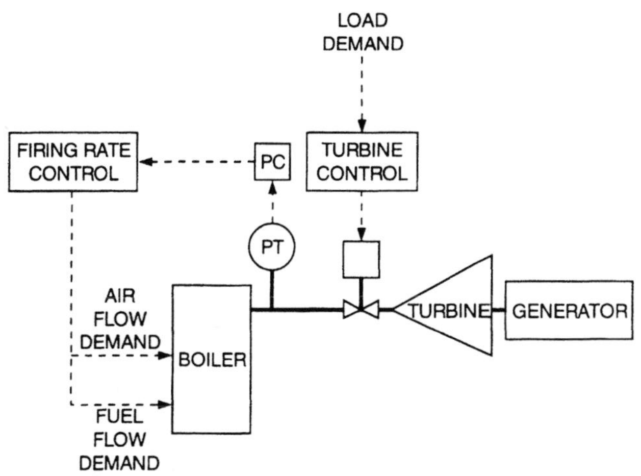

Fig. 18-4. Boiler follow scheme.

COORDINATED CONTROL. The Coordinated Control scheme came into being with the advent of once-through boilers in the early 1960s. The two-element Boiler Follow scheme was not adequate for once-through boilers. The Boiler Follow scheme utilized the stored energy in the boiler to supply steam during transients. Once-through boilers are drumless and do not have as much storage capacity as the drum-type boilers. The once-through boilers also have a startup bypass system that is not required for drum type boilers. The Coordinated Control scheme was developed to accommodate these and other characteristics of the once-through boilers. Since the 1970s, this control scheme has also been used to control drum type boilers because of the fast response and precise load control required of the individual generating units in the ever-growing interconnected systems. The basic principles of the Coordinated Control scheme are that load demand is provided as a feedforward signal to both the boiler and the turbine control system in parallel, and that closed loop controls are provided for both megawatts and throttle pressure. The objective is to use the load demand signal to simultaneously change the turbine valve position and the firing rate on load change while minimizing the interaction between the boiler and the turbine variables.

The Coordinated Control scheme, in its simplest form, is shown in Fig. 18-5. Many variations in method are used to implement the Coordinated Control concept. All major control system manufacturers have a particular design. A megawatt controller is provided to take the load demand signal and produce a control signal using the actual megawatt output of the generator to close the loop. The megawatt control signal is sent in parallel to the turbine and the boiler control system. The control signal to the turbine becomes a demand for steam flow, and a flow controller with steam flow feedback is used to close the loop. Without the flow controller, the signal to the turbine would in effect be a demand for valve position that is not linear with load. The insertion of the flow loop linearizes the relationship between the megawatt demand and the generator output and further enhances the response of the turbine valves to load demand changes.

The megawatt control signal is also the feedforward signal to the firing rate control and a pressure controller is provided to control the throttle pressure at its set point. With such an arrangement, both the turbine and the boiler control system immediately respond to a load change because of the effect of the feedforward signal. The steam flow loop further assists the turbine valves in reaching the desired valve position for the new load demand. The throttle pressure error resulting from the valve movement activates the pressure controller. The pressure error of the transient condition enables the system to utilize the stored boiler energy to temporarily satisfy the turbine steam demand. The pressure controller provides the signal to overfire or underfire during the process to restore the energy to the level corresponding to the prevailing boiler load, and ultimately, restore the pressure back to the set point.

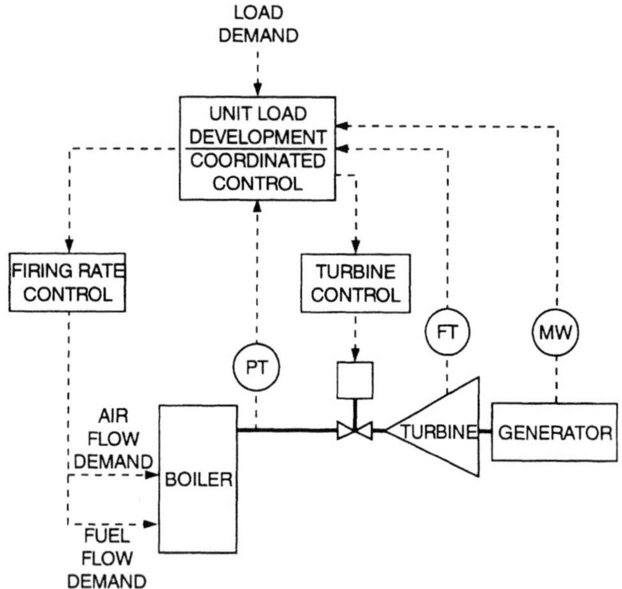

Fig. 18-5. Coordinated control scheme.

During steady-state operation, a change in the fuel's British thermal unit (Btu) content, feedwater temperature, or other variables related to boiler and turbine operation would also result in a pressure error, even when there is no change in load demand. Under such circumstances, the pressure controller would automatically sense the error and make the correction to restore the pressure to the normal set point.

The concept of coordinating the control actions of the turbine and boiler can be implemented in many ways. In one variation of the Coordinated scheme, the turbine and the boiler receive the load demand signal in parallel, but the turbine valves are used to control the throttle pressure and the firing rate is used to control megawatt error. This is more or less the opposite of the coordinated system just described and is sometimes referred to as the Integrated Control System.

TURBINE FOLLOW. A discussion on the load demand control of the Boiler Control system cannot be complete without mentioning the Turbine Follow scheme. The Turbine Follow scheme (Fig. 18-6) is the opposite of the Boiler Follow scheme. The load demand signal is sent directly to the boiler control system to regulate the boiler firing rate and the turbine control valves follow by controlling the throttle pressure.

On a load increase, for example, the load demand signal causes the firing rate to increase first. The throttle pressure goes up as the increased firing rate generates more steam.

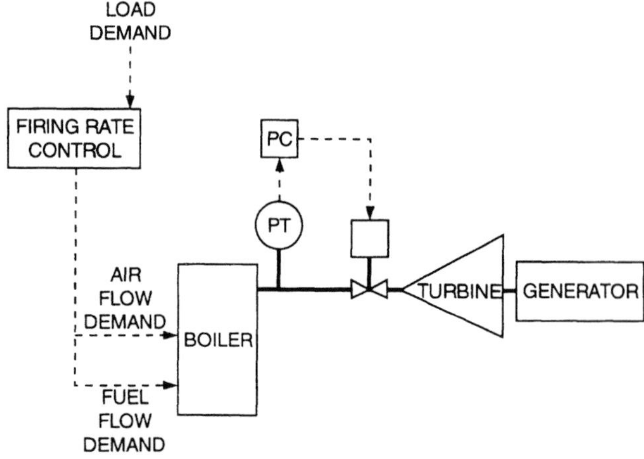

Fig. 18-6. Turbine follow scheme.

The turbine valves then open to admit more steam to the turbine and brings the pressure back to normal.

The Turbine Follow scheme provides a tight control on throttle pressure, but its response to load changes is very slow because the valves do not move until the firing rate has been adjusted to the new demand level. This scheme does not make use of any available stored energy from the boiler. Because of its slow response to load changes, there has rarely been any boiler control system with the Turbine Follow scheme alone. However, this system is mainly an offshoot of the Coordinated Control system and is often included in a Coordinated Control system as an automatically selected control mode to be used when the steam-generating capability of the boiler is temporarily limited because of loss of burners, fans, pumps, or pulverizers, or similar contingency operations.

A Coordinated Control scheme typically includes a unit load development program. The unit load development program is that part of the Coordinated Control scheme that receives the load demand from the load dispatcher and checks the demand signal against the allowable operating limits for the unit. The demand is subject to the maximum load that can be carried by the boiler–turbine–generator (BTG) unit and the minimum load desired. During online operation, the demand can also be automatically run back in the event of a loss of a major piece of plant auxiliary equipment such as losing one forced draft fan, one induced draft fan, or one boiler feed pump, which limits the load carrying capability of the unit. A maximum rate of change is imposed upon the load demand to ensure that unit load increases or decreases are executed within the limits that are safe for the operation of the BTG unit. The demand, after having been processed in this manner, is used as the load demand signal in the Coordinated Control scheme.

18.4.1.2 Firing Rate Control. Firing rate control is also known as combustion control. The function of this control loop is to adjust the fuel flow and the combustion air flow to the boiler to meet the firing rate demand from the load demand control system. A typical firing rate control scheme is shown in Fig. 18-7.

In general, the firing rate control loop employs two closed loop controllers: the total fuel flow controller and the total air flow controller.

The firing rate demand serves as the fuel flow set point for the fuel controller which uses the measured fuel flow as the feedback signal to close the loop. The controller output (fuel demand) regulates the fuel flow to the boiler. In the case of oil- or gas-fired boilers, the fuel flow is controlled by modulating the fuel flow control valves, and the fuel flow is measured by flow meters installed in the fuel supply line to the burners. For pulverized coal- or cyclone-fired boilers, the

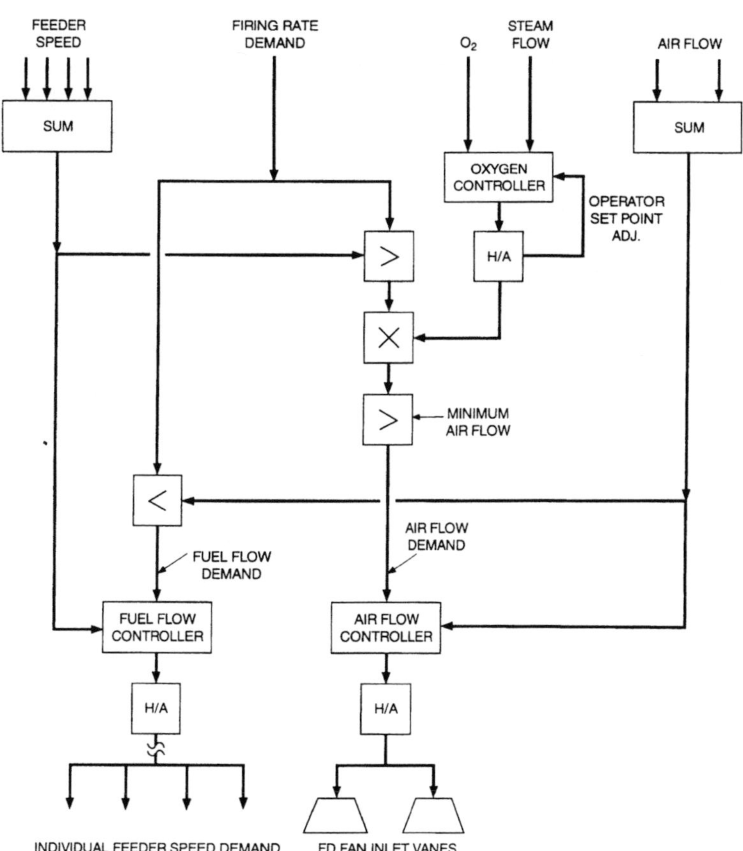

Fig. 18-7. Firing rate control.

coal flow is controlled by modulating the speed of all running coal feeders, and the feedback flow is represented by the sum of the speed of the running feeders.

The air flow set point is derived by modifying the firing rate demand signal with an oxygen control signal to account for the excess air required for optimum combustion. The optimum air flow is normally a function of the boiler load. The required excess air set point can be developed from a variable that is proportional to the load such as the boiler steam flow signal. An oxygen controller is used to compare the oxygen set point to the measured oxygen content in the flue gas and produce the oxygen control signal. The firing rate demand is multiplied by the oxygen control signal before it is used as the set point for the air flow controller. Thus, the air flow set point changes proportionally as the oxygen signal varies with load. This is why it is often said that the boiler air flow is controlled by fuel/air ratio. The air flow feedback signal is the sum of the air flows through the parallel air paths to the wind box and, in the case of coal-fueled boilers, the sum of all primary air flows to the individual pulverizers is added as a part of the feedback signal. The air flow controller output is the air demand signal which controls the boiler air flow by regulating the inlet vanes, the inlet control damper, or the actual speed of the forced draft fans.

There are two major considerations in controlling the boiler firing rate. First, the furnace must be kept air-rich under all operating conditions to ensure a safe combustion environment. This is accomplished by comparing the measured air flow to the fuel control signal, and also by comparing the measured fuel flow to the air flow control signal so that on load increase, the air flow increase will lead the fuel flow, and on load decrease, the fuel flow decrease will lead the air flow. The net result is that the boiler air flow is higher than the fuel flow at all times when the boiler is in operation.

Second, the air flow through the furnace must be maintained above a level that is sufficient to sweep off any unburned combustibles that could possibly accumulate in the air and gas flow path. A minimum air flow limiter is used to keep the air flow control signal from dropping below the minimum level before the air flow control signal is applied to the control drives.

18.4.1.3 Pulverizer Coal and Air Flow Control. For coal-fueled boilers, the coal flow is regulated by varying the speed of all coal feeders conveying coal to the pulverizers. The coal from each feeder is ground into fine powder in the pulverizer and carried into the coal burners by the primary air pushed through the pulverizer by the primary air fan. This necessitates a separate control loop for each pulverizer to control the volume of coal and air flow mixture sent to the furnace through the coal burners. The primary air is also used to dry the coal in each pulverizer. The air temperature must be high enough to ensure evaporation of moisture from the coal and low enough to prevent a pulverizer fire. For this reason, the temperature of the coal–air mixture must also be controlled. The method of implementing the flow and temperature control functions depends on the pulverizer type. All pulverizers can be classified as either low-storage or high-storage. In a low-storage pulverizer, the coal actually supplied to the burners is closely related to the coal feed rate to the pulverizer. There is very little time delay between the coal flow to the burners and the coal feed from the feeder. The opposite is true for the high-storage pulverizers. The following paragraphs describe the pulverizers most commonly used on utility boilers.

BALL AND RACE TYPE PULVERIZERS. In this low-storage type pulverizer, the fuel demand signal from the firing rate control loop regulates the speed of the feeder that supplies coal to the pulverizer (Fig. 18-8A). The same signal is used to position the primary air damper in the primary air pipe leading to the pulverizer. The speed control is open loop without using a speed feedback signal since the speed feedback is already used in the total coal flow controller in the firing rate control loop. To control the primary air flow to the pulverizer, the feeder speed demand is used as the flow set point with the measured air flow used as the feedback signal. The coal–air mixture temperature is controlled by modulating the hot and the cold primary air damper in opposite directions. The primary air is supplied from the primary air fans to a header before it is distributed to the pulverizers. The header pressure is controlled by regulating the control damper at the fan inlet in a closed pressure control loop.

BOWL AND ROLLER TYPE PULVERIZERS. This type of low-storage pulverizer can be operated under pressure or under suction. If it is operated under suction, an exhauster is installed at the outlet of the pulverizer, drawing the coal–air mixture to the burners. The coal–air flow is controlled by regulating in parallel the feeder speed and the exhauster damper in response to the demand signal from the firing rate system. No feedback measurement is used. The coal–air mixture temperature is controlled by modulating the hot air damper at the hot air inlet to the pulverizer. Cold air is introduced to the pulverizer through an atmospheric pressure damper that is mechanically modulated to maintain proper suction at the pulverizer inlet.

If the pulverizer is under pressure, the pulverizer is typically equipped with a hot and a cold air damper without a primary air damper. The coal flow is controlled by regulating the feeder speed, and the primary air flow is controlled by modulating the hot air and the cold air damper in parallel (Fig. 18-8B). The coal–air temperature is controlled by regulating the hot air and the cold air damper in opposite directions.

BALL AND TUBE TYPE PULVERIZERS. This type of pulverizer has a large coal storage capacity and is typically equipped with two feeders. Because of the large coal storage capacity in the pulverizer, the speed of the feeders is not a good indication of the coal flow carried by the primary air to the furnace. Instead, the amount of coal being supplied to the

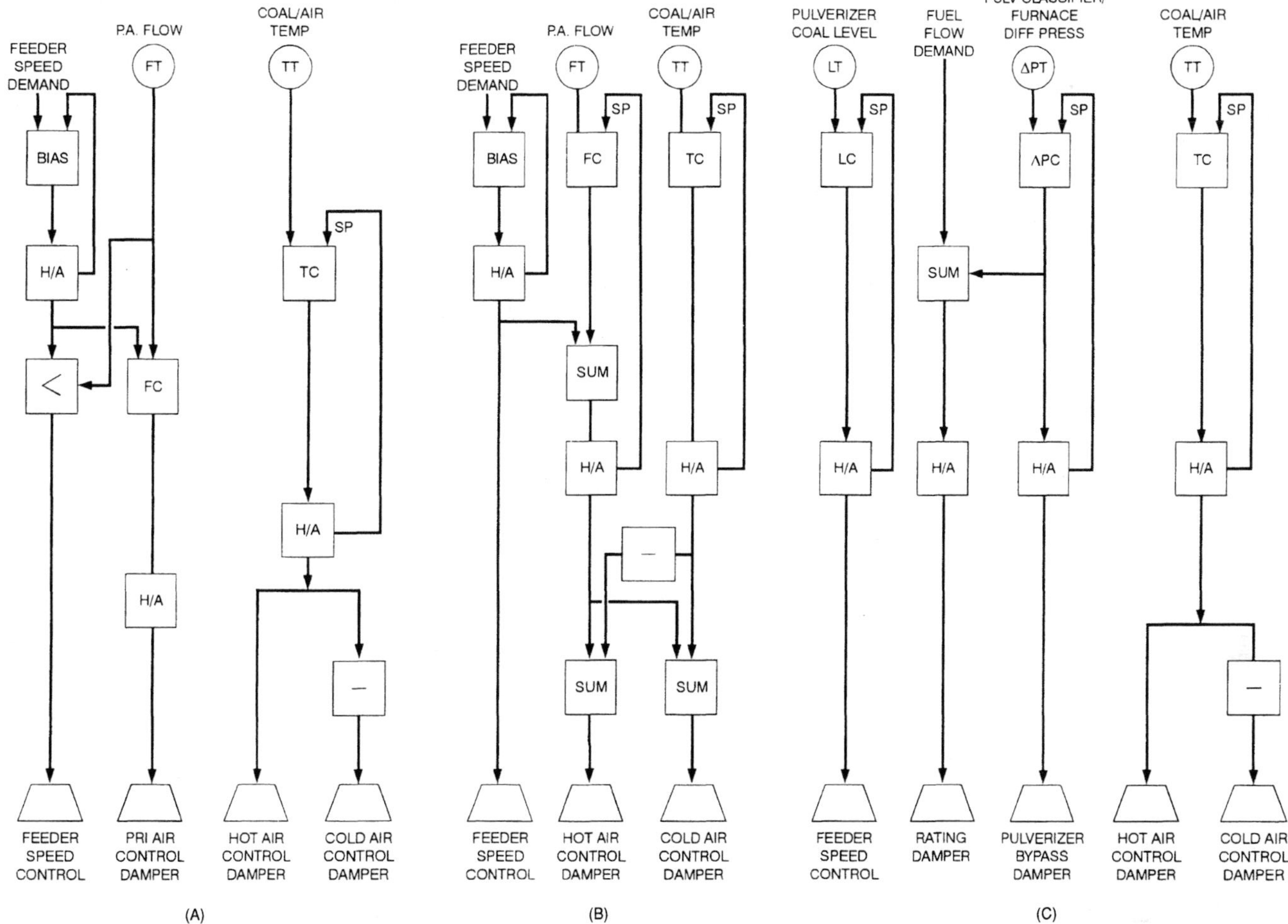

Fig. 18-8. Pulverizer coal/air flow control. (A) Ball and race type pulverizer. (B) Bowl and roller type pulverizer (pressurized). (C) Ball and tube type pulverizer.

burners is directly related to the velocity of primary air sweeping the pulverizer.

The fuel demand signal is sent to the pulverizer control loop to regulate the primary air flow to the pulverizer as load changes by modulating a rating damper located in the primary air duct to the pulverizer (Fig. 18-8C). The rating damper is equivalent to the primary air flow control damper of the ball and race type pulverizers. A pulverizer level control loop is included to control the speed of the two feeders in parallel, controlling the level in the pulverizer at a preset level, thus maintaining a constant coal inventory in the pulverizer.

A bypass damper is installed on each side of the pulverizer, diverting some of the primary air from the pulverizer air inlet to the coal–air outlet of the pulverizer. The bypass damper opens at low loads when the demand for coal is low, thus maintaining a minimum velocity of primary air to prevent any dropout and accumulation of coal in the pipes to the burners. The temperature of the coal–air mixture is controlled by modulating the cold and the hot primary air damper in the same manner as the other types of pulverizers.

The vast majority of utility boilers are equipped with more than one pulverizer. While the design of the feeders and the pulverizers for any given boiler may be identical, it is highly unlikely for the individual pieces of material handling equipment such as feeders and pulverizers to respond to the load demand in exactly the same manner. For the same demand signal at the same speed, one feeder may supply coal at a slightly higher or lower rate than the others. The same holds true for the coal–air flow out of the pulverizers. For this reason, the control system must have manual bias capability to adjust the automatic control signals to the primary air flow control dampers and the variable-speed drives of the feeders. All control systems have this characteristic in common, regardless of the type of pulverizer installed on a boiler. Manual bias adjustment is typically accomplished by using a bias adjustment knob on the hand/automatic control stations.

If the pulverizers of a boiler are operated under pressure, the primary air supply may be from an individual primary air fan installed for each pulverizer or from a common primary air system. In the latter case, the common primary air fans supply air to a common header for distribution to the pulverizers. The air pressure at the header is controlled by modulating the inlet control dampers of the primary air fans.

18.4.1.4 Secondary Air Flow Control. The forced draft fans supply combustion air through the air heaters to the wind box. For coal-fueled boilers, this is known as the secondary air as opposed to the primary air to the pulverizers. In a wall-fired furnace, the secondary air in the wind box enters the furnace through the openings around the burners. In recent years, some large wall-fired boilers have been equipped with a compartmentalized wind box for better air flow distribution and tighter control of the excess air at the burners. The compartments are typically arranged according to burner elevations with two compartments for each elevation of burners, one on each side of the furnace. The secondary air flow through each compartment is measured with air foils and regulated by a volume damper in the compartment.

The total air flow control in this case is slightly different from the scheme used on boilers without a compartmentalized wind box. The forced draft fans are used to control the secondary air supply pressure at the air duct to the wind box. A flow control loop is provided for each compartment to regulate the air flow according to the coal flow to the burners in the compartment.

In the case of a tangentially fired furnace, the secondary air enters the furnace through compartments at the burner corners where the burners are arranged vertically in alternate compartments. A damper is installed in each compartment. The damper in the burner compartment is called the fuel/air damper, and the damper in the compartment without a burner is called the auxiliary air damper. The fuel/air dampers are modulated in proportion to the fuel input to the burners, and the auxiliary air dampers are modulated to maintain a set wind box to furnace pressure differential to improve the air distribution through the four corners of the furnace.

18.4.1.5 Furnace Draft Control. On balanced draft boilers, the forced draft fans supply the combustion air to the furnace, and the induced draft fans remove products of combustion or the flue gas from the furnace. The furnace draft is controlled at a slightly negative pressure. Either the forced draft fans or the induced draft fans can be used to control the furnace pressure (furnace draft), and the other is used to control the air flow. The furnace pressure control and the air flow control are closely related and a change in one inevitably has an impact on the other.

To minimize the interaction between the two control loops, the selection of furnace draft and air flow control is made depending on how the air flow is measured. The furnace pressure is controlled by the forced draft fans if the air flow measurement is on the gas side of the boiler (such as the gas pressure differential across the convection pass). Conversely, the induced draft fans are used if the air flow is measured on the air side, for example, by flow elements such as piezometer ring at the forced draft fan discharge or air foils in the secondary air duct. Today, nearly all utility boilers use the induced draft fans for furnace pressure control because the air flow in modern boilers is typically measured on the clean air side of the furnace.

Furnace draft control is basically a two-element control loop using a furnace pressure controller plus a feedforward signal representing a variable that, when changed, has an impact on the furnace draft such as the boiler air flow or the forced draft fan inlet control vane position (Fig. 18-9).

The furnace of large boilers equipped with high-head and high-capacity induced draft fans presents a potential risk for furnace implosion. In most cases, the implosion is the result of the rapid decrease in gas pressure in the furnace following the decrease in gas temperature caused by a sudden collapse of the furnace flame on a master fuel trip. An implosion can also result from a malfunction of any draft equipment in the air and gas path and the ensuing imbalance between the air flow and the gas flow.

To minimize the potential for the occurrence of any such upset conditions, NFPA Code 85C (Standard for the Prevention of Furnace Explosion/Implosion in Multiple Burner Boiler-Furnaces) specifically recommends interlocking functions that must be incorporated into the operation of the forced draft and induced draft fans. The measures pertaining to the furnace pressure control loop include the following:

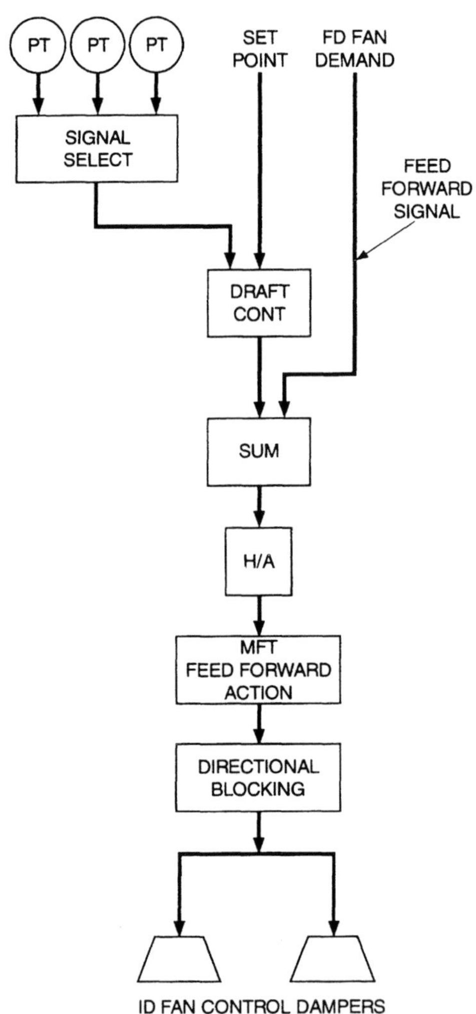

Fig. 18-9. Furnace draft control.

- Use three furnace pressure transmitters as a measure of redundancy.
- Use a feedforward signal to anticipate any demand changes.
- Apply directional blocking interlocks to the fans.
- Use feedforward action on a master fuel trip.

When three transmitters are used to provide input to the control loop, the measurements from the transmitters are continuously compared against one another. A voting logic, typically a median selector, is provided to select the valid signal as the feedback signal to the loop. The required feedforward signal is typically implemented in the two-element control system.

The purpose of directional blocking is to minimize pressure increases or decreases under unstable operating conditions. On an excessive furnace pressure excursion on the negative side, this interlock function prevents the forced draft fan control dampers from closing further and the induced draft fan control dampers from opening. Conversely, on excessive high furnace pressure, the interlock function blocks any further opening of the forced draft fan dampers and closing of the induced draft fan dampers.

Feedforward action on a master fuel trip is used mainly to runback the induced draft fan control damper to slow down the furnace pressure decay. The control system gradually dissipates this action after a time delay as the furnace pressure returns to the set point.

In addition to the measures applied to the furnace draft control loop, the same NFPA code also provides guidelines for operating the forced and induced draft fans. Protective interlocks must be used to ensure an open flow path from the forced draft fan inlet through the furnace, and the induced draft fans to the stack under all operating conditions, and the dampers on idle fans must not be closed unless other forced draft and induced draft fans are in operation. The forced draft and induced draft fans must be operated in pairs, and when a fan is tripped, its companion fan must also be tripped. The objective is to avoid any circumstance that could possibly cause an imbalance between the air and the gas flow through the furnace.

18.4.1.6 Superheat and Reheat Steam Temperature Control. Superheat steam temperature is controlled by regulating spray water flow to the superheater desuperheaters in nearly all utility boilers. The temperature control loop, in its simplest form, is a single-element control loop that measures the superheater outlet steam temperature and compares it to a set point. The controller output provides a corrective action to modulate the spray control valve that regulates the water to the desuperheater. This type of control is used mostly in small boilers where load upset is minimal.

An improvement to the single-element control is the two-element cascade-type control loop. Two controllers are provided: the superheater outlet temperature (primary) controller and the desuperheater outlet temperature (secondary) controller (Fig. 18-10A). The control output of the primary controller serves as the set point for the secondary controller which sends an output signal to modulate the spray water control valves. This arrangement allows the secondary controller to correct any disturbance in the desuperheater outlet temperature before the disturbance affects the superheater outlet temperature. Most utility boilers add a third element to the control system. The third element is an anticipatory signal indicative of any forthcoming load change that will upset the steam temperature in the boiler. The signal often used for this purpose is the boiler air flow signal. The superheat temperature can be controlled at its design point only if the boiler load is high enough to make superheat design temperature. The load point above which the steam temperature can be controlled is known as the control point. Below this point, the superheat outlet temperature drops and the spray control signal decreases until the flow control valve is automatically shut off.

Boiler reheat steam temperature, if not controlled, will drop below the design condition as boiler load decreases from full load. The temperature is raised to the design point by manipulating heat distribution within the boiler gas path. The method used to accomplish this varies from manufacturer to manufacturer. There are mainly three major ways to control boiler reheat steam temperature: gas recirculation, burner tilt, and damper positioning (refer to Chapter 7 for more information). In rare occasions, the amount of excess air is adjusted to control the reheat temperature. The control is essentially a two-element loop with a temperature controller and a feedforward signal representing boiler load such as boiler air flow. As the boiler load changes, the feedforward signal modulates the gas recirculation fan dampers, tilting burners, or the convection path gas dampers toward the desired position with additional adjustment of the output accomplished by the temperature controller based on the reheat temperature error (Fig. 18-10B).

Because of its detrimental effect on unit heat rate, reheat spray is designed to be used only during transient conditions when the reheat temperature excursion exceeds the desired set point by a preset margin, and in some cases, when operating the unit under overload conditions. Reheat spray at overload conditions may be required because boilers typically are designed without any action by burner tilt, gas recirculation, or damper positioning when operating at the maximum continuous rating (MCR). The reheat temperature will likely exceed the desired set point when the boiler load is increased above the MCR. Under such conditions, the reheat spray must be used to reduce the temperature to the set point value.

A spray controller is used for the reheat spray valves. The set point for the spray controller is set slightly higher than the set point for the heat distribution controller so that the control valves will start to function only when the temperature rises above the design temperature. As an alternative, a two-element cascade control loop may be used for the spray valves, with the second element being the steam temperature at the reheat desuperheater outlet similar to the scheme used for superheat temperature. However, this method is not often used because the spray valves are expected to open only under transient conditions and are closed during normal operation.

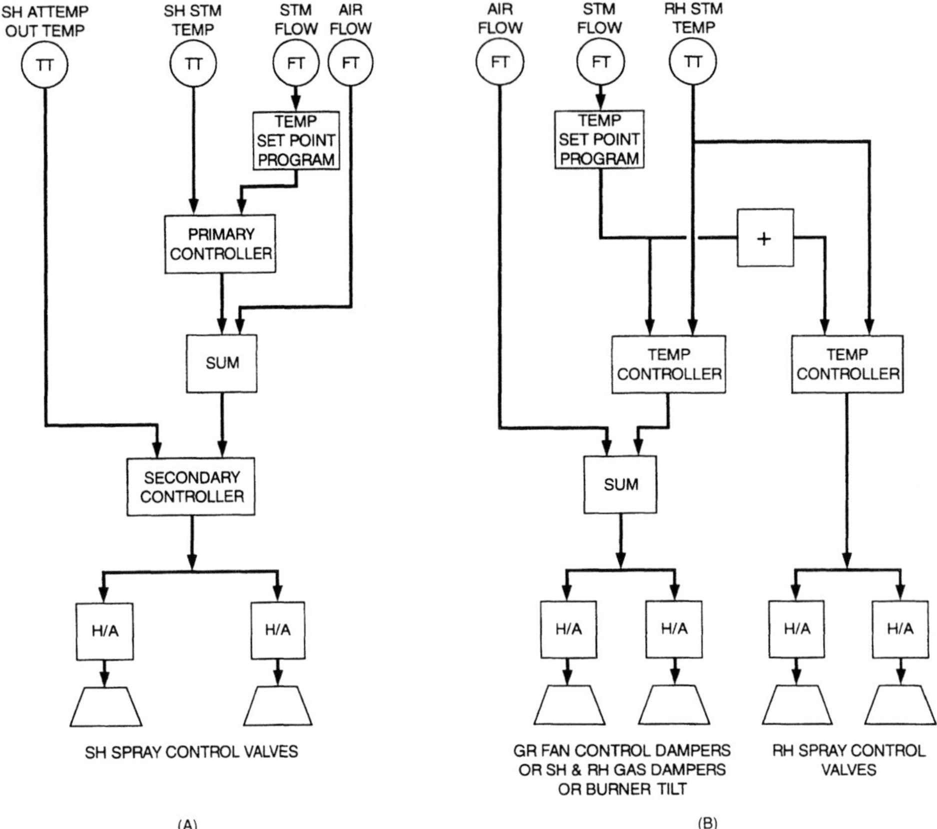

Fig. 18-10. Superheat and reheat temperature control.

Both the superheat and the reheat spray water flow represent a potential source of water induction to the turbine. To guard against this possibility, the spray valves are provided with tight shutoff valves that are interlocked so that the water flow is automatically blocked on a turbine trip. The valves remain closed during subsequent startup operation until the turbine load reaches a minimum level. As an added precaution, during normal operation, the shutoff valves are closed automatically whenever the spray water demand from the temperature control loop goes to zero.

18.4.1.7 Feedwater Flow Control. The purpose of the feedwater flow control is to supply feedwater to the boiler to make up for the steam supplied to the turbine. In drum type boilers, the feedwater flow is regulated to maintain the boiler drum level at a preset level. Any deviation of the drum level from the set point is an indication that the supply of feedwater to the boiler and the steam flow from the boiler to the turbine are out of balance, and a change in feedwater flow is necessary to restore the balance.

A feedwater flow control system that regulates the boiler feedwater flow by utilizing only the steam drum level is called a single-element control system. Single-element control is adequate for small boilers; however, with large boilers, drum level is not a good indication of whether the steam and feedwater flows are in balance because of the shrink and swell effect in the steam drum during load changes. On decreasing load, closing the turbine valves causes the steam in the steam line to back up to the drum, resulting in a pressure increase which reduces the size of the steam bubbles in the steam-generating tubes and "shrinks" the drum level. Conversely, on load increase, the opening of the turbine valves produces a drop in drum pressure and causes the water in the generating tubes to flash into steam; the steam bubbles produced serve to "swell" the drum level. In either case, the ensuing level deviation misleads the control system to change the feedwater flow in the wrong direction. The remedy for this is a three-element system, with the three elements being feedwater flow, steam flow, and drum level (Fig. 18-11).

This system is basically a cascade plus feedforward control loop. Steam flow is used as the feedforward signal which is summed with the drum level control signal and becomes the demand for feedwater flow. In general, the turbine first-stage pressure is used as the steam flow signal and the feedwater flow is measured at a point on the feedwater line between the boiler feedwater pump discharge and the economizer inlet. Feedwater flow measurement typically is temperature compensated to account for the change in water density because of variations in temperature.

The final control elements are the feedwater control valves installed in the feedwater line to the boiler for constant-speed motor-driven boiler feed pumps. Variable-speed motor-driven pumps are equipped with a hydraulic

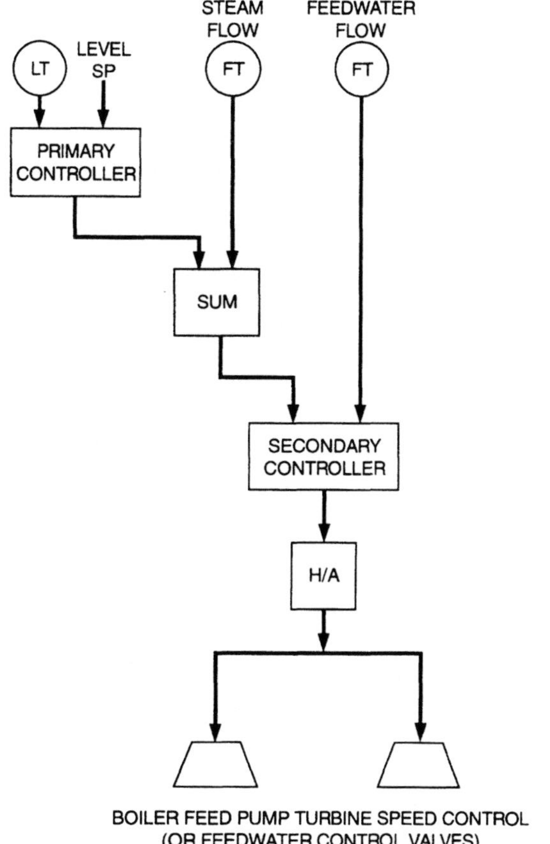

Fig. 18-11. Feedwater control.

coupling with a scoop tube that is positioned by a drive unit to change the speed of the pump.

With turbine-driven boiler feed pumps, the turbine speed is varied by positioning the turbine control valves. A typical turbine drive has a high-pressure and a low-pressure steam chest, each equipped with a hydraulic type control valve. The two valves are operated in sequence. The low-pressure valve opens first, taking extraction steam from the turbine. When the low-pressure valve has reached the maximum position, the high-pressure valve opens to admit additional steam to the turbine. The most common steam sources for the boiler feed pump turbines are the crossover steam to the low-pressure valve and main, cold reheat, or auxiliary steam to the high-pressure valve.

18.4.1.8 Superheater Bypass and Startup System Control. There are many types of bypass systems. The following describes the control loops for one particular bypass system being used primarily to facilitate boiler startup.

The major control loops involved in a boiler equipped with a superheater bypass system include low-load firing rate control, superheat attemperator control, and reheat attemperator control (Fig. 18-12).

A boiler with a superheater bypass system has a bypass valve from the boiler drum to the condenser and a set of division valves located immediately downstream of the primary superheater (see Chapter 7 for more information). During startup operation, the low-load firing rate control loop regulates the boiler firing rate to raise the gas temperature at the furnace exit and build up drum pressure. The drum pressure is maintained at a manually adjusted set point by modulating steam flow to the condenser through the bypass control valve. The steam pressure at the turbine inlet is raised and controlled through the throttling action of the division valves. After the steam line is heated and the steam condition matches the turbine metal temperature, the turbine valves are opened to admit steam to the turbine. As the steam enters the turbine, the generator may be synchronized. Once the generator is synchronized, the turbine load may be increased by releasing the stored energy in the drum through the division valves and pressurizing the superheaters downstream of the division valves without increasing the firing rate. Pressurizing the superheaters in this manner results in a fast and stable startup operation while matching steam temperature to the metal condition of the turbine as compared to pressurizing the superheater by increasing the firing rate as is the case of a boiler without a bypass system.

During this period of operation, both the superheat and

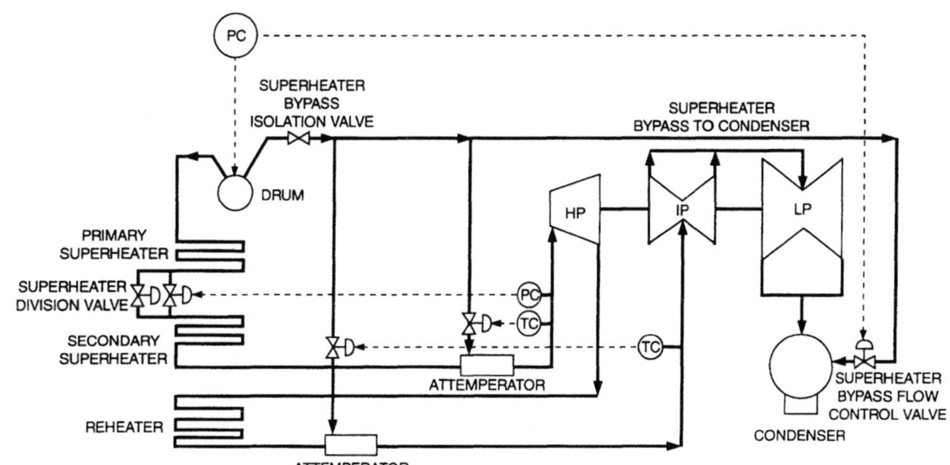

Fig. 18-12. Superheater bypass control.

the reheat attemperator control the superheat and the reheat steam temperature at a set point value that is either manually adjusted or developed from the turbine control system according to the prevailing metal condition in the turbine by regulating spray water flow to the superheat and the reheater attemperator, respectively.

18.4.2 Auxiliary Control Loops

The following paragraphs describe some auxiliary control loops commonly installed in power plants whose functions are not directly related to the boiler control loops.

18.4.2.1 Closed Cycle Cooling Water System. The closed cycle cooling water system supplies cooling water for various pieces of equipment (pump bearings, compressor cylinders, lube oil coolers, etc.) where clean water is required as the cooling medium. The cooling water system typically employs a set of cooling water pumps that supply water to the equipment via a water supply header. The return water is routed back to the pumps suction through a return header. The system consists mainly of three control loops: water supply, water temperature, and makeup water control.

A water bypass line with a control valve is installed between the two headers. The water supply loop ensures adequate cooling water flow through the system by modulating the bypass valve to maintain a constant pressure differential between the supply and the return header.

The water picks up heat as it passes through the various pieces of equipment in the cooling circuits. Therefore, the return water must be cooled in the closed cycle cooling water heat exchanger before it is pumped back to the supply header. The water temperature is controlled by modulating the cooling water supply to the heat exchanger. The heat exchanger cooling water is usually supplied from the plant's raw water system.

Any water loss in the closed cycle cooling system circuit is made up by adding makeup water from the plant's treated water system. This is done by regulating makeup water to maintain the water level in the closed cycle cooling water head tank.

18.4.2.2 Feedwater Heater Level Control. Feedwater heater level is controlled to keep the tube surface in the condensing zone of the heater from being submerged, to keep the drain cooler flooded, and most importantly, to prevent the level from rising into the extraction line. Two separate level control loops are provided for each heater, in accordance with the ASME Standard "Recommended Practice for the Prevention of Water Damage to Steam Turbines Used for Electric Power Generators." The first control loop controls the heater level at a normal set point by regulating the heater drain flow through a control valve to the shell side of the next lower pressure heater. An emergency drain valve to the condenser is also provided. The emergency control valve is controlled from the second control loop. The emergency valve is activated when the level rises above the normal level as a result of tube leak or other emergency conditions that will flood the heater.

Heater drain pumps are sometimes used for economic reasons. For high-pressure heater drains, the pump is used to pump the drain to the deaerator. For low-pressure heaters, the drain is normally pumped into the condensate stream. In either case, an emergency drain line is installed at the pump suction to dump the drains to the condenser on the occurrence of high level in the heater.

An additional interlocking is provided to automatically shut off all cascading drain valves and extraction line block valves leading into a heater in the event that the level in the heater rises above the set point of the second level control loop. In the case of heaters installed in the condenser neck where no block valves are installed on the extraction line to the heater, power-operated condensate inlet shutoff valve and heater bypass valve are interlocked to heater level. A high–high level condition automatically closes the inlet valve and opens the bypass valve, stopping water flow to the tube side of the heater altogether.

18.4.2.3 Deaerator Storage Tank and Condenser Hotwell Level Control Loops. The condensate system withdraws condensate from the condenser hotwell and pumps the condensate through low-pressure feedwater heaters to the deaerator. The condenser hotwell and the deaerator storage tank are the two reservoirs in the turbine cycle. There are three control loops for this system (Fig. 18-13).

The deaerator storage tank serves as a surge tank for the boiler feedwater. The tank level is controlled by a control valve on the condensate supply line to the deaerator. As with the boiler drum level control loop, a three-element scheme can be used for this application to enhance system response. In this case, the three elements are the tank level, feedwater flow out of the tank, and condensate flow into the deaerator. The condensate flow into the deaerator can be regulated by either modulating a control valve on the condensate line into the deaerator or by varying the speed of the condensate pumps. With variable-speed pumps, the pump discharge pressure varies with load. At low loads, the condensate pressure may be too low to be used for auxiliary services such as boiler feed pump injection water and turbine exhaust hood spray. In such instances, care must be taken to ensure that the pressure does not drop below the level required for such services.

The condenser hotwell serves as the reservoir for water in the turbine cycle. Periodically, cycle makeup water is needed because of water loss through water leakage, steam venting, or nonrecoverable steam usage during operation. The water lost must be replaced from an external source.

The condensate level in the hotwell is generally an indication of the amount of water in the turbine cycle. The hotwell level can be controlled within a safe range by modulating makeup and dump control valves. When the level reaches the high end of the level range, the dump valve opens to dump the excess condensate from the condenser hotwell to a con-

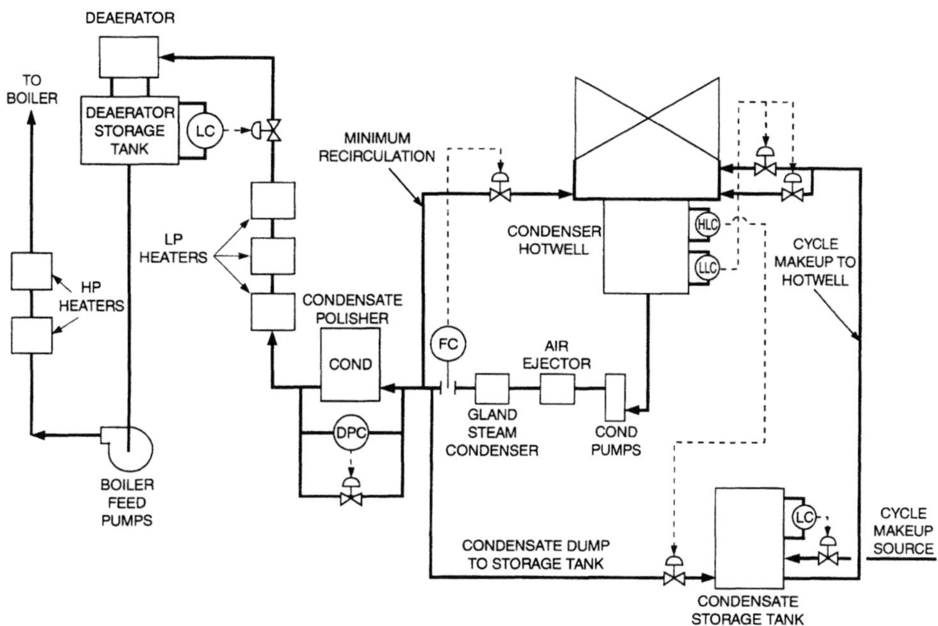

Fig. 18-13. Condensate system controls. HLC = Hotwell high-level controller; LLC = hotwell low-level controller.

densate storage tank. Loss of condensate from the turbine cycle is reflected by a low level in the hotwell. When the level reaches the low point, the makeup valves open to supply makeup water to the cycle. The makeup flow can vary drastically depending on the magnitude of the loss in condensate during operation. Because of the low-pressure differential between the condensate storage tank and the hotwell, the usual arrangement is to have two valves operating in sequence, using one valve for low-makeup flow condition and both valves on high-makeup flow.

As an alternative, the hotwell level can be controlled by regulating the condensate flow out of the hotwell, and the deaerator tank level can be controlled within a band (between an upper limit and a lower limit) by regulating the excess condensate flow and the cycle makeup flow. The condensate dump valve from the hotwell to the condensate storage tank opens when the deaerator tank level reaches the upper limit and the makeup valve from the condensate storage tank opens when the deaerator tank level reaches the lower limit. This scheme is seldom used because of its slow response to change in the deaerator tank level under transient conditions.

The gland steam condenser and air ejector are typically installed in the condensate stream before the condensate enters the low-pressure heaters. They require a minimum flow for proper operation. For pump protection, the condensate flow through the condensate pumps must also be kept above a minimum value. At low loads, the condensate flow required to maintain the deaerator tank level will fall short of the minimum flow required. Thus, a recirculation line from the condensate outlet of the gland steam condenser to the condenser is provided to keep the condensate flow above the minimum. The minimum recirculation control is a simple flow control loop, using the measured condensate flow to close the loop.

In systems equipped with a condensate polisher, a condensate bypass line with a control valve around the polisher is provided. When the polisher becomes dirty or the number of polisher vessels in service is reduced during operation, the condensate pressure drop across the polisher may exceed the design limit of the polisher. Under such conditions, the bypass valve opens to limit the pressure differential by regulating condensate flow bypassing the polisher.

18.4.2.4 Boiler Feed Pump Recirculation Control. Water passes through a running pump and picks up heat in the form of temperature rise; a high temperature rise can cause the water to flash into steam, and cavitation results when the steam bubbles collapse inside the pump. High temperature also causes rubbing because of the uneven expansion of the pump parts. For pump protection, the temperature rise must be minimized by maintaining a minimum flow through the pump at all times when the pump is running. For small pumps, this is usually accomplished by installing a permanent recirculation line from the pump discharge to the pump suction with an inline pressure breakdown orifice plate to restrict the recirculated flow. On large pumps, boiler-feed pumps for example, the flow is externally controlled because permanent recirculation consumes too much unneeded pumping power. The recirculation control valve of a boiler feed pump can be regulated to maintain either a minimum flow through the pump or control the water temperature rise as it passes through the pump.

If the flow measurement is used, the flow may be measured at the pump suction or the pump discharge. If the flow measuring element is located at the pump suction, it induces a pressure drop, thus reducing the net positive suction head available. A suction flow meter should be used only if the pressure loss is acceptable.

If the discharge flow is measured, the flow element must be installed at a point on the discharge line upstream of the recirculation line takeoff. This is not always practical because the pump discharge is usually routed through a congested area where the straight piping run required for the flow element is not available. If the flow element is located downstream of the recirculation line takeoff, the flow measured will be the difference between the pump through flow and the flow recirculated to the deaerator. In this case, a flow element on the recirculation line is required to provide the feedback for the control system to calculate the pump flow. The location of flow measurement is usually determined on a case-by-case basis.

If the temperature rise method is used, the temperature rise is determined by taking the difference between the suction flow temperature and the water temperature at the balancing head discharge of the pump. This method is direct, but the difficulty lies in taking an accurate reading of the temperature at the balancing head discharge because of the flow leakage of the pump seal injection water. The response of temperature sensors to temperature changes is also slow. For these reasons, this method is seldom used.

The recirculation valves are high-pressure drop valves. In the past, a pressure breakdown orifice was installed on the recirculation line at the outlet of the control valve to alleviate the burden imposed on the valve. In recent years, the improvements made on valve design have virtually eliminated this requirement.

The flow control loop may be on–off. The pump flow is measured with a high-flow and a low-flow set point. On increasing load, the valve stays open until the flow reaches the high set point. Above this point, the recirculation valve is closed. On decreasing load, the valve opens when the flow reaches the low set point. With this method, the recirculated flow is more than is required when operating between the two set points. The modulating control modulates the valve continuously to control the flow precisely at the minimum flow when the pump is running at low loads.

18.4.2.5 Air Heater Cold End Temperature Control. Boiler flue gas contains water vapor, sulfur, and other chemical mixtures. The gas temperature decreases as it passes through the air heater. If the cold end temperature of the air heater falls below the dew point of the flue gas, the water vapor condenses, leading to combination with the sulfur and sulfuric acid formation and causing corrosion of the air heater materials at the cold end. Thus, the cold end temperature must be maintained above this point. Two methods are used: air heater bypass and heating coils at the air heater air inlet.

The bypass method bypasses a portion of the incoming air from the forced draft fans around the air heater. This reduces the air flow through the air heater and raises the gas temperature leaving the air heater. The disadvantage of this method is that it reduces the heat picked up by the incoming air, thus decreasing the boiler efficiency. This method is not often used.

The heating medium for the heating coils is typically steam, water, or a glycol solution. For steam coils, the cold end temperature is controlled by regulating the steam supply from the turbine cycle to the coils. The condensate from the coils is collected in a drip tank and returned to the cycle. The water coils and the glycol coils are controlled in a similar manner.

In the case of glycol coils, the solution is recirculated in a closed loop, and the glycol, after passing through the coils, must be heated before it is returned to the coils. This is achieved by using a glycol heater with steam as the heating medium. The glycol solution temperature is raised to a set point as it flows through the heater by regulating the steam supply to the heater. The steam is condensed in the heater and returned to the cycle. The condensate level in the heater is controlled by regulating the drain flow from the heater (similar to the method used for feedwater heater level controls). The advantage of using glycol solution is that, unlike steam or water, it will not freeze even under very low ambient temperature conditions.

Because it is not practical to directly measure the actual cold end metal temperature, the feedback signal used for this loop is the average of the air temperature entering the air heater and the flue gas temperature leaving the air heater.

18.5 CONTROL SYSTEM HARDWARE AND SOFTWARE

When digital computer systems first appeared in power plants in the 1960s, they were installed as single central processor type computer systems and the major areas of application were data acquisition and supervisory control functions. Since then, as a result of the development of microprocessor technology and the rapid technological advances in recent years, the centralized digital computer system has evolved into a distributed system comprising numerous small computers, each with an assigned control or data acquisition function. Currently, all control hardware found in power plants (or other plants in industry) is known as "distributed" and "microprocessor-based." This type of hardware has, for all practical purposes, replaced the conventional type devices as the dominant tool to implement the control and monitoring functions necessary in operating a power plant. In this section, an overview of the "distributed" and "microprocessor-based" control systems and the essential elements that make up these systems is provided.

18.5.1 Microprocessor-Based Control Systems

The basic components in a microprocessor-based control system are the microprocessor, memory, and input/output modules and peripheral devices.

18.5.1.1 Microprocessors. A microprocessor-based control system is essentially a computer control system based on the latest computer hardware technology. As in any computer, there are four basic elements (Fig. 18-14) that must be present for the system to function:

- The main memory unit,
- The control unit,
- The arithmetic-logic unit (ALU), and
- The input/output (I/O) unit.

The control unit and the arithmetic-logic unit constitute the central processing unit (CPU), which is the "brain" of the computer. The memory unit is used to store programs and data. The program of a digital computer is a series of instructions that tell the computer how to do its work. The control unit is the conductor of the computer operation. It retrieves programmed instructions from the main memory and collects data (inputs) from the I/O unit for the ALU to do the work. It sends the result of the work done by the ALU (outputs) back to the I/O unit. The I/O unit shuttles the input and output data between the computer and the outside world. In the case of a control computer, the I/O unit provides the connection between the computer and the field-mounted process sensors and control drives. A computer system may have more than one I/O unit.

Some I/O units are also known as peripheral devices. Typical peripheral devices include cathode-ray tube (CRT) monitors, printers, plotters, keyboards, and auxiliary memories such as hard disks and magnetic tapes. Auxiliary memory supplements the main memory and is used to store instructions and data that are either too extensive for storage in the main memory or are not frequently used. Auxiliary memory also serves as the permanent offline storage place for all programs in the computer system.

The CPU was once the most expensive item in a computer system. With a structure as shown in Fig. 18-14, each system had only one CPU. The CPU executed the programmed instructions one at a time, generally in a prescribed sequence regardless of the size of the program and the number of programs stored in the system. All mainframe computer and

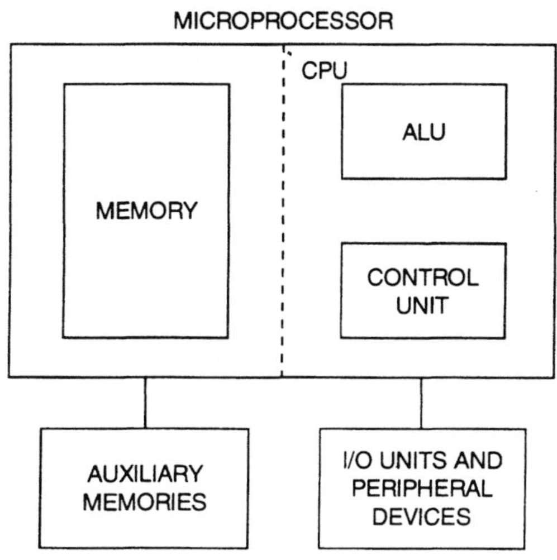

Fig. 18-14. The basic computer.

minicomputers operate this way. The advent of Integrated Circuit (IC) technology, followed by Large-Scale Integration (LSI), and in recent years Very Large Scale Integration (VLSI), has greatly reduced the size of computer components and drastically reduced the hardware costs. This has enabled the manufacturers to put all components of the ALU and the control unit on a single chip. The term microprocessor is used to denote a chip that contains all the components of a CPU. A microprocessor should not be confused with a computer. For a microprocessor to function as a computer, it requires the addition of one or more memory modules plus one or more interface chips to operate the various I/O devices. The term micro has been used to denote the small physical size of the components. The microprocessor technology thus makes it possible for control system manufacturers to put more than one CPU (or microprocessor) in a given system and distribute the system functions throughout the various microprocessors in the system, thus giving birth to the term "distributed control system."

18.5.1.2 Memory. Computer systems of the current generation use mostly large-scale integration or very large scale integration solid-state components to build the memories. Memories may be classified into RAM, ROM, PROM, EPROM, and EEPROM.

RAM is the acronym for *R*andom *A*ccess *M*emory with read/write capabilities. With RAM, the time it takes for the CPU to access any piece of information stored in the memory is always the same regardless of the memory location. This is in contrast to a serial (or sequential) access memory such as a magnetic tape, in which the access time varies with the location of the information to be accessed. With read/write capabilities, information stored in the RAM can be altered by writing new information at any memory location over the old information stored at the same spot. RAM is typically used to store program variables and I/O data. For a real-time process control computer, all I/O data that are constantly updated must be stored in the RAM. The solid-state RAM is volatile in that it will lose all stored contents (programs and data) on loss of electric power. Battery backup must be used to maintain the memory content on loss of power, if required.

ROM, PROM, EPROM, and EEPROM are nonvolatile random access type memories. The information stored in them can be read (retrieved) at will, but information cannot be written into the memory at will. ROM is *R*ead *O*nly *M*emory. PROM is *P*rogrammable *R*ead *O*nly *M*emory. ROM and PROM share one common characteristic: information can be written into the memory only once. Once it is written, the content can never be altered. The difference between the two lies in the method by which the information is written into the memory. For ROM, the information is written by a technique known as mask programming, which is done by the chip manufacturers at the time the memory chip is produced. For PROM, the programs are written into the memory through a programming unit known as the PROM burner after the chips are produced.

EPROM (erasable PROM) differs from the ROM and the PROM in that the content stored in the EPROM can be erased and overwritten with new information. This can be done by ultraviolet radiation or by electric pulses.

EEPROM is a variation of PROM (electronically erasable PROM). As in EPROM, the information stored in this type of memory can be altered, except that it can be accomplished without removing the memory from the computer assembly and without a separate programming unit.

The forerunner of the solid-state memory is the magnetic core memory. Magnetic core memory is similar to RAM except that it is slower, larger in physical size, and non-volatile. Nevertheless, magnetic core memory is no longer popular for the simple reason that it cannot compete from a cost standpoint with the solid-state memories.

Among the categories of the solid-state memories, ROMs are the cheapest, RAMs are the most expensive, and the costs of PROMs, EPROMs, and EEPROMs lie in between.

A distributed control system is composed of a set of processors, input modules, output modules, printers, and workstations that are interconnected by a communication network. A processor consists of a microprocessor-based CPU and a memory unit. In the memory unit, EPROM or ROM is typically used to store programs that are closely related to the operation of the processor; RAM is used to store application programs and I/O data, and also to serve as the scratch pad for the intermediate results of the computer calculations. Since RAM is volatile, its content is lost whenever the power is turned off. To avoid reloading the programs after a power loss, some manufacturers provide auxiliary batteries to serve as a backup power source during power outage. Some manufacturers do not provide battery backup for the RAM, but simply reload from information permanently stored on auxiliary memories connected to the communication network. Some systems use EEPROM instead of RAM with backup batteries. The drawback to EEPROM is that it does not run as fast as the RAM. PROM and EPROM are seldom used to store application programs because a special programming device is needed and the memory modules must be removed from the computer assembly to make changes to the programs.

18.5.1.3 I/O Modules and Workstations. The input and output modules provide the necessary interface between the distributed control system, and the field instrumentation and control devices. There are four basic types of I/O devices:

- Digital inputs,
- Digital outputs,
- Analog inputs, and
- Analog outputs.

Digital input devices sense the presence or absence of an electrical voltage, usually by means of a contact of an electrical switching device such as a pressure switch or a limit switch. The contact state is either open or closed, which is sensed by a voltage imposed on the contact. The voltage level is typically 120 V alternating current (ac) or 125 V direct current (dc), or lower voltage level such as 24- or 48-V dc. The digital input module converts the input voltage signal into a memory bit that is scanned by the processor. The input module filters signal noise and contact bouncing, and furnishes isolation to guard against voltage spikes external to the system. A digital output from the system is typically a solid-state switching device such as a triac, a power transistor, or solid-state relay that is switched on and off as dictated by the result of the processor calculations. The output contact switches the power, either internal or external, to the controlled equipment. In some cases, an interposing relay of the electromechanical type is used to either isolate the output device from the external system or to provide the current carrying-capacity required to handle the load current of the external device.

Analog signals take many forms: millivolt signals from thermocouples, milliampere or low-level voltage signals from transmitters, and signals from resistance temperature detectors (RTDs). The most commonly used analog signal from transmitters is a 4- to 20-milliamperes direct current (mAdc) signal. Analog input modules contain analog-to-digital (A/D) converters to digitize the analog signals from the field devices for use by the processor. The A/D computer changes the input signal to a digital value that can be stored in the processor's memory. Analog output modules contain digital-to-analog converters to convert the digital values produced by the processor into analog signals usable by the controlled equipment. Like the analog input signals, the most commonly used analog output signal is also 4 to 20 mAdc.

A workstation is an assembly of microprocessors and their peripheral devices. A workstation typically consists of one or more cathode ray tube (CRT) monitors, keyboards, pointing devices such as mouses and trackballs, and the associated electronics. The electronics mainly consist of processors with working memory and disk drives (hard drive and floppy disk). Because it is equipped with electronics, the workstation is itself a computer system with data processing and graphic display programs stored in memory or on the hard disk drive to support the CRT monitor. The workstation is the main interface point between the plant operation and maintenance personnel and the system. When used by plant personnel to monitor and control plant operation, it is called an operator workstation. When used by plant personnel to perform system programming functions, it becomes an engineering workstation. The CRT monitor is also called a visual display unit (VDU). Many manufacturers have specially designed keyboards with keys for specific functions to suit the system for the operator workstation. The keyboard for the engineer workstation can be either the standard typewriter keyboard or a keyboard specifically designed for programming functions, or a combination of both.

18.5.2 Communications Network

For the various components in a distributed system to perform effectively as an integrated unit, they need to exchange information and share resources (inputs, outputs, or processed data) by communicating with one another. This occurs through a communications line (sometimes known as the data highway) that connects the components to form a network.

The communications network is similar to the interconnecting wiring between the controllers in an analog electronic control system. But, unlike the analog system, the communications network handles the data in binary form. This makes it possible to scan a group of data signals from any component in the network and transmit them one signal at a time in series through communications lines as simple as two-wire cables. The type of signal transmission does not require a cable for each signal or a point-to-point connection between components in the system. A communications network can handle a large volume of data at a very high speed and saves a tremendous amount of field wiring. For this reason, the communications network has become indispensable to the distributed system. Such a network installed in a power plant is essentially a local area network (LAN) that is commonly employed to facilitate data transmission between computers in office buildings.

18.5.2.1 Network Topology. In a communications network, the point at which a component is connected to the network is generally known as a node, and the component is often referred to as a drop.

Network topology refers to the pattern in which the drops are connected to one another within the network. There are three basic types of network topology: star network, bus network, and ring network (Fig. 18-15).

In the star network topology, each component in the system is connected to a central computer by a point-to-point line. Any information exchange between two components is routed through the central computer. In such a scheme, the central computer serves as a traffic director and the communication channels are simple and structured. However, the network is very much dependent on the integrity of the central computer to keep the communications lines open. If the central device fails, all communications are lost.

The bus network topology is basically a single cable for transporting information between devices that are connected to the highway. Because this topology has only one transmission path for all system components, it requires a method for all components on the bus to share the use of the transmission line.

The ring network topology links system components in a continuous loop. Information is passed from drop to drop on the ring until it reaches its destination. The movement of information in the ring is usually unidirectional. It can be made bidirectional, but to do so requires more sophisticated software.

18.5.2.2 Signaling Method. Data signals between drops in the network are transmitted through cables that make up the data highway. In general, there are two basic signaling methods: baseband transmission and broadband transmission. The two methods refer to the way signals are imposed on the cable. The baseband method sends voltage pulses

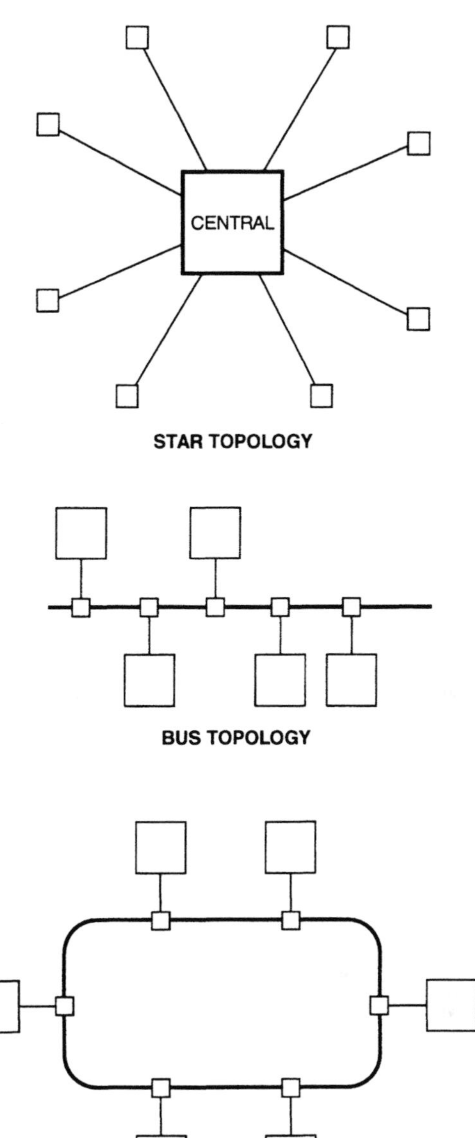

Fig. 18-15. Network topology.

representing the binary bits (ones and zeroes) through the cable. Sending signals by the broadband method is like radio or television broadcasting. The transmission has a wide bandwidth that is divided into channels of separate ranges of bandwidth with different frequencies. With this method, more than one signal can be sent simultaneously through the medium at different frequencies within the separate sections of the bandwidth simultaneously.

The baseband method has one transmission channel and can transmit data signals only one at a time. The equipment associated with this method is relatively inexpensive. The broadband method, because of its multiple channels, has a higher transmission capacity and can handle data, voice, and video signals on the same medium. However, the cost for equipment required for broadband transmission is relatively higher. It also requires the use of a modem (modulator–

demodulator). For data transmission, a modem converts the data signals from the binary form into signals that fall in the frequency band and propagate along the transmission lines (the medium) as electromagnetic waves. Another modem at the receiving end then converts the electromagnetic waves back to the original binary form. Both methods are used in power plant applications.

18.5.2.3 Transmission Media. Three types of transmission media are most often used for the communications network: twisted-pair wire, coaxial cable, and optical fiber.

The basic twisted-pair wire has two insulated copper conductors twisted together and sheathed with polyvinyl chloride (PVC). A large number of wire pairs can be encased in one sheath and made into cables. Twisted-pair wire is inexpensive, but it is susceptible to electromagnetic interference and noise from the environment. The signal attenuation of twisted wire pairs and the amount of cross-talk between adjacent pairs rise sharply with increasing frequency. Thus, its usage is confined to single-baseband applications.

The coaxial cable has a copper conductor in the core surrounded by insulation material (foam or plastic) that is covered by a semirigid metal conductor and encased in an outer shielding of PVC or Teflon. This type of cable reduces the interference from the environment and reduces the signal attenuation characteristics of the cable. Coaxial cables can be used for both baseband and broadband signal transmission without any limitation on the bandwidth. Coaxial cable costs more than the twisted-pair wire. In general, extra care is required to install the coaxial cables.

Optical fiber conductors are made of small-diameter glass fibers. Light signals can travel long distances without losing their signal strength. A fiber conductor has very precise light-propagation qualities. Optical fiber conductors have many advantages. They have large bandwidth and can be used for both baseband and broadband signal transmission. They are also immune to electromagnetic interference and corrosive agents. However, optical fiber conductors are the most expensive to use. They are difficult to install and, once installed, cannot be easily modified. However, they are gaining acceptance in the industry, and it is only a matter of time before they become well used in power plants.

18.5.2.4 Transmission Protocols. Protocols are sets of rules that govern the flow of data in a network. Their purpose is to ensure that data are correctly transmitted from point to point with procedures for automatic error detection and correction.

The data highway that forms the network is shared by all drops connected to the network. All drops must have access to the highway before they can communicate with each other. The access method is the protocol used to coordinate the data transmission activities among the drops and to avoid any conflict that may arise when two or more drops are attempting to access the medium at the same time. The access method, in general, can be classified into the contention and the noncontention methods.

In the contention method, the drops in the network compete with one another for access to the medium. One of the most commonly used contention systems is CSMA/CD.

CSMA/CD is the abbreviation for Carrier Sense Multiple Access with Collision Detection. In this method, each drop in the system "listens" to the traffic on the highway. If the drop has data to transmit to another drop, it "listens" and sends the data only if it senses no traffic. If more than one drop in the system is listening and transmitting information at exactly the same moment, a data collision occurs on the highway. The immediate result is that all data on the highway are garbled. This method permits the drops to recognize a data collision, and the drops involved back off immediately. Each drop then waits for a random period and tries again. The length of random period is different for each drop. This ensures that no collision happens on the second try. This method is also called the nondeterministic access method because the drops do not have a definite turn in which to transmit signals, and the time it takes for a signal to reach its destination is not predictable because of the potential for data collision and the retry. CSMA/CD is predominantly used in the bus topology. One of the most well-known users of CSMA/CD is Ethernet, which was developed by the Xerox Corporation.

In the noncontention method, the drops share the medium according to some predetermined schedule to avoid conflicts. One of the widely used noncontention methods is the token passing method. This method can be used for either the ring or bus topology.

The token ring was developed by IBM. In this method, a token is passed along the data highway from drop to drop. A token is a prescribed pattern of bits. Only the drop having possession of the token has the right to transmit data to any other drop on the highway. The receiving drop, on receipt of the data, sends the token back to the originating drop, which then passes the token to the next downstream drop on the ring. Token passing on a bus is similar to the procedure used for the token passing ring, except that the token passing sequence is prearranged with no regard to the physical arrangement of the drops on the bus. After the data are transmitted, the drop passes the token to the next drop in the sequence. The Manufacturing Automation Protocol (MAP) that is frequently used in factories was developed on the basis of a token bus.

The token passing technique allows predictable data transmission among the drops and potentially saves transmission time because a drop does not have to listen before transmission. On the other hand, the token may be damaged while passing through a drop and cause an interruption to the process. This system needs a monitor to oversee the transmission process and the monitor will step in whenever the token is lost. In this method, adding a drop to the highway has an impact on all drops in the system and requires changes in the communications software.

Another access method of the noncontention type is the polling method. The polling method requires a host station in the system that controls the information exchange activities. The host station continuously polls each drop in the system, asking if it has any information to transmit. If the answer is positive, the drop is given permission to send the information. The central controller then proceeds to the next drop and continues to make rounds. This method is most often used in the star topology.

IEEE 802 committee is dedicated to the development of communications network standards. The committee has established the following standards for the access methods:

- 802.3 CSMA/CD Bus Standard,
- 802.4 Token Bus Standard, and
- 802.5 Token Ring Standard.

These standards are followed by manufacturers who supply bus and ring networks, with some variations to suit the particulars of their systems.

18.5.2.5 RS232C Interface. In a discussion on communications network, mention must be made of the RS232C interface because, in the digital world, the RS232C interface is well known and the most widely used standard for direct point-to-point serial data transmission between two pieces of digital equipment (for example, a workstation and a printer). What is not well known is that this standard was originally developed to specify the interface between a computer and a modem.

RS232C is the recommended standard (RS) of the Electronics Industry Association (EIA), No. 232. The letter C refers to the revision level of the standard. The full name of RS232C is "Interface Between Data Terminal Equipment (DTE) and Data Communication Equipment (DCE) Employing Serial Binary Exchange." The standard specifies the interconnection and signal exchange between the DTE and the DCE (Fig. 18-16).

The DTE can be a computer, CRT, printer, or any digital equipment that sends or receives binary data in serial form, to and from another DTE at a separate location. The DCE is a modem that transforms the binary bit stream from the DTE into electromagnetic waves or vice versa. The standard is intended for serial data transmission, using pulses at a rate of up to about 20,000 bits per second over a distance of no more than 50 ft. The interface uses a 25-pin connector at both the DTE end and the DCE end for a maximum of 25 signal lines, with the function of each line defined for both the DTE and the DCE. Not all of the 25 lines need to be used to complete a

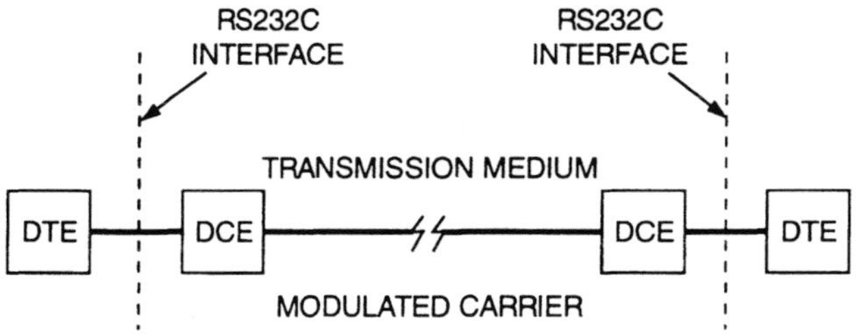

Fig. 18-16. RS232C interface.

transmission. In some cases, as few as three lines are sufficient to complete the interface and, for this reason, some manufacturers use a nine-pin connector for their equipment.

Since the 1970s, the computer industry has borrowed this standard and used it to facilitate serial data transmission between computers. It is commonplace to find an RS232C port on a computer or its peripherals. However, this standard was meant only for defining the interface between DTEs and DCEs. If it is used for data transmission between parties other than a DTE and a DCE (such as an interface between a workstation and printer or between two microprocessors), one party must play the role of the DTE and the other must play the role of the DCE, and how the pins are connected between the two ends depends on the roles of the two parties. This point sometimes causes confusion between the user and the computer manufacturer because equipment designed with a provision for an RS232C interface is not always identified clearly as DTE or DCE.

There are other standards similar to the RS232C for serial data transmission, such as RS422A and RS423A. The differences lie in the electrical and/or mechanical aspects of the signals and connections to allow for transmission at a higher rate and longer distance. By far, the RS232C is still the most popular standard in the computer industry.

18.5.3 Programmable Logic Controllers

The programmable logic controller was a digital control computer developed in 1969 for the automobile industry to replace the relay control logic in their factories. Originally, it was called programmable controller (PC), but the name has been changed to programmable logic controller (PLC) to avoid confusion with personal computers. At the time it was developed, the PLC had a few characteristics that made it different from other control computers:

- The PLC was solely used to perform on–off control functions, controlling machines on the factory floor.
- The PLC's hardware was designed for installation on the factory floor for operation in harsh ambient conditions.
- The software was made "easy-to-use" so that any programming and maintenance work could be handled by technicians and electricians who were accustomed to servicing the traditional hard-wired relay logic control systems.
- The PLC was designed to store only one control program that would run any time the PLC was turned on. This is unlike computers in other applications, which have many programs for various functions, and the programs run according to a certain schedule or when they are called on to run.

After its first appearance, the PLC was quickly adopted by other industries. Today, PLCs can be found anywhere a group of motors or solenoids need to be controlled. Today's PLCs retain the same characteristics of the original design except that they are faster and better, with many new features added over the years. Chief among the new features are remote I/Os, PID control algorithms, and operator workstations.

18.5.3.1 PLC System Configuration. A programmable logic controller, in its simplest and smallest form, consists of a processor, an input module, an output module, the necessary power supply, and the interconnecting wiring. The arrangement shown in Fig. 18-17 is a typical PLC control system. The processor consists of a CPU and a memory for program and data storage. Typically, battery-backed RAM is used to store the application software and the I/O data. However, manufacturers also offer EEPROM for this purpose. In addition, a programming device is needed to develop the necessary application program and to load the finished program into the processor.

In general, processor size is specified by the size of the user memory and the number of I/Os that can be accommodated. PLC manufacturers produce PLC modules of varying sizes with accompanying I/O modules to suit. There are ample varieties for a user to pick and choose to suit his applications. For each size of processor, the manufacturers often make provision for memory expansion. The inputs and outputs are usually composed into modules of 8, 16, or 32 similar types of inputs and outputs. The processor and I/O modules are designed for rack mounting and the racks can be stacked and mounted in cabinets of varying sizes.

In early PLC applications, all manual commands such as those to start and stop a motor or to open and close a valve were initiated by the plant operator through control switches or push buttons. Indicating lights showed the status of the controlled equipment. The switches and push buttons were wired to the I/O modules as digital inputs, and the indicating lights were driven by digital outputs from the PLC. In some instances, the status contacts from the controlled equipment were wired directly to the lights, bypassing the PLC altogether. The manual commands and status indications have been gradually replaced in recent years by CRT-based opera-

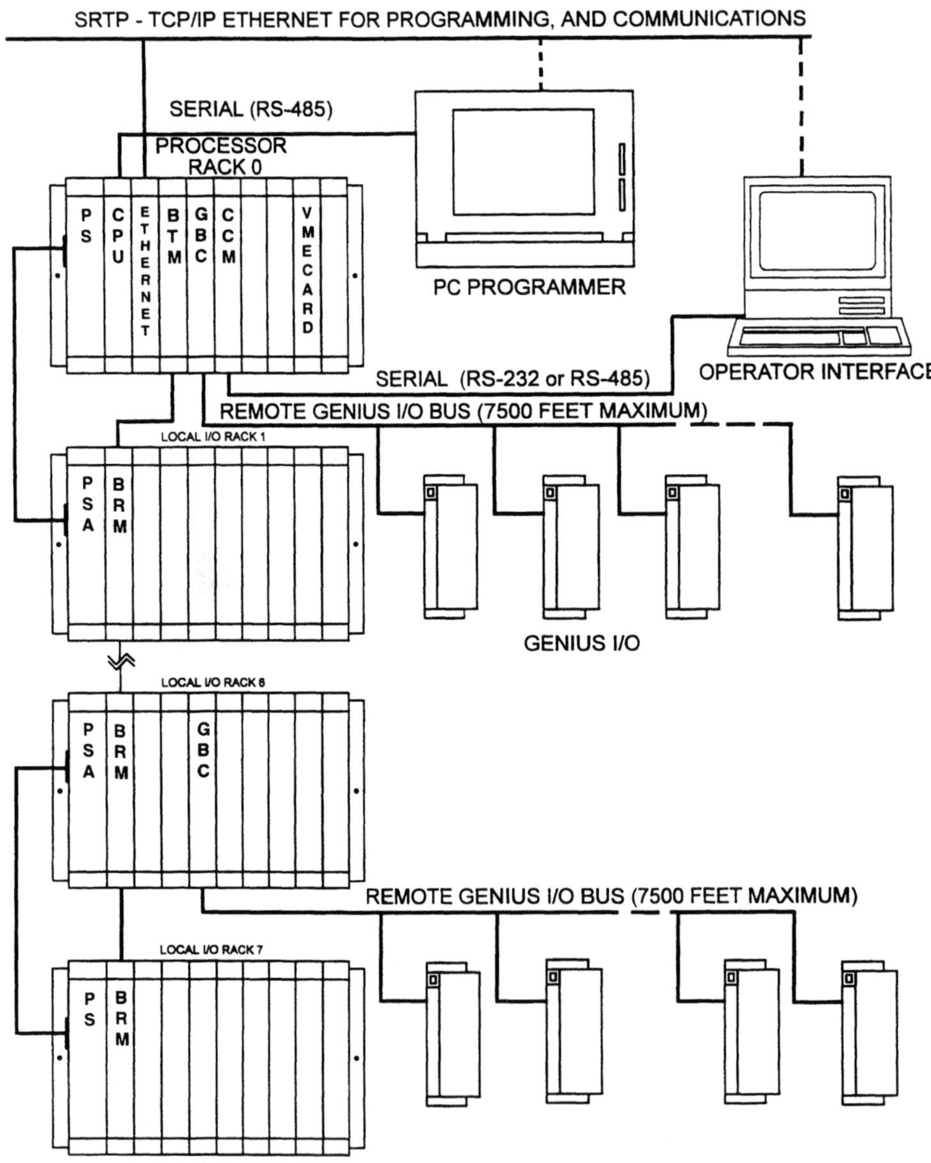

Fig. 18-17. Typical PLC system. (From GE Automation. Used with permission.)

tor workstations. PLC manufacturers have standard software display packages running on processors specially designed for this purpose. Communications networks connect the control processors, I/O modules, and display processors, thus forming a PLC system that performs specific control functions for plant equipment.

A PLC system configured in such a manner is in effect a distributed control system. The control programs can be distributed into more than one processor according to the nature of the tasks performed by the controlled equipment. The processors and their I/Os can be located throughout the plant near the controlled equipment or, alternatively, all processors can be installed at a centralized location and the I/O modules can be distributed geographically throughout the plant. In either case, extensive savings in wiring costs can be realized by locating the I/O modules near the controlled equipment. I/O modules distributed in this manner are known as remote I/Os.

18.5.3.2 Programming Language. Relay ladder logic diagrams have been identified with PLC programming from the very beginning of PLC application. The ladder diagram is derived from the schematic wiring diagram (sometimes referred to as elementary wiring diagram) developed by electrical system designers during the control system design process. The control logic is implemented by a network of contacts and relays. The logic outputs are either energized or deenergized by activating the various electric circuit paths. For example, a network of contacts, if arranged in series, will form an "AND" logic structure and, if arranged in parallel, will form an "OR" logic structure.

The schematic wiring diagram (Fig. 18-18), as the name

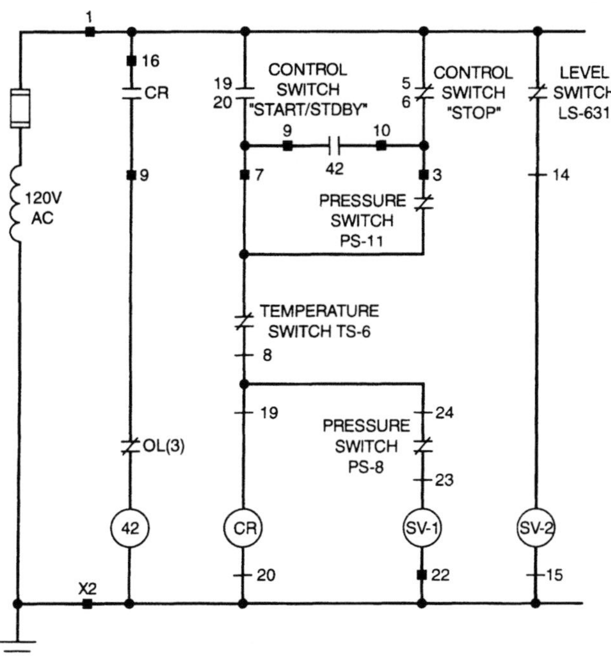

Fig. 18-18. Typical schematic diagram.

implies, shows how the contacts and relay coils are wired in relation to one another, and thus in effect becomes a diagram depicting the control logic of the system being designed.

The relay ladder diagram (Fig. 18-19) is identical to the schematic wiring diagram. Each horizontal line is called a rung. All contacts and relays are placed on the rung as if they are connected to form a circuit path. The PLC programming language was developed in the ladder diagram format because the operations and maintenance personnel in the factory were familiar with such documentation and, as a consequence, this would enable the factory to make the transition from the relay control to PLC control without retraining factory personnel.

There are two types of programming devices: hand-held and stand-alone. The stand-alone programming device typically is CRT based, consisting of a CRT monitor, a keyboard, and the necessary electronics. Programming is similar to the programming of any other type of computer. The finished programs are copied onto floppy disks or tapes and loaded into the processor. The programming device can also be used as a tool for troubleshooting. This can be done by displaying the ladder diagram on the monitor with the PLC operating online and checking the power flow of the contacts and relays on the rungs. Furthermore, the programming device has the capability for simulating input conditions by manually forcing the opening and closing operation of contacts on the rung. This provision allows technicians to check system logic without actually operating the field devices and thus greatly facilitates checkout activities during initial equipment startup.

The handheld programmer is like a portable calculator with a keypad and a liquid crystal display (LCD). Developing a PLC control program with the handheld device requires that it be plugged into the PLC processor, and the programmer enters the programming instructions on the keypad. The program is loaded directly into the processor's memory. This programming device is limited to programming the control logic for small PLCs.

PLC manufacturers are mainly hardware suppliers. Each manufacturer in this field is capable of furnishing a complete line of PLC equipment. However, PLC manufacturers typically do not design the control logic for the user's application and do not provide the related programming services. In the power industry, these activities are typically the responsibility of the equipment purchaser or designer.

18.5.3.3 Application in Power Plants. PLCs are used in power plants to implement on–off control functions. The applications can be divided into two categories: PLCs used in stand-alone processes and those used for balance-of-plant equipment.

In a power plant, stand-alone processes refer to those processes that are traditionally designed and supplied by manufacturers as a complete system with all process equipment and control equipment included. Systems in this category include water treatment, coal handling, boiler soot

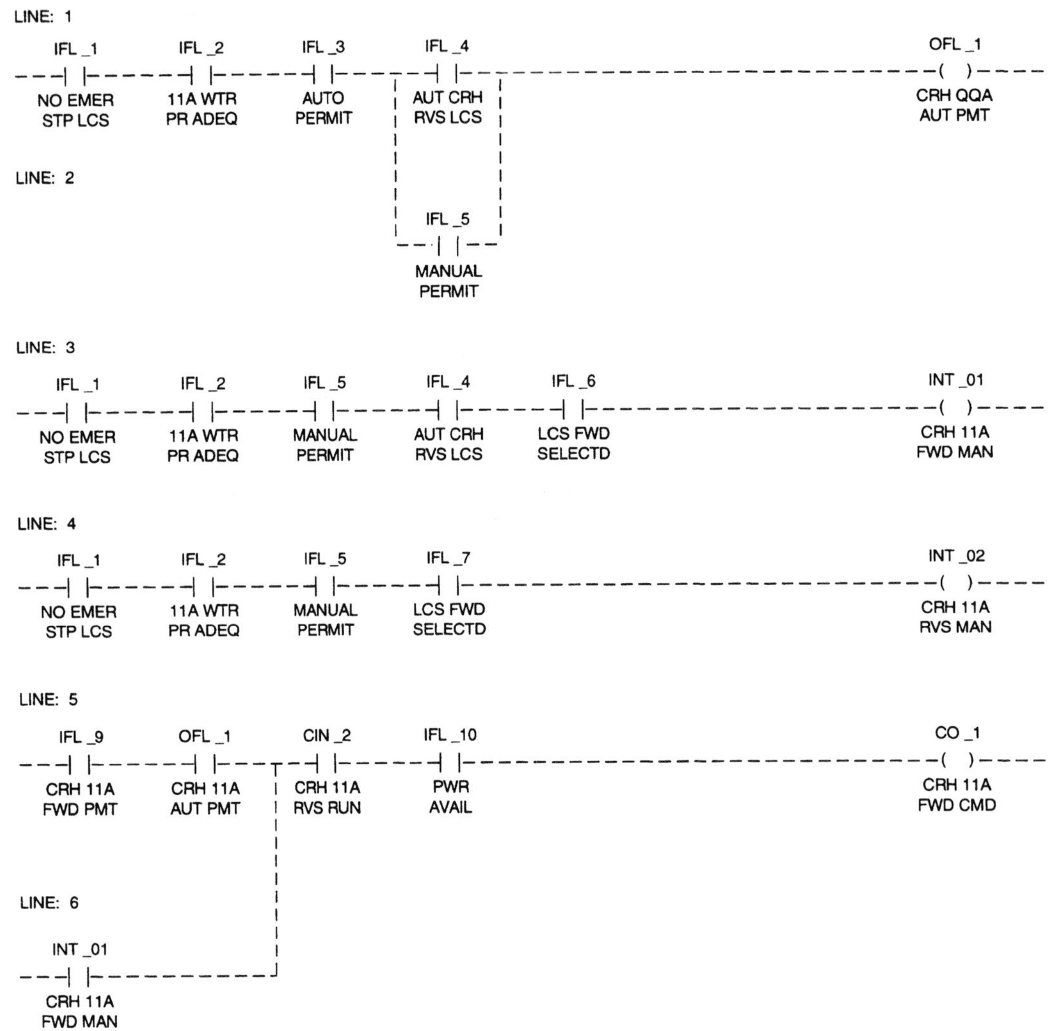

Fig. 18-19. Typical ladder diagram.

blowing, fly ash handling, bottom ash handling, and burner control. Except for the burner control system, changing plant load conditions do not have a significant impact on the operation of the systems. The controls deal mainly with the start/stop operations of motors, solenoid valves, and sequential operation of the equipment in the process. The systems are typically furnished with a control panel for installation near the system equipment for local operation. The PLC components are housed in the panel. The I/O modules may be installed in the panel or at the process equipment if the equipment is located at a distance from the panel. A small panel insert with switches and lights or a CRT-based panel may be supplied and installed in the central control room for remote operation from the central control room.

The balance-of-plant equipment listed below represents mechanical equipment associated with the boiler system, the turbine–generator system, and the plant auxiliary systems:

- Boiler draft equipment—induced draft fans, forced draft fans, primary air fans, and shutoff dampers;
- Condensate and feedwater equipment—condensate pumps, boiler feed pumps, and shutoff valves;
- Turbine extraction equipment—turbine extraction line block valves and steam line drain valves;
- Condenser circulating water equipment—circulating water pumps, cooling tower fans, and butterfly valves in the circulating water system;
- Condenser air removal equipment—condenser vacuum pumps and steam–air ejector valves;
- Accessory equipment for main turbine–generator, and large pumps and fans—lube oil pumps, turning gears, seal oil pumps, and related equipment;
- Plant auxiliary cooling water equipment—closed cycle cooling water pumps, river water pumps, and service water pumps; and
- Compressed air system—instrument air compressors and service air compressors.

PLC applications with respect to balance-of-plant systems and their equipment deal mainly with motor start/stop operations and associated protective interlock functions. Typically, the logic designer will use a number of PLCs for distribution of the control logic according to the process functions. The processors may be located in a centralized electronic equipment room or physically distributed throughout the plant. The distribution of I/O modules are handled in a similar manner. This is usually determined on a case-by-case basis depending on space availability, wiring and installation costs, user's philosophy on plant operation and maintenance activities, and the particular plant requirements.

It should be noted that current state-of-the-art PLCs also have the capability of performing modulating type control functions. However, at the present time, this type of control function in power plants typically is allocated to a larger dedicated plant distributed control system.

18.5.4 Distributed Control System

Strictly from the technical standpoint and in the broadest sense, any microprocessor-based control system can be made into a distributed control system as long as the control programs are stored in more than one processor. However, in the power industry, the term distributed control system (DCS) is generally applied to the system that implements boiler control and data acquisition functions of the power plant.

The DCS, when it first appeared on the market, was applied to the boiler controls which, until the late 1970s, were mostly of the analog electronic type. The data acquisition systems at that time were predominantly centralized computer systems. Since the early 1980s, with the advent of operator workstations and the increasing power of microprocessors, DCS manufacturers have expanded the functions performed by the DCS into the data acquisition area, and today, they have completely replaced the centralized computer system. Thus, the DCS has in effect become a power plant control and information management system. This is why the DCS is frequently referred to as a DCDAS (Distributed Control and Data Acquisition System), DCIS (Distributed Control and Information System), or by a similar description and abbreviation.

18.5.4.1 Configuration. As previously mentioned, a state-of-the-art distributed control system typically is composed of modularized microprocessor-based processing units, input modules, output modules, operator workstations, engineer workstations, printers, and other types of peripheral devices, all connected through a multiple level communications network. DCS manufacturers have standard modules for different functions. They generally fall into two categories: control modules and data processing modules.

The control modules are structured to perform a variety of control and computing tasks such as PID (proportional plus integral plus derivative) control, binary logic, and arithmetic functions. Some manufacturers have separate modules for modulating and on–off control functions, and others have combined the two into one module. For some manufacturers, each module is available in varying sizes to suit a user's needs.

The data processing modules manipulate data, performing such functions as data table, graphic display, variable trending, report generation, and long-term data storage and retrieval. Module structure is dependent largely on the particular manufacturer's design and, as a rule, auxiliary memory units such as hard disk drives are required if the processing function involves a large amount of data storage. For long-term data storage, magnetic tapes are the medium typically used. However, optical disks are also gaining acceptance.

Input and output modules are similar to those used in PLC applications including analog, digital, and pulse type inputs and outputs. Analog input modules accept signals from thermocouple, resistance temperature detectors, and pressure, flow, and level transmitters. Analog output signals are wired to the modulating control loops final control elements such as control valves and control dampers. The standard transmitter and analog output signals are 4 to 20 mAdc. Digital inputs and outputs are the same as those for PLC applications.

I/O modules can be either dumb or intelligent. A dumb module merely digitizes the input signals and sends them to the processors. An intelligent module is equipped with a microprocessor and has the capability of further processing the digitized signals such as conversion to engineering units, square root extraction, and alarm limit checking. Some manufacturers even use their digital I/O modules to perform ladder control logic functions, thus further decentralizing the functions of the DCS. As with the PLC I/O modules, the DCS I/O modules can also be housed in environmentally hardened cabinets located near the controlled equipment to shorten the field I/O wiring. Experience in recent design has demonstrated that considerable savings in wiring costs can be realized if the I/O cabinets are geographically distributed in such a manner.

System configuration for a DCS varies from manufacturer to manufacturer. At the lowest level are I/O modules. The I/O modules associated with a particular control loop or related loops may be grouped together. The I/O signals are bussed to a control module or a group of control modules. Some manufacturers designate control and I/O modules grouped in this way as a processing unit dedicated to a particular process system in a plant area. When applied to a power plant or a given boiler–turbine–generator unit in the plant, the DCS may comprise a number of processing units, each connected to a plantwide communications network. In addition, the data processing modules, operator and engineer workstations, and other peripheral devices are connected to the plant communication networks.

Figure 18-20 shows typical system configurations of DCS systems installed in a power plant.

18.5.4.2 Programming Language. Control functions can be programmed in any computer language as long as the language can be operated in a real-time environment (run-

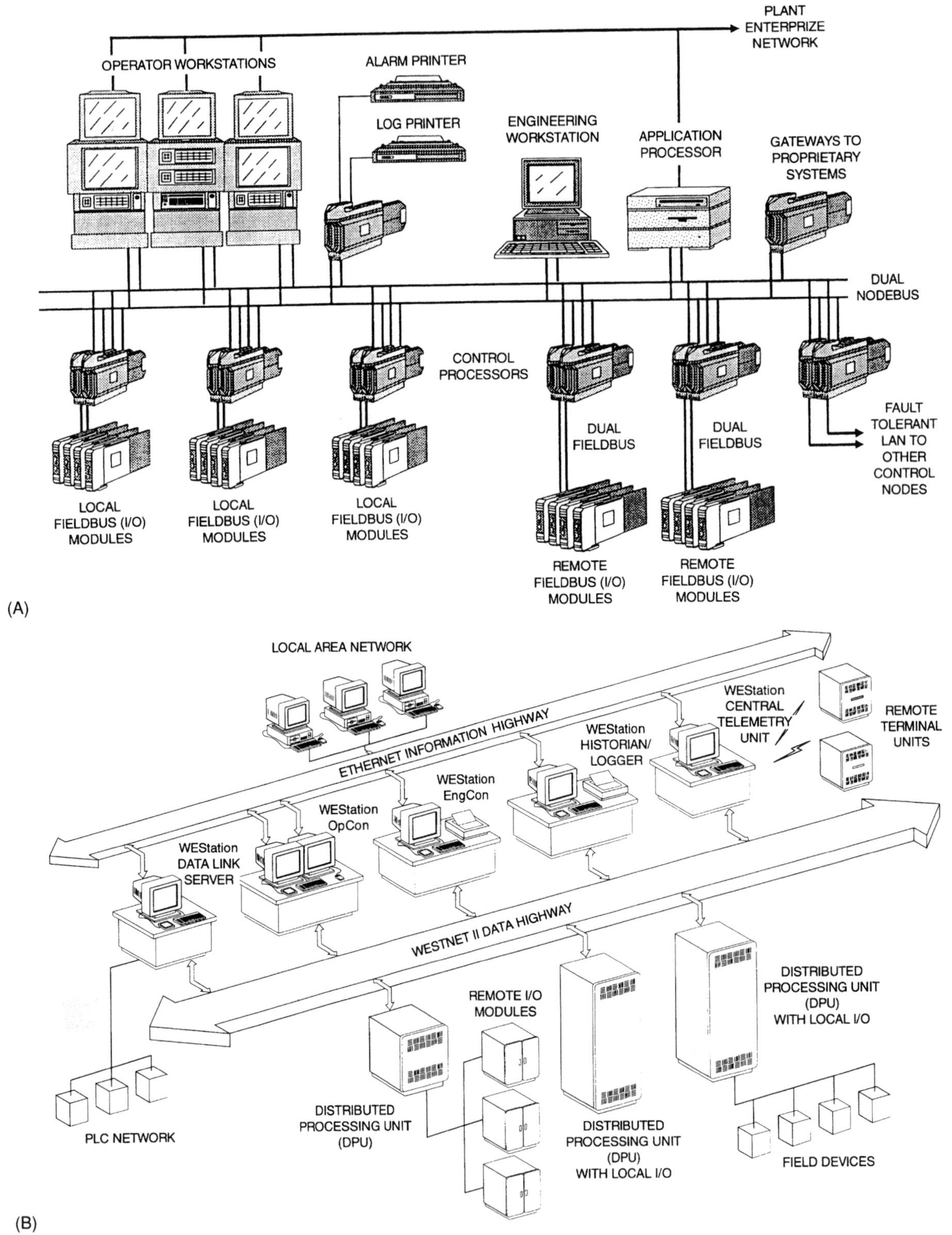

Fig. 18-20. (A) DCS configuration (From Foxboro. Used with permission.) (B) DCS configuration. (From Westinghouse. Used with permission.)

ning the program using live data from the I/O modules). Examples of the languages that can be adopted for this purpose are C, APL, FORTRAN, and BASIC. These are all high-level languages. The engineers who use them need to have some knowledge of computer programming.

However, the programming format that DCS manufacturers use most is what is known as "function blocks." A function block is like a subroutine consisting of a set of algorithms designed to execute a specific control or data manipulating function such as a PID control function or a high/low signal selecting function that is expected to be used repetitively in a control program. When developing a control program, the engineer selects the blocks required to fit the control strategy and strings the blocks together to form a control loop. The attributes of each block are then placed into the program through a fill-in-the-blank type data entry form. The manufacturers have a standard repertoire of function blocks typically constructed from control functions listed in the Instrument Society of America (ISA) standards.

A control program composed of function blocks can be used for both modulating and on–off control loops. The manufacturers also generally provide ladder diagram programs because of their popularity among technicians in power plants. The ladder logic can be accommodated in a DCS in many ways. Some manufacturers use a separate PLC-like control module, some use modules that are suitable for both function blocks and ladder diagrams, and some put the ladder logic into the I/O modules. Because the on–off control logic is essentially Boolean algebra, it can also be readily programmed in Boolean statements. Most of the control processors available have made provisions for programming the logic in this fashion.

In addition to the control functions, the DCS needs programs to implement all operator interface, report generation, and data storage and retrieval functions. The manufacturers generally divide these functions into separate packages, each with a specially structured program that serves as a platform for the user to develop graphic displays, other forms of data presentation, and format operating logs. In general, the programming functions are user-friendly and menu-driven so that they can be a programming tool that is easily understood by the user's personnel. Even though the programming packages are designed for use by technical personnel who have little knowledge of computer programming, it is not practical for the user to assume that his personnel can become proficient in developing DCS programs with absolutely no training.

Programming functions are conducted from the engineer workstation, usually with a full complement of CRT screen, keyboard, auxiliary memory, and floppy disks. The workstation is normally connected to the DCS communications network, and the programs developed from the workstation can be directly downloaded to the individual processor modules in the system.

18.5.4.3 DCS Application in Power Plants. DCS application in power plants typically covers the following areas:

- Boiler controls, including the combustion (firing rate), furnace draft, steam temperature, and feedwater control loops;
- Burner control and pulverizer control;
- Control loops in the plant auxiliary systems that need to be monitored and/or operated from the central control room;
- Alarm annunciation and recording functions;
- Monitoring functions for other separate stand-alone controllers or control systems;
- Remote indication and recording of plant operating parameters;
- Periodic reports and event logs; and
- Historical data storage and retrieval functions.

In nearly all power plants built in recent years, the monitoring and data processing tasks that the DCS is capable of handling have largely replaced the conventional mimic panels, annunciator light boxes, indicators, and recorders in the power plant control rooms. It should also be mentioned that DCS application in power plants has been expanding into motor controls for the balance-of-plant equipment (pumps, fans, etc.) which was once predominantly an area for PLC applications. At the present time, the choice between PLC and DCS for this application is largely a matter of cost and users' preference.

18.5.5 Personal Computer

The basic components of a personal computer (PC) include a processor with a hard disk, monitor, keyboard, and mouse or other pointing device. This is essentially equivalent to an operator workstation or an engineering workstation. The personal computers in an office building are connected to a network, and this is in effect a distributed control system. The only difference between a PC system and a DCS is that the PCs in the office do not operate in a real-time environment. They only receive input data from keyboards, mouses, and other PCs, and produce output data on the screen or on a printer.

A PC can be made to work as a control computer if it is connected to I/O modules that interface with the process with the PC's operating system revised to handle the I/Os in a real-time environment. However, PC applications in power plants so far have been limited to offline programming functions as a supplement to or replacement for the workstations of a DCS or PLC system.

A DCS or PLC manufacturer will include a PC in his system to be used as a workstation either directly connected to the network or installed in the user's engineering office for the user's personnel to monitor plant operations or to develop application programs for downloading to the control processors.

18.5.6 Stand-Alone Controllers

Stand-alone controllers are microprocessor-based controllers that have a microprocessor and an I/O module with a capacity for a limited number of I/Os of both the analog and digital type all contained in a single package. The package typically has the capacity for one or two PID control loops and has the appearance of a conventional panel-mounted hand/automatic station (Fig. 18-21) with pushbuttons and bar graph displays for automatic/manual mode selection, for raising and lowering the control output signal to the final control element when the loop is in the manual mode, and to generate a set point signal for the process controller. The control loop functions generally are programmed from a handheld programming device. In some designs, the pro-

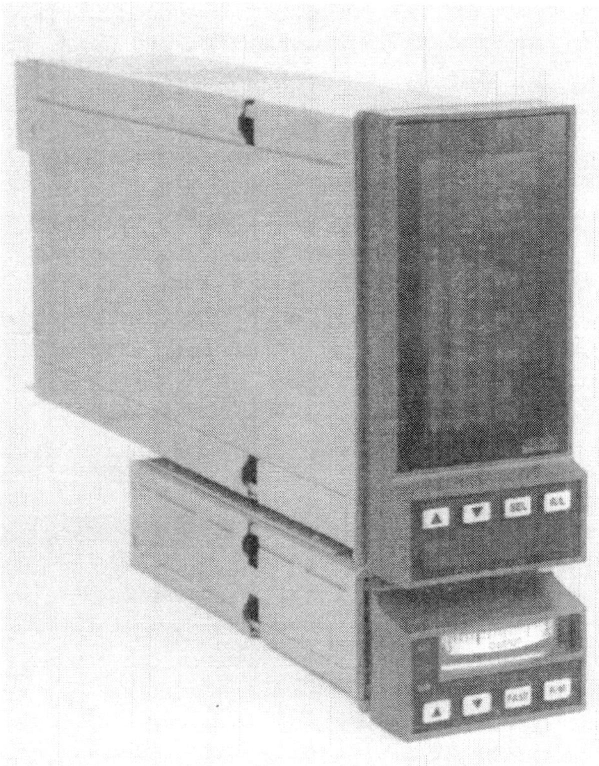

Fig. 18-21. Stand-alone controller. (From Foxboro. Used with permission.)

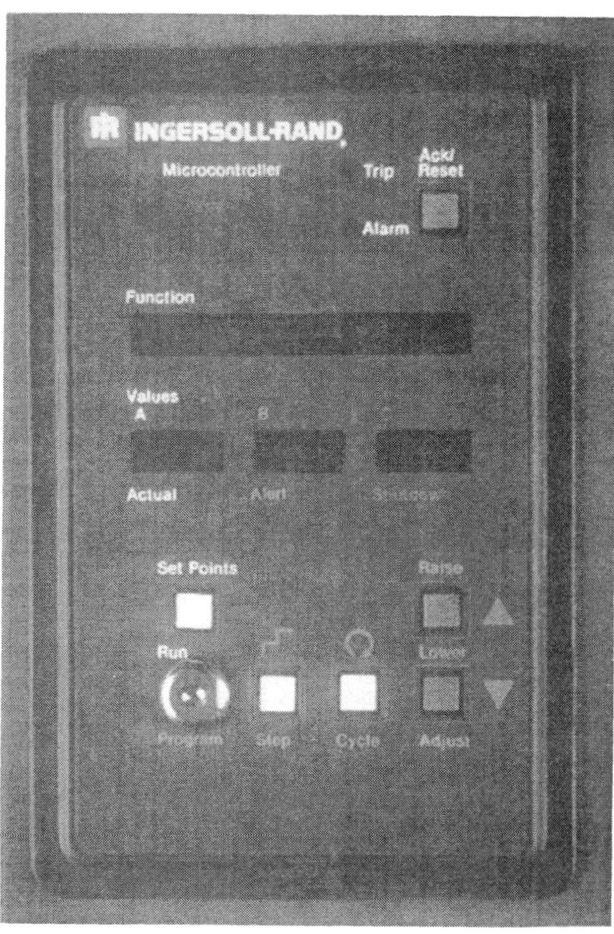

Fig. 18-22. Local stand-alone controller. (From Ingersoll-Rand. Used with permission.)

gramming function is built into the unit. The stand-alone controllers are often used in process applications that include only a few PID loops. The loops are generally in a local area of the plant separate from the central control room. One example might be local control of a small auxiliary boiler with a fuel/air control loop and a drum control loop.

There are also stand-alone PLCs. These are mini-PLCs with all the essential elements of a PLC contained in a small enclosure. Stand-alone PLCs perform binary logic functions equivalent to a handful of relays and are used in applications similar to the stand-alone PID controllers.

There are many other variations for utilization of stand-alone controllers. One example is the implementation of control and monitoring functions to operate a centrifugal air compressor. In this application, PID loops are needed for discharge air pressure control by regulating the compressor air inlet valve and for surge control by opening the compressor bypass valve. In addition, the compressor air pressure, lube oil temperature, bearing vibration, and other similar parameters are measured for monitoring the machine health. In this case, a stand-alone controller with a faceplate can be used to perform all these functions (Fig. 18-22). This type of stand-alone controller, as a rule, is furnished by the process equipment manufacturer as part of the equipment package and is custom designed. It should be recognized that, even though the stand-alone controllers can be operated independent of the plant DCS or PLC system, they are typically equipped with RS232C ports for either sending data to or receiving remote control signals from the plant distributed control system.

18.5.7 Hardware Redundancy

It is normal practice in power plant control system applications that critical control functions be implemented with redundant pieces of hardware. The redundant pieces are installed in the system to perform the same functions so that the system will suffer no loss of the control function if one piece fails during operation. The purpose is twofold. First, under normal operating conditions, redundancy serves to mitigate the risk of a total unit shutdown on the occurrence of a component failure in the control system. Second, it ensures that the unit trip or equipment trip logic will function even if there is a component failure in the control system. A control system that is capable of meeting these criteria is said to have been designed for "single component failure criteria" or "fault tolerance."

In a DCS or PLC system, the starting point for component redundancy is the control processor for critical control functions. For example, the processors for the boiler fuel/air control loops must have backups or a processor failure could result in an uncontrolled excursion boiler fuel or air flow, possibly resulting in a boiler trip.

When two redundant processors are installed, they can be arranged to operate in parallel, with the output of one processor continuously compared to the output of the other. Only one output is transmitted to the controlled equipment. The controlled equipment will be brought to a safe state if a discrepancy exists between the two outputs. An alternate to this method is to designate one processor as the primary controller that will normally perform the control functions, with the other processor designated as the backup controller that tracks the output of the primary controller. The latter will immediately take over the control function in the event that a failure in the former is detected.

The control system inputs can also be made redundant. For analog inputs, this means two or more sensors for one measuring point connected to separate input modules. For the most critical loops such as furnace draft and drum level controls, triple redundancy is typically used. In this case, the control processors use either the averaged value or the median value of the three process variable sensors. Should one of the sensors fail, the control mode will revert to the dual redundant scheme. For digital inputs, a voting logic is applied to the redundant inputs. The logic can be one out of two, two out of two, or in case of three inputs, two out of three, depending on the criticality of the application. For outputs, a redundant scheme similar to the arrangement of the inputs can also be provided.

Another redundant scheme is known as the triple modular redundancy (TMR). This measure uses three of each element in the control scheme (three inputs, three processors, and three outputs). Each of the three processors receives all three redundant inputs, executes the necessary computing functions, and produces an output signal. A voting logic or an averaging circuit is used for the three outputs before they are sent to drive the controlled equipment. As compared to the dual redundant scheme, the TMR provides an extra level of protection and is often employed for the most critical applications such as the unit protection logic of the plant.

18.5.8 Power Plant Simulators

Power plant simulators are used to train plant operators. The idea originated in the nuclear industry, where it is mandatory to conduct training classes using a simulator as one of the training tools.

A simulator comprises three parts. The first part is a computer that models the behavior of the plant equipment and process. The second part is the processors that duplicate the control functions of the actual plant control system. The last part is the hardware that emulates the instruments and control devices that the operator uses during actual online operation. The simulators used for fossil-fueled power plant application are available in several degrees of complexity and cost, ranging from simple generic simulators to simulators that include a mathematical replica of the plant itself and a physical replica of the operations control center.

A simple generic simulator has a computer model for the boiler or the turbine developed from the manufacturer's standard design. It also has a small console for the trainee to practice operating a typical boiler and turbine unit. On the other extreme, a high-fidelity, full-scope simulator includes a computer model that simulates the operation of all equipment and processes in the plant, a simulator room that is an exact copy of the real control room of the plant with all the operator instrument and control equipment, and tapes that generate the noise of a turbine running under various operating circumstances such as turbine on turning gear, turbine at full load, and turbine tripped. The design can be so realistic that a person standing in the simulator room will not be able to tell if it is the real thing or the simulation. The selection of what size simulator to use is a matter of economics and the particular needs of the user. Figure 18-23 shows a full-scope simulator console. In this case, the unit being simulated has a control room with a conventional control panel. The simulator console was sized so that the displays of the screens on the console duplicate the instruments and control devices on the actual control panel.

In addition to the computer model and the control console, simulators of medium and large size also have an instructor console for the instructor to operate the simulator. Major simulator functions on the instructor console include initial conditions, run and freeze, snapshot and backtrack, and malfunctions. Initial conditions represent the starting point of a training session. They can be a cold start condition where all plant equipment is at standstill, a hot start immediately after a turbine trip, or the turbine running at a certain load point. The conditions can be defined by the instructor, and once defined, can be saved for later use. The run and freeze function enables the instructor to stop the process at any time during the exercise for a coaching session with the trainee and then the process can be restarted. The simulator can take a snapshot of the plant status automatically at

Fig. 18-23. Full-sized simulator. (From ESSCOR. Used with permission.)

programmed intervals set by the instructor, and the backtrack function plays back the snapshots in slow motion to show the trainees the cause and effect of the simulated operation. Malfunctions are intended to simulate abnormal conditions that can occur during actual operation. During the exercise, the instructor can initiate a malfunction, such as a condensate pump trip or a feedwater control valve failure, and allow the trainee to cope with the aftermath of such an event.

The use of simulators is not widespread. The main reason is that for a medium- or large-sized simulator, the cost is relatively high as compared to the price for the control equipment of the plant, particularly if the user wants a custom-made model that closely resembles the plant equipment. This situation is likely to change as the cost of computer equipment comes down and as the technology advances.

A major advantage of training by simulation is that it allows an operator to experience emergency conditions that may be experienced infrequently in actual operation. Lessons learned in training can guide the operator to correct responses that can minimize loss of generation and/or damage to equipment, when and if the actual event occurs. Also, simulators can be used to test new control strategies before they are implemented on the actual equipment. Regardless of cost, experience shows that simulators are a useful tool for training plant operators.

18.5.9 Artificial Intelligence

A computer cannot think; it can do only what it is told to do. Therefore, the computer is only as good as the programs stored in its memory. Artificial intelligence is an attempt to enable the computer to emulate the mental reasoning capability of a human. A good example of this is to program a computer to play chess. This is done by having an expert chess player put his expertise on this subject into the computer program so that, when the computer plays the game against a human opponent, it responds to its opponent's moves and learns from its own mistakes like an expert.

Artificial intelligence for control system applications follows the same principle by building the expertise of experienced operators and engineers into the computer so the computer can provide solutions to problems encountered in operating and maintaining the plant equipment. The computer solutions can be presented in the form of guidance messages displayed on the CRT screens or, in some cases, as direct output to the final control elements. There are many branches of artificial intelligence. Among the popular ones are expert systems and neural networks.

An expert system has two parts: the knowledge base and the reasoning engine. The knowledge base is where the expertise of the system is stored. The reasoning engine is a set of decision-making rules structured in the form of if–then statements that represent the expert's problem-solving approaches under different circumstances. A problem is presented to the computer as inputs either from the process or manually entered by the plant personnel. The computer then consults the knowledge base and seeks the rules that match the inputs. The process continues until a solution or possible solutions are reached. The expert system essentially solves problems through a deductive process using the knowledge stored in the computer and the clues derived from the input data.

A neural network is a program that develops a pattern recognition process to allow the computer to emulate the way humans recognize patterns. When presented with a set of input data, the neural network program first tries to identify relationships among the input data and then attempts to match the inputs to the established patterns derived from the data stored in the computer. Once the program is put to use, the database is continuously updated as more data are received. In this respect, the neural network is similar to a human. It gets better as it gains working experience because it "learns" from its own experience. In control system applications, a neural network is seldom used by itself because pattern recognition, unaccompanied by other programs, will not produce a direct control output signal. The neural network is most useful in providing information to other programs. One such application is to calculate the energy balance of a process or to estimate the composition of a process medium such as a particular type of gas or liquid by inference from temperature, pressure or flow measurements.

A self-tuning controller is a controller that uses the neural network principle to automatically change the tuning parameters to improve the system response as the process load changes. A case in point is a stand-alone PID controller with a built-in pattern recognition capability that changes its tuning parameters such as proportional gain, reset time, and rate time according to the behavior of the process by continuously absorbing process information from input data. Self-tuning PID controllers have been used in the process industry with great success in recent years.

18.6 OPERATOR INTERFACE

The phrase "operator interface" is short for "operator–machine interface" or "man–machine interface." Interface, as defined by Webster dictionary means, "a point or means of interaction between two systems, disciplines, etc." Generally, operator interface refers to the interaction between the plant operator and the plant process or equipment, and operator interface devices are the tools that an operator uses to facilitate this interaction. Traditionally, for a control room operator, this encompasses all instrumentation mounted on the control panel that is normally encountered in operating a power plant. For a DCS or PLC system, this means devices associated with the operator workstation: CRT monitors, keyboards, mouses, trackballs, and printers. Figure 18-24 shows a typical CRT-based DCS control console.

A brief description of typical operator interface function follows.

Fig. 18-24. Typical CRT-based DCS control console.

18.6.1 Process Graphic Displays

The idea of graphic displays, simply put, is "a picture is worth a thousand words." The forerunners of process graphic displays were panel mimics that showed a given process (for example, a piping schematic for circulating water piping to and from the condenser with indicating lights for the valves). Panel mimics had their disadvantages. Panel spaces were limited and it was not possible to provide a mimic for every process system in the plant. Often, the space available was not sufficient to adequately depict all the flow paths in a system. The result was usually a schematic diagram cluttered with lines and lights.

CRT displays have revolutionized mimic design. With its nearly unlimited data storage capacity, an operator workstation can provide a piping schematic display for each system in the plant and at different levels of detail. The designer generally sets up a display hierarchy starting with a plant overview (showing the major plant areas and their interconnection at the top level). From the overview, the plant can be divided into process systems such as boiler air and gas, feedwater, condensate, and condenser circulating water which corresponds to the individual sections of a conventional boiler–turbine–generator control panel. Additional displays can be made to show details of each system or the equipment in the system. Instead of moving in front of the control panel to reach the panel section that the operator is controlling, the operator merely calls up the display needed and uses the display to access the data wanted at that particular moment by further calling up other types of displays (bar charts, trend displays, faceplates) to control the process. The process graphics (Fig. 18-25) have become the centerpiece of operator interface activities and they are the starting point from which the operator calls up all other displays to monitor plant operation and initiate control actions while performing duties in the control room.

18.6.2 Bar Charts and Trend Displays

A bar chart is an emulation of the panel-mounted conventional analog indicators on the CRT screen. They present an indication of variables that pertain to plant operation (such as steam pressure and unit load) and equipment condition (such as lube oil pressure and feedwater water heater level). The indicators were vertical, horizontal, or round dial types, all of which can be duplicated on a CRT display.

The CRT displays have several advantages. An indicator on a CRT display can be duplicated in different display groups as many times as the operator needs. For the conventional indicators, this requires additional hardware and panel space. The CRT display can also change color when the variable indicated is in the alarm state and, if needed, the value of the variable can also be displayed in the numeric form.

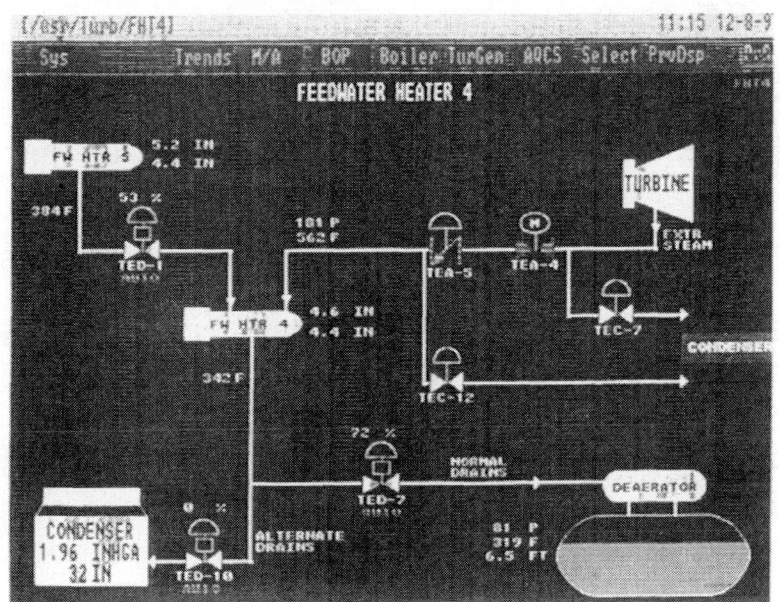

Fig. 18-25. Process graphics display.

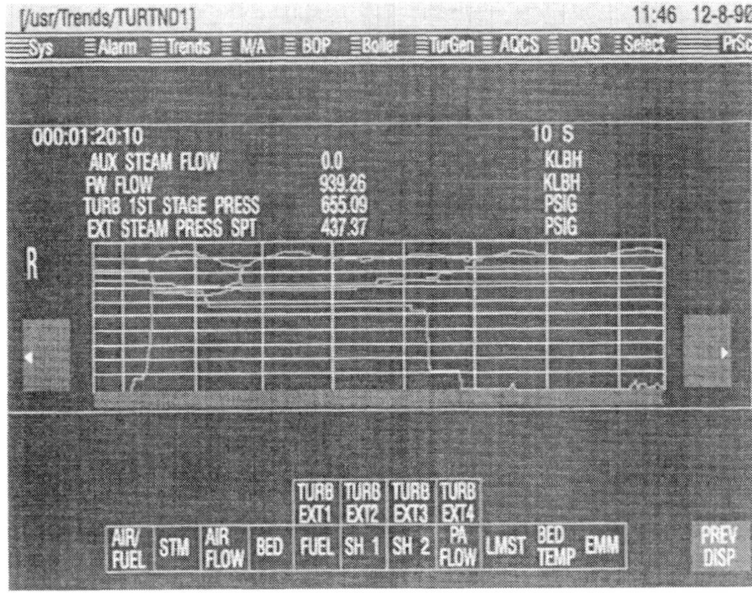

Fig. 18-26. Trend displays.

Trend displays (Fig. 18-26) may be considered an emulation of a strip chart recorder on a CRT screen. Typically, a trend display is capable of trending three to five selectable variables with a time scale that can be changed in predetermined increments. The data trended can be either real-time or historical data. Like the bar charts, the trended displays can also be reduced and grouped to emulate a recorder panel showing the trend of multiple variables. Any system data that can be displayed on a CRT screen can also be sent to a hardware strip chart recorder to create a permanent hard copy if desired.

18.6.3 Alarm Displays

Alarm displays (Fig. 18-27) show all points, analog or digital, in an alarm state in the form of written messages with associated information such as time of alarm, point status, variable value, and alarm set point value. The messages are generally displayed on the CRT screen in the order of occurrence. The alarms can also be prioritized with points in different priority levels displayed in different colors. The operator can be alerted automatically when a point first comes into the alarm state even though the alarm display is

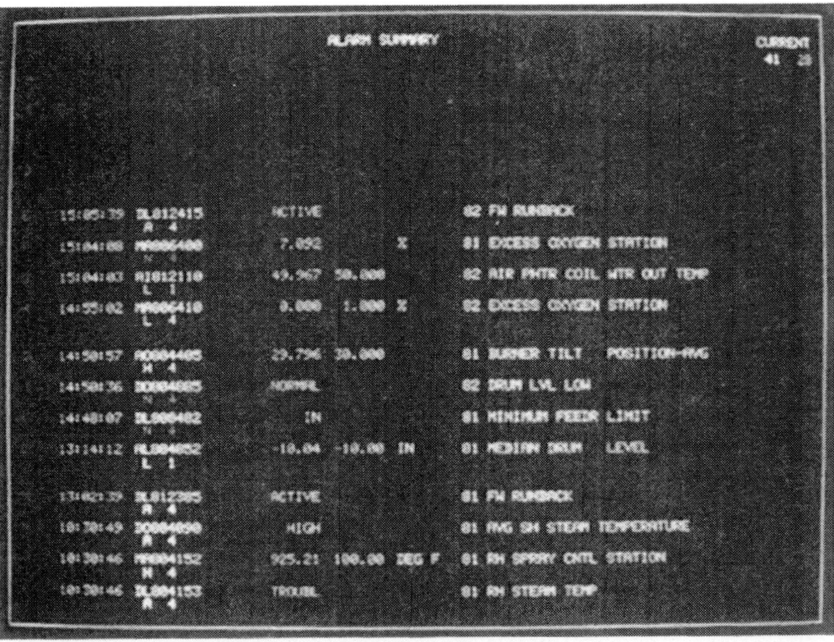

Fig. 18-27. Alarm display. (From Bailey Controls Company. Used with permission.)

not on a CRT screen at the time. This is accomplished either by reserving a line or a particular area on the screen for showing any new alarm messages or, in some systems, by using a small custom-made alarm panel with small indicator lights that are automatically lit on the occurrence of a new alarm to prompt the operator to call up the relevant alarm displays.

A new alarm message, when it first appears on the screen, will flash while accompanied by a horn. The message stops flashing and the horn is silenced as soon as the message is acknowledged by the operator with a push button. When the status of the point returns to normal, it is again accompanied by a flashing message and the horn until it is acknowledged by the operator. Such a sequence can be made identical to the annunciation sequences specified in ISA standards for conventional visual annunciators. Alarms are generally printed out as they occur and can be stored on tape or disk for long-term data storage.

In addition to the alarm message displays, the alarm messages displayed on the screen can be arranged in the same manner as the traditional annunciator windows to enable the operator to handle the alarms as if operating the unit with the traditional control panel arrangement.

18.6.4 Faceplates

Faceplates are CRT displays that emulate the conventional panel-mounted control stations. There are basically two types of faceplates.

The first type of faceplate emulates the functions of the conventional hard-wired hand/automatic station. Bar graphs of all relevant variables associated with the control loop or the final control element, including the set point, controlled variable, magnitude of the set point error, and output signal to the final element, are displayed. Figure 18-28A shows typical faceplate displays hand/automatic stations.

The second type of faceplate emulates the control devices used for on–off type control functions such as control switches and push buttons with indicating lights. Figure 18-28B shows such a faceplate with soft "back-lit" push buttons for motor manual start and stop operation. The appropriate buttons will be lit to indicate whether the motor is running or not running and both the start and stop buttons will be lit if the motor is tripped automatically after it has been started. The Maintenance Mode button is used to prohibit motor start when the motor is shut down for maintenance work. Such a faceplate completely emulates the control function of the most commonly used control switch (a three-position, spring-return-to-center motor control switch with a red and green target and three indicating lights).

For sequential type controls, the "soft" push buttons can also be arranged in pairs for start/stop operation of the various pieces of equipment controlled by the sequential control logic.

The faceplates typically are grouped according to the process system that they are related to, and both types can be mixed in one grouping in the same manner.

18.6.5 Printed Logs and Reports

The ability of DCS to generate logs and reports is nearly unlimited. Any data input to the DCS and any data processed by the DCS can be put in a programmed format and printed out on demand, automatically on the occurrence of a predetermined event, or periodically at a time designated by the operator.

18.6.5.1 Periodic Logs. These are logs automatically printed out hourly, daily, or at other fixed intervals. The logs typically contain data concerning the performance of the unit in the most recent period. The data printed can be instantaneous, averaged, or values accumulated in the period. These logs, with proper selection of the data contained, can be used to replace the manual data collection the operator routinely performs during each shift.

18.6.5.2 Trend Logs. A trend log prints out a selected group of analog data at predetermined intervals. This log can be used to follow the trend of a set of related operating parameters (for example, turbine metal temperatures during startup operations). These logs are used in place of the traditional multiple-point recorders.

18.6.5.3 Sequence of Events (SOE) Log. In a DCS, certain digital inputs designated as SOE points will be scanned by the DCS at a higher rate (on the order of one scan in one millisecond) than the other digital inputs (which are typically scanned once per second). An SOE log is triggered by a predetermined event such as a boiler trip and a turbine trip. On the occurrence of such an event, the SOE program stores the state of the SOE inputs and prints out the SOE log. This type of log is used by station operating personnel to review the sequence of events after the event to determine the cause of the trip and examine the system response after such an event.

18.6.5.4 Post Trip Review Log. This log continuously collects selected analog points and prints out the data for the period immediately before and after a unit trip or a similar event. This log is very useful for the purpose of reviewing any abnormal condition that might have caused the trip and for examining the behavior of the boiler and the turbine after the trip.

18.6.5.5 Event Log. An event log records actions taken by the operator through the operator workstation (for example, acknowledging a new alarm, initiating a motor start command, etc.), thus providing a record of what an operator has done during his shift.

Log programs provide the necessary infrastructure for the DCS to print out selected data on the basis of selection criteria entered by the user. In this regard, the types of logs and reports that can be generated are largely determined by the user. With proper criteria and programming, the user can produce as many logs and reports as needed to help operate and maintain the equipment installed in the plant.

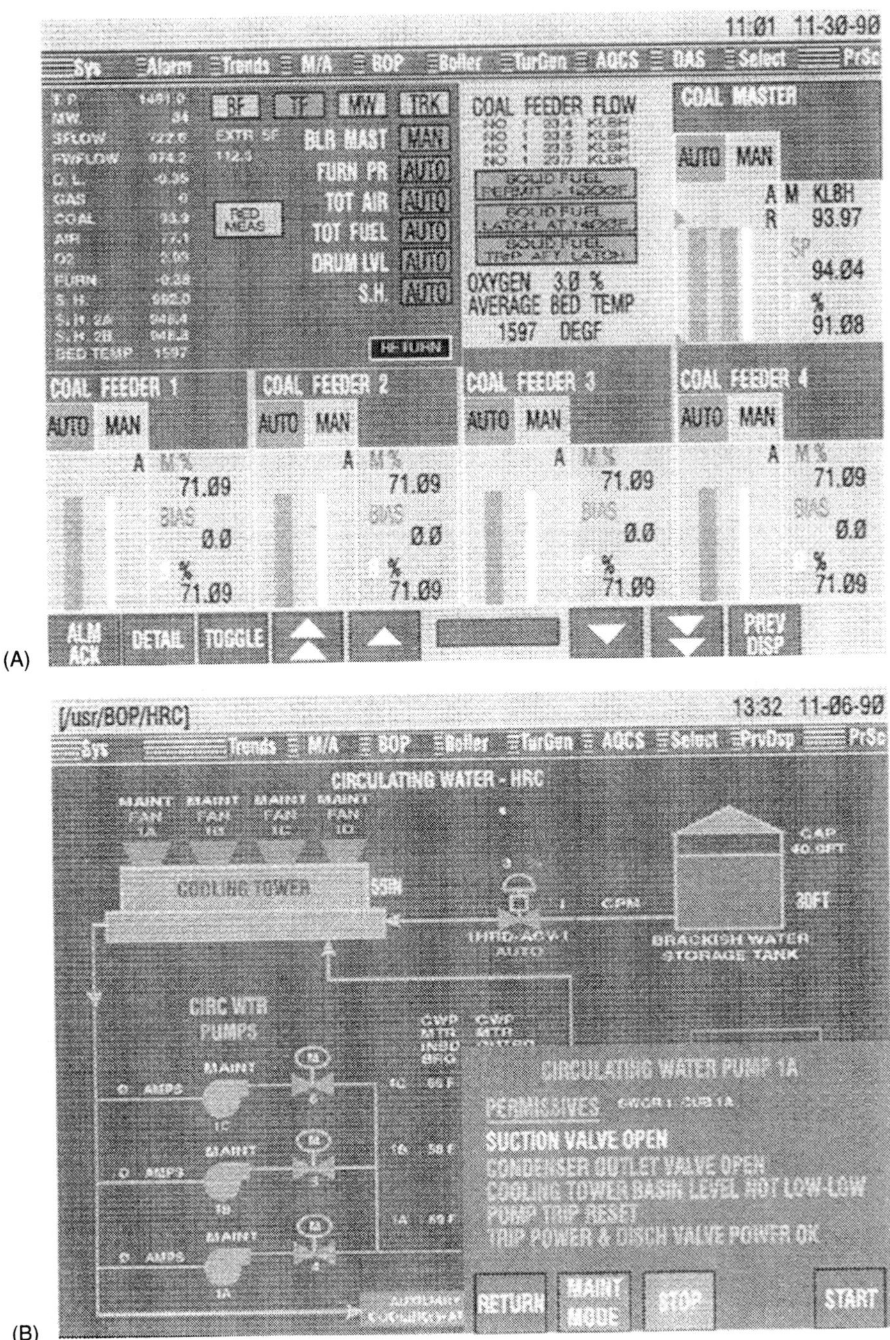

Fig. 18-28. Faceplate displays. (A) For hand/automatic stations. (B) On–off control.

18.6.6 Plant Operation from the Control Room

Operator interface devices are the tools that a control room operator uses to monitor and control plant operation. This section provides a brief rundown of the activities related to plant operation from cold start to full load that require the operator's attention and interaction with the plant equipment and systems.

From the standpoint of the operator in the central control room, the equipment and systems fall into two categories: those controlled by a local operator, and those controlled by the control room operator situated remote from the equipment. The former are mostly the plant systems that serve more than one or all units in the plant, and their operation is either intermittent or the outputs do not vary directly with the megawatt output of the plant. Those in the latter category are the unit-oriented equipment that must be placed into service when the unit is running and their operation must be regulated as the unit load changes.

18.6.6.1 Common Systems Operation. After a plant is built, it is rare that a unit has to be started from a very "cold" condition with every piece of equipment at standstill and every tank or vessel empty. However, if this is the plant condition at the time of a unit start, the operating personnel will have to get these systems into operation before doing anything else. The following steps should be taken:

1. Energize the plant auxiliary electrical distribution system by bringing in power from an outside source and powering up the control systems.
2. Start the compressed air systems (both the station service and the control air system).
3. Make water (plant service water and cycle makeup water) by first starting the plant raw water system (or well water system, as the case may be) with its associated water treatment system.
4. Prepare the coal handling and other material (such as limestone) handling systems for service.
5. Prepare the ignition fuel system, and if applicable, the auxiliary boiler fuel system for service.
6. Produce auxiliary steam by starting the auxiliary boiler when the required fuel, water, and compressed air are ready, or by taking steam from other sources if available in the plant.

Typically, plant common systems are installed with a control cabinet or a local control panel located in the vicinity of the system equipment (in some plants, the local control functions are selectively duplicated in the control room). While the systems are in operation, the major task for the central control room operator is to monitor the status of these systems. As a rule, abnormal operating conditions of each system are alarmed in the central control room in the form of a "system trouble" alarm for the control room operator to call on the local operator to investigate the trouble.

In addition, all equipment status information can be made available in the control room through the data highway of the equipment control systems (PLC or DCS). The key to this monitoring function is to constantly check the major parameters such as coal silo level, water tank level, and compressed air pressure to ensure that each of the common systems is running at a capacity sufficient to support the needs of the operating units in the plant.

18.6.6.2 Unit System Operation. A unit may be started as soon as the common service systems are operational. A myriad of activities will be conducted during a unit startup, some in parallel and some in sequence. The major ones are listed below.

1. Start the circulating water system.
2. Fill water tanks and vessels (such as the boiler drum, condensate storage tank, condenser hotwell, and deaerator storage tank) with treated water.
3. Place the condensate system into service by starting the condensate pumps. Because there is no demand for condensate at this time, the condensate flow will be recirculated from the pump discharge back to the condenser.
4. Use the auxiliary steam to peg the deaerator at a predetermined pressure.
5. Place the feedwater system into service by starting the boiler feed pumps. Because there is no demand for feedwater at this time, the feedwater flow will be recirculated from the pump discharge back to the deaerator.
6. Start the turbine lube oil system; put the turbine on turning gear and use auxiliary steam to heat the turbine and pull vacuum on the condenser.
7. Open all boiler dampers, start the air heaters, induced draft fans, and forced draft fans, and commence the furnace purge.
8. Initiate boiler firing after completion of the furnace purge by starting the required number of igniters for lightoff, and start the first pulverizer group following the successful lightoff.
9. Warm the steam lines as steam is produced from the boiler.
10. Open the turbine steam admission valves and roll the turbine off the turning gear as the boiler steam condition matches the turbine metal condition.
11. Increase the turbine speed to the generator synchronous speed (3,600 rpm) through the turbine control system and hold the speed for turbine heat soak if necessary.
12. Start the generator excitation system, synchronize the generator, and put the turbine–generator unit on minimum load.
13. Increase the unit load by increasing the boiler firing rate, and put additional burners (and pulverizers) and other auxiliary equipment (fans and pumps) into service as the unit load increases until the unit reaches full load.

Startup and online operations involve a variety of plant systems. The operating procedures are automated to a varying degree. The basic concept of CRT-based operator interface functions is to develop displays revolving around the individual plant systems, starting with a system overview followed by process graphic displays and supplemented with faceplate, bar graph, and trend displays of pertinent parameters. Similar displays can be made for large pieces of equipment such as a pulverizer, with a diagram showing the air and fuel flow paths in the pulverizer, measuring points, and damper drives, accompanied by bar chart, faceplate, and starting permissive displays. The supplementary displays can further be overlaid on the graphic display. The idea is to have the displays organized so that they can be easily and quickly called up as the need arises. With a CRT-based console, the operator remains in a seated position while performing his or her duties. All information and data gathered in any processor are available for the operator to make decisions about the control actions necessary during operation and particularly under emergency circumstances. Alarms are treated separately. New alarm messages take precedence over other displays and preempt the current displays if needed.

Printed logs also play a role in helping the operator. Typical usage includes periodic logs (hourly, daily, etc.) to collect routine information, alarm log to check alarms in recent past, trend log to trend variables (turbine metal temperatures during startup, for example), sequence of events log to determine the cause of a trip, and post-trip review log to examine the behavior of major equipment in the period immediately before and after a trip.

Data collected by the long-term data storage equipment are not just for the operator. The data represent the operating history of the various pieces of plant equipment (running time and number of starts, etc.) and the status of machine health (for example, trend of metal temperature or bearing vibration over time), and are used by the plant maintenance personnel as a reference to develop preventive maintenance programs for the plant equipment. This application will increase as the data collection and analyzing techniques of the DCS become more sophisticated.

18.7 THE PRESENT AND THE FUTURE OF THE PLANT CONTROL SYSTEM

As previously mentioned, the digital computer has completely revolutionized the concept of implementing control and monitoring functions in all industries. The use of microprocessor technology and the proliferation of the information highway further makes this revolution drastic and radical.

Microprocessors have become household items. A processor installed in a control system in a power plant works on the same principle as a control device found in a microwave oven. A workstation in a control room is nearly a carbon copy of a workstation in an accountant's office. The control system hardware is no longer a mystery. This makes things easier to any newcomer entering the control industry.

Furthermore, digital data transmission techniques have brought down the barrier between the plant control system and the outside world. The communications network that facilitates data transmission is universal. One can easily connect a processor installed in the control system of a power plant to the equipment manufacturer's factory or the engineer's office as long as a phone jack and a modem are available. Using this free flow of information exchange will certainly present a challenge to the engineers.

Lastly, control hardware of the previous generation was often a constraint for the implementation of complex and sophisticated control strategies. This is no longer the case. Microprocessors, with ever growing memory capacity, are able to do just about anything the engineers want them to do. This allows the engineers to spend their time improving the plant control system without worrying about hardware limitations.

The digital system has brought about a complete change of scenery in the control industry. At this point, no new concept is appearing on the far horizon that can possibly replace the digital concept. This system and its attendant hardware will be in use for years to come, and the drive, the urge, and the need to perfect the control systems of the future by utilizing even the known potentials of the digital system will be limited only by the imagination of the control engineers.

18.8 REFERENCES

ALBERT, C. L. and D. A. COGGAN, Eds. 1992. *Fundamentals of Industrial Control.* Instrument Society of America. Research Triangle Park, NC.

ASME. 1985. Recommended Practice for the Prevention of Water Damage to Steam Turbines used for Electric Power Generation. ANSI/ASME TDP-1.

BAILEY CONTROLS COMPANY. 29801 Euclid Ave., Wickliffe, OH 44092.

CLIFTON, CARL STEPHEN. 1986. *What Every Engineer Should Know About Data Communications.* Marcel Dekker, New York.

CONSIDINE, DOUGLAS M., P. G., Editor in Chief. 1993. *Process/Industrial Instruments and Controls Handbook.* McGraw-Hill, New York.

DUKELOW, SAM. G. 1991. *The Control of Boilers*, 2nd Ed. Instrument Society of America. Research Triangle Park, NC.

ELECTRONICS INDUSTRY ASSOCIATION (EIA). RS 232C. Interface Between Data Terminal Equipment (DTE) and Data Communication Equipment (DCE) Employing Serial Binary Exchange.

ESSCOR, INC. 512 Via dela Valle, Suite 311, Solana Beach, CA 92075.

FOXBORO COMPANY. 33 Commercial Street, Foxboro, MA 02035-2099.

FRIEND, GEORGE E., JOHN L. FIKE, H. CHARLES BAKER, and JOHN C. BELLAMY. 1988. *Understanding Data Communications.* Howard W. Sams & Company, a division of MacMillian, Inc. Indianapolis, IN.

GE FANUC AUTOMATION. P.O. Box 8106, Charlottesville, VA 22906.

GENERAL ELECTRIC COMPANY. 1994. *Generator Protection, Generator Instructions.* Publication GEK 75512F. General Electric Industrial and Power Systems. Revised 1994.

HENSHALL, JOHN, and SANDY SHAW. 1988. *OS2 Explained.* Ellis Harwood Limited. Chicester, West Sussex, England.

HOUSLEY, TREVOR. 1979. *Data Communications and Teleprocessing Systems.* Prentice-Hall, Englewood Cliffs, NJ.

HUBBY, R. N., LEEDS & NORTHRUP. 1989. Pulverized control: a tutorial review. Paper presented at Instrument Society of America Power Symposium. Phoenix, AZ, May 22–24, 1989.

IEEE POWER ENGINEERING SOCIETY. 1985. IEEE Guide for Protection, Interlocking, and Control of Fossil Fueled Unit-Connected Steam Stations. Institute of Electrical and Electronics Engineers. New York, New York.

INGERSOLL-RAND COMPANY. Hwy 45 South, Mayfield, KY 42066.

JONES, C. T., and L. A. BRYAN. 1983. *Programmable Controllers, Concepts, and Applications.* International Programmable Controls, and IPC/ASTEC Publication. Atlanta, GA.

KIBIRIGE, HARRY M. 1989. *Local Area Networks in Information Management.* Greenwood Press, Westport, CT.

LIPTAK, BELA G., Editor in Chief. 1985. *Instrument Engineers' Handbook*, revised edition. Chilton Book Company. Radnor, PA.

NFPA. 1993. *NFPA 85C—Standard for Prevention of Furnace Explosions/Implosions in Multiple Burner Boiler-Furnaces.* National Fire Protection Association. Quincy, MA.

SINGER, JOSEPH G., Editor. 1981. *Combustion Fossil Power Systems*, 3rd Edition. Combustion Engineering. Windsor, CT.

STULTZ, S. C., and J. B. KITTO, Eds. 1992. *Steam: Its Generation and Use*, 4th Edition. Babcock & Wilcox Company. New York, NY.

WESTINGHOUSE ELECTRIC CORPORATION. The Quadrangle. 4400 Alafaya Trail, Orlando, FL 32826-2399.

WESTINGHOUSE ELECTRIC CORPORATION. 1983. *Steam Turbine Generator Motoring and Overspeed Protection*. Steam Turbine-Generator Information. Revision 2, December 1983.

19

Site/Plant Arrangements

Roger M. Prewitt

The site arrangements and plant arrangements for an electricity generating power plant are drawings that provide the basis for the engineering and design of a power plant project. The site arrangements show the locations of major facilities and areas on the power plant site, the locations of the major utilities and facilities entering and exiting the site and interconnecting the major areas on the site, and traffic patterns for the site. The plant arrangements define the physical arrangement of the specific buildings on the site. This includes the location and spacing of support columns for the building; the location of exterior and interior walls; the locations of floors, doors, walkways, access ways, platforms, and roofs; and the location of the significant equipment within the building.

These drawings provide a primary means of communication and coordination between the various design engineering disciplines on a specific project, and offer a primary means of communication between the architect–engineer, client/owner, regulatory agencies, and other interested parties. Because the site and plant arrangement drawings are used in these ways, they must be developed at the very beginning of a project.

The location of a building relative to other buildings, facilities, and areas on the site impacts the physical arrangement of that building. Therefore, it is necessary to determine and develop the site arrangements before developing plant arrangements of the individual buildings. More importantly, the site arrangement drawings are essential for the licensing and permitting of a project, which occurs prior to the more detailed design associated with the plant arrangements.

19.1 SITE ARRANGEMENTS

19.1.1 Site Selection

The first step in producing the site arrangement for a power plant is selecting the site. The site must be carefully chosen, considering the economics of site development, plant lifetime costs, environmental factors, socioeconomic factors, and licensability factors. These requirements are frequently in conflict with one another and require careful screening to ensure the optimum location is selected.

A major consideration for site selection is the impact of federal, state, and local regulations on air quality, water allocation, water quality, wastewater treatment, noise, and solid waste disposal for the specific site. The National Environment Policy Act (NEPA) is the basic national charter for protecting the environment. NEPA establishes policy, sets goals, and provides the means for carrying out that policy. Federal agencies initiate the measures needed to direct their policies, plans, and programs so as to meet national environmental goals. Individual states typically adopt parallel programs for actions within their particular state.

The major federal statute regulating water quality and wastewater discharges is the Clean Water Act (CWA). The Environmental Protection Agency (EPA) is responsible for administering the CWA. Individual states determine if more stringent standards are required to meet their standards and they administer permits and review water treatment system designs.

The federal requirements for potable water supplies are established by the Safe Drinking Water Act (SDWA). Individual state agencies implement provisions of the SDWA and require the use of standard plumbing codes.

The Resource Conservation and Recovery Act (RCRA) is the controlling federal legislation for the regulation of solid waste disposal. The EPA and individual state agencies are responsible for implementing the provisions of the RCRA.

The Noise Control Act (NCA) provides the statutory guidance for federal noise control activities. The EPA and the Occupational Safety and Health Administration (OSHA) are responsible for determining noise emission standards and administering the requirements of the NCA. The impact of these regulations must be carefully considered during the site screening process.

The screening process used to select the site for a new power plant should involve the following major steps:

- Define the potential types and sizes of power plants to be placed on the site.
- Define the area from which the site is to be selected. This generally includes the owner's service territory, if the owner is a utility, or may include areas near a proposed load center or qualifying cogeneration host if the owner is an independent power producer, developer, or cogenerator.
- Characterize the siting area by reviewing USGS maps, aerial photographs, and transmission maps. These maps and photographs should be reviewed to identify major existing waterways, existing transmission routes, and potential sources for fuel.

- Define exclusion areas based on proximity to national or state parks, Prevention of Significant Deterioration (PSD) nonattainment areas, and other areas where a power generating facility could not be sited.
- Identify potential sites by further reviewing the maps; photographs; and available data on hydrology, soils, vegetation, and land use. The potential sites need to consider water sources, fuel sources, transportation access, access to adequate transmission lines, provisions for combustion waste disposal, topography, geology, land availability, environmental impacts, and availability of a work force.

Once the potential sites have been identified, a score or value should be determined for each of the factors noted previously for each site. Based on these scores, an overall ranking of the sites can be determined. From this ranking, several of the top ranking sites can be selected for further detailed evaluation. It is recommended that both a primary site and a secondary site be considered so that the plant can be relocated if insurmountable problems occur during land acquisition or permitting and licensing.

During detailed evaluation of the selected sites, site visits are mandatory to verify map work and to investigate existing site conditions. These site investigations can help verify the amount of wetlands, the likelihood of hazardous waste, or the types of foundation materials encountered. A factor that plays a large part in the determination of the site is the availability of water. Water sources and the long-term viability of these sources must be carefully evaluated. In some areas, an onsite reservoir may be necessary to store water for use by the plant during periods of drought. In addition, the site should be located so that two competitive sources of fuel are available. Fuel cost is one of the largest lifetime operating costs of a project, and in a deregulated environment, competition between sources of fuel and transportation companies or transportation modes is the surest guarantee for an economical delivered cost of fuel. For areas where selected sites are remote or isolated, the problems of housing and infrastructure for construction and operations staff must be addressed.

Following the site visits, site-specific evaluation criteria and parameters should be defined. These criteria will include fixed charges, rate of return, water sources, competitive fuel sources, transportation access, access to adequate transmission lines, provisions for combustion waste disposal, site topography, environmental considerations, licensability factors, and site development cost factors. The specific sites can then be evaluated and scored based on these criteria and parameters. By comparing the scores and evaluations, the sites can be ranked and a primary and secondary site selected.

It is sometimes necessary to perform sensitivity analyses for selected sites to determine the importance of one or more of the evaluation factors on the overall rankings of the sites. The sensitivity analyses are performed by varying the degree of importance for one or more of the evaluation factors with respect to the overall score for a site. The results of these analyses can be of help in determining how the evaluation factors affect the scoring and ranking of the sites.

19.1.2 Development of Site Arrangement

Once the site has been selected, the arrangement of the power plant on the site can be developed. The development of the site arrangement should be conducted in an orderly and logical fashion. To establish a site arrangement, the following items should be defined and identified, as appropriate for the type of power plants or sites:

- Fuel delivery modes (rail carriers, barge, truck, pipeline, etc.);
- Location of railroad connections;
- Location and type of water sources;
- Location of connection points to water and sewer lines;
- Amount of water to be stored on the site and the storage method (tanks or pond);
- Type of circulating water system (cooling tower or once-through);
- Location of public road connections;
- Location of connection point to electrical transmission system;
- Probable location of wastewater discharge points;
- Location of combustion waste disposal areas and amount of combustion wastes to be stored on the site;
- Direction of prevailing winds for winter and summer;
- Amount of fuel storage and flue gas desulfurization system additive storage to be maintained on site;
- Location of environmentally sensitive areas on the site or adjacent to the site that may affect the site arrangement; and
- Location of areas for construction facilities and construction laydown.

Of primary interest among these items are the major high unit cost utilities and facilities entering and exiting the central plant complex, and the traffic patterns throughout the site. The central plant complex includes the power generation facilities area, the control center area, and the air quality control equipment facilities. The major utilities and facilities and comparative installed unit costs typically include the following:

Major utility/facility entering or exiting the central plant complex	Comparative installed unit cost*
Plant access roads	$ X per foot
Railroad trackwork	$ X per foot
Natural gas pipeline	$ X per foot
Plant makeup water pipeline	$ X per foot
Electrical transmission lines	$2.5X per foot
Circulating water pipeline	$7X per foot
Coal conveyors	
On ground	$10X per foot
Overhead	$20X per foot

*The access roads, trackwork, and pipelines have a similar unit cost, denoted by the value "X." The comparative unit costs for the remaining items in the list are denoted as a multiple of "X" (2.5 time X, 7 times X, and so on).

Having established the major utilities/facilities and their unit costs, the "radial concept" of site arrangement can be

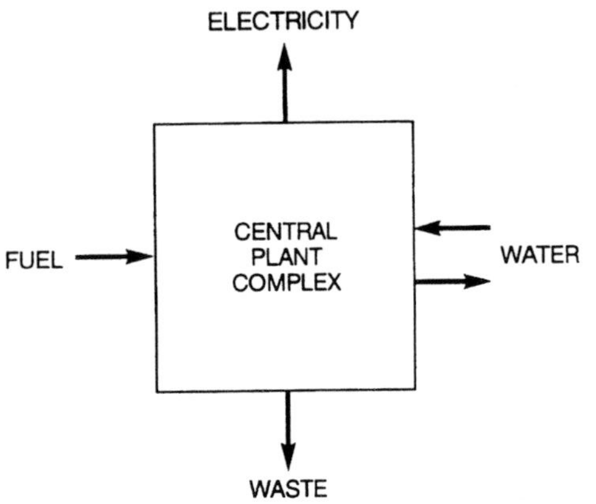

Fig. 19-1. Radial concept.

used to begin development of the arrangement. The radial concept involves arranging the major utilities/facilities in a radial manner around the central plant complex so that these major utilities/facilities do not interfere with one another. Figure 19-1 illustrates the radial concept.

The first step in developing the site arrangement is to analyze the shortest distance to bring the following major items from offsite to the central plant complex while adhering to the radial concept:

- Fuel supply,
- Transmission lines,
- Makeup water, and
- Access road.

After establishing these directions, at least three site arrangements should be developed for evaluation. Figures 19-2, 19-3, and 19-4 show three proposed arrangements for a multiple-unit, pulverized coal power plant site with a cooling tower circulating water system. For each arrangement, the fuel supply is delivered by rail from the south of the site, the transmission lines exit the site to the north, and the makeup water and plant access road enter the site from the north. Each of these arrangements was developed considering the appropriate items from the screening process and the radial concept, as follows.

The central plant complex is centrally located on the site in each arrangement to support the radial concept. This central location enables optimum use of the site for the remaining major areas (including coal storage area, combustion waste disposal area, cooling towers, recycle basin, and water supply storage pond) and allows for expansion of the storage and disposal areas. The central location also creates

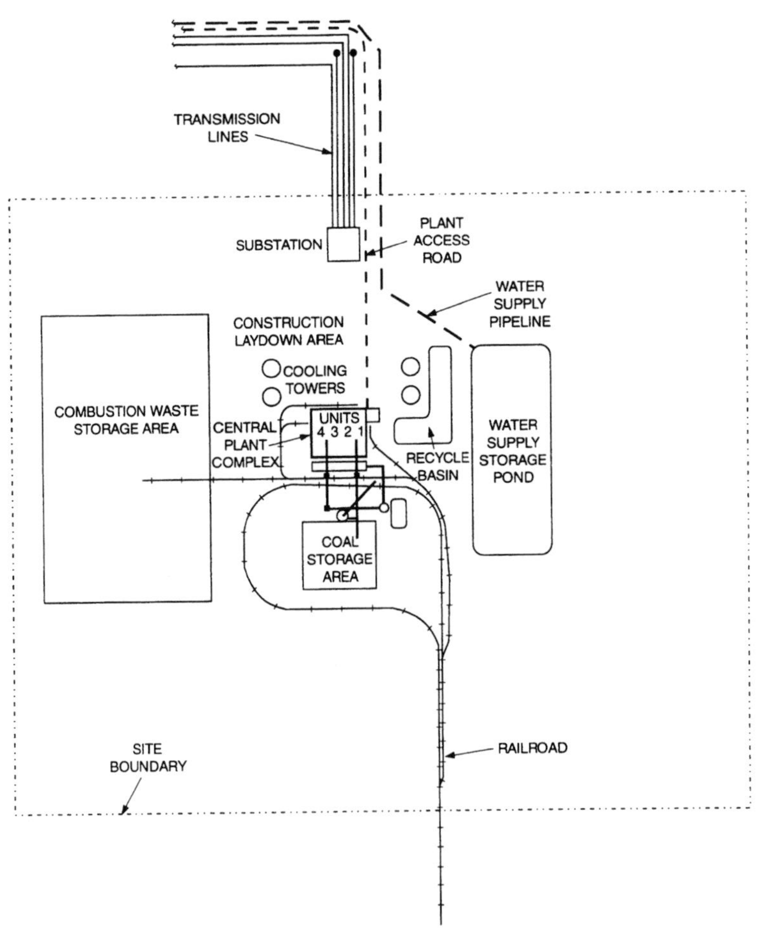

Fig. 19-2. Alternative site arrangement A.

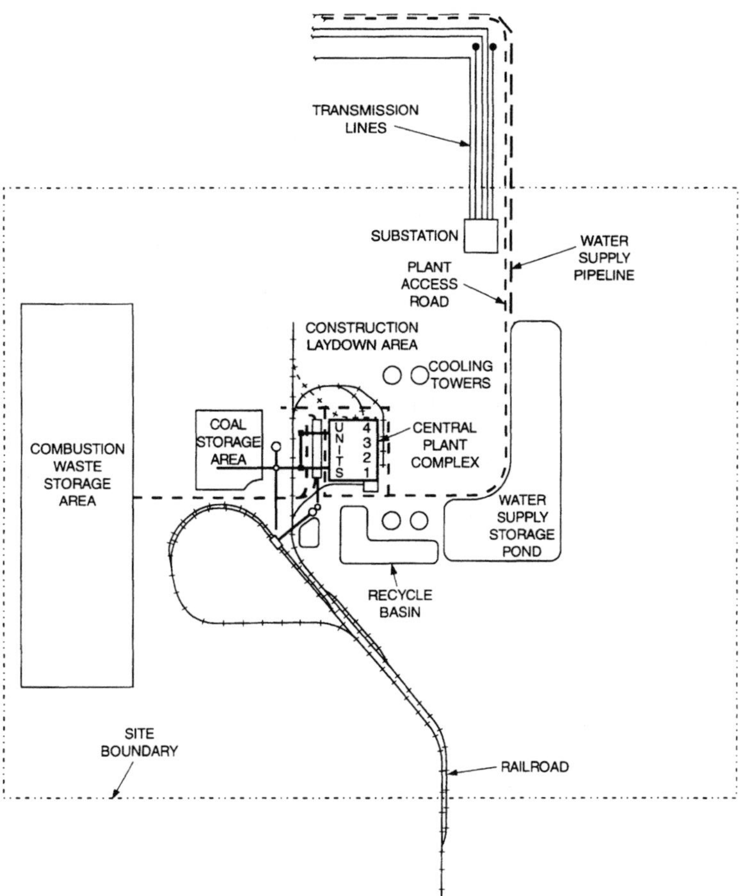

Fig. 19-3. Alternative site arrangement B.

open space, or "buffer zones," between the central plant and property adjacent to the site.

When the remaining major areas of the site are first arranged, the prevailing winter and summer winds must be taken into account. The drift from the cooling towers (discussed in Chapter 12) carried by the prevailing winds can create a major operating problem for the electrical substation equipment, transmission lines between the substation and central plant, and generator transformers located at the central plant. Mechanical draft cooling towers should be located so that the prevailing winter wind cannot blow the cooling tower drift onto the substation. Drift should also be prevented from blowing onto the coal storage area or the central plant complex, if possible. In any case, mechanical draft cooling towers should be located a minimum of 1,000 ft from a substation and 700 ft from a major structure such as the central plant complex. These location requirements somewhat dictate the length of the circulating water piping between the towers and the central plant, but this piping length should be kept as short as practical because of the high unit cost of this piping. Once the central plant and cooling towers have been located, the coal storage area should be located.

Coal storage requires a large, level space. The storage area should be large enough to contain sufficient coal to allow the plant to continue operation in the event that coal deliveries are interrupted. This storage capacity may range from as little as 15 days of operation to as many as 120 days, depending on the reliability of delivery system.

For rail delivery systems, space is required for a track loop for efficient unit train delivery for a high-capacity, 3,500 to 4,000 tons per hour unloading system. Unloading rates are determined by the length and net tonnage of the unit train and the free unloading time allowed by the railroad. A series of "ladder tracks" may be required for systems with random-car dumping systems. These systems may have longer free unloading times to allow for car switching. The railroad curves and grades required to reach the storage area and the length of the train to be used for delivery constrain the actual unloading point in the storage area. As noted previously, the unit cost of coal conveyors is the highest of any of the major utilities/facilities and this cost must be considered in the location and arrangement of the coal storage area relative to the central plant complex.

If coal is handled by truck, provisions should be made onsite to allow for a queue of trucks waiting to weigh or unload. Special treatment of the access road and its intersection with the public road system will likely be required for the increased truck traffic.

For barge delivery of coal, the distance between the barge dock, storage area, and the central plant should be kept to a

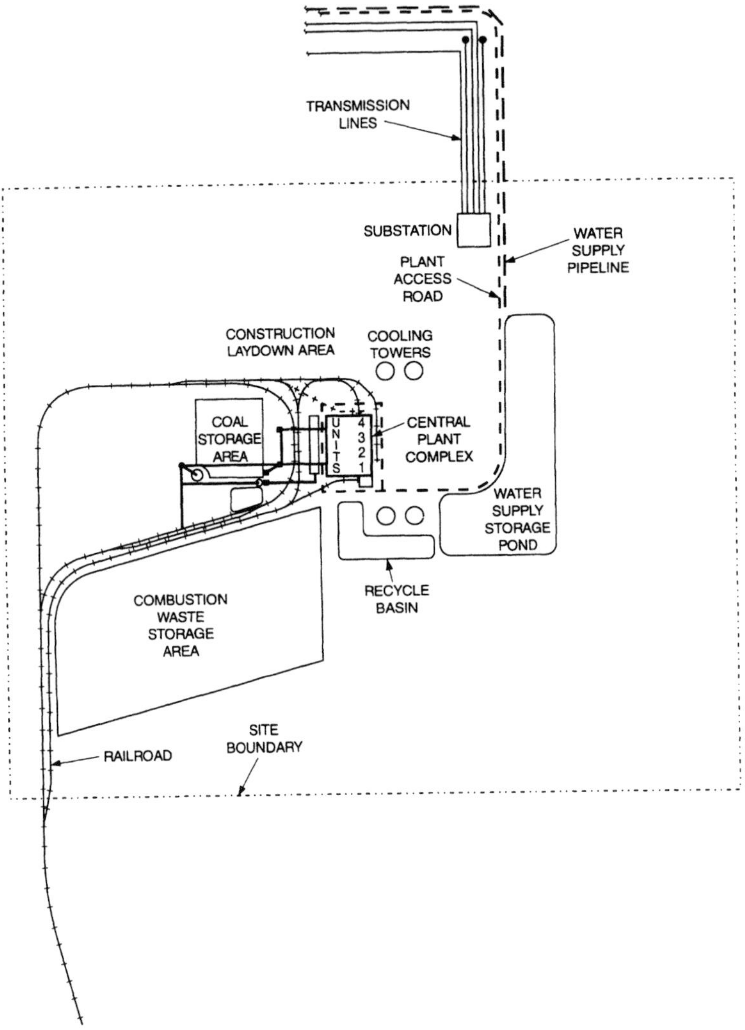

Fig. 19-4. Alternative site arrangement C.

minimum to minimize the conveyor length. Space must be provided adjacent to the coal storage area for a runoff pond to catch and treat rainfall runoff from the storage area. The prevailing seasonal wind directions must also be considered in the location of the coal storage area so that coal dust is not blown onto the central plant complex or onto the adjacent property. See Chapter 5 for more details on coal handling systems.

Onsite combustion waste disposal requires significant storage area. Typically, this area is sized to contain all the combustion waste generated over the life of the plant. For the examples shown in Figs. 19-2, 19-3, and 19-4, the combustion waste disposal area requires approximately 10% of the total site area. An alternative to onsite storage that should be investigated is to return the combustion wastes to the coal mine supplying the fuel for the plant. Combustion wastes generated in a pulverized coal power plant typically include bottom ash, fly ash, and flue gas desulfurization (FGD) system waste. Fly ash and FGD system wastes are typically handled dry. Short-term storage for these wastes, usually 3 days accumulation, is provided at the central plant, where trucks or other mobile equipment can pick up the wastes for transport to the disposal area. Bottom ash can be handled either wet or dry. When it is handled wet it is sluiced from the steam generator in the central plant to ponds in the waste disposal area. The sluice water is then decanted and returned to the plant for reuse.

Whether the combustion wastes are handled and stored dry, wet, or in a combination of the two, the site soil conditions and environmental requirements must be reviewed to determine if the ground in the disposal area must be lined. The disposal area should be developed such that only small areas are active at any one time. As with the coal storage area, space must be provided adjacent to the combustion waste disposal area for a pond to catch and treat any rainfall runoff from the disposal area.

For the water storage area, sufficient storage capacity should be provided to allow the plant to continue operation in the event that the supply of water to the site is interrupted or the water treatment equipment located at the plant site is out of service. This storage capacity is usually between 3 and 7 days of operation.

The recycle basin and other ponds designated for the containment of waste water should be sized to contain the runoff resulting from a 100-year, 24-hour storm.

With these major areas sized and located in alternative arrangements as shown in Figs. 19-2, 19-3, and 19-4, a comparative cost evaluation can be performed using the unit costs for the major utilities and facilities noted previously. In comparing Figs. 19-2, 19-3, and 19-4, the site arrangement shown in Fig. 19-2 has the lowest cost for transmission lines, access road, and coal and waste handling systems. It has a slightly higher cost for water supply pipeline and the cost for its railroad is higher than the arrangement in Fig. 19-3, but less than that in Fig. 19-4. Overall, the cost for the arrangement in Fig. 19-2 is approximately 20% less than the cost for the arrangement in Fig. 19-3, and approximately 30% less than the cost of the arrangement in Fig. 19-4.

The process of developing alternative site arrangements and performing cost evaluations should be repeated until an optimum arrangement is achieved.

The development of an optimum site arrangement is a detailed and complicated process that involves the consideration of a wide variety of factors. As explained and illustrated previously, in many cases, these factors will be in conflict with one another. However, if the development process is performed as described, an efficient and economical site arrangement can be determined.

19.2 PLANT ARRANGEMENTS

As stated previously, plant arrangements are drawings that define the physical arrangement of a building and the location and arrangement of the major equipment within that building. Major equipment is generally that equipment that has a significant effect on the building layout.

The physical arrangement of a building includes the overall length, width, and height of the building; the location and spacing of support columns; the location and heights of floors, platforms, and roofs; the location of exterior and interior walls; and the location of vertical and horizontal access ways, including stairs, elevators, access aisles, and doors.

Equipment within the building is typically located by the use of a three-coordinate system (X, Y, Z). The X and Y coordinates are horizontal dimensions off the building column rows. The Z coordinate is the elevation of the equipment. These coordinates are typically the dimensions to the centerline of the equipment. However, a piece of equipment may be composed of several subassemblies that each have a centerline. In this case, to avoid confusion, it is typical to dimension to a corner of the piece of equipment with the X and Y coordinates and to dimension to the bottom of the equipment with the Z coordinate.

When showing the arrangement of a piece of equipment on a plant arrangement drawing, it is not necessary to show every detail of the piece of equipment. What is important is to show enough detail to indicate the overall size and shape of the piece of equipment; its length, width, and height; the location of major piping connections; the space required for disassembly or inspection of the equipment (such as tube or shell pull space for a heat exchanger, or access door opening space for a cabinet); the arrangement of any significant appurtenances on the equipment that impacts the building layout; and enough detail to allow the construction contractor to properly set and install the equipment. This level of detail provides a significant amount of information without duplicating information that is available on other drawings or from other sources.

The primary purpose of the plant arrangement drawings is to provide the basis for engineering design. By defining the physical arrangement of a building and the location of equipment, the foundations, structural steel supports, platforms and floors, and roof and floor drain systems can be designed, and the architectural design can be developed. The plant arrangement drawings provide the basis for designing the piping systems that interconnect the mechanical equipment and provide the basis for routing these piping systems. The plant arrangement drawings also provide the basis for the design and layout of the electrical power and control cable systems and the design and layout of the raceway system for these power and control cable systems.

The plant arrangement drawings provide a primary means of coordination and communication among the various design engineering disciplines on a particular project. They also provide a primary means of communication between the architect–engineer, client/owner, regulatory agencies, and other interested parties.

19.2.1 Development of Plant Arrangements

Plant arrangement drawings typically are produced for every building of a power plant project. However, to illustrate the development process, only the plant arrangement drawings for a portion of the central plant complex are discussed in this chapter. The central plant complex includes the power generation facilities area, the control center area, and the air quality control facilities area. The power generation facilities area and the control center area compose what is commonly referred to as the Generation Building. The power generation facilities area is typically separated into the steam generator area and the steam turbine–generator area. For the purposes of this chapter, the development of the Generation Building plant arrangements for a 675-megawatt (MW) pulverized-coal-fueled power plant are discussed in detail. The methodology and procedures described are appropriate and applicable for any building of a power plant project.

The steps that should be followed in developing the plant arrangements for the Generation Building are as follows:

- Identify all the major equipment that will be located in the building and that will be shown on the plant arrangement drawings.
- Review and locate each piece of equipment with respect to its

requirements for maintainability, constructibility, and operability.

- Based on the equipment locations, determine the overall length, width, and height of the building; the location and spacing of building support column rows; the location and heights of floors, platforms, and roofs; the location of exterior and interior walls; and the location of vertical and horizontal accessways including stairs, elevators, access aisles, and doors.

An important consideration in the development of plant arrangements is the impact of federal, state, and local regulations on the arrangement. Common federal regulations to be considered include the Uniform Building Code (UBC), National Fire Protection Association (NFPA) standards, and Occupational Safety and Health Administration (OSHA) standards. Where these regulations directly affect the arrangements, they are referenced in the text. The individual states typically have construction codes and building codes that must be adhered to.

The first step in the development process is identifying the major equipment to be located in the Generation Building (Table 19-1). This is only a partial list, but it includes a variety of equipment with unique characteristics and location requirements.

Before beginning the second step in the process, the general criteria for maintainability, constructibility, and operability that apply specifically to the development of plant arrangements need to be reviewed before applying these concepts to the individual pieces of equipment. These three concepts are part of a larger concept commonly represented by the acronym MARCO—maintainability, availability, reliability, constructibility, and operability. The concepts of availability and reliability are more appropriate for system and equipment design and so are not included in this development of plant arrangements. The criteria for maintainability, constructibility, and operability that apply specifically to developing the plant arrangement drawings are discussed in the following sections.

19.2.1.1 Maintainability. Plant equipment should be designed to be maintained in place, if possible, with minimum disassembly of surrounding equipment and minimum usage of temporary scaffolding and handling equipment. Permanent maintenance platforms should be located where required to ensure safety and efficiency.

Plant building arrangements, piping routing, and cable tray locations should be designed for maximum equipment accessibility and to allow the following types of access:

- Space should be provided to allow plant personnel easy access to all equipment which may require maintenance.
- Space should be provided to give unobstructed access for maintenance tools and equipment required for maintenance on permanently installed equipment.
- Ample space should be provided to facilitate removal of any equipment that cannot be maintained in place or that may require replacement.

Permanent cranes and hoists should be provided for equipment that is not easily maintainable with mobile equipment. Lifting eyes and other portable equipment supports should be provided.

Structures and internal building spaces should be designed to minimize housecleaning requirements by avoiding dead pockets and difficult to reach horizontal surfaces.

Floor and equipment drains should be installed within buildings and structures and at material handling areas so that hose sprays can be used for washdown.

Special attention should be given to providing enclosures, curbs, drip guards and collection systems for fugitive water,

Table 19.1. Generation Building Equipment List

Steam Generator Area	Steam Turbine–Generator Area	Control Center Area
Steam generator	Steam turbine–generator	Main control room equipment
Coal pulverizers	Turbine lube oil reservoir	Water treatment equipment
Coal feeders	Hydrogen seal oil unit	HVAC equipment
Coal silos	Electro-hydraulic control unit	Battery and inverter equipment
Air heaters	Gland condenser	Water quality laboratory equipment
Primary air fans	Surface condenser	
Forced draft fans	Turbine–driven boiler feed pumps	
Bottom ash hopper	Motor-Driven startup boiler feed pump	
Ash sluice pumps	Deaerator	
Boiler blowdown tank	Condensate pumps	
	High-voltage switchgear	
	Secondary unit substations	
	Motor control centers	
	Generator transformer	
	Main auxiliary transformer	
	Reserve auxiliary transformers	
	High-pressure feedwater heaters	
	Low-pressure feedwater heaters	
	Closed cooling water heat exchangers	
	Turbine lube oil storage and purification equipment	

hose spray water, chemicals, oils, and coal dust so that equipment areas are kept as clean as possible.

19.2.1.2 Constructibility. Site traffic routes should be established to move material, personnel, and equipment to their required position with a minimum of effort and time, and in as short a distance as practical. The establishment of traffic routes determines access points to each building.

After access points are established at the buildings, vertical and horizontal access routes within the buildings should be provided.

Construction openings should be provided where equipment access route penetrate walls and slabs, and where equipment, materials, and personnel must use the route after the walls or slabs are placed.

Preassembly of equipment and materials should be considered to reduce onsite labor costs if it also contributes to reduced overall project cost or schedule benefits.

Piping and ductwork should be routed to minimize conflicts between contractors competing for the same work space.

19.2.1.3 Operability. Each piece of equipment must be located to ensure that it operates correctly and performs most efficiently. This can be illustrated with an example of a pump that is to be located inside a building and that is fed by gravity from a storage tank outdoors. Assuming that the storage tank is adjacent to the building and supported at grade elevation, the pump should be located near the outside wall adjacent to the tank and no higher than grade elevation in the building. This helps ensure a flooded suction for the pump and results in maximum utilization of the storage volume in the tank.

Systems and equipment should be located for easy operational access and logical operational sequences. Special attention should be given to adequate lighting, ventilation, and acoustic dampening of all operational spaces.

To the extent practical, all equipment that requires operator manipulation, including valves, switches, inspection ports, and dampers, should be located so that the operator can manipulate the equipment from a standing position without obstruction and without having to climb ladders or having to climb on top of equipment.

The second step in the development process is to review each piece of equipment with respect to its requirements for maintainability, constructibility, and operability. This step is very closely related to the third step, which is actually locating the equipment. To illustrate these two steps, each piece of equipment listed in Table 19-1 is reviewed here and its location identified. What becomes apparent is that maintainability and operability are the primary criteria for locating equipment. Where used in these descriptions, the general reference "access for maintenance" means there is no particular or unique requirement for that piece of equipment other than the general criteria previously noted.

The equipment locations are shown in Figs. 19-5 through 19-8, plant arrangement drawings for the Generation Building. Detailed information on the design and operation of the equipment listed is included in the chapter addressing that specific piece of equipment in this book.

- *Steam Generator Area*
  - —*Steam Generator.* The steam generator is a pulverized-coal-fueled drum-type reheating unit. The maintenance requirements consist of providing access to the drum, burners, access doors, instrumentation, and various associated components such as soot blowers. Access stairs and platforms must be provided as required by the steam generator manufacturer's design. Horizontal and vertical access ways must be provided for component removal. The operability of the steam generator is generally addressed by the manufacturer's design, but the location of related equipment, such as fans and coal pulverizers, does impact operation, and the requirements for this equipment are discussed. A more important requirement for locating the steam generator is its position relative to the steam turbine–generator. Since the steam generator is providing the high-pressure, high-temperature steam to the turbine, it is important that the relative locations of these two major pieces of equipment result in an economical arrangement of the steam piping connecting them. This arrangement is described in greater detail in the discussion of the fourth and final step of the process later in this chapter.

    On the basis of these requirements, the steam generator is located with its principal axis parallel to the flue gas flow and oriented in a north–south direction. The steam drum is arranged perpendicular to the principal axis, on the south side of the steam generator. This orientation provides a good arrangement of the steam piping to the turbine. The steam generator elevation must be established to provide space below it for the bottom ash hopper and related ash removal equipment.
  - —*Coal Pulverizers, Coal Feeders, and Coal Silos.* It is assumed that a 675-MW unit requires eight coal pulverizers, with a feeder and a silo associated with each pulverizer. Each pulverizer requires clear space around it and access space for journal maintenance. Each feeder requires access space around it for maintenance and inspections. For proper operation, the coal silo must be located above the feeder, and the feeder must be located above the pulverizer to allow for gravity feed of the coal from the silo to the feeder and from the feeder to the pulverizer. The pulverizers are very large rotating pieces of equipment and should be located at grade elevation so that their foundations can be designed to properly isolate the vibrations that occur during operation. The pulverizers are connected to the steam generator by coal feed pipes supplying a coal–air mixture to the steam generator burners. It is important that these coal feed pipes be as short as possible to keep pressure drops low and reduce material costs. There must also be sufficient space above the pulverizers for proper routing of the coal feed pipes.

    This coal feeding equipment can be positioned around the steam generator in several configurations to satisfy these requirements. The eight pulverizers can be positioned in a single line, parallel with the steam drum and between the steam generator and the turbine–generator areas. This location would result in more steam piping between the

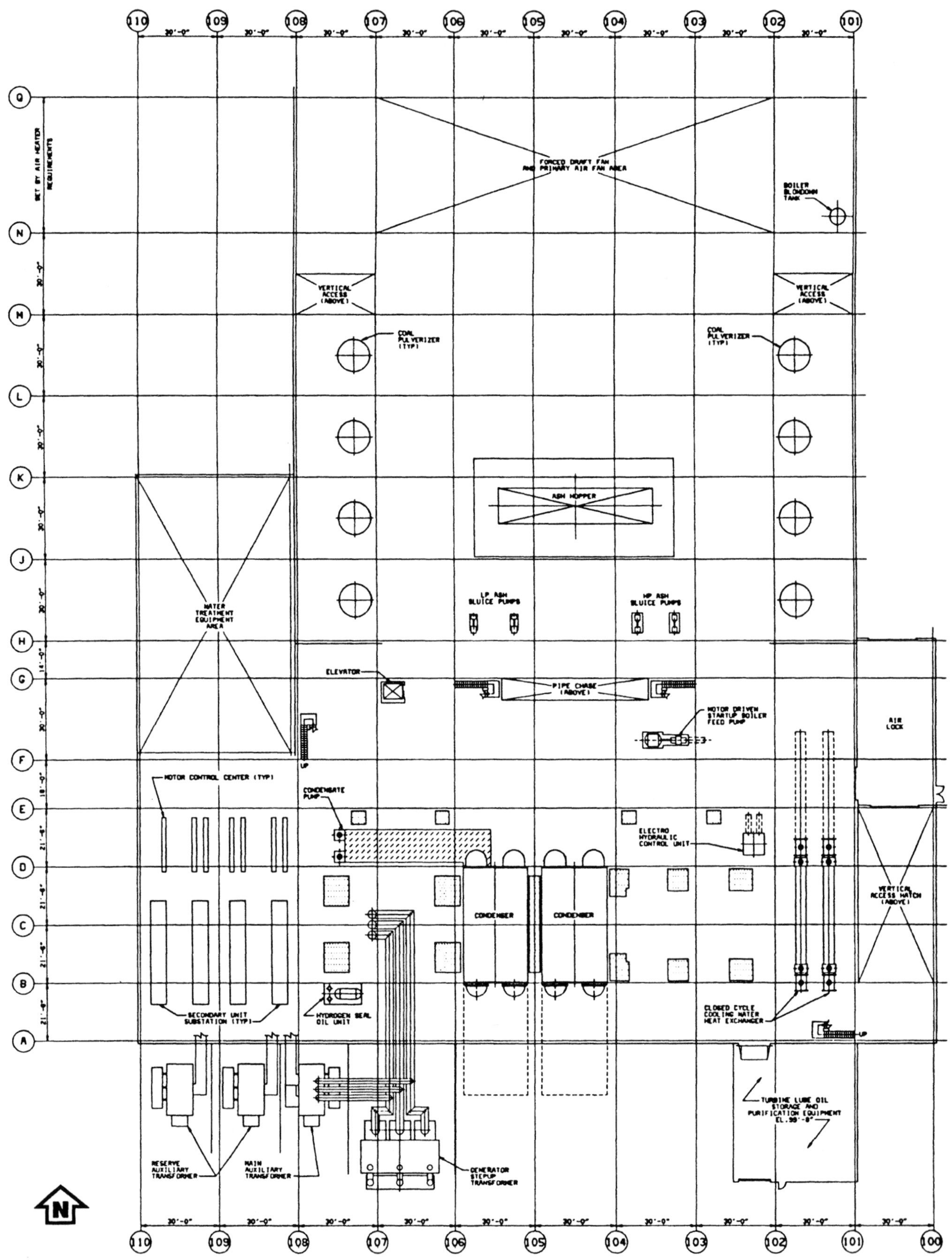

Fig. 19-5. Generation building arrangement: ground floor elevation 100′-0″.

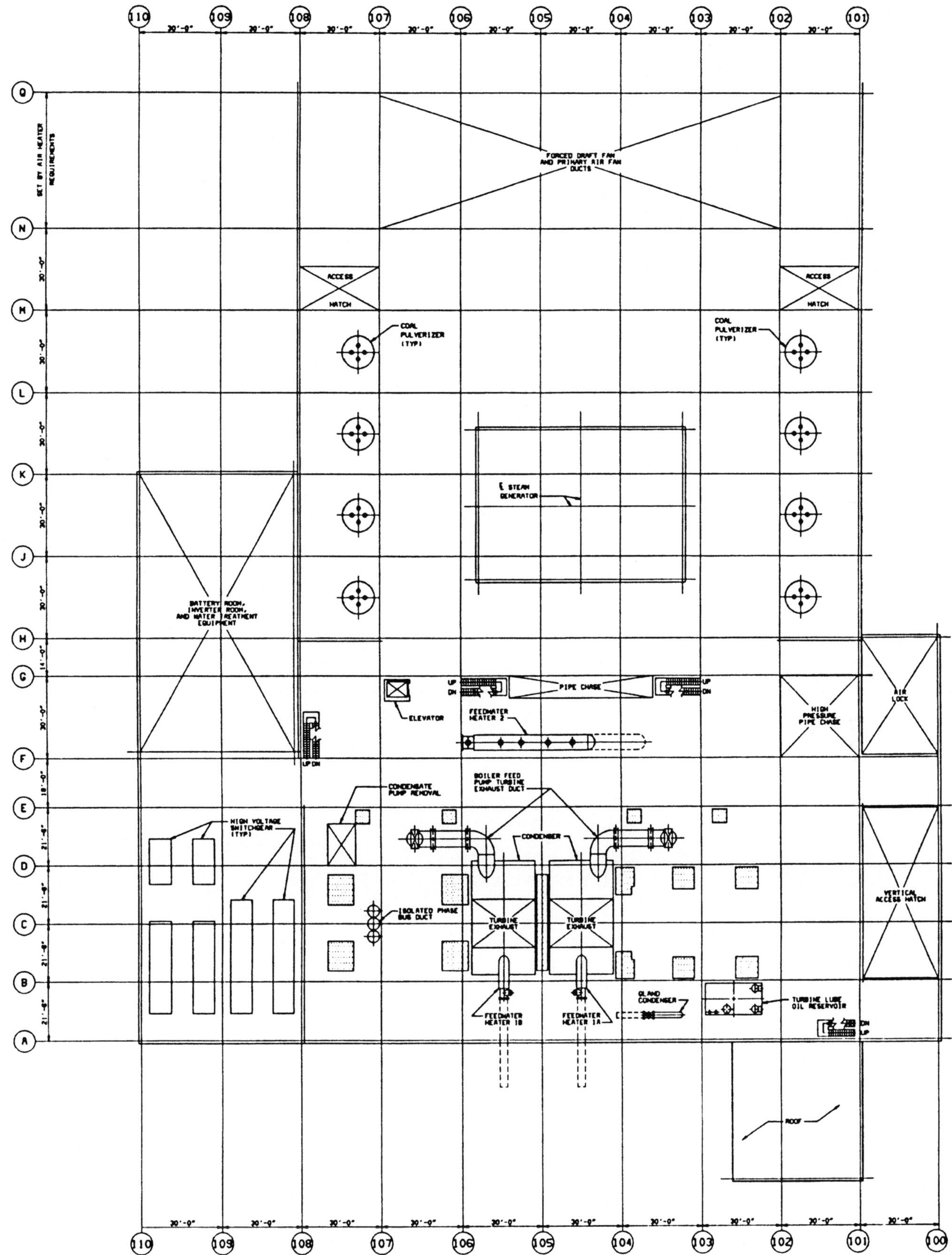

Fig. 19-6. Generation building arrangement: mezzanine floor elevation 121′-0″.

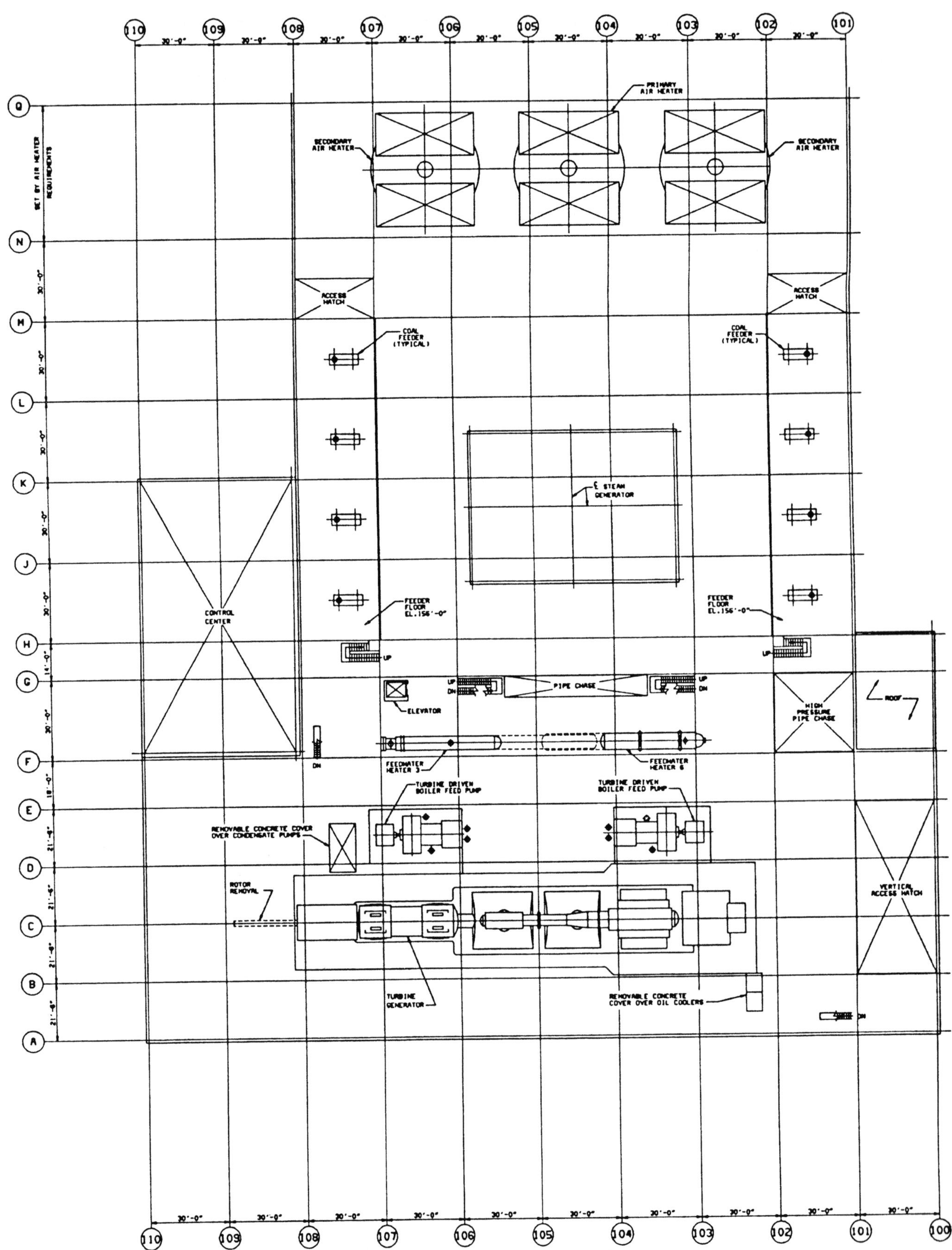

Fig. 19-7. Generation building arrangement: operating floor elevation 142'-0".

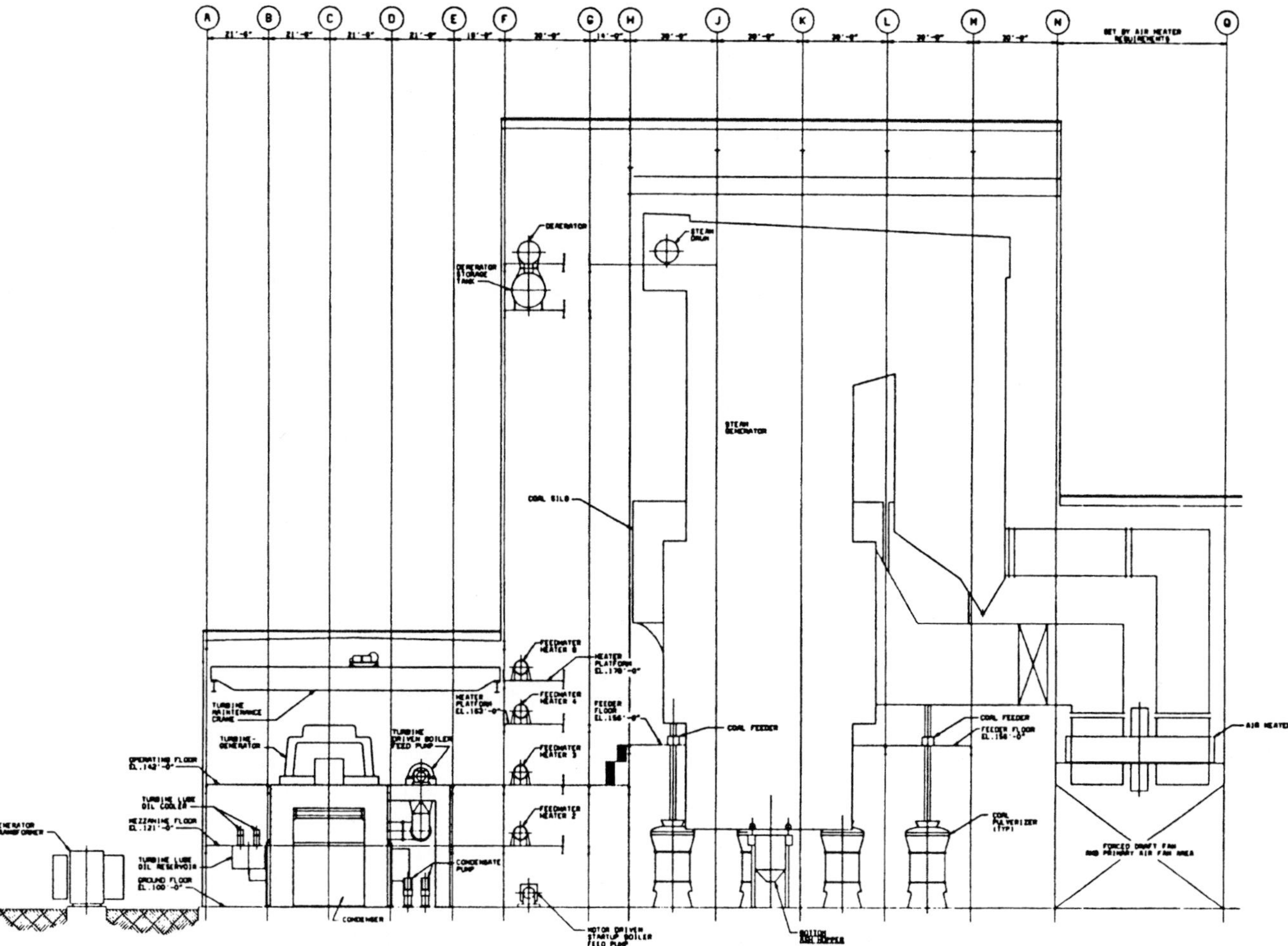

Fig. 19-8. Generation building arrangement: section elevation—east.

turbine and the steam generator. This is not economical because this steam piping is the most expensive in the plant. This location would also create obstructions for electrical tray and other piping which must be routed between the steam generator area and turbine area. A second choice is to position all the pulverizers on either the east side or the west side of the steam generator. This location is not preferred since it would require more coal feed piping because four of the pulverizers would be significantly farther from the burners. The optimum choice is to place four pulverizers along the east side of the steam generator and four along the west side of the steam generator. This arrangement offers the most economical and efficient routing of coal feed piping and meets the specified maintenance requirements.

Specifically, the pulverizers are located at the ground floor elevation and are spaced 30 ft apart. The coal feeders are approximately 56 ft above grade, directly above the pulverizers, to allow space for routing the coal pipes. The coal silos are directly above the feeders.

—*Primary Air Fans, Forced Draft Fans, and Air Heaters.* The 675-MW unit has two centrifugal primary air fans, two axial forced draft fans, a single primary air heater, and two secondary air heaters. These items of equipment are reviewed together because the air from the fans is discharged through ductwork to the air heaters. The air heaters require access for maintenance of motors, bearings, and lubricating oil systems. Vertical access is required adjacent to the air heaters for removal and replacement of cold-end baskets. Access and space are required around the fans for removal and replacement of rotors and drive motors. The air heaters typically are located and arranged by the steam generator manufacturer to be compatible with the steam generator design. With the orientation and expected design of the steam generator, the air heaters are located at the rear (north) of the furnace and elevated to minimize the interconnecting ductwork with the steam generator. The fans are large pieces of rotating equipment and, like the pulverizers, must be supported at grade elevation for vibration isolation. From an economical and efficiency perspective, it is important to place each fan as close to its associated air heater as possible to minimize the amount of interconnecting ductwork. On this basis and because of the noted maintenance requirements, the two primary air fans are located on the ground floor elevation directly below the primary air heater, and one forced draft fan is located on the ground floor elevation directly below each secondary air heater.

—*Bottom Ash Hopper and Ash Sluice Pumps.* The bottom ash hopper and the ash sluice pumps require access for maintenance and inspection. For proper operation, the bottom ash hopper must be directly below the steam generator furnace so that the bottom ash can flow by gravity from the furnace into the hopper. The ash sluice pumps supply water to sluice the bottom ash to waste storage and to sluice rejects from the pulverizers to waste storage. These pumps typically take suction by gravity from a storage tank, so it is important to locate them at grade elevation for proper operation. On the basis of these requirements, the bottom ash hopper must be directly below the furnace at the ground floor elevation. The steam generator elevation must be established to provide space for this hopper. Ideally, the ash sluice pumps should be near the perimeter of the building since they take suction from an outdoor storage tank. But in the steam generator area, it is more important that that area be used for locating the pulverizers and fans. Therefore, these pumps are located at the ground floor elevation adjacent to the ash hopper to limit the amount of piping from the pumps to the ash handling equipment.

—*Boiler Blowdown Tank.* The blowdown tank receives continuous blowdown from the steam drum and various other high-pressure and high-temperature blowdowns from the steam generator. The tank also receives drains by gravity from the steam generator when the steam generator is drained for maintenance. Access should be provided around the tank for inspection and maintenance. Vertical access must be provided above the tank because the tank must be vented to atmosphere for proper operation. The tank should be as near as possible to an outside wall so that the hot drains from the tank can be routed directly to an exterior drain manhole. The tank should be at grade elevation to allow for gravity drainage of the steam generator.

On the basis of these requirements, the boiler blowdown tank is located on the ground floor elevation on the east side of the steam generator and north of the pulverizers.

- *Steam Turbine–Generator Area*

—*Steam Turbine–Generator.* The steam turbine–generator is a tandem compound, four-flow machine. Because of the size and weight of its components, the turbine–generator must be accessible by an overhead crane for disassembly, maintenance, and reassembly. Sufficient clear space must be retained around the machine for laydown of the various components during disassembly. Space must be provided at the generator end of the machine for pulling the generator rotor. A properly sized vertical access way should be located adjacent to the machine for transferring components into and out of the building. The turbine–generator must be elevated so that the steam exhausts downward from the low-pressure turbine sections into the surface condensers. This elevation must be sufficient for the proper design and operation of the surface condensers. For proper support, the turbine–generator requires a reinforced concrete, pedestal-type foundation that is isolated from the rest of the building foundations and structure.

On the basis of these requirements, the turbine–generator is located at an elevation 42 ft above grade elevation. The longitudinal axis of the machine is oriented east–west, with the turbine to the east and the generator to the west. This orientation gives the most economical and efficient routing of the steam piping between the turbine and the steam generator.

—*Turbine–Generator Auxiliary Equipment.* Several items of equipment support operation of the turbine–generator and are furnished by the turbine–generator manufacturer. These items of equipment vary between manufacturers, but there are common items. Each manufacturer has specific requirements for the location of their equipment and these requirements must be adhered to for proper operation of the machine. Some of the common items of equipment, their maintenance requirements, and their locations according to typical manufacturer's requirements, are as follows:

(a) *Turbine Lube Oil Reservoir.* Access is required for inspection and maintenance of pumps and other equipment on top of the reservoir. Vertical access is required above the reservoir for removal of lube oil coolers. The top of the reservoir is at an elevation 21 ft above grade and adjacent to the south side of the turbine to allow gravity drainage of lube oil from the turbine–generator bearings to the reservoir.

(b) *Hydrogen Seal Oil Unit.* Access is required for inspection and maintenance of components on the unit. The unit is on the ground floor elevation and adjacent to the south side of the generator end of the machine.

(c) *Electro-Hydraulic Control Unit.* Access is required for inspection and maintenance of the components on the unit. Horizontal clear space is required for tube pull of heat exchangers located on the unit. The unit is on the ground floor elevation, on the north side of the turbine end of the machine.

(d) *Gland Condenser.* Access is required for inspection and maintenance. Horizontal clear space is required for tube pull from the gland condenser. The gland condenser is located at an elevation 21 ft above grade, adjacent to the south side of the turbine and west of the turbine lube oil reservoir. This location allows for gravity drainage of the turbine steam seal piping.

—*Surface Condenser.* The surface condenser consists of two condenser shells, one for each low-pressure section of the turbine–generator. The shells must be located directly below the low-pressure sections to accept the downward exhaust from these sections and to provide the most efficient operation of the unit. The condenser tubes are arranged perpendicular to the longitudinal axis of the turbine–generator. Horizontal clear space must be provided on the south side of the condenser for removal and reinstallation of tubes. The two hot wells are connected and provide condensate storage for suction to the condensate pumps.

On the basis of these requirements, the condensers are located on the ground floor elevation, between the turbine pedestal legs for the low-pressure sections.

—*Turbine-Driven Boiler Feed Pumps.* Two steam turbine driven, barrel-type boiler feed pumps are provided. Access is required for inspection and maintenance. Clear space is required for pump rotor removal and for disassembly of the steam turbine. For operability requirements, the turbines must be close to the surface condenser so that the turbine exhaust duct to the condenser is as short as possible.

On the basis of these requirements, the boiler feed pumps and drive turbines are located at an elevation 42 ft above grade, adjacent to the north side of the main turbine–generator. They are offset from the low-pressure turbines, with one located next to the generator and one located next to the high-pressure turbine. These locations offer several advantages. At this elevation, the turbine room crane can be used for maintenance, disassembly, and reassembly of the pumps and turbines. These locations allow for a direct and flexible routing of the turbine exhaust ducts from the bottoms of the turbines to the condensers, while avoiding the congestion with the low-pressure turbine extraction piping from the condenser necks. At this elevation and location, each pump will require a reinforced concrete pedestal type foundation.

—*Motor-Driven Startup Boiler Feed Pump*. This pump requires access for maintenance and clear space for pump rotor removal. This pump is designed only for initial plant startup and low-load operation. It takes suction from the deaerator storage tank. For proper operation, it should be located as low as possible to allow for the most net positive suction head (NPSH) available.

On the basis of these requirements, the pump is located on the ground floor elevation, north of the turbine end of the main turbine–generator.

—*Deaerator*. The deaerator consists of a deaerating section mounted on top of a storage tank section. Access is required around the deaerating section for removal and reinstallation of trays. Access is required around the storage tank for maintenance and inspection. The deaerator provides the required NPSH for the turbine driven boiler feed pumps and the startup boiler feed pump. The deaerator should be located as high as possible so that adequate NPSH is available at all times to ensure proper operation of the boiler feed pumps.

On the basis of these requirements, the deaerator is located at approximately the same elevation as the steam generator steam drum, on the south side of the steam generator. The actual elevation of the deaerator is dependent on the NPSH of the boiler feed pumps actually furnished.

—*Condensate Pumps*. Two vertical, multistage, centrifugal, can-type condensate pumps are provided. Access is required to the pumps for inspection and maintenance. Clear space is required above the pumps for removing the pumps from the cans. For proper operation, the pumps should be located as close to the condenser as possible. The pumps should be lower than the hot wells to allow for gravity flow of condensate from the hot wells to the pump cans and to allow for the most NPSH available.

On the basis of these requirements, the pumps are located at the ground floor elevation. The cans will be embedded in the foundation, below grade. The pumps are located to the west of the west boiler feed pump pedestal to allow for vertical pull of the pumps, clear of the pedestal, using the turbine room crane. The pump suction piping between the hot wells and the pumps is located below grade in a trench in the foundation.

—*High-Pressure and Low-Pressure Feedwater Heaters*. The feedwater heaters are horizontal, U-tube, closed type heaters. There are four low-pressure heaters and three high-pressure heaters. Access is required around the heaters for inspection and maintenance. Clear space is required at each heater for pulling the heater shell. In addition, an important consideration in locating the heaters is the economical routing of all the piping connected to and interconnecting the heaters.

Each heater has extraction steam piping from the turbine; condensate or boiler feed water piping; and, if applicable, heater drains piping from the next higher pressure heater connected to it.

On the basis of these requirements and considerations, the heaters are located as follows. The lowest pressure heater is actually supplied as two, half-capacity heaters with each one located in the condenser neck directly below a low-pressure turbine section. These heaters require the largest diameter extraction steam piping. Locating them in the condenser neck limits the length of this piping and avoids the use of floor space for the heaters.

These heaters are designed for tube pull, instead of shell pull, to the south of the condenser. The remaining heaters are located between the turbine–generator and the steam generator, below the deaerator. The low-pressure heaters get extraction steam from the low-pressure turbines and are located at the generator end.

The high-pressure heaters get extraction steam from the high-pressure and intermediate-pressure turbines and are located at the turbine end. This minimizes the extraction piping from the turbine to the heaters. In addition, the heaters are located with the succeeding extraction stage heaters placed at elevations below the preceding stage heaters. This means the heaters are stacked vertically, with a lower pressure heater located below the next higher pressure heater. This arrangement facilitates the heater drain piping layout and minimizes the required length of the larger diameter extraction steam piping associated with each succeeding heater by locating heaters with larger piping at a lower level. Also, this arrangement facilitates the layout of the condensate and boiler feed water piping since the condensate piping is routed from the condensate pumps through the low-pressure heaters to the deaerator, and the boiler feedwater piping is routed from the boiler feed pumps through the high-pressure heaters to the steam generator. On the basis of these requirements, the second low-pressure heater is located 21 ft above grade elevation with shell pull to the east.

The third low-pressure heater and the first high-pressure heater are located 42 ft above grade, with the low-pressure heater shell pull to the east and the high-pressure heater shell pull to the west. The fourth low-pressure heater and the second high-pressure heater are located approximately 63 ft above grade with shell pulls like the preceding heaters. This location typically is adjusted to be consistent with the elevation of the access platform for the first burner level on the steam generator. The third high-pressure heater is located approximately 78 ft above grade with shell pull to the west. Again, this location typically is adjusted to be consistent with the access to the second burner level on the steam generator.

—*Transformers*. The generator transformer, main auxiliary transformer, and reserve auxiliary transformers require access for inspection and maintenance. The transformers must be located outdoors for safety and to meet National Fire Protection Association (NFPA) code requirements. The generator transformer should be located as close to the main generator as possible to minimize the length of the isolated phase bus duct from the generator terminals to the transformer. The main and reserve auxiliary transformers should be located as close as possible to the generator transformer to minimize the length of bus duct interconnecting these transformers.

On the basis of these requirements, these transformers are located at grade elevation, south of the generator and outside the building. The main and reserve auxiliary transformers are located to the west of the generator transformer to avoid interference with the condenser tube pull space and to facilitate routing of bus duct to the high-voltage switchgear. The turbine area exterior wall adjacent to the transformers and the fire walls between the transformers are designed to meet NFPA code requirements.

—*High-Voltage Switchgear.* Access is required around the switchgear for breaker removal and maintenance. The switchgear should be located as close as possible to the main and reserve auxiliary transformers to minimize the length of bus duct interconnecting them.

On the basis of these requirements, the switchgear is located directly north of the transformers, inside the building, at an elevation 21 ft above grade. This elevation accommodates the location of the bus duct connections on the transformers and allows for connecting the bus duct to the bottom of the switchgear. A fire wall is required to isolate the switchgear from adjacent equipment to meet NFPA requirements.

—*Secondary Unit Substations (SUS) and Motor Control Centers (MCC).* The SUS require access for breaker removal and maintenance. The MCC require access for removal of starters and maintenance. The SUS are fed from the high-voltage switchgear and should be located as close to the switchgear as possible to minimize the interconnecting bus duct and cable. The MCC are fed from the SUS and should be as close to the SUS as possible to minimize the interconnecting cable.

On the basis of these requirements, the SUS are located on the ground floor elevation directly below the high-voltage switchgear. The MCC are located on the ground floor elevation directly north of the SUS.

—*Closed Cooling Water Heat Exchangers.* Two horizontal, shell and tube type heat exchangers are provided. Clear space is required for tube removal and replacement. The heat exchangers receive cooling water from the circulating water system and should be located as low as possible to minimize the amount of interconnecting piping with that system.

On the basis of these requirements; the heat exchangers are located on the ground floor elevation, west of the turbine foundation.

—*Turbine Lube Oil Storage and Purification Equipment.* This equipment is required for draining, storing, and purifying the main turbine lubricating oil. Access is required for inspection, maintenance, and delivery and removal of lube oil in barrels or tank truck. For oil spill containment, the equipment should be located as low as possible and should be enclosed by a curb or other means. The equipment should be located as close as possible to the turbine lube oil reservoir to minimize the length of interconnecting piping between the reservoir and the equipment.

On the basis of these requirements, the equipment is located in a separate room directly to the south of the reservoir. The floor of this room should be approximately 4 in. lower than the ground floor elevation for oil spill containment.

- *Control Center Area.* The arrangement of the facilities and equipment in the Control Center Area is typically very specific to the needs of the particular plant and client. Because of this, only general location requirements are discussed for the equipment to be located in this area. The Control Center Area is located adjacent to the generator end of the main turbine–generator and north of the high-voltage switchgear and SUS. This location provides the best access for control cable from the steam generator and turbine–generator areas and for power cable from the turbine–generator area.

Water treatment equipment, typically consisting of condensate polisher equipment, is located on the ground floor to minimize the length of interconnecting piping with the condensate pump discharge piping. The water quality laboratory, including water sampling and analysis and fuel analysis equipment, is located on the second floor. The battery and inverter rooms are also located on the second floor. The main control room and associated control equipment, and operating personnel facilities such as offices, toilets, kitchen, lunch room, and meeting rooms are on the third floor of the area.

The final step in the plant arrangement development process can be accomplished after the above listed equipment is located. This step is to determine the location and spacing of the building support column rows; the location and heights of floors, platforms, and roofs; the location of exterior and interior walls; and the location of vertical and horizontal access ways including stairs, elevators, access aisles, and doors. The plant arrangement drawings shown in Figs. 19-5 through 19-8 are the result of this development process and should be referred to during the following discussion.

—*Building Support Column Rows.* The east–west column row spacing for the Generation Building is based on the support requirements for the steam turbine–generator. Referring to Fig. 19-5, six column rows (102 through 108) spaced 30 ft apart most effectively accommodates the turbine–generator. A 30 ft wide bay to the east of the turbine end is provided for routing of the main steam, hot reheat, and cold reheat piping and for the closed cooling water heat exchangers. A second 30 ft wide bay is provided to the east for the vertical access opening required for transferring turbine–generator components from the turbine floor to the ground floor. An air lock is located directly north of the vertical access bay for staging equipment and components into and out of the building, and to isolate the rest of the building from this area during the transfer of components. Two 30-ft wide bays are provided to the west of the generator end to accommodate the SUS and switchgear and to accommodate generator rotor removal. Overall, the turbine–generator area is 300 ft long from column row 100 to column row 110.

This column row spacing accommodates the steam generator area and the control center area. With its north–south axis centered between rows 104 and 105, the steam generator can be located between the three 30-ft bays between rows 103 and 106. The 30-ft bays to the east and west of the steam generator (between rows 102 and 103 and rows 106 and 107) can effectively accommodate the routing of hot air ducts to the pulverizers and burners and the routing of coal piping from the pulverizers to the burners. The 30-ft bays between rows 101 and 102 and between 107 and 108 provide the location for the pulverizers and adequate space for maintenance access to the pulverizers. The space above the pulverizers is effectively used for locating the coal feeders and coal silos. Overall, the steam generator area is 210 ft long from column row 101 to column row 108. The control center area is located between column rows 108 and 110, which provides sufficient space for the equipment and facilities in this area and aligns the west column line of this area with turbine generator area.

The north–south column row spacing is keyed off the turbine–generator arrangement in the turbine area. From the longitudinal centerline of the turbine–generator, two 21-ft, 6-in. bays are located to the south to provide the necessary maintenance space on the turbine floor, and the necessary space on the lower levels for equipment, piping, and other facilities that must be located near the turbine generator. The turbine generator area south column line is designated column row A. The turbine lube oil storage and purification equipment room is located to the south of column row A between column rows 101 and 103. Two 21-ft, 6-in. bays are located to the north of the turbine generator centerline to provide space for the turbine driven boiler feed pumps on the turbine floor and space on the lower floors for equipment, piping, and other facilities. An 18 ft wide bay is provided between column rows E and F that runs the length of the turbine generator area. This bay is provided as a maintenance and access aisle to provide convenient access to all parts of the area. No equipment is located in this bay. The turbine generator area from column row A to column row F is 104 ft wide and accommodates a 100 ft wide turbine room crane located above the turbine–generator.

A 30-ft bay is provided between column rows F and G, and a 14-ft bay is provided between column rows G and H. These bays provide a transition area between the turbine–generator area and the steam generator area that is commonly referred to as the auxiliary bay area. The bay between rows F and G provides the location for the feedwater heaters and deaerator and the location for the boiler to turbine piping transition. The bay between rows G and H contains the steam generator south wall support wind bracing and provides a transition area for the unit to unit crossover of auxiliary services associated with the addition of future units. The auxiliary bay extends the full height of the steam generator.

As stated previously, a primary factor in determining the location of the steam generator relative to the turbine generator is the need for flexibility in the high-pressure/high-temperature piping that interconnects these components. This piping will be routed from the top, front south of the steam generator to the front east centerline of the turbine. The main steam and hot reheat steam piping will drop vertically from the top of the unit, along the front centerline, to a midpoint elevation crossover. This crossover from the steam generator centerline to the east end of the auxiliary bay will be followed by a second vertical drop to the ground floor level. At the lower level, the piping will run from the east end of the auxiliary bay to the front of the turbine. This down-crossover–down-across arrangement will effectively provide the required piping flexibility (while minimizing piping lengths) to accommodate pipe expansion. The auxiliary bay includes a vertical pipe chase dedicated to the accommodation of vertical piping runs including extraction steam, condensate, boiler feedwater, and other auxiliary piping systems that require significant elevation transitions. An elevator is also provided in the auxiliary bay.

Continuing on to the north, five 30 ft wide bays are provided to accommodate the steam generator and coal feed equipment. Two vertical access ways are provided between column rows 101 and 102, and between 107 and 108 at column row M. These access ways extend the full height of the steam generator and provide the location for transferring equipment and components up and down the building. The area to the north of column row N is not dimensioned, as this will depend on the actual arrangement of the air heaters. The ground floor level in this area will accommodate the primary air and forced draft fans and ductwork.

The control center area is located between column rows F and K to provide sufficient space for the equipment and facilities located in this area.

—*Building Floor Heights.* Figure 19-8 is the Plant Arrangement Sectional Elevation drawing for the Generation Building and identifies the major floor heights. The ground floor elevation is set at 100 ft for reference purposes. Based on the equipment locations and to allow sufficient spacing for structural steel, multilevel electrical raceway, and multilevel piping to be installed and maintained above headroom, the floor-to-floor spacing in the turbine generator area is 21 ft. The mezzanine floor elevation is 121 ft, and the operating floor elevation is 142 ft. These floor elevations are consistent throughout the turbine generator area and extend from column row A to column row G. The turbine room crane elevation is determined by the hook height required to allow removal of the largest turbine generator component (other than the generator stator). Typically, this height is set by the height required to lift a low-pressure turbine hood over the highest component on the turbine generator and is determined by the turbine generator actually furnished. The turbine generator area roof elevation is determined by the crane elevation and crane clearance requirements.

In the auxiliary bay area, a feedwater heater support floor is located at an elevation of approximately 163 ft. This elevation should be adjusted to be consistent with the access floor for the first burner elevation on the steam generator for efficient movement through the building. A second feedwater heater support floor is located at an elevation of approximately 178 ft. This elevation is typically adjusted to match the access floor for the second burner elevation on the steam generator. The deaerator elevation will be based on the actual NPSH requirements for the boiler feed pumps, and the deaerator should be no higher than necessary. For this example, the deaerator is located such that the deaerating heater section is at the same elevation as the steam drum, with the storage tank directly below. Access floors are provided for each. Additional floors will be provided in the auxiliary bay as necessary for access to other equipment.

The major floor in the steam generator area is the coal feeder floor located at approximately elevation 156 ft. The actual elevation of this floor is dependent on the size and arrangement of the pulverizers actually furnished. All remaining floors and platforms in this area will be located based on the access requirements for the steam generator and auxiliary equipment actually furnished. Access platforms or floors will be required around the ash hopper, around and above the pulverizers, around the air heaters, at all burner elevations, at all sootblower elevations, at all access doors on the steam generator and ductwork, at the coal unloading equipment elevation above the coal silos, at the steam drum, and at other locations as required for inspections and maintenance.

The floors in the control center area are set to match the mezzanine floor and operating floor elevations to accommodate the routing of control and power cable and piping interconnecting this area with the rest of the building, and to allow efficient movement through the building.

—*Walls and Access Ways*. Exterior walls in the turbine generator area are located along column rows A, 100, and 110. Interior walls are avoided as much as possible in this area because they hinder maintenance and access and they create dead spaces that affect cleanliness and ventilation. Interior walls are required for the air lock along column rows 101 and E from the ground floor up to the operating floor. A wall is also required along column row 108 at the mezzanine floor to isolate the high-voltage switchgear to meet NFPA code requirements.

Exterior walls for the control center area are located along column rows 110 and K. Interior walls are located along column rows 108 and F from the mezzanine floor up to provide a clean environment for the control room and other facilities located in this area. Additional interior walls will be provided as required to accommodate the facilities in the area.

The wall along row 110 is designed for expansion to a future unit. This design would allow the turbine generator areas and the control center areas of the two units to be connected. This would also allow the turbine room crane to be used for the second unit.

Exterior walls for the steam generator area are located along column rows 101 and 108. Interior walls are avoided in this area as much as possible unless required for noise or dust control or to meet code requirements. Partial interior walls are located along column row H between 101 and 102, and between 107 and 108 to isolate the pulverizers from the turbine–generator area for noise and dust control. These walls extend from the ground floor up to the feeder floor.

Access hatches are provided in the operating floor above the condensate pumps and above the turbine lube oil coolers to allow the turbine crane to be used for removal of the pumps and coolers.

Access stairs are located in the turbine generator area and auxiliary bay area as noted on the drawings. Access stairs will be provided in the steam generator area as required by the specific design of the steam generator, but typically would be located along column row M, adjacent to the two vertical access ways.

Large doors are provided at grade level at each end of the maintenance aisle between column rows E and F to facilitate moving components and equipment into and out of the building. Large doors are provided at grade elevation on the north and south sides of the air lock in the turbine area.

Large doors will be provided at grade elevation in the steam generator area near the pulverizers to accommodate pulverizer maintenance and journal removal.

19.3 SUMMARY

The procedures described in this chapter illustrate that the development of the site arrangements and plant arrangements for a power plant project is a complicated and difficult task involving significant time and the resources of a wide variety of technical disciplines. The procedures presented are based on significant experience with a broad range of power plant sites and technologies.

20

Gas Turbines

Jeffrey M. Smith

20.1 INTRODUCTION

When the gas turbine generator was introduced to the power generation industry in the late 1940s, it was a revolutionary self-contained fossil-fueled power plant. Twenty years later, gas turbines were established as an important means of meeting fast growing peak loads on utility systems. By the early 1990s, gas turbines in various application cycles had become a significant portion of the new power generation additions worldwide. In fewer than 50 years, what was originally a jet engine technology has been transformed into an essential high-technology solution to many power generation needs.

In a gas turbine, the working medium for transforming thermal energy into rotating mechanical energy is the hot combustion gas, hence the term "gas turbine." Gas turbines are also referred to as combustion turbines, or combustion gas turbines. Industry groups, users, and manufacturers use the terms interchangeably.

The gas turbine technology has many applications. The original jet engine technology was first made into a heavy-duty application for mechanical drive purposes. Pipeline pumping stations, gas compressor plants, and various modes of transportation have successfully used gas turbines. While the mechanical drive applications continue to have widespread use, the technology has advanced into larger gas turbine designs that are coupled to electric generators for power generation applications.

Gas turbine generators are self-contained packaged power plants. Air compression, fuel delivery, combustion, expansion of combustion gas through a turbine, and electricity generation are all accomplished in a compact combination of equipment, usually provided by a single supplier under a single contract.

In contrast to steam turbine-generators, which are designed for a particular application, the manufacturers of gas turbines have a defined product line, allowing for substantial standardization and assembly line manufacturing. The modular concept of the package power plants made gas turbines relatively quick and easy to install. Standardization and modularization combine to provide the product benefits of relatively low capital cost and fast installation of power generation capacity. The benefits of low capital cost and fast installation were initially offset by higher operating costs when compared to other installed capacity. Therefore, early utility applications of gas turbine–generators were strictly for peak load operation for a few hundred hours per year. Improvements in efficiency and reliability and the application of combined cycles have added to the economic benefits of the technology and now give gas turbine-based power plants a wider range of application on electric systems. An economic screening of power plants is shown in Fig. 20-1. Project specific comparisons will depend on fuel costs, capital costs, and maintenance requirements.

Gas turbine technology can be used in a variety of configurations for electric power generation. Conventional applications are simple cycle, combined cycle, or cogeneration. Of the conventional applications, simple cycle operation is generally the least efficient and is used primarily for peaking power generation. Combined cycles combine the gas turbine and steam turbine cycles into more efficient power plants by utilizing the gas turbine exhaust gas heat. Cogeneration cycles use the steam generated from the exhaust gas for process or heating requirements. Electric utility companies use gas turbines predominantly in simple cycle and combined cycle applications. Nonutility generator companies predominantly use combined cycles, either strictly for power generation or in conjunction with an industrial company as cogeneration power plants.

Most gas turbine applications rely on natural gas or No. 2 fuel oil for fuel. During the late 1980s and early 1990s, the availability and economics of natural gas made gas turbine-based power plants the preferred choice for the majority of new power generation additions in the United States. Fuel flexibility and efficient thermodynamic cycles have become important characteristics for gas turbine applications. Further improvements in the areas of fuel flexibility, cycle efficiency, and reduced exhaust gas emissions are the primary technical challenges to advancing gas turbine technology.

Gas turbine products and applications are changing rapidly. The technology continues to evolve through innovations in component cooling techniques, metallurgy improvements, and cycle innovations. Future enhancements in gas turbine technology can be expected from higher firing temperatures, increased pressure ratios, improved materials, multistaged combustion techniques, and emission control capability. The gas turbine–generator will continue to play an important role in meeting power generation requirements

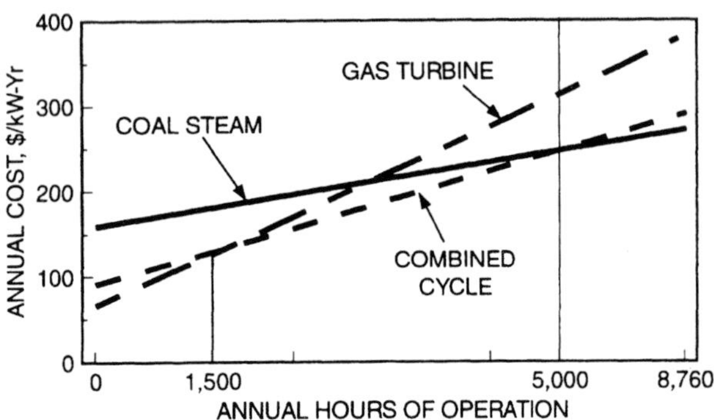

Fig. 20-1. Economic screening curve. (From Morris, et al. 1994.)

as technology advances and as the product and cycle designs respond to changes in fuel economics and allowable plant emissions.

This chapter describes the gas turbine applications for power generation. The basic gas turbine cycle is discussed, as are descriptions of more complex cycles. Components and auxiliary systems normally supporting the power plant packages are also discussed.

20.2 THERMODYNAMIC FUNDAMENTALS

A typical, simple or open cycle gas turbine system is shown schematically in Fig. 20-2. The gas turbine consists of a compressor section, combustor, and turbine section. As indicated in Fig. 20-2, ambient air is first compressed from state point 1 to state point 2 in the compressor section. Fuel is added and combusted in the combustor. The combustion products exit the combustor at state 3 and expand from state point 3 to state point 4 (atmosphere) in the turbine section. The expansion process produces useful shaft work. Typically, over 50% of the turbine shaft work produced is required to drive the compressor. The remaining work (W_{net}) is available to drive an electric generator or for mechanical drive applications.

Operation of an actual gas turbine approaches that of an idealized thermodynamic cycle called the air-standard Brayton cycle. Consequently, reviewing the characteristics of the Brayton cycle provides useful insights into the operating and performance characteristics of an actual gas turbine system.

Air, the working fluid of the Brayton cycle, is assumed to be an ideal gas. It is further assumed that the mass flow rate through the cycle is constant, the air is of fixed composition, and the specific heat of the air is constant.

Four thermodynamic processes make up the Brayton cycle: compression, heat addition at constant pressure, expansion, and heat rejection at constant pressure.

The compression and expansion portions of the cycle are internally reversible and adiabatic (isentropic) processes. Because the mass flow rate is assumed to be constant throughout the process, the direct combustion process of an actual gas turbine is conceptualized by a heat exchanger that transfers heat to the working fluid from an external source. The actual inlet and exhaust processes of the gas turbine are conceptualized by a heat exchanger that rejects heat at constant pressure to the surroundings to complete the cycle.

Figure 20-3 depicts the air-standard Brayton cycle. A comparison of Figs. 20-2 and 20-3 highlights the major differences between an open cycle gas turbine and the air-standard Brayton cycle.

Figure 20-4 shows the generalized pressure–volume (P–v) and temperature–entropy (T–s) diagrams for the air-standard Brayton cycle. The state point numbers in Fig. 20-4 correspond to the numbers shown in Fig. 20-3.

Actual gas turbine performance would approach that of the Brayton cycle if not for the following:

- Irreversibilities that occur in the compressor and turbine;
- Pressure losses that occur in the flow passages and the combustor;
- Heating of the compressed air by direct combustion rather than an indirect heat exchanger, thereby increasing the mass flow rate throughout the system downstream of the combustor;
- Passing of combustion products to atmosphere without completing the cycle; and
- Thermal, mechanical, and electrical losses.

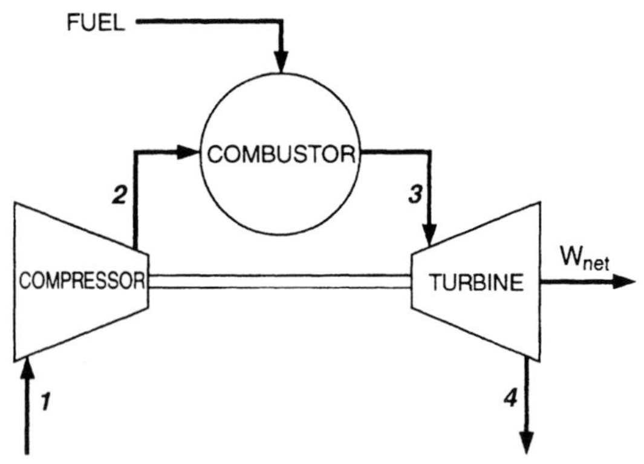

Fig. 20-2. Typical open-cycle gas turbine system.

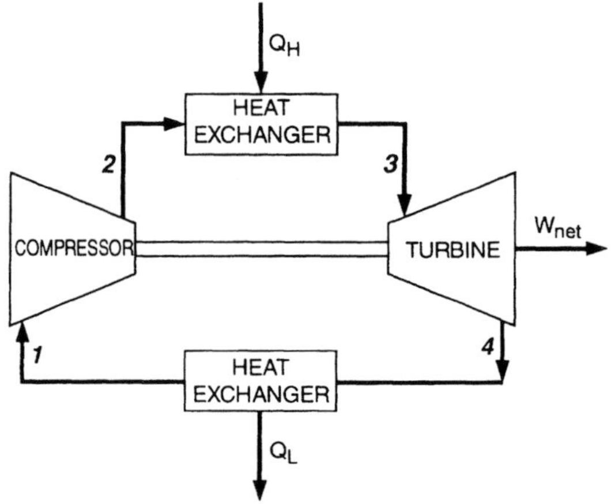

Fig. 20-3. Air-standard Brayton cycle.

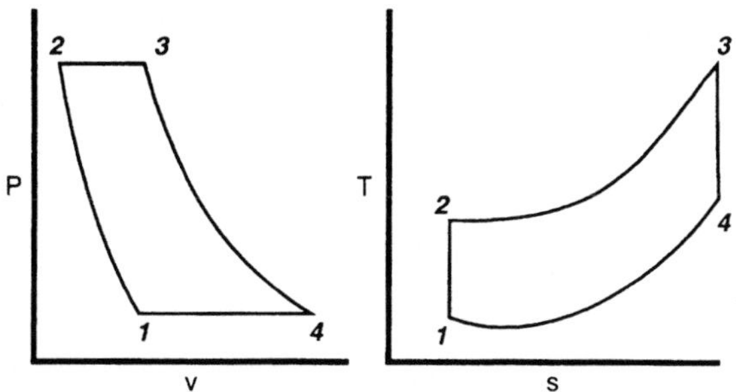

Fig. 20-4. *P–v* and *T–s* diagrams for the Brayton cycle.

Referring to Fig. 20-3, the thermal efficiency of the Brayton cycle is defined as follows:

$$\eta_{th} = \frac{W_{net}}{Q_H} = \frac{Q_H - Q_L}{Q_H} = 1 - \frac{Q_L}{Q_H} = 1 - \frac{C_p(T_4 - T_1)}{C_p(T_3 - T_2)} \quad (20\text{-}1)$$

where

W_{net} = useful work available,

Q_H = heat added to the system, and

Q_L = heat rejected from the system.

Since the specific heat is assumed constant, the expression may be simplified such that thermal efficiency is solely a function of cycle temperatures,

$$\eta_{th} = 1 - \frac{T_1\left(\frac{T_4}{T_1} = 1\right)}{T_2\left(\frac{T_3}{T_2} - 1\right)} \quad (20\text{-}2)$$

Referring to Fig. 20-4, and noting that,

$$\frac{P_3}{P_4} = \frac{P_2}{P_1} \quad (20\text{-}3)$$

and also noting that,

$$\frac{T_3}{T_4} = \frac{T_2}{T_1} \quad \text{or} \quad \frac{T_3}{T_2} = \frac{T_4}{T_1} \quad (20\text{-}4)$$

Consequently, Eq. (20-2) can be simplified and thermal efficiency can be written as follows:

$$\eta_{th} = 1 - \frac{T_1}{T_2} \quad \text{or} \quad \eta_{th} = 1 - \frac{T_4}{T_3} \quad (20\text{-}5)$$

For an ideal gas undergoing isentropic compression, the compressor pressure ratio is related to the temperature ratio by,

$$\frac{P_2}{P_1} = \left(\frac{T_2}{T_1}\right)^{k/(k-1)} \quad (20\text{-}6)$$

where

k = ratio of specific heats for air.

If Eq. (20-6) is solved for T_1/T_2 and the results submitted into Eq. (20-5), then the resulting equation defines thermal efficiency as a function of compressor pressure ratio.

$$\eta_{th} = 1 - \frac{1}{(P_2/P_1)^{(k-1)/k}} \quad (20\text{-}7)$$

If Eq. (20-7) is evaluated for various compressor pressure ratios, it is clear that cycle efficiency increases as the pressure ratio increases. Figure 20-5 shows the relationship of pressure ratio to cycle thermal efficiency of the Brayton cycle assuming a ratio of specific heats for air of 1.4.

By inspecting the generalized *T–s* diagram shown in Fig. 20-6 and referring to Eqs. (20-5) and (20-7) it can be seen that if turbine exhaust temperature, T_4, is constant, increasing the combustion temperature and the resultant turbine inlet temperature from T_3 to T_{3A} will increase the cycle's thermal efficiency. However, in actual gas turbines, these temperatures cannot be increased indefinitely because of temperature limitations of the materials. From this, it is quite understandable why many research and development efforts by gas turbine manufacturers continue to focus on improving hot gas path component metallurgy and cooling methods to increase the allowable turbine inlet temperatures and thereby increase cycle efficiency.

As previously mentioned, increasing the pressure ratio (P_2/P_1) increases the cycle efficiency. However, as pressure ratio changes, the net specific work per unit mass of working fluid also changes.

Both thermal efficiency and specific work are strongly influenced by pressure ratio and turbine inlet temperature. However, the maximum net specific work and the maximum thermal efficiency do not occur at the same pressure ratio. Therefore, in designing gas turbines, the design pressure ratio must be a compromise between the maximum thermal efficiency and the maximum specific work.

A typical performance curve for a real (non-ideal) turbine is shown in Fig. 20-7. As the compressor pressure ratio

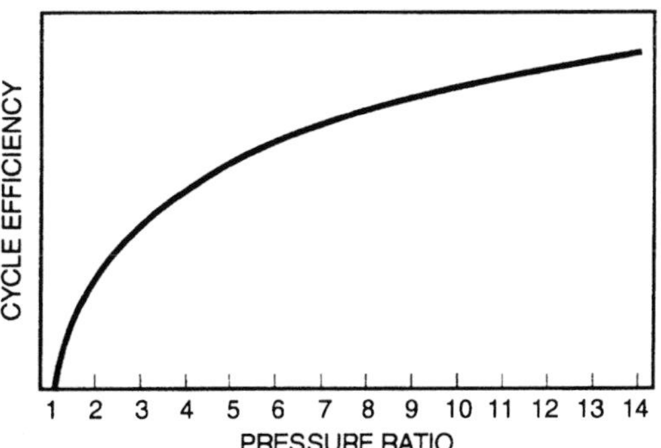

Fig. 20-5. Brayton cycle thermal efficiency versus pressure ratio at constant temperature.

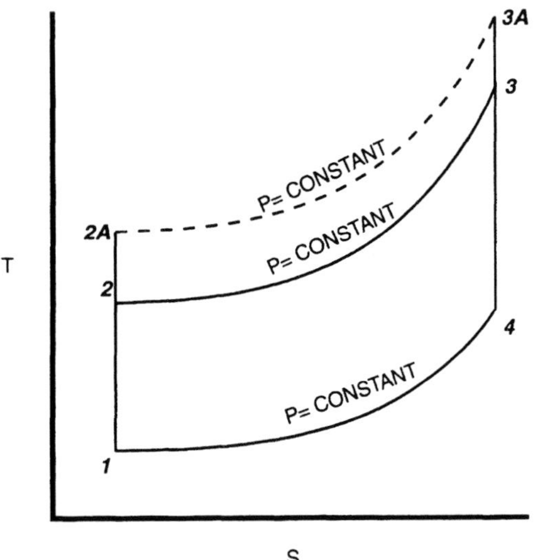

Fig. 20-6. Generalized *T–s* diagram.

(P_2/P_1) increases from 4 to 8, the specific work produced by the turbine section increases more rapidly than the specific work that the compressor consumes. The result is an increase in shaft specific work output and efficiency as pressure ratio increases.

As the compressor pressure ratio increases, the temperature rise across the compressor also increases. This in turn reduces the combustor temperature rise necessary to achieve a given turbine inlet temperature. As a result, the combustor heat input decreases and cycle efficiency increases. At the point corresponding to the maximum shaft specific work output, the turbine specific work produced and the compressor specific work consumed increase at the same rate. As pressure ratio increases further (beyond the point where the shaft specific work output is a maximum), the compressor

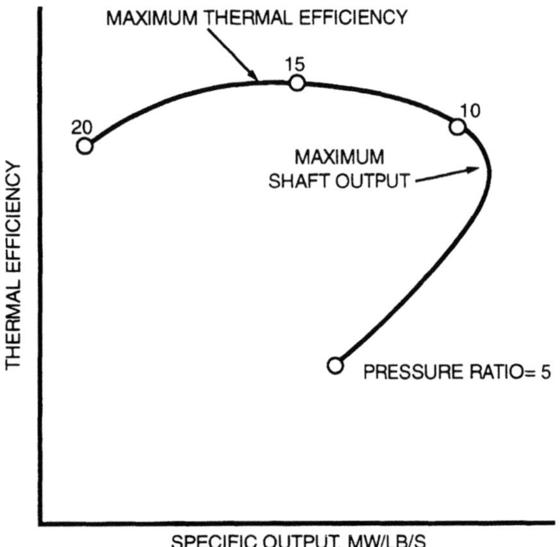

Fig. 20-7. Typical (non-ideal) real gas turbine simple cycle performance curve—efficiency vs. specific work at various pressure ratios at constant temperature.

specific work consumed increases at a greater rate than the turbine specific work produced and, as a result, the shaft specific work output actually decreases. However, because the combustor heat input continues to decrease, the efficiency continues to increase as pressure ratio increases. However, with further increases in pressure ratio, the thermal efficiency of real gas turbines will begin to decrease. The maximum shaft specific work for the example of Fig. 20-7 occurs at a pressure ratio of 8. As the pressure ratio is increased from 8 to 16, the shaft specific work output decreases because compressor specific work consumed is increasing faster than turbine specific work output. Therefore, shaft specific work output (turbine specific work produced minus compressor specific work consumed) decreases. However, because higher compressor outlet temperatures occur at the higher pressure ratios, fuel consumption is reduced and overall thermal efficiency increases. After a point, the increase in compressor specific work consumed more than offsets the advantage of higher compressor outlet temperature and overall efficiency begins to decrease. In Fig. 20-7, this occurs at pressure ratios greater than 16. The actual pressure ratio at which this occurs depends on the specific gas turbine considered.

The efficiencies of the compressor and turbine for an actual cycle are typically defined in relation to isentropic processes. Considering the state points for an isentropic and a nonisentropic cycle (Fig. 20-8), the compressor and turbine efficiencies are as follows:

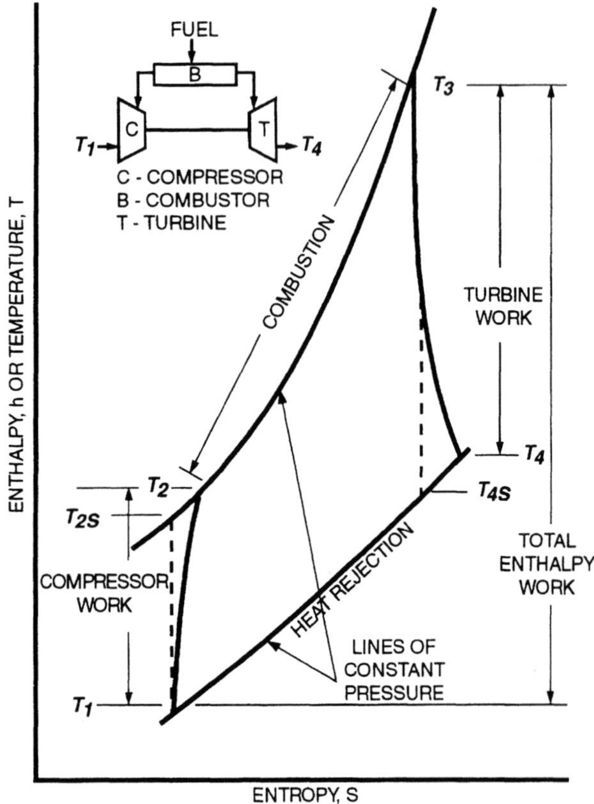

Fig. 20-8. *T–s* diagram for an isentropic and nonisentropic cycle.

$$\eta_{comp} = \frac{h_{2S} - h_1}{h_2 - h_1} \quad (20\text{-}8)$$

where

h_1 = enthalpy at compressor inlet,

h_{2S} = enthalpy at constant entropy and compressor discharge pressure, and

h_2 = actual enthalpy at compressor discharge pressure.

and

$$\eta_{turb} = \frac{h_3 - h_4}{h_3 - h_{4S}} \quad (20\text{-}9)$$

where

h_3 = enthalpy at turbine inlet,

h_{4S} = enthalpy at constant entropy and turbine exit pressure, and

h_4 = actual enthalpy at turbine exit pressure.

It is these process inefficiencies which cause real devices to depart from the ideal behavior described by Eq. (20-7).

20.3 GAS TURBINE APPLICATIONS

Gas turbines are frequently used in both simple cycle and combined cycle configurations. In simple cycle mode, the gas turbine is operated alone, without the benefit of recovering any of the energy in the hot exhaust gases. The exhaust gases are sent directly to the atmosphere. Figure 20-9 schematically represents a simple cycle configuration.

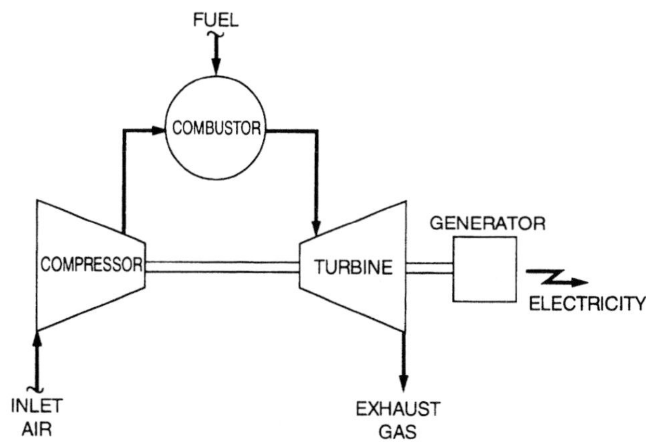

Fig. 20-9. Gas turbine generator cycles: simple cycle.

Combined cycle configurations vary, but generally consist of one or more gas turbines that exhaust into one or more heat recovery steam generators (HRSGs). The HRSG(s) generates steam that is normally used to power a steam turbine. Figure 20-10 is a schematic representation of one combined cycle arrangement for power generation.

20.3.1 Simple Cycle Configurations

Simple cycle gas turbines in electric utility systems are typically used for reserve or peaking capacity and are generally operated for a limited number of hours per year. Peaking operation is often defined as fewer than 2,000 hours of operation per year. In mechanical drive applications, and for some industrial power generation, simple cycle gas turbines are baseloaded and operate more than 5,000 hours per year.

Some plants are initially installed as simple cycle plants with provisions for future conversion to combined cycle. If

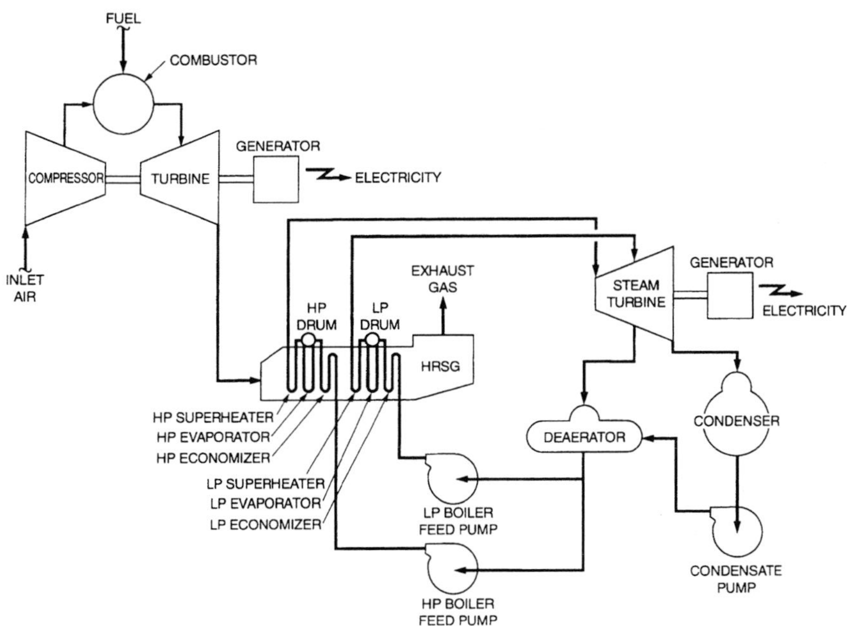

Fig. 20-10. Gas turbine generator cycles: combined cycle.

this conversion is anticipated, several provisions should be made in the design of the simple cycle plant. These provisions include sufficient space for the addition of the HRSG, steam turbine, condenser, and the remainder of the auxiliary systems. Another important consideration is the source of cooling water for the steam condenser, whether it be a river, a cooling tower, or other means.

Gas turbines typically have their own dedicated cooling, lubricating, and other service systems needed for simple cycle operation. This can eliminate the need to tie service systems into the combined cycle addition and will allow continued operation of the gas turbine during the conversion process and, with proper provisions, during periods when the combined cycle equipment is out of service.

If future simple cycle operation is desired, a bypass stack may be included with the connection of the HRSG. A typical method for providing this connection is to procure a diverter damper box at the outlet of the gas turbine. A bypass stack connected to the damper box is used in simple cycle operation. A separate connection on the damper box is used as the interface point for the ductwork to the HRSG. A blanking plate placed in this opening allows the erection of the future ductwork without disabling the gas turbine. If the damper is not purchased with the damper box, the gas turbine must be shut down while the diverter damper is installed.

20.3.2 Combined Cycle Configurations

Combined cycle plants are generally arranged with one or more gas turbines with each turbine driving a dedicated electrical generator. The exhaust gas from the gas turbine is directed through an HRSG that generates steam at one or more pressure levels. The steam is fed to a steam turbine that drives a dedicated electrical generator. With this arrangement, the gas turbine can be "decoupled" from the operation of the steam turbine, allowing for steam turbine shutdown with continued gas turbine operation. Gas turbine–generators, HRSGs, and steam turbine–generators can be arranged in many different combinations depending on the size of the gas turbine–generators, the electrical generation requirements of the project, and the project economics. Figure 20-10 shows a common multishaft combined cycle arrangement.

Another configuration that can be used is to install the one gas turbine, one steam turbine, and one generator using a single shaft. This arrangement can be lower in capital cost because one generator and one step-up transformer is eliminated and a single foundation can be utilized. However, operation is limited to concurrent operation of the gas turbine and steam turbine, unless the steam turbine can be decoupled from the generator through a clutch.

20.3.2.1 Gas Turbines in Combined Cycle. Heavy-duty industrial and aero-derivative gas turbines are both used in combined cycle applications. See Section 20.4 for a detailed description of the gas turbine types.

The exhaust gas temperature of heavy-duty machines is typically higher than that of aero-derivative machines. In addition, the exhaust flow per combustion turbine kilowatt is higher for the heavy-duty machines. In combined cycle mode, this allows more steam with higher superheat temperatures to be generated with the heavy-duty machines, which translates into more electrical output from the steam turbine. In general, for smaller ratings, the overall heat rate for a heavy-duty industrial gas turbine-based combined cycle is slightly higher than that for an aero-derivative based combined cycle plant of similar size. However, larger high-firing temperature heavy-duty based combined cycle plants will have lower heat rates compared to aero-derivative based combined cycle plants.

20.3.2.2 Heat Recovery Steam Generators. HRSGs are gas-to-water heat exchanger systems that extract energy from the gas turbine exhaust gases to generate steam. This steam can be generated over a wide range of pressures and temperatures for a variety of uses such as process or steam heating for industrial use or as high-pressure steam to drive a steam turbine. Many HRSG configurations may be considered when designing a combined cycle or process steam system. In many combined cycle applications, steam is generated at several pressures to make the most advantageous use of the energy available. Reheat cycles are also feasible because of the higher exhaust temperatures available from large heavy-duty advance technology gas turbines.

A typical combined cycle arrangement is shown schematically in Fig. 20-10. This typical arrangement has a two-pressure HRSG, with the high-pressure steam directed to the steam turbine throttle. The low-pressure steam can be directed to either the steam turbine and/or the deaerator. This arrangement utilizes an external deaerator. Other arrangements may include an integral deaerator that uses steam generated at the back end of the HRSG specifically for deaeration. More steam pressure sections can be designed into the HRSG, making it more complicated and more expensive but with improved recovery of the exhaust heat. The steam cycle design depends on the economics of the application and the project specific requirements.

20.3.3 Cogeneration

Cogeneration is generally defined as the production of two usable forms of energy, typically electricity and heat. The heat energy is usually recovered waste heat that is used in some manufacturing process or as heat for building or process heating. The recovery of waste heat displaces energy that would have been required from a conventional heat source, making better utilization of fuels. Figure 20-11 shows a typical arrangement of a gas turbine used in a cogeneration application.

Cogeneration is often achieved with combined cycles to produce electricity and process steam. When cogenerating with combined cycles, nearly 70% of the energy contained in the fuel can be used. The steam cycle adds flexibility to the system operation, enabling the operator to match the process steam requirements and maximize electricity pro-

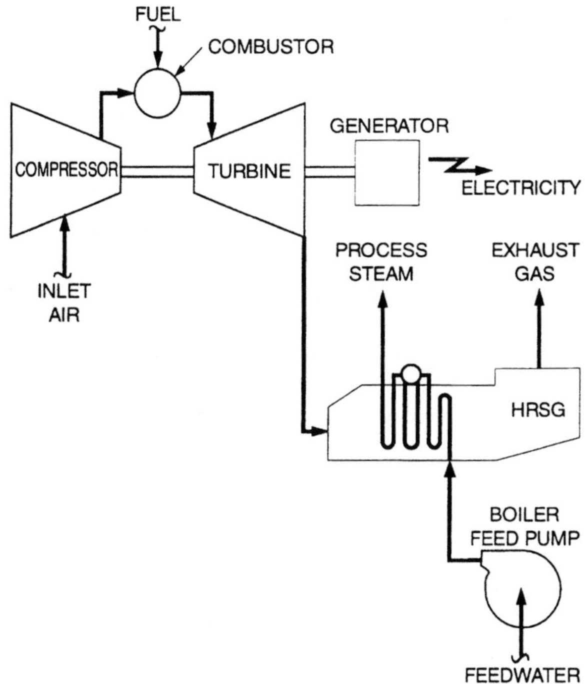

Fig. 20-11. Gas turbine generator cycles: cogeneration cycle.

duction. Process steam can be extracted from the steam turbine or taken directly from the HRSG at the conditions required for the required process.

20.3.4 Comparison with Conventional Coal-Fueled Power Plants

Gas turbine-based power plants have several distinct advantages over similarly sized and loaded coal-fueled power plants. Some of these advantages are listed below:

- Short schedule for design, installation, and startup;
- Higher overall efficiency, especially combined cycle plants;
- Good cycling capability;
- Fast starting and loading;
- Lower pollutant emissions;
- Lower installed cost;
- More compact site;
- Clean fuel source (natural gas or No. 2 fuel oil);
- No ash disposal; and
- No coal handling systems.

Conversely, gas turbine-based power plants do have some disadvantages when compared to conventional coal-fueled power plants. Some of these disadvantages are listed below:

- Higher fuel costs, due to use of natural gas or oil versus coal,
- Uncertain long-term fuel supply, and
- Output more dependent on ambient temperatures.

Gas turbine-based plants are relatively quick to design and erect because of the modular nature of the components. The gas turbine is assembled at the factory and mounted on a structural base plate or skid, minimizing the need for field assembly of the turbine. Other components and support systems such as cooling water and lubricating oil are modules that are easily erected and connected to the gas turbine skid. Fewer service systems are required for gas turbine operation than for a comparable coal-fueled plant. The gas turbine usually can be operated in simple cycle mode while the steam portion of the combined cycle is erected.

Combined cycle plants are the most fuel-efficient thermal power generating units available today. Combined cycle plants are capable of exceeding 58% efficiency based on the lower heating value of the fuel, mainly because of the increased efficiency of the gas turbines and the higher firing temperatures achieved, allowing the use of reheat steam cycles.

Gas turbines are capable of relatively quick starts. Heavy-duty industrial machines can achieve starting times as low as 10 minutes but usually no higher than 30 minutes from cold start to 100% load. Aero-derivative machines can achieve 100% load in 3 minutes or less. If equipped with bypass systems, the startup of the steam cycle portion of the combined cycle can be separate from the gas turbine. The gas turbine can be operating at full load while the steam turbine is warming up. The HRSG can be warmed up nearly as quickly as the gas turbine, with excess steam produced being bypassed to the condenser. The startup time of the gas turbine and the combined cycle plant is significantly less than the time required for a comparably sized coal-fueled power plant.

Gas turbines in simple cycle or combined cycle modes typically emit significantly fewer air pollutants than similarly sized coal-fueled power plants. This is primarily the result of fuel quality and combustion technology. Coal-fueled power plants have to deal with sulfur emissions.

Most gas turbines burn natural gas or No. 2 fuel oil. Natural gas is a particularly convenient energy source. It is clean and easily transported through pipelines. No. 2 fuel oil, though not as clean burning as natural gas, is easily transported by pipeline or truck. Neither of these fuel sources require scrubbing for sulfur dioxide (SO_2) removal. Sulfur is not usually a constituent of natural gas and is usually a small constituent in fuel oil. Therefore, scrubbing the gas turbine exhaust gas is not necessary. The primary emission concerns for gas turbines are nitrous oxides (NO_x) and carbon monoxide (CO). Both of these pollutants can be controlled through combustor techniques or with postcombustion systems in the HRSG. This is discussed further in Section 20.7.

Since gas turbines burn gaseous or liquid fuels, no ash is produced. This is in contrast to a typical coal plant where 5% to 10% of the coal is removed after combustion as ash.

Combined cycles and simple cycle gas turbines have lower installed costs than similarly sized coal-fueled power plants. This is because of the relative simplicity of the gas turbine-based power plant.

20.4 GAS TURBINE TYPES AND MANUFACTURERS

20.4.1 Gas Turbine Types

Gas turbine technology was first applied in jet engines for aircraft and eventually transformed into large, land-based prime movers, referred to as heavy-duty industrial units. Jet engines have been packaged into aero-derivative gas turbines. The heavy-duty industrial and aero-derivative machines are two separate and distinct product lines. By far, the majority of the gas turbines used in power generation are of the heavy-duty industrial type. These machines are generally less expensive per kilowatt and less efficient than the aero-derivative units.

The aero-derivative gas turbines use technology developed for the jet engine industry. The aero-derivative gas turbine features include lightweight components, compact design, and high efficiency. The heavy-duty machines are based on more rugged design and can use a much wider range of fuels than the aero-derivative machines. The exchange of technology has improved for both types of units. Advanced metallurgy and cooling techniques developed for jet engines have enabled heavy-duty machines to achieve higher firing temperatures and efficiency. Combustion emission controls developed for heavy-duty units are being applied to the aero-derivative units.

20.4.2 Gas Turbine Sizes

Gas turbine manufacturers have chosen to market discrete sizes of gas turbines to take advantage of the economies of standardized designs. Therefore, part of a manufacturer's long-range strategy must be to carefully evaluate market trends to establish sizes that will be most attractive to potential customers and that will maximize their ability to compete in the future. Because gas turbines come in discrete sizes, if a power producer dictates a narrow range of electrical output when specifying a gas turbine, some of the potential bidders will be precluded from bidding because they do not have an appropriately sized machine. Gas turbines are not customized to any appreciable extent. Auxiliary packages and accessories associated with the machine may be customized, but in general, the base machine is not.

20.4.3 Gas Turbine Manufacturers

Gas turbines are built in a wide range of output ratings, from < 0.5 megawatts (MW) to over 200 MW at ISO conditions (59° F, sea level). The number of available manufacturers and gas turbine models for the 0- to 50-MW range is much greater than the 50- to 100-MW or the 100-MW and greater ranges. The four major manufacturers of gas turbines over 50 MW include General Electric (GE), Asea Brown Boveri (ABB), Siemens Kraftwerk Union (KWU), and Westinghouse (W). A current listing of available gas turbines including performance from various manufacturers, packagers, and licensees may be found in the latest edition of the *Gas Turbine World Handbook.*

20.4.4 System Packagers

Gas turbine system packagers generally procure gas turbines or jet engines for packaging and resale. Prior to resale, the packagers complete the power train with a power turbine, gear drive, or generator and add additional auxiliary systems and controls as appropriate.

Some gas turbine suppliers, primarily GE, Westinghouse, and GM Allison, have developed contractual relationships with manufacturing licensees or business associates. These manufacturing licensees can manufacture all or part of the gas turbine using the purchased or licensed designs and specifications of the original manufacturer. A strict licensee can use the design but may not be assured of receiving improvements from the original manufacturer. It is not uncommon for the original manufacturer to compete with their own licensees for the sale of gas turbine equipment. In general, however, the original manufacturer maintains a technology edge in machine design.

20.5 GAS TURBINE COMPONENTS AND SYSTEMS

20.5.1 Major Components

Although various arrangements and features are available from the gas turbine manufacturers, this section describes the components of a typical simple cycle, single-shaft machine. A typical gas turbine consists of an inlet air system, compressor, combustion system, turbine, exhaust system, load gear (depending on turbine speed), and generator as illustrated in Fig. 20-12.

When the gas turbine is started, ambient air is drawn through the inlet air system, where it is filtered and then directed to the inlet of the compressor. The air is compressed by the compressor and directed to the combustion system. Inside the combustion system, the air is mixed with fuel and the mixture is ignited. The compressed and heated combustion gases then flow to the turbine. The combustion gases expand as they flow through the turbine, causing it to rotate. The rotating turbine drives the compressor and accessory equipment with available excess energy to produce shaft power which drives the generator. The gases exiting the turbine are exhausted to the atmosphere or directed to heat recovery equipment through an exhaust system. The number of stages within the compressor and turbine may vary, but there is always compression, followed by heating, followed by expansion of the working fluid (the combustion gas).

20.5.1.1 Inlet Air System. Two basic types of inlet air filtration systems are used most frequently: a two-stage high-efficiency inlet air filter system and a cartridge-type self-cleaning inlet air filter system. In addition, inlet air systems include a silencer and inlet ducting to the compressor. They

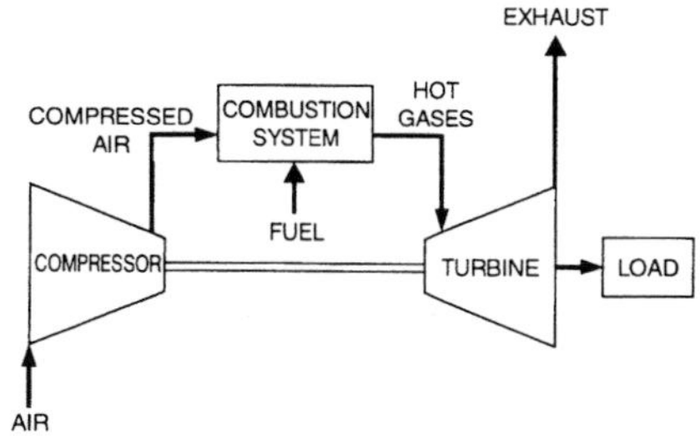

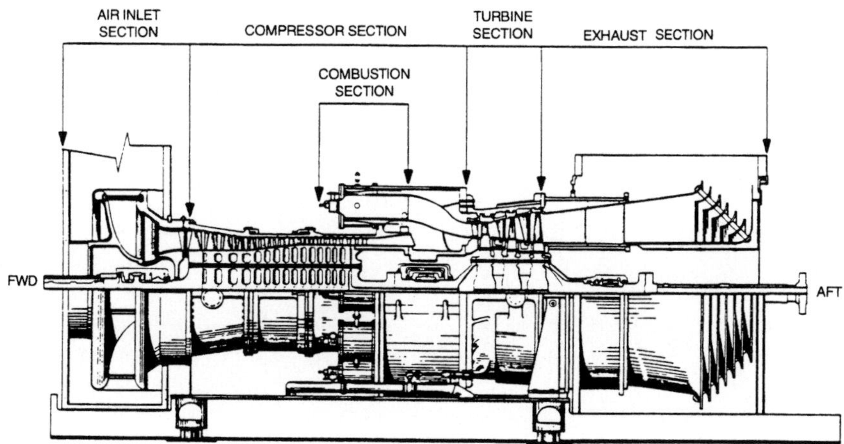

Fig. 20-12. Major sections of the MS-7000 gas turbine assembly. (From General Electric. Used with permission.)

may also include an evaporative cooler or chiller coils for lowering the temperature of the inlet air.

The multistage filtration systems are built around a high-efficiency filter. Depending on site conditions, the high-efficiency filter may be preceded by prefilters and/or inertial separator. The high-efficiency filter is a media type that can accumulate dust particles as small as one micron. The high-efficiency filter must be replaced when the pressure drop across the filter has reached a predetermined level. The life of the filter can be extended by installing an inertial separator and/or a prefilter upstream of the high-efficiency filter. The prefilter is disposable and more economical than the high-efficiency filter. The inertial separator is recommended where the airborne particles that need to be removed are large.

The self-cleaning cartridge inlet air filter system contains high-efficiency media filter cartridges that are cleaned automatically by reverse pulses of compressed air taken from an intermediate compressor stage as shown in Fig. 20-13. The filter cartridges are arranged in horizontal arrays within a module. A number of modules are arranged side by side to form a single level, with several levels stacked on top of the other as shown in Fig. 20-14. The clean air from the filter cartridges is collected by the air plenum and conveyed through ductwork to the compressor.

Cleaning the filter cartridges is accomplished by injecting compressed air backwards through the filter cartridges. The

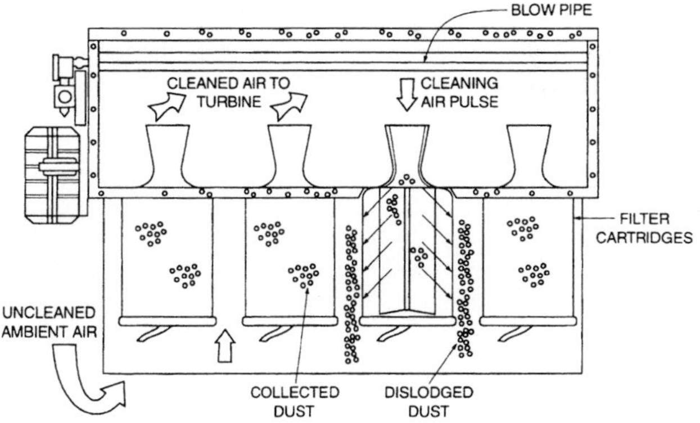

Fig. 20-13. Self-cleaning filter cleaning action. (From General Electric. Used with permission.)

pulse of air is controlled by a combination timer/sequencer that operates air valves in the supply air lines. The pressure drop is measured across the filters and the cleaning operation starts and stops when the pressure drop reaches preset levels. Since only a few modules are cleaned at one time, the turbine operation is not disrupted during cleaning.

In hot, dry climates, evaporative coolers may be installed downstream of either type filtration system to lower the inlet air temperature. Other means of inlet air cooling have also been used, including cold water or chiller systems. A lower inlet air temperature increases the density of the air, thereby

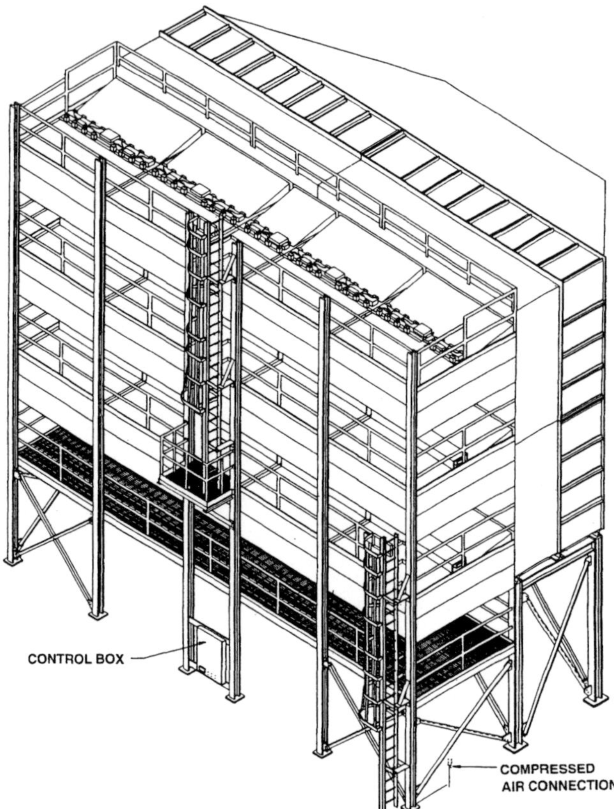

Fig. 20-14A. A self-cleaning inlet air filter. (From Donaldson Company, Inc. Used with permission.)

allowing a greater mass flow of air to be compressed. The increased mass flow allows higher fuel burn rates and results in higher gas turbine output.

20.5.1.2 Compressor. The inlet air is compressed in stages by a series of rotating and stationary air-foil-shaped blades. The rotating (rotor) blades supply the force to compress the air in each stage, and the stationary (stator) blades guide the air to the inlet of the next rotor stage. The compressor rotor blades are attached to wheels that are spun by the gas turbine shaft. The air in larger heavy-duty type machines typically is compressed to about 11 to 14 times normal atmospheric pressure. Aero-derivative gas turbines may have compressor pressure ratios as high as 30 to 1. Compressed air discharging from the compressor is directed to the combustion system. Air may also be extracted from the different compressor stages for turbine cooling, bearing sealing, and inlet air filter cleaning.

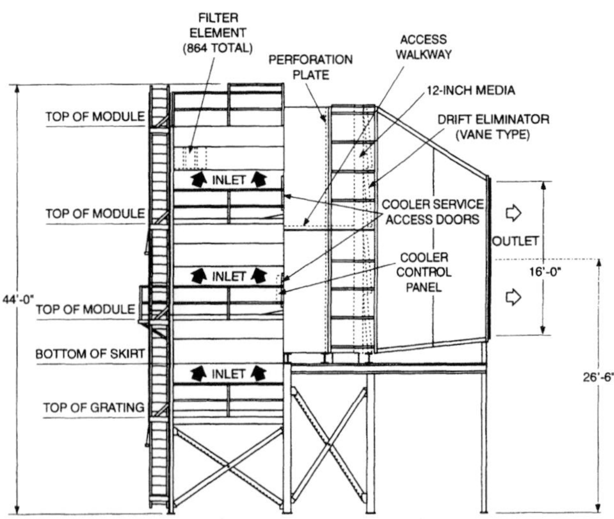

Fig. 20-14B. Self-cleaning inlet air filter. (From Donaldson Company, Inc. Used with permission.)

The compressor stator blades are attached to the inside wall of the compressor casing. The compressor casing supports the rotor at the bearing points. The casing typically is split horizontally in heavy-duty frame sized machines to facilitate maintenance. Aero-derivative type machines are split vertically at the transition of each major component (such as the compressor, the combustors, and the turbine).

The compressor rotor and stator blades are air-foil-shaped and designed to efficiently compress the air. Variable inlet guide vanes are provided at the inlet to the compressor or at midpoints within the compressor to control the quantity of compressor inlet air flow. The vanes are mechanically positioned by a hydraulic control system. The inlet guide vanes are used to control compressor surge during startup and when operating at low levels. They are also used to raise exhaust temperatures during part load operation of combined cycle units.

20.5.1.3 Combustion System. Three types of combustion systems are currently used: the silo type, the multiple-canular, and the annular ring. The combustion systems consist of combustion chamber(s) with integral fuel nozzles, atomizing air connections, spark plug ignition system, and flame detectors. Steam or water injection connections may also be included for emission control and power augmentation.

The differences in the combustion systems lie in the design of the combustion chambers. The silo type system uses a large single chamber or dual chambers that houses all of the nozzles, atomizing air connections, spark plug ignition systems, and flame detectors. The multiple-canular type system uses a number of smaller combustion chambers arranged in an annular configuration around the inlet of the turbine. Each combustion chamber, or can, contains its own nozzle(s) and atomizing air connections. The chambers are interconnected to ensure uniform combustion in each chamber. The annular ring system uses a single combustion chamber that is arranged in an annular configuration around the inlet of the turbine. The chamber contains all of the nozzles, atomizing air connections, spark plug ignition systems, and flame detectors.

Fuel is supplied to each combustion chamber through a nozzle assembly that is designed to disperse and mix the fuel with the proper amount of combustion air from the compressor. Fuel nozzles may be arranged with one or more fuel nozzles per combustion chamber.

Because firing temperatures have increased, most major manufacturers have developed multiple combustor arrangements to reduce NO_x emissions. Key features of the new combustor designs include adequate premixing of the air and fuel upstream of the combustors and stable control of the fuel and air mixture.

Combustion is initiated by a spark plug ignition system. Hot combustion gases from the reaction zone are diluted with sufficient cooling air to cool the gases to desired temperature before they are delivered to the turbine section.

During the starting sequence of a gas turbine, it is essential that the presence or absence of flame is known. A flame monitoring system is employed to establish the existence of a flame. If a flame is not established, the turbine is shut down.

20.5.1.4 Turbine. The action of the hot combustion gases expanding through the turbine section converts the energy of the hot gases into mechanical work. The hot exhaust gases are expanded through the turbine in stages by a series of rotating and stationary air-foil-shaped blades. As the high-energy gases expand through the stationary blades (nozzles) a portion of the thermal energy is converted into kinetic energy. The kinetic energy is then transferred to the rotating blades (turbine buckets) and converted into work. Approximately one-half of the work produced is available to produce electricity; the other half is used to drive the compressor.

The inlet end of the turbine rotor may be coupled to an accessory gear drive that has integral shafts for driving accessory equipment such as the main lube oil pump and the main hydraulic oil pump.

20.5.1.5 Exhaust System. After leaving the last stage of the turbine, the exhaust gases are either released to the atmosphere or directed through an exhaust system to heat recovery equipment. The exhaust system includes an exhaust plenum mounted on the outlet end of the turbine. The exhaust plenum directs the exhaust gas into the exhaust ductwork which transports the hot gas to the exhaust stack or into the inlet ductwork of the recovery equipment.

The exhaust system includes an expansion joint that allows for the thermal movements of the plenum and exhaust ductwork.

20.5.1.6 Generator. The function of the generator is to convert the mechanical energy of the turbine to electrical energy. A description of the design and operation of the generator is provided in Chapter 17.

Gas turbine generators are either air-cooled or hydrogen-cooled. Air-cooled generators are cooled by forced air provided by fans mounted on the rotor shaft. The air flow through the machine may be an open or closed design.

Open ventilated generators are cooled by ambient air drawn into the machine through filters and exhausted through outlet ducts to the atmosphere. Filters such as the self-cleaning pulse air type described for the gas turbine inlet air system are used to remove contaminants from the cooling air stream. Air silencers are installed in the inlet and exhaust ducts to reduce noise levels.

Closed air circuit generators are used where severe site conditions exist, such as in a desert, along the sea coast, or in other heavily contaminated environments. With this arrangement, the hot generator exhaust air is cooled by flowing through water-cooled heat exchangers before being returned to the generator inlet. This type of generator is referred to as a totally enclosed, water-to-air cooled (TEWAC) generator.

Hydrogen-cooled generators use a similar arrangement as the TEWAC generator. The cooling arrangement of the hydrogen-cooled generator uses hydrogen gas as the cooling medium instead of air. Because hydrogen is a combustible gas, the use of this gas requires that the generator be hermetically sealed from a combustion sources. This is accomplished with hydrogen seal housings at each end of the generator. Oil is pumped to the seal oil feed groove in the housings. The oil flows between the seal rings and along the shaft in both directions through the annular clearance between the rings and the shaft. By maintaining an oil pressure greater than that of the hydrogen, the oil forms the seal which prevents the hydrogen from escaping along the shaft. The hydrogen supply is typically maintained in tanks on a separate skid located near the gas turbine.

The use of hydrogen as the cooling medium improves the efficiency of the generator because of its lower density and higher thermal conductivity. The density of hydrogen is approximately one-fourteenth that of air; therefore, the use of hydrogen reduces the windage friction losses to a fraction of those when operating with air. For a high-speed machine, this accounts for a full load efficiency improvement of approximately 0.5%. In addition, the thermal conductivity of hydrogen is nearly seven times that of air, and hydrogen's ability to transfer heat through forced convection is about 50% better than that of air. This reduces the amount of active material required for a given output in the construction of the generator.

20.5.1.7 Starting System. Three starting systems are currently available to provide the breakaway torque required for initial rotation of the gas turbine and the torque necessary for acceleration to self-sustaining speed. The three types are an electric motor starting package, a diesel engine starting package, and a static starting system.

The electric motor and the diesel engine start the gas turbine in a similar fashion. The electric motor and the diesel engine are the power source for the rotation of the gas turbine. The system uses a clutch or torque converter as the means of speed control. The clutch or torque converter also allows for disengagement of the starting equipment once the machine reaches self-sustaining speed.

The static start system uses the generator to act similar to a synchronous motor. Once the gas turbine has reached a self-sustaining speed, the generator discontinues operation as a synchronous motor. The generator has adequate cooling (because of its design as a generator) so as to not require a cooling period between starts. The result is a starting system that can perform numerous consecutive start attempts without a delay for cooling. The system consists of an isolation transformer, static drive (variable-frequency converter system), series of motor-operated disconnect switches, and startup excitation transformer.

During unit shutdown, the turning gear provides for a slow roll of the gas turbine generator during cooldown of the unit to prevent bowing of the rotors.

20.5.2 Auxiliary Systems and Equipment

Auxiliary systems and equipment that support gas turbine operation include lube oil, hydraulic oil, cooling, fuel, NO_x control, fire protection, and compressor water wash.

20.5.2.1 Lube Oil System. Lubricating requirements of the gas turbine and generator are provided by the lube oil system. A portion of the lube oil is typically used in the hydraulic oil system for hydraulic control devices.

The lube oil system consists of reservoirs, pumps, coolers, heaters, filters, piping, valves, and various controls and instrumentation. Dual equipment is furnished for key components to allow for servicing and to allow for backup which increases reliability.

The lube oil reservoirs are usually fabricated carbon steel tanks. They may also be fabricated of stainless steel. Lube oil is pumped from the main reservoir to the bearing header, accessory gear drive, and hydraulic oil system. The pumps may be main shaft driven or electric motor driven. After the lube oil has lubricated the bearings, it flows back through gravity drain lines to the main reservoir or other supplementary reservoirs located near the bearings.

The lube oil is pumped through a cooler that rejects heat absorbed from the equipment bearings to the cooling system. Dual 100% capacity coolers are normally provided with a transfer valve to direct flow to either cooler, thereby allowing one cooler to be in service while the other cooler is available for maintenance.

Dual 100% capacity filters are normally provided to filter the lube oil. A transfer valve is similarly provided to direct the lube oil flow, which allows cleaning of the nonoperating filter.

During standby periods, the lube oil may be maintained at the proper viscosity by heaters installed in the main lube oil reservoir. Temperature controls cycle these heaters on and off as necessary to maintain the proper lube oil temperature.

A typical arrangement of pumps includes a main lube oil pump driven by an accessory gear off the main turbine shaft or by an electric motor (depending on the turbine manufacturer). Additional pumps may include an auxiliary alternating current (ac) motor-driven pump and an emergency direct current (dc) motor-driven pump. These are usually submerged, centrifugal driven type pumps.

The main lube oil pump operates during normal gas turbine operation. The auxiliary lube oil pump operates during gas turbine startup or shutdown or whenever sufficient pressure is not available in the lube oil system. The emergency lube oil pump starts automatically when ac power is not available or when the lube oil supply header pressure drops below the pressure setting.

Regulating valves are included to control lube oil system pressure. Instruments monitor the pressure and temperature of the system and provide alarms and automatic shutdown of the gas turbine as necessary.

20.5.2.2 Hydraulic Oil System. The fuel supply stop and control valves, variable inlet guide vanes, and other control and trip devices are normally powered by the hydraulic oil system. Filtered and regulated oil from the lube oil system is used as the high-pressure fluid to operate the hydraulically controlled equipment and devices.

The hydraulic oil system normally includes a main hydraulic oil supply pump, an auxiliary hydraulic oil supply pump, oil filters, an accumulator assembly, and hydraulic oil supply manifolds. The main pump is usually a variable-displacement pump driven by an accessory gear shaft or by an electric motor. The auxiliary pump is used to back up the main pump whenever the oil pressure is insufficient, such as during combustion turbine startup or low-speed conditions.

Oil from the lube oil system is pressurized by the hydraulic oil pumps. The pressure is controlled by compensators built into the pumps that vary the stroke of the pumps to maintain a set pressure at the pump discharge. Relief valves are also provided to relieve pressure in the event the pressure compensator fails. Filters prevent contaminants from plugging the control devices. The accumulator dampens any severe shock that may occur whenever a pump is started and supplies the transient demands of the system.

20.5.2.3 Cooling Air Systems. Internal cooling of turbine components has been an essential element of advancing gas turbine technology and performance. Impingement cooling and film cooling are used extensively in the advanced technology gas turbines to cool stationary and rotating components. Some of the cooling techniques originated as jet engine technology and have been used successfully in both heavy-duty and aero-derivative gas turbines. Extensive component cooling has kept internal metal temperatures at or below original design parameters while turbine inlet temperatures have been increased by as much as 15%.

The cooling air system provides air from compressor extractions to cool turbine internal parts, seal the turbine bearings, and supply air for operation of control valves in other combustion turbine auxiliary systems. Air extractions are also used to supply air for pulsing the self-cleaning filters in the inlet air system. Cooling air is also provided from external centrifugal type blowers to cool the turbine external casing.

Extraction connections from the compressor are piped to the turbine bearings to provide pressurized air to cool the bearings and help contain the lubricating fluid within the bearing area. Orifices in the air supply lines limit the air flow to the required amount.

Some gas turbine units require centrifugal blowers that are located external to the gas turbine. These blowers provide cooling air to the turbine shell and the exhaust system.

20.5.2.4 Fuel System. Most large gas turbines are designed to operate with both gas and liquid fuels. The two most common fuels are natural gas and No. 2 fuel oil. Control systems are provided to allow automatic changeover from one fuel system to another or to allow the burning of both fuels simultaneously.

The gas fuel system delivers gas fuel to the combustors at the correct pressure and flow rates for all operating conditions. A strainer and/or scrubber are normally provided on the inlet to the fuel gas supply connection to remove contaminants. Stop and control valves are provided to regulate the fuel gas flow. The stop valve shuts off the gas flow to the combustors whenever required for normal or emergency shutdown and controls the gas pressure in front of the control valve. The control valve meters the gas flow in accordance with the demands of the gas turbine.

An automatic vent valve is provided in vent piping between the stop and control valves. This valve relieves any fuel gas that accumulates between the stop and control valves to the atmosphere. Instruments and controls are supplied to monitor the fuel gas flow and pressure and to alarm abnormal conditions.

The liquid fuel system delivers fuel to the combustion chamber nozzles. Fuel oil is normally delivered to the liquid fuel system under low pressure. Oil filters are provided at the inlet to this system to remove contaminants in the fuel supply. A liquid fuel stop valve is provided to shut off the supply of liquid fuel during normal or emergency shutdown. This valve is normally controlled from the hydraulic oil system. Fuel oil pumps increase the liquid fuel pressure to that required by the combustion chambers. These pumps are sometimes driven by the turbine accessory gear.

A flow divider is provided in the supply lines to the combustion system to distribute the liquid fuel to the fuel nozzles. A pressure relief valve is also provided on the discharge side of the pumps to prevent overpressure of the fuel delivery components in the event of a malfunction.

Liquid fuel sprayed into the combustion chambers through the fuel nozzles forms large droplets that will not burn completely. A low-pressure air atomizing system is provided to break the liquid fuel stream into a fine mist. The atomizing air source is an extraction connection from the compressor. This high-temperature air must first be cooled to maintain a uniform temperature to the inlet of the atomizing air compressor. An atomizing air precooler uses water from the cooling water system to absorb the heat from the extraction air. The atomizing air compressors increase the extraction air pressure to the level required to atomize the liquid fuel. The main atomizing air compressor is driven by the turbine accessory gear. An electric motor-operated starting atomizing air compressor is provided to supplement the main compressor during gas turbine startup.

20.5.2.5 NO_x Control System. As with any combustion process, firing a fuel such as natural gas or No. 2 oil in a gas turbine produces emissions of NO_x, carbon monoxide, and unburned hydrocarbons. NO_x contributes to photochemical smog and can adversely affect human health and certain plants. NO_x also reacts in the presence of sunlight to form ozone. Stringent standards to regulate NO_x emissions have been established throughout the United States.

Historically, NO_x control has been achieved by the injection of diluent (water or steam) into the combustion zone. Diluent injection effectively limits the duration of time the gas is at the very high combustion temperatures where thermal NO_x is formed. Nitrous oxide formation is a function of the temperature of combustion and the duration of combustion at the higher temperatures.

Recent developments in combustion designs have produced dry low NO_x combustion systems. Thorough mixing of fuel, staged combustion, and very lean air/fuel ratios have been used to successfully limit the rate of NO_x production without the injection of diluent.

Massive diluent injection can be used for power augmentation in addition to emission control. This arrangement is particularly useful for cogeneration applications. When process steam demand is low, the excess steam can be injected back through the gas turbine to boost power and efficiency.

20.5.2.6 Fire Protection System. A carbon dioxide (CO_2) fire protection system is generally provided to extinguish fires in gas turbine compartments. The system works by reducing the oxygen content of the air in the compartment to a level insufficient to support combustion.

Carbon dioxide is supplied to the gas turbine compartments from a low-pressure liquid CO_2 storage tank. If a fire occurs, pilot-operated selector valves mounted in a CO_2 discharge manifold are automatically opened by an electric signal from fire detectors located in various compartments of the gas turbine. The system may also be manually activated. Two systems are normally provided: an initial discharge system that rapidly fills the gas turbine compartment with sufficient CO_2 to extinguish the fire, and an extended discharge system that maintains a concentration of CO_2 within the compartment for a prolonged period.

20.5.2.7 Compressor Wash System. Atmospheric air drawn in through the inlet air system can contain contaminants such as dirt, dust, insects, and oil fumes. Although a large amount of these contaminants is removed by the inlet air filters, some deposit on the internal surfaces of the compressor. Fouling of the compressor surfaces reduces the gas turbine performance. A decrease in unit performance is noted by a reduction in compressor pressure ratio and power output. When significant performance degradation has occurred, the compressor is cleaned to restore the unit's performance.

The preferred methods for cleaning the compressor are

offline or online liquid washing. These methods are suitable for removing most deposits from the compressor surfaces. In the past, dry abrasive cleaning methods were sometimes used to remove deposits. Abrasive cleaning used either organic (such as nutshells) or inorganic materials. These materials were fed into the compressor during operation to remove the deposits. This practice damaged compressor blades and blade coating and is no longer recommended by gas turbine manufacturers. Online washing consists of injecting demineralized water into the compressor inlet while the gas turbine is in operation.

The offline liquid cleaning solution consists of water mixed with detergent. The detergent is added to remove oily deposits. The water itself is demineralized or deionized water so that it does not cause fouling or corrosion. An offbase water wash system consisting of solution and mixing tanks, heaters, pumps, and piping prepares the water wash solution and delivers it to the gas turbine at the correct pressure, temperature, and flow rate.

Online cleaning occurs when the gas turbine is running at full speed and at some load. Offline cleaning is performed while the gas turbine is under no load and is being turned at cranking speed. Online washing is not as effective as offline cleaning and is therefore used to supplement offline cleaning rather than replace it.

20.5.3 Controls

20.5.3.1 Simple Cycle Gas Turbine Controls. Simple cycle gas turbine generators are furnished with a complete standalone control system by the manufacturer. The packaged control system controls all of the gas turbine and generator functions including protective relays and the auxiliary support equipment. The system has a panel-mounted computer monitor and keypad near the generating unit for equipment monitoring and control. It may also have a desk-mounted computer monitor and keyboard for use in a plant control room. A generator control panel is often included with control switches for the generator voltage, speed, exciter, and generator breaker. This serves as a hard-wired backup to the software driven controls in the main microprocessor controlled system. The system provides for automatic startup of the gas turbine unit from cold start to synchronization (closing the generator breaker) and loading the generator to a target load. The target load is selected manually by the unit operator or by a control signal from a remote load dispatch system. This system also provides for shutdown and tripping of the unit (controlled and uncontrolled).

20.5.3.2 Combined Cycle Unit Controls. A combined cycle unit is a steam power plant except that the "boiler" is a heat recovery steam generator (HRSG) and that all boiler heat comes from the gas turbine exhaust instead of firing fuel in the furnace. The "furnace" of an HRSG is an insulated exhaust duct with banks of tubes inside for heating water and steam. Combined cycle units are simpler than steam power plants with a fossil-fuel-fired boiler because there are no forced draft or induced draft fans, coal pulverizers, burner managements systems, ash disposal systems, or air pollution control systems such as precipitators and scrubbers. Boiler feed pumps are electric motor driven instead of steam turbine driven and there are no feedwater heaters except for a deaerator.

However, combined cycle units can have two components not always used in conventional steam power plants: a steam turbine bypass system and HRSG gas bypass system. The steam bypass system consists of control valves and piping to direct HRSG outlet steam flow directly to the condenser, bypassing the steam turbine. The HRSG gas bypass system is an isolation damper or diverter damper between the gas turbine generator and the HRSG that bypasses some or all of the gas turbine generator exhaust gas to an alternate exhaust stack (bypass stack) instead of the HRSG. This allows the HRSG to be completely isolated from the heat source so that the gas turbine generator can run independently in the event of an HRSG shutdown. Not all combined cycle units have an HRSG isolation damper because of cost. The rest of a combined cycle unit is basically the same as a conventional steam power plant with condensate, feedwater, circulating water systems, demineralizers, auxiliary equipment cooling systems, and auxiliary electric system.

The control system for a combined cycle unit is similar to the controls described in Chapter 18. A distributed control and information system (DCIS) coordinates the operation of the gas turbine generator, HRSG, steam turbine generator, and balance of plant pumps, valves, and motors. The control strategy for HRSG steam drum level, deaerator level, and condenser hotwell control is the same as described in Chapter 18. The control of HRSG firing rate is a direct function of the gas turbine generator load. That is, the more megawatts generated by the gas turbine generator, the more hot exhaust gas flow to the HRSG and the more steam generated.

The DCIS controls combined cycle unit load by calculating the power contribution from the gas turbine generators and the steam turbine so that the sum of the power from the gas turbine generators and steam turbine equals the desired combined cycle unit load. This is known as block load control. Generally, the steam turbine power will be about 50% of the power generated by each gas turbine generator so that two gas turbine generators operating at 100 MW each will provide enough energy for a steam turbine load of 100 MW and a total block load of 300 MW. The target block load can be set at the DCIS by the plant operator or from a signal to the DCIS from a remote load dispatch center. The DCIS sends a megawatt demand signal to each gas turbine generator control system to achieve the target block load. The remainder of the combined cycle control description will address control features unique to combined cycle plants.

STEAM TURBINE BYPASS CONTROLS. The bypass system provides a steam flow path to warm up the HRSG before the steam turbine is on line, a means of throttle pressure control for the steam turbine, and an HRSG steam energy sink in the event of a steam turbine trip or any condition requiring isolation of the steam turbine from the HRSG steam. Steam bypass systems can be designed to bypass up to 100% of the HRSG steam flow to the condenser. With a full 100% bypass system, tripping the steam turbine does not require tripping the HRSG, unlike the case for a conventional power plant boiler which must trip.

The steam bypass line branches off of the main steam line to the steam turbine and connects to the condenser above the hotwell water level. The bypass line has a motor-operated shutoff valve, a modulating control valve, and a desuperheating spray section which are all piped in series. The desuperheating spray section is closest to the condenser. The spray usually supplied from the condensate pump discharge serves to cool the bypass steam before entering the condenser. The condenser manufacturer defines a limiting superheat value for steam entering the condenser. The desuperheating spray is controlled to keep bypass steam within these limits (typically 30° F superheat). The DCIS measures the bypass steam absolute pressure and temperature at the condenser inlet and calculates the superheat content. It then calculates a spray demand control signal to modulate the spray valve at a superheat set point safely below the condenser limit. If the bypass steam exceeds the superheat limit, the DCIS commands the motor operated shutoff valve to close to protect the condenser.

The bypass line modulating control valve serves to control the pressure of the main steam line to the turbine. Closing the control valve raises pressure; opening reduces pressure. A pressure controller in the DCIS controls the bypass valve to maintain the pressure selected by the operator. This valve is used any time the main steam pressure must be set at a fixed value, especially during turbine startup when the pressure is maintained at a level recommended by the turbine manufacturer. Once the turbine is fully warmed, the bypass valve controller set point is switched from a fixed set point to a variable set point. The variable set point is the measured main steam pressure plus about 5 psig. In this control mode, the valve will fully close by trying to maintain a pressure slightly above measured main steam pressure. The turbine is then switched to a variable pressure mode so that the steam inlet control valve (part of the turbine) fully opens and the turbine takes all of the available energy from the HRSG. This is similar to the Turbine Following (Chapter 18) mode except the turbine control valve does not try to regulate main steam line pressure. Combined cycle steam turbines are designed for variable pressure or sliding pressure operation. That is, the turbine will accept any pressure available above a minimum. DCIS interlocks automatically fully open the bypass modulating control valves in the event of a turbine trip, allowing the steam to be developed to the condenser.

HRSG BYPASS DAMPER CONTROL. The HRSG bypass damper is controlled by a hydraulic or motor operated control drive. The hydraulic system consists of a pressure maintenance pump, a damper drive pump, and a number of hydraulic solenoids that are selectively energized to move the damper to a discrete position. Typical positions are 0 (HRSG fully isolated), 30, 45, 65, and 90 degrees (HRSG at full gas turbine gas flow). The damper is physically arranged so that both gravity and the gas turbine gas flow naturally isolate the HRSG on loss of hydraulic power. The damper is moved to the intermediate discrete positions only during HRSG warmup or to accommodate a limiting condition in the HRSG or steam generation process. During warmup, the damper is held at each position for a fixed time determined by the HRSG manufacturer. The DCIS turns the hydraulic drive pump on only when the damper is commanded to move. The smaller pressure maintenance pump cycles on and off to maintain hydraulic pressure at a level above damper position holding pressure. The DCIS should be programmed to isolate the HRSG on the following conditions:

- HRSG trouble: low drum water level, high drum water level, loss of cooling flow for HRSGs with forced circulation (loss of HRSG circulating water pumps);
- Steam turbine trip and no bypass path available (steam bypass shutoff valves closed or modulating bypass valves fail to open); and
- Condenser cooling system failure (circulating water pumps or cooling towers). See following sections on equipment limits and energy balances.

DEAERATOR PRESSURE CONTROL. The deaerator is generally the only feedwater heater in a combined cycle unit. The feedwater pumps pump deaerator water to the HRSG economizer which is the coolest section of the HRSG. The pressure of the deaerator must be maintained so that the temperature of the feedwater entering the economizer is above the dewpoint of the exhaust gas for the fuel fired in the gas turbine. If multiple fuels are fired in the gas turbine (such as gas or oil), the pressure of the deaerator steam supply must be adjusted to match the fuel fired. This is done by using a pressure control valve at the deaerator steam supply inlet. The set point for the DCIS pressure controller for that valve is programmed to a higher set point for oil and lower set point for gas.

EQUIPMENT LIMITS AND ENERGY BALANCES. For a simple cycle unit, all of the gas turbine exhaust energy goes up the stack and heats the outside air. When this heat energy is captured by an HRSG, the energy release path becomes water heated to steam and steam converted to power in the steam turbine or steam cooled by the condenser through the bypass system. The condenser in turn releases energy to the outside air via cooling towers or to a cooling body of water (river or lake). The HRSG, feedwater system, turbine, condenser, and condenser cooling system are all considered links in an energy release chain, and the heat input to that chain must be regulated to accommodate the weakest link in the chain. If the feedwater system, for example, becomes limited to 50% capacity because of a failure of one of its pumps, the heat input to the HRSG must be reduced to accommodate the limited feedwater supply. Otherwise, with a 100% heat input and 50% feedwater flow, the HRSG drums will soon be dry. Control system interlocks must be designed for this case to either reduce the gas turbine generator load to 50% or move the HRSG damper to a 50% heat input position.

For combined cycle units that have an HRSG isolation damper, any limits in the HRSG, feedwater, turbine, condenser, or condenser cooling system can be accommodated by partially or fully isolating the HRSG from the gas turbine so that the gas turbine generator can remain generating at full load. For combined cycle units that do not have an HRSG bypass damper, the control system must be designed to detect limits in the energy release chain of equipment and interlocks provided to automatically partially or completely reduce gas turbine generator load to accommodate the limiting condition.

20.6 TURBINE GENERATOR PERFORMANCE

20.6.1 Performance Parameters

Gas turbine generator performance is characterized by generator output and heat rate. Generator output is the electrical generation of the turbine generator and typically is measured at the generator terminals. Generator output is measured in units of kilowatts or megawatts.

Heat rate is the ratio of heat consumption to generator output and is usually expressed in units of British thermal units per kilowatt-hour (Btu/kWh). Heat consumption represents the thermal energy consumed by the gas turbine and can be calculated as the product of the fuel mass flow rate and the heating value of the fuel. For gas turbines, the convention is to use the lower heating value. Therefore, heat rates for gas turbines are typically expressed on a lower heating value basis.

In equation form, the heat rate for a combustion turbine generator is calculated as follows:

$$\text{Heat rate (Btu/kWh)} = \frac{\text{Heat consumption (Btu/h)}}{\text{Generator output (kW)}}$$

or

$$\text{Heat rate (Btu/kWh)} = \frac{[\text{Fuel consumption rate (lb/h)} \times \text{Fuel lower heating value (Btu/lb)}]}{\text{Generator output (kW)}}$$

Heat rates are an inverse form of thermal efficiency. A lower heat rate value indicates a higher thermal efficiency. Likewise, a higher heat rate value indicates a lower thermal efficiency.

Other principal gas turbine performance parameters are listed below, with their English units of measure in parentheses:

- Compressor inlet temperature (°F),
- Compression ratio,
- Turbine inlet temperature (°F),
- Exhaust temperature (°F),
- Exhaust gas flow (lb/h),
- Exhaust heat (Mbtu/h),
- Inlet pressure loss (in. H_2O), and
- Exhaust pressure loss (in. H_2O).

Each performance parameter is shown in Fig. 20-15 and is defined in the following paragraphs.

Compressor inlet temperature is the dry bulb temperature of the inlet air to the compressor section. When the gas turbine is not equipped with an evaporative cooler or an inlet chiller, the compressor inlet temperature is equal to the ambient dry bulb temperature. When the gas turbine is equipped with an evaporative cooler or an inlet chiller, the compressor inlet temperature is equal to the outlet temperature of the cooler or chiller.

The compression ratio or pressure ratio is a measure of the compressor discharge pressure to the compressor inlet pressure. Theoretically, higher compression ratios will result in higher thermal efficiencies. In practice, compression ratios are limited by the costs associated with the additional compressor stages required to achieve them. In addition, higher compression ratios result in higher temperatures of the compressed inlet air (because of the heat of compression). If compression ratios are too high, an intercooler may be required to reduce the temperature of the compressed inlet air.

For heavy-duty industrial gas turbines, compression ratios of 10 to 18 are most common. For aero-derivative gas turbine generators, compression ratios of 18 to 30 are most common. The trend for both heavy-duty and aero-derivative gas turbines is toward higher compression ratios to achieve lower heat rates.

The turbine inlet temperature, sometimes refered to as the turbine firing temperature, is the average temperature of the combustion gases entering the turbine section of the gas

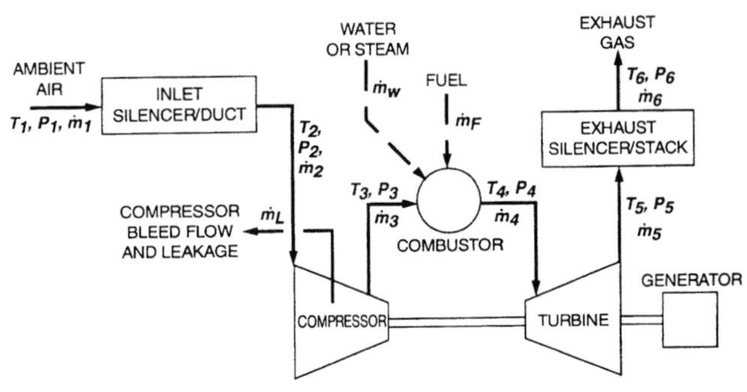

Fig. 20-15. Gas turbine generator performance parameters.

turbine. Firing temperature may be defined differently by various manufacturers.

Higher turbine inlet temperatures result in higher specific outputs (kilowatts per pound of fuel) and lower heat rates. In practice, higher turbine inlet temperatures are limited by the material properties of the turbine. In recent years, higher turbine inlet temperatures have been achieved by using ceramic and composite materials. In addition, improved turbine cooling technology has allowed increases in turbine inlet temperatures. Currently, the most advanced heavy-duty gas turbines have turbine inlet temperatures of approximately 2,350° F, an increase of about 350° F over the preceding generation of heavy-duty gas turbines.

The exhaust temperature is the average gas temperature leaving the turbine. For a given turbine inlet temperature, a lower exhaust temperature indicates higher thermal efficiency and higher specific output. Aero-derivative units generally have exhaust temperatures in the range of 800 to 950° F and are therefore more efficient in simple cycle. By contrast, current heavy-duty units generally have exhaust temperatures of 950 to 1,100° F.

Exhaust temperature is significant for heat recovery applications where an HRSG generates steam from gas turbine exhaust gases. The maximum temperature and pressure of steam generated in an HRSG are limited by the gas turbine exhaust temperature (assuming that the HRSG is not supplementally fired). Higher exhaust temperatures allow higher steam temperatures and pressures. This effect is critical for combined cycle applications where higher steam temperatures/pressures result in significantly lower combined cycle heat rates. Because of this effect, heavy-duty gas turbines are more commonly used in combined cycle units.

Exhaust gas flow is the mass flow rate of exhaust gases from the turbine. Since leakages are minimal and bleed air flow is ultimately redirected into the main turbine flow, the exhaust gas flow is essentially equal to the sum of the inlet air mass flow, the fuel mass flow, and any water/steam injection. The mass flow rate of steam that can be generated in an HRSG is directly proportional to the gas turbine exhaust gas flow.

Exhaust heat is a measure of the thermal energy contained in the exhaust gas flow. Exhaust heat is the product of the exhaust gas flow and the exhaust gas enthalpy. Exhaust heat can be estimated by calculating a heat balance around the gas turbine. By this method, exhaust heat is calculated as the heat input from fuel, inlet air, and injection water/steam, minus the thermal equivalent of the generator output and miscellaneous losses (including generator losses and turbine/generator bearing friction). Some manufacturers do not guarantee exhaust flow or temperature, but they do guarantee exhaust heat. This guarantee is of significance for heat recovery applications only.

Inlet pressure loss is the pressure loss associated with the inlet air flow through the inlet air filter, silencer, evaporative cooler (if used), and inlet ductwork to the compressor section of the gas turbine. Inlet pressure loss typically is measured in units of inches of water. For "new and clean" performance, inlet pressure loss values of 3.0 to 4.0 in. of water are common.

Exhaust pressure loss is the pressure loss associated with the exhaust gas flow from the exit of the turbine section of the gas turbine through the exhaust ductwork, silencer, and stack. In heat recovery applications, the exhaust gas flow through the HRSG results in additional exhaust pressure losses. For simple cycle applications, exhaust pressure losses of 4.0 to 5.0 in. of water are typical. For heat recovery applications, exhaust pressure losses of 10.0 to 17.0 in. of water are typical, depending on the HRSG configuration and the presence of emission-reduction or noise-abatement equipment.

20.6.2 Gas Turbine Generator Ratings

For purposes of comparison, it is common practice to report the performance characteristics of a gas turbine generator at a set of standard conditions. The set of standard conditions most commonly used in the power generation industry is ISO (International Standards Organization) conditions. ISO conditions have been established as an ambient dry bulb temperature of 59° F (15° C), a relative humidity of 60%, and an ambient barometric pressure of 14.7 psia (equivalent to average sea level conditions).

Performance values at ISO conditions are widely published, both by manufacturers and by trade journals. Perhaps the most complete publication of comparative performance is the annual *Gas Turbine World Handbook*.

Gas turbine users require information at a defined set of site conditions as a more realistic basis for comparing performance. Site conditions are usually defined at the average barometric pressure associated with the site elevation and at a certain site temperature and relative humidity. For a simple cycle peaking turbine, the site temperature and relative humidity will usually be based on extreme weather conditions coincident with the peak electric usage. For a baseload to intermediate load combined cycle or cogeneration unit, the site temperature and relative humidity are usually based on the average annual site weather conditions.

In addition to ISO and site ratings, gas turbines have both base and peak ratings. The base rating is indicative of the gas turbine generator performance operating at maximum continuous conditions. The peak rating represents a higher output which should be maintained for a limited period of time. The *Gas Turbine World Handbook* defines base rating as the output that can be maintained for 6,000 hours or more per year, and defines peak rating as the output that can be maintained for 2,000 hours or less per year.

In practice, the peak rating is achieved by operating at a higher turbine inlet temperature. The higher turbine inlet temperature of peak operation results in significantly more thermal wear and increased maintenance requirements. For this reason, many operators do not operate their units at the peak rating, unless the value of the on-peak electricity produced is greater than the increased maintenance costs.

20.6.3 Factors That Influence Performance

Factors that affect performance include the following:

- Ambient conditions,
- Inlet/exhaust pressure losses,
- Fuels, and
- Water/steam injection flow rates.

20.6.3.1 Ambient Conditions. Because gas turbines are open cycle internal combustion engines, their performance is significantly affected by ambient conditions. Ambient conditions that directly affect performance include dry-bulb temperature, specific humidity, and barometric pressure (or elevation).

Ambient dry-bulb temperature has a pronounced effect on electrical output. Lower air temperatures provide higher air densities which result in higher mass flows, and turbine output is directly proportional to higher mass flow. Higher air temperatures, which provide lower air densities and mass flows, result in lower output.

Lower ambient air temperatures also result in lower heat rates and better efficiency due to an increase in compressor pressure ratio. Likewise, higher ambient air temperatures result in higher gas turbine generator heat rates. A typical correction curve showing the effect of ambient air temperature on output, heat rate, and other parameters is shown in Fig. 20-16. Specific humidity affects gas turbine performance through the slight variance in water mass flow. For higher specific humidities, more moisture is contained in the inlet air flow stream, making the air less dense and resulting in a lower electrical output and a higher heat rate. Typical correction curves showing the effect of specific humidity on output and heat rate are shown in Fig. 20-17.

Barometric pressure or site elevation has an effect on output because of the variance in air density. Lower barometric pressure, which occurs at higher elevation, results in lower air density, lower inlet air mass flow rates, and lower electrical outputs. Barometric pressure has no effect on heat rate since the compressor pressure ratio and turbine inlet temperature are unaffected. A typical correction curve showing the affect of site elevation on output is shown in Fig. 20-18.

20.6.3.2 Inlet and Exhaust Pressure Losses. Higher inlet pressure losses reduce the air flow through the unit and thereby reduce output. A typical correction curve showing the effect of inlet pressure loss is shown in Fig. 20-19.

In practical terms, inlet pressure loss is a function of the inlet air system design and cleanliness of the inlet air filters. Lower inlet air pressure losses can be achieved by designing for lower inlet air velocities through the filter, silencer, and duct. The improved operating performance associated with a lower inlet air velocity design must be evaluated against the associated higher capital cost. A similar cost evaluation determines the optimum point that dirty air filters, which have higher pressure losses, should be changed out.

Higher exhaust pressure losses increase the turbine backpressure, thereby reducing power output and increasing heat rate. A typical correction curve showing the effect of exhaust pressure loss on performance is shown in Fig. 20-20.

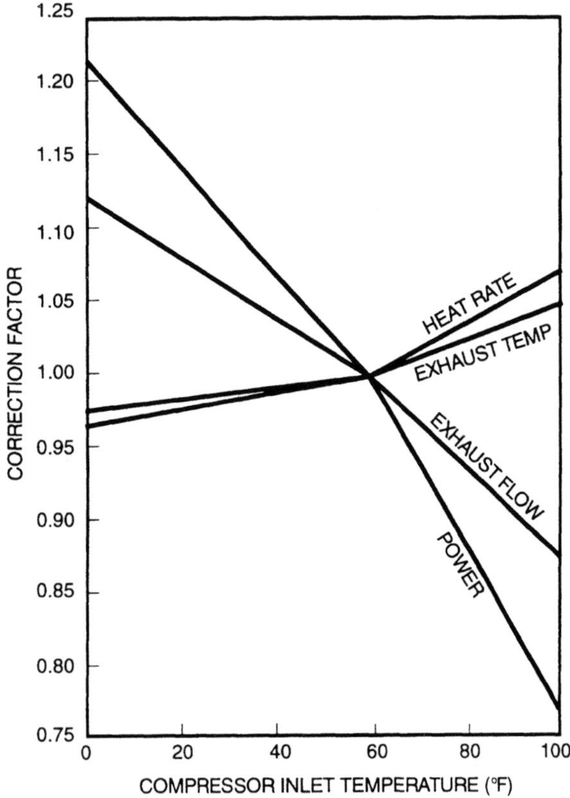

Fig. 20-16. Gas turbine generator correction curves for compressor inlet temperature.

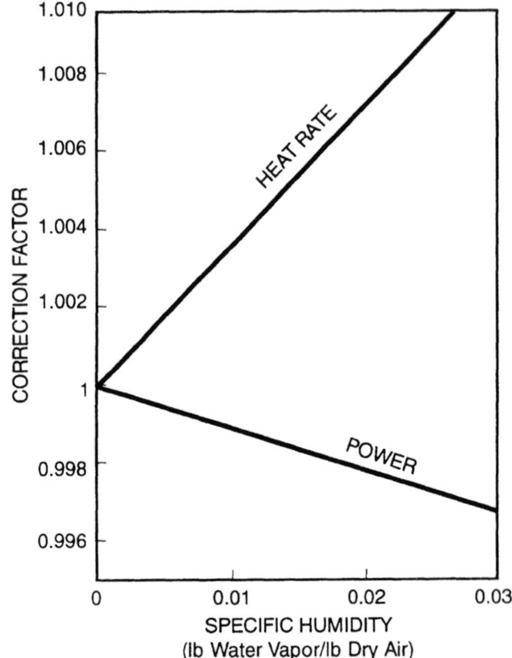

Fig. 20-17. Gas turbine generator correction curves for humidity.

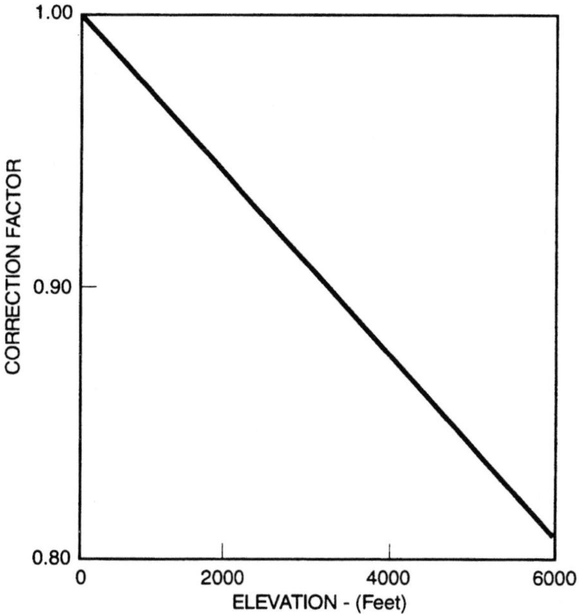

Fig. 20-18. Gas turbine generator correction curve for output and exhaust flow versus site elevation.

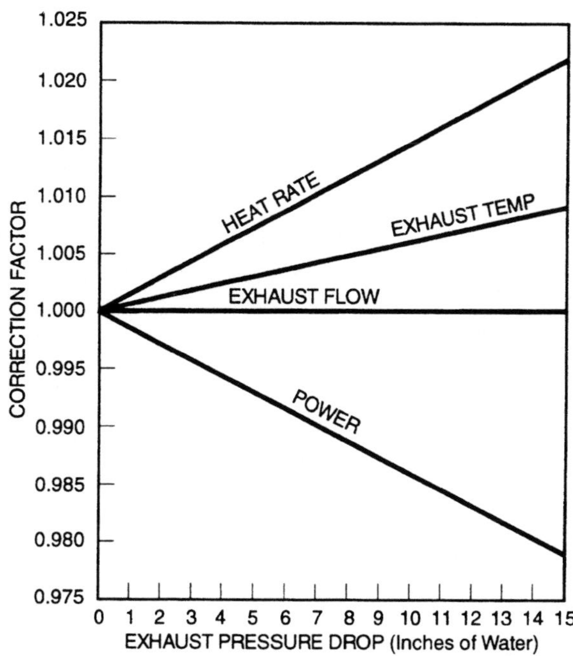

Fig. 20-20. Gas turbine generator correction curves for exhaust pressure drop.

Higher exhaust pressure losses are primarily a function of the exhaust system design. For simple cycle applications, the exhaust system typically consists of an exhaust duct, silencers, and a stack. Exhaust pressure losses of 4.0 to 5.0 in. of water are typical for simple cycle gas turbines. For combined cycle or cogeneration applications, the exhaust gases pass through an HRSG with the associated additional pressure loss. Exhaust pressure losses of 10.0 to 17.0 in. of water are typical for combined cycle and cogeneration applications depending on the complexity of the cycle arrangement, exhaust emission control, or noise-abatement.

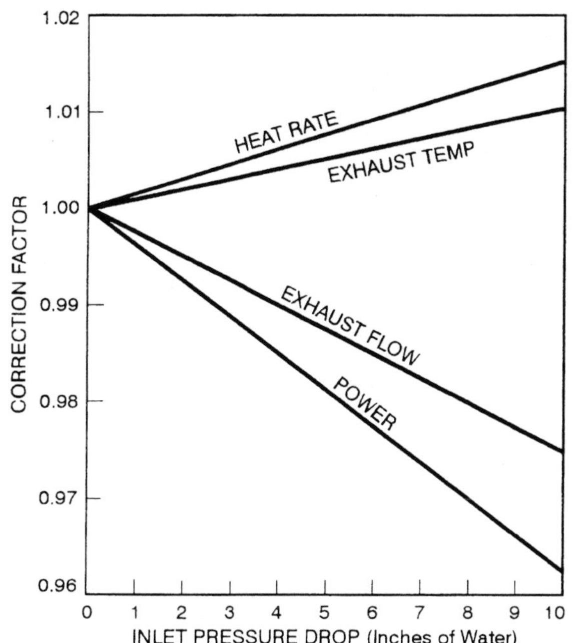

Fig. 20-19. Gas turbine generator correction curves for inlet pressure drop.

20.6.3.3 Fuels. Gas turbine output and heat rate are also affected by the type of fuel burned. The two most common fuels used today are natural gas and No. 2 fuel oil. Turbine output and heat rate are both slightly better for operation on natural gas than for operation on No. 2 fuel oil, assuming no water or steam injection is used for NO_x control or power augmentation.

Coal gasification combined cycle technology is being developed for future large-scale commercial use. The coal gasification process produces low to medium Btu gas which can be used as gas turbine fuel. Using the coal gas as the fuel results in higher fuel mass flow rates to achieve the same heat consumption rates as for conventional fuels. As a result of the higher fuel mass flow rates, higher electrical outputs can be generated, depending on the fuel heating value.

20.6.3.4 Water and Steam Injection. The most common method of controlling gas turbine NO_x emissions rates is to inject water or steam into the combustor to reduce the peak flame temperature. Besides reducing the rate of formation of NO_x, water or steam injection also affects the output and heat rate. Because water or steam injection results in higher mass flow rates through the turbine section, electrical output is increased. Water injection increases gas turbine generator heat rate because of the additional heat consumption required to vaporize the water. Steam injection improves gas turbine generator heat rate because of its higher energy entering the combustion zone.

Water or steam can be injected into the gas turbine at rates beyond what is required for NO_x control. For these applications, water or steam is injected for power augmentation (power boost). For power augmentation, the effect of water or steam injection on gas turbine generator output and heat rate is identical to the effect of water or steam injection for NO_x control.

A typical curve showing the effect of water injection on heat rate and output is given in Fig. 20-21. A typical curve showing the effect of steam injection on heat rate and output is given in Fig. 20-22.

20.6.4 Performance Degradation

The electrical output and heat rate of a gas turbine generator at the time of initial operation are referred to as the "new and clean" performance. Usually, the gas turbine generator manufacturer guarantees the gas turbine generator performance only under new and clean conditions. New and clean conditions are specified on a project-by-project basis, but typically are defined as performance during the first 20 to 100 hours of fired operation.

New and clean performance is significant because gas turbine–generators are subject to a high degree of performance degradation. As internal combustion engines operating in an open cycle, gas turbines pass very large quantities of air through the compressor, combustor, and turbine sections. Although inlet air is filtered, some contaminants in the ambient air pass through the compressor and turbine. These contaminants may include various salts, other corrosive agents, and particulate matter.

In addition, fuel is burned directly in the combustor section of the gas turbine. Any contaminant contained in the fuel passes through the combustor and turbine sections. Contaminants that may be contained in the fuel include sulfur, potassium, vanadium, and other corrosive chemicals. Also, combustion temperatures typically exceed 2,000° F, and the combination of corrosive chemicals and high temperatures produces a highly corrosive environment.

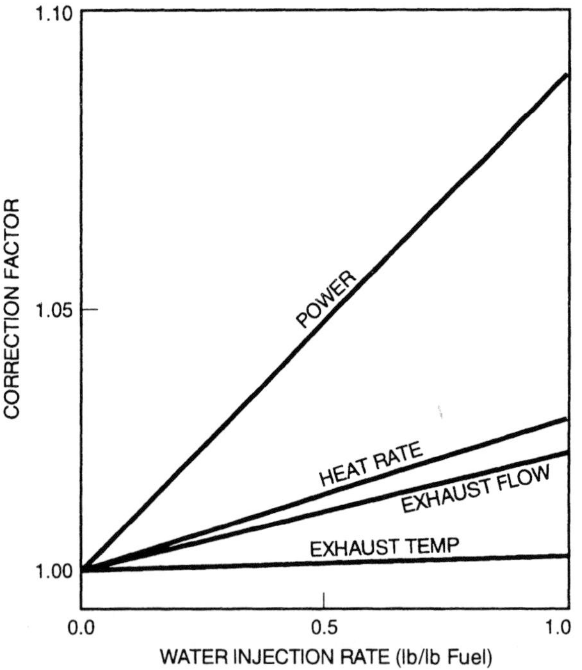

Fig. 20-21. Gas turbine generator correction curves for water injection.

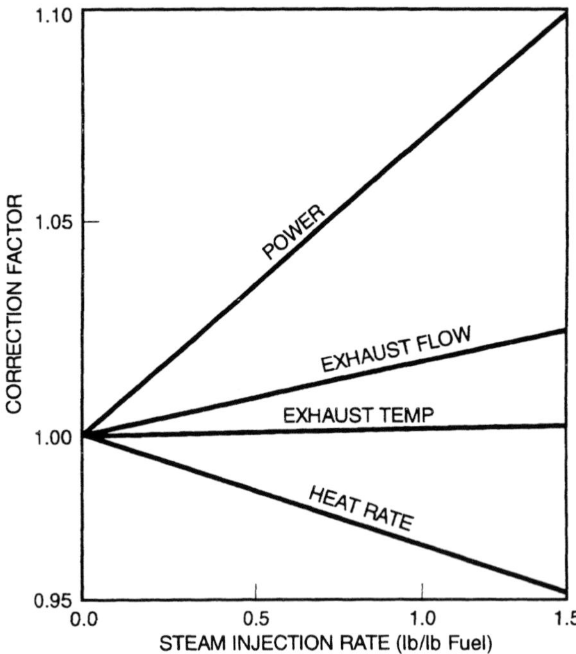

Fig. 20-22. Gas turbine generator correction curves for steam injection.

Contaminants in the inlet air steam and fuel can cause performance degradation by deposition, erosion, or corrosion of the hot gas path. Deposition takes place when particulate matter in inlet air or fuel is deposited and sticks to compressor or turbine blades and nozzles or other components. The deposits physically change the shape of the compressor or turbine blades and nozzles, reducing the aerodynamic performance. Deposits may also block cooling air flow ports and passages, resulting in thermal damage to turbine components.

Particulate matter may also cause erosion of compressor or turbine blades and nozzles. Again, erosion changes the aerodynamic shape of the blades and nozzles, which results in performance degradation.

Corrosion of air and gas path components occurs when salts and other corrosive agents are contained in the inlet air or fuel. Corrosion results in loss of material which degrades the aerodynamic performance. Corrosion also weakens metal components and may eventually lead to mechanical failure of the component.

Performance degradation can be partially reversed by cleaning. The cleaning operation removes deposits from the air and gas flow components, thus partially restoring the aerodynamic performance of the unit. Obviously, cleaning cannot reverse the effects of erosion or corrosion.

Cleaning (washing) can be performed with the unit in

service (online cleaning) or out of service (offline cleaning). Online cleaning typically consists of spraying demineralized water into the compressor inlet. The water dissolves and removes some deposits from the compressor section. Offline cleaning typically consists of spraying a heated demineralized water and detergent solution into the compressor inlet while the unit is slowly rotated. The detergent solution is followed by a rinse with demineralized water.

Performance degradation can also be reversed by overhauling or replacing gas path components. Manufacturers typically recommend periodic maintenance inspections and overhauls which are scheduled on the basis of hours of fired operation and type of fuel burned. On the basis of natural gas firing, one manufacturer generally recommends a combustion inspection at 8,000 hours, a hot gas path inspection at 24,000 hours, and a major overhaul at 48,000 hours of operation. Depending on the severity of performance degradation and other factors, components can be inspected and replaced during these inspections and overhauls with a corresponding recovery in performance.

A representative curve showing gas turbine performance degradation and the results of cleaning versus hours of operation is given in Fig. 20-23. The curve shows losses in electrical output and increases in heat rate from the original new and clean performance.

20.6.5 Performance Testing

Performance testing is conducted on gas turbine–generators at the time of initial operation to determine if a new unit meets its guaranteed electrical output and heat rate. Performance testing is also conducted on existing units to accurately determine current unit electrical output and heat rate.

The performance test code most commonly used for gas turbine testing in the United States is the American Society of Mechanical Engineers (ASME) Power Test Code 22 (ASME PTC-22). ASME PTC-22 specifies an input/output testing method in which fuel flow rate (input) and electrical generation (output) are directly measured to determine unit electrical output and heat rate. ASME PTC-22 also specifies test instrumentation guidelines, data reduction and correction methods, and test reporting requirements.

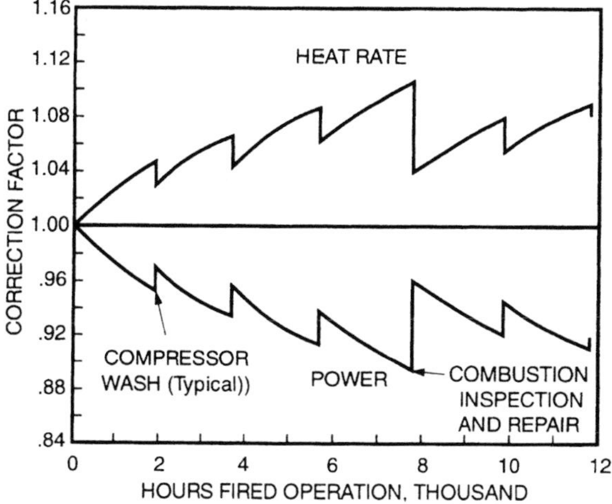

Fig. 20-23. Gas turbine generator performance degradation.

Some gas turbine manufacturers also have their own performance test procedures. These procedures typically are based on ASME PTC-22 and have relatively minor differences.

When performance tests are conducted for the purpose of verifying guaranteed performance, a number of parameters must be established prior to the test. These parameters, or guarantee conditions, should include the following:

- Ambient dry-bulb temperature;
- Ambient wet-bulb temperature (or relative humidity);
- Barometric pressure;
- Test fuel characteristics (including type of fuel, heating value, analysis, temperature, etc.);
- Water (or steam) injection rate, pressure, and temperature;
- Inlet and exhaust pressure loss;
- Point where electrical output is to be measured (for example, the generator terminals) and what auxiliary electrical loads are to be included;
- Condition of the gas turbine unit (if a new unit, a definition of new and clean); and
- Load point or points at which the test is to be conducted.

To the extent practicable, performance tests should be conducted at ambient conditions that closely match guarantee conditions. Minimizing the differences between guarantee conditions and test conditions reduces the inaccuracies of correcting the test results. However, some corrections from test conditions to guarantee conditions will be required. These corrections should be made using performance correction curves submitted by the manufacturer before the test and included in the performance test procedure.

20.7 EXHAUST EMISSIONS

20.7.1 Environmental Concerns

As with any combustion process, the byproduct or exhaust gas from a gas turbine contains elements that have been deemed as pollutants by the US Environmental Protection Agency (EPA). Several United States regulations limit the volume or rate of pollutants emitted from gas turbines. In addition to United States federal legislation, state and local regulations must be met for any new combustion sources of pollutants.

Historically, emissions of nitrous oxides (NO_x) have been the challenge for gas turbines. Regulations continue to change and combustion technology continues to improve, resulting in a steady decline in the allowable rate of NO_x emissions.

20.7.1.1 Trends in Best Available Control Technology (BACT). Under the federal Clean Air Act, BACT represents the maximum degree of pollutant reduction determined on a case-by-case basis considering technical, economic, energy, and environmental considerations. However, BACT cannot be less stringent than the emission limits established by any applicable New Source Performance Standards (NSPS).

A BACT analysis must follow the general requirements of EPA's "top down" BACT guidance document. This approach requires that the BACT analysis start by assuming the use of the lowest achievable emission rate (LAER) control alternative. In regions where LAER control technology is not specifically required, other less efficient emission control technologies are evaluated on the basis of technology, economics, energy, and environmental considerations.

20.7.2 Nitrogen Oxides Emissions and Control

20.7.2.1 Formation. Oxides of nitrogen (NO_x) emissions from gas turbines are composed of two types: fuel NO_x and thermal NO_x. Fuel NO_x is formed by the gas phase oxidation of char nitrogen (CN-compounds) in the fuel, and is largely independent of combustion temperature or the nature of the organic nitrogen compounds. Thermal NO_x is the result of the high temperature reaction between the nitrogen and oxygen in the combustion air. Formation of thermal NO_x is a function of combustion chamber design and turbine operating parameters, including flame temperature, residence time, combustion pressure, and fuel/air ratios in the primary combustion zone. Since natural gas is low in nitrogen, thermal NO_x is the primary concern for natural gas fired gas turbines. Formation of nitrogen oxides can be limited by lowering combustion temperatures and staging combustion (using a reducing atmosphere followed by an oxidizing atmosphere).

20.7.2.2 Water or Steam Injection. The use of water or steam injection in the combustion zones of a gas turbine is a commonly used method to limit the amount of NO_x formed. The injection of water or steam lowers combustion temperatures so thermal NO_x formation is inhibited. The degree of reduction in NO_x formation is somewhat proportional to the amount of water or steam injected into the turbine.

Since the gas turbine NSPS was last revised in 1982, gas turbines have significantly improved their tolerance to negative maintenance impacts from the water necessary to control NO_x emissions below the current NSPS level. Steam or water injection can be used to control gas turbine NO_x emission levels without significant impact on reliability or power output.

Due to the temperature-lowering mechanism of water or steam injection, it can be an effective NO_x control method for a turbine burning either natural gas or fuel oil. However, demineralized water must be used and additional capital cost is required because an onsite demineralization system must be built and the annual operation of this system can be a significant expense in unit operation.

20.7.2.3 Improved Low NO_x Combustion Chamber. Gas turbine manufacturers now offer burner designs that, when compared to earlier combustor technology, provide improved air/fuel mixing and reduced flame temperatures through water or steam injection. The lower combustion temperatures prevent thermal NO_x formation. Lower NO_x concentrations are formed when firing natural gas and No. 2 fuel oil in comparison to the earlier standard combustion chamber design (25 versus 42 ppmdv for natural gas, and 42 versus 65 ppmdv for No. 2 fuel oil).

Dry low NO_x technology is available for gas fueled turbines and uses a two-stage combustor. This combustor premixes a portion of the air and fuel in the first stage, while the remaining air and fuel are injected into the second stage, where the air and fuel are ignited. This two-stage process ensures thorough mixing of the air and fuel and minimizes the amount of air required. Current designs of the dry low-NO_x combustor are capable of NO_x emissions of 9 to 25 ppmdv at 15% O_2 while firing natural gas without the injection of steam or water. Burner manufacturers are developing a dry low-NO_x burner that will be effective when burning liquid fuels such as oil.

The use of dry low-NO_x burners eliminates the need for water demineralizing systems with the associated capital and operating costs. Because of this and the negligible differential cost, dry low-NO_x burners are becoming standard equipment for NO_x control on new natural gas fueled turbine units.

20.7.2.4 Selective Catalytic Reduction (SCR). SCR is a postcombustion method for NO_x emissions control. The SCR process combines vaporized ammonia with NO_x in the presence of a catalyst to form nitrogen and water. The vaporized ammonia is injected into the exhaust gases before it passes through the catalyst bed. The SCR process can achieve in excess of 90% NO_x reduction. A more detailed description of SCR systems is included in Chapter 14.

At NO_x inlet concentrations below approximately 150 ppm, a nonlinear relationship exists between removal and inlet concentration. This is a significant factor in sizing the catalyst because gas turbine exhaust frequently contains <150 ppm NO_x, especially when water injection or dry low-NO_x burners are also used. The NH_3/NO_x mole ratio, concentration of O_2 in the exhaust gas, and space velocity (inverse of residence time) are important factors in determining removal efficiency. The temperature specificity of a given catalyst is also important.

Most SCR systems on turbines in operation in the United States are on combined cycle units, and ammonia injection and SCR catalyst are integrated into the HRSG. Catalysts are usually located within the evaporator section of the HRSG because the temperature profile is in the range needed for the desired chemical reactions. The catalyst location is determined on a project-specific basis.

Exhaust gas temperatures from simple cycle gas turbines generally are too high for vanadium-based catalysts, so zeolite must be used. Zeolite catalyst will operate effectively at temperatures up to 1,100° F.

The most common commercial SCR catalysts for NO_x reduction are vanadium-based. Vanadium pentoxide (V_2O_5) generally is used as the active compound. It is normally supported by titanium or a mixture of titanium and silica. Oxides of tungsten and molybdenum are often incorporated as promoters.

Zeolites, which are crystalline aluminosilicate compounds, are another catalyst used in high temperature, simple cycle applications. Zeolite crystals are characterized by an interconnected system of pores, and adsorb only those compounds with molecular sizes comparable to the pore size. The small pore size creates a very high internal surface area. Zeolites are also called molecular sieves. Although operating experience with zeolite catalyst is currently limited to relatively small turbines, zeolite does offer a high temperature NO_x control alternative that has not been previously available.

Platinum is also used as the active compound in SCR catalysts. In NO_x reduction-only applications, platinum-based catalysts are not generally cost competitive with vanadium-based catalysts. However, in dual NO_x and carbon monoxide SCR applications, platinum is generally the preferred catalyst. The optimum temperature range for the CO reduction catalytic reaction is lower than for the NO_x reaction. At least one manufacturer has developed a dual application catalyst that is platinum-based.

In gas turbine applications, the main causes of catalyst deactivation are thermal degradation and poisoning. Thermal degradation occurs when maximum design temperatures are exceeded. Exhaust gases above the design temperature sinter the catalyst and cause a permanent loss of active surface area at temperatures above 800° F.

Primary poisons for SCR catalyst are arsenic, sulfur oxides, sodium, potassium, and calcium. Arsenic is generally not a problem for turbines firing natural gas or low sulfur oil, however, it can be present if refinery gas is burned. At low temperatures (<500° F), sulfur in the form of SO_3 (oxidized SO_2) can combine with catalytic compounds to form sulfates which block the catalyst's pores. The SO_3 can also react with the NH_3 to form ammonium sulfate and bisulfate which coats the catalyst surface. Ammonium sulfate and bisulfate deposition on downstream equipment can result in other problems such as corrosion and plugging.

These poisoning effects do not occur suddenly; they are a continual process over the life of the catalyst. As the catalyst becomes less active, the ratio of ammonia injected to NO_x removed must be increased to compensate for the lower activity level and maintain required NO_x emissions. The increase in stoichiometric ratio results in increased ammonia slip. In turn, the increased ammonia slip can cause ammonium sulfate or bisulfate deposition problems.

Eventually, the SCR catalyst will no longer meet performance requirements. However, it is unlikely that the entire array of catalyst will need to be replaced at any one time. The exception to this would be if severe mechanical damage resulted from a thermal excursion. The probability of such a failure is extremely low.

OTHER CONSIDERATIONS. The use of an SCR system could result in a negative environmental impact because of the release of unreacted ammonia to the atmosphere. Ammonia and a number of amine compounds are recognized hazardous air pollutants. Ammonia is also a hazardous material. Accordingly, this material must be handled and stored with extreme care.

If fuel oil is the primary fuel for the gas turbine, the use of SCR also has the potential to increase particulate emissions in the form of ammonia sulfate compounds. Because of the inherently higher sulfur content in fuel oil, SCR oxidizes more of the SO_2 to SO_3, and in the presence of unreacted ammonia, sulfate compounds are formed. Once the flue gas cools, the sulfate compounds precipitate out in the form of particulate and cause plugging and corrosion of downstream equipment.

20.7.3 Carbon Monoxide and Volatile Organic Compound Emissions

Typically, measures taken to minimize the formation of NO_x during combustion inhibit complete combustion, which in turn increases the emissions of CO and volatile organic compounds (VOCs).

CO is formed during the combustion process as a result of incomplete oxidation of the carbon in the fuel. CO and VOC formation are reduced by ensuring complete and efficient combustion of the fuel in the turbines. High combustion temperatures, adequate excess air, and good fuel/air mixing during combustion minimize CO and VOC emissions. Therefore, staging combustion and lowering combustion temperatures by water or steam injection or low-NO_x burners (which are used for NO_x emission control) can be counterproductive with regard to CO and VOC emissions.

20.7.3.1 Catalytic Oxidation. Catalytic oxidation is a postcombustion method for reduction of CO and VOC emissions. The process oxidizes CO to CO_2 with the use of a catalyst and oxidizes VOC hydrocarbons to CO_2 and H_2O. A CO control catalyst uses a precious metal base to promote the oxidation process. None of the catalyst components are considered toxic.

The optimum exhaust gas temperature range for CO and VOC catalyst operation is between 850 and 1,100° F. Exhaust gas from the gas turbine is typically between 950 and 1,100° F. Therefore, a CO and VOC catalyst, if used, would be installed at the discharge of the gas turbine. Oxidizing catalysts are capable of reducing CO and VOC emissions by approximately 95% and 50%, respectively.

If the fuel contains sulfur, then the use of an oxidizing CO catalyst in conjunction with SCR catalyst must be carefully examined. Oxidation catalyst can convert up to 60% of the SO_2 into SO_3. When this SO_3 comes in contact with the ammonia required for SCR operation, then ammonium sulfate and ammonium sulfite will be formed. Although the operating temperature is too high for these salts to deposit and foul the catalyst, they will condense and collect onto the cooler back-end of the HRSG. Also, the salts that do not condense can form particulate matter which will be emitted from the stack.

20.7.4 Sulfur Dioxide and Acid Gas Mist Emissions

The NSPS established by Environmental Protection Agency (EPA) for emissions from gas turbines sets a maximum SO_2 level in the flue gas of 150 ppmdv (at 15% O_2) and a maximum fuel sulfur content of 0.8% by weight (40 Code of Federal Regulations [CFR] 60.333). The EPA has not established a gas turbine NSPS for acid gas mist (H_2SO_4). The turbine manufacturers' emission data indicate that on average, approximately 3% of the SO_2 in the flue gas is oxidized to SO_3 which combines with water to form H_2SO_4.

Typically, natural gas has only a trace of sulfur (2,000 grains per million standard cubic feet or less), and no supplemental SO_2 or acid gas mist emission controls have been imposed on natural gas fired gas turbines. Permits for No. 2 fuel oil fired gas turbines typically have included limits on maximum allowable fuel sulfur contents.

As discussed in the previous section, the use of CO oxidation catalyst in the presence of sulfur bearing fuels will lead to an increase in SO_3 production. If oxidation catalyst is to be used, this increase must be included in the determination of acid mist emissions.

20.7.5 Particulate Matter Emissions

The emission of particulate matter from the gas turbine facility is controlled by ensuring complete combustion of the fuel. The NSPS for gas turbines do not establish an emission limit for particulate matter. The natural gas and No. 2 fuel oil fuels to be used in gas turbines will contain only trace quantities of noncombustible material. The manufacturers' standard gas turbine operating procedures ensure as complete combustion of the fuel as possible.

Particulate matter can also be formed by the use of postcombustion catalyst used on turbines fired by sulfur bearing fuels.

20.7.6 Summary of Emissions Control

The SCR process can achieve in excess of 90% NO_x reduction. Because flue gas temperature affects catalyst performance, installation of the catalyst in an area in which the flue gas is between 600 and 750° F is essential.

SCR catalysts are designed to provide optimum activity, selectivity, and chemical stability. Catalysts with high activity offer high NO_x conversion. Catalysts with optimum selectivity reduce the undesired chemical reactions. Chemically stable catalysts protect against catalyst deactivation and ensure acceptable service life.

Typically, measures taken to minimize formation of NO_x during combustion inhibit complete combustion and result in increased emissions of CO and VOCs. Catalytic oxidation is a postcombustion method that can be used to reduce CO and VOC emissions.

The control of SO_2 emissions from gas turbines is accomplished by using natural gas that contains only trace amounts of sulfur and by limiting the maximum allowable sulfur content in fuel oil. Particulate emissions are controlled by using proper operating procedures to ensure as complete combustion of the fuel as possible.

20.8 GAS TURBINE FUELS

Gas turbines can use a wide variety of liquid and gaseous fuels (conventional or nonconventional type). Conventional fuels used regularly at gas turbine installations in the power and process industries are natural gas and various grades of fuel oils, ranging from light petroleum naphthas to residual fuels. Nonconventional fuels used less frequently include crude oils, refinery gas, and propane. Other nonconventional fuels that can be used, and that are further characterized as synthetic fuels, include gaseous and liquid fuels derived, respectively, from coal gasification and liquefaction processes. In addition, research and development continues with regard to using coal–oil mixtures and coal–water mixtures as fuel for gas turbines.

20.8.1 Conventional Fuels

The fuel burned in the gas turbine has a decided effect on unit performance and fuel treatment requirements. A gas turbine, when firing natural gas, will have a higher output and a lower heat rate than the same gas turbine firing fuel oil.

20.8.1.1 Natural Gas. In general, natural gas is an ideal fuel for use in gas turbines because it is practically free from solid residue. Furthermore, natural gas usually has little inherent sulfur content and is environmentally acceptable because the resulting SO_2 emissions are inherently low. Methane (CH_4) and ethane (C_2H_6) are the principal combustible constituents of natural gas. Natural gas may contain significant quantities of nitrogen, N_2, and CO_2. The lower heating value (LHV) of natural gas can range from 300 to 1,500 Btu/ft^3 depending on composition, but the heating value for natural gas is normally in the range of 1,000 Btu/ft^3. Table 20-1 indicates typical ranges of constituents of natural gas.

It is not uncommon for gas turbine manufacturers to specify maximum allowable concentrations of H_2S, SO_2, and SO_3 in the natural gas fuel to ensure that they have taken proper precautions to prevent elevated temperature corrosion of turbine blading materials.

The distance sometimes associated with natural gas transmission and pipeline conditions can result in changes in composition from wellhead to end user. Thus, the as-fired natural gas analysis should be based on point of use rather than at the wellhead.

20.8.1.2 Liquid Fuel Oil. Liquid fuel oils burned in gas turbines include both true distillates and ash-bearing fuels. Light distillates generally do not require preheating because

Table 20-1. Composition of Typical Natural Gas[a]

	Natural Gas from Pipelines[a]				
Sample Number:	1214	1225	1249	1276	1358
Composition, mol%					
Methane	94.3	72.3	88.9	75.4	85.6
Ethane	2.1	5.9	6.3	6.4	7.8
Propane	0.4	2.7	1.8	3.6	1.4
Normal butane	0.2	0.3	0.2	1.0	0.0
Isobutane	0.0	0.2	0.1	0.6	0.1
Normal pentane	Tr	Tr	0.0	0.1	0.0
Isopentane	Tr	0.2	Tr	0.2	0.1
Cyclopentane	Tr	0.0	Tr	Tr	0.0
Hexanes plus	Tr	Tr	Tr	0.1	Tr
Nitrogen	0.0	17.8	2.2	12.0	4.7
Oxygen	Tr	Tr	Tr	Tr	Tr
Argon	0.0	Tr	0.0	Tr	Tr
Hydrogen	Tr	0.1	0.1	0.0	0.0
Carbon dioxide	2.8	0.1	0.1	0.1	0.2
Helium	Tr	0.4	0.1	0.4	0.1
Heating value[b]	1010	934	1071	1044	1051
Origin of sample	CO	KS	KS	OK	TX

[a]Analysis from Bureau of Mines Bulletin 617 (Tr = Trace).

[b]Calculated total Btu/ft^3, dry at 60° F and 30 in. Hg.

they have sufficiently low pour points under most ambient conditions. Heavy distillates can have high pour points because of the high wax content. Preheating heavy distillates is desirable to prevent filter plugging.

The American Society for Testing and Materials (ASTM) D2880 specifies five grades of combustion turbine liquid fuel for different classes of machines and types of service. These grades are as follows:

- Grade 0-GT includes naphtha and other light hydrocarbon liquids that have low flash points and low viscosities as compared to kerosene and fuel oils.
- Grade 1-GT is a light distillate fuel oil suitable for use in nearly all gas turbines.
- Grade 2-GT is a heavier distillate than Grade 1-GT and can be used in gas turbines not requiring the clean burning characteristics of Grade 1-GT.
- Grade 3-GT may be a heavier distillate than Grade 2-GT, a residual fuel oil that meets the low ash requirements, or a blend of distillate and residual oils that meets the low ash requirements.
- Grade 4-GT includes most residuals. Because of potentially wide ranging properties, the gas turbine manufacturer should be consulted on acceptable limits on properties.

Generally, fuel preheating is required for Grade 3-GT and Grade 4-GT and, in some cases, may be required for Grade 2-GT.

Maximum allowable limits are normally set by gas turbine manufacturers for trace metal contaminants such as vanadium, potassium, sodium, and lead because of their potential corrosive effects in the hot gas path parts of the gas turbine. Low-grade, heavy oil can contain sufficient amounts of vanadium that, if burned untreated, corrode the hot gas path components of the gas turbine. During combustion, the vanadium in the fuel forms vanadium pentoxide, an extremely corrosive compound. Vanadium pentoxide and sodium vanadate (also formed if sodium and vanadium are present in the fuel) are semimolten and corrosive at metal temperatures typical of gas turbine operation. If the vanadium is not inhibited and the sodium is not reduced to acceptable levels prior to introduction into the combustor, the gas turbine will have to be operated at firing temperatures lower than the corrosion threshold temperature (about 1,100 to 1,200° F) to minimize the corrosive effects. Maintaining firing temperatures at such a low level dramatically reduces output and efficiency of the gas turbine.

Typically, the corrosive effects of vanadium are inhibited by adding one of a variety of magnesium compounds to the fuel in the ratio of three parts of magnesium by weight to each part of vanadium. The vanadium pentoxide formed during combustion reacts with the magnesium to form magnesium orthovanadate. Using a magnesium compound as an inhibitor raises the melting point of the ash above gas turbine firing temperatures. The gas turbine firing temperature and, as a result, overall thermal efficiency, are not limited by the melting point of magnesium orthovanadate because it is above gas turbine firing temperature.

Although some deposition of the magnesium orthovanadate invariably will occur in the hot gas path parts, such deposits are minimal because magnesium orthovanadate has a melting point high enough to allow most of it to pass through the turbine. Furthermore, the ash produced from the inhibited fuel is dry and noncorrosive when compared to the molten ash produced if no inhibitor is added.

If excess magnesium compound is added to the fuel, ash deposits on the turbine blades will increase. Consequently, some manufacturers recommend that the weight ratio of magnesium to vanadium not exceed 3.5 to minimize the ash deposits on hot gas path components.

Sodium and potassium levels can be reduced to acceptable levels in the fuel by providing a system for washing the oil with water. Sodium and potassium levels generally should be established prior to using the fuel, particularly in the case of installations where the fuel is delivered by sea transport.

20.8.2 Nonconventional Fuels

20.8.2.1 Crude Oil. Crude oils can also be burned in gas turbines. However, the flash point for crude oils is low because of the light fractions that are present. These light fractions are otherwise normally driven off during distillation processes. Because of the low flash point, crude oils represent a greater fire hazard than true distillates and require special handling methods. Plants firing crude oils may require more extensive fire protection systems and alternate, more conventional fuel for startup and shut-down.

20.8.2.2 Refinery Gas. Refinery gas is produced during the conversion of crude oil to gasoline and other products.

Refinery gas generally has relatively high heating value and typically may be blended with lower heating value gas by-products prior to burning. Because of its nature, limited quantities of refinery gas are available.

20.8.2.3 Propane. Propane is normally used only as a backup fuel for instances where an interruption in the primary fuel supply (natural gas or oil) may occur. Propane also may be used as a startup fuel.

20.8.3 Synthetic Fuels

Because of the abundance of coal, synthetic fuels produced from coal by chemical or physical processes hold much promise for future, large-scale use. In addition to chemical processing, coal can be physically mixed with a liquid (water or oil) to produce a liquid mixture that can then be used as fuel in gas turbines.

20.8.3.1 Chemical Processes. Synthetic gas is produced by coal gasification and synthetic liquid fuel is produced by coal liquefaction. Coal gasification and coal liquefaction are chemical processes. At present, coal gasification is the more promising technology because it can be integrated with a conventional gas turbine combined cycle plant to produce power with high efficiency while passing relatively low levels of SO_2, NO_x, and particulates to the environment. Coal liquefaction is not competitive because less integration potential exists and the fuel by itself is not competitive with oil.

Coal gasification is a process that converts coal to a combustible gas which can then be used directly as fuel or, alternatively, may be used as a feedstock for synthetic natural gas (SNG) production or methanol production. In the coal gasification process, the coal is burned in an oxygen-deficient environment under specific temperature and pressure conditions. The composition of the product gas depends not only on stoichiometry, pressure, and temperature, but also on the type of oxidant used by the process. Either high-purity oxygen or air can be used in coal gasification processes.

If air is used as the oxidant, a low-Btu gas is produced. Low-Btu gases produced under such conditions typically have heating values below 150 Btu/scf (LHV) and contain approximately 50% nitrogen, by volume. Nitrogen in the air is required to sustain the gasification reaction. If high-purity oxygen (typically 95% to 99+%) is used, a medium-Btu gas is produced. The medium-Btu gas typically has a heating value of 200 to 500 Btu/scf (LHV), with a nitrogen content of about 1% to 2% by volume. The difference in nitrogen content accounts for most of the heating value difference between low- and medium-Btu gases. The combustible constituents of the low- and medium-Btu gases are primarily CO and H_2, with small quantities of methane, CH_4, and other hydrocarbons.

For an air-blown process, the coal feedstock is reacted with a mixture of steam or water and air. The combustible gas that results is composed primarily of CO, H_2, and CO_2. For the air-blown case, the gas also contains significant quantities of N_2. The CO_2 results from imperfect mixing as a result of localized stoichiometric or above stoichiometric concentrations of oxygen. The gas can contain small amounts of H_2O, CH_4, and C_2H_6 as well. The gas generally has a heating value range from 150 to 350 Btu/scf, depending on the final gas composition.

High-Btu gas or SNG may be produced by upgrading medium-Btu gas by catalytic conversion and methanation. High-Btu gas has a heating value of about 1,000 Btu/scf and consists almost entirely of methane. The first step in the upgrading process is called shift conversion, in which CO from the CO-rich gas is saturated with steam and passed through a catalytic reactor. Within the reactor, the catalyst converts a portion of the CO and H_2O into hydrogen and CO_2. The final step is the methanation process, during which the mixture of CO and H_2 available following shift conversion is converted to methane and water.

Coal liquefaction is the conversion of coal into a liquid fuel or liquid refinery feedstock. Liquid fuels are produced either by direct or indirect liquefaction methods. In indirect liquefaction, the coal is first gasified into a medium-Btu gas and then catalytically recombined at high pressures to yield a variety of liquid fuels.

The Fischer–Tropsch is an indirect liquefaction process developed during the 1940s that is still used commercially by South Africa's SASOL Corporation. As a part of the Fischer–Tropsch process, coal is gasified in the presence of steam in the manner previously described to produce CO and H_2. The gasification reaction is subsequently followed by catalytic reactions that yield methane and lighter hydrocarbons. From the product stream exiting the catalytic reactors, methane is extracted and used as pipeline gas and the remaining hydrocarbons are processed into liquid fuels with fuel properties ranging from those of petroleum distillate to residual fuel oil. Because the coal is gasified and the gas stream is treated to remove H_2S, the fuel products generally are free of sulfur as well as ash.

Most direct liquefaction methods use similar processing techniques. In general, coal is mixed with a paste or solvent, heated, then subjected to suitable conditions in the presence of hydrogen. Direct liquefaction yields fuels with properties ranging from light distillates to residual fuel oils. Because of the processes used in direct liquefaction, the fuel products are relatively free of ash but still contain most of the sulfur found in the coal.

20.8.3.2 Physical Processes. Physical processes include the use of slurry mixtures of coal and water or coal and oil. The fuel mixtures are burned directly in the gas turbines. Such processes require significant research and development effort before commercialization will be possible.

20.9 GAS TURBINE CYCLE VARIATIONS

There are several variations on the conventional gas turbine cycle that increase combustion turbine efficiency or power output. In each of the variations, efficiency or power output is increased by increasing mass flow through the power turbine, recovering heat from the waste gas, reducing compressor work, or increasing the average temperature of the heat source. Some of the cycle variations are as follows:

- Evaporative cooling cycle,
- Recuperative cycle,
- Intercooled cycle,
- Humid air turbine,
- Steam injected cycle,
- Reheat cycle,
- Intercooled steam injected cycle,
- Intercooled reheat steam injected cycle, and
- Intercooled reheat chemically recuperated cycle.

20.9.1 Evaporative Cooling Cycle

The evaporative cooling cycle enhances gas turbine performance primarily by cooling the compressor inlet air (Fig. 20-24). An evaporative cooler at the inlet of the compressor cools the combustion air by the evaporation of water. Evaporative coolers are limited to reducing the inlet air dry-bulb temperature by up to 90% of the difference between the dry-bulb and wet-bulb temperatures.

After evaporative cooling, the compressor inlet air is cooler and more dense. Since compressors are constant-volume machines, this results in a higher mass flow through the gas turbine and increases the power output. Evaporative coolers are normally used in hot, dry climates where the cooling effect is more pronounced.

Water used with evaporative coolers often contains dissolved solids such as sodium and potassium which, in combination with sulfur in the fuel, are principal ingredients in hot gas path corrosion. For this reason, water quality and the prevention of water carryover must be considered carefully when using evaporative cooling.

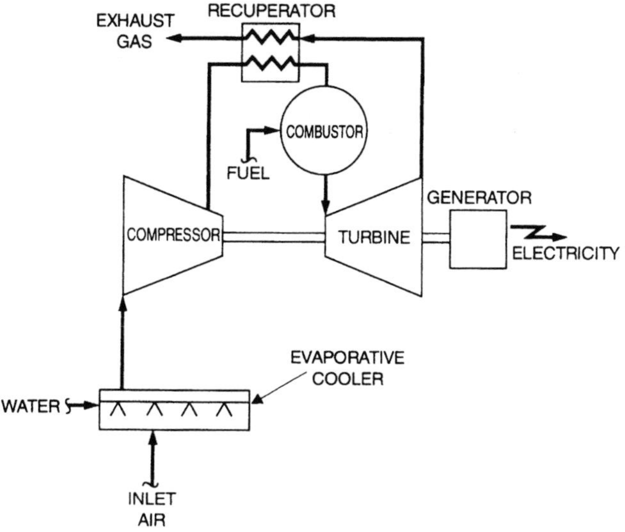

Fig. 20-24. Evaporative cooling and recuperative cycles.

20.9.2 Recuperative Cycle

In the recuperative (or regenerative) cycle, a counterflow heat exchanger transfers exhaust heat to the compressed air before it enters the combustor. This is shown schematically in Fig. 20-24. The amount of fuel needed to heat the air to combustion temperature is reduced by up to 25% (Javetski 1978).

20.9.3 Intercooled Cycle

In the intercooled cycle, a heat exchanger is placed in the air path between low- and high-pressure sections of the compressor, shown in Fig. 20-25. Compression of cool air requires less work than compression of warm air. This heat exchanger is designed to cool the air and reduce the amount of work required in the high-pressure section of the compressor, increasing both power output and efficiency. The intercooler can be a direct contact (evaporative) or an extended surface-type heat exchanger.

20.9.4 Steam-Injected Cycle

The steam-injected cycle is also known as the steam-injected gas turbine (STIG) or the Cheng cycle. Large volumes of steam are injected into the combustor and/or turbine section of a gas turbine to increase power and improve efficiency.

There is no steam turbine or associated equipment as in the combined cycle arrangement. The basic components for this cycle, shown in Fig. 20-26, are the gas turbine and HRSG. The HRSG produces steam for injection into the gas turbine or for optional process use. The typical application of the steam-injected cycle is in a cogeneration plant where increased power can be produced during low cogeneration demand and/or peak electrical demand periods.

Aero-derivative gas turbines are typically used in steam injection power augmentation applications. Increases in mass flow through typical heavy-duty gas turbines are limited by allowable loadings on turbine blades. However, several manufacturers offer heavy-duty machines capable of limited steam injection for power augmentation. These gas turbines are capable of injecting up to 2 pounds of steam per pound of fuel for NO_x control and/or power augmentation. These heavy-duty machines typically can produce

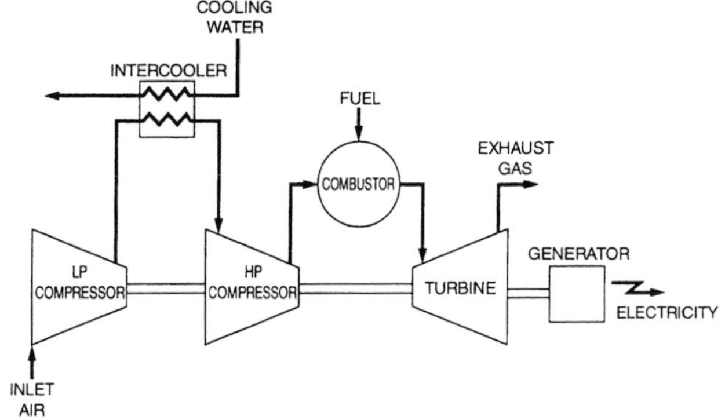

Fig. 20-25. Intercooled cycle.

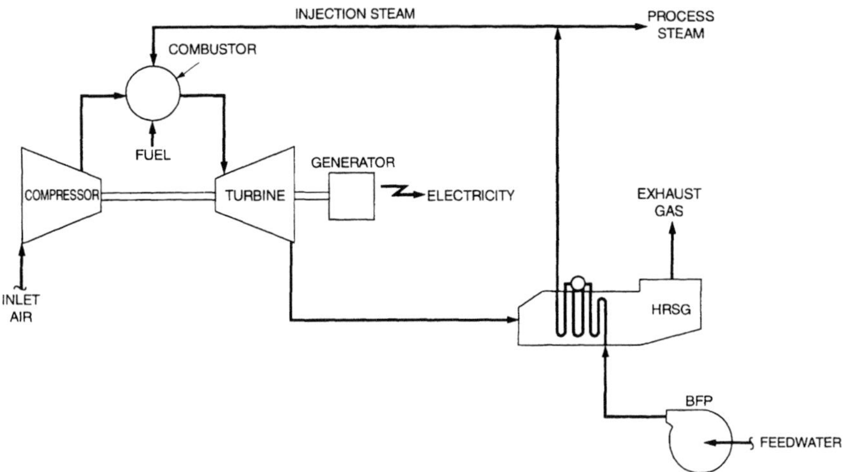

Fig. 20-26. Steam-injected cycle.

much more steam in an HRSG than can be used for injection steam.

20.9.5 Humid Air Turbine Cycle

An extension of steam-injected technology that is being investigated by several groups is called the humid air turbine cycle (Fig. 20-27). An intercooler and an aftercooler recover heat from the compressor (reducing the compressor work requirement) and transfer it to feedwater. The heated feedwater and compressor discharge air are combined in a saturator, which produces a mixture of steam and air. The steam and air mixture, or humid air, is heated in a recuperator which extracts heat from the gas turbine exhaust. The humid air is then fed to the combustor. Preliminary evaluations show the cycle to have 20% higher specific power output and an efficiency four percentage points higher than an equivalent combined cycle (Makanski 1990).

20.9.6 Reheat Cycle

The gas turbine reheat cycle is analogous to the steam turbine reheat cycle. The hot gases are partially expanded through the turbine, reheated by a second stage combustor, and returned to the turbine. This cycle is more complex than the standard gas turbine cycle. Figure 20-28 illustrates a schematic representation of this cycle.

- The reheat cycle is more efficient.
- To protect the reheat combustor, the high-pressure turbine is required to lower the temperature of the hot gases entering the reheat combustor.
- In general, it is necessary to increase fuel/air ratio to increase specific power. Metallurgy and cooling technology limit the turbine inlet temperature. The two-stage combustor allows more fuel to be injected without exceeding this temperature limit. This increases the exhaust temperature and makes the reheat combustion turbine more suitable for combined cycle.
- The reheat combustion turbine's higher specific power results in less residual oxygen in the exhaust gas. The inlet and exhaust sections are smaller than those for a simple cycle gas turbine of the same output.

20.9.7 Intercooled Steam-Injected Cycle

The intercooled steam-injected gas turbine (ISTIG) cycle is similar to the steam injected cycle. Inlet air is compressed in the low-pressure compressor and diverted through a water-cooled heat exchanger. The cooled air is then discharged to

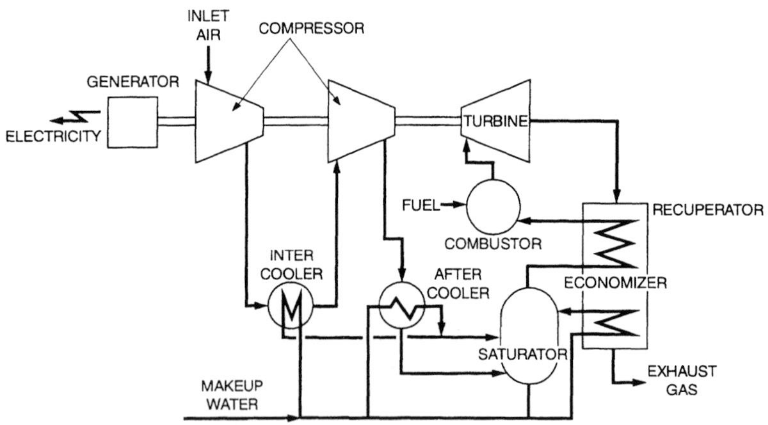

Fig. 20-27. Humid air turbine cycle.

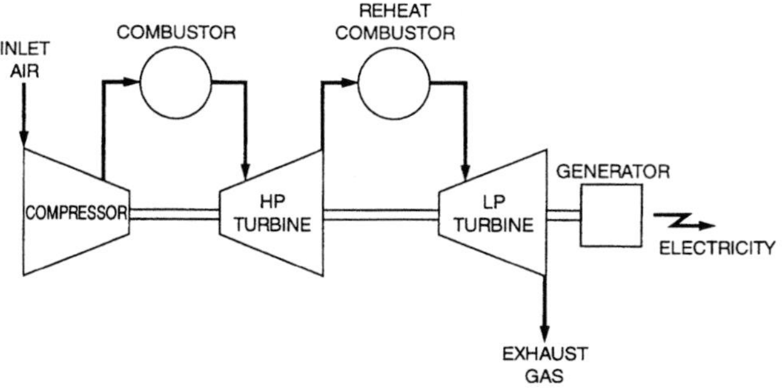

Fig. 20-28. Reheat cycle.

the high-pressure compressor inlet. The air temperature is lowered by intercooling to reduce the shaft work of the high-pressure compressor, thereby increasing power output and efficiency. The hot water from the heat exchanger can be used to preheat feedwater entering the HRSG. This further increases efficiency. The basic arrangement for the intercooled steam injected cycle is shown in Fig. 20-29.

20.9.8 Intercooled Reheat Steam-Injected Cycle

The intercooled reheat steam-injected gas turbine (IR-STIG) cycle employs compressor intercooling similar to the intercooled steam-injected cycle. The combustion products are refired in a reheat combustor after partial expansion in the turbine. This cycle increases the power output and efficiency.

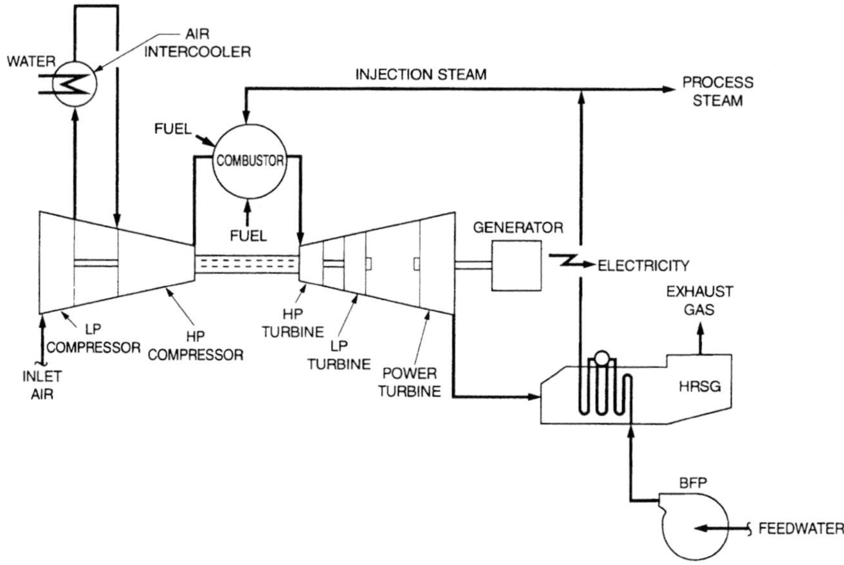

Fig. 20-29. Intercooled steam-injected cycle.

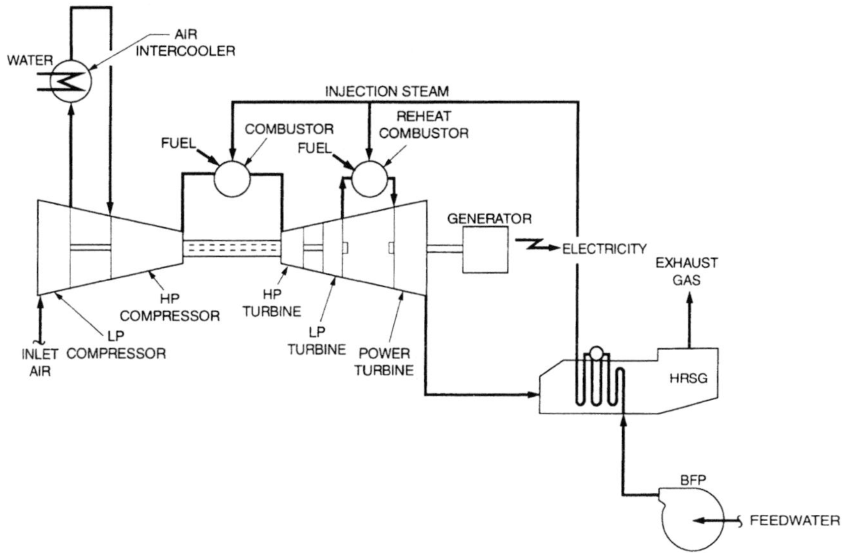

Fig. 20-30. Intercooled reheat steam-injected cycle.

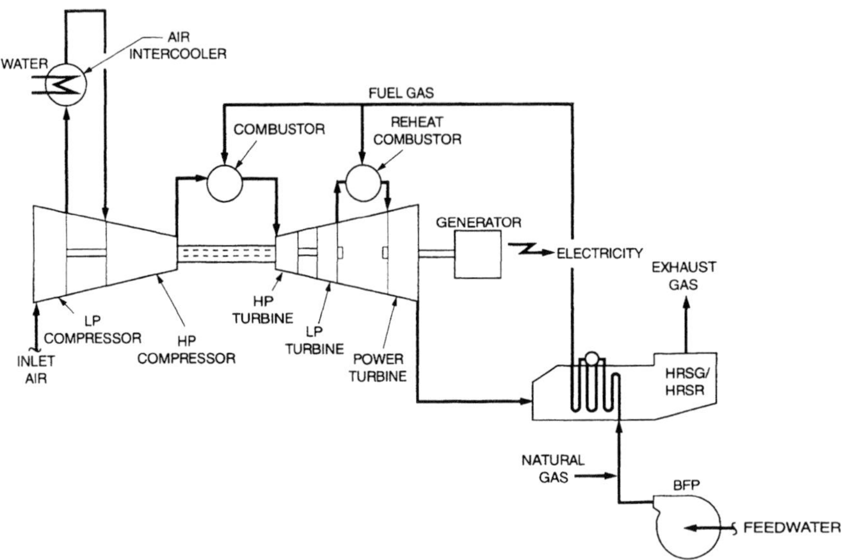

Fig. 20-31. Intercooled reheat chemically recuperated cycle.

Figure 20-30 shows the basic arrangement of the IR-STIG cycle.

20.9.9 Intercooled Reheat Chemically Recuperated Cycle

The intercooled reheat chemically recuperated gas turbine cycle uses a heat absorbing (endothermic) chemical reaction between steam and natural gas to recover exhaust heat. The chemical recuperation is through partial steam reforming of the natural gas. Gas turbine exhaust temperatures above 1,200° F are required for the chemical recuperation to be effective. The elevated exhaust temperatures can be achieved using a reheat intercooled gas turbine.

In the intercooled reheat chemically recuperated gas turbine cycle, steam in the HRSG is blended with desulfurized natural gas. Exhaust heat is absorbed by an endothermic chemical reaction between the steam and natural gas as the mixture passes over a nickel-based catalyst in the superheater section of the HRSG. In this cycle, the HRSG is basically converted to a heat recovery steam reformer. The gas produced by the reformer is a hydrogen-rich, low-Btu fuel gas that is fired in the gas turbine combustor and reheat combustor. The power output and efficiency of this cycle are higher than those of the intercooled steam injected cycle. Figure 20-31 shows a simplified flow diagram of the intercooled reheat chemically recuperated combustion turbine cycle.

20.10 REFERENCES

American Society of Mechanical Engineers (ASME). ASME PTC-22. Standard Specifications for Gas Turbine Fuel Oils.

American Society of Testing and Materials (ASTM). ASTM D2880. Gas Turbine Power Plants.

Environmental Protection Agency. 1985 and 1990. *BACT/LAER Clearinghouse—A Compilation of Control Technology Determinations.*

Donaldson Company, Inc. 1400 West 94th St., Bloomington, MN 55431.

Gas Turbine World Handbook. Pequot Publishing, Inc., Fairfield, CT.

General Electric Company. 1 River Road, Schenectady, NY 12345.

Javetski, John. 1978. A Special Report—The Changing World of Gas Turbines. Power. Vol. 122, No. 9. September 1978.

Makansi, Jason. 1990. Combined Cycle Power Plants. *Power.* June 1990.

May, Phillip A. et al., Radian Corporation. 1991. *Environmental and economic evaluation of combustion turbine SCR NO_x Control.* Paper presented at 1991 Joint EPA/EPRI Symposium on Stationary Combustion NO_x Control. Washington, D.C. March 1991.

Morris, Michael D., Chris K. Chidley, and Joseph W. Rogers, Black & Veatch. 1994. The advantages of small combined cycle power plants. Paper presented at the American Power Conference. April 27, Chicago, IL.

Schreiber, Henry, P. E. 1991. *Combustion NO_x controls for combustion turbines.* Electric Power Research Institute. Paper presented at 1991 Joint EPA/EPRI Symposium on Stationary Combustion NO_x Control. Washington, D.C. March 1991.

US Federal Bureau of Mines. Department of the Interior. 810 7th St. N.W., Washington, D.C. 20241.

21

Fluidized Bed Combustion

Kenneth E. Habiger

21.1 INTRODUCTION

During the mid to late 1980s, fluidized bed combustion (FBC) rapidly emerged as a viable option to stoker fired and pulverized coal-fueled units for the combustion of solid fuels. Initially used in the chemical and process industries, FBC was applied to the electric utility industry because of its perceived advantages over competing combustion technologies. Sulfur dioxide emissions could be controlled from FBC units without the use of external scrubbers, and nitrous oxide (NO_x) emissions from FBC units were inherently low. Furthermore, FBC units were touted as being "fuel flexible," with the capability of firing a wide range of solid fuels with varying heating value, ash content, and moisture content. Also, slagging and fouling tendencies were minimized in FBC units because of low combustion temperatures.

By 1991, the transition had been made from small industrial-sized boilers to several electrical utility boilers in operation in the size range from 100 to 165 megawatts (MW). Orders were in place for at least one large reheat boiler in the 250-MW size range and boiler manufacturers were willing to offer full commercial guarantees for units in the 250- to 300-MW size range (Makanski 1991).

21.1.1 Principles of Fluidized Bed Combustion Operation

A fluidized bed is composed of fuel (coal, coke, biomass, etc.) and bed material (ash, sand, and/or sorbent) contained within an atmospheric or pressurized vessel. The bed becomes fluidized when air or other gas flows upward at a velocity sufficient to expand the bed. The process is illustrated in Fig. 21-1. At low fluidizing velocities (3 to 10 ft/s), relatively high solids densities are maintained in the bed and only a small fraction of the solids are entrained from the bed. A fluidized bed that is operated in this velocity range is referred to as a bubbling fluidized bed (BFB). A schematic of a typical BFB combustor is illustrated in Fig. 21-2.

As the fluidizing velocity is increased, smaller particles are entrained in the gas stream and transported out of the bed. The bed surface, well-defined for a BFB combustor, becomes more diffuse and solids densities are reduced in the bed. A fluidized bed that is operated at velocities in the range of 13 to 22 ft/s is referred to as a circulating fluidized bed, or CFB. A schematic of a typical CFB combustor is illustrated in Fig. 21-3.

21.1.1.1 Advantages of Fluidized Bed Combustion. An environmentally attractive feature of FBC is that sulfur dioxide (SO_2) can be removed in the combustion process by adding limestone to the fluidized bed, eliminating the need for an external scrubber. The calcium oxide formed from the calcination of limestone reacts with SO_2 to form calcium sulfate, which is removed from the flue gas with a conventional particulate removal device.

In FBC, the combustion temperature is controlled at approximately 1,550 to 1,600° F as compared to approximately 3,000° F for conventional boilers. Combustion at the lower temperature has several benefits. First, the lower temperature minimizes sorbent (typically limestone) requirements because the required Ca/S molar ratio for a given SO_2 removal efficiency is minimized in this temperature range. Second, 1,550 to 1,600° F is well below the ash fusion temperatures of most fuels so the fuel ash never reaches its melting point. The slagging and fouling problems that are characteristic of pulverized coal units are significantly reduced, if not eliminated. Finally, the lower temperature reduces NO_x emissions.

Since combustion temperatures are below ash fusion temperatures, design of an FBC boiler is not as dependent on ash properties as is a conventional boiler. With proper design considerations, a FBC boiler can fire a wider range of fuels with less operating difficulty.

The advantages of FBC in comparison to conventional pulverized coal-fueled units can be summarized as follows:

- SO_2 can be removed in the combustion process by adding limestone to the fluidized bed, eliminating the need for an external desulfurization process.
- Fluidized bed boilers are inherently fuel flexible and, with proper design provisions, can burn a variety of fuels.
- Combustion in FBC units takes place at temperatures below the ash fusion temperatures of most fuels. Consequently, tendencies for slagging and fouling are reduced with FBC.
- Because of the reduced combustion temperatures, NO_x emissions are inherently low.

21.1.1.2 Heat Transfer. In a BFB unit, combustion heat is absorbed from the gas by heat transfer tubes immersed in the bed and by conventional waterwall surface. In-bed tubes can serve as either steam generation surface or superheat surface. Although an in-bed heat exchanger is a possible source of

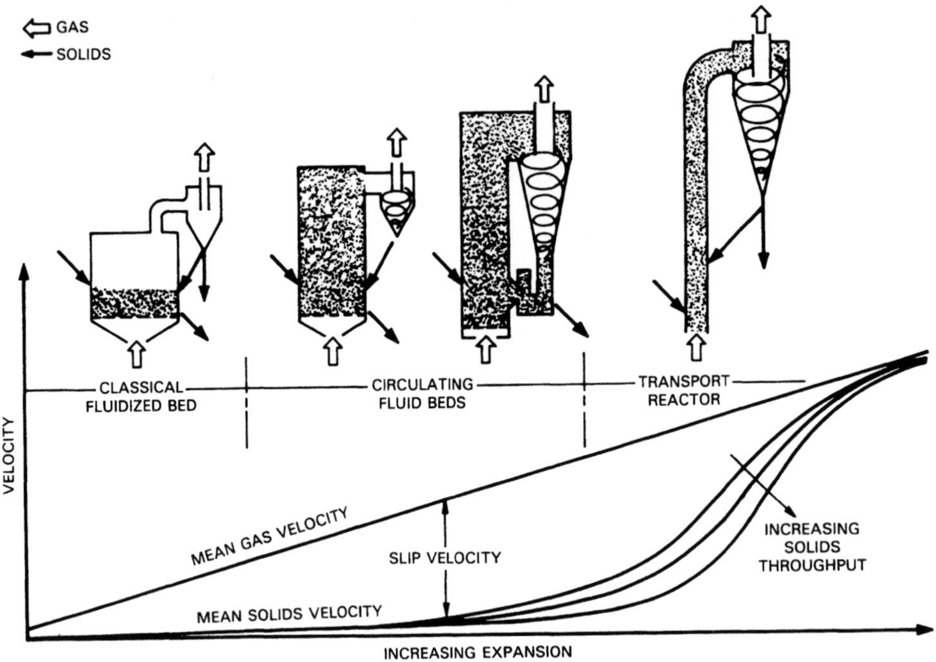

Fig. 21-1. Basic fluid bed systems. (From ABB Combustion Engineering Systems. Used with permission.)

erosion problems, the close contact with bed materials and excellent mixing provides high heat transfer. Heat transfer in the convection pass is similar to that for a conventional stoker-fired or pulverized coal-fueled steam generator.

In a CFB unit, combustion heat is absorbed from the gas by conventional waterwall surface, by platens located in the upper region of the combustor, or by heat transfer surface located in external heat exchangers. The velocity of the solids in a CFB is relatively lower than that of the gas. The high slip velocity (difference between mean gas velocity and the mean solids velocity), in combination with the intense mixing that occurs, results in high heat and mass transfer rates. The high inventory of circulated particles leads to consistently uniform temperatures throughout the combustor and the recycle system, as well as a long solids inventory residence time. The long residence and contact times, increased slip velocities, and intense mixing result in higher heat and mass transfer rates and higher combustion efficiency in a CFB than are available with a BFB.

21.1.1.3 CFB vs. BFB Technology. In comparison to BFBs, CFBs offer the following potential advantages:

- Greater fuel throughput per unit of cross-sectional area is possible because of higher combustion air velocities.
- Combustion efficiencies are higher.
- The calcium-to-sulfur ratio is lower for a given SO_2 removal rate.

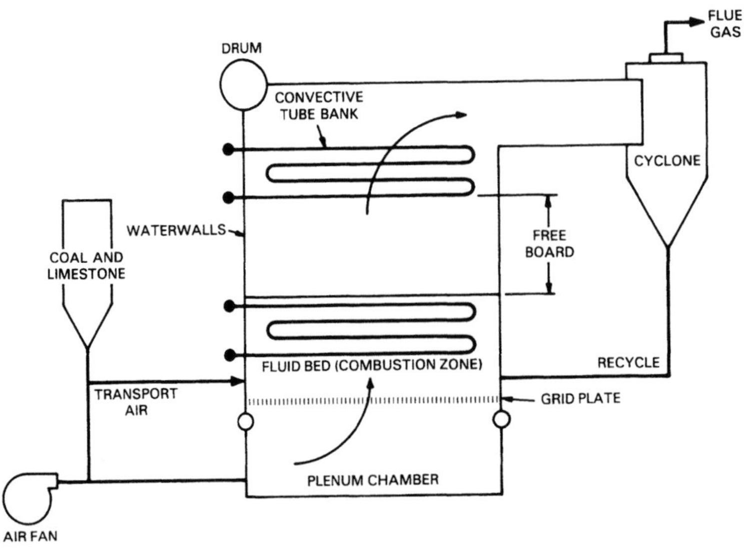

Fig. 21-2. Typical atmospheric bubbling bed combustor.

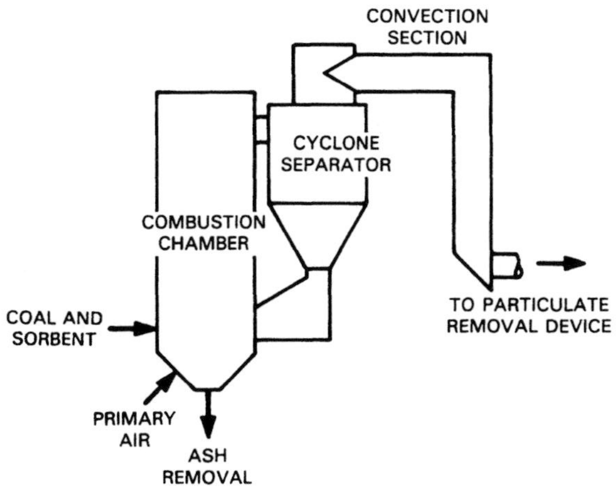

Fig. 21-3. Typical circulating bed combustor.

- Lower NO_x emissions are possible, primarily as a result of the intimate mixing in the combustion vessel and the introduction of secondary combustion air above the primary combustion zone.
- The fuel feeding system is simplified.

21.1.1.4 Categories of Fluidized Bed Combustion. Two categories of FBC systems are potentially applicable to power generation: atmospheric fluidized bed combustion (AFBC) and pressurized fluidized bed combustion (PFBC). AFBC systems are commercially available and are currently being applied to electric utility applications. In the AFBC option, the fluidized bed is used to generate steam for power production using a conventional Rankine cycle. A block flow diagram for power generation from an AFBC is presented in Fig. 21-4.

PFBC is under development, with several large demonstration units in operation or in the planning stage. PFBC units operate at 10 to 15 atmospheres and offer the potential advantages of smaller boilers in comparison to AFBC. The smaller equipment size provides opportunities for modular construction with PFBC units. Removal of particulate matter from the high-temperature, combustor discharge flue gas stream is an integral part of the PFBC process. However, gas cleanup at high temperatures (1,500 to 1,600° F, for example) is a technology that is not fully demonstrated. Because of the cleanup and other development issues, the PFBC technology is generally believed to lag the AFBC technology by 5 to 10 years.

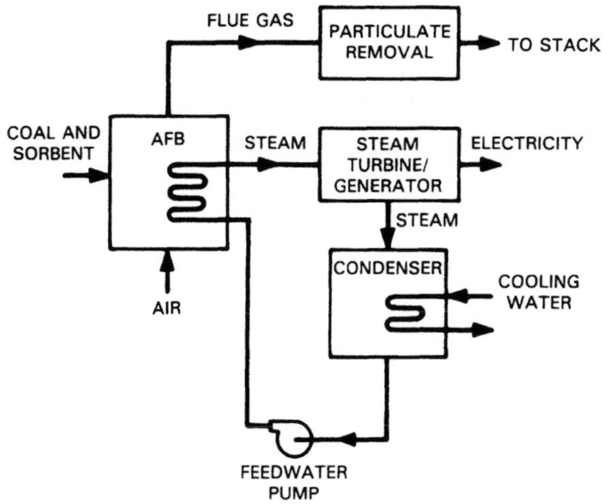

Fig. 21-4. Block diagram of AFBC power generation system.

21.2 ATMOSPHERIC FLUIDIZED BED COMBUSTION

21.2.1 Bubbling Fluidized Bed Combustors

The application of BFB technology to the electric utility industry in the United States had its beginnings in 1976 when an industrial-size FBC steam generator began operation at the Rivesville, West Virginia, plant of the Monongahela Power Company. This BFB type unit was designed by Foster Wheeler Energy Corporation, and generated 300,000 lb/h of steam at 1,350 psig and 925° F while firing bituminous coal in a fluidized bed of limestone.

The Rivesville unit was closely followed by test BFBs and pilot plants. In early 1978, Babcock & Wilcox, under joint sponsorship with the Electric Power Research Institute (EPRI), started up a 6 ft × 6 ft BFB in Alliance, Ohio, that was highly instrumented to provide data for scaling to commercial units. Much of the basic performance data available today for BFBs was generated by this unit. In 1979, a 100,000 lb/h BFB started up at Georgetown University. Designed by Foster Wheeler with funding from the US Department of Energy (DOE), this unit supplied the steam requirements for heating and air conditioning for campus buildings. Combustion Engineering, under a cost sharing program with the DOE, designed and constructed a 50,000 lb/h BFB at the Great Lakes Naval Base in North Chicago, Illinois. Initial operation of this demonstration unit was achieved in 1981.

Two important milestones for BFB technology were achieved in 1982. A 20-MW pilot plant began operation at Tennessee Valley Authority's (TVA) Shawnee Station. Designed by Babcock & Wilcox, the 20-MW TVA pilot plant operated successfully on a variety of coal types. Partially funded by the EPRI, this pilot plant provided much of the data used to scale to larger unit sizes. One noteworthy feature of the pilot plant was the inclusion of two complete feed systems: one for overbed feed and one for underbed feed.

Also, Northern States Power (NSP) completed a project to retrofit FBC to the 15-MW French Island Unit 2 plant. Located in La Crosse, Wisconsin, the French Island BFB was the first retrofit of wood-fueled FBC to an existing utility boiler dedicated entirely to electric power generation. The conversion was designed by Energy Products of Idaho, now JWP Energy Products. Although designed for firing wood, the unit successfully fired other fuels such as peat, refuse-derived fuel (RDF), and shredded tires.

An important milestone for BFB technology occurred in 1986 when NSP started up its Black Dog Station Unit 2, a pulverized coal unit that had been retrofitted with a 135-MW BFB. Unit 2 was originally rated at 100 MW, but was downrated to 85 MW when firing western subbituminous coal. With the retrofit, NSP not only recovered the lost generation, but increased capacity as well. Designed by Foster Wheeler, this unit was the largest operating FBC unit in the world when it began operation.

Continuing a trend of using BFB technology to increase the capacity of existing boilers, Montana Dakota Utilities (MDU) retrofitted a BFB to an existing 70-MW stoker-fired unit in 1987. Fueled by lignite, the original boiler was plagued by slagging problems and was unable to achieve full load. With the Babcock & Wilcox designed retrofit, MDU not only reached the original rated capacity of the unit, but achieved increased output as well.

The world's largest BFB unit in operation is the 160-MW unit at TVA's Shawnee Station. Designed by ABB Combustion Engineering, the BFB boiler was a replacement for the Unit 10 boiler and supplied steam to the existing Unit 10 turbine. The demonstration unit was designed to use high-sulfur Kentucky coal and locally available limestone. Design main steam conditions were 1,100,000 lb/h steam flow at 1,000° F and 1,800 lb/in.^{2}. Reheat steam design temperature was also 1,000° F (Jacobs 1986). The unit began operation in 1988.

The trend to larger unit sizes continues. It has been reported that a BFB retrofit is planned to an existing 350-MW oil-fueled unit in Japan (Makanski 1991).

21.2.1.1 General Configuration. A typical BFB arrangement is illustrated schematically in Fig. 21-5. Fuel and sorbent are introduced either above or below the fluidized bed. (Overbed feed is illustrated.) The bed, consisting of about 97% limestone or inert material and 3% burning fuel, is suspended by hot primary air entering the bottom of the combustion chamber. The bed temperature is controlled by heat transfer tubes immersed in the bed and by varying the quantity of coal in the bed. As the coal particle size decreases, as a result of either combustion or attrition, the particles are elutriated from the bed and carried out of the combustor. A portion of the particles elutriated from the bed are collected by a cyclone (or multiclone) collector downstream of the convection pass and returned to the bed to improve combustion efficiency.

Secondary air can be added above the bed to improve combustion efficiency and to achieve staged combustion, thus lowering NO_x emissions. Most of the early BFBs used tubular air heaters to minimize air leakage that could occur as a result of the relatively high primary air pressures required to suspend the bed. Recent designs have included regenerative type air heaters. The TVA 160-MW BFB shown in Fig. 21-6 shows such an arrangement.

The convection pass can be of the tower design located above the freeboard as illustrated in Fig. 21-6 or of the backpass type found in a conventional pulverized coal unit. Split backpass arrangements are normally used for reheat control.

21.2.1.2 Fuel Feed Systems. Two types of fuel feed systems are used with BFB units: overbed feed and underbed feed. With overbed feed, spreader feeders are used to distribute fuel above the bed. Overbed feed is less complex than underbed feed and minimizes the potential for erosion and plugging in the fuel feed system. However, overbed feed has a number of disadvantages. Partial combustion above the bed reduces the opportunity for sulfur capture, and fines in the feed can be entrained from the bed before combustion is complete. Typically, fuel feed size is restricted to 1¼ in. top size, with a maximum of 20% less than ¼ in. Even with fuel size restrictions and an ash recycle system, combustion efficiencies are lower with overbed feed than with underbed feed, particularly with nonreactive fuels. Also, limitations in spreader throwing distances preclude the use of overbed feed for very large units.

Underbed feed systems typically use pneumatic transport to deliver fuel and limestone to the combustor. The fuel is restricted in top size to ¼ in. to ½ in. to facilitate pneumatic

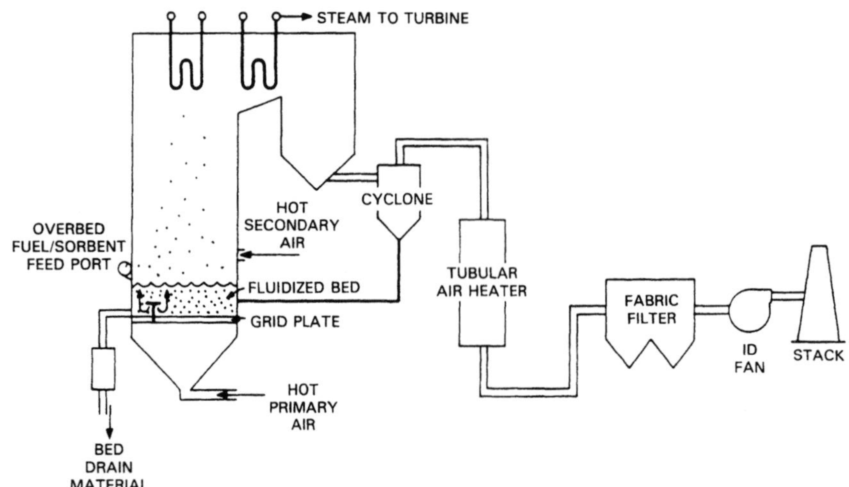

Fig. 21-5. Typical BFB arrangement.

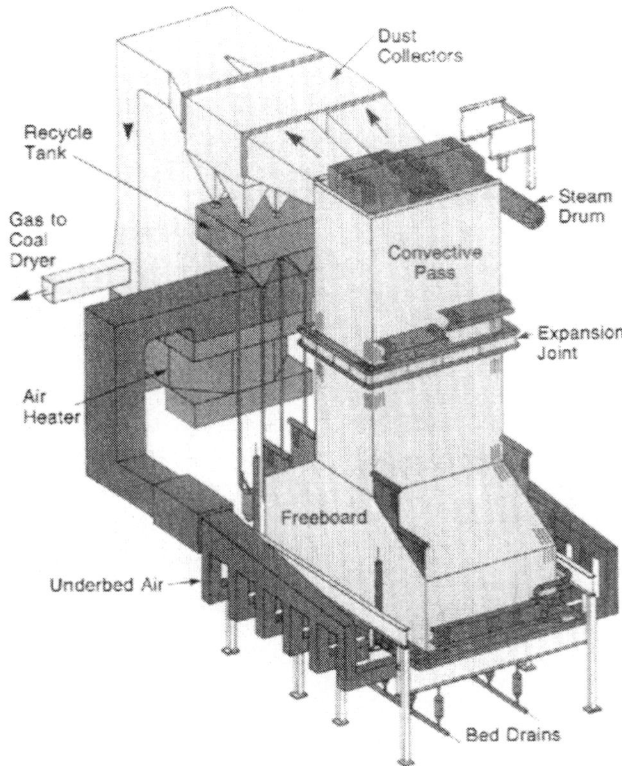

Fig. 21-6. Tennessee Valley Authority 160-MW BFB. (From ABB Combustion Engineering Systems. Used with permission.)

conveying and is introduced through nozzles located just above the air distributor (Singer 1991). With underbed feed, fines can be fired efficiently, and combustion efficiency and sulfur capture are usually greater than with overbed feed. Potential for pluggage and erosion is increased, however, and multiple feed ports are required. Crusher/dryers are usually required to control fuel topsize and moisture content. Fuel moisture content is restricted to about 6% to avoid plugging of pneumatic conveying lines. To achieve good fuel distribution, one feed port is typically required for every 9 to 20 ft^2 of bed area. Thus, while underbed feed provides better performance than overbed feed, it is more complex and has the potential for greater maintenance requirements.

The advantages and disadvantages of overbed versus underbed feed systems are summarized in Table 21-1.

21.2.1.3 Fuel Requirements. A BFB unit can accommodate a variety of fuels, including both low-grade and high-sulfur fuels. The size of feed materials depends on the reactivity of the fuel and the type of feed system used. Larger particle sizes can be used for highly reactive fuels such as lignite and some subbituminous coals. Smaller particle sizes are typically required for less reactive fuels such as anthracite and petroleum coke. As indicated in the previous section, fuel feed topsize is restricted to about 1¼ in. for overbed feed and about ½ in. for underbed feed. High-moisture fuels typically can be accommodated by overbed feed systems, while moisture content is restricted to about 6% with underbed feed systems.

Table 21-1. Advantages and Disadvantages of Overbed and Underbed Feed Systems

Advantages	Disadvantages
Overbed System	
1. Minimizes potential for plugging in the fuel feed system	1. Partial combustion above the bed can reduce sulfur capture
2. Capable of firing larger sized coal	2. Inadequate mixing of fuel and sorbent for sulfur capture
3. Fewer feed ports with conventional spreader feeders	3. Unable to use fines. (Fines tend to be elutriated from the bed, leading to excessive carbon losses and poor sulfur capture.)
Underbed System	
1. Fines can be fired efficiently	1. Large potential for pluggage in feed system
2. More uniform distribution of fuel and sorbent	2. Small fuel feed size; more crushing required
3. More complete combustion within the bed	3. Multiple feed ports needed in each section of the bed

21.2.1.4 Sorbent Requirements. The basic chemical compound used to capture sulfur in the fluidized bed is lime (calcium oxide), which is generally introduced into the system as limestone or dolomite. Once introduced into the bed, limestone is calcined to lime according to the following equation:

$$CaCO_3 \rightarrow CaO + CO_2 \quad (21\text{-}1)$$

The resulting lime reacts with SO_2 and oxygen to form calcium sulfate as follows:

$$CaO + SO_2 + \tfrac{1}{2}O_2 \rightarrow CaSO_4 \quad (21\text{-}2)$$

Dolomite, when introduced into the bed, calcines to form a solid mixture of CaO and MgO. The CaO forms calcium sulfate as does the limestone, but the magnesium oxide is unreactive.

Sorbent feed particle size affects the process as well as the design of the unit and auxiliary equipment. Stone elutriation, required Ca/S ratio, bed particle size, bed bubble phenomena, and bed voidage are among the major items affected or determined by a given sorbent size and fluidization velocity. Overall calcium use generally increases, up to a limit, with a decrease in sorbent feed size as a result of the increased surface area for sulfur absorption. With further decrease in sorbent size, the bed removal rate becomes excessive, thereby resulting in poor sorbent use. Sorbent use also depends on the porosity of the particle surface formed during calcination. Highly porous particles have larger surface areas exposed for the sulfation reaction. If the pores in the particle are too deep, however, they can become plugged by $CaSO_4$, thus stopping the sulfation reaction. The presence of dolomite usually enhances the reactivity of the sorbent. Apparently, porosity is increased by the calcination of $MgCO_3$.

Sorbent use in the sulfation process is difficult to predict. Most manufacturers require test burns with the fuel and sorbent to be used before they will offer guarantees on limestone usage.

Sorbent can be fed into the fluidized bed by mixing the sorbent with the fuel or by using separate feed systems similar to those used for the fuel. Separate feed systems are most often used for fuel and sorbent so that better process control of the Ca/S ratio can be achieved to improve control of SO_2 emissions.

21.2.1.5 Ash Systems. Solids are removed from the fluidized bed by elutriation and through drain ports located in each compartment of the bed. Solids that are elutriated from the bed and freeboard area of the combustor are captured by a mechanical dust collector (typically a multiclone collector). To improve combustion efficiency and maintain bed inventories, most of the solids are usually recycled to the combustor. Recycled material can be fed underbed or overbed. The recycle ratio, defined as the ratio of the recycle mass flow to fuel feed mass flow, usually ranges from 1:1 to 3:1 when firing bituminous coals (Singer 1991). Solids passing through the mechanical dust collector are collected by a conventional particulate removal device, such as an electrostatic precipitator or fabric filter.

A bed solids removal system is required to maintain an adequate bed level and inventory of bed material in the fluidized bed. The system also removes tramp material from the bed. It consists of bed drain lines with drain valves discharging bed material to a water-cooled ash cooler. To reduce carbon losses in the ash removed from the bed, ash classifiers are sometimes used. Fine material separated in the classifiers is returned to the bed, while the coarse materials are discharged to the ash removal system. In determining the required equipment capacities, consideration must be given to the maximum expected drain rates and maximum material discharge temperature.

The ash and other spent solids from a BFB are in a dry form that can be conveyed pneumatically and stored in silos. Disposal will generally be by truck transport to a landfill.

21.2.1.6 Heat Transfer. Although the overall steam cycle defines the heat that must be absorbed by the steam generating, superheating, and reheating surfaces, BFB manufacturers have some flexibility in the manner in which the heat transfer surfaces are arranged. Some manufacturers do not place the heat transfer surface in the bed, while others place horizontal or sloped tubes in the bed area. In addition, the inbed tubes may be either water or steam cooled. The design for the 160-MW TVA demonstration unit includes a waterwall combustor; superheater and steam generation tubes in the bed; and reheat, superheat, and economizer tubes in the convection section (Smith 1982).

The interrelationship of the cooling media, allowable pressure drops, heat transfer coefficients, bed temperature, fluidizing air velocity, surface requirements, and tube bundle geometry need to be taken into account in the design of the inbed surface arrangement. Steam flow and temperature imbalances should be minimized. In addition, the inbed tube bundle arrangement must effectively use the available bed volume dictated by the bed plan area, bed depth, and clearance requirements.

Heat transfer to surfaces immersed in a fluidized bed occurs by means of gas convection, radiation, and particle convection (Singer 1991). The gas convection contribution to the overall heat transfer is small. The radiation contribution is larger, but the overall heat transfer is dominated by particle convection. Relative magnitude heat transfer coefficients provided by Stevens and Begina of ABB Combustion Engineering (1991) for the three types of heat transfer to surfaces immersed in a fluidized bed are summarized as follows:

	Heat transfer coefficient (Btu/h-ft²-° F)
Gas convection	0–2
Radiation	8–12
Particle convection	30–50

ABB Combustion Engineering has found that the particle convective heat transfer coefficient correlates well with the Archimedes number and the Prandtl number as indicated by the following equation:

$$\mathrm{Nu} = \frac{h_c d_p}{k_g} = 0.31(\mathrm{Ar})^{.27}(\mathrm{Pr}) \qquad (21\text{-}3)$$

where

d_p = harmonic mean particle diameter,

h_c = total outside convective heat-transfer coefficient,

k_g = gas thermal conductivity,

Nu = Nusselt number,

Pr = Prandtl number, and

Ar = Archimedes number.

The Archimedes number is given by

$$\mathrm{Ar} = \frac{(d_p)^3 g \rho_g(\rho_p - \rho_g)}{\mu_g^{\,2}} \qquad (21\text{-}4)$$

where

d_p = mean particle diameter,

g = gravitational constant,

ρ_g = gas density,

ρ_p = particle density, and

μ_g = gas viscosity.

The Prandtl number is given by

$$\mathrm{Pr} = \frac{Cp_g \mu_g}{k_g} \tag{21-5}$$

where

Cp_g = gas heat capacity,

μ_g = gas viscosity, and

k_g = gas thermal conductivity.

The heat transfer surface in the convection pass is designed using methods and techniques proven through years of experience in conventional coal fueled steam generator applications. Allowance must be made for the higher convection pass dust loading in the BFB boilers, especially with the increased recycle needed to improve combustion efficiency and sulfur capture.

21.2.1.7 Air Heaters. Testing and operation at first-generation BFB facilities revealed that regenerative-type air heaters were not acceptable for fluidized bed service because of excessive air leakage. Higher air-side pressures required by the fluidized bed process dictated the need for tubular air heaters to minimize air in-leakage. With the development of advanced leakage control systems for regenerative air heaters, these types of air heaters have been selected for two large BFB applications. A low-leakage Ljungstrom regenerative-type air heater was used for the 135-MW Black Dog BFB (Wong 1987). In addition, the 160-MW TVA demonstration unit uses a single Ljungstrom regenerative air heater equipped with a special seal system to reduce leakage (Jacobs 1986).

21.2.1.8 Fans and Blowers. The combustion air systems for BFB boilers and conventional boilers are similar. Conventional design practices are suitable for the FBC air system, including ductwork, fans, and dampers. However, the higher air-side pressures associated with BFBs must be accounted for in the design.

Positive displacement or rotary blowers supply air at higher pressures for pneumatic conveying. Pneumatic conveying is used to convey fuel (underbed feed), limestone, and solid waste products.

21.2.1.9 Startup Burners. Startup burners are used to raise the temperature of the bed to the auto-ignition temperature of the fuel being fired. For coal-fueled BFBs, the auto-ignition temperature is approximately 1,200° F. Typically, the startup burners are placed in an air duct leading to the fluidized bed. Bed lances are sometimes used to supplement the heat input of the startup burners. Usually, startup is accomplished by heating one of several compartments in the bed. When ignition is accomplished in the first compartment, other compartments are gradually brought to temperature and placed into operation.

21.2.1.10 Operating Parameters. Typical BFB operating parameters are shown in Table 21.2. Bed temperature is normally selected to optimize sulfur removal while achieving acceptable carbon burnout. The optimum bed temperature for sulfur removal is approximately 1,550° F, as indicated in Fig. 21-7 (Kantesaria and Jukkola 1983). If sulfur removal is not a consideration or low-volatile fuels are being fired, higher bed temperatures can be used to increase combustion efficiency. Fluidizing air is required to provide the oxygen for combustion air as well as to lift and expand the bed. The air velocity dictates, to some extent, the height of the overall bed and affects the entrainment of smaller particles in the air leaving the bed. Typically, the fluidizing velocity ranges from 3 to 10 fps, with 8 fps being most common.

Bed depth and freeboard height (height from the top of the bed to the top of the combustor) are selected to match the fuel and type of feed system. Deeper beds are required when low-volatile fuels such as anthracite and petroleum coke are fired, while shallow beds can be used for more reactive fuels such as lignite and subbitimous coal. For a reactive fuel, a bed height of 2 ft in the slumped position is typical, which normally results in an expanded bed height of about 4 ft. Larger freeboard heights are usually selected for low-volatile fuels or when overbed feed is used. The increased freeboard provides additional residence time to complete the combustion process.

As discussed previously, fuel feed particle size is a function of the type of fuel feed system. In addition, larger fuel particle sizes can be used for reactive fuels, while smaller sizes are necessary when firing less reactive fuels. Care must be taken to minimize the amount of fines in the feed to a BFB, particularly if overbed feed is used. Typically, the amount of fines in the feed is restricted to <25% minus 16-mesh with overbed feed and <20% minus 30-mesh with underbed feed (Singer 1991).

Sorbent size is selected to maximize sorbent use and prevent excessive elutriation of unreacted sorbent from the combustor.

Ca/S ratios are selected to achieve the desired removal of

Table 21-2. Typical BFB Operating Parameters

Bed temperature, ° F	1,550–1,650
Superficial velocity, ft/s	3–10
Bed depth, ft	2–6
Freeboard height, ft	8–20
Coal feed particle size, in.	
Overbed feed	1¼ × 0
Underbed feed	½ × 0
Sorbent feed particle size, in.	⅛
Calcium/sulfur ratio	2.5–4.0
SO_2 removal, %	90
Combustion efficiency, %	90–98
Recycle ratio	0–5
Excess air, %	20–35
NO_x emissions, ppm	150–350

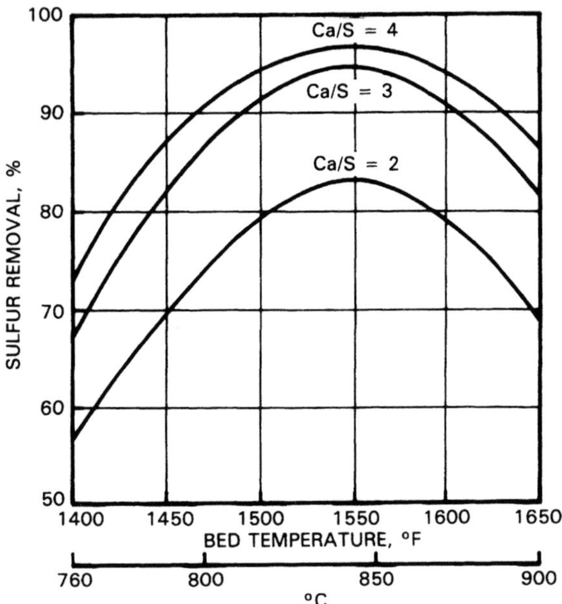

Fig. 21-7. Sulfur removal vs. temperature at various Ca/S mole ratios. (From ABB Combustion Engineering Systems. Used with permission.)

SO_2. The Ca/S ratio required to achieve a given SO_2 removal depends on the type of fuel feed system used, the reactivity of the sorbent, and the amount of solids recycled to the combustor (recycle ratio). Testing conducted at the TVA 20-MW pilot plant indicated that, with underbed feed, >95% sulfur removal could be achieved at Ca/S ratios ranging from about 2.6 to 3.1 if a recycle ratio of 1.5 was maintained. With overbed feed, sulfur retention was also sensitive to the amount of fines in the feed, but that sulfur retention values of 90% or greater could be achieved at Ca/S ratios ranging from 2.7 to 3.7 and recycle ratios ranging from 0 to 1.5, if the fuel feed contained <7% minus 30-mesh fines. The sulfur retention test results are shown in Fig. 21-8.

Combustion efficiency in a BFB usually ranges from 90% to 98% and is a function of bed temperature, type of fuel feed, fuel reactivity, bed depth, and recycle ratio. Results of underfeed tests at TVA's 20-MW pilot plant indicated combustion efficiencies ranging from about 90% to 97% for recycle ratios ranging from 0 to 2.0, respectively (Fig. 21-9). Testing at ABB Combustion Engineering's Great Lakes FBC Demonstration Plant indicated that combustion efficiency did not improve significantly when excess air levels were increased above the 10% to 20% level (Kantesaria and Jukkola 1983).

The mechanism of NO_x formation in a BFB is complicated and depends on the coal devolitization rate and its volatile content, excess air level, bed temperature, and other factors (Modrak, Tany, and Aulisio 1982). Generally, NO_x emissions in the range 150 to 350 ppm and lower can be achieved. According to Modrak, Tany, and Aulisio, however, NO_x emissions tend to increase as Ca/S levels increase. Thus, reducing SO_2 emissions may result in an increase in NO_x emissions.

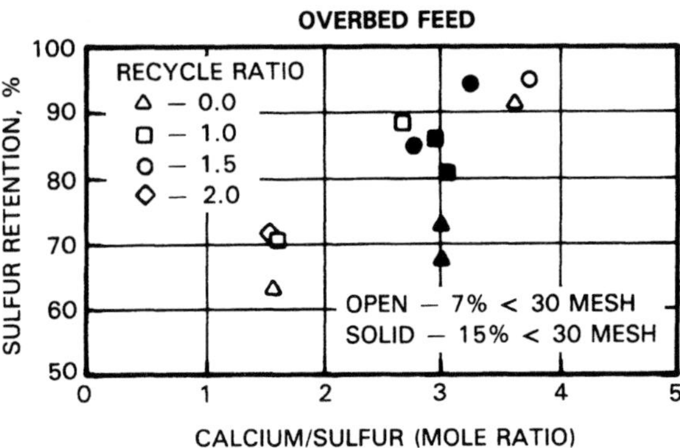

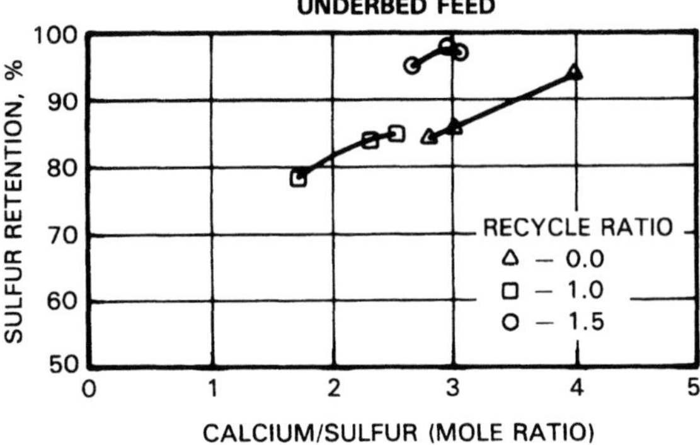

Fig. 21-8. Sulfur retention test results at TVA 20-MW pilot plant. (From Babcock & Wilcox. Used with permission.)

21.2.1.11 Control Philosophy. Two methods are commonly used to control load or steam flow with a BFB. The first involves velocity turndown whereby fuel and airflow are uniformly varied across the entire bed to control steam flow while maintaining bed temperature. The combustor vessel can be designed such that as fuel and airflow are reduced and the bed slumps, the in-bed heat transfer surface is uncovered, thus lowering the rate of heat transfer to steaming surface in the bed.

The second method of load control involves slumping the bed in separate zones of the combustor. To reduce load, fluidizing air to individual zones or compartments is reduced or isolated, thus slumping the beds in the affected compartments. This allows maintaining bed temperatures in the active compartments at the reduced loads. The boiler bed area of the 160-MW TVA BFB is divided into 12 compartments (Kopetz, O'Brien, and Tennant 1991). The division is accomplished by segmenting the air plenum below the distributor plate. Each compartment has its own secondary air damper, coal and limestone feed, and bed removal drain.

In practice, both methods of load control are usually used in conjunction with each other. Overall load turndown to 25% load is usually achievable when both methods of load control are used. Bed temperature control at a given fuel flow is a function of air flow and heat transfer in the bed area. It is desirable to maintain bed temperatures at near optimum levels over the entire load range to maximize sulfur capture. Although this goal usually is not achievable, a properly

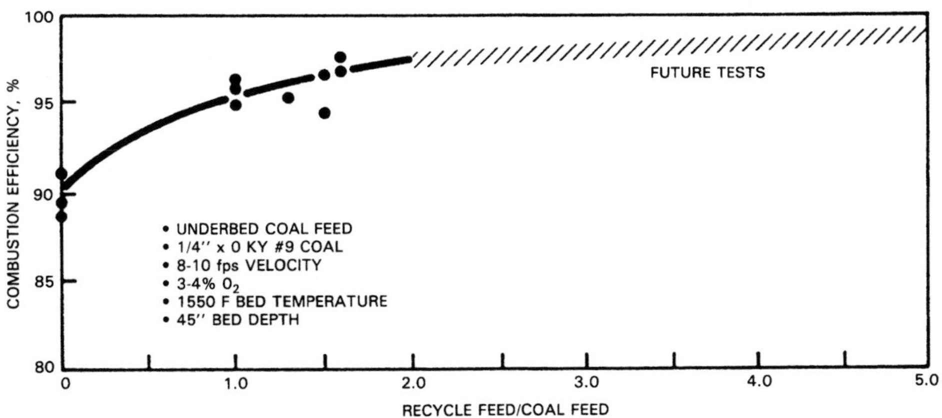

Fig. 21-9. Combustion efficiency at TVA's 20-MW pilot plant. (From Babcock & Wilcox. Used with permission.)

designed BFB combustor will allow effective bed temperature control over a wide load range. The designer can use bed zoning, bed slumping, bed inventory, and effective placement of the heat transfer surface to control bed temperatures. Once the BFB is constructed and placed into operation, airflow is the primary parameter that is varied to control bed temperature. Also, bed inventory can be varied somewhat to provide a secondary level of bed temperature control. Bed inventory is controlled by draining solids from the combustor to maintain a constant pressure differential across the bed.

Steam temperature control in a BFB is similar to that used for a conventional pulverized coal unit. Biasing control dampers in the convection pass are used for primary superheat/reheat steam temperature control. Secondary steam temperature control is achieved with desuperheater sprays.

21.2.1.12 BFB Applications. The primary use of BFB units in the electric utility industry has been in retrofit applications. Above steam flows of about 200,000 lb/h, CFB units are generally considered to be more economical than BFBs in new applications. Utility-sized BFBs require large plan areas to maintain acceptable fluidizing velocities and complicated underbed feed systems to maximize sorbent use and carbon burnout. Thus, most electric utility interest in FBC has been focused on CFB units.

BFBs can usually be retrofitted to an existing unit at a lower cost than can a CFB unit. Typically, more of the existing boiler can be preserved with a BFB, giving this technology a decided capital cost advantage. The lower capital cost is frequently sufficient to offset the greater operating efficiency and lower operating cost of CFB units.

21.2.1.13 Manufacturer's Offerings. ABB Combustion Engineering, Foster Wheeler, and Babcock & Wilcox all offer BFB units for large electric utility applications. The TVA 160-MW BFB, designed by ABB Combustion Engineering, is shown in Fig. 21-6. A schematic representation of Foster Wheeler's BFB unit for nonreheat retrofit applications is shown in Fig. 21-10. A cutaway view of TVA's 20-MW BFB

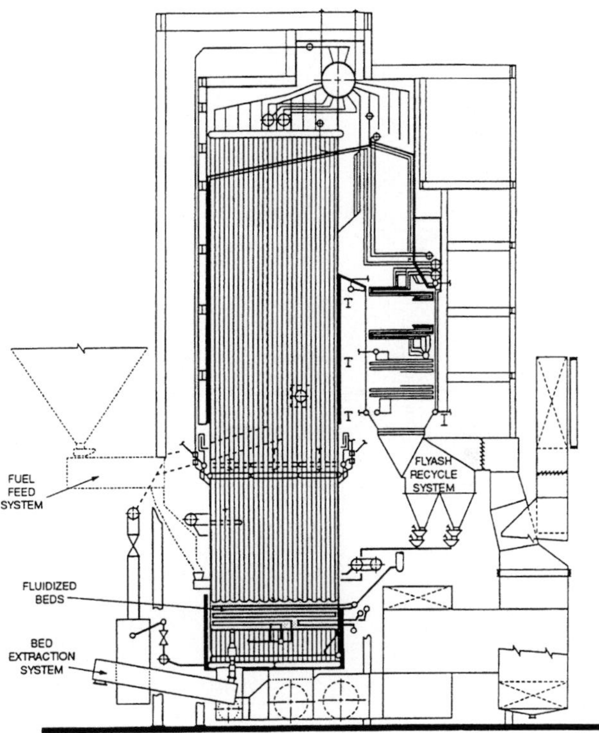

Fig. 21-10. BFB unit for nonreheat retrofit applications. (From Foster Wheeler. Used with permission.)

pilot plant, designed by Babcock & Wilcox, is shown in Fig. 21-11.

A number of manufacturers offer BFB units in smaller sizes and for industrial applications. A special report on fluidized bed boilers listed over 30 manufacturers of BFB units (Schwieger 1985).

21.2.2 Circulating Fluidized Bed Combustors

Development of CFB combustors began in about 1955 in Germany when Lurgi designed roasters for sulfide ores. In 1965, Lurgi developed the CFB technology for calcining aluminum hydroxide. The CFB technology was first applied

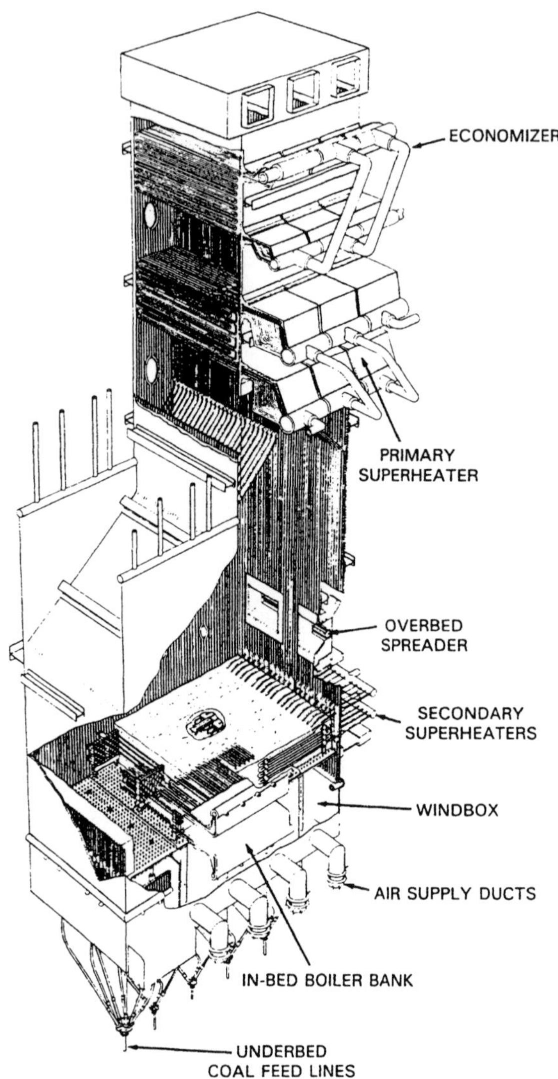

Fig. 21-11. 20-MW fluidized bed boiler. (From Babcock & Wilcox. Used with permission.)

to the combustion of coal by Lurgi in 1982 when a cogeneration unit was started up in Luenen, West Germany. This unit supplied molten salt as well as steam for the generation of electricity. The boiler had a thermal capacity of 84 MW, equivalent to a steam flow of about 250,000 lb/h, and delivered steam at 945 psig and 900° F.

Ahlstrom, a manufacturer headquartered in Finland, first applied the CFB technology in a board mill in Pihlava, Finland, in 1979. Retrofitted to an existing unit, this CFB was fueled with peat and woodwaste and generated 45,000 lb/h steam at 1,230 psig and 970° F. Ahlstrom's first coal-fueled CFB unit was installed at a paper mill in Kauttua, Finland. Started up in 1981, this cogeneration unit also fired peat and generated 20,000 lb/h steam at 1,235 psig and 930° F.

Lurgi and Ahlstrom dominated most of the early CFB experience in Europe, although other manufacturers, such as Studsvik in Germany, also offered CFB units. The first application of the CFB technology in the United States occurred at Uvalde, Texas, in 1981 when Struthers Wells Corporation started up a CFB that generated 50,000 lb/h steam at 2,450 psig and 665° F for enhanced oil recovery. This unit was closely followed by a CFB unit at Bakersfield, California, for Gulf Oil Company. Also applied for enhanced oil recovery, this unit was designed by Pyropower, a subsidiary company of Ahlstrom, and generated 50,000 lb/h of saturated steam at 2,500 psig.

The first application of reheat to a CFB unit occurred in 1985 at Duisberg, West Germany, where Lurgi supplied a 96-MW coal-fueled unit. A cogeneration unit, the CFB generated 595,000 lb/h steam at 2,100 psig and 995/995° F.

The first large application of CFB in the United States occurred at the Scott Paper Company in Chester, Pennsylvania. Under a technology agreement with Lurgi, ABB Combustion Engineering (CE) supplied a CFB that generated 650,000 lb/h steam at 1,450 psig and 950° F. The unit was designed to fire anthracite culm as well as alternate fuels. When this unit was placed in operation in 1986, it was the largest CFB operating in the world. The Scott Paper Company CFB was the forerunner for the AES Thames CFB installation, a coal-fueled cogeneration facility in Montville, Connecticut. The Thames installation included two 673,000 lb/h CFB units that supplied steam to a single-reheat turbine.

The first application of CFB to the electric utility industry in the United States occurred at Colorado-Ute Electric Association's Nucla Station, where three existing 12-MW turbines and one new 75-MW turbine were powered by a 925,000 lb/h Pyropower CFB. Started up in 1987, this CFB featured a split furnace design and produced steam at 1,510 psig and 1,005° F.

In 1990, a 150-MW CFB designed by CE/Lurgi was placed into operation for Texas-New Mexico Power company (TNP). Fired by Texas lignite, this unit generated 1,000,000 lb/h of steam at 1,800 psig and 1,005/1,005° F. When it was placed into operation, this unit was the largest operating CFB unit in the world.

Larger CFB units are being planned for future applications. CFB units as large as 250 MW have been proposed in two separate projects under the DOE's Clean Coal Technology Program (Makanski 1991). One of these projects will include a 250-MW reheat CFB manufactured by Foster Wheeler Energy Corp. The unit will produce a steam flow of 1,790,000 lb/h at 1,940 psig and 1,005/1,005° F.

21.2.2.1 General Configuration. A typical CFB arrangement is illustrated schematically in Fig. 21-12. In a CFB, primary air is introduced into the lower portion of the combustor, where the heavy bed material is fluidized and retained. The upper portion of the combustor contains the less dense material that is entrained from the bed. Secondary air typically is introduced at higher levels in the combustor to ensure complete combustion and to reduce NO_x emissions.

The combustion gas generated in the combustor flows upward with a considerable portion of the solids inventory entrained. These entrained solids are separated from the

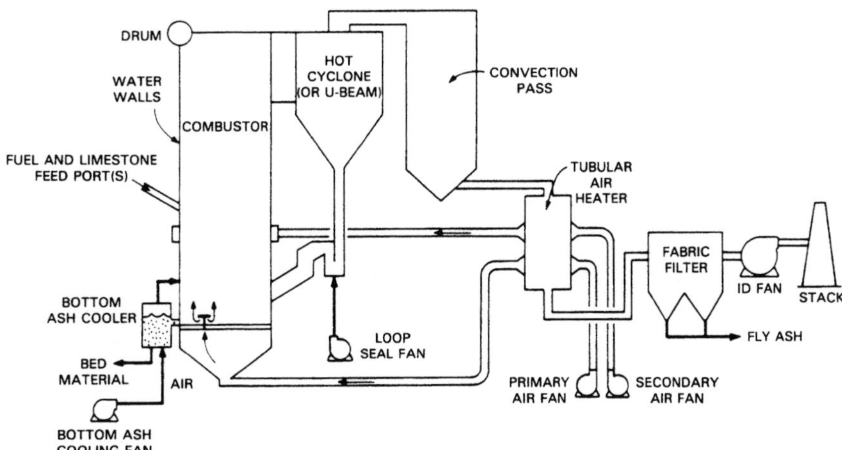

Fig. 21-12. Atmospheric circulating bed combustor.

combustion gas in hot cyclone-type dust collectors or in mechanical particle separators, and are continuously returned to the combustion chamber by a recycle loop.

The combustion chamber of a CFB unit for utility applications generally consists of membrane-type welded waterwalls to provide most of the evaporative boiler surface. The lower third of the combustor is refractory lined to protect the waterwalls from erosion in the high-velocity dense bed region. Several CFB designs offer external heat exchangers, which are unfired dense BFB units that extract heat from the solids collected by the dust collectors before it is returned to the combustor. The external heat exchangers are used to provide additional evaporative heat transfer surface as well as superheat and reheat surface, depending on the manufacturer's design.

The flue gas, after removal of more than 99% of the entrained solids in the cyclone or particle separator, exits the cyclone or separator to a convection pass. The convection pass designs are similar to those used with conventional coal-fueled units, and contain economizer, superheat, and reheat surface as required by the application.

21.2.2.2 Fuel Feed Systems. Fuel feed in a CFB is usually controlled with gravimetric feeders with rotary lock valves used to isolate the feeders from the combustor. Some CFB manufacturers have designed for negative pressure in the combustor at the fuel feed points and have eliminated the rotary lock valves. Fuel typically flows by gravity into the combustor, with some designs using an air assist in the fuel feed chute. Pneumatic fuel feed systems can also be used, including air swept crushers, and dense phase conveying systems are also available.

The number and location of feed points varies with the fuel being fired and with the manufacturer's design. Because of the thorough mixing that occurs in a CFB, typically a relatively few feed points are required in comparison to a BFB with underbed feed. Fuels with low bulk densities require more fuel feed points to physically accommodate the high volume flows. Feed points are usually in the front wall of the combustor or in the seal lines from the cyclone separator to the combustor, or both. High-moisture fuels are usually introduced in the seal lines from the cyclone, where they mix with hot recirculated solids to accomplish some drying before being introduced into the combustor.

The 150-MW Texas New Mexico Power CFB uses only four feed points to introduce lignite with more than 30% moisture to the CFB combustor (Stevens and Begina 1991). From gravimetric feeders, the lignite is conveyed by volumetric feeders to the siphon seal return lines. Reportedly, the TNP unit has achieved full capacity with three of four feed systems in operation.

The 110-MW CFB at the Nucla Station also uses four feed points (Haecock and Davis 1986). Coal is introduced into the front walls of the combustors through gravimetric feeders and rotary isolation valves.

Liquid fuels are usually dispersed in the combustor with bed lances, either stationary or retractable. Because of the short bed residence time with liquid fuels, typically more feed points are required with liquid fuels in comparison to solid fuels.

21.2.2.3 Fuel Requirements. In comparison to a BFB with underbed feed, fuel preparation requirements for a CFB are typically less severe. A wide variety of fuels can be accommodated, and fines in the fuel can be fired with no difficulty. While they can be fired satisfactorily in a CFB, high-moisture fuels and fuels with a large amount of fines can present severe materials handling difficulties. To prevent plugging, special design measures may be required, including the strategic placement of air blasters and the use of dryers.

To achieve acceptable combustion efficiencies, low-volatile fuels, such as anthracite, must be crushed to smaller topsize than higher volatile fuels. It is not unusual to see sizing specifications of 1/16 in. or less for low volatile fuels (Makanski 1991). Larger fuel sizes are acceptable for higher volatile fuels.

Lignite for the 150-MW TNP CFB is crushed to 3/8 in.

topsize prior to introduction into the combustor (Stevens and Begina 1991). Coal for the 100-MW Nucla unit is crushed to ¼ in. topsize (Heacock and Davis 1986).

21.2.2.4 Sorbent Requirements. Sorbent requirements for a CFB are similar to those for a BFB. Most manufacturers require test burns with the fuel and sorbent to be used before they will offer guarantees on limestone usage. Typically, limestone is crushed to a top size of 1,000 microns with an average size of 150 microns. Although the sorbent can be introduced into the combustor by mixing with the fuel, the more usual process is to introduce the sorbent into the combustor by separate feed systems. Most systems use day bins for temporary storage of sorbent, followed by rotary airlock feeders that drop the sorbent into pneumatic conveying lines for transport to the combustor. Gravimetric feeders can be used if more accurate control of sorbent feed is desired.

The 150-MW TNP CFB unit was designed to use a dense phase pneumatic system for introduction of limestone to the combustor. This system required extensive maintenance to avoid pluggage, and TNP planned to replace it with a dilute phase pneumatic system using gravimetric feeders to control limestone feed from the silos. Like the fuel, limestone is fed to each of the four syphon seal return lines.

The sorbent feed system at Nucla uses loss-in-weight type limestone feeders, in which the rate of feed is integrated over time. From the feeders, limestone is admitted to the pneumatic conveying system through rotary valves. Four limestone injection points are used for each of the two combustors. Two feed points are on the front wall of the combustor, one on the side wall, and one on the cyclone discharge of each combustor.

21.2.2.5 Ash Systems. Typically, ash is removed in three locations in a CFB installation: the combustor, the backpass or air heater hoppers, and the fly ash particulate removal device. The primary function of the combustor ash removal system is to control bed inventory. Bed material is removed as required to maintain the desired bed pressure drop. A secondary function of the combustor ash system is to remove oversized or tramp material that, if left to accumulate, could restrict airflow and defluidize a portion of the bed.

Two primary types of combustor ash removal systems are available for application to CFBs. The first type consists of water-cooled screw conveyors that cool the hot ash leaving the combustor to a temperature that can be accommodated by the ash handling system, similar to that used for BFBs. Usually, one or two bed drains are sufficient for a medium-to-large CFB.

The second type of combustor ash removal system consists of small fluidized bed heat exchangers, called strippers or classifiers, mounted on the sides of the combustor near the combustor floor. Air nozzles are located on the combustor floor to direct solids material to the inlet of the ash classifiers. Air flow is directed through the classifiers at sufficient velocities to fluidize the solids material. Fine materials are returned to the bed, while coarse materials are discharged from the classifiers for disposal. Ash strippers or classifiers typically include cooling coils immersed in the bed to cool the ash and recover heat for the feed heating cycle. An ash stripper design proposed by Foster Wheeler is shown in Fig. 21-13 (Goidich, McGee, and Richardson 1991).

Ash collected in the economizer, air heater hoppers, and the particulate removal device, such as a fabric filter or electrostatic precipitator, is removed with conventional dry ash removal systems.

Ash handling system capacities should be conservatively sized. The expected ash split between bed ash and fly ash is difficult to predict and the ash content of the fuel can vary significantly. A conservative design criterion for ash handling systems conveying capacity would be 100% of total solids capacity for the bed ash system and 100% of total solids capacity for the fly ash system.

21.2.2.6 Ash Recycle System. The combustion gas generated in the combustor of a CFB flows upward with a considerable portion of the solids inventory entrained. These entrained solids are separated from the combustion gas and are continuously returned to the combustion chamber by a recycle loop. Most CFBs use one or more hot cyclones to separate solids from the combustion gas. The cyclones are designed to remove all solid particles >100 microns in diameter, which normally results in an overall separation efficiency >99% (Singer 1991). The cyclones are constructed with steel plate lined with one or more layers of refractory. The hot face of the refractory is a dense erosion-resistant

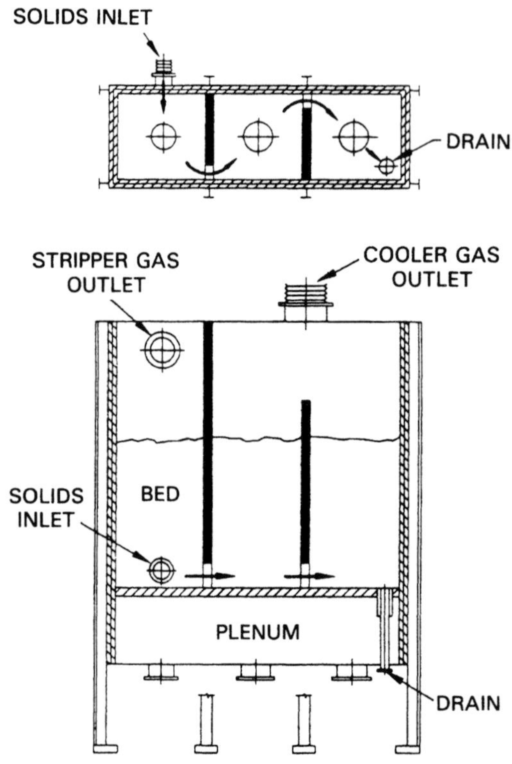

Fig. 21-13. Ash stripper/cooler. (From Foster Wheeler. Used with permission.)

material, while the inner layers are less dense insulating materials.

Cyclones may be uncooled, water cooled, or steam cooled. Uncooled cyclones use several layers of refractory with a total thickness in the range of 12 to 16 in. The outside surface temperature of uncooled cyclones is typically on the order of 250° F. With water or steam cooling, refractory can typically be reduced to a single layer 2 in. thick. Outside surface temperatures of cooled cyclones are nominally 130° F. The cooled cyclones have the advantages of less radiant heat loss, lighter weight, and more rapid startup times. Units with thick cyclone refractory typically require 12 to 16 hours for startup to avoid thermal stresses and spalling of the refractory. With cooled cyclones, startup times usually are not limited by refractory considerations.

Another type of particle separator used on CFBs, a U-beam collector, consists of a series of vertical channels that trap the particles in the flue gas and drop them into a particle storage hopper.

Solids collected by cyclone separators or U-beam collectors are returned to the combustor with a solids reinjection device often called a seal pot, L-valve, or J-valve. The solids reinjection device is essentially a nonmechanical loop seal that allows movement of solids against the combustor backpressure. The bottom of the seal is fluidized so that the solids in the seal can seek different levels on each side of the seal corresponding to the differential pressure. Flow of solids through the seal may or may not be controlled by a valve. On units with a fluidized bed heat exchanger (FBHE), a plug valve is typically used to regulate the flow of solids to the FBHE.

21.2.2.7 Fluidized Bed Heat Exchangers. Several circulating bed designs offer separate FBHEs, which are unfired dense BFBs that extract heat from the solids collected by the cyclone or U-beam dust collectors before it is returned to the combustor. The FBHEs are used as evaporative heat transfer surfaces as well as superheat and reheat surfaces, depending on the manufacturer's design. An FBHE can be designed for total recirculation of solids from the cyclones or U-beam separators, or for partial flow with the remainder of solids returned directly to the combustor. FBHEs use low fluidizing velocities (1 to 4 ft/s), and virtually no combustion occurs in the heat exchanger because of the low carbon content of the solids being circulated. Erosion/corrosion in the heat exchanger is not a significant concern because of the low fluidizing velocities and low levels of combustion.

The initial FBHEs were separate stand-alone vessels and were commonly called external heat exchangers. In more recent designs, the FBHEs have been incorporated into the design of the combustor, usually sharing a wall with the combustor. In the integrated FBHE design, the walls of the FBHE are formed by water-cooled membrane panels.

21.2.2.8 Heat Transfer. The flue gas, after removal of more than 99% of the entrained solids in the cyclone or U-beam particle separator, enters the convection pass. The convection pass designs are similar to those used with conventional coal-fueled units, and contain economizer, superheat, and reheat surface as required by the application. Depending on the manufacturer's design, the superheat surface is typically located in convection pass and combustion chamber, or the convection pass and external heat exchanger.

The location of the superheat surface in a combustion chamber is an area of concern because of the potential for erosion. Designs that offer a combustion chamber superheat surface typically use pendant sections, or wing walls that are located along the combustor walls, or tubes that traverse the combustor and that are protected from erosion.

The reheat surface is typically located in an external heat exchanger, in the convection pass and combustion chamber, or in the convection pass only.

The designs that use FBHEs for the reheater surface have several advantages over the other options. The FBHE can be isolated during boiler startup, thus protecting tubes from high temperatures. The FBHE also offers high heat transfer rates, which minimizes surface area requirements, and as previously noted, the erosion of tube surface in an FBHE is not a significant concern.

Designs that split the reheater surface between the combustion chamber and convection pass must incorporate some means to protect the combustion chamber reheater surface from high temperatures during startup. The placement of the reheater surface in the combustion chamber also has the disadvantage of possible erosion.

Designs that locate all of the reheat surface in the convection pass (similar to a conventional boiler) solve the startup tube protection problem, but the low flue gas temperatures mean a large surface area may be required.

21.2.2.9 Refractory. Refractory linings are typically used in CFBs to protect metal surfaces from erosion and/or to protect against corrosion in areas of a reducing atmosphere. The most common areas where refractory is used include the lower portion of the combustor, cyclone separators, seal pots, FBHEs, ash coolers, and solids return lines. Multiple layers of refractory are typically used for uncooled surfaces such as cyclone separators, in which the total thickness can range from 12 to 16 in. Dense erosion-resistant materials are used on the hot face, while lighter weight insulating materials are used on the inner face.

The refractory in a CFB can greatly increase the weight of the unit and increase startup times. As indicated previously, units with thick refractory layers can require 12 to 16 hours for startup to avoid thermal stresses and spalling of the refractory. In addition, refractory layers are susceptible to failure and can be a significant maintenance concern during operation. The three basic modes of refractory failure are thermal shock, erosion/abrasion, and corrosion. Once cracks appear in the refractory, foreign material quickly fills the void and accelerates the failure mechanism.

The various CFB manufacturers are striving to improve refractory performance. Water- or steam-cooled surfaces significantly reduce the required thickness of refractory and permit shop application of refractory where application can be more strictly controlled. New types of refractory are being developed for application to CFBs and greater quality control is being used in applying the refractory.

21.2.2.10 Air Heaters. Due to the high air pressures required in CFBs and the excessive leakage that could consequently occur with conventional regenerative air heaters, low-leakage-type air heaters, such as tubular air heaters and heat pipe air heaters, are used in CFB applications. Generally, separate primary and secondary air heaters are used, and an air heater also can be used for the FBHE fluidizing air.

With tubular air heaters, gas-over-tube/air-through-tube designs are usually used to facilitate cleaning with sootblowers. Tubular air heaters were used on almost all of the early CFB units. The tubular air heater sizes required with larger units and units that require gas outlet temperatures below 300° F present a problem because of excessive space requirements and pressure drops.

One alternative to tubular air heaters has been the recent application of heat pipe air heaters, which use a medium inside a sealed pipe to transfer heat from the hot flue gas to the cold combustion air. The heat transfer medium used inside the heat pipes is usually toluene or naphthalene. The heat pipe heat exchangers have no external power requirements and use soot blowers to keep the finned tube surfaces clean. The heat pipe air heaters require less space and pressure drop than tubular air heaters.

Cold-end protection by means of steam preheating, hot water preheating, or air bypass is usually required for air heaters in CFB applications (Singer 1991). There is the potential for reducing or eliminating cold-end protection, since SO_2/SO_3 levels at the air heater are low (from sorbent addition to the combustor). More research is required with various fuels, however, before taking this approach.

21.2.2.11 Fans and Blowers. The combustion air system for a CFB is similar to, and provides the same functions as, the air systems on a conventional boiler. Generally, separate centrifugal fans, arranged in parallel, are used for primary and secondary air. A series arrangement can also be used, with the second fan supplying the high-pressure primary air. Fluidizing air for ash coolers, FBHE, and seal pot is usually supplied by positive displacement blowers or centrifugal fans.

21.2.2.12 Startup Burners. The startup burner system for a CFB is similar to that for a BFB. In-duct burners are located in the primary air system and overbed burners are located in the walls of the combustion chamber. These burners are used to heat up the system and to raise the temperature of the bed to the auto-ignition temperature of the fuel. Bed lances are used in the lower combustor walls to provide additional heat input as required to maintain bed temperatures during abnormal operating conditions. The startup burner system on the Texas New Mexico Power 150-MW CFB includes two in-duct burners, each rated 55 MBtu/h; four retractable wall burners, each rated 80 MBtu/h; and eight bed lances, each rated 160 MBtu/h (Stevens and Begina 1991).

21.2.2.13 Ammonia Injection Systems. Although NO_x emissions are inherently low in a CFB, additional NO_x control measures are necessary to meet very stringent emission restrictions such as those encountered in California. Generally this has meant the application of selective noncatalytic reduction (SNR) using ammonia injection. The method and location of ammonia injection is thought to be critical in the overall NO_x reduction achieved. Injection points used by manufacturers have included the upper combustor area, cyclone separators, separators, and the overfire air duct. At least one manufacturer claims that its proprietary injection mechanism leads to almost undetectable NO_x emissions at low ammonia consumption rates.

21.2.2.14 Operating Parameters. Typical operating parameters for a CFB are shown in Table 21-3. As with BFBs, gas solid temperatures are usually maintained in the range of 1,550 to 1,650° F for optimum sulfur capture. Higher bed temperatures are used if sulfur capture is not a concern or if low-volatile fuels are being fired. The higher bed temperatures increase carbon burnout, but generally lead to higher NO_x emissions.

Typical superficial gas velocities in the combustor are much higher for CFBs compared to BFBs. Selection of the design velocity is usually a compromise between heat transfer rates, erosion, and turndown. Velocities at the higher end of the range promote good heat transfer and provide for large turndown ratios, but significantly increase the potential for erosion. Low velocities minimize the potential for erosion, but heat transfer and turndown suffer.

As indicated previously, fuel feed size is dictated by the type of fuel being fired. The smaller feed sizes are used when firing low-volatile, high-ash fuels, while the larger fuel feed sizes are used when firing reactive fuels. In comparison to BFBs, the fuel feed size for CFBs typically is much smaller.

Sorbent feed size in a CFB is about 1,000 microns. The management of both fuel and sorbent feed size is important to overall CFB performance (Singer 1991). Particle sizes that are too large will lead to excessive bed pressure drops and carbon losses when bed material is drained from the combus-

Table 21-3. Typical CFB Operating Parameters

Bed temperature, ° F	1,550–1,650
Superficial velocity, ft/s	15–30
Coal feed particle size, in.	1/16–3/8
Sorbent feed particle size, microns	1,000
Recycle ratio	10–100
Calcium/sulfur ratio	1.5–4.0
SO_2 removal, %	90–95
Combustion efficiency, %	98–99
NO_x emissions, ppm	10–100

tor. Particles that are too fine will be carried through the cyclone separator.

As with BFBs, the Ca/S ratios are selected to achieve the desired sulfur removal. The results of performance testing at the Nucla CFB are summarized in Fig. 21-14 (Friedman, Heller, and Boyd 1991). The correlation shown is for a combustor temperature of 1,500 to 1,550° F and indicates that a Ca/S ratio of 1.5 is required for 70% SO_2 removal, and a Ca/S ratio of 4.0 is required for 95% SO_2 removal. The results reported by Friedman indicate that SO_2 removal is quite temperature sensitive. At a bed temperature of 1,650° F, a Ca/S ratio of about 2.7 was required to achieve 70% SO_2 removal at Nucla.

Combustion efficiencies are typically quite good in a CFB. Results of testing at the Nucla Station indicated that combustion efficiency varied over a narrow range of 97.6% and 98.9% for loads ranging from 55 to 110 MW (Friedman, Heller, and Boyd 1991). No significant variation in combustion efficiency was noted when bed temperature, primary/secondary air distribution, excess air, and coal feed configuration were varied.

NO_x emissions reported by Friedman, Heller, and Boyd at the Nucla Station are summarized in Fig. 21-15. NO_x emissions for all tests were <0.35 lb/MBtu. As shown in Fig. 21-15, the NO_x emissions were strongly dependent on combustor temperature. At 1,425° F, emissions were <40 ppmv, but increased to about 240 ppmv at 1,725° F. NO_x emissions were also found to increase with increasing Ca/S ratio.

The impact of combustor temperature on emissions at the Nucla Station as reported by Friedman, Heller, and Boyd is summarized in Fig. 21-16. The plot emphasizes the importance of temperature control in minimizing gaseous emissions.

21.2.2.15 Control Philosophy. Turndown from full load operating conditions on a CFB is initiated by reducing fuel and air flow to the combustor. Initially, both fuel and air flow

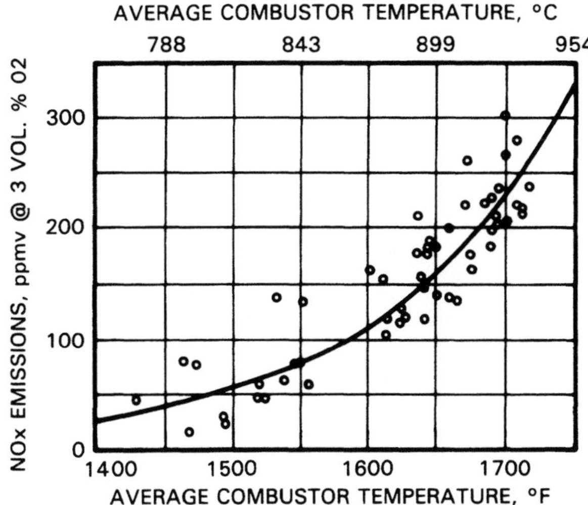

Fig. 21-15. NO_x emissions versus combustor operating temperature. (From Colorado UTE Electric Association. Used with permission.)

can be reduced and, thus, combustion temperatures can be maintained at full load values. When the load is reduced to a certain point, however, it is necessary to hold air flow constant to provide for adequate combustor performance and to prevent severe temperature maldistributions in the combustor. At loads below this point, combustor temperature drops. When the combustor temperature falls to a preselected value, startup burners must be placed in operation.

If the CFB is provided with a FBHE, turndown is achieved by first reducing solids flow to the FBHE. Since operation of the combustor is not initially impacted, combustor temperatures remain at optimum values. At reduced loads, solids flow to the FBHE are stopped. At loads below this point, fuel and air flow to the combustor must be decreased as discussed previously. One of the primary advantages of a FBHE, therefore, is that optimum combustor temperatures can be main-

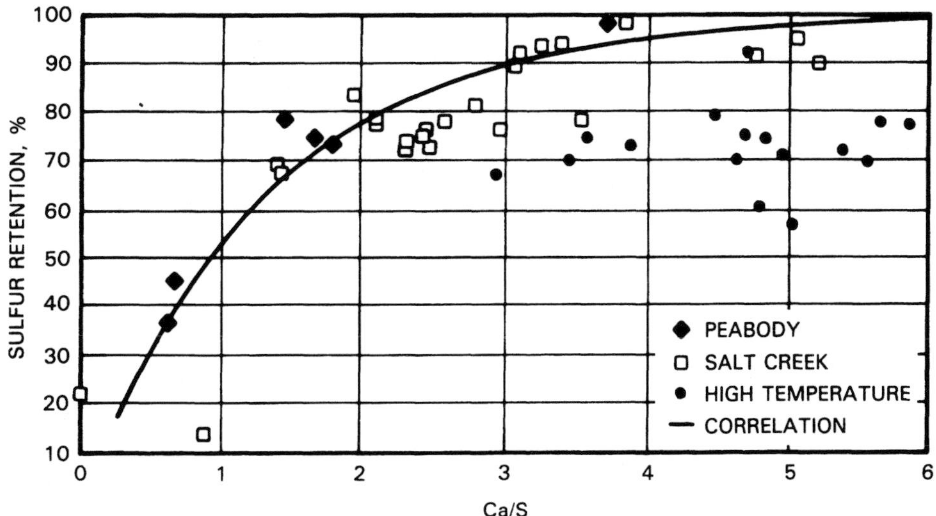

Fig. 21-14. Percent sulfur retention vs. Ca/S mole ratio. (From Colorado UTE Electric Association. Used with permission.)

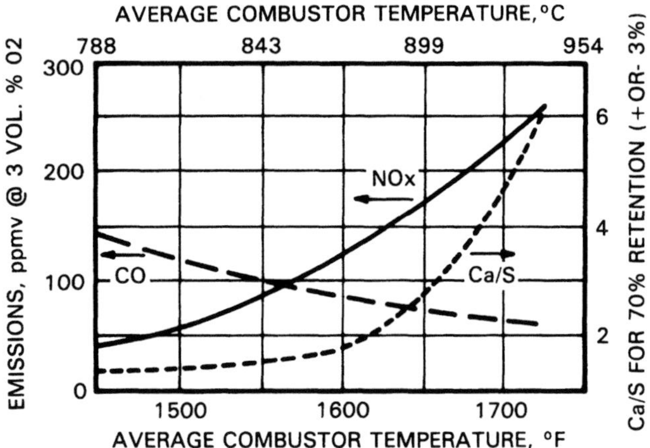

Fig. 21-16. Effect of combustor temperatures on gaseous emissions. (From Colorado UTE Electric Association. Used with permission.)

tained over a wider load range than for CFBs that do not have a FBHE.

As with BFBs, bed temperature control is maintained by varying airflow. Also, bed inventory is maintained by draining solids from the combustor to maintain a constant pressure differential across the bed. Steam temperature control in a CFB is similar to that used for a BFB or for a conventional pulverized coal unit.

21.2.2.16 CFB Applications. CFB units are considered commercially available and compete in the marketplace in sizes up to about 250 MW. To date, over 100 CFBs have been selected by independent power producers (IPPs) in unit sizes up to about 180 MW. Many of these have been cogeneration units. In electric utility applications, the largest operating units are the two 150-MW Texas New Mexico Power units in Texas. The largest CFB unit under construction is the Nova Scotia Power Corporation's 165-MW unit at Point Aconi in Nova Scotia.

Most new FBC units will likely be of the CFB type. The increased combustion efficiency, lower sorbent feed requirements, simplified feed, and reduced emissions with CFBs as compared to BFBs will be the factors influencing selection of an FBC type for new units. Whether or not CFBs will be able to compete in the marketplace with other technologies remains to be seen. Recent evaluations of pulverized coal units with scrubbers against CFBs have indicated that the conventional pulverized coal units are usually more economical than CFB units when good quality coal fuels are available. The impact of the Clean Air Act Amendments of 1990 will also be a factor that influences the decision of which technology to select. Although CFBs have a decided advantage over conventional pulverized coal units in minimizing NO_x emissions, SO_2 emissions from CFB units can be reduced to levels comparable to emissions from pulverized coal units with conventional scrubbers only with very high Ca/S ratios. Also, integrated gasification combined cycle units will compete with both pulverized coal and CFB units if gaseous emissions are severely restricted.

It appears most likely that CFB units will be selected for near term applications when waste fuels are burned or when fuel flexibility is a major consideration. Unlike pulverized coal units, CFB units can effectively burn waste fuels, and if properly designed, CFB units can burn a wide variety of fuels.

21.2.2.17 Manufacturers' Offerings. The most experienced manufacturers of CFB units are Lurgi/ABB Combustion Engineering Services and Pyropower. Manufacturers with fewer commercial units include Foster Wheeler, Riley Stoker, Babcock & Wilcox/Studsvik, Combustion Power, and others. An artist's rendition of the Lurgi/ABB-CES Texas New Mexico Power Company 150-MW CFB unit is shown in Fig. 21-17. A side view of the Pyropower CFB unit for the Argus Cogeneration Expansion Project is shown in Fig. 21-18. Foster Wheeler's CFB design for new reheat units is shown in Fig. 21-19.

21.3 PRESSURIZED FLUIDIZED BED COMBUSTION

21.3.1 History

Development of PFBC for electric utility generation began in the late 1960s for the express purpose of increasing cycle efficiency in coal-fueled plants (Drenker 1985). Numerous bench scale facilities were constructed and operated, primarily in Europe, to gather basic engineering data needed to scale to larger sizes. The bench scale facilities were quickly followed by small pilot plants. In the United States, Curtiss-Wright started up a small 1-MW unit in 1978 and followed with a 13-MW pilot facility in 1984. The Curtiss-Wright

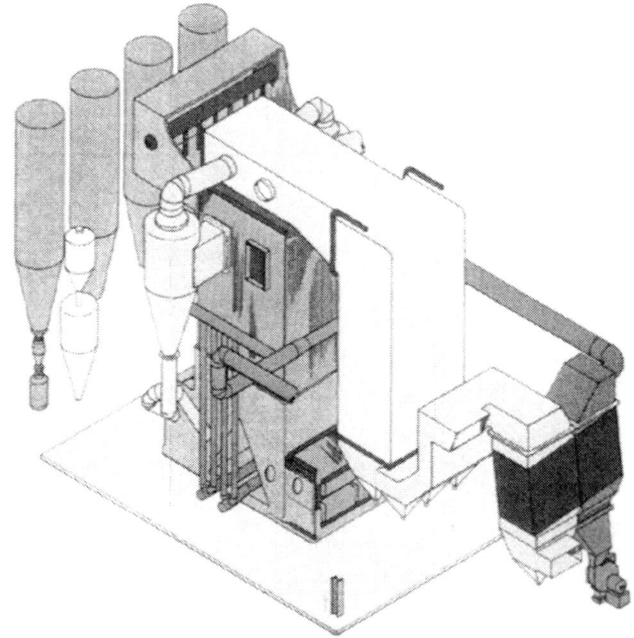

Fig. 21-17. 150-MW CFB units. (From ABB Combustion Engineering Systems. Used with permission.)

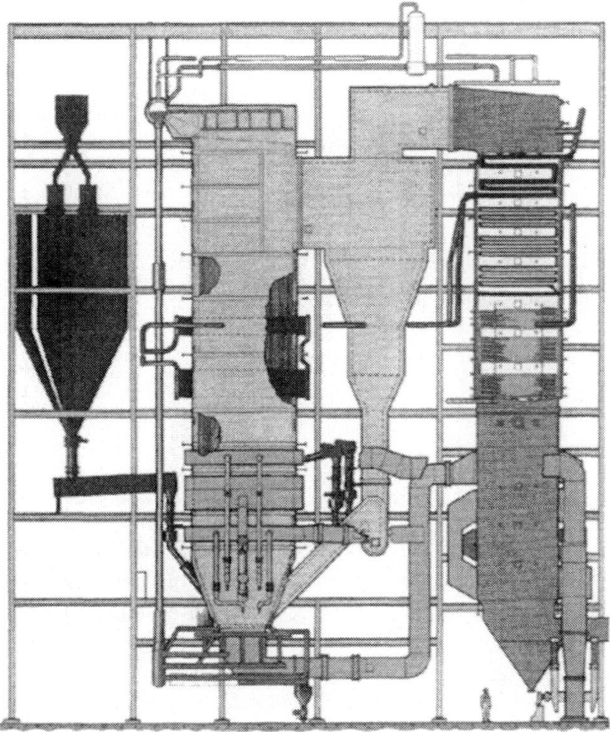

Fig. 21-18. CFB unit. (From Ahlstrom Pyropower Corporation. Used with permission.)

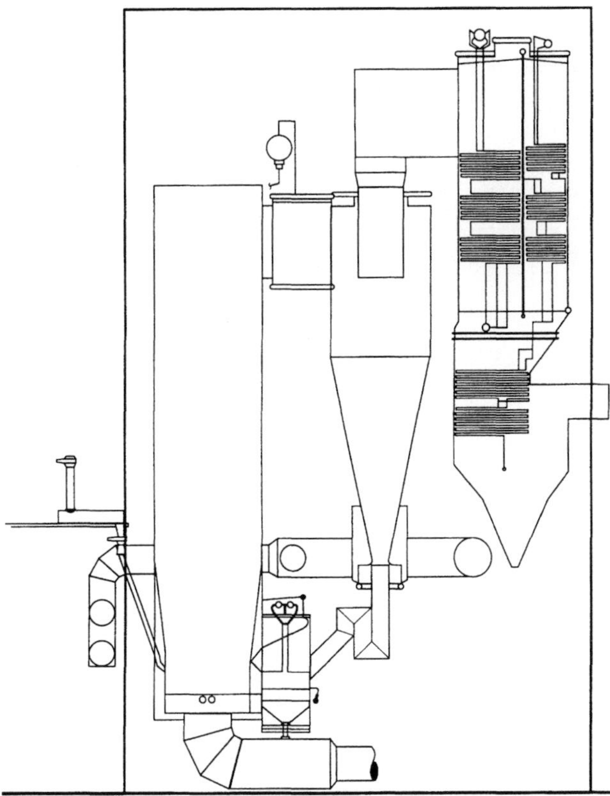

Fig. 21-19. CFB design for new reheat units. (From Foster Wheeler. Used with permission.)

facility was shut down just as it was being completed when DOE funding was terminated.

A 20-MW pilot facility, located in Grimethorpe, England, was built by the International Energy Agency in 1981. In 1984, the facility was taken over by Britain's National Coal Board and Central Electricity Generating Board. EPRI has worked closely with this group in supporting and planning test programs.

Sweden's Stal Laval and its parent, ASEA, jointly formed ASEA–PFBC to develop and market PFBC combined cycle power plants. In 1980, ASEA started construction on a 15-MW PFBC pilot plant that entered operation in 1983.

The PFBC commercialization process took a giant step forward in the United States when American Electric Power announced in the mid-1980s that it would retrofit a 74-MW PFBC/combined cycle unit at its Tidd Station on the Ohio River in Brilliant, Ohio. The bubbling bed PFBC repowered an existing steam turbine and added a new 16-MW turbine powered by flue gas from the PFBC. The unit started up in late 1990 and was scheduled to undergo 3 years of operational demonstration. In late 1991, the unit was shut down to correct design problems. Reportedly, the boiler heat transfer surface was understated by 25% (Coal & Synfuels Technology 1991).

AEP also planned to repower Units 3 and 4 of its Phillip Sporn plant in New Haven, West Virginia, with a 330-MW PFBC. Originally anticipated to be on line by 1996, it is now reported by Coal & Synfuel Technology (1991) that AEP has decided not to repower Sporn and is considering constructing a greenfield plant instead.

Two other PFBC pilot plants are operating outside the United States. One of these is the 79-MW ENDESA Plant in Escatron, Spain. This unit repowers an existing steam turbine and began initial operation in November 1990. Two additional PFBC units are in operation at the 135-MW Stockholm Energi Cogeneration Plant in Vartan, Sweden. Both units began operation in November 1990 and supply district heating to Stockholm (Anderson and Anderson 1991).

Both Alstrom and Deutsche Babcock have testing facilities operating. Alstrom has a 10-MW circulating bed PFBC in operation in Finland and Deutsche Babcock has a 15-MW bubbling bed PFBC testing facility in operation in Germany (Stallings, Boyd, and Brown 1991).

A 70-MW PFBC demonstration project will repower Iowa Power's Des Moines Energy Center. Cosponsored by Dairyland Power Cooperative, this Alstrom/Pyropower PFBC will be a scaleup of Alstrom's testing facility in Finland.

21.3.2 General Configuration

Development of PFBC has advanced to the stage where second-generation designs are now being considered. First-generation units were classified as either turbocharged or combined cycle units. In the turbocharged arrangement (Fig. 21-20), combustion gas from the PFBC boiler is cooled to

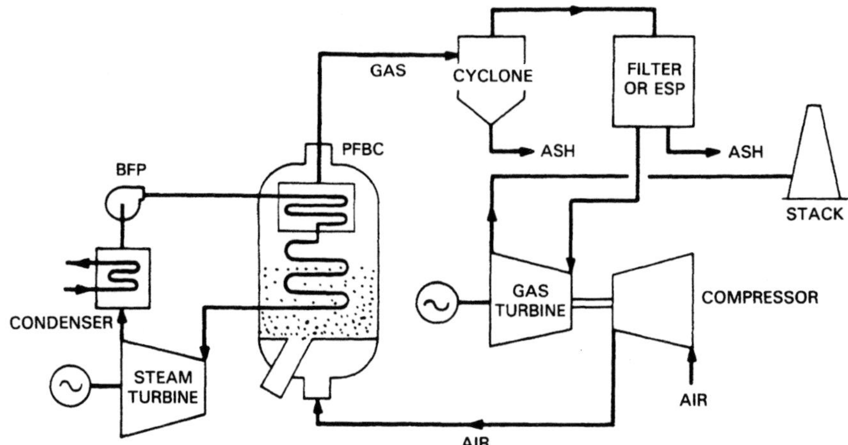

Fig. 21-20. PFBC turbocharged arrangement.

approximately 750° F and is used to drive a gas turbine. The gas turbine drives an air compressor, and there is little, if any, net gas turbine output. Electricity is produced by a turbine generator driven by steam generated in the PFBC boiler.

In the combined cycle arrangement (Fig. 21-21), 1,500 to 1,600° F combustion gas from the PFBC boiler is used to drive the gas turbine. About 20% of the net plant electrical output is provided by the gas turbine. With this arrangement, thermal efficiencies 2 to 3 percentage points higher than with the turbocharged cycle are feasible.

In second-generation units (Fig. 21-22), coal is first pyrolyzed in a carbonizer to produce low-Btu gas. This gas is mixed with the flue gases produced by the combustion of the char in the PFBC combustor and is burned in a topping combustor. The combustion of these gases in the topping combustor provides hot gases (potentially as hot as 2,300° F) for the gas turbine. These higher inlet gas temperatures allow for the use of higher efficiency gas turbines. Sulfur capture occurs in both the carbonizer and the PFBC combustor. Heat is recovered from the gas turbine exhaust in a heat recovery steam generation (HRSG) to heat the feedwater in an economizer and to superheat steam. Additional superheat is achieved in an external FBHE. Also, the FBHE is used to reheat steam from the HP steam turbine. Power generated by the steam turbine and the gas turbine is about equal, allowing for potential thermal efficiencies of over 45%. Steam cycles using superheated steam and double reheat can raise the cycle efficiency to approximately 47% (Hyre et al. 1991).

21.3.3 Circulating vs. Bubbling Bed

The PFB combustor can be either a circulating bed combustor or a bubbling bed combustor. Designs that use the circulating bed combustor usually incorporate an external FBHE that contains steam generation and reheat surface. Bubbling bed combustor designs may or may not incorporate an internal heat transfer surface.

First-generation PFBCs used bubbling bed combustors with in-bed heat transfer surface to generate steam. Relatively low (3 ft/s) fluidizing velocities were used to minimize

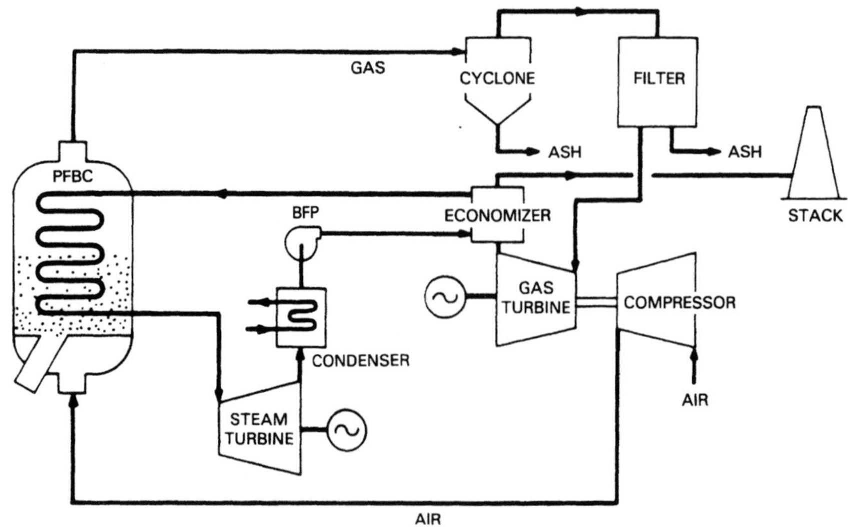

Fig. 21-21. PFBC combined cycle arrangement.

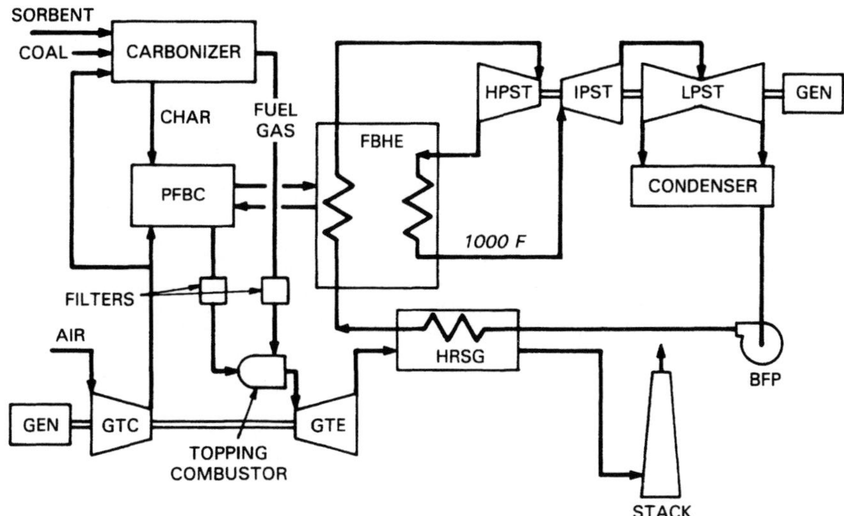

Fig. 21-22. Second generation PFBC plant—45% (HHV) efficiency.

erosion in the combustor. Although the fluidizing velocities were less than for AFBC units using bubbling beds, PFBC combustor sizes were reduced because of the higher operating pressures (12 to 15 atmospheres).

21.3.4 Hot Gas Cleanup

Cleanup of the hot flue gas prior to passing through the expander section of the combustion turbine is essential to avoid excessive erosion of turbine blades. In the first-generation turbocharged arrangement (Fig. 21-20), the flue gas temperature is reduced to approximately 750° F prior to gas cleanup. The reduced temperature allows the use of conventional electrostatic precipitator technology for particulate removal. Most turbocharged designs incorporate one or more cyclone separators in series for cleanup. Some designs added an ESP for additional particulate removal.

The first-generation combined cycle PFBCs and the second-generation PFBCs supply hot gases to the combustion turbine to achieve higher cycle efficiencies. In both cases, combustion gas temperatures are approximately 1,600° F. For the second-generation PFBC, particulates must be removed from both the fuel gas generated in the carbonizer and the flue gas from the PFBC. Both gas streams are at a temperature of about 1,600° F. A key technical issue for both designs is the development of reliable hot gas cleanup (HGCU) devices that can achieve the particulate removal required to protect downstream equipment and to meet New Source Performance Standards. With the second-generation designs, a second factor that must be considered is that the HGCU device must be compatible with the reducing gases from the carbonizer.

The AEP Tidd PFBC Plant will be used to test large-scale HGCU devices under DOE funding. The test facility will be designed to clean one-seventh of the gas generated by the PFBC combustor. The first particle filter tested will be Westinghouse's ceramic candle filter system. Testing was scheduled to begin in July 1992 (Mudd and Durner 1991).

21.3.5 Fuel and Sorbent Feed

Fuel and sorbent can be fed to the combustor or the carbonizer either in a dry form or as a pumpable paste. With dry feed systems, coal is dried and ground to an average particle size of approximately 50 microns and is carried to the combustor or carbonizer with flue gas through lock hoppers. The lock hoppers are required to seal against the pressure of the PFBC combustor. Sorbent (limestone) is ground to an average particle size of approximately 100 microns and can be either fed with the coal or in a separate stream.

With paste feed systems, coal is crushed to ¼ in. top size and mixed with water to about 25% water. The paste feed is pumped into the combustor with positive displacement-type pumps. Sorbent can be mixed with the coal or fed separately as with the dry feed system. When sorbent is mixed with the coal, a trim sorbent feed system is sometimes used to allow for more rapid response to variations in SO_2 emissions. Sorbent is typically crushed to 1,000 microns for feed in the trim systems.

One of the advantages of PFBC systems is that fewer feed points are required to achieve good mixing because pressurization allows reductions in combustor sizes.

21.3.6 Performance

Overall performance of PFBC units in terms of heat transfer, combustion efficiency, and emissions is expected to be better than for AFBC units. The higher operating pressure of PFBC units results in higher heat transfer rates and combustion efficiencies. In addition, because of the pressurization, NO_x emissions are less than for AFBC units (Carpenter et al. 1991).

Net plant heat rates for PFBC units depend on the design vintage (first or second) and the type of steam cycle used (subcritical or supercritical). New first-generation, subcritical PFBC units (2,400 psig, 1,000° F main steam and reheat steam) are expected to have a net plant heat rate of 8,570 Btu/kWh, a 40% thermal efficiency (Stallings, Boyd, and Brown 1991).

Both Foster Wheeler and Gilbert/Commonwealth have investigated the use of supercritical steam cycles and advanced gas turbines with second-generation PFBC units. One subcritical and three supercritical configurations were evaluated. The net plant heat rates for these configurations ranged from 7,600 Btu/kWh (45% thermal efficiency) for the subcritical configuration with single reheat to 6,780 Btu/kWh (50% thermal efficiency) for the developmental supercritical configuration (5,000 psig, 1,200° F steam) with double reheat (Hyre et al. 1991).

PFBC units, both bubbling bed and circulating bed, are expected to achieve low levels of emissions for SO_2, NO_x, and particulates. PFBC units operating in Escatron, Spain, and Vartan, Sweden, have achieved 90% SO_2 reduction, NO_x emissions of 0.12 to 0.50 lb/MBtu, and particulate emissions of <0.033 lb/MBtu (Stallings, Boyd, and Brown 1991).

Typical emissions for a second-generation PFBC firing Illinois bituminous coal are projected to be as follows (O'Donnell 1991):

Pollutant	Emissions, lb/MBtu	NSPS
SO_2	0.12	1.2
NO_x	0.026	0.60
CO	0.03	
Particulates	0.013	0.03

As shown, plant emissions are expected to be well below New Source Performance Standards for fossil-fueled power plants.

21.3.7 Future Developments

Several companies, including Foster Wheeler, M. W. Kellogg, Pyropower, and British Coal, are involved in the development of second-generation PFBC units. Foster Wheeler is constructing a PFBC pilot plant that will be used to test both the carbonizer and the circulating PFBC combustor. The carbonizer and its associated ceramic crossflow filter HGCU system will be tested first. The circulating PFBC combustor, burning coal and char from the carbonizer, will be tested second.

Westinghouse is developing and testing a multiannular swirl burner for use as the topping combustor.

In the design being developed by M. W. Kellogg, both the gasifier and the combustor unit will use transport reactors. These are refractory-lined pressure vessels that are connected by refractory-lined solids transfer pipes. The use of these transport reactors will allow for higher particle velocities than for typical circulating bed reactors (Carpenter et al. 1991).

British Coal has taken a different approach for their second-generation unit. To increase the probability of successful development, British Coal uses an atmospheric CFB combustor instead of a circulating PFBC combustor to burn the char from the carbonizer. A PFBC gasifier is used in the British Coal design.

21.4 REFERENCES

ABB COMBUSTION ENGINEERING SYSTEMS. ABB Power Plant Systems. P. O. Box 500, 1000 Prospect Hill Road, Windsor, CT 06095-0500.

AHLSTROM PYROPOWER CORPORATION. 8925 Rehco Road, P. O. Box 85480, San Diego, CA 92121-3269.

ANDERSON, J. and S. A. ANDERSON. 1991. Commissioning experience from three PFBC plants. Paper read at 1991 International Conference on Fluidized Bed Combustion. The American Society of Mechanical Engineers. April 1991, New York, NY, pp. 787–793.

BABCOCK & WILCOX. Power Generation Group. 20 S. Van Buren Avenue, P. O. Box 351, Barberton, OH 44203-0351.

CAPUANO, L., K. ATABAY, CE POWER SYSTEMS; and S. A. FOX. LURGI CORPORATION. 1984. Circulating fluid bed steam generation. Paper presented at Pacific Coast Electric Association Engineering and Operating Conference. San Francisco, CA, March 20–21.

CARPENTER, L., W. LANGAN, R. DELLEFIELD, G. NELKIN, and T. HAND (US DOE). 1991. Pressurized fluidized bed combustion: a commercially available clean coal technology. Paper read at 1991 International Conference on Fluidized Bed Combustion. The American Society of Mechanical Engineers. April 1991, New York, NY, pp. 467–474.

COAL AND SYNFUELS TECHNOLOGY. 1991. AEP spends $3 million to revamp Tidd. *Coal & Synfuels Technology.* Vol. 12, No. 41, p. 1.

COLORADO UTE ELECTRIC ASSOCIATION, INC. P. O. BOX 698, Nucla, CO 81424.

DRENKER, STEVEN G. 1985. *Proceedings: Pressurized Fluidized-Bed Combustion Power Plants.* Electric Power Research Institute, May 1985.

FOSTER WHEELER ENERGY CORPORATION. Perryville Corporate Park, Clinton, NJ 08809-4000.

FRIEDMAN, M. A., T. J. HELLER, and T. J. BOYD. 1991. Operational data from 110 MWe Nucla CFB. Paper read at 11th International Conference on Fluidized Bed Combustion. April 21–24, 1991, Montreal, Canada.

GOIDICH, S. J. (FOSTER WHEELER), A. R. MCGEE, and K. RICHARDSON. 1991. The NISCO Cogeneration Project 100-MWe circulating fluidized bed reheat steam generator. Paper read at the 1991 International Conference on Fluidized Bed Combustion. April 21–24, 1991, Montreal, Canada.

HAECOCK, F. A., JR. (COLORADO-UTE ELECTRIC ASSOCIATION), and C. M. DAVIS (PYROPOWER CORPORATION). 1986. Material handling systems for the Colorado-Ute Nucla CFB Demonstration Project. February 1986 (PP 86-1).

HYRE, M. R., D. A. HORAZAK, L. N. RUBOW (GILBERT/COMMONWEALTH), and A. ROBERTSON (FOSTER WHEELER). 1991. PFBC performance and cost improvements using state-of-the art and advanced topping/bottoming cycles. *Proceedings of the 1991 International Conference on Fluidized Bed Combustion.* April 21–24, 1991, Montreal, Canada.

JACOBS, R. V. (COMBUSTION ENGINEERING). 1986. Design of the TVA 160 MW atmospheric fluidized bed steam generator. Paper read at the 1986 ASME/IEEE Joint Power Generation Conference. October 12–23, 1986, Portland, OR.

KANTESARIA, P. P. and G. D. JUKKOLA (COMBUSTION ENGINEERING). 1983. Performance results from the Great Lakes Fluidized Bed Demonstration Plant. Paper read at the AIChE 1983 Spring National Meeting & Petro Expo '83. March 27–31, 1983, TIS-7275, Houston, TX.

KOPETZ, E. A., JR., W. B. O'BRIEN, and G. L. TENNANT (TENNESSEE VALLEY AUTHORITY). 1991. Startup and operating experience of TVA's 160-MW Shawnee atmospheric fluidized bed combustion demonstration plant: a 1991 update. Paper read at the 1991 International Conference on Fluidized Bed Combustion. April 21–24, 1991, Montreal, Canada.

LURGI/COMBUSTION ENGINEERING. *Circulating Fluid Bed Boilers.* 1985.

MAKANSKI, JASON. 1991. Fluidized bed boilers. *Power.* Vol. 135, No. 3, pp. 15–32.

MODRAK, T. M., J. T. TANY (BABCOCK & WILCOX), and C. J. AULISIO (EPRI). 1982. Sulfur capture and nitrogen oxide reduction on the $6' \times 6'$ atmospheric fluidized bed combustion test facility. Paper read at the Symposium on Combustion Chemistry, American Chemical Society, 29 March–2 April 1982, RDTPA 82-7, Las Vegas, NV.

MUDD, M. J. and M. W. DURNER. 1991. American Electric Power's PFBC hot gas clean up test program. Paper read at the 1991 International Conference on Fluidized Bed Combustion. The American Society of Mechanical Engineers. April 1991, New York, NY, pp. 953–957.

O'DONNELL, J. J. (M. W. KELLOGG). 1991. An advanced concept in pressurized fluid bed combustion. Paper presented at the 1991 International Conference on Fluidized Bed Combustion. The American Society of Mechanical Engineers. April 1991, New York, NY, pp. 183–191.

SCHWIEGER, BOB. 1985. Fluidized-bed boilers achieve commercial status worldwide. *Power.* Vol. 129, No. 2, pp. S.1–S.16.

SINGER, J. G., ED. 1991. *Combustion Fossil Power*, 4th edit. Combustion Engineering Inc., Windsor, CT.

SMITH, J. W. 1982. The Babcock & Wilcox atmospheric fluidized bed combustion development program. Paper read at the Southeastern Electric Exchange 1982 Annual Conference. April 21–21, 1982, Kissimmee, FL.

STALLINGS, J., T. BOYD, and R. BROWN. 1991. Update on status of fluidized-bed combustion technology. Electric Power Research Institute, Palo Alto, CA.

STEVENS, W. A. and T. F. BEGINA (ABB COMBUSTION ENGINEERING SYSTEMS). 1991. The design and commercial development of the 150 MW CFB unit for Texas-New Mexico Power Company (TIS 8544).

WHITNEY, S. A., ET AL. (BABCOCK & WILCOX). 1983. Utility fluidized bed operating experience. Paper read at the Joint Power Generation Conference. September 25–29, 1983, Indianapolis, IN. (BR-1244).

WONG, H. K., B. L. JENNESS, D. W. RENS, and E. KOLWALSKI (FOSTER WHEELER). 1987. Update of the Black Dog atmospheric fluidized bed combustion project. Paper read at the American Power Conference. April 1987. Chicago, IL.

22

Resource Recovery

Mitchell N. Bjeldanes and *Gordon V. Z. Beard*

22.1 INTRODUCTION

For the foreseeable future, the disposal of municipal and agricultural solid waste will continue to be an increasing problem. This waste is either composted, landfilled, recycled, or converted to some form of energy. It is the third option, also called Resource Recovery, with which this chapter is concerned.

There are currently 140 operating municipal solid waste combustion units in the United States. These plants process about 17% of the United States annual garbage output and generate 2,400 megawatts (MW) of electricity. This electric generation would power approximately 1.3 million homes or displace 32 million barrels of oil per year.

Resource recovery is waste volume reduction, and any monies gained from either recycled materials or energy recovered simply displace part of the cost of reducing the volume of garbage. The utility's role in the resource recovery approach is that of a community servant who assists the community and the environment by using this waste in either existing, converted, or new equipment. The amount of fuel this source represents to a utility can be relatively inconsequential, but the detriment to the utility from burning this fuel can be significant.

22.2 CLASSES OF WTEE TECHNOLOGY

This chapter broadly describes some of the various technologies used to generate power from recovered resources, and discusses aspects the utility must consider for the following three classes of waste to electric energy (WTEE) technology.[1]

- Unprocessed solid waste combustion,
- Processed solid waste combustion, and
- Solid waste liquid/gas conversion.

Within these classifications, 16 specific WTEE technologies have been identified, as listed below:

- Unprocessed solid waste combustion technologies
  - —Waterwall furnace,
  - —Refractory furnace,
  - —Rotary kiln furnace,
  - —Water-cooled rotary combustor furnace, and
  - —Controlled air furnace.
- Processed solid waste combustion technologies
  - —Spreader stoker-fired boiler,
  - —Suspension-fired boiler,
  - —Fluidized bed unit,
  - —Cyclone furnace unit, and
  - —Cofiring with other fuels.
- Solid waste liquid/gas conversion technologies
  - —Anaerobic digestion,
  - —Landfill gas recovery,
  - —Fermentation,
  - —Pyrolysis,
  - —Gasification, and
  - —Liquefaction.

These technologies are discussed in regard to their basic design, commercial availability, fuels, and cost and performance data.

22.2.1 Unprocessed Solid Waste Combustion Technologies (Mass Burning)

The mass burn technologies use municipal solid waste (MSW) in its as-discarded form. The MSW undergoes only limited processing to remove noncombustible and oversized items. Agricultural and other solid wastes are processed primarily to reduce gross size (for example, cutting trees into short logs).

Mass burn technologies include water-wall furnace, refractory furnace, rotary kiln furnace, water-cooled rotary combustor furnace, and controlled air furnace.

22.2.1.1 Waterwall Furnace. Typically solid waste is burned on a reciprocating grate inside the waterwall furnace. Heat is transferred from the burning waste and hot combustion gases to the water in the waterwalls, generating steam. This steam is used to power a steam turbine, generating electricity, or as heat for process energy. The waterwall

[1]Much of the information presented in this chapter was adapted from a Waste to Electric Energy Options Asseessment Study prepared by Black & Veatch for the Rural Electric Research program of the National Rural Electric Cooperative Association (NRECA).

furnace has the highest heat recovery efficiency of mass burn technologies and is preferred where the generation of electricity is important (Taylor 1984).

22.2.1.2 Refractory Furnace. In a typical refractory furnace, solid waste is burned on a reciprocating grate in the refractory-lined furnace. The combustion gases produced by the burning waste in the furnace travel through a convection-type heat recovery steam generator to generate steam. This technology has a lower heat recovery efficiency than the waterwall furnace because of the lack of radiant heat-absorbing, water-cooled furnace walls (Robinson 1986).

22.2.1.3 Rotary Kiln Furnace. The rotary kiln furnace is a variation of the refractory furnace. Solid waste is fed into a refractory-lined rotary kiln where the waste is mixed by tumbling while burning. Combustion of the waste and the flue gas is completed in a secondary refractory-lined combustion chamber. Steam is produced in a downstream convection-type heat recovery steam generator.

22.2.1.4 Water-Cooled Rotary Combustor Furnace. The water-cooled rotary combustor furnace is a combination of the water-wall furnace and the rotary kiln furnace. Solid waste is burned in a rotary kiln constructed of a membrane-type water-cooled shell. Combustion of the waste is completed in a radiant waterwall secondary combustion chamber. Steam is generated in both the waterwalls and a convection boiler section.

22.2.1.5 Controlled Air Furnace. The controlled air furnace is a variation of the refractory furnace. This furnace, which is suitable for smaller, modular units, can be either an excess air unit or a starved air unit. Both of these have two combustion chambers. The excess air unit uses excess primary combustion air in the first chamber to ensure complete combustion. The starved air unit uses less than stoichiometric (starved) primary combustion air to create a partial pyrolytic process in the first chamber. Auxiliary fuel is used in the secondary chamber to complete combustion of unburned gases.

22.2.2 Processed Solid Waste Combustion Technologies

The WTEE technologies identified in this subsection use refuse derived fuel (RDF). The solid wastes are subject to various processes to improve their physical and chemical properties. These processes include drying, comminution (reduction of particle size), densification, physical separation, and chemical modification (Solar Energy Research Institute Volume II 1979).

22.2.2.1 Spreader Stoker-Fired Boiler. In a spreader stoker-fired boiler, the solid waste is spread by multiple feeder–distributor units (spreaders) onto a grate system. The grate system most frequently used is a continuous-cleaning, traveling grate. The lighter portion of the waste burns in suspension above the grate, and the heavier portion burns in a bed on the grate. Underfire combustion air is used to cool the grate and to burn the waste on the grate, while overfire combustion air is used to cause mixing of gases and combustion above the grate. Steam is produced and superheated in a combination of waterwalls and boiler tube banks.

22.2.2.2 Suspension-Fired Boiler. In a suspension-fired boiler, most of the solid waste burns in suspension. Some of the waste falls down to dump grates at the bottom of the furnace, where combustion is completed. Air is used to convey the waste into the furnace through the burners in the furnace walls. In horizontally fired units, these burners are in the front and rear walls of the furnace. In tangentially fired boilers, the burners are in the corners of the furnace. The boiler arrangement for a suspension-fired boiler is very similar to that for a spreader stoker-fired boiler.

22.2.2.3 Fluidized Bed Unit. A fluidized bed unit burns solid waste in a bed of sand or limestone that is supported by upward-flowing air. These units are classified as either bubbling bed or circulating bed units. The distinction between the two technologies is the upward air velocities. In a bubbling bed unit, air velocities are maintained between 3 and 10 ft/s to minimize entrainment of the solids in the flue gas. In a circulating bed unit, air velocities are maintained between 13 and 20 ft/s, allowing entrainment of the bed material. A cyclone separator at the exit of the combustor separates this entrained material from the flue gas and returns the material to the bed. Steam is generated in a combination of in-bed tubes, waterwalls in the combustor, and in a waste heat convection boiler downstream of either the combustor or the cyclone.

Fluidized bed combustion (FBC) offers several advantages over conventional forms of combustion. First, these units have a greater ability to burn a wide variety of fuels, which is important considering the diverse nature of solid wastes. Second, the limestone in the bed combines with the SO_2 and acids to be removed as a solid. Finally, because of the lower combustion temperatures, fluidized bed units have inherently lower NO_x emissions.

22.2.2.4 Cyclone Furnace Unit. A cyclone furnace unit uses small cyclone furnaces 6 to 10 ft in diameter in conjunction with a typical waterwall furnace. Complete combustion takes place in the cyclone furnace. These are refractory-lined, water-cooled, horizontal chambers. High turbulence in the cyclone furnaces is provided by the cyclonic action of the tangentially supplied solid waste, primary air, and secondary air in the furnaces. Because of the low heat absorption rates and high heat release rates in the cyclone furnaces, high combustion temperatures (>3,000° F) are produced. The hot combustion gases are discharged from the cyclone furnaces into the boiler, where steam generation takes place in the waterwalls and convective tube sections.

22.2.2.5 Cofiring with Other Fuels. Solid waste can be burned simultaneously (cofired) with fossil fuels such as coal, oil, and natural gas, or in a more limited manner, one particular kind of solid waste can also be cofired with another solid waste. When the solid waste is cofired with fossil fuels, it can be treated as either the primary fuel or a supplementary fuel. If the solid waste is considered to be the primary fuel, the unit is designed for the combustion of solid waste with fossil fuels used as auxiliary fuels when the solid waste is not available (Bretz 1989). When the solid waste is the supplementary fuel, it is cofired with the fossil fuel in a unit designed to burn the fossil fuel. In this case, the amount of solid waste cofired is limited to about 15% to minimize the adverse effect on the boiler.

22.2.3 Solid Waste Liquid/Gas Conversion Technologies

These technologies convert solid waste into either liquid or gas fuels. The resulting fuel can then be burned in a conventional manner. The solid waste conversion technologies are divided into biological conversion processes and thermal conversion processes (not including combustion). Biological processes require wet, biodegradable solid waste or biomass, whereas thermal processes work best with solid waste with <50% moisture content (Reed and Das 1988). Biological conversion processes include anaerobic digestion, landfill gas recovery, and fermentation. Thermal conversion processes include pyrolysis, gasification, and liquefaction. These processes have been demonstrated, but the variability of the waste as a feed stock makes the process impractical at present.

22.2.3.1 Anaerobic Digestion. Anaerobic digestion is the decomposition of solid waste by bacteria in the absence of air. It produces methane from solid waste. Carbon dioxide is a major byproduct of this process. The typical anaerobic digestion process has three steps:

1. Decomposition of cellulose and other complex molecules to simple organic compounds by certain bacteria.
2. Conversion of organic compounds by acidogenic bacteria to simple organic acids ending with acetic acid (CH_3COOH) and hydrogen.
3. Conversion of acetic acid and hydrogen by methanogens to methane (CH_4) and carbon dioxide.

22.2.3.2 Landfill Gas Recovery. The recovery of landfill gas, which is primarily methane and carbon dioxide, for power generation can be applied either to existing landfills or new landfills. Landfill gas is generated by anaerobic digestion. Landfill gas either can be used directly from the landfill to generate steam in boilers or processed for pipeline use.

22.2.3.3 Fermentation. Fermentation is a biological conversion process used for the production of ethanol (ethyl alcohol). The most suitable feedstocks are wood, agricultural residues, grasses, and the organic portion of MSW. Robinson (1986) describes the typical fermentation process steps.

22.2.3.4 Pyrolysis. Pyrolysis is the decomposition of biomass or solid waste by heat in the absence of air. This process is an important first step for combustion and gasification processes and is accomplished at temperatures of 400 to 1,100° F. Pyrolysis generates three primary products: medium energy gas (primarily hydrogen, methane, CO, and CO_2); tarry liquids or pyrolysis oil, and charcoal.

Pyrolysis can be either slow or fast. Slow pyrolysis, which involves either slow heating rates or large pieces of solid waste, produces a high proportion of charcoal. Fast pyrolysis, which is the very rapid heating of fine solid waste particles, produces a high proportion of gas. This gas contains a large fraction of olefins, which are important feedstocks for the production of gasoline or alcohol.

22.2.3.5 Gasification. Gasification is the process of converting biomass or solid waste into combustible gases which typically contain 60% to 90% of the energy originally contained in the feedstock. Gasification involves pyrolysis as the first stage with subsequent, higher temperature reactions of the pyrolysis products to generate low molecular weight gases. There are four primary gasification processes:

- Pyrolytic gasification, in which the pyrolysis products of charcoal and oil are burned with air to produce higher yields of medium energy gas with a heating value of 300 to 500 Btu/scf;
- Air gasification, in which the biomass is burned with a limited supply of air to produce a low-energy gas (H_2 and CO, diluted with H_2O, CO_2, and N_2) with a heating value of 150 to 200 Btu/scf;
- Oxygen gasification, in which the biomass is burned with a limited supply of oxygen to produce a medium energy gas (H_2 and CO, diluted with H_2O and CO_2) with a heating value of 300 Btu/scf; and
- Hydrogasification, in which hydrogen is added under high pressure to a pyrolysis process to increase the yield of methane.

The air and oxygen gasification processes are direct gasification processes since they use air or oxygen in internal combustion processes to produce the required heat. The pyrolytic and hydrogasification processes are classified as indirect gasification processes because heat is produced externally (Reed and Das 1988).

Direct gasifiers are simpler and more suitable for small installations than indirect gasifiers. Direct gasifiers include fixed bed, fluidized bed, and suspended particle gasifiers. Fixed bed gasifiers are best for bulky fuels such as wood chips and corn cobs. Fluidized bed gasifiers use a wider variety of fuels, and suspended particle gasifiers use finely divided fuel particles. Fixed bed gasifiers, which are the only ones currently suitable for small-scale applications, include downdraft and updraft gasifiers. Downdraft gasifiers, in which air or oxygen flows downward through the downward-flowing fuel, produce a gas low in tar and oil. Updraft gasifiers, in which air or oxygen flows upward through the downward-flowing fuel, produce a gas high in tar and oil.

The medium energy gas is suitable for immediate combustion, limited pipeline distribution, or as synthesis-gas to produce methanol, gasoline, and methane. The low energy gas is suitable only for immediate combustion (Solar Energy Research Institute, Volume I, 1979).

22.2.3.6 Liquefaction. Liquefaction is the thermal conversion of biomass or solid waste into liquid fuels. Liquefaction processes include direct liquefaction, indirect liquefaction, and pyrolysis. In direct liquefaction, hydrogen is added to the biomass at high temperatures and pressures either directly or by a reaction of CO_2 and water. The reaction of hydrogen with the biomass produces hydrocarbon liquid fuels. Direct liquefaction can be performed either with or without a catalyst. Indirect liquefaction requires that the biomass or solid waste first be converted to synthesis-gas using one of the gasification processes. The synthesis-gas can then be converted to liquid fuels by catalytic processes (Singer 1981 and University of Oklahoma 1979).

22.3 SYSTEM ARRANGEMENTS

This section addresses four aspects of WTEE technologies:

- Commercial availability,
- General design requirements,
- Allowable fuels, and
- Cost and performance data.

No attempt has been made in this chapter to assess the environmental considerations associated with their use.

22.3.1 Unprocessed Solid Waste Combustion

Unprocessed solid waste combustion WTEE technologies use MSW, unprocessed agricultural waste, and other solid wastes. The solid waste is fed to a furnace that uses a combination of radiant and convective heat transfer surfaces to generate steam and may use a steam turbine to generate electricity.

22.3.1.1 Solid Waste Feeding to Furnace. Little processing other than gross size reduction or the removal of noncombustibles is done to the MSW or other solid waste before it is burned in the furnace. In one method (pit and crane), solid waste is dumped from trucks into a storage pit. After any noncombustible or oversize items are removed from the waste, it is transported by crane from the pit to an elevated charging hopper. In another method (tipping floor and inclined conveyor), waste is dumped from trucks onto a tipping floor. Front-end loaders are used to both separate the noncombustible or oversize items from the waste and to push the remaining waste onto an inclined conveyor. This conveyor then carries the waste to an elevated charging hopper. Both are suitable for large facilities; however, the pit and crane method is more prevalent. The third method is similar to the tipping conveyor method, except that the front-end loaders push the waste directly into floor-level charging hoppers. This latter method is suitable for smaller and modular batch-fed facilities (Robinson 1986 and Bechtel 1990). Once the waste is in the charging hoppers, it is fed onto the grate system in the furnace either by gravity feeding or by mechanical feeders.

22.3.1.2 Waterwall Furnace. The furnace enclosure for a waterwall furnace (Fig. 22-1) consists of water-filled boiler tubes joined together to form the furnace walls. This category includes those furnaces that combine refractory walls in the lower furnace area and waterwalls in the upper furnace area. Steam is generated in the waterwalls primarily from the radiant absorption. Figures 22-2 and 22-3 illustrate plant concepts.

The waterwall furnace is the most common WTEE technology. Hundreds of plants are operating with this technology worldwide. The United States plants either existing or in the advanced planning stages and burning MSW have the following capacity ranges:

- Minimum tons/day = 200,
- Maximum tons/day = 3,150, and
- Average tons/day = 1,200.

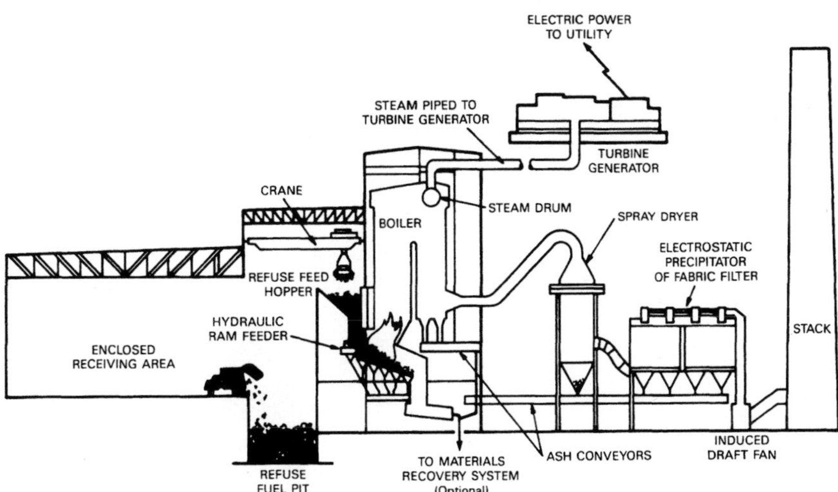

Fig. 22-1. Waterwall furnace WTEE plant.

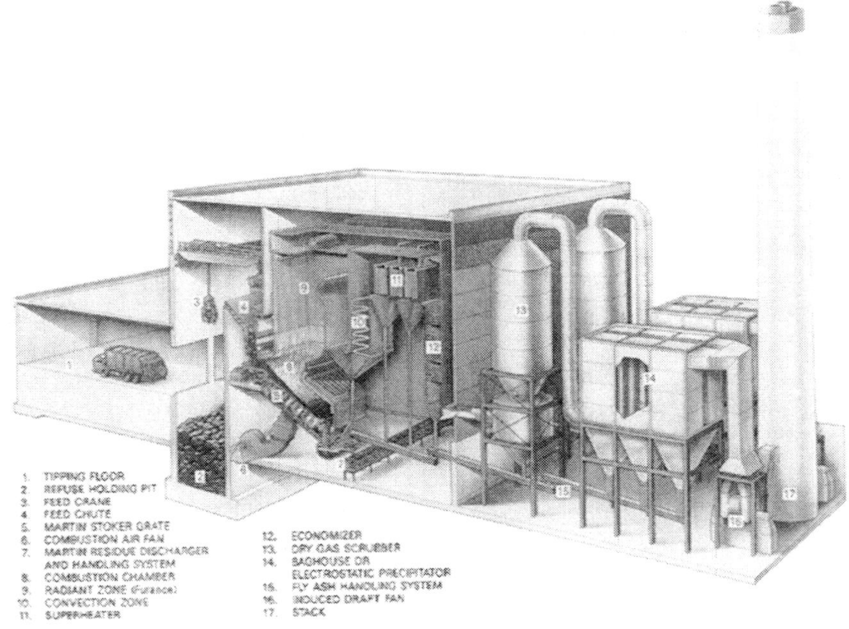

Fig. 22-2. Waterwall furnace—a typical Ogden Martin resource recovery facility. (From Ogden Martin. Used with permission.)

Figure 22-4 shows the capacity distribution of these plants (Berenyi 1988).

Waterwall furnace technology has been applied to the combustion of various other solid wastes. One such application is the 14-MW Modesto Power Plant in Westley, California, which uses tires for fuel. This plant went into operation in 1987 and burns 4.5 million tires per year (La Roe 1988 and Oxford 1988).

The general concepts for the major systems used in a waterwall furnace WTEE plant are listed below:

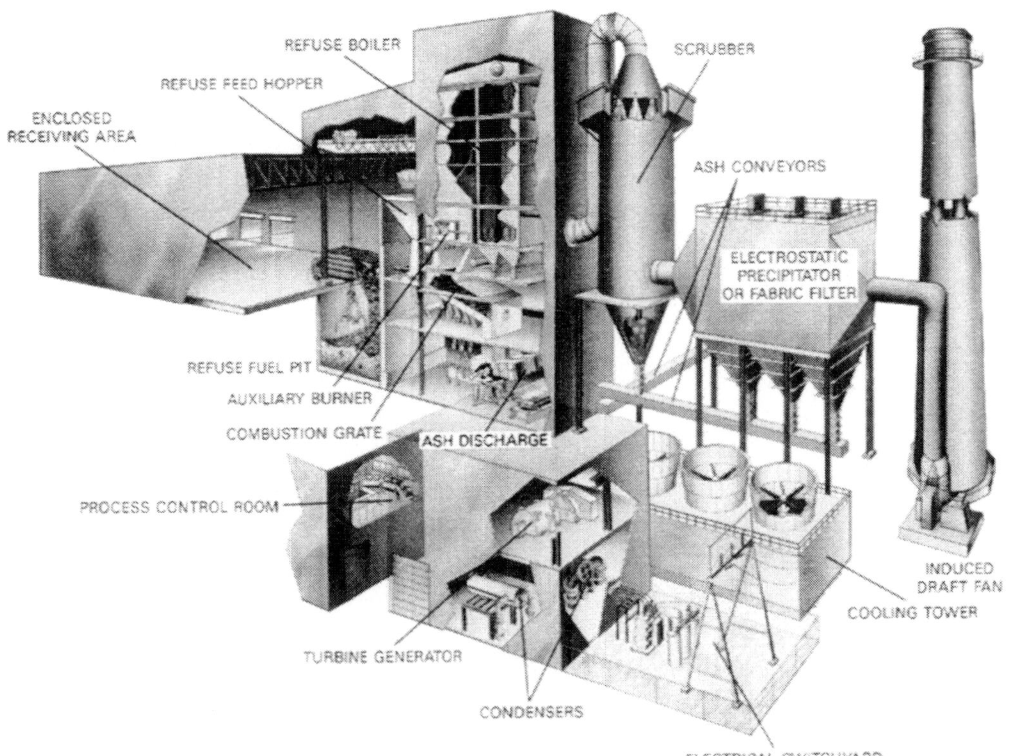

Fig. 22-3. Waterwall furnace—trash-to-energy system. (From Wheelabrator Environmental Systems, Inc. Used with permission.)

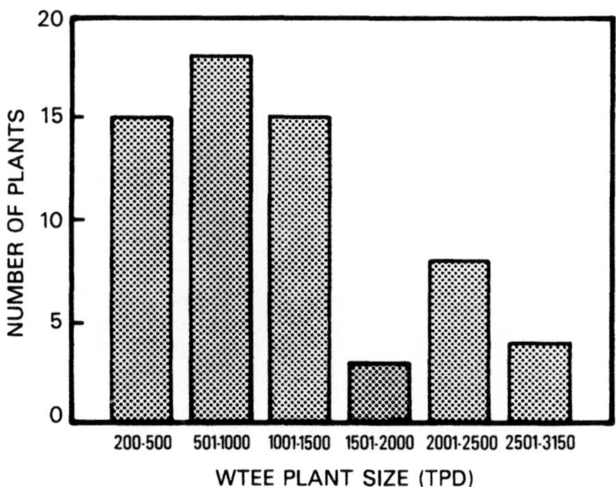

Fig. 22-4. Capacity distribution of United States waterwall furnace WTEE plants.

- Solid waste feeding equipment—The pit and crane method is typically used to feed MSW to the furnace (Robinson 1986). The Modesto Power Plant that burns whole tires uses conveyors to feed the waste to the furnace (Oxford 1988).
- Stoker grate system (Fig. 22-5)—Any grate system has four main functions (Bechtel 1990):
  1. Support burning waste.
  2. Convey waste from feed end to discharge end.
  3. Tumble waste to expose it to combustion air and flame.
  4. Allow combustion air, both underfire and overfire, to contact waste.

  The reciprocating grate is the most common equipment used for burning unprocessed solid waste, but chain, traveling, and water-cooled vibrating grates are also used. These grates all provide continuous ash removal. The Rocket Research Co. (1980) states that stationary pinhole grates, either flat or inclined, are suitable for high-moisture, low-ash fuels such as sugar cane bagasse and wood waste.

Fig. 22-5. Stoker grate system. Detroit Reciprograte® stoker for mass burning of municipal waste installed in a waterwall furnace for waste-to-energy service. (Courtesy of Detroit Stoker Company. Used with permission.)

- Furnace—Excess air of 80% to 100% percent is typically required to ensure complete combustion and reduce the corrosive effect of acid gases produced by combustion of solid waste (Bechtel [1990] and Matuny and Adkin [1989]). Furnaces burning MSW are designed to combat metal wastage through corrosion and abrasion from the combustion gases and the alternating oxidizing and reducing atmosphere. Measures for this include the following:
  1. Line the lower furnace waterwalls with refractory.
  2. Maintain lower steam temperatures.
  3. Use parallel-flow superheaters to minimize tube-metal temperatures.
  4. Design air systems for maximum mixing.
  5. Lower gas velocities through convection sections.
  6. Lower furnace heat release rates.

Robinson (1986) suggests that the use of refractory-lined waterwalls or refractory walls in the lower furnace maintains sufficient combustion temperatures, which is important when high-moisture or low-heating value fuels are being burned. Product brochures from Volund (1991) indicate that the use of refractory walls, however, typically increases the requirements for excess air to cool these walls.

Fuels burned in a waterwall furnace should be of non-uniform shape and size and have relatively high as-received heating value. The fuels that meet these criteria include the following:

- MSW (depending on composition);
- Whole waste tires;
- Corn stover (fodder);
- Rice straw;
- Wood wastes, especially forest residues; and
- Sugar cane bagasse.

Typical cost and performance data for a waterwall furnace WTEE plant are presented in Tables 22-1 and 22-2 for plants burning either MSW or wood waste. These data are shown for plants with input waste capacities ranging from 300 to 2,000 tons per day. Tables 22-1 and 22-2 are based on calculations made by Black & Veatch for the NRECA using a computer program developed by Bechtel (1990) for the Electric Power Research Institute (EPRI).

22.3.1.3 ***Refractory Furnace.*** The refractory furnace (Fig. 22-6) consists entirely of walls lined with a heat-resistant material. There is no heat recovery in the furnace. The combustion gases produced by the burning waste in the furnace travel through a convection-type heat recovery steam generator to generate steam.

The refractory furnace is a proven technology. It is a modification of the technology used for energy recovery from mass burning incinerator systems in the 1930s and 1940s. This technology is used infrequently because of its lower heat recovery efficiency. The refractory furnace does

Table 22-1. Waterwall Furnace WTEE Plant Burning MSW—Cost and Performance Data (Bechtel 1990)[a]

	Design Capacity, Tons/Day				
	300	500	1,000	1,500	2,000
Gross power output, MW	7.6	12.8	26.5	40.0	53.3
Auxiliary power, MW	0.9	1.5	3.0	4.5	5.9
Net power output, MW	6.7	11.3	23.5	35.5	47.4
Total direct cost, 1990 $/ton	112,333	100,000	85,800	78,200	73,350
Total direct cost, 1990 $/kW	5,030	4,425	3,651	3,304	3,095
Total O&M cost, 1990 $/ton	51.18	45.64	39.22	36.13	34.16
Fixed O&M cost, 1990 $/kW-yr	256.12	210.88	154.68	128.23	110.74
Variable O&M cost, 1990 $/MWh	64.31	58.76	51.33	48.82	47.56
Operating efficiency, kWh/ton	536.0	542.4	564.0	568.0	568.8
Net plant heat rate, Btu/kWh	17,780	17,570	16,897	16,778	16,755

[a]Table is based on calculations made by Black & Veatch for the NRECA using a computer program developed by Bechtel (1990) for EPRI.

not use the additional radiant transfer from the burning waste and combustion gas in the furnace. One such refractory furnace WTEE plant is the 225 ton/day Glen Cove Plant owned by the City of Glen Cove, New York. This plant has been in operation since 1983 and burns MSW and 25 tons of sewage sludge per day. It has a reciprocating grate and a convective waste heat boiler. Cost and performance data for this plant are provided in Table 22-3 (Berenyi 1988).

The general design considerations for the major systems used in a refractory furnace WTEE plant are as follows:

- Stoker grate system—The same as for the waterwall furnace.
- Furnace—Excess air of 130% to 170% is typically required to adequately cool the refractory-lined walls.

The hot face of the refractory enclosure should be maintained at a temperature of <1,800° F to limit slagging of the ash (Robinson 1986). Since the convection boiler is downstream of the furnace, corrosion of the boiler tubes resulting from high temperature gases, ash particles, and the oxidation/reduction atmosphere associated with the combustion process is greatly reduced (Peterson and Givonetti 1984). As with the waterwall furnace, the use of refractory walls in the furnace maintains sufficient combustion temperatures, which is important when burning high-moisture or low-heating value fuels (Robinson 1986).

Fuels burned in a refractory furnace should be of non-uniform shape and size, and should have relatively low as-received heating value. The low as-received heating value is necessary to protect refractory-lined walls from excessively high temperatures. Three fuels that meet these criteria are MSW (depending on composition), sugar cane bagasse, and wood wastes (wet), especially forest residues.

22.3.1.4 Rotary Kiln Furnace. The rotary kiln furnace (Fig. 22-7) offered by Volund USA Ltd. is a variation of the waterwall furnace. The reciprocating grate predries and ignites the waste, and the refractory-lined rotary kiln, located after the reciprocating grate, improves the burnout of the fuel (Volund 1991).

The rotary kiln furnace is a proven technology and is one of four basic heat-recovery incineration technologies used in industrial and institutional waste combustors (Energy 1987). Volund USA Ltd., the primary vendor of the rotary kiln grate system used in WTEE applications, has 20 rotary kiln WTEE

Table 22-2. Waterwall Furnace WTEE Plant Burning Wood Waste—Cost and Performance Data (Bechtel 1990)[a]

	Design Capacity, Tons/Day				
	300	500	1,000	1,500	2,000
Gross power output, MW	12.6	12.5	43.7	65.5	87.3
Auxiliary power, MW	1.4	2.3	4.6	6.8	9.1
Net power output, MW	11.2	19.2	39.1	58.7	78.2
Total direct cost, 1990 $/ton	112,333	100,000	85,800	78,200	73,350
Total direct cost, 1990 $/kW	3,009	2,604	2,194	1,998	1,876
Total O&M cost, 1990 $/ton	38.85	33.24	26.87	23.72	21.73
Fixed O&M cost, 1990 $/kW-yr	153.21	124.11	92.97	77.55	67.12
Variable O&M cost, 1990 $/MWh	23.98	20.42	17.00	15.63	14.89
Operating efficiency, kWh/ton	896.0	921.6	938.4	939.2	938.4
Net plant heat rate, Btu/kWh	16,297	15,844	15,561	15,547	15,561

[a]Table is based on calculations made by Black & Veatch for the NRECA using a computer program developed by Bechtel (1990) for EPRI.

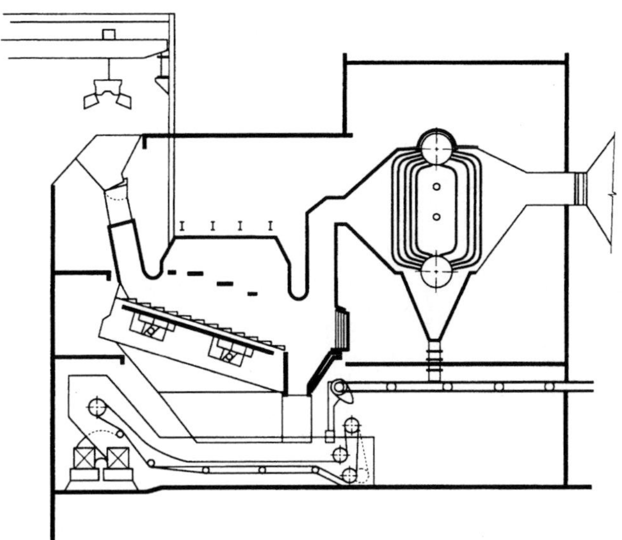

Fig. 22-6. Refractory furnace WTEE plant. (From William D. Robinson, P.E., editor, *The Solid Waste Handbook*. Copyright © 1986, John Wiley & Sons, Inc. Reprinted by permission of John Wiley & Sons, Inc.)

plants either operating or under construction in several foreign countries. These plants, all of which burn MSW, have the following capacity ranges:

- Minimum (tons/day) = 168,
- Maximum (tons/day) = 1,200, and
- Average (tons/day) = 530.

The 1,000 ton/day McKay Bay Refuse-to-Energy Facility owned by the City of Tampa, Florida, is a rotary kiln plant. The plant has been in operation since 1985 (Berenyi 1988). Cost and performance data for this facility are shown in Table 22-4.

The general design considerations for a rotary kiln WTEE plant are as follows:

- Rotary kiln grate system—Waste is typically predried and ignited on a reciprocating grate. Final combustion of the waste occurs in a slowly-rotating (0 to 12 revolutions per hour), inclined rotary kiln. The rotary kiln provides better carbon burnout for wastes with variable properties.
- Furnace—Same as for the waterwall furnace.

The rotary kiln furnace can burn the same fuels as the waterwall furnace. The rotary kiln is best suited for waste streams with wide variations in composition. Variations in waste composition might occur when a secondary waste fuel is infrequently mixed with the primary waste fuel.

22.3.1.5 Water-Cooled Rotary Combustor Furnace. The water-cooled rotary combustor furnace (Fig. 22-8) is a variation of the rotary kiln furnace. The rotary kiln and the reciprocating grate are replaced by the water-cooled rotary combustor.

The Westinghouse O'Conner combustor is designed to rotate at approximately 7 rph. The rotating action and incline cause the fuel to tumble over and down the axis of the combustor. Burning occurs in four stages within the combustor. The first two stages dry the fuel and initiate the burning. The third and fourth stages are the hottest sections in the combustor, where primary burning occurs. Temperatures can reach over 2,000° F. The rotating action of the combustor creates highly turbulent airflow that provides burnout with relatively little excess air required.

After about 30 minutes in the combustor, the remaining fuel and ash fall onto an afterburning grate where any combustibles still present are burned. Figure 22-9 depicts a rotary combustor with a conveyor feed system.

The water-cooled rotary combustor furnace is a proven technology that has been used in Japan and Thailand since the 1970s (TVA 1985). The 200 ton/day Sumner County plant in Gallatin, Tennessee, was the first United States plant to use this technology. It went into operation in 1981. United States plants, either existing or in the advanced planning stage and burning MSW, exhibit the following capacity ranges:

- Minimum (tons/day) = 48,
- Maximum (tons/day) = 2,688, and
- Average (tons/day) = 850.

Figure 22-10 shows the capacity distribution of these plants (Berenyi 1988).

The general design considerations for the major systems used in a water-cooled rotary furnace WTEE plant are the following:

- Water-cooled rotary combustor system—The rotary combustor has a slight incline of 6 degrees and rotates from 5 to

Table 22-3. Refractory Furnace WTEE Plant Burning MSW—Cost and Performance Data Glen Cove Plant (Berenyi 1988)

Item	Value	Item	Value
Design capacity, tons/day	225	Total O&M cost, 1990 $/ton	14.06
Gross power output, MW	2.5	Fixed O&M cost, 1990 $/kW-yr	305.51
Auxiliary power, MW	1.5	Variable O&M cost, 1990 $/MWh[b]	42.31
Net power output, MW	1.0	Operating efficiency, kWh/ton	107
Total capital cost, 1990 $/ton[a]	203,956	Net plant heat rate, Btu/kWh[b]	84,375
Total capital cost, 1990 $/kW[a]	45,890		

[a]Based on capital cost of $40,800,000 ($1987) escalated at 4.0%/yr for 3 years for capital cost of $45,890,000 ($1990).

[b]Assumed 4,500 Btu/lb heating value for MSW fuel.

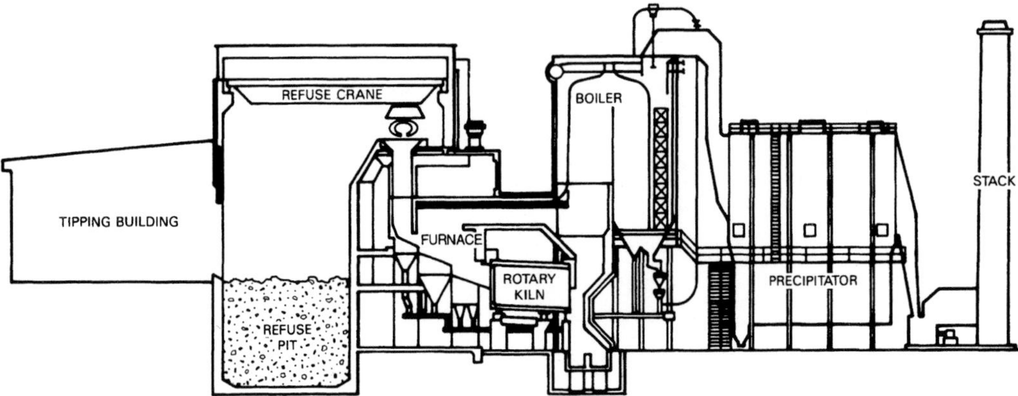

Fig. 22-7. Rotary kiln furnace waste-to-energy plant. (From Volund USA, Ltd. Used with permission.)

20 rph. Primary combustion air at 450° F is introduced into the combustor through holes in the steel plates. The preheated combustion air and mixing provides efficient combustion of high-moisture wastes (up to 65%) without the use of auxiliary fuel and it operates with only 50% excess air (TVA 1985 and McCarthy and Namburi 1989).

- Furnace—Secondary combustion air is fed into the waterwall furnace as overfire air for NO_x control (McCarthy and Namburi 1989). Corrosion in the combustor and waterwall furnace is reduced by the constant oxidizing atmosphere with turbulent mixing in the combustor and the maintaining of tube and steel surface temperatures above 350° F and below 600° F (Boilers 1988).

The primary criterion for fuels burned in a water-cooled rotary combustor furnace is that they be of nonuniform shape and size. This technology has more flexibility than the waterwall furnace in its ability to burn lower-heating value or higher-moisture wastes. Wood wastes, tires, sugar cane bagasse, MSW, corn stover, and rice hulls are burnable in a water-cooled rotary combustor.

Typical cost and performance data for a water-cooled rotary combustor furnace WTEE plant are presented in Table 22-5. These data for plants with input waste capacities ranging from 300 to 2,000 tons per day were generated using least-squares regression analysis of data from Berenyi (1988).

22.3.1.6 Controlled Air Furnace. The controlled air furnace (Fig. 22-11) is a variation of the refractory furnace. This furnace has two combustion chambers to fully burn solid waste, and the unit can be either excess air or starved air. When the first combustion chamber uses less-than-stoichiometric air, it is termed a starved air unit. The second combustion chamber completes combustion of unburned gases from the first chamber (Robinson 1986).

The controlled air furnace, both excess air and starved air units, is a proven technology, having been commercially available since the 1960s (Robinson 1986). United States plants, either existing or in the advanced planning stage and burning MSW, have the following capacity ranges:

- Minimum (tons/day) = 50,
- Maximum (tons/day) = 600, and
- Average (tons/day) = 250.

Figure 22-12 shows the capacity distribution of these plants (Berenyi 1988).

The general design considerations for the major systems used in a controlled air furnace WTEE plant are as follows:

- Solid waste feeding equipment—Front-end loaders and bulldozers push the waste from the tipping floor directly into floor-level charging hoppers (batch feeding).
- Furnace—Both the excess air and the starved air types consist of two refractory-lined combustion chambers: a primary chamber and a secondary chamber. In the primary chamber, waste is moved from the charging hopper and through the chamber by a series of transfer rams. In the starved air furnace, less-than-stoichiometric air allows partial combustion of the waste (primarily the fixed carbon), creating a low-Btu gas (volatile matter). In the excess air furnace, a slight amount of excess air is used for combustion of the waste.

Table 22-4. Rotary Kiln Furnace WTEE Plant Burning MSW—Cost and Performance Data, McKay Bay Refuse-to-Energy Facility (Berenyi 1988)

Item	Value	Item	Value
Design capacity, tons/day	1,000	Total O&M cost, 1990 $/ton	19.72
Gross power output, MW	17.0	Fixed O&M cost, 1990 $/kW-yr	242.61
Auxiliary power, MW	2.0	Variable O&M cost, 1990 $/MWh	14.07
Net power output, MW	15.0	Operating efficiency, kWh/ton	450
Total capital cost, 1990 $/ton[a]	82,160	Net plant heat rate, Btu/kWh[b]	20,000
Total capital cost, 1990 $/kW[a]	5,477		

[a]Based on capital cost of $73,040,000 ($1987) escalated at 4.0%/yr for 3 years for capital cost of $82,160,000 ($1990).

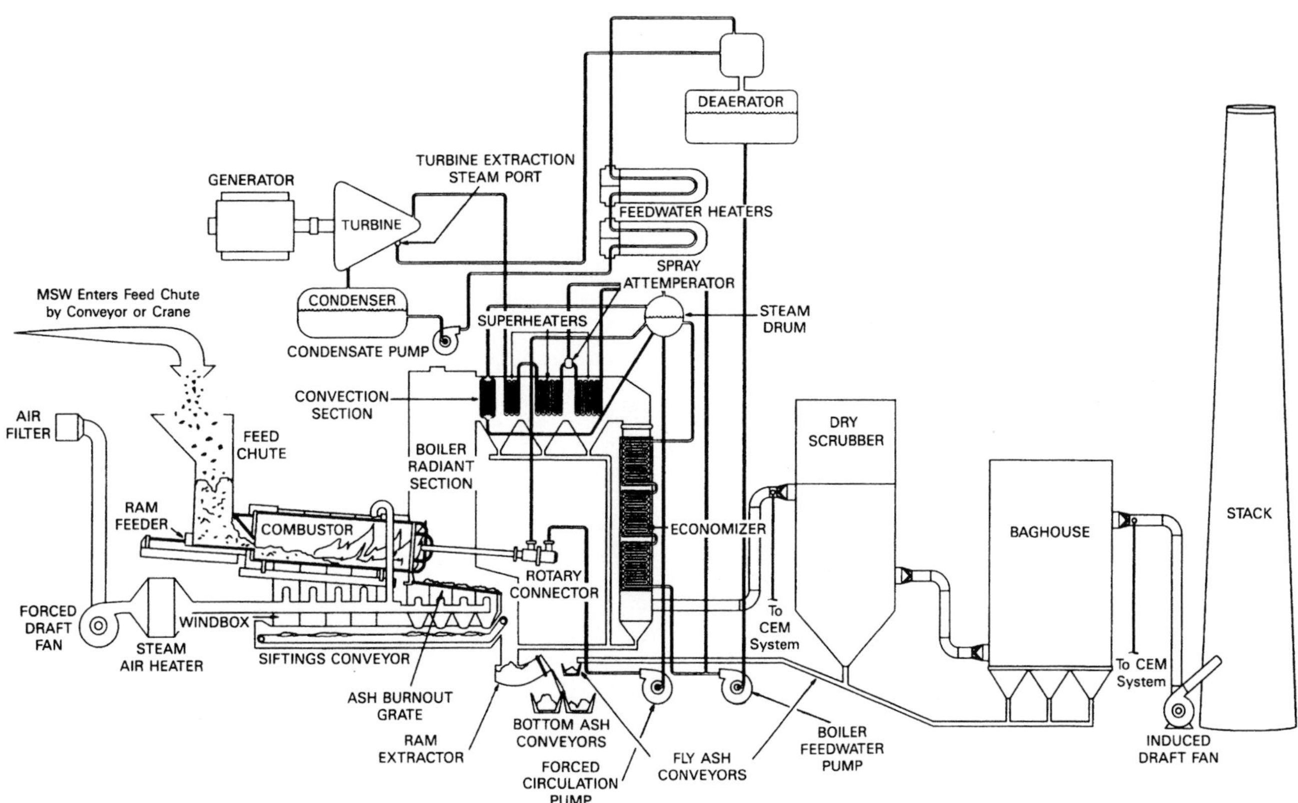

Fig. 22-8. Westinghouse O'Connor water-cooled rotary combustor system. (From Westinghouse Resource Energy Systems. Used with permission.)

In the secondary chamber, complete combustion consumes smoke and odors. Combustion temperatures range from 1,800 to 2,000° F. Auxiliary fuel burners are required below 1,600° F.

Fuels burned in a controlled air furnace should be of nonuniform shape and size. Some systems require a secondary criterion of moisture content of 30% or less. Five fuels meet these criteria: MSW, whole waste tires, corn stover, rice straw, and wood wastes, especially forest residues.

Typical cost and performance data for a controlled air furnace WTEE plant are presented in Table 22-6 for plants burning MSW. These data are shown for plants with input waste capacities ranging from 100 to 400 tons per day. Table 22-6 is based on calculations made by Black & Veatch for the NRECA using a computer program developed by Bechtel (1990) for EPRI.

22.3.2 Processed Solid Waste Combustion

Processed solid waste combustion WTEE technologies burn refuse-derived fuel (RDF) which has been processed from MSW, specific processed agricultural solid wastes, other processed solid wastes, and unprocessed agricultural and other solid wastes, such as sawdust, that are of a suitably

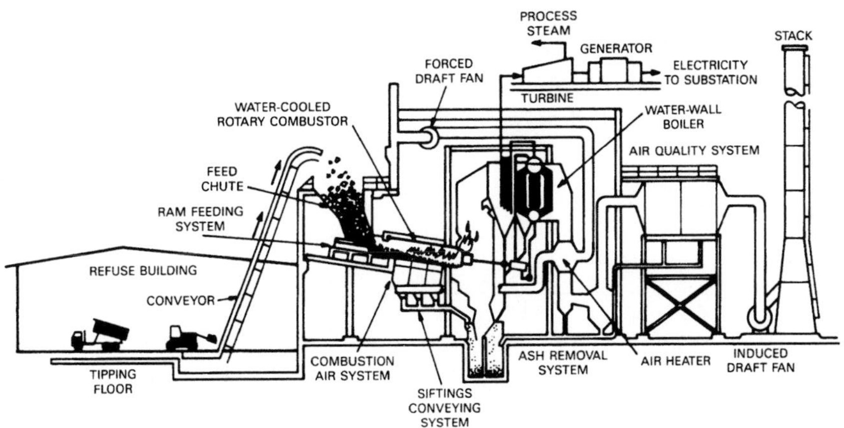

Fig. 22-9. Westinghouse water-cooled rotary combustor furnace WTEE plant. (From Westinghouse Resource Energy Systems. Used with permission.)

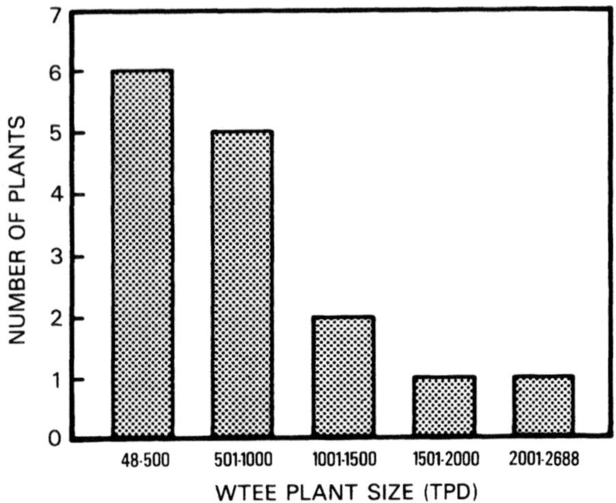

Fig. 22-10. Capacity distribution of United States water-cooled rotary combustor furnace WTEE plants.

uniform size for combustion. All of these technologies use a combination of radiant and convective heat transfer surfaces to generate steam and may use a steam turbine to generate electricity.

*22.3.2.1 **RDF Production.*** The American Society for Testing and Materials (ASTM) has developed seven category definitions of RDF based on the level of processing applied to the MSW. These seven categories are listed in Table 22-7. The primary categories of interest in this chapter are RDF-2 through RDF-5.

The primary process for producing RDF is the dry process, which involves the mechanical preparation of MSW into dry RDF and salable recovered materials. Table 22-8 presents typical dry process components and systems, as well as the characteristics and applications of the RDF produced by these processes. Figure 22-13 illustrates each of the processes listed in Table 22-8.

*22.3.2.2 **Processed Waste.*** Five major processes are available to improve the physical and/or chemical properties of agricultural and other solid wastes (Solar Energy Research Institute Vol. II 1979):

- Drying—Removal of physically bound water either thermally or mechanically in centrifuges and presses;
- Comminution—Reduction of particle size by shredding, cutting, grinding, or pulverization;
- Densification—Increase of bulk density of waste through pelletizing, cubing, briquetting, extrusion, or rolling-compressing;
- Physical separation—Removal of unwanted components from solid waste; and
- Chemical modification—Change of waste's chemical structure.

The major processes for agricultural and other solid wastes are drying, comminution, and densification.

An example of the combustion of agricultural wastes, such as rice hulls and straw, is the Floyd Myers Marsh Power Plant in California's Sacramento Valley, which has a suspension-fired boiler. Rice hulls are ground in one of four hammermills, each equipped with hard-faced hammers and chromed screens to accommodate the extreme abrasion caused by the rice hulls. These grinders double the density of the rice hulls from 10 lb/ft^3 to 20 lb/ft^3. Whole bales of rice straw are chopped in a tub-type grinder into pieces about 1½ to 2 in. long. A fifth, larger hammermill grinder then grinds the chopped straw.

An example of a facility that burns wood wastes is the 7-MW wood-waste-fueled cogeneration plant at Chapleau, Ontario, which has a spreader stoker-fired boiler. This plant receives wood waste from sawmills in the form of sawdust, bark, planer shavings, sawed-off tree trunks, and ends of trees. A wood shredder reduces the oversize material to <4 in. (Anthony 1989).

An example of a facility that burns coal screenings referred to as gob or culm is the anthracite culm-fired St. Nicholas Cogeneration Project at Mahanoy, Pennsylvania, which has a circulating fluidized bed boiler. The culm is first sized to <4 in. using a scalping screen. Two crushers reduce the remaining culm to 1 in. × 0 in. size and then to $\frac{1}{16}$ in. × 0 in. size. An oil-fired flash dryer dries the culm to 9% moisture (Thies and Heina 1990).

*22.3.2.3 **Spreader Stoker-Fired Boiler.*** A spreader stoker-fired boiler (Fig. 22-14) uses multiple feeder–distributor units (spreaders) to spread the solid waste onto a grate sys-

Table 22-5. Water-Cooled Rotary Combustor Furnace WTEE Plant Burning MSW—Cost and Performance Data (Bechtel 1988)

	Design Capacity, Tons/Day				
	300	500	1,000	1,500	2,000
Gross power output, MW	5.4	11.6	27.2	42.8	58.4
Auxiliary power, MW	0.8	1.5	3.1	4.8	6.5
Net power output, MW	4.5	10.1	24.1	38.0	51.9
Total capital cost, 1990 $/ton	84,309	83,261	81,860	81,052	80,483
Total capital cost, 1990 $/kW	5,566	4,113	3,402	3,199	3,099
Total O&M cost, 1990 $/ton	28.97	25.35	21.15	19.03	17.65
Fixed O&M cost, 1990 $/kW-yr	539.03	341.09	183.31	127.48	98.52
Variable O&M cost, 1990 $/MWh	24.89	19.37	13.79	11.30	9.81
Operating efficiency, kWh/ton	400.1	441.7	505.0	546.2	577.5
Net plant heat rate, Btu/kWh	23,251	21,451	19,230	18,039	17,239

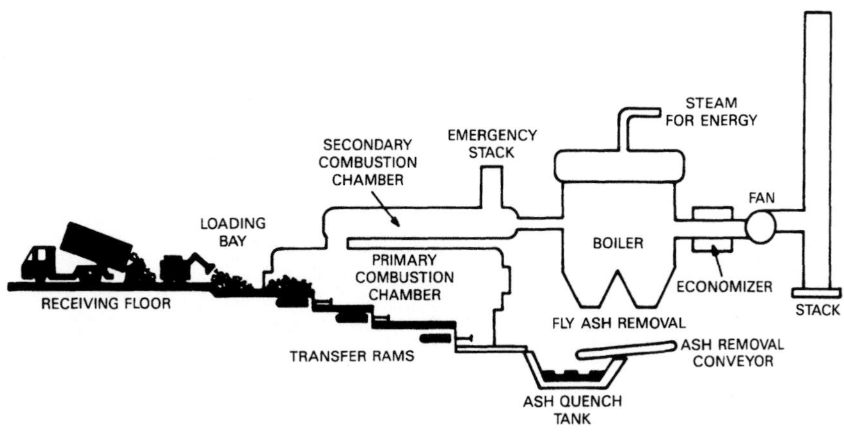

Fig. 22-11. Controlled air furnace WTEE plant.

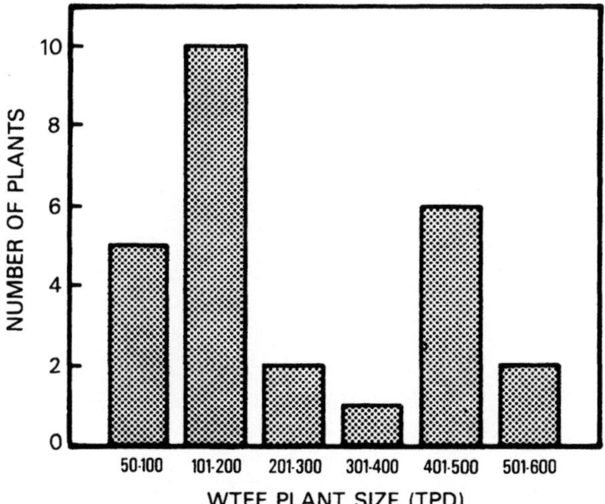

Fig. 22-12. Capacity distribution of United States controlled air furnace WTEE plants.

tem. The lighter portion of the waste burns in suspension above the grate, and the heavier portion burns in a bed on the grate. Underfire combustion air is used to support burning of the waste on the grate, while overfire air is used to complete the burning of the waste in suspension (Robinson 1986).

The spreader stoker-fired boiler is a proven technology, and has been used to burn bagasse and wood fibers since the 1930s and RDF since 1972 (Robinson 1986). United States plants, either existing or in the advanced planning stage and burning RDF, have the following capacity ranges:

- Minimum (tons/day) = 350,
- Maximum (tons/day) = 4,000, and
- Average (tons/day) = 1,600.

Figure 22-15 shows the capacity distribution of these plants (Berenyi 1988).

The spreader stoker-fired boiler technology has been applied to the combustion of various solid wastes. One such plant is the 7-MW cogeneration plant at Chapleau, Ontario. This plant went into operation in 1987 and burns 360 tons of wood waste per day (Anthony 1989). Another example is the 20-MW cogeneration plant that has been operating since 1981 at the Lihue Plantation Co.'s sugar factory on the island of Kauai, Hawaii. This plant burns 1,280 tons/day of sugar cane bagasse and is designed to also burn cane tops and leaves (Foster Wheeler 1991).

The general design considerations for the major systems used in a spreader stoker-fired boiler WTEE plant are as follows:

Table 22-6. Controlled Air Furnace WTEE Plant Burning MSW—Cost and Performance Data (Bechtel 1990)[a]

	Design Capacity, Tons/Day						
	100	150	200	250	300	350	400
Gross power output, MW	2.2	3.3	4.4	5.5	6.6	7.7	8.8
Auxiliary power, MW	0.3	0.5	0.6	0.8	1.0	1.1	1.3
Net power output, MW	1.9	2.8	3.8	4.7	5.6	6.6	7.5
Total direct cost, 1990 $/ton	94,479	87,218	82,500	78,800	76,000	73,414	72,000
Total direct cost, 1990 $/kW	4,942	4,590	4,342	4,191	4,071	3,909	3,840
Total O&M cost, 1990 $/ton	67.58	59.44	54.37	50.52	47.65	45.40	43.70
Fixed O&M cost, 1990 $/kW-yr	425.26	352.91	307.63	280.00	258.75	237.88	224.27
Variable O&M cost, 1990 $/MWh	92.28	84.52	79.29	75.80	73.09	69.87	68.53
Operating efficiency, kWh/ton	458.8	456.0	456.0	451.2	448.0	452.6	450.0
Net plant heat rate, Btu/kWh	20,772	20,897	20,899	21,121	21,272	21,056	21,178

[a]Table is based on calculations made by Black & Veatch for the NRECA using a computer program developed by Bechtel (1990) for EPRI.

Table 22-7. ASTM Refuse-Derived Fuel (RDF) Classification (Robinson 1986)

ASTM Classification	Form	Description
RDF-1	Raw	MSW used as fuel in as-discarded form with minimal processing to remove nonprocessible and oversized items such as appliances. Typically used in mass burn unit.
RDF-2	Coarse	MSW processed to coarse particle size with or without ferrous metal separation. Typically burned in a spreader stoker-fired boiler. Higher heating value of 4,500–5,000 Btu/lb and a yield of approximately 70–80%.
RDF-3	Fluff	MSW processed to a particle size that 95% by weight will pass through a 2-in. square mesh screen and from which most metals, glass, and other inorganic material have been removed. May be burned in spreader stoker-fired boiler or suspension-fired boilers equipped with dump grates. Typical higher heating value of about 6,000 Btu/lb and yield of approximately 65%. Used when cofiring.
RDF-4	Powder	MSW processed into powdered form, 95% by weight passes through a 10-mesh screen and from which most metals, glass, and other inorganic materials have been removed. Burned in suspension-fired boilers using pulverized coal-type burners. Typical higher heating value of 6,500 to 7,500 Btu/lb.
RDF-5	Densified	MSW that has been processed and densified (compressed) into the form of pellets, slugs, cubettes, or briquettes. Higher quality fuel with longer term storage capability, which reduces fuel transportation costs. May be burned in spreader stoker-fired boilers.
RDF-6	Liquid	MSW processed into liquid fuel.
RDF-7	Gas	MSW processed into gaseous fuel.

- Solid waste feeding/distribution equipment—Feeders, or metering devices, regulate the flow of waste to the furnace. The five types of devices are reciprocating plate or ram, variable-speed chain or belt conveyor, vibrating, rotating drum, and screw auger. Mechanical (rotating drum) or pneumatic (air-swept spout) distributors uniformly distribute the waste onto the grate. Waste particles should have a nominal size of 4 in. or less, depending on the density of the fuel. Applicable ASTM RDF classifications would be RDF-2 and RDF-3 (Robinson 1986).
- Stoker grate system—Continuously cleaning traveling grates are the most common, but continuously cleaning oscillating grates and water-cooled vibrating grates are also used (Oxford 1988).
- Furnace—Bechtel (1990) states that excess air of 40% to 50% is typically required to ensure complete combustion of the

Table 22-8. Typical Dry Process Components and Systems (Robinson 1986)

Process	Product Size Range	Yield RDF/ Raw Material[a]	Firing Method	Other Comments
Rotary shear + magnet	+ 8–12 in. strips (20–30 cm) 4–8 in. (10–20 cm) topsize	90–95%	Mass burn, fluid bed, pyrolysis, if no further processing	Frequent and erratic stop/start/reverse interruptions; maintenance and capacity uncertainty. In shakedown stages
Shred + magnet	90% 6 in. (15 cm)	+ 90%	Spreader stoker	RDF retains glass/grit and topsizes
Shred + magnet + screen	90–95% in. (10 cm) 60–70% 3 in. (7.5 cm)	50–60% if screen overs discarded	Spreader stoker	50–60% glass/grit and most onerous topsize removed if screen overs discarded
Trommel + shred + magnet	90% 6 in. (15 cm)	50–60% if trommel combustible unders discarded	Spreader stoker	90% Materials recovery, cans and glass + grit removal; reduced downstream wear
Trommel + shred + magnet + screen	90–95% 4 in. (10 cm) 60–70% 3 in. (7.5 cm)	50–60% if raw trommel combustible unders discarded	Full suspension, spreader stoker	
Flail + magnet + screen + shred + air knife	90% 2 in. (5 cm) 60–70% 1 in. (2.5 cm)	50–60%	Full suspension, spreader stoker	50–60% Glass/grit removal
A. Shred, magnets, screens, air knife B. + Screen overs to reshred	A. 90% 4 in. (10 cm) 75% 2 in. (5 cm) B. 90% 2 in. (5 cm) 75% ¾ in (1.9 cm)	85–90%	Spreader stoker or full suspension	80–85% Material recovery, primary shred option to 2-in. (51 mm)
[a]RDF preparation	(Typical pellet cubette sizes)	40–50%	Spreader; mass burn; coal-mixed then pulverized	90% Material recovery; clean, dry, and fine prepelletizing preparation required

[a]"Overs" and "unders" relate to materials greater or less than the optimum product size range.

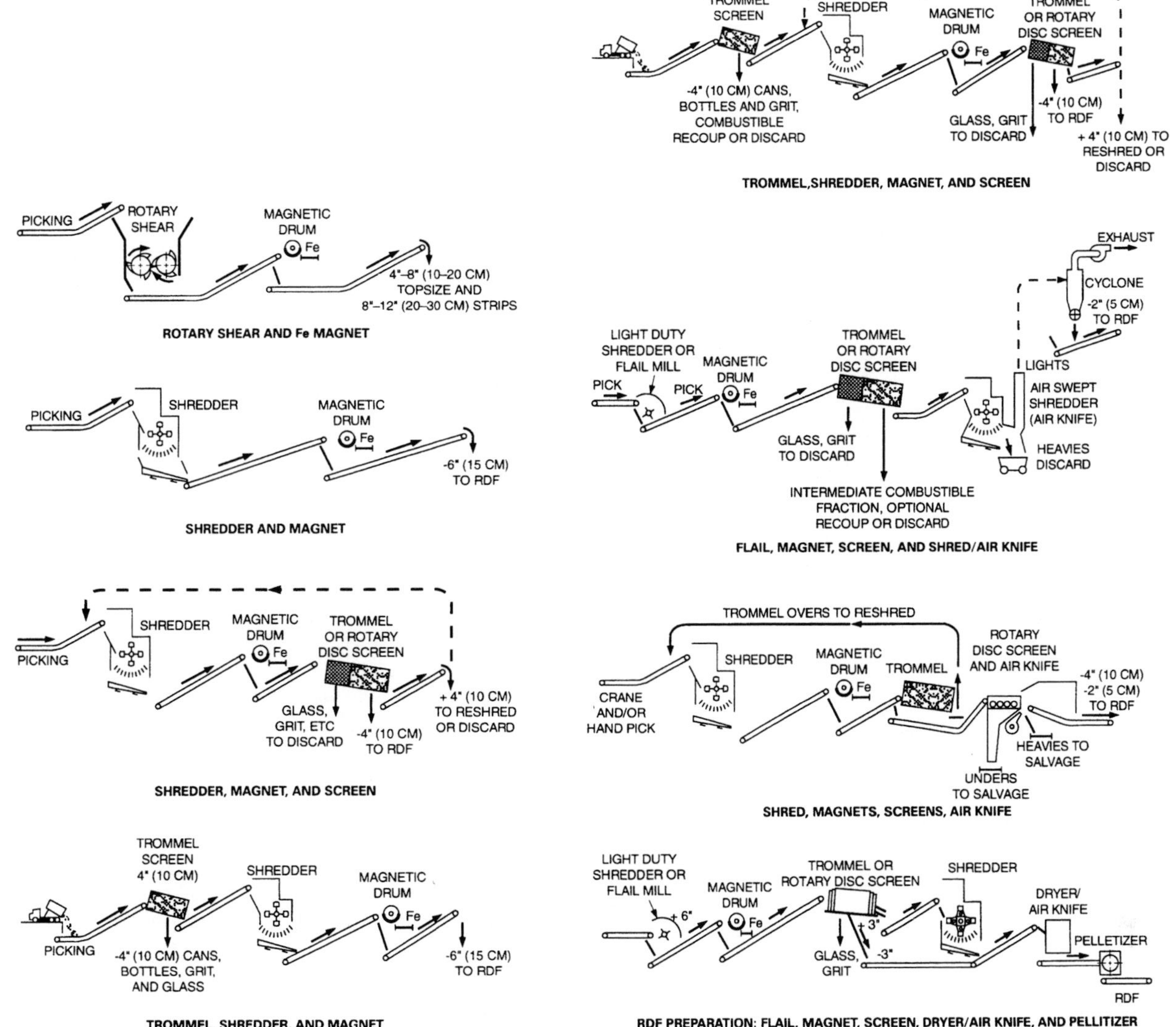

Fig. 22-13. Typical dry process components and systems. (From William D. Robinson, P. E., editor, *The Solid Waste Handbook*. Copyright © 1986, John Wiley & Sons, Inc. Reprinted by permission of John Wiley & Sons, Inc.)

> waste. Furnaces burning RDF should be designed to combat corrosion and erosion from the combustion gases by applying a metal overlay on boiler tubes, by using bimetallic composite tubes consisting of an inner carbon steel tube and an outer Inconel tube, and by keeping tube metal temperatures below about 800° F. Slagging in the furnace and fouling in the convection pass are concerns for certain fuels. A maximum gas velocity of 30 fps should be considered.

Fuels burned in a spreader stoker-fired boiler should be of uniform shape and relatively small size (dimensions <4 in. for RDF), either as-received or by processing, have moisture contents of less than 60%, and preferably have low to moderate slagging and fouling potential (Cheremisinoff et al. 1980). These six fuels meet these criteria: MSW processed into RDF-2 or RDF-3, shredded waste tires, wood wastes, peanut hulls, sugar cane bagasse, and corn stover.

Although documentation concerning the combustion of shredded waste tires or corn stover in a spreader stoker-fired boiler is limited, it is believed to be technically feasible to burn these fuels in a spreader stoker-fired boiler. However, the high heat content of tires, approximately 14,000 Btu/lb, requires special consideration.

Typical cost and performance data for a spreader stoker-fired boiler WTEE plant are presented in Tables 22-9 and 22-10 for plants burning either RDF or wood waste. These data are shown for plants with input waste capacities ranging from 300 to 3,300 tons per day.

22.3.2.4 Suspension-Fired Boiler. A suspension-fired boiler allows most of the solid waste to be burned in suspension, but some of the waste falls down to dump grates at the bottom of the furnace where burning is completed. Combustion air is used to convey the waste into the furnace.

The technology has been applied to the combustion of various agricultural wastes. One such plant is the 26-MW Floyd Myers Marsh Power Plant located near the city of

Fig. 22-14. Spreader stoker-fired boiler. Detroit rotograte stoker equipped with Detroit refuse metering feeder and arranged for RDF firing. (Courtesy of Detroit Stoker Company. Used with permission.)

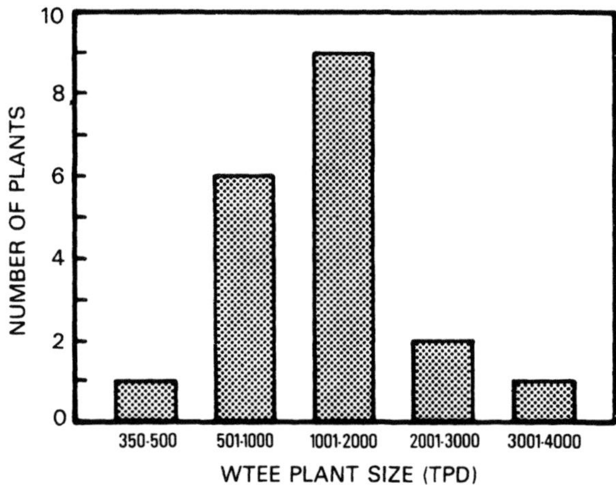

Fig. 22-15. Capacity distribution of United States spreader stoker-fired WTEE plants.

Williams, California. This plant went into operation in 1989 and burns 184,000 tons of waste rice hulls and straw per year. Another such plant is the 10.5-MW power plant at Lake Charles, Louisiana. This plant went into operation in 1984 and burns 90,000 tons of waste rice hulls per year (Schwieger 1985).

The general design considerations for the major systems used in a suspension-fired boiler WTEE plant are as follows:

- Burner system—RDF or other processed solid waste is conveyed to the burner wind boxes by a pneumatic transport system using heated combustion air. Two types of burner systems used are horizontally fired and tangentially fired. In the horizontally fired system, burners are located along rows on either the front wall or on both the front and back walls (opposed firing) of the furnace. Sufficient turbulence is provided by rotational motion imparted on the waste and air mixture by individual burners. For the tangentially fired system, burners are located along rows in the corners of the furnace. The waste and air mixture is directed by individual burners along a line tangent to a horizontal circle at the center of the furnace, providing sufficient turbulence. Wood waste and other dense wastes must be pulverized, while RDF and other less dense wastes must be sized <2 in. Applicable ASTM RDF classifications are RDF-3 and RDF-4. A dump grate at the bottom of the furnace is used to burn any waste particles that do not burn in suspension (Robinson 1986 and Singer 1981).
- Furnace—Singer (1981) states that this is the same as for the spreader stoker-fired boiler except the excess air requirement is only about 30%.

Fuels burned in a suspension-fired boiler should be of uniform shape and small size (dimensions <2 in. for RDF) either as-received or by processing (Makansi 1988). Wood waste and other dense wastes that must be pulverized should be dried to moisture contents of <20% and should have preferably low to moderate slagging and fouling potential. Fuels which meet these criteria include MSW processed into RDF-3 and RDF-4, wood wastes, rice straw and hulls, and sugar cane bagasse.

Table 22-9. Spreader Stoker-Fired Boiler WTEE Plant Burning RDF—Cost and Performance Data (Bechtel 1990)[a]

	Design Capacity, Tons/Day				
	300	500	1,000	1,500	2,000
Gross power output, MW	8.1	13.7	28.3	42.5	56.6
Auxiliary power, MW	1.2	2.0	4.0	5.9	7.9
Net power output, MW	6.9	11.7	24.3	36.6	48.7
Total direct cost, 1990 $/ton	163,000	136,600	107,500	93,400	84,600
Total direct cost, 1990 $/kW	7,087	5,838	4,424	3,828	3,474
Total O&M cost, 1990 $/ton	69.03	57.72	45.65	40.09	36.71
Fixed O&M cost, 1990 $/kW-yr	431.16	335.73	231.19	185.82	157.68
Variable O&M cost, 1990 $/MWh	66.87	57.76	47.68	44.16	42.44
Operating efficiency, kWh/ton	552.0	561.6	583.2	585.6	584.4
Net plant heat rate, Btu/kWh	19,587	19,252	18,539	18,463	18,501

[a]Table is based on calculations made by Black & Veatch for the NRECA using a computer program developed by Bechtel (1990) for EPRI.

Table 22-10. Spreader Stoker-Fired Boiler WTEE Plant Burning Wood Waste—Cost and Performance Data (Bechtel 1990)[a]

	Design Capacity, Tons/Day				
	300	500	1,000	1,500	2,000
Gross power output, MW	13.5	23.0	46.6	69.9	93.2
Auxiliary power, MW	1.8	3.1	6.1	9.2	12.3
Net power output, MW	11.7	19.9	40.5	60.7	80.9
Total direct cost, 1990 $/ton	153,000	128,300	100,950	87,700	79,450
Total direct cost, 1990 $/kW	3,923	3,224	2,493	2,167	1,964
Total O&M cost, 1990 $/ton	59.24	47.89	35.83	30.30	26.89
Fixed O&M cost, 1990 $/kW-yr	254.27	197.39	138.72	112.04	94.92
Variable O&M cost, 1990 $/MWh	28.42	23.13	17.97	16.02	14.90
Operating efficiency, kWh/ton	936.0	955.2	972.0	971.2	970.8
Net plant heat rate, Btu/kWh	15,600	15,287	15,023	15,035	15,041

[a]Table is based on calculations made by Black & Veatch for the NRECA using a computer program developed by Bechtel (1990) for EPRI.

22.3.2.5 Fluidized Bed Combustor. A fluidized bed combustor (FBC) burns solid waste in a preheated bed of inert material such as ash, sand, or limestone, that is supported by a column of upward-flowing air. Depending on the upward air velocities, these combustors are classified as either bubbling bed (Fig. 22-16) or circulating bed (Fig. 22-17).

The FBC is a commercially available technology, nearing the status of a fully proven technology. Table 22-11 shows US FBC plants that are either existing or in the advanced planning stage (Bechtel 1990; Berenyi 1988; Thies and Heina 1990; Wilhelm and Simbeck 1990).

The general design considerations for the major systems used in a fluidized bed combustor WTEE plant are as follows:

- FBC system—The bubbling bed type uses fluidizing air velocity maintained at 3 to 10 ft/s to prevent carryover of solids from the bed to the boiler. Fuel is fed either pneumatically from underneath the bed through multiple feed points, or mechanically or pneumatically from above the bed. Bed materials that are carried over are reinjected into the bed.

  The circulating bed maintains fluidizing air velocity at 13 to 20 ft/s to prevent a dense bed from forming and to encourage the carryover of solids from the bed. A cyclone separator recirculates the particles carried over from the furnace, and fuel is fed pneumatically into the combustor near the bottom of the combustor. The combustor must be designed to resist the abrasive and corrosive nature of both the bed material and the waste that is burned (Bechtel 1990 and Churgin 1989).

  Desired combustion temperature depends on the priority of

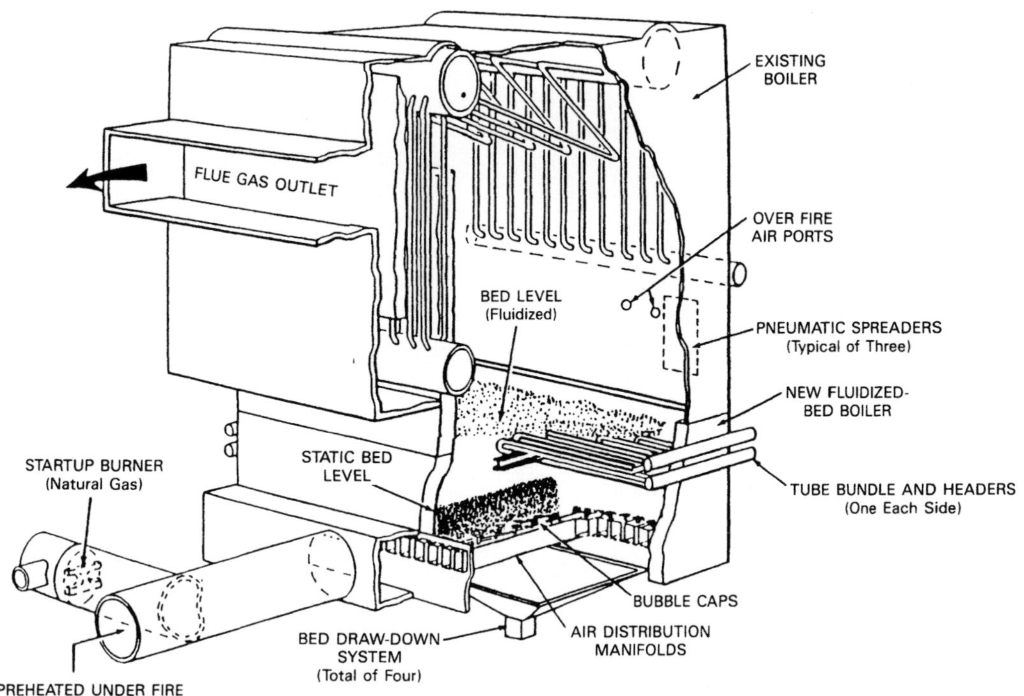

Fig. 22-16. Bubbling-bed fluidized bed combustor. (From Electric Power Research Institute and Government Institutes, Inc. Used with permission.)

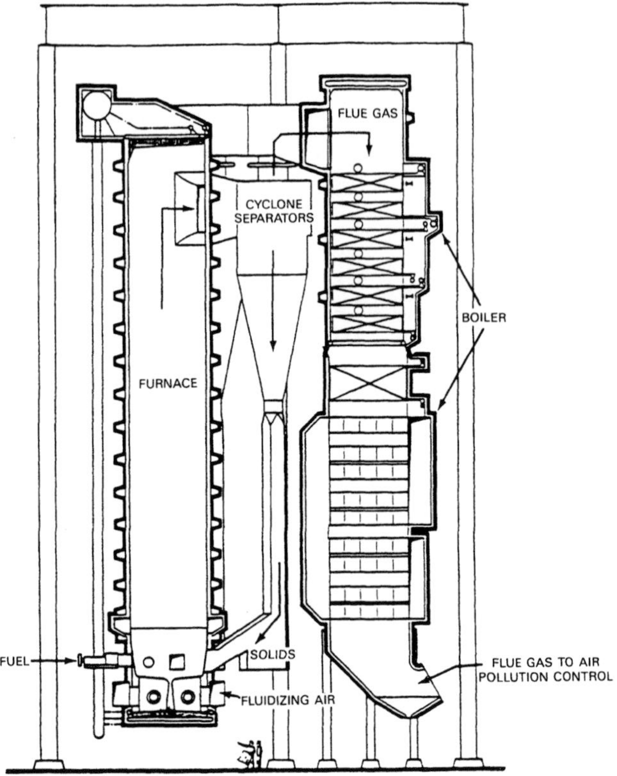

Fig. 22-17. Circulating-bed fluidized bed combustor. (From Electric Power Research Institute and Gotaverken Energy Systems AB. Used with permission.)

emissions control goals (Churgin 1989). Good mixing and complete combustion are required to destroy dioxins and furans. A temperature of 1,800° F is sometimes used as an indicator. To minimize production of NO_x, temperature should be kept below 1,600° F, and reduction of SO_2 through interaction with limestone is optimized at 1,450° F. Typical combustion temperatures are too high for neutralization of HCl in the combustor. Excess air of 50% is typically required (Bechtel 1990).

- Waterwall boiler—As was the case for the spreader stoker-fired boiler, fouling and corrosion in the boiler is a major concern, particularly with RDF.

Fuels burned in a FBC should be of uniform shape and small size (dimensions <2 in. for RDF), either as-received or by processing, and should not contain glass or metal fragments (Churgin 1989 and Bechtel 1990). Fuels meeting these criteria include MSW processed into RDF-3, shredded waste tires, wood wastes, gob, rice straw and hulls, corn stover, peanut hulls, sugar cane bagasse, broiler litter, and hog manure.

It is technically feasible to burn these fuels in a fluidized bed combustor. Since cattle manure has been burned in a California plant (Table 22-11), both broiler litter and hog manure are assumed burnable in a fluidized bed combustor (Wilhelm and Simbeck 1990).

Typical cost and performance data for a fluidized bed combustor WTEE plant are presented in Table 22-12 for plants burning either RDF, wood waste, or gob. These data are shown for plants with input waste capacities ranging from 300 to 2,000 tons per day.

22.3.2.6 Cyclone Furnace Unit. A cyclone furnace unit (Fig. 22-18) uses small cyclone furnaces 6 to 10 ft in diameter in conjunction with a typical waterwall furnace. Complete combustion takes place in the separate cyclone furnaces, which are refractory-lined, water-cooled, horizontal chambers. The hot combustion gases produced in the cyclone furnaces are then discharged into the boiler.

The cyclone furnace unit is a proven technology, in use commercially to burn coal since 1944. Currently, there are no existing or planned units using this technology that burn exclusively RDF. However, this technology is used to cofire RDF as a supplemental fuel with coal or oil in existing utility or industrial boilers (Robinson 1986).

Robinson (1986) and Babcock & Wilcox (1978) list the general design considerations for the major systems used in a cyclone furnace WTEE plant as the following:

- Cyclone furnace—RDF or other processed solid waste is conveyed to the burner ends of the individual cyclone furnaces by a pneumatic transport system using heated combustion air. Three types of combustion air are used to provide turbulence in the cyclone furnace. Primary air (20% of total) enters tangentially at burner end. Secondary air (75% of total) with a velocity of 300 ft/s enters tangentially at the roof of the main barrel of the cyclone furnace. Tertiary air (5% of total) enters through center of burner end. Excess air of 10% to 15% is typically required. To maintain the thin layer of slag on the cylinder walls necessary to hold the burning waste particles, the waste must have a sufficient ash content (4% to 25%, dry basis). Temperatures of over 3,000° F are required to complete combustion inside the cyclone furnaces. Tungsten carbide or other erosion-resistant wear liners are used inside the cyclone furnace to minimize the abrasive effects of the high-velocity waste particles on the inside walls of the furnace.
- Waterwall boiler cyclone furnaces are arranged on either a single wall or on two opposing walls in the lower portion of the boiler. The waterwalls in the lower portion of the boiler are lined with slag-resistant refractory to maintain the high temperatures to keep the excess molten slag leaving the cyclone furnaces in a liquid state, and to protect the waterwalls from the high-temperature (3,000° F) gases leaving the cyclone furnaces.

Fuels burned in a cyclone furnace unit should be of uniform shape and small size (dimensions <2 in. for RDF), either as-received or by processing. The fuel's proximate analysis should fall within the following ranges (Babcock & Wilcox 1978):

- Moisture—2% to 40%,
- Volatile matter—18% to 45% (dry basis),
- Fixed carbon—35% to 75% (dry basis), and
- Ash—4% to 25% (dry basis).

Cyclone furnaces are better suited to firing coal, rather than RDF or other wastes, because of the restrictive fuel analysis criterion. However, RDF-3 and wood wastes have

Table 22-11. United States Fluidized Bed Combustor WTEE Plants in Operation or Advanced Planning

Plant Name (Type)[a]	State	Owner	FBC Manufacturer	Design Capacity TPD/MW	Fuels (TPD)[b]
French Island (BB)	WI	Northern States Power	Energy Products of Idaho (EPI)	1,075/28	RDF (555), WW (520)
Tacoma (BB)	WA	City of Tacoma	EPI	1,630/43	RDF (310), WW (815), Coal (505)
Truckee Meadows (NA)	NV	Truckee Meadows	Wormser Engineering	1,000/40	RDF
Derry (NA)	NH	Power Recovery Systems	Power Recovery Systems	450/11	RDF
Robbins (CFB)	IL	Robbins Resource Recovery Company	Gotaverken Energy Systems	1,500/40	RDF
Erie (CFB)	PA	Erie Energy Recovery Company	Gotaverken Energy Systems	850/20	RDF
NEPCO Cogen. (CFB)	PA	Northeastern Power Company	Combustion Engineering, Inc.	1,300/50	AC (1160), Coal (170)
Gilberton Cogen. (CFB)	PA	Gilberton Power Co.,	Pyropower Corp.	NA/80	AC
Westwood (CFB)	PA	Westwood Energy Properties	Combustion Engineering, Inc.	NA/30	AC
St. Nicholas Cogen. (CFB)	PA	Schuylkill Energy Resources Inc.	Combustion Engineering, Inc.	5,000/80	AC
Cambria Cogen. (CFB)	PA	Cambria CoGen Co.	Pyropower Corp.	2,230/85	AC (2060), Coal (170)
Feather River (CFB)	CA	Sithe/Energies	B&W/Studsvik	NA/18	WW, CR
Fresno (CFB)	CA	Ultrasystems	C-E/Lurgi	NA/24	WW, CR
Rocklin (CFB)	CA	Ultrasystems	C-E/Lurgi	NA/25	WW, CR
Mecca (CFB)	CA	Colmac Energy Inc.	C-E/Lurgi	NA/47	WW, CR
Chinese Station (BB)	CA	Pacific Ultrapower	EPI	NA/20	WW
El Nido (BB)	CA	CAPCO	EPI	NA/10	CR
Chowchilla (BB)	CA	CAPCO	EPI	NA/10	CR
Madera (BB)	CA	CAPCO	EPI	NA/25	CR
Calexico (BB)	CA	South Valley Power	EPI	NA/5	Cattle manure
Delano (BB)	CA	Thermo Electron	EPI	NA/27	WW, CR
Mendota (CFB)	CA	Thermo Electron	Gotaverken Energy Systems	NA/26	WW, CR
Woodland (CFB)	CA	Thermo Electron	Gotaverken Energy Systems	NA/25	WW, CR

[a]Plant types:
- BB-bubbling bed fluidized bed combustor
- CFB-circulating bed fluidized bed combustor

[b]Fuel types:
- RDF—refuse-derived fuel
- WW—wood waste
- AC—anthracite culm/silt
- CR—crop residues, which include wheat straw, corn stalks, cotton stalks, rice straw, and rice hulls.

been cofired as a supplementary fuel to coal in a cyclone furnace WTEE plant.

22.3.2.7 Cofiring as a Supplemental Fuel with Fossil Fuels. Bechtel (1990) states that any of the five processed solid waste combustion technologies can be used to cofire solid waste as a supplemental fuel with fossil fuels.

The cofiring of solid wastes as a supplemental fuel with fossil fuels is a proven technology. Table 22-13 shows six operating US plants that cofire solid waste as a supplemental fuel with coal (Bechtel 1990 and Berenyi 1988).

Bechtel (1990), using guidelines published by the Electric Power Research Institute, lists the general design considerations particular to the cofiring of solid waste in a fossil-fueled generating station as the following:

- Burning RDF limits steam temperatures in the boiler to 800 to 850° F due to corrosion from HCl produced during combustion.
- RDF amounts are typically limited to 10% to 20% by heat input to minimize problems.
- Typical modifications to a fossil-fueled boiler include adding an RDF or solid waste feed point, adding a dump grate in a suspension-fired boiler, increasing the overfire air supply and adding overfire air nozzles, and enlarging the bottom ash handling system.

The primary fuels that have been cofired with coal are RDF and wood wastes. Typical cost and performance data for cofiring RDF in a coal-fueled power plant are presented in Table 22-14. These data are shown for RDF heat input percentages ranging from 5% to 20% for three sizes of power plants:

- Retrofit 50-MW power plant,
- New 210-MW power plant, and
- New 515-MW power plant.

22.3.3 Solid Waste Liquid/Gas Conversion

As stated previously, these processes have been demonstrated, but the variability of the waste as a feedstock makes the process impractical for the present. Because of this factor

Table 22-12. Fluidized Bed Combustor WTEE Plant—Cost and Performance Data (Bechtel 1990)[a]

	Design Capacity, Tons/Day				
RDF Fuel	300	500	1,000	1,500	2,000
Gross power output, MW	8.4	14.2	29.4	44.2	58.9
Auxiliary power, MW	1.5	2.5	4.9	7.4	9.8
Net power output, MW	6.9	11.7	24.5	36.8	49.1
Total direct cost, 1990 $/ton	158,333	137,200	113,100	101,000	93,250
Total direct cost, 1990 $/kW	6,884	5,863	4,616	4,117	3,798
Total O&M cost, 1990 $/ton	73.11	61.08	48.45	42.57	39.01
Fixed O&M cost, 1990 $/kW-yr	453.77	353.33	241.39	194.51	164.62
Variable O&M cost, 1990 $/MWh	71.25	61.42	50.47	46.89	44.96
Operating efficiency, kWh/ton	552.0	561.6	588.0	588.8	589.2
Net plant heat rate, Btu/kWh	19,587	19,252	18,388	18,363	18,350
	Design Capacity, Tons/Day				
Wood Waste Fuel	300	500	1,000	1,500	2,000
Gross power output, MW	14.0	23.9	48.4	72.5	96.7
Auxiliary power, MW	2.3	3.8	7.6	11.4	15.2
Net power output, MW	11.7	20.1	40.8	61.1	81.5
Total direct cost, 1990 $/ton	148,667	128,800	106,200	94,833	86,575
Total direct cost, 1990 $/kW	3,812	3,204	2,603	2,328	2,149
Total O&M cost, 1990 $/ton	63.31	51.36	38.67	32.85	29.23
Fixed O&M cost, 1990 $/kW-yr	267.61	205.67	144.95	117.15	99.18
Variable O&M cost, 1990 $/MWh	31.00	25.14	19.80	17.77	16.57
Operating efficiency, kWh/ton	936.0	964.8	979.2	977.6	978.0
Net plant heat rate, Btu/kWh	15,600	15,135	14,912	14,937	14,930
	Design Capacity, Tons/Day				
Gob Fuel	300	500	1,000	1,500	2,000
Gross power output, MW	5.3	8.9	18.3	28.0	37.4
Auxiliary power, MW	0.9	1.5	2.9	4.4	5.9
Net power output, MW	4.4	7.4	15.4	23.6	31.5
Total direct cost, 1990 $/ton	158,333	137,200	113,100	101,000	93,250
Total direct cost, 1990 $/kW	10,795	9,270	7,344	6,419	5,921
Total O&M cost, 1990 $/ton	97.81	85.86	73.27	67.42	63.83
Fixed O&M cost, 1990 $/kW-yr	711.59	558.65	384.03	303.31	256.60
Variable O&M cost, 1990 $/MWh	185.61	170.52	150.99	142.38	139.21
Operating efficiency, kWh/ton	352.0	355.2	369.6	377.6	378.0
Net plant heat rate, Btu/kWh	15,966	15,822	15,206	14,883	14,868

[a]Table is based on calculations made by Black & Veatch for the NRECA using a computer program developed by Bechtel (1990) for EPRI.

and space limitations, these technologies are not discussed further in this text.

22.4 FLUE GAS TREATMENT

The combustion of solid wastes produces both air pollutants and other solid wastes. The quantity and type of pollutants in the flue gas stream from resource recovery plants depends on source separation, combustion control, and flue gas treatment. There are three classes of air pollutants:

- Products of incomplete combustion,
- Particulates and trace metals, and
- Acid gases.

The products of incomplete combustion (PICs) form when carbon compounds in the solid waste are not completely combusted. The factors that contribute to the formation of PICs include fuel-rich conditions (too little oxygen), poor fuel-air mixing, low combustion temperature, and short combustion-zone residence time.

PICs include carbon monoxide (CO), long-chain hydrocarbons, chlorinated aromatic hydrocarbons, including dioxins and furans, and polychlorinated biphenyls (PCBs). The last two categories are those that cause the most concern because of their perceived toxicity (Li, Frillici, and Wilson 1990).

Particulates are formed during the combustion of solid waste from three sources:

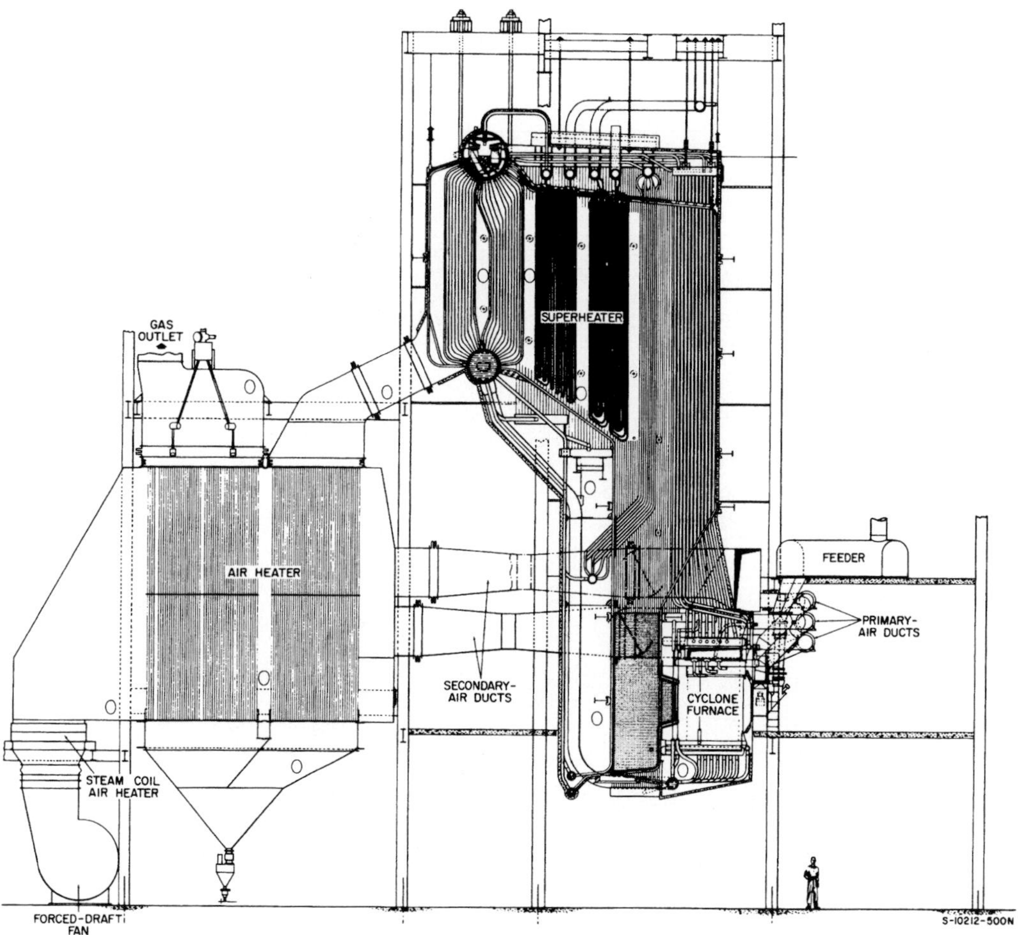

Fig. 22-18. Cyclone furnace WTEE plant. (From Babcock & Wilcox. Used with permission.)

- Inorganic material that does not burn;
- Organometallic compounds, including inorganic oxides and metal salts that form from high-temperature oxidation; and
- Unburned waste particles that form as a result of incomplete combustion.

Also of concern are trace metals that are present in the solid waste. Three trace metals controlled by EPA's Prevention of Significant Deterioration (PSD) regulations are lead, mercury, and beryllium. Other trace metals not covered by the PSD regulations include arsenic, cadmium, and chromium (Li, Frillici, and Wilson 1990).

Table 22-13. United States WTEE Plants That Cofire Solid Waste

Plant Name	State	Owner	Boiler Manufacturer	Design Capacity (TPD/MW)	Technology	Fuels (% heat input)
Ames	IA	City of Ames	Combustion Engineering, Inc. (Unit 7), Babcock & Wilcox (Unit 8)	200/100	Suspension	RDF (10–15)
Baltimore County	MD	Baltimore County, Baltimore Gas & Electric Company	Babcock & Wilcox	1,200/200	Cyclone furnace	RDF (15–20)
Lakeland	FL	City of Lakeland, Orlando Utilities Commission	Babcock & Wilcox	300/364	Suspension	RDF (10)
Madison	WI	City of Madison	Babcock & Wilcox	500/100	Suspension	RDF (10–15)
Rumford Cogen	ME	Rumford Cogeneration Co.	Pyropower Corp.	NA/85	Circulating fluidized bed	Wood wastes (30)
William J. Neal	ND	Basin Electric	Combustion Engineering, Inc.	240/50	Suspension	Sunflower seed hulls (30)

Table 22-14. RDF Cofiring WTEE Plant—Cost and Performance Data (Bechtel 1990)[a]

	Heat Input from RDF, %			
Retrofit 50-MW Plant	5	10	15	20
RDF capacity, tons/day	68	137	205	278
Incremental direct cost, 1990 $1,000	2,800	2,800	2,800	2,800
Incremental total O&M cost, 1990 $1,000	461	495	532	566
Incremental fixed O&M cost, 1990 $1,000	335	355	355	355
Incremental variable O&M cost, 1990 $1,000	106	140	177	211
Boiler efficiency—coal only, %	88.3	88.3	88.3	88.3
Boiler efficiency—cofiring, %	85.7	85.0	84.2	83.6
Coal usage savings, tons/yr	3,000	10,000	17,000	24,000
New 210-MW Plant				
RDF capacity, tons/day	291	586	886	1,190
Incremental direct cost, 1990 $1,000	4,500	4,500	4,500	4,500
Incremental total O&M cost, 1990 $1,000	1,019	1,175	1,323	1,473
Incremental fixed O&M cost, 1990 $1,000	731	731	731	731
Incremental variable O&M cost, 1990 $1,000	288	444	592	742
Boiler efficiency—coal only, %	88.6	88.6	88.6	88.6
Boiler efficiency—cofiring, %	87.1	86.4	85.8	85.1
Coal usage savings, tons/yr	23,000	52,000	82,000	113,000
New 515-MW Plant				
RDF capacity, tons/day	698	1,404	2,123	2,855
Incremental direct cost, 1990 $1,000	7,400	7,400	7,400	7,400
Incremental total O&M cost, 1990 $1,000	1,721	2,072	2,434	2,802
Incremental fixed O&M cost, 1990 $1,000	1,084	1,084	1,084	1,084
Incremental variable O&M cost, 1990 $1,000	637	988	1,350	1,718
Boiler efficiency—coal only, %	88.6	88.6	88.6	88.6
Boiler efficiency—Cofiring, %	87.1	86.4	85.8	85.1
Coal usage savings, tons/yr	55,000	126,000	198,000	271,000

[a]Table is based on calculation made by Black & Veatch for NRECA using a computer program developed by Bechtel (1990) for EPRI.

Acid gases that form during the combustion of solid waste include nitrogen oxides (NO_x), sulfur dioxide (SO_2), sulfur trioxide (SO_3), hydrogen chloride (HCl), and hydrogen fluoride (HF).

Nitrogen oxides are formed from the nitrogen in the waste (fuel NO_x) and during the combustion process (thermal NO_x). The sulfur oxides are created when the sulfur present in the waste is oxidized during combustion. Hydrogen chloride and hydrogen fluoride are both formed when elemental chlorine and fluorine, respectively, present in the waste react with hydrogen ions during combustion (Li, Frillici, and Wilson 1990).

When solid waste is burned on grates and not in suspension, the majority of ash produced during combustion is bottom ash. Bottom ash, which includes the unburnable inorganic material present in the waste, consists of relatively large particles and is considered nonhazardous because of its inert properties. Fly ash includes the particulates captured by and the reaction products generated by air pollution control systems. Fly ash is a special concern since it may contain high concentrations of heavy metals such as lead and cadmium. These and other toxic metals are more easily leachable from fly ash in landfills because of the small particle sizes and their correspondingly greater surface area/unit volume ratio (Energy from Wastes, *Power* 1988).

22.4.1 Source Separation

The public is moving toward recycling as a means of energy and resource conservation. Currently, the separation of newspaper, plastics, and aluminum is popular in residential areas. Iron is separated from the waste in prepared refuse burning plants and from the ash stream in some mass burning plants. The economic motivation to separate iron is regional, because it requires a locally available user.

Most of the objectionable elements in the waste stream are consumed in the combustion process or collected from the waste as particulate or condensed by reducing the temperature. Mercury is unique in that it is still in the gaseous state at the flue gas treatment temperatures. Other metals condense above these temperatures.

The primary sources of mercury in waste are the button-size batteries used in small electronic devices such as hearing aids and cameras. There is a small amount of mercury in most dry cell batteries. Source separation of batteries has been only marginally successful, probably because of the small size of the batteries and because they are of no value to the person doing the separation.

Industry is making progress in removing mercury from their manufacturing processes and designs. For example, mercury is no longer used in paints, and within 5 years batteries probably will not contain mercury.

A slow trend is developing for designers to consider final disposal when they design packaging. The packaging should be either recyclable or disposable without undesirable releases.

Unburned carbon in the ash is effective in collecting mercury in flue gas filter bags. It has been demonstrated that selenium filters effectively remove mercury from flue gas.

22.4.2 Combustion Control

In modern waste combustion facilities, proper combustion control eliminates the chlorinated aromatic hydrocarbons such as dioxin and furans.

It was previously noted that combustors that operate above 1,800° F were relatively free of dioxin and furans in the flue gas. Based on this, the criterion of 1,800° F for more than 2 seconds was previously the practice. There are two problems with this practice. First, some well-run combustors that did not operate at or above 1,800° F, such as fluidized beds, had very low dioxin/furan emissions. Second, high temperatures promote NO_x formation.

It has now been concluded that combustors operating with proper firing conditions including adequate excess air and turbulence will minimize dioxins and furans in the flue gas.

22.4.2.1 Flue Gas Treatment. Flue gas treatment systems consist of acid gas removal (HCl and SO_2 control) followed by particulate collection. They also may include selective catalytic reduction or ammonia injection for NO_x control.

The HCl and SO_x are controlled by alkali injection. Some combustors have had limited success adding the alkali to the fuel. Success has been better, however, with fluidized bed combustors. Alkali injection into the flue gas ducts after the combustion zone has been performed as an expedient, but this has been only slightly successful.

The wet and dry flue gas scrubber have been the most successful. The wet scrubber sprays the alkali slurry into the gas stream. The undesirable constituents in the flue gas are tied up in the slurry and removed. The drawback of this concept is that the sludge requires further disposal handling. One of the benefits is that the system reduces the gas temperature low enough (about 150° F) to condense out heavy metals.

The dry scrubber is an adaptation of spray drying technology. The alkali slurry is sprayed into the flue gas stream. The temperature must be controlled at high levels to ensure that the dry product will flow in the removal system. This could be on the order of 350° F.

The particulate collection equipment is either an electrostatic precipitator (ESP) or a baghouse filter. The ESP has a low pressure drop, but it may have to be four or five fields deep to accomplish the required particulate removal. If only particulate removal is required (no wet or dry scrubber precedes it), an ESP is used because small fires from burning char will not cause damage.

Baghouses have proven to be efficient particulate removal equipment in the utility industry, and have found to be very suitable in the resource recovery industry. The cake on the filter cloth acts as the filter. Charcoal in this cake, as either unburned charcoal or as injected carbon, removes metals. The unreacted alkali carried over reacts with the acidic compounds in the flue gas for further reduction of the effluents from the plant.

In 1987, the USEPA and Environment Canada joined to test atmospheric emissions from mass and prepared refuse combustors in the United States and Canada. This program concluded that efficient operation and current designs of air pollution control systems reduce emission to extremely low levels (Sawell and Constable 1991).

22.5 REFERENCES

ANTHONY, ROBERT J. 1989. Canadian Cogeneration Plant Converts Wood Waste to Power. *Solid Waste & Power.* April: S-36-40.

BABCOCK & WILCOX. 1978. *Steam: Its Generation and Use*, 39th ed. Babcock & Wilcox, New York, NY.

BABCOCK & WILCOX. Power Generation Group. 20 S. Van Buren Avenue, P. O. Box 351, Barberton, OH 44203-0351.

BECHTEL GROUP, INC. 1990. *Waste to Energy Screening Guide.* Draft Final Report, Project RP2190-5. Electric Power Research Institute, Palo Alto, CA.

BERENYI, EILEEN. 1988. *1988–1989 Resource Recovery Yearbook—Directory and Guide.* Robert Gould, Ed. Governmental Advisory Associates, New York, NY.

Boilers and Auxiliary Equipment. 1988. *Power.* June: B-14–40.

BRETZ, ELIZABETH. 1989. Energy from Wastes. *Power.* March: W-1–26.

CHEREMISINOFF, NICHOLAS P., PAUL N. CHEREMISINOFF, and FRED ELLERBUSCH. 1980. *Biomass: Applications, Technology, and Production.* Marcel Dekker, New York, NY.

CHURGIN, GENE S. 1989. Potential for Fluidized Bed Combustion in the Waste to Energy Market. Wastes to Energy '89: Competing in Tougher Markets. New Orleans, LA.

DETROIT STOKER COMPANY. Subsidiary of United Industrial Corporation. 1510 East First Street, P. O. Box 732, Monroe, MI 48161.

ELECTRIC POWER RESEARCH INSTITUTE. 1988. Guideline for Cofiring Refuse Derived Fuel in Electric Utility Boilers. EPRI CS-5754, 3 Vols. EPRI, Palo Alto, CA.

ELECTRIC POWER RESEARCH INSTITUTE. 3412 Hillview Avenue, Palo Alto, CA 94304.

Energy from Wastes. 1988. *Power.* March: W-1-36.

Energy from Waste: On-Site Heat Recovery Incineration. 1987. *Power.* March: W-1-15.

FOSTER WHEELER POWER SYSTEMS. 1991. Product Brochures provided by Roland A. Kiska. Foster Wheeler, New Jersey.

GERSHMAN, BRICKNER & BRATTON, INC. and SHIVE-HATTERY ENGINEERS. 1985. Potential for Energy Recovery from Municipal Solid Waste in Iowa. Iowa Energy Policy Council, Des Moines, IA.

LAROE, JOHN. 1988. Tired of Those Roadside Detractions? *Solid Waste & Power*. February: 20–25.

LI, RAMON, PAUL W. FRILLICI, and JOHN W. WILSON. 1990. Identifying and Controlling WTE Stack Emissions. *Solid Waste & Power*. February: 16–22.

MAKANSI, JASON. 1988. Developments to Watch: Directly Firing Pulverized Biomass in a Boiler. *Power*. February: 60–62.

MATUNY, MORTON G. and PATRICK ADKIN. 1989. Mass-Burn Plant Converts Solid Waste, Still Slashes Pollution. *Power*. April: S-41-44.

MCCARTHY, J. and J. P. NAMBURI. 1989. Feature Siting, Latest Advances at Resource Recovery Plant. *Power.* April: S-75-78.

NATIONAL RURAL ELECTRIC COOPERATIVE ASSOCIATION. 1800 Massachusetts Avenue, N.W., Washington, D.C. 20036.

OGDEN MARTIN. Ogden Projects, Inc. 40 Lane Road. P. O. Box 2615, Fairfield, NJ 07007-2615.

OXFORD MEETS PERFORMANCE GOALS FIRING WHOLE TIRES. 1988. *Power*. October: S-24-28.

PETERSON, CHARLES, and RAYMOND GIVONETTI. 1984. Municipal Solid Waste for Energy: A Technology Review. *Proceedings of the Eleventh Energy Technology Conference*. Richard F. Hill, Ed. Government Institutes, Inc., 1337-1356. Rockville, MD.

REED, THOMAS B. and AGUA DAS. 1988. *Handbook of Biomass Downdraft Gasifier Engine Systems*. Solar Energy Research Institute, SERI/SP-271-3022. Golden, CO.

ROBINSON, WILLIAM D., ED. 1986. *The Solid Waste Handbook—A Practical Guide*. John Wiley & Sons, New York, NY.

ROCKET RESEARCH COMPANY. 1980. *A Study of the Feasibility of Cogeneration Using Wood Waste as Fuel*. Technical Planning Study, TPS79-736-1, AP-1483. Electric Power Research Institute, Palo Alto, CA.

SAWELL, S. E. and T. W. CONSTABLE. 1991. *A Summary of the National Incinerator Testing Evaluation Program Ash Characterization and Solidification Studies*. 2nd International Conference on Municipal Waste Combustion at Tampa, Florida. Environment Canada, Conservation and Protection, Wastewater Technology Centre. Burlington, Ontario.

SCHWIEGER, BOB. 1985. Rice-Hulled-Fired Powerplant Burns a Nuisance Waste, Sells Electricity, Ash. *Power*. July: 86-91.

SINGER, JOSEPH G., ED. 1981. *Combustion—Fossil Power Systems*, 3rd Edition. Combustion Engineering, Inc., Windsor, CT.

SOLAR ENERGY RESEARCH INSTITUTE. 1979. *A Survey of Biomass Gasification: Volume I—Synopsis and Executive Summary*. SERI/TR-33-239. Solar Energy Research Institute, Golden, CO.

SOLAR ENERGY RESEARCH INSTITUTE. 1979. *A Survey of Biomass Gasification: Volume II—Principles of Gasification, SERI/TR-33-239*. Solar Energy Research Institute, Golden, CO.

TAYLOR, HUNTER F. 1984. *Energy Recovery from Municipal Solid Waste—A Feasibility Guide for Local Government in Virginia*. Temple Bayliss and Daniel T. Gallagher, Eds. Energy Division, Office of Emergency and Energy Services, Virginia.

TENNESSEE VALLEY AUTHORITY (TVA). 1985. *Sumner County Solid Waste Energy Recovery Facility—Volume 1: Feasibility Studies, Design, and Construction*. Research Project 2190-2, CS-4164. Electric Power Research Institute, Palo Alto, CA.

THIES, LAWRENCE and ROBERT HEINA. 1990. Burning Culm in CFB Boiler Minimizes Environmental Impact. *Power*. April: S-27-32.

UNIVERSITY OF OKLAHOMA. 1979. *Evaluation of the Potential for Producing Liquid Fuels from Biomaterials*. Final Report, EPRI AF-974. Electric Power Research Institute, Palo Alto, CA.

VOLUND USA. 1991. Volund Product Brochures, provided by George R. Kotynek. Volund USA LTD, Oak Brook, IL.

VOLUND USA LTD. Mountain Heights Center. 430 Mountain Avenue, New Providence, NJ 07974.

WESTINGHOUSE ELECTRIC CORPORATION. Resource Energy Systems Division. 2400 Ardmore Boulevard, Pittsburgh, PA 15221.

WESTINGHOUSE RESOURCE ENERGY SYSTEMS DIVISION. 1990. Westinghouse Product Brochures provided by J. J. Zebroski. Westinghouse Electric Corporation, Pittsburgh, PA.

WHEELABRATOR ENVIRONMENTAL SYSTEMS, INC. Liberty Lane, Hampton, NH 03842.

WILHELM, DONALD J. and DALE R. SIMBECK. 1990. The California FBC Boiler Story—A Status Report. Council of Industrial Boiler Owners Sixth Annual FBC Conference, Harrisburg, PA.

23

Nuclear Power

Lawrence F. Drbal

23.1 INTRODUCTION

The history of nuclear power is an evolution of scientific discoveries directed toward understanding the internal makeup of the atom. A key discovery was made in 1932 by Sir James Chadwick, an English physicist, who determined that a highly penetrating form of radiation found by other researchers was actually an uncharged particle with a mass approximately equal to that of a proton. This particle was named the neutron.

23.1.1 Discovery of Fissioning

Later, Enrico Fermi experimented with neutron bombardment of many different elements. He showed that many of these elements became radioactive when bombarded. In one series of experiments, he bombarded uranium salts with neutrons but was unable to interpret the set of events that resulted. In 1939, physicists Lise Meitner and Otto Frish correctly interpreted Fermi's experiment, finding that neutron bombardment of uranium actually split, or fissioned, the uranium nucleus into two parts with a large release of energy. Within months, other researchers confirmed Meitner and Frish's findings.

As the research continued, the release of neutrons from the fission process was quickly added to the discoveries, and the possibility of a chain reaction became apparent. Neutrons from the preceding fission process fissioned other uranium nuclei, producing still another generation of neutrons and continuing the fission process, releasing large amounts of energy. Fissioning was found to take place in uranium (U), thorium (Th), and protactinium (Pa), with the release of one, two, or three additional neutrons. The fission fragments produced in the fission process were unstable, radioactive, and varied in atomic number from 34 to 57. ${}^{235}_{92}U$ showed a higher probability of fissioning with lower energy neutrons, while ${}^{238}_{92}U$ and ${}^{232}_{90}Th$ fissioned with higher energy neutrons. Plutonium could be produced from neutron absorption of ${}^{238}_{92}U$ and was fissionable.

The potential of nuclear energy for use in a weapon of terrible destructive capability was evident, but its potential for use in peaceful energy applications was also recognized.

23.1.2 Chicago Pile

The first chain reaction was demonstrated in December 1942 at the University of Chicago beneath the stands of the Stagg Field Stadium. A team of scientists under the direction of Enrico Fermi designed and assembled a pile consisting of a large number of graphite blocks. The graphite acted as a moderator to reduce the neutron energy and as a reflector to minimize neutron leakage from the pile. Uranium in both metal and oxide form was inserted into holes in the graphite blocks. The chain reaction was demonstrated by gradually adding more graphite blocks and uranium to the pile, while measuring the neutron intensity. An exponential rise in neutron intensity showed the initiation of a chain reaction. Insertion of a cadmium control rod stopped the reaction. The heat generation rate was 2 kW, and the pile was air cooled.

23.1.3 Manhattan Project

The potential for development of an atomic bomb by Germany and the fear of Germany using this weapon against the Allies in World War II prompted several well-known scientists, including Albert Einstein, to meet with President Roosevelt. Following this meeting, President Roosevelt, in August 1942, established the Manhattan Engineer District of the US Army Corps of Engineers, commonly known as the Manhattan Project, to develop the atomic bomb. The Manhattan Project was successful, with the first detonation of the bomb in July 1945 in Nevada. Then, as history has so vividly recorded, atomic bombs were detonated at Hiroshima and Nagasaki, Japan, in August 1945, ending World War II shortly thereafter.

The Nevada and Nagasaki atomic bombs were made with ${}^{239}_{94}Pu$, while the Hiroshima bomb used ${}^{235}_{92}U$. Because of the uncertainty of relying on only one material for weapons, production of both ${}^{235}_{92}U$ and ${}^{239}_{94}Pu$ was pursued by the Manhattan Project. ${}^{239}_{94}Pu$ was produced from large-scale reactors in Hanford, Washington. The Hanford reactors were graphite moderated, water cooled, and fueled with natural uranium. Plutonium was chemically separated from uranium. Hanford continued plutonium production following World War II.

A parallel effort involved two separate processes—gaseous diffusion and mass spectrography—to produce ${}^{235}_{92}U$. ${}^{235}_{92}U$ occurs naturally in ${}^{238}_{92}U$ at about 0.7 wt % and requires further separation to obtain a higher enrichment of ${}^{235}_{92}U$. The favored method was the gaseous diffusion process developed at Oak Ridge, Tennessee. The gaseous diffusion method separated the 235 atoms from the 238 atoms by filtrating UF_6 gas through a series of porous barriers. The lighter 235 atoms passed through the barrier more readily than 238 atoms. A large number of diffusion barrier stages was required to produce almost pure $^{235}_{92}U$. Because this method required a longer development time, an alternate method was needed. The mass spectrography method was first used at Oak Ridge to produce the first quantities of $^{235}_{92}U$. This process involved passing a collimated beam of uranium ions through a magnetic field, which deflected the lighter 235 ions in larger numbers than the 238 ions.

23.1.4 Atomic Energy Commission

The Atomic Energy Commission (AEC) was created under the Atomic Energy Act established in 1946 by Congress. The AEC had oversight of all United States nuclear activities, including federal control and ownership of the production and use of fissile materials and control of civilian involvement in research and development activities.

The AEC was also given the responsibility for developing peaceful uses of nuclear energy. Most pertinent to the subject of this text were the assignments for developing reactor propulsion systems for submarines and for capitalizing on the breeding potential of $^{238}_{92}U$.

The latter efforts led to the building of the first experimental breeder reactor, the EBR-I, in 1951. EBR-I became the first reactor to produce electrical energy.

23.1.4.1 Submarine Propulsion. The development of a submarine reactor started with the construction of a land-based reactor called the S1W at the Arco Naval Reactor Test Station in Idaho. The S1W, which began operation in March 1953, was a pressurized water reactor (PWR) utilizing ordinary water for neutron energy moderating and heat removal. S1W became the prototype for PWRs. The success of the S1W program lead to the development and construction of the first nuclear-powered submarine, the *USS Nautilus*. Commissioned in 1955, the *Nautilus* was propelled by a small, high-power PWR using highly enriched U-235 fuel.

23.1.4.2 Shippingport. This successful PWR submarine technology was applied to build the first central station nuclear power plant near Shippingport, Pennsylvania. The Shippingport plant was built as part of the AEC prototype program approved by Congress in 1954. Placed in commercial operation in 1957, Shippingport was a 60 MWe PWR using a reactor water pressure of 2,000 psia (13.8 MPa). Steam production was at 490° F (254° C) and 600 psia (4.14 MPa).

23.1.4.3 Private Industry Involvement. With the help of the AEC, private companies started the commercial nuclear plant market in 1954. This involvement was a result of President Dwight Eisenhower's "Atoms for Peace" program. Congress, responding to the "Atoms for Peace" initiative, changed the AEC's role to allow civilian access and ownership of fissile (fissionable) material. The AEC was given the responsibility for the licensing and safety of all United States nuclear plants.

Three private contractors responded to AEC's offer of shared research costs and operational fuel costs for 7 years (Leclercq 1986):

- The Westinghouse Company built a 185-megawatt electric (MWe) PWR at Yankee Rowe, Massachusetts, for a group of electric utilities. Yankee Rowe began commercial operation in 1961.
- The Babcock and Wilcox Company supplied the nuclear steam supply system for Indian Point 1 near New York City. Indian Point 1 was owned by Consolidated Edison Company. Indian Point 1 was a 264-MWe PWR that began commercial operation in 1962 and is still operating.
- The General Electric Company built Dresden 1 for Commonwealth Edison Company. Dresden 1, which was located near Morris, Illinois, was a 200-MWe boiling water reactor (BWR). It went into commercial operation in 1960 and operated until 1984. Dresden 1 was based on technology of the Experimental Boiling Water Reactor (EBWR) built at Argonne National Laboratory and the Vallecitos test reactor built by General Electric.

These three commercial projects laid the foundation for private industry involvement. The AEC continued to subsidize larger power plants; however, the sale by General Electric of the 620-MWe Oyster Creek BWR marked the end of AEC funding (Leclercq 1986). Oyster Creek, which was owned by Jersey Central Power and Light Company, went into commercial operation in 1969.

In 1974, the AEC was divided into two separate agencies. The AEC regulatory responsibilities of licensing and reactor safety were given to the Nuclear Regulatory Commission (NRC). Promotion and research and development responsibilities were given to the Energy Research and Development Agency (ERDA), which later became part of the Department of Energy (DOE).

23.2 POWER REACTORS

Of the different reactor concepts examined over the last five decades, many were abandoned because of economics, material considerations, design deficiencies, or poor thermal efficiencies. Reactor types using light water[1] as a neutron energy moderator and heat transfer medium together with slightly enriched uranium fuel have become the dominant technology. There are two types of light water reactors (LWR): pressurized water (PWR) and boiling water (BWR)

[1]Light water is ordinary water. This terminology is used to differentiate light water from another reactor type that uses deuterium oxide, commonly called heavy water.

reactors. PWRs prohibit boiling in the core and use secondary heat exchangers or steam generators to produce steam, which powers the turbine–generator. BWRs produce steam directly in the reactor core and pass the steam to the turbine–generator. Other reactor technologies in worldwide use include the following:

- Heavy water reactors, which use deuterium oxide as the neutron energy moderator and heat transfer medium together with natural uranium fuel.
- Gas-cooled reactors, which use helium or carbon dioxide gas as the heat transfer medium and graphite as a neutron moderator. The fuel is either natural or enriched uranium.
- Breeder reactors, which use high-energy neutrons to fission Pu-239 and produce additional Pu-239 from neutron capture in U-238. Liquid sodium is the heat transfer medium.
- Water–graphite reactors, which use light water as the heat transfer medium and graphite as the neutron energy moderator.

23.2.1 Pressurized Water Reactors

Figure 23-1 shows a simplified schematic of a PWR. Light water, acting as the coolant and moderator, passes through the reactor core under high pressure (2,250 psia [15.5 MPa]) and removes the fission heat from the uranium fuel. Reactor coolant pumps move the water from the reactor to a primary heat exchanger or steam generator, and while passing through the primary side of the steam generator, the reactor coolant gives up the fission heat energy to water on the secondary side, producing steam at saturated conditions of 1,100 psi (7.6 MPa) and 556° F (291° C). The steam expands through a turbine coupled to an electrical generator and then is condensed back to liquid in the condenser. The water is then returned to the steam generator after preheating in several stages of feedwater heaters using turbine extraction steam.

The reactor core consists of a large number of fuel rods, housed in fuel assemblies located in a nearly cylindrical arrangement. A fuel assembly contains a square array of fuel rods—up to 17 × 17 or 289 fuel rods. Each fuel rod is a zirconium alloy-clad tube containing pellets of slightly enriched uranium dioxide (2% to 3% $^{235}_{92}U$) stacked end-to-end inside the tube to a height of 12 ft (3.6 m). The reactor is controlled by the use of control rods. The control rods enter from the top of the reactor and pass through the fuel assemblies. Each control rod is a stainless steel tube containing a neutron absorbing material such as boron carbide or an alloy of silver, indium, and cadmium. The reactor core is housed inside a larger cylindrical pressure vessel that is designed to withstand the high reactor coolant pressures. The pressure vessel is made of steel with a typical thickness of 6.5 to 7.5 in. (20 to 23 cm), a height of 45 ft (13.7 m), and a diameter of 15 ft (4.6 m).

Pressurized water reactors have two to four parallel independent steam generation loops consisting of a steam generator and reactor coolant pump. A pressurizer on one of the loops maintains the reactor water pressure at constant value.

23.2.2 Boiling Water Reactors

Figure 23-2 shows a simplified schematic of a BWR. Light water, which acts as the coolant and moderator, passes through the core where boiling takes place in the upper part of the core. The wet steam then passes through a bank of moisture separators and steam dryers in the upper part of the pressure vessel. The water that is not vaporized to steam is recirculated through the core with the entering feedwater using two recirculation pumps coupled to jet pumps (usually 10 or 12 per recirculation pump). The steam leaving the top of the pressure vessel is at saturated conditions of 1,040 psia (7.2 MPa) and 533° F (278° C).

The steam then expands through a turbine coupled to an electrical generator. After condensing to liquid in the condenser, the liquid is returned to the reactor as feedwater. Prior to entering the reactor, the feedwater is preheated in several stages of feedwater heaters. The balance of plant systems (e.g., turbine generator, feedwater heaters) are similar for both PWR and BWRs.

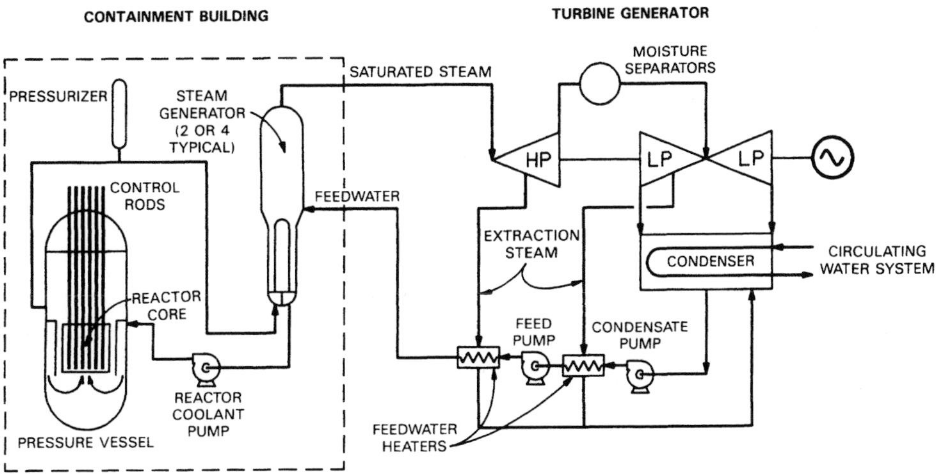

Fig. 23-1. Schematic for a pressurized water reactor.

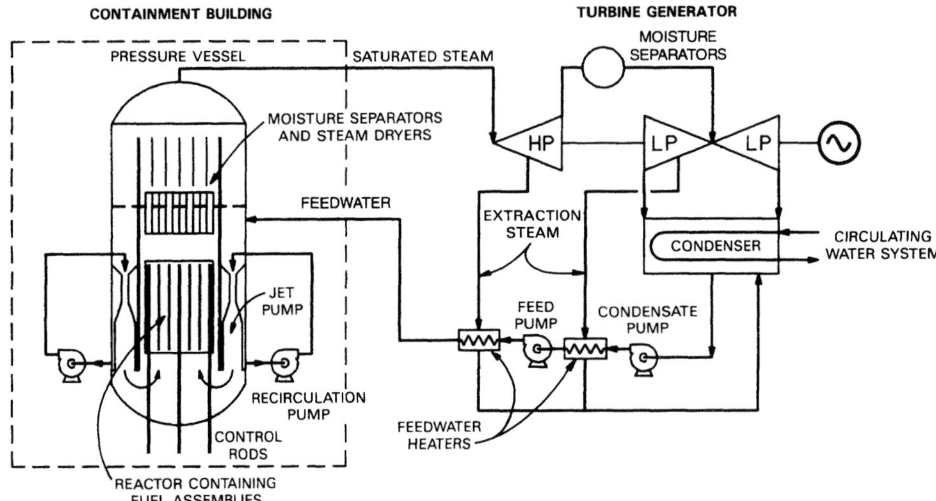

Fig. 23-2. Schematic for a boiling water reactor.

The BWR reactor core, like that in a PWR, consists of a large number of fuel rods housed in fuel assemblies in a nearly cylindrical arrangement. Each fuel assembly contains an 8×8 or 9×9 square array of 64 or 81 fuel rods (typically two of the fuel rods contain water rather than fuel) surrounded by a square Zircaloy channel box to ensure no coolant crossflow in the core. The fuel rods are similar to the PWR rods, although larger in diameter. Each fuel rod is a zirconium alloy-clad tube containing pellets of slightly enriched uranium dioxide (2% to 5% U-235) stacked end-to-end. The reactor is controlled by control rods housed in a cross-shaped, or cruciform, arrangement called a control element. The control elements enter from the bottom of the reactor and move in spaces between the fuel assemblies.

The BWR reactor core is housed in a pressure vessel that is larger than that of a PWR. A typical BWR pressure vessel, which also houses the reactor core, moisture separators, and steam dryers, has a diameter of 21 ft (6.4 m), with a height of 72 ft (22 m). Since a BWR operates at a nominal pressure of 1,000 psia (6.9 MPa), its pressure vessel is thinner than that of a PWR.

23.2.3 Heavy Water Reactors

Heavy water reactors (HWRs) use heavy water (deuterium oxide) rather than light water for neutron energy moderating and fission heat removal (coolant) functions. Since heavy water is a poorer neutron absorber than light water, it can be used as a coolant–moderator with natural uranium fuel. The HWRs were developed in Canada, where they are being operated successfully in a country that has abundant hydroelectric power for heavy-water production. Figure 23-3 shows a simplified schematic of an HWR.

The HWR design consists of a calandria reactor vessel that separates the heavy water coolant for fission heat removal from the heavy water moderator. The calandria reactor contains a large number of horizontal pressure tubes. Each Zircaloy pressure tube contains 12 fuel bundles of

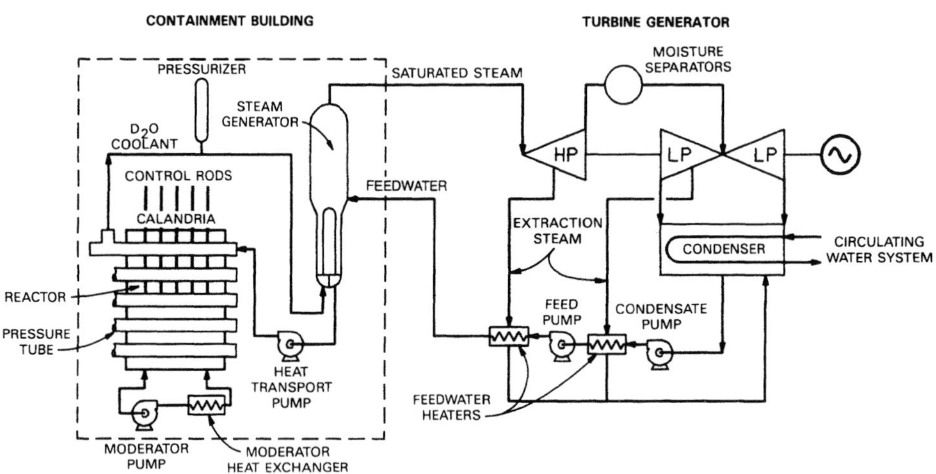

Fig. 23-3. Schematic for a heavy water reactor.

natural-uranium dioxide fuel. Each fuel bundle contains 28 or 37 Zircaloy-clad fuel elements. Heavy water is pumped through the fuel bundles within the pressure tubes (where the fission heat is removed) and then sent to two or more steam generators. The heavy water is not allowed to boil in the pressure tubes and is maintained at 1,315 psia (9.1 MPa). After passing through the steam generators, the heavy water coolant returns to the pressure tubes.

The steam generators produce steam on the secondary or shell side with ordinary water at saturated conditions of 683 psia (4.7 MPa) and 500° F (260° C). Steam is expanded through a turbine generator, condensed, preheated in several stages of feedwater heaters, then pumped back to the steam generators. The balance-of-plant systems for an HWR are similar to those for PWR and BWR saturated steam cycles.

The calandria vessel is filled with heavy water, which surrounds the fuel channels containing the pressure tubes and provides for neutron moderation and reflecting processes. Reactor control consists of vertical absorber rods that move between the fuel channels and provide for both shutdown and reactor power control.

HWRs are manufactured and marketed by Atomic Energy of Canada Limited (AECL). AECL currently markets its CANDU reactors (Canada deuterium uranium) in sizes from 500 to 900 MWe, and AECL recently has introduced its CANDU 3, which has a 450-MWe output. The majority of the CANDU designs are operating in Canada and countries such as Pakistan, India, Korea, and Argentina. Heavy water moderated reactors comprise approximately 7% of the world's nuclear capacity. There are no CANDUs operating in the United States since LWRs with enriched fuel have demonstrated more favorable economics.

23.2.4 Gas-Cooled Reactors

The largest program involving commercial gas-cooled reactors (GCRs) was developed in England. The first series of GCRs used graphite as a moderator and carbon dioxide as the coolant. The reactor consisted of a large number of graphite blocks housed in a steel pressure vessel. Numerous channels were drilled through the graphite blocks. The channels either contained fuel elements or were used for control rod insertion. The fuel elements consisted of natural uranium metal clad with Magnox, an alloy of magnesium. Carbon dioxide passed through the fuel channels, removing the fission heat. The gas was then circulated to heat exchangers that produced steam, and then returned to the reactor. The steam from the steam generators was expanded through a turbine generator, condensed, and returned to the steam generators.

The Magnox series operated successfully but had low thermal efficiency and low fuel lifetime because of the low operating temperature (780° F, 420° C) and radiation damage limits due to the metallic uranium fuel. Later designs integrated the heat exchangers around the core inside the pressure vessel which consisted of a prestressed concrete structure.

The advanced gas-cooled reactor (AGR) was developed to overcome the limitations of the Magnox series. The AGR design was similar to the Magnox except that the reactor fuel consisted of enriched uranium dioxide fuel pins clad with stainless steel and housed in fuel elements. Each fuel channel contained several fuel elements (for example, 8). Because of the high operating gas coolant temperature of 1,210° F (654° C) at the reactor outlet, superheat and reheat steam were produced.

23.2.5 High-Temperature Gas-Cooled Reactors

The high-temperature gas-cooled reactor (HTGR) is an American design that produces a higher gas temperature, and thus, a higher steam temperature and higher thermal efficiencies than those of the LWR or HWR. Thermal efficiencies are similar to those for modern pulverized coal plants. The HTGR, shown schematically in Fig. 23-4, uses helium gas as the coolant and graphite as the neutron energy moderator. This reactor consists of hexagonal graphite blocks in which cylindrical fuel rods containing small spherical fuel particles of enriched uranium and thorium are housed within fuel

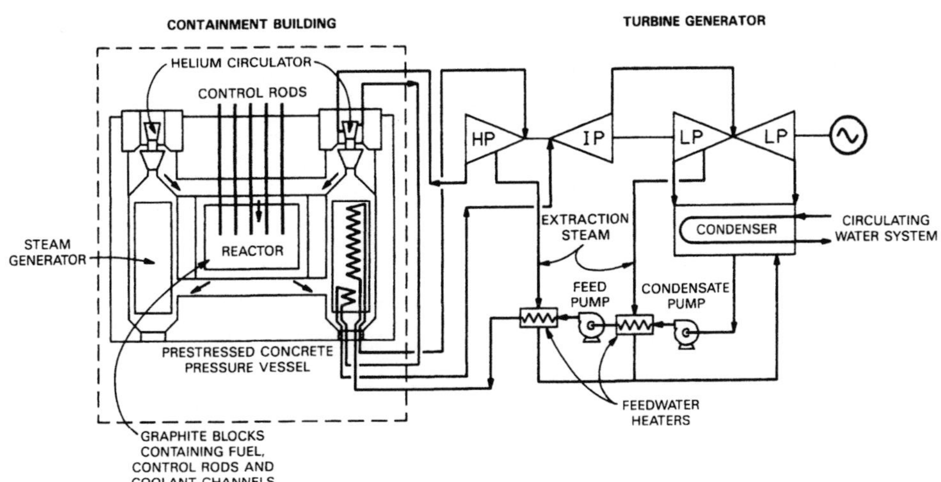

Fig. 23-4. Schematic for a high-temperature gas-cooled reactor.

holes interspersed with coolant holes for helium flow. The graphite blocks are stacked vertically to form the reactor core. Helium flows through the graphite blocks, removing the fission heat, and then passes to one or more steam generators that produce superheated and reheated steam. Superheated steam at 2,400 psig (15.4 MPag), 950° F (510° C) is sent to a high-pressure turbine where the steam exhaust is then reheated in the steam generators at 550 psia (3.8 MPa), 950° F (510° C) and sent back to the intermediate pressure turbine. The balance-of-plant systems (e.g., steam turbine generator, condensers, and feedwater heaters) are very similar to those of a modern pulverized coal plant.

The graphite core, steam generators, and helium circulators are located in a prestressed concrete pressure vessel. Control is provided by control rods of boron carbide, which enter from the top of the reactor core through channels in the graphite blocks.

23.2.6 Breeder Reactors

The breeder reactor has the capability of producing more fuel than it consumes through the breeding of uranium. The reactor core is surrounded by fertile material ${}^{238}_{92}U$, which captures the neutrons not used for fissioning, and through a series of nuclear decays produces ${}^{239}_{94}Pu$, a fissile fuel. Breeder reactors operate in the fast neutron energy range to take advantage of the higher number of neutrons produced per fission in uranium and plutonium fuel which result from the absorption of the high-energy neutrons. Breeder reactors can produce additional plutonium fuel to support several light water reactors, and thus have the potential to increase nuclear fuel reserves.

Figure 23-5 shows a simplified schematic of a breeder reactor. Liquid sodium is used as the coolant to remove the reactor fission energy and transfer the energy to steam generators. Sodium is used because of its good heat transfer properties, low neutron moderating characteristics, and low operating pressures. Since liquid sodium becomes radioactive in passing through the reactor, a primary heat exchanger (sodium to sodium) is used to prevent leakage of radioactive sodium into the steam cycle. An additional advantage of the primary heat exchangers is to prevent water getting into the nuclear core.

The breeder core consists of a number of fuel assemblies of stainless steel fuel rods that are packed with pellets of ${}^{238}_{92}U$ dioxide and ${}^{239}_{94}Pu$ dioxide material. The mixture is approximately 80% ${}^{238}_{92}U$ and 20% ${}^{239}_{94}Pu$. The core is also surrounded by a radial blanket of ${}^{238}_{92}U$ dioxide. Sodium passes through the breeder core and radial blanket, where it removes the fission energy and then flows to the primary heat exchanger, transferring the fission energy to the intermediate sodium coolant. The intermediate sodium coolant passes to a steam generator (sodium to water exchanger) that produces superheated steam at 1,535 psig (10.6 MPag), 906° F (486° C).

Superheated steam expands through the high- and low-pressure sections of the turbine generator, is condensed, and then returned to the sodium/water steam generator. The steam cycle is similar to a conventional steam cycle utilizing a 3,600 rpm nonreheat turbine generator.

23.2.7 Graphite-Moderated, Light Water-Cooled Reactors

A simplified schematic of a graphite-moderated, light water-cooled reactor (GWR) is shown in Fig. 23-6. Developed in the former Soviet Union, the GWR produced the first commercial electrical energy. This particular design was developed prior to the PWR since fabrication of major components such as pressure tubes could be accomplished at existing manufacturing plants and did not require specially built fabricating equipment such as that required for the PWR pressure vessel. The GWR received world attention from the Chernobyl accident. Chernobyl Unit 4 was a GWR. Later, the PWR became the dominant reactor manufactured and sold in the former Soviet Union.

The GWR reactor core consists of a large number of graphite blocks in a cylindrical configuration. A 1,000-MWe core has a diameter of 40 ft (12.2 m) and a height of 30 ft (7 m), and contains channels housing 1,661 pressure tubes.

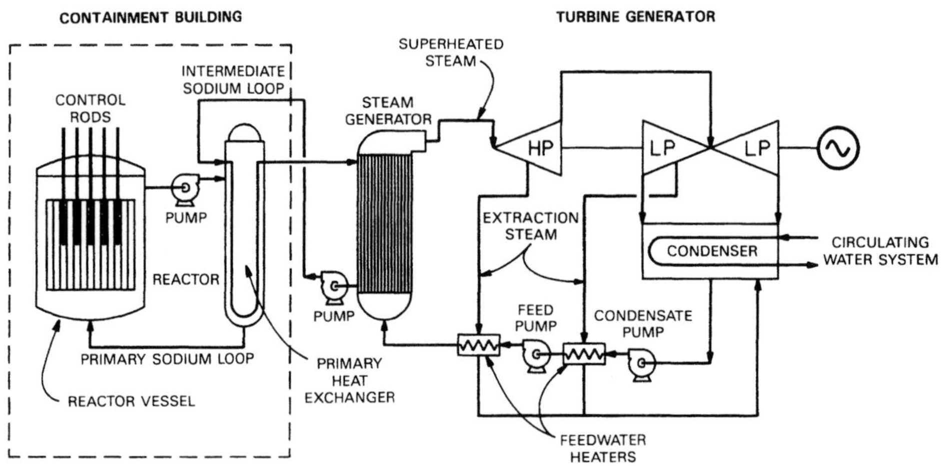

Fig. 23-5. Schematic for a breeder reactor.

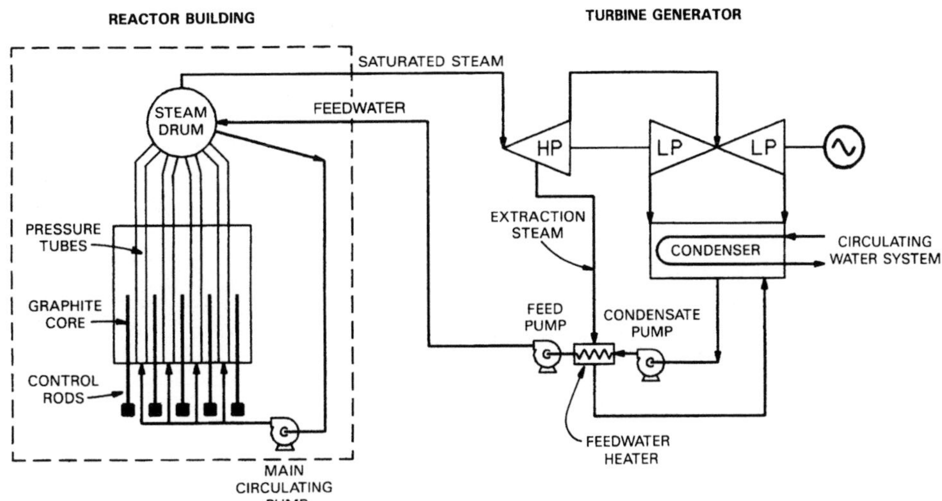

Fig. 23-6. Schematic for a graphite water reactor.

Each pressure tube is made of zirconium with steel ends and contains a fuel assembly holding two fuel subassemblies, one stacked on top of the other. Each subassembly consists of 18 zirconium-clad fuel elements containing slightly enriched (1.8% to 2% $^{235}_{92}U$) uranium dioxide fuel pellets.

Control rods enter the core in separate zirconium-clad channels from the bottom. The control rods consist of boron carbide in an aluminum alloy. Water is taken from one of four steam drums by the main circulating pumps and sent to the bottom of the core, passing through the pressure tubes where boiling takes place in the upper portion of the pressure tubes. The 14.5% vapor–liquid mixture from each pressure tube goes to a common steam drum (one of four) where the vapor is separated from the liquid. The steam leaving the steam drum is at saturated conditions of 1,015 psia (7.0 MPa) at 546° F, 286° C. The steam is expanded through high- and low-pressure sections of a turbine generator, condensed, preheated in feedwater heaters using extraction steam from the turbine, then returned to the steam drums.

23.3 THE ROLE OF NUCLEAR POWER

Nuclear power has steadily increased its share of the electrical generation in the United States. As shown in Table 23-1, in 1993, nuclear power plants generated 19.0% of the total United States electricity production, both utility and non-utility, making it the second major source of electricity generation. There are currently 108 nuclear power plants in operation in the United States with a net generation of almost 98,214 MWe. All are light water reactors, with pressurized water reactors representing two-thirds of the total and boiling water reactors the remaining third, as depicted in Table 23-2. It is expected that utility orders of advanced nuclear designs will restart in the late 1990s as additional electrical capacity is needed.

Globally, nuclear power plays a bigger role. In 1991, nuclear power comprised 17% of the world's electrical energy generation (US Council For Energy Awareness 1994). As depicted in Table 23-3, 33 countries have 415 operating units, with a net generation of 328,308 MWe. Another 91 units are under various stages of construction. Table 23-4 shows that of the 415 units globally, 238 units are PWRs, representing 57% of the world's total. BWRs follow, with 88 units representing 21% of the total. Gas-cooled, heavy water, and graphite water reactors follow PWRs and BWRs, representing 9%, 8%, and 4%, respectively, and liquid metal breeder reactors represent 1%.

Table 23-1. US Electricity Generation in 1993

Source	Billion kWh[a]	Percent
Coal	1,694	52
Nuclear	610	19
Gas	436	14
Hydro	276	9
Oil	103	3
Other	105	3
Total	3,224	100

Source: Nuclear Energy Institute.

[a]Indicates both utility and nonutility industrial generators.

Table 23-2. Nuclear Power Units in United States

	Operating		Total[a]	
Type	Number of Units	Net MWe	Number of Units	Net MWe
BWR	37	32,170	39	33,375
PWR	71	66,044	78	74,473
Total	108	98,214	117	107,848

Source: Nuclear News 1993.

[a]Includes plants under construction or planned as well as operating as of June 30, 1993.

Table 23-3. Nuclear Generation on a Global Basis

Nation	Operating No. Units	Operating Net MWe	Total[a] No. Units	Total[a] Net MWe
Argentina	2	935	3	1,627
Belgium	7	5,484	7	5,484
Brazil	1	626	3	3,084
Bulgaria	6	3,666	6	3,666
Canada	22	15,442	22	15,442
China	0	0	5	3,300
Cuba	0	0	2	834
Czech Republic	4	1,632	6	3,412
Finland	4	2,310	4	2,310
France	55	56,488	61	64,808
Germany	21	22,508	21	22,508
Hungary	4	1,729	4	1,729
India	9	1,834	16	3,874
Japan	43	33,171	54	43,716
Kazakhstan	1	135	1	135
Korea	9	7,220	16	13,083
Lithuania	2	2,760	2	2,760
Mexico	1	654	2	1,308
Netherlands	2	507	2	507
Pakistan	1	125	2	425
Philippines	0	0	1	605
Romania	0	0	5	3,100
Russia	25	19,799	40	33,424
Slovakia	4	1,632	8	3,296
Slovenia	1	620	1	620
South Africa	2	1,840	2	1,840
Spain	9	7,071	15	12,857
Sweden	12	10,002	12	10,002
Switzerland	5	2,936	5	2,936
Taiwan, China	6	4,884	6	4,884
Ukraine	14	12,095	20	17,794
United Kingdom	35	11,950	36	13,138
United States	108	98,214	116	107,848
Totals	415	328,308	506	406,357

Source: Nuclear News 1993.

[a]Includes plant operating, under construction, and planned, as of June 30, 1993.

Table 23-5 shows that, for 1993, 18 of the 29 countries outside the United States generated more than 20% of their electrical energy from nuclear power. Lithuania was the leader, with 87.2% of its electrical energy generation supplied from nuclear. Other countries include France with 77.7%, Belgium with 58.9%, Sweden with 42.0%, Switzerland with 37.9%, and Japan with 30.9%.

23.4 URANIUM FUEL CYCLE

The use of uranium as fuel is more complicated than use of coal. Coal essentially moves directly to the steam generator following mining and transportation. As shown in Fig. 23-7, uranium requires several processing steps prior to use in the reactor. The major steps in a uranium fuel cycle consist of mining and milling, conversion to UF_6, enrichment to several percent ${}^{235}_{92}U$, fuel fabrication, then energy production. Following energy production, the fuel is transferred to a spent fuel storage pool adjacent to the reactor in a separate spent fuel building.

From this point, the spent fuel is treated either in an open or a closed fuel cycle. The traditional concept is to use the closed fuel cycle in which the spent fuel is reprocessed to obtain the unused ${}^{235}_{92}U$ and reactor produced ${}^{239}_{94}Pu$. Typical fuels start with 2% to 5% ${}^{235}_{92}U$ and no ${}^{239}_{94}Pu$ and end up with

Table 23-4. Global Nuclear Generation by Reactor Type

Reactor Type	Operating No. Units	Operating Net MWe	Total[a] No. Units	Total[a] Net MWe
Pressurized water reactors	238	208,987	296	263,512
Boiling water reactors	88	72,166	99	83,140
Gas-cooled reactors	36	12,399	36	12,399
Heavy water reactors	34	18,793	50	26,688
Graphite water reactors	15	14,785	16	15,710
Liquid metal fast-breeder reactors	4	1,178	9	4,908
Totals	415	328,308	506	406,357

Source: Nuclear News 1993.

[a]As of June 30, 1993.

Table 23-5. Nuclear Capacity Outside The United States for 1993

	Capacity MWe[a]	Percent of Generation[a]
Argentina	935	14.2
Belgium	5,527	58.9
Brazil	626	0.2
Bulgaria	3,538	36.9
Canada	15,755	17.3
China	1,194	0.3
Czech, R.P.	1,648	29.2
Finland	2,310	32.4
France	59,033	77.7
Germany	22,559	29.7
Hungary	1,729	43.3
India	1,593	1.9
Japan	38,029	30.9
Kazakh.	70	0.5
Korea, Republic of (South)	7,220	40.3
Lithuania	2,370	87.2
Mexico	654	3.0
Netherlands, The	504	5.1
Pakistan	125	0.9
Russia	19,843	12.5
Slovak R.P.	1,632	53.6
Slovenia	632	43.3
South Africa, Republic of	1,842	4.5
Spain	7,101	36.0
Sweden	10,002	42.0
Switzerland	2,985	37.9
Taiwan, Republic of China	4,890	33.5
Ukrania	12,679	32.9
United Kingdom	11,909	26.3

Source: Nuclear Energy Institute.

[a]1993 data are preliminary.

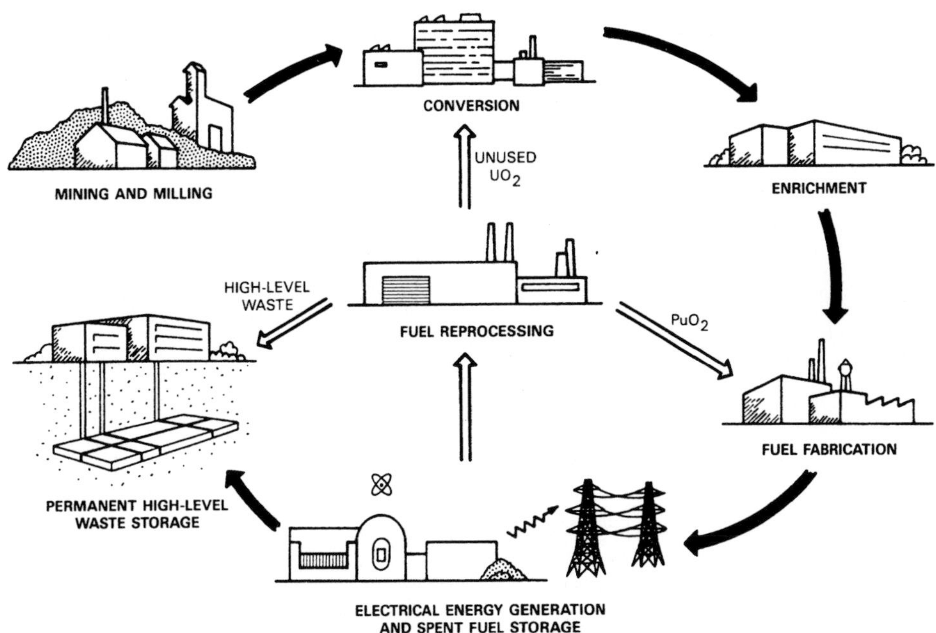

Fig. 23-7. Uranium fuel cycle.

1% to 2% ${}^{235}_{92}U$ and about 1% ${}^{239}_{94}Pu$ Since natural uranium contains about 0.7% ${}^{235}_{92}U$, this spent fuel is a rich resource. The reprocessed uranium is reenriched in ${}^{235}_{92}U$ and fabricated as new fuel. The Pu is fabricated as new fuel within a ${}^{239}_{94}PuO_2/{}^{238}_{92}UO_2$ mixture. The fission products remaining are high-level waste and are processed into a solid form for permanent repository storage. The current concept is to use the open cycle in which the spent fuel, after several years in water storage pools at the reactor site, will be placed in a federally owned underground repository. The first United States repository is not expected to be in operation until around 2010.

Some fuel was reprocessed in the United States until 1977, when the Carter Administration closed fuel reprocessing because of concerns about proliferation of Pu by hostile groups or foreign governments into weapons use. Other nations, such as England and France, continued to reprocess the spent fuel.

In the 1980s, the political picture again changed and fuel reprocessing was reopened tô commercial companies. No companies undertook the reprocessing, however, since uranium reserves were felt to be adequate. Due to reduced fuel reprocessing cost savings and concern over future political reversals away from fuel reprocessing, the United States now uses the open fuel cycle. England, France, and Japan, as well as some other nations, plan to reprocess and reuse their nuclear fuel.

Natural uranium contains approximately 0.7% by weight ${}^{235}_{92}U$ in ${}^{239}_{92}U$. Light water reactors require a higher enrichment of ${}^{235}_{92}U$ in ${}^{238}_{92}U$ to overcome the neutron absorption by the water molecules. Enrichments vary depending on the fuel lifetime in the reactor core, but typically are around 3 wt % ${}^{235}_{92}U$.

Several processes are used for enrichment, including gaseous diffusion, gas-centrifuge, and laser methods. The gaseous diffusion method was developed during the Manhattan Project and is the most widely used, including use within the United States, France, and the former Soviet Union. There are three gaseous diffusion plants in the United States, supplying approximately 50% of the world's enrichment demand.

A major problem for the nuclear industry has been the final, permanent storage of the spent fuel or fission product waste. Fission product or high-level waste is highly radioactive and thermally hot and must be stored safely in the environment away from human contact. Storage times to allow for radioactive decay to essentially background levels require thousands of years, depending on the radionuclide makeup of high level waste.

High-level wastes consist of spent fuel and commercial nuclear and defense wastes. More than 90% of the high-level waste activity content is contained in commercial spent fuel. By 2000, spent fuel is expected to reach 44,000 tons (40,000 tonnes) representing 557,599 ft^3 (15,786 m^3), while high-level defense waste is expected to reach about 11,690,000 ft^3 (331,000 m^3).

Many scientific studies have been conducted to determine the best solution for permanent storage. Deep underground storage appears to be the most acceptable solution, both technically and politically. Underground storage has two objectives: waste isolation from people, making accidental or intentional access highly unlikely; and waste isolation from circulating underground water. Underground water contamination involves the more probable method of population exposure. The United States took a major step toward a permanent storage solution with the passage of the Nuclear Waste Policy Act (NWPA) of 1982 (signed into law on January 7, 1983). This act required the DOE through OCRWM to accept the responsibility for the following activities:

1. Site, construct, and operate a deep, mined geologic repository.
2. Site, construct, and operate one monitored retrievable storage facility (MRS) as an interim solution under certain conditions.
3. Develop a system for transporting waste to the mined repository and MRS facility.

NWPA of 1982 established procedures for the DOE to select candidate sites, characterize the sites, and make recommendations to the President of the United States. The President would make the final decision, subject to approval by the Congress. Two national waste repositories were to be planned. The Act also established a Nuclear Waste Fund that required utilities to pay a fee of one mill (0.1 cent) per kilowatt-hour for nuclear generated electricity. This fee allowed the federal government to recover all of the costs of developing the repository and disposal of the spent fuel or other high-level wastes. The federal government would pay costs for its defense wastes.

The NWPA of 1982 was amended in 1987 and directed DOE to conduct site characterization studies at the Nevada Yucca Mountain site approximately 100 miles northwest of Las Vegas. Yucca Mountain was one of three candidate sites recommended earlier for characterization by DOE under NWPA 1982, but cost and political considerations led the US Congress to amend the Act to the Yucca Mountain Site only. Lack of rainfall, distance from population centers, and an uninhabitable area were key considerations for its selection. A second national waste repository, to be located in the eastern United States, was delayed indefinitely.

Site characterization includes determination of the capabilities of the underground site to permanently and safely contain the waste, including rock condition studies involving geology, hydrology, seismicity, and tectonic investigations. If the site is determined to be suitable, that location must be approved by the President and the host state. If the host state disapproves, the amended act requires Congress to pass a joint resolution to override the state's disapproval. A DOE application to the NRC for authorization to construct the repository is then required. Once authorization is received, construction can begin. The current timetable envisions site characterization starting in 1991/92 with completion in 2001. Construction would start in 2004 with commercial operation in 2010.

Figure 23-8 shows the conceptual arrangement of the permanent waste repository for the Yucca Mountain site. The repository consists of surface facilities to receive and prepare the waste. Spent fuel is received by either rail or truck in transportation casks. The spent fuel is received, consolidated and placed in canisters (with gas added to prevent oxidation), and then sealed. The canisters are checked for leakage, then placed in storage until transport to the final underground location.

The canisters, which are 15.5 ft (4.7 m) in length and 26 in. (0.7 m) in diameter, are then moved to the underground repository, approximately 1,000 ft (305 m) below the surface and placed in vertical bore holes of tuff (volcanic rock). The boreholes are located in a series of emplacement drifts and are approximately 25 ft (7.6 m) deep and 30 in. (0.8 m) in diameter. A metal plug of several inches of steel is added for shielding, then crushed tuff is packed around and on top of the canister. The borehole is then closed with a metal cover. The Yucca Mountain repository will accommodate about 70,000 tons (70,000 tonnes) of heavy metal over 2,095 acres (8,500 hectares). Current plans call for the utilization of about 1,380 acres (5,600 hectares). The active life of the repository is about 50 years, with the last 25 years following placement of the final canisters to be used as a monitoring or caretaker period to confirm the integrity of the repository. After 50 years, the repository is permanently sealed and the site returned to its natural state.

23.5 CURRENT GENERATION POWER REACTORS

The reactor workhorse is the pressurized water reactor, comprising 57% of the 415 power reactors operating worldwide. Boiling water reactors follow with 21% of the world's total. Gas-cooled, heavy water, and graphite water reactors follow with 9%, 7%, and 5% of the world's total, respectively. In the

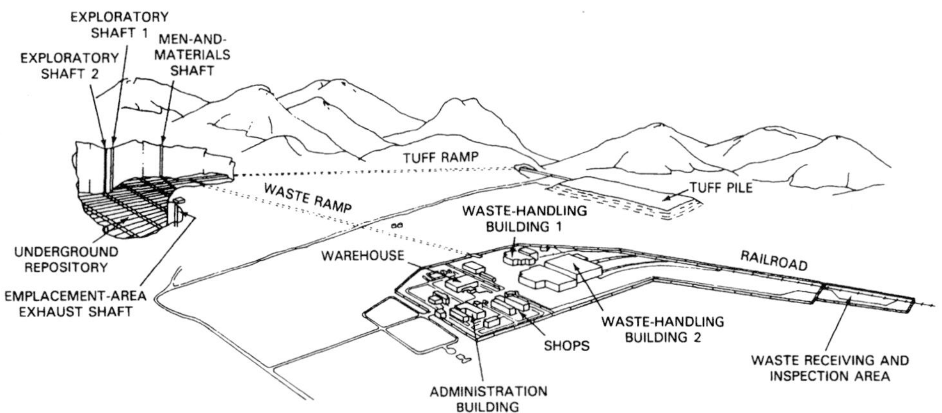

Fig. 23-8. Conceptual arrangement of the proposed high-level spent fuel repository at Yucca Mountain. (Source: Department of Energy.)

United States, 67% of the 108 operating reactors are pressurized water reactors followed by 33% for boiling water reactors.

23.5.1 Pressurized Water Reactors

Pressurized water reactors (PWRs) maintain a sufficiently high reactor coolant pressure to prohibit boiling in the reactor core. Figure 23-1 shows a simplified schematic of a PWR. The nuclear reactor is within a pressure vessel. Coupled to the pressure vessel are two or more reactor coolant system (RCS) loops, each containing a steam generator and one or more reactor coolant pumps. A pressurizer is connected to one of the loops to maintain reactor coolant pressure. Each loop represents a closed system. The reactor coolant containing radioactive fission products and activated corrosion products is separated from the secondary side which supplies steam to the turbine–generator.

In the United States, PWRs are manufactured and marketed by the Westinghouse Electric Corporation, the Babcock & Wilcox Company, and Combustion Engineering, Inc. Commercially available sizes range from 600 MWe to over 1,200 MWe. The majority of reactors purchased and constructed in the 1980s have been the larger sizes (1,000 to 1,200 MWe) to take advantage of economies of scale. The current Westinghouse nuclear steam supply system models consist of two, three, or four loops coupled to the pressure vessel, where each loop contains one reactor coolant pump and one U-tube steam generator. The number of loops increases with reactor size. For example, four-loop design is used for 1,150- and 1,280-MWe reactor sizes. The Combustion Engineering and Babcock & Wilcox designs use two loops coupled to the reactor, each loop consisting of a steam generator and two reactor coolant pumps. The Combustion Engineering and Westinghouse steam generators are a U-tube design, whereas the Babcock & Wilcox steam generator is once-through producing some superheat steam. The Westinghouse and Combustion Engineering steam generators produce steam at saturated conditions.

23.5.1.1 PWR Nuclear Steam Supply System. The heart of the nuclear steam supply system (NSSS) is the nuclear core. It provides the thermal energy which is transferred by the reactor coolant system to the steam generators for steam production. Figure 23-9 shows a typical PWR pressure vessel assembly, which includes the cylindrical pressure vessel with neutron shield, lower core support plate, core barrel, core baffle, nuclear core containing fuel assemblies, control rod cluster assemblies, and the upper core support structure. The upper core support structure includes the upper support plate, upper core plate, support columns, and control rod guide tubes. Core support structure components are fabricated from stainless steel.

Reactor coolant enters the pressure vessel through inlet nozzles, flows down between the core barrel and the vessel wall, then turns and flows up through the lower core plate and fuel assemblies, and out the pressure vessel through the outlet nozzles to the steam generators. The lower core plate supports the weight of the fuel assemblies and transfers the loads through the core barrel to the reactor vessel flange. The upper core support structure provides the alignment of the upper ends of the fuel assemblies and provides the control rod guide tubes and associated tube shafts which are part of the control rod drive mechanisms.

The neutron shield consists of steel panels attached to the lower areas of the core barrel subject to higher neutron flux. The neutron shield protects the pressure vessel from gamma and neutron radiation. Neutron detectors and temperature detectors (thermocouples) enter from the bottom head. The neutron detectors provide information necessary to monitor the neutron flux levels in the reactor and determine the fuel burnup of the fuel assemblies as well as the power produced. The thermocouples measure coolant temperature rise through the fuel assemblies.

The pressure vessel is a cylindrical container with a hemispherical bottom head and a removable upper head. The vessel is made of low-alloy steel, using stainless steel or Inconel cladding for surfaces in contact with water. For refueling operations, all internals inside the core panel above the upper core plate are removed starting with the upper pressure vessel head. The fuel assemblies are then removed or relocated by direct access. Table 23-6 shows typical reactor parameters for a large four-loop PWR design.

The reactor core consists of a series of fuel assemblies, arranged in a cylindrical shape, and supported by the lower core support plate. Figure 23-10 shows a typical fuel assembly with a control rod cluster. Each fuel assembly consists of a square array of fuel rods in a specific grid structure with the fuel rods held in place by spacers. The spacers are spring clipped grids located at several points along the length of the fuel assembly.

The bottom nozzle of the fuel assembly fits into the lower core support plate and controls the coolant flow distribution through the fuel assembly. The top nozzle provides upper fuel rod support and directs the hotter reactor coolant toward flow holes in the upper core plate. Depending on reactor size and manufacturers, fuel assemblies consist of 14×14, 15×15, 16×16, or 17×17 arrays of fuel rods. Reactor power is increased by adding fuel assemblies; for example, a 600-MWe reactor contains 121 fuel assemblies, whereas an 1,150-MWe reactor contains 193.

The control rod cluster as shown in Fig. 23-10 consists of a series of control rods (5 to greater than 20) fastened to a top spider. The control rods move through the full length of the fuel assembly in control rod guide thimbles, located in vacant fuel rod positions within a fuel assembly. Not all fuel assemblies contain control rod clusters. Each control rod contains a neutron absorber such as boron carbide or hafnium in a stainless steel clad tube. The control rod cluster is attached to a control rod drive shaft which moves through the control rod guide tube located in the upper support structure.

Electromagnetic mechanical control rod drive mechanisms on the top of the reactor vessel head raise or lower the

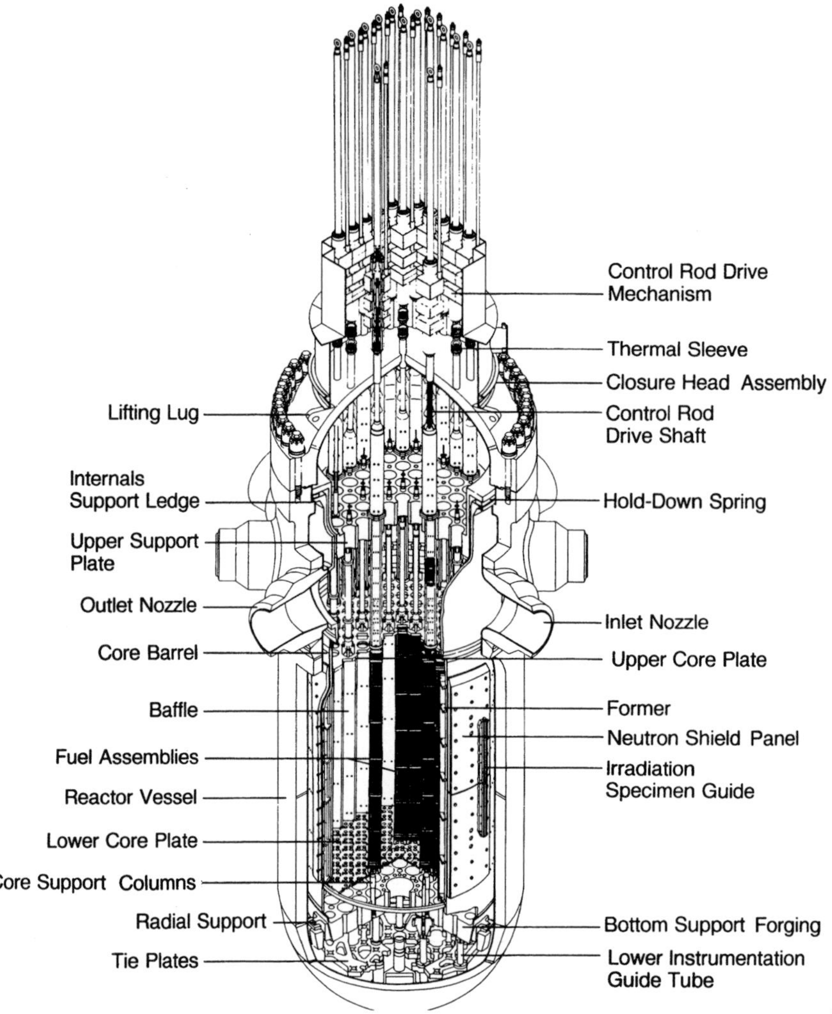

Fig. 23-9. Pressure vessel for a large PWR four-loop design. (From Westinghouse. Used with permission.)

control rod clusters through the attached control rod drive shafts for control of neutron flux and subsequent reactor power. Each control rod drive controls a control rod cluster and also contains a magnetic device that becomes deenergized during a reactor scram (rapid reactor shutdown), dropping the control rod clusters into the fuel assembly guide tubes by gravity. Control rod clusters are used for reactor startup and shutdown and for following more rapid changes in reactor power called load transients. The control rod clusters are located throughout the core in symmetrical group arrangements. A chemical shim system adjusts boric acid (boron is a neutron absorber) concentration in the reactor coolant to compensate for fuel consumption and fission product buildup over the fuel lifetime. Table 23-7 shows typical reactor fuel parameters for a large four-loop PWR design.

Table 23-6. Typical Reactor Parameters for a Large PWR Four-Loop Design

Parameter	Typical Values
Total heat output	3,411 MWt
System pressure	2,250 psia (15.5 MPa)
Coolant flow rate	138.4×10^6 lb/m (1.74×10^4 kg/s)
Coolant inlet temperature	557.5° F (291.9° C)
Average coolant temperature rise in vessel	61.0° F (33.9° C)
Coolant outlet temperature	618.5° F (325.8° C)
Equivalent core diameter	11.06 ft (3.38 m)
Core length (between fuel ends)	12.0 ft (3.66 m)
Fuel arrangement	17 × 17
Uranium core weight	89.8 tons (81.2 tonnes)
Number of fuel assemblies	193
Overall length of pressure vessel	44 ft-7 in. (13.6 m)

Source: Westinghouse 1984.

23.5.1.2 Reactor Coolant System Loop Components. The Reactor Coolant System (RCS) consists of two or more loops coupled to the reactor pressure vessel. Each loop contains at least one reactor coolant pump and a steam generator. A pressurizer is connected to one of the loops to maintain constant system pressure. The coolant loops form a closed system that contains the radioactive coolant.

The reactor coolant pumps move reactor coolant at high pressure and temperature through the pressure vessel, reactor core, steam generators (tube side), and reactor coolant piping. The pumps are centrifugal, single-stage, vertical type

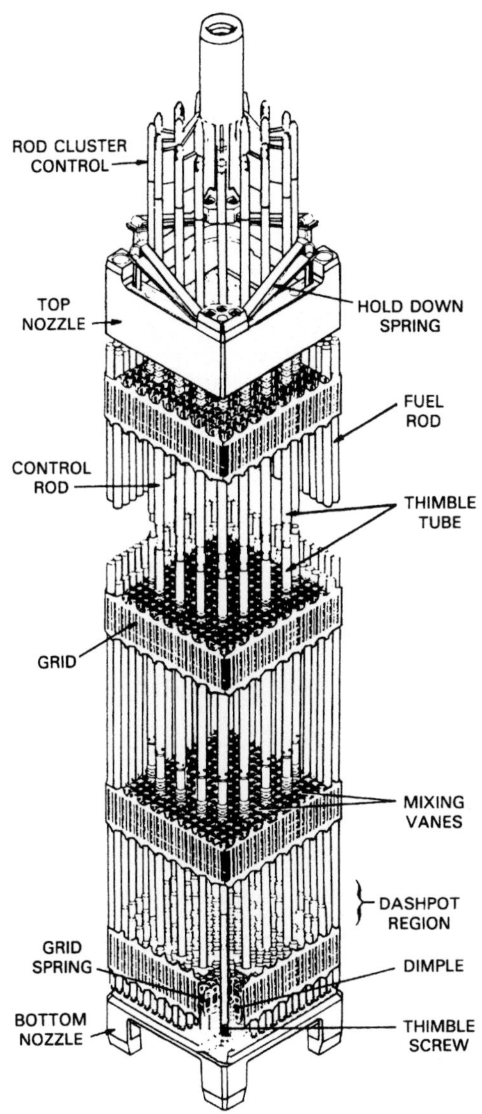

Fig. 23-10. 17 × 17 Fuel assembly with control rod cluster. (From Westinghouse. Used with permission.)

Table 23-7. Typical Reactor Fuel Parameters for a Large PWR Four-Loop Design

Parameter	Value
Fuel rod length	12 ft (3.66 m)
Outside fuel rod diameter	0.36 in. (0.914 cm)
Fuel rod cladding thickness	0.0225 in. (0.0572 cm)
Cladding material	Zircaloy 4
Fuel pellet diameter	0.3088 in. (0.7844 cm)
Lattice pitch	0.496 in. (1.26 cm)
Assembly fuel rod arrangement	17 × 17
Fuel rods in an assembly	264
Total fuel rods in a core	50,952
Number of control rod cluster assemblies	53

Source: Westinghouse 1984.

powered by a single-speed, air-cooled induction motor. The pumps are water cooled and use high-pressure injection water (greater than reactor coolant pressure) at the shaft seal to prevent the outleakage of radioactive reactor coolant.

Steam generators such as that shown in Fig. 23-11 produce steam for the turbine generator. The steam generator in Fig. 23-11 is a typical vertical mounted U-tube closed heat exchanger. The high-temperature reactor coolant enters the bottom of the steam generator at the reactor coolant inlet nozzle, passes through the U-tubes, and then out the steam generator at the bottom outlet. Feedwater from the high-pressure feedwater pumps enters the steam generator at the feedwater inlet, flows through a feedwater ring into the downcomer annulus (volume between the steam generator shell and the tube sheet wrapper), and enters the tube bundle near the tube sheet at the bottom of the steam generator. As the feedwater rises up the tube bundle, it boils. The wet steam then passes through moisture separators in the upper steam generator shell, where the entrained liquid is removed. The steam then passes through steam dryers to raise the steam quality to 99.75%. Table 23-8 shows typical parameters and other data for reactor coolant pumps and steam generators used in a large four-loop PWR design. The steam generator U-tubes are made of Inconel, and the shell and

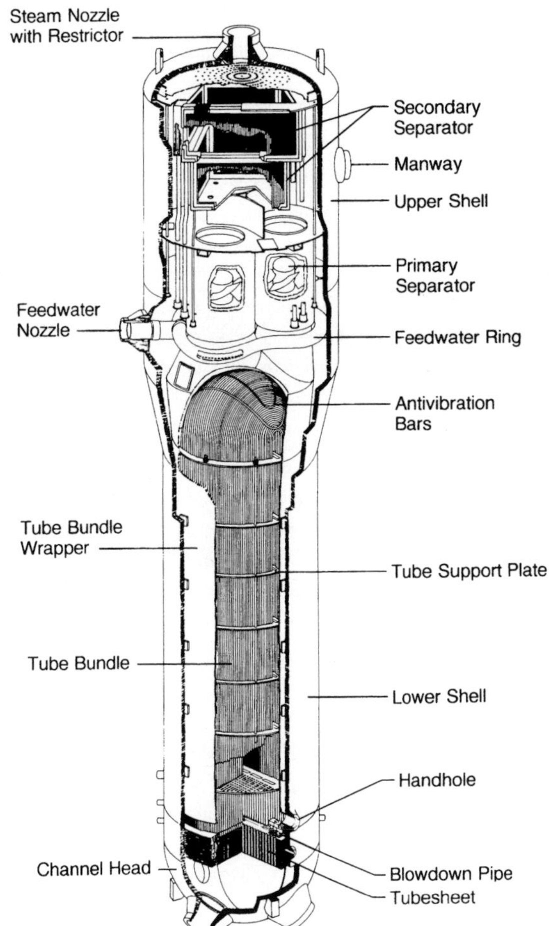

Fig. 23-11. Steam generator for a large PWR. (From Westinghouse. Used with permission.)

Table 23-8. Typical Design Parameters and Data for Components of a Large Four-Loop PWR

Reactor coolant pump	
Capacity	97,600 gpm (6.15 m^3/s)
Head	280 ft (85.3 m)
Pressure	2,500 psia (17.2 MPa)
Temperature	650° F (343° C)
Overall height	28 ft (8.5 m)
Operating rpm	1,189
Casing diameter	6ft-5 in. (196 cm)
Power	7,000 hp (5,220 kW)
Steam generator (U-tube)	
Overall height	67 ft-8 in. (20.6 m)
Upper shell diameter (outside)	14 ft-7¾ in. (4.5 m)
Lower shell diameter (outside)	11 ft-3 in. (3.4 m)
Design pressure (tube side)	2,500 psia (171.2 MPa)
Operating pressure (tube side)	2,250 psia (15.5 MPa)
Pressure (shell side)	1,000 psia (6.9 MPa)
Design pressure (shell side)	1,200 psia (8.3 MPa)
Steam flow rate	3.8×10^6 lb/h (480 kg/s)
Reactor coolant flow rate	35×10^6 lb/h (4,419 kg/s)
Reactor coolant inlet temperature	621° F (327° C)
Reactor coolant outlet temperature	558° F (292° C)
Dry weight	346 tons (314 tonnes)

Source: Westinghouse 1984.

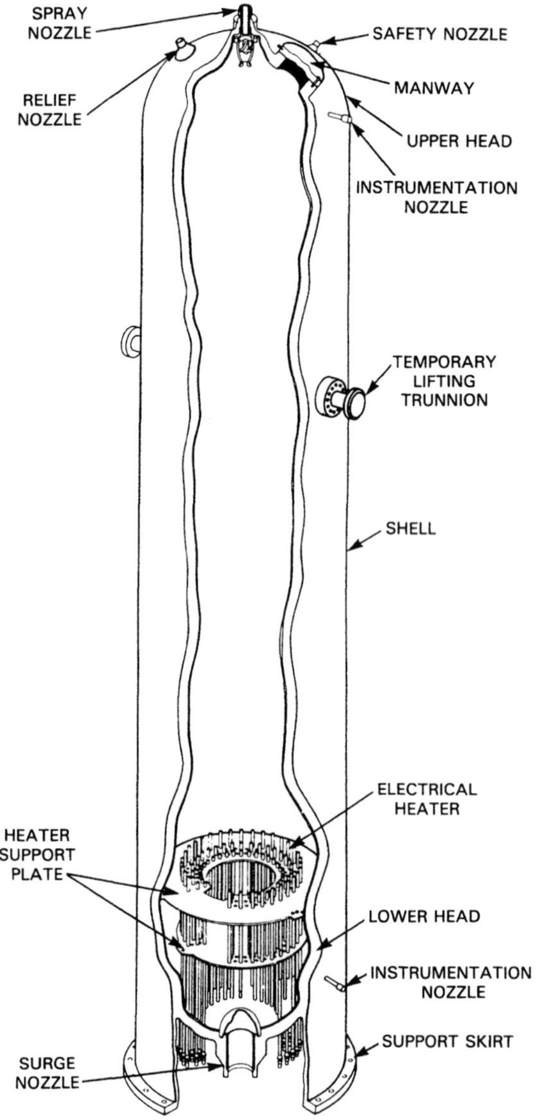

Fig. 23-12. Pressurizer for a large PWR design. (From Westinghouse. Used with permission.)

other pressure parts are carbon or low-alloy steel. Areas subject to reactor coolant contact are clad with stainless steel or Inconel.

Westinghouse and Combustion Engineering use U-tube steam generators. The steam generators used by Babcock & Wilcox are vertical, straight tube and shell exchanges. Reactor coolant enters at the top of the steam generator and slows downward through the Inconel tubes and exits at the bottom of the steam generator. Feedwater enters the steam generator near the middle and flows down an annulus between the steam generator shell and tube baffle and enters near the bottom. The water, as it passes up the steam generator, is converted to steam. The steam is superheated before leaving the steam generator. Since the steam is superheated, moisture separators and dryers are not used.

Increases or decreases in the generating plant electrical load cause fluctuations in the reactor coolant pressure. A pressurizer on one of the RCS loops is used to limit pressure changes during load transients and to maintain constant coolant system pressure during steady-state operation. Figure 23-12 shows a typical pressurizer. The pressurizer contains both a spray nozzle and electric immersion heaters. It also has self-actuating and power-operated relief valves (not shown in Fig. 23-12) which discharge through vent piping to a pressurizer relief tank.

An increase in reactor system pressure above the pressure setpoint opens spray valves that spray reactor coolant from a cold loop side into the pressurizer steam space, condensing steam and lowering the pressure. Likewise, a decrease in pressure or temperature below the system set points activates the immersion heaters, which increases the reactor system pressure to (or near) its setpoint value. The relief and safety valves are used in the event that the system pressure increases above the pressurizer's capability. Both relief and safety valves vent to a pressure relief tank that contains cooler water for condensing steam. A pressurizer for a large four-loop design has a typical height of approximately 53 ft (16.2 m), with an 8 ft (2.3 m) diameter, and contains approximately 75 immerser heaters with 1,800 kW of power. The pressurizer is low-alloy steel with stainless steel interior cladding. Its dry weight is 98 tons (88.7 tonnes).

23.5.1.3 PWR Auxiliary Systems. Numerous auxiliary systems are required to support a PWR. These systems provide important functions such as RCS coolant purification, RCS boron addition and removal, shutdown fission product decay heat removal, and emergency reactor core cooling for mitigation of accidents such as a reactor coolant line rupture. Key auxiliary systems include the following:

- Chemical Volume and Control System (CVCS) maintains RCS water purity and water inventory and provides for boric acid concentration adjustment.

- Boron Thermal Regeneration System (BTRS) maintains boron concentration in the RCS during normal, shutdown, and steady-state operations.
- Residual Heat Removal System (RHRS) removes fission product decay heat from reactor during shutdown and refueling.
- Steam Generator Blowdown Processing System (SGBDS) treats steam generator blowdown (secondary side) water and helps maintain secondary side water purity.
- Safety Injection System (SIS), an engineered safeguard system, provides emergency reactor core cooling from a loss-of-coolant accident (LOCA) such as a coolant line break.
- Emergency Feedwater System (EFWS) supplies feedwater to the shell side of the steam generators on loss of main feedwater (e.g., loss of feedwater pumps).
- Component Cooling Water System (CCWS) provides cooling water to reactor auxiliary systems containing radioactive fluids. The CCWS is an engineered safeguard system.
- Containment Spray System (CSS), an engineered safeguard system, limits containment pressure following a coolant line rupture.
- Containment Fan Cooling System (CFCS) provides fan cooling of the containment atmosphere utilizing cooling coil air handling units. The CFCS is an engineered safeguard system.

The Reactor Protection System (RPS) monitors the plant for abnormal conditions, minimizes spurious reactor trips, and provides for rapid, automatic reactor shutdown resulting from an unsafe or abnormal condition. The reactor protection system utilizes two-out-of-three and two-out-of-four logic of the particular trip parameter (e.g., reactor pressure) to activate reactor shutdown. Parameters vary according to manufacturer, but usually include parameters similar to the following (Lish 1972 and Westinghouse 1984):

- Manual trip,
- High neutron flux,
- High rate of power increase,
- High reactor power,
- High coolant temperature and pressure,
- Low coolant pressure,
- High pressurizer water level,
- Low or loss-of-coolant flow,
- Low stream generator water or feedwater flow,
- Turbine generator trip,
- Low steam generator pressure, and
- Safety injection initiation.

23.5.1.4 Containment. The containment structure that houses the reactor pressure vessel and reactor coolant system loops is a passive safety device. It is designed to withstand pressure loads during reactor loss-of-coolant accidents and failure of the safety injection systems and to contain gaseous fission product releases to the environment.[2] Several types of containment structures are in use, but the most common for large reactor systems consists of a post tensioned concrete structure with a steel plate liner. The steel liner acts as a leak-tight membrane prohibiting fission product leakage. The concrete structure is cylindrical and the concrete is prestressed by a post-tensioning system. The post-tensioning system uses a series of individual horizontal and vertical tendons that place the concrete structure in a compression mode. The foundation consists of a base slab concrete structure heavily reinforced with steel to support the weight of the reactor coolant system and the containment structure. A typical containment design pressure is 60 psig (0.4 MPag) which results from the peak pressure occurring from any rupture of the reactor coolant system (Westinghouse 1984). The containment volume for a large four-loop PWR is approximately 2,000,000 ft^3 (56,600 m^3). Its inside diameter is 124 ft (37.8 m) and its inside height is 205 ft (62.5 m) (Westinghouse 1984). Figure 23-13 shows a cutaway view of a typical containment structure, which houses a large four-loop PWR.

23.5.1.5 Power Conversion System. The Power Conversion System (PCS) uses steam from the steam generators to produce electrical energy. The PCS consists of the turbine generator, moisture separators, main condenser, condensate, feedwater, circulating water, and other auxiliary systems. The PCS using nuclear-generated steam is similar to a PCS using fossil-generated steam, except the nuclear PCS working fluid is saturated steam at approximately 1,000 psia (6.9 MPa). The turbine generator is usually an 1,800 rpm machine with a high-pressure section and two or more low-pressure sections. In contrast, modern fossil plants use steam conditions at 2,400 psig (16.6 MPag) and a 3,600 rpm turbine–generator. In a nuclear PCS, steam exiting the high-pressure section goes through a moisture separator–reheater to remove moisture and to provide one or two stages of reheating prior to passing through the low-pressure turbine. Removal of moisture from the high-pressure exhaust steam helps prevent low-pressure blading degradation. Extraction steam for the reheater stages is taken from a high-pressure turbine stage and from main steam.

Figure 23-14 shows a heat balance diagram at valves-wide-open (VWO). The turbine generator is a tandem-compound, six-flow, four-element, 1,800 rpm unit with one high-pressure and three low-pressure sections. Moisture separator–reheaters are utilized to dry and reheat the steam between the high- and low-pressure turbines. The PCS also contains a three-shell deaerating surface condenser, condenser evacuation system, two turbine-driven main feedwater pumps, three motor-driven condensate pumps, and seven stages of feedwater heating (four low-pressure and three high-pressure stages). Between the condensate pumps and low-pressure feedwater heaters is a full-flow condensate demineralizer system to maintain feedwater purity. Condenser cooling is provided by an open circulating water system taking water from a large constructed cooling lake.

[2]The containment at Three Mile Island retained the vast majority of radionuclides from the Unit 2 accident whereas at the Chernobyl Unit 4 accident, Unit 4 gaseous radionuclides were released to the environment because of a lack of a containment building.

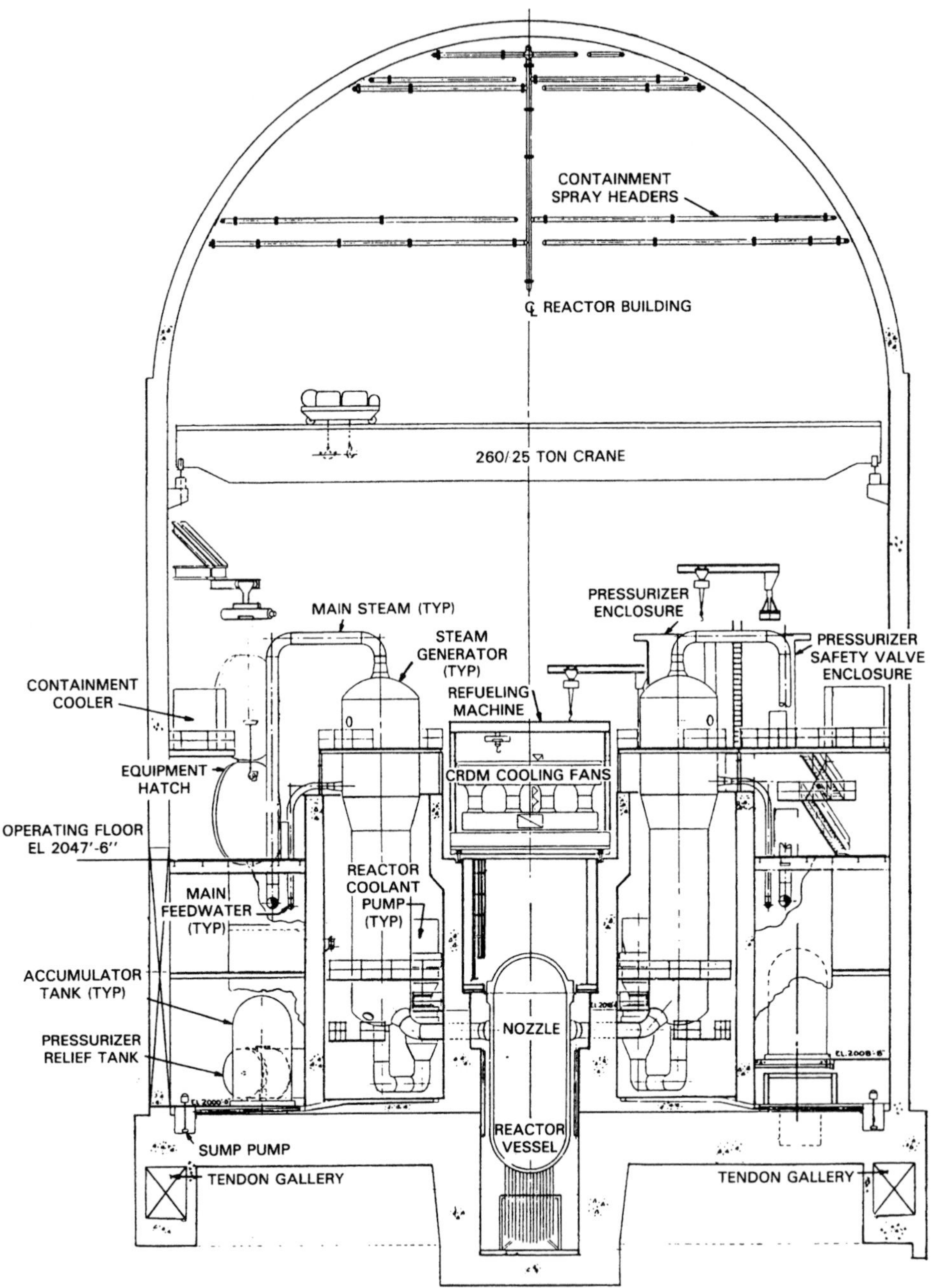

Fig. 23-13. Containment structure for a large four-loop PWR. (From Wolf Creek Nuclear Operating Corporation. Used with permission.)

23.5.2 Boiling Water Reactors

BWRs produce steam directly in the core. The BWR steam production method is simpler than that of a PWR, since separate steam generators are not required. Figure 23-2 shows a simplified schematic for a BWR. Steam produced in the nuclear core at saturated conditions of 1,000 psia (6.9 MPa) passes through moisture separators and steam dryers and then is sent to a turbine generator. Leaving the low-pressure sections of the turbine, the steam is condensed to liquid, preheated by several stages of feedwater heating, coolant pressure is increased to about 1,000 psia by the feedwater pumps, and sent back to the pressure vessel.

The sole manufacturer of BWRs in the United States is the General Electric Company. The BWR nuclear steam supply system consists of the nuclear reactor core contained within a large pressure vessel, auxiliary systems required for normal, shutdown, and accident operations, and the necessary instrumentation and controls. BWRs are available commercially in sizes from 600 MWe to 1,400 MWe. However, BWRs of the larger sizes (>1,000 MWe) were more commonly purchased in the 1970s to take advantage of economies of scale. General Electric's BWR 6 was introduced in 1972 and represents its latest product line currently in operation. The development of the next generation BWR, the Advanced BWR (ABWR), commenced in 1978. Construc-

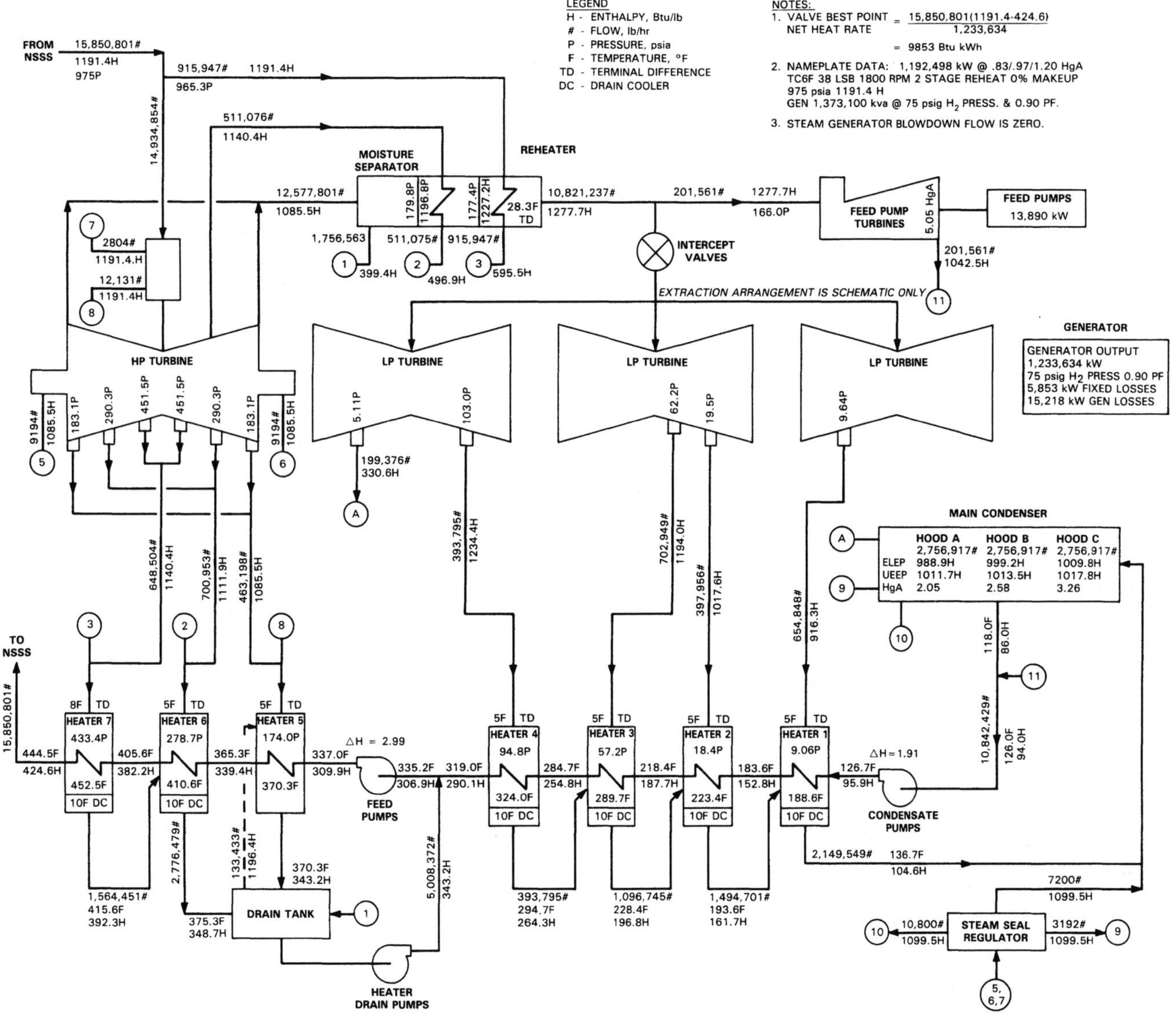

Fig. 23-14. Typical heat balance for a large (1,190 MWe) PWR. (From Wolf Creek Nuclear Operating Corporation. Used with permission.)

tion of the first ABWR began in 1991 and operation is scheduled for 1996.

23.5.2.1 BWR Nuclear Steam Supply System. The pressure vessel represents the dominant part of the steam supply system and is shown in Fig. 23-15. The pressure vessel is a large cylindrical container with a removable upper head. The vessel contains the nuclear core, shroud, top guide assembly, core plate assembly, steam separator and dryer assemblies, and jet pumps. Also included within the pressure vessel are the control rods, control rod drive housings, and the control rod drives. The pressure vessel is made of low-alloy steel with interior stainless steel cladding except for its removal head.

Feedwater enters the pressure vessel at the feedwater inlet and is distributed in an annulus between the stainless steel shroud and the pressure vessel wall. The feedwater is mixed with internal recirculating water within the annulus. Part of the coolant mixture (about two-thirds) enters the suction inlets of the jet pump assemblies. The remaining coolant mixture leaves the pressure vessel at the recirculation water outlet nozzle, is pumped through the recirculation loops (not shown on Fig. 23-15), and reenters the pressure vessel at the jet pump nozzle inlets. There, it mixes with the nonrecirculated coolant mixture within the jet pump diffusers and is discharged into the plenum below the core.

The coolant then flows up through the core where a portion is evaporated into steam. The steam–liquid mixture passes through the steam separator, separating the vapor from the liquid. The separator consists of an array of stand pipes. Each stand pipe includes a vaned separator that spins the mixture, and the resulting centrifugal forces separate the water from the steam. The separated water leaves the separator and enters the coolant pool that surrounds the standpipes

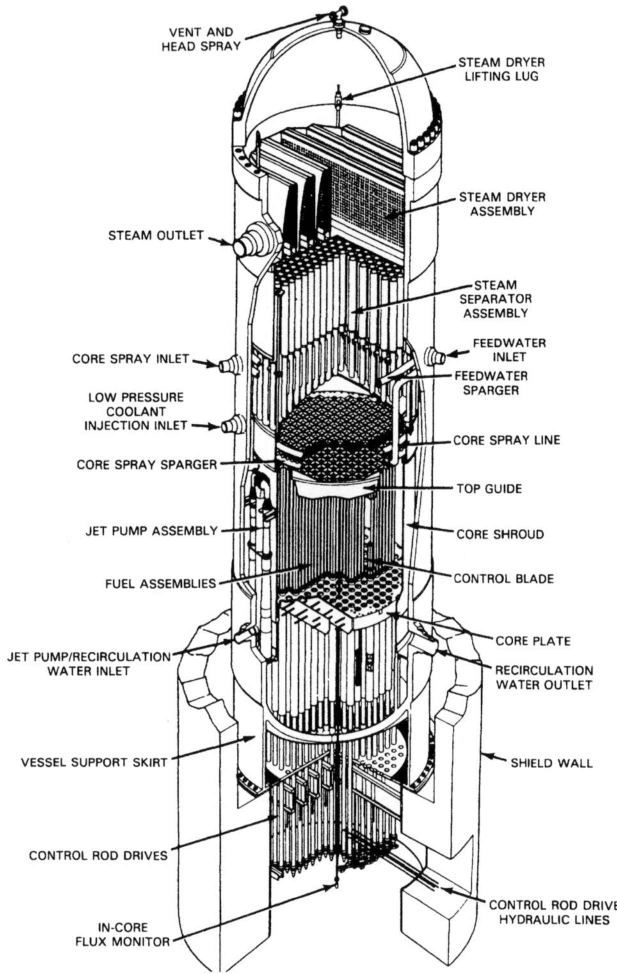

Fig. 23-15. Pressure vessel for a BWR. (From General Electric Company. Used with permission.)

and flows back into the annulus region where it mixes with entering feedwater.

The steam leaves the separators at the top flowing upwards and outwards through a series of drying vanes in the steam dryer where additional separated water is recovered. The steam then leaves the pressure vessel at the steam outlet nozzle and passes to the turbine–generator through the main steam lines. The steam separator assembly sits on the top flange of the core shroud and forms the top of the core discharge plenum. The steam dryer assembly is supported by pads extending from the pressure vessel wall above the steam separator. The pressure vessel head holds the steam dryer assembly in place during operation.

The nuclear core consists of a large number of fuel assemblies and control rods. The control rods move between the fuel assemblies. A typical fuel assembly is shown in Fig. 23-16. Each assembly consists of an 8 × 8 array of Zircaloy-clad fuel rods supported on the top and bottom by tie plates. The lower tie plate uses a nosepiece that fits into the lower fuel support plate assembly and distributes coolant flow to the fuel rods. The lower support plate assembly supports the fuel assembly loads during operation. The upper tie plate, which fits into the top fuel guide assembly, contains a handle for transferring the fuel assembly. Other advanced fuel designs manufactured by GE, Siemens, and ABB-CE contain 9 × 9 and 10 × 10 fuel rod arrays. Some of these advanced designs are currently being used in the United States and overseas.

Three types of fuel rods are used: tie rods, water rods, and standard fuel rods. The third and sixth fuel rods along the outer edge of the fuel assembly are tie rods that tie into the lower and upper tie plates and support the weight of the fuel assembly during fuel handling operations. Two rods in the fuel assembly interior are water rods, consisting of Zircaloy-clad tubes with top and bottom holes for water entry and exit. The water rods are for additional neutron moderating. The remaining 54 fuel rods are standard fuel rods containing slightly enriched UO_2 pellets. Each rod has end pins that fit into anchor holes in the tie plates. The fuel rods are larger for a BWR than for a PWR. Spacers with X springs are located along the fuel assembly length and maintain fuel rod spacing. Each fuel assembly is enclosed within a Zircaloy fuel channel. The fuel channel directs the coolant flow through the fuel rods and provides a guide for the control rods.

Control rods enter the nuclear core from the bottom. A typical control rod is shown in Fig. 23-17. It is cruciform in shape and contains 84 stainless steel tubes within a steel sheath. Each wing of the control rod contains 21 steel tubes with each tube containing boron carbide for neutron poisoning. Each control rod moves between four fuel assemblies. Control rods are lowered or raised from the bottom of the

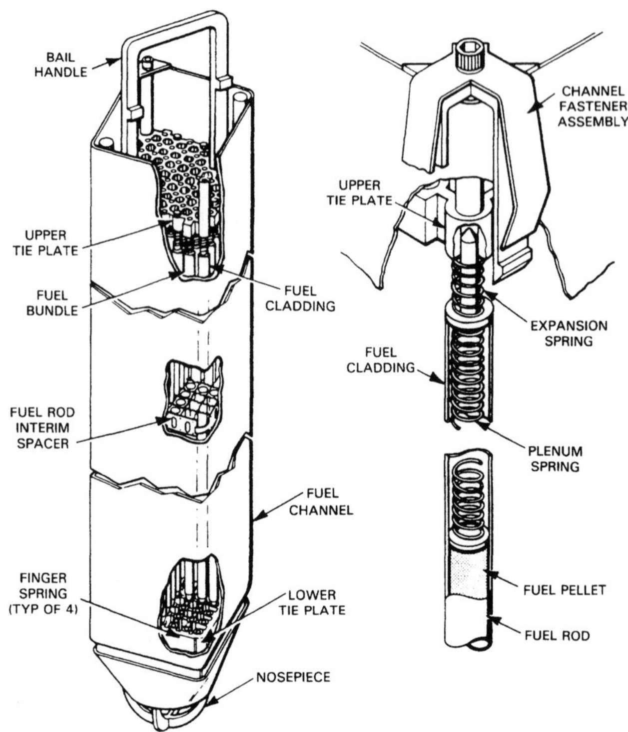

Fig. 23-16. 8 × 8 Fuel assembly for a BWR. (From General Electric Company. Used with permission.)

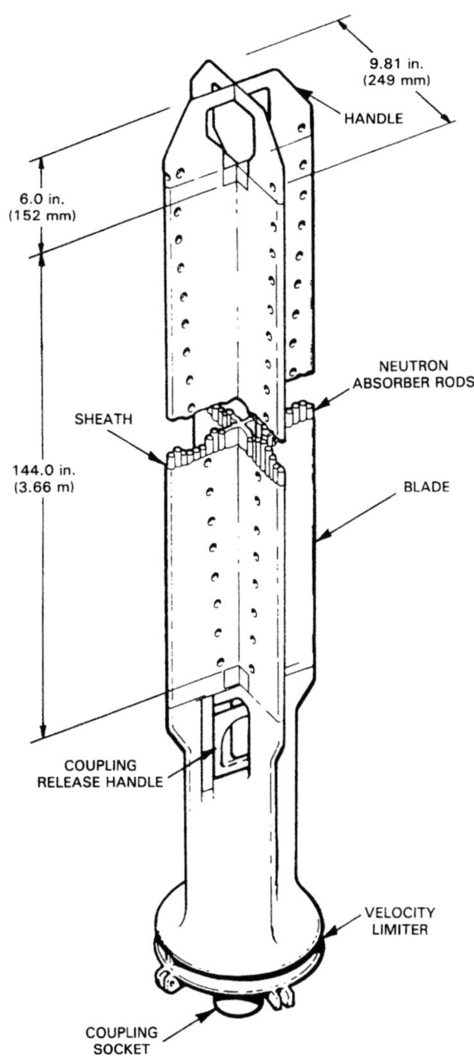

Fig. 23-17. Cruciform-shaped control rod for a BWR. (From General Electric Company. Used with permission.)

core through guide tubes during reactor operation. Each control rod is coupled to a control rod drive that enters the reactor vessel through housings welded to the bottom head of the reactor vessel. The drives are basically hydraulic cylinders, operated by hydraulic control units within the reactor building but outside of primary containment for maintenance access. Each control unit includes an accumulator, nitrogen tank, operating valves, and interconnecting piping. Hydraulic fluid, which is demineralized water, is supplied to the hydraulic control units from one of two full-capacity pumps at a pressure greater than reactor pressure (e.g., 1,750 psia, 12.1 MPa).

A series of withdraw and insert valves within the control unit supply high-pressure water to the top or bottom of the control rod drive piston in the pressure vessel control rod drive housings and lower or raise the control rod into the nuclear core. The drive mechanism can position the control rod at intermediate increments over the full core length. A control rod is held at a fixed position by a collet mechanism. The collet mechanism consists of collet fingers that engage notches in the control drive tube to prevent unintentional control rod withdrawal. Accumulators at each control unit under high pressure (1,750 psia, 12.1 MPa) are used to rapidly insert control rods into the core during an emergency reactor scram.

Fuel enrichments for a BWR vary depending on such factors as refueling cycle duration, initial core, and load requirements. The average enrichment for an initial core ranges from 1.7 wt % $^{235}_{92}$U to 2.0 wt % $^{235}_{92}$U (General Electric 1980). Fuel enrichments for reload fuel have average enrichments from 2.6 wt % to >3.0 wt %. In addition, different $^{235}_{92}$U enrichments may be used within the individual fuel assemblies to reduce local power peaking. Typically, lower enrichment rods are placed in the corner positions and in positions near the water rods. Table 23-9 shows typical reactor parameters for a BWR 6 design.

23.5.2.2 Recirculation System. Control rods are used to maintain and adjust the reactor power level and also to provide power distribution shaping within the core. A forced flow recirculation system called the reactor water recirculation system (RWRS) is also used to adjust the power level without changing the control rod positions. Power levels can be adjusted up to 40% of rated power using the reactor water recirculation system, providing the BWR with load following capability. The reactor water recirculation system for the BWR 6 design consists of two external loops coupled to the pressure vessel. Each loop contains a constant flow pump, an electric motor, a flow control valve, and two shutoff valves. Figure 23-18 shows a simplified drawing of the reactor water recirculation system. The recirculation pumps remove ap-

Table 23-9. Typical Reactor Parameters for a BWR 6 Design

Fuel assembly (8 × 8 arrangement)	
Fuel rod outside diameter	0.483 in. (1.23 cm)
Fuel rod clad thickness	0.032 in. (0.88 cm)
Fuel pellet outside diameter	0.410 in. (1.04 cm)
Number of fuel Rods	62
Channel inside diameter	5.215 in. (13.25 cm)
Active fuel length	12.5 ft (3.8 m)
Reactor core	
Active core height	12.5 ft (3.8 m)
Equivalent core diameter	14.1 ft (4.3 m)
Number of fuel assemblies	624
Uranium core weight	142.6 tons (129.6 tonnes)
Number of control rods	145
Total heat output	2,894 MWt
Pressure vessel	
Inside height	69.34 ft (21.1 m)
Inside diameter	18.17 ft (5.54 m)
Minimum thickness	5.4 in. (13.7 cm)
Reactor coolant	
Core coolant flow rate	84.5×10^6 lb/h (10,669 kg/s)
Feedwater flow rate	12.4×10^6 lb/h (1,566 kg/s)
Steam flow rate	12.4×10^6 lb/h (1,566 kg/s)
System pressure, steam dome	1,040 psia (7.2 MPa)
Core coolant inlet temperature	533° F (278° C)

Source: Gulf States Utilities Company.

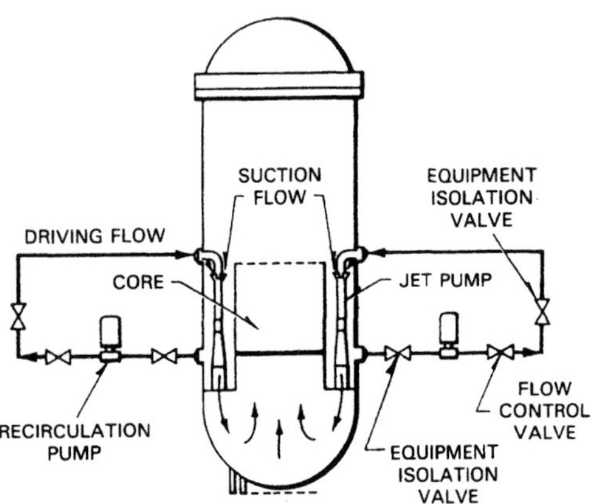

Fig. 23-18. Reactor water recirculation system. (From General Electric Company. Used with permission.)

proximately one-third of the recirculation coolant from the pressure vessel annulus region through the recirculation water outlet nozzle and circulate the coolant at a higher pressure to a manifold that distributes the coolant to the jet pumps through the jet pump/recirculation water inlet nozzles. The coolant flow is discharged through the jet pump nozzles, which provide additional suction and induce the surrounding recirculation water (remaining two-thirds of the coolant annulus recirculation) into the jet pump throat. The two streams mix in the diffuser section and then are discharged into the lower core plenum. The recirculation system adjusts the power level by temporarily reducing or increasing the void fraction in the core. For an increase in power level, the flow control valve increases the recirculation flow through the core, reducing the core void fraction. The reduced void fraction adds more moderating strength, increasing the thermal neutron population (additional reactivity) and correspondingly increasing the reactor thermal power. Once stabilized, the increased reactivity is balanced by the increased void fraction at the new power level.

*23.5.2.3 **Auxiliary Systems.*** A BWR requires numerous auxiliary systems for normal, abnormal, and emergency operations. Several of the auxiliary systems required for normal operation include the reactor water cleanup system (RWCS), residual heat removal (RHR) system in the shutdown mode, closed loop cooling system (CLCS), and the service water system (SWS). Since a large fraction of the reactor water is recirculated by the reactor water recirculation system, the reactor water cleanup system is used to maintain reactor water quality by removing fission and corrosion product contaminants in the coolant water. The shutdown cooling mode of the RHR provides for reactor cooling to remove fission product decay heat during reactor shutdown and before reactor core refueling. The closed loop cooling system provides closed loop cooling to specific reactor systems and equipment that may contain radioactivity (such as heat exchangers, pump and motor coolers), provides a second barrier between their systems and the service water system which removes the closed loop cooling system heat loads, and is the final heat sink.

Auxiliary systems required for abnormal operation primarily deal with standby systems used for reactor core backup cooling such as for hot standby. These systems include the reactor core isolation cooling (RCIC) system, the reactor steam condensing (hot standby) mode of the residual heat removal system, the closed loop cooling system, and the service water system. Both the closed loop cooling system and service water system provide for heat removal.

The last group of auxiliary systems are the safety systems, usually referred to as the emergency core cooling systems (ECCS). The ECCS is used in case of a loss of coolant accident to ensure that the reactor core is supplied with adequate cooling to mitigate core damage and the release of fission products to the environment. ECCS systems are designed for a large number of postulated accident scenarios, with the most severe thought to be a complete break of a main steamline or reactor recirculation line. ECCS systems include the high-pressure core spray system, low-pressure core spray system, low-pressure coolant injection mode of the RHR, automatic depressurization system, standby liquid control system, closed loop cooling system, and service water system. ECCS systems are combined into two independent, automatically activated emergency cooling systems with separate, independent ac (emergency diesel generators) and dc power sources (Divisions 1 and 2). Each of these independent systems is designed to supply cooling water to the nuclear core under all loss-of-coolant accident scenarios. The Division 1 system consists of the automatic depressurization, low-pressure core spray, low-pressure coolant injection mode of the RHR utilizing one RHR pump and RHR exchanger, RHR service water, emergency diesel generator (Division 1), and a dc source. The Division 2 system consists of the automatic depressurization, the low-pressure coolant injection mode of the RHR utilizing two RHR pumps and one RHR heat exchanger, RHR service water, emergency diesel generator (Division 2), and a dc power source. A third reactor core cooling system consisting of the high-pressure core spray system, a third emergency diesel generator (Division 3), and a dc power source provide backup to Division 1 and Division 2 systems in the event the automatic depressurization system does not function properly.

*23.5.2.4 **Containment.*** A passive containment structure is also provided with the BWR to prevent fission product releases to the environment in the event of a loss-of-coolant accident and failure of the ECCS systems to maintain core cooling. Figure 23-19 shows a simplified drawing of a typical reactor building for a General Electric, Mark III design.[3] The reactor building includes a shield building, a containment structure, and the drywell structure.

The drywell is a reinforced concrete cylindrical shell

[3]General Electric has also two earlier commercial containment designs, the Mark I and Mark II.

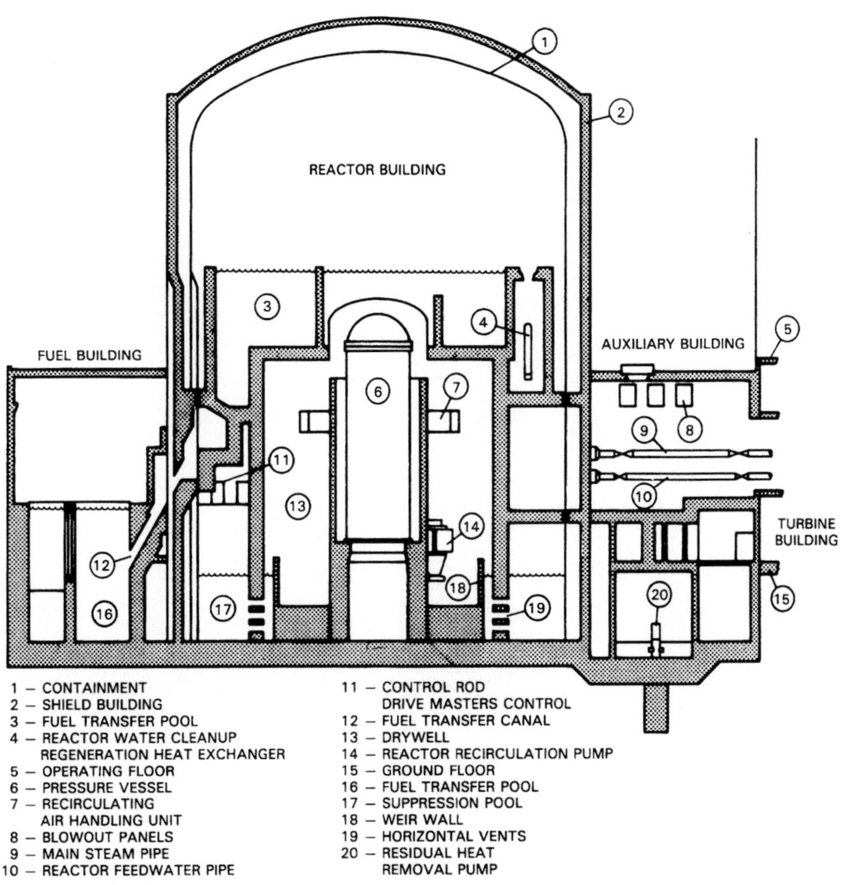

Fig. 23-19. Simplified view of a typical Mark III BWR reactor building.

structure that contains the reactor pressure vessel and reactor recirculation system. The drywell supports the upper containment pool, which is used for refueling operations and has a moveable drywell head cover for access to the reactor pressure vessel. The function of the drywell is to contain the transient blowdown pressure from a loss-of-coolant accident such as a recirculation line break, and to move the blowdown air–steam mixture through a series of large horizontal vent openings in the lower section of the drywell into the suppression pool. This condenses the air–steam mixture, reducing the drywell pressure. The drywell sits on the reactor building foundation mat.

The containment structure encloses the drywell and provides a leakage barrier for fission product leakage to the shield building. The containment structure for the Mark III design is a freestanding, vertical, cylindrical steel pressure vessel with an elliptical head and a flat-bottom steel liner plate. The steel shell is anchored into the reactor building foundation mat. Located at the lower area of the containment structure between the drywell and outer containment structure is a suppression pool. The suppression pool water level covers the horizontal vent openings in the drywell, thus maintaining a water seal between the containment and drywell air volumes. During a loss-of-coolant accident within the drywell, the high-pressure steam displaces the water in the vents and moves into the suppression pool, where it is condensed. The nuclear safety relief valves (SRVs) in the main steam lines also vent to the suppression pools (selected safety relief valves are part of the automatic depressurization system [ADS]).

The shield building houses the containment structure and provides an additional (secondary) barrier for fission product releases to the environment. The shield building provides missile protection for the containment structure. Fission product gases that might leak from the containment are collected in the annulus volume of the shield building and treated by the standby gas treatment system (SGTS) prior to release. The standby gas treatment system maintains a negative pressure in the annulus air volume and then recirculates the air through high-efficiency particulate (HEPA) filters for particulate removal and charcoal filters for iodine removal with a vent stream to the environment.

23.5.2.5 Power Conversion System. The power conversion system (PCS) for a BWR is similar to that for a PWR; both utilize main steam at saturated conditions of 1,000 psia (6.9 MPa). The main steam is radioactive, containing small amounts of fission products from fuel rod leakage (such as I-131, Xe-133), corrosion products activated in the core (such as Co-60), and water-activated gases (such as N-16). Since radioactivity is present in the PCS, special consideration as to equipment arrangement for maintenance access, shielding

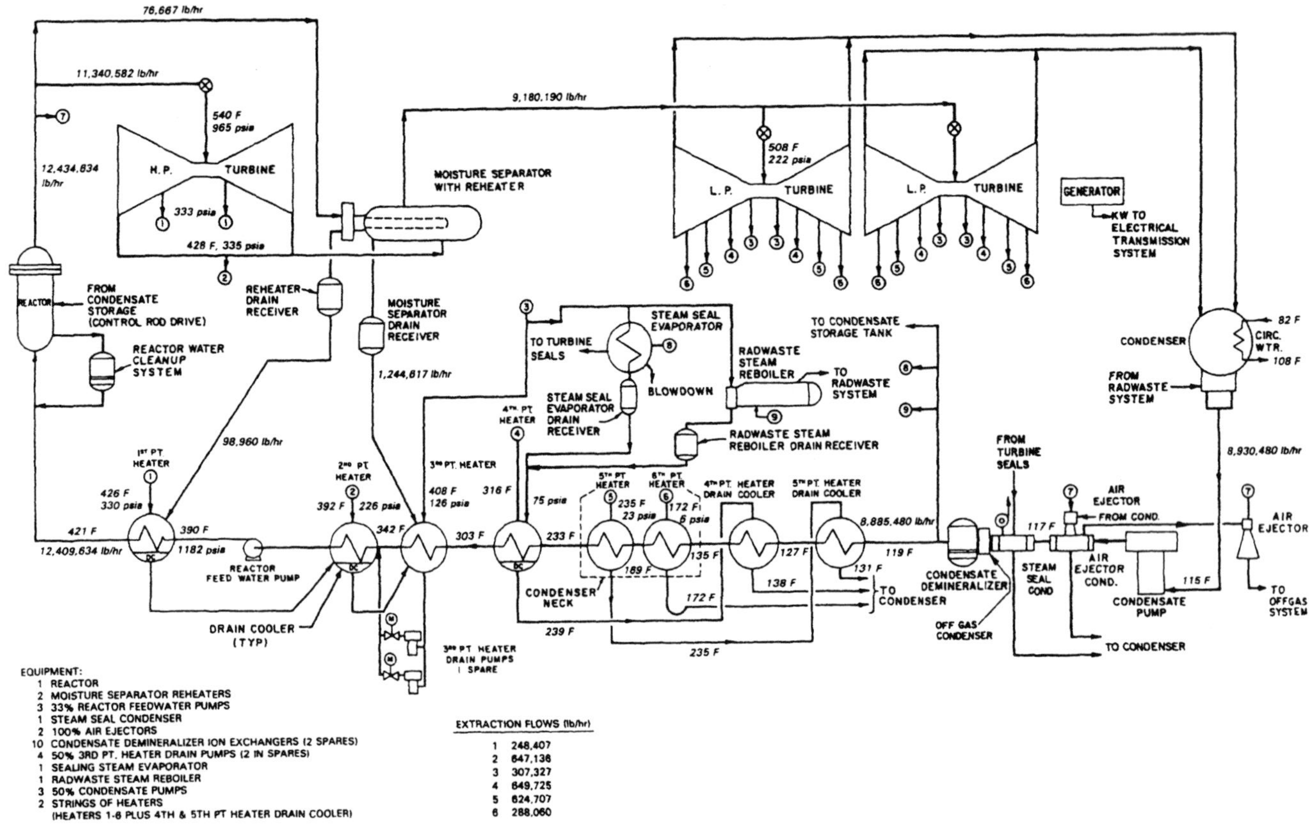

Fig. 23-20. Plant power cycle for a large BWR 6. (From Entergy Operations, Inc., USAR. Used with permission.)

material additions, and limited personnel access are required which are not needed for a PWR.

Figure 23-20 shows the plant power cycle for the River Bend PCS and is typical for a BWR. River Bend, located near Baton Rouge, Louisiana, is a BWR 6 producing main steam at the turbine throttle of 11.34×10^6 lb/h (1,432 kg/s), 965 psia (6.7 MPa), 540° F (287° C). The turbine–generator is an 1,800 rpm, tandem-compound, four-flow machine with 43-in. last-stage blades. The generator is a three-phase, 60 H, 22,000 V, 1,800 rpm hydrogen-cooled, synchronous generator rated at 1,151,100 kVA at 0.9 power factor. The output of the turbine–generator is 990 MWe at guarantee conditions with 3.0 in. Hg (abs) back pressure and 0% makeup. Steam leaves the reactor pressure vessel through main steam lines to the high-pressure section of the turbine–generator, and then passes through moisture separator/reheaters with one stage of reheating (using main steam) to the two two-flow low-pressure sections. Condensate from the condenser hot well is pumped by the condensate pumps through a full flow, deep bed condensate demineralizer system and then through five stages of low-pressure feedwater heating. Motor-driven feedwater pumps take suction from the last stage low-pressure heaters and move the feedwater through one high-pressure stage prior to entering the pressure vessel. Feedwater flow is controlled by a flow control valve.

Condenser cooling at River Bend is provided by a closed loop circulating water system utilizing mechanical draft cooling towers with makeup from the Mississippi River.

23.5.3 Gas-Cooled Reactors

The gas-cooled reactor (GCR) and its hybrid, the high-temperature gas-cooled reactor (HTGR), represent 9% of the world's total nuclear capacity. The largest gas-cooled reactor program is in the United Kingdom, where nearly all of its nuclear capacity are GCRs. The United Kingdom is also building its first PWR at Sizewell, and the PWR may well represent the United Kingdom's next generation of nuclear capacity. Gas-cooled reactors use graphite as a moderator and natural uranium or slightly enriched UO_2 as the fuel contained within metal-clad fuel elements. Carbon dioxide generally is used for the coolant gas. Reactor gas outlet temperatures ranged from 780° F (420° C) with metallic fuel to 1,210° F (654° C) with UO_2 fuel. The lower gas outlet temperatures with metallic fuel are due to the temperature limitations on the fuel and result in lower plant efficiencies.

23.5.4 High-Temperature Gas-Cooled Reactors

The HTGR operates at a much higher gas outlet temperature of 1,450° F (788° C) over the GCRs and provides sufficient energy for superheat and reheat steam to match conventional modern fossil unit conditions of 2,400 psig, 1,000° F, 1,000° F

(16.5 MPag, 538° C, 538° C), respectively. In the United States, the sole manufacturer of the HTGR is General Atomics. The first commercial HTGR was a 40-MWe unit at Peach Bottom, Pennsylvania, for Philadelphia Electric Company.

Peach Bottom 1, which was built under an AEC Power Reactor Demonstration Program, entered commercial operation in 1967 and ran successfully until 1974. Peach Bottom produced steam conditions of 1,450 psia (10 MPa) and 1,000° F (538° C).

General Atomics provided a 330-MWe HTGR to Public Service Company of Colorado for their Fort St. Vrain plant near Platteville, Colorado. Fort St. Vrain began operation in 1976, but its operating history was characterized by low availability and inconsistent energy production. The utility eventually took the unit out of service in 1989. Fuel cycle costs also became prohibitive since the Fort St. Vrain unit was the only HTGR supporting this type of fuel cycle technology. The plant experienced engineering problems and regulatory delays and generally suffered from the lack of an industry-wide experience base for resolution of specific plant problems (Fuller 1990).

One of the more notable problems involved water leakage into the coolant from the helium circulator water bearings during plant transients, which resulted in extended reactor shutdown for moisture removal. Under the present climate, the future of the HTGR design is uncertain. Perhaps the best hope for a commercial HTGR in the United States (and the world) lies with the advanced gas turbine modular HTGR design. Further discussion of the Fort St. Vrain design is warranted since the unit is representative of HTGR technology and lays the foundation for the advanced modular HTGR design discussed later.

Figure 23-21 shows the simplified flow diagram of Fort St. Vrain. As shown in the figure, helium coolant enters the reactor core from the top and passes through the core, entering the steam generators (one of 12) beneath the reactor core. Helium circulators (one of four) remove the helium from the steam generators and recirculate it through an annulus region between the reactor core and the pressure vessel back into the top of the reactor core.

The power conversion system is similar to that of a modern fossil unit, producing both superheat steam at 2,400 psig/1,000° F (16.5 MPag/538° C) and reheat steam at 1,000° F (538° C). Cold reheat steam drives the helium circulators which require a cold reheat turbine discharge pressure (960 psia, 6.6 MPa) significantly higher than for a conventional fossil cold reheat (540 psia, 3.7 MPa) system. Table 23-10 shows key design parameters for Fort St. Vrain.

Figure 23-22 shows the pressure vessel, which is a prestressed concrete reactor vessel (PCRV) housing the reactor core and core support structure, helium circulators, and the steam generators. Refueling is performed through penetrations in the PCRV top head. Control rod drives are also located in these PCRV penetrations. The PCRV bottom head has penetrations to accommodate service piping for the steam generators and the helium circulators as well as accommodating removal. The PCRV is constructed of concrete reinforced by bonded reinforcement steel and is prestressed by steel tendons. The PCRV primary functions are to contain the helium coolant operating pressure and to provide radia-

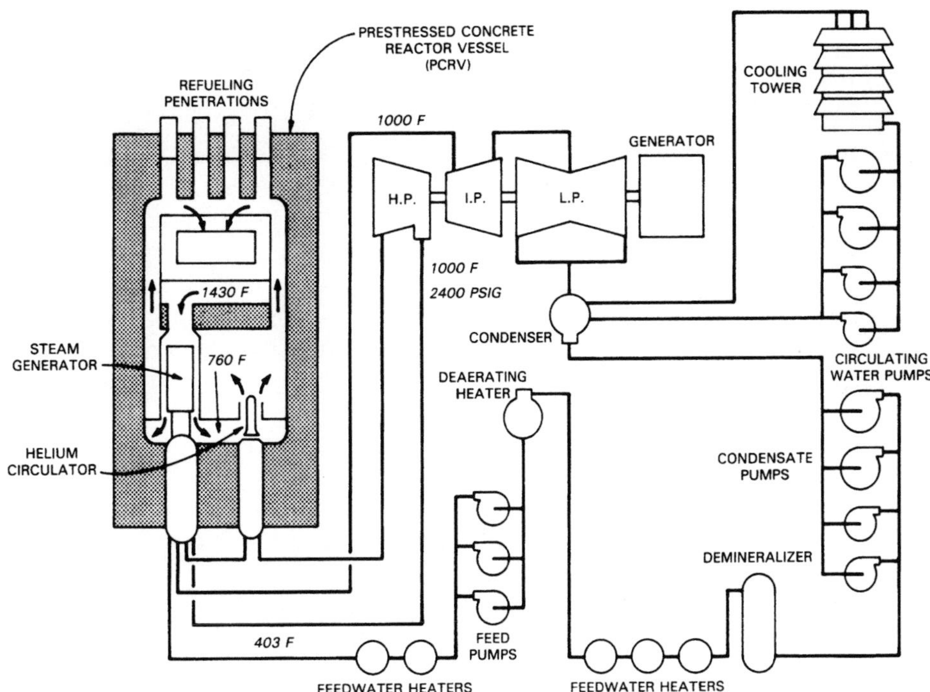

Fig. 23-21. Simplified flow diagram of Ft. St. Vrain HTGR. (From Public Service Company of Colorado. Used with permission.)

Table 23-10. Design and Performance Parameters for Fort St. Vrain

Output	
Thermal	842 MW
Gross electrical	342 MWe
Net electrical	330 MWe
Net plant efficiency	39%
Helium coolant system	
Core inlet	760° F (404° C)
Core outlet	1,430° F (777° C)
Circulator inlet pressure	686 psia (4.7 MPa)
Circulator outlet pressure	700 psia (4.8 MPa)
Helium flow rate	3.4 × 10⁶ lb/h (429 kg/s)
Helium circulators	
Horsepower (each)	5,500 hp (4,101 kW)
Speed (full power)	9,550 rpm
Main turbine inlet pressure	843 psia (5.8 MPa)
Main turbine outlet pressure	648 psia (4.5 MPa)
Steam generators	
Feedwater temperature	403° F (206° C)
Feedwater pressure	3,100 psia (21.4 MPa)
Feedwater flow rate (total)	2.31 × 10⁶ lb/h (292 kg/s)
Outlet steam pressure	2,512 psia (17.3 MPa)
Outlet steam temperature	1,005° F (541° C)
Reheat inlet pressure	649 psia (4.5 MPa)
Reheat outlet pressure	600 psia (4.1 MPa)
Reheat inlet temperature	670° F (354° C)
Output	
Reheat outlet temperature	1,002° F (984° C)
Main turbine–generator	
Type	Tandem compound, two-flow, 33.5 in last stage blade, 3,600 rpm
Rating	403 MVA at 0.85 PF
Main condenser	
Type	Two-pass, divided water box
Pressure	2.5 in. Hga
Feedwater heaters	
Type	Closed and deaerating
Stages	3—Low pressure 1—Deaerating 2—High pressure

Source: General Atomics (undated), Public Service Company's Nuclear Background.

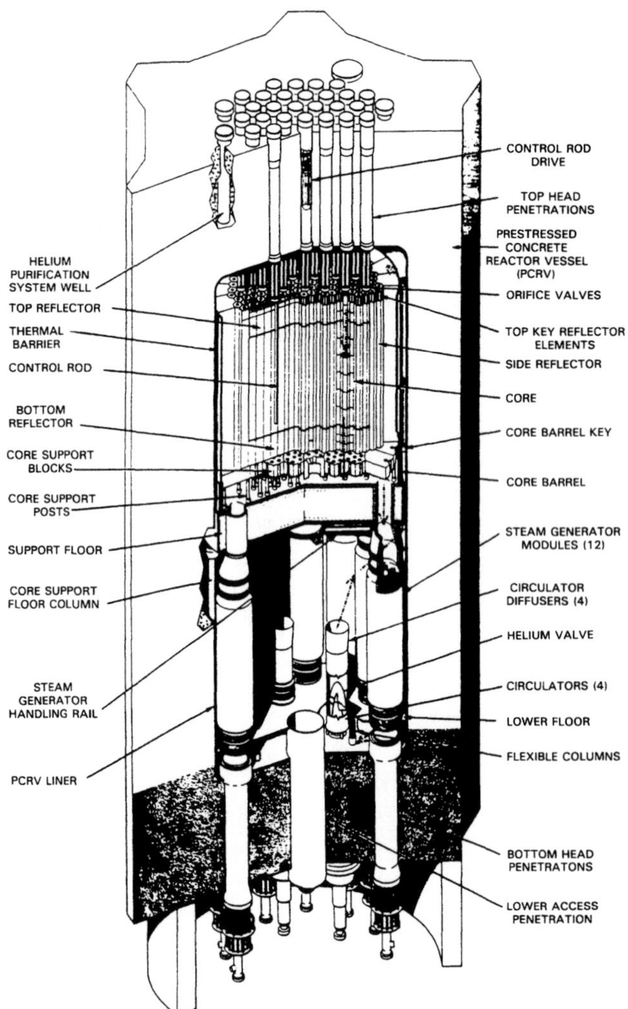

Fig. 23-22. Ft. St. Vrain HTGR pressure vessel. (From Public Service Company of Colorado. Used with permission.)

tion shielding. The main cavity that houses the reactor, steam generators, and helium circulators is in the form of a circular cylinder 31 ft (9.4 m) in diameter and 75 ft (22.9 m) high. The cavity has an insulated steel liner that provides a helium-tight membrane seal. The thermal barrier, in conjunction with water coolant tubes on the concrete side, provides for temperature control of the liner and concrete.

The reactor core is composed of a large number of hexagonal-shaped graphite fuel blocks stacked in vertical columns to form the active core. A series of vertically stacked graphite blocks surround the active core and act as a neutron reflector. The active core is 19.6 ft (6.0 m) in diameter by 15.6 ft (4.8 m) high. There are 74 control rods which enter from the top of the reactor and are controlled in pairs by 37 electric drives and cable drums in the PCRV top head penetrations.

The active core consists of 1,482 fuel blocks. Figure 23-23 shows a typical fuel block (also called a fuel element). Each fuel block is a hexagonal structure, 14 in. (35.6 cm) across the flats and 31 in. (78.7 cm) high, containing approximately 100 coolant passages and 210 cavities to hold the fuel rods. Each block has a central hole for handling and three graphite dowels on its top face to align the individual blocks within a column by matching three bottom recesses on the adjacent upper graphite block. Special fuel blocks also contain two additional holes or channels for control rod insertion, and a third channel for emergency insertion of boron carbide balls.

The fuel rods consist of coated fuel particles of uranium and thorium dicarbide, both encased with four separate coatings. The first coating is a porous pyrocarbon that permits fuel expansion and fission product gas accumulations. The second coating is a dense, isotropic pyrocarbon that provides a barrier for release of fission products. The third coating is silicon carbide, which retains metallic fission products. The fourth coating is a dense isotropic pyrocarbon, added in such a manner as to place the silicon carbide coating in a compression state. The $^{235}_{92}U$ carbide (UC) fuel particles are about

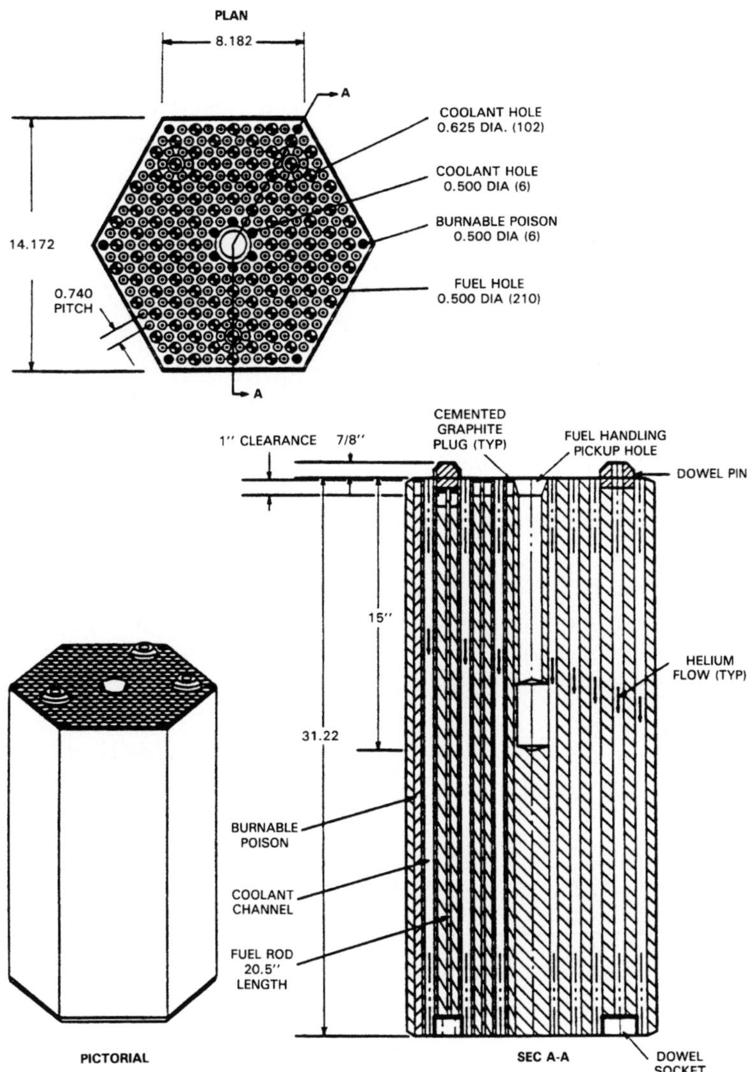

Fig. 23-23. HTGR fuel element. (From Public Service Company of Colorado. Used with permission.)

230 to 430 microns in size, whereas the $^{232}_{90}$U Th carbide (ThC) particles are around 430 to 730 microns in size (Walker and Johnston 1969). The fuel particles are cast with a carbon binder and formed into a rod approximately 1.5 in. (3.8 cm) long and 0.4 in. (1 cm) in diameter. The rods are placed in the graphite fuel block holes. The fuel blocks are grouped within the active core into 37 fuel regions for refueling and control and are also divided into several radial fuel zones. Each fuel zone contains a specific loading of uranium and thorium for uniform core temperature distribution.

The design average fuel burnup is 100,000 MWd/tonnes of U and Th with a 6-year fuel lifetime at equilibrium. The HTGR fuel consists of a mixture of highly enriched UC (93.5%) and ThC. Two different sized fuel particles for the UC and ThC permit easier separation during fuel reprocessing. The fissile particle contains 1 part $^{235}_{92}$U to 4.3 parts of $^{232}_{90}$Th. The fertile particle contains only $^{232}_{90}$Th. The $^{232}_{90}$Th is used to breed $^{233}_{92}$U, which is a source of fissile fuel during reactor operation and also can be recycled back to the core following reactor refueling, chemical separation, and fuel fabrication. The carbide fuel design allows for much higher temperature operation over light water reactors, since graphite is not temperature limited until 4,500° F (2,482° C).

The large amount of graphite (fuel, moderator, and reflector) acts as a heat sink because of the high heat capacity, offering additional operating flexibility during temperature transients. Thus, the need to supply immediate core emergency cooling in the event of a loss of helium cooling is not as great as for light water reactors. This HTGR feature represents an inherent safety advantage over light water reactors.

Each of the 37 fuel regions is located beneath a PCRV refueling penetration. The refueling penetrations also contain a control rod drive and controllable orificing mechanism to control helium flow into the reactor core. The central column of each fuel region contains two additional channels for control rod pair insertion and a third channel for emergency scram insertion of boron carbide spheres. Each control rod drive mechanism controls a pair of control rods. A control rod pair is supported from cables that are moved by control drives consisting of electric drives and cable drums. Control rods consist of boron carbide and graphite in metallic containers in an annulus-type arrangement for the cooling by the helium gas.

The helium coolant system actually comprises two loops. Each loop circulates once through a steam generator and two circulators. The steam generator consists of six identical modules, each module containing an economizer, evaporator, superheater, and reheat section. The tube bundles are helically coiled tubes. The feedwater, superheat, and cold and hot reheat lines enter the steam generator module from the bottom penetration of the PCRV. The helium circulators are single-stage axial flow compressors, with a main single-stage steam-driven turbine drive and an auxiliary single-stage water-turbine drive. Both the main and auxiliary drives are on the same shaft as the compressor. Under emergency conditions, the auxiliary turbine maintains helium flow in the core or auxiliary steam is used with the steam-driven turbine to maintain helium flow. Two of the four circulators are used for the shutdown cooling. Table 23-11 shows reactor design parameters for Fort St. Vrain.

23.6 ADVANCED DESIGNS

Construction of nuclear power plants in the United States is essentially complete for current LWR designs. No reactor orders have been placed from United States utilities for the past several years. Orders for new units are expected within this decade, particularly if licensing and construction times are reduced, progress toward a permanent high waste repository is made, and a more favorable economic picture returns. The public's perception of nuclear power as unsafe must also be changed.

Table 23-11. Reactor Parameters for Fort St. Vrain

Active core	
Thermal power	842 MWt
Effective core diameter	19.5 ft (5.9 m)
Active core height	15.6 ft (4.8 m)
Number of graphite fuel blocks	1,482
Number of fuel columns	247
Graphite reflector thickness	3.9 ft (1.2 m)
Number of fuel zones (refueling)	37
Number of control Rod Pans	37
Initial core loading	
Th-232	42,900 lb (19,500 kg)
U-235	1,194 lb (870 kg)
Average fuel burnup	100,000 MWd/tonne
Conversion ratio (equilibrium)	0.62
Prestressed concrete reactor vessel (PCRV)	
Overall height	106 ft (32.3 m)
Overall diameter	61 ft (18.6 m)
Internal height	75 ft (22.9 m)
Internal diameter	31 ft (9.4 m)

Source: General Atomics (undated), Public Service Company's Nuclear Background, and Walker and Johnson 1969.

Construction times increased dramatically for nuclear units coming on line in the 1980s with many units experiencing construction times of over 12 years from start of construction to commercial operation. The extended construction time also contributed to a rapid rise in capital costs. Units coming on line in early 1980 had average costs of $1,000 per kilowatt installed. By 1987, the average costs of units starting commercial operation had risen to over $4,000 kilowatt installed (Utility Data Institute 1988). Contributing factors to the longer construction times included additional license hearings, additional system and equipment modifications resulting from the lessons learned from the Three Mile Island accident, declining utility load growth, and a lessening of public acceptance of nuclear power safety. This last factor was evidenced in many ways, including the increased use of the judicial system by opponents to further delay plant startups (and in the case of the Shoreham unit, the cancellation and writeoff of a fully constructed 819-MWe BWR). Other nations such as Japan and France have been successful in keeping nuclear power economics in a favorable position over its primary rival, coal. The major reasons for their success are attributed to a stable licensing process and plant standardization.

The emphasis for new nuclear plants in the United States is directed toward advanced standardized designs. Advanced standardized designs are being pursued in two categories. The first category involves additional improvements to current designs based on lessons learned from experience and on utilities' future requirements. Utility future requirements are contained in an Advanced Light Water Reactor (ALWR) Utility Requirements Document (EPRI 1992) prepared by the Electric Power Research Institute (EPRI) which provides guidelines, cost and performance targets, and specific design concepts of what utilities want in advanced designs. Utilities, reactor manufacturers, and architect–engineers from the United States and other nations participated in preparation of the document. Utility expressed goals (Ryan 1989) include improvements over existing designs in areas such as the following:

- Decreasing reactor accident probabilities by a factor of 10;
- Increasing plant availability to 87% per year as compared to the current 70% per year;
- Reducing planned outages to 30 days/year. Increasing the refueling cycle to 24 months;
- Decreasing construction times (first concrete pour to fuel loading) to 4 years;
- Designing for a 60-year plant life;
- Reducing the personnel exposure rate to 100 person-rem per year or less;
- Reducing nuclear costs to a 30-year levelized life cycle cost of 6.5 cents per kilowatt-hour;
- Increasing thermal design margins;
- Designing for ease of operation with emphasis on human factors;
- Designing for high-quality level standardized design; and
- Establishing predictable licensable designs.

Advanced designs in this category include General Electric's advanced BWR (ABWR), Combustion Engineering's System 80+™,[4] and AECL's CANDU 3.

The second category involves designs using inherent/passive safety features to further reduce the risk of reactor accidents leading to core damage and fission product releases. The passive features replace existing active systems such as high-pressure and low-pressure injection pumps and emergency diesel generators in favor of inherent safety features such as natural circulation, gravity, and built-in heat sinks. Passive features also help to further demonstrate to the public that reactors can be made sufficiently safe. Other utility goals outlined in the ALWR requirements document for passive reactor designs include simplifying the designs, ease of operations utilizing state-of-the-art distributed digital control systems with emphasis on human factors, and modularity in construction methods for cost reduction. To take competitive advantage of passive design features, the reactor size must be reduced for passive and inherent systems to economically remove the reactor and coolant heat. Examples of passive designs include Westinghouse's AP 600, GE's simplified BWR (SBWR), and GA's gas turbine modular HTGR. The advanced liquid metal-cooled reactor, called PRISM (Power Reactor Innovative Small Module), is included in the passive category but is described in Section 23.7 on breeder reactors.

23.6.1 Advanced Reactors

23.6.1.1 Advanced Boiling Water Reactor. The advanced boiling water reactor (ABWR) is a General Electric design that was the result of a review effort by a multinational team of BWR vendors and architect–engineers under General Electric's direction with the goal of adding the best features and technology from all BWR designs. Objectives included simplified design, improved safety and reliability, reduced construction costs, better maintainability, reduced personnel radiation exposure, and reduced solid radwaste quantities. The first two ABWR units are scheduled for commercial operation in Japan in 1996 and 1997 (Wilkins 1990). Japan has selected the ABWR as its next generation standard BWR. In the United States, the ABWR is undergoing NRC review for certification as a standard plant. Once certified, the ABWR would be referenced by utilities for use on their sites without additional license hearings on design. The expected effect of standard certified designs is to reduce time and cost for licensing and construction. Certification is expected in the 1994/1995 time frame.

The ABWR is also undergoing First-of-a-Kind Engineering (FOAKE) as part of the FOAKE program. The object of the project is to complete engineering in the ABWR beyond design certification in sufficient detail to define a firm cost estimate and provide a detailed construction schedule. The project is sponsored by the Department of Energy, the Advanced Reactor Corporation, and a GE lead design team. Project completion is scheduled for September 1996.

The ABWR is a large BWR with a 3,926-megawatt thermal (MWt), 1,356-MWe rating. The reactor consists of 872 fuel assemblies of existing GE 8 × 8 design. Its power density is 50 kW/L. The more noticeable changes within the reactor assembly include the recirculation system and control rod drive mechanisms. Figure 23-24 shows the ABWR reactor assembly. The external recirculation loops are eliminated in favor of an internal recirculation system consisting of 10 internal wet-motor glandless pumps for recirculation control. Elimination of external recirculation piping has reduced the number of welds in safety-related pressure boundary piping and eliminated the large nozzle openings below the core, which results in lower pump capacity loads on the ECCS system. The internal recirculation system leads to better drywell space use and allows more access space for inspection and maintenance. The hydraulic control rod drives are replaced with electrical–hydraulic control rod drives for fine motion rod control and hydraulic scram utilization. Fine motion control provides for small power load changes and for fuel burnup reactivity compensation. The vessel is a standard BWR design, 23 ft (7 m) in diameter and 68.9 ft (21 m) in height. The annular space between the RPV shroud

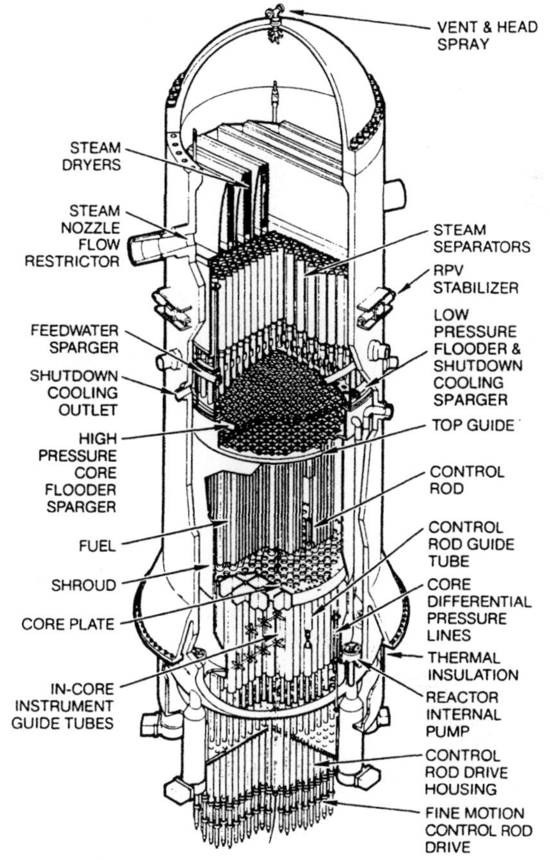

Fig. 23-24. Advanced BWR reactor pressure vessel assembly. (From General Electric Company. Used with permission.)

[4]System 80+™ is a trademark of Combustion Engineering. Used with permission.

and vessel wall was increased to house the internal recirculation pumps.

The ABWR has three complete divisions of emergency core cooling, each with its own dedicated cooling water and emergency diesel generator, resulting in a triple redundant system and a factor of 10 reduction in core damage accident probability over the BWR 6 design (Wilkins et al. 1986). Two ECCS divisions each utilize a traditional RHR pump for low-pressure coolant injection and a motor-driven high-pressure core flooder (HPCF) pump for coolant injection at high pressure. The third independent division combined the upgraded safety RCIC steam-driven turbine pump for high-pressure coolant injection and a third RHR pump for low-pressure coolant injection. The containment structure is a cylindrical shaped design that has integrated the containment walls with the reactor building to form the major structural part of the building. Figure 23-25 shows the ABWR reactor building. The containment structure is a cylindrical reinforced concrete containment vessel with the passive retaining walls lined with steel plates for leak tightness. The drywell structure that houses and supports the RPV utilizes a system of multiple horizontal vents in the lower drywell. These vents reduce pressure by transferring blowdown steam and water mixtures into the suppression pool. The ABWR reactor building volume is approximately 5,897,000 ft^3 (167,000 m^3). A 48-month construction schedule is believed attainable for the ABWR design (Wilkins et al. 1986).

The plant control system consists of a distributed digital control system (DCIS) using standardized microprocessor-based control and instrumentation modules (multiplexing units) and fiberoptic signal transmission. The ABWR has four separate divisions of safety system logic and control, including four separate redundant multiplexing networks to ensure plant safety. Displays of plant and control parameters are on touch-screen CRTs with full-color, hierarchical graphic displays.

23.6.1.2 Advanced Pressurized Water Reactors. Westinghouse and Combustion Engineering are working on advanced designs for a 1,300-MWe reactor based on the design

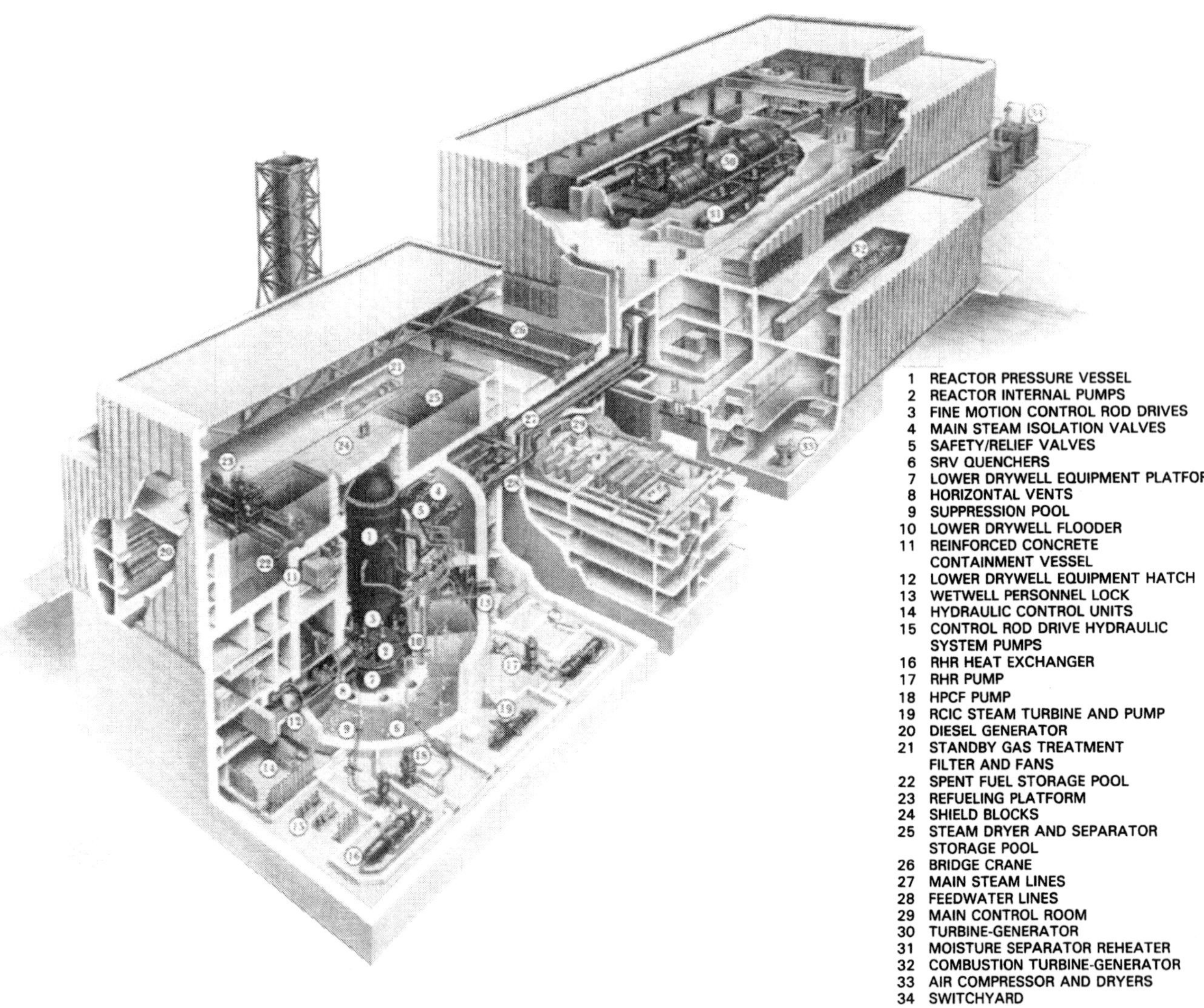

Fig. 23-25. View of the advanced BWR building. (From General Electric Company. Used with permission.)

of their present successful reactors. However, Westinghouse is placing more emphasis on marketing its passive PWR, discussed later. Combustion Engineering's advanced pressurized water reactor (APWR) is highlighted here as representative of APWR. Combustion Engineering's APWR, called the System 80+™, is based on the System 80®[5] design in operation at the Palo Verde Nuclear Generating Station. Combustion Engineering Inc., a subsidiary of Asea Brown Boveri, submitted its APWR design to the NRC for certification as a standard plant. Certification is expected in the 1994 to 1995 time frame. A 48-month construction schedule is envisioned (Combustion Engineering 1989). The APWR has incorporated many of the requirements of the EPRI ALWR requirements document.

Figure 23-26 shows the APWR reactor building, which consists of a spherical steel containment vessel (SCV) and an outer cylindrical concrete shield building (CSB). The steel containment vessel is a freestanding, low-leakage spherical welded steel pressure vessel with a 200 ft (61 m) diameter. It houses the nuclear steam supply system and also contains a large in-containment refueling water storage tank (IRWST) at the bottom of the containment in a toruslike configuration around the reactor vessel cavity. The containment sump also empties into the in-containment refueling water storage tank, which results in an additional safety feature. No switching from the out-of-containment refueling water storage tank (RWST) to the containment sump is required for safety injection pump suction once the refueling water storage tank is exhausted. The shield building is a cylindrical concrete building providing missile and other environmental protection (e.g., tornado, winds) for the steel containment vessel. The spherical containment shape provides a large floor area at the operating deck and adds more space for maintenance and refueling activities.

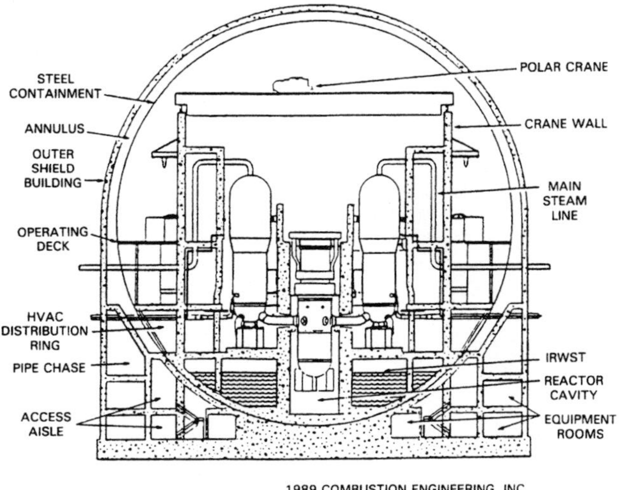

Fig. 23-26. Reactor building for an advanced PWR. (From ABB/CE Nuclear Power. Used with permission.)

The APWR is a 3,817-MWt (1,300-MWe) reactor with 241 standard CE fuel assemblies, 236 fuel rods per assembly. The reactor power density is 95.5 kW/L. There are 48 full-strength control rod assemblies, either four- or twelve-fingered. The 12 control rod assemblies provide for reactor scram protection while the four element control rod assemblies provide for power regulation during normal power operation. The reactor pressure vessel is essentially unchanged from the System 80+ design but is a ring-forged design, resulting in fewer welds and less complexity as compared to the older rolled and welded plate design.

The nuclear steam supply system consists of two parallel loops connected to the reactor pressure vessel; each loop contains a steam generator and two reactor coolant pumps. A pressurizer is connected by piping to one of the loops. The steam generators are of the vertical U-tube recirculating type. The steam generator secondary side water inventory is increased 25% over previous designs to provide additional margin for boilout time. The steam generator heat transfer surface area is 10% larger than previous designs and includes a 10% tube plugging margin. The steam generator also contains an integral axial flow economizer region in the steam generator lower shell. The pressurizer which maintains reactor coolant pressure is 33% larger over previous designs to improve transmit responses to events such as a turbine trip and loss of condenser vacuum.

The APWR has four trains of safety injection over the previous design of two trains. Each train feeds directly into the reactor vessel. The core damage probability is reduced by a factor of 10 (ABB/CE Nuclear Power 1991). Each train consists of a safety injection pump; safety injection tank; and associated piping, valves, and instrumentation. Each safety injection pump takes suction from the IRWST. Electrical power is provided by two independent electrical buses. The safety injection tanks supply treated water to the PRV following rapid reactor coolant system depressurization.

A safety depressurization system (SDS) was also added to vent reactor coolant through valves in the pressurizer to spargers in the in-containment refueling water storage tank. The safety depressurization system provides a means for controlled depressurization of the reactor coolant system following loss of feedwater (normal and emergency) which is then followed by initiation of the safety injection pumps to supply in-containment refueling water storage tank water to the reactor pressure vessel. The safety depressurization system/safety injection pump combination also provides a backup method for a decay heat removal using a feed and bleed approach.

Two separate emergency feedwater subsystems were added; each subsystem is dedicated but intertied to a steam generator. Each subsystem includes an emergency feedwater storage tank (EFST), 100% capacity motor-driven pump, and a 100% capacity steam-driven pump. Each pump (motor- and steam-driven) takes suction from an emergency feedwater

[5]System 80+™ is a registered trademark of Combustion Engineering Inc. Used with permission.

storage tank through separate lines that join inside the steel containment vessel.

The APWR has an advanced distributed digital control system (DCIS) with fiberoptic communication and video displays. The DCIS process information is directly usable by the operator for monitoring and control. The DCIS system greatly improves the man–machine interface, and operators have immediate access to reactor and plant data through CRTs. Alarms are computer validated and prioritized to aid the operator in determining control actions, and information overload is also greatly reduced. Essentially all plant status information is available for graphic display on CRT screens. Graphic presentations are hierarchical, allowing the operator menu access to go from system, to subgroup, to component level displays. Normal and accident displays are integrated.

23.6.1.3 Modular Pressurized Heavy Water Reactor The modular pressurized heavy water reactor (MPHWR), commonly called the CANDU 3, is a smaller version of AECL's CANDU design. The MPHWR technology and design are similar to that of larger CANDU reactors of the 500- to 900-MWe range currently in operation. The units are downsized, however, to take advantage of modular construction and to place the MPHWR in the commercial market for smaller sized reactors. The MPHWR has a 450-MWe rating. The MPHWR uses heavy water systems for moderating and removing reactor fission heat.

The discussion of the PHWR (Section 23.2.3) is also applicable to the MPHWR. The MPHWR reactor assembly is shown in Fig. 23-27. The reactor assembly includes the calandria shell and end shields, 232 fuel channel assemblies, control rod units, a shield tank assembly, and a reactivity mechanisms deck. The calandria shown in Fig. 23-27 is a horizontal cylindrical vessel with end shields that support the 232 horizonal tubes and tube extensions. A shield tank encloses the calandria vessel and contains light water for radioactive shielding. Attached to the shield tank is a shield tank extension that houses the vertical control rod components. The reactivity mechanisms deck is located on top of the shield tank extension. The calandria vessel, shield tank ves-

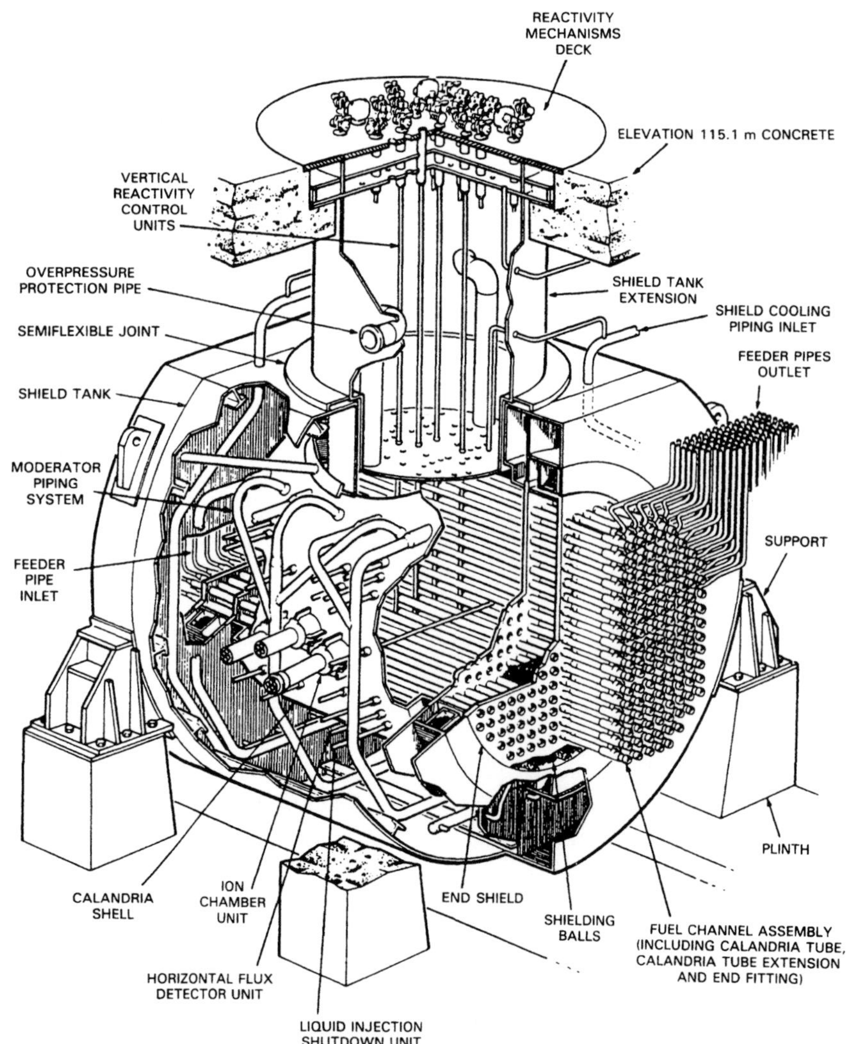

Fig. 23-27. MHPWR reactor assembly. (From AECL. Used with permission.)

sel, and other components are supported by the end walls of the shield tank which transmit the loads to the foundation mat.

Each horizontal calandria tube contains a pressure tube, insulated from the calandria tube by a CO_2-filled gas annular region. The pressure tube, calandria tube, and tube extension form the fuel channel assembly. The 232 fuel channel assemblies are factory assembled. Twelve fuel bundles are housed in each pressure tube. A fuel bundle consists of 37 Zircaloy-clad fuel elements held together by end plates. The fuel elements contain natural UO_2 fuel pellets. Each bundle as shown in Fig. 23-28 is 19.5 in. (495.3 mm) in length and about 4 in. (102 mm) in diameter. The reactor is refueled during power operation.

Reactor control consists of both horizontal and vertical control rods that move between the rows of fuel channels and are used for power adjustments, reactor startup, and reactor shutdown. Adjustor rods are used for short-term power flattening. There also is an independent liquid poison injection system for emergency shutdown. Long-term control of power distribution is achieved by online refueling (AECL 1989).

As shown in Fig. 23-27, the heavy water moderator is contained within the calandria and is separate from the fuel channel assemblies. The heavy water moderator is circulated through the calendria by a moderator system that provides heavy water cooling, purification, and circulation. A liquid poison system supplies a neutron poison to the moderator system for reactivity control during shutdown. Representative reactor data are given in Table 23-12.

The heavy water reactor coolant system or heat transport system is shown in Fig. 23-29 and consists of two U-tube steam generators with a preheater section, four motor-driven heat transport pumps with air/water-cooled squirrel cage induction motors, and four reactor inlet headers arranged such that two are located at each half of the reactor. The two reactor inlet headers on the same half of the reactor supply coolant flow to the fuel channels in the reactor half through individual feeder pipes. Each steam generator has two inlet pipes, both of which are connected to a reactor outlet header. Each reactor outlet header is connected to half of the fuel channels by feeder pipes.

The reactor heat transfer system is designed for unidirectional flow to facilitate refueling from one end of the reactor.

Table 23-12. MPHWR (CANDU 3) Data

Reactor	
Type	Horizontal pressure tube
Coolant	Pressurized heavy water
Moderator	Heavy water
Number of fuel channels	232
Effective core radius	96.8 in. (245.8 cm)
Core length	234 in. (594.4 cm)
Fuel	
Fuel	Compacted and sintered natural-UO_2 pellets
Form	Fuel bundle assembly of 37 elements
Length of bundle	19.5 in. (495.3 mm)
Outside diameter	4.0 in. (102.4 mm)
Bundle weight	51.7 lb (23.5 kg) (includes 42 lb (19.1 kg U)
Bundles per fuel channel	12
Heat transport system	
Number of steam generators	2
Steam generator type	Vertical U-tube with integral steam drum and preheater
Number of heat transport pumps	4
Heat transport pump type	Vertical, centrifugal, single suction, single discharge
Reactor outlet header pressure (gauge)	1,436 psi (9.9 MPa)
Reactor outlet temperature	590° F (310° C)
Reactor coolant flow	42.0×10^6 lb/h (5,300 kg/s)
Steam tempeature (nominal)	500° F (260° C)
Steam quality (minimum)	99.75%
Steam pressure (gauge)	667 psi (4.6 MPa)
Total fission heat	1,440.3 MWth
Net electrical output (nominal)	450 MWe

Source: AECL 1989. Used with permission.

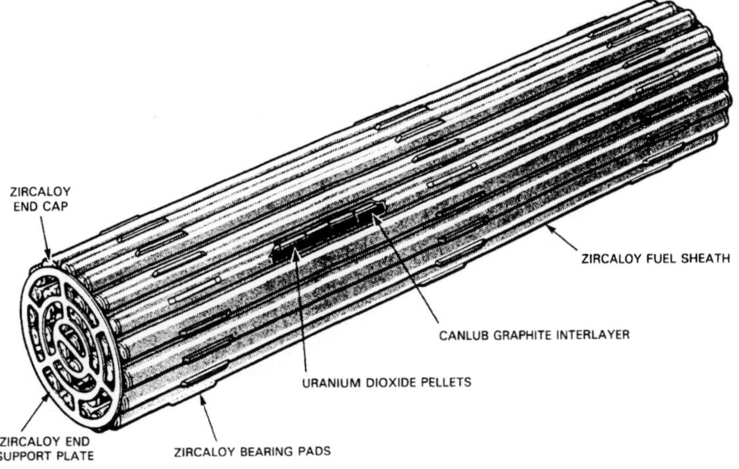

Fig. 23-28. MHPWR fuel bundle. (From AECL. Used with permission.)

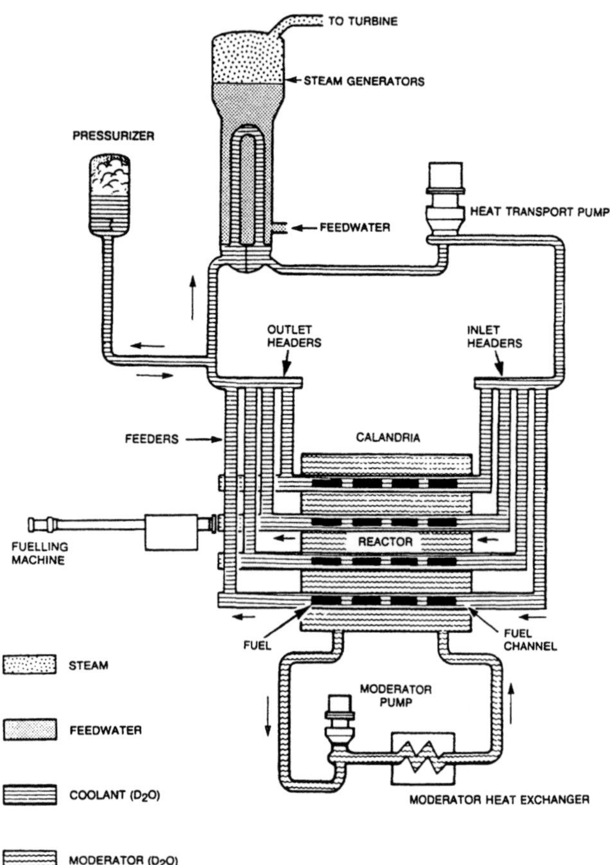

Fig. 23-29. MHPWR heat transport system. (From AECL. Used with permission.)

Both steam generators are located on opposite ends from the heat transport pumps by a single pump suction line to each pump. The suction to a pair of parallel pumps and inlet headers from a steam generator outlet feeds into the outlet header to the other steam generator. The coolant makes two core passes through each set of parallel pumps per cycle. The header arrangement results in reactor fuel channels in each vertical half of the core belonging to the same core pass. A pressurizer on one of the heat transfer system hot legs is used to maintain a constant coolant pressure at 1,436 psig (9.9 MPag). The steam generators supply saturated steam at 500° F (260° C) and 667 psig (4.6 MPag). The emergency core cooling system (ECCS) consists of two gas-pressurized accumulator water tanks, two ECC pumps, two heat exchangers, and one grade-level water tank. If a line break in the heat transport system occurs, ECCS operation involves an initial short-term high-pressure water injection by the gas-pressurized water accumulator tanks via the heat transport system to the reactor inlet and outlet headers. Following reactor depressurization and accumulator tank water depletion, the low-pressure ECC pumps supply water to the heat transport system from the grade-level water tank. Once the water tank is empty, the water supply is switched to the reactor building floor sump for long-term operation. The ECC pumps transport a mixture of light and heavy water from the reactor building floor sump through heat exchangers back into the heat transport system to the reactor inlet/outlet headers.

The ECCS system is classified as a Group 2 system. CANDU systems are grouped into either Group 1 or Group 2. Group 1 systems include those systems primarily involved with normal power production. Group 2 systems involve systems required for safety or safety support to maintain the plant in a safe configuration following loss of Group 1 systems or for accident mitigation. Group 1 and Group 2 systems are located for the most part in separate areas of the plant.

The reactor building as shown in Fig. 23-30 is a cylindrical reinforced concrete structure with an interior steel liner for leak tightness. The reactor building contains the reactor assembly, heat transfer system, moderator system, and certain components of the safety systems.

The power conversion system is similar to other conventional systems using main steam at saturated conditions. The turbine generator consists of one high-pressure double-flow section and two low-pressure double-flow sections. The feedwater system utilizes five stages of feedwater heating consisting of three low-pressure stages, one deaerator stage, and one high-pressure stage. Three 50% capacity motor-driven feedwater pumps supply feedwater from the deaerator to the steam generators. One 4% capacity diesel-driven auxiliary feedwater pump supplies feedwater to the steam generator in the event the feedwater pumps are not available. The plant control system uses a digital distributed control system for control and monitoring functions.

23.6.2 Advanced Passive Reactors

23.6.2.1 Simplified Boiling Water Reactors. The simplified boiling water reactor (SBWR) is General Electric's passive BWR, designed to meet utility requirements for a smaller, simpler, passively designed reactor. Key features of the SBWR include natural recirculation of reactor coolant within the core; passive emergency core cooling systems; simplified design with less piping, valves, and instrumentation and control; and modular construction with factory-made and installed systems/components. A 3-year construction schedule is visualized, and costs can be competitive with similar size coal-fueled units.

The SBWR conceptual design has been completed and involved a series of cooperative international programs under the sponsorship of DOE, EPRI, and Japan Atomic Power Company. The conceptual design effort was performed by General Electric working with United States, Japanese, Italian, and the Netherlands firms. The conceptual design establishes technical, economic, and licensing feasibilities (Wilkins 1990). Testing and development of key SBWR features is continuing, including full vertical scale testing of the gravity driven core cooling system and qualification testing of the high-capacity depressurization valve. Development and testing of the passive containment cooling system is

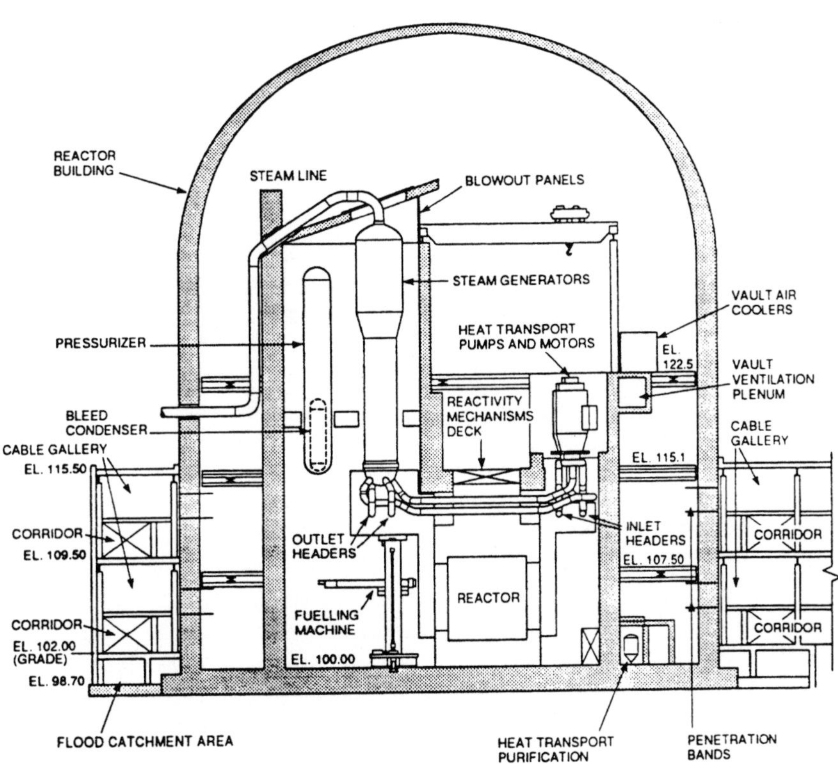

Fig. 23-30. MHPWR reactor building. (From AECL. Used with permission.)

being conducted in Japan by General Electric, Hitachi, and Toshiba. The Department of Energy and EPRI are sponsoring a detail design, development, and certification program for the SBWR using an international team of General Electric; United States, Italian, Japanese, and Netherlands vendors; architect–engineers; universities; utilities; and specialists. Standard plant certification by the NRC is expected in the mid to late 1990s.

The basic concept of the BWR has not changed for the SBWR. Water boils in the nuclear core from controlled nuclear fission heat. The mixture of water and steam rises from the core and then undergoes moisture removal and steam drying within the pressure vessel prior to steam exit. The saturated steam is sent to a conventional power conversion system that produces electrical energy from the turbine–generator and returns the steam as liquid feedwater back to the reactor.

Figure 23-31 shows the SBWR pressure vessel which houses the nuclear core, moisture separators, and steam dryers. Coolant flow through the reactor is provided by natural circulation rather than forced circulation as with current BWR designs. Natural circulation reduces the amount of pumps, piping, valves, instrumentation, and controls and simplifies the design. To accomplish the required coolant flow requirements, additional pressure vessel height was added. The space between the top of the core and steam separator assembly was also increased to provide a chimney effect.

The reactor core is rated at 3,000-MW thermal (650-MWe) with a lower power density (42 kW/L) smaller than with conventional designs (50 kW/L). The fuel assemblies are General Electric's standard 8 × 8 design containing slightly enriched UO_2 fuel; only the length has been reduced to 9 ft (2.7 m) to further enhance natural circulation. Control rods are conventional cruciform shape and enter from the bottom. The design of the control rods is similar to the ABWR design, being electrohydraulic fine motion rods. A 24-month refueling cycle is expected.

Like the ABWR design, the SBWR also uses a containment structure and reactor building integrated as a structural unit as shown in Fig. 23-32. The containment structure is a pressure suppression cylindrical reinforced concrete structure. All systems important to reactor safety area are contained within the reactor/containment structure.

Passive design systems are used for emergency responses including reactor isolation and loss-of-coolant (LOCA) events. As shown in Fig. 23-33, emergency responses utilize the suppression pool, gravity-driven cooling pool, and isolation condenser pool. For reactor isolation events, such as main steam line isolation valve closure, a series of isolation condensers tied to the main steam lines and located in the isolation condenser pool are used. On reactor isolation, the steam–water mixture from nuclear decay heat is collected and condensed in the isolation condensers. The liquid condensate is returned from the isolation condensers to the reactor. Activation of the safety/relief valves is not required. The decay heat is absorbed in the isolation condenser pool and eventually released to the environment as vapor (boil off).

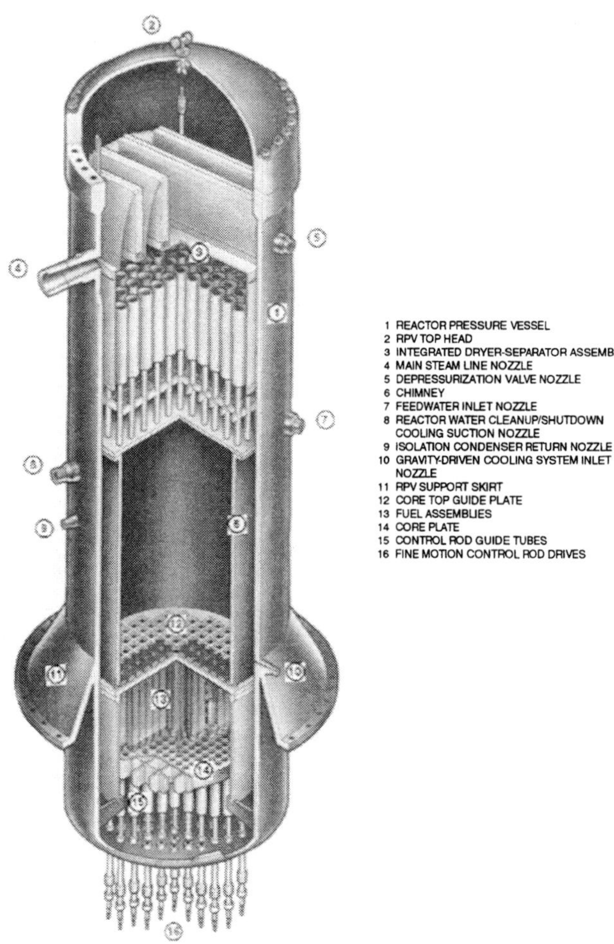

Fig. 23-31. SBWR pressure vessel. (From General Electric Company. Used with permission.)

For an event such as a LOCA, reactor core coverage is provided by a passive gravity-driven core cooling system as shown in Fig. 23-33. Reactor coolant depressurization occurs through the use of six depressurization valves on the main steam lines and discharges the moisture–steam mixture into the suppression pool. Once the pressure has been reduced, gravity flow from the gravity-driven cooling pool (GDCS) through separate lines into the RPV maintains water level above the reactor core. Large RPV openings near or below the core (such as recirculation piping in current designs) were eliminated, ensuring complete core coverage. The passive cooling function requires more water in the RPV above the core to reduce SRV lifting and for additional water capacity during depressurization. The large water supply in the RPV also increases transient time responses. During a LOCA event, passive containment cooling is provided by the passive containment cooling system. Core fission product decay heat is rejected as steam through the isolation condensers to the isolation condenser pool and the condensed water drains back to the GDCS. Vapor from pool heatup is vented to the atmosphere. The pool is sized for approximately 3 days of cooling capacity before operator action is required. Additional water can be added to maintain and continue the cooling needs. Sufficient water storage is available in the suppression pool and isolation condenser pool to completely flood the drywell to a water level above the core if needed.

The control room habitability area heating, ventilating, and air conditioning (HVAC) are also passively designed. A natural air circulation system is used with cooling capacity supplied by a refrigerated ice bed above the control room. A stored air supply is also used for oxygen replacement.

These passive features for safety systems have resulted in simpler design and fewer active components, such as active emergency core cooling system pumps, motors, valves, and emergency diesel generators, requiring less operator attention. Emergency electrical power for lighting and performance monitoring is supplied by plant dc batteries.

The power conversion system is similar to those for other nuclear systems using saturated steam. Refinements to reduce equipment and piping have been made. The General Electric turbine generator is 1,800 rpm tandem compound, two-flow design with 52 in. (1.3 m) last-stage buckets. One stage of reheat/moisture separation removes moisture from the high-pressure section discharge steam. The main condenser is single pressure with two cells. The condenser is located under and to the side of the turbine generator, thus reducing the turbine pedestal height. One single string of feedwater heaters is used, involving five or six stages of feedwater heating. Motor-driven variable- speed feedwater pumps improve system operability. The high-pressure feedwater drains return to the suction side of the feedwater pumps.

The plant control system consists of a distributed digital control system using standardized, microprocessor-based control and instrumentation modules, multiplexing units, and fiberoptic signal transmission. Displays of plant and control parameters are on touch-screen CRTs with full-color, hierarchical graphic displays. Typical SBWR plant data are shown in Table 23-13.

23.6.2.2 Advanced Passive Pressurized Water Reactor. The AP600 is Westinghouse's passive PWR, designed to meet EPRI's utility requirement document for a smaller, simpler, passively designed reactor. Key features of the AP600 include canned motor reactor coolant pumps mounted in the steam generator, a refueling water storage tank located in containment for passive reactor core cooling, passive safety systems for coolant injection, and a passively designed containment structure for long-term heat removal of reactor core decay heat and reactor coolant system sensible heat. The AP600 design also features modular construction with factory-made and installed systems/components. A 5-year lead time from plant order to commercial operation is thought to be realistic with a 36-month construction period (Vijuk and Bruschi 1988). Costs are competitive with similar sized coal units.

The AP600 conceptual design has been completed by Westinghouse and their subcontractors as part of EPRI and

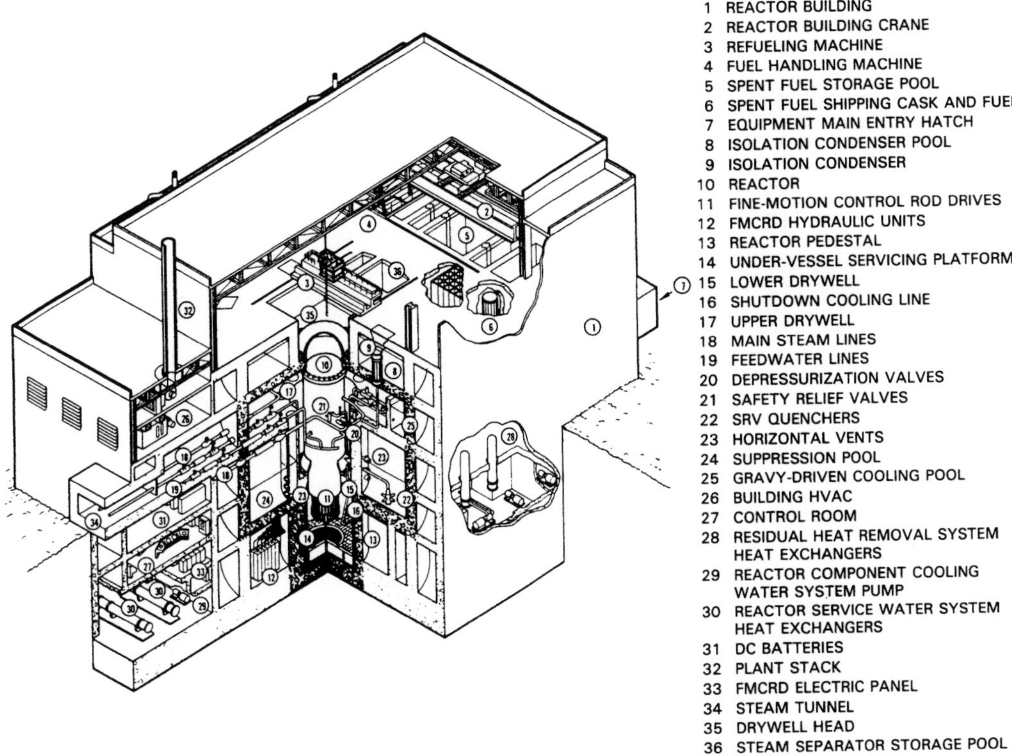

Fig. 23-32. SBWR reactor and auxiliary buildings. (From General Electric Company. Used with permission.)

DOE programs. Work is continuing on detail design and testing. A design certification application to the NRC as a standardized plant is planned for 1992 with NRC approval expected in the mid to late 1990s time period. Testing of the passive safety systems and the high-inertia rotor for the reactor coolant pumps is ongoing. The AP600 is also part of the FOAKE program with the objective of completing engineering beyond design certification in sufficient detail to define a firm cost estimate and provide a detailed construction schedule. The project is sponsored by the Department of Energy, the Advanced Reactor Corporation, and a Westinghouse lead design team.

The basic concept for the AP600 has not changed from the standard PWRs. Water is maintained under sufficient high pressure to prohibit boiling in the nuclear core. The reactor coolant pumps move water through the core, removing the fission heat, and sending the heated coolant to primary side (tube side) of the steam generators. Within the steam generator, heat energy is transferred to the secondary side (shell side), causing water to boil. The mixture of vapor

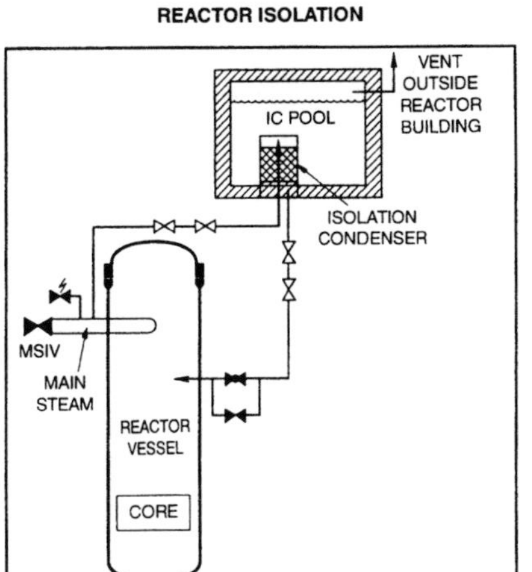

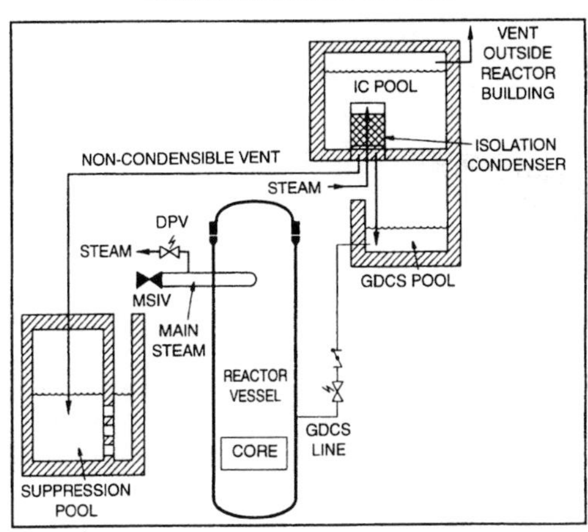

Fig. 23-33. SBWR decay heat removal systems. (From General Electric Company. Used with permission.)

Table 23-13. SBWR Plant Data

Plant	
Net output	~650 MWe
Gross reactor thermal power	2,000 MWt
Main steam flow	8.57×10^6 lb/h ($3{,}890 \times 10^3$ kg/h)
Turbine configuration	TC 2F - 52 in. last stage blade
Reheat stages	0
Design life	60 years
Pressure vessel	
Inner diameter	19.7 ft (6 m)
Height	80.21 ft (24.45 m)
Natural recirculation flow	5.2×10^7 lb/h (2.4×10^6 kg/h)
Reactor core	
Active fuel length	9.0 ft (2.7 m)
Equivalent core diameter	15.5 ft (4.7 m)
Power density	42.0 kW/L
Fuel assemblies	732 (8 × 8 type)
Fuel material	Enriched UO_2
Cladding material	Zircaloy 2
Control	
Control rod/drives	177
Neutron absorber	B_4C
Control rod form	Cruciform
Control rod drive	Electrohydraulic fine-motion
Containment	
Type	Pressure suppression
Configuration	Cylindrical reinforced concrete

Source: General Electric Company.

and water rises to the point at which it comes into contact with moisture separators and then steam dryers. The saturated steam leaving the steam generators is sent to the turbine generator of the power conversion system. The condensed steam is taken from the condenser and reheated using several stages of feedwater heating before returning to the steam generators for new steam production.

Figure 23-34 shows the AP600 reactor coolant system (RCS). The reactor coolant system consists of the pressure vessel housing the nuclear core with control assemblies, two steam generators, a pressurizer on one hot leg loop, and two reactor coolant pumps in the cold leg side of each steam generator. The pressure vessel and reactor internals are essentially unchanged from previous designs. The reactor core consists of 145 fuel assemblies of the 17 × 17 array design. The core design uses three low-enrichment regions and supports an 18- or 24-month refueling cycle. The active fuel length is 12 ft (3.65 m). The core uses a smaller power density than conventional designs. Top entering control rods are used for reactor shutdown and for load following and power regulation. Burnable poisons are used for burnup control of new fuel excess reactivity.

The steam generators are a U-tube/shell design consisting of tubes, shell, tube supports, baffles, separators, dryers, and so forth. Two canned motor reactor coolant pumps are mounted in an inverted direction to the primary side of the cold leg channel head. This arrangement eliminates the need for separate coolant pump foundations and essentially combines the steam generator and reactor coolant pumps as one structure. The combined structure's operating weight is 10% to 12% greater than that of a conventional steam generator. The canned motor pumps are hermetically sealed, requiring no external coolant sealing system. The canned motor pump has been modified to increase its rotating inertia for longer pump coast down. This modification is subject to an ongoing testing program and involves increasing the size and weight of the pump thrust runner with an addition of a heavy uranium metal disk (Vijuk and Bruschi 1988). The AP600 steam generator/canned motor pump arrangement offers a simplified support system and provides better access to the steam generators and coolant pumps for testing, inspection, and maintenance. The reduction in pipe whip restraints resulting from revised NRC criteria for leak-before-break was also used to simplify design and increase access.

As with the SBWR, the most significant design changes involved the addition of passive reactor cooling features for the plant's safety systems over the conventional active systems. The AP600 passive safety systems were designed to function for transient and accidental conditions such as emergency core decay heat removal, reactor coolant inventory control, loss-of-coolant safety injections, loss-of-coolant long-term recirculation, and long-term containment heat removal.

In the event that the normal core heat removal systems, such as steam generator heat removal, are not available, passive residual heat removal is accomplished through the

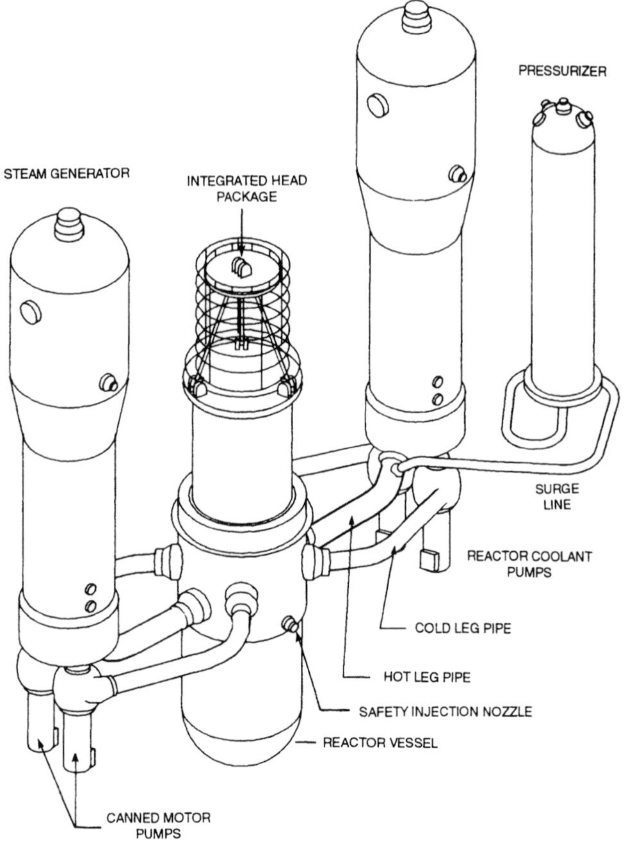

Fig. 23-34. AP600 reactor coolant system. (From Westinghouse. Used with permission.)

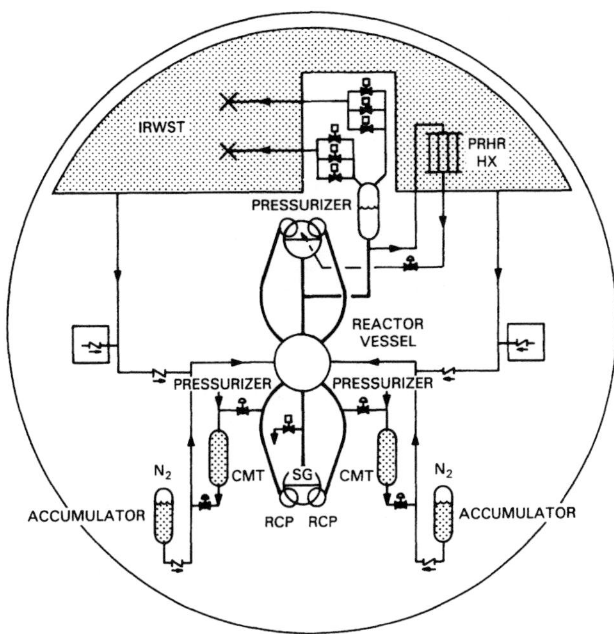

Fig. 23-35. AP600 passive decay heat removal systems. (From Westinghouse. Used with permission.)

passive residual heat removal heat exchanger (PRHR HX). The PRHR HX takes over the active conventional functions of the auxiliary feedwater and the residual heat removal systems. The residual heat removal function is shown in Fig. 23-35. The PRHR HX supply piping is connected to the reactor coolant hot leg loop piping, whereas the return piping is connected to the coolant cold leg loop piping. The PRHR HX is located above the reactor coolant system loop piping inside the in-containment refueling water storage tank (IRWST). A natural circulation flow cooling loop is established with flow coming from the coolant hot leg loop piping, through the PRHR HX, and then returning via the cold leg loop piping. The IRWST provides the heat sink for the PRHR HX and is designed for several hours of core decay heat removal before boiloff occurs at saturation temperature. Air-operated control valves control flow rate for reactor coolant system temperature control.

The reactor coolant makeup and safety injection functions are passively accomplished through two redundant systems, each consisting of a core makeup tank (CMT), an accumulator tank (AT), and a supply line from the IRWST. All are located above the reactor coolant loop piping. An automatic depressurization system (ADS) on the pressurizer with discharge to the IRWST is also included. Each system injects directly into the reactor through a nozzle in the reactor pressure vessel cold leg side shown in Fig. 23-35. The CMTs provide reactor coolant makeup following small leak pipe breaks, during transient events, or when the normal makeup system is not available. Each CMT is filled with 2,000 ft^3 (56.6 m^3) borated water and provides coolant water injection capability at any coolant pressure using gravity draining. A small pressure balance line connects the pressurizer steam space with the top of each CMT. System initiation occurs on low pressurizer water pressure or level and also includes stopping the reactor coolant pumps. The CMT discharge isolation valves open, allowing the CMTs to drain into the reactor vessel.

Safety injection following a loss-of-coolant accident is provided by the CMTs and ATs for the short-term makeup and by the IRWST following reactor coolant depressurization for the long-term makeup. A larger line running from the top of the CMT to the reactor coolant system cold loop leg is used for pressure balancing. The CMTs provide coolant makeup for smaller break area accidents. For the larger break area accidents, the ATs provide the additional higher flows for reactor core makeup. Each accumulator tank contains 1,700 ft^3 (48.1 m^3) of borated water under 300 ft^3 (8.5 m^3) of N_2 at 700 psig (4.8 MPag).

On exhaustion of the CMT and AT makeup water volumes, the IRSWT provides up to 10 hours of coolant makeup. To allow for gravity draining of the IRWST into the reactor pressure vessel, an ADS reduces the coolant pressure to below the static (elevational) head of the IRWST. The ADS consists of two redundant sets of valves on the pressurizer that provide phased discharge into the IRWST. On draining of the IRWST, the containment is flooded to a level above the reactor coolant loops. A long-term stable reactor core cooling and makeup system is established by water in containment draining by gravity through special vents on the reactor vessel injection lines back into the reactor pressure vessel. The passive containment cooling system provides the final heat sink by condensing reactor coolant steam on the containment shell. The condensed water drains back to the containment water pool for reactor coolant recirculation. The CMT/AT/IRWST passive system replaces the active low- and high-pressure safety injection systems and emergency diesel generators.

Figure 23-36 shows the reactor building which consists of a freestanding leak-tight steel cylindrical containment vessel surrounded by a concrete reactor shield building. The reactor shield building contains a passive containment cooling diffuser and a passive containment water cooling tank (CWCT) at the top of the shield building. The containment and shield building provide radioactivity containment, containment pressure suppression, containment structure missile protection, and passive containment cooling (PCC). The PCC system provides the ultimate heat sink for reactor coolant and core decay heat. Natural circulation air between the containment structure and the shield building provide for heat removal from the containment shell. This energy transfer from containment maintains a containment pressure below design pressure values. Water drainage by gravity from the overhead CWCT down and around the containment shell provides additional cooling for up to 3 days when higher heat removal because of a loss-of-coolant accident is required. Natural draft cooling by shield building air is sufficient for long-term containment cooling. This cooling process occurs as cooler air enters through the SB air inlet vents, passes

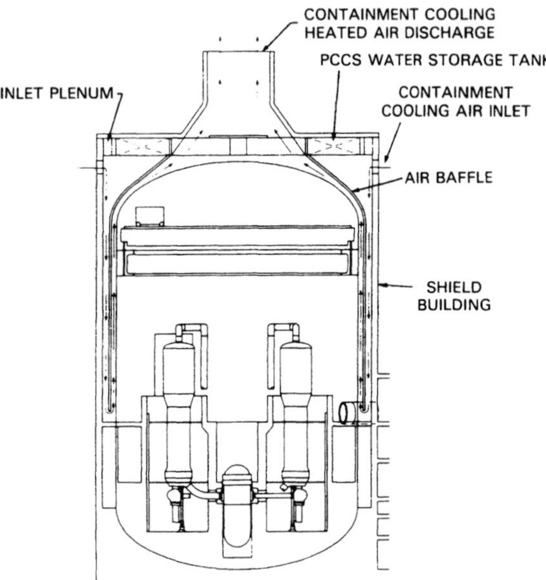

Fig. 23-36. AP600 reactor building. (From Westinghouse. Used with permission.)

around and up the outside containment shell by natural draft circulation, then discharges through the cooling diffuser.

All safety systems are within the containment, auxiliary, and fuel handling buildings. All reactor coolant containing systems are within containment.

The power conversion system is similar to other conventional saturated steam systems. The power conversion system uses a tandem compound, four-flow turbine with 44 in. (1.10 m) last-stage blades. There is one stage of steam reheat prior to entering the 2 two-flow pressure sections. Four stages of feedwater heating are used plus one deaerating stage. Table 23-14 summarizes representative plant data for the AP600.

Table 23-15 compares equipment and building volume reductions for the AP600 with an equivalent size conventional PWR. The major reduction in equipment was brought about through the addition of passive safety systems, which eliminates the high- and low-pressure injection systems, auxiliary feedwater system, safety RHR functions, and emergency service water. Expanded use of control rods for power maneuvers and a simplified chemical and volume control system also resulted in elimination of boron recycle systems. The safety functions of the cooling water systems, HVAC, and AC power systems have also been eliminated. The addition of a distributed digital control system also significantly reduced the amount of cable and cable trays.

Table 23-14. AP600 Plant Data

Net electrical output	600 MWe
Number of steam generators (vertical U-tubed)	2
Steam generator power (total)	1,940 MWt
Primary coolant flow rate	102,000 gpm (6.4 m^3/s)
Steam flow rate (total)	8.4×10^6 lb/h (1,047 kg/s)
Number of fuel assemblies	145 (17 × 17 array)
Fuel assembly height	12 ft (3.66 m)
Turbine–generator	TC 4F 47 in. (1.2 m) last stage blade
Stages of reheat	1
Feedwater heating stages	7 (1 deaerating stage)

Source: Vijuk and Bruschi 1988 and Westinghouse 1984.

23.6.2.3 Gas Turbine-Modular Helium Reactor. General Atomics is the principal designer of the gas turbine-modular helium reactor (GT-MHR) which represents the high-temperature gas-cooled reactor (HTGR) version of a smaller and simpler reactor system with passive safety. The GT-MHR has been developed based on previous HTGR designs, including Peach Bottom 1 and Fort St. Vrain. Development of a modularized HTGR started in 1984 with goals of providing a reactor with a high degree of safety in conformance with safety and regulatory requirements defined by the DOE and the NRC's Advanced Reactor Policy. Utility interests and needs are addressed through Gas-Cooled Reactor Associates (GCRA) and its utility members.

Technical support is provided by Oak Ridge National Laboratory. Reassessment of the nuclear Brayton cycle in 1992 led to the development of the GT-MHR. The GT-MHR

Table 23-15. AP600 Equipment Reduction Comparison

Item	Conventional Two-Loop	AP 600	Percent Reduction
Pumps			
Safety related	25	None	100
Non-safety-related	188	139	26
Valves (>2 in., 5 cm)			
Nuclear steam supply system	512	259	49
Balance of plant	2,041	1,530	25
Pipe (>2 in., 5 cm)			
Nuclear steam supply system >2 in. (5 cm)	44,300 ft (13,503 m)	11,042 ft (3,366 m)	75
Balance of plant > 2 in. (5 cm)	97,000 ft (29,566 m)	67,000 ft (20,421 m)	31
Diesel generators	2 (safety)	2 (non-safety)	100
Building volume			
Containment	2.7×10^6 ft^3 (7.6×10^4 m^3)	3.4×10^6 ft^3 (9.6×10^4 m^3)	46
Seismic	6.7×10^6 ft^3 (1.9×10^5 m^3)	1.7×10^6 ft^3 (4.8×10^4 m^3)	

Source: Vijuk and Bruschi 1988 and Westinghouse 1984.

reference plant consists of four identical reactor modules ranging in thermal power from 550 to 600 MWt. Each module is tied to a gas turbine and generator with electrical capacity up to 286 MWe. Conceptual design of the standardized plant is underway. A thorough technology development program of component testing is planned as the GT-MHR design develops. A non-nuclear integrated test of the first full-scale power conversion module is also planned prior to deployment of the first GT-MHR (Zgliczynski et al. 1994) The GT-MHR is also being evaluated for potential applications as a plutonium burner and multipurpose reactor.

Previous gas-cooled reactor designs transferred heat from the gas coolant to water via a steam generator. Steam was then used to generate electricity through a steam turbine, similar in nature to a PWR or BWR. The GT-MHR differs from the conventional LWR designs by using the heated helium gas coolant to directly turn a turbine, hence a Brayton cycle.

Figure 23-37 shows a simplified flow diagram for the GT-MHR. Each reactor unit is contained within three steel pressure vessels: a reactor vessel, a power conversion vessel, and a connecting cross-vessel. All three vessels are located underground in a concrete silo, which serves as an independent vented low-pressure containment structure (Zgliczynski et al. 1994). The GT-MHR steel reactor pressure vessel contains the active reactor core, core supports, top-entering control rods, a shutdown cooling system, and refueling access penetration. The power conversion vessel contains a turbomachine, three heat exchangers, a high-efficiency recuperator, and a water-cooled intercooler and precooler. The turbomachine consists of a turbine, generator, and two compressor sections mounted on a single shaft supported by magnetic bearings. Helium within the reactor vessel enters through the top of the core and passes through the reactor, where it removes the fission-generated heat energy. It then leaves through the lower reactor plenum and flows through the inner duct of the coaxial cross-vessel to the power conversion vessel. The heated helium coolant is expanded through the turbine, which directly drives the electrical generator and the high-pressure and low-pressure compressors. The helium exits the turbine and flows through the recuperator and precooler. An intercooled compressor system, which consists of a low-pressure compressor, intercooler, and high-pressure compressor, compresses the gas, which then passes through the recuperator, and finally returns to the reactor through the outer annulus of the coaxial cross-vessel. Key features of the GT-MHR are shown in Table 23-16.

The reactor core contains a large number of prismatic blocks stacked in columns and placed in an annular-ring arrangement. Helium coolant flows through coolant holes within the active core blocks. The center and outer core rings consist of unfueled graphite reflector blocks, while the inner ring of blocks contains the nuclear fuel. A steel core barrel assembly surrounds the reactor core and provides space between the barrel and the pressure vessel for helium returning from the recuperator for passage to the top of the core. The pressure vessel, which utilizes a high-temperature vessel material (9Cr-1 Mo-V), is uninsulated.

A shutdown cooling heat exchanger and shutdown cooling circulator are located at the bottom of the reactor vessel. Cooling water is supplied to the shutdown cooling heat

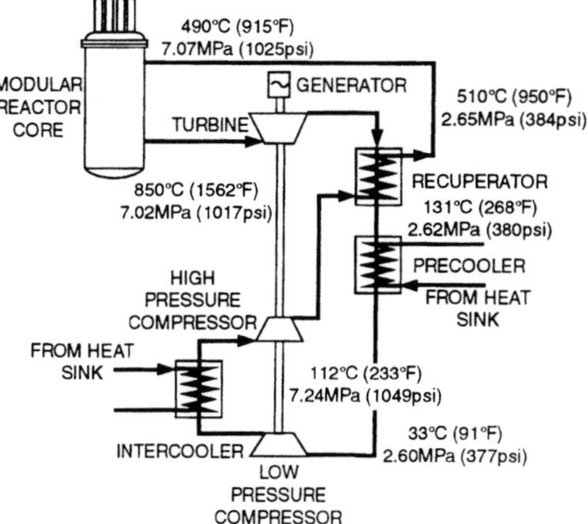

Fig. 23-37. GT-MHR flow diagram. (From General Atomics. Used with permission.)

Table 23-16. Key Features of the GT-MHR

Plant data	
Plant type	Gas turbine coupled with helium reactor
Constructon type	Modular
Reactor type	MHR
Core geometry	Annular
Fuel element type	Prismatic block
Power conversion system	Direct cycle helium gas turbine
Core thermal rating, MWt	Up to 600
Modular power output, MWe	Up to 286
Net efficiency, %	47.6
Thermal data	
Thermodynamic cycle	Recuperated/intercooled
Turbine inlet temperature, °C (°F)	850 (1,562)
Turbine inlet pressure, MPa (psia)	7.0 (1,017)
Compressor pressure ratio	2.8
Compressor efficiency, %	90
Turbine efficiency, %	92
Recuperator effectiveness, %	95
System pressure loss ($\Delta P/P$), %	6
Component design features	
Turbomachine	Single-shaft rotor
Compressor type (stages)	Multistage axial flow (11/15)
Turbine type (stages)	Multistage axial flow (9)
Generator type	Asynchronous 50/60 Hz
Bearing type	Active magnetic bearings
Recuperator type	Compact plate-fin modules
Precooler/intercooler type	Helical bundle
Pressure vessel(s)	Vertical steel vessels
Plant status	
Design status	Conceptual
Technology status	State-of-the-art
Deployment	Approximately 2005

Source: General Atomics 1994. Used with permission.

exchanger by the shutdown cooling water system, which rejects heat to the plant service water system. The shutdown cooling system is used when the gas turbine is unavailable. The shutdown cooling heat exchanger is a vertical shell and tube, cross-counter flow heat exchanger (Bechtel National, Inc. 1987). The shutdown cooling circulator is a single-stage radial compressor driven by an electric motor. Magnetic bearings and antifriction catcher bearings are also used.

The fuel consists of two types of fuel particles. The fissile fuel particles contain approximately 20% enriched uranium and natural uranium oxycarbide kernels. Each fuel particle consists of a fuel kernel coated with an initial porous graphite buffer, followed by three successive layers of pyrolytic carbon, silicon carbide, and pyrolytic carbon. This fuel coating (TRISO coating) prevents significant release of radionuclides, even at temperatures beyond those reached in severe accident conditions. The total outer fuel particle diameter is approximately 1 mm (0.039 in.) (General Atomics 1994). The fuel particles are bound together within graphite binder material, made into fuel rods, and placed in holes in the graphite fuel block. A 16- to 18-month refueling interval is expected. Control rods enter from the top of the core and pass through holes in the core and outer reflector regions. Control rods provide for reactor power control, startup, and shutdown functions. A backup shutdown mechanism is provided through the use of boron pellets placed in hoppers above selected channels in the reactor core.

Each GT-MHR unit consisting of interconnected reactor and power conversion modules is located in an underground vertical cylindrical concrete silo as shown in Fig. 23-38. Each silo is a vented, independent confinement structure. Independent safety systems for each unit are contained within individual concrete structures. Placing the confinement structure underground reduces seismic amplifications, provides for high-speed missile protection, and serves, along with the surrounding earth, as the final heat sink. The reference plant consists of four reactor structures and a reactor service area, all covered by a common enclosure to allow multiple use of auxiliary cranes and fuel handling equipment.

The GT-MHR has several unique features over the passive light water design. The large amount of graphite in the core results in slow temperature changes with the fuel even if a loss of helium cooling occurs. The graphite has a high heat capacity and is closely coupled with the fuel. Graphite

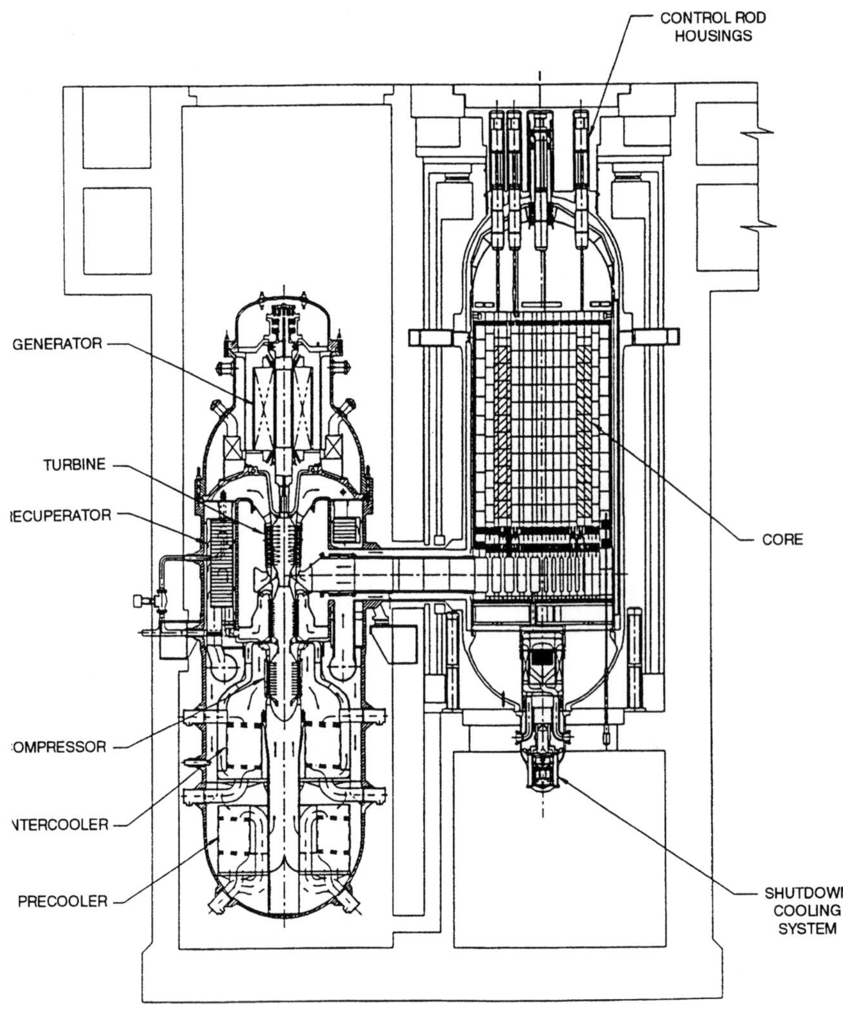

Fig. 23-38. GT-MHR plant arrangement. (From General Atomics. Used with permission.)

has good structural strength, and its structural strength and integrity are maintained at high temperatures far beyond achievable reactor conditions. Helium coolant is inert and does not react with the reactor core components. In addition, the radioactive isotopes of helium are very short lived (milliseconds).

The fuel is the most unique feature of the GT-MHR. Test and operating data show that the fuel particle coatings retain fission products if the fuel is maintained below 3,632° F (2,000° C) (General Atomics Undated). The core size, configuration, core power density, material selection, and total core power level are designed such that the core fuel temperature from decay heat production always is maintained below 2,912° F (1,600° C) by passive means. Even in the event of a total loss of helium flow, passive heat removal through conduction, radiation, and natural convection to soil surrounding the reactor silo will maintain the fuel temperature below 2,912° F (1,600° C) (Hamilton and Lommers 1990).

If both the normal reactor heat transport system and the shutdown cooling system are unavailable, the reactor cavity cooling system (RCCS) provides for adequate core cooling. Figure 23-39 shows the RCCS. Core decay heat is transferred by thermal conduction and radiation through the reactor vessel to the RCCS cooling panels outside the reactor enclosure. Air flow through the panels is by natural convection and provides adequate heat removal. The air flow to and from the panels is through ductwork to the intake/exhaust structure on the above-grade reactor structure.

The power conversion system, shown in Fig. 23-40, consists of a turbomachine, three heat exchangers, a recuperator, and water-cooled intercooler and precooler, tied to its own reactor module. This allows the entire conversion from heat

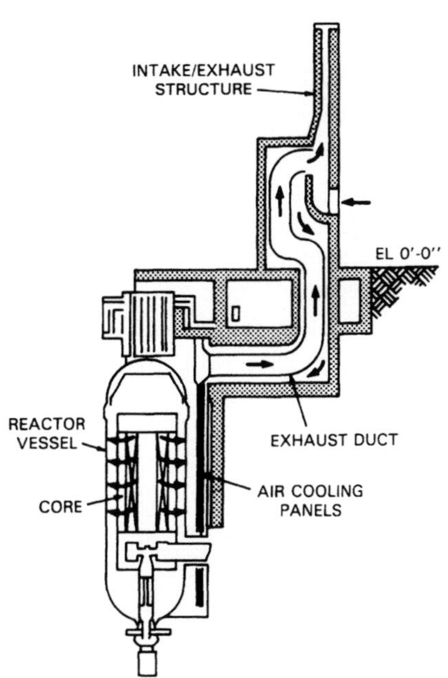

Fig. 23-39. GT-MHR passive reactor cavity cooling system. (From General Atomics. Used with permission.)

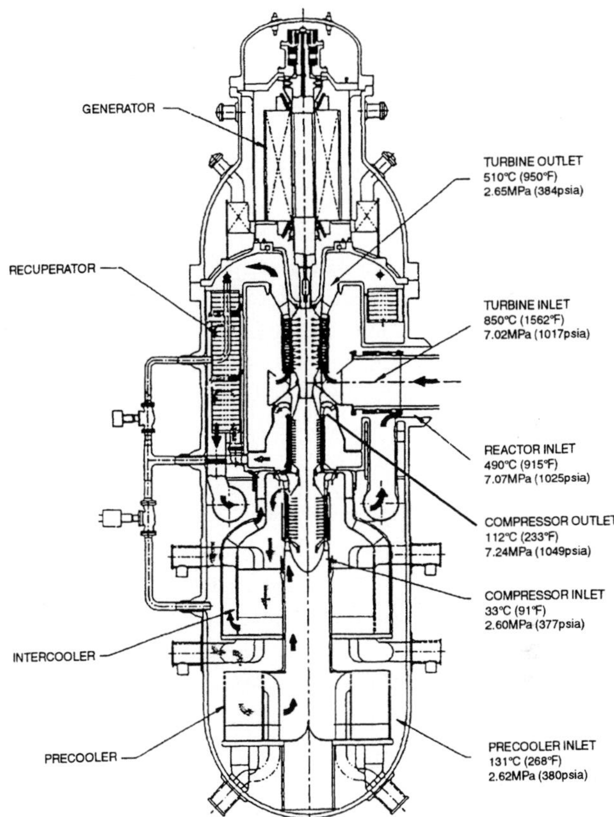

Fig. 23-40. GT-MHR power conversion vessel assembly. (From General Atomics. Used with permission.)

to electric power to take place in one single pressure vessel. The turbomachine, consisting of two compressor sections and the turbine, drives the generator with a single, vertical shaft. The rotational speed is 3,600 rpm, thus providing 60 Hz to the grid (General Atomics 1994). A six-module, compact, plate-fin recuperator combines a low-pressure loss with a high unit thermal density to achieve a 95% recuperator effectiveness. The precooler and intercooler are helical coil bundle, water-cooled, finned tube heat exchangers. The use of the Brayton single power conversion system instead of a typical Rankine steam cycle enables the GT-MHR to operate at a very high net plant efficiency of approximately 47%. This is nearly 50% greater than for conventional light water reactors, and nears the plant efficiency of natural gas power plants.

Research and development activities continue on the GT-MHR, and a Utility Review Team (Bliss et al. 1991) provided several recommendations regarding these activities. One recommendation was to increase the emphasis on fuel quality control process development. Billions of microspheres of UCO particles are contained in the fuel, and very stringent fuel quality and performance requirements must be applied. The technology to ensure viability of fuel and performance requirements on a commercial scale has previously been demonstrated for one Fort St. Vrain HTGR power station.

Another Utility Review Team recommendation suggests

the fuel test temperature range be broadened from 2,192° F (1,200° C) to 3,992° F (2,200° C) to better establish the high temperature fuel integrity limits. Tests show that significant fuel failure begins to occur at temperatures greater than 3,632° F (2,000° C) with the thermal decomposition of the silicon carbide coating. However, some tests show failure initiation in the 3,092 to 3,632° F (1,700 to 2,000° C) range. An important development need, therefore, is to quantify with greater confidence the safety margin between the calculated peak fuel temperature and the fuel particle failure temperature limit. Planned fuel performance tests should better reflect actual reactor operations including power increases and decreases for 150 load-following events per core.

NRC certification of the GT-MHR is expected to require a full-scale demonstration plant. Because of the unique features of the GT-MHR fuel and reactor design, very little or no fuel failure is expected under accident conditions even postulating the loss of the RCCS. A conventional LWR-type unvented, high-pressure, low-leakage containment structure has not been included for the GT-MHR. The Utility Review Team recommended a type of containment structure (not necessarily an LWR structure) be provided for additional fission product retention for the first GT-MHR unit. The Utility Review team felt results from the first plant operation and research and development work would better establish the need for a containment or some other barrier to environmental fission product releases.

The MHTGR has potential uses other than for electrical energy production. With a helium gas outlet temperature of about 1,292° F (700° C), the MHTGR produces steam at conditions (1,000° F, 538° C, 2,510 psig, 13.7 MPag) similar to those of a conventional fossil unit. This can be used for cogeneration applications such as for desalination; stimulating oil recovery from older fields, heavy oil, or shale oil deposits; or for coal liquifaction using the Lurgi process (Hamilton and Lommers 1990). Long term, the HTGR has the potential to operate at much higher helium gas outlet temperatures up to about 1,832° F (1,000° C). At these temperatures, a variety of different technologies such as coal gasification, synthesis gas production, and thermochemical water splitting are possible (Hamilton and Lommers 1990).

23.7 BREEDER REACTORS

23.7.1 The Role of Breeder Reactors

The world's reserves of uranium fuel are not unlimited and are being depleted by the 415 commercial power reactors in operation as well as the various countries' nuclear weapons programs. Further reductions also will occur from future LWR additions. Breeder reactors provide a method to greatly extend both the reserves of competitively priced fissile material and the fuel cycle itself. With breeding, fuel reserves may be extended approximately 55 times over current LWR usage (El-Wakil 1984). Breeder reactors produce more fissionable fuel than they consume, resulting in a conversion or breeding ratio >1.

The ability of nuclear power to breed new fuel was observed very early in its history. The first reactor to generate electricity was a United States breeder, the Experimental Breeder Reactor I (EBR I). EBR I operated from 1951 to 1963 and had a 1,400-KWt/2OO-KWe rating. Its coolant was a potassium–sodium alloy.

There are active breeder development programs in the United States, France, Germany, Japan, Russia, England, and India. France's program is the most advanced with two commercial units, the Phenix, a 250-MWe LMFBR that started operation in 1973, and the SuperPhenix, a 1,240-MWe LMFBR that started operation in 1986.

The status of the United States breeder program can best be described as in a state of declining emphasis and funding by the DOE. The current price of uranium, coupled with the decline in new nuclear power plant construction, has reduced the need for the breeder past the year 2000. One experimental breeder, the Experimental Breeder II (EBRI II), is in operation. EBR II, a 19.5-MWe LMFBR, has been in operation since 1965 and is the longest operating LMFBR in the United States. It demonstrated the feasibility of locating the reactor, fuel reprocessing facilities, and fuel fabrication facilities together. EBR-II is located at the Idaho National Engineering Laboratory and is currently operating with all metal fuel rods consisting of 90% uranium and 10% zirconium. EBR-II is currently supporting the DOE's Integral Fast Reactor (IFR)/Advanced Liquid Metal Reactor (ALMR) program.

The Fast Flux Test Facility (FFTE) started operation in 1980 but shut down in 1994 because of the lack of a clear mission. It was being used to test fuels and other materials, as well as to verify fast reactor design features (Glasstone and Sesonski 1981). FFTF was a 400-MWt reactor near Hanford, Washington.

Two commercial LMFBRs have been attempted in the United States. The first was the 60-MWe Enrico Fermi I LMFBR constructed by Detroit Edison in partnership with American and Japanese utilities (Leclercq 1986). It began operation in 1963 but was shut down in 1966 as a result of a loss of coolant incident that damaged part of its fuel. It was repaired but had little operation thereafter and was permanently shut down in 1971 for financial reasons. The other commercial LMFBR was the 300-MWe demonstration reactor, the Clinch River Breeder Reactor (CRBR), which was to be built near Clinch River, Tennessee. The CRBR became a political football between the US Congress and the Executive Branch; after $2 billion was spent for equipment and engineering, its funding was finally stopped in 1983 due to lack of perceived need. Current United States emphasis on breeder reactors is with the DOE's Advanced Liquid Metal Reactor (ALMR) program. The ALMR program involves conceptual design of a passive, smaller LMFBR with IFR features of onsite energy, fuel fabrication, and fuel reprocessing functions.

23.7.2 General Concepts

The principal fertile materials used in breeder reactors are either $^{238}_{92}U$ or $^{232}_{90}Th$. $^{238}_{92}U$ is the more common material. Figure 23-5 shows a simplified schematic of a breeder reactor. A breeder reactor consists of a primary sodium loop containing the reactor, sodium coolant pump, and primary heat exchanger (shell side) and an intermediate loop consisting of the primary heat exchanger (tube side), intermediate sodium pump, and steam generator. Liquid sodium is used in the primary and intermediate loops because of its good heat transfer properties and its low neutron moderating characteristics. Since sodium becomes radioactive from neutron capture of $^{23}_{11}Na$ (to $^{24}_{11}Na$ with 15-hour half-life) in the reactor core, a primary heat exchanger and intermediate sodium coolant loop are used to prevent radioactive Na-24 from leaking into the steam cycle. The intermediate cycle also prevents water from getting into the reactor core. Since sodium is a liquid metal, its pressure inside the primary and intermediate loops can be maintained above atmospheric temperatures but precludes a large depressurization event possible with LWRs. Sodium is a solid at room temperature with a melting point at 207.5° F (97.5° C) and requires heating to maintain a liquid state. Sodium is reactive with both water and air, and an inert gas blanket over the sodium is required.

Breeder reactor designs are divided into two types: loop and pool. For the loop design, the reactor vessel, primary heat exchanger, sodium pump, and interconnecting piping are separated within compartments and housed in a containment building. In a pool design, the reactor, primary heat exchanger, and sodium pump are located within a large reactor vessel, completely submerged in liquid sodium. Sodium coolant leaving the reactor passes through the primary heat exchanger and then is discharged directly into the reactor vessel. The submerged sodium pump moves the sodium from the primary heat exchanger's discharge into the reactor. Figure 23-41 shows a simplified schematic of the loop and pool designs.

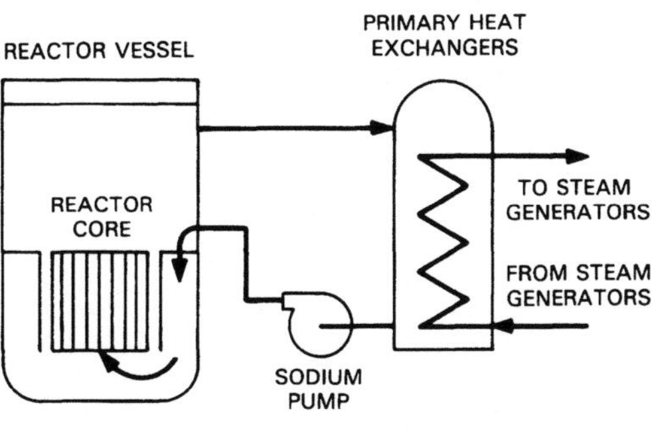

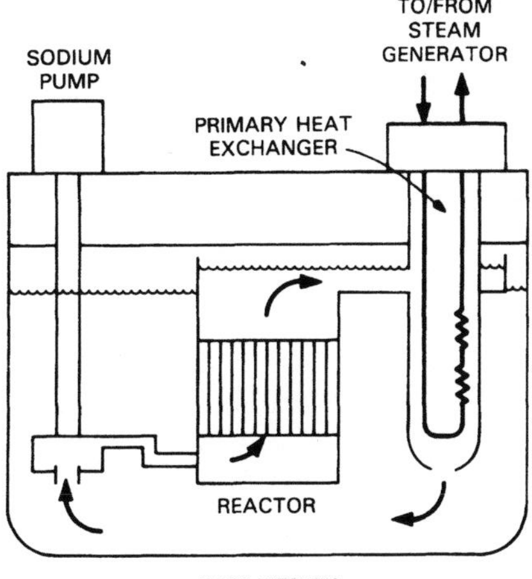

Fig. 23-41. Breeder loop versus pool design.

The pool design has a safety advantage over the loop design. It can absorb large amounts of core decay heat in the event of a pump failure because of the large pool of sodium with its high thermal inertia. The pool design is less sensitive than the loop design to sodium leaks in the primary system. The loop design represents a more straightforward mechanical design and has a large base of accumulated experience from the LWR designs. It also represents a separation of primary system components for better or easier maintenance and repair than the pool design. However, the multiple aspects of shielding the major components and interconnecting piping are a disadvantage with loop design. The large sodium-filled reactor vessel housing the reactor, sodium pump, and primary heat exchanger is more difficult to fabricate and poses additional maintenance difficulties over the loop design. The Clinch River Breeder Reactor was a loop design, while the EBR-II and France's Phenix designs are pool type.

23.7.3 Commercial Designs

The largest breeder reactor in operation is France's SuperPhenix. SuperPhenix is a 1,200-MWe plant of the pool design and represents an upgrade from Phenix I, an earlier 250-MWe pool type breeder reactor still in operation. SuperPhenix was built by NERSA, an international consortium of France, Italy, Germany, Netherlands, Belgium, and Britain (Leclercq 1986). SuperPhenix was first critical on September 1985 and connected to the grid in January 1986. Table 23-17 shows major design parameters for SuperPhenix. A more detailed description of SuperPhenix is found in El-Wakil (1984).

In Japan, there is an ongoing fast reactor program that has recently achieved a demonstration unit initial criticality (MONJU in April 1994) and has a larger unit under design (DFBR). The MONJU plant is 280 MWe which will be connected to the grid in December 1995. The goal is to have a commercial plant deployed by 2030 or before.

In Russia, there are two completed fast reactors, BNR 300 and BNR 600, which are connected to the grid. Units of 800 MWe are under construction.

23.7.4 Advanced Breeder Designs

In addition to passive LWR designs under development in the United States, an advanced liquid metal reactor (ALMR)

Table 23-17. Design Parameters for Superphenix Breeder Reactor

Parameter	Value
Power	3,000 MWt, 1,200 MWe (gross)
Fuel	PuO_2, UO_2 mixed oxide, enrichment 15% $^{239}_{94}Pu$
Fuel elements	364 assemblies, 271 rods per assembly
Fuel rods	Stainless steel clad, 8.9 ft (2.7 m) length, 2.2 in. (8.65 m) OD, 3.28 ft (1 m) length of Pu/UO_2 fuel with 1 ft (0.3 m) of fertile material top and bottom
Radial blanket	233 assemblies, 91 rods per assembly, depleted $^{238}_{92}U$
Control assemblies	21 stainless steel clad, 31 rods each
Reactor vessel	69 ft (21 m) ID, 64 ft (19.5 m) high, containing 3,500 tons (3,182 tonnes) sodium
Fuel burnup	70,000–100,000 MW/day tonne
Breeding rates	1.24
Primary sodium core temperatures	743° F (395° C) inlet 1,013° F (545° C) outlet
Primary loop	4 primary loops consisting of one sodium pump and 2 primary heat exchangers symmetrically located around the core within the reactor vessel
Intermediate loop	4 loops with one steam generator and intermediate sodium pump
Intermediate loop sodium temperatures	653° F (345° C) inlet 977° F (525° C) outlet
Steam generators–water temperature	455° F (235° C) inlet 909° F (487° C) outlet
Steam generator–water pressure	2,567 psia (17.7 MPa)
Steam flow rate	2,692,800 lb/h (340 kg/s) each steam generator
Turbine–generators	2–600 MW, 3,000 rpm
Plant efficiency	41.5% (gross) 8,227 Btu/kW-h (gross)

Source: El-Wakil 1984.

program is under development. The GE-led industrial team has completed the advanced conceptual design, which is an evolution of the Power Reactor Innovative Small Module (PRISM) privately developed by GE. This design can be used with either an oxide (Japan or Europe) or metal fuel (under development by Argonne National Laboratory). The ALMR (PRISM) utilizes innovative features such as passive safety and modular construction to achieve simplicity and economically viable performance. The ALMR (PRISM) has received a positive review by the NRC in a Preapplication Safety Evaluation Report (USNRC 1994). Its economic and technical feasibility have been reviewed by DOE committees and utility review panels.

The overall goal is to design and build a full-sized prototype reactor to demonstrate safety and performance features. Following prototype demonstration, NRC certification as a standard plant is planned after the year 2000.

The current commercial design consists of three identical 465-MWe blocks for a total commercial size of 1,395 MWe. Each 465-MWe block consists of three 471-MWt ALMRs tied to one 465-MWe turbine–generator. Site arrangement consists of the 465-MWe blocks arranged in parallel with one common control building. A 465-MWe block can be added sequentially as the need for capacity develops. The ALMR design uses factory-fabricated reactor modules with passive shutdown heat removal for loss-of-coolant events. The ALMR breeds additional $^{239}_{94}Pu$ with a 50-year compound system doubling time capability. The ALMR IFR features also provide the capability to reprocess the spent fuel and fabricate new $^{239}_{94}Pu$-enriched $^{238}_{92}U$ metal fuel onsite. Long-life actinide fuel cycle wastes are recycled with the new fuel and fissioned in the reactor to shorter-lived fission products, greatly reducing the fission product high-level waste decay period.

Figure 23-42 shows the general arrangement of the ALMR power block. The ALMR is a pool-type design. The reactor module, intermediate heat transport system, and part of the steam generator system are located below ground. The steam generation system is physically separated from the reactor modules or the nuclear part of the plant.

Figure 23-43 shows a closer view of the ALMR reactor module. The cylindrical shaped reactor includes two rings of hexagonal shaped fuel assemblies surrounded on each side by rings of similar shaped fertile blanket assemblies. The core arrangement starts with a center ring of internal blanket assemblies followed by an inner ring of fuel assemblies, then another ring of internal blanket assemblies interspersed with six control assemblies, an outer ring of fuel assemblies, and a final ring of radial blanket assemblies. There are 42 fuel assemblies, 24 internal blanket assemblies, and 33 radial blanket assemblies (Quinn et al. 1991).

A fuel assembly contains 331 0.25-in. (0.9-cm) fuel pins held in place by wire wrap. A metal channel or duct surrounds each assembly. A fuel pin contains slugs of enriched 63% $^{238}_{92}U$–27% $^{239}_{94}Pu$–10% Zr (Quinn et al. 1991). Uranium–Pu–Zr oxide fuel can also be used; however, the metal fuel is the more favored because of its superior inherent negative temperature feedback coefficient. The cladding and channel material is TH-9 ferritic stainless steel, chosen for its resistance to swelling over the long fuel burnup time expected with the 20-month refueling cycle. One-third of the fuel and one-fourth of the blanket assemblies are removed at each refueling (Quinn et al. 1991). The fertile blanket consists of assemblies of TH-9 steel clad pins of depleted 90% $^{238}_{92}U$–10% Zr (Quinn et al. 1991). The ALMR fuel technology is being developed by the Argonne National Laboratory as part of the Integral Fast Reactor (IFR) metal fuel cycle program.

The reactor vessel that contains the reactor also houses two intermediate heat exchangers[6] and four electromagnetic sodium coolant pumps submerged in a large pool of sodium with a helium cover gas. The two intermediate stainless steel heat exchangers supply intermediate hot sodium to one

[6]Primary and intermediate heat exchanger terminology is used interchangeably.

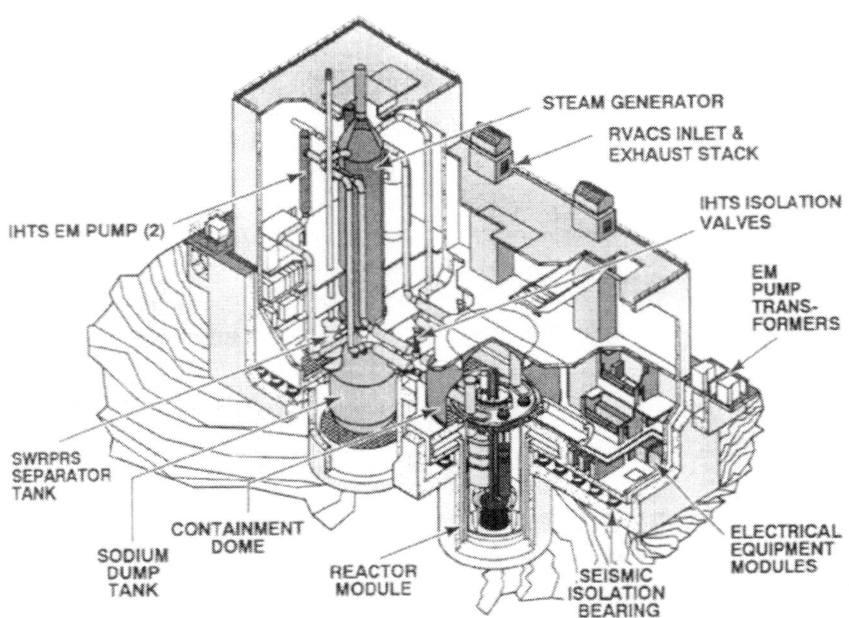

Fig. 23-42. ALMR power block. (From General Electric Company. Used with permission.)

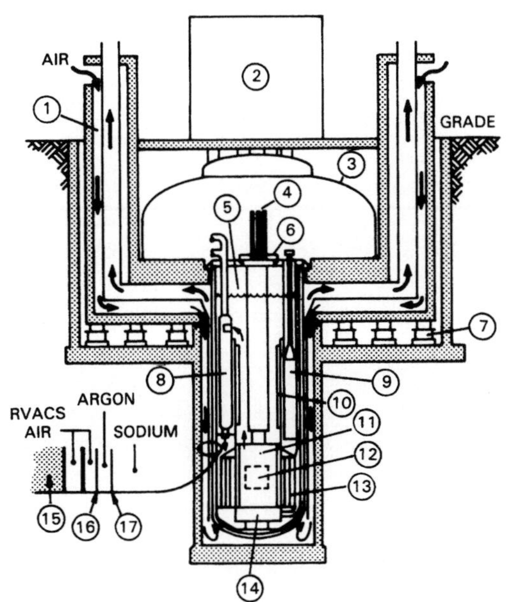

Fig. 23-43. ALMR reactor module. (From General Electric Company. Used with permission.)

steam generator. The primary sodium coolant boundary consists of the reactor vessel, reactor head closure with fittings, and the intermediate heat exchangers. The reactor module and safety system are seismically isolated in the horizontal direction by isolation bearings. The isolation bearings consist of an assembly of steel plates laminated with rubber. The steam generator supplies saturated steam at turbine throttle conditions of 955 psig (6.6 MPag). The balance of plant systems are similar to the saturated steam cycle. However, a recirculation system consisting of a recirculation pump and steam drum is added to steam cycle. Feedwater pumps supply feedwater to the steam drum. The recirculation pump taking suction from the steam drum moves the feedwater–steam mixture through the steam generator back to the drum. Steam exits the steam drum to the turbine generator following moisture removal and steam drying in the drum.

As shown in Fig. 23-42, the reactor vessel sits inside a lower containment vessel. The gap between the lower containment vessel and the reactor vessel is filled with argon at a pressure greater than that of the reactor sodium. Detectors monitor both sodium and argon pressure for leaks both from the reactor vessel and containment vessel. An upper containment dome encloses the reactor head closure. The lower containment vessel and upper containment dome provide a second low-leakage, pressure-retaining boundary. Isolation valves in the intermediate heat exchanger piping just outside the containment dome provide a final barrier between the nuclear safety-related reactor system and the non-safety-related steam generation and balance-of-plant systems. Table 23-18 provides data on the ALMR design.

Refueling is accomplished by a refueling enclosure that fits in place on the reactor facility deck directly over the upper containment dome. Refueling is accomplished using a

Table 23-18. ALMR Data

Number of reactors per power block	3
Net output per power block	465 MWe
Number of power blocks/net output	3/1,395 MWe
Reactor thermal power	471 MWe
Reactor sodium temperature	
Inlet	640° F (338° C)
Outlet	905° F (485° C)
Secondary loop	
Sodium inlet temperature to intermediate heat exchanger	540° F (282° C)
Sodium outlet temperature to intermediate heat exchanger	830° F (443° C)
Steam generator temperature—turbine inlet	539° F (282° C)
Steam generator pressure—turbine inlet	955 psi (6.6 MPa)
Reactor module dimensions	
Height	60 ft (18.3 m)
Diameter	20 ft (6 m)
Reactor fuel type	U–Pu–Zr metal
Refueling period	18 months
Compound system doubling time for breeding Pu (reference/capability)	100 years/50 years

Source: Berglund et al. 1990.

portable refueling cask connected to an adapter assembly. The adapter assembly is placed through a port in the upper containment dome and connected to the refueling port in the reactor head enclosure (Berglund et al. 1990).

The PRISM's primary means of core decay heat removal is through the intermediate heat exchangers, steam generator, condenser, and circulating water system. In the event the steam cycle is unavailable for decay heat rejection, an auxiliary cooling system (ACS) is used. The ACS provides for direct cooling of the steam generator using forced convection cooling with atmospheric air over the steam generator outside shell. If forced convection cooling is not available, natural convection can also be used. In the event the intermediate heat exchangers become unavailable, such as sodium pump failure, a passive design reactor vessel auxiliary cooling system (RVACS) is available. The RVACS shown in Fig. 23-43 removes core decay heat from the reactor vessel via the containment vessel by natural convection of outside atmospheric air around the containment vessel. Heat transfer calculations show the RVACS maintains sodium temperatures well within the design margins for vessel and core structural limits and below sodium boiling (Berglund et al. 1990). The RVACS coupled with the large core negative feedback temperature coefficient provides both passive and inherent design features for safe reactor shutdown and core decay heat removal, including the loss of all backup active heat removal systems.

A unique ALMR feature is the ability to reprocess the spent and fertile fuel and fabricate new fuel for new fissile and fertile fuel. This could be done at a reprocessing/fuel fabrication facility on site or at a centralized facility offsite that would support several ALMRs. As part of the Integral Fast Reactor (IFR) fuel cycle, long-lived actinides produced in core are processed with the plutonium and placed back in the ALMR core with the newly fabricated plutonium enriched fuel. These actinides fission with the fast-energy neutrons. The remaining actinide fission products, which have much shorter half-lives, are then retained as part of the high-level waste when the fuel is reprocessed. The potential result is to reduce the high-level waste decay time from tens of thousands of years to approximately 500 years and to reduce the biological risk factor below that of naturally occurring uranium ore (Berglund et al. 1990). The potential is there for ALMRs to significantly reduce LWR actinide quantities in spent fuel. ALMRs would also use the unused $^{239}_{94}Pu$ that is presently being discarded with the spent LWR fuel.

The ALMR is undergoing conceptual design and licensing assessment. It is quite likely that its present design configuration will change as the design develops. Utility reviews (EPRI 1990) of the ALMR, although favorable, have pointed to areas of additional study and development. These areas include utility acceptance of onsite fuel reprocessing; competitive busbar costs (¢/Kwh) in comparison with ALWR designs; inspection and maintenance of equipment immersed or containing liquid sodium; steam generator demonstrated reliability; ALMR metal fuel performance; and NRC acceptance of no independent low-leakage containment structure surrounding the reactor vessel similar to LWRs.

23.8 SUMMARY

Nuclear power is a complex technology requiring diligence in its design and operation by the thousands of engineers, technicians, operators, regulators and other professionals working in the nuclear industry. The future of nuclear power relies heavily on the continued safe operation of the existing commercial reactors and the emergence of economically competitive, standardized designed, and licensable advanced reactors. These advanced reactors must be viewed by the general public as safe, environmentally acceptable, and needed as an energy source.

In addition, the resolution of the high-level waste issue between political and other groups is needed. The opening of a commercial, high-level waste repository, thereby finally closing the nuclear fuel cycle, may also be required before nuclear power can enjoy widespread public acceptance. These mechanisms are being put into place, although more slowly than hoped, and a resurgence of nuclear power will occur within this decade.

Before leaving this chapter, two events that had a significant impact on nuclear power should be mentioned. The March 1979 accident at Unit 2 of the Three Mile Island Nuclear Station (TMI) resulted in renewed attention to the human element of nuclear power operation. Although equipment malfunctions initiated the accident, inappropriate operator actions turned the incident into a serious accident. The TMI accident resulted in significant damage to the reactor but caused negligible environment damage. Following TMI, the nuclear industry focused heavily on lessons learned from the accident. These lessons included redesign of control rooms for better man–machine interface, improved operator training including frequent training with control room simulators, improved emergency planning procedures and facilities, and numerous modifications to safety- and non-safety-related systems. Regulatory agencies increased their attention to the operations side, and the nuclear industry created its own organization, the Institute of Nuclear Power Operations, to promote excellence in nuclear power operation. The TMI Unit 2 cleanup program resulted in about 110 tons of damaged fuel and 55 tons of damaged reactor internals which have been collected, packaged, and shipped offsite for disposal. Unit 2 has been decontaminated in all areas of the plant except for the reactor building basement. The unit was placed in a monitored and secured storage condition and is scheduled for dismantlement with Unit 1, once Unit 1 reaches its end-of-life. The total cost of TMI-2 cleanup was approximately 1 billion dollars.

The April 1986 accident at Unit 4 of the Chernobyl Nuclear Power Station in the former Soviet Union was a serious accident with significant environmental damage. The Chernobyl accident was a result of both human error and reactor design deficiencies. There was little to be learned by the United States nuclear industry from this accident because of the significant difference in reactor types and containment structure. The accident did show the value of a passive containment building to hold fission products. This structure has always been part of the defense-in-depth approach of United States nuclear plant design.

23.9 REFERENCES

ABB Combustion Engineering Nuclear Power. 1000 Prospect Hill Road. P. O. Box 500, Windsor, CT 06095-0500.

ABB Combustion Engineering Nuclear Power. 1991. *System 80+™ 3817 MW(t) Standard Nuclear Power Plant Design Summary Description.* ASEA Brown Boveri.

Atomic Energy of Canada Limited (AECL). AECL Technologies. 9210 Corporate Boulevard, Suite 410, Rockville, MD 20850.

Atomic Energy of Canada Limited (AECL). September 1989. *Candu 3 Technical Outline.*

Bechtel National, Inc. et al. 1987. *Conceptual Design Summary Report, Modular HTGR Plant, Reference Modular High-Temperature Gas-Cooled Reactor Plant.* Issued by Bechtel National, Inc. under subcontract to Gas-Cooled Reactor Associates, For the Department of Energy Contract DE-AC03-78SF02034. DOE-HTGR-87-092. September.

Berglund, R. C., G. L. Gyorey, and F. E. Tippets. 1990. Progress in safety and performance design of the US advanced liquid metal reactor (ALMR). Paper presented at the ASME-AIEE Joint Power Generation Conference. October 21–25, 1990, Boston, MA.

Bliss, H. E. et al. 1991. *A Utility Assessment of the Modular High Temperature Gas-Cooled Reactor (MHTGR).* Final Report, EPRI NP-7135. Electric Power Research Institute, Palo Alto, CA. January.

Combustion Engineering, Inc. 1989. *System 80+™ Standard Design Your Nuclear Option for the 90's ... Today.*

Electric Power Research Institute. 1992. Advanced Light Water Reactor Utility Requirements Document, Revision 5, Palo Alto, CA. December.

El-Wakil, M. M. 1984. *Powerplant Technology.* McGraw-Hill, New York, NY.

Entergy Operations, Inc., USAR. River Bend Nuclear Station. P.O. Box 220, St. Francisville, LA 70775.

Fuller, Charles H. 1990. Fort St. Vrain Operational Experience. Paper presented at an IAEA Symposium, San Diego, CA.

General Atomics. 3550 General Atomics Court, San Diego, CA 92121-1194.

General Atomics. Undated. *Public Service Company's Nuclear Background.* San Diego, CA.

General Atomics. Undated. *The Modular High Temperature Gas-Cooled Reactor*, San Diego, CA.

General Atomics. 1994. Enabling Technologies That Minimize the Development Risks of the Gas Turbine Power Conversion System for the GT-MHR. GA White paper GT-003. San Diego, CA. February 9.

General Electric Company. 1 River Road, Schenectady, NY 12345.

General Electric Company. 1978. *BWR 6 General Description.* Nuclear Energy Divisions, San Jose, CA.

General Electric Company. 1980. *BWR 6 General Description of a Boiling Water Reactor.* Nuclear Energy Group, San Jose, CA.

Glasstone, S. and A. Sesonske. 1981. *Nuclear Reactor Engineering.* Van Nostrand Reinhold, New York, NY.

Gulf State Utilities. *Updated Safety Analysis Report* (USAR). Beaumont, TX.

Hamilton, C. J. and L. J. Lommers. 1990. Developing role of the modular high temperature gas-cooled reactor. Paper presented at the Combined ANS and ASME Conference on Excellent and Economic Nuclear Plant Performance. Newport, RI, September 1990.

Leclercq, Jacques. 1986. *The Nuclear Age.* Published by Le-Chêne and distributed by Hachette.

Lish, Kenneth C. 1972. *Nuclear Power Plant Systems and Equipment.* Industrial Press, New York, NY.

Nuclear Energy Institute. 1776 I Street NW, Suite 400, Washington, D.C. 20006-3708.

Nuclear News. 1993. World List of Nuclear Power Plants, Vol. 36, No. 11, pp 43–62. September.

Public Service Company of Colorado. P. O. Box 840, Denver, CO 80201-0840.

Quinn, J. E., M. L. Thompson, W. D. Burch, and J. J. Laidler. 1991. Balancing the fuel equation with the advanced liquid metal reactor system (ALMRs). Paper presented at the International Conference on Emergency Nuclear Energy Systems. Monterey, CA, June 16–21.

Ryan, Margaret L. et al. 1989. Outlook on Advanced Reactors. *Nucleonics Week.* March 30.

US Council for Energy Awareness (USCEA). 1994. *Energizer.* Washington D.C.: USCEA.

US Nuclear Regulatory Commission (USNRC). 1994. "Pre-application Safety Evaluation Report for the Power Reactor Innovative Small Module (PRISM) Liquid-Metal Reactor." Washington, D.C. February.

Utility Data Institute. 1988. *Construction Costs—US Steam-Electric Plants 1966–1987*. Washington, D.C.

Vijuk, Ronald and Howard Bruschi. 1988. AP600 Offers a Simpler Way to Greater Safety, Operability, and Maintainability. *Nuclear Engineering International*. November 1988.

Walker, R. E. and T. A. Johnston. 1969. Fort Saint Vrain Nuclear Power Station. *Nuclear Engineering International*, December 1969.

Warembourg, Don. 1969. Progress Report. Fort St. Vrain Nuclear Generating Station. Paper read at the 1969 Rocky Mountain Electric League Spring Conference. Greeley, CO, May 4–6.

Westinghouse Electric Corporation. 1984. *The Westinghouse Pressurized Water Reactor Nuclear Power Plant*. Water Reactor Divisions. Pittsburgh, PA.

Wilkins, D. R. 1990. GE advanced boiling water reactors. Paper presented at the American Power Conference. Chicago, IL, April 23–25.

Wilkins, D. R., J. F. Quirk, and R. J. McCandless. 1990. Status of advanced boiling water reactors. Paper presented at the American Nuclear Society 7th Pacific Basin Nuclear Conference. San Diego, CA, March 4–8.

Wilkins, D. R., T. Seko, S. Sugino, and H. Hashimoto. 1986. Advanced BWR: Design Improvements Build on Proven Technology. *Nuclear Engineering International*, June.

Wolf Creek Nuclear Operating Corporation. P. O. Box 411, Burlington, KS 66839.

Zgliczynski, J. B., F. A. Silady, and A. J. Neylan. 1994. The gas turbine-modular helium reactor (GT-MHR), high efficiency, cost competitive, nuclear energy for the next century. GA-A21661. Paper presented at the 1994 American Nuclear Society, International Topical Meeting on Advances in Reactor Safety. Pittsburgh, PA, April 17–21.

24

EMERGING TECHNOLOGIES

Larry E. Stoddard

Power generation technologies are continually being improved. In many cases, the improvements are small modifications of existing technologies; in other cases, the improvements come in the form of significantly different technological configurations. The impetus for change, whether a technology modification or technology revolution, is the improvement of cost and performance.

Specifically, an emerging technology may bring about one or more of the following power plant improvements:

- Reduced capital cost,
- Reduced nonfuel operating costs,
- Increased plant efficiency (resulting in reduced fuel costs),
- Increased reliability (resulting in improved availability and reduced maintenance costs),
- Use of alternate fuels (less expensive or renewable energy sources),
- Reduced emissions, or
- Production of saleable byproducts.

The impetus for new technology development is often provided by changing market or political forces. The oil crisis of the early 1970s led to heightened efforts to develop and commercialize technologies that use renewable sources of energy such as solar, wind, biomass, and various forms of ocean energy. At the same time, interest was high in developing technologies for producing fuels and chemicals from the enormous coal reserves. Subsequently, low natural gas prices led to the development of high-efficiency, high-reliability, natural-gas-fueled, combined cycle plants for intermediate and base load applications. Environmental concerns have also encouraged the development of clean coal technologies that reduce sulfur dioxide (SO_2), nitrogen oxides (NO_x), particulate matter, and trace element emissions while exhibiting high efficiencies, thereby reducing the emission of carbon dioxide (CO_2), a "greenhouse" gas.

New technologies emerge as the result of research and development (R&D) programs. R&D for a new technology is often internally funded and conducted by a manufacturer with the goal of patenting and marketing the technology. R&D is also conducted by colleges and universities, various laboratories, and research institutes. In addition to internal sources, funding for R&D may be provided by the federal government, for example the US Department of Energy, and state governments. Electric utilities within the United States fund a large number of technology R&D programs through the Electric Power Research Institute (EPRI); there are similar organizations in other countries. EPRI identifies electric utility power generation needs, assesses potential technologies for meeting those needs, and provides project management and funding of R&D activities to develop these potential technologies.

Figure 24-1 shows a typical progression for development of a technology; various organizations may use different terminology. In the concept stage, a new technology idea is assessed to identify potential benefits as well as any fundamental flaws that would make the technology unattractive. If the preliminary assessment warrants additional expenditure for further development, laboratory bench scale tests are conducted. The purpose of bench scale tests is to establish proof-of-concept for one or more critical steps in the overall process or system. They do not demonstrate operation of the complete system.

If the bench-scale tests prove successful, the next step may be the construction of a process development unit (PDU) to demonstrate process feasibility on a somewhat larger scale.

The next step after successful PDU operation is operation of a pilot plant. A pilot plant attempts to validate the fundamental design and operation of a complete system, though on a significantly smaller scale (possibly one tenth the capacity) than the corresponding commercial plant. A pilot plant is generally not sufficiently large to be operated economically. It is usually operated for only a few years to provide operational and performance data.

Following a successful pilot program, the next step in technology development is a demonstration-scale plant. A demonstration plant is generally about one third the capacity of a commercial scale plant. Demonstration plants may not be economically viable without some type of subsidy. They often operate for several years in a test or demonstration mode, perhaps operating on a variety of fuels, for example, different coals, and demonstrating various operating modes.

If the demonstration facility proves successful from performance and reliability perspectives, and if economics are favorable, the technology is ready for the first commercial plant. Typically, commercial plant costs follow a "learning curve" in which subsequent plants cost successively less than the previous plant. Developers often refer to a plant that

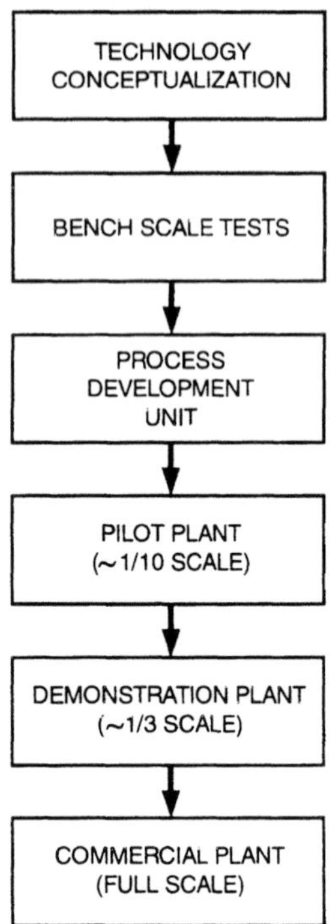

Fig. 24-1. Typical technology development steps. Note: At each level, the technology is assessed to determine if going to the next step is appropriate. At any step the development may be halted or may return to a prior step.

has reached mature plant cost and performance as the nth plant.

Throughout the development process, the conceptual designs of a new commercial plant are refined using data from the most recent developmental activity. Cost, performance, and reliability are evaluated to ascertain whether further development is warranted. Should insurmountable technical or economic barriers be identified, the development program is halted.

This chapter discusses emerging power generation technologies in four generic categories:

- Fossil-fueled
  —Gasification combined cycle,
  —Indirect fired combined cycle,
  —Magnetohydrodynamics, and
  —Fuel cells.
- Renewables
  —Solar,
  —Wind,
  —Biomass,
  —Ocean, and
  —Geothermal.
- Energy storage
  —Battery and
  —Compressed air.
- Fusion

The list of technologies discussed in this chapter is not comprehensive. The technologies included are those that have found fairly widespread interest within the power industry. Certainly some technologies are more attractive in a particular region. For example, interest in ocean technologies will be high in Hawaii and low in Kansas. Furthermore, technology development is dynamic, with new concepts continuously being formulated and tested.

24.1 FOSSIL-FUELED TECHNOLOGIES

Fossil-fueled technologies are those fired by coal, natural gas, or oil. Commercially available fossil-fueled technologies include oil- or gas-fueled steam plants, pulverized coal plants, atmospheric fluidized bed combustion plants, simple and combined cycle combustion turbine plants, and diesel engine plants. There are a large number of emerging fossil fueled technologies, of which the following are discussed in this section:

- Gasification combined cycle,
- Indirect fired combined cycle,
- Magnetohydrodynamics, and
- Fuel cells.

Coal is the primary fuel for the first three of these technologies. Fuel cells will initially be fueled by natural gas, with the possibility of later use in integrated coal gasification fuel cell applications.

Pressurized fluidized bed combustion is another prominent emerging coal-fueled technology; it is discussed in Chapter 21, Fluidized Bed Combustion.

24.1.1 Gasification Combined Cycle

Gasification combined cycle (GCC) is a very promising near-term power generation technology. A GCC system gasifies "dirty" carbonaceous material, producing a "clean" fuel gas, often called syngas, for a combined cycle power generation system. Syngas primarily consists of carbon monoxide (CO) and hydrogen (H_2); other gases are generally considered diluents. Typical feedstocks include lignite, subbituminous or bituminous coal, petroleum coke, heavy oil, and waste products such as municipal sewer sludge.

GCC systems promise relatively low heat rates (<8,500 Btu/kWh) while being very environmentally friendly. Sulfur removal for GCC systems is typically greater than 98%. Sulfur is often recovered in elemental form and may be a saleable product. NO_x emissions are comparable to those from natural gas fueled combined cycle plants. In slagging gasifiers, the ash is recovered as a nonleachable vitreous slag; this slag may be used as an aggregate material for construction. For many of the gasification concepts, most of the toxic elements, such as heavy metals, contained in the feed are encapsulated in the vitreous slag. Water usage is less than that of a pulverized coal or atmospheric fluidized bed plant.

Integrated gasification combined cycle (IGCC) plants are those in which steam production in the gasification system is integrated with that in the combined cycle portion of the system. Other forms of integration include extracting air from the combustion turbine for feed to the gasifier or air separation unit (ASU) and adding to the fuel gas surplus nitrogen from the ASU. Integration increases efficiency and reduces capital and operating costs. The combined cycle design in an IGCC plant is significantly different from that in a stand-alone unit; this reduces the attractiveness of IGCC for phased construction. Nonintegrated systems are more conducive to phased construction; the combined cycle plant is initially fueled by natural gas or oil, and the gasification plant is added later.

24.1.1.1 Gasification Concepts. Figure 24-2 illustrates the GCC system concept. The gasifier feed is dry, liquid, or a solids–water slurry, depending on the gasifier concept. The feed is reacted with oxygen and steam to produce syngas consisting primarily of carbon monoxide (CO) and hydrogen (H_2).

The gasifier operates in a reducing environment that converts most of the sulfur in the feed to hydrogen sulfide (H_2S). A small amount of the sulfur is in the form of carbonyl sulfide (COS). Other contaminants of concern in the syngas are ammonia (NH_3), hydrogen cyanide (HCN), chlorides (Cl), and entrained ash, which typically contains unconverted carbon.

Oxygen to sustain the gasification reactions is provided as air or as 95% pure oxygen produced by an air separation unit. The syngas heating value is "low Btu" (<200 Btu/scf, HHV basis) from air-blown gasifiers and "medium Btu" (200 to 500 Btu/scf) from oxygen-blown gasifiers. "Cold gas efficiency," the ratio of the HHV of the GCC product fuel gas to the gasifier feed HHV, is higher for oxygen systems than for air blown GCC systems because of the additional feed required to heat the nitrogen associated with the air. The additional heat input to the nitrogen in an air-blown gasifier is recovered, to a large extent, in steam generation produced by gas cooling or, in the case of hot gas cleanup, is contained in the hot fuel gas to the combustion turbine.

The raw syngas leaves the gasifier at a temperature ranging from 1,000° F (540° C) to about 3,000° F (1,650° C), depending on the gasifier concept. The syngas must have particulate, sulfur, and other contaminants removed. Particulate may be removed through dry filtering using ceramic filters or through wet scrubbing. Current sulfur removal technology requires the removal of ammonia, hydrogen cyanide, and chlorides prior to hydrogen sulfide removal; all three contaminants can be removed by wet scrubbing. Alternatives to wet scrubbing are catalytic conversion of ammonia and hydrogen cyanide and dry adsorption of chlorides.

Current sulfur removal technology requires cooling of the syngas to about 100° F (38° C). This cooling can be achieved using heat exchange equipment or through recycle gas quench or water quench. Typically, systems that use heat exchange equipment are more efficient than those using quench cooling; however, the capital cost of the systems with heat exchange is higher.

Oxygen-blown gasifiers are generally used with conventional, or "cold gas," cleanup systems. The high cost of downstream equipment caused by the increased mass flow of nitrogen in air blown systems offsets the cost of the air separation unit that produces the oxygen for an oxygen blown system.

Hot gas cleanup systems under development provide sulfur removal at about 1,000° F (540° C); this eliminates the need for much of the heat exchange equipment or quenching required by systems using cold gas cleanup. Ammonia, chloride, and fluoride removal requirements for hot gas cleanup systems are uncertain. Ammonia increases combustion turbine NO_x production and may require catalytic removal before the combustion turbine as ammonia or after the combustion turbine as NO_x. Chlorides and fluorides may exceed a concentration acceptable for the combustion turbine with respect to corrosion. The concentrations of ammonia, chlorides, and fluorides in the fuel gas are a function of the gasifier feed composition.

Hot gas cleanup systems hold the potential of increased system efficiency and lower system costs. Currently planned IGCC plants using hot gas cleanup generally will use air-blown gasifiers.

After sulfur removal, the syngas is transported to the combined cycle "power block," where it is used to fire the combustion turbine. Heat from the combustion turbine exhaust gases is used to generate steam in the heat recovery steam generator (HRSG). The steam is used to power a steam turbine–generator. Combined cycle systems are discussed in

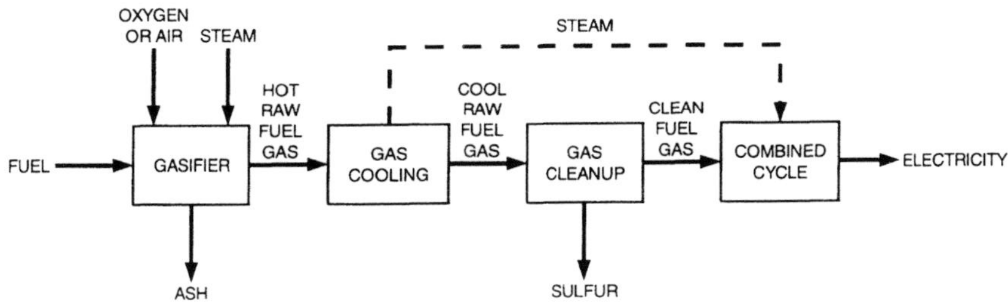

Fig. 24-2. Simplified gasification combined cycle (GCC) block flow diagram.

Chapter 20. The combined cycle power block for a GCC plant is similar to that for a natural gas combined cycle.

IGCC overall plant HHV efficiency ranges from 40% to 45% for an oxygen blown gasification technology with cold gas cleanup and an advanced combustion turbine. Efficiency is a function of the degree of plant integration; the ambient air temperature and relative humidity; the plant elevation; and the fuel's moisture, ash, and sulfur content. Future IGCC improvements such as hot gas cleanup and higher combustion turbine operating temperatures have the potential to increase efficiency above 45%.

GASIFIER TYPES. There are three fundamental types of gasifiers: moving bed gasifiers, fluidized bed gasifiers, and entrained bed gasifiers. These concepts are shown in Fig. 24-3 (Simbeck 1983).

MOVING BED GASIFIERS. With moving bed gasifiers, relatively large particles of coal (2 in. × ¼ in.) are pressurized in lock hoppers and fed into the top of the gasifier. The oxidant and steam are introduced near the bottom of the gasifier. The relatively large pieces of coal move slowly down through the bed and are heated by rising gas. There are five distinct processing zones in the bed. From the top, the first is the drying zone, where the coal is heated to 500° F (260° C) and moisture is driven out of the coal. The second is the pyrolysis or carbonization zone, where the coal is heated to 1,150° F (620° C) and small branches of the coal molecule thermally disassociate, producing ammonia, methane, naphtha, phenols, tars, and organic sulfur compounds. The slow heating rate and long residence time in this zone maximize the production of condensable hydrocarbons. The third is the gasification zone, where the coal is heated to near the combustion temperature of 2,200° F (1,200° C) for an air-blown gasifier and 2,700° F (1,480° C) for an oxygen-blown gasifier. The fourth is the combustion zone, where the small amount of remaining carbon is burned to generate the heat required to drive the endothermic gasification reactions. The fifth is the ash or slag zone. The ash flows down from the gasifier into a lockhopper, where it is periodically discharged into a sluice.

The gas leaves the gasifier at 500 to 1,000° F (260 to 540° C) depending on the coal feed moisture content. Tar, phenol, naphtha, and water are condensed as the gas is cooled. Removal of naphtha and organic sulfur compounds requires the Rectisol process, a refrigerated methanol wash, for acid gas removal. The British Gas Lurgi (BGL) process operates the gasifier in the slagging mode and reinjects all of the condensed organics into the gasifier combustion zone, where they are combusted or disassociated to syngas. The condensed organics cannot all be reinjected into the gasifier in the Lurgi dry bottom process. Instead they are burned in a boiler to produce the large amount of steam required by the dry bottom gasifier.

Coal fines, minus ¼ in., must be screened out of the gasifier feed. Coal fines are entrained in the gas leaving the gasifier and plug the downstream equipment, primarily the

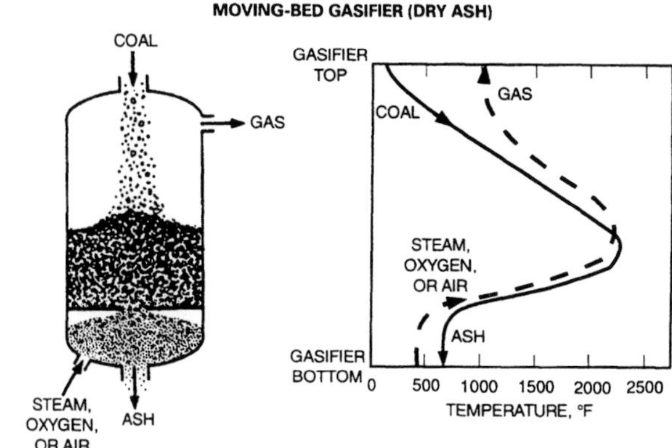

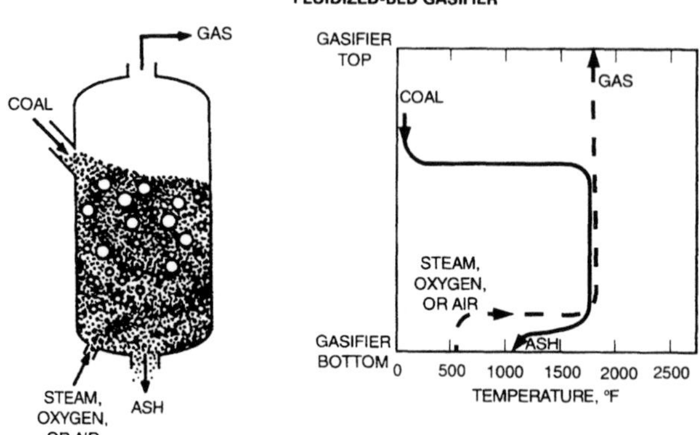

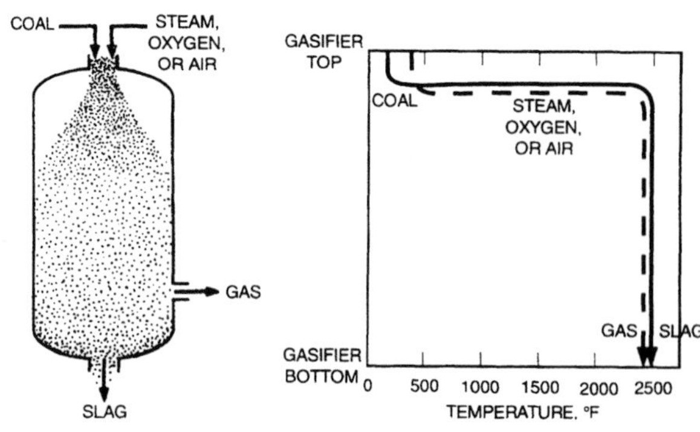

Fig. 24-3. Fundamental gasifier types. (From Electric Power Research Institute. Used with permission.)

level control valve on the first downstream waste heat exchanger. In the BGL process they are briquetted to a size larger than ¼ in. and fed to the gasifier.

The moving bed gasifier has demonstrated successful operation over a wide range of coals including high-moisture North Dakota lignite (38% by weight moisture) and high ash South African brown coal (30% by weight ash).

FLUIDIZED BED GASIFIERS. Fluidized bed gasifiers use the oxidant (air) to suspend the bed of coal particles, similar to fluidized bed combustion boilers. The feed coal is typically ¼ in. by 0 in. with a limit on the minus 100-mesh fraction. The small feed particles are rapidly heated to the bed temperature of 1,500 to 1,800° F (820 to 980° C). This results in a small amount of higher hydrocarbons in the syngas even though the bed residence time is less than that of a moving bed gasifier.

The small coal particle size results in significant entrainment in the gas leaving the gasifier. These particles are collected in cyclones and recycled to the gasifier. The spent particles, which are primarily ash, exit the bottom of the gasifier.

Fluidized bed gasifiers may have in situ sulfur removal through use of a limestone or dolomite sorbent as part of the bed material. The calcium sulfide (CaS) produced in the gasifier reducing atmosphere must be oxidized to calcium sulfate ($CaSO_4$) in the bottom of the gasifier or in a separate vessel prior to disposal.

Fluidized bed gasifiers typically operate at the lowest pressure required by the syngas user. For IGCC with hot gas cleanup this pressure is about 300 psig.

A key technical issue for fluidized bed gasifiers is achieving high carbon conversion. The coal feed to fluid bed gasifiers must be dry from surface moisture and, for IGCC, limited to about 25% (by weight) inherent moisture.

ENTRAINED FLOW GASIFIERS. Entrained flow gasifiers inject the feed, oxidant, and steam at the same location via a co-annular "burner." Figure 24-3 shows a single burner firing down at the top of the gasifier; however, some entrained flow gasifier concepts introduce the coal and steam near the bottom of the gasifier in multiple, horizontal-opposed burners. The finely ground feed (−200 mesh) is entrained in the oxygen and steam as it exits the burner and is rapidly converted to syngas at temperatures near 3,000° F (1,650° C). Any ash in the feed is melted. Most of this slag flows down the wall of the gasifier and exits via a slag tap at the bottom of the gasifier, where it is quenched in a pool of water. The slag freezes into small vitreous granules. Some flyash containing a small amount of unconverted carbon is entrained in the syngas leaving the gasifier.

The hot gas exiting the gasifier is cooled or quenched by radiant heat exchange, cool recycle gas, liquid water contact, or with a "second-stage" addition of feed. Typically, recycle gas quenching and radiant heat exchange are used together. Radiant heat exchange with some gas quenching and second-stage feed quench produce the highest IGCC efficiency. The water quench is less efficient but has a significantly lower capital cost.

Feed to entrained flow gasifiers can be dry solid–gas, solid–liquid slurry, or liquid. Dry solid–gas systems use nitrogen to convey the dry feed through the gasifier burner. Solid–liquid slurry feed systems can process slurries containing up to 65% by weight solids. Water is used to slurry coal and similar material. This water increases the syngas carbon dioxide concentration, which reduces IGCC efficiency. However, the operating cost penalty related to lower efficiency is essentially offset by lower equipment costs for the simpler slurry feed system.

24.1.1.2 Commercial Status. The first large-scale demonstration of an IGCC plant was the 1,150 tons of coal per day, 100-megawatt (MW) Cool Water plant near Daggett, California. Cool Water, which used an entrained flow Texaco gasifier, operated from 1984 through 1989, when the test program was completed.

The second major IGCC plant is the 2,400 tons per day, 160-MW Dow Syngas Project in Plaquemine, Louisiana. This plant, which uses a two-stage entrained flow Dow gasifier, has operated since 1987.

The Shell Development Company operated a 250 tons per day demonstration plant at Deer Park, Texas, from 1987 to 1991. This plant used an entrained flow Shell gasifier to produce syngas, which was burned in the Shell Oil Company Deer Park Refinery boilers.

The largest gasifier complex in the United States is the 13,000 tons per day Dakota Gasification Plant near Beulah, North Dakota. The plant, which is fueled with lignite, uses 14 Lurgi moving bed gasifiers. It produces synthetic natural gas (methane).

The largest gasifier complex in the world comprises the SASOL I, II, and III plants in South Africa. A total of 80,000 tons per day of coal are gasified in Lurgi gasifiers to produce a number of products, including transportation fuels.

A 500 tons per day BGL moving bed gasifier has been successfully demonstrated at Westfield, Scotland. Fuel gas from the gasifier powers a 27-MW combustion turbine.

It is anticipated that several IGCC plants in the 2,000 tons per day, 250-MW range will be operational in the mid to late 1990s. This size is consistent with capacities of combined cycle systems using advanced (2,300° F [1,260° C]) combustion turbines.

24.1.2 Indirect Fired Combined Cycle

The central concept of the indirect fired combined cycle technology is that heat generated by combustion of coal is transferred to air through a heat exchanger. The hot air is then expanded through a combustion turbine. Unlike pressurized fluidized bed combustion, where the hot flue gas from coal combustion is cleaned and expanded through the combustion turbine, the heat of combustion is transferred indirectly to the combustion turbine. As a result, the combustion turbine is not subjected to erosion or corrosion from products of combustion.

An indirect fired combined cycle concept is shown in Fig. 24-4. Compressed air from the combustion turbine air compressor is transported to a ceramic heat exchanger, where it is heated to above 2,000° F (1,090° C). The hot air is expanded through the combustion turbine. The exhaust air from the combustion turbine is used as combustion air. Flue gas from the combustor passes through a slag screen, which removes slag particles of a size that could foul the heat exchanger. The flue gas then passes through the ceramic heat exchanger, where it heats the air. Flue gas exiting the heat exchanger passes through a heat recovery steam generator (HRSG),

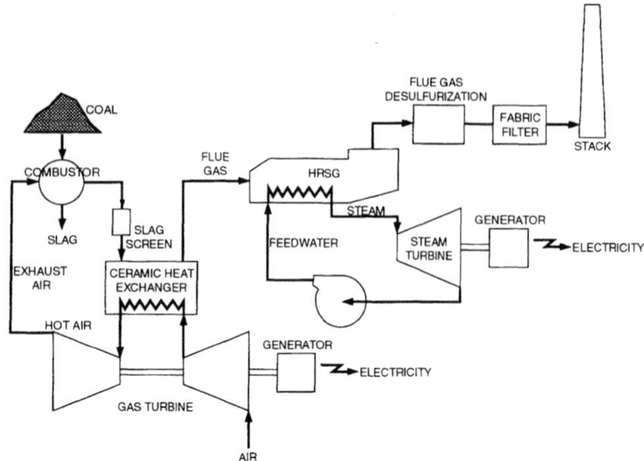

Fig. 24-4. Indirect coal-fired combined cycle.

which provides steam to the bottoming steam cycle. Flue gas desulfurization and particulate removal systems clean the flue gas after it exits the HRSG.

Indirect fired gas turbine concepts have been studied since the 1930s (Keller 1946). Keller and Gaehler reported applications of coal- and peat-fueled air heaters in 1961. However, early applications of the indirect fired concept were hampered by material limitations of metallic heat exchangers. Current indirect fired heat exchanger development makes use of high-temperature ceramic materials.

One indirect fired combined cycle program within the United States is the externally fired combined cycle (EFCC) being developed by the EFCC Consortium. The consortium is headed by Hague International and cofunded by the US Department of Energy's Morgantown Energy Technology Center (METC). Key systems of the EFCC will be tested at the Kennebunk Test Facility starting in 1994. The test will include a 7.5-megawatt thermal (MWt) low-NO_x coal-fueled combustor, a slag screen, a three-pass, 72-tube-string heat exchanger, and a 2-megawatts electric (MWe) combustion turbine (Vandervort et al. 1993). The exit flue gas from the heat exchanger will be cooled by a heat recovery steam generator; steam will be dumped to the atmosphere. The upper limit of air heating by the heat exchanger will be about 2,150° F (1,180° C).

Key developmental issues to be tested at the Kennebunk test facility are associated with the ceramic heat exchanger. Specific issues include achieving low leakage rates, preventing catastrophic sequential breakage of tubes, and development of ceramic materials that can withstand the corrosive environment associated with high-temperature coal combustion products. These issues have been previously addressed on a component basis, but the Kennebunk tests will be the first pilot scale, full system test of EFCC equipment.

Following successful testing of the EFCC systems at Kennebunk, Pennsylvania Electric Company plans to repower Unit 2 of its Warren Station with an EFCC system. This demonstration project will be cofunded by the US DOE under the Clean Coal Technology Demonstration Program, Round V. That project will replace existing pulverized coal boilers that power a 47-MW steam turbine. Including the combustion turbine output, the repowered plant will generate a net 66.0 MW, with a net plant heat rate of 9,600 Btu/kWh (HHV), a 29% improvement over the existing pulverized coal plant heat rate (LaHaye and Bary 1994).

More recently, two teams, one headed by United Technology and another headed by Foster Wheeler, have undertaken systems studies to define concepts and R&D plans under the Combustion 2000 program of the Pittsburgh Energy Technology Center (PETC). Proof-of-concept testing is expected to begin in the late 1990s (Ruth 1992).

Commercial applications of the indirect fired combined cycles are expected to include new coal-fueled plants as well as repowering of existing facilities. Heat rates for new facilities are projected to be significantly below 8,000 Btu/kWh (efficiencies significantly above 43%). The long-range goal is to reduce heat rates to below 7,300 Btu/kWh (efficiency above 47%).

24.1.3 Magnetohydrodynamics

Magnetohydrodynamics (MHD) is a power generation technology in which the electric generator is static (nonrotating) equipment. In the MHD concept, a fluid conductor flows through a static magnetic field, resulting in a dc electric flow perpendicular to the magnetic field. MHD/steam combined cycle power plants have the potential for very low heat rates (in the range of 6,500 Btu/kWh). SO_2 and NO_x emission levels from MHD plants are projected to be very low.

24.1.3.1 MHD Concept. The fundamental MHD concept is illustrated in Fig. 24-5 (MHD Program Plan 1990). The fluid conductor is typically an ionized flue gas resulting from combustion of coal or another fossil fuel. Potassium carbonate, called "seed," is injected during the combustion process to increase fluid conductivity. The fluid temperature is typically about 4,500 to 4,800° F (2,480 to 2,650° C).

The conductive fluid flows through the magnetic field, inducing an electric field by the Faraday effect. The electric field is orthogonal to both the fluid velocity and magnetic

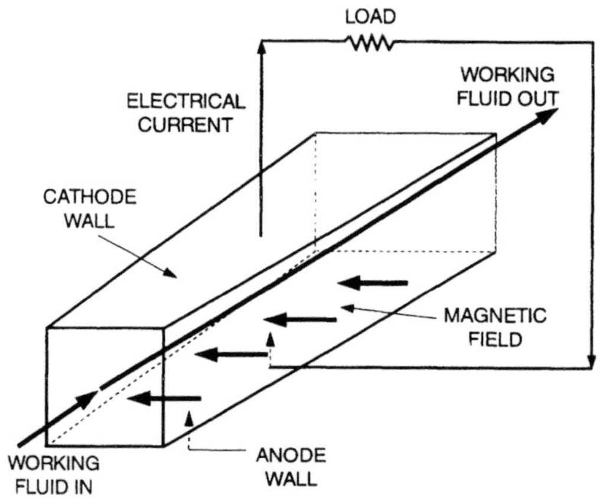

Fig. 24-5. Basic magnetohydrodynamic concept.

field vectors. As a result, a potential difference is developed between the two walls of the duct as shown in Fig. 24-5. The direct current (dc) generated is converted to alternating current (ac) by a solid-state inverter.

24.1.3.2 Commercial MHD Concept. The planned application of the MHD concept for utility scale electric power generation uses MHD as a topping cycle combined with a steam bottoming cycle, as shown in Fig. 24-6 (MHD Program Plan 1990). The topping cycle consists of the coal combustor, nozzle, MHD channel, magnet, power conditioning equipment (inverter), and a diffuser (Kessler et al. 1992). The bottoming cycle consists of a heat recovery/seed recovery unit, a particulate removal system, a steam turbine–generator system, cycle compressor, seed regeneration plant, and, for some concepts, an oxygen plant.

The combustor burns coal to produce a uniform product gas with a high electrical conductivity (about 10 mho/m). A typical MHD plant requires combustion gases of about 4,800° F (2,650° C) at a pressure of 5 to 10 atmospheres. The goal is to remove a large portion (50% to 70%) of the slag (molten ash) formed in the combustion process in the combustor. High ash carryover inhibits efficient seed recovery later in the process. Oxygen enriched air is used as the oxidant to achieve high flue gas temperatures.

Commercial-scale MHD plants will use superconducting magnets. Magnetic fields must be in the range of 4.5 to 6 tesla. To achieve superconducting properties, the magnets must be cooled to about 7° R (4K).

In addition to converting direct current to alternating current, the power conditioning system consolidates power from the electrode pairs and controls the electric field and current. Commercial power conditioning systems will use existing line-commutated solid state inverter technology.

The diffuser is the transition between the topping cycle and the bottoming cycle. The diffuser reduces the velocity of the hot gases from the MHD channel, partially converting kinetic energy into static pressure.

The heat recovery/seed recovery unit consists of radiative and convective heat transfer surfaces to generate and superheat steam. It also removes slag and the seed from the flue gas. In addition, the heat recovery/seed recovery preheats the oxidant supply. NO_x control may be achieved within the recovery unit by a second stage of combustion. The first stage of combustion within the MHD combustor is conducted in a fuel-rich environment. The second stage of combustion within the recovery unit takes place at a temperature above 2,800° F (1,540° C) with a residence time and cooling rate such that NO_x decomposes into N_2 and O_2.

Control of SO_x is intrinsic with the removal of the potassium seed from the flue gas. The potassium seed combines with the sulfur to form potassium sulfate, which condenses and is removed downstream by the particulate removal system. The recovered potassium sulfate is converted to potassium seed in the seed regeneration unit.

24.1.3.3 MHD Developmental Status. The MHD technology is still in the relatively early development stage, although test data indicate that there are no fundamental technical barriers to commercialization. A consensus development program, the Coal Fired MHD Preliminary Transition and Program Plan, was established by the US Department of Energy and Industry in 1984 (MHD Program Plan FY 1992). The goal of this plan was to demonstrate proof-of-concept of MHD plants by the early 1990s. In the plan, the topping cycle, bottoming cycle, and seed regeneration subsystems were to be fabricated and tested separately.

A 50-MWt topping cycle proof of concept test will be conducted at the Component Development and Integration Facility near Butte, Montana. The test equipment will include a coal-fired combustor, a nozzle, an MHD channel, a diffuser, and power conditioning and inverter equipment.

Integrated bottoming cycle tests will be conducted at the Coal Fired Flow Facility located near Tullahoma, Tennessee. Subsystems will include coal pulverizing and drying equipment, seed/ash handling equipment, and an intermediate temperature air heater.

A 28-MWt heat recovery/seed recovery unit is being tested at the Coal Fired Flow Facility. NO_x and SO_x emissions are well below New Source Performance Standards

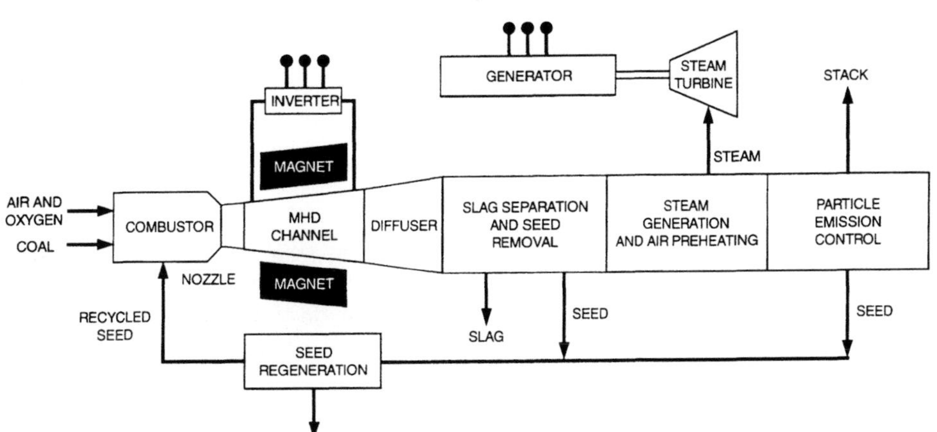

Fig. 24-6. Coal-fueled magnetohydrodynamics system concept.

(NSPS). Particulate emissions are also well within NSPS standards with the use of a baghouse or electrostatic precipitator.

24.1.4 Fuel Cells

Fuel cells are devices that convert fuel directly to electricity through electrochemical reactions. Although working fuel cells have been in existence since the 1930s, their primary use to date has been for space applications. Fuel cells for the power industry are still in the developmental and demonstration stage.

Fuel cells have several potential advantages for the power industry. Because fuel cells are galvanic devices, they are not limited by the Carnot efficiency. This fundamental difference offers significantly higher conversion efficiency than for technologies using thermal cycles, even at part load conditions. Fuel cells are inherently modular; therefore, relatively small units generating a few megawatts of power can be sited near centers of load growth. Fuel cells have low emissions, relatively small land area requirements, and relatively low profiles. Furthermore, they exhibit rapid response to load changes.

Advanced fuel cell concepts have the potential for integration with coal gasification plants and with steam bottoming cycles. These integrated coal gasification fuel cell plants show potential for very attractive heat rates.

A simplified hydrogen fuel cell concept is shown in Fig. 24-7. The fuel, in many cases, hydrogen, is fed to the anode; the oxidant, oxygen or air, is fed to the cathode. The chemical reaction at the anode releases electrons, which flow from the anode through the load to the cathode. Hydrogen ions are also released at the anode. These positive ions flow through the electrolyte and combine with negative hydroxyl ions from the cathode to form product water. The voltage produced by a single cell is on the order of 1 volt dc; the operating voltage of a fuel cell device is obtained by stacking electrodes in series. Therefore, the fundamental building block for fuel cell devices is referred to as a "stack." The output of the fuel cell stack is direct current electricity which is converted to alternating current by a solid-state inverter.

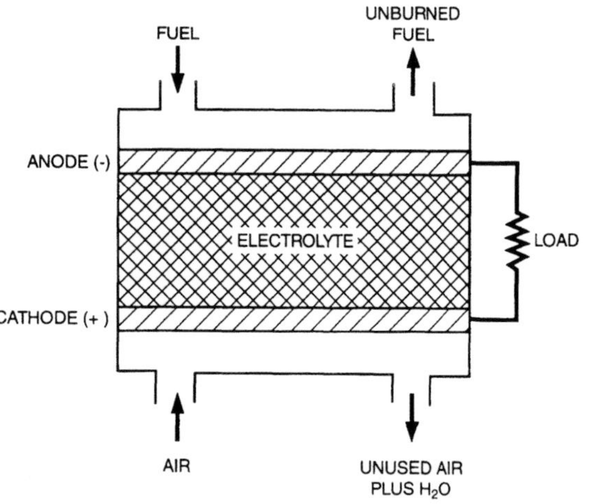

Fig. 24-7. Simplified fuel cell concept.

Development of fuel cells for the power industry has focused on three concepts: phosphoric acid fuel cells, molten carbonate fuel cells, and solid oxide fuel cells.

24.1.4.1 Phosphoric Acid Fuel Cells. The most developed fuel cell technology for power industry application is the phosphoric acid fuel cell (PAFC). PAFCs that are fueled by natural gas require external reforming of natural gas to supply a hydrogen-rich fuel source. In the reformer, natural gas and steam are reacted in the presence of a catalyst to form hydrogen and carbon monoxide. Carbon monoxide and steam are reacted to form hydrogen and carbon dioxide in a shift converter. This external reforming of natural gas requires process heat obtained by combustion of natural gas, thereby reducing the overall fuel-to-electricity conversion efficiency. The external reformer also adds significantly to equipment costs. PAFCs operate at approximately 400° F (205° C).

The electrolyte for PAFCs is concentrated phosphoric acid (H_3PO_4). Current transport through the electrolyte is by the H^+ ion. Electrodes typically are fabricated from finely dispersed platinum catalyst on carbon.

Tokyo Electric Power Company operates an 11-MW PAFC, as well as several smaller units (Douglas 1991). Stacks for the 11-MW unit were manufactured by International Fuel Cell Company in the United States. Typical PAFC fuel-to-electricity conversion efficiency is expected to be about 40% (8,500 Btu/kWh).

24.1.4.2 Molten Carbonate Fuel Cells. The second generation of utility application fuel cells is the molten carbonate fuel cell. Molten carbonate fuel cells operate at about 1,200° F (650° C) (Myles 1989). Because of this high temperature, it is possible to internally reform natural gas to hydrogen and carbon monoxide. Some manufacturers, although not all, have developed internally reforming molten carbonate fuel cells that eliminate the need for external reforming equipment. Molten carbonate fuel cells are expected to be more efficient, more reliable, and less expensive than phosphoric acid cells. It is expected that they will have fuel-to-electricity conversion efficiencies in the range of 54% to 60% (5,700 to 6,300 Btu/kWh).

The overall reaction for the molten carbonate fuel cell is as follows:

$$H_2 + \tfrac{1}{2}O_2 \rightarrow H_2O + \text{Electricity} + \text{Heat} \qquad (24\text{-}1)$$

The fuel cell stack consists of repeating layers of porous nickel anodes, electrolyte tile, and porous nickel oxide cathodes. The electrolyte consists of a mixture of alkali carbonates, commonly Li_2CO_3 and K_2CO_3, and is a molten liquid at the fuel cell operating temperature. The electrolyte is held in place by a pcrous lithium aluminate electrolyte matrix that forms the electrolyte tile.

Clean fuel gas at about 1,250° F (675° C) is fed to the

anode. Hydrogen reacts at the interface of the anode and the electrolyte according to the following:

$$H_2 + CO_3^{2-} \rightarrow H_2 + CO_2 + 2e^- \qquad (24\text{-}2)$$

Carbon monoxide in the fuel gas is also consumed in the anode via conversion to hydrogen by the water gas shift reaction.

$$CO + H_2O \rightarrow CO_2 + H_2 \qquad (24\text{-}3)$$

At the cathode, recycled carbon dioxide from the anode is converted to the carbonate ion:

$$CO_2 + 1/2O_2 + 2e^- \rightarrow CO_3^{2-} \qquad (24\text{-}4)$$

Small-scale (20 kW) molten carbonate fuel cell stacks have been successfully tested. Commercial plants include a 2-MW demonstration plant to begin operation at Santa Clara, California, in the mid 1990s with additional demonstration plants following shortly thereafter. Near-term plants for utility applications will be fueled by natural gas. Future plans are to fuel molten carbonate fuel cells with syngas (hydrogen and carbon monoxide) produced by coal gasification.

24.1.4.3 Solid Oxide Fuel Cells. Solid oxide fuel cells are a third-generation fuel cell technology being developed for utility-scale application. They have an yttrium-doped zirconium oxide ceramic electrolyte. Unlike phosphoric acid or molten carbonate electrolytes, the zirconium oxide ceramic is not corrosive. The anode and cathode are composed of thin layers of nickel–zirconia cermet and strontium–doped lanthanum manganite, respectively. Solid oxide fuel cells have the potential to be small. Although they are not projected to be as efficient as molten carbonate fuel cells, their high operating temperature (about 1,800° F) make solid oxide fuel cells good candidates to be used as topping cycles in conjunction with a conventional expansion turbine. Tests of these fuel cells have been conducted in Japan on two 20-kW units. One of these units has passed 6,500 hours at power. Tests of larger units are planned in the United States for 1996 or later (Douglas 1994).

24.2 RENEWABLE ENERGY SYSTEMS

Renewable energy systems are those power generation technologies in which the energy sources are not depleted by their use. Most renewable energy sources, such as solar, wind, biomass, and some ocean resources, have the sun as the ultimate source of energy. Wind energy, for example, is created primarily by global temperature gradients and the thermal interactions between land, sea, and air. Wave energy results from the interaction of winds with the sea surface. Ocean thermal energy conversion exploits temperature gradients in the ocean. The production of biomass feedstock depends on photosynthesis, a chemical process requiring sunlight. Exceptions include tidal energy, which results from the rotation of the earth coupled with the revolution of the moon about the earth, and geothermal energy, which uses the heat within the earth's interior.

Interest in renewable energy within the United States increased substantially during the 1970s as a result of the oil embargo by the Organization of Petroleum Exporting Countries (OPEC). As oil prices soared, the United States became increasingly aware of the vulnerability of the economy resulting from its dependence on foreign oil; therefore, research and development of renewable energy technologies were accelerated. However, dire predictions by economists of escalating oil prices were never realized when the embargo was lifted. Growing environmental concerns and the escalating trade deficit drive most of the current interest in renewable energy.

Renewable energy now provides 8% to 10% of the total energy generated within the United States (SERI/TP-260-3674 1990). The majority of this generation is from hydropower. Other renewable resources are still considered, in general, to be economically noncompetitive with fossil-fueled energy.

Two formidable hurdles exist for most of the renewable energy systems: high capital cost and the intermittent nature of most of the renewable resources. The economic feasibility of renewables, however, is improving as the technologies mature and as niche markets are found. The intermittent nature of most of the renewable resources can result in relatively low-capacity factors and in lack of availability when power is needed by the grid. Use of energy storage or supplementing with fossil fuels can help mitigate these concerns.

This section covers five primary renewable energy resources and the technologies used to exploit them: solar, wind, biomass, ocean, and geothermal.

24.2.1 Solar Energy

Solar systems are powered by energy from the sun. Two generic types of solar-electric systems are solar photovoltaics and solar thermal electric. Photovoltaics use a quantum process to convert solar radiation directly to electricity. A solar thermal electric system uses a working fluid that is heated by solar energy and expanded through a turbine or heat engine.

Solar radiation reaching the earth's surface is called insolation. There are two components of insolation at the earth's surface. Direct normal insolation (DNI), which normally is 80% of the total insolation, is that part of the radiation coming directly from the sun. Diffuse insolation is that part of the radiation that has been scattered by the atmosphere or is reflected off the ground or other surfaces. On a cloudy day, all of the insolation is diffuse. The vector sum of the diffuse and direct components of insolation on a surface is termed the global insolation; on a horizontal surface, the global insolation is often called the total horizontal insolation (THI). Systems in which the solar radiation is focused (more correctly, concentrated) use only direct normal insolation. Systems that do not concentrate the solar radiation use global radiation.

Insolation is typically given as a power density in units of kW/m^2, Btu/h-ft^2, or MJ/h-m^2. (The unit kW here is a unit of thermal power; 1 kW/m^2 = 317 Btu/h-ft^2 = 3.60 MJ/h-m^2.) Average daily insolation typically is given as an energy density in units of kWh/m^2-day, Btu/ft^2-day, or MJ/m^2-day. For much of the earth's area, the instantaneous insolation on a surface oriented toward the sun at noon on a very clear day typically is in the range of 0.9 kW/m^2 to 1.1 kW/m^2. The insolation may be decreased by clouds, atmospheric turbidity (dust or haze), or by the angle of the sun with respect to the surface of interest. The hourly variations in direct normal insolation and in total horizontal insolation on a typical clear day are shown in Fig. 24-8.

24.2.1.1 Solar Resource. The best information on insolation resource for a particular location is measured data at that location. Direct normal insolation is measured with a pyroheliometer, a device that tracks the sun. Global insolation is measured with a pyranometer, a device mounted at a fixed orientation. Algorithms are available for estimating global insolation for one orientation based on measured data for another orientation. The National Renewable Energy Laboratory (formerly Solar Energy Research Institute) in Golden, Colorado, has published average insolation resources for over 200 locations within the United States (SERI/TP-220-3880 1990). In addition, weather tapes giving hourly insolation data for typical meteorological years (TMYs) for these same locations are available from the National Climatic Center in Asheville, North Carolina. Data bases for these documents and tapes were derived from SOLMET weather data (including global insolation, but not direct normal insolation) taken at 26 sites, typically from 1952 to 1975. Direct normal insolation was estimated by empirically derived correlations.

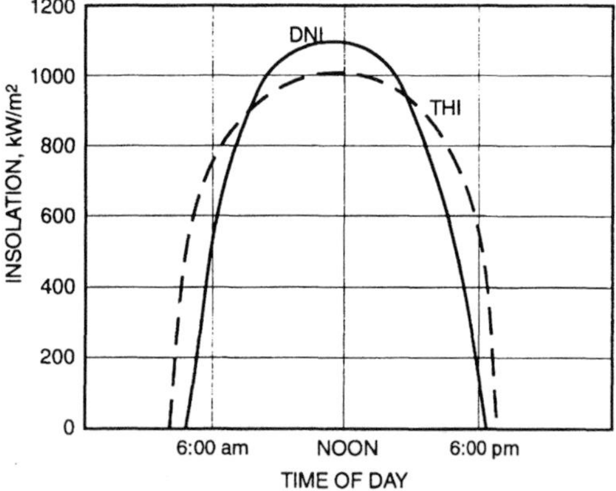

Fig. 24-8. Hourly variation of insolation on a typical clear day.

Although these insolation data are generally derived rather than direct measurements, they represent the best available insolation data for many locations within the United States. At present, a large network of insolation resource measurements is being taken in the United States. The National Renewable Energy Laboratory monitors these measurements and furnishes resource data for a particular location on request if such data are available.

Figure 24-9 is an isopleth map of average annual direct normal insolation for the United States. The highest insolation is available in the desert Southwest. Typically, regions near the coast have decreased insolation because of haze or cloud cover.

24.2.1.2 Solar Photovoltaic Systems. Solar photovoltaic (PV) systems convert solar radiant energy directly to electricity through the use of quantum effect solar cells. Solar cells are solid-state semiconductor devices, usually containing silicon. The cells typically have an n-type region and a p-type region as shown in Fig. 24-10. Sunlight striking the solar cell frees electrons from the valence bond to the conduction band. Internal potential gradients, formed by doping of the semiconductor material, separate the charge carriers producing an electric current. Metal contact lines or grids on the cell surfaces collect the charge carriers, and transmit the resulting electric current to the external load.

Solar cells are closely packed and electrically interconnected within an environmentally protected package called a module. Modules typically measure 1 to 4 ft (0.3 to 1.2 m) on a side and generate 10 to 100 watts of direct current (dc) power. Groups of modules are generally connected in parallel/series arrangements to achieve the system operating current and voltage. Such a group of modules is often referred to as a panel. Panels are interconnected to form arrays. For utility grid connection, the dc from the array is converted to alternating current (ac) using a solid-state inverter, often called a power conversion unit. The modular nature of PV systems allows installation in increments from tens to thousands of kilowatts.

The two basic configurations of PV systems are flat plate and concentrating systems. Flat plate systems do not concentrate the solar rays; insolation is absorbed evenly across the surface. Concentrating systems use reflective or refractive optics to concentrate sunlight on a line (multiple solar cells side-by-side) or to a point (single cell).

Flat plate collectors use both the direct and diffuse components of insolation. Flat plate collectors may be fixed (unchanging orientation), or they may track the sun about one or two axes to improve energy capture.

Concentrating collectors use only the direct component of insolation and must therefore track the sun. A line focus concentrating PV system tracks the sun about one axis; a point focus concentrating PV system must track the sun about two axes such that each module always points directly at the sun.

PV system power output is influenced by several factors:

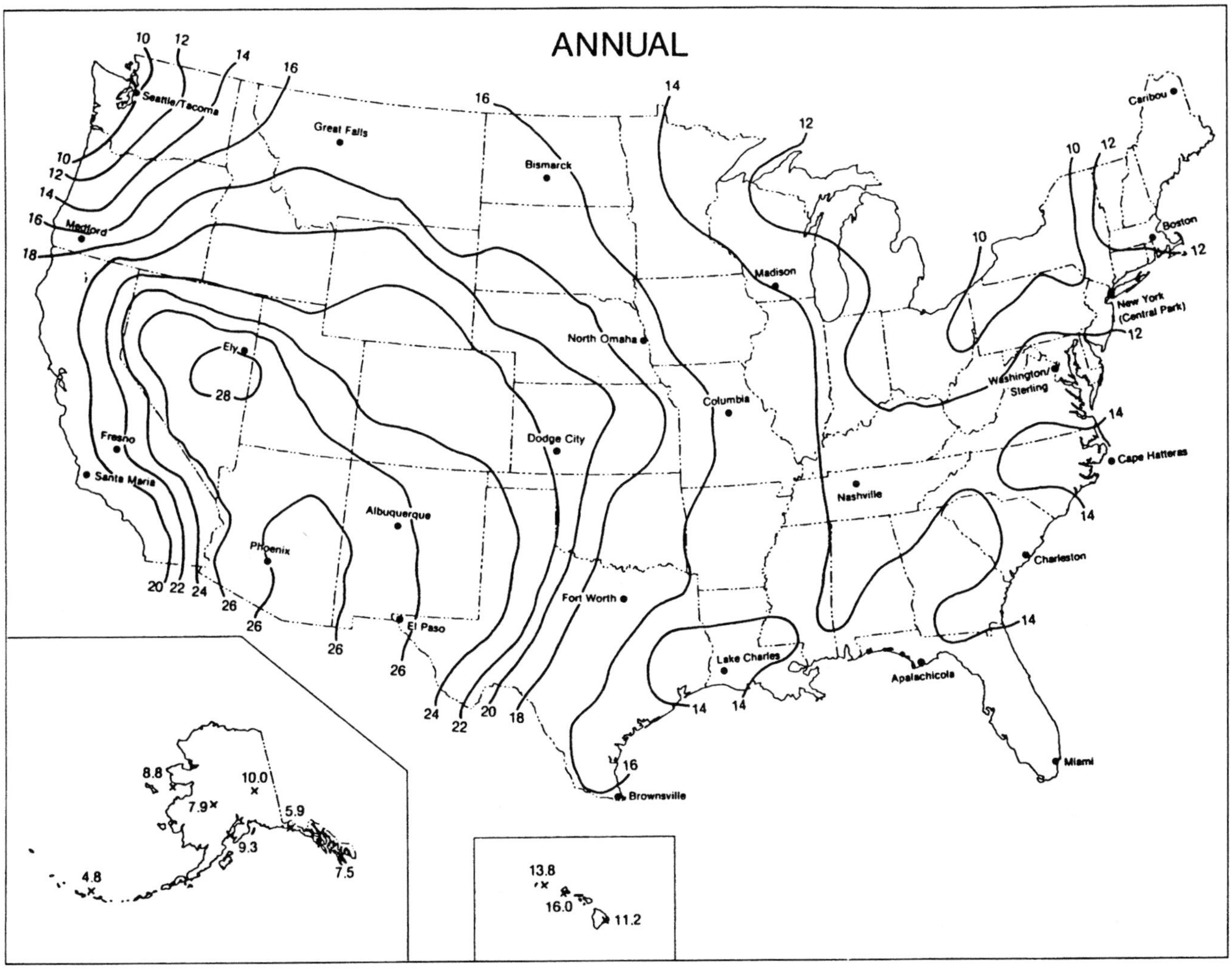

Fig. 24-9. Average daily direct normal solar radiation (MJ/m²). (From National Renewable Energy Research Laboratory. Used with permission.)

- Total active area of the modules (often termed aperture area), typically given in square meters.
- The amount of insolation striking the modules. This is the total global insolation for flat plate collectors and the direct normal insolation for concentration collectors. If the collector is not aimed directly at the sun (as in the case of a line focus system), the incident energy can be determined multiplying the direct normal insolation by the cosine of the angle between the collector normal and the sun angle.
- Efficiency of the modules. Temperature-dependent efficiency

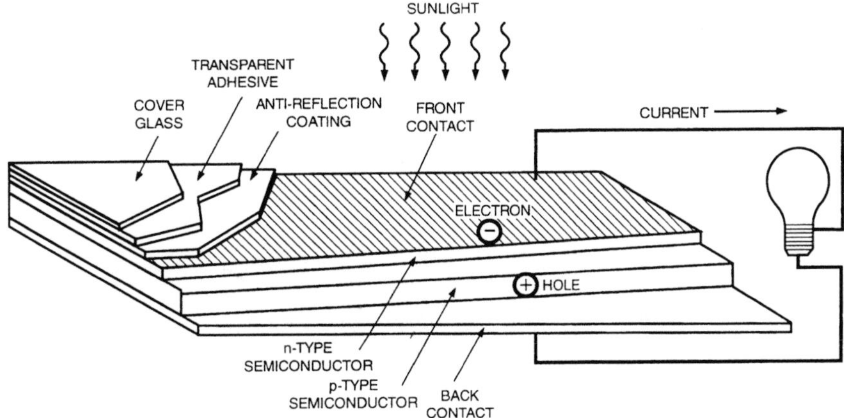

Fig. 24-10. Photovoltaic cell concept. (From National Renewable Energy Research Laboratory. Used with permission.)

data are provided by the manufacturer. Flat plate module efficiencies are typically about 10%, and concentrating module efficiencies are in the 15% to 20% range.

- A soiling factor, usually about 95% (a 5% soiling loss). Depending on the system location, rain may be adequate to clean the modules; in other locations, occasional washing may be required.
- A mismatch factor. Mismatch is nonuniformity of performance characteristics among modules. Tight specifications of module performance minimize mismatch. A typical mismatch factor is 98% (a 2% mismatch loss).
- Resistance (I^2R) losses in interconnecting wiring. This is a function of system design. I^2R factors typically are 99% or higher (1% or less loss).
- Power conversion unit efficiency. Power conversion unit efficiency is dependent on load. It is usually in the range of 95% to 97% at above 25% load but falls off sharply below that level.

Figure 24-11 is a typical PV system energy stairstep illustrating the cumulative effects of these factors.

Instantaneous system performance can easily be calculated manually given insolation data and module manufacturer data. Annual performance is best estimated using a performance computer model such as PVFORM (Menicucci and Fernandez). PVFORM is available from the PV Design Assistance Center at Sandia National Laboratories in Albuquerque, New Mexico. PVFORM reads insolation data (for example, TMY data) at hourly increments and computes temperature-corrected module efficiency and other efficiency factors based on user input. The code steps through each hour of the year and provides daily, monthly, and annual summaries as requested by the user.

Research and development on PV technology is aimed at increasing module efficiency and decreasing module cost. A chief contributor to PV module cost has been the semiconductor-grade crystalline silicon used in the manufacture of a large portion of the PV cells. Not only is the material expensive, but also traditional wafer slicing and trimming approaches have resulted in significant material waste. Many developmental efforts are aimed at crystal growing techniques to reduce the waste, at using lower grade materials, or at using thin film techniques.

Several demonstration multi-megawatt PV systems have been operated in the United States. The Sacramento Municipal Utility District owns and operates a 2.35-MW one-axis tracking flat plate system. ARCO Solar (which was purchased by Siemens Solar) owned and operated 6.5-MW and 860-kW two-axis tracking flat plate systems; these systems have been dismantled.

A national cooperative research and development project, PV for Utility Scale Applications, is installing eight 20-kW emerging module technology systems and three utility-scaleable systems totalling 800 kW. These systems are located near Davis, California. While large-scale commercialization of PV systems could occur in the 1990s, significant market penetration for central station power generation is more likely to occur in the 2000s.

24.2.1.3 Solar Thermal. Solar thermal technologies convert radiant energy from the sun to thermal energy. For low-temperature applications (typically below about 200° F [95° C]) such as domestic water heating, concentration of the sunlight is not required. To achieve the high temperatures required for generation of electrical power, the solar energy must be concentrated. The solar thermal technologies discussed in this chapter are all concentrating solar concepts that make use only of direct normal insolation.

All solar thermal electric technologies include a collector, which redirects and concentrates the insolation on a receiver. In the receiver, the solar energy is absorbed, heating a fluid that powers a heat engine to generate electricity. In some systems, a heat exchanger may be used to transfer heat from the receiver working fluid to the fluid used for power generation. An example of this is the heating of heat transfer oil in the solar receiver; heat in the oil is subsequently used to generate steam to power a turbine. In some cases, a thermal storage system is incorporated, allowing longer operating days or shifting of power generation from daytime hours to nighttime hours.

Three principal solar thermal concentrator concepts are currently under development for power generation: parabolic trough, central receiver, and parabolic dish. These technology concepts are illustrated in Fig. 24-12.

PARABOLIC TROUGH. A parabolic trough collector consists of a computer-controlled, movable parabolic-cylindrical mirror that focuses direct normal insolation on a receiver tube. The receiver tube is located along the focal axis of the

NOTE: ( ) = STEP EFFICIENCY
[ ] = CUMULATIVE EFFICIENCY

11.4 MWt	11.3 MWt	10.7 MWt	1.18 MWe	1.08 MWe	1.06 MWe	1.05 MWe	1.05 MWe
[100%]	[99.0%]	[94.0%]	[10.3%]	[9.5%]	[9.3%]	[9.2%]	[8.8%]
INCIDENT INSOLATION	SHADING LOSS	SOILING AND DEGRADATION LOSS	REFERENCE MODULE EFFICIENCY	MODULE EFFICIENCY TEMPERATURE DEGRADATION	LESS MISMATCH LOSS	LESS I^2R LOSS	LESS POWER CONVERSION UNIT AND AUXILIARY LOSSES
(100%)	(99.0%)	(95.0%)	(11.0%)	(92.0%)	(98.0%)	(99.0%)	(95.0%)

Fig. 24-11. Typical photovoltaic system efficiency stairstep.

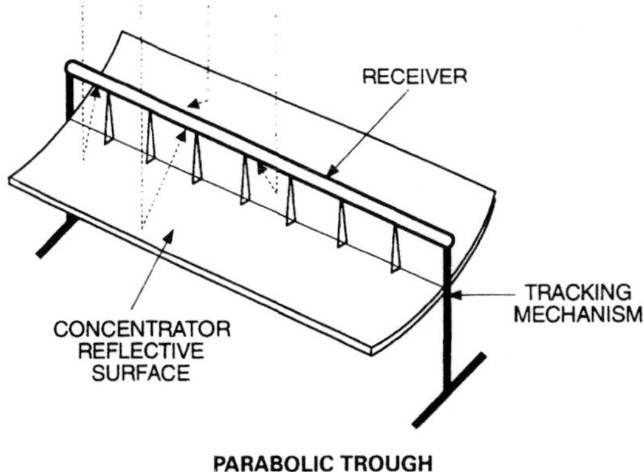

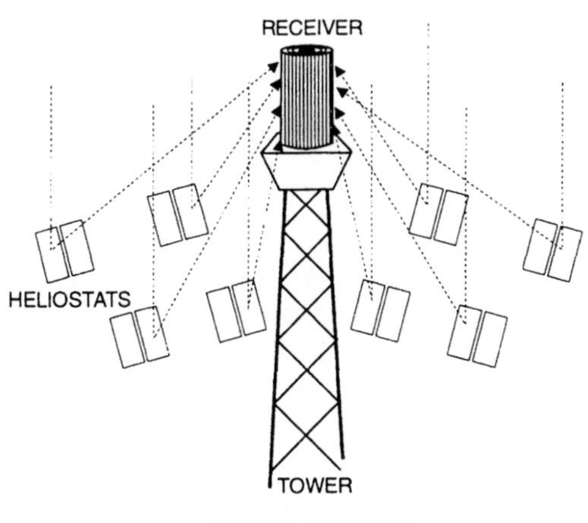

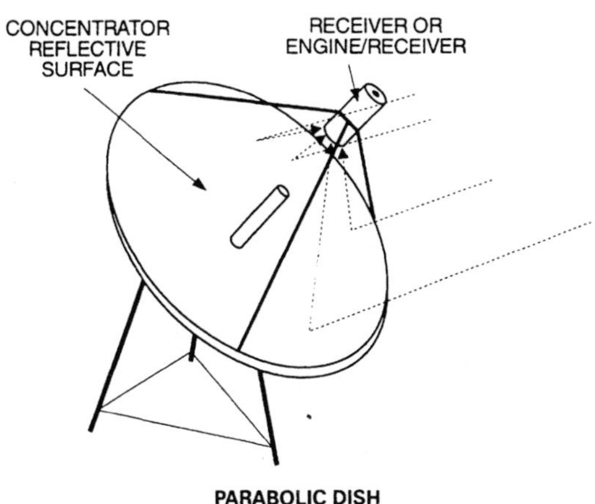

Fig. 24-12. Solar thermal electric generator concepts.

collector. The mirror rotates about its longitudinal axis, tracking the sun throughout the day. The solar energy focused on the receiver tube heats a fluid, typically a heat transfer oil such as diphenyl or silicone-type oil, to about 750° F (400° C). Parabolic trough systems usually consist of long rows of collectors placed end-to-end with each collector contributing to the temperature rise in the working fluid. A steam generator uses heat from the receiver working fluid to generate steam for powering a turbine. An alternate fluid such as toluene can be used instead of water as the power conversion medium.

Parabolic trough systems are commercially available; cost and application (solar resource and potential revenues) rather than technical issues determine the feasibility of their use. The principal commercial applications of parabolic troughs are the Luz Solar Electric Generation Systems (SEGS) in southern California. By 1990, 354 MW of generating capacity had been installed in the SEGS I through SEGS IX units (De Laquil et al. 1993). The SEGS units use a hybrid solar–fossil configuration, with 25% of their annual energy being generated with natural gas.

CENTRAL RECEIVER. In a solar thermal central receiver plant, direct normal insolation is redirected and concentrated by computer-controlled mirrors, called heliostats, onto a receiver located atop a tower. The solar thermal central receiver concept is sometimes referred to as the "power tower." Solar radiation striking the receiver heats a working fluid such as water–steam or molten salt. With a water–steam system, steam generated in the receiver is used directly to power a ground-level turbine generator. With a molten salt receiver, the incident energy heats the salt in the receiver, raising its temperature from about 550 to 1,050° F (290 to 565° C). The hot salt is transported to a storage tank at ground level and pumped from the storage tank to a steam generator, where steam is generated to drive the turbine. Figure 24-13 shows the central receiver concept with a molten salt receiver. An important advantage of molten salt systems is that thermal energy storage is less costly and more efficient than water–steam systems. Molten salt energy storage allows dispatch of power generation at peak hours, allows power generation during brief periods of cloud coverage, and decouples turbine–generator and receiver operation.

Among other things, the design of a solar thermal central receiver system requires a complex optimization of the number of heliostats, heliostat placement, tower height, and receiver dimensions. Design computer programs that facilitate optimization include DELSOL, developed by Sandia National Laboratories, and RCELL, developed by the University of Houston (Kistler 1986 and Laurence and Lipps 1984). A computer performance model for predicting annual energy generation is SOLERGY, which was developed by Sandia National Laboratories (Stoddard et al. 1987).

The water–steam solar thermal central receiver concept has been successfully demonstrated on a pilot scale at the 10-MW Solar One facility near Barstow, California (Holl and Barron 1989). Solar One started up in 1982, underwent a 2-year test and evaluation period, and ran in a power production phase from 1984 to 1988. It was shut down following its contracted demonstration period because operating costs exceeded revenues generated by electrical sales. A consortium headed by Southern California Edison and cofunded by the US Department of Energy plans to repower Solar One with a molten salt receiver and salt-to-steam generator.

Other small-scale water–steam, molten salt, and liquid

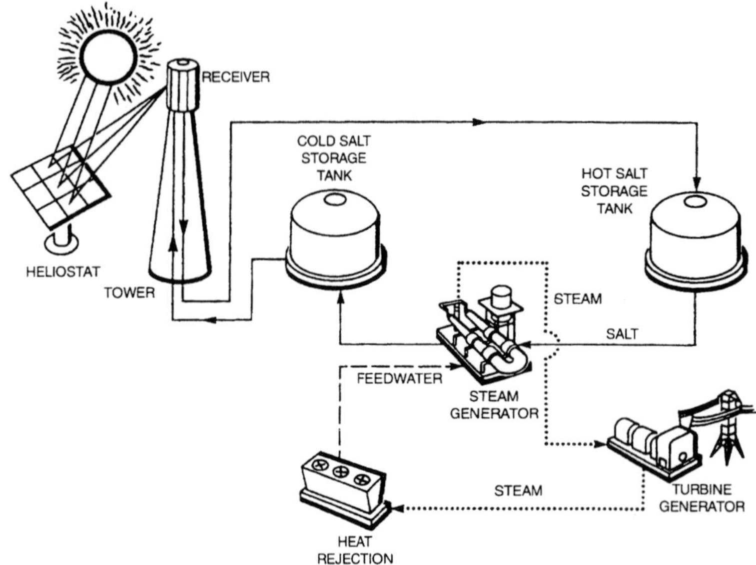

Fig. 24-13. Flow schematic of a molten salt central receiver system. (From Sandia National Laboratories. Used with permission.)

sodium solar thermal central receiver systems have been tested at various sites in the United States and in Europe (Holl and Barron 1989).

PARABOLIC DISH. A parabolic dish system consists of a reflective parabolic bowl that tracks the sun, redirecting and concentrating the solar energy on a receiver at the dish's focal point. In the receiver, heat is absorbed by a working fluid. The fluid may be piped to a central ground level power generation device, or it may power a prime mover, such as a toluene Rankine cycle turbine, a Brayton cycle turbine, or a Stirling cycle engine, located with the receiver at the focal point.

A 25-kW Stirling cycle engine system has demonstrated a solar-to-electric efficiency of 27.9% with an engine cycle efficiency of 39.5% (Holl and Barron 1989). Long-term reliability and availability of the Stirling cycle engine have not been demonstrated.

Two major dish system demonstrations have operated in the United States. The Solar Total Energy Project at Shenendoah, Georgia, produced 400 kW of electrical power and 2,000 kW of process heat. It operated between 1982 and 1990. The privately financed Warner Springs, California, SOLARPLANT I was designed to produce 4.9 MW of electricity. This facility, which began operation in 1985, experienced significant equipment and operational problems. Other small pilot facilities include the 25-kW (electric), 100-kW (thermal) White Cliffs project in Australia; the 100-kW (electric), 400-kW (thermal) Sulaibyah project in Kuwait; and the 25-kW (electric) Vanguard project in California.

24.2.2 Wind

Wind energy systems convert the wind's kinetic energy into mechanical or electrical energy. There are reports of windmills used to pump water in Japan and China as early as 2000 BC (Dunn 1986). Windmills played an important role in water pumping throughout the world; in the rural United States, water was pumped by wind power until the spread of rural electrification. Recent development of wind energy has concentrated on the generation of electricity.

Wind energy is the kinetic energy of a moving mass of air. Thus, the kinetic energy of a mass (m) of air moving at a velocity (v) is $\frac{1}{2}mv^2$. Considering a unit area perpendicular to the wind direction, the mass flow rate is ρAv, where ρ is the density of the air mass and A is the cross-sectional area perpendicular to the velocity. The energy flow rate (P) per unit area (A) is as follows:

$$P/A = \tfrac{1}{2}\rho v^3 \qquad (24\text{-}5)$$

The velocity-cubed dependence of wind power makes it extremely sensitive to the wind velocity resource. A doubling of the velocity results in an eightfold increase in available power. For example, a wind turbine that can generate 50 kW in a 10-mph wind theoretically can generate 400 kW in a 20-mph wind.

The wind velocity at a particular location varies with the height above ground level. This variation is often modeled by the following relationship (Dunn 1986):

$$v_1 = v_2[z_1/z_2]^x \qquad (24\text{-}6)$$

where

v_2 = wind velocity at a reference height z_2;

v_1 = wind velocity at the height of interest, z_1; and

x = constant which depends on the nature of the terrain.

Values of x vary from 0.1 for a sandy terrain to 0.3 for a rough forest. A typical value of x for a grassy terrain is 0.17. Wind velocities are generally measured at a height of 33 ft (10 m).

The actual amount of energy that can be extracted from the wind is less than the theoretical amount of energy available. Betz showed that the theoretical limit on power extracted from the wind is 59% (Dunn 1986). A typical efficiency for a wind turbine is about 40%; that is, about 40% of the power available in the area swept by the wind turbine blades is converted to electricity. Actual wind turbine design must consider not only high efficiency at rated wind speeds, but also reasonably high efficiencies at lower wind speeds. Therefore, a goal of wind turbine design is to achieve a high weighted average efficiency over the design wind velocity range.

Wind energy systems can be configured as horizontal axis machines or vertical axis machines, as shown in Fig. 24-14. The majority of operating wind systems are horizontal axis machines. Horizontal axis refers to the orientation of the axis of rotation, which is parallel to the wind flow as shown.

24.2.2.1 Wind Resource. The economic viability of a wind energy project is highly dependent on the available wind resource. On a global scale, winds result from temperature gradients between the equator and poles and between land and seas. On a smaller scale, thermal winds can be generated by local thermal effects. Local factors such as high altitude, unobstructed terrain, lofty air flow height, and natural wind tunneling features cause some areas to have inherently higher wind speeds.

Wind data are usually classified in wind power categories as listed in Table 24-1 (Vosburgh 1983). Wind power classifications vary from 1, the lowest power density, to 7, the highest power density; areas designated Class 5 or higher are generally considered suitable for cost-effective wind energy projects. Advanced wind technology is expected to make regions with wind classes of 3 and above cost effective.

Regional wind resource trends for the United States are catalogued in the *Wind Energy Resource Atlas of the United States* published by the National Renewable Energy Laboratory (formerly the Solar Energy Research Institute) (Elliot et al. 1986). The *Atlas* provides wind resource maps for the United States, Puerto Rico, and the Virgin Islands.

Table 24-1. Windpower Classes as Utilized by the Department of Energy and Pacific Northwest Laboratory

	10 m (33 ft)		50 m (164 ft)	
Wind Power Class	Wind Power Density, Watts/m^2	Mean Wind Speed m/s (mph)	Wind Power Density, Watts/m^2	Mean Wind Speed m/s (mph)
1	0	0	0	0
2	100	4.4 (9.8)	200	5.6 (12.5)
3	150	5.1 (11.5)	300	6.4 (14.3)
4	200	5.6 (12.5)	400	7.0 (15.7)
5	250	6.0 (13.4)	500	7.5 (16.8)
6	300	6.4 (14.3)	600	8.0 (17.9)
7	400	7.0 (15.7)	800	8.8 (19.7)
	1,000	9.4 (21.1)	2,000	11.9 (26.6)

Note: Based on Rayleigh windspeed distribution and standard sea level conditions. To maintain the same power density, speed increases approximately 5% per 5,000 ft (3% per 1,000 m) of elevation.

Source: Vosburgh 1983.

24.2.2.2 Wind Turbine Devices. Wind turbine technology has developed significantly in the last 25 years. Large horizontal wind turbine concepts were developed in the 1970s under the US National Aeronautics and Space Administration's Horizontal Axis Wind Turbine Program. That program was aimed at developing progressively larger wind machines, beginning with the 100-kW MOD-0, followed by the 200-kW MOD-0A, the 2-MW MOD-1, the 2.5-MW MOD-2, and the 3.2-MW MOD-5B (Vosburgh 1983). The MOD-5B, on Oahu in Hawaii, is the largest horizontal axis wind turbine with a 320 ft rotor diameter. A 3.6-MW vertical axis machine is in operation at Cap Chut, Quebec. This Darrius type machine has a 308 ft center column and blades that are 205 ft apart at their farthest separation.

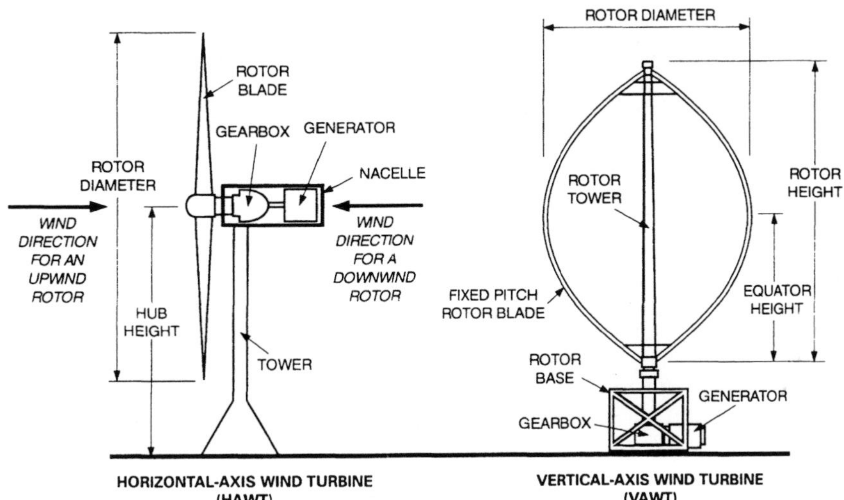

Fig. 24-14. Two basic wind turbine configurations.

Almost 80% of turbines worldwide are within the 50- to 150-kW range. These smaller machines were preferred because of material and joint strength, weight, and cost advantages. Recent advancements in technology have improved the reliability and efficiency of larger machines. Thus, the predominant choice of turbine size is expected to shift to medium-scale turbines, particularly those in the 250- to 500-kW range. Another advancement has been the development of a variable-speed wind turbine, which uses advanced power electronics to provide synchronized power to the grid while allowing variable rotor speed. It is projected that the variable speed capability together with advanced aerodynamic blades and a taller tower design should extract wind energy at least 10% more efficiently than current designs.

24.2.2.3 Wind Farms. Commercial or industrial wind power is usually provided by large groups of interconnected wind turbines, often called wind farms. Large wind farms are located in California at the Altamont Pass, San Gorgonio Pass, and the Tehachapi Mountains. There are over 16,000 wind turbines operating in California; these wind farms provide about 3% of the state's overall energy generation (SERI/SP-225-3555 1989). Wind farms have also been built or are being planned in Hawaii, Massachusetts, Maine, Oregon, Vermont, New Hampshire, Montana, and Minnesota.

24.2.3 Biomass Energy.

Biomass is regenerative (as opposed to fossil) organic material used for energy production. Sources for biomass fuel include terrestrial and aquatic vegetation, agricultural and forestry residues, and municipal and animal wastes. Like most renewable energy sources, biomass is a form of solar energy; however, biomass is distinct from thermal processes in that energy has been transformed by photosynthesis and stored as chemical energy of the organic material. Biomass can either be collected and burned directly or converted to other usable energy forms, such as gaseous or liquid fuels, or energy-intensive chemical feedstocks. The term biomass as a fuel category includes many of the waste categories discussed in Chapter 22, Resource Recovery. It also includes biomass farm concepts, where biomass is harvested solely as a fuel.

Biomass has a number of advantages over traditional fossil fuels. Its primary advantage is that it is renewable. It is also a low-cost fuel; many biomass sources are agricultural or industrial residues that, if not used for energy production, would result in disposal costs. Another advantage is the low sulfur content of biomass, typically $<0.1\%$ by weight compared to a range of about 1% to 5% for coal. Biomass also has a low (1%) ash content. This ash may be a saleable byproduct that can be returned to the soil as a fertilizer and soil conditioner. Furthermore, the growth and combustion cycle of biomass does not increase the atmospheric CO_2 level. (However, cultivation and harvest of biomass requires the use of fuels such as gasoline, which may increase the CO_2 level.) Biomass is much more reactive than coal, making it attractive as a feedstock for thermochemical gasification (Hall et al. 1993).

Biomass, however, also has a number of associated disadvantages as an energy source. Biomass is distributed in supply and has diverse physical and chemical characteristics. Some chemical constituents found in many biomass fuels can cause fouling or slagging problems or form unacceptable air pollutants. Biomass is generally bulky and expensive to transport; therefore, it must be produced, collected, and combusted or converted on a local basis. Furthermore, biomass used for energy can displace its use for food, materials, or chemicals. Even when the biomass feed stock plant is useful only for fuel, the production resources required, such as land and water, are fertile resources that are being diverted from the production of plants that could be used to produce food, animal feed, textiles, paper, construction products, or chemicals.

24.2.3.1 Biomass Resource. The biomass resource encompasses a large range of materials including waste products and agricultural and forestry products grown specifically for energy production. Because of the disperse nature of biomass, the most attractive biomass applications are those for which the biomass material is a waste from processes that gather biomass feedstock for other purposes. The practicality of using residues for energy is dependent on the sustainability of the primary industrial activity and the value of the residues for other uses such as fertilizer.

Dedicated biomass farms are another potential biomass resource. Typical biomass crops include short-rotation hardwoods, sorghum, napier grass, or types of cane. To provide a steady energy supply, biomass must either be produced year-round or harvested and stored. The most attractive regions of the United States for such dedicated farms appear to be the Gulf Coast, Great Lakes, and Pacific Northwest areas. Studies of currently planted cropland and idle cropland, with projections of future production rates and crop requirements for food and other purposes, indicate that biomass production for energy purposes could be as high as 26 quads (26×10^{15} Btus) per year (SERI/TP-260-3674 1990).

24.2.3.2 Biomass Energy Production Technologies. Biomass energy production generally uses direct combustion. However, biomass may also be converted to other solid, liquid, or gaseous fuel forms using one of two generic technology categories: thermochemical conversion, in which biomass is heated in the presence of reactive gases to produce fuel gases, oils, and char; and biochemical conversion, in which microbial processes break down the biomass into simple chemicals such as alcohols.

DIRECT COMBUSTION TECHNOLOGIES. In a direct combustion process, the biomass is burned to complete combustion, usually within a boiler. The thermal energy released is used to produce steam for process heating or for the generation of electricity. Currently, most biomass applications are fueled by waste products. Direct combustion technologies for these applications using unprocessed solid wastes include waterwall, refractory, rotary kiln, water-cooled rotary combustor, and controlled air furnaces; spreader stoker-fired boilers; suspension-fired boilers; fluidized bed boilers; and cyclone furnaces. These technologies are discussed in detail in Chapter 22, Resource Recovery.

THERMOCHEMICAL CONVERSION PROCESSES. Thermochemical processes for the conversion of biomass include pyrolysis and gasification. A principal difference between these processes is the amount of air or oxygen used relative to the quantity of fuel.

In pyrolytic conversion, biomass is reacted in an atmosphere containing <20% of the air theoretically required for complete combustion (that is, a substoichiometric environment). Gas, oil, and char are the products. The quantity and quality of the products depend on the type of reactor, type of biomass, and the conditions of reaction. The product gas, which contains CO and H_2 as well as other hydrocarbon gases, is a medium Btu gas (200 to 500 Btu/scf). The gas may be used directly as a fuel or it may be upgraded to pipeline quality substitute natural gas with a heating value of 1,000 Btu/scf. It may also be used as a chemical feedstock to synthesize other fuels such as methanol. The product oil is highly oxygenated and viscous with a somewhat lower heating value than that of residual oil. The char, primarily carbon, is also used as a fuel.

Biomass gasification takes place in an atmosphere containing 20% to 50% of the oxygen theoretically required for complete combustion. If air is used in the process, the product gas (containing H_2 and CO diluted with H_2O, CO_2, and N_2) is a low-Btu fuel (90 to 200 Btu/scf). If the process uses oxygen, the product gas is a medium-Btu fuel (200 to 500 Btu/scf) and does not contain N_2.

BIOCHEMICAL CONVERSION PROCESSES. Biochemical processes include anaerobic digestion and fermentation. Anaerobic digestion employs microbes to decompose biomass in the absence of oxygen to produce a medium-Btu gas comprising methane CH_4 and CO_2. The microbes, called anaerobes, live and grow in an oxygen-free environment by decomposing oxygen-containing compounds. The resultant medium-Btu gas may be used directly as a fuel, or it may be upgraded to pipeline quality substitute natural gas. Animal manure is a biomass form well suited as a feedstock for anaerobic digestion. Anaerobic digestion of lignocellulosic (woody) biomass is currently inefficient.

Fermentation employs microbes to decompose starches and sugars contained in biomass into ethanol. Ethanol is usable directly as a fuel or fuel additive, or it may be converted into higher alcohols and other organic fuels/chemicals.

Biochemical conversion technologies tend to be feedstock-limited, resulting in smaller capacity facilities compared to conversion plants proposed for fossil fuels, especially coal. Consequently, the cost of products produced from biomass tend to be more expensive than fossil-derived products because extreme economies of scale may not be achievable.

24.2.4 Ocean Energy Systems

The oceans of the world represent a vast resource for renewable energy. Three technologies for tapping this resource are ocean thermal energy conversion systems, wave energy systems, and tidal energy systems.

24.2.4.1 Ocean Thermal Energy Conversion. Ocean thermal energy conversion (OTEC) is an emerging technology concept for using the temperature difference between warm surface water (as high as 80° F [27° C]) and cold (as low as 40° F [4° C]) deep (3,000 ft [915 m]) water to generate electricity. Because of the relatively small temperature differences (only about 40° F [22° C]), OTEC plants exhibit very low efficiencies—in the range of 2.5%. An illustration of a floating OTEC plant is shown in Fig. 24-15.

OTEC concepts have been developed using two basic types of cycles. Closed cycle plants use a low boiling point working fluid such as ammonia. The working fluid is heated and vaporized by the warm surface water, expanded in a turbine generator, and condensed by deep cold water. Open cycle plants use seawater as the working fluid. The warm surface water is flashed to low-pressure steam, expanded in a turbine generator, and condensed by the deep cold water. The relatively small temperature difference between warm and cold seawater, and the resultant low system efficiency, require the pumping of enormous amounts of water for power generation. A 100-MW plant would require pumping about 75 million gpm of both warm and cold seawater. The cold water intake pipe for a 100-MW plant would range in length from 2,000 to 3,300 ft (Cavanagh et al. 1993).

OTEC may be shore-based in locations with steep offshore slopes. Shore-based OTEC plants can be tied to existing electric transmission grids. Offshore plants can be floating platforms or tower-mounted structures. Offshore OTEC plants require undersea power cables connecting to an onshore transmission grid, or they can produce a transportable product such as hydrogen.

The French scientist D'Arsonval first proposed the OTEC concept in 1881 (Cavanagh et al. 1993). An early system using the OTEC concept was built by George Claude, a French physicist, and his partner, Paul Boucherot, in 1926

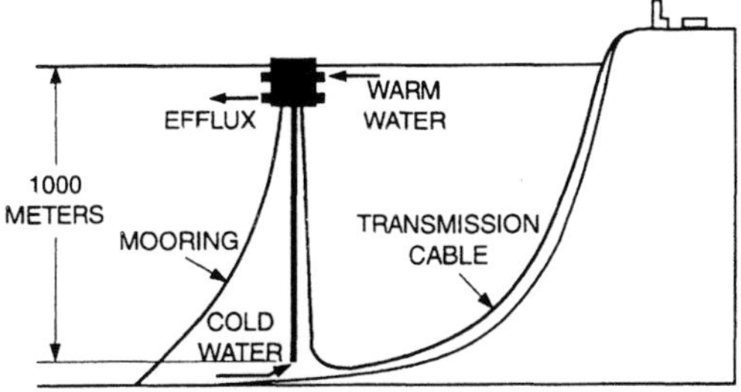

Fig. 24-15. Schematic illustration of an ocean thermal energy conversion (OTEC) floating platform. (Granted with permission from *Renewable Energy: Sources for Fuels and Electricity*, edited by Johannson, Kelly, Reddy, and Williams. Island Press, Washington, D.C., and Covelo, CA.)

(G. Claude 1930). That onshore plant generated 22 kW gross; however, auxiliary power requirements for the plant exceeded the gross power generation.

The closed cycle OTEC concept was evaluated in 1963 (Anderson and Anderson 1965) and developed further during the 1970s. A 50-kW demonstration unit, the MINI OTEC, was built in 1979 and tested for a short period off the coast of Hawaii. In 1981, the US Department of Energy funded OTEC-1, a deep ocean test facility. This unit, which was moored 12 miles off the coast of Hawaii, conducted tests on heat exchangers and biofouling prevention measures. The Japanese built and tested a 100-kW (gross) onshore OTEC facility in 1981. Another 50-kW unit was also built by the Japanese. Subsequent OTEC efforts have included conceptual design studies for large OTEC units.

The best potential OTEC sites are in tropical and subtropical areas because of the existence of higher temperature differences between surface and deep water. The ocean surface area that could support OTEC is about 23 million square miles (Cavanagh et al. 1993). Although there are very large resources available for OTEC, it appears that the large estimated capital cost may delay implementation.

24.2.4.2 Ocean Wave Energy. Wave energy systems convert kinetic energy and potential energy associated with ocean waves into electricity. The kinetic energy of waves is associated with the velocity of the water mass; the potential energy is associated with the displacement of water above or below the mean sea level. Wave velocity is related to wavelength—the greater the velocity, the larger the wavelength. The power of a wave is the rate at which energy is transferred across a unit length perpendicular to the wave direction; wave power is typically expressed in kW/m. Wave power remains relatively constant in deep water; however, in water shallower than half a wavelength, substantial power is lost through friction with the sea bed (Cavanagh et al. 1993).

The greatest resource for wave power typically occurs between 40 degrees and 60 degrees latitude in each hemisphere, although seas experiencing regular tradewinds at 30 degrees latitude can also exhibit suitable wave power for electrical power generation. The west coasts of the United States and Europe, and the coasts of New Zealand and Japan are considered suitable for wave energy extraction (Cavanagh et al. 1993).

Although devices for generating power from ocean waves have been conceptualized for more than 100 years, the increase in oil prices in 1973 intensified development of wave energy concepts; several of these are shown in Fig. 24-16. S. H. Salter described a wave energy conversion concept called the Salter Duck in 1974 (Salter 1974). The Salter Duck is a pitching device, as shown in Fig. 24-16. Measurements with small laboratory devices indicated a conversion efficiency of 80% of the wave energy.

Another wave-energy system concept is the British Sea Clam. The Sea Clam, which is a surge device as shown in Fig. 24-16, consists of a hollow, floating chamber and a collapsible bladder connected by a hollow duct containing an air turbine. An approaching wave collapses the bladder forcing air past the turbine and generating electricity.

Another concept, the oscillating water column, is typified by the Kvaerner wave-energy plant in Norway. The Kvaerner plant uses a water oscillation chamber blasted out of a rock cliff. The rise and fall of waves in this chamber cause air in the chamber to pass up and down through a Wells turbine, thereby generating electricity. The Wells turbine spins in the same direction while accepting air from either above or below the turbine.

24.2.4.3 Tidal Energy Conversion. Tidal energy conversion systems extract energy from the changes in the level of ocean water during tidal cycles. A tidal energy conversion plant typically consists of a tidal basin created by a dam, a turbogenerator, and a sluice gate in the dam to allow the tidal flow to enter or leave the tidal basin (Carmichael et al. 1986).

Tides are caused by the interacting forces of the gravitational pulls of the moon and sun. The forces involved are complex, but the moon is approximately twice as effective as the sun in causing tides. The rotation of the earth within the gravitational field of the moon results in the most apparent variable of the tidal cycle, its period. Times and amplitudes of high and low tide are predictable. They vary considerably with geographic location and season. In many of the coastal areas of the world, tides rise and fall each 12 hour and 35 minute period.

Tidal energy generation is approximately proportional to the square of the tidal amplitude. Tidal amplitude varies significantly from location to location. A 16 ft minimum tidal amplitude is considered necessary for a tidal energy conversion system to be economically feasible. Tidal amplitudes as high as 50 ft (17 m) exist in the Bay of Fundy on the east coast of Canada (Dunn).

Tidal energy using conventional hydroelectric technology has been demonstrated on a large scale. A 240-MW commercial plant has been operating across La Rance Estuary in northwest France since 1962 (Carmichael et al. 1986). At this site, the tide occurs twice daily and has a maximum amplitude of about 46 ft (14 m) and a mean amplitude of 28 ft (8.5 m). The Rance dam, which has a rated head of 18.5 ft (5.65 m), contains about 2 billion cubic ft (184 million m^3) of water with a mean basin area of 8.5 square miles (22 square km). It uses twenty-four 10 MW turbine–generators to generate 500 GWh of electricity annually. This provides a capacity factor of about 24%.

The first tidal power plant in North America is an 18-MW plant at Annapolis Royal on the Annapolis River on the Nova Scotia coast of the Bay of Fundy in Canada. This project uses an existing flood control dam on the Annapolis River.

24.2.5 Geothermal

Geothermal energy is thermal energy associated with the elevated temperature of the interior of the earth. There are four principal forms of geothermal energy resources:

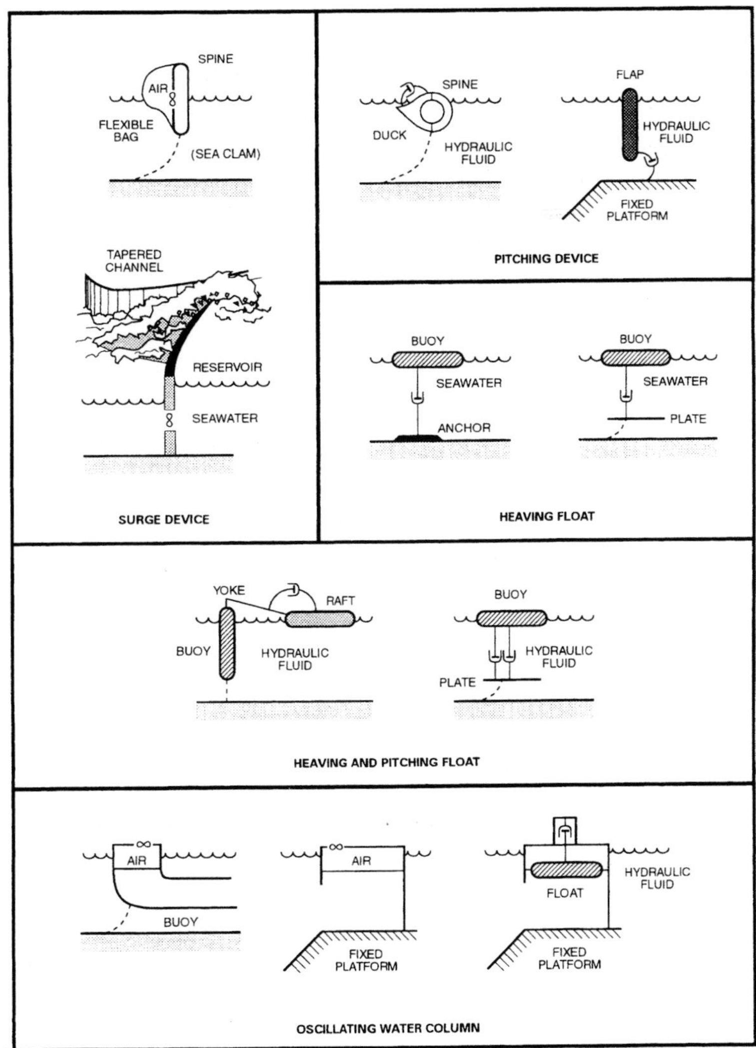

Fig. 24-16. Wave energy concepts. (Granted with permission from *Renewable Energy: Sources for Fuels and Electricity*, edited by Johannson, Kelly, Reddy, and Williams. Island Press, Washington, D.C. and Covello, CA.)

- Hydrothermal,
- Geopressured,
- Hot dry rock, and
- Magma energy.

Currently, only hydrothermal resources that have naturally occurring hot water or steam are being developed economically for electrical power generation. In vapor-dominated hydrothermal fields, dry steam exists in fractures, solution channels, and cavities in the reservoir rock. In liquid-dominated fields, a fractured porous reservoir is saturated with hot water.

Geopressured zones are reservoirs under extremely high pressure caused by the weight of overlying sediments. These reservoirs often contain mixtures of hot salt water and methane gas. Potentially, energy could be obtained from these zones by burning the methane and using the pressure and heat from the hot water.

Hot dry rock resources have no permeable aquifer overlying the heat source; therefore, there is no naturally occurring hot water or steam. Tapping the hot dry rock resource requires fracturing the rock and injecting water. About 75% of the geothermal resource in the United States is of the hot dry rock type.

Magma resources are associated with magmatic intrusions to within a few kilometers of the earth's surface, and are characterized by temperatures of 1,100 to 1,800° F (600 to 1,000° C). Tapping magma resources, as when tapping hot dry rock resources, requires water injection to extract the heat.

24.2.5.1 Geothermal Resource. Practical tapping of the geothermal resource is generally considered to be limited to a depth of about 30,000 ft (10 km). It is estimated that the worldwide geothermal resource to this depth, with temperatures greater than 185° F (85° C), is 100 million quads, where a quad is 10^{15} Btu; the United States resource is estimated to be 10 million quads (SERI/TP-260-3674). The US Geological Survey estimates that there are sufficient high-temperature (>300° F [150° C]) resources within the

United States to produce 23,000 MW over a 30-year period with only modest technology improvements. The estimated resource in the United States is 95,000 to 150,000 MW for 30 years with more advanced technologies capable of using lower temperature (>200° F [90° C]) geothermal resources (SERI/TP-260-3674 1990).

24.2.5.2 Conversion Technologies. The technologies used for power generation differ depending on the type of geothermal resource being tapped.

HYDROTHERMAL CONVERSION TECHNOLOGIES. Three technologies are typically used for production of electricity from hydrothermal resources: dry steam systems, single/double flash systems, and binary cycle systems. All technologies comprise three basic elements: a production well, which brings the geothermal resource to the surface; a conversion system, which converts the geothermal energy into electricity; and an injection well, which recycles the spent geothermal fluids into the reservoir. A perculation pond may be used instead of the injection well.

Dry steam systems are based on conventional steam power plant technologies that have been adapted for geothermal applications. A simplified schematic diagram of a dry steam system is shown in Fig. 24-17. Dry steam systems typically exploit geothermal reservoirs containing dry, superheated steam at temperatures exceeding 392° F (200° C). Slightly superheated steam from the production wells operate a condensing steam turbine–generator. The condensate is used as makeup water for a mechanical draft cooling tower. Cooling water blowdown is then reinjected back into the geothermal reservoir.

Dry steam systems are commercially available and economically competitive. The Geysers in California is the largest developed dry steam system in the world but only one of many that have been developed internationally. Dry steam power plants account for approximately 3,000 MW worldwide, with two thirds of that capacity operating at the Geysers (California Energy Commission).

Unfortunately, dry steam resources are relatively rare, and tapping other types of hydrothermal resources require more complex techniques. Flash systems are used for high temperature (>350° F [180° C]) liquid-dominated geothermal reservoirs. A simplified schematic of a single flash system is shown in Fig. 24-18. A two-phase mixture of liquid brine and steam is routed to a separator. Pressure is reduced in the separator, causing the fluid to flash into steam. The separated, scrubbed steam powers a turbine–generator. Geothermal fluid that does not flash is separated from the steam and is disposed of into an injection well or perculation pond.

A double flash system flashes the separated brine a second time to produce lower temperature steam for a separate steam turbine, or the steam may be introduced into the high-pressure turbine through a second admission port.

As of 1989, flash-steam plants accounted for approximately 1,500 MW of power generation worldwide (California Energy Commission 1991). Additional development is expected because liquid-dominated geothermal resources are more numerous than dry steam resources. However, the capital costs for single or double flash systems are higher than those for dry steam systems. Resource development costs must be reduced to make them economically attractive.

Binary cycle systems are the preferred technology for development of liquid-dominated reservoirs that do not have a sufficiently high temperature to flash steam. A simplified binary cycle system diagram is shown in Fig. 24-19.

Hot brine is pumped from the reservoir through counterflow heat exchangers. A working fluid such as a fluorocarbon or isobutane is vaporized in the heat exchangers. The vapor is piped to the turbine generator, expanded, condensed, and returned to the heat exchanger.

Worldwide, there are approximately 85 binary cycle plants with a total capacity of 126 MW (California Energy Commission 1991). Binary cycle systems potentially have broad applicability. About 80% of the hydrothermal resources in the United States are of moderate temperature appropriate for binary cycle system application. Likewise,

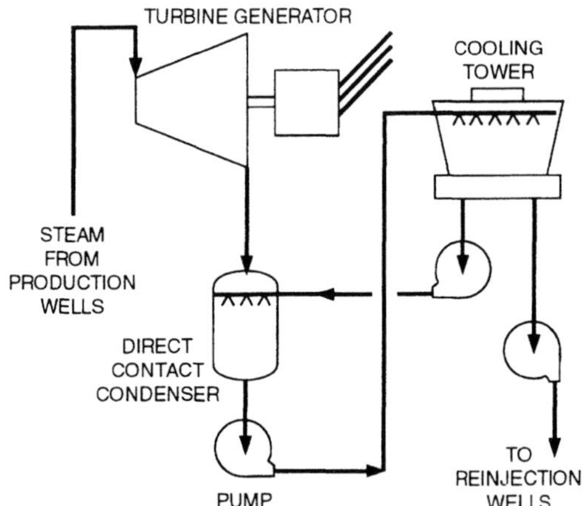

Fig. 24-17. Schematic of a dry steam system. (From American Society of Mechanical Engineers. Used with permission.)

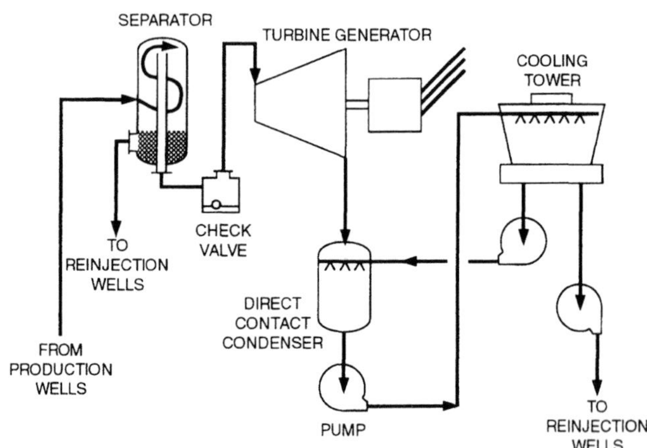

Fig. 24-18. Schematic of a single flash system. (From American Society of Mechanical Engineers. Used with permission.)

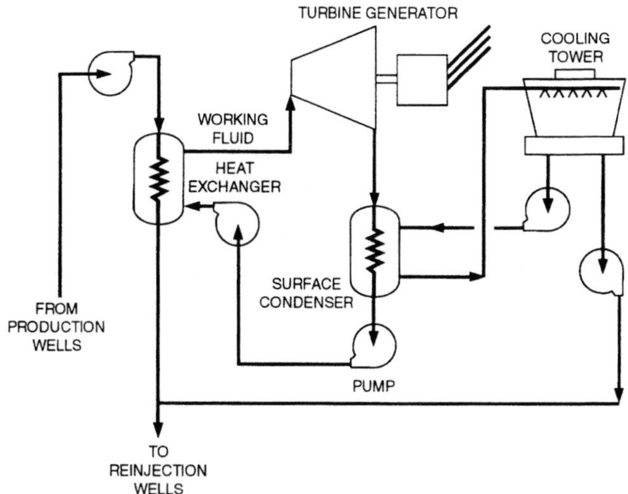

Fig. 24-19. Schematic of a binary cycle system. (From American Society of Mechanical Engineers. Used with permission.)

the binary systems are expected to have significant application for geopressure and hot dry rock reservoirs.

GEOPRESSURED CONVERSION TECHNOLOGY. Use of the geopressured resource requires separation of the methane and water mixture. The methane powers a heat engine or is sold commercially; the hot water is used in a binary cycle. Geopressured systems are expected to be used more for direct applications, such as enhanced oil recovery, than for power generation (Palmerini 1993).

HOT DRY ROCK CONVERSION TECHNOLOGY. To tap the energy contained in the underground hot dry rock, the rock must first be fractured by conventional oil and natural gas drilling techniques. Once a reservoir has been created in the hot dry rock, inlet and outlet water pipes connect the reservoir with surface power generation equipment. Circulating hot water is heated in this reservoir and generates power by one of the three hydrothermal conversion technologies presented previously.

MAGMA CONVERSION TECHNOLOGY. To exploit the energy contained in magma layers near the earth's surface, it will be necessary to tap into the magma chamber. Once penetrated, fluid would be circulated in the chamber, thus creating a type of down-hole heat exchanger.

24.2.5.3 Commercial Status. As of 1989, there were 5,892 MW of geothermal generating capacity installed worldwide, of which 2,770 MW were installed in the United States (Palmerini 1993). Dry steam systems were the first geothermal applications to be developed. The world's first dry steam system was initiated in 1904 with small scale experiments at Larderello, Italy. The Larderello system has grown to about 500 MW with a projected capacity of 900 MW in 1995.

Dry steam geothermal energy has been used for commercial generation of electricity in the United States since 1960 when Pacific Gas and Electric, with others, built a 12-MW power plant at the Geysers in northern California (Lacy 1988). This geothermal field supported plants generating over 1,800 MW of electricity, although production has declined in recent years. Another dry steam plant is the Sacramento Municipal Utility District Geothermal Plant No. 1, which is rated at 72 MW.

A liquid-dominated geothermal system has been in continuous operation at Wairakei, New Zealand, since 1956. Another system using the liquid-dominated resource was San Diego Gas & Electric's 46.6-MW Heber binary cycle plant; that plant is no longer in operation.

Applications of geopressured zones and hot dry rock technologies are still being developed. Research efforts are aimed at identifying potential sources and developing the necessary energy conversion technologies.

Environmental considerations for geothermal plants include possible emissions of H_2S and other gaseous pollutants as well as possible air toxics. Plant design must include equipment to control these emissions.

24.3 ENERGY STORAGE

Energy storage technologies allow utilities to "load level," that is, to use energy generated during off-peak hours to supply requirements during peak hours. Load leveling is facilitated by the stored energy when it is charged with off-peak, lower cost power and discharged during peak, higher cost power periods. In this type of application, storage concepts are economical when the cost of storage system construction, operation, and maintenance are offset by the differential between peaking and baseload energy costs or the avoidance of adding new capacity. If economically competitive, storage systems may also be useful in combination with intermittent energy sources, a common trait of many renewable energy resources. Furthermore, storage systems may provide additional system advantages, such as spinning reserve, and area frequency and voltage control.

Two energy storage technology concepts are discussed in this chapter: battery energy storage systems and compressed air energy storage. Pumped hydroelectric storage is a commercially available storage concept based on conventional hydroelectric technology, but it is not discussed in this book. Other developmental storage technologies include superconducting magnets and flywheels; these technologies are not discussed in this chapter.

24.3.1 Battery Storage

A battery storage system comprises the battery, dc switchgear, dc/ac converter/charger, transformer, ac switchgear, and a building to house these components. During the utility peak periods, when energy costs are the highest, the battery system discharges and supplies power to the utility ac power system. A typical discharge period is 4 to 5 hours. During off-peak hours, the battery system is recharged and then remains on "float" until the next peak period, thus completing one cycle of operation.

Battery energy storage systems have several advantages in addition to their load leveling capability. Because battery storage systems can be added in small increments, they offer a utility a means of matching peak load growth. Battery storage systems also have dynamic source benefits because they provide spinning reserve, area frequency and voltage control, and increased system reliability. Because of their small incremental sizes and because battery storage systems are environmentally compatible in virtually any area, they can be located near customer load centers. This results in a reduction of electrical transmission losses.

A primary disadvantage of battery energy storage systems is the relatively high initial cost. Furthermore, existing battery technology requires replacement every 8 to 10 years.

Currently, the only commercially available battery appropriate for utility use is the lead-acid battery, which uses lead electrodes and a sulfuric acid electrolyte. Advanced batteries such as the sodium–sulfur battery and the zinc–bromine battery are being developed for possible utility application.

Two lead-acid battery storage systems for supplying power to a utility ac power system have been operating successfully since 1986: a 17-MW, 14.4-megawatt-hour (MWh) facility in West Germany and a 1-MW, 4-MWh facility in Osaka, Japan. In 1988, a 10-MW, 40-MWh battery energy storage system began operation at the Southern California Edison Chino substation (Morris 1988). In 1993, Puerto Rico Electric Power Authority started up a 21-MW, 14-MWh battery system at its Sabana Llana substation near San Juan, Puerto Rico.

Typical round trip (ac to ac) efficiencies for battery storage systems are in the range of 72%. This efficiency includes battery round trip (dc to dc) of about 78% and power conditioning system efficiencies of about 94%.

24.3.2 Compressed Air Energy Storage

Compressed air energy storage (CAES) is a technology in which energy is stored in the form of compressed air in an underground cavern. Air is compressed during off-peak periods, stored in a cavern, and then used on demand during peak periods to generate power with a turbogeneration system.

A typical CAES unit consists of five basic components:

- Compressor train (including compressor, intercoolers, and after-cooler);
- Motor generator;
- Turbine expander train (including expanders and combustors);
- Recuperator; and
- Underground cavern.

Compression of the air is powered by a motor that uses off-peak electricity from the grid. During peak hours, the compressed air from the cavern is extracted and preheated in the recuperator. Once heated, the air is mixed with oil or gas, which is burned in the combustor. The hot gas is expanded in the turbine to generate electricity.

During the power generation (storage discharge) mode of operation, the heat rate for a CAES system is very low because no power is required for compression. An important performance parameter for a CAES system is the charging ratio, which is defined as the ratio of the electrical energy required to charge the storage versus the electrical energy generated during discharge (the number of kWh input in charging to produce 1 kWh output). A low charging ratio results in low off-peak electrical energy requirements during the charging cycle.

Fast startup is another advantage of CAES. A CAES plant can provide a startup time of about 9 minutes for an emergency start and about 12 minutes under normal conditions (EPRI 1991). By comparison, combustion turbine peaking plants typically require 20 to 30 minutes for a normal startup.

A significant contributor to the cost of a CAES plant is the construction of the underground cavern. Three types of geological formations used for compressed air storage are salt dome, aquifer, and rock caverns. Applications of these storage media are illustrated in Fig. 24-20. In addition to the geological formation classifications, there are two classes of cavern design concepts: constant volume (also called uncompensated) and constant pressure (also called compensated). In a constant volume cavern, the air pressure is allowed to drop as air is withdrawn from storage. In a constant pressure cavern, water from a surface reservoir displaces the compressed air to maintain constant pressure in the cavern.

The first commercial scale CAES plant in the world is the 290-MW Huntorf, Germany, plant operated by Nordwest Deutsche Kraftwerke (NWK) since 1978. The Huntorf plant runs on a daily cycle in which it charges the air storage for 8 hours and provides plan generation for 2 hours. The plant has reported high availability at 86% and a starting reliability of 98% (Gaul et al. 1993). The Huntorf plant has a salt cavern.

The second commercial scale CAES plant was built by the Alabama Electric Cooperative, Inc., in McIntosh, Alabama (Makonsi). This plant has the maximum existing CAES cavern capacity of 19 million ft^3. It began operation in 1991 and provides 110 MW of power generation. The cavern for the McIntosh plant was mined from a salt dome by dissolving salt with fresh water. The cavern, which is 230 ft in diameter, 1,000 ft tall, and 1,500 ft below grade, supplies compressed air supporting generation of 100 MW for 26 hours. The CAES plant has a full load net plant heat rate of 4,568 Btu/kWh (HHV) with a charging ratio of 0.82. At 25 MW, the performance is significantly reduced, with a heat rate of 5,253 Btu/kWh and a charging ratio of 1.3.

A schematic of the McIntosh plant is shown in Fig. 24-21. The McIntosh facility has high-pressure, intermediate-pressure, and low-pressure compressors; a motor generator; and high- and low-pressure expanders. The compressors and expanders are connected to the motor/generator by clutches. These clutches are used to decouple the compressors during

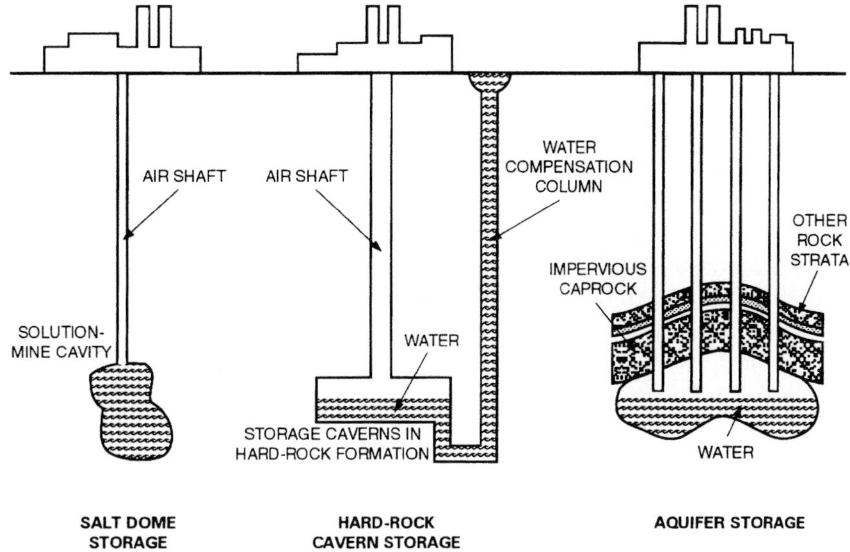

Fig. 24-20. Compressed air energy storage (CAES) cavern designs. (From Modern Power Systems. Used with permission.)

the generation cycle and to decouple the expanders during the compression cycle.

In addition to the NWK and McIntosh CAES facilities, a 35-MW CAES unit is under construction in Japan. Israel also has a 100-MW CAES unit under construction; this unit uses an aquifer cavern for storage.

24.4 FUSION

The potential of fusion energy is enormous because of the vast fuel reserves available from water or tritium obtained from reactions of lithium.

Fusion power involves the release of energy from the fusing of light nuclei such as deuterium (${}^{1}_{2}H$) and tritium (${}^{3}_{1}H$) into heavier nuclei, such as helium (${}^{4}_{2}He$). The heavier nuclei have a larger average binding energy per nucleon than their respective lighter nuclei parents. This fusing results in a more stable internal arrangement with the release of large amounts of energy. The fusion reactions of major interest include the following:

$$ {}^{2}_{1}H + {}^{3}_{1}H \rightarrow {}^{4}_{2}He + {}^{1}_{0}n + 17.6\ \text{MeV}^{1} \qquad (24\text{-}7)$$

$$ {}^{2}_{1}H + {}^{2}_{1}H \rightarrow {}^{3}_{2}He + {}^{1}_{0}n + 3.2\ \text{MeV} \qquad (24\text{-}8)$$

$$ {}^{2}_{1}H + {}^{2}_{1}H \rightarrow {}^{3}_{1}H + {}^{1}_{1}H + 4.0\ \text{MeV} \qquad (24\text{-}9)$$

$$ {}^{2}_{1}H + {}^{3}_{2}He \rightarrow {}^{4}_{2}He + {}^{1}_{0}p + 18.2\ \text{MeV} \qquad (24\text{-}10)$$

A large amount of energy is required to initiate fusion and overcome the strong electrostatic repulsion forces that tend

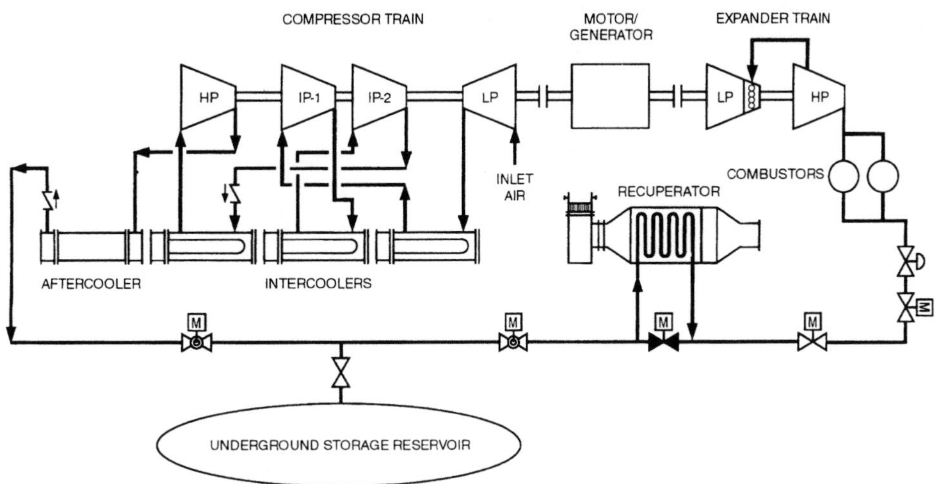

Fig. 24-21. Alabama Electric Cooperative (AEC) 100-MW compressed air energy storage (CAES) plant. (From American Society of Mechanical Engineers. Used with permission.)

---

[1]One million electron-volts (MeV) is equivalent to 1.5×10^{-16} Btu (1.6×10^{-13} joule).

to repel the positively charged light nuclei. Energy must be supplied to raise the kinetic energy of the light nuclei to overcome the repulsion forces. To sustain a deuterium–tritium reaction [Eq. (24-7)], a temperature greater than 50 million K is required. A temperature >500 million K is required for the deuterium–deuterium reaction [Eqs. (24-8) and (24-9)].

The deuterium–tritium reaction is the fusion process of major interest since it takes place at lower reaction temperature than the other fusion reactions and is a more attainable reaction. Approximately 80% of the 17.6-MeV reaction energy is imparted to the resulting neutrons (14.1 MeV), while the remaining 20% is contained by the helium-4 nucleus (3.5 MeV). At the high temperatures involved, the hydrogen isotope atoms are stripped of their electrons and a plasma consisting of positively charged ions and free negative electrons is formed. The plasma, as a whole, is neutrally charged since the positive charge ions and negative charged electrons balance.

Two concepts are under research and development for attaining the high fusion temperatures and thus sustaining a fusion reaction. The first concept, magnetic confinement, involves the use of magnetic forces to provide a magnetic bottle to confine and compress the light nuclei plasma. The second concept, inertial confinement, involves using a driver such as laser light or ion particle beams to impact, compress, and fuse small pellets of light nuclei fuel. Magnetic confinement is receiving the larger research effort.

24.4.1 Magnetic Confinement

Magnetic confinement involves trapping the heated plasma within magnetic lines of force, essentially creating a magnetic bottle. The leading concept for magnetic confinement is the tokamak,[2] which is a doughnut-shaped device in which magnetic field lines, twisting around and surrounding the torus, close on themselves to confine the plasma. The plasma consists of a series of charged particles that resist crossing the magnetic field lines. The major magnetic component fields consist of a toroidal field and a poloidal field. Figure 24-22 shows the two fields involved. The toroidal field is produced by passing an electric current through conducting metal rings located outside of the toroidal chamber. The poloidal field is produced by inducing current flow through the plasma which forms magnetic field lines as circles around the plasma.

24.4.1.1 Plasma Heating. Heating of the plasma to fusion temperatures involves several heating mechanisms. One mechanism uses the induced current itself. This current generates the poloidal field and increases the temperature by ohmic (resistance) heating. Resistance heating is limited to between 20 and 30 million °F (11 to 17 million °C) (Glasstone 1980). Another plasma heating mechanism involves neutral-beam heating, which is the injection of beams of

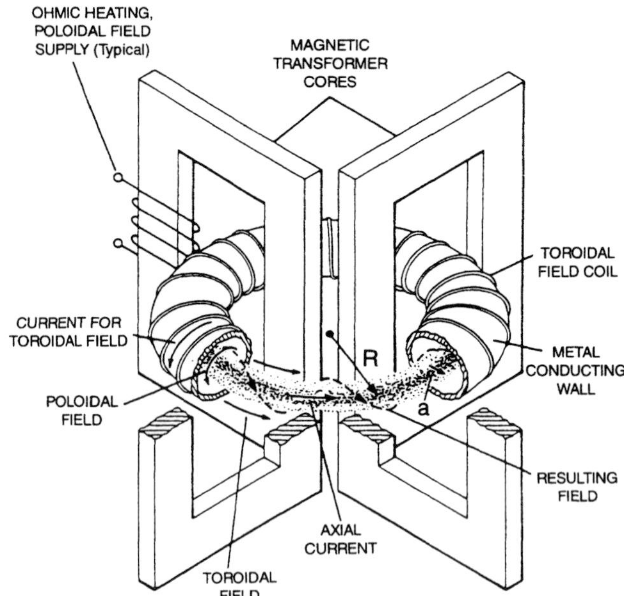

Fig. 24-22. Magnetic field containment concept.

high-energy neutral atoms directly into the plasma. The neutral atoms collide and transfer some of their energy to the plasma ions, increasing the plasma temperature.

Radiofrequency waves are also used for plasma heating. These waves are produced outside of the torus by an oscillator and directed to the plasma, transferring part of their energy to the plasma. The plasma particles collide with other particles, increasing the plasma kinetic energy and temperature.

Plasma compression is also used for plasma heating. The magnetic field strength is increased, compressing and heating the plasma.

24.4.1.2 Tokamak Fusion Test Reactor. The Tokamak Fusion Test Reactor (TFTR), which was completed in 1981 at the Princeton Plasma Physics Laboratory, is expected to be the first United States fusion device to demonstrate fusion break-even, where thermonuclear power generated equals or exceeds power input. Figure 24-23 shows the TFTR. Along with ohmic heating, the TFTR uses both neutral beam injection and radiofrequency waves for plasma heating. The magnets used for the toroidal and poloidal fields are superconducting, with cooling provided by liquid helium. TFTR has undergone physics and engineering testing with deuterium fuel for computer physics modeling and verification of the complex processes and interactions involved. If tests continue as planned and US Department of Energy funding continues, tritium will be added in 1994/95 leading to fusion energy break-even.

24.4.1.3 Tokamak Power Reactor. Figure 24-24 shows a conceptual tokamak reactor device for producing electrical energy. Coolant such as water or helium gas removes heat

[2]*Tokamak* is a Russian acronym for toroidal chamber with magnetic coils.

Fig. 24-23. Tokamak fusion test reactor (TFTR). (From Princeton Plasma Physics Laboratory. Used with permission.)

from the first wall and Li blanket material and transfers it to the steam generator. Superheated steam at conditions of 925 psig (6.4 MPag), 570° F (299° C) is produced by the steam generator to power a conventional turbine–generator. The generator net output is approximately 1,000 MWe. The plant efficiency is 39%. The auxiliary power load for plasma heating systems, coolant pumps, and other balance of plant loads is approximately 20% of gross generation.

24.4.1.4 Technological Maturity. Magnetic confinement devices such as the tokamak are still in research and development. Significant technical problems, including plasma confinement techniques, material qualifications, first wall thermal and radiation effects, optimum blanket and shield materials, and tritium control and handling, require more development. Commercial applications of this technology will not occur until well into the next century.

There are active fusion programs ongoing in the United States, Japan, Canada, Russia, and Europe. The United States has the TFTR, Japan the JT 60 tokamak, Russia the T15 tokamak, and Europe the Joint European Tokamak or JET. JET was constructed by a consortium of European nations. Like the TFTR, JET expects to demonstrate energy break-even in the mid-1990s.

The next tokamak test reactor to be built will most likely be the International Thermonuclear Engineering Test Reactor (ITER), shown in Fig. 24-25. It is being designed and constructed under a cost sharing program with the international community including the United States, Japan, Russian Republic, and the European community. Conceptual design activities have been completed and detailed design was started in 1992 with completion planned for the late 1990s. Construction is expected to be completed around 2005 if funding is available and the necessary national commitments are made following detailed design. ITER is intended to demonstrate the scientific and technological feasibility of fusion, including the demonstration of controlled ignition and extended burn of deuterium–tritium plasmas. The total cost of ITER for design, materials, and construction is approximately \$7 billion.

24.4.2 Inertial Confinement

Inertial confinement involves compressing a small deuterium–tritium pellet to the 180 million ° R (100 million K) reaction temperature such that the reaction time in the fuel sphere is shorter than the disassembly time. The disassembly time is calculated to be approximately 10^{-10} seconds (Glasstone 1980).

To achieve this type of compression, a driver such as a laser or particle beam is used. The laser or particle beam heats the pellet very rapidly, vaporizing the surface shell of the pellet. The surface shell is ablated (carried or blown off) rapidly at a high velocity in the outward direction. The remainder of the pellet undergoes an opposite reaction and is imploded and compressed in the radial direction. The compression raises the deuterium–tritium to its reaction temperature which initiates fusion.

The inertial confinement program (ICP) is centered around national defense activities for supporting nuclear weapons development and weapons effort testing. An objective of the program is achieving small-scale thermonuclear explosions

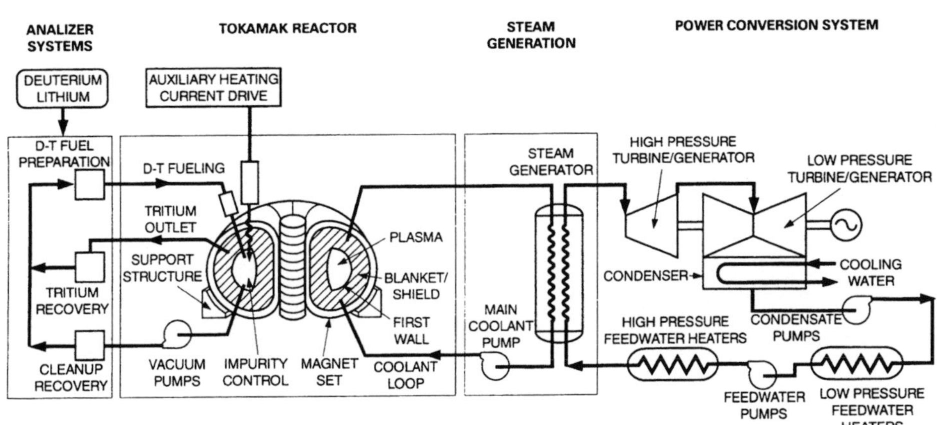

Fig. 24-24. Simplified schematic for a tokamak power reactor.

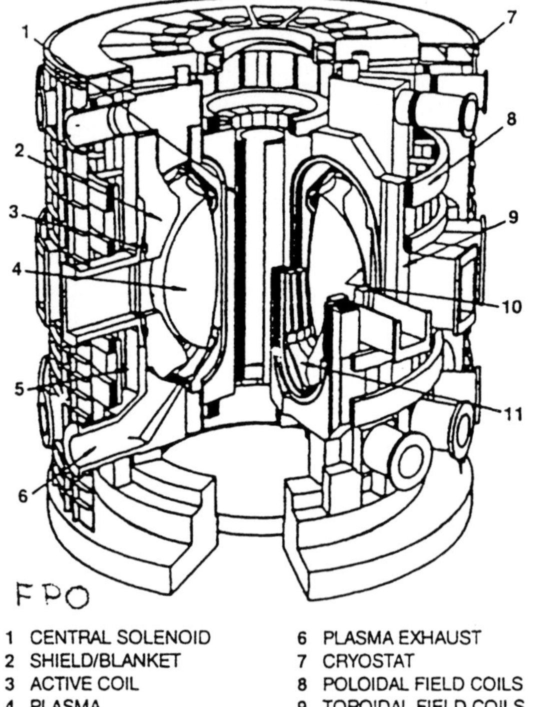

Fig. 24-25. International thermonuclear engineering test reactor (ITER).

in the laboratory. These efforts may lead to development of commercial fusion power generation.

It is difficult to project the direction of the civilian commercial program development following break-even demonstration. Construction of an engineering test reactor for ignition and burning demonstration as well as for electrical energy demonstration would be a minimum requirement prior to commercial reactors. Such a program would occur well into the next century.

24.4.3 Cold Fusion

In March 1989, Martin Fleischmann and Stanley Pons, researchers working at the University of Utah, announced the discovery of nuclear fusion in an arrangement of simple electrochemical cells. Each cell consisted of a palladium cathode surrounded by a spiral platinum anode immersed in a beaker containing heavy water (deuterium oxide) and lithium salts. The cells were connected to a battery. Fleischmann and Pons stated that the cells produced more than four times the power that was added to the cells and this energy release could only come from a deuterium–deuterium fusion reaction. Since the process took place at room temperatures, the process was called cold fusion.

It was postulated that the deuterium ions collecting on the palladium cathode migrated into the palladium crystal lattice, where they become very tightly packed and fused. Since the heat produced did not accompany the expected release of neutrons and radioactivity from the known fusion reactions of deuterium, it was postulated that the following, less likely reaction was occurring:

$$ {}^{2}_{1}H + {}^{2}_{1}H \rightarrow {}^{4}_{2}He + \gamma + 23.85 \text{ MeV} \qquad (24\text{-}11) $$

If this fusion process was correct and cold fusion could be harnessed for energy production, the potential impact could be enormous. Cold fusion could provide the world with an essentially inexhaustible fuel at a much lower capital cost than seen with magnetic or inertial fusion machines.

Following the Fleischmann and Pons announcement, researchers from around the world attempted to repeat their results; the majority of scientists were unable to do so. In November of 1989, a Department of Energy Research Advisory Board concluded that the present evidence of cold fusion was not persuasive and no special programs should be funded.

However, research has continued with more thorough, better instrumented experiments benefiting from previous lessons learned. Results have been mixed and not always repeatable; however, excess heat, neutrons, tritium, and protons have been observed in many of the various experiments, although not always in the same experiment (Storms 1991). Research suggests that some type of fusion mechanism (or multiple mechanisms) is occurring within the palladium lattice or on the metal surface. Once the mechanisms are understood, the question as to whether a useful energy source can be developed may be answered.

24.5 REFERENCES

American Society of Mechanical Engineers (ASME). 345 East 47th Street, New York, NY 10017-2392.

Anderson, J. H. and J. H. Anderson, Jr. 1965. Power from the sun by way of the sea. *Power*, Vol. 109. No. 2. February 1965, pp. 63–66.

California Energy Commission (CEC). 1991. *Energy Technology Status Report.* Appendix A, Vol. II. June 1991.

Carmichael, A. D., E. E. Adams, and M. A. Glucksman. 1986. *Ocean Energy Technologies: The State of the Art.* AP-4921, Research Project 1348-28. Massachusetts Institute of Technology (MIT), Final Report Prepared for Electric Power Research Institute, November 1986.

Cavallo, Alfred J., Susan M. Hock, and Don R. Smith. 1993. Wind energy: technology and economics. *Renewable Energy: Sources for Fuels and Electricity.* Island Press, Washington, D.C., pp. 121–156.

Cavanagh, James E., John H. Clarke, and Roger Price. 1993. Ocean energy systems. *Renewable Energy: Sources for Fuels and Electricity.* Island Press, Washington, D.C., pp. 513–547.

Claude, G. 1930. Power from the tropical seas. *Mechanical Engineering.* Vol. 52, pp. 1039–1044.

De Laquil, P., D. Kearney, M. Geyer, and R. Diver. 1993. Solar-thermal electric technology. *Renewable Energy: Sources for Fuels and Electricity*, Island Press, Washington, D.C.

Douglas, John. 1991. Fuel cells for urban power. *EPRI Journal*, September 1991, pp. 5–11.

Douglas, John. 1994. Solid futures in fuel cells. *EPRI Journal*, March 1994, pp. 6–13.

Dunn, P. D. 1986. *Renewable Energies: Sources, Conversions, and Application*, Peter Peregrinus Ltd., London.

Elliot, D. L., L. L. Wendell, and G. L. Gower. 1991. *An Assessment of the Available Windy Land Area and Wind Energy Potential in the Contiguous United States*. PNL-7789/UC-261, Prepared for the US Department of Energy under Contract DE-AC06-76RLO 1830 by Pacific Northwest by Battelle. August 1991.

Elliott, D. L. et al. 1986. *Wind Energy Resource Atlas of the United States*. DOE/CH 10093-4. Published by the Solar Technical Information Program. Solar Energy Research Institute, Golden, CO, October 1986.

Electric Power Research Institute (EPRI). 1991. Alabama cooperative generates power from air. *EPRI Journal*, December 1991.

Electric Power Research Institute (EPRI). 1993. Coal Gasification Guidebook: Status, Applications, and Technologies. EPRI TR- 102034. Prepared by SFA Pacific, Inc., Mountain View, CA.

Falcone, P. K. 1986. *A Handbook for Solar Central Receiver Design*. SAND86-8009. Sandia National Laboratories, Livermore, CA, December 1986.

Gaul, G. W. et al. 1993. Compressed air energy storage thermal performance improvements. ASME Paper 93-JPGC-GT-2. Presented at the Joint ASME/IEEE Power Generation Conference. Kansas City, MO, October 17–22, 1993.

Glasstone, S. 1980. *Fusion Energy*. US Department of Energy. Technical Information Center. US Government Printing Office.

Hall, D. O., Frank Roisillo-Calle, R. H. Williams, and J. Woods. 1993. Biomass for energy: supply prospects. *Renewable Energy: Sources for Fuels and Electricity*. Island Press, Washington, D.C.

Holl, R. J. and D. R. Barron. 1989. *Status of solar thermal electric technology*. EPRI GS-6573. Electric Power Research Institute, Palo Alto, CA, December 1989.

Island Press, 1718 Connecticut Avenue NW, Suite 300, Washington, D.C. 20009.

Johansson, T., H. Kelly, A. Reddy, and R. Williams. 1993. *Renewable Energy: Sources for Fuels and Electricity*. Island Press, Washington D.C. and Covelo, CA.

Keller, C. 1946. The Escher Wyss-AK closed cycle turbine, its actual development and future prospects. *Transactions of the ASME*, Vol. 68, pp. 791–882.

Keller, C. and W. Gaehler. 1961. Coal-burning gas turbine. Gas Turbine Power Conference. Washington, D.C., March 5–9, 1961.

Kessler, R. et al. 1992. Coal-burning magnetohydrodynamic power generation. *Proceedings of the American Power Conference*, Vol. 51–8, pp. 467–479.

Kistler, B. L. 1986. *A User's Manual for DELSOL3: A Computer Code for Calculating the Optical Performance and Optimal System Design for Solar Thermal Central Receiver Plants*. SAND86-8018. Sandia National Laboratories, November 1986.

Lacy, R. G. 1988. Conversion cycles for geothermal power generation. Geothermal Energy Symposium, Eleventh Annual Energy-Source Technology Conference and Exhibit. New Orleans, LA, January 10–13, 1988.

LaHaye, P. G. and M. R. Bary. 1994. Externally fired combustion cycle (EFCC), a DOE Clean Coal V project: effective means of rejuvenation for older coal fueled stations. Presented at the ASME Turbo Expo '94. The Hague, Netherlands, June 13–16, 1994.

Laurence, C. L. and F. W. Lipps. 1984. *User's Manual for the University of Houston Individual Heliostat Layout and Performance Code*. SAND84-8187. Sandia National Laboratories.

Makansi, J. 1992. Coop manages system load with compressed storage. *Power*, Vol. 136, No. 4, April 1992.

Menicucci, D. F. and J. P. Fernandez. 1988. *User's Manual for PVFORM: A Photovoltaic System Simulation Program for Stand-Alone and Grid-Interactive Applications*. SAND85-0376. Sandia National Laboratories, April, 1988.

MHD Program Plan: FY 1990, US Department of Energy. DOE/FE—0189P. Washington, D.C., October 1989.

MHD Program Plan: FY 1992, US Department of Energy. DOE/FE—0259P. Washington, D.C., October 1991.

Modern Power Systems. Wilmington Publishing Ltd. Wilmington House, Church Hill. Wilmington, Dartford DA2 7EF, UK.

Morris, D. 1988. The Chino battery facility. *EPRI Journal* Vol. 13, No. 2, March 1988, pp. 46–50.

Myles, K. M. et al. 1989. Status of molten carbonate fuel cell technology. Energy Technology Conference, pp. 197–204.

National Renewable Energy Laboratory (NREL). 1617 Cole Boulevard, Golden, CO 80401-3393.

Palmerini, C. G. 1993. Geothermal energy. *Renewable Energy: Sources for Fuels and Electricity*. Island Press, Washington, D.C.

Princeton University. Plasma Physics Laboratory. James Forrestral Campus, P. O. Box 451, Princeton, NJ 08544.

Probstein, R. F. and R. E. Hicks. 1982. *Synthetic Fuels*, McGraw-Hill, New York, NY.

Research Triangle Institute (RTI). 1991. *Biomass State-of-the-Art Assessment*, Final Report, Vol. 1. GS-7471, Research Project 2612-13. Prepared jointly for Virginia Power and Electric Power Research Institute. September 1991.

Ruth, Larry. 1992. Combustion 2000: planning and procuring an engineering development program. *PETC Review*. Pittsburgh Energy Technology Center. Issue 7. Winter 1992–1993.

Salter, S. H. 1974. Wave Power. *Nature*, Vol. 249, No. 5459, pp. 720–724.

Sandia National Laboratories. Albuquerque, NM 87185-5800. Livermore, CA 94551-0969.

SERI/TP-20-3880. 1990. *Insolation Data Manual and Direct Normal Solar Radiation Data Manual*. Solar Energy Research Institute, Golden, CO, July 1990.

SERI/SP-220-3555. 1989. *Wind Energy Technical Information Guide*. Solar Energy Research Institute. Golden, CO, December 1989.

SERI/TP-260-3674. 1990. The Potential of Renewable Energy: An Interlaboratory White Paper. Idaho National Engineering Laboratory, Los Alamos National Laboratory, Oak Ridge, National

Laboratory, Sandia National Laboratories, Solar Energy Research Institute, March 1990.

SIMBECK, D. R., R. L. DICKENSON, and E. D. OLIVER. 1983. *Coal Gasification Systems: A Guide to Status, Applications, and Economics*. Electric Power Research Institute, AP-3109.

SOLAR ENERGY RESEARCH INSTITUTE (SERI). 1988. *Photovoltaics Technical Information Guide*. Second Edition. SERI/SP-320-3328. DE88001184. UC270-276. Golden, CO, August 1988.

STODDARD, M. C., S. E. FAAS, C. J. CHIANG, and J. A. DIRKS. 1987. *Solergy: A Computer Code for Calculating the Annual Energy from Central Receiver Power Plants*. SAN86-8060. Sandia National Laboratories, Livermore, CA, May 1987.

STORMS, E. 1991. Review of experimental observations about the cold fusion effort. *Fusion Technology*, Vol. 20, No. 4, pp. 433–477.

US DEPARTMENT OF ENERGY. 1990. International Thermonuclear Experimental Reactor (ITER) Design and Technical Parameters. US Home Team Project Office. Lawrence Livermore National Laboratory.

VANDERVORT, C. L., M. R. BARY, L. E. STODDARD, and S. T. HIGGINS. 1993. Externally-fired combined cycle repowering of existing steam plants. Presented at the 38th ASME International Gas Turbine and Aeroengine Congress and Exposition. Cincinnati, OH, May 24–27, 1993.

VOSBURGH, P. N. 1983. *Commercial Applications of Wind Power*. Van Nostrand Reinhold, New York, NY.

25

Power Plant Planning and Design

Stephen M. Garrett

25.1 INTRODUCTION

The process of developing a new power plant from its inception to commercial operation is complex and dynamic. The power plant planning and design process described in this chapter is tailored to conventional fossil fueled power plants using oil, natural gas, or coal. The basic steps, shown in Fig. 25-1, are similar for alternative generating unit concepts.

The power plant design process changes depending on the unique financial, engineering, environmental, and other requirements for a specific plant. Even the designs of modular plants that incorporate standard components must consider the plant's specific application. One approach to the power plant design process is to design by function or system, purchase by component, construct by specialty contractor, and startup by system. Each of these steps is required in some form by all power plant designers.

It is vital that the goals, objectives, and constraints for each project be carefully defined in the planning and analysis stage. Project planning and analysis encompasses those strategic elements of a project that must be considered early in project development. Fuel supply studies, system planning studies, siting evaluation, transmission planning analyses, environmental feasibility analyses, and economic and financial feasibility analyses are integral to project planning and analysis for new power generation facilities.

Conceptual design engineering encompasses a broad variety of activities. The conceptual design process consists of systematically defining and evaluating the basic conditions and constraints applicable to a specific electric generating plant. Conceptual design engineering starts as part of the project planning and analysis activities, and is increasingly required in support of permitting and licensing activities.

Detailed design engineering includes determining the technical requirements for all plant components. It involves consideration of equipment sizing, reliability constraints, performance requirements, and codes and standards, all directed toward requirements for successful specification, construction, and startup.

As the detailed engineering is completed for systems and equipment, procurement and construction specifications are developed to delineate specific technical and commercial requirements consistent with the overall design objectives.

Finally, effective control of schedules, cost, design, and construction is essential to a successful power plant project. Project control activities include critical path scheduling of engineering and plant construction activities, cost control and cost risk assessment, design control, and construction control.

25.2 EVALUATING FUEL SUPPLY OPTIONS FOR POWER GENERATION

Among the most significant factors influencing the successful design and operation of new power generation facilities are the fuel supply and transportation arrangements. Successful fuel strategies for new and existing power generating facilities must meet the required operating requirements in an efficient, effective manner while maintaining adequate fuel flexibility and security.

Five factors must be considered in evaluations of alternative fuel supply options: potential fuel types, alternative fuel sources, transportation systems, fuel pricing, and design and operational impacts. These factors are investigated to varying degrees of detail depending on the stage of development of the power generation facilities. As a project develops from inception to design, the need for precise fuel supply sourcing information becomes progressively greater. The scope of fuel supply evaluations for broad planning assessments differs significantly from that for fuel contracting strategy evaluations that support project financing.

25.2.1 Potential Fuel Types

The mix of fuels for generation (such as hydroelectric plants, oil-fueled steam units, coal-fueled steam units, oil/gas simple cycle combustion turbine, or oil/gas-fueled combined cycle units) within a utility system dictates to a large extent which fuels are appropriate candidates for new plants. Power generation technologies and fuel types should be diversified to an extent consistent with cost-competitive power production and power production reliability. Environmental constraints also significantly influence fuel type selection.

25.2.2 Alternative Fuel Sources

Potential fossil fuel sources typically include natural gas, distillate fuel oil, residual fuel oil, lignite, bituminous coal,

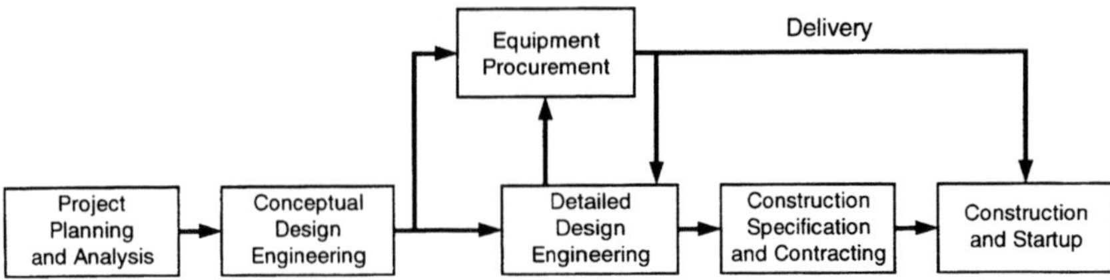

Fig. 25-1. Power plant design process.

and subbituminous coal. Less common sources include liquefied petroleum gas, liquefied natural gas, propane, butane, bitumen, coke, and other petroleum derivatives. Multiple fuel sources are desired to ensure pricing competition and thereby minimize plant operating costs. Factors considered in identification and evaluation of alternative fuel sources are proximity to power generation facilities, quantity, quality, reliability, and supplier commitments. An initial screening criterion is the location of candidate fuel sources relative to the power plant site. For petroleum and natural gas, the producing fields, refineries, processing plants, and major bulk terminals are identified and the chemical properties of each fuel are determined.

For coal fuels, the major producing coal field and their locations relative to the plant facilities are determined. For each coal-producing region, coal properties pertinent to power plant design are identified. These properties include typical and range values for proximate and ultimate analyses, mineral analysis of ash, ash fusion temperatures, and Hardgrove Grindability Index. Depending on the stage of project development, a detailed assessment of economically recoverable reserves and existing and planned production capability may be conducted.

25.2.3 Transportation Systems

Transportation systems from a fuel source to the power plant are integral to the ultimate reliability and delivered cost of a fuel supply. As such, transportation systems should be characterized according to system constraints and related upgrade needs, current and projected availability and source commitments considering competing transportation movements, and market conditions and associated impacts on availability and reliability.

For natural gas pipelines, existing capacity and current use of those lines is determined. Similarly, for fuel oils and other petroleum fuels, the location and condition of suitable product pipelines and bulk terminals are evaluated. Scheduled plans for pipeline expansion should be identified and considered in the evaluation.

For coal fuels, transportation alternatives (railroad, road, and conveyors) are evaluated for the specific power plant location. Ownership of railroad rights-of-way and their status, if applicable, are determined. The availability of right-of-way, condition of existing track, and potential socioeconomic and environmental impacts of a rail line can significantly impact a project's cost.

25.2.4 Fuel Pricing

Typically, fuel pricing for planning purposes is based on spot market conditions (free on board mine or wellhead) plus a transportation price component. Spot fuel price estimates are determined through market trends and surveys, published literature, and supplier contacts.

The maximum price for a specific fuel is often based on a netback basis assuming direct competition with other fuels of equivalent quality. Netback costs are calculated by subtracting transportation and utilization costs from a market price that has been benchmarked to the location of the power plant.

The minimum price is established by the fuel's production cost. Production costs are influenced by factors such as recovery and mining methods, recovery constraints, nature and extent of underground reserves, geological conditions, marketing operations, royalties, taxes, and fees. Recovery or mining operations are influenced by corporate organizational structures, scale of operations, capital structure, labor structure, recovery equipment age and condition, and beneficiation facilities. Netback and production costs vary with time as a function of technological changes.

Fuel transportation costs are determined considering the transport system logistics from the source to the end-use. This includes transport distances, transport equipment type, sizing, and transport constraints; fuel quantity and consignment sizing; transfer point capability and limitations; and destination point unloading capability and constraints. Competing transportation systems (such as multiple railroads) can have a significant effect on the delivered price of fuel.

Consideration of future generating costs is integral to the project planning process. Fuel price projections published by trade organizations, lending agencies, government agencies, and specialized private firms can be consolidated into a single estimate for fuel price escalation or established as a range of estimates. As an alternative, the cost of producing each fuel can be considered in forecasting future fuel prices. The cost structure associated with exploration, development, and fuel production establishes the key components of fuel price escalation. The escalation of each component can then be determined for incorporation into a consolidated rate. Historical fuel price cycles and seasonal price trends are often analyzed as part of a future trend projection.

25.2.5 Design and Operational Impacts

The properties of a fuel selected for a generating plant impact the facility's design and operation. For example, residual fuel oils can be burned in combustion turbines; however, the plant design must incorporate fuel treatment systems that would not be required for burning distillate fuel oils or natural gas. Also, burning residual oil in a combustion turbine increases nonfuel operations and maintenance (O&M) expenses.

The properties of a coal influence the conceptual and detailed plant design as well as plant operating characteristics. The rank of the coal has a significant effect on boiler sizing and initial capital costs. The physical size of a steam generator firing high moisture content lignitic fuels will be significantly greater than the boiler size required for burning high-rank bituminous coals. High ash content coals require larger ash removal and handling systems, thus increasing initial capital expenditures and annual operating and maintenance expenses. For a given sulfur dioxide emission limit, the costs for a desulfurization system increase for coals with relatively high sulfur content. There are numerous examples of coal quality impacts on plant design and operation. However, for most project planning scenarios, the impacts of fuel quality on design, initial capital cost, and annual O&M expenses are within the error band for the projection of delivered fuel cost.

25.3 SYSTEM PLANNING STUDIES

A system planning study defines the predicted timing and size for future capacity additions in a specific system and determines the most economical capacity expansion plan on the basis of specific study assumptions. The study may also evaluate the impact of demand-side management options for reducing electrical loads. "Integrated resource planning" and "least cost planning" are options that include evaluations of both supply-side and demand-side options. Each system planning study is tailored for a specific system and may meet specific requirements of a regulating agency or financial lending institution. However, several basic activities and tasks are common to most system planning studies.

25.3.1 Load Forecasting

Load forecasts can be made on an aggregate system basis for generation expansion planning or on a geographical basis for transmission planning. Several alternate methods are appropriate for forecasting electric loads. Forecasts for strategic planning require models with maximum explanatory power to produce moderate- to long-term projections of system energy (megawatt-hour, MWh) and peak demand (megawatts, MW). Strategic planning forecasts necessarily devote less attention to random short-term fluctuations to more correctly identify the major long-term relationships. For this reason, long-term relationships are likely to be less accurate in any 1 year than is a short-term forecast. However, year-to-year fluctuations from the forecast should be distributed randomly and handled operationally rather than allowed to influence long-term facilities planning.

Methods used for forecasting electric power needs fall into three categories: time series trend analysis, end-use, and econometric.

The time series trend analysis uses regression analysis of historical electrical loads to predict future trends. This analysis is easily performed and requires minimal data. However, it assumes no changes in the historical relationships among factors affecting electricity use, and it also assumes a continuation of the historical trend in these factors.

End-use forecasting models use very detailed electric power usage and consumption pattern data for residential, commercial, and industrial users. The electrical load characteristics of individual appliance and equipment types are determined by metering at the actual point of consumption. Daily load shapes can then be derived and summed to obtain total load curves by time-of-day. Once historical electric consumption and patterns are identified, the data are related to changes in socioeconomic conditions, weather conditions, and appliance saturation levels to predict future demand and energy use.

Econometric load forecasting models employ historical time series data and regression analysis to estimate macroscopic relationships between electricity sales, weather, and various economic and demographic factors. These economic and demographic factors include variables such as population, employment, personal income, and electricity prices. Econometric models reflect relationships as they have developed over time, making these models appropriate for forecasting uses. Specification of the relationships is based on economic theory and experimental testing.

Once a load forecast has been prepared, the capacity addition requirements are developed on the basis of factors such as system reserve requirements, existing generating capacity, and planned unit retirements.

25.3.2 Demand-Side Management

Demand-side management (also called load management) consists of techniques or methods to control the shape and growth of electric loads. Demand-side management can reduce the need for new supply-side generating plant additions by reducing energy requirements at the point of load. For example, switching off the power to air conditioners, electric heating systems, electric water heaters, and other selected equipment for a short period, the utility's peak electrical demand can be lowered. By using demand-side management, the installation of new generating units sometimes can be delayed, thus delaying capital expenditures.

Demand-side management programs are not without drawbacks. Incentives must be offered for customers to willingly allow the utility to control the operation of their appliances and equipment. The utility must also install control systems to shut off appliances and equipment at the appropriate time. If customers resist the program and do not elect to participate, the program is not economically successful.

25.3.3 Supply-Side Alternatives

Once capacity addition requirements have been established, a utility system planning study develops several supply-side alternatives. Power generation plants of commercially proven design and operation that use readily available fossil fuels are typically considered first to satisfy growing demand. For example, pulverized coal plants can satisfy baseload generation needs, and natural gas- or oil-fueled combustion turbine generators can satisfy peak demand needs.

The planner also evaluates nonconventional and advanced supply-side alternatives such as gasification combined cycle, pressurized fluidized bed, wind energy conversion technologies, solar systems, and compressed air energy storage. Obviously, the alternatives considered and selected for new power generation will change as technology develops, fuel costs fluctuate, environmental constraints tighten, and socioeconomic conditions change.

In a screening evaluation, many supply-side alternatives can be eliminated on the basis of technical maturity, unit performance, resource requirements, and environmental compliance and adaptability. Technological maturity examines the extent of commercial or demonstration experience supporting the technology. A technology that does not have current sufficient experience or projected development within the time frame of interest should be eliminated from further consideration. A technology with inadequate unit performance because of unavailability or equipment constraints is also eliminated. Unit resource requirements describe the types and amounts of resources critical to operate a unit that uses a particular technology. For example, solar systems require sufficient solar insolation and therefore are not appropriate for most northern latitudes. Environmental compliance and adaptability screen for the extent to which the technology can be designed and operated to satisfy current and potential environmental requirements. Technologies that cannot meet environmental constraints should be eliminated from further consideration.

The economics of supply-side alternatives can also be compared using screening curves which represent the comparative revenue requirements necessary for the construction and operation of the applicable supply-side generation options versus operating hours or capacity factor. An example of these screening curves is shown in Fig. 25-2. Revenue requirements are calculated as the sum of capital revenue requirements plus levelized O&M costs and fuel.

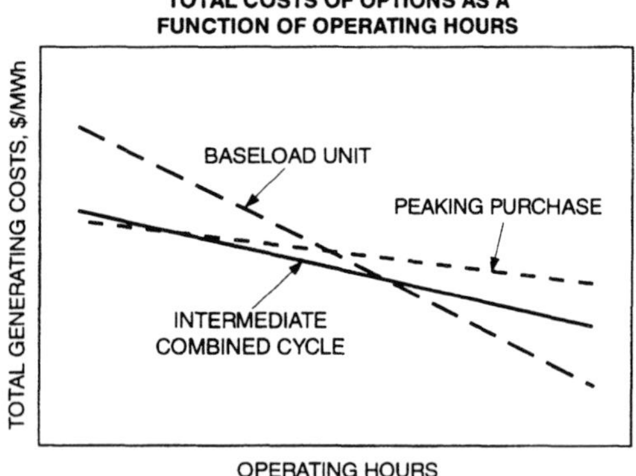

Fig. 25-2. Example screen curves for supply-side options.

25.3.4 Production Cost Analysis

Production cost models are used to evaluate the operational and cost impacts of adding various types of new generation to the utility's existing power supply system. The addition of new generating units is modeled taking into account the retirement of existing units, the repowering of older existing units, and other system changes. Production cost analyses also assess the effects caused by changes in projected load demand and the effects of implementing load management programs.

There are two basic types of production costing models: load duration models and chronological models. Both models determine the operating costs of a power supply system over a specified period, typically 10 to 20 years. The simulation is conducted in a sequence of time steps in which units are committed and dispatched to satisfy the projected load. System performance and production costs are then calculated in terms of generating unit electric power production (load factor, capacity factor, net energy to grid), heat rates, fuel consumption, fuel costs, and nonfuel O&M costs. The simulation determines the performance of individual units and the total system. Total system performance is determined for when new unit additions are scheduled for future installation, when existing units are assumed to be retired, or when other modifications are assumed to made to the current system. In this manner, different types and sizes of units can be compared on an operational and economic basis.

In both types of production cost models, generating units are characterized by their maximum and minimum capacity, fuel burn rate, fuel price, O&M costs, scheduled maintenance, and forced outage rates. The fuel burn rate takes the form of the following equation:

$$\mathrm{FBR} = C_0 + C_1 \times P + C_2 \times P^2 \tag{25-1}$$

where

FBR = fuel burn rate (MBtu/h),

P = generation level (MW), and

C_0, C_1, C_2 = fuel burn rate coefficients uniquely defined for each unit.

Generating units are selected for service, or committed, on the basis of full load operating costs. For example, the unit with the lowest operating cost is selected to run first, the second most cost-effective unit is selected to operate next, and so on. Within the production cost model, operating cost may be defined as simply the fuel cost or it may include variable O&M costs. The units are committed in order of increasing full load operating cost until the combined capacity is sufficient to meet demand and operating reserve requirements.

After the generating units are selected for service, the production of each unit is determined in a procedure that minimizes overall production cost. Cost minimization is obtained by dispatching all units to operate at the same incremental cost, subject to minimum and maximum capacity constraints. Incremental cost is the dollar per megawatt-hour cost to the system for the next megawatt-hour of generation produced from a power source. The cost minimized within a production cost model is fuel cost, or sometimes fuel cost plus variable O&M costs. Fuel cost, in dollars per hour, is given by the fuel burn rate equation multiplied by the fuel price in dollars per MBtu.

A numeric value for incremental cost is obtained by differentiating the fuel burn rate equation with respect to the generation level. Without consideration of the effect of variable O&M costs, the incremental cost is as follows:

$$\text{Incremental cost} = (C_1 + 2 \times C_2 \times P) \times \text{PR} \qquad (25\text{-}2)$$

where

PR = fuel price, \$/MBtu.

The unit dispatching process consists of determining the incremental cost level for which the sum of the generation level taken across all committed units is equal to the sum of system demand and operating reserves. Once the required incremental cost is determined, the production level and associated costs are calculated for each unit from specified operational characteristics and fuel prices.

A load duration based model is the simplest production cost model. This model assumes that system demand can be represented as a load duration tabulation or load duration curve of discrete load blocks. An example of a load duration curve is shown in Fig. 25-3. Each year is divided into periods for which there are discrete load blocks. Units are then assumed to be committed and dispatched to satisfy the demand for each period.

A chronological production cost model considers unit commitment and dispatch scheduling in an hour-by-hour simulation. Hourly simulation allows for increased accuracy of system production costs, because startup costs, startup times, ramp rates, minimum up and down times, and other characteristics are included for each specific generating unit.

25.3.5 Selection of a Least-Cost Plan

The production cost analysis yields a base plan that is the most economical mix of supply-side and demand-side alternatives relative only to system production costs. For a complete economic evaluation, additional costs for capital investment associated with the installation of facilities and fixed nonfuel O&M costs for each plan must be added to the production costs.

It is possible for an alternative power supply plan to be economically feasible yet not financially feasible. The final decision to implement a plan must consider the absolute magnitude of the cost impact. Assessment of the total cost impact is essential to ensure that the recommended option is financially feasible. In other words, the utility must be able to obtain the needed capital for a near-term capacity addition without causing unacceptable rate shock to the electric customers.

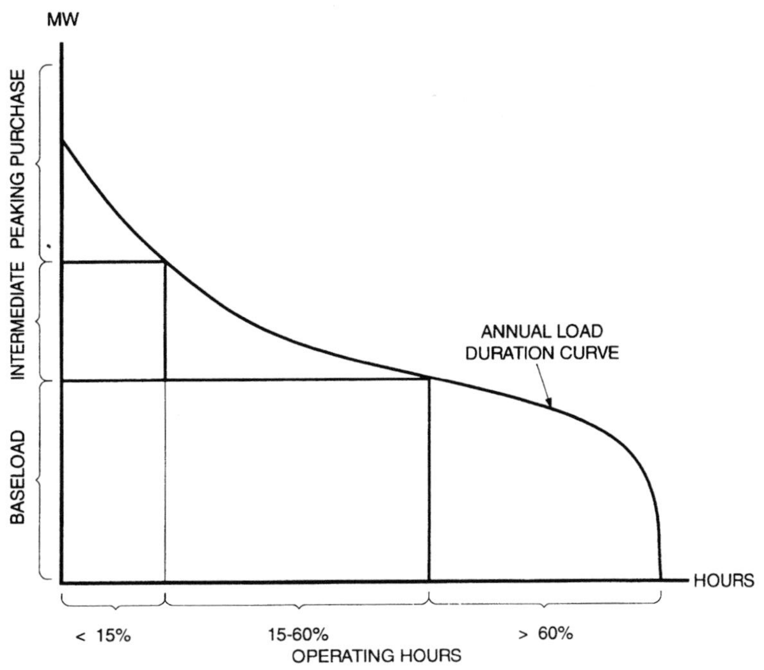

Fig. 25-3. Example load duration curve.

25.4 SITE EVALUATION

Effective power plant site selection (or "siting") involves a systematic process that applies technical, environmental, and land use criteria to identify, screen, and evaluate sites. The siting process for a comprehensive regional study can be separated into three basic screening levels. Smaller comparative studies include a similar procedure with one or two screening levels.

In the first screening level, the study's focus is narrowed by excluding major portions of a region because of requirements for specified resources or the avoidance of critical environmental or geological factors. The second screening level consists of identifying and evaluating potential sites within in the remaining siting areas to arrive at a list of candidate sites. In the third screening level, a preferred site, which is ranked as most suitable, and an alternate site are selected from the candidate sites.

25.4.1 Identification of Siting Areas

During this first screening level of the siting process, specific technical criteria associated with the development and operation of the proposed power plant are established. These criteria are then used in a sequential manner to eliminate certain geographical areas from further consideration. These technical criteria include access to fuel, transmission lines, and suitable water supplies as well as geological and environmental constraints. Geographical areas that satisfy the established requirements are carried forward into the next step as siting areas.

The sequencing of these criteria is flexible; however, it is usually advantageous to begin with the criterion that is most easily applied and will exclude significant areas.

Fuel is usually the first exclusionary criterion. Coal-fueled generating units must be sited for economic fuel delivery by rail, barges, or other means in the quantities required for a specific unit size. Similarly, natural gas-fueled units must be sited within economic distance of existing gas pipelines with sufficient capacity.

The second exclusionary criterion is the distance from suitable transmission lines. New transmission lines are expensive to install and are often subject to strict environmental constraints and right-of-way restrictions.

The third exclusionary criterion is access to reliable water supplies of suitable quality and quantity. The availability of ground and surface water should be investigated, and undependable or poor quality sources excluded. For river water supplies, a withdrawal of 10% of the flow volume at low flow conditions is considered reasonable.

The fourth exclusionary criterion is air quality sensitive areas. In the United States, areas designated as PSD Class I or nonattainment for any pollutant, as well as an appropriate buffer zone surrounding the areas, may be excluded from further consideration. More information on area classification can be found in Chapter 14.

The fifth exclusionary criterion assesses geological structures and features. As an example, areas having significant slope would require extensive earthwork so that site development may not be practical. In addition, areas with Karst/sinkhole potential or areas with unusually high seismic activity can be excluded from further consideration.

25.4.2 Selection of Candidate Sites

After siting areas are identified, the potential sites must be evaluated consistent with overall site development and environmental impacts. The sites are judged relative to project specific factors which typically include the following salient physical and cultural features:

- Proximity to linear facilities (road, rail, transmission corridors, water supply, and pipeline corridors) necessary to support plant development;
- Topographic factors;
- Air quality exclusion zones;
- Proximity to urbanized or unincorporated residential/commercial areas;
- Proximity to areas of potential use conflicts (such as airport instrument approaches and military security zones) or adverse community acceptance;
- Proximity to environmentally sensitive areas (such as a river system having a federal or state "wild" or "scenic" designation, or national or state parks);
- Aesthetic constraints (including noise, open space, and public view/scenic vista) affecting facilities placement or height;
- Cultural, terrestrial, or aquatic environmental constraints (including cemeteries, wetlands, or other sensitive habitats) impacting site licensing, development, or operation; and
- Geological factors.

Using information from onsite field reconnaissance and the previously developed knowledge of regional and site-specific information, comprehensive descriptions of each candidate site are developed. The descriptions must be sufficiently accurate to ensure that the engineering concepts are feasible and appropriate for the sites. Specific power plant systems and other items for which detailed site information must be defined include the plant technology, makeup water system, cooling water system, fuel supply and delivery, highway accessibility, transmission interconnection, and transmission system integration/stability.

Decision analysis methods are used to systematically evaluate each site in relation to the criteria listed above. An example of a weighted ranking decision analysis evaluation of a potential site is shown in Table 25-1. In the example, weights are assigned to the evaluation criteria based on the perceived relative importance. Each site is judged relative to the criterion and assigned a rank from 1 to 5 or 1 to 10; sites with the most favorable characteristics are given a value of 5 or 10. The weights are multiplied by the ranking for each criterion, then summed for a total site weighted ranking score. Those sites having the highest overall total weighted ranking score are selected as candidate sites.

Table 25-1. Example Weighted-Ranking Siting Evaluation[a]

Siting Criteria	Criteria Weight, percent	Site 1 Raw Score	Site 1 Weighted Score	Site 2 Raw Score	Site 2 Weighted Score	Site 3 Raw Score	Site 3 Weighted Score
Site development (60)							
Transmission	30	3	90	4	120	4	120
Water supply	8	3	24	3	24	4	32
Fuel delivery	10	3	30	3	30	4	40
Site access	6	2	12	4	24	2	12
Site topography/geology	6	0	0	2	12	3	18
Environmental/sociological (40)							
Air quality	15	2	30	2	30	3	45
Biological impacts	6	5	30	5	30	5	30
Land use	6	4	24	3	18	4	24
Socioeconomics	5	3	15	3	15	4	20
Cultural resources	4	2	8	2	8	2	8
Aesthetics	4	5	20	4	16	5	20
	100		283		327		369

[a]Based on a 1 to 5 rating, with 5 representing a site having favorable characteristics relative to the siting criteria.

25.4.3 Evaluation of Preferred and Alternative Sites

Within this phase of a siting study, a preferred site and an alternate candidate site are selected. This step includes further evaluations and refinement of the engineering and environmental characteristics of the candidate sites. Once this is accomplished, a preliminary conceptual design unique to that candidate site is prepared. The conceptual design includes development of a preliminary site arrangement and defines the following key plant design features:

- Existing site facilities,
- Steam generation system,
- Water supply and treatment system,
- Wastewater discharge,
- Fuel delivery systems,
- Air quality control systems,
- Heat rejection systems,
- Waste handling systems, and
- Transmission system.

The design concepts are prepared considering project specific requirements, environmental design criteria, site criteria, overall plant availability and reliability objectives, and economics. Preliminary site arrangement drawings are prepared to show the size and location of key plant facilities and structures; show spatial relationships of the facilities; and indicate roads, transmission lines, pipelines, and conveyors that link offsite systems. A linear facility map is often prepared to show the proposed connection routes to the offsite systems.

Differential capital and operating costs for the site-specific power plant are then estimated using the preliminary conceptual design definitions and site arrangement drawing. Key factors that result in significant differences in the initial plant cost include linear facilities (such as rail, roads, pipelines, and electrical transmission), site preparation, and local field labor rates for plant construction. The sites may also have differential annual operating costs that warrant review.

Decision analysis techniques similar to the weighted ranking method previously described and that incorporate differential capital and operating cost factors can then be used to select a preferred and alternate site. Sensitivity analyses, in which other possible weighting/ranking scenarios are considered, are used to identify the relative strengths and weaknesses of the selected sites.

25.5 TRANSMISSION PLANNING ANALYSIS

When new generating capacity is eliminated or introduced into a given electric power system, or other system modifications are made, the performance of the electric power system is affected. Transmission planning analysis establishes the nature and extent of the performance impacts and provides information regarding the equipment modifications or additions required to accommodate power flow within the system. Transmission planning analysis is composed of loadflow studies, stability studies, and horizon planning analysis.

25.5.1 Loadflow Studies

Loadflow studies demonstrate that a proposed upgrade or expansion of a specific system can actually deliver electric power as anticipated. A loadflow study determines the power flow (real and reactive) on each transmission line in the system, the voltage level at each bus, and the relative phase angle at each bus.

Generally, loadflow is evaluated for the following basic scenarios: Scenario 1—the present power system configuration, and Scenario 2—the system is modified to accommodate load growth, possible generating unit retirements, installation of new generating units, addition or upgrades to the transmission system, and other system alterations. Contingency cases are evaluated to determine the effects of system component outage. Loadflow modeling can range from a simple assessment of a radial system to a complex mathematical assessment of an integrated system network using coupled nonlinear equations. Network analysis requires a high degree of modeling sophistication for solution convergence.

Loadflow studies include determination and evaluation of transmission line characteristics for alternative scenarios including equipment capacity, voltage level, steady-state angle inspection, line losses, and system losses. Once adequate power flow is determined for alternative system upgrade scenarios, the options are evaluated on an economic basis. The capital costs for the improvements and the annual costs associated with capacity and energy replacement are determined for each upgrade scenario. The alternative having the lowest present worth cost is selected unless noncost factors dictate otherwise. The installation of supplemental transmission lines for system reliability (maintenance considerations plus single-line contingency criteria) is evaluated as a further refinement.

25.5.2 Stability Studies

An operating power system is considered stable if oscillations in the characteristics of the system damp out satisfactorily with time following a system disturbance. System disturbances cause changes in the generator voltages, rotor speeds, and rotor angles. Generally, transmission systems with limited power flow capability (high impedance, low voltage, or high load angles) are more likely to become unstable after transient events than systems that have more power flow capability. Stability studies are required when new generation or other system modifications are being considered.

Steady-state stability of a power system is assumed when small and gradual system disturbances can be accommodated without loss of synchronism between system elements. Steady-state instability occurs when no equilibrium condition (steady-state power transfer limit) exists. Voltage instability is characterized by the collapse of voltage levels in a specific area as a result of a system weakness, voltage-sensitive load, or demands of massive reactive transfer. Steady-state oscillatory instability can occur in some systems operating at high power flow levels. This condition is characterized by hunting (an undesired oscillation after external stimuli disappear) or steadily growing oscillations which might lead to a loss of synchronism. Transient (short-term condition after a periodic disturbance) stability indicates a system's ability to accommodate large disturbances such as generating unit trips, line faults (short circuits), or line loss. Also, imbalances in the amount of generation relative to loads can lead to frequency excursions and excessive power flows. This in turn can cause loss of transmission lines and generators (cascading).

Limits imposed on power transfer because of stability constraints vary with changes in the system conditions. As the operating conditions and configuration of the power system change as a result of factors such as generator unit additions, load growth or decline, or transmission line and substation additions, the system stability is impacted. Some stability problems become more limiting than others as a system is expanded.

The transient stability of a system after a disturbance is characterized by the first cycle of rotor angle swing response or "first-swing stability." On initiation of a transient, the synchronous machines in the system accelerate and decelerate with respect to each other. This phenomenon results in oscillations among the machines. The first-swing stability is analyzed to identify weaknesses in the system and the modifications or improvements required to ensure a stable system response.

Power system stability is analyzed using a deterministic mathematical approach to evaluate contingency modes that are most likely to occur. Examples of contingency modes investigated in a stability study include the following:

- A trip of a large generating unit,
- Loss of a large transmission intertie,
- Loss of a substation that interconnects several transmission lines,
- A permanent three-phase fault cleared using circuit breakers to remove the faulted equipment, and
- A single-phase fault in conjunction with a stuck (failed to open) breaker.

If the system is not stable for the specified contingency modes, modifications are required. Depending on the specific situation, these modifications include adding new transmission lines, adding shunt or series capacitors to improve the flow of power, static volt ampere reactive (VAR) controls, improvements in relay and breaker interrupt times, and generating unit modifications to improve transient response (high response exciters, application of power system stabilizers, or early valve actuation equipment for steam turbine generators).

25.5.3 Horizon Planning Analysis

Transmission systems are often analyzed for a specific "horizon" year that may be 10 to 20 years in the future. For each year up to the horizon year, power flow through the system is assessed. Specific criteria are established for the system, including equipment thermal constraints, voltage variation range limits, and outage requirements. If the system cannot adequately meet load demands while satisfying the horizon criteria, modifications or enhancements are made in the form of new transmission lines, substations, and VAR compensation. Least-cost alternatives are selected for each year up to the horizon year. A horizon planning analysis provides a detailed year-by-year plan to accommodate system growth.

25.6 ECONOMIC AND FINANCIAL FEASIBILITY

The economic and financial feasibility of a proposed power plant is the cornerstone of all project development activities. An economic feasibility assessment addresses a power sector development program against other projects that compete for scarce resources. Economic feasibility analyses are required by international lending agencies for power projects in developing countries but typically are not performed for domestic United States projects. A financial feasibility assessment determines whether a specific project can offer a sufficient return to investors to justify spending capital resources for development.

25.6.1 Economic Feasibility

Economic feasibility establishes a basis for comparing the expansion of the system's generation capability with other development programs that compete with the power sector for limited resources. The economic feasibility assessment examines the project's costs and benefits from a national perspective. Economic evaluations are based on real values and use border prices for both costs and benefits. Border prices refer to an equivalent export cost or benefit in the international market. Local taxes and duties are not included in this evaluation. Typically, economic feasibility is based on the determination of an internal rate of return for the project which is evaluated against criteria developed for the project. Project evaluation criteria generally are based on requirements developed by funding agencies such as the World Bank.

The primary activity in the economic feasibility assessment is the identification and valuation of project costs and benefits. Two scenarios are developed: the base case scenario, representing the conditions that would prevail in the absence of the project; and the project scenario, representing the anticipated conditions if the project is allowed to proceed. The differential benefits and costs between the two scenarios are quantified to determine the economic internal rate of return. The evaluation incorporates all costs and benefits that are readily quantified; benefits that cannot be quantified are addressed qualitatively. Depending on the project, benefits may be valued on the basis of existing electric tariffs or, as in the case of rural electrification projects, the cost of electric energy substitutes.

25.6.2 Financial Feasibility

Financial feasibility assesses the ability of the utility or developer to meet the financial obligations associated with the project. Financial costs, unlike economic costs, include import duties and taxes.

The projected expenditures and revenues associated with the project are modeled on the basis of the utility of developer's financial evaluation criteria. A variety of evaluation criteria can be used, including required rate of return, debt coverage ratios, and electric tariff impacts. Financial feasibility is determined by comparing project financial parameters with established evaluation criteria.

25.6.2.1 Financial Pro Forma Models. A financial pro forma model presents a projection of cash flows and balances throughout the economic life of a project. From these projected cash flows, net present values, and debt service coverage ratios, a judgment can be made regarding the project's financial feasibility. The cash flow analysis methodology used in financial pro forma models and the financial pro forma model input and results expected from the analysis are discussed in the following sections.

25.6.2.2 Cash Flow Analysis Methodology. The analysis period for a power plant is divided into construction and operation periods. The operation period begins at commercial operation (end of the construction period) and extends through the economic life of the project. Figure 25-4 illus-

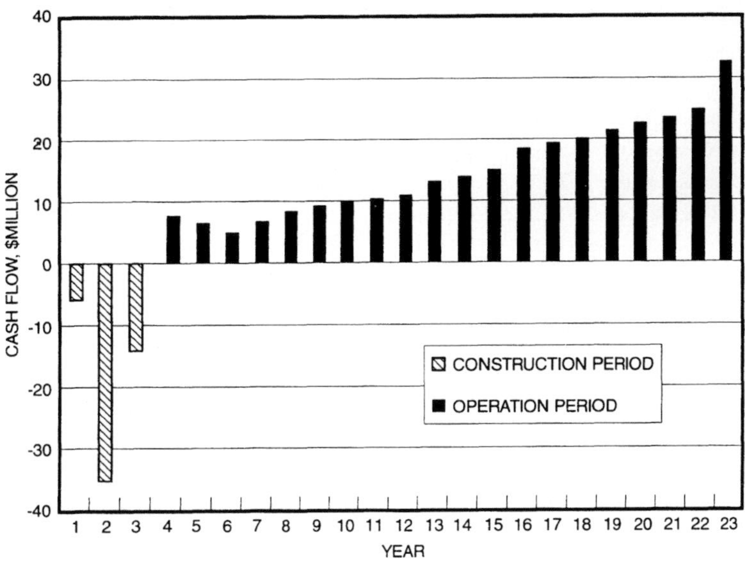

Fig. 25-4. Cash flow for an equity investor.

trates a sample cash flow for an equity investor. The equity investor invests cash during the construction period and receives cash distributions from the project during the operation period.

To maintain a positive cash balance during the construction period, the cash receipts should equal or slightly exceed the cash disbursements. Cash receipts consist of financing proceeds from various sources (such as long-term debt), and equity, leases, and interest income from any positive cash balances during the period. Cash disbursements include equipment and construction contract payments, cost of land, indirect construction costs, and interest during construction. The calculation of interest during construction depends on the accumulated monthly balances of the debt and lease proceeds. The primary results obtained from the construction period analysis include interest during construction, total capital cost, financing amounts, and the required equity investment.

During the operating period, cash flows are required to pay off the long-term debt and lease obligations and to provide an adequate return to the equity investors. Essentially all of the required cash flows are provided by revenues in excess of expenses that result from operating the proposed plant. Revenues are derived from a variety of sources, such as the sale of electricity, sale of process steam, and tipping fees for waste-to-energy projects. Expenses include fuel, nonfuel O&M, and other operating expenses.

25.6.2.3 Pro Forma Input Data. A meaningful financial pro forma analysis of a project requires a wide variety of input data. These input data can be grouped into the following major categories:

- General operating parameters/characteristics,
- Plant capital cost,
- Financing assumptions,
- Revenue data, and
- Expense data.

General operating parameters include the construction and operating periods as well as parameters particular to the operating period of the project. These latter parameters include electric operation period designation (on/off peak), capacity factors, and fuel specifications.

The plant capital cost can be expressed as one value, but a breakdown of the total capital cost is often necessary. The components of the total capital cost include equipment and construction costs, escalation, contingency, indirect construction costs, cost of land, and interest during construction. Also important are the distributions of these costs during the construction period.

The financing assumptions describe the financing method for the project and the structure of any lease or long-term debt repayments. The relative proportions of debt, lease, and equity financing must be defined. Other important data include interest rates, debt or lease terms, and number of debt or lease payments per year.

The revenue data include electric sales rates, steam sales rates, tipping fees for waste-to-energy plants, and other revenues. Electric sales rates are often divided into capacity charges and on- and off-peak energy rates.

Expense data include fuel prices, nonfuel O&M expenses, and other operating expenses. Nonfuel O&M expenses consist of costs for items such as operating labor, maintenance, and waste disposal.

Other cash flow parameters include depreciation assumptions and income tax assumptions. Various methods can be used to depreciate all or portions of the total capital cost. Income tax rates include federal, state, and local rates. Also, several options exist for the treatment of tax losses.

25.6.2.4 Pro Forma Analysis Results. Although each financial pro forma model usually has a unique presentation format, certain financial results are typical to most models:

- Debt service coverage ratios,
- Net present values,
- Internal rate of return, and
- Return on investment.

The debt service coverage ratio measures how well a project can cover its annual debt repayments. The ratio is calculated by dividing the cash available for debt service (revenues less expenses) for a given year by that year's debt service payment (principal and interest payment). This ratio must be at least 1.0 to cover the annual debt service payment, but it should be higher to ensure that the project can provide an adequate after-tax return to the equity investor. Also, for high-risk projects, the lender would require a higher minimum debt service coverage ratio.

A net present value (NPV) is usually reported for the after-tax cash flows distributed to the equity investor. The NPV is calculated by discounting the stream of after-tax cash flows to the beginning of the operating period and then subtracting the corresponding future value of the equity investment. A positive NPV indicates that the project is acceptable at the chosen discount rate.

An internal rate of return (IRR) can also be reported for the stream of after-tax cash flows. The IRR is that discount rate at which the NPV equals zero. An IRR higher than the chosen discount rate or "hurdle" rate indicates that the project is acceptable. One problem with the IRR method is that multiple IRRs may exist for a stream of cash flows if sign changes (alternating positive and negative cash flows) occur within that stream.

The return on investment (ROI) is a measure of how profitable a project is for a given year or over a fixed time. An annual ROI is calculated by dividing the after-tax cash flow distributed to the equity investor for a given year by the amount of the equity investment. This annual ROI can be shown on a nominal (undiscounted) basis or on a discounted basis. A cumulative ROI can also be calculated for a given year.

25.7 CONCEPTUAL DESIGN ENGINEERING

Conceptual design engineering is a systematic, objective, investigative process in which the basic technical requirements, operational characteristics, and constraints applicable to a specific power plant system are evaluated and defined. Conceptual design engineering provides the basis for selecting design concepts and equipment, and defines the key design features of the plant, functional systems and structure, system and equipment design constraint, plant performance, and plant costs. Conceptual design engineering lays the technical foundation for the follow-on detailed design process, equipment procurement, construction, and operation of the new facility.

25.7.1 Design Characterization

The first step in the conceptual design engineering process is design characterization. This activity focuses on developing the preliminary power plant technical and cost information to support project planning and analysis. As a project evolves, the engineering effort to characterize the project increases and the level of conceptual design becomes increasingly more comprehensive.

Three levels of conceptual design are defined in Table 25-2. The first level is a very broad engineering assessment for the project that generally supplies sufficiently accurate information to support system planning studies, preliminary economic and financial assessments, and plant siting evaluations. Estimates of capital costs at a conceptual level are typically prepared as part of this initial power plant characterization.

The second level of conceptual design characterization provides additional detail regarding the site layout, plant performance, and individual power plant systems. This level of design detail supports permitting and licensing activities.

The third level of design characterization is very specific with regard to system and equipment design and firmly lays the foundation for follow-on detail design activities and equipment specification and purchase.

25.7.2 The System Approach to Design

A power plant consists of numerous complex and unique processes and elements including buildings, structures, equip-

Table 25-2. Levels of Conceptual Design

Level 1 General Plant Description	Level 2 General Plant Description	Level 3 General Plant Description
Plant performance estimate	Plant performance estimate	Plant performance estimate
Capacity	Capacity	Capacity
Heat rate	Heat rate	Heat rate
Fuel consumption	Fuel consumption	Fuel consumption
	Solid waste production	Solid waste production
	Air pollutant emissions	Air pollutant emissions
	Consumable consumption rates	Consumable consumption rates
	Cooling requirements	Cooling requirements
	Water mass balance	Water mass balance
	Turbine cycle heat balance	Turbine cycle heat balance
		Other heat and material balances
	Boiler efficiency	Boiler efficiency
	Auxiliary power	Auxiliary power
	Site arrangement drawing	Site arrangement drawing
	System definitions	System definitions
		System design specification
		System description
		Performance requirements
		Design parameters and criteria
		List of major components
		Process flow diagram
		List of standards and codes
		Identification of interfacing systems
		Regulatory requirements
		Operating conditions
		Redundancy requirements
		Functional constraints
		Physical constraints
		Process control description
		Testing requirements
		Component design criteria
		Materials selection
		Plant arrangement drawings

ment, and controls. The systems approach is a design philosophy in which these elements are categorized into functional systems. This process is not unique and has been applied to the conceptual and detailed design of numerous power plants. The system approach is also used during the checkout, testing, and startup of plant equipment.

System categories for several types of fossil-fueled power plants are listed in Table 25-3. The table indicates which systems are generally applicable for a particular type of power plant, although the unique requirements of a project will likely result in a project specific system listing. Within each category shown on the table, several distinct systems can be defined. For example, the ash and scrubber solids category can be subdivided further into bottom ash and fly ash handling systems. Similarly, the cycle heat rejection category can be subdivided into a condensing system and circulating water system. The number of defined systems can exceed 100 for large coal-fueled power plants.

Table 25-3. List of Power Plant System Categories

	Generating Plant Type			
	Coal Steam	Gas/Oil Steam	Simple Cycle Combustion Turbine	Combined Cycle Combustion Turbine
Ash and scrubber solids	✓			
Auxiliary power supply	✓	✓	✓	✓
Auxiliary steam	✓	✓		✓
Buildings and structures	✓	✓	✓	✓
Bulk materials (other than coal)	✓			
Coal handling	✓			
Combustion gas	✓	✓		✓
Cleaning and exhaust				
Communication	✓	✓	✓	✓
Compressed air	✓	✓		✓
Compressed gas storage	✓	✓		✓
Control	✓	✓	✓	✓
Cycle heat rejection	✓	✓		✓
Drains and plumbing	✓	✓	✓	✓
Electrical	✓	✓	✓	✓
Equipment cooling	✓	✓		✓
Feedwater	✓	✓		✓
Fire protection	✓	✓	✓	✓
Fuel gas	✓	✓	✓	✓
Fuel oil	✓	✓	✓	✓
Generator terminal	✓	✓	✓	✓
Information	✓	✓	✓	✓
Lighting	✓	✓	✓	✓
Substation	✓	✓	✓	✓
Sampling and analysis	✓	✓	✓	✓
Space conditioning	✓	✓		✓
Steam generation	✓	✓		✓
Turbine extraction	✓	✓		✓
Turbine generator	✓	✓		✓
Waste collection and treatment	✓	✓	✓	✓
Water supply and storage	✓	✓	✓	✓
Water treatment	✓	✓	✓	✓

As a designer investigates a specific system, its conceptual design is defined by the following:

- Process flow arrangement,
- System operating requirements,
- Reliability and redundancy requirements,
- Equipment/component design criteria,
- Control philosophy, and
- Interfacing system requirements.

25.7.3 Evaluation of Alternative System Design Concepts

The majority of the decisions made during the power station design process are complex and involve several different but interacting components. These decisions also involve mechanical, electrical, chemical, control, electrical, and civil-structural engineering disciplines. System design tradeoff studies are performed during the project's conceptual design phase to select an optimum design concept from several possible alternatives. A design tradeoff study consists of an evaluation of technical design issues for a specific system coupled with an economic comparison of the alternatives. The study considers the entire system and all pertinent components, interfaces, and constraints that affect the final design. The results of the design tradeoff study establish the optimum design approach for implementation in the follow-on detailed design process. Tradeoff studies typically do not evaluate system alternatives from a financial perspective.

In some instances, selection of a specific system design concept from several alternatives can be made by an experienced designer without any formal technical or economic analysis. If low first cost is a requirement, evaluation of higher capital cost design alternatives is not required. If low life cycle costs are desired, then alternative design concepts having higher capital costs but lower annual operating costs achieved through a combination of enhanced efficiency characteristics, O&M savings, or lower replacement energy costs can be considered. Similarly, technical constraints may dictate the selection of a specific system design concept.

25.7.3.1 Cost–Benefit Analysis Methodology. Each design alternative is fully characterized with regard to its capital cost, performance and operational costs, overall system operational constraints and impacts, maintenance requirements, and equipment availability characteristics. The evaluation compares the initial capital cost of each alternative with the associated economic and operational benefits. The alternative that meets system operational requirements in the most cost-effective manner is selected for inclusion in the conceptual design. Following the completion of a tradeoff study, the specific design parameters for the system and individual components within the system serve as the basis for the follow-on detailed design.

The economic comparison of a low capital cost/low efficiency design option with a high capital cost/high efficiency design option is illustrated in Fig. 25-5. The economic comparison is presented on a cumulative present worth basis assuming conventional utility economic methods using fixed charge rates to annualize the equipment capital cost. Option A has the lowest initial capital cost, but its annual operating costs are more than those for Option B. Depending on the specific system being investigated, the annual operating costs may include fuel costs, replacement energy costs attributable to system unavailability, and O&M costs. At the end of the 30-year economic evaluation period, the cumulative present worth of the annual fixed costs plus annual operating costs is greater for Option A. However, it takes about 13 years for the enhanced operational efficiency associated with Option B to balance its increased capital cost with respect to Option A. Option A would be selected if low first cost is of utmost importance. Option B would be selected if the plant operator desires to minimize energy production costs over the life of the plant.

25.7.3.2 Availability Analysis Methodology. Availability analysis is used to measure the effective performance of generating units, systems, and major system components. Availability is defined mathematically as the sum of the durations the unit, system, or component was operational divided by the total period of interest. The difference between these two values is called downtime, or the period the

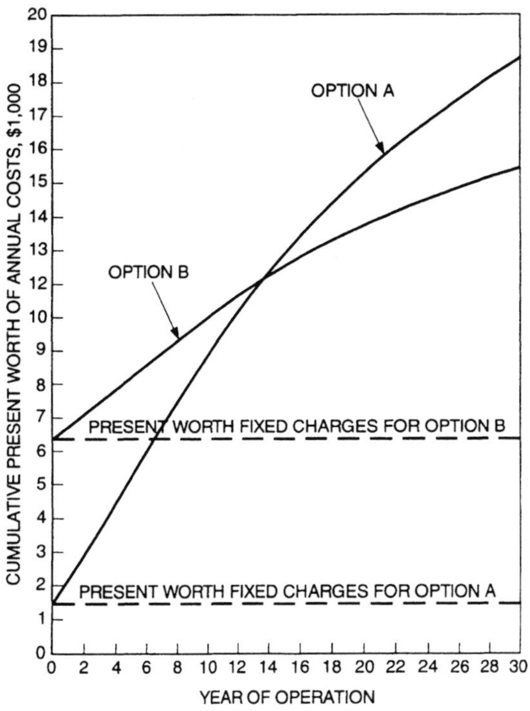

Fig. 25-5. Example tradeoff study economic comparison.

system or equipment is not operational. The two main reasons for downtime are planned maintenance or equipment failures.

Availability analysis involves the investigation of component failures, causes of failure, and the reasons for the time taken to return the component to service. The analysis normally considers the alternative methods of increasing the operational time of the component or of mitigating the effects of the component's downtime on the overall system. Operating plant improvements can often be justified by an availability analysis that assesses overall plant operation and determines the relative importance of system or component failures on plant productivity. Performing availability analyses allows the system designers to objectively assess alternatives.

Three principal availability analysis techniques are the Failure Mode and Effect Analysis, the Reliability Block Diagram (RBD), and the Fault Tree Analysis (FTA). All three techniques may be used for qualitative analysis to identify potential component failure modes and to determine the contribution of component failure to system availability. Only the RBD and FTA methods can reasonably be used for quantitative analysis, wherein the components or groups of components are characterized by average failure rates and mean times to restore.

FAILURE MODE AND EFFECTS ANALYSIS (FMEA). FMEA is a systematic procedure of documenting equipment failure modes and evaluating their consequences. The failure analysis is presented in tabular format and typically consists of information for each component such as name, identification number, function, failure modes, and the effects of failure. The major advantage of the FMEA method is that it permits a qualitative analysis with a minimum of training in reliability theory. Results adequate for most preliminary studies can be obtained without requiring failure data. The FMEA tabular format permits a reviewer to easily verify that all necessary equipment failure modes have been identified and their effects evaluated. A limitation of the FMEA is that it treats only single-event failures satisfactorily; double-event (and greater) failures are not detected conveniently.

RELIABILITY BLOCK DIAGRAM (RBD) METHOD. The RBD method gives a success-oriented, pictorial representation of a system that indicates normal system logic and functions. Major components or subsystems are represented by blocks interconnected by a series of parallel paths to represent the operating configuration of the system. A qualitative solution of an RBD yields information as to which components are essential to successful system operation. Multievent failure paths can be identified by this means. However, on a qualitative level, this task is more easily performed by Fault Tree Analysis. The major advantage of the RBD is in the calculation of system availability. The functional blocks of the diagram are characterized by the failure rate and mean time to restore the equipment performing the function. A series or parallel network can be simplified, using network reduction formulae, to a simple series network that can be solved for the value of availability for the total system.

If a calculated value of availability for a given design concept is compared to a target value for the system, the system design can be modified efficiently to improve availability. The intermediate results of the RBD simplification procedure guide the system designer to those areas that can be effectively improved by introducing equipment redundancy or alternative equipment designs. Simple system configurations often can be expressed in terms of an availability equation as a result of inspection. This approach bypasses the drawing of the block diagram but uses the same, or similar, theoretical principles to calculate the result.

An example RBD diagram for a series configuration 100% capacity boiler feed pump station is presented in Fig. 25-6. The boiler feed pump system availability is designated as A_{BFP} and is the product of the individual component availabilities, A_m, A_g, and A_p. The system unavailability is the mathematical complement of the availability and is designated as $\bar{A}_{BFP}$.

FAULT TREE ANALYSIS (FTA). FTA uses a fault tree to describe a system in terms of the events leading to failure. The "top event" is the failure mode being investigated (and this must be carefully defined), then a treelike structure is formed to represent all the sequences of failure events that result in this top event. The fault tree provides a graphical display of how the system can malfunction, enhancing the understanding of the failure mechanisms that control the system. A qualitative FTA can be performed on a system to identify the components that are essential to system operation. This approach can also highlight common mode failures that may not be evident without the analysis.

25.7.3.3 Alternative Design Concepts. Numerous design alternatives can be considered for virtually every power plant system. Large coal-fueled steam plants have a greater number of possible system design solutions than a natural gas-fueled simple cycle combustion turbine installation. Combustion turbine combined cycle plants have several design options from which to choose. Some typical system tradeoff studies that can be conducted as part of the conceptual design process are briefly described in the following paragraphs.

COAL DELIVERY. Alternative methods of transporting coal from the coal source to a power plant site are defined and

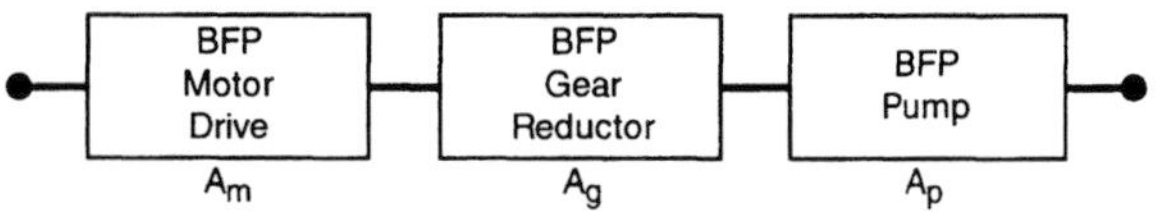

$$ABFP = (Am)\,(Ag)\,(Ap)$$

$$\bar{A}BFP = (1 - ABFP) = \bar{A}m \bullet Ag \bullet Ap + Am \bullet \bar{A}g \bullet Ap + Am \bullet Ag \bullet \bar{A}p + \bar{A}m \bullet \bar{A}g \bullet Ap + \ldots\ldots + \bar{A}m \bullet \bar{A}g \bullet \bar{A}p$$

Fig. 25-6. Example reliability block diagram series configuration boiler feed pump.

evaluated from a technical and economical perspective. Evaluation of commercial and institutional factors may also be required. The alternatives considered depend on the specific plant site location, plant size and fuel demand, site terrain, coal source location, regional infrastructure, and other site-specific local conditions. Some typical alternatives are truck delivery, coal conveyors, barge or ship delivery (inland or ocean-going movements), rail delivery, and possibly coal slurry pipelines.

Truck and conveyor systems are used for applications where the coal source is relatively close to the power plant site. Generally, the truck option requires a smaller up-front investment than a conveyor system, but annual fuel and maintenance expenses can be higher. Rail and coal slurry delivery systems are appropriate for relatively long movements. Coal movement by barge and ship is relatively inexpensive on an incremental ton-mile basis compared with the other coal delivery alternatives. However, a potentially significant investment may be required to develop the wharf and unloading facilities.

STEAM CYCLE ARRANGEMENT. The number, location, and design conditions for the regenerative feedwater heaters in the steam cycle are established in this system design tradeoff study. In addition, the locations for steam extraction off the steam turbine for feedwater heating, terminal temperature differences, and feedwater heater drain cooler approaches are established. Installing additional heaters in the steam cycle increases the initial capital cost for the plant, but lowers the annual fuel expenses because the plant heat rate is improved. If the fuel price is sufficiently high and projected to stay high, the differential capital cost associated with additional heaters can possibly be offset by savings in annual fuel costs. The number of years it takes the fuel savings to offset the higher capital cost would need to satisfy economic selection criteria established for the project.

BOILER FEED PUMP. The power required to drive the boiler feed pumps often represents the largest continuous auxiliary load in a power plant. This tradeoff study addresses the number and arrangement of the boiler feed pumps and the type of drive (electric motors or steam turbines). Electric motor drives are less expensive than steam turbine drives but require power that could be otherwise sold on the electric transmission grid. Steam turbine drives use high-pressure steam, and their operation impacts the plant's heat rate and power output. Steam turbine drives may offer less operational flexibility than electric motor drives. The availability of the system is dependent on the number of boiler feed pumps, the type of driver, and the configuration (series versus parallel arrangement). Replacement energy costs can be assigned to the alternative with a lower system availability.

CYCLE HEAT REJECTION. The most advantageous method of heat dissipation for the power plant is selected in this tradeoff study. Alternative heat rejection systems typically considered include mechanical draft evaporative cooling towers of either crossflow or counterflow design, natural draft cooling towers, air-cooled condensers, wet/dry cooling towers, and river water once-through cooling. For evaporative cooling towers, both fresh water and salt water towers may be considered if the site is located near the ocean. The choice of a plant cooling system may be readily apparent to the system designer on inspection of the site conditions. For example, an air-cooled condenser is the appropriate selection if the source of makeup water is severely restricted or water is totally unavailable.

Mechanical draft cooling towers are used extensively in power plant applications. Natural draft towers offer savings in operating costs because fans are not required to create an airflow across the cooling tower fill. However, the initial capital cost is higher for the natural draft cooling tower option. Once-through cooling can provide a high plant efficiency, but a significant investment may be required for intake and outfall structures. The use of once-through cooling is site dependent and may be an environmentally sensitive issue.

INDUCED DRAFT. This tradeoff study includes evaluating alternative induced draft fan arrangements, number of fans, type of fans, and fan drives. Options for fan types include centrifugal and axial flow. Multiple fan arrangements can offer enhanced reliability but the initial capital cost may be greater than for single fan options. Operating costs may be higher for a multiple fan configuration because of operation at off-design points. Fan drive options include electric motors and steam turbine drives. The electric motor option provides for a low first cost, but annual operating costs can be greater than for a steam turbine drive system. The tradeoff study would be conducted very early during the conceptual design phase in conjunction with establishment of steam generator design parameters. A similar tradeoff study can be performed for the combustion air system (forced draft fans and primary air fans).

AIR QUALITY CONTROL. Commercially available alternative air quality control systems for sulfur dioxide (SO_2) and particulate emission control are evaluated in this tradeoff study. The study establishes the installation that represents the best available technology for control of plant emissions and that also results in reasonable cost and reliability. Options for particulate emission control include conventional electrostatic precipitators and fabric filters.

For SO_2 emission control, over a hundred possible options presently exist. However, many of these options can be eliminated on the basis of factors such as size limitations, lack of commercial experience, and high costs. Typical scrubber options include spray dryer absorbers and wet limestone scrubbers. As part of this analysis, an initial water management plan may be developed to demonstrate feasible plans for reuse of plant wastewater streams. The study may also include an evaluation of alternate methods for handling and disposing of scrubber solids and fly ash.

HEAT RECOVERY STEAM GENERATOR PRESSURE LEVELS. This tradeoff study is applicable to combustion turbine combined cycle plants only. In this study, the optimum number of pressure levels for the heat recovery steam generator (HRSG) is determined. A greater number of pressure levels may increase initial capital costs but may also result in fuel savings associated with improved efficiency.

25.8 DETAILED DESIGN ENGINEERING

Detailed design engineering expands the information developed during the conceptual design stage into a form that permits construction of the power plant facilities. Typical construction drawings that are prepared during the power plant detailed design stage are listed in Table 25-4. Within the context of detailed design engineering, system designs are completed and equipment sized to a sufficient degree so that a contractor can clearly determine the manufacturing and construction work requirements. The construction drawings are prepared in a manner to minimize field engineering design and associated construction delays. Engineering calculations are performed to support design decisions, equipment and material selections, and preparation of construction drawings. Engineering lists are prepared to define specific design parameters of individual components, procurement

Table 25-4. Typical Power Plant Construction Drawings

Civil–Structural	Mechanical/Chemical	Electrical	Control
Site arrangement	Plant arrangements	One-line diagrams	Digital logic diagrams
Structural steel	Piping and instrument diagrams	Three-line diagrams	Analog logic diagrams
Platforms	Composite piping	Raceway	Instrument schematics
Stairs	Detailed piping	Arrangement	Control device layout
Suspended concrete slabs	Piping fabrication drawings	Grounding	Interface connection diagrams
Foundations	Pipe hangers	Cathodic protection	Instrument installation details
Materials handling diagrams		Raceway	
Architectural details		Schematics	
Underground utilities		Conduit schematics	
Construction facilities		Lighting	
Roads and walkways		Freeze protection	
Grading and fencing			
Railroad			

information, and installation information. Engineering lists are prepared for the following:

- Major equipment,
- Electrical loads,
- Electrical assembly,
- Electrical devices,
- Instrumentation,
- Control devices,
- Valves,
- Mechanical devices,
- Visual annunciators,
- Pipe hangers, and
- Recording annunciators.

Detailed design activities are defined by engineering discipline: civil–structural, mechanical, chemical, electrical, and control.

Civil–structural detailed design engineering includes analysis of geotechnical investigations and soil borings for foundation design; design engineering for structures in accordance with code requirements and seismic design criteria; design of structures, foundations, concrete, reinforcing steel, structural steel, ductwork, platforms, stairs, and enclosures; preparation of materials handling system flow diagrams; and architectural and landscaping design.

Detailed design of mechanical systems includes sizing and selection of equipment; piping and valve sizing; preparation of piping and instrument diagrams (P&IDs); routing of piping systems; preparation of piping spool drawings and pipe hanger details; pipe stress analysis and design for high-temperature critical systems; water hammer analysis; and preparation of heat and material balances.

Detailed electrical design activities include preparation of one-line and three-line diagrams for electrical systems (including metering, phasing, synchronizing, and protective relaying systems); preparation of arrangement drawings for electrical cable trays, wireway, and underground duct banks; preparation of grounding drawings, raceway, and circuit lists; preparation of raceway and conduit schematic drawings; and establishment of protective relay settings.

Instrumentation and control detailed design includes layout of control panels or computer screens; preparation of control system digital and analog logic diagrams for the power plant systems; preparation of control and instrument schematic and interconnection diagrams; and preparation of instrument installation drawings.

Detailed design of chemical systems focuses primarily on water systems throughout the plant. Detailed chemical engineering design includes preparation of water mass balance diagrams showing water usage requirements for plant systems; design of water treatment facilities; design of sanitary waste treatment systems; design and selection of materials for corrosive applications; equipment selection and sizing for chemical process systems; and design of water analysis laboratory facilities.

25.9 EQUIPMENT PROCUREMENT AND CONSTRUCTION CONTRACTING

Equipment procurement specifications and construction work packages are prepared as an integral part of the detailed design process. Early in the project, the number and type of procurement and construction specifications are identified based on the type and complexity of the power plant. The scope for each package is also determined. The procurement packages generally fall into one of the following categories:

- Procurement specification,
- Furnish and erect specification,
- Construction specification,
- Service contract, or
- Turnkey specification.

Simple cycle peaking installations which consist of modular combustion turbine–generator units require few procurement and construction specifications. In comparison, a coal-fueled power plant contains thousands of individual components and requires multiple procurement and erection specifications. A list of typical procurement specifications for a large power generating plant is presented in Table 25-5. A companion list of typical construction specifications is shown in Table 25-6. The lists shown in these tables would be tailored for the individual requirements of each project.

A procurement specification for equipment is a comprehensive document containing information on bidding procedures, contract conditions, bonding, and insurance. It also contains the technical design requirements for the supply of equipment and materials. An equipment manufacturer is responsible for supplying the equipment and related services for a well-defined fixed scope of work; technical specifications and detailed construction drawings are supplied by the engineer. Information from the conceptual and detailed design stages is incorporated in the technical portion of the procurement documents.

A furnish and erect specification may be desirable for equipment that has unique erection requirements that can best be accommodated by the original equipment manufacturer (such as a large field-erected steam generator). A construction specification contains the detailed information and drawings required to erect the individual equipment components (which have been purchased in separate procurement packages).

Another contract used for power plant construction is a service contract. Service contract cover specialized services such as site security and site cleanup.

A turnkey specification typically specifies performance guarantees with much, if not all, of the specific design details left to the discretion of the turnkey contractor. The turnkey approach requires one contractor to be responsible for plant design, equipment procurement, plant performance, cost, construction, and schedule. This type of specification is appropriate for independent power projects using project financing.

Table 25-5. Typical Procurement Contracts for a Large Coal-Fueled Power Plant

Civil-Structural	Electrical	Mechanical
Site buildings[a]	Cathodic protection	Air quality control equipment
Bulk materials handling	Communications	Compressed gas
Concrete batch plant	Conductor	Cooling towers
Cranes and hoists	Isolated phase bus	Fire protection system
Elevators	Generator and protective relaying	Steam turbine–generator
Enclosures	Motors	Emergency generator
Breeching and ducts	Raceway	Combustion turbine
Structural steel	Continuous ac power equipment	Heat exchangers
Chimney[a]	Motor starters	Condenser
	Switchgear and bus duct	Deaerator
	Secondary unit substation	Feedwater heaters
	Generator transformer	Maintenance equipment
	Auxiliary transformers	Materials handling

Chemical	Control	Mechanical (continued)
		Mechanical drives
Acid and inhibitor equipment	Control and information system	High-pressure pipe
Chemical feed equipment	Flow instrumentation	Low-pressure pipe
Chlorination equipment	Level instrumentation	Pipe supports
Demineralization system equipment	Pressure instrumentation	Pumps
Water pretreatment	Temperature instrumentation	Oil purification equipment
Laboratory furniture	Environmental monitoring equipment	Space conditioning equipment
Water quality control system	Panels, racks, and cabinets	Steam generator[a]
Special analytical equipment	Test equipment	Fans
Chemical mixers		Shop fabricated tanks
Sanitary waste treatment		Valves
		Insulation

[a]Furnish and erect specification.

25.10 PROJECT CONTROL

Effective control of schedules, cost, design, and construction is essential to achieve a successful project. Every project is unique, and planning and control activities are tailored accordingly. As a project evolves from the initial planning stage to engineering completion, the level of detail becomes more focused and specific. There are a variety of vehicles for planning and controliing a project. The methods described below highlight approaches used for planning and control of power plant design and construction.

25.10.1 Schedule Control

Small projects (usually those with fewer than 100 activities) may use bar charts, that are sometimes called Gantt charts, to schedule and control a project. Alternatively, a milestone schedule that combines the characteristics of a bar chart with key milestone dates can be developed for a project. An example of a typical milestone schedule for a new power generation facility is shown in Fig. 25-7. A bar chart milestone schedule is prepared early during the project and includes the most significant events in the project and the target dates for accomplishing each task.

With a bar chart or a milestone schedule, major tasks and activities are listed on the left of the chart and a time scale is drawn from left to right. The durations of the tasks are then indicated with a bar. The milestone schedule is easy to prepare and provides a ready reference for assessing, in a broad manner, the progress of small projects. It has the disadvantage of not allowing evaluation of resource constraints (such as manpower needs) relative to specific tasks. Also, it does not consider the sometimes complex interrelationships between tasks and the effects of schedule slippage.

Large design and construction projects, which may contain as many as 6,000 distinct activities, will most certainly use the Critical Path Method (CPM) or a similar technique for scheduling and project monitoring of design and construction. CPM is a systematic procedure derived from linear programming resource allocation algorithms that seeks the

Table 25-6. Typical Construction Contracts for a Large Coal-Fueled Power Plant

Civil and Structural	Electrical and Control	Mechanical	Chemical	Construction Services
Site development	Electrical construction	Major equipment erection	Chemical cleaning	Site services
Substructures	Control and instrumentation installation	Piping installation	Special protective coatings	Soil testing
Superstructures		Piping and equipment insulation		Security services
Painting				
HVAC construction				

LETTER OF INTENT
NOTICE TO PROCEED
START CONSTRUCTION
DRUM LIFT
COMPLETE BOILER HYDRO
SYNCHRONIZE UNIT
GUARANTEED COMPLETION

DESCRIPTION

-3 -2 -1 1 2 3 4 5 6 7 8 9 10 11 12 13 14 15 16 17 18 19 20 21 22 23 24 25 26 27 28 29 30 31 32 33

ENGINEERING
BOILER & TURBINE FINAL LOAD / ARRANGEMENTS
FINALIZE CONCEPTUAL DESIGN
DETAILED DESIGN
MAJOR PROCUREMENTS (NOTE 1)
STEAM GENERATOR
TURBINE GENERATOR
PRECIPITATOR
DEAERATOR
FEEDWATER HEATERS
BOILER FEED PUMPS
TRANSFORMERS
HIGH PRESSURE PIPE
CONSTRUCTION
MOBILIZE
SITE PREPARATION & CONSTRUCTION FACILITIES
BOILER AREA FOUNDATIONS
TURBINE - GENERATOR FOUNDATIONS & PEDESTAL
ERECT BOILER AREA STRUCTURAL STEEL
ERECT / ENCLOSE TURBINE HALL
ERECT BOILER
ERECT TURBINE GENERATOR
ERECT PRECIPITATOR
ERECT COAL HANDLING STRUCTURES & EQUIP.
INSTALL PIPING & VALVES
INSTALL ELECTRICAL SYSTEMS
STARTUP & COMMISSIONING
BACK FEED UNIT 2 SUBSTATION
ENERGIZE AUXILIARY POWER
MAKE & STORE DEMIN WATER
OPERATE CIRC WATER (RECIRC)
HYDROTEST & RESTORE BOILER
DRAFT PATH CHECKOUT
FIRING CHECKS ON OIL
BOILOUT & CHEM. CLEAN BOILER / PREBOILER SYS.
CHECKOUT COAL HANDLING OPERATION
FIRST FIRE ON COAL
TURBINE - GENERATOR LUBE OIL FLUSH
TURBINE ON TURNING GEAR
STEAM BLOW
ADMIT STEAM TO TURBINE
LOAD / CHECKOUT / SYNCHRONIZE UNIT
FINALIZE UNIT / PERFORMANCE TESTS

NOTE:
1. SHIPMENT TO THE SITE INCLUDES 2 MONTHS FOR SHIPMENT FROM THE FACTORY TO THE SITE. TIME AT WHICH THE EQUIPMENT IS AVAILABLE AT FACTORY IS NOTED WITH ▲ SYMBOL.

NO	DATE	REVISIONS & RECORD OF ISSUE	BY	CHK	APP	FLM
A	05-10-94	INITIAL ISSUE	DVB	MDS		

ENGINEER MDS DRAWN DVB
CHECKED DATE
MILESTONE SCHEDULE
DRAWING NUMBER 32151-010-F1100 REV A
CODE
AREA
05109401 54A 05/11/94 06:23:16

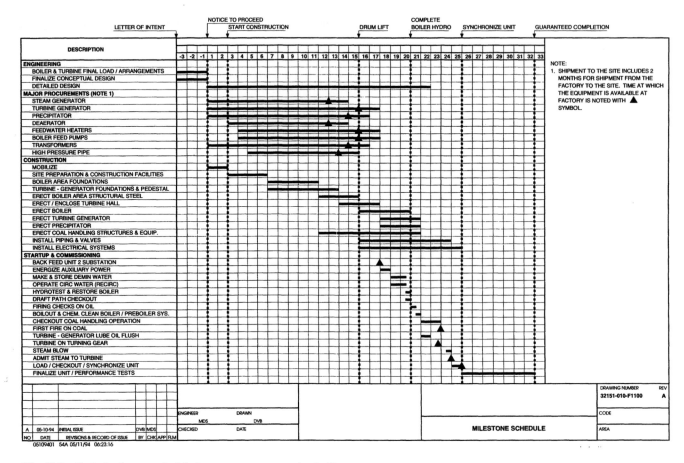

Fig. 25-7. Typical milestone schedule for a new power generation facility.

best allocation of limited resources to a number of competing activities. A similar scheduling and control method called Program Evaluation and Review Technique (PERT) is also used.

A CPM schedule is developed in essentially three steps. The first step is to identify all major project activities. These activities may include licensing, conceptual design, design engineering, procurement, construction, preoperational testing, plant startup, and initial operation. Second, the relationships between the activities are defined in a network diagram as illustrated in Fig. 25-8. The arrows represent activities and the circular nodes represent events. To develop a network diagram, planners define which activities must be completed before other activities can start. They also define which activities can occur simultaneously and which activities can't start until after completion of another activity. For the

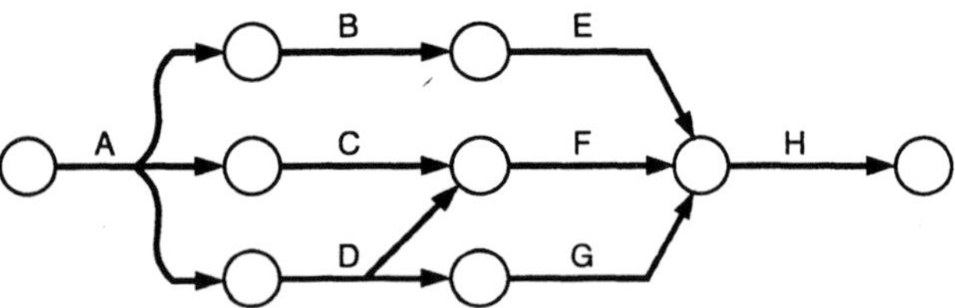

Fig. 25-8. Example critical path method network diagram.

construction effort, key interfaces between construction contractors are defined.

Once the logic of the network diagram is developed, the planner assigns time durations required for execution of each activity. Review of the network will reveal at least one path for which the early start time equals the late start time. This path is called the critical path and it establishes the shortest possible duration for the project schedule.

25.10.2 Cost Control

Controlling costs associated with large, capital intensive power plants requires significant dedication to the effort and a multifaceted approach. Methods for cost control include the following:

- Detailed study of design alternatives,
- Detailed design before procurement,
- Competitively bid fixed-price procurement,
- Coordination of procurement and design engineering,
- Proper management of construction, and
- Establishment of a cost monitoring system.

Cost control occurs early in project development. Technical and economic studies of design alternatives can be performed in advance of preparing the detailed plans and specifications. The objective is a careful balancing of capital costs and operating costs to develop an economical source of power generation.

By performing detailed design before procurement, equipment and construction needs are defined as explicitly as possible. The procurement process can then be based on competitively bid fixed prices. Designing before procurement also minimizes the costs associated with last-minute design changes. The coordination of design engineering with the design of purchased equipment is also important to avoid negotiated change orders. A knowledgeable construction management team is also vital to controlling plant costs.

Estimates of capital costs are prepared early during project development and are updated as design and procurement progresses. Preliminary estimates based on system concepts are less accurate than more definitive estimates prepared later during design. The current cost estimate, whether preliminary or very detailed, serves as the current basis for the project budget, which, with appropriate contingencies, is used to plan financing and revenue requirements. Design and construction performance are then measured against the budget. Establishing a cost monitoring system early during project development is important to control of capital expenditures.

25.10.3 Cost Risk

Costs for complex power plant projects with numerous activities and equipment procurements can be difficult to predict accurately. Variations from the anticipated average estimated values are inevitable. Traditionally, this concept has been dealt with through sensitivity analysis studies. These studies hypothetically raise or lower key cost components by some amount, then measure the change in overall project value. When the key cost components can be accurately identified and are very limited in number, this approach is satisfactory. However, as the number of uncertain cost elements expands, the sensitivity study approach becomes cumbersome and possibly incomprehensible.

An alternative approach is to use statistical methods to evaluate cost risk. In this method, a standard project cost estimate is developed, However, instead of simply estimating the most likely value for each cost element, a probability distribution is estimated to reflect the possible variation from the "target" or average value. The values within the range may be either uniformly distributed or centralized about the target. Completed project costs corresponding to various levels of risk or uncertainly can then be derived from the combined effect of variation in all pertinent cost components.

25.10.4 Design Control

Power station design can be effectively controlled by establishing a technically competent project team, implementing a usable quality assurance program, using a consistent set of engineering documents and drawings, and integrating a computerized database system into the design process. Design control must be a systematic process implemented early during project development.

A primary need for successful project design is to convey consistent information among the project designers. The information required varies from general site data and constraints in the conceptual design phase to highly specific information regarding systems, equipment, and structures in the later detailed design phases. Therefore, the design control process should establish methods for controlling design information from one stage of the design process to subsequent stages. Documents for controlling design information include the following:

- Regulatory requirements manual,
- Special studies and investigations,
- System analyses,
- Project design manual,
- System design specifications,
- Equipment procurement specifications,
- Construction specifications, and
- Project instructions.

Most of these documents are common in power station design work and are intended to reflect the increasing level of technical detail that develops as the project evolves.

25.10.5 Construction Control

In the construction phase, the system designs, drawings, and equipment, and structures procurement are brought together to create a functional power plant. Construction control requires the use of critical path method tools for progress assessment and controls for tracking material deliveries, storage, and erection. In addition, onsite construction managers assess the quality control procedures used by contractors and manage concrete tests, soil testing, structural steel checks, and various tests of mechanical boiler and piping systems.

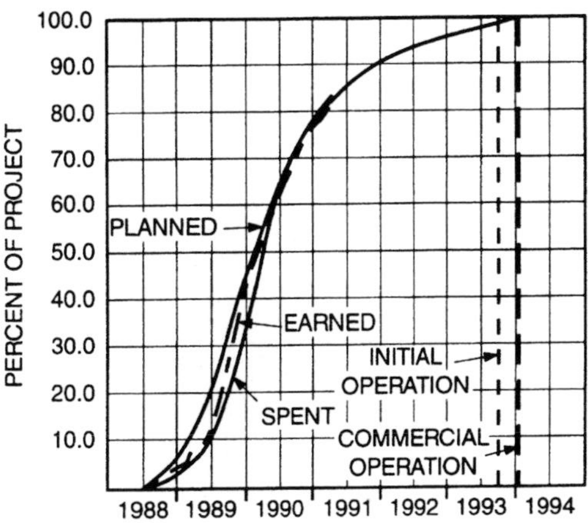

Fig. 25-9. Earned workday curve.

25.10.6 Progress Measurement Techniques

Critical path management can be used to monitor and control a project by comparing actual progress against planned progress. Engineering resources for design are allocated for activities within a critical path management workday S-curves. These curves are used to compare actual progress versus planned progress as a function of time. As activities are completed, they earn the workdays originally budgeted to them, regardless of the actual number of workdays spent on the activity. For activities that are only partially complete at a certain time, a reassessment is made of the remaining workdays for task completion. An example of an earned workday curve is illustrated in Fig. 25-9.

26

Permitting and Environmental Review Requirements

Anne F. Harris, Amy L. Carlson, Douglas C. Timpe, and *Lesley A. Wallingford*

26.1 INTRODUCTION

A critical first step in the development of a power generation facility is obtaining the requisite permits and approvals and satisfying the environmental review requirements applicable to the construction and operation of the proposed project. This chapter provides generic descriptions of the permits and approvals typically required in the United States for the construction and operation of power generation facilities.[1] Included in each description is an identification of the legislation requiring the approval, an overview of the permit application requirements, and a typical time frame for obtaining the permit. The permits identified are those typically required for coal-, gas-, and oil-fueled facilities. The specialized permit requirements applicable to hydroelectric and nuclear facilities are not addressed in this chapter.

The issuance of permits by federal agencies or the potential for impacts to state natural resources may also trigger the applicability of the National Environmental Policy Act (NEPA) and similar state environmental legislation. Obtaining information on baseline conditions, preparing environmental assessments or comparable documents, and complying with the procedural requirements associated with the environmental assessment process can add significant time and expense to project development. This chapter also describes the major legislation mandating such studies and offers general guidance for preparation of the documents required to comply with substantive and procedural environmental review requirements.

26.2 PRELIMINARY CONSIDERATIONS

Avoiding environmentally sensitive areas in site arrangements and in conceptual design, conducting site audits, and identifying specific technical information required for permit applications are important activities in project development. These activities are discussed in more detail in the following paragraphs.

26.2.1 Environmentally Sensitive Areas

The presence of environmentally sensitive areas on or adjacent to the project site can significantly impact project design and schedule. Every effort should be made to avoid or mitigate environmental impacts that are likely to result in project delays or the implementation of costly mitigation measures.

The criteria used in site selection methodologies to eliminate from further consideration those sites with air quality constraints or sensitive environmental features were discussed in Chapter 25. Once a site is selected, however, detailed baseline studies may reveal additional site features that should be considered in conceptual design to minimize or eliminate adverse environmental impacts. For example, if wetlands are present on the site, the wetland delineation and a survey of wetland boundaries will facilitate the development of a site arrangement that could minimize impacts to wetlands and the associated permit and mitigation requirements. Other environmental resources that should be considered in site development include threatened and endangered species, cultural resources listed or eligible for listing on the *National Register of Historic Places*, floodplains, wild and scenic rivers, coastal zones, sole source aquifers, critical habitats, and natural landmarks.

26.2.2 Site Audits

After site selection, one of the most important initial environmental activities associated with project development is conducting a Phase I site assessment to evaluate the possibility of existing pollution or contamination of the proposed site. Existing conditions that would require costly cleanup or ongoing treatment should be identified to enable developers, regulatory agency personnel, and investors to accurately evaluate the financial viability of the site.

A number of states have developed guidelines for Phase I site assessments. The American Society for Testing and Materials (ASTM) publications E 1527-93 and E 1528-93 provide guidance for those states that do not have guidelines. A Phase I assessment generally includes the following activities:

[1]International permitting requirements vary not only with the country in which the project will be located but also with the financing institution funding the project. Because of the variations of regulatory requirements in foreign countries, international permitting issues are not addressed in this chapter.

- A review of publicly available information and records to identify previous spills, citations, and corrective actions. This information is available from agencies such as the state and local departments of health and environment; the local planning, building permit, and fire departments; and regional pollution control agencies.
- Site reconnaissance, including a visual and physical inspection of buildings and structures.
- An evaluation of the present and past uses of both the subject and adjoining properties, including interviews with present and past property owners, neighbors, and employees.

A Phase II site assessment may be warranted if the results of the Phase I assessment indicate a significant potential for contamination, such as the presence of polychlorinated biphenyls (PCBs) or petroleum products in groundwater samples. Although no guidelines have been established for Phase II assessments, the nature and extent of the Phase II assessment will be dictated by the Phase I findings. Typical Phase II assessments include soil borings and analyses, ground water sampling and analyses, and the establishment of ground water monitoring wells to identify the nature and extent of contamination and to define the nature and extent of cleanup and/or treatment requirements.

26.2.3 Availability of Technical Information

The availability of existing technical information required to complete the requisite permit applications and the length of time required to obtain onsite information can have a major impact on the project schedule. The collection of baseline data may be weather or season dependent. For example, wetland surveys can be conducted only during certain months in the northeastern United States. These constraints should be taken into consideration when developing the overall project schedule once the environmental review requirements applicable to the project have been identified.

The application forms that must be submitted to federal, state, and local agencies typically require detailed engineering information which may be difficult to obtain at the early stages of project design. The technical information required to complete an air permit application for a typical combined cycle combustion turbine project is listed in Table 26-1. As the table indicates, the application must include details such as the annual capacity factor, hours of operation, combustion gas flows, and heat input. Other key information required at the outset of the permitting effort includes a project description, detailed site arrangement identifying the exact location of major project components (including construction lay-

Table 26-1. Information Typically Required for Air Permit Application for Combustion Turbine Facility

Site Information
Detailed site layouts/locations of structures (facility drawing)
Building dimensions (length, width, height)
Stack height and diameter
USGS topo. maps of the region
Facility fence boundary

Operational Characteristics
Annual capacity factor
Operational loads (at lowest load 50, 75, 100)
CTG output (kW)
CTG heat rate (Btu/kWh)
Fuel flow
Fuel composition/types/hours/consumption
Flow diagram showing operation
Date operation will begin
Heat input to CTG (at lowest load, 50, 75, 100)
Higher heating value of fuel
Lower heating value of fuel
Fuel tests/analyses
Number of startups/shutdowns/profiles

Material Handling
Fuel handling description/transfer points
Fuel storage information
Fuel transportation information
Storage of hazardous materials
Waste storage/disposal information
Length of all haul roads
Capacity and empty weight of each haul truck
Maximum design capacity of turbines
Amount of chemicals/substance delivered to site for emission control

Other Emission Structures
Type of cooling system (detail if cooling tower)
Auxiliary/backup equipment (detailed info as above)

Construction information
Construction start date
Construction end date
Construction equipment (e.g., unloader, scrapers)
Area of disturbed site (including laydown area)

Emission Inventory
PSD and NAAQS sources
Ambient air quality data

Stack Information (for each emission source)
Stack flue gas flow volume (in acfm)
Mass flow rate (lb/h) and molecular weight of flue gas
Stack exit temperature

Emission Information (for each source)
Control technologies
Removal efficiencies
Manufacturer's guaranteed levels for criteria pollutants (SO_2, NO_x, CO, PM, and VOCs)

Miscellaneous Information
Project costs
Cost of pollution control equipment
Owner/operator of facility

BACT Information
Design fuel analysis—ultimate and trace element (typical and ranges)
Design fuel higher heating value (typical and ranges)
Design fuel burn rate, MBtu/h (typical and ranges)
Design air margins (excess air and air heater leakage)
Air heater or combustor gas exit temperature
Design bottom ash/fly ash split
Uncontrolled emission rates (NO_x, CO, and nonmethane hydrocarbons)
Economic evaluation parameters of plant life

down areas), detailed water mass balance, and description of the proposed storm water management system.

Most permit application procedures include specific time frames for agency review, identified in the applicable regulations and triggered by the receipt of a "complete" application. As long as the reviewing agency continues to request specific information associated with the submittal, the formal review schedule cannot be initiated. Early coordination of the licensing and engineering teams and frequent consultation with the agencies with jurisdiction over the projects are critical to an expeditious application review process. Figure 26-1 is an example of a preliminary licensing schedule showing the major steps and lead times required to permit a baseload facility at the federal, state, and local levels.

26.3 FEDERAL PERMITS

Table 26-2 lists the federal, state, and local permits and approvals potentially necessary for construction and operation of a power generation facility. The table identifies the agency typically requiring the permit, the activity for which the permit is required, and the approximate time needed to obtain the permit if no significant environmental review is required by NEPA or comparable state environmental legislation.

26.3.1 US Army Corps of Engineers Permits

The US Army Corps of Engineers (COE) has been given jurisdiction over construction activities in or affecting most surface waters. The two major types of activities regulated by the COE are construction work or placement of structures in navigable waters, and discharges of dredge and fill material into waters of the United States.

26.3.1.1 Section 10 Permit. Section 10 of the Rivers and Harbors Act of 1899 requires a permit for the placement of structures in, over, or under, or for any work in or affecting navigable waters of the United States. Navigable waters are those subject to the ebb and flow of the tide, and rivers, lakes, and other waters used to transport interstate or foreign commerce (as defined in 33 Code of Federal Regulations [CFR] 329). COE regulations pertaining to Section 10 permits can be found at 33 CFR 320, 322, 329, and 330. Examples of activities that require a Section 10 Permit include installation of an outfall structure and piping in one of the Great Lakes, placement of pilings and piers, or a road crossing (bridge) or transmission lines built over a navigable river.

26.3.1.2 Section 404 Permit. Section 404 of the Clean Water Act (CWA) requires a permit for the discharge of dredge or fill material to waters of the United States. COE regulations pertaining to Section 404 permits can be found at 33 CFR 320, 325, 329, and 330. As defined in 33 CFR 328, waters of the United States comprise three categories: territorial seas, tidal waters, and nontidal waters. These categories include all intrastate and interstate waters (navigable or otherwise), wetlands, lakes, rivers and streams (including intermittent streams), impoundments, and cooling ponds as defined in 40 CFR 123.11. Waste treatment systems, including treatment ponds or lagoons designed to meet the requirements of the Clean Water Act, are *not* waters of the United States. Examples of activities requiring a Section 404 Permit include placing fill material for construction activities in a wetland area or installing utility lines in a wetland.

26.3.1.3 Types of Permits. Three types of permits—general, nationwide, and individual—are available from the COE for both the Section 10 and the Section 404 regulatory programs. Typically, general permits other than nationwide permits are not used for activities associated with constructing a power plant.

- General permits are available for activities causing only minimal individual and cumulative environmental impacts that are similar in nature or are controlled by other regulatory agencies.
- Nationwide permits, a type of general permit, are available for 40 predetermined activities identified in COE regulations. Examples of activities associated with the construction of power plants that would qualify for a nationwide permit include outfall structures permitted under the National Pollutant Discharge Elimination System (NPDES) permitting program and associated intake structures, and discharge of fill into wetlands of fewer than 10 acres. A nationwide permit verification typically requires about 3 months.
- Individual permits are required for activities that are not authorized by a nationwide or general permit or exempted from the COE's regulatory program. The issuance of an individual permit is a federal action that triggers review under the NEPA. Obtaining an individual permit that requires full compliance with NEPA requirements may take as long as 24 to 32 months. NEPA requirements are discussed in further detail in Section 26.6.1.

26.3.1.4 Permit Application Information Requirements. United States requirements for both predischarge notifications and permit applications are identified below.

PREDISCHARGE NOTIFICATION REQUIREMENTS. These are to be submitted prior to using certain prescribed nationwide permits, and include the following:

- Owner/operator and facility information;
- Project location, purpose, and description;
- A description of direct and indirect adverse environmental effects of the project;
- Wetlands delineation (when required);
- Confirmation of agency contact for threatened or endangered species and State Historic Preservation Office (SHPO) issues; and
- Any required fees.

SECTIONS 10 AND 404 APPLICATION REQUIREMENTS. Specific application requirements depend on the type of COE permit and the reason it is required. It is important to note that all information must be submitted in 8½ × 11 in. format. Enlarged copies of potentially illegible information should also

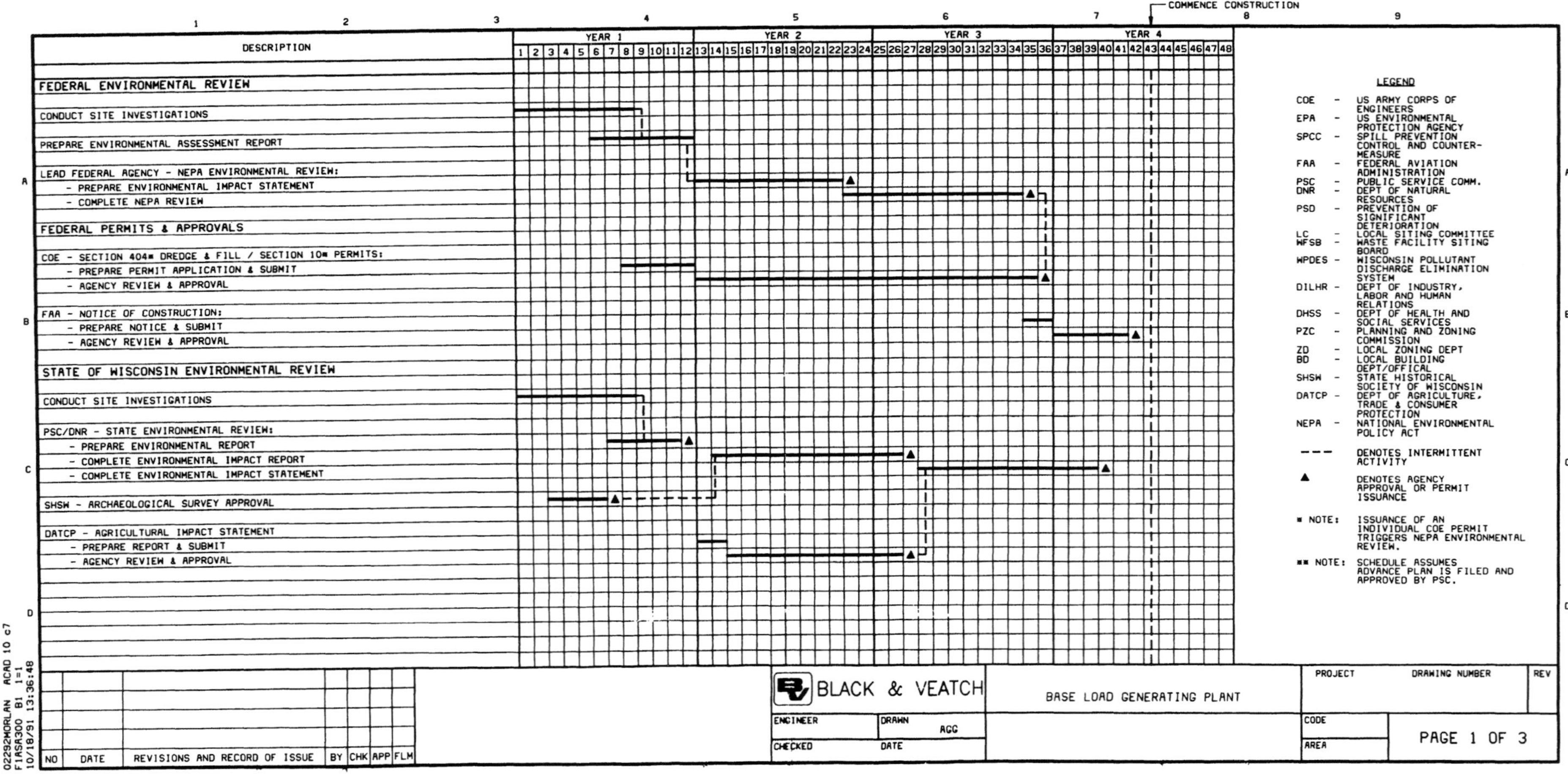

Fig. 26-1. Preliminary licensing schedule (page 1 of 3).

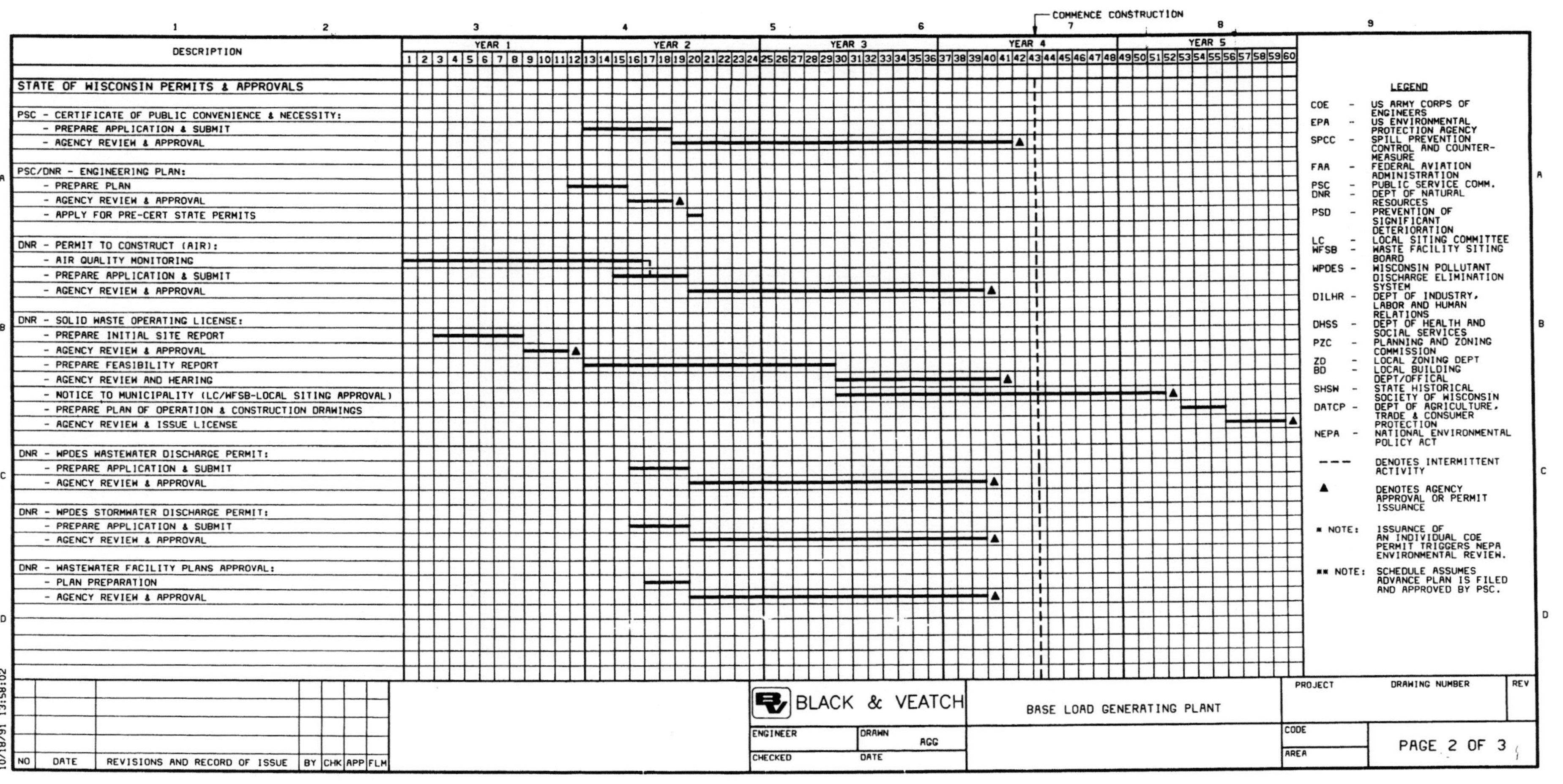

Fig. 26-1. Preliminary licensing schedule (page 2 of 3).

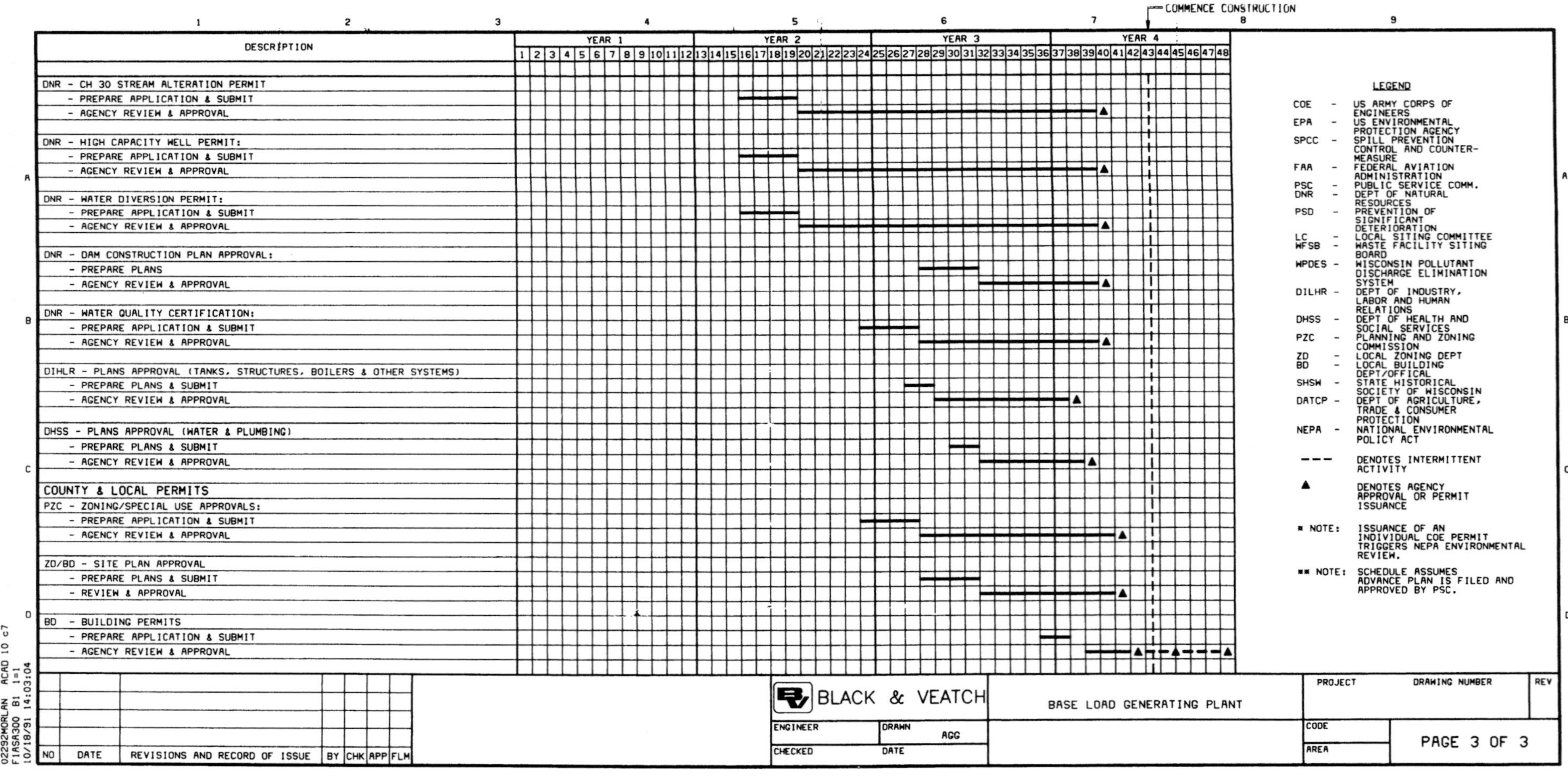

Fig. 26-1. Preliminary licensing schedule (page 3 of 3).

be provided. The application requirements for Sections 10 and 404 permits typically include the following:

- Site arrangement (including all ancillary facilities);
- Necessary drawings (e.g., intake/outfall structures, pipelines);
- 401 Water Quality Certification (or request, if certification not received);
- Coastal Zone Management Consistency Determination (if required);
- Project description;
- Wetlands information;
- Aquatic resources information;
- Cultural resources information;
- Threatened and endangered species information;
- Alternative site information;
- Impacts of construction (on all above issues);
- Impacts of operation (on all above issues); and
- Mitigation (on all above issues).

As indicated by the extensive information requirement listed above, detailed knowledge of the proposed project site is a prerequisite to preparing a COE application. Adequate time and budget should be allocated to ensure that site information is compiled.

26.3.2 Department of Energy

26.3.2.1 Alternate Fuels Capability Certification. The Powerplant and Industrial Fuel Use Act (FUA) prohibits the construction and operation of a new electric baseload power plant fueled with natural gas or other petroleum products that does not have the capability to use coal or another alternate fuel as a primary energy source. A baseload facility is defined as one whose annual generation in kilowatt-hours exceeds its design capacity multiplied by 3,500 hours (a 40% capacity factor). An example of a project requiring an Alternate Fuels Capability Certification is a baseload combined cycle combustion turbine unit firing natural gas. A natural gas-fueled peaking simple cycle unit does not require a certification.

The Department of Energy (DOE) Office of Fuels Programs is the agency authorized to administer the FUA. DOE regulations pertaining to Alternate Fuels Capability Certification are found at 10 CFR 500 through 516.

Developers of a new electric baseload power plant can either self-certify that the facility will have the capability to use coal or other alternate fuels, or can apply for a permanent exemption from meeting the statutory prohibition. A self-certification filing includes owner/operator information, plant location and configuration, design capacity, fuel type, name of the utility purchasing the energy, expected date of operation, and a notarized certification that the proposed facility meets the requirements described in the DOE regulations. Filing a self-certification with the DOE immediately triggers the applicant's compliance with the FUA. The DOE retains the right to collect additional information from the applicant within 60 days of submission, but the certification is effective on filing. To obtain a permanent exemption, a project must meet general and specific criteria and must provide detailed engineering and financial information in the petition for an exemption. Because of the liberal definition of "coal capable," compliance with this requirement is no longer a concern and can be handled with an appropriate self-certification.

26.3.2.2 Presidential Permit. Presidential Permits are required for electric and gas transmission lines that cross national borders. The DOE issues Presidential Permits for electric transmission lines; the Federal Energy Regulatory Commission (FERC) issues Presidential Permits for gas transmission lines. The United States Department of State also issues a Presidential Permit if a bridge or structure that would affect the integrity of United States borders is constructed. The US Department of State must approve all Presidential Permits issued by the DOE and FERC.

26.3.3 Environmental Protection Agency

26.3.3.1 Environmental Protection Agency—Prevention of Significant Deterioration (PSD) Permit to Construct. The Environmental Protection Agency (EPA) oversees permitting and new source review required by the Clean Air Act's (CAA) PSD program. A permit subject to PSD guidelines is required for the construction of any new combustion turbine facility located in an attainment area and that has the potential to emit more than 250 tons/year of any pollutant regulated under the CAA. For a combustion turbine facility, the major pollutants emitted could include nitrogen oxides (NO_x), sulfur dioxide (SO_2), carbon monoxide (CO), volatile organic compounds (VOCs), particulate matter (PM), and lead (Pb). Permit applicants are required to show source and pollutant applicability, provide a BACT demonstration, and perform a source impact analysis for each significant pollutant.

In many states, the state agency charged with protecting the environment has been delegated the authority or partial authority to administer the PSD program. In these states, the PSD permit is issued by the state. In cases where partial authority exists, a permit from both the EPA and the state agency may be required. To avoid duplication, the PSD program information is discussed in Section 26.4, State Permits and Approvals.

26.3.3.2 EPA—National Pollutant Discharge Elimination System Permit. The Federal CWA prohibits discharging wastewater and storm water into surface waters without first obtaining a permit from the EPA. The EPA regulates these discharges through the NPDES program.

In some states, the EPA has delegated the authority to administer the NPDES program to the state agency charged with protecting the environment. In those states, the NPDES permit is issued by the state. In cases where partial authority exists, a permit from both the EPA and the state agency may be required. To avoid duplication, the NPDES program in

Table 26-2. Potential Permitting Requirements for a Power Generating Facility

Agency	Permit/Approval	Reason Required	Approximate Time Frame[a]
Federal			
COE[b]	Section 10 Permit	Construction of outfall structure and associated intake structure in navigable waters of the United States	6–9 Months
	Section 404 Permit	Discharge of dredge or fill material into waters of the United States, including wetlands	6–9 Months
DOE	Alternate Fuels Capability Certification	Baseload facility using natural gas	3–6 Months
	Presidential Permit	Electric and gas transmission lines crossing national borders	6–12 Months
EPA	PSD Air Permit[c]	Construction of air pollution control equipment and emission sources subject to federal PSD program	12–15 Months if 1 year of baseline air quality data is not required
	Title IV Operating Permit[c]	Title IV CAAA of 1990	To be determined when regulations are finalized
	Title V Operating Permit[c]	Title V CAAA of 1990	To be determined when regulations are finalized
	NPDES Waste Water/Storm Water Discharge Permit	Wastewater discharge and storm water runoff during facility operation	9–12 Months
	NPDES Storm Water Discharge Permit	Construction activities	2 Months
	SPCC Plan	Onsite storage of petroleum products	3–6 Months
	Non-Transportation-Related Onshore facility Response Plan[d]	Onsite storage of petroleum products	6–9 Months
FAA	Notice of Proposed Construction or Alteration	Construction of an object that affects navigable airspace or is near an airport	3 Months
FERC	Qualify Facility (QF) Certification	Quality facility status	3–6 Months
	Exempt Wholesale Generator (EWG) Status	Selling electric energy at wholesale	9–12 Months
USCG	Marine Transportation-Related Facility Response Plan	Onsite storage of petroleum products	6–9 Months
State			
PUC	Certificate of Public Convenience and Necessity	Construction of an electric generation facility	6–9 Months
DEP, Division of Air	State Air Permit	Construction of air pollution control equipment and emission sources *not* subject to the federal PSD program	8–10 Months
DEP, Division of Water Quality	Section 401 Water Quality Certification	Impacts to state waters resulting from federal actions	3–6 Months
	Underground Storage Registration	Underground oil storage tank	N/A
DEP, Division of Water Resources	Surface/Ground Water Water Use/ Appropriation Permit	Water withdrawal	6–12 Months
DEP, Division of Solid Waste	Approval to Construct and Operate a Combustion Waste Landfill	Disposal of ash and other combustion wastes	36–60 Months

Table 26-2. *(Continued)*

Agency	Permit/Approval	Reason Required	Approximate Time Frame[a]
Local			
Planning Department	Site Plan Approval	Site development	6–12 Months
Zoning Department	Land Use: CUP/SUP, Rezoning Approval, Zoning Variances	Establishment of power generation and cogeneration plants as a permitted use	9–12 Months
Building Department	Building Permits	Construction of facility	2 Months
	Certificate of Occupancy	Commercial operation of facility	2 Months
Fire Marshal	Fire Safety Approval	Installation of fire protection system	2 Months

[a]Time frames represent projects that do not trigger applicability of the National Environmental Policy Act (NEPA).

[b]Certain prescribed federal and state actions may trigger the need for an environmental review or assessment of the project. Federal actions, such as REA financing or approval of the project, or the issuance of COE individual permits, require review under NEPA. Compliance with the substantive and procedural requirements of NEPA could add 1 to 2 years to the permit schedule. At a minimum, an Environmental Assessment discussing the environmental and socioeconomic impacts associated with construction and operation of the project will need to be prepared. Also, the state may require an environmental review of the project prior to issuing a Certificate of Public Convenience and Necessity or any environmental permits and approvals.

[c]Many states have been delegated either full or partial authority by EPA to issue this permit.

[d]Facilities that are adjacent to marine waters and require an EPA Response Plan may also require a Marine Transportation-Related Facility Response Plan from the US Coast Guard.

Abbreviations:

Federal
CAAA —Clean Air Act Amendments
COE —US Army Corps of Engineers
DOE —Department of Energy
EPA —Environmental Protection Agency
EWG —Exempt Wholesale Generator
FAA —Federal Aviation Administration
NEPA —National Environmental Policy Act
NPDES —National Pollutant Discharge Elimination System
PSD —Prevention of Significant Deterioration
REA —Rural Electrification Administration
SPCC —Spill Prevention Control and Countermeasure
USCG —US Coast Guard

State/Local
CUP —Conditional Use Permit
DEP —Department of Environmental Protection
PUC —Public Utility Commission
SUP —Special Use Permit

formation is discussed in Section 26.4, State Permits and Approvals.

26.3.3.3 Spill Prevention Control and Countermeasure Plan. Spill Prevention Control and Countermeasure (SPCC) plans were first mandated by the Federal Water Pollution Control Act Amendments of 1972, commonly known as the CWA, to establish procedures to prevent and contain discharges of oil into navigable waters of the United States. EPA published its Oil Pollution Prevention regulations at 40 CFR 112, effective in 1974, establishing requirements for SPCC plans for facilities storing large quantities of oil.

In 1991, EPA proposed changes to its existing regulations to implement the Oil Pollution Act of 1990, which amends the CWA. EPA considers these proposed regulations to be a clarification of the existing regulations and the best guidance available. According to these proposed regulations, SPCC plans for new facilities must be prepared and must be implemented *before* the facility initiates operations. SPCC plans for existing facilities must be reviewed and brought into compliance with the new regulations. Owners or operators of facilities failing to comply with these rules when finalized are liable for civil penalties of up to $5,000 for each day the violations continue.

To ensure compliance with the proposed rules, SPCC plans must include the following features:

- The entire containment system, including walls and floors, must be impervious to oil for 72 hours and must be constructed so that any discharge from a primary containment system, such as a tank or pipe, will not permeate, drain, infiltrate, or otherwise escape to surface waters before cleanup occurs.
- The engineer certifying the SPCC plan must be registered in the state in which the facility is located, and will be required to physically examine the facility and attest to that fact.

Plans must be amended *before* any change is made in facility design, construction, operation, or maintenance affecting the facility's potential for discharge of oil into waters of the United States.

The plan must also include the following information:

- Unit-by-unit storage capacity and type and quantity of oil stored;
- Estimates of quantities of oil that potentially could be discharged;
- Discussion of possible spill pathways;
- Procedures for routine handling of products and control of discharges;
- Provisions for disposal of recovered material;
- Detailed contact list with agency names and phone numbers for federal, state, and local personnel who must be contacted in case of an emergency, and essential information that must be provided to organizations on the contact list;
- Written commitment of manpower, equipment, and materials required to expeditiously control and remove any quantity of oil that may be discharged. EPA recommends that the facility

owner or operator consider factors such as financial capability in making the written commitments;

- Discussion of conducting self-inspections and regular maintenance on facility equipment; and
- Discussion of conducting annual training exercises, and ensuring that new employees are trained within their first week of work.

Additional requirements for events that occur during a period of flooding must also be addressed in the plan.

All in-plant drainage must be equipped with a diversion system that, in the event of an uncontrolled spill, will retain oil in the facility.

Integrity testing must be conducted every 10 years on aboveground tanks and when material repairs are conducted.

EPA anticipates that the proposed rules will be finalized in the near future.

26.3.3.4 Nontransportation-Related Onshore Facility Response Plan. Section 311 of the CWA and EPA regulations require nontransportation-related onshore facilities that could reasonably be expected to cause "substantial harm" to the environment by discharging oil into or on navigable waters to prepare a facility-specific plan for responding to such discharges and to submit the response plan to the EPA for review. EPA regulations for response plans are found at 40 CFR, codified at Part 112.

Response plan requirements apply to the following facility categories: (1) facilities with a maximum storage capacity greater than or equal to 42,000 gallons, if facility operations include over-water transfers of oil to and from vessels; and (2) facilities with a maximum storage capacity greater than or equal to one million gallons, if the facility lacks adequate secondary containment, is located near environmentally sensitive areas, is located near public drinking water intakes, or has experienced a spill greater than or equal to 10,000 gallons in the last 5 years.

In accordance with Appendix F of 40 CFR 112, response plans must contain detailed information on the following:

- Emergency response actions;
- Hazard evaluations;
- Small, medium, and worst-case discharge scenarios;
- Discharge detection systems;
- Response plan implementation;
- Procedures for facility self-inspections, mock alert drills; personnel training, and discharge prevention meetings; and
- Facility security systems.

Affected facilities were required to prepare and submit a response plan to the EPA by February 18, 1993, and to be operating in compliance with the submitted response plan by August 18, 1993, or the facility was restricted from handling, storing, or transporting oil. After August 18, 1993, a facility classified by the EPA as capable of causing "significant and substantial harm" is prohibited from handling, storing, or transporting oil unless the submitted response plan has been approved by the EPA. The facility is permitted, however, to operate without an approved response plan for up to 2 years, if it has certified in writing that it has ensured by contract or other approved means the availability of personnel and equipment necessary to respond to a worst-case discharge.

26.3.4 Federal Aviation Administration

Section 1101 of the Federal Aviation Act of 1958 requires filing with the Federal Aviation Administration (FAA) a Notice of Proposed Construction or Alteration (Notice) for construction or alteration of any structure over 200 ft above ground level. The FAA must also be notified of the construction or alteration of structures <200 ft within 20,000 ft of any airport. The FAA reviews the proposed construction for potential obstruction or hazard to air navigation and determines the appropriate marking and lighting standards for the structures. Examples of project components that may be subject to notification include stacks, coal silos, and evaporative draft cooling towers. Notice is also required for temporary construction equipment, such as construction cranes, that can reach a height of 200 ft or greater.

The Notice must include information on the location of the site and on the height, dimensions, and orientation of the proposed construction or alteration. The Notice must be submitted to the FAA at least 30 days before the start of construction. The FAA returns an Acknowledgement of Proposed Construction or Alteration within approximately 4 to 8 weeks. The acknowledgement details any special requirements, such as lighting or marking. A Supplemental Notice of Construction is required if changes are made in project plans that would affect airspace or if a construction project is abandoned or dismantled, and is also required at least 48 hours before the start of construction and within 5 days after the construction reaches its greatest height.

26.3.5 Federal Energy Regulatory Commission

Both the qualifying facility status and exempt wholesale generator status are granted by FERC. The processes by which projects receive FERC approval or certification are described in the following sections.

26.3.5.1 Qualifying Facility Certification. The Public Utility Regulatory Policies Act of 1978 (PURPA) affords certain protection and benefits to qualifying cogeneration facilities and small power producing facilities utilizing biomass, waste, geothermal, or renewable energy sources. As authorized by PURPA, FERC has promulgated regulations to implement the requirements and policies of PURPA. Facilities that meet the criteria identified in the regulations, such as specific operating and efficiency standards, are termed "qualifying facilities" (QFs).

Although achieving QF status is not a regulatory approval that is required for construction and operation of a new facility, the benefits afforded by PURPA often provide attractive contractual negotiating, financial, and operational advantages. PURPA benefits afforded to QFs include the following:

- An electric utility *must* purchase, either directly or indirectly, any energy and capacity made available from the QF.
- An electric utility must sell to any QF any energy and capacity requested, including supplementary, backup, maintenance, and interruptible power, at rates comparable to those for sales to other utility customers.
- An electric utility must make such interconnections with a QF as may be necessary to accomplish the required purchase or sale.
- An electric utility may transmit a QF's output to another utility rather than purchase that output, but the utility must obtain the QF's consent to do so.
- Each electric utility must offer to operate in parallel with a QF.

Eligible facilities may obtain QF status either through a simple self-certification process or through a formal FERC administrative process. In the self-certification process, facilities meeting the requirements need only provide notice to FERC of their qualifying status to be eligible for PURPA benefits. The notice contains general information about the applicant, a description of the facility, and a flow diagram demonstrating that the operating and efficiency criteria described in the regulations are met. FERC then publishes a notice of the self-certification in the *Federal Register*.

To obtain formal FERC certification, a detailed application describing the operating and efficiency standards of the facility and other technical data must be submitted to FERC. The application must also include a statement identifying the percentage of ownership any electric utility has in the facility. The application is reviewed under FERC procedural rules, which require publication of a notice of application in the *Federal Register* and public comment periods. Following verification by FERC technical staff that the facility complies with certification requirements, FERC issues a formal order verifying qualification for PURPA benefits. If the original application submitted to FERC contains the information required by FERC technical staff to evaluate the qualification of the facility, formal FERC certification can be obtained in 3 to 5 months.

FERC originally intended for project developers to use the self-certification process to obtain QF status. However, although FERC certification is more time consuming and costly, many financial institutions have been reluctant to finance projects without FERC certification. Further, some electric utilities and state regulatory agencies use FERC certification as a prerequisite for negotiating power sales contracts. Many project developers therefore request the formal FERC certification.

On January 25, 1995, FERC revised its rules under PURPA to include certain procedural modifications to encourage acceptance of the self-certification option among the financial institutions and to minimize the number of incomplete applications that were being submitted to FERC. FERC identified specific information needed to accelerate routine processing and evaluation of applications and provided additional specific application requirements for small power production facilities and cogeneration facilities. This information must be referred to prior to completing applications to obtain formal FERC certification. FERC also requires self certifiers to incorporate new application information into the notice of self-certification, and requires cogenerators and small power producers to provide copies of the notice of self-certification to each affected state commission and electric utility, thus providing greater assurance that a facility will meet the statutory and regulatory requirements for qualifying status.

26.3.6 Exempt Wholesale Generator Status

The Energy Policy Act of 1992 amended the Public Utility Holding Company Act of 1932 (PUHCA) to create a class of independent power producers known as exempt wholesale generators (EWGs). An EWG is defined as a person determined by FERC to be engaged directly, or indirectly and exclusively in the business of owning and/or operating an eligible facility and selling energy wholesale. Regulations governing procedures for implementing the amendments to PUHCA are found at 18 CFR Part 365. Under the amendments, EWGs are allowed to operate generating facilities under FERC regulation and sell electricity wholesale, without regard to geographical restrictions or regulation under PUHCA, thus providing a free wholesale market for independent power producers (IPPs). FERC rules and procedures establishing the requisite filing requirements for those seeking EWG status require those seeking EWG status to file an application with FERC and to provide the Security and Exchange Commission and affected state commissions with project information. FERC must render its determination within 60 days of receipt of the application.

26.3.7 US Coast Guard Marine Transportation-Related Facility Response Plan

Marine transportation-related (MTR) facilities, including certain Coast Guard-related onshore facilities, that could reasonably be expected to cause "substantial harm" to the environment by discharging oil into or on navigable waters of the United States must prepare a facility-specific plan for responding to such discharges and to submit the plan to the US Coast Guard (USCG) for review. USCG response plan requirements apply to MTR onshore facilities (including piping and any structures used for the transfer of oil to or from a vessel) capable of transferring oil, in bulk, to or from a vessel with a capacity of 250 barrels or more.

The authority to regulate transportation-related onshore facilities under Section 311 of the CWA has been delegated to the US Department of Transportation (DOT), and the authority for regulating MTR facilities has been further delegated to the USCG.

General response plans must contain detailed information on notification procedures, spill mitigation procedures, response activities, environmentally sensitive areas, disposal plans, training and drill procedures, plan review and update procedures, equipment, communications plans, and site-specific safety and health plans.

An affected MTR facility was to have prepared and submitted a response plan to the USCG by February 18, 1993, and was to be operating in compliance with its response plan by August 18, 1993, or the facility is prohibited from handling, storing, or transporting oil. After August 18, 1993, an MTR facility identified as being capable of causing "significant and substantial harm" may not handle, store, or transport oil unless the submitted response plan has been approved by the USCG. The facility is permitted, however, to operate without an approved response plan for up to 2 years, if it has certified in writing that it has ensured, by contract or other approved means, the availability of personnel and equipment necessary to respond to a worst-case discharge.

26.4 STATE PERMITS AND APPROVALS

26.4.1 Certificate of Public Convenience and Necessity

A certificate of public convenience and necessity (Certificate) from the state utility regulatory authority is typically required to commence site preparation for the construction of major utility power generation facilities and transmission facilities, such as electric and gas transmission lines. Depending on specific state requirements, underground gas and electric transmission lines may be exempt from the Certificate requirement. Although a power generation facility proposed by an IPP may also be exempt, a Certificate may be required for the associated gas and electric transmission lines.

Typically, a Certificate application requires a description of the transmission facility, including its location, and requires detailed planning information such as the following:

- A summary of any environmental impact studies for the project;
- A statement explaining the methodologies and analyses used to identify environmental concerns;
- A statement justifying the need for the facility;
- A description of any reasonable alternate locations for the proposed facility;
- A description of the comparative merits and detriments of each proposed location;
- A statement of the reasons why the primary proposed location is best suited for the facility; and
- A statement of detailed costs for the proposed facility.

Procedures associated with obtaining a Certificate typically include numerous public hearings in all counties through which the transmission facility will pass and review of the application by numerous state agencies such as the state's Departments of Environmental Protection, Transportation, Agriculture, and Commerce. The construction and operation of linear transmission facilities may also require numerous permits and approvals from federal, state, and local agencies. Preparing the required documentation and complying with the extensive procedural requirements may require 24 months or longer to complete.

26.4.2 Air Permits

Air permits are required for both construction and operation of power generation projects. The difficulty and feasibility of obtaining air permits are dependent on the type and location of the facility, applicability of federal/state/local air regulations, air emission control requirements, availability of existing representative meteorological and air quality data, and air dispersion modeling requirements. The PSD permit is the key air permit that a facility may need to obtain prior to construction. The CAA Title IV acid rain and Title V operating permits may be required prior to facility operation. The following subsections describe these air permits in further detail.

26.4.2.1 Prevention of Significant Deterioration Permit. PSD regulations were promulgated as a result of the 1977 CAA Amendments to ensure that air quality in existing attainment areas (areas in which air quality currently meets national ambient air quality standards [NAAQS]) does not significantly deteriorate or exceed AAQS, while providing a margin for future industrial growth. Sources subject to the PSD program are required to perform several analyses and obtain a permit to construct. The analyses include a BACT assessment, a PSD increment assessment, an NAAQS assessment, a preconstruction monitoring assessment, an additional impacts assessment, and public participation. The pollutant impacts are tracked by limiting PSD increment consumption. Increments are the amount of air quality degradation allowed after a "baseline" date which is set by regulations. Special consideration is given to pristine areas such as national parks and wilderness areas. These areas are referred to as PSD Class I areas. The PSD program is administered on a pollutant basis and is applicable only in areas where existing air quality is better than national ambient air quality standards.

ADMINISTRATION. The PSD program is administered by the Environmental Protection Agency (EPA). However, EPA can delegate all or portions of the program to state and local agencies that have approved permitting programs as part of their State Implementation Plan (SIP). The SIP is a record of specific state-mandated past compliance activities at facilities and a set of regulatory procedures implemented to bring nonattainment areas into attainment status.

APPLICABILITY. PSD regulations apply to new major stationary sources and major modifications under construction in areas designated as attainment or unclassifiable under Section 107 of the CAA for any criteria pollutant.

A new major stationary source is defined as any one of 28 source categories listed in 40 CFR 52.21 that emits, or has the potential to emit, 100 tons per year (tpy) or more of any of the regulated pollutants (Table 26-3). For all other categories of sources, a new major source is defined as one that has the potential to emit more than 250 tpy or more of a regulated pollutant. Steam-generating units firing fossil fuel (with >250 MBtu/h heat input) are considered one of the 28 source categories. Simple cycle combustion turbines are not

Table 26-3. Stationary Source Categories with a 100 tpy Threshold for Prevention of Significant Deterioration Major Source Status

1. Fossil-fueled steam electric plants of more than 250 MBtu per hour heat input
2. Coal cleaning plants (with thermal dryers)
3. Kraft pulp mills
4. Portland cement plants
5. Primary zinc smelters
6. Iron and steel mill plants
7. Primary aluminum ore reduction plants
8. Primary copper smelters
9. Municipal incinerators capable of charging more than 150 tons per day of refuse
10. Hydrofluoric acid plants
11. Sulfuric acid plants
12. Nitric acid plants
13. Petroleum refineries
14. Lime plants
15. Phosphate rock processing plants
16. Coke oven batteries
17. Sulfur recovery plants
18. Carbon black plants (furnace process)
19. Primary lead smelters
20. Fuel conversion plants
21. Sintering plants
22. Secondary metal production plants
23. Chemical process plants
24. Fossil-fuel steam generators (or combinations thereof) totaling more than 250 MBtu per hour heat input
25. Petroleum storage and transfer units with a total storage capacity exceeding 300,000 barrels
26. Taconite ore processing plants
27. Glass fiber processing plants
28. Charcoal production plants

one of the 28 source categories and are therefore not a major source unless they have the potential to emit 250 tpy or more of any of the regulated pollutants.

A major modification is defined as any physical change in or change in the method of operation of a major stationary source that would result in a significant emissions increase of any pollutant subject to regulation under the CAA, where source is defined as all emission units in the same industrial grouping (Standard Industrial Classification Code [SIC]), located on contiguous or adjacent properties and under common ownership. Net emission increases are generally calculated as the proposed emission increases minus any contemporaneous emission decrease at the source. Significant net emission increases of a pollutant are those that exceed the PSD significant emission rate for that pollutant. Certain activities are exempt from being considered modifications, such as routine maintenance, repair, and replacement activities. The exemptions are found at 40 CFR 52.21(b).

If the proposed source is determined to be a major new stationary source or modification to an existing stationary source, each pollutant must be compared to the new stationary PSD significant emission rates to determine PSD pollutant applicability. Specific PSD program review requirements are applicable only on a pollutant-by-pollutant basis. That is, the PSD requirements must be met individually for each pollutant that is potentially emitted in excess of the specified significant emission level for that pollutant.

Currently, six criteria pollutants are regulated under the PSD program: CO, NO_x, SO_2, PM_{10}, ozone (as volatile organic compounds [VOCs]), and Pb. Fluorides, sulfuric acid mist, total reduced sulfur, chlorofluorocarbons (CFCs), and halons are regulated as noncriteria pollutants. Criteria pollutants are those that have NAAQS established.

PSD REQUIREMENTS. Sources subject to the PSD requirements for a particular pollutant must perform the following analyses for that pollutant: ambient monitoring, air quality impact analysis, BACT determination, visibility and vegetation impacts and Class I area impacts. The applicant must also participate in a public comment process.

AMBIENT MONITORING. For each pollutant emitted in significant amounts, the applicant may be required to establish and operate a site-specific preapplication ambient air quality monitoring network for 12 months (a minimum of 4 months of air quality data will be required). If dispersion modeling results predict that the ambient impact from the proposed source will be less than certain threshold levels, the appropriate regulatory review agency may waive the monitoring requirement and may provide representative data on background concentrations of pollutants. Should preapplication ambient air quality modeling be required, the time required to obtain an air permit could be increased by 12 to 14 months.

Meteorological data are required as input to the dispersion modeling programs for the air quality impact analysis. The decision whether to conduct preapplication meteorological monitoring is made by the agency with jurisdiction over the permit and depends on the availability of representative National Weather Service data or historical onsite data and the type of modeling analysis required.

PSD INCREMENTS (CLASS I, CLASS II, CLASS III). The goal of the PSD program is to limit the degradation of air quality in those areas that currently are in attainment of the NAAQS. (These degradation limits are also known as the PSD increments.) All PSD areas are classified by the EPA as either Class I, Class II, or Class III. Class I areas are the most restrictive air quality regions, comprising areas of special national or regional value from a natural, scenic, recreational, or historical perspective. The PSD program provides special protection for such areas. Class II areas include areas where slightly more air quality degradation is allowed. Class III areas allow the greatest amount of air quality degradation. As of 1994, no Class III areas had been identified.

Increment levels are specific to each pollutant, ambient measurement averaging time, and PSD area classification. A new source subject to PSD requirements must perform a detailed air quality impact analysis to demonstrate that impacts of the new source emissions do not consume more than the remaining PSD increment. This requirement is generally waived if the new source has a predicted air quality impact that is less than the significant level as specified by the PSD program. Some states do not allow all of the remaining increments to be consumed by a single source. Those states typically permit only a percentage of the increment to be consumed.

The amount and complexity of air dispersion modeling associated with the PSD increment consumption analyses (and all other air impact analyses) are dependent on the project location, project emissions, and regulatory guideline requirements established for conducting the analyses. Greater air quality impacts, resulting in greater permitting difficulty, could be caused by the following factors:

- Complex terrain (terrain with elevations above stack top elevation)
- Shoreline fumigation
- Stack heights less than Good Engineering Practice (GEP) stack height. (GEP stack height is defined as the height at which air pollutants emitted from the stack are not unduly influenced by aerodynamic downwash from nearby structures or geographical features.)
- Number, size, and location of other pollutant emitting sources in the area.

Demonstrating compliance with the PSD Class II increments (and other air quality standards) may be difficult (i.e., may require extensive dispersion modeling and/or more stringent emission controls) if the proposed stacks are not constructed to GEP stack height. Emissions from non-GEP stacks may be affected by the formation of turbulent eddies on the lee side of buildings which can essentially "trap" the plume and bring it rapidly to ground level.

Demonstrating compliance with PSD Class II increments also depends on the number of sources and amount of emission increases there have been since the submittal of the first complete PSD application in an area. The likelihood of demonstrating compliance with PSD Class II increments is smaller with a greater number of large new pollutant emitting sources located near the facility.

Because of the lower SO_2, NO_x, and particulate matter emissions typically associated with natural gas firing, it would be less difficult to demonstrate compliance with the PSD Class II increments for a combustion turbine project firing natural gas than with oil- or coal-fueled technologies. However, the emissions resulting from use of the backup fuel (such as fuel oil) must also be considered.

NAAQS ANALYSIS. For each criteria pollutant emitted in significant amounts, an ambient air quality impact analysis must also be performed to demonstrate that the emissions will not cause or contribute to a violation of the applicable NAAQS. The analysis must take into account proposed source emissions, nearby source emissions, and ambient monitoring data (or background concentrations). Background concentrations represent the unquantified sources of air pollution and are typically added to source impacts predicted through dispersion modeling.

The requirement for the ambient air quality analysis is generally waived if the predicted impact from the new source is less than established significant concentrations. Some states may have ambient standards that are more stringent than NAAQS. Permit applicants must comply with the most stringent standards applicable to the project.

If the existing air quality of the area is "moderately" or "severely" polluted, demonstrating compliance with the NAAQS will be more difficult. Gas fired technologies in this case would have the higher probability of demonstrating compliance because of the lower emission rates.

BEST AVAILABLE CONTROL TECHNOLOGY. The PSD program requires new or modified sources subject to PSD review to use BACT to minimize the emission of each pollutant with a significant emission level. The BACT analysis will result in emission limits at least as stringent as required by any New Source Performance Standard (NSPS) or state emission regulations. The BACT requirements are discussed in further detail in Chapter 14.

ADDITIONAL IMPACTS ANALYSES. The PSD program requires that air quality impacts resulting from growth in the area of the project be assessed. Types of growth include the associated industrial, commercial, and residential growth that occurs as a result of construction and operation of the facility. In addition, the program requires that impacts on surrounding soils and vegetation be assessed. Furthermore, an assessment of potential visibility impairment in affected Class I areas must be conducted.

PUBLIC PARTICIPATION. The federal PSD program requires that adequate opportunity for public participation be a part of the permit review process (40 CFR 52.21). If there are significant public comments on the permit application, a public hearing may be required.

CLASS I ANALYSES. Class I areas are areas of special national or regional value from a natural, scenic, recreational, or historic perspective such as national parks or national wildlife refuges. The PSD program provides special protection for such areas by requiring that the applicant demonstrate that the PSD Class I increments will not be exceeded, nor will certain air quality-related values (including visibility) be adversely affected.

26.4.2.2 Title IV Acid Rain Operating Permit. One of the major new regulatory programs under the 1990 CAA Amendments (CAAA) concerns the control of SO_2 and NO_x, both precursors of acid deposition ("acid rain"). The centerpiece of Title IV of the CAAA is the establishment of an emissions allowance and trading program. The primary goal of Title IV is to reduce annual SO_2 emissions by 10 million tons below 1980 levels. This reduction will be achieved in two phases. Phase I begins in 1995 and affects 110 major existing electric utility stations with high SO_2 emissions. Phase II, which begins in the year 2000, imposes NO_x emission limitations and further tightens the SO_2 emission limits for the Phase I facilities and sets restrictions for other existing and new utility units.

The Title IV requirements are applicable only to "affected" units. (Under Title IV, a unit is defined as a fossil-fuel-fired combustion device. This includes combustion turbines.) Affected units include certain existing units and all new utility units. A utility unit is any unit that serves a generator that produces electricity for sale. A new unit is one which commenced operation after November 15, 1990.

Certain new units are exempted from being affected. These include cogeneration units that produce 25 MW or less of electricity for sale, and new utility units that produce 25 MW or less of electricity for sale *and* burn only fuels with $<0.05\%$ by weight of sulfur. A written petition must be submitted to obtain the latter exemption.

The Phase I permit applications were due to EPA by February 1993. In accordance with Section 408(d)(2) of the CAAA Title IV, a Phase II SO_2 permit application and compliance plan for each affected unit must be submitted to the permitting authority by January 1, 1996. A Phase II NO_x permit application and compliance plan for each affected unit must be submitted by January 1, 1998. Under Section 408(g) of the CAAA Title IV, a revised compliance plan can be submitted at any time. The compliance plan will be enforceable in lieu of a permit until a permit is issued. New units that are permitted after the state operating permit program becomes effective will generally have their acid rain permits incorporated into their Title V permits. Permit applications must include compliance plans and completed forms.

26.4.2.3 Title V Operating Permit. Title V of the CAAA, signed into law on November 15, 1990, adds new provisions to the CAA requiring all major sources (and some minor sources) of air pollution to obtain a permit to operate. The purpose of the operating permit program is to establish a program where all existing air permitting requirements can be centralized and enforced more efficiently. Also, states are to establish operating permit programs, based on EPA's minimum requirements, that will fund, through pollutant fees, the state permitting programs. The permit is intended to incorporate all applicable requirements such as NSPS, PSD, Hazardous Air Pollutants (HAPs), and existing National Emission Standards for Hazardous Air Pollutants (NESHAP) requirements. The permits required under the acid deposition control program (Title IV) may also be incorporated into the comprehensive operating permit.

The EPA's final regulations implementing Title V establish minimum requirements for an operating permit program that must be adopted by all states within a specified period of time. Within 1 year of receiving the state's program, the EPA is required to approve or disapprove it. Sources subject to the program are to submit complete permit applications within 1 year after the state program is approved by the EPA. The specific deadline date will vary from state to state, in accordance with the EPA's approval date. Even with approval delays, it is likely that all Title V operating permit applications will need to be submitted prior to the late 1990s.

Among other things, Title V permits must include emission limits (emissions inventories), compliance schedules, and monitoring and reporting requirements. In addition, the following information will generally need to be included in the permit application:

- Description and design parameters for all pollution controls and emission monitoring methods;
- Information required by all applicable regulations, including State Implementation Plan provisions;
- Data and calculations necessary to document emission determinations;
- Citations to all applicable regulations and test methods;
- Explanation of all exemptions and nonapplicable regulations;
- Information necessary to define alternate operating scenarios;
- Definition of present compliance status, a compliance plan and schedule, and certification of compliance by a responsible official; and
- Use of reporting, record keeping, and recording requirement.

Each permittee will have to submit semiannual reports and certify annually that it is in compliance. Permits are also to allow for inspection and entry to ensure compliance. Sources subject to the acid rain provisions of Title IV of the CAAA will be required to install and operate continuous emission monitors or some alternative method to provide information of the same type and accuracy.

Of particular note in the Title V provisions of the CAAA is a requirement that the state collect annual permit fees to cover the state's costs in administering the air regulatory program.

26.4.3 Water Quality Permits

Permits to discharge industrial wastewater, including storm waters, into waters of the United States are required during both construction and operational phases of power generation projects. The following subsections describe these permits and associated studies that may be required to obtain the permits.

26.4.3.1 Wastewater Discharge Permit. Section 402 of the CWA established the NPDES program which prohibits discharges of wastewater and storm water into surface waters without a permit from EPA. Although general NPDES permits are available for some activities, power generation facilities require "individual" (custom-designed) permits. The NPDES permit establishes effluent limitations, as well as specific requirements for monitoring, sampling and testing, recordkeeping, and reporting. The effluent limits are based on EPA's industry-wide effluent guidelines as applied to the quality of the receiving water.

EPA may delegate administration of the NPDES permitting program to the state agency charged with protecting the environment. In instances where the state has been delegated only partial authority to administer the NPDES program, a permit from both the EPA and the state agency may be required. As of June 1994, 40 states and territories have been delegated full or partial authority to administer the NPDES program.

EPA has prepared a number of application forms that should be used when submitting the permit applications for new or existing facilities. States may have developed additional forms to use. In addition to requiring general information about the facility, EPA requires the following technical information:

- Description of the receiving waters;
- A topographic map of the facility, extending at least 1 mile beyond property boundaries, showing the outline of facility; location of all intake and discharge structures; any hazardous waste treatment, storage, or disposal facilities; any injection wells; and all springs, rivers, and other surface water bodies;
- For each outfall, a description of all operations contributing wastewater to the effluent (including process wastewater, sanitary wastewater, cooling water, and storm water runoff), the average flow contributed by each operation, and the treatment received by the wastewater;
- Water mass balance; and
- An estimated amount of concentration and mass of any pollutant known or likely to be found in the effluent. (Note: Although estimated data are acceptable for new facilities, actual data must be submitted no later than 2 years after the facility has begun discharging.)

NPDES applications must be submitted at least 180 days before the discharge. In states that have delegated authority, other state agencies, such as a Wildlife Resource Commission, may assist in the review of the NPDES application. In states that do not have delegated authority, federal agencies, such as the US Army COE, US Fish and Wildlife Service, or the National Marine Fisheries Service may review the application. Typically, the permit application process includes 30 days for a public hearing, and the public has another 30 days to request an evidentiary hearing or legal review. The application process typically requires from 6 to 9 months. In states that do not have delegated authority, the NPDES permit may trigger the federal NEPA review requiring an environmental assessment (Section 26.6.1). Under these circumstances, it could take up to 24 months to obtain the permit. (Note: Facilities discharging wastewater into a municipal sewer system for treatment at a publicly owned treatment works [POTW] do not need an NPDES permit but will be required to comply with the POTW's pretreatment requirements before being allowed to discharge into the municipal system. The municipal facility must also have an EPA-approved pretreatment program to be eligible to authorize industrial discharges.)

New Source Performance Standards (NSPS) for wastewater discharges from steam electric power generating stations must be reviewed prior to design of the facility. NSPS have been identified for total suspended solids (TSS), total residual chlorine and free available chlorine (TRC and FAC), oil and grease, copper, iron, chromium, and zinc. Other pollutants that must be considered when the facility is designed include biochemical oxygen demand (BOD), chemical oxygen demand (COD), total organic carbon (TOC), flow, ammonia, thermal discharges (discussed below), and pH.

To improve or stabilize water quality standards, most states have identified additional pollutants of concern that must be addressed in the application. For example, in an attempt to eliminate the discharge of all phosphorus compounds into Lake Ontario, New York, is discouraging the use of any phosphate-containing scale or corrosion inhibitor in a generating facility's circulating water system. Because of its significant adverse impacts on aquatic biological systems, most states also prohibit or strictly limit discharges of chlorine to surface waters.

NPDES permits are issued for a term of 5 years. The renewal permit application process should be initiated at least 1 year before the expiration of the existing permit to ensure adequate time for application development.

26.4.3.2 Thermal Discharges. Section 316(a) of the CWA requires that applicants for an NPDES permit minimize adverse impacts to aquatic ecosystems from thermal discharges to surface waters. Heat is included in the definition of a "pollutant" (40 CFR 122.2) for which a facility must obtain an NPDES permit. The requirement for limitations in thermal discharge is a provision of the NPDES permitting procedures (40 CFR 122, 40 CFR 401). If the NPDES permit is administered by a state, the state's environmental department administers the requirement to limit thermal discharges.

Design features that reduce thermal impacts include the reduction of heat input to cooling waters, the use of cooling towers, or, in once-through cooling systems, lowering the temperature rise across the condenser by increasing cooling water flows. Documentation (i.e., monitoring, inspections, recordkeeping, and reporting) of the effects of thermal discharges on aquatic ecosystems may also be required. Reduction of thermal impacts should be considered early in the development of the conceptual design to minimize the potential for a requirement to conduct long-term aquatic monitoring which could significantly increase the project cost and schedule.

26.4.3.3 Cooling Water Intake Structure. Section 316(b) of the CWA requires applicants for a NPDES permit to minimize adverse impacts to aquatic ecosystems from cooling tower water intake structures (40 CFR 401.12 and 114). Significant aquatic impacts include impingement and entrainment of aquatic organisms such as fish and invertebrates. EPA regulations require implementation of the Best Available Technology (BAT) for cooling water intake structures. The location, design, construction, and capacity of cooling water intake structures are considered when evaluating BAT designs. Other critical factors include the shape of the shoreline, water velocity, size and placement of screens, and the use of biocides such as chlorine to control bacterial growth and mollusks such as zebra mussels. Because of concerns regarding cooling tower and intake structure impacts to aquatic organisms, natural resource agencies may require entrainment and impingement studies which can significantly increase project costs and schedules.

26.4.3.4 Storm Water Discharge Permit for Construction and Operation. As described previously, an NPDES permit must be obtained for all storm water runoff discharges associated with industrial activities, including runoff from steam electric generating facilities and construction sites disturbing over 5 acres. Storm water permits are issued as general, group, or individual permits, depending on the activity at the site that will be covered, whether the activity is covered under an existing permit, and whether an existing effluent limitation guideline for storm water for that particular pollutant has been established.

CONSTRUCTION. Typically, construction sites over 5 acres will require a general permit for storm water discharges. Storm water discharges associated with construction activities not authorized under a general permit that require an individual permit include the following:

- Nonstorm water discharges (except those listed in the general permit);
- Discharges already covered by an existing permit;
- Discharges that may contribute to a water quality violation; and
- Discharges that may affect a listed or proposed threatened or endangered species.

Storm water discharges that originate from the site after the completion of construction activities and after the site has undergone final stabilization will require an operational permit.

A Notice of Intent to use the General Permit for Construction must be submitted to the EPA (or delegated state agency) prior to initiating any construction activities at the site. A Pollution Prevention Plan, which describes the procedures that will be taken to reduce or eliminate pollutants in the storm water runoff and to minimize or prevent soil erosion, must be developed and implemented prior to submitting the Notice of Intent. The Notice of Intent is deemed approved on agency receipt unless and until the issuing authority subsequently determines that an individual permit is required.

OPERATION. Three different categories of permits are available to authorize storm water discharges during operation of the facility; group, general, and individual.

GROUP PERMITS. Group permits were developed for industries with similar discharges. Although no group permits were developed for the power plant industry, EPA drafted a group-general permit in November of 1993. The group-general permit would authorize storm water runoff from steam electric facilities that are subject to the effluent limitations for coal pile runoff. The final permit is expected to be issued in late 1994 or 1995.

GENERAL PERMITS. General permits are available for many activities associated with power generation facilities and can be obtained relatively quickly. The Notice of Intent to use the general permit for storm water discharges during facility operation must be submitted to the EPA (or delegated state agency) prior to commencing operation at the site. A Pollution Prevention Plan, which describes the procedures that will be taken to reduce or eliminate pollutants in the storm water runoff during facility operation, must be developed and implemented prior to submitting the Notice of Intent. The Notice of Intent is deemed approved upon agency receipt unless and until the issuing authority subsequently determines that an individual permit is required.

INDIVIDUAL PERMIT. Individual permit applications are the most costly and time consuming to obtain because they are specific for a particular facility. Storm water discharges that require an individual permit include the following:

- Nonstorm water discharges (except those listed in the general permit);
- Discharges already covered by an existing permit;
- Discharges that may contribute to a water quality violation; and
- Discharges that may affect a listed or proposed threatened and endangered species.

An application for an individual permit for storm water discharges during operation should be submitted with the application for a wastewater discharge (NPDES) permit. The agency generally reviews and issues the two permits concurrently. An individual storm water permit requires a Best Management Practices Plan, which is similar to the Pollution Prevention Plan required by a general permit. Application requirements for individual permits are similar to those described in the individual permits for wastewater discharges. New Source Performance Standards for pollutants in coal pile runoff are listed for total suspended solids. Other pollutants that will be considered during agency review include oil and grease, biochemical oxygen demand, chemical oxygen demand, total Kjeldahl nitrogen, nitrate plus nitrite nitrogen, and total phosphorus.

26.4.3.5 401 Water Quality Certificate. Section 401 of the CWA requires that a water quality certificate (Certificate) from the state or other appropriate authority be obtained or waived before a federal agency can issue a permit for an activity that may result in a discharge into navigable waters. The Certificate constitutes verification that the discharge will not result in an adverse impact on state water quality standards. Federal permit requirements that would trigger the requirement for a Certificate include Section 10 and/or Section 404 permits from the COE and an NPDES permit issued by the EPA.

26.4.3.6 Water Use/Appropriation. Almost all states impose certain administrative requirements and require permits to appropriate large amounts of water from surface water and ground water sources. These types of permits are generally issued for consumptive use of the water; quantities that are returned to the water source are typically not counted as quantities used. Recharge zones for ground water sources also must be considered to ensure that the waters are not withdrawn at a faster rate than they are replenished. States that rely on sole source aquifers for their drinking water use will have more stringent appropriation requirements.

In addition to state requirements, and depending on project location, federal reserved rights of water must be considered when water use is evaluated. Federal reserved rights are used to control the water necessary to continue activities on federally reserved lands and override state laws regarding the allocation of water. A dormant federal reserved right can be exercised at any time, potentially decreasing the amount of water available to the state to distribute.

26.4.3.7 Underground Storage Tank Registration. Underground storage tanks are regulated either by EPA through a combination of statutory and regulatory provisions or by states that have received delegated authority from EPA. Tanks to which these regulatory programs apply are those that contain a regulated substance under the Superfund Amendment and Reauthorization Act of 1987 or a liquid petroleum product with at least 10% of the tank volume, including connecting pipes, underground. Tank owners are required to notify the state that they possess such tanks and submit a complex tank registration form. EPA has established detection, prevention, and release detection requirements and performance standards. Underground storage tanks must also be installed according to industry codes, must be equipped with spill and overfill prevention and leak detection devices, and must include corrosion protection.

26.4.4 Landfill Permit

The Resource Conservation and Recovery Act (RCRA) was enacted in 1976 to ensure proper management and control of solid and hazardous waste. RCRA broadly defines hazardous waste and requires comprehensive management from "cradle to grave," including generation, treatment, storage, and disposal. RCRA exempts a number of waste types from regulation as a hazardous waste, such as those wastes produced in large quantities but that do not present a danger to health or the environment or that are covered by other environmental regulations. Exempted wastes include wastes from municipal resource recovery plants and wastes from oil and gas explorations, mining operations, and fossil fuel combustion. These exempted wastes are regulated as solid wastes under RCRA, Subtitle D.

Sources of solid waste from oil- and coal-fueled utility boilers include the following:

- Flyash—Noncombustible particles formed during coal combustion carried from the furnace by the flue gas and collected by air pollution control equipment such as precipitators and/or fabric filters;
- Bottom ash—The solid residue remaining from fuel combustion which falls to the bottom of the boiler furnace; and
- Pyrite—Iron sulfide separated from coal in the pulverizer and typically combined with bottom ash for disposal.

Although EPA continues to evaluate whether wastes from fossil fuel combustion should be more stringently controlled, the trend at the state and local level is to require increasing control and restrictions regarding the siting, construction, and operation of combustion waste landfills. Typical requirements include treatment such as dewatering, solid stabilization, and fixation. Most state regulations also require the implementation of leachate collection systems, single or double linings, and ground water monitoring.

Even though fossil fuel combustion wastes are not currently classified as hazardous wastes by the EPA, individual states may classify combustion wastes as hazardous under their own regulatory programs. Ash products resulting from fuel combustion therefore must be analyzed to determine whether they must be regulated as a solid or as a hazardous waste. Even if it does not qualify as a hazardous waste, most states will regulate combustion waste as a "special" or "industrial" waste rather than a "solid" or "municipal" waste.

Most states have comprehensive solid waste landfill permit application requirements which include the development of detailed engineering information and extensive agency and public review. A typical application must include the following:

- Detailed description of existing site conditions and the potential impacts to natural resources such as soils, geological and hydrologic conditions, flora, fauna, wetlands, aquatic resources, cultural resources, floodplains, and recreation and land use.
- Detailed plans and specifications prepared and approved by a professional engineer and addressing geological and hydrologic conditions and soils; control of impacts on ground and surface water quality; implementation of ground water monitoring; dust control; cover application; personnel protection; and operating and recordkeeping procedures. Supporting site borings will also be required.
- Evidence of financial responsibility to ensure sufficient funds for closure and post closure monitoring
- Ash analysis
- Closure and postclosure plans
- Prior solid waste operating compliance history
- Evidence of compliance with local zoning requirements
- Names and addresses of all adjoining landowners.

The extensive natural resource surveys and agency and public review requirements associated with the typical solid waste landfill permit process can take from 3 to 5 years in some states. Because of the potential impacts on the project schedule, it is critical to initiate the permit application process at the earliest stages of project development to ensure that the permit is obtained in a timely fashion. Typically a final permit or an alternative disposal plan would be required to obtain project financing.

26.5 LOCAL PERMITS AND APPROVALS

Although local permits and approvals do not carry the administrative burdens and penalties of federal and state approvals, they may present the most likely source of project delay or termination because of public concerns. For this reason, good public relations are essential to seeking local approvals.

26.5.1 Site Plan Approvals

Municipalities and counties generally require approval of a facility site plan before construction can begin. A site plan includes general information on facility components, location, dimensions, landscaping, lighting, parking, etc. Information typically required to be submitted in a site plan application includes design drawings of the facility and may include requests for variances or design waivers from certain requirements, such as yard setbacks, maximum heights, curbing requirements, parking spaces, sidewalk installation, water line looping, and noise limitations. Identification and status of local, state, and federal approvals may also be required. Descriptions of water usage, methods for treatment and disposal of wastes, provisions for monitoring existing air and water quality, and soil erosion control may also be addressed.

Typically, a planning commission, consisting of city and/or county representatives, has the responsibility of reviewing and approving the site plan application. Actions taken on the site plan may be limited to monthly meetings of the planning commission, which may inhibit a timely approval process. If the city does not have its own engineers, the planning commission may have to hire a consulting engineer to review the plan. Charges for independent engineer review will likely be transferred to the applicant. Site plan review and approval may take anywhere from 6 months to a year, depending on the size and type of the electrical generating facility, and the city and/or county in which it will be located. It may be prudent to have a local engineer interface with the planning commission on behalf of the applicant to assist with the application. Also, some states require that the applicant be represented at meetings by an attorney licensed in that state.

The site plan review and approval can result in facility component relocation (if considered aesthetically displeasing), a requirement for air and ground water quality monitoring during operation of the facility, landscaping requirements (for aesthetics and noise mitigation), sewer and/or water line improvements, and other city improvements such as the creation of a local park.

26.5.2 Land Use (Conditional Use Permit, Special Use Permit, Rezoning Request)

Typically, cities and counties adopt ordinances that allow them to regulate land use in a manner that encourages orderly development with the city or county's adopted land use policies, and that promotes and protects the public health, safety, and welfare of the community. Zoning district classifications are determined by the city or county's zoning commission, and generally include classifications such as residential, commercial, industrial, and agricultural.

Principal and conditional uses within each type of district are established by the zoning commission. Map designations are developed to assist the city in maintaining its land use policies and to assist developers in site selection. Power generating facilities typically are allowed only in industrial zoning districts. Agricultural and commercial districts could be considered as sites for power plants, but they may have to be rezoned. A less restrictive district, such as an industrial district, may limit the approved uses if it adjoins a more restrictive district.

Generally, districts that are less restrictive will admit the uses of the more restrictive ones. For example, a business district may allow "light" utilities, such as transmission lines and substations as a conditional use within that district, but would not allow "heavy" utilities, such as power generating facilities. An industrial district, however, that would allow power generating facilities as a principal use should also allow the lighter utilities.

26.5.2.1 Conditional Use Permit. If power generation is designated as a conditional use in a zoning district, approval to construct and operate a power generating facility in that district must be obtained from the city's zoning/planning department. The application for a Conditional Use Permit will likely include site plan arrangement and drawings and a description of surrounding land uses and property boundaries.

Typically, the zoning or planning department reviews the application and submits it to the zoning/planning commission with recommendations. Public notice of receipt of the application will be given, and a hearing before the zoning/planning commission will be scheduled and held. After the hearing, a Record of Decision, which will include certain conditions with which the applicant must comply, will be issued by the zoning/planning commission.

Applications for a Conditional Use Permit may be submitted and reviewed as part of the site plan approval process, or they may be considered independently. The main considerations of interest to the zoning/planning commission are the impact of the activity on neighboring business or persons, surrounding/adjacent land use, buffer zones (set backs), traffic/public access, utility support, noise, odors, lighting, height, and security and fencing.

26.5.2.2 Special Use Permit. When a proposed activity is not specifically identified in the zoning classifications, it may be possible to apply for a Special Use Permit, which authorizes a "one-time-only" activity within that district. Special use permits may be required for ancillary facilities, such as ash landfills or a coal storage yard. Special Use Permits may also be required for temporary construction-related activities. Application requirements and procedures are typically the same as for Conditional Use Permits.

26.5.2.3 Rezoning/Zoning Variance. If the zoning law does not allow for the proposed activity, the applicant could consider applying for a zoning variance or request that the area be rezoned. Rezoning requests and variances are typically more time consuming and difficult to obtain because they set a precedent for future zoning variance requests. Application requirements and procedures are typically the same as for Special Use Permits, although additional information may be required.

26.5.3 Building Permit and Certificate of Occupancy

Building permits of various types are required throughout the construction of the facility. Building permits are generally issued by the building department after the plans and specifications for the structures have been approved. Building permits are typically required for concrete foundation, electrical work, and plumbing work. After the entire facility has been built and inspected, a Certificate of Occupancy is issued by the building department.

26.5.4 Fire Safety Approval

As a part of the Certificate of Occupancy inspection, the local fire marshal will likely inspect the facility to ensure that sprinkler systems are in place and in operating condition. The location of hazardous substances and oil storage tanks at the facility site will also need to be identified. The fire marshal may have to approve tank design and the adequacy of fire fighting equipment.

26.6 ENVIRONMENTAL REVIEW REQUIREMENTS

Compliance with applicable environmental review requirements can require onsite studies of existing natural resources, evaluations of project impacts, and the development of mitigation measures, all of which can have significant impacts on project cost and schedules.

26.6.1 National Environmental Policy Act

The NEPA of 1969 became law on January 1, 1970. NEPA establishes a national environmental policy, sets environmental protection goals, and provides a means of achieving the goals. NEPA states a commitment to conduct federal activities in a manner to promote the general welfare and to be in harmony with the environment. NEPA ensures that federal agencies consider the environmental consequences of actions before they are taken. Federal actions include the issuance of a permit by a federal agency, such as an NPDES permit by EPA or a Section 404 permit by the COE.

26.6.1.1 Lead Agency System. Most major projects involve several federal agencies. In these instances, one agency is designated as the "lead agency" with the responsibility to prepare or supervise the preparation of an environmental impact statement (EIS). The selection of the lead agency is based on the following criteria:

- Magnitude of the agency's involvement,
- Project approval and disapproval authority,
- Duration of the agency's involvement, and
- Sequence of the agency's involvement.

Usually only one agency is designated as the lead agency. It is possible, however, that a combination of federal, state, or local agencies (including at least one federal agency) may act as joint lead agencies to prepare the environmental impact statement. If agreement cannot be reached concerning the designation of the lead agency, the Council on Environmental Quality will make the designation.

26.6.1.2 Procedural Regulations. Six general stages of NEPA implementation comprise the NEPA process:

- Agency guidance and categorical exclusions,
- Environmental assessments,
- Scoping process,
- Draft environmental impact statement,
- Final environmental impact statement, and
- Agency decision and record of decision,

These stages are discussed in more detail in the following paragraphs.

26.6.1.3 Agency Guidance and Categorical Exclusions. To simplify the NEPA process, agencies must establish specific criteria for actions that usually require an EIS, those that usually do not require an EIS nor environmental assessment (EA), and actions that normally require an EA but not necessarily an EIS. If neither an EA nor an EIS is required by regulation, the action usually can proceed without the need to comply with any further NEPA requirements.

26.6.1.4 Environmental Assessments. Federal actions that are neither excluded categorically nor clearly require an EIS must have an EA. The EA is a document developed by the lead agency, and provides a basis for determining the need for an EIS. The EA must include the need for the proposed action and alternatives, and a list of agencies and others consulted. The environmental topics addressed in a typical EA document are described in detail in Section 26.6.3.

After developing the EA, the lead agency may conclude that no significant environmental impacts will result from the proposed action. The agency then will issue a "finding of no significant impact" (FONSI), which briefly explains why an EIS is not necessary.

If the lead agency determines that an EIS is necessary, a notice of intent to prepare an EIS must be published (typically in the *Federal Register*), and the next stage required by Council of Environmental Quality regulations, the scoping process, must be initiated.

26.6.1.5 Actions Requiring an Environmental Impact Statement (EIS). NEPA Section 102(2)(c) requires that an EIS be prepared for all major federal actions that significantly affect the quality of the human environment. The following are the actions that normally require an EIS:

- Feasibility reports for authorization and construction of major federal projects,
- Proposed changes in federal projects that increase the size substantially or add additional purposes, and
- Proposed major changes in the operation or maintenance of completed federal projects.

26.6.1.6 Scoping Process for EIS. When it is determined that an EIS is necessary, an early and open process will be initiated to determine the scope of issues to be addressed, additional studies to be conducted, and significant issues related to the proposed action. This procedure is referred to as the "scoping process." The scoping process is the initial step in obtaining agency and public participation and focus on the EIS. The lead agency must determine within the scope of the EIS the extent of actions, alternatives, and impacts to be considered in the EIS.

The scoping process provides an early opportunity to shape the subject content of the EIS. Participants in the scoping process may bring subjects and issues to the lead agency's attention. Affected parties also may request time limits and page limits. Normally, the lead agency is required to set such limits if requested.

Participation in the scoping process generally includes invited affected federal, state, and local agencies. Any affected Indian tribe, the proponent of the action, and other interested persons (including those who might not be in accord with the action on environmental grounds) may also attend. It also is possible that citizen organizations may be formed for the purpose of opposing or supporting the proposed project.

26.6.1.7 Draft EIS. Following the scoping process, a draft EIS is prepared. The draft EIS must fulfill, to the extent possible, the requirements of the final EIS. The draft EIS is then submitted for review and comment to appropriate agencies, other interested organizations, and affected parties. The review period for the draft EIS must be at least 45 days.

26.6.1.8 Final EIS. After the review of the draft EIS, the agencies and public comments are resolved and incorporated into the final EIS. Several iterations of this document may be required to respond to all agency comments. In general, the final EIS must address the environmental impact of an action, any unavoidable adverse environmental impacts of the action, alternatives to the proposed action, the relationship between short-term uses and long-term productivity, any irreversible and unretrievable commitment of resources involved in the proposed action, and mitigation measures that will be taken to reduce or eliminate environmental impacts.

26.6.1.9 Decision and Record of Decision. In arriving at a decision on the proposed action, the final EIS is reviewed and evaluated by the lead agency. The lead agency then develops a Record of Decision on the proposed action, which is a concise statement of its decision and the alternatives and means employed to mitigate or minimize adverse environmental impacts. The decision on the proposed action may not be taken or recorded until the later of either 90 days after the notice of the draft EIS, or 30 days after notice of the final EIS.

26.6.1.10 Contents of an EIS. The EIS is a detailed statement developed through consultation with appropriate agencies. The document must consider the environmental impact of the proposed action and include an evaluation of unavoidable adverse environmental effects, alternatives, the relationship between local short-term uses of the environment and the maintenance and enhancement of long-term productivity, and irreversible and irretrievable commitments of resources. The generic outline for a comprehensive environmental analysis document presented in Appendix 26A includes the major topics that would typically be addressed in an EIS. The exact content of an EIS is influenced somewhat by the nature of the proposed action and the project specific issues identified during scoping.

26.6.2 State Environmental Quality Review Acts

State environmental review processes (known as "little NEPA" processes) are, in many ways, quite similar to the federal NEPA regulations. Obviously, the specific state agencies involved vary depending on the location of the project. Document review time, filing procedures, notice publication procedures, and other requirements can also vary. Thus, it is essential that the environmental review procedures for the state in which a project is located be thoroughly understood early in the design phase of the project to establish a compatible project design/licensing schedule.

26.6.3 Environmental Analysis

Environmental analyses are conducted in response to the applicable federal, state, and in some cases, local regulations to determine the extent of any project-related impacts on environmental resources. In most cases, the environmental analyses typically are based on field evaluations, literature searches, and information obtained from the resource agencies. The lead agency's environmental assessments and environmental impact statements are the documents that are the product of the environmental analysis for the project in question. These documents generally include descriptions of the proposed development features, the existing environment, the anticipated environmental impacts, and the proposed mitigation measures. The depth and breadth of issues that are addressed may vary depending on the regulations governing the proposed action, the input of the involved resource agencies, and the sensitivity of the affected resources environment to project impacts.

The environmental resources typically addressed during the environmental review process are described below. Appendix 26A includes a typical outline for an environmental analysis document that identifies the potential impacts to existing natural resources and proposed mitigation measures associated with the construction and operation of a power plant.

26.6.3.1 Air Quality. Air quality analyses begin with the characterization of local air quality and meteorological conditions. Regional and local air quality and meteorological conditions are first summarized from existing data. This information, together with project design data such as emission levels, stack height, and site configuration, is then used in determining potential project-related construction and operation impacts on air quality (AAQS analysis) and the required level of control (BACT analysis). Additional site-specific air quality/meteorological sampling is sometimes necessary if sufficient data are not available to satisfy the requirements defined during the agency scoping session (see Section 26.4.2).

26.6.3.2 Geology and Soils. Regional and site-specific geological and soil characteristics are described based on published literature sources, such as county soil surveys, and on the results of site geotechnical studies. Impacts to soils and geology (e.g., erosion) resulting from project construction and operation are also addressed.

26.6.3.3 Water Resources. Local streams and reservoirs often are the sources of makeup water, the receiving waterbody for plant wastewaters, and the intake/outfall points for the proposed sites. Wells may be used to supply potable water for the project. Existing literature and water quality records from sources such as the USEPA, US Geological Survey, state resource management agencies, and other relevant sources are typically used to assess the surface and ground water quality and quantity conditions, and hydrologic conditions, although additional site-specific studies may be required. Water quality parameters evaluated usually include temperature, pH, alkalinity, hardness, nutrients, dissolved oxygen, biochemical oxygen demand, silica, total dissolved solids, total suspended solids, coliforms, and heavy metals. This information is used to assess potential project impacts and mitigation measures on water quality, quantity, and hydrology.

26.6.3.4 Threatened and Endangered Species/Terrestrial Resources. The terrestrial resources (vegetation and wildlife) of the project site are briefly described based on existing information and information collected during site reconnaissance. Vegetative and wildlife communities onsite are typically described as well as federal and state protected animal and plant species. Common plant and animal species onsite are also listed. Project-related impacts on these resources are described and recommendations for mitigating such impacts are provided.

26.6.3.5 Aquatic Resources. The aquatic resources of the lakes and streams (plankton, benthic invertebrates, and fish) in the project area are characterized based on data extrapolated from published literature, information obtained from the resource agencies, and the results of any site-specific studies. Wetland communities are also described in terms of quantity and quality. Potential impacts of project construction and operation on these resources are described.

26.6.3.6 Transportation. The existing transportation networks (highways, rail, air) are described, in addition to project impacts and potential mitigation measures for plant construction and operation impacts.

The existing transportation environments are characterized from available information, site reconnaissance, and if necessary, specialized traffic studies. Existing traffic counts near the sites are used, if available, to describe the existing environment and assess project impacts. Project transportation impacts such as the number of workers and deliveries to the sites are estimated, and conclusions about project impacts and the need for mitigation measures are developed.

26.6.3.7 Socioeconomics/Land Use/Recreation. The objective of this task is to develop an economic, demographic, and land use profile of the project area, and to assess future trends and project impacts on these resources. The section on socioeconomics typically encompasses the anticipated construction period and the commercial operating period of the project. Areas of study include the following:

- Existing socioeconomic and demographic conditions (employment, population structure, personal income, etc.);
- Existing land use/zoning conditions, including recreational areas, recreational activities, and usage estimates;
- Future direct and indirect conditions with and without the project; and
- Mitigation measures, if required.

26.6.3.8 Cultural Resources. The objective of this task is to comply with the directives of Section 106 of the National Historic Preservation Act. Section 106 requires that historic properties potentially affected by a proposed project be identified and evaluated for listing on the *National Register of Historic Places.*

The State Historic Preservation Officer (SHPO) is typically consulted to identify any known resources or properties that may be affected by the project. The SHPO's opinion regarding the need for cultural resource investigations of the project site is solicited, as well as the identification of known resources or properties that may be affected by the proposed project. A Phase I cultural resources investigation of the project site is often required, particularly if the project is located in an undeveloped area. A Phase I survey generally includes a literature search and a basic walkover of the project area. The Phase I survey serves as a basis for evaluating project impacts on local cultural resources, and for developing mitigation measures for minimizing such impacts.

26.6.3.9 Aesthetics. A qualitative description of a site's current visual character is usually developed based on information available from existing aerial photographs and data collected from a field reconnaissance of the site area. The existing aesthetic setting is described in terms of general appearance, land cover diversity, land use, etc. Field studies, such as the raising of a tethered balloon to the height of the project's tallest structure and evaluating the visibility of the balloon from several vantage points, are sometimes conducted to simulate the project's visual impact on local aesthetic conditions. Sometimes more detailed studies of the visual impacts of particular project components such as cooling towers are required. An artist's rendering or superimposing the project on a photograph is also useful in enabling the public to visualize the proposed project within the existing environment.

26.6.3.10 Noise. Onsite ambient measurements sometimes have to be taken at the project site to accurately characterize the existing noise conditions. The noise impacts resulting from the operation of the proposed facility equipment are predicted for the areas surrounding the site based on the facility equipment design parameters, estimates of the equipment sound power levels, and computer modeling results of the facility noise impacts. It may be necessary to recommend mitigation measures for specific project components to minimize noise impacts.

The construction noise impact can also be modeled. The various phases and equipment operating locations are analyzed to predict the worst-case noise impact on the area and its expected duration.

Compliance with the pertinent noise regulations is always of concern during plant design. Current state and local noise regulations must be identified and taken into consideration when identifying major project components. In the absence of noise regulations, a project acoustic design standard is established in cooperation with the appropriate agencies using established community acceptability criteria. This acoustic design standard is then used to evaluate the facility noise emissions and establish mitigation requirements.

26.6.4 Mitigation

An additional component of the environmental analysis process is the development of the appropriate mitigation measures. The type and extent of mitigation can vary depending on the resource affected, the type and magnitude of impacts on the resource, the public and agency attitude toward such impacts, the established agency regulations and/or policies, and industry standards. Mitigation for a specific impact may be as minor as periodically checking the location of a known archaeological site near the project to ensure that the site is not being looted. Revising the site arrangement to avoid critical habitat areas or wetland areas is another example of a minor mitigation effort, particularly if it occurs early in the process before the more detailed designs for the major plant components are developed.

Conversely, mitigation may involve significant capital expenditures, such as with the installation and maintenance of a selective catalytic reduction system—a system designed to control air pollutant emissions—at a combustion turbine project or developing a plan for creating a wetland to replace wetlands lost or adversely affected by the project. In any case, it is essential for the environmental specialists and the design engineers to work together to ensure that project impacts are minimized to the extent feasible.

26.6.5 Other Environmental Legislation

Congress has enacted numerous statutes designed to afford protection to the nation's valuable natural resources. Those statutes with the greatest potential to impact the cost or schedule for the construction and operation of a power generation facility are briefly described in the following paragraphs.

26.6.5.1 Endangered Species Act. The purpose of the Endangered Species Act of 1973 (ESA) is to conserve threatened or endangered species and their habitats, to develop a program for the conservation of the endangered and threatened species, and to encourage Congress to take the necessary steps to achieve the purposes of the ESA. All federal departments and agencies are directed by the ESA to conserve endangered or threatened species and utilize their authority to enforce the provisions of the ESA. The policy of Congress as stated in the ESA is that federal agencies will cooperate with state and local agencies to resolve issues without sacrificing conservation of endangered species. Proposed project sites must be evaluated to determine whether any protected species occur onsite and to identify any mitigation requirements that may be necessary.

26.6.5.2 Fish and Wildlife Coordination Act. The Fish and Wildlife Coordination Act of 1958 (FWCA) authorizes the Secretary of the Interior to provide assistance to and cooperate with federal, state, and public or private agencies and organizations in the development, protection, rearing, and stocking of all species of wildlife, resources, and their habitat. The losses of wildlife from disease or other causes are to be controlled. The FWCA directs that the report and recommendations of the Secretary of the Interior regarding the wildlife aspects of a water resource project be included within any report prepared or submitted by any agency of the federal government responsible for engineering and construction of the project.

26.6.5.3 Wild and Scenic Rivers Act. The Wild and Scenic Rivers Act establishes a national policy of selecting certain rivers possessing remarkable scenic, recreational, geological, fish and wildlife, historic, cultural, or other similar values, to be preserved in free-flowing condition so that their immediate environments will be protected for the benefit and enjoyment of present and future generations. The Wild and Scenic Rivers System is managed by the Secretary of the Interior, but this governing authority is delegated to the individual states except where the river flows on federal lands (e.g., national parks). Water use and/or wastewater discharge to a designated wild and scenic river may be prohibited or strictly limited.

26.6.5.4 Coastal Zone Management Act. The Coastal Zone Management Act (CZMA) was passed to preserve, protect, develop, and where possible, restore or enhance the resources of the nation's coastal zone for this and succeeding generations. Responsibility for the management of programs to meet these objectives belongs to the individual states. CZMA provides funding for states with qualifying programs. Project development in or adjacent to coastal zones may be prohibited. Approved projects may be required to implement specific mitigation measures to minimize impacts on coastal resources.

26.6.5.5 National Historic Preservation Act. The National Historic Preservation Act of 1966 requires federal agencies which license or fund projects that would affect structures or sites listed or eligible for listing in the National Register of Historical Places (Register) to take into account the effects of the proposed project on such structures or sites. The Register, which is maintained by the National Park Service, is an official list of archaeological, historic, and architectural properties that have local, state, or national significance. Both states and the federal government nominate properties for inclusion in the Register. Federal agency consideration of the effects of a proposed project on resources listed or eligible for listing in the Register must include consultation with the applicable State Historic Preservation Officer. The developer may be required to identify and implement a management plan for listed resources impacted by project construction or operation.

26.6.6 Agency Consultation and Public Review

State and federal environmental review processes typically require the involvement of the appropriate federal, state, and sometimes local agencies, as well as the general public. Input from the agencies can be very useful in identifying the issues of concern and any additional studies that may be required to characterize the local environment. These agencies also may provide environmental data that already exist and guidance regarding acceptable mitigation for the project.

Developers should *never* underestimate the authority of the agencies and should avoid adversarial relationships if possible. This does not mean that developers should comply with every request or requirement imposed by the agency. Rather, developers should be selective in the issues they choose to contest. Issues on which the developers and the agencies disagree should be argued in a professional, unemotional manner, and should be backed up with sound, reasonably defendable information. This does not guarantee that the developer will win the argument, but it will significantly enhance the chances of permit issuance, and in most cases will not alienate the other party. An experienced consultant can be helpful in identifying these issues and assisting the developer with developing a reasonable argument or identifying mutually acceptable alternatives.

The developer should establish contacts with the agencies early, preferably prior to the formal initiation of the environmental review process. Such contacts should be maintained throughout the process to keep the involved agencies appraised of the progress of the project and to identify any changes in agency policies, procedures, or concerns that may affect the developer. Finally, honesty *is* the best policy when dealing with agencies. Obviously, a thorough knowledge of the subject matter is important when discussing project-related issues with agency personnel. However, an upfront disclosure of the proposed development plans and a sincere effort to work with the agencies and comply with the applicable regulatory requirements can be invaluable in completing a project in a cost-effective and expeditious manner.

The developer is also well advised to devote the appropriate attention to the concerns of the general public. Many times, local residents are opposed to project development, particularly if it occurs in an area that has little or no such development. Conversely, local residents can also be supportive of a project, especially if it will result in additional jobs, tax revenues, improved infrastructure, and other socio-economic benefits. The developer can enhance public support by distributing public relations materials describing the developer and the proposed project, holding public meetings, and meeting with public officials. An artist's rendering illustrating the location and size of the proposed project and its associated ancillary facilities is also a useful tool in acquainting the public with the project. Admittedly, public opinion can be the downfall of any project, regardless of the efforts and intentions of the developer. However, public opinion can also be the developer's strongest ally, even in the face of opposition by the resource agencies.

26.7 REFERENCES

AMERICAN SOCIETY OF TESTING AND MATERIALS (ASTM). 1993. *Standard Practice for Environmental Site Assessments: Phase I Environmental Site Assessment Process.* ASTM E 1527-93.

AMERICAN SOCIETY OF TESTING AND MATERIALS (ASTM). 1993. *Standard Practice for Environmental Site Assessments: Transaction Screen Process.* ASTM E 1528-93.

Certificate of Public Convenience and Necessity. Name and requirements of certificate differ from state to state (e.g., Certificate of Environmental Compatibility and Public Need).

Code of Federal Regulations. Published by the Office of the Federal Register National Archives and Rewards Administration. Washington, DC 20408.

Note: When a law is passed by Congress, it is compiled into a public law (pub. law), then codified in the United States Codes (USC). From these codified laws, various regulations are developed and passed by the government, then published in the Code of Federal Regulations (CFR).

Law/Regulations	Citations
BACT requirements in NSPS amended by CAAA of 1990	42 USC 7411 (b)(1)(B)

Law/Regulations	Citations
Clean Air Act Amendments (CAAA) of 1990	Pub. Law 101-549, 104 Stat. 2399
CAAA, Title IV	42 USC 7651 et seq.; 40 CFR 72, 40 CFR 73, 40 CFR 74, 40 CFR 75, 40 CFR 76, 40 CFR 77, 40 CFR 78
CAAA, Title V	42 USC 7661 et seq.
Clean Water Act (CWA) of 1978, sec. 11	33 USC 1251 et seq.; 40 CFR 112
Clean Water Act (CWA), sec. 404	42 USC 1344; 40 CFR 230, 40 CFR 231, 40 CFR 233, 33 CFR 330 40 CFR 70
Clean Water Act (NPDES program), sec. 402	42 USC 1342 et seq.; 40 CFR 122, 40 CFR 123, 40 CFR 125
Endangered Species Act of 1973 (ESA)	16 USC 1531 et seq.; 33 CFR 322
Energy Policy Act of 1992, sec. 711 (Amended the Public Utility Holding Company Act of 1932 (PUHCA))	42 USC 13201 et seq.
Federal Aviation Act of 1958, sec. 1101	42 USC 1101; 14 CFR 77
Federal Water Pollution Control Act Amendments (CWA) of 1972	33 USC 1251; 40 CFR 112
Fish and Wildlife Coordination Act of 1958 (FWCA)	16 USC 742 et seq.; 40 CFR 1500-1508
National Environmental Policy Act (NEPA) of 1969	42 USC 4371 et seq.; 42 USC 7609
National Historic Preservation Act of 1966	16 USC 470
New Source Performance Standards (NSPS) (for wastewater discharges from steam electric power generating stations): CWA, sec. 306	33 USC 1311 et seq., 33 USC 1361; 40 CFR 423.15
Powerplant & Industrial Fuel Use Act (FUA) of 1978	42 USC 8301 et seq.; 10 CFR 500 through 516
Prevention of Significant Deterioration (PSD) under CAA	33 USC 1326; 40 CFR 52.21
Public Utility Holding Company Act (PUHCA)	16 USC 79 et seq. Section 32—40 CFR 365
Public Utility Regulatory Policies Act of 1978 (PURPA)	16 USC 791, 16 USC 2601 et seq., 42 USC 7101, 31 USC 9701; 18 CFR 292
PURPA (Federal Register [FR]) Resource Conservation and Recovery Act (RCRA)	57 FR 55176 (11/24/92) 42 USC 6901 et seq.
Rivers and Harbors Act of 1899, sec. 10	42 USC 403; 33 CFR 322
Superfund Amendments & Reauthorization Act of 1987	26 USC 4611 et seq; 42 USC 9601 et seq.; Pub. Law 99-499
Wild and Scenic Rivers Act	16 USC 1278 et seq.

Law/Regulations Citations

Standard Industrial Classification (SIC) Code. *Standard Industrial Classification Manual*. National Technical Information Service. Executive Office of the President. Office of Management and Budget.

APPENDIX 26A: SUGGESTED OUTLINE FOR ENVIRONMENTAL ASSESSMENT (DETAILED ANALYSIS OF PREFERRED *AND* ALTERNATE SITES)

1.0 INTRODUCTION

2.0 NEED FOR THE PROJECT

3.0 PROJECT DESCRIPTION

- 3.1 Site Description
  - 3.1.1 Preferred Site
  - 3.1.2 Alternative Site
- 3.2 Description of Proposed Facility
  - 3.2.1 Site Layout
    - 3.2.1.1 Site Arrangement
    - 3.2.1.2 Water Supply Pipeline Corridor
  - 3.2.2 Power Generation Equipment
    - 3.2.2.1 Combustion Turbine Generator
    - 3.2.2.2 Heat Recovery Steam Generator
    - 3.2.2.3 Steam Turbine Generator
  - 3.2.3 Air Quality Control
  - 3.2.4 Heat Rejection System
  - 3.2.5 Fuel Supply and Storage
    - 3.2.5.1 Fuel Types and Quantities
    - 3.2.5.2 Fuel Supply Sources
    - 3.2.5.3 Fuel Storage Facilities

3.2.6 Water Supply and Treatment

3.2.6.1 Facility Water Uses and Quantities
3.2.6.2 Water Supply Pipeline and Intake Structure
3.2.6.3 Water Treatment

3.2.7 Chemical Supply and Storage

3.2.7.1 Chemical Types and Uses
3.2.7.2 Chemical Storage and Handling

3.2.8 Wastewater Treatment and Discharge

3.2.8.1 Wastewater Sources and Quantities
3.2.8.2 Wastewater Treatment and Disposal

3.2.9 Solid Waste Management

3.2.9.1 Solid Waste Sources and Types
3.2.9.2 Solid Waste Disposal

3.2.10 Electric Transmission Facilities

3.2.11 Transportation Facilities

3.2.12 Cost of Facility

3.3 Analysis of Alternatives

3.4 Regulatory Requirements

4.0 EXISTING ENVIRONMENT OF PREFERRED SITE

4.1 Air Quality

4.1.1 Existing Climatological Conditions
4.1.2 Existing Air Quality Conditions

4.2 Geology and Soils

4.3 Water Resources

4.3.1 Existing Water Supplies and Uses

4.3.1.1 Ground Water
4.3.1.2 Surface Water

4.3.2 Existing Water Quality

4.3.2.1 Ground Water
4.3.2.2 Surface Water

4.4 Terrestrial Resources

4.4.1 Botanical Resources

4.4.1.1 Species Composition and Community Types
4.4.1.2 Wetlands
4.4.1.3 Species of Special Concern

4.4.2 Wildlife Resources

4.4.2.1 Habitat Types
4.4.2.2 Species Composition
4.4.2.3 Species of Special Concern

4.5 Aquatic Resources

4.5.1 Aquatic Habitat
4.5.2 Fish Communities
4.5.3 Invertebrate Communities
4.5.4 Species of Special Concern

4.6 Transportation

4.6.1 Existing Trafficways
4.6.2 Existing Traffic Volumes and Patterns

4.7 Land Use

4.7.1 Existing Land Uses and Land Use Plans
4.7.2 Existing Zoning Restrictions

4.8 Recreation

4.8.1 Existing Facilities
4.8.2 Existing Usages

4.9 Socioeconomics

4.9.1 Demographics
4.9.2 Housing
4.9.3 Wages and Employment
4.9.4 Existing Public Facilities and Services

4.10 Cultural Resources
4.10.1 Archaeological Resources
4.10.2 Historic Resources

4.11 Aesthetics

4.12 Noise

4.13 Hazardous Wastes

5.0 ENVIRONMENTAL CONSEQUENCES

5.1 Air Quality

5.1.1 Construction
5.1.2 Operation

5.2 Geology and Soils

5.2.1 Construction
5.2.2 Operation

5.3 Water Resources

5.3.1 Construction
5.3.2 Operation

5.4 Terrestrial Resources

5.4.1 Botanical Resources

5.4.1.1 Construction
5.4.1.2 Operation

5.4.2 Wildlife Resources

5.4.2.1 Construction
5.4.2.2 Operation

5.5 Aquatic Resources

5.5.1 Construction
5.5.2 Operation

5.6 Transportation

5.6.1 Construction
5.6.2 Operation

5.7 Land Use

5.7.1 Construction
5.7.2 Operation

5.8 Recreation

5.8.1 Construction
5.8.2 Operation

5.9 Socieconomics

5.9.1 Construction
5.9.2 Operation

5.10 Cultural Resources

5.10.1 Construction
5.10.2 Operation

5.11 Aesthetics

5.11.1 Construction
5.11.2 Operation

5.12 Noise

5.12.1 Construction
5.12.2 Operation

5.13 Hazardous Wastes

6.0 MITIGATION

6.1 Air Quality

6.2 Geology and Soils

6.3 Water Resources

6.4 Terrestrial Resources

6.5 Aquatic Resources

6.6 Transportation

6.7 Land Use

6.8 Recreation

6.9 Socioeconomics

6.10 Cultural Resources

6.11 Aesthetics

6.12 Noise

6.13 Hazardous Wastes

7.0 UNAVOIDABLE ADVERSE IMPACTS

8.0 AGENCY CONSULTATION PROCESS

9.0 LITERATURE CITED

Appendix: Conversion Table

CLASSIFIED LIST OF UNITS[1]

To convert from	to	Multiply by
Acceleration		
ft/s^2	meter per second squared (m/s^2)	3.048 000*E−01
Angle		
degree	radian (rad)	1.745 329 E−02
Area		
acre	square meter (m^2)	4.046 873 E+03
ft^2	square meter (m^2)	9.290 304*E−02
mi^2 (U.S. statute)[(2)]	square meter (m^2)	2.589 998 E+06
Bending moment or torque		
dyne·cm	newton meter (N·m)	1.000 000*E−07
lbf·in[(3)]	newton meter (N·m)	1.129 848 E−01
lbf·ft	newton meter (N·m)	1.355 818 E+00
Electricity and magnetism		
ampere hour	coulomb (C)	3.600 000*E+03
faraday (chemical)	coulomb (C)	9.649 57 E+04
gauss	tesla (T)	1.000 000*E−04
maxwell	weber (Wb)	1.000 000*E−08
ohm centimeter	ohm meter (Ω·m)	1.000 000*E−02
Energy (Includes Work)		
British thermal unit (mean)	joule (J)	1.055 87 E+03
calorie (mean)	joule (J)	4.190 02 E+00
electronvolt	joule (J)	1.602 19 E−19
erg	joule (J)	1.000 000*E−07
ft-lbf	joule (J)	1.355 818 E+00
kilocalorie (mean)	joule (J)	4.190 02 E+03
kW·h	joule (J)	3.600 000*E+06
therm (U.S.)	joule (J)	1.054 804*E+08
Energy per unit area time		
Btu (International Table)/(ft^2·s)	watt per square meter (W/m^2)	1.135 653 E+04
Force		
dyne	newton (N)	1.000 000*E−05
kilogram-force	newton (N)	9.806 650*E+00
pound-force (lbf)	newton (N)	4.448 222 E+00
Heat		
Btu (International Table)·ft/(h·ft^2·°F)	watt per meter kelvin [W/(m·K)]	1.730 735 E+00
Btu (International Table)·in/(h·ft^2·°F) (thermal conductivity)	watt per meter kelvin [W/(m·K)]	1.442 279 E−01
Btu (International Table)·in/(s·ft^2·°F)	watt per meter kelvin [W/(m·K)]	5.192 204 E+02
Btu (International Table)/ft^2	joule per square meter (J/m^2)	1.135 653 E+04
Btu (International Table)/(h·ft^2·°F) (thermal conductance)	watt per square meter kelvin [W/(m^2·K)]	5.678 263 E+00
Btu (International Table)/(s·ft^2·°F)	watt per square meter kelvin [W/(m^2·K)]	2.044 175 E+04
Btu (International Table)/lb	joule per kilogram (J/kg)	2.326 000*E+03
Btu (International Table)/(lb·°F) (heat capacity)	joule per kilogram kelvin [J/(kg·K)]	4.186 800*E+03
Btu (International Table)/ft^3	joule per cubic meter (J/m^3)	3.725 895 E+04
°F·h·ft^2/Btu (International Table) (thermal resistance)	kelvin square meter per watt (K·m^2/W)	1.761 102 E−01
°F·h·ft^2/[Btu (International table)·in] (thermal resistivity)	kelvin meter per watt (K·m/W)	6.933 471 E+00
Length		
angstrom	meter (m)	1.000 000*E−10
foot	meter (m)	3.048 000*E−01
inch	meter (m)	2.540 000*E−02
micron (micrometer)	meter (m)	1.000 000*E−06

To convert from	to	Multiply by
Length (*continued*)		
mile (international nautical)	meter (m)	1.852 000*E+03
mile (U.S. nautical)	meter (m)	1.852 000*E+03
mile (international)	meter (m)	1.609 344*E+03
mile (U.S. statute[2])	meter (m)	1.609 347 E+03
yard	meter (m)	9.144 000*E−01
Light		
footcandle	lux (lx)	1.076 391 E+01
Mass		
grain	kilogram (kg)	6.479 891*E−05
gram	kilogram (kg)	1.000 000*E−03
kgf·s^2/m (mass)	kilogram (kg)	9.806 650*E+00
ounce (avoirdupois)	kilogram (kg)	2.834 952 E−02
pound (lb avoirdupois)	kilogram (kg)	4.535 924 E−01
ton (long, 2240 lb)	kilogram (kg)	1.016 047 E+03
ton (short, 2000 lb)	kilogram (kg)	9.071 847 E+02
tonne	kilogram (kg)	1.000 000*E+03
Mass per unit area		
lb/ft^2	kilogram per square meter (kg/m^2)	4.882 428 E+00
Mass per unit time		
lb/h	kilogram per second (kg/s)	1.259 979 E−04
lb/s	kilogram per second (kg/s)	4.535 924 E−01
ton (short)/h	kilogram per second (kg/s)	2.519 958 E−01
Mass per unit volume		
lb/ft^3	kilogram per cubic meter (kg/m^3)	1.601 846 E+01
lb/gal (U.S. liquid)	kilogram per cubic meter (kg/m^3)	1.198 264 E+02
ton (long)/yd^3	kilogram per cubic meter (kg/m^3)	1.328 939 E+03
ton (short/yd^3)	kilogram per cubic meter (kg/m^3)	1.186 553 E+03
Power		
Btu (International Table)/h	watt (W)	2.930 711 E−01
Btu (International Table)/s	watt (W)	1.055 056 E+03
ft lbf/min	watt (W)	2.259 697 E−02
horsepower (electric)	watt (W)	7.460 000*E+02
foot of refrigeration (= 12 000 Btu/h)	watt (W)	3.517 E+03
Pressure or stress (Force per unit area)		
atmosphere, standard	pascal (Pa)	1.013 250*E+05
foot of water (39.2° F)	pascal (Pa)	2.988 98 E+03
inch of mercury (60° F)	pascal (Pa)	3.376 85 E+03
inch of water (60° F)	pascal (Pa)	2.488 4 E+02
kip/in^2 (ksi)	pascal (Pa)	6.894 757 E+06
lbf/ft^2	pascal (Pa)	4.788 026 E+01
lbf/in^2 (psi)	pascal (Pa)	6.894 757 E+03
Radiation units		
curie	becquerel (Bq)	3.700 000*E+10
rad	gray (Gy)	1.000 000*E−02
rem	sievert (Sv)	1.000 000*E−02
roentgen	coulomb per kilogram (C/kg)	2.580 000*E−04
Temperature		
degree Celsius	kelvin (K)	$T_K = t_c + 273.15$
degree Fahrenheit	degree Celsius (° C)	$t_c = (t_f - 32)/1.8$
degree Fahrenheit	kelvin (K)	$T_k = (t_f + 459.67)/1.8$
degree Rankine	kelvin (K)	$T_K = T_R/1.8$
kelvin	degree Celsius (° C)	$t_c = T_K - 273.15$
Time		
day	second (s)	8.640 000*E+04
hour	second (s)	3.600 000*E+03
minute	second (s)	6.000 000*E+01
year (365 days)	second (s)	3.153 600*E+07
Velocity		
ft/h	meter per second (m/s)	8.466 667 E−05
ft/min	meter per second (m/s)	5.080 000*E−03
ft/s	meter per second (m/s)	3.048 000*E−01
knot (international)	meter per second (m/s)	5.144 444 E−01
mi/h (international)	meter per second (m/s)	4.470 400*E−01

To convert from	to	Multiply by
Viscosity		
centipoise (dynamic viscosity)	pascal second (Pa·s)	1.000 000*E−03
centistokes (kinematic viscosity)	square meter per second (m^2/s)	1.000 000*E−06
ft^2/s	square meter per second (m^2/s)	9.290 304*E−02
lb/(ft·h)	pascal second (Pa·s)	4.133 789 E−04
lb/(ft·s)	pascal second (Pa·s)	1.488 164 E+00
lbf·s/in^2	pascal second (Pa·s)	6.894 757 E+03
Volume		
acre foot	cubic meter (m^3)	1.233 489 E+03
barrel (oil, 42 gal)	cubic meter (m^3)	1.589 873 E−01
ft^3	cubic meter (m^3)	2.831 685 E−02
gallon (U.S. liquid)	cubic meter (m^3)	3.785 412 E−03
ounce (U.S. fluid)	cubic meter (m^3)	2.957 353 E−05
Volume per unit time		
ft^3/min	cubic meter per second (m^3/s)	4.719 474 E−04
gallon (U.S. liquid) per minute	cubic meter per second (m^3/s)	6.309 020 E−05

(1)ASTM Standard, E-380-93, "Standard Practice for Use of the International System of Units (SI) (the Modernized Metric System)," ASTM, Philadelphia, Pennsylvania (1994).

(2)1 statute mile = 5,280 feet.

(3)lbf = pound force, lb = pound mass.

INDEX